This book presents a systematic treatment of a broad area of modern optical physics dealing with coherence and fluctuations of light. It is a field that has largely developed since the first lasers became available in the 1960s.

The first three chapters cover various mathematical techniques that are needed later. A systematic account is then presented of optical coherence theory within the framework of classical optics, and this is applied to subjects that have not been treated systematically before, such as radiation from sources of different states of coherence, foundations of radiometry, effects of source coherence on the spectra of radiated fields, coherence theory of laser modes and scattering of partially coherent light by random media.

A semiclassical description of photoelectron detection precedes the treatment of field quantization and of the coherent states, and this is followed by a discussion of photon statistics, the quantum theory of photoelectric detection and applications to thermal light. This includes a discussion of correlation measurements and photon antibunching, and of the Einstein–Podolsky–Rosen locality paradox.

A chapter is devoted to the interaction between light and a two-level atom and the problem of resonance fluorescence, and this is followed by treatments of cooperative radiation effects. After describing some general techniques for analyzing interacting quantum systems, such as the regression theorem and master equations, the book goes on to treat the single-mode and the two-mode laser and the linear amplifier. The concluding chapters deal with squeezed states of light and their generation and detection, and with some quantum effects in nonlinear optics such as parametric down-conversion, phase conjugation and quantum non-demolition measurements. Each chapter concludes with a set of problems.

The authors are well-known scientists who have made substantial contributions to many of the topics treated in this book. Much of the book is based on various graduate courses given by them over the years.

This book is likely to become an indispensable aid to scientists and engineers concerned with new developments in modern optics, as well as to teachers and graduate students of physics and engineering.

Optical coherence and quantum optics

Optical coherence and quantum optics

LEONARD MANDEL

Lee DuBridge Professor of Physics and Optics
University of Rochester, Rochester, N. Y., USA

AND

EMIL WOLF

Wilson Professor of Optical Physics
University of Rochester, Rochester, N. Y., USA

CAMBRIDGE
UNIVERSITY PRESS

Published by the Press Syndicate of the University of Cambridge
The Pitt Building, Trumpington Street, Cambridge CB2 1RP
40 West 20th Street, New York, NY 10011-4211, USA
10 Stamford Road, Oakleigh, Melbourne 3166, Australia

First Published 1995

Printed in the United States of America

A catalogue record for this book is available from the British Library

Library of Congress cataloguing in publication data

Mandel, Leonard.
Optical coherence and quantum optics / by Leonard Mandel and Emil Wolf.
p. cm.
Includes bibliographical references and index.
ISBN 0 521 41711 2
1. Coherence (Optics) 2. Quantum optics. I. Wolf, Emil. II. Title.
QC403.M34 1995
535′.2–dc20 93-48873 CIP

ISBN 0 521 41711 2 hardback

KT

Dedicated to our wives

Jeanne and Marlies

in appreciation of their patience, understanding and help

Contents

Preface

Prior to the development of the first lasers in the 1960s, optical coherence was not a subject with which many scientists had much acquaintance, even though early contributions to the field were made by several distinguished physicists, including Max von Laue, Erwin Schrödinger and Frits Zernike. However, the situation changed once it was realized that the remarkable properties of laser light depended on its coherence. An earlier development that also triggered interest in optical coherence was a series of important experiments by Hanbury Brown and Twiss in the 1950s, showing that correlations between the fluctuations of mutually coherent beams of thermal light could be measured by photoelectric correlation and two-photon coincidence counting experiments. The interpretation of these experiments was, however, surrounded by controversy, which emphasized the need for understanding the coherence properties of light and their effect on the interaction between light and matter.

Undoubtedly it was the realization that the subject of optical coherence was not well understood that prompted the late Dr E. U. Condon to invite us, more than three decades ago, to prepare a review article on the subject of coherence and fluctuations of light for publication in the *Reviews of Modern Physics*, which he then edited. The article was well received and frequently cited, and this encouraged us to expand it into a book. Little did we know then how rapidly the subject would develop and that it would become the cornerstone of an essentially new field, now known as quantum optics. Also the first experiments dealing with non-classical states of light were reported in the 1970s, and they provided the impetus for the new quantum mechanical developments. As an indication of the growth of the field we note that the book *Principles of Optics*, by M. Born and E. Wolf, published in 1959, the year before the laser was invented, had a chapter of just over 60 pages on partially coherent light, which covered most of what was then known about the subject. It was based entirely on the classical wave theory; quantum optics barely existed at that time. By contrast, in the present book more than twice as much space is devoted to quantum as to classical phenomena. The book is perhaps unusual in covering both the classical and the quantum theory of fluctuating electromagnetic fields in some depth.

Despite the length of the book, we make no claim as to its completeness, especially with respect to the quantum mechanical sections, and several topics are treated only cursorily or not at all. For example, only a short section deals with the subject of laser cooling and trapping, which has grown to merit a book of its own, and the important new field of atom interferometry is not treated at all.

Although at first we tried to be consistent in the use of notation throughout the book, later we abandoned the attempt, in part because the size of the book made it impractical, and partly because the use of certain symbols has become standard in some subfields. As regards the much debated question of the best choice of units for electromagnetic quantities, we have demonstrated our open-mindedness by employing both Gaussian and SI units. However, SI units are always used in discussing experiments.

Much of the book is an outgrowth of lectures that we have both given over more than 30 years at the University of Rochester, New York, and elsewhere. In particular Section 3.2 of the book on the angular spectrum representation of wavefields is based on lectures first given by one of us (EW) at the former National Bureau of Standards in Gaithersburg, Maryland in 1979 and 1980. Part of the text was prepared by him during sabbatical leaves at the University of California in Berkeley and at the Schlumberger–Doll Research Laboratory in Ridgefield, Connecticut, and he wishes to acknowledge his indebtedness to Professor Sumner P. Davis and to Dr Robert P. Porter for providing congenial facilities for the work.

We wish to thank Mr K. J. Harper and Mrs P. T. Sulouff, the former and the present Head Librarians of the Physics, Optics and Astronomy library at the University of Rochester, for their assistance in tracing and checking some of the less accessible references.

Authors from academia are often fortunate in being able to call on their present and former graduate students for assistance, and we gratefully acknowledge the help we received from several generations of our students in checking various sections of the manuscript and making many valuable suggestions for improvement. We are particularly indebted to G. S. Agarwal, S. Bali, D. Branning, B. Cairns, F. C. Cheng, D. G. Fischer, A Fougères, A. Gamliel, T. P. Grayson, D. F. V. James, M. Kowarz, P. D. Lett, F. A. Narducci, J. W. Noh, J. R. Torgerson, L. J. Wang, W. Wang, and X. Y. Zou for their help. We are also grateful to Mr Fischer for preparing the subject index and to Dr W. Wang for having drawn some of the figures and for checking many of the references.

We are much indebted to our former and present secretaries, Mrs Ruth Andrus, Mrs Ellen Calkins, Ms Laura Gifford, and Mrs Jennifer Van Remmen, who patiently typed and retyped the greater part of the manuscript, and to Mrs Calkins also for preparing the author index.

We acknowledge with thanks the excellent cooperation we received from the staff of Cambridge University Press at all stages of the production of this book. In particular we wish to express our appreciation to Mrs Susan Bowring, the copy editor and to Mr Tony Tomlinson, the production manager, for the way in which they converted an imperfect manuscript into a fine looking book. Finally we wish to thank Dr Simon Capelin, the publishing director for physical sciences of Cambridge University Press, who went out of his way to accommodate our numerous wishes over the long period extending from the initial discussion to the final execution of the project.

University of Rochester
Rochester, New York
May 1995

Leonard Mandel and Emil Wolf

1

Elements of probability theory

1.1 Definitions

The concept of probability is of considerable importance in optics, as in any situation in which the outcome of a given trial or measurement is uncertain. Under these conditions it is desirable to be able to associate a measure with the likelihood of the outcome or the event in question; such a measure is called the *probability* of the event.

Several different definitions of probability have been adopted at various times in the past. The classical definition is based on an exhaustive enumeration of the possible outcomes of an experiment or trial. If the trial has N distinguishable, mutually exclusive outcomes, which are equally likely to occur, and if n out of these N possible outcomes have an attribute or characteristic that we call 'success', then the probability of success in any one trial is given by the ratio n/N. For example, if we roll a die, and if each of the six digits is equally likely to be on top when the die comes to rest, there are $N = 6$ distinguishable outcomes. If we identify success with an even number, for example, then since there are three different ways in which success can be achieved, it follows that the probability of success when the die is rolled is given by $3/6 = 1/2$. Unfortunately, an exhaustive enumeration of all possibilities is not always feasible.

Another common definition of probability is based on the notion of relative frequency of success. If in a large number of N independent trials the successful attribute appears n times, then the relative frequency of success is n/N. When N becomes very great, we identify this ratio with the *probability of success* in any one trial. However, n/N does not have a limit as $N \to \infty$ in the mathematical sense.

Alternatively, the concept of probability can be introduced in an axiomatic way, in which we simply associate measures $p(A)$, $p(B)$, $p(C)$, ... that we call probabilities with all possible outcomes or events A, B, C, ... of a trial. If the total event space is denoted by Ω, then $A \in \Omega$, $B \in \Omega$, etc. It is convenient to introduce the following notation which is illustrated geometrically by the Venn diagrams in Fig. 1.1:

$$A \cup B \Rightarrow A + B$$

denotes the *combination* or *union* of the two events A and B, which implies

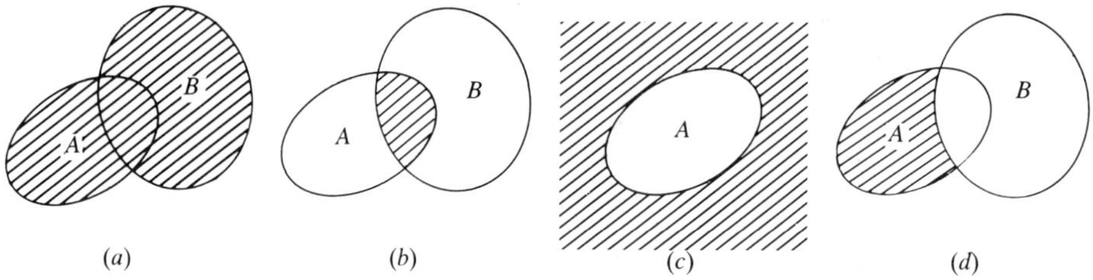

Fig. 1.1 Venn diagrams for certain combinations of events A and B. The shaded area illustrates the notion of (a) A or B or both; (b) both A and B; (c) not A; (d) A but not B.

either A, or B, or both (see Fig. 1.1(a));

$$A \cap B \Rightarrow A, B$$

denotes the *intersection* of the two events A and B, which implies both A and B (see Fig. 1.1(b));

$$\tilde{A} \Rightarrow -A$$

denotes the *complement* of the event A, which implies not A (see Fig. 1.1(c));

$$A \cap \tilde{B} \Rightarrow A - B$$

denotes the intersection of event A with the complement of B, which implies A but not B (see Fig. 1.1(d)). The null event $\varnothing$ is the complement of Ω. In all cases the notation on the right is the one customary in probability theory, and that on the left is the usual set theoretic notation. Figures 1.1(a) to 1.1(d) illustrate the notions of union of two events, intersection of two events, etc., etc., geometrically. Two events A and B are said to be *disjoint* or *mutually exclusive* if they do not overlap at all, or the intersection A, B is the null event $\varnothing$.

The following three axioms suffice to determine the properties of the probability $p(A)$ of a given event:

(a) $p(A) \geq 0,$ (1.1–1)
(b) $p(\Omega) = 1,$ (1.1–2)
(c) if $A_1, A_2, A_3, \ldots$ are mutually exclusive events, then

$$p(A_1 + A_2 + A_3 + \ldots) = p(A_1) + p(A_2) + p(A_3) + \ldots . \quad (1.1\text{–}3)$$

Equation (1.1–2) may be interpreted to mean that the probability of an outcome that is certain is unity. As $A + \tilde{A} = \Omega$, and A and $\tilde{A}$ are mutually exclusive, it follows from Eq. (1.1–3) that

$$p(A) + p(\tilde{A}) = p(A + \tilde{A}) = p(\Omega),$$

and from Eqs. (1.1–1) and (1.1–2)

$$0 \leq p(A) \leq 1. \quad (1.1\text{–}4)$$

The bounds on any probability are therefore zero from below and unity from above.

1.2 Properties of probabilities

Several important corollaries follow immediately from these relations. If the events $A_1, A_2, A_3, \ldots, A_N$ are mutually exclusive and represent the set of all possible outcomes, so that $A_1 + A_2 + \ldots + A_N = \Omega$, then from Eqs. (1.1–2) and (1.1–3) it follows that

$$\sum_{i=1}^{N} p(A_i) = p\left(\sum_{i=1}^{N} A_i\right) = p(\Omega) = 1. \tag{1.2–1}$$

Also, if A is a subset of B, or $A \subset B$, as illustrated by the Venn diagram in Fig. 1.2, then A and $B - A$ are mutually exclusive, and their union is B, so that by Eq. (1.1–3)

$$p(A) + p(B - A) = p(B),$$

or

$$p(A) \leqslant p(B) \quad \text{when } A \subset B. \tag{1.2–2}$$

Finally, we note that an event that cannot occur has probability zero, because $\varnothing + \Omega = \Omega$, so that

$$p(\varnothing) + p(\Omega) = p(\Omega),$$

and

$$p(\varnothing) = 0. \tag{1.2–3}$$

Thus if A and B are mutually exclusive, then the probability of both A and B $p(A, B) = 0$.

1.2.1 Joint probabilities

Events that are obtained by compounding other events are known as joint events, and the corresponding probabilities are joint probabilities. Thus $p(A, B)$ is the *joint probability* of both events A and B, or the probability of the intersection of A with B. The order in which the events A and B are listed is immaterial. As the compound event A, B is a subset of the event A (see Fig. 1.1(b)), it follows from Eq. (1.2–2) that

$$\left.\begin{aligned} &p(A, B) \leqslant p(A), \\ &\text{and similarly} \\ &p(A, B) \leqslant p(B). \end{aligned}\right\} \tag{1.2–4}$$

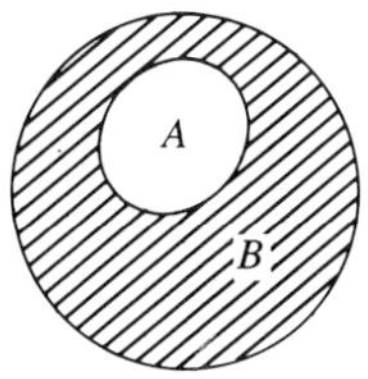

Fig. 1.2 Illustrating A as a subset of B.

The joint probability for two events is therefore always less than or equal to the probability for one of the events alone.

If $B_1, B_2, \ldots, B_M$ is a set of all possible mutually exclusive events, then

$$\sum_{i=1}^{M} p(A, B_i) = p(A, \Omega) = p(A). \tag{1.2–5}$$

This result follows immediately from the fact that A, B_1 $A, B_2 \ldots$ is also a set of mutually exclusive events spanning the whole space (see Eq. (1.1–3)). More generally, joint probabilities may involve more than two events. If $C_1, C_2, \ldots, C_N$ is a complete set of mutually exclusive events, then

$$\sum_{k=1}^{N} p(A, B, C_k, D, \ldots) = p(A, B, D, \ldots). \tag{1.2–6}$$

Let us now consider the situation in which two events A and B are not necessarily mutually exclusive (see Fig. 1.1(*a*)), and let us calculate the probability $p(A + B)$ of the union $A + B$. We cannot apply the summation law (1.2–6) to A and B directly. However, we note that the two events A and $B - A$ (cf. Figs. 1.1) are mutually exclusive, and their union is $A + B$. Then according to Eq. (1.1–3),

$$p(A) + p(B - A) = p(A + B). \tag{1.2–7}$$

Also, $B - A$ and A, B are mutually exclusive with union B, so that

$$p(B) = p(B - A) + p(A, B).$$

If we substitute for $p(B - A)$ in Eq. (1.2–7), we obtain immediately

$$p(A + B) = p(A) + p(B) - p(A, B), \tag{1.2–8}$$

so that

$$p(A + B) \leqslant p(A) + p(B). \tag{1.2–9}$$

Equation (1.2–8) is known as the *composition law* for two events that are not necessarily mutually exclusive. The relation is readily generalized to N events $A_1, A_2, \ldots, A_N$, for which it may be proved by induction that

$$p(A_1 + A_2 + \ldots + A_N) = \sum_{i=1}^{N} p(A_i) - \underset{\binom{N}{2}\text{ pairs}}{\sum_{i \neq j}^{N}\sum^{N}} p(A_i, A_j) + \underset{\binom{N}{3}\text{ triplets}}{\sum_{\substack{i \neq j \neq k \\ i \neq k}}^{N}\sum^{N}\sum^{N}} p(A_i, A_j, A_k)$$

$$- \ldots + (-1)^{N-1} p(A_1, A_2, \ldots, A_N). \tag{1.2–10}$$

Also, by repeated application of the inequality (1.2–9), one readily finds that

$$\begin{aligned} p(A_1 + A_2 + \ldots + A_N) &\leqslant p(A_1 + A_2 + \ldots + A_{N-1}) + p(A_N) \\ &\leqslant p(A_1 + A_2 + \ldots + A_{N-2}) + p(A_{N-1}) + p(A_N) \\ &\leqslant p(A_1) + p(A_2) + \ldots + p(A_N), \end{aligned} \tag{1.2–11}$$

with the equality sign holding for the special case of mutually exclusive events.

1.2.2 Conditional probabilities

The probability of some event A conditioned on some other event B is known as the *conditional probability of A given B*, and it is frequently denoted by $\mathcal{P}(A|B)$. It is given by the ratio

$$\mathcal{P}(A|B) = p(A, B)/p(B), \tag{1.2–12}$$

and it is, of course, defined only when B is not a null event. From Eq. (1.2–4) it follows immediately that

$$0 \leqslant \mathcal{P}(A|B) \leqslant 1, \tag{1.2–13}$$

so that a conditional probability is a true probability, with all the properties given earlier. If $A_1, A_2, \ldots, A_N$ is a complete set of mutually exclusive possible outcomes, then by virtue of the property (1.2–5) we have

$$\sum_{i=1}^{N} \mathcal{P}(A_i|B) = 1. \tag{1.2–14}$$

If the conditional probability of A given B is equal to the unconditional probability of A, or

$$\mathcal{P}(A|B) = p(A), \tag{1.2–15}$$

then it evidently does not matter whether event B occurs, or not, so far as event A is concerned. Events A and B are then described as being *statistically independent*. From Eqs. (1.2–15) and (1.2–12) we see that

$$p(A, B) = p(A)p(B) \tag{1.2–16}$$

whenever A and B are statistically independent, and this is sometimes taken to be the defining relation for statistical independence. More generally, the necessary and sufficient condition for N events $A_1, A_2, \ldots, A_N$ to be statistically independent is that the joint probability factorizes in the form

$$p(A_1, A_2, \ldots, A_N) = p(A_1)p(A_2) \ldots p(A_N). \tag{1.2–17}$$

A similar relation then holds for any subset of the N events. Needless to say, events that are mutually exclusive cannot be statistically independent, because the joint probability for mutually exclusive events is zero (except for the trivial case in which one or more of the events cannot happen at all).

A simple example may be helpful. Suppose that a die is rolled, and the number ending up on top is registered. We are interested in events of type A in which the number is divisible by 2, events B in which the number is divisible by 3, and events C in which the number is prime. These events are described by the following sets, with the indicated probabilities:

$$\left.\begin{aligned} A &= (2, 4, 6), & p(A) &= \tfrac{1}{2} \\ B &= (3, 6), & p(B) &= \tfrac{1}{3} \\ C &= (2, 3, 5), & p(C) &= \tfrac{1}{2}. \end{aligned}\right\} \tag{1.2–18}$$

The intersections among these sets are given by

$$\left.\begin{aligned} (A, B) &= (6), \quad p(A, B) = \tfrac{1}{6} \\ (A, C) &= (2), \quad p(A, C) = \tfrac{1}{6} \\ (B, C) &= (3), \quad p(B, C) = \tfrac{1}{6}. \end{aligned}\right\} \qquad (1.2\text{–}19)$$

It follows that

$$\left.\begin{aligned} p(A, B) &= p(A)p(B) \\ p(A, C) &\neq p(A)p(C) \\ p(B, C) &= p(B)p(C), \end{aligned}\right\} \qquad (1.2\text{–}20)$$

so that A and B are statistically independent, as are B and C, but A and C are not statistically independent.

1.2.3 Bayes' theorem on inverse probabilities

From the definition (1.2–12) of conditional probability the following two relations follow:

$$p(A, B) = \mathcal{P}'(A|B)p(B)$$

$$p(A, B) = \mathcal{P}(B|A)p(A).$$

On equating both expressions for $p(A, B)$ we obtain for mutually exclusive events A and B

$$\mathcal{P}'(A|B) = \frac{\mathcal{P}(B|A)p(A)}{p(B)} = \frac{\mathcal{P}(B|A)p(A)}{\sum_{\text{all}A} \mathcal{P}(B|A)p(A)}, \qquad (1.2\text{–}21)$$

where we have made use of Eq. (1.2–5) in the last expression. This relation is known as *Bayes' theorem*. If we call $\mathcal{P}(B|A)$ the conditional probability of B given A, we may think of $\mathcal{P}'(A|B)$ as the *inverse probability* of A given B. Bayes' theorem then allows the inverse probability to be determined from the forward conditional probability together with $p(A)$. In practice the theorem is often applied to experimental situations in which A is to be determined from measurements of B, but little or nothing is known about the a priori probability $p(A)$. Some assumption about the a priori characteristics of $p(A)$ then has to be made before Eq. (1.2–21) can be used, and this introduces a certain arbitrariness into the procedure, which has been criticized.

Let us illustrate the problem by a simple example. A vessel contains N balls, which are either black or white, in unknown proportion. A ball is picked at random and is found to be white. We wish to determine the inverse probability that the vessel contained n $(0 \leqslant n \leqslant N)$ white balls originally in the light of the experiment. Let $\mathcal{P}(1|n)$ be the conditional probability that a white ball is picked when the vessel actually contains n white balls $(n = 0, 1, \ldots, N)$. From the nature of the problem it is evident that $\mathcal{P}(1|n) = n/N$. Then from Eq. (1.2–21) the inverse probability $\mathcal{P}'(n|1)$ that the vessel originally contained n white balls, given that a white ball is picked, is given by

$$\mathcal{P}'(n|1) = \frac{\mathcal{P}(1|n)p(n)}{\sum_{n=0}^{N} \mathcal{P}(1|n)p(n)} = \frac{(n/N)p(n)}{\sum_{n=0}^{N}(n/N)p(n)}, \tag{1.2–22}$$

where $p(n)$ is the a priori probability that the vessel contains n white balls. Unfortunately, nothing is known about $p(n)$, so that, strictly speaking, Eq. (1.2–22) cannot be applied. However, in the absence of further information, if we arbitrarily assign equal weights to all values of n from 0 to N a priori, then $p(n) = 1/(N+1)$ and Eq. (1.2–22) leads to the solution

$$\mathcal{P}'(n|1) = \frac{n}{\sum_{n=0}^{N} n} = \frac{n}{\frac{1}{2}N(N+1)}. \tag{1.2–23}$$

By making some assignment to the a priori probabilities $p(n)$, we have been able to calculate the inverse probabilities $\mathcal{P}'(n|1)$. Although it may not be possible to give a formal justification for this procedure, it nevertheless leads to quantitative estimates that are often valuable.

1.3 Random variables and probability distributions

When the possible outcomes A of a trial or experiment are numbers, then the outcomes are automatically mutually exclusive. It is convenient to regard these numbers as the values of some variable x, which is known as a *random variable* or *variate*. If the possible values of x consist of the countable set of numbers x_1, x_2, x_3, . . . , then x is known as a *discrete random variable*, whereas if the possible values are any numbers in some interval (a, b) (which may be infinite), x is known as a *continuous random variable*. The set of all possible outcomes is known as the *ensemble* of x. Usually the random variable is taken to be real, but complex random variables $z = x + iy$, whose real and imaginary parts x, y are both random variables, will also be encountered.

With each of the possible outcomes or values x_1, x_2, . . . of the discrete variate we may associate a probability p_i $(i = 1, 2, \ldots)$, and as the different values are mutually exclusive, the corresponding probabilities must sum to unity, by virtue of Eq. (1.2–1),

$$\sum_{\text{all } i} p_i = 1. \tag{1.3–1}$$

A graph of the probability p_i versus x_i as in Fig. 1.3(a) consists of a series of points or lines, and illustrates the distribution of probability over the interval. In the special case in which one value x_0 is certain, and none of the other values x_1, x_2, . . . occurs, the form of p_i is

$$p_i = \delta_{i0}, \tag{1.3–2}$$

where δ_{ij} is the Kronecker delta symbol, i.e. $\delta_{ij} = 1$ if $i = j$ and $\delta_{ij} = 0$ otherwise.

If x is a continuous variate in the interval (a, b), it is convenient to associate a

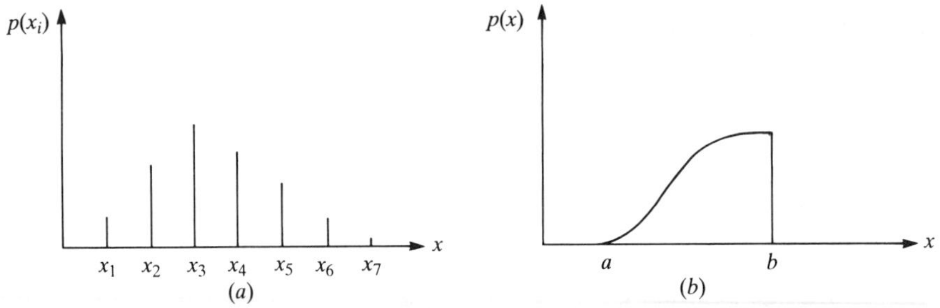

Fig. 1.3 Illustrating (*a*) discrete, (*b*) continuous probability distributions.

probability density $p(x)$ with the ensemble of x, such that $p(x)\,\mathrm{d}x$ gives the probability for x to be found in the infinitesimal interval from x to $x + \mathrm{d}x$. Then, corresponding to Eq. (1.3–1), we have the normalization condition

$$\int_a^b p(x)\,\mathrm{d}x = 1. \tag{1.3–3}$$

The form of $p(x)$ gives the probability distribution of the variate x (see Fig. 1.3(*b*)). The probability $P(x \leqslant X)$ for x to be equal to or below X ($a \leqslant X \leqslant b$) is given by the integral

$$P(x \leqslant X) = \int_a^X p(x)\,\mathrm{d}x. \tag{1.3–4}$$

Corresponding to Eq. (1.2–2) we have the relation

$$P(x \leqslant X_1) \leqslant P(x \leqslant X_2) \quad \text{when } X_1 \leqslant X_2. \tag{1.3–5}$$

$P(x \leqslant X)$ is therefore an increasing function of X that is bounded by unity, and its derivative is the probability density

$$\frac{\mathrm{d}P(x \leqslant X)}{\mathrm{d}X} = p(X). \tag{1.3–6}$$

The probability density $p(X)$ may not exist as an ordinary function when $P(x \leqslant X)$ is discontinuous, but it cannot be more singular than a Dirac delta function. If the continuous random variable x takes on the value x_0 with certainty, then $p(x)$ has the form

$$p(x) = \delta(x - x_0), \tag{1.3–7}$$

which can be compared with Eq. (1.3–2) for a discrete random variable. The need for delta functions to describe probability densities can be avoided by the use of Stieltjes integrals (Yaglom, 1962, Chap. 2, Sec. 9), but we shall not hesitate to use delta functions here. Indeed, by the introduction of delta functions we can incorporate the treatment of discrete variates in the treatment of continuous variates. If a discrete variate takes on the values $x_1, x_2, \ldots$ with probabilities $p_1, p_2, \ldots$, then we can formally describe this situation by a continuous variate x having the following probability density $p(x)$:

$$p(x) = \sum_i p_i \delta(x - x_i). \tag{1.3–8}$$

Because of this, and in order to avoid repetition, we shall henceforth formally regard x as a continuous variate.

A cautionary note regarding notation may be in order here. If x and y are two different random variables, their probability densities are sometimes denoted by $p(x)$ and $p(y)$, respectively, without any implication that the functional forms of the two probability distributions are equal. However, it is generally safer to use different symbols, e.g. $p(x)$ and $P(y)$, for two probability densities that are not necessarily equal.

1.3.1 Transformations of variates

Let x be a random variable defined on the interval (a, b) with probability density $p(x)$. It is sometimes necessary to make a transformation from x to a new variable y, where

$$y = f(x), \quad A \leqslant y \leqslant B, \tag{1.3–9}$$

and we wish to determine the probability density $P(y)$ of y. Let us first suppose that the transformation (1.3–9) has a single-valued inverse,

$$x = g(y). \tag{1.3–10}$$

Then if x and y correspond to each other, and the interval $\mathrm{d}x$ corresponds to the interval $\mathrm{d}y$, then evidently

$$P(y)|\mathrm{d}y| = p(x)|\mathrm{d}x|,$$

so that

$$\begin{aligned} P(y) &= p(x)\left|\frac{\mathrm{d}x}{\mathrm{d}y}\right| \\ &= p[g(y)]|g'(y)| \\ &= \frac{p[g(y)]}{|f'[g(y)]|}. \end{aligned} \tag{1.3–11}$$

More generally, if the inverse is multivalued, and to a given y there correspond several x's

$$\left.\begin{aligned} x_1 &= g_1(y) \\ x_2 &= g_2(y) \\ &\ldots\ldots, \end{aligned}\right\} \tag{1.3–12}$$

then we need to add the probabilities associated with these different, mutually exclusive x's, and we have in place of Eq. (1.3–11)

$$\begin{aligned} P(y) &= \sum_i p(x_i)\left|\frac{\mathrm{d}x}{\mathrm{d}y}\right|_{x=x_i} \\ &= \sum_i p[g_i(y)]|g_i'(y)| \\ &= \sum_i \frac{p[g_i(y)]}{|f'[g_i(y)]|}. \end{aligned} \tag{1.3–13}$$

The same result can also be formally expressed in the more compact form

$$P(y) = \int p(x)\delta[y - f(x)]\,dx, \tag{1.3–14}$$

if we expand the delta function in the usual way in terms of its zeros,

$$\delta[y - f(x)] = \sum_i \frac{\delta(x - x_i)}{|f'(x_i)|}. \tag{1.3–15}$$

Equation (1.3–14) can be interpreted to mean that the probability density of y is obtained by integrating the probability density $p(x)$ of x over all those values of x that correspond to y, i.e. those which are subject to the constraint $y = f(x)$.

As an example we consider the change of probability under the transformation $y = x^2$, which has the double-valued inverse

$$x = \pm\sqrt{y}$$

$$\frac{dx}{dy} = \pm\frac{1}{2\sqrt{y}}.$$

In this case Eq. (1.3–14) gives

$$P(y) = \frac{p(\sqrt{y})}{2\sqrt{y}} + \frac{p(-\sqrt{y})}{2\sqrt{y}}, \quad 0 \leqslant y. \tag{1.3–16}$$

Next we consider the more general situation in which we have a set of N variates $x_1, x_2, \ldots, x_N$, with joint probability density $p(x_1, x_2, \ldots, x_N)$, and we wish to transform to a new set $y_1, y_2, \ldots, y_N$, with probability density $P(y_1, y_2, \ldots, y_N)$. If the transformation

$$y_r = f_r(x_1, x_2, \ldots, x_N), \quad (r = 1, 2, \ldots, N)$$

has a single-valued inverse

$$x_r = g_r(y_1, y_2, \ldots, y_N), \quad (r = 1, 2, \ldots, N),$$

then

$$P(y_1, y_2, \ldots, y_N)|dy_1\,dy_2 \ldots dy_N| = p(x_1, x_2, \ldots, x_N)|\,dx_1\,dx_2 \ldots dx_N|, \tag{1.3–17a}$$

and

$$P(y_1, y_2, \ldots, y_N) = |J|p(x_1, x_2, \ldots, x_N), \tag{1.3–17b}$$

where J is the Jacobian of the transformation

$$|J| = \left|\frac{\partial(g_1, g_2, \ldots, g_N)}{\partial(y_1, y_2, \ldots, y_N)}\right|.$$

Once again we can express the transformation of the probability with the help of delta functions in the more compact form

$$P(y_1, y_2, \ldots, y_N) = \int p(x_1, x_2, \ldots x_N)\delta(y_1 - f_1)\delta(y_2 - f_2) \ldots \delta(y_N - f_N)\,dx_1\,dx_2 \ldots dx_N, \tag{1.3–18}$$

which also holds when the inverse is multivalued.

If the random variable z is complex, with $z = x + iy$, then we are effectively dealing with two real variates x, y. The probability density $P(z)$ of z is then simply the joint probability density of x and y, and the normalization condition on $P(z)$ becomes

$$\int P(z)\,d^2z = 1, \tag{1.3–19}$$

where d^2z is a shorthand notation for $dx\,dy$. Probability densities of complex variates can be treated in a manner that is a natural generalization of the foregoing treatment of real variates.

1.3.2 Expectations and moments

Perhaps the most important quantity associated with a random variable x is its *average* or *mean* or *expectation value*, which will be denoted by $\langle x \rangle$. The mean is obtained by weighting each value of x by the associated probability $p(x)\,dx$ for that value and integrating over the allowed range of x. Thus,

$$\langle x \rangle = \int xp(x)\,dx, \tag{1.3–20}$$

provided that the integral exists. More generally, if x is a random variable, any function $f(x)$ of x is itself a random variable, and its mean or expectation, if it exists, is given by

$$\langle f(x) \rangle = \int f(x)p(x)\,dx. \tag{1.3–21}$$

Among the functions of x that are of particular interest are the powers x^r ($r = 1, 2, \ldots$), for which

$$\nu_r \equiv \langle x^r \rangle = \int x^r p(x)\,dx. \tag{1.3–22}$$

ν_r or $\langle x^r \rangle$ is known as the *r'th moment* of x. The mean $\langle x \rangle$, which is of course the first moment ν_1, is often the most important moment, whereas the higher moments generally are progressively less important. It is worth noting that if $p(x)$ does not fall off sufficiently rapidly with x for large x, some of the moments may not exist. An example is provided by the Cauchy probability density

$$p(x) = \frac{a/\pi}{a^2 + (x - x_0)^2}, \quad (a > 0, -\infty < x < \infty), \tag{1.3–23}$$

whose moments all diverge.

When the random variable takes on only integer values $n = 0, 1, 2, \ldots$ with probabilities $p_0, p_1, p_2, \ldots$, we define the moments in a completely analogous manner by the formula

$$\nu_r \equiv \langle n^r \rangle = \sum_{n=0}^{\infty} n^r p_n. \tag{1.3–24}$$

However, another kind of moment defined by

$$\langle n^{(r)} \rangle \equiv \langle n(n-1)(n-2)\ldots(n-r+1) \rangle = \sum_{n=0}^{\infty} n(n-1)\ldots(n-r+1)p_n, \tag{1.3–25}$$

and known as the *r'th factorial moment of n*, sometimes turns out to be simpler for integer variates, as we shall see below. In the following definitions we limit ourselves to continuous variates, in order to avoid repetition.

Equation (1.3–22) defines the moments of x about the origin $x = 0$, but it is often more useful to deal with moments taken about some other value, such as the mean $\langle x \rangle$. In that case they are known as *central moments*, and we shall denote them by μ_r,

$$\mu_r \equiv \langle (x - \langle x \rangle)^r \rangle = \int (x - \langle x \rangle)^r p(x)\, dx. \tag{1.3–26}$$

The difference between x and its expectation $\langle x \rangle$ is known as the *deviation*

$$\Delta x \equiv x - \langle x \rangle,$$

so that we may write $\mu_r = \langle (\Delta x)^r \rangle$. The first central moment $\mu_1 = \langle \Delta x \rangle = 0$ from the definition, but the higher central moments are non-zero in general, although a probability density $p(x)$ that is symmetric about its mean will have zero odd central moments. The second central moment μ_2 is known as the *variance* or the *mean-squared deviation* or the *dispersion*. It is an important measure that determines the effective width of the probability density $p(x)$. The variance μ_2 is necessarily non-negative, and it vanishes only in the special case in which $p(x)$ is a delta function, i.e. when there is no uncertainty at all in the outcome x. The square root of the variance is known as the *root-mean-squared deviation* or *standard deviation* σ, and like the variance, it is a measure of dispersion, but is of the same dimensions as the mean $\langle x \rangle$. The standard deviation is sometimes used to normalize the higher moments of x. For example, the ratio

$$\alpha_3 \equiv \mu_3/\sigma^3, \tag{1.3–27}$$

which can be positive or negative, is known as the *coefficient of skewness*, and it provides a dimensionless measure of the asymmetry of the probability density $p(x)$ (see Fig. 1.4). Similarly, the dimensionless ratio

$$\alpha_4 \equiv \mu_4/\sigma^4, \tag{1.3–28}$$

which is known as the *kurtosis*, conveniently distinguishes probability distributions which are tall and thin from those that are short and wide (see Fig. 1.5). It

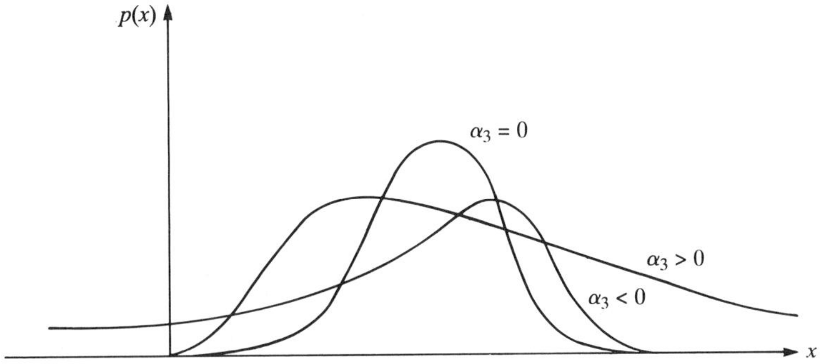

Fig. 1.4 Illustrating probability distributions with positive, zero and negative skewness.

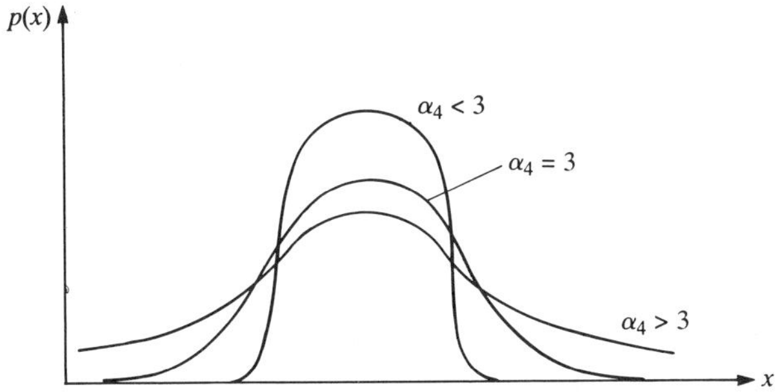

Fig. 1.5 Illustrating probability distributions with kurtosis greater than, equal to, and less than 3.

has the value 3 for a Gaussian distribution, which plays a central role in probability theory, as we shall see.

By making a binomial expansion under the integral in Eq. (1.3–26) and integrating term by term, we arrive at the following relation between the ordinary and the central moments,

$$\mu_r = \sum_{s=0}^{r} \binom{r}{s} \nu_s (-\nu_1)^{r-s}, \tag{1.3–29}$$

where $\nu_0 = 1$. In particular, the variance is given by

$$\mu_2 = \nu_2 - \nu_1^2. \tag{1.3–30}$$

The first and second moments characterize the most important features of the probability distribution. It is sometimes convenient to transform from x to a new variate y defined by

$$y \equiv (x - \langle x \rangle)/\sigma, \tag{1.3–31}$$

which has the property that its mean is zero, and its standard deviation is unity. The new variate y is said to be in *standard form*, and calculations are often simplified by this transformation.

Moments can also be defined jointly for several random variables $x, y, z, \ldots$ If

$p(x, y, z, \ldots)$ is the joint probability density of $x, y, z, \ldots$, then the central moment of order $l, m, n, \ldots$ is given by the expectation

$$\mu_{lmn\ldots} \equiv \langle (\Delta x)^l (\Delta y)^m (\Delta z)^n \ldots \rangle. \tag{1.3–32}$$

In particular, for two random variables there are three different variances: $\mu_{20} = \langle (\Delta x)^2 \rangle$, $\mu_{02} = \langle (\Delta y)^2 \rangle$, $\mu_{11} = \langle \Delta x \Delta y \rangle$; the last is known as the *covariance*. An alternative notation that is commonly used is to distinguish the different random variables by a suffix, so that they become $x_1, x_2, \ldots x_N$. The covariance between x_i and x_j is then written

$$\mu_{ij} \equiv \langle \Delta x_i \Delta x_j \rangle. \tag{1.3–33}$$

The covariances μ_{ij} can be regarded as the elements of a symmetric $N \times N$ matrix known as the *covariance matrix*, whose diagonal elements are the variances $\langle (\Delta x_1)^2 \rangle, \langle (\Delta x_2)^2 \rangle, \ldots, \langle (\Delta x_N)^2 \rangle$.

From the Schwarz inequality one can readily show that

$$|\mu_{ij}|^2 \leqslant \mu_{ii}\mu_{jj} = \sigma_i^2 \sigma_j^2, \tag{1.3–34}$$

where we have written $\sigma_i \equiv \langle (\Delta x_i)^2 \rangle^{1/2}$, etc. for the standard deviation of x_i. The normalized quantity

$$\rho_{ij} \equiv \mu_{ij}/\sigma_i \sigma_j \tag{1.3–35}$$

is known as the *correlation coefficient* and is evidently bounded by -1 and $+1$. Two variates whose correlation coefficient is $+1$ or -1 are said to be completely correlated or completely anticorrelated, respectively. For example, if $x_2 = ax_1 + b$, where a, b are real numbers, it is clear that x_2 and x_1 fluctuate up and down together when a is positive and in opposition when a is negative. As $\Delta x_2 = a \Delta x_1$, the correlation coefficient is readily found to be $\rho_{12} = a/|a|$, which is $+1$ or -1, so that the two variates x_1, x_2 are completely correlated or anticorrelated in this case. By contrast we note that $\rho_{12} = 0$ if x_1 and x_2 are statistically independent, because

$$\sigma_1 \sigma_2 \rho_{12} = \int \Delta x_1 \Delta x_2 p(x_1, x_2)\, dx_1\, dx_2 = \int \Delta x_1 p(x_1)\, dx_1 \int \Delta x_2 p(x_2)\, dx_2 = 0.$$

By a generalization of the same argument we can readily show that if $x_1, x_2, \ldots$ are statistically independent random variables, then

$$\langle \Delta x_i \Delta x_j \Delta x_k \ldots \rangle = 0 \quad \text{if } i, j, k, \ldots \text{ are all different.} \tag{1.3–36}$$

The covariance matrix μ_{ij} is then diagonal.

Linear combinations of statistically independent variates $x_1, x_2, \ldots$ are often encountered in practice. Let y be a new variate defined by

$$y = \sum_i a_i x_i, \tag{1.3–37}$$

where the coefficients $a_1, a_2, \ldots$ are arbitrary real numbers, and the $x_1, x_2, \ldots$ are independent. Then it follows immediately from the definition that

$$\langle y \rangle = \sum_i a_i \langle x_i \rangle, \tag{1.3–38}$$

and from Eq. (1.3–36) that

$$\langle(\Delta y)^2\rangle = \sum_i\sum_j a_i a_j \langle \Delta x_i \Delta x_j\rangle = \sum_i a_i^2 \sigma_i^2. \tag{1.3–39}$$

Finally we note that averages and moments can also be introduced for complex random variables by an obvious generalization of Eqs. (1.3–21) and (1.3–22). If $z_1, z_2, \ldots$ are complex random variables, the *covariance matrix* is the Hermitian matrix defined by

$$\mu_{ij} \equiv \langle \Delta z_i \Delta z_j^* \rangle. \tag{1.3–40}$$

The correlation coefficient ρ_{ij} given by Eq. (1.3–35) is then complex, but the standard deviations $\sigma_i = \langle |\Delta z_i|^2\rangle^{1/2}$ are real as before, and $|\rho_{ij}|$ is bounded by unity.

1.3.3 Chebyshev inequality

One sometimes wishes to know the probability that some randomly fluctuating quantity exceeds a certain threshold value. The Chebyshev inequality allows one to put an upper bound on this probability without detailed knowledge of the actual form of the probability distribution.

Let $p(x)$ be the probability density of some real random variable x, and let $g(x)$ be a non-negative real function of x. Suppose that $g(x) \geqslant K$, where K is some positive number, whenever x lies in the domain D. Then the probability $P[g(x) \geqslant K]$ that $g(x)$ exceeds K is evidently the probability that x lies within D, and is given by

$$P[g(x) \geqslant K] = \int_D p(x)\,\mathrm{d}x. \tag{1.3–41}$$

Let us now calculate the expectation of $g(x)$. By virtue of the non-negative character of $g(x)$ and $p(x)$, we have

$$\langle g(x)\rangle = \int g(x)p(x)\,\mathrm{d}x \geqslant \int_D g(x)p(x)\,\mathrm{d}x \geqslant K\int_D p(x)\,\mathrm{d}x = KP[g(x) \geqslant K],$$

so that

$$P[g(x) \geqslant K] \leqslant \langle g(x)\rangle / K. \tag{1.3–42}$$

This formula is known as the *Chebyshev inequality*. Note that it is sufficient for us to know the expectation of $\langle g(x)\rangle$ in order to determine an upper bound for $P[g(x) \geqslant K]$, although the upper bound is often rather weak.

As an example, let us calculate the probability that the value x departs from its mean $\langle x\rangle$ by more than η standard deviations σ. Let $g(x) = (\Delta x)^2$. Then the required probability is $P[(\Delta x)^2 \geqslant \eta^2\sigma^2]$. Now

$$\langle g(x)\rangle = \langle(\Delta x)^2\rangle = \sigma^2,$$

so that from Eq. (1.3–42), with $K = \eta^2\sigma^2$,

$$P[|\Delta x| \geqslant \eta\sigma] \leqslant \frac{1}{\eta^2}. \tag{1.3–43}$$

This is known as the *Bienaymé–Chebyshev inequality*. Thus, if $\eta = 3$, it tells us that the probability of encountering a fluctuation from the mean greater than three standard deviations is less than 1/9. Actually, if x were a Gaussian variate, the probability $P[|\Delta x| \geqslant 3\sigma]$ would be less than .003, so that the Bienaymé–Chebyshev inequality provides only a weak upper bound.

1.4 Generating functions

1.4.1 Moment generating function

It is sometimes possible to generate the moments more simply and conveniently from another function, known as the *generating function*, rather than to calculate them directly. For example, the function defined for real ξ by

$$M(\xi) \equiv \langle e^{x\xi} \rangle = \int e^{x\xi} p(x)\,dx, \qquad (1.4\text{–}1)$$

is known as the *moment generating function*, provided it exists. On expanding the exponential as a power series and integrating term by term, we find that

$$M(\xi) = \sum_{r=0}^{\infty} \frac{\nu_r \xi^r}{r!}, \qquad (1.4\text{–}2)$$

so that the moment ν_r can be obtained from $M(\xi)$ either by making a power series expansion in ξ or by differentiation, in which case we obtain

$$\nu_r = \left[\frac{d^r M(\xi)}{d\xi^r} \right]_{\xi=0}. \qquad (1.4\text{–}3)$$

For a discrete random variable, such as a random integer n, we simply replace the integral in Eq. (1.4–1) by a sum over n, and write

$$M(\xi) = \sum_{n=0}^{\infty} e^{n\xi} p(n). \qquad (1.4\text{–}4)$$

This equation generates the moments of n as before. It is also possible to derive the factorial moments $\langle n^{(r)} \rangle$ in an analogous manner. For that purpose we introduce the *factorial moment generating function* $F(\xi)$ defined by

$$F(\xi) \equiv \langle (1+\xi)^n \rangle = \sum_{n=0}^{\infty} \sum_{r=0}^{n} \frac{n(n-1)\ldots(n-r+1)}{r!} \xi^r p(n)$$

$$= \sum_{r=0}^{\infty} \frac{\langle n^{(r)} \rangle}{r!} \xi^r, \qquad (1.4\text{–}5)$$

which generates the factorial moments by power series expansion. It should be noted that the series is infinite if there is no upper bound on n. To avoid repetition we shall focus mainly on the continuous variate below.

1.4.2 Characteristic function

The moment generating function may not exist for every ξ. However, another generating function, defined by

$$C(\xi) \equiv \langle e^{ix\xi} \rangle = \int e^{ix\xi} p(x)\, dx, \tag{1.4–6}$$

and known as the *characteristic function* always exists as an ordinary function, because it is simply the Fourier transform of an absolutely integrable function (Goldberg, 1961, Chap. 2). In general $C(\xi)$ is complex. When $p(x)$ is square-integrable, Eq. (1.4–6) can be inverted to yield the probability density

$$p(x) = \frac{1}{2\pi} \int C(\xi)\, e^{-ix\xi}\, dx, \tag{1.4–7}$$

but even when $p(x)$ contains delta function contributions, Eqs. (1.4–6) and (1.4–7) are valid within the framework of generalized function theory. It is clear that the moments, if they exist, can be derived just as readily by differentiation from $C(\xi)$ as from $M(\xi)$ above. However, it is worth noting that the characteristic function $C(\xi)$ exists even when the moments do not. As an example we again consider the Cauchy distribution given by Eq. (1.3–23) above, which has no moments, for which

$$C(\xi) = \frac{a}{\pi} \int_{-\infty}^{\infty} \frac{e^{ix\xi}\, dx}{a^2 + (x - x_0)^2} = e^{-a|\xi|}\, e^{ix_0\xi}. \tag{1.4–8}$$

The right-hand side is not differentiable with respect to ξ at $\xi = 0$, and therefore cannot be used to generate the moments. Even when the moments exist, the probability density is not always uniquely determined by its moments, although counter-examples tend to be pathological and unphysical (Kendall, 1952, Chap. 4).

Characteristic functions are very rich in mathematical properties and whole books have been devoted to them (Lukacs, 1960). The following are a few of the more useful properties:

(a)
$$C(0) = 1, \tag{1.4–9}$$

as is obvious from the definition.

(b)
$$|C(\xi)| \leqslant C(0). \tag{1.4–10}$$

This follows from the fact that

$$|C(\xi)| \leqslant \int |e^{ix\xi} p(x)|\, dx = \int p(x)\, dx = 1.$$

The characteristic function can therefore never rise above its value at $\xi = 0$.

(c) $C(\xi)$ is continuous on the real axis, even if $p(x)$ has discontinuities. This can be seen from the inequality

$$|C(\xi+h)-C(\xi)| \leqslant \int |e^{ix(\xi+h)} - e^{ix\xi}|p(x)\,dx = 2\int |\sin(\tfrac{1}{2}xh)|p(x)\,dx \tag{1.4–11}$$

because the right-hand side tends to zero as $h \to 0$ for all ξ.

(d)
$$C(-\xi) = C^*(\xi), \tag{1.4–12}$$

where the asterisk denotes the complex conjugate. This follows immediately from the definition by virtue of the fact that $p(x)$ is necessarily real.

(e) $C(\xi)$ is non-negative definite, which means that for an arbitrary set of N real numbers $\xi_1, \xi_2, \ldots, \xi_N$ and N arbitrary complex numbers $a_1, a_2, \ldots, a_N$,

$$\sum_{i=1}^{N}\sum_{j=1}^{N} a_i^* a_j C(\xi_j - \xi_i) \geqslant 0. \tag{1.4–13}$$

This result follows from the inequality

$$\left\langle \left| \sum_{j=1}^{N} a_j e^{ix\xi_j} \right|^2 \right\rangle \geqslant 0$$

when the average on the left is written as an integral. We then have

$$\sum_{i=1}^{N}\sum_{j=1}^{N} \int a_i^* a_j e^{ix(\xi_j-\xi_i)} p(x)\,dx \geqslant 0$$

which, with the help of Eq. (1.4–6), leads immediately to the inequality (1.4–13).

There is an important theorem due to Bochner (1959, pp. 325–328; see also Goldberg, 1961, Chap. 5) which in its elementary form asserts that every non-negative definite function of a broad class has a non-negative Fourier transform and, conversely, that the Fourier transform of every non-negative function of a broad class is non-negative definite. This class includes functions which fall off sufficiently rapidly at infinity to ensure that their Fourier transforms are continuous functions. Absolutely integrable functions are of this kind. It follows that a complex function $C(\xi)$ that satisfies the conditions (1.4–13) and $C(0) = 1$ is a characteristic function.

From the definitions it is evident that the three generating functions $M(\xi)$, $F(\xi)$, $C(\xi)$, when they exist, are related as follows:

$$\left.\begin{aligned} M(i\xi) &= C(\xi) \\ F(e^{i\xi}-1) &= C(\xi) \\ F(\xi) &= M[\ln(1+\xi)]. \end{aligned}\right\} \tag{1.4–14}$$

Finally, if $z = x + iy$ is a complex random variable with probability density $p(z)$, so that

$$\int p(z)\,d^2z = 1, \quad (d^2z \equiv dx\,dy),$$

the characteristic function $C(u)$ is defined to be the two-dimensional Fourier transform of $p(z)$. This may be written

$$C(u) = \int e^{u^*z-uz^*} p(z)\, d^2z, \tag{1.4–15}$$

where u is a complex parameter, and $e^{u^*z-uz^*}$ is the two-dimensional Fourier kernel in x, y.

1.4.3 Cumulants

Another generating function that is sometimes useful is the logarithm of the moment generating function,

$$K(\xi) \equiv \ln[M(\xi)], \tag{1.4–16}$$

known as the *cumulant generating function*. It can be given a power series expansion in ξ with coefficients that are known as the *cumulants* κ_r,

$$K(\xi) = \sum_{r=1}^{\infty} \frac{\kappa_r \xi^r}{r!}. \tag{1.4–17}$$

There is no constant term in the expansion because $M(0) = 1$ and $K(0) = 0$. It can be shown that the series is either infinite or it terminates at the second term. This result is an immediate consequence of a theorem of Marcinkiewicz (see Lukacs, 1970, p. 213).

Cumulants have a number of interesting invariance properties, and they are particularly convenient when one is dealing with combinations of statistically independent variates. Let $x_1, x_2, \ldots, x_N$ be statistically independent, with

$$y = \sum_{i=1}^{N} c_i x_i, \tag{1.4–18}$$

where the coefficients c_i are constants. Let us first examine the relation between the moment generating functions $M_y(\xi)$ and $M_{x_i}(\xi)$ of y and x_i, respectively. By definition

$$M_y(\xi) = \langle e^{y\xi} \rangle = \left\langle \exp \sum_{i=1}^{N} c_i x_i \xi \right\rangle = \left\langle \prod_{i=1}^{N} e^{c_i x_i \xi} \right\rangle = \prod_{i=1}^{N} \langle e^{c_i x_i \xi} \rangle = \prod_{i=1}^{N} M_{x_i}(c_i \xi), \tag{1.4–19}$$

where the last step is a consequence of the assumed statistical independence of the x_i. It follows from Eq. (1.4–16) that the corresponding relation between cumulant generating functions is

$$K_y(\xi) = \sum_{i=1}^{N} K_{x_i}(c_i \xi), \tag{1.4–20}$$

and for the r'th cumulants

$$\kappa_r^{(y)} = \sum_{i=1}^{N} \kappa_r^{(x_i)} c_i^r. \tag{1.4–21}$$

In particular, if the constants c_i are all unity, the r'th cumulant of y is the sum of the r'th cumulants of the x_i's.

Cumulants behave particularly simply under translation of the random variable x. Consider the linear transformation from x to y,

$$y = \lambda(x - a), \quad (\lambda > 0) \tag{1.4–22}$$

that represents both a translation through a and a change of scale λ. From Eq. (1.3–11) the probability densities $p(x)$ of x and $P(y)$ of y are seen to be related by

$$P(y) = \frac{1}{\lambda} p\left(\frac{y}{\lambda} + a\right). \tag{1.4–23}$$

Hence, we have for the corresponding moment generating function

$$\begin{aligned} M_y(\xi) &= \int e^{y\xi} P(y)\,dy \\ &= \int e^{\lambda(x-a)\xi} p(x)\,dx \\ &= e^{-\lambda a \xi} M_x(\lambda\xi). \end{aligned} \tag{1.4–24}$$

We see that the effect of the translation on the moments is more complicated than the effect of the scale change. If $a = 0$, the moments are related very simply by

$$\langle y^r \rangle = \lambda^r \langle x^r \rangle, \tag{1.4–25}$$

but if $\lambda = 1$ and $a \neq 0$, there is no simple relation between $\langle y^r \rangle$ and $\langle x^r \rangle$. However, from Eq. (1.4–24), we have on taking logarithms,

$$K_y(\xi) = -\lambda a \xi + K_x(\lambda\xi), \tag{1.4–26}$$

so that the cumulants are connected by the relations

$$\left.\begin{aligned} \kappa_1^{(y)} &= \lambda \kappa_1^{(x)} - \lambda a \\ \kappa_r^{(y)} &= \lambda^r \kappa_r^{(x)}, \quad (r \geqslant 2). \end{aligned}\right\} \tag{1.4–27}$$

In particular, if there is no scale change, so that $\lambda = 1$, all the cumulants except the first remain completely unchanged. Cumulants (except for the first) are therefore invariant under translation.

In the special case in which $a = \langle x \rangle$ and $\lambda = 1$, the transformation (1.4–22) represents a translation by the mean, so that $y = \Delta x$ and the moments of y are the central moments of x. Then we have from Eq. (1.4–26)

$$K_{\Delta x}(\xi) = -\langle x \rangle \xi + K_x(\xi). \tag{1.4–28}$$

We can readily use the defining relation to establish a connection between the cumulants and the central moments. We start from the relation

$$K_{\Delta x}(\xi) = \ln[M_{\Delta x}(\xi)],$$

and differentiate both sides with respect to ξ. This gives

$$M_{\Delta x}(\xi) \frac{dK_{\Delta x}(\xi)}{d\xi} = \frac{dM_{\Delta x}(\xi)}{d\xi}.$$

On substituting the power series expansions (1.4–2) and (1.4–17) for $M_{\Delta x}(\xi)$ and $K_{\Delta x}(\xi)$, using Eq. (1.4–28) and recalling that the moments of Δx are the central moments $\mu_1, \mu_2, \ldots$, we obtain

$$\left[1 + \frac{\mu_2\xi^2}{2!} + \frac{\mu_3\xi^3}{3!} + \ldots\right]\left[-\langle x\rangle + \kappa_1 + \frac{\kappa_2\xi}{1!} + \frac{\kappa_3\xi^2}{2!} + \ldots\right] = \left[\mu_2\xi + \frac{\mu_3\xi^2}{2!} + \frac{\mu_4\xi^3}{3!} + \ldots\right]. \tag{1.4–29}$$

Equating coefficients of equal powers of ξ leads to the following relations between the cumulants of x and the central moments of x,

$$\left.\begin{aligned} \kappa_1 &= \langle x\rangle \\ \kappa_2 &= \mu_2 \\ \kappa_3 &= \mu_3 \\ \kappa_4 &= \mu_4 - 3\mu_2^2 \\ \kappa_5 &= \mu_5 - 10\mu_2\mu_3 \\ &\ldots\ldots\ldots \end{aligned}\right\} \tag{1.4–30}$$

The first cumulant is seen to be the mean, and all higher cumulants are independent of the choice of origin.

1.5 Some examples of probability distributions

1.5.1 Bernoulli or binomial distribution

Consider a sequence of N independent trials or observations, in which we focus on some particular feature (success) that has a probability β of showing up in any one trial. We wish to determine the probability $p_N(n)$ that there are n successes out of the N trials $(n \leqslant N)$. As all the trials are independent, the probability of encountering a particular sequence of n successes and $N - n$ failures is the product of the corresponding probabilities, which is $\beta^n(1-\beta)^{N-n}$, irrespective of the order in which the successes and failures occur. But there are $\binom{N}{n}$ different arrangements of the n successes and $N - n$ failures, each of which has the same probability. The total probability $p_N(n)$ is therefore the sum of the probabilities for these $\binom{N}{n}$ different arrangements, or

$$p_N(n) = \binom{N}{n}\beta^n(1-\beta)^{N-n}, \quad (0 \leqslant n \leqslant N). \tag{1.5–1}$$

This probability distribution is known as the *Bernoulli or binomial distribution*, the latter because $p_N(n)$ is a term in the binomial expansion of $[\beta + (1-\beta)]^N$.

We note that $p_N(n)$ contains two free parameters N, β, and that

$$\sum_{n=0}^{N} p_N(n) = [\beta + (1-\beta)]^N = 1,$$

as required. The form of the probability distribution is illustrated in Fig. 1.6 for the symmetric case $\beta = \frac{1}{2}$.

It is not difficult to calculate moments of n directly from Eq. (1.5–1), but we can determine the moment generating function equally easily. From Eq. (1.5–1) we have

$$\begin{aligned} M(\xi) &= \sum_{n=0}^{N} e^{\xi n} p_N(n) \\ &= \sum_{n=0}^{N} \binom{N}{n} (\beta e^{\xi})^n (1-\beta)^{N-n} \\ &= [1 + \beta(e^{\xi} - 1)]^N, \end{aligned} \tag{1.5–2}$$

from which the moments are calculated by making a power series expansion in ξ.

The factorial moments turn out to be much simpler than the ordinary moments, and we find for the generating function

$$\begin{aligned} F(\xi) &= \sum_{n=0}^{N} (1+\xi)^n p_N(n) \\ &= \sum_{n=0}^{N} \binom{N}{n} [\beta(1+\xi)]^n (1-\beta)^{N-n} \\ &= (1+\beta\xi)^N \\ &= \sum_{r=0}^{N} \binom{N}{r} \beta^r \xi^r. \end{aligned} \tag{1.5–3}$$

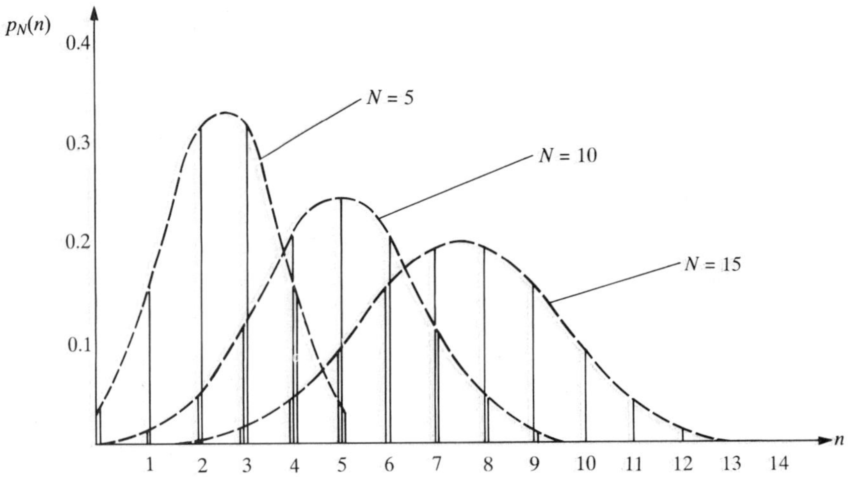

Fig. 1.6 Some forms of the Bernoulli distribution with $\beta = 1/2$.

The r'th factorial moment $\langle n^{(r)}\rangle$, which is the coefficient of $\xi^r/r!$, is therefore given by

$$\langle n^{(r)}\rangle = N(N-1)\ldots(N-r+1)\beta^r, \quad (r \leqslant N). \tag{1.5–4}$$

In particular, by choosing $r = 1, 2$, we obtain for the first two moments,

$$\left.\begin{aligned} \nu_1 &\equiv \langle n\rangle = N\beta \\ \nu_2 &\equiv \langle n^2\rangle = N\beta + N(N-1)\beta^2 \\ \mu_2 &= \nu_2 - \nu_1^2 = N\beta(1-\beta). \end{aligned}\right\} \tag{1.5–5}$$

The higher-order moments get progressively more complicated. The variance is greatest when $\beta = \frac{1}{2}$ for any given N, i.e. when the distribution is symmetric about $n = \frac{1}{2}N$. Although the width or standard deviation $\sigma = [N\beta(1-\beta)]^{1/2}$ increases with N, it should be noted that the relative width

$$\frac{\sigma}{\langle n\rangle} = \left(\frac{1-\beta}{N\beta}\right)^{1/2}$$

tends to zero as $N \to \infty$, so that in a relative sense the distribution becomes narrower with increasing N. The skewness α_3 and the kurtosis α_4 defined by Eqs. (1.3–27) and (1.3–28) are $(1-2\beta)/[N\beta(1-\beta)]^{1/2}$ and $3 - 6/N + 1/[N\beta(1-\beta)]$, respectively. The former becomes zero when $\beta = \frac{1}{2}$, and tends to zero as $N \to \infty$ irrespective of β, while the latter tends to 3 as $N \to \infty$.

1.5.2 *Poisson distribution*

When the number of trials N tends to infinity and the probability of success in any one trial β tends to zero in such a way that the expectation $\langle n\rangle = N\beta$ remains constant, the Bernoulli distribution simplifies. To see this we write

$$\begin{aligned} p_N(n) &= \frac{N(N-1)(N-2)\ldots(N-n+1)\beta^n(1-\beta)^{N-n}}{n!} \\ &= \frac{(1-1/N)(1-2/N)\ldots(1-(n-1)/N)(\beta N)^n(1-\beta)^{N-n}}{n!} \end{aligned}$$

and replace β by $\langle n\rangle/N$, so that

$$p_N(n) = \left[\prod_{r=1}^{n-1}\left(1-\frac{r}{N}\right)\right]\frac{\langle n\rangle^n}{n!}\frac{(1-\langle n\rangle/N)^N}{(1-\langle n\rangle/N)^n}.$$

As $N \to \infty$ while n remains finite, the product term tends to unity, as does the factor $(1-\langle n\rangle/N)^n$. However, the factor $(1-\langle n\rangle/N)^N$ tends to $e^{-\langle n\rangle}$, so that finally

$$\operatorname*{Lim}_{N\to\infty} p_N(n) \equiv p(n) = \frac{\langle n\rangle^n e^{-\langle n\rangle}}{n!}. \tag{1.5–6}$$

This is known as the *Poisson distribution* in n. It contains but a single parameter $\langle n\rangle$, and is useful for describing random events that occur at some known average rate. We shall see later in Chapter 9 that when light from a single-mode laser falls on a photodetector, photoelectric pulses are produced at random at an average rate proportional to the mean light intensity, and the number of pulses emitted within a given time interval therefore obeys a Poisson distribution.

Both the moment generating function and the factorial moment generating function are readily calculated from Eq. (1.5–6). Thus

$$\begin{aligned} M(\xi) &= \sum_{n=0}^{\infty} e^{\xi n} \frac{\langle n\rangle^n e^{-\langle n\rangle}}{n!} \\ &= \exp[\langle n\rangle(e^{\xi}-1)], \end{aligned} \tag{1.5–7}$$

and

$$\begin{aligned} F(\xi) &= \sum_{n=0}^{\infty} (1+\xi)^n \frac{\langle n\rangle^n e^{-\langle n\rangle}}{n!} \\ &= e^{\langle n\rangle \xi} \\ &= \sum_{r=0}^{n} \frac{\langle n\rangle^r \xi^r}{r!}. \end{aligned} \tag{1.5–8}$$

The factorial moments can be obtained by inspection of Eq. (1.5–8) and we have

$$\langle n^{(r)}\rangle = \langle n\rangle^r, \tag{1.5–9}$$

from which it follows that the variance

$$\mu_2 = \langle n^2\rangle - \langle n\rangle^2 = \langle n\rangle. \tag{1.5–10}$$

Like the Bernoulli distribution, the width of the Poisson distribution increases with $\langle n\rangle$, but the relative width becomes progressively smaller as $\langle n\rangle \to \infty$.

Finally, we calculate the cumulant generating function $K(\xi)$ from Eq. (1.5–7) and obtain

$$\begin{aligned} K(\xi) = \ln[M(\xi)] &= \langle n\rangle(e^{\xi}-1) \\ &= \langle n\rangle \sum_{r=1}^{\infty} \frac{\xi^r}{r!}. \end{aligned} \tag{1.5–11}$$

As the r'th cumulant κ_r is the coefficient of $\xi^r/r!$ in this expansion, we see immediately that

$$\kappa_r = \langle n\rangle, \quad r = 1, 2, \ldots \tag{1.5–12}$$

Having all cumulants equal to the mean is an important characteristic of a Poisson distribution. Finally, we note from the third and fourth moments that the skewness $\alpha_3 = \mu_3/\sigma^3 = 1/\sqrt{\langle n\rangle}$ is always positive because of the infinite tail of $p(n)$, and that the kurtosis is $\alpha_4 = \mu_4/\sigma^4 = 3 + 1/\langle n\rangle$, which tends to 3 as $\langle n\rangle \to \infty$. The limiting values of α_3 and α_4 as $\langle n\rangle \to \infty$ are characteristic of a Gaussian, as we show below.

One sometimes encounters situations in which a random variable n is the sum of N different, statistically independent Poisson variates $n_1, n_2, \ldots, n_N$. Then the moment generating function of n is given by

$$\begin{aligned}
M_n(\xi) &= \langle e^{(n_1+n_2+\cdots+n_N)\xi} \rangle \\
&= \prod_{i=1}^{N} \langle e^{n_i \xi} \rangle \\
&= \prod_{i=1}^{N} \exp\left[\langle n_i \rangle (e^{\xi} - 1)\right] \\
&= \exp\left[\sum_{i=1}^{N} \langle n_i \rangle (e^{\xi} - 1)\right].
\end{aligned} \tag{1.5–13}$$

This describes another Poisson distribution, whose parameter $\langle n \rangle$ is the sum of the parameters of the component distributions. Hence n is also a Poisson variate and the sum of independent Poisson variates is also a Poisson variate.

1.5.3 Bose–Einstein distribution

When the outcomes that enter into the Bernoulli distribution above are intrinsically indistinguishable, the combinatorial factor $\binom{N}{n}$ in Eq. (1.5–1) does not appear, and we have the simpler probability distribution

$$p_N(n) = K_N \beta^n (1 - \beta)^{N-n}, \quad 0 \leqslant n \leqslant N, \tag{1.5–14}$$

in which K_N is a normalization constant. Let us denote $\beta/(1 - \beta)$ by η and let $N \to \infty$ with the understanding that $\eta < 1$. Then

$$p_N(n) \to p(n) = K\eta^n, \quad 0 \leqslant n,$$

where K is another normalization constant. Summation over n shows immediately that $K = (1 - \eta)$, so that

$$p(n) = (1 - \eta)\eta^n. \tag{1.5–15}$$

This is known as the *Bose–Einstein distribution*. It describes, for example, the probability distribution of photons in one cell of phase space when the optical field is in thermal equilibrium.

The moment generating function is given by

$$\begin{aligned}
M(\xi) &= \sum_{n=0}^{\infty} (1 - \eta)\, e^{\xi n}\, \eta^n \\
&= \frac{1 - \eta}{1 - \eta e^{\xi}},
\end{aligned} \tag{1.5–16}$$

while the factorial moment generating function is

$$\begin{aligned}
F(\xi) &= \sum_{n=0}^{\infty} (1 + \xi)^n (1 - \eta)\eta^n \\
&= \frac{1 - \eta}{1 - \eta(1 + \xi)} \\
&= \frac{1}{1 - \eta\xi/(1 - \eta)}.
\end{aligned} \tag{1.5–17}$$

A power series expansion in ξ immediately leads to the first two moments:

$$\left.\begin{aligned}\langle n\rangle &= \frac{\eta}{1-\eta},\\ \langle n(n-1)\rangle &= 2\left(\frac{\eta}{1-\eta}\right)^2,\\ \langle(\Delta n)^2\rangle &= \langle n\rangle(1+\langle n\rangle).\end{aligned}\right\} \tag{1.5–18}$$

Equations (1.5–18) show that the fluctuations are greater than for a Poisson distribution with the same mean, which may be regarded as a consequence of the intrinsic indistinguishability of the particles being counted. Equation (1.5–18) enables us to re-express $p(n)$ in terms of $\langle n\rangle$, by substituting $\eta = (1 + 1/\langle n\rangle)^{-1}$, in the form

$$p(n) = \frac{1}{(1+\langle n\rangle)(1+1/\langle n\rangle)^n}. \tag{1.5–19}$$

1.5.4 *The weak law of large numbers*

Let us return to the Bernoulli distribution (1.5–1), for which the probability of success in any one trial is β, and use it to show that the ratio n/N of successes to trials tends towards β with probability unity when the number of trials N tends to infinity. Let ε be some arbitrarily small number. Then the probability that $|n/N - \beta|$ exceeds ε is given by

$$P\left(\left|\frac{n}{N} - \beta\right| \geqslant \varepsilon\right) = P(|n - N\beta| \geqslant N\varepsilon).$$

We note that $n - N\beta = \Delta n$ is the deviation of n from its expectation value. We now introduce a parameter θ such that $N\varepsilon = \theta\sigma$, where $\sigma = [N\beta(1-\beta)]^{1/2}$ is the standard deviation. Then $\theta = N^{1/2}\varepsilon/[\beta(1-\beta)]^{1/2}$,

$$P\left(\left|\frac{n}{N} - \beta\right| \geqslant \varepsilon\right) = P(|\Delta n| \geqslant \theta\sigma),$$

and by the Bienaymé–Chebyshev inequality (1.3–43),

$$P\left(\left|\frac{n}{N} - \beta\right| \geqslant \varepsilon\right) \leqslant \frac{1}{\theta^2} = \frac{\beta(1-\beta)}{N\varepsilon^2} \leqslant \frac{1}{4N\varepsilon^2}, \tag{1.5–20}$$

since $\beta(1-\beta) \leqslant \frac{1}{4}$. It follows that for any ε, no matter how small, $P[|(n/N) - \beta| \geqslant \varepsilon]$ can be made as small as we wish by choosing N to be sufficiently large, i.e.

$$\lim_{N\to\infty} P\left(\left|\frac{n}{N} - \beta\right| \geqslant \varepsilon\right) = 0, \tag{1.5–21a}$$

or

$$\lim_{N\to\infty} P\left(\left|\frac{n}{N} - \beta\right| < \varepsilon\right) = 1. \tag{1.5–21b}$$

Hence, as the number of trials tends to infinity, n/N tends to the constant β with probability unity. This result is known as the *weak law of large numbers*, and it provides the justification for a determination of the probability of success from a large number of independent trials. There is a stronger version of the law of large numbers that involves the higher moments, but we shall not consider it here.

1.5.5 Normal or Gaussian distribution

Let x be a continuous random variable defined on the infinite interval from $-\infty$ to ∞. The variable x is known as a *Gaussian random variable* if its probability density $p(x)$ is of the form

$$p(x) = \frac{1}{(2\pi)^{1/2}\sigma} \mathrm{e}^{-(x-\langle x\rangle)^2/2\sigma^2}. \tag{1.5–22}$$

The probability distribution $p(x)$ is illustrated in Fig. 1.7. It has two free parameters, which are the mean $\langle x\rangle$ and the standard deviation σ, and it is normalized so that

$$\int_{-\infty}^{\infty} p(x)\,\mathrm{d}x = 1.$$

The *normal* or *Gaussian* probability distribution (1.5–22) plays a central role in probability theory for several reasons. It has an especially simple structure, it is the limiting form of several other probability distributions, and because of the central limit theorem (see Section 1.5.6 below), it is a probability distribution that is encountered under a great variety of different conditions. We shall therefore examine the properties of the Gaussian distribution at some length.

The moment generating function $M(\xi)$ can readily be calculated by completing

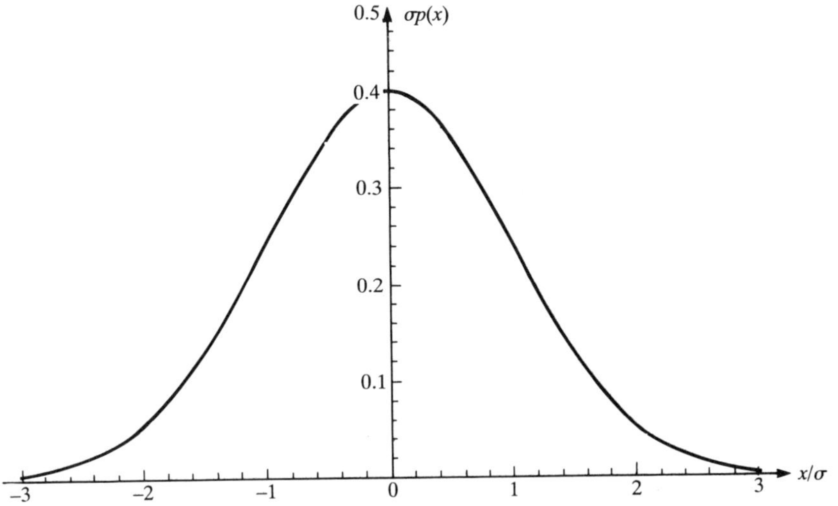

Fig. 1.7 The Gaussian probability distribution with mean $\langle x\rangle = 0$.

the square in the exponent. We obtain

$$\begin{aligned} M(\xi) &= \frac{1}{\sqrt{(2\pi)}\sigma}\int_{-\infty}^{\infty} e^{x\xi}\, e^{-(x-\langle x\rangle)^2/2\sigma^2}\, dx \\ &= \frac{1}{\sqrt{(2\pi)}\sigma} e^{\xi\langle x\rangle}\, e^{\sigma^2\xi^2/2}\int_{-\infty}^{\infty} e^{-[x-(\sigma^2\xi+\langle x\rangle)]^2/2\sigma^2}\, dx \\ &= e^{\xi\langle x\rangle}\, e^{\sigma^2\xi^2/2}. \end{aligned} \tag{1.5–23}$$

The characteristic function $C(\xi)$ follows immediately if we replace ξ by $i\xi$ in Eq. (1.5–23). Alternatively we may proceed from first principles by taking the Fourier transform of $p(x)$ and obtain,

$$C(\xi) = e^{i\xi\langle x\rangle}\, e^{-\sigma^2\xi^2/2}. \tag{1.5–24}$$

From the transformation law (1.4–24), the moment generating function and the characteristic function of Δx are

$$\left.\begin{aligned} M_{\Delta x}(\xi) &= e^{\sigma^2\xi^2/2} \\ C_{\Delta x}(\xi) &= e^{-\sigma^2\xi^2/2}. \end{aligned}\right\} \tag{1.5–25}$$

The central moments μ_r then follow immediately by expansion of $M_{\Delta x}(\xi)$ in a power series in ξ, because μ_r is just the coefficient of $\xi^r/r!$. We obtain the following expressions for the central moments

$$\mu_r = 0, \quad \text{if r is odd;} \tag{1.5–26a}$$

$$\begin{aligned} \mu_r &= \frac{(\tfrac{1}{2}\sigma^2)^{r/2} r!}{(\tfrac{1}{2}r)!} \\ &= \frac{\sigma^r r!}{r(r-2)(r-4)\ldots 4\cdot 2} \\ &= \sigma^r (r-1)(r-3)\ldots 5\cdot 3\cdot 1 \\ &= (r-1)!!\sigma^r, \quad \text{if } r \text{ is even,} \end{aligned} \tag{1.5–26b}$$

where $(r-1)!! \equiv (r-1)(r-3)\ \ldots\ 5\cdot 3\cdot 1$. The variance is σ^2, the skewness $\alpha_3 = \mu_3/\sigma^3 = 0$, and the kurtosis $\alpha_4 = \mu_4/\sigma^4 = 3$.

The cumulant generating function $K(\xi)$ is especially simple, and we find at once from Eq. (1.5–23) that

$$K(\xi) = \ln[M(\xi)] = \xi\langle x\rangle + \tfrac{1}{2}\sigma^2\xi^2. \tag{1.5–27}$$

There are therefore only two non-vanishing cumulants, the mean $\kappa_1 = \langle x\rangle$ and the variance $\kappa_2 = \sigma^2$. All higher cumulants are zero. The converse is also true: a probability density whose cumulants beyond the second order vanish is Gaussian. We shall not prove the result here, but it can be shown that there is no other probability distribution whose cumulant generating function is a polynomial in ξ. This theorem is essentially Marcinkiewicz' theorem, which we have already encountered (Section 1.4.3).

When the Gaussian distribution is put in standard form, with zero mean and unit variance, it simplifies to

$$p(x) = \frac{1}{\sqrt{(2\pi)}} e^{-x^2/2}, \tag{1.5–28}$$

and

$$\left.\begin{aligned} M(\xi) &= e^{\xi^2/2} \\ C(\xi) &= e^{-\xi^2/2} \\ K(\xi) &= \tfrac{1}{2}\xi^2. \end{aligned}\right\} \tag{1.5–29}$$

Many other probability distributions tend to the Gaussian form in a certain limit. As an example, we will show that the Poisson distribution tends to become Gaussian as $\langle n \rangle \to \infty$. From Eq. (1.5–11) the cumulant generating function of the Poisson variate n is given by

$$K_n(\xi) = \langle n \rangle (e^{\xi} - 1).$$

We now transform n to standard form by introducing the transformed variate

$$x \equiv \frac{n - \langle n \rangle}{\sigma} = \frac{n - \langle n \rangle}{\sqrt{\langle n \rangle}},$$

which may be regarded as effectively continuous for large $\langle n \rangle$. From the transformation law (1.4–26), the cumulant generating function of x is given by

$$\begin{aligned} K_x(\xi) &= -\sqrt{\langle n \rangle}\,\xi + K_n\left(\frac{\xi}{\sqrt{\langle n \rangle}}\right) \\ &= -\sqrt{\langle n \rangle}\,\xi + \langle n \rangle \left[\frac{\xi}{\sqrt{\langle n \rangle}} + \frac{1}{2!}\frac{\xi^2}{\langle n \rangle} + \frac{1}{3!}\frac{\xi^3}{\langle n \rangle^{3/2}} + \dots\right] \\ &= \tfrac{1}{2}\xi^2 + \frac{1}{\sqrt{\langle n \rangle}}\left[\frac{\xi^3}{3!} + \frac{\xi^4}{4!\langle n \rangle^{1/2}} + \dots\right]. \end{aligned} \tag{1.5–30}$$

For any given ξ, this expression tends to $\frac{1}{2}\xi^2$ as $\langle n \rangle \to \infty$, which is, by Eqs. (1.5–29), the cumulant generating function of a Gaussian variate in standard form.

A complex variate $z = x + iy$ is said to be Gaussian (for simplicity of zero mean) if x and y are both Gaussian variates of zero mean with the same standard deviation σ, and if they are statistically independent. Then the joint probability density $p(x, y)$ has the form

$$p(x, y) = \frac{1}{2\pi\sigma^2} e^{-(x^2+y^2)/2\sigma^2},$$

which can be written

$$p(z) = \frac{1}{2\pi\sigma^2} e^{-|z|^2/2\sigma^2} = \frac{1}{\pi\tilde{\sigma}^2} e^{-|z|^2/\tilde{\sigma}^2}, \tag{1.5–31}$$

where $\tilde{\sigma}^2 \equiv \langle |z|^2 \rangle = 2\sigma^2$ is the variance of the complex variate z. When z is not of zero mean, z has to be replaced by Δz in Eq. (1.5–31).

1.5.6 *The central limit theorem*

In many practical situations in which a fluctuating random variable y is being investigated, the fluctuations contain contributions from many independent causes. The central limit theorem asserts that, under rather general conditions, y will tend to become a Gaussian variate, as the number of causes becomes large.

Let $x_1, x_2, \ldots, x_N$ be a set of N statistically independent variates, all of which are in standard form, but with arbitrary probability densities. Then

$$y = \frac{1}{\sqrt{N}}(x_1 + x_2 + \ldots + x_N) \tag{1.5–32}$$

is a new variate in standard form, because its mean is also zero, and its variance

$$\langle y^2 \rangle = \frac{1}{N}\sum_{i=1}^{N} \langle x_i^2 \rangle = 1.$$

Since each x_i is in standard form, the corresponding cumulant generating function is

$$K_{x_i}(\xi) = \tfrac{1}{2}\xi^2 + \sum_{r=3}^{\infty} \kappa_r^{(x_i)} \frac{\xi^r}{r!} = \tfrac{1}{2}\xi^2 + O(\xi^3), \tag{1.5–33}$$

where O is the usual order symbol. By the composition rule (1.4–20), the cumulant generating function of y is given by

$$\begin{aligned} K_y(\xi) &= \sum_{i=1}^{N} K_{x_i}(\xi/\sqrt{N}) \\ &= \sum_{i=1}^{N} \left[\frac{1}{2}\frac{\xi^2}{N} + O\left(\frac{\xi^3}{N^{3/2}}\right) \right] \\ &= \frac{1}{2}\xi^2 + \sum_{i=1}^{N} O\left(\frac{\xi^3}{N^{3/2}}\right). \end{aligned} \tag{1.5–34}$$

As $N \to \infty$ the sum tends to zero because of the factor $N^{3/2}$ in the denominator, so that $K_y(\xi) \to \frac{1}{2}\xi^2$. Hence, y becomes Gaussian irrespective of the forms of the probability distributions of the x's. The central limit theorem is very powerful, and it holds under more general conditions than we have considered here.

1.5.7 *Gamma distribution*

The gamma distribution is another continuous one-parameter distribution for a random variable x defined on the interval $0 \leqslant x < \infty$. If the probability density $p(x)$ is given by

$$p(x) = \frac{x^{n-1}\,\mathrm{e}^{-x}}{\Gamma(n)}, \quad (n > 0), \tag{1.5–35}$$

then x is known as a *γ-variate of parameter n* (n need not be an integer), or as a $\gamma(n)$-variate. If n is an integer, then a $\gamma(n+1)$-variate has the probability distribution

$$p(x) = \frac{x^n \, \mathrm{e}^{-x}}{n!}. \tag{1.5–36}$$

This should not be confused with the Poisson distribution, which looks similar but is a probability distribution of n rather than of x. Some examples of the γ-distribution are shown in Fig. 1.8.

The moment generating function exists for $\xi < 1$, and we have

$$\begin{aligned} M(\xi) &= \frac{1}{\Gamma(n)} \int_0^\infty \mathrm{e}^{\xi x} x^{n-1} \, \mathrm{e}^{-x} \, \mathrm{d}x \\ &= \frac{1}{(1-\xi)^n} \\ &= \sum_{r=0}^{\infty} \binom{-n}{r} (-\xi)^r. \end{aligned} \tag{1.5–37}$$

The r'th moment is the coefficient of $\xi^r/r!$, and we find that

$$\nu_r \equiv \langle x^r \rangle = n(n+1)(n+2)\ldots(n+\mathrm{r}-1), \tag{1.5–38}$$

so that the mean is the parameter n,

$$\langle x \rangle = n, \tag{1.5–39}$$

and the variance is

$$\mu_2 = \langle x^2 \rangle - \langle x \rangle^2 = n \tag{1.5–40}$$

also. Although the dispersion increases with n, the relative width $\sigma/\langle x \rangle = 1/\sqrt{n}$ decreases with n. From Eq. (1.5–37) the cumulant generating function is

$$\begin{aligned} K(\xi) &= -n \ln(1-\xi) \\ &= n \sum_{r=1}^{\infty} \frac{\xi^r}{r}, \end{aligned} \tag{1.5–41}$$

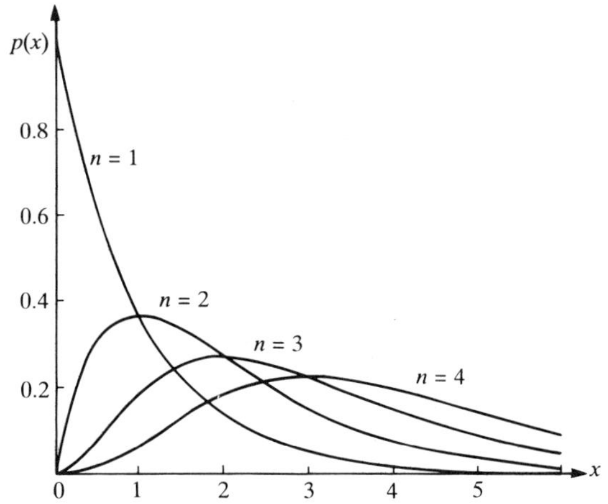

Fig. 1.8 Some examples of the γ-distribution for integer values of the parameter n.

which allows us to determine the cumulants by inspection:

$$\kappa_r = n(r-1)!. \tag{1.5–42}$$

With the help of Eq. (1.4–30) we find that the skewness is always positive,

$$\alpha_3 = \frac{\mu_3}{\sigma^3} = \frac{\kappa_3}{\sigma^3} = \frac{2}{\sqrt{n}}, \tag{1.5–43}$$

although it tends to zero as $n \to \infty$, while the kurtosis is

$$\alpha_4 = \frac{\mu_4}{\sigma^4} = \frac{(\kappa_4 + 3\kappa_2^2)}{\sigma^4} = 3 + \frac{6}{n}, \tag{1.5–44}$$

and tends to 3. This suggests that the γ-distribution tends to the Gaussian form as $n \to \infty$, as can be shown explicitly. If we transform x to standard form by putting

$$y = \frac{x-n}{\sqrt{n}},$$

we find immediately with the help of the transformation rule (1.4–26) that

$$\begin{aligned} K_y(\xi) &= \sqrt{(n)}\xi - n\ln\left(1 - \frac{\xi}{\sqrt{n}}\right) \\ &= \frac{\xi^2}{2} + \frac{1}{\sqrt{n}}\left(\frac{\xi^3}{3} + \frac{\xi^4}{4\sqrt{n}} + \ldots\right), \end{aligned} \tag{1.5–45}$$

and this tends to $\frac{1}{2}\xi^2$ as $n \to \infty$, which is characteristic of the cumulant generating function of a Gaussian variate in standard form.

Independent γ-variates can combine to form other γ-variates. To show this we consider a set of N independent γ-variates $x_1, x_2, \ldots, x_N$ with parameters $n_1, n_2, \ldots, n_N$, respectively, and let

$$y = x_1 + x_2 + \ldots + x_N.$$

From the general composition rule (1.4–20) it follows that the cumulant generating function of y is

$$\begin{aligned} K_y &= \sum_{i=1}^{N} K_{x_i}(\xi) \\ &= -\ln(1-\xi)\sum_{i=1}^{N} n_i \\ &= -n\ln(1-\xi), \end{aligned} \tag{1.5–46}$$

where $n \equiv \sum_{i=1}^{n} n_i$. Hence, y is a new γ-variate whose parameter is the sum of the parameters of the original γ-variates. This property is similar to that encountered earlier for Poisson variates and is sometimes known as the *reproducing property*.

Finally, we note that if x is a Gaussian variate with mean $\langle x\rangle$ and standard deviation σ, then

$$y = \frac{(x - \langle x\rangle)^2}{2\sigma^2} \tag{1.5–47}$$

is a gamma half-variate. To show this we use the general transformation rule (1.3–13), and we find for the probability density of y

$$\begin{aligned} P(y) &= \frac{1}{\sqrt{(2\pi)}\sigma} \mathrm{e}^{-y} \frac{\sigma}{\sqrt{(2y)}} + \frac{1}{\sqrt{(2\pi)}\sigma} \mathrm{e}^{-y} \frac{\sigma}{\sqrt{(2y)}} \\ &= \frac{1}{\sqrt{\pi}} y^{-1/2} \mathrm{e}^{-y} \\ &= \frac{1}{\Gamma(\frac{1}{2})} y^{-1/2} \mathrm{e}^{-y}, \end{aligned} \tag{1.5–48}$$

which is a γ-variate of the general form (1.5–35) with $n = \frac{1}{2}$. In the light of the composition law embodied in Eq. (1.5–46), it follows that if $x_1, x_2, \ldots, x_N$ are independent Gaussian variates, then

$$\sum_{i=1}^{N} \frac{(x_i - \langle x_i \rangle)^2}{2\sigma_i^2}$$

is a γ-variate of parameter $\frac{1}{2}N$.

1.6 Multivariate Gaussian distribution

Gaussian variates are often encountered in groups and they have to be treated jointly. Let us consider a number of Gaussian variates $x_1, x_2, \ldots, x_N$ with standard deviations $\sigma_1, \sigma_2, \ldots, \sigma_N$, respectively. Then each variate has a probability distribution of the form (1.5–22). If all the variates are statistically independent, we can immediately write down the joint probability distribution $p(x_1, x_2, \ldots, x_N)$, which takes the form of a product:

$$p(x_1, x_2, \ldots, x_N) = \frac{1}{(2\pi)^{N/2} \sigma_1 \sigma_2 \ldots \sigma_N} \exp\left[-\tfrac{1}{2} \sum_{i=1}^{N} \frac{(\Delta x_i)^2}{\sigma_i^2} \right]. \tag{1.6–1}$$

The x's are jointly Gaussian, but they are independent, so that this formula certainly does not describe the most general multivariate Gaussian distribution. The covariance matrix $\boldsymbol{\mu}$ of $x_1, x_2, \ldots x_N$ is diagonal, $\mu_{ij} = \delta_{ij} \sigma_i^2$, and its determinant is

$$\det \boldsymbol{\mu} = \sigma_1^2 \sigma_2^2 \sigma_3^3 \ldots \sigma_N^2. \tag{1.6–2}$$

The exponent in Eq. (1.6–1) can be written in matrix form. Let $\mathbf{x}$ be the column matrix

$$\mathbf{x} \equiv \begin{bmatrix} x_1 \\ x_2 \\ x_3 \\ \vdots \\ x_N \end{bmatrix}$$

and $\mathbf{x}^\dagger$ its adjoint row matrix. Similarly, let $\boldsymbol{\Delta}\mathbf{x}$ be the column matrix with elements Δx_i. Then we can express the $N \times N$ covariance matrix as

$$\boldsymbol{\mu} = \langle \boldsymbol{\Delta}\mathbf{x} \boldsymbol{\Delta}\mathbf{x}^\dagger \rangle, \tag{1.6–3}$$

and we can write the sum in Eq. (1.6–1) in the form

$$\sum_{i=1}^{N} \frac{(\Delta x_i)^2}{\sigma_i^2} = \mathbf{\Delta x}^\dagger \boldsymbol{\mu}^{-1} \mathbf{\Delta x}, \tag{1.6–4}$$

where $\boldsymbol{\mu}^{-1}$ is the reciprocal covariance matrix

$$\mu_{ij}^{-1} = \delta_{ij} \frac{1}{\sigma_i^2}.$$

This allows us to express Eq. (1.6–1) in the more compact form

$$p(x_1, x_2, \ldots x_N) = \frac{1}{(2\pi)^{N/2} (\det \boldsymbol{\mu})^{1/2}} \exp\left(-\tfrac{1}{2} \mathbf{\Delta x}^\dagger \boldsymbol{\mu}^{-1} \mathbf{\Delta x}\right). \tag{1.6–5}$$

We shall refer to this probability distribution as the general multivariate Gaussian distribution, irrespective of whether the covariance matrix $\boldsymbol{\mu}$ is diagonal or not.

We now consider a homogeneous linear transformation from $x_1, x_2, \ldots, x_N$ to $y_1, y_2, \ldots, y_N$, and show that this preserves the structure of Eq. (1.6–5), although the covariance matrix will no longer be diagonal in general. If $\mathbf{y}$ is the column matrix with elements $y_1, y_2, \ldots, y_N$, then it can always be expressed in the form

$$\mathbf{y} = \mathbf{U}\mathbf{x}, \tag{1.6–6}$$

where $\mathbf{U}$ is the $N \times N$ transformation matrix. The same transformation connects $\mathbf{\Delta y}$ and $\mathbf{\Delta x}$. For simplicity we shall take $\mathbf{U}$ to be orthogonal (i.e. unitary with real elements). Then the Jacobian of the transformation is unity, so that $\mathbf{U}\mathbf{U}^\dagger = \mathbf{1} = \mathbf{U}^\dagger\mathbf{U}$, and the joint probability density of $y_1, y_2, \ldots, y_N$ is (cf. Eq. (1.3–17))

$$P(y_1, y_2, \ldots, y_N) = p(x_1, x_2, \ldots, x_N) = \frac{1}{(2\pi)^{N/2} (\det \boldsymbol{\mu})^{1/2}} \exp\left(-\tfrac{1}{2} \mathbf{\Delta x}^\dagger \boldsymbol{\mu}^{-1} \mathbf{\Delta x}\right). \tag{1.6–7}$$

We now express $\mathbf{x}$ in terms of $\mathbf{y}$. By inversion of Eq. (1.6–6) we have

$$\mathbf{x} = \mathbf{U}^\dagger \mathbf{y},$$

so that

$$\mathbf{\Delta x}^\dagger \boldsymbol{\mu}^{-1} \mathbf{\Delta x} = \mathbf{\Delta y}^\dagger \mathbf{U} \boldsymbol{\mu}^{-1} \mathbf{U}^\dagger \mathbf{\Delta y}. \tag{1.6–8}$$

Also the covariance matrix $\widetilde{\boldsymbol{\mu}}$ of y is given by

$$\widetilde{\boldsymbol{\mu}} = \langle \mathbf{\Delta y} \mathbf{\Delta y}^\dagger \rangle = \mathbf{U} \langle \mathbf{\Delta x} \mathbf{\Delta x}^\dagger \rangle \mathbf{U}^\dagger = \mathbf{U} \boldsymbol{\mu} \mathbf{U}^\dagger, \tag{1.6–9}$$

and it is positive definite. In general it is no longer diagonal, and its reciprocal is

$$\widetilde{\boldsymbol{\mu}}^{-1} = \mathbf{U} \boldsymbol{\mu}^{-1} \mathbf{U}^\dagger, \tag{1.6–10}$$

as can be verified by direct multiplication. Hence, from Eqs. (1.6–8) and (1.6–10)

$$\mathbf{\Delta x}^\dagger \boldsymbol{\mu}^{-1} \mathbf{\Delta x} = \mathbf{\Delta y}^\dagger \widetilde{\boldsymbol{\mu}}^{-1} \mathbf{\Delta y}. \tag{1.6–11}$$

As the determinant of the Hermitian matrix $\boldsymbol{\mu}$ does not change under unitary transformation, $\det \boldsymbol{\mu} = \det \tilde{\boldsymbol{\mu}}$ and Eq. (1.6–7) becomes

$$P(y_1, y_2, \ldots, y_N) = \frac{1}{(2\pi)^{N/2}(\det \tilde{\boldsymbol{\mu}})^{1/2}} \exp(-\tfrac{1}{2}\boldsymbol{\Delta y}^\dagger \tilde{\boldsymbol{\mu}}^{-1} \boldsymbol{\Delta y}). \quad (1.6\text{–}12)$$

This has exactly the same structure as Eq. (1.6–5), except that $\tilde{\boldsymbol{\mu}}$, unlike $\boldsymbol{\mu}$, is not necessarily diagonal. The exponent is therefore not as simple as in Eq. (1.6–1), and instead of being quadratic it has the bilinear form

$$\tfrac{1}{2}\boldsymbol{\Delta y}^\dagger \tilde{\boldsymbol{\mu}}^{-1} \boldsymbol{\Delta y} = \tfrac{1}{2}\sum_{i=1}^{N}\sum_{j=1}^{N} \Delta y_i \tilde{\mu}_{ij}^{-1} \Delta y_j, \quad (1.6\text{–}13)$$

which is characteristic of the general multivariate Gaussian distribution.

If we integrate $P(y_1, y_2, \ldots, y_N)$ over $N - 1$ variates, say $y_2, y_3, \ldots y_N$, we obtain

$$P(y_1) = \int P(y_1, y_2, \ldots, y_N)\, \mathrm{d}y_2\, \mathrm{d}y_3 \ldots \mathrm{d}y_N = \frac{1}{\sqrt{(2\pi)}\sigma_{y_1}} \exp[-(\Delta y_1)^2/2\sigma_{y_1}^2], \quad (1.6\text{–}14)$$

which is the usual form of a Gaussian probability density in one variable. It is apparent from Eq. (1.6–12) that all higher-order moments and correlations of $y_1, y_2, \ldots, y_N$ must be expressible in terms of the second-order moments μ_{ij}, and that Gaussian variates remain Gaussian under linear transformation.

Let us apply these results to the special case of a bivariate Gaussian distribution $p(x_1, x_2)$, in which the covariance matrix $\boldsymbol{\mu}$ is not necessarily diagonal. With the help of the correlation coefficient $\rho_{12} = \mu_{12}/\sigma_1\sigma_2$ we can express the covariance matrix $\boldsymbol{\mu}$ in the form

$$\boldsymbol{\mu} = \begin{bmatrix} \sigma_1^2 & \rho_{12}\sigma_1\sigma_2 \\ \rho_{12}\sigma_1\sigma_2 & \sigma_2^2 \end{bmatrix}, \quad (1.6\text{–}15)$$

while its reciprocal is

$$\boldsymbol{\mu}^{-1} = \frac{1}{(1 - \rho_{12}^2)} \begin{bmatrix} 1/\sigma_1^2 & -\rho_{12}/\sigma_1\sigma_2 \\ -\rho_{12}/\sigma_1\sigma_2 & 1/\sigma_2^2 \end{bmatrix}. \quad (1.6\text{–}16)$$

Then

$$\det \boldsymbol{\mu} = \sigma_1^2\sigma_2^2(1 - \rho_{12}^2),$$

and we obtain for the bivariate Gaussian probability distribution the following form

$$p(x_1, x_2) = \frac{1}{2\pi\sigma_1\sigma_2(1 - \rho_{12}^2)^{1/2}} \exp\left[-\frac{1}{2(1 - \rho_{12}^2)}\left(\frac{\Delta x_1^2}{\sigma_1^2} - \frac{2\rho_{12}\Delta x_1 \Delta x_2}{\sigma_1\sigma_2} + \frac{\Delta x_2^2}{\sigma_2^2}\right)\right]. \quad (1.6\text{–}17)$$

It is apparent from this equation that, whenever two Gaussian variates are uncorrelated so that $\rho_{12} = 0$, the joint probability density factorizes into a product of the probability densities for the two separate variates. In other words, if two Gaussian variates are uncorrelated they are also statistically independent. This is not true of random variables in general.

1.6.1 The Gaussian moment theorem

Gaussian variates have the remarkable property that all higher-order correlations among them are expressible in terms of second-order correlations between pairs of variates. Let $x_1, x_2, \ldots$ be a set of Gaussian variates. Then for any set of n indices $i_1, i_2, \ldots, i_n$,

$$\left.\begin{aligned} \langle \Delta x_{i_1} \Delta x_{i_2} \ldots \Delta x_{i_n} \rangle &= 0, && \text{if } n \text{ is odd} \\ &= \sum_{\substack{\text{all}(n-1)!! \\ \text{pairings}}} \langle \Delta x_{i_1} \Delta x_{i_2} \rangle \langle \Delta x_{i_3} \Delta x_{i_4} \rangle \ldots \langle \Delta x_{i_{n-1}} \Delta x_{i_n} \rangle && \text{if } n \text{ is even.} \end{aligned}\right\} \tag{1.6–18}$$

This result is known as the *Gaussian moment theorem* (see for example, Mehta, 1965, Appendix A1).

To prove the theorem we start from a set of independent Gaussian variates y_{i_1}, $y_{i_2}, \ldots y_{i_n}$, and we note that

$$\langle \Delta y_{i_1} \Delta y_{i_2} \ldots \Delta y_{i_n} \rangle$$

vanishes if all the suffices are different, and more generally unless each suffix appears an even number of times. Then the number of factors has to be even, and the correlation is of the form

$$\begin{aligned} \langle (\Delta y_{i_1})^{2n_{i_1}} (\Delta y_{i_2})^{2n_{i_2}} \ldots \rangle &= \langle (\Delta y_{i_1})^{2n_{i_1}} \rangle \langle (\Delta y_{i_2})^{2n_{i_2}} \rangle \ldots \\ &= (2n_{i_1} - 1)!! \langle (\Delta y_{i_1})^2 \rangle^{n_{i_1}} (2n_{i_2} - 1)!! \langle (\Delta y_{i_2})^2 \rangle^{n_{i_2}} \ldots \end{aligned} \tag{1.6–19}$$

with the help of Eq. (1.5–26). But this is just the sum over all non-vanishing pairings of the original correlation, so that we can write formally for the correlation of an even number of independent Gaussian variates

$$\langle \Delta y_{i_1} \Delta y_{i_2} \ldots \Delta y_{i_n} \rangle = \sum_{\substack{\text{all} \\ \text{pairings}}} \langle \Delta y_{i_1} \Delta y_{i_2} \rangle \ldots \langle \Delta y_{i_{n-1}} \Delta y_{i_n} \rangle. \tag{1.6–20}$$

We now make a unitary transformation $\mathbf{U}$ from the y's to a set of variates $x_1, x_2, \ldots, x_n$, which are no longer independent in general,

$$\left.\begin{aligned} x_i &= U_{ij} y_j \\ y_i &= U^{\dagger}_{ij} x_j. \end{aligned}\right\} \tag{1.6–21}$$

Then

$$\langle \Delta x_{i_1} \Delta x_{i_2} \ldots \Delta x_{i_n} \rangle = U_{i_1 j_1} \ldots U_{i_n j_n} \langle \Delta y_{j_1} \Delta y_{j_2} \ldots \Delta y_{j_n} \rangle,$$

and application of Eq. (1.6–20) leads immediately to the general moment theorem (1.6–18).

We now consider two useful applications. If all the x_i in Eq. (1.6–18) are equal, we obtain immediately

$$\langle (\Delta x)^{2n} \rangle = \sum_{\substack{\text{all}(2n-1)!! \\ \text{pairings}}} \langle (\Delta x)^2 \rangle^n = (2n - 1)!! \sigma^{2n}, \tag{1.6–22}$$

which simply recovers the results given by Eq. (1.5–26b). For a less trivial application we consider two correlated Gaussian variates x, y. Then

$$\begin{aligned}\langle(\Delta x)^2(\Delta y)^2\rangle &= \langle(\Delta x)^2\rangle\langle(\Delta y)^2\rangle + \langle\Delta x\Delta y\rangle\langle\Delta x\Delta y\rangle + \langle\Delta x\Delta y\rangle\langle\Delta x\Delta y\rangle \\ &= \sigma_x^2\sigma_y^2(1 + 2\rho_{xy}^2), \qquad (1.6\text{–}23)\end{aligned}$$

so that $\langle(\Delta x)^2(\Delta y)^2\rangle$ always exceeds $\langle(\Delta x)^2\rangle\langle(\Delta y)^2\rangle$. We shall see later (in Chapter 8) that this formula finds application in the treatment of the intensity correlations of thermal light.

1.6.2 Moment generating function and characteristic function

We consider a multivariate Gaussian distribution $p(x_1, x_2, \ldots x_N)$, in which the variates $x_1, x_2, \ldots, x_N$ may be correlated. The moment generating function is given by

$$M(\xi_1, \xi_2, \ldots, \xi_N) = \left\langle \exp\left(\sum_{i=1}^{N}\xi_i x_i\right)\right\rangle = \langle e^{\boldsymbol{\xi}^\dagger \mathbf{x}}\rangle, \qquad (1.6\text{–}24)$$

where we have used $\boldsymbol{\xi}$ and $\mathbf{x}$ to denote the column matrices with elements ξ_i and x_i and $\boldsymbol{\xi}^\dagger$ is the adjoint row matrix. To determine $M(\xi_1, \xi_2, \ldots, \xi_N)$ we have to evaluate the integral

$$M(\xi_1, \xi_2, \ldots, \xi_N) = \frac{1}{(2\pi)^{N/2}(\det \boldsymbol{\mu})^{1/2}}\int \exp(\boldsymbol{\xi}^\dagger\mathbf{x} - \tfrac{1}{2}\Delta\mathbf{x}^\dagger\boldsymbol{\mu}^{-1}\Delta\mathbf{x})\,\mathrm{d}^N x. \qquad (1.6\text{–}25)$$

For this purpose we now consider the following bilinear form

$$\tfrac{1}{2}(\Delta\mathbf{x} - \boldsymbol{\mu}\boldsymbol{\xi})^\dagger\boldsymbol{\mu}^{-1}(\Delta\mathbf{x} - \boldsymbol{\mu}\boldsymbol{\xi}) = \tfrac{1}{2}\Delta\mathbf{x}^\dagger\boldsymbol{\mu}^{-1}\Delta\mathbf{x} + \tfrac{1}{2}\boldsymbol{\xi}^\dagger\boldsymbol{\mu}\boldsymbol{\xi} - \boldsymbol{\xi}^\dagger\Delta\mathbf{x}, \qquad (1.6\text{–}26)$$

in which we have made use of the properties that $\boldsymbol{\mu}^\dagger = \boldsymbol{\mu}$ because $\boldsymbol{\mu}$ is Hermitian and that $\boldsymbol{\xi}^\dagger\Delta\mathbf{x} = \Delta\mathbf{x}^\dagger\boldsymbol{\xi}$. This allows us to re-write the exponent in Eq. (1.6–25), and we obtain

$$M(\xi_1, \xi_2, \ldots, \xi_N) = e^{\boldsymbol{\xi}^\dagger\langle\mathbf{x}\rangle}\, e^{\boldsymbol{\xi}^\dagger\boldsymbol{\mu}\boldsymbol{\xi}/2}\left[\frac{1}{(2\pi)^{N/2}(\det\boldsymbol{\mu})^{1/2}}\int \exp[-\tfrac{1}{2}(\Delta\mathbf{x} - \boldsymbol{\mu}\boldsymbol{\xi})^\dagger\boldsymbol{\mu}^{-1}(\Delta\mathbf{x} - \boldsymbol{\mu}\boldsymbol{\xi})]\,\mathrm{d}^N x\right].$$

The integral becomes the usual Gaussian integral if we replace the mean $\langle x\rangle$ by $\langle\mathbf{x}\rangle - \boldsymbol{\mu}\boldsymbol{\xi}$, the factor in the square brackets then becomes unity, so that finally

$$\left.\begin{aligned}M(\xi_1, \xi_2, \ldots, \xi_N) &= e^{\boldsymbol{\xi}^\dagger\langle\mathbf{x}\rangle}\, e^{\xi^\dagger\boldsymbol{\mu}\xi/2} \\ &= \exp\left[\sum_{i=1}^{N}\xi_i\langle x_i\rangle + \tfrac{1}{2}\sum_{i=1}^{N}\sum_{j=1}^{N}\xi_i\mu_{ij}\xi_j\right].\end{aligned}\right\} \qquad (1.6\text{–}27)$$

Similarly, we may show that the characteristic function of the multivariate Gaussian distribution is given by

$$C(\xi_1, \xi_2, \ldots, \xi_N) = \langle e^{i\boldsymbol{\xi}^\dagger\mathbf{x}}\rangle = e^{i\boldsymbol{\xi}^\dagger\langle\mathbf{x}\rangle}\, e^{-\boldsymbol{\xi}^\dagger\boldsymbol{\mu}\boldsymbol{\xi}/2}, \qquad (1.6\text{–}28)$$

and the cumulant generating function follows from Eq. (1.6–27) on taking logarithms, namely

$$K(\xi_1, \xi_2, \ldots, \xi_N) = \sum_{i=1}^{N}\xi_i\langle x_i\rangle + \tfrac{1}{2}\sum_{i=1}^{N}\sum_{j=1}^{N}\xi_i\mu_{ij}\xi_j. \qquad (1.6\text{–}29)$$

Again we note that the series in ξ is a second-order polynomial in the N variables $\xi_1, \xi_2, \ldots, \xi_N$.

1.6.3 *Multiple complex Gaussian variates*

We have already seen that a complex Gaussian variate z of zero mean has the probability density (cf. Eq. (1.5–31))

$$p(z) = \frac{1}{\pi\mu} e^{-|z|^2/\mu},$$

where $\mu = \langle \Delta z \Delta z^* \rangle$ is the variance of z. If $z_1, z_2, \ldots, z_N$ is a set of N complex Gaussian variates, then the general multivariate Gaussian probability distribution has the form

$$p(z_1, z_2, \ldots, z_N) = \frac{1}{\pi^N \det \boldsymbol{\mu}} e^{-\boldsymbol{\Delta z}^\dagger \boldsymbol{\mu}^{-1} \boldsymbol{\Delta z}}, \tag{1.6–30}$$

in which $\boldsymbol{\mu} = \langle \boldsymbol{\Delta z \Delta z}^\dagger \rangle$ is the covariance matrix, $\boldsymbol{\Delta z}$ is the column matrix with elements $z_i - \langle z_i \rangle$, as before, and $\boldsymbol{\Delta z}^\dagger$ is its Hermitian adjoint. As each complex variate involves two real variates, $p(z_1, z_2, \ldots, z_N)$ has a structure that is reminiscent of the square of $p(x_1, x_2, \ldots, x_N)$. Just as for the real variate $\mathbf{x}$, $\boldsymbol{\mu}$ is positive definite.

The characteristic function can be obtained by taking the $2N$-dimensional Fourier transform of $p(z_1, z_2, \ldots, z_N)$. We introduce a set of complex parameters $u_1, u_2, \ldots u_N$ forming a column matrix $\mathbf{u}$. Then the matrix product $\mathbf{u}^\dagger \mathbf{z} - \mathbf{z}^\dagger \mathbf{u}$ is purely imaginary, and $e^{\mathbf{u}^\dagger \mathbf{z} - \mathbf{z}^\dagger \mathbf{u}}$ is the $2N$-dimensional Fourier kernel. Hence, we can express the characteristic function in the compact form

$$C(u_1, u_2, \ldots, u_N) = \int e^{\mathbf{u}^\dagger \mathbf{z} - \mathbf{z}^\dagger \mathbf{u}} p(z)\, d^{2N} z, \tag{1.6–31}$$

and an analysis like that used above to derive the moment generating function of a set of real Gaussian variates leads to the following expression for the characteristic function:

$$\begin{aligned} C(u_1, u_2, \ldots, u_N) &= e^{\mathbf{u}^\dagger \langle \mathbf{z} \rangle - \langle \mathbf{z} \rangle^\dagger \mathbf{u}} e^{-\mathbf{u}^\dagger \boldsymbol{\mu} \mathbf{u}} \\ &= \exp\left[\sum_{i=1}^{N} (u_i^* \langle z_i \rangle - \langle z_i \rangle^* u_i) - \sum_{i=1}^{N} \sum_{j=1}^{N} (u_i^* \mu_{ij} u_j)\right]. \end{aligned} \tag{1.6–32}$$

It should be noted that the single sum is purely imaginary and the double sum is real, just as for real variates.

The Gaussian moment theorem has a counterpart when the variates are complex. It may be shown that (Mehta, 1965, Appendix A2)

$$\begin{aligned} \langle \Delta z_{i_1}^* \Delta z_{i_2}^* \ldots \Delta z_{i_N}^* \Delta z_{j_M} \ldots \Delta z_{j_1} \rangle &= 0, \quad \text{if } N \neq M \\ &= \sum_{\substack{\text{all} N! \\ \text{pairings}}} \langle \Delta z_{i_1}^* \Delta z_{j_1} \rangle \langle \Delta z_{i_2}^* \Delta z_{j_2} \rangle \ldots \langle \Delta z_{i_N}^* \Delta z_{j_N} \rangle \\ &\quad \text{if } N = M. \end{aligned} \tag{1.6–33}$$

It follows from this, for example, that if $\langle z_1 \rangle = 0 = \langle z_2 \rangle$, then

$$\begin{aligned}\langle |z_1|^2 |z_2|^2 \rangle &= \langle z_1^* z_1 \rangle \langle z_2^* z_2 \rangle + \langle z_1^* z_2 \rangle \langle z_2^* z_1 \rangle \\ &= \langle |z_1|^2 \rangle \langle |z_2|^2 \rangle (1 + |\rho_{12}|^2). \qquad (1.6\text{–}34)\end{aligned}$$

Problems

1.1 ϕ is a random phase angle that is distributed uniformly over the range 0 to 2π, a is a constant, and

$$x = a \cos \phi, \quad y = a \sin \phi.$$

Calculate (a) the probability distributions of x and y; (b) the joint probability distribution of x and y; (c) the covariance of x and y. Are the variates x and y statistically independent?

1.2 Examine the question whether there exists a random variable x whose characteristic function $C(\xi)$ is real for real ξ and has the form illustrated below, and if so, for what values of a, b. What is the corresponding probability density? Calculate the moments of x.

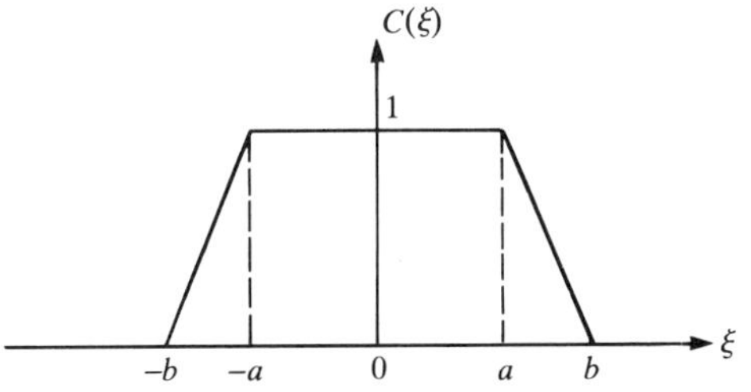

1.3 A certain experiment is performed repeatedly and independently, and it has probability α of being successful at any one time. What is the probability $P(m, n)$ of obtaining m failures and n successes in $m + n$ trials that end with the n'th success? Calculate the r'th factorial moment of m for a given n.

$$\left[\text{you may use the relation } \binom{m+n-1}{m} = (-1)^m \binom{-n}{m}.\right]$$

1.4 A line of unit length is divided into two parts by a point on the line chosen at random. Calculate (a) the average length of each part; (b) the average ratio of the shorter to the longer part; (c) the covariance matrix for the lengths of the two parts.

1.5 In a laboratory experiment to measure the acceleration of a body under gravity ($g = 981$ cm/s^2), which is capable of 1% accuracy, a student claims to have obtained the following four answers from four independent measurements: 980 cm/s^2, 981 cm/s^2, 983 cm/s^2, 981 cm/s^2. Has the student been cheating? Use the χ^2-test of hypothesis to give the level of confidence of your conclusion.

1.6 Show that the probability $P(n)$ that at least two out of any group of n people have the same birthday is given by

$$P(n) \approx 1 - \exp\left(\frac{n - n^2}{730}\right)$$

approximately, for moderate values of n. What is the smallest group of n for which $P(n) > \frac{1}{2}$?

(Take the year to have 365 days, with each day equally likely as someone's birthday.)

1.7 A certain system having only two possible states A or B is measured $2N$ different times independently, after being prepared in exactly the same way each time. It is found to be N times in state A and N times in state B. Assuming that nothing is known initially about the prior probability q of finding the system in state A, use Bayes' theorem to calculate the probability distribution of q in the light of the $2N$ different observations, and make a graph of the distribution for $N = 0, 2, 4, 6$.

1.8 A drunkard moves in one dimension by taking a step forward or backward at random, with probabilities α and $1 - \alpha$, respectively. He stands a distance of n steps from the edge of a cliff. Calculate the probability that he will go over the cliff as a function of n and α. What is the smallest value of α for which he is certain to fall off no matter where he starts? What is the asymptotic form of this probability as $n \to \infty$?

(Hint: try a composition law for constructing the probabilities.)

1.9 x, y are two real Gaussian random variables of zero mean with joint probability density

$$p(x, y) = \frac{1}{2\pi\sigma_x\sigma_y(1 - \rho^2)^{1/2}} \exp\left[-\frac{1}{2(1 - \rho^2)}\left(\frac{x^2}{\sigma_x^2} - \frac{2xy}{\sigma_x\sigma_y} + \frac{y^2}{\sigma_y^2}\right)\right]$$

in which $\sigma_x = \langle x^2\rangle^{1/2}$, $\sigma_y = \langle y^2\rangle^{1/2}$, $\rho = \langle xy\rangle/\sigma_x\sigma_y$. Show that the new real variables u, v defined by the transformation

$$u = x\cos\theta - y\sin\theta$$

$$v = x\sin\theta + y\cos\theta,$$

are also bivariate Gaussian and determine the angle θ for which u, v are statistically independent.

1.10 Show that any real characteristic function $C(h)$ obeys the inequality

$$1 - C(h) \geqslant \frac{1}{4^n}[1 - C(2^n h)] \quad \text{for } n = 0, 1, 2, 3, \ldots.$$

2

Random (or stochastic) processes

2.1 Introduction to statistical ensembles

The concept of a random or stochastic process or function represents a generalization of the idea of a set of random variables $x_1, x_2, \ldots$, when the set is no longer countable and the variables form a continuum. We therefore introduce a continuous parameter t, such as time, that labels the variates. We call $x(t)$ a *random process* or a *random function of* t if x does not depend on t in a deterministic way. Random processes are encountered in many fields of science, whenever fluctuations are present. Examples of a real random process $x(t)$ are the fluctuating voltage across an electrical resistor, and the coordinates of a particle undergoing Brownian motion. We shall see shortly that the optical field generated by any realistic light source must also be treated as a random function of position and time. Of course the parameter t may also stand for some quantity other than time, but for simplicity we shall take it to represent time. In our applications $x(t)$ will frequently represent a Cartesian component of the electric or magnetic field vector in a light beam. To begin with we shall take $x(t)$ to be real, but complex random processes will also be encountered.

2.1.1 The ensemble average

As x does not depend on t deterministically, we can only describe its values statistically, by some probability distribution or probability density. For every value of t, $x(t)$ is a random variable in some domain, with probability density $p[x(t)]$ or $p(x, t)$. After integrating over the domain, we have

$$\int p(x, t)\,\mathrm{d}x = 1,$$

as usual for a probability density. It should be noted that, because of the dependence on t, $p(x, t)$ now stands not for one, but for an *infinite family* of probability distributions. The totality of all variates x at all times t constitutes the random process $x(t)$. We may calculate the expectation of x at time t by using the probability density $p(x, t)$,

$$\langle x(t)\rangle = \int x p(x, t)\,\mathrm{d}x. \qquad (2.1\text{–}1a)$$

Alternatively, we may consider the set of all possible realizations or samples of the function $x(t)$, such as those illustrated in Fig. 2.1, as the random process. The countable collection of all possible realizations is known as the *ensemble* of $x(t)$. In an experiment a sample function may describe the outcome of a measurement as a function of time t, but repetition of the experiment will generally yield different sample functions or realizations, which may be labeled successively ${}^{(1)}x(t)$, ${}^{(2)}x(t)$, . . . , etc. We can then form the average or expectation of x at time t by averaging over the ensemble of all realizations

$$\langle x(t)\rangle = \lim_{N\to\infty} \frac{1}{N}\sum_{r=1}^{N} {}^{(r)}x(t). \tag{2.1–1b}$$

Equations (2.1–1a) and (2.1–1b) are equivalent definitions of the *ensemble average*.

The foregoing concepts and definitions are equally applicable to a complex random process $z(t)$, for which the random variable associated with a particular parameter t is complex. By writing $z(t) = x(t) + \mathrm{i}y(t)$, we see that a complex random process may be regarded as a pair of real random processes $x(t)$, $y(t)$. The probability density $p(z, t)$ is then the joint probability density of x, y at time t. The average of $z(t)$ may be calculated from $p(z, t)$ by an obvious generalization of Eq. (2.1–1a), which we may write as

$$\langle z(t)\rangle = \int z p(z, t)\, \mathrm{d}^2 z, \tag{2.1–2a}$$

with $\mathrm{d}^2 z \equiv \mathrm{d}x\, \mathrm{d}y$. This quantity is again equivalent to an average over the ensemble of all realizations ${}^{(r)}z(t)$,

$$\langle z(t)\rangle = \lim_{N\to\infty} \frac{1}{N}\sum_{r=1}^{N} {}^{(r)}z(t). \tag{2.1–2b}$$

For simplicity, we shall concentrate on real random processes in this section.

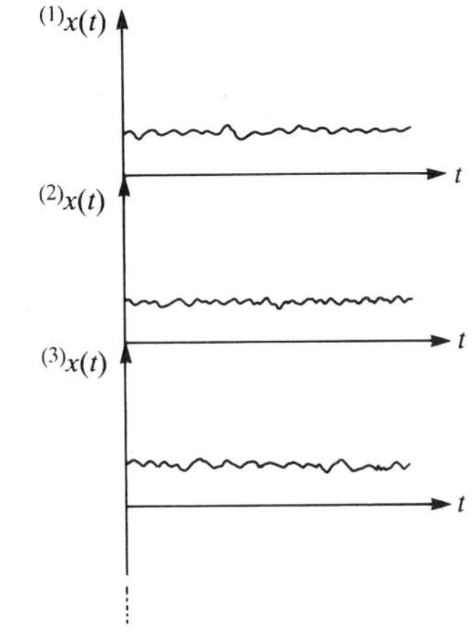

Fig. 2.1 The representation of a real random process $x(t)$ by an ensemble of realizations, or sample functions, ${}^{(j)}x(t)$, $(j = 1, 2, 3, \ldots)$.

2.1.2 *Joint probabilities and correlations*

Although $p(x, t)$ stands for an infinite family of probability densities, it still does not describe the random process completely. For example, it contains no information about possible correlations between $x(t_1)$ and $x(t_2)$ at two different times t_1 and t_2. Such information is provided by the joint or two-fold probability density of the variates at t_1, t_2, and it is usually expressed in the following equivalent ways,

$$p_2[x_2(t_2); x_1(t_1)] \quad \text{or} \quad p_2(x_2, t_2; x_1, t_1) \quad \text{or} \quad p_2(x_2, x_1; t_2, t_1),$$

which depends on two variates x_1, x_2 and two parameters t_1, t_2. The probability density $p_2(x_2, t_2; x_1, t_1)$ allows us to calculate two-time correlation functions such as

$$\Gamma(t_1, t_2) \equiv \langle x(t_1)x(t_2)\rangle = \int x_1 x_2 p_2(x_2, t_2; x_1, t_1)\,\mathrm{d}x_1\,\mathrm{d}x_2. \qquad (2.1\text{–}3)$$

The probability density $p_2(x_2, t_2; x_1, t_1)$ contains all the information carried by $p(x, t)$, which can also be written as $p_1(x, t)$, in addition to new information, for we have, from the usual property of joint probabilities, that

$$\int p_2(x_2, t_2; x_1, t_1)\,\mathrm{d}x_2 = p_1(x_1, t_1). \qquad (2.1\text{–}4)$$

It should be noted that the parameter t_2 must disappear in the process of integrating over all x_2 for reasons of compatibility, i.e. the internal consistency of the theory.

The quantity $\Gamma(t_1, t_2) \equiv \langle x(t_1)x(t_2)\rangle$ is known as the (two-time) *autocorrelation function* of the random process $x(t)$. After the mean $\langle x(t)\rangle$, it is the quantity next in order of importance in the description of the random process, because it yields information on how far correlations extend in time.

Although $p_2(x_2, t_2; x_1, t_1)$ contains more information than $p_1(x, t)$, it still does not allow us to calculate some expectation values, e.g. a three-time correlation function such as $\langle x(t_1)x(t_2)x(t_3)\rangle$, for which the three-fold joint probability density $p_3(x_3, t_3; x_2, t_2; x_1, t_1)$ is required. Evidently there exists an *infinite hierarchy of probability densities*,

$$\begin{aligned}
&p_1(x, t)\\
&p_2(x_2, t_2; x_1, t_1)\\
&p_3(x_3, t_3; x_2, t_2; x_1, t_1)\\
&\cdots\cdots\cdots\cdots\cdots\cdots\cdots\\
&p_n(x_n, t_n; x_{n-1}, t_{n-1}; \ldots ; x_1, t_1),
\end{aligned}$$

each of which contains more information than the preceding ones, and each of which encompasses all the information contained in the preceding ones. $p_n(x_n, t_n; x_{n-1}, t_{n-1}; \ldots ; x_1, t_1)\,\mathrm{d}x_1\,\mathrm{d}x_2 \ldots \mathrm{d}x_n$ is the n-fold joint probability that at time t_1 the random variable has a value between x_1 and $x_1 + \mathrm{d}x_1$, that at time t_2 it has a value between x_2 and $x_2 + \mathrm{d}x_2$, etc. The probability density $p_n(x_n, t_n; x_{n-1}, t_{n-1};$

$\ldots; x_1, t_1)$ must satisfy the consistency condition

$$\int p_n(x_n, t_n; x_{n-1}, t_{n-1}; \ldots; x_1, t_1)\, dx_{k+1}\, dx_{k+2} \ldots dx_n$$
$$= p_k(x_k, t_k; x_{k-1}, t_{k-1}; \ldots; x_1, t_1), \quad (2.1\text{–}5)$$

for any integer $k < n$. The time t_j associated with the variate x_j therefore disappears when we integrate over all x_j. Also $p_n(x_n, t_n; x_{n-1}, t_{n-1}; \ldots; x_1, t_1)$ is symmetric in the variates, which implies that

$$p_n(x_n, t_n; x_{n-1}, t_{n-1}; \ldots; x_1, t_1) = p_n\left(\prod[x_n, t_n; x_{n-1}, t_{n-1}; \ldots; x_1, t_1]\right), \quad (2.1\text{–}6)$$

where $\prod$ stands for any permutation of the indices 1 to n. Once p_n is known it can be used to calculate a multi-time correlation of the n'th order [or of order lower than n with the help of Eq. (2.1–5)]:

$$\langle x(t_1)x(t_2) \ldots x(t_n)\rangle =$$
$$\int x_1 x_2 \ldots x_n p_n(x_n, t_n; x_{n-1}, t_{n-1}; \ldots; x_1, t_1)\, dx_1\, dx_2 \ldots dx_n. \quad (2.1\text{–}7)$$

In general the higher-order correlation functions of a random process contain progressively more information, just as p_n contains more information than p_{n-1}. An exception to this arises in the case of a *Gaussian random process*, in which $p_n(x_n, t_n; x_{n-1}, t_{n-1}; \ldots; x_1, t_1)$ is a multivariate Gaussian probability density of the form given by Eq. (1.6–5). In that case, by virtue of the Gaussian moment theorem [see Eqs. (1.6–18)], we have for the n'th-order correlation function of a real Gaussian random process $x(t)$ of zero mean,

$$\begin{aligned}\langle x(t_1)x(t_2) \ldots x(t_n)\rangle &= 0 && \text{if } n \text{ is odd.}\\ &= \sum_{\substack{\text{all}(n-1)!!\\ \text{pairings}}} \langle x(t_1)x(t_2)\rangle\langle x(t_3)x(t_4)\rangle \ldots \langle x(t_{n-1})x(t_n)\rangle && \text{if } n \text{ is even,}\end{aligned} \quad (2.1\text{–}8)$$

where $(n-1)!! \equiv (n-1)(n-3) \ldots 5 \cdot 3 \cdot 1$. In this case the second-order auto-correlation function therefore already contains all the information about higher-order correlations.

2.1.3 The probability functional

Although p_n contains more information than p_{n-1}, there is no upper bound on n when t is a continuous parameter, and no p_n for finite n can, in general, describe the process completely. Only the joint probability density of infinite order at all times t, or the *probability functional* $p(\{x(t)\})$, in which $\{x(t)\}$ stands for the infinite set of all x's at all times t, contains the full statistical description of the random process. However, the explicit form of $p(\{x(t)\})$ is often not known.

Nevertheless, the concept of the *probability functional* is sometimes useful, for example to describe the expectation of some functional $f(\{x(t)\})$ of $\{x(t)\}$. One then writes

$$\langle f(\{x(t)\})\rangle = \int f(\{x(t)\})p(\{x(t)\})\,\mathrm{d}\{x(t)\}, \tag{2.1–9}$$

where $\mathrm{d}\{x(t)\}$ is a shorthand notation for $\prod_{\text{all}\,t}\mathrm{d}x(t)$. By analogy with the characteristic function for N variates, for example,

$$C(\xi_1, \xi_2, \ldots, \xi_N) = \left\langle \exp\left(\mathrm{i}\sum_{j=1}^{N} x_j\xi_j\right)\right\rangle,$$

one can associate a *characteristic functional* with the probability functional $p(\{x(t)\})$, defined by

$$\begin{aligned} C(\{\xi(t)\}) &= \left\langle \exp\left[\mathrm{i}\int x(t')\xi(t')\,\mathrm{d}t'\right]\right\rangle \\ &= \int \exp\left[\mathrm{i}\int x(t')\xi(t')\,\mathrm{d}t'\right]p(\{x(t)\})\,\mathrm{d}\{x(t)\}. \end{aligned} \tag{2.1–10}$$

In particular, for a Gaussian random process of zero mean one has, by an obvious generalization of Eq. (1.6–28), the characteristic functional

$$C(\{\xi(t)\}) = \exp\left[-\tfrac{1}{2}\iint_{-\infty}^{\infty} \xi(t')\xi(t'')\mu(t', t'')\,\mathrm{d}t'\,\mathrm{d}t''\right], \tag{2.1–11}$$

where $\mu(t', t'') \equiv \Gamma(t', t'') \equiv \langle x(t')x(t'')\rangle$ is the covariance matrix, which is also the autocorrelation function. In principle, multi-time moments or correlations of the random process $x(t)$ may be derived from $C(\{\xi(t)\})$ by functional differentiation‡, but we shall not use this technique.

2.2 Stationarity and ergodicity

Random functions of time frequently have the property that the character of the fluctuations does not change with time, even though any realization of the ensemble $x(t)$ changes continually in time. Such a process is said to be *statistically stationary*. An example is illustrated in Fig. 2.2. More precisely, we call a random process *stationary* if all the probability densities $p_1, p_2, p_3, \ldots$ governing the fluctuations are invariant under an arbitrary translation of the origin of time§, i.e. if

$$p_n(x_n, t_n; x_{n-1}, t_{n-1}; \ldots; x_1, t_1) = p_n(x_n, t_n + T; x_{n-1}, t_{n-1} + T; \ldots; x_1, t_1 + T) \quad \text{for all } T. \tag{2.2–1}$$

Under these circumstances, the expectation value of any function of $x(t_1)$, $x(t_2)$,

‡ For an account of the technique of functional differentiation required here see, for example, Beran (1968).

§ For a more thorough treatment of the mathematical foundations of the theory of stationary random processes see, for example, Yaglom (1962).

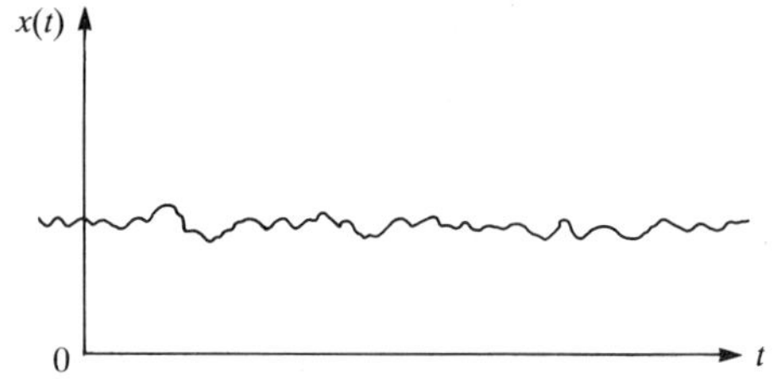

Fig. 2.2 Illustrating a realization of a stationary random process $x(t)$.

. . . is also invariant under time translation i.e.

$$\langle f[x(t_1), x(t_2), \ldots] \rangle = \langle f[x(t_1 + T), x(t_2 + T), \ldots] \rangle. \qquad (2.2\text{–}2)$$

It is clear from Eq. (2.2–1) that $p_1(x, t)$ cannot depend on t at all for a stationary process, and neither can the expectation $\langle x(t) \rangle$. For, by choosing $n = 1$ and $T = -t$, we have

$$\left.\begin{aligned} p_1(x, t) &= p_1(x, 0) \\ \langle x(t) \rangle &= \int x p_1(x, t)\,\mathrm{d}x = \int x p_1(x, 0)\,\mathrm{d}x. \end{aligned}\right\} \qquad (2.2\text{–}3)$$

Also by choosing $T = -t_1$, we have from Eq. (2.2–1),

$$p_n(x_n, t_n; x_{n-1}, t_{n-1}; \ldots; x_1, t_1) = p_n(x_n, t_n - t_1; x_{n-1}, t_{n-1} - t_1; \ldots; x_1, 0), \qquad (2.2\text{–}4)$$

so that the joint probabilities p_n can be expressed as functions only of the differences between one of the n time arguments and the remaining $n - 1$ time arguments. In particular, for the autocorrelation function we obtain

$$\begin{aligned} \Gamma(t_1, t_2) = \langle x(t_1) x(t_2) \rangle &= \int x_1 x_2 p_2(x_2, t_2; x_1, t_1)\,\mathrm{d}x_1\,\mathrm{d}x_2 \\ &= \int x_1 x_2 p_2(x_2, t + t_2 - t_1; x_1, t)\,\mathrm{d}x_1\,\mathrm{d}x_2 \\ &= \langle x(t) x(t + t_2 - t_1) \rangle \end{aligned} \qquad (2.2\text{–}5)$$

for all values of t, which depends only on the difference between the two time arguments. Therefore, $\Gamma(t_1, t_2)$ is frequently written as $\Gamma(t_2 - t_1)$. By replacing t in Eq. (2.2–5) by $t - t_2 + t_1$ we observe that $\Gamma(t_2 - t_1)$ is symmetric, i.e. that

$$\Gamma(t_2 - t_1) = \Gamma(t_1 - t_2) \qquad (2.2\text{–}6)$$

for a real stationary process. More generally, we can see by an argument similar to the one used in deriving Eq. (2.2–5) that for the N'th-order correlation function of a stationary random process

$$\Gamma^{(N)}(t_1, t_2, \ldots, t_N) \equiv \langle x(t_1) x(t_2) \ldots x(t_N) \rangle = \langle x(t_1 + T) x(t_2 + T) \ldots x(t_N + T) \rangle \qquad (2.2\text{–}7)$$

for all T. However, higher-order correlations are encountered less frequently.

When only the mean $\langle x(t) \rangle$ and the second-order correlation function $\Gamma(t_1, t_2)$

are of interest, a weaker form of stationarity is often invoked. Specifically when $x(t)$ is such that its mean $\langle x(t)\rangle$ is independent of t and its autocorrelation function $\Gamma(t_1, t_2)$ depends on t_1 and t_2 only through the difference of the two time arguments, the process is said to be *stationary in the wide sense*.

If instead of a real random process $x(t)$, we have a complex random process $z(t)$, the autocorrelation function $\Gamma(t_1, t_2)$ is defined by the equation

$$\Gamma(t_1, t_2) \equiv \langle z^*(t_1)z(t_2)\rangle \equiv \int z_1^* z_2 p_2(z_2, t_2; z_1, t_1)\, \mathrm{d}^2 z_1\, \mathrm{d}^2 z_2.$$

In this case, when the process is stationary, at least in the wide sense, $\langle z(t)\rangle$ is independent of t, $\Gamma(t_1, t_2) \equiv \Gamma(t_2 - t_1)$ and it obeys the Hermiticity condition

$$\Gamma(t_2 - t_1) = \Gamma^*(t_1 - t_2), \tag{2.2–8}$$

instead of the symmetry condition (2.2–6).

2.2.1 The time average of a stationary process

So far the means, or expectations, have been calculated by averaging over the ensemble of all realizations. However, sometimes only a single realization of the ensemble is available, say the k'th one ${}^{(k)}z(t)$, and one wishes to determine its average over a certain time interval T, or perhaps over all time. Let us define the finite time average of the k'th realization of a stationary random process $z(t)$, which may be complex, by

$$[{}^{(k)}z(t)]_T \equiv \frac{1}{T}\int_{t-T/2}^{t+T/2} {}^{(k)}z(t')\,\mathrm{d}t'. \tag{2.2–9}$$

Then $[{}^{(k)}z(t)]_T$ is itself a random process, and one might expect its fluctuations to become smaller as T increases. When $T \to \infty$, Eq. (2.2–9) gives the *time average* of ${}^{(k)}z(t)$

$$\overline{{}^{(k)}z} \equiv \operatorname*{Lim}_{T\to\infty} \frac{1}{T}\int_{t-T/2}^{t+T/2} {}^{(k)}z(t')\,\mathrm{d}t', \tag{2.2–10}$$

which no longer depends on t or T, but in general does depend on the particular realization k of the ensemble that we have chosen. In principle, there may be as many different time averages as there are elements of the ensemble.

More generally, if we are interested in multi-time correlations, such as the average of $z^*(t)z(t+\tau)$, for example, we simply construct a new random process $Z(t)$ from $z(t)$ by letting

$${}^{(k)}Z(t) = {}^{(k)}z^*(t)\,{}^{(k)}z(t+\tau)$$

for each realization k. We can then ask the same questions about ${}^{(k)}Z(t)$ that we have just asked about ${}^{(k)}z(t)$. For example, we can calculate the time average of ${}^{(k)}Z(t)$ and compare it with the ensemble average. Once again we find that the time average

$$\overline{{}^{(k)}Z} \equiv \operatorname*{Lim}_{T\to\infty} \frac{1}{T}\int_{t-T/2}^{t+T/2} {}^{(k)}Z(t')\,\mathrm{d}t'$$

may depend on the particular realization k.

2.2.2 *Ergodicity*

It frequently happens in practice that every realization of the ensemble carries the same statistical information about the stationary random process as every other realization. The different time averages $\overline{{}^{(k)}z}$ are then all equal and coincide with the ensemble average $\langle z \rangle$, and the same also applies to any other process $Z(t)$ that can be constructed from $z(t)$. The stationary random process $z(t)$ is then said to be an *ergodic process*.

It is not difficult to find a condition that ensures that the time average $\overline{{}^{(k)}z}$ coincides with the ensemble average $\langle z \rangle$ of a stationary random process $z(t)$. If we average $[{}^{(k)}z(t)]_T$ over the ensemble, we obtain

$$\begin{aligned}\langle [{}^{(k)}z(t)]_T \rangle &= \frac{1}{T}\int_{t-T/2}^{t+T/2} \langle {}^{(k)}z(t') \rangle \, \mathrm{d}t' \\ &= \frac{1}{T}\int_{t-T/2}^{t+T/2} \langle z \rangle \, \mathrm{d}t' \\ &= \langle z \rangle, \end{aligned} \tag{2.2–11}$$

so that the ensemble average of $[{}^{(k)}z(t)]_T$ coincides with the ensemble average of the random process $z(t)$ for all T. Next we calculate the dispersion of $[{}^{(k)}z(t)]_T$:

$$\begin{aligned}\langle |\Delta[{}^{(k)}z(t)]_T|^2 \rangle &\equiv \langle |[{}^{(k)}z(t)]_T - \langle z \rangle|^2 \rangle \\ &= \langle [{}^{(k)}z^*(t)]_T [{}^{(k)}z(t)]_T \rangle - |\langle z \rangle|^2 \\ &= \left\langle \left[\frac{1}{T}\int_{t-T/2}^{t+T/2} {}^{(k)}z^*(t')\,\mathrm{d}t'\right]\left[\frac{1}{T}\int_{t-T/2}^{t+T/2} {}^{(k)}z(t'')\,\mathrm{d}t''\right]\right\rangle - |\langle z \rangle|^2 \\ &= \frac{1}{T^2}\iint_{t-T/2}^{t+T/2} [\Gamma(t''-t') - |\langle z \rangle|^2]\,\mathrm{d}t'\,\mathrm{d}t'',\end{aligned}$$

where we have made use of the definition $\Gamma(t''-t') = \langle z^*(t')z(t'') \rangle$. By substituting $t' = t + t_1$, $t'' = t + t_2$ we can symmetrize the limits on the integral, so that

$$\langle |\Delta[{}^{(k)}z(t)]_T|^2 \rangle = \frac{1}{T^2}\iint_{-T/2}^{T/2} [\Gamma(t_2 - t_1) - |\langle z \rangle|^2]\,\mathrm{d}t_1\,\mathrm{d}t_2. \tag{2.2–12}$$

Because the integrand on the right of Eq. (2.2–12) depends only on the difference between the two time arguments t_1 and t_2 it can be converted into a single integral over the variable $\tau = t_2 - t_1$. This conversion can be done by a simple geometrical argument. Consider a contribution to the integral from a strip of length l along the line $\tau = t_2 - t_1 =$ constant, shown shaded in Fig. 2.3. Along this strip, $\Gamma(\tau)$ is constant and hence the contribution to the integral of this strip is $[\Gamma(\tau) - |\langle z \rangle|^2] l \,\mathrm{d}\tau/\sqrt{2}$. It follows from elementary geometry that $l = \sqrt{2}$ $(T - |\tau|)$. Hence the contribution from the strip is $[\Gamma(\tau) - |\langle z \rangle|^2](T - |\tau|)\,\mathrm{d}\tau$. Using this result we obtain the formula

$$\langle |\Delta[{}^{(k)}z(t)]_T|^2 \rangle = \frac{1}{T}\int_{-T}^{T}\left(1 - \frac{|\tau|}{T}\right)|\Gamma(\tau) - |\langle z \rangle|^2|\,\mathrm{d}\tau$$

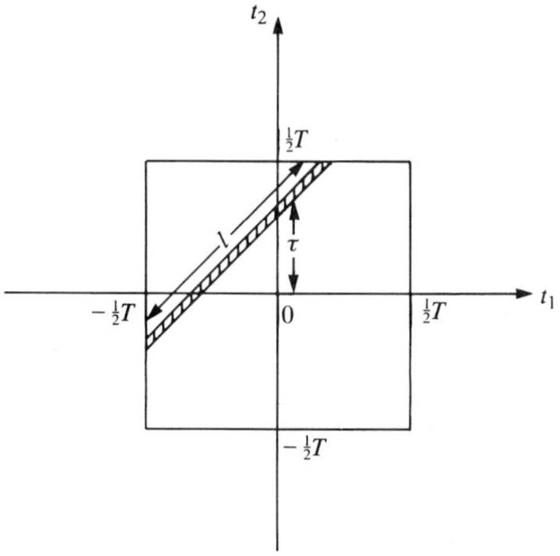

Fig. 2.3 Illustrating the geometry for the double integral in Eq. (2.2–12) leading to the inequality (2.2–13b).

$$\leqslant \frac{1}{T}\int_{-T}^{T} |\Gamma(\tau) - |\langle z \rangle|^2|\,\mathrm{d}\tau \qquad (2.2\text{–}13\text{a})$$

$$\therefore\ \langle|\Delta[^{(k)}z(t)]_T|^2\rangle \leqslant \frac{1}{T}\int_{-\infty}^{\infty} |\Gamma(\tau) - |\langle z \rangle|^2|\,\mathrm{d}\tau. \qquad (2.2\text{–}13\text{b})$$

Evidently with increasing T, the dispersion of $[^{(k)}z(t)]_T$ tends to zero whenever the integral is finite. But when $T \to \infty$, $[^{(k)}z(t)]_T$ becomes the time average $\overline{^{(k)}z}$. It follows that $\overline{^{(k)}z}$ coincides with the ensemble average $\langle z \rangle$ whenever

$$\int_{-\infty}^{\infty} |\Gamma(\tau) - |\langle z \rangle|^2|\,\mathrm{d}\tau < \infty. \qquad (2.2\text{–}14)$$

This condition is sufficient for the two averages to be the same, although it is not necessary, and a somewhat weaker but less convenient condition can be obtained from Eq. (2.2–13a). However, condition (2.2–14) still does not ensure that the process $z(t)$ is ergodic in the full sense, e.g. that the time average of another process $Z(t) \equiv z^*(t)z(t+\tau)$, constructed from $z(t)$, equals the ensemble average of $Z(t)$. For that purpose we would need to calculate the autocorrelation function of $Z(t)$ and to apply the same test to the random process $Z(t)$. In general an infinity of different criteria is needed to ensure the full ergodicity of the random process $z(t)$. However, for the special case of the Gaussian random process $z(t)$, for which all higher-order correlations are expressible in terms of second-order correlations (see Section 1.6), the criterion expressed by Eq. (2.2–14) is sufficient for full ergodicity.

Condition (2.2–14) has a simple physical interpretation. Since $\Delta z = z(t) - \langle z(t) \rangle$ and $\langle \Delta z^*(t)\Delta z(t+\tau) \rangle = \Gamma(\tau) - |\langle z \rangle|^2$, we see that the integral is finite if $\langle \Delta z^*(t)\Delta z(t+\tau) \rangle$ tends to zero sufficiently rapidly as $\tau \to \infty$. In other words, ergodicity holds if correlations of the random process die out sufficiently rapidly in time. In that case a sufficiently long record of a single realization $^{(k)}z(t)$ of the random process can be divided up into sections of shorter lengths which are uncorrelated, so that an ensemble can be constructed from a single realization. The average over this ensemble then equals the average over time, because a single realization of sufficient length already contains all the information about the ensemble. If the process is stationary and ergodic, all realizations of the random process look somewhat similar and differ only in detail.

Finally, we point out that, although the single condition (2.2–14) does not ensure ergodicity of the process $z(t)$, in general, if the integral in Eq. (2.2–14) diverges, then the process $z(t)$ is not ergodic, even though the possibility that ${}^{(k)}z = \langle z \rangle$ is not ruled out.

2.2.3 *Examples of random processes*

We now consider two simple examples of random processes, in order to illustrate the notions of stationarity and ergodicity we have been discussing.

(a) Consider the ensemble of complex functions

$$z(t) = \sum_{n=1}^{N} a_n \, \mathrm{e}^{-\mathrm{i}\omega_n t}, \tag{2.2–15}$$

in which $\omega_1, \omega_2, \ldots, \omega_N$ are fixed frequencies, but the coefficients $a_1, a_2, \ldots, a_N$ are random variables. As each set of possible values of the variates $a_1, a_2, \ldots, a_N$ defines a different realization, $z(t)$ is evidently a random process. Examples of possible realizations of the ensemble are illustrated in Fig. 2.4. Let us examine whether $z(t)$ is stationary in the wide sense.

From Eq. (2.2–15) we find for the expectation

$$\langle z(t) \rangle = \sum_{n=1}^{N} \langle a_n \rangle \, \mathrm{e}^{-\mathrm{i}\omega_n t}, \tag{2.2–16}$$

and for the autocorrelation function

$$\Gamma(t, t + \tau) = \langle z^*(t) z(t + \tau) \rangle = \sum_{n=1}^{N} \sum_{m=1}^{N} \langle a_n^* a_m \rangle \, \mathrm{e}^{\mathrm{i}(\omega_n - \omega_m)t} \, \mathrm{e}^{-\mathrm{i}\omega_m \tau}. \tag{2.2–17}$$

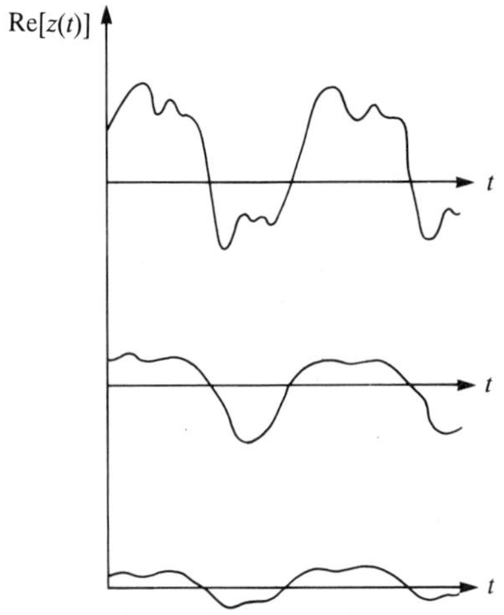

Fig. 2.4 Illustrating some possible realizations of the real part of the random process $z(t)$. This process is not ergodic.

In order that both $\langle z(t) \rangle$ and $\Gamma(t, t + \tau)$ be independent of t, we require that

$$\left.\begin{aligned} \langle a_n \rangle &= 0, \\ \langle a_n^* a_m \rangle &= \langle |a_n|^2 \rangle \delta_{nm}, \end{aligned}\right\} \tag{2.2–18}$$

where δ_{nm} is the Kronecker symbol, i.e. δ_{nm} has the value 1 or 0 according as $m = n$ or $m \neq n$. In other words, if the variates $a_1, a_2, \ldots, a_N$ are all of zero mean and if they are all uncorrelated, then $z(t)$ is stationary in the wide sense.

Assuming that this is the case, let us examine whether the process $z(t)$ is ergodic. According to the criterion given by Eq. (2.2–14), we need to calculate the integral

$$\int_{-\infty}^{\infty} \left| \sum_{n=1}^{N} \langle |a_n|^2 \rangle \, \mathrm{e}^{-\mathrm{i}\omega_n \tau} \right| \mathrm{d}\tau,$$

which evidently diverges, because correlations never die out. In this case, time averages of functions of the random process calculated from different realizations of the ensemble will, in general, not equal the ensemble average, and will differ also from each other. Therefore we do not have a single time average, but an infinity of them. Nevertheless, it so happens that $\overline{{}^{(k)}z} = 0 = \langle z \rangle$. However, we may readily show that for the random process $|z(t)|^2$ we obtain for the time average the result

$$\overline{|{}^{(k)}z|^2} = \sum_n |a_n^{(k)}|^2,$$

which evidently depends on the particular realization k of the ensemble. A glance at Fig. 2.4 shows why that is so: the different realizations of the ensemble are not statistically equivalent, and each one will yield a different answer; hence the process $z(t)$ is not ergodic.

(b) For our second example we consider the real random process $x(t)$ illustrated in Fig. 2.5, which is sometimes known as the *random telegraph signal* (Rice, 1944). The process $x(t)$ takes on two fixed values, a and $-a$, alternately and jumps instantaneously between them at completely random times, at an average rate R.

It is apparent from the construction of the process $x(t)$ that there is no preferred origin of time, so that $x(t)$ is stationary not only in the wide sense but

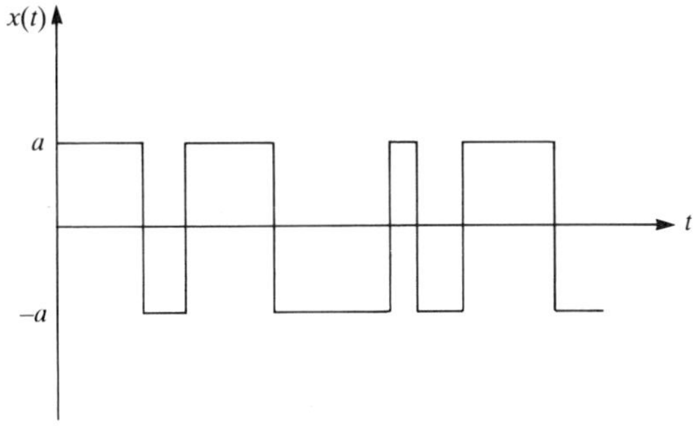

Fig. 2.5 Illustrating a realization of the random telegraph signal $x(t)$.

also in the strict sense. Moreover if the values a and $-a$ are equally probable, we have

$$\langle x(t)\rangle = 0. \tag{2.2–19}$$

Let us now calculate the autocorrelation function $\Gamma(\tau) = \langle x(t)x(t+\tau)\rangle$. The product $x(t)x(t+\tau)$ can take on only the two values a^2 and $-a^2$. If $x(t)$ has switched an even number of times in the interval from t to $t+\tau$ then $x(t)x(t+\tau) = a^2$, whereas the product yields $-a^2$ if there have been an odd number of switches. If $p(n, \tau)$ is the probability of n switches in the interval τ, it follows that

$$\begin{aligned}\Gamma(\tau) &= a^2 \sum_{n=0,2,4,\ldots}^{\infty} p(n, \tau) - a^2 \sum_{n=1,3,5,\ldots}^{\infty} p(n,\tau)\\ &= a^2 \sum_{n=0}^{\infty} (-1)^n p(n, \tau).\end{aligned} \tag{2.2–20}$$

As the switches occur at random at an average rate R, $p(n, \tau)$ is a Poisson distribution with parameter $\langle n\rangle = R\tau$, i.e.

$$p(n, \tau) = \frac{(R\tau)^n \mathrm{e}^{-R\tau}}{n!}. \tag{2.2–21}$$

When this expression is inserted in Eq. (2.2–20), it yields

$$\begin{aligned}\Gamma(\tau) &= a^2 \sum_{n=0}^{\infty} \frac{(-R\tau)^n \mathrm{e}^{-R\tau}}{n!}\\ &= a^2 \mathrm{e}^{-2R\tau} \quad (\tau \geqslant 0),\end{aligned}$$

and, since $\Gamma(-\tau) = \Gamma(\tau)$,

$$\Gamma(\tau) = a^2 \mathrm{e}^{-2R|\tau|} \tag{2.2–22}$$

for all values of τ.

The autocorrelation therefore decays exponentially to zero with increasing $|\tau|$ and $\Gamma(\tau)$ satisfies the criterion (2.2–14) for ergodicity. It is also apparent from the construction of the random process that any long realization will be statistically very similar to any other long realization, so that the process is ergodic. Finally we note that, despite the repeated discontinuities of the process $x(t)$, its autocorrelation function $\Gamma(\tau)$ is a continuous function of τ. This continuous behavior is a general property of all autocorrelations, as we show in the next section.

2.3 Properties of the autocorrelation function

Because of the importance of the autocorrelation function $\Gamma(\tau)$ for any real or complex stationary random process $z(t)$, we now summarize and examine some of its properties.

(a)
$$\Gamma(0) \geqslant 0. \tag{2.3–1}$$

This follows immediately from the definition $\Gamma(0) = \langle |z(t)|^2\rangle$. It is worth noting that $\Gamma(0)$ can vanish only in the trivial case in which the random process $z(t)$ is identically zero for all times t.

(b)
$$\Gamma(-\tau) = \Gamma^*(\tau). \tag{2.3–2}$$

This Hermiticity property follows from the fact that $z(t)$ is stationary, which allows us to make an arbitrary translation of the time origin. Consequently

$$\Gamma(\tau) = \langle z^*(t)z(t+\tau)\rangle = \langle z^*(t-\tau)z(t)\rangle = \Gamma^*(-\tau).$$

(c)
$$|\Gamma(\tau)| \leqslant \Gamma(0). \tag{2.3–3}$$

This property follows from the Schwarz inequality:

$$|\Gamma(\tau)|^2 = |\langle z^*(t)z(t+\tau)\rangle|^2 \leqslant \langle z^*(t)z(t)\rangle\langle z^*(t+\tau)z(t+\tau)\rangle = \Gamma^2(0),$$

which implies that $|\Gamma(\tau)|$ can never exceed its initial value $\Gamma(0)$, although it can fall below $\Gamma(0)$ and then return to $\Gamma(0)$.

(d) $\Gamma(\tau)$ is a non-negative definite function.

To establish this property we use the fact that for any positive integer N, for any N time-arguments $t_1, t_2, \ldots, t_N$ and for any N real or complex numbers $a_1, a_2, \ldots, a_N$, we necessarily have

$$\left\langle \left| \sum_{i=1}^{N} a_i z(t_i) \right|^2 \right\rangle \geqslant 0,$$

which implies, since $\langle z^*(t_i)z(t_j)\rangle = \Gamma(t_j - t_i)$, that

$$\sum_{i=1}^{N}\sum_{j=1}^{N} a_i^* a_j \Gamma(t_j - t_i) \geqslant 0. \tag{2.3–4a}$$

This inequality shows that the autocorrelation function $\Gamma(\tau)$ is non-negative definite. Alternatively an integral form of the non-negative definiteness property may be obtained by starting from the obvious inequality

$$\left\langle \left| \int_{T_1}^{T_2} f(t)z(t)\,\mathrm{d}t \right|^2 \right\rangle \geqslant 0,$$

where $f(t)$ is an arbitrary function and T_1 and T_2 are arbitrary times. We then obtain by a similar argument the alternative form of the non-negative definiteness condition,

$$\int_{T_1}^{T_2}\int_{T_1}^{T_2} f^*(t)f(t')\Gamma(t'-t)\,\mathrm{d}t\,\mathrm{d}t' \geqslant 0. \tag{2.3–4b}$$

(e) Let $t_1, t_2, \ldots, t_N$ be any N times and let $\Gamma(t_j - t_i) \equiv \Gamma_{ij}$. Then

$$\det \Gamma_{ij} \geqslant 0, \tag{2.3–5}$$

where $\det \Gamma_{ij}$ denotes the determinant of the $N \times N$ matrix Γ_{ij}. This inequality follows from the non-negative definiteness condition (2.3–4a), which can be written in matrix notation as

$$\mathbf{a}^\dagger \mathbf{\Gamma} \mathbf{a} \geqslant 0. \tag{2.3–6}$$

Here $\mathbf{a}$ is an arbitrary $N \times 1$ column matrix and $\mathbf{a}^\dagger$ is the conjugate $1 \times N$ row matrix. Now according to Eq. (2.3–2) Γ is Hermitian, and it can therefore be diagonalized by an $N \times N$ unitary transformation $\mathbf{U}$ (satisfying the conditions $\mathbf{U}\mathbf{U}^\dagger = 1 = \mathbf{U}^\dagger\mathbf{U}$), i.e.

$$\mathbf{U}\boldsymbol{\Gamma}\mathbf{U}^\dagger \equiv \widetilde{\boldsymbol{\Gamma}}, \tag{2.3–7}$$

where $\widetilde{\boldsymbol{\Gamma}}$ is a diagonal matrix. Hence we can rewrite the inequality (2.3–6) as

$$\mathbf{a}^\dagger\mathbf{U}^\dagger\mathbf{U}\boldsymbol{\Gamma}\mathbf{U}^\dagger\mathbf{U}\mathbf{a} \geqslant 0$$

or

$$\mathbf{b}^\dagger\widetilde{\boldsymbol{\Gamma}}\mathbf{b} \geqslant 0, \tag{2.3–8}$$

where $\mathbf{b} \equiv \mathbf{U}\mathbf{a}$ is another column matrix with arbitrary elements. If we choose $b_i = \delta_{ni}$, where δ_{ni} is the Kronecker symbol, Eq. (2.3–8) reduces to

$$\widetilde{\Gamma}_{nn} \geqslant 0, \tag{2.3–9}$$

$\widetilde{\Gamma}_{nn}$ being the n'th eigenvalue (diagonal element) of $\widetilde{\boldsymbol{\Gamma}}$. All eigenvalues are therefore non-negative, and so is the determinant det $\widetilde{\boldsymbol{\Gamma}}$, which is identical to the determinant det $\boldsymbol{\Gamma}$ of the original matrix and this evidently implies the inequality (2.3–5).

When the matrix Γ_{ij} is of order $N = 1$, Eq. (2.3–5) becomes identical to Eq. (2.3–1). When Γ_{ij} is of order $N = 2$, Eq. (2.3–5) reproduces the inequality (2.3–3). When $N = 3$ or greater a large number of new inequalities can be derived from Eq. (2.3–5), which we shall not explore here.

(f) If the stationary random process $z(t)$ is differentiable, then the 'velocity' $\xi(t) \equiv \dot{z}(t)$ (dot denotes differentiation with respect to time) of the process is a wide sense stationary random process of zero mean, and its autocorrelation function is

$$\Gamma_\xi(\tau) = -\ddot{\Gamma}_z(\tau). \tag{2.3–10}$$

To prove this result, we make use of the definition of the derivative and readily find that

$$\langle \xi(t) \rangle = \langle \dot{z}(t) \rangle = \operatorname*{Lim}_{h\to 0} \frac{\langle z(t+h) - z(t) \rangle}{h} = \operatorname*{Lim}_{h\to 0} \frac{\langle z \rangle - \langle z \rangle}{h} = 0. \tag{2.3–11}$$

Also

$$\begin{aligned}\Gamma_\xi(\tau) &= \langle \dot{z}^*(t)\dot{z}(t+\tau) \rangle \\ &= \operatorname*{Lim}_{\substack{h_1\to 0\\ h_2\to 0}} \left\langle \left[\frac{z^*(t+h_1) - z^*(t)}{h_1}\right]\left[\frac{z(t+\tau+h_2) - z(t+\tau)}{h_2}\right] \right\rangle \\ &= \operatorname*{Lim}_{\substack{h_1\to 0\\ h_2\to 0}} \left[\frac{\Gamma_z(\tau - h_1 + h_2) - \Gamma_z(\tau - h_1)}{h_2} - \frac{\Gamma_z(\tau + h_2) - \Gamma_z(\tau)}{h_2}\right]\frac{1}{h_1}\end{aligned}$$

$$= \lim_{h_1 \to 0} \frac{\dot{\Gamma}_z(\tau - h_1) - \dot{\Gamma}_z(\tau)}{h_1}$$

$$= -\ddot{\Gamma}_z(\tau).$$

$\xi(t)$ is therefore stationary, at least in the wide sense.

(g) If the stationary random process is differentiable, then $z(t)$ and its 'velocity' $\xi(t) \equiv \dot{z}(t)$ have the cross-correlation function $\dot{\Gamma}_z(\tau)$. Once again we have from first principles,

$$\langle z^*(t)\xi(t+\tau)\rangle = \lim_{h \to 0} \left\langle \frac{z^*(t)[z(t+\tau+h) - z(t+\tau)]}{h} \right\rangle$$

$$= \lim_{h \to 0} \frac{\Gamma_z(\tau + h) - \Gamma_z(\tau)}{h}$$

$$= \dot{\Gamma}_z(\tau). \tag{2.3–12}$$

As a corollary to this equation we conclude that a real, stationary differentiable random process $x(t)$ is uncorrelated with its derivative at the same time. For we find, using the symmetry property for the real process $x(t)$, namely

$$\Gamma(-\tau) = \Gamma(\tau),$$

that

$$\dot{\Gamma}(-\tau) = -\dot{\Gamma}(\tau)$$

$$\dot{\Gamma}(0) = -\dot{\Gamma}(0),$$

and therefore

$$\dot{\Gamma}(0) = 0. \tag{2.3–13}$$

Hence from Eq. (2.3–12)

$$\langle x(t)\dot{x}(t)\rangle = \dot{\Gamma}(0) = 0. \tag{2.3–14}$$

We conclude our discussion of the autocorrelation function by noting that its normalized version,

$$\gamma(\tau) = \frac{\Gamma(\tau)}{\Gamma(0)}, \tag{2.3–15}$$

is in its mathematical structure equivalent to a characteristic function. This is so because $\gamma(0) = 1$ and $\gamma(\tau)$, like $\Gamma(\tau)$, is non-negative definite [property (d) above] and therefore satisfies the requirements of Bochner's theorem (see Section 1.4.2). Consequently the Fourier transform of $\gamma(\tau)$ has all the properties of a probability density. As we shall see shortly, the Fourier transform of the un-normalized function $\Gamma(\tau)$ has an important physical significance.

The fact that the normalized autocorrelation function $\gamma(\tau)$ has the mathematical structure of a characteristic function is perhaps the single most important property of $\Gamma(\tau)$. It implies a wealth of results established in the field of probability theory (see, for example, Lukacs, 1970). The properties (a)–(g) that we have just discussed are consequences of this fact. Another important property of the autocorrelation function which can be deduced from this equivalence follows at once from Eq. (1.4–11), namely

(h) $\Gamma(\tau)$ is continuous.
This is so even when the random process $z(t)$ has jump discontinuities, such as, for example, the process illustrated in Fig. 2.5.

2.4 Spectral properties of a stationary random process

2.4.1 Spectral density and the Wiener–Khintchine theorem

One of the most important attributes of a stationary random process $z(t)$ is its spectrum. We could try to introduce the spectrum heuristically as follows. Let us formally represent $z(t)$ as a Fourier integral,

$$z(t) = \int_{-\infty}^{\infty} \tilde{z}(\omega)\,\mathrm{e}^{-\mathrm{i}\omega t}\,\mathrm{d}\omega, \tag{2.4–1a}$$

and let us assume that the integral exists and may be inverted, i.e. that

$$\tilde{z}(\omega) = \frac{1}{2\pi}\int_{-\infty}^{\infty} z(t)\,\mathrm{e}^{\mathrm{i}\omega t}\,\mathrm{d}t. \tag{2.4–1b}$$

We might then attempt to define the spectrum $S(\omega)$ of $z(t)$ by the expectation value of $|\tilde{z}(\omega)|^2$, i.e.

$$S(\omega) = \langle|\tilde{z}(\omega)|^2\rangle \tag{2.4–2}$$

so that $S(\omega)$ would be a measure of the strength of the fluctuations associated with a particular Fourier component of $z(t)$. However simple considerations show that the definition (2.4–2) is mathematically unsound. For if $z(t)$ is a stationary random process, it does not tend to zero as $t \to +\infty$ and $-\infty$, because the underlying probability densities that characterize the fluctuations of $z(t)$ are invariant with respect to the translation of the origin of time [see Eq. (2.2–1)]. Thus, in the statistical sense, $z(t)$ cannot behave any differently for large values of $|t|$ than it does for any other value of t. Consequently $z(t)$ is neither square-integrable nor absolutely integrable and hence the Fourier integral (2.4–1a) does not exist within the framework of the theory of ordinary functions.

The difficulty just noted was overcome by Wiener (1930) in a classic paper which was the origin of a whole new branch of mathematics, the so-called *generalized harmonic analysis*. Wiener considered a broad class of functions $z(t)$ [measurable in the sense of Lebesgue – see Titchmarsh (1939), Chapt. 10; Kestelman (1960), Chapt. 3] for which the integral

$$\Gamma(\tau) = \lim_{T\to\infty}\frac{1}{2T}\int_{-T}^{T} z^*(t)z(t+\tau)\,\mathrm{d}t \tag{2.4–3}$$

exists. Such functions evidently need not tend to zero as $T \to \pm\infty$. Wiener showed that the quantity

$$\sigma(\omega) = \frac{1}{2\pi}\int_{-\infty}^{\infty} \Gamma(\tau)\frac{\mathrm{e}^{\mathrm{i}\omega\tau}-1}{\mathrm{i}\tau}\,\mathrm{d}\tau, \tag{2.4–4}$$

which he called the spectrum of $z(t)$, also exists. For reasons that will soon become apparent, it is now more appropriate to refer to $\sigma(\omega)$ as the integrated spectrum. If $\sigma(\omega)$ is a differentiable function of ω and the order of differentiation and integration may be interchanged, it follows that

$$S(\omega) \equiv \frac{\mathrm{d}\sigma(\omega)}{\mathrm{d}\omega} = \frac{1}{2\pi}\int_{-\infty}^{\infty} \Gamma(\tau)\,\mathrm{e}^{\mathrm{i}\omega\tau}\,\mathrm{d}\tau. \tag{2.4–5}$$

The function $S(\omega)$ may be identified with the spectrum or, more precisely, with the *spectral density*, also called the *power spectrum*, of $z(t)$. Although Eq. (2.4–5) appears to have no resemblance to the formula (2.4–2) by which we introduced the spectrum heuristically, we shall soon see that the identification of the integral on the right of (2.4–5) with the spectral density of the random process $z(t)$ is quite appropriate.

Wiener's analysis pertained to a single function $z(t)$, rather than to an ensemble of functions, and he did not employ any statistical concepts in his analysis. However, when dealing with a stationary and ergodic ensemble of random functions, one can replace the autocorrelation function, defined by the time average (2.4–3), by the autocorrelation function

$$\Gamma(\tau) = \langle z^*(t)z(t+\tau)\rangle, \tag{2.4–6}$$

defined as an ensemble average, because the two averages are then equal.

Four years after the publication of Wiener's classic paper, Khintchine (1934) showed, with the help of Bochner's theorem (Section 1.4.2), that for a function $\Gamma(\tau)$ to be the autocorrelation function of a continuous stationary random process it must be expressible in the form of a Fourier–Stieltjes integral (Yaglom, 1962, Chapt. 2, Sec. 9)

$$\Gamma(\tau) = \int_{-\infty}^{\infty} \mathrm{e}^{-\mathrm{i}\omega\tau}\,\mathrm{d}\sigma(\omega), \tag{2.4–7}$$

where $\sigma(\omega)$ is a real, non-decreasing, bounded function. Although Khintchine's approach was entirely different from Wiener's, if one identifies Wiener's time-averaged function $\Gamma(\tau)$ with the autocorrelation function of an ergodic process, one may also identify Wiener's integrated spectrum with the distribution function $\sigma(\omega)$ in the Khintchine representation (2.4–7).

It is worth noting that both Wiener and Khintchine employed the notion of integrated spectrum rather than of spectral density, probably because the spectral density may become singular, though no more singular than the Dirac delta function.‡ In the following, as elsewhere in this book, we do not hesitate to use the Dirac delta function and we will, therefore, work with the spectral density. Its use can be justified rigorously within the framework of the theory of distributions or generalized function theory (Bremerman, 1965; Nussenzveig, 1972, Appendix A; Jones, 1982, and references cited in these texts). Actually, when one deals with a statistically stationary source and a stationary field, rather than with a

‡ This fact is evident from the formal analogy, noted at the end of Section 2.3, between the characteristic function and the autocorrelation function which, in turn, implies a formal analogy between the Fourier transforms of these quantities, i.e. between the probability density and the normalized spectral density. In particular, as noted in Section 1.3, the probability density cannot be more singular than the Dirac delta function and hence the same is true about the spectral density.

stationary random function, the use of singular functions and generalized functions may be avoided altogether, as has been shown by Wolf (1981, 1982). We will discuss this topic in Section 4.7.

Let us now examine why the integral in Eq. (2.4–5) may be identified with the spectral density. For this purpose we will use again the Fourier transform relations (2.4–1a) and (2.4–1b), regarding them now as symbolic formulas which, as just noted, can be given rigorous mathematical meaning, if one goes beyond ordinary function theory.

For each realization ${}^{(k)}z(t)$ of the stationary random process, the transform (2.4–1b) will generate a realization ${}^{(k)}\tilde{z}(\omega)$ and hence $\tilde{z}(\omega)$ is evidently also a random process, in which the frequency is the parameter rather than the time. Let us now consider the expectation, or ensemble average, of the product $\tilde{z}^*(\omega)\tilde{z}(\omega')$. From Eq. (2.4–1b) we have, if we interchange the operations of averaging and integration, that

$$\langle \tilde{z}^*(\omega)\tilde{z}(\omega')\rangle = \frac{1}{(2\pi)^2}\iint_{-\infty}^{\infty} \langle z^*(t)z(t')\rangle\, e^{i(\omega't'-\omega t)}\, dt\, dt'. \tag{2.4–8}$$

Since the process $z(t)$ is assumed to be stationary,

$$\langle z^*(t)z(t')\rangle = \Gamma(t'-t), \tag{2.4–9}$$

where Γ is the autocorrelation function of $z(t)$. On substituting from Eq. (2.4–9) into the integral in Eq. (2.4–8), and setting $t'-t=\tau$, we find that

$$\langle \tilde{z}^*(\omega)\tilde{z}(\omega')\rangle = \frac{1}{(2\pi)^2}\int_{-\infty}^{\infty} dt\, e^{i(\omega'-\omega)t}\int_{-\infty}^{\infty} d\tau\, \Gamma(\tau)\, e^{i\omega'\tau},$$

which implies that

$$\langle \tilde{z}^*(\omega)\tilde{z}(\omega')\rangle = \tilde{\Gamma}(\omega)\delta(\omega-\omega'), \tag{2.4–10}$$

where

$$\tilde{\Gamma}(\omega) = \frac{1}{2\pi}\int_{-\infty}^{\infty} \Gamma(\tau)\, e^{i\omega\tau}\, d\tau. \tag{2.4–11}$$

The formulas (2.4–10) and (2.4–11) are two very important relations. The first, Eq. (2.4–10), shows that the (generalized) Fourier components of a stationary random process belonging to different frequencies are uncorrelated, and that $\tilde{\Gamma}(\omega)$ is a measure of the strength of the fluctuations of the Fourier component at frequency ω, i.e. $\tilde{\Gamma}(\omega)$ may be identified with the spectral density $S(\omega)$ of $z(t)$:

$$S(\omega) \equiv \tilde{\Gamma}(\omega). \tag{2.4–12}$$

The singularity at $\omega'=\omega$ in Eq. (2.4–10) can be removed if we integrate both sides over a small ω'-range around ω. We then obtain, if we also use Eq. (2.4–12), the following expression for the spectral density:

$$S(\omega) = \lim_{\Delta\omega\to 0}\int_{\omega-\Delta\omega/2}^{\omega+\Delta\omega/2} \langle \tilde{z}^*(\omega)\tilde{z}(\omega')\rangle\, d\omega'. \tag{2.4–13}$$

The similarity between this expression for the spectral density and the naive definition (2.4–2) should be noted.

To bring out the significance of the formulas (2.4–10) and (2.4–11), we will re-write them in terms of $S(\omega)$ rather than $\widetilde{\Gamma}(\omega)$ by using Eq. (2.4–12):

$$\langle \tilde{z}^*(\omega)\tilde{z}(\omega')\rangle = S(\omega)\delta(\omega - \omega') \tag{2.4–14}$$

and

$$S(\omega) = \frac{1}{2\pi}\int_{-\infty}^{\infty} \Gamma(\tau)\,\mathrm{e}^{\mathrm{i}\omega\tau}\,\mathrm{d}\tau. \tag{2.4–15}$$

We may regard the formula (2.4–14) as *defining* the spectral density or power spectrum‡ $S(\omega)$ of a stationary random process $z(t)$. The expression (2.4–15) for the spectral density is seen to be in agreement with the (liberally interpreted) formula (2.4–5) based on Wiener's theory, provided that the function $z(t)$ in Wiener's definition (2.4–3) of $\Gamma(\tau)$ is regarded as a realization of a stationary ergodic random process.

The formula (2.4–15), together with its inverse

$$\Gamma(\tau) = \int_{-\infty}^{\infty} S(\omega)\,\mathrm{e}^{-\mathrm{i}\omega\tau}\,\mathrm{d}\omega, \tag{2.4–16}$$

are generally known as the *Wiener–Khintchine theorem*. Stated more explicitly the theorem asserts that *the autocorrelation function of a stationary random process and the spectral density (or power spectrum) of the process form a Fourier transform pair*§.

As an example we note that for the random telegraph signal treated in Section 2.2, whose autocorrelation function $\Gamma(\tau)$ is given by Eq. (2.2–22), $S(\omega)$ is the Lorentzian spectral density

$$S(\omega) = \frac{1}{2\pi}\frac{4Ra^2}{4R^2 + \omega^2}. \tag{2.4–17}$$

This function, illustrated in Fig. 2.6, is centered at zero frequency and has a width at half height of

$$\Delta\omega = 4R, \tag{2.4–18}$$

which is seen to be approximately the reciprocal of the range of $\Gamma(\tau)$.

‡ An alternative definition of the spectral density which is sometimes used, and which does not involve singular functions, generalized functions or Stieltjes integrals is based on the following truncation procedure. One defines the function $z_T(t)$ by

$$\begin{aligned} z_T(t) &= z(t) \quad \text{when } |t| < T \\ &= 0 \quad \text{when } |t| > T \end{aligned}$$

and its Fourier transform by

$$\tilde{z}(\omega; T) = \frac{1}{2\pi}\int_{-\infty}^{\infty} z_T(t)\,\mathrm{e}^{\mathrm{i}\omega\tau}\,\mathrm{d}\omega.$$

The spectral density may then be expressed in the form

$$S(\omega) = \operatorname*{Lim}_{T\to\infty}\frac{\langle \tilde{z}^*(\omega; T)\tilde{z}(\omega; T)\rangle}{2T}.$$

(cf. Goldman, 1953, Sec. 8.4; Middleton, 1960, Sec. 3.2).

§ It was not generally known until long after the publications of the papers by Wiener and Khintchine that the essential aspects of this theorem were discovered much earlier by Einstein (1914). For an English translation of this paper see Einstein (1987). An interesting commentary on Einstein's paper was given by Yaglom (1987).

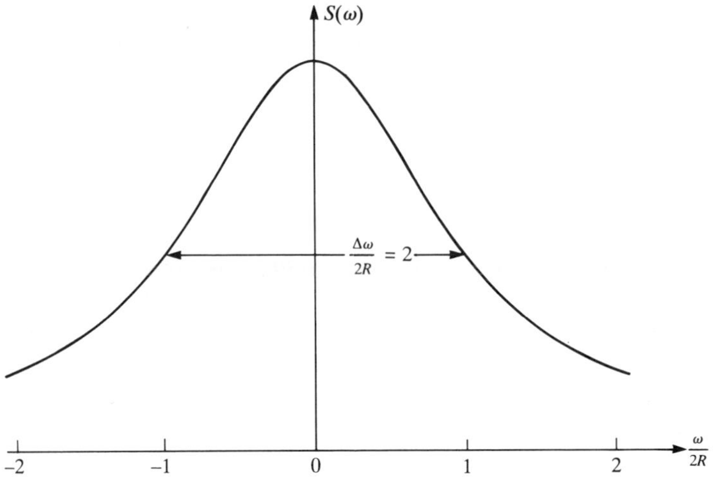

Fig. 2.6 Illustrating the form of the Lorentzian spectral density given by Eq. (2.4–17).

When the random process $z(t)$ describes an optical field, $S(\omega)$ will generally be peaked at optical frequencies of the order of $\omega = 10^{15}\ \mathrm{s}^{-1}$ and become vanishingly small outside the optical range.

2.4.2 Singularities of the spectral density

The spectral density $S(\omega)$, defined by Eq. (2.4–14), may contain delta function singularities if the mean is non-zero or if $z(t)$ contains oscillatory components. To illustrate how the singularities arise, we start by considering a stationary, ergodic random process $z_0(t)$ of zero mean, that satisfies the ergodicity condition (2.2–14). This condition ensures that the autocorrelation function $\Gamma_0(\tau)$ of $z_0(t)$ is absolutely integrable. Its Fourier transform $S_0(\omega)$ therefore exists and is a continuous function of ω. This last feature may be seen as follows. From Eq. (2.4–15) we have

$$|S_0(\omega + \delta\omega) - S_0(\omega)| = \frac{1}{2\pi}\left|\int_{-\infty}^{\infty} \Gamma_0(\tau)[e^{i(\omega+\delta\omega)\tau} - e^{i\omega\tau}]\,d\tau\right|$$
$$\leqslant \frac{1}{\pi}\int_{-\infty}^{\infty} |\Gamma_0(\tau)|\,|\sin[\tfrac{1}{2}(\delta\omega)\tau]|\,d\tau, \qquad (2.4\text{–}19)$$

and the right-hand side tends to zero as $\delta\omega \to 0$ for all ω. On the other hand, if correlations of $z(t)$ gradually die out, but so slowly that $\Gamma(\tau)$ is not absolutely integrable, then $S(\omega)$ may contain delta function singularities. To illustrate such a situation we consider a stationary random process $z(t)$ that differs from $z_0(t)$ only by a constant, which is of course the mean $\langle z \rangle$. Then the autocorrelation function $\Gamma(\tau)$ of $z(t)$ is related to $\Gamma_0(\tau)$ by

$$\Gamma(\tau) = \Gamma_0(\tau) + |\langle z \rangle|^2. \qquad (2.4\text{–}20)$$

It follows from the absolute integrability of $\Gamma_0(\tau)$, that $\Gamma(\tau) - |\langle z \rangle|^2$ is also absolutely integrable, and that the process $z(t)$ is ergodic, but evidently $\Gamma(\tau)$ is not absolutely integrable. If we take the Fourier transform of both sides of Eq. (2.4–20) and use Eq. (2.4–15), we obtain for the spectral density $S(\omega)$ of $z(t)$,

$$S(\omega) = S_0(\omega) + |\langle z \rangle|^2 \delta(\omega), \tag{2.4–21}$$

where $\delta(\omega)$ is the Dirac delta function. By virtue of the non-zero mean of the process $z(t)$, its spectral density now has a delta function singularity at zero frequency.

Finally, we suppose that the random process contains periodic contributions at various frequencies $\omega_1, \omega_2, \ldots$ as well. Then its autocorrelation function $\Gamma(\tau)$ will have the general form

$$\Gamma(\tau) = \Gamma_0(\tau) + |\langle z \rangle|^2 + \sum_j A_j \mathrm{e}^{-\mathrm{i}\omega_j \tau}. \tag{2.4–22}$$

Its spectral density, obtained by taking the Fourier transform of the expression (2.4–22), is given by

$$S(\omega) = S_0(\omega) + |\langle z \rangle|^2 \delta(\omega) + \sum_j A_j \delta(\omega - \omega_j), \tag{2.4–23}$$

and it is seen to have additional delta function singularities at each of the frequencies $\omega_1, \omega_2, \ldots$.

2.4.3 *Normalized correlations and normalized spectral densities*

It is sometimes convenient to introduce a normalized autocorrelation function $\gamma(\tau)$ and a normalized spectral density $\mathscr{s}(\omega)$ for a random process $z(t)$ (Davenport and Root, 1958, Chapts. 4 and 6). When $\langle z \rangle \neq 0$ we define $\gamma(\tau)$ by the formula

$$\gamma(\tau) = \frac{\Gamma(\tau) - |\langle z \rangle|^2}{\Gamma(0) - |\langle z \rangle|^2}. \tag{2.4–24}$$

Evidently $\gamma(0) = 1$. Also by virtue of the inequality (2.3–3) we have

$$0 \leqslant |\gamma(\tau)| \leqslant 1. \tag{2.4–25}$$

The subtractions of $|\langle z \rangle|^2$ in the definition ensure that $\gamma(\tau) \to 0$ as $\tau \to \infty$ when the process $z(t)$ is ergodic, because $\Gamma(\tau)$ must then tend to $|\langle z \rangle|^2$ as the correlations die out. As defined, $\gamma(\tau)$ is mathematically equivalent to a characteristic function, and its Fourier transform $\mathscr{s}(\omega)$, known as the normalized spectral density, has the properties of a true probability density. We therefore have

$$\mathscr{s}(\omega) = \frac{1}{2\pi} \int_{-\infty}^{\infty} \gamma(\tau) \, \mathrm{e}^{\mathrm{i}\omega\tau} \, \mathrm{d}\tau \tag{2.4–26}$$

and

$$\int_{-\infty}^{\infty} \mathscr{s}(\omega) \, \mathrm{d}\omega = 1. \tag{2.4–27}$$

By using the definition (2.4–24) of $\gamma(\tau)$ in Eq. (2.4–26), we can immediately relate $\mathscr{s}(\omega)$ to the un-normalized spectral density $S(\omega)$, and we find

$$\mathscr{s}(\omega) = \frac{S(\omega) - |\langle z \rangle|^2 \delta(\omega)}{\Gamma(0) - |\langle z \rangle|^2}. \tag{2.4–28}$$

We note that the singularity in $S(\omega)$ when $\langle z \rangle$ is non-zero is subtracted out. Therefore, $\mathfrak{s}(\omega)$ is a continuous function of ω for an ergodic process $z(t)$, whether $z(t)$ is of zero mean or not.

Because $\gamma(0) = 1$ and $\gamma(\tau)$ is absolutely integrable for an ergodic process [cf. Eq. (2.2–14)], the area under the curve $|\gamma(\tau)|^2$ is always finite and is a convenient measure of the range over which correlations of $z(t)$ extend in time. We can therefore define a *correlation time* T_c for the random process $z(t)$ by

$$T_c \equiv \int_{-\infty}^{\infty} |\gamma(\tau)|^2 \, d\tau. \tag{2.4–29}$$

By Parseval's theorem connecting the Fourier conjugates $\gamma(\tau)$ and $\mathfrak{s}(\omega)$, T_c is also given by

$$T_c = 2\pi \int_{-\infty}^{\infty} \mathfrak{s}^2(\omega) \, d\omega. \tag{2.4–30}$$

The reciprocal of T_c provides a convenient measure of the bandwidth of $\mathfrak{s}(\omega)$.

Other measures of coherence time and bandwidth (e.g. based on moments of time with respect to the correlation function or moments of frequency with respect to the spectral density) can be constructed. Some are discussed in Section 4.3.3, in connection with the correlation properties of an optical field.

2.4.4 *Cross-correlations and cross-spectral densities*

The cross-correlation function of two real random processes $x_1(t)$ and $x_2(t)$ is defined by the average product $\langle x_1(t_1)x_2(t_2)\rangle$ at two different times t_1, t_2, in complete analogy with the definition of the autocorrelation. If the two processes $z_1(t)$ and $z_2(t)$ are complex, we define the cross-correlation function by the formula

$$\Gamma_{12}(t, t + \tau) = \langle z_1^*(t) z_2(t + \tau)\rangle. \tag{2.4–31}$$

If $z_1(t)$ and $z_2(t)$ are jointly stationary, then $\Gamma_{12}(t, t+\tau)$, like $\Gamma(t, t+\tau)$, is a function of τ only, and one may then denote it by $\Gamma_{12}(\tau)$. It obeys the condition

$$\Gamma_{21}(\tau) = \Gamma_{12}^*(-\tau), \tag{2.4–32}$$

which can readily be derived from the definition of Γ_{12} by making a translation of the origin of time. However, many of the interesting properties of the autocorrelation function $\Gamma(\tau)$ are not shared by the cross-correlation function $\Gamma_{12}(\tau)$. For example, in general $\Gamma_{12}(\tau)$ does not have the properties of a characteristic function.

Let $z_1(t), z_2(t), \ldots, z_N(t)$ be a set of N different, jointly stationary random processes. Then

$$\Gamma_{ij}(\tau) \equiv \langle z_i^*(t) z_j(t + \tau)\rangle, \quad (i, j = 1, 2, \ldots, N) \tag{2.4–33}$$

is an $N \times N$ matrix known as the *cross-correlation matrix*. If the means $\langle z_i(t)\rangle$ for $i = 1, 2, \ldots, N$ are all zero, $\Gamma_{ij}(\tau)$ is often called the *covariance matrix* of the set of random processes. It is not difficult to show that $\Gamma_{ij}(\tau)$ obeys a kind of non-negative definiteness condition that is reminiscent of the conditions (2.3–4) obeyed by the autocorrelation function. To derive this condition we observe that for an arbitrary set of n times $t_1, t_2, \ldots, t_n$ $(n \leqslant N)$ and for an arbitrary set of n complex numbers $a_1, a_2, \ldots, a_n$,

$$\left|\sum_{i=1}^{n} a_i z_i(t_i)\right|^2 \geqslant 0.$$

On taking expectations of both sides of this inequality and using Eq. (2.4–33) we obtain the inequality

$$\sum_{i=1}^{n}\sum_{j=1}^{n} a_i^* a_j \Gamma_{ij}(t_j - t_i) \geqslant 0. \tag{2.4–34}$$

By analogy with the definition (2.4–14) of the spectral density $S(\omega)$, we may define the so-called *cross-spectral density* (or cross power spectrum), $W_{ij}(\omega)$, of the jointly stationary random processes $z_i(t)$ and $z_j(t)$ by the formula

$$\langle \tilde{z}_i^*(\omega)\tilde{z}_j(\omega')\rangle = W_{ij}(\omega)\delta(\omega - \omega'). \tag{2.4–35}$$

In this formula $\tilde{z}_i(\omega)$ is the (generalized) Fourier transform of $z_i(t)$ [cf. Eq. (2.4–1b)]. Formula (2.4–35) shows that (generalized) Fourier components that belong to different frequencies are uncorrelated. The cross-spectral density $W_{ij}(\omega)$ is evidently a measure of the correlations between the fluctuations of different components at the same frequency. When $j = i$ the cross-spectral density reduces to the spectral density, i.e.

$$W_{ii}(\omega) = S_i(\omega), \tag{2.4–36}$$

where $S_i(\omega)$ is the spectral density of the random process $z_i(t)$.

By an argument similar to that used in deriving Eq. (2.4–15), one can readily show that

$$W_{ij}(\omega) = \frac{1}{2\pi}\int_{-\infty}^{\infty} \Gamma_{ij}(\tau)\,\mathrm{e}^{\mathrm{i}\omega\tau}\,\mathrm{d}\tau, \tag{2.4–37}$$

i.e. the cross-spectral density function of $z_i(t)$ and $z_j(t)$ is the Fourier transform of their cross-correlation function. Equation (2.4–37), together with its Fourier inverse

$$\Gamma_{ij}(\tau) = \int_{-\infty}^{\infty} W_{ij}(\omega)\,\mathrm{e}^{-\mathrm{i}\omega\tau}\,\mathrm{d}\omega, \tag{2.4–38}$$

may evidently be regarded as a *generalized Wiener–Khintchine theorem*.

We note some properties of the cross-spectral density. $W_{ij}(\omega)$, just like $\Gamma_{ij}(\tau)$, is an element of an $N \times N$ matrix. However, unlike $\Gamma_{ij}(\tau)$, $W_{ij}(\omega)$ is Hermitian, for it follows from Eqs. (2.4–32) and (2.4–37) that

$$W_{ji}(\omega) = W_{ij}^*(\omega). \tag{2.4–39}$$

Although, unlike the spectral density, the cross-spectral density is neither real nor necessarily positive in general, it may be shown to satisfy a non-negative definiteness condition somewhat similar to the condition (2.4–34) for the cross-correlation matrix, namely:

$$\sum_{i=1}^{n}\sum_{j=1}^{n} a_i^* a_j W_{ij}(\omega) \geqslant 0. \tag{2.4–40}$$

Here again n is a positive integer and $a_1, a_2, \ldots, a_n$ $(n \leqslant N)$ are arbitrary complex numbers. To establish the inequality (2.4–40) we start with the obvious inequality

$$\left\langle \left| \int_{\omega_0-\varepsilon_1}^{\omega_0+\varepsilon_2} \left[\sum_{i=1}^{n} a_i \tilde{z}_i(\omega) \right] \mathrm{d}\omega \right|^2 \right\rangle \geqslant 0, \tag{2.4–41}$$

where ω_0 is any chosen frequency and $\varepsilon_1 > 0$, $\varepsilon_2 > 0$ are arbitrary. It follows from this inequality that

$$\left\langle \left| \int_{\omega_0-\varepsilon_1}^{\omega_0+\varepsilon_2} \mathrm{d}\omega \int_{\omega_0-\varepsilon_1}^{\omega_0+\varepsilon_2} \mathrm{d}\omega' \left[\sum_{i=1}^{n} \sum_{j=1}^{n} a_i^* a_j \tilde{z}_i^*(\omega) \tilde{z}_j(\omega') \right] \right| \right\rangle \geqslant 0.$$

If we interchange the orders of averaging and of the integrations and summations, we may re-write the inequality in the form

$$\int_{\omega_0-\varepsilon_1}^{\omega_0+\varepsilon_2} \mathrm{d}\omega \int_{\omega_0-\varepsilon_1}^{\omega_0+\varepsilon_2} \mathrm{d}\omega' \left[\sum_{i=1}^{n} \sum_{j=1}^{n} a_i^* a_j \langle \tilde{z}_i^*(\omega) \tilde{z}_j(\omega') \rangle \right] \geqslant 0. \tag{2.4–42}$$

Next we substitute for the expectation value in this inequality from Eq. (2.4–35) and carry out the trivial integration with respect to ω'. We then obtain the simpler inequality

$$\int_{\omega_0-\varepsilon_1}^{\omega_0+\varepsilon_2} \left[\sum_{i=1}^{n} \sum_{j=1}^{n} a_i^* a_j W_{ij}(\omega) \right] \mathrm{d}\omega \geqslant 0. \tag{2.4–43}$$

Since ε_1 and ε_2 are arbitrary non-negative quantities if the $W_{ij}(\omega)$'s are continuous functions of ω, the non-negative definiteness condition (2.4–40) follows at once from this formula.

In the special case when $n = 1$, the inequality (2.4–40), together with Eq. (2.4–36), implies that

$$W_{11}(\omega) \equiv S_1(\omega) \geqslant 0, \tag{2.4–44}$$

i.e. that the spectral density is real and non-negative, as we learned earlier. If we take $n = 2$, the inequality (2.4–40) and Eq. (2.4–44) imply that the determinant

$$\begin{vmatrix} S_1(\omega) & W_{12}(\omega) \\ W_{21}(\omega) & S_2(\omega) \end{vmatrix} \geqslant 0,$$

or, if we also use the Hermiticity of $\mathbf{W}$, expressed by Eq. (2.4–39),

$$|W_{12}(\omega)| \leqslant [S_1(\omega) S_2(\omega)]^{1/2}. \tag{2.4–45}$$

We may define a normalized cross-correlation function $\gamma_{ij}(\tau)$ by the formula

$$\begin{aligned} \gamma_{ij}(\tau) &\equiv \frac{\Gamma_{ij}(\tau) - \langle z_i^* \rangle \langle z_j \rangle}{[\Gamma_{ii}(0) - |\langle z_i \rangle|^2]^{1/2} [\Gamma_{jj}(0) - |\langle z_j \rangle|^2]^{1/2}} \\ &= \frac{\langle \Delta z_i^*(t) \Delta z_j(t+\tau) \rangle}{[\langle |\Delta z_i|^2 \rangle \langle |\Delta z_j|^2 \rangle]^{1/2}}, \end{aligned} \tag{2.4–46}$$

which will be recognized as the correlation coefficient of $\Delta z_i(t)$ and $\Delta z_j(t+\tau)$ [cf.

Eq. (1.3–35)]. It is normalized so that

$$0 \leqslant |\gamma_{ij}(\tau)| \leqslant 1, \tag{2.4–47}$$

with the upper bound of unity achievable only when the two processes $z_i(t)$ and $z_j(t)$ are completely correlated. In general $\gamma_{ij}(0) \neq 1$.

Let us denote the Fourier transform of $\gamma_{ij}(\tau)$ by $w_{ij}(\omega)$:

$$w_{ij}(\omega) = \frac{1}{2\pi}\int_{-\infty}^{\infty} \gamma_{ij}(\tau)\, e^{i\omega\tau}\, d\tau. \tag{2.4–48}$$

This quantity is sometimes known as the normalized cross-spectral density, although it is not normalized in any obvious way. Another normalized cross-spectral density function is often more useful for problems in which the random process is of zero mean. It is normalized like a correlation coefficient [cf. Eq. (1.3–35)], and is given by (Wolf, 1982, 1986; see also Carter and Wolf, 1975; Wolf and Carter, 1975, 1976; Mandel and Wolf, 1976, 1981)

$$\mu_{ij}(\omega) = \frac{W_{ij}(\omega)}{[W_{ii}(\omega)]^{1/2}[W_{jj}(\omega)]^{1/2}} \tag{2.4–49a}$$

or, with the help of Eq. (2.4–36),

$$\mu_{ij}(\omega) = \frac{W_{ij}(\omega)}{[S_i(\omega)]^{1/2}[S_j(\omega)]^{1/2}}. \tag{2.4–49b}$$

$\mu_{ij}(\omega)$ measures the degree of correlation between two Fourier components of $z_i(t)$ and $z_j(t)$ at the frequency ω. From the inequality (2.4–45) it follows immediately that

$$0 \leqslant |\mu_{ij}(\omega)| \leqslant 1. \tag{2.4–50}$$

Normalized correlation and spectral density functions will be discussed in more detail within the context of optics, in Section 4.3 below.

2.5 Orthogonal representation of a random process

In their study of the statistical properties of blackbody radiation, Einstein and Hopf (1910) suggested that, in any finite time interval, the field may be represented in the form of a Fourier series with coefficients which are normally distributed, independent random variables with zero mean. Shortly afterwards von Laue (1915a) questioned the correctness of this suggestion and a polemic ensued (Einstein, 1915; von Laue, 1915b), but no agreement was reached. Blackbody radiation may be shown to be a stationary Gaussian random process (see Section 2.1.2) and the Einstein–Hopf representation was later used for the representation of such processes in spite of the unresolved disagreements.

The question of expanding a random process in a Fourier series was later studied by Davis (1953) and Root and Pitcher (1955). It follows from these papers that, for the Einstein–Hopf representation to exist, the process must be not only Gaussian but also periodic with period T, i.e. if one has for all t, $x(t + T) = x(t)$, with probability 1. Blackbody radiation obviously does not satisfy the periodicity requirement, so that, as von Laue had claimed, the Einstein–Hopf expansion cannot be strictly valid. However it follows from the analysis of Root and Pitcher that, if in an arbitrary interval $0 \leq t \leq T$ one represents a stationary Gaussian random process $x(t)$ as a Fourier series with random coefficients, the correlation between the coefficients tends to zero as $T \to \infty$. In this limit, the Fourier expansion does not exist, of course, but the result implies that if T is large enough, the assumption of independent coefficients, though not strictly correct, is in general, a reasonable approximation.

In 1947 Kac and Siegert introduced a new type of expansion for a stationary Gaussian random process, in which the expansion coefficients are *strictly independent*. At about the same time Karhunen (1946) showed that, whether or not a process is stationary or Gaussian, an 'orthogonal expansion' of the process is, in general, possible. Expansions of the Karhunen–Loéve type have been finding increasing use in treatments of random processes. We will, therefore, briefly discuss such expansions.

2.5.1 The Karhunen–Loéve expansion

Let us consider a complex random process $z(t)$, $(-T \leq t \leq T)$, which is not necessarily stationary. For the sake of simplicity we assume that

$$\langle z(t) \rangle = 0. \tag{2.5–1}$$

We consider the possibility of expanding each realization of the process in the form

$$z(t) = \sum_n c_n \psi_n(t), \tag{2.5–2}$$

where the functions $\psi_n(t)$ form an orthonormal set on the interval $-T \leq t \leq T$, i.e.

$$\int_{-T}^{T} \psi_n^*(t) \psi_m(t)\, \mathrm{d}t = \delta_{nm}, \tag{2.5–3}$$

δ_{nm} being the Kronecker symbol. The coefficients c_n are uncorrelated random variables, i.e.

$$\langle c_n^* c_m \rangle = \lambda_n \delta_{nm}, \tag{2.5–4}$$

and the λ_n's are non-negative constants.

Let us for the moment assume that such an expansion exists and let us examine whether we can determine the functions $\psi_n(t)$ and the constants λ_n. For this purpose we form the autocorrelation function $\Gamma(t_1, t_2)$ of the process. Evidently

$$\begin{aligned} \Gamma(t_1, t_2) &= \langle z^*(t_1) z(t_2) \rangle \\ &= \sum_n \sum_m \langle c_n^* c_m \rangle \psi_n^*(t_1) \psi_m(t_2) \\ &= \sum_m \lambda_m \psi_m^*(t_1) \psi_m(t_2), \end{aligned} \tag{2.5–5}$$

when Eqs. (2.5–2) and (2.5–4) are used. If we multiply both sides of Eq. (2.5–5) by $\psi_n(t_1)$, integrate with respect to t_1 over the range $-T \leq t_1 \leq T$, and use the orthogonality condition (2.5–3), we obtain the relation

$$\int_{-T}^{T} \Gamma(t_1, t_2)\psi_n(t_1)\,\mathrm{d}t_1 = \lambda_n \psi_n(t_2). \tag{2.5–6}$$

Equation (2.5–6) is a homogeneous Fredholm integral equation for the ψ_n's. More precisely, the ψ_n's are seen to be the eigenfunctions of an integral operator, whose kernel, assumed to be continuous, is the autocorrelation function $\Gamma(t_1, t_2)$ of the process, and the λ_n's are the corresponding eigenvalues. Now the kernel $\Gamma(t_1, t_2)$ obeys the Hermiticity condition $\Gamma(t_2, t_1) = \Gamma^*(t_1, t_2)$, and we shall also assume that it is a Hilbert–Schmidt kernel, i.e. that

$$\int_{-T}^{T}\int_{-T}^{T} |\Gamma(t_1, t_2)|^2\,\mathrm{d}t_1\,\mathrm{d}t_2 < \infty. \tag{2.5–7}$$

Under these conditions one may draw the following conclusions about the eigenfunctions and eigenvalues of our integral equation (Riesz and Nagy, 1955, Sec. 97; Pogorzelski, 1966, Chapt. 5; Smithies, 1970, Chapt. 7):

(a) The integral equation (2.5–6) has at least one non-zero eigenvalue.
(b) Each eigenvalue has at most a finite degeneracy.
(c) There is a (finite or infinite) orthonormal sequence of eigensolutions of Eq. (2.5–6) with the property that any function $F(t)$ that is square-integrable on the interval $-T \leqslant t \leqslant T$, can be represented, in the sense of convergence in the mean‡, in the form

$$F(t) = h(t) + \sum_n f_n \psi_n(t), \tag{2.5–8}$$

where $h(t)$ is a function such that

$$\int_{-T}^{T} \Gamma(t_1, t_2)h(t_1)\,\mathrm{d}t_1 = 0, \tag{2.5–9}$$

and

$$f_n = \int_{-T}^{T} \psi_n^*(t)F(t)\,\mathrm{d}t. \tag{2.5–10}$$

(d) The kernel $\Gamma(t_1, t_2)$ of the integral equation (2.5–6), i.e. the autocorrelation function of the process, may be expanded in the form of Eq. (2.5–5), with the series converging to $\Gamma(t_1, t_2)$ in the mean.

The results that we have just stated hold under the assumption that $\Gamma(t_1, t_2)$ is a Hermitian, Hilbert–Schmidt kernel, i.e. that it obeys the condition (2.5–7) (cf. Riesz and Nagy, 1955, p. 365). However, one can readily show that Γ is also non-negative definite, i.e. that for any N real or complex numbers $a_1, a_2, \ldots, a_N$,

$$\sum_{i=1}^{N}\sum_{j=1}^{N} a_i^* a_j \Gamma(t_i, t_j) \geqslant 0. \tag{2.5–11}$$

This inequality can be established in a way strictly similar to that used in deriving

‡ A series $\sum_{n=1}^{N} s_n(t)$ is said to converge in the mean to $S(t)$ if $\mathrm{Lim}_{N\to\infty} \int_{-\infty}^{\infty} |S(t) - \sum_{n=1}^{N} s_n(t)|^2\,\mathrm{d}t = 0$.

the corresponding inequality (2.3–4a) for the autocorrelation function of a stationary random process. With this property of Γ one can show that

(e) The expansion (2.5–5) is uniformly convergent (Mercer's theorem).
(f) The eigenvalues are non-negative real numbers.

Finally we note that, if the non-negative definiteness condition on Γ may be replaced by the stronger condition of positive definiteness, i.e. if the equality sign in Eq. (2.5–11), or, more precisely, in the equivalent integral form [see Eq. (2.3–4b)], cannot be attained except in the trivial case when $\Gamma(t) \equiv 0$, then Eq. (2.5–9) can hold only if $h(t) \equiv 0$. In this case, as is evident from Eq. (2.5–8), the eigenfunctions form a *complete* set in the Hilbert space of square-integrable functions.

It follows that, if the sample functions of our random process $z(t)$ are square-integrable on the interval $-T \leqslant t \leqslant T$ and the autocorrelation function $\Gamma(t_1, t_2)$ of the process obeys the convergence condition (2.5–7) and is positive definite (rather than just non-negative definite), an expansion of the form (2.5–2) exists, subject to conditions (2.5–3) and (2.5–4). This representation is known as the *Karhunen and Loéve expansion* of the random process $z(t)$.

2.5.2 *The limit $T \to \infty$; an alternative approach to the Wiener–Khintchine theorem*

Suppose that the random process $z(t)$, again assumed to be of zero mean, is stationary, at least in the wide sense, so that its autocorrelation function is of the form

$$\Gamma(t_1, t_2) = \Gamma(t_2 - t_1). \tag{2.5–12}$$

We will consider, for this case, the limiting form of the Karhunen–Loéve expansion as $T \to \infty$, and we will show that it leads to the Wiener–Khintchine theorem.

By analogy with many eigenvalue problems, we may expect tha in the limit as $T \to \infty$, the spectrum of the integral operator $\int_{-T}^{T} \Gamma(t_2 - t_1) \ldots \mathrm{d}t_1$ will be continuous rather than discrete. For example, the eigenvalue problem associated with a harmonic oscillator on a finite interval leads to a Fourier series, whereas on an infinite interval it leads to a Fourier integral. Hence, instead of a series representation of the form given by Eq. (2.5–2), we may now expect an integral representation of the form

$$z(t) = \int_{-\infty}^{\infty} c(\omega)\psi(t; \omega)\,\mathrm{d}\omega, \tag{2.5–13}$$

and, in place of Eqs. (2.5–3) and (2.5–4), we may expect orthogonality conditions of the form

$$\frac{1}{2\pi}\int_{-\infty}^{\infty} \psi^*(t; \omega)\psi(t; \omega')\,\mathrm{d}t = \delta(\omega - \omega') \tag{2.5–14}$$

$$\langle c^*(\omega)c(\omega')\rangle = \lambda(\omega)\delta(\omega - \omega'), \tag{2.5–15}$$

where $\delta(\omega)$ is the Dirac delta function. We will try to determine the functions $\psi(t; \omega)$ and $\lambda(\omega)$ that obey these conditions.

We will proceed as we did before for finite T. We assume that a representation of the form (2.5–13) exists, subject to the requirements (2.5–14) and (2.5–15). It then follows that

$$\Gamma(t_2 - t_1) = \langle z^*(t_1) z(t_2) \rangle$$
$$= \iint_{-\infty}^{\infty} \langle c^*(\omega_1) c(\omega_2) \rangle \psi^*(t_1; \omega_1) \psi(t_2; \omega_2) \, d\omega_1 \, d\omega_2,$$

and hence, with the help of Eq. (2.5–15),

$$\Gamma(t_2 - t_1) = \int_{-\infty}^{\infty} \lambda(\omega) \psi^*(t_1; \omega) \psi(t_2; \omega) \, d\omega. \tag{2.5–16}$$

We now multiply both sides of the equation by $\psi(t_1; \omega')$ and integrate with respect to t_1 over the range $-\infty$ to ∞. We then readily find, after rearranging the order of the terms on the right-hand side, that

$$\int_{-\infty}^{\infty} \Gamma(t_2 - t_1) \psi(t_1; \omega') \, dt_1 = \int_{-\infty}^{\infty} d\omega \lambda(\omega) \psi(t_2; \omega) \int_{-\infty}^{\infty} dt_1 \psi^*(t_1; \omega) \psi(t_1; \omega'),$$

which implies that

$$\frac{1}{2\pi} \int_{-\infty}^{\infty} \Gamma(t_2 - t_1) \psi(t_1; \omega') \, dt_1 = \lambda(\omega') \psi(t_2; \omega'). \tag{2.5–17}$$

The last line follows immediately with the help of Eq. (2.5–14). The relation (2.5–17) corresponds to the integral equation (2.5–6) derived for the case when the time interval is finite and the process is not necessarily stationary.

If the assumed representation (2.5–13) exists, the functions $\psi(t; \omega)$ must be the eigenfunctions and the $\lambda(\omega)$'s the eigenvalues of the integral equation (2.5–17). We must, therefore, examine whether there exist non-trivial solutions to this equation. We will do so using Fourier transform techniques. We represent the autocorrelation function $\Gamma(\tau)$ and the eigenfunctions $\psi(t; \omega)$, if they exist, as Fourier integrals:

$$\Gamma(\tau) = \int_{-\infty}^{\infty} S(\eta) \, e^{i\eta\tau} \, d\eta, \tag{2.5–18}$$

$$\psi(t; \omega) = \int_{-\infty}^{\infty} \tilde{\psi}(\eta; \omega) \, e^{-i\eta t} \, d\eta. \tag{2.5–19}$$

Then, if we make use of the convolution theorem on Fourier transforms, we obtain from Eq. (2.5–17) the relation

$$S(\eta) \tilde{\psi}(\eta; \omega) = \lambda(\omega) \tilde{\psi}(\eta; \omega). \tag{2.5–20}$$

Now if we attempt to cancel the common factor $\tilde{\psi}(\eta; \omega)$ [which cannot be identically zero under our assumption that a non-trivial eigensolution of Eq. (2.5–17) exists], we are led to the conclusion that $S(\eta) = \lambda(\omega)$, which is impossible if $S(\eta)$ is not a constant. Thus, we conclude that Eq. (2.5–20) does not admit solutions for $\tilde{\psi}(\eta; \omega)$ that are ordinary functions. However, it is not difficult to see that the equation has singular solutions, a result that can be established rigorously by means of the theory of generalized functions. As elsewhere in this book, we avoid the use of generalized function theory, and instead we make formal use of the Dirac delta function. The singular solutions of (2.5–20) may then be expressed in the form

$$\tilde{\psi}(\eta;\, \omega) = \delta(\eta - \omega) \tag{2.5–21}$$

$$\lambda(\omega) = S(\omega). \tag{2.5–22}$$

These solutions must be understood in the sense that if we substitute them into Eq. (2.5–20), multiply each side of the equation by an arbitrary test function $F(\eta)$, and integrate both sides with respect to η from $-\infty$ to ∞, we obtain a consistent result.

Let us now examine the implication of such singular solutions. From Eqs. (2.5–21) and (2.5–22) it follows at once that the eigenfunctions of the integral equation (2.5–17) are the functions

$$\psi(t;\, \omega) = \mathrm{e}^{-\mathrm{i}\omega t}, \tag{2.5–23}$$

i.e. they are periodic exponentials, *irrespective* of the nature of the autocorrelation function $\Gamma(\tau)$ of the process. The orthogonal representation (2.5–13) then becomes

$$z(t) = \int_{-\infty}^{\infty} c(\omega)\, \mathrm{e}^{-\mathrm{i}\omega t}\, \mathrm{d}\omega, \tag{2.5–24}$$

which is just the Fourier integral representation of $z(t)$. From what has been said above it is clear that this equation must also be interpreted in the sense of generalized function theory. Moreover, according to Eqs. (2.5–15) and (2.5–22),

$$\langle c^*(\omega) c(\omega') \rangle = S(\omega)\delta(\omega - \omega'). \tag{2.5–25}$$

Thus, we have indeed found a limiting form, as $T \to \infty$, of the Karhunen–Loéve expansion for a random process that is stationary, at least in the wide sense.

Comparison of Eqs. (2.5–24) and (2.5–25) with Eqs. (2.4–1a) and (2.4–14) shows that the eigenvalue $\lambda(\omega) = S(\omega)$ of the integral equation (2.5–17) is the spectral density of the random process $z(t)$. Moreover, from Eq. (2.5–18) $S(\omega)$ is the Fourier transform of the autocorrelation function $\Gamma(\tau)$. This conclusion is nothing else than a statement of the Wiener–Khintchine theorem (cf. Section 2.4.1), which has been derived here from the Karhunen–Loéve expansion of the stationary random process $z(t)$ in the limit $T \to \infty$. Moreover, this expansion on the infinite interval is seen to be just the generalized Fourier integral representation of $z(t)$.

2.6 Time development and classification of random processes

2.6.1 Conditional probability densities

We have already seen that there exists a whole hierarchy of joint probability densities $p_1(x, t)$, $p_2(x_2, t_2; x_1, t_1)$, . . . , in which the higher-order p_n generally contain progressively more and more information about the random process $x(t)$ and how it develops in time (see, for example Oppenheim, Shuler and Weiss 1977, Chapt. 2). Because the future time development of $x(t)$ is in general influenced by its past behavior, it is often convenient to introduce *conditional probability densities* $\mathcal{P}_{n,k}$, giving the n-fold joint probability density that x has the value x_{k+1} at time t_{k+1}, x_{k+2} at time t_{k+2}, . . . , and x_{k+n} at time t_{k+n}, conditioned on the values x_k at the earlier time t_k, x_{k-1} at time t_{k-1}, . . . , x_1 at time t_1. In analogy with Eq. (1.2–12), $\mathcal{P}_{n,k}$ is expressible as the ratio of two probability densities

$$\mathcal{P}_{n,k}(x_{k+n}, t_{k+n}; \ldots ; x_{k+1}, t_{k+1}|x_k, t_k; \ldots ; x_1, t_1)$$
$$= \frac{p_{k+n}(x_{k+n}, t_{k+n}; x_{k+n-1}, t_{k+n-1}; \ldots ; x_1, t_1)}{p_k(x_k, t_k; x_{k-1}, t_{k-1}; \ldots ; x_1, t_1)}, \quad (2.6\text{–}1)$$

and the times are now understood to be ordered so that $t_1 \leqslant t_2 \leqslant t_3 \leqslant \ldots$ $\mathcal{P}_{n,k}$ is normalized to unity in the usual way when it is integrated over all $x_{k+1}, x_{k+2}, \ldots, x_{k+n}$.

If we multiply both sides of Eq. (2.6–1) by p_k, put $k = 1 = n$ and integrate both sides with respect to x_1, we obtain

$$\int p_2(x_2, t_2; x_1, t_1)\,\mathrm{d}x_1 = p_1(x_2, t_2) = \int \mathcal{P}_{1,1}(x_2, t_2|x_1, t_1)p_1(x_1, t_1)\,\mathrm{d}x_1. \quad (2.6\text{–}2)$$

This allows the probability distribution of x at a later time to be obtained from the distribution at an earlier time, when the conditional probability density $\mathcal{P}_{1,1}$ is known. However $\mathcal{P}_{1,1}$ may be governed by the past history of the process. In the special case in which the two times t_1, t_2 are equal, $\mathcal{P}_{1,1}$ must reduce to a delta function,

$$\mathcal{P}_{1,1}(x_2, t|x_1, t) = \delta(x_2 - x_1). \quad (2.6\text{–}3)$$

How strongly the values of x at earlier times influence $x(t)$ at later times t depends on the dynamics of the random process. Some random processes evolve in ways that are only weakly influenced by their history, and in other cases $x(t)$ is strongly influenced by the past. In the following we introduce a natural classification, based on the influence of the past.

2.6.2 Completely random or separable process

This is a process $x(t)$ in which successive values of x are completely independent of past values. It follows that if the times are all different, then the conditional probability density equals the unconditional probability density, or

$$\mathcal{P}_{n,k}(x_{k+n}, t_{k+n}; \ldots ; x_{k+1}, t_{k+1}|x_k, t_k; \ldots ; x_1, t_1) =$$
$$p_n(x_{k+n}, t_{k+n}; \ldots ; x_{k+1}, t_{k+1}). \quad (2.6\text{–}4)$$

With the help of Eq. (2.6–1), and on putting $k = 1$, we obtain

$$p_{n+1}(x_{n+1}, t_{n+1}; \ldots ; x_1, t_1) = p_n(x_{n+1}, t_{n+1}; \ldots ; x_2, t_2)p_1(x_1, t_1),$$

which is a recursion formula that allows p_{n+1} to be expressed in terms of p_n. Repeated application leads to the result,

$$p_n(x_n, t_n; \ldots ; x_1, t_1) = p_1(x_n, t_n)p_1(x_{n-1}, t_{n-1}) \ldots p_1(x_1, t_1), \quad (2.6\text{–}5)$$

which shows that the variates $x_1, x_2, \ldots, x_n$ at different times are all statistically independent of each other. The random process $x(t)$ is therefore also known as *separable*. An example is a delta-correlated Gaussian random process, for which the two-time correlation function $\langle x(t_1)x(t_2)\rangle$ is proportional to $\delta(t_2 - t_1)$, because the lack of correlation of the variates at different times implies their statistical independence.

2.6.3 First-order Markov process

After considering a process which is independent of its history, we now consider one which is influenced only by its immediate or most recent past, for which the conditional probability density satisfies the relation

$$\mathcal{P}_{n,k}(x_{k+n}, t_{k+n}; \ldots; x_{k+1}, t_{k+1}|x_k, t_k; \ldots; x_1, t_1)$$
$$= \mathcal{P}_{n,1}(x_{k+n}, t_{k+n}; \ldots; x_{k+1}, t_{k+1}|x_k, t_k). \quad (t_1 \leqslant t_2 \leqslant t_3 \leqslant \ldots) \quad (2.6\text{–}6)$$

In other words, only the most recent past governs the time evolution of the process. We can express this by saying that the future of $x(t)$ is influenced by the present, but not by its earlier history. Such a process is known as a *Markov process*, or more precisely as a *first-order Markov process*, to distinguish it from higher-order Markov processes described below.

It follows from Eq. (2.6–1) that

$$p_n(x_n, t_n; \ldots; x_1, t_1) =$$
$$\mathcal{P}_{1,n-1}(x_n, t_n|x_{n-1}, t_{n-1}; \ldots; x_1, t_1)p_{n-1}(x_{n-1}, t_{n-1}; \ldots; x_1, t_1),$$

and with the help of the Markovian property (2.6–6) we have

$$p_n(x_n, t_n; \ldots; x_1, t_1) = \mathcal{P}_{1,1}(x_n, t_n|x_{n-1}, t_{n-1})p_{n-1}(x_{n-1}, t_{n-1}; \ldots; x_1, t_1).$$

This is in the form of a recursion relation, and repeated application leads to the result

$$p_n(x_n, t_n; \ldots; x_1, t_1) =$$
$$\mathcal{P}_{1,1}(x_n, t_n|x_{n-1}, t_{n-1})\mathcal{P}_{1,1}(x_{n-1}, t_{n-1}|x_{n-2}, t_{n-2}) \ldots \mathcal{P}_{1,1}(x_2, t_2|x_1, t_1)p_1(x_1, t_1).$$
$$(2.6\text{–}7)$$

For a Markov process the conditional probability density $\mathcal{P}_{1,1}(x_n, t_n|x_{n-1}, t_{n-1})$ is known as the *transition probability density*. It is governed only by the dynamics of the random process, and is independent of the behavior at times earlier than t_{n-1}. The joint probability density p_n of any order n is then completely determined by p_1 together with the transition probability density $\mathcal{P}_{1,1}$, whose suffices (1, 1) are often dropped because they are understood. This makes the treatment of the first-order Markov process particularly simple.

The transition probability density $\mathcal{P}(x_n, t_n|x_{n-1}, t_{n-1})$ obeys a simple integral relation for a Markov process (Smoluchowski, 1906; Chapman, 1916; Kolmogoroff, 1931). We start from the general consistency relation with $t_3 \geqslant t_2 \geqslant t_1$

$$p_2(x_3, t_3; x_1, t_1) = \int p_3(x_3, t_3; x_2, t_2; x_1, t_1)\, dx_2,$$

which holds for any process $x(t)$, and we use Eq. (2.6–7) to substitute for p_3 under the integral. We then obtain

$$p_2(x_3, t_3; x_1, t_1) = \int \mathcal{P}(x_3, t_3|x_2, t_2)\mathcal{P}(x_2, t_2|x_1, t_1)p_1(x_1, t_1)\,\mathrm{d}x_2,$$

and division of both sides by $p_1(x_1, t_1)$ leads to the integral relation

$$\mathcal{P}(x_3, t_3|x_1, t_1) = \int \mathcal{P}(x_3, t_3|x_2, t_2)\mathcal{P}(x_2, t_2|x_1, t_1)\,\mathrm{d}x_2,$$

$$(t_3 \geqslant t_2 \geqslant t_1) \quad (2.6\text{–}8)$$

which is known as the *Smoluchowski–Chapman–Kolmogorov relation*. It is a necessary, although not sufficient, condition for the process to be a first-order Markov process.

2.6.4 Higher-order Markov process

The next stage in the categorization of random processes is reached when the conditional probability density $\mathcal{P}_{n,k}$ depends only on the two most recent values of x, i.e.

$$\mathcal{P}_{n,k}(x_{k+n}, t_{k+n}; \ldots x_{k+1}, t_{k+1}|x_k, t_k; \ldots; x_1, t_1)$$
$$= \mathcal{P}_{n,2}(x_{k+n}, t_{k+n}; \ldots; x_{k+1}, t_{k+1}|x_k, t_k; x_{k-1}, t_{k-1}). \quad (2.6\text{–}9)$$

Such a process is known as a *second-order Markov process*. With the help of the same kind of argument as that leading to Eq. (2.6–7) one finds from Eq. (2.6–9)

$$p_n(x_n, t_n; \ldots; x_1, t_1)$$
$$= \mathcal{P}_{1,2}(x_n, t_n|x_{n-1}, t_{n-1}; x_{n-2}, t_{n-2})\mathcal{P}_{1,2}(x_{n-1}, t_{n-1}|x_{n-2}, t_{n-2}; x_{n-3}, t_{n-3})$$
$$\times \ldots \mathcal{P}_{1,2}(x_3, t_3|x_2, t_2; x_1, t_1)p_2(x_2, t_2; x_1, t_1), \quad (2.6\text{–}10)$$

so that all higher-order joint probability densities are determined by $\mathcal{P}_{1,2}$ and p_2. From this equation we obtain the following generalization of the Smoluchowski–Chapman–Kolmogorov relation (2.6–8):

$$\mathcal{P}_{1,2}(x_4, t_4|x_2, t_2; x_1, t_1) = \int \mathcal{P}_{1,2}(x_4, t_4|x_3, t_3; x_2, t_2)\mathcal{P}_{1,2}(x_3, t_3|x_2, t_2; x_1, t_1)\,\mathrm{d}x_3,$$

$$(t_4 \geqslant t_3 \geqslant t_2 \geqslant t_1). \quad (2.6\text{–}11)$$

It is clear that we can introduce a hierarchy of higher-order Markov processes by proceeding in this manner. A third-order Markov process would be one in which the evolution of the random process is governed by the three most recent values of $x(t)$, and all fourth- and higher-order joint probability densities are expressible in terms of $\mathcal{P}_{1,3}$ and p_3, and so on. The essential element in a Markov process of any order is that memory about the past history of the process does not persist indefinitely, but eventually dies out.

Finally, there exist random processes whose time evolution depends on their entire history. In that case no finite-order transition probability density $\mathcal{P}_{n,k}$ can completely describe the evolution.

2.7 Master equations in integro-differential form

We shall now use the general integral relation (2.6–2) that connects the probability density $p_1(x_2, t_2)$ with $p_1(x_1, t_1)$ at an earlier time to obtain the time derivative of $p_1(x, t)$. If we put $t_1 = t$ and $t_2 = t + \delta t$ in Eq. (2.6–2) we obtain (after dropping the suffix 1 on $p_1(x, t)$ for simplicity)

$$p(x, t + \delta t) = \int \mathcal{P}_{1,1}(x, t + \delta t | x_1, t) p(x_1, t)\, dx_1. \tag{2.7–1}$$

By making use of the normalization of $\mathcal{P}_{1,1}$ $(x_1, t + \delta t | x, t)$ when it is integrated with respect to x_1 we may also write

$$p(x, t) = \int \mathcal{P}_{1,1}(x_1, t + \delta t | x, t) p(x, t)\, dx_1. \tag{2.7–2}$$

It then follows from first principles that

$$\begin{aligned}
\frac{\partial p(x, t)}{\partial t} & \\
&= \lim_{\delta t \to 0} \frac{1}{\delta t}[p(x, t + \delta t) - p(x, t)] \\
&= \lim_{\delta t \to 0} \frac{1}{\delta t}\int [\mathcal{P}_{1,1}(x, t + \delta t | x_1, t) p(x_1, t) - \mathcal{P}_{1,1}(x_1, t + \delta t | x, t) p(x, t)]\, dx_1 \\
&= \lim_{\delta t \to 0} \frac{1}{\delta t}\int \{[\mathcal{P}_{1,1}(x, t + \delta t | x_1, t) - \mathcal{P}_{1,1}(x, t | x_1, t)] p(x_1, t) \\
&\qquad - [\mathcal{P}_{1,1}(x_1, t + \delta t | x, t) - \mathcal{P}_{1,1}(x_1, t | x, t)] p(x, t)\}\, dx_1,
\end{aligned} \tag{2.7–3}$$

where we have made explicit use of the result [cf. Eq. (2.6–3)]

$$\mathcal{P}_{1,1}(x, t | x_1, t) = \delta(x - x_1) = \mathcal{P}_{1,1}(x_1, t | x, t).$$

We now introduce the following notation for the rate of change of $\mathcal{P}_{1,1}$,

$$\mathcal{A}(x, x_1, t) \equiv \lim_{\delta t \to 0} \frac{1}{\delta t}[\mathcal{P}_{1,1}(x, t + \delta t | x_1, t) - \mathcal{P}_{1,1}(x, t | x_1, t)]. \tag{2.7–4}$$

When $\mathcal{P}_{1,1}$ is a transition probability density, $\mathcal{A}(x, x_1, t)$ is known as the *transition rate*. It satisfies a number of simple relations, which follow immediately from the definition. First

$$\mathcal{A}(x, x_1, t) \geqslant 0 \quad \text{if } x \neq x_1, \tag{2.7–5}$$

because $\mathcal{P}_{1,1}(x, t | x_1, t)$ is then zero, whereas $\mathcal{P}_{1,1}(x, t + \delta t | x_1, t)$ is non-negative. Also

$$\int \mathcal{A}(x, x_1, t)\, dx = 0, \tag{2.7–6}$$

because both conditional probability densities in Eq. (2.7–4) integrate to unity. However, $\mathcal{A}(x, x_1, t)$ can be a highly singular function in some cases, as we shall see in the example in Section 2.10 below. When the definition (2.7–4) is used in Eq. (2.7–3), we immediately obtain the equation of motion

$$\frac{\partial p(x,t)}{\partial t} = \int [\mathcal{A}(x, x_1, t)p(x_1, t) - \mathcal{A}(x_1, x, t)p(x, t)]\, dx_1. \qquad (2.7\text{–}7)$$

This equation is sometimes known as the *Pauli master equation*, after Pauli (1928, p. 30) who obtained a similar equation for a quantum system, with the help of certain assumptions that are not needed here (see Section 17.3 below). It has the character of a rate equation, and it tells us that $p(x, t)$ increases at a rate governed by transition rates from other values x_1 to x, and decreases at a rate governed by transition rates from x to other values x_1.

By virtue of the property (2.7–6), the second term in Eq. (2.7–7) integrates to zero, and the relation may be written as

$$\frac{\partial p(x,t)}{\partial t} = \int \mathcal{A}(x, x_1, t)p(x_1, t)\, dx_1, \qquad (2.7\text{–}8)$$

although the more symmetric form (2.7–7) is sometimes preferable. In the special case when the transition rate is symmetric,

$$\mathcal{A}(x, x_1, t) = \mathcal{A}(x_1, x, t), \qquad (2.7\text{–}9)$$

Eq. (2.7–7) simplifies to

$$\frac{\partial p(x,t)}{\partial t} = \int \mathcal{A}(x, x_1, t)[p(x_1, t) - p(x, t)]\, dx_1. \qquad (2.7\text{–}10)$$

Finally we note that for a random process taking on only the discrete values x_1, $x_2, x_3, \ldots$ the master equation (2.7–7) takes the form

$$\frac{\partial p(x_i,t)}{\partial t} = \sum_{j\neq i} [\mathcal{A}(x_i, x_j, t)p(x_j, t) - \mathcal{A}(x_j, x_i, t)p(x_i, t)]. \qquad (2.7\text{–}11)$$

In Eq. (2.7–11) $p(x_i, t)$ is the probability rather than probability density.

Although the master equation we have derived, holds in principle for any random process $x(t)$, it is generally useful only for a first-order Markov process. The reason is that only in that case is $\mathcal{P}_{1,1}(x, t + \delta t|x_1, t)$ determined entirely by the dynamics of the process and independent of all other probabilities, and similarly for the transition rate $\mathcal{A}(x, x_1, t)$. In the more general situation, $\mathcal{A}(x, x_1, t)$ depends on the probabilities of the process at earlier times. It is then preferable to deal with the time evolution of the joint probability density p_n, which involves transition rates conditioned on earlier values of the random process (Oppenheim, Shuler and Weiss, 1977, chapt. 3; Srinivas and Wolf, 1977). We shall not deal with this more general case here.

2.8 Master equations in differential form

We shall now derive an equation of motion for $p(x, t)$ in the form of a partial differential equation, which is often more useful than the integro-differential equation (2.7–7).

2.8.1 *The Kramers–Moyal differential equation*

Our starting point is again Eq. (2.7–1) for $p(x, t+\delta t)$, which allows us to express the rate of change of $p(x, t)$ in the form

$$\begin{aligned}\frac{\partial p(x,t)}{\partial t} &= \lim_{\delta t\to 0}\frac{1}{\delta t}[p(x, t+\delta t) - p(x,t)] \\ &= \lim_{\delta t\to 0}\frac{1}{\delta t}\left[\int \mathcal{P}(x, t+\delta t|x_1, t)p(x_1, t)\,\mathrm{d}x_1 - p(x,t)\right].\end{aligned}$$

We have discarded the suffices 1, 1 of $\mathcal{P}_{1,1}$, with the understanding that the relation is really useful only when $\mathcal{P}_{1,1}$ is a true transition probability density. We now make the change of variable $x - x_1 \equiv \Delta x$, and write

$$\frac{\partial p(x,t)}{\partial t} = \lim_{\delta t\to 0}\frac{1}{\delta t}\left[\int \mathcal{P}(x - \Delta x + \Delta x, t+\delta t|x - \Delta x, t)p(x-\Delta x, t)\,\mathrm{d}\Delta x - p(x,t)\right]. \tag{2.8–1}$$

The integrand can be regarded as a function $f(y, z)$ of the two variables $y = x - \Delta x$ and $z = \Delta x$. With respect to the variable y, let us expand $f(y, z)$ in a Taylor series about $y = x$, keeping z constant,

$$f(x - \Delta x, z) = \sum_{r=0}^{\infty}\frac{(-1)^r}{r!}(\Delta x)^r\frac{\partial^r}{\partial x^r}f(x, z). \tag{2.8–2}$$

When this expansion is substituted under the integral in Eq. (2.8–1) we obtain the equation

$$\frac{\partial p(x,t)}{\partial t} = \lim_{\delta t\to 0}\frac{1}{\delta t}\left[\int \sum_{r=0}^{\infty}\frac{(-\Delta x)^r}{r!}\frac{\partial^r}{\partial x^r}\mathcal{P}(x + \Delta x, t+\delta t|x, t)p(x, t)\,\mathrm{d}\Delta x - p(x,t)\right]. \tag{2.8–3}$$

We now interchange the order of summation and integration and integrate term by term. The term $r = 0$ integrates to $p(x, t)$, which cancels the last term. For the remaining terms we define the quantities

$$D_r(x,t) \equiv \lim_{\delta t\to 0}\frac{1}{\delta t}\int (\Delta x)^r\mathcal{P}(x+\Delta x, t+\delta t|x, t)\,\mathrm{d}\Delta x, \quad r = 1, 2, \ldots, \tag{2.8–4a}$$

which are known as the *transition moments* of the random process $x(t)$. The transition moment $D_r(x, t)$ is proportional to the r'th moment of the change of the process in a short time δt, subject to the initial value x at time t. This is sometimes expressed in the form

$$D_r(x,t) \equiv \lim_{\delta t\to 0}\frac{\langle(\Delta x)^r\rangle_{x,t}}{\delta t}. \tag{2.8–4b}$$

With the help of the definition (2.8–4a), Eq. (2.8–3) can be re-written in the more compact form

$$\frac{\partial p(x,t)}{\partial t} = \sum_{r=1}^{\infty} \frac{(-1)^r}{r!} \frac{\partial^r}{\partial x^r}[D_r(x,t)p(x,t)], \tag{2.8–5}$$

which is known as the *Kramers–Moyal differential equation* (Kramers, 1940; Moyal, 1949). In general, it is a partial differential equation of infinite order, although the order may become finite in certain cases, as we shall see.

The effect of the first term on the right involving D_1 on the change of $p(x, t)$ in a short time δt is given by

$$-\delta t \frac{\partial}{\partial x}[D_1(x,t)p(x,t)],$$

and this is illustrated in Fig. 2.7(a) for the case in which D_1 is a positive constant. The gradient of $p(x, t)$ determines the change of $p(x, t)$ and the term involving D_1 clearly causes the probability density to drift to the right when it is positive. $D_1(x, t)$ is therefore known as the *drift coefficient*.

The effect of the term involving D_2 on the change of $p(x, t)$ is given by

$$\tfrac{1}{2}\delta t \frac{\partial^2}{\partial x^2}[D_2(x,t)p(x,t)],$$

and this is illustrated in Fig. 2.7(b) for the case D_2 is constant and positive. Evidently D_2 causes the probability density to spread out, and it is known as the *diffusion coefficient*. The higher transition moments have no well-established names.

Finally, we point out once again that, although no explicit Markovian assumption was made in the derivation of Eq. (2.8–5), because Eq. (2.7–1) that forms the starting point holds quite generally, it is nevertheless only for a first-order Markov process that the results are really useful. The reason is that only when $\mathcal{P}(x, t|x_0, t_0)$ is a transition probability density are the transition moments determined by the dynamics of the random process, and independent of the probabilities of x at earlier times.

When $x(t)$ is a first-order Markov process, it is not difficult to see that the transition probability density $\mathcal{P}(x, t|x_0, t_0)$ must also obey a partial differential

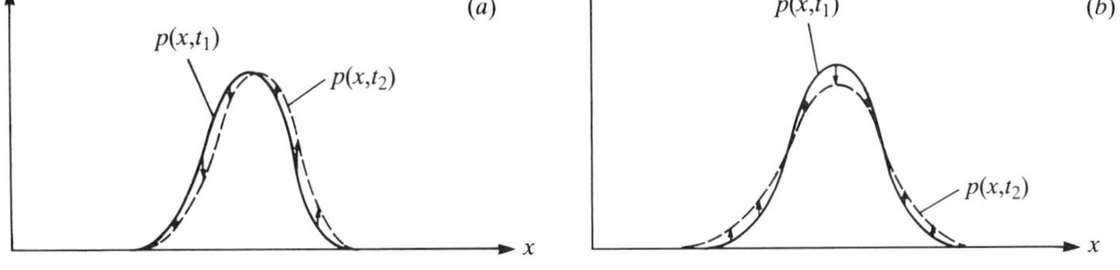

Fig. 2.7 Effect of (a) the drift term and (b) the diffusion term on the change of the probability density $p(x, t)$ in a short time interval from t_1 to t_2.

equation like Eq. (2.8–5). From the Smoluchowski–Chapman–Kolmogorov relation (2.6–8) we have

$$\mathcal{P}(x, t + \delta t|x_0, t_0) = \int \mathcal{P}(x, t + \delta t|x_1, t_1)\mathcal{P}(x_1, t_1|x_0, t_0)\,\mathrm{d}x_1, \quad (t + \delta t \geqslant t_1 \geqslant t_0). \tag{2.8–6}$$

This relation is completely analogous to Eq. (2.7–1), except that $p(x, t)$ on both sides of the equation is replaced by the conditional or transition probability density $\mathcal{P}(x, t|x_0, t_0)$. As the value $x_0(t_0)$ plays no role in the differentiation or integration, the argument leading to Eq. (2.8–5) can be repeated step by step, and we obtain instead of Eq. (2.8–5)

$$\frac{\partial \mathcal{P}(x, t|x_0, t_0)}{\partial t} = \sum_{r=1}^{\infty} \frac{(-1)^r}{r!} \frac{\partial^r}{\partial x^r}[D_r(x, t)\mathcal{P}(x, t|x_0, t_0)]. \tag{2.8–7}$$

$\mathcal{P}(x, t|x_0, t_0)$ and $p(x, t)$ therefore obey the same equation of motion for a first-order Markov process.

2.8.2 *Vector random process*

One sometimes needs to deal with a random process that has several different components, and is most conveniently treated as a vector $\mathbf{x}(t)$. For example, the complex random process $z(t)$, with real and imaginary components $x_1(t)$ and $x_2(t)$, may be regarded as a real, two-dimensional vector process $\mathbf{x}(t)$. For a vector process Eq. (2.7–1) still holds, except that the integration is with respect to the vector $\mathbf{x}_1$, and we write

$$p(\mathbf{x}, t + \delta t) = \int \mathcal{P}(\mathbf{x}, t + \delta t|\mathbf{x}_1, t)p(\mathbf{x}_1, t)\,\mathrm{d}\mathbf{x}_1 \tag{2.8–8}$$

The argument leading to the partial differential equation (2.8–5) can now be repeated with only trivial changes. The Taylor series (2.8–2) becomes an expansion of a scalar function of a vector, and we have instead

$$f(\mathbf{x} - \mathbf{\Delta x}) = f(x) - \Delta x_i \frac{\partial}{\partial x_i} f(\mathbf{x}) + \frac{1}{2!}\Delta x_i \Delta x_j \frac{\partial^2}{\partial x_i \partial x_j} f(\mathbf{x}) - \ldots, \tag{2.8–9}$$

where summation over repeated indices is implied. This leads to the partial differential equation

$$\frac{\partial p(\mathbf{x}, t)}{\partial t} = -\frac{\partial}{\partial x_i}[D_i(\mathbf{x}, t)p(\mathbf{x}, t)] + \frac{1}{2!}\frac{\partial^2}{\partial x_i \partial x_j}[D_{ij}(\mathbf{x}, t)p(\mathbf{x}, t)] - \ldots, \tag{2.8–10}$$

in which the r'th transition moment is a Cartesian tensor of rank r,

$$D_{ijk\ldots}(\mathbf{x}, t) = \lim_{\delta t \to 0} \frac{1}{\delta t} \int \Delta x_i \Delta x_j \Delta x_k \ldots \mathcal{P}(\mathbf{x} + \mathbf{\Delta x}, t + \delta t|\mathbf{x}, t)\,\mathrm{d}\mathbf{\Delta x}. \tag{2.8–11}$$

Except for this feature, the equation of motion is similar to Eq. (2.8–5) and an analogous relation can also be derived for $\mathcal{P}(\mathbf{x}, t|\mathbf{x}_0, t_0)$ for a first-order vector Markov process $\mathbf{x}(t)$. The first-order moment $D_i(\mathbf{x}, t)$ is known as the *drift vector*, and the second-order moment $D_{ij}(\mathbf{x}, t)$ as the *diffusion tensor*.

2.8.3 *The order of the Kramers–Moyal differential equation*

Although the Kramers–Moyal differential equation (2.8–5) is formally of infinite order, the order can become finite if all transition moments D_r for r greater than some value N vanish. We now use an argument of Lax (1966) to show that there are severe constraints on the order N.

From the Schwarz inequality we have for any integers r_1, r_2,

$$\left[\int (\Delta x)^{r_1}(\Delta x)^{r_2}\mathcal{P}(x+\Delta x, t+\delta t|x, t)\,\mathrm{d}\Delta x\right]^2$$

$$\leqslant \int (\Delta x)^{2r_1}\mathcal{P}(x+\Delta x, t+\delta t|x, t)\,\mathrm{d}\Delta x$$

$$\times \int (\Delta x)^{2r_2}\mathcal{P}(x+\Delta x, t+\delta t|x, t)\,\mathrm{d}\Delta x,$$

and with the help of the definition (2.8–4) of the transition moment D_r this leads to the inequality

$$D^2_{r_1+r_2} \leqslant D_{2r_1}D_{2r_2}. \tag{2.8–12}$$

As this relation holds for any integer values r_1, r_2 we may, in particular, put $r_1 = 1$, $r_2 = N - 1$, and obtain

$$D^2_N \leqslant D_2 D_{2N-2}. \tag{2.8–13}$$

Now let us suppose that all transition moments D_r above some value $r = N$ vanish. Then the Kramers–Moyal equation becomes an N'th-order partial differential equation. If $N \geqslant 3$, then $2N - 2 \geqslant N + 1$, so that both $D_{N+1} = 0$ and $D_{2N-2} = 0$. But from Eq. (2.8–13) we see that if $D_{2N-2} = 0$, then $D_N = 0$. It follows that if the transition moments vanish above order N, they also vanish above order $N - 1$.

This argument can be repeated recursively, to show that all transition moments above $N - 2$, $N - 3$, etc. vanish also, so long as the order remains greater than the second. When $N = 2$ or $N = 1$, the argument breaks down, because it is then no longer true that $2N - 2 \geqslant N + 1$. It follows that, if the Kramers–Moyal differential equation is of finite order, then it must be of the first or of the second order. In that case the equation is known as the *Fokker–Planck equation*. A similar result, known as Marcinkiewicz's theorem (see e.g. Lukacs, 1970, p. 213), which we encountered in Section 1.4.3, holds in connection with the theory of characteristic functions.

2.9 Langevin equation and Fokker–Planck equation

Many processes $\mathbf{x}(t)$ in nature are governed by an equation of motion of the general form

$$\frac{\mathrm{d}\mathbf{x}}{\mathrm{d}t} = \mathbf{A}(\mathbf{x}, t) + \mathbf{q}(t), \tag{2.9–1}$$

Brownian motion, the left-hand side is the rate of change of momentum and the right-hand side represents the fluctuating force on the particle. This force has been divided into its short-term average $\mathbf{A}(\mathbf{x}, t)$ together with a rapidly fluctuating part $\mathbf{q}(t)$ representing the departure from the average. As $\mathbf{q}(t)$ is a random process, it follows that $\mathbf{x}(t)$ is also a random process in general, and the equation of motion (2.9–1) is known as the *stochastic equation*.

In order to determine the statistical properties of $\mathbf{x}(t)$ we have to specify the statistics of $\mathbf{q}(t)$. We shall take $\mathbf{q}(t)$ to be a Gaussian random process of zero mean

$$\langle \mathbf{q}(t) \rangle = 0, \tag{2.9–2}$$

with extremely rapid fluctuations. We shall therefore approximate the two-time correlation function by a delta function, and write

$$\langle q_i(t) q_j(t') \rangle = g_{ij}(t) \delta(t - t'). \tag{2.9–3}$$

Under these conditions, Eq. (2.9–1) is usually known as the *Langevin equation*.‡ As the random process $\mathbf{q}(t)$ evolves in its own way so as to drive $\mathbf{x}(t)$, it is apparent that $\mathbf{x}(t_1)$ will be uncorrelated with $\mathbf{q}(t_2)$ at a later time t_2, or

$$\langle x_i(t_1) q_j(t_2) \rangle = 0, \quad t_2 > t_1. \tag{2.9–4}$$

Moreover, the future evolution of $\mathbf{x}(t)$ through Eq. (2.9–1) is governed only by the present, and the past history of $\mathbf{x}(t)$ plays no role, so that $\mathbf{x}(t)$ is Markovian of the first order. The statistics of $\mathbf{x}(t)$ are therefore completely determined by the probability density $p(\mathbf{x}, t)$ and by the transition probability density $\mathcal{P}(\mathbf{x}, t | \mathbf{x}_0, t_0)$, which obey Fokker–Planck equations, as we now show.

2.9.1 Transition moments for the Langevin process

We start by calculating the first-order transition moment. After integrating Eq. (2.9–1) over a very short time interval δt, we have

$$\Delta x_i(t) \equiv x_i(t + \delta t) - x_i(t) = A_i(\mathbf{x}, t)\delta t + \int_t^{t+\delta t} q_i(t')\,\mathrm{d}t'. \tag{2.9–5}$$

Although δt is very small, the integral in Eq. (2.9–5) cannot be replaced by $q_i(t)\delta t$, because $q_i(t')$, being delta-correlated, can fluctuate even in an infinitesimal interval. If we calculate the average of both sides of this equation, subject to the constraint that $\mathbf{x}(t)$ has some given value, divide by δt and proceed to the limit $\delta t \to 0$, we obtain the drift vector $D_i(\mathbf{x}, t)$ of the random process [cf. Eq. (2.8–4b)]. Now

$$\begin{aligned} \langle x_i(t + \delta t) - x_i(t) \rangle_{\mathbf{x},t} &= A_i(\mathbf{x}, t)\delta t + \int_t^{t+\delta t} \langle q_i(t') \rangle\,\mathrm{d}t' \\ &= A_i(\mathbf{x}, t)\,\delta t, \end{aligned}$$

by virtue of Eq. (2.9–2), so that

$$D_i(\mathbf{x}, t) = A_i(\mathbf{x}, t). \tag{2.9–6}$$

‡ For a more complete discussion of Langevin and Fokker–Planck equations see Risken, 1984.

We have therefore shown that the average forcing term in the Langevin equation is the drift vector of the random process. This is also apparent from the fact that $A_i(\mathbf{x}, t)$ is the conditional average of the velocity of $\mathbf{x}(t)$.

Next we calculate the diffusion tensor from Eq. (2.9–5). This requires that we average the product $\Delta x_i \Delta x_j$ subject to the constraint (indicated by subscript $\mathbf{x}$, t) that $\mathbf{x}$ has a given value at time t. We obtain

$$D_{ij}(\mathbf{x}, t) = \operatorname*{Lim}_{\delta t\to 0} \frac{1}{\delta t}\left\langle\left[A_i(\mathbf{x}, t)\delta t + \int_t^{t+\delta t} q_i(t')\,\mathrm{d}t'\right]\left[A_j(\mathbf{x}, t)\delta t + \int_t^{t+\delta t} q_j(t')\,\mathrm{d}t'\right]\right\rangle_{\mathbf{x},t}$$

$$= \operatorname*{Lim}_{\delta t\to 0}\left[A_i(\mathbf{x}, t)A_j(\mathbf{x}, t)\delta t + \left\langle A_i(\mathbf{x}, t)\int_t^{t+\delta t} q_j(t')\right\rangle_{\mathbf{x},t}\mathrm{d}t'\right.$$

$$\left. + \left\langle A_j(\mathbf{x}, t)\int_t^{t+\delta t} q_i(t')\right\rangle_{\mathbf{x},t}\mathrm{d}t' + \frac{1}{\delta t}\int_t^{t+\delta t}\int_t^{t+\delta t}\langle q_i(t')q_j(t'')\rangle\,\mathrm{d}t'\,\mathrm{d}t''\right].$$

The first term on the right vanishes in the limit $\delta t \to 0$, and the second and third terms average to zero by virtue of the fact that $\mathbf{x}(t)$ is independent of $\mathbf{q}(t')$ at later times t' [cf. Eq. (2.9–4)]. For the fourth term we make use of Eq. (2.9–3), and obtain finally

$$\begin{aligned} D_{ij}(\mathbf{x}, t) &= \operatorname*{Lim}_{\delta t\to 0} \frac{1}{\delta t}\int_t^{t+\delta t}\int_t^{t+\delta t} g_{ij}(t')\delta(t' - t'')\,\mathrm{d}t'\,\mathrm{d}t'' \\ &= g_{ij}(t). \end{aligned} \qquad (2.9\text{–}7)$$

It follows that the strength of the Langevin noise yields the diffusion tensor.

A similar argument can now be used to calculate any higher-order transition moment $D_{ijk}(\mathbf{x}, t)$, etc., and it will be found that all higher-order moments vanish in the limit $\delta t \to 0$. It follows from Eq. (2.8–10) that $p(\mathbf{x}, t)$ obeys the second-order Fokker–Planck equation

$$\frac{\partial p(\mathbf{x}, t)}{\partial t} = -\frac{\partial}{\partial x_i}[A_i(\mathbf{x}, t)p(\mathbf{x}, t)] + \frac{1}{2}\frac{\partial^2}{\partial x_i \partial x_j}[g_{ij}(\mathbf{x}, t)p(\mathbf{x}, t)], \qquad (2.9\text{–}8)$$

and similarly also for $\mathcal{P}(\mathbf{x}, t|\mathbf{x}_0, t_0)$. The Langevin process obeying the stochastic equation (2.9–1) is therefore a Fokker–Planck process.

2.9.2 Steady-state solution of the Fokker–Planck equation

The Fokker–Planck equation (2.9–8) can always be formally re-written in the form

$$\frac{\partial p(\mathbf{x}, t)}{\partial t} + \frac{\partial}{\partial x_i} j_i(\mathbf{x}, t) = 0, \qquad (2.9\text{–}9)$$

(summation over repeated indices being implied) in which

$$j_i(\mathbf{x}, t) \equiv A_i(\mathbf{x}, t)p(\mathbf{x}, t) - \frac{1}{2}\frac{\partial}{\partial x_k}[g_{ik}(\mathbf{x}, t)p(\mathbf{x}, t)] \qquad (2.9\text{–}10)$$

is known as the *probability current density*. Equation (2.9–9) will be recognized as the conservation law for probability.

We shall examine the steady-state solution of Eq. (2.9–9) when A_i and g_{ij} are independent of time, and when g_{ij} is diagonal, i.e. when

$$g_{ij}(\mathbf{x}) = \delta_{ij} D(\mathbf{x}). \tag{2.9–11}$$

In the steady state $\partial p/\partial t = 0$, so that from Eq. (2.9–9) the divergence of the probability current density $\mathbf{j}$ vanishes also. In one dimension this would imply that the scalar j is a constant, and as the probability density and its derivatives vanish at infinity, it then follows from Eq. (2.9–10) that the constant must be zero. However, in several dimensions it is by no means obvious that

$$\mathbf{j}_s(\mathbf{x}) = 0 \tag{2.9–12}$$

is the most general steady-state solution (the subscript s stands for steady state), because any divergence-free vector current density satisfies Eq. (2.9–9) in the steady state. In the following we shall nevertheless adopt Eq. (2.9–12) as the starting point for obtaining the steady-state solution of the Fokker–Planck equation (2.9–9).

If $p_s(\mathbf{x})$ is the solution in the steady state, then Eq. (2.9–12) implies that

$$\mathbf{A} p_s - \tfrac{1}{2}\mathbf{\nabla}(D p_s) = 0,$$

so that

$$\tfrac{1}{2} D \mathbf{\nabla} p_s = p_s[\mathbf{A} - \tfrac{1}{2}\mathbf{\nabla} D]$$

or

$$\frac{1}{p_s}\mathbf{\nabla} p_s = \mathbf{\nabla}(\ln p_s) = \frac{2\mathbf{A}}{D} - \frac{1}{D}\mathbf{\nabla} D. \tag{2.9–13}$$

Because the vector on the right is the gradient of the scalar field $\ln p_s$, it must be irrotational, and therefore

$$\frac{\partial}{\partial x_i}\left(\frac{2A_j}{D} - \frac{1}{D}\frac{\partial D}{\partial x_j}\right) = \frac{\partial}{\partial x_j}\left(\frac{2A_i}{D} - \frac{1}{D}\frac{\partial D}{\partial x_i}\right)$$

or

$$\frac{\partial}{\partial x_i}\left(\frac{2A_j}{D}\right) = \frac{\partial}{\partial x_j}\left(\frac{2A_i}{D}\right).$$

Hence $2\mathbf{A}/D$ is also irrotational, and is therefore expressible as the gradient of a scalar. Let us put

$$\frac{2\mathbf{A}(\mathbf{x})}{D(\mathbf{x})} \equiv -\mathbf{\nabla} U(\mathbf{x}), \tag{2.9–14}$$

where $U(\mathbf{x})$ is a scalar potential that may be expressed as a line integral from some reference point $\mathbf{x}_0$ to $\mathbf{x}$ along any path,

$$U(\mathbf{x}) = -\int_{\mathbf{x}_0}^{\mathbf{x}} \frac{2\mathbf{A}(\mathbf{x})}{D(\mathbf{x})}\cdot d\mathbf{x} + \text{constant}. \tag{2.9–15}$$

Then, on combining Eq. (2.9–13) with Eq. (2.9–14) we have

$$\begin{aligned}\nabla U &= -\frac{1}{D}\nabla D - \frac{1}{p_s}\nabla p_s \\ &= -\frac{1}{Dp_s}\nabla(Dp_s) \\ &= -\nabla \ln(Dp_s).\end{aligned}$$

This equation can be integrated at once, and we obtain

$$U(\mathbf{x}) = -\ln(Dp_s) + \text{constant}$$

or

$$p_s(\mathbf{x}) = \frac{K}{D(\mathbf{x})}\mathrm{e}^{-U(\mathbf{x})}, \qquad (2.9\text{–}16)$$

where K is a normalization constant. As $U(\mathbf{x})$ is defined only up to a constant, the normalization is frequently included in the definition of the potential $U(\mathbf{x})$, and the constant K can then be set equal to unity.

2.9.3 Time-dependent solution of the Fokker–Planck equation

In order to obtain the general time-dependent solution of Eq. (2.9–9) with A_i and g_{ij} constant we can attempt to make a separation of variables, by writing

$$p(\mathbf{x}, t) = f(\mathbf{x})\Theta(t), \qquad (2.9\text{–}17)$$

and then substituting back into the Fokker–Planck equation. If $\mathscr{L}$ stands for the differential operator

$$\mathscr{L} \equiv \frac{\partial}{\partial x_i}A_i - \frac{1}{2}\frac{\partial^2}{\partial x_i \partial x_i}D,$$

(summation over repeated indices being again implied), the resulting equation has the form

$$-\frac{1}{\Theta(t)}\frac{\mathrm{d}\Theta(t)}{\mathrm{d}t} = \frac{1}{f(\mathbf{x})}\mathscr{L}f(\mathbf{x}).$$

As the left-hand side depends only on t, whereas the right-hand side depends only on $\mathbf{x}$, each expression must be a constant λ. We are therefore led to the two differential equations

$$\mathscr{L}f(\mathbf{x}) = \lambda f(\mathbf{x}) \qquad (2.9\text{–}18)$$

$$\frac{\mathrm{d}\Theta(t)}{\mathrm{d}t} = -\lambda\Theta(t). \qquad (2.9\text{–}19)$$

Equation (2.9–18) is an eigenvalue equation with an infinity of eigensolutions $f_{\{n\}}(\mathbf{x})$ and eigenvalues $\lambda_{\{n\}}$ that are labeled by an infinite set of integers $\{n\}$. The dimensionality of the set is the dimensionality of the random process $\mathbf{x}(t)$, i.e., in the case of the three-dimensional process $\mathbf{x}(t)$, $\{n\}$ stands for n_1, n_2, n_3. Equation (2.9–19) can be integrated immediately, and the general solution for

$p(\mathbf{x}, t)$ is then a linear combination of all possible eigensolutions, i.e., it is of the form

$$p(\mathbf{x}, t) = \sum_{\{n\}} C_{\{n\}} f_{\{n\}}(\mathbf{x}) \exp(-\lambda_{\{n\}} t), \tag{2.9–20}$$

where the constants $C_{\{n\}}$ are determined by the initial conditions, and the $\lambda_{\{n\}} \geqslant 0$.

It is possible to transform Eq. (2.9–18) to a Sturm–Liouville equation by a suitable transformation (see, for example, Morse and Feshbach, 1953, p. 719 *et sec.*), but we shall not go into the general procedure here. The method will be illustrated in Chapters 18 and 19, in connection with the solution of the Fokker–Planck equation for the laser.

2.10 The Wiener process (or one-dimensional random walk)

2.10.1 The random walk problem

As an illustrative example of a simple random process, we now consider an idealized model of a body performing a one-dimensional random walk on a lattice, which has been discussed by Einstein (1906). It is often taken as a model of *Brownian motion*, and is also known as the drunkard's walk.

We consider a particle (or drunkard) located initially at the origin, that takes successive unit steps either forwards or backwards with equal probability. From the nature of the problem it is apparent that any future position is governed by the most recent past position without reference to the more distant past, so that the random process is Markovian. We wish to determine the probability $P(n, N)$ that the particle will be located at position n ($n = 0, \pm 1, \pm 2, \ldots$) after N steps. Let n_+, n_- be the number of steps taken forwards and backwards, respectively. Then evidently

$$\left.\begin{aligned} n_+ + n_- &= N \\ n_+ - n_- &= n, \end{aligned}\right\} \tag{2.10–1}$$

so that

$$\left.\begin{aligned} n_+ &= \tfrac{1}{2}(N + n) \\ n_- &= \tfrac{1}{2}(N - n). \end{aligned}\right\} \tag{2.10–2}$$

It is obvious from the definition that N and n must have the same parity, because only an even position can be reached after an even number of steps, etc. The probability $P(n, N)$ is given by a Bernoulli distribution in which n_+ is the number of 'successes' out of N trials, each of which occurs with probability $\frac{1}{2}$ [cf. Eq. (1.5–1)]. Hence

$$\begin{aligned} P(n, N) &= \binom{N}{n_+}\left(\frac{1}{2}\right)^{n_+}\left(\frac{1}{2}\right)^{n_-} \\ &= \frac{N!}{n_+! n_-!}\left(\frac{1}{2}\right)^N. \end{aligned} \tag{2.10–3}$$

We now substitute for n_+ and n_- from Eq. (2.10–2), and assume that the numbers are large, so that all factorials can be well approximated by Stirling's theorem

$$N! \sim (2\pi)^{1/2} N^{N+1/2} \mathrm{e}^{-N}. \tag{2.10–4}$$

We then obtain

$$\begin{aligned}
P(n, N) &= \frac{N!}{\left(\dfrac{N+n}{2}\right)!\left(\dfrac{N-n}{2}\right)!2^N} \\
&= \frac{N^{N+1/2}\mathrm{e}^{-N}}{(2\pi)^{1/2}2^N[(N+n)/2]^{(N+n+1)/2}[(N-n)/2]^{(N-n+1)/2}\mathrm{e}^{-N}} \\
&= \frac{2}{(2\pi N)^{1/2}}\frac{1}{(1+n/N)^{(N+n+1)/2}(1-n/N)^{(N-n+1)/2}} \\
&= \frac{2}{(2\pi N)^{1/2}}\exp\left[-\frac{1}{2}(N+n+1)\ln\left(1+\frac{n}{N}\right)-\frac{1}{2}(N-n+1)\ln\left(1-\frac{n}{N}\right)\right].
\end{aligned}$$

We expand each logarithm in a power series, and make use of the fact that when n_+ and n_- are large they tend to become approximately equal by symmetry, so that $n/N \ll 1$. We therefore discard terms of order $(n/N)^2$ in the exponent. After collecting terms we obtain

$$P(n, N) = \frac{2}{(2\pi N)^{1/2}}\exp\left[-\frac{1}{2}\frac{n^2}{N} + O\left(\frac{n}{N}\right)^2\right]. \tag{2.10–5}$$

We now introduce the position coordinate x and the elapsed time t, by writing

$$\left.\begin{aligned} x &= na \\ t &= NT, \end{aligned}\right\} \tag{2.10–6}$$

where a is the size of each step and T is the interval between successive steps. When a and T are small and when the numbers n and N are sufficiently large, we are justified in considering x and t as essentially continuous variables. If $p(x, t)$ is the probability density of x, then

$$p(x, t)\delta x = \tfrac{1}{2}P(n, N)\delta n,$$

with $\delta x/\delta n = a$. The factor $\frac{1}{2}$ appears because of the parity restriction on n. Hence from Eq. (2.10–5)

$$p(x, t) = \frac{1}{(2\pi t/T)^{1/2}a}\exp\left(-\frac{1}{2}\frac{x^2T}{a^2t}\right).$$

If we now let $a \to 0$ and $T \to 0$ in such a way that $a^2/T \equiv D$ remains constant, then

$$p(x, t) = \frac{1}{(2\pi Dt)^{1/2}}\mathrm{e}^{-x^2/2Dt}. \tag{2.10–7}$$

This is a Gaussian probability distribution in x, with zero mean and dispersion

$$\langle(\Delta x)^2\rangle = Dt. \qquad (2.10\text{–}8)$$

The form of $p(x, t)$ is illustrated in Fig. 2.8 for several different times. Evidently the probability density is non-stationary and steadily diffuses and consequently $x(t)$ is known as a *diffusion process*. Another name is the *Wiener process*, after Wiener (1923, 1930), who studied its features.

It is not difficult to obtain an expression for the conditional probability density $\mathcal{P}(x, t|x_0, t_0)$. Let us first note that $p(x, t)$ given by Eq. (2.10–7) is also $\mathcal{P}(x, t|0, 0)$, because it was assumed that $x(0) = 0$. But from the nature of the process it is apparent that its evolution is independent of where or when it starts; $\mathcal{P}(x, t|x_0, t_0)$ only depends on the difference $x - x_0$ and on the time interval $t - t_0$ $(t > t_0)$. It follows that $\mathcal{P}(x, t|x_0, t_0)$ can be obtained immediately from Eq. (2.10–7), and must be of the form

$$\mathcal{P}(x, t|x_0, t_0) = \frac{1}{[2\pi D(t - t_0)]^{1/2}} \mathrm{e}^{-(x-x_0)^2/2D(t-t_0)}. \qquad (2.10\text{–}9)$$

It is a true transition probability density, independent of any other probability.

2.10.2 *Joint probabilities and autocorrelation*

Once $p(x, t)$ and $\mathcal{P}(x, t|x_0, t_0)$ are known, we can construct any joint probability density $p_n(x_n, t_n; \ldots; x_1, t_1)$ by making use of the Markovian property (2.6–7). Thus, for $n = 2$ we find that

$$p_2(x_2, t_2; x_1, t_1) = \frac{1}{[2\pi D(t_2 - t_1)]^{1/2}} \mathrm{e}^{-(x_2-x_1)^2/2D(t_2-t_1)} \frac{1}{[2\pi D t_1]^{1/2}} \mathrm{e}^{-x_1^2/2Dt_1}, \quad (t_2 > t_1) \qquad (2.10\text{–}10)$$

which allows us to calculate the autocorrelation function of $x(t)$ in the usual way,

$$\Gamma(t_1, t_2) \equiv \langle x(t_1)x(t_2)\rangle = \iint_{-\infty}^{\infty} x_1 x_2 p_2(x_2, t_2; x_1, t_1)\,\mathrm{d}x_1\,\mathrm{d}x_2.$$

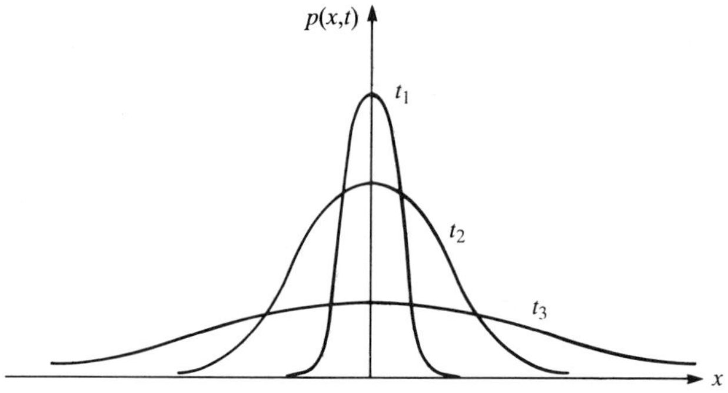

Fig. 2.8 The time evolution of $p(x, t)$ for the random walk at three different times $(t_1 < t_2 < t_3)$.

With the help of Eq. (2.10–10) and the substitution $x_2 - x_1 = x_3$, we obtain

$$\Gamma(t_1, t_2) = \iint_{-\infty}^{\infty} x_1(x_1 + x_3)\frac{1}{[2\pi D(t_2 - t_1)]^{1/2}}\,\mathrm{e}^{-x_3^2/2D(t_2-t_1)}\frac{1}{(2\pi D t_1)^{1/2}}\,\mathrm{e}^{-x_1^2/2Dt_1}\,\mathrm{d}x_1\,\mathrm{d}x_3$$
$$= Dt_1 \quad (t_2 \geqslant t_1), \tag{2.10–11}$$

because the second integral involving the product x_1x_3 vanishes. The answer is unusual in that it is independent of the later time t_2, because the change of x is independent of the initial value of x. Had t_2 been less than t_1 we would have obtained Dt_2. The two answers can be combined with the help of the unit step function

$$\begin{aligned}\theta(\tau) &= 1 \quad \text{for } \tau > 0\\ &= \tfrac{1}{2} \quad \text{for } \tau = 0\\ &= 0 \quad \text{for } \tau < 0,\end{aligned}$$

in the form

$$\Gamma(t_1, t_2) = D[t_1\theta(t_2 - t_1) + t_2\theta(t_1 - t_2)]. \tag{2.10–12}$$

2.10.3 Equation of motion of the Wiener process

Once the transition probability density $\mathcal{P}(x, t|x_0, t_0)$ is known, the transition rate $\mathcal{A}(x, x_0, t)$ can be calculated from Eq. (2.7–4), according to which

$$\mathcal{A}(x, x_0, t) = \operatorname*{Lim}_{\delta t\to 0}\frac{1}{\delta t}[\mathcal{P}(x, t + \delta t|x_0, t) - \delta(x - x_0)]. \tag{2.10–13}$$

We now make use of the Gaussian character of $\mathcal{P}(x, t + \delta t|x_0, t)$, to express the transition probability density as a Fourier integral of the characteristic function [cf. Eq. (1.5–24)], in the form

$$\mathcal{P}(x, t + \delta t|x_0, t) = \frac{1}{2\pi}\int_{-\infty}^{\infty} \mathrm{e}^{-\xi^2 D\delta t/2}\,\mathrm{e}^{-\mathrm{i}(x-x_0)\xi}\,\mathrm{d}\xi. \tag{2.10–14}$$

The delta function can also be represented by a Fourier integral,

$$\delta(x - x_0) = \frac{1}{2\pi}\int_{-\infty}^{\infty} \mathrm{e}^{-\mathrm{i}(x-x_0)\xi}\,\mathrm{d}\xi. \tag{2.10–15}$$

If we use Eqs. (2.10–14) and (2.10–15) in Eq. (2.10–13), expand $\exp(-\xi^2 D\delta t/2)$ as a power series in δt, and proceed to the limit $\delta t \to 0$, we immediately obtain

$$\begin{aligned}\mathcal{A}(x, x_0, t) &= \frac{1}{2\pi}\int_{-\infty}^{\infty}(-\tfrac{1}{2}\xi^2 D)\,\mathrm{e}^{-\mathrm{i}(x-x_0)\xi}\,\mathrm{d}\xi\\ &= \frac{1}{2}D\frac{\partial^2}{\partial x^2}\frac{1}{2\pi}\int_{-\infty}^{\infty}\mathrm{e}^{-\mathrm{i}(x-x_0)\xi}\,\mathrm{d}\xi\\ &= \frac{1}{2}D\frac{\partial^2}{\partial x^2}\delta(x - x_0).\end{aligned} \tag{2.10–16}$$

This is a highly singular function. Nevertheless, $\mathscr{A}(x, x_0, t)$ satisfies the two general conditions (2.7–5) and (2.7–6) for transition rates, and viewed as a distribution and used under an integral, it leads to reasonable results.

Thus, from the general master equation (2.7–8), we obtain with the help of Eq. (2.10–16) the following equation of motion for $p(x, t)$:

$$\frac{\partial p(x,t)}{\partial t} = \int_{-\infty}^{\infty} \frac{1}{2} D \left[\frac{\partial^2}{\partial x^2} \delta(x - x_0) \right] p(x_0, t)\, \mathrm{d}x_0$$
$$= \frac{1}{2} D \frac{\partial^2}{\partial x^2} p(x, t). \qquad (2.10\text{–}17)$$

This will be recognized as the diffusion equation, which might of course have been expected for the process illustrated in Fig. 2.8.

The Wiener process is also sometimes represented by the stochastic Langevin equation with zero drift, namely

$$\frac{\mathrm{d}x(t)}{\mathrm{d}t} = q(t), \qquad (2.10\text{–}18)$$

in which $q(t)$ is a δ-correlated Gaussian noise of zero mean,

$$\left.\begin{aligned} \langle q(t) \rangle &= 0 \\ \langle q(t) q(t') \rangle &= D\delta(t - t'). \end{aligned}\right\} \qquad (2.10\text{–}19)$$

It then follows from the general relationship between Langevin and Fokker–Planck equations (cf. Section 2.9), that $p(x, t)$ obeys the equation of motion (2.10–17). However, it was already pointed out by Einstein (1906) that the stochastic equation (2.10–18) leads to an internal contradiction, because the velocity really does not exist. If we attempt to define a r.m.s. velocity for the Wiener process through the ratio $\langle [x(t + \delta t) - x(t)]^2 \rangle^{1/2} / \delta t$, we obtain with the help of Eq. (2.10–12)

$$\frac{1}{\delta t} \langle [x(t + \delta t) - x(t)]^2 \rangle^{1/2} = \frac{1}{\delta t} [\langle x^2(t + \delta t) \rangle + \langle x^2(t) \rangle - 2 \langle x(t) x(t + \delta t) \rangle]^{1/2}$$
$$= \frac{1}{\delta t} [D(t + \delta t) + Dt - 2Dt]^{1/2}$$
$$= \left(\frac{D}{\delta t} \right)^{1/2}, \qquad (2.10\text{–}20)$$

and this has no limit as $\delta t \to 0$. The stochastic equation (2.10–18) is therefore not strictly meaningful.

Problems

2.1 Consider the complex random process

$$z(t) = u(t) + v,$$

in which $u(t)$ is a complex stationary random process and v is a complex time-independent random variable that is statistically independent of $u(t)$.

Determine whether $z(t)$ is (a) stationary in the wide sense and (b) ergodic with respect to its mean.

2.2 Show that the characteristic functional $C(\{y\})$ of the real Gaussian random process $x(t)$ of zero mean is given by

$$C(\{y\}) = \exp\left[-\frac{1}{2}\iint_{-\infty}^{\infty} \mathrm{d}t_1\,\mathrm{d}t_2\, y(t_1)y(t_2)\Gamma(t_1, t_2)\right],$$

where $\Gamma(t_1, t_2) = \langle x(t_1)x(t_2)\rangle$.

2.3 If $x(t)$ is a real, wide-sense, stationary, random process and $y(t)$ is a *local time average* of $x(t)$ over the interval $2T$ defined by

$$y(t) = \frac{1}{2T}\int_{t-T}^{t+T} x(t')\,\mathrm{d}t'$$

show that the autocorrelation functions $\Gamma_{xx}(\tau)$ and $\Gamma_{yy}(\tau)$ of $x(t)$ and $y(t)$ are related by the equation

$$\Gamma_{yy}(\tau) = \frac{1}{2T}\int_{-2T}^{2T}\left(1 - \frac{|\tau'|}{2T}\right)\Gamma_{xx}(\tau - \tau')\,\mathrm{d}\tau'.$$

2.4 $x(t)$ is a real stationary random process of zero mean and autocorrelation function

$$\Gamma(\tau) = \mathrm{e}^{-\alpha|\tau|}/2\alpha.$$

Find the Karhunen–Loéve expansion of $x(t)$ in the time interval $-T < t < T$.

2.5 A real random process $x(t)$ consists of a sequence of constant segments alternating between positive and negative values. The process $x(t)$ switches from one value to the next at random, at an average rate R per second. The positive values x_+ and the negative values x_- of $x(t)$ have different probability distributions, with dispersions σ_+ and σ_- but with $\langle x_+\rangle = -\langle x_-\rangle$, and each value is independent of every other value. Calculate the spectral density of $x(t)$.

2.6 A one-dimensional random process $x(t)$ obeys the stochastic differential equation

$$\frac{\mathrm{d}x}{\mathrm{d}t} = Ax - x^3 q(t),$$

in which A is a constant, and $q(t)$ is a Gaussian white noise with

$$\langle q(t)\rangle = 0,$$
$$\langle q(t)q(t')\rangle = D\delta(t - t').$$

Calculate the probability density of x in the steady state.

2.7 Consider the complex random process defined by the equation

$$z(t) = a\,\mathrm{e}^{-\mathrm{i}\omega_0 t}\xi(t) + b|\xi(t)|^2,$$

in which a, b, ω_0 are fixed numbers, and $\xi(t)$ is a complex Gaussian, stationary and ergodic random process, with spectral density $S(\omega)$. Calculate the spectral density of $z(t)$. Is $z(t)$ stationary in the wide sense? Is $z(t)$ ergodic?

2.8 A certain discrete random process $x(t)$ takes on only one of three possible values x_1, x_2, x_3 and the value always changes after an interval T. The conditional probabilities $\mathcal{P}(x_i,\ t+T|x_j,\ t;\ x_k,\ t-T)$ are given in the following table:

	$j=1, k=2$	$j=1, k=3$	$j=2, k=1$	$j=2, k=3$	$j=3, k=1$	$j=3, k=2$
$i=1$	0	0	0	$\frac{1}{2}$	0	$\frac{1}{2}$
$i=2$	$\frac{1}{2}$	$\frac{1}{2}$	0	0	1	$\frac{1}{2}$
$i=3$	$\frac{1}{2}$	$\frac{1}{2}$	1	$\frac{1}{2}$	0	0

By considering two different sets of initial probabilities $p_2(x_j,\ t;\ x_k,\ t-T)$ in turn, namely

	$k=1$	$k=2$	$k=3$
$j=1$	0	$\frac{1}{6}$	$\frac{1}{6}$
$j=2$	$\frac{1}{6}$	0	$\frac{1}{6}$
$j=3$	$\frac{1}{6}$	$\frac{1}{6}$	0

and

	$k=1$	$k=2$	$k=3$
$j=1$	0	$\frac{1}{12}$	$\frac{1}{3}$
$j=2$	$\frac{1}{4}$	0	$\frac{1}{6}$
$j=3$	$\frac{1}{6}$	0	0

show explicitly that the equation

$$p(x_i,\ t+T) = \sum_j \mathcal{P}(x_i,\ t+T|x_j,\ t)p(x_j,\ t)$$

is satisfied in both cases, but that $\mathcal{P}(x_j,\ t+T|x_j,\ t)$ is not a transition probability.

2.9 The following complex random process is occasionally used to represent the field of an ideal two-mode laser:

$$z(t) = a\exp\{-\mathrm{i}[\omega_1 t + \phi_1(t)]\} + a\exp\{-\mathrm{i}[\omega_2 t + \phi_2(t)]\}.$$

Here a is a complex constant, ω_1 and ω_2 are fixed frequencies, and the two phases $\phi_1(t)$ and $\phi_2(t)$ are independent Wiener processes; i.e. each performs a one-dimensional random walk with diffusion constant D. Calculate the spectral density of the random process $|z(t)|^2$. Is $|z(t)|^2$ stationary in the wide sense?

2.10 A stationary, ergodic, complex Gaussian random process $z(t)$ of zero mean has normalized spectral density $\mathfrak{s}_1(\omega)$. Calculate the normalized spectral density $\mathfrak{s}_2(\omega)$ of the random process $|z(t)|^2$, and derive the relation between the mean-squared widths $\langle(\Delta\omega)^2\rangle$ of $\mathfrak{s}_1(\omega)$ and $\mathfrak{s}_2(\omega)$.

$$\left(\text{Use the definition } \langle \omega^n \rangle_{s_1} \equiv \frac{1}{2\pi} \int_{-\infty}^{\infty} \omega^n s_1(\omega)\, \mathrm{d}\omega \right).$$

Show that $\mathscr{s}_2(\omega)$ is symmetric about $\omega = 0$, no matter what the form of $\mathscr{s}_1(\omega)$.

2.11 A complex stationary, ergodic random process of zero mean has the normalized autocorrelation function $\gamma(t)$. Show that for any three instants of time t_1, t_2, t_3,

$$|\gamma(t_3 - t_2) - \gamma(t_1 - t_2)\gamma(t_3 - t_1)|^2 \leqslant [1 - |\gamma(t_1 - t_2)|^2][1 - |\gamma(t_3 - t_1)|^2],$$

and for any τ

$$|\gamma(\tau)|^2 + |\gamma(2\tau)|^2 + |\gamma(3\tau)|^2 \leqslant 1 + 2\,\mathrm{Re}\,[\gamma(\tau)\gamma(2\tau)\gamma^*(3\tau)],$$

where Re denotes the real part.

2.12 Consider a complex stationary Gaussian random process $z(t)$ of zero mean. Let

$$\begin{aligned} z_T(t) &= z(t) \quad \text{when } |t| \leqslant T \\ &= 0 \qquad \text{when } |t| > T, \end{aligned}$$

and let $\xi(\nu, T)$ be the Fourier inverse of $z_T(t)$. Further let

$$S_T(\nu) = \frac{\xi^*(\nu, T)\xi(\nu, T)}{2T}$$

and

$$S(\nu) = \operatorname*{Lim}_{T\to\infty} \langle S_T(\nu) \rangle.$$

Show that at any frequency ν for which $S(\nu) \neq 0$, the variance of the process $S_T(\nu)$ does not tend to zero as $T \to \infty$. What is the implication of this result for the problem of determining the spectral density function of the process from one of its sample functions?

3

Some useful mathematical techniques

3.1 The complex analytic signal

The basic variables in optics, e.g. the current and charge densities and the electric and magnetic fields, are real functions of position and time. In the study of their correlation properties it is, nevertheless, useful to represent them by certain complex functions. The representation is a natural generalization of one that is frequently employed in connection with real monochromatic signals and, as we will see later (Section 11.12), it has a counterpart in the description of the electromagnetic field on the basis of quantum field theory.

In this section we will discuss this complex representation for deterministic functions and examine some of its main properties. In later chapters we will make use of it in connection with random functions.

3.1.1 Definition and basic properties of analytic signals

Let $x(t)$ be a real function of a real variable t, defined on the range $-\infty < t < \infty$ and let us assume that it is square-integrable, i.e. that

$$\int_{-\infty}^{\infty} x^2(t)\,\mathrm{d}t < \infty. \tag{3.1–1}$$

We may represent $x(t)$ as a Fourier integral

$$x(t) = \int_{-\infty}^{\infty} \tilde{x}(\nu)\,\mathrm{e}^{-2\pi i\nu t}\,\mathrm{d}\nu, \tag{3.1–2a}$$

where

$$\tilde{x}(\nu) = \int_{-\infty}^{\infty} x(t)\,\mathrm{e}^{2\pi i\nu t}\,\mathrm{d}t. \tag{3.1–2b}$$

Since $x(t)$ is real, the (generally complex) spectral amplitudes $\tilde{x}(\nu)$ obey the relation

$$\tilde{x}(-\nu) = \tilde{x}^*(\nu). \tag{3.1–3}$$

We see from Eq. (3.1–3) that the negative frequency components ($\nu < 0$) do not carry any information not already contained in the positive frequency components ($\nu > 0$), and hence there will be no loss of generality if, in place of $x(t)$, we employ a function $z(t)$ which is obtained from the Fourier integral (3.1–2a) by suppressing the negative frequency components:

$$z(t) = \int_{-\infty}^{\infty} \tilde{z}(\nu)\,\mathrm{e}^{-2\pi i\nu t}\,\mathrm{d}\nu, \tag{3.1–4}$$

where

$$\begin{aligned} \tilde{z}(\nu) &= \tilde{x}(\nu) \quad \text{when } \nu \geqslant 0 \\ &= 0 \qquad\quad \text{when } \nu < 0. \end{aligned} \tag{3.1–4a}$$

The function $z(t)$, defined by Eq. (3.1–4), is evidently a complex function of the real variable t. For reasons that will become apparent shortly it is known as the *complex analytic signal*[‡] associated with the real signal $x(t)$.

Since we assumed that $x(t)$ is square-integrable, its Fourier transform $\tilde{x}(\nu)$ is, according to Plancherel's theorem, also square-integrable. It then follows from Eq. (3.1–4) that the same is true about $z(t)$, i.e. that

$$\int_{-\infty}^{\infty} |z(t)|^2\,\mathrm{d}t < \infty. \tag{3.1–5}$$

With the help of Eq. (3.1–3), $x(t)$ may be represented in the alternative form

$$x(t) = \int_0^{\infty} a(\nu)\cos\left[\varphi(\nu) - 2\pi\nu t\right]\mathrm{d}\nu, \tag{3.1–6}$$

where

$$a(\nu)\,\mathrm{e}^{i\varphi(\nu)} = 2\tilde{z}(\nu), \quad [a(\nu),\ \varphi(\nu)\ \text{real}] \tag{3.1–6a}$$

and the analytic signal $z(t)$ may then be expressed in the form

$$z(t) = \tfrac{1}{2}\int_0^{\infty} a(\nu)\,\mathrm{e}^{i[\varphi(\nu)-2\pi\nu t]}\,\mathrm{d}\nu. \tag{3.1–7}$$

The transition from Eq. (3.1–6) to Eq. (3.1–7) shows that the complex analytic signal is a natural generalization of the complex representation that is frequently used in connection with real monochromatic wavefields.

It is clear from Eqs. (3.1–6) and (3.1–7) that

$$x(t) = z(t) + z^*(t), \tag{3.1–8a}$$

i.e.

$$x(t) = 2\,\mathrm{Re}\,z(t), \tag{3.1–8b}$$

where Re denotes the real part. Equation (3.1–8b) shows that the real part of the complex analytic signal $z(t)$ is one-half of the real signal $x(t)$ from which it was derived.

The analytic signal representation has many interesting properties. Its most

‡ This concept is due to Gabor (1946). Gabor's original definition differs trivially from ours, by the presence of a multiplicative factor 2 in front of the integral (3.1–4). For our purposes the present definition is preferable since, as we shall see later (Section 11.11), it leads to a more elegant correspondence between the classical and the quantum theory of optical coherence.

Some applications of analytic signals to problems of communication engineering were discussed by Gabor, *op. cit.*, Ville (1948, 1950) and by Oswald (1956).

important feature becomes apparent when one examines the possibility of extending the definition of $z(t)$ to complex valued arguments. Let us introduce the complex variable

$$w = t + \mathrm{i}s, \quad (t, s \text{ real}) \tag{3.1–9}$$

and let us try to continue $z(t)$ analytically from the real t-axis into the complex w-plane, by means of Eq. (3.1–4). We then have, formally at any rate,

$$z(w) = \int_0^{\infty} f(\nu, w)\, \mathrm{d}\nu, \tag{3.1–10}$$

where

$$f(\nu, w) = \tilde{x}(\nu)\, \mathrm{e}^{-\mathrm{i}2\pi\nu w}. \tag{3.1–11}$$

We see that for each value of ν, the integrand $f(\nu, w)$ in Eq. (3.1–10) is an entire analytic function of w, i.e. it is analytic and regular in the whole complex w-plane. Now according to a well-known mathematical theorem, a sum (and in an appropriate limit also the integral) of such functions is itself analytic and regular provided that certain continuity and convergence requirements are satisfied.‡ Now from Eqs. (3.1–10) and (3.1–11) we see that the integral (3.1–10) will converge if $s < 0$. This heuristic argument suggests that if $\tilde{x}(\nu)$ is sufficiently well behaved *$z(w)$ is an analytic and regular function of w in the whole lower half of the complex w-plane*. Or, put in a slightly different way, *the function $z(t)$, defined by Eq. (3.1–4) is then the boundary value on the real t-axis of a function that is analytic and regular in the lower half of the complex w-plane*.

It can be shown that the result that we have just stated is true under somewhat more general conditions than our heuristic derivation might suggest. In fact the square-integrability requirement (3.1–5) is both a necessary and a sufficient condition for its validity.§

According to Eq. (3.1–8b), the real part of $z(t)$ is $\frac{1}{2}x(t)$. Let us denote the imaginary part of $z(t)$ by $\frac{1}{2}y(t)$, i.e.

$$z(t) = \tfrac{1}{2}[x(t) + \mathrm{i}y(t)], \tag{3.1–12}$$

where $y(t)$ is also real. The analytic property of $z(t)$ that we have just discussed may be shown to imply that $x(t)$ and $y(t)$ form a *Hilbert transform pair* (known also as conjugate pair), i.e. that

$$y(t) = \frac{1}{\pi} P\int_{-\infty}^{\infty} \frac{x(t')}{t' - t}\, \mathrm{d}t', \tag{3.1–13a}$$

$$x(t) = -\frac{1}{\pi} P\int_{-\infty}^{\infty} \frac{y(t')}{t' - t}\, \mathrm{d}t', \tag{3.1–13b}$$

where P denotes the Cauchy principal value taken at $t' = t$. For a rigorous derivation of these relations we must refer elsewhere (Titchmarsh, 1948, p. 128, Theorem 95 and p. 125, Theorem 93). However, if we assume that $x(t)$ is

‡ Sufficiency requirements are discussed, for example, in Copson (1935), p. 110.

§ cf. Paley and Wiener, (1934), p. 8, Theorem V. This result is closely related to the so-called *Titchmarsh theorem* of which an excellent account is given in Nussenzveig, (1972), pp. 27–28.

not only square-integrable but is also continuous on the real time axis $(-\infty < t < \infty)$, a simple derivation of these relations can be obtained by the use of the extended Cauchy integral formula of the theory of analytic functions. This theorem states (Copson, 1935, p. 66 and p. 134) that if a function $z(w)$ of a complex variable $w = t + is$, $(t,\ s$ real), is analytic and regular throughout a closed domain D of the w-plane and is continuous on the boundary C of D, then (Morse and Feshbach, 1953, p. 368)

$$\oint_C \frac{z(w)}{w - w_0}\,dw = 2\pi i\, z(w_0) \quad \text{if } w_0 \in D \tag{3.1–14a}$$

$$= \pi i\, z(w_0) \quad \text{if } w_0 \in C \tag{3.1–14b}$$

$$= 0 \quad \text{otherwise,} \tag{3.1–14c}$$

where the integration is taken around the curve C in the counterclockwise sense. In the case when w_0 lies on the boundary curve C the integral must be interpreted as the Cauchy principal value.

Let us apply this theorem to the analytic signal and let us choose as the contour C the curve consisting of the portion $-T < t < T$ of the real axis and a semi-circle Γ of radius T in the lower half plane, centered at the origin (see Fig. 3.1). Let $w_0 = t$ be a point on the segment of the real axis in the interval $-T < t < T$. It then follows from the formula (3.1–14b) that in the limit as $T \to \infty$

$$P\int_{+\infty}^{-\infty} \frac{z(t')}{t' - t}\,dt' + \lim_{T\to\infty} \int_\Gamma \frac{z(w)}{w - t}\,dw = \pi i\, z(t). \tag{3.1–15}$$

Now it may be deduced from Eqs. (3.1–10) and (3.1–11), using the square-integrability of $\tilde{x}(\nu)$, that

$$\lim_{T\to\infty} \int_\Gamma \frac{z(w)}{w - t}\,dw = 0 \tag{3.1–16}$$

and hence Eq. (3.1–15) implies that

$$P\int_{-\infty}^{\infty} \frac{z(t')}{t' - t}\,dt' = -\pi i\, z(t). \tag{3.1–17}$$

On substituting from Eq. (3.1–12) into this formula and equating the real and the imaginary parts, the Hilbert transform relations (3.1–13) follow.

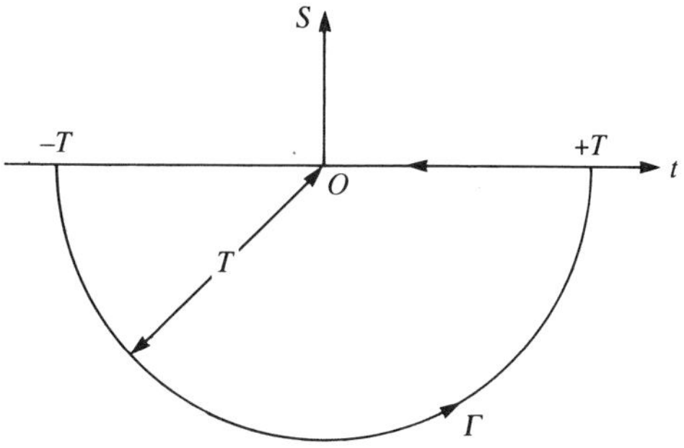

Fig. 3.1 Notation relating to formula (3.1–15). $w = t + is$.

Relations of the form (3.1–13), but involving frequencies rather than time, play an important role in physics, where they are often referred to as *dispersion relations*. This terminology has its origin in the fact that such relations were first found to arise in the theory of dispersion of light in material media. Dispersion relations usually appear in physical theories as a consequence of *causality* (Nussenzveig, 1972, Chapt. 1). If one deals with a linear system whose response is characterized by a function $K(t)$, (e.g. the dielectric susceptibility of a linear medium responding to an incident electric field), then causality demands that $K(t) = 0$ for $t < 0$ (no output before an input is applied). This condition may be shown to impose certain relations between the real and the imaginary parts of the Fourier transform $k(\nu)$ of $K(t)$, of which the Hilbert transforms are the simplest. In the present problem, the appearance of relations of this form is not due to causality, because these relations hold in the time domain and not in the frequency domain. The *mathematical* origin of these relations is, however, the same in both cases. They arise essentially from the fact that the Fourier transform $K(t)$ of $k(\nu)$ and the Fourier transform $\tilde{z}(\nu)$ of $z(t)$ [see Eq. (3.1–4a)] vanish identically for negative values of their arguments.

It is sometimes useful to express the transition from a real signal $x(t)$ to the associated complex analytic signal $z(t)$ by means of one of the well-known singular functions of field theory, namely the so-called negative frequency part of the Dirac delta function‡, defined symbolically by the expression

$$\delta_-(t) = \int_0^\infty e^{2\pi i\nu t}\,d\nu \tag{3.1–18a}$$

$$= \frac{1}{2}\left[\delta(t) - \frac{1}{\pi i}P\frac{1}{t}\right], \tag{3.1–18b}$$

where P denotes, as before, the Cauchy principal value. It follows from Eqs. (3.1–18a) and (3.1–2), that

$$\begin{aligned}\int_{-\infty}^{\infty} x(t')\delta_-(t'-t)\,dt' &= \int_{-\infty}^{\infty} dt'x(t')\int_0^\infty d\nu\, e^{2\pi i\nu(t'-t)} \\ &= \int_0^\infty d\nu\, e^{-2\pi i\nu t}\int_{-\infty}^{\infty} dt'x(t')\,e^{2\pi i\nu t'} \\ &= \int_0^\infty \tilde{x}(\nu)\,e^{-2\pi i\nu t}\,d\nu.\end{aligned} \tag{3.1–19}$$

Now according to Eq. (3.1–4), the right-hand side of Eq. (3.1–19) is precisely the analytic signal $z(t)$ associated with $x(t)$. Hence

$$z(t) = \int_{-\infty}^{\infty} x(t')\delta_-(t'-t)\,dt'. \tag{3.1–20}$$

This formula shows that $z(t)$ is a linear transform of $x(t)$, the kernel of the transform being the singular function δ_-. If we use the representation (3.1–18b)

‡ This terminology derives from the following decomposition of the Dirac delta function: $\delta(t) = \int_{-\infty}^{\infty} \exp(-2\pi i\nu t)\,d\nu = \delta_-(t) + \delta_+(t)$, where $\delta_-(t) = \int_{-\infty}^{0} \exp(-2\pi i\nu t)\,d\nu = \int_0^\infty \exp(2\pi i\nu t)\,d\nu$ and $\delta_+(t) = \int_0^\infty \exp(-2\pi i\nu t)\,d\nu$. For a fuller discussion of these singular functions see Heitler (1954), Sec. 8. See also Appendix A4.1.

of the δ_- function, Eq. (3.1–20) implies that

$$z(t) = \frac{1}{2}\left[x(t) + \frac{\mathrm{i}}{\pi} P \int_{-\infty}^{\infty} \frac{x(t')}{t' - t} \mathrm{d}t'\right], \tag{3.1–21}$$

in agreement with Eqs. (3.1–12) and (3.1–13).

Finally we note the following relations that may readily be derived by the use of formulas (3.1–2a), (3.1–4), (3.1–12) and (3.1–3):

$$\int_{-\infty}^{\infty} x^2(t)\,\mathrm{d}t = \int_{-\infty}^{\infty} y^2(t)\,\mathrm{d}t = 2\int_{-\infty}^{\infty} |z(t)|^2\,\mathrm{d}t = 2\int_0^{\infty} |\tilde{z}(\nu)|^2\,\mathrm{d}\nu \tag{3.1–22}$$

and

$$\int_{-\infty}^{\infty} x(t) y(t)\,\mathrm{d}t = 0. \tag{3.1–23}$$

3.1.2 *Quasi-monochromatic signals and their envelopes*

We will now show that the analytic signal provides an unambiguous definition of the envelope of a real fluctuating signal.

In practice one often encounters real signals whose Fourier transforms (Fourier spectra) are effectively confined to the frequency range

$$\nu_0 - \tfrac{1}{2}\Delta\nu \leqslant |\nu| \leqslant \nu_0 + \tfrac{1}{2}\Delta\nu \quad (\nu_0 > 0,\ \Delta\nu > 0) \tag{3.1–24}$$

around some frequencies $\pm\nu_0$, where

$$\frac{\Delta\nu}{\nu_0} \ll 1. \tag{3.1–25}$$

More explicitly, if we represent $x(t)$ in the form of a Fourier integral [Eq. (3.1–2a)] then $|\tilde{x}(\nu)|$ is essentially zero outside the ranges defined by the inequalities (3.1–24) (see Fig. 3.2). Such a signal is said to be *quasi-monochromatic*.

According to Eqs. (3.1–4) and (3.1–4a), the analytic signal associated with $x(t)$ may be expressed in the form

$$z(t) = \int_{-\infty}^{\infty} \tilde{z}(\nu)\, \mathrm{e}^{-2\pi \mathrm{i}\nu t}\,\mathrm{d}\nu, \tag{3.1–26}$$

where

$$\left.\begin{aligned} \tilde{z}(\nu) &= \tilde{x}(\nu) \quad \text{when } \nu \geqslant 0 \\ &= 0 \qquad\ \ \text{when } \nu < 0 \end{aligned}\right\}. \tag{3.1–27}$$

For a quasi-monochromatic signal whose spectral amplitude distribution is indicated in Fig. 3.2(*a*), $|\tilde{z}(\nu)|$ will have the form indicated in Fig. 3.2(*b*).

Were the signal *strictly monochromatic*, of frequency ν_0, one would have

$$x(t) = \xi_0\, \mathrm{e}^{-2\pi \mathrm{i}\nu_0 t} + \xi_0^* \mathrm{e}^{2\pi \mathrm{i}\nu_0 t} \tag{3.1–28a}$$

$$= 2\,\mathrm{Re}\,(\xi_0\, \mathrm{e}^{-2\pi \mathrm{i}\nu_0 t}) \tag{3.1–28b}$$

$$= A_0 \cos(\Phi_0 - 2\pi\nu_0 t), \tag{3.1–28c}$$

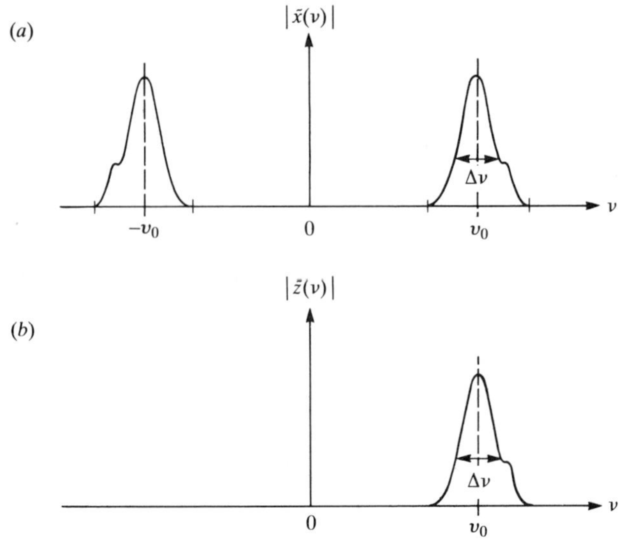

Fig. 3.2 An example of (*a*) a Fourier spectrum of a quasi-monochromatic signal $x(t)$ and of (*b*) the associate analytic signal $z(t)$. Only the absolute values of the Fourier spectra are shown.

where A_0 and Φ_0 are real constants and

$$\xi_0 = \tfrac{1}{2}A_0\, e^{i\Phi_0}. \tag{3.1–29}$$

In this case the Fourier spectrum of $x(t)$ clearly is

$$\xi(\nu) = \xi_0^*\delta(\nu + \nu_0) + \xi_0\delta(\nu - \nu_0), \tag{3.1–30}$$

where δ denotes the Dirac delta function, and the analytic signal associated with the monochromatic signal (3.1–28) is

$$\begin{aligned} z(t) &= \xi_0\, e^{-2\pi i\nu_0 t} \\ &= \tfrac{1}{2}A_0\, e^{i[\Phi_0 - 2\pi\nu_0 t]}. \end{aligned} \tag{3.1–31}$$

If instead of being strictly monochromatic the signal is quasi-monochromatic, with its Fourier spectrum occupying the small frequency range given by the inequalities (3.1–24), we may represent it in a form resembling Eq. (3.1–28c), namely

$$x(t) = A(t)\cos[\Phi(t) - 2\pi\nu_0 t], \tag{3.1–32}$$

where the 'amplitude' A and the 'phase' Φ are, however, no longer constants but are functions of time. The representation (3.1–32), although often written in this way, is clearly not unique. For there are many ways of choosing the two functions $A(t)$ and $\Phi(t)$ so that the right-hand side of Eq. (3.1–32) is equal to $x(t)$. However, a unique choice of A and Φ may be made (subject to the constraint $0 \leqslant \Phi(t) < 2\pi$) if we demand that the analytic signal $z(t)$ associated with $x(t)$ is of a form analogous to the representation (3.1–31), namely

$$\begin{aligned} z(t) &= \tfrac{1}{2}[x(t) + iy(t)] \\ &= \tfrac{1}{2}A(t)\, e^{i[\Phi(t) - 2\pi\nu_0 t]}. \end{aligned} \tag{3.1–33}$$

For we now have, according to Eq. (3.1–13a), that

$$y(t) = \frac{1}{\pi} P \int_{-\infty}^{\infty} \frac{x(t')}{t' - t} \, dt' \tag{3.1–34}$$

and, according to Eqs. (3.1–33) and (3.1–12),

$$y(t) = A(t) \sin [\Phi(t) - 2\pi\nu_0 t]. \tag{3.1–35}$$

If $x(t)$ is given, $y(t)$ may be determined from Eq. (3.1–34), and Eqs. (3.1–32) and (3.1–35) then give, if Eq. (3.1–33) is also used,

$$A(t) = [x^2(t) + y^2(t)]^{1/2} = 2|z(t)|, \tag{3.1–36a}$$

$$\Phi(t) = 2\pi\nu_0 t + \chi(t), \tag{3.1–36b}$$

where

$$\cos \chi(t) = \tfrac{1}{2}\frac{x(t)}{|z(t)|}, \quad \sin \chi(t) = \tfrac{1}{2}\frac{y(t)}{|z(t)|}. \tag{3.1–36c}$$

It is seen from Eq. (3.1–36a) that $A(t)$ is independent of the exact choice of the frequency ν_0 and Eqs. (3.1–36b) and (3.1–36c) imply that $\Phi(t)$ depends on ν_0 only through the simple additive factor $2\pi\nu_0 t$.

Let us examine the behavior of $A(t)$ and $\Phi(t)$ more closely. According to Eqs. (3.1–33) and (3.1–26)

$$\tfrac{1}{2}A(t)\, e^{i\Phi(t)} = \int_{-\infty}^{\infty} \tilde{z}(\nu)\, e^{-2\pi i(\nu-\nu_0)t} \, d\nu. \tag{3.1–37}$$

If we set

$$\nu - \nu_0 = \mu \tag{3.1–38}$$

and recall that according to Eq. (3.1–27) $\tilde{z}(\nu) = 0$ when $\nu < 0$, Eq. (3.1–37) implies that

$$\tfrac{1}{2}A(t)\, e^{i\Phi(t)} = \int_{-\nu_0}^{\infty} \zeta(\mu)\, e^{-2\pi i\mu t} \, d\mu, \tag{3.1–39}$$

where

$$\zeta(\mu) = \tilde{z}(\nu_0 + \mu), \quad \mu \geqslant -\nu_0. \tag{3.1–40}$$

Thus $\zeta(\mu)$ is of the same form as $\tilde{z}(\nu)$ but is shifted by the amount ν_0 in the negative ν-direction (see Fig. 3.3).

Since the spectral amplitudes $|\tilde{z}(\nu)|$ have appreciable values only in the range defined by the inequalities (3.1–24), $|\zeta(\mu)|$ will be appreciable only for values of μ in the range

$$-\tfrac{1}{2}\Delta\nu \leqslant \mu \leqslant \tfrac{1}{2}\Delta\nu. \tag{3.1–41}$$

Consequently, the integral on the right of Eq. (3.1–39) consists of low-frequency components only. Moreover, in view of the quasi-monochromatic assumption expressed by the inequalities (3.1–24), it is clear from Eq. (3.1–37) that $A(t)$ and $\Phi(t)$ will vary much more slowly with t than $\cos 2\pi\nu_0 t$ and $\sin 2\pi\nu_0 t$. In fact, $A(t)$ and $\Phi(t)$ will be essentially constant over any time interval for which the term

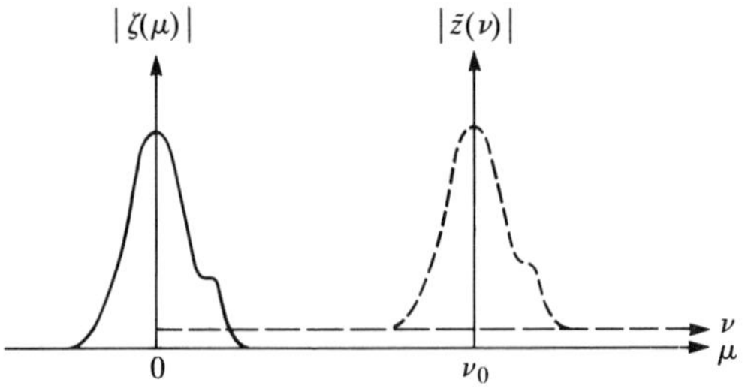

Fig. 3.3 The absolute value of the shifted Fourier spectrum $\zeta(\mu) = \tilde{z}(\nu_0 + \mu)$.

$2\pi\mu t$, $(\frac{1}{2}\Delta\nu \leqslant \mu \leqslant \frac{1}{2}\Delta\nu)$, in the exponential in Eq. (3.1–39) is small compared to 2π, i.e. over any time interval of duration Δt such that

$$\Delta t \ll \frac{1}{\Delta\nu}. \tag{3.1–42}$$

Hence the two real signals, given by the expressions (3.1–32) and (3.1–35), are nearly periodic functions of t, with frequency ν_0, slowly modulated in amplitude by $A(t)$ and in phase by $\Phi(t)$. The modulation is negligible over any time interval which is appreciably smaller than the inverse bandwidth $1/\Delta\nu$ (see Fig. 3.4). Consequently, the following relations follow at once from Eqs. (3.1–32), (3.1–35) and (3.1–33):

$$\widetilde{x^2(t)} = \widetilde{y^2(t)} = \tfrac{1}{2}A^2(t) = 2|z(t)|^2, \tag{3.1–43}$$

where the curly lines indicate a 'short-time' average, i.e. an average over a time interval of duration that is short compared to $1/\Delta\nu$ but long compared to $1/\nu_0$.

Since $A(t)$ and $\Phi(t)$ remain essentially constant over any time interval which is much shorter than $1/\Delta\nu$, it follows from Eqs. (3.1–32) and (3.1–35) that

$$y(t) \approx -x(t - 1/4\nu_0) \tag{3.1–44}$$

and also that

$$y(t) \approx \frac{1}{2\pi\nu_0}\frac{\mathrm{d}x(t)}{\mathrm{d}t}, \tag{3.1–45a}$$

$$x(t) \approx -\frac{1}{2\pi\nu_0}\frac{\mathrm{d}y(t)}{\mathrm{d}t}. \tag{3.1–45b}$$

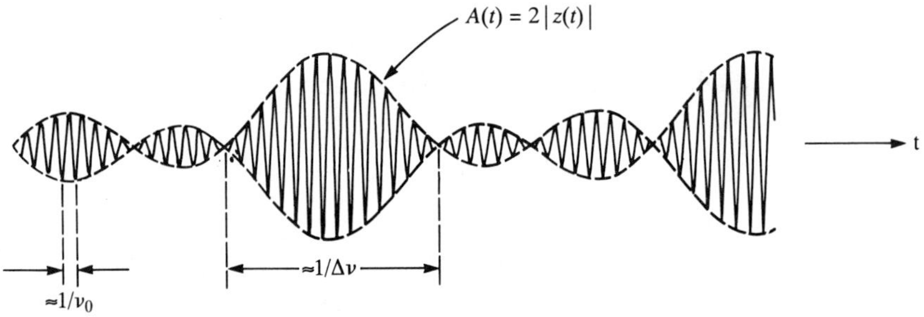

Fig. 3.4 Illustrating the behavior of a quasi-monochromatic signal $x(t)$ and of its real envelope $A(t)$.

From Fig. 3.4 and Eq. (3.1–33) it is obvious that the function

$$A(t)\,e^{i\Phi(t)} = 2z(t)\,e^{2\pi i \nu_0 t} \tag{3.1–46}$$

may be regarded as representing the *complex envelope*‡ of the real quasi-monochromatic signal $x(t)$. The envelope changes slowly with t in comparison with the periodic factor.

There are several interesting theorems concerning the envelope that we have just defined. Suppose that the real signal $x(t)$ is strictly bandlimited to the frequency range $\nu_0 - \frac{1}{2}\alpha \leqslant |\nu| \leqslant \nu_0 + \frac{1}{2}\alpha$ i.e. that the Fourier spectrum $\tilde{x}(\nu)$ of $x(t)$ is zero everywhere outside this range. Let us now consider the Fourier spectrum $\sigma(\nu)$ of the squared amplitude $A^2(t) = 4|z(t)|^2$ of the complex envelope of $x(t)$. It is given by the formula

$$\sigma(\nu) = 4\int_{-\infty}^{\infty} |z(t)|^2\, e^{2\pi i \nu t}\, dt \tag{3.1–47}$$

or, using Eq. (3.1–4) and the convolution theorem on Fourier transforms,

$$\sigma(\nu) = 4\int_{0}^{\infty} \tilde{x}^*(\nu')\tilde{x}(\nu + \nu')\, d\nu'. \tag{3.1–48}$$

Since $\tilde{x}(\nu)$ vanishes outside the ranges defined by Eq. (3.1–24) it is clear from Eq. (3.1–48) that $\sigma(\nu)$ will vanish when $|\nu| > \alpha$. Hence we have established the following *theorem* due to Dugundji (1958), illustrated in Fig. 3.5: *If $x(t)$ is strictly*

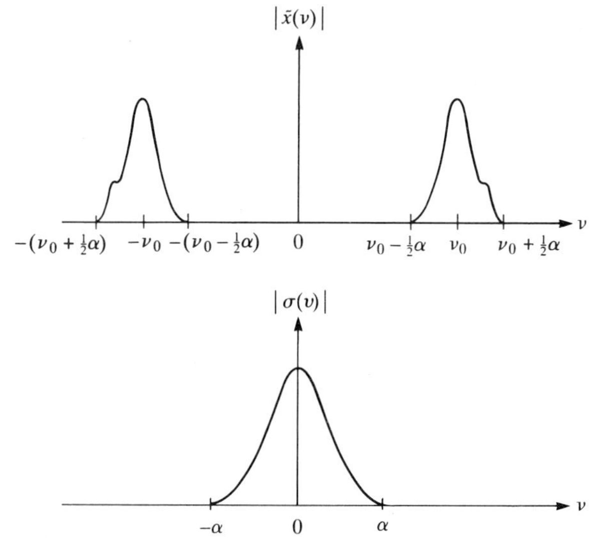

Fig. 3.5 Illustrating Dugundji's theorem. If the signal $x(t)$ is bandlimited to the range $\nu_0 - \frac{1}{2}\alpha \leqslant |\nu| \leqslant \nu_0 + \frac{1}{2}\alpha$, its square amplitude $A^2(t) = 4|z(t)|^2$ is necessarily bandlimited to the range $-\alpha \leqslant \nu \leqslant \alpha$. The term $\tilde{x}(\nu)$ is the Fourier spectrum of $x(t)$ and $\sigma(\nu)$ is the Fourier spectrum of $A^2(t)$.

‡ A formally different definition of an envelope for a certain class of narrow-band signals was given by Rice (1944, Sec. 3.7). It was shown by Dugundji (1958) that, whenever Rice's definition is applicable, it is equivalent to the present one.

bandlimited to the range $\nu_0 - \frac{1}{2}\alpha \leqslant \nu \leqslant \nu_0 + \frac{1}{2}\alpha$ *then the squared amplitude* $A^2(t)$ *of its complex envelope is bandlimited to the range* $-\alpha \leqslant \nu \leqslant \alpha$.

As we already noted, the envelope representation (3.1–32) of a real quasi-monochromatic signal is not unique. The question then arises as to the properties that distinguish the definition based on the analytic signal from other possible ones. Within the context of stationary ensembles of real, quasi-monochromatic signals this was answered by Mandel (1967). His result may be stated as follows. Suppose that with each member $x(t)$ of the ensemble we associate another signal $y(t)$, which is a linear transform of $x(t)$, with a real kernel $K(t)$:

$$y(t) = \int_{-\infty}^{\infty} K(t - t')x(t')\,\mathrm{d}t'. \tag{3.1–49}$$

Let us now define the complex function

$$z(t) = \tfrac{1}{2}[x(t) + \mathrm{i}y(t)], \tag{3.1–50}$$

and let

$$z(t) = z_0(t)\,\mathrm{e}^{-2\pi \mathrm{i}\nu_0 t}. \tag{3.1–51}$$

Then the expectation

$$\rho = \left\langle \left| \frac{\mathrm{d}z_0(t)}{\mathrm{d}t} \right|^2 \right\rangle \tag{3.1–52}$$

will be a *minimum* when the mean-squared deviation

$$\int_{-\infty}^{\infty} (\nu - \nu_0)^2 S(\nu)\,\mathrm{d}\nu = \text{minimum}$$

with

$$\nu_0 = \frac{\displaystyle\int_{-\infty}^{\infty} \nu S(\nu)\,\mathrm{d}\nu}{\displaystyle\int_{-\infty}^{\infty} S(\nu)\,\mathrm{d}\nu}, \tag{3.1–53}$$

where $S(\nu)$ is the power spectrum (see Sec. 2.4.1) of $z(t)$ and the minimum is attained when the kernel $K(t)$ of the transform (3.1–49) is the Hilbert transform kernel, i.e. when $z(t)$ is the analytic signal associated with the real signal $x(t)$. This result implies that *of all possible envelopes defined by means of linear transforms on the real signal, the one defined by means of the analytic signal exhibits the slowest rate of fluctuations of the complex envelope in the mean-squared sense*.

3.1.3 Relationships between correlation functions of real and associated complex analytic random processes

We will now establish a number of useful relations involving cross-correlation functions of real stationary random processes, the conjugate processes, and the associated complex analytic processes.

Let $x_1(t)$ and $x_2(t)$ represent two real jointly wide-sense stationary random processes of zero mean and let $z_1(t)$ and $z_2(t)$ be the associated analytic signals. Since the sample functions of a stationary random process are not square-integrable [see Section 2.4], the transition from $x_j(t)$ to $z_j(t)$, $(j = 1, 2)$, has to be interpreted with some caution. To avoid mathematical complexity we will formally define the analytic signals in the same way as we did in Section 3.1.1; precise justification of the formulas can, however, only be given within the framework of generalized function theory. Thus if we write

$$x_j(t) = \int_{-\infty}^{\infty} \tilde{x}_j(\nu)\, e^{-2\pi i \nu t}\, d\nu, \quad (j = 1, 2), \tag{3.1–54}$$

then

$$z_j(t) = \int_{-\infty}^{\infty} \tilde{z}_j(\nu)\, e^{-2\pi i \nu t}\, d\nu, \tag{3.1–55}$$

where

$$\left.\begin{aligned} \tilde{z}_j(\nu) &= \tilde{x}_j(\nu) && \text{when } \nu \geqslant 0 \\ &= 0 && \text{when } \nu < 0. \end{aligned}\right\} \tag{3.1–56}$$

Further

$$z_j(t) = \tfrac{1}{2}[x_j(t) + i y_j(t)], \quad (j = 1, 2), \tag{3.1–57}$$

where the (real) functions $x_j(t)$ and $y_j(t)$ are related by Hilbert transforms [Eqs. (3.1–13)].

Since $x_j(t)$ was assumed to be of zero mean, i.e. since

$$\langle x_j(t) \rangle = 0, \quad (j = 1, 2), \tag{3.1–58}$$

and since $y_j(t)$ and $z_j(t)$ are homogeneous linear transforms of $x_j(t)$, they both are also of zero mean, i.e.

$$\langle y_j(t) \rangle = 0, \quad (j = 1, 2), \tag{3.1–59}$$

$$\langle z_j(t) \rangle = 0, \quad (j = 1, 2). \tag{3.1–60}$$

Moreover, just like $x_j(t)$ the processes $y_j(t)$ and $z_j(t)$ are stationary, at least in the wide sense.

Let us now consider the cross-correlation function

$$\Gamma_{12}(\tau) = \langle z_1^*(t) z_2(t + \tau) \rangle \tag{3.1–61}$$

of two complex processes. According to the generalized Wiener–Khintchine theorem [Eq. (2.4–38)]

$$\Gamma_{12}(\tau) = \int_{-\infty}^{\infty} W_{12}(\nu)\, e^{-2\pi i \nu \tau}\, d\nu, \tag{3.1–62}$$

where $W_{12}(\nu)$ is the cross-spectral density of the two processes defined by the formula

$$\langle \tilde{z}_1^*(\nu) \tilde{z}_2(\nu') \rangle = W_{12}(\nu)\delta(\nu - \nu'). \tag{3.1–63}$$

Now according to Eq. (3.1–56), $\tilde{z}_j(\nu) = 0$, $(j = 1, 2)$, when $\nu < 0$ and hence Eq. (3.1–63) implies that

$$W_{12}(\nu) = 0, \quad \text{when } \nu < 0. \tag{3.1–64}$$

Consequently Eq. (3.1–62) may be written as

$$\Gamma_{12}(\tau) = \int_0^\infty W_{12}(\nu)\, e^{-2\pi i\nu\tau}\, d\nu. \tag{3.1–65}$$

Since the integration on the right extends over positive frequency only, $\Gamma_{12}(\tau)$ is an analytic signal. If we assume that $\Gamma_{12}(\tau)$ is a square-integrable function of τ and use the basic property of analytic signals expressed by Eqs. (3.1–13), we have the following theorem:

Theorem I: *If $z_1(t)$ and $z_2(t)$ are the complex analytic signal representations of two real, jointly wide-sense stationary random processes of zero means, then the cross-correlation function (assumed to be square-integrable) $\Gamma_{12}(\tau) = \langle z_1^*(t) z_2(t+\tau)\rangle$ is also an analytic signal and its real part $\Gamma_{12}^{(r)}(\tau)$ and its imaginary part $\Gamma_{12}^{(i)}(\tau)$, $[\Gamma_{12}(\tau) = \Gamma_{12}^{(r)}(\tau) + i\Gamma_{12}^{(i)}(\tau)]$, form a Hilbert transform pair, i.e.*

$$\Gamma_{12}^{(i)}(\tau) = \frac{1}{\pi} P\int_{-\infty}^{\infty} \frac{\Gamma_{12}^{(r)}(\tau')}{\tau' - \tau}\, d\tau', \tag{3.1–66a}$$

$$\Gamma_{12}^{(r)}(\tau) = -\frac{1}{\pi} P\int_{-\infty}^{\infty} \frac{\Gamma_{12}^{(i)}(\tau')}{\tau' - \tau}\, d\tau'. \tag{3.1–66b}$$

In order to derive another useful theorem involving $z_1(t)$ and $z_2(t)$ let us consider the cross-correlation function

$${}'\Gamma_{12}(\tau) = \langle {}'z_1^*(t) z_2(t+\tau)\rangle, \tag{3.1–67}$$

where

$${}'z_1(t) = z_1^*(t). \tag{3.1–68}$$

According to Eqs. (3.1–55) and (3.1–56) one readily finds that the function ${}'z_1(t)$ has the Fourier representation

$${}'z_1(t) = \int_{-\infty}^{\infty} {}'\tilde{z}_1(\nu)\, e^{-2\pi i\nu t}\, d\nu, \tag{3.1–69}$$

where

$$\left.\begin{aligned} {}'\tilde{z}_1(\nu) &= 0 && \text{when } \nu > 0 \\ &= \tilde{x}_1^*(\nu) && \text{when } \nu \leq 0. \end{aligned}\right\} \tag{3.1–70}$$

The cross-spectral density function ${}'W_{12}(\nu)$ of ${}'z_1(t)$ and $z_2(t)$ is defined by the formula [see Eq. (2.4–35)]

$$\langle {}'\tilde{z}_1^*(\nu)\tilde{z}_2(\nu')\rangle = {}'W_{12}(\nu)\delta(\nu - \nu'). \tag{3.1–71}$$

Now according to Eq. (3.1–70), ${}'\tilde{z}_1(\nu)$ is zero for all positive frequencies and according to Eq. (3.1–56) $\tilde{z}_1(\nu)$ is zero for all negative frequencies. It therefore follows from Eq. (3.1–71) that

$$'W_{12}(\nu) = 0 \quad \text{for all } \nu \neq 0. \tag{3.1–72}$$

Further, according to the generalized Wiener–Khintchine theorem [Eq. 2.4–38], the cross-correlation function (3.1–67) has the Fourier representation

$$'\Gamma_{12}(\tau) = \int_{-\infty}^{\infty} {'W_{12}(\nu)}\, e^{-2\pi i\nu\tau}\, d\nu, \tag{3.1–73}$$

and, using Eq. (3.1–72), we see that

$$'\Gamma_{12}(\tau) \equiv 0. \tag{3.1–74}$$

Hence we have established the following theorem.

Theorem II: *Under the same conditions as stated in Theorem* I,

$$'\Gamma_{12}(\tau) \equiv \langle z_1(t) z_2(t+\tau)\rangle = 0 \tag{3.1–75}$$

for all values of τ.

From Theorems I and II several other interesting results readily follow. On substituting from Eq. (3.1–57) into Eq. (3.1–75) and on equating real and imaginary parts, we obtain at once the following theorem:

Theorem III: *If* $x_1(t)$ *and* $x_2(t)$ *are two real, jointly wide-sense stationary random processes of zero means and* $y_1(t)$ *and* $y_2(t)$ *are the corresponding conjugate processes (i.e. their Hilbert transforms), then*

$$\langle x_1(t) x_2(t+\tau)\rangle = \langle y_1(t) y_2(t+\tau)\rangle \tag{3.1–76a}$$

and

$$\langle x_1(t) y_2(t+\tau)\rangle = -\langle y_1(t) x_2(t+\tau)\rangle. \tag{3.1–76b}$$

In the special case when $x_2(t) = x_1(t)$ we have, of course, $y_2(t) = y_1(t)$ and we obtain at once from Eqs. (3.1–76), if we take $\tau = 0$ and drop the (common) suffices,

$$\langle x^2(t)\rangle = \langle y^2(t)\rangle \tag{3.1–77a}$$

and

$$\langle x(t) y(t)\rangle = 0. \tag{3.1–77b}$$

Since according to Eqs. (3.1–58) and (3.1–59) we also have $\langle x(t)\rangle = \langle y(t)\rangle = 0$, Eqs. (3.1–77) imply:

Theorem IV: *The variance* $\langle (x - \langle x\rangle)^2\rangle$ *of a real, wide-sense stationary random process* $x(t)$ *of zero mean is equal to the variance* $\langle (y - \langle y\rangle)^2\rangle$ *of its conjugate process* $y(t)$ *and, at each instant of time, the two processes are uncorrelated.*

Let us express the complex cross-correlation function $\Gamma_{12}(\tau)$, defined by Eq. (3.1–61), in terms of the real processes $x_j(t)$ and $y_j(t)$ by the use of Eq. (3.1–57).

If we also make use of the relations (3.1–76) we obtain the formula

$$\Gamma_{12}(\tau) = \tfrac{1}{2}[\langle x_1(t)x_2(t+\tau)\rangle + \mathrm{i}\langle x_1(t)y_2(t+\tau)\rangle]. \tag{3.1–78}$$

If, as before, we denote the real and imaginary parts of $\Gamma_{12}(\tau)$ by $\Gamma_{12}^{(\mathrm{r})}(\tau)$ and $\Gamma_{12}^{(\mathrm{i})}(\tau)$, respectively, and equate the real and imaginary parts in Eq. (3.1–78) we obtain

Theorem V: *Under the same conditions as stated in Theorem III, the real and imaginary parts of the cross-correlation function $\Gamma_{12}(\tau)$ of the complex analytic signals $z_1(t)$ and $z_2(t)$ associated with the real signals $x_1(t)$ and $x_2(t)$ are given by the formulas*

$$\Gamma_{12}^{(\mathrm{r})}(\tau) = \tfrac{1}{2}\langle x_1(t)x_2(t+\tau)\rangle, \tag{3.1–79a}$$

$$\Gamma_{12}^{(\mathrm{i})}(\tau) = \tfrac{1}{2}\langle x_1(t)y_2(t+\tau)\rangle. \tag{3.1–79b}$$

There is an interesting consequence of this theorem and of Theorem I for a quasi-monochromatic, wide-sense stationary, real random process $x(t)$. The power spectrum‡ of such a process $x(t)$ is appreciable only in a spectral range whose width is small compared to the mean frequency. Now according to Eq. (3.1–79a) and Theorem I, $\langle x(t)x(t+\tau)\rangle = 2\Gamma^{(\mathrm{r})}(\tau) = 2\,\mathrm{Re}\,\Gamma(\tau)$, where $\Gamma(\tau)$ is the autocorrelation function of the analytic signal $z(t)$ associated with $x(t)$ and Re again denotes the real part. Now we have learned in Section 3.1.2 that $\Gamma(\tau)\exp(2\pi\mathrm{i}\nu_0\tau)$ is the complex envelope of $\Gamma^{(\mathrm{r})}(\tau)$ and hence, according to Eq. (3.1–79a), it is also the complex envelope of $\tfrac{1}{2}\langle x(t)x(t+\tau)\rangle$. The general behavior of the real autocorrelation function $\langle x(t)x(t+\tau)\rangle$ and of $2|\Gamma(\tau)|$ is shown in Fig. 3.6. It is clear that the effective width of $|\Gamma(\tau)|$ is a measure of the correlation time τ_c of the real process $x(t)$, i.e. of the time interval τ over which $x(t)$ and $x(t+\tau)$ are effectively correlated. In optics, where $x(t)$ usually represents a fluctuating real field, this correlation time is known as the *coherence time* of the field. It will be defined more precisely later (Section 4.3.3).

3.1.4 Statistical properties of the analytic signal associated with a real Gaussian random process

Let $x(t)$ be a real, wide-sense stationary Gaussian random process of zero mean. Its probability density at any instant of time is given by the formula

$$p(x) = \frac{1}{\sigma\sqrt{(2\pi)}}\mathrm{e}^{-x^2/2\sigma^2}, \tag{3.1–80}$$

where $\sigma^2 = \langle x^2\rangle$ is the variance of x. The process $y(t)$, conjugate to $x(t)$, is a linear transform of $x(t)$ [given by Eq. (3.1–13a)] and hence, according to a well-known theorem, (Davenport and Root, 1958, Sec. 8.4) is also a wide-sense stationary Gaussian random process of zero mean. Moreover, according to Theorem IV [Eq. (3.1–77)], the variance of y is equal to the variance σ of x.

‡ We stress that this assumption now refers to the *power spectrum* of a random signal and *not* to the *Fourier spectrum* of a deterministic signal, as was the case before.

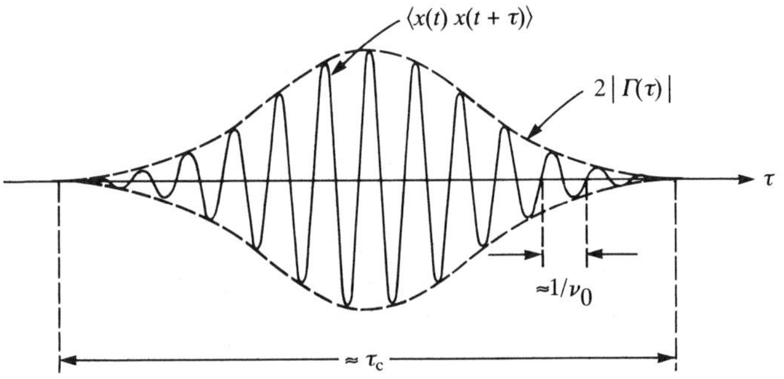

Fig. 3.6 Illustrating the relation between the correlation functions of a real quasi-monochromatic signal $x(t)$ and of the associated complex analytic signal.

Hence

$$p(y) = \frac{1}{\sigma\sqrt{(2\pi)}} \mathrm{e}^{-y^2/2\sigma^2}. \tag{3.1-81}$$

According to Eq. (3.1–77b) the random processes x and y are uncorrelated and consequently their joint probability $p(x, y)$ is just the product of $p(x)$ and $p(y)$; hence

$$p(x, y) = \frac{1}{2\pi\sigma^2} \mathrm{e}^{-(x^2+y^2)/2\sigma^2}, \tag{3.1-82}$$

i.e. it is a two-dimensional Gaussian distribution. Since the loci of constant values of the distribution (3.1–82) are circles, such a Gaussian distribution is said to be *circular*.

Let us now consider the joint probability density‡ $p(A, \varphi)$ of the modulus (amplitude) $A(t)$ and the argument (phase) $\varphi(t)$ of the associated analytic signal [Eqs. (3.1–12) and (3.1–33)]

$$z(t) = \tfrac{1}{2}[x(t) + \mathrm{i}y(t)] = \tfrac{1}{2}A(t)\,\mathrm{e}^{\mathrm{i}\varphi(t)}. \tag{3.1-83}$$

According to the elementary law for transformation of probabilities

$$p(A, \varphi)\,\mathrm{d}A\,\mathrm{d}\varphi = p(x, y)\,\mathrm{d}x\,\mathrm{d}y. \tag{3.1-84}$$

Since $x = A\cos\varphi$ and $y = A\sin\varphi$, we readily find that $\mathrm{d}x\,\mathrm{d}y = A\,\mathrm{d}A\,\mathrm{d}\varphi$ and hence Eqs. (3.1–84) and (3.1–82) give

$$p(A, \varphi) = \frac{A}{2\pi\sigma^2} \mathrm{e}^{-A^2/2\sigma^2}. \tag{3.1-85}$$

Since φ does not appear on the right-hand of Eq. (3.1–85), it is clear that all values of the phase ($0 \leqslant \varphi \leqslant 2\pi$) are equally probable, i.e. the probability density of φ is

$$p(\varphi) = \frac{1}{2\pi}. \tag{3.1-86}$$

The probability density of the amplitude is obtained by integrating Eq. (3.1–85)

‡ The symbol p stands here collectively for 'probability density of' rather than for a specific functional form.

with respect to φ over the range $0 \leqslant \varphi < 2\pi$ and evidently is

$$p(A) = \frac{A}{\sigma^2} e^{-A^2/2\sigma^2}, \tag{3.1-87}$$

which is the *Rayleigh distribution*. It is illustrated in Fig. 3.7.

Because the joint probability density $p(A, \varphi)$ [Eq. (3.1–85)] is the product of the probability densities $p(\varphi)$ and $p(A)$ given by Eqs. (3.1–86) and (3.1–87), the phase $\varphi(t)$ and the amplitude $A(t)$ of the analytic signal are, at each instant of time t, statistically independent. However, the random processes $\varphi(t)$ and $A(t)$ are, in general, statistically dependent.

Finally, let us consider the probability density of the square of the instantaneous amplitude. We will refer to it, for short, as the *instantaneous intensity* and denote it by $I(t)$:

$$I(t) = A^2(t). \tag{3.1-88}$$

Again using the elementary law for transformation of probabilities (see Section 1.3.1),

$$p(I)\,dI = p(A)\,dA, \tag{3.1-89}$$

we readily find from Eqs. (3.1–87)–(3.1–89) that

$$p(I) = \frac{1}{\langle I \rangle} e^{-I/\langle I \rangle}, \quad I \geqslant 0, \tag{3.1-90}$$

where $\langle I \rangle = 2\sigma^2$. Equation (3.1–90) shows that the instantaneous intensity has an *exponential distribution*. It is illustrated in Fig. 3.8. The nth moment $\langle I^n \rangle$ of this distribution may be shown to be $n!\langle I \rangle^n$.

We have only considered some of the simplest statistical properties of the complex process associated, via the analytic signal representation, with a real, wide-sense stationary Gaussian random process. The much more difficult problem of determining, from knowledge of the statistical properties of any real fluctuating field, the statistical properties of the associated complex analytic field was studied by Agarwal and Wolf (1972).

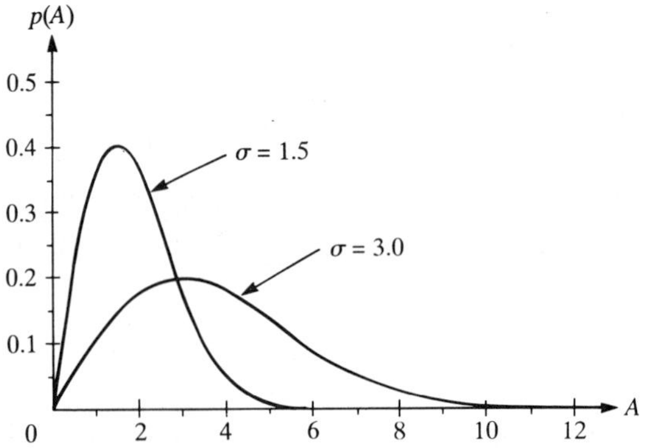

Fig. 3.7 The Rayleigh distribution $p(A) = \frac{A}{\sigma^2} e^{-A^2/2\sigma^2}$.

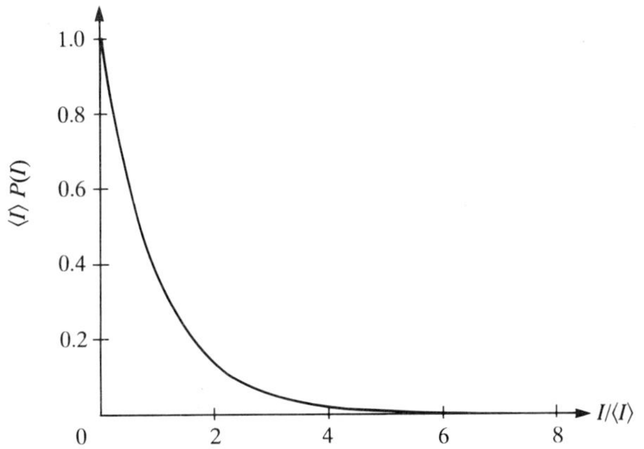

Fig. 3.8 The (scaled) exponential distribution $p(I) = \dfrac{1}{\langle I \rangle} e^{-I/\langle I \rangle}$.

3.2 The angular spectrum representation of wavefields

In this section we will describe a useful mathematical technique for studying the properties of wavefields in homogeneous media. It is based on a certain integral representation, known as the angular spectrum of plane waves. In its simplest form the representation applies to deterministic fields, but it can be generalized (as will be done in Section 5.6.3) to random fields. On the other hand it is restricted to wavefields in a domain that is either a half-space or that is bounded by two mutually parallel planes. Apart from its simplicity, the usefulness of this method lies in its strong intuitive appeal, which often makes it possible to obtain a qualitative understanding of various physical phenomena without carrying out detailed calculations. We begin by deriving the angular spectrum representation in a slab geometry.

3.2.1 The angular spectrum of a wavefield in a slab geometry

Consider a monochromatic scalar wavefield

$$V(\mathbf{r}, t) = U(\mathbf{r})\, e^{-i\omega t} \tag{3.2–1}$$

in a slab D (see Fig. 3.9) occupying the region

$$0 \leqslant z \leqslant Z, \tag{3.2–2}$$

in a homogeneous medium of refractive index $n(\omega)$. We assume that the sources of the field are located outside D. Then throughout D, the space-dependent part $U(\mathbf{r})$ of $V(\mathbf{r}, t)$ will satisfy the Helmholtz equation

$$(\nabla^2 + k^2)U(\mathbf{r}) = 0, \tag{3.2–3}$$

where

$$k = n(\omega)k_0, \tag{3.2–4a}$$

$$k_0 = \omega/c, \tag{3.2–4b}$$

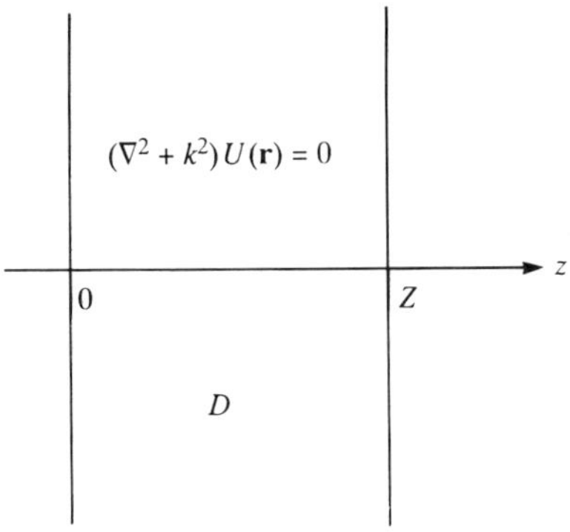

Fig. 3.9 Notation relating to the angular spectrum representation of a wave-field in a slab geometry.

(c being the speed of light in vacuo) are the wave numbers in the medium and in free space respectively, associated with frequency ω.

Let us assume that in any plane $z =$ constant in the slab D, the field may be represented as a Fourier integral, namely

$$U(x, y, z) = \iint_{-\infty}^{\infty} \mathcal{U}(u, v; z)\, \mathrm{e}^{\mathrm{i}(ux+vy)}\, \mathrm{d}u\, \mathrm{d}v, \tag{3.2–5}$$

($\mathbf{r} = x, y, z$). On substituting from Eq. (3.2–5) into Eq. (3.2–3) and on interchanging the operation $(\nabla^2 + k^2)$ and integration, we obtain the formula

$$\iint_{-\infty}^{\infty} [\nabla^2 + k^2][\mathcal{U}(u, v; z)\, \mathrm{e}^{\mathrm{i}(ux+vy)}]\, \mathrm{d}u\, \mathrm{d}v = 0. \tag{3.2–6}$$

After carrying out the differentiation under the integral sign we find that

$$\iint_{-\infty}^{\infty} \left[(-u^2 - v^2 + k^2)\mathcal{U}(u, v; z) + \frac{\partial^2 \mathcal{U}(u, v; z)}{\partial z^2}\right] \mathrm{e}^{\mathrm{i}(ux+vy)}\, \mathrm{d}u\, \mathrm{d}v = 0. \tag{3.2–7}$$

Since Eq. (3.2–7) holds for all values of x and y, the term in the brackets under the integral sign must be zero. Hence the function $\mathcal{U}(u, v; z)$ satisfies the differential equation

$$\frac{\partial^2 \mathcal{U}(u, v; z)}{\partial z^2} + w^2 \mathcal{U}(u, v; z) = 0, \tag{3.2–8}$$

where

$$w^2 = k^2 - u^2 - v^2. \tag{3.2–9}$$

For the sake of definiteness we denote by w the root of Eq. (3.2–9) defined as

$$\left.\begin{aligned} w &= +(k^2 - u^2 - v^2)^{1/2} \quad \text{when } u^2 + v^2 \leqslant k^2 \\ &= +\mathrm{i}(u^2 + v^2 - k^2)^{1/2} \quad \text{when } u^2 + v^2 > k^2. \end{aligned}\right\} \tag{3.2–10}$$

The general solution of the differential equation (3.2–8) is

$$\mathcal{U}(u, v; z) = A(u, v)\, \mathrm{e}^{\mathrm{i}wz} + B(u, v)\, \mathrm{e}^{-\mathrm{i}wz}, \tag{3.2–11}$$

where $A(u, v)$ and $B(u, v)$ are arbitrary functions. On substituting from Eq. (3.2–11) into the Fourier integral (3.2–5) we obtain the following expression for the wavefield within the slab:

$$U(x, y, z) = \iint_{-\infty}^{\infty} A(u, v)\, e^{i(ux+vy+wz)}\, du\, dv + \iint_{-\infty}^{\infty} B(u, v)\, e^{i(ux+vy-wz)}\, du\, dv. \tag{3.2–12}$$

This formula represents the wavefield within the slab as a linear superposition of contributions of the form $\exp(i\mathbf{K}^{+} \cdot \mathbf{r})$ and $\exp(i\mathbf{K}^{-} \cdot \mathbf{r})$, where $\mathbf{K}^{\pm} = (u, v, \pm w)$. According to Eqs. (3.2–9) and (3.2–4) we see that $(\mathbf{K}^{\pm})^2 \equiv u^2 + v^2 + w^2 = k^2$ $[= k_0 n(\omega)]^2$. This result can be readily seen to imply that each term of the integrands in Eq. (3.2–12) satisfies the same differential equation, namely the Helmholtz equation (3.2–3), as does the field $U(\mathbf{r})$. Consequently the exponential terms in the integrals (3.2–12) are *modes* of that equation and we may say that the formula (3.2–12) is a *mode representation* of the wavefield in a slab geometry. It should not be confused with a Fourier representation which it superficially resembles. Unlike the mode representation (3.2–12), a Fourier representation of a function of three real variables would involve three-folded, not two-folded integrals and, because $U(x, y, z)$ was assumed to be known only in the strip $0 \leqslant z \leqslant Z$ rather than throughout all space, the Fourier decomposition, unlike the decomposition (3.2–12), is not unique and is not a mode representation of the wavefield.

We will now turn our attention to the physical significance of the formula (3.2–12) under the assumption that the medium within the slab is *non-absorbing*. The refractive index $n(\omega)$ and, consequently, the wave number k are then real quantities. The formula (3.2–12) then represents the wavefield in terms of contributions from four types of plane-wave modes:

(a) $$e^{i(ux+vy+wz)} \quad \text{with } w = +(k^2 - u^2 - v^2)^{1/2}, \quad u^2 + v^2 \leqslant k^2. \tag{3.2–13a}$$

These are clearly homogeneous plane waves‡ that propagate from the boundary plane $z = 0$ towards the boundary plane $z = Z > 0$.

(b) $$e^{i(ux+vy+wz)} \quad \text{with } w = +i(u^2 + v^2 - k^2)^{1/2}, \quad u^2 + v^2 > k^2. \tag{3.2–13b}$$

Since in this case $e^{iwz} = e^{-|w|z}$ the surfaces of constant amplitude of such waves are given by z = constant, whereas the surfaces of constant phase are given by $ux + vy$ = constant. These waves are obviously inhomogeneous. Their amplitudes decay exponentially on propagation from the plane $z = 0$ towards the plane $z = Z > 0$ of the slab.

(c) $$e^{i(ux+vy-wz)} \quad \text{with } w = +(k^2 - u^2 - v^2)^{1/2}, \quad u^2 + v^2 \leqslant k^2. \tag{3.2–13c}$$

Evidently these waves are homogeneous plane waves that propagate from the boundary plane $z = Z > 0$ towards the boundary plane $z = 0$.

(d) $$e^{i(ux+vy-wz)} \quad \text{with } w = i(u^2 + v^2 - k^2)^{1/2}, \quad u^2 + v^2 > k^2. \tag{3.2–13d}$$

‡ A wave is said to be homogeneous if its surfaces of constant amplitude and constant phase coincide. If that is not so, the wave is said to be inhomogeneous.

Since now $e^{-iwz} = e^{|w|z}$ such waves are inhomogeneous and are similar to those discussed under (b) above, except that their amplitudes decay exponentially as the wave propagates from the plane $z = Z$ towards the plane $z = 0$ of the slab.

It is clear that modes of all the four types that we just discussed are needed, in general, to represent the wavefield uniquely in a domain $0 \leqslant z < Z$. Because of the physical significance of these modes as plane waves, the representation (3.2–12) is said to be a representation of the wavefield as an *angular spectrum of plane waves*.

3.2.2 *The angular spectrum of a wavefield in a half-space*

In many wave-propagation problems, e.g. in the analysis of diffraction at an aperture in a plane screen or in studies of beam propagation from a laser, one deals with wavefields in a half-space, say $z \geqslant 0$. In such cases it is often useful to make use of a mode representation of a wavefield in that half-space. Such a representation may be derived from the main result that we have just obtained for a slab geometry [Eq. (3.2–12)] by proceeding to the limit $Z \to \infty$.

We will assume that $n(\omega) = 1$ throughout the half-space $z \geqslant 0$ and that the field is outgoing at infinity; more specifically that if $\mathbf{s} \equiv (s_x, s_y, s_z \geqslant 0)$ is any fixed unit vector pointing into the half-space $z > 0$, then asymptotically (see Sec. 3.3.4),

$$U(r\mathbf{s}) \sim F(\mathbf{s})\frac{e^{ik_0 r}}{r} \quad \text{as } k_0 r \to \infty. \tag{3.2–14}$$

The function $F(\mathbf{s})$ is known as the *radiation pattern* of the field.

Since the half-space $z \geqslant 0$ may be regarded as the limit of the slab $0 \leqslant z \leqslant Z$ as $Z \to \infty$, it is clear that the wavefield in the half-space may also be represented in the form given by Eq. (3.2–12). However, the assumed outgoing behavior of the wavefield and some simple physical considerations lead to a simplification, as we will now show. For this purpose it is useful to separate the second integral on the right-hand side of Eq. (3.2–12) into contributions from homogeneous and inhomogeneous waves:

$$\iint_{-\infty}^{\infty} B(u, v)\, e^{i(ux+vy-wz)}\, du\, dv$$

$$= \iint_{u^2+v^2 \leqslant k_0^2} B(u, v)\, e^{i(ux+vy-|w|z)}\, du\, dv + \iint_{u^2+v^2 > k_0^2} B(u, v)\, e^{i(ux+vy)}\, e^{|w|z}\, du\, dv, \tag{3.2–15}$$

where w is given by Eq. (3.2–10). Since $|w| = (u^2 + v^2 - k_0^2)^{1/2}$ when $u^2 + v^2 > k_0^2$, the amplitudes of all the plane-wave modes in the second integral on the right-hand side of Eq. (3.2–15) increase without limit as z increases. For any realizable field such modes will obviously be absent, which implies that

$$B(u, v) = 0 \quad \text{when } u^2 + v^2 > k_0^2. \tag{3.2–16a}$$

Next let us consider the first integral on the right-hand side of Eq. (3.2–15). It represents contributions from homogeneous plane waves which propagate towards the plane $z = 0$ and it seems obvious that their combined effect will represent a field that is incoming rather than outgoing, as required by Eq. (3.2–14). That this is indeed so can be shown rigorously by the use of the principle of stationary phase (see also Miyamoto and Wolf, 1962, p. 615, Appendix) that will be discussed in Section 3.3. We may, therefore, conclude that in addition to requirement (3.2–16a) we must also have

$$B(u, v) = 0 \quad \text{when } u^2 + v^2 \leqslant k_0^2. \tag{3.2–16b}$$

It follows from Eqs. (3.2–12) and (3.2–16) that any wavefield in the half-space $z \geqslant 0$ which is square-integrable in any plane‡ $z =$ const. and which is outgoing at infinity in that half-space may be represented in the form

$$U(x, y, z) = \iint_{-\infty}^{\infty} A(u, v)\, \mathrm{e}^{\mathrm{i}(ux+vy+wz)}\, \mathrm{d}u\, \mathrm{d}v, \tag{3.2–17}$$

where w is given by Eqs. (3.2–10).

For later purposes it will be useful to change variables by setting

$$u = k_0 p, \quad v = k_0 q, \quad w = k_0 m, \tag{3.2–18}$$

where k_0 is, as before, the free-space wave number associated with the temporal frequency ω [Eq. (3.2–4b)]. Then the formula (3.2–17) becomes

$$U(x, y, z) = \iint_{-\infty}^{\infty} a(p, q)\, \mathrm{e}^{\mathrm{i}k_0(px+qy+mz)}\, \mathrm{d}p\, \mathrm{d}q, \tag{3.2–19}$$

where

$$a(p, q) = k_0^2 A(k_0 p, k_0 q) \tag{3.2–20}$$

and

$$m = +(1 - p^2 - q^2)^{1/2} \quad \text{when } p^2 + q^2 \leqslant 1 \tag{3.2–21a}$$

$$= +\mathrm{i}(p^2 + q^2 - 1)^{1/2} \quad \text{when } p^2 + q^2 > 1. \tag{3.2–21b}$$

The formulas (3.2–19) and (3.2–17) are equivalent versions of the representation of a monochromatic wavefield in the half-space $z \geqslant 0$, which is outgoing at infinity in that half-space, in the form of an angular spectrum of plane waves. The plane-wave modes are of two kinds. Those for which $p^2 + q^2 \leqslant 1$ are homogeneous waves that propagate into the half-space $z \geqslant 0$. And those for which $p^2 + q^2 > 1$ are inhomogeneous waves that decay exponentially in amplitude with increasing distance z from the boundary plane $z = 0$ of the half-space and are known as *evanescent waves*.

We have postulated that the field is outgoing at infinity, i.e. that it has the asymptotic behavior indicated by Eq. (3.2–14). That the angular spectrum representation (3.2–19) has indeed this behavior will be seen later [Eq. (3.2–34)

‡ It follows from Eq. (3.2–17) and an elementary inequality on integrals that this requirement will be satisfied whenever the boundary value $U(x, y, 0)$ is square-integrable. Some other sufficiency conditions for the validity of the angular spectrum representation of wavefields in a half-space were discussed by Lalor (1968).

and also Eq. (3.3–95)]. According to the later formula, one has, in general,‡

$$U(r\mathbf{s}) \equiv U(x, y, z) \sim -\frac{2\pi \mathrm{i}}{k_0}\left(\frac{z}{r}\right)a\left(\frac{x}{r}, \frac{y}{r}\right)\frac{\mathrm{e}^{\mathrm{i}k_0 r}}{r} \quad \text{as } k_0 r \to \infty \qquad (3.2\text{–}22)$$

in any fixed direction $\mathbf{s}$ $(s_x = x/r, s_y = y/r, s_z = z/r \geqslant 0)$, with $r = (x^2 + y^2 + z^2)^{1/2}$, pointing into the half-space $z \geqslant 0$ (Fig. 3.10).

The formula (3.2–22) shows that the (generally complex) spectral amplitude of one and only one plane wave of the angular spectrum representation contributes to the asymptotic behavior of the field in a fixed direction $(x/r, y/r, z/r)$; namely the one that is labeled by the parameters $p = x/r$, $q = y/r$. Since $(x/r)^2 + (y/r)^2 \leqslant 1$ this plane wave is necessarily homogeneous and is precisely that plane wave that propagates in the $\mathbf{s}$-direction for which the asymptotic (far-zone) limit is being considered. The physical reason for this result will become evident later (Sec. 3.3), when the asymptotic evaluation of certain integrals, based on the method of stationary phase, will be discussed.

We have just seen that there is an intimate relationship between the far field and the spectral amplitudes of the homogeneous plane-wave modes in the angular spectrum representation of the field in the half-space $z \geqslant 0$. We will now show that there is also a simple relationship between the spectral amplitudes of all the plane-wave modes and the boundary values of the field in the plane $z = 0$; and that this relationship provides a simple physical picture of the information that each plane-wave mode carries. For this purpose let us first represent the field in the plane $z = 0$ as a two-dimensional Fourier integral

$$U(x, y, 0) = \iint_{-\infty}^{\infty} \tilde{U}_0(u, v)\, \mathrm{e}^{\mathrm{i}(ux+vy)}\, \mathrm{d}u\, \mathrm{d}v, \qquad (3.2\text{–}23)$$

where, in the notation used in Eq. (3.2–5), $\tilde{U}_0(u, v) \equiv \mathcal{U}(u, v; 0)$. In physical terms Eq. (3.2–23) represents the boundary values of the field in the plane $z = 0$

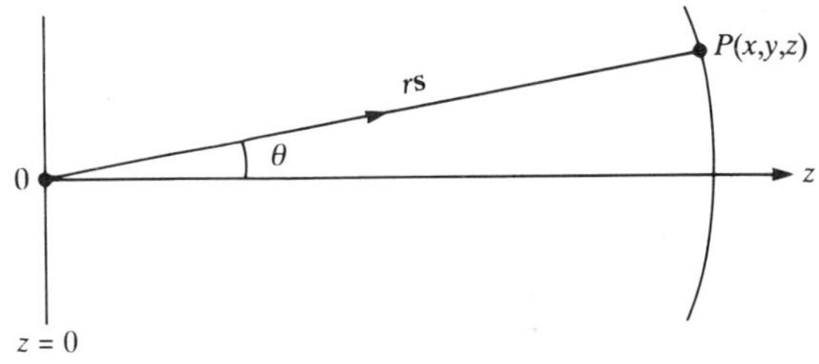

Fig. 3.10 Illustrating the asymptotic (far-zone) behavior of the angular spectrum representation (3.2–19) of a wavefield in the half-space $z \geqslant 0$. The field at a point $P(r\mathbf{s})$ in the far zone is given by formula (3.2–22), namely

$$U(r\mathbf{s}) \sim -\frac{2\pi \mathrm{i}}{k_0} a(s_x, s_y) \frac{\mathrm{e}^{\mathrm{i}k_0 r}}{r} \cos\theta,$$

where s_x, s_y, $s_z = \cos\theta$ are the Cartesian components (direction cosines) of the unit vector $\mathbf{s}$.

‡ Under special circumstances there may be points in the far field where the asymptotic formula (3.2–22) does not apply (Sherman, Stamnes and Lalor, 1976).

in terms of all possible two-dimensional spatial periodic components, labeled by two-dimensional spatial frequencies (u, v) $[-\infty < u < \infty,\ -\infty < v < \infty]$. The (generally complex) Fourier amplitudes $\widetilde{U}_0(u, v)$, which are given by the Fourier inverse of $U(x, y, 0)$,

$$\widetilde{U}_0(u, v) = \frac{1}{(2\pi)^2} \iint_{-\infty}^{\infty} U(x, y, 0)\, \mathrm{e}^{-\mathrm{i}(ux+vy)}\, \mathrm{d}x\, \mathrm{d}y, \tag{3.2–24}$$

represents the strengths with which the spatial-frequency component (u, v) contributes to the boundary value of the field in the plane $z = 0$.

It follows at once from the angular spectrum representation (3.2–19) that

$$U(x, y, 0) = \iint_{-\infty}^{\infty} a(p, q)\, \mathrm{e}^{\mathrm{i}k_0(px+qy)}\, \mathrm{d}p\, \mathrm{d}q, \tag{3.2–25}$$

or, if we change the variables of integration from (p, q) to (u, v) according to the first two relations in Eq. (3.2–18), the formula (3.2–25) takes the form

$$U(x, y, 0) = \frac{1}{k_0^2} \iint_{-\infty}^{\infty} a\left(\frac{u}{k_0}, \frac{v}{k_0}\right) \mathrm{e}^{\mathrm{i}(ux+vy)}\, \mathrm{d}u\, \mathrm{d}v. \tag{3.2–25a}$$

A comparison of this equation with Eq. (3.2–23) shows that

$$\widetilde{U}_0(u, v) = \frac{1}{k_0^2} a\left(\frac{u}{k_0}, \frac{v}{k_0}\right), \tag{3.2–26}$$

which implies that

$$a(p, q) = k_0^2 \widetilde{U}_0(k_0 p, k_0 q). \tag{3.2–27}$$

This relation shows that *the spectral amplitude $a(p, q)$ of each plane-wave mode in the angular spectrum representation of the field is uniquely specified by one and only one spatial-frequency component (Fourier component) of the boundary value of the field $U_0(x, y) \equiv U(x, y, 0)$ in the plane $z = 0$; namely the one labeled by the spatial frequency*

$$u = k_0 p, \quad v = k_0 q. \tag{3.2–28}$$

Some additional insight into the intimate relationship that exists between the angular spectrum modes and the boundary values of the field in the plane $z = 0$ may be obtained if we introduce the spatial periods Δx, Δy $[0 \leqslant \Delta x < \infty$, $0 \leqslant \Delta y < \infty]$ associated with the spatial frequencies (u, v) via the relations

$$\Delta x = \frac{2\pi}{|u|}, \quad \Delta y = \frac{2\pi}{|v|}. \tag{3.2–29}$$

Since, according to Eq. (3.2–18) $u = k_0 p = 2\pi p/\lambda_0$ and $v = k_0 q = 2\pi q/\lambda_0$, where $\lambda_0 = 2\pi/k_0$ is the wavelength of the monochromatic field, Eqs. (3.2–29) imply that

$$\Delta x = \frac{\lambda_0}{|p|}, \quad \Delta y = \frac{\lambda_0}{|q|}. \tag{3.2–30}$$

Since for homogeneous waves $p^2 + q^2 \leqslant 1$ and for evanescent waves $p^2 + q^2 > 1$ it follows at once from Eqs. (3.2–30) that the homogeneous waves are associated

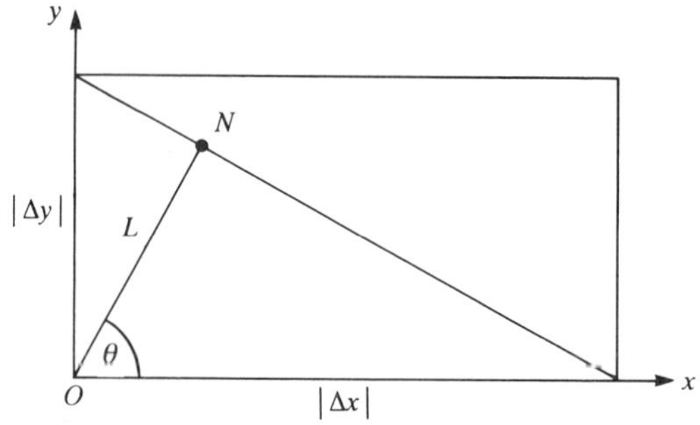

Fig. 3.11 Illustrating the meaning of the distance L, defined by Eq. (3.2–32).

with spatial periods for which

$$\frac{1}{(\Delta x)^2} + \frac{1}{(\Delta y)^2} \leqslant \frac{1}{\lambda_0^2}, \tag{3.2–31a}$$

and the evanescent waves are associated with spatial periods for which

$$\frac{1}{(\Delta x)^2} + \frac{1}{(\Delta y)^2} > \frac{1}{\lambda_0^2}. \tag{3.2–31b}$$

To see clearly the physical significance of the above inequalities let us introduce a length L, which represents the distance $\overline{ON}$ of the perpendicular dropped from a corner O of a rectangle with sides Δx and Δy onto the opposite diagonal of the rectangle (Fig. 3.11). If θ is the angle that $\overline{ON}$ makes with the x-axis one obviously has $L = \Delta x \cos\theta = \Delta y \sin\theta$ and hence

$$\frac{1}{(\Delta x)^2} + \frac{1}{(\Delta y)^2} = \frac{1}{L^2}. \tag{3.2–32}$$

From what we have just learned it is clear that spatial-frequency components of $U(x, y, 0)$ for which $L \geqslant \lambda_0$ give rise to homogeneous waves, whereas those for which $L < \lambda_0$ give rise to evanescent waves.

We note that in the special case when $\Delta y \to \infty$ (one-dimensional boundary field) the length L is just equal to Δx. There is no such simple correspondence for a two-dimensional boundary field but we may still regard L as a measure of the size of the periodic component with which it is associated via Eq. (3.2–32). With this understanding the inequalities (3.2–31a) and (3.2–31b) imply, roughly speaking, that the homogeneous waves carry information about periodic details of the field in the plane $z = 0$ which are larger than the wavelength, whereas the evanescent waves carry information about periodic details that are smaller than the wavelength. A precise delimitation between the two Δx, Δy domains is shown in Fig. 3.12.

It is clear from Eqs. (3.2–18) and (3.2–21) that homogeneous waves are associated with those spatial frequencies for which

$$u^2 + v^2 \leqslant k_0^2 \tag{3.2–33a}$$

whereas the evanescent waves are associated with spatial frequencies for which

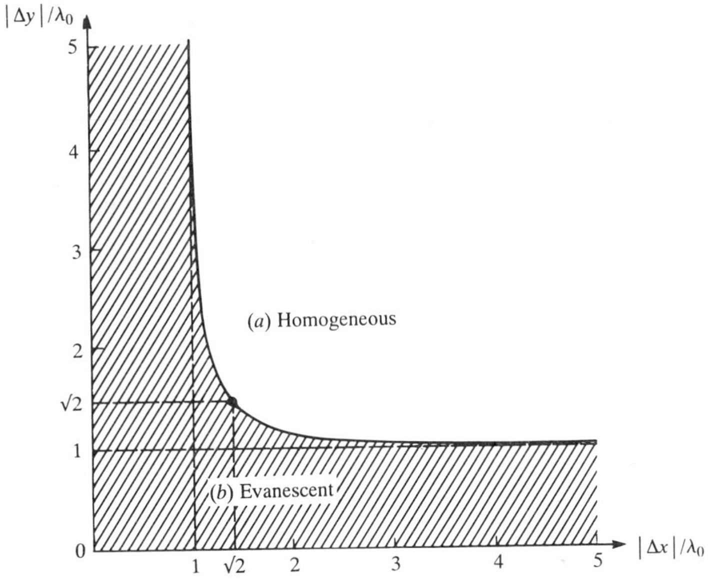

Fig. 3.12 The spatial periodicities of the field in the plane $z = 0$ about which information is carried by (a) homogeneous waves and (b) evanescent waves.

$$u^2 + v^2 > k_0^2. \tag{3.2–33b}$$

Spatial frequencies that satisfy the inequality (3.2–33a) are often said to be *low spatial frequencies* whereas those that satisfy the inequality (3.2–33b) are said to be *high spatial frequencies*.

Let us now briefly return to the formula (3.2–22) for the far field. If we substitute for the spectral amplitude function the expression (3.2–27), equation (3.2–22) becomes

$$U(r\mathbf{s}) \equiv U(x, y, z) \sim -2\pi \mathrm{i} k_0 \tilde{U}_0\left(k_0\frac{x}{r}, k_0\frac{y}{r}\right)\frac{\mathrm{e}^{\mathrm{i}k_0 r}}{r}\cos\theta \quad \text{as } k_0 r \to \infty \tag{3.2–34}$$

along any fixed θ direction ($\cos\theta = z/r$) which points into the half-space $z \geqslant 0$. Since $(x/r)^2 + (y/r)^2$ cannot exceed unity, $(k_0x/r)^2 + (k_0y/r)^2 \leqslant k_0^2$ and the formula (3.2–34) shows that the far field is specified entirely by the low spatial-frequency components of the field in the plane $z = 0$. This was to be expected, because, as we have just learned, the high spatial frequencies give rise to evanescent waves and since the amplitude of such waves decays exponentially with the distance z from the boundary plane $z = 0$, they do not, in general contribute to the far field.‡ The expression (3.2–34), which will be derived more directly in Section 3.2.5 [Eq. (3.2–88)], is a precise version of the well-known Fraunhofer formula of elementary diffraction theory (cf. Goodman, 1968, p. 61).

‡ Wavefields in a half-space, which do not contain evanescent waves at all in their angular spectrum representation, have a number of interesting properties, that were first discussed by Sherman (1969). Such fields are often very good approximations to fields encountered in practice, except close to sources and boundaries of objects with which the field may interact.

3.2.3 *An example: diffraction by a semi-transparent object*‡

To illustrate the physical insight that the angular spectrum representation can provide, we will consider the transmission of light through a thin, semi-transparent, weakly scattering object. Suppose that the object is placed on a plane $z = 0$ and is illuminated by a monochromatic plane wave that propagates along the positive z-direction (see Fig. 3.13). The space-dependent part of the wave can then be represented as

$$U^{(i)}(x, y, z) = C\,e^{ik_0 z}, \tag{3.2–35}$$

where C is a constant. If the refractive index changes slowly throughout the object, we may assume that the field U that emerges from it is, to a good approximation, given by

$$U(x, y, 0) = T(x, y)U^{(i)}(x, y, 0). \tag{3.2–36}$$

If we allow only the light that passes through the object to reach the half-space $z \geqslant 0$ then evidently $T(x, y) = 0$ at all points (x, y) that are located outside the area of the plane $z = 0$ occupied by the object. The function $T(x, y)$ is called *the transmission function of the object* or *the amplitude transmittance* (for normal incidence, at the given temporal frequency of the incident light). The modulus of T is a measure of how much light the object absorbs on transmission and the phase of T is a measure of the optical thickness of the object. It is this kind of information that one often seeks. Typically one would place an optical imaging system, e.g. a microscope, to the right of the object (i.e. in the half-space $z > 0$) and one would try to deduce the properties of the object from the analysis of its image.

Assuming that Eq. (3.2–36) describes sufficiently accurately the change in the incident wave due to the presence of the object it is clear that the maximum information the imaging system can provide about the object is its transmission function $T(x, y)$. We will examine how much information about $T(x, y)$ is contained in the transmitted wave before it reaches the imaging system.

It is clear from Eqs. (3.2–36) and (3.2–35) that in the plane $z = 0$, the field

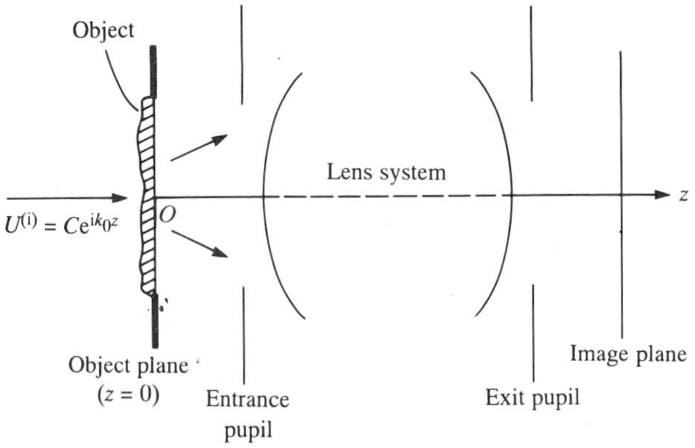

Fig. 3.13 Imaging of a thin semi-transparent object.

‡ This example and our method of analysis is essentially due to Gabor (1961).

emerging from the object is given by

$$U(x, y, 0) = CT(x, y). \tag{3.2–37}$$

In the absence of the imaging system, Eq. (3.2–37) represents the boundary values of a field that propagates freely in the half-space $z > 0$ and is outgoing at infinity in that half-space. Hence, according to Eqs. (3.2–19), (3.2–27), and (3.2–37), the field throughout that half-space may be represented in the form

$$U(x, y, z) = \iint_{-\infty}^{\infty} Ck_0^2\tilde{T}(k_0p, k_0q)\, \mathrm{e}^{\mathrm{i}k_0(px+qy+mz)}\, \mathrm{d}p\, \mathrm{d}q, \tag{3.2–38}$$

where

$$\tilde{T}(u, v) = \frac{1}{(2\pi)^2}\iint_{-\infty}^{\infty} T(x, y)\, \mathrm{e}^{-\mathrm{i}(ux+vy)}\, \mathrm{d}x\, \mathrm{d}y, \tag{3.2–39}$$

it being understood that $T(x, y) \equiv 0$ outside the region of the object plane occupied by the object.

According to Eq. (3.2–38) the amplitude of a typical plane wave, labeled by (p, q) in the angular spectrum representation of the transmitted field, is

$$a(p, q) = Ck_0^2\tilde{T}(k_0p, k_0q). \tag{3.2–40}$$

We see that the wave carries information about the Fourier components of the transmission function labeled by the (two-dimensional) spatial frequency $u = k_0p$, $v = k_0q$ that corresponds to the spatial periods Δx, Δy that are related to p and q by the formulas (3.2–30), which we will re-write as

$$|p| = \frac{\lambda_0}{\Delta x}, \quad |q| = \frac{\lambda_0}{\Delta y}. \tag{3.2–41}$$

Obviously if Δx and Δy are large enough, $|p|^2 + |q|^2 \ll 1$, i.e. the wave will propagate in a direction which makes a small angle θ with the z-axis (see Fig. 3.10). With decreasing values of Δx and Δy, θ will increase and it will eventually become so large that the corresponding homogeneous waves will not enter the imaging system at all. Consequently this information will be lost. This simple argument clearly shows the true origin of the loss of resolution in optical image formation.

Next let us consider the information about very small details, indicated by the inequality (3.2–31b), of the object structure, which is carried by evanescent waves. Since the amplitude of an evanescent wave decays exponentially with increasing distance from the object plane this information will obviously be harder and harder to discern the further one is away from the object plane. As an example consider a periodic element with periods

$$\Delta x = \lambda_0/5, \quad \Delta y = \infty. \tag{3.2–42}$$

(The second formula in Eq. (3.2–42) implies that the object is one-dimensional, extending only in the x-direction.) According to Eqs. (3.2–41) and (3.2–42) information about this element is carried by evanescent plane-wave modes for which

$$|p| = 5, \quad |q| = 0. \tag{3.2–43}$$

For this wave, we have, according to Eqs. (3.2–43) and (3.2–21b),

$$m = \mathrm{i}(25 - 1)^{1/2} \approx 4.9\mathrm{i}, \tag{3.2–44}$$

and hence [cf. Eq. (3.2–38)] the space-dependent part of the wave is given by

$$\begin{aligned} U(x, y, z) &= Ck_0^2\widetilde{T}(k_0p, 0)\,\mathrm{e}^{\mathrm{i}k_0px}\,\mathrm{e}^{-k_0|m|z} \\ &= Ck_0^2\widetilde{T}(k_0p, 0)\,\mathrm{e}^{\mathrm{i}5k_0x}\,\mathrm{e}^{-4.9k_0z}. \end{aligned} \tag{3.2–45}$$

Since

$$\mathrm{e}^{-4.9k_0z} \approx \mathrm{e}^{-31z/\lambda_0}, \tag{3.2–46}$$

the amplitude of this wave will decay by a factor e^{-31} in a distance of only one wavelength from the object plane and by a factor e^{-310} in a distance of ten wavelengths! Thus for all practical purposes information about such fine detail of the object is lost on propagation.

3.2.4 The Weyl representation of a spherical wave

In a well-known paper dealing with the propagation of electromagnetic waves over a conducting sphere Weyl (1919) derived a new representation of a spherical wave. It may be regarded as the angular spectrum representation of a wavefield in free space, generated by a point source located at the origin. Because the spherical wave is the Green's function of the Helmholtz operator, the Weyl representation has found many useful applications in connection with analyses of radiation, diffraction and scattering problems by the use of the angular spectrum techniques.

Because the spherical wave has a singularity at its center, a rigorous derivation of the Weyl representation requires some mathematical sophistication. We will ignore the underlying subtleties and give a purely formal derivation. Rigorous treatments may be found elsewhere (see, for example, Baños, 1966, Sec. 2.13).

Let us consider the diverging spherical wave (the time-dependent factor $\mathrm{e}^{-\mathrm{i}\omega t}$ being omitted)

$$G(\mathbf{r}) = \frac{\mathrm{e}^{\mathrm{i}k_0r}}{r}, \tag{3.2–47}$$

where $r = |\mathbf{r}|$ is the distance of the field point from the center of the wave. At any point except at the origin $\mathbf{r} = 0$, $G(\mathbf{r})$ satisfies the Helmholtz equation.

$$(\nabla^2 + k_0^2)G(\mathbf{r}) = 0, \quad (\mathbf{r} \neq 0). \tag{3.2–48}$$

Hence it seems reasonable to assume that, if we choose a rectangular Cartesian coordinate system with the origin at the source point, we might be able to represent $G(\mathbf{r})$ in each of the half-spaces $z > 0$ and $z < 0$ in the form of an angular spectrum of plane waves. Let us consider first the representation $G(\mathbf{r})$ in the half-space $z > 0$. We will then have, formally at any rate,

$$\frac{\mathrm{e}^{\mathrm{i}k_0r}}{r} = \iint_{-\infty}^{\infty} a(p, q)\,\mathrm{e}^{\mathrm{i}k_0(px+qy+mz)}\,\mathrm{d}p\,\mathrm{d}q, \quad (z > 0), \tag{3.2–49}$$

where m is given by the expressions (3.2–21), namely

$$m = +(1 - p^2 - q^2)^{1/2} \quad \text{when } p^2 + q^2 \leqslant 1 \tag{3.2–50a}$$

$$= +\mathrm{i}(p^2 + q^2 - 1)^{1/2} \quad \text{when } p^2 + q^2 > 1. \tag{3.2–50b}$$

Suppose that a function $a(p, q)$ exists which admits the representation (3.2–49) and, moreover, that the formula (3.2–49) remains valid as $z \to 0$, except at the origin (the singularity of the spherical wave). Then, in the limit as $z \to 0$,

$$\frac{\exp\left[\mathrm{i}k_0(x^2 + y^2)^{1/2}\right]}{(x^2 + y^2)^{1/2}} = \iint_{-\infty}^{\infty} a(p, q) \exp\left[\mathrm{i}k_0(px + qy)\right] \mathrm{d}p\, \mathrm{d}q. \tag{3.2–51}$$

On making a trivial change of the variables of integration and applying the Fourier inversion formula, we obtain the following expression for the spectral amplitude function $a(p, q)$:

$$a(p, q) = \left(\frac{k_0}{2\pi}\right)^2 \iint_{-\infty}^{\infty} \frac{\exp\left[\mathrm{i}k_0(x^2 + y^2)^{1/2}\right]}{(x^2 + y^2)^{1/2}} \exp\left[-\mathrm{i}k_0(px + qy)\right] \mathrm{d}x\, \mathrm{d}y. \tag{3.2–52}$$

The integral on the right-hand side of Eq. (3.2–52) may be evaluated in a closed form. For this purpose we set

$$x = R \cos\psi, \quad y = R \sin\psi, \tag{3.2–53a}$$

$$p = \rho \cos\chi, \quad q = \rho \sin\chi. \tag{3.2–53b}$$

Then Eq. (3.2–52) becomes

$$a(p, q) = \left(\frac{k_0}{2\pi}\right)^2 \int_0^{\infty}\int_0^{2\pi} \mathrm{e}^{\mathrm{i}k_0 R}\, \mathrm{e}^{-\mathrm{i}k_0 R\rho\cos(\psi - \chi)}\, \mathrm{d}R\, \mathrm{d}\psi. \tag{3.2–54}$$

The integration with respect to ψ can be carried out at once and gives [Watson, 1944, p. 20, Eq. (5), with an obvious substitution]

$$\int_0^{2\pi} \mathrm{e}^{-\mathrm{i}k_0 R\rho\cos(\psi - \chi)}\, \mathrm{d}\psi = 2\pi J_0(k_0 R\rho), \tag{3.2–55}$$

where $J_0(x)$ is the Bessel function of the first kind and zero order. Hence Eq. (3.2–54) reduces to

$$a(p, q) = \frac{k_0^2}{2\pi} \int_0^{\infty} \mathrm{e}^{\mathrm{i}k_0 R} J_0(k_0 R\rho)\, \mathrm{d}R. \tag{3.2–56}$$

The integral on the right may be evaluated by using two well-known Hankel-transform formulas [Erdelyi, 1954, p. 7, Sec. 8.2, formulas (5) and (6)]

$$\int_0^{\infty} \frac{\cos ax}{\sqrt{x}} J_0(xy)\sqrt{(xy)}\, \mathrm{d}x = 0 \qquad \text{when } 0 < y < a$$

$$= \left(\frac{y}{y^2 - a^2}\right)^{1/2} \qquad \text{when } a < y < \infty$$

$$\int_0^{\infty} \frac{\sin ax}{\sqrt{x}} J_0(xy)\sqrt{(xy)}\, \mathrm{d}x = \left(\frac{y}{a^2 - y^2}\right)^{1/2} \qquad \text{when } 0 < y < a$$

$$= 0 \qquad \text{when } a < y < \infty.$$

On taking the complex combination of these expressions and canceling some common factors we obtain the formula

$$\left.\begin{aligned}\int_0^{\infty} \mathrm{e}^{\mathrm{i}ax} J_0(xy)\,\mathrm{d}x &= \frac{\mathrm{i}}{(a^2 - y^2)^{1/2}} \quad \text{when } 0 < y < a \\ &= \frac{1}{(y^2 - a^2)^{1/2}} \quad \text{when } y < a < \infty.\end{aligned}\right\} \tag{3.2–57}$$

If we set $a = k_0$, $x = R$, $y = k_0\rho$ in these formulas and use the fact that, according to Eq. (3.2–53b), $\rho^2 = p^2 + q^2$, we find that

$$\left.\begin{aligned}\int_0^{\infty} \mathrm{e}^{\mathrm{i}k_0R} J_0(k_0R\rho)\,\mathrm{d}R &= \frac{\mathrm{i}}{k_0(1 - p^2 - q^2)^{1/2}} \quad \text{when } p^2 + q^2 < 1 \\ &= \frac{1}{k_0(p^2 + q^2 - 1)^{1/2}} \quad \text{when } p^2 + q^2 > 1.\end{aligned}\right\} \tag{3.2–58}$$

On substituting from the formula (3.2–58) into Eq. (3.2–56) and recalling the definition (3.2–50) of the quantity m, we obtain the following simple expression for the angular spectrum amplitude:

$$a(p, q) = \frac{\mathrm{i}k_0}{2\pi}\frac{1}{m}. \tag{3.2–59}$$

On substituting from Eq. (3.2–59) into Eq. (3.2–49) we finally obtain the required representation of the spherical wave:

$$\frac{\mathrm{e}^{\mathrm{i}k_0r}}{r} = \frac{\mathrm{i}k_0}{2\pi}\iint_{-\infty}^{\infty} \frac{1}{m}\mathrm{e}^{\mathrm{i}k_0(px+qy+mz)}\,\mathrm{d}p\,\mathrm{d}q, \quad (z > 0). \tag{3.2–60}$$

More refined analysis shows that this formula is indeed valid not only throughout the half-space $z > 0$, but, as we assumed, also on the plane $z = 0$ except at the origin.

In a strictly similar manner, or by using a symmetry argument, one can show that when $z < 0$ Eq. (3.2–60) holds with z replaced by $|z|$. Combining the two formulas we have, for all z,

$$\frac{\mathrm{e}^{\mathrm{i}k_0r}}{r} = \frac{\mathrm{i}k_0}{2\pi}\iint_{-\infty}^{\infty} \frac{1}{m}\mathrm{e}^{\mathrm{i}k_0(px+qy+m|z|)}\,\mathrm{d}p\,\mathrm{d}q. \tag{3.2–61}$$

The expression on the right-hand side of Eq. (3.2–61) is one form of the *Weyl representation of a diverging spherical wave*. It expresses the spherical wave in each half-space $z \leqslant 0$ and $z \geqslant 0$ as an angular spectrum of plane waves, with the homogeneous waves propagating into the far zone and with the evanescent waves decaying exponentially in amplitude with increasing distance $|z|$ from the plane $z = 0$. We note that the angular spectral amplitude function of each of the plane waves, given by Eq. (3.2–59), is singular when $m = 0$, i.e. when $p^2 + q^2 = 1$. This is precisely the circle in the p, q-plane, where the nature of the plane-wave modes changes, the modes being homogeneous when $p^2 + q^2 \leqslant 1$ and evanescent when $p^2 + q^2 > 1$. The singular behavior of the spectral amplitude $a(p, q)$ on this circle is a reflection of the singularity of the spherical wave at the origin. These singularities in the integrand in the Weyl representation (3.2–61) are, however, integrable, except, of course, when $x = y = z = 0$. Another reflection of the presence of the singularity of the spherical wave is the non-analytic nature of the integrand [the appearance of $|z|$ rather than z in Eq. (3.2–61)].

Before proceeding further let us briefly consider the corresponding *Weyl representation of a converging spherical wave*. It can be obtained at once by taking the complex conjugate of Eq. (3.2–61). This gives, if we also change the variables of integration from p and q to $-p$ and $-q$,

$$\frac{e^{-ik_0 r}}{r} = -\frac{ik_0}{2\pi}\iint_{-\infty}^{\infty} \frac{1}{m^*} e^{ik_0(px+qy-m^*|z|)}\,dp\,dq, \tag{3.2–62}$$

where the asterisk denotes the complex conjugate. If we recall that according to Eq. (3.2–50) $m^* = m$ for homogeneous waves and $m^* = -m$ for evanescent waves, we can readily see that the formula (3.2–62) represents the converging spherical wave in each half-space $z > 0$ and $z < 0$ as superposition of homogeneous waves that propagate from the far zone towards the origin and of evanescent waves whose amplitudes decay exponentially with increasing $|z|$. Thus, as compared with the case of a diverging wave, each homogeneous wave propagates in the reverse direction but the evanescent waves are unchanged‡.

A somewhat different version of the Weyl formula may readily be obtained from Eq. (3.2–61) by introducing different variables of integration. For this purpose we set

$$p = \sin\alpha\cos\beta, \quad q = \sin\alpha\sin\beta, \quad m = \cos\alpha. \tag{3.2–63}$$

Recalling that $p^2 + q^2 + m^2 = 1$, [a condition which ensures that each term under the integral sign is a mode of the Helmholtz equation $(\nabla^2 + k_0^2)U(\mathbf{r}) = 0$], it is clear that for homogeneous waves ($p^2 + q^2 \leqslant 1$) α and β are just the spherical polar angles of their directions of propagation, with the polar axis being along the positive z-direction. However, since for evanescent waves ($p^2 + q^2 > 1$) m is complex, the transformation (3.2–63) requires that for such waves α itself be complex. It may readily be verified that for homogeneous waves

$$0 \leqslant \alpha < \pi/2, \quad 0 \leqslant \beta < 2\pi, \tag{3.2–64}$$

whereas for evanescent waves

$$\alpha = \frac{\pi}{2} + i\alpha', \quad 0 \leqslant \beta < 2\pi, \tag{3.2–65}$$

with $-\infty < \alpha' < 0$. In Fig. 3.14(a) the portion of the α-contour associated with

‡ If we subtract Eq. (3.2–62) from Eq. (3.2–61), the contributions from the evanescent waves cancel out. We then obtain, after simple algebraic manipulations, the well-known formula (see, for example, Courant and Hilbert, 1962, p. 195)

$$\frac{\sin k_0 r}{k_0 r} = \frac{1}{4\pi}\int_{4\pi} \exp(ik_0\mathbf{s}\cdot\mathbf{r})\,d\Omega.$$

Here $\mathbf{s}$ is a real unit vector, $d\Omega$ is an element of solid angle [cf. Eq. (3.2–66)] and the integration is taken over the whole 4π-solid angle generated by $\mathbf{s}$. In physical terms the function $\sin(k_0 r)/k_0 r$ may be regarded as arising from the superposition of homogeneous plane waves, all of the same amplitude $1/4\pi$, propagating in all possible directions.

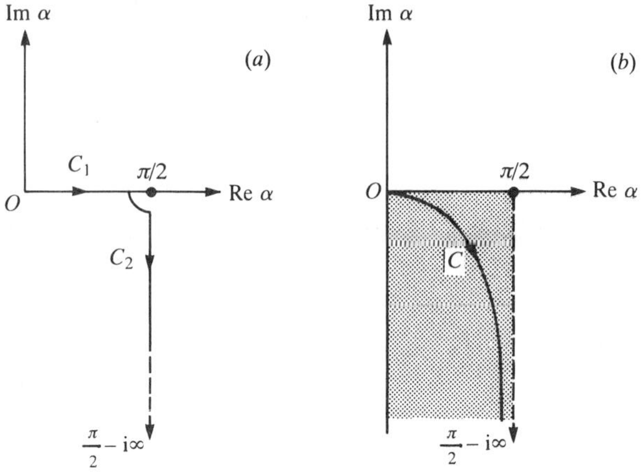

Fig. 3.14 The α-contours used in the Weyl representations of a spherical wave. In (a) each point on the horizontal segment C_1 is associated with a homogeneous plane wave and each point on the vertical line C_2 is associated with an evanescent wave in the formula (3.2–61). The curve C in (b) is the contour used in the formula (3.2–67).

homogeneous waves is denoted by C_1 and the portion associated with evanescent waves is denoted by C_2. One can readily show by calculating the Jacobian of the transformation $p, q \to \alpha, \beta$ that in both cases

$$\frac{\mathrm{d}p\,\mathrm{d}q}{m} = \sin\alpha\,\mathrm{d}\alpha\,\mathrm{d}\beta. \tag{3.2–66}$$

Evidently for real values of α, the expression (3.2–66) represents an element of solid angle formed by the (real) directions of propagation of the homogeneous plane waves in the integrand in Eq. (3.2–61). On making use of the relation (3.2–66) in Eq. (3.2–61) we obtain the following alternative representation of a diverging spherical wave:

$$\frac{\mathrm{e}^{\mathrm{i}k_0 r}}{r} = \frac{\mathrm{i}k_0}{2\pi}\int_0^{2\pi}\mathrm{d}\beta\int_{C_1+C_2}\mathrm{e}^{\mathrm{i}k_0(px+qy+m|z|)}\sin\alpha\,\mathrm{d}\alpha. \tag{3.2–67}$$

Here p, q and m are, of course, given by Eqs. (3.2–63).

By applying well-known rules relating to integration in the complex plane, the contour $C_1 + C_2$ may be distorted into a contour such as that denoted by C in Fig. 3.14(b). It is any curve in the complex α-plane that begins at the origin, approaches asymptotically the point $\pi/2 - \mathrm{i}\infty$, and lies entirely within the shaded strip. The formula (3.2–67) with $C_1 + C_2$ replaced by C is essentially the original version of Weyl's representation of a diverging spherical wave.

Let us examine the behavior of the mode functions along such a contour C. Since each of the three parameters p, q and m takes on complex values at every point on C except at the origin, we will set

$$p = p_1 + \mathrm{i}p_2, \quad q = q_1 + \mathrm{i}q_2, \quad m = m_1 + \mathrm{i}m_2, \tag{3.2–68}$$

where p_1, p_2, q_1, q_2, m_1 and m_2 are real. We then obviously have

$$\mathrm{e}^{\mathrm{i}k_0(px+qy+m|z|)} = \mathrm{e}^{\mathrm{i}k_0(p_1x+q_1y+m_1|z|)}\,\mathrm{e}^{-k_0(p_2x+q_2y+m_2|z|)}, \tag{3.2–69}$$

showing that, except for special values of the parameters, the surfaces of constant amplitude and of constant phase are distinct. Hence these modes are inhomogeneous plane waves which propagate in directions whose direction cosines are proportional to p_1, q_1, $\pm m_1$ and whose amplitudes decay exponentially along the directions p_2, q_2, $\pm m_2$. Here the upper or the lower signs are taken in front of m_1 and m_2 according as $z > 0$ or $z < 0$. These modes are obviously generalizations of the evanescent waves that we encountered earlier.

3.2.5 *The Rayleigh diffraction formulas*

We will now show how, with the help of the Weyl representation of the spherical wave, one may readily solve the Dirichlet and the Neumann boundary value problems of the Helmholtz equation for a half-space.

Let us first consider the Dirichlet problem, i.e. the problem of determining the solution of the Helmholtz equation (3.2–3) for the wavefield $U(x, y, z)$, valid throughout the half-space $z > 0$, from knowledge of the boundary values $U(x, y, 0)$ on the plane $z = 0$. We will assume that $U(x, y, z)$ is outgoing at infinity in the half-space $z > 0$.

We begin with the angular spectrum representation (3.2–19) of the field throughout the half-space, namely

$$U(x, y, z) = \iint_{-\infty}^{\infty} a(p, q)\, \mathrm{e}^{\mathrm{i}k_0(px+qy+mz)}\, \mathrm{d}p\, \mathrm{d}q, \tag{3.2–70}$$

where, according to Eqs. (3.2–27) and (3.2–24),

$$a(p, q) = \left(\frac{k_0}{2\pi}\right)^2 \iint_{-\infty}^{\infty} U(x', y', 0)\, \mathrm{e}^{-\mathrm{i}k_0(px'+qy')}\, \mathrm{d}x'\, \mathrm{d}y'. \tag{3.2–71}$$

If we substitute from Eq. (3.2–71) into Eq. (3.2–70) and interchange the order of the integrations, we obtain the following expression for the field in terms of the boundary values:

$$U(x, y, z) = \iint_{-\infty}^{\infty} U(x', y', 0) G_{\mathrm{D}}(x - x', y - y', z)\, \mathrm{d}x'\, \mathrm{d}y', \tag{3.2–72}$$

where

$$G_{\mathrm{D}}(x - x', y - y', z) = \left(\frac{k_0}{2\pi}\right)^2 \iint_{-\infty}^{\infty} \mathrm{e}^{\mathrm{i}k_0[p(x-x')+q(y-y')+mz]}\, \mathrm{d}p\, \mathrm{d}q. \tag{3.2–73}$$

Now we have according to Weyl's formula (3.2–60) with $\mathbf{r} \equiv (x, y, z > 0)$, $\mathbf{r}' \equiv (x, y, 0)$

$$\frac{\mathrm{e}^{\mathrm{i}k_0|\mathbf{r}-\mathbf{r}'|}}{|\mathbf{r} - \mathbf{r}'|} = \frac{\mathrm{i}k_0}{2\pi} \iint_{-\infty}^{\infty} \frac{1}{m}\, \mathrm{e}^{\mathrm{i}k_0[p(x-x')+q(y-y')+mz]}\, \mathrm{d}p\, \mathrm{d}q. \tag{3.2–74}$$

If we differentiate this formula with respect to z and interchange the order of differentiation and integration on the right we find that

$$\frac{\partial}{\partial z}\left[\frac{e^{ik_0|\mathbf{r}-\mathbf{r}'|}}{|\mathbf{r}-\mathbf{r}'|}\right] = -\frac{k_0^2}{2\pi}\iint_{-\infty}^{\infty} e^{ik_0[p(x-x')+q(y-y')+mz]}\,dp\,dq. \qquad (3.2\text{–}75)$$

On comparing Eqs. (3.2–73) and (3.2–75) we see that

$$G_D(x-x', y-y', z) = -\frac{1}{2\pi}\frac{\partial}{\partial z}\left(\frac{e^{ik_0R}}{R}\right), \qquad (3.2\text{–}76)$$

where

$$R = |\mathbf{r}-\mathbf{r}'| = [(x-x')^2 + (y-y')^2 + z^2]^{1/2}. \qquad (3.2\text{–}77)$$

Finally on substituting from Eq. (3.2–76) into Eq. (3.2–72) we obtain the required solution to the Dirichlet problem:

$$U(x, y, z) = -\frac{1}{2\pi}\iint_{-\infty}^{\infty} U(x', y', 0)\frac{\partial}{\partial z}\left(\frac{e^{ik_0R}}{R}\right)dx'\,dy'. \qquad (3.2\text{–}78)$$

This formula was first derived by Rayleigh‡ (1897) and is sometimes called the *Rayleigh diffraction formula of the first kind*.

The solution of the corresponding Neumann problem, i.e. the problem of determining the (outgoing) solution of the Helmholtz equation throughout the half-space $z > 0$ from knowledge of the boundary values of the derivative $\partial U(x, y, z)/\partial z$ on the plane $z = 0$, may be obtained in a somewhat similar manner. We have, on differentiating Eq. (3.2–70) with respect to z and on interchanging the order of differentiation and integration,

$$\left[\frac{\partial U(x, y, z)}{\partial z}\right]_{z=0} = ik_0\iint_{-\infty}^{\infty} m a(p, q)\, e^{ik_0(px+qy)}\,dp\,dq. \qquad (3.2\text{–}79)$$

On taking the Fourier inverse of this formula (after making a trivial change of the variables of integration) we obtain the following expression for the spectral amplitude function of the field:

$$a(p, q) = \left(\frac{k_0}{2\pi}\right)^2\frac{1}{ik_0m}\iint_{-\infty}^{\infty}\left[\frac{\partial U(x', y', z)}{\partial z}\right]_{z=0} e^{-ik_0(px'+qy')}\,dx'\,dy'. \qquad (3.2\text{–}80)$$

Next we substitute $a(p, q)$ from this equation into Eq. (3.2–70) and interchange the order of the integrals. We then find that

$$U(x, y, z) = \iint_{-\infty}^{\infty}\left[\frac{\partial U(x', y', z)}{\partial z}\right]_{z=0} G_N(x-x', y-y', z)\,dx'\,dy', \qquad (3.2\text{–}81)$$

where

$$G_N(x-x', y-y', z) = -\frac{ik_0}{(2\pi)^2}\iint_{-\infty}^{\infty}\frac{1}{m}e^{ik_0[p(x-x')+q(y-y')+mz]}\,dp\,dq. \qquad (3.2\text{–}82)$$

‡ A different derivation of Rayleigh's formula (based on the method of images) valid under somewhat broader conditions, was given by Luneburg (1964), Sec. 45. For a very simple derivation of this formula and also of the second Rayleigh formula [Eq. (3.2–84) below], using the Helmholtz–Kirchhoff integral theorem, see Baker and Copson (1950, pp. 157–158).

On comparing this formula with the Weyl representation (3.2–60) of the spherical wave we see at once that

$$G_N(x - x', y - y', z) = -\frac{1}{2\pi}\frac{e^{ik_0R}}{R}, \qquad (3.2\text{–}83)$$

where R is again given by Eq. (3.2–77). Finally on substituting from Eq. (3.2–83) into Eq. (3.2–81) we find that

$$U(x, y, z) = -\frac{1}{2\pi}\iint_{-\infty}^{\infty}\left[\frac{\partial U(x', y', z)}{\partial z}\right]_{z=0}\frac{e^{ik_0R}}{R}\,dx'\,dy'. \qquad (3.2\text{–}84)$$

This formula is the required solution of the Neumann problem. It is sometimes referred to as the *Rayleigh diffraction formula of the second kind* (Rayleigh, 1897).

We conclude this section by considering the behavior of the far field on the basis of the Rayleigh diffraction formula of the first kind [Eq. (3.2–78)]. For this purpose we note that when r is large enough,

$$R \sim r - \mathbf{s}\cdot\mathbf{r}', \qquad (3.2\text{–}85)$$

where $\mathbf{s}$ is the unit vector in the direction of $\mathbf{r}$ (see Fig. 3.15). Hence

$$\frac{e^{ik_0R}}{R} \sim \frac{e^{ik_0r}}{r}e^{-ik_0\mathbf{s}\cdot\mathbf{r}'} \quad \text{as } k_0r \to \infty, \qquad (3.2\text{–}86)$$

with $\mathbf{s}$ being kept fixed. On differentiating Eq. (3.2–86) with respect to z we readily find that in this limit

$$\frac{\partial}{\partial z}\left(\frac{e^{ik_0R}}{R}\right) \sim ik_0\left(\frac{z}{r}\right)\frac{e^{ik_0r}}{r}e^{-ik_0\mathbf{s}\cdot\mathbf{r}'}. \qquad (3.2\text{–}87)$$

On substituting from Eq. (3.2–87) into Eq. (3.2–78) we obtain the formula

$$U(r\mathbf{s}) \equiv U(x, y, z) \sim -2\pi i k_0\left(\frac{z}{r}\right)\tilde{U}_0\left(k_0\frac{x}{r}, k_0\frac{y}{r}\right)\frac{e^{ik_0r}}{r} \quad \text{as } k_0r \to \infty, \qquad (3.2\text{–}88)$$

where $\tilde{U}_0(u, v)$ is the Fourier transform of $U(x, y, 0)$, defined by Eq. (3.2–24). This formula for the far field is in agreement with Eq. (3.2–34), which was obtained in a different manner.

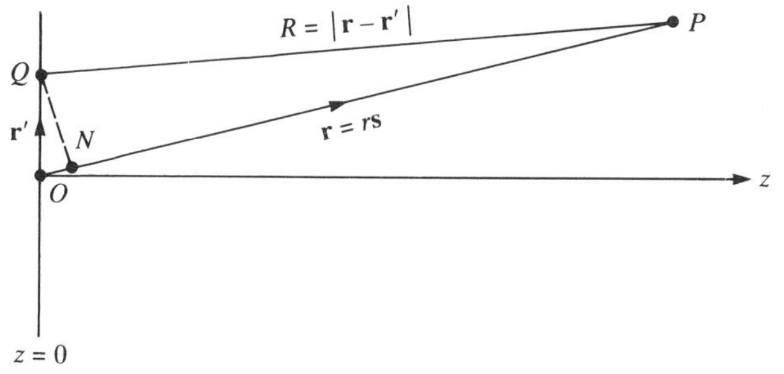

Fig. 3.15 Illustrating the far-zone approximation (3.2–85). $Q(\mathbf{r}')$, $[\mathbf{r}' = (x', y', 0)]$, and $P(\mathbf{r})$, $[\mathbf{r} = (x, y, z)]$, are points in the plane $z = 0$ and in the far zone respectively. When r is sufficiently large $\overline{QP} \sim \overline{OP} - \overline{ON}$, where N is the foot of the perpendicular dropped from Q onto the line OP, i.e. $R \sim r - \mathbf{s}\cdot\mathbf{r}'$.

3.3 The method of stationary phase

A very useful procedure for obtaining approximations to various integrals that frequently occur in wave theory is based on the so-called method (or the principle) of stationary phase. It provides an asymptotic approximation to integrals for large values of an appropriate parameter. Before explaining the essence of the method it will be useful to explain what is meant by an asymptotic expansion.

3.3.1 *Definition of an asymptotic expansion*

Suppose that

$$F(\xi) = G(\xi)\left[\sum_{n=1}^{N} \frac{a_n}{\xi^n} + R_N(\xi)\right] \tag{3.3–1}$$

where, for *all* N,

$$\xi^N R_N(\xi) \to 0 \tag{3.3–2}$$

as $|\xi| \to \infty$ within some range of $\arg \xi$, the a_n's being constants and $G(\xi)$ is some function of ξ. One then writes

$$F(\xi) \sim G(\xi)\left[a_0 + \frac{a_1}{\xi} + \frac{a_2}{\xi^2} + \ldots\right] \tag{3.3–3}$$

and one says that the right-hand side of the formula (3.3–3) is the *asymptotic expansion* of $F(\xi)$ for the particular range of $\arg \xi$. The formula (3.3–2) implies that when $|\xi|$ is large enough the absolute value of the difference between $F(\xi)/G(\xi)$ and the sum of the finite series $\sum_{n=1}^{N} a_n/\xi^n$ is of the order of $1/|\xi|^{N+1}$.

Asymptotic expansions may behave very differently from ordinary series. In particular they may fail to converge for some or all values of ξ. Nevertheless they often provide excellent approximations for sufficiently large values of $|\xi|$. In fact, when $|\xi|$ is large enough, the first term $a_0 G(\xi)$ on the right of Eq. (3.3–3) alone may provide a very good approximation to $F(\xi)$. If only this first term (or the first few terms) of the asymptotic expansion is used, one sometimes speaks of an *asymptotic approximation* to $F(\xi)$.

A discussion of the general properties of asymptotic series is outside the scope of this book‡, as is a rigorous treatment of the method of stationary phase. We will only briefly explain the essence of the method and make its validity plausible by an argument that has a strong physical appeal. We will use the method later to obtain a formal derivation of the asymptotic approximation to some integrals that frequently occur in optics.

3.3.2 *Method of stationary phase for single integrals*

Let us consider an integral of the form

$$F(k) = \int_a^b f(x)\, e^{ikg(x)}\, dx, \tag{3.3–4}$$

‡ For detailed and rigorous treatments of the subject see, for example, Whittaker and Watson (1940), Chap. 8.

where $f(x)$ and $g(x)$ are real, well-behaved functions of a real variable x and a and b are real constants. The parameter k is also assumed to be real. Without loss of generality, we may assume k to be positive. In physical applications $F(k)$ often represents the combined effect of waves of amplitudes $f(x)$, phases $g(x)$, all having the same wave number k.

To obtain some idea about the behavior of the integral (3.3–4) as a function of k, we consider a simple example. We take

$$f(x) \equiv 1, \quad g(x) = x^2, \tag{3.3–5}$$

and consider the real part (denoted by Re) of the integral:

$$\text{Re}\, F_1(k) = \int_a^b \cos(kx^2)\,\mathrm{d}x, \quad (a < 0,\, b > 0). \tag{3.3–6}$$

Let us compare the behavior of the integrand

$$G(x, k) = \cos kx^2 \tag{3.3–7}$$

in Eq. (3.3–6) for different values of k. With fixed k, $G(x, k)$ will oscillate between the values $+1$ and -1, with the rate of the oscillations depending on the value of k. Since, with k fixed, the zeros of $G(x, k)$ are given by

$$x = \pm[(n + \tfrac{1}{2})\pi/k]^{1/2}, \quad n = 0, 1, 2, 3, \ldots,$$

it is clear that if one takes k larger and larger, $G(x, k)$ will oscillate more and more rapidly (see Fig. 3.16).

Suppose now that we drop the assumption that $f(x) = 1$, but still take $g(x) = x^2$. Instead of Eq. (3.3–6) we then have the integral

$$\text{Re}\, F_2(k) = \int_a^b f(x)\cos(kx^2)\,\mathrm{d}x. \tag{3.3–8}$$

The factor $f(x)$ will give rise to 'amplitude modulation' of the cosine term; but it is clear that irrespective of the exact form of $f(x)$, if only k is large enough, the integrand of the integral on the right-hand side of Eq. (3.3–8) will again oscillate very rapidly and *there will be a tendency for the positive and negative contributions of the integrand to cancel out*. Moreover this cancellation can be expected to take place irrespective of whether $g(x) = x^2$ (as in our example) or whether it has some other form. However, for sufficiently larger values of k, the cancellation will not be complete in the neighborhood of points (if any) where $g(x)$ is stationary within the interval of integration, i.e. where

$$\frac{\mathrm{d}g(x)}{\mathrm{d}x} = 0, \tag{3.3–9}$$

(the origin $x = 0$ in the above examples), and also at the end points

$$x = a \quad \text{and} \quad x = b. \tag{3.3–10}$$

These special points are called *critical points* of the integrand in Eq. (3.3–4). Those that satisfy Eq. (3.3–9) are said to be *critical points of the first kind*, and the end points (3.3–10) are said to be *critical points of the second kind*. Of course, in special cases several critical points may coincide or one or both of the end points may themselves be stationary points of $g(x)$. Other complications may

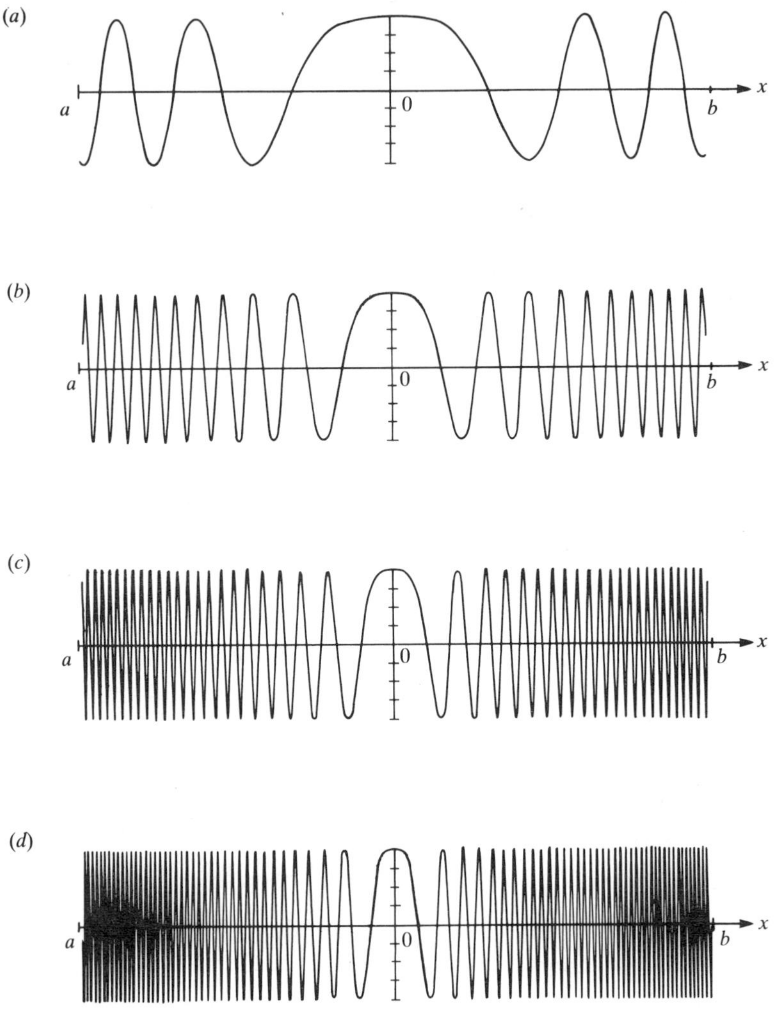

Fig. 3.16 Illustrating the principle of stationary phase. Comparison of the behavior of the functions (*a*) $G(x, 1) = \cos x^2$; (*b*) $G(x, 4) = \cos 4x^2$; (*c*) $G(x, 9) = \cos 9x^2$; (*d*) $G(x, 16) = \cos 16x^2$.

arise, for example when $g(x)$ or $f(x)$ have singular behavior. However, excluding these more complicated situations one can show that the asymptotic behavior as $k \to \infty$ of an integral of the form (3.3–4) is determined entirely by the behavior of the integrand at the critical points, and, moreover, the leading term in the asymptotic expansion of $F(k)$ often depends on the critical points of the first kind, i.e. on interior points in the range of integration where $g(x)$ is stationary. This fact is the essence of the *principle of stationary phase*.

In wave-theoretical problems, the absence of contributions from the whole range of integration except from the critical points may be regarded as a manifestation of destructive wave-interference. In fact, the method of stationary phase was originally introduced in connection with problems involving water waves (in 1887 by W. Thompson, who later became Lord Kelvin). It should be contrasted with another well-known asymptotic technique, the method of steepest descent. (See, for example, Copson, 1967, Sec. 7; Dennery and Krzywicki, 1967, p. 87 *et seq*.) Although the method of steepest descent has a formal similarity with the method of stationary phase, it has quite a different physical significance, in that in its wave-theoretical context it may be interpreted in terms of amplitude decay, rather than in terms of phase interference. Also from the mathematical standpoint there are basic differences between the two methods, because the method of steepest descent makes use of the analytic properties of the functions $f(x)$ and $g(x)$, considered as functions of a complex rather than a real variable, whereas the method of stationary phase may be developed entirely within the framework of functions of a real variable‡. For this reason the method of stationary phase, unlike the method of steepest descent, may readily be extended to two-dimensional integrals (see Section 3.3.3). Moreover, unlike the method of stationary phase, the method of steepest descent cannot be readily used when the integral has finite limits.

We will now use the principle of stationary phase to determine the asymptotic behavior of the integral (3.3–4) for large values of k. Our derivation will be heuristic, but the results may be justified by rigorous analysis. (See, for example, van der Corput, 1934–1935, 1936; Focke, 1954; Erdelyi 1955, Sec. 2.9; Braun, 1956).

(a) *Contribution from critical points of the first kind*

Let us assume that in the integral (3.3–4) $f(x)$ is continuous and $g(x)$ is twice continuously differentiable in the interval $a < x < b$. Suppose, to begin with, that there is one and only one critical point of the first kind, i.e. that there is one point x_1 and no other point in the interval at which

$$g'(x_1) = 0, \tag{3.3–11}$$

where the prime denotes differentiation with respect to x. We also assume that the second derivative of $g(x)$, i.e. $g''(x)$, is not zero at $x = x_1$:

$$g''(x_1) \neq 0. \tag{3.3–12}$$

Then at points x in the immediate neighborhood of x_1,

$$g(x) \approx g(x_1) + \tfrac{1}{2}(x - x_1)^2 g''(x_1). \tag{3.3–13}$$

Since, according to the principle of stationary phase, the asymptotic approximation to the integral for large values of k comes from the immediate neighborhood of x_1, we have from Eqs. (3.3–4) and (3.3–13)

$$F(k) \sim \int_a^b f(x_1)\, \mathrm{e}^{\mathrm{i}kg(x_1)}\, \mathrm{e}^{\mathrm{i}k(x-x_1)^2 g''(x_1)/2}\, \mathrm{d}x. \tag{3.3–14}$$

For the same reason we may extend the range of integration from (a, b) to $(-\infty, +\infty)$ and we obtain the formula

$$F(k) \sim f(x_1)\, \mathrm{e}^{\mathrm{i}kg(x_1)} \int_{-\infty}^{\infty} \mathrm{e}^{\mathrm{i}k(x-x_1)^2 g''(x_1)/2}\, \mathrm{d}x. \tag{3.3–15}$$

‡ That the method of stationary phase should not be regarded as a step-child of the method of steepest descent was stressed particularly by van Kampen (1958).

If we change the variable of integration from x to $u = x - x_1$ and use the symmetry of the integrand, the formula (3.3–15) becomes

$$F(k) \sim 2f(x_1)\,\mathrm{e}^{\mathrm{i}kg(x_1)}\int_0^{\infty}\mathrm{e}^{\mathrm{i}kg''(x_1)u^2/2}\,\mathrm{d}u. \qquad (3.3\text{–}16)$$

The integral on the right side is the well-known Fresnel integral, whose value is (Gradshteyn and Ryzhik, 1980, 395, #3.691 1)

$$\int_0^{\infty}\mathrm{e}^{\mathrm{i}au^2}\,\mathrm{d}u = \frac{1}{2}\left(\frac{\pi}{|a|}\right)^{1/2}\mathrm{e}^{\pm\mathrm{i}\pi/4}, \qquad (3.3\text{–}17)$$

where the upper or the lower sign is taken according as $a \gtrless 0$. With the help of this result, Eq. (3.3–16) becomes

$$F^{(1)}(k) \sim \left[\frac{2\pi}{k|g''(x_1)|}\right]^{1/2} f(x_1)\,\mathrm{e}^{\mathrm{i}kg(x_1)}\,\mathrm{e}^{\pm\mathrm{i}\pi/4} \quad \text{as } k \to \infty, \qquad (3.3\text{–}18)$$

where the upper or lower sign is taken, according as‡ $g''(x_1) \gtrless 0$ and we have written $F^{(1)}(k)$ rather than $F(k)$ to stress that the expression on the right-hand side of Eq. (3.3–18) is a contribution from a critical point of the first kind.

In deriving the formula (3.3–18) we assumed that the integrand had only one critical point of the first kind. If it has several such points, $x_1, x_2, \ldots, x_n$ say, the corresponding asymptotic approximation to $F(k)$ is obtained by summing the contributions, given by expressions of the form (3.3–18), from all of them, i.e. one then has

$$F^{(1)}(k) \sim \left(\frac{2\pi}{k}\right)^{1/2}\sum_{j=1}^{n}\frac{\varepsilon_j}{|g''(x_j)|^{1/2}}f(x_j)\,\mathrm{e}^{\mathrm{i}kg(x_j)}, \qquad (3.3\text{–}19)$$

where

$$\varepsilon_j = \mathrm{e}^{\pm\mathrm{i}\pi/4} \quad \text{according as } g''(x_j) \gtrless 0, \qquad (3.3\text{–}20)$$

it being assumed, of course, that $g''(x_j) \neq 0$, $(j = 1, 2, \ldots, n)$.

(b) *Contributions from critical points of the second kind*

Next we will determine the contributions from critical points of the second kind, namely end points of the interval of integration. We will assume that the end points are not stationary points of $g(x)$, i.e. that

$$g'(a) \neq 0, \quad g'(b) \neq 0.$$

We re-write the integral (3.3–4) in the form

$$\int_a^b f(x)\,\mathrm{e}^{\mathrm{i}kg(x)}\,\mathrm{d}x = \frac{1}{\mathrm{i}k}\int_a^b\left[\frac{\mathrm{d}}{\mathrm{d}x}\,\mathrm{e}^{\mathrm{i}kg(x)}\right]\frac{f(x)}{g'(x)}\,\mathrm{d}x.$$

‡ The term $|g''(x_1)|$ has a simple geometrical meaning. According to elementary differential geometry the radius of curvature $\rho(x)$ at a typical point of the curve $y = g(x)$ is given by $\rho(x) = |(1 + y'^2)^{3/2}/y''|$. Hence at a point $x = x_1$ where y is stationary, $\rho(x_1) = 1/|y''(x_1)|$ and we see that the factor $1/|g''(x_1)|^{1/2}$ which appears to the right-hand side of Eq. (3.3–18) is just the square-root of the radius of curvature of the phase function $g(x)$ at the stationary point.

The other factor, $\mathrm{e}^{\pm\mathrm{i}\pi/4}$ in the formula (3.3–18) often has an interesting interpretation also, for example, in connection with the well-known phase anomaly of waves near focus (see Born and Wolf, 1980, Sec. 8.8.4).

Integrating by parts on the right-hand side we obtain the formula

$$\int_a^b f(x)\,\mathrm{e}^{\mathrm{i}kg(x)}\,\mathrm{d}x = \frac{1}{\mathrm{i}k}\left[\frac{f(b)}{g'(b)}\mathrm{e}^{\mathrm{i}kg(b)} - \frac{f(a)}{g'(a)}\mathrm{e}^{\mathrm{i}kg(a)}\right] - \frac{1}{\mathrm{i}k}\int_a^b \frac{\mathrm{d}}{\mathrm{d}x}\left[\frac{f(x)}{g'(x)}\right]\mathrm{e}^{\mathrm{i}kg(x)}\,\mathrm{d}x. \tag{3.3–21}$$

Again integrating by parts, we find at once that the third term on the right-hand side of Eq. (3.3–21) is of the order of $1/k^2$ and hence can be neglected in comparison with the two other terms. Thus we may conclude that the contribution from critical points of the second kind is

$$F^{(2)}(k) \sim \frac{1}{\mathrm{i}k}\left[\frac{f(b)}{g'(b)}\mathrm{e}^{\mathrm{i}kg(b)} - \frac{f(a)}{g'(a)}\mathrm{e}^{\mathrm{i}kg(a)}\right]. \tag{3.3–22}$$

The preceding analysis indicates that when k is large enough

$$F(k) \sim F^{(1)}(k) + F^{(2)}(k), \tag{3.3–23}$$

where $F^{(1)}(k)$ is given by the formula (3.3–19) and $F^{(2)}(k)$ by the formula (3.3–22). Rigorous mathematical analysis shows that under the conditions that we assumed, the expression (3.3–19) for $F^{(1)}(k)$ and the expression (3.3–22) for $F^{(2)}(k)$ indeed represent the leading terms in the asymptotic expansions‡ of the contributions of the critical points of the first and of the second kind, respectively, to the integral $F(k)$, as $k \to \infty$.

Finally we note that, since for large values of k, $F^{(1)}(k)$ is of the order of $1/\sqrt{k}$ whereas $F^{(2)}(k)$ is of the order of $1/k$, the contributions of the interior stationary points, if there are any, are more important than the contributions of the end points.

3.3.3 *Method of stationary phase for double integrals*

Next we consider the asymptotic approximation as $k \to \infty$ to a double integral of the form

$$F(k) = \iint_D f(x, y)\,\mathrm{e}^{\mathrm{i}kg(x,y)}\,\mathrm{d}x\,\mathrm{d}y, \tag{3.3–24}$$

where $f(x, y)$ and $g(x, y)$ are real, well-behaved functions of two real variables x and y and D is a two-dimensional, simply-connected closed domain. We assume that the curve, C say, bounding D is smooth, i.e. that the tangent to C varies in a continuous manner along the curve.

Similar considerations as those that we have just outlined in connection with the one-dimensional integral lead to the conclusion that, for sufficiently large values of k, the term $\exp[\mathrm{i}kg(x, y)]$ in Eq. (3.3–24) oscillates so rapidly as the point (x, y) explores the domain of integration that the various contributions cancel out, except for contributions from the immediate neighborhood of certain special points, again called *critical points*. This is the essence of the *method of stationary phase for double integrals* of the form (3.3–24).

‡ The complete asymptotic expansions are given by Focke (1954). See also Stamnes (1986), Sec. 8.

The most important critical points are: points (if any) within the domain of integration, at which the function $g(x, y)$ is stationary, i.e. points (x_1, y_1) within the domain D where

$$\left.\frac{\partial g(x, y)}{\partial x}\right|_{x_1, y_1} = \left.\frac{\partial g(x, y)}{\partial y}\right|_{x_1, y_1} = 0; \tag{3.3–25}$$

and points on the boundary curve C for which $g(x, y)$ is stationary with respect to a small displacement $\mathrm{d}l$ along C, i.e. points on C where

$$\frac{\partial g}{\partial l} = 0. \tag{3.3–26}$$

Points within the domain D which satisfy the requirement (3.3–25) are called *critical points of the first kind* and those on the curve C which satisfy the requirement (3.3–26) are called *critical points of the second kind* (see Fig. 3.17).

We will now give a heuristic derivation of the contributions of these types of critical points to the asymptotic behavior of the double integral (3.3–24) for large values of k‡.

(a) Contributions from critical points of the first kind

We assume that in the double integral (3.3–24), $f(x, y)$ is continuous and that $g(x, y)$ has continuous second-order partial derivatives throughout the domain D. We also assume, to begin with, that there is only one critical point $P_1(x_1, y_1)$ of the first kind in D. Throughout the immediate neighborhood of that point,

$$g(x, y) \approx g(x_1, y_1) + \tfrac{1}{2}[(x - x_1)^2 g_{xx} + 2(x - x_1)(y - y_1)g_{xy} + (y - y_1)^2 g_{yy}], \tag{3.3–27}$$

where $g_{xx} = \partial^2 g/\partial x^2$ etc., evaluated at P_1. We assume that at P_1

$$g_{xx}g_{yy} - g_{xy}^2 \neq 0. \tag{3.3–28}$$

On substituting from Eq. (3.3–27) into Eq. (3.3–24) and by using a strictly similar argument based on the principle of stationary phase as we made before in

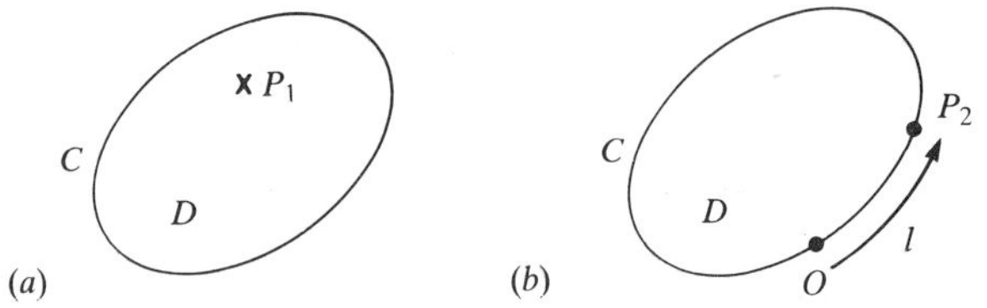

Fig. 3.17 (*a*) A critical point P_1 of the *first* kind: an interior stationary point of $g(x, y)$. (*b*) A critical point P_2 of the *second* kind: point on the boundary C of the domain of integration where $g(x, y)$ is stationary with respect to a displacement $\mathrm{d}l$ along the boundary, measured from some fixed point O on C.

‡ More rigorous and more complete treatments are given by Focke (1954), Jones and Kline (1958), and Chako (1965).

connection with one-dimensional integrals, we have, when k is large enough,

$$F(k) \sim f(x_1, y_1)\,\mathrm{e}^{ikg(x_1,y_1)} \iint\limits_{-\infty}^{\infty} \exp\left[\tfrac{1}{2}\mathrm{i}k(g_{xx}\xi^2 + 2g_{xy}\xi\eta + g_{yy}\eta^2)\right] \mathrm{d}\xi\,\mathrm{d}\eta, \tag{3.3–29}$$

where we set

$$\xi = x - x_1, \quad \eta = y - y_1. \tag{3.3–30}$$

The term in the exponent of the integrand in Eq. (3.3–29) may be simplified by a suitable rotation of the ξ-, η-axes about the origin $\xi = \eta = 0$. For that purpose we set

$$u = \xi\cos\theta + \eta\sin\theta, \tag{3.3–31a}$$

$$v = -\xi\sin\theta + \eta\cos\theta, \tag{3.3–31b}$$

and choose θ so that the quadratic form in the integrand reduces to the sum of squares, i.e. so that

$$g_{xx}\xi^2 + 2g_{xy}\xi\eta + g_{yy}\eta^2 \equiv Au^2 + Bv^2, \tag{3.3–32}$$

where A and B are some constants, which we will determine later. Then Eq. (3.3–29) becomes

$$F(k) \sim f(x_1, y_1)\,\mathrm{e}^{ikg(x_1,y_1)} I_A(k) I_B(k), \tag{3.3–33}$$

where

$$I_A(k) = 2\int_0^{\infty} \exp\left(\tfrac{1}{2}\mathrm{i}kAu^2\right) \mathrm{d}u, \quad I_B(k) = 2\int_0^{\infty} \exp\left(\tfrac{1}{2}\mathrm{i}kBv^2\right) \mathrm{d}v. \tag{3.3–34}$$

The integrals (3.3–34) are given by the formula (3.3–17) and using it, the expression (3.3–33) becomes

$$F(k) \sim \frac{2\pi\varepsilon_A\varepsilon_B}{k(|A|\,|B|)^{1/2}} f(x_1, y_1)\,\mathrm{e}^{ikg(x_1,y_1)}, \tag{3.3–35}$$

where

$$\left.\begin{aligned} \varepsilon_A &= \mathrm{e}^{\pm \mathrm{i}\pi/4} \quad \text{according as } A \gtrless 0, \\ \varepsilon_B &= \mathrm{e}^{\pm \mathrm{i}\pi/4} \quad \text{according as } B \gtrless 0. \end{aligned}\right\} \tag{3.3–36}$$

Next we express the factor $(|A|\,|B|)^{1/2}$ in the formula (3.3–35) in terms of the second-order partial derivatives of the phase function $g(x, y)$. To do so we use the fact that, under rotation of axes expressed by the formulas (3.3–31), a quadratic form in two variables has two invariants, namely the trace

$$\Sigma = g_{xx} + g_{yy} \tag{3.3–37a}$$

and the determinant‡

$$\Delta = g_{xx}g_{yy} - g_{xy}^2. \tag{3.3–37b}$$

‡ The quantity Δ may be shown to be proportional to the Gaussian curvature of the surface $g(x, y) = \text{constant}$ [Eisenhart, 1947, p. 225, Eq. (40.10)].

In the present case [see Eq. (3.3–32)]

$$\Sigma = A + B, \tag{3.3–38a}$$

and

$$\Delta = AB. \tag{3.3–38b}$$

It will be convenient to set

$$\varepsilon_A \varepsilon_B = \mathrm{i}\sigma. \tag{3.3–39}$$

If we use Eqs. (3.3–37) and (3.3–38) we readily find that

$$\left.\begin{aligned} &A > 0,\ B > 0 \text{ implies that } \Delta > 0,\ \Sigma > 0, \\ &A < 0,\ B < 0 \text{ implies that } \Delta > 0,\ \Sigma < 0, \\ &A > 0,\ B < 0 \text{ or } A < 0,\ B > 0 \text{ implies that } \Delta < 0. \end{aligned}\right\} \tag{3.3–40}$$

With the help of these inequalities the formula (3.3–35) becomes

$$F^{(1)}(k) \sim \frac{2\pi \mathrm{i}\sigma}{k\sqrt{|\Delta|}} f(x_1, y_1)\, \mathrm{e}^{\mathrm{i}kg(x_1,y_1)}, \tag{3.3–41}$$

where

$$\left.\begin{aligned} \sigma &= 1 && \text{when } \Delta > 0,\ \Sigma > 0, \\ &= -1 && \text{when } \Delta > 0,\ \Sigma < 0, \\ &= -\mathrm{i} && \text{when } \Delta < 0. \end{aligned}\right\} \tag{3.3–42}$$

On the left of Eq. (3.3–41) we have written $F^{(1)}(k)$ rather than $F(k)$ to stress that the expression represents the contribution to the asymptotic approximation from a critical point of the first kind, assuming there is such a critical point, and that the requirement (3.3–28) is satisfied.

If the integrand of the integral (3.3–24) has several critical points of the first kind, the corresponding asymptotic approximation to $F(k)$ is obtained by adding together their individual contributions, each of which is given by an expression of the form (3.3–41).

(b) Contribution from critical points of the second kind

We will now give a heuristic derivation for the contribution from a critical point of the second kind, i.e. for the contribution from a point $P_2(x_2, y_2)$ on the boundary C of the domain D, where the phase function $g(x, y)$ is stationary with respect to a small displacement along C [see Eq. (3.3–26)].

Let O be some fixed origin on C and let l be the distance from O to a typical point P_2 on C, taken as positive in the counter-clockwise sense. Let α be the angle between the tangent to C at that point and the positive x-direction, as shown in Fig. 3.18. Denoting differentiation with respect to l by prime we have, at the critical point P_2,

$$g' \equiv g_x x' + g_y y' = 0. \tag{3.3–43}$$

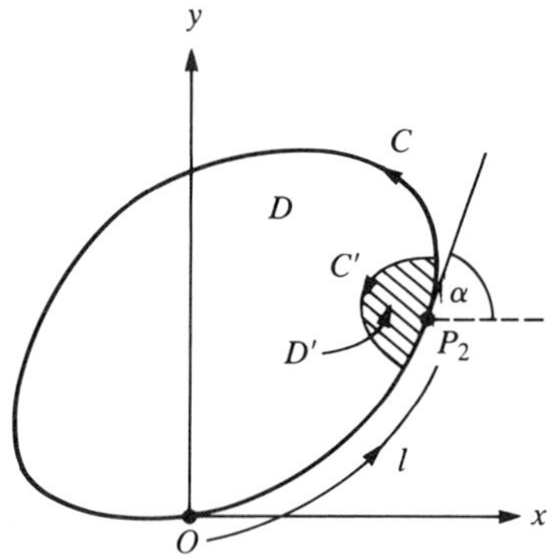

Fig. 3.18 Notation relating to evaluation of the contribution from a critical point P_2 of the second kind to the two-dimensional integral (3.3–24), in the asymptotic limit $k \to \infty$.

Hence at P_2

$$\tan\alpha = \frac{y'}{x'} = -\frac{g_x}{g_y}. \tag{3.3–44}$$

We first assume that at P_2

$$x' \neq 0 \quad \text{and} \quad g_y \neq 0. \tag{3.3–45}$$

It is understood that all the quantities in Eqs. (3.3–43)–(3.3–45) are evaluated at P_2.

For later purposes we note the following formulas:

$$x' = \cos\alpha = \pm\frac{1}{(1+\tan^2\alpha)^{1/2}} = \pm\frac{g_y}{(g_x^2+g_y^2)^{1/2}}, \tag{3.3–46a}$$

$$y' = \sin\alpha = \mp\frac{\tan\alpha}{(1+\tan^2\alpha)^{1/2}} = \mp\frac{g_x}{(g_x^2+g_y^2)^{1/2}}, \tag{3.3–46b}$$

or, if we introduce a factor η defined as

$$\eta = \pm 1 \quad \text{according as} \quad x' g_y \gtrless 0, \tag{3.3–47a}$$

we obtain [noting that according to Eqs. (3.3–43), $y' g_y = -x' g_x$]

$$x' = \cos\alpha = \frac{\eta g_y}{(g_x^2+g_y^2)^{1/2}}, \tag{3.3–48a}$$

$$y' = \sin\alpha = \frac{-\eta g_x}{(g_x^2+g_y^2)^{1/2}}. \tag{3.3–48b}$$

We note that according to Eq. (3.3–43) we have, at P_2,

$$x' y' (g_x x' + g_y y') = 0$$

i.e.

$$(g_x y') x'^2 + (g_y x') y'^2 = 0.$$

Hence in place of Eq. (3.3–47a) we may also write

$$\eta = \pm 1 \quad \text{according as } y' g_x \lessgtr 0. \tag{3.3–47b}$$

In order to evaluate the contribution from P_2, we restrict the domain of integration, in agreement with the method of stationary phase, to a small neighborhood D' lying within D (see Fig. 3.18) and make use of Green's theorem:

$$\iint_{D'} \left(\frac{\partial Q}{\partial x} - \frac{\partial P}{\partial y} \right) \mathrm{d}x\,\mathrm{d}y = \int_{C'} [P\,\mathrm{d}x + Q\,\mathrm{d}y], \tag{3.3–49}$$

where C' is the boundary of D', a part of which coincides with a portion of C and the rest of which lies wholly within D and is described in the counter-clockwise sense. It will be sufficient to consider the special case when $Q \equiv 0$. Then the identity (3.3–49) becomes

$$-\iint_{D'} \frac{\partial P}{\partial y} \mathrm{d}x\,\mathrm{d}y = \int_{C'} P\,\mathrm{d}x. \tag{3.3–50}$$

Let us take $P = f\,\mathrm{e}^{\mathrm{i}kg}$. Then $\partial P/\partial y = \mathrm{i}kg_y f\,\mathrm{e}^{\mathrm{i}kg} + f_y\,\mathrm{e}^{\mathrm{i}kg}$ and Eq. (3.3–50) gives

$$-\mathrm{i}k \iint_{D'} g_y f\,\mathrm{e}^{\mathrm{i}kg}\,\mathrm{d}x\,\mathrm{d}y - \iint_{D'} f_y\,\mathrm{e}^{\mathrm{i}kg}\,\mathrm{d}x\,\mathrm{d}y = \int_{C'} f\,\mathrm{e}^{\mathrm{i}kg}\,\mathrm{d}x. \tag{3.3–51}$$

Now when k is sufficiently large, the second term on the left may evidently be neglected compared with the first one. Further, since the domain D' may be taken to be arbitrarily small, we may replace the factor g_y in the integrand of the first integral by its value at the critical point $P_2(x_2, y_2)$. Moreover, we may also replace $\mathrm{d}x$ on the right by $x'\,\mathrm{d}l$. We then obtain the formula

$$-\mathrm{i}kg_y \iint_{D'} f\,\mathrm{e}^{\mathrm{i}kg}\,\mathrm{d}x\,\mathrm{d}y = \int_{C'} f\,\mathrm{e}^{\mathrm{i}kg} x'\,\mathrm{d}l. \tag{3.3–52}$$

Now in the integral on the right we may replace f and x' by the values these quantities take at $P_2(x_2, y_2)$, the latter being given by Eq. (3.3–48a). We then find, after dividing both sides by $-\mathrm{i}kg_y(x_2, y_2)$, that

$$\iint_{D'} f\,\mathrm{e}^{\mathrm{i}kg}\,\mathrm{d}x\,\mathrm{d}y = \frac{\mathrm{i}}{k} \frac{f}{(g_x^2 + g_y^2)^{1/2}} \int_{C'} \eta\,\mathrm{e}^{\mathrm{i}kg}\,\mathrm{d}l, \tag{3.3–53}$$

with g in the last integral being regarded as a function of the arclength l of C'.

Next we expand the phase function about the critical point P_2 which, for simplicitly, we now take to be at the origin O from which the arclength l is measured. We then have, for points on C',

$$g(l) \approx g(0) + \tfrac{1}{2} g''(0) l^2 \tag{3.3–54}$$

and hence

$$\begin{aligned} \int_{C'} \eta\,\mathrm{e}^{\mathrm{i}kg}\,\mathrm{d}l &\approx \mathrm{e}^{\mathrm{i}kg(0)} \int_{C'} \eta\,\mathrm{e}^{\mathrm{i}kg''(0)l^2/2}\,\mathrm{d}l \\ &\approx \mathrm{e}^{\mathrm{i}kg(0)} \int_{-\infty}^{\infty} \eta\,\mathrm{e}^{\mathrm{i}kg''(0)l^2/2}\,\mathrm{d}l \end{aligned} \tag{3.3–55}$$

for sufficiently large values of k. Using Eq. (3.3–17) again, and the symmetry of the integrand in Eq. (3.3–55), we find that

$$e^{ikg(0)}\int_{-\infty}^{\infty} \eta\, e^{ikg''(0)l^2/2}\, dl = \left(\frac{2\pi}{k|g''(0)|}\right)^{1/2} \eta\, e^{i[kg(0)\pm\pi/4]}, \tag{3.3–56}$$

with the upper or lower sign being taken according as $g''(0) \gtrless 0$. On substituting from Eqs. (3.3–56) and (3.3–55) into Eq. (3.3–53) we see at once that

$$\iint_{D'} f\, e^{ikg}\, dx\, dy \sim \frac{i\eta\varepsilon\sqrt{(2\pi)}}{k^{3/2}}\left[\frac{f\, e^{ikg}}{|g''|^{1/2}(g_x^2 + g_y^2)^{1/2}}\right]_{P_2}, \tag{3.3–57}$$

where

$$\varepsilon = e^{\pm i\pi/4} \quad \text{according as } g''(0) \gtrless 0.$$

Next we will express the quantity $g''(0)$ (i.e. the value of g'' at the critical point P_2) in a more explicit form. We have at any point on C,

$$g' = g_x x' + g_y y', \tag{3.3–58}$$

$$g'' = (g_{xx}x' + g_{xy}y')x' + (g_{yx}x' + g_{yy}y')y' + g_x x'' + g_y y''. \tag{3.3–59}$$

We also have, at any point on C,

$$x' = \cos\alpha, \quad y' = \sin\alpha, \tag{3.3–60}$$

$$x'' = -(\sin\alpha)\alpha', \quad y'' = (\cos\alpha)\alpha', \tag{3.3–61}$$

where

$$\alpha' = \frac{d\alpha}{dl} = \frac{1}{\rho}, \tag{3.3–62}$$

ρ being the radius of curvature of the curve C at the critical point P_2, taken as positive or negative (i.e. α increasing or decreasing), according as the center of curvature lies on the same side as the domain D' or on the opposite side, with respect to the boundary element at P_2 [ρ is positive in Fig. (3.18)]. From Eqs. (3.3–60) to (3.3–62),

$$x'' = -\frac{y'}{\rho}, \quad y'' = \frac{x'}{\rho}, \tag{3.3–63}$$

and using these expressions in Eq. (3.3–59), we obtain for g'' the expression

$$g'' = g_{xx}x'^2 + 2g_{xy}x'y' + g_{yy}y'^2 + \frac{1}{\rho}(-g_x y' + g_y x'). \tag{3.3–64}$$

Now the values of x' and y' at the critical points are given by Eqs. (3.3–48) and, hence Eq. (3.3–64) gives

$$g''(0) = \frac{g_{xx}g_y^2 - 2g_{xy}g_x g_y + g_{yy}g_x^2}{g_x^2 + g_y^2} + \frac{\eta}{\rho}(g_x^2 + g_y^2)^{1/2}. \tag{3.3–65}$$

Finally, on substituting from Eq. (3.3–65) into Eq. (3.3–57) we obtain the required contribution from a critical point $P_2(x_2, y_2)$ of the second kind to the asymptotic approximation to the double integral (3.3–24) as $k \to \infty$:

$$F^{(2)}(k) \sim \frac{\sqrt{(2\pi)}\mathrm{i}\varepsilon\eta}{k^{3/2}\sqrt{|\Omega|}} f(x_2, y_2)\,\mathrm{e}^{\mathrm{i}kg(x_2,y_2)}, \tag{3.3–66}$$

where

$$\Omega = \left[g_{xx}g_y^2 - 2g_{xy}g_xg_y + g_{yy}g_x^2 + \left(\frac{\eta}{\rho}\right)(g_x^2 + g_y^2)^{3/2} \right]_{x_2,y_2} \tag{3.3–67a}$$

$$\eta = \pm 1 \quad \text{according as } (x'g_y)_{x_2,y_2} \gtrless 0 \tag{3.3–67b}$$

and

$$\varepsilon = \mathrm{e}^{\pm \mathrm{i}\pi/4} \quad \text{according as } \Omega \gtrless 0. \tag{3.3–67c}$$

Finally we will examine whether the formula (3.3–66) remains valid when $x' = 0$ and $g_y = 0$ at P_2, a case which we explicitly excluded in the preceding analysis [see Eq. (3.3–45)]. If $x' \equiv \cos\alpha = 0$, then $y' = \sin\alpha = \pm 1$ and it follows from Eq. (3.3–43) that at P_2, $g_y = 0$. We may assume that $g_x \neq 0$ at P_2, for otherwise P_2 would also be a critical point of the first kind, a situation that we exclude. In place of Eq. (3.3–50) we now use Green's theorem in the form

$$\iint_{D'} \frac{\partial Q}{\partial x}\,\mathrm{d}x\,\mathrm{d}y = \int_{C'} Q\,\mathrm{d}y. \tag{3.3–68}$$

If we take $Q = f\,\mathrm{e}^{\mathrm{i}kg}$ we obtain, in a manner strictly similar to that used to obtain Eq. (3.3–52), the formula

$$\mathrm{i}kg_x \iint_{D'} f\,\mathrm{e}^{\mathrm{i}kg}\,\mathrm{d}x\,\mathrm{d}y = \int_{C'} f\,\mathrm{e}^{\mathrm{i}kg} y'\,\mathrm{d}l. \tag{3.3–69}$$

On replacing f and y' on the right by the values which these quantities take at P_2 and on using Eqs. (3.3–48b) and (3.3–47b) we obtain the formula (3.3–53) again, where now $(g_x^2 + g_y^2)^{1/2} = |g_x|$, because $g_y = 0$. The rest of the calculation is the same as before and leads again to the formula (3.3–66), except that η is now given by Eq. (3.3–47b) rather than by Eq. (3.3–67b).

In general, there will be several critical points of the second kind. Their combined contribution to the asymptotic approximation to $F(k)$ is obtained by just adding their individual contributions, each of which is given by the formula that we have just derived.

Finally we note that according to Eq. (3.3–41) the contribution from critical points of the first kind to the asymptotic approximation of the double integral (3.3–24) is of the order of $1/k$, whereas according to Eq. (3.3–66) the contribution from critical points of the second kind is of the order of $1/k^{3/2}$. Hence if the integrand has critical points of the first kind, their contributions will be, in general, more important than the contributions from critical points of the second kind.

We assumed throughout that the boundary curve C of the domain D is smooth. If this is not so, e.g. if there are points on C where the slope of the tangent has a discontinuity, (a corner of a square domain, for example), such points will also contribute to the asymptotic approximation. Points of this kind are called *critical points of the third kind* and may be shown to provide a contribution of the order of $1/k^2$ (see, for example, Stamnes, 1986, Sec. 9.1.4).

3.3.4 An example: far-zone behavior of the angular spectrum representation of wavefields

As an example of the usefulness of the formula (3.3–41) for the asymptotic approximation of certain types of double integrals, we will now derive an expression for the far-zone behavior of a wavefield which is represented in the form of an angular spectrum of plane waves.

We consider a wavefield $U(x, y, z)\,\mathrm{e}^{-\mathrm{i}\omega t}$ in the half-space $z \geqslant 0$ (assumed to be free space), outgoing at infinity in that half-space. According to Eq. (3.2–19) the space-dependent part of the wavefield in that domain has an angular spectrum representation of the form

$$U(x, y, z) = \iint_{-\infty}^{\infty} a(p, q)\,\mathrm{e}^{\mathrm{i}k_0(px+qy+mz)}\,\mathrm{d}p\,\mathrm{d}q, \qquad (3.3\text{–}70)$$

where

$$m = +(1 - p^2 - q^2)^{1/2} \quad \text{when } p^2 + q^2 \leqslant 1 \qquad (3.3\text{–}71\text{a})$$

$$= +\mathrm{i}(p^2 + q^2 - 1)^{1/2} \quad \text{when } p^2 + q^2 > 1 \qquad (3.3\text{–}71\text{b})$$

and $k_0 = \omega/c$, c being the speed of light in vacuo.

Since the evanescent waves, i.e. the plane waves in the integrand of Eq. (3.3–70) for which Eq. (3.3–71b) applies, decay exponentially in amplitude with increasing distance from the plane $z = 0$ they will not, in general, contribute to the field in the far zone. Hence, for the purpose of determining the far-zone behavior of the field we may, instead of the integral (3.3–70), consider the integral

$$U_{\mathrm{H}}(x, y, z) = \iint_{p^2+q^2\leqslant 1} a(p, q)\,\mathrm{e}^{\mathrm{i}k_0(px+qy+mz)}\,\mathrm{d}p\,\mathrm{d}q, \qquad (3.3\text{–}72)$$

which contains contributions from homogeneous waves only.

Let us consider the behavior of U_{H} at a point $P(x, y, z > 0)$, in the far zone, in a direction specified by a unit vector $\mathbf{s}$ $(s_x, s_y, s_z > 0)$. Then

$$s_x = \frac{x}{r}, \quad s_y = \frac{y}{r}, \quad s_z = \frac{z}{r} > 0, \qquad (3.3\text{–}73)$$

where

$$r = (x^2 + y^2 + z^2)^{1/2} \qquad (3.3\text{–}74)$$

is the distance of P from the origin (see Fig. 3.10). On substituting from Eq. (3.3–73) into Eq. (3.3–72) for x, y and z we obtain for U_{H} the expression

$$U_{\mathrm{H}}(x, y, z) = \iint_{p^2+q^2\leqslant 1} a(p, q)\,\mathrm{e}^{\mathrm{i}\kappa g(p,q;s_x,s_y)}\,\mathrm{d}p\,\mathrm{d}q, \qquad (3.3\text{–}75)$$

where

$$\kappa = k_0 r, \qquad (3.3\text{–}76)$$

$$g(p, q; s_x, s_y) = ps_x + qs_y + ms_z, \tag{3.3–77}$$

with m being given by Eq. (3.3–71a) and

$$s_z = (1 - s_x^2 - s_y^2)^{1/2}. \tag{3.3–78}$$

We are interested in the far-zone behavior of U_H; more specifically in the asymptotic behavior of the double integral on the right of Eq. (3.3–75), as $\kappa \to \infty$, with s_x and s_y being kept fixed. The integral is of the form (3.3–24) and hence the required asymptotic approximation to U_H is given by the formula (3.3–41) (with obvious change in the notation), provided the 'phase function' $g(p, q; s_x, s_y)$ has a single stationary point i.e. a single critical point of the first kind within the domain of integration. Assuming for the moment that this is so, the formula (3.3–41), applied to the present case, gives

$$U_H(s_x r, s_y r, s_z r) \sim \frac{2\pi i \sigma}{\kappa\sqrt{|\Delta|}} a(p_1, q_1)\, e^{i\kappa g(p_1, q_1; s_x, s_y)}, \tag{3.3–79}$$

where (p_1, q_1) is the stationary point, i.e. the point where

$$g_p = g_q = 0 \tag{3.3–80}$$

$(g_p = \partial g/\partial p,\ g_q = \partial g/\partial q)$. Further

$$\sigma = +1 \quad \text{when } \Delta > 0, \Sigma > 0, \tag{3.3–81a}$$

$$= -1 \quad \text{when } \Delta > 0, \Sigma < 0, \tag{3.3–81b}$$

$$= -i \quad \text{when } \Delta < 0, \tag{3.3–81c}$$

where

$$\Delta = (g_{pp} g_{qq} - g_{pq}^2)_1 \tag{3.3–82}$$

$$\Sigma = (g_{pp} + g_{qq})_1 \tag{3.3–83}$$

$(g_{pp} = \partial^2 g/\partial p^2$ etc.) and the subscript 1 following the parentheses indicates that the expression in the parentheses is evaluated at (p_1, q_1). We also assume that the requirement (3.3–28) is satisfied, i.e. that $\Delta \neq 0$.

Let us now examine whether, in the present case, the phase function has a stationary point. We have, on differentiating the expression (3.3–77),

$$g_p = s_x + m_p s_z. \tag{3.3–84}$$

Now from Eq. (3.3–71a)

$$m_p = -\frac{p}{m} \tag{3.3–85}$$

and, with the help of this expression, Eq. (3.3–84) becomes

$$g_p = s_x - \frac{p}{m} s_z. \tag{3.3–86a}$$

Similarly

$$g_q = s_y - \frac{q}{m} s_z. \tag{3.3–86b}$$

Hence g will be stationary [Eqs. (3.3–80) will be satisfied] when $p = p_1$, $q = q_1$,

with

$$\frac{p_1}{m_1} = \frac{s_x}{s_z}, \quad \frac{q_1}{m_1} = \frac{s_y}{s_z}, \tag{3.3–87a}$$

where [see Eq. (3.3–71a)]

$$m_1 = +(1 - p_1^2 - q_1^2)^{1/2}. \tag{3.3–87b}$$

Since the ratios s_x/s_z and s_y/s_z are fixed we see from Eq. (3.3–87a) that the phase function has indeed one and only one stationary point. Moreover since **s** is a unit vector,

$$s_z = +(1 - s_x^2 - s_y^2)^{1/2} \tag{3.3–88}$$

and it follows from Eqs. (3.3–87) and (3.3–88) that

$$p_1 = s_x, \quad q_1 = s_y, \quad m_1 = s_z. \tag{3.3–89}$$

In view of what we have already learned about the physical significance of the principle of stationary phase this result implies that, *in general, one and only one plane wave in the angular spectrum representation of the field contributes to the far field at a point located in the direction of the unit vector* **s**; *namely the wave that propagates in that particular direction*, the effect of the other waves being canceled by destructive interference.

To determine the asymptotic approximation to the field in the far zone in the **s**-direction we must, according to Eqs (3.3–80) and (3.3–82), also evaluate the second derivatives of the phase function. We have, on differentiating Eq. (3.3–84),

$$g_{pp} = -s_z\left(\frac{m - pm_p}{m^2}\right)$$

or, using Eq. (3.3–85),

$$g_{pp} = -\frac{s_z}{m}\left(1 + \frac{p^2}{m^2}\right).$$

Hence, at the stationary point p_1, q_1 given by Eqs. (3.3–89), we have

$$(g_{pp})_1 = -\left[1 + \left(\frac{s_x}{s_z}\right)^2\right]. \tag{3.3–90a}$$

In a similar way one finds that

$$(g_{qq})_1 = -\left[1 + \left(\frac{s_y}{s_z}\right)^2\right], \tag{3.3–90b}$$

$$(g_{pq})_1 = -\left(\frac{s_x s_y}{s_z^2}\right). \tag{3.3–90c}$$

With these expressions and using the identity (3.3–88), the quantities Δ and Σ defined by Eqs. (3.3–82) and (3.3–83) are readily found to have the value

$$\Sigma = -\left[1 + \frac{1}{s_z^2}\right], \quad \Delta = \frac{1}{s_z^2}. \tag{3.3–91}$$

Hence according to Eq. (3.3–81b) we have, in the present case,

$$\sigma = -1. \tag{3.3–92}$$

We must also evaluate $a(p, q)$ and $g(p, q)$ at the stationary point (p_1, q_1), given by Eqs. (3.3–89). Evidently,

$$a(p_1, q_1) = a(s_x, s_y) \tag{3.3–93}$$

and on substituting from Eq. (3.3–89) into Eq. (3.3–77)

$$g(p_1, q_1; s_x, s_y) = s_x^2 + s_y^2 + s_z^2 = 1. \tag{3.3–94}$$

Finally, on substituting from Eq. (3.3–76) and Eqs. (3.3–91)–(3.3–94) into the formula (3.3–79) and recalling that, in general, the far-zone behavior of U and of U_{H} are the same, we obtain for the far field the asymptotic formula

$$U(x, y, z) \sim -\frac{2\pi\mathrm{i}}{k_0}\left(\frac{z}{r}\right)a\left(\frac{x}{r}, \frac{y}{r}\right)\frac{\mathrm{e}^{ik_0 r}}{r} \quad \text{as } k_0 r \to \infty, \tag{3.3–95}$$

with the limit being taken along any fixed direction $s_x = x/r$, $s_y = y/r$, $s_z = z/r > 0$ $(r^2 = x^2 + y^2 + z^2)$ pointing into the half-space $z > 0$.

We note that in view of the elementary relation (3.2–26) between the Fourier transform $\widetilde{U}_0(u, v)$ of the field in the boundary plane $z = 0$ and the angular spectrum amplitude function $a(p, q)$, the asymptotic formula (3.3–95) is in agreement with Eq. (3.2–88), derived from the Rayleigh diffraction integral of the first kind.

Problems

3.1 Determine the Hilbert transforms of the following functions:

(i) $\sin \omega x$,

(ii) $\dfrac{\cos \omega x}{\omega x}$,

(iii) $\dfrac{1}{1 + (x/a)^2}$,

where ω and a are real non-zero constants.

3.2 A scalar wavefield $U(\mathbf{r})$ obeys the Helmholtz equation

$$(\nabla^2 + k^2)U(\mathbf{r}) = 0$$

throughout the half-space $z > 0$ and behaves as an outgoing wave at infinity in that half-space. In addition it has the property that in two planes $z = z_1 > 0$ and $z = z_2 > z_1$

$$U(x, y, z_2) = U(x, y, z_1)$$

for all values of x and y. Show that:

(a) the field distribution in the plane $z = 0$ can only have spatial frequency components (u, v) that satisfy the condition

$$u^2 + v^2 = k^2 - \left(\frac{2\pi\mu}{z_2 - z_1}\right)^2, \quad (\mu = 0, 1, 2, \ldots);$$

(b) the field distributions in the half-space $z > 0$ are identical in all planes that are at distances $N(z_2 - z_1)$ from the plane $z = z_1$, where N is any integer.

3.3 $U^{(1)}(x, y, z)$ and $U^{(2)}(x, y, z)$ represent two scalar wavefields which propagate into the half-space $z > 0$. Both fields are bandlimited to the spatial frequency domain $u^2 + v^2 \leqslant k^2$ in a plane $z = \text{constant} \geqslant 0$. Show that if in some plane $z = z_1 > 0$ the fields are complex conjugates of each other, i.e. if

$$U^{(2)}(x, y, z_1) = [U^{(1)}(x, y, z_1)]^*$$

for all values of x and y, then

$$U^{(2)}(x, y, z_1 + d) = [U^{(1)}(x, y, z_1 - d)]^*$$

for all values of d such that $|d| \leqslant z_1$ and for all values of x and y.

3.4 Consider a monochromatic wavefield, $U(x, y, z)\,\mathrm{e}^{-\mathrm{i}\omega t}$, which propagates into the half-space $z > 0$ and whose angular spectrum representation contains only homogeneous waves.

Derive an expression for the field $U(x, y, 0)$ in the plane $z = 0$ from the knowledge of the field in the plane $z = z_1 > 0$ in a form analogous to the Rayleigh diffraction formula.

3.5 Consider a monochromatic wavefield $U(x, y, z)\,\mathrm{e}^{-\mathrm{i}\omega t}$, propagating into the half-space $z > 0$. Let

$$C(\zeta, \eta; z) = \iint_{-\infty}^{\infty} U(\zeta + x, \eta + y, z)U^*(x, y, z)\,\mathrm{d}x\,\mathrm{d}y$$

be the spatial autocorrelation function of the field in a particular z-plane.

(a) Express $C(\zeta, \eta; z)$ in terms of the complex spectral amplitude function $a(p, q)$ of the angular spectrum representation of the field.

(b) Show that it is possible for $C(\zeta, \eta; z)$ to be independent of z. Find the general condition which the field in the boundary plane $z = 0$ must satisfy in this case. Can this condition be satisfied if the field in the boundary plane is of finite support? Justify your answer.

3.6 Show that there are solutions of the Helmholtz equation representing propagation into the half-space $z > 0$ which are of the form

$$U(x, y, z) = \mathrm{e}^{\mathrm{i}\beta z} f(x, y),$$

where β is a real constant.

Derive a general expression for the function $f(x, y)$ when $0 < \beta \leqslant k$ and discuss the structure of the angular spectrum representation of such fields. Why are solutions of this kind known as non-diffracting beams?

3.7 Solutions to a wide class of two-dimensional monochromatic wave-propagation problems are expressible in the form

$$\psi(kr, \theta) = \int_{-1}^{+1} f(t)\, e^{ikr[t \sin\theta + (1-t^2)^{1/2}\cos\theta]}\, dt,$$

where (r, θ), $0 \leqslant \theta < \pi/2$, are polar coordinates of a field point and k is a positive constant (wave number).

Obtain an asymptotic approximation for $\psi(kr, \theta)$ as $kr \to \infty$, with θ fixed.

3.8 Starting from the integral representation

$$J_n(x) = \frac{1}{2\pi}\int_{-\pi}^{\pi} e^{i(x \sin t - nt)}\, dt$$

of the Bessel function of the first kind, derive an asymptotic approximation for J_n, valid for large values of x $(x \gg n)$.

3.9 In the theory of diffraction of waves at a circular aperture, the following integral arises:

$$J = \int_0^{2\pi} \frac{e^{ika\sigma}}{\sigma^2}(1 - \rho\cos\phi)\, d\phi,$$

where ρ is a positive constant and

$$\sigma = (1 - 2\rho\cos\phi + \rho^2)^{1/2}.$$

Further $k = 2\pi/\lambda$ is the wave number and a is the radius of the aperture.

Assuming that $\rho \neq 1$ find the asymptotic approximation to J, when the radius of the aperture is large compared to the wavelength λ $(ka \gg 1)$.

4

Second-order coherence theory of scalar wavefields

4.1 Introduction

We will now study the properties of fluctuating electromagnetic fields, paying attention mainly to the optical region of the electromagnetic spectrum. It seems hardly necessary to stress that every electromagnetic field found in nature has some fluctuations associated with it. Even though these fluctuations are, as a rule, much too rapid to be observed directly, one can deduce their existence from suitable experiments that provide information about correlations between the fluctuations at two or more space-time points.

The simplest manifestations of correlations in optical fields are the well-known interference effects that arise when two light beams that originate from the same source are superposed. With the availability of modern light detectors and electronic circuitry of very short resolving time, other types of correlations in optical fields began to be studied in more recent times. These investigations, as well as the development of lasers and other novel types of light sources, led to a systematic classification of optical correlation phenomena and the complete statistical description of optical fields. The area of optics concerned with such questions is now generally known as *optical coherence theory*.

The first investigations of coherence phenomena are due to Verdet (1865, 1869) and von Laue (1907a, b). Some early investigations of Stokes (1852) and Michelson (1890, 1891a, b, c, 1892, 1920) although not explicitly mentioning coherence – because this concept is of a much later origin – have also contributed to the clarification and development of this subject. These investigations were carried further chiefly by Wiener (1927–1928, 1929, 1930), van Cittert (1934) Zernike (1938), Hopkins (1951, 1953, 1957), Wolf (1954a,b, 1955, 1959, 1981a, b, 1982, 1986); Blanc-Lapierre and Dumontet (1955), Pancharatnam (1956, 1957, 1963a, b, 1975), and Agarwal and Wolf (1993). The main outcome of these researches‡ was the introduction of a precise measure of the correlations between the fluctuating field variables at two space-time points and the formulation of dynamical laws which the correlation functions (in general a set of second-rank

‡ For a fuller discussion of the historical development of coherence theory, see Born and Wolf, (1980), Sec. 10.1. Some of the basic papers on this subject are reprinted in *Selected Papers on Coherence and Fluctuations of Light*, L. Mandel and E. Wolf eds., (Dover, New York, 1970), Vol. 1 (1950–1960), Vol. II (1961–1966); reprinted by SPIE Optical Engineering Press Milestone Series, MS 19, Parts I and II (Bellingham, WA, 1990). These two volumes also contain a comprehensive bibliography.

correlation tensors involving the electric and the magnetic fields) obey in free space. This 'second-order' theory provides a unified treatment of all the well-known interference and polarization phenomena of traditional optics.

Soon after this stage of development of the subject was completed, some quite new optical correlation phenomena were discovered that required for their elucidation the analysis of correlation properties of a wider class (which will be discussed in Chapter 6). Studies of such effects have eventually led to the formulation of a general theory of optical correlations and to the complete statistical description of optical fields. This general formulation has been obtained within the framework of both the classical and the quantum theory.

In this and in the next two chapters we will present an account of optical coherence theory on the basis of the classical theory of the optical field. An account based on quantum theory will be presented in later chapters.

4.2 Some elementary concepts and definitions

It will be useful to begin with a few elementary concepts and definitions that are customarily employed in discussions of simple optical correlation experiments.

4.2.1 Temporal coherence and the coherence time

Consider a beam of light from a small source σ. We assume that the light is quasi-monochromatic, i.e. that its bandwidth $\Delta\nu$ is small compared to its mean frequency $\bar{\nu}$ and that it is macroscopically steady.‡ Suppose that the beam is divided into two beams in a Michelson interferometer at a point P_1 and that the two beams are re-united after a path difference $\Delta l = c\Delta t$ (c being the speed of light in vacuum) has been introduced between them (Fig. 4.1). If the path difference Δl is sufficiently small, interference fringes are formed in the plane of observation $\mathscr{B}$. The formation of the fringes is said to be a manifestation of *temporal coherence* between the two beams, since their ability to form the fringes may be explained as arising from correlations that exist between them under conditions where a time delay Δt has been introduced between the beams. It is a well-known experimental fact that the interference fringes will be formed only if the time delay Δt is such that

$$\Delta t \Delta\nu \lesssim 1, \tag{4.2–1}$$

where $\Delta\nu$ is the bandwidth of the light. The time delay

$$\Delta t \sim \frac{1}{\Delta\nu} \tag{4.2–2}$$

is known as the *coherence time* of the light and the corresponding path difference

$$\Delta l = c\Delta t \sim \frac{c}{\Delta\nu} \tag{4.2–3}$$

‡ By 'macroscopically steady' we mean that it does not exhibit fluctuations on a *macroscopic* time-scale. In the more precise language of the theory of random processes, the fluctuations can be represented as a stationary random process (Section 2.2), whose mean period and correlation time are much shorter than the averaging interval needed to make an observation.

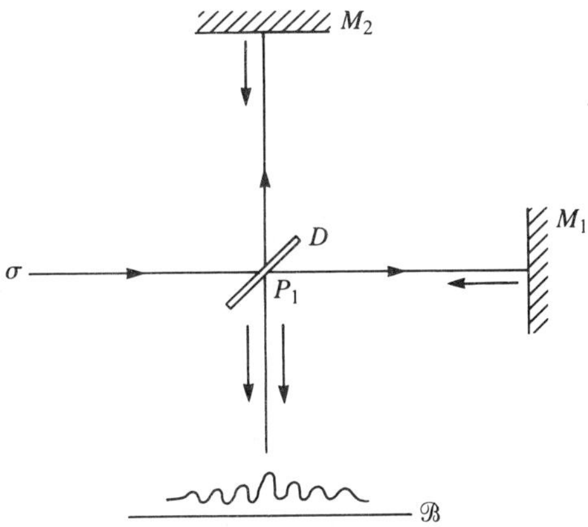

Fig. 4.1 Temporal coherence illustrated by means of an interference experiment using a Michelson interferometer. σ = source, D = beam divider, M_1, M_2 = mirrors, $\mathcal{B}$ = plane of observation. (For the sake of simplicity, a compensating plate and a collimating lens system are not shown.)

is called the *coherence length*, or more precisely, the *longitudinal coherence length* of the light. Since $\nu = c/\lambda$, where λ is the wavelength, $\Delta\nu \sim c\Delta\lambda/\bar{\lambda}^2$, the expression for the coherence length may also be expressed in the form

$$\Delta l \sim \left(\frac{\bar{\lambda}}{\Delta\lambda}\right)\bar{\lambda}, \qquad (4.2\text{–}4)$$

where $\bar{\lambda}$ is the mean wavelength.

A rough understanding of this phenomenon may be obtained as follows. The fringes in the observation plane $\mathcal{B}$ may be considered to arise from the addition of spatially periodic distributions, each of them being formed by a frequency component present in the spectrum of the light. Now the periodic distributions formed by light of different frequency components will have different spatial periodicities. Hence, with increasing time delay between the two beams, their addition will lead to a less and less well-defined fringe pattern, because the maxima of the various monochromatic contributions will get more and more out of step. For a sufficiently long time delay, the periodic intensity distributions will get so much out of step that the superposed pattern will no longer exhibit any pronounced intensity maxima and minima, i.e. no fringe pattern will be formed. Simple calculations show that with increasing time delay the fringes will disappear when Δt reaches a value that is of the order of magnitude indicated by the relation (4.2–2).

Alternatively we may gain some understanding of these effects from the following considerations, which utilize the concept of correlations. According to Section 3.1.2 a sample function of a quasi-monochromatic light disturbance, regarded as a stationary random process, may be pictured as a succession of slowly modulated wave trains, whose mean frequency coincides with the mean frequency of the light and whose duration is of the order of the reciprocal bandwidth of the light, i.e. of the order of the coherence time [Eq. (4.2–2)]. The beam divider D of the Michelson interferometer splits each wave train into two wave trains of the same general form but of reduced amplitudes. At the plane of observation $\mathscr{B}$ the wave trains of the two partial beams become superposed after a time delay has been introduced between them, i.e. after the corresponding wave trains in the two beams have been shifted relative to each other. Clearly there will be a strong correlation between the fluctuations in the two beams at $\mathscr{B}$ if the time delay introduced is short compared to $1/\Delta\nu$, i.e. short compared to the coherence time of the light, and there will be effectively no correlation between them when the time delay is much greater than $1/\Delta\nu$. The formation or the absence of interference fringes in the observation plane $\mathscr{B}$ is thus directly related to the *correlation* or the *lack of correlation*, respectively, between the fluctuations of the two partial beams reaching $\mathscr{B}$.

In Section 7.3 we will discuss this phenomenon in a broader context and from a more rigorous standpoint, in connection with Michelson's well-known method for determining the energy distribution in spectral lines from two-beam interference experiments.

We conclude this elementary discussion by two simple examples which illustrate the order of magnitude of the coherence time and of the coherence length in typical cases. For light generated by thermal sources (incandescent matter, gas discharge) with a high degree of monochromaticity (i.e. a narrow spectral width), the bandwidth $\Delta\nu$ is typically of the order of $10^8\,\mathrm{s}^{-1}$ or greater. The corresponding coherence time Δt is of the order of $10^{-8}\,\mathrm{s}$ and the coherence length $\Delta l \sim 3 \times 10^{10}\,\mathrm{cm\,s}^{-1} \times 10^{-8}\,\mathrm{s} \sim 3\,\mathrm{m}$. On the other hand, for a well-stabilized laser $\Delta\nu$ can be as large as $10^4\,\mathrm{s}^{-1}$ (10 kHz), so that light generated by such a laser has a coherence time of the order of $10^{-4}\,\mathrm{s}$ and a coherence length $\Delta l \sim 3 \times 10^{10}\,\mathrm{cm\,s}^{-1} \times 10^{-4}\,\mathrm{s} \sim 30\,\mathrm{km}$.

4.2.2 Spatial coherence and the coherence area

Next let us briefly consider another type of interference experiment, the Young's interference experiment with quasi-monochromatic light from an extended source σ (Fig. 4.2). We assume that σ is a *thermal* source, such as incandescent matter or gas discharge. We consider a symmetrical arrangement for simplicity, with the source having the form of a square of sides Δs. If the pinholes P_1 and P_2 are sufficiently close to the axis of symmetry, interference fringes will be observed in

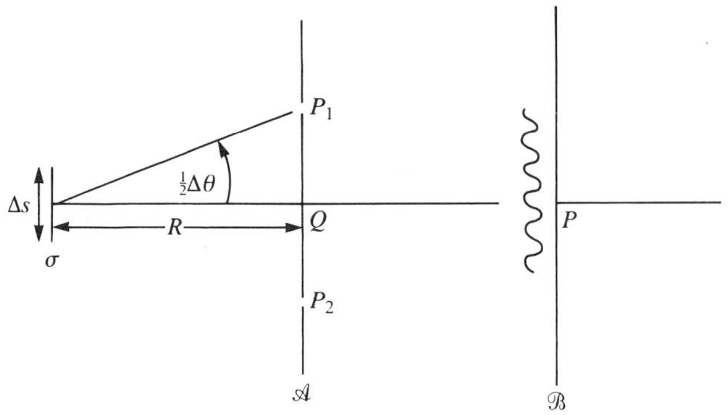

Fig. 4.2 Spatial coherence illustrated by means of Young's interference experiment with light from a thermal source σ.

the neighborhood of the axial point P at the plane of observation $\mathscr{B}$. The appearance of the fringes is said to be a manifestation of *spatial coherence* between the two light beams reaching P from the two pinholes P_1 and P_2, because the ability of the two beams to form fringes may be explained as arising from the correlation that exists between them under conditions where a spatial separation (the distance P_1P_2) has been introduced. In performing an interference experiment of this kind it is found that if the separation between the source and the plane $\mathscr{A}$ containing the pinholes is large enough, interference fringes will be formed near P only if

$$\Delta\theta\Delta s \lesssim \bar{\lambda}, \tag{4.2–5}$$

where $\Delta\theta$ is the angle that the distance P_1P_2 of separation between the pinholes subtends at the source and $\bar{\lambda} = c/\bar{\nu}$ is a mean wavelength of the light. If R denotes the distance between the plane of the source and the plane containing the pinholes, the foregoing result implies that, in order to observe fringes in the neighborhood of P, the two pinholes must be situated within a region around the axial point Q in the plane $\mathscr{A}$, whose area ΔA is given by the order of magnitude relation

$$\Delta A \sim (R\Delta\theta)^2 \sim \frac{R^2\bar{\lambda}^2}{S}, \tag{4.2–6}$$

where $S = (\Delta s)^2$ is the area of the source. This region is said to be the *coherence area* of the light in the plane $\mathscr{A}$ around the point Q and the square-root of the coherence area is sometimes called the *transverse coherence length*. It should be noted that according to Eq. (4.2–6), the coherence area becomes larger and larger with increasing R. However, there is an invariant quantity associated with the coherence area, independent of the distance R, namely the solid angle $\Delta A/R^2$. According to Eq. (4.2–6) this solid angle is given by the expression

$$\Delta\Omega \sim \bar{\lambda}^2/S. \tag{4.2–7}$$

It is sometimes useful to express the coherence area in an alternative form, that involves the solid angle $\Delta\Omega'$ that the source subtends at Q. Since $S = R^2\Delta\Omega'$ we obtain at once from (4.2–6) the expression

$$\Delta A \sim \bar{\lambda}^2/\Delta\Omega' \tag{4.2–8}$$

for the coherence area.

A rough elementary derivation of Eq. (4.2–5) may be obtained as follows. Each source point gives rise to an interference pattern in the plane of observation. Since fluctuations in the light from different points of a thermal source may be assumed to be mutually independent and hence have no fixed phase relationship to each other, the intensity distribution in the plane $\mathscr{B}$ is obtained by adding together, at each point, the intensities of the individual patterns generated by the different source points. Now the maxima of these patterns will be displaced with respect to each other. If the source and the positions of the planes $\mathscr{A}$ and $\mathscr{B}$ are kept fixed but the separation between the two pinholes is gradually increased from near coincidence to larger and larger values, i.e. if the angle $\Delta\theta$ in Fig. 4.2 is gradually increased, the individual patterns will get more and more out of step and will eventually give rise to an essentially uniform distribution near the axial point in the plane of observation. A simple calculation shows that this happens when $\Delta\theta \sim \bar{\lambda}/\Delta s$, in agreement with Eq. (4.2–5).

As in the case of temporal coherence, we may gain a somewhat deeper understanding of the phenomenon of spatial coherence if we analyze the experiment in terms of the concept of correlations. This will be done in Section 4.3. It is however possible, as we will now show, to obtain a qualitative understanding of the origin of correlations in a field, generated by an extended thermal source, from very simple considerations.

The essence of the phenomenon may be readily understood by considering, instead of the extended source σ, two point sources S_1 and S_2 (Fig. 4.3). Let us assume that the light emitted by these sources is quasi-monochromatic, with the same mean frequency $\bar{\nu}$ and the same effective spectral range $\Delta\nu$ and that the sources are statistically independent, so that there are no correlations between the two light fields generated by them. Let us now consider the light disturbances at the two points P_1 and P_2 in the space surrounding the source. If we ignore, for the sake of simplicity, the polarization properties of the field, we may represent the light disturbances reaching P_1 from the source points S_1 and S_2 by complex analytic scalar signals $V_1(t)$ and $V_2(t)$, respectively. Similarly we may represent the light disturbances reaching P_2 from the two source points by complex analytic signals $V_1'(t)$ and $V_2'(t)$.

If the difference between the distances $R_{11} = \overline{S_1P_1}$ and $R_{12} = \overline{S_1P_2}$ is small compared to the coherence length ($\sim c/\Delta\nu$) of the light, one obviously has, except for a deterministic phase factor,

$$V_1'(t) = V_1(t). \tag{4.2–9a}$$

Similarly if the difference between $R_{21} = \overline{S_2P_1}$ and $R_{22} = \overline{S_2P_2}$ is small compared to the coherence length of the light, one has, except for a deterministic phase factor,

$$V_2'(t) = V_2(t). \tag{4.2–9b}$$

The total field at P_1 arises from the superposition of the two fields generated by each of the two point sources (see Fig. 4.3) and hence is given by

$$V(P_1, t) = V_1(t) + V_2(t); \tag{4.2–10a}$$

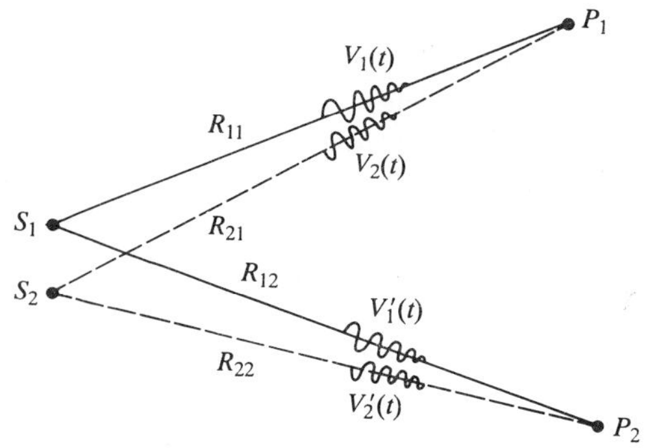

Fig. 4.3 Illustrating the origin of spatial coherence at two points P_1 and P_2, generated by two uncorrelated point sources S_1 and S_2.

similarly, the total field at P_2 is given by

$$V(P_2, t) = V_1'(t) + V_2'(t). \tag{4.2–10b}$$

Now since $V_1(t)$ and $V_2(t)$ are generated by the source points S_1 and S_2 which are statistically independent, these two disturbances will not be correlated. For the same reason $V_1'(t)$ and $V_2'(t)$ will also not be correlated. However the two *sums* $V_1(t) + V_2(t)$ and $V_1'(t) + V_2'(t)$ will evidently be correlated because of the relations (4.2–9). This conclusion is illustrated graphically in Fig. 4.3, where the (essentially identical) wave trains V_1 and V_1' arriving at the points P_1 and P_2 from S_1 are drawn in solid lines, and the (essentially identical) wave trains V_2 and V_2' arriving at P_1 and P_2 from S_2 are shown in dashed lines. Clearly, although the solidly drawn wave trains and the wave trains drawn in dashed lines may be of completely different forms, the *sum* of the two wave trains arriving at P_1 and the *sum* of the two wave trains arriving at P_2 will be similar to each other. Thus the fields at P_1 and P_2, represented by equations (4.2–10), will indeed be strongly *correlated*. Hence we see that, *even though the two sources S_1 and S_2 are statistically independent, they will give rise to correlations in the field and these correlations are generated in the process of propagation and superposition.*

The model that we just discussed applies only to a particular geometrical situation ($R_{11} \sim R_{12}$, $R_{21} \sim R_{22}$). It is clear that, when the geometrical conditions are relaxed, the situation becomes somewhat more complicated. Instead of a high degree of correlation between the total fields at the points P_1 and P_2, one then evidently finds a somewhat lower degree of correlation, depending on the exact location of the two points. We will return to this problem in Section 4.4.4 in a more general context.

We will illustrate our elementary analysis relating to spatial coherence by a few examples. Suppose that the linear dimension of the thermal source σ (Fig. 4.2) is $\Delta s = 1$ mm and that the source emits quasi-monochromatic light of mean wavelength $\bar{\lambda} = 5000$ Å. Let the plane $\mathscr{A}$ of the pinholes be situated at a distance $R = 2$ meters from the plane containing the source. Then, according to Eq. (4.2–6), the coherence area in the plane $\mathscr{A}$ of the pinholes is

$$\Delta A = \left(\frac{2 \times 10^2}{10^{-1}}\right)^2 (5 \times 10^{-5})^2 \text{ cm}^2 = 1 \text{ mm}^2, \tag{4.2–11}$$

i.e. its linear dimensions are of the order of 1 mm.

As a second example let us estimate the coherence area in a beam of sunlight, illuminating the surface of the earth. To satisfy our assumption of quasi-monochromaticity, we would first have to pass the sunlight through a filter with a narrow passband, say around the wavelength $\bar{\lambda} \sim 5000$ Å. Now the angular radius that the sun's disk subtends at the surface of the earth is approximately $\alpha = 0°16' \sim 0.00465$ radians. Hence, neglecting limb darkening, the solid angle $\Delta\Omega'$ that the sun's disk subtends at the earth's surface is $\Delta\Omega' \approx \pi\alpha^2 \sim 3.14 \times (4.65 \times 10^{-3})^2$ sr. $\sim 6.81 \times 10^{-5}$ sr. and thus, according to Eq. (4.2–8) the coherence area is‡

$$\Delta A \sim \frac{(5 \times 10^{-5})^2}{6.81 \times 10^{-5}} \text{ cm}^2 \approx 3.67 \times 10^{-3} \text{ mm}^2. \tag{4.2–12}$$

‡ This result for sunlight appears to be the first estimate of the size of a region of coherence that can be found in the literature. It was obtained by Verdet (1865, 1869).

Hence the linear dimension of the coherence area on the surface of the earth of filtered sunlight is of the order of $(3.67 \times 10^{-3})^{1/2}$ mm ~ 0.061 mm.

It is instructive to compare the coherence area of light reaching the earth from the sun with the coherence area of light that arrives from a more distant star. For this purpose we first note that, according to Eq. (4.2–8), the coherence area varies inversely with the solid angle that the source subtends at the axial point Q of the plane where the coherence area is to be estimated. Now when viewed from the earth's surface, the angular diameter of a typical star will be many orders of magnitude smaller than the angular diameter of the sun. Hence the coherence area of the star light reaching the earth's surface must be very much larger than that of sunlight. Consider, for example, the star Betelgeuse (α Orionis). This actually is the first star whose angular diameter was determined by an interferometric technique (described in Section 7.2) and it was found to have the value $2\alpha \sim 0.047$ seconds of arc $\sim 2.3 \times 10^{-7}$ radians. The solid angle $\Delta\Omega'$ that this star subtends at the surface of the earth is, therefore, $\Delta\Omega' \approx \pi\alpha^2 \sim 4.15 \times 10^{-14}$ sr. Hence the coherence area of light from Betelgeuse on the earth's surface, after the light has been passed through a filter that transmits a narrow band around the wavelength $\bar{\lambda} = 5000$ Å, is

$$\Delta A \sim \frac{(5 \times 10^{-5}\ \mathrm{cm})^2}{4.15 \times 10^{-14}} \sim 6\ \mathrm{m}^2. \tag{4.2–13}$$

This result implies the existence of appreciable correlations between light vibrations reaching the earth's surface from Betelgeuse at two points on earth up to a maximum separation of about $\sqrt{6}$ meters ~ 2.45 meters ~ 8 feet. Actually many stars have angular diameters that are appreciably smaller than that of Betelgeuse, so that a high degree of correlation exists over even greater areas in the light reaching the earth's surface from such stars.

Our analysis also gives a rough indication as to why stellar images formed by well-correlated telescopes have, on good observing nights (i.e. in the absence of appreciable atmospheric tremor), the appearance of a diffraction pattern, very familiar from the theory of imaging with strictly coherent light [see, for example, Born and Wolf (1980), Sec. 8.5]. For as we just saw, there exists a high degree of correlation in the star light entering the aperture of the telescope over areas which are generally much larger than that of the aperture. Consequently, the secondary wavelets entering the telescope and propagating to the image plane will superpose essentially in the same way as wavelets in a completely *coherent* beam.

We introduced the concept of spatial coherence of light generated by a thermal source that directly illuminates a distant plane in free space. It is, however, evident that this concept applies much more generally, irrespective of the nature of the source and of the surrounding medium. The existence of spatial coherence in the field may be revealed by interference effects in a two-pinhole experiment. The appearance of the interference fringes is again a manifestation of the correlations between light vibrations at the two pinholes. Naturally the coherence area will not, in general, be given by the simple formulas (4.2–6) and (4.2–8). We will learn later on (Section 4.4) how the degree of correlation may be determined for light generated by any type of source.

4.2.3 Coherence volume and the degeneracy parameter

We will now introduce two other concepts which are also useful in gaining intuitive insight into the coherence properties of light.

Consider an optical field consisting of a nearly plane, quasi-monochromatic, linearly polarized wave. The right-angle cylinder, whose base is the coherence area ΔA in a plane perpendicular to the direction of propagation of the wave and whose height is the *longitudinal coherence length* (Δl) (see Fig. 4.4), is known as the *coherence volume*. It clearly occupies a domain of volume

$$\Delta V = \Delta A \Delta l. \tag{4.2–14}$$

For a quasi-monochromatic plane wave, the coherence volume would be independent of the specific location of the volume. In the more realistic case when the wavefield is only approximately a plane wave, ΔA, and consequently ΔV, will depend on the location, and hence it is then more appropriate to speak about the *coherence volume around a particular point in the field*.

Suppose that the field is generated by a thermal source σ that has the form of a square of area S and that emits quasi-monochromatic light of mean wavelength $\bar{\lambda}$. Then, according to Eq. (4.2–6), the coherence area around the axial point Q in a plane $\mathscr{A}$ parallel to the source and at a large distance R from it (see Fig. 4.2) is given by $\Delta A \sim (R^2/S)\bar{\lambda}^2$. According to Eq. (4.2–4) the coherence length is given by $\Delta l \sim \bar{\lambda}^2/\Delta\lambda$. Hence Eq. (4.2–14) leads to the following expression for the coherence volume around the point Q:

$$\Delta V \sim \left(\frac{R^2}{S}\right)\left(\frac{\bar{\lambda}}{\Delta\lambda}\right)\bar{\lambda}^3. \tag{4.2–15a}$$

Since $S/R^2 = \Delta\Omega'$ is the solid angle that the source area S subtends at the point Q, we may also express ΔV as

$$\Delta V \sim \frac{1}{\Delta\Omega'}\left(\frac{\bar{\lambda}}{\Delta\lambda}\right)\bar{\lambda}^3. \tag{4.2–15b}$$

Let us estimate the coherence volume for the three examples that we considered in Section 4.2.2 in connection with the coherence area, assuming that in each case the effective wavelength range of the filtered light $\Delta\lambda = 10^{-7}\bar{\lambda} \sim 5 \times 10^{-4}$ Å, with $\bar{\lambda} \sim 5000$ Å as before. In this case, the coherence length,

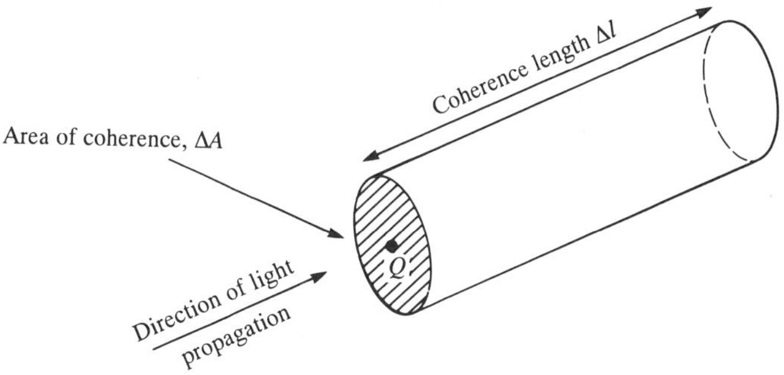

Fig. 4.4 Illustrating the concept of the coherence volume.

according to Eq. (4.2–4) is $\Delta l = 5$ m. For a two-dimensional thermal source of area 1 mm^2, we found that the coherence area around the axial point Q in a plane parallel to the source and at distance $R = 2$ m from it has size $\Delta A \sim 1$ mm^2 [Eq. (4.2–11)]. Hence, according to (4.2–14) the coherence volume around Q is

$$\Delta V \sim 1\ \text{mm}^2 \times 5\ \text{m} = 5\ \text{cm}^3. \tag{4.2–16}$$

For filtered sunlight reaching the earth's surface, we found [Eq. (4.2–12)] that $\Delta A \sim 3.67 \times 10^{-3}$ mm^2, so that the coherence volume of sunlight on the earth's surface is

$$\Delta V \sim 3.67 \times 10^{-3}\ \text{mm}^2 \times 5\ \text{m} \sim 18\ \text{mm}^3. \tag{4.2–17}$$

For filtered light from the star Betelgeuse, we found that $\Delta A \sim 6$ m^2 [Eq. (4.2–13)] so that

$$\Delta V \sim 6\ \text{m}^2 \times 5\ \text{m} = 30\ \text{m}^3. \tag{4.2–18}$$

We derived the expressions (4.2–15) for the coherence volume on the assumption that we are dealing with radiation far from a thermal source. However, the concept of a coherence volume clearly applies much more generally. Consider, for example, light from a common type laboratory He:Ne laser. Let as assume that the cross-section of the laser beam is 1 mm^2 and that the mean wavelength of the light is $\bar{\lambda} = 6 \times 10^{-5}$ cm ($\bar{\nu} \sim 5 \times 10^{14}$ Hz). Over a short time interval, of the order of a few seconds, one can easily achieve stability that ensures a narrow bandwidth $\Delta\nu \sim 10^6$ Hz, implying an effective wavelength range $\Delta\lambda \sim 1.2 \times 10^{-13}$ cm. According to Eq. (4.2–3) the coherence length during such a short time interval will be of the order $\Delta l \sim 3 \times 10^{10} \times 10^{-6}$ cm $\sim 3 \times 10^4$cm. Assuming that the laser beam is spatially coherent over its whole cross-section (which would be the case if the laser operated on a single mode), Eq. (4.2–14) may clearly be used to provide an estimate of the coherence volume ΔV and one finds that

$$\Delta V \sim 10^{-2} \times 3 \times 10^4\ \text{cm}^3 = 300\ \text{cm}^3. \tag{4.2–19}$$

The concept of coherence volume has an interesting interpretation in the context of quantum mechanics, when the light is analyzed in terms of photons. To see this let us again assume, for the sake of simplicity, that the field is that of a nearly plane, quasi-monochromatic wave and let $\mathbf{p}(p_x, p_y, p_z)$ be the momentum of a photon that is located[‡] in the neighborhood of a point $\mathbf{r}(x, y, z)$. We may associate with the field a six-dimensional phase space of the photons, with coordinates x, y, z, p_x, p_y, p_z. Now x and p_x cannot be measured simultaneously with accuracy greater than allowed by the Heisenberg uncertainty relation $\Delta x \Delta p_x \geqslant \hbar/2$ ($\hbar = h/2\pi$, h being Planck's constant) and similarly for the other conjugate pairs Δy, Δp_y and Δz, Δp_z. It is thus natural to imagine the phase space to be divided into cells, each of which has the size

$$\Delta x \Delta y \Delta z \Delta p_x \Delta p_y \Delta p_z = h^3. \tag{4.2–20}$$

Photons of the same polarization, which belong to a region of the phase space

‡ Caution must be exercised in speaking about the position of a photon, since a photon cannot be localized more closely than to a distance of the order of the wavelength (see Section 12.11).

whose size is not greater than indicated by Eq. (4.2–20), are *intrinsically indistinguishable* from each other.

The coherence volume that we introduced from considerations based on classical theory alone may now readily be shown, at least in typical cases of practical interest, to be precisely the volume $\Delta x \Delta y \Delta z$ of ordinary space, given by the formula (4.2–20), subject to the constraints imposed on the product $\Delta p_x \Delta p_y \Delta p_z$ by the geometry and by the bandwidth of the light.‡ In other words *the coherence volume is that region of space throughout which the photons in the field are intrinsically indistinguishable from each other*. To justify this statement let us first estimate the uncertainty in the components of the momentum of a photon in the far zone of a field generated by a plane, quasi-monochromatic thermal source σ of linear dimensions Δs. Let 2ϕ denote the angle that the source subtends at a point Q, assumed for simplicity to lie on the normal to σ, at distance R from σ in the far zone (Fig. 4.5). According to the de Broglie relation (Born and Wolf, 1980, Sec. 3 of Appendix II), the momentum $\mathbf{p}$ of the photon is related to its wavelength λ by the formula

$$\mathbf{p} = \frac{h}{\lambda}\mathbf{s}, \tag{4.2–21}$$

where $\mathbf{s}$ is the unit vector in the direction of $\mathbf{p}$. The uncertainties in the x- and y-components of the momentum arise from our ignorance of the exact point of the source from which the photon was emitted and is clearly given by the projections of the momentum vector onto the x- and y-axes, assumed to lie in the plane of the source. Hence

$$\Delta p_x = \Delta p_y \sim 2p \sin\phi$$

or, using Eq. (4.2–21),

$$\Delta p_x = \Delta p_y \sim \frac{2h}{\bar{\lambda}} \sin\phi \sim \frac{h}{\bar{\lambda}} \frac{\Delta s}{R}, \tag{4.2–22}$$

where $\bar{\lambda}$ is the mean wavelength of the light. With ϕ assumed to be sufficiently small, the uncertainty in the z-component of the momentum arises principally from the uncertainty in the wavelength. If $\Delta\lambda$ is the effective wavelength range of the light, we have from Eq. (4.2–21)

$$\Delta p_z \sim \frac{h}{\bar{\lambda}^2} \Delta\lambda. \tag{4.2–23}$$

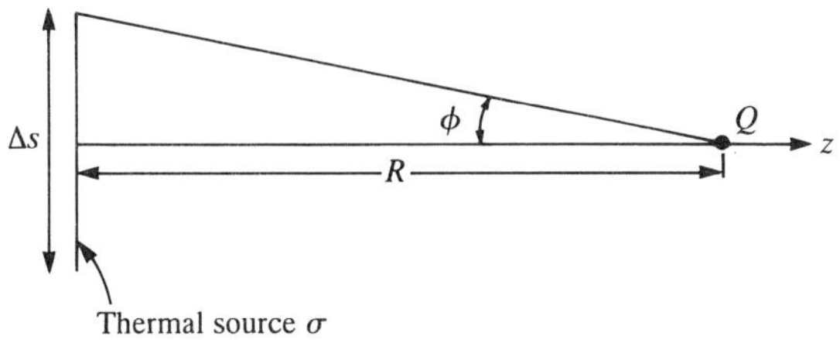

Fig. 4.5 Illustrating the analysis leading to Eqs. (4.2–22) and (4.2–23) for the uncertainties in the components of the momentum of a photon emitted by an extended thermal source.

‡ This result was first noted by Hanbury Brown and Twiss (1957), Appendix I, p. 321.

If follows from Eqs. (4.2–22) and (4.2–23) that in the present case

$$\Delta p_x \Delta p_y \Delta p_z = h^3 \frac{\Delta\lambda}{\bar{\lambda}^4} \frac{S}{R^2}, \tag{4.2–24}$$

where again $S = (\Delta s)^2$ is the order of magnitude of the source size. On substituting from (4.2–24) into the expression (4.2–20) for the size of a cell in phase space, we see that the volume of space around the point Q, throughout which photons emitted by our source are intrinsically indistinguishable, is

$$\Delta x \Delta y \Delta z = \left(\frac{R^2}{S}\right)\left(\frac{\bar{\lambda}}{\Delta\lambda}\right)\bar{\lambda}^3. \tag{4.2–25}$$

Comparison of Eq. (4.2–25) with Eq. (4.2–15a) shows that this value is precisely the coherence volume derived by considerations based on classical theory. Thus we have justified, in this special but important practical case, our earlier assertion as to the quantum-mechanical significance of the coherence volume.

As another example let us consider blackbody radiation in a large thermally insulated enclosure. It is well known that photons in a region of volume V inside the enclosure, with energies in the range $\Delta E = h\Delta\nu$ around the mean energy $\bar{E} = h\bar{\nu}$, will belong to the same cell of phase space if

$$\frac{8\pi V}{c^3}\bar{\nu}^2\Delta\nu \leqslant 1. \tag{4.2–26}$$

This result implies that the maximum size of any region inside the cavity, throughout which the photons are indistinguishable, is given by $V = c^3/(8\pi\bar{\nu}^2\Delta\nu)$ or, using the relation $\bar{\nu} = c/\bar{\lambda}$,

$$V = \frac{1}{8\pi}\left(\frac{\bar{\lambda}}{\Delta\lambda}\right)\bar{\lambda}^3. \tag{4.2–27}$$

This formula is precisely of the form of Eq. (4.2–15b) with the angle $\Delta\Omega'$ replaced by the factor 8π. Now $\Delta\Omega'$ in Eq. (4.2–15b) represents the solid angle that the source subtends at the point Q, i.e. the solid angle formed by all the directions along which radiation from the source can reach the coherence volume. Blackbody radiation inside a thermally insulated cavity can be regarded as a mixture of plane waves (with an appropriate statistical distribution of amplitudes and phases) which propagate in all possible directions. Hence in this case $\Delta\Omega' = 4\pi$. The extra factor $1/2$ can also be readily understood; it arises from the fact that the blackbody radiation is unpolarized and may thus be regarded as consisting of two independent polarized states (e.g. linear or circular) for each direction of propagation. Thus Eq. (4.2–27), derived from considerations of indistinguishability of photons, is again in complete agreement with the expression for the coherence volume, based on classical wave theory.

It is of interest also to examine the average number of photons of the same spin-state that are contained in a coherence volume for typical optical fields. This quantity is known as the degeneracy parameter of the field (Mandel, 1961a). In the language of quantum statistics, the degeneracy parameter represents the expectation value of the number of photons that are in the same quantum state. We will see later (Section 14.6) that this parameter plays an important role in connection with photoelectric detection of light fluctuations. The values of the degeneracy parameter can be significantly different for thermal light and for laser light as we will now show.

For blackbody radiation at equilibrium temperature T, the value of the degeneracy parameter δ at frequency ν is known from early work of Einstein (1912) (see also Bothe, 1927 and Fürth, 1928) to be given by

$$\delta = \frac{1}{e^{h\nu/k_B T} - 1}, \tag{4.2–28}$$

(h is Planck's constant, k_B is Boltzmann's constant). For light of frequency $\nu = 5 \times 10^{14}$ Hz from a typical incandescent source at temperature $T = 3000$ K, one obtains from (4.2–28) the value

$$\delta \sim 3 \times 10^{-4}, \tag{4.2–29}$$

implying that such light is highly non-degenerate ($\delta \ll 1$). In order that $\delta \sim 1$ at this frequency it is necessary for the temperature to be of the order $T \sim 3 \times 10^4$ K, as can be readily deduced from Eq. (4.2–28).

The situation is quite different for laser light. Consider, for example, a common type laboratory He:Ne laser with a 1 milliwatt power output, generating a beam of 1 mm^2 cross-section, of mean wavelength $\bar{\lambda} = 6 \times 10^{-5}$ cm ($\bar{\nu} \sim 5 \times 10^{14}$ Hz). The number of photons per unit volume, i.e. the energy per unit volume expressed in terms of energy of a single photon, in a light beam generated by such a laser is

$$\begin{aligned} \rho &= \frac{10^{-3}}{10^{-2} \times (6.67 \times 10^{-34}) \times (3 \times 10^{14}) \times 3 \times 10^{10}} \text{ photon/cm}^3 \\ &\sim 10^7 \text{ photon/cm}^3. \end{aligned} \tag{4.2–30}$$

We have seen earlier [Eq. (4.2–19)] that over a short enough time interval, the stability of the output is such that the laser light has a coherence volume $\Delta V \sim 300$ cm^3. Hence in this case the degeneracy parameter has the value

$$\delta = \rho \cdot \Delta V \sim 1 \times 10^7 \times 3 \times 10^2 = 3 \times 10^9. \tag{4.2–31}$$

Such light is therefore highly degenerate ($\delta \gg 1$). Comparison of Eq. (4.2–31) with Eq. (4.2–29) shows a difference of 13 orders of magnitude in the degeneracies of blackbody radiation and of the laser beam.

4.3 Interference of two stationary light beams as a second-order correlation phenomenon

In the preceding section we introduced rough criteria which indicate conditions under which simple interference effects may be expected to take place. We also briefly noted that such phenomena depend on *correlations* that exist between the light fluctuations in the interfering beams. Correlations may, of course, be analyzed by means of the mathematical techniques of the theory of stochastic processes, described in Chapter 2. A general treatment along these lines will be presented in Chapter 8. However, when one is concerned with the average intensity, as is often the case, one only needs to take into account second-order correlations (correlations between the light vibrations at two space-time points). We will now introduce a precise measure for such correlations from a detailed analysis of a simple interference experiment. This measure, which is classical in this context, will later be shown (in Chapter 12) also to correspond to a measure defined quantum mechanically.

4.3.1 The laws of interference. The mutual coherence function and the complex degree of coherence

In order to bring out the essential aspects of the theory we will ignore polarization phenomena throughout this chapter. These will be treated in Chapter 6.

Let $V^{(r)}(\mathbf{r}, t)$ denote a real field variable at a point represented by a position vector $\mathbf{r}$, at time t. This function may represent, for example, a Cartesian component of the electric field or of the vector potential. We purposely do not specify the nature of $V^{(r)}$ more closely at this stage, since the main analysis is independent of any particular choice of the field variable and different choices may be best suited for describing different experimental situations. In the case of photoelectric detection, it is appropriate, as will be shown in Section 14.1, to consider the vector potential (more precisely the analytic signal associated with it) as the basic field variable. But there are other detection processes for which other choices may be more suitable.

For any realistic light beam, $V^{(r)}$ will be a fluctuating function of time, which may be regarded as a typical member of an ensemble consisting of all possible realizations of the field. There are several reasons why $V^{(r)}$ will fluctuate. When the light is produced by a thermal source the fluctuations arise mainly because $V^{(r)}$ consists of a large number of contributions that are effectively independent of each other, so that their superposition gives rise to a fluctuating field which can only be described in statistical terms. But even light from a well-stabilized source, such as a laser, will exhibit some random fluctuations, since the effect of spontaneous emission is never entirely absent. In addition there will be other sources of irregular fluctuations, for example, vibrations of the mirrors at the end of the resonant cavity.

It is convenient to carry out the analysis not in terms of the real field variable $V^{(r)}(\mathbf{r}, t)$, but rather in terms of the associated analytic signal $V(\mathbf{r}, t)$, which we discussed in detail in Chapter 3. We will see later (Section 14.2) that the complex field $V(\mathbf{r}, t)$ appears naturally in the theory of photoelectric detection of light fluctuations as an eigenvalue of the operator which represents the annihilation of a photon at the space-time point $(\mathbf{r}, t)$.

Consider now a quasi-monochromatic light represented by a statistically stationary ensemble of analytic signals $V(\mathbf{r}, t)$. By quasi-monochromatic light we mean, as mentioned earlier, that the effective bandwidth of the light, i.e. the effective width $\Delta\nu$ of its power spectrum at each point $\mathbf{r}$, is small compared with its mean frequency $\bar{\nu}$:

$$\frac{\Delta\nu}{\bar{\nu}} \ll 1. \qquad (4.3\text{–}1)$$

One may regard such a field as being represented at each point by an ensemble of quasi-monochromatic signals (see Section 3.1.2) centered at the frequency $\bar{\nu}$.

Because of the high frequency of optical vibrations, V cannot be measured as a function of time with commonly available optical detectors. Optical periods are of the order of 10^{-15} s, whereas photoelectric detectors have typically resolving times of the order of 10^{-9} s, though special techniques exist by means of which still shorter resolving times may be achieved. However, although one cannot study the rapid time variations of the field, one can make measurements of the *correlations* of the field at two or more space-time points. Let us consider such measurements when the optical field is a well-collimated quasi-monochromatic light beam.

Suppose that the light vibrations at points $P_1(\mathbf{r}_1)$ and $P_2(\mathbf{r}_2)$ in the beam are isolated by placing an opaque screen $\mathscr{A}$ across the beam, with pinholes at the two points, and that we observe the intensity distribution resulting from the superposition of the light emerging from the two pinholes, on a screen $\mathscr{B}$ at a distance d from $\mathscr{A}$ (see Fig. 4.6). We assume that d is large compared with optical wavelengths. The instantaneous field at a point P on the screen $\mathscr{B}$ is, to a good approximation, given by

$$V(\mathbf{r}, t) = K_1 V(\mathbf{r}_1, t - t_1) + K_2 V(\mathbf{r}_2, t - t_2), \qquad (4.3\text{–}2)$$

where

$$t_1 = R_1/c, \quad t_2 = R_2/c, \qquad (4.3\text{–}2a)$$

are the times needed for the light to travel from P_1 to P and from P_2 to P respectively, c is the speed of light in vacuo and K_1 and K_2 are constant factors that depend on the size of the pinholes and on the geometry. It follows from elementary diffraction theory[‡] that K_1 and K_2 are purely imaginary numbers.

The instantaneous intensity $I(\mathbf{r}, t)$, at the point $P(\mathbf{r})$ at time t may be defined

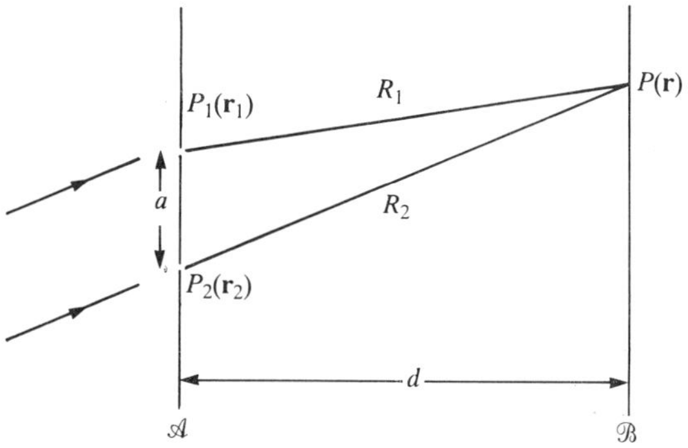

Fig. 4.6 Notation relating to Young's interference experiment from which the second-order correlation functions of a light beam may be determined.

‡ See, for example, Born and Wolf (1980), Sec. 8.3.
When the beam is incident on the plane $\mathscr{A}$ of the pinholes along or close to the direction normal to $\mathscr{A}$ and the angles that the diffracted directions P_1P and P_2P make with the normal to $\mathscr{A}$ are also small, then $K_1 \approx K_2 \approx -\mathrm{i}(\mathrm{d}\mathscr{A})/(\bar{\lambda}R)$, where R is the distance $\overline{P_1P} \approx \overline{P_2P}$ and $\mathrm{d}\mathscr{A}$ is the area of each aperture.

by the formula§

$$I(\mathbf{r}, t) = V^*(\mathbf{r}, t)V(\mathbf{r}, t). \tag{4.3–3}$$

From Eqs. (4.3–2) and (4.3–3) it follows that

$$I(\mathbf{r}, t) = |K_1|^2 I_1(\mathbf{r}_1, t - t_1) + |K_2|^2 I_2(\mathbf{r}_2, t - t_2) \\ + 2\,\mathrm{Re}\,\{K_1^* K_2 V^*(\mathbf{r}_1, t - t_1)V(\mathbf{r}_2, t - t_2)\}, \tag{4.3–4}$$

where Re denotes the real part. If we take the average of $I(\mathbf{r}, t)$ over an ensemble of different realizations of the field and denote this ensemble average by $\langle \ldots \rangle_{\mathrm{e}}$, we obtain the formula

$$\langle I(\mathbf{r}, t)\rangle_{\mathrm{e}} = |K_1|^2 \langle I(\mathbf{r}_1, t - t_1)\rangle_{\mathrm{e}} + |K_2|^2 \langle I(\mathbf{r}_2, t - t_2)\rangle_{\mathrm{e}} \\ + 2\,\mathrm{Re}\,\{K_1^* K_2 \Gamma(\mathbf{r}_1, \mathbf{r}_2, t - t_1, t - t_2)\}, \tag{4.3–5}$$

where

$$\Gamma(\mathbf{r}_1, \mathbf{r}_2; t_1, t_2) = \langle V^*(\mathbf{r}_1, t_1)V(\mathbf{r}_2, t_2)\rangle_{\mathrm{e}} \tag{4.3–6}$$

and

$$\langle I(\mathbf{r}_j, t_j)\rangle_{\mathrm{e}} = \langle V^*(\mathbf{r}_j, t_j)V(\mathbf{r}_j, t_j)\rangle_{\mathrm{e}} = \Gamma(\mathbf{r}_j, \mathbf{r}_j, t_j, t_j), \quad (j = 1, 2). \tag{4.3–7}$$

The function $\Gamma(\mathbf{r}_1, \mathbf{r}_2, t_1, t_2)$, defined by Eq. (4.3–6) will be recognized as the cross-correlation function of the random processes $V(\mathbf{r}_1, t)$ and $V(\mathbf{r}_2, t)$ (see Section 2.4.4). It represents, in the present context, the correlation that exists between the light vibrations at the pinholes at P_1 and P_2, at times t_1 and t_2 respectively. The quantity $\langle I(\mathbf{r}_j, t_j)\rangle_{\mathrm{e}}$ represents the (ensemble) averaged intensity of the light at the pinhole P_j at the time t_j $(j = 1, 2)$. We shall see later [Eq. (4.3–19)] that, under usual circumstances, the third term on the right-hand side of Eq. (4.3–5) gives rise to a sinusoidal modulation of the averaged intensity $\langle I(\mathbf{r}, t)\rangle_{\mathrm{e}}$ with $\mathbf{r}$.

Usually one is concerned with *stationary fields*, in which case all the ensemble averages are independent of the origin of time; moreover the field is as a rule also *ergodic*. Under these circumstances, as we have learned in Section 2.2.2, the ensemble averages become time-independent and may be replaced by the corresponding time averages.

Let us denote the time average of a stationary random process $f(t)$ by $\langle f(t)\rangle_{\mathrm{t}}$, i.e.

$$\langle f(t)\rangle_{\mathrm{t}} = \operatorname*{Lim}_{T\to\infty} \frac{1}{2T}\int_{-T}^{T} f(t)\,\mathrm{d}t. \tag{4.3–8}$$

Then the 'ensemble cross-correlation function' $\Gamma(\mathbf{r}_1, \mathbf{r}_2; \mathrm{t}_1, \mathrm{t}_2)$ may be replaced by the corresponding temporal cross-correlation function and this function depends

§ The instantaneous intensity $I(\mathbf{r}, t)$, defined by Eq. (4.3–3), is not strictly proportional to the square of a typical realization of the real field variable $V^{(\mathrm{r})}(\mathbf{r}, t)$. However, with the help of the envelope representation of real quasi-monochromatic signals the average of the square of a realization of $V^{(\mathrm{r})}(\mathbf{r}, t)$, taken over a time interval of a few mean periods of the light vibrations, may readily be shown to be equal to $2I(\mathbf{r}, t)$[cf. Eq. (3.1–43)].

on the two time arguments only through their difference $t_2 - t_1$. Hence if we set

$$\Gamma(\mathbf{r}_1, \mathbf{r}_2, \tau) = \langle V^*(\mathbf{r}_1, t)V(\mathbf{r}_2, t+\tau)\rangle_t = \lim_{T\to\infty} \frac{1}{2T}\int_{-T}^{T} V^*(\mathbf{r}_1, t)V(\mathbf{r}_2, t+\tau)\,dt, \tag{4.3-9}$$

the expression (4.3–5) for the averaged intensity at P becomes, under the assumption of stationarity and ergodicity,

$$\langle I(\mathbf{r}, t)\rangle = |K_1|^2\langle I(\mathbf{r}_1, t)\rangle + |K_2|^2\langle I(\mathbf{r}_2, t)\rangle + 2\,\mathrm{Re}\,\{K_1^*K_2\Gamma(\mathbf{r}_1, \mathbf{r}_2, t_1 - t_2)\}, \tag{4.3-10}$$

where we have omitted the subscripts t or e for the two types of averages, since it is now unnecessary to distinguish between them.

We note that, if the last term on the right-hand side of Eq. (4.3–10) does not vanish, the averaged intensity $\langle I(\mathbf{r}, t)\rangle$ is not equal to the sum of the (averaged) intensities of the two beams which reach the point P of observation from the two pinholes. It differs from their sum by the term $2\,\mathrm{Re}\,\{K_1^*K_2\Gamma(\mathbf{r}_1, \mathbf{r}_2, t_1 - t_2)\}$. Since $K_1 \neq 0$, $K_2 \neq 0$ it follows that if $\Gamma \neq 0$, the superposition of the two beams will give rise to *interference*.

The cross-correlation function $\Gamma(\mathbf{r}_1, \mathbf{r}_2, \tau)$ is known as the *mutual coherence function* (Wolf, 1955) and is the central quantity of the elementary theory of optical coherence. From the definition (4.3–3) of the instantaneous intensity $I(\mathbf{r}, t)$ and from the definition (4.3–9) of the mutual coherence function $\Gamma(\mathbf{r}_1, \mathbf{r}_2, \tau)$, it follows at once that $\Gamma(\mathbf{r}, \mathbf{r}, 0)$ represents the averaged intensity at the point $\mathbf{r}$:

$$\langle I(\mathbf{r}, t)\rangle = \langle V^*(\mathbf{r}, t)V(\mathbf{r}, t)\rangle = \Gamma(\mathbf{r}, \mathbf{r}, 0). \tag{4.3-11}$$

It is convenient to normalize the mutual coherence function by setting

$$\gamma(\mathbf{r}_1, \mathbf{r}_2, \tau) = \frac{\Gamma(\mathbf{r}_1, \mathbf{r}_2, \tau)}{[\Gamma(\mathbf{r}_1, \mathbf{r}_1, 0)]^{1/2}[\Gamma(\mathbf{r}_2, \mathbf{r}_2, 0)]^{1/2}} \tag{4.3-12a}$$

$$= \frac{\Gamma(\mathbf{r}_1, \mathbf{r}_2, \tau)}{[\langle I(\mathbf{r}_1, t)\rangle]^{1/2}[\langle I(\mathbf{r}_2, t)\rangle]^{1/2}}. \tag{4.3-12b}$$

For reasons that will become apparent shortly, $\gamma(\mathbf{r}_1, \mathbf{r}_2, \tau)$ is called the *complex degree of coherence* of the light vibrations at the points $P_1(\mathbf{r}_1)$ and $P_2(\mathbf{r}_2)$. According to the inequality (2.4–47), which the cross-correlation function of any two jointly stationary random processes satisfies, we have

$$0 \leq |\gamma(\mathbf{r}_1, \mathbf{r}_2, \tau)| \leq 1 \tag{4.3-13}$$

for all values of the arguments $\mathbf{r}_1$, $\mathbf{r}_2$, and τ of γ.

The first two terms on the right-hand side of Eq. (4.3–10) have simple meanings. To see this suppose that the pinhole P_2 is closed, so that only the light from the pinhole P_1 reaches the plane $\mathcal{B}$ of observation. In this case $K_2 \equiv 0$ and it is clear from Eq. (4.3–10) that in this case

$$|K_1|^2\langle I(\mathbf{r}_1, t)\rangle \equiv \langle I^{(1)}(\mathbf{r}, t)\rangle \tag{4.3-14a}$$

represents the averaged intensity of the light at the point $P(\mathbf{r})$, which reaches this point from the pinhole P_1, only. Similarly

$$|K_2|^2\langle I(\mathbf{r}_2, t)\rangle \equiv \langle I^{(2)}(\mathbf{r}, t)\rangle \tag{4.3–14b}$$

represents the averaged intensity of the light at the point $P(\mathbf{r})$ which reaches this point from the pinhole P_2 only.

The last term on the right-hand side of (4.3–10) may readily be expressed in terms of $\langle I^{(1)}\rangle$, $\langle I^{(2)}\rangle$ and γ. We have from Eqs. (4.3–12b), (4.3–14a) and (4.3–14b), if we also use the formula (4.3–2) and recall that the factors K_1 and K_2 are purely imaginary,

$$K_1^*K_2\Gamma(\mathbf{r}_1, \mathbf{r}_2, t_1 - t_2) = [\langle I^{(1)}(\mathbf{r}, t)\rangle]^{1/2}[\langle I^{(2)}(\mathbf{r}, t)\rangle]^{1/2}\gamma[\mathbf{r}_1, \mathbf{r}_2, (R_1 - R_2)/c].$$

Using this relation and Eqs. (4.3–14a) and (4.3–14b) in Eq. (4.3–10), we finally obtain the following expression for the averaged intensity of the light at P when the light reaches the plane of observation $\mathcal{B}$ via both the pinholes:

$$\langle I(\mathbf{r}, t)\rangle = \langle I^{(1)}(\mathbf{r}, t)\rangle + \langle I^{(2)}(\mathbf{r}, t)\rangle + 2[\langle I^{(1)}(\mathbf{r}, t)\rangle]^{1/2}[\langle I^{(2)}(\mathbf{r}, t)\rangle]^{1/2}\,\mathrm{Re}\,\gamma[\mathbf{r}_1, \mathbf{r}_2, (R_1 - R_2)/c]. \tag{4.3–15}$$

We see at once from Eq. (4.3–15) that measurements of the averaged intensities $\langle I(\mathbf{r}, t)\rangle$, $\langle I^{(1)}(\mathbf{r}, t)\rangle$ and $\langle I^{(2)}(\mathbf{r}, t)\rangle$ make it possible to determine the real part of the complex degree of coherence $\gamma(\mathbf{r}_1, \mathbf{r}_2, \tau)$. Moreover, it is apparent from Eq. (4.3–12b) that if, in addition, measurements of the averaged intensities $\langle I(\mathbf{r}_1, t)\rangle$ and $\langle I(\mathbf{r}_2, t)\rangle$ of the light at the two pinholes are made, the real part of the mutual coherence function $\Gamma(\mathbf{r}_1, \mathbf{r}_2, \tau)$ may then be determined.

Although, as we just saw, direct measurements of the averaged intensities provide information about the real parts of the correlation functions Γ and γ only, the imaginary parts could, in principle, be determined from the knowledge of their real parts for all values of the parameter τ. This may be seen as follows. Since $V(\mathbf{r}_1, t)$ and $V(\mathbf{r}_2, t)$ are analytic signals, it follows from Theorem I of Section 3.1.3 that their cross-correlation function, i.e. the mutual coherence function $\Gamma(\mathbf{r}_1, \mathbf{r}_2, \tau)$, is also an analytic signal and hence the real part, $\mathrm{Re}\,\Gamma$, and the imaginary part $\mathrm{Im}\,\Gamma$, of Γ are coupled by the Hilbert transform relations‡

$$\left.\begin{aligned}\mathrm{Im}\,\Gamma(\mathbf{r}_1, \mathbf{r}_2, \tau) &= \frac{1}{\pi}P\int_{-\infty}^{\infty}\frac{\mathrm{Re}\,\Gamma(\mathbf{r}_1, \mathbf{r}_2, \tau')}{\tau' - \tau}\,\mathrm{d}\tau',\\ \mathrm{Re}\,\Gamma(\mathbf{r}_1, \mathbf{r}_2, \tau) &= -\frac{1}{\pi}P\int_{-\infty}^{\infty}\frac{\mathrm{Im}\,\Gamma(\mathbf{r}_1, \mathbf{r}_2, \tau')}{\tau' - \tau}\,\mathrm{d}\tau',\end{aligned}\right\} \tag{4.3–16}$$

where P denotes the Cauchy principal value of the integrals at $\tau' = \tau$. Moreover,

‡ According to Theorems III and V, of Section 3.1.3, the real and imaginary parts of Γ may be expressed in terms of the real field $V^{(r)}(\mathbf{r}, t)$ and its conjugate field $V^{(i)}(\mathbf{r}, t)$ (the Hilbert transform of $V^{(r)}$) as follows.

$$\mathrm{Re}\,\Gamma = \tfrac{1}{2}\langle V^{(r)}(\mathbf{r}_1, t)V^{(r)}(\mathbf{r}_2, t + \tau)\rangle = \tfrac{1}{2}\langle V^{(i)}(\mathbf{r}_1, t)V^{(i)}(\mathbf{r}_2, t + \tau)\rangle,$$

$$\mathrm{Im}\,\Gamma = \tfrac{1}{2}\langle V^{(r)}(\mathbf{r}_1, t)V^{(i)}(\mathbf{r}_2, t + \tau)\rangle = -\tfrac{1}{2}\langle V^{(i)}(\mathbf{r}_1,t)V^{(r)}(\mathbf{r}_2, t + \tau)\rangle.$$

Since $V = \frac{1}{2}[V^{(r)} + \mathrm{i}V^{(i)}]$ it follows at once from these relations that $\langle V^{(r)2}(\mathbf{r}, t)\rangle = \langle V^{(i)2}(\mathbf{r}, t)\rangle = 2\langle V^*(\mathbf{r}, t)V(\mathbf{r}, t)\rangle = 2\langle I(\mathbf{r}, t)\rangle$ and that $\langle V^{(r)}(\mathbf{r}, t)V^{(i)}(\mathbf{r}, t)\rangle = 0$.

since the complex degree of coherence γ differs from the mutual coherence function Γ by a multiplicative factor that does not depend on τ, γ is also an analytic signal and hence its real and imaginary parts are likewise coupled by the Hilbert transform relations.

However, it is the absolute value of the complex degree of coherence γ, rather than its real part, that is a true measure of the 'sharpness' of the interference effects to which superposition of the two beams give rise. To see this let us examine more closely the expression (4.3–15) for the averaged intensity $\langle I(\mathbf{r}, t)\rangle$ of the light in the plane $\mathscr{B}$ of observation. Let us set

$$\gamma(\mathbf{r}_1, \mathbf{r}_2, \tau) = |\gamma(\mathbf{r}_1, \mathbf{r}_2, \tau)|e^{i[\alpha(\mathbf{r}_1,\mathbf{r}_2,\tau)-2\pi\bar{\nu}\tau]}, \tag{4.3–17}$$

where

$$\alpha(\mathbf{r}_1, \mathbf{r}_2, \tau) = \arg\gamma(\mathbf{r}_1, \mathbf{r}_2, \tau) + 2\pi\bar{\nu}\tau. \tag{4.3–18}$$

On substituting from Eq. (4.3–17) into Eq. (4.3–15), we obtain the following expression for $\langle I(\mathbf{r}, t)\rangle$:

$$\begin{aligned}\langle I(\mathbf{r}, t)\rangle = {} & \langle I^{(1)}(\mathbf{r}, t)\rangle + \langle I^{(2)}(\mathbf{r}, t)\rangle \\ & + 2[\langle I^{(1)}(\mathbf{r}, t)\rangle]^{1/2}[\langle I^{(2)}(\mathbf{r}, t)\rangle]^{1/2}|\gamma[\mathbf{r}_1, \mathbf{r}_2, (R_1 - R_2)/c]| \\ & \times \cos\{\alpha[\mathbf{r}_1, \mathbf{r}_2, (R_1 - R_2)/c] - \delta\},\end{aligned} \tag{4.3–19}$$

where

$$\delta = \frac{2\pi\bar{\nu}}{c}(R_1 - R_2) = \bar{k}(R_1 - R_2), \tag{4.3–20}$$

with

$$\bar{k} = \frac{2\pi\bar{\nu}}{c} = \frac{2\pi}{\bar{\lambda}}, \tag{4.3–21}$$

$\bar{\lambda}$ denoting the mean wavelength of the light. Now since we assumed that the plane of observation $\mathscr{B}$ is many wavelengths away from the plane $\mathscr{A}$ of the pinholes, the averaged intensities $\langle I^{(1)}\rangle$ and $\langle I^{(2)}\rangle$ of the two beams will change slowly with the position $P(\mathbf{r})$, on the screen $\mathscr{B}$. Moreover, since we also assumed that the light is quasi-monochromatic, it follows from the properties of the envelope representation (see Section 3.1.2) that $|\gamma|$ and α will also change slowly over any part of the plane of observation $\mathscr{B}$ for which the change in the distance $R_1 - R_2$ (the path delay $[PP_1] - [PP_2]$ introduced between the light emerging from the two pinholes) is small compared to the coherence length of the light. Hence the changes in $|\gamma|$ and α due to the changes in the argument $(R_1 - R_2)/c$ on the right-hand side of Eq. (4.3–19) may be neglected, provided that

$$||R_1 - R_2|_{P'} - |R_1 - R_2|_P| \ll \frac{c}{\Delta\nu}, \tag{4.3–22}$$

where $|R_1 - R_2|_P$ represents the difference in the distances of the points P from the two pinholes, $|R_1 - R_2|_{P'}$ represents this difference for a neighboring point P' in the plane $\mathscr{B}$ and $\Delta\nu$ denotes the effective bandwidth of the light. However, the cosine term on the right-hand side of Eq. (4.3–19) will change rapidly with the position $\mathbf{r}$ of the point P on the screen $\mathcal{B}$ because of the presence of the term δ. According to (4.3–20) this term is inversely proportional to the (small) mean wavelength $\bar{\lambda}$ of the light. Hence *over a sufficiently small portion of the plane of observation* $\mathcal{B}$, *the averaged intensity* $\langle I(\mathbf{r}, t)\rangle$ *will vary nearly sinusoidally with position, provided that* $|\gamma| \neq 0$.

The usual measure of the sharpness of interference fringes is the so-called *visibility*, a concept introduced by Michelson (1890). The visibility $\mathcal{V}(\mathbf{r})$ at a point $P(\mathbf{r})$ in an interference pattern is defined by the formula

$$\mathcal{V}(\mathbf{r}) = \frac{\langle I\rangle_{\text{max}} - \langle I\rangle_{\text{min}}}{\langle I\rangle_{\text{max}} + \langle I\rangle_{\text{min}}}, \tag{4.3–23}$$

where $\langle I\rangle_{\text{max}}$ and $\langle I\rangle_{\text{min}}$ represent the maximum and the minimum values that the averaged intensity assumes in the immediate neighborhood of P. Now from Eq. (4.3–19) we have, to a good approximation,

$$\langle I\rangle_{\text{max}} = \langle I^{(1)}(\mathbf{r}, t)\rangle + \langle I^{(2)}(\mathbf{r}, t)\rangle + 2[\langle I^{(1)}(\mathbf{r}, t)\rangle]^{1/2}[\langle I^{(2)}(\mathbf{r}, t)\rangle]^{1/2}|\gamma[\mathbf{r}_1, \mathbf{r}_2, (R_1 - R_2)/c]|, \tag{4.3–24a}$$

$$\langle I\rangle_{\text{min}} = \langle I^{(1)}(\mathbf{r}, t)\rangle + \langle I^{(2)}(\mathbf{r}, t)\rangle - 2[\langle I^{(1)}(\mathbf{r}, t)\rangle]^{1/2}[\langle I^{(2)}(\mathbf{r}, t)\rangle]^{1/2}|\gamma[\mathbf{r}_1, \mathbf{r}_2, (R_1 - R_2)/c]|, \tag{4.3–24b}$$

and hence (4.3–23) becomes

$$\mathcal{V}(\mathbf{r}) = 2\left[\eta(\mathbf{r}) + \frac{1}{\eta(\mathbf{r})}\right]^{-1} |\gamma[\mathbf{r}_1, \mathbf{r}_2, (R_1 - R_2)/c]|, \tag{4.3–25}$$

where

$$\eta(\mathbf{r}) = \left[\frac{\langle I^{(1)}(\mathbf{r}, t)\rangle}{\langle I^{(2)}(\mathbf{r}, t)\rangle}\right]^{1/2}. \tag{4.3–26}$$

In particular when the averaged intensities of the two beams at P are equal, as is frequently the case, then $\eta = 1$ and Eq. (4.3–25) reduces to

$$\mathcal{V}(\mathbf{r}) = |\gamma[\mathbf{r}_1, \mathbf{r}_2, (R_1 - R_2)/c]|, \tag{4.3–25a}$$

i.e. $|\gamma|$ is then just equal to the visibility of the fringes. The behavior of the averaged intensity in the plane of observation is shown in Fig. 4.7, under the assumption that the averaged intensities of the two interfering beams are equal. According to Eq. (4.3–13), $0 \leqslant |\gamma| \leqslant 1$. We see from the figure that, in the extreme case $|\gamma| = 1$, the averaged intensity around any point P in the fringe pattern then undergoes the greatest possible periodic variation, between the values $4\langle I^{(1)}\rangle$ and zero. In the other extreme case, $\gamma = 0$, no interference fringes are formed at all; the averaged intensity distribution in the neighborhood of P is then essentially uniform. These are the cases that traditionally are said to represent *complete coherence* (more precisely complete second-order coherence) and *complete* (second-order) *incoherence*, respectively. The intermediate values ($0 < |\gamma| < 1$) characterize *partial coherence*; the averaged intensity distribution in the fringe pattern around P then exhibits periodic variation between the values $2(1 + |\gamma|)\langle I^{(1)}\rangle$ and $2(1 - |\gamma|)\langle I^{(1)}\rangle$.

The argument (phase) of γ also has a simple significance. It follows from Eqs.

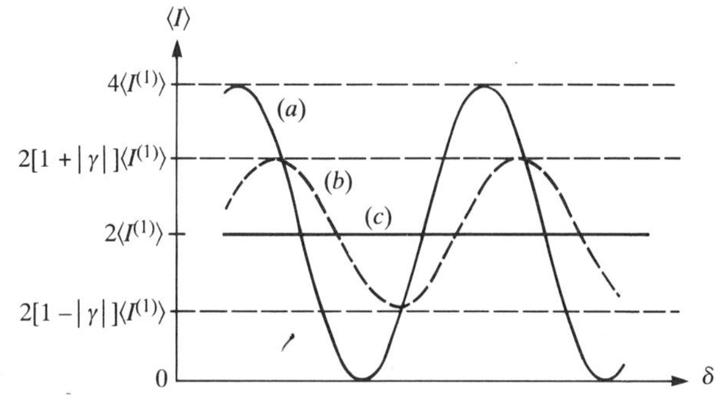

Fig. 4.7 Behavior of the averaged intensity $\langle I \rangle$ around a point $P(\mathbf{r})$ in the plane of observation $\mathscr{B}$, produced by the superposition of two quasi-monochromatic beams of equal averaged intensity $\langle I^{(1)} \rangle$ in Young's interference experiment illustrated in Fig. 4.6:

$$\langle I(\mathbf{r}, t) \rangle = 2\langle I^{(1)}(\mathbf{r}, t)\rangle[1 + |\gamma[\mathbf{r}_1, \mathbf{r}_2, (R_1 - R_2)/c]| \cos\{\alpha[\mathbf{r}_1, \mathbf{r}_2, (R_1 - R_2)/c] - \delta\}].$$

The curves illustrate the following three cases: (a) $|\gamma| = 1$: complete (second-order) coherence, (b) $0 < |\gamma| < 1$: partial (second-order) coherence, (c) $\gamma = 0$: complete (second-order) incoherence.

(4.3–18) and (4.3–19), that the locations of the maxima of the averaged intensity in the fringe pattern are, to a high degree of accuracy, given by

$$\begin{aligned} \arg \gamma[\mathbf{r}_1, \mathbf{r}_2, (R_1 - R_2)/c] &\equiv \alpha[\mathbf{r}_1, \mathbf{r}_2, (R_1 - R_2)/c] - 2\pi\bar{\nu}(R_1 - R_2)/c \\ &= 2m\pi \qquad (m = 0, \pm 1, \pm 2, \ldots). \end{aligned} \tag{4.3–27}$$

The positions of the maxima, given by (4.3–27), coincide with those which would be obtained if the two pinholes were illuminated by strictly monochromatic light of wavelength $\bar{\lambda}$ and the phase of the vibrations of the light at P_1 were retarded with respect to the phase of the light vibrations at P_2 by the amount $\alpha[\mathbf{r}_1, \mathbf{r}_2, (R_1 - R_2/c]$. Hence, for the purpose of describing the interference effects near the point P, we may regard $\alpha[\mathbf{r}_1, \mathbf{r}_2, (R_1 - R_2)/c]$ as representing the 'effective retardation' of the light at P_1 with respect to the light at P_2. Equation (4.3–27) shows that the argument of γ may be determined by measuring the position of the maxima of the fringe pattern. More precisely, let us use the fact that, with monochromatic illumination of the two pinholes, a phase retardation of amount 2π leads to a displacement of the interference pattern in the direction parallel to P_1P_2 by an amount $a\bar{\lambda}/d$, [Born and Wolf, 1980, Sec. 10.4, Eq. (6)], a being the distance between P_1 and P_2 and d the distance between the planes $\mathscr{A}$ and $\mathscr{B}$. We see that *the fringes* in our experiment *are displaced relative to the fringes that would be formed with monochromatic light of wavelength* $\bar{\lambda}$ *and with co-phasal illumination of the pinholes* by an amount

$$\begin{aligned} x &= \frac{\bar{\lambda}}{2\pi}\left(\frac{a}{d}\right)\alpha[\mathbf{r}_1, \mathbf{r}_2, (R_1 - R_2)/c] \\ &= \frac{a}{d}\left\{(R_1 - R_2) + \frac{\bar{\lambda}}{2\pi} \arg \gamma[\mathbf{r}_1, \mathbf{r}_2, (R_1 - R_2)/c]\right\} \end{aligned} \tag{4.3–28}$$

in the direction parallel to the line joining the two pinholes.

We have seen that, on the one hand, γ is a measure of the correlation of the complex field at two points P_1 and P_2 [Eq. (4.3–12)]. On the other hand, it is a measure of the sharpness of the fringes [Eq. (4.3–25)], obtained by superposing the beams propagated from these points, and it also specifies the location of the fringe maxima [Eq. (4.3–28)].‡ Since the sharpness of the fringes may be regarded as a manifestation of coherence between interfering beams, it is evident that the term '*complex degree of coherence*' of the field at points P_1 and P_2 conveys the significance of γ in interference experiments. This term is, however, somewhat ambiguous, since γ depends not only on the location of the two points P_1 and P_2 but also on the delay $\tau = (R_1 - R_2)/c$. It would be more appropriate to reserve the term 'complex degree of coherence' for the quantity $\gamma(\mathbf{r}_1, \mathbf{r}_2, \tau_0)$, where τ_0 is that value of τ which maximizes $|\gamma(\mathbf{r}_1, \mathbf{r}_2, \tau)|$, with $\mathbf{r}_1$ and $\mathbf{r}_2$ kept fixed. However, when the light is quasi-monochromatic, as we assumed, and when the interference fringes are observed in a region in the plane $\mathscr{B}$ where the visibility is maximum, as is usually the case, this distinction is not very significant. For, as is evident from the discussion following Eq. (4.3–21), $|\gamma(\mathbf{r}_1, \mathbf{r}_2, \tau)|$ and $\alpha(\mathbf{r}_1, \mathbf{r}_2, \tau) = \arg \gamma(\mathbf{r}_1, \mathbf{r}_2, \tau) + 2\pi\bar{\nu}\tau$ change very slowly with τ and remain sensibly constant over any τ-range which is short compared to the coherence time $(1/\Delta\nu)$ of the light. Hence

$$\gamma(\mathbf{r}_1, \mathbf{r}_2, \tau_1) \approx \gamma(\mathbf{r}_1, \mathbf{r}_2, \tau_2)\, \mathrm{e}^{-2\pi i\bar{\nu}(\tau_1-\tau_2)} \qquad (4.3\text{–}29)$$

for any two values of τ_1 and τ_2 such that

$$|\tau_1 - \tau_2| \ll \frac{1}{\Delta\nu}. \qquad (4.3\text{–}30)$$

Thus over a τ-range that satisfies the condition (4.3–30) γ (and also Γ) is effectively periodic in τ, with period equal to the mean period $T = 1/\bar{\nu}$ of the light; this result applies, in particular, to the τ-range around the value $\tau_{\max}$ that maximizes $|\gamma|$.

In practice one is often interested in interference effects under conditions close to geometrical symmetry, so that only those values of the correlation functions $\Gamma(\mathbf{r}_1, \mathbf{r}_2, \tau)$ and $\gamma(\mathbf{r}_1, \mathbf{r}_2, \tau)$ play a role for which the τ-argument has values close to zero. According to Eq. (4.3–29) and the analogous relation for Γ, we may then approximate the correlation functions by the expression:

$$\Gamma(\mathbf{r}_1, \mathbf{r}_2, \tau) \approx J(\mathbf{r}_1, \mathbf{r}_2)\, \mathrm{e}^{-2\pi i\bar{\nu}\tau}, \qquad (4.3\text{–}31)$$

$$\gamma(\mathbf{r}_1, \mathbf{r}_2, \tau) \approx j(\mathbf{r}_1, \mathbf{r}_2)\, \mathrm{e}^{-2\pi i\bar{\nu}\tau}, \qquad (4.3\text{–}32)$$

provided that

$$|\tau| \ll 1/\Delta\nu. \qquad (4.3\text{–}33)$$

The quantities $J(\mathbf{r}_1, \mathbf{r}_2)$ and $j(\mathbf{r}_1, \mathbf{r}_2)$ are the '*equal-time correlation functions*'

$$J(\mathbf{r}_1, \mathbf{r}_2) \equiv \Gamma(\mathbf{r}_1, \mathbf{r}_2, 0) = \langle V^*(\mathbf{r}_1, t) V(\mathbf{r}_2, t)\rangle, \qquad (4.3\text{–}34)$$

‡ The experimental arrangement (essentially that of a Young's interference experiment) by means of which we elucidated the significance of the complex degree of coherence $\gamma(\mathbf{r}_1, \mathbf{r}_2, \tau)$ is not appropriate when $\mathbf{r}_2 = \mathbf{r}_1$. In that case one deals with the *complex degree of self-coherence* $\gamma(\mathbf{r}, \mathbf{r}, \tau)$ whose significance may be elucidated by means of the Michelson interferometer. This situation will be discussed in Section 7.3.

$$j(\mathbf{r}_1, \mathbf{r}_2) \equiv \gamma(\mathbf{r}_1, \mathbf{r}_2, 0) = \frac{J(\mathbf{r}_1, \mathbf{r}_2)}{[J(\mathbf{r}_1, \mathbf{r}_1)]^{1/2}[J(\mathbf{r}_2, \mathbf{r}_2)]^{1/2}}. \qquad (4.3\text{–}35)$$

The equal-time correlation function $J(\mathbf{r}_1, \mathbf{r}_2)$ is called the *mutual intensity* of the light vibrations at the points $P_1(\mathbf{r}_1)$ and $P_2(\mathbf{r}_2)$ and $j(\mathbf{r}_1, \mathbf{r}_2)$, just like $\gamma(\mathbf{r}_1, \mathbf{r}_2, \tau)$, is usually known as their *complex degree of coherence*. These two less general correlation functions are adequate for the analysis of many coherence problems of instrumental optics.

Since the correlation effects that we discussed in this section are characterized by a correlation function that depends only on *two* space-time points, we will refer to them as *coherence effects of the second order*. A general classification of coherence effects will be given in Chapter 8.

It is evident that the phenomena of temporal and spatial coherence that we briefly discussed in rough qualitative terms in Sections 4.2.1 and 4.2.2 are characterized by $\Gamma(\mathbf{r}_1, \mathbf{r}_1, \tau)$ (known sometimes as the *self-coherence function*) and by $\Gamma(\mathbf{r}_1, \mathbf{r}_2, 0)$ [or, more generally, by $\Gamma(\mathbf{r}_1, \mathbf{r}_2, \tau_0)$ where τ_0 is a constant], respectively. In the first case the dependence of the correlation on the parameter τ is crucial, with the points P_1 and P_2 being coincident and kept fixed; in the second case the dependence on the position of the two points is crucial, while the time delay τ is kept essentially fixed; more precisely it is restricted to a range that is short compared to $1/\Delta\nu$. However, only in very simple cases can one sharply distinguish between temporal and spatial coherence. In general these two types of coherence phenomena are not independent of each other, since, as we will learn in Sections 4.4 and 4.6, the dependence of the mutual coherence function $\Gamma(\mathbf{r}_1, \mathbf{r}_2, \tau)$ on the position variables $\mathbf{r}_1$ and $\mathbf{r}_2$ and on the temporal variable τ is coupled. We will soon learn (Section 4.3.2) that the mutual coherence function also provides information about two-point correlations in the field for any particular frequency component of the light.

In relating the correlation functions Γ and γ to results of measurements we have, of course, implicitly assumed that the detecting apparatus measures the average of the instantaneous intensity $I(\mathbf{r}, t) = V^*(\mathbf{r}, t)V(\mathbf{r}, t)$. In practice this will almost certainly be the case if V is identified with the appropriate field variable and if the detector performs a time average over a time interval that is long compared to the time scales of the fluctuating field, i.e. long compared to the mean period and to the coherence time of the light. (Alternatively, the ensemble average of the intensity may be found from a succession of measurements, whether the measurement times are long or short). Under these conditions the measured time average may be assumed to differ negligibly from the average over an infinitely long time span, defined by Eq. (4.3–8). If these conditions are not satisfied, other types of interference effects (transient interference) may take place. Such effects are discussed, for example, in Mandel and Wolf (1965, Sec. 7).

We conclude this section by noting a number of properties of the mutual coherence function, which follow at once from the general results relating to the cross-correlation function of two jointly stationary complex random processes. Corresponding to Eq. (2.4–32) we have the property that

$$\Gamma(\mathbf{r}_2, \mathbf{r}_1, \tau) = \Gamma^*(\mathbf{r}_1, \mathbf{r}_2, -\tau). \qquad (4.3\text{–}36)$$

The non-negative definiteness condition (2.4–34) implies that, with any n pairs of values of the space-time variables $(\mathbf{r}_1, t_1)$, $(\mathbf{r}_2, t_2)$, . . . , $(\mathbf{r}_n, t_n)$, where n is any arbitrary positive integer, and with any n real or complex numbers a_1, a_2, . . . , a_n,

$$\sum_{j=1}^{n}\sum_{k=1}^{n} a_j^* a_k \Gamma(\mathbf{r}_j, \mathbf{r}_k, t_k - t_j) \geqslant 0. \tag{4.3–37}$$

In particular, if we choose $n = 1$, Eq. (4.3–37) implies that $\Gamma(\mathbf{r}_1, \mathbf{r}_1, 0) \geqslant 0$ for any point $\mathbf{r}_1$, a result that is obvious from the definition of the mutual coherence function. If we choose $n = 2$, Eq. (4.3–37) implies that $|\Gamma(\mathbf{r}_1, \mathbf{r}_2, \tau)| \leqslant [\Gamma(\mathbf{r}_1, \mathbf{r}_1, 0)]^{1/2}[\Gamma(\mathbf{r}_2, \mathbf{r}_2, 0)]^{1/2}$, a result that we have used in normalizing the mutual coherence function [Eqs. (4.3–12) and (4.3–13)].

4.3.2 Second-order correlations in the space-frequency domain. The cross-spectral density and the spectral degree of coherence

In our general study of stochastic processes we have learned (Section 2.4) that an important concept is that of the cross-spectral density [Eq. (2.4–35)]. We will now consider the cross-spectral density in the context of optical coherence theory.

Let $V(\mathbf{r}, t)$ again denote the analytic signal, representing the fluctuating optical field at the space-time point $(\mathbf{r}, t)$. We assume that the optical field is stationary, at least in the wide sense, and ergodic, and we represent $V(\mathbf{r}, t)$ as a Fourier integral (in the sense of the theory of generalized functions) with respect to the time variable:

$$V(\mathbf{r}, t) = \int_0^\infty \widetilde{V}(\mathbf{r}, \nu)\, \mathrm{e}^{-2\pi i \nu t}\, \mathrm{d}\nu. \tag{4.3–38}$$

The cross-spectral density function $W(\mathbf{r}_1, \mathbf{r}_2, \nu)$ (the cross-power spectrum) of the light disturbances at points $\mathbf{r}_1$ and $\mathbf{r}_2$ at frequency ν may be defined by the equation [cf. Eq. (2.4–35)]

$$\langle \widetilde{V}^*(\mathbf{r}_1, \nu)\widetilde{V}(\mathbf{r}_2, \nu')\rangle = W(\mathbf{r}_1, \mathbf{r}_2, \nu)\delta(\nu - \nu'), \tag{4.3–39}$$

where the (ensemble) average on the left-hand side of Eq. (4.3–39) is taken over the different realizations of the field and the δ on the right-hand side is the Dirac delta function. It is clear from Eq. (4.3–39) that the cross-spectral density function is a measure of the correlation between the spectral amplitudes of any particular frequency component of the light vibrations at the points $\mathbf{r}_1$ and $\mathbf{r}_2$.

According to the generalized Wiener–Khintchine theorem [Eqs. (2.4–37) and (2.4–38)], the mutual coherence function and the cross-spectral density function form a Fourier transform pair:

$$\Gamma(\mathbf{r}_1, \mathbf{r}_2, \tau) = \int_0^\infty W(\mathbf{r}_1, \mathbf{r}_2, \nu)\, \mathrm{e}^{-2\pi i \nu \tau}\, \mathrm{d}\nu, \tag{4.3–40a}$$

$$W(\mathbf{r}_1, \mathbf{r}_2, \nu) = \int_{-\infty}^\infty \Gamma(\mathbf{r}_1, \mathbf{r}_2, \tau)\, \mathrm{e}^{2\pi i \nu \tau}\, \mathrm{d}\tau. \tag{4.3–40b}$$

In the special case when the two points $\mathbf{r}_1$ and $\mathbf{r}_2$ coincide, the cross-spectral density function becomes a function of the location of only one point and of the frequency and, according to Eqs. (2.4–14) and (2.4–15), this function represents the *spectral density* (the power spectrum) of the light. We will denote it by $S(\mathbf{r}, \nu)$:

$$S(\mathbf{r}, \nu) = W(\mathbf{r}, \mathbf{r}, \nu). \tag{4.3–41}$$

Using Eqs. (4.3–40a) and (4.3–41) we have

$$\Gamma(\mathbf{r}, \mathbf{r}, \tau) = \int_0^\infty S(\mathbf{r}, \nu)\,\mathrm{e}^{-2\pi i\nu\tau}\,\mathrm{d}\nu, \tag{4.3–42a}$$

$$S(\mathbf{r}, \nu) = \int_{-\infty}^\infty \Gamma(\mathbf{r}, \mathbf{r}, \tau)\,\mathrm{e}^{2\pi i\nu\tau}\,\mathrm{d}\tau. \tag{4.3–42b}$$

We note a few properties of the cross-spectral density function. In view of Eqs. (4.3–36) and (4.3–40b) it is clearly Hermitian in the sense that

$$W(\mathbf{r}_2, \mathbf{r}_1, \nu) = W^*(\mathbf{r}_1, \mathbf{r}_2, \nu). \tag{4.3–43}$$

Moreover, according to Eq. (2.4–40), it is a non-negative definite function: for any n points $\mathbf{r}_1, \mathbf{r}_2, \ldots, \mathbf{r}_n$ (with n being an arbitrary positive integer), for any n real or complex numbers $a_1, a_2, \ldots, a_n$ and for any frequency ν,

$$\sum_{j=1}^{n}\sum_{k=1}^{n} a_j^* a_k W(\mathbf{r}_j, \mathbf{r}_k, \nu) \geqslant 0. \tag{4.3–44}$$

In particular, with the choice $n = 1$, Eqs. (4.3–44) and (4.3–41) imply that the spectral density is non-negative.

$$S(\mathbf{r}, \nu) \geqslant 0, \tag{4.3–45}$$

which, of course, is to be expected from its significance as a measure of the average energy density at the point $\mathbf{r}$, at frequency ν. With the choice $n = 2$, Eq. (4.3–44) implies that

$$|W(\mathbf{r}_1, \mathbf{r}_2, \nu)| \leqslant [W(\mathbf{r}_1, \mathbf{r}_1, \nu)]^{1/2}[W(\mathbf{r}_2, \mathbf{r}_2, \nu)].^{1/2} \tag{4.3–46}$$

It is useful to normalize the cross-spectral density function be setting

$$\mu(\mathbf{r}_1, \mathbf{r}_2, \nu) = \frac{W(\mathbf{r}_1, \mathbf{r}_2, \nu)}{[W(\mathbf{r}_1, \mathbf{r}_1, \nu)]^{1/2}[W(\mathbf{r}_2, \mathbf{r}_2, \nu)]^{1/2}} \tag{4.3–47a}$$

$$= \frac{W(\mathbf{r}_1, \mathbf{r}_2, \nu)}{[S(\mathbf{r}_1, \nu)]^{1/2}[S(\mathbf{r}_2, \nu)]^{1/2}}. \tag{4.3–47b}$$

In view of the inequality (4.3–46) we have

$$0 \leqslant |\mu(\mathbf{r}_1, \mathbf{r}_2, \nu)| \leqslant 1 \tag{4.3–48}$$

for all values of the arguments $\mathbf{r}_1$, $\mathbf{r}_2$ and ν. We will refer to $\mu(\mathbf{r}_1, \mathbf{r}_2, \nu)$ as the *spectral degree of coherence at frequency* ν, of the light at the points $P_1(\mathbf{r}_1)$ and $P_2(\mathbf{r}_2)$. It is sometimes also referred to as the complex degree of spatial (or spectral) coherence at frequency ν (see Wolf and Carter, 1975, 1976, Mandel and Wolf, 1976 and Bastiaans, 1977).

It may be worthwhile to note that, in spite of some formal similarity between the definitions of the complex degrees of coherence $\gamma(\mathbf{r}_1, \mathbf{r}_2, \tau)$ [Eqs. (4.3–12)] and $\mu(\mathbf{r}_1, \mathbf{r}_2, \tau)$ [Eqs. (4.3–47)] and the fact that $\Gamma(\mathbf{r}_1, \mathbf{r}_2, \tau)$ and $W(\mathbf{r}_1, \mathbf{r}_2, \nu)$ form a Fourier transform pair, γ and μ are, in general, *not* Fourier transforms of each other. The relationship between the two degrees of coherence is discussed in a note by Friberg and Wolf (1995).

Let us now again consider the two-beam interference experiment which we discussed in Section 4.3.1 and which is illustrated in Fig. 4.6. We will examine the relationship between the spectral densities of the light emerging from the pinholes and the light reaching the plane $\mathscr{B}$ of observation. For this purpose we first derive a generalization of Eq. (4.3–10), namely an expression for the self-coherence function,

$$\Gamma(\mathbf{r}, \mathbf{r}, \tau) = \langle V^*(\mathbf{r}, t)V(\mathbf{r}, t + \tau)\rangle \tag{4.3–49}$$

rather than for the averaged intensity at a typical point $P(\mathbf{r})$ in the plane of observation. On substituting for $V(\mathbf{r}, t)$ from Eq. (4.3–2) into Eq. (4.3–49) we find that

$$\begin{aligned}\Gamma(\mathbf{r}, \mathbf{r}, \tau) = {} & |K_1|^2\langle V^*(\mathbf{r}_1, t - t_1)V(\mathbf{r}_1, t + \tau - t_1)\rangle \\ & + |K_2|^2\langle V^*(\mathbf{r}_2, t - t_2)V(\mathbf{r}_2, t + \tau - t_2)\rangle \\ & + K_1^*K_2\langle V^*(\mathbf{r}_1, t - t_1)V(\mathbf{r}_2, t + \tau - t_2)\rangle \\ & + K_2^*K_1\langle V^*(\mathbf{r}_2, t - t_2)V(\mathbf{r}_1, t + \tau - t_1)\rangle.\end{aligned} \tag{4.3–50}$$

If we use the fact that the field is stationary, at least in the wide sense, we have the formula $\langle V^*(\mathbf{r}_1, t - t_1)V(\mathbf{r}_1, t + \tau - t_1)\rangle = \langle V^*(\mathbf{r}_1, t)V(\mathbf{r}_1, t + \tau)\rangle$ etc. and Eq. (4.3–50) may be expressed in the form

$$\begin{aligned}\Gamma(\mathbf{r}, \mathbf{r}, \tau) = {} & |K_1|^2\Gamma(\mathbf{r}_1, \mathbf{r}_1, \tau) + |K_2|^2\Gamma(\mathbf{r}_2, \mathbf{r}_2, \tau) \\ & + K_1^*K_2\Gamma(\mathbf{r}_1, \mathbf{r}_2, \tau + t_1 - t_2) + K_2^*K_1\Gamma(\mathbf{r}_2, \mathbf{r}_1, \tau + t_2 - t_1).\end{aligned} \tag{4.3–51}$$

Next we multiply both sides of Eq. (4.3–51) by $e^{2\pi i\nu\tau}$ and integrate with respect to τ from $-\infty$ to $+\infty$. In carrying out the integrations on the right-hand side of (4.3–51) we may neglect the weak dependence of the factors K_1 and K_2 on the frequency, because the light is assumed to be quasi-monochromatic. If we recall Eq. (4.3–40b), we obtain the following expression for the spectral density at the point $P(\mathbf{r})$ in the plane of observation:

$$\begin{aligned}W(\mathbf{r}, \mathbf{r}, \nu) = {} & |K_1|^2W(\mathbf{r}_1, \mathbf{r}_1, \nu) + |K_2|^2W(\mathbf{r}_2, \mathbf{r}_2, \nu) \\ & + K_1^*K_2W(\mathbf{r}_1, \mathbf{r}_2, \nu)\,e^{-2\pi i\nu(t_1-t_2)} + K_2^*K_1W(\mathbf{r}_2, \mathbf{r}_1, \nu)\,e^{-2\pi i\nu(t_2-t_1)}.\end{aligned} \tag{4.3–52}$$

By an argument similar to that given in connection with Eq. (4.3–14a), it follows that the first term on the right-hand side of (4.3–52), namely

$$|K_1|^2W(\mathbf{r}_1, \mathbf{r}_1, \nu) \equiv W^{(1)}(\mathbf{r}, \mathbf{r}, \nu), \tag{4.3–53a}$$

represents the spectral density at frequency ν of the light that reaches the point $P(\mathbf{r})$ from the pinhole P_1 only. Similarly the second term on the right-hand side of

(4.3–52), namely

$$|K_2|^2 W(\mathbf{r}_2, \mathbf{r}_2, \nu) \equiv W^{(2)}(\mathbf{r}, \mathbf{r}, \nu), \tag{4.3–53b}$$

represents the spectral density of the light that reaches the point $P(\mathbf{r})$ from the pinhole P_2 only.

The last two terms on the right-hand side of (4.3–52) may readily be expressed in terms of $W^{(1)}$, $W^{(2)}$, and μ. We have from Eqs. (4.3–47), (4.3–53a) and (4.3–53b), if we also use the fact that the factors K_1 and K_2 are purely imaginary,

$$K_1^* K_2 W(\mathbf{r}_1, \mathbf{r}_2, \nu) = [W^{(1)}(\mathbf{r}, \mathbf{r}, \nu)]^{1/2}[W^{(2)}(\mathbf{r}, \mathbf{r}, \nu)]^{1/2}\mu(\mathbf{r}_1, \mathbf{r}_2, \nu).$$

Next we make use of this relation and of Eqs. (4.3–53) and (4.3–43) in the formula (4.3–52). If we also recall the fact that, when its two spatial arguments are equal, the cross-spectral density reduces to the spectral density [Eq. (4.3–41)] and also that $t_1 = R_1/c$, and $t_2 = R_2/c$ (see Fig. 4.6), the formula (4.3–52) gives the following expression for the spectral density of the light at the point $P(\mathbf{r})$ on the plane $\mathscr{B}$ of observation:

$$\begin{aligned} S(\mathbf{r}, \nu) &= S^{(1)}(\mathbf{r}, \nu) + S^{(2)}(\mathbf{r}, \nu) \\ &\quad + 2[S^{(1)}(\mathbf{r}, \nu)]^{1/2}[S^{(2)}(\mathbf{r}, \nu)]^{1/2}\,\mathrm{Re}\,[\mu(\mathbf{r}_1, \mathbf{r}_2, \nu)\,\mathrm{e}^{-2\pi i\nu(R_1-R_2)/c}]. \end{aligned} \tag{4.3–54}$$

The formula (4.3–54), sometimes called the *spectral interference law*, shows that, in general, the spectral density $S(\mathbf{r}, \nu)$ of the light at P is not just the sum of the spectral densities $S^{(1)}(\mathbf{r}, \nu)$ and $S^{(2)}(\mathbf{r}, \nu)$ of the two beams reaching P from the two pinholes, but differs from it by the last term on the right-hand side. This term depends on the spectral degree of coherence $\mu(\mathbf{r}_1, \mathbf{r}_2, \nu)$ of the light at the two pinholes. In particular it follows from Eq. (4.3–54) that even when the two beams have the same spectral distributions, i.e. when $S^{(2)}(\mathbf{r}, \nu) = S^{(1)}(\mathbf{r}, \nu)$, the spectral distribution of the light obtained by superposing the two beams will, in general, be different. Spectral changes produced on interference will be discussed in Section 5.8.

We note that the expression (4.3–54) for the spectral density at a typical point in the plane $\mathscr{B}$ has a similar form as the expression (4.3–15) for the averaged intensity. The similarity suggests that, by analogy with the expression (4.3–23), it may be useful to introduce the concept of *spectral visibility* and we will do so later [Eq. (4.5–18)]. Here we only mention that the spectral interference law and the concept of spectral visibility have found useful applications (see, for example, Heiniger, Herden and Tschudi, 1983 and James, Kandpal and Wolf, 1995).

The cross-spectral density function $W(\mathbf{r}_1, \mathbf{r}_2, \nu)$ and the spectral density $S(\mathbf{r}, \nu)$ can, in principle, be determined from measurements of the mutual coherence function $\Gamma(\mathbf{r}_1, \mathbf{r}_2, \tau)$ and the self-coherence function $\Gamma(\mathbf{r}, \mathbf{r}, \tau)$ by the use of the Fourier transform relationship (4.3–40b). The spectral degree of coherence $\mu(\mathbf{r}_1, \mathbf{r}_2, \nu)$ can then be obtained from the formula (4.3–47b). Alternatively it may also be determined with the help of narrow-band filters, as we will now show, following an analysis due to Wolf (1983).

As we learned earlier, one may determine the mutual coherence function $\Gamma(\mathbf{r}_1, \mathbf{r}_2, \tau)$ of the light at the two pinholes in a Young's interference experiment from measurements of the (averaged) intensities at the pinholes and the visibility and the location of intensity maxima in the fringe pattern. Now the mutual coherence function is, according to Eq. (4.3–40a), just the Fourier transform of the cross-spectral density $W(\mathbf{r}_1, \mathbf{r}_2, \nu)$ which may be defined by Eq. (4.3–39).

Suppose that we place identical narrow-band filters behind each pinhole. If $T(\nu)$ is the complex amplitude transmission function of each filter, the cross spectral density $W^{(+)}(\mathbf{r}_1, \mathbf{r}_2, \nu)$ of the light emerging from the filters is given by a formula of the form (4.3–39), but with $\widetilde{V}(\mathbf{r}, \nu)$ replaced by $T(\nu)\widetilde{V}(\mathbf{r}, \nu)$, i.e. by the formula

$$\langle T^*(\nu)\widetilde{V}^*(\mathbf{r}_1, \nu)T(\nu')\widetilde{V}(\mathbf{r}_2, \nu')\rangle = W^{(+)}(\mathbf{r}_1, \mathbf{r}_2, \nu)\delta(\nu - \nu'). \quad (4.3\text{–}55)$$

Since $T(\nu)$ is a deterministic function, it may be taken outside the averaging brackets in Eq. (4.3–55) and, if we make use of Eq. (4.3–39), we obtain the following relation between the cross-spectral densities of the filtered and the unfiltered light at the two pinholes:

$$W^{(+)}(\mathbf{r}_1, \mathbf{r}_2, \nu) = |T(\nu)|^2 W(\mathbf{r}_1, \mathbf{r}_2, \nu). \quad (4.3\text{–}56)$$

It follows at once from Eqs. (4.3–56) and (4.3–40a) that the mutual coherence function of the filtered light at the two pinholes is given by

$$\Gamma^{(+)}(\mathbf{r}_1, \mathbf{r}_2, \tau) = \int_0^\infty |T(\nu)|^2 W(\mathbf{r}_1, \mathbf{r}_2, \nu)\,\mathrm{e}^{-2\pi i\nu\tau}\,\mathrm{d}\nu. \quad (4.3\text{–}57)$$

Suppose now that each filter has a passband of effective width $\Delta\nu$, centered on a frequency ν_0. If the cross-spectral density $W(\mathbf{r}_1, \mathbf{r}_2, \nu)$ of the light incident on the pinholes is a continuous function of ν, as we now assume, and if $\Delta\nu$ is so small that $W(\mathbf{r}_1, \mathbf{r}_2, \nu)$ does not appreciably change across the effective passband $\nu_0 - \frac{1}{2}\Delta\nu \leqslant \nu \leqslant \nu_0 + \frac{1}{2}\Delta\nu$ of the filters (see Fig. 4.8), we may replace $W(\mathbf{r}_1, \mathbf{r}_2, \nu)$ in the integral in Eq. (4.3–57) by $W(\mathbf{r}_1, \mathbf{r}_2, \nu_0)$. We then obtain the following

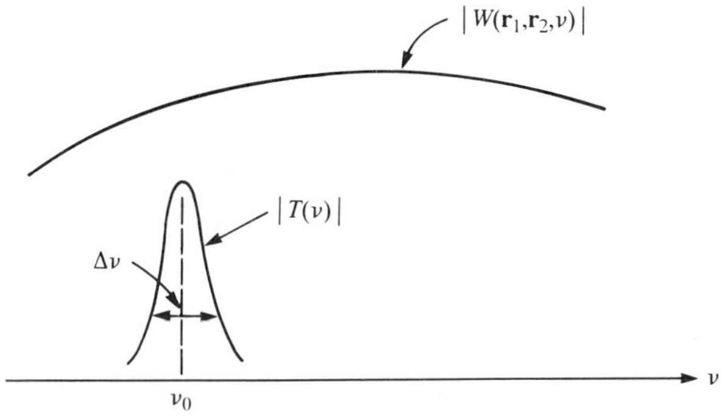

Fig. 4.8 Schematic illustration of the relative frequency dependence of the modulus of the complex amplitude transmission function, $T(\nu)$, of the filters placed in front of the pinholes in Young's interference experiments, and of the absolute value of the cross-spectral density $W(\mathbf{r}_1, \mathbf{r}_2, \nu)$ of the unfiltered light at the pinholes. The effective passbands $\nu_0 - \frac{1}{2}\Delta\nu \leqslant \nu \leqslant \nu_0 + \frac{1}{2}\Delta\nu$ of the filters are assumed to be so narrow that with $\mathbf{r}_1$ and $\mathbf{r}_2$ fixed, both the modulus and the phase (not shown) of $W(\mathbf{r}_1, \mathbf{r}_2, \nu)$ [and also of $W(\mathbf{r}_1, \mathbf{r}_1, \nu)$ and $W(\mathbf{r}_2, \mathbf{r}_2, \nu)$] are approximately constant over the passbands.

expression for $\Gamma^{(+)}$:

$$\Gamma^{(+)}(\mathbf{r}_1, \mathbf{r}_2, \tau) = W(\mathbf{r}_1, \mathbf{r}_2, \nu_0)\int_0^\infty |T(\nu)|^2 \, \mathrm{e}^{-2\pi i\nu\tau} \, \mathrm{d}\nu. \tag{4.3–58}$$

In a similar way it follows that if the spectral densities $W(\mathbf{r}_1, \mathbf{r}_1, \nu)$ and $W(\mathbf{r}_2, \mathbf{r}_2, \nu)$ at the two pinholes are continuous functions of the frequency ν and if $\Delta\nu$ is small enough then, to a good approximation,

$$\Gamma^{(+)}(\mathbf{r}_j, \mathbf{r}_j, \tau) = S(\mathbf{r}_j, \nu_0)\int_0^\infty |T(\nu)|^2 \, \mathrm{e}^{-2\pi i\nu\tau} \, \mathrm{d}\nu, \quad (j = 1, 2). \tag{4.3–59}$$

The formulas (4.3–58) and (4.3–59) imply that the cross-spectral density $W(\mathbf{r}_1, \mathbf{r}_2, \nu)$ and the spectral densities $S(\mathbf{r}_j, \nu)$, $(j = 1, 2)$, of the light incident on the pinholes may be determined from measurements of the mutual coherence function and the self-coherence function, respectively, of the filtered light and from the knowledge of the transmission function $T(\nu)$ of the filters.

On substituting from Eqs. (4.3–58) and (4.3–59) into Eq. (4.3–12a) we obtain the following expressions for the complex degree of coherence of the filtered light emerging from the two pinholes:

$$\gamma^{(+)}(\mathbf{r}_1, \mathbf{r}_2, \tau) = \mu(\mathbf{r}_1, \mathbf{r}_2, \nu_0)\theta(\tau), \tag{4.3–60}$$

where

$$\mu(\mathbf{r}_1, \mathbf{r}_2, \nu_0) = \frac{W(\mathbf{r}_1, \mathbf{r}_2, \nu_0)}{[S(\mathbf{r}_1, \nu_0)]^{1/2}[S(\mathbf{r}_2, \nu_0)]^{1/2}} \tag{4.3–61}$$

and

$$\theta(\tau) = \frac{\displaystyle\int_0^\infty |T(\nu)|^2 \, \mathrm{e}^{-2\pi i\nu\tau} \, \mathrm{d}\nu}{\displaystyle\int_0^\infty |T(\nu)|^2 \, \mathrm{d}\nu}. \tag{4.3–62}$$

The factor $\mu(\mathbf{r}_1, \mathbf{r}_2, \nu_0)$, defined by Eq. (4.3–61), which appears on the right-hand side of Eq. (4.3–60), is precisely the spectral degree of coherence at frequency ν_0 [Eq. (4.3–47b)] of the unfiltered light at the two pinholes. It is also equal to the spectral degree of coherence $\mu^{(+)}(\mathbf{r}_1, \mathbf{r}_2, \nu_0)$ of the filtered light, as is readily seen on substituting from Eq. (4.3–56) into Eq. (4.3–47b), with W replaced by $W^{(+)}$, and on comparing the resulting expression with Eq. (4.3–61). Hence the *spectral degree of coherence remains unchanged by filtering*. On the other hand the complex degree of coherence $\gamma(\mathbf{r}_1, \mathbf{r}_2, \tau)$ of the unfiltered and the filtered light may readily be shown to differ from each other. The other factor, $\theta(\tau)$, which appears on the right-hand side of Eq. (4.3–60) is, according to Eq. (4.3–62), the normalized Fourier transform of the squared modulus of the transmission function $T(\nu)$ of each filter.

It is evident from Eq. (4.3–60) that, considered as a function of τ, $|\gamma^{(+)}|$ attains its maximum when $|\theta(\tau)|$ reaches it maximum, i.e. when $\tau = 0$. Moreover, since $\theta(0) = 1$ it follows that one then has

$$\gamma^{(+)}(\mathbf{r}_1, \mathbf{r}_2, 0) = \mu(\mathbf{r}_1, \mathbf{r}_2, \nu_0). \tag{4.3–63}$$

This formula implies that *the 'equal-time' complex degree of coherence* $j^{(+)}(\mathbf{r}_1, \mathbf{r}_2) \equiv \gamma^{(+)}(\mathbf{r}_1, \mathbf{r}_2, 0)$ [Eq. (4.3–35)] *of the filtered light at the two pinholes is equal to the spectral degree of coherence* $\mu(\mathbf{r}_1, \mathbf{r}_2, \nu_0)$ *of the unfiltered (and also of the filtered) light for the mid-frequency* ν_0 *of the filters*. From this result it is clear that both the modulus and the phase of the spectral degree of coherence can be determined from the interference experiments discussed in Section 4.3.1, provided identical filters with sufficiently narrow passbands are placed behind the pinholes.

Since, with the positions of the pinholes fixed, $|\gamma^{(+)}(\mathbf{r}_1, \mathbf{r}_2, \tau)|$ attains its maximum when $\tau = 0$, the interference fringes formed by the filtered light will be sharpest near the center of the fringe pattern, i.e. in the neighborhood of the point which is equidistant from both pinholes. Now according to Eq. (4.3–48) the modulus of the spectral degree of coherence can take on any value between zero and unity, with both the extrema being attainable. Hence it is clear from Eq. (4.3–63), (4.3–60) and from the relation (4.3–25) between the fringe visibility $\mathcal{V}$ and $|\gamma|$ that the *maximum visibility of the interference fringes formed by the filtered light will not, in general, tend to unity as the passband of the filters decrease.* It is, of course, assumed that measurements are made on a time scale that involves averaging over a time interval which is long compared with the reciprocal bandwidth of the filters and that the detector is sensitive enough to measure the reduced intensities of the filtered light. It is actually not difficult to realize conditions under which the maximum visibility $\mathcal{V}_{max}$ will have any chosen value in the range $0 \leqslant \mathcal{V}_{max} \leqslant 1$, irrespective of how narrow the effective bandwidth $\Delta\nu$ of the filters may be. One only needs to illuminate the pinholes by a distant, spatially incoherent, uniform, circular, quasi-monochromatic source of mean frequency in the neighborhood of $\nu = \nu_0$, placed symmetrically with respect to the pinholes, and to separate the pinholes by an appropriate distance that may be readily calculated from the van Cittert–Zernike theorem (Section 4.4.4).

Although under conditions when Eq. (4.3–60) applies, the maximum fringe visibility will not change if the bandwidth $\Delta\nu$ of the filters is further reduced, the interference pattern will, nevertheless, be modified. If $\Delta\nu$ is decreased but the transmissivity $|T(\nu_0)|$ of the filters at the mid-frequency ν_0 is kept fixed, the effective width of $|\theta(\tau)|$ will increase, as can be readily deduced from Eq. (4.3–62). Consequently the effective width of $|\gamma^{(+)}(\mathbf{r}_1, \mathbf{r}_2, \tau)|$ (with $\mathbf{r}_1$, $\mathbf{r}_2$ fixed) will also increase, as is evident from Eq. (4.3–60). This implies that the visibility of the fringes in the plane of observation will then fall off more slowly with distance from the center of the fringe pattern.

In the formula (4.3–60) for the complex degree of coherence of the filtered light, the spatial coherence properties, characterized by $\mu(\mathbf{r}_1, \mathbf{r}_2, \nu_0)$, and the temporal coherence properties, characterized by $\theta(\tau)$, are completely separated. This separation is a manifestation of the so-called cross-spectral purity of the filtered light, which will be discussed in Section 4.5.1.

4.3.3 Coherence time and bandwidth

In our elementary discussion of temporal coherence given in Section 4.2.1 we introduced the concept of the coherence time. This quantity is a measure of the time interval in which appreciable amplitude correlations and phase correlations of the light vibrations at a particular point P in a fluctuating optical field will persist. We introduced the coherence time Δt from considerations of a simple interference experiment and we gave a rough estimate for it, namely

$$\Delta t = 1/\Delta \nu, \tag{4.3–64}$$

where $\Delta \nu$ is the effective spectral width of the light at P. We will now define the coherence time in a more precise way and will also introduce a corresponding definition of the bandwidth.

We learned in Section 4.3.1 that a measure of the correlations of the light vibrations at a point P in a stationary optical field at time instants separated by a time interval τ is the self-coherence function

$$\Gamma(\tau) = \langle V^*(P, t)V(P, t + \tau)\rangle, \tag{4.3–65}$$

where $V(P, t)$ is the complex analytic signal representation of the field at the point P. (For the sake of simplicity we do not display from now on the dependence of the self-coherence function and of the spectral density function on position.) It seems, therefore, natural and mathematically convenient to define the coherence time Δt at P as the normalized root-mean square (r.m.s.) width of the squared modulus of $\Gamma(\tau)$, i.e. by the formula[‡]

$$(\Delta\tau)^2 = \frac{\int_{-\infty}^{\infty} \tau^2|\Gamma(\tau)|^2\,\mathrm{d}\tau}{\int_{-\infty}^{\infty} |\Gamma(\tau)|^2\,\mathrm{d}\tau}. \tag{4.3–66}$$

Similarly, we may define the effective spectral width $\Delta\nu$ (the bandwidth) of the light at the point P as the normalized r.m.s. width $\Delta\nu$ of the spectral density function $S(\nu)$,

$$S(\nu) = \int_{-\infty}^{\infty} \Gamma(\tau)\,\mathrm{e}^{2\pi i\nu\tau}\,\mathrm{d}\tau, \tag{4.3–67}$$

by the formula

$$(\Delta\nu)^2 = \frac{\int_0^{\infty} (\nu - \bar{\nu})^2 S^2(\nu)\,\mathrm{d}\nu}{\int_0^{\infty} S^2(\nu)\,\mathrm{d}\nu}, \tag{4.3–68}$$

where

$$\bar{\nu} = \frac{\int_0^{\infty} \nu S^2(\nu)\,\mathrm{d}\nu}{\int_0^{\infty} S^2(\nu)\,\mathrm{d}\nu} \tag{4.3–69}$$

may be identified with the mean frequency of the light.

‡ The average value $\bar{\tau} = \int_{-\infty}^{\infty}\tau|\Gamma(\tau)|^2\,\mathrm{d}\tau / \int_{-\infty}^{\infty}|\Gamma(\tau)|^2\,\mathrm{d}\tau$ of τ is zero because $|\Gamma(\tau)|$ is an even function of τ. It is assumed here, of course, that the various moments appearing in these formulas exist.

The expressions (4.3–66) and (4.3–68) for the coherence time and the effective spectral width may be re-written in a slightly different form, from which a basic inequality relating to their product may readily be deduced. We set

$$\left.\begin{aligned} \Phi(\xi) &= S(\bar{\nu}+\xi) \qquad \text{when } \xi \geqslant -\bar{\nu} \quad \text{(a)} \\ &= 0 \qquad\qquad\quad \text{when } \xi < -\bar{\nu} \quad \text{(b)} \\ \psi(\tau) &= \Gamma(\tau)\,\mathrm{e}^{2\pi i\bar{\nu}\tau}. \qquad\qquad\qquad \text{(c)} \end{aligned}\right\} \tag{4.3–70}$$

We shall assume that $\Phi(\xi)$ is a square-integrable and continuous function of ξ $(-\infty < \xi < \infty)$. The continuity requirement implies, in particular, that $\Phi(-\bar{\nu}) = S(0) = 0$.‡ From Eq. (4.3–67) and our assumption of square-integrability it follows that the functions $\psi(\tau)$ and $\Phi(\xi)$ form a Fourier transform pair:

$$\psi(\tau) = \int_{-\infty}^{\infty} \Phi(\xi)\,\mathrm{e}^{-2\pi i\xi\tau}\,\mathrm{d}\xi, \quad \Phi(\xi) = \int_{-\infty}^{\infty} \psi(\tau)\,\mathrm{e}^{2\pi i\xi\tau}\,\mathrm{d}\tau \tag{4.3–71}$$

and the expressions for $\Delta\tau$ and $\Delta\nu$ become

$$(\Delta\tau)^2 = \frac{1}{N}\int_{-\infty}^{\infty} \tau^2|\psi(\tau)|^2\,\mathrm{d}\tau, \quad (\Delta\nu)^2 = \frac{1}{N}\int_{-\infty}^{\infty} \xi^2\Phi^2(\xi)\,\mathrm{d}\xi, \tag{4.3–72}$$

where

$$N = \int_{-\infty}^{\infty} |\psi(\tau)|^2\,\mathrm{d}\tau = \int_{-\infty}^{\infty} \Phi^2(\xi)\,\mathrm{d}\xi. \tag{4.3–73}$$

The equality of the two integrals in Eq. (4.3–73) follows from Parseval's theorem on Fourier transforms.

Next we express the integral that appears in the expression for $\Delta\nu$ in Eq. (4.3–72) in terms of ψ. We have, on using the second expression in Eq. (4.3–71),

$$\begin{aligned} (\Delta\nu)^2 &= \frac{1}{N}\int_{-\infty}^{\infty} \xi^2\Phi(\xi)\,\mathrm{d}\xi\int_{-\infty}^{\infty} \psi(\tau)\,\mathrm{e}^{2\pi i\xi\tau}\,\mathrm{d}\tau \\ &= \frac{1}{N}\int_{-\infty}^{\infty} \mathrm{d}\tau\,\psi(\tau)\left(\frac{1}{2\pi i}\right)^2\frac{\mathrm{d}^2}{\mathrm{d}\tau^2}\int_{-\infty}^{\infty} \Phi(\xi)\,\mathrm{e}^{2\pi i\xi\tau}\,\mathrm{d}\xi \\ &= -\frac{1}{4\pi^2}\frac{1}{N}\int_{-\infty}^{\infty} \psi(\tau)\frac{\mathrm{d}^2}{\mathrm{d}\tau^2}\psi^*(\tau)\,\mathrm{d}\tau \\ &= \frac{1}{4\pi^2}\frac{1}{N}\int_{-\infty}^{\infty}\left|\frac{\mathrm{d}\psi}{\mathrm{d}\tau}\right|^2\mathrm{d}\tau. \end{aligned} \tag{4.3–74}$$

In passing from the second to the third line, the first relation in Eq. (4.3–71) was used, together with the fact that, since $\Phi(\xi)$ is the 'shifted spectral density' [see Eqs. (4.3–70a) and (4.3–70b)], it is necessarily real. In passing from the third to the fourth line we integrated by parts and used the fact that $\psi(\tau) \to 0$ as $\tau \to \pm\infty$; this is so because $\psi(\tau)$ was assumed to be square-integrable.

It follows from the first relation in Eq. (4.3–72), from Eq. (4.3–74) and from

‡ This condition is necessary for the integral in the numerator of Eq. (4.3–68) to converge (see Wolf, 1958). The more general case when $S(0) \neq 0$ was investigated in a somewhat different context by Kay and Silverman (1957). See also Kharkevich (1960), Sec. 12.

the first relation in (4.3–73) that

$$(\Delta\tau)^2(\Delta\nu)^2 = \frac{1}{16\pi^2}\left[\frac{4\left(\int_{-\infty}^{\infty}\tau^2|\psi(\tau)|^2\,\mathrm{d}\tau\right)\left(\int_{-\infty}^{\infty}\left|\frac{\mathrm{d}\psi}{\mathrm{d}\tau}\right|^2\mathrm{d}\tau\right)}{\left(\int_{-\infty}^{\infty}|\psi(\tau)|^2\,\mathrm{d}\tau\right)^2}\right]. \tag{4.3–75}$$

Now there is a mathematical lemma (Weyl, 1950, pp. 393–394; Born and Wolf, 1980, Appendix VIII), according to which the term in the large brackets on the right-hand side of (4.3–75) is greater than or equal to unity for any function $\psi(\tau)$ for which the integrals exist. Hence with the definitions given by Eqs. (4.3–66) and (4.3–68) we obtain the following *reciprocity inequality*:

$$\Delta\tau\,\Delta\nu \geqslant 1/4\pi. \tag{4.3–76}$$

One can readily show that the bracketed term on the right-hand side of Eq. (4.3–75) will equal unity, and hence the equality sign in the formula (4.3–76) will apply, if and only if $\psi(\tau)$ is a Gaussian function (Born and Wolf, 1980, Appendix VIII). Now $\Phi(\xi)$ is a Fourier transform of $\psi(\tau)$ and, since the Fourier transform of a Gaussian function is again a Gaussian function, $\Phi(\xi)$ is also Gaussian in this case. Hence it will not vanish for any value of its argument $(-\infty < \xi < \infty)$. However, as is seen from Eq. (4.3–70b), (which is a consequence of the fact that $\Gamma(\tau)$ is an analytic signal), $\Phi(\xi)$, as defined by Eqs. (4.3–70), does not obey this requirement. Thus in the present context the equality sign in the formula (4.3–76) never strictly holds. However, when the spectral density $S(\nu)$ is approximately of Gaussian form and its mean frequency $\bar{\nu}$ is large compared to the effective spectral range, then the inequality sign in Eq. (4.3–76) may clearly be replaced by the order of magnitude relation

$$\Delta\tau\,\Delta\nu \sim 1/4\pi. \tag{4.3–77}$$

The definitions of the coherence time and of the effective spectral range that we have just discussed are useful when the light is quasi-monochromatic and when its spectrum has a single reasonably well-defined peak. It is more difficult to provide useful definitions of these quantities when the spectrum has several peaks (as is the case with multimode laser light) or when the absolute value of the mutual coherence function is multiply peaked (as, for example, is the case with white-light Brewster fringes, where it is doubly peaked). If the spectrum consists of two lines, whose widths are much smaller than their separation, the concept of an effective spectral range evidently no longer has an unambiguous meaning.

Different definitions of the coherence time and of the spectral range from those that we discussed may be introduced from considerations of the extent of a unit cell of photon phase space (see Mandel, 1959). The coherence time $(\Delta\tau)'$ is then defined by the formula

$$(\Delta\tau)' = \int_{-\infty}^{\infty}|\gamma(\tau)|^2\,\mathrm{d}\tau, \tag{4.3–78}$$

where

$$\gamma(\tau) = \frac{\Gamma(\tau)}{\Gamma(0)} = \frac{\langle V^*(P,t)V(P,t+\tau)\rangle}{\langle V^*(P,t)V(P,t)\rangle} \tag{4.3–79}$$

is the complex degree of 'self-coherence'; and the effective spectral range is then defined as

$$(\Delta\nu)' = \frac{1}{\displaystyle\int_0^\infty \mathscr{s}^2(\nu)\,\mathrm{d}\nu}, \tag{4.3–80}$$

where

$$\mathscr{s}(\nu) = \int_{-\infty}^{\infty} \gamma(\tau)\,\mathrm{e}^{2\pi i\nu\tau}\,\mathrm{d}\tau \tag{4.3–81}$$

is the normalized spectral density [see Eqs. (2.4–26) and (2.4–27)]. From Eqs. (4.3–78) and (4.3–79) it follows that $(\Delta\tau)'$ may be expressed as

$$(\Delta\tau)' = \frac{1}{[\Gamma(0)]^2}\int_{-\infty}^{\infty} |\Gamma(\tau)|^2\,\mathrm{d}\tau \tag{4.3–82}$$

and from Eqs. (4.3–80), (4.3–81), (4.3–79) and (4.3–67), with the use of Parseval's theorem, it follows that $(\Delta\nu)'$ is expressible in the form

$$(\Delta\nu)' = \frac{[\Gamma(0)]^2}{\displaystyle\int_0^\infty S^2(\nu)\,\mathrm{d}\nu}. \tag{4.3–83}$$

If now we take the product of the expression (4.3–82) and (4.3–83) and again use Parseval's theorem we obtain the relation

$$(\Delta\tau)'(\Delta\nu)' = 1. \tag{4.3–84}$$

Thus, with these definitions, the coherence time is always the reciprocal of the effective spectral width.

It may be shown (Mandel and Wolf, 1962; Mehta, 1963) that, with simple types of spectral profiles, $\Delta\nu \approx (\Delta\nu)'$ and $\Delta\tau \approx (\Delta\tau)'$. However, if the spectral density distribution is of a more complicated form, the two sets of definitions may lead to results of quite different orders of magnitude. Hence in some cases caution must be exercised in applying a particular definition.

4.4 Propagation of correlations

In Section 4.2.2 we indicated by a simple argument that the state of coherence of light may be appreciably changed in the process of propagation. More specifically, we showed that, even if the light originates from uncorrelated point sources, the field at points sufficiently far away from the sources may be highly correlated. From the standpoint of the general theory of partial coherence, this change in the state of coherence may be understood as a consequence of the fact that the mutual coherence function obeys two precise propagation laws. In free space they are just two wave equations. We will now derive these equations and also two related differential equations for the cross-spectral density.

4.4.1 *Differential equations for the propagation of the mutual coherence and of the cross-spectral density in free space*

Let $V^{(r)}(\mathbf{r}, t)$ represent as before, a sample function of the real random process that characterizes the fluctuating light disturbance at the point $\mathbf{r}$, at time t. If we identify $V^{(r)}$ with a Cartesian component of the electric field or of the vector potential, it will obey, in free space, the wave equation

$$\nabla^2 V^{(r)}(\mathbf{r}, t) = \frac{1}{c^2}\frac{\partial^2 V^{(r)}(\mathbf{r}, t)}{\partial t^2}, \tag{4.4–1}$$

(c, as before, is the speed of light in vacuo). It can readily be shown that the complex analytic signal $V(\mathbf{r}, t)$ associated with $V^{(r)}(\mathbf{r}, t)$ also obeys this equation. To demonstrate this fact, let us represent $V^{(r)}(\mathbf{r}, t)$ as a (generalized) Fourier integral

$$V^{(r)}(\mathbf{r}, t) = \int_{-\infty}^{\infty} \tilde{V}(\mathbf{r}, \nu)\,\mathrm{e}^{-2\pi i\nu t}\,\mathrm{d}\nu. \tag{4.4–2}$$

On taking the Fourier transform of Eq. (4.4–1) we obtain the well-known result that $\tilde{V}(\mathbf{r}, \nu)$ obeys the Helmholtz equation

$$\nabla^2 \tilde{V}(\mathbf{r}, \nu) + k^2 \tilde{V}(\mathbf{r}, \nu) = 0, \tag{4.4–3}$$

where

$$k = 2\pi\nu/c. \tag{4.4–4}$$

The analytic signal $V(\mathbf{r}, t)$ associated with our real field $V^{(r)}(\mathbf{r}, t)$ is obtained by suppressing the negative frequency components on the right-hand side of Eq. (4.4–2):

$$V(\mathbf{r}, t) = \int_{0}^{\infty} \tilde{V}(\mathbf{r}, \nu)\,\mathrm{e}^{-2\pi i\nu t}\,\mathrm{d}\nu. \tag{4.4–5}$$

If we apply the operator $\nabla^2 - (1/c^2)\partial^2/\partial t^2$ to both sides of Eq. (4.4–5) and interchange, on the right-hand side, the order of this operator and of the integral operator, we find that

$$\nabla^2 V(\mathbf{r}, t) - \frac{1}{c^2}\frac{\partial^2 V(\mathbf{r}, t)}{\partial t^2} = \int_{0}^{\infty} [\nabla^2 \tilde{V}(\mathbf{r}, \nu) + k^2 \tilde{V}(\mathbf{r}, \nu)]\,\mathrm{e}^{-2\pi i\nu t}\,\mathrm{d}\nu. \tag{4.4–6}$$

If we make use of Eq. (4.4–3), we see that the integral on the right-hand side vanishes identically. Hence it follows that, in free space, the complex disturbance $V(\mathbf{r}, t)$ indeed satisfies the wave equation

$$\nabla^2 V(\mathbf{r}, t) = \frac{1}{c^2}\frac{\partial^2 V(\mathbf{r}, t)}{\partial t^2}. \tag{4.4–7}$$

Let us now take the complex conjugate of Eq. (4.4–7) and write $\mathbf{r}_1$ in place of $\mathbf{r}$ and t_1 in place of t. We then obtain the equation

$$\nabla_1^2 V^*(\mathbf{r}_1, t_1) = \frac{1}{c^2}\frac{\partial^2 V^*(\mathbf{r}_1, t_1)}{\partial t_1^2}, \tag{4.4–8}$$

where ∇_1^2 is the Laplacian operator taken with respect to the point $\mathbf{r}_1$. Next we multiply both sides of Eq. (4.4–8) by $V(\mathbf{r}_2, t_2)$. Since the differential operators in Eq. (4.4–8) are taken with respect to the variables $\mathbf{r}_1$, t_1, we may place $V(\mathbf{r}_2, t_2)$ under the operator signs and we obtain the equation

$$\nabla_1^2[V^*(\mathbf{r}_1, t_1)V(\mathbf{r}_2, t_2)] = \frac{1}{c^2}\frac{\partial^2}{\partial t_1^2}[V^*(\mathbf{r}_1, t_1)V(\mathbf{r}_2, t_2)]. \qquad (4.4\text{–}9)$$

Let us take the ensemble average of Eq. (4.4–9), over the different realizations of the field. If we interchange the order of the ensemble average and of the differential operators and recall the definition (4.3–6) of the second-order correlation function, we obtain the equation

$$\nabla_1^2\Gamma(\mathbf{r}_1, \mathbf{r}_2; t_1, t_2) = \frac{1}{c^2}\frac{\partial^2}{\partial t_1^2}\Gamma(\mathbf{r}_1, \mathbf{r}_2; t_1, t_2). \qquad (4.4\text{–}10a)$$

In a strictly similar way we can also derive the equation

$$\nabla_2^2\Gamma(\mathbf{r}_1, \mathbf{r}_2; t_1, t_2) = \frac{1}{c^2}\frac{\partial^2}{\partial t_2^2}\Gamma(\mathbf{r}_1, \mathbf{r}_2; t_1, t_2), \qquad (4.4\text{–}10b)$$

where ∇_2^2 is the Laplacian operator taken with respect to the point $\mathbf{r}_2$. Thus we have established the important result that *in free space the second-order correlation function $\Gamma(\mathbf{r}_1, \mathbf{r}_2; t_1, t_2)$ of an optical field obeys the two wave equations (4.4–10a) and (4.4–10b).*

Suppose now that the ensemble that represents the statistical properties of the field is stationary, at least in the wide sense, and is also ergodic. The correlation function $\Gamma(\mathbf{r}_1, \mathbf{r}_2; t_1, t_2)$ will then depend on the two time variables only through their difference $t_2 - t_1 = \tau$ and, moreover, it is then immaterial whether the correlation function is defined by means of an ensemble average or a time average. Also it is clear from the definition of τ that the operators $\partial^2/\partial t_1^2$ and $\partial^2/\partial t_2^2$ on the right-hand sides of Eqs. (4.4–10) may each be replaced by $\partial^2/\partial\tau^2$. Hence Eqs. (4.4–10) reduce to the following two wave equations, which the mutual coherence function of the field in free space necessarily obeys:

$$\nabla_1^2\Gamma(\mathbf{r}_1, \mathbf{r}_2, \tau) = \frac{1}{c^2}\frac{\partial^2}{\partial \tau^2}\Gamma(\mathbf{r}_1, \mathbf{r}_2, \tau), \qquad (4.4\text{–}11a)$$

$$\nabla_2^2\Gamma(\mathbf{r}_1, \mathbf{r}_2, \tau) = \frac{1}{c^2}\frac{\partial^2}{\partial \tau^2}\Gamma(\mathbf{r}_1, \mathbf{r}_2, \tau). \qquad (4.4\text{–}11b)$$

These two wave equations were first obtained by Wolf (1955).

Each of the two wave equations (4.4–11) describes the changes in the mutual coherence function as one of the two points ($\mathbf{r}_1$ or $\mathbf{r}_2$) is kept fixed, whilst the other one and the parameter τ vary. Now τ represents the difference between the two time instants t_1, t_2 at which the correlation between the light vibrations at the two points $P_1(\mathbf{r}_1)$ and $P_2(\mathbf{r}_2)$ is being considered. In experiments τ only appears in the expressions through the difference between the geometrical path lengths, $R_2 - R_1 = c\tau$ [cf. Fig. 4.6 and Eq. (4.3–2a)]. The actual time makes, therefore, no appearance in the basic formulas of coherence theory of stationary fields.

We noted earlier, that spatial coherence is characterized by the dependence of the mutual coherence function on $\mathbf{r}_1$ and $\mathbf{r}_2$, and that temporal coherence is characterized by its dependence on τ. Since Eqs. (4.4–11) couple the dependence of Γ on all the variables $\mathbf{r}_1$, $\mathbf{r}_2$ and τ, it is clear that, in general, spatial and temporal coherence properties of light are not independent of each other. A good example of such a dependence is the fact, discussed from an elementary standpoint in Section 4.2.2 (cf. Fig. 4.3), that light generated by a spatially incoherent source acquires a finite degree of spatial coherence in the process of propagation. Our rough elementary explanation of this phenomenon was in terms of finite wave trains emitted by different source points. The finite duration of the wave trains implies that the emitted light has a finite, non-zero bandwidth and hence possesses some temporal coherence. Because of the coupling implied by Eqs. (4.4–11) between spatial and temporal coherence, spatial coherence can be generated in the field produced by the spatially incoherent source. The two wave equations (4.4–11) provide a basis for a precise quantitative description of this phenomenon. We will analyze it more fully in Sections 4.4.2–4.4.5.

Since the cross-spectral density function $W(\mathbf{r}_1, \mathbf{r}_2, \nu)$ is the Fourier transform of the mutual coherence function [Eq. (4.3–40b)], we obtain at once, on taking the Fourier transform of Eqs. (4.4–11), the following two Helmholtz equations that W must satisfy in free space:

$$\nabla_1^2 W(\mathbf{r}_1, \mathbf{r}_2, \nu) + k^2 W(\mathbf{r}_1, \mathbf{r}_2, \nu) = 0, \qquad (4.4\text{–}12a)$$

$$\nabla_2^2 W(\mathbf{r}_1, \mathbf{r}_2, \nu) + k^2 W(\mathbf{r}_1, \mathbf{r}_2, \nu) = 0, \qquad (4.4\text{–}12b)$$

where $k = 2\pi\nu/c$.

It is clear that the problem of determining the mutual coherence function and the cross-spectral density at any pair of points in a domain in free space, bounded by a closed surface, is now reduced to a standard problem in the theory of partial differential equations. Numerous techniques, both exact and approximate, for solving such problems are well known. We will consider in detail only two problems of this kind, which are of practical interest.

4.4.2 Propagation of correlations from a plane

Consider a fluctuating optical field $V(\mathbf{r}, t)$, which propagates into the half-space $z > 0$. We will derive a formula for the mutual coherence function of the light vibrations at any pair of points $P_1(\mathbf{r}_1)$, $P_2(\mathbf{r}_2)$ in the field in terms of the values of the mutual coherence function and of some of its derivatives at all pairs of points $S_1(\mathbf{r}_1')$ and $S_2(\mathbf{r}_2')$ in the plane $z = 0$ (Fig. 4.9).

Since the cross-spectral density function obeys the Helmholtz equations (4.4–12), which are simpler than the wave equations for the mutual coherence function, we consider first the propagation of the cross-spectral density. For this purpose we make use of the Green's functions appropriate to the Dirichlet boundary value problem for the Helmholtz equation in a half-space. One must, however, take care to correctly represent the behavior of $W(\mathbf{r}_1, \mathbf{r}_2, \nu)$, far away from the origin in the half-space $z > 0$. Since the field propagates into the half-space $z > 0$, it is clear from Eq. (4.3–39) that $W(\mathbf{r}_1, \mathbf{r}_2, \nu)$ behaves asymptotically as an *outgoing* wave with respect to $\mathbf{r}_2$, but as an *incoming* wave with respect to $\mathbf{r}_1$.

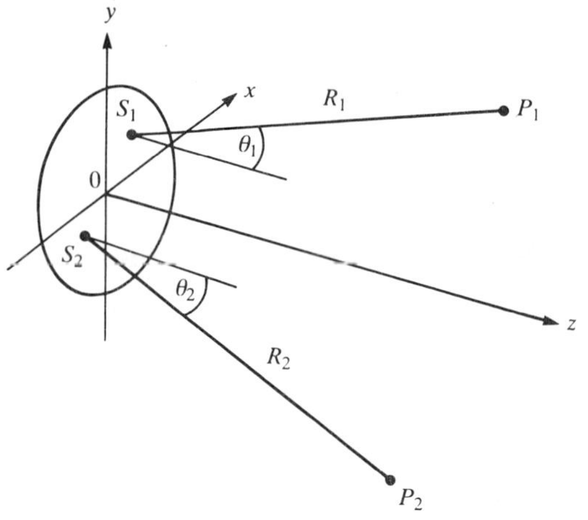

Fig. 4.9 Illustrating the notation relating to the propagation of the cross-spectral density and of the mutual coherence from the plane $z = 0$ into the half-space $z > 0$.

Let us first determine the solution of Eq. (4.4–12b) for $W(\mathbf{r}_1, \mathbf{r}_2, \nu)$ in terms of $W(\mathbf{r}_1, \mathbf{r}'_2, \nu)$, with $\mathbf{r}'_2$ constrained to be located in the plane $z = 0$ and with $\mathbf{r}_1$ kept fixed for the moment. We have, according to Rayleigh's first diffraction formula [Eq. (3.2–78)]

$$W(\mathbf{r}_1, \mathbf{r}_2, \nu) = -\frac{1}{2\pi}\int_{(z=0)} W(\mathbf{r}_1, \mathbf{r}'_2, \nu)\frac{\partial}{\partial z_2}\left(\frac{e^{ikR_2}}{R_2}\right) d^2r'_2, \qquad (4.4\text{–}13)$$

where $R_2 = |\mathbf{r}_2 - \mathbf{r}'_2|$ and $\partial/\partial z_2$ indicates differentiation along the positive z-direction. In a similar way we obtain the formula

$$W(\mathbf{r}_1, \mathbf{r}'_2, \nu) = -\frac{1}{2\pi}\int_{(z=0)} W(\mathbf{r}'_1, \mathbf{r}'_2, \nu)\frac{\partial}{\partial z_1}\left(\frac{e^{-ikR_1}}{R_1}\right) d^2r'_1, \qquad (4.4\text{–}14)$$

where $R_1 = |\mathbf{r}_1 - \mathbf{r}'_1|$. On substituting from Eq. (4.4–14) into Eq. (4.4–13) we obtain the formula

$$W(\mathbf{r}_1, \mathbf{r}_2, \nu) = \frac{1}{(2\pi)^2}\iint_{(z=0)} W(\mathbf{r}'_1, \mathbf{r}'_2, \nu)\left[\frac{\partial}{\partial z_1}\left(\frac{e^{-ikR_1}}{R_1}\right)\right]\left[\frac{\partial}{\partial z_2}\left(\frac{e^{ikR_2}}{R_2}\right)\right] d^2r'_1\, d^2r'_2. \qquad (4.4\text{–}15)$$

This formula expresses the cross-spectral density function of the field in the half-space $z > 0$ in terms of its boundary values on the plane $z = 0$. We will re-write it in a more explicit form. We have

$$\frac{\partial}{\partial z_1}\left(\frac{e^{-ikR_1}}{R_1}\right) = -\left(ik + \frac{1}{R_1}\right)\frac{e^{-ikR_1}}{R_1}\cos\theta_1, \qquad (4.4\text{–}16a)$$

$$\frac{\partial}{\partial z_2}\left(\frac{e^{-ikR_2}}{R_2}\right) = -\left(-ik + \frac{1}{R_2}\right)\frac{e^{ikR_2}}{R_2}\cos\theta_2, \qquad (4.4\text{–}16b)$$

where θ_1 and θ_2 are the angles that the lines S_1P_1 and S_2P_2 make with the positive z-direction. On substituting from Eqs. (4.4–16) into Eq. (4.4–15) we obtain the formula

$$W(\mathbf{r}_1, \mathbf{r}_2, \nu) = \left(\frac{k}{2\pi}\right)^2 \iint_{(z=0)} W(\mathbf{r}_1', \mathbf{r}_2', \nu) \times \left[1 + \frac{\mathrm{i}}{k}\left(\frac{1}{R_2} - \frac{1}{R_1}\right) + \frac{1}{k^2}\frac{1}{R_1 R_2}\right] \frac{\mathrm{e}^{\mathrm{i}k(R_2 - R_1)}}{R_1 R_2} \cos\theta_1 \cos\theta_2 \, \mathrm{d}^2 r_1' \, \mathrm{d}^2 r_2'. \tag{4.4–17}$$

Equation (4.4–17) is the required expression for the propagation of the cross-spectral density.

Before proceeding further we note an approximate form of Eq. (4.4–17) valid when the points $P_1(\mathbf{r}_1)$ and $P_2(\mathbf{r}_2)$ are many wavelengths away from the plane $z = 0$. Since we then have $R_1 \gg \lambda$, $R_2 \gg \lambda$, it follows that $1/R_1 \ll k$, $1/R_2 \ll k$. Under these circumstances Eq. (4.4–17) reduces to

$$W(\mathbf{r}_1, \mathbf{r}_2, \nu) \approx \left(\frac{k}{2\pi}\right)^2 \iint_{(z=0)} W(\mathbf{r}_1', \mathbf{r}_2', \nu) \frac{\mathrm{e}^{\mathrm{i}k(R_2 - R_1)}}{R_1 R_2} \cos\theta_1 \cos\theta_2 \, \mathrm{d}^2 r_1' \, \mathrm{d}^2 r_2'. \tag{4.4–18}$$

From Eq. (4.4–17) we can readily obtain the required law for the propagation of the mutual coherence function. We multiply both sides of Eq. (4.4–17) by $\mathrm{e}^{-2\pi\mathrm{i}\nu\tau}$ and integrate over the range $0 < \nu < \infty$. The left-hand side then represents the mutual coherence function $\Gamma(\mathbf{r}_1, \mathbf{r}_2, \tau)$ [see Eq. (4.3–40a)]. On the right-hand side we use the fact that

$$\Gamma(\mathbf{r}_1', \mathbf{r}_2', \tau) = \int_0^\infty W(\mathbf{r}_1', \mathbf{r}_2', \nu)\, \mathrm{e}^{-2\pi\mathrm{i}\nu\tau} \, \mathrm{d}\nu, \tag{4.4–19}$$

and we have, on interchanging the orders of integrations and differentiation,

$$\frac{1}{c}\frac{\partial \Gamma(\mathbf{r}_1', \mathbf{r}_2', \tau)}{\partial \tau} = -\int_0^\infty (\mathrm{i}k) W(\mathbf{r}_1', \mathbf{r}_2', \nu)\, \mathrm{e}^{-2\pi\mathrm{i}\nu\tau} \, \mathrm{d}\nu, \tag{4.4–19a}$$

and

$$\frac{1}{c^2}\frac{\partial^2 \Gamma(\mathbf{r}_1', \mathbf{r}_2', \tau)}{\partial \tau^2} = \int_0^\infty (\mathrm{i}k)^2 W(\mathbf{r}_1', \mathbf{r}_2', \nu)\, \mathrm{e}^{-2\pi\mathrm{i}\nu\tau} \, \mathrm{d}\nu, \tag{4.4–19b}$$

where use was made of the relation $2\pi\nu/c = k$. We then obtain the following formula, essentially due to Parrent (1959), for the mutual coherence function of the light at any two points in the half-space $z > 0$:

$$\Gamma(\mathbf{r}_1, \mathbf{r}_2, \tau) = \frac{1}{(2\pi)^2} \iint_{(z=0)} \frac{\cos\theta_1 \cos\theta_2}{R_1^2 R_2^2} \mathscr{D}\Gamma\left(\mathbf{r}_1', \mathbf{r}_2', \tau - \frac{R_2 - R_1}{c}\right) \mathrm{d}^2 r_1' \mathrm{d}^2 r_2', \tag{4.4–20}$$

where $\mathscr{D}$ is the differential operator

$$\mathscr{D} = 1 + \frac{R_2 - R_1}{c}\frac{\partial}{\partial \tau} - \frac{R_1 R_2}{c^2}\frac{\partial^2}{\partial \tau^2}. \tag{4.4–20a}$$

Equation (4.4–20) shows that the values of the mutual coherence function, for all pairs of points $P_1(\mathbf{r}_1)$ and $P_2(\mathbf{r}_2)$ in the half-space $z > 0$ into which the light propagates, and for all τ-values, may be determined from the knowledge of the mutual coherence function and its first and second derivatives with respect to τ at all pairs of points $S_1(\mathbf{r}_1')$ and $S_2(\mathbf{r}_2')$ in the plane $z = 0$.

If the points $P_1(\mathbf{r}_1)$ and $P_2(\mathbf{r}_2)$ are so far away from the plane $z = 0$ that $R_1 \gg \lambda$, $R_2 \gg \lambda$ for all points $S_1(\mathbf{r}_1')$ and $S_2(\mathbf{r}_2')$ in this plane and for all wavelengths $\lambda = c/\nu = 2\pi/k$ for which the absolute value $|W(\mathbf{r}_1', \mathbf{r}_2', \nu)|$ of the cross-spectral density function is appreciable, the formula (4.4–20) takes a simpler form. We may obtain this form by simply taking the Fourier transform of Eq. (4.4–18) and making use of Eq. (4.4–19b). We then find that

$$\Gamma(\mathbf{r}_1, \mathbf{r}_2, \tau) \approx -\frac{1}{(2\pi c)^2}\iint_{(z=0)} \frac{\cos\theta_1 \cos\theta_2}{R_1 R_2}\Gamma''\left(\mathbf{r}_1', \mathbf{r}_2', \tau - \frac{R_2 - R_1}{c}\right) \mathrm{d}^2 r_1'\, \mathrm{d}^2 r_2', \tag{4.4–21}$$

where Γ'' represents the second derivative of Γ with respect to τ.

With the help of the preceding formulas, one may derive a number of useful results that have a bearing on the coherence properties of optical images, on problems encountered in radiometry in connection with determining the angular distribution of the radiant intensity from sources of different states of coherence, etc. We will consider such questions later.

4.4.3 Propagation of correlations from finite surfaces

The formulas that we have just derived, relating to the propagation of the cross-spectral density function and the mutual coherence function from a plane surface, are exact consequences of the differential equations satisfied by these correlation functions. One can also derive corresponding formulas for propagation from curved surfaces. However, such formulas are of limited use as they require knowledge of the Green's functions associated with such surfaces and, except for the simplest geometrical shapes, the Green's functions cannot be obtained in a closed form. It is possible, however, to derive an approximate propagation law that is quite adequate for many situations frequently encountered in practice. The approximate law, which we will now derive, may, in a sense, be regarded as a counterpart of the well-known Huygens–Fresnel principle for propagation of monochromatic light.

Let us consider a light wave which emerges from an optical system and let $\mathscr{A}$ be a (fictitious) open surface intercepting the wave (see Fig. 4.10). Suppose that the values of one of the second-order correlation functions (the mutual coherence function or the cross-spectral density function) are known for all pairs of points $S_1(\mathbf{r}_1')$ and $S_2(\mathbf{r}_2')$ on the surface $\mathscr{A}$. We wish to determine the values of this correlation function at all pairs of points $P_1(\mathbf{r}_1)$, $P_2(\mathbf{r}_2')$ on that side of the surface $\mathscr{A}$ into which the light propagates.

Let $V(\mathbf{r}_1', t)$ and $V(\mathbf{r}_2', t)$ be the complex light disturbances at $S_1(\mathbf{r}_1')$ and $S_2(\mathbf{r}_2')$ and let $\widetilde{V}(\mathbf{r}_1', \nu)$ and $\widetilde{V}(\mathbf{r}_2', \nu)$ be their corresponding spectral amplitudes. More precisely these four quantities refer to a typical realization of the underlying

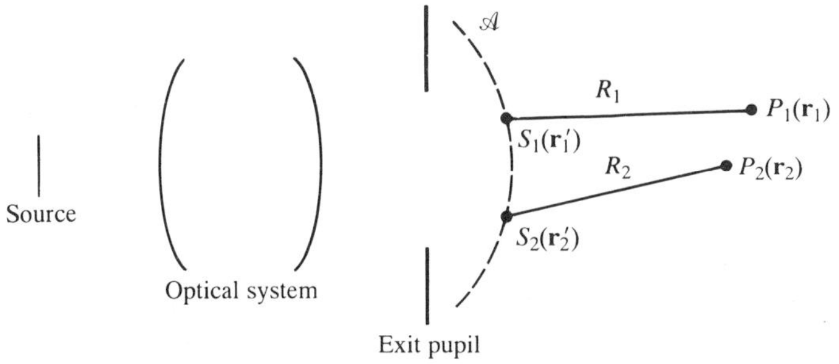

Fig. 4.10 Illustrating the notation relating to the approximate propagation laws for the cross-spectral density and for the mutual intensity.

stochastic process that characterizes the fluctuating field. We will assume that the process is stationary, at least in the wide sense, and is also ergodic. The spectral amplitudes must then be interpreted as generalized functions. The complex amplitudes at the points $P_1(\mathbf{r}_1)$ and $P_2(\mathbf{r}_2)$ may be expressed in terms of the complex amplitudes at all points on the surface $\mathscr{A}$ in the following approximate form, which is the mathematical expression of the Huygens–Fresnel principle (Born and Wolf, 1980, Sec. 8.2):

$$\widetilde{V}(\mathbf{r}_1, \nu) = \int_{\mathscr{A}} \widetilde{V}(\mathbf{r}_1', \nu) \frac{e^{ikR_1}}{R_1} \Lambda_1 \, d^2 r_1', \tag{4.4–22a}$$

$$\widetilde{V}(\mathbf{r}_2, \nu) = \int_{\mathscr{A}} \widetilde{V}(\mathbf{r}_2', \nu) \frac{e^{ikR_2}}{R_2} \Lambda_2 \, d^2 r_2', \tag{4.4–22b}$$

Here R_1 and R_2 are the distances $\overline{S_1 P_1}$ and $\overline{S_2 P_2}$, respectively, and Λ_1 and Λ_2 are inclination factors. If the angles of diffraction at the exit pupil are sufficiently small, one has approximately

$$\Lambda_2(k) \approx \Lambda_1(k) \approx \frac{ik}{2\pi}. \tag{4.4–23}$$

From Eqs. (4.4–22) it follows that

$$\widetilde{V}^*(\mathbf{r}_1, \nu)\widetilde{V}(\mathbf{r}_2, \nu') \approx \int_{\mathscr{A}}\int_{\mathscr{A}} \widetilde{V}^*(\mathbf{r}_1', \nu)\widetilde{V}(\mathbf{r}_2', \nu') \frac{e^{i(k'R_2 - kR_1)}}{R_1 R_2} \Lambda_1^*(k)\Lambda_2(k') \, d^2 r_1' \, d^2 r_2' \tag{4.4–24}$$

where, of course, $k = 2\pi\nu/c$ and $k' = 2\pi\nu'/c$. If now we take the ensemble average of both sides of Eq. (4.4–24) over the different realizations, interchange averaging and integration on the right-hand side and recall the definition (4.3–39) of the cross-spectral density function, we obtain the following propagation law:

$$W(\mathbf{r}_1, \mathbf{r}_2, \nu) = \int_{\mathscr{A}}\int_{\mathscr{A}} W(\mathbf{r}_1', \mathbf{r}_2', \nu) \frac{e^{ik(R_2 - R_1)}}{R_1 R_2} \Lambda_1^*(k)\Lambda_2(k) \, d^2 r_1' \, d^2 r_2'. \tag{4.4–25}$$

If we assume that the light is quasi-monochromatic around a mean frequency $\bar{\nu}$, we may derive from Eq. (4.4–25) a propagation law for the mutual coherence function. For this purpose we first neglect the weak dependence of Λ_1 and Λ_2 on the frequency, i.e. we replace these quantities by the values $\overline{\Lambda}_1$ and $\overline{\Lambda}_2$, which they take at the mean frequency. Next we multiply both sides of Eq. (4.4–25) by $e^{-2\pi i\nu\tau}$, integrate over ν $(0<\nu<\infty)$, and recall the relation (4.3–40a) between the cross-spectral density function and the mutual coherence function. We then obtain the required propagation law:

$$\Gamma(\mathbf{r}_1, \mathbf{r}_2, \tau) = \int_{\mathscr{A}}\int_{\mathscr{A}} \frac{\Gamma[\mathbf{r}_1', \mathbf{r}_2', \tau - (R_2 - R_1)/c]}{R_1 R_2} \bar{\Lambda}_1^* \bar{\Lambda}_2 \, d^2 r_1' \, d^2 r_2'. \quad (4.4\text{–}26)$$

In practice the path difference $|R_2 - R_1|$ is usually small compared to the coherence length ($\sim c/\Delta\nu$) of the light; i.e. the retardation term on the right-hand side of Eq. (4.4–26) is small compared to the coherence time:

$$\frac{|R_2 - R_1|}{c} \ll \frac{1}{\Delta\nu}. \quad (4.4\text{–}27)$$

Under these circumstances we have, from the properties of the envelope representation of real signals (Section 3.1.2),

$$\Gamma[\mathbf{r}_1', \mathbf{r}_2', \tau - (R_2 - R_1)/c] \approx \Gamma(\mathbf{r}_1', \mathbf{r}_2', \tau)\, e^{i\bar{k}(R_2 - R_1)}, \quad (4.4\text{–}28)$$

$(\bar{k} = 2\pi\bar{\nu}/c)$, and the formula (4.4–26) gives

$$\Gamma(\mathbf{r}_1, \mathbf{r}_2, \tau) \approx \int_{\mathscr{A}}\int_{\mathscr{A}} \Gamma(\mathbf{r}_1', \mathbf{r}_2', \tau) \frac{e^{i\bar{k}(R_2 - R_1)}}{R_1 R_2} \bar{\Lambda}_1^* \bar{\Lambda}_2 \, d^2 r_1' \, d^2 r_2'. \quad (4.4\text{–}29)$$

More generally, if the region between the surface $\mathscr{A}$ and the field points $P_1(\mathbf{r}_1)$ and $P_2(\mathbf{r}_2)$ is not free space, one would clearly have, in place of Eq. (4.4–25), the formula

$$W(\mathbf{r}_1, \mathbf{r}_2, \nu) = \int_{\mathscr{A}}\int_{\mathscr{A}} W(\mathbf{r}_1', \mathbf{r}_2', \nu) K^*(\mathbf{r}_1, \mathbf{r}_1', \nu) K(\mathbf{r}_2, \mathbf{r}_2', \nu) \, d^2 r_1' \, d^2 r_2'. \quad (4.4\text{–}30)$$

In this formula the function $K(\mathbf{r}, \mathbf{r}', \nu)$ is a transmission function (the impulse response function) that characterizes the propagation in the domain beyond the surface $\mathscr{A}$, i.e. it represents the complex disturbance at $\mathbf{r}$ due to a monochromatic point source of frequency ν, and of unit strength and zero phase, situated at $\mathbf{r}'$ on the surface $\mathscr{A}$. If the effective band of the light is sufficiently narrow, one obtains from Eq. (4.4–30), in a similar way to that employed in deriving Eq. (4.4–29), the following approximate generalized propagation law for the mutual coherence function:

$$\Gamma(\mathbf{r}_1, \mathbf{r}_2, \tau) \approx \int_{\mathscr{A}}\int_{\mathscr{A}} \Gamma(\mathbf{r}_1', \mathbf{r}_2', \tau) K^*(\mathbf{r}_1, \mathbf{r}_1', \bar{\nu}) K(\mathbf{r}_2, \mathbf{r}_2', \bar{\nu}) \, d^2 r_1' \, d^2 r_2'. \quad (4.4\text{–}31)$$

4.4.4 The van Cittert–Zernike theorem

One of the central theorems of the elementary theory of partial coherence was formulated by van Cittert (1934) and later, in a more general form, by Zernike (1938). It expresses the field correlations at two points in the field, generated by a spatially incoherent, quasi-monochromatic, planar source.

The van Cittert–Zernike theorem may be readily derived by starting from our formula (4.4–29). We set $\tau = 0$ in Eq. (4.4–29) and recall that, according to Eq. (4.3–34), $\Gamma(\mathbf{r}_1, \mathbf{r}_2, 0)$ is just the mutual intensity $J(\mathbf{r}_1, \mathbf{r}_2)$. If we also make in Eq.

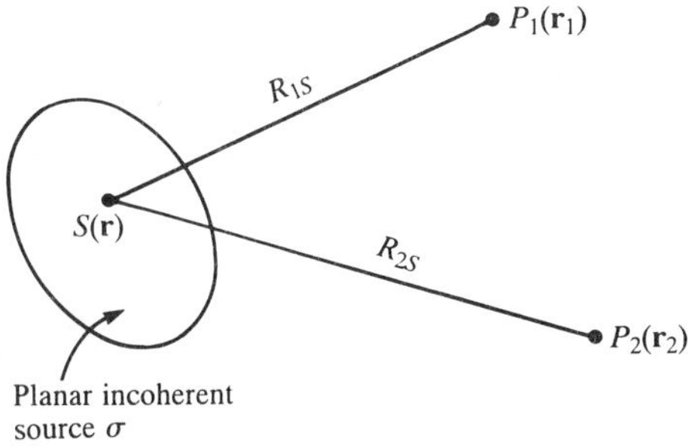

Fig. 4.11 Illustrating the notation relating to the van Cittert–Zernike theorem.

(4.4–29) the small angle approximation indicated by Eq. (4.4–23), we obtain the following formula, sometimes called *Zernike's propagation law* for the mutual intensity:

$$J(\mathbf{r}_1, \mathbf{r}_2) = \left(\frac{\bar{k}}{2\pi}\right)^2 \int_{\mathscr{A}} \int_{\mathscr{A}} J(\mathbf{r}_1', \mathbf{r}_2') \frac{e^{i\bar{k}(R_2 - R_1)}}{R_1 R_2} \, d^2 r_1' \, d^2 r_2'. \tag{4.4–32}$$

Suppose now that the open surface $\mathscr{A}$ coincides with the radiating surface σ of a spatially incoherent, planar, quasi-monochromatic secondary source. Then, for any two points $S_1(\mathbf{r}_1')$ and $S_2(\mathbf{r}_2')$ on σ,

$$J(\mathbf{r}_1', \mathbf{r}_2') = I(\mathbf{r}_1')\delta^{(2)}(\mathbf{r}_2' - \mathbf{r}_1'), \tag{4.4–33}$$

where $I(\mathbf{r}')$ is a measure of the intensity at $\mathbf{r}'$ and $\delta^{(2)}$ is the two-dimensional Dirac delta function. The presence of the delta function on the right-hand side of Eq. (4.4–33) expresses the fact that any two elements of the source are assumed to be mutually uncorrelated. This, of course, is an idealization. Any source found in nature will be correlated over distances that are at least of the order of the mean wavelength of the light. This is true even for sources that are generally regarded as spatially completely incoherent, e.g. blackbody sources, as we will see later (Section 13.1). However, if the correlations extend over distances that are not larger than about a mean wavelength and if the linear dimensions of such a source are large compared to the mean wavelength, as we now assume, the idealization expressed by Eq. (4.4–33) usually leads to a good approximation for the mutual intensity of the field.

On substituting from Eq. (4.4–33) into Eq. (4.4–32), we obtain the formula

$$J(\mathbf{r}_1, \mathbf{r}_2) = \left(\frac{\bar{k}}{2\pi}\right)^2 \int_{\sigma} I(\mathbf{r}') \frac{e^{i\bar{k}(R_{2S} - R_{1S})}}{R_{1S} R_{2S}} \, d^2 r', \tag{4.4–34}$$

where R_{1S} and R_{2S} are the distances from a typical point $S(\mathbf{r}')$ on the source to the points $P_1(\mathbf{r}_1)$ and $P_2(\mathbf{r}_2)$, respectively (see Fig. 4.11).

If we normalize Eq. (4.4–34) according to Eq. (4.3–35) we obtain the following expression for the (equal-time) complex degree of coherence of the field generated by our spatially incoherent source σ:

$$j(\mathbf{r}_1, \mathbf{r}_2) = \frac{1}{[I(\mathbf{r}_1)]^{1/2}[I(\mathbf{r}_2)]^{1/2}} \left(\frac{\bar{k}}{2\pi}\right)^2 \int_{\sigma} I(\mathbf{r}') \frac{e^{i\bar{k}(R_{2S} - R_{1S})}}{R_{1S} R_{2S}} \, d^2 r', \tag{4.4–35}$$

where

$$I(\mathbf{r}_j) = J(\mathbf{r}_j, \mathbf{r}_j) = \left(\frac{\bar{k}}{2\pi}\right)^2 \int_\sigma \frac{I(\mathbf{r}')}{R_{jS}^2} \mathrm{d}^2 r', \quad (j = 1, 2), \tag{4.4–36}$$

is the intensity at the field point $P_j(\mathbf{r}_j)$.

The formula (4.4–35) is the mathematical formulation of the van Cittert–Zernike theorem. It expresses the equal-time degree of coherence at two points $P_1(\mathbf{r}_1)$ and $P_2(\mathbf{r}_2)$ in the field generated by a planar, spatially incoherent, quasi-monochromatic source σ in terms of the intensity distribution $I(\mathbf{r}')$ across the source and the intensities $I(\mathbf{r}_1)$ and $I(\mathbf{r}_2)$ at the two field points. We stress that Eq. (4.4–35) was derived on the assumption that the path differences $|R_{2S} - R_{1S}|$ are small compared to the coherence length of the light and that the angles which the lines SP_1 and SP_2 make with the normal to the source are small.

The integral that appears in the expression (4.4–35) is of the same form as one which is frequently encountered in quite a different connection, namely in calculations of diffraction patterns on the basis of the Huygens–Fresnel–Kirchhoff diffraction theory (Born and Wolf, 1980, Sec. 8.2 and Sec. 8.3). More specifically the van Cittert–Zernike theorem, expressed by the formula (4.4–35), implies that, under the conditions stated, *the equal-time degree of coherence* $j(\mathbf{r}_1, \mathbf{r}_2)$, *is equal to the normalized complex amplitude in a certain diffraction pattern: namely, the normalized complex amplitude at a point* $P_2(\mathbf{r}_2)$ *in the diffraction pattern formed by a monochromatic spherical wave of frequency* $\bar{k}c/2\pi$ *converging towards the point* $P_1(\mathbf{r}_1)$ *and diffracted at an aperture in an opaque screen, of the same size, shape and location as the incoherent source* σ, *with the amplitude distribution across the aperture being proportional to the intensity distribution across the source.*

In many cases of practical interest the field points $P_1(\mathbf{r}_1)$ and $P_2(\mathbf{r}_2)$ are situated in the far-zone of the source. Under these circumstances the van Cittert–Zernike theorem takes a simpler form, which we will now derive.

Let us take the origin O of the position vectors to be in the source region σ, and let $\mathbf{s}_1$ and $\mathbf{s}_2$ be the unit vectors pointing in the directions from O to the field points $P_1(\mathbf{r}_1)$ and $P_2(\mathbf{r}_2)$ (see Fig. 4.12). Then

$$\mathbf{r}_1 = r_1\mathbf{s}_1, \quad \mathbf{r}_2 = r_2\mathbf{s}_2. \tag{4.4–37}$$

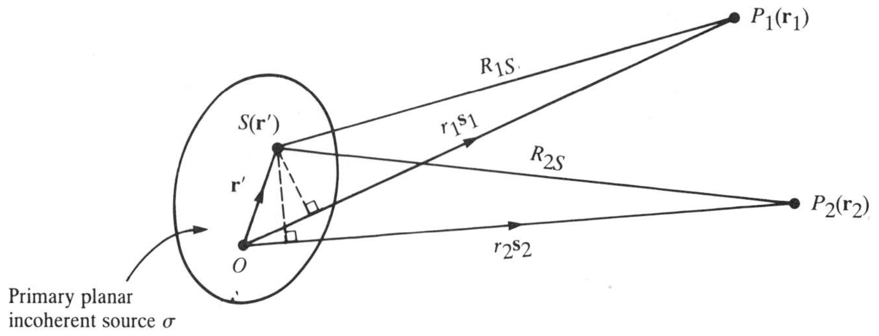

Fig. 4.12 Illustrating the notation relating to the far-zone form of the van Cittert–Zernike theorem.

We will consider the asymptotic behavior of the integral in Eq. (4.4–35) as $\bar{k}r_1 \to \infty$, $\bar{k}r_2 \to \infty$, with the two directions specified by the unit vectors $\mathbf{s}_1$ and $\mathbf{s}_2$ fixed. When r_1 and r_2 are sufficiently large

$$R_{1S} \sim r_1 - \mathbf{s}_1 \cdot \mathbf{r}', \quad R_{2S} \sim r_2 - \mathbf{s}_2 \cdot \mathbf{r}'. \tag{4.4–38}$$

If we substitute from Eq. (4.4–38) into the formulas (4.4–34) to (4.4–36) and neglect in the denominators the terms $\mathbf{s}_1 \cdot \mathbf{r}'$ and $\mathbf{s}_2 \cdot \mathbf{r}'$ in comparison with r_1 and r_2 respectively, the formulas simplify to

$$J(\mathbf{r}_1, \mathbf{r}_2) = \left(\frac{\bar{k}}{2\pi}\right)^2 \frac{e^{i\bar{k}(r_2-r_1)}}{r_1 r_2} \int_\sigma I(\mathbf{r}')\, e^{-i\bar{k}(\mathbf{s}_2-\mathbf{s}_1)\cdot\mathbf{r}'}\, d^2r', \tag{4.4–39}$$

and

$$j(\mathbf{r}_1, \mathbf{r}_2) = e^{i\bar{k}(r_2-r_1)} \frac{\displaystyle\int_\sigma I(\mathbf{r}')\, e^{-i\bar{k}(\mathbf{s}_2-\mathbf{s}_1)\cdot\mathbf{r}'}\, d^2r'}{\displaystyle\int_\sigma I(\mathbf{r}')\, d^2r'}. \tag{4.4–40}$$

We see that when the field points $P_1(\mathbf{r}_1)$ and $P_2(\mathbf{r}_2)$ are in the far zone of the source, the mutual intensity $J(\mathbf{r}_1, \mathbf{r}_2)$ and the equal-time degree of coherence $j(\mathbf{r}_1, \mathbf{r}_2)$ are expressible in terms of the Fourier transform of the intensity $I(\mathbf{r}')$ across the source. We may refer to the formula (4.4–40) as the *far-zone form of the van Cittert–Zernike theorem*.

As an example let us consider an incoherent, quasi-monochromatic, circular source σ, of radius a, centered at O and of uniform intensity $I(\mathbf{r}') = \text{constant}$. For the sake of simplicity we take the points $P_1(\mathbf{r}_1)$ and $P_2(\mathbf{r}_2)$ in the far zone to be situated at the same distance $r\ (= r_1 = r_2)$ from the origin O and close to the normal direction. The expression (4.4–40) then reduces to

$$j(\mathbf{r}_1, \mathbf{r}_2) = \frac{\displaystyle\int_{r'\leqslant a} e^{-i\bar{k}(\mathbf{s}_2-\mathbf{s}_1)\cdot\mathbf{r}'}\, d^2r'}{\displaystyle\int_{r'\leqslant a} d^2r'}. \tag{4.4–41}$$

To evaluate the integral in the numerator of Eq. (4.4–41) we set

$$\mathbf{r}' = (\rho\cos\theta, \rho\sin\theta), \quad \mathbf{s}_{2\perp} - \mathbf{s}_{1\perp} \equiv (w\cos\psi, w\sin\psi), \tag{4.4–42}$$

where $\mathbf{s}_{1\perp}$ and $\mathbf{s}_{2\perp}$ are the projections, considered as two-dimensional vectors, of the three-dimensional unit vectors $\mathbf{s}_1$ and $\mathbf{s}_2$ onto the source plane. The formula (4.4–41) then becomes

$$j(\mathbf{r}_1, \mathbf{r}_2) = \frac{1}{\pi a^2} \int_0^a\!\!\int_0^{2\pi} e^{-i\bar{k}\rho w\cos(\theta-\psi)} \rho\, d\rho\, d\theta. \tag{4.4–43}$$

The integral that appears on the right-hand side is well known in the theory of Fraunhofer diffraction at a circular aperture (see, for example, Born and Wolf, 1980, Sec. 8.5.2). On substituting its value into Eq. (4.4–43) one finds that

$$j(\mathbf{r}_1, \mathbf{r}_2) = \frac{2J_1(v)}{v}. \tag{4.4–44}$$

Here

$$v = \bar{k}a|\mathbf{s}_{2\perp} - \mathbf{s}_{1\perp}| \tag{4.4–45}$$

and $J_1(x)$ is the Bessel function of the first kind and first order. We note that, since

$$\mathbf{s}_{1\perp} = \left(\frac{x_1}{r}, \frac{y_1}{r}, 0\right), \quad \mathbf{s}_{2\perp} = \left(\frac{x_2}{r}, \frac{y_2}{r}, 0\right), \tag{4.4–46}$$

the variable v may also be expressed as

$$v = \bar{k}\left(\frac{a}{r}\right)d_{12} = \bar{k}\left(\frac{a}{r}\right)[(x_1 - x_2)^2 + (y_1 - y_2)^2]^{1/2}, \tag{4.4–47}$$

where evidently d_{12} is the distance between the points $P_1(\mathbf{r}_1)$ and $P_2(\mathbf{r}_2)$.

The behavior of the function $2J_1(v)/v$, which according to Eq. (4.4–44) represents the equal-time degree of coherence in the present example, is shown in Fig. 4.13. We see that it decreases steadily from the value 1 for $v = 0$ to the value 0 for $v = 3.83$. Hence, with increasing separation of the two points $P_1(\mathbf{r}_1)$ and $P_2(\mathbf{r}_2)$ the (equal-time) degree of coherence $j(\mathbf{r}_1, \mathbf{r}_2)$ decreases steadily at first and there is complete incoherence $[j(\mathbf{r}_1, \mathbf{r}_2) = 0]$ when

$$d_{12} = \frac{3.83}{\bar{k}}\left(\frac{r}{a}\right) = \frac{0.61r\bar{\lambda}}{a}. \tag{4.4–48}$$

A further increase in the separation leads to a re-introduction of a small amount of coherence, but the degree of coherence remains, in absolute value, smaller than 0.14, and there is further complete incoherence at $v = 7.02$. Since the function $2J_1(v)/v$ changes sign as v passes through each of its zeros, the phase of the complex degree of coherence changes there by π.

The function $2J_1(v)/v$ decreases steadily from the value 1 for $v = 0$ to the value 0.88 for $v = 1$, i.e. for the separation

$$d_{12} = 0.16\bar{\lambda}r/a. \tag{4.4–49}$$

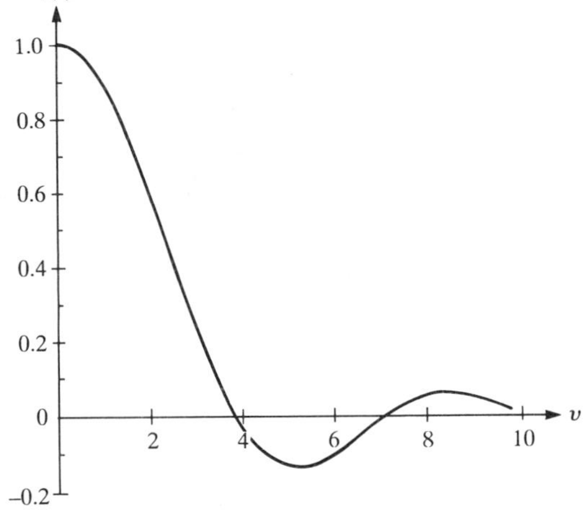

Fig. 4.13 The function $2J_1(v)/v$.

Now, in practice, a drop from the ideal value of unity, which does not exceed about 12%, is not regarded as very significant. Therefore, *in the far zone and close to the direction normal to the source plane and parallel to it, the light produced by a spatially incoherent, quasi-monochromatic, uniform, circular source of radius a is approximately coherent over a circular area* ΔA *whose diameter is* $0.16\bar{\lambda}/\alpha$, *where* $\alpha = a/r$ *is the angular radius subtended by the source, when viewed from* ΔA. We note that $\Delta A = \pi(0.16\bar{\lambda}/2\alpha)^2$, i.e. that

$$\Delta A = 0.063 r^2 \bar{\lambda}^2 / S, \qquad (4.4\text{–}50)$$

where $S = \pi a^2$ is the area of the source. This expression is seen to be in agreement with the order of magnitude relation (4.2–6) which we derived earlier by a rough intuitive argument.

4.4.5 Propagation of correlations from primary sources

In the preceding sections we considered the propagation of second-order correlations in free space, without paying any special attention to the way in which the fluctuating field was generated in the first place. The field fluctuations are, of course, a reflection of the fact that the sources of the field emit radiation by a mechanism that is never strictly deterministic and can only be described in statistical terms. We will now examine how the correlations of a field are related to the correlations of its source.

Let $Q^{(\mathrm{r})}(\mathbf{r}, t)$ represent a realization of a (real) fluctuating scalar source, assumed to occupy a finite domain D in free space, and let $V^{(\mathrm{r})}(\mathbf{r}, t)$ be the field generated by the source. Then $V^{(\mathrm{r})}(\mathbf{r}, t)$ and $Q^{(\mathrm{r})}(\mathbf{r}, t)$ are related by the inhomogeneous wave equation

$$\nabla^2 V^{(\mathrm{r})}(\mathbf{r}, t) - \frac{1}{c^2}\frac{\partial^2 V^{(\mathrm{r})}(\mathbf{r}, t)}{\partial t^2} = -4\pi Q^{(\mathrm{r})}(\mathbf{r}, t). \qquad (4.4\text{–}51)$$

Let us associate with $V^{(\mathrm{r})}$ and $Q^{(\mathrm{r})}$ the analytic signals V and Q, respectively. By a straightforward extension of the analysis that led to Eq. (4.4–7), it follows that $V(\mathbf{r}, t)$ obeys the inhomogeneous wave equation

$$\nabla^2 V(\mathbf{r}, t) - \frac{1}{c^2}\frac{\partial^2 V(\mathbf{r}, t)}{\partial t^2} = -4\pi Q(\mathbf{r}, t). \qquad (4.4\text{–}52)$$

Let us write $\mathbf{r}_1$ in place of $\mathbf{r}$ and t_1 in place of t and then take the complex conjugate of the equation. This gives

$$\nabla_1^2 V^*(\mathbf{r}_1, t_1) - \frac{1}{c^2}\frac{\partial^2 V^*(\mathbf{r}_1, t_1)}{\partial t_1^2} = -4\pi Q^*(\mathbf{r}_1, t_1), \qquad (4.4\text{–}53)$$

where ∇_1^2 is the Laplacian operator taken with respect to $\mathbf{r}_1$. Next we write $\mathbf{r}_2$ in place of $\mathbf{r}$ and t_2 in place of t in Eq. (4.4–52), and we obtain the equation

$$\nabla_2^2 V(\mathbf{r}_2, t_2) - \frac{1}{c^2}\frac{\partial^2 V(\mathbf{r}_2, t_2)}{\partial t_2^2} = -4\pi Q(\mathbf{r}_2, t_2), \qquad (4.4\text{–}54)$$

where ∇_2^2 is, of course, the Laplacian operator taken with respect to $\mathbf{r}_2$. If we multiply the left-hand side of Eq. (4.4–53) by the left-hand side of Eq. (4.4–54) and also multiply together the right-hand sides, we find at once that

$$\left(\nabla_1^2 - \frac{1}{c^2}\frac{\partial^2}{\partial t_1^2}\right)\left(\nabla_2^2 - \frac{1}{c^2}\frac{\partial^2}{\partial t_2^2}\right)V^*(\mathbf{r}_1, t_1)\mathbf{V}(r_2, t_2) = (4\pi)^2 Q^*(\mathbf{r}_1, t_1)Q(\mathbf{r}_2, t_2). \tag{4.4–55}$$

If we take averages of both sides of Eq. (4.4–55) over the source ensemble, we find that

$$\left(\nabla_1^2 - \frac{1}{c^2}\frac{\partial^2}{\partial t_1^2}\right)\left(\nabla_2^2 - \frac{1}{c^2}\frac{\partial^2}{\partial t_2^2}\right)\Gamma_V(\mathbf{r}_1, \mathbf{r}_2; t_1, t_2) = (4\pi)^2 \Gamma_Q(\mathbf{r}_1, \mathbf{r}_2; t_1, t_2), \tag{4.4–56}$$

where

$$\Gamma_V(\mathbf{r}_1, \mathbf{r}_2; t_1, t_2) \equiv \langle V^*(\mathbf{r}_1, t_1)V(\mathbf{r}_2, t_2)\rangle, \tag{4.4–57}$$

$$\Gamma_Q(\mathbf{r}_1, \mathbf{r}_2; t_1, t_2) \equiv \langle Q^*(\mathbf{r}_1, t_1)Q(\mathbf{r}_2, t_2)\rangle. \tag{4.4–58}$$

We see that the correlation functions of the field and of the source are coupled by a fourth-order differential equation.

Suppose that the statistical ensemble which represents the fluctuations of the source is stationary, at least in the wide sense. The source correlation function $\Gamma_Q(\mathbf{r}_1, \mathbf{r}_2;\ t_1, t_2)$ will then depend on its two time arguments only through the time difference $t_2 - t_1$; and, since the relationship between the field variable V and the source variable Q is, according to Eq. (4.4–52), linear, the same is true for the field correlation function $\Gamma_V(\mathbf{r}_1, \mathbf{r}_2;\ t_1, t_2)$. We may, therefore, set

$$\Gamma_V(\mathbf{r}_1, \mathbf{r}_2; t_1, t_2) \equiv \Gamma_V(\mathbf{r}_1, \mathbf{r}_2, t_2 - t_1), \tag{4.4–59}$$

$$\Gamma_Q(\mathbf{r}_1, \mathbf{r}_2; t_1, t_2) \equiv \Gamma_Q(\mathbf{r}_1, \mathbf{r}_2, t_2 - t_1). \tag{4.4–60}$$

The function $\Gamma_V(\mathbf{r}_1, \mathbf{r}_2, \tau)$ is, of course, the mutual coherence function of the field that we denoted before by $\Gamma(\mathbf{r}_1, \mathbf{r}_2, \tau)$, where $\tau = t_2 - t_1$. The operators $\partial^2/\partial t_1^2$ and $\partial^2/\partial t_2^2$ on the right-hand side of Eq. (4.4–56) may each be replaced by $\partial^2/\partial \tau^2$. It follows that when the source is stationary, at least in the wide sense, the second-order correlation functions of the field and of the source are related by the fourth-order differential equation

$$\left(\nabla_1^2 - \frac{1}{c^2}\frac{\partial^2}{\partial \tau^2}\right)\left(\nabla_2^2 - \frac{1}{c^2}\frac{\partial^2}{\partial \tau^2}\right)\Gamma_V(\mathbf{r}_1, \mathbf{r}_2, \tau) = (4\pi)^2 \Gamma_Q(\mathbf{r}_1, \mathbf{r}_2, \tau). \tag{4.4–61}$$

Let us represent Γ_V and Γ_Q in terms of the cross-spectral densities W_V and W_Q of the field and of the source, via the Fourier transform relations

$$\Gamma_V(\mathbf{r}_1, \mathbf{r}_2, \tau) = \int_0^\infty W_V(\mathbf{r}_1, \mathbf{r}_2, \nu)\, e^{-2\pi i\nu\tau}\, d\nu, \tag{4.4–62a}$$

$$\Gamma_Q(\mathbf{r}_1, \mathbf{r}_2, \tau) = \int_0^\infty W_Q(\mathbf{r}_1, \mathbf{r}_2, \nu)\, e^{-2\pi i\nu\tau}\, d\nu. \tag{4.4–62b}$$

On substituting from Eqs. (4.4–62) into Eq. (4.4–61) and interchanging the order of differentiation and integration we obtain the following fourth-order differential equation, which relates the two cross-spectral densities:

$$(\nabla_1^2 + k^2)(\nabla_2^2 + k^2)W_V(\mathbf{r}_1, \mathbf{r}_2, \nu) = (4\pi)^2 W_Q(\mathbf{r}_1, \mathbf{r}_2, \nu), \qquad (4.4\text{–}63)$$

where

$$k = 2\pi\nu/c \qquad (4.4\text{–}64)$$

is the wave number associated with the frequency ν.

Equation (4.4–61) or (4.4–63) may be taken as the starting point in studying the properties of fields generated by steady-state primary sources of any state of coherence. We will consider such problems later (Section 5.2). Here we will only derive a useful formula which we will then need later. It provides an explicit solution of Eq. (4.4–63) for W_V in terms of W_Q, when the source radiates in free space. We will obtain the solution in two stages. First we re-write Eq. (4.4–63) in the form

$$(\nabla_2^2 + k^2)[(\nabla_1^2 + k^2)W_V(\mathbf{r}_1, \mathbf{r}_2, \nu)] = (4\pi)^2 W_Q(\mathbf{r}_1, \mathbf{r}_2, \nu) \qquad (4.4\text{–}65)$$

and solve this equation for $(\nabla_1^2 + k^2)W_V(\mathbf{r}_1, \mathbf{r}_2, \nu)$, whilst keeping $\mathbf{r}_1$ fixed. The solution of such a 'reduced wave equation' is well known to be (see, for example, Papas, 1965, Sec. 2.1)

$$(\nabla_1^2 + k^2)W_V(\mathbf{r}_1, \mathbf{r}_2, \nu) = -4\pi\int_D W_Q(\mathbf{r}_1, \mathbf{r}_2', \nu)\frac{e^{ikR_2}}{R_2}\,d^3r_2', \qquad (4.4\text{–}66)$$

where (see Fig. 4.14)

$$R_2 = |\mathbf{r}_2 - \mathbf{r}_2'| \qquad (4.4\text{–}67)$$

and the integration extends throughout the source region D. Equation (4.4–66) is of the same general form as Eq. (4.4–65) and so we may obtain its solution also at once. However, we must bear in mind that $V(\mathbf{r}_1)$ and $V(\mathbf{r}_2)$ enter the definition of W_V as complex conjugates. Hence the corresponding Green's function will be $\exp(-ikR_1)/R_1$ rather than $\exp(ikR_2)/R_2$. We then obtain the required solution of Eq. (4.4–63):

$$W_V(\mathbf{r}_1, \mathbf{r}_2, \nu) = \int_D\int_D W_Q(\mathbf{r}_1', \mathbf{r}_2', \nu)\frac{e^{ik(R_2-R_1)}}{R_1R_2}\,d^3r_1'\,d^3r_2', \qquad (4.4\text{–}68)$$

where

$$R_j = |\mathbf{r}_j - \mathbf{r}_j'|, \quad (j = 1, 2). \qquad (4.4\text{–}69)$$

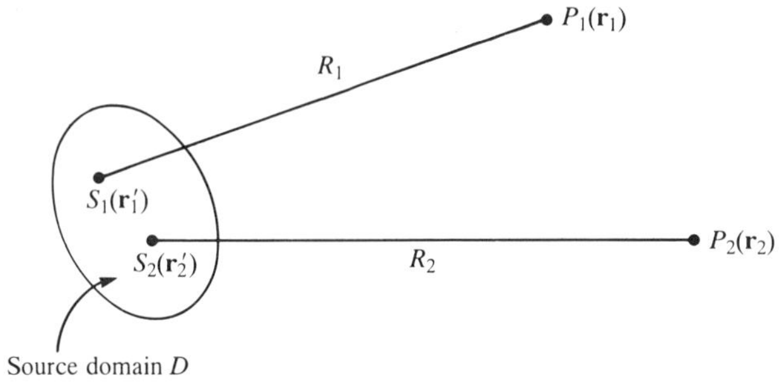

Fig. 4.14 Illustrating the notation relating to Eqs. (4.4–66) to (4.4–70).

The corresponding solution of Eq. (4.4–61) may be obtained in a similar way, starting with the well-known retarded Green's function solution of the inhomogeneous wave equation; or, more simply, by just taking the Fourier transform of Eq. (4.4–68) and using Eqs. (4.4–62). The result is

$$\Gamma_V(\mathbf{r}_1, \mathbf{r}_2, \tau) = \int_D \int_D \frac{\Gamma_Q[\mathbf{r}_1', \mathbf{r}_2', \tau - (R_2 - R_1)/c]}{R_1 R_2} \, \mathrm{d}^3 r_1' \, \mathrm{d}^3 r_2'. \qquad (4.4\text{–}70)$$

4.5 Fields of special types

4.5.1 Cross-spectrally pure light

We have seen in Section 4.3.2 that, if light of the same spectral distribution emanates from two pinholes at P_1 and P_2 and one examines the superposed light at a screen $\mathcal{B}$, some distance away from the plane containing the pinholes, one will find that, in general, it will have different spectral compositions at different points on the screen. We will now show that, under certain circumstances which are often encountered in practice, the spectral composition of the superposed light depends on the spectral composition of the two interfering beams in a relatively simple manner.

According to the spectral interference law (4.3–54) the spectral density $S(\mathbf{r}, \nu)$ of the light at the point $P(\mathbf{r})$ is related to the spectral densities $S^{(1)}(\mathbf{r}, \nu)$ and $S^{(2)}(\mathbf{r}, \nu)$ of the light reaching P from the two pinholes P_1 and P_2, respectively, by the formula

$$S(\mathbf{r}, \nu) = S^{(1)}(\mathbf{r}, \nu) + S^{(2)}(\mathbf{r}, \nu) + 2[S^{(1)}(\mathbf{r}, \nu)]^{1/2}[S^{(2)}(\mathbf{r}, \nu)]^{1/2} \operatorname{Re}[\mu(\mathbf{r}_1, \mathbf{r}_2, \nu)\, \mathrm{e}^{-2\pi i \nu \tau}], \qquad (4.5\text{–}1)$$

where $\mu(\mathbf{r}_1, \mathbf{r}_2, \nu)$ is the spectral degree of coherence of the light at the two pinholes $P_1(\mathbf{r}_1)$ and $P_2(\mathbf{r}_2)$ and

$$\tau = \frac{[P_1 P] - [P_2 P]}{c} \qquad (4.5\text{–}2)$$

is the difference in the times needed for light to travel from P_1 to P and from P_2 to P respectively (see Fig. 4.6).

Suppose now that the relative spectral distributions of the light at the two pinholes are the same, i.e. that

$$S(\mathbf{r}_2, \nu) = C_{12} S(\mathbf{r}_1, \nu), \qquad (4.5\text{–}3)$$

where C_{12} is a positive proportionality factor which is independent of the frequency. The spectral densities of the light reaching the point P in the plane $\mathcal{B}$ of observation from the two pinholes are, according to Eqs. (4.3–53) and (Eq. (4.3–41), given by

$$S^{(1)}(\mathbf{r}, \nu) = |K_1|^2 S(\mathbf{r}_1, \nu), \qquad (4.5\text{–}4a)$$

$$S^{(2)}(\mathbf{r}, \nu) = |K_2|^2 S(\mathbf{r}_2, \nu) = C_{12} |K_2|^2 S(\mathbf{r}_1, \nu). \qquad (4.5\text{–}4b)$$

On substituting from Eqs. (4.5–4) into Eq. (4.5–1) we obtain the following expression for $S(\mathbf{r}, \nu)$:

$$S(\mathbf{r}, \nu) = S(\mathbf{r}_1, \nu)\{|K_1|^2 + C_{12}|K_2|^2 + 2\surd(C_{12})|K_1|\,|K_2| \operatorname{Re}[\mu(\mathbf{r}_1, \mathbf{r}_2, \nu)\,\mathrm{e}^{-2\pi i\nu\tau}]\}. \tag{4.5–5}$$

Suppose now that there is a point, $P_0(\mathbf{r}_0)$ say, on the screen $\mathscr{B}$ where the spectral density of the light is the same as that at the pinholes, except for a proportionality factor that is independent of the frequency, and let τ_0 be the corresponding τ-value, i.e.

$$\tau_0 = \frac{[P_1P_0] - [P_2P_0]}{c}. \tag{4.5–6}$$

It is clear that, for this to be the case, the term in the curly brackets in Eq. (4.5–5), with τ replaced by τ_0, must be independent of the frequency. Now the factors K_1 and K_2 are each inversely proportional to the wavelength, i.e. directly proportional to the frequency (see footnote on p. 161). If we assume that the spectrum of the light at each pinhole is not too broad, we may neglect the frequency dependence of these two factors. The last term on the right-hand side of Eq. (4.5–5), (with τ replaced by τ_0) will be independent of the frequency if

$$\mu(\mathbf{r}_1, \mathbf{r}_2, \nu)\,\mathrm{e}^{-2\pi i\nu\tau_0} = f(\mathbf{r}_1, \mathbf{r}_2, \tau_0), \tag{4.5–7}$$

where $f(\mathbf{r}_1, \mathbf{r}_2, \tau_0)$ is some function of $\mathbf{r}_1$, $\mathbf{r}_2$, and τ_0 only.

When a region exists around a point P_0 in the plane of observation $\mathscr{B}$ such that the spectral distribution of the light in this region is of the same form as the spectral distribution of the light at the pinholes, we say that the light at the pinholes is *cross-spectrally pure*.‡ This will evidently be the case when Eq. (4.5–7) holds. The time delay τ_0, given by Eq. (4.5–6), then specifies the location of the point P_0.

We will now show that cross-spectrally pure light has a number of interesting properties. According to Eqs. (4.3–40a) and (4.3–47b), the mutual coherence function $\Gamma(\mathbf{r}_1, \mathbf{r}_2, \tau)$ may be expressed in the form

$$\Gamma(\mathbf{r}_1, \mathbf{r}_2, \tau) = \int_0^\infty [S(\mathbf{r}_1, \nu)]^{1/2}[S(\mathbf{r}_2, \nu)]^{1/2}\mu(\mathbf{r}_1, \mathbf{r}_2, \nu)\,\mathrm{e}^{-2\pi i\nu\tau}\,\mathrm{d}\nu. \tag{4.5–8}$$

On substituting from Eqs. (4.5–3) and (4.5–7) into Eq. (4.5–8) we obtain the following expression for $\Gamma(\mathbf{r}_1, \mathbf{r}_2, \tau)$:

$$\Gamma(\mathbf{r}_1, \mathbf{r}_2, \tau) = \surd(C_{12})f(\mathbf{r}_1, \mathbf{r}_2, \tau_0)\int_0^\infty S(\mathbf{r}_1, \nu)\,\mathrm{e}^{-2\pi i\nu(\tau-\tau_0)}\,\mathrm{d}\nu. \tag{4.5–9}$$

Now according to Eqs. (4.3–40a) and (4.3–41), the integral on the right-hand side of Eq. (4.5–9) is precisely $\Gamma(\mathbf{r}_1, \mathbf{r}_1, \tau - \tau_0)$ and hence Eq. (4.5–9) may be written as

$$\Gamma(\mathbf{r}_1, \mathbf{r}_2, \tau) = \surd(C_{12})f(\mathbf{r}_1, \mathbf{r}_2, \tau_0)\Gamma(\mathbf{r}_1, \mathbf{r}_1, \tau - \tau_0). \tag{4.5–10}$$

From this equation we can readily see the physical significance of the function

‡ This concept is due to Mandel (1961b), who also established some of the special properties that such light exhibits. The present treatment is largely based on the analysis of Mandel and Wolf (1976).

$f(\mathbf{r}_1, \mathbf{r}_2, \tau_0)$ that we introduced rather formally via Eq. (4.5–7). For this purpose we first note that, because of Eq. (4.5–3), we now have, from Eq. (4.3–40a), $\Gamma(\mathbf{r}_2, \mathbf{r}_2, \tau) = C_{12}\Gamma(\mathbf{r}_1, \mathbf{r}_1, \tau)$ and hence

$$[\Gamma(\mathbf{r}_1, \mathbf{r}_1, 0)]^{1/2}[\Gamma(\mathbf{r}_2, \mathbf{r}_2, 0)]^{1/2} = \sqrt{(C_{12})}\Gamma(\mathbf{r}_1, \mathbf{r}_1, 0). \qquad (4.5\text{–}11)$$

On dividing both sides of Eq. (4.5–10) by this factor and recalling the definition (4.3–12a) of the complex degree of coherence γ, we obtain the relation

$$\gamma(\mathbf{r}_1, \mathbf{r}_2, \tau) = f(\mathbf{r}_1, \mathbf{r}_2, \tau_0)\gamma(\mathbf{r}_1, \mathbf{r}_1, \tau - \tau_0). \qquad (4.5\text{–}12)$$

Finally if we set $\tau = \tau_0$ in Eq. (4.5–12) and recall that $\gamma(\mathbf{r}_1, \mathbf{r}_1, 0) \equiv 1$, we obtain the result that

$$f(\mathbf{r}_1, \mathbf{r}_2, \tau_0) = \gamma(\mathbf{r}_1, \mathbf{r}_2, \tau_0). \qquad (4.5\text{–}13)$$

Thus the function $f(\mathbf{r}_1, \mathbf{r}_2, \tau_0)$ is precisely the complex degree of coherence of the light vibrations at the two pinholes, for time delay τ_0. On making this identification in Eq. (4.5–7) we see that our *condition for cross-spectrally pure light* is

$$\mu(\mathbf{r}_1, \mathbf{r}_2, \nu) = \gamma(\mathbf{r}_1, \mathbf{r}_2, \tau_0)\, e^{2\pi i\nu\tau_0}. \qquad (4.5\text{–}14)$$

Strictly speaking this relation must hold for spectral components of every frequency ν that is contained in the spectrum of the light at P_1 and P_2. However, to a good approximation, it may clearly be relaxed by demanding that it be satisfied only for those frequency components that contribute significantly to the average intensity at P_1 and P_2.

In physical terms the condition (4.5–14) implies that *when the light at P_1 and P_2 is cross-spectrally pure, the absolute value of the spectral degree of coherence $\mu(\mathbf{r}_1, \mathbf{r}_2, \nu)$ is the same for every frequency ν present in the spectrum of the light at P_1 and P_2* and is equal to the absolute value of the degree of coherence $\gamma(\mathbf{r}_1, \mathbf{r}_2, \tau_0)$; and that, moreover, the phases of μ and of γ only differ by an additive term $2\pi\nu\tau_0$. Of course, for the light at P_1 and P_2 to be cross-spectrally pure, it must satisfy, in addition to (4.5–14), the requirement (4.5–3) that expresses the fact that the spectra of the light at P_1 and P_2 are the same, except possibly for a proportionality constant.

The condition (4.5–14) for cross-spectrally pure light has an immediate implication with regards to the form of the spectral distribution of the light at any point $P(\mathbf{r})$ in the plane of observation. On substituting from Eq. (4.5–14) into the general expression (4.5–5) for the spectral density we find that for cross-spectrally pure light

$$S(\mathbf{r}, \nu) = S(\mathbf{r}_1, \nu)\left\{A + B|\gamma(\mathbf{r}_1, \mathbf{r}_2, \tau_0)| \cos\left[\arg \gamma(\mathbf{r}_1, \mathbf{r}_2, \tau_0) - 2\pi\nu\left(\frac{R_1 - R_2}{c} - \tau_0\right)\right]\right\}, \qquad (4.5\text{–}15)$$

where R_1 and R_2 are the distances from the pinholes P_1 and P_2 to P, and

$$A = |K_1|^2 + C_{12}|K_2|^2, \quad B = 2\sqrt{(C_{12})}|K_1|\,|K_2|. \qquad (4.5\text{–}16)$$

It is seen from Eq. (4.5–15) that, as the point $P(\mathbf{r})$ explores the plane of observation, the spectral density $S(\mathbf{r}, \nu)$, with the frequency ν fixed, varies sinusoidally between the values

$$\left.\begin{aligned} S_{\max}(\mathbf{r}, \nu) &= S(\mathbf{r}_1, \nu)[A + B|\gamma(\mathbf{r}_1, \mathbf{r}_2, \tau_0)|] \\ S_{\min}(\mathbf{r}, \nu) &= S(\mathbf{r}_1, \nu)[A - B|\gamma(\mathbf{r}_1, \mathbf{r}_2, \tau_0)|] \end{aligned}\right\}. \tag{4.5–17}$$

By analogy with the definition (4.3–23) of the visibility of interference fringes we may define the *spectral visibility* $\mathcal{V}(\mathbf{r}, \nu)$ at frequency ν, at a point $P(\mathbf{r})$ in the interference pattern, by the formula

$$\mathcal{V}(\mathbf{r}, \nu) = \frac{S_{\max}(\mathbf{r}, \nu) - S_{\min}(\mathbf{r}, \nu)}{S_{\max}(\mathbf{r}, \nu) + S_{\min}(\mathbf{r}, \nu)}. \tag{4.5–18}$$

On substituting from Eqs. (4.5–17) into Eq. (4.5–18) we find that‡

$$\mathcal{V}(\mathbf{r}, \nu) = \frac{B}{A}|\gamma(\mathbf{r}_1, \mathbf{r}_2, \tau_0)|. \tag{4.5–19}$$

Since both the factors $|K_1|$ and $|K_2|$ are proportional to frequency, we see from (4.5–16) that the ratio B/A is frequency-independent and hence *the spectral visibility of interference fringes formed by two cross-spectrally pure beams is independent of the frequency*. Moreover, since $|K_1|$ and $|K_2|$ and hence the ratio B/A depend rather weakly on the position of the point P, the spectral visibility (with P_1 and P_2 fixed) is effectively constant over a portion of the plane of observation.

The sinusoidal variation of the spectral density across the interference pattern as given by Eq. (4.5–15), with the spectral visibility independent of the frequency, expresses the spectral modulation law for cross-spectrally pure beams.

We have noted several times already that the temporal and the spatial coherence properties of light are, in general, not independent of each other. However, cross-spectrally pure light is rather exceptional in this regard, as is seen at once if we substitute from Eq. (4.5–13) into Eq. (4.5–12). We then obtain the relation

$$\gamma(\mathbf{r}_1, \mathbf{r}_2, \tau) = \gamma(\mathbf{r}_1, \mathbf{r}_2, \tau_0)\gamma(\mathbf{r}_1, \mathbf{r}_1, \tau - \tau_0). \tag{4.5–20}$$

We see that the complex degree of coherence $\gamma(\mathbf{r}_1, \mathbf{r}_2, \tau)$ is now expressible as the product of two factors: one factor characterizes spatial coherence at the two pinholes for time delay τ_0 and the other characterizes temporal coherence at one of the pinholes. The formula (4.5–20) is usually referred to as *the reduction formula for cross-spectrally pure light*. It may also be expressed in the equivalent form

$$\Gamma(\mathbf{r}_1, \mathbf{r}_2, \tau) = \Gamma(\mathbf{r}_1, \mathbf{r}_2, \tau_0)\gamma(\mathbf{r}_1, \mathbf{r}_1, \tau - \tau_0), \tag{4.5–20a}$$

‡ More generally, whether or not the beams are cross-spectrally pure one can readily show by the use of Eqs. (4.5–18) and (4.3–54) that [cf. Eqs. (4.3–25) and (4.3–26)]

$$\mathcal{V}(\mathbf{r}, \nu) = 2[\eta(\mathbf{r}, \nu) + 1/\eta(\mathbf{r}, \nu)]^{-1}|\mu(\mathbf{r}_1, \mathbf{r}_2, \nu)|,$$

where $\eta(\mathbf{r}, \nu) = [S^{(1)}(\mathbf{r}, \nu)/S^{(2)}(\mathbf{r}, \nu)]^{1/2}$. Further one can also readily deduce from Eq. (4.3–54) that the location of the maxima of the spectral density $S(\mathbf{r}, \nu)$, with the frequency ν fixed, is given by the formula [cf. Eq. (4.3–27)] $\alpha(\mathbf{r}_1, \mathbf{r}_2, \nu) - 2\pi\nu(R_1 - R_2)/c = 2m\pi$, ($m = 0, \pm 1, \pm 2, \ldots$), where $\alpha(\mathbf{r}_1, \mathbf{r}_2, \nu)$ is the argument (phase) of the spectral degree of coherence: $\mu(\mathbf{r}_1, \mathbf{r}_2, \nu) = |\mu(\mathbf{r}_1, \mathbf{r}_2, \nu)| \exp[i\alpha(\mathbf{r}_1, \mathbf{r}_2, \nu)]$.

which follows at once from Eq. (4.5–20) on multiplying both sides by the product $[\Gamma(\mathbf{r}_1, \mathbf{r}_1, 0)\Gamma(\mathbf{r}_2, \mathbf{r}_2, 0)]^{1/2}$ and recalling the definition (4.3–12a) of the complex degree of coherence in terms of the mutual coherence function.

Conversely one can show that, when the reduction formula (4.5–20) holds with some value of τ_0, and the spectral densities of the light at the points P_1 and P_2 are equal to each other, except perhaps for a frequency-independent proportionality factor, then the light at these two points is cross-spectrally pure.

Cross-spectrally pure light can be generated, for example, by linear filtering of the light that emerges from the two pinholes in a Young interference experiment (see, for example, Kandpal, Saxena, Mehta, Vaishya and Joshi, 1993). We have shown in Section 4.3.2 that if the filters have sufficiently narrow passbands, the degree of coherence of the filtered light is given by Eq. (4.3–63), namely $\gamma^{(+)}(\mathbf{r}_1, \mathbf{r}_2, 0) = \mu(\mathbf{r}_1, \mathbf{r}_2, \nu_0)$, where ν_0 is the mid-frequency of the filters. This relation is seen to be in agreement with the condition (4.5–14) for cross-spectrally pure light. Other examples of such light are given by Mandel (1961b).

4.5.2 *Coherent light in the space-time domain*‡

We showed earlier that the absolute values of the complex degrees of coherence $\gamma(\mathbf{r}_1, \mathbf{r}_2, \tau)$ and $\mu(\mathbf{r}_1, \mathbf{r}_2, \nu)$ have upper bounds equal to unity. We saw that, when these upper bounds are attained, one can produce interference fringes of maximum possible visibility (unity) in suitable interference experiments. These limiting cases represent *complete second-order coherence*.

Complete second-order coherence in the space-time domain implies certain functional forms of the complex degree of coherence $\gamma(\mathbf{r}_1, \mathbf{r}_2, \tau)$ and of the mutual coherence function $\Gamma(\mathbf{r}_1, \mathbf{r}_2, \tau)$; and complete coherence in the space-frequency domain implies certain functional forms of the spectral degree of coherence $\mu(\mathbf{r}_1, \mathbf{r}_2, \nu)$ and of the cross-spectral density $W(\mathbf{r}_1, \mathbf{r}_2, \nu)$.

In the present section we consider the implications of complete second-order coherence in the space-time domain. The corresponding results relating to complete second-order coherence in the space-frequency domain will be discussed in the subsequent section (Section 4.5.3).

We begin with the non-negative definiteness condition satisfied by the complex degree of coherence $\gamma(\mathbf{r}_1, \mathbf{r}_2, \tau)$, namely the inequality

$$\sum_{j=1}^{n}\sum_{k=1}^{n} a_j^* a_k \gamma(\mathbf{r}_j, \mathbf{r}_k, \tau_k - \tau_j) \geqslant 0, \tag{4.5–21}$$

valid with any positive integer n, any set of points $\mathbf{r}_1, \mathbf{r}_2, \ldots, \mathbf{r}_n$, any set of temporal parameters $\tau_1, \tau_2, \ldots, \tau_n$ and any real or complex constants $a_1, a_2, \ldots, a_n$. The inequality (4.5–21) follows from Eqs. (4.3–37) and (4.3–12a). If we chose $n = 3$ it implies that

$$\begin{vmatrix} 1 & \gamma(\mathbf{r}_1, \mathbf{r}_2, \tau_2 - \tau_1) & \gamma(\mathbf{r}_1, \mathbf{r}_3, \tau_3 - \tau_1) \\ \gamma(\mathbf{r}_2, \mathbf{r}_1, \tau_1 - \tau_2) & 1 & \gamma(\mathbf{r}_2, \mathbf{r}_3, \tau_3 - \tau_2) \\ \gamma(\mathbf{r}_3, \mathbf{r}_1, \tau_1 - \tau_3) & \gamma(\mathbf{r}_3, \mathbf{r}_2, \tau_2 - \tau_3) & 1 \end{vmatrix} \geqslant 0 \tag{4.5–22}$$

‡ The analysis presented in this section is based on an investigation of Mehta, Wolf and Balachandran (1966).

for all real values of τ_1, τ_2, τ_3 and for any set of points $\mathbf{r}_1$, $\mathbf{r}_2$, $\mathbf{r}_3$ in the domain D in which the complex degree of coherence $\gamma(\mathbf{r}_j, \mathbf{r}_k, \tau)$ is defined. On evaluating the determinant in Eq. (4.5–22) and on using the Hermiticity relation

$$\gamma(\mathbf{r}_j, \mathbf{r}_k, \tau_k - \tau_j) = \gamma^*(\mathbf{r}_k, \mathbf{r}_j, \tau_j - \tau_k), \tag{4.5–23}$$

which follows from Eqs. (4.3–36) and (4.3–12a) of Section 4.3, we obtain the following inequality valid for all values of τ_1, τ_2 and τ_3:

$$|\gamma(\mathbf{r}_3, \mathbf{r}_2, \tau_2 - \tau_3) - \gamma(\mathbf{r}_1, \mathbf{r}_2, \tau_2 - \tau_1)\gamma(\mathbf{r}_3, \mathbf{r}_1, \tau_1 - \tau_3)|^2$$
$$\leqslant [1 - |\gamma(\mathbf{r}_1, \mathbf{r}_2, \tau_2 - \tau_1)|^2][1 - |\gamma(\mathbf{r}_3, \mathbf{r}_1, \tau_1 - \tau_3)|^2]. \tag{4.5–24}$$

With the help of this inequality we may readily derive a number of theorems relating to complete second-order coherence in the space-time domain.

(a) Complete self-coherence at a fixed point

Suppose that the light at some particular point $\mathbf{r} = \mathbf{R}$ in the domain D is completely coherent in the sense that

$$|\gamma(\mathbf{R}, \mathbf{R}, \tau)| = 1. \tag{4.5–25}$$

for all values of τ. This assumption implies that

$$\gamma(\mathbf{R}, \mathbf{R}, \tau) = \mathrm{e}^{\mathrm{i}\phi(\tau)}, \tag{4.5–26}$$

where $\phi(\tau)$ is some real function of the real variable τ and also perhaps of $\mathbf{R}$ (not displayed). Since γ satisfies the relation (4.5–23), $\phi(\tau)$ must obey the relation

$$\phi(-\tau) = -\phi(\tau) + 2m\pi, \tag{4.5–27}$$

where m is an integer.

Let us set $\mathbf{r}_1 = \mathbf{r}_2 = \mathbf{R}$ and $\mathbf{r}_3 = \mathbf{r} \in D$ in the inequality (4.5–24) and let us also make use of Eq. (4.5–25). We then obtain the inequality

$$|\gamma(\mathbf{r}, \mathbf{R}, \tau_2 - \tau_3) - \gamma(\mathbf{R}, \mathbf{R}, \tau_2 - \tau_1)\gamma(\mathbf{r}, \mathbf{R}, \tau_1 - \tau_3)|^2 \leqslant 0. \tag{4.5–28}$$

Since the left-hand side cannot be negative, it is clear that the inequality (4.5–28) can only hold if the left-hand side vanishes, i.e. if

$$\gamma(\mathbf{r}, \mathbf{R}, \tau_2 - \tau_3) = \gamma(\mathbf{R}, \mathbf{R}, \tau_2 - \tau_1)\gamma(\mathbf{r}, \mathbf{R}, \tau_1 - \tau_3). \tag{4.5–29}$$

If we choose $\tau_1 = \tau_3 = \tau_2 - \tau$ it follows that

$$\gamma(\mathbf{r}, \mathbf{R}, \tau) = \gamma(\mathbf{R}, \mathbf{R}, \tau)\gamma(\mathbf{r}, \mathbf{R}, 0). \tag{4.5–30}$$

Next let us set $\mathbf{r} = \mathbf{R}$ in Eq. (4.5–29) and make use of Eqs. (4.5–26) and (4.5–27). We then obtain the following functional equation for ϕ:

$$\phi(\tau_1 - \tau_2) + \phi(\tau_2 - \tau_3) + \phi(\tau_3 - \tau_1) = 2m\pi, \tag{4.5–31}$$

where m is any integer.

To solve the functional equation (4.5–31) we will make use of the fact that it holds for all values of τ_1, τ_2 and τ_3. Let us differentiate it with respect to τ_1 and then set $\tau_1 = \tau_3 = \tau_2 + \tau$. This gives the following differential equation for $\phi(\tau)$:

$$\frac{\mathrm{d}\phi(\tau)}{\mathrm{d}\tau} = -2\pi\nu_0, \tag{4.5–32}$$

where $\nu_0 = -(\mathrm{d}\phi/\mathrm{d}\tau)_{\tau=0}/2\pi$ is a constant. On integrating Eq. (4.5–32) with respect to τ we see that $\phi(\tau)$ must have the form

$$\phi(\tau) = -2\pi\nu_0\tau + \beta, \tag{4.5–33}$$

where β is a constant. From Eqs. (4.5–33) and (4.5–26) it follows that

$$\gamma(\mathbf{R}, \mathbf{R}, \tau) = \mathrm{e}^{\mathrm{i}(\beta - 2\pi\nu_0\tau)}. \tag{4.5–34}$$

Now $\gamma(\mathbf{R}, \mathbf{R}, 0) = 1$. Hence the constant β in Eq. (4.5–34) must be an integral multiple of 2π. Moreover, since γ, just like Γ, is an analytical signal which contains only positive frequency components, the constant ν_0 must be positive. Hence

$$\gamma(\mathbf{R}, \mathbf{R}, \tau) = \mathrm{e}^{-2\pi\mathrm{i}\nu_0\tau}, \quad (\nu_0 > 0). \tag{4.5–35}$$

On substituting from Eq. (4.5–35) into Eq. (4.5–30) we obtain for $\gamma(\mathbf{r}, \mathbf{R}, \tau)$ the expression

$$\gamma(\mathbf{r}, \mathbf{R}, \tau) = \gamma(\mathbf{r}, \mathbf{R}, 0)\,\mathrm{e}^{-2\pi\mathrm{i}\nu_0\tau}, \quad (\nu_0 > 0). \tag{4.5–36}$$

From this formula and the relation (4.5–23) we also obtain at once the expression

$$\gamma(\mathbf{R}, \mathbf{r}, \tau) = \gamma(\mathbf{R}, \mathbf{r}, 0)\,\mathrm{e}^{-2\pi\mathrm{i}\nu_0\tau}, \quad (\nu_0 > 0). \tag{4.5–37}$$

We may summarize the main result that we just obtained by saying that *if the light is completely self-coherent at some point* $\mathbf{R} \in D$ *in the sense expressed by Eq. (4.5–25), then* $\gamma(\mathbf{R}, \mathbf{R}, \tau)$, $\gamma(\mathbf{r}, \mathbf{R}, \tau)$ *and* $\gamma(\mathbf{R}, \mathbf{r}, \tau)$ *with* $\mathbf{r} \in D$ *are, for all values of* τ, *necessarily of the form given by Eqs. (4.5–35) to (4.5–37)*, i.e. they are periodic in τ.

(b) Complete mutual coherence at two fixed points

Next let us examine the case when the light is mutually completely coherent at two points $\mathbf{r}_1 = \mathbf{R}_1$, $\mathbf{r}_2 = \mathbf{R}_2$, both of which are located in the domain D containing the optical field. Expressed mathematically, we assume that

$$|\gamma(\mathbf{R}_1, \mathbf{R}_2, \tau)| = 1 \quad \text{for all } \tau \tag{4.5–38}$$

with the points $\mathbf{R}_1$ and $\mathbf{R}_2$ fixed. To see the implications of Eq. (4.5–38) we proceed in a similar way as before. We set $\mathbf{r}_1 = \mathbf{R}_1$, $\mathbf{r}_2 = \mathbf{R}_2$ and $\mathbf{r}_3 = \mathbf{r} \in D$ in Eq.(4.5–24) and use Eq. (4.5–38). We then obtain the inequality

$$|\gamma(\mathbf{r}, \mathbf{R}_2, \tau_2 - \tau_3) - \gamma(\mathbf{R}_1, \mathbf{R}_2, \tau_2 - \tau_1)\gamma(\mathbf{r}, \mathbf{R}_1, \tau_1 - \tau_3)|^2 \leqslant 0, \tag{4.5–39}$$

which can evidently only be satisfied if the relation

$$\gamma(\mathbf{r}, \mathbf{R}_2, \tau_2 - \tau_3) = \gamma(\mathbf{R}_1, \mathbf{R}_2, \tau_2 - \tau_1)\gamma(\mathbf{r}, \mathbf{R}_1, \tau_1 - \tau_3) \tag{4.5–40}$$

holds for all values of τ_1, τ_2 and τ_3. On taking the absolute values of both sides of

Eq. (4.5–40), setting $\mathbf{r} = \mathbf{R}_1$ and on using Eq. (4.5–38) once more, it follows that

$$|\gamma(\mathbf{R}_1, \mathbf{R}_1, \tau)| = 1 \quad \text{for all } \tau. \tag{4.5–41}$$

In a similar way, with $\mathbf{r} = \mathbf{R}_2$ one finds that

$$|\gamma(\mathbf{R}_2, \mathbf{R}_2, \tau)| = 1 \quad \text{for all } \tau. \tag{4.5–42}$$

Now Eqs. (4.5–41) and (4.5–42) imply that the field is completely self-coherent at each of the points $\mathbf{R}_1$ and $\mathbf{R}_2$ and hence, according to the results that we obtained earlier [Eqs. (4.5–35) to (4.5–37)],

$$\gamma(\mathbf{R}_1, \mathbf{R}_1, \tau) = \gamma(\mathbf{R}_2, \mathbf{R}_2, \tau) = \mathrm{e}^{-2\pi \mathrm{i}\nu_0\tau}, \quad (\nu_0 > 0), \tag{4.5–43}$$

and

$$\gamma(\mathbf{R}_1, \mathbf{R}_2, \tau) = \gamma(\mathbf{R}_1, \mathbf{R}_2, 0)\,\mathrm{e}^{-2\pi \mathrm{i}\nu_0\tau}, \quad (\nu_0 > 0), \tag{4.5–44}$$

However, since in view of Eq. (4.5–38), $\gamma(\mathbf{R}_1, \mathbf{R}_2, 0)$ is unimodular, Eq. (4.5–44) implies that

$$\gamma(\mathbf{R}_1, \mathbf{R}_2, \tau) = \mathrm{e}^{\mathrm{i}(\beta - 2\pi\nu_0\tau)}, \tag{4.5–45}$$

where β is a real constant. Finally, setting $\tau_1 = \tau_2 = \tau_3 + \tau$ in Eq. (4.5–40) and making use of Eq. (4.5–45), we obtain the formula

$$\gamma(\mathbf{r}, \mathbf{R}_2, \tau) = \gamma(\mathbf{r}, \mathbf{R}_1, \tau)\,\mathrm{e}^{\mathrm{i}\beta}, \tag{4.5–46}$$

valid for all τ.

We have thus shown that *if the light at two points $\mathbf{R}_1 \in D$, $\mathbf{R}_2 \in D$ is mutually completely coherent in the sense expressed by Eq. (4.5–38), then $\gamma(\mathbf{R}_1, \mathbf{R}_1, \tau)$, $\gamma(\mathbf{R}_2, \mathbf{R}_2, \tau)$ and $\gamma(\mathbf{R}_1, \mathbf{R}_2, \tau)$ are, for all values of τ, necessarily of the forms given by Eqs.* (4.5–43) *and* (4.5–44); *and that, moreover, if $\mathbf{r}$ is any other point in D, $\gamma(\mathbf{r}, \mathbf{R}_2, \tau)$ and $\gamma(\mathbf{r}, \mathbf{R}_1, \tau)$ are related by* Eq. (4.5–46), *where β is a real constant.*

(c) Complete coherence throughout a volume

Finally we will consider the case when the light is completely coherent throughout a volume D, in the sense that

$$|\gamma(\mathbf{r}_1, \mathbf{r}_2, \tau)| = 1 \quad \text{for all } \tau \tag{4.5–47}$$

and for all $\mathbf{r}_1 \in D$, $\mathbf{r}_2 \in D$. We start once again with the inequality (4.5–24), where we now set $\tau_1 = \tau_2 = \tau_3$ and make use of the assumption (4.5–47). We then obtain the relation

$$\gamma(\mathbf{r}_3, \mathbf{r}_2, 0) = \gamma(\mathbf{r}_1, \mathbf{r}_2, 0)\gamma(\mathbf{r}_3, \mathbf{r}_1, 0). \tag{4.5–48}$$

Equation (4.5–47) implies that

$$\gamma(\mathbf{r}_j, \mathbf{r}_k, 0) = \mathrm{e}^{\mathrm{i}\phi(\mathbf{r}_j, \mathbf{r}_k, 0)}, \; (j,\, k = 1, 2, 3), \tag{4.5–49}$$

where ϕ is real. From Eqs. (4.5–48) and (4.5–49) it follows that the phase function ϕ satisfies the functional equation

$$\phi(\mathbf{r}_1, \mathbf{r}_2, 0) + \phi(\mathbf{r}_3, \mathbf{r}_1, 0) = \phi(\mathbf{r}_3, \mathbf{r}_2, 0) + 2m\pi, \tag{4.5–50}$$

where m is an integer. Since the origin in D is arbitrary, we may take $\mathbf{r}_3 = 0$, and we obtain from Eq. (4.5–50) the relation

$$\phi(\mathbf{r}_1, \mathbf{r}_2, 0) = \alpha(\mathbf{r}_2) - \alpha(\mathbf{r}_1) + 2m\pi, \tag{4.5–51}$$

where $\alpha(\mathbf{r}) = \phi(0, \mathbf{r}, 0)$ is a function of $\mathbf{r}$ only.

From Eqs. (4.5–49) and (4.5–51) it follows that

$$\gamma(\mathbf{r}_1, \mathbf{r}_2, 0) = \mathrm{e}^{\mathrm{i}[\alpha(\mathbf{r}_2) - \alpha(\mathbf{r}_1)]}. \tag{4.5–52}$$

Now under the assumption of complete second-order coherence expressed by Eq. (4.5–47), the formula (4.5–44) applies with $\mathbf{R}_1 \to \mathbf{r}_1$, $\mathbf{R}_2 \to \mathbf{r}_2$ and hence, if we substitute from Eq. (4.5–52) into Eq. (4.5–44), we obtain the following expression for $\gamma(\mathbf{r}_1, \mathbf{r}_2, \tau)$:

$$\gamma(\mathbf{r}_1, \mathbf{r}_2, \tau) = \mathrm{e}^{\mathrm{i}[\alpha(\mathbf{r}_2) - \alpha(\mathbf{r}_1) - 2\pi\nu_0\tau]}, \quad (\nu_0 > 0). \tag{4.5–53}$$

Hence we have established the following result: *If a field is completely coherent throughout a volume D, i.e. if* $|\gamma(\mathbf{r}_1, \mathbf{r}_2, \tau)| = 1$ *for all points* $\mathbf{r}_1 \in D$, $\mathbf{r}_2 \in D$ *and for all values of* τ, *then the complex degree of coherence has necessarily the form given by* Eq. (4.5–53), *with* $\alpha(\mathbf{r})$ *being a real function of position.* If we recall the definition of the complex degree of coherence [Eq. (4.3–12a)], it follows that, *under these circumstances, the mutual coherence function has the 'factorized'* τ*-periodic form*

$$\Gamma(\mathbf{r}_1, \mathbf{r}_2, \tau) = U^*(\mathbf{r}_1)U(\mathbf{r}_2)\,\mathrm{e}^{-2\pi\mathrm{i}\nu_0\tau}, \quad (\nu_0 > 0), \tag{4.5–54}$$

where

$$U(\mathbf{r}) = [\Gamma(\mathbf{r}, \mathbf{r}, 0)]^{1/2}\,\mathrm{e}^{\mathrm{i}\alpha(\mathbf{r})}. \tag{4.5–55}$$

$\Gamma(\mathbf{r}, \mathbf{r}, 0) = \langle V^*(\mathbf{r}, t)V(\mathbf{r}, t)\rangle$ is, of course, the average intensity at the point $\mathbf{r}$, $\alpha(\mathbf{r})$ is a phase factor and ν_0 is a positive constant.

If the domain D is free space, the mutual coherence function $\Gamma(\mathbf{r}_1, \mathbf{r}_2, \tau)$ satisfies the two wave equations (4.4–11). On substituting from Eq. (4.5–54) into these equations it follows that the function $U(\mathbf{r})$ then necessarily satisfies the Helmholtz equation

$$\nabla^2 U(\mathbf{r}) + k_0^2 U(\mathbf{r}) = 0, \tag{4.5–56}$$

where

$$k_0 = 2\pi\nu_0/c. \tag{4.5–57}$$

c being the speed of light in vacuum.

We have shown in this section that complete coherence in the space-time domain always implies strictly periodic τ-dependence of the degree of coherence and of the mutual coherence function. Hence the spectral- and cross-spectral densities of such fields are represented by Dirac delta functions, with their singularities at some positive frequency ν_0. Clearly such fields cannot occur in nature. They should be regarded as representing a mathematical limit, rather than a realizable optical field.

We will next consider complete coherence in the space-frequency domain.

4.5.3 *Coherent light in the space–frequency domain*‡

We begin with the non-negative definiteness condition satisfied by the spectral degree of coherence, $\mu(\mathbf{r}_1, \mathbf{r}_2, \nu)$, namely the inequality

$$\sum_{j=1}^{n}\sum_{k=1}^{n} a_j^* a_k \mu(\mathbf{r}_j, \mathbf{r}_k, \nu) \geqslant 0, \tag{4.5–58}$$

which is valid with any positive integer n, any sets of points $\mathbf{r}_1, \mathbf{r}_2, \ldots, \mathbf{r}_n$ and any set of real or complex constants $a_1, a_2, \ldots, a_n$. This inequality follows at once from Eqs. (4.3–44) and (4.3–47).

Let $\mathbf{r}_1, \mathbf{r}_2, \mathbf{r}_3$ be any three points in the domain D containing the optical field. If we choose $n = 3$ in Eq. (4.5–58) the inequality implies that

$$\begin{vmatrix} \mu_{11} & \mu_{12} & \mu_{13} \\ \mu_{21} & \mu_{22} & \mu_{23} \\ \mu_{31} & \mu_{32} & \mu_{33} \end{vmatrix} \geqslant 0, \tag{4.5–59}$$

where we used the abbreviation

$$\mu_{jk} = \mu(\mathbf{r}_j, \mathbf{r}_k, \nu). \tag{4.5–60}$$

Now according to Eqs. (4.3–43) and (4.3–47) the spectral degree of coherence is Hermitian, i.e.

$$\mu_{kj} = \mu_{jk}^*. \tag{4.5–61}$$

Moreover, from its definition [Eq. (4.3–47)] we see that

$$\mu_{jj} = 1. \tag{4.5–62}$$

Hence the inequality (4.5–59) implies that

$$\begin{vmatrix} 1 & \mu_{12} & \mu_{13} \\ \mu_{12}^* & 1 & \mu_{23} \\ \mu_{13}^* & \mu_{23}^* & 1 \end{vmatrix} \geqslant 0. \tag{4.5–63}$$

On evaluating the determinant we obtain the inequality

$$2\,\mathrm{Re}\,(\mu_{12}\mu_{23}\mu_{13}^*) \geqslant |\mu_{12}|^2 + |\mu_{23}|^2 + |\mu_{13}|^2 - 1, \tag{4.5–64}$$

where Re denotes the real part.

Suppose now that the field is spatially fully coherent at some particular frequency ν_0, throughout a volume D, i.e. that

$$|\mu(\mathbf{r}_1, \mathbf{r}_2, \nu_0)| = 1, \quad \nu_0 > 0 \text{ fixed}, \tag{4.5–65}$$

for all $\mathbf{r}_1 \in D$, $\mathbf{r}_2 \in D$. Then, in particular, $|\mu_{12}|^2 = |\mu_{23}|^2 = |\mu_{31}|^2 = 1$ and the inequality (4.5–64) reduces to

$$\mathrm{Re}\,(\mu_{12}\mu_{23}\mu_{13}^*) \geqslant 1, \tag{4.5–66}$$

where the frequency argument of the μ_{jk} $(j, k = 1, 2, 3)$ is understood to be ν_0. In view of the assumption (4.5–65), μ_{jk} is of the form

$$\mu_{jk} = \mathrm{e}^{\mathrm{i}\psi_{jk}}, \tag{4.5–67}$$

‡ The analysis of this section is largely based on an investigation of Mandel and Wolf (1981).

where $\psi_{jk} = \psi(\mathbf{r}_j, \mathbf{r}_k, \nu_0)$ is real. Because of the Hermiticity relation (4.5–61), we also have

$$\psi_{kj} = -\psi_{jk}. \tag{4.5–68}$$

If we make use of Eqs. (4.5–67) and (4.5–68), the inequality (4.5–66) is seen to imply that

$$\cos(\psi_{12} - \psi_{32} + \psi_{31}) \geqslant 1. \tag{4.5–69}$$

Since $|\cos\theta|$ cannot exceed unity for real values of θ and is equal to unity when $\theta = 2m\pi$ where m is any integer, the inequality (4.5–69) implies that

$$\psi(\mathbf{r}_1, \mathbf{r}_2, \nu_0) + \psi(\mathbf{r}_3, \mathbf{r}_1, \nu_0) = \psi(\mathbf{r}_3, \mathbf{r}_2, \nu_0) + 2m\pi, \tag{4.5–70}$$

where m is an integer. Since the origin in D is arbitrary, we may take $\mathbf{r}_3 = 0$ and we then obtain from Eq. (4.5–70) the relation

$$\psi(\mathbf{r}_1, \mathbf{r}_2, \nu_0) = \beta(\mathbf{r}_2; \nu_0) - \beta(\mathbf{r}_1; \nu_0) + 2m\pi, \tag{4.5–71}$$

where $\beta(\mathbf{r}; \nu_0) = \psi(0, \mathbf{r}, \nu_0)$.

From Eqs. (4.5–67) and (4.5–71) it follows that the spectral degree of coherence must have the form

$$\mu(\mathbf{r}_1, \mathbf{r}_2, \nu_0) = e^{i[\beta(\mathbf{r}_2;\nu_0)-\beta(\mathbf{r}_1;\nu_0)]}. \tag{4.5–72}$$

Hence we have established the following result. *If a field is spatially fully coherent at some particular frequency* ν_0 *throughout a volume* D*, i.e if* $|\mu(\mathbf{r}_1, \mathbf{r}_2, \nu_0)| = 1$ *for all* $\mathbf{r}_1 \in D$, $\mathbf{r}_2 \in D$, *then the spectral degree of coherence has necessarily the form given by Eq. (4.5–72), where* $\beta(\mathbf{r}; \nu_0)$ *is a real function of position*. If we recall the definition (4.3–47) of the spectral degree of coherence, it follows that, *under these circumstances, the cross-spectral density at frequency* ν_0 *has the factorized form*

$$W(\mathbf{r}_1, \mathbf{r}_2, \nu_0) = \mathcal{U}^*(\mathbf{r}_1, \nu_0)\mathcal{U}(\mathbf{r}_2, \nu_0), \tag{4.5–73}$$

where

$$\mathcal{U}(\mathbf{r}, \nu_0) = [S(\mathbf{r}, \nu_0)]^{1/2} e^{i\beta(\mathbf{r},\nu_0)}, \tag{4.5–74}$$

with $S(\mathbf{r}, \nu_0)$ representing the spectral density at frequency ν_0 at the point $\mathbf{r}$ and $\beta(\mathbf{r}; \nu_0)$ is a phase factor.

If the domain D is free space, the cross-spectral density $W(\mathbf{r}_1, \mathbf{r}_2, \nu)$ satisfies two Helmholtz equations [Eqs. (4.4–12)]. On substituting from Eq. (4.5–73) into these equations we readily find that the function $\mathcal{U}(\mathbf{r}, \nu_0)$ necessarily satisfies the Helmholtz equation

$$\nabla^2\mathcal{U}(\mathbf{r}, \nu_0) + k_0^2\mathcal{U}(\mathbf{r}, \nu_0) = 0 \tag{4.5–75}$$

where

$$k_0 = 2\pi\nu_0/c. \tag{4.5–76}$$

We considered in this section a field that is completely coherent, at some particular frequency ν_0, throughout a volume D. Let us now suppose that the field is completely coherent throughout D at *each* frequency ν $(0 < \nu < \infty)$. Then, according to the result that we just established, the cross-spectral density at each frequency ν of such a field must necessarily have the factorized form (4.5–73). Consequently, since the mutual coherence function and the cross-spectral density form a Fourier transform pair (4.3–40), it follows that *the mutual coherence function of a field that is completely coherent at all frequencies must necessarily be expressible in the form*

$$\Gamma(\mathbf{r}_1, \mathbf{r}_2, \tau) = \int_0^\infty \mathcal{U}^*(\mathbf{r}_1, \nu)\mathcal{U}(\mathbf{r}_2, \nu)\, \mathrm{e}^{-2\pi i\nu\tau}\, \mathrm{d}\nu. \tag{4.5–77}$$

We note that a mutual coherence function of the form (4.5–77) will, in general, *not* be periodic in τ and hence the field that it represents will have a spectral density

$$S(\mathbf{r}, \nu) \equiv W(\mathbf{r}, \mathbf{r}, \nu) = |\mathcal{U}(\mathbf{r}, \nu)|^2 \tag{4.5–78}$$

of non-vanishing width. In view of this fact and the discussion following Eq. (4.5–57), it would seem that complete coherence in the space-frequency domain is a more realistic concept than complete coherence in the space-time domain.

An example of a field which is completely coherent in the space-frequency domain occurs in connection with a single-mode laser resonator, as is shown in Section 7.4. Such a field is, however, not fully coherent in the space-time domain.

4.6 Free fields of any state of coherence

All fields encountered in nature are generated by sources located at finite distances from the observer and the fields generally interact with material media (lenses, mirrors, screens, etc.). Nevertheless, for reasons that we will note shortly, it is of interest to consider idealized fields that have no sources, except perhaps at infinity and that do not interact with any material objects. Such fields are known as *free fields*. We will consider free fields of any state of coherence.

Although free fields represent an idealization, the concept of such fields is useful, because most measurements are performed at distances that are many optical wavelengths away from the sources of the fields and from objects with which the fields interact. Under these circumstances the field in the region of interest may usually be well approximated by a free field, which differs from the actual field by the absence of evanescent waves (see Section 3.2). Such waves decay exponentially in amplitude with increasing distance from their sources. The contributions of the evanescent waves in regions where measurements on the field are made are usually negligible. Naturally, free fields are easier to analyze mathematically than more general fields.

We have seen in Section 4.4.1 that the space-time correlation function of a field in free space satisfies two wave equations, i.e. differential equations which are second-order in time and space variables. Sudarshan (1969) showed that, if the domain is the whole space, i.e. if one deals with a free field, the second-order space-time correlation function also obeys equations which are first-order in time, but are non-local in space. This result has a number of interesting physical consequences.

In this section we will derive Sudarshan's equations for propagation of correlation functions of free fields and will then briefly discuss some of their consequences.

4.6.1 Sudarshan's equations for the propagation of second-order correlation functions of free fields

Let $V(\mathbf{r}, t)$ be a realization of a statistical ensemble of free fields. We take $V(\mathbf{r}, t)$ to be the complex analytical signal associated with the real field variable. It satisfies the wave equation (4.4–7)

$$\nabla^2 V(\mathbf{r}, t) = \frac{1}{c^2}\frac{\partial^2 V(\mathbf{r}, t)}{\partial t^2} \qquad (4.6\text{–}1)$$

throughout the whole space. The general formal solution of Eq. (4.6–1), valid throughout the whole space, may be expressed as superposition of plane-wave modes, namely

$$V(\mathbf{r}, t) = \int a(\mathbf{K})\,\mathrm{e}^{\mathrm{i}(\mathbf{K}\cdot\mathbf{r}-Kct)}\,\mathrm{d}^3K + \int b(\mathbf{K})\,\mathrm{e}^{\mathrm{i}(\mathbf{K}\cdot\mathbf{r}+Kct)}\,\mathrm{d}^3K, \qquad (4.6\text{–}2)$$

where $K = |\mathbf{K}|$ and the integrations extend over the whole three-dimensional $\mathbf{K}$-space.

The representation (4.6–2) of a free field may be readily established as follows. We express $V(\mathbf{r}, t)$ as a (possibly generalized) three-dimensional Fourier integral,

$$V(\mathbf{r}, t) = \int \widetilde{V}(\mathbf{K}, t)\,\mathrm{e}^{\mathrm{i}\mathbf{K}\cdot\mathbf{r}}\,\mathrm{d}^3K, \qquad (4.6\text{–}3)$$

and substitute this expression into the differential equation (4.6–1). On interchanging the order of differentiation and integration and taking the Fourier inverse of the resulting equation, we readily find that

$$\left(-K^2 - \frac{1}{c^2}\frac{\partial^2}{\partial t^2}\right)\widetilde{V}(\mathbf{K}, t) = 0. \qquad (4.6\text{–}4)$$

The general solution of Eq. (4.6–4) is

$$\widetilde{V}(\mathbf{K}, t) = a(\mathbf{K})\,\mathrm{e}^{-\mathrm{i}Kct} + b(\mathbf{K})\,\mathrm{e}^{\mathrm{i}Kct}, \qquad (4.6\text{–}5)$$

where $a(\mathbf{K})$ and $b(\mathbf{K})$ are arbitrary functions of $\mathbf{K}$. On substituting for $\widetilde{V}(\mathbf{K}, t)$ from Eq. (4.6–5) into Eq. (4.6–3) the representation (4.6–2) follows.

Since we are representing the field by a complex analytic signal, it cannot contain any negative frequency components and hence $b(\mathbf{K}) \equiv 0$ in Eq. (4.6–2). Consequently the field is represented by an ensemble of realizations of the form

$$V(\mathbf{r}, t) = \int a(\mathbf{K})\,\mathrm{e}^{\mathrm{i}(\mathbf{K}\cdot\mathbf{r}-Kct)}\,\mathrm{d}^3K. \qquad (4.6\text{–}6)$$

If we differentiate both sides of Eq. (4.6–6) with respect to t, we obtain the equation

$$\frac{\partial V}{\partial t} = -\mathrm{i}c\int Ka(\mathbf{K})\,\mathrm{e}^{\mathrm{i}(\mathbf{K}\cdot\mathbf{r}-Kct)}\,\mathrm{d}^3K. \qquad (4.6\text{–}7)$$

The integral on the right-hand side of this formula may be expressed in terms of a certain non-local operator acting on $V(\mathbf{r}, t)$. This operator, usually denoted by $\sqrt{(-\nabla^2)}$, is well known in particle physics and in quantum field theory and may be defined as follows:

Let $F(\mathbf{r})$ be any scalar function of $\mathbf{r}$ that has a Fourier integral representation

$$F(\mathbf{r}) = \int \tilde{F}(\mathbf{K})\,\mathrm{e}^{\mathrm{i}\mathbf{K}\cdot\mathbf{r}}\,\mathrm{d}^3K. \tag{4.6–8}$$

Then

$$\nabla^2 F(\mathbf{r}) = -\int K^2 \tilde{F}(\mathbf{K})\,\mathrm{e}^{\mathrm{i}\mathbf{K}\cdot\mathbf{r}}\,\mathrm{d}^3K. \tag{4.6–9}$$

This formula suggests the use of the symbolic notation $\sqrt{(-\nabla^2)}F(\mathbf{r})$ to denote the integral that appears on the right-hand side of Eq. (4.6–9):

$$\sqrt{(-\nabla^2)}F(\mathbf{r}) = \int K \tilde{F}(\mathbf{K})\,\mathrm{e}^{\mathrm{i}\mathbf{K}\cdot\mathbf{r}}\,\mathrm{d}^3K. \tag{4.6–10}$$

This formula *defines* the operator $\sqrt{(-\nabla^2)}$. We may express the right-hand side of Eq. (4.6–10) directly in terms of $F(\mathbf{r})$ rather than in terms of $\tilde{F}(\mathbf{K})$ by substituting for $\tilde{F}(\mathbf{K})$ the expression

$$\tilde{F}(\mathbf{K}) = \frac{1}{(2\pi)^3}\int F(\mathbf{r}')\,\mathrm{e}^{-\mathrm{i}\mathbf{K}\cdot\mathbf{r}'}\,\mathrm{d}^3r', \tag{4.6–11}$$

which is the Fourier inverse of Eq. (4.6–8). One then obtains after simple algebraic manipulation the formula

$$\sqrt{(-\nabla^2)}F(\mathbf{r}) = \int F(\mathbf{r}')H(\mathbf{r} - \mathbf{r}')\,\mathrm{d}^3r', \tag{4.6–12}$$

where

$$H(\mathbf{R}) = \frac{1}{(2\pi)^3}\int K\mathrm{e}^{\mathrm{i}\mathbf{K}\cdot\mathbf{R}}\,\mathrm{d}^3K. \tag{4.6–13}$$

It is shown in Appendix 4.1 that the kernel $H(\mathbf{R})$, given by Eq. (4.6–13), of the integral transform in Eq. (4.6–12) may be symbolically expressed in the form

$$H(\mathbf{R}) = \frac{\mathrm{i}}{2\pi R}\frac{\mathrm{d}^2}{\mathrm{d}R^2}[\delta^{(+)}(R) - \delta^{(-)}(R)], \tag{4.6–14}$$

where $R = |\mathbf{R}|$ and $\delta^{(+)}(R)$ and $\delta^{(-)}(R)$ are the positive and the negative frequency parts of the Dirac delta function, defined in Appendix 4.1 [Eq. (A4.1–4)]. Equation (4.6–12) shows that the operator $\sqrt{(-\nabla^2)}$ is non-local, i.e. the value of $\sqrt{(-\nabla^2)}F(\mathbf{r})$ depends not only on the value of F at the point $\mathbf{r}$ but on the values of F at all points in space.

Let us now return to Eq. (4.6–7). We note that the integral which appears on the right-hand side of that equation is just $-\mathrm{i}c\sqrt{(-\nabla^2)}V(\mathbf{r}, t)$, and hence we see that the analytic signal representation of any free field satisfies the equation

$$\sqrt{(-\nabla^2)}V(\mathbf{r}, t) = -\frac{1}{\mathrm{i}c}\frac{\partial V(\mathbf{r}, t)}{\partial t}. \tag{4.6–15}$$

This equation was first derived by Sudarshan (1969).

Because Eq. (4.6–15) is first-order in time, it is possible to determine the values of $V(\mathbf{r}, t)$ for all $\mathbf{r}$ and all t from the knowledge of $V(\mathbf{r}, t')$ at any particular instant of time t'. An explicit formula for this time evolution can be obtained as follows. We first take the Fourier inverse of Eq. (4.6–6). This gives

$$a(\mathbf{K}) = \frac{1}{(2\pi)^3} \mathrm{e}^{\mathrm{i}Kct'} \int V(\mathbf{r}', t') \, \mathrm{e}^{-\mathrm{i}\mathbf{K}\cdot\mathbf{r}'} \, \mathrm{d}^3 r'. \tag{4.6–16}$$

Next we substitute from Eq. (4.6–16) into Eq. (4.6–6) and interchange the order of integrations. We then obtain the formula

$$V(\mathbf{r}, t) = \int V(\mathbf{r}', t') G(\mathbf{r} - \mathbf{r}', t - t') \, \mathrm{d}^3 r', \tag{4.6–17}$$

where

$$G(\mathbf{R}, T) = \frac{1}{(2\pi)^3} \int \mathrm{e}^{\mathrm{i}(\mathbf{K}\cdot\mathbf{R} - KcT)} \, \mathrm{d}^3 K. \tag{4.6–18}$$

It is shown in Appendix 4.2 that $G(\mathbf{R}, T)$ may be expressed in the form

$$G(\mathbf{R}, T) = -\frac{1}{2\pi R}\frac{\partial}{\partial R}[\delta^{(+)}(R - cT) + \delta^{(-)}(R + cT)], \tag{4.6–19}$$

where $R = |\mathbf{R}|$.

The formula (4.6–17) for the time-development of the analytic signal is valid not only for all values of $t > t'$ but also for all values of $t < t'$. This fact is a reflection of the requirement that in order to construct the complex analytic signal associated with a real signal the values of the signal must be known for all values of t $(-\infty < t < \infty)$ [see Eq. (3.1–21)].

Now, in the present context, $V(\mathbf{r}, t)$ is a realization (sample function) of a fluctuating free field, whose second-order correlation function may be defined in the usual way as

$$\Gamma(\mathbf{r}_1, t_1; \mathbf{r}_2, t_2) = \langle V^*(\mathbf{r}_1, t_1) V(\mathbf{r}_2, t_2) \rangle. \tag{4.6–20}$$

Let $\sqrt{(-\nabla_1^2)}$ and $\sqrt{(-\nabla_2^2)}$ denote the operator $\sqrt{(-\nabla^2)}$, acting with respect to spatial variables $\mathbf{r}_1$ and $\mathbf{r}_2$, respectively. Let us apply the operator $\sqrt{(-\nabla_j^2)}$, $(j = 1, 2)$, to Eq. (4.6–20) and interchange the order of the operations on the right-hand side of the resulting equations. This gives

$$\sqrt{(-\nabla_j^2)}\Gamma(\mathbf{r}_1, t_1; \mathbf{r}_2, t_2) = \langle \sqrt{(-\nabla_j^2)} V^*(\mathbf{r}_1, t_1) V(\mathbf{r}_2, t_2) \rangle, \quad (j = 1, 2). \tag{4.6–21}$$

On making use of Eq. (4.6–15) on the right-hand side of Eq. (4.6–21) and interchanging again the order of the operations, we obtain Sudarshan's equations (Sudarshan, 1969)

$$\sqrt{(-\nabla_1^2)}\Gamma(\mathbf{r}_1, t_1; \mathbf{r}_2, t_2) = \frac{1}{\mathrm{i}c}\frac{\partial}{\partial t_1}\Gamma(\mathbf{r}_1, t_1; \mathbf{r}_2, t_2), \tag{4.6–22a}$$

$$\sqrt{(-\nabla_2^2)}\Gamma(\mathbf{r}_1, t_1; \mathbf{r}_2, t_2) = -\frac{1}{\mathrm{i}c}\frac{\partial}{\partial t_2}\Gamma(\mathbf{r}_1, t_1; \mathbf{r}_2, t_2), \tag{4.6–22b}$$

which must be satisfied by the second-order correlation function $\Gamma(\mathbf{r}_1, t_1; \mathbf{r}_2, t_2)$ of any ensemble of free fields.

We note that unlike the wave equations (4.4–10) for Γ, Sudarshan's equations are first-order in the temporal variables and are non-local in the spatial variables. This fact implies that the knowledge of the correlation function $\Gamma(\mathbf{r}'_1, t'_1; \mathbf{r}'_2, t'_2)$ for all values of $\mathbf{r}'_1$ and $\mathbf{r}'_2$, but with t'_1, t'_2 fixed, determines $\Gamma(\mathbf{r}_1, t_1; \mathbf{r}_2, t_2)$ for all values of its arguments. We will derive shortly the solution to this 'initial value' problem and discuss some of its physical consequences.

Since most optical fields that one encounters in nature are stationary, at least in the wide sense, we will specialize Eqs. (4.6–22) to fields of this kind. For such fields $\Gamma(\mathbf{r}_1, t_1; \mathbf{r}_2, t_2)$ will depend on its two time arguments only through the difference $\tau = t_2 - t_1$ and becomes the mutual coherence function $\Gamma(\mathbf{r}_1, \mathbf{r}_2, \tau)$. Since now $-\partial/\partial t_1 = \partial/\partial t_2 = \partial/\partial \tau$, Eqs. (4.6–22) become

$$\sqrt{(-\nabla_1^2)}\Gamma(\mathbf{r}_1, \mathbf{r}_2, \tau) = -\frac{1}{ic}\frac{\partial}{\partial \tau}\Gamma(\mathbf{r}_1, \mathbf{r}_2, \tau), \tag{4.6–23a}$$

$$\sqrt{(-\nabla_2^2)}\Gamma(\mathbf{r}_1, \mathbf{r}_2, \tau) = -\frac{1}{ic}\frac{\partial}{\partial \tau}\Gamma(\mathbf{r}_1, \mathbf{r}_2, \tau). \tag{4.6–23b}$$

Finally, if we take the Fourier transform of Eqs. (4.6–23) and recall that the Fourier transform of the mutual coherence function is just the cross-spectral density $W(\mathbf{r}_1, \mathbf{r}_2, \nu)$ [Eq. (4.3–40b)], we obtain the equations

$$\sqrt{(-\nabla_1^2)}W(\mathbf{r}_1, \mathbf{r}_2, \nu) - kW(\mathbf{r}_1, \mathbf{r}_2, \nu) = 0, \tag{4.6–24a}$$

$$\sqrt{(-\nabla_2^2)}W(\mathbf{r}_1, \mathbf{r}_2, \nu) - kW(\mathbf{r}_1, \mathbf{r}_2, \nu) = 0, \tag{4.6–24b}$$

where $k = 2\pi\nu/c$.

4.6.2 Time evolution of the second-order correlation functions of free fields‡

Let us again consider an ensemble of free fields which, to begin with, we do not restrict to being stationary. The time evolution of each realization $V(\mathbf{r}, t)$ is given by Eq. (4.6–17). On substituting from Eq. (4.6–17) into Eq. (4.6–20) that defines the correlation function Γ, we obtain the formula

$$\Gamma(\mathbf{r}_1, t_1; \mathbf{r}_2, t_2) =$$

$$\iint \Gamma(\mathbf{r}'_1, t'_1; \mathbf{r}_2, t'_2)G^*(\mathbf{r}_1 - \mathbf{r}'_1, t_1 - t'_1)G(\mathbf{r}_2 - \mathbf{r}'_2, t_2 - t'_2)\,\mathrm{d}^3r'_1\,\mathrm{d}^3r'_2, \tag{4.6–25}$$

where $G(\mathbf{R}, T)$ is the Green's function (4.6–19). This formula, which is valid for all values of t_1 and t_2 ($t_1 \gtrless t'_1$, $t_2 \gtrless t'_2$) describes, in a closed form, the time evolution of the correlation function Γ of a free field. It makes it possible to determine $\Gamma(\mathbf{r}_1, t_1; \mathbf{r}_2, t_2)$ for values of all its arguments from the knowledge of $\Gamma(\mathbf{r}'_1, t'_1; \mathbf{r}'_2, t'_2)$, where $\mathbf{r}'_1$ and $\mathbf{r}'_2$ take on all possible values, but t'_1 and t'_2 are any two (fixed) temporal arguments.

Next we will specialize the formula (4.6–25) to free fields which are stationary, at least in the wide sense. For this purpose we set $t_2 - t_1 = \tau$, $t'_2 - t'_1 = \tau'$ and eliminate t_2 and t'_2 from Eq. (4.6–25) by means of these relations. We now also have $\Gamma(\mathbf{r}_1, t_1; \mathbf{r}_2, t_2) \equiv \Gamma(\mathbf{r}_1, \mathbf{r}_2, \tau)$, $\Gamma(\mathbf{r}'_1, t'_1; \mathbf{r}'_2, t'_2) \equiv \Gamma(\mathbf{r}'_1, \mathbf{r}'_2, \tau')$, and we obtain

‡ The main results of this section are due to Wolf, Devaney and Foley (1981), p. 123.

the formula

$$\Gamma(\mathbf{r}_1, \mathbf{r}_2, \tau) =$$

$$\iint \Gamma(\mathbf{r}_1', \mathbf{r}_2', \tau')G^*(\mathbf{r}_1 - \mathbf{r}_1', t_1 - t_1')G(\mathbf{r}_2 - \mathbf{r}_2', \tau + t_1 - \tau' - t_1')\,\mathrm{d}^3r_1'\,\mathrm{d}^3r_2'. \tag{4.6–26}$$

If we set $t_1 = t_1' = 0$ in Eq. (4.6–26) we obtain the formula

$$\Gamma(\mathbf{r}_1, \mathbf{r}_2, \tau) = \iint \Gamma(\mathbf{r}_1', \mathbf{r}_2', \tau')G^*(\mathbf{r}_1 - \mathbf{r}_1', 0)G(\mathbf{r}_2 - \mathbf{r}_2', \tau - \tau')\,\mathrm{d}^3r_1'\,\mathrm{d}^3r_2'. \tag{4.6–27}$$

It follows from Eq. (4.6–18) and from the Fourier integral representation of the Dirac delta function that

$$G^*(\mathbf{R}, 0) = \delta^{(3)}(\mathbf{R}). \tag{4.6–28}$$

If we substitute from Eq. (4.6–28) into Eq. (4.6–27), we obtain the following formula for the time evolution of the mutual coherence function of any free field which is stationary, at least in the wide sense:

$$\Gamma(\mathbf{r}_1, \mathbf{r}_2, \tau) = \int \Gamma(\mathbf{r}_1, \mathbf{r}_2', \tau')G(\mathbf{r}_2 - \mathbf{r}_2', \tau - \tau')\,\mathrm{d}^3r_2', \tag{4.6–29}$$

where $\tau'(\gtrless \tau)$ is arbitrary.

4.6.3 A relationship between temporal and spatial coherence properties of free fields

The formula (4.6–29) that we have just derived has a number of interesting consequences which we will briefly consider in this and in the next section.

Since the parameter τ' in Eq. (4.6–29) is arbitrary we may, in particular, set it equal to zero. The correlation function that then enters the integrand of Eq. (4.6–29) is the 'equal-time' coherence function $\Gamma(\mathbf{r}_1, \mathbf{r}_2, 0)$, i.e. the mutual intensity that we encountered earlier [Eq. (4.3–34)] and denoted by $J(\mathbf{r}_1, \mathbf{r}_2)$. We thus obtain the formula

$$\Gamma(\mathbf{r}_1, \mathbf{r}_2, \tau) = \int J(\mathbf{r}_1, \mathbf{r}_2')G(\mathbf{r}_2 - \mathbf{r}_2', \tau)\,\mathrm{d}^3r_2', \tag{4.6–30}$$

which expresses the mutual coherence function (i.e. an unequal-time correlation function) in terms of the mutual intensity (i.e. an equal-time correlation function).

If we take $\mathbf{r}_2 = \mathbf{r}_1$ ($= \mathbf{r}$ say) and denote the variable of integration by $\mathbf{r}'$ rather than $\mathbf{r}_2'$, Eq. (4.6–30) becomes

$$\Gamma(\mathbf{r}, \mathbf{r}, \tau) = \int J(\mathbf{r}, \mathbf{r}')G(\mathbf{r} - \mathbf{r}', \tau)\,\mathrm{d}^3r'. \tag{4.6–31}$$

The function on the left is the self-coherence [see the discussion that follows Eq. (4.3–35)], which is a measure of the temporal coherence of the field at the point $\mathbf{r}$, whereas the function $J(\mathbf{r}, \mathbf{r}')$ on the right is a measure of the spatial coherence at the two points $\mathbf{r}$ and $\mathbf{r}'$. Hence Eq. (4.6–31) expresses the temporal coherence properties of a (stationary) free field in terms of its spatial coherence properties.

4.6.4 *A relationship between spectral properties and spatial coherence properties of free fields*‡

If we take the Fourier τ-transform of Eq. (4.6–30) and use Eq. (4.3–40b), we obtain the relation

$$W(\mathbf{r}_1, \mathbf{r}_2, \nu) = \int J(\mathbf{r}_1, \mathbf{r}_2')\widetilde{G}(\mathbf{r}_2 - \mathbf{r}_2', \nu)\, \mathrm{d}^3 r_2', \tag{4.6–32}$$

where $W(\mathbf{r}_1, \mathbf{r}_2, \nu)$ is the cross-spectral density of the field and $\widetilde{G}(\mathbf{R}, \nu)$ is the Fourier transform of the Green's function (4.6–19), namely

$$\widetilde{G}(\mathbf{R}, \nu) = \int_{-\infty}^{\infty} G(\mathbf{R}, T)\, \mathrm{e}^{2\pi i \nu T}\, \mathrm{d}T. \tag{4.6–33}$$

This Fourier transform is evaluated in Appendix 4.2 and is found to be

$$\left.\begin{aligned} \widetilde{G}(\mathbf{R}, \nu) &= \frac{k^2}{\pi c}\left(\frac{\sin kR}{kR}\right) && \text{when } \nu > 0 \\ &= 0 && \text{when } \nu < 0, \end{aligned}\right\} \tag{4.6–34}$$

where $k = 2\pi\nu/c$. On substituting from Eq. (4.6–34) into Eq. (4.6–32) we obtain another interesting relation, namely

$$W(\mathbf{r}_1, \mathbf{r}_2, \nu) = \frac{k^2}{\pi c}\int J(\mathbf{r}_1, \mathbf{r}_2')\frac{\sin k|\mathbf{r}_2 - \mathbf{r}_2'|}{k|\mathbf{r}_2 - \mathbf{r}_2'|}\, \mathrm{d}^3 r_2' \quad \text{when } \nu > 0 \tag{4.6–35a}$$

$$= 0 \quad \text{when } \nu < 0. \tag{4.6–35b}$$

The fact that $W(\mathbf{r}_1, \mathbf{r}_2, \nu) = 0$ when $\nu < 0$ was, of course, to be expected, since $\Gamma(\mathbf{r}_1, \mathbf{r}_2, \tau)$ is an analytic signal [see discussion that follows Eq. (4.3–15)]. The formula (4.6–35a) expresses the cross-spectral density of a free, stationary field in terms of its mutual intensity or, to put it somewhat differently, it expresses its spectral properties in terms of its spatial coherence properties.

If we choose $\mathbf{r}_2 = \mathbf{r}_1$ ($= \mathbf{r}$ say), the left-hand side of Eqs. (4.6–35) reduces to the spectral density $S(\mathbf{r}, \nu) = W(\mathbf{r}, \mathbf{r}, \nu)$ of the light at the point $\mathbf{r}$. If we denote the variable of integration on the right of Eq. (4.6–35a) by $\mathbf{r}'$ rather than $\mathbf{r}_2'$, we obtain the following expression for the spectral density of the field in terms of the mutual intensity:

$$S(\mathbf{r}, \nu) = \frac{k^2}{\pi c}\int J(\mathbf{r}, \mathbf{r}')\frac{\sin k|\mathbf{r} - \mathbf{r}'|}{k|\mathbf{r} - \mathbf{r}'|}\, \mathrm{d}^3 \mathbf{r}'. \tag{4.6–36}$$

Some implications of the formulas (4.6–35) and (4.6–36) for one-dimensional, statistically stationary, homogeneous free fields were discussed by Wolf, Devaney and Gori (1983).

4.7 Coherent-mode representation and ensemble representation of sources and fields in the space-frequency domain

We have noted earlier (Section 2.4.1) that a stationary random function, $V(t)$ say, does not admit a Fourier integral representation within the framework of

‡ The results in this section are due to Wolf and Devaney (1981).

ordinary function theory, because the sample functions do not tend to zero as $t \to \pm\infty$. In the previous section of this chapter we ignored this difficulty and we formally employed the Fourier integral representation in order to provide a heuristic description of fluctuating fields in the space-frequency domain, rather than in the space-time domain. The main results that we obtained by this non-rigorous approach can be justified by more sophisticated mathematical techniques, such as generalized harmonic analysis and the theory of generalized functions.

In spite of the impossibility of defining the Fourier spectrum of a stationary random function, one can obtain, as first shown by Wolf (1982; see also 1981a, b, 1986 and Agarwal and Wolf, 1993), a mathematically rigorous representation of stationary random sources and of stationary random fields in the space-frequency domain, in terms of ensembles of ordinary functions, at least within the framework of second-order theory. Naturally this representation does not use as a basis the time harmonic Fourier kernels $e^{-2\pi i\nu t}$ but rather, as we will see, the eigenfunctions of an integral equation whose kernel is the cross-spectral density. In this section we will present an outline of this theory. We will make considerable use of the theory in Chapters 5 and 7.

4.7.1 Coherent-mode representation of partially coherent fields in free space

Let us consider a stationary optical field $V(\mathbf{r}, t)$ in some finite closed domain D in free space and let $\Gamma(\mathbf{r}_1, \mathbf{r}_2, \tau)$ be its mutual coherence function. We assume that $\Gamma(\mathbf{r}_1, \mathbf{r}_2, \tau)$ falls off sufficiently rapidly with τ as $|\tau| \to \infty$ to ensure that, for all points $\mathbf{r}_1 \in D$ and $\mathbf{r}_2 \in D$, the mutual coherence function is absolutely integrable with respect to τ, i.e. that

$$\int_{-\infty}^{\infty} |\Gamma(\mathbf{r}_1, \mathbf{r}_2, \tau)| \, d\tau < \infty. \tag{4.7–1}$$

It then follows from a well-known theorem of Fourier integral analysis (Goldberg, 1965, p. 6) that $\Gamma(\mathbf{r}_1, \mathbf{r}_2, \tau)$ has a Fourier frequency transform

$$W(\mathbf{r}_1, \mathbf{r}_2, \nu) = \int_{-\infty}^{\infty} \Gamma(\mathbf{r}_1, \mathbf{r}_2, \tau) \, e^{2\pi i\nu\tau} \, d\tau \tag{4.7–2}$$

and that this transform, namely the cross-spectral density function $W(\mathbf{r}_1, \mathbf{r}_2, \nu)$, is a continuous function of ν.

It also readily follows from Eq. (4.3–12a) and the inequality (4.3–13) involving the cross-correlation function of a stationary random function that absolute integrability of the mutual coherence function Γ implies its square-integrability:

$$\int_{-\infty}^{\infty} |\Gamma(\mathbf{r}_1, \mathbf{r}_2, \tau)|^2 \, d\tau < \infty. \tag{4.7–3}$$

Consequently by Parseval's relation (Goldberg, 1965, p. 51) the cross-spectral density is square-integrable with respect to ν:

$$\int_{0}^{\infty} |W(\mathbf{r}_1, \mathbf{r}_2, \nu)|^2 \, d\nu < \infty, \tag{4.7–4}$$

and Eq. (4.7–2) may be inverted:

$$\Gamma(\mathbf{r}_1, \mathbf{r}_2, \tau) = \int_0^\infty W(\mathbf{r}_1, \mathbf{r}_2, \nu)\, \mathrm{e}^{-2\pi i\nu\tau}\, \mathrm{d}\nu. \qquad (4.7\text{–}5)$$

On the right-hand side of Eq. (4.7–5) the lower limit of integration is 0 rather than $-\infty$ because the mutual coherence function is an analytic signal.

Equations (4.7–2) to (4.7–5) and the continuity of the cross-spectral density as a function of ν are consequences of the single assumption expressed by Eq. (4.7–1). We will now also assume, that the cross-spectral density is a continuous function of $\mathbf{r}_1$ and $\mathbf{r}_2$ throughout the domain D. Then $|W(\mathbf{r}_1, \mathbf{r}_2, \nu)|^2$ will necessarily be bounded in D. Consequently

$$\int_D \int_D |W(\mathbf{r}_1, \mathbf{r}_2, \nu)|^2\, \mathrm{d}^3 r_1\, \mathrm{d}^3 r_2 < \infty. \qquad (4.7\text{–}6)$$

We have seen earlier [Eq. (4.3–43)] that the cross-spectral density satisfies the condition

$$W(\mathbf{r}_2, \mathbf{r}_1, \nu) = W^*(\mathbf{r}_1, \mathbf{r}_2, \nu). \qquad (4.7\text{–}7)$$

It is also a non-negative definite function in the sense that

$$\int_D \int_D W(\mathbf{r}_1, \mathbf{r}_2, \nu) f^*(\mathbf{r}_1) f(\mathbf{r}_2)\, \mathrm{d}^3 r_1\, \mathrm{d}^3 r_2 \geqslant 0, \qquad (4.7\text{–}8)$$

where $f(\mathbf{r})$ is any square-integrable function. This inequality, which may be regarded as a 'continuous analogue' of the inequality (4.3–44), may be established by a similar argument as was used in Section 2.4.4 in proving the non-negative definiteness of the cross-spectral density matrix $W_{jk}(\nu)$, but replacing the sum in Eq. (2.4–40) by an integral [cf. Eqs. (2.3–4a) and (2.3–4b)].

The condition (4.7–6) implies that the cross-spectral density function is a Hilbert–Schmidt kernel and Eqs. (4.7–7) and (4.7–8) show that the kernel is Hermitian and non-negative definite. By the multi-dimensional version of Mercer's theorem discussed in Section 2.5.1 it follows that the cross-spectral density, assumed to be continuous, may be expressed in the form

$$W(\mathbf{r}_1, \mathbf{r}_2, \nu) = \sum_n \alpha_n(\nu) \psi_n^*(\mathbf{r}_1, \nu) \psi_n(\mathbf{r}_2, \nu), \qquad (4.7\text{–}9)$$

the series on the right being absolutely and uniformly convergent. The functions $\psi_n(\mathbf{r}, \nu)$ are the eigenfunctions and the coefficients $\alpha_n(\nu)$ are the eigenvalues of the integral equation

$$\int_D W(\mathbf{r}_1, \mathbf{r}_2, \nu) \psi_n(\mathbf{r}_1, \nu)\, \mathrm{d}^3 r_1 = \alpha_n(\nu) \psi_n(\mathbf{r}_2, \nu), \qquad (4.7\text{–}10)$$

which is seen to be a homogeneous Fredholm integral equation of the second kind. As in the one-dimensional version of Eq. (4.7–10), discussed in Section 2.5, the Hermiticity of W ensures that the integral equation (4.7–10) has at least one non-zero eigenvalue, and the Hermiticity and the non-negative definiteness of W ensures that all the eigenvalues are real and non-negative, i.e. that

$$\alpha_n(\nu) \geqslant 0. \qquad (4.7\text{–}11)$$

Moreover one can again assume, without loss of generality, that the eigenfunctions form an orthonormal set, i.e. that

$$\int_D \psi_n^*(\mathbf{r}, \nu)\psi_m(\mathbf{r}, \nu)\, \mathrm{d}^3 r = \delta_{nm}, \tag{4.7–12}$$

where δ_{nm} is the Kronecker symbol.

In the present context the Mercer expansion (4.7–9) has an interesting physical significance. To see this we re-write it in the form

$$W(\mathbf{r}_1, \mathbf{r}_2, \nu) = \sum_n \alpha_n(\nu) W^{(n)}(\mathbf{r}_1, \mathbf{r}_2, \nu), \tag{4.7–13}$$

where

$$W^{(n)}(\mathbf{r}_1, \mathbf{r}_2, \nu) = \psi_n^*(\mathbf{r}_1, \nu)\psi_n(\mathbf{r}_2, \nu). \tag{4.7–14}$$

An expression of the form (4.7–14) represents the cross-spectral density function of a field that is completely coherent in the space-frequency domain, as is clear from a theorem derived in Section 4.5.3 [see Eq. (4.5–73)]; or more directly, by noting that the corresponding spectral degree of coherence

$$\mu^{(n)}(\mathbf{r}_1, \mathbf{r}_2, \nu) = \frac{W^{(n)}(\mathbf{r}_1, \mathbf{r}_2, \nu)}{[W^{(n)}(\mathbf{r}_1, \mathbf{r}_1, \nu)]^{1/2}[W^{(n)}(\mathbf{r}_2, \mathbf{r}_2, \nu)]^{1/2}} \tag{4.7–15}$$

is unimodular [$|\mu^{(n)}(\mathbf{r}_1, \mathbf{r}_2, \nu)| = 1$ for all $\mathbf{r}_1 \in D$, $\mathbf{r}_2 \in D$].

We have shown earlier [Eqs. (4.4–12)] that in free space the cross-spectral density obeys the two Helmholtz equations

$$\nabla_1^2 W(\mathbf{r}_1, \mathbf{r}_2, \nu) + k^2 W(\mathbf{r}_1, \mathbf{r}_2, \nu) = 0, \tag{4.7–16a}$$

$$\nabla_2^2 W(\mathbf{r}_1, \mathbf{r}_2, \nu) + k^2 W(\mathbf{r}_1, \mathbf{r}_2, \nu) = 0, \tag{4.7–16b}$$

where

$$k = 2\pi\nu/c \tag{4.7–17}$$

is the wave number associated with the frequency ν. If we substitute the expansion (4.7–9) for W into Eq. (4.7–16b), multiply the equation by $\psi_n(\mathbf{r}_1, \nu)$, integrate both sides of the equation with respect to $\mathbf{r}_1$ over the domain D, interchange the order of summation and integration and use the orthonormality relations (4.7–12), we readily find that

$$\nabla^2 \psi_n(\mathbf{r}, \nu) + k^2 \psi_n(\mathbf{r}, \nu) = 0 \tag{4.7–18}$$

throughout D. From Eqs. (4.7–14) and (4.7–18) one then immediately finds that $W^{(n)}$ satisfies the two Helmholtz equations

$$\nabla_1^2 W^{(n)}(\mathbf{r}_1, \mathbf{r}_2, \nu) + k^2 W^{(n)}(\mathbf{r}_1, \mathbf{r}_2, \nu) = 0, \tag{4.7–19a}$$

$$\nabla_2^2 W^{(n)}(\mathbf{r}_1, \mathbf{r}_2, \nu) + k^2 W^{(n)}(\mathbf{r}_1, \mathbf{r}_2, \nu) = 0. \tag{4.7–19b}$$

Thus, for each n, $W^{(n)}(\mathbf{r}_1, \mathbf{r}_2, \nu)$ satisfies the same equations as does the cross-spectral density of the field and hence we may regard $W^{(n)}(\mathbf{r}_1, \mathbf{r}_2, \nu)$ as representing a *mode* of the field. Thus we have shown that *the expansion (4.7–13) represents the cross-spectral density of the field as a superposition of modes that are completely coherent in the space-frequency domain*. For this reason the expansion (4.7–9) is sometimes referred to as *the coherent-mode representation* of the cross-spectral density.

It can also be readily shown from the definition (4.7–14) of $W^{(n)}$, by the use of the orthonormality relations (4.7–12), that the modes are mutually orthonormal; more precisely that

$$\int_D\int_D [W^{(n)}(\mathbf{r}_1, \mathbf{r}_2, \nu)]^* W^{(m)}(\mathbf{r}_1, \mathbf{r}_2, \nu)\, \mathrm{d}^3 r_1\, \mathrm{d}^3 r_2 = \delta_{nm}. \tag{4.7–20}$$

Let us now briefly consider the spectral density of the field. According to Eq. (4.3–41), the spectral density $S(\mathbf{r}, \nu)$ of the light at the point $\mathbf{r}$ is just the 'diagonal element' of the cross-spectral density, i.e.

$$S(\mathbf{r}, \nu) \equiv W(\mathbf{r}, \mathbf{r}, \nu). \tag{4.7–21}$$

On substituting from Eq. (4.7–13) into Eq. (4.7–21) we obtain for the spectral density the expression

$$S(\mathbf{r}, \nu) = \sum_n \alpha_n(\nu) S^{(n)}(\mathbf{r}, \nu), \tag{4.7–22}$$

where

$$S^{(n)}(\mathbf{r}, \nu) \equiv W^{(n)}(\mathbf{r}, \mathbf{r}, \nu) = |\psi_n(\mathbf{r}, \nu)|^2. \tag{4.7–23}$$

If we integrate Eq. (4.7–22) throughout D, interchange the order of summation and integration and use the formula

$$\int_D S^{(n)}(\mathbf{r}, \nu)\, \mathrm{d}^3 r = 1, \tag{4.7–24}$$

which follows at once from the definition (4.7–23) of $S^{(n)}$ and from the orthonormality relation (4.7–12), we obtain the relation

$$\int_D S(\mathbf{r}, \nu)\, \mathrm{d}^3 r = \sum_n \alpha_n(\nu). \tag{4.7–25}$$

Equation (4.7–22), together with Eq. (4.7–23), shows that the contribution of the mode labeled by the index n to the spectral density at the point $\mathbf{r}$ is $\alpha_n(\nu)|\psi_n(\mathbf{r}, \nu)|^2$; and Eq. (4.7–25) shows that its contribution to the integral of the spectral density throughout the domain D, which is a measure of the total energy in D at frequency ν, is precisely $\alpha_n(\nu)$.

Finally we note that according to Eqs. (4.7–5) and (4.7–13), the mutual coherence function of the field may be expressed in the form

$$\Gamma(\mathbf{r}_1, \mathbf{r}_2, \tau) = \sum_n \Gamma^{(n)}(\mathbf{r}_1, \mathbf{r}_2, \tau), \tag{4.7–26}$$

where $\Gamma^{(n)}(\mathbf{r}_1, \mathbf{r}_2, \tau)$, the mutual coherence function of the mode labeled by the index n, is given by

$$\Gamma^{(n)}(\mathbf{r}_1, \mathbf{r}_2, \tau) = \int_0^\infty \alpha_n(\nu) W^{(n)}(\mathbf{r}_1, \mathbf{r}_2, \nu)\, \mathrm{e}^{-2\pi i\nu\tau}\, \mathrm{d}\nu \tag{4.7–27a}$$

$$= \int_0^\infty \alpha_n(\nu) \psi_n^*(\mathbf{r}_1, \nu) \psi_n(\mathbf{r}_2, \nu)\, \mathrm{e}^{-2\pi i\nu\tau}\, \mathrm{d}\nu. \tag{4.7–27b}$$

From Eq. (4.7–27a) one can readily show, with the help of the orthogonality relations (4.7–20), that the $\Gamma^{(n)}$'s are also orthogonal in the sense that

$$\int_{-\infty}^{\infty} \mathrm{d}\tau \int_D \mathrm{d}^3 r_1 \int_D \mathrm{d}^3 r_2 [\Gamma^{(n)}(\mathbf{r}_1, \mathbf{r}_2, \tau)]^* \Gamma^{(m)}(\mathbf{r}_1, \mathbf{r}_2, \tau) = \delta_{nm} \int_0^{\infty} \alpha_n^2(\nu)\, \mathrm{d}\nu. \quad (4.7\text{–}28)$$

However, unlike the spectral degree of coherence $\mu^{(n)}(\mathbf{r}_1, \mathbf{r}_2, \nu)$ the complex degree of coherence $\gamma^{(n)}(\mathbf{r}_1, \mathbf{r}_2, \tau)$ of each mode, obtained on substituting from Eq. (4.7–27b) into Eq. (4.3–12a) is not, in general, unimodular. Hence the modes $\Gamma^{(n)}(\mathbf{r}_1, \mathbf{r}_2, \tau)$ are not, in general, coherent in the space-time domain. This conclusion was to be expected, since the expressions (4.7–27) are not of the form given by Eq. (4.5–54), which follows from the requirement of complete coherence in that domain.

4.7.2 Rigorous representation of the cross-spectral density as a correlation function

With the help of the coherent-mode representation that we have just discussed we will now construct an ensemble of strictly monochromatic wave functions $\{U(\mathbf{r}, \nu)\, \mathrm{e}^{-2\pi i \nu t}\}$, all of the same frequency ν, such that the cross-spectral density $W(\mathbf{r}_1, \mathbf{r}_2, \nu)$ is equal to their cross-correlation function. The mathematical analysis that underlies the construction is similar, in part, to that which we encountered earlier in connection with the Karhunen–Loéve orthogonal expansion of a random process (Section 2.5.1), but the physical consequences of the results go beyond it.

Let us consider a set of functions $\{U(\mathbf{r}, \nu)\}$, each member of which is a linear superposition of the eigenfunctions $\psi_n(\mathbf{r}, \nu)$ of the integral equation (4.7–10):

$$U(\mathbf{r}, \nu) = \sum_n a_n(\nu) \psi_n(\mathbf{r}, \nu). \quad (4.7\text{–}29)$$

In this expansion a_n are random coefficients, whose properties we will specify shortly. The cross-correlation of $U(\mathbf{r}_1, \nu)$ and $U(\mathbf{r}_2, \nu)$ at two points $\mathbf{r}_1$ and $\mathbf{r}_2$ in the field is then given by

$$\langle U^*(\mathbf{r}_1, \nu) U(\mathbf{r}_2, \nu) \rangle_\nu = \sum_n \sum_m \langle a_n^*(\nu) a_m(\nu) \rangle_\nu \psi_n^*(\mathbf{r}_1, \nu) \psi_m(\mathbf{r}_2, \nu), \quad (4.7\text{–}30)$$

where the angle brackets, with suffix ν, denote the average over the ensemble of the frequency-dependent functions $U(\mathbf{r}, \nu)$ or, equivalently, over the ensemble of the random coefficients $a_n(\nu)$. Suppose now that we choose the random coefficient $a_n(\nu)$ so that

$$\langle a_n^*(\nu) a_m(\nu) \rangle_\nu = \alpha_n(\nu) \delta_{nm}, \quad (4.7\text{–}31)$$

where the α_n's are the eigenvalues of the integral equation (4.7–10) and δ_{nm} is the Kronecker symbol. We also demand that, for each realization,

$$\sum_n |a_n(\nu)|^2 < \infty. \quad (4.7\text{–}32)$$

Such an ensemble $\{a_n(\nu)\}$ can always be chosen, for example, by taking

$$a_n(\nu) = [\alpha_n(\nu)]^{1/2}\, \mathrm{e}^{i\theta_n}, \quad (4.7\text{–}33)$$

where, for each, n, θ_n is a real random variable that is uniformly distributed in the interval $0 \leq \theta_n < 2\pi$ and θ_n and θ_m are statistically independent when $n \neq m$. With this choice, the requirement (4.7–31) is evidently satisfied. The requirement expressed by Eq. (4.7–32) is then also satisfied because we have

$$\sum_n \langle |a_n(\nu)|^2 \rangle_\nu = \sum_n \alpha_n(\nu) \qquad (4.7\text{–}34)$$

and the sum of the eigenvalues of the integral equation (4.7–10) with a continuous Hilbert–Schmidt kernel is known to be finite [Tricomi, 1957, Sec. 3.12, Eq. (8), with $\alpha_n = 1/\lambda_n$]. That the sum on the right-hand side of Eq. (4.7–34) is finite is also obvious from the relation (4.7–25).

Each of the sample functions $U(\mathbf{r}, \nu)$ given by an expansion of the form (4.7–29), subject to the constraint (4.7–32), may be readily seen to be square-integrable over the domain D. To show this we integrate the squared modulus of $U(\mathbf{r}, \nu)$ throughout D. If then we substitute in the integrand the expansion (4.7–29) for $U(\mathbf{r}, \nu)$ and interchange the order of integration and summation, we obtain the formula

$$\int_D |U(\mathbf{r}, \nu)|^2 \, \mathrm{d}^3r = \sum_n \sum_m a_n^*(\nu) a_m(\nu) \int_D \psi_n^*(\mathbf{r}, \nu) \psi_m(\mathbf{r}, \nu) \, \mathrm{d}^3r. \qquad (4.7\text{–}35)$$

Since according to Eq. (4.7–12), the ψ_n's form an orthonormal set, Eq. (4.7–35) reduces to

$$\int_D |U(\mathbf{r}, \nu)|^2 \, \mathrm{d}^3r = \sum_n |a_n(\nu)|^2, \qquad (4.7\text{–}36)$$

and the right-hand side of this equation is finite, because of the requirement (4.7–32).

Let us now substitute from Eq. (4.7–31) into Eq. (4.7–30). We then obtain the formula

$$\langle U^*(\mathbf{r}_1, \nu) U(\mathbf{r}_2, \nu) \rangle_\nu = \sum_n \alpha_n(\nu) \psi_n^*(\mathbf{r}_1, \nu) \psi_n(\mathbf{r}_2, \nu). \qquad (4.7\text{–}37)$$

On comparing the right-hand side of this equation with the right-hand side of the Mercer-type expansion (4.7–9) for the cross-spectral density, we see that they are equal. The left-hand sides must, therefore, also be equal and hence we obtain the formula

$$W(\mathbf{r}_1, \mathbf{r}_2, \nu) = \langle U^*(\mathbf{r}_1, \nu) U(\mathbf{r}_2, \nu) \rangle_\nu. \qquad (4.7\text{–}38)$$

Thus we have constructed *an ensemble* $\{U(\mathbf{r}, \nu)\}$ *of random fields that provide a representation of the cross-spectral density* $W(\mathbf{r}_1, \mathbf{r}_2, \nu)$ *of the given field as the cross-correlation function over this ensemble.*

Since according to Eq. (4.7–18) each term in the expansion (4.7–29) for $U(\mathbf{r}, \nu)$ satisfies the Helmholtz equation, it follows that each member of our ensemble $\{U(\mathbf{r}, \nu)\}$ will also satisfy this equation, i.e. that

$$\nabla^2 U(\mathbf{r}, \nu) + k^2 U(\mathbf{r}, \nu) = 0, \qquad (4.7\text{–}39)$$

where $k = 2\pi\nu/c$ is the wave number associated with frequency ν.

We note that the result expressed by Eq. (4.7–38) has some resemblance to the formula (4.3–39) for the cross-spectral density in terms of $\widetilde{V}(\mathbf{r}, \nu)$. Moreover, $\widetilde{V}(\mathbf{r}, \nu)$, just like $U(\mathbf{r}, \nu)$, satisfies the Helmholtz equation [Eq. (4.4–3)] in free space. However, all our earlier derivations of results relating to stationary fields, which involved $\widetilde{V}(\mathbf{r}, \nu)$, must be regarded as being heuristic, because $\widetilde{V}(\mathbf{r}, \nu)$ was formally introduced as the Fourier transform of the random function $V(\mathbf{r}, t)$; and, as we have noted several times already, the Fourier transform of a stationary random function does not exist within the framework of ordinary function theory. It can only be rigorously interpreted as a generalized function. On the other hand, the functions $U(\mathbf{r}, \nu)$ of the present theory are ordinary functions. They may be used to provide a rigorous formulation of many of the results that we obtained earlier by plausibility arguments.

Apart from greater mathematical rigor, the formula (4.7–38) for the cross-spectral density and its 'diagonal' form

$$S(\mathbf{r}, \nu) = \langle U^*(\mathbf{r}, \nu) U(\mathbf{r}, \nu) \rangle_\nu \tag{4.7–40}$$

for the spectral density of the field at a point $\mathbf{r}$ agree well with the intuitive physical meanings of these quantities. For, in view of Eqs. (4.7–38) and (4.7–39), we may regard each member $U(\mathbf{r}, \nu)$ of our ensemble as the time-independent part of a monochromatic wave function.

$$V(\mathbf{r}, t) = U(\mathbf{r}, \nu)\, \mathrm{e}^{-2\pi i \nu t}; \tag{4.7–41}$$

and the spectrum and the cross-spectrum of the field are then expressed as averages of quantities that are quadratic in their complex amplitudes.

4.7.3 Natural modes of oscillations of partially coherent primary sources and a representation of their cross-spectral density as a correlation function

Representations strictly similar to those given by Eqs. (4.7–9) and (4.7–38) for the cross-spectral density of a stationary field can also be obtained for the cross-spectral density of a stationary source. The 'source-analogues' to these two formulas, which we will derive in this section, lead to the concept of natural modes of oscillations of a partially coherent source and they provide a useful method for the study of the properties of fields generated by such sources in terms of coherent modes.

Let us consider a primary fluctuating scalar source localized in some finite domain σ in free space. Let $Q(\mathbf{r}, t)$ be the source distribution, which we take to be the analytic signal associated with a real source distribution $Q^{(\mathrm{r})}(\mathbf{r}, t)$. We assume that the fluctuations are characterized by a statistical ensemble that is stationary, at least in the wide sense, and we denote by $\Gamma_Q(\mathbf{r}_1, \mathbf{r}_2, \tau)$ the cross-correlation function of Q, namely

$$\Gamma_Q(\mathbf{r}_1, \mathbf{r}_2, \tau) = \langle Q^*(\mathbf{r}_1, t) Q(\mathbf{r}_2, t + \tau) \rangle. \tag{4.7–42}$$

We assume, by analogy with the assumption expressed by Eq. (4.7–1), that Γ_Q is absolutely integrable with respect to τ, and hence similar results will hold for the source distribution as those expressed by Eqs. (4.7–2) to (4.7–5) for the optical field. In particular the cross-spectral density of the source distribution, namely

$$W_Q(\mathbf{r}_1, \mathbf{r}_2, \nu) = \int_{-\infty}^{\infty} \Gamma_Q(\mathbf{r}_1, \mathbf{r}_2, \tau)\,e^{2\pi i\nu\tau}\,d\tau, \tag{4.7–43}$$

will exist and will be a continuous function of ν. We also assume that $W_Q(\mathbf{r}_1, \mathbf{r}_2, \nu)$ is a continuous function of $\mathbf{r}_1$ and $\mathbf{r}_2$ throughout the source domain σ. Then W_Q will be a non-negative definite Hilbert–Schmidt kernel and it follows, by analogy with Eq. (4.7–9), that it can be represented in the form of Mercer's expansion, namely

$$W_Q(\mathbf{r}_1, \mathbf{r}_2, \nu) = \sum_n \beta_n(\nu)\phi_n^*(\mathbf{r}_1, \nu)\phi_n(\mathbf{r}_2, \nu), \tag{4.7–44}$$

which is absolutely and uniformly convergent. The ϕ_n's are the eigenfunctions and the β_n's are the eigenvalues of the integral equation

$$\int_\sigma W_Q(\mathbf{r}_1, \mathbf{r}_2, \nu)\phi_n(\mathbf{r}_1, \nu)\,d^3r_1 = \beta_n(\nu)\phi_n(\mathbf{r}_2, \nu). \tag{4.7–45}$$

Again the eigenvalues are necessarily real and non-negative,

$$\beta_n(\nu) \geqslant 0, \tag{4.7–46}$$

and the eigenfunctions may be taken to be orthonormalized:

$$\int_\sigma \phi_n^*(\mathbf{r}, \nu)\phi_m(\mathbf{r}, \nu)\,d^3r = \delta_{nm}. \tag{4.7–47}$$

The expansion (4.7–44) represents the cross-spectral density of the source as a linear superposition of the cross-spectral densities

$$W_Q^{(n)}(\mathbf{r}_1, \mathbf{r}_2, \nu) = \phi_n^*(\mathbf{r}_1, \nu)\phi_n(\mathbf{r}_2, \nu), \tag{4.7–48}$$

which are seen to factorize with respect to the two spatial variables $\mathbf{r}_1$ and $\mathbf{r}_2$. This fact implies that the associated spectral degree of coherence

$$\mu_Q^{(n)}(\mathbf{r}_1, \mathbf{r}_2, \nu) = \frac{W_Q^{(n)}(\mathbf{r}_1, \mathbf{r}_2, \nu)}{[W_Q^{(n)}(\mathbf{r}_1, \mathbf{r}_1, \nu)]^{1/2}[W_Q^{(n)}(\mathbf{r}_2, \mathbf{r}_2, \nu)]^{1/2}} \tag{4.7–49}$$

is unimodular. Hence the expansion (4.7–44) represents the cross-spectral density of the source as a linear superposition of elementary sources which are completely coherent in the space-frequency domain. We may regard these elementary sources ϕ_n, or rather the products $\phi_n(\mathbf{r}, \nu)\,e^{-2\pi i\nu t}$, as representing *natural modes of oscillations* of the given source. By analogy with Eq. (4.7–20) one can also show that the cross-spectral densities of the different modes are mutually orthogonal.

Of special interest for applications of this mode representation is an anologue to Eq. (4.7–38), namely the formula

$$W_Q(\mathbf{r}_1, \mathbf{r}_2, \nu) = \langle U_Q^*(\mathbf{r}_1, \nu)U_Q(\mathbf{r}_2, \nu)\rangle_\nu. \tag{4.7–50}$$

It represents *the cross-spectral density of the source distribution as a correlation function in the space-frequency domain*. In Eq. (4.7–50) the U_Q's are sample functions of a statistical ensemble, of which each member has the form

$$U_Q(\mathbf{r}, \nu) = \sum_n b_n(\nu)\phi_n(\mathbf{r}, \nu), \tag{4.7–51}$$

where the $b_n(\nu)$'s are random variables satisfying the requirements

$$\langle b_n^*(\nu) b_m(\nu)\rangle_\nu = \beta_n(\nu)\delta_{nm} \tag{4.7–52}$$

and

$$\sum_n \langle |b_n(\nu)|^2\rangle_\nu < \infty. \tag{4.7–53}$$

The average in (4.7–50) is, of course, taken over the ensemble of the U_Q or, equivalently, over the ensemble of the random coefficients b_n. The formula (4.7–50) may be derived by the same argument as employed in the derivation of the corresponding formula (4.7–38) for the cross-spectral density of the field.

Let us now consider the field $V(\mathbf{r}, t)$ generated by the fluctuating source and let $W_V(\mathbf{r}_1, \mathbf{r}_2, \nu)$ be the cross-spectral density of this field. According to Eq. (4.4–63), W_V is related to the cross-spectral density $W_Q(\mathbf{r}_1, \mathbf{r}_2, \nu)$ of the source by the equation

$$(\nabla_1^2 + k^2)(\nabla_2^2 + k^2)W_V(\mathbf{r}_1, \mathbf{r}_2, \nu) = (4\pi)^2 W_Q(\mathbf{r}_1, \mathbf{r}_2, \nu). \tag{4.7–54}$$

Now as we have already noted, the expansion (4.7–44) represents W_Q as a linear combination of contributions from completely coherent elementary sources. Each such elementary source ϕ_n may be expected to generate a field ψ_n that is an outgoing solution of the Helmholtz equation

$$(\nabla^2 + k^2)\psi_n(\mathbf{r}, \nu) = -4\pi\phi_n(\mathbf{r}, \nu). \tag{4.7–55}$$

Assuming, for the sake of simplicity, that the region outside the source domain σ is unbounded free space, the outgoing solution of Eq. (4.7–55) is

$$\psi_n(\mathbf{r}, \nu) = \int_\sigma \phi_n(\mathbf{r}', \nu)\frac{\exp(ik|\mathbf{r} - \mathbf{r}'|)}{|\mathbf{r} - \mathbf{r}'|}\,d^3r'. \tag{4.7–56}$$

If, on the other hand, the region outside the source domain σ contains material objects that respond linearly to an incident field, or if the region is bounded, one would have, in place of Eq. (4.7–56), a formula of the same form, but with the outgoing spherical wave function $\exp(ik|\mathbf{r} - \mathbf{r}'|)/|\mathbf{r} - \mathbf{r}'|$ replaced by the appropriate Green's function.

If we now construct the ensemble of functions

$$U_V(\mathbf{r}, \nu) = \sum_n b_n(\nu)\psi_n(\mathbf{r}, \nu), \tag{4.7–57}$$

where the ψ_n's are given by Eq. (4.7–56) and the b_n's are the same expansion coefficients as those that appear in Eq. (4.7–51). Equations (4.7–57), (4.7–55) and (4.7–51) imply that $U_V(\mathbf{r}, \nu)$ is an outgoing solution of the reduced wave equation

$$(\nabla^2 + k^2)U_V(\mathbf{r}, \nu) = -4\pi U_Q(\mathbf{r}, \nu). \tag{4.7–58}$$

From Eqs. (4.7–58) and (4.7–50) it readily follows that

$$(\nabla_1^2 + k^2)(\nabla_2^2 + k^2)\langle U_V^*(\mathbf{r}_1, \nu)U_V(\mathbf{r}_2, \nu)\rangle_\nu = (4\pi)^2 W_Q(\mathbf{r}_1, \mathbf{r}_2, \nu), \tag{4.7–59}$$

which implies, in view of Eq. (4.7–54), that

$$W_V(\mathbf{r}_1, \mathbf{r}_2, \nu) = \langle U_V^*(\mathbf{r}_1, \nu) U_V(\mathbf{r}_2, \nu) \rangle_\nu. \tag{4.7–60}$$

This formula expresses *the cross-spectral density of the field generated by the primary source as a correlation function in the space-frequency domain.*

Finally if we substitute from Eq. (4.7–57) into Eq. (4.7–60) and use Eq. (4.7–52), we obtain the following expansion for W_V:

$$W_V(\mathbf{r}_1, \mathbf{r}_2, \nu) = \sum_n \beta_n(\nu) \psi_n^*(\mathbf{r}_1, \nu) \psi_n(\mathbf{r}_2, \nu). \tag{4.7–61}$$

Now each term in this series is factorized with respect to the two spatial variables $\mathbf{r}_1$ and $\mathbf{r}_2$ and consequently its associated spectral degree of coherence is unimodular. Hence *Eq. (4.7–61) represents the cross-spectral density of the field produced by the primary source as a linear combination of elementary fields, each generated by a natural source mode which is completely coherent in the space-frequency domain.* However, unlike the natural modes of the source, these partial fields are not, in general, mutually orthogonal but, in view of Eq. (4.7–52), their contributions $\beta_n(\nu)\psi_n^*(\mathbf{r}_1, \nu)\psi_n(\mathbf{r}_2, \nu)$ to the cross-spectral density W_V of the field are *uncorrelated*.

Appendix 4.1

The kernel $H(\mathbf{R})$ of the integral transform representation of the operator $\surd(-\nabla^2)$ [Eq. (4.6–14)]

In this Appendix we obtain a formal evaluation of the integral (4.6–13), namely

$$H(\mathbf{R}) = \frac{1}{(2\pi)^3} \int K \, e^{i\mathbf{K}\cdot\mathbf{R}} \, d^3K, \tag{A4.1–1}$$

where the integration extends over the whole **K**-space.

We introduce spherical polar coordinates (K, θ, ϕ) in **K**-space, with the polar axis $\theta = 0$ along **R**. The formula (A4.1–1) then becomes

$$H(\mathbf{R}) = \frac{1}{(2\pi)^3} \int_0^\infty dK K^3 \int_0^{2\pi} d\phi \int_0^\pi d\theta \sin\theta \, e^{iKR\cos\theta}. \tag{A4.1–2}$$

The integration with respect to ϕ gives 2π and, if we set $x = \cos\theta$, Eq. (A4.1–2) reduces to

$$\begin{aligned}
H(\mathbf{R}) &= \frac{1}{(2\pi)^2} \int_0^\infty dK K^3 \int_{-1}^1 dx \, e^{iKRx} \\
&= \frac{1}{(2\pi)^2} \int_0^\infty K^3 \frac{e^{iKR} - e^{-iKR}}{iKR} dK \\
&= -\frac{i}{2\pi R}\left[\frac{1}{2\pi}\int_0^\infty K^2 e^{iKR} \, dK - \frac{1}{2\pi}\int_0^\infty K^2 e^{-iKR} \, dK\right] \\
&= -\frac{i}{2\pi R}\left\{-\frac{d^2}{dR^2}\left[\frac{1}{2\pi}\int_0^\infty e^{iKR} \, dK\right] + \frac{d^2}{dR^2}\left[\frac{1}{2\pi}\int_0^\infty e^{-iKR} \, dK\right]\right\},
\end{aligned}$$

i.e.

$$H(\mathbf{R}) = \frac{\mathrm{i}}{2\pi R}\frac{\mathrm{d}^2}{\mathrm{d}R^2}[\delta^{(+)}(R) - \delta^{(-)}(R)], \tag{A4.1–3}$$

where

$$\delta^{(+)}(\xi) = \frac{1}{2\pi}\int_0^\infty \mathrm{e}^{\mathrm{i}K\xi}\,\mathrm{d}K, \qquad \delta^{(-)}(\xi) = \frac{1}{2\pi}\int_0^\infty \mathrm{e}^{-\mathrm{i}K\xi}\,\mathrm{d}K. \tag{A4.1–4}$$

These positive and negative frequency parts of the Dirac delta function are trivially related to the singular functions δ_- and δ_+ which we encountered in Section 3.1. The reason for the slightly different notation used here is due to the difference in the signs of the arguments of the exponential kernels which appear in the temporal and the spatial Fourier transforms. The functions $\delta^{(+)}(\xi)$ and $\delta^{(-)}(\xi)$ may also be represented in the form [Heitler (1984), Sec. 8]

$$\delta^{(\pm)}(\xi) = \frac{1}{2}\left[\delta(\xi) \mp \frac{1}{\pi \mathrm{i}}P\frac{1}{\xi}\right], \tag{A4.1–5}$$

where P denotes the Cauchy principal value.

The formula (A4.1–3) is Eq. (4.6–14) of the text.

Appendix 4.2

The Green's function $G(\mathbf{R}, T)$ for the time evolution of the analytic signal representation of free fields and its Fourier transform $\tilde{G}(\mathbf{R}, \nu)$ [Eqs. (4.6–19) and (4.6–34)].

The Green's function $G(\mathbf{R}, T)$ which appears in Eq. (4.6–17) for the time evolution of the analytic signal representation of free fields is defined by the formula (4.6–18), namely

$$G(\mathbf{R}, T) = \frac{1}{(2\pi)^3}\int \mathrm{e}^{\mathrm{i}(\mathbf{K}\cdot\mathbf{R} - KcT)}\,\mathrm{d}^3K, \tag{A4.2–1}$$

where the integration extends over the whole **K**-space.

We proceed in a similar manner as we did in Appendix 4.1 in evaluating the kernel $H(\mathbf{R})$. We introduce spherical polar coordinates (K, θ, ϕ) in **K**-space, with the polar axis $\theta = 0$ along **R**, integrate with respect to ϕ and set $x = \cos\theta$. The expression (A4.2–1) for the Green's function then becomes

$$\begin{aligned} G(\mathbf{R}, T) &= \frac{1}{(2\pi)^2}\int_0^\infty \mathrm{d}KK^2\int_{-1}^{1}\mathrm{d}x\,\mathrm{e}^{\mathrm{i}K(Rx - cT)} \\ &= \frac{1}{(2\pi)^2}\int_0^\infty K^2\,\mathrm{e}^{-\mathrm{i}KcT}\left[\frac{\mathrm{e}^{\mathrm{i}KR} - \mathrm{e}^{-\mathrm{i}KR}}{\mathrm{i}KR}\right]\mathrm{d}K \\ &= -\frac{1}{2\pi R}\left[\frac{1}{2\pi}\int_0^\infty \mathrm{i}K\,\mathrm{e}^{\mathrm{i}K(R - cT)}\,\mathrm{d}K - \frac{1}{2\pi}\int_0^\infty \mathrm{i}K\,\mathrm{e}^{-\mathrm{i}K(R + cT)}\,\mathrm{d}K\right], \end{aligned}$$

i.e.

$$G(\mathbf{R}, T) = -\frac{1}{2\pi R}\frac{\partial}{\partial R}[\delta^{(+)}(R - cT) + \delta^{(-)}(R + cT)], \qquad \text{(A4.2–2)}$$

where $\delta^{(+)}$ and $\delta^{(-)}$ are the functions defined by Eqs. (A4.1–4). The formula (A4.2–2) is Eq. (4.6–19) of the text.

We will next determine the Fourier transform

$$\widetilde{G}(\mathbf{R}, \nu) = \int_{-\infty}^{\infty} G(\mathbf{R}, T)\, e^{2\pi i \nu T}\, dT \qquad \text{(A4.2–3)}$$

of $G(\mathbf{R}, T)$. We can obtain it almost at once by starting from the second expression in Eq. (A4.2–2) above, which we re-write in the form

$$G(\mathbf{R}, T) = \int_0^{\infty} g(k)\, e^{-ikcT}\, dk, \qquad \text{(A4.2–4)}$$

where

$$g(k) = \frac{2k^2}{(2\pi)^2}\frac{\sin kR}{kR}. \qquad \text{(A4.2–5)}$$

If in the integral in Eq. (A4.2–4) we change the variable of integration from k to ν, where

$$kc = 2\pi\nu \qquad \text{(A4.2–6)}$$

we obtain for $G(\mathbf{R}, T)$ the expression

$$G(\mathbf{R}, T) = \frac{2\pi}{c}\int_0^{\infty} g(2\pi\nu/c)\, e^{-2\pi i \nu T}\, d\nu. \qquad \text{(A4.2–7)}$$

On taking the Fourier inverse of Eq. (A4.2–7) and comparing it with Eq. (A4.2–3) we find that

$$\left.\begin{aligned} \widetilde{G}(\mathbf{R}, \nu) &= \frac{2\pi}{c} g(2\pi\nu/c) \quad && \text{when } \nu > 0 \\ &= 0 && \text{when } \nu < 0. \end{aligned}\right\} \qquad \text{(A4.2–8)}$$

On substituting from Eq. (A4.2–5) into Eq. (A4.2–8) and again using the relation (A4.2–6) we finally obtain for $\widetilde{G}(\mathbf{R}, \nu)$ the expression

$$\left.\begin{aligned} \widetilde{G}(\mathbf{R}, \nu) &= \frac{k^2}{\pi c}\left(\frac{\sin kR}{kR}\right) \quad && \text{when } \nu > 0 \\ &= 0 && \text{when } \nu < 0, \end{aligned}\right\} \qquad \text{(A4.2–9)}$$

which is Eq. (4.6–34) of the text.

Problems

4.1 The mutual intensity function for all pairs of points on a surface intercepting a certain stationary quasi-monochromatic light beam is of the form

$$J(P_1, P_2) = f(P_1)g(P_2),$$

where $f(P)$ and $g(P)$ are known functions of position on the surface. Show that
(a) $g(P) = \alpha f^*(P)$ where α is a real constant, and the asterisk denotes the complex conjugate.
(b) The light in the space into which it propagates is completely spatially coherent within the framework of second-order coherence theory.

4.2 The cross-spectral density function of a planar, secondary source, at points specified by position vectors $\boldsymbol{\rho}_1$ and $\boldsymbol{\rho}_2$, has the factorized form

$$W(\boldsymbol{\rho}_1, \boldsymbol{\rho}_2, \omega) = F\left(\frac{\boldsymbol{\rho}_1 + \boldsymbol{\rho}_2}{2}, \omega\right)G(\boldsymbol{\rho}_2 - \boldsymbol{\rho}_1, \omega).$$

Show that in order that $F(\boldsymbol{\rho}, \omega)$ represents the spectral density and $G(\boldsymbol{\rho}, \omega)$ the spectral degree of coherence of light across the source, the spectral density has to satisfy a certain functional equation. Show further that the functional equation is satisfied by any spectral density distribution whose spatial dependence has the form

$$S^{(0)}(\boldsymbol{\rho}, \omega) \equiv S^{(0)}(x, y, \omega) = S^{(0)}(0, 0; \omega)\,\mathrm{e}^{(\beta_1 x + \beta_2 y)},$$

where β_1 and β_2 are constants.

4.3 The mutual coherence function of a stationary optical field in free space is of the form

$$\Gamma(\mathbf{r}_1, \mathbf{r}_2, \tau) = F(\mathbf{r}_1, \mathbf{r}_2)G(\tau).$$

Show that the function $F(\mathbf{r}_1, \mathbf{r}_2)$ must satisfy two Helmholtz equations and determine the most general form of $G(\tau)$.

4.4 Consider a statistically stationary field in free space. Show that the complex degree of coherence $\gamma(\mathbf{r}_1, \mathbf{r}_2, \tau)$ and the spectral degree of coherence $\mu(\mathbf{r}_1, \mathbf{r}_2, \nu)$ of the field are related by the formula

$$\gamma(\mathbf{r}_1, \mathbf{r}_2, \tau) = \int_0^\infty \sqrt{\mathscr{s}(\mathbf{r}_1, \nu)}\sqrt{\mathscr{s}(\mathbf{r}_2, \nu)}\,\mu(\mathbf{r}_1, \mathbf{r}_2, \nu)\,\mathrm{e}^{-2\pi i\nu\tau}\,\mathrm{d}\nu,$$

where

$$\mathscr{s}(\mathbf{r}, \nu) = \frac{S(\mathbf{r}, \nu)}{\int_0^\infty S(\mathbf{r}, \nu)\,\mathrm{d}\nu}$$

is the normalized spectral density at $\mathbf{r}$.

Suppose next that the normalized spectral densities at two points $\mathbf{r}_1'$ and $\mathbf{r}_2'$ are equal to each other and that

$$\mu(\mathbf{r}_1', \mathbf{r}_2', \nu) = \eta(\mathbf{r}_1', \mathbf{r}_2')\,\mathrm{e}^{2\pi i\nu\tau(\mathbf{r}_1', \mathbf{r}_2')},$$

with $\eta(\mathbf{r}_2', \mathbf{r}_1') = \eta^*(\mathbf{r}_1', \mathbf{r}_2')$ and $\tau(\mathbf{r}_1', \mathbf{r}_2')$ is real and satisfies the relation $\tau(\mathbf{r}_2', \mathbf{r}_1') = -\tau(\mathbf{r}_1', \mathbf{r}_2')$. Show that under these circumstances $\gamma(\mathbf{r}_1', \mathbf{r}_2', \tau)$ obeys the reduction formula for cross-spectrally pure beams, namely

$$\gamma(\mathbf{r}_1', \mathbf{r}_2', \tau) = \gamma(\mathbf{r}_1', \mathbf{r}_2', \tau_{12})\gamma(\mathbf{r}_1', \mathbf{r}_1', \tau - \tau_{12}),$$

where τ_{12} depends on $\mathbf{r}_1'$ and $\mathbf{r}_2'$.

4.5 Let $r\mathbf{s}$ ($\mathbf{s}^2 = 1$) be the position vector (referred to an origin in the source domain) of a point in the far zone of a planar, secondary, statistically stationary source occupying a finite region σ in the plane $z = 0$. Show that the 'longitudinal' spectral degree of coherence $\mu^{(\infty)}(r_1\mathbf{s}, r_2\mathbf{s}, \nu)$ of the far field is unimodular, i.e. that the field at any two points in the far zone, located along the same outward direction from the source, is spatially fully coherent at each frequency ν.

Using the relation given by the first formula of problem 4.4, show that the complex degree of coherence $\gamma^{(\infty)}(r_1\mathbf{s}, r_2\mathbf{s}, \tau)$ is not unimodular in general, but that $\gamma^{(\infty)}(r_1\mathbf{s}, r_2\mathbf{s}, \tau) = 1$ for a particular value τ_0 of τ. Determine τ_0 and interpret the result in physical terms.

4.6 Consider the integral equation

$$\int_{-1}^{1} K(x, x')\phi(x')\, \mathrm{d}x' = \lambda\phi(x)$$

with the non-Hermitian kernel

$$K(x, x') = (1 + \mathrm{i}\sqrt{3}x)(1 + \mathrm{i}\sqrt{3}x').$$

(a) Show that the integral equation has no solution with $\lambda \neq 0$.
(b) Show that if one of the i's in the expression for $K(x, x')$ is changed to $-\mathrm{i}$ (so that the kernel becomes Hermitian), the equation has a solution, with the eigenvalue $\lambda = 4$.

4.7 A monochromatic plane wave of frequency ω, propagating in the positive z-direction, is incident on a diffuser which moves with constant speed v in the positive y-direction (see figure). An opaque screen, pierced by small

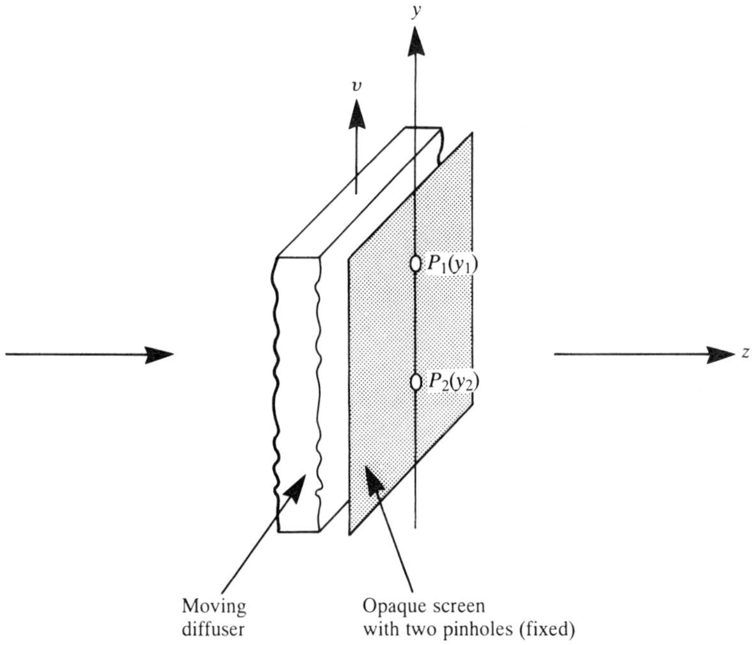

pinholes at points $P_1(y_1)$ and $P_2(y_2)$ is placed immediately behind the diffuser. The amplitude transmission function of the diffuser, assumed to be statistically homogeneous and non-absorbing, is $t(y)$.

(a) Determine the complex degree of coherence $\gamma_{12}(\tau)$ of the light at P_1 and P_2.

(b) Determine the spectral degree of coherence $\mu_{12}(\omega)$ of the light at P_1 and P_2.

(c) Estimate the coherence time of the light emerging from one of the pinholes for the case when the correlation function of the diffuser is $\exp(-a^2y^2)$, a being a positive constant.

4.8 In the theory of atmospheric turbulence one frequently deals with real fields $f(\mathbf{r})$ (e.g. the refractive index function), which are random functions of the position vector $\mathbf{r}$. One defines its correlation function as

$$R(\mathbf{r}_1, \mathbf{r}_2) = \langle f(\mathbf{r}_1) f(\mathbf{r}_2) \rangle,$$

the angular brackets denoting ensemble average. If $R(\mathbf{r}_1, \mathbf{r}_2)$ depends only on the 'difference vector' $\mathbf{r} = \mathbf{r}_2 - \mathbf{r}_1$, e.g. if $R(\mathbf{r}_1, \mathbf{r}_2) \equiv R(\mathbf{r}_2 - \mathbf{r}_1)$ the random field $f(\mathbf{r})$ is said to be statistically *homogeneous*.

(a) The spectrum $S(\mathbf{k})$ of a homogeneous random field $f(\mathbf{r})$ may be identified with the Fourier inverse of its correlation function, i.e.

$$S(\mathbf{k}) = \frac{1}{(2\pi)^3} \int R(\mathbf{r})\, \mathrm{e}^{-\mathrm{i}\mathbf{k}\cdot\mathbf{r}}\, \mathrm{d}^3 r.$$

Assuming that $f(\mathbf{r})$ is also statistically *isotropic*, i.e. that $R(\mathbf{r})$ depends only on the magnitude $r = |\mathbf{r}|$, show that $S(\mathbf{k})$ is a function of the magnitude $k = |\mathbf{k}|$ only and that

$$S(k) = \frac{1}{2\pi^2 k} \int_0^\infty r R(r) \sin kr\, \mathrm{d}r$$

and

$$R(r) = \frac{4\pi}{r} \int_0^\infty k S(k) \sin kr\, \mathrm{d}k.$$

(b) For such a field it is also convenient to introduce a related spectral function $V(k)$, as the one-dimensional Fourier inverse of $R(r)$, i.e.

$$V(k) = \frac{1}{2\pi} \int_{-\infty}^{\infty} R(r)\, \mathrm{e}^{-\mathrm{i}kr}\, \mathrm{d}r,$$

where, for negative values of its arguments, R is formally defined by the relation $R(-r) = R(r)$. Express $S(k)$ in terms of $V(k)$.

5

Radiation from sources of any state of coherence

5.1 Introduction

In this chapter we will apply the correlation theory of scalar wavefields developed in the previous chapter to study radiation from localized primary and secondary scalar sources of any state of coherence. In general both the spectral density and the coherence properties of the source will determine the nature of the field. Among the topics which we will consider in some detail are partially coherent optical beams, the role of coherence in the radiometric model of energy transport and the influence of coherence properties of sources on the spectra of the emitted radiation.

It is convenient and mathematically simpler to employ the space-frequency rather than the space-time description in treating these problems, i.e. to characterize the correlation properties of the sources and fields by the cross-spectral density function rather than by the mutual coherence function. Corresponding results in the space-time and space-frequency domains are, of course, related via the basic Fourier transform relations (4.3–40).

5.2 Radiation from three-dimensional primary sources

5.2.1 General formulas

Let us consider radiation from a fluctuating primary source, which occupies a finite domain D. We assume that the fluctuations may be described by an ensemble which is stationary, at least in the wide sense. According to Eq. (4.4–68), the cross-spectral density function $W(\mathbf{r}_1, \mathbf{r}_2, \nu)$ of the field which the source generates at any two field points $P_1(\mathbf{r}_1)$ and $P_2(\mathbf{r}_2)$ is expressible in terms of the cross-spectral density function $W_Q(\mathbf{r}_1', \mathbf{r}_2', \nu)$ of the source distribution by the formula

$$W(\mathbf{r}_1, \mathbf{r}_2, \nu) = \int_D \int_D W_Q(\mathbf{r}_1', \mathbf{r}_2', \nu) \frac{\mathrm{e}^{ik(R_2 - R_1)}}{R_1 R_2} \mathrm{d}^3 r_1' \, \mathrm{d}^3 r_2', \qquad (5.2\text{–}1)$$

where $\mathbf{r}_1'$ and $\mathbf{r}_2'$ are the position vectors of any two source points S_1 and S_2,

$$R_j = |\mathbf{r}_j - \mathbf{r}_j'|, \quad (j = 1, 2), \qquad (5.2\text{–}2)$$

and $k = 2\pi\nu/c$ (c being the speed of light in vacuum) is the wave number associated with the frequency ν.

One is frequently interested in the behavior of the field in the far zone of the source. The formula (5.2–1) then appreciably simplifies, as we will now show. We take the origin O of the position vectors $\mathbf{r}_1$, $\mathbf{r}_2$, $\mathbf{r}'_1$, $\mathbf{r}'_2$ to be at some fixed point in the source region and set (see Fig. 5.1)

$$\mathbf{r}_1 = r_1\mathbf{s}_1, \quad \mathbf{r}_2 = r_2\mathbf{s}_2, \tag{5.2–3}$$

where $r_1 = |\mathbf{r}_1|$, $r_2 = |\mathbf{r}_2|$ and $\mathbf{s}_1$ and $\mathbf{s}_2$ are unit vectors pointing from the origin O to the two field points. If the field points are sufficiently far away from the origin, we may make, in the exponent of the integrand in Eq. (5.2–1), the approximation

$$R_1 \equiv |\mathbf{r}_1 - \mathbf{r}'_1| \approx r_1 - \mathbf{r}'_1 \cdot \mathbf{s}_1, \quad R_2 \equiv |\mathbf{r}_2 - \mathbf{r}'_2| \approx r_2 - \mathbf{r}'_2 \cdot \mathbf{s}_2, \tag{5.2–4}$$

whilst in the denominator we may just make the approximation $R_1 \approx r_1$, $R_2 \approx r_2$. We then obtain the following expression for the cross-spectral density:

$$W^{(\infty)}(\mathbf{r}_1, \mathbf{r}_2, \nu) = \frac{\mathrm{e}^{\mathrm{i}k(r_2-r_1)}}{r_1 r_2}\int_D\int_D W_Q(\mathbf{r}'_1, \mathbf{r}'_2, \nu)\,\mathrm{e}^{-\mathrm{i}k(\mathbf{s}_2\cdot\mathbf{r}'_2-\mathbf{s}_1\cdot\mathbf{r}'_1)}\,\mathrm{d}^3r'_1\,\mathrm{d}^3r'_2. \tag{5.2–5}$$

We wrote $W^{(\infty)}$ rather than W on the left-hand side of Eq. (5.2–5) to stress that the formula applies to the far field.

The right-hand side of Eq. (5.2–5) is the product of two factors. The first one depends only on the distances r_1 and r_2 from the origin O of the points P_1 and P_2 in the far zone. The second depends only on the directions, specified by the unit vectors $\mathbf{s}_1$ and $\mathbf{s}_2$, pointing from the origin O to P_1 and P_2. It is useful to emphasize this directional dependence and for this reason we rewrite Eq. (5.2–5) as

$$W^{(\infty)}(r_1\mathbf{s}_1, r_2\mathbf{s}_2, \nu) = L(\mathbf{s}_1, \mathbf{s}_2, \nu)\frac{\mathrm{e}^{\mathrm{i}k(r_2-r_1)}}{r_1 r_2}, \tag{5.2–6}$$

where

$$L(\mathbf{s}_1, \mathbf{s}_2, \nu) = \int_D\int_D W_Q(\mathbf{r}'_1, \mathbf{r}'_2, \nu)\,\mathrm{e}^{-\mathrm{i}k(\mathbf{s}_2\cdot\mathbf{r}'_2-\mathbf{s}_1\cdot\mathbf{r}'_1)}\,\mathrm{d}^3r'_1\,\mathrm{d}^3r'_2. \tag{5.2–7}$$

The function $L(\mathbf{s}_1, \mathbf{s}_2, \nu)$ is known as the *radiant cross-intensity* of the radiated field. It will be seen later [Eq. (5.6–52)] that the radiant cross-intensity is also a measure of correlations between the plane-wave modes of frequency ν of the field which propagate in directions specified by the unit vectors $\mathbf{s}_1$ and $\mathbf{s}_2$.

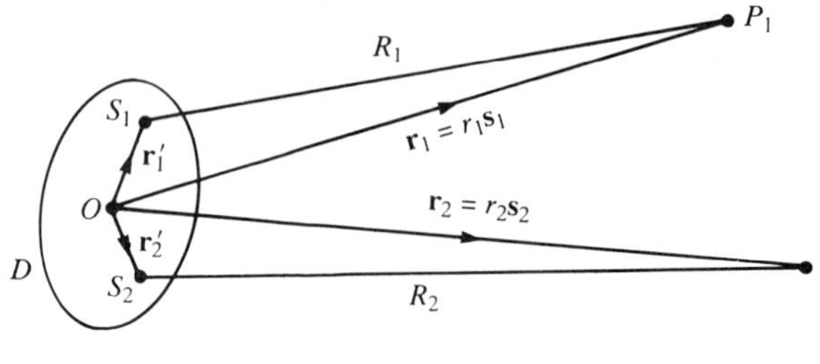

Fig. 5.1 Illustrating the notation relating to the formula (5.2–5) for the cross-spectral density of the far field generated by a three-dimensional primary source.

We derived Eq. (5.2–6) using several approximations but it may be shown, for example with the help of the principle of stationary phase, that it is the exact asymptotic formula for $W(r_1\mathbf{s}_1, r_2\mathbf{s}_2, \nu)$ as $kr_1 \to \infty$ and $kr_2 \to \infty$, with $\mathbf{s}_1$ and $\mathbf{s}_2$ being kept fixed. This fact may be verified by the use of Eqs. (4.7–38), (3.3–95) and (3.2–27).

Formulas of the form given by Eq. (5.2–6) are known also to apply to other situations. For example, the cross-spectral density in the far zone of a field scattered by a localized object may be shown to have this form. Of course, the radiant cross-intensity is then no longer given by Eq. (5.2–7).

Since the cross-spectral density $W_Q(\mathbf{r}_1', \mathbf{r}_2', \nu)$ of the source has zero value when either $\mathbf{r}_1'$ or $\mathbf{r}_2'$ or both represent points outside the source domain D, we may extend the integrations in Eq. (5.2–7) formally over all space. We then obtain for the radiant cross-intensity the expression

$$L(\mathbf{s}_1, \mathbf{s}_2, \nu) = (2\pi)^6 \widetilde{W}_Q(-k\mathbf{s}_1, k\mathbf{s}_2, \nu), \tag{5.2–8}$$

where

$$\widetilde{W}_Q(\mathbf{K}_1, \mathbf{K}_2, \nu) = \frac{1}{(2\pi)^6} \iint W_Q(\mathbf{r}_1', \mathbf{r}_2', \nu)\, \mathrm{e}^{-\mathrm{i}(\mathbf{K}_1\cdot\mathbf{r}_1' + \mathbf{K}_2\cdot\mathbf{r}_2')}\, \mathrm{d}^3 r_1'\, \mathrm{d}^3 r_2' \tag{5.2–9}$$

is the six-dimensional spatial Fourier transform of the cross-spectral density of the source.

Equation (5.2–6), together with the expression (5.2–8) for the radiant cross-intensity, is the basic formula relating to radiation from fluctuating 'steady-state' (more precisely wide-sense stationary) three-dimensional sources. We see from Eq. (5.2–8) that not all the spatial Fourier components of W_Q contribute to the far field. Only those contribute that are labeled by the pairs of three-dimensional vectors

$$\mathbf{K}_1 = -k\mathbf{s}_1, \quad \mathbf{K}_2 = k\mathbf{s}_2. \tag{5.2–10}$$

Since $\mathbf{s}_1$ and $\mathbf{s}_2$ are unit vectors, the magnitudes of the spatial-frequency vectors that enter the expression (5.2–8) are

$$|\mathbf{K}_1| = |\mathbf{K}_2| = k = 2\pi\nu/c, \tag{5.2–11}$$

i.e. their magnitudes are equal to the free-space wave number k, associated with the temporal frequency ν.

In the special case when the points P_1 and P_2 in the far zone coincide, we obtain at once from Eq. (5.2–6) the formula

$$S^{(\infty)}(r\mathbf{s}, \nu) = J(\mathbf{s}, \nu)/r^2. \tag{5.2–12}$$

Here

$$S^{(\infty)}(r\mathbf{s}, \nu) \equiv W^{(\infty)}(r\mathbf{s}, r\mathbf{s}, \nu), \tag{5.2–13}$$

i.e. the 'diagonal elements' of $W^{(\infty)}(r_1\mathbf{s}_1, r_2\mathbf{s}_2, \nu)$, represent the spectral density of the field at the point P specified by the position vector $r\mathbf{s}$ and

$$J(\mathbf{s}, \nu) \equiv L(\mathbf{s}, \mathbf{s}, \nu). \tag{5.2–14}$$

The function $J(\mathbf{s}, \nu)$ is known as the *radiant intensity* of the field.

We have introduced the radiant intensity here from considerations based on the far-zone behavior of the field. We will learn later, in connection with radiation from a planar source [Eq. (5.7–35)], that the radiant intensity also represents the rate at which the source radiates energy, at frequency ν, per unit solid angle around the **s**-direction.

It follows from Eqs. (5.2–14) and (5.2–8) that the radiant intensity may be calculated from knowledge of the cross-spectral density of the source by means of the formula

$$J(\mathbf{s}, \nu) = (2\pi)^6 \widetilde{W}_Q(-k\mathbf{s}, k\mathbf{s}, \nu). \tag{5.2–15}$$

The use of this expression for determining the radiant intensity makes it necessary to evaluate a six-fold Fourier transform. However, because the two spatial-frequency arguments on the right-hand side of Eq. (5.2–15) only differ from each other in sign, the formula may be expressed in a simpler form, as we will now show.

From Eq. (5.2–9) it follows that

$$\widetilde{W}_Q(-k\mathbf{s}, k\mathbf{s}, \nu) = \frac{1}{(2\pi)^6} \iint W_Q(\mathbf{r}_1', \mathbf{r}_2', \nu)\, \mathrm{e}^{-ik\mathbf{s}\cdot(\mathbf{r}_2'-\mathbf{r}_1')}\, \mathrm{d}^3 r_1'\, \mathrm{d}^3 r_2'. \tag{5.2–16}$$

Let us change the variables of integration from $\mathbf{r}_1'$ and $\mathbf{r}_2'$ to $\mathbf{r}'$ and $\mathbf{r}$ by making the transformation

$$\mathbf{r}_2' - \mathbf{r}_1' = \mathbf{r}', \quad \tfrac{1}{2}(\mathbf{r}_1' + \mathbf{r}_2') = \mathbf{r}, \tag{5.2–17a}$$

whose inverse is

$$\mathbf{r}_1' = \mathbf{r} - \tfrac{1}{2}\mathbf{r}', \quad \mathbf{r}_2' = \mathbf{r} + \tfrac{1}{2}\mathbf{r}'. \tag{5.2–17b}$$

The Jacobian of the transformation (5.2–17b) can readily be shown to be unity and hence we obtain from Eq. (5.2–16) the following expression for $\widetilde{W}_Q(-k\mathbf{s}, k\mathbf{s}, \nu)$:

$$\widetilde{W}_Q(-k\mathbf{s}, k\mathbf{s}, \nu) = \frac{1}{(2\pi)^6} \iint W_Q(\mathbf{r} - \tfrac{1}{2}\mathbf{r}', \mathbf{r} + \tfrac{1}{2}\mathbf{r}', \nu)\, \mathrm{e}^{-ik\mathbf{s}\cdot\mathbf{r}'}\, \mathrm{d}^3 r\, \mathrm{d}^3 r', \tag{5.2–18}$$

where the integration extends formally over all possible values of the variables of integration.

It is convenient to set

$$\int W_Q(\mathbf{r} - \tfrac{1}{2}\mathbf{r}', \mathbf{r} + \tfrac{1}{2}\mathbf{r}', \nu)\, \mathrm{d}^3 r = C_Q(\mathbf{r}', \nu). \tag{5.2–19}$$

We will refer to the function $C_Q(\mathbf{r}', \nu)$ as the *source-integrated cross-spectral density*.

It follows from Eqs. (5.2–18) and (5.2–19) that

$$\widetilde{W}_Q(-k\mathbf{s}, k\mathbf{s}, \nu) = \frac{1}{(2\pi)^3} \widetilde{C}_Q(k\mathbf{s}, \nu), \tag{5.2–20}$$

where

$$\widetilde{C}_Q(\mathbf{f}, \nu) = \frac{1}{(2\pi)^3} \int C_Q(\mathbf{r}', \nu)\, \mathrm{e}^{-i\mathbf{f}\cdot\mathbf{r}'}\, \mathrm{d}^3 r' \tag{5.2–21}$$

is the three-dimensional spatial Fourier transform of the source-integrated cross-spectral density. Finally on substituting from Eq. (5.2–20) into Eq. (5.2–15) we obtain the following alternative expression for the radiant intensity:

$$J(\mathbf{s}, \nu) = (2\pi)^3 \widetilde{C}_Q(k\mathbf{s}, \nu). \tag{5.2–22}$$

Let us now consider the spectral degree of coherence [cf. Eq. (4.3–47b)] of the far field, namely

$$\mu^{(\infty)}(r_1\mathbf{s}_1, r_2\mathbf{s}_2, \nu) = \frac{W^{(\infty)}(r_1\mathbf{s}_1, r_2\mathbf{s}_2, \nu)}{[S^{(\infty)}(r_1\mathbf{s}_1, \nu)]^{1/2}[S^{(\infty)}(r_2\mathbf{s}_2, \nu)]^{1/2}}. \tag{5.2–23}$$

We may express the right-hand side of Eq. (5.2–23) in terms of the radiant cross-intensity and the radiant intensity by making use of Eqs. (5.2–6), (5.2–13) and (5.2–12) and we then obtain the following expression for $\mu^{(\infty)}$:

$$\mu^{(\infty)}(r_1\mathbf{s}_1, r_2\mathbf{s}_2, \nu) = \frac{L(\mathbf{s}_1, \mathbf{s}_2, \nu)}{[J(\mathbf{s}_1, \nu)]^{1/2}[J(\mathbf{s}_2, \nu)]^{1/2}} \mathrm{e}^{ik(r_2-r_1)}. \tag{5.2–24}$$

If we substitute for the radiant cross-intensity and for the radiant intensity from Eqs. (5.2–8) and (5.2–15), we obtain the following expression for the spectral degree of coherence of the far field in terms of the Fourier transform of the cross-spectral density of the source:

$$\mu^{(\infty)}(r_1\mathbf{s}_1, r_2\mathbf{s}_2, \nu) = \frac{\widetilde{W}_Q(-k\mathbf{s}_1, k\mathbf{s}_2, \nu)}{[\widetilde{W}_Q(-k\mathbf{s}_1, k\mathbf{s}_1, \nu)]^{1/2}[\widetilde{W}_Q(-k\mathbf{s}_2, k\mathbf{s}_2, \nu)]^{1/2}} \mathrm{e}^{ik(r_2-r_1)}. \tag{5.2–25}$$

When the two points in the far zone are at equal distances from the origin (i.e. when $r_2 = r_1$), one often speaks of *transverse coherence*. The spectral degree of coherence of the far field at such a pair of points is given by Eq. (5.2–25), with the exponential factor replaced by unity. Clearly, it is independent of the common distance ($r_2 = r_1$) from the origin.

When the two points in the far zone are located in the same direction ($\mathbf{s}_2 = \mathbf{s}_1$) one speaks of *longitudinal coherence*. The spectral degree of coherence of the far field at such a pair of points is, according to Eq. (5.2–25), given by

$$\mu^{(\infty)}(r_1\mathbf{s}, r_2\mathbf{s}, \nu) = \mathrm{e}^{ik(r_2-r_1)}. \tag{5.2–26}$$

Now the modulus of this expression is unity for all values r_1 and r_2 associated with points in the far zone. Hence the field at any two points in the far zone, which lie along the same direction when viewed from the source, is completely coherent at each temporal frequency ν, irrespective of the distance $r_2 - r_1$ between the two points.

5.2.2 *Radiation from some model sources*

We will illustrate the general results that we have just obtained by considering radiation from a class of model sources which represent rather well many sources encountered in nature or used in the laboratory.

Let $W_Q(\mathbf{r}_1, \mathbf{r}_2, \nu)$ be again the cross-spectral density of a source which occupies a finite domain D and whose statistical behavior is characterized by an ensemble which is stationary, at least in the wide sense. Then

$$S_Q(\mathbf{r}, \nu) \equiv W_Q(\mathbf{r}, \mathbf{r}, \nu) \qquad (5.2\text{–}27)$$

represents the spectral density of the source distribution and

$$\mu_Q(\mathbf{r}_1, \mathbf{r}_2, \nu) = \frac{W_Q(\mathbf{r}_1, \mathbf{r}_2, \nu)}{[S_Q(\mathbf{r}_1, \nu)]^{1/2}[S_Q(\mathbf{r}_2, \nu)]^{1/2}} \qquad (5.2\text{–}28)$$

is its spectral degree of coherence.

We will assume that $\mu_Q(\mathbf{r}_1, \mathbf{r}_2, \nu)$ depends on $\mathbf{r}_1$ and $\mathbf{r}_2$ only through the difference $\mathbf{r}_2 - \mathbf{r}_1$, i.e. that it is of the form

$$\mu_Q(\mathbf{r}_1, \mathbf{r}_2, \nu) \equiv g_Q(\mathbf{r}_2 - \mathbf{r}_1, \nu) \qquad (5.2\text{–}29)$$

for each effective frequency ν present in the source spectrum. Sources of this kind are known as (primary) *Schell-model sources*, because such sources (actually their analogue for two-dimensional secondary sources) appear to have been first considered by Schell, (1961, 1967). It is clear from Eqs. (5.2–29) and (5.2–28) that the cross-spectral density of a Schell-model source is of the form

$$W_Q(\mathbf{r}_1, \mathbf{r}_2, \nu) = [S_Q(\mathbf{r}_1, \nu)]^{1/2}[S_Q(\mathbf{r}_2, \nu)]^{1/2} g_Q(\mathbf{r}_2 - \mathbf{r}_1, \nu). \qquad (5.2\text{–}30)$$

We will consider a particular class of Schell-model sources, namely those for which the spectral density $S_Q(\mathbf{r}, \nu)$, considered as a function of $\mathbf{r}$, varies so slowly with position that it is approximately constant over distances across the source that are of the order of the correlation length Δ (the effective width of $|g_Q(\mathbf{r}', \nu)|$ – see Fig. 5.2). It is customary to describe this situation by saying that $S_Q(\mathbf{r}, \nu)$ is a *slow* function of $\mathbf{r}$ and that $g_Q(\mathbf{r}', \nu)$ is a *fast* function of $\mathbf{r}'$. In addition we also assume that the linear dimensions of the source are large compared with the wavelength $\lambda = c/\nu$ and with the correlation length Δ. Schell-model sources of this kind are known as *quasi-homogeneous sources* and, as we will see, they generate fields which are relatively simple to analyze mathematically and yet are rich enough in their properties to represent a variety of situations of practical interest.

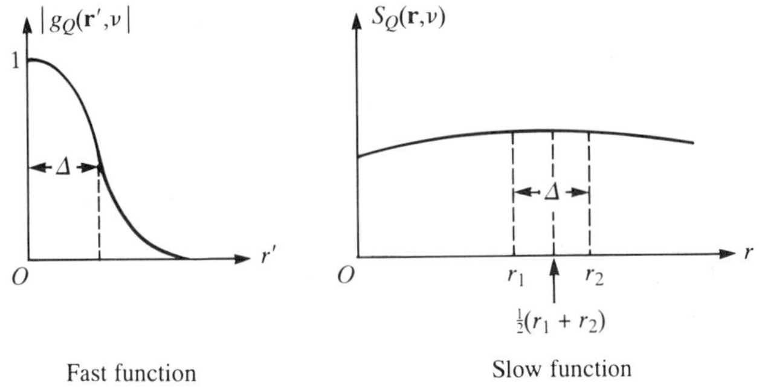

Fig. 5.2 Illustrating the concept of a quasi-homogeneous source. The modulus $|g_Q(\mathbf{r}', \nu)|$ of the spectral degree of coherence of the source distribution changes much more rapidly with $\mathbf{r}'$ than its spectral density $S_Q(\mathbf{r}, \nu)$ changes with $\mathbf{r}$. For the purpose of illustration, the source is taken to be one-dimensional.

When Δ is sufficiently small, of the order of or smaller than the wavelength λ, such a source may be said to be 'locally' spatially incoherent at the frequency ν. When Δ is of the order of many wavelengths, the source may be said to be locally spatially coherent. This distinction is sometimes useful, because, as we will see later, the fields generated by quasi-homogeneous sources belonging to these two categories have somewhat different structure. Quasi-homogeneous sources are, of course, always spatially rather incoherent in the 'global' sense, because their linear dimensions are large compared with the correlation length Δ.

Since for a quasi-homogeneous source the spectral density $S_Q(\mathbf{r}, \nu)$ is assumed to vary slowly with position over the effective width Δ of $|g_Q(\mathbf{r}', \nu)|$, we may make on the right-hand side of Eq. (5.2–30) the approximation

$$S_Q(\mathbf{r}_1, \nu) \approx S_Q(\mathbf{r}_2, \nu) \approx S_Q[\tfrac{1}{2}(\mathbf{r}_1 + \mathbf{r}_2), \nu]. \qquad (5.2\text{–}31)$$

Using this approximation in Eq. (5.2–30) we see that the cross-spectral density $W_Q(\mathbf{r}_1, \mathbf{r}_2, \nu)$ of a quasi-homogeneous source may be expressed, to a good approximation, in the form

$$W_Q(\mathbf{r}_1, \mathbf{r}_2, \nu) = S_Q[\tfrac{1}{2}(\mathbf{r}_1 + \mathbf{r}_2), \nu]g_Q(\mathbf{r}_2 - \mathbf{r}_1, \nu). \qquad (5.2\text{–}32)$$

The six-fold Fourier transform of the cross-spectral density of the source, which enters the expressions (5.2–8) and (5.2–15) for the radiant cross-intensity and the radiant intensity, reduces in this case to the product of two three-dimensional Fourier transforms. To see this we substitute from Eq. (5.2–32) into Eq. (5.2–9) and change the variables of integration from $\mathbf{r}_1'$ and $\mathbf{r}_2'$ to $\mathbf{r}$ and $\mathbf{r}'$ by the use of the transformation (5.2–17b). We then find at once that

$$\widetilde{W}_Q(\mathbf{K}_1, \mathbf{K}_2, \nu) = \tilde{S}_Q(\mathbf{K}_1 + \mathbf{K}_2, \nu)\tilde{g}_Q[\tfrac{1}{2}(\mathbf{K}_2 - \mathbf{K}_1), \nu], \qquad (5.2\text{–}33)$$

where $\tilde{S}_Q(\mathbf{K}, \nu)$ and $\tilde{g}_Q(\mathbf{K}', \nu)$ are the three-dimensional spatial Fourier transforms of $S_Q(\mathbf{r}, \nu)$ and $g_Q(\mathbf{r}', \nu)$ respectively, namely

$$\tilde{S}_Q(\mathbf{K}, \nu) = \frac{1}{(2\pi)^3}\int S_Q(\mathbf{r}, \nu)\,\mathrm{e}^{-\mathrm{i}\mathbf{K}\cdot\mathbf{r}}\,\mathrm{d}^3r, \qquad (5.2\text{–}34)$$

$$\tilde{g}_Q(\mathbf{K}', \nu) = \frac{1}{(2\pi)^3}\int g_Q(\mathbf{r}', \nu)\,\mathrm{e}^{-\mathrm{i}\mathbf{K}'\cdot\mathbf{r}'}\,\mathrm{d}^3r'. \qquad (5.2\text{–}35)$$

Thus we see that, when the cross-spectral density $W_Q(\mathbf{r}_1, \mathbf{r}_2, \nu)$ factorizes in the form (5.2–32) appropriate to a quasi-homogeneous source, its six-dimensional spatial Fourier transform $\widetilde{W}_Q(\mathbf{K}_1, \mathbf{K}_2, \nu)$ also factorizes, in the form given in Eq. (5.2–33). Moreover, since for a quasi-homogeneous source $S_Q(\mathbf{r}, \nu)$ is a slow function of $\mathbf{r}$ and $g_Q(\mathbf{r}', \nu)$ is a fast function of $\mathbf{r}'$, it follows from the reciprocity relation involving the effective widths of Fourier transform pairs [Eq. (4.3–76)] that the first factor $\tilde{S}_Q(\mathbf{K}, \nu)$ on the right-hand side of Eq. (5.2–33) is a fast function of $\mathbf{K}$ and the second factor $\tilde{g}_Q(\mathbf{K}', \nu)$ is a slow function of $\mathbf{K}'$. We will make use of these properties shortly.

On substituting from Eq. (5.2–33) into Eqs. (5.2–8) and (5.2–15) we obtain the following expressions for the radiant cross-intensity and the radiant intensity of the field radiated by a three-dimensional, primary, quasi-homogeneous source:

$$L(\mathbf{s}_1, \mathbf{s}_2, \nu) = (2\pi)^6\tilde{S}_Q[k(\mathbf{s}_2 - \mathbf{s}_1), \nu]\tilde{g}_Q[\tfrac{1}{2}k(\mathbf{s}_1 + \mathbf{s}_2), \nu] \qquad (5.2\text{–}36)$$

and

$$J(\mathbf{s}, \nu) = (2\pi)^6 \tilde{S}_Q(0, \nu)\tilde{g}_Q(k\mathbf{s}, \nu), \qquad (5.2\text{–}37)$$

with $k = 2\pi\nu/c$. Further, on substituting from Eqs. (5.2–36) and (5.2–37) into the formula (5.2–24), we find that the spectral degree of coherence of the far field, generated by the source, is given by the expression

$$\mu^{(\infty)}(r_1\mathbf{s}_1, r_2\mathbf{s}_2, \nu) = G_Q(k\mathbf{s}_1, k\mathbf{s}_2, \nu)\frac{\tilde{S}_Q[k(\mathbf{s}_2 - \mathbf{s}_1), \nu]}{\tilde{S}_Q(0, \nu)} \mathrm{e}^{ik(r_2-r_1)}, \qquad (5.2\text{–}38)$$

where

$$G_Q(k\mathbf{s}_1, k\mathbf{s}_2, \nu) = \frac{\tilde{g}_Q[\frac{1}{2}k(\mathbf{s}_1 + \mathbf{s}_2), \nu]}{[\tilde{g}_Q(k\mathbf{s}_1, \nu)]^{1/2}[\tilde{g}_Q(k\mathbf{s}_2, \nu)]^{1/2}}. \qquad (5.2\text{–}39)$$

Now as we have already noted, $\tilde{S}_Q(\mathbf{K}, \nu)$ is a fast function of $\mathbf{K}$, whereas $\tilde{g}_Q(\mathbf{K}', \nu)$ is a slow function of $\mathbf{K}'$. It is, therefore, clear that, for arguments $k(\mathbf{s}_2 - \mathbf{s}_1)$ for which the factor $\tilde{S}_Q[k(\mathbf{s}_2 - \mathbf{s}_1), \nu]/\tilde{S}_Q(0, \nu)$ in Eq. (5.2–38) differs appreciably from zero, $\mathbf{s}_2$ may be replaced by $\mathbf{s}_1$ on the right-hand side of Eq. (5.2–39). Hence we may set in Eq. (5.2–38)

$$G_Q(k\mathbf{s}_1, k\mathbf{s}_2, \nu) \approx G_Q(k\mathbf{s}_1, \mathrm{k}\mathbf{s}_1, \nu) = 1. \qquad (5.2\text{–}40)$$

On making use of this approximation in Eq. (5.2–38) we obtain the following expression for the spectral degree of coherence of the far field generated by a three-dimensional, primary, quasi-homogeneous source:

$$\mu^{(\infty)}(r_1\mathbf{s}_1, r_2\mathbf{s}_2, \nu) \approx \frac{\tilde{S}_Q[k(\mathbf{s}_2 - \mathbf{s}_1), \nu]}{\tilde{S}_Q(0, \nu)} \mathrm{e}^{ik(r_2-r_1)}. \qquad (5.2\text{–}41)$$

The formulas (5.2–37) and (5.2–41) bring into evidence the following two interesting *reciprocity relations* that pertain to radiation generated by sources of this type:

(a) The angular distribution of the radiant intensity $J(\mathbf{s}, \nu)$ is proportional to the three-dimensional spatial Fourier transform of the spectral degree of coherence of the source [Eq. (5.2–37)].

(b) The spectral degree of coherence of the far field is, apart from a simple geometrical phase factor, equal to the normalized three-dimensional spatial Fourier transform of the spectral density of the source [Eq. (5.2–41)].

Thus we see that the influences of the spatial distributions of the spectral density of the source and of its spectral degree of coherence on the far field are quite distinct. The result expressed by the theorem (b) above [Eq. (5.2–41)] may be regarded as an analogue for three-dimensional, primary, quasi-homogeneous sources of the far-zone form of the van Cittert–Zernike theorem [Eq. (4.4–40)].

We will illustrate these results by a simple example. Let us consider a three-dimensional, primary, isotropic, quasi-homogeneous source, whose spatial distributions of the spectral density and of the spectral degree of coherence are both Gaussian:

$$S_Q(\mathbf{r}, \nu) = [A(\nu)]^2 \,\mathrm{e}^{-r^2/2\sigma_S^2(\nu)}, \quad g_Q(\mathbf{r}', \nu) = \mathrm{e}^{-r'^2/2\sigma_g^2(\nu)}, \qquad (5.2\text{–}42)$$

where $A(\nu)$, $\sigma_S(\nu)$, and $\sigma_g(\nu)$ are positive quantities, $r = |\mathbf{r}|$ and $r' = |\mathbf{r}'|$. The assumption that the source is quasi-homogeneous at a frequency ν implies that

$$\sigma_S(\nu) \gg \sigma_g(\nu). \tag{5.2–43}$$

The three-dimensional spatial Fourier transforms of the expressions (5.2–42) are readily found to be [if we no longer display the frequency dependence of $A(\nu)$, $\sigma_S(\nu)$, and $\sigma_g(\nu)$],

$$\tilde{S}_Q(\mathbf{K}, \nu) = A^2\left(\frac{\sigma_S}{\sqrt{(2\pi)}}\right)^3 e^{-(K\sigma_S)^2/2}, \quad \tilde{g}_Q(\mathbf{K}', \nu) = \left(\frac{\sigma_g}{\sqrt{(2\pi)}}\right)^3 e^{-(K'\sigma_g)^2/2}. \tag{5.2–44}$$

On substituting from Eq. (5.2–44) into Eq. (5.2–37) we obtain the following expression for the radiant intensity generated by the source.

$$J(\mathbf{s}, \nu) = (2\pi\sigma_S\sigma_g)^3 A^2 e^{-(k\sigma_g)^2/2}. \tag{5.2–45}$$

The radiant intensity is now seen to be independent of direction (characterized by the unit vector **s**) as it must be, because the source was assumed to be isotropic; it is also seen to be proportional to the effective volume $(4\pi/3)\sigma_S^3$ of the source. We note that when the effective correlation distance σ_g of the source is smaller than about the wavelength $\lambda = 2\pi/k$ associated with the frequency ν, i.e. when the source is locally incoherent, the expression (5.2–45), for the radiant intensity reduces to

$$J(\mathbf{s}, \nu) \approx (2\pi\sigma_S\sigma_g)^3 A^2. \tag{5.2–46}$$

Thus the radiant intensity is now proportional not only to the effective volume of the source, but also to its effective volume of coherence, $(4\pi/3)\sigma_g^3$.

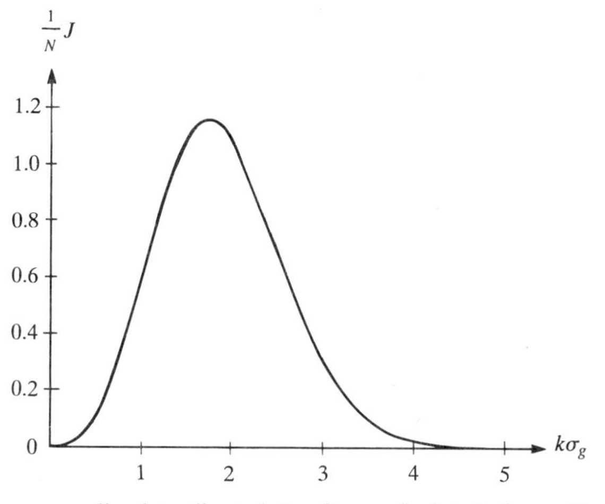

Fig. 5.3 The normalized radiant intensity, calculated from Eq. (5.2–45), $J(\mathbf{s}, \nu)/N = (k\sigma_g)^3 \exp[-\frac{1}{2}(k\sigma_g)^2]$, with $N = (2\pi/k)^3 A^2 \sigma_S^3$, as function of the normalized effective correlation distance $k\sigma_g$, generated by a three-dimensional, Gaussian, quasi-homogeneous source characterized by Eqs. (5.2–42). Note that the normalization constant N is proportional to the effective source volume $(4\pi/3)\sigma_S^3$. (After Carter and Wolf, 1981b.)

The dependence of the radiant intensity on the effective correlation distance σ_g of the source, computed from Eq. (5.2–45), is shown in Fig. 5.3.

Let us now briefly also examine the coherence properties of the far field, generated by the type of source under consideration. The spectral degree of coherence is obtained on substituting for $\tilde{S}_Q(\mathbf{K}, \nu)$ from Eq. (5.2–44) into Eq. (5.2–41). We then find that

$$\mu^{(\infty)}(r_1\mathbf{s}_1, r_2\mathbf{s}_2, \nu) = e^{-(k\sigma_S)^2|\mathbf{s}_2-\mathbf{s}_1|^2/2}\, e^{ik(r_2-r_1)}. \tag{5.2–47}$$

Now since $\mathbf{s}_1$ and $\mathbf{s}_2$ are unit vectors we have (see Fig. 5.4)

$$|\mathbf{s}_2 - \mathbf{s}_1|^2 = [2\sin(\theta/2)]^2, \tag{5.2–48}$$

where θ is the angle between the unit vectors $\mathbf{s}_1$ and $\mathbf{s}_2$. On substituting from Eq. (5.2–48) into Eq. (5.2–47) we obtain the following expression for $\mu^{(\infty)}$:

$$\mu^{(\infty)}(r_1\mathbf{s}_1, r_2\mathbf{s}_2, \nu) = e^{-2(k\sigma_S)^2\sin^2(\theta/2)}\, e^{ik(r_2-r_1)}. \tag{5.2–49}$$

It follows at once from this formula that the angular separation θ, for which the spatial coherence at frequency ν in the far zone is appreciable, must satisfy the order of magnitude relation

$$\sin^2(\theta/2) \lesssim \frac{1}{2(k\sigma_S)^2}. \tag{5.2–50}$$

Let us examine two extreme cases. If the effective linear dimension σ_S of the source is much smaller than the wavelength [subject, of course, to the (now rather idealized) assumption (5.2–43)], then $k\sigma_S \ll 1$ and the inequality (5.2–50) is satisfied for all possible angular separations between the unit vectors $\mathbf{s}_1$ and $\mathbf{s}_2$, i.e. for $0 \leqslant \theta \leqslant \pi$. The far field is then spatially completely coherent at the frequency ν. If, on the other hand, the effective linear dimension σ_S of the source is much larger than the wavelength, $k\sigma_S \gg 1$ and the value of $\sin^2(\theta/2)$, which satisfies the inequality (5.2–50), may then be approximated by $(\theta/2)^2$. It follows that, under these circumstances, spatial coherence at frequency ν in the far zone of the source extends over an angular region for which

$$\theta \lesssim \frac{\sqrt{2}}{k\sigma_S}. \tag{5.2–51}$$

The behavior of the absolute value of the spectral degree of coherence of the far field, as function of the angular separation θ, computed from Eq. (5.2–49), is shown in Fig. 5.5 for a few selected values of the normalized effective linear dimension $k\sigma_S$ of the source.

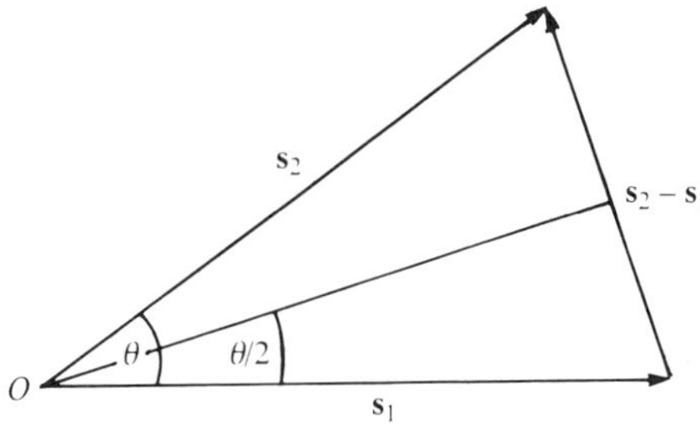

Fig. 5.4 Illustrating the formula (5.2–48).

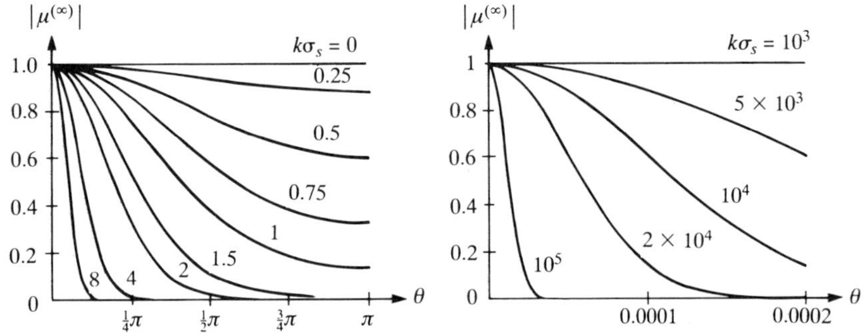

Fig. 5.5 The behavior of the absolute value of the spectral degree of coherence $\mu^{(\infty)}(r_1\mathbf{s}_1, r_2\mathbf{s}_2, \nu)$ given by Eq. (5.2–49), of the far field generated by a three-dimensional, Gaussian, quasi-homogeneous source, characterized by Eqs. (5.2–42) and (5.2–43), as a function of the angle θ between $\mathbf{s}_1$ and $\mathbf{s}_2$. (After Carter and Wolf, 1981b.)

In this section we have illustrated the general results derived in Section 5.2.1 with reference to quasi-homogeneous sources. Radiation from other types of sources, including completely coherent and completely incoherent sources, as well as Gaussian Schell-model sources are discussed in two papers by Carter and Wolf (1981a,b).

5.3 Radiation from planar, secondary sources

Many sources employed in the laboratory are secondary planar sources. A source of this kind is usually an aperture in an opaque planar screen, illuminated either directly or via an optical system by a primary source. In this section we will study radiation fields produced by such sources.‡

We will take the primary source to be located on one side, say $z < 0$, of the aperture plane and to radiate into the half-space $z > 0$ on the other side of it.

5.3.1 General formulas

Within the framework of correlation theory, a secondary planar source may be characterized by the cross-spectral density $W^{(0)}(\boldsymbol{\rho}_1, \boldsymbol{\rho}_2, \nu)$ of the fluctuating field $V(\boldsymbol{\rho}, t)$ in the source plane $z = 0$. We assume that the fluctuations of $V(\boldsymbol{\rho}, t)$ may be represented by an ensemble which is stationary, at least in the wide sense.

It follows from Eq. (4.4–18) that the cross-spectral density $W(\mathbf{r}_1, \mathbf{r}_2, \nu)$ of the field at two points $P_1(\mathbf{r}_1)$ and $P_2(\mathbf{r}_2)$, located at distances from the secondary source σ which are large compared with the wavelength $\lambda = 2\pi/k = c/\nu$, is given by

$$W(\mathbf{r}_1, \mathbf{r}_2, \nu) = \left(\frac{k}{2\pi}\right)^2 \int_\sigma\int_\sigma W^{(0)}(\boldsymbol{\rho}_1, \boldsymbol{\rho}_2, \nu) \frac{e^{ik(R_2-R_1)}}{R_1 R_2} \cos\theta_1' \cos\theta_2' \, d^2\rho_1 \, d^2\rho_2. \tag{5.3–1}$$

‡ Radiation from *primary* planar sources of any state of coherence was discussed by Wolf and Carter (1978).

Here

$$R_1 = |\mathbf{r}_1 - \boldsymbol{\rho}_1|, \quad R_2 = |\mathbf{r}_2 - \boldsymbol{\rho}_2| \tag{5.3–2}$$

are the distances from the source points $S_1(\boldsymbol{\rho}_1)$ and $S_2(\boldsymbol{\rho}_2)$ to the field points $P_1(\mathbf{r}_1)$ and $P_2(\mathbf{r}_2)$, respectively, and θ_1' and θ_2' are the angles that the lines S_1P_1 and S_2P_2 make with the positive z-axis (Fig. 5.6).

Suppose now that the field points are situated in the far zone of the source. If we denote by $\mathbf{s}_{1\perp}$ and $\mathbf{s}_{2\perp}$ the projections, considered as two-dimensional vectors, of the unit vectors $\mathbf{s}_1$ and $\mathbf{s}_2$ respectively onto the source plane $z = 0$ we may then make the approximation

$$R_1 \approx r_1 - \boldsymbol{\rho}_1 \cdot \mathbf{s}_{1\perp}, \quad R_2 \approx r_2 - \boldsymbol{\rho}_2 \cdot \mathbf{s}_{2\perp} \tag{5.3–3}$$

($R_1 = |\mathbf{R}_1|$ etc.) in the exponent of the integral. In the denominator of the integral in Eq. (5.3–1) we may make the approximation $R_1 \approx r_1$, $R_2 \approx r_2$. We also replace the angles θ_1' and θ_2' by the fixed angles θ_1 and θ_2 which the lines OP_1 and OP_2 make with the normal to the source plane. The formula (5.3–1) then becomes (again writing $W^{(\infty)}$ for the far-zone value of W)

$$W^{(\infty)}(r_1\mathbf{s}_1, r_2\mathbf{s}_2, \nu) = L(\mathbf{s}_1, \mathbf{s}_2, \nu)\frac{e^{ik(r_2-r_1)}}{r_1 r_2}, \tag{5.3–4}$$

where

$$L(\mathbf{s}_1, \mathbf{s}_2, \nu) = \left(\frac{k}{2\pi}\right)^2 \cos\theta_1 \cos\theta_2 \int_\sigma\int_\sigma W^{(0)}(\boldsymbol{\rho}_1, \boldsymbol{\rho}_2, \nu) \exp[-ik(\mathbf{s}_{2\perp}\cdot\boldsymbol{\rho}_2 - \mathbf{s}_{1\perp}\cdot\boldsymbol{\rho}_1)]\, d^2\rho_1\, d^2\rho_2. \tag{5.3–5}$$

Since the secondary source occupies a finite region σ of the plane $z = 0$, $W^{(0)}(\boldsymbol{\rho}_1, \boldsymbol{\rho}_2, \nu)$ will have zero values when $\boldsymbol{\rho}_1$ or $\boldsymbol{\rho}_2$ represent points in the source plane which are located outside the source region. We may, therefore, extend the integrals on the right-hand side of Eq. (5.3–5) over the whole plane $z = 0$ and the formula (5.3–5) may then be rewritten in a more compact form as

$$L(\mathbf{s}_1, \mathbf{s}_2, \nu) = (2\pi k)^2 \widetilde{W}^{(0)}(-k\mathbf{s}_{1\perp}, k\mathbf{s}_{2\perp}, \nu)\cos\theta_1\cos\theta_2, \tag{5.3–6}$$

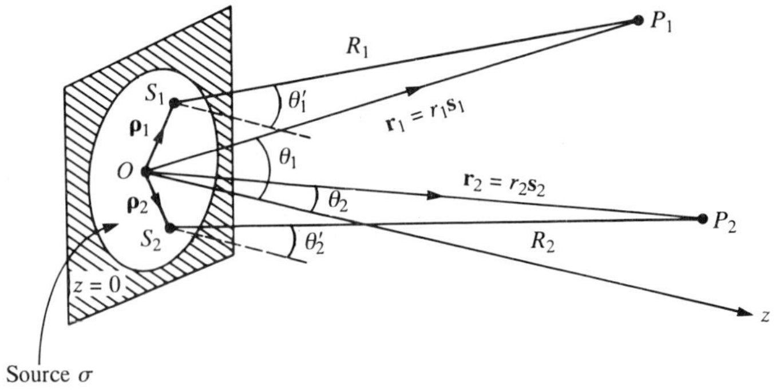

Fig. 5.6 Illustrating the notation relating to the calculation of the cross-spectral density of the far field generated by a planar source. S_1 and S_2 are two source points, P_1 and P_2 are two points in the far zone.

where

$$\widetilde{W}^{(0)}(\mathbf{f}_1, \mathbf{f}_2, \nu) = \frac{1}{(2\pi)^4} \iint W^{(0)}(\boldsymbol{\rho}_1, \boldsymbol{\rho}_2, \nu)\, e^{-i(\mathbf{f}_1 \cdot \boldsymbol{\rho}_1 + \mathbf{f}_2 \cdot \boldsymbol{\rho}_2)}\, d^2\rho_1\, d^2\rho_2 \qquad (5.3\text{–}7)$$

is the four-dimensional spatial Fourier transform of $W^{(0)}(\boldsymbol{\rho}_1, \boldsymbol{\rho}_2, \nu)$. Since $\mathbf{s}_{1\perp}$ and $\mathbf{s}_{2\perp}$ are the projections of unit vectors, $|\mathbf{s}_{1\perp}| \leq 1$, $|\mathbf{s}_{2\perp}| \leq 1$. Therefore it is clear from Eq. (5.3–6) that only those Fourier components, labeled by spatial-frequency vectors $\mathbf{f}_1$ and $\mathbf{f}_2$ for which $|\mathbf{f}_1| \leq k$ and $|\mathbf{f}_2| \leq k$, contribute to the correlation properties of the far field.

It follows at once from Eq. (5.3–6) that the radiant intensity $J(\mathbf{s}, \nu) = L(\mathbf{s}, \mathbf{s}, \nu)$ is given by the formula

$$J(\mathbf{s}, \nu) = (2\pi k)^2 \widetilde{W}^{(0)}(-k\mathbf{s}_\perp, k\mathbf{s}_\perp, \nu) \cos^2\theta, \qquad (5.3\text{–}8)$$

where θ is the angle which the vector $\overrightarrow{OP} = r\mathbf{s}$ makes with the normal to the source plane (Fig. 5.7).

As in the case of radiation from three-dimensional primary sources, it is sometimes convenient to express the radiant intensity in an alternative form, in terms of the *source-integrated cross-spectral density*

$$C^{(0)}(\boldsymbol{\rho}', \nu) = \int W^{(0)}(\boldsymbol{\rho} - \tfrac{1}{2}\boldsymbol{\rho}', \boldsymbol{\rho} + \tfrac{1}{2}\boldsymbol{\rho}', \nu)\, d^2\rho. \qquad (5.3\text{–}9)$$

From Eqs. (5.3–8), (5.3–7) and (5.3–9) it follows that

$$J(\mathbf{s}, \nu) = k^2 \widetilde{C}^{(0)}(k\mathbf{s}_\perp, \nu) \cos^2\theta, \qquad (5.3\text{–}10)$$

where $\widetilde{C}^{(0)}(\mathbf{f}, \nu)$ is the two-dimensional spatial Fourier transform of $C^{(0)}(\boldsymbol{\rho}', \nu)$, namely

$$\widetilde{C}^{(0)}(\mathbf{f}, \nu) = \frac{1}{(2\pi)^2} \int C^{(0)}(\boldsymbol{\rho}', \nu)\, e^{-i\mathbf{f} \cdot \boldsymbol{\rho}'}\, d^2\boldsymbol{\rho}'. \qquad (5.3\text{–}11)$$

The spectral degree of coherence of the far field is given by the equation [cf. Eq. (5.2–24)]

$$\mu^{(\infty)}(r_1\mathbf{s}_1, r_2\mathbf{s}_2, \nu) = \frac{L(\mathbf{s}_1, \mathbf{s}_2, \nu)}{[J(\mathbf{s}_1, \nu)]^{1/2}[J(\mathbf{s}_2, \nu)]^{1/2}}\, e^{ik(r_2 - r_1)}. \qquad (5.3\text{–}12)$$

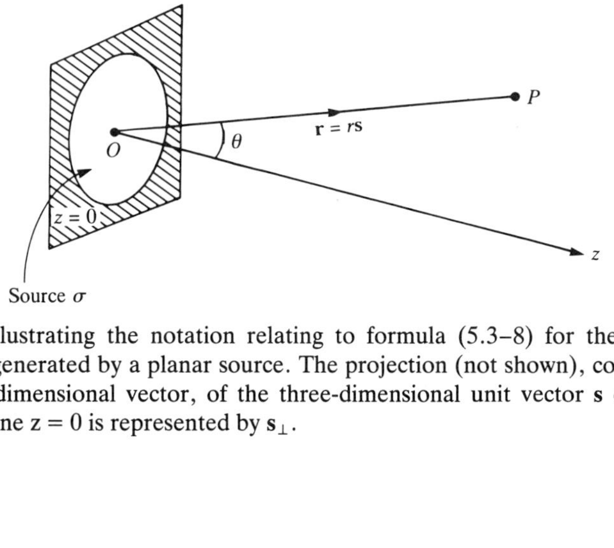

Fig. 5.7 Illustrating the notation relating to formula (5.3–8) for the radiant intensity generated by a planar source. The projection (not shown), considered as a two-dimensional vector, of the three-dimensional unit vector $\mathbf{s}$ onto the source plane $z = 0$ is represented by $\mathbf{s}_\perp$.

On substituting Eqs. (5.3–6) and (5.3–8) into Eq. (5.3–12) we obtain for $\mu^{(\infty)}$ the formula

$$\mu^{(\infty)}(r_1\mathbf{s}_1, r_2\mathbf{s}_2, \nu) = \frac{\widetilde{W}^{(0)}(-k\mathbf{s}_{1\perp}, k\mathbf{s}_{2\perp}, \nu)}{[\widetilde{W}^{(0)}(-k\mathbf{s}_{1\perp}, k\mathbf{s}_{1\perp}, \nu)]^{1/2}[\widetilde{W}^{(0)}(-k\mathbf{s}_{2\perp}, k\mathbf{s}_{2\perp}, \nu)]^{1/2}} \mathrm{e}^{ik(r_2-r_1)}. \tag{5.3–13}$$

Similar conclusions can be drawn from Eq. (5.3–13) regarding transverse and longitudinal coherence of the far field as those obtained in connection with the corresponding formula (5.2–25) pertaining to radiation from three-dimensional primary sources.

5.3.2 Radiation from planar, secondary, quasi-homogeneous sources

Among the variety of fields to which the general formulas that we have just derived apply, those generated by certain types of sources are of particular interest in practice. These are secondary, planar sources analogous to the three-dimensional primary sources discussed in Section 5.2.2. Chief amongst them are *Schell-model sources* characterized by the property that their spectral degree of coherence $\mu^{(0)}(\boldsymbol{\rho}_1, \boldsymbol{\rho}_2, \nu)$ in the source plane depends on $\boldsymbol{\rho}_1$ and $\boldsymbol{\rho}_2$ only through the difference $\boldsymbol{\rho}_2 - \boldsymbol{\rho}_1$, i.e. it is of the form

$$\mu^{(0)}(\boldsymbol{\rho}_1, \boldsymbol{\rho}_2, \nu) = g(\boldsymbol{\rho}_2 - \boldsymbol{\rho}_1, \nu). \tag{5.3–14}$$

Recalling the definition (4.3–47b) of the spectral degree of coherence it follows that the cross-spectral density function of a secondary, planar Schell-model source has the form

$$W^{(0)}(\boldsymbol{\rho}_1, \boldsymbol{\rho}_2, \nu) = [S^{(0)}(\boldsymbol{\rho}_1, \nu)]^{1/2}[S^{(0)}(\boldsymbol{\rho}_2, \nu)]^{1/2} g^{(0)}(\boldsymbol{\rho}_1 - \boldsymbol{\rho}_2, \nu), \tag{5.3–15}$$

where $S^{(0)}(\boldsymbol{\rho}, \nu)$ is the spectral density of the light at a typical point in the source plane.

Frequently the spectral degree of coherence $g^{(0)}(\boldsymbol{\rho}', \nu)$ of the light in the source plane varies much more rapidly with $\boldsymbol{\rho}'$ than the spectral density $S^{(0)}(\boldsymbol{\rho}, \nu)$ varies with $\boldsymbol{\rho}'$ for each frequency component ν present in the light. Moreover, the linear dimensions of the source are usually large compared with the wavelength $\lambda = c/\nu$ and with the spectral correlation length of the light [the effective spatial width of $g^{(0)}(\boldsymbol{\rho}', \nu)$] in the source plane. The source is then said to be a *quasi-homogeneous*, planar, secondary source. Just like the three-dimensional, primary, quasi-homogeneous sources which we studied in Section 5.2.2, planar, secondary, quasi-homogeneous sources are also rather incoherent in the 'global' sense because their linear dimensions are large compared to their effective spectral correlation lengths. Again it is useful to distinguish between 'locally coherent' and 'locally incoherent' quasi-homogeneous sources. For the former, the spectral correlation length of the light in the source plane is large compared to the wavelength $\lambda = c/\nu$; for the latter it is of the order of, or smaller than, the wavelengths. As we will see in Section 5.6.3, locally coherent, Gaussian-correlated, quasi-homogeneous sources, unlike the locally incoherent ones, can generate beams.

Because the spectral density of a quasi-homogeneous source is a 'slow function' of $\boldsymbol{\rho}$, whereas its spectral degree of coherence is a 'fast function' of the difference variable $\boldsymbol{\rho}' = \boldsymbol{\rho}_2 - \boldsymbol{\rho}_1$, the expression (5.3–15) for its cross-spectral density may be well-approximated by the expression

$$W^{(0)}(\boldsymbol{\rho}_1, \boldsymbol{\rho}_2, \nu) = S^{(0)}[\tfrac{1}{2}(\boldsymbol{\rho}_1 + \boldsymbol{\rho}_2), \nu]g^{(0)}(\boldsymbol{\rho}_2 - \boldsymbol{\rho}_1, \nu). \qquad (5.3\text{–}16)$$

The four-dimensional spatial Fourier transform, defined by Eq. (5.3–7), of the expression (5.3–16) is given by the formula

$$\widetilde{W}^{(0)}(\mathbf{f}_1, \mathbf{f}_2, \nu) = \tilde{S}^{(0)}[\mathbf{f}_1 + \mathbf{f}_2, \nu]\tilde{g}^{(0)}[\tfrac{1}{2}(\mathbf{f}_2 - \mathbf{f}_1), \nu], \qquad (5.3\text{–}17)$$

where $\tilde{S}^{(0)}(\mathbf{f}, \nu)$ and $\tilde{g}^{(0)}(\mathbf{f}', \nu)$ are the two-dimensional spatial Fourier transforms of $S^{(0)}(\boldsymbol{\rho}, \nu)$ and $g^{(0)}(\boldsymbol{\rho}', \nu)$ respectively, defined by formulas

$$\tilde{S}^{(0)}(\mathbf{f}, \nu) = \frac{1}{(2\pi)^2}\int S^{(0)}(\boldsymbol{\rho}, \nu)\, \mathrm{e}^{-\mathrm{i}\mathbf{f}\cdot\boldsymbol{\rho}}\, \mathrm{d}^2\rho, \qquad (5.3\text{–}18)$$

$$\tilde{g}^{(0)}(\mathbf{f}', \nu) = \frac{1}{(2\pi)^2}\int g^{(0)}(\boldsymbol{\rho}', \nu)\, \mathrm{e}^{-\mathrm{i}\mathbf{f}'\cdot\boldsymbol{\rho}'}\, \mathrm{d}^2\rho'. \qquad (5.3\text{–}19)$$

On substituting from Eq. (5.3–17) into Eqs. (5.3–6) and (5.3–8) we obtain the following expressions for the radiant cross-intensity and the radiant intensity of the field generated by a quasi-homogeneous, secondary planar source:

$$L(\mathbf{s}_1, \mathbf{s}_2, \nu) = (2\pi k)^2\tilde{S}^{(0)}[k(\mathbf{s}_{2\perp} - \mathbf{s}_{1\perp}), \nu]\tilde{g}^{(0)}[\tfrac{1}{2}k(\mathbf{s}_{1\perp} + \mathbf{s}_{2\perp}), \nu]\cos\theta_1\cos\theta_2, \qquad (5.3\text{–}20)$$

$$J(\mathbf{s}, \nu) = (2\pi k)^2\tilde{S}^{(0)}(0, \nu)\tilde{g}^{(0)}(k\mathbf{s}_\perp, \nu)\cos^2\theta. \qquad (5.3\text{–}21)$$

An expression for the spectral degree of coherence $\mu^{(\infty)}(r_1\mathbf{s}_1, r_2\mathbf{s}_2, \nu)$ of the far field is also readily obtained on substituting from Eqs. (5.3–17) into Eq. (5.3–13). If we then make a similar approximation to that made in passing from Eq. (5.2–38) to Eq. (5.2–41), we find that

$$\mu^{(\infty)}(r_1\mathbf{s}_1, r_2\mathbf{s}_2, \nu) = \frac{\tilde{S}^{(0)}(k(\mathbf{s}_{2\perp} - \mathbf{s}_{1\perp}), \nu)}{\tilde{S}^{(0)}(0, \nu)}\,\mathrm{e}^{\mathrm{i}k(r_2 - r_1)}. \qquad (5.3\text{–}22)$$

We note that Eqs. (5.3–21) and (5.3–22) are very similar to the corresponding formulas (5.2–37) and (5.2–41) for radiation from three-dimensional, primary quasi-homogeneous sources.

The formulas (5.3–21) and (5.3–22) bring into evidence the following two reciprocity relations due to Carter and Wolf (1977; see also Goodman, 1965, Appendix A and Goodman, 1979, Sec. 4C) for radiation from planar, secondary, quasi-homogeneous sources:

(a) The angular distribution of the radiant intensity $J(\mathbf{s}, \nu)$ is proportional to the product of the two-dimensional spatial Fourier transform of the spectral degree of coherence of the field in the source plane and the square of the cosine of the angle which the $\mathbf{s}$-direction makes with the normal to the source plane [Eq. (5.3–21)].

(b) The spectral degree of coherence of the far field is equal, apart from a simple geometrical phase factor, to the normalized two-dimensional spatial Fourier transform of the spectral density of the field in the source plane [Eq. (5.3–22)].

The reciprocity relation (a) implies that a quasi-homogeneous source may generate radiant intensity which is rotationally symmetric about the normal of the source plane, irrespective of the shape of the source and irrespective of the spatial distribution of the spectral density across it. Some examples of this fact are discussed in a paper by Li and Wolf (1982).

The reciprocity relation (b) is essentially a generalization for quasi-homogeneous sources of the far-zone form of the van Cittert–Zernike theorem [Eq. (4.4–40)]. However the van Cittert–Zernike theorem in its traditional formulation applies to the equal-time degree of coherence $j(r_1\mathbf{s}_1, r_2\mathbf{s}_2)$, whereas the formula (5.3–22) pertains to the spectral degree of coherence $\mu(r_1\mathbf{s}_1, r_2\mathbf{s}_2, \nu)$. This distinction is often insignificant when the light is quasi-monochromatic.

We will illustrate the reciprocity relations by considering radiation from a planar, secondary, quasi-homogeneous source of radius $a \gg \lambda$ whose spatial distributions of the spectral density and the spectral degree of coherence are both Gaussian,[‡] i.e.

$$\left.\begin{aligned} S^{(0)}(\boldsymbol{\rho}, \nu) &= A^2(\nu)\,\mathrm{e}^{-\rho^2/2\sigma_S^2(\nu)} && \text{when } \rho \leqslant a \\ &= 0 && \text{when } \rho > a \end{aligned}\right\} \tag{5.3–23}$$

and

$$g^{(0)}(\boldsymbol{\rho}', \nu) = \mathrm{e}^{-\rho'^2/2\sigma_g^2(\nu)}, \tag{5.3–24}$$

where A, σ_S and σ_g are real and positive. Since the source is assumed to be quasi-homogeneous we must have

$$\sigma_S \gg \sigma_g. \tag{5.3–25}$$

To simplify the calculations we will also assume that

$$\sigma_S \ll a. \tag{5.3–26}$$

The two-dimensional spatial Fourier transforms, defined by formulas (5.3–18) and (5.3–19), of the Gaussian distributions (5.3–23) and (5.3–24) are

$$\tilde{S}^{(0)}(\mathbf{f}, \nu) \approx \frac{A^2\sigma_S^2}{2\pi}\,\mathrm{e}^{-\sigma_S^2 f^2/2}, \tag{5.3–27}$$

$$\tilde{g}^{(0)}(\mathbf{f}', \nu) = \frac{\sigma_g^2}{2\pi}\,\mathrm{e}^{-\sigma_g^2 f'^2/2}. \tag{5.3–28}$$

In calculating $\tilde{S}^{(0)}$ use was made of the assumption (5.3–26), which allowed the replacement of the truncated Gaussian distribution (5.3–23) by the complete Gaussian distribution in determining its Fourier transform.

On substituting from Eqs. (5.3–27) and (5.3–28) into Eq. (5.3–21) and making use of the relation $\mathbf{s}_\perp^2 = \sin^2\theta$, we obtain the following expression for the radiant intensity:

$$J(\mathbf{s}, \nu) = J^{(0)}(\nu)\cos^2\theta\,\mathrm{e}^{-[(k\sigma_g)^2\sin^2\theta]/2}, \tag{5.3–29}$$

‡ The radiant intensity and the far-zone coherence properties of model sources of other kinds were studied by Baltes, Steinle and Antes (1976); Steinle and Baltes (1977) and Baltes and Steinle (1977). See also Section 5.4.2. below. Far-zone range criteria for radiation generated by planar, secondary, quasi-homogeneous sources were given by Leader (1978).

where

$$J^{(0)}(\nu) = (kA\sigma_g\sigma_S)^2. \tag{5.3–30}$$

Let us briefly consider two limiting cases. When $k\sigma_g \to 0$, the source becomes spatially incoherent (zero correlation length) at frequency ν and Eq. (5.3–29) shows that in this case

$$\left(\frac{J(\mathbf{s}, \nu)}{J^{(0)}(\nu)}\right)_{\text{incoh.}} \to \cos^2\theta; \tag{5.3–31}$$

hence the radiant intensity falls off with θ as $\cos^2\theta$, i.e. more rapidly than with a Lambertian source [cf. Eq. (5.3–45) below]. In the limiting case when $k\sigma_g \to \infty$ [but retaining the assumption (5.3–25) in taking this limit] the source becomes locally coherent and Eq. (5.3–29) and (5.3–30) imply that one now has

$$\left.\begin{aligned}\left(\frac{J(\mathbf{s}, \nu)}{J^{(0)}(\nu)}\right)_{\text{coh.}} &= 0 \quad \text{when } \theta \neq 0\\ &= 1 \quad \text{when } \theta = 0,\end{aligned}\right\} \tag{5.3–32}$$

i.e. in this limit the source radiates only in the direction $\theta = 0$ normal to the source plane.

The behavior of the normalized distribution of the radiant intensity, calculated from Eq. (5.3–29) for selected values of the spectral correlation width σ_g, is shown on a polar diagram in Fig. 5.8(*a*).

Next let us consider the spectral degree of coherence $\mu^{(\infty)}(r_1\mathbf{s}_1, r_2\mathbf{s}_2, \nu)$ of the far field. We find at once, on substituting from Eq. (5.3–27) into Eq. (5.3–22), that

$$\mu^{(\infty)}(r_1\mathbf{s}_1, r_2\mathbf{s}_2, \nu) = \mathrm{e}^{-(k\sigma_S)^2u_{12}^2/2}\,\mathrm{e}^{ik(r_2-r_1)}, \tag{5.3–33}$$

where

$$u_{12} = |\mathbf{s}_{2\perp} - \mathbf{s}_{1\perp}|. \tag{5.3–34}$$

The behavior of the spectral degree of coherence $\mu^{(\infty)}$, given by Eq. (5.3–33), is shown in Fig. 5.8(*b*) for different values of the effective linear source size σ_S. It is not difficult to show that, when the points $P_1(r_1\mathbf{s}_1)$ and $P_2(r_2\mathbf{s}_2)$ in the far zone are sufficiently close to the forward direction and are situated in the same meridional plane (i.e. a plane containing the z-axis), the variable u_{12} represents the angular separation of the two points as viewed from the source. Figure 5.8(*b*) shows that as the effective linear size σ_S of the source increases, the angular separation of the points at which the far field has appreciable correlation then becomes smaller and smaller.

5.3.3 An inverse problem for planar, secondary, quasi-homogeneous sources

In the preceding section (Section 5.3.2) we obtained expressions for the radiant intensity and for the spectral degree of coherence of the far field generated by a planar, secondary, quasi-homogeneous source. We will now consider the inverse problem, namely that of determining the spectral density and the spectral degree of coherence of the source from far-field measurements.

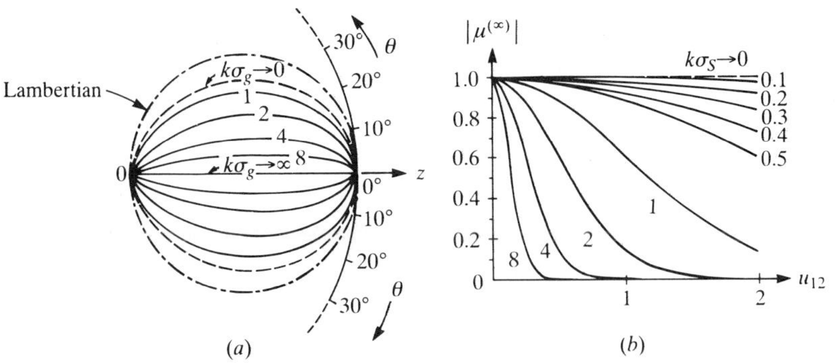

Fig. 5.8 (*a*) Polar diagram of the normalized radiant intensity, given by Eq. (5.3–29),and (*b*) the behavior of the absolute value of the spectral degree of coherence, given by Eq. (5.3–33), of the far field generated by a Gaussian, quasi-homogeneous, planar, secondary source. The variable u_{12} is defined by Eq. (5.3–34). (Adapted from Carter and Wolf, 1977.)

We have from Eq. (5.3–21),

$$\tilde{g}^{(0)}(k\mathbf{s}_\perp, \nu) = \frac{J(\mathbf{s}, \nu)}{(2\pi k)^2 \tilde{S}^{(0)}(0, \nu)\cos^2\theta}. \tag{5.3–35}$$

This formula gives the two-dimensional spatial Fourier component of the spectral degree of coherence of the source, labeled by the spatial-frequency vector $\mathbf{f} = k\mathbf{s}_\perp$, in terms of the radiant intensity of the field associated with one particular direction; namely with the direction specified by the unit vector $\mathbf{s} \equiv (s_x, s_y, s_z)$, where s_x, s_y are the Cartesian components of the two-dimensional vector $\mathbf{s}_\perp$ and $s_z = +(1 - \mathbf{s}_\perp^2)^{1/2} = \cos\theta$. However, since $|\mathbf{s}_\perp| \leq 1$, only the low-frequency spatial Fourier components, i.e. those for which $|\mathbf{f}| \leq k$, can be determined from the knowledge of the radiant intensity. The high-frequency spatial Fourier components, i.e. those for which $|\mathbf{f}| > k$, are associated with evanescent waves (see Section 3.2.2). In many cases of practical interest, the high-frequency spatial Fourier components are negligible. Under these circumstances we may, to a good approximation, omit them. We then obtain from Eq. (5.3–35), if we use the relation $\cos\theta = +(1 - \mathbf{s}_\perp^2)^{1/2}$, the following approximate expression for $g^{(0)}$:

$$g^{(0)}(\boldsymbol{\rho}', \nu) \approx \frac{1}{(2\pi k)^2 \tilde{S}^{(0)}(0, \nu)} \int_{\mathbf{s}_\perp^2 \leq 1} (1 - \mathbf{s}_\perp^2)^{-1} J(\mathbf{s}, \nu)\, e^{ik\mathbf{s}_\perp \cdot \boldsymbol{\rho}'}\, d^2(ks_\perp). \tag{5.3–36}$$

Now, since the spectral degree of coherence $g^{(0)}(\boldsymbol{\rho}', \nu)$ has the value unity when $\boldsymbol{\rho}' = 0$, it follows from Eq. (5.3–36) that

$$\tilde{S}^{(0)}(0, \nu) \approx \frac{1}{(2\pi k)^2} \int_{\mathbf{s}_\perp^2 \leq 1} (1 - \mathbf{s}_\perp^2)^{-1} J(\mathbf{s}, \nu)\, d^2(ks_\perp). \tag{5.3–37}$$

Further from Eqs. (5.3–36) and (5.3–37),

$$g^{(0)}(\boldsymbol{\rho}', \nu) = \frac{\int_{\mathbf{s}_\perp^2 \leq 1} (1 - \mathbf{s}_\perp^2)^{-1} J(\mathbf{s}, \nu)\, e^{ik\mathbf{s}_\perp \cdot \boldsymbol{\rho}'}\, d^2 s_\perp}{\int_{\mathbf{s}_\perp^2 \leq 1} (1 - \mathbf{s}_\perp^2)^{-1} J(\mathbf{s}, \nu)\, d^2 s_\perp}. \tag{5.3–38}$$

This formula provides an approximate solution to the problem of determining the spectral degree of coherence of a planar, secondary, quasi-homogeneous source from measurements of the angular distribution of the radiant intensity produced by the source.

To determine the distribution of the spectral density across the source from far-field measurements it will be convenient to first re-write the formula (5.3–22) as

$$\mu^{(\infty)}(r_1\mathbf{s}_1, r_2\mathbf{s}_2, \nu) = g^{(\infty)}[k(\mathbf{s}_{2\perp} - \mathbf{s}_{1\perp}), \nu]\, \mathrm{e}^{ik(r_2 - r_1)}, \tag{5.3–39}$$

where

$$g^{(\infty)}[k(\mathbf{s}_{2\perp} - \mathbf{s}_{1\perp}), \nu] = \frac{\tilde{S}^{(0)}[k(\mathbf{s}_{2\perp} - \mathbf{s}_{1\perp}), \nu]}{\tilde{S}^{(0)}(0, \nu)}. \tag{5.3–40}$$

We note that when $r_1 = r_2$ ($= r$ say), i.e. when the two field points $P_1(r_1\mathbf{s}_1)$ and $P_2(r_2\mathbf{s}_2)$ in the far zone are at equal distances from the origin, $\mu^{(\infty)}$ and $g^{(\infty)}$ become equal, i.e.

$$\mu^{(\infty)}(r\mathbf{s}_1, r\mathbf{s}_2, \nu) = g^{(\infty)}[k(\mathbf{s}_{2\perp} - \mathbf{s}_{1\perp}), \nu]. \tag{5.3–41}$$

The formula (5.3–40) expresses some of the normalized spatial Fourier components, labeled by spatial-frequency vectors

$$\mathbf{f} = k(\mathbf{s}_{2\perp} - \mathbf{s}_{1\perp}) \tag{5.3–42}$$

of the spectral density distribution $S^{(0)}(\boldsymbol{\rho}, \nu)$, in terms of the spectral degree of coherence $g^{(\infty)}[k(\mathbf{s}_{2\perp} - \mathbf{s}_{1\perp}), \nu]$ of the far field, at equidistant points from the origin, in directions $\mathbf{s}_1$ and $\mathbf{s}_2$. Since $|\mathbf{s}_{1\perp}| \leqslant 1$ and $|\mathbf{s}_{2\perp}| \leqslant 1$, we see that only those spatial-frequency components of the normalized spectral density distribution across the source for which $|\mathbf{f}| < 2k$ may be obtained from measurements of the spectral degree of coherence of the far field. For the same reason as given earlier in connection with determining the spectral degree of coherence of the source from measurements of the radiant intensity, we will neglect spatial-frequency components for which $|\mathbf{f}| > 2k$. We then obtain, on taking the inverse Fourier transform of Eq. (5.3–40), the following expression for the spectral density of the source:

$$S^{(0)}(\boldsymbol{\rho}, \nu) \approx \tilde{S}^{(0)}(0, \nu) \int_{|\mathbf{f}| \leqslant 2k} g^{(\infty)}(\mathbf{f}, \nu)\, \mathrm{e}^{i\mathbf{f}\cdot\boldsymbol{\rho}}\, \mathrm{d}^2 f, \tag{5.3–43}$$

where $\mathbf{f}$ and the unit vectors $\mathbf{s}_1$ and $\mathbf{s}_2$ are related by Eq. (5.3–42) and $\tilde{S}^{(0)}(0, \nu)$ is given by Eq. (5.3–37). It follows at once from the definition of the Fourier transform [See Eq. (5.3–18)] that

$$\tilde{S}^{(0)}(0, \nu) = \frac{1}{(2\pi)^2} \int_\sigma S^{(0)}(\boldsymbol{\rho}, \nu)\, \mathrm{d}^2\rho, \tag{5.3–44}$$

showing that $\tilde{S}^{(0)}(0, \nu)$ is proportional to the integral of the spectral density taken over the source.

The formulas (5.3–38) and (5.3–43) make it possible to determine the spectral degree of coherence and the normalized spectral density of a quasi-homogeneous, secondary, planar source from far-field measurements.

We will illustrate the use one of the inversion formulas by determining the spectral degree of coherence of a quasi-homogeneous, secondary, planar, Lambertian source. The angular distribution of the radiant intensity generated by a Lambertian source obeys *Lambert's law*

$$J(\mathbf{s}, \nu) = J^{(0)}(\nu) \cos\theta, \tag{5.3–45}$$

where $J^{(0)}(\nu) > 0$. On substituting from Eq. (5.3–45) into the formula (5.3–38) and recalling that $\mathbf{s}_\perp^2 = s_x^2 + s_y^2$ we find that

$$g^{(0)}(\boldsymbol{\rho}', \nu) = \frac{\mathscr{G}^{(0)}(\boldsymbol{\rho}', \nu)}{\mathscr{G}^{(0)}(0, \nu)}, \tag{5.3–46}$$

where

$$\mathscr{G}^{(0)}(\boldsymbol{\rho}', \nu) = \iint_{s_x^2+s_y^2 \leq 1} \frac{1}{[1 - s_x^2 - s_y^2]^{1/2}} \mathrm{e}^{\mathrm{i}k(s_x x' + s_y y')} \, \mathrm{d}s_x \, \mathrm{d}s_y, \tag{5.3–47}$$

x', y' being the Cartesian components of $\boldsymbol{\rho}'$.

To evaluate the integral on the right-hand side of Eq. (5.3–47) we set

$$s_x = \tau \cos\chi, \quad s_y = \tau \sin\chi, \tag{5.3–48a}$$

$$x' = \rho' \cos\theta, \quad y' = \rho' \sin\theta. \tag{5.3–48b}$$

The formula (5.3–47) then becomes

$$\mathscr{G}^{(0)}(\boldsymbol{\rho}', \nu) = \int_0^{2\pi} \mathrm{d}\chi \int_0^1 \frac{1}{[1 - \tau^2]^{1/2}} \mathrm{e}^{\mathrm{i}k\tau\rho' \cos(\chi - \theta)} \, \tau \, \mathrm{d}\tau. \tag{5.3–49}$$

The integration with respect to χ may be carried out at once by recalling a well-known integral representation of the Bessel function $J_0(x)$ of the first kind and zero order [Watson, 1966, p. 20, Eq. (5)], namely

$$\frac{1}{2\pi} \int_0^{2\pi} \mathrm{e}^{\mathrm{i}k\tau\rho' \cos(\chi - \theta)} \, \mathrm{d}\chi = J_0(k\tau\rho'). \tag{5.3–50}$$

By use of this formula, Eq. (5.3–49) reduces to

$$\mathscr{G}^{(0)}(\boldsymbol{\rho}', \nu) = 2\pi \int_0^1 \frac{1}{[1 - \tau^2]^{1/2}} J_0(k\tau\rho') \, \tau \, \mathrm{d}\tau. \tag{5.3–51}$$

This integral may be evaluated in closed form (see, for example, Gradshteyn and Ryzhik, 1980, p. 682, Formula 2 of 6.554) and one then obtains the following expression for $\mathscr{G}^{(0)}$:

$$\mathscr{G}^{(0)}(\boldsymbol{\rho}', \nu) = 2\pi \left(\frac{\sin k\rho'}{k\rho'} \right). \tag{5.3–52}$$

It follows from Eqs. (5.3–46) and (5.3–52), since evidently $\mathscr{G}^{(0)}(0, \nu) = 2\pi$, that

$$g^{(0)}(\boldsymbol{\rho}_2 - \boldsymbol{\rho}_1, \nu) = \frac{\sin k|\boldsymbol{\rho}_2 - \boldsymbol{\rho}_1|}{k|\boldsymbol{\rho}_2 - \boldsymbol{\rho}_1|}. \tag{5.3–53}$$

We thus conclude that *all quasi-homogeneous, secondary, planar, Lambertian sources have the same degree of spatial coherence, given by Eq. (5.3–53),*‡ provided that contributions from its high-frequency spatial Fourier components ($|\mathbf{f}| > k$) are negligible. The behavior of the function $\sin k\rho'/k\rho'$ is shown in Fig. 5.9. We see that the field correlations across a Lambertian source extend over a distance $\Delta\rho'$ such that $k\Delta\rho' \lessapprox \pi/2$, i.e. over distances $\Delta\rho' \lessapprox \lambda/4$, where $\lambda = c/\nu$ is the wavelength. Thus a Lambertian source is spatially not strictly incoherent, although its correlation widths, at optical frequencies, are very small, being of the order of the wavelength for each spectral component.

One encounters generalizations of the Lambertian distribution of radiant intensity, given by Eq. (5.3–45), when one studies the correlation properties of a quasi-homogeneous source which generates an arbitrary rotationally symmetric distribution $J(\mathbf{s}, \nu)$. We may then represent $J(\mathbf{s}, \nu)$ as a series involving powers of $\cos\theta$:

$$J(\mathbf{s}, \nu) = \sum_{n=0}^{\infty} a_n(\nu) \cos^n \theta. \tag{5.3–54}$$

Because of the linearity of the inverse formula (5.3–38), the cross-spectral density of the quasi-homogeneous source that gives rise to $J(\mathbf{s}, \nu)$ is a linear combination of cross-spectral densities produced by the more elementary sources (all of which are also assumed to be quasi-homogeneous), each of which generates a radiant intensity of the form [see Fig. 5.10(*a*)]

$$J^{(n)}(\mathbf{s}, \nu) = a_n(\nu) \cos^n \theta. \tag{5.3–55}$$

The inverse problem associated with the radiant intensity distribution (5.3–55) was considered in several papers (Antes, Baltes and Steinle, 1976; Baltes, Steinle and Antes, 1976; Carter, 1984; see also Baltes, 1977, Secs. 3.3 and 3.4 and

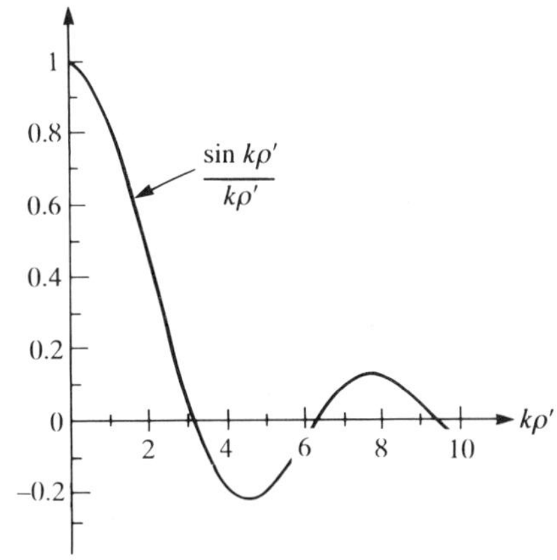

Fig. 5.9 The spectral degree of coherence $g^{(0)}(\boldsymbol{\rho}', \nu) = \sin(k\rho')/(k\rho')$, where $\boldsymbol{\rho}' = |\boldsymbol{\rho}_2 - \boldsymbol{\rho}_1|$, of a planar, secondary, quasi-homogeneous Lambertian source [Eq. (5.3–53)].

‡ This result is consistent with known correlation properties of blackbody sources. In this connection see Carter and Wolf (1975), Sec. II.

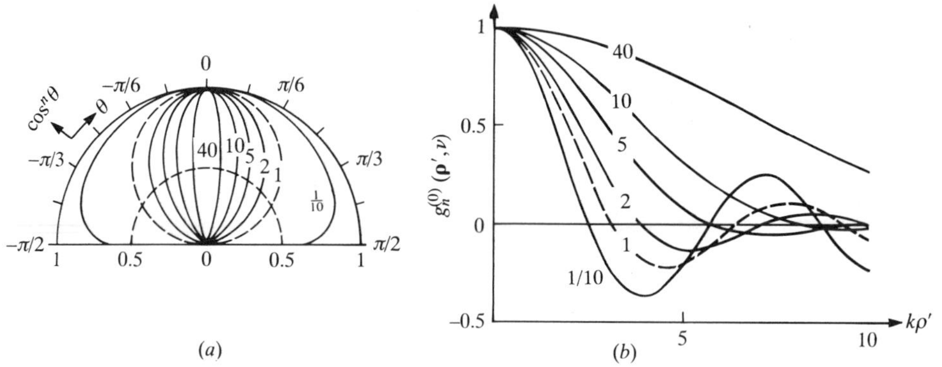

Fig. 5.10 The polar diagram (a) illustrating the angular distribution $J^{(n)}(\mathbf{s}, \nu)/a_n(\nu)$ [Eq. (5.3–55)] of the normalized radiant intensity, generated by a Bessel-correlated, planar, secondary, quasi-homogeneous source. Figure (b) shows the behavior of the spectral degree of coherence of the source [Eq. (5.3–56)]. The numbers on the curves represent the values of the parameter n. The dashed curve (for $n = 1$) corresponds to a Lambertian source. (Adapted from Baltes, Steinle and Antes, 1976.)

McGuire, 1979). It was found that the spectral degree of coherence $g_n^{(0)}(\boldsymbol{\rho}', \nu)$ of a quasi-homogeneous secondary source which generates the radiant intensity (5.3–55) is given by the formula

$$g_n^{(0)}(\boldsymbol{\rho}', \nu) = 2^{n/2}\Gamma\left(1 + \frac{n}{2}\right)\frac{J_{n/2}(k\rho')}{(k\rho')^{n/2}}, \tag{5.3–56}$$

where $J_{n/2}$ is a Bessel function of the first kind and Γ is the gamma function, it being assumed that its high-frequency spatial Fourier components ($|\mathbf{f}| > k$) may be neglected. The formula (5.3–56) is valid for all values of $n \geqslant 0$, not just for non-negative integral values and, as may readily be verified, it reduces correctly to the formula (5.3–53) when $n = 1$.

Sources whose spectral degree of coherence is given by Eq. (5.3–56) are sometimes said to be *Bessel correlated*. In Fig. 5.10(b) the spectral degree of coherence of such sources is plotted as a function of $k\rho'$ for selected values of the parameter n.

5.4 Equivalence theorems for planar sources which generate the same radiant intensity

One of the very useful properties of laser sources is their ability to generate optical beams, i.e. to produce fields that are highly directional. The radiant intensity $J(\mathbf{s}, \nu)$ of such fields is concentrated in a very narrow solid angle; more precisely, $J(\mathbf{s}, \nu)$, considered as function of $\mathbf{s}$, is sharply peaked around some particular direction $\mathbf{s}_0$. Because a laser source is a highly coherent source, one might assume that only coherent sources can generate beams. Actually this is not so. It was found that sources with entirely different coherence properties may generate identical distributions of the radiant intensity. In particular it was shown that quasi-homogeneous sources which, as we learned earlier, are rather incoherent in the global sense can produce the same distribution of radiant intensity as a fully spatially coherent laser source. This discovery triggered off investigations that led to the formulation of a number of 'equivalence theorems'. These theorems provide conditions under which sources of different states of coherence and different spatial distributions of spectral density will generate fields which have the same radiant intensity. The coherence properties of the light generated by two equivalent sources will, of course, be different. This fact suggests a number of potential applications. For example, beams of low spatial coherence will not produce the detrimental speckle effects which are usually present when highly coherent laser beams are used; this property is useful, for example in lithography where very high resolution is required.

In the present section we will formulate the main equivalence theorems and we will illustrate them by numerical examples and by some experimental results.

5.4.1 An equivalence theorem for planar sources

According to Eq. (5.3–8) the radiant intensity generated by a planar, secondary source is given by the formula

$$J(\mathbf{s}, \nu) = (2\pi k)^2 \widetilde{W}^{(0)}(-k\mathbf{s}_\perp, k\mathbf{s}_\perp, \nu) \cos^2\theta, \tag{5.4–1}$$

where $\widetilde{W}^{(0)}(\mathbf{f}_1, \mathbf{f}_2, \nu)$ is the four-dimensional spatial Fourier transform, defined by Eq. (5.3–7), of the cross-spectral density $W^{(0)}(\boldsymbol{\rho}_1, \boldsymbol{\rho}_2, \nu)$ of the field distribution $V^{(0)}(\boldsymbol{\rho}, t)$ in the source plane, namely

$$\widetilde{W}^{(0)}(\mathbf{f}_1, \mathbf{f}_2, \nu) = \frac{1}{(2\pi)^4} \iint W^{(0)}(\boldsymbol{\rho}_1, \boldsymbol{\rho}_2, \nu)\, \mathrm{e}^{-\mathrm{i}(\mathbf{f}_1\cdot\boldsymbol{\rho}_1 + \mathbf{f}_2\cdot\boldsymbol{\rho}_2)}\, \mathrm{d}^2\rho_1\, \mathrm{d}^2\rho_2, \tag{5.4–2}$$

where the integrations extend formally over the whole source plane.

According to Eq. (5.4–1) the radiant intensity is fully determined by certain spatial Fourier components of the cross-spectral density $W^{(0)}$, namely by those components labeled by pairs of spatial-frequency vectors $\mathbf{f}_1$ and $\mathbf{f}_2$ which satisfy the relation $\mathbf{f}_2 = -\mathbf{f}_1$. The corresponding Fourier components of $W^{(0)}$ are said to be *anti-diagonal* components. Moreover as we have already noted, and as is evident at once from Eq. (5.4–1), only the low spatial-frequency Fourier components, i.e. those for which $|\mathbf{f}_1| \leqslant k$, $|\mathbf{f}_2| \leqslant k$, appear in the expression for the radiant intensity. Hence we have the following *equivalence theorem for radiant intensity*, due to Collett and Wolf (1978; see also Collett and Wolf, 1979; Saleh, 1979; Saleh and Irshid, 1982): *Two planar secondary sources, whose cross-spectral density functions $W^{(0)}(\boldsymbol{\rho}_1, \boldsymbol{\rho}_2, \nu)$ have the same low-frequency anti-diagonal spatial Fourier components, generate the same distribution of the radiant intensity.*

It is important to appreciate that, even though the anti-diagonal low-frequency elements of the two sources may be identical, the other low-frequency elements (for which $\mathbf{f}_2 \neq -\mathbf{f}_1$) may be entirely different. Consequently the two sources will, in general, have different coherence properties and different spatial distributions of their spectral densities. This in turn implies, as can immediately be seen from Eqs. (5.3–4) and (5.3–6), that, although the two sources will generate the same radiant intensity, the far fields which they produce will have different coherence properties.

It is not difficult to understand the physical reasons why entirely different sources may generate the same distribution of the radiant intensity. For this purpose let us first express the cross-spectral density of the source in terms of its spectral density and its spectral degree of coherence [Eq. (4.3–47b)]:

$$W^{(0)}(\boldsymbol{\rho}_1, \boldsymbol{\rho}_2, \nu) = [S^{(0)}(\boldsymbol{\rho}_1, \nu)]^{1/2}[S^{(0)}(\boldsymbol{\rho}_2, \nu)]^{1/2}\mu^{(0)}(\boldsymbol{\rho}_1, \boldsymbol{\rho}_2, \nu). \quad (5.4\text{–}3)$$

On substituting from Eq. (5.4–3) into Eq. (5.4–2) and setting $\mathbf{f}_1 = -k\mathbf{s}_\perp$, $\mathbf{f}_2 = k\mathbf{s}_\perp$ we obtain the following expression for the low-frequency anti-diagonal Fourier components of the cross-spectral density of the source:

$$\widetilde{W}^{(0)}(-k\mathbf{s}_\perp, k\mathbf{s}_\perp, \nu) = \frac{1}{(2\pi)^4}\iint [S^{(0)}(\boldsymbol{\rho}_1, \nu)]^{1/2}[S^{(0)}(\boldsymbol{\rho}_2, \nu)]^{1/2}\mu^{(0)}(\boldsymbol{\rho}_1, \boldsymbol{\rho}_2, \nu) \times \mathrm{e}^{-ik\mathbf{s}_\perp\cdot(\boldsymbol{\rho}_2-\boldsymbol{\rho}_1)}\,\mathrm{d}^2\rho_1\,\mathrm{d}^2\rho_2. \quad (5.4\text{–}4)$$

We see that both the spectral density and the spectral degree of coherence of the source appear under the integral sign on the right-hand side of Eq. (5.4–4). It is, therefore, possible that $\widetilde{W}^{(0)}(-k\mathbf{s}_\perp, k\mathbf{s}_\perp, \nu)$ and hence also the radiant intensity, given by Eq. (5.4–1), may be identical, even when the two sources are quite different. Stated in another way, it is possible to replace $S^{(0)}(\boldsymbol{\rho}, \nu)$ and $\mu^{(0)}(\boldsymbol{\rho}_1, \boldsymbol{\rho}_2, \nu)$ by different functions $'S^{(0)}(\boldsymbol{\rho}, \nu)$ and $'\mu^{(0)}(\boldsymbol{\rho}_1, \boldsymbol{\rho}_2, \nu)$ respectively, without changing the value of the integral in Eq. (5.4–4). Hence *one can make a 'trade-off' between coherence and the spatial distribution of the spectral density of the source, without affecting the distribution of the radiant intensity which the source generates*.

Another formulation of the equivalence theorem may be based on the formula (5.3–10), namely

$$J(\mathbf{s}, \nu) = k^2\widetilde{C}^{(0)}(k\mathbf{s}_\perp, \nu)\cos^2\theta, \quad (5.4\text{–}5)$$

rather than on Eq. (5.4–1). Here $\widetilde{C}^{(0)}(\mathbf{f}, \nu)$ is the two-dimensional spatial Fourier transform of the source-integrated cross-spectral density

$$C^{(0)}(\boldsymbol{\rho}', \nu) = \int W^{(0)}(\boldsymbol{\rho} - \tfrac{1}{2}\boldsymbol{\rho}', \boldsymbol{\rho} + \tfrac{1}{2}\boldsymbol{\rho}', \nu)\,\mathrm{d}^2\rho. \quad (5.4\text{–}6)$$

The formula (5.4–5) brings at once into evidence the following alternative formulation of the equivalence theorem for the radiant intensity: *two planar sources, whose source-integrated cross-spectral density $C^{(0)}(\boldsymbol{\rho}', \nu)$ has the same low-frequency ($|\mathbf{f}| \leqslant k$) Fourier components, will generate the same distribution of the radiant intensity*.

An interesting consequence of the equivalence theorem which we have just discussed is the possibility of producing sources which are globally rather incoherent but which can nevertheless generate fields that are highly directional. In particular such sources can generate the same distribution of radiant intensity as a fully coherent laser source. Examples of such equivalent sources will be given in the next two sub-sections (5.4.2 and 5.4.3).

5.4.2 Example: equivalent Gaussian Schell-model sources

We will illustrate the equivalence theorem that we just discussed for a particular class of Schell-model sources. We recall that such a source is characterized by the property that its spectral degree of coherence $\mu^{(0)}(\boldsymbol{\rho}_1, \boldsymbol{\rho}_2, \nu)$ depends on $\boldsymbol{\rho}_1$ and $\boldsymbol{\rho}_2$ only through the difference $\boldsymbol{\rho}' = \boldsymbol{\rho}_2 - \boldsymbol{\rho}_1$. Consequently the cross-spectral density of a Schell-model source has the form indicated by (5.3–15), namely

$$W^{(0)}(\boldsymbol{\rho}_1, \boldsymbol{\rho}_2, \nu) = [S^{(0)}(\boldsymbol{\rho}_1, \nu)]^{1/2}[S^{(0)}(\boldsymbol{\rho}_2, \nu)]^{1/2} g^{(0)}(\boldsymbol{\rho}_2 - \boldsymbol{\rho}_1, \nu). \quad (5.4\text{–}7)$$

On substituting from Eq. (5.4–7) into (5.4–6) one finds that for a Schell-model source

$$C^{(0)}(\boldsymbol{\rho}', \nu) = g^{(0)}(\boldsymbol{\rho}', \nu) H^{(0)}(\boldsymbol{\rho}', \nu), \quad (5.4\text{–}8)$$

where

$$H^{(0)}(\boldsymbol{\rho}', \nu) = \int [S^{(0)}(\boldsymbol{\rho} - \tfrac{1}{2}\boldsymbol{\rho}', \nu)]^{1/2}[S^{(0)}(\boldsymbol{\rho} + \tfrac{1}{2}\boldsymbol{\rho}', \nu)]^{1/2}\, \mathrm{d}^2\rho. \quad (5.4\text{–}9)$$

Suppose now that the spatial distributions of the spectral density and of the spectral degree of coherence are both Gaussian, namely

$$S^{(0)}(\boldsymbol{\rho}, \nu) = A^2(\nu)\, \mathrm{e}^{-\rho^2/2\sigma_S^2(\nu)}, \quad (5.4\text{–}10)$$

$$g^{(0)}(\boldsymbol{\rho}', \nu) = \mathrm{e}^{-\rho'^2/2\sigma_g^2(\nu)}, \quad (5.4\text{–}11)$$

where A, σ_S and σ_g are all positive. We will refer to such a source as a *Gaussian Schell-model source*.

On substituting from Eq. (5.4–10) into Eq. (5.4–9) and evaluating the resulting integral we readily find that in this case

$$H^{(0)}(\boldsymbol{\rho}', \nu) = 2\pi A^2(\nu)\sigma_S^2(\nu)\, \mathrm{e}^{-\rho'^2/8\sigma_S^2(\nu)}; \quad (5.4\text{–}12)$$

and if we substitute from Eqs. (5.4–12) and (5.4–11) into Eq. (5.4–8) we find that for a Gaussian Schell-model source

$$C^{(0)}(\boldsymbol{\rho}', \nu) = 2\pi A^2(\nu)\sigma_S^2(\nu)\, \mathrm{e}^{-\rho'^2/2\delta^2(\nu)}, \quad (5.4\text{–}13)$$

where

$$\frac{1}{\delta^2(\nu)} = \frac{1}{4\sigma_S^2(\nu)} + \frac{1}{\sigma_g^2(\nu)}. \quad (5.4\text{–}14)$$

The Fourier transform, defined by Eq. (5.3–11), of the expression (5.4–13) is readily found to be

$$\tilde{C}^{(0)}(\mathbf{f}, \nu) = (A\sigma_S\delta)^2\, \mathrm{e}^{-f^2\delta^2/2}. \quad (5.4\text{–}15)$$

On the right-hand side of this formula and also in some of the formulas which follow we do not show explicitly the dependence of some of the quantities on the frequency ν.

It follows from the second form of the equivalence theorem stated below Eq. (5.4–6) and from Eqs. (5.4–15) and (5.4–14) that *two Gaussian Schell-model sources, for which the quantities*

(a) $$\frac{1}{4\sigma_S^2} + \frac{1}{\sigma_g^2}$$

have the same value and also the products

(b) $$A\sigma_S$$

are the same, generate the same distribution of the radiant intensity‡.

The radiant intensity $J(\mathbf{s}, \nu)$ produced by such a rotationally symmetric source, in a direction making an angle θ with the normal to the source plane, is obtained at once by substituting for $\tilde{C}^{(0)}$ from Eq. (5.4–15) into Eq. (5.4–5) and by making use of the relation $\mathbf{s}_\perp^2 = \sin^2\theta$. One then finds that

$$J(\mathbf{s}, \nu) = J^{(0)}(\nu)\cos^2\theta\, e^{-[(k\delta)^2\sin^2\theta]/2}, \tag{5.4–16}$$

where

$$J^{(0)}(\nu) = (kA\sigma_S\delta)^2 \tag{5.4–17}$$

is the radiant intensity in the direction normal to the source plane ($\theta = 0$).

The expression (5.4–16), together with (5.4–17), shows explicitly that Gaussian Schell-model sources for which the parameters δ and also the products $A\sigma_S$ have the same values will indeed generate the same distribution of the radiant intensity. The dependence of the parameter δ, defined by Eq. (5.4–14), on σ_S and σ_g illustrates the 'trade-off' between the state of coherence of the source (characterized by σ_g) and the suitably normalized spatial distribution of spectral density (characterized by σ_S), which is at the root of the equivalence theorem. This fact is also illustrated in Figs. 5.11 and 5.12.

We note two extreme cases. In the quasi-homogeneous limit, characterized by the condition

$$\sigma_S \gg \sigma_g, \tag{5.4–18}$$

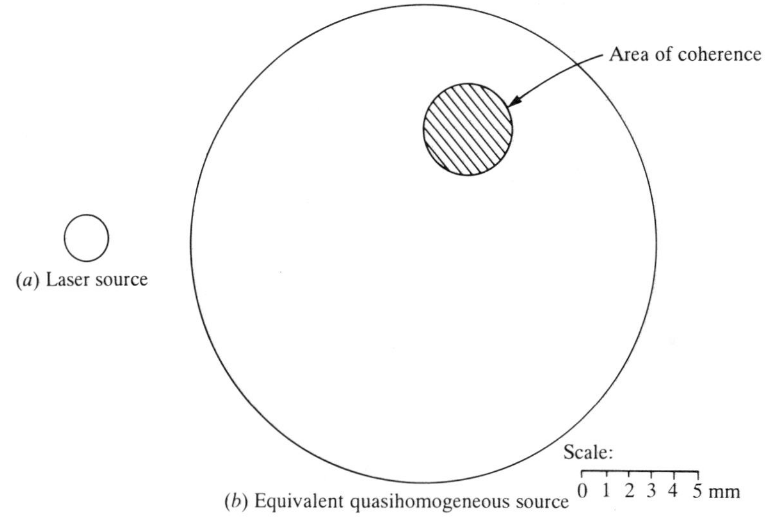

Fig. 5.11 Illustrating the effective sizes of (*a*) a laser source and of (*b*) an 'equivalent' quasi-homogeneous source. The coherence area of the quasi-homogeneous source is shown shaded in Fig. (*b*). (After Wolf, 1978.)

‡ An equivalence theorem for Gaussian Schell-model sources which generate fields with the same distribution of the *spectral degree of coherence* in the far zone was formulated by Kandpal and Wolf (1994).

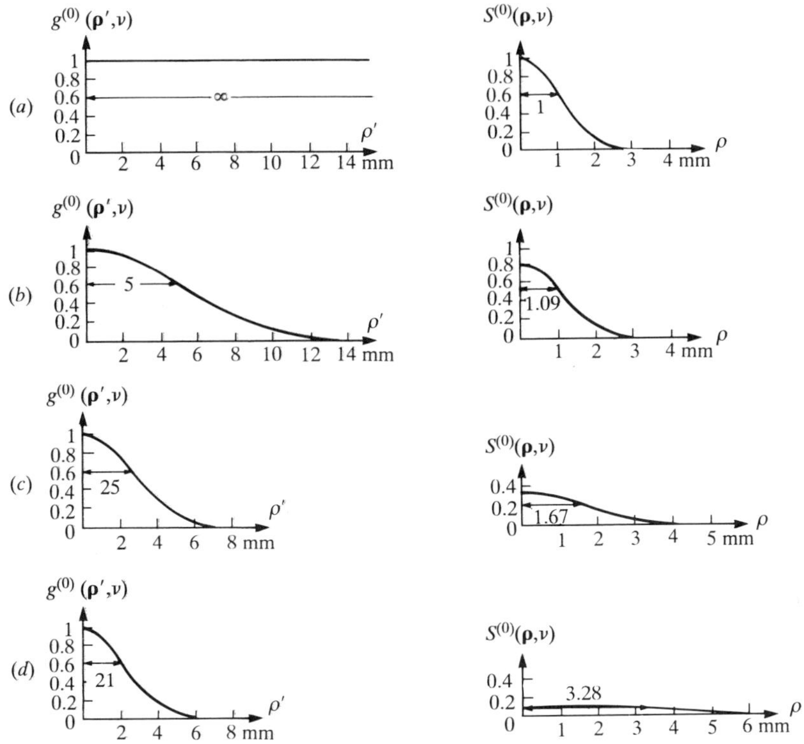

Fig. 5.12 The spectral degree of coherence $g^{(0)}(\boldsymbol{\rho}', \nu)$ and the distribution of spectral density $S^{(0)}(\boldsymbol{\rho}, \nu)$ across four planar, secondary, Gaussian, Schell-model sources which generate identical distributions of the radiant intensity. The curves in (*a*) pertain to a completely (spatially) coherent source (e.g. a single-mode laser) and the curves in (*d*) to a rather incoherent source. The parameters characterizing the four sources are: (*a*) $\sigma_g = \infty$, $\sigma_S = 1$ mm, $A = 1$ (arbitrary units); (*b*) $\sigma_g = 5$ mm, $\sigma_S = 1.09$ mm, $A = 0.84$; (*c*) $\sigma_g = 2.5$ mm, $\sigma_S = 1.67$ mm, $A = 0.36$; (*d*) $\sigma_g = 2.1$ mm, $\sigma_S = 3.28$ mm, $A = 0.09$. The normalized radiant intensity generated by all these sources is given by the expression (5.4–16), namely $J(\theta)/J(0) = \cos^2\theta \exp[-\frac{1}{2}(k\delta)^2 \sin^2\theta]$, with $\delta =$ 2 mm. (After Wolf and Collett, 1978.)

the Gaussian Schell-model source becomes globally rather incoherent and we see from Eq. (5.4–14) that in this case

$$\delta \approx \sigma_g. \tag{5.4–19}$$

In this limit the expression (5.4–16) for the radiant intensity correctly reduces to Eq. (5.3–29).

In the other extreme case, when

$$\sigma_S \ll \sigma_g, \tag{5.4–20}$$

the source is essentially spatially coherent in the global sense and Eq. (5.4–14) then gives

$$\delta \approx 2\sigma_S. \tag{5.4–21}$$

It follows from Eqs. (5.4–16), (5.4–17) and (5.4–21) that the radiant intensity is now, to a good approximation, independent of the exact value of the effective correlation length σ_g of the source.

The limit $\sigma_g \to \infty$ characterizes, of course, a Gaussian Schell-model source which is spatially completely coherent at frequency ν, i.e. for which

$$g^{(0)}(\boldsymbol{\rho}', \nu) \equiv 1 \quad \text{for all } \boldsymbol{\rho}'. \tag{5.4–22}$$

Instead of the the approximate relation (5.4–21) we then have from Eq. (5.4–14) the exact relation $\delta = 2\sigma_S$. Examples of such sources are lasers with flat output mirrors, each of which operates in its lowest-order mode, provided effects arising from diffraction at the edge of the mirror are neglected. Let us re-write Eq. (5.4–10), appropriate to such a source, in the form

$$S^{(0)}(\boldsymbol{\rho}, \nu) = A_{\mathrm{L}}^2 \mathrm{e}^{-2\rho^2/\delta_{\mathrm{L}}^2}, \tag{5.4–23}$$

where $\delta_{\mathrm{L}} = 2(\sigma_S)_{\mathrm{L}}$, the subscript L indicating that the parameters now pertain to a laser $[(\sigma_g)_{\mathrm{L}} \to \infty]$. It is clear from the equivalence theorem which we just discussed that *Gaussian Schell-model sources for which* $\delta = \delta_{\mathrm{L}} \equiv 2(\sigma_S)_{\mathrm{L}}$ *and* $A\sigma_S = A_{\mathrm{L}}(\sigma_S)_{\mathrm{L}}$ *will generate the same angular distribution of radiant intensity as the laser source*. Written out more explicitly the first condition implies, according to Eq. (5.4–14), that

$$\frac{1}{4(\sigma_S)_{\mathrm{L}}^2} = \frac{1}{4\sigma_S^2} + \frac{1}{\sigma_g^2}, \tag{5.4–24}$$

from which we see that

$$\sigma_S \geqslant (\sigma_S)_{\mathrm{L}} \tag{5.4–25}$$

and

$$\sigma_g \geqslant 2(\sigma_S)_{\mathrm{L}}. \tag{5.4–26}$$

Hence a partially coherent Gaussian Schell-model source which is 'equivalent' to the fully coherent laser source must have a larger effective size than the laser and must be correlated over distances that are at least twice as large as the effective linear dimensions of the laser source. Those with the smallest correlation lengths $[\sigma_g \approx 2(\sigma_S)_{\mathrm{L}}]$ are quasi-homogeneous sources because, according to Eq. (5.4–24), σ_S must then be necessarily much greater than σ_g. A comparison of a laser source with such an 'equivalent' quasi-homogeneous source is indicated in Fig. 5.11. Figure 5.12 shows the degree of coherence and the distribution of the spectral density across some other Gaussian Schell-model sources which generate the same distribution of radiant intensity as a laser (see also Gori and Palma, 1978; De Santis, Gori and Palma, 1979; Gori, 1980a,b).

5.4.3 An experimental test of the equivalence theorem

An experimental test of the equivalence theorem for radiant intensities was made by De Santis, Gori, Guattari and Palma (1979), using an optical system shown schematically in Fig. 5.13. A Gaussian spot of laser light was produced via a lens

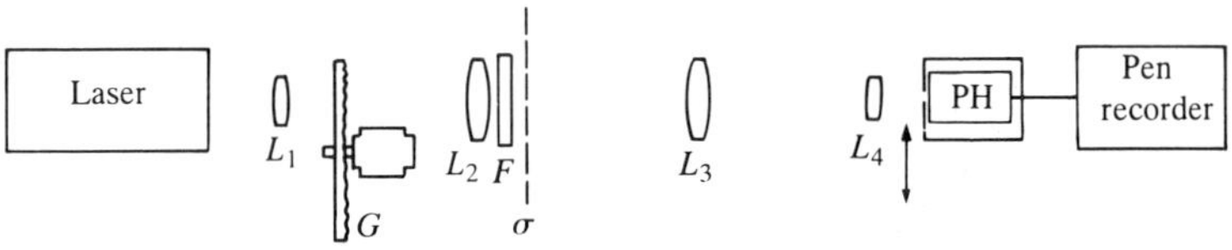

Fig. 5.13 The experimental set up used for testing the equivalence theorem for radiant intensity. The meanings of the symbols are explained in the text. (Adapted from De Santis, Gori, Guattari and Palma, 1979.)

L_1 on a ground glass plate G. The plate was located in the focal plane of a lens L_2, which was followed by an amplitude transmission filter F that has a Gaussian transmission function. As was first demonstrated by Martienssen and Spiller (1964), if the plate is continuously rotated, the cross-section of the beam that emerges from the plate is essentially a spatially incoherent secondary source, provided that the spot size is large compared with the inhomogeneity scale of the ground glass plate. Using the van Cittert–Zernike theorem and the propagation law for the cross-spectral density one can show that the light in a plane σ which follows the filter F is a Gaussian Schell-model source. The r.m.s. widths of the two distributions (5.4–10) and (5.4–11) which characterize this source are found to depend on:

(a) The r.m.s. width of the surface roughness correlation function of the ground glass plate;
(b) The r.m.s. width of the transmission function of the amplitude filter F;
(c) The focal length of the lens L_2.

By choosing these three parameters appropriately one can produce Gaussian Schell-model sources with desired parameters.

The angular distribution of the radiant intensity generated by such sources may be studied with the help of an optical system formed by lenses L_3 and L_4. The image produced by this system was scanned by a photodetector PH.

In these experiments the angular distribution produced by the Gaussian laser beam in the absence of both the ground glass plate G and the amplitude filter F was first measured. Next, corresponding measurements were made with a ground glass plate and an amplitude filter in place. The r.m.s. widths of the ground glass plate correlation function and of the filter transmission function were chosen so that the coherent laser source and the partially coherent Schell-model source satisfied the conditions of the equivalence theorem.

In Fig. 5.14 the observed optical intensities in the cross-section of the coherent laser beam (a) and of the beam just emerging from the ground glass plate when it is not rotating (b) are shown. The average size of the speckles in Fig. 5.14(b) is a rough measure of the r.m.s. width of the transverse correlations (of the order of σ_g) of the light which emerges from the plate when it is rotated. In spite of the considerable difference between the two sources, the far-zone intensity distributions were found to be essentially the same (Fig. 5.15), except for a scale factor which was left somewhat arbitrary in these experiments.

A different arrangement for testing some of the theoretical predictions relating

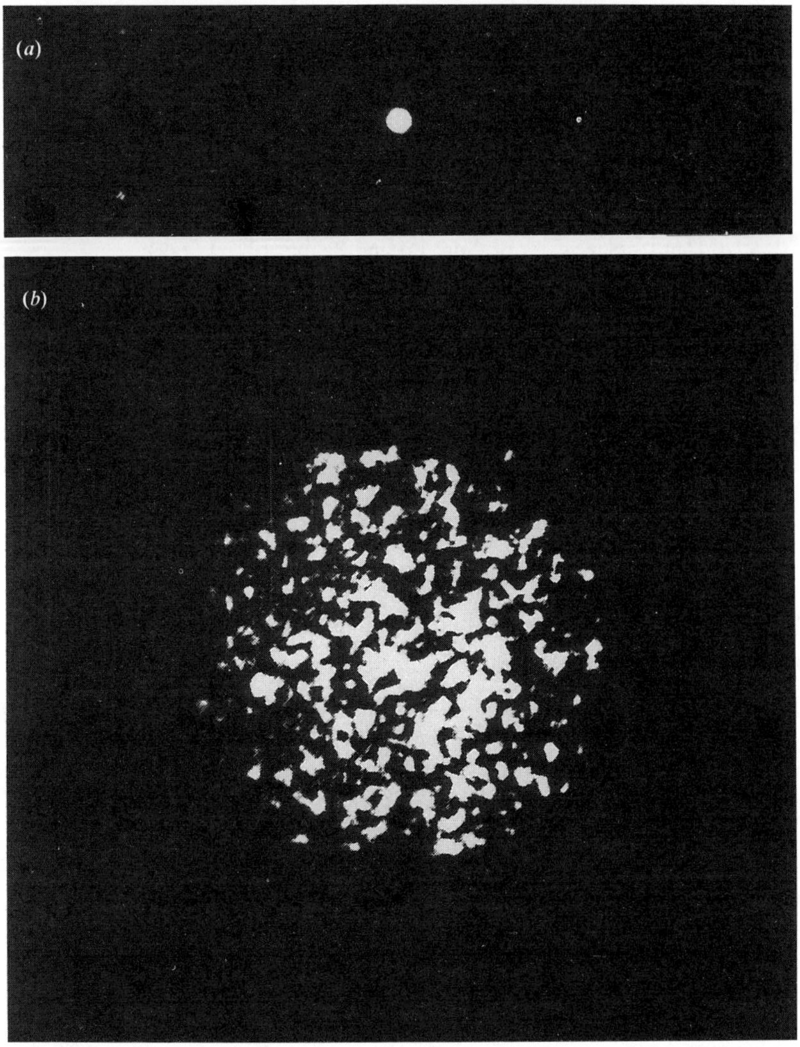

Fig. 5.14 Intensity distribution across a coherent laser source (a), and across an 'equivalent' partially coherent source σ (b). (Adapted from De Santis, Gori, Guattari and Palma, 1979.)

to radiation from partially coherent sources was employed by Farina, Narducci and Collett (1980). It will be described later, in Section 5.6.4, in connection with Gaussian Schell-model beams. Here we only mention that the secondary source used in these experiments was obtained by placing, in the path of a laser beam, a rotating phase screen created by spraying a finishing mist on a clear glass blank. With appropriate coating a quasi-homogeneous source was produced and it was shown that such a source can indeed generate highly directional beams.

Other methods of producing sources with different coherence properties have been described in the literature. One of them utilizes liquid crystals to scatter a laser beam (Scudieri, Bertolotti and Bartolino, 1974; Carter and Bertolotti, 1978).

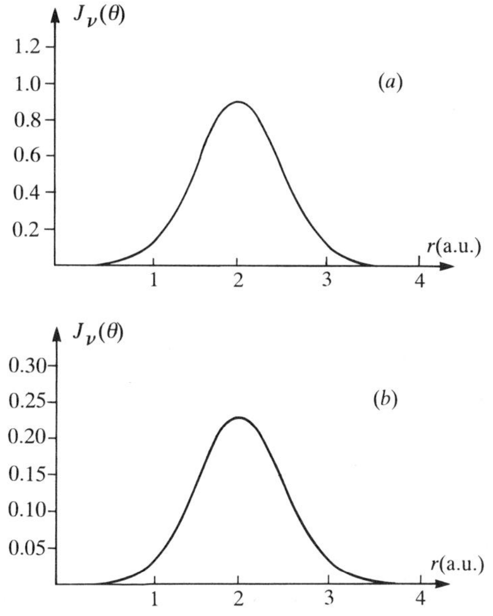

Fig. 5.15 The measured angular distributions of the radiant intensity $J_\nu(\theta)$ (in arbitrary but identical units) generated by the two sources illustrated in Fig. 5.14. (Adapted from De Santis, Gori, Guattari and Palma, 1979.)

By applying a d.c. electric field to the liquid crystal, the coherence properties of the scattered light were modified. Other methods utilize various transmission filters or holographic filters to change the coherence properties of the light which passes through them (Courjon and Bulabois, 1979; Courjon, Bulabois and Carter, 1981). Methods for modifying the coherence of light by interaction with sound waves have also been developed (Ohtsuka and Imai, 1979; Imai and Ohtsuka, 1980; Turunen, Tervonen, Friberg, 1990; Tervonen, Friberg and Turunen, 1992). A method has also been proposed for generating Gaussian Schell-model sources from primary quasi-homogeneous sources by the use of feedback systems (Dechamps, Courjon and Bulabois, 1983). Sources of controlled degree of coherence, constructed by means of these and other techniques, are finding useful applications in fields such as microdensitometry (Kinzly, 1972; Reynolds and Smith, 1973), line width measurements (Nyyssonen, 1977, 1979), and lithography (Oldham, Subramanian and Neureuther, 1981), where high spatial coherence of light has sometimes detrimental effects, because it gives rise to speckled images which make it difficult to obtain good resolution. An example showing how reduction of spatial coherence may improve resolution is shown in Fig. 5.16.

5.5 Coherent-mode representation of Gaussian Schell-model sources

Throughout this chapter we have illustrated some of our analysis with reference to Schell-model sources (which include quasi-homogeneous sources in an appropriate limit). Because of the importance of model sources of this type and also in order to obtain better insight into some basic differences between globally coherent and globally incoherent sources, we will consider the coherent-mode representation, described in Section 4.7, of Gaussian Schell-model sources. To keep the analysis as simple as possible we will consider only one-dimensional sources of this class.

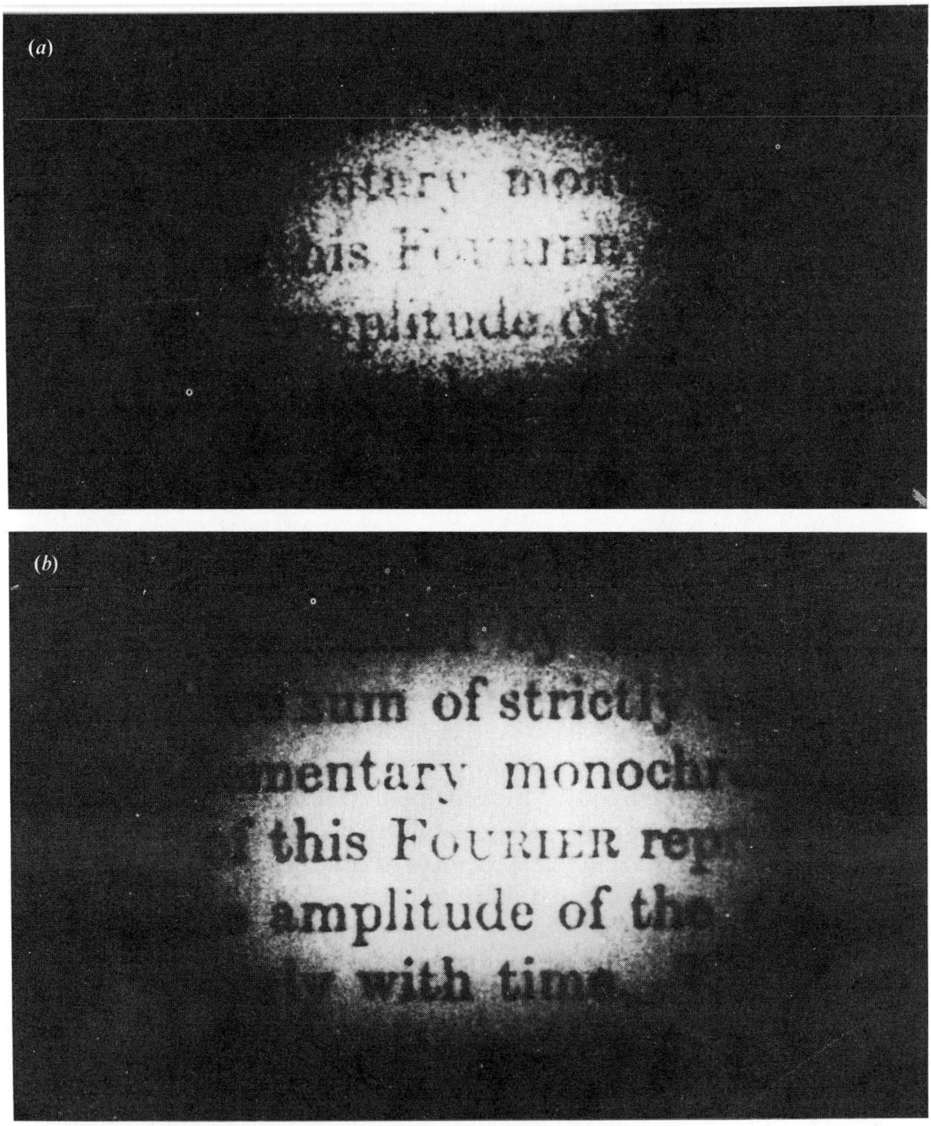

Fig. 5.16 Illustrating the improvement in resolution by changing the spectral degree of coherence of light. (*a*): Photograph of text obtained with spatially coherent He:Ne laser light. The speckles that are produced obscure the image, making the words nearly unreadable. (*b*): Photograph of the same text obtained with light from a quasi-homogeneous (i.e. globally incoherent) source. The speckles have disappeared and the text has become readable. (Courtesy of L. M. Narducci and J. D. Farina.)

A one-dimensional, secondary, Gaussian Schell-model source is characterized by a cross-spectral density of the form [cf. Eqs. (5.3–15), (5.3–23) and (5.3–24)]

$$W^{(0)}(x_1, x_2, \nu) = [S^{(0)}(x_1, \nu)]^{1/2}[S^{(0)}(x_2, \nu)]^{1/2} g^{(0)}(x_2 - x_1, \nu), \quad (5.5\text{–}1)$$

where the spatial distribution of the spectral density $S^{(0)}(x, \nu)$ and the spectral degree of coherence $g^{(0)}(x', \nu)$ are given by expressions of the form

$$S^{(0)}(x, \nu) = A^2(\nu)\, \mathrm{e}^{-x^2/2\sigma_S^2(\nu)}, \quad (5.5\text{–}2)$$

$$g^{(0)}(x', \nu) = \mathrm{e}^{-x'^2/2\sigma_g^2(\nu)}, \quad (5.5\text{–}3)$$

where $A(\nu)$, $\sigma_S(\nu)$ and $\sigma_g(\nu)$ are positive.

According to the one-dimensional analogue of Eq. (4.7–44), the cross-spectral density $W^{(0)}(x_1, x_2, \nu)$ may be expressed in the form

$$W^{(0)}(x_1, x_2, \nu) = \sum \beta_n(\nu) \phi_n^*(x_1, \nu) \phi_n(x_2, \nu), \quad (5.5\text{–}4)$$

where $\beta_n(\nu)$ are the eigenvalues and $\phi_n(x, \nu)$ are the (orthonormalized) eigenfunctions of the homogeneous Fredholm integral equation

$$\int_{-\infty}^{\infty} W^{(0)}(x_1, x_2, \nu) \phi_n(x_1, \nu)\, \mathrm{d}x_1 = \beta_n(\nu) \phi_n(x_2, \nu). \quad (5.5\text{–}5)$$

In physical terms Eq. (5.5–4) represents the cross-spectral density function of the secondary source as a linear superposition of elementary cross-spectral density modes, each of which is completely spatially coherent at each frequency.

The integral equation (5.5–5), with the kernel given by Eqs. (5.5–1) to (5.5–3) may be solved in a closed form. The (normalized) eigenfunctions and the eigenvalues are found to be (see, for example, Gori, 1980b, or Starikov and Wolf, 1982)

$$\phi_n(x, \nu) = \left(\frac{2c}{\pi}\right)^{1/4} \frac{1}{(2^n n!)^{1/2}} H_n[x(2c)^{1/2}]\, \mathrm{e}^{-cx^2}, \quad (5.5\text{–}6)$$

$$\beta_n(\nu) = A^2 \left(\frac{\pi}{a + b + c}\right)^{1/2} \left(\frac{b}{a + b + c}\right)^n, \quad (5.5\text{–}7)$$

where $H_n(x)$ are the Hermite polynomials and

$$a(\nu) = \frac{1}{4\sigma_S^2(\nu)}, \quad b(\nu) = \frac{1}{2\sigma_g^2(\nu)}, \quad (5.5\text{–}8a)$$

$$c(\nu) = [a^2(\nu) + 2a(\nu)b(\nu)]^{1/2}. \quad (5.5\text{–}8b)$$

From Eq. (5.5–7), which gives the eigenvalues for a Gaussian Schell-model source, we can at once obtain a simple expression for the relative weights with which the different modes contribute to the cross-spectral density of the source. The ratio of the eigenvalue β_n to the lowest eigenvalue β_0 is evidently given by

$$\frac{\beta_n}{\beta_0} = \left(\frac{b}{a + b + c}\right)^n. \quad (5.5\text{–}9)$$

We may express Eq. (5.5–9) in a physically more significant form by dividing the numerator and the denominator on the right by the factors b^n. We then obtain the following simple expression for the ratio β_n/β_0:

$$\frac{\beta_n}{\beta_0} = \left[\frac{1}{(q^2/2) + 1 + q[(q/2)^2 + 1]^{1/2}}\right]^n, \tag{5.5–10}$$

where the parameter q is the ratio of the r.m.s. width of the spectral degree of coherence to that of the spectral density function:

$$q(\nu) = \frac{\sigma_g(\nu)}{\sigma_S(\nu)}. \tag{5.5–11}$$

Clearly this parameter may be regarded as a measure of the *degree of global coherence* of a Gaussian Schell-model source. Let us consider two extreme cases.

When $\sigma_g \gg \sigma_S$ the source is spatially very *coherent in the global sense*. According to Eq. (5.5–11) we have in this case $q \gg 1$ and it then follows from Eq. (5.5–10) that

$$\frac{\beta_n}{\beta_0} \approx \frac{1}{q^{2n}}. \tag{5.5–12}$$

This formula implies that, for all $n \neq 0$, $\beta_n \ll \beta_0$ and hence in the coherent limit the behavior of the source is well approximated by the lowest-order mode.

In the other extreme case, when $\sigma_g \ll \sigma_S$ the source is spatially very *incoherent in the global sense* and belongs to the class of quasi-homogeneous sources which we considered earlier (Sec. 5.3.2). According to Eq. (5.5–11) we now have $q \ll 1$ and it then follows from Eq. (5.5–10) that

$$\frac{\beta_n}{\beta_0} \approx 1 - nq. \tag{5.5–13}$$

This formula implies that a large number of modes (of the order $1/q$) is now needed to represent the source adequately.

In Fig. 5.17 the behavior of the ratio β_n/β_0 as a function of n is shown, for various selected values of the parameter q. It is seen from the figure that as we proceed from highly (globally) coherent sources to incoherent ones, more and more modes are needed to represent the spatial correlation properties. The contributions of the different modes to the intensity in the far field have been studied by Starikov and Wolf (1982). The contributions of the different modes to the cross-spectral density in any transverse cross-section z = constant were obtained by Gori (1983).

The eigenfunctions, given by Eq. (5.5–6), and sometimes called *Hermite–Gaussian functions*, are well known in the theory of laser resonators. They represent the x-dependent parts of the transverse modes of confocal laser resonators with spherical mirrors and rectangular boundaries along the x- and y-directions, when diffraction produced by the edges of the mirrors is neglected (see, for example, Kogelnik and Li, 1966; Siegman, 1971, Sec. 8.4). We will study the structure of the lowest-order transverse mode in Section 5.6.2.

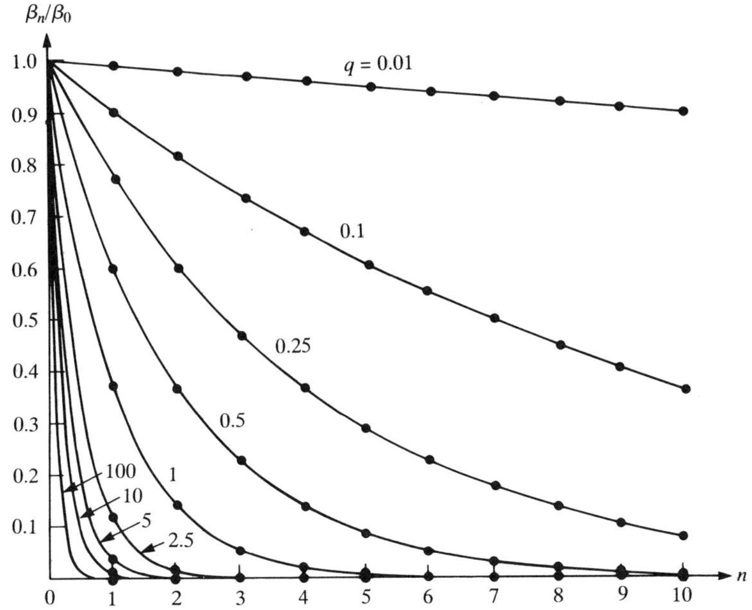

Fig. 5.17 The ratio of the $(n+1)$th eigenvalue β_n to the lowest-order eigenvalue β_0 of a one-dimensional, secondary, Gaussian, Schell-model source as a function of n, [Eq. (5.5–10)], for selected values of the degree of global coherence $q = \sigma_g/\sigma_S$. Values $q \gg 1$ and $q \ll 1$ characterize sources that are spatially coherent and spatially incoherent in the global sense, respectively. (After Starikov and Wolf, 1982.)

5.6 Optical beams

Up to now we have been mainly concerned with properties of fields in the far zones of fluctuating sources. We will now consider fields at arbitrary distances from the sources and we will study the changes which the fields undergo on propagation. Because of the complexity of the problem, we will restrict ourselves to highly directional fields, i.e. to beams.

We will first derive conditions under which a monochromatic planar source generates a beam. We will then obtain two useful representations of monochromatic beams. One of them is in terms of the angular spectrum of plane waves; the other is an integral-transform formulation, which utilizes an appropriate Green's function. We will apply the results to study the structure of a monochromatic Gaussian beam. We will then generalize the results to partially coherent beams, generated by planar sources of any state of spatial coherence. Finally we will study the main properties of beams produced by Gaussian Schell-model sources.

5.6.1 Monochromatic beams

Let us consider a monochromatic wave field of frequency ν,

$$V(x, y, z, t) = U(x, y, z; \nu)\,\mathrm{e}^{-2\pi i \nu t}, \qquad (5.6\text{–}1)$$

propagating into the half-space $z > 0$, which is assumed to be free space. As we have learned in Section 3.2.2, under usual circumstances, the space-dependent part $U(x, y, z; \nu)$ may be expressed throughout the half-space $z \geqslant 0$ as a superposition of plane waves, all with the same wave number

$$k = 2\pi\nu/c, \tag{5.6–2}$$

namely [Eq. (3.2–19) with k written now in place of k_0]

$$U(x, y, z; \nu) = \iint_{-\infty}^{\infty} a(p, q; \nu)\, \mathrm{e}^{\mathrm{i}k(px+qy+mz)}\, \mathrm{d}p\, \mathrm{d}q, \tag{5.6–3}$$

where

$$m = +[1 - p^2 - q^2]^{1/2} \quad \text{when } p^2 + q^2 \leqslant 1 \tag{5.6–4a}$$

$$= +\mathrm{i}[p^2 + q^2 - 1]^{1/2} \quad \text{when } p^2 + q^2 > 1. \tag{5.6–4b}$$

As we noted in Section 3.2.2, the plane waves for which $p^2 + q^2 \leqslant 1$ are ordinary homogeneous plane waves which propagate in directions specified by the unit vector (p, q, m). Those for which $p^2 + q^2 > 1$ represent evanescent plane waves, whose amplitudes decay exponentially with increasing z.

We will need the following two results derived in Section 3.2.2:

(a) The spectral amplitude function $a(p, q; \nu)$ can be expressed in terms of the boundary values of the field $U(x, y, z; \nu)$ in the plane $z = 0$ as [Eq. (3.2–27)]

$$a(p, q; \nu) = k^2 \tilde{U}^{(0)}(kp, kq; \nu), \tag{5.6–5}$$

where

$$\tilde{U}^{(0)}(u, v; \nu) = \frac{1}{(2\pi)^2} \iint_{-\infty}^{\infty} U(x, y, 0; \nu)\, \mathrm{e}^{-\mathrm{i}(ux+vy)}\, \mathrm{d}x\, \mathrm{d}y \tag{5.6–6}$$

is the two-dimensional spatial Fourier transform of $U(x, y, 0; \nu)$.

(b) Let

$$r = [x^2 + y^2 + z^2]^{1/2} \tag{5.6–7}$$

be the distance of the field point $P(x, y, z)$ from the origin. Then as P recedes from the origin in any fixed direction specified by a unit vector $\mathbf{s} \equiv s_x, s_y, s_z > 0$, one has, in the asymptotic limit as $kr \to \infty$, [Eq. (3.2–22)]

$$U(r\mathbf{s}, \nu) \sim U^{(\infty)}(r\mathbf{s}, \nu) = -\frac{2\pi\mathrm{i}}{k} \cos\theta\, a(s_x, s_y; \nu) \frac{\mathrm{e}^{\mathrm{i}kr}}{r}. \tag{5.6–8}$$

Here θ is the angle which the unit vector $\mathbf{s}$ makes with the positive z-axis (see Fig. 5.18).

The preceding relations are valid quite generally. We will now specialize them to the case when the field is beam-like. Clearly in that case the amplitudes $U^{(\infty)}(r\mathbf{s}; \nu)$ of the far field will be negligible, except for $\mathbf{s}$-directions which are close to the beam axis. If we take the beam axis to be the z-direction, this result implies, according to Eq. (5.6–8), that the absolute values $|a(s_x, s_y; \nu)|$ of the spectral amplitudes will be negligible unless $s_x^2 + s_y^2 \ll 1$. This result means, in

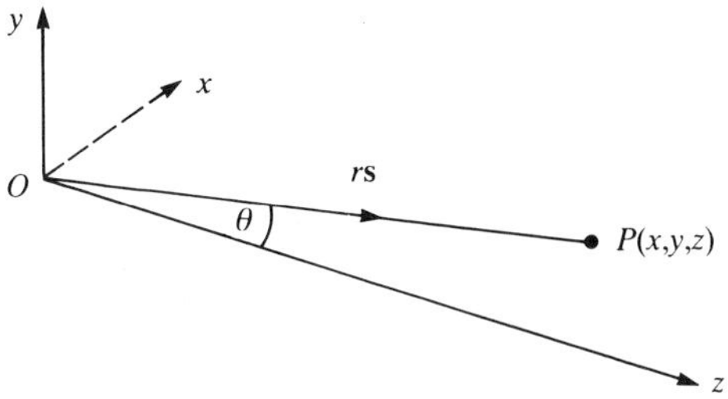

Fig. 5.18 Illustrating the notation relating to Eq. (5.6–8). **s** is a unit vector in the direction OP.

physical terms, that *for a beam which propagates close to the positive z-direction, only those plane waves will be present with non-negligible amplitudes in its angular spectrum representation (5.6–3), whose directions fall within a narrow solid angle around the z-axis*, as one would intuitively expect. From this result and from the relation (5.6–5) we obtain at once a necessary and sufficient condition for a field distribution $U(x, y, 0; \nu)\,e^{-2\pi i\nu t}$ in the plane $z = 0$ to generate, in the half-space $z > 0$, a beam which propagates close to the positive z-direction: *$U(x, y, 0; \nu)$ must essentially contain only very low spatial-frequency components, i.e. the modulus of its two-dimensional spatial Fourier transform $\widetilde{U}^{(0)}(u, v; \nu)$ must be negligible unless $u^2 + v^2 \ll k^2$.*

Since, as we just saw, for a beam which propagates close to the z-direction one has

$$|a(p, q; \nu)| \approx 0 \quad \text{unless } p^2 + q^2 \ll 1 \tag{5.6–9}$$

only the first of the two expressions (5.6–4) for m now applies; and that expression can be approximated by the first two terms of its binomial expansion, i.e. by

$$m \approx 1 - \tfrac{1}{2}(p^2 + q^2). \tag{5.6–10}$$

If we substitute from Eq. (5.6–10) into Eq. (5.6–3) we obtain the following representation of a monochromatic beam (with the periodic time-dependent factor $e^{-2\pi i\nu t}$ omitted) *which propagates close to the z-axis into the half-space $z > 0$*:

$$U(x, y, z; \nu) = e^{ikz} \iint_{-\infty}^{\infty} a(p, q; \nu)\, e^{ik(px+qy)}\, e^{-ik(p^2+q^2)z/2}\, dp\, dq, \tag{5.6–11}$$

it being understood that the (generally complex) amplitude function $a(p, q; \nu)$ satisfies the constraint (5.6–9).

If we use in Eq. (5.6–11) the simple relation Eq. (5.6–5) between $a(p, q; \nu)$ and $\widetilde{U}^{(0)}(kp, kq; \nu)$ and change the variables of integration from p and q to $u = kp$, $v = kq$, we obtain at once the following representation of the beam in terms of the Fourier transform of its boundary values in the plane $z = 0$:

$$U(x, y, z; \nu) = e^{ikz} \iint_{-\infty}^{\infty} \widetilde{U}^{(0)}(u, v; \nu)\, e^{i(ux+vy)}\, e^{-iz(u^2+v^2)/2k}\, du\, dv, \tag{5.6–12}$$

it is being understood that

$$|\widetilde{U}^{(0)}(u, v; \nu)| \approx 0 \quad \text{unless } u^2 + v^2 \ll k^2. \tag{5.6–13}$$

We may also represent the beam throughout the half-space $z > 0$ directly in terms of the boundary values of the field in the plane $z = 0$. For this purpose we substitute for $\widetilde{U}^{(0)}(u, v; \nu)$ from Eq. (5.6–6) into Eq. (5.6–12) and interchange the order of the integrations. We then obtain the following formula for the beam [writing now $U(x, y, z)$ in place of $U(x, y, z; \nu)$, etc.]:

$$U(x, y, z) = \mathrm{e}^{ikz} \iint_{-\infty}^{\infty} U(x', y', 0) G(x - x', y - y', z)\, \mathrm{d}x'\, \mathrm{d}y', \tag{5.6–14}$$

where

$$G(x - x', y - y', z) = \frac{1}{(2\pi)^2} \iint_{-\infty}^{\infty} \mathrm{e}^{-\mathrm{i}z(u^2+v^2)/2k}\, \mathrm{e}^{\mathrm{i}[u(x-x')+v(y-y')]}\, \mathrm{d}u\, \mathrm{d}v. \tag{5.6–15}$$

The double integral on the right is a product of two simple integrals, each of which is a one-dimensional Fourier transform of a Gaussian distribution with an imaginary variance. Its value and also the values of several other integrals that we will encounter later can readily be obtained from the formula [Gradshteyn and Ryzhik, 1980, p. 307, Eq. (2) of Sec. 3.323, together with the formulas 9.253 and 8.956(1)]

$$\int_{-\infty}^{\infty} \mathrm{e}^{-\beta^2 t^2}\, \mathrm{e}^{-\mathrm{i}qt}\, \mathrm{d}t = \frac{\sqrt{\pi}}{\beta} \mathrm{e}^{-q^2/4\beta^2}, \quad (\mathrm{Re}\, \beta > 0). \tag{5.6–16}$$

We then readily find that

$$G(x, y, z) = -\frac{\mathrm{i}k}{2\pi z} \mathrm{e}^{\mathrm{i}k(x^2+y^2)/2z}. \tag{5.6–17}$$

Instead of using the angular spectrum to obtain a representation of a beam, one often employs an alternative approach which we will now briefly discuss. Let us set

$$U(x, y, z) = \psi(x, y, z)\, \mathrm{e}^{\mathrm{i}kz}. \tag{5.6–18}$$

For a strictly uni-directional beam (i.e. a homogeneous plane wave) which propagates in the positive z-direction, ψ will be constant. For any physically realizable beam that propagates close to the z-axis, $\psi(x, y, z)$ will, of course, vary with x, y and z, but if the angular spread of the beam is sufficiently small, it may be expected to vary very slowly with z. Let us assume that this variation is so slow that $|\partial^2\psi/\partial z^2| \ll 2k|\partial\psi/\partial z|$; or, what may readily be shown to amount to essentially the same thing, that the change in $|\partial\psi/\partial z|$ in a z-interval of the order of a wavelength is negligible compared with the value of $|\partial\psi/\partial z|$ itself. On substituting from Eq. (5.6–18) into the Helmholtz equation for U [Eq. (4.7–39)] and on neglecting the term $\partial^2\psi/\partial z^2$ we find that, to a good approximation, $\psi(x, y, z)$ satisfies the so-called *paraxial equation*

$$\frac{\partial^2\psi}{\partial x^2} + \frac{\partial^2\psi}{\partial y^2} + 2\mathrm{i}k\frac{\partial\psi}{\partial z} = 0. \tag{5.6–19}$$

By applying to the paraxial equation (5.6–19) the same procedure that we applied in Section 3.2 to the Helmholtz equation, it is not difficult to show that the most general solution of Eq. (5.6–19), which propagates into the half-space $z > 0$, may be expressed in the form

$$\psi(x, y, z) = \iint_{-\infty}^{\infty} a(p, q)\,\mathrm{e}^{-\mathrm{i}k(px+qy)}\,\mathrm{e}^{-\mathrm{i}k(p^2+q^2)z/2}\,\mathrm{d}p\,\mathrm{d}q, \qquad (5.6\text{–}20)$$

where $a(p, q)$ is an arbitrary function of the parameters p and q, and that the solution of Eq. (5.6–19) which takes on the prescribed boundary values $\psi(x, y, 0)$ on the plane $z = 0$ is[‡]

$$\psi(x, y, z) = \iint_{-\infty}^{\infty} \psi(x', y', 0) G(x - x', y - y', z)\,\mathrm{d}x'\,\mathrm{d}y', \qquad (5.6\text{–}21)$$

where $G(x - x', y - y', z)$ is the Green's function (5.6–17). To derive this formula one needs only to set $z = 0$ in Eq. (5.6–20), invert the resulting formula to determine $a(p, q)$ in terms of $\psi(x, y, 0)$ and then substitute this expression for $a(p, q)$ into Eq. (5.6–20).

In view of the relationship (5.6–18) between $\psi(x, y, z)$ and $U(x, y, z)$ the formulas (5.6–20) and (5.6–21) are seen to be in agreement with the formulas (5.6–11) and (5.6–14). However, it should be noted that the solution (5.6–20) of the paraxial equation (5.6–19) will not represent a beam, unless the function $a(p, q)$ satisfies the requirement (5.6–9). Nor will the solution (5.6–21) of the paraxial equation represent a beam, unless $\psi(x, y, 0)$ is effectively bandlimited to a circle in the spatial-frequency plane, whose radius is much smaller than the wave number k, i.e. unless the modulus of the Fourier transform $\tilde{\psi}(u, v, 0)$ of $\psi(x, y, 0)$ is negligible, except when $u^2 + v^2 \ll k^2$. These results imply that, whilst every well-behaved solution of the paraxial equation may be expressed in the forms (5.6–20) and (5.6–21), not all solutions of that equation represent beams. It is for this reason that we based our representation of beams on the concept of the angular spectrum of plane waves rather than on the paraxial equation. Such an approach also provides a clear insight into many of the physically significant features of wavefields, whether or not they have a beam-like structure.

5.6.2 *Example: monochromatic Gaussian beams*

We will now make use of one of the beam representations derived in Section 5.6.1 to elucidate the main properties of Gaussian beams. Such beams are produced by many commonly used lasers.

A Gaussian beam is a very directional field which is generated by the field distribution

$$U(x, y, 0) = A\,\mathrm{e}^{-(x^2+y^2)/w_0^2} \qquad (5.6\text{–}22)$$

‡ Apart from normalization and scaling factors, the integral transform that appears on the right-hand side of Eq. (5.6–21), with the kernel G given by Eq. (5.6–17), is known as the (two-dimensional) Fresnel transform. Some of the main properties of Fresnel transforms are discussed, for example, by Gori (1981, 1994).

in some plane $z = 0$. In Eq. (5.6–22) A and w_0 are positive constants. We note that when the distance

$$\rho = [x^2 + y^2]^{1/2} \tag{5.6–23}$$

is equal to w_0, U drops to $1/\mathrm{e}$ times its axial value $U(0, 0, 0) = A$. Hence w_0 is the effective radius of the circularly symmetric distribution in the plane $z = 0$. For this reason the parameter w_0 is usually called the *minimum spot size*.

The Gaussian distribution (5.6–22) is often an approximation to the true field in the plane $z = 0$ for distances ρ which do not exceed some value $\rho = a$. If a is significantly larger than w_0, the effect of approximating the field in the plane $z = 0$ by the full Gaussian distribution is, as a rule, negligible in determining the field throughout the half-space $z > 0$ into which the beam propagates.

In order to represent the field generated by the planar secondary source with boundary values given by Eq. (5.6–22) in the form of an angular spectrum of plane waves [Eq. (5.6–3)] we must first determine its spectral-amplitude function $a(p, q)$. It is obtained by substituting for $U(x, y, 0)$ from Eq. (5.6–22) into Eq. (5.6–6) and using the relation (5.6–5). We then obtain for $a(p, q)$ the expression

$$a(p, q) = A\left(\frac{k}{2\pi}\right)^2 \iint_{-\infty}^{\infty} \mathrm{e}^{-(x^2+y^2)/w_0^2}\, \mathrm{e}^{-ik(px+qy)}\, \mathrm{d}x\, \mathrm{d}y. \tag{5.6–24}$$

The double integral on the right-hand side of Eq. (5.6–24) may be evaluated at once using the formula (5.6–16) and we find that

$$a(p, q) = \frac{A(kw_0)^2}{4\pi}\, \mathrm{e}^{-(kw_0)^2(p^2+q^2)/4}. \tag{5.6–25}$$

In order that the expression (5.6–25) is the spectral amplitude function of a beam which propagates close to the z-axis, the condition (5.6–9) must also be satisfied, i.e. $|a(p, q)|$ must only have appreciable values when $p^2 + q^2 \ll 1$. Now the expression (5.6–25) attains its maximum when $p = q = 0$ and decreases to $1/\mathrm{e}$ of its maximum value when $\frac{1}{4}(kw_0)^2(p^2 + q^2) = 1$. Hence $|a(p, q)|$ has appreciable values only when

$$p^2 + q^2 \lesssim \frac{4}{(kw_0)^2}. \tag{5.6–26}$$

It follows that the beam condition (5.6–9) will be satisfied in the present case provided that $4/(kw_0)^2 \ll 1$, i.e. provided that $w_0 \gg 2/k$. Since $k = 2\pi/\lambda$ where λ is the wavelength, this condition is equivalent to the requirement that

$$w_0 \gg \lambda/\pi. \tag{5.6–27}$$

Hence, provided that the minimum spot size w_0 is much larger than the wavelength, the Gaussian distribution (5.6–22) in the plane $z = 0$ will generate a beam, known as a *Gaussian beam*.

According to Eqs. (5.6–11) and (5.6–25) a Gaussian beam has the following angular spectrum representation throughout the half-space $z \geqslant 0$:

$$U(x, y, z) = \frac{A(kw_0)^2}{4\pi}\, \mathrm{e}^{ikz} \iint_{-\infty}^{\infty} \mathrm{e}^{-[(kw_0)^2/4 + ikz/2](p^2+q^2)}\, \mathrm{e}^{ik(px+qy)}\, \mathrm{d}p\, \mathrm{d}q. \tag{5.6–28}$$

The double integral on the right-hand side of Eq. (5.6–28) may be readily evaluated by the use of the formula (5.6–16). If we write $U(\boldsymbol{\rho}, z)$ in place of $U(x, y, z)$, where

$$\boldsymbol{\rho} \equiv (x, y) \tag{5.6–29}$$

is the 'transverse vector' which, together with z, specifies the location of the field point (x, y, z), we obtain, after evaluating the integral in Eq. (5.6–28), the following expression for the beam:

$$U(\boldsymbol{\rho}, z) = \frac{A(kw_0)^2}{(kw_0)^2 + 2ikz} \mathrm{e}^{-\{(k\rho)^2/[(kw_0)^2+2ikz]\}} \mathrm{e}^{ikz}. \tag{5.6–30}$$

We next express the right-hand side of Eq. (5.6–30) in a form that shows explicitly the amplitude and the phase of $U(\boldsymbol{\rho}, z)$. After long but straightforward calculation one finds that

$$\frac{1}{(kw_0)^2 + 2ikz} = \frac{1}{(kw)^2}\left\{1 - i\left[\left(\frac{w}{w_0}\right)^2 - 1\right]^{1/2}\right\}, \tag{5.6–31}$$

$$\frac{(kw_0)^2}{(kw_0)^2 + 2ikz} = \left(\frac{w_0}{w}\right)\mathrm{e}^{i\psi}, \tag{5.6–32}$$

$$\mathrm{e}^{-\{(k\rho)^2/[(kw_0)^2+2ikz]\}} = \mathrm{e}^{-\rho^2/w^2}\,\mathrm{e}^{ik\rho^2/2R}, \tag{5.6–33}$$

where the quantities, w, ψ and R, each of which is a function of z, are defined by the formulas

$$w(z) = w_0\left[1 + \left(\frac{2z}{kw_0^2}\right)^2\right]^{1/2}, \tag{5.6–34a}$$

$$\cos\psi(z) = \frac{w_0}{w(z)}, \tag{5.6–34b}$$

$$\sin\psi(z) = -\left[1 - \left(\frac{w_0}{w(z)}\right)^2\right]^{1/2}, \tag{5.6–34c}$$

$$R(z) = z\left[1 + \left(\frac{kw_0^2}{2z}\right)^2\right]. \tag{5.6–34d}$$

The behavior of the quantities $w(z)$, $\psi(z)$ and $R(z)$, defined by these formulas, is illustrated in Fig. 5.19.

On substituting from Eqs. (5.6–32) and (5.6–33) into Eq. (5.6–30) we finally obtain the following expression for a monochromatic Gaussian beam (with the periodic time-dependent factor $\mathrm{e}^{-2\pi i\nu t}$ omitted):

$$U(\boldsymbol{\rho}, z) = A\left[\frac{w_0}{w(z)}\right]\mathrm{e}^{-\rho^2/[w(z)]^2}\,\mathrm{e}^{i\{k[z+\rho^2/2R(z)]+\psi(z)\}}. \tag{5.6–35}$$

Let us examine some implications of this expression.

We see that, as the beam propagates from the plane $z = 0$ into the half-space $z > 0$, the field amplitude $|U(\boldsymbol{\rho}, z)|$ remains Gaussian in each transverse cross-section $z =$ constant, but its width increases with increasing values of z. The

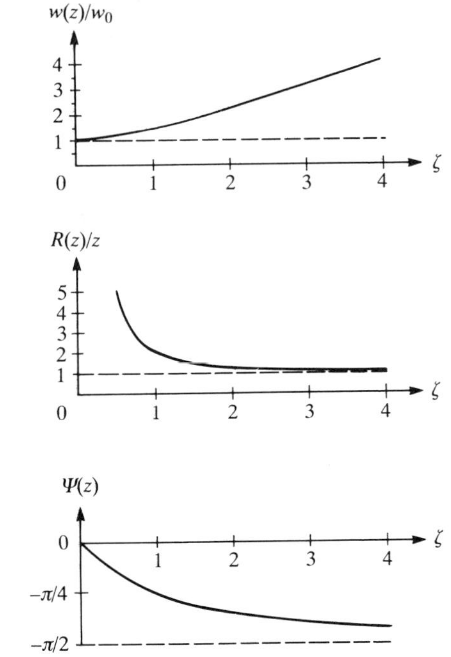

Fig. 5.19 The behavior of the quantities $w(z)$, $\psi(z)$ and $R(z)$ [defined by Eqs. (5.6–34)], associated with a monochromatic Gaussian beam as functions of the normalized distance $\zeta = 2z/(kw_0^2) = \lambda z/(\pi w_0^2)$ from the waist $z = 0$ of the beam.

amplitude in each transverse cross-section falls off to $1/e$ times its axial value at distance $\rho = w(z)$ from the axis. For this reason $w(z)$ is known as the *spot size at distance* z. The smallest beam radius is, according to Eq. (5.6–34a), just the minimum spot size w_0 and the plane in which this minimum spot size occurs is called the *waist of the beam*.

Next let us examine the phase, $\phi(\boldsymbol{\rho}, z)$ say, of the Gaussian beam. We see from Eq. (5.6–35) that it changes from its initial value $\phi(\boldsymbol{\rho}, 0) = 0$ at the waist to the value

$$\phi(\boldsymbol{\rho}, z) = k\left[z + \frac{\rho^2}{2R(z)}\right] + \psi(z), \qquad (5.6\text{–}36)$$

after propagating a distance z. The first term on the right represents the distance from the source plane, expressed in units of $1/k = \lambda/2\pi$. The second term, $k\rho^2/2R(z)$, represents the distance, expressed in the same units, between a spherical surface of radius $R(z)$ and the corresponding z-plane, at height ρ from the axis. According to Eq. (5.6–34d) the spherical surface is centered on the beam axis at distance

$$d(z) = \frac{k^2 w_0^4}{2z} \qquad (5.6\text{–}37)$$

behind the source plane (i.e. in the half-space $z < 0$) (see Fig. 5.20). The last term $\psi(z)$, on the right of Eq. (5.6–36), represents an additional phase shift which is

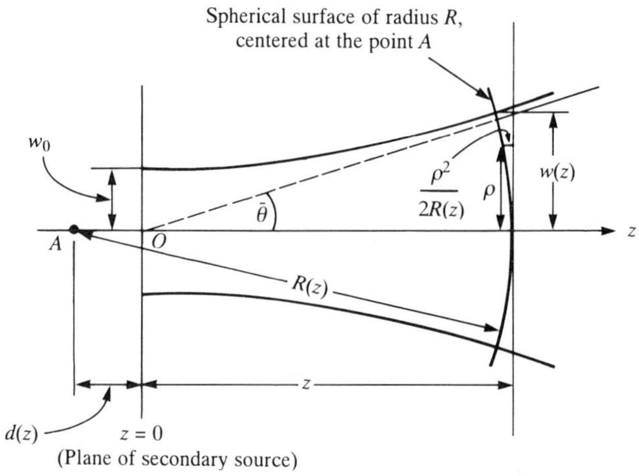

Fig. 5.20 Illustrating the geometrical significance of some of the parameters that characterize a monochromatic Gaussian beam. The quantities $w(z)$, $R(z)$, $d(z)$ and $\bar{\theta}$ are given by Eqs. (5.6–34a), (5.6–34d), (5.6–37) and (5.6–40) respectively.

associated with the so-called phase anomaly near focus (Born and Wolf, 1980, Sec. 8.8.4).

Let us next examine the behavior of the beam in the far zone of the secondary source. The simplified form of the expression (5.6–35), appropriate to the far field, may be obtained by substituting into Eq. (5.6–35) the asymptotic values, as $kz \to \infty$, of the quantities $w(z)$, $\psi(z)$ and $R(z)$, obtained from the formulas (5.6–34). Alternatively, the far-zone expression for the beam may be obtained by substituting into the general asymptotic formula (5.6–8) the spectral amplitude function (5.6–25) for a Gaussian beam and making use of the fact that, for a beam-field, $kr \sim kz$ in the far zone. One then obtains the following expression for the far field $U^{(\infty)}(\theta, z)$:

$$U^{(\infty)}(\theta, z) = -\tfrac{1}{2}\mathrm{i}kAw_0^2\,\mathrm{e}^{-(kw_0)^2\theta^2/4}\,\mathrm{e}^{\mathrm{i}kz(\theta^2/2)}\,\frac{\mathrm{e}^{\mathrm{i}kz}}{z}, \tag{5.6–38}$$

where

$$\theta \approx \frac{\rho}{z} = \frac{(x^2 + y^2)^{1/2}}{z}, \tag{5.6–39}$$

and we have also used the small angle approximation $\cos\theta \approx 1$, $\sin\theta \approx \theta$.

We see from Eq. (5.6–38) that the amplitude of the Gaussian beam in the far zone drops to $1/\mathrm{e}$ of its value on the beam axis $\theta = 0$ when $\theta = \bar{\theta}$, where $(kw_0)^2\bar{\theta}^2/4 = 1$, i.e. when

$$\bar{\theta} = \frac{2}{kw_0} = \frac{\lambda}{\pi w_0}. \tag{5.6–40}$$

$\bar{\theta}$ is known as the *angular spread* of the Gaussian beam.

Finally let us determine the radiant intensity $J(\mathbf{s}, \nu)$ of the Gaussian beam.

According to Eq. (5.2–12), the radiant intensity is given by the general formula

$$J(\mathbf{s}, \nu) = r^2 S^{(\infty)}(r\mathbf{s}, \nu), \tag{5.6–41}$$

where $S^{(\infty)}(r\mathbf{s}, \nu)$ is the spectral density at a point P in the far zone, specified by position vector $r\mathbf{s}$, which makes an angle θ with the positive z-axis. But $S^{(\infty)}(r\mathbf{s}, \nu)$ is also equal to the squared amplitude of the field at the point P. Hence for a Gaussian beam we have from Eq. (5.6–38)

$$S^{(\infty)}(r\mathbf{s}, \nu) = |U^{(\infty)}(\theta, z)|^2 = \left(\frac{kAw_0^2}{2z}\right)^2 \mathrm{e}^{-(kw_0)^2\theta^2/2}. \tag{5.6–42}$$

On substituting from Eq. (5.6–42) into Eq. (5.6–41) and on making the approximation $kr \sim kz$ appropriate for a beam we obtain the following expression for the radiant intensity of a monochromatic Gaussian beam:

$$J(\mathbf{s}, \nu) = (\tfrac{1}{2}kAw_0^2)^2\, \mathrm{e}^{-(kw_0)^2\theta^2/2}. \tag{5.6–43}$$

Except for a trivial difference in notation, this formula is in agreement with the coherent limit ($k\sigma_g \to \infty$) of the expression (5.4–16) for the radiant intensity generated by a Gaussian Schell-model source, specialized to a beam ($\sin\theta \approx \theta$, $\cos\theta \approx 1$).

5.6.3 Partially coherent beams

We learned in Section 5.5 that for a source to generate a beam, it is not necessary that it be spatially fully coherent. In fact, as we have seen, even sources that are globally rather incoherent can generate fields of this type. We will now discuss the mathematical representation of partially coherent beams. In Section 5.6.4 we will illustrate the results by applying them to beams produced by Gaussian Schell-model sources.

In our study of monochromatic beams in Section 5.6.1 we found it useful to start with the angular spectrum representation of monochromatic fields. When we deal with a fluctuating field whose statistical properties can be described by an ensemble which is stationary, at least in the wide sense, the basic quantity to consider is the mutual coherence function $\Gamma(\mathbf{r}_1, \mathbf{r}_2, \tau)$ or its Fourier transform, the cross-spectral density $W(\mathbf{r}_1, \mathbf{r}_2, \nu)$. Now according to Eq. (4.7–38), the cross-spectral density can always be represented as a correlation function,

$$W(\mathbf{r}_1, \mathbf{r}_2, \nu) = \langle U^*(\mathbf{r}_1, \nu) U(\mathbf{r}_2, \nu)\rangle, \tag{5.6–44}$$

where the average [previously denoted by $\langle \ldots \rangle_\nu$] is taken over an ensemble of monochromatic wavefields $\{U(\mathbf{r}, \nu)\exp(-2\pi i\nu t)\}$, all of the same frequency ν.

Starting from Eq. (5.6–44) we may readily generalize the angular spectrum representation to fields of any state of coherence that propagate into the half-space $z > 0$. For this purpose we first express each member $U(\mathbf{r}, \nu)$ of the statistical ensemble which represents the partially coherent field in the form of an angular spectrum of the plane waves [Eqs. (5.6–3)], namely

$$U(\mathbf{r}, \nu) = \iint_{-\infty}^{\infty} a(p, q; \nu)\, \mathrm{e}^{\mathrm{i}k(px+qy+mz)}\, \mathrm{d}p\, \mathrm{d}q, \tag{5.6–45}$$

where

$$m = +(1 - p^2 - q^2)^{1/2} \quad \text{when } p^2 + q^2 \leqslant 1, \tag{5.6–46a}$$

$$= +\mathrm{i}(p^2 + q^2 - 1)^{1/2} \quad \text{when } p^2 + q^2 > 1, \tag{5.6–46b}$$

$\mathbf{r} \equiv (x, y, z)$ and $k = 2\pi\nu/c$, c being the speed of light in vacuo. It is to be noted that, since $U(\mathbf{r}, \nu)$ is now a random function, so is the spectral-amplitude function $a(p, q; \nu)$. On substituting from Eq. (5.6–45) into Eq. (5.6–44) and interchanging the order of averaging and integration we obtain the following "double angular spectrum representation" of the field at the points $[\mathbf{r}_1 \equiv (x_1, y_1, z_1 \geqslant 0), \mathbf{r}_2 \equiv (x_2, y_2, z_2 \geqslant 0)]$, in the half-space $z \geqslant 0$:

$$W(\mathbf{r}_1, \mathbf{r}_2, \nu) = \iiiint_{-\infty}^{\infty} \mathscr{A}(p_1, q_1; p_2, q_2; \nu) \times \mathrm{e}^{\mathrm{i}k(p_2x_2 + q_2y_2 + m_2z_2 - p_1x_1 - q_1y_1 - m_1^*z_1)}\, \mathrm{d}p_1\, \mathrm{d}q_1\, \mathrm{d}p_2\, \mathrm{d}q_2 \tag{5.6–47}$$

where

$$\mathscr{A}(p_1, q_1; p_2, q_2; \nu) = \langle a^*(p_1, q_1; \nu) a(p_2, q_2; \nu)\rangle. \tag{5.6–48}$$

The formula (5.6–47) represents the cross-spectral density of the field throughout the half-space $z \geqslant 0$ as a superposition of correlated pairs of plane waves, both homogeneous ($p^2 + q^2 \leqslant 1$) and evanescent ($p^2 + q^2 > 1$), which propagate into that half-space. The plane waves forming each pair are, in general, correlated and their correlation is characterized by the function $\mathscr{A}(p_1, q_1; p_2, q_2; \nu)$, known as the *angular correlation function of the field* (Marchand and Wolf, 1972). We will see shortly that it is related in a very simple way to the radiant cross-intensity of the field.

If we set $z_1 = z_2 = 0$ in Eq. (5.6–47) and take the Fourier inverse of the resulting formula, we obtain the following expression for the angular correlation function in terms of the Fourier transform of the cross-spectral density of the field in the plane $z = 0$:

$$\mathscr{A}(p_1, q_1; p_2, q_2; \nu) = k^4 \widetilde{W}^{(0)}(-kp_1, -kq_1; kp_2, kq_2; \nu), \tag{5.6–49}$$

where

$$\widetilde{W}^{(0)}(u_1, v_1; u_2, v_2; \nu) = \frac{1}{(2\pi)^4} \iiiint_{-\infty}^{\infty} W(x_1, y_1, 0; x_2, y_2, 0; \nu) \times \mathrm{e}^{-\mathrm{i}(u_1x_1 + v_1y_1 + u_2x_2 + v_2y_2)}\, \mathrm{d}x_1\, \mathrm{d}y_1\, \mathrm{d}x_2\, \mathrm{d}y_2. \tag{5.6–50}$$

When the field points $P_1(\mathbf{r}_1)$ and $P_2(\mathbf{r}_2)$ are in the far zone of the source, the expression (5.6–47) for the cross-spectral density takes a much simpler form that can be most easily derived with the help of the asymptotic formula (5.6–8). Suppose that $P_1(\mathbf{r}_1)$ and $P_2(\mathbf{r}_2)$ recede to infinity in fixed directions specified by the unit vectors $\mathbf{s}_1 \equiv (s_{1x}, s_{1y}, s_{1z} > 0)$, $\mathbf{s}_2 \equiv (s_{2x}, s_{2y}, s_{2z} > 0)$ respectively. In the asymptotic (i.e. far-zone) limit as $kr_1 \to \infty$ and $kr_2 \to \infty$, we obtain, on substituting from Eq. (5.6–8) into Eq. (5.6–44) and on using Eq. (5.6–48), the following expression for the cross-spectral density $W^{(\infty)}$:

$$W^{(\infty)}(r_1\mathbf{s}_1, r_2\mathbf{s}_2, \nu) = \left(\frac{2\pi}{k}\right)^2 \mathscr{A}(\mathbf{s}_{1\perp}, \mathbf{s}_{2\perp}, \nu) \frac{\mathrm{e}^{\mathrm{i}k(r_2 - r_1)}}{r_1 r_2} \cos\theta_1 \cos\theta_2. \tag{5.6–51}$$

Here $\mathbf{s}_{1\perp} \equiv (s_{1x}, s_{1y}, 0)$ and $\mathbf{s}_{2\perp} \equiv (s_{2x}, s_{2y}, 0)$ are the projections, considered as two-dimensional vectors, of the unit vectors $\mathbf{s}_1$ and $\mathbf{s}_2$ on the plane $z = 0$, $\mathscr{A}(\mathbf{s}_{1\perp}, \mathbf{s}_{2\perp}, \nu) \equiv \mathscr{A}(s_{1x}, s_{1y}; s_{2x}, s_{2y}; \nu)$, and θ_1 and θ_2 are the angles which the unit vectors $\mathbf{s}_1$ and $\mathbf{s}_2$ make with the positive z-axis (see Fig. 5.6).

On comparing the formula (5.6–51) with Eq. (5.3–4) we see that the angular correlation function and the radiant cross-intensity are related by the simple formula:

$$L(\mathbf{s}_1, \mathbf{s}_2, \nu) = \left(\frac{2\pi}{k}\right)^2 \mathscr{A}(\mathbf{s}_{1\perp}, \mathbf{s}_{2\perp}, \nu) \cos\theta_1 \cos\theta_2. \tag{5.6–52}$$

The radiant intensity $J(\mathbf{s}, \nu)$ of the field is, according to Eq. (5.2–14), just the 'diagonal element' of the radiant cross-intensity. Hence it follows from Eq. (5.6–52) that the radiant intensity is related to the 'diagonal element' of the angular correlation function by the formula

$$J(\mathbf{s}, \nu) = \left(\frac{2\pi}{k}\right)^2 \mathscr{A}(\mathbf{s}_\perp, \mathbf{s}_\perp, \nu) \cos^2\theta. \tag{5.6–53}$$

If we substitute from Eqs. (5.6–52) and (5.6–53) into Eq. (5.3–12), we obtain the following expression for the spectral degree of coherence of the far field in terms of the angular correlation function:

$$\mu^{(\infty)}(r_1\mathbf{s}_1, r_2\mathbf{s}_2, \nu) = \frac{\mathscr{A}(\mathbf{s}_{1\perp}, \mathbf{s}_{2\perp}, \nu)}{[\mathscr{A}(\mathbf{s}_{1\perp}, \mathbf{s}_{1\perp}, \nu)]^{1/2}[\mathscr{A}(\mathbf{s}_{2\perp}, \mathbf{s}_{2\perp}, \nu)]^{1/2}} \mathrm{e}^{\mathrm{i}k(r_2 - r_1)}. \tag{5.6–54}$$

In view of Eq. (5.6–48), the first factor on the right of Eq. (5.6–54) is evidently a quantitative measure of the correlation that exists between the plane-wave modes of the field which propagate in directions specified by the unit vectors $\mathbf{s}_1$ and $\mathbf{s}_2$. For this reason this factor is sometimes called the *degree of angular correlation* of the field.

So far we have not imposed any restrictions on the fluctuating field, except that it is represented by a statistical ensemble that is stationary, at least in the wide sense, and that it propagates into the half-space $z > 0$. Suppose now that the field is beam-like and that it propagates close to the z-direction. The radiant intensity $J(\mathbf{s}, \nu)$ of such a field will then be appreciable only for those $\mathbf{s}$-directions that lie within a narrow solid angle around the z-axis. This implies, according to Eq. (5.6–53), that $|\mathscr{A}(\mathbf{s}_\perp, \mathbf{s}_\perp, \nu)|$ will only be appreciable when $\mathbf{s}_\perp^2 \ll 1$. Consequently, in view of the inequality

$$|\mathscr{A}(\mathbf{s}_{1\perp}, \mathbf{s}_{2\perp}, \nu)| \leqslant [\mathscr{A}(\mathbf{s}_{1\perp}, \mathbf{s}_{1\perp}, \nu)]^{1/2}[\mathscr{A}(\mathbf{s}_{2\perp}, \mathbf{s}_{2\perp}, \nu)]^{1/2}, \tag{5.6–55}$$

which follows at once from Eq. (5.6–54) and from the fact that the absolute value of the spectral degree of coherence cannot exceed unity‡ [Eq. (4.3–48)], $|\mathscr{A}(\mathbf{s}_{1\perp}, \mathbf{s}_{2\perp}, \nu)|$ will be only appreciable when $|\mathbf{s}_{1\perp}| \ll 1$ and $|\mathbf{s}_{2\perp}| \ll 1$. Hence Eq. (5.6–47)

‡ This argument establishes the inequality (5.6–55) only when $\mathbf{s}_{1\perp}$ and $\mathbf{s}_{2\perp}$ are associated with homogeneous waves of the angular spectrum representation of the field, because in Eq. (5.6–54) $|\mathbf{s}_{1\perp}| \leqslant 1$ and $|\mathbf{s}_{2\perp}| \leqslant 1$. This is the range of interest here. However the inequality (5.6–55) holds in the more general form $|\mathscr{A}(p_1, q_1; p_2, q_2; \nu)| \leqslant [\mathscr{A}(p_1, q_1; p_1, q_1; \nu)]^{1/2}[\mathscr{A}(p_2, q_2; p_2\, q_2; \nu)]^{1/2}$ for all real values of the variables p_1, q_1, p_2, q_2, whether associated with homogeneous waves ($p^2 + q^2 \leqslant 1$) or with evanescent waves ($p^2 + q^2 > 1$). This general relation may be derived directly from the definition (5.6–48) of the angular correlation function by the use of the Schwarz inequality.

is a representation of the cross-spectral density of a beam which propagates close to the z-direction, provided that

$$|\mathscr{A}(p, q; p, q; \nu)| \approx 0 \quad \text{unless } p^2 + q^2 \ll 1. \tag{5.6–56}$$

If we make use of Eq. (5.6–49) we may readily express the condition for a partially coherent beam in terms of a constraint on the boundary value of the cross-spectral density in the plane $z = 0$, namely: $W^{(0)}(x_1, y_1, 0; x_2, y_2, 0; \nu)$ must contain only low spatial-frequency anti-diagonal components, i.e.

$$|\widetilde{W}^{(0)}(-u, -v; u, v; \nu)| \approx 0 \quad \text{unless } u^2 + v^2 \ll k^2. \tag{5.6–57}$$

It follows from Eq. (5.6–46a) that when $p_1^2 + q_1^2 \ll 1$ and $p_2^2 + q_2^2 \ll 1$, m_1 and m_2 may be approximated as follows:

$$m_1 \approx 1 - \tfrac{1}{2}(p_1^2 + q_1^2), \quad m_2 \approx 1 - \tfrac{1}{2}(p_2^2 + q_2^2). \tag{5.6–58}$$

Making use of these approximations in Eq. (5.6–47) we see that the cross-spectral density of a beam, which propagates close to the $+z$-direction, may be expressed in the form

$$W(\mathbf{r}_1, \mathbf{r}_2, \nu) = e^{ik(z_2 - z_1)} \iiiint\limits_{-\infty}^{\infty} \mathscr{A}(p_1, q_1; p_2, q_2; \nu)\, e^{ik(p_2x_2 + q_2y_2 - p_1x_1 - q_1y_1)} \times e^{ik[(p_1^2+q_1^2)z_1 - (p_2^2+q_2^2)z_2]/2}\, dp_1\, dq_1\, dp_2\, dq_2. \tag{5.6–59}$$

We note in passing that the cross-spectral density (5.6–59) of a beam of any state of coherence is identical to the cross-spectral density of an *ensemble of monochromatic beams*, all of the same frequency ν, whose spectral amplitudes $a(p, q, \nu)$ are random variables, with correlation function $\mathscr{A}(p_1, q_1; p_2, q_2; \nu) = \langle a^*(p_1, q_1; \nu) a(p_2, q_2; \nu)\rangle$. This result follows at once on substituting from Eq. (5.6–11) into Eq. (5.6–44), interchanging the operations of averaging and integration and comparing the resulting expression with Eq. (5.6–59).

If we substitute on the right-hand side of Eq. (5.6–59) the expression (5.6–49) for the angular correlation function and change the variables of integration from p_1, q_1, p_2, q_2 to $u_1 = -kp_1$, $v_1 = -kq_1$, $u_2 = kp_2$, $v_2 = kq_2$, we obtain the following representation for the cross-spectral density of the beam throughout the half-space $z > 0$, in terms of the Fourier transform of its boundary values in the plane $z = 0$:

$$W(\mathbf{r}_1, \mathbf{r}_2, \nu) = e^{ik(z_2 - z_1)} \iiiint\limits_{-\infty}^{\infty} \widetilde{W}^{(0)}(u_1, v_1; u_2, v_2; \nu)\, e^{i(u_1x_1 + v_1y_1 + u_2x_2 + v_2y_2)} \times e^{[i/(2k)][(u_1^2+v_1^2)z_1 - (u_2^2+v_2^2)z_2]}\, du_1\, dv_1\, du_2\, dv_2. \tag{5.6–60}$$

We may also readily represent the cross-spectral density of the beam throughout the half-space $z > 0$ in terms of the boundary values of the cross-spectral density itself in the plane $z = 0$, rather than in terms of its Fourier transform. For this purpose we substitute for $\widetilde{W}^{(0)}$ on the right-hand side of Eq. (5.6–60) from Eq. (5.6–50) and interchange the order of integrations. We then obtain the formula

$$W(\mathbf{r}_1, \mathbf{r}_2, \nu) = e^{ik(z_2 - z_1)} \iiiint_{-\infty}^{\infty} W(x_1', y_1', 0; x_2', y_2', 0; \nu) G^*(x_1 - x_1', y_1 - y_1', z_1) \times G(x_2 - x_2', y_2 - y_2', z_2)\, dx_1' \, dy_1' \, dx_2' \, dy_2', \tag{5.6–61}$$

where G is the Green's function (5.6–15), whose explicit form is given by Eq. (5.6–17).

5.6.4 *Gaussian Schell-model beams*

We will now apply some of the general results that we have just established to study the properties of beams generated by planar, secondary, Gaussian, Schell-model sources. Such sources have cross-spectral densities of the form given by Eq. (5.4–7), namely

$$W^{(0)}(\boldsymbol{\rho}_1, \boldsymbol{\rho}_2, \nu) = [S^{(0)}(\boldsymbol{\rho}_1, \nu)]^{1/2}[S^{(0)}(\boldsymbol{\rho}_2, \nu)]^{1/2} g^{(0)}(\boldsymbol{\rho}_2 - \boldsymbol{\rho}_1, \nu), \tag{5.6–62}$$

with

$$S^{(0)}(\boldsymbol{\rho}, \nu) = A^2(\nu)\, e^{-\rho^2/2\sigma_S^2(\nu)}, \tag{5.6–63}$$

$$g^{(0)}(\boldsymbol{\rho}', \nu) = e^{-\rho'^2/2\sigma_g^2(\nu)}, \tag{5.6–64}$$

representing the spectral density and the spectral degree of coherence of the light in the source plane, A, σ_S and σ_g being positive quantities. We have shown in Section 5.4.2 that, with a suitable choice of σ_S and σ_g, such a source can generate a field whose radiant intensity $J(\mathbf{s}, \nu)$ has appreciable values only within a cone of narrow solid angle, i.e. it can generate a beam. Such beams are called *Gaussian Schell-model beams*.

On substituting from Eqs. (5.6–63) and (5.6–64) into Eq. (5.6–62) we obtain for $W^{(0)}$ the expression

$$W^{(0)}(\boldsymbol{\rho}_1, \boldsymbol{\rho}_2, \nu) = A^2\, e^{-(a\rho_1^2 + a\rho_2^2 - 2b\boldsymbol{\rho}_1 \cdot \boldsymbol{\rho}_2)}, \tag{5.6–65}$$

where (if we omit the explicit dependence of some of the quantities on the frequency ν)

$$a = \frac{1}{4\sigma_S^2} + \frac{1}{2\sigma_g^2}, \quad b = \frac{1}{2\sigma_g^2}. \tag{5.6–66}$$

Since $W^{(0)}$ is rotationally symmetric about the origin in the source plane $z = 0$, it is clear that the cross-spectral density $W(\mathbf{r}_1, \mathbf{r}_2, \nu)$ of the field in the half-space $z > 0$ into which the source is assumed to radiate will have rotational symmetry about the z-axis. It is, therefore, convenient to specify the location of each point $\mathbf{r}(x, y, z)$ in the half-space by the variables $\boldsymbol{\rho}$, z, where $\boldsymbol{\rho}$ is a two-dimensional transverse vector with components (x, y). The cross-spectral density $W(\mathbf{r}_1, \mathbf{r}_2, \nu)$ will then become a function of $\boldsymbol{\rho}_1$, z_1, $\boldsymbol{\rho}_2$, z_2 and ν and we will denote it by $W(\boldsymbol{\rho}_1, z_1; \boldsymbol{\rho}_2, z_2; \nu)$. We will denote the four-dimensional Fourier transform, defined by Eq. (5.6–50), of the cross-spectral density function of the field across the source plane by $\widetilde{W}^{(0)}(\mathbf{f}_1, \mathbf{f}_2, \nu)$, rather than by $\widetilde{W}^{(0)}(u_1, v_1; u_2, v_2; \nu)$, where $\mathbf{f}_1 = (u_1, v_1)$ and $\mathbf{f}_2 = (u_2, v_2)$ are two-dimensional spatial-frequency vectors. With this notation the expression (5.6–60) for the cross-spectral density of the field, in the approximation appropriate for a beam, takes the form

$$W(\boldsymbol{\rho}_1, z_1; \boldsymbol{\rho}_2, z_2; \nu) = \\ \mathrm{e}^{\mathrm{i}k(z_2-z_1)}\iint \widetilde{W}^{(0)}(\mathbf{f}_1, \mathbf{f}_2, \nu)\,\mathrm{e}^{\mathrm{i}(\mathbf{f}_1\cdot\boldsymbol{\rho}_1+\mathbf{f}_2\cdot\boldsymbol{\rho}_2)}\,\mathrm{e}^{[\mathrm{i}/(2k)](f_1^2z_1-f_2^2z_2)}\,\mathrm{d}^2f_1\,\mathrm{d}^2f_2, \tag{5.6–67}$$

where, according to Eq. (5.6–50),

$$\widetilde{W}^{(0)}(\mathbf{f}_1, \mathbf{f}_2, \nu) = \frac{1}{(2\pi)^4}\iint W^{(0)}(\boldsymbol{\rho}_1', \boldsymbol{\rho}_2', \nu)\,\mathrm{e}^{-\mathrm{i}(\mathbf{f}_1\cdot\boldsymbol{\rho}_1'+\mathbf{f}_2\cdot\boldsymbol{\rho}_2')}\,\mathrm{d}^2\rho_1'\,\mathrm{d}^2\rho_2'. \tag{5.6–68}$$

The integrations on the right-hand side of Eq. (5.6–67) extend over the complete $\mathbf{f}_1$, $\mathbf{f}_2$-planes and in Eq. (5.6–68) they extend twice independently over the source plane $z = 0$.

On substituting from Eq. (5.6–65) into Eq. (5.6–68) we obtain the following expression for the four-dimensional Fourier transform of the cross-spectral density of the field distribution across a Gaussian Schell-model source:

$$\widetilde{W}^{(0)}(\mathbf{f}_1, \mathbf{f}_2, \nu) = \frac{A^2}{(2\pi)^4}\iint \mathrm{e}^{-(a\rho_1'^2+a\rho_2'^2-2b\boldsymbol{\rho}_1'\cdot\boldsymbol{\rho}_2')}\,\mathrm{e}^{-\mathrm{i}(\mathbf{f}_1\cdot\boldsymbol{\rho}_1'+\mathbf{f}_2\cdot\boldsymbol{\rho}_2')}\,\mathrm{d}^2\rho_1'\,\mathrm{d}^2\rho_2'. \tag{5.6–69}$$

This four-dimensional Fourier transform may be evaluated by a long but straightforward calculation with the help of the formula (5.6–16). The result is:

$$\widetilde{W}^{(0)}(\mathbf{f}_1, \mathbf{f}_2, \nu) = \frac{A^2}{(4\pi)^2(a^2-b^2)}\,\mathrm{e}^{-(\alpha f_1^2+\alpha f_2^2+2\beta\mathbf{f}_1\cdot\mathbf{f}_2)}, \tag{5.6–70}$$

where

$$\alpha = \frac{a}{4(a^2-b^2)}, \qquad \beta = \frac{b}{4(a^2-b^2)}. \tag{5.6–71}$$

We have so far not imposed any restrictions on the parameters to ensure that the Gaussian Schell-model source generates a beam. The required condition is obtained on substituting from Eq. (5.6–70), with $\mathbf{f}_1 = (-u, -v)$, $\mathbf{f}_2 \equiv (u, v)$ into the general beam condition expressed by Eq. (5.6–57). We then find that, for a Gaussian Schell-model source to generate a beam, we must have

$$\mathrm{e}^{-2(\alpha-\beta)f^2} \approx 0 \quad \text{unless } f^2 \ll k^2. \tag{5.6–72}$$

The exponential term in Eq. (5.6–72) has the value unity when $f = 0$ and decreases to the value $1/\mathrm{e}$ when $2(\alpha-\beta)f^2 = 1$. Hence the exponential term will have appreciable value only when $f^2 \lesssim 1/2(\alpha-\beta)$. The requirement (5.6–72) will therefore be satisfied if

$$\frac{1}{2(\alpha-\beta)} \ll k^2. \tag{5.6–72a}$$

If we substitute for α and β from Eqs. (5.6–71) and recall the definitions (5.6–66) of the parameters a and b and also use the relation $k = 2\pi/\lambda$, where λ is the wavelength we obtain the following necessary and sufficient condition for a planar, secondary, Gaussian Schell-model source to generate a beam:

$$\frac{1}{(2\sigma_S)^2} + \frac{1}{\sigma_g^2} \ll \frac{2\pi^2}{\lambda^2}. \tag{5.6–73}$$

We note two extreme cases:

(a) When $\sigma_g \gg \sigma_S$ the source is essentially spatially coherent and the beam condition (5.6–73) then implies that

$$\sigma_S \gg \frac{\lambda}{2\pi\sqrt{2}}. \tag{5.6–74}$$

Hence, roughly speaking, the effective linear dimensions of the source must now be large compared with the wavelength.

(b) When $\sigma_g \ll \sigma_S$ the source is quasi-homogeneous (globally essentially spatially incoherent – cf. Section 5.3.2) and the beam condition implies that

$$\sigma_g \gg \frac{\lambda}{\pi\sqrt{2}}. \tag{5.6–75}$$

Hence the correlation length of the light across the source must now be large compared with the wavelength, i.e. in order to generate a beam, a quasi-homogeneous source must be locally rather coherent.

Let us now return to the more general case. Assuming that the beam condition (5.6–73) is satisfied we obtain at once, on substituting from Eq. (5.6–70) into Eq. (5.6–67), the following integral representation of the cross-spectral density of a beam generated by a Gaussian Schell-model source:

$$W(\boldsymbol{\rho}_1, z_1; \boldsymbol{\rho}_2, z_2; \nu) = \frac{A^2}{(4\pi)^2(a^2 - b^2)} \mathrm{e}^{\mathrm{i}k(z_2 - z_1)} \times \iint \mathrm{e}^{-(\gamma_1 f_1^2 + \gamma_2 f_2^2 + 2\beta \mathbf{f}_1 \cdot \mathbf{f}_2)} \mathrm{e}^{\mathrm{i}(\mathbf{f}_1 \cdot \boldsymbol{\rho}_1 + \mathbf{f}_2 \cdot \boldsymbol{\rho}_2)} \mathrm{d}^2 f_1 \, \mathrm{d}^2 f_2, \tag{5.6–76}$$

where

$$\gamma_1 = \alpha - \frac{\mathrm{i}z_1}{2k}, \quad \gamma_2 = \alpha + \frac{\mathrm{i}z_2}{2k}. \tag{5.6–77}$$

The four-fold Fourier transform on the right-hand side of Eq. (5.6–76) may again be evaluated with the help of the formula (5.6–16) and one then obtains the following expression for the cross-spectral density:

$$W(\boldsymbol{\rho}_1, z_1; \boldsymbol{\rho}_2, z_2; \nu) = \frac{A^2}{16(a^2 - b^2)(\gamma_1\gamma_2 - \beta^2)} \mathrm{e}^{\mathrm{i}k(z_2 - z_1)} \exp\left[-\frac{1}{4(\gamma_1\gamma_2 - \beta^2)}(\gamma_2\rho_1^2 + \gamma_1\rho_2^2 - 2\beta\boldsymbol{\rho}_1 \cdot \boldsymbol{\rho}_2)\right]. \tag{5.6–78}$$

Here the constants α, β, γ_1 and γ_2 are defined by Eqs. (5.6–71) and (5.6–77) with the constants a and b related to the source parameters σ_S and σ_g by Eqs. (5.6–66).

We will now discuss some implications of the formula (5.6–78). In particular we will study the correlations at pairs of points in transverse cross-sections of the beam and also the distribution of the spectral density throughout the beam.

Let us set $z_2 = z_1 = z$ in Eq. (5.6–78) and let us use the abbreviated notation $W(\boldsymbol{\rho}_1, \boldsymbol{\rho}_2, z; \nu)$ for $W(\boldsymbol{\rho}_1, z; \boldsymbol{\rho}_2, z; \nu)$. The formula (5.6–78) then becomes

$$W(\boldsymbol{\rho}_1, \boldsymbol{\rho}_2, z; \nu) = \frac{A^2}{16(a^2 - b^2)(\gamma\gamma^* - \beta^2)} \exp\left[-\frac{1}{4(\gamma\gamma^* - \beta^2)}(\gamma\rho_1^2 + \gamma^*\rho_2^2 - 2\beta\boldsymbol{\rho}_1 \cdot \boldsymbol{\rho}_2)\right], \tag{5.6–79}$$

where

$$\gamma = \alpha + \frac{\mathrm{i}z}{2k} \tag{5.6–80}$$

and γ^* is, of course, the complex conjugate of γ. After a long but straightforward calculation the expression (5.6–79) may be recast into a physically more meaningful form. We will only indicate the main steps of the calculations.

From the definitions (5.6–66) it follows that

$$a^2 - b^2 = \frac{1}{4\sigma_S^2\delta^2}, \tag{5.6–81}$$

where

$$\frac{1}{\delta^2} = \frac{1}{(2\sigma_S)^2} + \frac{1}{\sigma_g^2}. \tag{5.6–82}$$

We note that the parameter δ represents the same quantity which enters the beam condition (5.6–73); we also encountered it earlier [Eq. (5.4–14)] in connection with the equivalence theorem for the radiant intensity.

We deduce from Eqs. (5.6–80), (5.6–71) and (5.6–81) that

$$\gamma\gamma^* - \beta^2 = [\tfrac{1}{2}\sigma_S\delta\Delta(z)]^2, \tag{5.6–83}$$

where

$$\Delta(z) = +[1 + (z/k\sigma_S\delta)^2]^{1/2}. \tag{5.6–84}$$

For reasons that will become apparent shortly the quantity $\Delta(z)$ is sometimes called the *expansion coefficient* of the beam. We note that $\Delta(z) > \Delta(0)$. We find at once from Eqs. (5.6–80), (5.6–83) and (5.6–81) that

$$\frac{A^2}{16(a^2 - b^2)(\gamma\gamma^* - \beta^2)} = \frac{A^2}{[\Delta(z)]^2}. \tag{5.6–85}$$

Also, if we use the expressions (5.6–80) and (5.6–71) for γ, α and β and make use of Eq. (5.6–83), we find that

$$\frac{\gamma\rho_1^2 + \gamma^*\rho_2^2 - 2\beta\boldsymbol{\rho}_1 \cdot \boldsymbol{\rho}_2}{4(\gamma\gamma^* - \beta^2)} = \frac{1}{[\sigma_S\delta\Delta(z)]^2}\left[\frac{a\rho_1^2 + a\rho_2^2 - 2b\boldsymbol{\rho}_1 \cdot \boldsymbol{\rho}_2}{4(a^2 - b^2)} + \frac{\mathrm{i}z}{2k}(\rho_1^2 - \rho_2^2)\right]. \tag{5.6–86}$$

The expression on the right of Eq. (5.6–86) can be expressed in a simpler form. Let us consider first its real part. If we substitute for the parameters a and b from Eq. (5.6–66) and make use of Eq. (5.6–81) we find after a long calculation that

$$\frac{1}{[\sigma_S\delta\Delta(z)]^2}\,\frac{a\rho_1^2 + a\rho_2^2 - 2b\boldsymbol{\rho}_1\cdot\boldsymbol{\rho}_2}{4(a^2 - b^2)} = \frac{(\boldsymbol{\rho}_1 + \boldsymbol{\rho}_2)^2}{8\sigma_S^2[\Delta(z)]^2} + \frac{(\boldsymbol{\rho}_2 - \boldsymbol{\rho}_1)^2}{2\delta^2[\Delta(z)]^2}. \tag{5.6–87}$$

Next let us consider the imaginary part of the expression on the right of Eq. (5.6–86). If we recall the definition of $\Delta(z)$ given by Eq. (5.6–84) we readily find that

$$\frac{iz}{2k[\sigma_S\delta\Delta(z)]^2}(\rho_1^2 - \rho_2^2) = \frac{ik(\rho_1^2 - \rho_2^2)}{2R(z)}, \tag{5.6–88}$$

where

$$R(z) = z\left[1 + \left(\frac{k\sigma_S\delta}{z}\right)^2\right]. \tag{5.6–89}$$

On adding Eqs. (5.6–87) and (5.6–88) and using Eq. (5.6–86) we obtain the following expression for the exponent in the last factor on the right-hand side of Eq. (5.6–79):

$$\frac{\gamma\rho_1^2 + \gamma^*\rho_2^2 - 2\beta\boldsymbol{\rho}_1\cdot\boldsymbol{\rho}_2}{4(\gamma\gamma^* - \beta^2)} = \frac{(\boldsymbol{\rho}_1 + \boldsymbol{\rho}_2)^2}{8\sigma_S^2[\Delta(z)]^2} + \frac{(\boldsymbol{\rho}_2 - \boldsymbol{\rho}_1)^2}{2\delta^2[\Delta(z)]^2} + \frac{ik(\rho_2^2 - \rho_1^2)}{2R(z)}. \tag{5.6–90}$$

Finally, on substituting from Eqs. (5.6–90) and (5.6–85) into Eq. (5.6–79), we obtain the following expression, derived previously by Friberg and Sudol (1982; see also Friberg and Sudol, 1983), for the cross-spectral density in any transverse cross-section of the beam generated by a planar, secondary, Gaussian, Schell-model source:

$$W(\boldsymbol{\rho}_1, \boldsymbol{\rho}_2, z; \nu) = \frac{A^2}{[\Delta(z)]^2}\exp\left[-\frac{(\boldsymbol{\rho}_1 + \boldsymbol{\rho}_2)^2}{8\sigma_S^2[\Delta(z)]^2}\right]\exp\left[-\frac{(\boldsymbol{\rho}_2 - \boldsymbol{\rho}_1)^2}{2\delta^2[\Delta(z)]^2}\right]\exp\left[\frac{ik(\rho_2^2 - \rho_1^2)}{2R(z)}\right]. \tag{5.6–91}$$

As a partial check of Eq. (5.6–91) let us consider its limiting form as $z \to 0$. We see from Eqs. (5.6–84) and (5.6–89) that $\Delta(0) = 1$ and $1/R(0) = 0$ and hence

$$W(\boldsymbol{\rho}_1, \boldsymbol{\rho}_2, 0; \nu) = A^2\exp\left[-\frac{(\boldsymbol{\rho}_1 + \boldsymbol{\rho}_2)^2}{8\sigma_S^2}\right]\exp\left[-\frac{(\boldsymbol{\rho}_2 - \boldsymbol{\rho}_1)^2}{2\delta^2}\right]. \tag{5.6–92}$$

If we recall the definition (5.6–82) of the parameter δ, we can readily verify that Eq. (5.6–92) agrees with the assumed expression (5.6–65) for the cross-spectral density in the source plane $z = 0$.

We see that $W(\boldsymbol{\rho}_1, \boldsymbol{\rho}_2, 0; \nu)$ is the product of three terms, namely the constant factor A^2 and two Gaussian distributions, one being a function of the sum of $\boldsymbol{\rho}_1$ and $\boldsymbol{\rho}_2$ and the other of their difference. On comparing Eqs. (5.6–91) and (5.6–92) we see that, as the light propagates from the source plane $z = 0$ to any transverse plane $z = \text{const.} > 0$, the cross-spectral density retains the same form except that it acquires a phase factor‡ $k(\rho_2^2 - \rho_1^2)/2R(z)$. The first factor A^2 is reduced to $A^2/[\Delta(z)]^2$. The second and the third factor again represent Gaussian distributions of the sum and the difference of the transverse variables, but their effective widths increase with z, σ_S having been replaced by $\sigma_S\Delta(z)$ and δ by $\delta\Delta(z)$. It is for this reason that we referred to $\Delta(z)$ as the expansion coefficient of the beam. The behavior of $\Delta(z)$ and of $R(z)$ for several values of the parameter $q = \sigma_g/\sigma_S$ [Eq. (5.5–11)] that characterizes the degree of global coherence is shown in Fig. 5.21.

Next let us briefly consider the limiting form of the formula (5.6–91), when the Gaussian Schell-model source is spatially completely coherent, i.e. when its spectral degree of coherence $g^{(0)}(\boldsymbol{\rho}') \equiv 1$. In this case, as we see from Eq. (5.6–64), $\sigma_g \to \infty$ and the parameter δ, defined by Eq. (5.6–82), has the value (with subscript c denoting the coherent limit)

$$\delta_c = 2\sigma_S; \tag{5.6–93}$$

and the expressions (5.6–84) and (5.6–89) become

$$\Delta_c(z) = \left[1 + \left(\frac{z}{2k\sigma_S^2}\right)^2\right]^{1/2}, \quad R_c(z) = z\left[1 + \left(\frac{2k\sigma_S^2}{z}\right)^2\right]. \tag{5.6–94}$$

The formula (5.6–91) for the cross-spectral density now reduces to

$$W_c(\boldsymbol{\rho}_1, \boldsymbol{\rho}_2, z; \nu) = \frac{A^2}{[\Delta_c(z)]^2} \exp\left[-\frac{\rho_1^2 + \rho_2^2}{4\sigma_S^2[\Delta_c(z)]^2}\right] \exp\left[\frac{ik(\rho_2^2 - \rho_1^2)}{2R_c(z)}\right]. \tag{5.6–95}$$

We note that this formula may be expressed in the factorized form

$$W_c(\boldsymbol{\rho}_1, \boldsymbol{\rho}_2, z; \nu) = U_c^*(\boldsymbol{\rho}_1, z; \nu) U_c(\boldsymbol{\rho}_2, z; \nu), \tag{5.6–96}$$

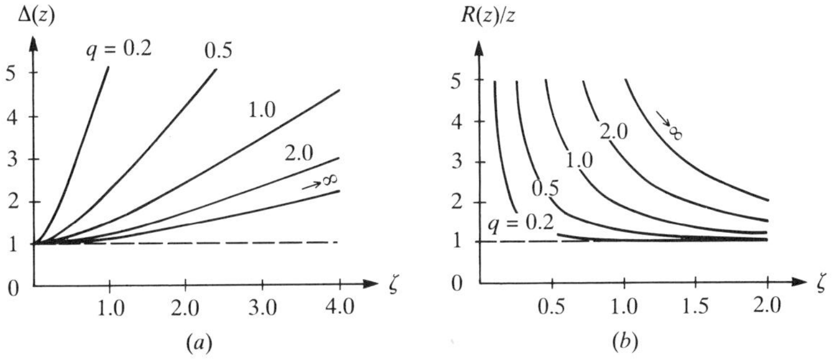

Fig. 5.21 The behavior of (*a*) the expansion coefficient $\Delta(z)$, defined by Eq. (5.6–84), and of (*b*) the quantity $R(z)/z$, defined by Eq. (5.6–89), for a Gaussian Schell-model beam as functions of the normalized distance $\zeta = z/k\sigma_S^2$ from the source plane, for selected values of the degree of global coherence $q = \sigma_g/\sigma_S$. Values $q \gg 1$ and $q \ll 1$ are associated with beams generated by spatially coherent and spatially incoherent sources respectively.

‡ The significance of this phase factor can be understood by considering the coherent-mode decomposition of the beam. One finds (Gori, 1983) that each coherent mode contains a phase factor $k\rho^2/2R(z)$ which is associated with the curvature of its wave front. This, in turn, implies that the cross-spectral density of the beam, in any transverse cross-section, will contain the phase factor $k(\rho_2^2 - \rho_1^2)/2R(z)$.

where

$$U_c(\boldsymbol{\rho}, z; \nu) = \frac{A}{\Delta_c(z)} \exp\left[-\frac{\rho^2}{4\sigma_S^2[\Delta_c(z)]^2}\right] \exp\left[\frac{ik\rho^2}{2R_c(z)}\right] \exp[i\phi_c(z)], \quad (5.6\text{–}97)$$

with $\phi_c(z)$ being some function of z. This function could be determined by considering the coherent limit of the cross-spectral density $W(\boldsymbol{\rho}_1, z_1; \boldsymbol{\rho}_2, z_2; \nu)$ at points $(\boldsymbol{\rho}_1, z_1)$, $(\boldsymbol{\rho}_2, z_2)$ in two different cross-sections, rather than at points $(\boldsymbol{\rho}_1, z)$, $(\boldsymbol{\rho}_2, z)$ in the same cross-section, because when $z_2 = z_1 = z$ the exponential term $\exp[i\phi_c(z)]$ cancels out in the product $U_c^*(\boldsymbol{\rho}_1, z_1; \nu)U_c(\boldsymbol{\rho}_2, z_2; \nu)$. The fact that the expression (5.6–91) now factorizes implies that the beam is spatially completely coherent at frequency ν (see Section 4.5.3) in any transverse cross-section $z = \text{const.} > 0$, as one would expect. Moreover, it may be readily verified that, with an appropriate choice of $\phi_c(z)$, the expression (5.6–97) agrees (except for notation) with our earlier expression (5.6–35) for a monochromatic Gaussian beam.

Let us now return to the more general case of a beam generated by a Gaussian Schell-model source of any state of coherence. If we set $\boldsymbol{\rho}_1 = \boldsymbol{\rho}_2 = \boldsymbol{\rho}$ in Eq. (5.6–91) we obtain the following expression for the spectral density (i.e. the optical intensity at frequency ν) of the beam field:

$$S(\boldsymbol{\rho}, z; \nu) = \frac{A^2}{[\Delta(z)]^2} \exp\left[-\frac{\rho^2}{2\sigma_S^2[\Delta(z)]^2}\right]. \quad (5.6\text{–}98)$$

We see that, in any transverse cross-section of the beam, the spectral density at frequency ν has a Gaussian profile and has the value $A^2/[\Delta(z)]^2$ on the axis $\rho = 0$ of the beam. If we define the *beam radius* $\bar{\rho}_S(z)$ in a cross-section $z = \text{const.} > 0$ as the value of ρ at which the spectral density at frequency ν drops to $1/e$ of its axial value, we see at once from Eq. (5.6–98) that

$$\bar{\rho}_S(z) = \sigma_S \Delta(z) \sqrt{2}. \quad (5.6\text{–}99)$$

Since according to Eq. (5.6–84) $\Delta(0) = 1$,

$$\bar{\rho}_S(0) = \sigma_S \sqrt{2}, \quad (5.6\text{–}100)$$

and it follows from Eqs. (5.6–99), (5.6–100) and (5.6–84) that the beam radius changes with distance z according to the 'beam expansion law'

$$\bar{\rho}_S(z) = [\bar{\rho}_S^2(0) + \bar{\theta}_S^2 z^2]^{1/2}, \quad (5.6\text{–}101)$$

where, if Eq. (5.6–82) is also used,

$$\bar{\theta}_S^2 = \frac{2}{k^2\delta^2} = \frac{2}{k^2}\left[\frac{1}{(2\sigma_S)^2} + \frac{1}{\sigma_g^2}\right]. \quad (5.6\text{–}102)$$

We see from Eq. (5.6–101) that as $z \to \infty$

$$\frac{\bar{\rho}_S(z)}{z} \sim \bar{\theta}_S, \quad (5.6\text{–}103)$$

so that $\bar{\theta}_S$ represents the *angular spread of the beam*.‡

‡ It should be noted that we define the angular spread of the (partially coherent) beam in terms of a $1/e$ drop in the spectral density, whereas for the monochromatic Gaussian beam [Eq. (5.6–40)] we defined it in terms of a $1/e$ drop in field amplitude, as is customary.

In Fig. 5.22 the increase in the beam radius with distance z from the source is shown for selected values of the parameters σ_S and σ_g which, according to Eqs. (5.6–63) and (5.6–64), characterize the effective source size and the effective spectral coherence width of the source. Figure 5.22(a) shows that, for beams with the same initial beam radii, those which are more coherent, i.e. those for which σ_g is larger, are more directional. Figure 5.22(b) shows that, for beams with the same spectral degree of coherence, those which have smaller initial radii are less directional. Figure 5.23 shows the z-dependence of the beam radii for four beams that satisfy the conditions of the equivalence theorem for radiant intensity (Section 5.4.2). We see that, with increasing distance from the sources, the radii of the beams tend to the same value, in spite of the fact that the sources have different effective sizes and different spectral coherence widths.

In addition to the angular spread of the beam we may also determine the entire

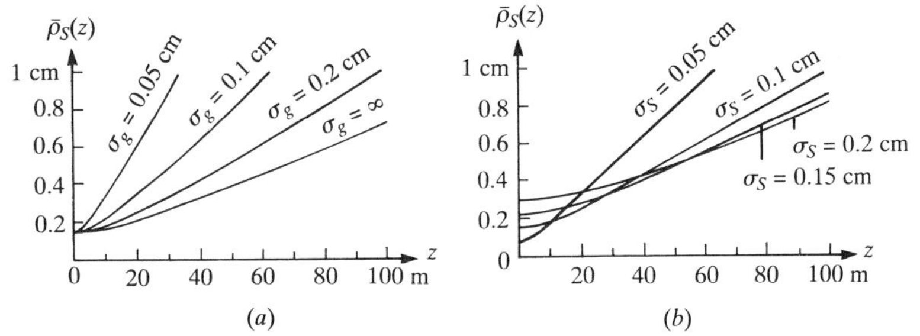

Fig. 5.22 (a) The beam radii for Gaussian Schell-model beams with the same initial radii ($\sigma_S = 0.1$ cm), but with different degrees of coherence, as a function of distance z. The wavelength of each beam was taken to be 6328 Å. (b) The beam radii for beams with the same initial spectral degree of coherence ($\sigma_g = 0.2$ cm), but with different initial radii. The wavelength of each beam was again taken to be 6328 Å. (Adapted from Foley and Zubairy, 1978.)

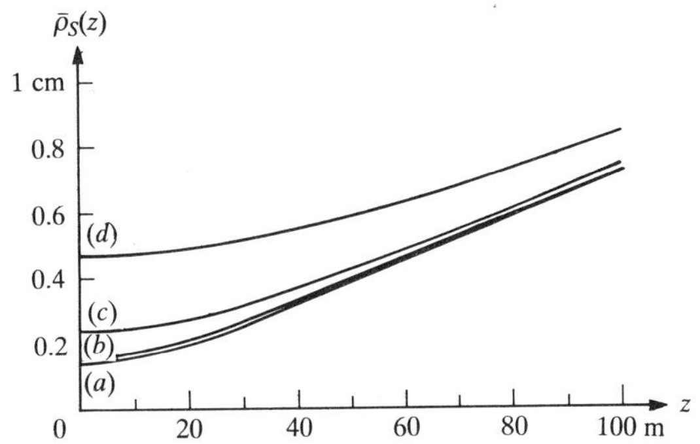

Fig. 5.23 The beam radii of Gaussian Schell-model beams as functions of the distance z from four sources which satisfy the conditions of the equivalence theorem for radiant intensity (cf. Sec. 5.4.2) and hence have the same angular spread $\bar{\theta}_S$.

The parameters for the sources generating these beams are: (a) $\sigma_S = 0.1$ cm and $\sigma_g = \infty$, (b) $\sigma_S = 0.109$ cm and $\sigma_g = 0.5$ cm, (c) $\sigma_S = 0.167$ cm and $\sigma_g = 0.25$ cm, (d) $\sigma_S = 0.328$ cm and $\sigma_g = 0.21$ cm. The wavelength for each beam is 6328 Å. (Adapted from Foley and Zubairy, 1978.)

angular distribution of the radiant intensity. To do so we make use of Eq. (5.2–12) and the fact that, when the field is beam-like, we may replace r by z in the far-zone limit, i.e.

$$J(\mathbf{s}, \nu) = \lim_{kz\to\infty} z^2 S(\boldsymbol{\rho}, z; \nu), \tag{5.6–104}$$

where the direction $\mathbf{s}$, and consequently the angle $\theta = \rho/z$, are fixed. Now according to Eq. (5.6–84),

$$\Delta(z) \sim \frac{z}{k\sigma_S\delta} \quad \text{as } kz \to \infty \tag{5.6–105}$$

and, if we substitute from Eq. (5.6–105) into Eq. (5.6–98) and then substitute the resulting expression into Eq. (5.6–104), we find that

$$J(\mathbf{s}, \nu) \sim (kA\sigma_S\delta)^2 \, \mathrm{e}^{-(k\delta)^2\theta^2/2}. \tag{5.6–106}$$

This formula is in agreement with the 'beam-limit' ($\cos\theta \approx 1$, $\sin\theta \approx \theta$) of Eq. (5.4–16) that we derived earlier in a different manner, in connection with the equivalence theorem for the radiant intensity generated by Gaussian Schell-model sources.

Let us now turn our attention to the spectral degree of coherence of the beam. One finds on substituting from Eqs. (5.6–91) into Eq. (4.3–47) and making use of Eq. (5.6–82) that the spectral degree of coherence of the beam in any transverse cross-section is given by

$$\mu(\boldsymbol{\rho}_1, \boldsymbol{\rho}_2, z; \nu) = \exp\left[-\frac{(\boldsymbol{\rho}_2 - \boldsymbol{\rho}_1)^2}{2\sigma_g^2[\Delta(z)]^2}\right]\exp\left[\frac{ik(\rho_2^2 - \rho_1^2)}{2R(z)}\right]. \tag{5.6–107}$$

We see that the absolute value of μ is given by a Gaussian distribution. Since according to Eq. (5.6–84) $\Delta(z)$ increases with z, the coherence area of the light in a transverse cross-section of the beam also increases with z. If we define the *spectral coherence width* $\bar{\rho}_\mu(z)$ of the beam as that separation $|\boldsymbol{\rho}_2 - \boldsymbol{\rho}_1|$ of points in a transverse cross-section at which $|\mu|$ drops from its maximum value unity (for $|\boldsymbol{\rho}_2 - \boldsymbol{\rho}_1| = 0$) to the value $1/\mathrm{e}$, we see at once from Eq. (5.6–107) that

$$\bar{\rho}_\mu(z) = \sigma_g \Delta(z)\sqrt{2}. \tag{5.6–108}$$

On comparing this expression for the spectral coherence width with the formula (5.6–99) for the beam radius, we see that both obey the same 'expansion law' and that

$$\frac{\bar{\rho}_\mu(z)}{\bar{\rho}_S(z)} = \frac{\sigma_g}{\sigma_S}. \tag{5.6–109}$$

This formula implies that *the ratio of the spectral coherence width of the light in any transverse cross-section of the beam to the beam radius in that cross-section is constant on propagation*, i.e. it is the same for all cross-sections. Moreover this constant ratio is just the degree of global coherence of the source [Eq. (5.5–11)]. More generally, the ratio $\bar{\rho}_\mu(z')/\bar{\rho}_S(z')$ may be interpreted as the degree of global coherence of the light in the plane $z = z'$. Hence Eq. (5.6–109) implies that *the degree of global coherence of the light in any transverse cross-section of a Gaussian Schell-model beam is invariant on propagation*.

If we substitute into Eq. (5.6–108) the expression (5.6–84) for $\Delta(z)$ and also make use of the expression (5.6–82), we obtain the following explicit formula for the dependence of the spectral coherence width (at frequency ν) on z:

$$\bar{\rho}_\mu(z) = [\bar{\rho}_\mu^2(0) + \bar{\theta}_\mu^2 z^2]^{1/2}, \tag{5.6–110}$$

where

$$\bar{\rho}_\mu(0) = \sigma_g \sqrt{2} \tag{5.6–111}$$

and

$$\bar{\theta}_\mu^2 = \frac{2}{k^2}\left(\frac{\sigma_g}{\sigma_S}\right)^2 \frac{1}{\delta^2} = \frac{2}{k^2}\left(\frac{\sigma_g}{\sigma_S}\right)^2 \left[\frac{1}{(2\sigma_S)^2} + \frac{1}{\sigma_g^2}\right]. \tag{5.6–112}$$

It follows from Eq. (5.6–110) that as $z \to \infty$

$$\frac{\bar{\rho}_\mu(z)}{z} \sim \bar{\theta}_\mu. \tag{5.6–113}$$

Hence $\bar{\theta}_\mu$ represents the semi-angle of the solid-angle cone, with vertex at the origin, within which the light in any transverse cross-section in the far zone is essentially spatially coherent at frequency ν. For this reason $\bar{\theta}_\mu$ is sometimes called the *far-zone coherence angle*.

It follows at once from Eqs. (5.6–111) and (5.6–102) that

$$\bar{\rho}_\mu(0)\bar{\theta}_S = \frac{1}{k}\left[4 + \left(\frac{\sigma_g}{\sigma_S}\right)^2\right]^{1/2}, \tag{5.6–114}$$

and from Eqs. (5.6–100) and (5.6–112) that

$$\bar{\rho}_S(0)\bar{\theta}_\mu = \frac{1}{k}\left[4 + \left(\frac{\sigma_g}{\sigma_S}\right)^2\right]^{1/2}. \tag{5.6–115}$$

These formulas bring into evidence two *reciprocity relations* between a Gaussian Schell-model source which generates a beam and the far-zone behavior of the beam:‡ The first [Eq. (5.6–114)] shows that the angular beam spread is inversely proportional to the spectral coherence width of the light in the source plane. The other [Eq. (5.6–115)] shows that the far-zone coherence angle is inversely proportional to the effective linear size of the source.

If we substitute in Eqs. (5.6–114) and (5.6–115) for $\bar{\rho}_\mu(0)$ and $\bar{\rho}_S(0)$ from Eqs. (5.6–111) and (5.6–100) and recall that $k = 2\pi/\lambda$, the reciprocity relations may be expressed in the form

$$\bar{\theta}_S = \left(\frac{\lambda}{2\pi\sqrt{2}}\right)\frac{1}{\sigma_g}\left[4 + \left(\frac{\sigma_g}{\sigma_S}\right)^2\right]^{1/2}, \tag{5.6–116a}$$

$$\bar{\theta}_\mu = \left(\frac{\lambda}{2\pi\sqrt{2}}\right)\frac{1}{\sigma_S}\left[4 + \left(\frac{\sigma_g}{\sigma_S}\right)^2\right]^{1/2}. \tag{5.6–116b}$$

We note that, in the special case when the source is globally very coherent

‡ Reciprocity relations that apply to a broader class of partially coherent sources and the far fields that they generate were formulated by Friberg and Wolf (1983).

($\sigma_g \gg \sigma_S$), Eqs. (5.6–116a) and (5.6–116b) reduce to

$$\bar{\theta}_S \approx \left(\frac{\lambda}{2\pi\sqrt{2}}\right)\frac{1}{\sigma_S}, \quad \bar{\theta}_\mu \approx \left(\frac{\lambda}{2\pi\sqrt{2}}\right)\left(\frac{\sigma_g}{\sigma_S}\right)\frac{1}{\sigma_S}. \tag{5.6–117}$$

When the source is globally rather incoherent ($\sigma_g \ll \sigma_S$), i.e. when it is a quasi-homogeneous source, Eqs. (5.6–116a) and (5.6–116b) reduce to

$$\bar{\theta}_S \approx \left(\frac{\lambda}{\pi\sqrt{2}}\right)\frac{1}{\sigma_g}, \quad \bar{\theta}_\mu \approx \left(\frac{\lambda}{\pi\sqrt{2}}\right)\frac{1}{\sigma_S}. \tag{5.6–118}$$

Some of the theoretical results discussed in this section were tested experimentally by Farina, Narducci and Collett (1980). A secondary Gaussian quasi-homogeneous source was obtained by illuminating a suitable phase screen with a helium-neon (He:Ne) laser beam which was broadened and collimated by a beam-expanding telescope. The phase screen was produced by spraying a finishing mist on a clear glass substrate. The spray gave rise to a uniform rough coating

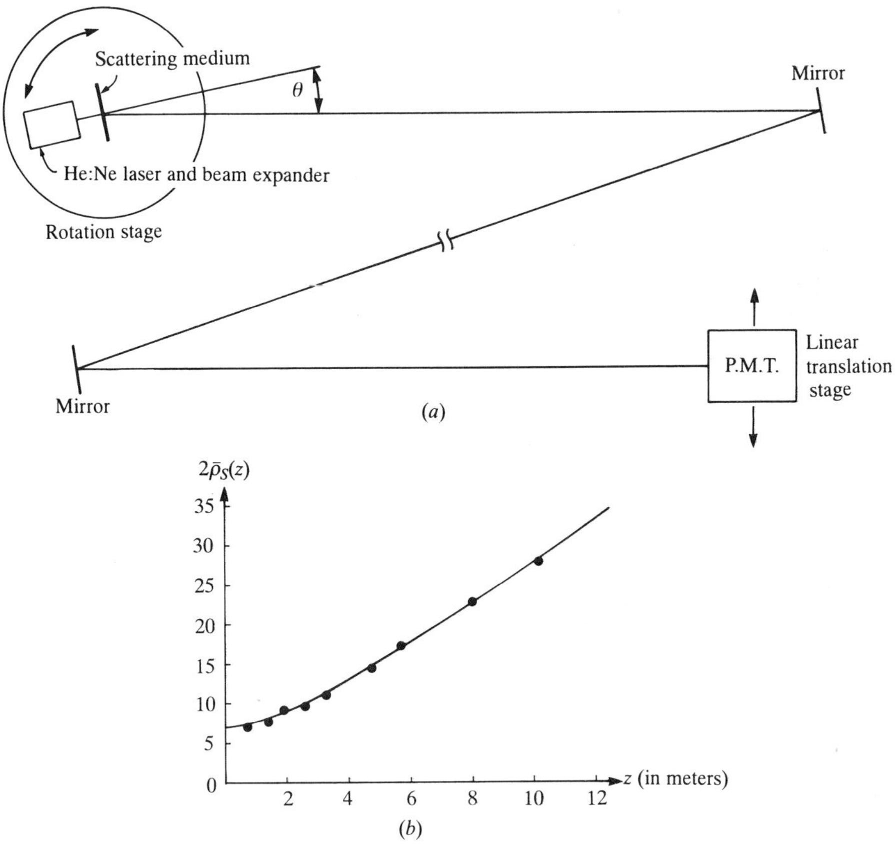

Fig. 5.24 (*a*) The experimental set up for measurement of the angular distribution of the radiant intensity generated by planar, secondary, Gaussian, quasi-homogeneous sources. (*b*) The effective beam diameter $2\bar{\rho}_S(z)$ as a function of the distance z from the source. The solid line in (*b*) was calculated from theory. The dots represent results of measurements (Adapted from Farina, Narducci and Collett, 1980.)

whose thickness could be controlled by varying the amount of deposition. The coated plate was mounted on the rotating shaft of a synchronous motor, which turned the plate without introducing significant vibrations. When the rotating plate was illuminated by the laser, the light in the plane immediately behind and parallel to the plate was a secondary source with the required properties. This source generated a beam which was allowed to propagate different path lengths with the help of arrays of mirrors [Fig. 5.24(*a*)]. Measurements on the beam were then made with a cooled photomultiplier mounted on a linear translation stage. Using this device the beam diameters were determined at different distances from the secondary source. The results, shown in Fig. 5.24(*b*), were found to be in good agreement with the theory. Other theoretical predictions, relating to beams generated by secondary Gaussian quasi-homogeneous sources, were also verified with this experimental set up.

5.7 Foundations of radiometry

Up to now we have studied the properties of fields generated by fluctuating sources within the framework of second-order classical coherence theory. The basic ingredients of that theory are the wave equation of physical optics and some elementary statistical considerations. There is, however, a much older model that is sometimes used for describing and analyzing the radiation characteristics of sources, known as radiometry.[‡] It centers around the notion that, in a region of space which contains radiation, energy propagates along geometrical trajectories, i.e. along light rays. This model is mainly used in connection with thermal sources and has been very successful in the analysis of problems which are encountered, for example, in illumination engineering. A generalization of radiometry, *the theory of radiative energy transfer*, extends the radiometric model to propagation in both deterministic and random media.[§] This theory is a basic tool of astrophysics and has numerous applications in other fields.

Radiometry and the theory of radiative energy transfer have developed quite independently of modern theories of radiation and, in spite of their long history, their foundations have not yet been fully clarified (cf. Wolf, 1978). In this section we will discuss results of investigations that clarify this question to some extent. We will largely confine ourselves to elucidating the connection between some of the basic concepts of these older disciplines and those of optical coherence theory.

We begin with a brief discussion of energy conservation in scalar wavefields.

5.7.1 Energy density, energy flux and the energy conservation law in scalar wavefields

Two important quantities that are associated with any electromagnetic field are the energy density and the energy flux density, the first of which is a scalar while

‡ An account of the early work in this field is given in an article by Trotter (1919). Radiometry appears to have been systematized in the early years of the twentieth century, in connection with the theory of heat radiation. Max Planck's *The Theory of Heat Radiation* (1959), the first edition of which was published in 1906, remains to this day one of the most comprehensive accounts of this subject.

§ For accounts of the theory of radiative energy transfer see, for example, Hopf (1934) or Chandrasekhar (1960). Some of the basic papers on this subject are reprinted in Menzel (1966).

the second is a vector. They are related by a well-known conservation law, sometimes known as Poynting's theorem (see, for example Born and Wolf, 1980, Sec. 1.1.4). With scalar wavefields one can also associate scalar and vector densities, which obey a similar conservation law and which may be regarded as analogues of the energy density and the energy flux density of an electromagnetic field. In this section we will introduce these quantities for scalar wavefields, derive the corresponding conservation law and discuss some of its implications. We will consider both deterministic and random wavefields.

Let $V(\mathbf{r}, t)$ be a complex deterministic wavefield in a region in free space. It satisfies the wave equation

$$\nabla^2 V(\mathbf{r}, t) = \frac{1}{c^2}\frac{\partial^2 V(\mathbf{r}, t)}{\partial t^2}, \tag{5.7–1}$$

c being the speed of light in vacuum. General field-theoretical considerations suggest (see, for example, Wentzel, 1949, Sec. 8) that the (real) scalar

$$H(\mathbf{r}, t) = \alpha\left[\frac{1}{c^2}\frac{\partial V^*}{\partial t}\frac{\partial V}{\partial t} + \boldsymbol{\nabla} V^* \cdot \boldsymbol{\nabla} V\right] \tag{5.7–2}$$

can be identified with the energy density and the (real) vector

$$\mathbf{F}(\mathbf{r}, t) = -\alpha\left[\frac{\partial V^*}{\partial t}\boldsymbol{\nabla} V + \frac{\partial V}{\partial t}\boldsymbol{\nabla} V^*\right] \tag{5.7–3}$$

with the energy flux density vector. Here α is a positive constant, whose value depends on the choice of units.

If we take the time derivative of H and the divergence of $\mathbf{F}$ and use Eq. (5.7–1), we readily find that

$$\frac{\partial H(\mathbf{r}, t)}{\partial t} + \boldsymbol{\nabla} \cdot \mathbf{F}(\mathbf{r}, t) = 0. \tag{5.7–4}$$

Let us integrate Eq. (5.7–4) throughout a domain D. This gives

$$\int_D \frac{\partial H}{\partial t}\,\mathrm{d}^3 r + \int_D \boldsymbol{\nabla} \cdot \mathbf{F}\,\mathrm{d}^3 r = 0. \tag{5.7–5}$$

If in the first term we interchange the operations of integration and differentiation and apply Gauss' theorem to the second term, Eq. (5.7–5) becomes

$$\frac{\mathrm{d}}{\mathrm{d}t}\int_D H(\mathbf{r}, t)\,\mathrm{d}^3 r + \int_\sigma \mathbf{F}(\mathbf{r}, t) \cdot \mathbf{n}\,\mathrm{d}\sigma = 0, \tag{5.7–6}$$

where the second integral is taken over the surface σ bounding the volume D and $\mathbf{n}$ is the outward unit normal to D at the element $\mathrm{d}\sigma$. The formula (5.7–6) may be given the following interpretation. The rate of increase (or decrease) of the energy contained in D at any given instant t is equal to the rate at which energy enters (or leaves) D through the boundary σ. With this interpretation Eq. (5.7–6) represents the *energy conservation law* of the field. The formula (5.7–4) is the differential form of this law.

The physical significance of the energy flux vector $\mathbf{F}(\mathbf{r}, t)$ must be interpreted with some caution, as can be seen from the following considerations. According to elementary vector calculus one has the identity $\nabla \cdot (\nabla \times \mathbf{f}) \equiv 0$, where $\mathbf{f}(\mathbf{r}, t)$ is any vector function which is twice continuously differentiable with respect to the spatial variables. Hence Eqs. (5.7–4) and (5.7–6) remain unchanged if one adds to $\mathbf{F}$ the curl of any sufficiently well-behaved vector field $\mathbf{f}$, showing that the flux density vector $\mathbf{F}$, defined by Eq. (5.7–3), is not the only vector that is consistent with energy conservation. These remarks indicate that one cannot regard $\mathbf{F}(\mathbf{r}, t)$ as representing the rate of energy flow at the point $\mathbf{r}$, at time t. It is only the integral of the normal component of the energy flux vector, taken over any closed surface in the region of space containing the field, that has an unambiguous physical meaning.‡

Let us next assume that the complex function $V(\mathbf{r}, t)$ is an analytic signal which represents a monochromatic field of frequency ν, i.e.

$$V(\mathbf{r}, t) = U(\mathbf{r}, \nu)\, \mathrm{e}^{-2\pi i \nu t}. \tag{5.7–7}$$

The expressions (5.7–2) and (5.7–3) for the energy density and for the energy flux density vector now become, (with $k = 2\pi\nu/c$),

$$H_\nu(\mathbf{r}) = \alpha k^2 \left[U^* U + \frac{1}{k^2} \nabla U^* \cdot \nabla U \right], \tag{5.7–8}$$

$$\mathbf{F}_\nu(\mathbf{r}) = -i\alpha k c [U^* \nabla U - U \nabla U^*], \tag{5.7–9}$$

where we have written $H_\nu(\mathbf{r})$ and $\mathbf{F}_\nu(\mathbf{r})$ in place of $H(\mathbf{r}, t)$ and $\mathbf{F}(\mathbf{r}, t)$ to stress that these quantities are now independent of t but depend on the frequency ν. The differential form [Eq. (5.7–4)] of the energy conservation law now becomes

$$\nabla \cdot \mathbf{F}_\nu(\mathbf{r}) = 0. \tag{5.7–10}$$

So far we have considered the field variable to be deterministic. Suppose now that it is random and that it represents a fluctuating optical field. Assuming that the fluctuations may be represented by a statistical ensemble which is stationary, at least in the wide sense, the cross-spectral density of the field may be expressed in the form [cf. Eq. (4.7–38)]

$$W(\mathbf{r}_1, \mathbf{r}_2, \nu) = \langle U^*(\mathbf{r}_1, \nu) U(\mathbf{r}_2, \nu) \rangle. \tag{5.7–11}$$

Here the angular brackets denote the expectation value, taken over an appropriate ensemble of strictly monochromatic wave fields, all of frequency ν. The average energy density and energy flux density at frequency ν are then given by the expectation values of the expressions (5.7–8) and (5.7–9), namely by

$$\langle H_\nu(\mathbf{r}) \rangle = \alpha k^2 \left[\langle U^* U \rangle + \frac{1}{k^2} \langle \nabla U^* \cdot \nabla U \rangle \right] \tag{5.7–12}$$

and

$$\langle \mathbf{F}_\nu(\mathbf{r}) \rangle = -i\alpha k c [\langle U^* \nabla U \rangle - \langle U \nabla U^* \rangle]. \tag{5.7–13}$$

Furthermore on taking the ensemble average of Eq. (5.7–10) and interchanging

‡ In this connection the following passage from Lorentz (1909, p. 25) seems worth quoting: 'in general it will not be possible to trace the path of parts or elements of energy in the same sense in which we can follow in their course the ultimate particles of which energy is made up.'

the order of averaging and differentiation we obtain the differential form of the conservation law for random, statistical stationary wavefields:

$$\nabla \cdot \langle \mathbf{F}_\nu(\mathbf{r}) \rangle = 0. \tag{5.7–14}$$

If we make use of Eq. (5.7–11) we may express $\langle H_\nu \rangle$ and $\langle \mathbf{F}_\nu \rangle$ in terms of the cross-spectral density rather than in terms of the ensemble $\{U\}$, and we then find that

$$\langle H_\nu(\mathbf{r}) \rangle = \alpha k^2 \left[W(\mathbf{r}, \mathbf{r}, \nu) + \frac{1}{k^2} \operatorname*{Lim}_{\mathbf{r}_1 \to \mathbf{r}} \operatorname*{Lim}_{\mathbf{r}_2 \to \mathbf{r}} \nabla_2 \cdot \nabla_1 W(\mathbf{r}_1, \mathbf{r}_2, \nu) \right], \tag{5.7–15}$$

and

$$\langle \mathbf{F}_\nu(\mathbf{r}) \rangle = -\mathrm{i}\alpha k c \operatorname*{Lim}_{\mathbf{r}_1 \to \mathbf{r}} \operatorname*{Lim}_{\mathbf{r}_2 \to \mathbf{r}} [\nabla_2 W(\mathbf{r}_1, \mathbf{r}_2, \nu) - \nabla_1 W(\mathbf{r}_1, \mathbf{r}_2, \nu)]. \tag{5.7–16}$$

An equivalent expression for the average energy flux density vector which involves one rather than two limiting processes is sometimes useful. One can derive it in the following way. We set

$$\mathbf{r} = \tfrac{1}{2}(\mathbf{r}_1 + \mathbf{r}_2), \quad \mathbf{r}' = \mathbf{r}_2 - \mathbf{r}_1. \tag{5.7–17}$$

We then have $W(\mathbf{r}_1, \mathbf{r}_2, \nu) = W(\mathbf{r} - \tfrac{1}{2}\mathbf{r}', \mathbf{r} + \tfrac{1}{2}\mathbf{r}', \nu)$ where

$$W(\mathbf{r} - \tfrac{1}{2}\mathbf{r}', \mathbf{r} + \tfrac{1}{2}\mathbf{r}', \nu) = \langle U^*(\mathbf{r} - \tfrac{1}{2}\mathbf{r}', \nu) U(\mathbf{r} + \tfrac{1}{2}\mathbf{r}', \nu) \rangle. \tag{5.7–18}$$

If we apply to both sides of Eq. (5.7–18) the gradient operator ∇', taken with respect to the 'difference variable' $\mathbf{r}'$, interchange the orders of differentiation and averaging and then proceed to the limit $\mathbf{r}' \to 0$, we obtain the formula

$$\operatorname*{Lim}_{\mathbf{r}' \to 0} \nabla' \{W(\mathbf{r} - \tfrac{1}{2}\mathbf{r}', \mathbf{r} + \tfrac{1}{2}\mathbf{r}', \nu)\} = \tfrac{1}{2} \langle U^*(\mathbf{r}) \nabla U(\mathbf{r}) - U(\mathbf{r}) \nabla U^*(\mathbf{r}) \rangle.$$

On comparing this formula with Eq. (5.7–13) we obtain an alternative expression for the average energy flux density vector:

$$\langle \mathbf{F}_\nu(\mathbf{r}) \rangle = -2\mathrm{i}\alpha k c \operatorname*{Lim}_{\mathbf{r}' \to 0} \nabla' W(\mathbf{r} - \tfrac{1}{2}\mathbf{r}', \mathbf{r} + \tfrac{1}{2}\mathbf{r}', \nu). \tag{5.7–19}$$

There is also the following alternative expression for $\langle H_\nu(\mathbf{r}) \rangle$, involving a single limiting process, which can be derived from Eq. (5.7–15):

$$\langle H_\nu(\mathbf{r}) \rangle = -2\alpha \operatorname*{Lim}_{\mathbf{r}' \to 0} \nabla'^2 W(\mathbf{r} - \tfrac{1}{2}\mathbf{r}', \mathbf{r} + \tfrac{1}{2}\mathbf{r}', \nu). \tag{5.7–20}$$

Let us now suppose that the field point $P(\mathbf{r})$ is in the far zone of a planar radiating source which occupies a finite portion of the plane $z = 0$ and radiates into the half-space $z > 0$. We will show that there exist simple relations between the energy density, the energy flux density vector and the spectral density at the point P. To demonstrate this we first represent each member of the statistical ensemble $\{U(\mathbf{r}, \nu)\}$ that express the field throughout the half-space $z > 0$ in the form of an angular spectrum of plane waves [Eq. (3.2–19), with k written in place of k_0 and with other trivial changes in notation], namely

$$U(\mathbf{r}, \nu) = \int a(\mathbf{s}'_\perp, \nu)\, \mathrm{e}^{\mathrm{i}k\mathbf{s}' \cdot \mathbf{r}}\, \mathrm{d}^2 \mathbf{s}'_\perp, \tag{5.7–21}$$

where $\mathbf{s}' \equiv (s'_x, s'_y, s'_z)$, $\mathbf{s}'_\perp \equiv (s'_x, s'_y, 0)$ and

$$s'_z = +[1 - s'^2_x - s'^2_y]^{1/2} \quad \text{when } s'^2_x + s'^2_y \leqslant 1 \tag{5.7–22a}$$

$$= +\,\mathrm{i}[s'^2_x + s'^2_y - 1]^{1/2} \quad \text{when } s'^2_x + s'^2_y > 1. \tag{5.7–22b}$$

It follows at once from Eq. (5.7–21) that

$$\nabla U(\mathbf{r}, \nu) = \mathrm{i}k \int \mathbf{s}' a(\mathbf{s}'_\perp, \nu)\, \mathrm{e}^{\mathrm{i}k\mathbf{s}'\cdot\mathbf{r}}\, \mathrm{d}^2\mathbf{s}'_\perp. \tag{5.7–23}$$

We have shown earlier [Eq. (3.2–22)] that the far-zone form of the expression (5.7–21), as the point $\mathbf{r} = r\mathbf{s}$, $(\mathbf{s}^2 = 1)$ recedes to the far zone $(kr \to \infty)$ in a fixed direction specified by a unit vector $\mathbf{s}$, is

$$U^{(\infty)}(r\mathbf{s}, \nu) = -\frac{2\pi \mathrm{i}}{k} \cos\theta\, a(\mathbf{s}_\perp, \nu) \frac{\mathrm{e}^{\mathrm{i}kr}}{r}, \tag{5.7–24}$$

From this expression, or by applying the principle of stationary phase to the integral (5.7–23), we find that

$$\nabla U^{(\infty)}(r\mathbf{s}, \nu) = 2\pi \cos\theta\, a(\mathbf{s}_\perp, \nu) \frac{\mathrm{e}^{\mathrm{i}kr}}{r} \mathbf{s}. \tag{5.7–25}$$

In these formulas θ is the angle which the unit vector $\mathbf{s}$ makes with the positive z-axis and $\mathbf{s}_\perp$ is the projection, considered as a two-dimensional vector, of the unit vector $\mathbf{s}$ onto the source plane $z = 0$ (see Fig. 5.18). On substituting from Eqs. (5.7–24) and (5.7–25) into the formulas (5.7–12) and (5.7–13) we obtain the following expressions for the average energy density and the average energy flux density vector in the far zone of the source:

$$\langle H^{(\infty)}_\nu(\mathbf{r})\rangle = 2(2\pi)^2 \alpha \mathscr{A}(\mathbf{s}_\perp, \mathbf{s}_\perp, \nu) \frac{\cos^2\theta}{r^2}, \tag{5.7–26}$$

$$\langle \mathbf{F}^{(\infty)}_\nu(\mathbf{r})\rangle = 2(2\pi)^2 \alpha c \mathscr{A}(\mathbf{s}_\perp, \mathbf{s}_\perp, \nu) \frac{\cos^2\theta}{r^2} \mathbf{s}. \tag{5.7–27}$$

In these formulas

$$\mathscr{A}(\mathbf{s}_\perp, \mathbf{s}_\perp, \nu) = \langle a^*(\mathbf{s}_\perp, \nu) a(\mathbf{s}_\perp, \nu)\rangle \tag{5.7–28}$$

is the 'diagonal element' of the angular correlation function of the radiated field [cf. Eq. (5.6–48)]. It is related to the cross-spectral density of the field in the source plane by Eq. (5.6–49). Now it follows from Eqs. (5.6–53) and (5.2–12) that the spectral density of the far field, $S^{(\infty)}(r\mathbf{s}, \nu) = W^{(\infty)}(r\mathbf{s}, r\mathbf{s}, \nu) = J(\mathbf{s}, \nu)/r^2$, is expressible in the form

$$S^{(\infty)}(r\mathbf{s}, \nu) = \left(\frac{2\pi}{k}\right)^2 \mathscr{A}(\mathbf{s}_\perp, \mathbf{s}_\perp, \nu) \frac{\cos^2\theta}{r^2}. \tag{5.7–29}$$

So far the constant α has been arbitrary. Let us now make the choice

$$\alpha = \frac{1}{2ck^2} = \frac{c}{8\pi^2\nu^2}. \tag{5.7–30}$$

We deduce at once from Eqs. (5.7–26), (5.7–27) and (5.7–29) that with this choice

$$\langle H_\nu^{(\infty)}(r\mathbf{s})\rangle = \frac{1}{c}S^{(\infty)}(r\mathbf{s}, \nu) \tag{5.7–31}$$

and

$$\langle \mathbf{F}_\nu^{(\infty)}(r\mathbf{s})\rangle = S^{(\infty)}(r\mathbf{s}, \nu)\mathbf{s}. \tag{5.7–32}$$

The formulas (5.7–31) and (5.7–32) imply that in the far zone we may picture energy transport as propagation of energy along straight lines in the outward radial directions (i.e. in directions pointing from the origin to the field point) with the vacuum speed of light. We stress that this simple model is, in general, only appropriate for energy transport in the far zone of the radiating source.

From the significance of the flux density vector it is clear that the rate at which energy at frequency ν is radiated by the source into the half-space $z > 0$ is given by

$$\mathscr{F}_\nu = \int_{(2\pi)} \langle \mathbf{F}_\nu^{(\infty)}(r\mathbf{s})\rangle \cdot \mathbf{s} r^2 \, \mathrm{d}\Omega, \tag{5.7–33}$$

where the integration extends over the 2π-solid angle generated by all the unit vectors $\mathbf{s}$ pointing into the half-space $z > 0$. If we substitute into this formula the expression (5.7–32) for the averaged flux density vector, we obtain for $\mathscr{F}_\nu$ the expression

$$\mathscr{F}_\nu = \int_{(2\pi)} r^2 S^{(\infty)}(r\mathbf{s}, \nu)\, \mathrm{d}\Omega. \tag{5.7–34}$$

Now according to Eq. (5.2–12), the integrand on the right-hand side of Eq. (5.7–34) is just the radiant intensity $J(\mathbf{s}, \nu)$. Hence $\mathscr{F}_\nu$ may also be expressed in the form

$$\mathscr{F}_\nu = \int_{(2\pi)} J(\mathbf{s}, \nu)\, \mathrm{d}\Omega. \tag{5.7–35}$$

This formula shows that the radiant intensity, which we introduced in Section 5.2.1 from considerations of the far-zone behavior of the spectral density, represents the rate at which the source radiates energy at frequency ν per unit solid angle around the $\mathbf{s}$-direction.

5.7.2 Basic concepts of radiometry

A basic assumption of radiometry may be expressed by the following elementary radiometric law. The rate $\mathrm{d}^2\mathscr{F}_\nu$ at which energy at frequency ν is radiated into an element $\mathrm{d}\Omega$ of solid angle by an element $\mathrm{d}\sigma$ of a planar steady-state source σ is given by the expression

$$\mathrm{d}^2\mathscr{F}_\nu = B_\nu^{(0)}(\boldsymbol{\rho}, \mathbf{s}) \cos\theta \, \mathrm{d}\sigma \, \mathrm{d}\Omega. \tag{5.7–36}$$

Here $\boldsymbol{\rho}$ is the (two-dimensional) position vector of the point Q in the source plane at which the element $\mathrm{d}\sigma$ is located, $\mathbf{s}$ is the unit vector along the axis of the element $\mathrm{d}\Omega$ of solid angle and $\theta(0 \leqslant \theta \leqslant \pi/2)$ is the angle which the unit vector $\mathbf{s}$ makes with the normal to the source plane (Fig. 5.25). The function $B_\nu^{(0)}(\boldsymbol{\rho}, \mathbf{s})$ is

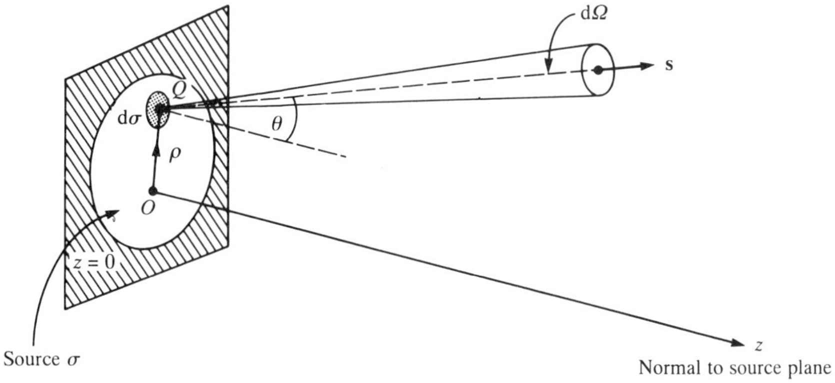

Fig. 5.25 Illustrating the notation relating to the definition of the radiance $B_\nu(\boldsymbol{\rho}, \mathbf{s})$ [Eq. (5.7–36)].

called the (spectral) *radiance* or *brightness*. In view of its apparent physical significance this quantity must evidently satisfy the following two requirements:

$$B_\nu^{(0)}(\boldsymbol{\rho}, \mathbf{s}) \geqslant 0 \quad \text{when } \boldsymbol{\rho} \in \sigma \tag{5.7–37a}$$

and

$$B_\nu^{(0)}(\boldsymbol{\rho}, \mathbf{s}) = 0 \quad \text{when } \boldsymbol{\rho} \notin \sigma. \tag{5.7–37b}$$

The formula (5.7–36) implies that the rate $\mathrm{d}\mathscr{F}_\nu$ at which the source radiates energy at frequency ν into the element $\mathrm{d}\Omega$ of solid angle is given by

$$\mathrm{d}\mathscr{F}_\nu = \cos\theta \int_\sigma B_\nu^{(0)}(\boldsymbol{\rho}, \mathbf{s})\,\mathrm{d}^2\rho \tag{5.7–38}$$

and that the total rate $\mathscr{F}_\nu$ at which the source radiates energy at that frequency is expressible in the form

$$\mathscr{F}_\nu = \int_{(2\pi)} \mathrm{d}\Omega \cos\theta \int_\sigma \mathrm{d}^2\rho B_\nu^{(0)}(\boldsymbol{\rho}, \mathbf{s}). \tag{5.7–39}$$

The first integral on the right of Eq. (5.7–39) extends over the 2π-solid angle subtended by a hemisphere in the half-space into which the source radiates, the center of the hemisphere being in the source region.

Comparing Eqs. (5.7–39) and (5.7–35) we see that the elementary radiometric law expressed by Eq. (5.7–36) will be consistent with physical optics if the radiant intensity is expressible in terms of the radiance by the formula

$$J(\mathbf{s}, \nu) = \cos\theta \int_\sigma B_\nu^{(0)}(\boldsymbol{\rho}, \mathbf{s})\,\mathrm{d}^2\rho. \tag{5.7–40}$$

Now we derived earlier the following expression for the radiant intensity in terms of the cross-spectral density $W^{(0)}(\boldsymbol{\rho}_1, \boldsymbol{\rho}_2, \nu)$ of the light in the source plane [Eqs. (5.3–8) and (5.3–7)]:

$$J(\mathbf{s}, \nu) = \left(\frac{k}{2\pi}\right)^2 \cos^2\theta \int_\sigma\int_\sigma W^{(0)}(\boldsymbol{\rho}_1, \boldsymbol{\rho}_2, \nu)\,\mathrm{e}^{-ik\mathbf{s}_\perp\cdot(\boldsymbol{\rho}_2-\boldsymbol{\rho}_1)}\,\mathrm{d}^2\rho_1\,\mathrm{d}^2\rho_2. \tag{5.7–41}$$

On comparing Eqs. (5.7–41) and (5.7–40) we see that the following relation must hold for all real unit vectors $\mathbf{s} \equiv (s_x, s_y, s_z \geqslant 0)$, with $\mathbf{s}_\perp \equiv (s_x, s_y, 0)$:

$$\int_\sigma B_\nu^{(0)}(\boldsymbol{\rho}, \mathbf{s})\, \mathrm{d}^2\rho = \left(\frac{k}{2\pi}\right)^2 \cos\theta \int_\sigma\int_\sigma W^{(0)}(\boldsymbol{\rho}_1, \boldsymbol{\rho}_2, \nu)\, \mathrm{e}^{-\mathrm{i}k\mathbf{s}_\perp\cdot(\boldsymbol{\rho}_2-\boldsymbol{\rho}_1)}\, \mathrm{d}^2\rho_1\, \mathrm{d}^2\rho_2. \tag{5.7–42}$$

It turns out that many functions $B_\nu^{(0)}(\boldsymbol{\rho}, \mathbf{s})$ will satisfy the requirement (5.7–42). However, as was shown by Friberg (1979), none of them can be expressed as a linear transform of the cross-spectral density $W^{(0)}$ of the field distribution across the source and also satisfy the constraints (5.7–37) for all possible sources. However, if one relaxes some of the requirements on the radiance function, one can obtain a kind of generalization of the radiance function that is useful in some applications. In this connection one should bear in mind that traditional radiometry has developed in connection with radiation from thermal sources, which are spatially rather incoherent. We will show in Section 5.7.3 that for such sources and, in fact, for the broader class of quasi-homogeneous sources, a generalized radiance function can be introduced which has the properties usually attributed to the radiance and which is consistent with physical optics.

Let us drop the requirement expressed by Eq. (5.7–37b) and let us extend the $\boldsymbol{\rho}$-integrations in Eqs. (5.7–40) and (5.7–42) throughout the entire source plane, which we take to be the plane $z = 0$. Since the cross-spectral density $W^{(0)}(\boldsymbol{\rho}_1, \boldsymbol{\rho}_2, \nu)$ has zero value whenever $\boldsymbol{\rho}_1$ or $\boldsymbol{\rho}_2$ (or both) represent points in the plane $z = 0$ which are located outside the source region σ, we can formally extend both double integrals on the right-hand side of Eq. (5.7–42) over the whole plane $z = 0$. We then obtain the relation

$$\int \mathcal{B}_\nu^{(0)}(\boldsymbol{\rho}, \mathbf{s})\, \mathrm{d}^2\rho = \left(\frac{k}{2\pi}\right)^2 \cos\theta \int\int W^{(0)}(\boldsymbol{\rho}_1, \boldsymbol{\rho}_2, \nu)\, \mathrm{e}^{-\mathrm{i}k\mathbf{s}_\perp\cdot(\boldsymbol{\rho}_2-\boldsymbol{\rho}_1)}\, \mathrm{d}^2\rho_1\, \mathrm{d}^2\rho_2, \tag{5.7–43}$$

where all the integrals are taken over the entire source plane $z = 0$. Let us now change the variables of integration on the right-hand side of Eq. (5.7–43) from $\boldsymbol{\rho}_1, \boldsymbol{\rho}_2$ to $\boldsymbol{\rho}, \boldsymbol{\rho}'$, where

$$\boldsymbol{\rho} = \tfrac{1}{2}(\boldsymbol{\rho}_1 + \boldsymbol{\rho}_2), \quad \boldsymbol{\rho}' = \boldsymbol{\rho}_2 - \boldsymbol{\rho}_1. \tag{5.7–44}$$

Equation (5.7–43) then becomes

$$\int \mathcal{B}_\nu^{(0)}(\boldsymbol{\rho}, \mathbf{s})\, \mathrm{d}^2\rho = \left(\frac{k}{2\pi}\right)^2 \cos\theta \int\int W^{(0)}(\boldsymbol{\rho} - \tfrac{1}{2}\boldsymbol{\rho}', \boldsymbol{\rho} + \tfrac{1}{2}\boldsymbol{\rho}', \nu)\, \mathrm{e}^{-\mathrm{i}k\mathbf{s}_\perp\cdot\boldsymbol{\rho}'}\, \mathrm{d}^2\rho\, \mathrm{d}^2\rho'. \tag{5.7–45}$$

This relation is obviously satisfied with the choice

$$\mathcal{B}_\nu^{(0)}(\boldsymbol{\rho}, \mathbf{s}) = \left(\frac{k}{2\pi}\right)^2 \cos\theta \int W^{(0)}(\boldsymbol{\rho} - \tfrac{1}{2}\boldsymbol{\rho}', \boldsymbol{\rho} + \tfrac{1}{2}\boldsymbol{\rho}', \nu)\, \mathrm{e}^{-\mathrm{i}k\mathbf{s}_\perp\cdot\boldsymbol{\rho}'}\, \mathrm{d}^2\rho'. \tag{5.7–46}$$

This expression for the generalized radiance was introduced by Walther (1968) and is frequently used in the analysis of radiometric problems which involve partially coherent sources (see, for example, Marchand and Wolf, 1974a). In view of our earlier remarks it is not surprising that it can take on negative values

(Marchand and Wolf, 1974a,b), in violation of the implicit postulate (5.7–37a). However, it can be shown that it is always real.

Another possible (generally non-equivalent) definition of a generalized radiance is provided by the formula

$$'\mathscr{B}_\nu^{(0)}(\boldsymbol{\rho}, \mathbf{s}) = \left(\frac{k}{2\pi}\right)^2 \cos\theta\, \mathrm{e}^{\mathrm{i}k\mathbf{s}_\perp\cdot\boldsymbol{\rho}} \int W^{(0)}(\boldsymbol{\rho}, \boldsymbol{\rho}', \nu)\, \mathrm{e}^{-\mathrm{i}k\mathbf{s}_\perp\cdot\boldsymbol{\rho}'}\, \mathrm{d}^2\rho'. \quad (5.7\text{–}47)$$

In order to deduce some of the properties of $'\mathscr{B}_\nu^{(0)}(\boldsymbol{\rho}, \mathbf{s})$ we will express it in a somewhat different form. For this purpose we represent $W^{(0)}$ in the form given by Eq. (5.7–11), namely

$$W^{(0)}(\boldsymbol{\rho}, \boldsymbol{\rho}', \nu) = \langle U^{(0)*}(\boldsymbol{\rho}, \nu) U^{(0)}(\boldsymbol{\rho}', \nu)\rangle. \quad (5.7\text{–}48)$$

On substituting from Eq. (5.7–48) into Eq. (5.7–47) we find that

$$'\mathscr{B}_\nu^{(0)}(\boldsymbol{\rho}, \mathbf{s}) = k^2 \cos\theta \langle [U^{(0)}(\boldsymbol{\rho}, \nu)]^* \widetilde{U}^{(0)}(k\mathbf{s}_\perp, \nu)\rangle\, \mathrm{e}^{\mathrm{i}k\mathbf{s}_\perp\cdot\boldsymbol{\rho}}, \quad (5.7\text{–}49)$$

where

$$\widetilde{U}^{(0)}(\mathbf{f}, \nu) = \frac{1}{(2\pi)^2} \int U^{(0)}(\boldsymbol{\rho}', \nu)\, \mathrm{e}^{-\mathrm{i}\mathbf{f}\cdot\boldsymbol{\rho}'}\, \mathrm{d}^2\rho' \quad (5.7\text{–}50)$$

is the two-dimensional spatial Fourier transform of $U^{(0)}(\boldsymbol{\rho}', \nu)$.

It is clear from Eq. (5.7–49) that $'\mathscr{B}_\nu^{(0)}(\boldsymbol{\rho}, \mathbf{s})$ is, in general, a complex quantity. However, its integral with respect to $\boldsymbol{\rho}$, taken over the entire source plane $z = 0$, is necessarily real and non-negative. This fact follows at once from Eq. (5.7–49) which yields, on integration,

$$\int {'\mathscr{B}_\nu^{(0)}}(\boldsymbol{\rho}, \mathbf{s})\, \mathrm{d}^2\rho = (2\pi k)^2 \cos\theta \langle [U^{(0)}(k\mathbf{s}_\perp, \nu)]^* \widetilde{U}^{(0)}(k\mathbf{s}_\perp, \nu)\rangle \geqslant 0. \quad (5.7\text{–}51)$$

Because this integral is real-valued it remains unchanged if $'\mathscr{B}_\nu^{(0)}(\boldsymbol{\rho}, \mathbf{s})$ is replaced by

$$''\mathscr{B}_\nu^{(0)}(\boldsymbol{\rho}, \mathbf{s}) = \mathrm{Re}\, {'\mathscr{B}_\nu^{(0)}}(\boldsymbol{\rho}, \mathbf{s}), \quad (5.7\text{–}52)$$

where Re denotes the real part. Hence it is clear that in calculating the radiant intensity via the radiometric formula (5.7–40), any of the expressions (5.7–46), (5.7–47) or (5.7–52) for the radiance function could be used. A definition of the generalized radiance, essentially equivalent to $''\mathscr{B}_\nu^{(0)}(\mathbf{r}, \mathbf{s})$ was first proposed by Walther (1973; see also Walther, 1978a,b).‡ However, just like the radiance defined by Eq. (5.7–46), $''\mathscr{B}_\nu^{(0)}(\mathbf{r}, \mathbf{s})$ may sometimes take on negative values (Marchand and Wolf, 1974b; Walther, 1974).

The mathematical structure of the formulas (5.7–46) and (5.7–47) contains a hint about the reasons for the difficulties encountered when one tries to define the radiance function on the basis of physical optics. These formulas have the same mathematical structure as expressions for generalized phase-space distribution functions, also known as quasi-probabilities, which are sometimes employed in calculations of the expectation values of quantum mechanical operators by methods similar to those used in classical statistical mechanics [cf. Section 11.8].

‡ Many of the basic papers concerning the foundations of radiometry are reprinted in *Selected Papers on Coherence and Radiometry*, ed. A. T. Friberg (1993).

Because such generalized distribution functions are functions of c-number representatives of non-commuting operators, they are not true probabilities[‡] and consequently they may become negative or even complex. In particular, the formula (5.7–46) resembles an expression for the so-called Wigner distribution function (Wigner, 1932),[§] whilst the formula (5.7–47) resembles a phase-space distribution function introduced by Margenau and Hill (1961).

Whichever expression for the generalized radiance function is adopted, it must always satisfy the requirement (5.7–39), which ensures that it yields the correct value for the rate at which the source radiates energy into the half-space $z > 0$. The formula (5.7–39) may be expressed in two alternative, but equivalent forms, namely

$$\mathscr{F}_\nu = \int_{(2\pi)} J(\mathbf{s}, \nu)\,\mathrm{d}\Omega \tag{5.7–53}$$

and (again extending the $\boldsymbol{\rho}$-integration over the entire source plane)

$$\mathscr{F}_\nu = \int E_\nu(\boldsymbol{\rho})\,\mathrm{d}^2\rho. \tag{5.7–54}$$

In the first formula, which is just Eq. (5.7–35), $J(\mathbf{s}, \nu)$ is the radiant intensity, which is expressible in terms of the generalized radiance by Eq. (5.7–40). The radiant intensity represents the rate at which the source radiates energy at the frequency ν per unit solid angle around the $\mathbf{s}$-direction. In the formula (5.7–54)

$$E_\nu(\boldsymbol{\rho}) = \int_{(2\pi)} B_\nu^{(0)}(\boldsymbol{\rho}, \mathbf{s})\cos\theta\,\mathrm{d}\Omega. \tag{5.7–55}$$

$E_\nu(\boldsymbol{\rho})$ is another radiometric quantity, known as the *radiant emittance*. According to Eq. (5.7–54) it may be regarded as representing the rate at which the source radiates energy at frequency ν per unit area of the source as a function of the position vector $\boldsymbol{\rho}$. We will now derive an expression for the (generalized) radiant emittance in terms of the cross-spectral density.

Since, as we have already noted, the generalized radiance is not uniquely determined by the cross-spectral density, it is evident from Eq. (5.7–55) that the same is true about the generalized radiant emittance. We will derive an expression for the radiant emittance associated with the generalized radiance defined by Eq. (5.7–46). On substituting on the right-hand side of Eq. (5.7–55) from Eq. (5.7–46), interchanging the orders of integration and making use of the expression $\mathrm{d}\Omega = \mathrm{d}s_x\,\mathrm{d}s_y/\cos\theta$ for the element of the solid angle we readily find that

$$\mathscr{E}_\nu(\boldsymbol{\rho}) = \int W^{(0)}(\boldsymbol{\rho} - \tfrac{1}{2}\boldsymbol{\rho}', \boldsymbol{\rho} + \tfrac{1}{2}\boldsymbol{\rho}', \nu)K_\nu(\boldsymbol{\rho}')\,\mathrm{d}^2\rho', \tag{5.7–56}$$

where

$$K_\nu(\boldsymbol{\rho}') = \left(\frac{k}{2\pi}\right)^2 \iint_{s_x^2+s_y^2\leqslant 1} \cos\theta\,\mathrm{e}^{-ik\mathbf{s}_\perp\cdot\boldsymbol{\rho}'}\,\mathrm{d}s_x\,\mathrm{d}s_y, \tag{5.7–57}$$

‡ The impossibility of introducing phase–space distribution functions which would obey all the postulates of probability theory was demonstrated by Wigner (1971), p. 25.

§ For an excellent discussion of the Wigner distribution function see Imre, Özizmir, Rosenbaum and Zweifel (1967).

$\mathbf{s}_\perp \equiv (s_x, s_y)$ and $\cos\theta = [1 - s_x^2 - s_y^2]^{1/2}$. The integral on the right-hand side of Eq. (5.7–57) may readily be evaluated by changing the variables of integration to polar coordinates and making use of the integral representation (5.3–50) of the Bessel function $J_0(x)$ and of the spherical Bessel function, namely,

$$j_1(x) = \frac{1}{x}\left(\frac{\sin x}{x} - \cos x\right). \tag{5.7–58}$$

One then obtains for the kernel $K_\nu(\boldsymbol{\rho}')$ of the integral transform (5.7–56) the expression (cf. Marchand and Wolf, 1974a)

$$K_\nu(\boldsymbol{\rho}') = \frac{k^2}{2\pi}\left[\frac{j_1(k\rho')}{k\rho'}\right]. \tag{5.7–59}$$

On substituting from Eq. (5.7–59) into Eq. (5.7–56) we obtain the following expression for the generalized radiant emittance:

$$\mathscr{E}_\nu(\boldsymbol{\rho}) = \frac{k^2}{2\pi}\int W^{(0)}(\boldsymbol{\rho} - \tfrac{1}{2}\boldsymbol{\rho}', \boldsymbol{\rho} + \tfrac{1}{2}\boldsymbol{\rho}', \nu)\left[\frac{j_1(k\rho')}{k\rho'}\right]\mathrm{d}^2\rho'. \tag{5.7–60}$$

It may be shown that the generalized radiant emittance given by Eq. (5.7–60) can take on negative values indicating that, just like the radiance, it is not in general a measurable quantity.

5.7.3 *Radiance function of a planar, secondary, quasi-homogeneous source*‡

We mentioned earlier that radiometry has largely developed in connection with problems involving thermal sources, i.e. blackbody sources or sources whose output may be derived from blackbody sources by linear filtering [cf. Section 13.2]. Sources of this kind usually belong to the class of quasi-homogeneous sources. It is, therefore, of interest to examine the forms of the expressions for the basic radiometric quantities for sources of this kind.

For a planar, secondary, quasi-homogeneous source the cross-spectral density has the form [Eq. (5.3–16)]

$$W^{(0)}(\boldsymbol{\rho}_1, \boldsymbol{\rho}_2, \nu) = S^{(0)}[\tfrac{1}{2}(\boldsymbol{\rho}_1 + \boldsymbol{\rho}_2), \nu]g^{(0)}(\boldsymbol{\rho}_2 - \boldsymbol{\rho}_1, \nu), \tag{5.7–61}$$

where the spectral density $S^{(0)}(\boldsymbol{\rho}, \nu)$ varies much more slowly with $\boldsymbol{\rho}$ than the spectral degree of coherence $g^{(0)}(\boldsymbol{\rho}', \nu)$ varies with $\boldsymbol{\rho}'$. Moreover the linear dimensions of such a source are large compared with its spectral coherence width, at frequency ν, of the light distribution across the source, i.e. compared with the effective width of $|g^{(0)}(\boldsymbol{\rho}', \nu)|$ (cf. Fig. 5.2).

If we substitute from Eq. (5.7–61) into Eq. (5.7–46), we readily obtain the following expression for one of the generalized radiance functions of a quasi-homogeneous source:

$$\mathscr{B}_\nu^{(0)}(\boldsymbol{\rho}, \mathbf{s}) = k^2 S^{(0)}(\boldsymbol{\rho}, \nu)\tilde{g}^{(0)}(k\mathbf{s}_\perp, \nu)\cos\theta, \tag{5.7–62}$$

‡ The discussion presented in this section is largely based on an investigation of Carter and Wolf (1977; see also Friberg, 1981). For some analogous calculations relating to Gaussian Schell-model sources see Baltes and Steinle (1977) and Baltes, Steinle and Antes (1978).

where

$$\tilde{g}^{(0)}(\mathbf{f}, \nu) = \frac{1}{(2\pi)^2} \int g^{(0)}(\boldsymbol{\rho}', \nu)\, \mathrm{e}^{-\mathrm{i}\mathbf{f}\cdot\boldsymbol{\rho}'}\, \mathrm{d}^2\rho' \qquad (5.7\text{–}63)$$

is the two-dimensional spatial Fourier transform of $g^{(0)}(\boldsymbol{\rho}')$.

We see that the radiance function, given by Eq. (5.7–62), has a factorized form, being at each frequency the product of a function of $\boldsymbol{\rho}$ and a function of $\mathbf{s}$. The first factor $k^2 S^{(0)}(\boldsymbol{\rho}, \nu)$ is, of course, non-negative. The second factor, $\tilde{g}^{(0)}(k\mathbf{s}_\perp, \nu)$ is also non-negative, because it is the Fourier transform of a correlation coefficient which is necessarily non-negative definite, and hence by Bochner's theorem (Section 1.4.2)

$$\tilde{g}^{(0)}(\mathbf{f}, \nu) \geqslant 0 \qquad (5.7\text{–}64)$$

for all two-dimensional vectors $\mathbf{f}$. Moreover, since $0 \leqslant \theta \leqslant \pi/2$, the last factor $\cos\theta$ is also non-negative. Thus Eq. (5.7–62) shows that *the radiance function (5.7–46) of a quasi-homogeneous source is necessarily non-negative*:

$$\mathcal{B}_\nu^{(0)}(\boldsymbol{\rho}, \mathbf{s}) \geqslant 0. \qquad (5.7\text{–}65)$$

Moreover, since the spectral density $S^{(0)}(\boldsymbol{\rho}, \nu)$ vanishes at all points in the source plane outside the region σ occupied by the source,

$$\mathcal{B}_\nu^{(0)}(\boldsymbol{\rho}, \mathbf{s}) \equiv 0 \quad \text{when } \boldsymbol{\rho} \notin \sigma. \qquad (5.7\text{–}66)$$

Thus we see that *the radiance function, defined by Eq. (5.7–46), of a quasi-homogeneous source satisfies the postulates (5.7–37a) and (5.7–37b) of traditional radiometry*. Further, if we substitute from Eq. (5.7–61) into the two other definitions of the radiance function [Eqs. (5.7–47) and (5.7–52)] and use the fact that $S^{(0)}(\boldsymbol{\rho}, \nu)$ varies much more slowly with $\boldsymbol{\rho}$ than $g^{(0)}(\boldsymbol{\rho}', \nu)$ varies with $\boldsymbol{\rho}'$ we find that, to a good approximation, each is equal to the expression (5.7–62), so that

$$\mathcal{B}_\nu^{(0)}(\boldsymbol{\rho}, \mathbf{s}) \approx {}'\mathcal{B}_\nu^{(0)}(\boldsymbol{\rho}, \mathbf{s}) = {}''\mathcal{B}_\nu^{(0)}(\boldsymbol{\rho}, \mathbf{s}), \qquad (5.7\text{–}67)$$

i.e. *for a quasi-homogeneous source the three definitions of the radiance function are essentially equivalent to each other*.

Let us now consider the radiant emittance. If we substitute from Eq. (5.7–61) into the formula (5.7–60) we obtain at once the following expression for the radiant emittance of a quasi-homogeneous source:

$$\mathscr{E}_\nu(\boldsymbol{\rho}) = C_g(\nu) S^{(0)}(\boldsymbol{\rho}, \nu), \qquad (5.7\text{–}68)$$

where

$$C_g(\nu) = \frac{k^2}{2\pi} \int g^{(0)}(\boldsymbol{\rho}', \nu) \left[\frac{j_1(k\rho')}{k\rho'}\right] \mathrm{d}^2\rho'. \qquad (5.7\text{–}69)$$

The formulas (5.7–68) and (5.7–69) show that the *generalized radiant emittance* [based on the expression (5.7–46) for the radiance] *of a quasi-homogeneous source is proportional to the spectral density $S^{(0)}(\boldsymbol{\rho}, \nu)$, the proportionality factor $C_g(\nu)$ depending on the spectral degree of coherence of the light distribution across the source*. For reasons that will become apparent shortly, the factor $C_g(\nu)$ is sometimes called the *efficiency factor* of the source. It can be expressed in a somewhat different form, from which its main properties may be readily derived. If we substitute from Eq. (5.7–62) into Eq. (5.7–55) and compare the resulting expression with Eq. (5.7–68), we find that

$$C_g(\nu) = k^2 \int_{(2\pi)} \tilde{g}^{(0)}(k\mathbf{s}_\perp, \nu) \cos^2\theta \, \mathrm{d}\Omega. \tag{5.7–70}$$

Since according to Eq. (5.7–64) $\tilde{g}^{(0)}(k\mathbf{s}_\perp, \nu) \geqslant 0$ for all values of its arguments, we see from Eq. (5.7–70) that

$$C_g(\nu) \geqslant 0. \tag{5.7–71}$$

If we make use in Eq. (5.7–68) of this result and also use the fact that the spectral density $S^{(0)}(\boldsymbol{\rho}, \nu)$ is non-negative, it follows that

$$\mathscr{E}_\nu(\boldsymbol{\rho}) \geqslant 0. \tag{5.7–72}$$

Hence *the generalized radiant emittance of a quasi-homogeneous source*, associated with the generalized radiance (5.7–46), *is necessarily non-negative*. (In view of Eq. (5.7–67) the same is true if, instead of the expression (5.7–46), one uses either of the alternative expressions (5.7–47) or (5.7–52) for the radiance). Moreover, since the spectral density $S^{(0)}(\boldsymbol{\rho}, \nu)$ is zero when the point $\boldsymbol{\rho}$ is situated outside the source area σ, it follows from Eq. (5.7–68) that

$$\mathscr{E}_\nu(\boldsymbol{\rho}) \equiv 0 \quad \text{when } \boldsymbol{\rho} \notin \sigma. \tag{5.7–73}$$

From Eq. (5.7–69) we may also obtain an upper bound on $C_g(\nu)$. For this purpose we take the Fourier inverse of Eq. (5.7–63), set $\boldsymbol{\rho}' = 0$ and use the fact that $g^{(0)}(0, \nu) = 1$. We then find that

$$\int_{f^2 \leqslant k^2} \tilde{g}^{(0)}(\mathbf{f}, \nu) \, \mathrm{d}^2 f + \int_{f^2 > k^2} \tilde{g}^{(0)}(\mathbf{f}, \nu) \, \mathrm{d}^2 f = 1. \tag{5.7–74}$$

If we set $\mathbf{f} = k\mathbf{s}_\perp = (ks_x, ks_y)$ and use the inequality (5.7–64), it follows that

$$k^2 \iint_{s_x^2 + s_y^2 \leqslant 1} \tilde{g}^{(0)}(k\mathbf{s}_\perp, \nu) \, \mathrm{d}s_x \, \mathrm{d}s_y \leqslant 1. \tag{5.7–75}$$

Since $\mathrm{d}\Omega = \mathrm{d}s_x \, \mathrm{d}s_y / \cos\theta$ it follows at once from Eqs. (5.7–75) and (5.7–70) that

$$C_g(\nu) \leqslant 1. \tag{5.7–76}$$

We see from the inequality (5.7–76) and from Eq. (5.7–68) that *the radiant emittance of a quasi-homogeneous source at any source point is smaller or equal to the spectral density at that point*.

The behavior of the efficiency factor C_g for a secondary, planar, Gaussian correlated, quasi-homogeneous source as a function of its effective spectral coherence width σ_g is shown in Fig. 5.26. We see that C_g increases monotonically with σ_g, from the value zero for a completely incoherent source ($\sigma_g \ll \lambda$) to the value unity for a locally coherent source ($\sigma_g \gg \lambda$). For a Lambertian quasi-homogeneous source the value of C_g was shown to be 1/2 (Wolf and Carter, 1978a).

The radiant intensity $J(\mathbf{s}, \nu)$ generated by a quasi-homogeneous source may be obtained on substituting from Eq. (5.7–62) into Eq. (5.7–40). We then find at

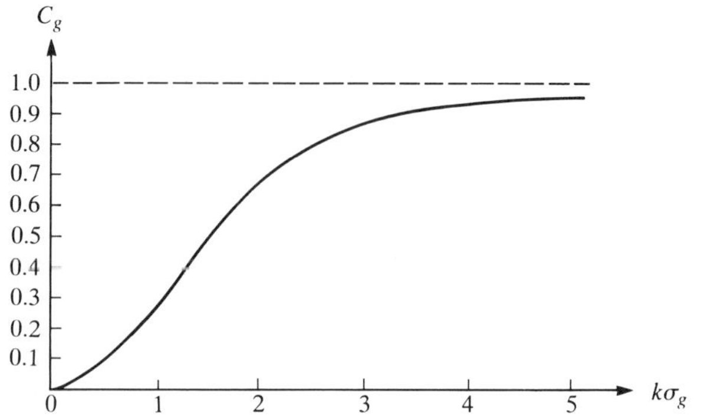

Fig. 5.26 The efficiency factor C_g of a planar, secondary, Gaussian-correlated $[g^{(0)}(\boldsymbol{\rho}') = \exp(-\rho'^2/2\sigma_g^2)]$, quasi-homogeneous source, as a function of its normalized effective correlation width $k\sigma_g$ (Adapted from Wolf and Carter, 1978a.)

once that

$$J(\mathbf{s}, \nu) = (2\pi k)^2 \tilde{S}^{(0)}(0, \nu)\tilde{g}^{(0)}(k\mathbf{s}_\perp, \nu)\cos^2\theta, \qquad (5.7\text{–}77)$$

where $\tilde{S}^{(0)}(0, \nu)$ is the Fourier transform of the spectral density distribution across the source, evaluated for the spatial frequency $\mathbf{f} = 0$, i.e.

$$\tilde{S}^{(0)}(0, \nu) = \frac{1}{(2\pi)^2}\int S^{(0)}(\boldsymbol{\rho}, \nu)\, d^2\rho. \qquad (5.7\text{–}78)$$

The expression (5.7–77) is in agreement with the formula (5.3–21), which was derived in a different manner.

Finally we note that the rate at which a quasi-homogeneous source radiates energy at frequency ν is, according to Eqs. (5.7–54), (5.7–68) and (5.7–78) equal to

$$\mathscr{F}_\nu = (2\pi)^2 C_g(\nu)\tilde{S}^{(0)}(0, \nu). \qquad (5.7\text{–}79)$$

We see from this formula and from Eq. (5.7–78) that, for a quasi-homogeneous source, $\mathscr{F}_\nu$ is proportional to the source-integrated spectral density at frequency ν. Now according to Eq. (5.7–69) the proportionality factor $C_g(\nu)$, $(0 \leq C_g(\nu) \leq 1)$, depends only on the spectral degree of coherence of the source and hence is a measure of the radiation 'efficiency' with which sources, having the same spectral density but different spectral degrees of coherence, generate radiation. It is for this reason that $C_g(\nu)$ is referred to as the efficiency factor.

We have demonstrated in this section that the radiometric model, when applied to radiation from quasi-homogeneous sources, is consistent with physical optics, at least as regards properties of the basic radiometric quantities [cf. Eqs. (5.7–65), (5.7–66), (5.7–72) and (5.7–73)]. This is a rather satisfactory conclusion, since thermal sources, around which traditional radiometry has been developed and to which it is mainly applied, usually belong to this class of sources. For other types of sources the radiometric model has to be interpreted with caution, because the radiance and the radiant emittance may not possess all the properties that are commonly attributed to them. Nevertheless the concepts of radiance and of radiant emittance can always be used in calculating measurable radiometric quantities, such as radiant flux and radiant intensity.

5.7.4 Radiative energy transfer model

The theory of radiative energy transfer is a generalization of the radiometry that we discussed in the preceding sections. The simple radiometric model, expressed by Eq. (5.7–36) for the rate at which energy is radiated from an element of a planar source, is broadened in the sense that the radiance is generalized to a field quantity. The surface element $d\sigma$ may now be any portion of a (generally fictitious) surface in a region of space which contains radiation. Instead of radiance one then usually speaks of the *specific intensity* of the radiation. We will denote the specific intensity by $I_\nu(\mathbf{r}, \mathbf{s})$, where $\mathbf{r}$ denotes the position vector of the element $d\sigma$ and $\mathbf{s}$ denotes a unit vector specifying a direction. According to the radiative transfer model one may regard energy as being transported across the surface element $d\sigma$ along pencils of rays. The rate $d^2\mathscr{F}_\nu$ at which energy at frequency ν is transported, in a steady-state field, across $d\sigma$ into an element $d\Omega$ of solid angle about the $\mathbf{s}$-direction is assumed to be given by

$$d^2\mathscr{F}_\nu = I_\nu(\mathbf{r}, \mathbf{s}) \cos\theta \, d\sigma \, d\Omega, \tag{5.7–80}$$

where θ is the angle between the unit vector $\mathbf{s}$ and the unit vector normal $\mathbf{n}$ to $d\sigma$ (see Fig. 5.27). In the special case when $d\sigma$ coincides with an element of a planar radiating surface, the specific intensity becomes just the radiance $B_\nu(\mathbf{r}, \mathbf{s})$ and the formula (5.7–80) reduces to the basic radiometric law, expressed by Eq. (5.7–36). Since according to Eq. (5.7–80) the specific intensity represents the rate at which energy at frequency ν is transported per unit projected area (projected onto a plane perpendicular to $\mathbf{s}$) per unit solid angle, it is necessarily non-negative, i.e.

$$I_\nu(\mathbf{r}, \mathbf{s}) \geqslant 0 \tag{5.7–81}$$

for all values of its arguments.

One usually introduces two other basic quantities in the phenomenological theory of radiative energy transfer, which are defined in terms of the specific intensity: the so-called *space density of radiation*, $'H_\nu(\mathbf{r})$, and the *net flux* $'\mathbf{F}_\nu(\mathbf{r})$,

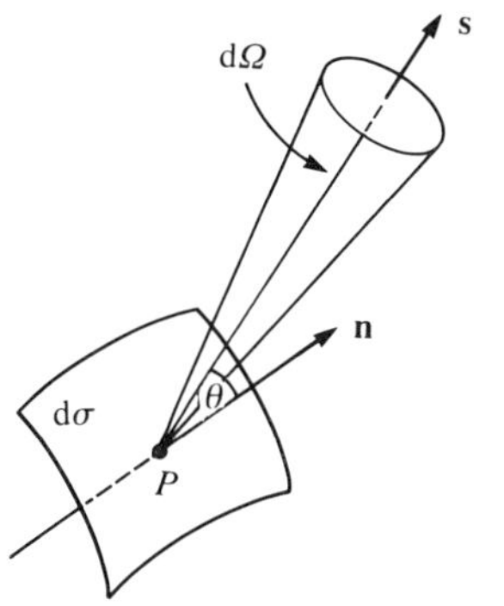

Fig. 5.27 Illustrating the notation relating to the definition (5.7–80) of the specific intensity of radiation $I_\nu(\mathbf{r}, \mathbf{s})$.

at the point $\mathbf{r}$ in the field, at frequency ν. They are defined by the formulas (Planck, 1959, Sec. 22; Chandrasekhar, 1960, Sec. 2.2 and 2.3)

$$'H_\nu(\mathbf{r}) = \frac{1}{c}\int_{(4\pi)} I_\nu(\mathbf{r}, \mathbf{s})\,\mathrm{d}\Omega \tag{5.7–82}$$

and

$$'\mathbf{F}_\nu(\mathbf{r}) = \int_{(4\pi)} I_\nu(\mathbf{r}, \mathbf{s})\mathbf{s}\,\mathrm{d}\Omega. \tag{5.7–83}$$

In Eq. (5.7–82) c denotes the speed of light in vacuum and the integrations in Eqs. (5.7–82) and (5.7–83) extend over the complete 4π-solid angle generated by the unit vector $\mathbf{s}$. It is generally taken for granted that the space density of radiation, $'H_\nu(\mathbf{r})$ may be identified with the expectation value of energy density and $'\mathbf{F}_\nu(r)$ with the expectation value of the energy flux density vector at the point $\mathbf{r}$, at frequency ν.

By elementary considerations involving nothing more than simple quasi-geometric arguments regarding energy conservation, one is led to an integro-differential equation for the propagation of the specific intensity in any isotropic medium, known as the *equation of radiative energy transfer*. This equation may be written in the form (Hopf, 1934, Sec. 2)

$$\mathbf{s}\cdot\boldsymbol{\nabla}I_\nu(\mathbf{r}, \mathbf{s}) = -\alpha_\nu(\mathbf{r}, \mathbf{s})I_\nu(\mathbf{r}, \mathbf{s}) + \int_{(4\pi)} \beta_\nu(\mathbf{r}, \mathbf{s}, \mathbf{s}')I_\nu(\mathbf{r}, \mathbf{s}')\,\mathrm{d}\Omega' + D_\nu(\mathbf{r}, \mathbf{s}). \tag{5.7–84}$$

The functions $\alpha_\nu(\mathbf{r}, \mathbf{s})$, $\beta_\nu(\mathbf{r}, \mathbf{s}, \mathbf{s}')$, and $D_\nu(\mathbf{r}, \mathbf{s})$ are known as the *extinction coefficient*, the *differential scattering coefficient* and the *source function*, respectively. The left-hand side of Eq. (5.7–84) represents the rate of change of the specific intensity in the direction $\mathbf{s}$. On the right-hand side, the first term represents the rate of decrease in energy due to absorption along the $\mathbf{s}$-direction, the second term (with the integration extending over the 4π-solid angle of directions generated by the unit vector $\mathbf{s}'$) represents the rate of increase in energy along the $\mathbf{s}$-direction due to scattering from all $\mathbf{s}'$-directions and the last term represents the rate at which energy is generated by sources of the field.

In spite of the extensive use of the theory of radiative energy transfer, no satisfactory derivation of its basic equation, Eq. (5.7–84), from electromagnetic theory or even from scalar wave theory has been obtained up to now, except in some special cases.

Solutions of the equation (5.7–84) for two special cases are worth noting. When scattering is negligible ($\beta_\nu \approx 0$) and no sources are present ($D_\nu = 0$), the equation reduces to

$$\mathbf{s}\cdot\boldsymbol{\nabla}I_\nu(\mathbf{r}, \mathbf{s}) = -\alpha_\nu(\mathbf{r}, \mathbf{s})I_\nu(\mathbf{r}, \mathbf{s}). \tag{5.7–85}$$

It follows from this equation that the specific intensity $I_\nu(\mathbf{r}, \mathbf{s})$ at any point $\mathbf{r}$, connected to the point $\mathbf{r}_0$ by a vector in the $\mathbf{s}$-direction, is related to the specific intensity $I_\nu(\mathbf{r}_0, \mathbf{s})$ at the point $\mathbf{r}_0$ by the formula

$$I_\nu(\mathbf{r}, \mathbf{s}) = I_\nu(\mathbf{r}_0, \mathbf{s})\exp\left\{-\int_{\mathbf{r}_0}^{\mathbf{r}} \alpha_\nu(\mathbf{r}, \mathbf{s})\,\mathrm{d}l\right\}, \tag{5.7–86}$$

where the integration is taken along the line joining $\mathbf{r}_0$ to $\mathbf{r}$ [(Fig. 5.28)]. The formula (5.7–86) is known as *Beer's law*.

In free space ($\alpha_\nu = \beta_\nu = D_\nu = 0$) the equation of radiative energy transfer takes on the simple form

$$\mathbf{s}\cdot\nabla I_\nu(\mathbf{r},\mathbf{s}) = 0. \tag{5.7–87}$$

It will be seen shortly that Eq. (5.7–87) may be regarded as a basic equation of radiometry in free space. It implies that, along a line in the **s**-direction joining the points $\mathbf{r}_0$ and $\mathbf{r}$,

$$I_\nu(\mathbf{r},\mathbf{s}) = I_\nu(\mathbf{r}_0,\mathbf{s}). \tag{5.7–88}$$

This result is usually expressed by saying that in free space the specific intensity is constant along a ray.

5.7.5 *Radiometry as a short wavelength limit of statistical wave theory with quasi-homogeneous sources*‡

In Section 5.7.3 we introduced a generalized radiance function for a planar, secondary, quasi-homogeneous source. We will now show that a natural extension of the definition of one such function is suitable to represent the generalized radiance not only for points in the source plane but also for all points throughout the half-space $z > 0$. We will show further that, in the short-wavelength limit (more precisely in the asymptotic limit as the wave number $k \to \infty$), this function obeys all the basic postulates of traditional radiometry and satisfies the free-space form (5.7–87) of the equation (5.7–84) of radiative energy transfer.

Let us return to the definition (5.7–49) of the generalized radiance, namely

$$'\mathcal{B}_\nu(\boldsymbol{\rho},\mathbf{s}) = k^2 s_z \langle [U^{(0)}(\boldsymbol{\rho},\nu)]^* \tilde{U}^{(0)}(k\mathbf{s}_\perp,\nu)\rangle\, \mathrm{e}^{ik\mathbf{s}_\perp\cdot\boldsymbol{\rho}}, \tag{5.7–89a}$$

where we have made use of the relation $\cos\theta = s_z$ and have written U_0 and $\tilde{U}_0$ rather than U and $\tilde{U}$ to stress that these quantities now pertain to points in the source plane $z = 0$. According to Eq. (3.2–27) we may express the formula (5.7–89a) in the alternative form

$$'\mathcal{B}_\nu(\boldsymbol{\rho},\mathbf{s}) = s_z \langle [U^{(0)}(\boldsymbol{\rho},\nu)]^* a(\mathbf{s}_\perp,\nu)\rangle\, \mathrm{e}^{ik\mathbf{s}_\perp\cdot\boldsymbol{\rho}}, \tag{5.7–89b}$$

where $a(\mathbf{s}_\perp,\nu)$ is the angular spectrum amplitude of the (random) field $U(\mathbf{r},\nu)$ generated by the source in the half-space $z > 0$.

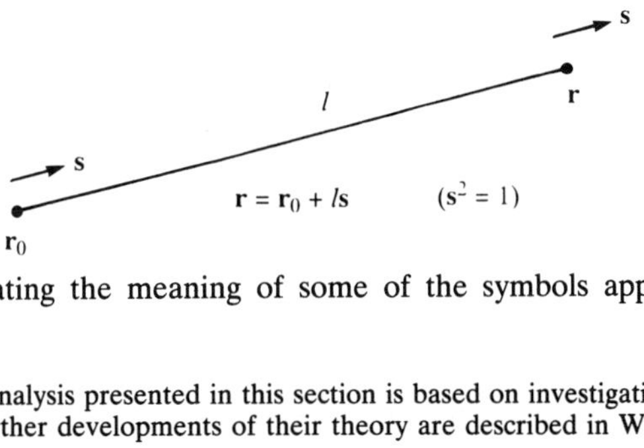

Fig. 5.28 Illustrating the meaning of some of the symbols appearing in Eq. (5.7–86).

‡ The main part of the analysis presented in this section is based on investigations by Foley and Wolf (1985, 1991). Some further developments of their theory are described in Wolf (1994) and in Foley and Wolf (1995).

The structure of the formula (5.7–89b) suggests the following extension, denoted by $\mathscr{B}_\nu(\mathbf{r}, \mathbf{s})$ of the function $'\mathscr{B}_\nu^{(0)}(\boldsymbol{\rho}, \mathbf{s})$, to points at arbitrary locations $\mathbf{r} \equiv (x, y, z)$ in the half-space $z > 0$ into which the source radiates:

$$\mathscr{B}_\nu(\mathbf{r}, \mathbf{s}) = s_z \langle U^*(\mathbf{r}, \nu) a(\mathbf{s}_\perp, \nu) \rangle \, \mathrm{e}^{ik\mathbf{s}\cdot\mathbf{r}}, \tag{5.7–90}$$

where $\mathbf{s} \equiv (s_x, s_y, s_z > 0)$ is, as usual, a real unit vector and $\mathbf{s}_\perp \equiv (s_x, s_y, 0)$. Evidently

$$\lim_{z\to+0} \mathscr{B}_\nu(\mathbf{r}, \mathbf{s}) = {'\mathscr{B}_\nu^{(0)}}(\boldsymbol{\rho}, \mathbf{s}) \quad [\boldsymbol{\rho} \equiv (x, y, 0)]. \tag{5.7–91}$$

We next show that the extension (5.7–90) of the definition (5.7–89b) is physically meaningful, for radiation generated by quasi-homogeneous sources, at least in the asymptotic limit as $k \to \infty$. For this purpose we first express the field variable $U(\mathbf{r}, \nu)$ in terms of its boundary values $U^{(0)}(\boldsymbol{\rho}, \nu)$ in the plane $z = 0$ by the use of the Rayleigh diffraction formula of the first kind [Eq. (3.2–78)], namely

$$U(\mathbf{r}, \nu) = \int_\sigma U^{(0)}(\boldsymbol{\rho}, \nu) G(R, \nu) \, \mathrm{d}^2\rho, \tag{5.7–92}$$

where

$$G(R, \nu) = -\frac{1}{2\pi}\frac{\partial}{\partial z}\left(\frac{\mathrm{e}^{ikR}}{R}\right), \tag{5.7–93}$$

$$R = |\mathbf{r} - \boldsymbol{\rho}| \tag{5.7–94}$$

and the integration extends over the source domain σ. On substituting from Eq. (5.7–92) into Eq. (5.7–90), interchanging the order of averaging and integration and making use of Eq. (5.7–89b) we find that

$$\mathscr{B}_\nu(\mathbf{r}, \mathbf{s}) = \mathrm{e}^{ik\mathbf{s}\cdot\mathbf{r}} \int_\sigma {'\mathscr{B}_\nu^{(0)}}(\boldsymbol{\rho}, \mathbf{s}) G^*(R, \nu) \, \mathrm{e}^{-ik\mathbf{s}_\perp\cdot\boldsymbol{\rho}} \, \mathrm{d}^2\rho. \tag{5.7–95}$$

In general this quantity is complex and hence cannot represent a true radiance function. However, the situation is different when the source is quasi-homogeneous and the wavelength of the radiated field is short enough, as we will now show.

For a planar, secondary, quasi-homogeneous source the generalized radiance function in the source plane is given by Eq. (5.7–62) [see also Eq. (5.7–67)]. On substituting that expression into Eq. (5.7–95) we obtain for $\mathscr{B}_\nu(\mathbf{r}, \mathbf{s})$ the following expression:

$$\mathscr{B}_\nu(\mathbf{r}, \mathbf{s}) = k^2 s_z \tilde{g}^{(0)}(k\mathbf{s}_\perp, \nu) C_\nu^*(\mathbf{r}, \mathbf{s}_\perp) \, \mathrm{e}^{ik\mathbf{s}\cdot\mathbf{r}}, \tag{5.7–96}$$

where

$$C_\nu(\mathbf{r}, \mathbf{s}_\perp) = \int_\sigma G(R, \nu) S^{(0)}(\boldsymbol{\rho}, \nu) \, \mathrm{e}^{ik\mathbf{s}_\perp\cdot\boldsymbol{\rho}} \, \mathrm{d}^2\rho. \tag{5.7–97}$$

In the formula (5.7–96) $\tilde{g}^{(0)}(\mathbf{f}, \nu)$ is again the two-dimensional Fourier transform [defined by Eq. (5.7–63)] of the spectral degree of coherence of the source and $S^{(0)}(\boldsymbol{\rho}, \nu)$ is the spectral density at a typical source point.

Let us now consider the behavior of the expression (5.7–96) for very short wavelengths λ or, more precisely, in the asymptotic limit as $k \equiv 2\pi/\lambda \to \infty$, keeping, however, the source unaffected by this limiting process. For this purpose we first carry out the differentiation on the right-hand side of Eq. (5.7–93) and substitute the resulting expression for the Green's function $G(R, \nu)$ into Eq. (5.7–97). We then find that

$$C_\nu(\mathbf{r}, \mathbf{s}_\perp) = C_\nu^{(1)}(\mathbf{r}, \mathbf{s}_\perp) + C_\nu^{(2)}(\mathbf{r}, \mathbf{s}_\perp), \tag{5.7–98}$$

where

$$C_\nu^{(1)}(\mathbf{r}, \mathbf{s}_\perp) = \frac{kz}{2\pi \mathrm{i}} \int_\sigma S^{(0)}(\boldsymbol{\rho}, \nu) \frac{\mathrm{e}^{\mathrm{i}k\Phi(R,\boldsymbol{\rho})}}{R^2} \mathrm{d}^2\rho, \tag{5.7–99}$$

$$C_\nu^{(2)}(\mathbf{r}, \mathbf{s}_\perp) = \frac{z}{2\pi} \int_\sigma S^{(0)}(\boldsymbol{\rho}, \nu) \frac{\mathrm{e}^{\mathrm{i}k\Phi(R,\boldsymbol{\rho})}}{R^3} \mathrm{d}^2\rho \tag{5.7–100}$$

and

$$\Phi(R, \boldsymbol{\rho}) \equiv \Phi(\mathbf{r}, \mathbf{s}_\perp, \boldsymbol{\rho}) = |\mathbf{r} - \boldsymbol{\rho}| + \mathbf{s}_\perp \cdot \boldsymbol{\rho}. \tag{5.7–101}$$

As k becomes larger and larger, the exponential term in the above integrals will, in general, oscillate more and more rapidly as $\boldsymbol{\rho}$ explores the domain of integration (the source domain) and can then be evaluated by the use of the principle of stationary phase for double integrals (Section 3.3.3). Moreover as is evident by comparing the two expressions on the right-hand sides of Eqs. (5.7–99) and (5.7–100), $C_\nu^{(2)}$ is of a lower order in k as $k \to \infty$. Hence, except perhaps for some special cases, (e.g. where the field points are located on an axis of symmetry), we can omit the contribution $C_\nu^{(2)}$ in the asymptotic evaluation of C_ν and we then have

$$C_\nu(\mathbf{r}, \mathbf{s}_\perp) \sim C_\nu^{(1)}(\mathbf{r}, \mathbf{s}_\perp) \quad \text{as } k \to \infty. \tag{5.7–102}$$

The asymptotic evaluation of $C_\nu^{(1)}(\mathbf{r}, \mathbf{s}_\perp)$ is carried out in Appendix 5.1 and one finds that

$$\left.\begin{aligned} C_\nu^{(1)}(\mathbf{r}, \mathbf{s}_\perp) &\sim S^{(0)}(\boldsymbol{\rho}_0, \nu)\, \mathrm{e}^{\mathrm{i}k\mathbf{s}\cdot\mathbf{r}} && \text{if } \boldsymbol{\rho}_0 \in \sigma \\ &\sim O\left(\frac{1}{k^{1/2}}\right) && \text{if } \boldsymbol{\rho}_0 \notin \sigma \end{aligned}\right\} \tag{5.7–103}$$

(O being the order symbol) as $k \to \infty$, where $\boldsymbol{\rho}_0$ (denoted by $\boldsymbol{\rho}_0'$, in Appendix 5.1) is a point in the source plane, given by the formula

$$\boldsymbol{\rho}_0 = \mathbf{r}_\perp - \frac{z}{s_z}\mathbf{s}_\perp. \tag{5.7–104}$$

On substituting from Eq. (5.7–103) into Eq. (5.7–96) we find, to the leading order in k in the asymptotic approximation as $k \to \infty$,

$$\left.\begin{aligned} \mathscr{B}_\nu(\mathbf{r}, \mathbf{s}) &\sim k^2 s_z S^{(0)}(\boldsymbol{\rho}_0, \nu)\tilde{g}^{(0)}(k\mathbf{s}_\perp, \nu) && \text{if } \boldsymbol{\rho}_0 \in \sigma \\ &\sim 0 && \text{if } \boldsymbol{\rho}_0 \notin \sigma. \end{aligned}\right\} \tag{5.7–105}$$

On comparing the expression (5.7–105) with the expression (5.7–62) for the radiance function of a quasi-homogeneous source, namely

$$\mathcal{B}_\nu^{(0)}(\boldsymbol{\rho}, \mathbf{s}) = k^2 s_z S_\nu^{(0)}(\boldsymbol{\rho}, \nu)\tilde{g}_\nu^{(0)}(k\mathbf{s}_\perp, \nu), \quad (\boldsymbol{\rho} \in \sigma), \qquad (5.7\text{–}106)$$

and recalling that $s_z = \cos\theta$ we see that in the asymptotic limit as $k \to \infty$,

$$\left.\begin{aligned} \mathcal{B}_\nu(\mathbf{r}, \mathbf{s}) &= B_\nu(\boldsymbol{\rho}_0, \mathbf{s}) && \text{if } \boldsymbol{\rho}_0 \in \sigma \\ &= 0 && \text{if } \boldsymbol{\rho}_0 \notin \sigma, \end{aligned}\right\} \qquad (5.7\text{–}107)$$

with the position vector $\boldsymbol{\rho}_0$ of a point in the source plane related to a typical field point $\mathbf{r}$ by the formula (5.7–104).

To see more clearly the physical significance of the result which we have just obtained, let us denote the points with position vectors $\boldsymbol{\rho}_0$ and $\mathbf{r}$, related by Eq. (5.7–104), by Q_0 and by P respectively. One can then show by elementary geometry that Q_0 is precisely the point at which the line through P, in the direction specified by the unit vector $\mathbf{s}$, intersects the source plane $z = 0$ (see Fig. 5.29). Hence Eqs. (5.7–107) and (5.7–106) imply that

$$\left.\begin{aligned} \mathcal{B}_\nu(P, \mathbf{s}) &= k^2 s_z S^{(0)}(Q_0, \nu)\tilde{g}^{(0)}(k\mathbf{s}_\perp, \nu) && \text{if } \mathbf{s} \in \Omega_P \\ &= 0 && \text{if } \mathbf{s} \notin \Omega_P, \end{aligned}\right\} \qquad (5.7\text{–}108)$$

where Ω_P denotes the solid angle generated by the vectors pointing from all the source points to P.

From the simple relation (5.7–107) and from Eqs. (5.7–65) and (5.7–66) it follows that, in the asymptotic limit as $k \to \infty$, the function $\mathcal{B}_\nu$ is necessarily non-negative and that it vanishes at all points in the source plane which are located outside the source region. Hence it obeys the two basic postulates of traditional radiometry. Moreover we see at once from Eqs. (5.7–104) and (5.7–107) that

$$\mathcal{B}_\nu(P, \mathbf{s}) = \mathcal{B}_\nu^{(0)}(Q_0, \mathbf{s}) \qquad (5.7\text{–}109)$$

i.e. $\mathcal{B}_\nu(P, \mathbf{s})$ *is constant along each straight line in the half-space* $z > 0$. The formula (5.7–109), which is formally identical with Eq. (5.7–88) for the specific intensity of radiation, implies that in free space the function $\mathcal{B}_\nu(P, \mathbf{s}) \equiv \mathcal{B}_\nu(\mathbf{r}, \mathbf{s})$

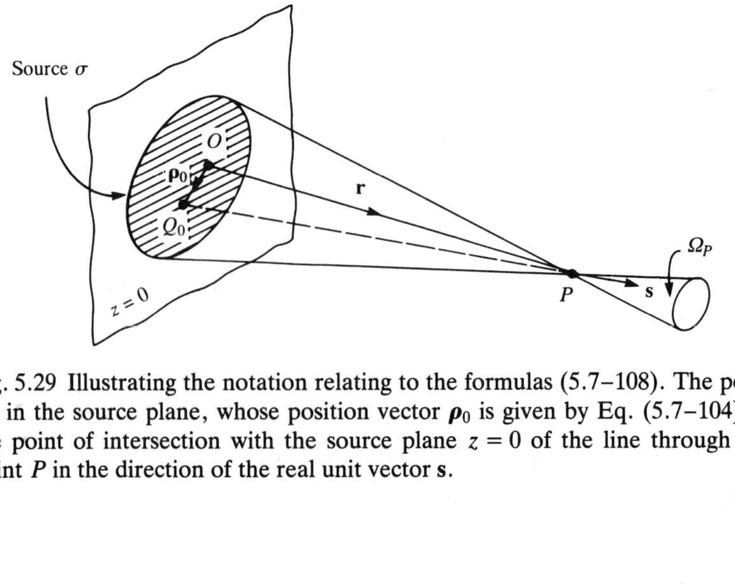

Fig. 5.29 Illustrating the notation relating to the formulas (5.7–108). The point Q_0 in the source plane, whose position vector $\boldsymbol{\rho}_0$ is given by Eq. (5.7–104), is the point of intersection with the source plane $z = 0$ of the line through the point P in the direction of the real unit vector $\mathbf{s}$.

satisfies the equation (5.7–87) of radiative energy transfer, namely

$$\mathbf{s}\cdot\nabla\mathcal{B}_\nu(\mathbf{r},\mathbf{s}) = 0. \tag{5.7–110}$$

Evidently in this case $\mathcal{B}_\nu(\mathbf{r},\mathbf{s})$ may be identified with the specific intensity $I_\nu(\mathbf{r},\mathbf{s})$ (Section 5.7.4) and also with the radiance function $B_\nu(\mathbf{r},\mathbf{s})$ of traditional radiometry. For this reason we will denote this limiting form by $B_\nu(\mathbf{r},\mathbf{s})$.

We may summarize the main results that we have derived in this section by the following statement: *In the asymptotic limit as* $k = 2\pi/\lambda \equiv 2\pi\nu/c \to \infty$ *the radiance function given by* Eq. (5.7–105) or, more explicitly [if Eq. (5.7–104) is also used], *by the formula*

$$B_\nu(\mathbf{r},\mathbf{s}) = k^2 s_z[S^{(0)}(\mathbf{r}_\perp - (z/s_z)\mathbf{s}_\perp, \nu)]\tilde{g}^{(0)}(k\mathbf{s}_\perp, \nu) \tag{5.7–111}$$

correctly describes the behavior of the field generated by any planar, secondary, quasi-homogeneous source. We stress that the formula (5.7–111) shows that the radiance depends not only on the distribution of the spectral density $S^{(0)}(\boldsymbol{\rho},\nu)$ across the source but also on the degree of coherence $g^{(0)}(\mathbf{r}_2 - \mathbf{r}_1, \nu)$ of the source.

In Figure 5.30, polar diagrams calculated from Eq. (5.7–105) are shown, illustrating the behavior of the radiance function of fields generated by some quasi-homogeneous sources. They show, as was to be expected, that the coherence properties of a source can considerably affect the radiance function.

The starting point of the preceding analysis was the definition (5.7–49) of the generalized radiance function in the plane of the secondary source. The same results were shown to follow from the alternative definition (5.7–46) (Kim and Wolf, 1987). In fact there are some indications that in the asymptotic limit as $k \to \infty$ many other possible definitions of generalized radiances will yield the same result (Agarwal, Foley and Wolf, 1987; Friberg, Agarwal, Foley and Wolf, 1992).

5.8 Effects of spatial coherence of a source on the spectrum of radiated fields

In Sections 5.2 and 5.3 we considered radiation from sources of arbitrary state of coherence and we learned how coherence properties of a source affect the *spatial* distribution of the radiated energy at any chosen frequency. In this section we will examine how coherence properties of the source affect the *spectral* distribution of energy, i.e. how they affect the spectrum of the emitted radiation, considered as a function of frequency. It turns out that the changes in the state of spatial coherence of the source can lead to appreciable changes in the spectrum of the radiated field.

We begin by considering a very simple example which illustrates the essence of this phenomenon.

5.8.1 Spectrum of the field generated by two partially correlated sources‡

We consider the field generated by two fluctuating sources, located at points P_1 and P_2. Let $\{Q(P_1,\omega)\}$ and $\{Q(P_2,\omega)\}$ be the ensembles that represent the

‡ The analysis in this section follows closely that given by Wolf (1987c).

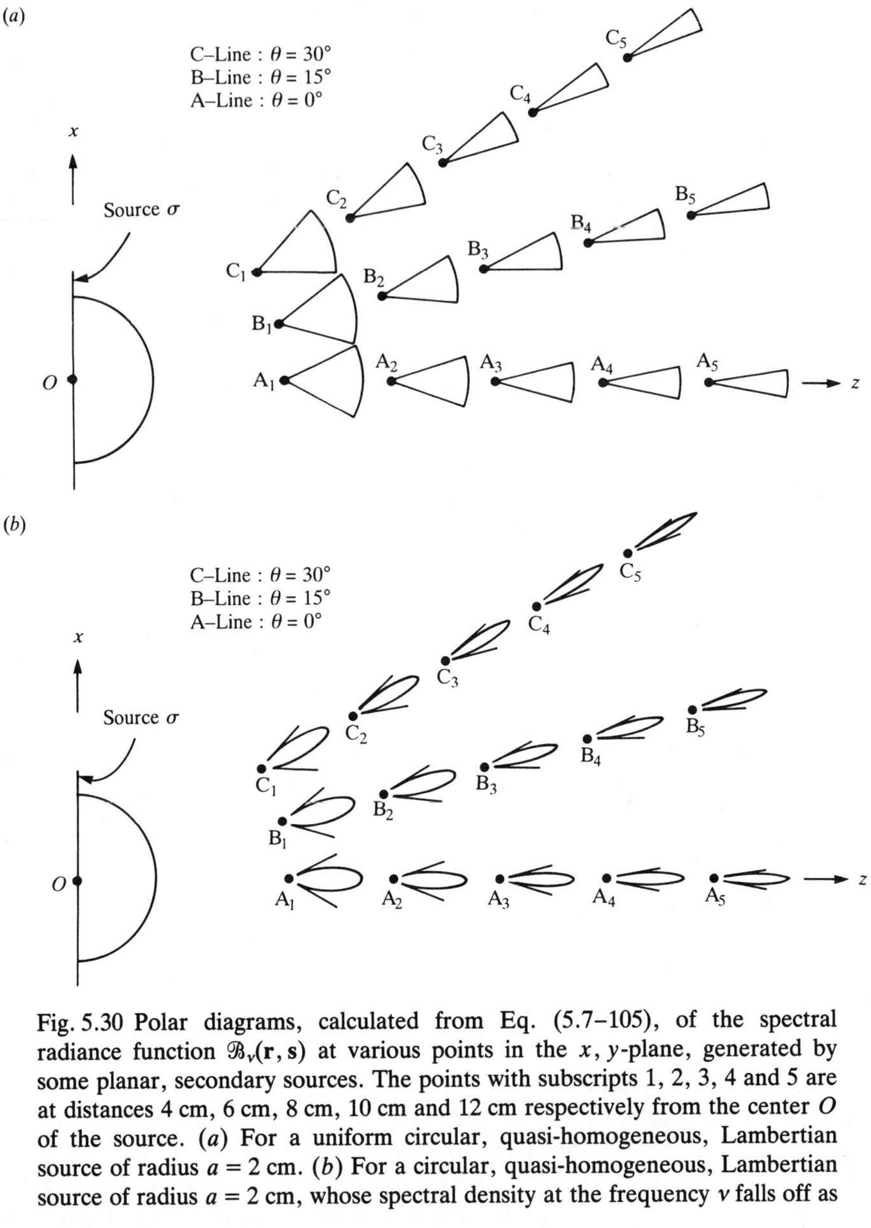

Fig. 5.30 Polar diagrams, calculated from Eq. (5.7–105), of the spectral radiance function $\mathscr{B}_\nu(\mathbf{r}, \mathbf{s})$ at various points in the x, y-plane, generated by some planar, secondary sources. The points with subscripts 1, 2, 3, 4 and 5 are at distances 4 cm, 6 cm, 8 cm, 10 cm and 12 cm respectively from the center O of the source. (*a*) For a uniform circular, quasi-homogeneous, Lambertian source of radius $a = 2$ cm. (*b*) For a circular, quasi-homogeneous, Lambertian source of radius $a = 2$ cm, whose spectral density at the frequency ν falls off as

source fluctuations, assumed to be statistically stationary, and let $\{U(P, \omega)\}$ be the ensemble that represents the field generated by the sources at a point P (Fig. 5.31).[§] If the sources are sufficiently small, each realization $U(P, \omega)$ may be expressed in the form

$$U(P, \omega) = Q(P_1, \omega)\frac{e^{ikR_1}}{R_1} + Q(P_2, \omega)\frac{e^{ikR_2}}{R_2}, \qquad (5.8\text{–}1)$$

§ Following the notation employed in most of the publications referred to in this section, we now use the angular frequency ω rather than the circular frequency $\nu = \omega/2\pi$.

(c)

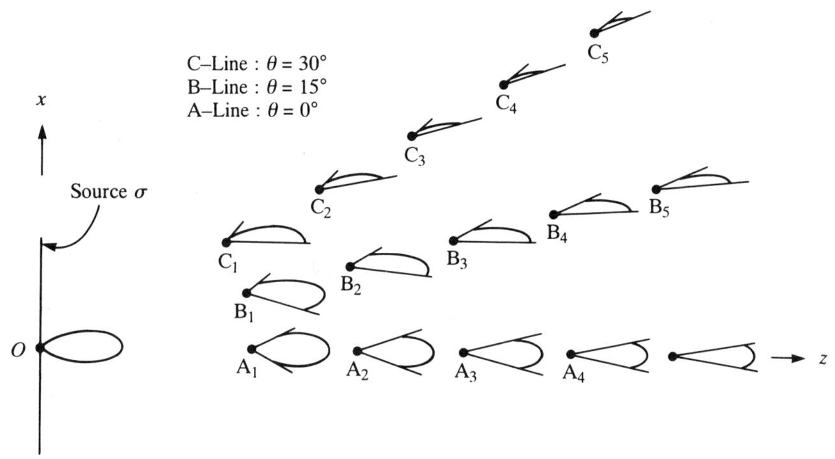

(d)

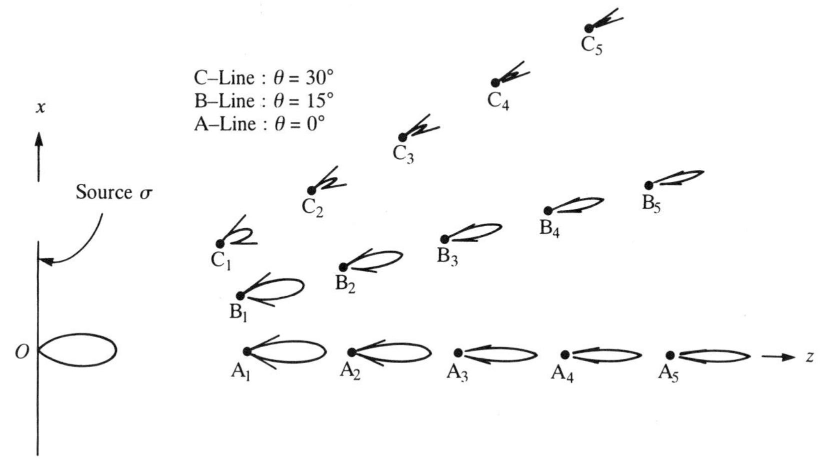

Caption for Fig. 5.30 (*cont.*).
a Gaussian function (Eq. (5.3–23)] with r.m.s. width $\sigma_S(\nu) = 1$ cm. (*c*) For a uniform, circular, quasi-homogeneous, Gaussian-correlated source [Eq. (5.3–24)] of radius $a = 2$ cm, with $\sigma_g(\nu) = 0.5\lambda$. (*d*) For a circular, Gaussian-correlated source of radius $a = 2$ cm, $\sigma_g(\nu) = 0.5\lambda$, whose spectral density at frequency ν falls off as a Gaussian function, with $\sigma_S(\nu) = 1$ cm. (After Foley and Wolf, 1991.)

where R_1 and R_2 are the distances from P_1 to P and from P_2 to P respectively and $k = \omega/c$, c being the speed of light in vacuo. The spectrum of the field at the point P is given by the expression

$$S(P, \omega) = \langle U^*(P, \omega)U(P, \omega)\rangle, \qquad (5.8\text{–}2)$$

where, as usual, the angular brackets denote the ensemble average. On substituting from Eq. (5.8–1) into Eq. (5.8–2) one readily finds that

$$S(P, \omega) = \left(\frac{1}{R_1^2} + \frac{1}{R_2^2}\right)S_Q(\omega) + \left[W_Q(P_1, P_2, \omega)\frac{e^{ik(R_2-R_1)}}{R_1R_2} + \text{c.c.}\right]. \qquad (5.8\text{–}3)$$

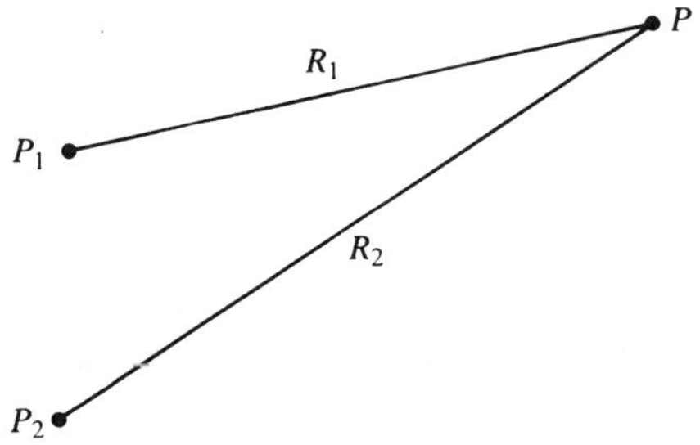

Fig. 5.31 Geometry and notation relating to the determination of the spectrum $S(P, \omega)$ of the field at a point P, produced by two small sources with identical spectra $S_Q(\omega)$, located at points P_1 and P_2.

Here

$$S_Q(\omega) = \langle Q^*(P_1, \omega)Q(P_1, \omega)\rangle = \langle Q^*(P_2, \omega)Q(P_2, \omega)\rangle \tag{5.8–4}$$

is the spectrum, assumed to be the same, of each source,

$$W_Q(P_1, P_2, \omega) = \langle Q^*(P_1, \omega)Q(P_2, \omega)\rangle \tag{5.8–5}$$

is the cross-spectral density of the two sources and c.c. denotes the complex conjugate.

Let us introduce the degree of correlation of the two sources by the formula

$$\mu_Q(P_1, P_2, \omega) = \frac{W_Q(P_1, P_2, \omega)}{S_Q(\omega)} \tag{5.8–6}$$

[cf. Eq. (4.3–47b)]. On substituting from Eq. (5.8–6) for W_Q into Eq. (5.8–3) we find that

$$S(P, \omega) = S_Q(\omega)\left\{\frac{1}{R_1^2} + \frac{1}{R_2^2} + \left[\mu_Q(P_1, P_2, \omega)\frac{e^{ik(R_2-R_1)}}{R_1R_2} + \text{c.c.}\right]\right\}. \tag{5.8–7}$$

For the sake of simplicity, let us choose the field point P to be located on the perpendicular bisector of the line joining P_1 and P_2. Then $R_1 = R_2$ (= R say) and the formula (5.8–7) reduces to

$$S(P, \omega) = \frac{2}{R^2}S_Q(\omega)[1 + \operatorname{Re}\mu_Q(P_1, P_2, \omega)], \tag{5.8–8}$$

where Re denotes the real part.

We see from Eq. (5.8–8) that in the special case when $\mu_Q(P_1, P_2, \omega) = 0$ for all frequencies present in the source spectrum $S_Q(\omega)$, i.e. when the two sources are uncorrelated, the field spectrum $S(P, \omega)$ is proportional to the source spectrum $S_Q(\omega)$. However, *in general, the two spectra S and S_Q will not be proportional to each other*, because of the presence of the degree of correlation $\mu_Q(P_1, P_2, \omega)$ in the expression (5.8–8). Moreover, as can be seen at once from Eq. (5.8–7), the field spectrum $S(P, \omega)$ will be different, in general, at different points in space. This simple example shows clearly that source correlations can modify the spectrum of the emitted field in a non-trivial way.

The modified field-spectrum may take on many different forms, depending on the nature of the frequency-dependence of the correlation coefficient $\mu_Q(P_1, P_2, \omega)$. Consider, for example, the situation when

$$\mu_Q(P_1, P_2, \omega) = a\mathrm{e}^{-(\omega-\omega_1)^2/2\delta_1^2} - 1, \tag{5.8–9}$$

where a, ω_1 and $\delta_1 \ll \omega_1$ are positive constants. In order that the expression (5.8–9) represents a correlation coefficient, it must satisfy the inequality $0 \leqslant |\mu_Q| \leqslant 1$, which implies that $a \leqslant 2$.

Suppose also that the spectrum $S_Q(\omega)$ of each of the two sources consists of a single spectral line of Gaussian profile:

$$S_Q(\omega) = A\,\mathrm{e}^{-(\omega-\omega_0)^2/2\delta_0^2}, \tag{5.8–10}$$

where A, ω_0 and $\delta_0 \ll \omega_0$ are also positive constants.

On substituting from Eqs. (5.8–9) and (5.8–10) into the formula (5.8–8) we obtain the following expression for the spectrum of the field at the point P:

$$S(P, \omega) = \frac{2Aa}{R^2}\mathrm{e}^{-(\omega-\omega_0)^2/2\delta_0^2}\,\mathrm{e}^{-(\omega-\omega_1)^2/2\delta_1^2}. \tag{5.8–11}$$

By a straightforward calculation, performed most simply by using the so-called product theorem for Gaussian functions (see Appendix 5.2), one can show that this expression may be re-written in the form

$$S(P, \omega) = A'\,\mathrm{e}^{-(\omega-\omega_0')^2/2\delta_0'^2}, \tag{5.8–12}$$

where

$$A' = \left(\frac{2Aa}{R^2}\right)\mathrm{e}^{-(\omega_1-\omega_0)^2/2(\delta_0^2+\delta_1^2)}, \tag{5.8–13}$$

$$\omega_0' = (\delta_1^2\omega_0 + \delta_0^2\omega_1)/(\delta_0^2 + \delta_1^2) \tag{5.8–14}$$

and

$$\frac{1}{\delta_0'^2} = \frac{1}{\delta_0^2} + \frac{1}{\delta_1^2}. \tag{5.8–15}$$

Equation (5.8–12) shows that the spectrum of the field at the point P consists of a single line of Gaussian profile. However this line is not centered, in general, at the center frequency ω_0 of the source spectrum $S_Q(\omega)$ but is centered at the frequency ω_0' given by Eq. (5.8–14).

If the two sources were uncorrelated ($\mu_Q = 0$), we would have, according to Eqs. (5.8–8) and (5.8–10),

$$[S(P, \omega)]_{\text{uncorr.}} = \left(\frac{2A}{R^2}\right)\mathrm{e}^{-(\omega-\omega_0)^2/2\delta_0^2}. \tag{5.8–16}$$

Comparison of this expression with Eq. (5.8–12) shows that although both lines have Gaussian profiles, they differ from each other. Since according to Eq. (5.8–15) $\delta_0' < \delta_0$, the spectral line of the field produced by the correlated sources is narrower than the spectral line produced by the uncorrelated ones. Further we can readily deduce from Eq. (5.8–14) that $\omega_0' \gtreqless \omega_0$ according as $\omega_1 \gtreqless \omega_0$. Hence if $\omega_1 < \omega_0$ the spectral line (5.8–12) of the field is centered on a lower frequency than the spectral line (5.8–16) produced by two uncorrelated sources or, as one says, it is *redshifted*; and if $\omega_1 > \omega_0$, the spectral line (5.8–12) is centered on a higher frequency and it is then said to be *blueshifted*. These results are illustrated in Fig. 5.32. Examples of other kinds of spectral changes produced by suitable source correlations are given in Figs. 5.33 and 5.34. (See also Gamliel and Wolf, 1988; Gamliel, 1988.)

The prediction that correlations between radiating sources can give rise to shifts of spectral lines was first demonstrated, with acoustical rather than optical waves, by Bocko, Douglass and Knox (1987). Two partially correlated sources (speakers) with the same spectrum were obtained with the help of two independent generators which produced random signals $X(t)$ and $Y(t)$. The combined signals

$$Q(P_1, t) = X(t) + Y(t) \tag{5.8–17a}$$

and

$$Q(P_2, t) = Y(t) - X(t) \tag{5.8–17b}$$

drove the two speakers, located in the neighborhood of the points P_1 and P_2 (Fig. 5.35). It follows from Eqs. (5.8–17) that

$$\begin{aligned}\langle Q^*(P_1, t)Q(P_1, t + \tau)\rangle &= \langle Q^*(P_2, t)Q(P_2, t + \tau)\rangle \\ &= \langle X^*(t)X(t + \tau)\rangle + \langle Y^*(t)Y(t + \tau)\rangle, \end{aligned} \tag{5.8–18}$$

where we have used the fact that

$$\langle X^*(t)Y(t + \tau)\rangle = \langle Y^*(t)X(t + \tau)\rangle = 0, \tag{5.8–19}$$

because the two generators were independent. On taking the Fourier transform of

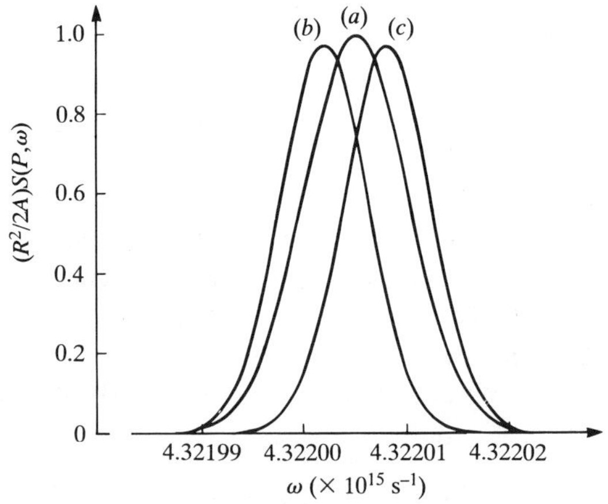

Fig. 5.32 Redshift and blueshift of a spectral line as predicted by formulas (5.8–12) to (5.8–15). The spectrum $S_Q(\omega)$ of both of the two source distributions is a single line of a Gaussian profile given by Eq. (5.8–10) with $A = 1$, $\omega_0 = 4.32201 \times 10^{15}\ \text{s}^{-1}$ (Hg line $\lambda = 4358.33$ Å), $\delta_0 = 5 \times 10^9\ \text{s}^{-1}$. (*a*) The field spectrum $S(P, \omega)$ at P when the two sources are uncorrelated [$\mu_Q = 0$]. (*b*) The field spectra at P when the two sources are correlated in accordance with Eq. (5.8–9), with $a = 1.8$, $\delta_1 = 7.5 \times 10^9\ \text{s}^{-1}$, and with $\omega_1 = \omega_0 - 2\delta_0$ (redshifted line) and (*c*) $\omega_1 = \omega_0 + 2\delta_0$ (blueshifted line). (After Wolf, 1987c.)

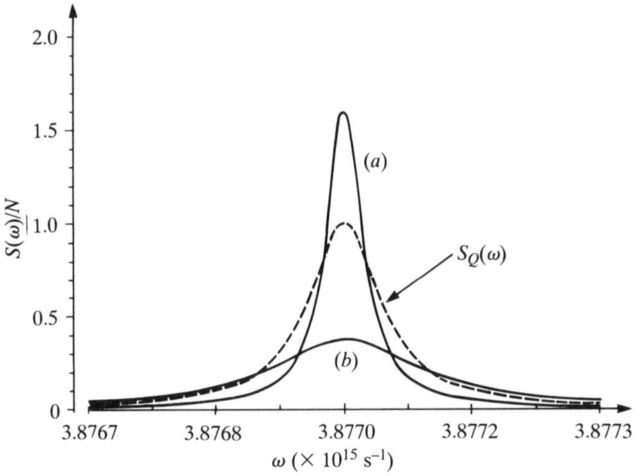

Fig. 5.33 Line narrowing (a) and line broadening (b) produced by source correlations, calculated from Eq. (5.8–8). The source spectrum is a single line of Lorentzian profile $S_Q(\omega) = A_0[(\omega - \omega_0)^2 + \Gamma_0^2]^{-1}$ and the spectral degree of coherence is

$$\mu_Q(\omega) = \frac{A_j(\omega - \omega_0)^2 + \Gamma_0^2}{A_0(\omega - \omega_j)^2 + \Gamma_j^2} - 1,$$

with $\omega_0 = 3.77 \times 10^{15}\,\text{s}^{-1}$, $\Gamma_0 = 5 \times 10^{10}\,\text{s}^{-1}$. In case ($a$), $j = 1$, $\Gamma_1 = 2.5 \times 10^{10}\,\text{s}^{-1}$, $A_1/A_0 = 0.4$. In case (b), $j = 2$, $\Gamma_2 = 10^{11}\,\text{s}^{-1}$, $A_2/A_0 = 1.5$. The normalization factor $N = A_0/\Gamma_0^2$.

Eqs. (5.8–18) and making use of the Wiener–Khintchine theorem [Eqs. (2.4–15) and (2.4–16)] we see at once that the spectra of the two sources are the same, each being given by

$$S_Q(\omega) = S_X(\omega) + S_Y(\omega) \tag{5.8–20}$$

where $S_X(\omega)$ and $S_Y(\omega)$ are the (power) spectra of $X(t)$ and $Y(t)$ respectively.

It further follows from Eqs. (5.8–17) that the cross-correlation function of the two sources is given by the expression

$$\langle Q^*(P_1, t)Q(P_2, t + \tau)\rangle = -\langle X^*(t)X(t + \tau)\rangle + \langle Y^*(t)Y(t + \tau)\rangle. \tag{5.8–21}$$

On taking the Fourier transform of this formula and using the Wiener–Khintchine theorem again we obtain for the cross-spectral density of the two sources the expression

$$W_Q(P_1, P_2, \omega) = -S_X(\omega) + S_Y(\omega). \tag{5.8–22}$$

From Eqs. (5.8–22) and (5.8–20) it follows that the degree of correlation $\mu_Q(P_1, P_2, \omega)$ of the two sources, defined by Eq. (5.8–6), is given by the formula

$$\mu_Q(P_1, P_2, \omega) = \frac{-S_X(\omega) + S_Y(\omega)}{S_X(\omega) + S_Y(\omega)}. \tag{5.8–23}$$

In this experiment a microphone M was placed at a location equidistant from the two acoustic sources. The spectra were measured at the microphone one at a

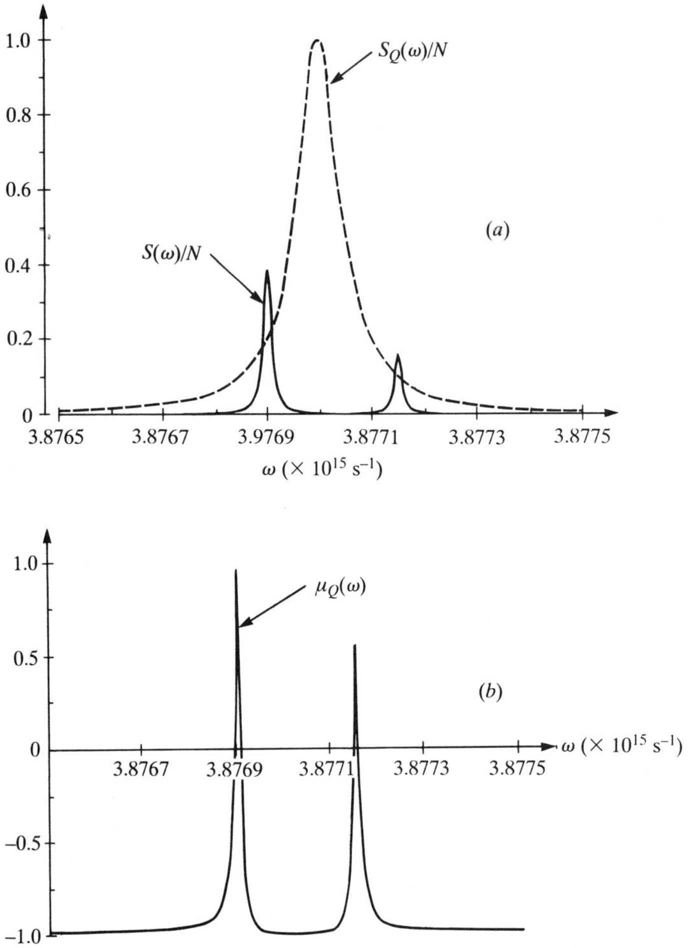

Fig. 5.34 Generation of two spectral lines, calculated from Eq. (5.8–8), from a single spectral line of Lorentzian profile $S_Q(\omega) = A_0[(\omega - \omega_0)^2 + \Gamma_0^2]^{-1}$ and spectral degree of coherence

$$\mu_Q(\omega) = \sum_{j=1}^{2} \frac{A_j}{A_0} \frac{(\omega - \omega_0)^2 + \Gamma_0^2}{(\omega - \omega_j)^2 + \Gamma_j^2} - 1,$$

with $\omega_0 = 3.87700 \times 10^{15}\,\mathrm{s}^{-1}$, $\Gamma_0 = 5 \times 10^{10}\,\mathrm{s}^{-1}$, $\omega_1 = 3.87690 \times 10^{15}\,\mathrm{s}^{-1}$, $\Gamma_1 = 8 \times 10^9\,\mathrm{s}^{-1}$, $A_1/A_0 = 0.01$, $\omega_2 = 3.87715 \times 10^{15}\,\mathrm{s}^{-1}$, $\Gamma_2 = 8 \times 10^9\,\mathrm{s}^{-1}$, $A_2/A_0 = 0.004$. The normalization factor $N = A_0/\Gamma_0^2$. Figure (*a*) shows the spectral lines and figure (*b*) shows the behavior of the spectral degree of coherence $\mu_Q(\omega)$.

time, with one source switched off and then the other. The spectra were found to have the same shape as the spectra of the two sources. Next, with both sources switched on, the spectrum $S(M, \omega)$ of the sound at the microphone M was measured. According to Eq. (5.8–8), with $S_Q(\omega)$ given by Eq. (5.8–20) and $\mu_Q(P_1, P_2, \omega)$ given by Eq. (5.8–23), one finds that

$$S(M, \omega) = \frac{4}{R^2} S_Y(\omega). \tag{5.8–24}$$

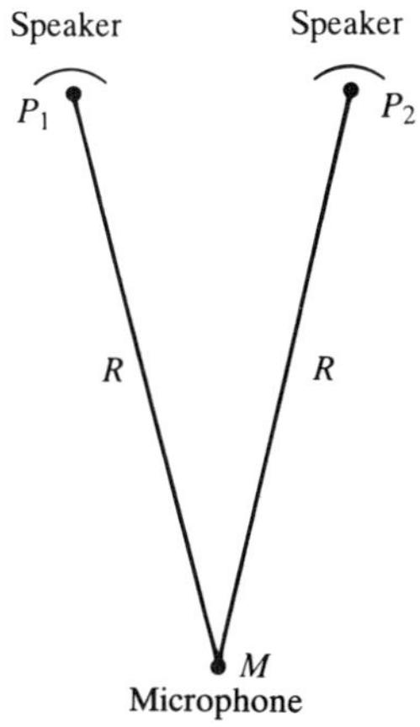

Fig. 5.35 The configuration of the acoustical experiment which was used to demonstrate the generation of frequency shifts of spectral lines by source correlations. (Adapted from Bocko, Douglass and Knox, 1987.)

This spectrum is evidently not proportional to the spectrum $S_Q(\omega)$ [Eq. (5.8–20)] of each of the two sources, unless $S_X(\omega)$ is proportional to $S_Y(\omega)$. Depending on the forms of $S_X(\omega)$ and $S_Y(\omega)$ the peak of the field spectrum $S(M, \omega)$ can be displaced with respect to the peak of the source spectrum S_Q and this was observed in the experiment. With suitable choices of the spectra of the signals $X(t)$ and $Y(t)$, both redshift [Fig. 5.36(*a*)] and blueshift [Fig. 5.36(*b*)] were observed.

An optical experiment illustrating shifts of spectral lines produced by two partially correlated sources was performed by Gori, Guattari, Palma and Padovani (1988) soon afterwards. Secondary sources P_1 and P_2 were produced by

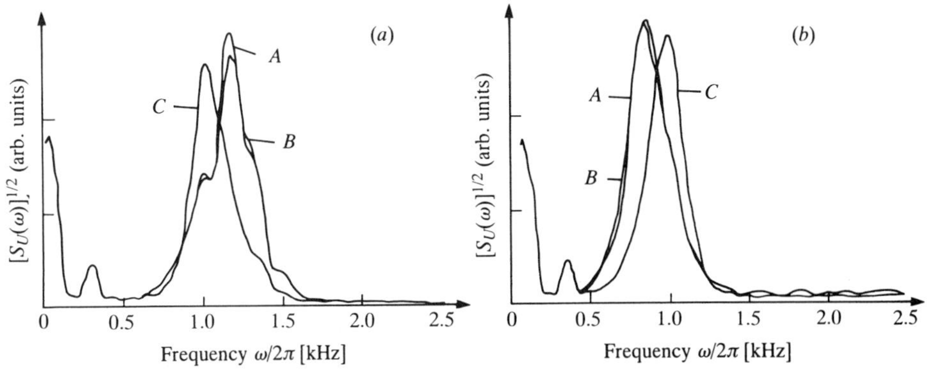

Fig. 5.36 Square-root of the spectral density of the measured lines in the acoustical experiment outlined in Fig. 5.35. The curves A and B are the spectral lines measured at the microphone M when only one of the sources was switched on. The curves C represent the measured spectral lines at M when both sources were switched on. The line is seen to be redshifted with respect to A and B in case (a) and blueshifted in case (b), depending on the choice of the correlation coefficient μ_Q. (Reproduced from Bocko, Douglass and Knox, 1987.)

simultaneously illuminating two pinholes with two coherent but mutually independent beams B_α, B_β, using a beam splitter BS. (Fig. 5.37). The correlation between the sources P_1 and P_2 depends on the orientation of the beam splitter and can be varied by changing the inclination of the beam splitter with respect to the normal to the plane $\mathscr{A}$ which contains the two pinholes.

The spectrum of the field produced by the secondary sources P_1 and P_2 was examined by a spectrum analyzer SA, located in the vicinity of a point, at the same distance R from each source. Both redshifted and blueshifted spectra were observed with this arrangement (Fig. 5.38).

An interesting technique for producing two small sources which can be correlated in any prescribed manner was described by Faklis and Morris (1992).

Another conceptually simple way to illustrate the effect of coherence on the

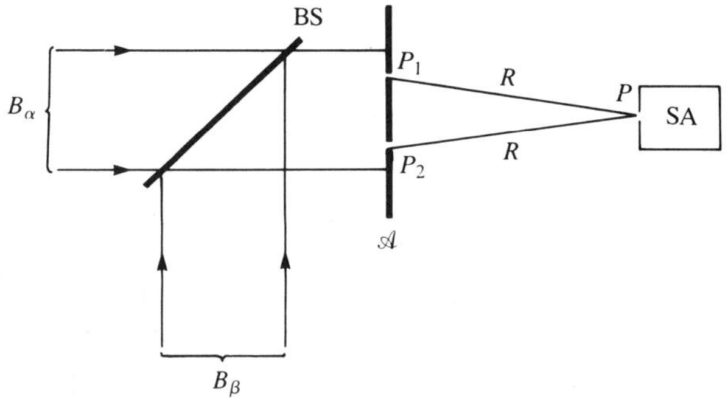

Fig. 5.37 The lay-out of an optical experiment used to demonstrate the generation of frequency shifts of spectral lines due to source correlations. (After Gori, Guattari, Palma and Padovani, 1988.)

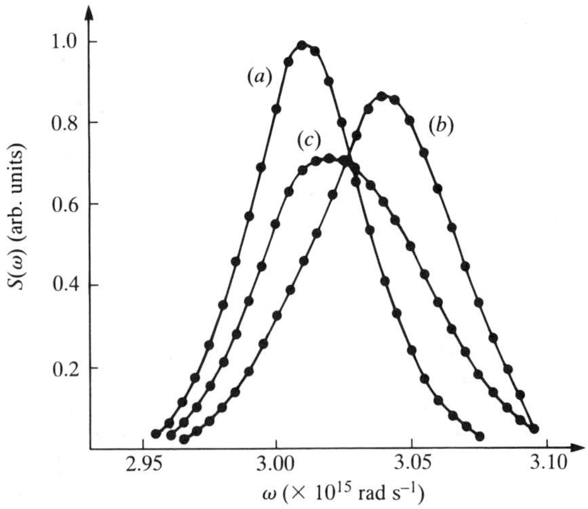

Fig. 5.38 Results of measurements (denoted by heavy dots), demonstrating shifts of spectral lines due to correlation between two small optical sources, with the arrangement of Fig. 5.37. For the sake of clarity the experimental data points have been connected by lines. Redshifted spectrum (*a*), blueshifted spectrum (*b*) and the source spectrum, multiplied by a factor 2 (*c*). (After Gori, Guattari, Palma and Padovani, 1988.)

spectrum of light is by means of a Young's interference experiment. If the pinholes are illuminated by partially coherent light, the spectrum of the light in the fringe pattern will differ from the spectrum of the light incident on the pinholes for two reasons: because of diffraction at the pinholes and because of the correlation between the light at the two pinholes. However, the change arising from partial coherence is only appreciable when the bandwidth of the incident light is sufficiently broad, as was first shown by James and Wolf (1991a,b). We will present a brief account of the main part of their analysis.

Suppose the two pinholes are illuminated by a distant, planar, incoherent, uniform circular source, as shown in Fig. 5.39. We assume a symmetric arrangement, with the planes of the pinholes being parallel to the source plane. It readily follows from one of the reciprocity theorems for quasi-homogeneous sources [Eq. (5.3–22)] that the spectral degree of coherence of the light incident on the pinholes is given by

$$\mu(P_1, P_2, \omega) = \frac{2J_1(d\omega\alpha/c)}{d\omega\alpha/c}, \tag{5.8–25}$$

where J_1 is the Bessel function of the first kind and first order, 2α is the angular diameter which the source subtends at the midpoint O between the two pinholes and d is the distance between them, c being the speed of light in vacuum. Let us for simplicity consider the spectrum $S(P_0, \omega)$ at a point P_0 some distance behind the plane containing the pinholes, on the normal CO to the source plane. It readily follows from Eqs. (4.3–54) and (5.8–25) and an elementary formula for diffraction at a pinhole that

$$S(P_0, \omega) = 2\left(\frac{A}{2\pi cr}\right)^2 \omega^2 \left[1 + \frac{2J_1(d\omega\alpha/c)}{d\omega\alpha/c}\right] S^{(i)}(\omega), \tag{5.8–26}$$

($\omega = 2\pi\nu$), where $S^{(i)}(\omega)$ is the spectrum of the light incident on the two pinholes, A is the area of each pinhole (assumed to be sufficiently small) and r is the distance $\overline{P_1P_0}$ ($= \overline{P_2P_0}$). Figure 5.40 shows the form of the spectrum $S(P_0, \omega)$, calculated from Eq. (5.8–26) for different separations d of the pinholes, when the spectrum of the incident light is the Planck spectrum

$$S^{(i)}(\omega) = \frac{A'\omega^3}{\exp(\hbar\omega/k_B T) - 1}. \tag{5.8–27}$$

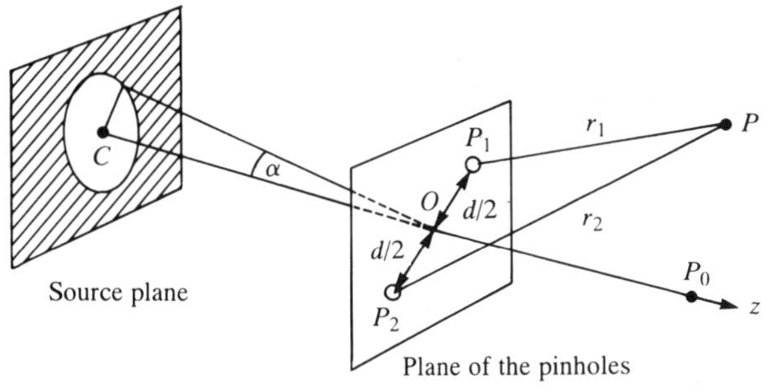

Fig. 5.39 Illustrating the geometry and notation relating to Eq. (5.8–25). (After James and Wolf, 1991a.)

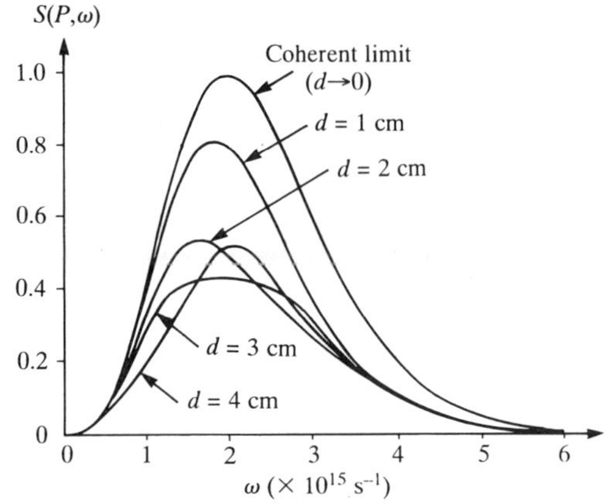

Fig. 5.40 The changes produced at the axial point P_0 (Fig. 5.39) in the Planck spectrum by interference, with different pinhole separations d. The source was assumed to be at temperature $T = 3000$ K and to subtend an angular semi-diameter $\alpha = 2.96 \times 10^{-5}$ radians at O. The units on the vertical axis are arbitrary. (After James and Wolf, 1991a.)

Here A' is a constant, $\hbar$ is the Planck's constant divided by 2π, k_B is the Boltzmann constant and T is the temperature. The figure shows that there is quite a drastic change in the spectrum at P_0 as the two pinholes are separated by larger and larger distances, i.e. as the spectral degree of coherence $\mu(P_1, P_2, \omega)$ of the light incident on the two pinholes changes with their separation distance d. This effect was demonstrated experimentally by Kandpal, Vaishya, Chander, Saxena, Mehta and Joshi (1992) and by Santarsiero and Gori (1992).

We see from Eq. (5.8–25) that the spectral degree of coherence depends on ω and d only through their product. If we write $\mu(d, \omega; \alpha)$ in place of $\mu(P_1, P_2, \omega)$ this result implies that

$$\mu(d, \omega; \alpha) = \mu(\beta d, \omega/\beta; \alpha) \tag{5.8–28}$$

where β is any positive number. This formula shows that, under the circumstances for which the formula (5.8–25) applies, the spectral degree of coherence at frequency ω, of the light at points separated by a distance d, is the same as the spectral degree of coherence at frequency ω/β of the light at points separated by a distance βd. Thus there exists a 'trade off' between frequency and separation distance. This result is an example of the so-called *interferometric equivalence principle* for certain types of fields (James and Wolf, 1991c). Such fields are produced, for example, by all planar, quasi-homogeneous sources, which have the same normalized spectrum at all source points. This principle is of interest in connection with attaining high resolution in radio astronomy using only two radio telescopes with fixed separation.

5.8.2 *Spectrum of the far field generated by planar, secondary, quasi-homogeneous sources*

We have just seen that when light from two small correlated sources with the same spectrum is superposed the spectrum will, in general, change. It might be expected that spectral changes will also occur when light is produced by an extended partially coherent or by a completely coherent source, i.e. that the spectrum of light produced by such sources will, in general, not be invariant on propagation. In this section we will briefly consider such changes when the light originates in a planar, secondary source σ, which we will assume to be quasi-homogeneous.

We begin with Eq. (5.3–21) for the radiant intensity produced by such a source, namely

$$J(\mathbf{s}, \omega) = (2\pi k)^2 \tilde{S}^{(0)}(0, \omega) \tilde{g}^{(0)}(k\mathbf{s}_\perp, \omega) \cos^2\theta. \qquad (5.8\text{–}29)$$

Here

$$\tilde{S}^{(0)}(0, \omega) = \frac{1}{(2\pi)^2} \int_\sigma S^{(0)}(\boldsymbol{\rho}, \omega)\, \mathrm{d}^2\rho, \qquad (5.8\text{–}30)$$

$S^{(0)}(\boldsymbol{\rho}, \omega)$ being the spectral density of the field at a typical source point and

$$\tilde{g}^{(0)}(\mathbf{f}', \omega) = \frac{1}{(2\pi)^2} \int g^{(0)}(\boldsymbol{\rho}', \omega)\, \mathrm{e}^{-\mathrm{i}\mathbf{f}' \cdot \boldsymbol{\rho}'}\, \mathrm{d}^2\rho' \qquad (5.8\text{–}31)$$

is the two-dimensional spatial Fourier transform of the spectral degree of coherence $g^{(0)}(\boldsymbol{\rho}_2 - \boldsymbol{\rho}_1, \omega)$ of the light in the source plane. Further

$$k = \omega/c \qquad (5.8\text{–}32)$$

denotes, as usual, the wave number associated with frequency ω, θ is the angle which the unit vector $\mathbf{s}$ makes with the normal to the source plane (see Fig. 5.41) and $\mathbf{s}_\perp$ is the projection, considered as a two-dimensional vector, of the unit vector $\mathbf{s}$ onto the source plane.

Since we are interested in the spectrum $S^{(\infty)}(r\mathbf{s}, \omega)$ in the far zone rather than in the radiant intensity $J(\mathbf{s}, \omega)$, we make use of the simple relation (5.2–12), namely

$$S^{(\infty)}(r\mathbf{s}, \omega) = J(\mathbf{s}, \omega)/r^2, \qquad (5.8\text{–}33)$$

and we then obtain from Eq. (5.8–33) and (5.8–29) the following formula for the far-zone spectrum:

$$S^{(\infty)}(r\mathbf{s}, \omega) = \left(\frac{2\pi k}{r}\right)^2 \tilde{S}^{(0)}(0, \omega) \tilde{g}^{(0)}(k\mathbf{s}_\perp, \omega) \cos^2\theta. \qquad (5.8\text{–}34)$$

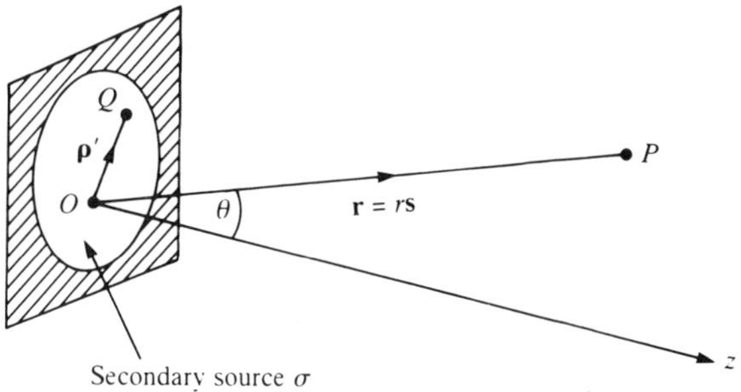

Fig. 5.41 Notation relating to the derivation of Eq. (5.8–34) for the far-zone spectrum produced by a planar, secondary, quasi-homogeneous source.

Suppose now that the normalized spectrum, $\mathfrak{s}^{(0)}(\omega)$ say, is the same at every source point, i.e. that

$$S^{(0)}(\boldsymbol{\rho}, \omega) = I^{(0)}(\boldsymbol{\rho})\mathfrak{s}^{(0)}(\omega) \tag{5.8–35}$$

where

$$I(\boldsymbol{\rho}) = \int_0^\infty S^{(0)}(\boldsymbol{\rho}, \omega)\, \mathrm{d}\omega. \tag{5.8–36}$$

Evidently

$$\int_0^\infty \mathfrak{s}^{(0)}(\omega)\, \mathrm{d}\omega = 1. \tag{5.8–37}$$

The function $I(\boldsymbol{\rho})$, defined by Eq. (5.8–36), represents the optical intensity of the light at the source point $Q(\boldsymbol{\rho})$. On substituting from Eq. (5.8–35) into Eq. (5.8–34) we obtain the following expression for the far-zone spectrum:‡

$$S^{(\infty)}(r\mathbf{s}, \omega) = \left(\frac{2\pi k}{r}\right)^2 \tilde{I}(0)\mathfrak{s}^{(0)}(\omega)\tilde{g}^{(0)}(k\mathbf{s}_\perp, \omega)\cos^2\theta, \tag{5.8–38}$$

where

$$\tilde{I}(0) = \frac{1}{(2\pi)^2}\int_\sigma I(\boldsymbol{\rho})\, \mathrm{d}^2\rho. \tag{5.8–39}$$

The formula (5.8–38) shows that, apart from simple geometrical factors, the spectrum of the light in the far zone differs, in general, from that in the source plane for two reasons:

(a) Because of the presence of the proportionality factor $k^2 = (\omega/c)^2$, which is reminiscent of the proportionality factor that appears in formulas which describe the effect of diffraction of light at an aperture.
(b) Because of the presence of the frequency-dependent factor $\tilde{g}^{(0)}(k\mathbf{s}_\perp, \omega)$, which depends on the correlation properties of the light in the source plane.

If, in particular, the source spectrum $\mathfrak{s}^{(0)}(\omega)$ consists of a single spectral line centered at frequency ω_0 and if $\tilde{g}^{(0)}$ is peaked at a different frequency ω_1, the far-field spectrum will be centered at some frequency which differs from ω_0 by an amount depending on ω_1. Thus, just as in the simple case discussed in Section 5.8.1, of a radiating system consisting of two correlated sources, correlations of the field across a planar, secondary, quasi-homogeneous source can produce changes in the spectrum of the radiated field and these changes may appear essentially as a shift of the spectral line. It is not a true shift because the displaced line is slightly distorted. Moreover, because of the presence of the quantity $k\mathbf{s}_\perp$ in the argument of $\tilde{g}^{(0)}$, spectral changes will, in general, also depend on the direction of observation, **s**.

As an example let us consider the situation when the normalized spectrum of the secondary source is just a single line of Gaussian profile. Then

$$\mathfrak{s}^{(0)}(\omega) = \frac{1}{\delta_0\sqrt{(2\pi)}}\,\mathrm{e}^{-(\omega-\omega_0)^2/2\delta_0^2}, \quad (\delta_0 \ll \omega_0), \tag{5.8–40}$$

‡ An analogous formula for the far-zone spectrum of the radiation generated by a primary, three-dimensional, quasi-homogeneous source was derived by Wolf (1987a). See also Wolf (1987b).

where δ_0 and ω_0 are positive constants. We assume that at each effective frequency ω, the spectral degree of coherence of the light in the source plane is given by a Gaussian function, i.e. that

$$g^{(0)}(\boldsymbol{\rho}', \omega) = e^{-\rho'^2/2\sigma_g^2}, \qquad (5.8\text{–}41)$$

where σ_g is much smaller than the scale length of variation of the intensity function $I(\boldsymbol{\rho})$, defined by Eq. (5.8–36).

The two-dimensional spatial Fourier transform of the expression (5.8–41) is

$$\tilde{g}^{(0)}(\mathbf{f}, \omega) = \frac{\sigma_g^2}{2\pi} e^{-\sigma_g^2 f^2/2}. \qquad (5.8\text{–}42)$$

On substituting from Eqs. (5.8–40) and (5.8–42) into the formula (5.8–38) and using Eq. (5.8–32) and the fact that $\mathbf{s}_\perp^2 = \sin^2\theta$, we obtain the following expression for the far-zone spectrum:

$$S^{(\infty)}(r\mathbf{s}, \omega) = \frac{A}{r^2}\omega^2 e^{-(\omega-\omega_0)^2/2\delta_0^2} e^{-\omega^2/2\alpha^2(\theta)} \cos^2\theta, \qquad (5.8\text{–}43)$$

where

$$A = \frac{\sqrt{(2\pi)}\sigma_g^2 \tilde{I}(0)}{\delta_0 c^2}, \qquad (5.8\text{–}44)$$

$$\frac{1}{\alpha^2(\theta)} = \frac{\sigma_g^2}{c^2}\sin^2\theta. \qquad (5.8\text{–}45)$$

The product of the two Gaussian functions on the right-hand side of Eq. (5.8–43) can be expressed as a single Gaussian function, for example by the use of the product theorem for Gaussian functions (see Appendix 5.2). One then finds that

$$e^{-(\omega-\omega_0)^2/2\delta_0^2} e^{-\omega^2/2\alpha^2} = e^{-\omega_0^2/2(\delta_0^2+\alpha^2)} e^{-(\omega-\tilde{\omega})^2/2\tilde{\delta}^2}, \qquad (5.8\text{–}46)$$

where

$$\tilde{\omega}(\theta) = \frac{\omega_0\alpha^2(\theta)}{\delta_0^2 + \alpha^2(\theta)}, \qquad (5.8\text{–}47a)$$

$$\frac{1}{\tilde{\delta}^2(\theta)} = \frac{1}{\delta_0^2} + \frac{1}{\alpha^2(\theta)}. \qquad (5.8\text{–}47b)$$

On substituting from Eq. (5.8–46) into Eq. (5.8–43) we finally obtain the following expression for the far-zone spectrum:

$$S^{(\infty)}(r\mathbf{s}, \omega) = \left[\frac{A}{r^2} e^{-\omega_0^2/2[\delta_0^2+\alpha^2(\theta)]}\right]\omega^2 e^{-[\omega-\tilde{\omega}(\theta)]^2/2\tilde{\delta}^2(\theta)} \cos^2\theta. \qquad (5.8\text{–}48)$$

The formula (5.8–48) shows that the far-field spectrum is proportional to the product of ω^2 and a Gaussian distribution. The factor ω^2 produces a slight distortion of the Gaussian. Except when $\theta = 0$ [see Eq. (5.8–49) below], this Gaussian is not centered at the mean frequency ω_0 of the source spectrum but at a lower frequency, $\tilde{\omega}(\theta)$, given by Eq. (5.8–47a). Further, according to Eq. (5.8–47b) $\tilde{\delta}(\theta) < \delta_0$ (except when $\theta = 0$); i.e. the far-zone spectrum is narrower than the spectral line of the source. For $\theta = 0$, i.e. for the case when the direction of observation is perpendicular to the source plane, Eqs. (5.8–45), (5.8–47a) and (5.8–47b) give $1/\alpha(0) = 0$, $\tilde{\omega}(0) = \omega_0$ and $\tilde{\delta}(0) = \delta_0$, and the expression (5.8–48) for the far-zone spectrum reduces to

$$[S^{(\infty)}(r\mathbf{s}, \omega)]_{\theta=0} = \frac{A}{r^2}\omega^2 \, e^{-(\omega-\omega_0)^2/2\delta_0^2}. \qquad (5.8\text{–}49)$$

Some examples of spectral changes, calculated from Eq. (5.8–48), are shown in Fig. 5.42.

We have implicitly assumed in the preceding discussion that $\sigma_g \neq 0$, i.e. that the field in the source plane has a non-zero correlation length. Let us briefly consider the extreme case when the source is spatially completely incoherent, i.e. when

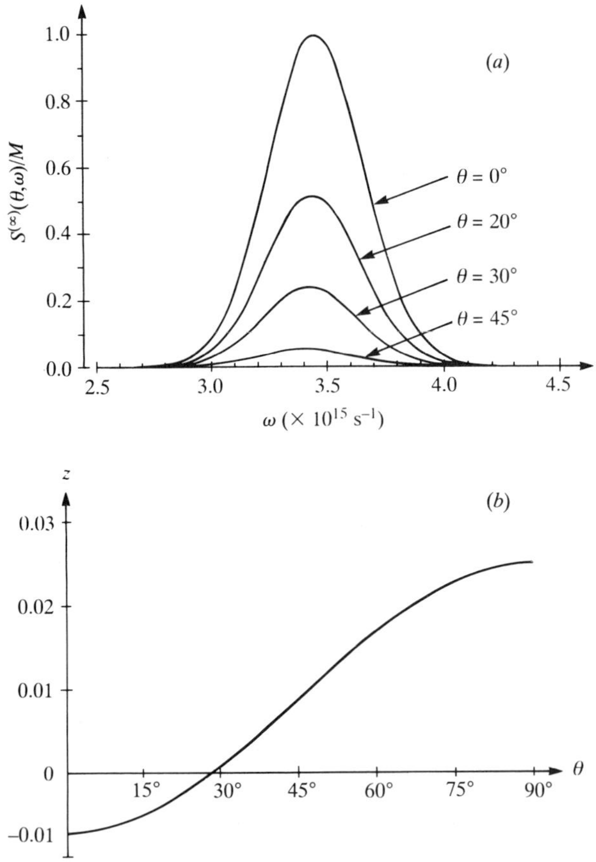

Fig. 5.42 (*a*) Far-zone spectra $S^{(\infty)}(\theta, \omega) \equiv S^{(\infty)}(r\mathbf{s}, \omega)$ (given by Eq. 5.8–48) in units of $M = A\omega_0^2/r^2$, as functions of the direction of observation θ, produced by a Gaussian-correlated [Eq. (5.8–41)], planar, secondary, quasi-homogeneous source, whose normalized spectrum is a line of Gaussian profile [Eq. (5.8–40)]. The constants were taken as $\sigma_g = 3\bar{\lambda} = 3\lambda_0/2\pi$, $\lambda_0 = 5500$ Å ($\omega_0 = 3.43 \times 10^{15}\ \text{s}^{-1}$) and $\delta_0 = 6 \times 10^{-2}\ \omega_0$. (*b*) The relative shift of the center frequency $z(\theta) = [\omega_0 - \tilde{\omega}(\theta)]/\tilde{\omega}(\theta) = [\tilde{\lambda}(\theta) - \lambda_0]/\lambda_0$, $\tilde{\lambda}(\theta) = 2\pi c/\tilde{\omega}(\theta)$.

$\sigma_g \to 0$. This limit is to be understood in the sense that $\sigma_g^2 \tilde{I}(0)$ remains finite as $\sigma_g \to 0$ to ensure that the factor A, defined by Eq. (5.8–44), has a non-zero value. We see from Eq. (5.8–45) that in this case $1/\alpha \to 0$, irrespective of the angle θ of observation. The formulas (5.8–47) now give $\tilde{\omega}(\theta) = \omega_0$ and $\tilde{\delta}(\theta) = \delta_0$ for all values of θ in the range $0 < \theta \leqslant \pi/2$, and Eq. (5.8–48) reduces to the right-hand side of Eq. (5.8–49), i.e. we have, with a spatially incoherent source (subscript 'incoh.'),

$$[S^{(\infty)}(r\mathbf{s}, \omega)]_{\text{incoh.}} = \frac{A}{r^2}\omega^2 \, \mathrm{e}^{-(\omega-\omega_0)^2/2\delta_0^2} \cos^2\theta. \tag{5.8–50}$$

We note the far-zone spectrum generated by the incoherent source depends on the direction of observation, its peak value falling off as $\cos^2\theta$.

We see that, in the two cases represented by Eqs. (5.8–49) and (5.8–50), the far-zone spectra are proportional to the product of the normalized source spectrum $\mathscr{s}^{(0)}(\omega)$ [Eq. (5.8–40)] and the factor ω^2, which produces a very slight distortion of the line and a small amount of blueshift.

We have so far only considered the effects of source correlations on the spectra of radiated fields in free space. In practice diffracting and sometimes also dispersive elements will be interposed between the source and the detector.‡ A diffracting aperture will, in general, change the coherence properties of the light which passes through it, even in the absence of dispersion. For example, if partially coherent light is diffracted by an aperture whose linear dimensions are of the order of or smaller than the width of the transverse correlations of the incident light, the light which emerges from the aperture will be essentially spatially coherent. However, if the size of the aperture is sufficiently large, the emerging light will be partially coherent. Consequently one can expect the spectrum of the light which is transmitted by the aperture to be, in general, affected by the size of the aperture. This effect was first noted by Kandpal, Vaishya and Joshi (1989).

Systematic studies of the effect of aperture size on the spectrum of the light diffracted by an aperture were made by Foley (1990, 1991). He considered a simple system, illustrated in Fig. 5.43. An uniform, spatially incoherent, planar, circular source σ of radius a_S was located in the front focal plane of a thin lens L of focal length f. An opaque screen, containing a circular aperture $\mathscr{A}$ of radius a, was placed in the back focal plane of the lens. Foley found that the spectrum of the diffracted light in the far zone differs from the spectrum of the light in the aperture by a factor which depends chiefly on the ratio of the aperture radius to the effective spectral correlation length

$$l(\omega) = \frac{0.61cf}{\omega a_S} \tag{5.8–51}$$

of the light in the aperture. An example of such a spectral change is illustrated in Fig. 5.44.

Returning to Eq. (5.8–34) it is clear that with suitable choice of the spectral

‡ A general expression relating to spectral changes due to the propagation of light from a source of any state of coherence through a linear, time-invariant system was derived by Wolf and Fienup (1991).

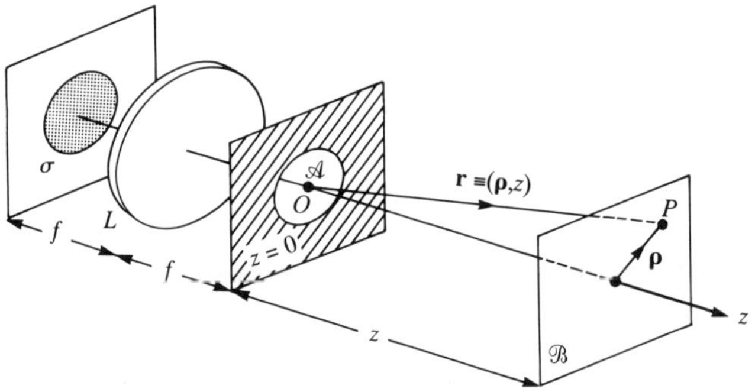

Fig. 5.43 An arrangement for studying the effect of the aperture size on the spectrum of partially coherent light. σ is a uniform, spatially incoherent, planar, circular source of radius a_S, located in the front focal plane of a thin lens L. A circular aperture $\mathscr{A}$ of radius a is placed in the back focal plane of the lens. The plane of observation $\mathscr{B}$ is in the far zone of the aperture. (Adapted from Foley, 1990.)

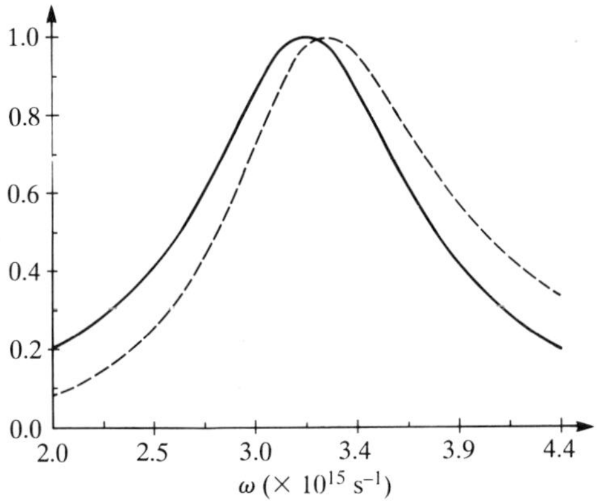

Fig. 5.44 Illustrating the effect of an aperture on the spectrum of light, for the system shown in Fig. 5.43. Normalized on-axis far-zone spectrum (dashed line) and normalized spectrum in the aperture (solid line), for the case when the normalized spectrum in the aperture has a Lorentzian profile $\mathscr{s}^{(\mathrm{o})}(\omega) = A\Gamma^2/[(\omega - \bar{\omega})^2 + \Gamma^2]$, ($A$ = const.), with $\bar{\omega} = 3.2 \times 10^{15}\ \mathrm{s}^{-1}$, $\Gamma = 0.6 \times 10^{15}\ \mathrm{s}^{-1}$ and $a/l = 0.25$, l being the effective correlation width of the light in the aperture, at frequency $\bar{\omega}$. (After Foley, 1990.)

degree of coherence of the source, one can produce changes more drastic than line shifts, as is also evident from the theoretical predictions illustrated in Figs. 5.33 and 5.34. A system for synthesizing sources (not necessarily quasi-homogeneous) with prescribed coherence properties has been developed by Indebetouw (1989). It is shown schematically in Fig. 5.45. The input is a thin slit illuminated by spatially incoherent polychromatic light of uniform spectral density. The slit is imaged by a two-lens afocal system (lenses L_1 and L_2 with focal lengths f) onto a mask in plane 3. The pupil of this imaging system (plane 2 in Fig. 5.45) contains an aperture and a prism which shift the image of the slit along the y_3-axis by an

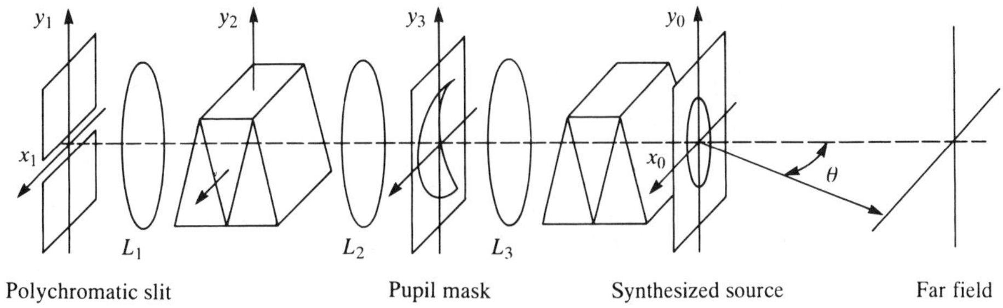

Fig. 5.45 An optical system used for synthesizing planar, secondary sources with prescribed coherence properties. (After Indebetouw, 1989.)

amount which depends on frequency. Under suitable experimental conditions the plane 3 acts as an intermediate secondary source which is spatially incoherent along the x_3-axis and spectrally dispersed along the y_3-axis. A pupil mask with complex amplitude transmission function $T(x_3, \omega)$ is used to produce the desired spectral degree of coherence in a plane 0 behind another prism, which is identical to the first. The purpose of this second prism is to eliminate a certain unwanted phase term.

A detailed calculation shows that, in the plane 0, one obtains a secondary source whose spectral density is the same at each point and whose spectral degree of coherence varies along the x-direction and is proportional to the Fourier transform of the squared modulus of the pupil mask transmission function $T(x_3, \omega)$ which modulates the dispersed image of the slit.

An example of a pupil mask used to synthesize a source with a particular degree of coherence is shown in Fig. 5.46. The synthesized source had an approximately uniform spectral density, $S(\omega) = S_0 = \text{const.}$, over a certain frequency range, and the spectral density was zero outside this range. The far-field spectra, produced by this source in three selected directions of observation, are shown in Fig. 5.47.

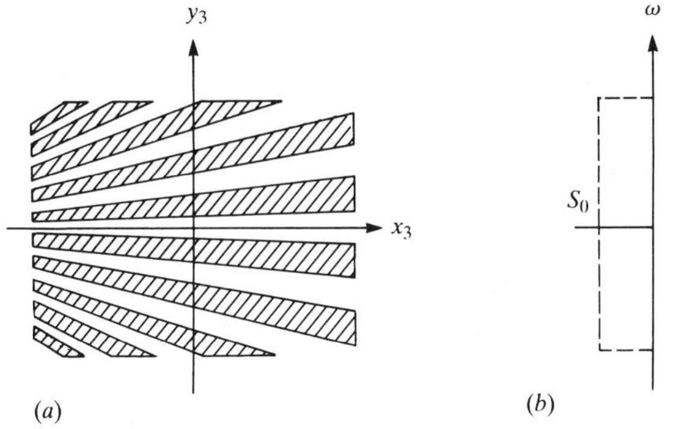

Fig. 5.46 A pupil mask (*a*) used to synthesize a source whose spectrum is uniform (*b*) which produces the far-field spectra shown in Fig. 5.47. (After Indebetouw, 1989.)

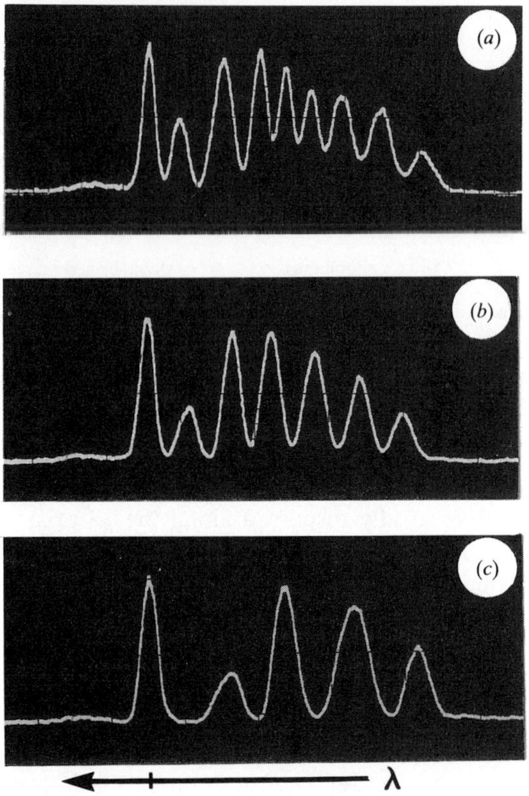

Fig. 5.47 Far-field spectra observed in directions (a) $\theta = 3°$, (b) $\theta = 0°$ and (c) $\theta = -3°$, produced by the synthesized, spectrally uniform source, generated by means of the pupil mask shown in Fig. 5.46. (After Indebetouw, 1989.)

Finally it should be noted that, in all the situations leading to spectral changes which we have discussed so far, no new frequency components were generated by the correlation mechanism. This fact is evident, for example, from Eq. (5.8–38) which shows that $S^{(\infty)}(r\mathbf{s}, \omega) = 0$ whenever $\mathscr{s}^{(0)}(\omega) = 0$. Source correlations produce essentially enhancements or reductions in the strengths of the various frequency components contained in the original spectrum, but they do not create new ones. Consequently the largest spectral shifts which can be produced in this way are of the order of the effective width of the original spectral line. Similar remarks apply to spectral changes induced by the somewhat analogous mechanism of scattering on partially correlated, static, random media [cf. Section 7.6.4(c), especially Eq. (7.6–59)]. However for dynamic scattering or, more specifically, for scattering from random media whose responses to the incident field vary not only in space, but also in time, spectral components may be generated in the scattered light which are not present in the incident light. In particular such a process can give rise to large line shifts. Under certain circumstances the spectral changes which are produced in this manner can imitate the Doppler effect in its main features, even though the source, the scatterer and the observer are at rest with respect to each other. (Wolf, 1989; James, Savedoff and Wolf, 1990; James and Wolf, 1990, 1994). An example is given in Fig. 7.9.

5.8.3 *A condition for spectral invariance: the scaling law for planar, secondary, quasi-homogeneous sources*

We showed in the preceding section that the far-zone spectrum of the light generated by a planar, quasi-homogeneous, secondary source differs, in general, from the source spectrum even on propagation in free space. We may express this fact by saying that the spectrum of the light is, in general, not invariant on propagation. On the other hand, such spectral changes have not been observed until relatively recently. This suggests that the usual laboratory sources have some rather special coherence properties. We will see shortly that this is indeed the case.

Let us again consider the field produced by a planar, secondary, quasi-homogeneous source, whose normalized source spectrum $\mathcal{s}^{(0)}(\omega)$ [cf. Eq. (5.8–35)] is the same at every source point. We will first derive a condition which the spectral degree of coherence of the source must satisfy in order for the normalized spectrum of the far field produced by the source to be independent of the direction $\mathbf{s}$ of observation. For this purpose we make use of two inversion formulas which we derived in Section 5.3. The first of them, [Eq.(5.3–37)] becomes, if we make use of Eq. (5.8–33),

$$\tilde{S}^{(0)}(0, \omega) = \frac{r^2}{(2\pi)^2}\int_{\mathbf{s}_\perp^2 \leq 1} (1 - \mathbf{s}_\perp^2)^{-1} S^{(\infty)}(r\mathbf{s}, \omega)\,\mathrm{d}^2 s_\perp, \tag{5.8–52}$$

where we have assumed that the high spatial-frequency components ($f^2 > k^2$) of the spectral degree of coherence of the light in the source plane are negligible. Now if the normalized spectrum $\mathcal{s}^{(0)}$ is the same at every source point, we have, on taking the two-dimensional spatial Fourier transform of Eq. (5.8–35) and evaluating it at the spatial frequency $\mathbf{f} = 0$,

$$\tilde{S}^{(0)}(0, \omega) = \tilde{I}(0)\mathcal{s}^{(0)}(\omega), \tag{5.8–53}$$

where $\tilde{I}(0)$ is proportional to the source-integrated intensity [Eqs. (5.8–39) and (5.8–36)]. Using this relation in Eq. (5.8–52) we obtain the following formula for the reconstructed (subscript 'rec') normalized source spectrum:

$$[\mathcal{s}^{(0)}(\omega)]_{\text{rec}} = \frac{r^2}{(2\pi)^2\tilde{I}(0)}\int_{\mathbf{s}_\perp^2 \leq 1} (1 - \mathbf{s}_\perp^2)^{-1} S^{(\infty)}(r\mathbf{s}, \omega)\,\mathrm{d}^2 s_\perp. \tag{5.8–54}$$

If we integrate both sides of this equation over the complete frequency range and use the normalization condition (5.8–37), we see at once that

$$\frac{(2\pi)^2\tilde{I}(0)}{r^2} = \int_0^\infty \mathrm{d}\omega \int_{\mathbf{s}_\perp^2 \leq 1} (1 - \mathbf{s}_\perp^2)^{-1} S^{(\infty)}(r\mathbf{s}, \omega)\,\mathrm{d}^2 s_\perp. \tag{5.8–55}$$

On substituting from this equation into Eq. (5.8–54) we finally obtain the following expression for the normalized source spectrum in terms of the far-zone spectrum:

$$[\mathscr{s}^{(0)}(\omega)]_{\mathrm{rec}} = \frac{\displaystyle\int_{\mathbf{s}_\perp^2 \leqslant 1} (1 - \mathbf{s}_\perp^2)^{-1} S^{(\infty)}(r\mathbf{s}, \omega)\, \mathrm{d}^2 s_\perp}{\displaystyle\int_0^\infty \mathrm{d}\omega \int_{\mathbf{s}_\perp^2 \leqslant 1} (1 - \mathbf{s}_\perp^2)^{-1} S^{(\infty)}(r\mathbf{s}, \omega)\, \mathrm{d}^2 s_\perp}. \qquad (5.8\text{–}56)$$

Further, on making use of the simple relation (5.8–33) in the second inversion formula [Eq. (5.3–38)], for quasi-homogeneous sources we obtain the corresponding expression for the spectral degree of coherence of the source:

$$[g^{(0)}(\boldsymbol{\rho}', \omega)]_{\mathrm{rec}} = \frac{\displaystyle\int_{\mathbf{s}_\perp^2 \leqslant 1} (1 - \mathbf{s}_\perp^2)^{-1} S^{(\infty)}(r\mathbf{s}, \omega) \exp(\mathrm{i}k\mathbf{s}_\perp \cdot \boldsymbol{\rho}')\, \mathrm{d}^2 s_\perp}{\displaystyle\int_{\mathbf{s}_\perp^2 \leqslant 1} (1 - \mathbf{s}_\perp^2)^{-1} S^{(\infty)}(r\mathbf{s}, \omega)\, \mathrm{d}^2 s_\perp}. \qquad (5.8\text{–}57)$$

The formulas (5.8–56) and (5.8–57) bring into evidence an interesting fact; namely that, for the class of sources under consideration, the far-zone spectrum allows the determination of both the normalized source spectrum and the spectral degree of coherence of the source, provided that the high spatial-frequency components ($|\mathbf{f}| > k$) of the spectral degree of coherence are negligible, as is usually the case. Examples of such reconstructions from normalized far-zone spectra, [defined by Eq. (5.8–58) below] are presented in Fig. 5.49, for the case when both the normalized source spectrum and the spectral degree of coherence have Gaussian profiles [see Eqs. (5.8–40) and (5.8–41)]. The far-zone spectra are then given by Eq. (5.8–48) and are shown in Fig. 5.48.

We see from Fig. 5.49 that with sources to which these calculations pertain the reconstructed source spectra and the reconstructed spectral degree of coherence of the source reproduce very closely the assumed source data provided that the r.m.s. width of the spectral degree of coherence of the source is of the order of or greater than $0.5\lambda_0$. Assuming that the high spatial-frequency components of the spectral degree of coherence are negligible, we evidently need not distinguish between the true and the reconstructed data and we will, therefore, drop the subscript 'rec' from now on.

Let us introduce the normalized far-zone spectrum

$$\mathscr{s}^{(\infty)}(r\mathbf{s}, \omega) = \frac{S^{(\infty)}(r\mathbf{s}, \omega)}{I^{(\infty)}(r\mathbf{s})}, \qquad (5.8\text{–}58)$$

where

$$I^{(\infty)}(r\mathbf{s}) = \int_0^\infty S^{(\infty)}(r\mathbf{s}, \omega)\, \mathrm{d}\omega. \qquad (5.8\text{–}59)$$

Evidently this normalization ensures that

$$\int_0^\infty \mathscr{s}^{(\infty)}(r\mathbf{s}, \omega)\, \mathrm{d}\omega = 1. \qquad (5.8\text{–}60)$$

It is clear from the relation (5.8–33) between the far-zone spectrum and the

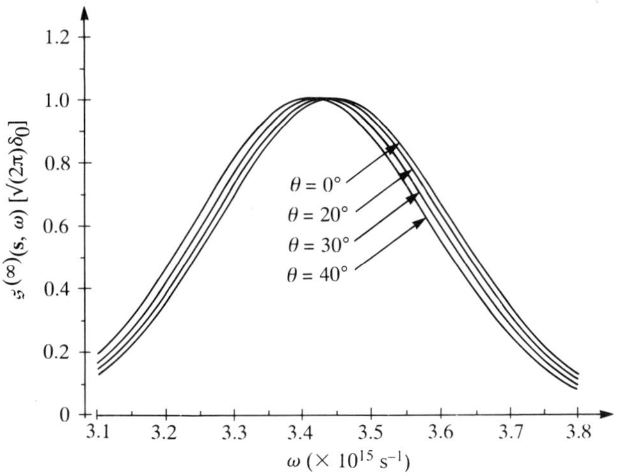

Fig. 5.48 Normalized far-zone spectra

$$\mathfrak{s}^{(\infty)}(r\mathbf{s}, \omega) = S^{(\infty)}(r\mathbf{s}, \omega) \Big/ \int_0^\infty S^{(\infty)}(r\mathbf{s}, \omega)\, d\omega,$$

calculated from Eq. (5.8–48), in units of $[\delta_0\sqrt{(2\pi)}]^{-1}$, produced by a planar, secondary, quasi-homogeneous source, whose normalized source spectrum is a single line of Gaussian profile and whose spectral degree of coherence is also Gaussian, with $\omega_0 = 3.43 \times 10^{15}$ Hz, $\delta_0/\omega_0 = 1/20$, $\sigma_g = 0.5\lambda_0$. (After Wolf, 1992.)

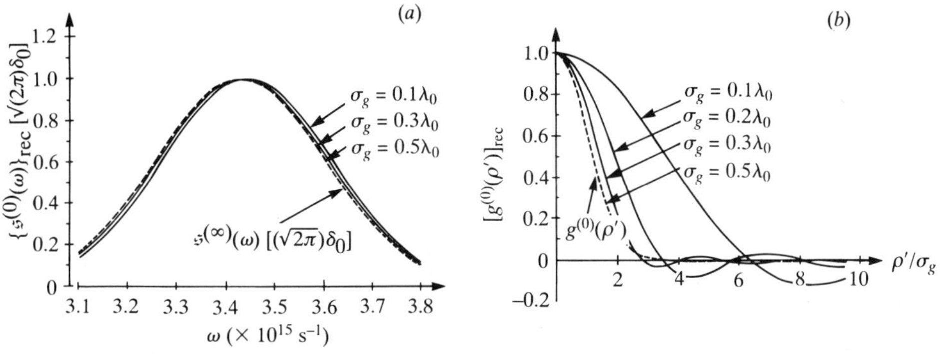

Fig. 5.49 (*a*) The reconstructed normalized source spectra $[\mathfrak{s}^{(0)}(\omega)]_{rec}$ in units of $[\delta_0\sqrt{(2\pi)}]^{-1}$ and (*b*) the reconstructed spectral degree of coherence $[g^{(0)}(\boldsymbol{\rho}', \omega_0)]_{rec}$ (with the argument ω_0 suppressed in the figure), calculated from the far-zone spectra shown in Fig. 5.48, using Eqs. (5.8–56) and (5.8–57). (After Wolf, 1992.)

radiant intensity and from the definition (5.8–58) that $\mathfrak{s}^{(\infty)}(r\mathbf{s}, \omega)$ is, in fact, independent of the distance r of the point of observation in the far zone from the origin. Suppose now that it is also independent of the direction $\mathbf{s}$ of observation. We may then write $\mathfrak{s}^{(\infty)}(\omega)$ in place of $\mathfrak{s}^{(\infty)}(r\mathbf{s}, \omega)$ and Eq. (5.8–58) implies that

$$S^{(\infty)}(r\mathbf{s}, \omega) = \mathfrak{s}^{(\infty)}(\omega) I^{(\infty)}(r\mathbf{s}). \tag{5.8–61}$$

On substituting from Eq. (5.8–61) into the first inversion formula (5.8–56), making use of the normalization condition (5.8–60) and neglecting contributions from the high spatial-frequency components, we obtain the result that

$$\mathscr{g}^{(\infty)}(\omega) = \mathscr{g}^{(0)}(\omega). \tag{5.8–62}$$

We have thus established the following theorem.

Theorem: *Consider a planar, secondary, quasi-homogeneous source with the same normalized spectrum at each source point. If the normalized spectrum of the field generated by such a source is the same throughout the far zone, it is necessarily equal to the normalized source spectrum.*

Further if we substitute from Eq. (5.8–61) into the second inversion formula [Eq. (5.8–57)] and again neglect the contributions from high spatial-frequency components, we obtain the following expression for the spectral degree of coherence of the source:

$$g^{(0)}(\boldsymbol{\rho}', \omega) = \frac{\displaystyle\int_{\mathbf{s}_\perp^2 \leq 1} (1 - \mathbf{s}_\perp^2)^{-1} I^{(\infty)}(r\mathbf{s}) \exp(\mathrm{i}k\mathbf{s}_\perp \cdot \boldsymbol{\rho}') \, \mathrm{d}^2 s_\perp}{\displaystyle\int_{\mathbf{s}_\perp^2 \leq 1} (1 - \mathbf{s}_\perp^2)^{-1} I^{(\infty)}(r\mathbf{s}) \, \mathrm{d}^2 s_\perp}. \tag{5.8–63}$$

We see that $g^{(0)}$ now has a particular functional form, namely (writing $\boldsymbol{\rho}_2 - \boldsymbol{\rho}_1$ in place of $\boldsymbol{\rho}'$)

$$g^{(0)}(\boldsymbol{\rho}_2 - \boldsymbol{\rho}_1, \omega) = h[k(\boldsymbol{\rho}_2 - \boldsymbol{\rho}_1)], \quad (k = \omega/c = 2\pi/\lambda), \tag{5.8–64}$$

i.e. it is a function of the variable

$$\zeta = k(\boldsymbol{\rho}_2 - \boldsymbol{\rho}_1) = 2\pi \frac{\boldsymbol{\rho}_2 - \boldsymbol{\rho}_1}{\lambda} \tag{5.8–65}$$

only. For obvious reasons a spectral degree of coherence which has the functional form (5.8–64) is said to obey the *scaling law*.

The result which we have just established, together with that expressed by Eq. (5.8–62), may be summarized in the following theorem, first formulated by Wolf (1986, 1992):

Theorem: *Consider the field produced by a planar, secondary, quasi-homogeneous source which has the same normalized spectrum at each source point. In order that the normalized spectrum of the light generated by the source is the same throughout the far zone as at the source itself, the spectral degree of coherence of the light in the source plane must satisfy the scaling law, i.e. it must have the functional form (5.8–64).*

Examples of sources which satisfy the scaling law are all planar, secondary, quasi-homogeneous *Lambertian* sources, because according to Eq. (5.3–53) all such sources have the spectral degree of coherence

$$g^{(0)}(\boldsymbol{\rho}_2 - \boldsymbol{\rho}_1, \omega) = \frac{\sin k|\boldsymbol{\rho}_2 - \boldsymbol{\rho}_1|}{k|\boldsymbol{\rho}_2 - \boldsymbol{\rho}_1|}, \tag{5.8–66}$$

which evidently obeys the scaling law (5.8–64). Because quasi-homogeneous Lambertian sources are so often encountered in the laboratory and because, as we have just learned, the normalized spectrum of the light throughout the far zone generated by such a source is the same as the normalized source spectrum, 'spectral invariance' was until recently taken for granted. However, as we have just learned, such invariance is, in fact, not a general property of light. This result was first predicted by Wolf (1986) and was soon afterwards confirmed by Morris and Faklis (1987) by means of experiments which we will now briefly describe.

Morris and Faklis illuminated an aperture in plane I (see Fig. 5.50) by light from a broadband, essentially incoherent thermal source (a tungsten lamp), located directly in front of the aperture. A planar, secondary, quasi-homogeneous source was produced in plane II by means of (*a*) an ordinary lens of focal length f and (*b*) an achromatic Fourier transform lens. The spectra of the light produced by these two sources was measured in the far zone (plane III), for different directions of observation θ. These spectra were compared with the spectra of the secondary sources in plane II.

Provided that the aperture in plane II is sufficiently large, the spectral degree of coherence of the secondary source in plane II, formed by the lens in arrangement (*a*), may be shown to satisfy the scaling law. On the other hand the spectral degree of coherence of the light in plane II formed by the achromatic Fourier transform lens may be shown not to obey the scaling law. According to the theoretical analysis presented earlier in this section we can expect that, with the experimental set up shown in Fig. 5.50(*a*), the normalized spectrum of the far field (plane III) will be the same for all directions θ and will be equal to the normalized source spectrum; and that, for the arrangement shown in Fig. 5.50(*b*), the normalized far-zone spectrum will be different for different θ-directions and consequently, except perhaps for some special directions, it will also differ from the normalized source spectrum. This is indeed what the experiments of Morris and Faklis demonstrated. Their main results are shown in Fig. 5.51.

We stress that the scaling law is a condition for spectral invariance in the *far zone* of a planar, secondary, quasi-homogeneous source. Closer to the source, the

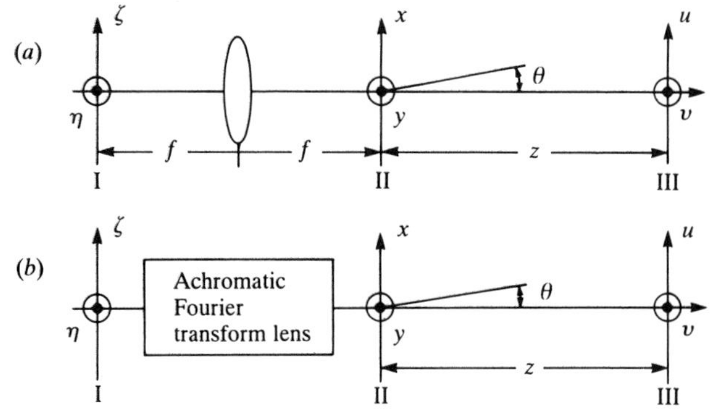

Fig. 5.50 The experimental configurations for realization of a planar, secondary, quasi-homogeneous source in plane II, which (*a*) obeys the scaling law and (*b*) does not obey it. (After Morris and Faklis, 1987.)

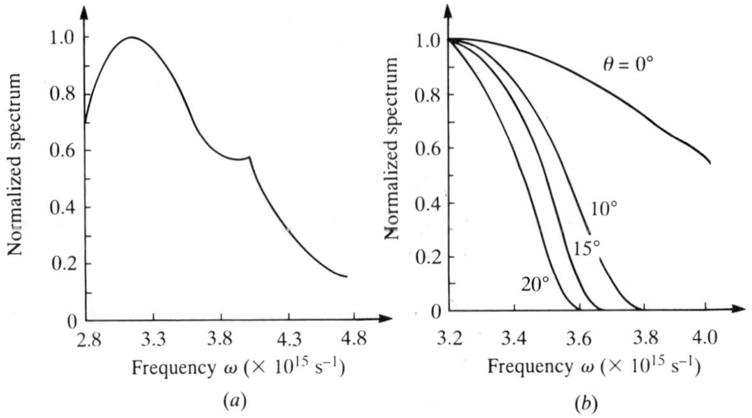

Fig. 5.51 Measured values in different directions θ of the normalized spectrum in the far zone produced by a source (a) which satisfies the scaling law and (b) which does not satisfy it, obtained by using the experimental arrangements shown in Fig. 5.50. For the scaling-law source (a) all the normalized far-zone spectra, as well as the source spectrum, were found to be the same; hence only one curve is shown in that figure. (Adapted from Morris and Faklis, 1987.)

field spectrum may differ both from the far-zone spectrum and from the source spectrum.

Appendix 5.1

Derivation of the asymptotic approximation (5.7–103)‡

We have, according to Eq. (5.7–99)

$$C_\nu^{(1)}(\mathbf{r}, \mathbf{s}_\perp) = \frac{kz}{2\pi i} N_\nu(\mathbf{r}, \mathbf{s}_\perp), \tag{A5.1–1}$$

where, (if we write $\boldsymbol{\rho}'$ in place of $\boldsymbol{\rho}$),

$$N_\nu(\mathbf{r}, \mathbf{s}_\perp) = \int_\sigma S^{(0)}(\boldsymbol{\rho}', \nu) \frac{\mathrm{e}^{ik\Phi(\mathbf{r}, \mathbf{s}_\perp, \boldsymbol{\rho}')}}{R^2} \mathrm{d}^2\rho' \tag{A5.1–2}$$

and

$$\Phi(\mathbf{r}, \mathbf{s}, \boldsymbol{\rho}') = R + \mathbf{s}_\perp \cdot \boldsymbol{\rho}', \tag{A5.1–3}$$

with

$$R = |\mathbf{r} - \boldsymbol{\rho}'|. \tag{A5.1–4}$$

Let

$$\mathbf{r} \equiv (x, y, z), \boldsymbol{\rho}' = (x', y', 0). \tag{A5.1–5}$$

‡ After Foley and Wolf (1991).

The formula (A5.1–2) then becomes

$$N_\nu(\mathbf{r}, \mathbf{s}_\perp) = \int_\sigma f(x', y'; \mathbf{r}; \nu) e^{ikg(x',y';\mathbf{r};\mathbf{s}_\perp)} \, dx' \, dy', \tag{A5.1–6}$$

where

$$f(x', y'; \mathbf{r}; \nu) = \frac{1}{R^2} S^{(0)}(x', y'; \nu), \tag{A5.1–7a}$$

$$g(x', y'; \mathbf{r}; \mathbf{s}_\perp) = s_x x' + s_y y' + R, \tag{A5.1–7b}$$

with

$$R = [(x' - x)^2 + (y' - y)^2 + z^2]^{1/2}. \tag{A5.1–8}$$

The asymptotic behavior as $k \to \infty$ of the integral (A5.1–6) may be determined, using the principle of stationary phase for double integrals discussed in Section 3.3.3. In this limiting process the source is taken to be unchanged, i.e. the spectrum $S^{(0)}(x', y', \nu)$ in the integral is kept fixed. We must first determine the location of the critical points of the first kind (if any) of the integrand. These are points at which the function $g(x', y'; \mathbf{r}; s_x, s_y)$ is stationary within the source region σ, with respect to variations of (x', y'), i.e. where

$$g_{x'} = g_{y'} = 0 \tag{A5.1–9}$$

within σ, with $g_{x'}$ and $g_{y'}$ denoting the first partial derivatives of g with respect to x' and y' respectively. Now from Eqs. (A5.1–7b) and (A5.1–8)

$$g_{x'} = s_x + \frac{x' - x}{R}, \quad g_{y'} = s_y + \frac{y' - y}{R}. \tag{A5.1–10}$$

From Eqs. (A5.1–10) and (A5.1–8) it follows that, if a point $x' = x_0'$, $y' = y_0'$ is a critical point of the first kind, it must satisfy the equations

$$x_0' - x = -s_x [(x_0' - x)^2 + (y_0' - y)^2 + z^2]^{1/2} \tag{A5.1–11a}$$

and

$$y_0' - y = -s_y [(x_0' - x)^2 + (y_0' - y)^2 + z^2]^{1/2}. \tag{A5.1–11b}$$

On squaring both sides of these equations we obtain the simultaneous equations for the quantities $(x_0' - x)^2$ and $(y_0' - y)^2$. They can be readily solved and give

$$x_0' = x - \frac{s_x}{s_z} z, \quad y_0' = y - \frac{s_y}{s_z} z, \tag{A5.1–12a}$$

or, more explicitly in vector form, with $\boldsymbol{\rho}_0' \equiv (x_0', y_0', 0)$, $\mathbf{r}_\perp \equiv (x, y, 0)$,

$$\boldsymbol{\rho}_0' = \mathbf{r}_\perp - \frac{z}{s_z} \mathbf{s}_\perp. \tag{A5.1–12b}$$

The formulas (A5.1–12) show that the function $g(x', y'; \mathbf{r}; s_x, s_y)$ has one and only one stationary point in the source plane $z = 0$. This point will be a critical point of the first kind of the integrand on the right-hand side of Eq. (A5.1–2) only if it is located within the source region σ. The geometrical significance of this point is illustrated in Fig. 5.29 of the text, where it is denoted by Q_0.

According to the general formula (3.3–41) the asymptotic approximation to the integral (A5.1–6) is given by [if we ignore the dependence of the function $f(x', y', \mathbf{r}; \nu)$ on ν, which is justified for reasons explained in Section 5.7.5]

$$N_\nu(\mathbf{r}, \mathbf{s}_\perp) \sim \frac{2\pi i \alpha}{k\sqrt{|\Delta|}} f(x_0', y_0'; \mathbf{r}; \nu) e^{ikg(x_0', y_0'; \mathbf{r}; \mathbf{s}_\perp)} \quad \text{as } k \to \infty, \qquad \text{(A5.1–13)}$$

where

$$\Delta = (g_{x'x'} g_{y'y'} - g^2_{x'y'})|_{x_0', y_0'} \qquad \text{(A5.1–14)}$$

and

$$\left.\begin{aligned} \alpha &= +1 \quad \text{when } \Delta > 0, \Sigma > 0 \\ &= -1 \quad \text{when } \Delta > 0, \Sigma < 0 \\ &= -i \quad \text{when } \Delta < 0, \end{aligned}\right\} \qquad \text{(A5.1–15)}$$

$$\Sigma = (g_{x'x'} + g_{y'y'})|_{x_0', y_0'}, \qquad \text{(A5.1–16)}$$

it being assumed that $\Delta \neq 0$. The term $g_{x'x'}$ denotes the second partial derivative of g with respect to x', etc. These derivatives are readily obtained by differentiating the expressions (A5.1–10). One then obtains the following formulas:

$$\left.\begin{aligned} g_{x'x'} &= \frac{1}{R}\left[1 - \frac{(x' - x)^2}{R^2}\right], \\ g_{y'y'} &= \frac{1}{R}\left[1 - \frac{(y' - y)^2}{R^2}\right], \\ g_{x'y'} &= -\frac{(x' - x)(y' - y)}{R^3}. \end{aligned}\right\} \qquad \text{(A5.1–17)}$$

The values of these quantities at the critical point $x' = x_0'$, $y' = y_0'$, can be obtained at once on substituting from Eqs. (A5.1–12a) into Eq. (A5.1–17). One finds that

$$\left.\begin{aligned} g_{x'x'}|_{x_0', y_0'} &= \frac{s_z}{z}(1 - s_x^2), \\ g_{y'y'}|_{x_0', y_0'} &= \frac{s_z}{z}(1 - s_y^2), \\ g_{x'y'}|_{x_0', y_0'} &= -\frac{s_x s_y s_z}{z}, \end{aligned}\right\} \qquad \text{(A5.1–18)}$$

where we used the fact that

$$R|_{x_0', y_0'} = \frac{z}{s_z}, \qquad \text{(A5.1–19)}$$

which readily follows from Eqs. (A5.1–8) and (A5.1–12a).

On substituting from Eqs. (A5.1–18) into the formulas (A5.1–14), (A5.1–15) and (A5.1–16) we find that

$$\Delta = \left(\frac{s_z^2}{z}\right)^2, \quad \alpha = 1, \quad \text{and } \Sigma > 0. \tag{A5.1–20}$$

Furthermore, on substituting from Eqs. (A5.1–12a) into Eqs. (A5.1–7), and making use of Eq. (A5.1–19), we obtain for $f(x_0', y_0'; \mathbf{r}, \nu)$ and $g(x_0', y_0'; \mathbf{r}; \mathbf{s}_\perp)$ the expressions

$$f(x_0', y_0'; \mathbf{r}, \nu) = \left(\frac{s_z}{z}\right)^2 S^{(0)}(x_0', y_0', \nu), \tag{A5.1–21}$$

$$g(x_0', y_0'; \mathbf{r}, \mathbf{s}_\perp) = xs_x + ys_y + zs_z = \mathbf{r} \cdot \mathbf{s}. \tag{A5.1–22}$$

On substituting from Eqs. (A5.1–20), (A5.1–21) and (A5.1–22) into the general formula (A5.1–13) we find that

$$N_\nu(\mathbf{r}, \mathbf{s}_\perp) \sim \frac{2\pi \mathrm{i}}{k} \frac{1}{z} S^{(0)}(\boldsymbol{\rho}_0', \nu) \mathrm{e}^{\mathrm{i}k\mathbf{s} \cdot \mathbf{r}} \quad \text{as } k \to \infty, \tag{A5.1–23}$$

provided that the point $\boldsymbol{\rho}_0'$, specified by Eq. (A5.1–12b), lies within the domain of integration (the source domain). If it does not, the contributions to the asymptotic approximation to $N_\nu(\mathbf{r}, \mathbf{s}_\perp)$ comes from critical points of the second kind, which are located on the boundary curve of σ [cf. Section 3.3.2(b)] and which give (except in some special cases)

$$N_\nu(\mathbf{r}, \mathbf{s}_\perp) \sim O\left(\frac{1}{k^{3/2}}\right) \quad \text{as } k \to \infty. \tag{A5.1–24}$$

Finally on substituting from Eqs. (A5.1–23) and (A5.1–24) into Eq. (A5.1–1) we see that, as $k \to \infty$,

$$\left.\begin{aligned} C_\nu^{(1)}(\mathbf{r}, \mathbf{s}_\perp) &\sim S^{(0)}(\boldsymbol{\rho}_0', \nu)\mathrm{e}^{\mathrm{i}k\mathbf{s}\cdot\mathbf{r}} && \text{when } \boldsymbol{\rho}_0' \in \sigma \\ &\sim O\left(\frac{1}{k^{1/2}}\right) && \text{when } \boldsymbol{\rho}_0' \notin \sigma, \end{aligned}\right\} \tag{A5.1–25}$$

where $\boldsymbol{\rho}_0'$ is given by Eq. (A5.1–12b). Equation (A5.1–25) is the formula (5.7–103) used in the text.

Appendix 5.2

Product theorem for Gaussian functions‡

In this appendix we will establish the following theorem, known as the *Product theorem for Gaussian functions*:

If $G(\omega - \omega_\mathrm{j}; \delta_\mathrm{j})$ represents the Gaussian function

$$G(\omega - \omega_\mathrm{j}; \delta_\mathrm{j}) = \exp[-(\omega - \omega_j)^2/2\delta_j^2], \tag{A5.2–1}$$

then the product of two such functions is proportional to another Gaussian

‡ After Wolf, Foley and Gori (1989).

function, given by

$$G(\omega - \omega_1; \delta_1)G(\omega - \omega_2; \delta_2) = G[\omega_1 - \omega_2; (\delta_1^2 + \delta_2^2)^{1/2}]G(\omega - \tilde{\omega}; \tilde{\delta}), \tag{A5.2–2}$$

where

$$\tilde{\omega} = \frac{\omega_1\delta_2^2 + \omega_2\delta_1^2}{\delta_1^2 + \delta_2^2} \tag{A5.2–3}$$

and

$$\frac{1}{\tilde{\delta}^2} = \frac{1}{\delta_1^2} + \frac{1}{\delta_2^2}. \tag{A5.2–4}$$

To establish this theorem let us multiply together two Gaussian functions and express the product in the form

$$G(\omega - \omega_1; \delta_1)G(\omega - \omega_2; \delta_2) = \exp[-g(\omega)]. \tag{A5.2–5}$$

Then

$$\begin{aligned} g(\omega) &= \frac{1}{2\delta_1^2\delta_2^2}[\delta_2^2(\omega - \omega_1)^2 + \delta_1^2(\omega - \omega_2)^2] \\ &= \frac{1}{2\delta_1^2\delta_2^2}(a^2\omega^2 - 2b\omega + c), \end{aligned} \tag{A.5.2–6}$$

where

$$a^2 = \delta_1^2 + \delta_2^2, \tag{A5.2–7a}$$

$$b = \omega_1\delta_2^2 + \omega_2\delta_1^2, \tag{A5.2–7b}$$

$$c = \omega_1^2\delta_2^2 + \omega_2^2\delta_1^2. \tag{A5.2–7c}$$

On completing the square in Eq. (A5.2–6) we find that

$$\begin{aligned} g(\omega) &= \frac{a^2}{2\delta_1^2\delta_2^2}\left(\omega - \frac{b}{a^2}\right)^2 + \frac{1}{2\delta_1^2\delta_2^2}\left(c - \frac{b^2}{a^2}\right) \\ &= \frac{1}{2\tilde{\delta}^2}(\omega - \tilde{\omega})^2 + \frac{1}{2\delta_1^2\delta_2^2}\left(c - \frac{b^2}{a^2}\right), \end{aligned} \tag{A5.2–8}$$

where

$$\frac{1}{\tilde{\delta}^2} = \frac{a^2}{\delta_1^2\delta_2^2} = \frac{1}{\delta_1^2} + \frac{1}{\delta_2^2}, \tag{A5.2–9}$$

and

$$\tilde{\omega} = \frac{b}{a^2} = \frac{\omega_1\delta_2^2 + \omega_2\delta_1^2}{\delta_1^2 + \delta_2^2}. \tag{A5.2–10}$$

By using Eqs. (A5.2–7a) to (A5.2–7c) one can readily show that

$$c - \frac{b^2}{a^2} = \frac{\delta_1^2\delta_2^2}{\delta_1^2 + \delta_2^2}(\omega_1 - \omega_2)^2. \tag{A5.2–11}$$

On substituting from Eq. (A5.2–11) into Eq. (A5.2–8) and from Eq. (A5.2–8) into Eq. (A5.2–5) we find, if we recall the definition (A5.2–1), that

$$
\begin{aligned}
G(\omega - \omega_1; \delta_1)G(\omega - \omega_2; \delta_2) &= \exp\left[-\frac{(\omega_1 - \omega_2)^2}{2(\delta_1^2 + \delta_2^2)}\right]\exp\left[-\frac{(\omega - \tilde{\omega})^2}{2\tilde{\delta}^2}\right] \\
&= G[\omega_1 - \omega_2; (\delta_1^2 + \delta_2^2)^{1/2}]G(\omega - \tilde{\omega}; \tilde{\delta}),
\end{aligned}
\tag{A5.2–12}
$$

as asserted by the theorem.

Some consequences of this theorem are discussed in the last part of Appendix A of Wolf, Foley and Gori (1989).

Problems

5.1 Consider a three-dimensional, statistically stationary, finite source.
(a) Show that when the source is spatially completely incoherent, the radiant intensity of the field radiated by the source is independent of direction.
(b) What form does the expression for the radiant intensity take when the source is spatially completely coherent?
(c) Consider a coherent, uniform, equiphasal, spherical source of radius a. Show that for certain values of a, the source will not radiate.

5.2 A fluctuating source $Q(\mathbf{r}, t)$ occupies a finite, three-dimensional domain D in free space. The fluctuations are characterized by a stationary statistical ensemble. Show that if the source is spatially coherent at some frequency ω, the field generated by the source is also spatially coherent at the same frequency.

5.3 The cross-spectral density function of a fluctuating, statistically stationary source distribution $Q(\mathbf{r}, t)$, occupying a three-dimensional domain D, has the form

$$
\begin{aligned}
W_Q(\mathbf{r}_1, \mathbf{r}_2, \omega) = {} & p(\omega)[f^*(\mathbf{r}_1, \omega)f(\mathbf{r}_2, \omega) + g^*(\mathbf{r}_1, \omega)g(\mathbf{r}_2, \omega)] \\
& + q(\omega)[f^*(\mathbf{r}_1, \omega)g(\mathbf{r}_2, \omega) + g^*(\mathbf{r}_1, \omega)f(\mathbf{r}_2, \omega)],
\end{aligned}
$$

where $p(\omega)$ and $q(\omega)$ are real, $p(\omega)$ is positive and

$$|q(\omega)| \leqslant p(\omega),$$

$$\int_D f^*(\mathbf{r}, \omega)g(\mathbf{r}, \omega)\, \mathrm{d}^3r = 0,$$

$$\int_D |f(\mathbf{r}, \omega)|^2\, \mathrm{d}^3r = \int_D |g(\mathbf{r}, \omega)|^2\, \mathrm{d}^3r = 1.$$

Determine the coherent-mode representation of $W_Q(\mathbf{r}_1, \mathbf{r}_2, \omega)$.

5.4 A planar, secondary, quasi-homogeneous source of finite extent radiates into the half-space $z > 0$. The total flux which the source radiates at frequency ω is given by the expression

$$F_{\omega} = \int J_{\omega}(\mathbf{s})\, \mathrm{d}\Omega,$$

where $J_{\omega}(\mathbf{s})$ is the radiant intensity generated by the source and the integration extends over the 2π solid angle subtended at the source by the hemisphere at infinity in the half-space $z > 0$. Show that

$$F_{\omega} \leqslant \int_{\sigma} S^{(0)}(\boldsymbol{\rho}, \omega)\, \mathrm{d}^2 r,$$

where $S^{(0)}(\boldsymbol{\rho}, \omega)$ is the spectral density across the source and σ is the area occupied by the source.

5.5 A monochromatic field propagates into the half-space $z > 0$. Show that:
(a) the evanescent waves do not contribute to the total flux of energy crossing any plane $z = \text{const.} > 0$;
(b) the energy is conserved in the sense that the total flux entering the half-space from the plane $z = 0$ is equal to the total energy flux crossing the hemisphere at infinity in that half-space.

5.6 Consider two identical small sources, separated by a distance d. The spectrum of each source is $S_Q(\omega)$ and the correlation between them is characterized by $\mu_Q(\omega)$.
Derive an expression for the total power radiated by the two sources and discuss the limiting cases $d \ll \lambda$ and $d \gg \lambda$, $(\lambda = 2\pi c/\omega)$. Comment on the implications of the result for the case $d \ll \lambda$ on the overall behavior of the spectrum of the far field.

5.7 Consider a linear, time-invariant optical system. The light in the input plane $z = z_0$ (see figure) is statistically stationary and has the same source spectrum $S_0(\omega)$ at each point in that plane and its spectral degree of coherence is $\mu_0(\boldsymbol{\rho}'_1, \boldsymbol{\rho}'_2, \omega)$. Further $K(\boldsymbol{\rho}, \boldsymbol{\rho}', \omega)$ is the impulse response function of the

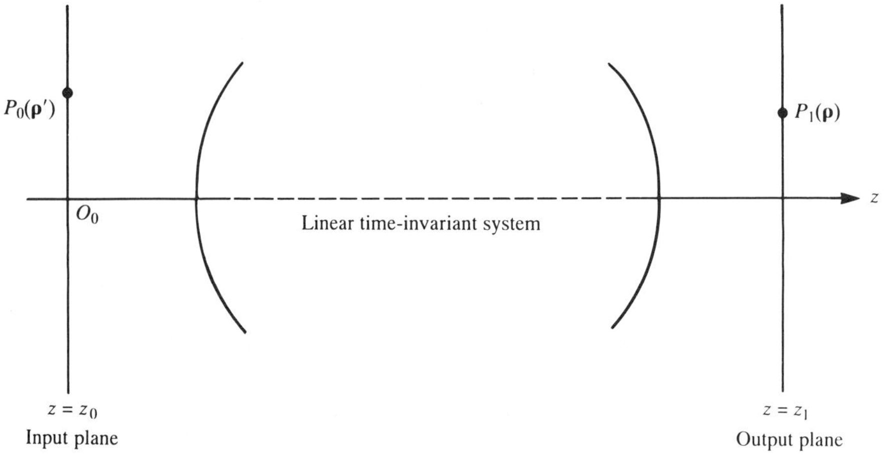

system for transmission of monochromatic light from point $\boldsymbol{\rho}'$ in the input plane $z = z_0$ to a point $\boldsymbol{\rho}$ in the output plane $z = z_1$.

(a) Derive an expression for the spectrum $S_1(\boldsymbol{\rho}, \omega)$ in the output plane.

(b) Consider the special case when light from a partially coherent planar secondary source occupies a finite region σ of the input plane $z = z_0$ and radiates to the far zone in free space. Obtain an expression $S_1(\boldsymbol{\rho}, \omega)$ for this case.

(c) Simplify the expression obtained in (b) for the case when the source is quasi-homogeneous and show that the result is in agreement with one of the reciprocity relations for radiation generated by quasi-homogeneous sources [Section 5.3.2].

(d) What form does the expression for $S_1(\boldsymbol{\rho}, \omega)$ take when the input is a polychromatic plane wave which is incident in the positive z-direction?

5.8 Consider a three-dimensional, spatially incoherent, primary source, occupying a finite domain D. The fluctuations of the source are statistically stationary and the spectrum of the source is the same at every source point.

(a) Show that the normalized spectrum of the field generated by the source is the same at every point outside the source region D.

(b) Derive an expression for the spectral degree of coherence of the field in the far zone.

5.9 A three-dimensional primary, quasi-homogeneous source, whose spectrum $S_Q(\omega)$ is the same at every source point, has a spectral degree of coherence which obeys the three-dimensional scaling law, i.e. it is of the form

$$\mu_Q(\mathbf{r}', \omega) = h(k\mathbf{r}'), \quad (k = \omega/c).$$

Show that the normalized spectral density $\mathscr{s}_V(r\mathbf{u}, \omega)$ of the field generated by the source is the same throughout the far zone and find a relation between the normalized far-zone spectrum $\mathscr{s}_V$ and the source spectrum S_Q.

5.10 Consider *paraxial* propagation of light from a partially coherent, statistically stationary, planar, secondary source that is located in the plane $z = 0$ and that radiates into the half-space $z > 0$. Let $\mathbf{r} \equiv (x, y, z) \equiv (\boldsymbol{\rho}, z)$, $[\boldsymbol{\rho} = (x, y)]$, be a position vector of a typical point in that half-space.

Show that if the normalized spectrum $\mathscr{s}(\boldsymbol{\rho}, z; \omega)$ of the field is independent of $\boldsymbol{\rho}$ across every plane $z = \text{cons.} > 0$, then the normalized spectrum of the field is invariant on propagation, i.e. it is the same throughout the whole half-space $z \geqslant 0$.

6

Second-order coherence theory of vector electromagnetic fields

6.1 Introduction

In the preceding chapters we studied second-order correlation effects in fluctuating scalar wavefields. We have seen that the simplest manifestations of second-order correlations in such fields are the well-known interference phenomena. We have shown that a careful analysis of these leads to the introduction of scalar correlation functions (the mutual coherence function or its Fourier transform, the cross-spectral density function), in terms of which all second-order correlation effects in scalar wavefields may be analyzed.

We will now turn our attention to vector fields, more precisely to fluctuating electromagnetic fields. We will begin by first analyzing a well-known phenomenon that brings into evidence second-order correlations in electromagnetic fields, namely partial polarization. We will see that an analysis of partial polarization can be conveniently carried out in terms of certain 2×2 correlation matrices. Their natural generalizations are 3×3 correlation matrices which are associated with a set of second-rank correlation tensors. Using these tensors one obtains a unified description of all second-order correlation phenomena in electromagnetic fields.

Just as in the scalar case, we will find it convenient to employ the analytic signal representation (Section 3.1.1). Consider a single realization of a fluctuating field and let $\mathbf{E}^{(\mathrm{r})}(\mathbf{r}, t)$, $\mathbf{H}^{(\mathrm{r})}(\mathbf{r}, t)$ denote the (real) electric and magnetic vectors respectively, at a point specified by a position vector $\mathbf{r}$ at time t. We express each field vector as a Fourier integral,

$$\mathbf{E}^{(\mathrm{r})}(\mathbf{r}, t) = \int_{-\infty}^{\infty} \tilde{\mathbf{E}}^{(\mathrm{r})}(\mathbf{r}, \nu)\, \mathrm{e}^{-2\pi i\nu t}\, \mathrm{d}\nu, \quad \mathbf{H}^{(\mathrm{r})}(\mathbf{r}, t) = \int_{-\infty}^{\infty} \tilde{\mathbf{H}}^{(\mathrm{r})}(\mathbf{r}, \nu)\, \mathrm{e}^{-2\pi i\nu t}\, \mathrm{d}\nu. \quad (6.1\text{–}1)$$

The associated complex analytic signals are then given by the integrals

$$\mathbf{E}(\mathbf{r}, t) = \int_{0}^{\infty} \tilde{\mathbf{E}}^{(\mathrm{r})}(\mathbf{r}, \nu)\, \mathrm{e}^{-2\pi i\nu t}\, \mathrm{d}\nu, \quad \mathbf{H}(\mathbf{r}, t) = \int_{0}^{\infty} \tilde{\mathbf{H}}^{(\mathrm{r})}(\mathbf{r}, \nu)\, \mathrm{e}^{-2\pi i\nu t}\, \mathrm{d}\nu. \quad (6.1\text{–}2)$$

For fields which are statistically stationary the Fourier decomposition (6.1–1) will not exist in the sense of ordinary function theory (see Section 2.4) and has to be interpreted in the sense of the theory of generalized functions. It follows from Eq. (3.1–12), when applied to each Cartesian component of the complex electric and magnetic field, that

$$\mathbf{E}(\mathbf{r}, t) = \tfrac{1}{2}[\mathbf{E}^{(\mathrm{r})}(\mathbf{r}, t) + \mathrm{i}\mathbf{E}^{(\mathrm{i})}(\mathbf{r}, t)]. \quad (6.1\text{–}3a)$$

and

$$\mathbf{H}(\mathbf{r}, t) = \tfrac{1}{2}[\mathbf{H}^{(r)}(\mathbf{r}, t) + \mathrm{i}\mathbf{H}^{(i)}(\mathbf{r}, t)], \tag{6.1–3b}$$

where the real and imaginary parts of each of these two complex field vectors form a Hilbert transform pair.

It will be convenient for later purposes to derive expressions for the average electric and magnetic energy densities and for the average Poynting vector of an electromagnetic field in free space in terms of the complex field vectors **E** and **H**. Assuming that the field fluctuations are stationary (at least in the wide sense), ergodic and have zero mean we have, according to Eqs. (3.1–77), that

$$\langle [\mathbf{E}^{(r)}(\mathbf{r}, t)]^2 \rangle = \langle [\mathbf{E}^{(i)}(\mathbf{r}, t)]^2 \rangle, \tag{6.1–4a}$$

and

$$\langle \mathbf{E}^{(r)}(\mathbf{r}, t) \cdot \mathbf{E}^{(i)}(\mathbf{r}, t) \rangle = 0. \tag{6.1–4b}$$

Now in the Gaussian system of units‡ the average electric energy density $\langle w_e(\mathbf{r}, t) \rangle$ is given by the expression

$$\langle w_e(\mathbf{r}, t) \rangle = \frac{1}{8\pi} \langle [\mathbf{E}^{(r)}(\mathbf{r}, t)]^2 \rangle. \tag{6.1–5}$$

If we make use of the relations (6.1–3a) and (6.1–4) we readily find that in terms of the complex field $\mathbf{E}(\mathbf{r}, t)$,

$$\langle w_e(\mathbf{r}, t) \rangle = \frac{1}{4\pi} \langle \mathbf{E}^*(\mathbf{r}, t) \cdot \mathbf{E}(\mathbf{r}, t) \rangle. \tag{6.1–6}$$

In a strictly similar manner it follows that the average magnetic energy density $\langle w_m(\mathbf{r}, t) \rangle$ may be expressed in the form

$$\langle w_m(\mathbf{r}, t) \rangle = \frac{1}{4\pi} \langle \mathbf{H}^*(\mathbf{r}, t) \cdot \mathbf{H}(\mathbf{r}, t) \rangle. \tag{6.1–7}$$

The average Poynting vector is given by the expression

$$\langle \mathbf{S}(\mathbf{r}, t) \rangle = \frac{c}{4\pi} \langle \mathbf{E}^{(r)}(\mathbf{r}, t) \times \mathbf{H}^{(r)}(\mathbf{r}, t) \rangle. \tag{6.1–8}$$

Now according to Eqs. (6.1–3),

$$\begin{aligned} \mathrm{Re}\, \langle \mathbf{E}^* \times \mathbf{H} \rangle &= \tfrac{1}{4} \mathrm{Re}\, [\langle (\mathbf{E}^{(r)} - \mathrm{i}\mathbf{E}^{(i)}) \times (\mathbf{H}^{(r)} + \mathrm{i}\mathbf{H}^{(i)}) \rangle] \\ &= \tfrac{1}{4}[\langle \mathbf{E}^{(r)} \times \mathbf{H}^{(r)} \rangle + \langle \mathbf{E}^{(i)} \times \mathbf{H}^{(i)} \rangle], \end{aligned} \tag{6.1–9}$$

where Re denotes the real part. If we make use of the relation (3.1–77a) we readily find that the two terms on the right-hand side of Eq. (6.1–9) are equal to each other and hence

$$\mathrm{Re}\, \langle \mathbf{E}^* \times \mathbf{H} \rangle = \tfrac{1}{2} \langle \mathbf{E}^{(r)} \times \mathbf{H}^{(r)} \rangle. \tag{6.1–10}$$

‡ Both the Gaussian and SI units are used in this book. Generally Gaussian units are used only in the formal development presented in this chapter and the next, whereas magnitudes of electromagnetic quantities are always given in SI units.

On substituting from this relation into Eq. (6.1–8) we obtain the required expression for the average Poynting vector:

$$\langle \mathbf{S}(\mathbf{r}, t)\rangle = \frac{c}{2\pi}\,\mathrm{Re}\,\langle \mathbf{E}^*(\mathbf{r}, t) \times \mathbf{H}(\mathbf{r}, t)\rangle. \tag{6.1–11}$$

6.2 The 2 × 2 equal-time coherence matrix of a well-collimated, uniform, quasi-monochromatic light beam‡

Consider a well-collimated, uniform, quasi-monochromatic light beam of mean frequency $\bar{\nu}$. Let us choose a right-hand Cartesian coordinate system of axes x, y, z, with the z-axis along the effective direction of propagation of the beam. Let

$$E_x(t) = a_1(t)\,\mathrm{e}^{\mathrm{i}[\alpha_1(t)-2\pi\bar{\nu}t]}, \quad E_y(t) = a_2(t)\,\mathrm{e}^{\mathrm{i}[\alpha_2(t)-2\pi\bar{\nu}t]}, \tag{6.2–1}$$

$[a_1(t) \geqslant 0,\ a_2(t) \geqslant 0]$ be the components of the complex electric field vector in the x- and y-directions respectively, at a point (x, y) in some transverse plane $z = z_0$ at time t. If the point (x, y) is within the beam and not too close to its edge, $E_x(t)$ and $E_y(t)$ will be effectively independent of x and y, because the beam is assumed to be uniform; however it will depend on z_0 (which we do not show explicitly in the arguments of E_x and E_y).

It follows from the properties of the envelope representation of narrow-band signals, discussed in Section 3.1.2, that the complex amplitude factors $a_1(t)\,\mathrm{e}^{\mathrm{i}\alpha_1(t)}$ and $a_2(t)\,\mathrm{e}^{\mathrm{i}\alpha_2(t)}$ in the expression (6.2–1) will vary slowly with t in comparison with $\cos 2\pi\bar{\nu}t$ and $\sin 2\pi\bar{\nu}t$, remaining effectively constant over time intervals that are short compared to the reciprocal bandwidth of light. The amplitudes $a_1(t)$ and $a_2(t)$, just like $E_x(t)$ and $E_y(t)$, are random functions.

In general, $E_x(t)$ and $E_y(t)$ will be correlated. We will now show that some information about this correlation may be obtained from simple experiments. Suppose that the beam is passed through a compensator and then through a polarizer. Let ε_1 and ε_2 denote the phase changes produced in E_x and E_y respectively by the compensator, on passage of the beam from a plane $z = z_0$ to a plane $z = z_1$ (See Fig. 6.1). Further let θ denote the angle which the direction of vibration of the electric field emerging from the polarizer makes with the x-axis. We denote by $\mathbf{i}_\theta$ the unit vector, with components $\cos\theta$, $\sin\theta$, along the x- and y-directions respectively. Apart from an unessential constant phase factor, the complex electric field at any point in a cross-section $z = z_2$ of the beam that emerges from this two-component device is given by (if reflection losses at the surfaces of the compensator and the polarizer are ignored)

$$\mathbf{E}'(t; \varepsilon_1, \varepsilon_2, \theta) = [E_x(t)\,\mathrm{e}^{\mathrm{i}\varepsilon_1}\cos\theta + E_y(t)\,\mathrm{e}^{\mathrm{i}\varepsilon_2}\sin\theta]\mathbf{i}_\theta. \tag{6.2–2}$$

The average electric energy density $\langle w'_\mathrm{e}\rangle$ at a typical point in the illuminated part of the plane $z = z_2$ is, according to Eq. (6.1–6), given by the expression

$$\langle w'_\mathrm{e}(\varepsilon_1, \varepsilon_2, \theta)\rangle = \frac{1}{4\pi}\langle \mathbf{E}'^*(t; \varepsilon_1, \varepsilon_2, \theta)\cdot\mathbf{E}'(t; \varepsilon_1, \varepsilon_2, \theta)\rangle, \tag{6.2–3}$$

which is independent of time, because of our assumption that the field is

‡ The analysis presented in this section is largely based on a paper by Wolf (1959).

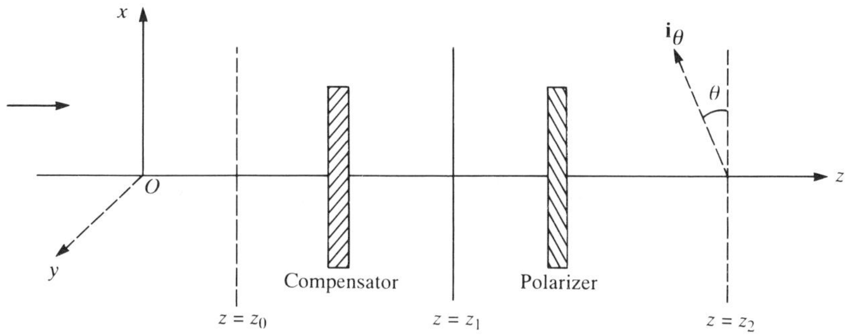

Fig. 6.1 Illustrating an arrangement for determining the elements of the 2 × 2 equal-time coherence matrix of a uniform quasi-monochromatic light beam.

statistically stationary. On substituting from Eq. (6.2–2) into Eq. (6.2–3) we obtain for $\langle w'_e \rangle$ the expression

$$\langle w'_e(\delta, \theta) \rangle = \frac{1}{4\pi}[J_{xx}\cos^2\theta + J_{yy}\sin^2\theta + J_{xy}\,e^{i\delta}\sin\theta\cos\theta + J_{yx}\,e^{-i\delta}\cos\theta\sin\theta], \tag{6.2–4}$$

where

$$\delta = \varepsilon_2 - \varepsilon_1 \tag{6.2–5}$$

and J_{kl}, (k, l stand for x or y), are the elements of the covariance matrix

$$\mathbf{J} \equiv \begin{bmatrix} \langle E_x^*(t)E_x(t)\rangle & \langle E_x^*(t)E_y(t)\rangle \\ \langle E_y^*(t)E_x(t)\rangle & \langle E_y^*(t)E_y(t)\rangle \end{bmatrix}. \tag{6.2–6}$$

Since the phase changes ε_1 and ε_2 enter the right-hand side of Eq. (6.2–4) only through the difference $\delta = \varepsilon_2 - \varepsilon_1$, we have written $\langle w'_e(\delta, \theta)\rangle$ rather than $\langle w'_e(\varepsilon_1, \varepsilon_2, \theta)\rangle$ on the left-hand side of that equation. The covariance matrix **J** defined by Eq. (6.2–6) may be called the 2 × 2 *equal-time coherence matrix* or, for short, the *coherence matrix* of the beam. For reasons which will become evident in Section 6.3 it is sometimes also known as the *polarization matrix*. It should be mentioned that the term 'coherence matrix' is used in optical coherence theory in several different senses, some of which we will encounter later on.

Let tr **J** denote the trace of the coherence matrix. We see that

$$\begin{aligned} \operatorname{tr}\mathbf{J} &\equiv J_{xx} + J_{yy} \\ &= \langle E_x^*(t)E_x(t)\rangle + \langle E_y^*(t)E_y(t)\rangle \\ &= 4\pi\langle w_e \rangle, \end{aligned} \tag{6.2–7}$$

i.e. it is proportional to the average electric energy density $\langle w_e \rangle$ of the light at any point in the cross-section $z = z_0$ of the beam incident on the device. The off-diagonal elements J_{xy} and J_{yx} are complex numbers, in general, and they represent the 'equal-time' correlation between the x- and y-components of the complex electric field in the plane $z = z_0$. It is clear that J_{xy} and J_{yx} are complex conjugates of each other, i.e. that

$$J_{yx} = J_{xy}^*, \tag{6.2–8}$$

showing that the coherence matrix is Hermitian. Moreover, it is non-negative definite, i.e. with any arbitrary complex numbers c_1 and c_2,

$$c_1^* c_1 J_{xx} + c_2^* c_2 J_{yy} + c_1^* c_2 J_{xy} + c_2^* c_1 J_{yx} \geqslant 0. \tag{6.2–9}$$

This result follows at once from the obvious fact that $\langle |c_1 E_x + c_2 E_y|^2 \rangle \geqslant 0$. One can readily deduce from the inequality (6.2–9) or from the Schwarz inequality and the Hermiticity relation (6.2–8) that the determinant of the coherence matrix, which we will denote by $\det \mathbf{J}$, is non-negative, i.e. that

$$\det \mathbf{J} \equiv J_{xx} J_{yy} - J_{xy} J_{yx} \geqslant 0. \tag{6.2–10}$$

We see from Eq. (6.2–4) that both the diagonal and the off-diagonal elements of the coherence matrix are, in general, needed to calculate the change in the average electric energy density of the light transmitted through the device. In their mathematical structure the off-diagonal elements are analogous to the mutual intensity of the scalar theory [Eq. (4.3–34)]. We will normalize the off-diagonal element J_{xy} like the mutual intensity, by setting

$$j_{xy} \equiv |j_{xy}|\, e^{i\beta_{xy}} \equiv \frac{J_{xy}}{(J_{xx})^{1/2}(J_{yy})^{1/2}}. \tag{6.2–11}$$

It follows from Eqs. (6.2–11), (6.2–10) and (6.2–8) that

$$0 \leqslant |j_{xy}| \leqslant 1. \tag{6.2–12}$$

It is clear that j_{xy} may be considered to be a measure of the degree of correlation that exists between the components $E_x(t)$ and $E_y(t)$ of the complex electric field vector at a typical point in the plane $z = z_0$ at any particular instant of time.

Let us now return to the expression (6.2–4) for the average electric energy density at points in the plane $z = z_2$. If we use Eqs. (6.2–8) and (6.2–11), Eq. (6.2–4) may be expressed in the form

$$\langle w_e'(\delta, \theta) \rangle = \frac{1}{4\pi}[J_{xx} \cos^2\theta + J_{yy} \sin^2\theta + 2(J_{xx})^{1/2}(J_{yy})^{1/2}|j_{xy}| \cos\theta \sin\theta \cos(\beta_{xy} + \delta)] \tag{6.2–13}$$

or, if we set

$$\frac{1}{4\pi} J_{xx} \cos^2\theta = \langle I^{(1)}(\theta) \rangle, \quad \frac{1}{4\pi} J_{yy} \sin^2\theta = \langle I^{(2)}(\theta) \rangle, \tag{6.2–13a}$$

the formula (6.2–13) may be expressed in more compact form

$$\langle w_e'(\delta, \theta) \rangle = \langle I^{(1)}(\theta) \rangle + \langle I^{(2)}(\theta) \rangle + 2[\langle I^{(1)}(\theta) \rangle]^{1/2}[\langle I^{(2)}(\theta) \rangle]^{1/2} |j_{xy}| \cos(\beta_{xy} + \delta). \tag{6.2–14}$$

Equation (6.2–14) is analogous to the law (4.3–19) of the scalar theory, for interference with partially coherent light.

With the help of Eq. (6.2–13) the elements of the coherence matrix may be determined from measurements which employ a compensator and a polarizer.

One only needs to measure the values of the average electric energy density $\langle w'_e(\delta, \theta)\rangle$ at any point in the plane $z = z_2$ for a number of selected pairs of values of δ and θ and solve Eq. (6.2–13) for the elements J_{xx}, J_{yy} and $J_{xy} = J^*_{yx}$.

Let $\{\delta, \theta\}$ denote the measurement in the plane $z = z_2$ of the average value of the electric energy density, which corresponds to a particular pair of values δ, θ. A convenient set of measurements, for example, is

$$\{0, 0°\}, \{0, 45°\}, \{0, 90°\}, \{0, 135°\}, \{\pi/2, 45°\}, \{\pi/2, 135°\}.$$

We then readily find from Eq. (6.2–4) that

$$J_{xx} = 4\pi\langle w'_e(0, 0°)\rangle, \tag{6.2–15a}$$

$$J_{yy} = 4\pi\langle w'_e(0, 90°)\rangle, \tag{6.2–15b}$$

$$J_{xy} = J^*_{yx} = 2\pi\{[\langle w'_e(0, 45°)\rangle - \langle w'_e(0, 135°)\rangle] + \mathrm{i}[\langle w'_e(\pi/2, 45°)\rangle - \langle w'_e(\pi/2, 135°)\rangle]\}. \tag{6.2–15c}$$

The three formulas give the elements of the coherence matrix in terms of the values of the averaged energy density determined from the six experiments.‡

Let us examine more closely how the average electric density $\langle w'_e(\delta, \theta)\rangle$ changes with δ and θ. For this purpose we re-write Eq. (6.2–13) in a more convenient form, by the use of the trigonometric identities $\cos^2\theta = (1 + \cos 2\theta)/2$, $\sin^2\theta = (1 - \cos 2\theta)/2$ and $2\sin\theta\cos\theta = \sin 2\theta$. Equation (6.2–13) then becomes, if we make use of Eq. (6.2–11) again,

$$\langle w'_e(\delta, \theta)\rangle = \frac{1}{8\pi}[(J_{xx} + J_{yy}) + (J_{xx} - J_{yy})\cos 2\theta + 2|J_{xy}|\sin 2\theta\cos(\beta_{xy} + \delta)]. \tag{6.2–17}$$

The second and the third term on the right-hand side of Eq. (6.2–17) may be combined with the help of an elementary trigonometric identity and we then obtain for the average electric density in the plane $z = z_2$ the expression

$$\langle w'_e(\delta, \theta)\rangle = \frac{1}{8\pi}\{\mathrm{tr}\,\mathbf{J} + R(\delta)\cos[2\theta - \phi(\delta)]\}, \tag{6.2–18}$$

where

$$R(\delta) = [(J_{xx} - J_{yy})^2 + 4|J_{xy}|^2\cos^2(\beta_{xy} + \delta)]^{1/2}, \tag{6.2–19}$$

$$\cos\phi(\delta) = \frac{J_{xx} - J_{yy}}{R(\delta)}, \quad \sin\phi(\delta) = \frac{2|J_{xy}|}{R(\delta)}\cos(\beta_{xy} + \delta). \tag{6.2–20}$$

‡ Actually four properly chosen experiments are sufficient to determine the elements of the coherence matrix, e.g. the experiments involving the measurements

$$\{0, 0°\}, \{0, 45°\}, \{0, 135°\}, \{\pi/2, 45°\}.$$

In terms of the four measured values of the average electric energy density obtained from these experiments, the elements of the coherence matrix may readily be shown to be given by

$$J_{xx} = 4\pi\langle w'_e(0, 0°)\rangle, \tag{6.2–16a}$$

$$J_{yy} = 4\pi[\langle w'_e(0, 45°)\rangle + \langle w'_e(0, 135°)\rangle - \langle w'_e(0, 0°)\rangle], \tag{6.2–16b}$$

$$J_{xy} = J^*_{yx} = 2\pi\{[\langle w'_e(0, 45°)\rangle - \langle w'_e(0, 135°)\rangle] - \mathrm{i}[\langle w'_e(0, 45°)\rangle + \langle w'_e(0, 135°)\rangle - 2\langle w'_e(\pi/2, 45°)\rangle]\}. \tag{6.2–16c}$$

Suppose that we keep θ fixed but vary δ. We see from Eq. (6.2–17) that $\langle w'_e(\delta, \theta)\rangle$ now varies sinusoidally between the values

$$\langle w'_e(\delta, \theta)\rangle_{\max(\delta)} = \frac{1}{8\pi}[\operatorname{tr}\mathbf{J} + (J_{xx} - J_{yy})\cos 2\theta + 2|J_{xy}|\sin 2\theta], \quad (6.2\text{–}21a)$$

and

$$\langle w'_e(\delta, \theta)\rangle_{\min(\delta)} = \frac{1}{8\pi}[\operatorname{tr}\mathbf{J} + (J_{xx} - J_{yy})\cos 2\theta - 2|J_{xy}|\sin 2\theta], \quad (6.2\text{–}21b)$$

provided that $\sin 2\theta \geqslant 0$. These two expressions may be rewritten in the form

$$\langle w'_e(\delta, \theta)\rangle_{\max(\delta)} = \frac{1}{8\pi}[\operatorname{tr}\mathbf{J} + R'M] \quad (6.2\text{–}22a)$$

and

$$\langle w'_e(\delta, \theta)\rangle_{\min(\delta)} = \frac{1}{8\pi}[\operatorname{tr}\mathbf{J} + R'N], \quad (6.2\text{–}22b)$$

where

$$R' = [(\operatorname{tr}\mathbf{J})^2 - 4\det\mathbf{J}]^{1/2}, \quad (6.2\text{–}23)$$

$$M = \max[\cos(2\theta + \phi'), \cos(2\theta - \phi')], \quad (6.2\text{–}24a)$$

$$N = \min[\cos(2\theta + \phi'), \cos(2\theta - \phi')], \quad (6.2\text{–}24b)$$

$$\cos\phi' = \frac{J_{xx} - J_{yy}}{R'}, \quad \sin\phi' = \frac{2|J_{xy}|}{R'}. \quad (6.2\text{–}25)$$

We see from Eq. (6.2–17) that maxima occur when

$$\delta = -\beta_{xy} \pm 2m\pi, \quad (m = 0, 1, 2, \ldots) \quad (6.2\text{–}26a)$$

and the minima occur when

$$\delta = -\beta_{xy} \pm (2m + 1)\pi, \quad (m = 0, 1, 2, \ldots). \quad (6.2\text{–}26b)$$

Next suppose that we keep δ fixed and vary θ. It follows from Eq. (6.2–18) that $\langle w'_e(\delta, \theta)\rangle$ again varies sinusoidally, between the values

$$\langle w'_e(\delta, \theta)\rangle_{\max(\delta)} = \frac{1}{8\pi}[\operatorname{tr}\mathbf{J} + R(\delta)], \quad (6.2\text{–}27a)$$

and

$$\langle w'_e(\delta, \theta)\rangle_{\min(\delta)} = \frac{1}{8\pi}[\operatorname{tr}\mathbf{J} - R(\delta)]. \quad (6.2\text{–}27b)$$

From Eq. (6.2–18) it is clear that the maxima with respect to θ occur when

$$\theta = \tfrac{1}{2}\phi \pm m\pi, \quad (m = 0, 1, 2, \ldots) \quad (6.2\text{–}28a)$$

and the minima with respect to θ occur when

$$\theta = \tfrac{1}{2}\phi \pm (m + \tfrac{1}{2})\pi, \quad (m = 0, 1, 2, \ldots). \quad (6.2\text{–}28b)$$

Finally let us determine the extrema of the average electric energy density with respect to both δ and θ. They can be obtained either by taking the maximum of Eq. (6.2–22a) and the minimum of Eq. (6.2–22b) with respect to θ or the maximum of Eq. (6.2–27a) and the minimum of Eq. (6.2–27b) with respect to δ. The results are

$$\langle w_e'(\delta, \theta)\rangle_{\max(\delta,\theta)} = \frac{1}{8\pi}\{\mathrm{tr}\,\mathbf{J} + [(\mathrm{tr}\,\mathbf{J})^2 - 4\det\mathbf{J}]^{1/2}\}, \tag{6.2–29a}$$

$$\langle w_e'(\delta, \theta)\rangle_{\min(\delta,\theta)} = \frac{1}{8\pi}\{\mathrm{tr}\,\mathbf{J} - [(\mathrm{tr}\,\mathbf{J})^2 - 4\det\mathbf{J}]^{1/2}\}, \tag{6.2–29b}$$

The values of the pairs (δ, θ) for which the maxima occur are given by Eqs. (6.2–26a) and (6.2–28a) and those for which the minima occur are given by Eqs. (6.2–26b) and (6.2–28b).

It follows from Eqs. (6.2–29) that

$$\frac{\langle w_e'(\delta, \theta)\rangle_{\max(\delta,\theta)} - \langle w_e'(\delta, \theta)\rangle_{\min(\delta,\theta)}}{\langle w_e'(\delta, \theta)\rangle_{\max(\delta,\theta)} + \langle w_e'(\delta, \theta)\rangle_{\min(\delta,\theta)}} = \left[1 - \frac{4\det\mathbf{J}}{(\mathrm{tr}\,\mathbf{J})^2}\right]^{1/2}. \tag{6.2–30}$$

The expression on the left-hand side of Eq. (6.2–30) is analogous to the expression for the visibility of interference fringes in Young's interference experiment, which we discussed in Section 4.3.1. We showed there that the visibility was proportional to the absolute value of the degree of coherence of the light at the pinholes [Eq. (4.3–25)]. We will see later that the expression on the right-hand side of Eq. (6.2–30) also has a simple physical significance.

In defining the coherence matrix we employed a fixed but arbitrary (x, y)-coordinate system in a plane perpendicular to the direction of propagation of the beam (the z-axis). We will now briefly consider how the coherence matrix transforms when the x-, y-axes are rotated through an angle Θ about the z-axis. In terms of E_x and E_y, the components $E_{x'}$ and $E_{y'}$ of the complex electric vector, referred to the new axes x', y' (see Fig. 6.2), are

$$\left.\begin{aligned} E_{x'} &= E_x\cos\Theta + E_y\sin\Theta, \\ E_{y'} &= -E_x\sin\Theta + E_y\cos\Theta. \end{aligned}\right\} \tag{6.2–31}$$

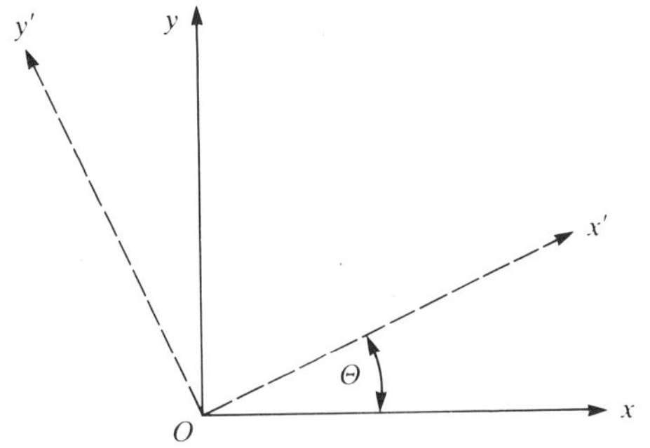

Fig. 6.2 Illustrating the notation relating to determining the change in the coherence matrix **J** under rotation of axes about the direction of propagation of a beam.

The elements of the coherence matrix $\mathbf{J}'$, referred to the rotated axes are

$$J_{k'l'} = \langle E_{k'}^* E_{l'} \rangle, \tag{6.2–32}$$

where k', l' each take on the values x' and y'. From Eqs. (6.2–32) and (6.2–31) we readily find that

$$\mathbf{J}' = \begin{bmatrix} J_{xx}c^2 + J_{yy}s^2 + (J_{xy} + J_{yx})cs & (J_{yy} - J_{xx})cs + J_{xy}c^2 - J_{yx}s^2 \\ (J_{yy} - J_{xx})cs + J_{yx}c^2 - J_{xy}s^2 & J_{xx}s^2 + J_{yy}c^2 - (J_{xy} + J_{yx})cs \end{bmatrix}, \tag{6.2–33}$$

where

$$c = \cos\Theta, \quad s = \sin\Theta. \tag{6.2–34}$$

It follows at once from Eq. (6.2–33) that

$$\operatorname{tr}\mathbf{J}' = J_{xx} + J_{yy} = \operatorname{tr}\mathbf{J} \tag{6.2–35}$$

and that

$$\det\mathbf{J}' = J_{xx}J_{yy} - J_{xy}J_{yx} = \det\mathbf{J}. \tag{6.2–36}$$

Equations (6.2–35) and (6.2–36) show that the trace of the coherence matrix and its determinant each remain unchanged under any rotation of the x-, y-axes about the z-direction, a result which also follows from well-known theorems on matrices.

Instead of using a coherence (polarization) matrix to represent correlations in quasi-monochromatic beams of elementary radiation, one can employ a closely related and much older representation due to Stokes (1852) (which actually predates Maxwell's electromagnetic theory). Because Stokes' representation is widely used, we will now briefly describe it and we will show how it is related to the modern characterization of the state of polarization of a light wave in terms of coherence matrices.

Let us again represent the components $E_x(t)$ and $E_y(t)$ of the complex electric field in two mutually orthogonal directions perpendicular to the direction of propagation of the beam in the form given by Eqs. (6.2–1), namely

$$E_x(t) = a_1(t)\,e^{i[\alpha_1(t) - 2\pi\bar{\nu}t]}, \quad E_y(t) = a_2(t)\,e^{i[\alpha_2(t) - 2\pi\bar{\nu}t]}. \tag{6.2–37}$$

One then defines the so-called *Stokes parameters* of the beam by the expressions

$$\left.\begin{aligned} s_0 &= \langle a_1^2 \rangle + \langle a_2^2 \rangle, \\ s_1 &= \langle a_1^2 \rangle - \langle a_2^2 \rangle, \\ s_2 &= 2\langle a_1 a_2 \cos\Delta(t) \rangle, \\ s_3 &= 2\langle a_1 a_2 \sin\Delta(t) \rangle, \end{aligned}\right\} \tag{6.2–38}$$

where

$$\Delta(t) = \alpha_2(t) - \alpha_1(t). \tag{6.2–39}$$

On comparing Eqs. (6.2–38) with Eq. (6.2–6) and making use of Eqs. (6.2–39) and (6.2–37) we readily find that the four Stokes parameters and the elements of the coherence matrix are related by the formulas

$$\left.\begin{aligned} s_0 &= J_{xx} + J_{yy}, \\ s_1 &= J_{xx} - J_{yy}, \\ s_2 &= J_{xy} + J_{yx}, \\ s_3 &= \mathrm{i}(J_{yx} - J_{xy}). \end{aligned}\right\} \quad (6.2\text{–}40a) \qquad \left.\begin{aligned} J_{xx} &= \tfrac{1}{2}(s_0 + s_1), \\ J_{yy} &= \tfrac{1}{2}(s_0 - s_1), \\ J_{xy} &= \tfrac{1}{2}(s_2 + \mathrm{i}s_3), \\ J_{yx} &= \tfrac{1}{2}(s_2 - \mathrm{i}s_3). \end{aligned}\right\} \quad (6.2\text{–}40b)$$

The relations (6.2–40) may be expressed in a more compact form in terms of the Pauli spin matrices which we take to be

$$\boldsymbol{\sigma}_1 = \begin{bmatrix} 1 & 0 \\ 0 & -1 \end{bmatrix}, \quad \boldsymbol{\sigma}_2 = \begin{bmatrix} 0 & 1 \\ 1 & 0 \end{bmatrix}, \quad \boldsymbol{\sigma}_3 = \begin{bmatrix} 0 & \mathrm{i} \\ -\mathrm{i} & 0 \end{bmatrix} \qquad (6.2\text{–}41)$$

and the unit matrix

$$\boldsymbol{\sigma}_0 = \begin{bmatrix} 1 & 0 \\ 0 & 1 \end{bmatrix}. \qquad (6.2\text{–}42)$$

One readily finds that the relations (6.2–40a) and (6.2–40b) are equivalent to the formulas

$$s_j = \mathrm{tr}\,[\mathbf{J}\boldsymbol{\sigma}_j], \qquad j = 0, 1, 2, 3 \qquad (6.2\text{–}43a)$$

and

$$\mathbf{J} = \tfrac{1}{2}\sum_{j=0}^{3} s_j \boldsymbol{\sigma}_j. \qquad (6.2\text{–}43b)$$

In view of these relations it is clear that any property of the coherence matrix may be translated into a property of the Stokes parameters. For example, if we make use of Eqs. (6.2–40a) we readily find that the non-negativity of the determinant of the coherence matrix [Eq. (6.2–10)] implies that the four Stokes parameters satisfy the inequality

$$s_0^2 \geqslant s_1^2 + s_2^2 + s_3^2. \qquad (6.2\text{–}44)$$

Just like the elements of the coherence matrix, the four Stokes parameters of a quasi-monochromatic beam may be determined from simple experiments that involve a compensator and a polarizer. Suppose that $\langle w_\mathrm{e}'(\delta, \theta) \rangle$ denotes, as before, the average electric energy density of the beam after it passes through a compensator which introduces a relative phase change $\delta = \varepsilon_2 - \varepsilon_1$ between the y- and x-components of the complex electric field vector. It then passes through a polarizer which transmits only the component of the electric field which makes an angle θ with the x direction. Then it follows from Eqs. (6.2–15) and (6.2–40a) that the Stokes parameters of the beam incident on this two-component device are given by

$$\left.\begin{aligned} s_0 &= 4\pi\{\langle w_\mathrm{e}'(0, 0^\circ)\rangle + \langle w_\mathrm{e}'(0, 90^\circ)\rangle\}, \\ s_1 &= 4\pi\{\langle w_\mathrm{e}'(0, 0^\circ)\rangle - \langle w_\mathrm{e}'(0, 90^\circ)\rangle\}, \\ s_2 &= 4\pi\{\langle w_\mathrm{e}'(0, 45^\circ)\rangle - \langle w_\mathrm{e}'(0, 135^\circ)\rangle\}, \\ s_3 &= 4\pi\{\langle w_\mathrm{e}'(\pi/2, 45^\circ)\rangle - \langle w_\mathrm{e}'(\pi/2, 135^\circ)\rangle\}. \end{aligned}\right\} \qquad (6.2\text{–}45)$$

6.3 Completely unpolarized and completely polarized light. The degree of polarization

According to Eq. (6.2–12), the normalized, generally complex, off-diagonal element j_{xy} of the coherence matrix, which is a measure of the correlation between the components of the complex electric vector along the x- and y-directions, is bounded by zero and unity in absolute value. We will now consider the form of the coherence matrix in the extreme cases when $|j_{xy}|$ attains one or other of these limiting values.

6.3.1 *Unpolarized light (natural light)*

Let us first consider a beam for which

$$j_{xy} = 0, \tag{6.3–1}$$

irrespective of the particular choice of the x-, y-axes. According to Eqs. (6.2–11) and (6.2–8) this requirement implies that, irrespective of the choice of the x-, y-axes,

$$J_{xy} = J_{yx} = 0. \tag{6.3–2}$$

In view of the transformation law given by Eq. (6.2–33) this implies that irrespective of the choice of Θ, $(J_{yy} - J_{xx}) \cos\Theta \sin\Theta = 0$, i.e. with any choice of the x-, y-axes,

$$J_{xx} = J_{yy}. \tag{6.3–3}$$

It follows, in view of Eqs. (6.3–2) and (6.3–3), that the coherence matrix now has the form

$$\mathbf{J} = J_{xx}\begin{bmatrix} 1 & 0 \\ 0 & 1 \end{bmatrix}, \tag{6.3–4}$$

i.e. *it is proportional to the unit matrix.* The proportionality factor J_{xx} is also independent of the choice of the axes because, when Eqs. (6.3–2) and (6.3–3) hold, the leading diagonal term $J_{x'x'}$ in the 'transformed' coherence matrix $\mathbf{J}'$ [Eq. (6.2–33)] is then independent of Θ. Stated more explicitly, we have shown that when the x- and y-components of the electric field vector are uncorrelated for all pairs of directions [Eq. (6.3–1)], the average $J_{xx} = \langle E_x^*(t)E_x(t)\rangle$ has the same value for every direction which is perpendicular to the direction of propagation of the beam.

If we make use of the relations (6.3–2) and (6.3–3) in Eq. (6.2–4) we find at once that

$$\langle w_e'(\delta, \theta)\rangle = \frac{1}{4\pi}J_{xx}. \tag{6.3–5}$$

Since the right-hand side of this equation is a constant, the left-hand side must be independent of θ and δ. Hence if a beam which is characterized by a coherence matrix of the form given by Eq. (6.3–4) is transmitted through a compensator and a polarizer, the average electric energy density of the transmitted beam is unaffected by the retardation introduced by the compensator and by the orientation of the polarizer. The same result can readily be seen to apply also to the magnetic energy density of the transmitted field. Light with this property is said to be *unpolarized*. It is also called *natural light*, because such light is generated by sources that are commonly encountered in nature (e.g. many stellar sources). Since $\langle w'_e(\delta, \theta)\rangle$ is now independent of δ and θ it follows at once that the expression on the left-hand side of Eq. (6.2–30) takes on its smallest possible value, namely zero, i.e.

$$\frac{\langle w'_e(\delta, \theta)\rangle_{\max(\delta,\theta)} - \langle w'_e(\delta, \theta)\rangle_{\min(\delta,\theta)}}{\langle w'_e(\delta, \theta)\rangle_{\max(\delta,\theta)} + \langle w'_e(\delta, \theta)\rangle_{\min(\delta,\theta)}} = 0. \tag{6.3–6}$$

6.3.2 Completely polarized light

Let us now consider the other extreme case, namely the case when

$$|j_{xy}| = 1. \tag{6.3–7}$$

This requirement implies that the components of the complex electric vector in the x- and y-directions are now completely correlated. It follows at once from Eq. (6.3–7) and from the definition (6.2–11) of the correlation coefficient j_{xy} that we now have

$$|J_{xy}| = (J_{xx})^{1/2}(J_{yy})^{1/2} \tag{6.3–8}$$

or, using also the Hermiticity condition (6.2–8), it follows that

$$\det \mathbf{J} = 0. \tag{6.3–9}$$

Conversely Eq. (6.3–9) may readily be shown to imply Eq. (6.3–7).

We recall that the determinant of the coherence matrix is invariant with respect to rotation of the x-, y-axes about the direction of propagation [Eq. (6.2–36)]. Hence if Eq. (6.3–9) and, consequently also Eq. (6.3–7), hold for one particular set of x-, y-axes, these equations hold for all of them. In physical terms this result implies that if two components of the complex electric vector $\mathbf{E}(t)$ along any pair of (mutually orthogonal) x- and y-directions are completely correlated, they are correlated for all such pairs of directions.

In view of the relation (6.3–8) and the Hermiticity condition (6.2–8) the coherence matrix $\mathbf{J}$ now has the form

$$\mathbf{J} = \begin{bmatrix} J_{xx} & (J_{xx})^{1/2}(J_{yy})^{1/2}\,\mathrm{e}^{\mathrm{i}\alpha} \\ (J_{xx})^{1/2}(J_{yy})^{1/2}\,\mathrm{e}^{-\mathrm{i}\alpha} & J_{yy} \end{bmatrix}, \tag{6.3–10}$$

where α is a (real) phase factor.

Light for which the condition (6.3–7) or, equivalently, the condition (6.3–9) holds is said to be *completely polarized*. This terminology arises from the fact that a deterministic monochromatic wave, which is necessarily completely polarized in the conventional sense (cf. Born and Wolf, 1980, Secs. 1.4.2 and 1.4.3) may be regarded as a deterministic analogue of a wave of this kind. This can readily be seen by considering a plane, monochromatic, electromagnetic wave which propagates along the positive z-direction. Let

$$E_x(z, t) = e_1\,\mathrm{e}^{\mathrm{i}(k_0 z - 2\pi\nu_0 t)}, \quad E_y(z, t) = e_2\,\mathrm{e}^{\mathrm{i}(k_0 z - 2\pi\nu_0 t)} \tag{6.3–11}$$

be the components of its complex electric field vector in two mutually orthogonal directions perpendicular to the z-direction. In Eq. (6.3–11), e_1 and e_2 are (generally complex) constants and $k_0 = 2\pi\nu_0/c$, c being the speed of light in vacuo. The elements of the coherence matrix of such a wave are $J_{kl} = E_k^* E_l = e_k^* e_l\ (k, l = x, y)$, i.e. the coherence matrix has now the form

$$\mathbf{J} = \begin{bmatrix} e_1^* e_1 & e_1^* e_2 \\ e_2^* e_1 & e_2^* e_2 \end{bmatrix}, \tag{6.3–12}$$

No averaging symbol appears in the expressions for the elements of the coherence matrix (6.3–12), because we have chosen to represent the wave by a deterministic ensemble, i.e. an ensemble in which all the sample functions are identical (with probability unity). We see that the determinant of the coherence matrix (6.3–12) has zero value, just as for a fluctuating wave for which the requirement (6.3–9) holds.

It is clear that our 'canonical' experiment, which involves only a compensator and a polarizer, cannot distinguish between a quasi-monochromatic wave whose coherence matrix has a determinant with zero value and the strictly deterministic monochromatic plane wave given by Eq. (6.3–11), for which

$$e_1 = (J_{xx})^{1/2}\, \mathrm{e}^{\mathrm{i}\beta_1}, \quad e_2 = (J_{yy})^{1/2}\, \mathrm{e}^{\mathrm{i}\beta_2}, \tag{6.3–13}$$

where β_1 and β_2 are arbitrary constants.

Because a completely polarized wave is characterized by the requirement that $\det \mathbf{J} = 0$, the expression on the right-hand side of Eq. (6.2–30) is unity, i.e.

$$\frac{\langle w'_{\mathrm{e}}(\delta, \theta)\rangle_{\max(\delta,\theta)} - \langle w'_{\mathrm{e}}(\delta, \theta)\rangle_{\min(\delta,\theta)}}{\langle w'_{\mathrm{e}}(\delta, \theta)\rangle_{\max(\delta,\theta)} + \langle w'_{\mathrm{e}}(\delta, \theta)\rangle_{\min(\delta,\theta)}} = 1, \tag{6.3–14}$$

showing that for a completely polarized wave, the expression on the left attains its greatest possible value. Equation (6.3–14) implies that, in this case,

$$\langle w'_{\mathrm{e}}(\delta, \theta)\rangle_{\min(\delta,\theta)} = 0. \tag{6.3–15}$$

6.3.3 The degree of polarization

We shall now show that any coherence matrix $\mathbf{J}$ can be uniquely expressed as the sum of two matrices, one of which represents completely unpolarized light and the other completely polarized light. Expressed mathematically this means that one can always express $\mathbf{J}$ in the form

$$\mathbf{J} = \mathbf{J}^{\mathrm{unpol}} + \mathbf{J}^{\mathrm{pol}}, \tag{6.3–16}$$

where

$$\mathbf{J}^{\mathrm{unpol}} = A\begin{bmatrix} 1 & 0 \\ 0 & 1 \end{bmatrix}, \quad \mathbf{J}^{\mathrm{pol}} = \begin{bmatrix} B & D \\ D^* & C \end{bmatrix}, \tag{6.3–17}$$

with [cf. Eqs. (6.3–4) and (6.3–10)]

$$A \geqslant 0, \quad B \geqslant 0, \quad C \geqslant 0 \tag{6.3–18a}$$

and

$$BC - DD^* = 0. \tag{6.3–18b}$$

To show this let J_{kl} $(k, l = x, y)$ again be the elements of the coherence matrix $\mathbf{J}$. We have, according to Eqs. (6.3–16) and (6.3–17)

$$A + B = J_{xx}, \qquad A + C = J_{yy}, \tag{6.3–19}$$

$$D = J_{xy}, \qquad D^* = J_{yx}. \tag{6.3–20}$$

On substituting into Eq. (6.3–18b) for B and C from Eqs. (6.3–19) and for D and D^* from Eq. (6.3–20) we obtain the following equation for the element A:

$$(J_{xx} - A)(J_{yy} - A) - J_{xy}J_{yx} = 0. \tag{6.3–21}$$

This equation shows that A is an eigenvalue of the coherence matrix $\mathbf{J}$. The solutions of Eq. (6.3–21) are readily found to be (Wolf, 1959)

$$A = \tfrac{1}{2}\{\operatorname{tr}\mathbf{J} \pm [(\operatorname{tr}\mathbf{J})^2 - 4\det\mathbf{J}]^{1/2}\}. \tag{6.3–22}$$

Since the coherence matrix $\mathbf{J}$ is Hermitian and non-negative definite [Eqs. (6.2–8) and (6.2–9)] both the eigenvalues A are necessarily non-negative, as may also be readily verified by direct calculation.

Let us consider first the root of A given by the expression (6.3–22) with the negative sign in front of the square root. On substituting that root into Eqs. (6.3–19) we obtain the following expressions for the elements B and C:

$$B = \tfrac{1}{2}(J_{xx} - J_{yy}) + \tfrac{1}{2}[(\operatorname{tr}\mathbf{J})^2 - 4\det\mathbf{J}]^{1/2}, \tag{6.3–23a}$$

$$C = \tfrac{1}{2}(J_{yy} - J_{xx}) + \tfrac{1}{2}[(\operatorname{tr}\mathbf{J})^2 - 4\det\mathbf{J}]^{1/2}. \tag{6.3–23b}$$

Now if we make use of the Hermiticity relation $J_{yx} = J^*_{xy}$ [Eq. (6.2–8)] we find at once that

$$[(\operatorname{tr}\mathbf{J})^2 - 4\det\mathbf{J}]^{1/2} = [(J_{xx} - J_{yy})^2 + 4|J_{xy}|^2]^{1/2} \geqslant |J_{xx} - J_{yy}| \tag{6.3–24}$$

and hence it is clear that the matrix elements B and C, given by Eqs. (6.3–23), are necessarily non-negative, as required by two of the inequalities in (6.3–18a). On the other hand, the other expression for A, given by Eq. (6.3–22) with the positive sign in front of the square root, may readily be shown to yield negative values for the elements B and C and, therefore, they do not satisfy two of the inequalities in (6.3–18a). Hence we have shown that there is a unique decomposition of any coherence matrix $\mathbf{J}$ in the form (6.3–16).

It follows from the expression for the coherence matrix $\mathbf{J}^{\text{pol}}$ of the polarized portion of the light [the second matrix in Eq. (6.3–17)] and from Eqs. (6.3–23) that its trace is given by

$$\operatorname{tr}\mathbf{J}^{\text{pol}} = [(\operatorname{tr}\mathbf{J})^2 - 4\det\mathbf{J}]^{1/2}, \tag{6.3–25}$$

and hence the ratio

$$P \equiv \frac{\operatorname{tr}\mathbf{J}^{\text{pol}}}{\operatorname{tr}\mathbf{J}} = \left[1 - \frac{4\det\mathbf{J}}{(\operatorname{tr}\mathbf{J})^2}\right]^{1/2}. \tag{6.3–26}$$

Now according to Eq. (6.2–7) the trace of the coherence matrix of a wave is proportional to the average electric energy density and we may, therefore, regard the trace as a measure of the intensity of the wave. Hence the ratio $\mathrm{tr}\,\mathbf{J}^{\mathrm{pol}}/\mathrm{tr}\,\mathbf{J}$, which we denoted by the letter P in the formula (6.3–26), represents the ratio of the intensity of the 'polarized portion' of the wave to its total intensity. For this reason P is called the *degree of polarization* of the wave represented by the coherence matrix $\mathbf{J}$. In view of the inequality (6.3–24) it is clear that the degree of polarization, given by the expression on the right-hand side of Eq. (6.3–26), is a non-negative number bounded by zero and unity, i.e.

$$0 \leqslant P \leqslant 1. \tag{6.3–27}$$

It follows from Eq. (6.3–26) that when $P = 1$, the determinant of the coherence matrix $\mathbf{J}$ has zero value, which is precisely the condition (6.3–9) for the light to be completely *polarized*. When $P = 0$, we have, according to Eq. (6.3–26), the relation $(\mathrm{tr}\,\mathbf{J})^2 = 4\det\mathbf{J}$ which, written out more explicitly, implies that

$$(J_{xx} - J_{yy})^2 + 4J_{xy}J_{yx} = 0. \tag{6.3–28}$$

Since according to Eq. (6.2–8) $J_{yx} = J^*_{xy}$, the left-hand side of Eq. (6.3–28) is the sum of two squares, and each of the two terms must necessarily be zero, i.e. $J_{xx} = J_{yy}$ and $J_{xy} = J_{yx} = 0$. These are precisely the conditions (6.3–2) and (6.3–3) for the light to be *unpolarized*. In all other cases $(0 < P < 1)$ the light is said to be *partially polarized*.

We note that the expression (6.3–26) for the degree of polarization is identical with the expression on the right-hand side of Eq. (6.2–30) and hence

$$\frac{\langle w'_{\mathrm{e}}(\delta, \theta)\rangle_{\max(\delta,\theta)} - \langle w'_{\mathrm{e}}(\delta, \theta)\rangle_{\min(\delta,\theta)}}{\langle w'_{\mathrm{e}}(\delta, \theta)\rangle_{\max(\delta,\theta)} + \langle w'_{\mathrm{e}}(\delta, \theta)\rangle_{\min(\delta,\theta)}} = P. \tag{6.3–29}$$

Thus the left-hand side of Eq. (6.3–29) which, as we already noted, is analogous to the expression for the visibility of interference fringes in Young's interference experiment, is precisely the degree of polarization of the light beam. In view of Eq. (6.3–29) the equations (6.3–6) and (6.3–14) now acquire a clear meaning.

It is evident from Eq. (6.3–26) that the degree of polarization is independent of the choice of the x-, y-axes, because it is expressed in terms of invariants of the coherence matrix under rotation of the axes about the direction of propagation.

The degree of polarization may also be readily expressed in terms of the eigenvalues A_1 and A_2 of the coherence matrix $\mathbf{J}$, i.e. in terms of the two roots of Eq. (6.3–21), given by Eq. (6.3–22). For this purpose we recall that every Hermitian matrix may be diagonalized by a unitary transformation and that its trace and its determinant are invariant under such a transformation. (In general, this unitary transformation will not represent a real rotation of the x-, y-axes about the direction of propagation of the wave). Hence

$$\mathrm{tr}\,\mathbf{J} = A_1 + A_2 \tag{6.3–30a}$$

and

$$\det\mathbf{J} = A_1 A_2. \tag{6.3–30b}$$

On substituting from Eqs. (6.3–30) into Eq. (6.3–26) we obtain the required alternative expression for the degree of polarization, namely

$$P = \left| \frac{A_1 - A_2}{A_1 + A_2} \right|. \qquad (6.3\text{–}31)$$

Finally let us express the degree of polarization in terms of the four Stokes parameters rather than in terms of the elements of the coherence matrix. We find at once from the relations (6.2–40b) that

$$\det \mathbf{J} = \tfrac{1}{4}(s_0^2 - s_1^2 - s_2^2 - s_3^2), \qquad (6.3\text{–}32a)$$

and that

$$\operatorname{tr} \mathbf{J} = s_0. \qquad (6.3\text{–}32b)$$

On substituting from these formulas into Eq. (6.3–26) we obtain the required expression for the degree of polarization:

$$P = \frac{(s_1^2 + s_2^2 + s_3^2)^{1/2}}{s_0}. \qquad (6.3\text{–}33)$$

It immediately follows from this expression that for a completely polarized beam ($P = 1$)

$$s_1^2 + s_2^2 + s_3^2 = s_0^2, \qquad (6.3\text{–}34)$$

whereas for a completely unpolarized beam ($P = 0$)

$$s_1 = s_2 = s_3 = 0. \qquad (6.3\text{–}35)$$

The states of polarization of a completely polarized wave can be represented geometrically on a sphere known as the Poincaré sphere. There are also interesting geometrical representations for waves of arbitrary state of polarization. Some relations involve the degree of polarization P, the degree of correlation j_{xy} and the parameters that characterize the polarization ellipse of the polarized portion of the beam. A discussion of some of these topics may be found, for example, in Born and Wolf (1980, Sec. 10.8.2), Wolf (1959) and Parrent and Roman (1960).

6.4 Transmission of a quasi-monochromatic beam through linear, non-image-forming devices‡

We will now study how the coherence matrix changes when a quasi-monochromatic light beam is transmitted through some of the commonly used linear, non-image-forming optical devices. For this purpose it is useful to combine the components $E_x(t)$ and $E_y(t)$ of the complex electric vector at some point in a plane $z = z_0$ transverse to the direction of propagation of the beam [Eq. (6.2–1)] into a row vector

$$\mathscr{E}(t) = [E_x(t) \quad E_y(t)]. \qquad (6.4\text{–}1)$$

‡ The analysis of this section is largely based on a paper by Parrent and Roman (1960). It may be regarded, in some respects, as generalization to quasi-monochromatic light of any state of polarization of a matrix calculus introduced by Jones (1941a; see also Jones 1941b, 1942, 1947a, b, 1948, 1956; Hurwitz and Jones, 1941) for transmission of monochromatic light through linear systems. All these papers are reprinted in Swindell (1975).

Its Hermitian conjugate (Hermitian adjoint) is the column vector

$$\mathscr{E}^\dagger(t) = \begin{bmatrix} E_x^*(t) \\ E_y^*(t) \end{bmatrix}. \tag{6.4–1a}$$

The coherence matrix $\mathbf{J}$ of the incident beam may then be expressed in the form

$$\mathbf{J} = \langle \mathscr{E}^\dagger(t)\mathscr{E}(t) \rangle. \tag{6.4–2}$$

If the beam passes through a linear device such as a compensator, an absorber, a rotator or a polarizer, the vector $\mathscr{E}(t)$ is transformed into another vector which in an 'output plane' $z = z_1$ may be represented by the row vector

$$\mathscr{E}'(t) = \mathscr{E}(t)\mathbf{T}, \tag{6.4–3}$$

where $\mathbf{T}$ is a 2×2 matrix which characterizes the device. We will call $\mathbf{T}$ the *transmission matrix* of the device.

The coherence matrix of the complex electric field of the beam which emerges from the device, at a point in the output plane $z = z_1$, is given by

$$\mathbf{J}' = \langle \mathscr{E}'^\dagger(t)\mathscr{E}'(t) \rangle. \tag{6.4–4}$$

On substituting from Eq. (6.4–3) into Eq. (6.4–4) and omitting the argument t of $\mathscr{E}(t)$ we find at once that

$$\mathbf{J}' = \langle \mathbf{T}^\dagger \mathscr{E}^\dagger \mathscr{E} \mathbf{T} \rangle = \mathbf{T}^\dagger \langle \mathscr{E}^\dagger \mathscr{E} \rangle \mathbf{T}$$

or, if we make use of Eq. (6.4–2),

$$\mathbf{J}' = \mathbf{T}^\dagger \mathbf{J} \mathbf{T}, \tag{6.4–5}$$

where $\mathbf{T}^\dagger$ is the Hermitian adjoint of $\mathbf{T}$.

We note an immediate consequence of Eq. (6.4–5). If we take the determinant of both sides of that equation and use the well-known theorem that the determinant of a product of matrices is equal to the product of their determinants (Aitken, 1944, Sec. 34), it follows at once from Eq. (6.4–5) that

$$\begin{aligned} \det \mathbf{J}' &= \det \mathbf{T}^\dagger \times \det \mathbf{J} \times \det \mathbf{T} \\ &= \det \mathbf{J} \, |\det \mathbf{T}|^2. \end{aligned} \tag{6.4–6}$$

Suppose now that the incident beam is completely polarized ($P = 1$). Then according to Eq. (6.3–26) we have $\det \mathbf{J} = 0$ and Eq. (6.4–6) implies that $\det \mathbf{J}' = 0$; consequently the degree of polarization P' of the beam transmitted by the device is also equal to unity, i.e. the transmitted beam is also fully polarized. This result might have been expected, since it is intuitively clear, and can be readily verified, that in order for a fully polarized beam to become *depolarized*, i.e. in order for its degree of polarization to change from the value unity to a smaller value, it is necessary that the device introduces some randomness. Such a device or a medium (e.g. the atmosphere) cannot be represented by a single transmission matrix $\mathbf{T}$ but rather it must be characterized by a statistical ensemble of such transmission matrices (see, for example, Kim, Mandel and Wolf, 1987).

An expression for the average electric energy density $\langle w'_e \rangle$ of the transmitted beam is obtained on substituting from Eq. (6.4–5) into the formula (6.2–7), applied now to the beam that emerges from the linear device. One finds at once that

$$\langle w_e' \rangle = \frac{1}{4\pi} \operatorname{tr}\{\mathbf{T}^\dagger \mathbf{J} \mathbf{T}\}. \tag{6.4–7}$$

We will now determine the form of the transformation matrix $\mathbf{T}$ for some common linear non-image-forming devices.

6.4.1 A compensator

Let ε_1 and ε_2 denote the phase changes produced in the components E_x and E_y respectively by a compensator, as the beam propagates through it from a plane $z = z_0$ to a plane $z = z_1$. We assume that the relative phase difference

$$\delta = \varepsilon_2 - \varepsilon_1, \tag{6.4–8}$$

which is, in general, a function of frequency, is small compared to $2\pi l/\bar{\lambda}$, where l is the coherence length of the light and $\bar{\lambda}$ is the mean wavelength. We also assume that losses due to reflection and absorption introduced by the compensator are negligible.

It is clear that the compensator changes $\mathscr{E}$ into the row vector

$$\mathscr{E}' = [E_x\, \mathrm{e}^{\mathrm{i}\varepsilon_1} \quad E_y\, \mathrm{e}^{\mathrm{i}\varepsilon_2}] = \mathscr{E}\begin{bmatrix} \mathrm{e}^{\mathrm{i}\varepsilon_1} & 0 \\ 0 & \mathrm{e}^{\mathrm{i}\varepsilon_2} \end{bmatrix}. \tag{6.4–9}$$

Since only the phase difference between the two Cartesian components of the electric field is now significant, we may express the relation between $\mathscr{E}$ and $\mathscr{E}'$ in the form

$$\mathscr{E}' = \mathscr{E}\begin{bmatrix} \mathrm{e}^{-\mathrm{i}\delta/2} & 0 \\ 0 & \mathrm{e}^{\mathrm{i}\delta/2} \end{bmatrix}. \tag{6.4–9a}$$

On comparing Eq. (6.4–9) with the general expression (6.4–3) we see that the transmission matrix of the compensator, which we will denote by $\mathbf{T}_{\mathrm{C}}$, is

$$\mathbf{T}_{\mathrm{C}} = \begin{bmatrix} \mathrm{e}^{-\mathrm{i}\delta/2} & 0 \\ 0 & \mathrm{e}^{\mathrm{i}\delta/2} \end{bmatrix}. \tag{6.4–10}$$

One can readily verify that $\mathbf{T}_{\mathrm{C}}^\dagger \mathbf{T}_{\mathrm{C}} = \mathbf{T}_{\mathrm{C}} \mathbf{T}_{\mathrm{C}}^\dagger = \mathbf{1}$, where $\mathbf{1}$ is the unit matrix. Hence the transmission matrix of our (somewhat idealized) compensator is unitary. Using this fact and recalling that the trace of a product of matrices is unchanged under their cyclic permutation we conclude from Eqs. (6.4–7) and (6.4–10) that

$$\begin{aligned} \langle w_e' \rangle &= \frac{1}{4\pi} \operatorname{tr}\{\mathbf{J} \mathbf{T}_{\mathrm{C}} \mathbf{T}_{\mathrm{C}}^\dagger\} \\ &= \frac{1}{4\pi} \operatorname{tr} \mathbf{J} \\ &= \langle w_e \rangle, \end{aligned} \tag{6.4–11}$$

which shows that the average electric energy density of the beam is unchanged by passage through the compensator.

6.4.2 *An absorber*

For an absorber the components E_x and E_y will be attenuated. Let $e^{-\alpha_1}$ and $e^{-\alpha_2}$ ($\alpha_1 > 0$, $\alpha_2 > 0$) be the attenuation factors for the x- and y-components of the complex electric field vector, on propagating through the absorber from the plane $z = z_0$ to the plane $z = z_1$. We assume that the variation of α_1 and α_2 over the narrow spectral range of the quasi-monochromatic beam and also the reflection losses are negligible. Then

$$\mathscr{E}' = [E_x \, e^{-\alpha_1} \quad E_y \, e^{-\alpha_2}] = \mathscr{E} \begin{bmatrix} e^{-\alpha_1} & 0 \\ 0 & e^{-\alpha_2} \end{bmatrix}. \tag{6.4–12}$$

Hence the transmission matrix, $\mathbf{T}_A$, of the absorber is

$$\mathbf{T}_A = \begin{bmatrix} e^{-\alpha_1} & 0 \\ 0 & e^{-\alpha_2} \end{bmatrix}. \tag{6.4–13}$$

We note that $\mathbf{T}_A^\dagger = \mathbf{T}_A$ showing that the transmission matrix of the absorber is Hermitian. Using this fact we readily find from Eq. (6.4–7) that the average electric density of the transmitted beam is given by the expression

$$\begin{aligned} \langle w_e' \rangle &= \frac{1}{4\pi} \operatorname{tr} \{\mathbf{T}_A^\dagger \mathbf{J} \mathbf{T}_A\} \\ &= \frac{1}{4\pi} \operatorname{tr} \{\mathbf{J} \mathbf{T}_A \mathbf{T}_A^\dagger\} \\ &= \frac{1}{4\pi} \operatorname{tr} \{\mathbf{J} \mathbf{T}_A^2\} \\ &= \frac{1}{4\pi} [J_{xx} \, e^{-2\alpha_1} + J_{yy} \, e^{-2\alpha_2}]. \end{aligned} \tag{6.4–14}$$

6.4.3 *A rotator*

Various materials and physical devices produce a rotation of the electric field vector about the direction of propagation of the beam. Such a device is called a rotator. Let χ be the angle of rotation of the complex electric vector about the direction of propagation, as the light is transmitted through the rotator from a plane $z = z_0$ to a plane $z = z_1$. If we assume that χ is effectively independent of the frequency over the narrow spectral range of the quasi-monochromatic beam, and neglect reflection and absorption losses, the rotator changes $\mathscr{E}$ into

$$\mathscr{E}' = \mathscr{E} \begin{bmatrix} \cos\chi & \sin\chi \\ -\sin\chi & \cos\chi \end{bmatrix}. \tag{6.4–15}$$

Hence the transmission matrix $\mathbf{T}_R$ of the rotator is given by

$$\mathbf{T}_R = \begin{bmatrix} \cos\chi & \sin\chi \\ -\sin\chi & \cos\chi \end{bmatrix}. \tag{6.4–16}$$

It is readily seen that $\mathbf{T}_R$ is a real, orthogonal and hence an unitary matrix so that its transpose $\mathbf{T}_R^T$, its Hermitian adjoint $\mathbf{T}_R^\dagger$ and its inverse $\mathbf{T}_R^{-1}$ are equal to each other ($\mathbf{T}_R^T = \mathbf{T}_R^\dagger = \mathbf{T}_R^{-1}$). It follows from Eq. (6.4–7) and the unitarity of $\mathbf{T}_R$ that the average electric energy density of the transmitted beam is given by the expression

$$\begin{aligned}\langle w'_e \rangle &= \frac{1}{4\pi} \operatorname{tr}\{\mathbf{T}_R^\dagger \mathbf{J} \mathbf{T}_R\} \\ &= \frac{1}{4\pi} \operatorname{tr}\{\mathbf{J} \mathbf{T}_R \mathbf{T}_R^\dagger\} \\ &= \frac{1}{4\pi} \operatorname{tr} \mathbf{J} \\ &= \langle w_e \rangle, \end{aligned} \tag{6.4–17}$$

i.e. the average electric energy density is unchanged by passage through the rotator.

6.4.4 A polarizer

Let us next consider a polarizer which transmits the component of the electric field in a direction that makes an angle θ with the x-direction. The component $E(\theta)$ of the incident field in that direction is given by

$$E(\theta) = E_x \cos\theta + E_y \sin\theta. \tag{6.4–18}$$

Hence, if reflection and transmission losses are neglected, the row vector $\mathscr{E}'$ which represents the transmitted field is given by

$$\begin{aligned}\mathscr{E}' &= [E(\theta)\cos\theta \quad E(\theta)\sin\theta] \\ &= [(E_x\cos\theta + E_y\sin\theta)\cos\theta \quad (E_x\cos\theta + E_y\sin\theta)\sin\theta] \\ &= \mathscr{E}\begin{bmatrix} \cos^2\theta & \cos\theta\sin\theta \\ \sin\theta\cos\theta & \sin^2\theta \end{bmatrix}. \end{aligned} \tag{6.4–19}$$

The transmission matrix $\mathbf{T}_P$ of the polarizer is, therefore,

$$\mathbf{T}_P = \begin{bmatrix} \cos^2\theta & \cos\theta\sin\theta \\ \sin\theta\cos\theta & \sin^2\theta \end{bmatrix}. \tag{6.4–20}$$

This matrix can readily be seen to satisfy the idempotency condition $\mathbf{T}_P\mathbf{T}_P = \mathbf{T}_P$, i.e. $\mathbf{T}_P$ represents a (real symmetric) projection operator. This, of course, expresses the fact that the electric field which emerges from the polarizer is not affected by the passage through another identical polarizer.

The energy density of the transmitted beam is, according to Eq. (6.4–7), given by

$$\begin{aligned}\langle w'_e \rangle &= \frac{1}{4\pi} \operatorname{tr}\{\mathbf{T}_P^\dagger \mathbf{J} \mathbf{T}_P\} \\ &= \frac{1}{4\pi} \operatorname{tr}\{\mathbf{J} \mathbf{T}_P \mathbf{T}_P^\dagger\} \\ &= \frac{1}{4\pi} \operatorname{tr}\{\mathbf{J} \mathbf{T}_P \mathbf{T}_P\} \\ &= \frac{1}{4\pi} \operatorname{tr}\{\mathbf{J} \mathbf{T}_P\}. \end{aligned} \tag{6.4–21}$$

6.4.5 A cascaded system

Suppose that a quasi-monochromatic light beam is passed through a succession of non-imaging devices $D_1, D_2, \ldots, D_n$, with their individual axes aligned along the direction of propagation of the beam (Fig. 6.3). We will derive an expression for the transmission matrix of the whole system in terms of the transmission matrices of the individual components.

Let $\mathscr{E}_0$ be the row vector representing a realization of the complex electric field of the incident beam in a plane $z = z_0$. After transmission through D_1 the complex electric field in a plane $z = z_1$ located between D_1 and D_2 will be given by

$$\mathscr{E}_1 = \mathscr{E}_0 \mathbf{T}_1, \tag{6.4–22}$$

where $\mathbf{T}_1$ is the transmission matrix of D_1. After transmission through the second device D_2 the complex electric field in a plane $z = z_2$ located between D_2 and D_3 will be represented by a row vector $\mathscr{E}_2$, where

$$\mathscr{E}_2 = \mathscr{E}_1 \mathbf{T}_2, \tag{6.4–23}$$

$\mathbf{T}_2$ being the transmission matrix of D_2. It is clear that the successive transmissions between planes $z = z_j$ and $z = z_{j+1}$ are expressible by the set of recurrence relations

$$\mathscr{E}_j = \mathscr{E}_{j-1} \mathbf{T}_j, \quad j = 1, 2, \ldots, n. \tag{6.4–24}$$

It follows at once from these relations that

$$\mathscr{E}_n = \mathscr{E}_0 \mathbf{T}, \tag{6.4–25}$$

where

$$\mathbf{T} = \mathbf{T}_1 \mathbf{T}_2 \ldots \mathbf{T}_n \tag{6.4–26}$$

is the transmission matrix that characterizes the propagation of the beam from the 'input plane' $z = z_0$ to the 'output plane' $z = z_n$ (see Fig. 6.3).

We see from Eqs. (6.4–5) and (6.4–26) that if $\mathbf{J}$ is the coherence matrix in the input plane $z = z_0$ of a beam incident on a cascaded system formed by elements $D_1, D_2, \ldots, D_n$, the coherence matrix of the emerging beam in the output plane $z = z_n$ is given by

$$\mathbf{J}' = (\mathbf{T}_1 \mathbf{T}_2 \ldots \mathbf{T}_n)^\dagger \mathbf{J} (\mathbf{T}_1 \mathbf{T}_2 \ldots \mathbf{T}_n), \tag{6.4–27}$$

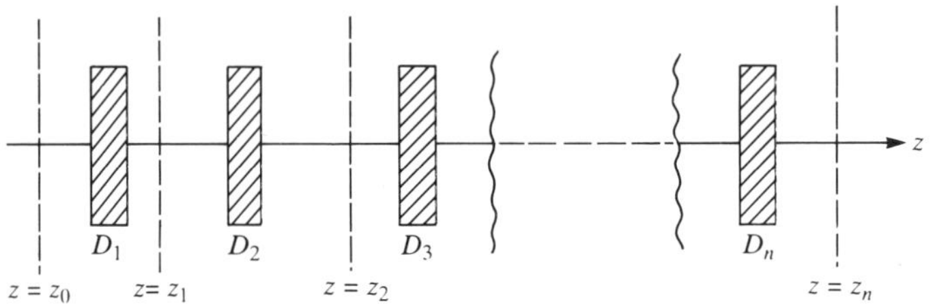

Fig. 6.3 Notation relating to determination of the transmission matrix of a cascaded system.

which implies that

$$\mathbf{J}' = \mathbf{T}_n^\dagger \mathbf{T}_{n-1}^\dagger \ldots \mathbf{T}_1^\dagger \mathbf{J} \mathbf{T}_1 \mathbf{T}_2 \ldots \mathbf{T}_n. \tag{6.4–28}$$

We will illustrate the use of this formula by applying it to a problem which we treated in Section 6.2 in a more elementary way; namely to determine the average electric energy density of a beam transmitted by a two-component system consisting of a compensator and a polarizer. For this case the formula (6.4–28) gives

$$\mathbf{J}' = \mathbf{T}_\mathrm{P}^\dagger \mathbf{T}_\mathrm{C}^\dagger \mathbf{J} \mathbf{T}_\mathrm{C} \mathbf{T}_\mathrm{P}, \tag{6.4–29}$$

where $\mathbf{T}_\mathrm{C}$ and $\mathbf{T}_\mathrm{P}$ are the transmission matrices of the compensator and of the polarizer, given by Eqs. (6.4–10) and (6.4–20), respectively.

The average energy density $\langle w_\mathrm{e}' \rangle$ of the beam emerging from this system is, according to Eqs. (6.4–29) and (6.2–7), given by the expression

$$\begin{aligned}\langle w_\mathrm{e}' \rangle &= \frac{1}{4\pi} \operatorname{tr} \{\mathbf{T}_\mathrm{P}^\dagger \mathbf{T}_\mathrm{C}^\dagger \mathbf{J} \mathbf{T}_\mathrm{C} \mathbf{T}_\mathrm{P}\} \\ &= \frac{1}{4\pi} \operatorname{tr} \{\mathbf{T}_\mathrm{C} \mathbf{T}_\mathrm{P} \mathbf{T}_\mathrm{P}^\dagger \mathbf{T}_\mathrm{C}^\dagger \mathbf{J}\}.\end{aligned} \tag{6.4–30}$$

We noted earlier that $\mathbf{T}_\mathrm{P}$ is a real and symmetric idempotent matrix so that $\mathbf{T}_\mathrm{P}\mathbf{T}_\mathrm{P}^\dagger = \mathbf{T}_\mathrm{P}$. Hence Eq. (6.4–30) may be expressed in the form

$$\langle w_\mathrm{e}' \rangle = \frac{1}{4\pi} \operatorname{tr} \{\mathbf{K}\mathbf{J}\}, \tag{6.4–31}$$

where

$$\mathbf{K} = \mathbf{T}_\mathrm{C} \mathbf{T}_\mathrm{P} \mathbf{T}_\mathrm{C}^\dagger. \tag{6.4–32}$$

On substituting from Eqs. (6.4–10) and (6.4–20) into Eq. (6.4–32) one readily finds that

$$\mathbf{K} = \begin{bmatrix} \cos^2\theta & \cos\theta \sin\theta \, \mathrm{e}^{-\mathrm{i}\delta} \\ \sin\theta\cos\theta \, \mathrm{e}^{\mathrm{i}\delta} & \sin^2\theta \end{bmatrix}. \tag{6.4–33}$$

Finally on substituting from Eq. (6.4–33) into Eq. (6.4–31) it follows that

$$\langle w_\mathrm{e}' \rangle = \frac{1}{4\pi}[J_{xx}\cos^2\theta + J_{yy}\sin^2\theta + J_{xy}\,\mathrm{e}^{+\mathrm{i}\delta}\sin\theta\cos\theta + J_{yx}\,\mathrm{e}^{-\mathrm{i}\delta}\cos\theta\sin\theta], \tag{6.4–34}$$

in agreement with the formula (6.2–4).

The transmission of the beam through various devices may, of course, also be described in terms of the Stokes parameters rather than in terms of the coherence matrices. To derive the corresponding transformation law let us denote by s_j and s_j', $j = 1, 2, 3$, the Stokes parameters of the beam in the input and the output planes respectively. We then have, according to Eqs. (6.2–43)

$$s_j = \operatorname{tr}[\mathbf{J}\boldsymbol{\sigma}_j], \tag{6.4–35a}$$

$$\mathbf{J} = \tfrac{1}{2} s_j \boldsymbol{\sigma}_j, \tag{6.4–35b}$$

where $\boldsymbol{\sigma}_1$, $\boldsymbol{\sigma}_2$, $\boldsymbol{\sigma}_3$ are the Pauli spin matrices defined by Eq. (6.2–41) and $\boldsymbol{\sigma}_0$ is the 2×2 unit matrix (6.2–42). On the right-hand side of Eq. (6.4–35b) and in what follows, Einstein's summation convention is used, i.e. summation over repeated dummy indices is implied. Now according to Eq. (6.4–5) the coherence matrices $\mathbf{J}$ and $\mathbf{J}'$ in the input and output planes are related by the formula

$$\mathbf{J}' = \mathbf{T}^{\dagger}\mathbf{J}\mathbf{T}. \tag{6.4–36}$$

On substituting from Eq. (6.4–36) into Eq. (6.4–35a) we find at once that

$$s'_j = \operatorname{tr}[\mathbf{T}^{\dagger}\mathbf{J}\mathbf{T}\boldsymbol{\sigma}_j], \tag{6.4–37}$$

or, making use of Eq. (6.4–35b),

$$\begin{aligned} s'_j &= \tfrac{1}{2}\operatorname{tr}[\mathbf{T}^{\dagger}s_k\boldsymbol{\sigma}_k\mathbf{T}\boldsymbol{\sigma}_j] \\ &= \tfrac{1}{2}\operatorname{tr}[\mathbf{T}\boldsymbol{\sigma}_j\mathbf{T}^{\dagger}s_k\boldsymbol{\sigma}_k], \end{aligned}$$

i.e.

$$s'_j = M_{jk}s_k, \tag{6.4–38}$$

where

$$M_{jk} = \tfrac{1}{2}\operatorname{tr}[\mathbf{T}\boldsymbol{\sigma}_j\mathbf{T}^{\dagger}\boldsymbol{\sigma}_k]. \tag{6.4–39}$$

The formula (6.4–38) is the required law for the transformation of the Stokes parameters and it shows that each Stokes parameter of the field in the output plane is a linear combination of the Stokes parameters of the field in the input plane. This linear relationship is seen to be characterized by a 4×4 matrix $\mathbf{M} = [M_{jk}]$, generally known as the *Mueller matrix*.‡ The formula (6.4–39) expresses the elements of the Mueller matrix of the system in terms of the elements of its transmission matrix.

Partly for historical reasons, the Stokes parameters have been used much more frequently than the coherence matrices. It should be noted, however, that the influence of a linear non-image-forming, non-depolarizing deterministic device is characterized much more simply by using the coherence matrix rather than by the Stokes parameters description. For in the coherence matrix representation the device is characterized by a 2×2 matrix, whereas in the Stokes parameter representation 4×4 matrices are needed. Evidently for a non-depolarizing device the elements of a Mueller matrix satisfy an appreciable number of constraints (see, for example, Barakat, 1981; Simon, 1982).

‡ Named after H. Mueller who employed such matrices in lectures on phenomenological optics that he gave at the Massachusetts Institute of Technology during 1945–1948. Mueller did not publish his results in the open literature, except in a brief abstract (Mueller, 1948; see also Parke, 1949; Shurcliff, 1962, p. 117). Such matrices were, however, used previously, at least in some special cases, by Soleilett (1929) and Perrin (1942). The relationship between the Mueller matrices and the matrices used in the Jones calculus (see footnote at the beginning of Section 6.4) were discussed in several papers, e.g. Barakat (1981), Simon (1982), Kim, Mandel and Wolf (1987).

6.5 The general second-order coherence matrices and coherence tensors of a stationary electromagnetic field

6.5.1 *The electric, magnetic and mixed coherence matrices (tensors)*

The coherence matrix J, defined by Eq. (6.2–6) characterizes the simplest correlation properties of an uniform, quasi-monochromatic beam of electromagnetic radiation. In particular, as we saw, it is adequate to describe the state of polarization of the beam.

To characterize correlation properties of an electromagnetic field which is not necessarily quasi-monochromatic or beam-like, more general coherence matrices are needed. They may be defined as follows. Consider a fluctuating field whose statistical properties are characterized by a stationary ergodic ensemble and let $\mathbf{E}(\mathbf{r}, t)$ and $\mathbf{H}(\mathbf{r}, t)$ be the complex analytic signal representations of the electric and the magnetic field vectors respectively, at a space-time point $(\mathbf{r}, t)$. We assume, for simplicity, that they have zero mean values:

$$\langle \mathbf{E}(\mathbf{r}, t)\rangle = \langle \mathbf{H}(\mathbf{r}, t)\rangle = 0. \tag{6.5–1}$$

The coherence matrices (introduced by Wolf, 1954) are then defined by the formulas:

$$\mathsf{E}(\mathbf{r}_1, \mathbf{r}_2, \tau) \equiv [\mathscr{E}_{jk}(\mathbf{r}_1, \mathbf{r}_2, \tau)] = [\langle E_j^*(\mathbf{r}_1, t)E_k(\mathbf{r}_2, t+\tau)\rangle], \tag{6.5–2a}$$

$$\mathsf{H}(\mathbf{r}_1, \mathbf{r}_2, \tau) \equiv [\mathscr{H}_{jk}(\mathbf{r}_1, \mathbf{r}_2, \tau)] = [\langle H_j^*(\mathbf{r}_1, t)H_k(\mathbf{r}_2, t+\tau)\rangle], \tag{6.5–2b}$$

$$\mathsf{M}(\mathbf{r}_1, \mathbf{r}_2, \tau) \equiv [\mathscr{M}_{jk}(\mathbf{r}_1, \mathbf{r}_2, \tau)] = [\langle E_j^*(\mathbf{r}_1, t)H_k(\mathbf{r}_2, t+\tau)\rangle], \tag{6.5–2c}$$

$$\mathsf{N}(\mathbf{r}_1, \mathbf{r}_2, \tau) \equiv [\mathscr{N}_{jk}(\mathbf{r}_1, \mathbf{r}_2, \tau)] = [\langle H_j^*(\mathbf{r}_1, t)E_k(\mathbf{r}_2, t+\tau)\rangle], \tag{6.5–2d}$$

where j and k each stand for the subscripts x, y, z that label components taken with respect to a fixed Cartesian rectangular right-handed coordinate system of axes. E and H are called the *electric* and the *magnetic coherence matrices* respectively and M and N are called *mixed coherence matrices*. These matrices may, of course, be regarded as representing second-rank Cartesian tensors. In the subsequent discussion we will often find it convenient to use the language and the notation of tensor analysis rather than of matrix algebra.

We note certain obvious relations that follow at once from the above definitions and from the assumption of stationarity, namely,

$$\mathscr{E}_{kj}(\mathbf{r}_1, \mathbf{r}_2, \tau) = \mathscr{E}_{jk}^*(\mathbf{r}_2, \mathbf{r}_1, -\tau), \tag{6.5–3a}$$

$$\mathscr{H}_{kj}(\mathbf{r}_1, \mathbf{r}_2, \tau) = \mathscr{H}_{jk}^*(\mathbf{r}_2, \mathbf{r}_1, -\tau), \tag{6.5–3b}$$

$$\mathscr{M}_{kj}(\mathbf{r}_1, \mathbf{r}_2, \tau) = \mathscr{N}_{jk}^*(\mathbf{r}_2, \mathbf{r}_1, -\tau). \tag{6.5–3c}$$

The coherence matrices also obey various non-negative definiteness conditions, which may be derived as follows. We have the obvious inequality

$$\left\langle \left| \int \mathrm{d}^3 r \int \mathrm{d}t \{ f_j(\mathbf{r}, t)E_j(\mathbf{r}, t) + g_j(\mathbf{r}, t)H_j(\mathbf{r}, t)\} \right|^2 \right\rangle \geqslant 0, \tag{6.5–4}$$

where $f_j(\mathbf{r}, t)$ and $g_j(\mathbf{r}, t)$, $(j = x, y, z)$, are arbitrary well-behaved functions of $\mathbf{r}$,

t and the integration extends over an arbitrary space-time region containing the field. The Einstein summation convention is again implied. We readily obtain from the inequality (6.5–4) and from the defining equations (6.5–2) the following non-negative definiteness condition [originally derived in the context of the quantized electromagnetic field by Mehta and Wolf (1967a)]:

$$\begin{aligned}\int d^3r_1\int dt_1\int d^3r_2\int dt_2\{&f_j^*(\mathbf{r}_1, t_1)f_k(\mathbf{r}_2, t_2)\mathscr{E}_{jk}(\mathbf{r}_1, \mathbf{r}_2, t_2 - t_1)\\ &+ g_j^*(\mathbf{r}_1, t_1)g_k(\mathbf{r}_2, t_2)\mathscr{H}_{jk}(\mathbf{r}_1, \mathbf{r}_2, t_2 - t_1)\\ &+ f_j^*(\mathbf{r}_1, t_1)g_k(\mathbf{r}_2, t_2)\mathscr{M}_{jk}(\mathbf{r}_1, \mathbf{r}_2, t_2 - t_1)\\ &+ g_j^*(\mathbf{r}_1, t_1)f_k(\mathbf{r}_2, t_2)\mathscr{N}_{jk}(\mathbf{r}_1, \mathbf{r}_2, t_2 - t_1)\} \geqslant 0. \quad (6.5\text{–}5)\end{aligned}$$

Here we have also made use of the stationarity of the field.

Two paticular cases of this inequality are of special interest. If we choose $g_j(\mathbf{r}, t) \equiv 0$ in the inequality (6.5–5) we obtain the following non-negative definiteness condition obeyed by the electric coherence matrix:

$$\int d^3r_1\int dt_1\int d^3r_2\int dt_2\{f_j^*(\mathbf{r}_1, t_1)f_k(\mathbf{r}_2, t_2)\mathscr{E}_{jk}(\mathbf{r}_1, \mathbf{r}_2, t_2 - t_1)\} \geqslant 0, \quad (6.5\text{–}6)$$

where Einstein's summation convention is again used. Similarly, if we set $f_j(\mathbf{r}, t) \equiv 0$, we obtain a corresponding non-negative definiteness condition satisfied by the magnetic coherence matrix:

$$\int d^3r_1\int dt_1\int d^3r_2\int dt_2\{g_j^*(\mathbf{r}_1, t_1)g_k(\mathbf{r}_2, t_2)\mathscr{H}_{jk}(\mathbf{r}_1, \mathbf{r}_2, t_2 - t_1)\} \geqslant 0. \quad (6.5\text{–}7)$$

The non-negative definiteness conditions (6.5–5) to (6.5–7) imply alternative forms of these conditions, involving summations rather than integrations. To show this let us choose

$$f_j(\mathbf{r}, t) = \sum_{p=1}^{P}\sum_{m=1}^{M} a_{jpm}\delta^{(3)}(\mathbf{r} - \mathbf{r}_p)\delta(t - t_m), \quad (6.5\text{–}8)$$

$$g_j(\mathbf{r}, t) = \sum_{p=1}^{P}\sum_{m=1}^{M} b_{jpm}\delta^{(3)}(\mathbf{r} - \mathbf{r}_p)\delta(t - t_m), \quad (6.5\text{–}9)$$

where P and M are arbitrary positive integers, a_{jpm}, b_{jpm} ($j = x, y, z,\ \ p = 1, 2, \ldots, P,\ \ m = 1, 2, \ldots, M$) are arbitrary (real or complex) constants, $\mathbf{r}_p$ are arbitrary points and t_m are arbitrary instants of time in the space-time region over which the integration in Eq. (6.5–5) is carried out. We then obtain from the inequalities (6.5–5) to (6.5–7) the following non-negative definiteness conditions:

$$\begin{aligned}&a_{jpm}^*a_{kqn}\mathscr{E}_{jk}(\mathbf{r}_p, \mathbf{r}_q, t_n - t_m) + b_{jpm}^*b_{kqn}\mathscr{H}_{jk}(\mathbf{r}_p, \mathbf{r}_q, t_n - t_m)\\ &\quad + a_{jpm}^*b_{kqn}\mathscr{M}_{jk}(\mathbf{r}_p, \mathbf{r}_q, t_n - t_m) + b_{jpm}^*a_{kqn}\mathscr{N}_{jk}(\mathbf{r}_p, \mathbf{r}_q, t_n - t_m) \geqslant 0, \quad (6.5\text{–}10)\end{aligned}$$

$$a_{jpm}^*a_{kqn}\mathscr{E}_{jk}(\mathbf{r}_p, \mathbf{r}_q, t_n - t_m) \geqslant 0, \quad (6.5\text{–}11)$$

$$b_{jpm}b_{kqn}\mathscr{H}_{jk}(\mathbf{r}_p, \mathbf{r}_q, t_n - t_m) \geqslant 0. \quad (6.5\text{–}12)$$

We may readily express the average electric and magnetic energy densities

$\langle w_e(\mathbf{r}, t)\rangle$ and $\langle w_m(\mathbf{r}, t)\rangle$ of a stationary electromagnetic field in free space in terms of the electric and magnetic coherence matrices. We have at once from Eqs. (6.1–6) and (6.1–7) and from Eqs. (6.5–2a) and (6.5–2b) that

$$\langle w_e(\mathbf{r}, t)\rangle = \frac{1}{4\pi}\,\mathrm{tr}\,\mathsf{E}(\mathbf{r}, \mathbf{r}, 0), \tag{6.5–13}$$

and

$$\langle w_m(\mathbf{r}, t)\rangle = \frac{1}{4\pi}\,\mathrm{tr}\,\mathsf{H}(\mathbf{r}, \mathbf{r}, 0). \tag{6.5–14}$$

Although the time argument formally appears on the left-hand sides of Eqs. (6.5–13) and (6.5–14) [and also in Eqs. (6.5–15) below], the expressions are actually independent of time as is seen from the right-hand sides of these equations. This fact is a consequence of our earlier assumption that the electromagnetic field is statistically stationary.

We may also readily express the averaged Poynting vector, $\langle S(\mathbf{r}, t)\rangle$, of such a field in terms of the mixed coherence matrices. According to Eq. (6.1–11) and Eqs. (6.5–2c) and (6.5–2d) the Cartesian components of $\langle \mathbf{S}(\mathbf{r}, t)\rangle$ are given by the expressions

$$\begin{aligned}\langle S_j(\mathbf{r}, t)\rangle &= \frac{c}{2\pi}\varepsilon_{jkl}\,\mathrm{Re}\,\{\langle E_k^* H_l\rangle - \langle E_k^* H_l\rangle\} \\ &= \frac{c}{2\pi}\varepsilon_{jkl}\,\mathrm{Re}\,\{\mathcal{M}_{kl}(\mathbf{r}, \mathbf{r}, 0) - \mathcal{M}_{lk}(\mathbf{r}, \mathbf{r}, 0)\}\end{aligned} \tag{6.5–15a}$$

or, if we make use of the relation (6.5–3c),

$$\langle S_j(\mathbf{r}, t)\rangle = \frac{c}{2\pi}\varepsilon_{jkl}\,\mathrm{Re}\,\{\mathcal{M}_{kl}(\mathbf{r}, \mathbf{r}, 0) - \mathcal{N}_{kl}^*(\mathbf{r}, \mathbf{r}, 0)\}. \tag{6.5–15b}$$

In these formulas the subscripts j, k, l represent the indices x, y, z or their cyclic permutations.

6.5.2 *First-order differential equations for the propagation of the coherence tensors*[‡]

Since the electric and the magnetic fields are related by Maxwell's equations, the correlation matrices, or, equivalently, the correlation tensors which such matrices represent, are not independent of each other. We shall now derive the main relations which exist between them.

To begin with we will express Maxwell's equations in tensor form. For this purpose we recall that the vector product of any two vectors **a** and **b** may be written as

$$(\mathbf{a} \times \mathbf{b})_j = \varepsilon_{jkl} a_k b_l, \tag{6.5–16}$$

where ε_{jkl} is the completely antisymmetric unit tensor of Levi–Civita, i.e. $\varepsilon_{jkl} = +1$ or -1 according to whether the subscripts j, k, l are an even or odd

[‡] The analysis presented in this section is based largely on investigations of Wolf (1956) and Roman and Wolf (1960a).

permutation of the integers 1, 2, 3 and $\varepsilon_{jkl} = 0$ when two (or more) of the subscripts are equal. Cartesian components are now labeled by the suffices 1, 2, 3 rather than by x, y, z. If in the identity (6.5–16) **a** represents the vector operator $\boldsymbol{\nabla} = (\partial/\partial x_1, \partial/\partial x_2, \partial/\partial x_3) \equiv \partial/\partial \mathbf{r} \equiv \boldsymbol{\partial}$ we have

$$(\boldsymbol{\nabla} \times \mathbf{b})_j = \varepsilon_{jkl}\partial_k b_l. \tag{6.5–17}$$

Since we will be dealing with functions which depend on the position of the two points $\mathbf{r}_1$, $\mathbf{r}_2$ we must distinguish between two differential operators. We will use superscript 1 or 2 according as the operator acts with respect to the coordinates of $\mathbf{r}_1$ or $\mathbf{r}_2$, i.e.

$$\boldsymbol{\partial}^1 \equiv \partial/\partial \mathbf{r}_1, \quad \boldsymbol{\partial}^2 \equiv \partial/\partial \mathbf{r}_2. \tag{6.5–18}$$

With this notation Maxwell's equations in free space, for each realization of the ensemble of the complex field, may be expressed in the form‡

$$\varepsilon_{jkl}\partial^1_k E_l(\mathbf{r}_1, t_1) + \frac{1}{c}\frac{\partial}{\partial t_1} H_j(\mathbf{r}_1, t_1) = 0, \tag{6.5–19}$$

$$\varepsilon_{jkl}\partial^1_k H_l(\mathbf{r}_1, t_1) - \frac{1}{c}\frac{\partial}{\partial t_1} E_j(\mathbf{r}_1, t_1) = 0, \tag{6.5–20}$$

$$\partial^1_j E_j(\mathbf{r}_1, t_1) = 0, \tag{6.5–21}$$

$$\partial^1_j H_j(\mathbf{r}_1, t_1) = 0. \tag{6.5–22}$$

Here we again employ Einstein's summation convention, i.e. the repetition of a dummy suffix implies summation over all possible values of the suffix.

Let us take the complex conjugate of Eq. (6.5–19) and multiply it by $E_m(\mathbf{r}_2, t_2)$. We then obtain the relation

$$\varepsilon_{jkl}\partial^1_k E^*_l(\mathbf{r}_1, t_1)E_m(\mathbf{r}_2, t_2) + \frac{1}{c}\frac{\partial}{\partial t_1} H^*_j(\mathbf{r}_1, t_1)E_m(\mathbf{r}_2, t_2) = 0. \tag{6.5–23}$$

Next we set $t_2 = t_1 + \tau$ and keep t_2 fixed. Then $\partial/\partial t_1 = -\partial/\partial\tau$ and we find from Eq. (6.5–23), on taking the ensemble average, and finally writing t in place of t_1

$$\varepsilon_{jkl}\partial^1_k\langle E^*_l(\mathbf{r}_1, t)E_m(\mathbf{r}_2, t+\tau)\rangle - \frac{1}{c}\frac{\partial}{\partial\tau}\langle H^*_j(\mathbf{r}_1, t)E_m(\mathbf{r}_2, t+\tau)\rangle = 0, \tag{6.5–24}$$

or, in terms of the coherence tensors $\mathscr{E}$ and $\mathscr{N}$, represented by the matrices [Eqs. (6.5–2a) and (6.5–2d)],

$$\varepsilon_{jkl}\partial^1_k \mathscr{E}_{lm}(\mathbf{r}_1, \mathbf{r}_2, \tau) - \frac{1}{c}\frac{\partial}{\partial\tau}\mathscr{N}_{jm}(\mathbf{r}_1, \mathbf{r}_2, \tau) = 0. \tag{6.5–25}$$

In a similar manner we may also derive from Maxwell's equation (6.5–19) the following equation which relates the coherence tensors $\mathscr{M}$ and $\mathscr{H}$ [Eqs. (6.5–2b)

‡ We use here the fact that not only the real fields $\mathbf{E}^{(r)}$, $\mathbf{H}^{(r)}$ but also their associated complex analytic signal representations **E**, **H** satisfy Maxwell's equations. That this is so may readily be verified, for example, by use of the procedure employed in Section 4.4.1 to show that if a real function $V^{(r)}(\mathbf{r}, t)$ satisfies the wave equation, so does the associated complex analytic signal $V(\mathbf{r}, t)$.

and (6.5–2c)]:

$$\varepsilon_{jkl}\partial_k^1 \mathcal{M}_{lm}(\mathbf{r}_1, \mathbf{r}_2, \tau) - \frac{1}{c}\frac{\partial}{\partial \tau}\mathcal{H}_{jm}(\mathbf{r}_1, \mathbf{r}_2, \tau) = 0. \tag{6.5–26}$$

From the second Maxwell equation (6.5–20), we may derive, in the same manner, the following two equations that relate the tensors $\mathcal{N}$ and $\mathcal{E}$ and $\mathcal{H}$ and $\mathcal{M}$ respectively:

$$\varepsilon_{jkl}\partial_k^1 \mathcal{N}_{lm}(\mathbf{r}_1, \mathbf{r}_2, \tau) + \frac{1}{c}\frac{\partial}{\partial \tau}\mathcal{E}_{jm}(\mathbf{r}_1, \mathbf{r}_2, \tau) = 0, \tag{6.5–27}$$

and

$$\varepsilon_{jkl}\partial_k^1 \mathcal{H}_{lm}(\mathbf{r}_1, \mathbf{r}_2, \tau) + \frac{1}{c}\frac{\partial}{\partial \tau}\mathcal{M}_{jm}(\mathbf{r}_1, \mathbf{r}_2, \tau) = 0. \tag{6.5–28}$$

Next we will consider some consequences of the divergence equations (6.5–21) and (6.5–22). If we take the complex conjugate of Eq. (6.5–21), multiply it by $E_k(\mathbf{r}_2, t_2)$, take the ensemble average and use the fact that the field is stationary, we obtain the equation

$$\partial_j^1 \mathcal{E}_{jk}(\mathbf{r}_1, \mathbf{r}_2, \tau) = 0. \tag{6.5–29}$$

In a strictly similar way we can also derive from Eq. (6.5–21) the equation

$$\partial_j^1 \mathcal{M}_{jk}(\mathbf{r}_1, \mathbf{r}_2, \tau) = 0. \tag{6.5–30}$$

From the divergence equation (6.5–22) we may derive, by the same procedure, the two equations

$$\partial_j^1 \mathcal{N}_{jk}(\mathbf{r}_1, \mathbf{r}_2, \tau) = 0, \tag{6.5–31}$$

and

$$\partial_j^1 \mathcal{H}_{jk}(\mathbf{r}_1, \mathbf{r}_2, \tau) = 0. \tag{6.5–32}$$

The equations (6.5–25) to (6.5–32) are basic equations for the propagation of the electromagnetic field correlation tensors in free space. There is another set of similar equations which involve the operator ∂^2 rather than ∂^1 and which can be derived in a strictly analogous manner. These two sets of equations are not independent. Each may be derived from the other with the help of the relations (6.5–3).

Various consequences of these equations and some generalizations have been discussed in the literature. In particular, conservation laws involving field correlations have been derived (Roman and Wolf, 1960b; Roman, 1961b) and the equations were generalized to include the effects of random charges and random currents (Roman, 1961a; Beran and Parrent, 1962). Explicit expressions for the coherence tensors of blackbody radiation were obtained and some of their consequences were studied by Bourret (1960) and by Mehta and Wolf (1964a, b).

6.5.3 Wave equations for propagation of the coherence tensors

The first-order differential equations that we have just derived relate the four second-order coherence tensors of a stationary electromagnetic field in free space.

We will now show that these equations imply that each tensor satisfies two wave equations.

Let us apply the operator $(1/c)\partial/\partial\tau$ to Eq. (6.5–27). We then obtain the equation

$$\varepsilon_{jkl}\partial^1_k \frac{1}{c}\frac{\partial}{\partial\tau}\mathcal{N}_{lm} + \frac{1}{c^2}\frac{\partial^2}{\partial\tau^2}\mathcal{E}_{jm} = 0. \tag{6.5–33}$$

Next we substitute on the left-hand side of this equation from Eq. (6.5–25) and find that

$$\varepsilon_{jkl}\varepsilon_{lab}\partial^1_k\partial^1_a\mathcal{E}_{bm} + \frac{1}{c^2}\frac{\partial^2}{\partial\tau^2}\mathcal{E}_{jm} = 0. \tag{6.5–34}$$

We now make use of the identity [Jeffreys, 1931, p. 15, Eq. (55)]

$$\varepsilon_{jkl}\varepsilon_{lab} = \varepsilon_{jkl}\varepsilon_{abl} = \delta_{ja}\delta_{kb} - \delta_{jb}\delta_{ka}, \tag{6.5–35}$$

where δ_{mn} is the Kronecker symbol. The equation (6.5–34) then reduces to

$$\partial^1_b\partial^1_j\mathcal{E}_{bm} - \partial^1_k\partial^1_k\mathcal{E}_{jm} + \frac{1}{c^2}\frac{\partial^2}{\partial\tau^2}\mathcal{E}_{jm} = 0. \tag{6.5–36}$$

Now according to Eq. (6.5–29) the first term on the left vanishes. Also $\partial^1_k\partial^1_k \equiv \boldsymbol{\nabla}^2_1$ where $\boldsymbol{\nabla}^2_1$ is the Laplacian operator taken with respect to the coordinates of the first point $\mathbf{r}_1$. Hence, if we also replace the dummy suffix m by k, Eq. (6.5–36) reduces to the equation

$$\boldsymbol{\nabla}^2_1\mathcal{E}_{jk}(\mathbf{r}_1, \mathbf{r}_2, \tau) = \frac{1}{c^2}\frac{\partial^2}{\partial\tau^2}\mathcal{E}_{jk}(\mathbf{r}_1, \mathbf{r}_2, \tau), \tag{6.5–37}$$

showing that in free space the electric coherence tensor satisfies the wave equation.

In strictly similar manner one may show that the other coherence tensors also satisfy this equation, namely

$$\boldsymbol{\nabla}^2_1\mathcal{H}_{jk}(\mathbf{r}_1, \mathbf{r}_2, \tau) = \frac{1}{c^2}\frac{\partial^2}{\partial\tau^2}\mathcal{H}_{jk}(\mathbf{r}_1, \mathbf{r}_2, \tau), \tag{6.5–38}$$

$$\boldsymbol{\nabla}^2_1\mathcal{M}_{jk}(\mathbf{r}_1, \mathbf{r}_2, \tau) = \frac{1}{c^2}\frac{\partial^2}{\partial\tau^2}\mathcal{M}_{jk}(\mathbf{r}_1, \mathbf{r}_2, \tau), \tag{6.5–39}$$

$$\boldsymbol{\nabla}^2_1\mathcal{N}_{jk}(\mathbf{r}_1, \mathbf{r}_2, \tau) = \frac{1}{c^2}\frac{\partial^2}{\partial\tau^2}\mathcal{N}_{jk}(\mathbf{r}_1, \mathbf{r}_2, \tau), \tag{6.5–40}$$

Further, on making use of Eqs. (6.5–3), one readily finds from Eqs. (6.5–37) to (6.5–40) that each of the four coherence tensors satisfies a wave equation in which the Laplacian operator acts with respect to the coordinates of the second point $\mathbf{r}_2$ (in which case we denote the Laplacian operator by $\boldsymbol{\nabla}^2_2$), i.e. each coherence tensor also satisfies a wave equation of the form

$$\boldsymbol{\nabla}^2_2\mathcal{E}_{jk}(\mathbf{r}_1, \mathbf{r}_2, \tau) = \frac{1}{c^2}\frac{\partial^2}{\partial\tau^2}\mathcal{E}_{jk}(\mathbf{r}_1, \mathbf{r}_2, \tau). \tag{6.5–41}$$

We conclude this section by stressing that we have defined coherence matrices (coherence tensors) of stationary electromagnetic fields and derived dynamical equations which they obey in free space. These definitions and equations provide the mathematical framework for a unified treatment of many problems involving second-order temporal coherence, spatial coherence and the state of polarization of the field.

6.6 The second-order cross-spectral density tensors of a stationary electromagnetic field

6.6.1 *The electric, magnetic and mixed cross-spectral density tensors*

We have seen earlier, e.g. in our discussion of the foundations of radiometry (Section 5.7) and in our study of the effects of source correlations on the spectra of the emitted radiation (Section 5.8), that in some cases it is more natural and more convenient to use the space-frequency rather than the space-time description of a fluctuating wavefield. It is also advantageous to use the space-frequency description in the analysis of light propagation in dispersive media and in some other situations involving the interaction of a fluctuating wavefield with matter. It is, therefore, appropriate to discuss the space-frequency analogues to some of the results involving the space–time coherence tensors which we just considered.

The basic mathematical quantities in the space-frequency description are the so-called cross-spectral density matrices, or, equivalently, the cross-spectral density tensors. They are the natural generalizations of the scalar cross-spectral density function which we frequently encountered in the preceding chapters.

Let $\widetilde{\mathbf{E}}(\mathbf{r}, \nu)$ and $\widetilde{\mathbf{H}}(\mathbf{r}, \nu)$ be the (generalized) Fourier transforms of the fluctuating complex electric and magnetic fields respectively, assumed to be stationary and of zero mean. The cross-spectral tensors may then be introduced by the formulas [cf. Eq. (4.3–39)]

$$\langle \widetilde{\mathbf{E}}_j^*(\mathbf{r}_1, \nu)\widetilde{\mathbf{E}}_k(\mathbf{r}_2, \nu')\rangle = W_{jk}^{(e)}(\mathbf{r}_1, \mathbf{r}_2, \nu)\delta(\nu - \nu'), \tag{6.6–1a}$$

$$\langle \widetilde{\mathbf{H}}_j^*(\mathbf{r}_1, \nu)\widetilde{\mathbf{H}}_k(\mathbf{r}_2, \nu')\rangle = W_{jk}^{(h)}(\mathbf{r}_1, \mathbf{r}_2, \nu)\delta(\nu - \nu'), \tag{6.6–1b}$$

$$\langle \widetilde{\mathbf{E}}_j^*(\mathbf{r}_1, \nu)\widetilde{\mathbf{H}}_k(\mathbf{r}_2, \nu')\rangle = W_{jk}^{(m)}(\mathbf{r}_1, \mathbf{r}_2, \nu)\delta(\nu - \nu'), \tag{6.6–1c}$$

$$\langle \widetilde{\mathbf{H}}_j^*(\mathbf{r}_1, \nu)\widetilde{\mathbf{E}}_k(\mathbf{r}_2, \nu')\rangle = W_{jk}^{(n)}(\mathbf{r}_1, \mathbf{r}_2, \nu)\delta(\nu - \nu'), \tag{6.6–1d}$$

where the subscripts j, $k = x$, y, z again label Cartesian components, δ is the Dirac delta function and $\nu \geqslant 0$. Each of the four quantities $W_{jk}^{(e)}$, $W_{jk}^{(h)}$, $W_{jk}^{(m)}$ and $W_{jk}^{(n)}$ are components of second-rank tensors. $W_{jk}^{(e)}$ and $W_{jk}^{(h)}$ are known as the *electric* and the *magnetic cross-spectral density tensors* respectively and $W_{jk}^{(m)}$ and $W_{jk}^{(n)}$ are called *mixed cross-spectral density tensors.* The nine components of each of these four tensors, may, of course, be arranged in the form of 3×3 matrices known as the electric, magnetic or mixed *cross-spectral density matrices* respectively.

According to the generalized Wiener–Khintchine theorem [Eqs. (2.4–37) and (2.4–38)], each coherence tensor, defined by Eqs. (6.5–2), and the corresponding cross-spectral density tensor form a Fourier transform pair:

$$\mathscr{E}_{jk}(\mathbf{r}_1, \mathbf{r}_2, \tau) = \int_0^\infty W_{jk}^{(e)}(\mathbf{r}_1, \mathbf{r}_2, \nu)\,\mathrm{e}^{-2\pi i\nu\tau}\,\mathrm{d}\nu, \tag{6.6–2a}$$

$$\mathscr{H}_{jk}(\mathbf{r}_1, \mathbf{r}_2, \tau) = \int_0^\infty W_{jk}^{(h)}(\mathbf{r}_1, \mathbf{r}_2, \nu)\,\mathrm{e}^{-2\pi i\nu\tau}\,\mathrm{d}\nu, \tag{6.6–2b}$$

$$\mathscr{M}_{jk}(\mathbf{r}_1, \mathbf{r}_2, \tau) = \int_0^\infty W_{jk}^{(m)}(\mathbf{r}_1, \mathbf{r}_2, \nu)\,\mathrm{e}^{-2\pi i\nu\tau}\,\mathrm{d}\nu, \tag{6.6–2c}$$

$$\mathscr{N}_{jk}(\mathbf{r}_1, \mathbf{r}_2, \tau) = \int_0^\infty W_{jk}^{(n)}(\mathbf{r}_1, \mathbf{r}_2, \nu)\,\mathrm{e}^{-2\pi i\nu\tau}\,\mathrm{d}\nu. \tag{6.6–2d}$$

The lower limits on the integrals in these formulas are 0 rather than $-\infty$ because we use the analytic signal representation of the fields. Taking the Fourier inverse of Eqs. (6.6–2) we have

$$W_{jk}^{(e)}(\mathbf{r}_1, \mathbf{r}_2, \nu) = \int_{-\infty}^\infty \mathscr{E}_{jk}(\mathbf{r}_1, \mathbf{r}_2, \tau)\,\mathrm{e}^{2\pi i\nu\tau}\,\mathrm{d}\tau, \quad (\nu \geqslant 0), \tag{6.6–3}$$

etc.

From the defining equations (6.6–1a) and (6.6–1b) we see at once that the cross-spectral density tensors of the electric and the magnetic fields each satisfies a relation of the form

$$W_{kj}^{(\alpha)}(\mathbf{r}_1, \mathbf{r}_2, \nu) = [W_{jk}^{(\alpha)}(\mathbf{r}_2, \mathbf{r}_1, \nu)]^*, \tag{6.6–4}$$

where α stands for either of the two superscripts e or h. Further from Eq. (6.6–1c) and (6.6–1d) the following relation involving the mixed cross-spectral density tensors evidently also holds:

$$W_{kj}^{(m)}(\mathbf{r}_1, \mathbf{r}_2, \nu) = [W_{jk}^{(n)}(\mathbf{r}_2, \mathbf{r}_1, \nu)]^*. \tag{6.6–5}$$

The cross-spectral density tensors satisfy a number of non-negative definiteness conditions, which may be derived as follows. We start from the obvious inequality

$$\left\langle \left| \int_{\nu-\varepsilon_1}^{\nu+\varepsilon_2} \mathrm{d}\nu' \int \mathrm{d}^3 r \{f_j(\mathbf{r})\widetilde{E}_j(\mathbf{r}, \nu') + g_j(\mathbf{r})\widetilde{H}_j(\mathbf{r}, \nu')\} \right|^2 \right\rangle \geqslant 0. \tag{6.6–6}$$

Here $f_j(\mathbf{r})$ and $g_j(\mathbf{r})$, $j = 1, 2, 3$, are arbitrary well-behaved functions of position, ν is an arbitrary positive frequency and ε_1 and ε_2 are arbitrary positive numbers, with $\nu - \varepsilon_1 > 0$. The volume integration extends over an arbitrary region of space containing the field. As before, summation over a repeated dummy index is implied.

From the inequality (6.6–6) we readily find, if we recall Eqs. (6.6–1), that

$$\begin{aligned}\int_{\nu-\varepsilon_1}^{\nu+\varepsilon_2} \mathrm{d}\nu' \int \mathrm{d}^3 r_1 \int \mathrm{d}^3 r_2 \{&f_j^*(\mathbf{r}_1)f_k(\mathbf{r}_2)W_{jk}^{(e)}(\mathbf{r}_1, \mathbf{r}_2, \nu') + g_j^*(\mathbf{r}_1)g_k(\mathbf{r}_2)W_{jk}^{(h)}(\mathbf{r}_1, \mathbf{r}_2, \nu') \\ &+ f_j^*(\mathbf{r}_1)g_k(\mathbf{r}_2)W_{jk}^{(m)}(\mathbf{r}_1, \mathbf{r}_2, \nu') + g_j^*(\mathbf{r}_1)f_k(\mathbf{r}_2)W_{jk}^{(n)}(\mathbf{r}_1, \mathbf{r}_2, \nu')\} \geqslant 0.\end{aligned} \tag{6.6–7}$$

Since this inequality holds for integration over an arbitrarily small frequency range it implies the following non-negative definiteness condition involving the four cross-spectral density tensors, originally derived in the context of the theory of quantized electromagnetic fields by Mehta and Wolf (1967b, Appendix III).

$$\int d^3r_1 \int d^3r_2 \{f_j^*(\mathbf{r}_1) f_k(\mathbf{r}_2) W_{jk}^{(e)}(\mathbf{r}_1, \mathbf{r}_2, \nu) + g_j^*(\mathbf{r}_1) g_k(\mathbf{r}_2) W_{jk}^{(h)}(\mathbf{r}_1, \mathbf{r}_2, \nu) + f_j^*(\mathbf{r}_1) g_k(\mathbf{r}_2) W_{jk}^{(m)}(\mathbf{r}_1, \mathbf{r}_2, \omega) + g_j^*(\mathbf{r}_1) f_k(\mathbf{r}_2) W_{jk}^{(n)}(\mathbf{r}_1, \mathbf{r}_2, \nu)\} \geqslant 0. \tag{6.6–8}$$

We note two special cases. If in Eq. (6.6–8) we choose $g_j(\mathbf{r}) \equiv 0$, $(j = x, y, z)$, we obtain the following non-negative definiteness condition obeyed by the electric cross-spectral density tensor:

$$\int d^3r_1 \int d^3r_2 f_j^*(\mathbf{r}_1) f_k(\mathbf{r}_2) W_{jk}^{(e)}(\mathbf{r}_1, \mathbf{r}_2, \nu) \geqslant 0. \tag{6.6–9}$$

Similarly, if in Eq. (6.6–8) we choose $f_j(\mathbf{r}) \equiv 0$, $(j = x, y, z)$, we obtain a corresponding non-negative definiteness condition satisfied by the magnetic cross-spectral density tensor:

$$\int d^3r_1 \int d^3r_2 g_j^*(\mathbf{r}_1) g_k(\mathbf{r}_2) W_{jk}^{(h)}(\mathbf{r}_1, \mathbf{r}_2, \nu) \geqslant 0. \tag{6.6–10}$$

Just as the corresponding inequalities involving the second-order coherence tensors, the inequalities (6.6–8) to (6.6–10) imply alternative forms of these conditions, involving summations rather than integrations. To show this let us choose

$$f_j(\mathbf{r}) = a_{jp} \delta^{(3)}(\mathbf{r} - \mathbf{r}_p), \tag{6.6–11}$$

$$g_j(\mathbf{r}) = b_{jp} \delta^{(3)}(\mathbf{r} - \mathbf{r}_p), \tag{6.6–12}$$

where P is an arbitrary positive integer, a_{jp}, b_{jp}, $j = x, y, z$, $p = 1, 2, 3, \ldots, P$, are arbitrary (real or complex) constants and $\mathbf{r}_p$ are arbitrary points within the domain over which the spatial integrations in Eq. (6.6–8) are taken. We then obtain from the inequalities (6.6–8) to (6.6–10) the following non-negative definiteness conditions:

$$a_{jp}^* a_{kq} W_{jk}^{(e)}(\mathbf{r}_p, \mathbf{r}_q, \nu) + b_{jp}^* b_{kq} W_{jk}^{(h)}(\mathbf{r}_p, \mathbf{r}_q, \nu) + a_{jp}^* b_{kq} W_{jk}^{(m)}(\mathbf{r}_p, \mathbf{r}_q, \nu) + b_{jp}^* a_{kq} W_{jk}^{(n)}(\mathbf{r}_p, \mathbf{r}_q, \nu) \geqslant 0, \tag{6.6–13}$$

$$a_{jp}^* a_{kq} W_{jk}^{(e)}(\mathbf{r}_p, \mathbf{r}_q, \nu) \geqslant 0, \tag{6.6–14}$$

$$b_{jp}^* b_{kq} W_{jk}^{(h)}(\mathbf{r}_p, \mathbf{r}_q, \nu) \geqslant 0. \tag{6.6–15}$$

From Eqs. (6.1–6), (6.1–7) and (6.1–11) and from Eqs. (6.6–2) we readily obtain the following expressions for the average electric and magnetic energy densities and the average Poynting vector in terms of the cross-spectral density tensors:

$$\langle w_e(\mathbf{r}, t) \rangle = \frac{1}{4\pi} \int_0^\infty \operatorname{tr} \mathbf{W}^{(e)}(\mathbf{r}, \mathbf{r}, \nu)\, d\nu, \tag{6.6–16}$$

$$\langle w_m(\mathbf{r}, t) \rangle = \frac{1}{4\pi} \int_0^\infty \operatorname{tr} \mathbf{W}^{(m)}(\mathbf{r}, \mathbf{r}, \nu)\, d\nu, \tag{6.6–17}$$

$$\langle S_j(\mathbf{r}, t) \rangle = \frac{c}{2\pi} \varepsilon_{jkl} \operatorname{Re} \left\{ \int_0^\infty [W_{kl}^{(m)}(\mathbf{r}, \mathbf{r}, \nu) - W_{lk}^{(m)}(\mathbf{r}, \mathbf{r}, \nu)]\, d\nu \right\} \tag{6.6–18a}$$

$$= \frac{c}{2\pi} \varepsilon_{jkl} \operatorname{Re} \left\{ \int_0^\infty [W_{kl}^{(m)}(\mathbf{r}, \mathbf{r}, \nu) - W_{kl}^{(n)}(\mathbf{r}, \mathbf{r}, \nu)]\, d\nu \right\}. \tag{6.6–18b}$$

Although the time argument appears on the left-hand side of Eqs. (6.6–16) to (6.6–18), the averages are time independent, for the same reasons as noted earlier in connection with Eqs. (6.5–13) to (6.5–15).

Explicit expressions for the cross-spectral density tensors of blackbody radiation have been derived by Mehta and Wolf (1967c).

6.6.2 *First-order differential equations for the propagation of the cross-spectral density tensors*

The four cross-spectral density tensors are related by a set of differential equations. Their free-space form may be obtained at once by taking the Fourier transform of Eqs. (6.5–25) to (6.5–32) which relate the coherence tensors and by making use of relations (6.6–3). One then obtains the required equations:

$$\varepsilon_{jkl}\partial^1_k W^{(e)}_{lm}(\mathbf{r}_1, \mathbf{r}_2, \nu) + \mathrm{i}\frac{2\pi\nu}{c} W^{(n)}_{jm}(\mathbf{r}_1, \mathbf{r}_2, \nu) = 0, \tag{6.6–19}$$

$$\varepsilon_{jkl}\partial^1_k W^{(m)}_{lm}(\mathbf{r}_1, \mathbf{r}_2, \nu) + \mathrm{i}\frac{2\pi\nu}{c} W^{(h)}_{jm}(\mathbf{r}_1, \mathbf{r}_2, \nu) = 0, \tag{6.6–20}$$

$$\varepsilon_{jkl}\partial^1_k W^{(n)}_{lm}(\mathbf{r}_1, \mathbf{r}_2, \nu) - \mathrm{i}\frac{2\pi\nu}{c} W^{(e)}_{jm}(\mathbf{r}_1, \mathbf{r}_2, \nu) = 0, \tag{6.6–21}$$

$$\varepsilon_{jkl}\partial^1_k W^{(h)}_{lm}(\mathbf{r}_1, \mathbf{r}_2, \nu) - \mathrm{i}\frac{2\pi\nu}{c} W^{(m)}_{jm}(\mathbf{r}_1, \mathbf{r}_2, \nu) = 0, \tag{6.6–22}$$

$$\partial^1_j W^{(e)}_{jk}(\mathbf{r}_1, \mathbf{r}_2, \nu) = 0, \tag{6.6–23}$$

$$\partial^1_j W^{(m)}_{jk}(\mathbf{r}_1, \mathbf{r}_2, \nu) = 0, \tag{6.6–24}$$

$$\partial^1_j W^{(n)}_{jk}(\mathbf{r}_1, \mathbf{r}_2, \nu) = 0, \tag{6.6–25}$$

$$\partial^1_j W^{(h)}_{jk}(\mathbf{r}_1, \mathbf{r}_2, \nu) = 0, \tag{6.6–26}$$

There is a similar set of equations, which involves the operator ∂^2 rather than ∂^1. It may be obtained from Eqs. (6.6–19) to (6.6–26) by making use of the relations (6.6–4).

6.6.3 *Helmholtz equations for propagation of the cross-spectral density tensors*

In free space each cross-spectral density tensor obeys the Helmholtz equation. To show this, one can either eliminate the tensors $\mathbf{W}^{(n)}$ or $\mathbf{W}^{(e)}$ between Eqs. (6.6–19) and (6.6–21) or eliminate $\mathbf{W}^{(m)}$ or $\mathbf{W}^{(h)}$ between Eqs. (6.6–20) and (6.6–22). A simpler way to obtain the same result is to take the Fourier transforms of Eqs. (6.5–37)–(6.5–40) and use the fact that the cross-spectral density tensors are Fourier transforms of the coherence tensors [Eq. (6.6–3)]. One then obtains at once the Helmholtz equations

$$\nabla^2_1 W^{(\alpha)}_{jk}(\mathbf{r}_1, \mathbf{r}_2, \nu) + \left(\frac{2\pi\nu}{c}\right)^2 W^{(\alpha)}_{jk}(\mathbf{r}_1, \mathbf{r}_2, \nu) = 0, \tag{6.6–27}$$

where ∇_1^2 is again the Laplacian operator taken with respect to the coordinates of the first point $\mathbf{r}_1$ and α stands for any of the four superscripts e, h, m or n.

If we take the complex conjugate of Eq. (6.6–27), use the relation (6.6–4) and interchange the arguments $\mathbf{r}_1$ and $\mathbf{r}_2$ we obtain a second set of Helmholtz equations

$$\nabla_2^2 W_{jk}^{(\alpha)}(\mathbf{r}_1, \mathbf{r}_2, \nu) + \left(\frac{2\pi\nu}{c}\right)^2 W_{jk}^{(\alpha)}(\mathbf{r}_1, \mathbf{r}_2, \nu) = 0, \qquad (6.6\text{–}28)$$

where ∇_2^2 is the Laplacian operator taken with respect to the coordinates of the second point $\mathbf{r}_2$.

Problems

6.1 Determine the Mueller matrices of a compensator, of an absorber, a rotator and a polarizer.

6.2 Explain how one could use a combination of the four non-image forming devices discussed in Sections 6.4.1–6.4.4 to convert unpolarized light into circularly polarized light. Can one accomplish this by using only the lossless devices, i.e. a compensator and a rotator?

6.3 Show, by an example or otherwise, that it is possible for the degree of polarization of a well-collimated, uniform, quasi-monochromatic light beam to decrease after passage through one of the four non-image forming devices discussed in Sections 6.4.1–6.4.4.

6.4 The cross spectral tensors $W_{ij}^{(\alpha)}(\mathbf{r}_1, \mathbf{r}_2, \nu)$, $(\alpha = e, h, n, m;\ i, j = 1, 2, 3)$ of an electromagnetic field that is statistically homogeneous and isotropic depend on $\mathbf{r}_1$ and $\mathbf{r}_2$ only through the difference $\mathbf{r}' = \mathbf{r}_2 - \mathbf{r}_1$ and are invariant under rotation of coordinate axes.

Show that any cross-spectral tensor of the form

$$C_{ij}(\mathbf{r}', \nu) = F(r', \nu)x_i'x_j' + G(r', \nu)\delta_{ij},$$

where $\mathbf{r}' \equiv (x_1', x_2', x_3')$, δ_{ij} is the Kronecker δ symbol and $r' = |\mathbf{r}'|$, is invariant under rotation.

6.5 Let $\mathbf{E}(\mathbf{r}, \nu)$ and $\mathbf{H}(\mathbf{r}, \nu)$ be space-frequency realizations of an ensemble of stationary free electromagnetic fields in vacuum. They may be represented as angular spectra

$$\mathbf{E}(\mathbf{r}, \nu) = \int_{(4\pi)} \mathbf{e}(\mathbf{s}, \nu)\, e^{ik\mathbf{s}\cdot\mathbf{r}}\, d\Omega,$$

$$\mathbf{H}(\mathbf{r}, \nu) = \int_{(4\pi)} \mathbf{h}(\mathbf{s}, \nu)\, e^{ik\mathbf{s}\cdot\mathbf{r}}\, d\Omega,$$

where $\mathbf{s}$ are real unit vectors and the integrations extend over the whole 4π solid angle generated by $\mathbf{s}$.

Show that the average energy density and the average Poynting vector of the field may be expressed in the form (in Gaussian systems of units)

$$\langle W_\nu(\mathbf{r})\rangle = \int_{(4\pi)} Q_\nu(\mathbf{r}, \mathbf{s})\,\mathrm{d}\Omega,$$

and

$$\langle \mathbf{S}_\nu(\mathbf{r})\rangle = \int_{(4\pi)} \mathbf{R}_\nu(\mathbf{r}, \mathbf{s})\,\mathrm{d}\Omega,$$

where

$$Q_\nu(\mathbf{r}, \mathbf{s}) = \frac{1}{4\pi}[\langle \mathbf{E}^*(\mathbf{r}, \nu)\cdot \mathbf{e}(\mathbf{s}, \nu)\rangle + \langle \mathbf{H}^*(\mathbf{r}, \nu)\cdot \mathbf{h}(\mathbf{s}, \nu)\rangle]\,\mathrm{e}^{ik\mathbf{s}\cdot\mathbf{r}}$$

and

$$\mathbf{R}_\nu(\mathbf{r}, \mathbf{s}) = \frac{c}{4\pi}[\langle \mathbf{E}^*(\mathbf{r}, \nu)\times \mathbf{h}(\mathbf{s}, \nu)\rangle - \langle \mathbf{H}^*(\mathbf{r}, \nu)\times \mathbf{e}(\mathbf{s}, \nu)\rangle]\,\mathrm{e}^{ik\mathbf{s}\cdot\mathbf{r}}.$$

6.6 The quantities $Q_\nu(\mathbf{r}, \mathbf{s})$ and $\mathbf{R}_\nu(\mathbf{r}, \mathbf{s})$ defined in Problem 6.5 may be regarded as 'angular components' of the average energy density and of the average energy flux vector. Show that they are related by the formula

$$cQ_\nu(\mathbf{r}, \mathbf{s}) = \mathbf{s}\cdot\mathbf{R}_\nu(\mathbf{r}, \mathbf{s}),$$

where c is the speed of light in vacuum. Show further that they satisfy the differential equations

$$\mathbf{s}\cdot\nabla Q_\nu(\mathbf{r}, \mathbf{s}) - \frac{1}{2ik}\nabla^2 Q_\nu(\mathbf{r}, \mathbf{s}) = 0$$

and

$$\mathbf{s}\cdot\nabla \mathbf{R}_\nu(\mathbf{r}, \mathbf{s}) - \frac{1}{2ik}\nabla^2 \mathbf{R}_\nu(\mathbf{r}, \mathbf{s}) = 0,$$

where $k = 2\pi\nu/c$.

6.7 Consider the electromagnetic field generated by a fluctuating charge-current source distribution localized, for all times, in a finite domain D. The fluctuations are statistically stationary. If $\mathbf{W}^{(e)}(\mathbf{r}_1, \mathbf{r}_2, \nu)$ and $\mathbf{W}^{(h)}(\mathbf{r}_1, \mathbf{r}_2, \nu)$ are the cross-spectral tensors of the complex electric and magnetic fields generated by the source, show that in the *far zone*,

$$\mathrm{tr}\,\mathbf{W}^{(e)}(r\mathbf{s}, r\mathbf{s}, \nu) = \mathrm{tr}\,\mathbf{W}^{(h)}(r\mathbf{s}, r\mathbf{s}, \nu), \quad (\mathbf{s}^2 = 1).$$

7

Some applications of second-order coherence theory

7.1 Introduction

In Chapter 5 we discussed some applications of second-order coherence theory to problems involving radiation from localized sources of any state of coherence. In the present chapter we will describe some other applications of the second-order theory. The first two concern classic interferometric techniques for determining the angular diameters of stars and the energy distribution in spectral lines. Both techniques were introduced by Albert Michelson many years ago and the underlying principles were explained by him without the use of any concept of coherence theory (which was formulated later). However, second-order coherence theory provided a deeper understanding of the physical principles involved and also suggested useful modifications of these techniques, some of which will be discussed in Section 9.10.

Another application which will be described in this chapter concerns the determination of the angular and the spectral distribution of energy in optical fields scattered from fluctuating linear media. The analysis will be based on the second-order coherence theory of the full electromagnetic field, which we developed in Chapter 6.

7.2 Stellar interferometry

As is well known, the angular diameters that stars subtend at the surface of the earth are so small that no available telescopes can resolve them. In the focal plane of a telescope, the star light gives rise to a diffraction pattern which is indistinguishable from that which would be produced by light from a point source, diffracted at the aperture of the telescope and degraded by the passage of the light through the earth's atmosphere.

Michelson (1890, 1920) showed that the angular diameter of a star and, in some cases, also the intensity distribution across the star may be determined from measurements made with a system consisting of an interferometer mounted on a telescope. A schematic diagram of such a system is shown in Fig. 7.1.[‡] Light from a star is incident on the two outer mirrors M_1 and M_2 of the interferometer, is reflected at the two inner mirrors M_3 and M_4 and is brought to the back focal plane $\mathscr{F}$ of a telescope to which the interferometer is attached. The inner mirrors M_3 and M_4 are fixed, while the outer mirrors M_1 and M_2 can be separated

[‡] A detailed description of two types of Michelson stellar interferometers is given in articles by Michelson and Pease (1921) and Pease (1925, 1930, 1931).

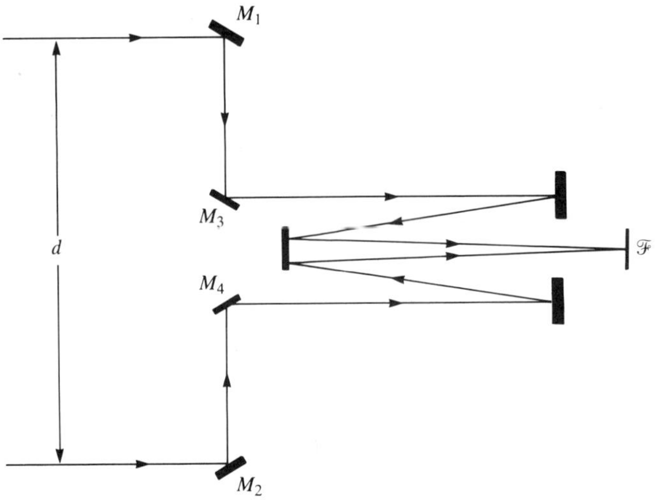

Fig. 7.1 A schematic diagram of the Michelson stellar interferometer.

symmetrically in the direction joining M_3 and M_4. In the focal plane $\mathcal{F}$ of the telescope one then observes the diffraction image of the star on which the telescope is focused, crossed by fringes formed by the two interfering beams.

The visibility of the fringes in the focal plane $\mathcal{F}$ depends on the separation of the mirrors M_1 and M_2. Michelson showed by an elementary argument that, from measurements of the variation of the visibility of the fringes with the separation of the two mirrors, one may obtain information about the intensity distribution across the star, at least in cases where one may assume that the distribution is rotationally symmetric. In particular Michelson showed that, if the stellar disk is circular and uniform, the visibility, considered as a function of the separation d of the two outer mirrors M_1 and M_2, will have zeros for certain separations and that the smallest separation for which the visibility has zero value is given by the expression

$$d_0 = \frac{0.61\bar{\lambda}}{\alpha}, \tag{7.2–1}$$

where $\bar{\lambda}$ is the mean wavelength and α is the angular radius of the star. Thus from the measurement of d_0, the angular diameter of the star may be determined. Angular diameters of several stars, down to about 0.02 seconds of arc, were first determined in this way with a system designed and built by Michelson and Pease and completed in 1920. The two moveable mirrors were suported by a 20 foot (~ 6 meters) long beam, which was mounted on the 100 inch (~ 2.5 meters) telescope at the Mount Wilson Observatory in California. The results obtained with this instrument are summarized in the article by Pease (1931).

From the standpoint of second-order coherence theory the principles of the method can be readily understood. To keep the analysis as simple as possible we will idealize the situation by considering the star to be a planar uncorrelated source σ, perpendicular to the axis of the telescope, and the mirrors to be situated in a plane parallel to the source plane. The absence of correlations across the source implies that the mutual intensity $J(\mathbf{r}_1', \mathbf{r}_2')$ [Eq. (4.3–34)] is δ-correlated, i.e. for any two points $\mathbf{r}_1'$ and $\mathbf{r}_2'$ on σ (see Fig. 7.2),

$$J(\mathbf{r}_1', \mathbf{r}_2') = I(\mathbf{r}_1')\delta^{(2)}(\mathbf{r}_2' - \mathbf{r}_1'), \qquad (7.2\text{–}2)$$

where $I(\mathbf{r}_1')$ is a measure of the average intensity at $\mathbf{r}_1'$ and $\delta^{(2)}$ is the two-dimensional Dirac delta function. According to the far-zone form of the van Cittert–Zernike theorem, the light that reaches the outer mirrors M_1 and M_2 of the interferometer will be partially coherent. The equal-time complex degree of coherence $j(r_1\mathbf{s}_1, r_2\mathbf{s}_2)$ of the light incident on the two mirrors is given by Eq. (4.4–40), namely

$$j(r_1\mathbf{s}_1, r_2\mathbf{s}_2) = e^{i\bar{k}(r_2-r_1)} \frac{\int_\sigma I(\mathbf{r}')\, e^{-i\bar{k}(\mathbf{s}_2-\mathbf{s}_1)\cdot\mathbf{r}'}\, d^2r'}{\int_\sigma I(\mathbf{r}')\, d^2r'}. \qquad (7.2\text{–}3)$$

In this formula r_1 and r_2 are the distances of the mirrors M_1 and M_2 from a chosen point O' on the stellar disk σ, $\mathbf{s}_1$ and $\mathbf{s}_2$ are unit vectors along the directions $O'M_1$ and $O'M_2$ and $\bar{k} = 2\pi/\bar{\lambda}$ is the mean wave-number of the light. In the integrals on the right-hand side of Eq. (7.2–3) the intensity distribution across σ is regarded as a function of the position vector $\mathbf{r}'$, specifying the location of a typical source point S.

Let us rewrite the right-hand side of Eq. (7.2–3) in a more explicit form. We take a Cartesian rectangular coordinate system OX, OY, OZ with the origin O at the mid-point between the two mirrors and with the axis OZ along the telescope axis and pointing into the telescope (Fig. 7.2). We choose as O' the point at which the (negative) z-axis intersects the stellar disk and denote the distance $\overline{O'O}$ by R. If $(x_1, y_1, 0)$ and $(x_2, y_2, 0)$ are the coordinates of the centers

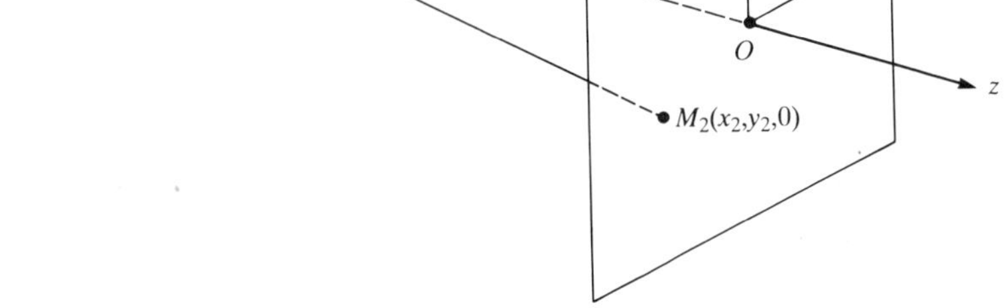

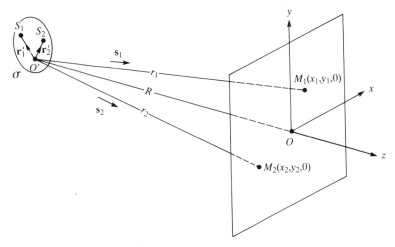

Fig. 7.2 Notation illustrating the meaning of some of the symbols in Eqs. (7.2–3) to (7.2–7). $\mathbf{s}_1$ and $\mathbf{s}_2$ are unit vectors in the directions $O'M_1$ and $O'M_2$ respectively.

of the mirrors M_1 and M_2 and (ξ, η) are the x- and y-components of a typical point $\mathbf{r}'$ on the stellar disk, we have

$$\mathbf{s}_l = \left(\frac{x_l}{r_l}, \frac{y_l}{r_l}, \frac{R}{r_l}\right), \tag{7.2–4a}$$

$$r_l = (x_l^2 + y_l^2 + R^2)^{1/2}, \tag{7.2–4b}$$

$(l = 1, 2)$. The expression (7.2–3) now becomes

$$j(x_1 - x_2, y_1 - y_2) = e^{i\bar{k}(r_2 - r_1)} \frac{\int_\sigma I(\xi, \eta) \exp\left\{i\bar{k}\left[\frac{x_1 - x_2}{R}\xi + \frac{y_1 - y_2}{R}\eta\right]\right\} d\xi\, d\eta}{\int_\sigma I(\xi, \eta)\, d\xi\, d\eta}, \tag{7.2–5}$$

where we have written $j(x_1 - x_2, y_1 - y_2)$ in place of $j(x_1, y_1, x_2, y_2)$, because the four coordinates that specify the location of the mirrors M_1 and M_2 enter the right-hand side of Eq. (7.2–5) only through the differences $x_1 - x_2$ and $y_1 - y_2$. We have also written $I(\xi, \eta)$ in place of $I(\mathbf{r}')$.

It may readily be seen that the factor $\exp[i\bar{k}(r_2 - r_1)]$ on the right-hand side of Eq. (7.2–5) may be replaced by unity. We have, according to Eq. (7.2–4b),

$$\begin{aligned} r_l &= R\left[\left(\frac{x_l}{R}\right)^2 + \left(\frac{y_l}{R}\right)^2 + 1\right]^{1/2} \\ &= R\left[1 + \frac{1}{2}\frac{x_l^2 + y_l^2}{R^2} + O\left(\frac{x_l^2 + y_l^2}{R^2}\right)^2\right], \quad (l = 1, 2), \end{aligned} \tag{7.2–6}$$

and hence with R large enough,

$$\bar{k}(r_2 - r_1) \approx \frac{\pi[(x_2^2 + y_2^2) - (x_1^2 + y_1^2)]}{\lambda R}. \tag{7.2–7}$$

Since the quantity on the right of Eq. (7.2–7) contains in the denominator the astronomically large distance R of the star from the earth's surface, $\bar{k}(r_2 - r_1)$ will be negligible in comparison to unity and hence $\exp[i\bar{k}(r_2 - r_1)] \approx 1$. From now on we will, therefore, neglect the first exponential factor on the right-hand side of Eq. (7.2–5).

The ratios

$$\frac{\xi}{R} = p, \quad \frac{\eta}{R} = q \tag{7.2–8}$$

may evidently be identified with the angular coordinates of the point $(\xi, \eta, -R)$ on the stellar disk when viewed from the midpoint between the outer mirrors of the interferometer. If we also set

$$I(\xi, \eta) \equiv I(pR, qR) \equiv i(p, q), \tag{7.2–9}$$

so that $i(p, q)$ is a measure of the intensity across the stellar disk as a function of the angular variables p and q, the formula (7.2–5) becomes

$$j(x_1 - x_2, y_1 - y_2) = \frac{\int_{\sigma'} i(p, q) \exp\{\mathrm{i}\bar{k}[(x_1 - x_2)p + (y_1 - y_2)q]\} \,\mathrm{d}p\,\mathrm{d}q}{\int_{\sigma'} i(p, q) \,\mathrm{d}p\,\mathrm{d}q}, \tag{7.2–10}$$

where σ' is the domain of the p, q-plane that corresponds to the source domain σ of the ξ, η-plane.

Equation (7.2–10) shows that the equal-time complex degree of coherence of the light incident on the two outer mirrors of the interferometer is the normalized Fourier transform of the intensity distribution across the stellar disk. Now according to Eqs. (4.3–25a) and (4.3–35) the absolute value $|j|$ of the equal-time degree of coherence j is equal to the visibility of the interference fringes in the central portion of the fringe pattern, formed by light reaching the focal plane $\mathscr{F}$ of the telescope. Moreover, according to Eq. (4.3–28) the phase of j may be determined from the location of the intensity maxima in the fringe pattern. We see, on taking the Fourier inverse of Eq. (7.2–10), that the normalized intensity distribution across the stellar disk may, in principle, be determined from measurements of the visibility and the location of the intensity maxima of the fringe pattern formed in the focal plane $\mathscr{F}$ of the telescope. Measurements of the location of the intensity maxima are, however, almost impossible to carry out in practice, because of the disturbing effects of the atmosphere, which produce irregular motion of the fringes. However, if the intensity distribution $i(p, q)$ across the stellar disk may be assumed to have rotational symmetry about the z-axis, elementary calculations relating to the Fourier inverse of Eq. (7.2–10) show that $i(p, q)$ may be determined even without knowledge of the phase of the equal-time degree of coherence except for certain ambiguities that arise if j vanishes for some values of its arguments, because j then becomes real. Such ambiguities can, in some cases, be removed by plausibility considerations.

In Fig. 7.3 the variations of the fringe visibility with the distance d of separation between the two outer mirrors M_1 and M_2 of the Michelson stellar interferometer are shown for some simple model sources.

If one is interested in determining only the angular size of the star rather than the intensity distribution across it the analysis greatly simplifies. Suppose, for example, that one may assume the stellar disk to be circularly symmetric about O' and of uniform intensity, i.e. $i(p, q)$ is constant across the disk. If the disk subtends an angular diameter 2α at the interferometer we then have, according to Eq. (4.4–44),

$$j(x_1 - x_2, y_1 - y_2) = \frac{2J_1(v)}{v}, \tag{7.2–11}$$

where

$$v = \frac{2\pi\alpha}{\bar{\lambda}} d, \qquad d = [(x_1 - x_2)^2 + (y_1 - y_2)^2]^{1/2}, \tag{7.2–12}$$

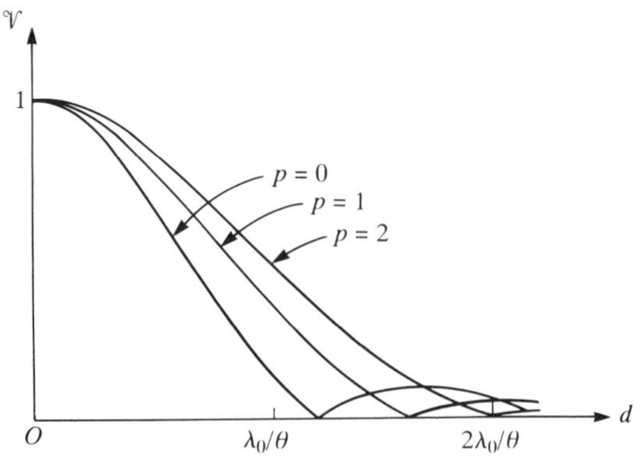

Fig. 7.3 Examples of the variation of the fringe visibility $\mathcal{V}$ with the separation d between the outer mirrors M_1 and M_2 of the Michelson stellar interferometer. The source is assumed to be spatially incoherent and circular of angular diameter 2α, with the intensity distribution across the disk given by $I(\theta) = I_0(\alpha^2 - \theta^2)^p$, θ denoting angular distance from the center of the disk, and I_0 is a constant.

J_1 is the Bessel function of the first kind and first order and $\bar{\lambda} = 2\pi/\bar{k}$ is the mean wavelength. The behavior of the function $2J_1(v)/v$ is shown in Fig. 4.13. It has the value unity when $v = 0$ and it decreases steadily to zero at $v = 3.83$; as v increases further, $2J_1(v)/v$ oscillates with diminishing amplitude. If d_0 denotes the smallest value of the separation d of the outer mirrors M_1 and M_2 at which the equal-time degree of coherence j vanishes, we have $3.83 = 2\pi\alpha d_0/\bar{\lambda}$, i.e.

$$\alpha = \frac{0.61\bar{\lambda}}{d_0}. \tag{7.2–13}$$

Thus by determining the smallest distance d_0 of separation between the outer mirrors for which the fringes in the focal plane $\mathcal{F}$ of the telescope disappear, the angular radius α of the star may be determined from Eq. (7.2–13). This formula is in agreement with Eq. (7.2–1), derived by Michelson by a different argument.

In more recent times various interferometers have been constructed which utilize the principles introduced by Michelson. For example, at Sydney University, Australia, an 11.4 m baseline prototype stellar interferometer was constructed (Davis and Tango, 1985, 1986) as a first step towards building a very long baseline instrument of that kind. It is located at the grounds of the Australian National Measurement Laboratory in West Lindsfield near Sydney. Another stellar interferometer, known as the Infrared Spatial Interferometer, is located on Mount Wilson. It is designed to operate at infrared wavelengths around 10 μm and employs a pair of movable telescopes, each consisting of a flat mirror of 2 m diameter and a parabolic mirror of 1.65 m diameter, and it uses a heterodyne signal detection system (Bester, Danchi and Townes, in Breckinridge, 1990, p. 40).

Many other interferometer techniques for use in optical astronomy have been developed. Notable among them is the so-called speckle interferometry due to Labeyrie (1970, see also Dainty, 1984) and stellar intensity interferometry introduced by Hanbury Brown and Twiss, which will be discussed in Sections 9.10 and 14.6.1. A review of these and other high resolution techniques used in optical astronomy can be found, for example, in Labeyrie (1976) and Breckinridge (1990).

We have confined our discussion to interferometers used in the visible region of the optical spectrum. However, the same principles are being used with great success in radio astronomy where instruments analogous to the Michelson stellar interferometer have, in fact, become basic research tools (see, for example, Rohlfs, 1986; Thompson, Moran and Swenson, 1986).

7.3 Interference spectroscopy

7.3.1 General principles

We will now consider the other method due to Michelson which we referred to at the beginning of this chapter, for determining the energy distribution in spectral lines (Michelson, 1891, 1892, 1927). This method is used to determine the energy distribution in spectral lines and is capable of resolving lines that are too narrow to analyze by conventional prism or grating spectrographs. A light beam is divided into two beams in the neighborhood of a point P_1 in a Michelson interferometer (see Fig. 4.1). The beams are superposed after a path difference $c\tau$ has been introduced between them. One then determines the visibility $\mathcal{V}(\tau) \equiv \mathcal{V}(P_1, \tau)$ of the resulting interference fringes as a function of τ. Michelson showed that from knowledge of the visibility one may obtain information about the energy distribution in the spectrum of the light and that, in particular, when the spectrum is symmetric about some frequency ν_0, the spectral profile is the Fourier transform of the visibility, except for certain ambiguities that arise if the visibility has zeros for some values of its argument. Such ambiguities may sometimes be removed by a plausibility argument. By using this technique Michelson found that certain spectral lines which appeared to be singlets when the light was analyzed by conventional techniques were, in fact, doublets or multiplets and he was able to estimate their widths in some cases.

The principle of this method may again be readily understood from second-order coherence theory. If we assume that the two beams have the same mean intensities then, according to Eq. (4.3–25a), the visibility of the fringes in the plane of observation $\mathcal{B}$ is related to the complex degree of coherence of the light at the beam divider D, located at the point $P_1(\mathbf{r}_1)$ (see Fig. 4.1), by the formula

$$\mathcal{V}(\tau) = |\gamma(\tau)|, \tag{7.3–1}$$

where $\gamma(\tau) \equiv \gamma(\mathbf{r}_1, \mathbf{r}_1, \tau)$ is the complex degree of self-coherence of the light at the point P_1. Let us represent $\gamma(\tau)$ as a Fourier integral, namely as

$$\gamma(\tau) = \int_0^\infty \mathfrak{s}(\nu)\, \mathrm{e}^{-2\pi i \nu\tau}\, \mathrm{d}\nu. \tag{7.3–2}$$

According to Eq. (4.3–12a) and Eq. (4.3–42a), the quantity

$$\mathfrak{s}(\nu) = \frac{S(\nu)}{\displaystyle\int_0^\infty S(\nu)\, \mathrm{d}\nu} \tag{7.3–3}$$

represents the normalized spectral density of the light at P_1.

Let us consider first the case when the normalized spectral density is symmetric about some frequency, ν_0 say.‡ It is then convenient to introduce the 'shifted' normalized spectrum

$$\begin{aligned} \mathscr{s}_{sh}(\mu) &= \mathscr{s}(\nu_0 + \mu) \quad \text{when } \mu \geqslant -\nu_0 \\ &= 0 \quad \text{when } \mu < -\nu_0. \end{aligned} \qquad (7.3\text{–}4)$$

If in the integral on the right-hand side of Eq. (7.3–2) we change the variable of integration from ν to $\mu = \nu - \nu_0$ and use Eq. (7.3–4), we find at once that

$$\gamma(\tau) = \gamma_{sh}(\tau)\, e^{-2\pi i \nu_0 \tau}, \qquad (7.3\text{–}5)$$

where

$$\gamma_{sh}(\tau) = \int_{-\infty}^{\infty} \mathscr{s}_{sh}(\mu)\, e^{-2\pi i \mu \tau}\, d\mu. \qquad (7.3\text{–}6)$$

Since $\mathscr{s}_{sh}(\mu)$ is an even function of μ, Eq. (7.3–6) implies that

$$\gamma_{sh}(\tau) = 2\int_0^{\infty} \mathscr{s}_{sh}(\mu) \cos(2\pi\mu\tau)\, d\mu. \qquad (7.3\text{–}7)$$

On taking the Fourier inverse of Eq. (7.3–7) we obtain the following expression for $\mathscr{s}_{sh}(\mu)$ in terms of $\gamma_{sh}(\tau)$:

$$\mathscr{s}_{sh}(\mu) = 2\int_0^{\infty} \gamma_{sh}(\tau) \cos(2\pi\mu\tau)\, d\tau. \qquad (7.3\text{–}8)$$

Now the spectral density $\mathscr{s}(\nu)$, and hence also its 'shifted' form $\mathscr{s}_{sh}(\mu)$ [Eq. (7.3–4)] are necessarily real. According to Eq. (7.3–7), $\gamma_{sh}(\tau)$ must, therefore, be also real and hence

$$\gamma_{sh}(\tau) = \varepsilon(\tau)|\gamma_{sh}(\tau)|, \qquad (7.3\text{–}9)$$

where $\varepsilon(\tau)$ can only take on the values $+1$ and -1. Assuming that $\gamma_{sh}(\tau)$ and consequently also $|\gamma_{sh}(\tau)|$ are continuous functions of τ, which we may expect to be the case in practice, we see from Eq. (7.3–9) that $\varepsilon(\tau)$ will also be continuous, except possibly for the values of τ, if any, at which $\gamma_{sh}(\tau) = 0$. Only at these special values could $\varepsilon(\tau)$ switch from $+1$ to -1 or vice versa. Moreover, since the normalization of the complex degree of coherence $\gamma(\tau)$ ensures that $\gamma(0) = 1$, we have from Eqs. (7.3–5) and (7.3–9) that

$$\varepsilon(0) = +1. \qquad (7.3\text{–}10)$$

Let us assume first that $\gamma_{sh}(\tau)$ has no zeros at all, in which case, as we see from Eqs. (7.3–5) and (7.3–1), $\gamma(\tau)$ and $\mathscr{V}(\tau)$ will not have zeros either. Then, since the value of $\varepsilon(\tau)$ cannot change except at a zero of $\gamma_{sh}(\tau)$, we must have $\varepsilon(\tau) \equiv \varepsilon(0) = +1$ and Eq. (7.3–9) simplifies to

$$\gamma_{sh}(\tau) = |\gamma_{sh}(\tau)|. \qquad (7.3\text{–}11)$$

‡ Because $\gamma(\tau)$ is an analytic signal, $\mathscr{s}(\nu) = 0$ for $\nu < 0$. Hence the assumption of symmetry implies that unless $\mathscr{s}(\nu) = 0$ for $\nu > 2\nu_0$ we are ignoring the 'tail' of $\mathscr{s}(\nu)$ for $\nu < 0$. Such an approximation is clearly a reasonable one when the light is quasi-monochromatic, because $2\nu_0$ is then very much greater than the effective bandwidth of the light.

If we again recall Eqs. (7.3–5) and (7.3–1) it follows that we now have $\gamma_{sh}(\tau) = \mathscr{V}(\tau)$ and Eq. (7.3–8) gives

$$\mathscr{s}_{sh}(\mu) = 2\int_0^\infty \mathscr{V}(\tau)\cos(2\pi\mu\tau)\,d\tau. \tag{7.3–12}$$

Thus we have shown that *if the spectral density is symmetric about some frequency* ν_0 *and the complex degree of self-coherence of the light, and consequently also the visibility, has no zeros, the normalized spectral profile function* $\mathscr{s}_{sh}(\mu)$ *is equal to twice the Fourier cosine transform of the fringe visibility function*. This result is illustrated by an example in Fig. 7.4.

Next let us consider the more complicated case when the spectral density is still symmetric about some frequency ν_0, but $\gamma_{sh}(\tau)$ becomes zero for some τ values. Suppose that the zeros of $\gamma_{sh}(\tau)$ on the positive τ-axis occur when $\tau_1, \tau_2, \ldots, \tau_N$, $(0 < \tau_1 < \tau_2 < \ldots < \tau_N)$. In this case we obviously obtain from Eqs. (7.3–8) and (7.3–9), if we also use Eqs. (7.3–5) and (7.3–1), the following expression for the normalized spectral profile function:

$$\mathscr{s}_{sh}(\mu) = 2\int_0^{\tau_1} \mathscr{V}(\tau)\cos 2\pi\mu\tau\,d\tau + 2\sum_{n=1}^{N-1}\int_{\tau_n}^{\tau_{n+1}} \varepsilon_n\mathscr{V}(\tau)\cos 2\pi\mu\tau\,d\tau$$
$$+ 2\int_{\tau_N}^\infty \varepsilon_n\mathscr{V}(\tau)\cos 2\pi\mu\tau\,d\tau, \tag{7.3–13}$$

where each of the quantities ε_n has the value $+1$ or -1. If all the zeros τ_n, $(n = 1, 2, \ldots, N)$, are simple zeros of $\gamma_{sh}(\tau)$, continuity demands that $\gamma_{sh}(\tau)$ changes sign as τ passes through each zero, which implies that $\varepsilon_1 = -1$, $\varepsilon_2 = +1$, $\varepsilon_3 = -1$, etc. If the zeros are not simple but are of second or higher order, the preceding argument has to be modified in an obvious way. Unfortunately, the (necessarily somewhat inaccurate) knowledge of the visibility function $\mathscr{V}(\tau) = |\gamma_{sh}(\tau)|$ obtained from measurement does not clearly indicate a possible multiplicity of the zeros; but nevertheless the values of the ε_n's and consequently of $\mathscr{s}_{sh}(\mu)$ can sometimes be determined from plausibility considerations.

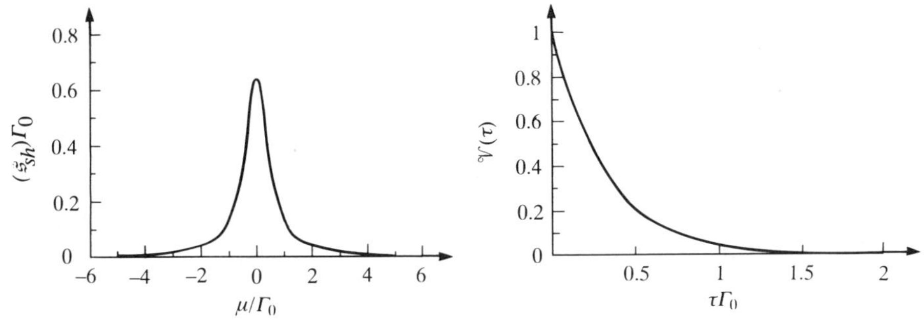

Fig. 7.4 Example of a normalized spectral profile $\mathscr{s}_{sh}(\mu)$ and the corresponding visibility function $\mathscr{V}(\tau)$. The spectral profile is that of a Lorentzian line $\mathscr{s}_{sh}(\mu) = (\Gamma_0/2\pi)/(\mu^2 + \frac{1}{4}\Gamma_0^2)$, $(\mu \geqslant \nu_0)$, with $2\pi\Gamma_0 = 1.877 \times 10^9\ s^{-1}$, $\nu_0 = 2.466 \times 10^{15}\ s^{-1}$ (Lyman line). The corresponding visibility function $\mathscr{V}(\tau) = \exp(-\pi\Gamma_0\tau)$.

7.3.2 The phase problem

Let us now briefly consider the general case when the spectrum is not symmetric. In such a case the argument leading from Eq. (7.3–2) to Eq. (7.3–8) no longer applies. Instead we then proceed in the following way. We first take the Fourier inverse of Eq. (7.3–2) and obtain the formula

$$\mathfrak{s}(\nu) = \int_{-\infty}^{\infty} \gamma(\tau)\, e^{2\pi i \nu\tau}\, d\tau$$

$$= \int_{-\infty}^{0} \gamma(\tau)\, e^{2\pi i \nu\tau}\, d\tau + \int_{0}^{\infty} \gamma(\tau)\, e^{2\pi i \nu\tau}\, d\tau. \qquad (7.3\text{–}14)$$

In the first integral on the right-hand side of Eq. (7.3–14) let us change the variable of integration from τ to $-\tau$ and make use of the relation

$$\gamma(-\tau) = \gamma^*(\tau), \qquad (7.3\text{–}15)$$

which follows from the definition (4.3–12a) of γ and from the relation (4.3–36). We then obtain for the normalized spectral density the expression

$$\mathfrak{s}(\nu) = 2\,\mathrm{Re} \int_{0}^{\infty} \gamma(\tau)\, e^{2\pi i \nu\tau}\, d\tau, \qquad (7.3\text{–}16)$$

where Re denotes the real part. Now $\gamma(\tau)$ is, in general, complex. Let $\psi(\tau)$ be the phase of $\gamma(\tau)$, i.e.

$$\gamma(\tau) = |\gamma(\tau)| e^{i\psi(\tau)}. \qquad (7.3\text{–}17)$$

If we substitute from Eq. (7.3–17) into Eq. (7.3–16) and use the fact that, according to Eq. (7.3–1), $|\gamma(\tau)|$ is equal to the visibility function $\mathcal{V}(\tau)$, we obtain the formula

$$\mathfrak{s}(\nu) = 2\,\mathrm{Re} \int_{0}^{\infty} \mathcal{V}(\tau)\, e^{i[\psi(\tau) + 2\pi\nu\tau]}\, d\tau. \qquad (7.3\text{–}18)$$

This formula shows that the normalized spectral density function $\mathfrak{s}(\nu)$ (no longer assumed to be symmetric) may be calculated from knowledge of the fringe visibility function $\mathcal{V}(\tau)$ and the phase $\psi(\tau)$ of the complex degree of self-coherence of the light. We have learned earlier [cf. Eq. (4.3–28)] that $\psi(\tau)$ may be determined from measurements of the location of the maxima in the interference fringes. Such measurements are, however, much harder to perform than measurements of the visibility function.

It has been usually assumed, following a discussion of this question by Rayleigh (1892), that measurements of the visibility and of the position of the intensity maxima provide two *independent* sets of information, so that both have to be performed in order to determine asymmetrical spectral distributions. However, more recent investigations based on coherence theory (Wolf, 1962; Roman and Marathay, 1963; Marathay and Roman, 1964; Dialetis and Wolf, 1967; Dialetis, 1967; Nussenzveig, 1967) have shown that the analytic properties of the complex degree of coherence impose certain restrictions on the phase function $\psi(\tau)$ which can be associated with a given modulus $|\gamma(\tau)|$ and consequently with a given visibility curve $\mathcal{V}(\tau)$. Suppose that $\gamma(\tau)$ is square-integrable, obeys the Paley–

Wiener condition (Paley and Wiener, 1934, p. 16, Theorem XII)

$$\int_0^\infty \frac{|\ln|\gamma(\tau)||}{\tau^2+1}\,\mathrm{d}\tau < \infty, \tag{7.3–19}$$

and is a continuous function of τ. It can be shown that $\psi(\tau)$ is then necessarily of the form (Toll, 1956)

$$\psi(\tau) = \psi_{\min}(\tau) + \psi_{\mathrm{B}}(\tau) - 2\pi\nu_0\tau. \tag{7.3–20}$$

Here ν_0 is an arbitrary non-negative constant,

$$\psi_{\min}(\tau) = \frac{2\tau}{\pi} P\int_0^\infty \frac{\ln|\gamma(\tau')|}{\tau'^2-\tau^2}\,\mathrm{d}\tau', \tag{7.3–21}$$

P denoting the Cauchy principal value of the integral taken at $\tau' = \tau$, and

$$\psi_{\mathrm{B}}(\tau) = \sum_j \arg\frac{\tau-\tau^{(j)}}{\tau-\tau^{(j)*}}, \tag{7.3–22}$$

the $\tau^{(j)}$'s being the zeros of $\gamma(\tau)$, considered as a function of complex τ in the lower half, Π^-, of the complex τ-plane ($\mathrm{Im}\,\tau \leqslant 0$).

The result expressed by Eq. (7.3–20) is difficult to establish rigorously‡ and we will not derive it here. We will just briefly indicate the origin of the three terms in the expression (7.3–20) for the phase $\psi(\tau)$ of the complex degree of self-coherence $\gamma(\tau)$.

We learned earlier that because $\gamma(\tau)$ is an analytic signal, it is the boundary value on the real τ-axis of a function which is analytic and regular in the lower half of the τ-plane, Π^- say (cf. Section 3.1.1). Consequently the function

$$\ln\gamma(\tau) \equiv \ln|\gamma(\tau)| + \mathrm{i}\psi(\tau), \tag{7.3–23}$$

when considered as a function of complex τ, will also be analytic in that half-plane, but it will have branch points at the points $\tau^{(j)}$, ($j = 1, 2, \ldots$), if any, at which $\gamma(\tau)$ has zero values. If $\gamma(\tau)$ has no zeros in Π^-, nor on the real axis $\mathrm{Im}\,\tau = 0$, the analyticity of $\ln\gamma(\tau)$ may be shown, with the help of Cauchy's integral formula, to imply that the phase of $\gamma(\tau)$ is given by the expression (7.3–21) (cf. Page, 1955, p. 224) for real values of τ. This contribution is known as the *minimum phase* associated with $|\gamma(\tau)|$, a terminology related to the fact that the contribution $\psi_{\mathrm{B}}(\tau)$ to the total phase in Eq. (7.3–20) may readily be shown to be non-negative.

If $\gamma(\tau)$, considered as a function of complex τ, has zeros in the half-plane Π^- or on the real axis $\mathrm{Im}\,\tau = 0$, each such zero may be shown to contribute an additive term to the total phase of $\gamma(\tau)$. If the zeros do not lie on the real axis, their contribution is given by the expression under the summation sign on the right-hand side of Eq. (7.3–22). This contribution is known as the *Blaschke phase*, because it is the phase of the so-called *Blaschke factor*

$$B_j(\tau) = \frac{\tau-\tau^{(j)}}{\tau-\tau^{(j)*}}, \qquad (\mathrm{Im}\,\tau^{(j)} < 0). \tag{7.3–24}$$

‡ A rigorous, though not quite complete, proof was given by Toll (1956). See also Weaver and Pao (1981), and Burge, Fiddy, Greenaway and Ross (1976).

Such a factor is clearly a function which is analytic and regular in the lower half of the complex τ-plane, has a single zero in that half-plane, at the point $\tau = \tau^{(j)}$, and is unimodular on the real τ-axis. The third additive term on the right-hand side of Eq. (7.3–20), which is linear in τ, affects only the absolute location of the spectral energy distribution and not its shape (the spectral profile), as can readily be seen from the 'shift' theorem for Fourier transforms.

If $\gamma(\tau)$ has zeros on the real axis, our earlier assumption about the continuity of $\ln \gamma(\tau)$ no longer holds. In that case one can show that the phase $\psi(\tau)$ is given by the sum of the expression (7.3–20) and an expression that represents the contribution from the axial zeros. These contributions can be determined by using contour integration techniques.

It is evident from the preceeding discussion that the phase function $\psi(\tau)$ of the complex degree of self-coherence, which is needed for unambiguous reconstruction of the spectral profile by Michelson's technique, is uniquely specified by $|\gamma(\tau)|$ (which is equivalent to the knowledge of the visibility curve) and the location of the zeros of the analytic continuation of $\gamma(\tau)$ into the lower half of the complex τ-plane (including the real axis $\text{Im}\,\tau = 0$). The location of the zeros other than those situated on the real τ-axis cannot, unfortunately, be determined from knowledge of $|\gamma(\tau)|$ for real τ-values.

A number of indirect methods have been proposed for determining the phase of $\gamma(\tau)$ and the normalized spectrum $\mathfrak{s}(\nu)$ from measurements of the modulus of $\gamma(\tau)$. In one method the spectrum is modified by passing the light through a filter with an exponential frequency transmittance (Mehta, 1965) and in another method a sharp, nearly monochromatic component at some known frequency is added (Gamo, 1963, p. 801). We will now briefly discuss the principles underlying these methods.

The method that employs exponential filters utilizes the fact that since $\gamma(\tau) \equiv \gamma(\tau_r, \tau_i)$, considered as a function of the complex variable $\tau = \tau_r + i\tau_i$, (τ_r, τ_i real), is analytic and regular in the lower half Π^- of the complex τ-plane ($\tau_i < 0$), its modulus and phase are related by the following differential equations, which are consequences of the Cauchy–Riemann equations:

$$\frac{\partial|\gamma(\tau_r, \tau_i)|}{\partial \tau_r} = -|\gamma(\tau_r, \tau_i)|\frac{\partial \psi(\tau_r, \tau_i)}{\partial \tau_i}, \quad (\tau_i < 0), \tag{7.3–25a}$$

$$\frac{\partial|\gamma(\tau_r, \tau_i)|}{\partial \tau_i} = |\gamma(\tau_r, \tau_i)|\frac{\partial \psi(\tau_r, \tau_i)}{\partial \tau_r}, \quad (\tau_i < 0). \tag{7.3–25b}$$

Hence if both $|\gamma(\tau)|$ and its derivative $\partial|\gamma(\tau)|/\partial\tau_i$ are known along the real axis, the phase $\psi(\tau)$ along the real axis can be determined by integrating Eq. (7.3–25b). One then obtains for ψ the expression

$$\begin{aligned}\psi(\tau_r, 0) &= \int_0^{\tau_r} \frac{1}{|\gamma(\tau_r', 0)|}\left[\frac{\partial|\gamma(\tau_r', \tau_i)|}{\partial\tau_i}\right]_{\tau_i \to -0} d\tau_r' \\ &= \left[\int_0^{\tau_r} \frac{\partial \ln|\gamma(\tau_r', \tau_i)|}{\partial \tau_i} d\tau_r'\right]_{\tau_i \to -0}.\end{aligned} \tag{7.3–26}$$

The problem is thus reduced to determining experimentally $|\gamma(\tau_r, \tau_i)|$ off the real axis; from this information the limiting values of the derivative $\partial|\gamma(\tau_r', \tau_i)|/\partial\tau_i$ as $\tau_i \to -0$ can be estimated. To carry out such measurements one can proceed as follows. Suppose that a filter with exponential transmittance $e^{-2\pi\nu T}$, where T is a positive constant, is available and that a light beam, whose normalized spectral distribution $\mathscr{s}(\nu)$ is to be determined, is passed through such a filter. The emerging light has the modified spectral distribution

$$\mathscr{s}_m(\nu) = K\mathscr{s}(\nu)\,e^{-2\pi\nu T}, \tag{7.3–27}$$

where K is a normalizing constant that represents the ratio of the original to the modified mean light intensity. The complex degree of self-coherence of the transmitted light is then given by

$$\gamma_m(\tau_r, 0) = \int_0^\infty \mathscr{s}_m(\nu)\,e^{-2\pi i\nu\tau_r}\,d\nu, \tag{7.3–28}$$

or, with the help of Eq. (7.3–27),

$$\begin{aligned}\gamma_m(\tau_r, 0) &= K\int_0^\infty \mathscr{s}(\nu)\,e^{-2\pi i\nu(\tau_r - iT)}\,d\nu \\ &= K\gamma(\tau_r, -T), \quad (T > 0).\end{aligned} \tag{7.3–29}$$

This formula shows that the measurement of $|\gamma_m|$ along the real axis is equivalent to measurement of $|\gamma|$ off the real τ-axis, along the line $\mathrm{Im}\,\tau = -T$.

The preceding analysis shows that with exponential filters, i.e. filters which modify the spectrum in accordance with Eq. (7.3–27), one can determine the phase ψ of the complex degree of self-coherence along the real axis from Eq. (7.3–26).

Next let us consider the other method mentioned earlier for determining the spectrum from measurements of the absolute value of the complex degree of self-coherence. Suppose that we generate another light beam whose bandwidth, centered on some frequency ν_1, is much smaller than the effective width of $\mathscr{s}(\nu)$. We will approximate the spectrum of this beam by the expression $I_1\delta(\nu - \nu_1)$, where $\delta(\nu)$ is the Dirac delta function. We assume that the frequency ν_1 lies outside the effective spectral range of $\mathscr{s}(\nu)$. If this light beam (to be referred to as the reference beam) is superposed on the light beam whose spectrum $\mathscr{s}(\nu)$ is to be determined, the (normalized) spectral density of the resulting beam is

$$\mathscr{s}_{Tot}(\nu) = [\mathscr{s}(\nu) + \beta\delta(\nu - \nu_1)]/(1 + \beta), \tag{7.3–30}$$

where β is the ratio $I_1/\langle I\rangle$ of the mean light intensities of the two constituent beams. The complex degree of self-coherence $\gamma_{Tot}(\tau)$ of the mixture is, therefore,

$$\begin{aligned}\gamma_{Tot}(\tau) &= \int_0^\infty \mathscr{s}_{Tot}(\nu)\,e^{-2\pi i\nu\tau}\,d\nu \\ &= \frac{\gamma(\tau) + \beta e^{-2\pi i\nu_1\tau}}{1 + \beta}.\end{aligned} \tag{7.3–31}$$

It follows from Eqs. (7.3–30) and (7.3–31) that

$$\int_{-\infty}^{\infty} |\gamma_{\text{Tot}}(\tau)|^2 \, e^{2\pi i \nu \tau} \, d\tau$$

$$= \int_0^{\infty} \mathscr{s}_{\text{Tot}}(\nu') \mathscr{s}_{\text{Tot}}(\nu' + \nu) \, d\nu'$$

$$= \frac{1}{(1+\beta)^2} \left\{ \int_0^{\infty} \mathscr{s}(\nu') \mathscr{s}(\nu' + \nu) \, d\nu' + \beta^2 \delta(\nu) + \beta \mathscr{s}(\nu_1 + \nu) + \beta \mathscr{s}(\nu_1 - \nu) \right\}$$

$$= \frac{1}{(1+\beta)^2} \left\{ \int_{-\infty}^{\infty} |\gamma(\tau)|^2 \, e^{2\pi i \nu \tau} \, d\tau + \beta^2 \delta(\nu) + \beta \mathscr{s}(\nu_1 + \nu) + \beta \mathscr{s}(\nu_1 - \nu) \right\}. \tag{7.3–32}$$

If both $|\gamma_{\text{Tot}}(\tau)|$ and $|\gamma(\tau)|$ are determined, one can extract the sum of the last two terms from Eq. (7.3–32). These two terms represent the unknown spectral distribution, displaced by $-\nu_1$, and the image of this distribution reflected about the origin. Hence $\mathscr{s}(\nu)$ can be derived from $|\gamma_{\text{Tot}}(\tau)|$. Without the reference beam there would exist a fundamental ambiguity in the reconstruction, because the degrees of self-coherence associated with any spectral distribution $\mathscr{s}(\nu)$ and its mirror image in some line $\nu = \nu'$ have the same modulus.

The two reconstruction procedures which we just described have been tested experimentally with some success (Kohler and Mandel, 1973); and also in connection with inverse problems involving a spatial variable rather than the temporal variable τ.

Other methods for determining the phase of the complex degree of coherence have been proposed in the context of both interference spectroscopy and stellar interferometry. One of them is somewhat similar to the second method which we have just described, but instead of superposing an essentially monochromatic light beam on the light beam whose spectral density is to be determined, one superposes on it a beam of arbitrary but known spectrum (Mehta, 1968). Other proposed methods involve measurements of correlations of order higher than second (e.g. Gamo, 1963; Beard, 1969).

The problem of obtaining phase information from other data, especially from intensity measurements, occurs not only in optical coherence theory but in many other fields, e.g. in x-ray crystallography, electron microscopy and image reconstruction. Some of the techniques used for solving such phase problems are reviewed in articles by Ferwerda (1978) and Millane (1990).

Michelson's method for determining the energy distribution in spectral lines from visibility curves has, in more recent times, been superseded to some extent by another interferometric technique, called Fourier spectroscopy (also known as the interferogram method) (Fellgett, 1958a,b; Jacquinot, 1958, 1960; Strong and Vanasse, 1959; Connes, 1961a,b,c,d; Vannasse and Sakai, 1967). This method, which is used mainly in the infrared region of the spectrum, allows the determination of the real part $\gamma^{(r)}(\tau)$ of the complex degree of self-coherence $\gamma(\tau)$. From knowledge of $\gamma^{(r)}(\tau)$ as a function of τ, the normalized spectral density may, in principle, be completely determined. An example is shown in Fig. 7.5.

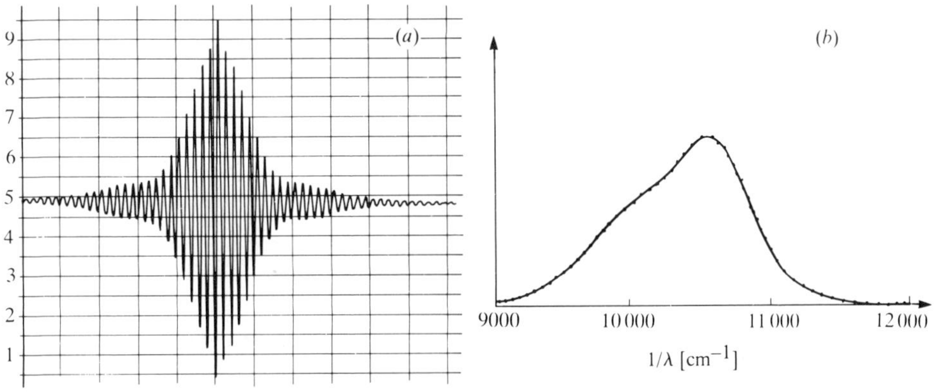

Fig. 7.5 Determination of an infrared spectrum of mean wavelength $\bar{\lambda} \approx 1\ \mu m$, from an interferogram: (*a*) The interferogram, (*b*) The spectrum. (After Connes, 1961a.)

7.4 Coherence of transverse laser resonator modes

The theory of resonant modes in a laser cavity has its origin in classic papers of Fox and Li (1961) and Boyd and Gordon (1961). It was subsequently extended and refined by numerous authors. The theory has played a central role in laser physics and laser engineering, but being based on a monochromatic model it did not elucidate the coherence properties of the modes. Early attempts to clarify this question were made by Wolf (1963), Streifer (1966), Allen, Gatehouse and Jones (1971) and Gori (1980). A more complete analysis was later carried out by Wolf and Agarwal (1984), based on coherence theory in the space-frequency domain. In this section we will present this analysis of the coherence properties of transverse laser resonator modes.

7.4.1 Steady-state condition for the cross-spectral density of light at a resonator mirror

Let us consider an empty resonator consisting of two mirrors A and B. Suppose that light with some initial distribution at the mirror A is launched into the cavity. We assume that the light fluctuations are characterized by an ensemble that is stationary, at least in the wide sense. We denote by $W_0(\boldsymbol{\rho}_1, \boldsymbol{\rho}_2, \nu)$ the cross-spectral density of the initial light distribution, at points specified by position vectors $\boldsymbol{\rho}_1$ and $\boldsymbol{\rho}_2$, on the mirror A (Fig. 7.6). The light propagates to the mirror B, where it is reflected and diffracted, and some of it returns to the mirror A. Let $W_1(\boldsymbol{\rho}_1, \boldsymbol{\rho}_2, \nu)$ be the cross-spectral density of the light that returns to A. After reflection and diffraction at that mirror some of the light propagates towards the mirror B again and this process continues. We denote by $W_j(\boldsymbol{\rho}_1, \boldsymbol{\rho}_2, \nu)$ the cross-spectral density of the light across the mirror A after the completion of j cycles $A \to B \to A$.

According to Eq. (4.7–38) the cross-spectral density W_j may be expressed in the form

$$W_j(\boldsymbol{\rho}_1, \boldsymbol{\rho}_2, \nu) = \langle U_j^*(\boldsymbol{\rho}_1, \nu) U_j(\boldsymbol{\rho}_2, \nu) \rangle, \tag{7.4–1}$$

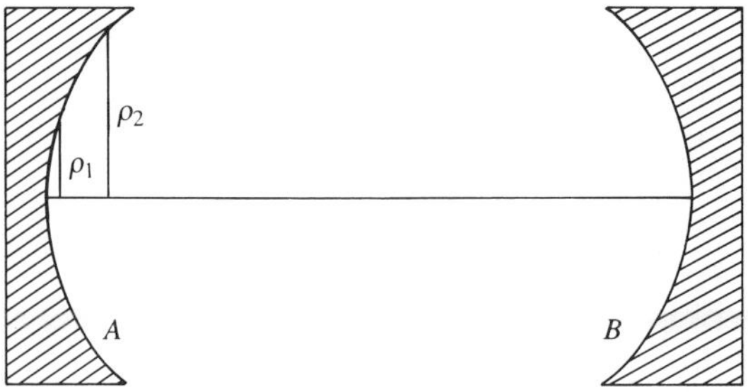

Fig. 7.6 Illustrating the geometry and notation relating to the determination of the steady-state condition for the cross-spectral density of light in a laser resonator.

where the average is taken over an ensemble of random functions $\{U_j(\boldsymbol{\rho}, \nu)\}$. Because each member $U_j(\boldsymbol{\rho}, \nu)$ of this ensemble is a boundary value of a field that satisfies the Helmholtz equation

$$(\nabla^2 + k^2)U_j(\mathbf{r}, \nu) = 0 \tag{7.4–2}$$

$(k = 2\pi\nu/c$, c being the speed of light in vacuo) in the space between the two mirrors, U_{j+1} and U_j will be related by a linear transformation, i.e.

$$U_{j+1}(\boldsymbol{\rho}, \nu) = \int_A L(\boldsymbol{\rho}, \boldsymbol{\rho}', \nu)U_j(\boldsymbol{\rho}', \nu)\,\mathrm{d}^2\rho', \tag{7.4–3}$$

$(j = 0, 1, 2, \ldots)$. Here $L(\boldsymbol{\rho}, \boldsymbol{\rho}', \nu)$ is a propagator, for propagation of monochromatic light of frequency ν from the point $\boldsymbol{\rho}'$ on the mirror A to the point $\boldsymbol{\rho}$ on the same mirror, after a single reflection and diffraction at the other mirror, B. Explicit approximate expressions for this propagator may be obtained with the help of the Huygens–Fresnel principle (see, for example Boyd and Gordon, 1961). On replacing the subscript j by $j + 1$ in Eq. (7.4–1), then substituting for U_{j+1} from Eq. (7.4–3) and interchanging the orders of averaging and integration, we obtain the following relation between the cross-spectral densities W_j and W_{j+1}:

$$W_{j+1}(\boldsymbol{\rho}_1, \boldsymbol{\rho}_2, \nu) = \int_A\int_A L^*(\boldsymbol{\rho}_1, \boldsymbol{\rho}_1', \nu)L(\boldsymbol{\rho}_2, \boldsymbol{\rho}_2', \nu)W_j(\boldsymbol{\rho}_1', \boldsymbol{\rho}_2', \nu)\,\mathrm{d}^2\rho_1'\,\mathrm{d}^2\rho_2'. \tag{7.4–4}$$

It seems reasonable to assume that, after a sufficient number of transits between the two mirrors, a steady state is reached in the sense that $W_{j+1}(\boldsymbol{\rho}_1, \boldsymbol{\rho}_2, \nu)$ will be equal to $W_j(\boldsymbol{\rho}_1, \boldsymbol{\rho}_2, \nu)$, up to some proportionality factor, $\sigma(\nu)$. This factor represents the loss of light due to diffraction and absorption in one complete round-trip $(A \to B \to A)$. More explicitly, the steady state is characterized by the requirement that for sufficiently large values of j,

$$W_{j+1}(\boldsymbol{\rho}_1, \boldsymbol{\rho}_2, \nu) = \sigma(\nu)W_j(\boldsymbol{\rho}_1, \boldsymbol{\rho}_2, \nu). \tag{7.4–5}$$

Since the 'diagonal' values $W_j(\boldsymbol{\rho}, \boldsymbol{\rho}, \nu)$ and $W_{j+1}(\boldsymbol{\rho}, \boldsymbol{\rho}, \nu)$ represent spectral densities, they are necessarily real and positive and hence

$$\sigma(\nu) > 0. \tag{7.4–6}$$

On substituting for W_{j+1} from Eq. (7.4–5) into Eq. (7.4–4) and suppressing the suffix j we obtain the equation

$$\int_A\int_A W(\boldsymbol{\rho}_1', \boldsymbol{\rho}_2', \nu)L^*(\boldsymbol{\rho}_1, \boldsymbol{\rho}_1', \nu)L(\boldsymbol{\rho}_2, \boldsymbol{\rho}_2', \nu)\,\mathrm{d}^2\rho_1'\,\mathrm{d}^2\rho_2' = \sigma(\nu)W(\boldsymbol{\rho}_1, \boldsymbol{\rho}_2, \nu). \quad (7.4\text{–}7)$$

This equation is an integral equation for the boundary values of the modes $W(\boldsymbol{\rho}_1, \boldsymbol{\rho}_2, \nu)$ of the cross-spectral density on the mirror A that the resonator can sustain, i.e. it is an equation for the boundary values of the cross-spectral densities associated with the laser resonator modes.

We have considered in detail the light distribution on the mirror A but strictly similar results apply, of course, to light on the other mirror, B.

7.4.2 Nature of the solutions of the integral equation (7.4–7)

To elucidate the nature of the solutions of the integral equation (7.4–7) we expand the (non-Hermitian) propagator $L(\boldsymbol{\rho}_1, \boldsymbol{\rho}_2, \nu)$ into a bi-orthogonal series (Morse and Feshbach, 1953, pp. 884–886).‡

$$L(\boldsymbol{\rho}_1, \boldsymbol{\rho}_2, \nu) = \sum_n \alpha_n(\nu)\phi_n(\boldsymbol{\rho}_1, \nu)\chi_n^*(\boldsymbol{\rho}_2, \nu). \quad (7.4\text{–}8)$$

Here $\alpha_n(\nu)$ and $\phi_n(\boldsymbol{\rho}, \nu)$ are the eigenvalues and the eigenfunctions respectively of the Fredholm integral equation

$$\int_A L(\boldsymbol{\rho}_1, \boldsymbol{\rho}_2, \nu)\phi_n(\boldsymbol{\rho}_2, \nu)\,\mathrm{d}^2\rho_2 = \alpha_n(\nu)\phi_n(\boldsymbol{\rho}_1, \nu) \quad (7.4\text{–}9)$$

and $\beta_n(\nu)$ and $\chi_n(\boldsymbol{\rho}, \nu)$ are the eigenvalues and the eigenfunctions of an associated Fredholm integral equation whose kernel is adjoint to L, namely

$$\int_A L^*(\boldsymbol{\rho}_2, \boldsymbol{\rho}_1, \nu)\chi_n(\boldsymbol{\rho}_2, \nu)\,\mathrm{d}^2\rho_2 = \beta_n(\nu)\chi_n(\boldsymbol{\rho}_1, \nu). \quad (7.4\text{–}10)$$

The kernel $L(\boldsymbol{\rho}_1, \boldsymbol{\rho}_2, \nu)$ for laser resonators is defined on a finite domain A and is, for each ν, a continuous function in both the spatial variables $\boldsymbol{\rho}_1$ and $\boldsymbol{\rho}_2$. It is known that with such kernels the following theorems hold:§

(a) To each eigenvalue α_n of Eq. (7.4–9) there corresponds an eigenvalue of Eq. (7.4–10) and

$$\beta_n = \alpha_n^*. \quad (7.4\text{–}11)$$

Moreover, the ranks (degrees of degeneracy) of α_n and of β_n are the same.

(b) The corresponding eigenfunctions of the two equations are, with suitable normalization, orthonormal over the domain A, i.e.

$$\int_A \phi_n^*(\boldsymbol{\rho}, \nu)\chi_m(\boldsymbol{\rho}, \nu)\,\mathrm{d}^2\rho = \delta_{nm}, \quad (7.4\text{–}12)$$

where δ_{nm} is the Kronecker symbol.

‡ The series (7.4–8) should not be confused with the so-called Schmidt expansion, which is bi-orthogonal in a different sense, which has also been used in the theory of laser resonator modes.

§ For proofs of one-dimensional versions of these results see Smithies (1970), theorems 6.7.3 and 6.7.4.

Let us substitute from Eq. (7.4–8) into Eq. (7.4–7) and interchange the order of the integrations and summations. We then obtain the relation

$$\sum_n\sum_m \alpha_n^*(\nu)\alpha_m(\nu)w_{nm}(\nu)\phi_n^*(\boldsymbol{\rho}_1,\nu)\phi_m(\boldsymbol{\rho}_2,\nu) = \sigma(\nu)W(\boldsymbol{\rho}_1,\boldsymbol{\rho}_2,\nu), \quad (7.4\text{–}13)$$

where

$$w_{nm} = \int_A\int_A W(\boldsymbol{\rho}_1',\boldsymbol{\rho}_2',\nu)\chi_n(\boldsymbol{\rho}_1',\nu)\chi_m^*(\boldsymbol{\rho}_2',\nu)\,\mathrm{d}^2\rho_1'\,\mathrm{d}^2\rho_2'. \quad (7.4\text{–}14)$$

Next let us multiply both sides of Eq. (7.4–13) by $\chi_N(\boldsymbol{\rho}_1,\nu)\chi_M^*(\boldsymbol{\rho}_2,\nu)$, integrate with respect to $\boldsymbol{\rho}_1$ and $\boldsymbol{\rho}_2$ and make use of the bi-orthogonality relations (7.4–12) and of the expression (7.4–14) for w_{nm}. One then obtains the relation

$$\sum_n\sum_m \alpha_n^*(\nu)\alpha_m(\nu)w_{nm}(\nu)\delta_{nN}\delta_{mM} = \sigma(\nu)w_{NM}(\nu), \quad (7.4\text{–}15)$$

which implies that

$$[\sigma(\nu) - \alpha_N^*(\nu)\alpha_M(\nu)]w_{NM}(\nu) = 0 \quad \text{(no summation).} \quad (7.4\text{–}16)$$

Equation (7.4–16) implies that either $w_{NM}(\nu) = 0$ or that $\sigma(\nu) = \alpha_N^*(\nu)\alpha_M(\nu)$. The first case ($w_{NM} = 0$) is of no interest here because the corresponding term, with $n = N$, $m = M$, does not contribute to the double sum on the left-hand side of Eq. (7.4–13). The other solution implies that the eigenvalues of the integral equation (7.4–7) are

$$\sigma_{NM}(\nu) = \alpha_N^*(\nu)\alpha_M(\nu). \quad (7.4\text{–}17)$$

We must now distinguish two situations.

(a) $\sigma_{NM}(\nu)$ is non-degenerate

Suppose first that $\sigma_{NM}(\nu)$ is non-degenerate in the sense that there is no other pair $\alpha_{N'}(\nu)$ and $\alpha_{M'}(\nu)$ of eigenvalues of the integral equation (7.4–9) for which $\alpha_{N'}^*(\nu)\alpha_{M'}(\nu) = \alpha_N^*(\nu)\alpha_M(\nu)$, i.e. we assume that

$$\alpha_N^*\alpha_M \neq \alpha_{N'}^*\alpha_{M'} \quad \text{for all } \alpha_{N'} \neq \alpha_N,\ \alpha_{N'} \neq \alpha_M,\ \alpha_{M'} \neq \alpha_M,\ \alpha_{M'} \neq \alpha_N. \quad (7.4\text{–}18)$$

The formula (7.4–16) then implies that with a particular choice

$$\sigma_{kl}(\nu) = \alpha_k^*(\nu)\alpha_l(\nu) \quad (7.4\text{–}19)$$

of a non-degenerate eigenvalue of the basic integral equation (7.4–7), $w_{NM} = 0$ unless $k = N$ and $l = M$. The expansion (7.4–13) then reduces to the single term

$$W(\boldsymbol{\rho}_1,\boldsymbol{\rho}_2,\nu) = w_{kl}\phi_k^*(\boldsymbol{\rho}_1,\nu)\phi_l(\boldsymbol{\rho}_2,\nu). \quad (7.4\text{–}20)$$

Since the cross-spectral density is necessarily Hermitian [Eq. (4.3–43)] it follows at once from Eq. (7.4–20) that

$$w_{kl}\phi_k^*(\boldsymbol{\rho}_2,\nu)\phi_l(\boldsymbol{\rho}_1,\nu) = w_{kl}^*\phi_k(\boldsymbol{\rho}_1,\nu)\phi_l^*(\boldsymbol{\rho}_2,\nu), \quad (7.4\text{–}21)$$

i.e. that

$$\frac{\phi_l(\boldsymbol{\rho}_1, \nu)}{\phi_k(\boldsymbol{\rho}_1, \nu)} = \frac{w^*_{kl}}{w_{kl}} \frac{\phi^*_l(\boldsymbol{\rho}_2, \nu)}{\phi^*_k(\boldsymbol{\rho}_2, \nu)}. \tag{7.4–22}$$

For each frequency ν, the left-hand side of Eq. (7.4–22) is a function of $\boldsymbol{\rho}_1$ only, whereas the right-hand side is a function of $\boldsymbol{\rho}_2$ only. This is only possible if each side is independent of the spatial variables. If we denote each side of Eq. (7.4–22) by $\Omega_{kl}(\nu)$, it follows that

$$\phi_l(\boldsymbol{\rho}, \nu) = \Omega_{kl}(\nu)\phi_k(\boldsymbol{\rho}, \nu), \tag{7.4–23}$$

and using this relation, the expression (7.4–20) becomes

$$W(\boldsymbol{\rho}_1, \boldsymbol{\rho}_2, \nu) = w_{kl}\Omega_{kl}\phi^*_k(\boldsymbol{\rho}_1, \nu)\phi_k(\boldsymbol{\rho}_2, \nu). \tag{7.4–24}$$

Next we substitute from Eq. (7.4–24) into Eq. (7.4–14). We then obtain the following expression for w_{nm}:

$$w_{nm} = w_{kl}\Omega_{kl}\int_A \phi^*_k(\boldsymbol{\rho}'_1, \nu)\chi_n(\boldsymbol{\rho}'_1, \nu)\,\mathrm{d}^2\rho'_1\int_A \phi_k(\boldsymbol{\rho}'_2, \nu)\chi^*_m(\boldsymbol{\rho}'_2, \nu)\,\mathrm{d}^2\rho'_2. \tag{7.4–25}$$

If we make use of the bi-orthonormality relation (7.4–12), the expression (7.4–25) for w_{nm} reduces at once to

$$w_{nm} = w_{kl}\Omega_{kl}\delta_{kn}\delta_{km}, \tag{7.4–26}$$

which implies that

$$w_{nm} = 0 \quad \text{unless } n = m = k \tag{7.4–27}$$

and that

$$w_{kk} = w_{kl}\Omega_{kl}. \tag{7.4–28}$$

If we make use of the relation (7.4–28) in Eq. (7.4–24), we see that the solutions of the integral equation (7.4–7) are given by

$$W^{(k)}(\boldsymbol{\rho}_1, \boldsymbol{\rho}_2, \nu) = \lambda^{(k)}(\nu)\phi^*_k(\boldsymbol{\rho}_1, \nu)\phi_k(\boldsymbol{\rho}_2, \nu), \tag{7.4–29}$$

with

$$\lambda^{(k)}(\nu) = w_{kk}(\nu); \tag{7.4–30}$$

and if we make use of Eq. (7.4–19), we readily find that the corresponding eigenvalues are

$$\sigma^{(k)}(\nu) = \alpha^*_k(\nu)\alpha_k(\nu). \tag{7.4–31}$$

The proportionality factor $\lambda^{(k)}$ on the right-hand side of Eq. (7.4–29) depends on normalization. Let us normalize the ϕ_k so that

$$\int_A |\phi_k(\boldsymbol{\rho}, \nu)|^2\,\mathrm{d}^2\rho = 1. \tag{7.4–32}$$

The right-hand side of Eq. (7.4–29) may then be identified with the Mercer expansion [cf. Eq. (4.7–9)] of $W^{(k)}$, which now consists of a single term. This result implies that each solution of our integral equation is also a mode in the sense of the theory of coherent-mode representation of statistical wavefields which we discussed in Section 4.7. The physical significance of $\lambda^{(k)}$ can readily be understood from the following considerations. Since $W^{(k)}(\boldsymbol{\rho}, \boldsymbol{\rho}, \nu)$ represents the spectral density of the kth mode at the point $\boldsymbol{\rho}$ on the mirror A, we have, according to Eq. (7.4–29),

$$S^{(k)}(\boldsymbol{\rho}, \nu) \equiv W^{(k)}(\boldsymbol{\rho}, \boldsymbol{\rho}, \nu) = \lambda^{(k)}(\nu)|\phi_k(\boldsymbol{\rho}, \nu)|^2. \qquad (7.4\text{–}33)$$

On integrating both sides of Eq. (7.4–33) over the mirror A and on using Eq. (7.4–32) we obtain the relation

$$\int_A S^{(k)}(\boldsymbol{\rho}, \nu)\, \mathrm{d}^2\rho = \lambda^{(k)}(\nu). \qquad (7.4\text{–}34)$$

Hence $\lambda^{(k)}(\nu)$ represents the spectrum of the kth mode, integrated over the mirror A and is a measure of the rate at which energy at frequency ν propagates in steady state from the mirror A into the resonator cavity.

The integral equation (7.4–9) for the functions $\phi_n(\boldsymbol{\rho}, \nu)$ which enter the expression (7.4–29) is precisely the integral equation of the usual (monochromatic) theory of laser resonator modes of Fox and Li (1961) and Boyd and Gordon (1961), to which we referred at the beginning of this section. We may, therefore, call the functions $\phi_n(\boldsymbol{\rho}, \nu)$ the *Fox–Li modes* and we see from the preceding analysis that they have a broader significance than would appear from the manner in which they were originally introduced.

Since each solution (7.4–29) of the integral equation (7.4–7) for the steady-state distribution of the cross-spectral density on the mirror A factorizes with respect to the two spatial variables $\boldsymbol{\rho}_1$ and $\boldsymbol{\rho}_2$, its spectral degree of coherence at frequency ν [Eq. (4.3–47)], namely

$$\mu^{(k)}(\boldsymbol{\rho}_1, \boldsymbol{\rho}_2, \nu) = \frac{W^{(k)}(\boldsymbol{\rho}_1, \boldsymbol{\rho}_2, \nu)}{[W^{(k)}(\boldsymbol{\rho}_1, \boldsymbol{\rho}_1, \nu)]^{1/2}[W^{(k)}(\boldsymbol{\rho}_2, \boldsymbol{\rho}_2, \nu)]^{1/2}}, \qquad (7.4\text{–}35)$$

is necessarily unimodular. This result implies (see Section 4.5.3) that each solution (7.4–29) is the cross-spectral density of a field distribution that is spatially completely coherent at frequency ν over the whole surface of the mirror A. Thus we have shown that if there is no degeneracy, i.e. if the requirement (7.4–18) is satisfied, the integral equation (7.4–7) only admits solutions that represent light which is spatially completely coherent at each frequency ν.

According to the analysis of Section 4.5.3, the factorization of the cross-spectral density into a product of two functions, each of which depends only on one of the two spatial variables, is both a necessary and a sufficient condition for complete spatial coherence in the space-frequency domain. Hence if a laser operates in more than one transverse mode, the output cannot be spatially fully coherent across the mirror surface, a conclusion which is supported by experiment (Bertolotti, Daino, Gori and Sette, 1965).

Up to now the frequency ν was left somewhat arbitrary. It seems reasonable to assume that when a laser operates in the steady state, the integral equation (7.4–7) will be satisfied for every frequency component that is present in the laser output. It then follows on taking the Fourier transform of Eq. (7.4–29) that the mutual coherence function [Eq. (4.3–40a)] of a laser resonator mode is given by

$$\Gamma^{(k)}(\boldsymbol{\rho}_1, \boldsymbol{\rho}_2, \tau) = \int_0^\infty \lambda^{(k)}(\nu)\phi_k^*(\boldsymbol{\rho}_1, \nu)\phi_k(\boldsymbol{\rho}_2, \nu)\, \mathrm{e}^{-2\pi i\nu\tau}\, \mathrm{d}\nu. \qquad (7.4\text{–}36)$$

It should be noted that, unlike the complex degree of spatial coherence

$\mu^{(k)}(\boldsymbol{\rho}_1, \boldsymbol{\rho}_2, \nu)$, the complex degree of coherence in the space-time domain $\gamma^{(k)}(\boldsymbol{\rho}_1, \boldsymbol{\rho}_2, \tau)$ [Eq. (4.3–12a)] of each mode, namely

$$\gamma^{(k)}(\boldsymbol{\rho}_1, \boldsymbol{\rho}_2, \tau) = \frac{\Gamma^{(k)}(\boldsymbol{\rho}_1, \boldsymbol{\rho}_2, \tau)}{[\Gamma^{(k)}(\boldsymbol{\rho}_1, \boldsymbol{\rho}_1, 0)]^{1/2}[\Gamma^{(k)}(\boldsymbol{\rho}_2, \boldsymbol{\rho}_2, 0)]^{1/2}}, \tag{7.4–37}$$

is not unimodular for all values of its arguments. In fact $\gamma^{(k)}(\boldsymbol{\rho}_1, \boldsymbol{\rho}_2, \tau) \to 0$ as $|\tau| \to \infty$, because of ergodicity (cf. Section 2.2.2) of the optical field. However, under usual circumstances, when $|\tau|$ is much smaller than the coherence time of the light, $|\gamma^{(k)}|$ will generally differ from unity by negligible amounts for all positions of the two points, specified by $\boldsymbol{\rho}_1$ and $\boldsymbol{\rho}_2$, on the mirrors of the resonator.

Let us now briefly consider the situation that we have excluded so far, namely when:

(b) $\sigma_{NM}(\nu)$ *is degenerate*

In this case there will be eigenvalues $\alpha_{N'}$ and $\alpha_{M'}$ such that

$$\alpha_N^*\alpha_M = \alpha_{N'}^*\alpha_{M'}, \quad \alpha_{N'} \neq \alpha_N, \quad \alpha_{N'} \neq \alpha_M, \quad \alpha_{M'} \neq \alpha_M, \quad \alpha_{M'} \neq \alpha_N. \tag{7.4–38}$$

In place of Eq. (7.4–20) we now have

$$W(\boldsymbol{\rho}_1, \boldsymbol{\rho}_2, \nu) = \sum_k\sum_l c_{kl}(\nu)\phi_k^*(\boldsymbol{\rho}_1, \nu)\phi_l(\boldsymbol{\rho}_2, \nu), \tag{7.4–39}$$

where the c_{kl} are constants. These constants are arbitrary except for the constraint $c_{lk}(\nu) = c_{kl}^*(\nu)$, which must be satisfied because the cross-spectral density is necessarily Hermitian [Eq. (4.3–43)]. Let us diagonalize the matrix $\mathbf{C} = [c_{kl}(\nu)]$ by a unitary matrix $\mathbf{U} = [u_{kl}]$. Then $\mathbf{C} = \mathbf{U}^\dagger\mathbf{\Lambda}\mathbf{U}$, where $\mathbf{\Lambda} = [\Lambda_k]$ is a diagonal matrix. We also normalize the ϕ_k's in accordance with Eq. (7.4–32). The degenerate solution (7.4–39) of our basic integral equation (7.4–7) takes the form of its Mercer expansion

$$W(\boldsymbol{\rho}_1, \boldsymbol{\rho}_2, \nu) = \sum_k \Lambda_k(\nu)f_k^*(\boldsymbol{\rho}_1, \nu)f_k(\boldsymbol{\rho}_2, \nu), \tag{7.4–40}$$

where

$$f_k(\boldsymbol{\rho}, \nu) = \sum_l u_{kl}(\nu)\phi_l(\boldsymbol{\rho}, \nu). \tag{7.4–41}$$

Because the solution (7.4–40) is not factorized with respect to the spatial variables $\boldsymbol{\rho}_1$ and $\boldsymbol{\rho}_2$, the field across the mirror A is no longer spatially completely coherent.‡

From Eq. (7.4–40) we now obtain, for the spectral density of the degenerate mode at the point $\boldsymbol{\rho}$ of the mirror A, the following expression in place of Eq. (7.4–33):

$$S(\boldsymbol{\rho}, \nu) \equiv W(\boldsymbol{\rho}, \boldsymbol{\rho}, \nu) = \sum_k \Lambda_k(\nu)|f_k(\boldsymbol{\rho}, \nu)|^2; \tag{7.4–42}$$

‡ The spatial coherence properties of such degenerate solutions for an ideal plane-parallel Fabry–Perot resonator have been studied by Friberg and Turunen (1994).

and in place of Eq. (7.4–36) we have the following expression for its mutual coherence function:

$$\Gamma(\boldsymbol{\rho}_1, \boldsymbol{\rho}_2, \tau) = \sum_k \int_0^\infty \Lambda_k(\nu) f_k^*(\boldsymbol{\rho}_1, \nu) f_k(\boldsymbol{\rho}_2, \nu)\, \mathrm{e}^{-2\pi i\nu\tau}\, \mathrm{d}\nu. \tag{7.4–43}$$

7.5 Dielectric response and the spectrum of induced polarization in a fluctuating medium‡

As another application of second-order coherence theory we will now derive some formulas relating to the macroscopic response of a time-fluctuating medium (deterministic or random) to an electric field. In Section 7.6 we will make use of them in connection with the theory of scattering of electromagnetic waves.

7.5.1 Medium whose macroscopic properties do not change in time

Let us begin by recalling some standard results relating to an electromagnetic field in a linear, spatially non-dispersive medium§ whose macroscopic properties do not depend on time. At each point **r** in the medium (for simplicity **r** is not explicitly shown as an argument of the various quantities – e.g. **E**, **P**, η, N and α), the Fourier transforms

$$\tilde{\mathbf{E}}(\omega) = \frac{1}{2\pi}\int_{-\infty}^{\infty} \mathbf{E}(t)\, \mathrm{e}^{i\omega t}\, \mathrm{d}t, \tag{7.5–1a}$$

and

$$\tilde{\mathbf{P}}(\omega) = \frac{1}{2\pi}\int_{-\infty}^{\infty} \mathbf{P}(t)\, \mathrm{e}^{i\omega t}\, \mathrm{d}t, \tag{7.5–1b}$$

of the real electric field $\mathbf{E}(t)$ and of the induced real polarization $\mathbf{P}(t)$ are connected by a constitutive relation of the form

$$\tilde{\mathbf{P}}(\omega) = \tilde{\eta}(\omega)\tilde{\mathbf{E}}(\omega), \tag{7.5–2}$$

where $\tilde{\eta}(\omega)$ is the dielectric susceptibility (which, in general, is a tensor). A similar relation holds, of course, between the Fourier time-transform of the magnetic field and of the induced magnetization. We will confine our discussion to the response of the medium to the electric field only. For the sake of simplicity we will also restrict ourselves to an isotropic medium. The dielectric susceptibility can then be replaced by a scalar.

It is known from the microscopic theory that for many media $\tilde{\eta}(\omega)$ may be expressed in terms of the average number N of molecules per unit volume and the mean polarizability $\alpha(\omega)$ of each molecule by the formula

$$\tilde{\eta}(\omega) = N\alpha(\omega)\left[1 - \frac{4\pi}{3}N\alpha(\omega)\right]^{-1}. \tag{7.5–3}$$

‡ The analysis in this section follows largely a treatment due to Mandel and Wolf (1973).

§ A spatially non-dispersive medium is a medium whose response to an applied electromagnetic field is local, i.e. the induced polarization **P** and the induced magnetization **M** at a point **r** in the medium depend on the values of the electric field **E** and of the magnitude field **H** at that point only.

This is the well-known Lorentz–Lorenz relation, more commonly written in the alternative but equivalent form (Born and Wolf, 1980, Sec. 2.3.3)

$$\frac{4\pi}{3} N\alpha(\omega) = \frac{n^2(\omega) - 1}{n^2(\omega) + 2}, \tag{7.5–3a}$$

which involves the refractive index of the medium, $n(\omega)$, rather than the dielectric susceptibility $\tilde{\eta}(\omega)$. The equivalence follows from the relation $n^2(\omega) = 1 + 4\pi\tilde{\eta}(\omega)$. The frequency dependence of the polarizability $\alpha(\omega)$ may be calculated from the Lorentz oscillator model of the medium (Born and Wolf, 1980, Sec. 2.3.4).

Let us take the Fourier transform of Eq. (7.5–2) and make use of the fact that the response of the medium is necessarily causal. We then obtain the following constitutive relation valid in the time domain:

$$\mathbf{P}(t) = \frac{1}{2\pi}\int_{-\infty}^{t} \eta(t - t')\mathbf{E}(t')\,\mathrm{d}t' \tag{7.5–4a}$$

or, equivalently,

$$\mathbf{P}(t) = \frac{1}{2\pi}\int_{0}^{\infty} \eta(t')\mathbf{E}(t - t')\,\mathrm{d}t', \tag{7.5–4b}$$

where

$$\eta(\tau) = \int_{-\infty}^{\infty} \tilde{\eta}(\omega)\,\mathrm{e}^{-\mathrm{i}\omega\tau}\,\mathrm{d}\omega. \tag{7.5–5}$$

The fact that in Eq. (7.5–4a) η depends on t and t' only through the difference $t - t'$ is a reflection of the assumption that the macroscopic properties of the medium are independent of time. Since both $\mathbf{P}(t)$ and $\mathbf{E}(t)$ are real, $\eta(\tau)$ is necessarily also real and $\eta(\tau) = 0$ for $\tau < 0$ because of the assumption of causality.

7.5.2 Medium whose macroscopic properties depend on time in a deterministic manner

Suppose now that the macroscopic properties of the medium change in the course of time, but that the response remains linear and spatially non-dispersive. For the moment we will also assume that these changes are deterministic. This situation would arise, for example, if a vessel, filled with liquid at rest, was traversed by a monochromatic compression wave. Then in place of Eq. (7.5–4b) we would obtain the general linear causal relationship

$$\mathbf{P}(t) = \frac{1}{2\pi}\int_{0}^{\infty} \eta(t; t')\mathbf{E}(t - t')\,\mathrm{d}t', \tag{7.5–6}$$

in which the response function $\eta(t; t')$ depends on two arguments. Its dependence on the second argument, t', characterizes the response of the medium to a sufficiently short pulse of electromagnetic radiation.

To obtain some insight into the generalized susceptibility function $\eta(t;\, t')$, let us take its Fourier transform with respect to its second argument:

$$\hat{\eta}(t;\, \omega') = \frac{1}{2\pi}\int_0^{\infty} \eta(t;\, t')\, \mathrm{e}^{\mathrm{i}\omega' t'}\, \mathrm{d}t'. \tag{7.5–7}$$

Now when the response of the medium is time independent, the ordinary susceptibility $\tilde{\eta}(\omega)$ is related to the (time-independent) number density N by Eq. (7.5–3). When the medium fluctuates, N will become a function of time, $N(t)$ say, and if the fluctuations are not too rapid, we may expect that the right-hand side of Eq. (7.5–3) will represent the time-dependent generalization $\hat{\eta}(t;\, \omega')$ of $\tilde{\eta}(\omega)$, i.e. that

$$\hat{\eta}(t;\, \omega') = N(t)\alpha(\omega')\left[1 - \frac{4\pi}{3}N(t)\alpha(\omega')\right]^{-1}. \tag{7.5–8}$$

Although the precise range of validity of this equation can only be determined from microscopic considerations, these remarks indicate the physical origin of the generalized dielectric susceptibility which depends on two rather than one temporal argument.

As an illustration of the preceding results let us consider the situation when a monochromatic field of frequency ω_0 interacts with the medium. Then

$$\mathbf{E}(t) = \tfrac{1}{2}[\mathbf{a}\, \mathrm{e}^{-\mathrm{i}\omega_0 t} + \mathbf{a}^*\, \mathrm{e}^{\mathrm{i}\omega_0 t}], \tag{7.5–9}$$

where the vector $\mathbf{a}$ is independent of time. It then follows from Eqs. (7.5–9), (7.5–6) and (7.5–7) that the induced polarization $\mathbf{P}(t)$ is given by the expression

$$\mathbf{P}(t) = \tfrac{1}{2}[\hat{\eta}(t;\, \omega_0)\mathbf{a}\, \mathrm{e}^{-\mathrm{i}\omega_0 t} + \hat{\eta}^*(t;\, \omega_0)\mathbf{a}^*\, \mathrm{e}^{\mathrm{i}\omega_0 t}], \tag{7.5–10}$$

where the relation

$$\hat{\eta}(t;\, -\omega_0) = \hat{\eta}^*(t;\, \omega_0), \tag{7.5–11}$$

which is a consequence of the reality of $\eta(t;\, t')$, was used. From Eq. (7.5–10) it readily follows that the Fourier transform $\tilde{\mathbf{P}}(\omega)$ of the polarization $\mathbf{P}(t)$ is given by

$$\tilde{\mathbf{P}}(\omega) = \tfrac{1}{2}[\bar{\eta}(\omega - \omega_0;\, \omega_0)\mathbf{a} + \bar{\eta}^*(-\omega - \omega_0;\, \omega_0)\mathbf{a}^*], \tag{7.5–12}$$

where

$$\bar{\eta}(\omega;\, \omega_0) = \frac{1}{2\pi}\int_{-\infty}^{\infty} \hat{\eta}(t;\, \omega_0)\, \mathrm{e}^{\mathrm{i}\omega t}\, \mathrm{d}t, \tag{7.5–13}$$

or, with the help of Eq. (7.5–7),

$$\bar{\eta}(\omega;\, \omega_0) = \frac{1}{(2\pi)^2}\int_{-\infty}^{\infty} \mathrm{d}t\, \mathrm{e}^{\mathrm{i}\omega t}\int_0^{\infty} \mathrm{d}t'\eta(t;\, t')\, \mathrm{e}^{\mathrm{i}\omega_0 t'}. \tag{7.5–14}$$

Equation (7.5–12) shows that when the electric field is monochromatic and of frequency ω_0, the Fourier spectrum of the induced polarization consists of two separate contributions from the susceptibility spectrum, centered at frequencies $\omega = \pm\omega_0$.

Finally we note the following relation which follows from Eq. (7.5–13) and the relation (7.5–11):

$$\bar{\eta}(-\omega, -\omega_0) = \bar{\eta}^*(\omega, \omega_0). \tag{7.5–15}$$

7.5.3 *Medium whose macroscopic properties change randomly in time*

We will now consider the situation when both the incident electric field and the physical properties of the medium fluctuate and the fluctuations are not deterministic. The fluctuations in the electric field may arise, for example, from motion or collisions of the radiating atoms which generate the field. The fluctuations in the physical properties of the medium may be due to variations of temperature or pressure, or other causes. We will assume that $\mathbf{E}(t)$ and $\hat{\eta}(t;\omega')$ are random processes, which are stationary, at least in the wide sense. We also assume that these two processes are statistically independent, as will be the case, to a good approximation, whenever the effect of the electric field $\mathbf{E}(t)$ on the susceptibility fluctuations is sufficiently small. In most cases of practical interest the time scales of the variation of $\eta(t;\ t')$ with respect to t and t' are quite different. The t-variations, if associated with thermal fluctuations, are usually confined within a bandwidth of some MHz (often kHz) about zero frequency, whereas the t'-variations, associated with optical transitions between atomic states, are centered at frequencies in the range of about 10^{14}–10^{15} Hz and may have bandwidths of the order of 100 MHz. However, for microwave transitions and rapidly fluctuating media the two frequency ranges may approach each other or even overlap.

We will now examine how the spectral density tensors of the electric field and of the induced polarization are related. This can be done most simply as follows. We first represent the electric field vector in Eq. (7.5–6) as a (generalized) Fourier integral and make use of Eq. (7.5–7). We then obtain the following expression for the induced polarization:

$$\mathbf{P}(t) = \int_{-\infty}^{\infty} \hat{\eta}(t;\ \omega')\tilde{\mathbf{E}}(\omega')\, \mathrm{e}^{-\mathrm{i}\omega' t}\, \mathrm{d}\omega'. \tag{7.5–16}$$

It can be readily deduced from this formula or, more simply, from Eq. (7.5–6), that the random process $\mathbf{P}(t)$, just like $\hat{\eta}(t;\ \omega')$ and $\mathbf{E}(t)$, is also stationary, at least in the wide sense.

We note that if the incident field is quasi-monochromatic of center frequency ω_0, and if $\eta(t;\ \omega')$ varies slowly with ω' in the neighborhood of $\omega' = \omega_0$, Eq. (7.5–16) may be approximated by

$$\mathbf{P}(t) = \hat{\eta}(t;\ \omega_0)\mathbf{E}(t). \tag{7.5–17}$$

This relation is frequently used without taking into account the fact that $\hat{\eta}$ depends not only on time but also on frequency. This can be justified when the effective frequencies of the electric field are not too close to the resonance region of the medium. If this is not so, the dependence of $\hat{\eta}$ on frequency must be taken into account (cf. Wolf and Foley, 1989, Sec. II).

From Eq. (7.5–16) it follows at once that the correlation tensor of the induced polarization is given by the formula

$$\langle P_j(t)P_k(t+\tau)\rangle = \iint_{-\infty}^{\infty} \langle \hat{\eta}^*(t;\,\omega_1')\hat{\eta}(t+\tau;\,\omega_2')\rangle \times \langle \tilde{E}_j^*(\omega_1')\tilde{E}_k(\omega_2')\rangle\, e^{i\omega_1' t - i\omega_2'(t+\tau)}\, d\omega_1'\, d\omega_2', \qquad (7.5\text{–}18)$$

where the suffices $j,\ k,\ (j = 1, 2, 3,\ k = 1, 2, 3)$, label Cartesian components and the angular brackets denote ensemble averages. We have made use here of the assumed statistical independence of the random processes $\hat{\eta}(t;\,\omega)$ and $\mathbf{E}(t)$. We now recall the relation [Eq. (6.6–1a) in a slightly different notation]

$$\langle \tilde{E}_j^*(\omega_1')\tilde{E}_k(\omega_2')\rangle = W_{jk}^{(E)}(\omega_1')\delta(\omega_1' - \omega_2'), \qquad (7.5\text{–}19)$$

which defines the spectral density tensor, $W_{\mathrm{jk}}^{(E)}$, of the electric field. If we use this relation in Eq. (7.5–18), the expression for the correlation tensor of the induced polarization reduces to

$$\langle P_j(t)P_k(t+\tau)\rangle = \int_{-\infty}^{\infty} \langle \hat{\eta}^*(t;\,\omega')\hat{\eta}(t+\tau,\,\omega')\rangle W_{jk}^{(E)}(\omega')\, e^{-i\omega'\tau}\, d\omega', \qquad (7.5\text{–}20)$$

where we have omitted the suffix 1 on ω_1'.

Let us now introduce the spectral density $W^{(\hat{\eta})}(\omega;\,\omega')$ of $\hat{\eta}(t;\,\omega')$ and the spectral density tensor $W_{jk}^{(P)}(\omega)$ of $\mathbf{P}(t)$ via the Wiener–Khintchine theorem [Eqs. (2.4–15) and (2.4–16)]

$$W^{(\hat{\eta})}(\omega;\,\omega') = \frac{1}{2\pi}\int_{-\infty}^{\infty} \langle \hat{\eta}^*(t;\,\omega')\hat{\eta}(t+\tau;\,\omega')\rangle\, e^{i\omega\tau}\, d\tau, \qquad (7.5\text{–}21)$$

$$W_{jk}^{(P)}(\omega) = \frac{1}{2\pi}\int_{-\infty}^{\infty} \langle P_j(t)P_k(t+\tau)\rangle\, e^{i\omega\tau}\, d\tau. \qquad (7.5\text{–}22)$$

As a consequence of the condition (7.5–11) and using the fact that $\hat{\eta}(t;\,\omega')$ is a wide sense stationary random process, one can readily deduce from Eq. (7.5–21) that the spectral density $W^{(\hat{\eta})}$ satisfies the following relation which we will need shortly:

$$W^{(\hat{\eta})}(-\omega;\,-\omega') = W^{(\hat{\eta})}(\omega;\,\omega'). \qquad (7.5\text{–}23)$$

On taking the Fourier transform of Eq. (7.5–20) with respect to τ, interchanging the order of integrations, and making use of Eqs. (7.5–21) and (7.5–22) we obtain the following expression for the spectral density tensor of the induced polarization:

$$W_{jk}^{(P)}(\omega) = \int_{-\infty}^{\infty} W_{jk}^{(E)}(\omega')W^{(\hat{\eta})}(\omega-\omega';\,\omega')\, d\omega'. \qquad (7.5\text{–}24)$$

This formula shows that if $W^{(\hat{\eta})}$ were independent of the second argument, the spectral density of the induced polarization would be just the convolution of the spectral densities of $\mathbf{E}(t)$ and of $\hat{\eta}(t;\,\omega)$. However, because the second argument is present, which is a consequence of the dispersive properties of the medium, the integral in Eq.(7.5–24) is more complicated than a convolution.

The tensorial properties of the spectral density tensor $W_{jk}^{(P)}$ are sometimes unimportant. For example, if the incident light is fully polarized, then the spectral properties are adequately represented by scalar functions and the tensorial indices in Eq. (7.5–24) can be suppressed. If, in addition, $W^{(E)}(\omega)$ is peaked at frequencies $\pm\omega_0$ and is sufficiently narrow compared with the effective width of the spectral density function $W^{(\tilde{\eta})}(\omega, \omega')$ in both ω and ω', one may often approximate $W^{(E)}(\omega')$ under the integral sign in Eq. (7.5–24) by the expression

$$W^{(E)}(\omega') = \langle I \rangle[\delta(\omega' - \omega_0) + \delta(\omega' + \omega_0)], \tag{7.5–25}$$

where $\delta(\omega)$ is the Dirac delta function and $\langle I \rangle$ denotes the expectation value (the average) of the intensity of the electric field. On substituting from Eq. (7.5–25) into Eq. (7.5–24) and making use of the relation (7.5–23), we obtain the following simple expression for the spectral density of the induced polarization in terms of the spectral density of $\tilde{\eta}$:

$$W^{(P)}(\omega) = \langle I \rangle[W^{(\tilde{\eta})}(\omega - \omega_0; \omega_0) + W^{(\tilde{\eta})}(-\omega - \omega_0; \omega_0)]. \tag{7.5–26}$$

This formula shows that the spectral density of the induced polarization **P** reflects in a simple manner the spectral density of the fluctuations of $\tilde{\eta}(t; \omega_0)$. In particular, if for a given frequency ω_0, $W^{(\tilde{\eta})}(\omega, \omega_0)$ has the form of narrow lines centered at frequencies $\omega = \pm\omega_1$ $(\omega_1 < \omega_0)$ as, for example, is often the case for acoustical waves in solids or liquids, the spectrum $W^{(P)}(\omega)$ of the induced polarization will, according to Eq. (7.5–26), consist of doublets centered at frequencies $\pm\omega_0$, with the two lines in each doublet being separated by $2\omega_1$. The well-known Brillouin doublets in the spectrum of light scattered from such media are a reflection of this result, as will be shown explicitly in Section 7.6.5.

Finally, we stress that the formula (7.5–26) does not apply to all situations, even with laser light. Most lasers generate light whose bandwidths are of the order of hundreds of kHz, or even larger, and if this bandwidth is comparable to or greater than the bandwidth of the fluctuation spectrum of $\tilde{\eta}$, the approximate formula (7.5–26) then no longer applies (cf. Mandel, 1969). The more general expression (7.5–24) is relevant to many scattering experiments. We will now turn our attention to the theory underlying such experiments.

7.6 Scattering from random media

With the extensive use of laser light in scattering experiments, a good deal of information has been obtained about thermal fluctuations in gases and liquids, about lattice vibrations in solids, about phase transitions, etc.‡ In this section we will derive some of the main formulas which elucidate the nature of the scattered field without, however, restricting ourselves to narrow-band light, and we will illustrate the results by a few examples. We begin by deriving some of the more important formulas relevant to scattering from deterministic media.

7.6.1 Basic equations for deterministic scattering

Let $\mathbf{E}^{(i)}(\mathbf{r}, t)$ and $\mathbf{H}^{(i)}(\mathbf{r}, t)$ denote the real electric and magnetic field vector, respectively, of a deterministic electromagnetic field incident on a deterministic

‡ For a review of some of this research see, for example, Cummins (1969), Cummins and Swinney (1970), Chu (1974), Crosignani, Di Porto and Bertolotti (1975) and Berne and Pecora (1976).

medium which occupies a finite volume V in free space. As usual $\mathbf{r}$ denotes the position vector of a point in space and t denotes the time. As a consequence of the interaction of the incident field with the scatterer, a new field $\mathbf{E}(\mathbf{r}, t)$, $\mathbf{H}(\mathbf{r}, t)$ will be generated. Let $\mathbf{P}(\mathbf{r}, t)$ and $\mathbf{M}(\mathbf{r}, t)$ be the polarization and magnetization, respectively, in the scattering medium, induced by the incident wave. These two vector fields will be some functionals of $\mathbf{E}$ and $\mathbf{H}$. We will assume, as is usually the case, that $\mathbf{P}$ depends only on $\mathbf{E}$ and $\mathbf{M}$ depends only on $\mathbf{H}$:

$$\mathbf{P} = f(\mathbf{E}), \quad \mathbf{M} = g(\mathbf{H}). \tag{7.6–1}$$

The exact form of these *constitutive relations* depends, of course, on the nature of the scattering medium.

Let

$$\mathbf{D} = \mathbf{E} + 4\pi\mathbf{P}, \quad \mathbf{B} = \mathbf{H} + 4\pi\mathbf{M} \tag{7.6–2}$$

be the electric displacement vector and the magnetic induction vector, respectively.

We will assume that the scattering medium is non-conducting and is free of external charges and currents. If we eliminate $\mathbf{D}$ and $\mathbf{H}$ from Maxwell's equations by the use of Eq. (7.6–2), we obtain the following set of equations (in the Gaussian system of units):

$$\boldsymbol{\nabla} \times \mathbf{B} - \frac{1}{c}\dot{\mathbf{E}} = \frac{4\pi}{c}[\dot{\mathbf{P}} + c\boldsymbol{\nabla} \times \mathbf{M}], \tag{7.6–3a}$$

$$\boldsymbol{\nabla} \times \mathbf{E} + \frac{1}{c}\dot{\mathbf{B}} = 0, \tag{7.6–3b}$$

$$\boldsymbol{\nabla} \cdot \mathbf{E} = -4\pi\boldsymbol{\nabla} \cdot \mathbf{P}, \tag{7.6–3c}$$

$$\boldsymbol{\nabla} \cdot \mathbf{B} = 0, \tag{7.6–3d}$$

the dot denoting differentiation with respect to time. The particular solutions of these equations which represent the electromagnetic field obtained by scattering of the incident field $\mathbf{E}^{(i)}$, $\mathbf{H}^{(i)}$ by the medium are given by

$$\mathbf{E}(\mathbf{r}, t) = \mathbf{E}^{(i)}(\mathbf{r}, t) + \mathbf{E}^{(s)}(\mathbf{r}, t), \tag{7.6–4a}$$

$$\mathbf{B}(\mathbf{r}, t) = \mathbf{B}^{(i)}(\mathbf{r}, t) + \mathbf{B}^{(s)}(\mathbf{r}, t), \tag{7.6–4b}$$

where the field vectors $\mathbf{E}^{(s)}$, $\mathbf{B}^{(s)}$ of the scattered wave, which behave as outgoing waves at infinity, satisfy the coupled equations (Born and Wolf, 1980, Sec. 2.2.2)

$$\mathbf{E}^{(s)}(\mathbf{r}, t) = -\frac{1}{c}\boldsymbol{\nabla} \times \dot{\boldsymbol{\Pi}}_{\mathrm{m}}(\mathbf{r}, t) + \boldsymbol{\nabla} \times [\boldsymbol{\nabla} \times \boldsymbol{\Pi}_{\mathrm{e}}(\mathbf{r}, t)] - 4\pi\mathbf{P}(\mathbf{r}, t), \tag{7.6–5a}$$

$$\mathbf{B}^{(s)}(\mathbf{r}, t) = \frac{1}{c}\boldsymbol{\nabla} \times \dot{\boldsymbol{\Pi}}_{\mathrm{e}}(\mathbf{r}, t) + \boldsymbol{\nabla} \times [\boldsymbol{\nabla} \times \boldsymbol{\Pi}_{\mathrm{m}}(\mathbf{r}, t)]. \tag{7.6–5b}$$

In these equations $\boldsymbol{\Pi}_{\mathrm{e}}$ and $\boldsymbol{\Pi}_{\mathrm{m}}$ are the electric and magnetic Hertz vectors, respectively, given by the formulas

$$\boldsymbol{\Pi}_{\mathrm{e}}(\mathbf{r}, t) = \int_V \frac{\mathbf{P}(\mathbf{r}', t - |\mathbf{r} - \mathbf{r}'|/c)}{|\mathbf{r} - \mathbf{r}'|}\,\mathrm{d}^3r', \tag{7.6–6a}$$

$$\boldsymbol{\Pi}_{\mathrm{m}}(\mathbf{r}, t) = \int_V \frac{\mathbf{M}(\mathbf{r}', t - |\mathbf{r} - \mathbf{r}'|/c)}{|\mathbf{r} - \mathbf{r}'|} \, \mathrm{d}^3 r'. \tag{7.6–6b}$$

7.6.2 *Scattering from deterministic media in the first-order Born approximation*

If the medium is linear, isotropic and non-magnetic the constitutive relations are of the form [cf. Eq. (7.5–6)]

$$\mathbf{P}(\mathbf{r}, t) = \frac{1}{2\pi} \int_0^\infty \eta(\mathbf{r}, t; t') \mathbf{E}(\mathbf{r}, t - t') \, \mathrm{d}t', \tag{7.6–7a}$$

$$\mathbf{M}(\mathbf{r}, t) = 0, \tag{7.6–7b}$$

where $\eta(\mathbf{r}; t; t')$ is the generalized dielectric susceptibility, introduced in Section 7.5.2.

We will again express the total field $(\mathbf{E}, \mathbf{B})$ as the sum of the incident field $(\mathbf{E}^{(\mathrm{i})}, \mathbf{B}^{(\mathrm{i})})$ and the scattered field $(\mathbf{E}^{(\mathrm{s})}, \mathbf{B}^{(\mathrm{s})})$. The expression (7.6–7a) for the induced polarization then becomes

$$\mathbf{P}(\mathbf{r}, t) = \mathbf{P}_1(\mathbf{r}, t) + \mathbf{P}_2(\mathbf{r}, t), \tag{7.6–8}$$

where

$$\mathbf{P}_1(\mathbf{r}, t) = \frac{1}{2\pi} \int_0^\infty \eta(\mathbf{r}, t; t') \mathbf{E}^{(\mathrm{i})}(\mathbf{r}, t - t') \, \mathrm{d}t', \tag{7.6–9a}$$

$$\mathbf{P}_2(\mathbf{r}, t) = \frac{1}{2\pi} \int_0^\infty \eta(\mathbf{r}, t; t') \mathbf{E}^{(\mathrm{s})}(\mathbf{r}, t - t') \, \mathrm{d}t'. \tag{7.6–9b}$$

Suppose that the scattering medium is weak in the sense that, for all values of its arguments, the polarization induced by the scattered field is much smaller than the polarization induced by the incident field, i.e. that

$$|\mathbf{P}_2(\mathbf{r}, t)| \ll |\mathbf{P}_1(\mathbf{r}, t)|, \tag{7.6–10a}$$

and also that

$$\left| \frac{\partial \mathbf{P}_2(\mathbf{r}, t)}{\partial t} \right| \ll \left| \frac{\partial \mathbf{P}_1(\mathbf{r}, t)}{\partial t} \right| \tag{7.6–10b}$$

and

$$|\boldsymbol{\nabla} \cdot \mathbf{P}_2(\mathbf{r}, t)| \ll |\boldsymbol{\nabla} \cdot \mathbf{P}_1(\mathbf{r}, t)|. \tag{7.6–10c}$$

Under these circumstances the basic equations (7.6–5a) and (7.6–5b) reduce to

$$\mathbf{E}^{(\mathrm{s})}(\mathbf{r}, t) = \boldsymbol{\nabla} \times [\boldsymbol{\nabla} \times \boldsymbol{\Pi}(\mathbf{r}, t)] - 4\pi \mathbf{P}_1(\mathbf{r}, t), \tag{7.6–11a}$$

$$\mathbf{B}^{(\mathrm{s})}(\mathbf{r}, t) = \frac{1}{c} \boldsymbol{\nabla} \times [\dot{\boldsymbol{\Pi}}(\mathbf{r}, t)], \tag{7.6–11b}$$

where

$$\boldsymbol{\Pi}(\mathbf{r}, t) = \int_V \frac{\mathbf{P}_1(\mathbf{r}', t - |\mathbf{r} - \mathbf{r}'|/c)}{|\mathbf{r} - \mathbf{r}'|} \, \mathrm{d}^3 r'. \tag{7.6–12}$$

Only the electric Hertz vector appears in Eqs. (7.6–11), because $\boldsymbol{\Pi}_m(\mathbf{r}, t) \equiv 0$, since the medium was assumed to be non-magnetic. For this reason we have now dropped the subscript 'e' on $\boldsymbol{\Pi}_e$.

We note that Eqs. (7.6–11) imply that

$$\nabla \cdot \mathbf{E}^{(s)} = -4\pi \nabla \cdot \mathbf{P}_1 \tag{7.6–11c}$$

and

$$\nabla \cdot \mathbf{B}^{(s)} = 0, \tag{7.6–11d}$$

Equations (7.6–11), together with Eq. (7.6–12), describe the scattered electromagnetic field with the same kind of accuracy as is implied by the use of the first-order Born approximation in quantum collision theory. If the generalized dielectric susceptibility function $\eta(\mathbf{r}, t; t')$ of the scattering medium is known, $\mathbf{P}_1(\mathbf{r}, t)$ may be calculated from Eq. (7.6–9a) and the Hertz vector $\boldsymbol{\Pi}(\mathbf{r}, t)$ may be determined by the use of Eq. (7.6–12). The scattered field at all points outside the medium may then be obtained from Eqs. (7.6–11). For purposes of later analysis it will be useful to employ the Fourier transforms of the field vectors rather than with the field vectors themselves. We will use the same definition of Fourier transforms as in Eq. (7.5–1), namely

$$\tilde{\mathbf{E}}^{(s)}(\mathbf{r}, \omega) = \frac{1}{2\pi}\int_{-\infty}^{\infty} \mathbf{E}^{(s)}(\mathbf{r}, t)\, e^{i\omega t}\, dt, \tag{7.6–13}$$

$$\tilde{\mathbf{P}}_1(\mathbf{r}, \omega) = \frac{1}{2\pi}\int_{-\infty}^{\infty} \mathbf{P}_1(\mathbf{r}, t)\, e^{i\omega t}\, dt, \tag{7.6–14}$$

etc. If we substitute for $\mathbf{P}_1(\mathbf{r}, t)$ under the integral sign in Eq. (7.6–14) the expression (7.6–9a), we find after a straightforward calculation that

$$\tilde{\mathbf{P}}_1(\mathbf{r}, \omega) = \int_{-\infty}^{\infty} \bar{\eta}(\mathbf{r}, \omega - \omega'; \omega')\tilde{\mathbf{E}}^{(i)}(\mathbf{r}, \omega')\, d\omega', \tag{7.6–15}$$

where

$$\bar{\eta}(\mathbf{r}, \omega; \omega') = \frac{1}{2\pi}\int_{-\infty}^{\infty} \hat{\eta}(\mathbf{r}, t; \omega')\, e^{i\omega t}\, dt \tag{7.6–16a}$$

$$= \frac{1}{(2\pi)^2}\int_{-\infty}^{\infty} dt\, e^{i\omega t}\int_0^{\infty} dt'\, \eta(\mathbf{r}, t; t')\, e^{i\omega' t'}. \tag{7.6–16b}$$

The quantities $\bar{\eta}(\mathbf{r}, \omega; \omega')$ and $\hat{\eta}(\mathbf{r}, t; \omega')$ are, of course, generalizations of the quantities $\bar{\eta}(\omega; \omega')$ and $\hat{\eta}(t, \omega')$ that we encountered earlier [Eqs. (7.5–13) and (7.5–14)].

On taking the Fourier transform of Eqs. (7.6–11) we readily obtain the following expressions for the frequency components of the scattered field:

$$\tilde{\mathbf{E}}^{(s)}(\mathbf{r}, \omega) = \nabla \times [\nabla \times \tilde{\boldsymbol{\Pi}}(\mathbf{r}, \omega)] - 4\pi\tilde{\mathbf{P}}_1(\mathbf{r}, \omega), \tag{7.6–17a}$$

$$\tilde{\mathbf{B}}^{(s)}(\mathbf{r}, \omega) = -ik\nabla \times \tilde{\boldsymbol{\Pi}}(\mathbf{r}, \omega), \tag{7.6–17b}$$

where

$$k = \omega/c \tag{7.6–18}$$

is the wave number associated with frequency ω and

$$\tilde{\boldsymbol{\Pi}}(\mathbf{r}, \omega) = \int_V \tilde{\mathbf{P}}_1(\mathbf{r}', \omega)\frac{e^{ik|\mathbf{r}-\mathbf{r}'|}}{|\mathbf{r}-\mathbf{r}'|}\,d^3r'. \tag{7.6–19}$$

We will now consider the scattered field at points $\mathbf{P}(\mathbf{r})$ in the far zone of the scattering medium. Let us set

$$\mathbf{r} = r\mathbf{s}, \tag{7.6–20}$$

where the origin of the position vector $\mathbf{r}$ is taken at some point in the scattering volume and $\mathbf{s}$ is a unit vector. When r is large enough we evidently have (see Fig. 7.7)

$$|\mathbf{r}-\mathbf{r}'| \sim r - \mathbf{s}\cdot\mathbf{r}' \tag{7.6–21}$$

($r = |\mathbf{r}|$) and hence

$$\frac{e^{ik|\mathbf{r}-\mathbf{r}'|}}{|\mathbf{r}-\mathbf{r}'|} \sim \frac{e^{ikr}}{r}e^{-ik\mathbf{s}\cdot\mathbf{r}'}, \quad (kr\to\infty, \mathbf{s} \text{ fixed}). \tag{7.6–22}$$

Making use of the asymptotic approximation (7.6–22) in Eq. (7.6–19) we obtain the following expression for the Hertz vector $\tilde{\boldsymbol{\Pi}}$:

$$\tilde{\boldsymbol{\Pi}}(r\mathbf{s}, \omega) = (2\pi)^3\mathcal{P}_1(k\mathbf{s}, \omega)\frac{e^{ikr}}{r}, \tag{7.6–23}$$

where

$$\mathcal{P}_1(\mathbf{K}, \omega) = \frac{1}{(2\pi)^3}\int_V \tilde{\mathbf{P}}_1(\mathbf{r}', \omega)\,e^{-i\mathbf{K}\cdot\mathbf{r}'}\,d^3r' \tag{7.6–24}$$

is the three-dimensional spatial Fourier transform of $\tilde{\mathbf{P}}_1(\mathbf{r}, \omega)$. Since $\tilde{\mathbf{P}}_1(\mathbf{r}, \omega)$ is the one-dimensional temporal Fourier transform of $\mathbf{P}_1(\mathbf{r}, t)$, we may express $\mathcal{P}_1(\mathbf{K}, \omega)$ directly in terms of $\mathbf{P}_1(\mathbf{r}, t)$ in the form

$$\mathcal{P}_1(k\mathbf{s}, \omega) = \frac{1}{(2\pi)^4}\int_V\int_{-\infty}^{\infty}\mathbf{P}_1(\mathbf{r}', t')\,e^{i(\omega t' - k\mathbf{s}\cdot\mathbf{r}')}\,d^3r'\,dt'. \tag{7.6–25}$$

Alternatively we may express $\mathcal{P}_1(k\mathbf{s}, \omega)$ in the form

$$\mathcal{P}_1(k\mathbf{s}, \omega) = \frac{1}{(2\pi)^3}\int_V d^3r'\,e^{-ik\mathbf{s}\cdot\mathbf{r}'}\int_{-\infty}^{\infty}d\omega'\,\tilde{\eta}(\mathbf{r}', \omega-\omega'; \omega')\tilde{\mathbf{E}}^{(i)}(\mathbf{r}', \omega'), \tag{7.6–26}$$

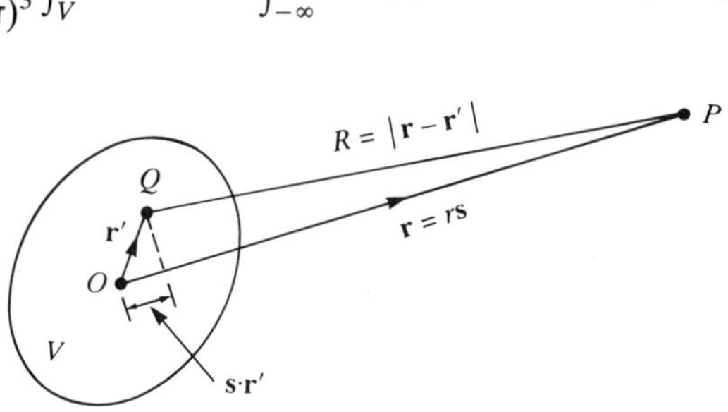

Fig. 7.7 Illustrating the notation relating to the far-zone approximation (7.6–21). In the figure Q is a typical point in the source domain V and P is a point in the far zone.

obtained at once on substituting from Eq. (7.6–15) into Eq. (7.6–24) and setting $\mathbf{K} = k\mathbf{s}$.

Finally, on substituting from Eq. (7.6–23) into Eqs. (7.6–17), noting that $\tilde{\mathbf{P}}_1(\mathbf{r}, \omega) = 0$ when $\mathbf{r}$ represents a point outside the scattering volume, one can show, with the help of vector identities, dyadic Green's functions (Papas, 1965, Sec. 2.2), or by using the angular spectrum representation of wavefields [Friberg and Wolf, 1983, Eq. (4.5)], that at points in the far zone,

$$\tilde{\mathbf{E}}^{(\mathrm{s})}(r\mathbf{s}, \omega) \sim -(2\pi)^3 k^2 \mathbf{s} \times [\mathbf{s} \times \boldsymbol{\mathscr{P}}_1(k\mathbf{s}, \omega)] \frac{e^{ikr}}{r}, \qquad (7.6\text{–}27a)$$

$$\tilde{\mathbf{B}}^{(\mathrm{s})}(r\mathbf{s}, \omega) \sim (2\pi)^3 k^2 \mathbf{s} \times \boldsymbol{\mathscr{P}}_1(k\mathbf{s}, \omega) \frac{e^{ikr}}{r}. \qquad (7.6\text{–}27b)$$

7.6.3 Scattering from random media in the first-order Born approximation‡

We will now consider the more difficult problem of scattering from random media. Because of the complexity of the problem, we will make the following simplifying assumptions about the scattering medium:

(a) The scattering medium is linear, isotropic, non-magnetic, statistically homogeneous, and its temporal fluctuations are stationary, at least in the wide sense, and are approximately of zero mean. The assumptions of homogeneity§ and stationarity imply that the correlation function of the dielectric susceptibility $\hat{\eta}(\mathbf{r}, t; \omega')$, at pairs of points $\mathbf{r}_1$ and $\mathbf{r}_2$ in the medium and at times t_1 and t_2, depends only on the differences $\mathbf{r}_2 - \mathbf{r}_1$ and $t_2 - t_1$:

$$\langle \hat{\eta}^*(\mathbf{r}_1, t_1; \omega')\hat{\eta}(\mathbf{r}_2, t_2; \omega')\rangle = G(\mathbf{r}_2 - \mathbf{r}_1, t_2 - t_1; \omega'). \qquad (7.6\text{–}28)$$

The angular brackets denote here an average over an ensemble of realizations of the random medium.

(b) The spatial extent of the susceptibility correlations is small compared to the size of the scattering volume V, i.e. the separations $R = |\mathbf{R}| = |\mathbf{r}_2 - \mathbf{r}_1|$ for which $|G(\mathbf{R}, T; \omega)|$ has appreciable values are much smaller than the linear dimensions of V.

(c) The scattering is sufficiently weak so that it may be treated within the framework of the first-order Born approximation.

As regards the incident (real) electric field, we will assume that it is statistically homogeneous and stationary, at least in the wide sense. Its correlation tensor will then also depend on the spatial and temporal variables only through the differences $\mathbf{r}_2 - \mathbf{r}_1$ and $t_2 - t_1$, i.e. it will have the form

$$\langle [E_l^{(\mathrm{i})}(\mathbf{r}_1, t_1)]^* E_m^{(\mathrm{i})}(\mathbf{r}_2, t_2)\rangle = \mathscr{E}_{lm}^{(\mathrm{i})}(\mathbf{r}_2 - \mathbf{r}_1, t_2 - t_1), \qquad (7.6\text{–}29)$$

‡ The analysis of Sections 7.6.3 and 7.6.4 follows closely that given by Wolf and Foley (1989).

§ Strict homogeneity demands that the medium occupies all space. However, when the linear dimensions of the scattering volume are large compared both with the scale of the correlation of the dielectric susceptibility [the effective $\mathbf{r}'$-range of $G(\mathbf{r}', T; \omega')$] and with the wavelength $\lambda' = 2\pi c/\omega'$, the assumption of an infinite medium does not introduce any appreciable errors and leads to a much simpler analysis. Similar remarks apply to the assumption of stationarity.

where l and m label Cartesian components. The angular brackets on the left denote the average taken over the ensemble that characterizes the statistical properties of the incident field.

In the subsequent analysis we will also need the cross-spectral density tensor $W_{lm}^{(i)}(\mathbf{r}_2 - \mathbf{r}_1, \omega)$ of the incident electric field. It may be defined formally by the equation [cf. Eq. (6.6–1a)]

$$\langle [\widetilde{E}_l^{(i)}(\mathbf{r}_1, \omega)]^* \widetilde{E}_m^{(i)}(\mathbf{r}_2, \omega') \rangle = W_{lm}^{(i)}(\mathbf{r}_2 - \mathbf{r}_1, \omega)\delta(\omega - \omega'), \qquad (7.6\text{–}30)$$

where $\widetilde{E}_l^{(i)}$ and $\widetilde{E}_m^{(i)}$ are the (generalized) Fourier transforms of $E_l^{(i)}$ and $E_m^{(i)}$, respectively, and δ is the Dirac delta function. According to the generalized Wiener–Khintchine theorem [Eqs. (2.4–37) and (2.4–38)], the electric cross-spectral density tensor is the Fourier transform of the electric correlation tensor, i.e.

$$W_{lm}^{(i)}(\mathbf{R}, \omega) = \frac{1}{2\pi}\int_{-\infty}^{\infty} \mathscr{E}_{lm}^{(i)}(\mathbf{R}, T)\, \mathrm{e}^{\mathrm{i}\omega T}\, \mathrm{d}T. \qquad (7.6\text{–}31)$$

It is to be noted that the expectation values in Eqs. (7.6–28) and (7.6–29) are taken over two different statistical ensembles. In Eq. (7.6–28) it is taken over the ensemble that characterizes the fluctuations of the dielectric susceptibility, whereas in Eq. (7.6–29) it is taken over the ensemble that characterizes the fluctuations of the incident field. We will assume that these two kinds of fluctuations are statistically independent. This assumption is reasonable if the incident field is not too strong.

We will now derive expressions for the angular and spectral distributions of the energy of the scattered field in the far zone. For this purpose we first note that because both $\widetilde{\eta}$ and $\widetilde{\mathbf{E}}^{(i)}$ are random processes, so is the quantity $\boldsymbol{\mathcal{P}}_1(\mathbf{K}, \omega)$, given by Eq. (7.6–26). Hence the (generalized) Fourier transforms $\widetilde{\mathbf{E}}^{(s)}$ and $\widetilde{\mathbf{B}}^{(s)}$ of the scattered field in the far zone, given by Eqs. (7.6–27), are also random processes. For each realization we have from Eq. (7.6–27a)

$$[\widetilde{\mathbf{E}}^{(s)}(r\mathbf{s}, \omega)]^* \cdot \widetilde{\mathbf{E}}^{(s)}(r\mathbf{s}, \omega') =$$

$$(2\pi)^6 \frac{k^2 k'^2}{r^2} \mathrm{e}^{\mathrm{i}(k'-k)r} \{\mathbf{s} \times [\mathbf{s} \times \boldsymbol{\mathcal{P}}_1^*(\mathbf{k}, \omega)]\} \cdot \{\mathbf{s} \times [\mathbf{s} \times \boldsymbol{\mathcal{P}}_1(\mathbf{k}', \omega')]\} \quad (7.6\text{–}32)$$

where

$$\mathbf{k} = k\mathbf{s} = \frac{\omega}{c}\mathbf{s}, \qquad \mathbf{k}' = k'\mathbf{s} = \frac{\omega'}{c}\mathbf{s}. \qquad (7.6\text{–}33)$$

The right-hand side of Eq. (7.6–32) may be simplified by the use of the vector identity $(\mathbf{A} \times \mathbf{B}) \cdot (\mathbf{C} \times \mathbf{D}) = (\mathbf{A} \cdot \mathbf{C})(\mathbf{B} \cdot \mathbf{D}) - (\mathbf{A} \cdot \mathbf{D})(\mathbf{B} \cdot \mathbf{C})$, and one then finds that (cf. Wolf and Foley, 1989, p. 579),

$$[\widetilde{\mathbf{E}}^{(s)}(r\mathbf{s}, \omega)]^* \cdot \widetilde{\mathbf{E}}^{(s)}(r\mathbf{s}, \omega') = (2\pi)^6 \frac{k^2 k'^2}{r^2} \mathrm{e}^{\mathrm{i}(k-k')r} (\delta_{lm} - s_l s_m) \mathcal{P}_{1l}^*(\mathbf{k}, \omega) \mathcal{P}_{1m}(\mathbf{k}', \omega'), \qquad (7.6\text{–}34)$$

where s_l and s_m are Cartesian components of the unit vector $\mathbf{s}$, δ_{lm} denotes the Kronecker symbol and the summation is taken over repeated suffices.

Next we take the average of Eq. (7.6–34) over the ensembles of the incident field and the fluctuating medium. Denoting this double average by double angular brackets, we have at once from Eq. (7.6–34) that

$$\langle\langle[\tilde{\mathbf{E}}^{(s)}(r\mathbf{s}, \omega)]^* \cdot \tilde{\mathbf{E}}^{(s)}(r\mathbf{s}, \omega')\rangle\rangle = (2\pi)^6 \frac{k^2 k'^2}{r^2} e^{i(k'-k)r}(\delta_{lm} - s_l s_m)\langle\langle \mathscr{P}^*_{1l}(\mathbf{k}, \omega)\mathscr{P}_{1m}(\mathbf{k}', \omega')\rangle\rangle. \quad (7.6\text{–}35)$$

By a rather long but straightforward calculation, carried out in Appendix 7.1 one finds that the double average which appears on the right-hand side of Eq. (7.6–35) may be expressed, to a good approximation, in the form

$$\langle\langle \mathscr{P}^*_{1l}(\mathbf{k}, \omega)\mathscr{P}_{1m}(\mathbf{k}', \omega')\rangle\rangle \approx \frac{V\delta(\omega - \omega')}{(2\pi)^6}\int_V d^3 R\, e^{-i\mathbf{k}\cdot\mathbf{R}}\int_{-\infty}^{\infty} d\omega_1\, \bar{G}(\mathbf{R}, \omega - \omega_1; \omega_1)W^{(i)}_{lm}(\mathbf{R}, \omega_1), \quad (7.6\text{–}36)$$

where

$$\bar{G}(\mathbf{R}, \Omega; \omega') = \frac{1}{2\pi}\int_{-\infty}^{\infty} G(\mathbf{R}, T; \omega')\, e^{i\Omega T}\, dT \quad (7.6\text{–}37)$$

is the temporal Fourier transform of the correlation function (7.6–28) of the generalized susceptibility function $\hat{\eta}(\mathbf{r}, t; \omega')$ of the scattering medium. The approximation sign in Eq. (7.6–36) arises from the use of our assumption that the spatial correlation length of the dielectric susceptibility fluctuations is small compared with the linear dimensions of the scattering volume.

Finally on substituting from Eq. (7.6–36) into Eq. (7.6–35) we see that

$$\langle\langle[\tilde{\mathbf{E}}^{(s)}(r\mathbf{s}, \omega)]^* \cdot \tilde{\mathbf{E}}^{(s)}(r\mathbf{s}, \omega')\rangle\rangle = S^{(s)}(r\mathbf{s}, \omega)\delta(\omega - \omega'), \quad (7.6\text{–}38)$$

where

$$S^{(s)}(r\mathbf{s}, \omega) = \frac{Vk^4}{r^2}(\delta_{lm} - s_l s_m)\int_{-\infty}^{\infty} d\omega_1 \int_V d^3 R\, \bar{G}(\mathbf{R}, \omega - \omega_1; \omega_1)W^{(i)}_{lm}(\mathbf{R}, \omega_1)\, e^{-i\mathbf{k}\cdot\mathbf{R}} \quad (7.6\text{–}39)$$

and k and $\mathbf{k}$ are defined by Eq. (7.6–33).

The quantity $S^{(s)}(r\mathbf{s}, \omega)$, defined by Eq. (7.6–38) and given by the expression (7.6–39), is proportional to the expectation value of the electric energy density at frequency ω of the scattered electric field, at a typical point $\mathbf{r} = r\mathbf{s}$ in the far zone. According to well-known properties of the far field [see, for example Friberg and Wolf, 1983, Eqs. (4.10) and (4.11)], it is also proportional to the expectation value of the magnetic energy density and of the magnitude of the Poynting vector at frequency ω in the far zone. We may, therefore, regard $S^{(s)}(r\mathbf{s}, \omega)$ as representing the spectral density of the scattered field. The formula (7.6–39) expresses it as a linear transform of the cross-spectral density $W^{(i)}_{lm}$, defined by Eq. (7.6–30), of the incident fluctuating electric field. The kernel of the transform is seen to be proportional to the temporal Fourier transform (7.6–37) of the two-point correlation function G, defined by Eq. (7.6–28), of the generalized dielectric susceptibility of the scattering medium. This two-point correlation function is analogous to the so-called Van Hove time-dependent two-particle correlation function (Van Hove, 1954; Glauber, 1962, Sec. 5), also known as the pair distribution function, which is frequently encountered in plasma physics and in the theory of neutron scattering.

7.6.4 Some special cases

(a) Scattering of a linearly polarized polychromatic field

Suppose that the incident field is a fluctuating polychromatic plane wave, incident in the direction specified by a real unit vector $\mathbf{s}_0$, with its electric vector linearly polarized along a direction specified by a unit vector $\mathbf{e}_0$ ($\mathbf{s}_0 \cdot \mathbf{e}_0 = 0$) (Fig. 7.8). Then each realization of the incident field may be represented in the form

$$\mathbf{E}^{(\mathrm{i})}(\mathbf{r}, t) = \mathbf{e}_0 \int_{-\infty}^{\infty} A(\omega)\, \mathrm{e}^{\mathrm{i}(k\mathbf{s}_0 \cdot \mathbf{r} - \omega t)}\, \mathrm{d}\omega. \tag{7.6–40}$$

The Fourier transform of $\mathbf{E}^{(\mathrm{i})}(\mathbf{r}, t)$ with respect to t is evidently

$$\tilde{\mathbf{E}}^{(\mathrm{i})}(\mathbf{r}, \omega) = A(\omega)\, \mathrm{e}^{\mathrm{i}k\mathbf{s}_0 \cdot \mathbf{r}}\, \mathbf{e}_0, \tag{7.6–41}$$

where $A(\omega)$ is, for each frequency ω, a random variable. On substituting from Eq. (7.6–41) into Eq. (7.6–30) we see that

$$\langle A^*(\omega)A(\omega')\rangle\, \mathrm{e}^{\mathrm{i}\mathbf{s}_0 \cdot (k'\mathbf{r}_2 - k\mathbf{r}_1)} e_{0l} e_{0m} = W^{(\mathrm{i})}_{lm}(\mathbf{r}_2 - \mathbf{r}_1, \omega)\delta(\omega - \omega'), \tag{7.6–42}$$

where e_{0l} and e_{0m} are Cartesian components of the unit vector $\mathbf{e}_0$. Now the spectral density $S^{(\mathrm{i})}(\omega)$ of the incident field is just the trace of $W^{(\mathrm{i})}_{lm}(0, \omega)$, and hence it follows from Eq. (7.6–42) that

$$\langle A^*(\omega)A(\omega')\rangle = S^{(\mathrm{i})}(\omega)\delta(\omega - \omega'). \tag{7.6–43}$$

If we use this relation in Eq. (7.6–42), we see that the cross-spectral density tensor of the incident field is given by the expression

$$W^{(\mathrm{i})}_{lm}(\mathbf{R}, \omega) = S^{(\mathrm{i})}(\omega)\, \mathrm{e}^{\mathrm{i}k\mathbf{s}_0 \cdot \mathbf{R}} e_{0l} e_{0m}. \tag{7.6–44}$$

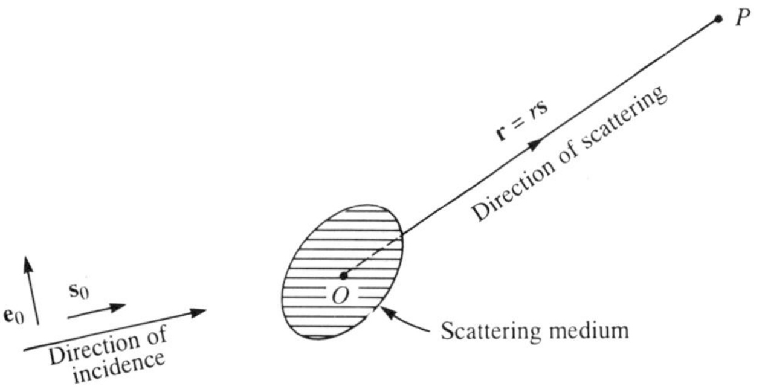

Fig. 7.8 Scattering of a linearly polarized plane, polychromatic, electromagnetic wave. The vectors $\mathbf{s}_0$ and $\mathbf{s}$ are real unit vectors in the direction of propagation of the incident field $\mathbf{E}^{(\mathrm{i})}$, $\mathbf{B}^{(\mathrm{i})}$ and the scattered field $\mathbf{E}^{(\mathrm{s})}$, $\mathbf{B}^{(\mathrm{s})}$, respectively, while $\mathbf{e}_0$ is a real unit vector in the direction of polarization of the incident electric field. The point P is a point in the far zone.

On substituting from Eq. (7.6–44) into the more general formula (7.6–39) we obtain the following expression for the expectation value of the spectral density of the scattered field:

$$S^{(s)}(r\mathbf{s}, \omega) = \frac{Vk^4 \sin^2 \psi}{r^2} \int_{-\infty}^{\infty} d\omega' \int_V d^3R \, \bar{G}(\mathbf{R}, \omega - \omega'; \omega') S^{(i)}(\omega') \, e^{-i(\mathbf{k}-\mathbf{k}_0')\cdot\mathbf{R}}, \tag{7.6–45}$$

where

$$\mathbf{k} = k\mathbf{s} = \frac{\omega}{c}\mathbf{s} \tag{7.6–46}$$

is the wave vector of the ω-component of the scattered field,

$$\mathbf{k}_0' = k'\mathbf{s}_0 = \frac{\omega'}{c}\mathbf{s}_0 \tag{7.6–47}$$

is the wave vector of the ω'-component of the incident field and ψ is the angle between the direction $\mathbf{s}$ of scattering and the direction of polarization, $\mathbf{e}_0$, of the incident electric field, i.e. $\cos\psi = \mathbf{s}\cdot\mathbf{e}_0$. In deriving Eq. (7.6–45) we also made use of the identity

$$(\delta_{lm} - s_l s_m)e_{0l}e_{0m} = 1 - (\mathbf{s}\cdot\mathbf{e}_0)^2 = 1 - \cos^2\psi = \sin^2\psi. \tag{7.6–48}$$

It is useful at this stage to introduce a function $\mathscr{S}(\mathbf{K}, \Omega; \omega')$ defined by the formula

$$\mathscr{S}(\mathbf{K}, \Omega; \omega') = \frac{1}{(2\pi)^3}\int_V \bar{G}(\mathbf{R}, \Omega; \omega') \, e^{-i\mathbf{K}\cdot\mathbf{R}} \, d^3R. \tag{7.6–49}$$

The expression (7.6–45) for the spectral density of the scattered field in the far zone then takes the form

$$S^{(s)}(r\mathbf{s}, \omega) = \frac{(2\pi)^3 Vk^4 \sin^2\psi}{r^2}\int_{-\infty}^{\infty} \mathscr{S}(\mathbf{k} - \mathbf{k}_0', \omega - \omega'; \omega') S^{(i)}(\omega') \, d\omega'. \tag{7.6–50}$$

The function $\mathscr{S}(\mathbf{K}, \Omega; \omega')$, given by Eq. (7.6–49), has a simple meaning, as is readily seen by substituting for $\bar{G}$ in Eq. (7.6–49) from Eq. (7.6–37). One then finds that

$$\mathscr{S}(\mathbf{K}, \Omega; \omega') = \frac{1}{(2\pi)^4}\int_V d^3R \int_{-\infty}^{\infty} dT \, G(\mathbf{R}, T; \omega') \, e^{-i(\mathbf{K}\cdot\mathbf{R} - \Omega T)} \tag{7.6–51a}$$

or, more explicitly, if we recall the definition (7.6–28) of the function $G(\mathbf{R}, T; \omega')$,

$$\mathscr{S}(\mathbf{K}, \Omega; \omega') = \frac{1}{(2\pi)^4}\int_V d^3R \int_{-\infty}^{\infty} dT \, \langle \hat{\eta}^*(\mathbf{r}, t; \omega')\hat{\eta}(\mathbf{r} + \mathbf{R}, t + T; \omega')\rangle \, e^{-i(\mathbf{K}\cdot\mathbf{R} - \Omega T)}. \tag{7.6–51b}$$

We see that $\mathscr{S}(\mathbf{K}, \Omega; \omega')$ is the four-dimensional space-time Fourier transform of the two-point correlation function of the dielectric susceptibility $\hat{\eta}$ of the scattering medium; we will refer to it as the *generalized structure function* of the medium. Under circumstances when $\hat{\eta}$ is effectively independent of ω' (i.e. when ω' is outside the resonance region of the medium) it is analogous to the so-called dynamical structure factor (Forster, 1975) for particle-density fluctuations.

The formula (7.6–50), which is the main result of this section, shows that, when the incident field is a linearly polarized polychromatic plane wave, the spectral density of the scattered field in the far zone is (apart from simple geometrical factors), equal to a 'weighted integral' taken over the spectrum of the incident field. This result holds, of course, only within the accuracy of the first Born approximation. The weighting factor is the generalized structure function of the medium. This weighted integral should not be confused with a convolution, because its kernel depends on the frequencies ω and ω' in a more complicated way than through their difference.

The formula (7.6–50) has been applied to study the changes in line spectra generated by scattering from some model media, whose physical properties vary randomly in both space and time (Foley and Wolf, 1989; Wolf, 1989; James, Savedoff and Wolf, 1990; James and Wolf, 1990, 1994). It was found that in some cases the changes can imitate the Doppler effect in its essential features, even though the source, the scatterer and the observer are at rest with respect to each other. An example is given in Fig. 7.9.

(b) *Scattering of a linearly polarized monochromatic field*

Often, especially when the field incident on the scatterer is generated by a laser, one idealizes the situation by assuming the field to be monochromatic. The corresponding simplified formula for the spectrum of the scattered field

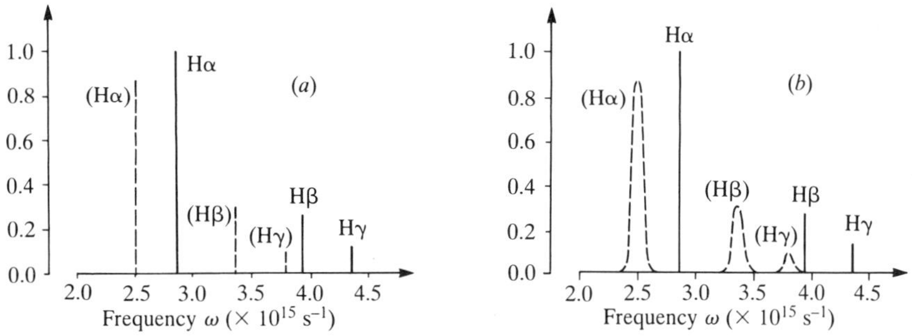

Fig. 7.9 The three narrow hydrogen Balmer lines Hα, $2.8705 \times 10^{15}\ \mathrm{s}^{-1}$; Hβ, $3.875 \times 10^{15}\ \mathrm{s}^{-1}$ and Hγ, $4.3402 \times 10^{15}\ \mathrm{s}^{-1}$, (*a*) as they would appear in the rest frame of their source (solid lines), and Doppler-shifted (dashed lines) due to motion at velocity $0.145c$ away from the observer. (*b*) The same lines (shown dashed), shifted and broadened by scattering on a medium at rest with respect to the source and the observer, whose correlation function (7.6–28) is the Gaussian function

$$G(\mathbf{R}, T; \omega') = G_0 \exp\left[-\tfrac{1}{2}\left(\frac{X^2}{\sigma_x^2} + \frac{Y^2}{\sigma_y^2} + \frac{Z^2}{\sigma_z^2} + \frac{c^2T^2}{\sigma_\tau^2}\right)\right],$$

($\mathbf{R} = X, Y, Z$), with $\sigma_x = \sigma_y = 34.21$ nm, $\sigma_z = 8.731$ μm, $\sigma_\tau = 1.376$ μm, $s_{0z} = 0.800$, $s_z = 0.921$, $\mathbf{s} \cdot \mathbf{s}_0 = 0.965$. (After James and Wolf, 1990.)

$S^{(s)}(r\mathbf{s}, \omega)$, may then be obtained by setting

$$S^{(i)}(\omega') = \tfrac{1}{2}I_0[\delta(\omega' - \omega_0) + \delta(\omega' + \omega_0)] \tag{7.6–52}$$

in Eq. (7.6–50), with I_0 being a positive constant. One then obtains for $S^{(s)}(r\mathbf{s}, \omega)$ the expression

$$S^{(s)}(r\mathbf{s}, \omega) = \frac{(2\pi)^3 k^4 I_0 V \sin^2 \psi}{2r^2}[\mathscr{S}(k\mathbf{s} - k_0\mathbf{s}_0, \omega - \omega_0; \omega_0) + \mathscr{S}(k\mathbf{s} + k_0\mathbf{s}_0, \omega + \omega_0; -\omega_0)], \tag{7.6–53}$$

where $k_0 = \omega_0/c$. If, as is usually the case, the fluctuations of the dielectric susceptibility are slow on the scale of the optical period $2\pi/\omega_0$ we may make the approximation

$$\mathscr{S}(k\mathbf{s} + k_0\mathbf{s}_0, \omega + \omega_0; -\omega_0) \approx 0 \quad \text{when } \omega > 0, \tag{7.6–54}$$

and the expression (7.6–53) for the spectrum of the scattered field reduces to‡

$$S^{(s)}(r\mathbf{s}, \omega) \approx \frac{(2\pi)^3 k^4 I_0 V \sin^2 \psi}{2r^2}\mathscr{S}(k\mathbf{s} - k_0\mathbf{s}_0, \omega - \omega_0; \omega_0). \tag{7.6–55}$$

The formula (7.6–55) also applies when $\omega < 0$ because $S^{(s)}(r\mathbf{s}, -\omega) = S^{(s)}(r\mathbf{s}, \omega)$ as is evident from Eq. (7.6–38), because $\widetilde{\mathbf{E}}^{(s)}(r\mathbf{s}, \omega)$ is the (generalized) Fourier transform of a real function.

Equation (7.6–55) shows that the spectral density $S^{(s)}(r\mathbf{s}, \omega)$ is now directly proportional to (a) the scattering volume V, (b) the square of the inverse distance r from the scatterer, (c) the square of the sine of the angle ψ between the direction of polarization of the incident wave and the direction of scattering and (d) the value of the generalized structure function $\mathscr{S}(\mathbf{K}, \Omega; \omega_0)$ with arguments $\mathbf{K} = k\mathbf{s} - k_0\mathbf{s}_0$ and $\Omega = \omega - \omega_0$.

It is clear from Eq. (7.6–55) and the definition (7.6–51a) of the generalized structure function that the measurement of the spectral density in some particular direction $\mathbf{s}$ and for a chosen frequency ω depends on a particular space-time Fourier component of the space-time correlation function G of $\hat{\eta}$ [Eq. (7.6–28)]. The dependence of the spectral density of the scattered light on the difference $k\mathbf{s} - k_0\mathbf{s}_0$ of the two wave vectors reflects the spatial behavior of the fluctuations of the dielectric susceptibility $\hat{\eta}$, whereas its dependence on $\Omega = \omega - \omega_0$ reflects the temporal behavior of these fluctuations.

(c) The static limit

Suppose now that the scattering medium exhibits randomness in its physical properties in the spatial but not in the temporal domain, i.e. its dielectric susceptibility $\hat{\eta}(\mathbf{r}, t; \omega')$ is a random function of $\mathbf{r}$ but not of t. This is the case, for example, when one employs the so-called 'frozen atmosphere' model in studies of atmospheric scattering. The time dependence of the fluctuations of the

‡ Somewhat similar expressions for the spectrum of the scattered field have been derived in different manners by Komarov and Fisher (1962), van Kampen (1969), p. 235, Eq. (7a) and Pecora (1964), Eq. (30).

refractive index of the atmosphere is then ignored, because its variations are much slower than the fluctuations of an optical field, and only the spatial fluctuations of the refractive index are then taken into account. We refer to such a situation as *static scattering* and we will then write $\tilde{\eta}(\mathbf{r}, \omega')$ in place of $\hat{\eta}(\mathbf{r}, t; \omega')$. The space-time correlation function $G(\mathbf{R}, T; \omega')$, defined by Eq. (7.6–28), will become independent of T and we will denote it by $g(\mathbf{R}; \omega')$:

$$g(\mathbf{R}, \omega') = \langle \tilde{\eta}^*(\mathbf{r}, \omega')\tilde{\eta}(\mathbf{r} + \mathbf{R}, \omega') \rangle . \qquad (7.6\text{–}56)$$

Equation (7.6–37) now gives

$$\bar{G}(\mathbf{R}, \Omega; \omega') = g(\mathbf{R}, \omega')\delta(\Omega). \qquad (7.6\text{–}57)$$

On substituting from this equation into the general formula (7.6–39) we obtain the following expression for the spectral density of the scattered field in the static limit:

$$S^{(s)}(r\mathbf{s}, \omega) = \frac{Vk^4}{r^2}(\delta_{lm} - s_l s_m)\int_V g(\mathbf{R}, \omega) W^{(i)}_{lm}(\mathbf{R}, \omega)\, \mathrm{e}^{-\mathrm{i}\mathbf{k}\cdot\mathbf{R}}\, \mathrm{d}^3 R. \qquad (7.6\text{–}58)$$

Two special cases of the formula (7.6–58) are of particular interest. When the incident field is a linearly polarized, polychromatic plane wave [Eq. (7.6–40)] the cross-spectral density tensor of the incident field is given by Eq. (7.6–44) and, if we make also use of the identity (7.6–48), Eq. (7.6–58) reduces to

$$S^{(s)}(r\mathbf{s}, \omega) = \frac{(2\pi)^3 Vk^4 \sin^2\psi}{r^2}\tilde{g}(\mathbf{k} - \mathbf{k}_0, \omega) S^{(i)}(\omega), \qquad (7.6\text{–}59)$$

where, as before, $\mathbf{k} = (\omega/c)\mathbf{s}$, $\mathbf{k}_0 = (\omega/c)\mathbf{s}_0$ and $\tilde{g}(\mathbf{K}, \omega)$ is the three-dimensional Fourier transform of the correlation function $g(\mathbf{R}, \omega)$, i.e.

$$\tilde{g}(\mathbf{K}, \omega) = \frac{1}{(2\pi)^3}\int_V g(\mathbf{R}, \omega)\, \mathrm{e}^{-\mathrm{i}\mathbf{K}\cdot\mathbf{R}}\, \mathrm{d}^3 R \qquad (7.6\text{–}60)$$

or, more explicitly, from Eq. (7.6–56),

$$\tilde{g}(\mathbf{K}, \omega) = \frac{1}{(2\pi)^3}\int_V \langle \tilde{\eta}^*(\mathbf{r}, \omega)\tilde{\eta}(\mathbf{r} + \mathbf{R}, \omega) \rangle\, \mathrm{e}^{-\mathrm{i}\mathbf{K}\cdot\mathbf{R}}\, \mathrm{d}^3 R. \qquad (7.6\text{–}61)$$

The formula (7.6–59) shows that even in static scattering the spectrum of the scattered light differs, in general, from the spectrum of the incident light. The change arises from the proportionality factor $k^4 = (\omega/c)^4$ and from the dependence on frequency of the Fourier transform $\tilde{g}$ of the static correlation function g. It also depends on the direction of observation. The formula (7.6–59) resembles in its main features the formula (5.8–29) for the spectrum of the field generated by a quasi-homogeneous source in free space. Spectral changes induced by static scattering were studied by the use of an analogue of formula (7.6–59) derived from scalar theory rather than from full electromagnetic theory by Wolf, Foley and Gori [1989, Eq. (3.7)].

Let us consider another special case, namely scattering of *monochromatic*, linearly polarized light of frequency ω_0 on a static medium. In this case the spectrum of the incident field is given by Eq. (7.6–52) and, for $\omega > 0$, Eq.

(7.6–59) reduces to

$$S^{(s)}(r\mathbf{s}, \omega) = \frac{(2\pi)^3 I_0 V k^4 \sin^2 \psi}{2r^2} \tilde{g}(\mathbf{k} - \mathbf{k}_0, \omega)\delta(\omega - \omega_0). \qquad (7.6\text{–}62)$$

A formula of this kind was first derived by Einstein (1910) in a well-known investigation that was the starting point of the statistical theory of light scattering. We note that according to Eq. (7.6–62) the spectral density of the light scattered in a direction specified by the unit vector **s** is proportional to a three-dimensional spatial Fourier component of the static correlation function $g(\mathbf{R}, \omega)$ of the dielectric susceptibility; namely the component labeled by the vector

$$\mathbf{K}_0 = k_0(\mathbf{s} - \mathbf{s}_0). \qquad (7.6\text{–}63)$$

Since **s** and $\mathbf{s}_0$ are unit vectors, it follows that

$$|\mathbf{K}_0| \leqslant 2k_0 = 4\pi/\lambda_0, \qquad (7.6\text{–}64)$$

where λ_0 is the wavelength of the incident light. Now the vector $\mathbf{K}_0$ is associated with spatial periodic components $(\Delta x, \Delta y, \Delta z)$ of the correlation function $g(\mathbf{R}, \omega)$ by the relations

$$\Delta x = \frac{2\pi}{|K_{0x}|}, \quad \Delta y = \frac{2\pi}{|K_{0y}|}, \quad \Delta z = \frac{2\pi}{|K_{0z}|}, \qquad (7.6\text{–}65)$$

where K_{0x}, K_{0y} and K_{0z} are the Cartesian components of $\mathbf{K}_0$. Hence the inequality (7.6–64) may be expressed in the form [analogous to Eq. (3.2–31a) which we encountered earlier in connection with the angular spectrum representation of wavefields]

$$\frac{1}{(\Delta x)^2} + \frac{1}{(\Delta y)^2} + \frac{1}{(\Delta z)^2} \leqslant \left(\frac{2}{\lambda_0}\right)^2. \qquad (7.6\text{–}66)$$

This inequality implies, roughly speaking, that only those periodic details of $g(\mathbf{R}, \omega)$ which are greater than about half a wavelength of the incident light contribute to the scattered field. Conversely, from measurements of the spectral density of the scattered field for different directions of scattering and with different directions of incidence one may, in principle, obtain information about the correlations of the dielectric susceptibility of some solids down to details of the order of the wavelength λ_0 of the incident light. In ordinary gases and liquids, however, the correlation lengths are usually much smaller than the wavelength. Scattering experiments of this kind are then less useful, although they can provide information about the variance of the density fluctuations from measurements of the intensity of the light scattered in the forward direction $\mathbf{s} = \mathbf{s}_0$ (cf. Van Kampen, 1969, Sec. 3).

7.6.5 Scattering from a simple fluid

We now examine the case of the scattering of monochromatic light by a medium exhibiting random fluctuations in its physical properties, we will consider a classic problem involving scattering by density fluctuations in a simple fluid, i.e. a fluid comprised of only one kind of molecule.

When light passes through a transparent medium, some of it is scattered even when the medium appears to be homogeneous. Smoluchowski (1908) suggested that, even in an apparently homogeneous medium, the density is not strictly constant but fluctuates about an equilibrium value and that the observed scattering is caused by such fluctuations. Soon afterwards Einstein (1910) carried out quantitative calculations based on this suggestion. Actually at that time, little was known about the time dependence of the density fluctuations and consequently Einstein confined his analysis to calculating effects that arise from spatial inhomogeneities only. In spite of this limitation his investigation provided the first real insight into the origin of critical opalescence. This effect occurs when the temperature of a transparent scattering medium approaches the critical value at which the phase transition takes place, and results in a dramatic increase in the intensity of the scattered light and a milky appearance of the scattering medium.

The first indication of the origin of the temporal fluctuations of the density was provided by Debye's theory of specific heats (Debye, 1912), according to which the fluctuations arise from thermally excited sound waves that continuously traverse the medium. This model was later used by Brillouin (1914, 1922) to predict the spectrum of light produced by scattering of monochromatic light by the sound waves. His calculations indicated that the spectrum of the scattered light consists of two lines, now known as the *Brillouin doublet*, located symmetrically about the frequency of the incident light. Later experiments by Gross (1930a,b,c; 1932) showed that the spectrum of the scattered light contains, in addition, a line centered at the original frequency, now known as the *Rayleigh line*. An explanation of the origin of this line was given by Landau and Placzek (1934; see also Landau and Lifshitz, 1960, Sec. 94), who showed that it arises from a non-propagating contribution (the so-called isobaric entropy fluctuations) to the fluctuating density.

Since these pioneering investigations were carried out, numerous contributions to the understanding of light scattering have been made, both theoretical and experimental, triggered largely by the development of the laser (see, for example, Fabelinskii, 1968). These researches have provided valuable information, for example about the dynamics of the fluctuations and about energy and momentum transfer in gases, liquids and solids.

We will now apply the formula (7.6–50) to the classic problem of scattering by density fluctuations. Specifically we consider the scattering of a plane, monochromatic, linearly polarized electromagnetic wave of frequency ω_0 and wave vector $k_0\mathbf{s}_0$ ($\mathbf{s}_0^2 = 1$) by a monatomic, non-absorbing one-component fluid under equilibrium conditions. Let

$$\Delta\hat{\eta}(\mathbf{r}, t; \omega_0) = \hat{\eta}(\mathbf{r}, t; \omega_0) - \langle\hat{\eta}\rangle_{\text{eq}} \tag{7.6–67}$$

represent the fluctuation of the dielectric susceptibility $\hat{\eta}$ from its equilibrium value $\langle\hat{\eta}\rangle_{\text{eq}}$. We assume that $|\langle\hat{\eta}\rangle_{\text{eq}}|$ is so small that its contribution to the scattered field can be neglected. Were this not so, the first Born approximation, on which our analysis is based, would not adequately describe the scattered field.

According to Eq. (7.6–55) the spectral density of the scattered light is given by

$$S^{(\text{s})}(r\mathbf{s}, \omega) = \frac{(2\pi)^3 k^4 I_0 V \sin^2\psi}{2r^2}\mathscr{S}^{(\Delta\hat{\eta})}(k\mathbf{s} - k_0\mathbf{s}_0, \omega - \omega_0; \omega_0), \tag{7.6–68}$$

where, according to Eqs. (7.6–51a) and (7.6–28),

$$\mathcal{S}^{(\Delta\hat{\eta})}(\mathbf{K}, \Omega; \omega_0) = \frac{1}{(2\pi)^4}\int_V \mathrm{d}^3R \int_{-\infty}^{\infty} \mathrm{d}T\, G^{(\Delta\hat{\eta})}(\mathbf{R}, T; \omega_0)\, \mathrm{e}^{-\mathrm{i}(\mathbf{K}\cdot\mathbf{R}-\Omega T)} \quad (7.6\text{–}69)$$

and

$$G^{(\Delta\hat{\eta})}(\mathbf{R}, T; \omega_0) = \langle \Delta\hat{\eta}^*(\mathbf{r}, t; \omega_0)\Delta\hat{\eta}(\mathbf{r}+\mathbf{R}, t+T; \omega_0)\rangle. \quad (7.6\text{–}70)$$

Assuming that the fluctuations of the dielectric susceptibility arise from fluctuations in the density ρ and in the temperature T, we have

$$\Delta\hat{\eta} = \left(\frac{\partial\hat{\eta}}{\partial\rho}\right)_T \Delta\rho + \left(\frac{\partial\hat{\eta}}{\partial T}\right)_\rho \Delta T. \quad (7.6\text{–}71)$$

Under usual circumstances $(\partial\hat{\eta}/\partial\rho)_T \gg (\partial\hat{\eta}/\partial T)_\rho$ and the second term on the right-hand side of Eq. (7.6–71) may then be neglected. We can then write,

$$\Delta\hat{\eta}(\mathbf{r}, t; \omega_0) = \left(\frac{\partial\hat{\eta}}{\partial\rho}\right)_{\mathrm{eq}} \Delta\rho(\mathbf{r}, t), \quad (7.6\text{–}72)$$

where the derivative $(\partial\hat{\eta}/\partial\rho)_{\mathrm{eq}}$, being an equilibrium quantity, is independent of $\mathbf{r}$ and t; it does, however, depend on the frequency ω_0 of the incident light.

It follows from Eqs. (7.6–70) and (7.6–72) that

$$G^{(\Delta\hat{\eta})}(\mathbf{R}, T; \omega_0) = \left(\frac{\partial\hat{\eta}}{\partial\rho}\right)^2_{\mathrm{eq}} G^{(\Delta\rho)}(\mathbf{R}, T), \quad (7.6\text{–}73)$$

where

$$G^{(\Delta\rho)}(\mathbf{R}, T) = \langle \Delta\rho(\mathbf{r}, t)\Delta\rho(\mathbf{r}+\mathbf{R}, t+T)\rangle \quad (7.6\text{–}74)$$

is the correlation function of the density fluctuations. On substituting from Eq. (7.6–73) into Eq. (7.6–69) we obtain at once the following expression for $\mathcal{S}^{(\Delta\hat{\eta})}$:

$$\mathcal{S}^{(\Delta\hat{\eta})}(\mathbf{K}, \Omega) = \left(\frac{\partial\hat{\eta}}{\partial\rho}\right)^2_{\mathrm{eq}} \mathcal{S}^{(\Delta\rho)}(\mathbf{K}, \Omega), \quad (7.6\text{–}75)$$

where

$$\mathcal{S}^{(\Delta\rho)}(\mathbf{K}, \Omega) = \frac{1}{(2\pi)^4}\int_V \mathrm{d}^3R \int_{-\infty}^{\infty} \mathrm{d}T\, G^{(\Delta\rho)}(\mathbf{R}, T)\, \mathrm{e}^{-\mathrm{i}(\mathbf{K}\cdot\mathbf{R}-\Omega T)} \quad (7.6\text{–}76)$$

is the structure function of the medium, expressed in terms of the correlation function (7.6–74) of the density fluctuations.

On substituting from Eq. (7.6–75) into Eq. (7.6–68) we obtain the following expression for the spectral density of the scattered light:

$$S^{(\mathrm{s})}(r\mathbf{s}, \omega) = \frac{(2\pi)^3 k^4 I_0 V \sin^2\psi}{2r^2}\left(\frac{\partial\hat{\eta}}{\partial\rho}\right)^2_{\mathrm{eq}} \mathcal{S}^{(\Delta\rho)}(k\mathbf{s} - k_0\mathbf{s}_0, \omega - \omega_0). \quad (7.6\text{–}77)$$

This equation is a slight generalization of a basic formula employed in the theory of light scattering from density fluctuations under equilibrium conditions.

The determination of the structure function $\mathcal{S}^{(\Delta\rho)}$ is a rather complicated problem that can be solved only approximately, on the basis of linearized hydrodynamic equations of irreversible thermodynamics (Mountain, 1966). This subject does not fall within the context of this book and we will, therefore, only state the result. For a simple fluid under conditions of local thermodynamic equilibrium and at temperatures not too close to the critical points of the fluid‡,

$$\mathcal{S}^{(\Delta\rho)}(\mathbf{K}, \Omega) =$$

$$g_0\left\{a_1\frac{\gamma_1 K^2}{(\gamma_1 K^2)^2 + \Omega^2} + a_2\left[\frac{\gamma_2 K^2}{(\gamma_2 K^2)^2 + (\Omega + vK)^2} + \frac{\gamma_2 K^2}{(\gamma_2 K^2)^2 + (\Omega - vK)^2}\right]\right\}, \tag{7.6–78}$$

where g_0, a_1, a_2, γ_1, and γ_2 are constants and v is the speed of sound in the fluid (taken to be constant).

Now according to Eq. (7.6–77) we need the values of the structure function (7.6–78) for the arguments $\mathbf{K} = k\mathbf{s} - k_0\mathbf{s}_0$ and $\Omega = \omega - \omega_0$. We have

$$K^2 = |k\mathbf{s} - k_0\mathbf{s}_0|^2 = \left(\frac{\omega}{c}\right)^2 + \left(\frac{\omega_0}{c}\right)^2 - 2\left(\frac{\omega}{c}\right)\left(\frac{\omega_0}{c}\right)\cos\theta, \tag{7.6–79}$$

where θ is the angle between the direction $\mathbf{s}$ of scattering and the direction $\mathbf{s}_0$ of incidence. If the frequency shift introduced on scattering is small enough, as is usually the case, we may replace ω by ω_0 on the right-hand side of Eq. (7.6–79) and we then obtain for K the expression

$$K = |k\mathbf{s} - k_0\mathbf{s}_0| \approx 2\left(\frac{\omega_0}{c}\right)\sin\left(\frac{\theta}{2}\right). \tag{7.6–80}$$

On substituting from Eq. (7.6–80) into Eq. (7.6–78) we find that

$$\begin{aligned}\mathcal{S}^{(\Delta\rho)}(k\mathbf{s} - k_0\mathbf{s}_0, \omega - \omega_0) = g_0\{&a_1 L[\omega - \omega_0; \Gamma_R(\theta)] \\ &+ a_2 L[\omega - \omega_0 + \omega_B(\theta); \Gamma_B(\theta)] \\ &+ a_2 L[\omega - \omega_0 - \omega_B(\theta); \Gamma_B(\theta)]\},\end{aligned} \tag{7.6–81}$$

where $L[\omega - \omega_0; \Gamma]$ is the Lorentzian function

$$L[\omega - \omega_0; \Gamma] = \frac{\Gamma}{(\omega - \omega_0)^2 + \Gamma^2}, \tag{7.6–82}$$

and

$$\Gamma_R(\theta) = 4\gamma_1\left(\frac{\omega_0}{c}\right)^2 \sin^2\left(\frac{\theta}{2}\right), \tag{7.6–83}$$

$$\Gamma_B(\theta) = 4\gamma_2\left(\frac{\omega_0}{c}\right)^2 \sin^2\left(\frac{\theta}{2}\right), \tag{7.6–84}$$

$$\omega_B(\theta) = 2\left(\frac{v}{c}\right)\omega_0 \sin\left(\frac{\theta}{2}\right). \tag{7.6–85}$$

It follows from Eqs. (7.6–77) and (7.6–81) that the spectrum of the scattered

‡ Near the critical point the factor g_0 in Eq. (7.6–78) becomes a function of K, reflecting a long-range correlation of the density fluctuations which then occurs.

light in any fixed direction of scattering consists of a Lorentzian line centered at the frequency ω_0 of the incident light and two Lorentzian lines located symmetrically with respect to it, at frequencies $\omega_0 \pm \omega_B(\theta)$. The central line, which has the width $\Gamma_R(\theta)$, is the Rayleigh line and the two other lines, each of width $\Gamma_B(\theta)$, form the Brillouin doublet. An example of such an observed Rayleigh–Brillouin spectrum is shown in Fig. 7.10.

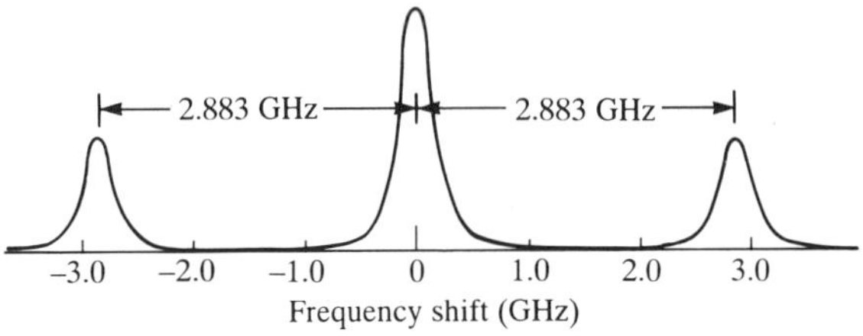

Fig. 7.10 An observed Rayleigh–Brillouin spectrum of liquid argon, at temperature $T = 84.97$ K for angle of scattering $\theta = 90° \, 14'$, with the wavelength of incident laser light $\lambda_0 = 5145$ Å (After Fleury and Boon, 1969.)

On a more elementary level the Brillouin doublet may be regarded, roughly speaking, as arising from a Doppler shift of the incident light wave on reflection from moving sound waves (Benedek, 1968). Its origin can also be understood quantum mechanically as a consequence of conservation of energy and momentum between an incident photon, a scattered photon and an acoustical phonon. The Brillouin line centered at the lower frequency is also known as the *Stokes line*, because of an earlier observation by Stokes that luminescence was generally observed at a frequency that is lower than that of the exciting light. The other line, centered at the higher frequency, is often referred to as the *anti-Stokes line*.

Appendix 7.1

Evaluation of the expectation value $\langle\langle \mathcal{P}^*_{1l}(\mathbf{k}, \omega)\mathcal{P}_{1m}(\mathbf{k}', \omega')\rangle\rangle$ [Eq. (7.6–36)][‡]

It follows from Eq. (7.6–26) that

$$\langle\langle \mathcal{P}^*_{1l}(\mathbf{k}, \omega)\mathcal{P}_{1m}(\mathbf{k}', \omega')\rangle\rangle = \frac{1}{(2\pi)^6}\int_V \mathrm{d}^3 r_1 \int_V \mathrm{d}^3 r_2 \int_{-\infty}^{\infty} \mathrm{d}\omega_1 \int_{-\infty}^{\infty} \mathrm{d}\omega_2 \, \mathrm{e}^{\mathrm{i}(\mathbf{k}\cdot\mathbf{r}_1 - \mathbf{k}'\cdot\mathbf{r}_2)}$$
$$\times \langle \tilde{\eta}^*(\mathbf{r}_1, \omega - \omega_1; \omega_1)\tilde{\eta}(\mathbf{r}_2, \omega' - \omega_2; \omega_2)\rangle \langle [\tilde{\mathbf{E}}^{(\mathrm{i})}_l(\mathbf{r}_1, \omega_1)]^* \tilde{\mathbf{E}}^{(\mathrm{i})}_m(\mathbf{r}_2, \omega_2)\rangle, \quad \text{(A7.1–1)}$$

where we have made use of the assumption that the fluctuations of the medium and of the incident field are statistically independent. The second expectation value that appears on the right-hand side of Eq. (A7.1–1) is, according to Eq. (7.6–30), given by

$$\langle [\tilde{\mathbf{E}}^{(\mathrm{i})}_l(\mathbf{r}_1, \omega)]^* \tilde{\mathbf{E}}^{(\mathrm{i})}_m(\mathbf{r}_2, \omega')\rangle = W^{(\mathrm{i})}_{lm}(\mathbf{r}_2 - \mathbf{r}_1, \omega)\delta(\omega - \omega'). \quad \text{(A7.1–2)}$$

‡ After Wolf and Foley (1989).

On substituting from Eq. (A7.1–2) into Eq. (A7.1–1) and carrying out the trivial integration with respect to ω_2 we find that

$$\langle\langle \mathcal{P}^*_{1l}(\mathbf{k}, \omega)\mathcal{P}_{1m}(\mathbf{k}', \omega')\rangle\rangle = \frac{1}{(2\pi)^6}\int_V \mathrm{d}^3 r_1 \int_V \mathrm{d}^3 r_2 \int_{-\infty}^{\infty} \mathrm{d}\omega_1\, \mathrm{e}^{\mathrm{i}(\mathbf{k}\cdot\mathbf{r}_1 - \mathbf{k}'\cdot\mathbf{r}_2)}$$
$$\times \langle [\bar{\eta}(\mathbf{r}_1, \omega - \omega_1; \omega_1)]^* \bar{\eta}(\mathbf{r}_2, \omega' - \omega_1; \omega_1)\rangle W^{(\mathrm{i})}_{lm}(\mathbf{r}_2 - \mathbf{r}_1, \omega_1). \quad \text{(A7.1–3)}$$

The expectation value on the right-hand side of Eq. (A7.1–3), which involves the dielectric susceptibility, may be expressed in the following form which follows from Eqs. (7.6–16a) and (7.6–28):

$$\langle [\bar{\eta}(\mathbf{r}_1, \Omega; \omega_1)]^* \bar{\eta}(\mathbf{r}_2, \Omega', \omega_1)\rangle = \bar{G}(\mathbf{r}_2 - \mathbf{r}_1, \Omega; \omega_1)\delta(\Omega - \Omega'), \quad \text{(A7.1–4)}$$

where, according to Eq. (7.6–37),

$$\bar{G}(\mathbf{R}, \Omega; \omega') = \frac{1}{2\pi}\int_{-\infty}^{\infty} G(\mathbf{R}, T; \omega')\, \mathrm{e}^{\mathrm{i}\Omega T}\, \mathrm{d}T, \quad \text{(A7.1–5)}$$

$G(\mathbf{R}, T; \omega')$ being the correlation function, defined by Eq. (7.6–28), of the dielectric susceptibility, namely

$$G(\mathbf{R}, T; \omega') = \langle \hat{\eta}^*(\mathbf{r}, t; \omega')\hat{\eta}(\mathbf{r} + \mathbf{R}, t + T; \omega')\rangle. \quad \text{(A7.1–6)}$$

On substituting from Eq. (A7.1–4) into Eq. (A7.1–3) we obtain the formula

$$\langle\langle \mathcal{P}^*_{1l}(\mathbf{k}, \omega)\mathcal{P}_{1m}(\mathbf{k}', \omega')\rangle\rangle = \frac{\delta(\omega - \omega')}{(2\pi)^6}\int_V \mathrm{d}^3 r_1 \int_V \mathrm{d}^3 r_2\, \mathrm{e}^{-\mathrm{i}\mathbf{k}\cdot(\mathbf{r}_2 - \mathbf{r}_1)}$$
$$\times \int_{-\infty}^{\infty} \mathrm{d}\omega_1\, \bar{G}(\mathbf{r}_2 - \mathbf{r}_1, \omega - \omega_1; \omega_1) W^{(\mathrm{i})}_{lm}(\mathbf{r}_2 - \mathbf{r}_1, \omega_1). \quad \text{(A7.1–7)}$$

Because of our assumption that the spatial correlation length of the dielectric susceptibility fluctuations is small compared with the linear dimensions of the scattering volume, the expression (A7.1–7) may readily be shown to reduce to

$$\langle\langle \mathcal{P}^*_{1l}(\mathbf{k}, \omega)\mathcal{P}_{1m}(\mathbf{k}', \omega')\rangle\rangle \approx \frac{V\delta(\omega - \omega')}{(2\pi)^6}\int_V \mathrm{d}^3 R\, \mathrm{e}^{-\mathrm{i}\mathbf{k}\cdot\mathbf{R}}$$
$$\times \int_{-\infty}^{\infty} \mathrm{d}\omega_1\, \bar{G}(\mathbf{R}, \omega - \omega_1; \omega_1) W^{(\mathrm{i})}_{lm}(\mathbf{R}, \omega_1), \quad \text{(A7.1–8)}$$

where V is the volume occupied by the scattering medium. This formula is Eq. (7.6–36) of the text.

Problems

7.1 A double star consists of two components having the same angular diameter 2α and angular separation 2β. The stars emit light of the same mean wavelength and may be modeled as circular areas of constant intensities. The ratio of the intensities of the two components is $1{:}b$. The light received is passed through a filter so as to become quasi-monochromatic.

(a) Derive an expression for the equal-time degree of coherence of the star light in the observation plane of a Michelson stellar interferometer.
(b) If $\beta \gg \alpha$ show how β may be determined from the visibility curve.

7.2 A spectrum of light consists of N lines of the same profile but of different intensities, centered on frequencies $\nu_1, \nu_2, \ldots, \nu_N$. The light is analyzed by means of a Michelson interferometer. Derive an expression for the visibility curve.

7.3 The spectrum of a certain light beam consists of two lines of identical Gaussian profiles, centered on frequencies ν_1 and ν_2. Assuming that the separation $\Delta\nu = |\nu_1 - \nu_2|$ of the two lines is large compared to their effective widths, show how one may determine $\Delta\nu$ from the visibility curve when the light is passed through a Michelson interferometer.

7.4 A Michelson interferometer is illuminated by a quasi-monochromatic light beam having a rectangular spectral distribution of width $\Delta\nu$, centered on frequency $\bar{\nu}$. The interference fringes first vanish when one of the mirrors is displaced by a distance d_0 from its 'symmetric' position with respect to the other mirror. Determine the width $\Delta\nu$ of the spectral distribution.

7.5 Determine the generalized structure function $S(\mathbf{K}, \Omega; \omega')$ [Eq. (7.6–51a)] of a scattering medium, whose space-time correlation function is given by the expression

$$G(\mathbf{R}, T; \omega') = \frac{B}{(2\pi\sigma^2)^{3/2}} \, \mathrm{e}^{-R^2/2\sigma^2} \, \mathrm{e}^{-T^2/2\tau^2},$$

where B, σ and τ are positive parameters which may depend on the frequency ω' and σ is small compared to the dimensions of the scattering volume.

7.6 Consider a polychromatic, linearly polarized, plane, electromagnetic wave whose spectrum is a line with Gaussian profile, of r.m.s. width Γ_0, centered on frequency ω_0, i.e.

$$S^{(\mathrm{i})}(\omega) = \frac{A}{\sqrt{(2\pi)}\Gamma_0}[g(\omega - \omega_0; \Gamma_0) + g(\omega + \omega_0; \Gamma_0)],$$

where A, ω_0 and Γ_0 are positive constants and

$$g(\omega \pm \omega_0; \Gamma_0) = \exp[-(\omega \pm \omega_0)^2/2\Gamma_0^2].$$

The wave is incident on a statistically homogeneous medium, whose space-time correlation function is given by the expression stated in Problem 7.5. Assuming that over the effective spectral range of the incident light the frequency-dependence of B, σ and τ may be neglected, determine, within the accuracy of the first Born approximation, the spectrum of the scattered field in the far zone of the medium.

7.7 A linearly polarized, polychromatic, plane wave is incident on a weak, homogeneous, static scatterer, whose spatial correlation function [defined by Eq. (7.6–56) of the text] has the functional form

$$g_\eta(\mathbf{R}, \omega) = f(\omega)h(kR), \quad (k = \omega/c).$$

Show that the normalized spectrum of the scattered field is the same throughout the far zone but is not, in general, equal to the normalized spectrum of the wave incident on the scatterer.

8

Higher-order correlations in optical fields

8.1 Introduction

In the preceding chapters we have been largely concerned with the simplest coherence effects of optical fields, namely those which depend on the correlation of the field variable at two space-time points $(\mathbf{r}_1, t_1)$ and $(\mathbf{r}_2, t_2)$. As we have seen, these effects include the most elementary coherence phenomena involving interference, diffraction and radiation from fluctuating sources.

In this chapter we present an extension of the theory to cover more complicated situations, which have to be described by correlations of higher order, i.e. by correlations of the field variables at more than two space-time points or the expectation values involving various powers and products of the field variables.‡ Situations of this kind have become of considerable importance since the development of the laser and of nonlincar optics. Thc basic difference between the statistical properties of thermal light and laser light can, in fact, only be understood by going beyond the elementary second-order correlation theory.

Many of the higher-order coherence phenomena are most clearly manifest in the photoelectric detection process, which can only be adequately described by the quantum theory of detection or by taking into account the quantum features of the field, both of which will be studied in the succeeding chapters. However, because the field is still described classically in the semi-classical theory of light detection, and also because the classical field description provides a natural stepping stone to the quantum description of field correlations, we will now discuss the general description of field correlations of all orders on the basis of the classical theory of the fluctuating wavefield. The usefulness of this description will become evident in Chapter 12.

8.2 Space–time correlation functions of arbitrary order

Let us begin by recalling the definition (4.3–6) of the second-order correlation function of a fluctuating scalar wavefield, represented by an analytic signal $V(\mathbf{r}, t)$, namely

$$\Gamma(\mathbf{r}_1, \mathbf{r}_2, t_1, t_2) = \langle V^*(\mathbf{r}_1, t_1) V(\mathbf{r}_2, t_2) \rangle, \tag{8.2–1}$$

‡ Within the framework of the classical theory of fluctuating wavefields, higher-order correlation functions were introduced by Wolf (1963, 1964) and Mandel (1964).

where the angle brackets denote the ensemble average. More explicitly,

$$\Gamma(\mathbf{r}_1, \mathbf{r}_2, t_1, t_2) = \iint V_1^* V_2\, p_2(V_1, V_2; \mathbf{r}_1, t_1; \mathbf{r}_2, t_2)\, \mathrm{d}^2V_1\, \mathrm{d}^2V_2, \quad (8.2\text{–}2)$$

where p_2 is the joint probability density of the fluctuating field at the two space-time points $(\mathbf{r}_1, t_1)$ and $(\mathbf{r}_2, t_2)$. Although, as we saw in the preceding sections, this function is very useful for analyzing various coherence phenomena it cannot, in general, provide any information about effects that involve probability densities of third and higher orders (cf. Section 2.1.2).

Among the most important quantities that are needed to elucidate coherence phenomena which depend on probability densities of order higher than the second are the correlation functions

$$\begin{aligned}\Gamma^{(M,N)}&(\mathbf{r}_1, \mathbf{r}_2, \dots, \mathbf{r}_{M+N}; t_1, t_2, \dots, t_{M+N})\\ &= \langle V^*(\mathbf{r}_1, t_1)V^*(\mathbf{r}_2, t_2) \dots V^*(\mathbf{r}_M, t_M)\\ &\quad \times V(\mathbf{r}_{M+1}, t_{M+1})V(\mathbf{r}_{M+2}, t_{M+2}) \dots V(\mathbf{r}_{M+N}, t_{M+N})\rangle \quad (8.2\text{–}3)\end{aligned}$$

or, more explicitly,

$$\begin{aligned}\Gamma^{(M,N)}&(\mathbf{r}_1, \mathbf{r}_2, \dots, \mathbf{r}_{M+N}; t_1, t_2, \dots, t_{M+N})\\ &= \underbrace{\int \dots \int}_{(M+N)} V_1^* V_2^* \dots V_M^* V_{M+1} V_{M+2} \dots V_{M+N}\\ &\quad \times p_{M+N}(V_1, V_2, \dots, V_{M+N}; \mathbf{r}_1, t_1; \mathbf{r}_2, t_2; \dots; \mathbf{r}_{M+N}, t_{M+N})\\ &\qquad \mathrm{d}^2V_1\, \mathrm{d}^2V_2 \dots \times \mathrm{d}^2V_{M+N}. \quad (8.2\text{–}4)\end{aligned}$$

We will refer to $\Gamma^{(M,N)}$ as the *(space-time) cross-correlation function* of order (M, N) of the random field $V(\mathbf{r}, t)$. For reasons that will become apparent later, correlation functions for which $N = M$ are particularly useful. They are often referred to as correlation functions of order $2M$. This terminology is in agreement with that used in some of the preceding sections of this book but it is not uniform; some authors refer to $\Gamma^{(M,M)}$ as a correlation function of order M rather than $2M$.

To simplify the notation we set

$$\left.\begin{aligned}\mathbf{r}_m, t_m &= x_m, & m &= 1, 2, \dots, M,\\ \mathbf{r}_{M+n}, t_{M+n} &= y_n, & n &= 1, 2, \dots, N.\end{aligned}\right\} \quad (8.2\text{–}5)$$

The formula (8.2–3) may then be re-written as

$$\begin{aligned}\Gamma^{(M,N)}&(x_1, x_2, \dots, x_M; y_1, y_2, \dots, y_N)\\ &= \langle V^*(x_1)V^*(x_2) \dots V^*(x_M)V(y_1)V(y_2) \dots V(y_N)\rangle. \quad (8.2\text{–}6)\end{aligned}$$

We note a number of properties of these correlation functions. First, we see at once from their definition that

$$\begin{aligned}[\Gamma^{(M,N)}&(x_1, x_2, \dots, x_M; y_1, y_2, \dots, y_N)]^* =\\ &\Gamma^{(N,M)}(y_1, y_2, \dots, y_N; x_1, x_2, \dots, x_M). \quad (8.2\text{–}7)\end{aligned}$$

Further one can readily show with the help of the Schwarz inequality that

$$|\Gamma^{(M,N)}(x_1, x_2, \ldots, x_M; y_1, y_2, \ldots, y_N)|^2 \leqslant \Gamma^{(M,M)}(x_1, x_2, \ldots, x_M; x_1, x_2, \ldots, x_M) \times \Gamma^{(N,N)}(y_1, y_2, \ldots, y_N; y_1, y_2, \ldots, y_N). \quad (8.2\text{–}8)$$

The expressions on the right-hand side of the inequality (8.2–8) have simple meanings, as can be seen at once by taking $M = N$ and $y_j = x_j$, $(j = 1, 2, \ldots, M)$, in Eq. (8.2–6). One then finds that

$$\Gamma^{(M,M)}(x_1, x_2, \ldots, x_M; x_1, x_2, \ldots, x_M) = \langle I(x_1)I(x_2) \ldots I(x_M)\rangle, \quad (8.2\text{–}9)$$

where

$$I(x_j) = V^*(x_j)V(x_j), \quad (j = 1, 2, \ldots, M), \quad (8.2\text{–}10)$$

is the instantaneous intensity at the space-time point x_j.

The correlation functions (8.2–6) satisfy non-negative definiteness conditions, which may be derived as follows. Consider the expression

$$F \equiv \sum_{i=1}^{r} a_i V(x_1^{(i)})V(x_2^{(i)}) \ldots V(x_M^{(i)}), \quad (8.2\text{–}11)$$

where r is an arbitrary positive integer, $x_m^{(i)}$ $(m = 1, 2, \ldots, M; i = 1, 2, \ldots, r)$ are arbitrary space-time points and the a_i's are arbitrary (real or complex) constants. Since obviously $\langle F^*F\rangle \geqslant 0$ it follows that

$$\sum_{i=1}^{r}\sum_{j=1}^{r} a_i^* a_j \Gamma^{(M,M)}(x_1^{(i)}, x_2^{(i)}, \ldots, x_M^{(i)}; x_1^{(j)}, x_2^{(j)}, \ldots, x_M^{(j)}) \geqslant 0. \quad (8.2\text{–}12)$$

When the field is stationary, all the correlation functions $\Gamma^{(M,N)}$ are, of course, invariant with respect to translation of the origin of time. They then depend on only $M + N - 1$ time arguments, which we may take to be

$$\tau_2 = t_2 - t_1, \quad \tau_3 = t_3 - t_1, \ldots, \tau_{M+N} = t_{M+N} - t_1. \quad (8.2\text{–}13)$$

So far we have restricted our considerations to scalar fields. For a vector field $\mathbf{V}(\mathbf{r}, t)$, such as the electric or the magnetic field, the analogous space-time correlation functions will be tensors defined by the formula

$$\Gamma^{(M,N)}_{j_1,j_2,\ldots,j_M;k_1,k_2,\ldots,k_N}(x_1, x_2, \ldots, x_M; y_1, y_2, \ldots, y_N) = \langle V^*_{j_1}(x_1)V^*_{j_2}(x_2) \ldots V^*_{j_M}(x_M)V_{k_1}(y_1)V_{k_2}(y_2) \ldots V_{k_N}(y_N)\rangle, \quad (8.2\text{–}14)$$

where the subscripts $j_1, j_2, \ldots, k_1, k_2, \ldots$ label Cartesian components of the vector field $\mathbf{V}$. A further generalization relating to the electromagnetic field, which we will not discuss, would involve Cartesian components of both the electric and the magnetic fields in an expression of the form appearing on the right-hand side of Eq. (8.2–14). We encountered second-order correlation tensors of this kind in Section 6.5.

In practice the correlation functions with $M = N$, sometimes called even-order correlation functions, are of more importance than the 'odd-order ones' for which $N \neq M$. This is so because frequently the fluctuating field behaves as a Gaussian random process, and for such a process, the odd-order correlation functions are all zero [Eq. (8.4–2a) below]. Moreover, when the field is quasi-monochromatic and the statistical ensemble which characterizes the fluctuations is stationary, but not necessarily Gaussian, the odd-order correlations are zero except for very large values of the order indices M and N, as we will demonstrate in the next Section.

8.3 Space-frequency correlation functions of arbitrary order‡

The correlation functions which we discussed in the two preceding sections characterize field correlations in the space-time domain. We will now turn our attention to correlations in the space-frequency domain.

Let us again consider a fluctuating scalar field represented by an analytic signal $V(\mathbf{r}, t)$, which is not necessarily stationary and let us represent it as a Fourier integral:

$$V(\mathbf{r}, t) = \int_0^\infty \widetilde{V}(\mathbf{r}, \nu)\, e^{-2\pi i\nu t}\, d\nu. \tag{8.3–1}$$

If we substitute from Eq. (8.3–1) into Eq. (8.2–3) we obtain for $\Gamma^{(M,N)}$ the expression

$$\begin{aligned}\Gamma^{(M,N)}&(\mathbf{r}_1, \mathbf{r}_2, \ldots, \mathbf{r}_{M+N}; t_1, t_2, \ldots, t_{M+N})\\ &= \int_0^\infty d\nu_1 \int_0^\infty d\nu_2 \ldots \int_0^\infty d\nu_{M+N}\, \Phi^{(M,N)}(\mathbf{r}_1, \mathbf{r}_2, \ldots, \mathbf{r}_{M+N}; \nu_1, \nu_2, \ldots, \nu_{M+N})\\ &\quad\times \prod_{j=1}^{M} \exp(2\pi i\nu_j t_j) \prod_{k=M+1}^{M+N} \exp(-2\pi i\nu_k t_k),\end{aligned} \tag{8.3–2}$$

where

$$\begin{aligned}\Phi^{(M,N)}&(\mathbf{r}_1, \mathbf{r}_2, \ldots, \mathbf{r}_{M+N}; \nu_1, \nu_2, \ldots, \nu_{M+N})\\ &= \langle \widetilde{V}^*(\mathbf{r}_1, \nu_1)\widetilde{V}^*(\mathbf{r}_2, \nu_2), \ldots, \widetilde{V}^*(\mathbf{r}_M, \nu_M)\\ &\quad\times \widetilde{V}(\mathbf{r}_{M+1}, \nu_{M+1})\widetilde{V}(\mathbf{r}_{M+2}, \nu_{M+2}), \ldots, \widetilde{V}(\mathbf{r}_{M+N}, \nu_{M+N})\rangle.\end{aligned} \tag{8.3–3}$$

We will refer to $\Phi^{(M,N)}$, defined by Eq. (8.3–3), as *the spectral cross-correlation function of order* (M, N) *of the field at the points* $\mathbf{r}_1, \mathbf{r}_2, \ldots, \mathbf{r}_{M+N}$.

If we take the Fourier inverse of Eq. (8.3–2) we obtain the following expression for the spectral cross-correlation functions in terms of the space-time correlation functions $\Gamma^{(M,N)}$:

$$\begin{aligned}\Phi^{(M,N)}&(\mathbf{r}_1, \mathbf{r}_2, \ldots, \mathbf{r}_{M+N}; \nu_1, \nu_2, \ldots, \nu_{M+N})\\ &= \int_{-\infty}^\infty dt_1 \int_{-\infty}^\infty dt_2 \ldots \int_{-\infty}^\infty dt_{M+N}\, \Gamma^{(M,N)}(\mathbf{r}_1, \mathbf{r}_2, \ldots, \mathbf{r}_{M+N}; t_1, t_2, \ldots, t_{M+N})\\ &\quad\times \prod_{j=1}^{M} \exp(-2\pi i\nu_j t_j) \prod_{k=M+1}^{M+N} \exp(2\pi i\nu_k t_k).\end{aligned} \tag{8.3–4}$$

Suppose now that the ensemble representing the fluctuating field is *stationary*.

‡ Some of the relations presented in this section were first obtained by Mehta and Mandel (1967).

The space-time correlation functions $\Gamma^{(M,N)}$ will then be invariant with respect to translation of the origin of time. In particular if, as before, we set

$$\tau_l = t_l - t_1, \quad (l = 2, 3, \ldots, M + N), \tag{8.3–5}$$

$\Gamma^{(M,N)}$ will be independent of t_1 and we then have

$$\Gamma^{(M,N)} \equiv \Gamma^{(M,N)}(\mathbf{r}_1, \mathbf{r}_2, \ldots, \mathbf{r}_{M+N}; \tau_2, \ldots, \tau_{M+N}). \tag{8.3–6}$$

On substituting from Eq. (8.3–6) into the multiple integral on the right-hand side of Eq. (8.3–4) and carrying out the (trivial) integration with respect to t_1 we obtain the following expression for the spectral cross-correlation function $\Phi^{(M,N)}$:

$$\begin{aligned}\Phi^{(M,N)}&(\mathbf{r}_1, \mathbf{r}_2, \ldots, \mathbf{r}_{M+N}; \nu_1, \nu_2, \ldots, \nu_{M+N}) \\ &= \delta(\nu_1 + \nu_2 + \ldots + \nu_M - \nu_{M+1} - \nu_{M+2} - \ldots - \nu_{M+N}) \\ &\quad \times W^{(M,N)}(\mathbf{r}_1, \mathbf{r}_2, \ldots, \mathbf{r}_{M+N}; \nu_2, \nu_3, \ldots, \nu_{M+N}),\end{aligned} \tag{8.3–7}$$

where $\delta(\nu)$ is the Dirac delta function and

$$\begin{aligned}W^{(M,N)}&(\mathbf{r}_1, \mathbf{r}_2, \ldots, \mathbf{r}_{M+N}; \nu_2, \nu_3, \ldots, \nu_{M+N}) \\ &= \int_{-\infty}^{\infty} \mathrm{d}\tau_2 \int_{-\infty}^{\infty} \mathrm{d}\tau_3 \ldots \int_{-\infty}^{\infty} \mathrm{d}\tau_{M+N} \Gamma^{(M,N)}(\mathbf{r}_1, \mathbf{r}_2, \ldots, \mathbf{r}_{M+N}; \tau_2, \tau_3, \ldots, \tau_{M+N}) \\ &\quad \times \prod_{j=2}^{M} \exp(-2\pi \mathrm{i} \nu_j \tau_j) \prod_{k=M+1}^{M+N} \exp(2\pi \mathrm{i} \nu_k \tau_k).\end{aligned} \tag{8.3–8}$$

We have implicitly assumed in this derivation that $M > 1$. When $M = 1$ one readily finds that the first product term (with index j) on the right-hand side in Eq. (8.3–8) has to be replaced by unity.

The formulas (8.3–3) and (8.3–7) imply that any $M + N$ (generalized) Fourier components $\widetilde{V}(\mathbf{r}_j, \nu_j)$, $(j = 1, 2, \ldots, M + N)$, of a stationary random field are δ-correlated, in the sense that the spectral cross-correlation function

$$\Phi^{(M,N)}(\mathbf{r}_1, \mathbf{r}_2, \ldots, \mathbf{r}_{M+N}; \nu_1, \nu_2, \ldots, \nu_{M+N}) = 0 \tag{8.3–9}$$

unless

$$\nu_1 + \nu_2 + \ldots + \nu_M - \nu_{M+1} - \nu_{M+2} - \ldots - \nu_{M+N} = 0. \tag{8.3–10}$$

When the $M + N$ frequencies satisfy Eq. (8.3–10) the components $\widetilde{V}(\mathbf{r}_j, \nu_j)$ will in general be correlated, their correlation then being characterized by the function $W^{(M,N)}$.

Let us eliminate $\Phi^{(M,N)}$ between Eqs. (8.3–3) and (8.3–7). One then obtains the equation

$$\begin{aligned}\langle \widetilde{V}^*(\mathbf{r}_1, \nu_1)\widetilde{V}^*(\mathbf{r}_2, \nu_2) \ldots \widetilde{V}^*(\mathbf{r}_M, \nu_M)\widetilde{V}(\mathbf{r}_{M+1}, \nu_{M+1})\widetilde{V}(\mathbf{r}_{M+2}, \nu_{M+2}) \ldots \\ \widetilde{V}(\mathbf{r}_{M+N}, \nu_{M+N})\rangle \\ = W^{(M,N)}(\mathbf{r}_1, \mathbf{r}_2, \ldots, \mathbf{r}_{M+N}; \nu_2, \ldots, \nu_{M+N}) \\ \times \delta(\nu_1 + \nu_2 + \ldots + \nu_M - \nu_{M+1} - \nu_{M+2} - \ldots - \nu_{M+N}).\end{aligned} \tag{8.3–11}$$

This formula, which evidently is a generalization of Eq. (4.3–39) (pertaining to the case when $M = N = 1$) may be regarded as defining *the cross-spectral density function of order* (M, N) *of a stationary optical field* $V(\mathbf{r}, t)$.

If we take the Fourier transform of Eq. (8.3–8) with respect to the frequencies $\nu_2, \nu_3, \ldots \nu_{M+N}$, we obtain the following expression for $\Gamma^{(M,N)}$ in terms of $W^{(M,N)}$:

$$\Gamma^{(M,N)}(\mathbf{r}_1, \mathbf{r}_2, \ldots, \mathbf{r}_{M+N}; \tau_2, \tau_3, \ldots, \tau_{M+N})$$
$$= \int_0^\infty \mathrm{d}\nu_2 \int_0^\infty \mathrm{d}\nu_3 \ldots \int_0^\infty \mathrm{d}\nu_{M+N} W^{(M,N)}(\mathbf{r}_1, \mathbf{r}_2, \ldots, \mathbf{r}_{M+N}; \nu_2, \nu_3, \ldots, \nu_{M+N})$$
$$\times \prod_{j=2}^{M} \exp(2\pi i \nu_j \tau_j) \prod_{k=M+1}^{M+N} \exp(-2\pi i \nu_k \tau_k). \tag{8.3–12}$$

The pair of formulas (8.3–8) and (8.3–12) for higher-order correlations is analogous to the generalized *Wiener–Khintchine theorem* [Eqs. (2.4–37), and 2.4–38].

We will show that, when the light is quasi-monochromatic, $W^{(M,N)}$ has zero value when $M \neq N$, unless M and N are large numbers. It follows from Eq. (8.3–7) that in order that the spectral correlation of order (M, N) be non-zero, the condition (8.3–10) must be satisfied. Now for quasi-monochromatic light, the (generalized) Fourier amplitudes $|\tilde{V}(\mathbf{r}_j, \nu_j)|$ [see Eq. (8.3–1)] differ appreciably from zero only when the ν_j's are confined to the interval $\nu_0 - \frac{1}{2}\Delta\nu \leqslant \nu_j \leqslant \nu_0 + \frac{1}{2}\Delta\nu$, where ν_0 is a mid-frequency and $\Delta\nu \ll \nu_0$ is the effective bandwidth of the light.

If we set

$$\nu_j = \nu_0 + \delta\nu_j, \qquad (j = 1, 2, \ldots, M + N), \tag{8.3–13}$$

then

$$\nu_1 + \nu_2 + \ldots + \nu_M = M\nu_0 + (\delta\nu_1 + \delta\nu_2 + \ldots + \delta\nu_M) \tag{8.3–14a}$$

and

$$\nu_{M+1} + \nu_{M+2} + \ldots + \nu_{M+N} = N\nu_0 + (\delta\nu_{M+1} + \delta\nu_{M+2} + \ldots + \delta\nu_{M+N}). \tag{8.3–14b}$$

Since $|\delta\nu_j| \leqslant \frac{1}{2}\Delta\nu_j$ Eq. (8.3–14a) implies that

$$M(\nu_0 - \tfrac{1}{2}\Delta\nu) \leqslant \nu_1 + \nu_2 + \ldots + \nu_M \leqslant M(\nu_0 + \tfrac{1}{2}\Delta\nu), \tag{8.3–15a}$$

and Eq. (8.3–14b) implies that

$$N(\nu_0 - \tfrac{1}{2}\Delta\nu) \leqslant \nu_{M+1} + \nu_{M+2} + \ldots + \nu_{M+N} \leqslant N(\nu_0 + \tfrac{1}{2}\Delta\nu). \tag{8.3–15b}$$

If now $\nu_1, \nu_2, \ldots, \nu_{M+N}$ are $M + N$ frequencies that satisfy the requirement (8.3–10), it follows at once from the inequalities (8.3–15) that

$$|N - M|\nu_0 - \tfrac{1}{2}(N + M)\Delta\nu \leqslant 0 \leqslant |N - M|\nu_0 + \tfrac{1}{2}(N + M)\Delta\nu. \tag{8.3–16}$$

This inequality can readily be seen to imply that

$$\frac{\Delta\nu}{\nu_0} \geqslant \frac{2|M - N|}{M + N}. \tag{8.3–17}$$

Clearly, for quasi-monochromatic light ($\Delta\nu/\nu_0 \ll 1$), this inequality can only be satisfied if either $M = N$ or if M and N are both large compared to unity. Hence we conclude that for quasi-monochromatic light $W^{(M,N)}$, and consequently also $\Gamma^{(M,N)}$, can only be non-zero under these circumstances.

8.4 Correlation functions of fields obeying Gaussian statistics

When the field fluctuations can be described as a Gaussian random process, the field is often said to obey Gaussian statistics. All the correlation functions of the field can then be expressed in terms of the lowest-order ones, by the use of the moment theorem for such processes [Eq. (1.6–33)]. We will now briefly consider some of the important formulas which hold for fields of this kind.

8.4.1 Space-time domain

When the field $V(\mathbf{r}, t)$ at each point is of zero mean, i.e. when

$$\langle V(\mathbf{r}, t)\rangle = 0 \tag{8.4–1}$$

for all $\mathbf{r}$, it follows at once from the definition (8.2–3) of the space-time correlation function $\Gamma^{(N,M)}$ and from the moment theorem for a Gaussian random process [Eq. (1.6–33)] that

$$\Gamma^{(M,N)}(x_1, x_2, \ldots, x_M; y_1, y_2, \ldots, y_N) = 0 \quad \text{if } N \neq M \tag{8.4–2a}$$

and

$$\begin{aligned}\Gamma^{(M,M)}&(x_1, x_2, \ldots, x_M; y_1, y_2, \ldots, y_M)\\ &= \sum_{\pi}\Gamma^{(1,1)}(x_{i_1}, y_{j_1})\Gamma^{(1,1)}(x_{i_2}, y_{j_2}) \ldots \Gamma^{(1,1)}(x_{i_M}, y_{j_M}),\end{aligned} \tag{8.4–2b}$$

where the subscripts i_p and j_q, $(1 \leqslant i_p \leqslant M,\ 1 \leqslant j_q \leqslant M)$, are integers and $\sum_\pi$ denotes summation over all the $M!$ possible permutations of the subscripts.

Some consequences of the formula (8.4–2b) for the case when $M = N = 2$ are useful in practice and we will, therefore, briefly consider this special case in some detail.

When $M = 2$, Eq. (8.4–2b) gives

$$\Gamma^{(2,2)}(x_1, x_2; x_3, x_4) = \Gamma^{(1,1)}(x_1, x_3)\Gamma^{(1,1)}(x_2, x_4) + \Gamma^{(1,1)}(x_1, x_4)\Gamma^{(1,1)}(x_2, x_3). \tag{8.4–3}$$

Let us set $x_3 = x_1$, $x_4 = x_2$. Since, according to Eq. (8.2–9), $\Gamma^{(1,1)}(x_j, x_j) = \langle I(x_j)\rangle$ is the average value of the intensity at the space-time point x_j and since, according to Eq. (8.2–7), $\Gamma^{(1,1)}(x_2, x_1) = [\Gamma^{(1,1)}(x_1, x_2)]^*$, Eq. (8.4–3) gives

$$\Gamma^{(2,2)}(x_1, x_2; x_1, x_2) = \langle I(x_1)\rangle\langle I(x_2)\rangle + |\Gamma^{(1,1)}(x_1, x_2)|^2. \tag{8.4–4}$$

If, in addition, the field is stationary, $\Gamma^{(2,2)}(x_1, x_2;\ x_1,\ x_2)$ and $\Gamma^{(1,1)}(x_1, x_2)$ depend on the time arguments only through the difference $t_2 - t_1$ and we then write

$$\Gamma^{(2,2)}(x_1, x_2; x_1, x_2) \equiv \Gamma^{(2,2)}(\mathbf{r}_1, \mathbf{r}_2, t_2 - t_1),$$
$$\Gamma^{(1,1)}(x_1, x_2) \equiv \Gamma^{(1,1)}(\mathbf{r}_1, \mathbf{r}_2, t_2 - t_1).$$

Equation (8.4–4) then reduces to

$$\Gamma^{(2,2)}(\mathbf{r}_1, \mathbf{r}_2, t_2 - t_1) = \langle I(\mathbf{r}_1, t)\rangle\langle I(\mathbf{r}_2, t)\rangle[1 + |\gamma^{(1,1)}(\mathbf{r}_1, \mathbf{r}_2, t_2 - t_1)|^2], \tag{8.4–5}$$

where

$$\gamma^{(1,1)}(\mathbf{r}_1, \mathbf{r}_2, \tau) = \frac{\Gamma^{(1,1)}(\mathbf{r}_1, \mathbf{r}_2, \tau)}{[\langle I(\mathbf{r}_1, t)\rangle]^{1/2}[\langle I(\mathbf{r}_2, t)\rangle]^{1/2}} \tag{8.4–6}$$

is just the usual (second-order) complex degree of coherence [Eq. (4.3–12b)]. The average values $\langle I(\mathbf{r}_j, t)\rangle$, $(j = 1, 2)$, of the intensities are now, of course, independent of time, because the field has been assumed to be stationary.

Let us introduce the intensity fluctuations

$$\Delta I_j = I(\mathbf{r}_j, t_j) - \langle I(\mathbf{r}_j, t_j)\rangle. \tag{8.4–7}$$

Then

$$\langle \Delta I_1 \Delta I_2\rangle \equiv \langle I(\mathbf{r}_1, t_1)I(\mathbf{r}_2, t_2)\rangle - \langle I(\mathbf{r}_1, t_1)\rangle\langle I(\mathbf{r}_2, t_2)\rangle. \tag{8.4–8}$$

Now

$$\langle I(\mathbf{r}_1, t_1)I(\mathbf{r}_2, t_2)\rangle = \Gamma^{(2,2)}(\mathbf{r}_1, \mathbf{r}_2, t_2 - t_1) \tag{8.4–9}$$

and it follows from Eqs. (8.4–8), (8.4–9) and (8.4–5) that

$$\langle \Delta I_1 \Delta I_2\rangle = \langle I(\mathbf{r}_1, t_1)\rangle\langle I(\mathbf{r}_2, t_2)\rangle|\gamma^{(1,1)}(\mathbf{r}_1, \mathbf{r}_2, t_2 - t_1)|^2. \tag{8.4–10}$$

In particular, if $\mathbf{r}_1 = \mathbf{r}_2 = \mathbf{r}$, $t_1 = t_2 = t$ and we make use of the fact that $\gamma(\mathbf{r}, \mathbf{r}, 0) = 1$, Eq. (8.4–10) reduces to

$$\langle(\Delta I)^2\rangle = \langle I(\mathbf{r}, t)\rangle^2, \tag{8.4–11}$$

where $\Delta I \equiv \Delta I(\mathbf{r}, t)$.

Up to now we have considered a scalar field $V(\mathbf{r}, t)$ only, so that the preceding formulas cannot take into account the polarization properties of the electromagnetic field. We will now generalize the formula (8.4–10) to include polarization effects.

Let $\mathbf{V}(\mathbf{r}, t)$ represent a vector field, e.g. the electric field vector of a plane wave, whose statistical properties are described by a stationary Gaussian random process of zero mean. Let $V_x(\mathbf{r}, t)$ and $V_y(\mathbf{r}, t)$ be the components of $\mathbf{V}(\mathbf{r}, t)$, represented by complex analytic signals, in two mutually orthogonal directions, perpendicular to the direction of propagation of the wave. The instantaneous intensity is then given by the expression

$$I(\mathbf{r}, t) = I_x(\mathbf{r}, t) + I_y(\mathbf{r}, t), \tag{8.4–12}$$

where $I_x = V_x^* V_x$ and $I_y = V_y^* V_y$. The fluctuation of the intensity, i.e. the quantity

$\Delta I(\mathbf{r}, t) = I(\mathbf{r}, t) - \langle I(\mathbf{r}, t)\rangle$ is then given by the simple formula

$$\Delta I(\mathbf{r}, t) = \Delta I_x(\mathbf{r}, t) + \Delta I_y(\mathbf{r}, t), \tag{8.4–13}$$

where $\Delta I_j(\mathbf{r}, t) = I_j(\mathbf{r}, t) - \langle I_j(\mathbf{r}, t)\rangle$, $(j = x, y)$, represents the intensity fluctuations of the two components of $\mathbf{V}(\mathbf{r}, t)$. Using Eq. (8.4–13) we deduce at once that the correlation between the intensity fluctuations of the wave at any two space-time points is given by the formula

$$\langle \Delta I(\mathbf{r}_1, t)\Delta I(\mathbf{r}_2, t+\tau)\rangle = \sum_i\sum_j \langle \Delta I_i(\mathbf{r}_1, t)\Delta I_j(\mathbf{r}_2, t+\tau)\rangle, \quad (i = x, y; j = x, y). \tag{8.4–14}$$

The formula (8.4–14) holds generally. Let us now specialize it to the case when the statistical properties of $\mathbf{V}(\mathbf{r}, t)$ are those of a Gaussian random process. We then readily find, by an argument similar to one that we employed earlier in connection with the scalar field, that each term under the summation signs in Eq. (8.4–14) is given by a formula that is analogous to Eq. (8.4–10). Consequently Eq. (8.4–14) implies that

$$\langle \Delta I(\mathbf{r}_1, t)\Delta I(\mathbf{r}_2, t+\tau)\rangle = \sum_i\sum_j \langle I_i(\mathbf{r}_1, t)\rangle\langle I_j(\mathbf{r}_2, t)\rangle |\gamma_{ij}^{(1,1)}(\mathbf{r}_1, \mathbf{r}_2, \tau)|^2, \tag{8.4–15}$$

where

$$\gamma_{ij}^{(1,1)}(\mathbf{r}_1, \mathbf{r}_2, \tau) = \frac{\Gamma_{ij}^{(1,1)}(\mathbf{r}_1, \mathbf{r}_2, \tau)}{[\langle I_i(\mathbf{r}_1, t)\rangle]^{1/2}[\langle I_j(\mathbf{r}_2, t)\rangle]^{1/2}} \tag{8.4–16}$$

and

$$\Gamma_{ij}^{(1,1)}(\mathbf{r}_1, \mathbf{r}_2, \tau) = \langle V_i^*(\mathbf{r}_1, t)V_j(\mathbf{r}_2, t+\tau)\rangle. \tag{8.4–17}$$

Equation (8.4–15) is the required generalization of Eq. (8.4–10).

Although the formula (8.4–15) appears to depend on the choice of the x- and y-directions it must be independent of it, because the quantity on the left-hand side is a scalar, as may be readily verified by expressing the right-hand side of that equation in a slightly different form. For this purpose we note that each term under the summation sign in Eq. (8.4–15) is, according to Eq. (8.4–16), equal to $|\Gamma_{ij}^{(1,1)}(\mathbf{r}_1, \mathbf{r}_2, \tau)|^2$. Consequently the formula (8.4–15) may be expressed in the form

$$\langle \Delta I(\mathbf{r}_1, t)\Delta I(\mathbf{r}_2, t+\tau)\rangle = \mathrm{tr}\,[\mathbf{J}(\tau)\mathbf{J}^\dagger(\tau)], \tag{8.4–18}$$

where

$$\mathbf{J}(\tau) = \begin{bmatrix} \Gamma_{xx}^{(1,1)}(\mathbf{r}_1, \mathbf{r}_2, \tau) & \Gamma_{xy}^{(1,1)}(\mathbf{r}_1, \mathbf{r}_2, \tau) \\ \Gamma_{yx}^{(1,1)}(\mathbf{r}_1, \mathbf{r}_2, \tau) & \Gamma_{yy}^{(1,1)}(\mathbf{r}_1, \mathbf{r}_2, \tau) \end{bmatrix}. \tag{8.4–19}$$

If the vector field $\mathbf{V}(\mathbf{r}, t)$ is chosen to be the electric field vector, this matrix is evidently a generalization of the 2×2 equal-time coherence (or polarization) matrix (6.2–6) and $\mathbf{J}^\dagger$ is its Hermitian conjugate (adjoint). Since the right-hand side of Eq. (8.4–18) is the trace of a matrix it is indeed independent of the choice of the x, y-directions.

We will now briefly consider some special cases of Eq. (8.4–15) that are of practical interest. Suppose first that the wave is *linearly polarized* in the x-direction, say. Then $\langle I_y(\mathbf{r}, t)\rangle \equiv 0$ and Eq. (8.4–15) reduces to

$$\langle \Delta I(\mathbf{r}_1, t)\Delta I(\mathbf{r}_2, t+\tau)\rangle = \langle I_x(\mathbf{r}_1, t)\rangle\langle I_x(\mathbf{r}_2, t)\rangle|\gamma_{xx}^{(1,1)}(\mathbf{r}_1, \mathbf{r}_2, \tau)|^2. \quad (8.4\text{–}20)$$

This formula is in agreement with Eq. (8.4–10) for a scalar field, as was to be expected.

Next, suppose that the wave is completely *unpolarized*. The x- and y-components of the field are then uncorrelated (cf. Section 6.3.1), i.e.

$$\gamma_{xy}^{(1,1)}(\mathbf{r}_1, \mathbf{r}_2, \tau) = 0 \quad (8.4\text{–}21)$$

and

$$\langle I_x(\mathbf{r}, t)\rangle = \langle I_y(\mathbf{r}, t)\rangle = \tfrac{1}{2}\langle I(\mathbf{r}, t)\rangle. \quad (8.4\text{–}22)$$

Moreover, since for an unpolarized wave the x- and y-directions are completely equivalent, we also have

$$\gamma_{xx}^{(1,1)}(\mathbf{r}_1, \mathbf{r}_2, \tau) = \gamma_{yy}^{(1,1)}(\mathbf{r}_1, \mathbf{r}_2, \tau). \quad (8.4\text{–}23)$$

On making use of Eqs. (8.4–21), (8.4–22) and (8.4–23) in Eq. (8.4–15) and dropping the subscripts on $\gamma_{xx}^{(1,1)}$ and $\gamma_{yy}^{(1,1)}$ we obtain the formula

$$\langle \Delta I(\mathbf{r}_1, t)\Delta I(\mathbf{r}_2, t+\tau)\rangle = \tfrac{1}{2}\langle I(\mathbf{r}_1, t)\rangle\langle I(\mathbf{r}_2, t)\rangle|\gamma^{(1,1)}(\mathbf{r}_1, \mathbf{r}_2, \tau)|^2. \quad (8.4\text{–}24)$$

On comparing Eqs. (8.4–24) and (8.4–20) we see that the correlation between the intensity fluctuations of the unpolarized wave is just one half of that of the polarized wave of the same averaged intensity and with the same degree of coherence.

The formula (8.4–24) shows that from knowledge of the correlations of the intensity fluctuations of an unpolarized, statistically stationary wave, whose fluctuations are governed by Gaussian statistics, one can determine the absolute value of the second-order degree of coherence of the wave. This result is the basis of correlation (or intensity) interferometry, which we will discuss in Sections 9.9 and 9.10.

As another example let us consider the variance of the intensity fluctuations in a partially polarized wave. We have from the general formula (8.4–15), with $\mathbf{r}_1 = \mathbf{r}_2 = \mathbf{r}$, $\tau = 0$,

$$\langle[\Delta I(\mathbf{r}, t)]^2\rangle = \sum_i\sum_j\langle I_i(\mathbf{r}, t)\rangle\langle I_j(\mathbf{r}, t)\rangle|\gamma_{ij}^{(1,1)}(\mathbf{r}, \mathbf{r}, 0)|^2 \quad (8.4\text{–}25)$$

or, since the normalization of the correlation coefficient is such that $\gamma_{xx}^{(1,1)}(\mathbf{r}, \mathbf{r}, 0) = \gamma_{yy}^{(1,1)}(\mathbf{r}, \mathbf{r}, 0) = 1$, the formula (8.4–25) reduces to

$$\langle[\Delta I(\mathbf{r}, t)]^2\rangle = \langle I_x(\mathbf{r}, t)\rangle^2 + \langle I_y(\mathbf{r}, t)\rangle^2 + 2\langle I_x(\mathbf{r}, t)\rangle\langle I_y(\mathbf{r}, t)\rangle|\gamma_{xy}^{(1,1)}(\mathbf{r}, \mathbf{r}, 0)|^2. \quad (8.4\text{–}26)$$

Now one can readily show by the use of the transformation (6.2–33) of the elements of the coherence matrix under a rotation of the x-, y-axis about the direction of propagation of the wave that the axes may be so chosen that $\langle I_x \rangle = \langle I_y \rangle$; and Eq. (6.3–26) implies that under these circumstances $|\gamma_{xy}^{(1,1)}(\mathbf{r}, \mathbf{r}, 0)|$ is just equal to the degree of polarization $P(\mathbf{r})$ of the wave. Hence Eq. (8.4–26) implies that

$$\langle [\Delta I(\mathbf{r}, t)]^2 \rangle = \tfrac{1}{2}[1 + P^2(\mathbf{r})]\langle I(\mathbf{r}, t)\rangle^2, \tag{8.4–27}$$

where Eq. (8.4–22) was used.

8.4.2 *Space-frequency domain*

We have seen in Section 8.4.1 that when the field fluctuations can be described as a Gaussian random process, the space-time correlation functions $\Gamma^{(M,N)}$ are expressible in terms of $\Gamma^{(1,1)}$. From this result it readily follows that the spectral cross-correlation functions $\Phi^{(M,N)}$ are expressible in terms of $\Phi^{(1,1)}$. To see this we only need to substitute from Eq. (8.4–2) into Eqs. (8.3–4) and we then readily find that

$$\Phi^{(M,N)}(\mathbf{r}_1, \mathbf{r}_2, \ldots, \mathbf{r}_M; \mathbf{r}'_1, \mathbf{r}'_2, \ldots, \mathbf{r}'_N; \nu_1, \nu_2, \ldots, \nu_{M+N}) = 0 \quad \text{if } N \neq M \tag{8.4–28a}$$

and

$$\Phi^{(M,M)}(\mathbf{r}_1, \mathbf{r}_2, \ldots, \mathbf{r}_M; \mathbf{r}'_1, \mathbf{r}'_2, \ldots, \mathbf{r}'_M; \nu_1, \nu_2, \ldots, \nu_{2M})$$
$$= \sum_{\pi} \Phi^{(1,1)}(\mathbf{r}_{i_1}, \mathbf{r}'_{j_1}; \nu_{i_1}, \nu_{j_1})\Phi^{(1,1)}(\mathbf{r}_{i_2}, \mathbf{r}'_{j_2}; \nu_{i_2}, \nu_{j_2}) \ldots \Phi^{(1,1)}(\mathbf{r}_{i_M}, \mathbf{r}'_{j_M}; \nu_{i_M}, \nu_{j_M}). \tag{8.4–28b}$$

As an example, let us consider the case when $M = N = 2$. For this case Eq. (8.4–28b) gives, if we write $\mathbf{r}_3$ in place of $\mathbf{r}'_1$ and $\mathbf{r}_4$ in place of $\mathbf{r}'_2$:

$$\Phi^{(2,2)}(\mathbf{r}_1, \mathbf{r}_2, \mathbf{r}_3, \mathbf{r}_4; \nu_1, \nu_2, \nu_3, \nu_4)$$
$$= \Phi^{(1,1)}(\mathbf{r}_1, \mathbf{r}_3; \nu_1, \nu_3)\Phi^{(1,1)}(\mathbf{r}_2, \mathbf{r}_4; \nu_2, \nu_4) + \Phi^{(1,1)}(\mathbf{r}_1, \mathbf{r}_4; \nu_1, \nu_4)\Phi^{(1,1)}(\mathbf{r}_2, \mathbf{r}_3; \nu_2, \nu_3). \tag{8.4–29}$$

If the field is stationary one readily obtains from Eqs. (8.4–28) and Eq. (8.3–7) corresponding formulas for the higher-order cross-spectral densities in terms of the lowest-order ones. In the special case when $M = N = 2$ we obtain, for example,

$$W^{(2,2)}(\mathbf{r}_1, \mathbf{r}_2, \mathbf{r}_3, \mathbf{r}_4; \nu_2, \nu_3, \nu_4)\delta(\nu_1 + \nu_2 - \nu_3 - \nu_4)$$
$$= W^{(1,1)}(\mathbf{r}_1, \mathbf{r}_3; \nu_3)W^{(1,1)}(\mathbf{r}_2, \mathbf{r}_4; \nu_4)\delta(\nu_3 - \nu_1)\delta(\nu_4 - \nu_2)$$
$$+ W^{(1,1)}(\mathbf{r}_1, \mathbf{r}_4; \nu_4)W^{(1,1)}(\mathbf{r}_2, \mathbf{r}_3; \nu_3)\delta(\nu_4 - \nu_1)\delta(\nu_3 - \nu_2). \tag{8.4–30}$$

8.5 Coherent-mode representation of cross-spectral densities of arbitrary order‡

8.5.1 General expressions

In Section 4.7 we showed that the cross-spectral density of the lowest order ($M = N = 1$), which we denoted by $W(\mathbf{r}_1, \mathbf{r}_2, \nu)$, of a stationary optical field in a finite domain D in free space, may be expressed in the following form [Eq. (4.7–9)]:

$$W(\mathbf{r}_1, \mathbf{r}_2, \nu) = \sum_n \alpha_n(\nu)\psi_n^*(\mathbf{r}_1, \nu)\psi_n(\mathbf{r}_2, \nu), \qquad (8.5\text{–}1)$$

sometimes called the coherent-mode representation of the cross-spectral density. Here ψ_n are the (orthonormalized) eigenfunctions and α_n the eigenvalues of the integral equation

$$\int_D W(\mathbf{r}_1, \mathbf{r}_2, \nu)\psi_n(\mathbf{r}_1, \nu)\, \mathrm{d}^3 r_1 = \alpha_n(\nu)\psi_n(\mathbf{r}_2, \nu). \qquad (8.5\text{–}2)$$

We also showed that the cross-spectral density $W(\mathbf{r}_1, \mathbf{r}_2, \nu)$ can be expressed as a correlation function, in the form [Eq. (4.7–38)]

$$W(\mathbf{r}_1, \mathbf{r}_2, \nu) = \langle U^*(\mathbf{r}_1, \nu)U(\mathbf{r}_2, \nu)\rangle_\nu. \qquad (8.5\text{–}3)$$

Here $\{U(\mathbf{r}, \nu)\exp(-2\pi i\nu t)\}$ is an ensemble of monochromatic fields, all of frequency ν, and the average on the right-hand side of Eq. (8.5–3) is taken over this ensemble. The fields $U(\mathbf{r}, \nu)$ can be generated from the coherent mode $\psi_n(\mathbf{r}, \nu)$ by means of the formula

$$U(\mathbf{r}, \nu) = \sum_n a_n(\nu)\psi_n(\mathbf{r}, \nu), \qquad (8.5\text{–}4)$$

where the $a_n(\nu)$ are random coefficients such that

$$\langle a_n^*(\nu)a_m(\nu)\rangle_\nu = \alpha_n(\nu)\delta_{nm}, \qquad (8.5\text{–}5)$$

δ_{nm} being the Kronecker symbol.

The representations (8.5–1) and (8.5–3) have proved to be very useful in treatments of various problems of statistical optics, as is evident, for example, from the analysis of some of the propagation problems with partially coherent light considered in Chapter 5 and in connection with the theory of laser resonator modes discussed in Section 7.4.

In this section we will extend the space-frequency formulation to correlations of arbitrary order. However, unlike the situation with the lowest-order correlations, it appears that a rigorous formulation of the higher-order theory requires the use of generalized functions. As elsewhere in this book we do not employ the rather complicated apparatus of generalized function theory but use instead heuristic arguments and the Dirac delta function.

We begin by formally expanding the (generalized) Fourier transform $\widetilde{V}(\mathbf{r}, \nu)$ of the fluctuating field variable $V(\mathbf{r}, t)$ [Eq. (4.3–38)] in terms of the eigenfunctions

‡ The analysis presented in this section is based on an investigation of Agarwal and Wolf (1993).

$\psi_n(\mathbf{r}, \nu)$ of the integral Eq. (8.5–2):

$$\widetilde{V}(\mathbf{r}, \nu) = \sum_n b_n(\nu)\psi_n(\mathbf{r}, \nu). \qquad (8.5\text{–}6)$$

Because the functions ψ_n form an orthonormal set over the domain D it follows at once from the expansion (8.5–6) that in terms of $\widetilde{V}$, the expansion coefficients $b_n(\nu)$ are given by the formula

$$b_n(\nu) = \int_D \widetilde{V}(\mathbf{r}, \nu)\psi_n^*(\mathbf{r}, \nu)\,\mathrm{d}^3 r. \qquad (8.5\text{–}7)$$

It is important to appreciate that two different statistical ensembles appear in our analysis: the ensemble of the random functions $U(\mathbf{r}, \nu)$ and the ensemble of the random functions $\widetilde{V}(\mathbf{r}, \nu)$. The former is characterized by an ensemble of the expansion coefficients $a_n(\nu)$, the latter by an ensemble of expansion coefficients $b_n(\nu)$.

We can readily establish a connection between correlations of the a_n's and the b_n's. We have at once, on making use of the formula (8.5–7) and interchanging the order of averaging and integration, that

$$\langle b_n^*(\nu)b_m(\nu')\rangle = \int_D\int_D \langle \widetilde{V}^*(\mathbf{r}_1, \nu)\widetilde{V}(\mathbf{r}_2, \nu')\rangle \psi_n(\mathbf{r}_1, \nu)\psi_m^*(\mathbf{r}_2, \nu')\,\mathrm{d}^3 r_1\,\mathrm{d}^3 r_2. \qquad (8.5\text{–}8)$$

The angular brackets on the left-hand side represent, of course, the average over the ensemble of the b_n's. This average must be distinguished from the average over the ensemble of the a_n's, which is denoted by angular brackets with subscript ν, as in Eq. (8.5–5).

Let us now make use of Eq. (4.3–39) by means of which the cross-spectral density function $W(\mathbf{r}_1, \mathbf{r}_2, \nu)$ was introduced:

$$\langle \widetilde{V}^*(\mathbf{r}_1, \nu)\widetilde{V}(\mathbf{r}_2, \nu')\rangle = W(\mathbf{r}_1, \mathbf{r}_2, \nu)\delta(\nu - \nu'). \qquad (8.5\text{–}9)$$

On substituting from Eq. (8.5–9) into Eq. (8.5–8) we obtain the formula

$$\langle b_n^*(\nu)b_m(\nu')\rangle = \delta(\nu - \nu')\int_D\int_D W(\mathbf{r}_1, \mathbf{r}_2, \nu)\psi_n(\mathbf{r}_1, \nu)\psi_m^*(\mathbf{r}_2, \nu')\,\mathrm{d}^3 r_1\,\mathrm{d}^3 r_2. \qquad (8.5\text{–}10)$$

According to Eq. (8.5–2), the integral with respect to $\mathbf{r}_1$ is just equal to $\alpha_n(\nu)\psi_n(\mathbf{r}_2, \nu)$ and hence Eq. (8.5–10) reduces to

$$\langle b_n^*(\nu)b_m(\nu')\rangle = \alpha_n(\nu)\delta(\nu - \nu')\int_D \psi_n(\mathbf{r}_2, \nu)\psi_m^*(\mathbf{r}_2, \nu)\,\mathrm{d}^3 r_2. \qquad (8.5\text{–}11)$$

Since the eigenfunctions ψ_n form an orthonormal set, the integral on the right is just equal to the Kronecker symbol δ_{nm} and hence

$$\langle b_n^*(\nu)b_m(\nu')\rangle = \alpha_n(\nu)\delta_{nm}\delta(\nu - \nu'). \qquad (8.5\text{–}12)$$

In view of Eq. (8.5–5), Equation (8.5–12) leads to the following relation between second-order correlations of the random expansion coefficients b_n and a_n:

$$\langle b_n^*(\nu) b_m(\nu') \rangle = \langle a_n^*(\nu) a_m(\nu) \rangle_\nu \delta(\nu - \nu'). \tag{8.5–13}$$

We may readily express cross-spectral densities of arbitrary order in terms of the mode functions ψ_n of the second-order theory. We then have, on substituting on the left-hand side of Eq. (8.3–11) for $\widetilde{V}$ the expansion (8.5–6) and interchanging the order of summation and averaging,

$$\begin{aligned} W^{(M,N)}&(\mathbf{r}_1, \mathbf{r}_2, \ldots, \mathbf{r}_{M+N}; \nu_2, \nu_3, \ldots, \nu_{M+N}) \\ &\times \delta(\nu_1 + \nu_2 + \ldots + \nu_M - \nu_{M+1} - \nu_{M+2} - \ldots - \nu_{M+N}) \\ &= \sum_{n_j} \mathcal{B}^{(M,N)}_{n_1, n_2, \ldots, n_{M+N}}(\nu_1, \nu_2, \ldots, \nu_{M+N}) \\ &\times \prod_{j=1}^{M} \psi_{n_j}^*(\mathbf{r}_j, \nu_j) \prod_{k=M+1}^{M+N} \psi_{n_k}(\mathbf{r}_k, \nu_k), \end{aligned} \tag{8.5–14}$$

where $\sum_{n_j}$ denotes summations over all possible values of the integers n_1, n_2, $\ldots$, n_{M+N} and

$$\mathcal{B}^{(M,N)}_{n_1, n_2, \ldots, n_{M+N}}(\nu_1, \nu_2, \ldots, \nu_{M+N}) = \left\langle \prod_{j=1}^{M} b_{n_j}^*(\nu_j) \prod_{k=M+1}^{M+N} b_{n_k}(\nu_k) \right\rangle \tag{8.5–15}$$

are the moments of the coefficients $b_n(\nu)$ in the expansion (8.5–6). It is evident from the structure of Eq. (8.5–14) that the moments are zero unless the constraint (8.3–10) is satisfied and when it is satisfied they have delta-function singularities.

The formula (8.5–14) may readily be inverted to obtain expressions for the moments $\mathcal{B}^{(M,N)}$ in terms of the cross-spectral density function $W^{(M,N)}$. For this purpose we substitute from Eq. (8.5–7) into the expression (8.5–15) and make use of Eq. (8.3–11). We then obtain for $\mathcal{B}^{(M,N)}$ the required formula:

$$\begin{aligned} \mathcal{B}^{(M,N)}_{n_1, n_2, \ldots, n_{M+N}}&(\nu_1, \nu_2, \ldots, \nu_{M+N}) \\ = \delta(\nu_1 + \nu_2 + \ldots &+ \nu_M - \nu_{M+1} - \nu_{M+2} - \ldots - \nu_{M+N}) \\ \int_D \mathrm{d}^3 r_1 \int_D \mathrm{d}^3 r_2 \ldots &\int_D \mathrm{d}^3 r_{M+N} W^{(M,N)}(\mathbf{r}_1, \mathbf{r}_2, \ldots, \mathbf{r}_{M+N}; \nu_2, \nu_3, \ldots, \nu_{M+N}) \\ \times \prod_{j=1}^{M} \psi_{n_j}(\mathbf{r}_j, \nu_j) &\prod_{k=M+1}^{M+N} \psi_{n_k}^*(\mathbf{r}_k, \nu_k). \end{aligned} \tag{8.5–16}$$

8.5.2 A single-mode field

Let us now briefly consider the special case of a single-mode field, i.e one whose cross-spectral density is given by a single term in the expansion (8.5–1):

$$W(\mathbf{r}_1, \mathbf{r}_2, \nu) = \alpha_0(\nu) \psi_0^*(\mathbf{r}_1, \nu) \psi_0(\mathbf{r}_2, \nu). \tag{8.5–17}$$

For such a field the spectral degree of coherence $\mu(\mathbf{r}_1, \mathbf{r}_2, \nu)$ [Eq. (4.3–47)] is unimodular, and hence the field is completely coherent in the space-frequency domain. In this case the expression (8.5–14) becomes

$$W^{(M,N)}(\mathbf{r}_1, \mathbf{r}_2, \ldots, \mathbf{r}_{M+N}; \nu_2, \nu_3, \ldots, \nu_{M+N}) \times \delta(\nu_1 + \nu_2 + \ldots + \nu_M - \nu_{M+1} - \nu_{M+2} - \ldots - \nu_{M+N}) = \mathcal{B}^{(M,N)}_{0,0,\ldots,0}(\nu_1, \nu_2, \ldots, \nu_{M+N}) \times \prod_{j=1}^{M} \psi_0^*(\mathbf{r}_j, \nu_j) \prod_{k=M+1}^{M+N} \psi_0(\mathbf{r}_k, \nu_k) \qquad (8.5\text{–}18)$$

showing that *for a single-mode field the cross-spectral densities of all orders factorize*.

8.5.3 Fields obeying Gaussian statistics

Finally let us consider the special case when the field obeys Gaussian statistics. In this case all the moments $\mathcal{B}^{(M,N)}$ may be expressed in terms of the second-order moments by making use of the moment theorem for a complex Gaussian random process [Eq. (1.6–33)]. Moreover, according to Eq. (8.5–12), the second-order moments may be expressed in terms of the eigenvalues of the integral equation (8.5–2). For example when $M = N = 2$,

$$\langle b_j^*(\nu_1) b_k^*(\nu_2) b_l(\nu_3) b_m(\nu_4)\rangle = \langle b_j^*(\nu_1) b_l(\nu_3)\rangle\langle b_k^*(\nu_2) b_m(\nu_4)\rangle + \langle b_j^*(\nu_1) b_m(\nu_4)\rangle\langle b_k^*(\nu_2) b_l(\nu_3)\rangle. \qquad (8.5\text{–}19)$$

From the definition (8.5–15) and the formula (8.5–12), Eq. (8.5–19) is seen to imply that

$$\begin{aligned}\mathcal{B}^{(2,2)}_{jklm}(\nu_1, \nu_2, \nu_3, \nu_4) &= \alpha_j(\nu_1)\delta_{jl}\delta(\nu_1 - \nu_3)\alpha_k(\nu_2)\delta_{km}\delta(\nu_2 - \nu_4) \\ &\quad + \alpha_j(\nu_1)\delta_{jm}\delta(\nu_1 - \nu_4)\alpha_k(\nu_2)\delta_{kl}\delta(\nu_2 - \nu_3) \\ &= \alpha_j(\nu_1)\alpha_k(\nu_2)[\delta_{jl}\delta_{km}\delta(\nu_1 - \nu_3)\delta(\nu_2 - \nu_4) \\ &\quad + \delta_{jm}\delta_{kl}\delta(\nu_1 - \nu_4)\delta(\nu_2 - \nu_3)]. \end{aligned} \qquad (8.5\text{–}20)$$

With the help of the formal identity[‡]

$$\delta(\nu_1 - \nu_3)\delta(\nu_2 - \nu_4) = \delta(\nu_1 + \nu_2 - \nu_3 - \nu_4)\delta(\nu_2 - \nu_4), \qquad (8.5\text{–}21)$$

the expression (8.5–20) for $\mathcal{B}^{(2,2)}_{jklm}$ becomes

$$\mathcal{B}^{(2,2)}_{jklm}(\nu_1, \nu_2, \nu_3, \nu_4) = \alpha_j(\nu_1)\alpha_k(\nu_2)\delta(\nu_1 + \nu_2 - \nu_3 - \nu_4) \times [\delta(\nu_2 - \nu_4)\delta_{jl}\delta_{km} + \delta(\nu_2 - \nu_3)\delta_{jm}\delta_{kl}]. \qquad (8.5\text{–}22)$$

Problems

8.1 $V(t)$ is the analytic signal representation of a linearly polarized, stationary, thermal field, at some fixed point in space. If $I(t) = V^*(t)V(t)$ is the instantaneous intensity, show that the spectrum $S_{\Delta I}(\nu)$ of the intensity

‡ This identity may be verified by multiplying both sides of Eq. (8.5–21) by a test function $f(\nu_1, \nu_2)$ and integrating with respect to ν_1 and ν_2.

fluctuations $\Delta I(t) = I(t) - \langle I(t) \rangle$ may be expressed in the form

$$S_{\Delta I}(\nu) = \int_0^\infty S_V(\nu') S_V(\nu' + \nu)\, \mathrm{d}\nu',$$

where $S_V(\nu)$ is the spectral density of the light. Discuss qualitatively the general behavior of $S_{\Delta I}(\nu)$ when $S_V(\nu)$ represents a spectral line.

8.2 Consider a well-collimated uniform light beam generated by a thermal source. Making use of the fact that thermal light can be modeled as a Gaussian random process, show that the degree of polarization of the beam may be expressed in the form

$$P = \left[1 - \frac{4 \det \sigma}{(\mathrm{tr}\, \sigma)^2}\right]^{1/2},$$

where $\det \sigma$ and $\mathrm{tr}\, \sigma$ are the determinant and the trace, respectively, of the matrix $\sigma_{ij} = (\mathscr{E}_{ij} \mathscr{E}_{ji})^{1/2}$, $\mathscr{E}_{ij}$ being the 2×2 equal-time coherence matrix discussed in Section 6.2.

8.3 $\mathbf{E}(\mathbf{r}, t)$ is the analytic signal representation of the electric field of a plane wave of thermal light which propagates in the z-direction. If $I(\mathbf{r}, t) = \mathbf{E}^*(\mathbf{r}, t) \cdot \mathbf{E}(\mathbf{r}, t)$ is the instantaneous intensity of the wave and $\Delta I(\mathbf{r}, t) = I(\mathbf{r}, t) - \langle I(\mathbf{r}, t) \rangle$ is the fluctuation of the intensity about its mean at the space-time point $(\mathbf{r}, t)$, show that the correlation of the intensity fluctuations at any two space-time points may be expressed in the form

$$\langle \Delta I(\mathbf{r}_1, t_1) \Delta I(\mathbf{r}_2, t_2) \rangle = \mathrm{tr}\, (\mathscr{E}^\dagger \mathscr{E}),$$

where $\mathscr{E}$ is the coherence matrix of the electric field, i.e.

$$\mathscr{E}_{ij} = \langle E_i^*(\mathbf{r}_1, t_1) E_j(\mathbf{r}_2, t_2) \rangle, \qquad (i, j = x, y),$$

and $\mathscr{E}^\dagger$ is the Hermitian adjoint of $\mathscr{E}$.

8.4 Derive an expression for the intensity correlation function

$$\langle I(\mathbf{r}_1, t) I(\mathbf{r}_2, t + \tau_2) \ldots I(\mathbf{r}_N, t + \tau_N) \rangle$$

at N points $\mathbf{r}_1, \mathbf{r}_2, \ldots, \mathbf{r}_N$ in a statistically stationary scalar wavefield, in terms of the cross-spectral density $W^{(N,N)}(\mathbf{r}_1, \mathbf{r}_2, \ldots, \mathbf{r}_{2N}; \nu_2, \ldots, \nu_{2N})$. Simplify the expression for the case when $N = 2$ and the field obeys Gaussian statistics.

9

Semiclassical theory of photoelectric detection of light

9.1 Introduction

It has been known since the nineteenth century that when light falls on certain metallic surfaces, electrons are sometimes released from the metal. This is known as the *photoelectric effect*, and the emitted particles are called *photoelectrons*. If a positively charged electrode is placed near the photoemissive cathode so as to attract the photoelectrons, an electric current can be made to flow in response to the incident light. The device thereby becomes a *photoelectric detector* of the optical field, and it has proved to be one of the most important of all photometric instruments. Various means exist for amplifying the photoelectric current. In one important device, known as the *photomultiplier* and shown schematically in Fig. 9.1, the photoelectrons are accelerated sufficiently that on striking the positive electrode they cause the release of several secondary electrons for each incident primary electron, and these electrons are then accelerated in turn to strike other secondary emitting surfaces. After 10 or more similar stages of amplification, the emission of each photoelectron from the cathode results in a pulse of millions of electrons at the anode, which is large enough to be registered by an electronic counter. By counting these photoelectric pulses we have an extremely sensitive detector of light.

It has been found experimentally that photoelectric emission from a given surface occurs only if the frequency of the incident light is high enough to exceed a certain threshold value (see Fig. 9.2). Once this critical frequency is exceeded, the number of photoelectrons released per second is proportional to the intensity of the incident light, whereas the average kinetic energy of the photoelectrons is independent of the light intensity. This was not easy to understand within the

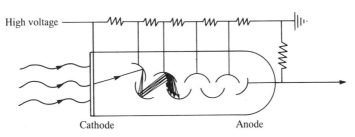

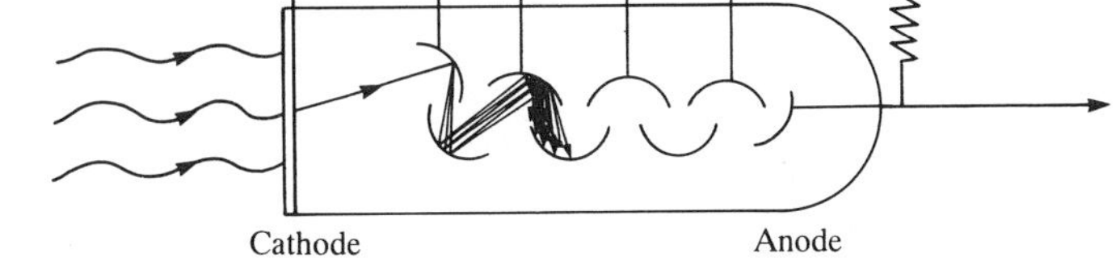

Fig. 9.1 Schematic of a photomultiplier.

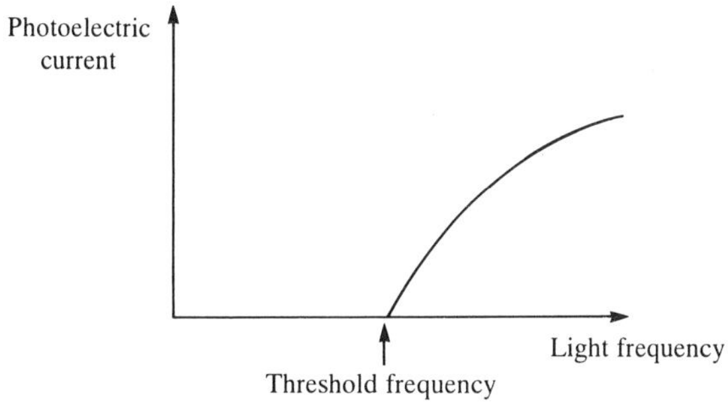

Fig. 9.2 Response of a photocathode to light of different frequencies.

classical framework of physics. It seemed to Einstein (1905) to be explicable most readily if the electrons within the metal were bound within some potential well with minimum binding energy E_0, and if the light were regarded as consisting of discrete particles or *photons*, of energy $\hbar\omega$ at frequency ω ($\hbar$ is Planck's constant divided by 2π). The flux of photons is proportional to the intensity of the light or to the power flow. When a photon is absorbed at the photoelectric surface it may transfer its energy $\hbar\omega$ to an electron, but unless $\hbar\omega > E_0$ this is insufficient to release the electron. Only when $\omega > E_0/\hbar$ does photoelectric emission take place, and then the number of electrons released per second is expected to be proportional to the photon flux or to the light intensity.

Although this simple 'quantum' picture accounted for some features of the photoelectric effect, it remained for a more completely developed theory of quantum mechanics and quantum electrodynamics to describe the effect in detail, and to yield expressions for the probability of photoelectric emission at various times. We shall take up the fully quantized treatment of the problem in Chapter 14. It turns out, however, that for many purposes the quantization of the electromagnetic field is not necessary at all, and the response of the photodetector can be understood even if we continue to picture the field in terms of classical electromagnetic waves, provided the photoelectrons are treated by quantum mechanics. The field then simply behaves as an external potential that perturbs the bound electrons of the photocathode. Such an approach to the problem is sometimes known as *semiclassical* (Mandel, Sudarshan and Wolf, 1964; Lamb and Scully, 1969, p. 363), and it is substantially simpler than the fully quantized treatment. Of course, it has certain limitations, and if pushed too far the semiclassical treatment will reveal some internal contradictions. However, that does not detract from its usefulness in many circumstances. As we shall see, for those electromagnetic fields for which an adequate classical description exists, the semiclassical and the fully quantized treatments of the photodetection problem yield virtually identical answers. Our approach in this chapter will be based largely on that adopted by Mandel, Sudarshan and Wolf (1964).

9.2 Review of elementary quantum mechanics

We assume that the photodetector contains a large number of bound electrons, a relatively few of which may be emitted under the influence of the incident electromagnetic field. We assume that the different emissions do not directly influence each other, and we start by focusing our attention on a single electron.

According to quantum mechanics the state of an electron is characterized by a certain state vector $|\psi\rangle$ in a Hilbert space. For simplicity we shall here ignore states relating to internal coordinates like spin. The dynamical variables like the energy, momentum, etc. that characterize the electron are represented by Hermitian operators $\hat{H}$, $\hat{\mathbf{p}}$, etc. that operate on the state vector. These operators need not commute, and we distinguish them from c-numbers by the caret $\hat{}$. In the quantum mechanical interaction picture that we shall adopt here,‡ the state $|\psi(t)\rangle$ is a function of time, that evolves according to the Schrödinger equation

$$\frac{\partial}{\partial t}|\psi(t)\rangle = \frac{1}{\mathrm{i}\hbar}\hat{H}_{\mathrm{I}}(t)|\psi(t)\rangle, \tag{9.2–1}$$

where $\hat{H}_{\mathrm{I}}(t)$ is the energy of interaction of the electron, and in general depends on time also. In the interaction picture any dynamical variable, such as $\hat{H}_{\mathrm{I}}(t)$ or the momentum $\hat{\mathbf{p}}(t)$, evolves in time according to the general rule

$$\hat{\mathbf{p}}(t) = \exp[\mathrm{i}\hat{H}_0(t - t_0)/\hbar]\hat{\mathbf{p}}(t_0)\exp[-\mathrm{i}\hat{H}_0(t - t_0)/\hbar], \tag{9.2–2}$$

where $\hat{H}_0$ is the non-interacting or free part of the electron energy, and t_0 is the time when the interaction is assumed to be turned on. The form of the interaction $\hat{H}_{\mathrm{I}}$ will be discussed in more detail in Section 14.1. For a non-relativistic electron of charge e, mass m and momentum $\hat{\mathbf{p}}(t)$ interacting with an electromagnetic wave of vector potential $\mathbf{A}(\mathbf{r}, t)$, the interaction energy can be written (cf. Section 14.1) in SI units

$$\hat{H}_{\mathrm{I}}(t) = -\frac{e}{m}\hat{\mathbf{p}}(t)\cdot\mathbf{A}(\mathbf{r}, t), \tag{9.2–3}$$

for an electron starting in a tightly bound state at position $\mathbf{r}$.

Like all other dynamical variables, $\hat{H}_0$ has a complete orthonormal set of eigenstates $|E\rangle$ and associated eigenvalues E, satisfying the relation

$$\hat{H}_0|E\rangle = E|E\rangle, \tag{9.2–4}$$

which may be continuous or discrete, or partly one and partly the other. The eigenvalues E represent the possible outcomes of a measurement of the non-interacting electron energy $\hat{H}_0$. For simplicity in writing general relations, we shall treat the set $|E\rangle$ formally as if it were discrete, so that we can express the orthonormality of the states by the scalar product relation

$$\langle E'|E\rangle = \delta_{\mathrm{EE'}}, \tag{9.2–5}$$

and the completeness by the sum over projectors

$$\sum_E |E\rangle\langle E| = 1. \tag{9.2–6}$$

‡ For a review of quantum mechanics, including the interaction picture, see for example, Schiff (1955), Messiah (1961) and Cohen-Tannoudji, Diu and Laloe (1977). Throughout this book we distinguish Hilbert space operators from c-numbers by the caret $\hat{}$.

The probability that when the electron is in the quantum state $|\psi(t)\rangle$ its energy is found to be E is then given by the squared projection $|\langle E|\psi(t)\rangle|^2$.

In order to determine $|\psi(t)\rangle$, given the initial state $|\psi(t_0)\rangle$, we need to integrate Eq. (9.2–1) between times t_0 and t. Formal integration converts the differential equation to an integral one, and yields

$$|\psi(t)\rangle = |\psi(t_0)\rangle + \frac{1}{i\hbar}\int_{t_0}^{t} \hat{H}_I(t_1)|\psi(t_1)\rangle\, dt_1, \tag{9.2–7}$$

which is a Volterra-type integral equation that can be solved by iteration. Thus, we may look on the initial state $|\psi(t_0)\rangle$ as being a zero-order approximation to $|\psi(t_1)\rangle$, and substitute it for $|\psi(t_1)\rangle$ under the integral. This gives

$$|\psi(t)\rangle = |\psi(t_0)\rangle + \frac{1}{i\hbar}\int_{t_0}^{t} \hat{H}_I(t_1)|\psi(t_0)\rangle\, dt_1,$$

which may be used as a first-order approximation to $|\psi(t_1)\rangle$ under the integral in Eq. (9.2–7), and substituting this back yields the second-order approximation

$$|\psi(t)\rangle = |\psi(t_0)\rangle + \frac{1}{i\hbar}\int_{t_0}^{t} \hat{H}_I(t_1)|\psi(t_0)\rangle\, dt_1 + \frac{1}{(i\hbar)^2}\int_{t_0}^{t} dt_1\int_{t_0}^{t_1} dt_2 \hat{H}_I(t_1)\hat{H}_I(t_2)|\psi(t_0)\rangle. \tag{9.2–8}$$

By proceeding in this way we obtain an infinite series, which, in principle, is the solution of Eq. (9.2–7). However, if the interaction time interval $t - t_0$ is sufficiently short, and particularly if we are interested in the projection of $|\psi(t)\rangle$ onto some other state, it may be sufficient if the series is terminated at the first non-vanishing term. For example, if we wish to calculate the probability of a transition from the initial state $|\psi(t_0)\rangle$ to some new state $|\Phi\rangle$ at time t, and if $|\Phi\rangle$ is orthogonal to $|\psi(t_0)\rangle$, then

$$\text{Transition probability} = |\langle\Phi|\psi(t)\rangle|^2 = \frac{1}{\hbar^2}\left|\int_{t_0}^{t} \langle\Phi|\hat{H}_I(t_1)|\psi(t_0)\rangle\, dt_1\right|^2 \tag{9.2–9}$$

to the lowest non-vanishing order.

9.3 The differential photodetection probability

We now apply these considerations to the problem of an initially bound electron that is interacting with a quasi-monochromatic electromagnetic wave. We suppose that the initial state $|\psi(t_0)\rangle$ of the electron is an eigenstate $|E_0\rangle$ of the unperturbed electron energy $\hat{H}_0$,

$$\hat{H}_0|E_0\rangle = E_0|E_0\rangle, \tag{9.3–1}$$

with an eigenvalue E_0 that is negative for the bound electron and of order -1 eV, as is typical of many photoemissive surfaces. Alternatively, if we write $|E_0| = \hbar\omega_0$, then ω_0 is some optical frequency of order $10^{15}\,\text{s}^{-1}$. Under the influence of the electromagnetic wave represented by its vector potential $\mathbf{A}(\mathbf{r}, t)$, the electron may be induced to make a transition from the bound state $|E_0\rangle$ to one of a continuum of positive energy eigenstates $|E\rangle$,

$$\hat{H}_0|E\rangle = E|E\rangle, \quad \text{with } E > 0, \tag{9.3–2}$$

with the emission of the electron. We shall suppose that after some amplification the emitted electron is registered by a counter. From Eq. (9.2–9) the desired transition probability to the free state in a short time interval Δt takes the form

$$\text{Transition probability} = \frac{1}{\hbar^2}\left|\int_{t_0}^{t_0+\Delta t}\langle E|\hat{H}_\mathrm{I}(t_1)|E_0\rangle\,\mathrm{d}t_1\right|^2, \tag{9.3–3}$$

with the interaction energy $\hat{H}_\mathrm{I}(t_1)$ given by

$$\hat{H}_\mathrm{I}(t_1) = -\frac{e}{m}\exp[\mathrm{i}\hat{H}_0(t_1 - t_0)/\hbar]\hat{\mathbf{p}}(t_0)\exp[-\mathrm{i}\hat{H}_0(t_1 - t_0)/\hbar]\cdot\mathbf{A}(\mathbf{r}, t_1)$$

according to Eqs. (9.2–2) and (9.2–3). Hence

$$\begin{aligned}\langle E|\hat{H}_\mathrm{I}(t_1)|E_0\rangle &\\ &= -\frac{e}{m}\langle E|\exp[\mathrm{i}\hat{H}_0(t_1 - t_0)/\hbar]\hat{\mathbf{p}}(t_0)\exp[-\mathrm{i}\hat{H}_0(t_1 - t_0)/\hbar]|E_0\rangle\cdot\mathbf{A}(\mathbf{r}, t_1)\\ &= -\frac{e}{m}\mathrm{e}^{\mathrm{i}(E-E_0)(t_1-t_0)/\hbar}\langle E|\hat{\mathbf{p}}(t_0)|E_0\rangle\cdot\mathbf{A}(\mathbf{r}, t_1).\end{aligned} \tag{9.3–4}$$

We have made use of the fact, embodied in Eqs. (9.3–1) and (9.3–2), that the states $|E\rangle$, $|E_0\rangle$ are both left and right eigenstates of $\hat{H}_0$. If the field is quasi-monochromatic and centered on frequency ω, and if the interval Δt is chosen to be much longer than the period $2\pi/\omega$ but still short compared with the coherence time of the light, then it is legitimate to represent $\mathbf{A}(\mathbf{r}, t_1)$ under the integral in the form

$$\mathbf{A}(\mathbf{r}, t_1) = \mathbf{V}(\mathbf{r}, t_0)\,\mathrm{e}^{-\mathrm{i}\omega(t_1-t_0)} + \mathbf{V}^*(\mathbf{r}, t_0)\,\mathrm{e}^{\mathrm{i}\omega(t_1-t_0)}. \tag{9.3–5}$$

Here $\mathbf{V}(\mathbf{r}, t)$ is the analytic signal representation of $\mathbf{A}(\mathbf{r}, t)$ (cf. Section 3.1), and $\mathbf{V}(\mathbf{r}, t)\,\mathrm{e}^{\mathrm{i}\omega t}$ is a slowly varying envelope function that does not change appreciably in a time short compared with the coherence time, so that for small δt

$$\mathbf{V}(\mathbf{r}, t_0 + \delta t) = \mathbf{V}(\mathbf{r}, t_0)\,\mathrm{e}^{-\mathrm{i}\omega\delta t}. \tag{9.3–6}$$

Hence, with the help of Eq. (9.3–5) we obtain

$$\begin{aligned}\int_{t_0}^{t_0+\Delta t}\mathrm{d}t_1\langle E|\hat{H}_\mathrm{I}(t_1)|E_0\rangle &= -\frac{e}{m}\langle E|\hat{\mathbf{p}}(t_0)|E_0\rangle\\ &\cdot\int_0^{\Delta t}\mathrm{d}t'\left\{\mathrm{e}^{\mathrm{i}[(E-E_0)/\hbar-\omega]t'}\mathbf{V}(\mathbf{r}, t_0) + \mathrm{e}^{\mathrm{i}[(E-E_0)/\hbar+\omega]t'}\mathbf{V}^*(\mathbf{r}, t_0)\right\}.\end{aligned} \tag{9.3–7}$$

Now we have taken E to be positive, and $-E_0/\hbar$ and ω to be optical frequencies of order $10^{15}\,\mathrm{s}^{-1}$, so that the term proportional to $\mathbf{V}^*(\mathbf{r}, t_0)$ oscillates at a very high frequency for all positive energies E. Its contribution to the integral will therefore by very small if $\Delta t \gg 10^{-15}$ s. The term in $\mathbf{V}(\mathbf{r}, t_0)$, on the other hand, oscillates much more slowly, and it makes an appreciable contribution whenever $(E - E_0)/\hbar \sim \omega$. We are therefore justified in discarding the $\mathbf{V}^*(\mathbf{r}, t_0)$ term in Eq.

(9.3–7) and we obtain after integration,

$$\int_{t_0}^{t_0+\Delta t} \mathrm{d}t_1 \langle E|\hat{H}_{\mathrm{I}}(t_1)|E_0\rangle \approx -\frac{e}{m}\langle E|\hat{\mathbf{p}}(t_0)|E_0\rangle \cdot \mathbf{V}(\mathbf{r}, t_0) \times \mathrm{e}^{\mathrm{i}[(E-E_0)/\hbar-\omega]\Delta t/2}\left\{\frac{\sin\frac{1}{2}[(E-E_0)/\hbar-\omega]\Delta t}{\frac{1}{2}[(E-E_0)/\hbar-\omega]}\right\},$$

so that from Eq. (9.3–3)

$$\text{Transition probability} = \left(\frac{e}{m\hbar}\right)^2 \left|\langle E|\hat{\mathbf{p}}(t_0)|E_0\rangle \cdot \mathbf{V}(\mathbf{r}, t_0)\right|^2 \times \left\{\frac{\sin\frac{1}{2}[(E-E_0)/\hbar-\omega]\Delta t}{\frac{1}{2}[(E-E_0)/\hbar-\omega]}\right\}^2. \tag{9.3–8}$$

This formula gives the probability that the photoelectron makes a transition from the bound state $|E_0\rangle$ to the free state $|E\rangle$ in the time Δt. If we are interested only in the probability that the electron becomes free, no matter what its final state, we need to sum this expression over all positive energies E. In practice, it is customary to replace such a sum by an integral with respect to E with the help of the density of states $\sigma(E)$, such that $\sigma(E)\,\mathrm{d}E$ gives the number of electron states lying within the energy interval $\mathrm{d}E$. Moreover, the detection process may favor electrons with certain energies over those with other energies, and we can allow for this possibility by multiplying by some response function $g(E)$ before performing the integration. We then obtain for the probability that the photoelectron is detected, the formula

$$\text{Photodetection probability} = \left(\frac{e}{m\hbar}\right)^2 \int_0^\infty \sigma(E)g(E)|\langle E|\hat{\mathbf{p}}(t_0)|E_0\rangle \cdot \mathbf{V}(\mathbf{r}, t_0)|^2 \times \left\{\frac{\sin\frac{1}{2}[(E-E_0)/\hbar-\omega]\Delta t}{\frac{1}{2}[(E-E_0)/\hbar-\omega]}\right\}^2 \mathrm{d}E.$$

The behavior of the integrand is dominated by the last factor, which has a peak at $E = E_0 + \hbar\omega$ and falls rapidly towards zero on either side of the peak. Indeed this factor has some of the characteristics of the delta function when $(E_0 + \hbar\omega)\Delta t \gg 1$, and to a first approximation it can be replaced by $\delta(E - E_0 - \hbar\omega)$. Provided $E_0 + \hbar\omega > 0$, so that the peak falls within the range of integration, and provided $\sigma(E)$, $g(E)$, $\langle E|\hat{\mathbf{p}}(t_0)|E_0\rangle$ vary slowly in the neighborhood of $E = E_0 + \hbar\omega$, these factors can be treated as having the approximately constant values $\sigma(E_0 + \hbar\omega)$, $g(E_0 + \hbar\omega)$, $\langle E_0 + \hbar\omega|\hat{\mathbf{p}}(t_0)|E_0\rangle$ under the integral. The integration can then be carried out, and we obtain finally

Photodetection probability

$$\begin{aligned} &\approx \left(\frac{e}{m\hbar}\right)^2 \sigma(E_0 + \hbar\omega)g(E_0 + \hbar\omega) \\ &\quad \times |\langle E_0 + \hbar\omega|\hat{\mathbf{p}}(t_0)|E_0\rangle \cdot \boldsymbol{\varepsilon}|^2 2\pi\hbar I(\mathbf{r}, t_0)\Delta t \quad \text{if } E_0 + \hbar\omega > 0 \\ &\approx 0 \qquad\qquad\qquad\qquad\qquad\qquad\qquad\qquad\quad \text{if } E_0 + \hbar\omega < 0. \end{aligned} \tag{9.3–9}$$

When $E_0 + \hbar\omega < 0$, the detection probability is very small because the important peak of the delta-function-like factor falls outside the range of integration. In Eq. (9.3–9) we have written $\mathbf{V}(\mathbf{r}, t) = \boldsymbol{\varepsilon} V(\mathbf{r}, t)$, where $\boldsymbol{\varepsilon}$ is a unit polarization vector characterizing the polarization of the incident light, and $I(\mathbf{r}, t)$ is the instantaneous light intensity [cf. Eq. (3.1–88)]

$$I(\mathbf{r}, t) \equiv |\mathbf{V}(\mathbf{r}, t)|^2.$$

The probability of photodetection in a short time Δt is therefore proportional to the light intensity and to Δt, with the remaining factors independent of the strength of the optical field, although they may depend on its frequency and on its polarization. If the illuminated photoelectric surface contains N bound electrons and if the light is in the form of a plane wave falling normally on the photocathode, if the bound states are not significantly depleted, and if different photoemissions do not influence each other, then it is to be expected that the photodetection probability $P(t)\Delta t$ within Δt is given by

$$P(t)\Delta t \sim \eta I(\mathbf{r}, t)\Delta t. \tag{9.3–10}$$

$\mathbf{r}$ is now any point within the illuminated photocathode surface, and

$$\begin{aligned} \eta &= \left(\frac{e}{m\hbar}\right)^2 N\sigma(E_0 + \hbar\omega)g(E_0 + \hbar\omega) \\ &\quad \times |\langle E_0 + \hbar\omega|\hat{\mathbf{p}}(t)|E_0\rangle \cdot \boldsymbol{\varepsilon}|^2 2\pi\hbar \quad \text{if } E_0 + \hbar\omega > 0 \\ &= 0 \qquad\qquad\qquad\qquad\qquad\qquad\quad \text{otherwise} \end{aligned} \tag{9.3–11}$$

is a constant characterizing the detector efficiency for a particular frequency and a particular polarization. We shall see in Chapter 14 that a somewhat more detailed, fully quantized treatment leads to an essentially similar answer. Of course it should be understood that $P(t)\Delta t$ given by Eq. (9.3–10) is a differential probability, which is meaningful only so long as $\eta I(\mathbf{r}, t)\Delta t \ll 1$. If the light intensity $I(\mathbf{r}, t)$ is so high that this condition is not satisfied, even if we choose Δt to be only an order of magnitude longer than the optical period, then the perturbation expansion that we used to solve Eq. (9.2–7) obviously cannot be terminated at the lowest non-vanishing term. In that case there are significant higher-order contributions to the interaction, and the detection probability is not simply proportional to the instantaneous light intensity $I(t)$. We shall henceforth discount this possibility, which requires extremely high light intensities.

A number of features of the simple solution (9.3–10) are worth pointing out:

(a) The Einstein photoelectric condition $\hbar\omega > -E_0$, implying a minimum threshold frequency for photoemission, emerges naturally even from a semiclassical analysis, without explicit introduction of the photon concept.

(b) When the condition $\hbar\omega > -E_0$ is satisfied, there is a non-vanishing probability for photoelectron emission to occur as soon as the detector is exposed to the light, no matter how weak the field may be.

(c) The detection probability is proportional to Δt after a sum over final states is performed, so that one can speak of an instantaneous rate of photoelectric emission.

(d) Although we started with a characterization of the electromagnetic field by the real vector potential $\mathbf{A}(\mathbf{r}, t)$, the *analytic signal* representation (see Section 3.1) $\mathbf{V}(\mathbf{r}, \mathrm{t})$ emerges naturally in the course of the treatment. The fully quantized analysis in Chapter 14 will reveal that this is so because $\mathbf{V}(\mathbf{r}, t)$ is closely connected with the quantum operator corresponding to the absorption of a photon.

9.4 Joint probabilities of multiple photodetections

Having derived an expression for the differential probability $P(t)\Delta t$ for a photoelectric detection to occur at a certain place and time t within Δt, we may readily generalize this to a situation in which several detections are registered at several different times, and possibly by different detectors. For simplicity we start with two detectors located at different positions in an optical field, but in such a way that the light falls normally on each photoelectric surface (see Fig. 9.3). Moreover, we suppose that each illuminated photocathode is small enough that the field over the surface can be regarded approximately as a plane wave. If $\mathbf{r}_1$, $\mathbf{r}_2$ are the centers of the two photocathodes, then the differential probability for the first detector at $\mathbf{r}_1$ to register a photodetection at time t_1 within Δt_1 is

$$P(\mathbf{r}_1, t_1)\Delta t_1 = \eta_1 I(\mathbf{r}_1, t_1)\Delta t_1, \qquad (9.4\text{–}1)$$

and the differential probability for the other detector at $\mathbf{r}_2$ to register a photodetection at time t_2 within Δt_2 is

$$P(\mathbf{r}_2, t_2)\Delta t_2 = \eta_2 I(\mathbf{r}_2, t_2)\Delta t_2. \qquad (9.4\text{–}2)$$

The two constants η_1, η_2 characterize the sensitivities of the two detectors. Now if the two photoelectric emissions do not influence each other in any way, then the joint or two-fold differential probability $P_2(\mathbf{r}_1, t_1; \mathbf{r}_2, t_2)\Delta t_1 \Delta t_2$ for both detections to occur is the product of the two separate detection probabilities, or

$$P_2(\mathbf{r}_1, t_1; \mathbf{r}_2, t_2)\Delta t_1 \Delta t_2 = \eta_1 \eta_2 I(\mathbf{r}_1, t_1) I(\mathbf{r}_2, t_2)\Delta t_1 \Delta t_2. \qquad (9.4\text{–}3)$$

This argument can evidently be extended to any larger number of photoelectric detections, so that we have for N multiple detections,

$$P_N(\mathbf{r}_1, t_1; \ldots; \mathbf{r}_N, t_N)\Delta t_1 \ldots \Delta t_N = \eta_1 \ldots \eta_N I(\mathbf{r}_1, t_1) \ldots I(\mathbf{r}_N, t_N)\Delta t_1 \ldots \Delta t_N. \qquad (9.4\text{–}4)$$

Moreover, it is not necessary for all the detections to be associated with different detectors. For example, the differential probability that the same detector located

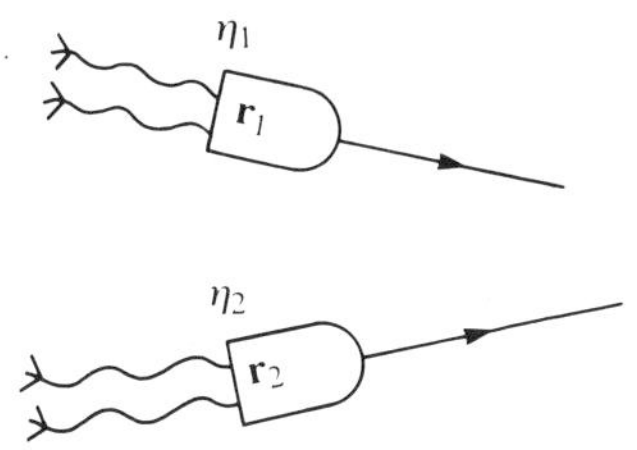

Fig. 9.3 The geometry for multiple photoelectric detections with two detectors.

at $\mathbf{r}$ registers detections at time t_1 within Δt_1 and at time t_2 within Δt_2 is given by

$$P_2(\mathbf{r}, t_1; \mathbf{r}, t_2)\Delta t_1 \Delta t_2 = \eta_1^2 I(\mathbf{r}, t_1) I(\mathbf{r}, t_2)\Delta t_1 \Delta t_2. \tag{9.4–5}$$

9.5 Integral detection probabilities

So far we have dealt only with detections in one or more differential time intervals Δt, and we have found that the probability of detection within Δt is $\eta I(\mathbf{r}, t)\Delta t$. This same differential quantity is also the expected number of detections (well below unity) within Δt. However, in practice we may be interested in all the n detections occurring in some finite time interval from t to $t + T$, which is not necessarily short compared with the coherence time of the light. It is then to be expected that the light intensity will vary over the time interval T. For the moment we shall ignore any randomness in this variation, and we look on it as prescribed or deterministic. In the language of the theory of random processes, we consider a single realization of a possible ensemble of optical fields with light intensity $I(\mathbf{r}, t)$.

We first divide the finite time interval from t to $t + T$ into a large number of differential intervals of width Δt. The expected or average number of detections in each differential interval is known and given by Eq. (9.3–10). The average number of detections $\langle n \rangle$ occurring within the finite time interval t to $t + T$ is then obtained by adding together the averages for all the differential intervals, and this clearly yields

$$\langle n \rangle = \eta \int_t^{t+T} I(\mathbf{r}, t')\, dt'. \tag{9.5–1}$$

Needless to say, this average number can be much greater than 1, and it cannot be interpreted as a probability.

However, it is not difficult to write down the probability $p(n, t, T)$ for n photoelectric detections to occur within the interval from t to $t + T$ when the expected number $\langle n \rangle$ is given by Eq. (9.5–1). If the events are independent, as we have assumed, the probability $p(n, t, T)$ must be a Poisson distribution in n with parameter $\langle n \rangle$, and we may write immediately

$$p(n, t, T) = \frac{1}{n!}\left[\eta \int_t^{t+T} I(\mathbf{r}, t')\, dt'\right]^n \exp\left[-\eta \int_t^{t+T} I(\mathbf{r}, t')\, dt'\right]. \tag{9.5–2}$$

The derivation of $p(n, t, T)$ can be made more explicit (Mandel, 1963, p. 181; see also Loudon, 1983, Section 6.6), but we shall not enter into the details here. The problem is treated again by full quantization in Chapter 14 below. We emphasize, however, that the Poisson distribution holds only so long as the time evolution of $I(\mathbf{r}, t)$ is deterministic and there are no ensembles, although the variation can have any form. Once ensembles or time averages are introduced the conclusion changes significantly, as we now show.

9.6 Photoelectric detection in a fluctuating field

We now apply the foregoing results to the more realistic situation in which a randomly fluctuating electromagnetic field impinges on one or more photodetectors. In that case we have to introduce an ensemble of realizations of the optical field, and we regard all the foregoing equations as applying to just one realization of the ensemble. However, physically meaningful results are obtained only after we perform an average over all members of the ensemble. If we denote the ensemble average by $\langle\ \rangle$, then from Eq. (9.3–10) we can write for the differential photelectric detection probability at $\mathbf{r}$ at time t within Δt of a fluctuating optical field

$$P_1(\mathbf{r}, t)\Delta t = \eta\langle I(\mathbf{r}, t)\rangle\Delta t. \tag{9.6–1}$$

If the field is stationary, the average $\langle I(\mathbf{r}, t)\rangle$ is of course independent of time t, and so is the detection probability.

Similarly, after averaging over the ensemble we obtain from Eq. (9.5–1) for the average number of photoelectric detections in a finite time interval from t to $t + T$,

$$\langle n\rangle = \eta\int_t^{t+T}\langle I(\mathbf{r}, t')\rangle\,\mathrm{d}t'. \tag{9.6–2}$$

Once again, if the field is stationary, the average $\langle I(\mathbf{r}, t')\rangle$ does not depend on t' and may by taken outside the integral, so that

$$\langle n\rangle = \eta\langle I(\mathbf{r})\rangle T, \tag{9.6–3}$$

which is strictly proportional to T.

From Eq. (9.4–4) for the joint N-fold photoelectric detection probability, we arrive at

$$P_N(\mathbf{r}_1, t_1; \ldots; \mathbf{r}_N, t_N)\Delta t_1 \ldots \Delta t_N = \eta_1 \ldots \eta_N\langle I(\mathbf{r}_1, t_1) \ldots \times I(\mathbf{r}_\mathrm{N}, t_\mathrm{N})\rangle\Delta t_1 \ldots \Delta t_N, \tag{9.6–4}$$

and we note that this probability involves an intensity correlation function of the N'th order. In the steady state, it depends only on the difference between the various time arguments. In particular, when $N = 2$ we have

$$P_2(\mathbf{r}_1, t_1; \mathbf{r}_2, t_2) = \eta_1\eta_2\langle I(\mathbf{r}_1, t_1)I(\mathbf{r}_2, t_2)\rangle. \tag{9.6–5}$$

Now in general, for a fluctuating field

$$\langle I(\mathbf{r}_1, t_1)I(\mathbf{r}_2, t_2)\rangle \neq \langle I(\mathbf{r}_1, t_1)\rangle\,\langle I(\mathbf{r}_2, \mathrm{t}_2)\rangle,$$

from which it follows that in general

$$P_2(\mathbf{r}_1, t_1; \mathbf{r}_2, t_2) \neq P_1(\mathbf{r}_1, t_1)P_1(\mathbf{r}_2, t_2). \tag{9.6–6}$$

The different photoelectric detections in general are therefore not independent either. At first sight this appears to contradict the assumption that was made in Section 9.4, that no influence of one photoemission on another occurs. However, there is no contradiction. Although one photoemission does not influence another, the two are correlated in general by the fluctuations of the common electromagnetic field.

In the special case when the radiation field is stationary, polarized and Gaussian, it follows with the help of the Gaussian moment theorem that [cf. Eq. (8.4–5)]

$$\langle I(\mathbf{r}_1, t_1)I(\mathbf{r}_2, t_2)\rangle = \langle I(\mathbf{r}_1)\rangle\,\langle I(\mathbf{r}_2)\rangle[1 + |\gamma(\mathbf{r}_1, \mathbf{r}_2, t_2 - t_1)|^2], \tag{9.6–7}$$

where $|\gamma(\mathbf{r}_1, \mathbf{r}_2, t_2 - t_1)|$ is the degree of coherence of the light at position $\mathbf{r}_1$ at time t_1 and at position $\mathbf{r}_2$ at time t_2 (see Section 4.3–1). The photoelectric emissions at detectors placed at positions $\mathbf{r}_1$ and $\mathbf{r}_2$ in an optical field will therefore be correlated at times when the degree of coherence at these points is non-zero. This was first demonstrated experimentally by Brown and Twiss (1956a) in measurements with two photodetectors (see Fig. 9.6 below). They later applied the same correlation technique to determine $|\gamma(\mathbf{r}_1, \mathbf{r}_2, 0)|$ for the light from a distant star, from which the angular size of the star can be deduced (Brown and Twiss, 1958a). The original experiment and the method, known as *stellar intensity correlation interferometry*, will be described in more detail in Sections 9.9 and 9.10 below.

9.6.1 Photoelectric bunching

One consequence of these correlations is the phenomenon of *photoelectric bunching*, whereby photoelectric emissions from a photodetector are more likely to occur close together than far apart in time. Let us focus our attention on two detections at times t and $t + \tau$ by a single photodetector at $\mathbf{r}$. We drop the position label and obtain from Eq. (9.6–5)

$$P_2(t, t + \tau) = \eta^2 \langle I(t) I(t + \tau) \rangle. \tag{9.6–8}$$

Once again we note that for a stationary field the correlation function on the right depends only on the time difference τ and not on t. Now from the Schwarz inequality and stationarity it follows that

$$\langle I(t) I(t + \tau) \rangle \leqslant \langle I^2(t) \rangle,$$

so that

$$P_2(t, t + \tau) \leqslant P_2(t, t). \tag{9.6–9}$$

Two photodetections are therefore more likely to occur at the same time, or very close together, than farther apart. Moreover, as $\tau \to \infty$ it follows for an ergodic process that $\langle I(t) I(t + \tau) \rangle \to \langle I \rangle^2$, or

$$P_2(t, t + \tau) \to P_1^2(t) \quad \text{as } \tau \to \infty, \tag{9.6–10}$$

so that the detections become independent when they are far separated in time.

A plot of the joint photoelectric detection probability density $P_2(t, t + \tau)$ against the time interval τ must therefore be of the general form shown in Fig. 9.4. The value of $P_2\ (t, t + \tau)$ is greatest when $\tau = 0$, and it falls to the constant value $P_1^2(t)$ for large τ, although not necessarily monotonically. The precise shape is of course determined by the statistical character of the optical field. The enhanced tendency of photoelectric emissions to occur close together, within a time of order of the intensity correlation time T_c, is known as *photoelectric bunching*. It occurs only with fluctuating fields, such as those produced by a thermal equilibrium light source, for example, and it has been observed with thermal light (Twiss, Little and Brown, 1957; Rebka and Pound, 1957; Morgan and Mandel, 1966; Scarl, 1966; Arecchi, Gatti and Sona, 1966). The phenomenon has been attributed to the boson character of the thermal photons, which tend to

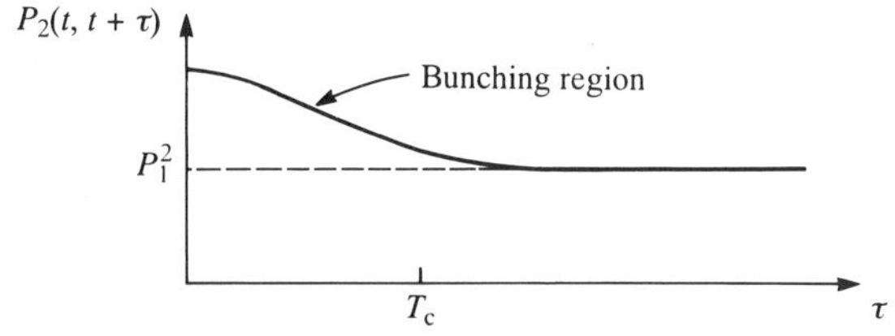

Fig. 9.4 The expected form of the joint probability density of photoelectric detection at two different times, as a function of the time interval τ between them.

bunch in this way (see Sections 12.10 and 12.1). However, as the field is here described as a classical electromagnetic wave, the photoelectric bunching evidently has an alternative explanation in terms of intensity fluctuations. Photoelectric emissions are more likely to occur whenever the instantaneous light intensity is high than when it is low. In Section 14.7, where we discuss the bunching phenomenon again in more detail and in a somewhat more rigorous manner in terms of a quantum field, we shall find that photoelectric antibunching can also occur in certain circumstances. However, this effect contradicts Eq. (9.6–9), which is valid only within the framework of classical optics.

9.7 Photoelectric counting statistics of a fluctuating field

When the optical field fluctuates and has to be treated as a random process, the probability $p(n, t, T)$ of detecting n photoelectric emissions in a finite time interval from t to $t + T$ is no longer a Poisson distribution in general. To derive the appropriate probability we start from Eq. (9.5–2), which holds for a single realization, and again average over the ensemble. We then obtain

$$p(n, t, T) = \left\langle \frac{1}{n!}\left[\eta\int_t^{t+T} I(\mathbf{r}, t')\,\mathrm{d}t'\right]^n \exp\left[-\eta\int_t^{t+T} I(\mathbf{r}, t')\,\mathrm{d}t'\right]\right\rangle,$$

which can be written more compactly as

$$p(n, t, T) = \left\langle \frac{1}{n!} W^n \,\mathrm{e}^{-W}\right\rangle, \tag{9.7–1}$$

with

$$W \equiv \eta\int_t^{t+T} I(\mathbf{r}, t')\,\mathrm{d}t'. \tag{9.7–2}$$

Here W, which is η times the integrated light intensity, can be regarded as a new random variable whose probability density $\mathcal{P}(W)$ can, in principle, be derived from the known statistics of the optical field. The averaging operation in Eq. (9.7–1) can then be formally expressed as an integral, and we can write

$$p(n, t, T) = \int_0^\infty \frac{1}{n!} W^n \,\mathrm{e}^{-W}\mathcal{P}(W)\,\mathrm{d}W. \tag{9.7–3}$$

This formula, which was first derived in the 1950s (Mandel, 1958, 1959), shows clearly that statistics enters in two different ways into the integral detection probability $p(n, t, T)$: in the randomness of the individual photoelectric emissions, which may be attributed to the underlying quantum mechanics of the photoelectric process, and in the fluctuations of the radiation field itself. But even in the absence of the latter, it is impossible to predict when and how many photo-electric emissions will occur, as is already apparent from Eqs. (9.3–10) and (9.5–2).

The act of averaging over the fluctuations of W, i.e. over the fluctuations of the electromagnetic field, in general causes $p(n, t, T)$ to depart from a Poisson distribution, despite the formal Poisson structure of Eqs. (9.7–1) and (9.7–3). This is most apparent if we examine the first two moments of n. We find for the average number of photoelectric detections

$$\begin{aligned} \langle n \rangle &= \sum_{n=1}^{\infty} np(n, t, T) \\ &= \int_0^{\infty} \mathrm{d}W \sum_{n=1}^{\infty} n \frac{W^n}{n!} \mathrm{e}^{-W} \mathcal{P}(W) \\ &= \int_0^{\infty} W \mathcal{P}(W) \, \mathrm{d}W \\ &= \langle W \rangle, \end{aligned} \tag{9.7–4}$$

and for the second factorial moment of n,

$$\begin{aligned} \langle n(n-1) \rangle &= \sum_{n=2}^{\infty} n(n-1) p(n, t, T) \\ &= \int_0^{\infty} \mathrm{d}W \sum_{n=2}^{\infty} n(n-1) \frac{W^n}{n!} \mathrm{e}^{-W} \mathcal{P}(W) \\ &= \int_0^{\infty} W^2 \mathcal{P}(W) \, \mathrm{d}W \\ &= \langle W^2 \rangle, \end{aligned} \tag{9.7–5}$$

so that the variance of n is given by

$$\begin{aligned} \langle (\Delta n)^2 \rangle &= \langle n(n-1) \rangle + \langle n \rangle - \langle n \rangle^2 \\ &= \langle n \rangle + \langle (\Delta W)^2 \rangle. \end{aligned} \tag{9.7–6}$$

This shows clearly that the number of photoelectric detections n does not obey a Poisson distribution whenever W fluctuates, because the variance of n exceeds the mean $\langle n \rangle$. Although Eq. (9.7–6) may suggest that the variance $\langle (\Delta n)^2 \rangle$ can never fall below $\langle n \rangle$, we shall find in Chapter 14 that this can happen for certain quantum states of the field, although it cannot happen in a classical field.

Two limiting cases of the general formula (9.7–1) are worth examining. Let us suppose first that the measurement interval T is very great compared with the intensity correlation time of the light, so that $I(\mathbf{r}, t')$ under the integral in Eq.

(9.7–2) goes through many changes in time. Then the time integral divided by T can be approximated by a time average, and for an ergodic process the time average equals the average over the ensemble. Hence

$$W = \eta\int_t^{t+T} I(\mathbf{r}, t')\,\mathrm{d}t' \approx \eta\langle I(\mathbf{r}, t)\rangle T, \tag{9.7–7}$$

which implies that W fluctuates very little, and W/T not at all in the limit $T \to \infty$. Hence $\mathcal{P}(W)$ can be approximated by a delta function under the integral in Eq. (9.7–3), and we obtain

$$p(n, t, T) \approx \frac{1}{n!}(\eta\langle I\rangle T)^n \exp(-\eta\langle I\rangle T) \tag{9.7–8}$$

for sufficiently large T. The distribution is Poissonian because the fluctuations have averaged out.

Next we consider the other extreme, in which the interval T is made short compared with the correlation time of the light intensity $I(\mathbf{r}, t)$. Then $I(\mathbf{r}, t')$ in Eq. (9.7–2) does not vary appreciably over the integration interval, and we may approximate W by writing

$$W \approx \eta I(\mathbf{r}, t)T. \tag{9.7–9}$$

Hence W is proportional to the instantaneous light intensity $I(\mathbf{r}, t)$, and the probability distribution of W becomes the probability distribution of I, apart from a scale factor. In place of Eq. (9.7–3) we may therefore write, for sufficiently short time intervals T,

$$p(n, t, T) = \int_0^\infty \frac{1}{n!}(\eta IT)^n \mathrm{e}^{-\eta IT} P(I)\,\mathrm{d}I. \tag{9.7–10}$$

where $P(I)$ is the probability density of the light intensity I.

As an illustration, we now apply this formula to the situation in which a beam of polarized light from a thermal equilibrium source is received by a photodetector. From the central limit theorem (see Section 1.5.6) we would expect the probability distribution of the analytic signal V representing this optical field to be Gaussian, because the field at each point is the sum of fields contributed by many independent sources. Hence the distribution $P(I)$ of $I = |V|^2$ is exponential, i.e.

$$P(I) = \frac{1}{\langle I\rangle}\exp\left(-\frac{I}{\langle I\rangle}\right). \tag{9.7–11}$$

When this is inserted in Eq. (9.7–10) we obtain

$$\begin{aligned} p(n, t, T) &= \frac{1}{n!\langle I\rangle}\int_0^\infty (\eta IT)^n \exp[-I(\eta T + 1/\langle I\rangle)]\,\mathrm{d}I \\ &= \frac{(\eta T)^n}{\langle I\rangle[\eta T + 1/\langle I\rangle]^n} \\ &= \frac{1}{[1 + \eta\langle I\rangle T][1 + 1/\eta\langle I\rangle T]^n}, \end{aligned} \tag{9.7–12}$$

which is known as the *Bose–Einstein distribution* with parameter $\langle n\rangle = \eta\langle I\rangle T$.

This shows quite explicitly that $p(n, t, T)$ need not be Poissonian, despite the formal Poisson structure of Eqs. (9.7–3) and (9.7–10).

The form of the probability distribution (9.7–12) is sometimes considered as reflecting the fluctuation properties of thermal equilibrium photons, which obey a similar probability distribution [cf. Eq. (13.1–10) below]. However, we emphasize once again that we are here regarding the incident light as a classical electromagnetic wave, that gives rise to random emissions of photoelectrons. It would therefore be wrong to draw any conclusions from our equations that are applicable to a quantum field. The remarkable agreement between the predictions of this semiclassical treatment and the fully quantized treatment has led to speculation that quantum electrodynamics may not be necessary even for so-called photon counting experiments. However, we shall see later that there are results of measurements for which the semiclassical analysis cannot account, whereas the fully quantized one can.

From Eqs. (9.7–1) and (9.7–2) the generating functions of n and of the integrated light intensity W are simply related. For example, the factorial moment generating function defined by (cf. Section 1.4.1)

$$F(\xi) = \langle (1+\xi)^n \rangle = \sum_{r=0}^{\infty} \frac{\xi^r}{r!} \langle n^{(r)} \rangle, \tag{9.7–13}$$

is expressible in the form

$$\begin{aligned} \langle (1+\xi)^n \rangle &= \sum_{n=0}^{\infty} (1+\xi)^n p(n, t, T) \\ &= \left\langle \sum_{n=0}^{\infty} (1+\xi)^n \frac{W^n \mathrm{e}^{-W}}{n!} \right\rangle \\ &= \langle \mathrm{e}^{W\xi} \rangle. \end{aligned} \tag{9.7–14}$$

If we put $\xi = \mathrm{i}x$ the term on the right hand side becomes $\langle \mathrm{e}^{\mathrm{i}xW} \rangle$, which is the characteristic function $C_W(x)$ of W, and if $1 + \mathrm{i}x = \mathrm{e}^{\mathrm{i}y}$ the term on the left becomes $\langle \mathrm{e}^{\mathrm{i}yn} \rangle$, which is the characteristic function $C_n(y)$ of n. Hence we have the relation

$$C_n(y) = C_W(x), \tag{9.7–15}$$

and the statistics of W are derivable, in principle, from the statistics of n.

9.8 Photoelectric current fluctuations

We have seen that when light falls on a photoemissive detector it causes some electrons to be emitted from the photoelectric surface. After some amplification each electron may result in a measurable current pulse, and we have studied the statistics of these pulses. However, in some experiments no attempt is made to resolve the individual current pulses and one treats the photoelectric current as a continuous random process $J(t)$. This was the case in the original correlation experiments of Brown and Twiss, 1956a, 1957a,b). Let us therefore examine the statistics of $J(t)$ and relate them to the light fluctuations.

For simplicity we shall suppose that every photoelectron emitted at time t' gives rise to a definite photoelectric current pulse $k(t - t')$ (see Fig. 9.5) which vanishes for $t < t'$. Then

$$J(t) = \sum_j k(t - t_j), \tag{9.8–1}$$

where the sum is to be taken over the various random emission times t_j. Let $J_{n,T}(t)$ be the photocurrent due to the emission of exactly n current pulses within some long time interval T,

$$J_{n,T}(t) = \sum_{j=1}^{n} k(t - t_j) \quad (-T/2 \leq t_j \leq T/2). \tag{9.8–2}$$

We now use the standard arguments, as in the derivation of the generalized Campbell theorem (Rice, 1944, Sec. 1.2), to study the correlations of $J(t)$. For a stationary process the probability that photoelectron j is emitted at time t_j within $\mathrm{d}t_j$ is $\mathrm{d}t_j/T$, and we can use this to calculate the expectation of $J_{n,T}(t)$ conditioned on a fixed number n within T. Thus

$$\langle J_{n,T}(t)\rangle_n = \sum_{j=1}^{n}\int_{-T/2}^{T/2} k(t - t_j)(\mathrm{d}t_j/T) = \frac{1}{T}\sum_{j=1}^{n}\int_{t-T/2}^{t+T/2} k(t')\,\mathrm{d}t' \approx \frac{1}{T}\sum_{j=1}^{n} Q = \frac{nQ}{T}, \tag{9.8–3}$$

where

$$Q \equiv \int_{-\infty}^{\infty} k(t')\,\mathrm{d}t' \tag{9.8–4}$$

is the total charge delivered by any one current pulse. If $p(n, T)$ is the probability that n electrons are emitted within the interval T, then the mean photocurrent $\langle J(t)\rangle$ is obtained from $\langle J_{n,T}(t)\rangle_n$ by averaging over n in the limit of large T. Thus we have

$$\langle J(t)\rangle = \sum_{n=0}^{\infty} p(n, T)\frac{nQ}{T} = \langle n\rangle\frac{Q}{T} = \eta\langle I\rangle Q. \tag{9.8–5}$$

We have made use of the fact that $\langle n\rangle/T$ is the average incident light intensity

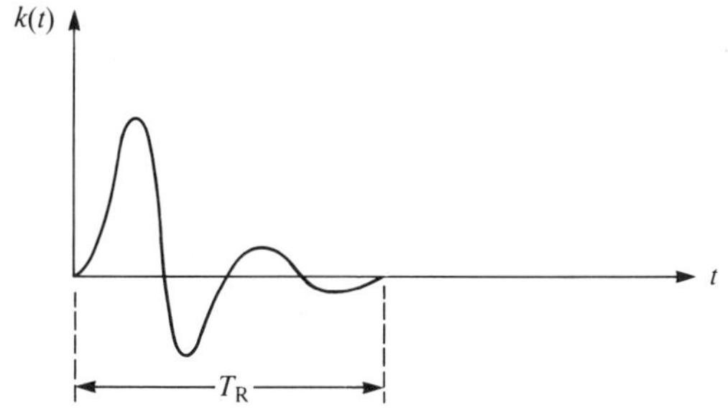

Fig. 9.5 Example of a typical photoelectric current pulse. T_{R} is the response time of the photodetector.

$\langle I \rangle$ expressed in units of photons per second multiplied by the detector quantum efficiency η.

Next we calculate the autocorrelation function $\langle J(t)J(t+\tau)\rangle$ of the photocurrent. From Eq. (9.6–5) it follows that, when n photoelectrons are emitted within the time interval T, the joint probability that electron i is emitted at time t_i within dt_i and electron j is emitted at time t_j within dt_j is given by

$$P_2(t_i, t_j) = \langle I(t_i)I(t_j)\rangle \Big/ \iint_{-T/2}^{T/2} \langle I(t_i)I(t_j)\rangle \, dt_i \, dt_j. \tag{9.8–6}$$

This expression allows for the possibility of two-time photoelectric correlations and it is normalized to unity over the interval T. We now use Eq. (9.8–6) to calculate the autocorrelation $\langle J_{n,T}(t)J_{n,T}(t+\tau)\rangle_n$ conditioned on a given n within T. Then we have

$$\langle J_{n,T}(t)J_{n,T}(t+\tau)\rangle_n = \left\langle \sum_{i=1}^{n}\sum_{j=1}^{n} k(t-t_i)k(t-t_j+\tau)\right\rangle_n .$$

It is convenient to divide the double sum into two parts, corresponding to a single detection $(i = j)$ and two separate detections $(i \neq j)$. Then

$$\begin{aligned}\langle J_{n,T}(t)J_{n,T}(t+\tau)\rangle_n &= \left\langle \sum_{j=1}^{n} k(t-t_j)k(t-t_j+\tau)\right\rangle \\ &\quad + \left\langle \sum_{i\neq j}^{n}\sum^{n} k(t-t_i)k(t-t_j+\tau)\right\rangle \\ &= \sum_{j=1}^{n}\int_{-T/2}^{T/2} k(t-t_j)k(t-t_j+\tau)\frac{dt_j}{T} \\ &\quad + \frac{\displaystyle\sum_{i\neq j}^{n}\sum^{n}\iint_{-T/2}^{T/2} k(t-t_i)k(t-t_j+\tau)\langle I(t_i)I(t_j)\rangle \, dt_i\, dt_j}{\displaystyle\iint_{-T/2}^{T/2}\langle I(t_i)I(t_j)\rangle \, dt_i\, dt_j}.\end{aligned} \tag{9.8–7}$$

When T is much longer than the range T_R of the current pulse $k(t)$, this simplifies to

$$\begin{aligned}\langle J_{n,T}(t)J_{n,T}(t+\tau)\rangle_n &= \frac{n}{T}\int_{-\infty}^{\infty} k(t')k(t'+\tau)\,dt' \\ &\quad + \frac{\displaystyle n(n-1)\iint_{-\infty}^{\infty} k(t')k(t'')\langle I(t)I(t+t'-t''+\tau)\rangle \, dt'\, dt''}{\displaystyle\iint_{-T/2}^{T/2}\langle I(t')I(t'')\rangle \, dt'\, dt''}\end{aligned}$$

when we make use of the stationarity of $I(t)$. Finally, we average over the number n within T as before. Then

$$\langle J(t)J(t+\tau)\rangle = \frac{\langle n\rangle_T}{T}\int_{-\infty}^{\infty} k(t')k(t'+\tau)\,dt' + \frac{\langle n(n-1)\rangle_T \iint_{-\infty}^{\infty} k(t')k(t'')\langle I(t)I(t+t'-t''+\tau)\rangle\,dt'\,dt''}{\iint_{-T/2}^{T/2} \langle I(t')I(t'')\rangle\,dt'\,dt''}. \tag{9.8–8}$$

But it follows from Section 9.7 that [cf. Eq. (9.7–4)]

$$\langle n\rangle_T = \eta\langle I\rangle T$$

and [cf. Eq. (9.7–5)]

$$\langle n(n-1)\rangle_T = \eta^2 \iint_{-T/2}^{T/2} \langle I(t')I(t'')\rangle\,dt'\,dt'',$$

and when these are substituted in Eq. (9.8–8) we obtain

$$\langle J(t)J(t+\tau)\rangle = \eta\langle I\rangle\int_{-\infty}^{\infty} k(t')k(t'+\tau)\,dt' + \eta^2\iint_{-\infty}^{\infty} k(t')k(t'')\langle I(t)I(t+t'-t''+\tau)\rangle\,dt'\,dt''. \tag{9.8–9}$$

We now subtract $\langle J(t)\rangle^2$ from both sides of the equation with the help of Eq. (9.8–5), and we arrive at the autocorrelation of the current fluctuations

$$\langle \Delta J(t)\Delta J(t+\tau)\rangle = \eta\langle I\rangle\int_{-\infty}^{\infty} k(t')k(t'+\tau)\,dt' + \eta^2\iint_{-\infty}^{\infty} k(t')k(t'')\langle \Delta I(t)\Delta I(t+t'-t''+\tau)\rangle\,dt'\,dt''. \tag{9.8–10}$$

The first term is attributable to the shot noise of the photocurrent, or the discreteness of the electronic charge, whereas the second one is associated with the fluctuations of the incident optical field. Which term is larger is determined by the nature of the optical field and by the detection efficiency η. For light from familiar thermal sources the first term is always dominant.

Finally, we can make use of similar arguments to calculate the cross-correlation between two photoelectric currents $J_1(t)$, $J_2(t)$ from two photodetectors illuminated by light of intensities $I_1(t)$, $I_2(t)$. Each current is given by an expression like that in Eq. (9.8–1). Let us suppose that the two photodetectors are identical except perhaps for their quantum efficiencies η_1, η_2. We may then proceed as in the derivation of Eq. (9.8–7), starting from

$$\langle J_{1n,T}(t)J_{2n,T}(t+\tau)\rangle_n = \left\langle \sum_{i=1}^{n}\sum_{j=1}^{n} k(t-t_{1i})k(t-t_{2j}+\tau)\right\rangle_n,$$

except that with two detectors the double sum can no longer be separated into terms involving only one photodetection and terms involving two detections, as before. The first term on the right of Eq. (9.8–7) is therefore absent. In other respects the calculation proceeds as before and we obtain in place of Eq. (9.8–10)

$$\langle \Delta J_1(t)\Delta J_2(t+\tau)\rangle = \eta_1\eta_2 \iint_{-\infty}^{\infty} k(t')k(t'')\langle \Delta I_1(t)\Delta I_2(t+t'-t''+\tau)\rangle \,\mathrm{d}t'\,\mathrm{d}t''. \tag{9.8–11}$$

It follows that the fluctuations of the two photoelectric currents $J_1(t)$, $J_2(t)$ are correlated in general when the light intensity fluctuations at the two photodetectors are correlated.

9.8.1 Special cases

We now consider a few special cases in which the foregoing relations simplify.

(a) Fast response photodetector

If the response time T_{R} (see Fig. 9.5) of the photodetector is very short compared with any correlation times of the incident light, we are justified in approximating the current pulse $k(t)$ under the time integral by a $\delta(t)$ function multiplied by Q with respect to slower test functions. Then Eqs. (9.8–10) and (9.8–11) reduce to

$$\langle \Delta J(t)\Delta J(t+\tau)\rangle \approx \eta\langle I\rangle Q\delta(\tau) + \eta^2 Q^2\langle \Delta I(t)\Delta I(t+\tau)\rangle \tag{9.8–12}$$

$$\langle \Delta J_1(t)\Delta J_2(t+\tau)\rangle \approx \eta_1\eta_2 Q^2\langle \Delta I_1(t)\Delta I_2(t+\tau)\rangle. \tag{9.8–13}$$

(b) Slow response photodetector

If the detector response is slow compared with the intensity fluctuations of the incident field, which is often the case, we may treat the intensity correlation function under the time integral as being approximately proportional to a delta function. Then after putting $\langle \Delta I(t)\Delta I(t+\tau)\rangle = \langle (\Delta I)^2\rangle T_{\mathrm{C}}\delta(\tau)$, we obtain from Eq. (9.8–10)

$$\langle \Delta J(t)\Delta J(t+\tau)\rangle = \eta\langle I\rangle \int_{-\infty}^{\infty} k(t')k(t'+\tau)\,\mathrm{d}t' + \eta^2\langle (\Delta I)^2\rangle T_{\mathrm{c}}\int_{-\infty}^{\infty} k(t')k(t'+\tau)\,\mathrm{d}t', \tag{9.8–14}$$

where T_{c} is a measure of the intensity correlation time of the light and $T_{\mathrm{R}} \gg T_{\mathrm{c}}$. By a similar argument we have from Eq. (9.8–11)

$$\langle \Delta J_1(t)\Delta J_2(t+\tau)\rangle = \eta_1\eta_2\langle \Delta I_1\Delta I_2\rangle T_{\mathrm{c}}\int_{-\infty}^{\infty} k(t')k(t'+\tau)\,\mathrm{d}t'. \tag{9.8–15}$$

9.8.2 *Thermal light*

Let us now apply the latter formulas for slow detectors to unpolarized thermal light. It was shown in Section 8.4 that in this case [see Eq. (8.4–24)]

$$\langle(\Delta I)^2\rangle = \tfrac{1}{2}\langle I\rangle^2 \tag{9.8–16}$$

and

$$\langle\Delta I_1 \Delta I_2\rangle = \tfrac{1}{2}\langle I_1\rangle\langle I_2\rangle|\gamma_{12}|^2, \tag{9.8–17}$$

where $|\gamma_{12}| \equiv |\gamma(\mathbf{r}_1, \mathbf{r}_2, 0)|$ is the modulus of the equal-time degree of coherence between the optical fields at the two photodetectors. When these results are inserted, Eqs. (9.8–14) and (9.8–15) become

$$\langle\Delta J(t)\Delta J(t+\tau)\rangle = (\eta\langle I\rangle + \tfrac{1}{2}\eta^2\langle I\rangle^2 T_c)\int_{-\infty}^{\infty} k(t')k(t'+\tau)\,dt' \tag{9.8–18}$$

and

$$\langle\Delta J_1(t)\Delta J_2(t+\tau)\rangle = \tfrac{1}{2}\eta_1\eta_2\langle I_1\rangle\langle I_2\rangle|\gamma_{12}|^2 T_c\int_{-\infty}^{\infty} k(t')k(t'+\tau)\,dt'. \tag{9.8–19}$$

The last equation describes what has come to be known as the Brown–Twiss effect, which was first observed with thermal light (Brown and Twiss, 1956a, 1957b). It shows that the correlation between the two photoelectric currents is proportional to the square of the degree of coherence of the light fields at the two detectors. In Eq. (9.8–18) for the autocorrelation function, the first or shot noise term is the dominant one whenever $\eta\langle I\rangle T_c \ll 1$, i.e. when the mean number of photons detected within the coherence T_c is much less than unity. This is almost invariably the case when the temperature of the thermal source is well below 10^5 K.

9.8.3 *Spectral density of the photocurrent*

Finally let us return to the more general formula (9.8–10), and take the Fourier transform with respect to τ of each term in the equation. As is well known, the Fourier transform of the photocurrent autocorrelation function is the spectral density $\chi(\omega)$ of the current fluctuations (cf. Section 2.4). After introducing the following Fourier transforms,

$$\int_{-\infty}^{\infty} \langle\Delta J(t)\Delta J(t+\tau)\rangle\, e^{i\omega\tau}\,d\tau \equiv \chi(\omega) \tag{9.8–20}$$

$$\int_{-\infty}^{\infty} k(t)\, e^{i\omega t}\,dt \equiv K(\omega) \tag{9.8–21}$$

$$\int_{-\infty}^{\infty} \lambda(\tau)\, e^{i\omega\tau}\,d\tau \equiv \psi(\omega), \tag{9.8–22}$$

where

$$\lambda(\tau) \equiv \langle\Delta I(t)\Delta I(t+\tau)\rangle/\langle I\rangle^2 \tag{9.8–23}$$

is the normalized autocorrelation of the light intensity fluctuations, we readily obtain from Eq. (9.8–10)

$$\chi(\omega) = \eta\langle I\rangle|K(\omega)|^2 + \eta^2\langle I\rangle^2|K(\omega)|^2\psi(\omega)$$
$$= \eta\langle I\rangle|K(\omega)|^2[1 + \eta\langle I\rangle\psi(\omega)]. \tag{9.8–24}$$

This shows that a spectral analysis of the photocurrent fluctuations in principle yields information about the spectrum of the light intensity fluctuations. The spectral function $\psi(\omega)$ is centered on zero frequency rather than on a high optical frequency. The function $K(\omega)$ can be regarded as the frequency response of the photodetector and the associated electronics.

In the special case in which we are dealing with polarized thermal light it was shown in Section 8.4 (cf Eq. 8.4–20) that

$$\lambda(\tau) \equiv \langle\Delta I(t)\Delta I(t+\tau)\rangle/\langle I\rangle^2 = |\gamma(\tau)|^2, \tag{9.8–25}$$

where $\gamma(\tau)$ is the normalized second-order autocorrelation function of the optical field. Then

$$\psi(\omega) = \tfrac{1}{2}\int_{-\infty}^{\infty} |\gamma(\tau)|^2\,\mathrm{e}^{\mathrm{i}\omega\tau}\,\mathrm{d}\tau = \frac{1}{2\pi}\int_{-\infty}^{\infty}\phi(\omega')\phi(\omega+\omega')\,\mathrm{d}\omega', \tag{9.8–26}$$

where

$$\phi(\omega) \equiv \int_{-\infty}^{\infty}\gamma(\tau)\,\mathrm{e}^{\mathrm{i}\omega\tau}\,\mathrm{d}\tau \tag{9.8–27}$$

is the normalized spectral density of the optical field [denoted by $\mathscr{s}(\omega)$, except for a possible factor 2π, in Sections 2.4.3, 4.3.3 and 5.8.2]. In this case, with phase and amplitude fluctuations both contributing equally to the light spectrum, it is possible to extract information about the spectrum of the incident light from the photocurrent fluctuations.

9.9 The Hanbury Brown–Twiss effect (semi-classical treatment)

The first experimental evidence for the existence of correlations between the outputs of two photoelectric detectors illuminated by partially correlated light waves was obtained in some classic experiments performed by Brown and Twiss in the 1950s, with the apparatus shown in Fig. 9.6(*a*) (Brown and Twiss, 1956, 1957a,b). A secondary light source was formed by a circular pinhole on which the image of a mercury arc was focused by a lens. The beam of light from the pinhole was divided by a semi-transparent mirror to illuminate the cathodes of photomultipliers P_1 and P_2. The photomultiplier P_2 was mounted on a horizontal slide which could be traversed in a direction normal to the direction of propagation of the incident light. Thus the cathode apertures, as viewed from the pinhole, could be superimposed or separated by any amount up to several times their widths. The fluctuations $\Delta J_1(t)$ and $\Delta J_2(t)$ in the anode currents of the photomultipliers were transmitted to the correlator through coaxial cables of equal lengths. In each case a high-pass filter was inserted between the anode and the input to the cable, to remove the steady direct-current component. The normal-

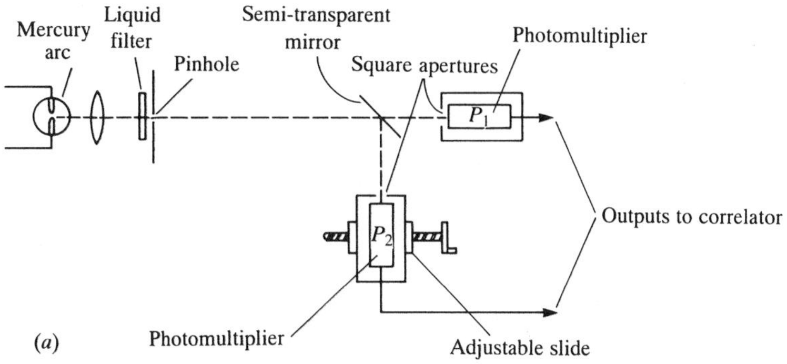

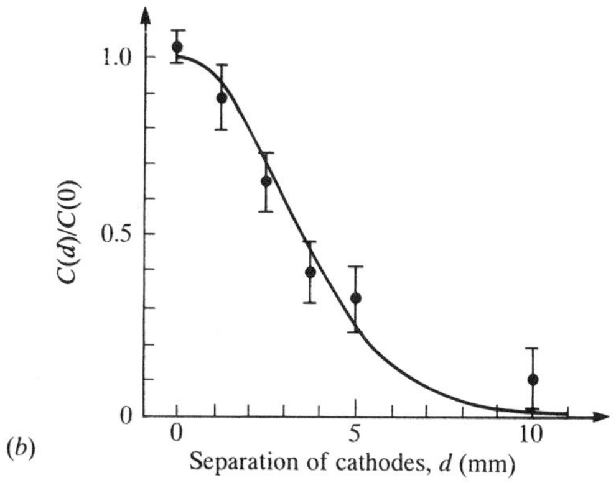

Fig. 9.6(*a*) A simplified outline of the apparatus for measuring correlations between the fluctuating current outputs of two photodetectors illuminated by partially coherent thermal light. (*b*) Experimental and theoretical values of the normalized correlation coefficient $C(d)/C(0)$ [Eq. (9.9–1)] for different effective separations d between the photocathodes. The experimental values are shown as points, with the vertical bars indicating the range of probable errors. (After Brown and Twiss, 1957b.)

ized correlation coefficient

$$C(d) = \frac{\langle \Delta J_1(t) \Delta J_2(t) \rangle}{(\langle [\Delta J_1(t)]^2 \rangle)^{1/2} (\langle [\Delta J_2(t)]^2 \rangle)^{1/2}} \tag{9.9–1}$$

of the photocurrent fluctuations was measured as a function of the effective separation $d = |\mathbf{r}_1 - \mathbf{r}_2|$ between the photocathodes, with the results shown in Fig. 9.6(*b*). It can be seen that the normalized correlation coefficient $C(d)$ has a maximum value when the photocathodes are optically superposed ($d = 0$) and that it decreases with increasing separation d.

The principle of this effect, now known as the *Hanbury Brown–Twiss effect*, can readily be understood from the theory developed in Section 9.8. Since the (unpolarized) incident light is produced by a statistically stationary thermal source and the field fluctuations are very rapid compared with the detector response, the correlations in the current fluctuations are given by Eqs. (9.8–18) and (9.8–19).

We therefore have

$$\langle(\Delta J_l(t))^2\rangle = \eta_l\langle I_l\rangle[1 + \tfrac{1}{2}\eta_l\langle I_l\rangle T_c]\int_0^\infty k^2(t')\,dt' \quad (l = 1, 2) \tag{9.9–2}$$

$$\langle\Delta J_1(t)\Delta J_2(t)\rangle = \tfrac{1}{2}\eta_1\eta_2\langle I_1\rangle\langle I_2\rangle|\gamma(\mathbf{r}_1, \mathbf{r}_2, 0)|^2 T_c\int_0^\infty k^2(t')\,dt', \tag{9.9–3}$$

from which it follows that the normalized cross-correlation coefficient $C(d)$ defined by Eq. (9.9–1) is given by

$$C(d) = \frac{\tfrac{1}{2}(\eta_1\eta_2\langle I_1\rangle\langle I_2\rangle)^{1/2}T_c}{(1 + \tfrac{1}{2}\eta_1\langle I_1\rangle T_c)^{1/2}(1 + \tfrac{1}{2}\eta_2\langle I_2\rangle T_c)^{1/2}}|\gamma(\mathbf{r}_1, \mathbf{r}_2, 0)|^2. \tag{9.9–4}$$

The quantity $\frac{1}{2}\eta_l\langle I_l\rangle T_c \equiv \delta$ $(l = 1, 2)$ is the average number of photoelectric counts due to light of one polarization registered by the detector in the correlation time T_c. For thermal light sources of temperature well below 10^5 K, δ is always much less than unity (see Section 13.1.4), and it was well below 1 in the experiments of Brown and Twiss. With $\eta_1\langle I_1\rangle \approx \eta_2\langle I_2\rangle$ we can therefore simplify Eq. (9.9–4) further and write

$$C(d) \approx \delta|\gamma(\mathbf{r}_1, \mathbf{r}_2, 0)|^2. \tag{9.9–5}$$

This formula shows that the normalized correlation coefficient of the fluctuations in the photoelectric current outputs is proportional to the squared modulus of the second-order equal-time degree of coherence of the light incident on the two detectors. Now for a circular, uniform spatially incoherent source such as was used in the experiments of Brown and Twiss, the equal time degree of coherence $|j(P_1, P_2)| \equiv |\gamma(\mathbf{r}_1, \mathbf{r}_2, 0)|$ was calculated in Section 4.4.4 by the use of the van Cittert–Zernike theorem, with the result given by Eq. (4.4–44). Its squared modulus has the form of the correlation coefficient $C(d)$ measured by Brown and Twiss and displayed in Fig. 9.6(b). The presence of the very small factor δ in Eq. (9.9–5) accounts for the relatively large error bars in the results shown in Fig. 9.6(b).

9.10 Stellar intensity interferometry

In Section 7.2 we described Michelson's classic technique for determining stellar diameters. It is based on measurements of the fringe visibility in the interference pattern formed by light from the star that reaches the detector plane via two mirrors $M_1(\mathbf{r}_1)$ and $M_2(\mathbf{r}_2)$ or equivalently on measurements of the absolute value of the equal-time degree of coherence $j(M_1, M_2) \equiv \gamma(\mathbf{r}_1, \mathbf{r}_2, 0)$ of the light reaching the two mirrors. According to the van Cittert–Zernike theorem the value of $\gamma(\mathbf{r}_1, \mathbf{r}_2, 0)$ for light at pairs of points in the far zone of an incoherent source is proportional to the Fourier transform of the intensity distribution across the source. Therefore measurements of γ provide information about the angular diameter of the star. Unfortunately, very large separations are necessary if stars of very small angular size are to be resolved, and the interferometry becomes progressively more difficult. The slightest flexing of the telescope arms can produce path-difference changes of many wavelengths, which cause the interference fringes to move around. Atmospheric turbulence can produce even larger fringe variations. It is significant that the smallest stellar angular diameters (of the order of 0.01 second of arc) measured with this method have not improved greatly in the years since Michelson's measurements.

The photoelectric correlation technique that we have just discussed provides an alternative method, with certain advantages, for measuring the absolute value of the degree of coherence $|\gamma(\mathbf{r}_1, \mathbf{r}_2, 0)|$ [see Eq. (9.9–5)], and therefore makes it possible to determine stellar diameters in a different way. A schematic diagram of a system, known as the *stellar intensity interferometer*, first constructed by Brown and Twiss (1956b, 1958a,b) is shown in Fig. 9.7. Two parabolic mirrors were aimed at a distant star, and the light collected by each one was directed to a photomultiplier at each focus. The photoelectric signals were amplified and correlated as before. A delay in one of the cables compensated for the fact that the light reached one photodetector before the other one. The mirror separation was then varied and the correlation was measured as a function of the separation. The principal advantage of this technique over the Michelson method lies in the fact that the light beams falling on the two mirrors do not need to interfere. Only the two photoelectric signals are actually brought together. As a result, the optical path difference does not have to be maintained strictly constant, and slight mirror movements and atmospheric turbulence have a very small effect. Much larger baseline separations are therefore feasible. Moreover, the light-collecting mirrors need not be of very high optical quality, as their only function is to direct the light to a photodetector at the focus, and the use of very large collectors therefore becomes possible also. As a result, Brown and Twiss were able to achieve better angular resolution than Michelson with much cruder optical equipment. The results of their first published measurements are shown in Fig. 9.8. The data correspond to the values of the normalized correlation coefficient $|\gamma(\mathbf{r}_1, \mathbf{r}_2, 0)|^2$ of light from the star Sirius, from which its angular diameter was determined to be $0.0068'' \pm 0.0005''$, in reasonable agreement with the values obtained by the Michelson interferometry technique.

A large stellar intensity interferometer was later constructed at Narrabri, Australia[‡] (Brown, Davis and Allen, 1967a,b; 1974) (see Fig. 9.9). It makes use

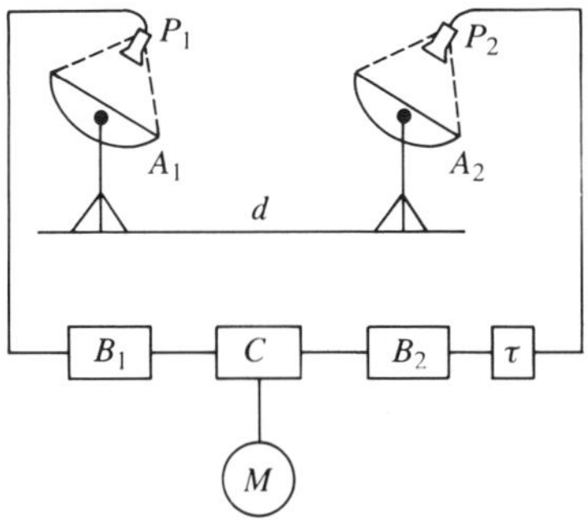

Fig. 9.7 Schematic diagram of the stellar intensity interferometer (A, mirrors; B, amplifiers; C, multiplier; M, integrator; P, phototubes; τ, delay line). (After Brown and Twiss, 1958a.)

[‡] A delightful account of the building and operation of the Narrabri interferometer, with a wealth of related useful information, is given in Brown (1974).

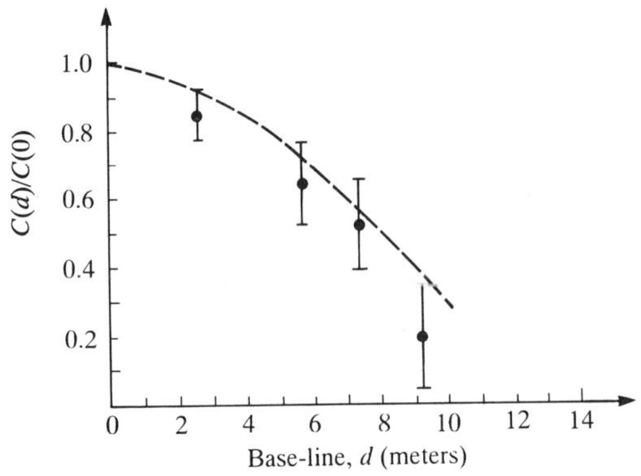

Fig. 9.8 Comparison between the measured values of the normalized correlation coefficient $C(d)/C(0)$ $|\gamma(r_1, r_2, 0)|^2$ of light from the star Sirius and the theoretical values for a star of angular diameter 0.0063″. The vertical bars indicate the range of probable error of observation. (Adapted from Brown and Twiss, 1956b.)

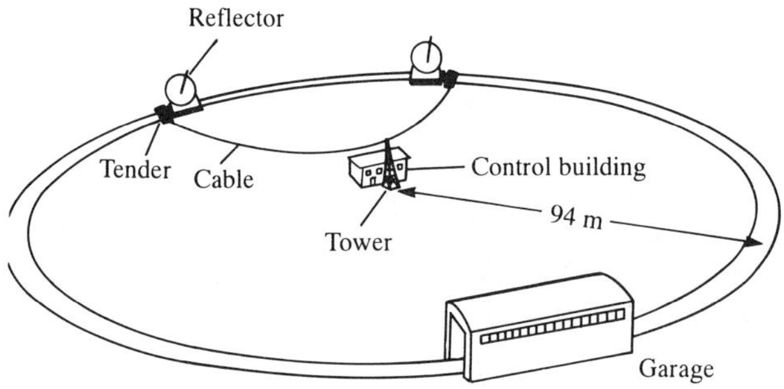

Fig. 9.9 The general layout of the intensity interferometer at the Narrabri Observatory. (After Brown, Davis and Allen, 1967a.)

of two 6.5 m diameter mirrors mounted on two computer-controlled trolleys, that move on a circular track of 188 m diameter. This allows baselines up to $d = 188$ m and has made it possible to measure angular diameters of stars as small as 0.0004″ arc, with resolutions of about 0.000 03″. Despite the fact that it permits large mirrors to be used, the technique is ultimately limited by the amount of light collected, because, as can be seen from Eq. (9.9–3), the measured correlation is proportional to the square of the light intensity.

9.11 Fluctuation spectroscopy

The technique of measuring correlations of photoelectric current fluctuations has other applications. For narrow-band light beams from thermal sources it provides an opportunity to obtain information about the spectral distribution of the light without the use of any dispersive elements (cf. also Section 7.3). As is apparent from Eq. (9.8–24), by making a spectral analysis of the photoelectric current fluctuations at the output of the photodetector, we can derive the function $\psi(\omega)$, which is the spectral density of the light intensity fluctuations. Of course, in general, intensity and phase fluctuations need not be correlated, so that there may not be any simple relation between the spectral densities of the light intensity and of the optical field. However, in the special case of thermal light, $\psi(\omega)$ is simply related to the normalized spectral density $\phi(\omega)$ of the optical field (denoted by $\mathfrak{s}(\omega)$ in previous chapters) by the convolution in Eq. (9.8–26). We then have the interesting possibility of obtaining information about the spectrum at optical frequencies from measurements made at very much lower electric current frequencies. This is because $\psi(\omega)$ given by Eq. (9.8–26) is centered on $\omega = 0$, whereas the spectral density $\phi(\omega)$ of the optical field is centered on some high frequency of order 10^{15} Hz.

In general, there is no unique procedure for inverting Eq. (9.8–26) and deriving $\phi(\omega)$ from measurements of $\psi(\omega)$. Nevertheless, some features of the spectrum $\phi(\omega)$ of the optical field can be derived from the measurements of $\psi(\omega)$. For example, when $\phi(\omega)$ is bandlimited within a frequency band $\Delta\omega$ centered on ω_0, i.e. if $\phi(\omega) = 0$ outside the range $\omega_0 - \frac{1}{2}\Delta\omega < \omega < \omega_0 + \frac{1}{2}\Delta\omega$, we see at once from Eq. (9.8–26) that $\psi(\omega) = 0$ when $|\omega| > \Delta\omega$ so that ψ is bandlimited to the range $-\Delta\omega < \omega < \Delta\omega$. If the light is quasi-monochromatic ($\Delta\omega/\omega_0 \ll 1$), then ψ contains only low-frequency components, extending over the effective bandwidth $\Delta\omega$ of the light. In Fig. 9.10 the relationship between the normalized spectra $\phi(\omega)$ of the light and $\psi(\omega)$ of the photocurrent fluctuations is illustrated graphically for three spectral profiles. We see that the effective widths of ϕ and ψ are in each case of the same order of magnitude.

This technique for obtaining information about optical spectra, which is known as *fluctuation spectroscopy*, has been used most successfully on scattered laser light (Ford and Benedek, 1965), particularly with heterodyning, when some of the unscattered laser light is superposed on the scattered field (Cummins, Knable and

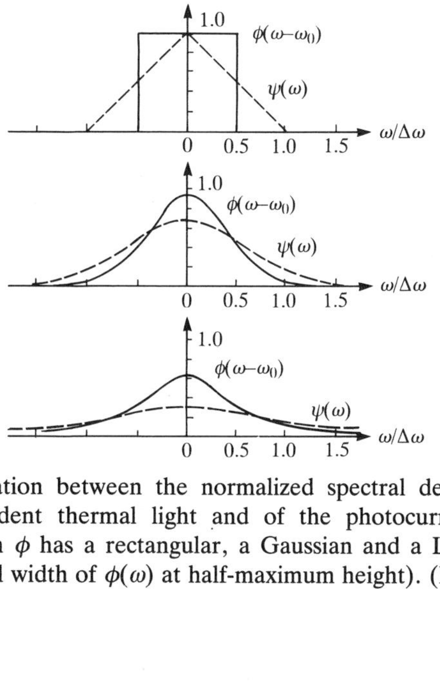

Fig. 9.10 The relation between the normalized spectral densities $\phi(\omega)$ and $\psi(\omega)$ of the incident thermal light and of the photocurrent fluctuations, respectively, when ϕ has a rectangular, a Gaussian and a Lorentzian profile ($\Delta\omega$ is the spectral width of $\phi(\omega)$ at half-maximum height). (Reproduced from Mandel, 1963.)

Yeh, 1964; Lastovka and Benedek, 1966). The subject has been reviewed by Cummins and Swinney (1970, p. 133).

Problems

9.1 A laser is oscillating simultaneously with equal amplitudes in two longitudinal modes of frequencies ω_1 and ω_2. Both modes are randomly phased and independent of each other. Calculate the probability density $p(\tau)$ for one photoelectric detection at time t and one at time $t + \tau$ when the laser light falls on a photodetector of surface area S.

9.2 If the phase $\varphi(t)$ of each laser mode in problem 9.1 performs a one-dimensional random walk with $\langle[\varphi(t + \tau) - \varphi(t)]^2\rangle = 2D\tau$, where D is the phase diffusion constant, calculate $p(\tau)$.

9.3 A quasi-monochromatic light beam with Gaussian statistics represented by $V_1(t)$ and a parallel strictly monochromatic light beam $V_2(t) = A\,\mathrm{e}^{-\mathrm{i}(\omega_2 t+\varphi)}$, in which A is constant and the phase ϕ is randomly distributed over the range 0 to 2π, are superimposed and fall on a photodetector. Both light fields are similarly polarized. Calculate the variance of the number of photoelectric counts registered by the detector in some counting time interval T in terms of the mean intensities $\langle I_1\rangle$, $\langle I_2\rangle$, the frequency ω_2 and the normalized spectral density of the Gaussian field, when T is much longer than the coherence time of the Gaussian field.

9.4 From the known differential photodetection probability $P(t)\Delta t$, together with the composition property

$$p(n, t, t + T + \Delta T) = p(n - 1, t, t + T)P(t + T)\Delta T + p(n, t, t + T)[1 - P(t + T)\Delta T],$$

derive a differential equation for $p(n, t, t + T)$ and solve it for $n = 0$.

9.5 Prove by induction or in any other way that the probability $p(n, t, t + T)$ given by Eq. (9.7–1) follows from the solution for $p(0, t, t + T)$ obtained in the previous problem.

10

Quantization of the free electromagnetic field

10.1 Introduction

Up to now the electromagnetic field has been treated as a classical field, describable by c-number functions. The great success of classical electromagnetic theory in accounting for a variety of optical phenomena, particularly those connected with wave propagation, interference and diffraction, amply justifies the classical approach. Moreover, as we have seen in the preceding chapters, in some cases the classical wave theory also gives a good account of itself in the treatment of the interaction of electromagnetic fields. For example, it is able to describe such seemingly non-classical effects as photoelectric bunching and the photoelectric counting statistics. It might almost seem that there is little justification for going beyond the domain of classical wave theory in optics.

On the other hand, it can be argued that optics lies well and truly in the quantum domain, in the sense that we often encounter situations in which very few quanta or photons are present. In the microwave region of the electromagnetic spectrum, and at still longer wavelengths, the number of photons in each mode of the field is usually very large, and we are justified in treating the system classically. However, in the optical region the situation is usually just the opposite. As we show in Section 13.1, for light produced by practically all sources other than lasers, the average number of photons per mode is typically much less than unity. It might seem, therefore, that a classical description would be hopelessly inadequate, and that optics must always be regarded as a branch of quantum electrodynamics. Indeed, as we shall see in Section 10.7, the very idea of an oscillatory field of definite phase, as is implied by writing $v\,\mathrm{e}^{-i\omega t}$ for the optical disturbance, is meaningless according to the quantum theory of radiation when there are so few photons.

Why then does classical optics work so well in many cases? One answer to this question is that we rarely attempt to measure the non-classical features of light, like the wildly fluctuating absolute phase of the wave. Phase differences are measured interferometrically, but they do not present the same problem. For many purposes it suffices to deal with the light intensity, and this is frequently well described by classical wave optics. Indeed, classical optics is able to account for some phenomena even at extremely low light levels.

However classical optics does not always work. There are some optical phenomena, usually, but not always, involving small photon numbers, for which the field needs to be treated quantum mechanically, and quantum electrodynamics plays an essential role. The quantum theory of electromagnetic radiation is the most successful and all-embracing theory of optics, and to date none of its predictions has been contradicted by experiment.

In the following sections we shall therefore take up the problem of quantizing the free or non-interacting electromagnetic field and investigate some of its properties. After discussing some of the principal non-classical features, we shall go on to study the coherent states of the quantum field, and show that it is possible to establish a remarkably close correspondence between certain coherence properties of the field given by the classical wave theory and by the quantum theory.

10.2 Classical field Hamiltonian and the canonical equations of motion

Let us consider a classical electromagnetic field in empty space, in the absence of any sources such as charges or currents. The field therefore satisfies the homogeneous Maxwell equations and, in SI units, these take the form[‡]

$$\boldsymbol{\nabla} \times \mathbf{E}(\mathbf{r}, t) = -\frac{\partial}{\partial t}\mathbf{B}(\mathbf{r}, t), \tag{10.2–1}$$

$$\boldsymbol{\nabla} \times \mathbf{B}(\mathbf{r}, t) = \frac{1}{c^2}\frac{\partial}{\partial t}\mathbf{E}(\mathbf{r}, t), \tag{10.2–2}$$

$$\boldsymbol{\nabla} \cdot \mathbf{E}(\mathbf{r}, t) = 0, \tag{10.2–3}$$

$$\boldsymbol{\nabla} \cdot \mathbf{B}(\mathbf{r}, t) = 0. \tag{10.2–4}$$

$\mathbf{E}(\mathbf{r}, t)$ and $\mathbf{B}(\mathbf{r}, t)$ are the usual electric and magnetic field vectors at the space–time point $(\mathbf{r}, t)$. It is often convenient to represent the free electromagnetic field by the transverse[§] vector potential $\mathbf{A}(\mathbf{r}, t)$ in the Coulomb gauge, which satisfies the homogeneous wave equation

$$\nabla^2\mathbf{A}(\mathbf{r}, t) - \frac{1}{c^2}\frac{\partial^2}{\partial t^2}\mathbf{A}(\mathbf{r}, t) = 0, \tag{10.2–5}$$

and the divergence condition

$$\boldsymbol{\nabla} \cdot \mathbf{A}(\mathbf{r}, t) = 0. \tag{10.2–6}$$

In terms of $\mathbf{A}(\mathbf{r}, t)$, the electric and magnetic fields $\mathbf{E}(\mathbf{r}, t)$ and $\mathbf{B}(\mathbf{r}, t)$ are given by

$$\mathbf{E}(\mathbf{r}, t) = -\frac{\partial}{\partial t}\mathbf{A}(\mathbf{r}, t), \tag{10.2–7}$$

$$\mathbf{B}(\mathbf{r}, t) = \boldsymbol{\nabla} \times \mathbf{A}(\mathbf{r}, t). \tag{10.2–8}$$

‡ Both SI and Gaussian units are used in this book, but measured quantities are always expressed in SI units.

§ We recall that the transverse part of the vector potential is gauge invariant.

10.2.1 Plane wave expansions

In order to obtain the Hamiltonian equations of motion, it is useful first to make a Fourier decomposition of $\mathbf{A}(\mathbf{r}, t)$ with respect to its space variables x, y, z. This can be done either in the form of a Fourier integral or a Fourier series. Although there is no particular advantage in either one at this stage, we shall find it a little easier eventually to deal with the quantized field in terms of a discrete decomposition of $\mathbf{A}(\mathbf{r}, t)$. Accordingly, we imagine the electromagnetic field to be contained in a very large cube of side L and we impose periodic boundary conditions on the field. At a suitable stage in the calculation we then allow L to tend to infinity.[‡] Needless to say, any physically meaningful results should not depend on the magnitude of L.[§]

We now write the three-dimensional Fourier expansion of $\mathbf{A}(\mathbf{r}, t)$ in terms of plane-wave modes in the form

$$\mathbf{A}(\mathbf{r}, t) = \frac{1}{\varepsilon_0^{1/2} L^{3/2}} \sum_{\mathbf{k}} \mathcal{A}_{\mathbf{k}}(t)\, \mathrm{e}^{\mathrm{i}\mathbf{k}\cdot\mathbf{r}}, \tag{10.2–9}$$

where the vector $\mathbf{k}$ has components

$$\left.\begin{aligned} k_1 &= 2\pi n_1/L, \quad n_1 = 0, \pm1, \pm2, \ldots \\ k_2 &= 2\pi n_2/L, \quad n_2 = 0, \pm1, \pm2, \ldots \\ k_3 &= 2\pi n_3/L, \quad n_3 = 0, \pm1, \pm2, \ldots \end{aligned}\right\} \tag{10.2–10}$$

forming a discrete set, and the sum $\sum_{\mathbf{k}}$ is understood to be a sum over the integers n_1, n_2, n_3. The factor $\varepsilon_0^{1/2}L^{3/2}$, where ε_0 is the vacuum dielectric constant, is introduced for later convenience.

In view of the transversality condition (10.2–6), we have

$$\frac{\mathrm{i}}{\varepsilon_0^{1/2} L^{3/2}} \sum_{\mathbf{k}} \mathbf{k} \cdot \mathcal{A}_{\mathbf{k}}(t)\, \mathrm{e}^{\mathrm{i}\mathbf{k}\cdot\mathbf{r}} = 0$$

for all $\mathbf{r}$, which requires that

$$\mathbf{k} \cdot \mathcal{A}_{\mathbf{k}}(t) = 0. \tag{10.2–11}$$

In addition, the reality of $\mathbf{A}(\mathbf{r}, t)$ leads to the condition

$$\mathcal{A}_{-\mathbf{k}}(t) = \mathcal{A}^*_{\mathbf{k}}(t). \tag{10.2–12}$$

Since $\mathbf{A}(\mathbf{r}, t)$ satisfies the homogeneous wave equation (10.2–5), it follows

‡ Actually it can be shown that the introduction of periodic boundary conditions is not strictly necessary in the limit $L \to \infty$. This question has been discussed by Ledermann (1944) and Peierls (1954) in connection with the normal modes of a crystal.

§ If we let $L \to \infty$, any discrete sum over $\mathbf{k}$-vectors $\sum_{\mathbf{k}}(\,)$ becomes an integral according to the rule $\sum_{\mathbf{k}}(\,) \to (L/2\pi)^3 \int(\,)\,\mathrm{d}^3k$, where $(L/2\pi)^3$ is the density of modes. This density can be derived as follows. From Eq. (10.2–10) $\delta k_i = (2\pi/L)\delta n_i$ $(i = 1, 2, 3)$, so that the number of modes corresponding to δk_i is δn_i. Hence the number of modes corresponding to the three-dimensional interval $\delta k_1 \delta k_2 \delta k_3$ is $\delta n_1 \delta n_2 \delta n_3 = (L/2\pi)^3 \delta k_1 \delta k_2 \delta k_3$. These steps can be avoided by a direct, continuous decomposition of the field vectors. The continuous mode representation is briefly discussed in Section 10.10 below.

immediately that

$$\frac{1}{\varepsilon_0^{1/2}L^{3/2}}\sum_{\mathbf{k}}\left(-k^2-\frac{1}{c^2}\frac{\partial^2}{\partial t^2}\right)\mathcal{A}_{\mathbf{k}}(t)\,\mathrm{e}^{\mathrm{i}\mathbf{k}\cdot\mathbf{r}}=0$$

for all $\mathbf{r}$, so that $\mathcal{A}_{\mathbf{k}}(t)$ satisfies the equation of motion

$$\left(\frac{\partial^2}{\partial t^2}+\omega^2\right)\mathcal{A}_{\mathbf{k}}(t)=0. \tag{10.2–13}$$

We have introduced the angular frequency $\omega_{\mathbf{k}} = ck$, and abbreviated it by writing ω. The general solution of this equation, which also obeys the condition (10.2–12), is given by

$$\mathcal{A}_{\mathbf{k}}(t)=\mathbf{c}_{\mathbf{k}}\,\mathrm{e}^{-\mathrm{i}\omega t}+\mathbf{c}^*_{-\mathbf{k}}\,\mathrm{e}^{\mathrm{i}\omega t}. \tag{10.2–14}$$

10.2.2 Unit polarization vectors

It is advantageous to resolve the vector $\mathbf{c}_{\mathbf{k}}$ into two orthogonal components, which are chosen so that Eq. (10.2–11) is satisfied automatically. We can do this most easily by selecting a pair of orthonormal real base vectors $\boldsymbol{\varepsilon}_{\mathbf{k}1}$, $\boldsymbol{\varepsilon}_{\mathbf{k}2}$ that obey the conditions

$$\left.\begin{aligned}&\mathbf{k}\cdot\boldsymbol{\varepsilon}_{\mathbf{k}s}=0,\quad (s=1,2)\\&\boldsymbol{\varepsilon}^*_{\mathbf{k}s}\cdot\boldsymbol{\varepsilon}_{\mathbf{k}s'}=\delta_{ss'}\quad (s,s'=1,2)\\&\boldsymbol{\varepsilon}_{\mathbf{k}1}\times\boldsymbol{\varepsilon}_{\mathbf{k}2}=\mathbf{k}/k\equiv\boldsymbol{\kappa},\end{aligned}\right\} \tag{10.2–15}$$

which signify transversality, orthonormality and right-handedness, respectively, and then putting

$$\mathbf{c}_{\mathbf{k}}=\sum_{s=1}^{2}c_{\mathbf{k}s}\boldsymbol{\varepsilon}_{\mathbf{k}s}. \tag{10.2–16}$$

The complex conjugation in the second Eq. (10.2–15) is of course unnecessary so long as the $\boldsymbol{\varepsilon}_{\mathbf{k}s}$ are real, but has been introduced for later convenience. The two real base vectors $\boldsymbol{\varepsilon}_{\mathbf{k}1}$, $\boldsymbol{\varepsilon}_{\mathbf{k}2}$ represent two states of orthogonal linear polarization, and Eq. (10.2–16) corresponds to a resolution of the field amplitude into two orthogonal linear polarizations. The conditions (10.2–15) that we imposed on the two base vectors do not actually determine them, but still leave them undefined up to a rotation about the wave vector $\mathbf{k}$.

However, it is sometimes more useful to resolve the field in other ways, for example into two orthogonal components of elliptic polarization. These polarization states are represented by complex unit base vectors $\boldsymbol{\varepsilon}_{\mathbf{k}1}$ and $\boldsymbol{\varepsilon}_{\mathbf{k}2}$ in general, and we shall impose the same conditions (10.2–15) on them. When the $\boldsymbol{\varepsilon}_{\mathbf{k}s}$ are complex, Eq. (10.2–16) represents a more general expansion of the field into two orthogonal polarization components.

It is not difficult to construct complex base vectors $\boldsymbol{\varepsilon}_{\mathbf{k}1}$, $\boldsymbol{\varepsilon}_{\mathbf{k}2}$ from real ones denoted by $\boldsymbol{\varepsilon}^{(\mathrm{R})}_{\mathbf{k}1}$, $\boldsymbol{\varepsilon}^{(\mathrm{R})}_{\mathbf{k}2}$. Let us arrange each pair of base vectors in the form of a column matrix,

$$\boldsymbol{\varepsilon}_{\mathbf{k}}^{(\mathrm{R})} \equiv \begin{bmatrix} \boldsymbol{\varepsilon}_{\mathbf{k}1}^{(\mathrm{R})} \\ \boldsymbol{\varepsilon}_{\mathbf{k}2}^{(\mathrm{R})} \end{bmatrix}, \quad \boldsymbol{\varepsilon}_{\mathbf{k}} \equiv \begin{bmatrix} \boldsymbol{\varepsilon}_{\mathbf{k}1} \\ \boldsymbol{\varepsilon}_{\mathbf{k}2} \end{bmatrix}.$$

Then any 2×2 unitary transformation matrix $\mathbf{U}$, with $\mathbf{U}\mathbf{U}^\dagger = \mathbf{1} = \mathbf{U}^\dagger\mathbf{U}$, acting on the column matrix $\boldsymbol{\varepsilon}_{\mathbf{k}}^{(\mathrm{R})}$ in general generates a pair of complex base vectors,

$$\boldsymbol{\varepsilon}_{\mathbf{k}} = \mathbf{U}\boldsymbol{\varepsilon}_{\mathbf{k}}^{(\mathrm{R})}. \tag{10.2–17a}$$

Moreover if the original base vectors satisfy conditions (10.2–15), then so do the new ones, as may be readily verified. For example, if the original matrix scalar product is the unit matrix,

$$\boldsymbol{\varepsilon}_{\mathbf{k}}^{(\mathrm{R})} \cdot \boldsymbol{\varepsilon}_{\mathbf{k}}^{(\mathrm{R})\dagger} = \mathbf{1} \equiv \begin{bmatrix} 1 & 0 \\ 0 & 1 \end{bmatrix},$$

which expresses the orthonormality of the original base vectors, then from Eq. (10.2–17a) the corresponding matrix scalar product for the new base vectors is

$$\boldsymbol{\varepsilon}_{\mathbf{k}} \cdot \boldsymbol{\varepsilon}_{\mathbf{k}}^\dagger = \mathbf{U}\boldsymbol{\varepsilon}_{\mathbf{k}}^{(\mathrm{R})} \cdot \boldsymbol{\varepsilon}_{\mathbf{k}}^{(\mathrm{R})\dagger}\mathbf{U}^\dagger = \mathbf{U}\mathbf{U}^\dagger = \mathbf{1}$$

also. $\mathbf{U}$ may be written explicitly in the form

$$\mathbf{U} = \frac{1}{(1 + |d|^2)^{1/2}} \begin{bmatrix} 1 & d \\ -d^* & 1 \end{bmatrix}, \tag{10.2–17b}$$

in which d is an arbitrary complex number. Hence

$$\left.\begin{aligned} \boldsymbol{\varepsilon}_{\mathbf{k}1} &= \frac{1}{(1 + |d|^2)^{1/2}}[\boldsymbol{\varepsilon}_{\mathbf{k}1}^{(\mathrm{R})} + d\boldsymbol{\varepsilon}_{\mathbf{k}2}^{(\mathrm{R})}] \\ \boldsymbol{\varepsilon}_{\mathbf{k}2} &= \frac{1}{(1 + |d|^2)^{1/2}}[-d^*\boldsymbol{\varepsilon}_{\mathbf{k}1}^{(\mathrm{R})} + \boldsymbol{\varepsilon}_{\mathbf{k}2}^{(\mathrm{R})}]. \end{aligned}\right\} \tag{10.2–17c}$$

By choosing different complex numbers d we generate different pairs of base vectors, corresponding to different polarization states.

Let us show that for an arbitrary complex number d, the transformation $\mathbf{U}$ given above generates elliptic polarization states. For that purpose we examine the motion of the real part of $\boldsymbol{\varepsilon}_{\mathbf{k}s}\,\mathrm{e}^{-\mathrm{i}\omega t}$ in time. Thus from Eq. (10.2–17c) we have, on decomposing $d = d_\mathrm{r} + \mathrm{i}d_\mathrm{i}$ into its real and imaginary parts,

$$\left.\begin{aligned} \mathrm{Re}\,[\boldsymbol{\varepsilon}_{\mathbf{k}1}\,\mathrm{e}^{-\mathrm{i}\omega t}] &= \mathrm{Re}\,\frac{1}{(1 + |d|^2)^{1/2}}(\boldsymbol{\varepsilon}_{\mathbf{k}1}^{(\mathrm{R})} + d\boldsymbol{\varepsilon}_{\mathbf{k}2}^{(\mathrm{R})})\,\mathrm{e}^{-\mathrm{i}\omega t} \\ &= \frac{1}{(1 + |d|^2)^{1/2}}[(\boldsymbol{\varepsilon}_{\mathbf{k}1}^{(\mathrm{R})} + d_\mathrm{r}\boldsymbol{\varepsilon}_{\mathbf{k}2}^{(\mathrm{R})})\cos\omega t + d_\mathrm{i}\boldsymbol{\varepsilon}_{\mathbf{k}2}^{(\mathrm{R})}\sin\omega t] \\ \text{and} \\ \mathrm{Re}\,[\boldsymbol{\varepsilon}_{\mathbf{k}2}\,\mathrm{e}^{-\mathrm{i}\omega t}] &= \frac{1}{(1 + |d|^2)^{1/2}}[(-d_\mathrm{r}\boldsymbol{\varepsilon}_{\mathbf{k}1}^{(\mathrm{R})} + \boldsymbol{\varepsilon}_{\mathbf{k}2}^{(\mathrm{R})})\cos\omega t + d_\mathrm{i}\boldsymbol{\varepsilon}_{\mathbf{k}1}^{(\mathrm{R})}\sin\omega t]. \end{aligned}\right\} \tag{10.2–18}$$

The right-hand sides of these equations represent motion on an ellipse in a plane perpendicular to the wave vector $\mathbf{k}$, because the amplitudes and directions of the

$\cos\omega t$ and $\sin\omega t$ oscillations are different in general. Moreover, one ellipse is traced out in the clockwise and the other in the counter-clockwise direction.

In the special case in which d is real and $d_i = 0$, the motions reduce to two cosine oscillations in the directions of the two orthogonal real unit vectors

$$\frac{\boldsymbol{\varepsilon}_{\mathbf{k}1}^{(R)} + d\boldsymbol{\varepsilon}_{\mathbf{k}2}^{(R)}}{(1+d^2)^{1/2}} \quad \text{and} \quad \frac{-d\boldsymbol{\varepsilon}_{\mathbf{k}1}^{(R)} + \boldsymbol{\varepsilon}_{\mathbf{k}2}^{(R)}}{(1+d^2)^{1/2}},$$

and these evidently represent two orthogonal linear polarization states. The new vectors $\boldsymbol{\varepsilon}_{\mathbf{k}1}$, $\boldsymbol{\varepsilon}_{\mathbf{k}2}$ are rotated through an angle $\arctan(d)$ about the $\mathbf{k}$-axis relative to the old base vectors. On the other hand, when $d = \mathrm{i}$ and $d_r = 0$, we have from Eqs. (10.2–18)

$$\mathrm{Re}\,[\boldsymbol{\varepsilon}_{\mathbf{k}1}\,\mathrm{e}^{-\mathrm{i}\omega t}] = \frac{1}{\sqrt{2}}[\boldsymbol{\varepsilon}_{\mathbf{k}1}^{(R)}\cos\omega t + \boldsymbol{\varepsilon}_{\mathbf{k}2}^{(R)}\sin\omega t]$$

$$\mathrm{Re}\,[\boldsymbol{\varepsilon}_{\mathbf{k}2}\,\mathrm{e}^{-\mathrm{i}\omega t}] = \frac{1}{\sqrt{2}}[\boldsymbol{\varepsilon}_{\mathbf{k}2}^{(R)}\cos\omega t + \boldsymbol{\varepsilon}_{\mathbf{k}1}^{(R)}\sin\omega t].$$

These equations describe two counter-rotating motions on a circle, and therefore represent orthogonal states of circular polarization. More generally, one can show from Eqs. (10.2–18) that the squares of the semi-major axis a and of the semi-minor axis b of the polarization ellipse are given by

$$a^2 = \frac{2d_i^2}{(1+|d|^2)\{1+|d|^2 - [(|d|^2-1)^2 + 4d_r^2]^{1/2}\}}$$

$$b^2 = \frac{2d_i^2}{(1+|d|^2)\{1+|d|^2 + [(|d|^2-1)^2 + 4d_r^2]^{1/2}\}},$$

while the difference is

$$a^2 - b^2 = \frac{[(|d|^2-1)^2 + 4d_r^2]^{1/2}}{|d|^2+1} \leqslant 1.$$

When $b = 0$, the ellipse reduces to a line, and the difference takes its largest value 1 whenever d is real, corresponding to linear polarization. On the other hand, the ellipse reduces to a circle with $a = b$ when $d = \mathrm{i}$, corresponding to circular polarization.

It is sometimes useful to be able to identify a set of base vectors associated with a particular wave vector $\mathbf{k}$ whose polar and azimuthal angles are θ and ϕ, respectively. It may readily be shown that the two real orthogonal base vectors

$$\left.\begin{aligned} \boldsymbol{\varepsilon}_{\mathbf{k}1} &= \cos\theta\cos\phi\,\mathbf{x}_1 + \cos\theta\sin\phi\,\mathbf{y}_1 - \sin\theta\,\mathbf{z}_1 \\ \boldsymbol{\varepsilon}_{\mathbf{k}2} &= -\sin\phi\,\mathbf{x}_1 + \cos\phi\,\mathbf{y}_1 \end{aligned}\right\} \tag{10.2–19a}$$

satisfy the conditions (10.2–15) and therefore represent two orthogonal linear polarizations associated with the wave vector $\mathbf{k}$. The unit vectors $\mathbf{x}_1$, $\mathbf{y}_1$, $\mathbf{z}_1$ point in the directions of the coordinate axes. On the other hand, the complex base vectors

$$\left.\begin{aligned}\boldsymbol{\varepsilon}_{\mathbf{k}1} &= \frac{1}{\sqrt{2}}[(\cos\theta\cos\phi - \mathrm{i}\sin\phi)\mathbf{x}_1 + (\cos\theta\sin\phi + \mathrm{i}\cos\phi)\mathbf{y}_1 - \sin\theta\,\mathbf{z}_1] \\ \boldsymbol{\varepsilon}_{\mathbf{k}2} &= \frac{1}{\sqrt{2}}[(\mathrm{i}\cos\theta\cos\phi - \sin\phi)\mathbf{x}_1 + (\mathrm{i}\cos\theta\sin\phi + \cos\phi)\mathbf{y}_1 - \mathrm{i}\sin\theta\,\mathbf{z}_1]\end{aligned}\right\} \tag{10.2–19b}$$

represent right and left circular polarizations, respectively, associated with the wave vector $\mathbf{k}$.

As we shall see, from time to time one encounters the sum over the two components $s = 1$ and $s = 2$ of the Cartesian tensor combination $(\boldsymbol{\varepsilon}^*_{\mathbf{k}s})_i(\boldsymbol{\varepsilon}_{\mathbf{k}s})_j$. In order to evaluate the sum we first observe that, since the basis of polarizations is arbitrary up to a unitary transformation, we can choose $\boldsymbol{\varepsilon}_{\mathbf{k}1}$, $\boldsymbol{\varepsilon}_{\mathbf{k}2}$ to be real, orthogonal unit vectors. Together with unit vector $\boldsymbol{\kappa}$ in the direction of the wave vector $\mathbf{k}$, these unit vectors $(\boldsymbol{\varepsilon}_{\mathbf{k}1}, \boldsymbol{\varepsilon}_{\mathbf{k}2}, \boldsymbol{\kappa})$ form a right-hand, orthogonal, Cartesian basis. The components $(\boldsymbol{\varepsilon}_{\mathbf{k}1})_i$, $(\boldsymbol{\varepsilon}_{\mathbf{k}2})_i$, κ_i are the three direction cosines of the i-axis in this Cartesian basis. By the well-known properties of the direction cosines, we may write for the cosine of the angle between the i-axis and the j-axis

$$\delta_{ij} = (\boldsymbol{\varepsilon}_{\mathbf{k}1})_i(\boldsymbol{\varepsilon}_{\mathbf{k}1})_j + (\boldsymbol{\varepsilon}_{\mathbf{k}2})_i(\boldsymbol{\varepsilon}_{\mathbf{k}2})_j + \kappa_i\kappa_j$$

or

$$\sum_s(\boldsymbol{\varepsilon}_{\mathbf{k}s})_i(\boldsymbol{\varepsilon}_{\mathbf{k}s})_j = \delta_{ij} - \kappa_i\kappa_j. \tag{10.2–19c}$$

As the expansion (10.2–16) is valid no matter which base vectors $\boldsymbol{\varepsilon}_{\mathbf{k}s}$ are chosen, it is often an advantage to leave them deliberately undefined in an analysis, until a stage is reached when it becomes clear that one particular set of base vectors simplifies the calculation. At that stage the base vectors can be chosen appropriately.

On substituting from Eq. (10.2–16) into Eq. (10.2–14) and using the result in Eq. (10.2–9) we arrive at the expansion

$$\begin{aligned}\mathbf{A}(\mathbf{r}, t) &= \frac{1}{\varepsilon_0^{1/2}L^{3/2}}\sum_{\mathbf{k}}\sum_s[c_{\mathbf{k}s}\boldsymbol{\varepsilon}_{\mathbf{k}s}\,\mathrm{e}^{-\mathrm{i}\omega t} + c^*_{-\mathbf{k}s}\boldsymbol{\varepsilon}^*_{-\mathbf{k}s}\,\mathrm{e}^{\mathrm{i}\omega t}]\,\mathrm{e}^{\mathrm{i}\mathbf{k}\cdot\mathbf{r}} \\ &= \frac{1}{\varepsilon_0^{1/2}L^{3/2}}\sum_{\mathbf{k}}\sum_s[c_{\mathbf{k}s}\boldsymbol{\varepsilon}_{\mathbf{k}s}\,\mathrm{e}^{\mathrm{i}(\mathbf{k}\cdot\mathbf{r}-\omega t)} + c^*_{\mathbf{k}s}\boldsymbol{\varepsilon}^*_{\mathbf{k}s}\,\mathrm{e}^{-\mathrm{i}(\mathbf{k}\cdot\mathbf{r}-\omega t)}] \\ &= \frac{1}{\varepsilon_0^{1/2}L^{3/2}}\sum_{\mathbf{k}}\sum_s[u_{\mathbf{k}s}(t)\boldsymbol{\varepsilon}_{\mathbf{k}s}\,\mathrm{e}^{\mathrm{i}\mathbf{k}\cdot\mathbf{r}} + u^*_{\mathbf{k}s}(t)\boldsymbol{\varepsilon}^*_{\mathbf{k}s}\,\mathrm{e}^{-\mathrm{i}\mathbf{k}\cdot\mathbf{r}}],\end{aligned} \tag{10.2–20}$$

where we have written

$$u_{\mathbf{k}s}(t) = c_{\mathbf{k}s}\,\mathrm{e}^{-\mathrm{i}\omega t}. \tag{10.2–21}$$

We regard Eq. (10.2–20) as an expansion of $\mathbf{A}(\mathbf{r}, t)$ in terms of the fundamental vector mode functions $\boldsymbol{\varepsilon}_{\mathbf{k}s}\,\mathrm{e}^{\mathrm{i}\mathbf{k}\cdot\mathbf{r}}$, with complex amplitudes $u_{\mathbf{k}s}(t)$. Each mode is labeled by a wave vector $\mathbf{k}$ and a polarization index s, and the corresponding mode function evidently satisfies the Helmholtz equation

$$(\nabla^2 + k^2)\boldsymbol{\varepsilon}_{\mathbf{k}s}\,\mathrm{e}^{\mathrm{i}\mathbf{k}\cdot\mathbf{r}} = 0, \tag{10.2–22}$$

while the corresponding mode amplitude $u_{\mathbf{k}s}(t)$ satisfies the same harmonic oscillator equation of motion (10.2–13) as $\mathscr{A}_{\mathbf{k}}(t)$.

We can immediately make use of the relation (10.2–20) to write corresponding mode expansions for the $\mathbf{E}(\mathbf{r}, t)$ and $\mathbf{B}(\mathbf{r}, t)$ vectors with the help of Eqs. (10.2–7), (10.2–8). Thus

$$\mathbf{E}(\mathbf{r}, t) = \frac{\mathrm{i}}{\varepsilon_0^{1/2}L^{3/2}}\sum_{\mathbf{k}}\sum_{s}\omega[u_{\mathbf{k}s}(t)\boldsymbol{\varepsilon}_{\mathbf{k}s}\,\mathrm{e}^{\mathrm{i}\mathbf{k}\cdot\mathbf{r}} - \text{c.c.}], \tag{10.2–23}$$

and

$$\mathbf{B}(\mathbf{r}, t) = \frac{\mathrm{i}}{\varepsilon_0^{1/2}L^{3/2}}\sum_{\mathbf{k}}\sum_{s}[u_{\mathbf{k}s}(t)(\mathbf{k}\times\boldsymbol{\varepsilon}_{\mathbf{k}s})\,\mathrm{e}^{\mathrm{i}\mathbf{k}\cdot\mathbf{r}} - \text{c.c.}]. \tag{10.2–24}$$

10.2.3 Energy of the electromagnetic field

Let us now use these equations to evaluate the energy H of the field, which is given by

$$H = \frac{1}{2}\int_{L^3}\left[\varepsilon_0\mathbf{E}^2(\mathbf{r}, t) + \frac{1}{\mu_0}\mathbf{B}^2(\mathbf{r}, t)\right]\mathrm{d}^3r. \tag{10.2–25}$$

The integration extends over the space contained within the box of volume L^3. On substituting for $\mathbf{E}(\mathbf{r}, t)$ and $\mathbf{B}(\mathbf{r}, t)$ and performing the integration over space with the help of the relations

$$\int_{L^3}\mathrm{e}^{\mathrm{i}(\mathbf{k}-\mathbf{k}')\cdot\mathbf{r}}\,\mathrm{d}^3r = L^3\delta^3_{\mathbf{k}\mathbf{k}'},$$

$$(\mathbf{k}\times\boldsymbol{\varepsilon}^*_{\mathbf{k}s})\cdot(\mathbf{k}\times\boldsymbol{\varepsilon}_{\mathbf{k}s'}) = k^2\boldsymbol{\varepsilon}^*_{\mathbf{k}s}\cdot\boldsymbol{\varepsilon}_{\mathbf{k}s'} = k^2\delta_{ss'},$$

we obtain the compact expression

$$H = 2\sum_{\mathbf{k}}\sum_{s}\omega^2|u_{\mathbf{k}s}(t)|^2, \tag{10.2–26}$$

which expresses the energy as a sum over the modes.

For the purpose of field quantization it is desirable to write H in Hamiltonian form, which we do by introducing a pair of real canonical variables $q_{\mathbf{k}s}(t)$ and $p_{\mathbf{k}s}(t)$ defined by

$$q_{\mathbf{k}s}(t) = [u_{\mathbf{k}s}(t) + u^*_{\mathbf{k}s}(t)], \tag{10.2–27}$$

$$p_{\mathbf{k}s}(t) = -\mathrm{i}\omega[u_{\mathbf{k}s}(t) - u^*_{\mathbf{k}s}(t)]. \tag{10.2–28}$$

In view of the time dependence of $u_{\mathbf{k}s}(t)$ given by Eq. (10.2–21), both $q_{\mathbf{k}s}(t)$ and $p_{\mathbf{k}s}(t)$ oscillate sinusoidally in time at frequency ω, and we have the relations

$$\frac{\partial}{\partial t}q_{\mathbf{k}s}(t) = p_{\mathbf{k}s}(t), \tag{10.2–29}$$

$$\frac{\partial}{\partial t}p_{\mathbf{k}s}(t) = -\omega^2 q_{\mathbf{k}s}(t). \tag{10.2–30}$$

In terms of $q_{\mathbf{k}s}(t)$ and $p_{\mathbf{k}s}(t)$ the expression (10.2–26) for the energy becomes

$$H = \tfrac{1}{2}\sum_{\mathbf{k}}\sum_{s}[p^2_{\mathbf{k}s}(t) + \omega^2 q^2_{\mathbf{k}s}(t)]. \tag{10.2–31}$$

This will be recognized as the energy of *a system of independent harmonic oscillators*, one for each $\mathbf{k}$, s mode of the electromagnetic field. The state of the classical radiation field is specified by the set of all canonical variables $q_{\mathbf{k}s}(t)$, $p_{\mathbf{k}s}(t)$. The set is infinite, but because we are dealing with a finite volume and a discrete set of modes, it is countably infinite. In terms of the canonical variables, the canonical equations of motion are

$$\frac{\partial H}{\partial p_{\mathbf{k}s}} = \frac{\partial q_{\mathbf{k}s}}{\partial t}, \tag{10.2–32}$$

and

$$\frac{\partial H}{\partial q_{\mathbf{k}s}} = -\frac{\partial p_{\mathbf{k}s}}{\partial t}, \tag{10.2–33}$$

and these are equivalent to Eqs. (10.2–29) and (10.2–30), respectively.

In terms of the canonical variables, the expansions (10.2–20), (10.2–23) and (10.2–24) for the field vectors $\mathbf{A}(\mathbf{r}, t)$, $\mathbf{E}(\mathbf{r}, t)$, $\mathbf{B}(\mathbf{r}, t)$ become

$$\mathbf{A}(\mathbf{r}, t) = \frac{1}{2\varepsilon_0^{1/2}L^{3/2}}\sum_{\mathbf{k}}\sum_{s}\left\{\left[q_{\mathbf{k}s}(t) + \frac{\mathrm{i}}{\omega}p_{\mathbf{k}s}(t)\right]\boldsymbol{\varepsilon}_{\mathbf{k}s}\,\mathrm{e}^{\mathrm{i}\mathbf{k}\cdot\mathbf{r}} + \text{c.c.}\right\}, \tag{10.2–34}$$

$$\mathbf{E}(\mathbf{r}, t) = \frac{\mathrm{i}}{2\varepsilon_0^{1/2}L^{3/2}}\sum_{\mathbf{k}}\sum_{s}\{[\omega q_{\mathbf{k}s}(t) + \mathrm{i}p_{\mathbf{k}s}(t)]\boldsymbol{\varepsilon}_{\mathbf{k}s}\,\mathrm{e}^{\mathrm{i}\mathbf{k}\cdot\mathbf{r}} - \text{c.c.}\}, \tag{10.2–35}$$

$$\mathbf{B}(\mathbf{r}, t) = \frac{\mathrm{i}}{2\varepsilon_0^{1/2}L^{3/2}}\sum_{\mathbf{k}}\sum_{s}\left\{\left[q_{\mathbf{k}s}(t) + \frac{\mathrm{i}}{\omega}p_{\mathbf{k}s}(t)\right]\mathbf{k}\times\boldsymbol{\varepsilon}_{\mathbf{k}s}\,\mathrm{e}^{\mathrm{i}\mathbf{k}\cdot\mathbf{r}} - \text{c.c.}\right\}. \tag{10.2–36}$$

10.3 Canonical quantization of the transverse field

In order to describe the electromagnetic field in quantum mechanics, we have to associate Hilbert space operators with the dynamical variables, which do not all commute in general. As before we shall adopt the convention of denoting Hilbert space operators by the same symbols as the corresponding classical variables, but with the caret $\hat{}$; e.g. the operators corresponding to $q_{\mathbf{k}s}(t)$ and $p_{\mathbf{k}s}(t)$ will be denoted by $\hat{q}_{\mathbf{k}s}(t)$ and $\hat{p}_{\mathbf{k}s}(t)$. According to the postulates of quantum mechanics, each pair of canonically conjugate operators $\hat{q}_{\mathbf{k}s}(t)$, $\hat{p}_{\mathbf{k}s}(t)$ has the non-zero commutator $i\hbar$. As the classical variables associated with two different modes are uncoupled, the corresponding Hilbert space operators commute. We may therefore write the following set of commutation relations:‡

$$[\hat{q}_{\mathbf{k}s}(t), \hat{p}_{\mathbf{k}'s'}(t)] = i\hbar\delta^3_{\mathbf{k}\mathbf{k}'}\delta_{ss'} \tag{10.3–1}$$

$$[\hat{q}_{\mathbf{k}s}(t), \hat{q}_{\mathbf{k}'s'}(t)] = 0, \tag{10.2–2}$$

$$[\hat{p}_{\mathbf{k}s}(t), \hat{p}_{\mathbf{k}'s'}(t)] = 0. \tag{10.3–3}$$

The state of the quantum mechanical system, i.e. of the electromagnetic field, is now described by a state vector or ket $|\psi\rangle$ in Hilbert space. The result of a measurement of a physical observable, O say, will be one of the eigenvalues of the Hilbert space operator $\hat{O}$ corresponding to that observable. But unless the state $|\psi\rangle$ is an eigenstate of the observable $\hat{O}$, the outcome of such a measurement is uncertain and only the probabilities of various outcomes can be predicted. The quantum expectation value of the observable $\hat{O}$ is given by the scalar product of the new vector $\hat{O}|\psi\rangle$ with the Hermitian conjugate vector $\langle\psi|$, i.e. by $\langle\psi|\hat{O}|\psi\rangle$. The same formula also allows us to calculate the probability of a particular outcome, say O_i, of a measurement of the observable $\hat{O}$. In order to compute this probability, we have to find the eigenstate $|O_i\rangle$ of $\hat{O}$ belonging to the eigenvalue O_i, form the projection operator $|O_i\rangle\langle O_i|$, and calculate its expectation value $|\langle\psi|O_i\rangle|^2$ in the state $|\psi\rangle$. Since physical observables are always associated with Hermitian operators the corresponding eigenvalues are all real. But whereas the spectrum of eigenvalues is continuous for $\hat{q}_{\mathbf{k}s}(t)$, $\hat{p}_{\mathbf{k}s}(t)$ and therefore for the various field operators $\hat{\mathbf{A}}(\mathbf{r}, t)$, $\hat{\mathbf{E}}(\mathbf{r}, t)$, $\hat{\mathbf{B}}(\mathbf{r}, t)$ etc., which are linear functions of $\hat{q}_{\mathbf{k}s}(t)$ and $\hat{p}_{\mathbf{k}s}(t)$, we shall see that the spectrum of eigenvalues of other operators, like the energy $\hat{H}$, is discrete. Apart from the fact that the dynamical variables are to be regarded as Hilbert space operators which do not all commute, all the previous expansions like (10.2–20), (10.2–23), (10.2–24), and the equations of motion like (10.2–13), (10.2–29), (10.2–30) remain valid as operator equations. Thus the Hamiltonian of the quantized radiation field is§

$$\hat{H} = \frac{1}{2}\sum_{\mathbf{k}}\sum_{s}[\hat{p}^2_{\mathbf{k}s}(t) + \omega^2\hat{q}^2_{\mathbf{k}s}(t)]. \tag{10.3–4}$$

It should be noted that $\hat{q}_{\mathbf{k}s}(t)$, $\hat{p}_{\mathbf{k}s}(t)$ and the field vectors $\hat{\mathbf{A}}(\mathbf{r}, t)$, $\hat{\mathbf{E}}(\mathbf{r}, t)$, $\hat{\mathbf{B}}(\mathbf{r}, t)$, etc., are the dynamical variables, while the space-time variables $\mathbf{r}$, t only play the role of parameters.

For many purposes it is more convenient to deal, not with the real dynamical variables or Hermitian operators $\hat{q}_{\mathbf{k}s}(t)$ and $\hat{p}_{\mathbf{k}s}(t)$, but with a set of non-Hermitian operators defined by

$$\hat{a}_{\mathbf{k}s}(t) = \frac{1}{(2\hbar\omega)^{1/2}}[\omega\hat{q}_{\mathbf{k}s}(t) + i\hat{p}_{\mathbf{k}s}(t)], \tag{10.3–5}$$

‡ In spirit we follow the procedure that was first described by Dirac (1927). See also Dirac (1958), Heitler (1954), Louisell (1973), Power (1964).

§ The association of Hilbert space operators with classical dynamical variables can be ambiguous when the Hilbert space operators do not commute. Thus, it may not be obvious whether the classical variable qp should be associated with $\hat{q}\hat{p}$, or $\hat{p}\hat{q}$, or $(\hat{q}\hat{p} + \hat{p}\hat{q})/2$, or some other form. However, such ambiguities do not arise in the case of the Hamiltonian given by Eq. (10.3–4).

$$\hat{a}^\dagger_{\mathbf{k}s}(t) = \frac{1}{(2\hbar\omega)^{1/2}}[\omega\hat{q}_{\mathbf{k}s}(t) - \mathrm{i}\hat{p}_{\mathbf{k}s}(t)], \tag{10.3–6}$$

the second of which is the Hermitian conjugate of the first. These equations can immediately be inverted, so that all $\hat{q}_{\mathbf{k}s}(t)$ and $\hat{p}_{\mathbf{k}s}(t)$ operators are expressible in terms of $\hat{a}_{\mathbf{k}s}(t)$ and $\hat{a}^\dagger_{\mathbf{k}s}(t)$ operators:

$$\hat{q}_{\mathbf{k}s}(t) = (\hbar/2\omega)^{1/2}[\hat{a}_{\mathbf{k}s}(t) + \hat{a}^\dagger_{\mathbf{k}s}(t)], \tag{10.3–7}$$

$$\hat{p}_{\mathbf{k}s}(t) = \mathrm{i}(\hbar\omega/2)^{1/2}[\hat{a}^\dagger_{\mathbf{k}s}(t) - \hat{a}_{\mathbf{k}s}(t)]. \tag{10.3–8}$$

With the help of Eqs. (10.3–1) to (10.3–3), we readily find the corresponding commutation relations for the $\hat{a}_{\mathbf{k}s}(t)$ and $\hat{a}^\dagger_{\mathbf{k}s}(t)$ operators to be

$$[\hat{a}_{\mathbf{k}s}(t), \hat{a}^\dagger_{\mathbf{k}'s'}(t)] = \delta^3_{\mathbf{k}\mathbf{k}'}\delta_{ss'} \tag{10.3–9}$$

$$[\hat{a}_{\mathbf{k}s}(t), \hat{a}_{\mathbf{k}'s'}(t)] = 0, \tag{10.3–10}$$

$$[\hat{a}^\dagger_{\mathbf{k}s}(t), \hat{a}^\dagger_{\mathbf{k}'s'}(t)] = 0. \tag{10.3–11}$$

Apart from a factor $(\hbar/2\omega)^{1/2}$, the $\hat{a}_{\mathbf{k}s}(t)$, $\hat{a}^\dagger_{\mathbf{k}s}(t)$ operators evidently correspond to the complex amplitudes $u_{\mathbf{k}s}(t)$, $u^*_{\mathbf{k}s}(t)$, and they also have the same time dependence (cf. Eq. 10.2–21),

$$\hat{a}_{\mathbf{k}s}(t) = \hat{a}_{\mathbf{k}s}(0)\,\mathrm{e}^{-\mathrm{i}\omega t}, \tag{10.3–12}$$

$$\hat{a}^\dagger_{\mathbf{k}s}(t) = \hat{a}^\dagger_{\mathbf{k}s}(0)\,\mathrm{e}^{\mathrm{i}\omega t}. \tag{10.3–13}$$

The operator products $\hat{a}_{\mathbf{k}s}(t)\hat{a}^\dagger_{\mathbf{k}s}(t)$ and $\hat{a}^\dagger_{\mathbf{k}s}(t)\hat{a}_{\mathbf{k}s}(t)$ are therefore time independent. But, because of the non-commutativity of the $\hat{a}_{\mathbf{k}s}(t)$, $\hat{a}^\dagger_{\mathbf{k}s}(t)$ operators, it is not immediately obvious what the operator form of a c-number equation like Eq. (10.2–26) for the energy should be. However, if we use the Hamiltonian form given by Eq. (10.3–4), and substitute for $\hat{q}_{\mathbf{k}s}(t)$ and $\hat{p}_{\mathbf{k}s}(t)$ from Eqs. (10.3–7) and (10.3–8), we obtain

$$\hat{H} = \frac{1}{2}\sum_{\mathbf{k}}\sum_{s}\hbar\omega[\hat{a}_{\mathbf{k}s}(t)\hat{a}^\dagger_{\mathbf{k}s}(t) + \hat{a}^\dagger_{\mathbf{k}s}(t)\hat{a}_{\mathbf{k}s}(t)], \tag{10.3–14}$$

in which the operators appear in symmetrized form with respect to their order. Alternatively, we may use the commutation relation (10.3–9) to express $\hat{H}$ in the normally ordered form

$$\hat{H} = \sum_{\mathbf{k}}\sum_{s}\hbar\omega[\hat{a}^\dagger_{\mathbf{k}s}(t)\hat{a}_{\mathbf{k}s}(t) + \tfrac{1}{2}], \tag{10.3–15}$$

in which $\hat{a}^\dagger_{\mathbf{k}s}(t)$ operators stand to the left of $\hat{a}_{\mathbf{k}s}(t)$ operators.

The contribution $\frac{1}{2}\hbar\omega$ to the energy of each $\mathbf{k}, s$ oscillator mode is the so-called *zero point contribution*. It is a reflection of the fact that, according to the uncertainty principle, a quantum-mechanical harmonic oscillator can never come to rest, not even in the ground state. Although the zero point energy contribution is therefore to be expected, it has the unfortunate consequence for an unbounded set of modes of giving an infinite contribution to the energy. This is a difficulty of Q.E.D. that has never been resolved satisfactorily. It may be argued that infinitely large wave numbers are unphysical, and that, in all physically meaningful problems, the sum in Eq. (10.3–15) should really be finite. This argument becomes quite compelling when we note that, for sufficiently large ω, one term $\frac{1}{2}\hbar\omega$ in the sum may exceed the energy of the whole universe. In working with a finite sum, we sometimes find that the effect of the zero point terms cancels, and that the solution remains finite and well behaved when the number of modes is allowed to become infinite at a later stage of the calculation.

Alternatively, it is possible to take the point of view that the zero point energy is a consequence of the direct association of quantum mechanical operators $\hat{q}$, $\hat{p}$ with classical variables q, p, when the only guide for such an association is the correspondence principle. (See second footnote on p. 474.) This principle demands only that the results given by the quantum theory agree with those of the classical theory in the classical limit, when the excitations become very great. In this limit the $\frac{1}{2}$ may be regarded as negligible compared with the $\hat{a}^\dagger_{\mathbf{k}s}(t)\hat{a}_{\mathbf{k}s}(t)$ term in Eq. (10.3–15), so that the correspondence principle allows us to write

$$\hat{H} = \sum_{\mathbf{k}}\sum_{s}\hbar\omega\hat{a}^\dagger_{\mathbf{k}s}(t)\hat{a}_{\mathbf{k}s}(t) \qquad (10.3\text{–}16)$$

for the energy. In the following we shall sometimes find it convenient to adopt this simpler expression, which is free from the foregoing problems. Unless we are concerned with the time dependence of an operator, we shall generally suppress the time argument, with the understanding that all operators are evaluated at the same time.

10.4 Spectrum of the energy; photons

Important differences between the classical and quantum theories of radiation appear immediately when we examine the spectrum of the energy operator given by Eqs. (10.3–4) or (10.3–16) (Messiah, 1961, Chap. 12). Whereas the corresponding classical expression, given by Eq. (10.2–31), admits all possible non-negative values for the energy, the same is not true of the operator expression for $\hat{H}$. The Hermitian operator $\hat{a}^\dagger_{\mathbf{k}s}\hat{a}_{\mathbf{k}s}$ that appears in Eq. (10.3–16) is a particularly important one and will be denoted by $\hat{n}_{\mathbf{k}s}$. We shall see that its spectrum is the set of integers, 0, 1, 2, ... From the commutation relations (10.3–9) and (10.3–10), we immediately find

$$\begin{aligned}[\hat{a}_{\mathbf{k}s}, \hat{n}_{\mathbf{k}'s'}] &= \hat{a}_{\mathbf{k}s}\hat{a}^\dagger_{\mathbf{k}'s'}\hat{a}_{\mathbf{k}'s'} - \hat{a}^\dagger_{\mathbf{k}'s'}\hat{a}_{\mathbf{k}'s'}\hat{a}_{\mathbf{k}s} \\ &= [\hat{a}_{\mathbf{k}s}, \hat{a}^\dagger_{\mathbf{k}'s'}]\hat{a}_{\mathbf{k}'s'} \\ &= \hat{a}_{\mathbf{k}s}\delta^3_{\mathbf{k}\mathbf{k}'}\delta_{ss'}, \end{aligned} \qquad (10.4\text{–}1)$$

and similarly

$$[\hat{a}^\dagger_{\mathbf{k}s}, \hat{n}_{\mathbf{k}'s'}] = -\hat{a}^\dagger_{\mathbf{k}'s'}\delta^3_{\mathbf{k}\mathbf{k}'}\delta_{ss'}. \qquad (10.4\text{–}2)$$

Let us now examine the eigenvalues of $\hat{n}_{\mathbf{k}s}$, which will immediately lead us to the eigenvalues of $\hat{H}$. If $n_{\mathbf{k}s}$ is an eigenvalue of $\hat{n}_{\mathbf{k}s}$, and if the corresponding eigenstate, normalized to unity, is denoted by $|n_{\mathbf{k}s}\rangle$, we have

$$\hat{n}_{\mathbf{k}s}|n_{\mathbf{k}s}\rangle = n_{\mathbf{k}s}|n_{\mathbf{k}s}\rangle. \qquad (10.4\text{–}3)$$

Since $\hat{n}_{\mathbf{k}s}$ is a Hermitian operator, the number $n_{\mathbf{k}s}$ is of course real.

Next we consider the state $\hat{a}^\dagger_{\mathbf{k}s}|n_{\mathbf{k}s}\rangle$. With the help of Eq. (10.4–2) we may write

$$\hat{n}_{\mathbf{k}s}\hat{a}^\dagger_{\mathbf{k}s}|n_{\mathbf{k}s}\rangle = \hat{a}^\dagger_{\mathbf{k}s}(\hat{n}_{\mathbf{k}s} + 1)|n_{\mathbf{k}s}\rangle,$$

and, on using Eq. (10.4–3), we find

$$\hat{n}_{\mathbf{k}s}\hat{a}^\dagger_{\mathbf{k}s}|n_{\mathbf{k}s}\rangle = (n_{\mathbf{k}s} + 1)\hat{a}^\dagger_{\mathbf{k}s}|n_{\mathbf{k}s}\rangle. \tag{10.4–4}$$

Comparison of Eqs. (10.4–4) and (10.4–3) shows that, if $|n_{\mathbf{k}s}\rangle$ is an eigenstate of $\hat{n}_{\mathbf{k}s}$ belonging to the eigenvalue $n_{\mathbf{k}s}$, then $\hat{a}^\dagger_{\mathbf{k}s}|n_{\mathbf{k}s}\rangle$ is also an eigenstate of $\hat{n}_{\mathbf{k}s}$, which belongs to the eigenvalue $(n_{\mathbf{k}s} + 1)$. Apart from some normalization constant $g_{\mathbf{k}s}$ we therefore have

$$\hat{a}^\dagger_{\mathbf{k}s}|n_{\mathbf{k}s}\rangle = g_{\mathbf{k}s}|n_{\mathbf{k}s} + 1\rangle.$$

This shows that the effect of the $\hat{a}^\dagger_{\mathbf{k}s}$ operator acting to the right on an eigenstate of $\hat{n}_{\mathbf{k}s}$ is to increase the corresponding eigenvalue by unity. The value of the normalization constant $g_{\mathbf{k}s}$ is readily found. We take the norms of both sides of the last equation,

$$\langle n_{\mathbf{k}s}|\hat{a}_{\mathbf{k}s}\hat{a}^\dagger_{\mathbf{k}s}|n_{\mathbf{k}s}\rangle = |g_{\mathbf{k}s}|^2\langle n_{\mathbf{k}s} + 1|n_{\mathbf{k}s} + 1\rangle = |g_{\mathbf{k}s}|^2,$$

and use the commutation relation (10.3–9) together with Eq. (10.4–3). We then find

$$|g_{\mathbf{k}s}|^2 = \langle n_{\mathbf{k}s}|\hat{n}_{\mathbf{k}s} + 1|n_{\mathbf{k}s}\rangle = n_{\mathbf{k}s} + 1,$$

or

$$|g_{\mathbf{k}s}| = (n_{\mathbf{k}s} + 1)^{1/2}.$$

Hence, apart from a unimodular factor, we have

$$\hat{a}^\dagger_{\mathbf{k}s}|n_{\mathbf{k}s}\rangle = (n_{\mathbf{k}s} + 1)^{1/2}|n_{\mathbf{k}s} + 1\rangle. \tag{10.4–5}$$

This argument can, of course, be repeated and it follows that, if $n_{\mathbf{k}s}$ is an eigenvalue of $\hat{n}_{\mathbf{k}s}$, so is $n_{\mathbf{k}s} + 1$, $n_{\mathbf{k}s} + 2$, $n_{\mathbf{k}s} + 3, \ldots$, etc., and moreover

$$(\hat{a}^\dagger_{\mathbf{k}s})^r|n_{\mathbf{k}s}\rangle = [(n_{\mathbf{k}s} + 1)\ldots(n_{\mathbf{k}s} + r + 1)]^{1/2}|n_{\mathbf{k}s} + r\rangle, \quad r = 1, 2, 3, \text{etc.} \tag{10.4–6}$$

The spectrum of eigenvalues of $\hat{n}_{\mathbf{k}s}$ is therefore unbounded from above.

Next let us consider the state $\hat{a}_{\mathbf{k}s}|n_{\mathbf{k}s}\rangle$. On using Eq. (10.4–1) we have

$$\hat{n}_{\mathbf{k}s}\hat{a}_{\mathbf{k}s}|n_{\mathbf{k}s}\rangle = \hat{a}_{\mathbf{k}s}(\hat{n}_{\mathbf{k}s} - 1)|n_{\mathbf{k}s}\rangle,$$

and with the help of Eq. (10.4–3) we find

$$\hat{n}_{\mathbf{k}s}\hat{a}_{\mathbf{k}s}|n_{\mathbf{k}s}\rangle = (n_{\mathbf{k}s} - 1)\hat{a}_{\mathbf{k}s}|n_{\mathbf{k}s}\rangle. \tag{10.4–7}$$

This shows that $\hat{a}_{\mathbf{k}s}|n_{\mathbf{k}s}\rangle$ is also an eigenstate of $\hat{n}_{\mathbf{k}s}$, belonging to the eigenvalue $n_{\mathbf{k}s} - 1$, and therefore

$$\hat{a}_{\mathbf{k}s}|n_{\mathbf{k}s}\rangle = d_{\mathbf{k}s}|n_{\mathbf{k}s} - 1\rangle,$$

where $d_{\mathbf{k}s}$ is another normalization constant. As before we can obtain $d_{\mathbf{k}s}$ by taking norms of both sides, and we find

$$|d_{\mathbf{k}s}| = \sqrt{n_{\mathbf{k}s}},$$

so that, apart from a unimodular factor,

$$\hat{a}_{\mathbf{k}s}|n_{\mathbf{k}s}\rangle = \sqrt{(n_{\mathbf{k}s})}|n_{\mathbf{k}s} - 1\rangle. \qquad (10.4\text{–}8)$$

Moreover, by repeated application of the $\hat{a}_{\mathbf{k}s}$ operator, we see that $n_{\mathbf{k}s} - 1$, $n_{\mathbf{k}s} - 2, \ldots$, etc., are also eigenvalues of $\hat{n}_{\mathbf{k}s}$, and that

$$(\hat{a}_{\mathbf{k}s})^r|n_{\mathbf{k}s}\rangle = [n_{\mathbf{k}s}(n_{\mathbf{k}s} - 1) \ldots (n_{\mathbf{k}s} - r + 1)]^{1/2}|n_{\mathbf{k}s} - r\rangle, \quad r = 1, 2, 3, \text{etc.} \qquad (10.4\text{–}9)$$

Now the sequence of numbers $n_{\mathbf{k}s}$, $n_{\mathbf{k}s} - 1$, $n_{\mathbf{k}s} - 2, \ldots$, etc., must eventually become negative, yet it is easy to show that eigenvalues of $\hat{n}_{\mathbf{k}s}$ cannot in fact be negative. For, if $|m_{\mathbf{k}s}\rangle$ is any eigenstate of $\hat{n}_{\mathbf{k}s}$ belonging to the eigenvalue $m_{\mathbf{k}s}$, we have for the norm of the vector $\hat{a}_{\mathbf{k}s}|m_{\mathbf{k}s}\rangle$,

$$0 \leq \langle m_{\mathbf{k}s}|\hat{a}^\dagger_{\mathbf{k}s}\hat{a}_{\mathbf{k}s}|m_{\mathbf{k}s}\rangle = \langle m_{\mathbf{k}s}|\hat{n}_{\mathbf{k}s}|m_{\mathbf{k}s}\rangle = m_{\mathbf{k}s}. \qquad (10.4\text{–}10)$$

The sequence of eigenvalues $\ldots, n_{\mathbf{k}s} - 1, n_{\mathbf{k}s}, n_{\mathbf{k}s} + 1, \ldots$ is therefore bounded from below and cannot become negative. This requirement can only be reconciled with Eqs. (10.4–8) and (10.4–9) if the lowest eigenvalue is zero, and we see from Eq. (10.4–10) that $\hat{a}_{\mathbf{k}s}|0_{\mathbf{k}s}\rangle$ is the null vector

$$\hat{a}_{\mathbf{k}s}|0_{\mathbf{k}s}\rangle = 0, \qquad (10.4\text{–}11)$$

so that the sequence terminates automatically. The spectrum of $\hat{n}_{\mathbf{k}s}$ is therefore the set of non-negative integers $0, 1, 2, \ldots$, etc. The operator $\hat{n}_{\mathbf{k}s}$ is known as the *number operator* for the $\mathbf{k}$, s mode.

10.4.1 Fock states

So far we have discussed only one mode of the radiation field. Together all the $\hat{n}_{\mathbf{k}s}$ operators form a complete set of commuting observables for the field. Since the operators corresponding to different modes $\mathbf{k}, s$ operate on different subspaces of Hilbert space, we can form a state vector characterizing the entire field by taking the direct product of $|n_{\mathbf{k}s}\rangle$ state vectors over all the modes:

$$\prod_{\mathbf{k},s}|n_{\mathbf{k}s}\rangle.$$

Such a state is known as a *Fock state* of the radiation field, and it is characterized by the (infinite) set of occupation numbers $n_{\mathbf{k}_1s_1}, n_{\mathbf{k}_2s_2}, \ldots$ for all the modes. We shall use the short-hand notation $\{n\}$ for the set of all $n_{\mathbf{k}s}$ and write

$$|\{n\}\rangle = \prod_{\mathbf{k},s}|n_{\mathbf{k}s}\rangle. \qquad (10.4\text{–}12)$$

It follows immediately that the Fock state $|\{n\}\rangle$ is an eigenstate of the number operator for the mode $\mathbf{k}, s$, i.e.

$$\hat{n}_{\mathbf{k}s}|\{n\}\rangle = n_{\mathbf{k}s}|\{n\}\rangle. \tag{10.4–13}$$

If we define a *total number operator* $\hat{n}$ by summing $\hat{n}_{\mathbf{k}s}$ over all modes,

$$\hat{n} \equiv \sum_{\mathbf{k},s} \hat{n}_{\mathbf{k}s}, \tag{10.4–14}$$

we find

$$\hat{n}|\{n\}\rangle = \left(\sum_{\mathbf{k},s} n_{\mathbf{k}s}\right)|\{n\}\rangle \equiv n|\{n\}\rangle, \tag{10.4–15}$$

so that the Fock state $|\{n\}\rangle$ is also an eigenstate of $\hat{n}$, with an eigenvalue which is the total occupation number n summed over all modes.

The state $|\{0\}\rangle$ for which all the occupation numbers are zero has the lowest eigenvalue of $\hat{n}$ and is known as the *vacuum state* $|\text{vac}\rangle$. We can form any Fock state by repeated application of the $\hat{a}^\dagger_{\mathbf{k}s}$ operators on the vacuum. Thus, we have immediately from Eq. (10.4–6)

$$|\{m\}\rangle = \prod_{\mathbf{k},s}\left[\frac{(\hat{a}^\dagger_{\mathbf{k}s})^{m_{\mathbf{k}s}}}{\sqrt{(m_{\mathbf{k}s})!}}\right]|\text{vac}\rangle. \tag{10.4–16}$$

Since the energy of the field given by Eq. (10.3–16) is a linear combination of the set of operators $\hat{n}_{\mathbf{k}s}$, it follows at once that the Fock states are also eigenstates of the energy $\hat{H}$, namely

$$\hat{H}|\{n\}\rangle = \left(\sum_{\mathbf{k},s} n_{\mathbf{k}s}\hbar\omega\right)|\{n\}\rangle, \tag{10.4–17}$$

and that the energy eigenvalues are given by $\sum_{\mathbf{k},s} n_{\mathbf{k}s}\hbar\omega$. These eigenvalues are degenerate, and there will generally be many Fock states corresponding to any eigenvalue. Indeed in the limit in which the modes become continuous, the degeneracy becomes infinite.

The discrete excitations or quanta of the electromagnetic field, corresponding to the occupation numbers $\{n\}$, are usually known as *photons*. Thus a state $|\ldots 0, 0, 1_{\mathbf{k}s}, 0, 0, \ldots\rangle$ is described as a state with one photon of wave vector $\mathbf{k}$ and polarization s. We shall sometimes find it convenient to label this state $|1_{\mathbf{k}s}, \{0\}\rangle$, in the sense that the mode $\mathbf{k}, s$, which is displayed explicitly, has occupation number 1, whereas the other modes are all empty. As the operators $\hat{a}_{\mathbf{k}s}$ and $\hat{a}^\dagger_{\mathbf{k}s}$ lower and raise the photon occupation number of a state by unity, they are known as the lowering or photon *annihilation operator*, and the raising or photon *creation operator*, respectively. The eigenvalues of the photon number operator $\hat{n}_{\mathbf{k}s}$ are unbounded, so that an arbitrarily large number of photons may be found in the same quantum state, which means that the photons are bosons and obey Bose–Einstein statistics. The energy eigenvalue $\sum_{\mathbf{k},s} n_{\mathbf{k}s}\hbar\omega$ in Eq. (10.4–17) can be interpreted to mean that each of the $n_{\mathbf{k}s}$ photons belonging to the mode $\mathbf{k}, s$ carries an energy $\hbar\omega$. This energy is independent of both the polarization s of the photon and the direction of the wave vector $\mathbf{k}$; it depends only on the frequency ω.

10.4.2 Approximately localized photons

We have defined photons as quantum excitations of the normal modes of the electromagnetic field, and we have associated them with plane waves of definite wave vector $\mathbf{k}$ and definite polarization s.

Now a plane wave has no localization in space or time, and therefore the excitation in the one-photon state $|1_{\mathbf{k}s}, \{0\}\rangle$ must be regarded as distributed over all space-time. However, it is sometimes necessary to deal with excitations of the electromagnetic field, or photons, that are localized to a certain extent and propagate with the velocity of light. Such localized photons may be visualized as particles in a sense, although this interpretation is not without its dangers, for photons lack some of the properties of particles. For example, a photon has no precise position no matter what the state may be (Newton and Wigner, 1949; Amrein, 1969); the concept of position is meaningful only in a restricted sense to be examined in Section 12.11.

Nevertheless, it is not difficult to introduce a state corresponding to an approximately localized photon at some given time. As an example, let us consider the state

$$|\psi\rangle = C\sum_{\mathbf{k}} \mathrm{e}^{-(\mathbf{k}-\mathbf{k}_0)^2/2\sigma^2}\,\mathrm{e}^{-\mathrm{i}\mathbf{k}\cdot\mathbf{r}_0}|1_{\mathbf{k}s}, \{0\}\rangle,$$

which is a linear superposition of one-photon Fock states with different wave vectors $\mathbf{k}$, suitably normalized by proper choice of the constant C. It can be seen at once that $|\psi\rangle$ is an eigenstate of the total photon number operator $\hat{n}$ with eigenvalue unity, so that it is also a one-photon state. However, it is a state in which the wave vector $\mathbf{k}$ has no definite value, but a Gaussian spread about $\mathbf{k}_0$. We may regard the corresponding photon as being approximately localized in the form of a wave packet centered at position $\mathbf{r}_0$ at the given time.

10.4.3 Fock states as a basis

Since the Fock states $|\{n\}\rangle$ are eigenstates of a complete set of commuting observables, they form a complete set, which serves as a basis for the representation of arbitrary states and operators. The completeness is expressed by the resolution of the identity operator in terms of Fock state projectors:

$$1 = \sum_{\{n\}}|\{n\}\rangle\langle\{n\}|. \tag{10.4–18}$$

In addition, since the states $|\{n\}\rangle$ are eigenstates of the Hermitian operators $\hat{n}_{\mathbf{k}s}$, it follows that the different Fock states are orthonormal, so that

$$\langle\{n\}|\{m\}\rangle = \prod_{\mathbf{k},s}\delta_{n_{\mathbf{k}s}m_{\mathbf{k}s}}. \tag{10.4–19}$$

By making use of these relations, we can obtain a representation of any state of the field and of any field operator.

As an example, let us consider the density operator $\hat{\rho}$ of the field. This is the natural object for describing the state of a system that is not in a pure quantum state, but is describable by an ensemble of quantum states $|\psi_i\rangle$, with probabilities p_i. The operator $\hat{\rho}$ is the mean projection operator averaged over the ensemble, or

$$\hat{\rho} = \sum_i p_i|\psi_i\rangle\langle\psi_i|. \tag{10.4–20}$$

Since the states of the radiation fields encountered in practice are almost never pure states, we shall usually find it appropriate to describe them by their density operators. In order to make a representation of $\hat{\rho}$ in the basis of Fock states we simply multiply $\hat{\rho}$ on both sides by the identity operator in the form of Eq. (10.4–18). We then obtain

$$\hat{\rho} = \sum_{\{n\}}\sum_{\{m\}} \langle\{n\}|\hat{\rho}|\{m\}\rangle|\{n\}\rangle\langle\{m\}|. \tag{10.4–21}$$

As another example, we consider the Fock representation of the annihilation and creation operators $\hat{a}_{\mathbf{k}'s'}$ and $\hat{a}^\dagger_{\mathbf{k}'s'}$. Again we multiply $\hat{a}_{\mathbf{k}'s'}$ by the identity operator in the form of Eq. (10.4–18) on the right and on the left, and use Eq. (10.4–8) to obtain

$$\hat{a}_{\mathbf{k}'s'} = \sum_{\{n\}}(n_{\mathbf{k}'s'})^{1/2}|n_{\mathbf{k}'s'} - 1\rangle\langle n_{\mathbf{k}'s'}| \prod_{\mathbf{k},s\neq\mathbf{k}',s'} |n_{\mathbf{k}s}\rangle\langle n_{\mathbf{k}s}|. \tag{10.4–22}$$

Similarly, with the help of Eq. (10.4–5), we have

$$\hat{a}^\dagger_{\mathbf{k}'s'} = \sum_{\{n\}}(n_{\mathbf{k}'s'} + 1)^{1/2}|n_{\mathbf{k}'s'} + 1\rangle\langle n_{\mathbf{k}'s'}| \prod_{\mathbf{k},s\neq\mathbf{k}',s'} |n_{\mathbf{k}s}\rangle\langle n_{\mathbf{k}s}|. \tag{10.4–23}$$

We note that these representations are entirely off-diagonal in the $\mathbf{k}'$, s' mode, so that the expectation values of $\hat{a}_{\mathbf{k}s}$ and $\hat{a}^\dagger_{\mathbf{k}s}$ vanish for a Fock state,

$$\langle\{n\}|\hat{a}_{\mathbf{k}s}|\{n\}\rangle = 0 = \langle\{n\}|\hat{a}^\dagger_{\mathbf{k}s}|\{n\}\rangle. \tag{10.4–24}$$

By adding and subtracting the expansions (10.4–22) and (10.4–23) for $\hat{a}_{\mathbf{k}'s'}$ and $\hat{a}^\dagger_{\mathbf{k}'s'}$ we immediately obtain the Fock representations of the $\hat{q}_{\mathbf{k}'s'}$ and $\hat{p}_{\mathbf{k}'s'}$ operators with the help of Eqs. (10.3–7) and (10.3–8). We again find that the expectation values vanish for the Fock state,

$$\langle\{n\}|\hat{q}_{\mathbf{k}s}|\{n\}\rangle = 0 = \langle\{n\}|\hat{p}_{\mathbf{k}s}|\{n\}\rangle. \tag{10.4–25}$$

This is a reflection of the fact that the eigenvalues of $\hat{q}_{\mathbf{k}s}$ and $\hat{p}_{\mathbf{k}s}$ are distributed over positive and negative values with equal probability when the field is in a Fock state. The second moments do not of course vanish. From Eqs. (10.3–7) and (10.3–8) and the commutation rule (10.3–9) we have

$$\begin{aligned}\langle\{n\}|\hat{q}^2_{\mathbf{k}s}|\{n\}\rangle &= (\hbar/2\omega)\langle\{n\}|\hat{a}^2_{\mathbf{k}s} + \hat{a}^{\dagger 2}_{\mathbf{k}s} + 2\hat{n}_{\mathbf{k}s} + 1|\{n\}\rangle \\ &= (\hbar/\omega)(n_{\mathbf{k}s} + \tfrac{1}{2}),\end{aligned} \tag{10.4–26}$$

and

$$\begin{aligned}\langle\{n\}|\hat{p}^2_{\mathbf{k}s}|\{n\}\rangle &= (\hbar\omega/2)\langle\{n\}| - \hat{a}^{\dagger 2}_{\mathbf{k}s} - \hat{a}^2_{\mathbf{k}s} + 2\hat{n}_{\mathbf{k}s} + 1|\{n\}\rangle \\ &= \hbar\omega(n_{\mathbf{k}s} + \tfrac{1}{2}),\end{aligned} \tag{10.4–27}$$

so that the product of the standard deviations $[\langle(\Delta\hat{q}_{\mathbf{k}s})^2\rangle]^{1/2}$ and $[\langle(\Delta\hat{p}_{\mathbf{k}s})^2\rangle]^{1/2}$ is given by

$$[\langle(\Delta\hat{q}_{\mathbf{k}s})^2\rangle]^{1/2}[\langle(\Delta\hat{p}_{\mathbf{k}s})^2\rangle]^{1/2} = [\langle\hat{q}^2_{\mathbf{k}s}\rangle]^{1/2}[\langle\hat{p}^2_{\mathbf{k}s}\rangle]^{1/2}$$
$$= \hbar(n_{\mathbf{k}s} + \tfrac{1}{2}). \qquad (10.4\text{–}28)$$

The uncertainty product therefore increases with the excitation of the state, and is a minimum for zero excitation, when it reaches its lowest possible value $\hbar/2$ for any state. But it is worth noting that $\hat{q}_{\mathbf{k}s}$ and $\hat{p}_{\mathbf{k}s}$ fluctuate, in the sense of not having well-defined values, even in the ground state.

10.4.4 The q-representation of the Fock state

More generally, we may obtain the probability density $P(q_{\mathbf{k}s})$ of the eigenvalues $q_{\mathbf{k}s}$ for a Fock state having a set of photon occupation numbers $\{n\}$ by noting that this is given by the expectation value of the projector for the state $|q_{\mathbf{k}s}\rangle$ with eigenvalue $q_{\mathbf{k}s}$:

$$P(q_{\mathbf{k}s}) = \langle n_{\mathbf{k}s}|q_{\mathbf{k}s}\rangle\langle q_{\mathbf{k}s}|n_{\mathbf{k}s}\rangle = |\langle q_{\mathbf{k}s}|n_{\mathbf{k}s}\rangle|^2. \qquad (10.4\text{–}29)$$

The probability density is therefore known once the scalar product $\langle q_{\mathbf{k}s}|n_{\mathbf{k}s}\rangle$ has been determined. This scalar product is also the amplitude of the state $|n_{\mathbf{k}s}\rangle$ in the q-representation, or the so-called Schrödinger wave amplitude.

In order to derive it we first note that, from Eqs. (10.3–5) and (10.4–11),

$$\langle q_{\mathbf{k}s}|\hat{a}_{\mathbf{k}s}|0_{\mathbf{k}s}\rangle = 0 = (\omega/2\hbar)^{1/2}\langle q_{\mathbf{k}s}|[\hat{q}_{\mathbf{k}s} + (\mathrm{i}/\omega)\hat{p}_{\mathbf{k}s}]|0_{\mathbf{k}s}\rangle$$
$$= (\hbar/2\omega)^{1/2}[(\omega/\hbar)q_{\mathbf{k}s} + \partial/\partial q_{\mathbf{k}s}]\langle q_{\mathbf{k}s}|0_{\mathbf{k}s}\rangle, \qquad (10.4\text{–}30)$$

where we have made use of a well-known property of the canonically conjugate operators $\hat{q}_{\mathbf{k}s}$, $\hat{p}_{\mathbf{k}s}$, which allows us to write for any state vector $|\psi\rangle$ in the Hilbert space associated with these operators,

$$\langle q_{\mathbf{k}s}|\hat{p}_{\mathbf{k}s}|\psi\rangle = (\hbar/\mathrm{i})(\partial/\partial q_{\mathbf{k}s})\langle q_{\mathbf{k}s}|\psi\rangle. \qquad (10.4\text{–}31)$$

We see that $\langle q_{\mathbf{k}s}|0\rangle$ can be obtained from Eq. (10.4–30) by solving a simple linear, first-order differential equation. The result is

$$\langle q_{\mathbf{k}s}|0\rangle = \left(\frac{\omega}{\pi\hbar}\right)^{1/4} \exp(-\tfrac{1}{2}q^2_{\mathbf{k}s}\omega/\hbar), \qquad (10.4\text{–}32)$$

when the normalization constant is chosen so as to ensure that

$$\int_{-\infty}^{\infty} |\langle q_{\mathbf{k}s}|0\rangle|^2 \, \mathrm{d}q_{\mathbf{k}s} = 1.$$

Equation (10.4–32) gives the amplitude, or Schrödinger wave function, of the ground state in the q-representation, which is seen to be of Gaussian form. We may now find $\langle q_{\mathbf{k}s}|n_{\mathbf{k}s}\rangle$ by noting that, from Eqs. (10.4–16), (10.4–31), (10.4–32) and (10.3–6),

$$\langle q_{\mathbf{k}s}|n_{\mathbf{k}s}\rangle = \frac{1}{\sqrt{(n_{\mathbf{k}s}!)}}\langle q_{\mathbf{k}s}|(\hat{a}^\dagger_{\mathbf{k}s})^{n_{\mathbf{k}s}}|0\rangle$$

$$= \frac{1}{\sqrt{(n_{\mathbf{k}s}!)}}\left(\frac{\hbar}{2\omega}\right)^{n_{\mathbf{k}s}/2}\left[\frac{\omega}{\hbar}q_{\mathbf{k}s} - \frac{\partial}{\partial q_{\mathbf{k}s}}\right]^{n_{\mathbf{k}s}}\langle q_{\mathbf{k}s}|0\rangle$$

$$= \frac{1}{\sqrt{(n_{\mathbf{k}s}!)}}\left(\frac{\hbar}{2\omega}\right)^{n_{\mathbf{k}s}/2}\left[\frac{\omega}{\hbar}q_{\mathbf{k}s} - \frac{\partial}{\partial q_{\mathbf{k}s}}\right]^{n_{\mathbf{k}s}}\left(\frac{\omega}{\pi\hbar}\right)^{1/4}\exp\left(-\tfrac{1}{2}q_{\mathbf{k}s}^2\omega/\hbar\right). \tag{10.4–33}$$

The $n_{\mathbf{k}s}$-fold application of the differential operator $[(\omega/\hbar)q_{\mathbf{k}s} - \partial/\partial q_{\mathbf{k}s}]$ to the Gaussian function $\exp(-\frac{1}{2}q_{\mathbf{k}s}^2\omega/\hbar)$ leads to $n_{\mathbf{k}s}$'th order Hermite functions, and we obtain finally

$$\langle q_{\mathbf{k}s}|n_{\mathbf{k}s}\rangle = \frac{1}{(2^{n_{\mathbf{k}s}}n_{\mathbf{k}s}!)^{1/2}}\left(\frac{\omega}{\pi\hbar}\right)^{1/4}H_{n_{\mathbf{k}s}}[\sqrt{(\omega/\hbar)}q_{\mathbf{k}s}]\exp\left(-\tfrac{1}{2}q_{\mathbf{k}s}^2\omega/\hbar\right), \tag{10.4–34}$$

in which $H_n(x)$ is the n'th-order Hermite polynomial in x.

10.4.5 Time dependence of the field operators

Finally, let us examine the equations of motion of $\hat{a}_{\mathbf{k}s}(t)$ and $\hat{a}^\dagger_{\mathbf{k}s}(t)$, and formally confirm that the time dependence is given by Eqs. (10.3–12) and (10.3–13). The Heisenberg equation of motion for any field operator $\hat{O}$ which is not explicitly time dependent is

$$\frac{\mathrm{d}\hat{O}}{\mathrm{d}t} = \frac{1}{\mathrm{i}\hbar}[\hat{O}, \hat{H}], \tag{10.4–35}$$

where $\hat{H}$ is given by Eq. (10.3–15) or Eq. (10.3–16). It follows immediately that the number operators $\hat{n}_{\mathbf{k}s}$ and $\hat{n}$ are time independent, since they commute with $\hat{H}$. For the annihilation and creation operators $\hat{a}_{\mathbf{k}s}(t)$ and $\hat{a}^\dagger_{\mathbf{k}s}(t)$ we have from Eqs. (10.4–1) and (10.4–2)

$$\frac{\mathrm{d}\hat{a}_{\mathbf{k}s}(t)}{\mathrm{d}t} = -\mathrm{i}\omega\hat{a}_{\mathbf{k}s}(t), \tag{10.4–36}$$

$$\frac{\mathrm{d}\hat{a}^\dagger_{\mathbf{k}s}(t)}{\mathrm{d}t} = \mathrm{i}\omega\hat{a}^\dagger_{\mathbf{k}s}(t), \tag{10.4–37}$$

from which the solutions (10.3–12) and (10.3–13) follow immediately. From the c-number Eqs. (10.2–34)–(10.2–36), the expansions for the field operators $\hat{\mathbf{A}}(\mathbf{r}, t)$, $\hat{\mathbf{E}}(\mathbf{r}, t)$, $\hat{\mathbf{B}}(\mathbf{r}, t)$ can therefore be written [cf. Eqs. (10.2–20), (10.2–23), (10.2–24)]

$$\hat{\mathbf{A}}(\mathbf{r}, t) = \frac{1}{L^{3/2}}\sum_{\mathbf{k}}\sum_{s}\left(\frac{\hbar}{2\omega\varepsilon_0}\right)^{1/2}[\hat{a}_{\mathbf{k}s}(0)\varepsilon_{\mathbf{k}s}\,\mathrm{e}^{\mathrm{i}(\mathbf{k}\cdot\mathbf{r}-\omega t)} + \text{h.c.}] \tag{10.4–38}$$

$$\hat{\mathbf{E}}(\mathbf{r}, t) = \frac{1}{L^{3/2}}\sum_{\mathbf{k}}\sum_{s}\left(\frac{\hbar\omega}{2\varepsilon_0}\right)^{1/2}[\mathrm{i}\hat{a}_{\mathbf{k}s}(0)\varepsilon_{\mathbf{k}s}\,\mathrm{e}^{\mathrm{i}(\mathbf{k}\cdot\mathbf{r}-\omega t)} + \text{h.c.}] \tag{10.4–39}$$

$$\hat{\mathbf{B}}(\mathbf{r}, t) = \frac{1}{L^{3/2}}\sum_{\mathbf{k}}\sum_{s}\left(\frac{\hbar}{2\omega\varepsilon_0}\right)^{1/2}[\mathrm{i}\hat{a}_{\mathbf{k}s}(0)(\mathbf{k}\times\varepsilon_{\mathbf{k}s})\,\mathrm{e}^{\mathrm{i}(\mathbf{k}\cdot\mathbf{r}-\omega t)} + \text{h.c.}]. \tag{10.4–40}$$

Here h.c. stands for the Hermitian conjugate of the preceding term. It is not difficult to show directly from the Heisenberg equations of motion for $\hat{\mathbf{E}}(\mathbf{r}, t)$, $\hat{\mathbf{B}}(\mathbf{r}, t)$, and from the commutation relations among the different field operators $\hat{\mathbf{E}}(\mathbf{r}, t)$, $\hat{\mathbf{B}}(\mathbf{r}, t)$, $\hat{\mathbf{A}}(\mathbf{r}, t)$ to be discussed in Section 10.8, that the quantized fields are coupled by the same Maxwell equations as the classical fields from which they are derived, i.e.

$$\nabla \times \hat{\mathbf{E}}(\mathbf{r}, t) = -\frac{\partial \hat{\mathbf{B}}(\mathbf{r}, t)}{\partial t} \tag{10.4–41}$$

$$\nabla \times \hat{\mathbf{B}}(\mathbf{r}, t) = \frac{1}{c^2}\frac{\partial \hat{\mathbf{E}}(\mathbf{r}, t)}{\partial t}. \tag{10.4–42}$$

The same conclusion also follows immediately from the mode expansions (10.4–38) to (10.4–40). Finally we note that the first term on the right of each of the Eqs. (10.4–38) to (10.4–40) is sometimes denoted by $\hat{\mathbf{A}}^{(+)}(\mathbf{r}, t)$, $\hat{\mathbf{E}}^{(+)}(\mathbf{r}, t)$, $\hat{\mathbf{B}}^{(+)}(\mathbf{r}, t)$, respectively, because it depends only on positive frequencies. These positive frequency parts of the field correspond to the complex analytic signals that were introduced in Section 3.1.

10.5 Momentum of the quantized field

In classical electromagnetic theory, the total linear momentum $\mathbf{P}$ of the field is proportional to the volume integral of the Poynting vector, which we can express in terms of the $\mathbf{E}(\mathbf{r}, t)$ and $\mathbf{B}(\mathbf{r}, t)$ vectors in SI units in the form

$$\mathbf{P} = \varepsilon_0 \int_{L^3} \mathbf{E}(\mathbf{r}, t) \times \mathbf{B}(\mathbf{r}, t)\, \mathrm{d}^3 r. \tag{10.5–1}$$

In the quantized theory the dynamical variables have to be replaced by Hermitian Hilbert space operators. We have already encountered the operator forms (10.4–39) and (10.4–40) of the electric and magnetic fields. Unfortunately, the replacement of $\mathbf{E}(\mathbf{r}, t)$ and $\mathbf{B}(\mathbf{r}, t)$ by Hilbert space operators in Eq. (10.5–1) does not lead to a Hermitian form, since $\hat{\mathbf{E}}(\mathbf{r}, t)$ and $\hat{\mathbf{B}}(\mathbf{r}, t)$ do not commute (see Section 10.8). However, a simple Hermitian form can be constructed from Eq. (10.5–1) by symmetrization. We therefore adopt the following expression for the momentum of the quantized field

$$\hat{\mathbf{P}} = \tfrac{1}{2}\varepsilon_0 \int_{L^3} [\hat{\mathbf{E}}(\mathbf{r}, t) \times \hat{\mathbf{B}}(\mathbf{r}, t) - \hat{\mathbf{B}}(\mathbf{r}, t) \times \hat{\mathbf{E}}(\mathbf{r}, t)]\, \mathrm{d}^3 r. \tag{10.5–2}$$

With the help of the expansions (10.4–39) and (10.4–40) for $\hat{\mathbf{E}}(\mathbf{r}, t)$ and $\hat{\mathbf{B}}(\mathbf{r}, t)$, we obtain

$$\hat{\mathbf{P}} = \frac{1}{2L^3}\sum_{\mathbf{k},s}\sum_{\mathbf{k}',s'}\left(\frac{\hbar\omega}{2}\right)^{1/2}\left(\frac{\hbar}{2\omega'}\right)^{1/2}\int_{L^3}[\mathrm{i}\hat{a}_{\mathbf{k}s}\boldsymbol{\varepsilon}_{\mathbf{k}s}\,\mathrm{e}^{\mathrm{i}(\mathbf{k}\cdot\mathbf{r}-\omega t)} - \mathrm{i}\hat{a}^\dagger_{\mathbf{k}s}\boldsymbol{\varepsilon}^*_{\mathbf{k}s}\,\mathrm{e}^{-\mathrm{i}(\mathbf{k}\cdot\mathbf{r}-\omega t)}]$$
$$\times\,[\mathrm{i}\hat{a}_{\mathbf{k}'s'}(\mathbf{k}' \times \boldsymbol{\varepsilon}_{\mathbf{k}'s'})\,\mathrm{e}^{\mathrm{i}(\mathbf{k}'\cdot\mathbf{r}-\omega' t)} - \mathrm{i}\hat{a}^\dagger_{\mathbf{k}'s'}(\mathbf{k}' \times \boldsymbol{\varepsilon}^*_{\mathbf{k}'s'})\,\mathrm{e}^{-\mathrm{i}(\mathbf{k}'\cdot\mathbf{r}-\omega' t)}]\,\mathrm{d}^3 r + \text{h.c.}$$

The volume integrals are easily evaluated, and they yield $L^3\delta^3_{-\mathbf{k}\mathbf{k}'}$ or $L^3\delta^3_{\mathbf{k}\mathbf{k}'}$. Next we expand the vector triple products, which reduce to

$$\boldsymbol{\varepsilon}_{\mathbf{k}s} \times (\mathbf{k} \times \boldsymbol{\varepsilon}^*_{\mathbf{k}s'}) = (\boldsymbol{\varepsilon}_{\mathbf{k}s} \cdot \boldsymbol{\varepsilon}^*_{\mathbf{k}s'})\mathbf{k} = \delta_{ss'}\mathbf{k}, \tag{10.5–3}$$

etc., because $\mathbf{k}$ is perpendicular to $\boldsymbol{\varepsilon}_{\mathbf{k}s}$. The terms in $\hat{a}_{\mathbf{k}s}\hat{a}_{-\mathbf{k}s'}$ and in $\hat{a}^\dagger_{\mathbf{k}s}\hat{a}^\dagger_{-\mathbf{k}s'}$ vanish, because they change sign when we introduce the change of variables $\mathbf{k} \to -\mathbf{k}$, $s \to s'$, $s' \to s$. Finally we combine terms with the help of Eq. (10.5–3) to obtain

$$\hat{\mathbf{P}} = \tfrac{1}{2}\sum_{\mathbf{k},s}\hbar\mathbf{k}(\hat{a}^\dagger_{\mathbf{k}s}\hat{a}_{\mathbf{k}s} + \hat{a}_{\mathbf{k}s}\hat{a}^\dagger_{\mathbf{k}s}), \tag{10.5–4}$$

which is to be compared with Eq. (10.3–14) for the energy. Apart from the appearance of the wave vector $\mathbf{k}$ instead of the angular frequency ω, the expressions for $\hat{\mathbf{P}}$ and $\hat{H}$ are similar. By using the commutation rule (10.3–9) we may also express $\hat{\mathbf{P}}$ in the normally ordered form

$$\hat{\mathbf{P}} = \sum_{\mathbf{k},s}\hbar\mathbf{k}(\hat{a}^\dagger_{\mathbf{k}s}\hat{a}_{\mathbf{k}s} + \tfrac{1}{2}). \tag{10.5–5}$$

This time the zero point contribution to the momentum vanishes, because there is a $-\tfrac{1}{2}\hbar\mathbf{k}$ term for every $\tfrac{1}{2}\hbar\mathbf{k}$ term when we sum over all modes. We may therefore express the momentum in the simpler form

$$\hat{\mathbf{P}} = \sum_{\mathbf{k},s}\hbar\mathbf{k}\hat{n}_{\mathbf{k}s}. \tag{10.5–6}$$

This shows that $\hat{\mathbf{P}}$, like $\hat{H}$, is expressible as a function of the number operators $\hat{n}_{\mathbf{k}s}$, and that $\hat{\mathbf{P}}$ and $\hat{H}$ commute. The Fock states $|\{n\}\rangle$ are therefore also eigenstates of $\hat{\mathbf{P}}$, and the eigenvalues are given by the equation

$$\hat{\mathbf{P}}|\{n\}\rangle = \left(\sum_{\mathbf{k}}\sum_{s}\hbar\mathbf{k}n_{\mathbf{k}s}\right)|\{n\}\rangle. \tag{10.5–7}$$

The eigenvalues of $\hat{\mathbf{P}}$, like those of $\hat{H}$, are degenerate. The eigenvalue $\sum_{\mathbf{k}}\sum_s n_{\mathbf{k}s}\hbar\mathbf{k}$ can readily be interpreted to mean that each of the $n_{\mathbf{k}s}$ photons belonging to the $\mathbf{k}$, s mode of the field carries a momentum $\hbar\mathbf{k}$, which is independent of the polarization. As the Heisenberg equation of motion for any operator $\hat{O}$ that is not explicitly time dependent is of the form

$$\frac{\mathrm{d}\hat{O}}{\mathrm{d}t} = \frac{1}{\mathrm{i}\hbar}[\hat{O}, \hat{H}], \tag{10.5–8}$$

the commutation of $\hat{\mathbf{P}}$ with $\hat{H}$ implies that the momentum is a constant of the motion for a free field.

10.6 Angular momentum of the quantized field

In classical electromagnetic theory the total angular momentum $\mathbf{J}(\mathbf{r}_0)$ of the electromagnetic field in vacuum with respect to the point $\mathbf{r}_0$ is given by the integrated moment of the momentum density about $\mathbf{r}_0$,

$$\begin{aligned}\mathbf{J}(\mathbf{r}_0) &= \varepsilon_0\int_{L^3}(\mathbf{r} - \mathbf{r}_0) \times [\mathbf{E}(\mathbf{r}, t) \times \mathbf{B}(\mathbf{r}, t)]\,\mathrm{d}^3r \\ &= \varepsilon_0\int_{L^3}\mathbf{r} \times [\mathbf{E}(\mathbf{r}, t) \times \mathbf{B}(\mathbf{r}, t)]\,\mathrm{d}^3r - \mathbf{r}_0 \times \varepsilon_0\int_{L^3}\mathbf{E}(\mathbf{r}, t) \times \mathbf{B}(\mathbf{r}, t)\,\mathrm{d}^3r \\ &= \mathbf{J}(0) - \mathbf{r}_0 \times \mathbf{P}.\end{aligned} \tag{10.6–1}$$

Here $\mathbf{P}$ is the total electromagnetic momentum that we have just discussed. Equation (10.6–1) shows that $\mathbf{J}(\mathbf{r}_0)$ depends on $\mathbf{r}_0$ in general. When the field is quantized, and $\hat{\mathbf{E}}(\mathbf{r}, t)$ and $\hat{\mathbf{B}}(\mathbf{r}, t)$ are non-commuting Hilbert space operators, we can define a quantum angular momentum operator $\hat{\mathbf{J}}(\mathbf{r}_0)$ in an analogous manner, provided we symmetrize the operators as before, in order to ensure that $\hat{\mathbf{J}}(\mathbf{r}_0)$ is Hermitian. We then write

$$\begin{aligned}\hat{\mathbf{J}}(\mathbf{r}_0) &= \tfrac{1}{2}\varepsilon_0\int_{L^3}(\mathbf{r}-\mathbf{r}_0)\times[\hat{\mathbf{E}}(\mathbf{r},t)\times\hat{\mathbf{B}}(\mathbf{r},t)-\hat{\mathbf{B}}(\mathbf{r},t)\times\hat{\mathbf{E}}(\mathbf{r},t)]\,\mathrm{d}^3r\\ &= \hat{\mathbf{J}}(0)-\mathbf{r}_0\times\hat{\mathbf{P}},\end{aligned}\tag{10.6–2}$$

where $\hat{\mathbf{P}}$ is the momentum operator given by Eqs. (10.5–2) and (10.5–6).

10.6.1 Angular momentum as a constant of the motion

Let us first examine the question whether the total angular momentum $\hat{\mathbf{J}}(\mathbf{r}_0)$ is a constant of the motion for the quantized field. As we have already shown that $\hat{\mathbf{P}}$ is a constant of the motion, it suffices if we can demonstrate the same thing for $\hat{\mathbf{J}}(0)$. From the definition, and with the help of the Maxwell equations (10.4–41) and (10.4–42), we have

$$\begin{aligned}\frac{\mathrm{d}\hat{\mathbf{J}}(0)}{\mathrm{d}t} &= \frac{1}{2}\varepsilon_0\int_{L^3}\mathbf{r}\times\left[\frac{\partial\hat{\mathbf{E}}}{\partial t}\times\hat{\mathbf{B}}+\hat{\mathbf{E}}\times\frac{\partial\hat{\mathbf{B}}}{\partial t}-\frac{\partial\hat{\mathbf{B}}}{\partial t}\times\hat{\mathbf{E}}-\hat{\mathbf{B}}\times\frac{\partial\hat{\mathbf{E}}}{\partial t}\right]\mathrm{d}^3r\\ &= \frac{1}{2}\int_{L^3}\mathbf{r}\times\left[\frac{1}{\mu_0}(\boldsymbol{\nabla}\times\hat{\mathbf{B}})\times\hat{\mathbf{B}}-\frac{1}{\mu_0}\hat{\mathbf{B}}\times(\boldsymbol{\nabla}\times\hat{\mathbf{B}})\right.\\ &\qquad\left.-\,\varepsilon_0\hat{\mathbf{E}}\times(\boldsymbol{\nabla}\times\hat{\mathbf{E}})+\varepsilon_0(\boldsymbol{\nabla}\times\hat{\mathbf{E}})\times\hat{\mathbf{E}}\right]\mathrm{d}^3r\\ &= -\int_{L^3}\mathbf{r}\times\left[\frac{1}{\mu_0}\hat{\mathbf{B}}\times(\boldsymbol{\nabla}\times\hat{\mathbf{B}})+\varepsilon_0\hat{\mathbf{E}}\times(\boldsymbol{\nabla}\times\hat{\mathbf{E}})\right]\mathrm{d}^3r,\end{aligned}\tag{10.6–3}$$

where ε_0, μ_0 are the vacuum dielectric constant and the magnetic permeability, respectively. The last line follows from the previous one because all $\hat{\mathbf{B}}(\mathbf{r}, t)$ operators commute with each other at the same time, and similarly for $\hat{\mathbf{E}}(\mathbf{r}, t)$, as we shall show explicitly in Section 10.8. We expand the vector triple products by writing

$$\begin{aligned}\hat{\mathbf{B}}\times(\boldsymbol{\nabla}\times\hat{\mathbf{B}}) &= \hat{B}_i\boldsymbol{\nabla}\hat{B}_i-\hat{\mathbf{B}}\cdot\boldsymbol{\nabla}\hat{\mathbf{B}}\\ &= \tfrac{1}{2}\boldsymbol{\nabla}(\hat{\mathbf{B}}\cdot\hat{\mathbf{B}})-\boldsymbol{\nabla}\cdot\hat{\mathbf{B}}\hat{\mathbf{B}},\end{aligned}$$

since $\boldsymbol{\nabla}\cdot\hat{\mathbf{B}}=0$. Hence

$$\begin{aligned}\mathbf{r}\times(\hat{\mathbf{B}}\times(\boldsymbol{\nabla}\times\hat{\mathbf{B}})) &= \tfrac{1}{2}(\mathbf{r}\times\boldsymbol{\nabla})\hat{\mathbf{B}}\cdot\hat{\mathbf{B}}-\mathbf{r}\times(\boldsymbol{\nabla}\cdot\hat{\mathbf{B}}\hat{\mathbf{B}})\\ &= -\tfrac{1}{2}\boldsymbol{\nabla}\times(\mathbf{r}\hat{\mathbf{B}}\cdot\hat{\mathbf{B}})-\boldsymbol{\nabla}\cdot(\hat{\mathbf{B}}\mathbf{r}\times\hat{\mathbf{B}}),\end{aligned}$$

and a similar expansion can be written for the term in $\hat{\mathbf{E}}(\mathbf{r}, t)$ in Eq. (10.6–3). We now insert these under the integral in Eq. (10.6–3), and apply the generalized form of Gauss' theorem to convert the volume integrals to integrals over the surface of the normalization cube. We then have

$$\frac{\mathrm{d}\hat{\mathbf{J}}(0)}{\mathrm{d}t} = \frac{1}{2}\oint \mathrm{d}\mathbf{s} \times \mathbf{r}\left(\frac{\hat{\mathbf{B}}^2}{\mu_0} + \varepsilon_0 \hat{\mathbf{E}}^2\right) + \oint \mathrm{d}\mathbf{s} \cdot \left(\frac{\hat{\mathbf{B}}\mathbf{r} \times \hat{\mathbf{B}}}{\mu_0} + \varepsilon_0 \hat{\mathbf{E}}\mathbf{r} \times \hat{\mathbf{E}}\right). \tag{10.6–4}$$

The first integral vanishes, as becomes evident if we combine contributions from the surface elements at $(-\frac{1}{2}L, y, z)$ and $(\frac{1}{2}L, y, z)$, for example. For it follows at once from the periodic boundary conditions that $\hat{\mathbf{B}}(-\frac{1}{2}L, y, z) = \hat{\mathbf{B}}(\frac{1}{2}L, y, z)$ and $\hat{\mathbf{E}}(-\frac{1}{2}L, y, z) = \hat{\mathbf{E}}(\frac{1}{2}L, y, z)$, etc., whereas the vectors $\mathrm{d}\mathbf{s} \times \mathbf{r}$ are equal and opposite at $(-\frac{1}{2}L, y, z)$ and $(\frac{1}{2}L, y, z)$, etc. We are therefore left with

$$\frac{\mathrm{d}\hat{\mathbf{J}}(0)}{\mathrm{d}t} = \oint \mathbf{r} \times \left[\frac{1}{\mu_0}\hat{\mathbf{B}}\hat{\mathbf{B}} + \varepsilon_0 \hat{\mathbf{E}}\hat{\mathbf{E}}\right] \cdot \mathrm{d}\mathbf{s}, \tag{10.6–5}$$

in which the products $\hat{\mathbf{B}}(\mathbf{r}, t)\hat{\mathbf{B}}(\mathbf{r}, t)$ and $\hat{\mathbf{E}}(\mathbf{r}, t)\hat{\mathbf{E}}(\mathbf{r}, t)$ are dyadics, and this term is not zero. However, the integral is to be taken over the boundary of the normalization volume, which is ultimately made arbitrarily large.

In classical electromagnetic theory such surface terms are usually discarded, when we are dealing with locally produced fields, on the grounds that the field at the boundary may be taken as negligibly small. But such an argument has to be treated with great caution when the fields are Hilbert space operators. We have already encountered the zero point contribution that arises in certain operator products, whose expectation value is independent of the quantum state of the field. Let us write Eq. (10.6–5) in component form

$$\frac{\mathrm{d}\hat{J}_l(0)}{\mathrm{d}t} = \epsilon_{lmn}\oint r_m\left(\frac{1}{\mu_0}\hat{B}_n\hat{B}_p + \varepsilon_0 \hat{E}_n\hat{E}_p\right)\mathrm{d}s_p, \tag{10.6–6}$$

with summation on repeated indices understood, where ϵ_{lmn} is the completely antisymmetric unit tensor of Levi Civita, i.e. $\epsilon_{lmn} = 1$ or -1 according as the subscripts l, m, n are an even or odd permutation of x, y, z and $\epsilon_{lmn} = 0$ when two suffices are equal. We see at once that the terms with $m \neq p$ also vanish from considerations of symmetry at the boundary of the normalization box. But if $m = p$ in Eq. (10.6–6), then the only non-vanishing contributions must come from terms in which $n \neq p$, because of the presence of the antisymmetric tensor ϵ_{lmn}.

Let us examine an operator product like $\hat{E}_n\hat{E}_p$ more closely. It can be seen from the mode expansion (10.4–39) that $\hat{\mathbf{E}}(\mathbf{r}, t)$ can always be expressed as the sum of two operators $\hat{\mathbf{E}}^{(+)}(\mathbf{r}, t)$ and $\hat{\mathbf{E}}^{(-)}(\mathbf{r}, t)$ that are conjugates of each other, such that $\hat{\mathbf{E}}^{(+)}(\mathbf{r}, t)$ involves only annihilation operators and $\hat{\mathbf{E}}^{(-)}(\mathbf{r}, t)$ involves only creation operators. These configuration space creation and annihilation operators $\hat{\mathbf{E}}^{(-)}(\mathbf{r}, t)$ and $\hat{\mathbf{E}}^{(+)}(\mathbf{r}, t)$ are discussed in more detail in Chapters 11 and 12. The product $\hat{E}_n(\mathbf{r}, t)\hat{E}_p(\mathbf{r}, t)$ can therefore be written

$$\begin{aligned}\hat{E}_n(\mathbf{r}, t)\hat{E}_p(\mathbf{r}, t) &= \hat{E}_n^{(+)}(\mathbf{r}, t)\hat{E}_p^{(+)}(\mathbf{r}, t) + \hat{E}_n^{(-)}(\mathbf{r}, t)\hat{E}_p^{(-)}(\mathbf{r}, t) \\ &\quad + \hat{E}_n^{(-)}(\mathbf{r}, t)\hat{E}_p^{(+)}(\mathbf{r}, t) + \hat{E}_n^{(+)}(\mathbf{r}, t)\hat{E}_p^{(-)}(\mathbf{r}, t).\end{aligned} \tag{10.6–7}$$

Of the four terms on the right of this equation, the first three are in *normal order*, with any creation operators to the left of any annihilation operators. Their expectation values therefore vanish for any state $|\psi\rangle$ of the field for which the photons are localized far from $\mathbf{r}$, t, with $\mathbf{r}$ lying on the boundary of the normalization volume, so that $\hat{\mathbf{E}}^{(+)}(\mathbf{r}, t)|\psi\rangle = 0$. The last term in Eq. (10.6–7) is not in normal order. However, it can be shown from the mode expansions that $\hat{E}_n^{(+)}(\mathbf{r}, t)$ and $\hat{E}_p^{(-)}(\mathbf{r}, t)$ commute when $n \neq p$, so that the term may be written in normal order $\hat{E}_p^{(-)}(\mathbf{r}, t)\hat{E}_n^{(+)}(\mathbf{r}, t)$. Similar remarks also apply to the $\hat{B}_n\hat{B}_p$ term in Eq. (10.6–6). We have therefore shown that $\mathrm{d}\hat{\mathbf{J}}(0)/\mathrm{d}t$ vanishes except for some normally ordered terms at the boundary of the normalization volume, whose matrix elements are zero between all states that correspond to excitations localized within the volume. In this restricted sense the total angular momentum of the quantized field can be considered to be a constant of the motion (Lenstra and Mandel, 1982).

10.6.2 Decomposition of the total angular momentum

In classical electromagnetic theory it is found that the total angular momentum $\mathbf{J}(\mathbf{r}_0)$ of the field can always be decomposed into the sum of two terms, one of which is identifiable as the orbital and the other as the intrinsic part of the total angular momentum (cf. for example, Gottfried, 1966 and Jackson, 1975). In order to make the classical decomposition we start by writing the n'th component of the Poynting vector in the form

$$\varepsilon_0(\mathbf{E} \times \mathbf{B})_n = \varepsilon_0[\mathbf{E} \times (\boldsymbol{\nabla} \times \mathbf{A})]_n$$

By expanding the vector triple product, and using the fact that $\boldsymbol{\nabla} \cdot \mathbf{E} = 0$, we can re-write this as

$$\begin{aligned} \varepsilon_0(\mathbf{E} \times \mathbf{B})_n &= \varepsilon_0(E_i \nabla_n A_i - E_i \nabla_i A_n) \\ &= \varepsilon_0 E_i \nabla_n A_i - \varepsilon_0 \nabla_i (E_i A_n), \end{aligned}$$

in which summation over repeated indices is implied. If we make use of the completely antisymmetric tensor ϵ_{lmn} to write out the components of the vector product we have

$$\begin{aligned} \varepsilon_0[(\mathbf{r} - \mathbf{r}_0) \times (\mathbf{E} \times \mathbf{B})]_l &= \varepsilon_0 \epsilon_{lmn}(r_m - r_{0m})(\mathbf{E} \times \mathbf{B})_n \\ &= \varepsilon_0 \epsilon_{lmn}(r_m - r_{0m})[E_i \nabla_n A_i - \nabla_i E_i A_n] \\ &= \varepsilon_0 \epsilon_{lmn}(r_m - r_{0m}) E_i \nabla_n A_i - \varepsilon_0 \epsilon_{lmn} \nabla_i (r_m - r_{0m}) E_i A_n \\ &\quad + \varepsilon_0 \epsilon_{lmn}(\nabla_i (r_m - r_{0m})) E_i A_n. \end{aligned}$$

Now $\nabla_i(r_m - r_{0m}) = \delta_{im}$, so that we obtain

$$\begin{aligned} \varepsilon_0[(\mathbf{r} - \mathbf{r}_0) \times (\mathbf{E} \times \mathbf{B})]_l &= \varepsilon_0 E_i[(\mathbf{r} - \mathbf{r}_0) \times \boldsymbol{\nabla}]_l A_i \\ &\quad - \varepsilon_0 \boldsymbol{\nabla} \cdot \mathbf{E}[(\mathbf{r} - \mathbf{r}_0) \times \mathbf{A}]_l + \varepsilon_0(\mathbf{E} \times \mathbf{A})_l. \end{aligned}$$

Equation (10.6–1) for the total angular momentum $\hat{\mathbf{J}}(\mathbf{r}_0)$ can therefore be written

$$\mathbf{J}(\mathbf{r}_0) = \varepsilon_0 \int_{L^3} \{E_i[(\mathbf{r} - \mathbf{r}_0) \times \boldsymbol{\nabla}]A_i - \boldsymbol{\nabla} \cdot \mathbf{E}((\mathbf{r} - \mathbf{r}_0) \times \mathbf{A})\}\, \mathrm{d}^3 r + \varepsilon_0 \int_{L^3} (\mathbf{E} \times \mathbf{A})\, \mathrm{d}^3 r$$

$$= \varepsilon_0 \int_{L^3} E_i[(\mathbf{r} - \mathbf{r}_0) \times \boldsymbol{\nabla}] A_i \, d^3r - \varepsilon_0 \oint \mathbf{E}[(\mathbf{r} - \mathbf{r}_0) \times \mathbf{A}] \cdot d\mathbf{s} + \varepsilon_0 \int_{L^3} (\mathbf{E} \times \mathbf{A}) \, d^3r, \tag{10.6–8}$$

where we have used Gauss' theorem to transform the second volume integral into an integral over the surface of the normalization box of side L. In the limit of sufficiently large L, and for electromagnetic fields that are localized well inside the box, the surface term is often discarded, and the total angular momentum splits into two parts,

$$\mathbf{J}(\mathbf{r}_0) = \mathbf{J}_o(\mathbf{r}_0) + \mathbf{J}_s, \tag{10.6–9}$$

where

$$\mathbf{J}_s = \varepsilon_0 \int_{L^3} (\mathbf{E} \times \mathbf{A}) \, d^3r \tag{10.6–10}$$

and

$$\mathbf{J}_o(\mathbf{r}_0) = \varepsilon_0 \int_{L^3} E_i[(\mathbf{r} - \mathbf{r}_0) \times \boldsymbol{\nabla}] A_i \, d^3r. \tag{10.6–11a}$$

The part $\mathbf{J}_s$ is independent of the point $\mathbf{r}_0$ or of the choice of origin, and therefore represents the *intrinsic or spin angular momentum* of the electromagnetic field. The part $\mathbf{J}_o(\mathbf{r}_0)$ depends on the chosen point $\mathbf{r}_0$, and is seen to involve the familiar differential operator $(\mathbf{r} - \mathbf{r}_0) \times \boldsymbol{\nabla}$ that we associate with the angular momentum about $\mathbf{r}_0$; it therefore represents the *orbital angular momentum* of the field. Comparison of Eqs. (10.6–1) and (10.6–11a) shows that the orbital angular momentum can always be decomposed in the form

$$\mathbf{J}_o(\mathbf{r}_0) = \mathbf{J}_o(0) - \mathbf{r}_0 \times \mathbf{P}. \tag{10.6–11b}$$

In dealing with the quantized electromagnetic field we can decompose the total angular momentum in a similar manner. We define the corresponding Hermitian spin and orbital angular momentum operators by taking the symmetrized expressions

$$\hat{\mathbf{J}}_s = \tfrac{1}{2}\varepsilon_0 \int_{L^3} (\hat{\mathbf{E}} \times \hat{\mathbf{A}} - \hat{\mathbf{A}} \times \hat{\mathbf{E}}) \, d^3r, \tag{10.6–12}$$

and

$$\hat{\mathbf{J}}_o(\mathbf{r}_0) = \hat{\mathbf{J}}_o(0) - \mathbf{r}_0 \times \hat{\mathbf{P}}, \tag{10.6–13a}$$

with

$$\hat{\mathbf{J}}_o(0) = \tfrac{1}{2}\varepsilon_0 \int_{L^3} \{\hat{E}_i(\mathbf{r} \times \boldsymbol{\nabla})\hat{A}_i + [(\mathbf{r} \times \boldsymbol{\nabla})\hat{A}_i]\hat{E}_i\} \, d^3r. \tag{10.6–13b}$$

We have discarded the contribution to $\hat{\mathbf{J}}_o(0)$ of the surface term for the same reason as before, in connection with $d\hat{\mathbf{J}}/dt$, because this term can be written as a normally ordered operator whose expectation vanishes in any state in which the excitations of the field are localized inside the normalization cube. The side L of this cube is eventually allowed to become infinite.

10.6.3 Intrinsic (or spin) angular momentum

Let us examine the intrinsic or spin angular momentum $\hat{\mathbf{J}}_{\mathrm{s}}$ in a little more detail. With the help of the mode expansions (10.4–38) and (10.4–39) we find

$$\hat{\mathbf{J}}_{\mathrm{s}} = \frac{1}{2L^3}\sum_{\mathbf{k},s}\sum_{\mathbf{k}',s'}\left(\frac{\hbar\omega}{2}\right)^{1/2}\left(\frac{\hbar}{2\omega'}\right)^{1/2}\int_{L^3}[\mathrm{i}\hat{a}_{\mathbf{k}s}\boldsymbol{\varepsilon}_{\mathbf{k}s}\,\mathrm{e}^{\mathrm{i}(\mathbf{k}\cdot\mathbf{r}-\omega t)} + \mathrm{h.c.}]$$
$$\times\,[\hat{a}_{\mathbf{k}'s'}\boldsymbol{\varepsilon}_{\mathbf{k}'s'}\,\mathrm{e}^{\mathrm{i}(\mathbf{k}'\cdot\mathbf{r}-\omega' t)} + \mathrm{h.c.}]\,\mathrm{d}^3r + \mathrm{h.c.}$$

The volume integrals yield $L^3\delta^3_{-\mathbf{k}\mathbf{k}'}$ and $L^3\delta^3_{\mathbf{k}\mathbf{k}'}$. The terms in $\delta^3_{-\mathbf{k}\mathbf{k}'}$ then vanish, as can be seen by making the change of variables $\mathbf{k}\to-\mathbf{k}$, $s\to s'$, $s'\to s$. With the help of the commutation relations (10.3–9) to (10.3–11) among the creation and annihilation operators we then have

$$\hat{\mathbf{J}}_{\mathrm{s}} = \mathrm{i}\hbar\sum_{\mathbf{k},s}\sum_{s'}(\hat{a}^\dagger_{\mathbf{k}s'}\hat{a}_{\mathbf{k}s} + \tfrac{1}{2}\delta_{ss'})(\varepsilon_{\mathbf{k}s}\times\varepsilon^*_{\mathbf{k}s'}). \qquad (10.6\text{–}14)$$

This is a general expression that holds for any choice of basis for the two complex polarization vectors $\varepsilon_{\mathbf{k}s}$ $(s = 1, 2)$. Naturally, it is desirable to select a basis in which the vector products $\varepsilon_{\mathbf{k}s}\times\varepsilon^*_{\mathbf{k}s'}$ have a particularly simple form, in order to simplify the expression for $\hat{\mathbf{J}}_s$.

Let us therefore examine two such choices of basis. If $\varepsilon_{\mathbf{k}1}$, $\varepsilon_{\mathbf{k}2}$ are real orthogonal unit vectors, corresponding to orthogonal linear polarizations, then

$$\varepsilon_{\mathbf{k}s}\times\varepsilon^*_{\mathbf{k}s'} = \pm\boldsymbol{\kappa}(1-\delta_{ss'}), \qquad (10.6\text{–}15)$$

where $\boldsymbol{\kappa}$ is the unit vector in the direction of the wave vector $\mathbf{k}$. By substituting from Eq. (10.6–15) in Eq. (10.6–14) we obtain

$$\hat{\mathbf{J}}_{\mathrm{s}} = \mathrm{i}\sum_{\mathbf{k}}\hbar\boldsymbol{\kappa}(\hat{a}^\dagger_{\mathbf{k}2}\hat{a}_{\mathbf{k}1} - \hat{a}^\dagger_{\mathbf{k}1}\hat{a}_{\mathbf{k}2}), \qquad (10.6\text{–}16)$$

which shows that $\hat{\mathbf{J}}_{\mathrm{s}}$ is entirely off-diagonal in terms of Fock states in the basis of orthogonal linear polarizations. Moreover, each component of $\mathbf{J}_{\mathrm{s}}$ points in the direction of the wave vector.

The question immediately arises whether there exists a basis of polarizations in which $\hat{\mathbf{J}}_{\mathrm{s}}$ is diagonal in terms of Fock states, which requires the vector product $\varepsilon_{\mathbf{k}s}\times\varepsilon^*_{\mathbf{k}s'}$ to be proportional to $\delta_{ss'}$. Such a basis is provided by the orthogonal circular polarization vectors $\varepsilon_{\mathbf{k}\lambda}$ $(\lambda = \pm1)$, defined by

$$\left.\begin{aligned}\varepsilon_{\mathbf{k},+1} &= \frac{1}{\sqrt{2}}(\varepsilon_{\mathbf{k}1} + \mathrm{i}\varepsilon_{\mathbf{k}2})\\ \varepsilon_{\mathbf{k},-1} &= \frac{1}{\sqrt{2}}(\mathrm{i}\varepsilon_{\mathbf{k}1} - \varepsilon_{\mathbf{k}2}),\end{aligned}\right\} \qquad (10.6\text{–}17)$$

in terms of the real orthogonal unit vectors $\varepsilon_{\mathbf{k}1}$, $\varepsilon_{\mathbf{k}2}$. Here $\varepsilon_{\mathbf{k},+1}$ and $\varepsilon_{\mathbf{k},-1}$ represent right-hand and left-hand circularly polarized light, respectively. These vectors satisfy the condition

$$\varepsilon_{\mathbf{k},\lambda}\times\varepsilon^*_{\mathbf{k},\lambda'} = -\mathrm{i}\lambda\boldsymbol{\kappa}\delta_{\lambda\lambda'},\ (\lambda,\ \lambda' = \pm1), \qquad (10.6\text{–}18)$$

which allows us to simplify Eq. (10.6–14) so that it becomes

$$\hat{\mathbf{J}}_s = \sum_{\mathbf{k},\lambda} \hbar\boldsymbol{\kappa}\lambda[\hat{a}^\dagger_{\mathbf{k}\lambda}\hat{a}_{\mathbf{k}\lambda} + \tfrac{1}{2}]$$

$$= \sum_{\mathbf{k}} \hbar\boldsymbol{\kappa}[\hat{n}_{\mathbf{k},+1} - \hat{n}_{\mathbf{k},-1}]. \tag{10.6–19}$$

This shows that $\hat{\mathbf{J}}_s$ is diagonal in the photon number in the basis of circular polarizations, and that the intrinsic angular momentum associated with any one wave vector $\mathbf{k}$ is $\hbar\boldsymbol{\kappa}$ times the difference between the number of right-hand and left-hand circularly polarized photons. Thus, for a plane wave of light, the intrinsic angular momentum lies entirely along the direction of propagation. For a one-photon state corresponding to a plane wave with wave vector $\mathbf{k}$, the component of the intrinsic angular momentum in the direction of propagation has eigenvalues $\pm\hbar\boldsymbol{\kappa}$. The projection of the intrinsic angular momentum on the direction of propagation is known as the *helicity*, and the quantum number $\lambda = \pm 1$ is known as the *spin* of the photon. The two eigenvalues of the helicity are characteristic of a spin 1 particle of zero mass. The validity of these relations has been demonstrated, and the angular momentum of a circularly polarized light beam has been measured in an experiment in which the beam fell on a doubly refracting plate that was free to rotate about the beam direction (Beth, 1936).

According to Eq. (10.6–19) the spin angular momentum $\hat{\mathbf{J}}_s$, like $\hat{\mathbf{P}}$ and $\hat{H}$, can be written as a function of the number operators $\hat{n}_{\mathbf{k}s}$ in the basis of circular polarizations. Therefore, $\hat{\mathbf{J}}_s$ commutes with $\hat{H}$ and is another constant of the motion for a free electromagnetic field.

10.6.4 Orbital angular momentum

We have already noted that the total angular momentum $\hat{\mathbf{J}}(\mathbf{r}_0)$ is a constant of the motion in a weak sense, and we have just shown that $\hat{\mathbf{J}}_s$ is strictly a constant. As $\hat{\mathbf{J}}(\mathbf{r}_0)$ is expressible as the sum of intrinsic and orbital parts $\hat{\mathbf{J}}_s$ and $\hat{\mathbf{J}}_0(\mathbf{r}_0)$, it follows that the orbital angular momentum of a free field must also be a constant of the motion in the weak sense.

We can try to use the mode expansions (10.4–38) and (10.4–39) to express $\hat{\mathbf{J}}_0(0)$ given by Eq. (10.6–13b) as a sum over modes, but the result is not simple when a discrete mode decomposition is used. This is because our box quantization with discrete modes imposes preferred directions in space. The consequences of this have been examined in some detail (Lenstra and Mandel, 1982). In Section 10.10 we briefly discuss the use of a continuous mode decomposition of the field operators, in which $\mathbf{k}$ is a continuous variable and no box normalization is introduced. If we write

$$\hat{\mathscr{A}}(\mathbf{k}, t) = \sum_s \hat{a}(\mathbf{k}, s)\boldsymbol{\varepsilon}(\mathbf{k}, s)\,\mathrm{e}^{-i\omega t} \tag{10.6–20}$$

in this continuous Fock space, it may be shown (Simmons and Guttmann, 1970, Appendix VI) that the orbital angular momentum is expressible as

$$\hat{\mathbf{J}}_0(0) = \frac{-i\hbar}{2}\int [\hat{\mathscr{A}}_i^\dagger(\mathbf{k}, t)(\mathbf{k} \times \nabla_{\mathbf{k}})\hat{\mathscr{A}}_i(\mathbf{k}, t) - \text{h.c.}]\,\mathrm{d}^3 k, \tag{10.6–21}$$

where $\nabla_{\mathbf{k}}$ is the differential gradient operator with respect to $\mathbf{k}$.

10.7 Phase operators for the quantized field

We have already seen that each $\mathbf{k}$, s mode of the free electromagnetic field behaves as a harmonic oscillator. For a classical harmonic oscillator of complex amplitude $v_{\mathbf{k}s}$, the canonical variable $q_{\mathbf{k}s}(t)$ is given by (cf. 10.2–27)

$$\begin{aligned} q_{\mathbf{k}s}(t) &= v_{\mathbf{k}s}\,\mathrm{e}^{-\mathrm{i}\omega t} + v^*_{\mathbf{k}s}\,\mathrm{e}^{\mathrm{i}\omega t} \\ &= 2|v_{\mathbf{k}s}|\cos(\omega t - \phi_{\mathbf{k}s}), \end{aligned} \tag{10.7–1}$$

where the phase $\phi_{\mathbf{k}s}$ is obtainable from the complex amplitude $v_{\mathbf{k}s}$ via the relation

$$v_{\mathbf{k}s} = |v_{\mathbf{k}s}|\,\mathrm{e}^{\mathrm{i}\phi_{\mathbf{k}s}}. \tag{10.7–2}$$

We now wish to examine the question whether there exists an analogous quantum mechanical observable for the phase, that is given by some Hermitian operator $\hat{\phi}_{\mathbf{k}s}$. It is tempting to try the same approach as in the classical argument above in order to find the observable. Thus, by analogy with Eq. (10.7–2), we might attempt to decompose the annihilation operator by writing

$$\hat{a} = \hat{g}\,\mathrm{e}^{\mathrm{i}\hat{\phi}}. \tag{10.7–3}$$

In the following we shall be concerned with only one mode of the field, and we therefore discard the mode label $\mathbf{k}$, s temporarily. If the decomposition is to lead to observables $\hat{g}$ and $\hat{\phi}$, the corresponding operators must be Hermitian. Now, in fact, it turns out to be impossible to generate a Hermitian phase operator in this way, but the recognition that this cannot be done appears to be relatively recent.‡

10.7.1 First attempts to construct a phase operator

Let us suppose for a moment that the decomposition (10.7–3), with $\hat{g}$ and $\hat{\phi}$ Hermitian, is valid, and let us examine some of its implications.

By taking the Hermitian conjugate of Eq. (10.7–3)

$$\hat{a}^\dagger = \mathrm{e}^{-\mathrm{i}\hat{\phi}}\hat{g}, \tag{10.7–4}$$

and multiplying Eqs. (10.7–3) and (10.7–4), we obtain

$$\hat{a}\hat{a}^\dagger = \hat{n} + 1 = \hat{g}^2,$$

or

$$\hat{g} = (\hat{n} + 1)^{1/2}.$$

Hence

$$\hat{a}^\dagger\hat{a} = \hat{n} = \mathrm{e}^{-\mathrm{i}\hat{\phi}}(\hat{n} + 1)\,\mathrm{e}^{\mathrm{i}\hat{\phi}}, \tag{10.7–5}$$

‡ A phase operator $\hat{\phi}$ of the type given by Eq. (10.7–3) was first introduced by Dirac (1927); see also Heitler (1954). The non-Hermitian character of $\hat{\phi}$ was pointed out by Louisell (1963). Hermitian operators for sine and cosine of the phase were first introduced by Susskind and Glogower (1964) and Carruthers and Nieto (1965, 1968). Phase operators and their measurement are treated in more detail in the book by Vogel and Welsch (1994).

or

$$[\hat{n}, e^{i\hat{\phi}}] = -e^{i\hat{\phi}}. \tag{10.7–6}$$

Moreover, on making use of the general operator expansion theorem‡

$$\exp(\hat{A})\,\hat{B}\exp(-\hat{A}) = \hat{B} + [\hat{A}, \hat{B}] + \frac{1}{2!}[A, [\hat{A}, \hat{B}]] + \frac{1}{3!}[\hat{A}, [\hat{A}, [\hat{A}, \hat{B}]]] + \ldots, \tag{10.7–7}$$

we have from Eq. 10.7–5

$$\begin{aligned}\hat{n} - 1 &= e^{-i\hat{\phi}}\,\hat{n}\,e^{i\hat{\phi}} \\ &= \hat{n} - i[\hat{\phi}, \hat{n}] + \frac{(-i)^2}{2!}[\hat{\phi}, [\hat{\phi}, \hat{n}]] + \ldots,\end{aligned}$$

which requires that

$$[\hat{n}, \hat{\phi}] \stackrel{?}{=} i. \tag{10.7–8}$$

Therefore, $\hat{n}$ and $\hat{\phi}$ appear to behave like canonically conjugate observables, and by the usual argument, Eq. (10.7–8) implies the existence of an uncertainty relation connecting the dispersions of $\hat{n}$ and $\hat{\phi}$:

$$\langle(\Delta\hat{n})^2\rangle^{1/2}\langle(\Delta\hat{\phi})^2\rangle^{1/2} \stackrel{?}{\geqslant} \tfrac{1}{2}. \tag{10.7–9}$$

One should be suspicious at this stage, because the phase is usually defined modulo 2π, so that its uncertainty cannot be arbitrarily large as required by Eq. (10.7–9) when $\langle(\Delta\hat{n})^2\rangle^{1/2}$ is very small. In fact, because $\hat{n}$ has a discrete spectrum whereas ϕ is continuous and bounded, Eq. (10.7–8) is an impossible relation, and the contradiction implicit in it becomes obvious if we examine the matrix elements obtainable from the equation by multiplying by the number eigenstate $|n\rangle$ on the right and by $\langle m|$ on the left. We are then led to

$$(m - n)\langle m|\hat{\phi}|n\rangle \stackrel{?}{=} i\delta_{mn},$$

which is obviously wrong, for the left side vanishes when $m = n$, whereas the right side becomes equal to i. It follows that both Eqs. (10.7–8) and (10.7–9) are actually wrong, although each step in the derivation was self-consistent. The difficulties stem from the fact that $e^{i\hat{\phi}}$ and $e^{-i\hat{\phi}}$, despite their appearance, are not unitary operators, so that $\hat{\phi}$ is not a Hermitian phase operator.

The non-unitarity of $e^{i\hat{\phi}}$, which we shall denote by $\hat{\mathscr{E}}$, can be demonstrated as follows. From the definition $\hat{a} = (\hat{n} + 1)^{1/2}\hat{\mathscr{E}}$ it follows that

$$\begin{aligned}\hat{\mathscr{E}} &= \frac{1}{(\hat{n} + 1)^{1/2}}\hat{a} \\ \hat{\mathscr{E}}^\dagger &= \hat{a}^\dagger\frac{1}{(\hat{n} + 1)^{1/2}}.\end{aligned}$$

‡ This theorem may readily be proved by writing $\hat{f}(x) = \exp(\hat{A}x)\hat{B}\exp(-\hat{A}x)$ and making a Taylor expansion of $\hat{f}(x)$ about $x = 0$. The derivatives of $\hat{f}(x)$ generate the commutators. See Section 10.11 for a derivation of the theorem.

Hence

$$\hat{\mathscr{E}}\hat{\mathscr{E}}^\dagger = \frac{1}{(\hat{n}+1)^{1/2}}\hat{a}\hat{a}^\dagger\frac{1}{(\hat{n}+1)^{1/2}} = \frac{1}{(\hat{n}+1)^{1/2}}(\hat{n}+1)\frac{1}{(\hat{n}+1)^{1/2}} = 1, \tag{10.7–10a}$$

as required for a unitary operator. But

$$\hat{\mathscr{E}}^\dagger\hat{\mathscr{E}} = \hat{a}^\dagger\frac{1}{\hat{n}+1}\hat{a}, \tag{10.7–10b}$$

and this is not unity because it gives zero when it operates either to the left or to the right on the vacuum state. By taking matrix elements of $\hat{\mathscr{E}}^\dagger\hat{\mathscr{E}}$ with respect to Fock states we may readily show that $\hat{\mathscr{E}}^\dagger\hat{\mathscr{E}} = 1 - |0\rangle\langle 0|$, so that $\hat{\mathscr{E}}$ is not unitary and $\hat{\phi}$ is not Hermitian.

10.7.2 Cosine and sine operators

As was shown by Susskind and Glogower (1964) and by Carruthers and Nieto (1965, 1968), a consistent way out of the difficulty is to avoid the $\hat{\phi}$ operators, and instead introduce operators $\hat{C}$ and $\hat{S}$ that are the analogs of the cosine and sine of $\hat{\phi}$. Let us put

$$\left.\begin{aligned} \hat{C} &\equiv \frac{1}{2}[\widehat{\exp(\mathrm{i}\phi)} + \widehat{\exp(\mathrm{i}\phi)}^\dagger] \\ \hat{S} &\equiv \frac{1}{2\mathrm{i}}[\widehat{\exp(\mathrm{i}\phi)} - \widehat{\exp(\mathrm{i}\phi)}^\dagger], \end{aligned}\right\} \tag{10.7–11}$$

where $\widehat{\exp(\mathrm{i}\phi)}$ is defined by $\hat{\mathscr{E}} = [1/(\hat{n}+1)^{1/2}]\hat{a}$ as before. Both $\hat{C}$ and $\hat{S}$ are manifestly Hermitian, despite the fact that $\widehat{\exp(\mathrm{i}\phi)}$ is not unitary, and $\hat{C} + \mathrm{i}\hat{S} = \widehat{\exp(\mathrm{i}\phi)}$.

From the commutation rules (10.4–1) and (10.4–2), we find, on substituting $\hat{a} = (\hat{n}+1)^{1/2}(\hat{C} + \mathrm{i}\hat{S})$ and $\hat{a}^\dagger = (\hat{C} - \mathrm{i}\hat{S})(\hat{n}+1)^{1/2}$,

$$[(\hat{n}+1)^{1/2}(\hat{C} + \mathrm{i}\hat{S}), \hat{n}] = (\hat{n}+1)^{1/2}(\hat{C} + \mathrm{i}\hat{S}),$$

and, after multiplying by $1/(\hat{n}+1)^{1/2}$ on the left to cancel out the square root, we have

$$[\hat{C} + \mathrm{i}\hat{S}, \hat{n}] = \hat{C} + \mathrm{i}\hat{S}. \tag{10.7–12}$$

Similarly we may show that

$$[\hat{C} - \mathrm{i}\hat{S}, \hat{n}] = -(\hat{C} - \mathrm{i}\hat{S}). \tag{10.7–13}$$

By adding and subtracting these equations we obtain

$$[\hat{C}, \hat{n}] = \mathrm{i}\hat{S} \tag{10.7–14}$$

$$[\hat{S}, \hat{n}] = -\mathrm{i}\hat{C}, \tag{10.7–15}$$

and these commutation relations immediately lead to the uncertainty relations

$$\langle(\Delta\hat{n})^2\rangle^{1/2}\langle(\Delta\hat{C})^2\rangle^{1/2} \geqslant \tfrac{1}{2}|\langle\hat{S}\rangle| \tag{10.7–16}$$

$$\langle(\Delta\hat{n})^2\rangle^{1/2}\langle(\Delta\hat{S})^2\rangle^{1/2} \geqslant \tfrac{1}{2}|\langle\hat{C}\rangle|, \tag{10.7–17}$$

which take the place of Eq. (10.7–9). Equation (10.7–12) also shows that Eq. (10.7–6) is in fact correct, despite the fact that $\widehat{\exp(\mathrm{i}\phi)}$ is a non-unitary operator.

The operators $\hat{C}$ and $\hat{S}$ can be shown to have continuous spectra between -1 and 1, as expected for cosine and sine operators. However, $\hat{C}$ and $\hat{S}$ do not commute, so there appear to be two different phase operators

$$\left.\begin{aligned}\hat{\phi}_C &= \arccos\hat{C}\\ \hat{\phi}_S &= \arcsin\hat{S},\end{aligned}\right\} \tag{10.7–18}$$

which do not commute either.

Despite the non-commutativity of $\hat{n}$ and $\hat{C}$, $\hat{S}$ and $\widehat{\exp(\mathrm{i}\phi)}$ and the consequent difficulty of defining the absolute phase, one can argue that, according to the formalism, phase differences between two modes of the field can be determined with arbitrary accuracy, in principle, even when the total photon number is well defined, so long as $\hat{n}_1$ and $\hat{n}_2$ are not determined separately. Let us think of the two modes as corresponding to two plane electromagnetic waves propagating at an angle with respect to each other, in which case the phase difference can be determined from an interference experiment. We associate suffices 1 and 2 with the two modes, and recall that operators belonging to different modes commute. Consider the operator $\exp(\widehat{\mathrm{i}\phi_1})[\exp(\widehat{\mathrm{i}\phi_2})]^\dagger$ that is the analog of $\exp\mathrm{i}(\phi_1-\phi_2)$. Its commutator with the total photon number $\hat{n}_1+\hat{n}_2$ is given by

$$\begin{aligned}[\exp(\widehat{\mathrm{i}\phi_1})[\exp(\widehat{\mathrm{i}\phi_2})]^\dagger,\ \hat{n}_1+\hat{n}_2] &= [(\hat{C}_1+\mathrm{i}\hat{S}_1)(\hat{C}_2-\mathrm{i}\hat{S}_2),\ \hat{n}_1+\hat{n}_2]\\ &= 0\end{aligned} \tag{10.7–19}$$

when we use Eqs. (10.7–12) and (10.7–13).

10.7.3 Phase operator based on the phase state projector

The failure of $\hat{C}$ and $\hat{S}$ to commute, and the absence of a ready, intuitive interpretation for their eigenvalues, have led a number of workers to explore other possible forms of phase operator (Lerner, 1968; Zak, 1969; Turski, 1972; Paul, 1974; Schubert and Vogel, 1978; Barnett and Pegg, 1986, 1989; Sanders, Barnett and Knight, 1986; Lynch, 1986, 1987; Pegg and Barnett, 1988, 1989; Shapiro, Shepard and Wong, 1989; Schleich, Horowicz and Varro, 1989; Bandilla, Paul and Ritze, 1991; Vogel and Schleich, 1991; Noh, Fougères and Mandel 1991, 1992a,b). In particular, Pegg and Barnett (1988) attempted to construct a Hermitian phase operator via a limiting procedure based on the state $|\theta\rangle$ of definite phase θ in a truncated Hilbert space, namely the $(s+1)$-dimensional subspace of Fock space. The state $|\theta\rangle$, defined by the expansion in terms of the Fock states

$$|\theta\rangle = \frac{1}{(s+1)^{1/2}}\sum_{n=0}^{s} \mathrm{e}^{\mathrm{i}n\theta}|n\rangle, \tag{10.7–20}$$

behaves in some ways as a state of definite phase θ when s is large. We recall that if the state of the free field in the Schrödinger picture at time $t=0$ is $|\psi(0)\rangle$, then at a later time t,

$$|\psi(t)\rangle = \exp(-\mathrm{i}\hat{H}_0 t/\hbar)|\psi(0)\rangle.$$

Let us now take the initial state $|\psi(0)\rangle$ to be the state $|\theta\rangle$ given by Eq. (10.7–20). Then with $\hat{H}_0 = \hbar\omega\hat{n}$, we obtain

$$\begin{aligned} |\psi(t)\rangle &= \mathrm{e}^{-\mathrm{i}\omega t\hat{n}}|\theta\rangle \\ &= \frac{1}{(s+1)^{1/2}}\sum_{n=0}^{s} \mathrm{e}^{\mathrm{i}n(\theta-\omega t)}|n\rangle \\ &= |\theta-\omega t\rangle. \end{aligned} \tag{10.7–21}$$

It is then natural to try and introduce a Hermitian phase operator $\hat{\phi}$ in terms of projectors of the type $|\theta\rangle\langle\theta|$. However, strictly speaking the limit $s \to \infty$ in Eq. (10.7–20) does not exist and $|\theta\rangle$ is not normalizable in the limit. The different phase states are not orthogonal and form an overcomplete set.

Pegg and Barnett (1988) overcome the problem of overcompleteness by introducing a discrete, orthonormal and complete set of phase states $|\theta_m\rangle$, with θ_m defined by

$$\theta_m = \theta_0 + \frac{2\pi m}{s+1}, \quad m = 0, 1, 2, \ldots, s, \tag{10.7–22}$$

where θ_0 is any fixed reference phase. From the definition one readily finds that

$$\begin{aligned} \langle\theta_m|\theta_{m'}\rangle &= \frac{1}{s+1}\sum_{n=0}^{s}\sum_{n'=0}^{s} \mathrm{e}^{\mathrm{i}(n'\theta_{m'}-n\theta_m)}\langle n|n'\rangle \\ &= \frac{1}{s+1}\sum_{n=0}^{s} \mathrm{e}^{\mathrm{i}n(\theta_{m'}-\theta_m)} \\ &= \frac{1-\mathrm{e}^{\mathrm{i}2\pi(m'-m)}}{(s+1)[1-\mathrm{e}^{\mathrm{i}2\pi(m'-m)/(s+1)}]} \\ &= \delta_{mm'}, \end{aligned}$$

as required. They attempt to overcome the limit problem by working with finite s and letting $s \to \infty$ only after expectation values have been calculated. With this understanding, they treat only physically realizable states whose photon occupation numbers are bounded and smaller than s. The phase operator $\hat{\phi}$ is then taken to be

$$\hat{\phi} = \sum_{m=0}^{s} \theta_m|\theta_m\rangle\langle\theta_m|. \tag{10.7–23}$$

Evidently the states $|\theta_m\rangle$ are eigenstates of $\hat{\phi}$ with eigenvalue θ_m.

The probability distribution $P(\theta_m)$ of the discrete phase angle θ_m can be obtained from the density operator $\hat{\rho}$ by writing

$$P(\theta_m) = \langle\theta_m|\hat{\rho}|\theta_m\rangle, \tag{10.7–24a}$$

with normalization $\sum_{m=0}^{s} P(\theta_m) = 1$. Alternatively, one can construct a continuous probability density $p(\phi)$ in terms of the continuous phase states $|\phi\rangle$ by writing

$$p(\phi) = \frac{s+1}{2\pi}\langle\phi|\hat{\rho}|\phi\rangle$$
$$= \frac{1}{2\pi}\sum_{n=0}^{s}\sum_{n'=0}^{s} e^{i(n'-n)\phi}\langle n|\hat{\rho}|n'\rangle, \qquad (10.7\text{–}24b)$$

which is normalized so that $\int_{-\pi}^{\pi} p(\phi)\,d\phi = 1$. The results of calculations based on the phase operator $\hat{\phi}$ agree in some cases with the Susskind and Glogower approach.

10.7.4 Operationally defined phase operators

An entirely different approach to the phase problem has been adopted by Noh, Fougères and Mandel (1991, 1992a,b), who start by analyzing what is usually measured in an experiment and then introduce operators that represent the measurement, based on the correspondence with classical optics. As measurements always involve the difference between two phases, and as an interference or homodyne experiment usually yields the cosine or sine of the phase difference, they introduce measured operators $\hat{C}_M$ and $\hat{S}_M$ for the cosine and sine of the phase difference that correspond to a particular measurement scheme. As an example, we consider the measurement scheme illustrated in Fig. 10.1, where two input fields labeled 1 and 2 are mixed by beam splitters, and the four detectors D_3, D_4, D_5, D_6 at the output ports of the interferometer count the emerging photons. With the help of a 90° phase shifter inserted in one interferometer arm it is possible to obtain values for the cosine and sine of the phase difference simultaneously from the differences between the photon counts n_3 and n_4, n_5 and n_6 registered by detectors D_3 and D_4, and D_5 and D_6, respectively. The appropriate measured cosine and sine operators are taken to be

$$\left.\begin{aligned}\hat{C}_M &= (\hat{n}_4 - \hat{n}_3)[(\hat{n}_4 - \hat{n}_3)^2 + (\hat{n}_6 - \hat{n}_5)^2]^{-1/2}\\ \hat{S}_M &= (\hat{n}_6 - \hat{n}_5)[(\hat{n}_4 - \hat{n}_3)^2 + (\hat{n}_6 - \hat{n}_5)^2]^{-1/2},\end{aligned}\right\} \qquad (10.7\text{–}25)$$

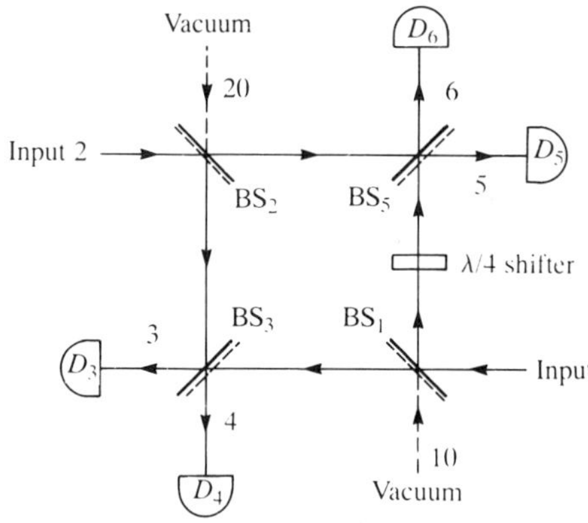

Fig. 10.1 Outline of an interferometer for measuring the phase difference between light beams entering at input ports 1 and 2. (Reproduced from Noh, Fougères and Mandel, 1991.)

and these can be expressed in terms of the field operators at the input ports of the apparatus in Fig. 10.1. It can be shown that $\hat{C}_M$ and $\hat{S}_M$ commute, which is a reflection of the fact that the cosine and sine measurements are compatible and do not disturb each other. All averages and moments of $\hat{C}_M$ and $\hat{S}_M$ are then calculated from these equations. However, a different measurement scheme in general leads to different operators, which may not commute.

From Eq. (10.7–25) it can be shown that for fields with $\langle \hat{n}_1 \rangle$, $\langle \hat{n}_2 \rangle \ll 1$, the dispersions of $\hat{C}_M$ and $\hat{S}_M$ obey the inequalities

$$\left.\begin{aligned} \frac{\langle (\Delta \hat{C}_M)^2 \rangle^{1/2}}{\langle \hat{C}_M \rangle} &\geqslant 1 \\ \frac{\langle (\Delta \hat{S}_M)^2 \rangle^{1/2}}{\langle \hat{S}_M \rangle} &\geqslant 1, \end{aligned}\right\} \tag{10.7–26}$$

so that both the cosine and sine of the phase difference are then ill defined.

If the formalism given by Eqs. (10.7–25) is sound, various moments of $\hat{C}_M$ and $\hat{S}_M$ derived from the equations should then agree with measurements in which the quadruplet of numbers n_3, n_4, n_5, n_6 is registered repeatedly and used to calculate C_M and S_M. Experimental quadruplets with $n_4 = n_3$ and $n_6 = n_5$ have to be discarded, because they do not yield values of C_M, S_M, and this requires renormalizing the theoretical moments for purposes of comparison with experiment.

Figures 10.2 and 10.3 show the expected variation of $\langle \hat{C}_M \rangle$ and $\langle (\Delta \hat{C}_M)^2 \rangle + \langle (\Delta \hat{S}_M)^2 \rangle$ with $\langle \hat{n}_1 \rangle$ for different ratios $\langle \hat{n}_2 \rangle / \langle \hat{n}_1 \rangle$ and for different phase differences $\theta_2 - \theta_1$, when the input field is in the form of a two-mode coherent state $|v_1\rangle$, $|v_2\rangle$ (see Chapter 11), with $v_1 = \langle \hat{n}_1 \rangle^{1/2} e^{i\theta_1}$, $v_2 = \langle \hat{n}_2 \rangle^{1/2} e^{i\theta_2}$. Experimentally measured quantities, obtained with two He:Ne laser beams as input, are

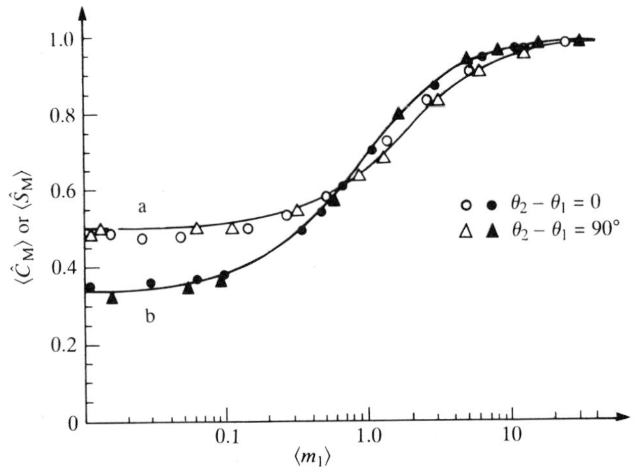

Fig. 10.2 Experimental results for the mean of the measured cosine $\hat{C}_M$ and sine $\hat{S}_M$ of the phase difference as a function of average number of photon counts $\langle m_1 \rangle$ superimposed on the theoretical curves based on Eqs. (10.7–25). ○, ●: $\theta_2 - \theta_1 = 0°$; △, ▲: $\theta_2 - \theta_1 = 90°$, (After Noh, Fougères and Mandel, 1991.)

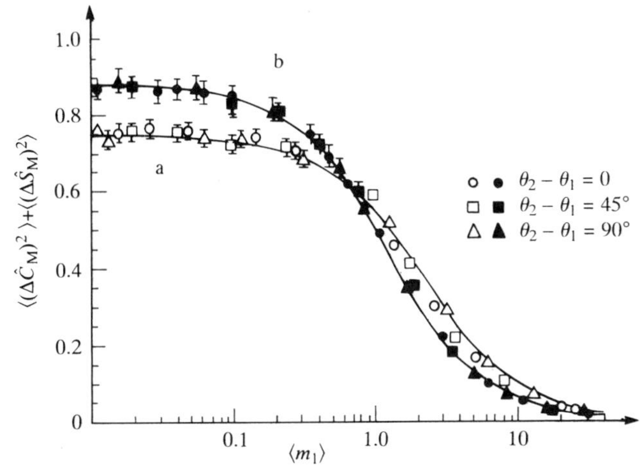

Fig. 10.3 Experimental results for the sum of the dispersions $\langle(\Delta\hat{C}_M)^2\rangle + \langle(\Delta\hat{S}_M)^2\rangle$ as a function of average number of photon counts $\langle m_1 \rangle$ superimposed on the theoretical curves based on Eqs. (10.7–25). ○, ●: $\theta_2 - \theta_1 = 0°$; □, ■: $\theta_2 - \theta_1 = 45°$; △, ▲: $\theta_2 - \theta_1 = 90°$. (After Noh, Fougères and Mandel, 1991.)

shown superimposed on the curves (Noh, Fougères and Mandel, 1991, 1992). They exhibit good agreement between theory based on Eqs. (10.7–25) and experiment.

Although this procedure yields the means and higher-order moments of $\hat{C}_M$ and $\hat{S}_M$, it is possible to derive the whole probability distribution $p(\phi_2 - \phi_1)$ of the phase difference by a modified experimental procedure (Noh, Fougères and Mandel, 1993). For this purpose a phase shift θ is deliberately imposed on the field at input port 2 of the interferometer in Fig. 10.1, and measurements are repeated for many different values of θ in the range $-\pi$ to π. Then the characteristic function of the phase difference $\phi_2 - \phi_1$, conditioned on the shift θ, is given by

$$C(x|\theta) = \langle(\hat{C}_M + i\hat{S}_M)^x\rangle', \tag{10.7–27}$$

where $\langle\ \rangle'$ denotes the quantum expectation in the input state modified by the phase shift θ. Hence, by Fourier inversion, we have for the probability density $p(\phi_2 - \phi_1|\theta)$ conditioned on the shift θ,

$$p(\phi_2 - \phi_1|\theta) = \frac{1}{2\pi}\int_{-\infty}^{\infty} C(x|\theta)\, e^{-i(\phi_2-\phi_1-\theta)x}\, dx. \tag{10.7–28}$$

When this is averaged over all values of θ it leads to the desired probability density

$$P(\phi_2 - \phi_1) = \frac{1}{2\pi}\int_{-\pi}^{\pi} p(\phi_2 - \phi_1|\theta)\, d\theta. \tag{10.7–29}$$

Figure 10.4 shows a histogram of the experimentally measured probability

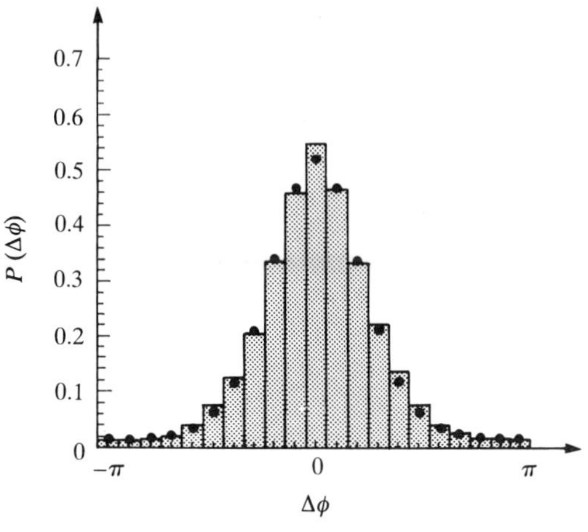

Fig. 10.4 The measured probability distribution of the phase difference for a two-mode coherent state $|v_1 = 1\rangle|v_2 = 2.36\rangle$ at the interferometer inputs. The black dots show the theoretically expected values given by Eq. (10.7–30). (Reproduced from Noh, Fougères and Mandel, 1993.)

distribution

$$P_N \equiv \int_{(N-\frac{1}{2})B}^{(N+\frac{1}{2})B} P(\phi_2 - \phi_1)\,\mathrm{d}(\phi_2 - \phi_1) \tag{10.7–30}$$

within bins of width $B = 18°$ for a two-mode coherent state $|v_1 = 1\rangle_1|v_2 = 2.36\rangle_2$ at the interferometer input. The black dots are theoretical values based on Eq. (10.7–30). There is reasonable agreement between theory and experiment, whereas there are distinct discrepancies between experiment and the theoretical approaches given in Sections 10.7.2 and 10.7.3 above.

10.8 Space-time commutation relations

Since, as we have seen, the non-Hermitian operators $\hat{a}_{\mathbf{k}s}$ and $\hat{a}^\dagger_{\mathbf{k}s}$ do not commute, it follows that the field operators $\hat{\mathbf{A}}(\mathbf{r}, t)$, $\hat{\mathbf{E}}(\mathbf{r}, t)$, $\hat{\mathbf{B}}(\mathbf{r}, t)$, will not in general commute either. We can evaluate any of the field commutators, which will of course be tensors, by making use of the series expansions (10.4–38) to (10.4–40), together with the commutation relations (10.3–9) to (10.3–11). Thus, we find for example,

$$\begin{aligned}
&[\hat{E}_i(\mathbf{r}_1, t_1), \hat{E}_j(\mathbf{r}_2, t_2)] \\
&\quad = \frac{1}{\varepsilon_0 L^3}\sum_{\mathbf{k},\mathbf{k}'}\sum_{s,s'}\left(\frac{\hbar\omega}{2}\frac{\hbar\omega'}{2}\right)^{1/2}[\hat{a}_{\mathbf{k}s}, \hat{a}^\dagger_{\mathbf{k}'s'}] \\
&\qquad \times \{(\varepsilon_{\mathbf{k}s})_i(\varepsilon^*_{\mathbf{k}'s'})_j \exp[\mathrm{i}(\mathbf{k}\cdot\mathbf{r}_1 - \mathbf{k}'\cdot\mathbf{r}_2 - \omega t_1 + \omega' t_2)] - \text{c.c.}\} \\
&\quad = \frac{1}{\varepsilon_0 L^3}\sum_{\mathbf{k}}\sum_{s}\frac{\hbar\omega}{2}(\varepsilon_{\mathbf{k}s})_i(\varepsilon^*_{\mathbf{k}s})_j[\exp\mathrm{i}[\mathbf{k}\cdot(\mathbf{r}_1 - \mathbf{r}_2) - \omega(t_1 - t_2)] - \text{c.c.}].
\end{aligned} \tag{10.8–1}$$

For the sum over s of the tensor $(\boldsymbol{\varepsilon}_{\mathbf{k}s})_i(\boldsymbol{\varepsilon}^*_{\mathbf{k}s})_j$ we refer to Section 10.2.2, where it was shown that

$$\sum_s (\boldsymbol{\varepsilon}_{\mathbf{k}s})_i(\boldsymbol{\varepsilon}_{\mathbf{k}s})_j = \delta_{ij} - \kappa_i\kappa_j. \tag{10.8–2}$$

With the help of this result, Eq. (10.8–1) becomes

$$[\hat{E}_i(\mathbf{r}_1, t_1), \hat{E}_j(\mathbf{r}_2, t_2)] = \frac{1}{\varepsilon_0 L^3}\sum_{\mathbf{k}}\left(\frac{\hbar\omega}{2}\right)(\delta_{ij} - \kappa_i\kappa_j) \times \{\exp \mathrm{i}[\mathbf{k}\cdot(\mathbf{r}_1 - \mathbf{r}_2) - \omega(t_1 - t_2)]\} - \text{c.c.} \tag{10.8–3}$$

As the size L^3 of the normalization volume becomes very great, and the mode spacing becomes very small, the sum may evidently be approximated by an integral. From the expressions (10.2–10) for the components of the wave vector $\mathbf{k}$, it follows that the number of modes contained within the wave number range $\Delta k_1 \Delta k_2 \Delta k_3$ is approximately $(L/2\pi)^3 \Delta k_1 \Delta k_2 \Delta k_3$, so that the density of modes is $(L/2\pi)^3$. We may therefore replace the sum over the discrete variables k_1, k_2, k_3 in Eq. (10.8–3) by an integral with respect to continuous variables, with the density of modes as weight, according to the general rule

$$\sum_{\mathbf{k}} \to \left(\frac{L}{2\pi}\right)^3 \int \mathrm{d}^3k. \tag{10.8–4}$$

Equation (10.8–3) therefore becomes

$$\begin{aligned}
[\hat{E}_i(\mathbf{r}_1, t_1), \hat{E}_j(\mathbf{r}_2, t_2)] \\
&= \frac{1}{(2\pi)^3}\int \frac{\hbar\omega}{2\varepsilon_0}(\delta_{ij} - \kappa_i\kappa_j)\{\exp \mathrm{i}[\mathbf{k}\cdot(\mathbf{r}_1 - \mathbf{r}_2) - \omega(t_1 - t_2)] - \text{c.c.}\}\,\mathrm{d}^3k \\
&= \frac{-\mathrm{i}}{(2\pi)^3}\int \frac{\hbar\omega}{\varepsilon_0}(\delta_{ij} - \kappa_i\kappa_j)\exp[\mathrm{i}\mathbf{k}\cdot(\mathbf{r}_1 - \mathbf{r}_2)]\sin\omega(t_1 - t_2)\,\mathrm{d}^3k,
\end{aligned} \tag{10.8–5}$$

where the second line follows from the first with the help of the change of variable $\mathbf{k} \to -\mathbf{k}$ in the complex conjugate term. It will be seen that the answer is indeed independent of the size of the normalization volume L^3, as expected for a physically meaningful result. Needless to say, this integral form of the commutator would have been obtained directly, had we chosen to make an integral rather than a series expansion of the field operator.

It is sometimes convenient to express Eq. (10.8–5) in another form, by introducing the singular function defined by‡

$$D(\mathbf{r}, t) \equiv -\frac{1}{(2\pi)^3}\int e^{\mathrm{i}\mathbf{k}\cdot\mathbf{r}} \sin\omega t \frac{\mathrm{d}^3k}{k}. \tag{10.8–6}$$

‡ For a more complete discussion of the singular D-function see Heitler (1954) Chap. 2.

By transforming to spherical polar coordinates according to the relations

$$\left.\begin{aligned} k_1 &= k\sin\theta\cos\phi, \\ k_2 &= k\sin\theta\sin\phi, \\ k_3 &= k\cos\theta, \end{aligned}\right\} \tag{10.8–7}$$

and carrying out the integrations over θ and ϕ, we may readily show that the function can also be expressed in the form

$$\begin{aligned} D(\mathbf{r}, t) &= \frac{1}{8\pi^2 r}\int_{-\infty}^{\infty}[\mathrm{e}^{\mathrm{i}k(r+ct)} - \mathrm{e}^{\mathrm{i}k(r-ct)}]\,\mathrm{d}k \\ &= \frac{1}{4\pi r}[\delta(r+ct) - \delta(r-ct)], \end{aligned} \tag{10.8–8}$$

where $r \equiv |\mathbf{r}|$. This shows that $D(\mathbf{r}, t)$ vanishes off the light cone $r = \pm ct$. With the help of this function, the commutator in Eq. (10.8–5) may be written

$$[\hat{E}_i(\mathbf{r}_1, t_1), \hat{E}_j(\mathbf{r}_2, t_2)] = \frac{\mathrm{i}\hbar c}{\varepsilon_0}\left[\frac{\delta_{ij}}{c^2}\frac{\partial^2}{\partial t_1 \partial t_2} - \frac{\partial^2}{\partial r_{1i}\partial r_{2j}}\right]D(\mathbf{r}_1 - \mathbf{r}_2, t_1 - t_2). \tag{10.8–9}$$

Since $D(\mathbf{r}_1 - \mathbf{r}_2, t_1 - t_2)$ is zero off the light cone connecting $\mathbf{r}_1, t_1$ and $\mathbf{r}_2, t_2$, this indicates that the electric fields at two different space-time points can be measured simultaneously only if the points cannot be connected by a light signal. If the two space-time points can be so connected, then one measurement may influence the outcome of the other. In particular, we note that the electric fields commute at any two fixed points in space *at the same time*.

Similarly by starting from the series expansion (10.4–40) for the magnetic field $\hat{\mathbf{B}}(\mathbf{r}, t)$ and proceeding as before, we may readily show that the magnetic fields obey an almost identical commutation relation,

$$[\hat{B}_i(\mathbf{r}_1, t_1), \hat{B}_j(\mathbf{r}_2, t_2)] = \frac{\mathrm{i}\hbar}{\varepsilon_0 c}\left[\frac{\delta_{ij}}{c^2}\frac{\partial^2}{\partial t_1 \partial t_2} - \frac{\partial^2}{\partial r_{1i}\partial r_{2j}}\right]D(\mathbf{r}_1 - \mathbf{r}_2, t_1 - t_2), \tag{10.8–10}$$

while the commutator of the electric and magnetic field may be written

$$\left.\begin{aligned} [\hat{E}_i(\mathbf{r}_1, t_1), \hat{B}_j(\mathbf{r}_2, t_2)] &= \frac{\hbar}{\varepsilon_0}\epsilon_{ijl}\frac{1}{(2\pi)^3}\int k_l \exp[\mathrm{i}\mathbf{k}\cdot(\mathbf{r}_1 - \mathbf{r}_2)]\cos\omega(t_1 - t_2)\,\mathrm{d}^3k \\ &= -\frac{\mathrm{i}\hbar}{\varepsilon_0 c}\epsilon_{ijl}\frac{\partial^2}{\partial t_1 \partial r_{1l}}D(\mathbf{r}_1 - \mathbf{r}_2, t_1 - t_2). \end{aligned}\right\} \tag{10.8–11}$$

Once again we see that the fields only fail to commute at space-time points which can be connected by a light signal. Even at such points the components of $\hat{\mathbf{E}}(\mathbf{r}_1, t_1)$ and $\hat{\mathbf{B}}(\mathbf{r}_2, t_2)$, resolved in the same direction, commute. However, the equal-time commutator between the electric and the magnetic field does not vanish, but can be seen to reduce to

$$[\hat{E}_i(\mathbf{r}_1, t), \hat{B}_j(\mathbf{r}_2, t)] = \frac{\hbar}{\mathrm{i}\varepsilon_0}\epsilon_{ijl}\frac{\partial}{\partial r_{1l}}\delta^3(\mathbf{r}_1 - \mathbf{r}_2). \qquad (10.8\text{–}12)$$

Finally, by the same procedure, we find for the commutator of the vector potential in the Coulomb gauge,

$$[\hat{A}_i(\mathbf{r}_1, t_1), \hat{A}_j(\mathbf{r}_2, t_2)] = \frac{\mathrm{i}\hbar}{\varepsilon_0}\frac{1}{(2\pi)^3}\int\frac{1}{\omega}\left(\delta_{ij} - \frac{k_ik_j}{k^2}\right)\exp[\mathrm{i}\mathbf{k}\cdot(\mathbf{r}_1 - \mathbf{r}_2)]\sin\omega(t_1 - t_2)\,\mathrm{d}^3k, \qquad (10.8\text{–}13)$$

which vanishes for equal times $t_1 = t_2$, whereas the mixed commutator is given by

$$[\hat{E}_i(\mathbf{r}_1, t_1), \hat{A}_j(\mathbf{r}_2, t_2)] = \frac{\mathrm{i}\hbar}{\varepsilon_0}\frac{1}{(2\pi)^3}\int\left(\delta_{ij} - \frac{k_ik_j}{k^2}\right)\exp[\mathrm{i}\mathbf{k}\cdot(\mathbf{r}_1 - \mathbf{r}_2)]\cos\omega(t_1 - t_2)\,\mathrm{d}^3k. \qquad (10.8\text{–}14)$$

In the special case in which the times t_1, t_2 are equal, the integral becomes one of the representations of the transverse delta function $\delta^{\mathrm{T}}_{ij}(\mathbf{r}_1 - \mathbf{r}_2)$, and we may write

$$[\hat{E}_i(\mathbf{r}_1, t), \hat{A}_j(\mathbf{r}_2, t)] = \frac{\mathrm{i}\hbar}{\varepsilon_0}\delta^{\mathrm{T}}_{ij}(\mathbf{r}_1 - \mathbf{r}_2). \qquad (10.8\text{–}15)$$

Note that this is non-zero even off the light cone, when the events $\mathbf{r}_1, t$ and $\mathbf{r}_2, t$ are disjoint. This reflects the fact that the vector potential is not a physically meaningful local observable. In a similar manner we find for the equal-time commutator connecting the vector potential $\hat{\mathbf{A}}(\mathbf{r}, t)$ and the magnetic field $\hat{\mathbf{B}}(\mathbf{r}, t)$,

$$[\hat{A}_i(\mathbf{r}_1, t), \hat{B}_j(\mathbf{r}_2, t)] = 0. \qquad (10.8\text{–}16)$$

10.8.1 Equations of motion for $\hat{\mathbf{E}}$ *and* $\hat{\mathbf{B}}$

Finally, we note that the commutator (10.8–12) allows us to simplify the Heisenberg equations of motion for the electromagnetic field vectors $\hat{\mathbf{E}}(\mathbf{r}, t)$ and $\hat{\mathbf{B}}(\mathbf{r}, t)$. From the quantized Hamiltonian for the field

$$\hat{H} = \frac{1}{2}\int\left[\varepsilon_0\hat{\mathbf{E}}^2(\mathbf{r}, t) + \frac{1}{\mu_0}\hat{\mathbf{B}}^2(\mathbf{r}, t)\right]\mathrm{d}^3r, \qquad (10.8\text{–}17)$$

we find

$$\begin{aligned}
\frac{\partial\hat{E}_i(\mathbf{r}, t)}{\partial t}
&= \frac{1}{\mathrm{i}\hbar}\left[\hat{E}_i(\mathbf{r}, t), \frac{1}{2}\int\left\{\varepsilon_0\hat{\mathbf{E}}^2(\mathbf{r}', t) + \frac{1}{\mu_0}\hat{\mathbf{B}}^2(\mathbf{r}', t)\right\}\mathrm{d}^3r'\right]\\
&= \frac{1}{2\mathrm{i}\hbar\mu_0}\int\{[\hat{E}_i(\mathbf{r}, t), \hat{B}_j(\mathbf{r}', t)]\hat{B}_j(\mathbf{r}', t) + \hat{B}_j(\mathbf{r}', t)[\hat{E}_i(\mathbf{r}, t), \hat{B}_j(\mathbf{r}', t)]\}\,\mathrm{d}^3r'\\
&= -\frac{1}{\varepsilon_0\mu_0}\int\epsilon_{ijk}\left[\frac{\partial}{\partial r_k}\delta^3(\mathbf{r} - \mathbf{r}')\right]\hat{B}_j(\mathbf{r}', t)\,\mathrm{d}^3r'\\
&= -c^2\epsilon_{ijk}\frac{\partial}{\partial r_k}\hat{B}_j(\mathbf{r}, t)\\
&= c^2(\boldsymbol{\nabla}\times\hat{\mathbf{B}}(\mathbf{r}, t))_i,
\end{aligned} \qquad (10.8\text{–}18)$$

and

$$\frac{\partial \hat{B}_i(\mathbf{r}, t)}{\partial t} = \frac{1}{i\hbar}\left[\hat{B}_i(\mathbf{r}, t), \frac{1}{2}\int\left\{\varepsilon_0 \hat{\mathbf{E}}^2(\mathbf{r}', t) + \frac{1}{\mu_0}\hat{\mathbf{B}}^2(\mathbf{r}', t)\right\} \mathrm{d}^3 r'\right]$$

$$= \frac{\varepsilon_0}{2i\hbar}\int\{[\hat{B}_i(\mathbf{r}, t), \hat{E}_j(\mathbf{r}', t)]\hat{E}_j(\mathbf{r}', t) + \hat{E}_j(\mathbf{r}', t)[\hat{B}_i(\mathbf{r}, t), \hat{E}_j(\mathbf{r}', t)]\} \mathrm{d}^3 r'$$

$$= \int \epsilon_{jik}\left[\frac{\partial}{\partial r'_k}\delta^3(\mathbf{r}' - \mathbf{r})\right]\hat{E}_j(\mathbf{r}', t)\, \mathrm{d}^3 r'$$

$$= -(\nabla \times \hat{\mathbf{E}}(\mathbf{r}, t))_i. \tag{10.8–19}$$

These are just the Maxwell equations for the free field, and they coincide with the Heisenberg equations of motion.

10.9 Vacuum fluctuations

We have already encountered several significant differences between the predictions of the quantized field theory and those of the classical theory. We found that, for each mode of the quantum field, the eigenvalues of the energy, momentum and angular momentum take on only certain discrete values, and that some components of the fields at certain space-time separations may not be simultaneously measurable. An even more striking manifestation of a quantized electromagnetic field is provided by the phenomenon of vacuum fluctuations.

The vacuum state $|\text{vac}\rangle$ of the field is the state of lowest energy, and we showed in Section 10.4 (Eq. 10.4–24) that the expectation values of both $\hat{a}_{\mathbf{k}s}$ and $\hat{a}^\dagger_{\mathbf{k}s}$ vanish in the vacuum state, because

$$\hat{a}_{\mathbf{k}s}|\text{vac}\rangle = 0 = \langle\text{vac}|\hat{a}^\dagger_{\mathbf{k}s}. \tag{10.9–1}$$

Hence for any field vector $\hat{\mathbf{F}}(\mathbf{r}, t)$, which may be the electric or magnetic field or the vector potential, having a mode expansion of the general form [cf. Eqs. (10.4–38) to (10.4–40)]

$$\hat{\mathbf{F}}(\mathbf{r}, t) = \frac{1}{L^{3/2}}\sum_{\mathbf{k},s} l(\omega)\hat{a}_{\mathbf{k}s}\boldsymbol{\varepsilon}_{\mathbf{k}s}\, e^{i(\mathbf{k}\cdot\mathbf{r}-\omega t)} + \text{h.c.} \tag{10.9–2}$$

where $l(\omega)$ is some slowly varying function of frequency which is different for each field vector, we have by reason of Eq. (10.9–1)

$$\langle\text{vac}|\hat{\mathbf{F}}(\mathbf{r}, t)|\text{vac}\rangle = 0. \tag{10.9–3}$$

However, the expectation of the square of the field operator does not vanish, as we may readily show. This implies that there are fluctuations of the electromagnetic field, even in its state of lowest energy.

If we use the mode expansion (10.9–2) and make use of the fact that

$$\langle\text{vac}|\hat{a}^\dagger_{\mathbf{k}s}\hat{a}_{\mathbf{k}'s'}|\text{vac}\rangle = 0$$

$$\langle\text{vac}|\hat{a}^\dagger_{\mathbf{k}s}\hat{a}^\dagger_{\mathbf{k}'s'}|\text{vac}\rangle = 0$$

$$\langle\text{vac}|\hat{a}_{\mathbf{k}s}\hat{a}_{\mathbf{k}'s'}|\text{vac}\rangle = 0,$$

we find immediately that

$$\langle \text{vac} | \hat{\mathbf{F}}^2(\mathbf{r}, t) | \text{vac} \rangle = \frac{1}{L^3} \sum_{\mathbf{k}s} \sum_{\mathbf{k}'s'} l(\omega) l^*(\omega') \langle \text{vac} | \hat{a}_{\mathbf{k}s} \hat{a}^\dagger_{\mathbf{k}'s'} | \text{vac} \rangle (\varepsilon_{\mathbf{k}s} \cdot \varepsilon^*_{\mathbf{k}'s'}) \times e^{i[(\mathbf{k}-\mathbf{k}')\cdot\mathbf{r} - (\omega - \omega')t]}.$$

With the help of the commutation relation (10.3–9) we have

$$\langle \text{vac} | \hat{a}_{\mathbf{k}s} \hat{a}^\dagger_{\mathbf{k}'s'} | \text{vac} \rangle = \langle \text{vac} | \hat{a}^\dagger_{\mathbf{k}'s'} \hat{a}_{\mathbf{k}s} + \delta^3_{\mathbf{k}\mathbf{k}'} \delta_{ss'} | \text{vac} \rangle = \delta^3_{\mathbf{k}\mathbf{k}'} \delta_{ss'},$$

so that

$$\langle \text{vac} | \hat{\mathbf{F}}^2(\mathbf{r}, t) | \text{vac} \rangle = \frac{1}{L^3} \sum_{\mathbf{k},s} |l(\omega)|^2 = \frac{2}{L^3} \sum_{\mathbf{k}} |l(\omega)|^2 \to \frac{2}{(2\pi)^3} \int |l(\omega)|^2 \, d^3k. \tag{10.9–4}$$

This is clearly non-zero, and indeed is infinite for an unbounded set of modes. As $\langle \text{vac} | (\Delta \hat{\mathbf{F}})^2 | \text{vac} \rangle = \langle \text{vac} | \hat{\mathbf{F}}^2 | \text{vac} \rangle$, where $\Delta \hat{\mathbf{F}} \equiv \hat{\mathbf{F}} - \langle \hat{\mathbf{F}} \rangle$ is the deviation from the mean, we see that the field fluctuates in the vacuum state. At first sight the infinite fluctuations appear to be totally unphysical, and indeed may cast doubt on the validity of the whole quantization procedure. However, Eq. (10.9–4) shows clearly that the infinity is a consequence of the infinitely high frequencies and wave numbers in the mode expansion, and these are not observable in practice.

Leaving aside the question of the infinities for the moment, let us briefly discuss why there should be field fluctuations at all. By virtue of the commutation relations (10.4–1) and (10.4–2), the field vector $\hat{\mathbf{F}}(\mathbf{r}, t)$ does not commute with the photon occupation number $\hat{n}_{\mathbf{k}s}$, and therefore does not commute with the energy $\hat{H}$. It follows that in any state of definite energy, of which the vacuum state is an example, there must be a dispersion of $\hat{\mathbf{F}}(\mathbf{r}, t)$, so that $\langle (\Delta \hat{\mathbf{F}})^2 \rangle_{\text{vac}}$ is non-zero. We recall that the canonical variables of a quantum harmonic oscillator continue to fluctuate in the ground state, and each $\mathbf{k}$, s mode of the field therefore exhibits its quantum nature.

10.9.1 Fluctuations of locally averaged fields

In order to arrive at a physically meaningful measure of the magnitude of the fluctuations we note that in any measurement there are always bounds on the $\mathbf{k}$, s modes to which the apparatus responds. In the space-time domain one may argue that no measurement of $\mathbf{F}(\mathbf{r}, t)$ can actually be made at a point in space and at an instant of time. The field is defined in terms of its effect on a test charge, which implies an effect averaged over a finite region of space and time. But, in averaging over a finite region, we shall find that the high-frequency contributions integrate to zero, so that a finite answer is obtained.

We can make these considerations more quantitative by introducing a real, non-negative, integrable spread or averaging function $g(\mathbf{r}, t)$, which is effectively zero outside a small space-time region, of linear dimensions $l_1 l_2 l_3 t_0$ (see Fig. 10.5). We may suppose that $g(\mathbf{r}, t)$ is normalized so that

$$\int g(\mathbf{r}, t) \, d^3r \, dt = 1. \tag{10.9–5}$$

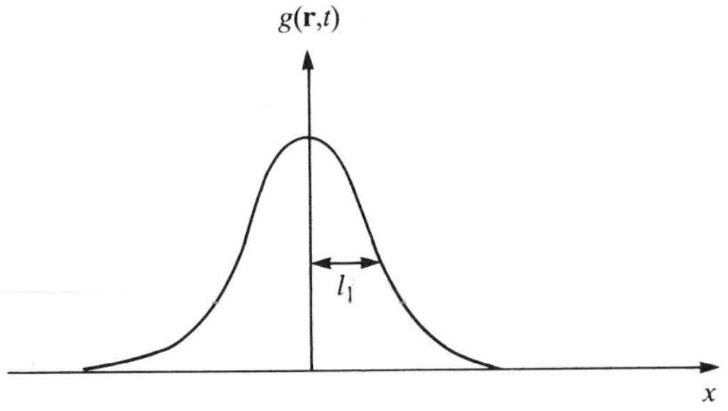

Fig. 10.5 Possible variation of the spread function $g(\mathbf{r}, t)$ with x.

The precise form of $g(\mathbf{r}, t)$ will be left arbitrary for the moment. With the help of the spread function $g(\mathbf{r}, t)$, we can express the physical effect of the field in the neighborhood of the space-time point $\mathbf{r}$, t by the convolution

$$\hat{\hat{\mathbf{F}}}(\mathbf{r}, t) = \int \hat{\mathbf{F}}(\mathbf{r} + \mathbf{r}', t + t')g(\mathbf{r}', t')\,\mathrm{d}^3r'\,\mathrm{d}t', \tag{10.9–6}$$

which represents a kind of local averaging of $\hat{\mathbf{F}}(\mathbf{r}, t)$ over a small region. With the help of the expansion (10.9–2), it then follows that

$$\hat{\hat{\mathbf{F}}}(\mathbf{r}, t) = \frac{1}{L^{3/2}}\sum_{\mathbf{k}}\sum_{s}\left[l(\omega)\hat{a}_{\mathbf{k}s}\boldsymbol{\varepsilon}_{\mathbf{k}s}\,\mathrm{e}^{\mathrm{i}(\mathbf{k}\cdot\mathbf{r}-\omega t)}\int \mathrm{e}^{\mathrm{i}(\mathbf{k}\cdot\mathbf{r}'-\omega t')}\,g(\mathbf{r}', t')\,\mathrm{d}^3r'\,\mathrm{d}t' + \mathrm{h.c.}\right]$$

$$= \frac{1}{L^{3/2}}\sum_{\mathbf{k}}\sum_{s}[l(\omega)\hat{a}_{\mathbf{k}s}\boldsymbol{\varepsilon}_{\mathbf{k}s}\,\mathrm{e}^{\mathrm{i}(\mathbf{k}\cdot\mathbf{r}-\omega t)}G(\mathbf{k}) + \mathrm{h.c.}], \tag{10.9–7}$$

where $G(\mathbf{k})$ is the four-dimensional Fourier transform of $g(\mathbf{r}, t)$, defined by

$$G(\mathbf{k}) \equiv \int g(\mathbf{r}, t)\,\mathrm{e}^{\mathrm{i}(\mathbf{k}\cdot\mathbf{r}-\omega t)}\,\mathrm{d}^3r'\,\mathrm{d}t \quad (\omega = ck). \tag{10.9–8}$$

Thus, the effect of calculating a local space-time average of $\hat{\mathbf{F}}(\mathbf{r}, t)$ with the spread function $g(\mathbf{r}, t)$, is to weight the amplitudes of the corresponding Fourier components of $\hat{\mathbf{F}}(\mathbf{r}, t)$ by $G(\mathbf{k})$.

It follows immediately from Eq. (10.9–7) that the vacuum expectation value of $\hat{\hat{\mathbf{F}}}(\mathbf{r}, t)$, like that of $\hat{\mathbf{F}}(\mathbf{r}, t)$, vanishes, i.e.

$$\langle \mathrm{vac}\,|\,\hat{\hat{\mathbf{F}}}\,|\,\mathrm{vac}\rangle = 0, \tag{10.9–9}$$

and by comparison with Eq. (10.9–4) we have

$$\langle \mathrm{vac}\,|\,\hat{\hat{\mathbf{F}}}{}^2|\,\mathrm{vac}\rangle = \frac{2}{(2\pi)^3}\int |l(\omega)|^2|G(\mathbf{k})|^2\,\mathrm{d}^3k. \tag{10.9–10}$$

We now observe that if $G(\mathbf{k})$ falls off sufficiently rapidly with $\mathbf{k}$ for large k, the vacuum expectation value of $\hat{\hat{\mathbf{F}}}{}^2$, unlike that of $\hat{\mathbf{F}}^2$, is finite. Roughly speaking, the less sharply the spread function $g(\mathbf{r}, t)$ falls off with $\mathbf{r}$, t, the more rapidly will $G(\mathbf{k})$ fall off for large $\mathbf{k}$. In fact, there is very little constraint on the allowed form of $g(\mathbf{r}, t)$. Even the very sharply varying function

$$\left.\begin{aligned} g(\mathbf{r}, t) &= 1/(\xi_1\xi_2\xi_3 t_0), \quad \text{for } |x| \leqslant \tfrac{1}{2}\xi_1,\ |y| \leqslant \tfrac{1}{2}\xi_2,\ |z| \leqslant \tfrac{1}{2}\xi_3,\ |t| \leqslant \tfrac{1}{2}t_0, \\ &= 0, \quad \text{otherwise,} \end{aligned}\right\} \tag{10.9–11}$$

whose four-dimensional Fourier transform is given by

$$G(\mathbf{k}) = \prod_{j=1,2,3}\left[\frac{\sin\left(\frac{1}{2}k_j\xi_j\right)}{\frac{1}{2}k_j\xi_j}\right]\left(\frac{\sin\frac{1}{2}\omega t_0}{\frac{1}{2}\omega t_0}\right), \tag{10.9–12}$$

leads to a finite value of $\langle \hat{\mathbf{F}}^2\rangle$. A less sharply varying function would lead to an even better convergence of the integral. Loosely speaking, if the averaging function $g(\mathbf{r}, t)$ has a range of order ξ in each direction of space and a range ξ/c in time, then $G(\mathbf{k})$ has a range of order $1/\xi$ in each variable k_1, k_2, k_3.

In practice there are usually constraints imposed on the wave numbers k_1, k_2, k_3 or frequency ω, so that it is more convenient to look on $G(\mathbf{k})$ as the fundamental quantity characteristic of some filter, and $g(\mathbf{r}, t)$ as being derived from it. We may regard $G(\mathbf{k})$ as the response function of the receiver.

10.9.2 Order of magnitude of vacuum fluctuations

It is interesting to explore the question whether the vacuum – or zero point – fluctuations of the field, say the electric field, would be important if they were measured, or whether they are completely negligible. In practice, the electric field is not usually measured in the optical domain, but such measurements are quite common with microwaves.

Let us then suppose that a receiver antenna has been set up to measure a quasi-monochromatic incoming optical field of midfrequency ω_0. The receiver will of course have a certain acceptance bandwidth and a certain acceptance solid angle, determined by the longitudinal and transverse coherence lengths of the incoming light, because all portions of the accepted field should be coherent with each other. Suppose that the light is traveling in the z-direction but with some small angular spread. Then the frequencies being accepted lie in the range $\omega_0 - c\delta k_3/2$ to $\omega_0 + c\delta k_3/2$, and the accepted directions are characterized by $-\delta k_1/2$ to $\delta k_1/2$ and $-\delta k_2/2$ to $\delta k_2/2$. It is then natural to take Eq. (10.9–10) as the fundamental relation for the mean square field, with $G(\mathbf{k})$ rather than $g(\mathbf{r}, t)$ playing the primary role. For simplicity we approximate $|G(\mathbf{k})|$ by the rectangular form

$$\left.\begin{aligned} |G(\mathbf{k})| &= 1, \quad \text{when } -\tfrac{1}{2}\delta k_1 \leqslant k_1 \leqslant \tfrac{1}{2}\delta k_1,\ -\tfrac{1}{2}\delta k_2 \leqslant k_2 \leqslant \tfrac{1}{2}\delta k_2, \\ &\qquad\qquad \omega_0/c - \tfrac{1}{2}\delta k_3 \leqslant k_3 \leqslant \omega_0/c + \tfrac{1}{2}\delta k_3 \\ &= 0, \quad \text{otherwise.} \end{aligned}\right\} \tag{10.9–13}$$

If the field is accepted by the receiver and it is coherent over the accepted range, then $2\pi/\delta k_3$ must be approximately equal to the longitudinal coherence length l_3 of the incoming light, and $2\pi/\delta k_1$, $2\pi/\delta k_2$ must be equal to the transverse coherence lengths l_1, l_2, respectively. The product $l_1 l_2 l_3 = V_c$, where V_c is the coherence volume, and l_1, l_2, l_3 are all large compared with the mid-wavelength $2\pi c/\omega_0$. After putting $|l(\omega)|^2 = \hbar\omega/2\varepsilon_0$ when $\hat{\mathbf{F}}$ in Eq. (10.9–10) is identified with the electric field $\hat{\mathbf{E}}$, and using Eq. (10.9–13), we obtain for the fluctuations of the measured $\hat{\widetilde{\mathbf{E}}}(\mathbf{r}, t)$ within the interval $(\delta k_1)(\delta k_2)(\delta k_3)$,

$$\begin{aligned}
\langle \text{vac} | \hat{\widetilde{\mathbf{E}}}^2(\mathbf{r}, t) | \text{vac} \rangle &\approx \frac{2}{(2\pi)^3} \int_{-\delta k_1/2}^{\delta k_1/2} \mathrm{d}k_1 \int_{-\delta k_2/2}^{\delta k_2/2} \mathrm{d}k_2 \int_{\omega_0/c - \delta k_3/2}^{\omega_0/c + \delta k_3/2} \frac{\hbar\omega}{2\varepsilon_0} \mathrm{d}k_3 \\
&\approx \frac{2}{(2\pi)^3} \frac{\hbar\omega_0}{2\varepsilon_0} (\delta k_1)(\delta k_2)(\delta k_3) \\
&\approx \frac{\hbar\omega_0}{\varepsilon_0} \frac{1}{V_c}. \qquad (10.9\text{–}14)
\end{aligned}$$

Let us compare this with the measured mean square of the electric field $\langle \widetilde{E}^2_{\text{light}} \rangle$ produced by the incoming quasi-monochromatic light. Let $\mathcal{N}$ be the average density of photons within the light beam of midfrequency ω_0. Then by equating energy densities we have

$$\varepsilon_0 \langle \widetilde{E}^2_{\text{light}} \rangle = \mathcal{N} \hbar\omega_0. \qquad (10.9\text{–}15)$$

The ratio of $\langle \widetilde{E}^2_{\text{light}} \rangle$ to $\langle \text{vac} | \hat{\widetilde{\mathbf{E}}}^2 | \text{vac} \rangle$ provides us with a rough indication of the signal-to-noise ratio for the measurement as a result of vacuum fluctuations, and we obtain from Eqs. (10.9–14) and (10.9–15)

$$\frac{\langle \widetilde{E}^2_{\text{light}} \rangle}{\langle \text{vac} | \hat{\widetilde{\mathbf{E}}}^2 | \text{vac} \rangle} \approx \mathcal{N} V_c. \qquad (10.9\text{–}16)$$

This shows that the average number of photons within a coherence volume is the key parameter determining the relative importance of vacuum fluctuations. So long as this number is very great, we are dealing with fields that are describable in classical terms, for which vacuum effects are negligible.

It was pointed out before, and will be proved in Section 13.1, that for radiation from all familiar thermal sources, the mean photon number in a coherence volume is usually much less than unity in the visible region. Indeed, it is only since the development of the laser that optical fields with large values of this parameter have been produced. It follows from Eq. (10.9–16) that the vacuum fluctuations, far from being a small effect, would dominate the output of an apparatus designed to detect the electric field of a light beam from most ordinary sources. Of course our argument is over-simplified, and a proper discussion of the problem really should include the measuring apparatus and its interaction with the field.

Despite the ubiquitous character of the electromagnetic vacuum, we shall see that there are quantum states in which one quadrature component of the field fluctuates less than in the vacuum state, at the cost of an increase in the fluctuations of the other field quadrature. These so-called squeezing effects have been observed experimentally and will be discussed in more detail in Chapter 21.

10.9.3 The Casimir force between conductors

Certain physical effects are sometimes attributed to vacuum fluctuations. One of the more striking examples is the attractive force between a pair of parallel, uncharged, conducting plates in vacuum (see Fig. 10.6). This force is also referred

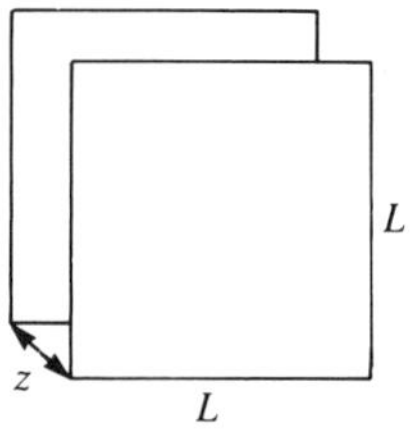

Fig. 10.6 The vacuum field trapped between two parallel, uncharged, conducting plates with separation z results in a force of attraction per unit area of order $\hbar c/z^4$.

to as a Van der Waals attraction and has been calculated by Casimir (1949; Casimir and Polder, 1948). It is interesting to note that one can account for it, and obtain an approximate value of its magnitude, by assuming that the force is a consequence of the separation-dependent vacuum field energy trapped between the two plates (Power 1964, Sec. 7.5). If the plates are squares of side L and are separated by a distance z, we may suppose that the system constitutes a 'cavity' that supports modes with wave number k down to about $1/z$. The vacuum field energy trapped between the plates may therefore be written approximately as

$$U = \sum_{\substack{\mathbf{k},s \\ k>1/z}} \tfrac{1}{2}\hbar\omega \approx (L^2 z)\int_{1/z}^{K} \hbar ck\, k^2\, \mathrm{d}k$$

$$\approx \tfrac{1}{4}L^2\hbar c\left[zK^4 - \frac{1}{z^3}\right] = U_{\text{upper}} - U_{\text{lower}}. \qquad (10.9\text{–}17)$$

We have replaced the sum by an integral in the usual way, and we have introduced a high frequency cut-off K to make the energy finite. We can think of the negative rate of change of the lower cut-off energy U_{lower} with separation z as constituting a force of attraction, whose magnitude F per unit area is given by

$$F = -\frac{1}{L^2}\frac{\mathrm{d}U_{\text{lower}}}{\mathrm{d}z} \sim \frac{\hbar c}{z^4}. \qquad (10.9\text{–}18)$$

A more careful treatment leads to a scale factor $\pi/480$ in front of $\hbar c/z^4$ in Eq. (10.9–18). It is interesting to note from the structure of Eq. (10.9–18) that the force is proportional to $\hbar$ and is therefore quantum mechanical, but apparently has nothing to do with the force between charged or polarized particles, since the electronic charge does not appear. The magnitude of the force, which has been measured, (Derjaguin, Abrikosova and Lifshitz, 1956; Kitchener and Prosser, 1957) is not insignificant for small separations z, and the validity of the force formula has been confirmed.

10.9.4 The Lamb shift

In 1948 Welton succeeded in accounting for the *Lamb shift* between the s and p energy levels of atomic hydrogen in terms of the perturbation of the electronic orbit brought about by vacuum fluctuations (Welton, 1948; see also Series, 1957, Chap. 9). A perturbation $\boldsymbol{\delta}\mathbf{r}$ in electronic position in general causes a change of potential energy $\boldsymbol{\delta} V$ given by

$$\begin{aligned}\boldsymbol{\delta} V &= V(\mathbf{r}+\boldsymbol{\delta}\mathbf{r}) - V(\mathbf{r})\\ &= \boldsymbol{\nabla} V\cdot\boldsymbol{\delta}\mathbf{r} + \frac{1}{2}\left(\frac{\partial}{\partial r_i}\frac{\partial}{\partial r_j}V\right)\boldsymbol{\delta} r_i\boldsymbol{\delta} r_j + \ldots .\end{aligned}$$

When we average this over the random displacements $\boldsymbol{\delta}\mathbf{r}$, the term in $\langle(\boldsymbol{\delta}\mathbf{r})^2\rangle$ is the leading non-zero term and we find that

$$\langle\boldsymbol{\delta} V\rangle = \tfrac{1}{6}(\nabla^2 V)\langle(\boldsymbol{\delta}\mathbf{r})^2\rangle. \tag{10.9–19}$$

In order to calculate the value of $\langle(\boldsymbol{\delta}\mathbf{r})^2\rangle$ resulting from the fluctuations of the vacuum field, we observe that, under the influence of an electric field $\mathbf{E}_\omega$ of frequency ω, the electronic position $\mathbf{r}$ obeys the equation of motion

$$m\ddot{\mathbf{r}} = -e\mathbf{E}_\omega\cos\omega t,$$

and this results in a mean squared displacement about its equilibrium value of

$$\langle(\boldsymbol{\delta}\mathbf{r}_\omega)^2\rangle = \frac{\frac{1}{2}e^2}{m^2\omega^4}\langle E_\omega^2\rangle_{\text{vac}}.$$

Let us identify the vacuum expectation value $\langle E_\omega^2\rangle_{\text{vac}}$ of the electric field at frequency ω with $\hbar\omega/\varepsilon_0 L^3$. On summing over all modes with frequencies in excess of the atomic frequency ω_0, and replacing the sum by an integral in the usual way, we arrive at a total mean squared displacement

$$\begin{aligned}\langle(\boldsymbol{\delta}\mathbf{r})^2\rangle &= \frac{\hbar e^2}{(2\pi)^3\varepsilon_0 m^2}\int\frac{\mathrm{d}^3k}{\omega^3}\\ &= \frac{\hbar e^2}{2\pi^2\varepsilon_0 m^2 c^3}\int_{\omega_0}^{\Omega}\frac{\mathrm{d}\omega}{\omega}.\end{aligned} \tag{10.9–20}$$

The integral diverges logarithmically at the upper end, and it has to be provided with a cut-off Ω, which is usually chosen to be of order $mc^2/\hbar$. When this expression for $\langle(\boldsymbol{\delta}\mathbf{r})^2\rangle$ is inserted in Eq. (10.9–19), and we average $\nabla^2 V(\mathbf{r})$ over the electronic orbit with the help of the wave function $\psi(\mathbf{r})$, we obtain finally for the perturbation of the atomic energy level,

$$\langle\delta V\rangle = \frac{\hbar e^2}{12\pi^2\varepsilon_0 m^2 c^3}\int\mathrm{d}^3 r\,\nabla^2 V(\mathbf{r})|\psi(\mathbf{r})|^2\int_{\omega_0}^{\Omega}\frac{\mathrm{d}\omega}{\omega}. \tag{10.9–21}$$

If we take the potential energy $V(\mathbf{r})$ to be $-e^2/4\pi\varepsilon_0 r$, then $\nabla^2 V(\mathbf{r}) = (e^2/\varepsilon_0)\delta^3(\mathbf{r})$, and the volume integral reduces to $(e^2/\varepsilon_0)|\psi(0)|^2$. This vanishes for a p-state but gives a finite value for an s-state. The difference between the s and p energy levels is therefore

$$\Delta E = \frac{\hbar e^4}{12\pi^2\varepsilon_0^2 m^2 c^3}|\psi_s(0)|^2\ln(mc^2/\hbar\omega_0). \tag{10.9–22}$$

This leads to $\Delta E/\hbar \sim 1040$ MHz for the 2s-state of hydrogen, and is in reasonable agreement with measurements by Lamb and Retherford (1947). The answer is similar to Bethe's expression (Bethe, 1947), based on the idea of mass renormalization, which does not directly invoke the electromagnetic vacuum field. The question whether the vacuum fluctuations are 'really' responsible, at least in part, for the Lamb shift has been discussed by Ackerhalt, Knight and Eberly (1973), Milonni, Ackerhalt and Smith (1973), Senitzky (1973), Dalibard, Dupont-Roc and Cohen-Tannoudji (1982), and Milonni (1988). We shall approach the Lamb shift in another way in Section 15.5 below.

10.9.5 Vacuum effects in the beam splitter

For another illustration of the important role that is sometimes played by the vacuum field let us examine the beam splitter illustrated in Fig. 10.7. After decomposing all fields into plane-wave modes in the usual way, we consider a single incident mode labeled 1, which gives rise to a reflected mode 2 and a transmitted mode 3. If r, t are the complex amplitude reflectivity and transmissivity for light incident from one side and r', t' for light coming from the other side, and if there are no losses in the beam splitter, then these parameters must obey the following reciprocity relations (due to Stokes, 1849):

$$\left.\begin{aligned} |r| = |r'|, \quad |t| = |t'| \\ |r|^2 + |t|^2 = 1 \\ rt^* + r'^*t' = 0. \end{aligned}\right\} \tag{10.9–23}$$

It follows that an incoming classical wave of complex amplitude v_1 gives rise to a reflected wave v_2, and a transmitted wave v_3 such that

$$v_2 = rv_1 \tag{10.9–24}$$

$$v_3 = tv_1. \tag{10.9–25}$$

From these relations it follows immediately that

$$|v_2|^2 + |v_3|^2 = (|t|^2 + |r|^2)|v_1|^2 = |v_1|^2, \tag{10.9–26}$$

so that the incoming energy is conserved.

Now suppose that we wish to apply a similar argument to the treatment of a quantum field. Then v_1, v_2, v_3 have to be replaced by the complex amplitude operators $\hat{a}_1$, $\hat{a}_2$, $\hat{a}_3$, which obey the commutation relations

$$[\hat{a}_j, \hat{a}_j^\dagger] = 1, \quad j = 1, 2, 3, \tag{10.9–27}$$

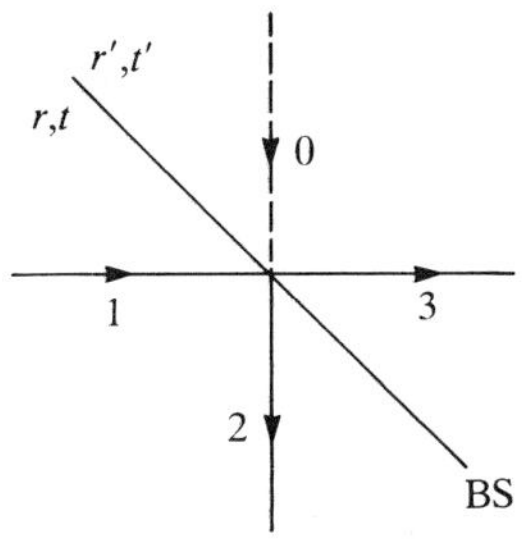

Fig. 10.7 Input and output modes for a beam splitter (BS).

and

$$[\hat{a}_2, \hat{a}_3^\dagger] = 0. \tag{10.9–28}$$

The last equation reflects the fact that one output can be measured without disturbing the other. However, if we simply replace v_1, v_2, v_3 in Eqs. (10.9–24) and (10.9–25) by the operators $\hat{a}_1$, $\hat{a}_2$, $\hat{a}_3$, we readily find that Eqs. (10.9–27) and (10.9–28) do not hold for $\hat{a}_2$, $\hat{a}_3$. Instead we obtain:

$$[\hat{a}_2, \hat{a}_2^\dagger] = |r|^2[\hat{a}_1, \hat{a}_1^\dagger] = |r|^2$$
$$[\hat{a}_3, \hat{a}_3^\dagger] = |t|^2[\hat{a}_1, \hat{a}_1^\dagger] = |t|^2$$
$$[\hat{a}_2, \hat{a}_3^\dagger] = rt^*[\hat{a}_1, \hat{a}_1^\dagger] = rt^*.$$

The reason for the discrepancy is that we have ignored the fourth beam splitter input port in Fig. 10.7, which is justifiably ignored in the classical treatment because no light enters that way. However, even if no energy is flowing through the mode labeled 0, in a quantized field treatment there is a vacuum field that enters here and contributes to the two output modes. Accordingly, in place of Eqs. (10.9–24) and (10.9–25) we need to write

$$\hat{a}_2 = r\hat{a}_1 + t'\hat{a}_0 \tag{10.9–29}$$

$$\hat{a}_3 = t\hat{a}_1 + r'\hat{a}_0, \tag{10.9–30}$$

where $\hat{a}_0$ obeys the commutation relation (10.9–27) for $j = 0$, and operators associated with modes 0 and 1 commute. We then find that

$$\begin{aligned}[\hat{a}_2, \hat{a}_2^\dagger] &= |r|^2[\hat{a}_1, \hat{a}_1^\dagger] + |t|^2[\hat{a}_0, \hat{a}_0^\dagger] \\ &= |r|^2 + |t|^2 \\ &= 1,\end{aligned} \tag{10.9–31}$$

and similarly for $\hat{a}_3$, whereas with the help of Eq. (10.9–23) we have

$$\begin{aligned}[\hat{a}_2, \hat{a}_3^\dagger] &= rt^*[\hat{a}_1, \hat{a}_1^\dagger] + r'^*t'[\hat{a}_0, \hat{a}_0^\dagger] \\ &= rt^* + r'^*t' \\ &= 0,\end{aligned} \tag{10.9–32}$$

as required.

Once again we see that the vacuum field plays a fundamental role and is required for internal consistency. The vacuum has certain consequences in quantum electrodynamics that have no counterpart in the classical domain and it cannot be ignored. The beam splitter will be treated in more detail in Section 12.12 below.

10.10 Continuous Fock space

Up to now we have dealt entirely with discrete modes of the electromagnetic field, to which we were led by the artifice of enclosing the field in a fictitious cube of side L and imposing periodic boundary conditions. No physically meaningful results actually depend on the size of the cube, which is allowed to become infinite at a suitable stage of the calculation. However, this procedure is not essential for the quantization of the field, and it is quite possible to deal directly with the infinite space domain. In some ways it would seem more natural and elegant to do so, although the continuous representation in fact turns out to be less compact. Before leaving the formal development of the quantum theory of radiation, let us briefly examine the situation that arises when the field is decomposed into a continuous set of modes,‡ and the field vectors are represented by Fourier integrals rather than by Fourier series as in Eqs. (10.4–38) to (10.4–40).

The wave vector $\mathbf{k}$ is now a continuous vector index, whose components range over all values from $-\infty$ to ∞, although the polarization index s still takes on only two values by virtue of the transversality of the field vectors. We then write

$$\hat{\mathbf{A}}(\mathbf{r}, t) = \frac{1}{\varepsilon_0^{1/2}(2\pi)^{3/2}} \sum_s \int \left(\frac{\hbar}{2\omega}\right)^{1/2} [\hat{a}(\mathbf{k}, s)\boldsymbol{\varepsilon}(\mathbf{k}, s)\, e^{i(\mathbf{k}\cdot\mathbf{r}-\omega t)} + \text{h.c.}]\, d^3k, \tag{10.10–1}$$

$$\hat{\mathbf{E}}(\mathbf{r}, t) = \frac{i}{\varepsilon_0^{1/2}(2\pi)^{3/2}} \sum_s \int \left(\frac{\hbar\omega}{2}\right)^{1/2} [\hat{a}(\mathbf{k}, s)\boldsymbol{\varepsilon}(\mathbf{k}, s)\, e^{i(\mathbf{k}\cdot\mathbf{r}-\omega t)} - \text{h.c.}]\, d^3k, \tag{10.10–2}$$

$$\hat{\mathbf{B}}(\mathbf{r}, t) = \frac{i}{\varepsilon_0^{1/2}(2\pi)^{3/2}} \sum_s \int \left(\frac{\hbar}{2\omega}\right)^{1/2} [\hat{a}(\mathbf{k}, s)(\mathbf{k} \times \boldsymbol{\varepsilon}(\mathbf{k}, s))\, e^{i(\mathbf{k}\cdot\mathbf{r}-\omega t)} - \text{h.c.}]\, d^3k. \tag{10.10–3}$$

The unit polarization vectors $\boldsymbol{\varepsilon}(\mathbf{k}, s)$ obey the same orthonormality relations (10.2–15) as before. The terms $\hat{a}(\mathbf{k}, s)$ and $\hat{a}^\dagger(\mathbf{k}, s)$ again play the role of annihilation and creation operators for photons of wave vector $\mathbf{k}$ and polarization s, but the commutation rules (10.3–9) to (10.3–11) must now by replaced by

$$\left.\begin{aligned} &[\hat{a}(\mathbf{k}, s), \hat{a}^\dagger(\mathbf{k}', s')] = \delta^3(\mathbf{k} - \mathbf{k}')\delta_{ss'}, \\ &[\hat{a}(\mathbf{k}, s), \hat{a}(\mathbf{k}', s')] = 0, \\ &[\hat{a}^\dagger(\mathbf{k}, s), \hat{a}^\dagger(\mathbf{k}', s')] = 0. \end{aligned}\right\} \tag{10.10–4}$$

As is evident from these relations, $\hat{a}(\mathbf{k}, s)$ and $\hat{a}^\dagger(\mathbf{k}, s)$ are not dimensionless operators, as are the corresponding operators in the discrete Fock space. With the help of the commutation relations it may readily be shown that the commutators of $\hat{\mathbf{E}}(\mathbf{r}, t)$, $\hat{\mathbf{B}}(\mathbf{r}, t)$, etc. have the values found in Section 10.8.

As before, we may introduce Fock states of the field, corresponding to a discrete number of excitations or photons. However, the concept of the number of excitations of one mode is no longer strictly meaningful, since one mode of the continuum has zero measure. The Fock states are now labeled by displaying the wave vector and polarization index of each excitation explicitly. For example, a

‡ For a fuller discussion of continuous Fock space, see for example Schweber (1961), Chap. 7.

typical n-photon Fock state might be

$$|\mathbf{k}_1 s_1, \mathbf{k}_2 s_2, \ldots, \mathbf{k}_n s_n\rangle,$$

in which the mode labels may be repeated. Because of the continuous nature of the modes, the states are not normalizable in the ordinary sense, but they can be delta-normalized. Orthonormality is expressed by the condition

$$\langle \mathbf{k}'_m s'_m, \ldots, \mathbf{k}'_1 s'_1 | \mathbf{k}_1 s_1, \ldots, \mathbf{k}_n s_n\rangle = \delta_{nm}\frac{1}{n!}\sum_P \delta^3(\mathbf{k}_1 - \mathbf{k}'_1)\delta_{s_1 s'_1}\delta^3(\mathbf{k}_2 - \mathbf{k}'_2)\delta_{s_2 s'_2}\ldots \tag{10.10–5}$$

where $\sum_P$ denotes the sum over all $n!$ pairings of mode labels in the ket vector with those in the bra vector. The scalar product vanishes unless all the mode labels coincide, but the different state vectors are no longer dimensionless.

As before we can form Fock states by the repeated action of creation operators on the vacuum state, which is labelled $|\,\mathrm{vac}\rangle$. For example

$$|\mathbf{k}_1 s_1, \ldots, \mathbf{k}_n s_n\rangle = \frac{1}{\sqrt{(n!)}}\hat{a}^\dagger(\mathbf{k}_n, s_n)\ldots\hat{a}^\dagger(\mathbf{k}_1, s_1)|\,\mathrm{vac}\rangle. \tag{10.10–6}$$

The action of a creation or annihilation operator on an arbitrary Fock state is given by the relations

$$\hat{a}^\dagger(\mathbf{k}, s)|\mathbf{k}_1 s_1, \ldots, \mathbf{k}_n s_n\rangle = (n+1)^{1/2}|\mathbf{k}s, \mathbf{k}_1 s_1, \ldots, \mathbf{k}_n s_n\rangle, \tag{10.10–7}$$

and

$$\begin{aligned}\hat{a}(\mathbf{k}, s)|\mathbf{k}_1 s_1, \ldots, \mathbf{k}_n s_n\rangle &= \frac{1}{\sqrt{n}}\sum_{i=1}^{n}\delta^3(\mathbf{k} - \mathbf{k}_i)\\ &\quad\times \delta_{ss_i}|\mathbf{k}_1 s_1, \ldots, \mathbf{k}_{i-1}s_{i-1}, \mathbf{k}_{i+1}s_{i+1}, \ldots, \mathbf{k}_n s_n\rangle,\end{aligned} \tag{10.10–8}$$

which should be compared with Eqs. (10.4–5) and (10.4–8), respectively. The more complicated Eq. (10.10–8) arises because the possible coincidence of the $\mathbf{k}$, s mode with each of the set $\mathbf{k}_1 s_1, \ldots, \mathbf{k}_n s_n$ has to be tested. The last equation can be derived by writing the Fock state $|\mathbf{k}_1 s_1, \ldots, \mathbf{k}_n s_n\rangle$ in the form of Eq. (10.10–6), applying the commutation relations (10.10–4) to move the $\hat{a}(\mathbf{k}, s)$ operator repeatedly to the right, and finally using the result

$$\hat{a}(\mathbf{k}, s)|\,\mathrm{vac}\rangle = 0.$$

When $\mathbf{k}$, s does not coincide with any modes of the set $\mathbf{k}_1 s_1, \ldots, \mathbf{k}_n s_n$, we have

$$\hat{a}(\mathbf{k}, s)|\mathbf{k}_1 s_1, \ldots, \mathbf{k}_n s_n\rangle = 0. \tag{10.10–9}$$

The number operator $\hat{n}_s(D)$, corresponding to the number of photons of polarization s whose wave vectors fall within some domain D, is defined by

$$\hat{n}_s(D) = \int_D a^\dagger(\mathbf{k}, s)\hat{a}(\mathbf{k}, s)\,\mathrm{d}^3 k. \tag{10.10–10}$$

This obeys the commutation relation

$$[\hat{a}(\mathbf{k}, s), \hat{n}_{s'}(D)] = \hat{a}(\mathbf{k}, s)\delta_{ss'}U(\mathbf{k}, D), \tag{10.10–11}$$

where $U(\mathbf{k}, D)$ is unity if $\mathbf{k} \in D$ and zero otherwise. When $\hat{n}_s(D)$ acts on the Fock state $|\mathbf{k}_1 s_1, \ldots, \mathbf{k}_n s_n\rangle$, it reproduces the state multiplied by the eigenvalue equal to the number of photons with polarization s and wave vectors lying within the domain D.

The continuous Fock states form a complete set, so that we can make a resolution of the unit operator in the usual way by taking the sum over all possible Fock state projectors. We then have

$$\begin{aligned}1 = |\,\mathrm{vac}\rangle\langle\mathrm{vac}\,| &+ \sum_{s_1}\int |\mathbf{k}_1 s_1\rangle\langle\mathbf{k}_1 s_1|\,\mathrm{d}^3k_1 \\ &+ \sum_{s_1}\sum_{s_2}\iint |\mathbf{k}_1 s_1, \mathbf{k}_2 s_2\rangle\langle\mathbf{k}_2 s_2, \mathbf{k}_1 s_1|\,\mathrm{d}^3k_1\,\mathrm{d}^3k_2 \\ &+ \sum_{s_1}\sum_{s_2}\sum_{s_3}\iiint |\mathbf{k}_1 s_1, \mathbf{k}_2 s_2, \mathbf{k}_3 s_3\rangle\langle\mathbf{k}_3 s_3, \mathbf{k}_2 s_2, \mathbf{k}_1 s_1|\,\mathrm{d}^3k_1\,\mathrm{d}^3k_2\,\mathrm{d}^3k_3 \\ &+ \ldots . \end{aligned} \tag{10.10–12}$$

This expansion can be used to make a Fock representation of any state or operator of the electromagnetic field, if we multiply by 1 in the form given by Eq. (10.10–12). It should be evident that, in general, the resulting expression will be much less compact than the corresponding discrete Fock representation, as given by Eq. (10.4–21) for example. It is partly for this reason that we prefer to work with the discrete set of modes in the following sections.

10.11 Some theorems on operator algebra

Calculations in quantum electrodynamics frequently require the manipulation of certain combinations of Hilbert space operators. We therefore collect together in this section some frequently used operator theorems.‡ Some of them have already been encountered, and others will be used in later chapters.

10.11.1 The operator expansion theorem

Let $\hat{A}$ and $\hat{B}$ be two operators that do not necessarily commute, and let

$$\hat{f}(x) \equiv \exp(x\hat{A})\hat{B}\exp(-x\hat{A}).$$

We expand $\hat{f}(x)$ as a Taylor series in x about the origin. From the definition

$$\hat{f}'(x) = \exp(x\hat{A})(\hat{A}\hat{B} - \hat{B}\hat{A})\exp(-x\hat{A}),$$

so that

$$\hat{f}'(0) = [\hat{A}, \hat{B}].$$

‡ For a more complete discussion of several operator theorems see Louisell (1973) Chapter 3. We follow his derivations in proving many of the theorems given here.

Similarly

$$\hat{f}''(x) = \exp(x\hat{A})(\hat{A}[\hat{A}, \hat{B}] - [\hat{A}, \hat{B}]\hat{A})\exp(-x\hat{A}),$$

so that

$$\hat{f}''(0) = [\hat{A}, [\hat{A}, \hat{B}]],$$

etc. We now write the Taylor series for $\hat{f}(x)$,

$$\hat{f}(x) = \hat{f}(0) + x\hat{f}'(0) + \frac{x^2}{2!}\hat{f}''(0) + \ldots,$$

and substitute for $\hat{f}^{(n)}(0)$, which immediately leads to the series

$$\exp(x\hat{A})\hat{B}\exp(-x\hat{A}) = \hat{B} + x[\hat{A}, \hat{B}] + \frac{x^2}{2!}[\hat{A}, [\hat{A}, \hat{B}]] + \ldots. \quad (10.11\text{–}1)$$

This is the so-called *operator expansion theorem*, which is especially useful in connection with unitary transformations.

In the special case in which the commutator $[\hat{A}, \hat{B}] = c$ is a c-number, the series terminates after the second term and we have

$$\exp(x\hat{A})\hat{B}\exp(-x\hat{A}) = \hat{B} + cx. \quad (10.11\text{–}2)$$

In that case $\exp(x\hat{A})$ acts as a translation operator with respect to $\hat{B}$. For example, if $\hat{A}$ and $\hat{B}$ are canonical conjugates, and $\hat{B} = \hat{q}$ while $\hat{A} = \mathrm{i}\hat{p}/\hbar$, then $[\hat{A}, \hat{B}] = 1$, and

$$\mathrm{e}^{\mathrm{i}\hat{p}x/\hbar}\,\hat{q}\,\mathrm{e}^{-\mathrm{i}\hat{p}x/\hbar} = \hat{q} + x. \quad (10.11\text{–}3)$$

Another important example of Eq. (10.11–2) occurs if we put $x = 1$ and $\hat{A} = -v\hat{a}^\dagger + v^*\hat{a}$, where $\hat{a}$, $\hat{a}^\dagger$ are the usual annihilation and creation operators, and v is any complex number. Then, if we choose $\hat{B} = \hat{a}$ or $\hat{B} = \hat{a}^\dagger$, we find $[\hat{A}, \hat{B}] = v$ or $[\hat{A}, \hat{B}] = v^*$, respectively, and

$$\left.\begin{aligned} \mathrm{e}^{-v\hat{a}^\dagger + v^*\hat{a}}\,\hat{a}\,\mathrm{e}^{v\hat{a}^\dagger - v^*\hat{a}} &= \hat{a} + v \\ \mathrm{e}^{-v\hat{a}^\dagger + v^*\hat{a}}\,\hat{a}^\dagger\,\mathrm{e}^{v\hat{a}^\dagger - v^*\hat{a}} &= \hat{a}^\dagger + v^*. \end{aligned}\right\} \quad (10.11\text{–}4)$$

The operator $\exp(v\hat{a}^\dagger - v^*\hat{a})$ is known as the *displacement operator*, and will be discussed in more detail in Section 11.3.

When $\hat{A}$ in Eq. (10.11–1) is the number operator $\hat{n} = \hat{a}^\dagger\hat{a}$ and $\hat{B}$ is identified with $\hat{a}$ or $\hat{a}^\dagger$, we effect a scaling transformation of $\hat{a}$ or $\hat{a}^\dagger$. As

$$[\hat{n}, \hat{a}] = -\hat{a},$$

and as all higher-order commutators also give $\hat{a}$ with alternating signs, it follows from the operator expansion theorem that

$$\begin{aligned} \mathrm{e}^{x\hat{n}}\,\hat{a}\,\mathrm{e}^{-x\hat{n}} &= \hat{a} - x\hat{a} + \frac{x^2}{2!}\hat{a} - \frac{x^3}{3!}\hat{a} + \ldots \\ &= \hat{a}\,\mathrm{e}^{-x}, \end{aligned} \quad (10.11\text{–}5)$$

and similarly

$$\mathrm{e}^{x\hat{n}}\,\hat{a}^\dagger\,\mathrm{e}^{-x\hat{n}} = \hat{a}^\dagger\,\mathrm{e}^{x}. \quad (10.11\text{–}6)$$

The operator expansion theorem also applies to differential operators. For example, if $\hat{A} = \frac{1}{2}q^2$ and $\hat{B} = \mathrm{d}/\mathrm{d}q$, then

$$[\hat{A}, \hat{B}] = -q,$$

and from Eq. (10.11–2)

$$\mathrm{e}^{xq^2/2}\frac{\mathrm{d}}{\mathrm{d}x}\mathrm{e}^{-xq^2/2} = \frac{\mathrm{d}}{\mathrm{d}q} - xq. \qquad (10.11\text{–}7)$$

We shall encounter other examples of the operator expansion theorem in later sections.

10.11.2 Theorems on similarity transformation

If $\hat{A}$, $\hat{B}$ are two operators that do not necessarily commute, and n is a positive integer, then by direct multiplication

$$\begin{aligned}[\exp(x\hat{A})\hat{B}\exp(-x\hat{A})]^n &= \exp(x\hat{A})\hat{B}\exp(-x\hat{A})\exp(x\hat{A})\hat{B}\exp(-x\hat{A}) \\ &\quad \ldots \exp(x\hat{A})\hat{B}\exp(-x\hat{A}) \\ &= \exp(x\hat{A})\hat{B}^n\exp(-x\hat{A}). \end{aligned} \qquad (10.11\text{–}8)$$

It follows immediately that for any operator function $f(\hat{B})$ having a power series expansion in $\hat{B}$,

$$\exp(x\hat{A})f(\hat{B})\exp(-x\hat{A}) = f(\exp(x\hat{A})\hat{B}\exp(-x\hat{A})). \qquad (10.11\text{–}9)$$

By combining Eqs. (10.11–2) and (10.11–9), we find for any pair of operators whose commutator is a c-number c,

$$\exp(x\hat{A})f(\hat{B})\exp(-x\hat{A}) = f(\hat{B} + cx). \qquad (10.11\text{–}10)$$

We can generalize Eqs. (10.11–8) and (10.11–9) somewhat by replacing $\exp(x\hat{A})$ by some other operator $\hat{G}$, that has an inverse. We then obtain

$$(\hat{G}\hat{B}\hat{G}^{-1})^n = \hat{G}\hat{B}^n\hat{G}^{-1}$$

and

$$\hat{G}f(\hat{B})\hat{G}^{-1} = f(\hat{G}\hat{B}\hat{G}^{-1}). \qquad (10.11\text{–}11)$$

By combining Eqs. (10.11–7) and (10.11–9), for example, we find

$$\mathrm{e}^{xq^2/2}f\left(\frac{\mathrm{d}}{\mathrm{d}q}\right)\mathrm{e}^{-xq^2/2} = f\left(\frac{\mathrm{d}}{\mathrm{d}q} - qx\right) \qquad (10.11\text{–}12)$$

and from Eqs. (10.11–4) and (10.11–9)

$$\mathrm{e}^{-v\hat{a}^\dagger + v^*\hat{a}}f(\hat{a}, \hat{a}^\dagger)\mathrm{e}^{v\hat{a}^\dagger - v^*\hat{a}} = f(\hat{a} + v, \hat{a}^\dagger + v^*). \qquad (10.11\text{–}13)$$

Also

$$\left.\begin{aligned}\mathrm{e}^{-v\hat{a}^\dagger}f(\hat{a}, \hat{a}^\dagger)\mathrm{e}^{v\hat{a}^\dagger} &= f(\hat{a}+v, \hat{a}^\dagger) \\ \mathrm{e}^{v^*\hat{a}}f(\hat{a}, \hat{a}^\dagger)\mathrm{e}^{-v^*\hat{a}} &= f(\hat{a}, \hat{a}^\dagger + v^*),\end{aligned}\right\} \qquad (10.11\text{–}14)$$

and by combining Eqs. (10.11–5) and (10.11–6) with Eq. (10.11–9) we obtain

$$e^{x\hat{n}} f(\hat{a}, \hat{a}^\dagger) e^{-x\hat{n}} = f(\hat{a} e^{-x}, \hat{a}^\dagger e^{x}). \tag{10.11–15}$$

10.11.3 Derivative theorems

Let $f(\hat{a}, \hat{a}^\dagger)$ be some function of creation and annihilation operators having a power series expansion. From Eqs. (10.11–14) we have, when v, v^* are very small,

$$\begin{aligned}\frac{\partial f(\hat{a}, \hat{a}^\dagger)}{\partial \hat{a}} &= \lim_{\delta v \to 0} \frac{f(\hat{a} + \delta v, \hat{a}^\dagger) - f(\hat{a}, \hat{a}^\dagger)}{\delta v} \\ &= \lim_{\delta v \to 0} \frac{(1 - \delta v \hat{a}^\dagger) f(\hat{a}, \hat{a}^\dagger)(1 + \delta v \hat{a}^\dagger) - f(\hat{a}, \hat{a}^\dagger)}{\delta v} \\ &= -[\hat{a}^\dagger, f(\hat{a}, \hat{a}^\dagger)],\end{aligned} \tag{10.11–16}$$

and similarly,

$$\frac{\partial f(\hat{a}, \hat{a}^\dagger)}{\partial \hat{a}^\dagger} = [\hat{a}, f(\hat{a}, \hat{a}^\dagger)]. \tag{10.11–17}$$

As an immediate application of these relations we note that if $f(\hat{a}, \hat{a}^\dagger) = \hat{a}^n$ or $f(\hat{a}, \hat{a}^\dagger) = \hat{a}^{\dagger n}$, $n = 1, 2, \ldots$, then

$$\left.\begin{aligned}[\hat{a}^\dagger, \hat{a}^n] &= -n\hat{a}^{n-1} \\ [\hat{a}, \hat{a}^{\dagger n}] &= n\hat{a}^{\dagger n-1},\end{aligned}\right\} \tag{10.11–18}$$

and if $f(\hat{a}, \hat{a}^\dagger) = e^{v\hat{a}^\dagger}$ or $e^{-v^*\hat{a}}$, we obtain

$$\left.\begin{aligned}[\hat{a}, e^{v\hat{a}^\dagger}] &= v\, e^{v\hat{a}^\dagger} \\ [\hat{a}^\dagger, e^{-v^*\hat{a}}] &= v^*\, e^{-v^*\hat{a}}.\end{aligned}\right\} \tag{10.11–19}$$

10.11.4 Normal and antinormal ordering

Some of the foregoing relations are occasionally useful for putting certain products of creation and annihilation operators into normal or antinormal order. In *normal order* all the creation operators $\hat{a}^\dagger$ stand to the left of all annihilation operators $\hat{a}$, and the opposite is true in *antinormal order*. Thus, suppose that the function $f^{(N)}(\hat{a}, \hat{a}^\dagger)$ is already in normal order. Then

$$e^{v^*\hat{a}} f^{(N)}(\hat{a}, \hat{a}^\dagger)$$

is not in normal order. However, we can put it in normal order by multiplying both sides of the second Eq. (10.11–14) by $e^{v^*\hat{a}}$ on the right, and we obtain

$$e^{v^*\hat{a}} f^{(N)}(\hat{a}, \hat{a}^\dagger) = f^{(N)}(\hat{a}, \hat{a}^\dagger + v^*) e^{v^*\hat{a}}, \tag{10.11–20}$$

where the right-hand side is now in normal order. Similarly

$$e^{-v\hat{a}^\dagger} f^{(A)}(\hat{a}, \hat{a}^\dagger) = f^{(A)}(\hat{a} + v, \hat{a}^\dagger) e^{-v\hat{a}^\dagger}, \tag{10.11–21}$$

and the right-hand side is in antinormal order if $f^{(A)}(\hat{a}, \hat{a}^\dagger)$ is a function in antinormal order.

The product

$$\hat{a} f^{(N)}(\hat{a}, \hat{a}^\dagger)$$

is also not in normal order. However, from Eq. (10.11–17) we obtain

$$\hat{a} f^{(N)}(\hat{a}, \hat{a}^\dagger) = f^{(N)}(\hat{a}, \hat{a}^\dagger)\hat{a} + \frac{\partial}{\partial \hat{a}^\dagger} f^{(N)}(\hat{a}, \hat{a}^\dagger), \tag{10.11–22}$$

and the right-hand side has been put in normal order if $f^{(N)}(\hat{a}, \hat{a}^\dagger)$ is in normal order. The opposite, or antinormal, ordering is achieved if we use Eq. (10.11–16) to write, for any antinormally ordered function $f^{(A)}(\hat{a}, \hat{a}^\dagger)$,

$$\hat{a}^\dagger f^{(A)}(\hat{a}, \hat{a}^\dagger) = f^{(A)}(\hat{a}, \hat{a}^\dagger)\hat{a}^\dagger - \frac{\partial}{\partial \hat{a}} f^{(A)}(\hat{a}, \hat{a}^\dagger). \tag{10.11–23}$$

More generally,

$$\hat{a}^n f^{(N)}(\hat{a}, \hat{a}^\dagger) = :\left(\hat{a} + \frac{\partial}{\partial \hat{a}^\dagger}\right)^n f^{(N)}(\hat{a}, \hat{a}^\dagger):, \quad n = 1, 2, \ldots . \tag{10.11–24}$$

where : : is a symbol that re-arranges every operator within it in normal order, without use of commutation relations. For example $:\hat{a}\hat{a}^\dagger: = \hat{a}^\dagger\hat{a}$. We can prove Eq. (10.11–24) most easily by induction, by assuming that it holds for some n. Then from Eq. (10.11–22)

$$\begin{aligned} \hat{a}^{n+1} f^{(N)}(\hat{a}, \hat{a}^\dagger) &= \hat{a}^n \hat{a} f^{(N)}(\hat{a}, \hat{a}^\dagger) \\ &= \hat{a}^n :\left(\hat{a} + \frac{\partial}{\partial \hat{a}^\dagger}\right) f^{(N)}(\hat{a}, \hat{a}^\dagger):, \end{aligned}$$

and since $\hat{a}^n$ is multiplied on the right by a normally ordered function, we have from Eq. (10.11–24)

$$\hat{a}^{n+1} f^{(N)}(\hat{a}, \hat{a}^\dagger) = :\left(\hat{a} + \frac{\partial}{\partial \hat{a}^\dagger}\right)^{n+1} f^{(N)}(\hat{a}, \hat{a}^\dagger):.$$

Hence Eq. (10.11–24) also holds when n is replaced by $n + 1$, and since it holds for $n = 1$, it holds for all positive integers n.

10.11.5 The Campbell–Baker–Hausdorff theorem

Let $\hat{A}$, $\hat{B}$ be two operators that do not necessarily commute, but whose commutator $[\hat{A}, \hat{B}]$ commutes with both $\hat{A}$ and $\hat{B}$, so that

$$[\hat{A}, [\hat{A}, \hat{B}]] = 0 = [\hat{B}, [\hat{A}, \hat{B}]]. \tag{10.11–25}$$

Then

$$\begin{aligned} \exp[x(\hat{A} + \hat{B})] &= \exp(x\hat{A})\exp(x\hat{B})\exp(-x^2[\hat{A}, \hat{B}]/2) \\ &= \exp(x\hat{B})\exp(x\hat{A})\exp(x^2[\hat{A}, \hat{B}]/2). \end{aligned} \tag{10.11–26}$$

This is known as the *Campbell–Baker–Hausdorff theorem* (Campbell, 1898;

Baker, 1902, 1903; Hausdorff, 1906). It can be proved as follows.‡ Let us write

$$\exp(x\hat{A})\exp(x\hat{B}) \equiv \hat{C}(x),$$

and differentiate both sides with respect to x. Then

$$\begin{aligned}\frac{\mathrm{d}\hat{C}(x)}{\mathrm{d}x} &= \hat{A}\exp(x\hat{A})\exp(x\hat{B}) + \exp(x\hat{A})\hat{B}\exp(x\hat{B}) \\ &= \{\hat{A} + \exp(x\hat{A})\hat{B}\exp(-x\hat{A})\}\hat{C}(x),\end{aligned}$$

and with the help of the operator expansion theorem (10.11–1),

$$\frac{\mathrm{d}\hat{C}(x)}{\mathrm{d}x} = \{(\hat{A} + \hat{B}) + x[\hat{A}, \hat{B}]\}\hat{C}(x), \tag{10.11–27}$$

because the higher-order commutators vanish. As $(\hat{A} + \hat{B})$ commutes with $[\hat{A}, \hat{B}]$, the two terms within { } can be treated effectively like c-numbers, and the order of operators does not matter.

But we also have from the definition of $\hat{C}(x)$

$$\begin{aligned}\frac{\mathrm{d}\hat{C}(x)}{\mathrm{d}x} &= \exp(x\hat{A})\hat{A}\exp(x\hat{B}) + \exp(x\hat{A})\exp(x\hat{B})\hat{B} \\ &= \exp(x\hat{A})\exp(x\hat{B})[\exp(-x\hat{B})\hat{A}\exp(x\hat{B}) + \hat{B}] \\ &= \hat{C}(x)\{\hat{A} + \hat{B} + x[\hat{A}, \hat{B}]\}.\end{aligned} \tag{10.11–28}$$

Comparison of Eqs. (10.11–27) and (10.11–28) shows that $\hat{C}(x)$ and $\hat{A} + \hat{B} + x[\hat{A}, \hat{B}]$ commute. Hence the equation can be integrated with respect to x like an ordinary differential equation, to yield

$$\hat{C}(x) = \exp\{x(\hat{A} + \hat{B}) + x^2[\hat{A}, \hat{B}]/2\} = \exp[x(\hat{A} + \hat{B})]\exp(x^2[\hat{A}, \hat{B}]/2), \tag{10.11–29}$$

as may be confirmed by direct differentiation. On substituting for $\hat{C}(x)$, we arrive at Eq. (10.11–26).

Problems

10.1 From the mode expansion show that the Heisenberg equations of motion for the free $\hat{\mathbf{E}}(\mathbf{r}, t)$ and $\hat{\mathbf{B}}(\mathbf{r}, t)$ fields yield two of Maxwell's equations.

10.2 Show that the components $\hat{E}_i(\mathbf{r}_1, t_1)$ and $\hat{E}_j(\mathbf{r}_2, t_2)$ of the quantized electric field operators commute if the events $\mathbf{r}_1$, t_1 and $\mathbf{r}_2$, t_2 are not connected by a light cone.

10.3 Show that the quantized field operators $\hat{E}_i(\mathbf{r}_1, t_1)$ and $\hat{A}_j(\mathbf{r}_2, t_2)$ do not commute in general, even if the events $\mathbf{r}_1$, t_1 and $\mathbf{r}_2$, t_2 have a space-like separation.

‡ Our proof follows that given by Healy (1982), Appendix D.

10.4 Calculate the mean squared dispersion of the locally averaged electric field $\widehat{\tilde{\mathbf{E}}}(\mathbf{r}, t)$ in the vacuum state when the spread function $g(\mathbf{r}, t)$ has the form:

(a) $g(\mathbf{r}, t) \propto e^{-r^2/2l^2}\delta(t)$;

(b) $g(\mathbf{r}, t) \propto \delta^{(3)}(\mathbf{r})\, e^{-t^2/2T^2}$.

10.5 Calculate and compare the vacuum expectations of the locally averaged squared electric field when the spread function $g(\mathbf{r})$ depends only on position and is of the form

(a) $g(\mathbf{r}) \propto e^{-r/D}, \quad D > 0,$

(b) $g(\mathbf{r}) \propto \dfrac{1}{(r^2 + D^2)^2}.$

10.6 Starting from the plane-wave expansions of the field vectors, show that the orbital angular momentum of the electromagnetic field about the point $\mathbf{r}_0$ differs from the orbital angular momentum about the origin by the term $\mathbf{r}_0 \times \sum_{\mathbf{k}s} \hbar \mathbf{k} \hat{n}_{\mathbf{k}s}$.

10.7 Use the Fock representation to determine if the creation operator for a single-mode electromagnetic field has a right eigenstate.

10.8 Show that in the limit $s \to \infty$ the phase states $|\theta\rangle$ defined by Eq. (10.7–20) form a complete orthonormal set.

11

Coherent states of the electromagnetic field

11.1 Introduction

In our discussion of the classical electromagnetic field we found it convenient to describe the field by a complex amplitude, both in the frequency and in the time domains. The complex representation is convenient partly because it contains information about the magnitude and about the phase of the electromagnetic disturbance, and partly because of its analytic properties. These features turn out to be particularly useful for the description of the optical coherence properties of the field.

We shall see that an analogous quantum state of the field exists, leading to an interesting representation that is also particularly useful for the treatment of optical coherence. This coherent-state representation leads to a close correspondence between the quantum and classical correlation functions. The coherent states of the field come as close as possible to being classical states of definite complex amplitude. We shall find that coherent states turn out to be particularly appropriate for the description of the electromagnetic fields generated by coherent sources, like lasers and parametric oscillators; indeed it turns out that the field produced by any deterministic current source is in a coherent state. In the following sections we shall examine some of the properties of coherent states, and then go on to discuss representations based on these states.

Coherent states were first discovered in connection with the quantum harmonic oscillator by Schrödinger (1926), who referred to them as states of minimum uncertainty product. They remained largely a curiosity until comparatively recent times, when their properties were further investigated by Klauder (1960). Later Bargmann (1961) introduced a functional representation of quantum states, having many features in common with the coherent-state representation. The recognition that coherent states are particularly important and appropriate for the quantum treatment of optical coherence and their adoption in quantum optics are due largely to the work of Glauber (1963a,b 1965), who coined the name 'coherent state'. Many of the properties of these states to be examined in this chapter were first treated by Glauber in the papers cited above.

11.2 Fock representation of the coherent state

In the introduction to the coherent states it will be convenient to confine our attention at first to just one **k**, s mode of the electromagnetic field. We shall therefore simplify the notation by discarding the mode label for the time being, with the understanding that we are considering only one mode.

Suppose that the annihilation operator $\hat{a}$ has a right eigenstate. Because $\hat{a}$ is not Hermitian, its eigenvalue will in general be some complex number v, and we label the corresponding eigenstate $|v\rangle$ by the same complex number v. We may therefore write the equation

$$\hat{a}|v\rangle = v|v\rangle. \tag{11.2–1}$$

The conjugate state $\langle v|$ is obviously a left eigenstate of the creation operator $\hat{a}^\dagger$, as can be seen by taking the Hermitian conjugate of Eq. (11.2–1),

$$\langle v|\hat{a}^\dagger = v^*\langle v|. \tag{11.2–2}$$

The state $|v\rangle$ so introduced will be called the coherent state, although the justification for this name will not appear until later. Equation (11.2–1) and its conjugate are entirely formal so far, and we have no proof that non-trivial states satisfying these equations exist. However, it is not difficult to find an explicit representation for $|v\rangle$ in the basis of Fock states $|n\rangle$ labelled by the occupation number n. Since the Fock states form a complete set [see Eq. (10.4–18)], we may use them to represent $|v\rangle$ in the form

$$|v\rangle = \sum_{n=0}^{\infty} c_n|n\rangle, \tag{11.2–3}$$

in which the c_n are complex numbers to be determined. On substituting the expansion (11.2–3) in Eq. (11.2–1), and using Eqs. (10.4–8) and (10.4–11), we obtain

$$\sum_{n=1}^{\infty} c_n\sqrt{n}\,|n-1\rangle = v\sum_{n=0}^{\infty} c_n|n\rangle.$$

As the $|n\rangle$ $(n = 0, 1, 2, \ldots)$ are a set of orthogonal state vectors, this equation is satisfied only if the coefficients of corresponding Fock state vectors on both sides are equal. We therefore have, on equating coefficients of $|n-1\rangle$,

$$c_n = \frac{v}{\sqrt{n}}c_{n-1}. \tag{11.2–4}$$

This is a recursion formula connecting c_n and c_{n-1} from which we obtain by repeated application

$$c_n = \frac{v^2}{[n(n-1)]^{1/2}}c_{n-2} = \ldots = \frac{v^n}{\sqrt{(n!)}}c_0, \tag{11.2–5}$$

so that

$$|v\rangle = c_0\sum_{n=0}^{\infty}\frac{v^n}{\sqrt{(n!)}}|n\rangle. \tag{11.2–6}$$

$|c_0|$ can be determined from the requirement that the state $|v\rangle$ be normalized to unity, which implies

$$\langle v|v\rangle = 1 = |c_0|^2 \sum_{n=0}^{\infty} \sum_{m=0}^{\infty} \frac{v^{*n} v^m}{\sqrt{(n!)}\sqrt{(m!)}} \langle n|m\rangle$$

$$= |c_0|^2 \sum_{n=0}^{\infty} \frac{|v|^{2n}}{n!}$$

$$= |c_0|^2 \, \mathrm{e}^{|v|^2}, \tag{11.2–7}$$

when we make use of the orthogonality (10.4–19) of the Fock states. Hence

$$|c_0| = \mathrm{e}^{-|v|^2/2}, \tag{11.2–8}$$

and, apart from a unimodular factor, we have

$$|v\rangle = \mathrm{e}^{-|v|^2/2} \sum_{n=0}^{\infty} \frac{v^n}{\sqrt{(n!)}} |n\rangle, \tag{11.2–9}$$

and

$$\langle v| = \mathrm{e}^{-|v|^2/2} \sum_{n=0}^{\infty} \frac{v^{*n}}{\sqrt{(n!)}} \langle n|. \tag{11.2–10}$$

It is interesting to note that for every complex number v, other than zero, the coherent state $|v\rangle$ has a non-zero projection on every Fock state $|n\rangle$. Thus

$$\langle n|v\rangle = \mathrm{e}^{-|v|^2/2} \frac{v^n}{\sqrt{(n!)}}. \tag{11.2–11}$$

When $v = 0$, the coherent state $|v\rangle$ becomes the vacuum state $|\text{vac}\rangle$, which may be regarded as either a coherent state or a Fock state. The squared modulus of the projection of $|v\rangle$ onto $|n\rangle$ gives the probability $p(n)$ that n excitations or photons will be found in the coherent state $|v\rangle$. We therefore have

$$p(n) = |\langle n|v\rangle|^2 = \frac{|v|^{2n}}{n!} \mathrm{e}^{-|v|^2}, \tag{11.2–12}$$

which will be recognized as a Poisson distribution in n, with parameter $|v|^2$. The mean number of photons present when the state is a coherent state $|v\rangle$ is therefore given by

$$\sum_{n=0}^{\infty} np(n) = |v|^2 = \langle v|\hat{a}^\dagger \hat{a}|v\rangle, \tag{11.2–13}$$

and this is large or small according as $|v|$ is a large or small number. But, no matter how small $|v|$ may be (with the exception of $v = 0$), there is some non-zero probability $p(n)$ that any number of photons n, however large, is present in the field. In a certain sense the photon number is as random as possible in a coherent state, subject to a certain mean number $|v|^2$. It is a remarkable feature of the combination of Fock states given by Eq. (11.2–9) that the state is unchanged when acted on by the annihilation operator $\hat{a}$. A possible connection between the coherent state of the quantum field and a classical field is suggested by the fact that it is possible to absorb photons from an electromagnetic field in a coherent state repeatedly, without changing the state in any way.

At first sight it might seem that, because the coherent state is not the eigenstate of any observable, it does not correspond to any readily measurable feature of the electromagnetic field. However, this is not so, because in practice most measurements of the field in the optical domain are based on the process of photoelectric detection. This includes instruments such as the photomultiplier, the photoconductor, the photographic plate and the eye. These instruments function by the absorption of photons, and it is therefore the absorption operator which is most closely associated with the measurement of the field. It is partly because the coherent states are eigenstates of the absorption operator, that these states prove to be particularly convenient for the description of many properties of the field encountered in photoelectric measurements.

11.3 The coherent state as a displaced vacuum state – the displacement operator

By combining the expansion (11.2–9) with the representation (10.4–16) of the Fock state $|n\rangle$, we can write

$$\begin{aligned} |v\rangle &= e^{-|v|^2/2} \sum_{n=0}^{\infty} \frac{v^n \hat{a}^{\dagger n}}{n!} |\text{vac}\rangle \\ &= e^{-|v|^2/2} e^{v\hat{a}^\dagger} |\text{vac}\rangle, \end{aligned} \tag{11.3–1}$$

which shows that the coherent state $|v\rangle$ may also be regarded as a displaced vacuum state (Glauber, 1963b).

We can re-express this in a somewhat more symmetric form, by inserting the operator $e^{-v^*\hat{a}}$ between $e^{v\hat{a}^\dagger}$ and $|\text{vac}\rangle$ in Eq. (11.3–1). Thus

$$|v\rangle = e^{-|v|^2/2} e^{v\hat{a}^\dagger} e^{-v^*\hat{a}} |\text{vac}\rangle. \tag{11.3–2}$$

The insertion of $e^{-v^*\hat{a}}$ can be justified by expansion of the exponential operator, which leads to the result

$$\begin{aligned} e^{-v^*\hat{a}} |\text{vac}\rangle &= \left[1 - v^*\hat{a} + \frac{(v^*\hat{a})^2}{2!} - \dots\right] |\text{vac}\rangle \\ &= |\text{vac}\rangle. \end{aligned} \tag{11.3–3}$$

We now make use of the Campbell–Baker–Hausdorff operator identity (see Section 10.11) for two operators $\hat{A}$, $\hat{B}$

$$\exp(\hat{A} + \hat{B}) = \exp\hat{A} \exp\hat{B} \exp(-[\hat{A}, \hat{B}]/2), \tag{11.3–4a}$$

provided that

$$[\hat{A}, [\hat{A}, \hat{B}]] = 0 = [\hat{B}, [\hat{A}, \hat{B}]]. \tag{11.3–4b}$$

The condition (11.3–4b) is obviously satisfied for any pair of operators $\hat{A}$, $\hat{B}$ whose commutator $[\hat{A}, \hat{B}]$ is a c-number. If we now put $\hat{A} = v\hat{a}^\dagger$, $\hat{B} = -v^*\hat{a}$ and make use of the commutation relation Eq. (10.3–9), we obtain $[\hat{A}, \hat{B}] = |v|^2$ and

$$e^{v\hat{a}^\dagger - v^*\hat{a}} = e^{v\hat{a}^\dagger} e^{-v^*\hat{a}} e^{-|v|^2/2}, \qquad (11.3\text{–}5)$$

which allows us to re-write Eq. (11.3–2) in the more compact form

$$|v\rangle = \hat{D}(v)|0\rangle, \qquad (11.3\text{–}6)$$

where

$$\hat{D}(v) \equiv e^{v\hat{a}^\dagger - v^*\hat{a}}. \qquad (11.3\text{–}7)$$

$\hat{D}(v)$ is the *displacement operator* that creates the coherent state $|v\rangle$ from the vacuum state $|0\rangle = |\text{vac}\rangle$. It is clear by inspection of Eq. (11.3–7) that $\hat{D}(v)$ is a unitary operator such that

$$\hat{D}^\dagger(v)\hat{D}(v) = 1 = \hat{D}(v)\hat{D}^\dagger(v), \qquad (11.3\text{–}8)$$

and

$$\hat{D}^\dagger(v) = \hat{D}(-v). \qquad (11.3\text{–}9)$$

11.3.1 Properties of the displacement operator

Because of its general usefulness, we shall now present some of the principal properties of the displacement operator, which were obtained by Glauber (1963b):

(a) Unitary transformation of $\hat{a}$ or $\hat{a}^\dagger$ leads to the complex displacement v or v^*, respectively. Thus

$$\hat{D}^\dagger(v)\hat{a}\hat{D}(v) = \hat{a} + v \qquad (11.3\text{–}10)$$

and

$$\hat{D}^\dagger(v)\hat{a}^\dagger\hat{D}(v) = \hat{a}^\dagger + v^*. \qquad (11.3\text{–}11)$$

These properties follow immediately by use of the operator expansion theorem (10.7–7), when we observe that the commutators

$$\left.\begin{aligned}[-v\hat{a}^\dagger + v^*\hat{a}, \hat{a}] &= v \\ [-v\hat{a}^\dagger + v^*\hat{a}, \hat{a}^\dagger] &= v^*\end{aligned}\right\} \qquad (11.3\text{–}12)$$

reduce to c-numbers. We then have

$$e^{-v\hat{a}^\dagger + v^*\hat{a}}\,\hat{a}\,e^{v\hat{a}^\dagger - v^*\hat{a}} = \hat{a} + v.$$

(b) It follows from Eqs. (11.3–10) and (11.3–11) that, for any function $f(\hat{a}, \hat{a}^\dagger)$ of the operators $\hat{a}$, $\hat{a}^\dagger$ having a power series expansion,

$$\hat{D}^\dagger(v)f(\hat{a}, \hat{a}^\dagger)\hat{D}(v) = f(\hat{a} + v, \hat{a}^\dagger + v^*). \qquad (11.3\text{–}13)$$

To prove this result we make a power series expansion of $f(\hat{a}, \hat{a}^\dagger)$, and introduce the unit operator $\hat{D}(v)\hat{D}^\dagger(v)$ between every pair of neighboring operators. Thus, for a typical element $\hat{a}^2\hat{a}^\dagger$ in the expansion we have

$$\hat{D}^\dagger(v)\hat{a}^2\hat{a}^\dagger\hat{D}(v) = \hat{D}^\dagger(v)\hat{a}\hat{D}(v)\hat{D}^\dagger(v)\hat{a}\hat{D}(v)\hat{D}^\dagger(v)\hat{a}^\dagger\hat{D}(v)$$
$$= (\hat{a} + v)(\hat{a} + v)(\hat{a}^\dagger + v^*).$$

Eq. (11.3–13) then follows immediately.

(c) The product of two displacement operators is another displacement operator, apart from a phase factor, whose total displacement is the sum of the two separate displacements.
To prove this we note that from the Campbell–Baker–Hausdorff identity (11.3–4) we have

$$\begin{aligned}\hat{D}(v)\hat{D}(v') &= \mathrm{e}^{v\hat{a}^\dagger - v^*\hat{a}}\,\mathrm{e}^{v'\hat{a}^\dagger - v'^*\hat{a}} \\ &= \mathrm{e}^{(v+v')\hat{a}^\dagger - (v^*+v'^*)\hat{a}}\,\mathrm{e}^{(vv'^* - v^*v')/2} \\ &= \mathrm{e}^{(vv'^* - v^*v')/2}\hat{D}(v + v'). \qquad (11.3\text{–}14)\end{aligned}$$

Since $vv'^* - v^*v'$ is purely imaginary, the factor in front of $\hat{D}(v + v')$ is simply a phase factor. It follows as an immediate corollary from this result that the effect of the displacement operator $\hat{D}(v')$ acting on the coherent state $|v\rangle$ is to 'displace' the state further to $|v + v'\rangle$. For, from Eq. (11.3–14) we have,

$$\begin{aligned}\hat{D}(v')|v''\rangle &= \hat{D}(v')\hat{D}(v'')|\mathrm{vac}\rangle \\ &= \mathrm{e}^{(v'v''^* - v'^*v'')/2}\hat{D}(v' + v'')|\mathrm{vac}\rangle \\ &= \mathrm{e}^{(v'v''^* - v'^*v'')/2}|v' + v''\rangle. \qquad (11.3\text{–}15)\end{aligned}$$

(d) Two different displacement operators $\hat{D}(v)$ and $\hat{D}(v')$ are orthogonal in the sense that

$$\mathrm{Tr}[\hat{D}(v)\hat{D}^\dagger(v')] = \pi\delta^2(v - v'), \qquad (11.3\text{–}16)$$

where $\delta^2(v)$ is a shorthand notation for $\delta(\mathrm{Re}[v])\,\delta(\mathrm{Im}[v])$.

To prove this we can make use of Eqs. (10.3–5) and (10.3–6) in order to express $\hat{a}$, $\hat{a}^\dagger$ in terms of the operators $\hat{q}$, $\hat{p}$. We then note that

$$\mathrm{Tr}[\hat{D}(v)] = \int \langle q|\exp\left\{\frac{1}{(2\hbar\omega)^{1/2}}[(v - v^*)\omega\hat{q} - \mathrm{i}(v + v^*)\hat{p}]\right\}|q\rangle\,\mathrm{d}q, \qquad (11.3\text{–}17)$$

where $|q\rangle$ is the eigenstate of the $\hat{q}$-operator. With the help of the Campbell–Baker–Hausdorff identity, together with the commutator given by Eq. (10.3–1), this can be re-written as

$$\mathrm{Tr}[\hat{D}(v)] =$$
$$\mathrm{e}^{(v^2 - v^{*2})/4}\int \langle q|\exp\left[\left(\frac{\omega}{2\hbar}\right)^{1/2}(v - v^*)\hat{q}\right]\exp\left[\frac{-\mathrm{i}}{(2\hbar\omega)^{1/2}}(v + v^*)\hat{p}\right]|q\rangle\,\mathrm{d}q.$$

After recalling that $\mathrm{e}^{\mathrm{i}x\hat{p}/\hbar}$ is the displacement operator for the $|q\rangle$ state, so that $\mathrm{e}^{\mathrm{i}x\hat{p}/\hbar}|q\rangle = |q + x\rangle$, we obtain

$$\begin{aligned}
\operatorname{Tr}[\hat{D}(v)] &= \mathrm{e}^{(v^2-v^{*2})/4}\int \exp\left[\left(\frac{\omega}{2\hbar}\right)^{1/2}(v-v^*)q\right]\langle q|q-(v+v^*)\left(\frac{\hbar}{2\omega}\right)^{1/2}\rangle\mathrm{d}q \\
&= \mathrm{e}^{(v^2-v^{*2})/4}\left(\frac{\omega}{2\hbar}\right)^{1/2}\delta(\operatorname{Re}[v])\int \exp\left[\left(\frac{\omega}{2\hbar}\right)^{1/2}(v-v^*)q\right]\mathrm{d}q \\
&= \mathrm{e}^{(v^2-v^{*2})/4}\left(\frac{\omega}{2\hbar}\right)^{1/2}\delta(\operatorname{Re}[v])2\pi\left(\frac{\hbar}{2\omega}\right)^{1/2}\delta(\operatorname{Im}[v]) \\
&= \pi\delta^2(v).
\end{aligned} \tag{11.3–18}$$

Equation (11.3–16) then follows immediately from this with the help of the product rule (11.3–14). The same result can be proved in a more elegant way with the help of the completeness property for coherent states to be derived in Section 11.6.

11.4 *q*-representation of the coherent state

To find the q-representation, also known as the coordinate representation or the Schrödinger wave function, $\psi_v(q)$ of the coherent state $|v\rangle$, we have to evaluate the matrix element $\langle q|v\rangle$, where $|q\rangle$ is the eigenstate of the $\hat{q}$ operator. From Eq. (11.2–1) we have

$$\langle q|\hat{a}|v\rangle = v\langle q|v\rangle \equiv v\psi_v(q). \tag{11.4–1}$$

After substituting for $\hat{a}$ in terms of the $\hat{q}$ and $\hat{p}$ operators from Eq. (10.3–5), and making use of the differential form of $\hat{p}$, we have from Eq. (11.4–1)

$$\frac{1}{(2\hbar\omega)^{1/2}}\left[q\omega+\hbar\frac{\partial}{\partial q}\right]\psi_v(q) = v\psi_v(q), \tag{11.4–2}$$

which is a first-order differential equation for the wave function $\psi_v(q)$. The general solution may be written

$$\psi_v(q) = A\exp\left\{\frac{-\omega}{2\hbar}\left[q-\left(\frac{2\hbar}{\omega}\right)^{1/2}v\right]^2\right\}, \tag{11.4–3}$$

where A is a normalization constant to be chosen so that

$$\int|\psi_v(q)|^2\,\mathrm{d}q = 1. \tag{11.4–4}$$

It will be seen that $\psi_v(q)$ has the formal structure of a Gaussian function of q, in which the peak of the Gaussian is displaced by the complex distance $(2\hbar/\omega)^{1/2}v$ from the origin. Since $v = 0$ yields the vacuum state, we see once again that the coherent state appears as a displacement from the vacuum state. The normalization condition (11.4–4) yields

$$|A| = \left(\frac{\omega}{\pi\hbar}\right)^{1/4}\mathrm{e}^{(\operatorname{Im} v)^2},$$

so that, up to a phase factor,

$$\psi_v(q) = \left(\frac{\omega}{\pi\hbar}\right)^{1/4} e^{(\mathrm{Im}\,v)^2} \exp\left\{\frac{-\omega}{2\hbar}\left[q - \left(\frac{2\hbar}{\omega}\right)^{1/2} v\right]^2\right\}. \tag{11.4–5}$$

We shall not have much occasion to use this representation of the coherent state.

11.5 Time evolution and uncertainty products

In the Schrödinger picture any state $|\psi(t)\rangle$ evolves in time according to the rule

$$|\psi(t)\rangle = \exp(-\mathrm{i}\hat{H}t/\hbar)|\psi(0)\rangle,$$

where $\exp(-\mathrm{i}\hat{H}t/\hbar)$ is the time evolution operator. When $|\psi(0)\rangle$ is an eigenstate of $\hat{H}$, the time evolution operator can be replaced by a simple phase factor, and the state does not change. In general, however, a state evolves into different states as time passes. Let us consider the case in which the initial state $|\psi(0)\rangle$ is the coherent state $|v\rangle$. As the coherent state is not an eigenstate of $\hat{H}$, it is expected to evolve into other states in time. When we make use of the explicit form of the Hamiltonian $\hat{H}$ given by Eq. (10.3–15), and confine our attention to just one mode of the field, as before, so that $\hat{H} = \hbar\omega(\hat{n} + \frac{1}{2})$, we have

$$|\psi(t)\rangle = e^{-\mathrm{i}\omega t/2}\, e^{-\mathrm{i}\omega t\hat{n}}|v\rangle, \tag{11.5–1}$$

where ω is the frequency of the mode. The effect of the time evolution is most easily calculated by making use of the Fock representation (11.2–9) for $|v\rangle$, and we find that

$$\begin{aligned}
|\psi(t)\rangle &= e^{-\mathrm{i}\omega t/2}\, e^{-|v|^2/2} \sum_{n=0}^{\infty} \frac{v^n}{\sqrt{(n!)}}\, e^{-\mathrm{i}\omega t\hat{n}}|n\rangle \\
&= e^{-\mathrm{i}\omega t/2}\, e^{-|v|^2/2} \sum_{n=0}^{\infty} \frac{(v\,e^{-\mathrm{i}\omega t})^n}{\sqrt{(n!)}}|n\rangle \\
&= e^{-\mathrm{i}\omega t/2}|v\,e^{-\mathrm{i}\omega t}\rangle.
\end{aligned} \tag{11.5–2}$$

Apart from the phase factor $e^{-\mathrm{i}\omega t/2}$, this is just another coherent state belonging to the complex eigenvalue $v\,e^{-\mathrm{i}\omega t}$. Thus the coherent state evolves into other coherent states continuously and periodically in time, such that the cycle is repeated at time intervals $2\pi/\omega$.

Let us now consider the time dependence of the expectation values of some elementary field operators, when the field is in a coherent state. Here it is generally a little more convenient to work in the Heisenberg picture. Thus, with the help of Eqs. (10.3–12) and the defining relation (11.2–1), which we take to hold at time $t = 0$, we have

$$\langle v|\hat{a}(t)|v\rangle = \langle v|\hat{a}(0)|v\rangle\, e^{-\mathrm{i}\omega t} = v\,e^{-\mathrm{i}\omega t}. \tag{11.5–3}$$

This result could, of course, have been obtained by inspection of Eq. (11.5–2). Similarly, we find for the expectation value of the creation operator $\hat{a}^\dagger(t)$,

$$\langle v|\hat{a}^\dagger(t)|v\rangle = \langle v|\hat{a}^\dagger(0)|v\rangle\, e^{\mathrm{i}\omega t} = v^*\, e^{\mathrm{i}\omega t}. \tag{11.5–4}$$

By combining these two results, and making use of Eqs. (10.3–7) and (10.3–8) for the canonically conjugate operators $\hat{q}(t)$ and $\hat{p}(t)$, we can also write down the expectation values of $\hat{q}(t)$ and $\hat{p}(t)$ in a coherent state. It is worth noting that these operators are the analogs of the canonically conjugate variables $\hat{A}(t)$ and $-\varepsilon_0\hat{E}(t)$ for our single-mode electromagnetic field. We immediately find that

$$\langle v|\hat{q}(t)|v\rangle = \left(\frac{\hbar}{2\omega}\right)^{1/2}[v\,\mathrm{e}^{-\mathrm{i}\omega t} + v^*\,\mathrm{e}^{\mathrm{i}\omega t}] = \left(\frac{2\hbar}{\omega}\right)^{1/2}|v|\cos(\omega t - \arg v), \tag{11.5–5}$$

and

$$\langle v|\hat{p}(t)|v\rangle = \mathrm{i}\left(\frac{\hbar\omega}{2}\right)^{1/2}[-v\,\mathrm{e}^{-\mathrm{i}\omega t} + v^*\,\mathrm{e}^{\mathrm{i}\omega t}] = -(2\hbar\omega)^{1/2}|v|\sin(\omega t - \arg v). \tag{11.5–6}$$

These results are reminiscent of the motion of a classical harmonic oscillator of frequency ω, having a well-defined complex amplitude v. Moreover, the dispersions $\langle v|(\Delta\hat{a}(t))^2|v\rangle$ and $\langle v|(\Delta\hat{a}^\dagger(t))^2|v\rangle$ of $\hat{a}(t)$ and $\hat{a}^\dagger(t)$ vanish in a coherent state, by virtue of the relations (11.2–1) and (11.2–2). It therefore seems that, when the field is in a coherent state $|v\rangle$, it behaves rather like a classical wave of definite amplitude and phase, and that $\hat{a}(t)$ is just the complex amplitude operator.

11.5.1 Canonical uncertainty product

However, the classical analogy must not be pushed too far, for our quantum oscillator in a coherent state has certain features that are entirely unlike any classical ones. If the complex amplitude of a classical oscillator has the definite value $v\,\mathrm{e}^{-\mathrm{i}\omega t}$ at any time t, then its real amplitude is equally well-defined by $|v|\cos(\omega t - \arg v)$. But, whereas the complex amplitude $\hat{a}(t)$ of a quantum oscillator in a coherent state $|v\rangle$ has a definite value $v\,\mathrm{e}^{-\mathrm{i}\omega t}$, the same is not true of the real amplitude $\hat{q}(t) = (\hbar/2\omega)^{1/2}(\hat{a}(t) + \hat{a}^\dagger(t))$, for the state $|v\rangle$ is not an eigenstate of $\hat{q}(t)$. When we make use of the commutation rule (10.3–9), we readily find from Eq. (10.3–7) that

$$\begin{aligned}\hat{q}^2(t) &= \frac{\hbar}{2\omega}[\hat{a}^2(t) + \hat{a}^{\dagger 2}(t) + \hat{a}(t)\hat{a}^\dagger(t) + \hat{a}^\dagger(t)\hat{a}(t)]\\ &= \frac{\hbar}{2\omega}[\hat{a}^2(t) + \hat{a}^{\dagger 2}(t) + 2\hat{a}^\dagger(t)\hat{a}(t) + 1],\end{aligned} \tag{11.5–7}$$

so that

$$\langle v|\hat{q}^2(t)|v\rangle = \frac{\hbar}{2\omega}[v^2\,\mathrm{e}^{-2\mathrm{i}\omega t} + v^{*2}\,\mathrm{e}^{2\mathrm{i}\omega t} + 2v^*v + 1]. \tag{11.5–8}$$

From Eqs. (11.5–8) and (11.5–5) we therefore have for the dispersion of $\hat{q}(t)$,

$$\begin{aligned}\langle v|(\Delta\hat{q}(t))^2|v\rangle &= \langle v|\hat{q}^2(t)|v\rangle - \langle v|\hat{q}(t)|v\rangle^2\\ &= \frac{\hbar}{2\omega}.\end{aligned} \tag{11.5–9}$$

Similarly we may show for the canonically conjugate variable $\hat{p}(t) = \mathrm{i}(\hbar\omega/2)^{1/2}[\hat{a}^\dagger(t) - \hat{a}(t)]$ that

$$\langle v|(\Delta\hat{p}(t))^2|v\rangle = \frac{\hbar\omega}{2}. \tag{11.5–10}$$

Hence the real (Hermitian) and imaginary (antiHermitian) parts of the complex amplitude $\hat{a}(t)$ do not have well-defined values in a coherent state, even though the complex amplitude itself is well defined. However, the Hermitian canonical variables $\hat{q}(t)$ and $\hat{p}(t)$, which correspond to the vector potential and the electric field in our single-mode problem, are as well defined as quantum mechanics allows, because the product of the uncertainties is given by

$$\langle v|(\Delta\hat{q}(t))^2|v\rangle^{1/2}\langle v|(\Delta\hat{p}(t))^2|v\rangle^{1/2} = \tfrac{1}{2}\hbar, \tag{11.5–11}$$

which is the lowest possible value. The real field therefore behaves as nearly like a classical field as is possible when the state is a coherent state. It is worth noting that, although the dispersions of both the canonical variables are non-zero, they are independent of the eigenvalue v that identifies the coherent state. Whether $\langle(\Delta\hat{q}(t))^2\rangle$ is appreciable or not compared with $\langle\hat{q}(t)\rangle^2$ evidently depends on the magnitude $|v|$, as inspection of Eqs. (11.5–5) and (11.5–9) shows. The departure from classical behavior is therefore unimportant when $|v| \gg 1$, but is quite significant when $|v| \lesssim 1$, especially for the vacuum state.

It is worth emphasizing that these properties of coherent states, namely that the states correspond to classical states of well-defined complex amplitude, and that the real amplitude is as well defined as possible, hold irrespective of the mean number of photons present. The coherent states should not be associated with the classical limit $n \to \infty$ of the electromagnetic field, although, as we shall see in Section 11.12, they are associated with classical current sources. The field produced by a single-mode laser operating far above its threshold also comes close to being in a coherent state, no matter how strong an attenuator may be placed in front of the source.

11.5.2 More general states of minimum uncertainty product

It is not difficult to show that the coherent states are members of a wider class of states having the property that the product of the dispersions of $\hat{q}$ and $\hat{p}$ is a minimum. In order to demonstrate this we introduce the unitary operator (Stoler, 1970, 1971; Yuen, 1976)

$$\hat{U}(\theta) = \mathrm{e}^{\theta(\hat{a}^2 - \hat{a}^{\dagger 2})/2}, \tag{11.5–12}$$

in which θ is some real number, and use it to operate on the coherent state $|v\rangle$. The resulting state

$$|v, \theta\rangle \equiv \hat{U}(\theta)|v\rangle \tag{11.5–13}$$

is also a state of minimum uncertainty product.

Let us calculate the moments of

$$\hat{q} \equiv \left(\frac{\hbar}{2\omega}\right)^{1/2}(\hat{a} + \hat{a}^\dagger)$$

in the state $|v, \theta\rangle$. From Eq. (11.5–13) we have

$$\langle v, \theta|\hat{a}|v, \theta\rangle = \langle v|\hat{U}^\dagger(\theta)\hat{a}\hat{U}(\theta)|v\rangle. \tag{11.5–14}$$

Now by use of the operator expansion theorem we find that

$$\begin{aligned}\hat{U}^\dagger(\theta)\hat{a}\hat{U}(\theta) &= e^{-\theta(\hat{a}^2-\hat{a}^{\dagger 2})/2}\hat{a}\, e^{\theta(\hat{a}^2-\hat{a}^{\dagger 2})/2} \\ &= \hat{a}\left(1 + \frac{\theta^2}{2!} + \frac{\theta^4}{4!} + \ldots\right) - \hat{a}^\dagger\left(\theta + \frac{\theta^3}{3!} + \frac{\theta^5}{5!} + \ldots\right) \\ &= \hat{a}\cosh\theta - \hat{a}^\dagger\sinh\theta,\end{aligned} \tag{11.5–15}$$

and after the conjugate is added, we obtain

$$\begin{aligned}\hat{U}^\dagger(\theta)(\hat{a} + \hat{a}^\dagger)\hat{U}(\theta) &= (\hat{a} + \hat{a}^\dagger)(\cosh\theta - \sinh\theta) \\ &= (\hat{a} + \hat{a}^\dagger)e^{-\theta}.\end{aligned} \tag{11.5–16}$$

Hence

$$\begin{aligned}\langle v, \theta|\hat{q}|v, \theta\rangle &= \left(\frac{\hbar}{2\omega}\right)^{1/2}\langle v|\hat{U}^\dagger(\theta)(\hat{a} + \hat{a}^\dagger)\hat{U}(\theta)|v\rangle \\ &= e^{-\theta}\langle v|\hat{q}|v\rangle.\end{aligned} \tag{11.5–17}$$

Similarly we have for the second moment from Eq. (11.5–16)

$$\begin{aligned}\langle v, \theta|\hat{q}^2|v, \theta\rangle &= \frac{\hbar}{2\omega}\langle v|\hat{U}^\dagger(\theta)(\hat{a} + \hat{a}^\dagger)\hat{U}(\theta)\hat{U}^\dagger(\theta)(\hat{a} + \hat{a}^\dagger)\hat{U}(\theta)|v\rangle \\ &= e^{-2\theta}\langle v|\hat{q}^2|v\rangle,\end{aligned} \tag{11.5–18}$$

so that

$$\langle v, \theta|(\Delta\hat{q})^2|v, \theta\rangle = e^{-2\theta}\langle v|(\Delta\hat{q})^2|v\rangle. \tag{11.5–19}$$

In a similar manner we may show for the canonical conjugate $\hat{p} = i(\hbar\omega/2)^{1/2}(\hat{a}^\dagger - \hat{a})$ that

$$\langle v, \theta|(\Delta\hat{p})^2|v, \theta\rangle = e^{2\theta}\langle v|(\Delta\hat{p})^2|v\rangle. \tag{11.5–20}$$

It follows from Eqs. (11.5–19) and (11.5–20) that the uncertainty product is given by

$$\begin{aligned}\langle v, \theta|(\Delta\hat{q})^2|v, \theta\rangle^{1/2}\langle v, \theta|(\Delta\hat{p})^2|v, \theta\rangle^{1/2} &= \langle v|(\Delta\hat{q})^2|v\rangle^{1/2}\langle v|(\Delta\hat{p})^2|v\rangle^{1/2} \\ &= \tfrac{1}{2}\hbar.\end{aligned} \tag{11.5–21}$$

Hence the states $|v, \theta\rangle$ are also states of minimum uncertainty product. However, their canonical dispersions are not independent of the state, as for coherent states.

As equations (11.5–19) and (11.5–20) indicate, the dispersion of either $\hat{q}$ or $\hat{p}$ can be made arbitrarily small by suitable choice of θ, at the cost of a corresponding increase in the dispersion of the other canonical variable. Such states are examples of what are called 'squeezed states' or 'two-photon coherent states', which are discussed in more detail in Chapter 21.

11.6 Coherent states as a basis; non-orthogonality and over-completeness

Up to now we have examined some of the properties of the coherent states by representing them in terms of other, more familiar states. But one of the main reasons for the importance of coherent states in quantum optics is that they themselves form a basis for the representation of arbitrary quantum states. Because the coherent states are eigenstates of a non-Hermitian operator, the coherent-state representation has some unusual features that we shall now examine.

Let us consider the scalar product of two coherent states $|v_1\rangle$ and $|v_2\rangle$. With the help of the Fock expansions (11.2–9) and (11.2–10) we have

$$\begin{aligned}
\langle v_2|v_1\rangle &= \mathrm{e}^{-|v_1|^2/2}\,\mathrm{e}^{-|v_2|^2/2}\sum_n\sum_m \frac{v_2^{*n} v_1^m}{\sqrt{(n!)}\sqrt{(m!)}}\langle n|m\rangle \\
&= \mathrm{e}^{-|v_1|^2/2}\,\mathrm{e}^{-|v_2|^2/2}\sum_n \frac{(v_2^* v_1)^n}{n!} \\
&= \mathrm{e}^{-(|v_1|^2+|v_2|^2-2v_2^* v_1)/2} \\
&= \mathrm{e}^{-|v_1-v_2|^2/2}\,\mathrm{e}^{(v_2^* v_1 - v_2 v_1^*)/2},
\end{aligned} \tag{11.6–1}$$

where we have written the answer in the form of a product of a real number $\mathrm{e}^{-|v_1-v_2|^2/2}$ and a unimodular phase factor. Evidently $\mathrm{e}^{-|v_1-v_2|^2/2}$ is the modulus of $\langle v_2|v_1\rangle$, and we have

$$|\langle v_2|v_1\rangle|^2 = \mathrm{e}^{-|v_1-v_2|^2}. \tag{11.6–2}$$

As there are no values of v_1 and v_2 for which this term vanishes, it follows that no two coherent states are ever orthogonal, in contrast to the more usual situation with eigenstates of a Hermitian operator. When $v_1 = v_2$, the right side of Eq. (11.6–1) becomes unity, as required by the normalization condition.

Although the scalar product of two coherent states $|v_1\rangle$ and $|v_2\rangle$ never vanishes, it can become very small for even quite modest differences between the complex numbers v_1 and v_2. As an example, let us consider the two coherent states with eigenvalues $v_1 = 23 + 20\mathrm{i}$ and $v_2 = 21 + 23\mathrm{i}$. From the point of view of the photon statistics, the two states $|v_1\rangle$ and $|v_2\rangle$ are quite close, for the mean photon numbers and the low-order moments of the photon number are rather similar. Yet the magnitude of the scalar product $|\langle v_2|v_1\rangle| \approx 10^{-3}$, so that the states are almost orthogonal. This is a reflection of the fact that, for large v_1 and v_2, the function $\mathrm{e}^{-|v_1-v_2|^2/2}$ takes on some of the character of the $\delta^2(v_1 - v_2)$ function.

Despite their non-orthogonality, the coherent states span the whole Hilbert space of state vectors and form a convenient basis for the representation of other states. In order to show this, we shall make a resolution of the identity operator 1 in terms of coherent state projectors. Consider the integral

$$\frac{1}{\pi}\int |v\rangle\langle v|\,\mathrm{d}^2 v,$$

taken over the entire complex v-plane. We use the symbol $\mathrm{d}^2 v$ as an abbreviation for d(Re v) d(Im v). By writing $v = r\,\mathrm{e}^{\mathrm{i}\theta}$, so that $\mathrm{d}^2v = r\,\mathrm{d}r\,\mathrm{d}\theta$, and making use of the expansions (11.2–9) and (11.2–10) we have

$$\frac{1}{\pi}\int |v\rangle\langle v|\,\mathrm{d}^2v = \frac{1}{\pi}\int_0^\infty \mathrm{d}r \int_0^{2\pi}\mathrm{d}\theta \sum_{n=0}^{\infty}\sum_{m=0}^{\infty} \mathrm{e}^{-r^2}\frac{r^{n+m+1}}{\sqrt{(n!)}\sqrt{(m!)}}\mathrm{e}^{\mathrm{i}(n-m)\theta}|n\rangle\langle m|. \tag{11.6–3}$$

If we now formally interchange the orders of summation and integration,‡ and carry out the integration over θ, we obtain a factor $2\pi\delta_{nm}$, which reduces the double summation to a single one. The equation therefore simplifies to

$$\begin{aligned}\frac{1}{\pi}\int |v\rangle\langle v|\,\mathrm{d}^2v &= \sum_{n=0}^{\infty}\frac{1}{n!}|n\rangle\langle n|\int_0^\infty 2\,\mathrm{e}^{-r^2}r^{2n+1}\,\mathrm{d}r \\ &= \sum_{n=0}^{\infty}|n\rangle\langle n| \\ &= 1,\end{aligned} \tag{11.6–4}$$

in view of the completeness of the set of Fock states. We have therefore shown that the coherent states also satisfy a completeness relation, so that they form a basis for the representation of other states.

Thus, if we have an arbitrary state $|\psi\rangle$, and multiply it on the left by the unit operator in the form given in Eq. (11.6–4), we obtain

$$|\psi\rangle = \frac{1}{\pi}\int |v\rangle\langle v|\psi\rangle\,\mathrm{d}^2v. \tag{11.6–5}$$

This is an expansion of $|\psi\rangle$ in terms of coherent states $|v\rangle$ with amplitudes $(1/\pi)\langle v|\psi\rangle$. In the following we shall usually take it for granted that, when the limits are not specified, the integral is to be taken over the entire complex v-plane.

11.6.1 Linear dependence of coherent states

The peculiarity inherent in the use of a non-orthogonal, but complete (actually over-complete) set of states as a basis becomes evident if we identify $|\psi\rangle$ in Eq. (11.6–5) with one of the coherent states, say $|v'\rangle$. We then have

$$\begin{aligned}|v'\rangle &= \frac{1}{\pi}\int |v\rangle\langle v|v'\rangle\,\mathrm{d}^2v \\ &= \frac{1}{\pi}\int |v\rangle\,\mathrm{e}^{-|v-v'|^2/2}\,\mathrm{e}^{(v^*v'-vv'^*)/2}\,\mathrm{d}^2v,\end{aligned} \tag{11.6–6}$$

with the help of Eq. (11.6–1). We have therefore obtained an expansion of one of the coherent states in terms of all of them. Such an expansion would of course be impossible, in principle, with a complete orthogonal set of states.

‡ For a more rigorous proof and a more detailed discussion, see Klauder (1960) and Klauder and Sudarshan (1968) Chap. 8.

It is clear from Eq. (11.6–6) that the different coherent states are not linearly independent. Moreover, we may formally use the equation to make a resolution of zero in the form

$$\int |v\rangle \left[\frac{1}{\pi} \mathrm{e}^{-|v-v'|^2/2} \, \mathrm{e}^{(v^*v'-vv'^*)/2} - \delta^2(v-v') \right] \mathrm{d}^2 v = 0, \qquad (11.6\text{–}7)$$

and since v' is an arbitrary complex number, there exists an infinity of different integral representations of zero in terms of coherent states. Any one of these integrals can be added to the right-hand side of Eq. (11.6–5) without changing its validity. It is evident, therefore, that an expansion of the type (11.6–5) in terms of coherent states is not unique. Moreover, Eq. (11.6–7) is not the only possible resolution of zero. Consider, for example, the integral $\int v^n |v\rangle \, \mathrm{d}^2 v$, where $n = 1, 2, 3$, etc. By expanding the coherent state $|v\rangle$ in terms of Fock states, changing to polar coordinates $v = r\,\mathrm{e}^{\mathrm{i}\theta}$, and carrying out the integration over θ, we see at once that the integral vanishes. It follows that, for an arbitrary function $f(|v|)$ for which the integral exists,

$$\int f(|v|) v^n |v\rangle \, \mathrm{d}^2 v = 0, \quad \text{for } n = 1, 2, 3, \ldots \qquad (11.6\text{–}8)$$

which again illustrates the linear dependence of the coherent states.

11.6.2 Over-completeness

The set of coherent states is usually said to be *over-complete*, in the sense that the states form a basis and yet are expressible in terms of each other. It should be emphasized, however, that the set of states cannot be made exactly complete by subtraction of a countable number of coherent states from it (Cahill, 1965). Nor would such a subtraction be particularly desirable, for it is just the over-completeness which gives the coherent-state representation some of its most interesting and important features, as we shall see.

The non-orthogonality and over-completeness of the set of coherent states, and the non-uniqueness of the representation, are sometimes a source of bewilderment when they are first encountered, and a simple geometrical analogy may be helpful. Consider the problem of representing the position $\mathbf{r}$ of a point P in a plane (see Fig. 11.1). One way of proceeding is to set up a pair of orthogonal axes as shown in Fig. 11.1(a), and to resolve $\mathbf{r}$ in terms of unit vectors $\mathbf{e}_1$ and $\mathbf{e}_2$ in the directions of the two axes. We then have

$$\mathbf{r} = c_1\mathbf{e}_1 + c_2\mathbf{e}_2,$$

where the pair of numbers (c_1, c_2) is representative of the point P. This representation is based on the complete orthogonal, linearly independent set of vectors $\mathbf{e}_1$, $\mathbf{e}_2$, and it is unique.

Alternatively, we may use a coordinate system as shown in Fig. 11.1(b), in which the axes are not orthogonal. Once again we have a representation based on a pair of numbers (c_1, c_2) such that

$$\mathbf{r} = c_1\mathbf{e}_1 + c_2\mathbf{e}_2$$

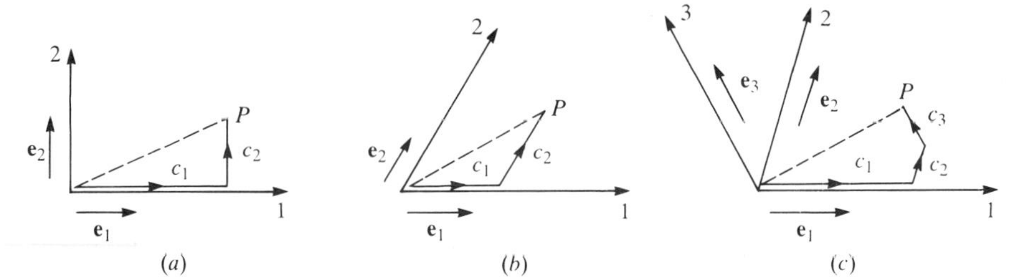

Fig. 11.1 Examples of the representation of the position of a point P in a plane by components of the position vector. (a) The basis is complete and orthogonal. (b) The basis is complete and non-orthogonal. (c) The basis is over-complete and non-orthogonal.

where $\mathbf{e}_1$, $\mathbf{e}_2$ are unit vectors in the directions of the axes. This time the set of base vectors is non-orthogonal, but the vectors are still linearly independent and the set is still complete, for any vector can be represented in this way. Moreover, the representation is again unique.

Alternatively, we may choose to use the coordinate system with three axes shown in Fig. 11.1(c). This time the position of the point P is represented by the triplet of numbers (c_1, c_2, c_3) such that

$$\mathbf{r} = c_1\mathbf{e}_1 + c_2\mathbf{e}_2 + c_3\mathbf{e}_3,$$

where $\mathbf{e}_1$, $\mathbf{e}_2$, $\mathbf{e}_3$ are unit vectors in the directions of the three axes, respectively. This representation is based on a non-orthogonal and over-complete set of basis vectors, any one of which may be represented in terms of some combination of the others. The base vectors $\mathbf{e}_1$, $\mathbf{e}_2$, $\mathbf{e}_3$ are no longer linearly independent, for the condition

$$l_1\mathbf{e}_1 + l_2\mathbf{e}_2 + l_3\mathbf{e}_3 = 0$$

does not imply $l_1 = l_2 = l_3 = 0$. Moreover, as a result of the linear dependence and the over-completeness, the representation of P by the triplet of numbers (c_1, c_2, c_3) is no longer unique, as is obvious by inspection of the figure. Of course, in this example the basis is finite, and the over-completeness can be removed by removal of one of the base vectors. The situation is not quite so simple with the (doubly) infinite basis formed by the set of coherent states.

In dealing with a non-orthogonal set of states, we also have to observe some caution in the interpretation of projections, or scalar products, like $\langle v|\psi\rangle$, between some state $|\psi\rangle$ and a coherent state $|v\rangle$. Let us recall that the corresponding scalar product $\langle n|\psi\rangle$, between the state $|\psi\rangle$ and a Fock state $|n\rangle$, is usually referred to as the probability amplitude of $|\psi\rangle$ in the n-representation, on the grounds that $|\langle n|\psi\rangle|^2$ gives the probability of finding n photons. As the Fock states are orthogonal, the probabilities $|\langle n|\psi\rangle|^2$ are mutually exclusive for different n, and they are normalized to unity,

$$\sum_{n=0}^{\infty} |\langle n|\psi\rangle|^2 = \sum_{n=0}^{\infty} \langle \psi|n\rangle\langle n|\psi\rangle = \langle \psi|\psi\rangle = 1.$$

On the other hand, the non-orthogonality of the coherent states $|v\rangle$ is reflected in the fact that the squared scalar products $|\langle v|\psi\rangle|^2$ do not represent mutually exclusive probabilities (or rather probability densities) and do not integrate to unity. Instead we have

$$\int |\langle v|\psi\rangle|^2 \,\mathrm{d}^2 v = \langle\psi|\int |v\rangle\langle v|\,\mathrm{d}^2 v|\psi\rangle = \pi, \qquad (11.6\text{–}9)$$

by virtue of Eq. (11.6–4). If $|\langle v|\psi\rangle|^2$ is interpreted as the probability density of finding the complex amplitude v, then the differential probabilities are evidently not mutually exclusive, and the right-hand side of Eq. (11.6–9) gives some idea of the degree of overlap. However, as we have seen, the non-orthogonality is significant mainly for neighboring coherent states; two states belonging to two appreciably different eigenvalues are almost orthogonal. In the same way, the non-exclusiveness of the probability densities $|\langle v|\psi\rangle|^2$ is associated mainly with neighboring coherent states, and if v_1 and v_2 are appreciably different, $|\langle v_1|\psi\rangle|^2\,\mathrm{d}^2v/\pi$ and $|\langle v_2|\psi\rangle|^2\,\mathrm{d}^2v/\pi$ play the role of almost mutually exclusive differential probabilities.

11.6.3 Representation of operators in terms of coherent states

The completeness relation (11.6–4) can also be used to represent operators in terms of coherent states. If we have an arbitrary operator $\hat{A}$ and formally multiply it by the unit operator given by Eq. (11.6–4) both on the right and on the left, we obtain

$$\hat{A} = \frac{1}{\pi^2}\iint \langle v'|\hat{A}|v''\rangle |v'\rangle\langle v''|\,\mathrm{d}^2v'\,\mathrm{d}^2v'', \qquad (11.6\text{–}10)$$

which is an integral representation of $\hat{A}$ in terms of coherent states. For the same reasons as before, this representation is not unique, for we may add an arbitrary integral of the form

$$\iint f(|v'|)g(|v''|)v'^{n}v''^{*m}|v'\rangle\langle v''|\,\mathrm{d}^2v'\,\mathrm{d}^2v'',$$

$$n, m = 1, 2, 3, \ldots,$$

to the right-hand side of Eq. (11.6–10) without changing its validity.

11.6.4 Evaluation of matrix elements of operators in normal order

The problem of evaluating the matrix elements $\langle v'|\hat{A}|v''\rangle$ appearing in Eq. (11.6–10) becomes particularly simple for operators for which the coherent states are right or left eigenstates. For example, by virtue of Eqs. (11.2–1) and (11.2–2), we have immediately

$$\langle v'|\hat{a}|v''\rangle = v''\langle v'|v''\rangle = v''\,\mathrm{e}^{-|v'-v''|^2/2}\,\mathrm{e}^{(v'^*v''-v'v''^*)/2}, \qquad (11.6\text{–}11)$$

$$\langle v'|\hat{a}^\dagger|v''\rangle = v'^*\langle v'|v''\rangle = v'^*\,\mathrm{e}^{-|v'-v''|^2/2}\,\mathrm{e}^{(v'^*v''-v'v''^*)/2}, \qquad (11.6\text{–}12)$$

and

$$\langle v'|\hat{n}|v''\rangle = \langle v'|\hat{a}^\dagger\hat{a}|v''\rangle = v'^*v''\,\mathrm{e}^{-|v'-v''|^2/2}\,\mathrm{e}^{(v'^*v''-v'v''^*)/2}. \qquad (11.6\text{–}13)$$

It is clear from these results that the evaluation of the matrix elements is very straightforward whenever the operator $\hat{A}$ is expressible as a sum of products of creation and annihilation operators, with the creation operators always standing to the left of the annihilation operators. An operator arranged in this way is said to be in *normal order*. Examples of normally ordered operators are $\hat{n}$, $\hat{a}^{\dagger 2}\hat{a}^3$, $\hat{n} + 2\hat{a}^{\dagger 2}$, $\hat{a}^\dagger \hat{n}$, $\sum_{n=0}^{\infty} \hat{a}^{\dagger n}\hat{a}^n x^n/n!$. We shall find that normally ordered operators appear naturally in the description of photoelectric measurements of the electromagnetic field and are therefore of particular importance in quantum optics. If $f^{(\mathrm{N})}(\hat{a}, \hat{a}^\dagger)$ is any function of the $\hat{a}$, $\hat{a}^\dagger$ operators which is in normal order and is expandable as a power series, we evidently have, by application of the same arguments as in the derivation of Eqs. (11.6–11) to (11.6–13),

$$\langle v'|f^{(\mathrm{N})}(\hat{a}, \hat{a}^\dagger)|v''\rangle = f^{(\mathrm{N})}(v'', v'^*)\,\mathrm{e}^{-|v'-v''|^2/2}\,\mathrm{e}^{(v'^*v''-v'v''^*)/2}. \qquad (11.6\text{–}14)$$

In particular, if $v' = v''$, we obtain

$$\langle v|f^{(\mathrm{N})}(\hat{a}, \hat{a}^\dagger)|v\rangle = f^{(\mathrm{N})}(v, v^*), \qquad (11.6\text{–}15)$$

so that the expectation value of a normally ordered operator in a coherent state is obtained by replacing all creation and annihilation operators by their left and right eigenvalues, respectively.

11.7 Representation of states and operators by entire functions

We have seen that the coherent-state expansion of an arbitrary state $|\psi\rangle$ in the form (11.6–5) is not unique, and the same objection applies to the expansion of some operator $\hat{A}$ in the form (11.6–10). But it is not difficult to arrive at a unique representation in terms of coherent states.

Consider the matrix element $\langle v|\psi\rangle$, which is representative of the state $|\psi\rangle$. By making use of the Fock expansion (11.2–10) of $\langle v|$, we can write

$$\langle v|\psi\rangle = \mathrm{e}^{-|v|^2/2}\sum_{n=0}^{\infty}\frac{v^{*n}}{\sqrt{(n!)}}\langle n|\psi\rangle \equiv \mathrm{e}^{-|v|^2/2}F_\psi(v^*). \qquad (11.7\text{–}1)$$

As $|\langle n|\psi\rangle| \leqslant 1$, the series is absolutely convergent for all $|\psi\rangle$ and all v^*. Now the function $F_\psi(v^*)$ is itself a representation of the state $|\psi\rangle$, and may be regarded as the element of a Hilbert space. Moreover, $F_\psi(v^*)$ is an entire analytic function of v^*. We have therefore arrived at a simple representation of the state $|\psi\rangle$ in terms of entire functions that is unique. It is closely related to representations that were introduced by Bargmann (1961, 1962).

We immediately encounter one of the remarkable features of the coherent-state representation. In general, a state $|\psi\rangle$ is determined by its representation with respect to all elements of the basis. For example, in the Fock representation, $|\psi\rangle$ is determined only when $\langle n|\psi\rangle$ is given for all integer values of n from 0 to ∞, and in the q-representation it is determined by $\langle q|\psi\rangle$ for all values of q from $-\infty$ to ∞. But an entire function is determined by its values within some finite range, no matter how small. It follows that, in the coherent-state representation, $|\psi\rangle$ *is completely determined by its matrix elements* $\langle v|\psi\rangle$ *within some arbitrarily small but finite range of* v^*. We see at once that no state can be orthogonal to a group of coherent states lying within some small range.‡ For if the entire analytic function $\langle v|\psi\rangle\,e^{|v|^2/2}$ vanishes for some v^* it must vanish for all v^*, and this is possible only for a null vector $|\psi\rangle$. Needless to say, this result is partly a reflection of the over-completeness of the set of coherent states. The power of the representation, and the virtue of the over-completeness, now begin to be evident.

Let us now consider the corresponding problem of representing a traceable, positive definite Hermitian operator $\hat{A}$ in terms of coherent states. Once again we make use of the Fock expansions (11.2–9) and (11.2–10) to write

$$\begin{aligned}\langle v|\hat{A}|v'\rangle &= e^{-|v|^2/2}\,e^{-|v'|^2/2}\sum_{n=0}\sum_{m=0}\frac{v^{*n}}{\sqrt{(n!)}}\frac{v'^m}{\sqrt{(m!)}}\langle n|\hat{A}|m\rangle \\ &\equiv e^{-(|v|^2+|v'|^2)/2}F_A(v^*, v'). \qquad (11.7\text{–}2)\end{aligned}$$

Now for a Hermitian, traceable, non-negative definite operator $\hat{A}$, $|\langle n|\hat{A}|m\rangle|$ has an upper bound. Perhaps the easiest way to see this is to expand $\hat{A}$ in terms of its own eigenstates $|\lambda_i\rangle$, where λ_i is the associated eigenvalue which is real and non-negative. Then we have from the expansion

$$\hat{A} = \sum_i \lambda_i|\lambda_i\rangle\langle\lambda_i|,$$

$$\langle n|\hat{A}|m\rangle = \sum_i \lambda_i\langle n|\lambda_i\rangle\langle\lambda_i|m\rangle,$$

so that

$$|\langle n|\hat{A}|m\rangle| \leqslant \sum_i \lambda_i|\langle n|\lambda_i\rangle|\,|\langle\lambda_i|m\rangle| \leqslant \sum_i \lambda_i = \mathrm{Tr}\,\hat{A}.$$

It follows that the double series in Eq. (11.7–2) is absolutely convergent for all v^* and all v', so that $F_A(v^*, v')$ is an entire analytic function of the two variables v^*, v' and, once again, we may regard the function $F_A(v^*, v')$ as a unique representation of the operator $\hat{A}$.

Because of the analyticity of $F_A(v^*, v')$, it is sufficient for the function to be known over some arbitrarily small range of v^* and v' for it to be known everywhere, and for it to determine the operator $\hat{A}$. In particular, it is sufficient for the diagonal matrix element $\langle v|\hat{A}|v\rangle$ to be known for some v^*, v, for the entire operator $\hat{A}$ to be determined (Jordan, 1964). In order to make a representation of an operator and study its properties, we may therefore confine our attention to the diagonal matrix elements alone. For, let $\hat{A}$ and $\hat{B}$ be two operators for which the diagonal matrix elements $\langle v|\hat{A}|v\rangle$ and $\langle v|\hat{B}|v\rangle$ coincide for some range of v. Then, for this range

$$\langle v|\hat{A} - \hat{B}|v\rangle = 0, \qquad (11.7\text{–}3)$$

and since, according to Eq. (11.7–2), $\langle v|\hat{A} - \hat{B}|v'\rangle\,e^{(|v|^2+|v'|^2)/2}$ is an entire function of v^* and v', it must vanish for all v^*, v' if it vanishes for some range of v^*, v'. We therefore conclude that

$$\hat{A} = \hat{B}, \qquad (11.7\text{–}4)$$

‡ It is of course possible for a state $|\psi\rangle$ that is not a coherent state to be orthogonal to any one coherent state $|v\rangle$. For example if $|\psi\rangle = (1/v^*)|1\rangle - (\sqrt{2}/v^{*2})\langle 2|$, then $\langle v|\psi\rangle = 0$.

which shows that *the operators are determined by their diagonal matrix elements alone*, within some arbitrarily small but finite range of v. This is a remarkable conclusion, which once again emphasizes the power of the coherent-state representation. The result has no counterpart in any representation based on a complete set of states.

It is not difficult to derive the matrix elements of the operator $\hat{A}$ in some other representation explicitly from $\langle v|\hat{A}|v\rangle$. For example, if we formally treat v, v^* as independent variables, we have from Eq. (11.7–2)

$$\frac{1}{\sqrt{(n!)}\sqrt{(m!)}}\left[\frac{\partial^{n+m}\langle v|\hat{A}|v\rangle\, \mathrm{e}^{|v|^2}}{\partial v^{*n}\partial v^{m}}\right]_{\substack{v^*=0\\ v=0}} = \langle n|\hat{A}|m\rangle. \tag{11.7–5}$$

The diagonal matrix element $\langle v|\hat{A}|v\rangle$ is therefore sufficient to generate all the matrix elements $\langle n|\hat{A}|m\rangle$, and, moreover, it is sufficient for $\langle v|\hat{A}|v\rangle$ to be known merely in the neighborhood of $v = 0$. Once $\langle n|\hat{A}|m\rangle$ has been determined, we can substitute it in Eq. (11.7–2) and derive any off-diagonal matrix element $\langle v'|\hat{A}|v''\rangle$.

11.8 Diagonal coherent-state representation of the density operator (Glauber–Sudarshan *P*-representation)

The previous Section 11.7 suggests that it may be possible to make a representation of a traceable operator, such as the density operator $\hat{\rho}$ describing the state of the system, entirely in terms of coherent-state projectors $|v\rangle\langle v|$, rather than in the lengthier form given by Eq. (11.6–10). Thus we write

$$\hat{\rho} = \int \phi(v)|v\rangle\langle v|\, \mathrm{d}^2 v, \tag{11.8–1}$$

in which $\phi(v)$ is some real function of v. Because $\mathrm{d}^2 v \equiv \mathrm{d}(\mathrm{Re}\, v)\,\mathrm{d}(\mathrm{Im}\, v)$, this is still a double integral representation, but it involves only projection operators and is substantially simpler than the four-fold integral in Eq. (11.6–10). A formal derivation of Eq. (11.8–1) will be given in Sections 11.8.4 and 11.10 below.

Such a representation was first introduced independently by Sudarshan (1963)[‡] and by Glauber (1963c, 1970), and it was later reformulated in a mathematically rigorous form by Klauder and co-workers (Klauder, McKenna and Currie, 1965; Klauder, 1966; see also Rocca, 1966; Miller and Mishkin, 1967; Klauder and Sudarshan, 1968). The surface element $\mathrm{d}^2 v = \mathrm{d}(\mathrm{Re}\, v)\,\mathrm{d}(\mathrm{Im}\, v)$ is sometimes referred to as an element of phase space. The integral in Eq. (11.8–1) is then an integral over all phase space, in which $\phi(v)$ plays the role of a phase space density or weight function. Equation (11.8–1) is also known as the *P-representation* of the state $\hat{\rho}$, and the weight function is then denoted by $P(v)$. The representation suggests that the state of the electromagnetic field may be regarded as a mixture

‡ Sudarshan (1963) expressed $\phi(v)$ in terms of the matrix elements $\langle n|\hat{\rho}|m\rangle$ of $\hat{\rho}$ in the Fock representation in the form of an infinite series of highly singular functions,

$$\phi(v) = \sum_{n=0}^{\infty}\sum_{m=0}^{\infty}\langle n|\hat{\rho}|m\rangle \frac{\sqrt{(n!)}\sqrt{(m!)}}{(n+m)!}\frac{1}{2\pi r}\,\mathrm{e}^{r^2-\mathrm{i}\theta(n-m)}\left(-\frac{\partial}{\partial r}\right)^{n+m}\delta(r),$$

where $v = r\,\mathrm{e}^{\mathrm{i}\theta}$, and the delta function is normalized so that $\int_0^\infty \delta(r)\,\mathrm{d}r = 1$.

of coherent states, with relative weight $P(v)$ or $\phi(v)$.‡ Because $\hat{\rho}$ is Hermitian, $\phi(v)$ is real. The requirement that $\hat{\rho}$ have unit trace implies that $\phi(v)$ is normalized to unity, for

$$1 = \mathrm{Tr}\,\hat{\rho} = \mathrm{Tr}\int \phi(v)|v\rangle\langle v|\,\mathrm{d}^2v = \int \phi(v)\,\mathrm{d}^2v. \qquad (11.8\text{–}2)$$

Both these properties are in accord with the intuitive notion that $\phi(v)$ plays the role of a probability density in phase space.

We define a classical state of light to be one in which $\phi(v)$ is a probability density. However, the interpretation of $\phi(v)$ is more complicated in general. In the first place, the different coherent states are not orthogonal, so that even if $\phi(v)$ behaved like a true probability density, it would not describe probabilities of mutually exclusive states. In fact, there exist states of the field, called non-classical states, for which $\phi(v)$ is less well behaved than a probability density, which has to be non-negative and cannot be more singular than a delta function. It is not difficult to find examples in which $\phi(v)$ takes on negative values and becomes more singular than a delta function, as we shall see below. As an example of a phase space density that is not always a probability density let us consider the Wigner distribution (Wigner, 1932).

11.8.1 Quasi-probability densities; the Wigner distribution

In 1932 Wigner introduced a distribution function $W(q, p)$, now known as the Wigner distribution, to characterize the state $|\psi\rangle$ of a quantum system in phase space. In one dimension, if $\hat{\rho}$ is the density operator of the system and $|q\rangle$ and $|p\rangle$ denote a position and a momentum state, respectively, then the Wigner distribution is defined by

$$W(q, p) \equiv \frac{1}{2\pi}\int_{-\infty}^{\infty} \langle q + \tfrac{1}{2}\hbar x|\hat{\rho}|q - \tfrac{1}{2}\hbar x\rangle\,\mathrm{e}^{\mathrm{i}xp}\,\mathrm{d}x. \qquad (11.8\text{–}3\mathrm{a})$$

In the special case in which the state is the pure state $|\psi\rangle$, $\hat{\rho} = |\psi\rangle\langle\psi|$ and $\psi(q) = \langle q|\psi\rangle$ is the Schrödinger wave function, so that

$$\langle q + \tfrac{1}{2}\hbar x|\hat{\rho}|q - \tfrac{1}{2}\hbar x\rangle = \psi^*(q - \tfrac{1}{2}\hbar x)\psi(q + \tfrac{1}{2}\hbar x). \qquad (11.8\text{–}3\mathrm{b})$$

$W(q, p)$ so defined is real and normalized to unity. It can be shown that the expectation of a symmetric or Weyl-ordered function $F^{(\mathrm{w})}(\hat{q}, \hat{p})$ of $\hat{q}$, $\hat{p}$ is given by§

$$\langle F^{(\mathrm{w})}(\hat{q}, \hat{p})\rangle = \int F^{(\mathrm{w})}(q, p)W(q, p)\,\mathrm{d}q\,\mathrm{d}p. \qquad (11.8\text{–}4)$$

The right-hand side of this equation has the same structure as a classical ensemble average, in which q, p are random variables with probability density or phase space density $W(q, p)$.

Despite the formal similarity between $W(q, p)$ and a joint probability density,

‡ The symbol $P(v)$, which was first used by Glauber (1963c), is the origin of the name P-representation. However, because P is already used so widely for other quantities, we shall use $\phi(v)$ instead.

§ A Weyl-ordered function of $\hat{q}$, $\hat{p}$ is one that is symmetric with respect to the order of operators. Examples of Weyl-ordered functions of $\hat{q}$, $\hat{p}$ are $\hat{q}\hat{p} + \hat{p}\hat{q}$, $\hat{q}^2\hat{p} + \hat{q}\hat{p}\hat{q} + \hat{p}\hat{q}^2$.

$W(q, p)$ does not have all the characteristics of a probability density, and it may become negative. It is known as a *quasi-probability density*. Of course, $W(q, p)$ does not correspond to any directly measurable quantity, because the joint probability of a pair of canonically conjugate variables cannot be measured, and indeed has no meaning in quantum mechanics. However, the marginal integral

$$\begin{aligned}\int W(q, p)\,\mathrm{d}p &= \frac{1}{2\pi}\int \mathrm{d}p \int \mathrm{d}x\, \psi^*(q - \tfrac{1}{2}\hbar x)\psi(q + \tfrac{1}{2}\hbar x)\,\mathrm{e}^{\mathrm{i}xp} \\ &= \int \psi^*(q - \tfrac{1}{2}\hbar x)\psi(q + \tfrac{1}{2}\hbar x)\delta(x)\,\mathrm{d}x \\ &= \psi^*(q)\psi(q)\end{aligned}$$

is the true probability density of q. Similarly, one may show that

$$\int W(q, p)\,\mathrm{d}q = \tilde{\psi}^*(p)\tilde{\psi}(p),$$

is the probability density of p, where $\tilde{\psi}(p)$ is the wave function in p-space. The Wigner distribution therefore has some, but not all, of the attributes of a probability density. This is a reflection of the fact that quantum mechanics admits states having no classical counterpart.

Similarly, the phase space density $\phi(v)$, that we encounter in the diagonal representation of the density operator $\hat{\rho}$, is a quasi-probability density. Because there are no measurements of the field that yield the weight function $\phi(v)$ directly, $\phi(v)$ need not have all the characteristics of a probability density. As we shall see, for certain states of the field the phase space function $\phi(v)$ can be much less well behaved and much more singular than the Wigner distribution. However, $\phi(v)$ does have one very valuable advantage over all other phase space densities; it allows us to see at a glance whether the state of the field has a classical description or not.

11.8.2 Two virtues of the diagonal representation

As the coherent state $|v\rangle$ is analogous to a classical field of complex amplitude v, it follows that when $\phi(v)$ is a true probability density, then Eq. (11.8–1) represents an ensemble of coherent states and therefore corresponds to an ensemble of classical fields with probability density $\phi(v)$. On the other hand, when $\phi(v)$ is not a probability density, because it is negative for some v or more singular than a delta function, then there is no classical ensemble that corresponds to the quantum state represented by Eq. (11.8–1). We are then dealing with a state which is purely quantum mechanical and without a classical analog. Nevertheless, integrals involving the function $\phi(v)$ may be perfectly meaningful and well defined. We shall see that the light emitted by lasers and many familiar thermal sources has a classical analog, because in those cases $\phi(v)$ is a probability density.

The fact that $\phi(v)$ is non-negative for a classical state allows us to construct some simple tests for recognizing a non-classical state. For example, if $f(v)$ is some real positive function of v, then

$$\int \phi(v) f(v)\, d^2 v$$

can vanish only if $\phi(v)$ represents a non-classical state. Thus, if $p(n)$ is the probability for n photons in the quantum state $\hat{\rho}$, then

$$p(n) = \mathrm{Tr}\,(\hat{\rho}|n\rangle\langle n|),$$

and with the help of Eqs. (11.8–1) and (11.2–12) we find

$$p(n) = \int \phi(v)\, e^{-|v|^2} \frac{|v|^{2n}}{n!}\, d^2 v. \tag{11.8–5}$$

Because $e^{-|v|^2}|v|^{2n}/n! > 0$ when $v \neq 0$, it follows that $p(n)$ cannot be zero for any n when $\phi(v)$ is a probability density, with the exception of the vacuum state for which $\phi(v) = \delta^2(v)$ and $p(0) = 1$. Except for the vacuum state, any state for which $p(n) = 0$ has no classical analog and is purely quantum mechanical. This is a reflection of the fact that the coherent state $|v\rangle$ contains contributions from Fock states with all possible occupation numbers.

A second important feature of the representation (11.8–1) is that it allows us to calculate expectation values of normally ordered operators (in which the creation operators always stand to the left of the annihilation operators) very simply, in a way which is completely analogous to the way an average is calculated in classical probability theory (see Section 11.9).

11.8.3 Diagonal representation of $\hat{\rho}$ by a sequence of functions

As was shown by Klauder and co-workers (Klauder, McKenna and Currie, 1965; Klauder, 1966; see also Rocca, 1966), it is possible to avoid the use of singular functions $\phi(v)$ by introducing instead a sequence of density operators $\hat{\rho}_1, \hat{\rho}_2, \ldots$ together with the sequence of phase space functions $\phi_1(v), \phi_2(v), \ldots$ such that

$$\hat{\rho}_N = \int \phi_N(v)|v\rangle\langle v|\, d^2 v, \quad N = 1, 2, \ldots.$$

Here $\phi_1(v), \phi_2(v), \ldots$ may be a sequence of well-behaved functions, or a sequence of tempered distributions, or some other sequence. For each of the $\hat{\rho}_1$, $\hat{\rho}_2, \ldots$ it is possible to calculate $\mathrm{Tr}\,(\hat{\rho}_N \hat{O})$, which is the expectation value of the operator $\hat{O}$ in the state $\hat{\rho}_N$. Then the sequence $\phi_1(v), \phi_2(v), \ldots$ converges to a phase space representation of $\hat{\rho}$ if

$$\lim_{N\to\infty} \mathrm{Tr}\,(\hat{\rho}_N \hat{O}) = \mathrm{Tr}\,(\hat{\rho}\hat{O}) \tag{11.8–6}$$

for any operator $\hat{O}$ whose expectation value exists. The character of the sequence $\phi_1(v), \phi_2(v), \ldots$ determines the character of the convergence of $\hat{\rho}_N$ to $\hat{\rho}$, and the range of operators allowed. In particular, it was shown by Klauder (1966) that there exists a sequence of infinitely differentiable functions $\phi_1(v), \phi_2(v), \ldots$ falling off at infinity faster than any inverse power, such that $\hat{\rho}_N$ converges to $\hat{\rho}$ in the trace-class norm,‡ and that the foregoing equation is satisfied for any bounded operator $\hat{O}$.

At first sight, the restriction to bounded operators appears to be a severe one, which limits the usefulness of the representation in general. However, by making a representation of the operator in terms of its own eigenstates, we may calculate the expectation value of an unbounded operator $\hat{O}$ also. For example, suppose that $\hat{O}$ may be represented in the form

$$\hat{O} = \sum_n \beta_n |\beta_n\rangle\langle\beta_n|, \tag{11.8–7a}$$

where β_n, $|\beta_n\rangle$, $n = 1, 2, \ldots$, is a complete set of eigenvalues and eigenstates of $\hat{O}$. Now the projection operators $|\beta_n\rangle\langle\beta_n|$ are all bounded by unity, and their expectation values $p_n \equiv \mathrm{Tr}\,(\hat{\rho}|\beta_n\rangle\langle\beta_n|)$ may therefore be calculated with the help of the diagonal representation. We can then use the relation

$$\langle\hat{O}\rangle = \mathrm{Tr}\,(\hat{\rho}\hat{O}) = \sum_n \beta_n p_n \tag{11.8–7b}$$

for an operator $\hat{O}$ that is not necessarily bounded, but whose expectation value exists. This procedure avoids the introduction of singular functions, but it does not have the virtue of indicating at a glance the presence of a non-classical state. In the following we shall not use the representation based on sequences.

We now present several procedures for determining the phase space density $\phi(v)$ in the diagonal coherent-state representation (11.8–1). We shall see that singular functions are sometimes encountered, but that does not prevent relationships like that in Eq. (11.8–5) from being useful in practice.

11.8.4 Diagonal representation of the density operator in anti-normal order

Suppose that the density operator $\hat{\rho}$ is expressible as a function of the annihilation and creation operators of the form

$$\hat{\rho} = \rho^{(\mathrm{A})}(\hat{a}, \hat{a}^\dagger) = \sum_{n,m} C_{nm}\hat{a}^n\hat{a}^{\dagger m}, \tag{11.8–8}$$

in which the annihilation operators in each term stand to the left of the creation operators. This order of the operators is known as the *anti-normal order*, and is indicated by the superscript A in $\rho^{(\mathrm{A})}(\hat{a}, \hat{a}^\dagger)$. In order to make a diagonal coherent state representation of $\hat{\rho}$ we merely insert the unit operator in the form (11.6–4) between the annihilation and creation operators in each term in Eq. (11.8–8). We then find, with the help of Eqs. (11.2–1) and (11.2–2), on interchanging orders of summation and integration,

$$\begin{aligned}\hat{\rho} = \rho^{(\mathrm{A})}(\hat{a}, \hat{a}^\dagger) &= \sum_{n,m} C_{nm}\frac{1}{\pi}\int \hat{a}^n|v\rangle\langle v|\hat{a}^{\dagger m}\,\mathrm{d}^2v \\ &= \int \frac{1}{\pi}\sum_{n,m} C_{nm}v^n v^{*m}|v\rangle\langle v|\,\mathrm{d}^2v\end{aligned}$$

‡ For a discussion of the convergence, including the definition of the trace-class norm, see Klauder and Sudarshan (1968), Sec. 8.4.

$$= \int \frac{1}{\pi} \rho^{(A)}(v, v^*) |v\rangle\langle v| \, d^2v. \tag{11.8–9}$$

We have therefore succeeded in making a diagonal representation of $\hat{\rho}$, in which the phase space function $\phi(v) = (1/\pi)\rho^{(A)}(v, v^*)$, and this can be regarded as a formal derivation of the representation (11.8–1). The same procedure can be applied to other operators, although the integral on the right of Eq. (11.8–9) may not always converge.

11.8.5 *Integral representation of* $\phi(v)$

An alternative procedure for obtaining $\phi(v)$ is to express $\phi(v)$ as a Fourier integral in terms of a characteristic function (Mehta and Sudarshan, 1965). The following approach is due to Mehta (1967). Let us suppose that $\hat{\rho}$ can be written in the diagonal form (11.8–1), and let us multiply both sides of the equation by the coherent state $\langle -u|$ on the left and by the coherent state $|u\rangle$ on the right, where u is some complex number. We then obtain

$$\langle -u|\hat{\rho}|u\rangle = \int \phi(v) \langle -u|v\rangle\langle v|u\rangle \, d^2v,$$

and with the help of Eq. (11.6–1) for the scalar product of two coherent states, this becomes

$$\langle -u|\hat{\rho}|u\rangle = e^{-|u|^2} \int \phi(v) \, e^{-|v|^2} \, e^{uv^* - u^*v} \, d^2v. \tag{11.8–10}$$

We now observe that Eq. (11.8–10) expresses $\langle -u|\hat{\rho}|u\rangle \, e^{|u|^2}$ as the Fourier transform of $\phi(v) \, e^{-|v|^2}$, because the kernel $e^{uv^* - u^*v}$ is just the two-dimensional Fourier kernel. This is perhaps more obvious if we write $u = p + iq$, $v = P + iQ$, so that $e^{uv^* - u^*v} = e^{2i(qP - pQ)}$. If we now formally invert the Fourier integral in Eq. (11.8–10), without paying too much attention to questions of convergence, we obtain

$$\phi(v) \, e^{-|v|^2} = \frac{1}{\pi^2} \int e^{|u|^2} \langle -u|\hat{\rho}|u\rangle \, e^{-uv^* + u^*v} \, d^2u. \tag{11.8–11}$$

We have therefore formally solved the problem of determining the phase space function $\phi(v)$ for a given density operator $\hat{\rho}$. However, the possibility of encountering mathematical convergence difficulties is at once apparent from inspection of Eq. (11.8–11), when we see the factor $e^{|u|^2}$ in the integrand. Unless the matrix element $\langle -u|\hat{\rho}|u\rangle$ decreases with u for large u more rapidly than $e^{-|u|^2}$, the integral will not exist in the ordinary sense of function theory, and can only be interpreted as a generalized function. Indeed, it is not difficult to find examples of states for which it does not exist. However, the integral converges for all optical fields for which a classical analog exists, except for a possible δ-function singularity, because then $\phi(v)$ is a probability density. Another integral representation of $\phi(v)$ will be given in the next section (Eq. (11.9–12)).

When $\phi(v)$ does not exist as an ordinary function, we may look on it as a generalized function defined by Eq. (11.8–11). It is straightforward to show that

$\langle -u|\hat{\rho}|u\rangle$ exists and is bounded by unity. For we can always express the density operator as a linear combination of projection operators in the form

$$\hat{\rho} = \sum_r p_r|\psi_r\rangle\langle\psi_r|,$$

with p_r real and non-negative, and $\sum_r p_r = 1$, so that

$$|\langle -u|\hat{\rho}|u\rangle| \leqslant \sum_r p_r|\langle -u|\psi_r\rangle\langle\psi_r|u\rangle| \leqslant \sum_r p_r = 1. \tag{11.8–12}$$

Moreover, $\langle -u|\hat{\rho}|u\rangle$, and therefore $e^{|u|^2}\langle -u|\hat{\rho}|u\rangle$, may be differentiated any number of times with respect to u and $-u^*$. To illustrate the integral procedure for determining the phase space density $\phi(v)$ we now consider a few examples.

11.8.6 Examples of $\phi(v)$

(a) Coherent state

Suppose that the field is in a coherent state $|v'\rangle$, so that $\hat{\rho} = |v'\rangle\langle v'|$. Then from Eq. (11.6–1), the matrix element

$$\begin{aligned}\langle -u|\hat{\rho}|u\rangle &= \langle -u|v'\rangle\langle v'|u\rangle \\ &= e^{-|u|^2-|v'|^2}\, e^{uv'^*-u^*v'},\end{aligned}$$

and when this expression is used in Eq. (11.8–11), we obtain

$$\begin{aligned}\phi(v)\, e^{-|v|^2} &= \frac{1}{\pi^2}\int e^{-|v'|^2}\, e^{u(v'^*-v^*)-u^*(v'-v)}\, d^2u \\ &= e^{-|v'|^2}\delta^2(v-v'),\end{aligned}$$

so that

$$\phi(v) = \delta^2(v-v'). \tag{11.8–13}$$

The phase space function for the coherent state is therefore a two-dimensional delta function, and we have

$$|v'\rangle\langle v'| = \int \delta^2(v-v')|v\rangle\langle v|\, d^2v.$$

This result could, of course, have been written down directly, but the calculation illustrates the procedure. In this case the function $\phi(v)$ has the character of a classical probability density. Indeed, a classical oscillator of complex amplitude v' would be assigned the same phase space function in a classical integral representation.

(b) The randomly phased laser model

As we shall show later (in Chapter 18) the optical field of a laser that is oscillating in a single mode far above threshold is in a state that is close in many respects to the coherent state $|v\rangle$. The intensity of the field is almost free from fluctuations and is held constant by the intrinsic saturation properties of the laser. On the other hand, the phase of the oscillation may be subject to fluctuations and, in general, drifts randomly in time. In a succession of measurements performed on the same laser, the state of the field appears to be a mixture of coherent states $|v = r_0\,\mathrm{e}^{\mathrm{i}\theta}\rangle$ having the same value of $|v|$, but the phases θ of the complex amplitude v are distributed uniformly over the range 0 to 2π. The following phase space function $\phi(v)$ is sometimes taken to be a better approximation to the laser field than the pure coherent state [cf. Eq. (18.5–16) below]:

$$\phi(v) = \frac{1}{2\pi r_0}\delta(|v| - r_0). \tag{11.8–14}$$

Since θ does not appear explicitly in this expression, θ is distributed uniformly over the whole range. The factor $1/2\pi r_0$ ensures that the probability density of the phase is given by

$$\int \phi(v)|v|\,\mathrm{d}|v| = \frac{1}{2\pi},$$

as required. This phase space function is identical with the probability distribution that would be assigned to a randomly phased mixture of classical oscillators. It is worth noting that in the state given by Eq. (11.8–14), unlike the pure coherent state of the field, the expectation values of the annihilation and creation operators $\hat{a}$ and $\hat{a}^\dagger$ vanish, for

$$\begin{aligned}\langle\hat{a}\rangle &= \int v\frac{1}{2\pi|v|}\delta(|v| - r_0)\,\mathrm{d}^2v \\ &= \frac{1}{2\pi}\int_0^{2\pi} r_0\,\mathrm{e}^{\mathrm{i}\theta}\,\mathrm{d}\theta \\ &= 0,\end{aligned}$$

and similarly for the expectation of $\hat{a}^\dagger$.

It is interesting to compare the probability distributions of the real part of the complex amplitude v in the coherent state $|v'\rangle$ and in the mixed state given by Eq. (11.8–14).‡ If we write $v = x + \mathrm{i}y$, the probability distribution of x, which we denote by $p(x)$, is given by

$$p(x) = \int \phi(x, y)\,\mathrm{d}y.$$

For the coherent state $|v'\rangle$ with $v' = x' + \mathrm{i}y'$ and $\phi(v)$ given by Eq. (11.8–13), we have

$$\begin{aligned}p(x) &= \int_{-\infty}^{\infty} \delta(x - x')\delta(y - y')\,\mathrm{d}y \\ &= \delta(x - x'),\end{aligned} \tag{11.8–15}$$

so that x has a well-defined value x'. However, the situation is quite different for

‡ Note that this is not the same as the distribution of the real – or Hermitian – part of the dynamical variable $\hat{a}$, which was discussed in Section 11.5.

the mixture represented by Eq. (11.8–14). In that case we find

$$\begin{aligned} p(x) &= \frac{1}{2\pi}\int_{-\infty}^{\infty} \frac{\delta[(x^2+y^2)^{1/2}-r_0]}{(x^2+y^2)^{1/2}}\,\mathrm{d}y \\ &\left.\begin{aligned} &= \frac{1}{\pi(r_0^2-x^2)^{1/2}}, \quad \text{for } |x|<r_0, \\ &= 0, \quad\quad\quad\quad\quad\quad \text{for } |x|>r_0. \end{aligned}\right\} \end{aligned} \tag{11.8–16}$$

We have made use of the general property of the delta function that

$$\delta[f(y)] = \sum_i \delta(y-y_i)\bigg/\left|\frac{\mathrm{d}f}{\mathrm{d}y}\right|_{y=y_i}, \tag{11.8–17}$$

where y_i, $i = 1, 2, \ldots$, are the zeros of the function $f(y)$. The function $p(x)$ given by Eq. (11.8–16) is the curve (a) illustrated in Fig. 11.2. It has the form expected for the classical probability distribution of $\mathrm{Re}\,(r_0\,\mathrm{e}^{\mathrm{i}\theta})$ when θ is a random phase angle, and corresponds approximately to the distribution of the real amplitude of a single-mode laser.

(c) *Thermal light*

We shall show in Section 13.1 that for blackbody radiation, and more generally for light in the form of a plane wave derived from any thermal source, each mode of the field has the phase space distribution

$$\phi(v) = \frac{1}{\pi\langle n\rangle}\,\mathrm{e}^{-|v|^2/\langle n\rangle}, \tag{11.8–18}$$

where $\langle n\rangle$ is the mean number of photons in that mode. This is a Gaussian distribution in the complex variable v, and is seen to be identical in form with the corresponding probability distribution in the classical description of thermal light. The distribution $p(x)$ of the real part of v follows immediately from Eq. (11.8–18) when we integrate over $\mathrm{Im}\,(v)$, and we obtain

$$p(x) = \frac{1}{(\pi\langle n\rangle)^{1/2}}\,\mathrm{e}^{-x^2/\langle n\rangle}. \tag{11.8–19}$$

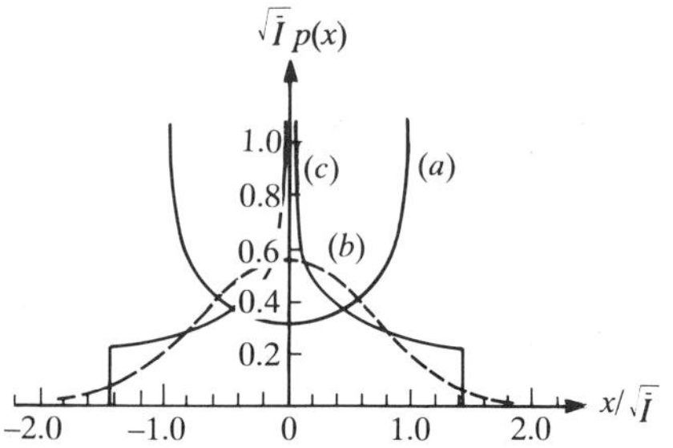

Fig. 11.2 The expected form of the probability distribution of $x = \mathrm{Re}\,v$, corresponding to (a) the randomly phased single-mode laser [Eq. (11.8–16)], (b) thermal light [Eq. (11.8–19)], (c) the superposition of two single-mode lasers [Eq. (11.8–29)]. (Reproduced from Mandel, 1965.)

For comparison, this function is also shown in Fig. 11.2 [curve (*b*)]. These examples all show that when the state of the field has a classical analog, $\phi(v)$ coincides with the corresponding classical probability density.

(d) The Fock state

As a less trivial example we now consider the diagonal coherent state representation of the density operator for a field that is in a Fock state $|n\rangle$, in which case $\hat{\rho} = |n\rangle\langle n|$. The matrix element $\langle -u|\hat{\rho}|u\rangle$ is readily evaluated with the help of Eq. (11.2–11), and we find that

$$\begin{aligned}\langle -u|\hat{\rho}|u\rangle &= \langle -u|n\rangle\langle n|u\rangle \\ &= \frac{\mathrm{e}^{-|u|^2}(-|u|^2)^n}{n!}.\end{aligned}$$

When this result is used in Eq. (11.8–11) we obtain

$$\phi(v)\,\mathrm{e}^{-|v|^2} = \frac{1}{n!}\frac{1}{\pi^2}\int(-|u|^2)^n\,\mathrm{e}^{-uv^*+u^*v}\,\mathrm{d}^2u,$$

which is a divergent integral that does not exist as an ordinary function. Moreover, the integral is more singular than a delta function when $n \neq 0$. However if we disregard the singular behavior, and proceed in a completely formal way, we may differentiate under the integral and write

$$\begin{aligned}\phi(v) &= \frac{\mathrm{e}^{|v|^2}}{n!}\frac{\partial^{2n}}{\partial v^{*n}\partial v^n}\frac{1}{\pi^2}\int\mathrm{e}^{-uv^*+u^*v}\,\mathrm{d}^2u \\ &= \frac{\mathrm{e}^{|v|^2}}{n!}\frac{\partial^{2n}}{\partial v^{*n}\partial v^n}\delta^2(v).\end{aligned} \tag{11.8–20}$$

Hence $\phi(v)$ is expressible as the $2n$'th derivative of a delta function, and it is much more singular than any classical probability density; it falls into the class of tempered (or temperate) distributions (Bremerman, 1965, Sec. 8.8; Nussenzveig, 1972, Appendix A.10). This is a reflection of the fact that the Fock state $|n\rangle$ is a quantum state of the field having no classical counterpart. As we have already pointed out, in general $\phi(v)$ is not a probability density for states that have no classical analogue. Of course, in the special case in which the Fock state is the vacuum state, $n = 0$ and Eq. (11.8–20) yields

$$\phi(v) = \delta^2(v),$$

in agreement with Eq. (11.8–13), because the vacuum is also an example of a coherent state.

(e) Superposition of two coherent states

Next we consider an example of a still more singular phase space density $\phi(v)$. Let the state $|\psi\rangle$ be a linear superposition of coherent states $|v'\rangle$ and $|v''\rangle$,

$$|\psi\rangle = c'|v'\rangle + c''|v''\rangle,$$

where the complex numbers c', c'' are appropriately chosen to ensure the normalization of $|\psi\rangle$. Although each coherent state $|v'\rangle$ or $|v''\rangle$ separately can be given a classical interpretation in terms of an optical field with a certain complex amplitude v' or v'', the superposition state $|\psi\rangle$ has no classical interpretation. It does not represent the field obtained by the physical superposition or interference of two optical fields that we discuss below, but represents a characteristically quantum mechanical state. The density operator $\hat{\rho} = |\psi\rangle\langle\psi|$, and the matrix element is given by

$$\begin{aligned}\langle -u|\hat{\rho}|u\rangle &= (c'\langle -u|v'\rangle + c''\langle -u|v''\rangle)(c'^*\langle v'|u\rangle + c''^*\langle v''|u\rangle)\\ &= |c'|^2 \mathrm{e}^{-(|u|^2+|v'|^2)}\,\mathrm{e}^{uv'^*-u^*v'} + |c''|^2 \mathrm{e}^{-(|u|^2+|v''|^2)}\,\mathrm{e}^{uv''^*-u^*v''}\\ &\quad + \mathrm{e}^{-(|u|^2+|v'|^2/2+|v''|^2/2)}[c'c''^*\,\mathrm{e}^{uv''^*-u^*v'} + c''c'^*\,\mathrm{e}^{uv'^*-u^*v''}].\end{aligned}$$

We therefore have from Eq. (11.8–11) for the phase space density

$$\begin{aligned}\phi(v) = \frac{\mathrm{e}^{|v|^2}}{\pi^2}\int\Big[&|c'|^2\,\mathrm{e}^{-|v'|^2}\,\mathrm{e}^{u(v'^*-v^*)-u^*(v'-v)} + |c''|^2\,\mathrm{e}^{-|v''|^2}\,\mathrm{e}^{u(v''^*-v^*)-u^*(v''-v)}\\ &+ c'c''^*\,\mathrm{e}^{-(|v'|^2+|v''|^2)/2}\,\mathrm{e}^{u(v''^*-v'^*)/2+u^*(v''-v')/2}\,\mathrm{e}^{u(v'^*/2+v''^*/2-v^*)-u^*(v'/2+v''/2-v)}\\ &+ c''c'^*\,\mathrm{e}^{-(|v'|^2+|v''|^2)/2}\,\mathrm{e}^{u(v'^*-v''^*)/2+u^*(v'-v'')/2}\,\mathrm{e}^{u(v'^*/2+v''^*/2-v^*)-u^*(v'/2+v''/2-v)}\Big]\,\mathrm{d}^2u.\end{aligned}$$

The first two terms can be expressed as delta functions, as before. However, each of the remaining two terms in the integrand consists of a Fourier kernel multiplied by a *real* exponential function of u, u^*, and for some values of the arguments v', v'' this grows more rapidly than any power of u, u^*. The integrals are therefore even more singular than those we encountered in making a representation of the Fock state. Still, if we are willing to proceed in a purely formal way, expand the exponential factors $\mathrm{e}^{u(v''^*-v'^*)/2}$, $\mathrm{e}^{u^*(v''-v')/2}$, etc. as power series in u, u^*, and then represent each power by differentiating the Fourier kernel under the integral, we arrive at

$$\begin{aligned}\phi(v) &= |c'|^2\delta^{(2)}(v-v') + |c''|^2\delta^{(2)}(v-v'')\\ &\quad + c'c''^*\exp\left(|v|^2 - \tfrac{1}{2}|v'|^2 - \tfrac{1}{2}|v''|^2\right)\exp\left[\tfrac{1}{2}(v'^*-v''^*)\frac{\partial}{\partial(v^*-\frac{1}{2}v'^*-\frac{1}{2}v''^*)}\right]\\ &\quad \times\exp\left[\tfrac{1}{2}(v''-v')\frac{\partial}{\partial(v-\frac{1}{2}v'-\frac{1}{2}v'')}\right]\delta^{(2)}(v-\tfrac{1}{2}v'-\tfrac{1}{2}v'')\\ &\quad + c''c'^*\exp\left(|v|^2 - \tfrac{1}{2}|v'|^2 - \tfrac{1}{2}|v''|^2\right)\exp\left[\tfrac{1}{2}(v''^*-v'^*)\frac{\partial}{\partial(v^*-\frac{1}{2}v'^*-\frac{1}{2}v''^*)}\right]\\ &\quad \times\exp\left[\tfrac{1}{2}(v'-v'')\frac{\partial}{\partial(v-\frac{1}{2}v'-\frac{1}{2}v'')}\right]\delta^{(2)}(v-\tfrac{1}{2}v'-\tfrac{1}{2}v''). \qquad (11.8\text{–}21)\end{aligned}$$

This expression involves derivatives of the delta function of infinitely high order, and lies outside the domain of the usually recognized classes of distributions. Nevertheless, when it is associated with a sufficiently well-behaved (infinitely differentiable) test function, it yields meaningful, finite answers by formal application of the rules of differentiation and integration.

As an example, we consider the following integral, which has a simple physical interpretation,

$$\int |v|^2 \phi(v)\, d^2v,$$

with $\phi(v)$ given by Eq. (11.8–21). The contributions of the first two terms in Eq. (11.8–21) can be evaluated by inspection, and they are $|c'|^2|v'|^2$ and $|c''|^2|v''|^2$. To evaluate the term in $c'c''^*$ we expand each exponential operator as a power series and integrate by parts. It is convenient to make the change of variable $v - \frac{1}{2}v' - \frac{1}{2}v'' = V$. Then

$$\text{3rd term} = c'c''^* \int \left\{ \exp\left[-\tfrac{1}{2}(v'^* - v''^*)\frac{\partial}{\partial V^*}\right] \exp\left[-\tfrac{1}{2}(v'' - v')\frac{\partial}{\partial V}\right] e^A \right\} \times |V + \tfrac{1}{2}v' + \tfrac{1}{2}v''|^2 \delta^2(V)\, d^2V,$$

with

$$A \equiv |V|^2 + \tfrac{1}{2}V(v'^* + v''^*) + \tfrac{1}{2}V^*(v' + v'') + \tfrac{1}{4}(-|v'|^2 - |v''|^2 + v'v''^* + v'^*v'').$$

The term in the { } brackets can be simplified with the help of the Campbell–Baker–Hausdorff identity in the form

$$e^B e^A = e^{[B,A]} e^A e^B,$$

applied to differential operators, provided that both B and A commute with $[B, A]$. With $B = -\frac{1}{2}(v'' - v')\partial/\partial V$ we easily find that

$$[B, A] \equiv C = -\tfrac{1}{2}(v'' - v')V^* - \tfrac{1}{4}(v'' - v')(v'^* + v''^*),$$

so that $[C, B] = 0 = [C, A]$, and

$$\text{3rd term} = c'c''^* \int \left\{ \exp\left[-\tfrac{1}{2}(v'^* - v''^*)\frac{\partial}{\partial V^*}\right] e^{C+A} \exp\left[-\tfrac{1}{2}(v'' - v')\frac{\partial}{\partial V}\right] \right\} \times |V + \tfrac{1}{2}v' + \tfrac{1}{2}v''|^2 \delta^2(V)\, d^2V.$$

A second application of the Campbell–Baker–Hausdorff identity yields the result

$$\text{3rd term} = c'c''^* \int e^{(v''^* - v'^*)(V + v')/2}\, e^{C+A} \exp\left[-\tfrac{1}{2}(v'^* - v''^*)\frac{\partial}{\partial V^*}\right] \times \exp\left[-\tfrac{1}{2}(v'' - v')\frac{\partial}{\partial V}\right] \times |V + \tfrac{1}{2}v' + \tfrac{1}{2}v''|^2 \delta^2(V)\, d^2V.$$

The integrand can now be easily evaluated with the help of a power series expansion of the exponential operators. Only the first two terms in each expansion make a contribution, and when they are combined with the delta function under the integral we obtain

$$\text{3rd term} = c'c''^*(v'v''^*)\, e^{-|v'|^2/2 - |v''|^2/2 + v'v''^*}.$$

The term in c'^*c'' can be derived in a similar manner, so that we have finally, when all four terms are combined,

$$\int |v|^2 \phi(v)\, d^2v = |c'|^2|v'|^2 + |c''|^2|v''|^2 + (c'c''^* v'v''^*\, e^{-|v'|^2/2 - |v''|^2/2 + v'v''^*} + \text{c.c.}) \tag{11.8–22}$$

This is a perfectly definite, and well-behaved solution to the problem, despite the extremely singular character of $\phi(v)$. It will become clear in the next section that the integral we have evaluated is nothing but the expectation value of the photon number operator $\hat{n}$ in the state $|\psi\rangle$. This can, of course, be calculated directly much more simply.

(f) *Interfering fields*

In optics we frequently encounter situations in which two or more light beams are superposed to produce interference effects. Then the total electromagnetic field at each point in space and time is the sum of the fields contributed by the separate beams, and similarly, the total excitation of each mode is the sum of the separate excitations contributed by each of the components of the total field.

Let us now consider the phase space representation of the state of the superposed field. We suppose that one beam on its own may be regarded as an excitation of the field to the coherent state $|v_1\rangle = \hat{D}(v_1)|0\rangle$, while the other one on its own may be regarded as the excitation to the coherent state $|v_2\rangle = \hat{D}(v_2)|0\rangle$. When both beams or both excitations are present at once, the total excitation is characterized by the product $\hat{D}(v_2)\hat{D}(v_1)$ of the two displacement operators acting on the vacuum. Now the product of two displacement operators is another displacement operator $\hat{D}(v_1 + v_2)$, apart from a phase factor [see Eq. (11.3–14)], so that the density operator of the superposed field is given by

$$\begin{aligned}\hat{\rho} &= \hat{D}(v_2)\hat{D}(v_1)|0\rangle\langle 0|\hat{D}^\dagger(v_1)\hat{D}^\dagger(v_2)\\ &= \hat{D}(v_1 + v_2)|0\rangle\langle 0|\hat{D}^\dagger(v_1 + v_2)\\ &= |v_1 + v_2\rangle\langle v_1 + v_2|.\end{aligned}$$

Now consider the more general situation in which the states in question are not necessarily pure coherent states, but are arbitrary. We can then make a diagonal coherent state representation of $\hat{\rho}$ by combining the coherent state projectors above and writing

$$\hat{\rho} = \int \phi_{12}(v_1, v_2)|v_1 + v_2\rangle\langle v_1 + v_2|\, d^2v_1\, d^2v_2, \tag{11.8–23}$$

in which $\phi_{12}(v_1, v_2)$ is some joint phase space density of v_1 and v_2. Of course, this representation in terms of a function of two separate complex variables v_1 and v_2 is meaningful only in so far as it is possible for the two excitations to exist separately, for example, if one or the other of the light beams can be blocked out in turn. In that case the two separate fields can be represented by density operators $\hat{\rho}_1$ and $\hat{\rho}_2$ such that

$$\begin{aligned}\hat{\rho}_1 &= \int \phi_{12}(v_1, v_2)|v_1\rangle\langle v_1|\, d^2v_1\, d^2v_2\\ &= \int \phi_1(v_1)|v_1\rangle\langle v_1|\, d^2v_1,\end{aligned}$$

with

$$\phi_1(v_1) \equiv \int \phi_{12}(v_1, v_2)\,\mathrm{d}^2 v_2,$$

and

$$\begin{aligned}\hat{\rho}_2 &= \int \phi_{12}(v_1, v_2)|v_2\rangle\langle v_2|\,\mathrm{d}^2 v_1\,\mathrm{d}^2 v_2 \\ &= \int \phi_2(v_2)|v_2\rangle\langle v_2|\,\mathrm{d}^2 v_2,\end{aligned}$$

with

$$\phi_2(v_2) \equiv \int \phi_{12}(v_1, v_2)\,\mathrm{d}^2 v_1.$$

By making the change of variable $v_1 + v_2 = v$, we can rewrite Eq. (11.8–23) in the form

$$\begin{aligned}\hat{\rho} &= \int \phi_{12}(v_1, v - v_1)|v\rangle\langle v|\,\mathrm{d}^2 v_1\,\mathrm{d}^2 v \\ &= \int \phi(v)|v\rangle\langle v|\,\mathrm{d}^2 v,\end{aligned}$$

where

$$\phi(v) \equiv \int \phi_{12}(v_1, v - v_1)\,\mathrm{d}^2 v_1, \qquad (11.8\text{–}24)$$

which is the conventional form of the diagonal coherent state representation.

In the special case in which the two excitations of the field are independent, the joint phase space function $\phi_{12}(v_1, v_2)$ in Eq. (11.8–23) factorizes into the product of the separate phase space functions for the separate excitations, and we have

$$\phi_{12}(v_1, v_2) = \phi_1(v_1)\phi_2(v_2). \qquad (11.8\text{–}25)$$

In that case Eq. (11.8–24) reduces to

$$\phi(v) = \int \phi_1(v_1)\phi_2(v - v_1)\,\mathrm{d}^2 v_1, \qquad (11.8\text{–}26)$$

which shows that the diagonal representation of the superposed field is a convolution of the diagonal representations of the separate fields.

As an obvious application of this equation, we observe that if each excitation is in the form of a coherent state, with $\phi_1(v) = \delta^2(v - v')$ and $\phi_2(v) = \delta^2(v - v'')$, then

$$\begin{aligned}\phi(v) &= \int \delta^2(v_1 - v')\delta^2(v - v_1 - v'')\,\mathrm{d}^2 v_1 \\ &= \delta^2(v - v' - v''),\end{aligned} \qquad (11.8\text{–}27)$$

which corresponds to the coherent state $|v' + v''\rangle$, as expected.

A less trivial application of Eq. (11.8–26) is provided by the situation in which the light beams from two single-mode lasers are superposed (Mandel, 1965). Let us suppose that the two lasers are identical but independent, and that each phase space distribution is given by Eq. (11.8–14). Then from Eq. (11.8–26) the phase space function of the resultant field is given by

$$\phi(v) = \frac{1}{(2\pi r_0)^2}\int \delta(|v'| - r_0)\delta(|v - v'| - r_0)\,\mathrm{d}^2v'.$$

If we put $v' = |v'|\,\mathrm{e}^{\mathrm{i}\theta'}$, $v = |v|\,\mathrm{e}^{\mathrm{i}\theta}$, we can carry out the integration over $|v'|$ and obtain

$$\begin{aligned}\phi(v) &= \frac{1}{(2\pi)^2 r_0}\int_0^{2\pi}\delta[[|v|^2 + r_0^2 - 2|v|r_0\cos(\theta - \theta')]^{1/2} - r_0]\,\mathrm{d}\theta' \\ &= \left.\begin{aligned}&\frac{1}{\pi^2|v|(4r_0^2 - |v|^2)^{1/2}}, && \text{for } |v| < 2r_0, \\ &= 0, && \text{for } |v| > 2r_0,\end{aligned}\right\}\end{aligned} \tag{11.8–28}$$

when we make use of the property (11.8–17) of the delta function. We note that $|v|$ for the superposed field no longer has a well-defined value, but is distributed over the range 0 to $2r_0$. Indeed the distribution of $|v|$ has the same form as the distribution of the real part of v given by Eq. (11.8–16) for a single laser beam (except that $|v|$ is necessarily limited to the half range $|v| \geqslant 0$). This result may be understood in the following way. For each laser beam the absolute amplitude $|v| = r_0$ is constant, but the random phase causes $\mathrm{Re}\,v$ to be distributed over the range $-r_0$ to r_0 with a density given by Eq. (11.8–16). When the two beams are superposed interference takes place, and the resultant absolute amplitude varies with the phase difference between the two beams within the range 0 to $2r_0$. The random phases now lead to a similar distribution for the resultant absolute amplitude.

By writing $v = x + \mathrm{i}y$ and integrating $\phi(x, y)$ given by Eq. (11.8–28) over all y, we arrive at the following expression for the distribution of x alone,

$$\left.\begin{aligned}p(x) &= \frac{1}{\pi^2}\int_{-(4r_0^2-x^2)^{1/2}}^{(4r_0^2-x^2)^{1/2}}\frac{\mathrm{d}y}{(x^2 + y^2)^{1/2}(4r_0^2 - x^2 - y^2)^{1/2}}, && \text{for } |x| \leqslant 2r_0 \\ &= 0, && \text{otherwise.}\end{aligned}\right\}$$

The integral is expressible as a complete elliptic integral K of the first kind and we have

$$\left.\begin{aligned}p(x) &= \frac{1}{\pi^2 r_0}K\left(1 - \frac{x^2}{4r_0^2}\right)^{1/2}, && \text{for } |x| \leqslant 2r_0, \\ &= 0, && \text{otherwise.}\end{aligned}\right\} \tag{11.8–29}$$

This distribution is also illustrated in Fig. 11.2 [curve (c)]. It is singular for $x = 0$, but shares with the Gaussian distribution the feature that the most probable value of x is zero. When several such independent laser beams are superposed, the distribution of the real part of the resultant complex amplitude tends quite rapidly towards a Gaussian form, as expected from the central limit theorem.

11.9 The optical equivalence theorem for normally ordered operators

So far the diagonal coherent state representation of the density operator $\hat{\rho}$ has appeared only as a compact representation of $\hat{\rho}$. Its major virtues become apparent when we use it to evaluate the expectations of operators, particularly those that are of greatest importance in quantum optics.

As we shall see in Section 12.2, most laboratory measurements of the optical field correspond to the expectation values of certain simple products of creation and annihilation operators in normal order, i.e. products with all creation operators standing to the left of all annihilation operators. The expectation values of such operators can be evaluated with the help of the diagonal representation in a particularly convenient way (Sudarshan, 1963; Glauber, 1963b; Klauder, 1966; Klauder and Sudarshan, 1968, Sect. 8.4). Moreover, this method emphasizes the character of $\phi(v)$ in the expansion of $\hat{\rho}$ as a phase space weighting function or quasi-probability density.

Consider any normally ordered function $g^{(\mathrm{N})}(\hat{a}, \hat{a}^\dagger)$ of the annihilation and creation operators $\hat{a}$, $\hat{a}^\dagger$, having a power series expansion of the form

$$g^{(\mathrm{N})}(\hat{a}, \hat{a}^\dagger) = \sum_{n,m} c_{nm} \hat{a}^{\dagger n} \hat{a}^m. \tag{11.9–1}$$

The expectation value of $g^{(\mathrm{N})}(\hat{a}, \hat{a}^\dagger)$ in the state characterized by the density operator $\hat{\rho}$ is given by

$$\langle g^{(\mathrm{N})}(\hat{a}, \hat{a}^\dagger)\rangle = \mathrm{Tr}[\hat{\rho} g^{(\mathrm{N})}(\hat{a}, \hat{a}^\dagger)]. \tag{11.9–2}$$

Let us now introduce the diagonal representation (11.8–1) of $\hat{\rho}$ explicitly in Eq. (11.9–2), together with the expansion (11.9–1) for $g^{(\mathrm{N})}(\hat{a}, \hat{a}^\dagger)$. We then find that

$$\begin{aligned}\langle g^{(\mathrm{N})}(\hat{a}, \hat{a}^\dagger)\rangle &= \mathrm{Tr} \int \phi(v) \sum_{n,m} c_{nm} |v\rangle\langle v| \hat{a}^{\dagger n} \hat{a}^m \,\mathrm{d}^2 v \\ &= \int \phi(v) \sum_{n,m} c_{nm} \langle v| \hat{a}^{\dagger n} \hat{a}^m |v\rangle \,\mathrm{d}^2 v,\end{aligned}$$

when we recall the identity $\mathrm{Tr}\,|A\rangle\langle B| = \langle B|A\rangle$. Now $\langle v|$ is the left eigenstate of $\hat{a}^\dagger$ and $|v\rangle$ is the right eigenstate of $\hat{a}$. Hence

$$\langle v| \hat{a}^{\dagger n} \hat{a}^m |v\rangle = v^{*n} v^m \tag{11.9–3}$$

and

$$\begin{aligned}\langle g^{(\mathrm{N})}(\hat{a}, \hat{a}^\dagger)\rangle &= \int \phi(v) \sum_{n,m} c_{nm} v^{*n} v^m \,\mathrm{d}^2 v \\ &= \int \phi(v) g^{(\mathrm{N})}(v, v^*) \,\mathrm{d}^2 v.\end{aligned} \tag{11.9–4}$$

The expectation value of the normally ordered operator $g^{(\mathrm{N})}(\hat{a}, \hat{a}^\dagger)$ can therefore be found by replacing the operators $\hat{a}$, $\hat{a}^\dagger$ by their right and left eigenvalues v, v^*, respectively, and 'averaging' the resulting c-number function $g^{(\mathrm{N})}(v, v^*)$ over the entire complex v-plane with $\phi(v)$ as weighting function. The expectation value of the quantum mechanical operator is therefore determined in the same manner as the expectation value of the corresponding c-number function in classical optics. If we think of v as a complex random variable with 'probability density' $\phi(v)$, then the expectation value of the function $g^{(N)}(v, v^*)$ may properly be denoted by

$$\int \phi(v) g^{(N)}(v, v^*)\, d^2v \equiv \langle g^{(N)}(v, v^*)\rangle_\phi, \tag{11.9–5}$$

where $\langle\ \rangle_\phi$ denotes the ensemble average with respect to the weighting function $\phi(v)$. Equation (11.9–4) can therefore be written in the compact form

$$\langle g^{(N)}(\hat{a}, \hat{a}^\dagger)\rangle = \langle g^{(N)}(v, v^*)\rangle_\phi. \tag{11.9–6}$$

Note that the angular brackets on the left of the equation signify the quantum mechanical expectation, whereas those on the right denote the ensemble average with respect to the phase space density $\phi(v)$.

This result, which was obtained by Sudarshan (1963), has been called the *optical equivalence theorem* by Klauder (1966), for it implies the formal equivalence between expectations of normally ordered operators in quantum optics and expectations of the corresponding c-number function in classical optics. Needless to say, as was emphasized by Glauber (1963b), the theorem does not imply the equivalence of quantum electrodynamics and classical optics. As we have seen, $\phi(v)$ refers to overlapping quantum states and it need not always have the character of a classical probability density. Still, in those cases in which $\phi(v)$ does behave like a classical probability density, there is indeed little difference between the evaluation of quantum expectations and the corresponding classical expectations.

The function $\phi(v)$ is not the only phase space density that allows expectations of quantum operators to be calculated like classical averages. We have already encountered the Wigner distribution in Section 11.8.1. However, $\phi(v)$ is distinguished from other phase space densities by the fact that it coincides with the corresponding classical probability density whenever a classical description of the state of the field exists. The rather undisciplined behavior of $\phi(v)$ for states that are strongly non-classical is the price we have to pay for the correspondence with classical optics.

As a simple example of the application of the optical equivalence theorem, let us calculate the mean number of photons in the field. As the number operator $\hat{n} = \hat{a}^\dagger \hat{a}$, we have immediately

$$\langle \hat{n}\rangle = \langle |v|^2\rangle_\phi. \tag{11.9–7}$$

The integral that we evaluated in Eq. (11.8–22) was just of this form. In particular, for a coherent state of the field, for which $\phi(v)$ has the form (11.8–13), we obtain

$$\begin{aligned}\langle \hat{n}\rangle &= \int |v|^2 \delta^2(v - v')\, d^2v \\ &= |v'|^2,\end{aligned} \tag{11.9–8}$$

in agreement with Eq. (11.2–13).

11.9.1 *Quantum characteristic functions*

Among the normally ordered operators that will be of particular interest to us in quantum optics is the generator

$$\mathrm{e}^{u\hat{a}^\dagger}\,\mathrm{e}^{-u^*\hat{a}},$$

whose expectation value is the characteristic function $C_{\mathrm{N}}(u, u^*)$ that is associated with normal ordering. Thus

$$C_{\mathrm{N}}(u, u^*) \equiv \langle \mathrm{e}^{u\hat{a}^\dagger}\,\mathrm{e}^{-u^*\hat{a}} \rangle = \mathrm{Tr}\,[\hat{\rho}\,\mathrm{e}^{u\hat{a}^\dagger}\,\mathrm{e}^{-u^*\hat{a}}]. \qquad (11.9\text{–}9)$$

The derivatives of this characteristic function with respect to u and u^* generate normally ordered moments of creation and annihilation operators.‡ For example,

$$\left[\frac{\partial^{n+m} C_{\mathrm{N}}(u, u^*)}{\partial u^n (\partial(-u^*))^m}\right]_{\substack{u=0\\ u^*=0}} = \langle \hat{a}^{\dagger n} \hat{a}^m \rangle. \qquad (11.9\text{–}10)$$

With the help of the equivalence theorem we now have

$$C_{\mathrm{N}}(u, u^*) = \langle \mathrm{e}^{uv^* - u^*v} \rangle_\phi = \int \phi(v)\, \mathrm{e}^{uv^* - u^*v}\, \mathrm{d}^2 v. \qquad (11.9\text{–}11)$$

Since $\mathrm{e}^{uv^*-u^*v}$ is the two-dimensional Fourier kernel, we see that $C_{\mathrm{N}}(u, u^*)$ is also the characteristic function associated with the phase space density $\phi(v)$ in the usual sense of probability theory, [except for the possible non-classical character of $\phi(v)$]. By Fourier inversion of Eq. (11.9–11) we obtain another integral representation of $\phi(v)$ in the form

$$\phi(v) = \frac{1}{\pi^2} \int C_{\mathrm{N}}(u, u^*)\, \mathrm{e}^{-uv^* + u^*v}\, \mathrm{d}^2 u. \qquad (11.9\text{–}12)$$

Needless to say, the existence of this integral in the ordinary sense, or the positive-definiteness of $\phi(v)$, are no more assured by this expression than by Eq. (11.8–11).

In order to explore the behavior of $C_{\mathrm{N}}(u, u^*)$ a little further, we note that the characteristic function $C_{\mathrm{N}}(u, u^*)$ is only one of a whole set of possible quantum characteristic functions, which are defined by different operator orderings. For example, when the operators are arranged in anti-normal order, we obtain the characteristic function

$$C_{\mathrm{A}}(u, u^*) \equiv \langle \mathrm{e}^{-u^*\hat{a}}\, \mathrm{e}^{u\hat{a}^\dagger} \rangle. \qquad (11.9\text{–}13)$$

The two functions $C_{\mathrm{N}}(u, u^*)$ and $C_{\mathrm{A}}(u, u^*)$ are simply related. For, from the Campbell–Baker–Hausdorff identity (11.3–4) it follows that

$$\mathrm{e}^{-|u|^2/2}\, \mathrm{e}^{u\hat{a}^\dagger}\, \mathrm{e}^{-u^*\hat{a}} = \mathrm{e}^{u\hat{a}^\dagger - u^*\hat{a}} = \mathrm{e}^{|u|^2/2}\, \mathrm{e}^{-u^*\hat{a}}\, \mathrm{e}^{u\hat{a}^\dagger}, \qquad (11.9\text{–}14)$$

so that

$$C_{\mathrm{N}}(u, u^*) = \mathrm{e}^{|u|^2} C_{\mathrm{A}}(u, u^*). \qquad (11.9\text{–}15)$$

‡ We have written $C_{\mathrm{N}}(u, u^*)$ rather than $C_{\mathrm{N}}(u)$ in order to emphasize that u, u^* may be treated as independent variables with respect to differentiation.

Clearly $C_N(u, u^*)$ is a function that grows much more rapidly with u than $C_A(u, u^*)$.

We are led to an interesting expression for $C_A(u, u^*)$ if we insert the unit operator in the form of Eq. (11.6–4) in the definition (11.9–13). We then have

$$C_A(u, u^*) = \mathrm{Tr} \int \frac{1}{\pi} \hat{\rho}\, e^{-u^*\hat{a}} |v\rangle\langle v| \, e^{u\hat{a}^\dagger} \, d^2v.$$

We now observe that the $\hat{a}$ and $\hat{a}^\dagger$ operators stand to the left and right of their respective eigenstates under the integral, so that they may be replaced by their eigenvalues v and v^*. On taking the trace we then arrive at

$$C_A(u, u^*) = \int \frac{1}{\pi} \langle v|\hat{\rho}|v\rangle \, e^{-u^*v + uv^*} \, d^2v, \tag{11.9–16}$$

which shows that $C_A(u, u^*)$ is the Fourier transform of $(1/\pi)\langle v|\hat{\rho}|v\rangle$, which is an absolutely integrable function.

Another possible characteristic function is the symmetrized, or Weyl-ordered, characteristic function

$$C_W(u, u^*) \equiv \langle e^{u\hat{a}^\dagger - u^*\hat{a}} \rangle = \langle \hat{D}(u) \rangle, \tag{11.9–17}$$

which is just the expectation value of the displacement operator $\hat{D}(u)$ that we discussed in Section 11.3. As $\hat{D}(u)$ is unitary we have

$$|C_W(u, u^*)| \leq 1. \tag{11.9–18}$$

With the help of the Campbell–Baker–Hausdorff identities (11.3–4) we find

$$e^{|u|^2/2} C_A(u, u^*) = C_W(u, u^*) = e^{-|u|^2/2} C_N(u, u^*), \tag{11.9–19}$$

and this yields the following bounds on $C_N(u, u^*)$ and $C_A(u, u^*)$:

$$\left.\begin{aligned} |C_A(u, u^*)| &\leq e^{-|u|^2/2} \\ |C_N(u, u^*)| &\leq e^{|u|^2/2}. \end{aligned}\right\} \tag{11.9–20}$$

We see immediately that the bound on $C_N(u, u^*)$ is very weak, and in general too weak to ensure that the integral (11.9–12) representing the phase space function $\phi(v)$ exists as an ordinary function in all cases. Fortunately, however, $C_N(u, u^*)$ grows much more slowly with u than is indicated by the upper bound $e^{|u|^2/2}$ in many cases that are important in practice.

11.10 More general phase space representations

11.10.1 Introduction

We now turn to a more general, unified treatment of phase space representations and operator ordering, from which many of the previous results emerge as special cases. We start by recalling the so-called optical equivalence theorem expressed by Eq. (11.9–4) for any function $g^{(N)}(\hat{a}, \hat{a}^\dagger)$ of annihilation and creation operators in normal order. By combining the formula (11.8–9) for the phase space density with Eq. (11.9–4) we may write

$$\langle g^{(\mathrm{N})}(\hat{a}, \hat{a}^\dagger)\rangle = \int \frac{1}{\pi}\rho^{(\mathrm{A})}(v, v^*)g^{(\mathrm{N})}(v, v^*)\,\mathrm{d}^2 v, \qquad (11.10\text{–}1)$$

where the function $\rho^{(\mathrm{A})}(v, v^*)$ is obtained from the density operator $\hat{\rho}$ by expressing $\hat{\rho}$ in anti-normal order and then replacing $\hat{a}$, $\hat{a}^\dagger$ by v, v^*, respectively. We may therefore evaluate the quantum expectation of the normally ordered operator $g^{(\mathrm{N})}(\hat{a}, \hat{a}^\dagger)$ by evaluating a c-number integral with a certain weight function or phase space density.

This result has an interesting generalization for other operator orderings Ω. Let $\bar{\Omega}$ denote the order which is reciprocal to Ω in a sense to be made more precise below. It can be shown that for the Ω-ordered operator $g^{(\Omega)}(\hat{a}, \hat{a}^\dagger)$

$$\langle g^{(\Omega)}(\hat{a}, \hat{a}^\dagger)\rangle = \int \frac{1}{\pi}\rho^{(\bar{\Omega})}(v, v^*)g^{(\Omega)}(v, v^*)\,\mathrm{d}^2 v. \qquad (11.10\text{–}2)$$

Our optical equivalence theorem (11.10–1) is then just a special case of Eq. (11.10–2) in which Ω is identified with the normal order N and the reciprocal order $\bar{\Omega}$ with the anti-normal order A.

The subject of generalized phase space distributions and the associated operator ordering evolved out of the work of Wigner (1932) to which we referred in Section 11.8.1. Moyal (1949) further developed the concept and showed that the Wigner distribution is connected with the evaluation of expectations of operators in symmetric or Weyl order. Important generalizations applicable to other operator orderings were made by Lax (1968), Agarwal and Wolf (1968, 1970a,b,c) and Cahill and Glauber (1969a,b). The following analysis is based largely on the work of Agarwal and Wolf [for a review, see also Peřina (1985, Chap. 16 and 1991, Sec. 4.8)].

11.10.2 Operator ordering

We start by introducing the idea of an ordered operator delta function, which played a major role in the development. The two-dimensional Fourier integral

$$\frac{1}{\pi^2}\int \mathrm{e}^{\beta(\hat{a}^\dagger - v^*) - \beta^*(\hat{a} - v)}\,\mathrm{d}^2\beta$$

can be identified with the delta function $\delta^2(\hat{a} - v)_\mathrm{W}$, with the operators $\hat{a}$, $\hat{a}^\dagger$ in the symmetric or Weyl order (W), by virtue of the fact that a power series expansion of the exponential generates operator products in symmetrized order. On the other hand, as we saw in Section 11.9.1 in the discussion of characteristic functions for different orderings, we have the relations

$$\mathrm{e}^{-|\beta|^2/2}\,\mathrm{e}^{\beta(\hat{a}^\dagger - v^*) - \beta^*(\hat{a} - v)} = \mathrm{e}^{-\beta^*(\hat{a} - v)}\,\mathrm{e}^{\beta(\hat{a}^\dagger - v^*)}$$

$$\mathrm{e}^{|\beta|^2/2}\,\mathrm{e}^{\beta(\hat{a}^\dagger - v^*) - \beta^*(\hat{a} - v)} = \mathrm{e}^{\beta(\hat{a}^\dagger - v^*)}\,\mathrm{e}^{-\beta^*(\hat{a} - v)}.$$

We may therefore write more generally for the Ω-ordered delta function,

$$\frac{1}{\pi}\int \Omega(\beta, \beta^*)\,\mathrm{e}^{\beta(\hat{a}^\dagger - v^*) - \beta^*(\hat{a} - v)}\,\mathrm{d}^2\beta \equiv \pi\delta^2(\hat{a} - v)_\Omega \equiv \hat{\Delta}(v, \Omega), \qquad (11.10\text{–}3)$$

with the understanding that Ω stands for the Weyl order (W) when $\Omega(\beta, \beta^*) = 1$,

Ω stands for the normal order (N) when $\Omega(\beta, \beta^*) = e^{|\beta|^2}$, and Ω stands for the anti-normal order (A) when $\Omega(\beta, \beta^*) = e^{-|\beta|^2/2}$. Other operator orderings associated with different $\Omega(\beta, \beta^*)$ were identified by Agarwal and Wolf (1968, 1970a,b,c), but we shall not consider them here. The order $\bar{\Omega}$ is said to be reciprocal to order Ω if $\bar{\Omega}(\beta, \beta^*)\Omega(\beta, \beta^*) = 1$. The normal and anti-normal orders are therefore reciprocals of each other, and the Weyl order is its own reciprocal.

A number of properties of the operator $\hat{\Delta}$-functions follow immediately from the definition, when we recall the displacement operator $\hat{D}(v)$ and its properties treated in Section 11.3.1. For example, from Eq. (11.10–3), we have

$$\begin{aligned}\mathrm{Tr}[\hat{\Delta}(v, \Omega)] &= \mathrm{Tr}\frac{1}{\pi}\int \hat{D}(\beta)\, e^{-\beta v^* + \beta^* v}\,\Omega(\beta, \beta^*)\, d^2\beta \\ &= \int \delta^2(\beta)\, e^{-\beta v^* + \beta^* v}\,\Omega(\beta, \beta^*)\, d^2\beta \\ &= \Omega(0, 0) \\ &= 1, \end{aligned} \tag{11.10–4}$$

and similarly,

$$\begin{aligned}&\mathrm{Tr}[\hat{\Delta}(\alpha, \Omega)\hat{\Delta}(\gamma, \bar{\Omega})] \\ &= \mathrm{Tr}\frac{1}{\pi^2}\int \hat{D}(\beta)\hat{D}(\beta')\, e^{-\beta\alpha^* + \beta^*\alpha}\, e^{-\beta'\gamma^* + \beta'^*\gamma}\,\Omega(\beta, \beta^*)\,\bar{\Omega}(\beta', \beta'^*)\, d^2\beta\, d^2\beta' \\ &= \frac{1}{\pi}\int \delta^2(\beta + \beta')\, e^{-\beta(\alpha^* - \gamma^*) + \beta^*(\alpha - \gamma)}\,\Omega(\beta, \beta^*)\,\bar{\Omega}(-\beta, -\beta^*)\, d^2\beta\, d^2\beta' \\ &= \frac{1}{\pi}\int e^{-\beta(\alpha^* - \gamma^*) + \beta^*(\alpha - \gamma)}\, d^2\beta \\ &= \pi\delta^2(\alpha - \gamma). \end{aligned} \tag{11.10–5}$$

Also,

$$\begin{aligned}\frac{1}{\pi}\int \hat{\Delta}(v, \Omega)\, d^2v &\equiv \frac{1}{\pi^2}\int \Omega(\beta, \beta^*)\, e^{\beta\hat{a}^\dagger - \beta^*\hat{a}}\, e^{\beta^* v - \beta v^*}\, d^2v\, d^2\beta \\ &= \int \Omega(\beta, \beta^*)\, e^{\beta\hat{a}^\dagger - \beta^*\hat{a}}\,\delta^2(\beta)\, d^2\beta \\ &= \Omega(0, 0) \\ &= 1. \end{aligned} \tag{11.10–6}$$

We may use the ordered operator $\hat{\Delta}$-function to put any operator $\hat{A} = A(\hat{a}, \hat{a}^\dagger)$ into the order Ω. Let us consider the integral

$$\frac{1}{\pi}\int \mathrm{Tr}[\hat{\Delta}(\beta, \bar{\Omega})A(\hat{a}, \hat{a}^\dagger)]\hat{\Delta}(\beta, \Omega)\, d^2\beta,$$

and use Eqs. (11.10–3) and (11.10–4). Then we obtain

$$\frac{1}{\pi}\int \mathrm{Tr}\,[\hat{\Delta}(\beta,\,\bar{\Omega})A(\hat{a},\,\hat{a}^\dagger)]\hat{\Delta}(\beta,\,\Omega)\,\mathrm{d}^2\beta = \int \mathrm{Tr}\,[\delta^2(\hat{a}-\beta)_{\bar{\Omega}}A(\hat{a},\,\hat{a}^\dagger)]\hat{\Delta}(\beta,\,\Omega)\,\mathrm{d}^2\beta$$

$$= \int A^{(\bar{\Omega})}(\beta,\,\beta^*)\delta^2(\hat{a}-\beta)_\Omega\,\mathrm{d}^2\beta$$

$$= A^{(\Omega)}(\hat{a},\,\hat{a}^\dagger). \tag{11.10–7}$$

The operation on $\hat{A}$ represented by the integral on the left therefore reproduces $\hat{A}$, but in the Ω-ordered form. Similarly, if $A(\hat{a},\,\hat{a}^\dagger)$, $B(\hat{a},\,\hat{a}^\dagger)$ are two operators and we use Eq. (11.10–3) and Eq. (11.10–5), we find by the same argument

$$\mathrm{Tr}\,[A(\hat{a},\,\hat{a}^\dagger)B(\hat{a},\,\hat{a}^\dagger)]$$

$$= \frac{1}{\pi^2}\mathrm{Tr}\int \mathrm{Tr}\,[\hat{\Delta}(v,\,\bar{\Omega})A(\hat{a},\,\hat{a}^\dagger)]\,\mathrm{Tr}\,[\hat{\Delta}(v',\,\Omega)B(\hat{a},\,\hat{a}^\dagger)]\hat{\Delta}(v,\,\Omega)\hat{\Delta}(v',\,\bar{\Omega})\,\mathrm{d}^2v\,\mathrm{d}^2v'$$

$$= \frac{1}{\pi}\int \mathrm{Tr}\,[\hat{\Delta}(v,\,\bar{\Omega})A(\hat{a},\,\hat{a}^\dagger)]\,\mathrm{Tr}\,[\hat{\Delta}(v,\,\Omega)B(\hat{a},\,\hat{a}^\dagger)]\,\mathrm{d}^2v$$

$$= \frac{1}{\pi}\int A^{(\bar{\Omega})}(v,\,v^*)B^{(\Omega)}(v,\,v^*)\,\mathrm{d}^2v. \tag{11.10–8}$$

11.10.3 Application to quantum expectations and the diagonal coherent-state representation

Let us apply Eq. (11.10–8) to the special case when $\hat{A}$ is the density operator $\hat{\rho}$, so that $A(\hat{a},\,\hat{a}^\dagger) = \rho(\hat{a},\,\hat{a}^\dagger)$. Then

$$\mathrm{Tr}\,(\hat{\rho}\hat{B}) = \langle\hat{B}\rangle = \frac{1}{\pi}\int \rho^{(\bar{\Omega})}(v,\,v^*)B^{(\Omega)}(v,\,v^*)\,\mathrm{d}^2v. \tag{11.10–9}$$

When Ω stands for normal ordering N, this will be recognized as the optical equivalence theorem in which the phase space density $\phi(v)$ has been expressed in the form $(1/\pi)\rho^{(\mathrm{A})}(v,\,v^*)$ [cf. Eq. (11.8–9)]. Evidently the theorem applies to other kinds of orderings also.

Finally, we shall use Eq. (11.10–7) to derive the diagonal coherent-state representation Eq. (11.8–1). Let us first identify Ω with anti-normal ordering A. Then, by definition,

$$\hat{\Delta}(v,\,\Omega^{(\mathrm{A})}) = \frac{1}{\pi}\int \Omega^{(\mathrm{A})}(\beta,\,\beta^*)\,\mathrm{e}^{\beta\hat{a}^\dagger-\beta^*\hat{a}}\,\mathrm{e}^{-v^*\beta+v\beta^*}\,\mathrm{d}^2\beta$$

$$= \frac{1}{\pi}\int \mathrm{e}^{-|\beta|^2/2}\,\mathrm{e}^{\beta\hat{a}^\dagger-\beta^*\hat{a}}\,\mathrm{e}^{-v^*\beta+v\beta^*}\,\mathrm{d}^2\beta$$

$$= \frac{1}{\pi}\int \mathrm{e}^{-\beta^*\hat{a}}\,\mathrm{e}^{\beta\hat{a}^\dagger}\,\mathrm{e}^{-v^*\beta+v\beta^*}\,\mathrm{d}^2\beta.$$

We now insert the unit operator in the form

$$1 = \frac{1}{\pi}\int |v'\rangle\langle v'|\,\mathrm{d}^2v'$$

between the factors $e^{-\beta^*\hat{a}}$ and $e^{\beta\hat{a}^\dagger}$ and make use of the fact that $|v'\rangle$ is the right eigenstate of $\hat{a}$. Then we obtain

$$\begin{aligned}\hat{\Delta}(v, \Omega^{(A)}) &= \frac{1}{\pi^2}\int e^{\beta(v'^*-v^*)-\beta^*(v'-v)}|v'\rangle\langle v'|\,d^2\beta\,d^2v' \\ &= \int \delta^2(v'-v)|v'\rangle\langle v'|\,d^2v' \\ &= |v\rangle\langle v|. \end{aligned} \tag{11.10–10}$$

When this result is used in Eq. (11.10–7) with the operator $\hat{A}$ identified with the density operator $\hat{\rho}$ and with $\Omega \rightarrow \Omega^{(A)}$, we find

$$\begin{aligned}\hat{\rho} &= \int \frac{1}{\pi}\operatorname{Tr}[\hat{\Delta}(v, \Omega^{(N)})\hat{\rho}]|v\rangle\langle v|\,d^2v \\ &= \int \frac{1}{\pi}\rho^{(A)}(v, v^*)|v\rangle\langle v|\,d^2v. \end{aligned} \tag{11.10–11}$$

This can be regarded as a formal derivation of the diagonal or P-representation of the density operator $\hat{\rho}$.

11.11 Multimode fields

Until now we have confined our discussion of coherent states, and the representations based on them, to a single mode of the electromagnetic field. It is now time to return to the more general situation, when there is an unbounded set of modes labelled by the wave vector $\mathbf{k}$ and polarization index s.

A coherent state in this larger Hilbert space is now labeled by a set of complex numbers $v_{\mathbf{k}s}$, one for each mode, and it is given by the direct product of states $|v_{\mathbf{k}s}\rangle$ over all modes,

$$\prod_{\mathbf{k},s}|v_{\mathbf{k},s}\rangle.$$

This obviously includes the special case in which only one mode of the field is occupied, while the others are empty. It is again convenient to make use of the notation $\{v\}$ for the set of complex numbers $v_{\mathbf{k}s}$ over all modes. Then we may label the coherent state by $|\{v\}\rangle$ and write

$$|\{v\}\rangle = \prod_{\mathbf{k},s}|v_{\mathbf{k}s}\rangle, \tag{11.11–1a}$$

$$= \prod_{\mathbf{k},s} e^{(v_{\mathbf{k}s}\hat{a}^\dagger_{\mathbf{k}s} - v^*_{\mathbf{k}s}\hat{a}_{\mathbf{k}s})}|0_{\mathbf{k}s}\rangle, \tag{11.11–1b}$$

$$= \prod_{\mathbf{k},s} e^{-|v_{\mathbf{k}s}|^2/2}\sum_{n_{\mathbf{k}s}}\frac{(v_{\mathbf{k}s})^{n_{\mathbf{k}s}}}{\sqrt{(n_{\mathbf{k}s}!)}}|n_{\mathbf{k}s}\rangle. \tag{11.11–1c}$$

The state $|\{v\}\rangle$ and its conjugate $\langle\{v\}|$ are still right and left eigenstates of $\hat{a}_{\mathbf{k}s}$ and $\hat{a}^\dagger_{\mathbf{k}s}$ so that we have

$$\hat{a}_{\mathbf{k}s}|\{v\}\rangle = v_{\mathbf{k}s}|\{v\}\rangle, \tag{11.11–2}$$

and

$$\langle\{v\}|\hat{a}^{\dagger}_{\mathbf{k}s} = v^{*}_{\mathbf{k}s}\langle\{v\}|. \tag{11.11–3}$$

From Eq. (11.6–1) the scalar product of two coherent states is given by

$$\langle\{v'\}|\{v''\}\rangle = \prod_{\mathbf{k},s} \mathrm{e}^{-|v'_{\mathbf{k}s}-v''_{\mathbf{k}s}|^2/2}\, \mathrm{e}^{(v'^{*}_{\mathbf{k}s}v''_{\mathbf{k}s}-v'_{\mathbf{k}s}v''^{*}_{\mathbf{k}s})/2}, \tag{11.11–4}$$

so that

$$|\langle\{v'\}|\{v''\}\rangle|^2 = \prod_{\mathbf{k},s} \mathrm{e}^{-|v'_{\mathbf{k}s}-v''_{\mathbf{k}s}|^2}$$

$$= \exp\left(-\sum_{\mathbf{k},s}|v'_{\mathbf{k}s} - v''_{\mathbf{k}s}|^2\right). \tag{11.11–5}$$

It is at once clear that these expressions will be non-vanishing, just as for single-mode fields, only if

$$\sum_{\mathbf{k},s}|v'_{\mathbf{k}s} - v''_{\mathbf{k}s}|^2 < \infty. \tag{11.11–6}$$

Since

$$\sum_{\mathbf{k},s}|v'_{\mathbf{k}s} - v''_{\mathbf{k}s}|^2 \leqslant \sum_{\mathbf{k},s}(|v'_{\mathbf{k}s}| + |v''_{\mathbf{k}s}|)^2 \leqslant 2\sum_{\mathbf{k},s}(|v'_{\mathbf{k}s}|^2 + |v''_{\mathbf{k}s}|^2),$$

condition (11.11–6) will be satisfied if, for each coherent state,

$$\sum_{\mathbf{k}s}|v_{\mathbf{k}s}|^2 < \infty, \tag{11.11–7}$$

which implies, according to Eq. (11.2–13), that the mean number of photons is finite.

We can make a resolution of the unit operator in terms of the coherent states $|\{v\}\rangle$, by taking products over all modes of the operators in Eq. (11.6–4). We then obtain

$$1 = \prod_{\mathbf{k},s}\left[\frac{1}{\pi}\int|v_{\mathbf{k}s}\rangle\langle v_{\mathbf{k}s}|\, \mathrm{d}^2 v_{\mathbf{k}s}\right]. \tag{11.11–8}$$

It is convenient to introduce the notation

$$\mathrm{d}\{v\} \equiv \prod_{\mathbf{k},s}[\mathrm{d}^2 v_{\mathbf{k}s}] \tag{11.11–9a}$$

for the product of the differentials and the notation,

$$\mathrm{d}\mu\{v\} \equiv \prod_{\mathbf{k},s}\left[\frac{1}{\pi}\mathrm{d}^2 v_{\mathbf{k}s}\right], \tag{11.11–9b}$$

for the product of scaled differentials, which allows us to write Eq. (11.11–8) in the compact form

$$1 = \int|\{v\}\rangle\langle\{v\}|\, \mathrm{d}\mu\{v\}, \tag{11.11–10}$$

to express the completeness of coherent states.

A diagonal coherent state-representation of the density operator $\hat{\rho}$ can again be given, and it takes the form

$$\hat{\rho} = \int \phi(\{v\})|\{v\}\rangle\langle\{v\}|\,\mathrm{d}\{v\}, \tag{11.11–11}$$

in which ϕ is now a functional of the set $\{v\}$, known as the phase space functional.

An optical equivalence theorem that is strictly analogous to Eq. (11.9–4) can be given for any functional $g^{(\mathrm{N})}(\{\hat{a}\}, \{\hat{a}^\dagger\})$ of the annihilation and creation operators in normal order, and we write

$$\begin{aligned}\langle g^{(\mathrm{N})}(\{\hat{a}\}, \{\hat{a}^\dagger\})\rangle &= \int \phi(\{v\}) g^{(\mathrm{N})}(\{v\}, \{v^*\})\,\mathrm{d}\{v\} \\ &= \langle g^{(\mathrm{N})}(\{v\}, \{v^*\})\rangle_\phi. \end{aligned} \tag{11.11–12}$$

For example let us consider the normally ordered characteristic functional

$$C_\mathrm{N}(\{u\}) \equiv \left\langle \prod_{\mathbf{k},s}[\mathrm{e}^{u_{\mathbf{k}s}\hat{a}^\dagger_{\mathbf{k}s}}\,\mathrm{e}^{-u^*_{\mathbf{k}s}\hat{a}_{\mathbf{k}s}}]\right\rangle = \left\langle \prod_{\mathbf{k},s}[\mathrm{e}^{u_{\mathbf{k}s}v^*_{\mathbf{k}s}-u^*_{\mathbf{k}s}v_{\mathbf{k}s}}]\right\rangle_{\phi\{v\}}, \tag{11.11–13}$$

which is formally a multiple Fourier transform of the phase space functional. Fourier inversion of Eq. (11.11–13) then yields the phase space functional,

$$\phi(\{v\}) = \int C_\mathrm{N}(\{u\}) \prod_{\mathbf{k},s}\left[\frac{1}{\pi^2}\,\mathrm{e}^{u^*_{\mathbf{k}s}v_{\mathbf{k}s}-u_{\mathbf{k}s}v^*_{\mathbf{k}s}}\,\mathrm{d}^2u_{\mathbf{k}s}\right]. \tag{11.11–14}$$

As before, the Fourier transform may not exist in the sense of ordinary function theory, but as a generalized functional.

The various electromagnetic field operators like $\hat{\mathbf{E}}(\mathbf{r}, t)$, $\hat{\mathbf{B}}(\mathbf{r}, t)$, $\hat{\mathbf{A}}(\mathbf{r}, t)$ have particularly simple expectation values when the field is in a coherent state. From the mode expansions (10.4–38) to (10.4–40) and the eigenvalue relations (11.11–2) and (11.11–3) we find immediately,

$$\langle\{v\}|\hat{\mathbf{A}}(\mathbf{r}, t)|\{v\}\rangle = \frac{1}{\varepsilon_0^{1/2}L^{3/2}}\sum_{\mathbf{k},s}\left(\frac{\hbar}{2\omega}\right)^{1/2}[v_{\mathbf{k}s}\boldsymbol{\varepsilon}_{\mathbf{k}s}\,\mathrm{e}^{\mathrm{i}(\mathbf{k}\cdot\mathbf{r}-\omega t)} + \mathrm{c.c.}] \tag{11.11–15}$$

$$\langle\{v\}|\hat{\mathbf{E}}(\mathbf{r}, t)|\{v\}\rangle = \frac{\mathrm{i}}{\varepsilon_0^{1/2}L^{3/2}}\sum_{\mathbf{k},s}\left(\frac{\hbar\omega}{2}\right)^{1/2}[v_{\mathbf{k}s}\boldsymbol{\varepsilon}_{\mathbf{k}s}\,\mathrm{e}^{\mathrm{i}(\mathbf{k}\cdot\mathbf{r}-\omega t)} - \mathrm{c.c.}] \tag{11.11–16}$$

$$\langle\{v\}|\hat{\mathbf{B}}(\mathbf{r}, t)|\{v\}\rangle = \frac{\mathrm{i}}{\varepsilon_0^{1/2}L^{3/2}}\sum_{\mathbf{k},s}\left(\frac{\hbar}{2\omega}\right)^{1/2}[v_{\mathbf{k}s}\mathbf{k}\times\boldsymbol{\varepsilon}_{\mathbf{k}s}\,\mathrm{e}^{\mathrm{i}(\mathbf{k}\cdot\mathbf{r}-\omega t)} - \mathrm{c.c.}]. \tag{11.11–17}$$

Each of these expansions will be seen to be just the mode expansion of the corresponding classical field. However, as we showed in the case of a single mode, the variance of the field vectors does not vanish when the field is in a coherent state, even though the expectation value of each vector suggests a strong resemblance to a classical field of well-defined complex amplitude. This is another example of the difference between the classical and quantum descriptions of the electromagnetic field.

In Section 11.8 we drew attention to the fact that the field of a laser that is oscillating in a single mode far above threshold, may be regarded as a mixture of single mode coherent states $|v\rangle$ with the same value of $|v|$ but randomly distributed values of $\arg v$. Let us consider the corresponding phase space functional $\phi(\{v\})$ in the multimode notation. If $\mathbf{k}'$, s' is the excited mode of the field and all other modes are unexcited, and if the excitation of the mode is characterized by Eq. (11.8–14), then we evidently have

$$\phi(\{v\}) = \frac{1}{2\pi|v_{\mathbf{k}'s'}|}\delta(|v_{\mathbf{k}'s'}| - r_0) \prod_{\mathbf{k},s\neq\mathbf{k}',s'} \delta^2(v_{\mathbf{k},s}). \tag{11.11–18}$$

This representation of the state still oversimplifies the actual physical situation, because, in practice, the frequency of the light is not constant but drifts within some narrow range $c\Delta k'$. However, in a succession of randomly timed measurements, each of which takes a time interval short compared with $1/\Delta k'$, the frequency drift is not observable. The experimental situation may then be quite well described by the phase space functional in Eq. (11.11–18).

11.11.1 Coherent states in the continuous mode representation

When the field vectors are given Fourier integral expansions, as in Eqs. (10.10–1) to (10.10–3), and the $\mathbf{k}$ modes form a continuum, we can still introduce coherent states as eigenstates of the annihilation operator $\hat{a}(\mathbf{k}, s)$ (Kibble, 1968). The states $|v(\mathbf{k}, s)\rangle$ are now functionals of the continuous function $v(\mathbf{k}, s)$ of $\mathbf{k}$, and we write

$$\hat{a}(\mathbf{k}, s)|v(\mathbf{k},s)\rangle = v(\mathbf{k}, s)|v(\mathbf{k}, s)\rangle. \tag{11.11–19}$$

By making use of the completeness relation (10.10–12) for the continuous Fock states, we can make a continuous Fock representation of the coherent state $|v(\mathbf{k}, s)\rangle$, corresponding to that given by Eq. (11.11–1c). A calculation similar to the one carried out in Section 11.2 leads to the result

$$\begin{aligned}|v(\mathbf{k}, s)\rangle = \exp&\left[-\tfrac{1}{2}\sum_s\int|v(\mathbf{k}, s)|^2\,\mathrm{d}^3k\right]\\ &\times\Bigg\{|\mathrm{vac}\rangle + \frac{1}{\sqrt{1!}}\sum_{s_1}\int v(\mathbf{k}_1, s_1)|\mathbf{k}_1, s_1\rangle\,\mathrm{d}^3k_1\\ &+ \frac{1}{\sqrt{2!}}\sum_{s_1}\sum_{s_2}\int\!\!\int v(\mathbf{k}_1, s_1)v(\mathbf{k}_2, s_2)|\mathbf{k}_1, s_1, \mathbf{k}_2, s_2\rangle\,\mathrm{d}^3k_1\,\mathrm{d}^3k_2\\ &+ \ldots\Bigg\}\end{aligned} \tag{11.11–20}$$

Evidently coherent states are a little easier to handle in terms of discrete modes.‡

11.12 Positive-frequency and negative-frequency field operators

We have already encountered the mode expansions of the electromagnetic field operators in Section 10.4 [see Eqs. (10.4–38), (10.4–39), (10.4–40)]. All of them

‡ For a more rigorous discussion of the approach to the multimode case, see Klauder and Sudarshan (1968), Sec. 7.4.

are of the general form

$$\frac{1}{L^{3/2}}\sum_{\mathbf{k},s}[l(\omega)\hat{a}_{\mathbf{k}s}\boldsymbol{\varepsilon}_{\mathbf{k}s}\,\mathrm{e}^{\mathrm{i}(\mathbf{k}\cdot\mathbf{r}-\omega t)} + l^*(\omega)\hat{a}^\dagger_{\mathbf{k}s}\boldsymbol{\varepsilon}^*_{\mathbf{k}s}\,\mathrm{e}^{-\mathrm{i}(\mathbf{k}\cdot\mathbf{r}-\omega t)}],$$

in which $\boldsymbol{\varepsilon}_{\mathbf{k}s}$ is a unit polarization vector orthogonal to $\mathbf{k}$, and $l(\omega)$ is some simple function of ω, such as $(\hbar/2\omega\varepsilon_0)^{1/2}$ for the vector potential. It is often convenient to combine these expansions by writing

$$\hat{\mathbf{F}}(\mathbf{r}, t) = \hat{\mathbf{F}}^{(+)}(\mathbf{r}, t) + \hat{\mathbf{F}}^{(-)}(\mathbf{r}, t), \tag{11.12–1}$$

where $\hat{\mathbf{F}}(\mathbf{r}, t)$ is any one of the field vectors, and

$$\hat{\mathbf{F}}^{(+)}(\mathbf{r}, t) = \frac{1}{L^{3/2}}\sum_{\mathbf{k},s} l(\omega)\hat{a}_{\mathbf{k}s}\boldsymbol{\varepsilon}_{\mathbf{k}s}\,\mathrm{e}^{\mathrm{i}(\mathbf{k}\cdot\mathbf{r}-\omega t)}. \tag{11.12–2}$$

By assigning different values to $l(\omega)$ we may define a whole range of linearly related field vectors in this way. The decomposition emphasizes the fact that each field $\hat{\mathbf{F}}$ may be regarded as the sum of two operators which are conjugates of each other, each of which is a combination of annihilation operators only, or creation operators only. To a certain extent the non-Hermitian operators $\hat{\mathbf{F}}^{(+)}(\mathbf{r}, t)$ and $\hat{\mathbf{F}}^{(-)}(\mathbf{r}, t)$ play the role of configuration space annihilation and creation operators at the space-time point $(\mathbf{r}, t)$, as we shall see more explicitly in Chapter 14. Moreover, $\hat{\mathbf{F}}^{(+)}(\mathbf{r}, t)$ is the positive frequency part of the field $\hat{\mathbf{F}}(\mathbf{r}, t)$, in the sense that it depends on time only through the factors $\mathrm{e}^{-\mathrm{i}\omega t}$ in the expansion, whereas $\hat{\mathbf{F}}^{(-)}(\mathbf{r}, t)$ is the negative frequency part of the field, whose time dependence is governed by the factors $\mathrm{e}^{\mathrm{i}\omega t}$. Evidently $\hat{\mathbf{F}}^{(+)}(\mathbf{r}, t)$ and its conjugate $\hat{\mathbf{F}}^{(-)}(\mathbf{r}, t)$ correspond very closely to the analytic signals that were used in the classical treatment of optical coherence (Chapters 4, 6), except that they are Hilbert space operators. We shall see shortly that the correspondence is more than purely formal, and the interpretation of $\hat{\mathbf{F}}^{(+)}(\mathbf{r}, t)$ and $\hat{\mathbf{F}}^{(-)}(\mathbf{r}, t)$ as annihilation and creation operators provides another explanation for the appearance of analytic signals in the treatment of optical coherence.

It is worth recalling that the real field is often difficult to measure in the optical domain and beyond. Most observations in optics are based on the absorption of light, either through the use of a photodetector, or a photographic plate or even the eye. We shall therefore not be surprised to find that the absorption operator $\hat{\mathbf{F}}^{(+)}(\mathbf{r}, t)$, rather than the real field $\hat{\mathbf{F}}(\mathbf{r}, t)$, plays the dominant role in the description of the experiments.

We now observe that, because of Eqs. (11.11–2) and (11.11–3), the coherent state is also a right or left eigenstate of $\hat{\mathbf{F}}^{(+)}(\mathbf{r}, t)$ or $\hat{\mathbf{F}}^{(-)}(\mathbf{r}, t)$, respectively. For

$$\begin{aligned}\hat{\mathbf{F}}^{(+)}(\mathbf{r}, t)|\{v\}\rangle &= \frac{1}{L^{3/2}}\sum_{\mathbf{k},s} l(\omega)v_{\mathbf{k}s}\boldsymbol{\varepsilon}_{\mathbf{k}s}\,\mathrm{e}^{\mathrm{i}(\mathbf{k}\cdot\mathbf{r}-\omega t)}|\{v\}\rangle \\ &= \mathbf{F}^{(+)}(\mathbf{r}, t)|\{v\}\rangle,\end{aligned} \tag{11.12–3}$$

where

$$\mathbf{F}^{(+)}(\mathbf{r}, t) \equiv \frac{1}{L^{3/2}}\sum_{\mathbf{k},s} l(\omega)v_{\mathbf{k}s}\boldsymbol{\varepsilon}_{\mathbf{k}s}\,\mathrm{e}^{\mathrm{i}(\mathbf{k}\cdot\mathbf{r}-\omega t)}, \tag{11.12–4}$$

and similarly,

$$\langle\{v\}|\hat{\mathbf{F}}^{(-)}(\mathbf{r}, t) = \mathbf{F}^{(-)}(\mathbf{r}, t)\langle\{v\}|. \qquad (11.12\text{–}5)$$

Here $\mathbf{F}^{(+)}(\mathbf{r}, t)$ and $\mathbf{F}^{(-)}(\mathbf{r}, t)$ are just the positive frequency and negative frequency parts of the classical c-number fields having the set of Fourier amplitudes $\{v\}$. Evidently, $\mathbf{F}^{(+)}(\mathbf{r}, t)$ and $\mathbf{F}^{(-)}(\mathbf{r}, t)$ are both analytic signals in the sense of Section 3.1; they are analytic in opposite halves of the complex t-plane.

11.12.1 Commutation relations

We see that the analytic signals, which are used so widely in the classical treatment of optical coherence, may be regarded as the right and left eigenvalues of the positive frequency and negative frequency parts of the field operators, belonging to the appropriate coherent state. We shall therefore not be surprised to find a rather close correspondence between classical correlation functions of the electromagnetic field expressed in terms of analytic signals and quantum correlation functions expressed in terms of positive and negative frequency parts of the field operators. The quantum correlations will simply be expectations of products of the $\hat{\mathbf{F}}^{(+)}(\mathbf{r}, t)$ and $\hat{\mathbf{F}}^{(-)}(\mathbf{r}, t)$ operators, taken at different space-time points. However, because of the non-commutativity of the $\hat{\mathbf{F}}^{(+)}(\mathbf{r}, t)$ and $\hat{\mathbf{F}}^{(-)}(\mathbf{r}, t)$ operators, a wide variety of different correlation functions can be defined for the same set of operators.

The commutator of $\hat{\mathbf{F}}^{(+)}(\mathbf{r}, t)$ and $\hat{\mathbf{F}}^{(-)}(\mathbf{r}, t)$ is readily evaluated with the help of the commutation relation (10.3–9). From the expansion (11.12–2) we find

$$\begin{aligned}
&[\hat{F}_i^{(+)}(\mathbf{r}_1, t_1), \hat{F}_j^{(-)}(\mathbf{r}_2, t_2)] \\
&\quad = \frac{1}{L^3}\sum_{\mathbf{k},s}\sum_{\mathbf{k}',s'} l(\omega)l^*(\omega')[\hat{a}_{\mathbf{k}s}, \hat{a}^\dagger_{\mathbf{k}'s'}](\boldsymbol{\varepsilon}_{\mathbf{k}s})_i(\boldsymbol{\varepsilon}^*_{\mathbf{k}'s'})_j\, \mathrm{e}^{\mathrm{i}(\mathbf{k}\cdot\mathbf{r}_1-\mathbf{k}'\cdot\mathbf{r}_2-\omega t_1+\omega' t_2)} \\
&\quad = \frac{1}{L^3}\sum_{\mathbf{k}s}|l(\omega)|^2(\boldsymbol{\varepsilon}_{\mathbf{k}s})_i(\boldsymbol{\varepsilon}^*_{\mathbf{k}s})_j\, \mathrm{e}^{\mathrm{i}[\mathbf{k}\cdot(\mathbf{r}_1-\mathbf{r}_2)-\omega(t_1-t_2)]} \\
&\quad = \frac{1}{L^3}\sum_{\mathbf{k}}|l(\omega)|^2\left(\delta_{ij} - \frac{k_i k_j}{k^2}\right)\mathrm{e}^{\mathrm{i}[\mathbf{k}\cdot(\mathbf{r}_1-\mathbf{r}_2)-\omega(t_1-t_2)]} \\
&\quad \to \frac{1}{(2\pi)^3}\int|l(\omega)|^2\left(\delta_{ij} - \frac{k_i k_j}{k^2}\right)\mathrm{e}^{\mathrm{i}[\mathbf{k}\cdot(\mathbf{r}_1-\mathbf{r}_2)-\omega(t_1-t_2)]}\,\mathrm{d}^3k, \qquad (11.12\text{–}6)
\end{aligned}$$

when we make use of the tensor relation (10.8–2), together with the rule (10.8–4) for replacing the sum by an integral. Of course, for a free field the different $\hat{\mathbf{F}}^{(+)}(\mathbf{r}, t)$ operators commute among themselves, by virtue of the commutation of the $\hat{a}_{\mathbf{k}s}$ operators, and so do the different $\hat{\mathbf{F}}^{(-)}(\mathbf{r}, t)$.

11.12.2 Normally ordered correlations

Of all the correlation functions of the field that can be written down we shall find that the normally ordered ones of the general type

$$\Gamma^{(N,M)} \equiv$$

$$\langle \hat{F}_{i_1}^{(-)}(\mathbf{r}_1, t_1)\hat{F}_{i_2}^{(-)}(\mathbf{r}_2, t_2)\ldots\hat{F}_{i_N}^{(-)}(\mathbf{r}_N, t_N)\hat{F}_{j_M}^{(+)}(\mathbf{r}'_M, t'_M)\ldots\hat{F}_{j_2}^{(+)}(\mathbf{r}'_2, t'_2)\hat{F}_{j_1}^{(+)}(\mathbf{r}'_1, t'_1)\rangle$$

are of most interest in practice. With the help of the optical equivalence theorem we now obtain

$$\begin{aligned}\Gamma^{(N,M)} &\equiv \langle \hat{F}_{i1}^{(-)}(\mathbf{r}_1, t_1)\hat{F}_{i2}^{(-)}(\mathbf{r}_2, t_2)\ldots \hat{F}_{iN}^{(-)}(\mathbf{r}_N, t_N)\hat{F}_{jM}^{(+)}(\mathbf{r}'_M, t'_M)\ldots \\ &\quad\times \hat{F}_{j2}^{(+)}(\mathbf{r}'_2, t'_2)\hat{F}_{j1}^{(+)}(\mathbf{r}'_1, t'_1)\rangle \\ &= \langle F_{i1}^{(-)}(\mathbf{r}_1, t_1)F_{i2}^{(-)}(\mathbf{r}_2, t_2)\ldots F_{iN}^{(-)}(\mathbf{r}_N, t_N)F_{jM}^{(+)}(\mathbf{r}'_M, t'_M)\ldots \\ &\quad\times F_{j2}^{(+)}(\mathbf{r}'_2, t'_2)F_{j1}^{(+)}(\mathbf{r}'_1, t'_1)\rangle_\phi, \end{aligned} \tag{11.12–7}$$

which emphasizes the close correspondence between classical and quantum correlations (cf. also Wolf, 1963).

Finally, let us evaluate this correlation function explicitly when the field is in the coherent state $|\{v'\}\rangle$. If $\mathbf{F}'^{(+)}(\mathbf{r}, t)$ is the eigenvalue of $\hat{\mathbf{F}}^{(+)}(\mathbf{r}, t)$ belonging to the state $|\{v'\}\rangle$, it follows immediately from the fact that all operators in $\Gamma^{(N,M)}$ can be replaced by their right and left eigenvalues that

$$\begin{aligned}&\Gamma^{(N,M)} = \\ &F_{i1}^{'(-)}(\mathbf{r}_1, t_1)F_{i2}^{'(-)}(\mathbf{r}_2, t_2)\ldots F_{iN}^{'(-)}(\mathbf{r}_N, t_N)F_{jM}^{'(+)}(\mathbf{r}'_M, t'_M)\ldots F_{j2}^{'(+)}(\mathbf{r}'_2, t'_2)F_{j1}^{'(+)}(\mathbf{r}'_1, t'_1).\end{aligned} \tag{11.12–8}$$

Thus, the correlation function factorizes into a product of complex functions, and we recall from Section 4.5 dealing with classical correlations that this is a characteristic of a completely coherent field. The justification for the name 'coherent state of the field' is now apparent, for all normally ordered correlations factorize in the same manner. In particular, we find when we evaluate the degree of second-order (normally ordered) coherence at any two space-time points $\mathbf{r}_1$, t_1 and $\mathbf{r}_2$, t_2 in the coherent state $|\{v'\}\rangle$,

$$\frac{|\langle\{v'\}|\hat{F}_i^{(-)}(\mathbf{r}_1, t_1)\hat{F}_j^{(+)}(\mathbf{r}_2, t_2)|\{v'\}\rangle|}{\langle\{v'\}|\hat{F}_i^{(-)}(\mathbf{r}_1, t_1)\hat{F}_i^{(+)}(\mathbf{r}_1, t_1)|\{v'\}\rangle^{1/2}\langle\{v'\}|\hat{F}_j^{(-)}(\mathbf{r}_2, t_2)\hat{F}_j^{(+)}(\mathbf{r}_2, t_2)|\{v'\}\rangle^{1/2}} = 1, \tag{11.12–9}$$

as expected for a coherent field.

11.13 The field produced by a classical current

So far we have been discussing the coherent states of the electromagnetic field in an entirely mathematical way. We have shown that they have interesting properties and form a useful representation. However, as yet we have no indication how such states might be produced, if indeed they are ever encountered in practice. Before concluding our discussion of coherent states we shall remedy this situation by considering at least one example in which a source of radiation gives rise to a coherent state of the field. We shall show that a classical, prescribed current distribution always produces an electromagnetic field in a coherent state; this was first shown by Glauber (1963b). Although the subject of the interaction of radiation with matter belongs properly in later chapters, the calculation is a very simple one and provides a good illustration of the formalism we have been developing.

Let us consider a quantum electromagnetic field of vector potential $\hat{\mathbf{A}}(\mathbf{r}, t)$ that is interacting with a classical electric current described by the vector current density $\mathbf{j}(\mathbf{r}, t)$. Then, as is well known from electromagnetic theory, the energy $\hat{H}_I(t)$ of the interaction is given by the volume integral over all space

$$\hat{H}_I(t) = -\int_{L^3} \mathbf{j}(\mathbf{r}, t) \cdot \hat{\mathbf{A}}(\mathbf{r}, t)\, d^3r. \tag{11.13–1}$$

We shall treat the current as a classical, c-number current that is not affected by its interaction with the field. The field, on the other hand, will be regarded as a quantized system in which the field vectors are Hilbert space operators. It will be convenient to work in the interaction picture in which the field operators evolve in time just as they would in a non-interacting system. With the help of the expansion (10.4–38) for $\hat{\mathbf{A}}(\mathbf{r}, t)$, we write for $\hat{H}_I(t)$

$$\hat{H}_I(t) = -\frac{1}{\varepsilon_0^{1/2}L^{3/2}}\sum_{\mathbf{k},s}\left(\frac{\hbar}{2\omega}\right)^{1/2}[\boldsymbol{\varepsilon}_{\mathbf{k}s}\cdot\mathbf{J}(\mathbf{k}, t)\hat{a}_{\mathbf{k}s}\,e^{-i\omega t} + \boldsymbol{\varepsilon}^*_{\mathbf{k}s}\cdot\mathbf{J}^*(\mathbf{k}, t)\hat{a}^\dagger_{\mathbf{k}s}\,e^{i\omega t}], \tag{11.13–2}$$

where

$$\mathbf{J}(\mathbf{k}, t) \equiv \int_{L^3} \mathbf{j}(\mathbf{r}, t)\, e^{i\mathbf{k}\cdot\mathbf{r}}\, d^3r \tag{11.13–3}$$

may be regarded as the Fourier transform of the current density. Now the infinitesimal time-evolution operator in the interaction picture, governing the evolution of the state from time t to the slightly later time $t + \Delta t$, is given by

$$\hat{U}(t + \Delta t, t) = \exp(-i\hat{H}_I(t)\Delta t/\hbar). \tag{11.13–4}$$

By making use of the expansion (11.13–2) for $\hat{H}_I$ and introducing the abbreviation

$$-\frac{i}{\hbar}\frac{1}{\varepsilon_0^{1/2}L^{3/2}}\left(\frac{\hbar}{2\omega}\right)^{1/2}\boldsymbol{\varepsilon}_{\mathbf{k}s}\cdot\mathbf{J}(\mathbf{k}, t)\,e^{-i\omega t} \equiv u^*_{\mathbf{k}s}(t), \tag{11.13–5}$$

we can express $\hat{U}(t + \Delta t, t)$ in the compact form

$$\begin{aligned}\hat{U}(t + \Delta t, t) &= \exp\left(\Delta t\sum_{\mathbf{k},s}[-u^*_{\mathbf{k}s}(t)\hat{a}_{\mathbf{k}s} + u_{\mathbf{k}s}(t)\hat{a}^\dagger_{\mathbf{k}s}]\right)\\ &= \prod_{\mathbf{k},s}\exp[\Delta t(-u^*_{\mathbf{k}s}(t)\hat{a}_{\mathbf{k}s} + u_{\mathbf{k}s}(t)\hat{a}^\dagger_{\mathbf{k}s})]\\ &= \prod_{\mathbf{k},s}\hat{D}[\Delta t\, u_{\mathbf{k}s}(t)],\end{aligned} \tag{11.13–6}$$

where $\hat{D}$ is the displacement operator defined by Eq. (11.3–7) that we encountered earlier. It follows that the density operator $\hat{\rho}(t + \Delta t)$ at time $t + \Delta t$ is related to the density operator $\hat{\rho}(t)$ at the earlier time t by

$$\hat{\rho}(t + \Delta t) = \prod_{\mathbf{k},s}\hat{D}[\Delta t\, u_{\mathbf{k}s}(t)]\hat{\rho}(t)\prod_{\mathbf{k},s}\hat{D}^\dagger[\Delta t\, u_{\mathbf{k}s}(t)]. \tag{11.13–7}$$

Now we have already shown that the state resulting from the action of the displacement operator $\hat{D}(v)$ on the vacuum $|\text{vac}\rangle$ is the coherent state $|v\rangle$, and that the state resulting from the action of $\hat{D}(v)$ on the coherent state $|v'\rangle$ is the coherent state $|v+v'\rangle$, apart from a phase factor [see Eq. (11.3–15)]. On applying these results to the multimode situation in Eq. (11.13–7), we see at once that, if the initial state of the electromagnetic field at time $t=0$ is the vacuum,

$$\hat{\rho}(0) = |\text{vac}\rangle\langle\text{vac}|,$$

and if the classical current is then turned on, then at time Δt

$$\hat{\rho}(\Delta t) = \prod_{\mathbf{k},s} |\Delta t u_{\mathbf{k}s}(0)\rangle\langle\Delta t u_{\mathbf{k}s}(0)|. \tag{11.13–8}$$

By repeated application of the differential displacement operators, we find that after a finite time t the density operator will be of the form

$$\hat{\rho}(t) = |\{v(t)\}\rangle\langle\{v(t)\}|, \tag{11.13–9}$$

where $|\{v(t)\}\rangle$ is the multimode coherent state, and $v_{\mathbf{k}s}(t)$ is given by

$$v_{\mathbf{k}s}(t) \equiv \int_0^t u_{\mathbf{k}s}(t')\,\mathrm{d}t'. \tag{11.13–10}$$

As the current changes, the state changes continually, but it remains a coherent state at all times when the source is a prescribed classical current.

We see therefore that the coherent state of the field is not merely a mathematical construct with interesting properties, but is, in principle, realizable in a rather direct way. Moreover, this example shows again that the coherent state is a quantum state that is rather intimately connected with classical fields through the classical sources. If the initial current is not completely prescribed except in statistical terms, in such a way that each realization of the ensemble may be regarded as a prescribed classical current, then the density operator $\hat{\rho}(t)$ will be in the form of a mixture of coherent states

$$\hat{\rho}(t) = \int \phi(\{v(t)\})|\{v(t)\}\rangle\langle\{v(t)\}|\,\mathrm{d}\{v\}, \tag{11.13–11}$$

in which the weighting functional $\phi(\{v(t)\})$ is now a true classical probability functional. The reason is, of course, that it reflects the uncertainty in the description of the classical current, rather than any quantum uncertainties resulting from the interaction of coupled quantum systems.

Problems

11.1 By making use of the analytic properties of the projection of a coherent state on another state, determine if the photon creation operator $\hat{a}^\dagger$ for a single-mode field has a right eigenstate.

11.2 The following expression was given by Sudarshan (1963) for the phase space density, or the P-representation $\phi(v)$,

$$\phi(v) = \frac{1}{2\pi r}\sum_{n=0}^{\infty}\sum_{m=0}^{\infty}\rho_{nm}\frac{\sqrt{(n!)}\sqrt{(m!)}}{(m+n)!}\exp[r^2+\mathrm{i}\theta(m-n)]\left(-\frac{\partial}{\partial r}\right)^{n+m}\delta(r).$$

Here $\rho_{nm} = \langle n|\hat{\rho}|m\rangle$, $v = r\,\mathrm{e}^{\mathrm{i}\theta}$, and the radial delta function is normalized so that $\int_0^\infty \delta(r)\,\mathrm{d}r = 1$. Show that formally this is a solution to the problem of making a diagonal coherent state representation of the density operator $\hat{\rho}$.

11.3 A laser is oscillating simultaneously in two independent axial modes of frequencies ω_1, ω_2 having the same linear polarization characterized by the unit vector $\boldsymbol{\varepsilon}$, and the same energy density. Consider a reduced Hilbert space with just these two modes, and assume that each mode can be treated as a randomly phased single-mode laser. What is the P-representation of the laser field? Calculate the characteristic function $\langle \exp(\mathrm{i}\boldsymbol{\varepsilon}\cdot\hat{\mathbf{E}}\,\xi)\rangle$ of the Hermitian electric field operator $\hat{\mathbf{E}}$.

11.4 Determine all the right eigenvalues of the operator $\hat{a}\cosh\theta + \hat{a}^\dagger\sinh\theta$ for real values of θ.

11.5 An electromagnetic field has the following Heisenberg equation of motion for the photon annihilation operator $\hat{a}_l(t)$ corresponding to the l'th mode of the field,

$$\frac{\mathrm{d}}{\mathrm{d}t}\hat{a}_l(t) = f_l(\{\hat{a}(t)\}, t).$$

Here f_l is some functional of the set $\{\hat{a}(t)\}$ and of the time t. Show that for this field a state in the Schrödinger picture that is coherent at time $t = 0$ will remain coherent for all t.

11.6 A multimode optical field is in the quantum state

$$|\psi\rangle = \frac{1}{\sqrt{2}}(|\{n'\}\rangle + |\{n''\}\rangle),$$

where $|\{n'\}\rangle$, $|\{n''\}\rangle$ are two different Fock states with total photon numbers

$$N' = \sum_{\mathbf{k},s} n'_{\mathbf{k}s}, \qquad N'' = \sum_{\mathbf{k},s} n''_{\mathbf{k}s}.$$

Show that $|\psi\rangle$ is non-classical if $(N' - N'')^2/(N' + N'') < 2$. Can $|\psi\rangle$ be non-classical even if this condition does not hold?

11.7 A single-mode optical field has the characteristic function

$$\langle \mathrm{e}^{-u^*\hat{a}}\,\mathrm{e}^{u\hat{a}^\dagger}\rangle = \mathrm{e}^{-|u|^2(1+m)}\,\mathrm{e}^{uv^*G^* - u^*vG},$$

where G, m are real numbers and u, v are complex. Calculate the diagonal coherent-state representation (P-representation) of the field.

11.8 Calculate the expectation of the operator $\hat{A}^\dagger\hat{A}$ in the coherent state $|v\rangle$, when $\hat{A} \equiv (\hat{n} + 1)^{-1/2}\hat{a}$.

11.9 Consider a single-mode field in the state

$$|\psi\rangle = c_1|v_1\rangle + ic_2|v_2\rangle,$$

which is a linear superposition of two different coherent states $|v_1\rangle$, $|v_2\rangle$, with $\langle\psi|\psi\rangle = 1$. All four numbers c_1, c_2, v_1, v_2 are real. Can the photon statistics in the state $|\psi\rangle$ be sub-Poissonian? Under what conditions is the variance of the photons Poissonian?

11.10 Construct a single-mode quantum state that is orthogonal to the two coherent states $|v'\rangle$ and $|v''\rangle$ when $v' \neq v''$.

12

Quantum correlations and photon statistics

12.1 Introduction

We have now prepared the way for a quantum treatment of problems that are of interest in optics. We have examined some basic properties of the quantized field and have explored the coherent-state representation and some of its virtues. Until now our discussions have been essentially formal; we have not concerned ourselves with what is actually measured in the laboratory, nor with questions concerning the interactions of the quantized field with other systems.

In the following sections we shall turn our attention to some of the quantum field theoretic quantities that are of particular interest in quantum optics and are of significance in the measurement of the field. Our discussions in this chapter will still be somewhat formal, in that we shall not yet come to grips with the problem of the interaction of the field with the apparatus. That problem will be tackled in Chapter 14. For the moment we shall continue to oversimplify and treat the field as a free field. However, within the idealized context of the free field we shall examine some of the questions encountered in measurements in quantum optics. We shall see that normally ordered correlations play a dominant role and examine some of their properties. Anti-normally ordered correlations will be seen to arise in much less common and less useful measurements of the field. We shall also examine the sense in which a photon may be said to be localized in a measurement in which a photoelectric emission is registered. Finally, we shall discuss the subject of locality violations in quantum optics and optical tests of the Bell inequalities.

12.2 Photoelectric measurement of the optical field; normal ordering

Most measurements of the electromagnetic field in the optical domain are based on the absorption of photons via the photoelectric effect. This is true not only insofar as photodiodes, photomultipliers, etc., are used, but also for such homely devices as the photographic plate and the eye. These all function by absorption of photons. Although we shall deal with the problem of photoelectric detection more carefully and in more detail in Chapter 14, it is possible to arrive at an expression for the detection probability by a simple heuristic argument that was first given by Glauber (1963a).

We have already encountered the configuration space photon absorption operator $\hat{\mathbf{F}}^{(+)}(\mathbf{r}, t)$ defined by Eq. (11.12–2), and its conjugate $\hat{\mathbf{F}}^{(-)}(\mathbf{r}, t)$, the creation operator,

$$\hat{\mathbf{F}}^{(+)}(\mathbf{r}, t) = \frac{1}{L^{3/2}} \sum_{\mathbf{k},s} l(k) \hat{a}_{\mathbf{k}s} \boldsymbol{\varepsilon}_{\mathbf{k}s} \, \mathrm{e}^{\mathrm{i}(\mathbf{k}\cdot\mathbf{r}-\omega t)}, \tag{12.2–1}$$

$$\hat{\mathbf{F}}^{(-)}(\mathbf{r}, t) = \frac{1}{L^{3/2}} \sum_{\mathbf{k},s} l^*(k) \hat{a}^\dagger_{\mathbf{k}s} \boldsymbol{\varepsilon}^*_{\mathbf{k}s} \, \mathrm{e}^{-\mathrm{i}(\mathbf{k}\cdot\mathbf{r}-\omega t)}, \quad (\omega = ck) \tag{12.2–2}$$

These operators are the positive frequency part and the negative frequency part of the corresponding Hermitian field vectors, respectively. We shall show in Section 12.11 that the state $\hat{\mathbf{V}}^\dagger(\mathbf{r}, t)|\mathrm{vac}\rangle$ $[\hat{\mathbf{V}}^\dagger = \hat{\mathbf{F}}^{(-)}$ with $l^*(k) = 1$ in Eq. (12.2–2)] is one in which a photon is approximately localized‡ at the space-time point $\mathbf{r}$, t, so that $\hat{\mathbf{F}}^{(-)}(\mathbf{r}, t)$ has the effect of creating a photon in the vicinity of $\mathbf{r}$, t out of the vacuum. Similarly $\hat{\mathbf{F}}^{(+)}(\mathbf{r}, t)$ corresponds to the absorption of a photon in the neighborhood of $\mathbf{r}$, t.

Now let us suppose that we have a photodetector located at the point $\mathbf{r}$ in an optical field, and that it registers an absorption, by emission of a photoelectron, at time t. We refer to this event as a photon detection and ask for the probability of this process. Of course, in practice, photodetectors are devices of substantial size, that can hardly be located at a point, but for the purposes of this argument, we may suppose that the photocathode has been effectively stopped down by use of a small aperture. Moreover, we suppose further that a polarizer has been placed in front of the aperture, so that the detector responds to light of only one polarization, say of type s. Then the operator corresponding to the detection of a photon of polarization s at $\mathbf{r}$, t will be

$$\hat{\mathbf{F}}^{(+)}(s, \mathbf{r}, t) = \frac{1}{L^{3/2}} \sum_{\mathbf{k}} l(k) \hat{a}_{\mathbf{k}s} \boldsymbol{\varepsilon}_{\mathbf{k}s} \, \mathrm{e}^{\mathrm{i}(\mathbf{k}\cdot\mathbf{r}-\omega t)}, \quad (\omega = ck). \tag{12.2–3}$$

For the moment we deliberately leave the factor $l(k)$ and the nature of the corresponding field vector $\hat{\mathbf{F}}^{(+)}(s, \mathbf{r}, t)$ unspecified. If the detector electrons interact with the field through an interaction of the form $-\hat{\mathbf{p}} \cdot \hat{\mathbf{A}}$ (see Section 14.1), then $\hat{\mathbf{F}}^{(+)}(s, \mathbf{r}, t)$ would be the positive frequency part of the vector potential operator $\hat{\mathbf{A}}$, etc. It is worth noting that, in the special case in which the basis vectors $\boldsymbol{\varepsilon}_{\mathbf{k}s}$ correspond to orthogonal linear polarizations, and they lie along two of the coordinate axes, $\hat{\mathbf{F}}^{(+)}(s, \mathbf{r}, t)$ may correspond to one of the Cartesian components of $\hat{\mathbf{F}}^{(+)}(\mathbf{r}, t)$.

Let us discuss the photoelectric detection of the optical field within the quantum mechanical interaction picture. If the electromagnetic field is initially in some quantum state $|\psi_1\rangle$, and finally, following the detection, it is in a quantum state $|\psi_2\rangle$, then the probability amplitude for the process (Glauber, 1963a) in which a photon is detected at $\mathbf{r}$, t is the matrix element $\langle\psi_2|\hat{\mathbf{F}}^{(+)}(s, \mathbf{r}, t)|\psi_1\rangle$, while the probability for the transition is proportional to $|\langle\psi_2|\hat{\mathbf{F}}^{(+)}(s, \mathbf{r}, t)|\psi_1\rangle|^2$. If we are not interested in the final state, and we wish to know the probability of

‡ No position operator exists for the photon. A photon cannot therefore be localized at a point in space-time. However, an approximate localization is possible within a region whose linear dimensions are large compared with the wavelength. This point is discussed in Section 12.11

photodetection irrespective of the final state of the field, we have to sum this probability over a complete set of final states $|\psi_2\rangle$, namely

$$\sum_{\text{all } \psi_2} |\langle \psi_2 | \hat{\mathbf{F}}^{(+)}(s, \mathbf{r}, t) | \psi_1 \rangle|^2.$$

More generally, if the initial state is not a pure state $|\psi_1\rangle$, but is some ensemble of states characterized by the density operator $\hat{\rho}$, so that

$$\hat{\rho} = \sum_{\psi_1} p(\psi_1) |\psi_1\rangle\langle\psi_1|, \tag{12.2–4}$$

where $p(\psi_1)$ is the probability associated with state $|\psi_1\rangle$, we have to average the foregoing expression over the ensemble of all ψ_1 with weight $p(\psi_1)$. We may then write

$$\text{Rate of photodetection} = C_1 \sum_{\psi_1} p(\psi_1) \sum_{\text{all } \psi_2} |\langle \psi_2 | \hat{\mathbf{F}}^{(+)}(s, \mathbf{r}, t) | \psi_1 \rangle|^2, \tag{12.2–5}$$

where C_1 is some constant characteristic of the detector. For a short time interval Δt over which the rate of detection does not vary appreciably, we have the differential probability

$$\begin{aligned} &\text{Probability of photodetection for} \\ &\quad \text{polarization } s \text{ at } \mathbf{r}, t \text{ within } \Delta t \equiv P_1(s, \mathbf{r}, t)\Delta t \\ &\qquad = C_1 \Delta t \sum_{\psi_1} p(\psi_1) \sum_{\text{all } \psi_2} |\langle \psi_2 | \hat{\mathbf{F}}^{(+)}(s, \mathbf{r}, t) | \psi_1 \rangle|^2, \end{aligned} \tag{12.2–6}$$

with the understanding that Δt is sufficiently short to ensure that the right side of Eq. (12.2–6) is much less than unity. The suffix 1 in $P_1(s, \mathbf{r}, t)$ emphasizes that we are dealing with a single photodetection event. If we now expand the square in Eq. (12.2–6) by writing

$$P_1(s, \mathbf{r}, t)\Delta t = C_1 \Delta t \sum_{\psi_1} p(\psi_1) \sum_{\text{all } \psi_2} \langle \psi_1 | \hat{F}_i^{(-)}(s, \mathbf{r}, t) | \psi_2 \rangle \langle \psi_2 | \hat{F}_i^{(+)}(s, \mathbf{r}, t) | \psi_1 \rangle,$$

where summation over repeated Cartesian indices is understood, we see that the right-hand side contains the sum over a complete set of final state projectors $|\psi_2\rangle\langle\psi_2|$. By definition, this sum is the unit operator, and the equation reduces to

$$\begin{aligned} P_1(s, \mathbf{r}, t)\Delta t &= C_1 \Delta t \sum_{\psi_1} p(\psi_1) \langle \psi_1 | \hat{F}_i^{(-)}(s, \mathbf{r}, t) \hat{F}_i^{(+)}(s, \mathbf{r}, t) | \psi_1 \rangle \\ &= C_1 \Delta t \,\mathrm{Tr}\,[\hat{\rho} \hat{F}_i^{(-)}(s, \mathbf{r}, t) \hat{F}_i^{(+)}(s, \mathbf{r}, t)] \\ &= C_1 \Delta t \langle \hat{F}_i^{(-)}(s, \mathbf{r}, t) \hat{F}_i^{(+)}(s, \mathbf{r}, t) \rangle. \end{aligned} \tag{12.2–7}$$

The probability of detection is therefore proportional to the expectation value of the normally ordered scalar product of $\hat{\mathbf{F}}^{(-)}(s, \mathbf{r}, t)$ with $\hat{\mathbf{F}}^{(+)}(s, \mathbf{r}, t)$. The normal ordering arises naturally here, as a consequence of the fact that the basic detection process is an absorption, characterized by the absorption operator $\hat{\mathbf{F}}^{(+)}(s, \mathbf{r}, t)$.

As in the classical treatment of detection, it is convenient to introduce the notion of *intensity* of the field, defined by the scalar product

$$\hat{I}(s, \mathbf{r}, t) \equiv \hat{\mathbf{F}}^{(-)}(s, \mathbf{r}, t) \cdot \hat{\mathbf{F}}^{(+)}(s, \mathbf{r}, t). \tag{12.2–8}$$

The operator $\hat{I}(s, \mathbf{r}, t)$ so defined is of course the intensity associated with the s polarization component of the field vector $\hat{\mathbf{F}}$. Similarly, we may associate a total intensity operator $\hat{I}(\mathbf{r}, t)$ with the total operator $\hat{\mathbf{F}}^{(+)}(\mathbf{r}, t)$, by writing

$$\hat{I}(\mathbf{r}, t) \equiv \hat{\mathbf{F}}^{(-)}(\mathbf{r}, t) \cdot \hat{\mathbf{F}}^{(+)}(\mathbf{r}, t). \tag{12.2–9}$$

We note also that there exists a whole range of different intensities, each of which is associated with a different field vector $\hat{\mathbf{F}}(\mathbf{r}, t)$ according to the choice of $l(k)$ in Eq. (12.2–1). As in the classical case, it is often convenient to leave the actual choice undefined at this stage, in order to deal with a variety of situations. By introducing the intensity operator, we can re-write Eq. (12.2–7) in the more compact form

$$P_1(s, \mathbf{r}, t)\Delta t = C_1\Delta t\langle \hat{I}(s, \mathbf{r}, t)\rangle = C_1\Delta t\langle I(s, \mathbf{r}, t)\rangle_\phi. \tag{12.2–10}$$

The last expression follows from the optical equivalence theorem, with $I(s, \mathbf{r}, t) \equiv \mathbf{F}^{(-)}(s, \mathbf{r}, t) \cdot \mathbf{F}^{(+)}(s, \mathbf{r}, t)$, where $\mathbf{F}^{(+)}(s, \mathbf{r}, t)$ and $\mathbf{F}^{(-)}(s, \mathbf{r}, t)$ are the right and left eigenvalues of $\hat{\mathbf{F}}^{(+)}(s, \mathbf{r}, t)$ and $\hat{\mathbf{F}}^{(-)}(s, \mathbf{r}, t)$, respectively. When the polarizer is removed and the detector is free to respond to all polarizations, we obtain the probability of detection $P_1(\mathbf{r}, t)\Delta t$ by a similar argument in which the total absorption operator $\hat{\mathbf{F}}^{(+)}(\mathbf{r}, t)$ is used, and we arrive at

$$P_1(\mathbf{r}, t) = C_1\Delta t\langle \hat{I}(\mathbf{r}, t)\rangle = C_1\Delta t\langle I(\mathbf{r}, t)\rangle_\phi. \tag{12.2–11}$$

where C_1 is not necessarily the same constant as in Eq. (12.2–10).

The similarity of this result to Eq. (9.6–1), which was obtained by a semi-classical calculation, is at once evident. Of course, we must not lose sight of the fact that the answer was derived by a simple, heuristic argument, but we shall find in Section 14.5 that an identical expression is obtained when the interaction of the quantized field with the detector is treated explicitly. This simple argument does not yield the value of the constant C_1 in Eqs. (12.2–10) and (12.2–11), although it is not difficult to guess its dependence on the geometry of the detector. We shall return to this question in Section 12.3.

12.2.1 Multiple photodetections; higher-order correlation functions

So far we have considered a single photodetection event only. Let us now examine the more general situation in which a number of photoelectric detections are observed, and we wish to know the joint probability of all the events. This joint probability evidently contains information about possible correlations among the photoelectric emissions. The different photodetection events may be registered by the same detector at different times, or by different detectors located at different points in space. For the sake of generality, let us consider a number of photo-detectors located at points $\mathbf{r}_1$, $\mathbf{r}_2$, etc., with polarizers s_1, s_2, etc., in front of them as shown in Fig. 12.1. We ask for the joint probability that detections are registered at position $\mathbf{r}_1$ and at time t_1 within a short time interval Δt_1, at position

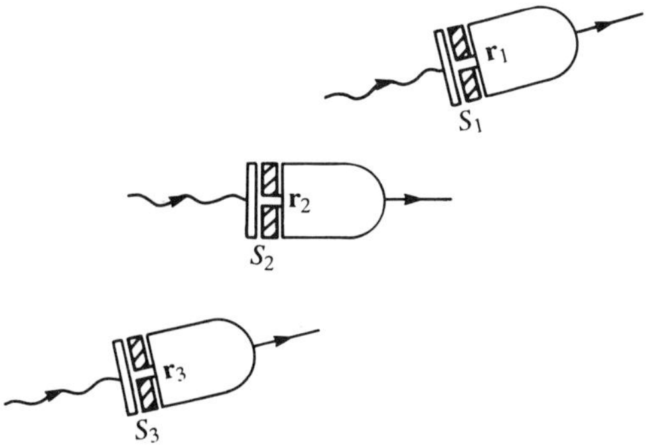

Fig. 12.1 Illustrating multiple photodetections with three detectors.

$\mathbf{r}_2$ and at time t_2 within Δt_2, etc. If there are N detections, ordered so that

$$t_1 < t_2 < \ldots < t_N,$$

the probability amplitude for the transition from the initial state of the field $|\psi_1\rangle$ to the final state $|\psi_2\rangle$ following the detections is now

$$\langle\psi_2|\hat{\mathbf{F}}^{(+)}(s_N, \mathbf{r}_N, t_N)\ldots\hat{\mathbf{F}}^{(+)}(s_2, \mathbf{r}_2, t_2)\hat{\mathbf{F}}^{(+)}(s_1, \mathbf{r}_1, t_1)|\psi_1\rangle,$$

in which the absorption operator is applied N times in succession. The direct product of vector operators is to be understood as a tensor operator. To arrive at the differential joint N-fold probability of photodetection we proceed exactly as before. We multiply the probability amplitude by its conjugate and sum over a complete orthogonal set of final states $|\psi_2\rangle$. If the initial state is not necessarily a pure state, but is characterized by the density operator $\hat{\rho}$ in Eq. (12.2–4), we also sum over ψ_1 with weight function $p(\psi_1)$. We then find

Joint N-fold probability of photodetection
for polarization s_1 at $\mathbf{r}_1$, t_1 within Δt_1,
for polarization s_2 at $\mathbf{r}_2$, t_2 within Δt_2,
⋮
for polarization s_N at $\mathbf{r}_N$, t_N within Δt_N

$$\begin{aligned}
&\equiv P_N(s_1, \mathbf{r}_1, t_1; s_2, \mathbf{r}_2, t_2; \ldots; s_N, \mathbf{r}_N, t_N)\Delta t_1 \Delta t_2 \ldots \Delta t_N \\
&= C_N \Delta t_1 \ldots \Delta t_N \sum_{\psi_1} p(\psi_1) \sum_{\text{all } \psi_2} \langle\psi_1|\hat{F}_{i_1}^{(-)}(s_1, \mathbf{r}_1, t_1)\ldots\hat{F}_{i_N}^{(-)}(s_N, \mathbf{r}_N, t_N)|\psi_2\rangle \\
&\quad \times \langle\psi_2|\hat{F}_{i_N}^{(+)}(s_N, \mathbf{r}_N, t_N)\ldots\hat{F}_{i_1}^{(+)}(s_1, \mathbf{r}_1, t_1)|\psi_1\rangle \\
&= C_N \Delta t_1 \ldots \Delta t_N \sum_{\psi_1} p(\psi_1)\langle\psi_1|\hat{F}_{i_1}^{(-)}(s_1, \mathbf{r}_1, t_1)\ldots\hat{F}_{i_N}^{(-)}(s_N, \mathbf{r}_N, t_N)\hat{F}_{i_N}^{(+)}(s_N, \mathbf{r}_N, t_N) \\
&\quad \times \ldots\hat{F}_{i_1}^{(+)}(s_1, \mathbf{r}_1, t_1)|\psi_1\rangle \\
&= C_N \Delta t_1 \ldots \Delta t_N \operatorname{Tr}[\hat{\rho}\hat{F}_{i_1}^{(-)}(s_1, \mathbf{r}_1, t_1)\ldots\hat{F}_{i_N}^{(-)}(s_N, \mathbf{r}_N, t_N)\hat{F}_{i_N}^{(+)}(s_N, \mathbf{r}_N, t_N) \\
&\quad \times \ldots\hat{F}_{i_1}^{(+)}(s_1, \mathbf{r}_1, t_1)] \\
&= C_N \Delta t_1 \ldots \Delta t_N \langle\hat{F}_{i_1}^{(-)}(s_1, \mathbf{r}_1, t_1)\ldots\hat{F}_{i_N}^{(-)}(s_N, \mathbf{r}_N, t_N)\hat{F}_{i_N}^{(+)}(s_N, \mathbf{r}_N, t_N) \\
&\quad \times \ldots\hat{F}_{i_1}^{(+)}(s_1, \mathbf{r}_1, t_1)\rangle,
\end{aligned} \tag{12.2–12}$$

where summation over repeated Cartesian indices is again understood, and C_N is some constant characteristic of the N detectors.

We now see that the calculation of the detection probability $P_N \Delta t_1 \ldots \Delta t_N$ leads to a quantum mechanical, normally ordered correlation function in which the arguments are repeated. It is worth repeating that the normal ordering appears because the fundamental measuring process is a photon absorption, and that the time ordering arises naturally as a forward ordering for the creation operators, and a backward ordering of the annihilation operators. However, for as long as it is meaningful to treat the field as a free field, since the different $\hat{F}_i^{(+)}(s, \mathbf{r}, t)$ commute among themselves for a free field, as do the different $\hat{F}_i^{(-)}(s, \mathbf{r}, t)$, the time ordering may be rearranged in the operator product. We note once again that normally ordered correlation functions are intimately related to photoelectric measurements of the field.

12.2.2 Ordering symbols and ordering operators

The expression for P_N can once again be shortened by introduction of the intensity operator. However, because of the normal and time ordering, the operator product in Eq. (12.2–12) cannot be written as a simple product of intensity operators. Several different elements of notation have been introduced to meet this situation. A pair of colons surrounding an operator (e.g. $:\hat{0}:$) is sometimes used as a normal ordering symbol that has the effect of rearranging the operator in normal order, without the use of commutation relations. For example, $:\hat{a}\hat{a}^\dagger:$ stands for the operator $\hat{a}^\dagger\hat{a}$, whereas $:\hat{a}^\dagger\hat{a}: = \hat{a}^\dagger\hat{a}$, since $\hat{a}^\dagger\hat{a}$ is already in normal order. Also $:\hat{n}^2: = \hat{a}^{\dagger 2}\hat{a}^2$, etc.

We shall denote by $\mathcal{T}$ the time ordering symbol that rearranges creation operators in forward time order and annihilation operators in backward time order, so that if $t_1 < t_2 < t_3 < t_4 < \ldots$, we have, for example,

$$\mathcal{T}\hat{F}_i^{(-)}(\mathbf{r}_2, t_2)\hat{F}_j^{(-)}(\mathbf{r}_1, t_1)\hat{F}_k^{(+)}(\mathbf{r}_3, t_3)\hat{F}_l^{(+)}(\mathbf{r}_4, t_4)$$
$$= \hat{F}_j^{(-)}(\mathbf{r}_1, t_1)\hat{F}_i^{(-)}(\mathbf{r}_2, t_2)\hat{F}_l^{(+)}(\mathbf{r}_4, t_4)\hat{F}_k^{(+)}(\mathbf{r}_3, t_3).$$

With the help of this notation we may rewrite Eq. (12.2–12) in the more compact form

$$\begin{aligned} P_N(s_1, \mathbf{r}_1, t_1; \ldots, s_N; \mathbf{r}_N, t_N)\Delta t_1 \ldots \Delta t_N &= C_N \Delta t_1 \ldots \Delta t_N \langle \mathcal{T}:\hat{I}(s_1, \mathbf{r}_1, t_1) \ldots \hat{I}(s_N, \mathbf{r}_N, t_N):\rangle \\ &= C_N \Delta t_1 \ldots \Delta t_N \langle I(s_1, \mathbf{r}_1, t_1) \ldots I(s_N, r_N, t_N)\rangle_\phi, \end{aligned} \qquad (12.2\text{–}13)$$

which shows that the joint detection probability is proportional to the normally ordered, time ordered intensity correlation function of the optical field. The last expression holds for a free field and again follows from the optical equivalence theorem. If each detector responds to all polarizations, we obtain the joint N-fold probability by making use of the total absorption operators, so that

$$P_N(\mathbf{r}_1, t_1, \ldots, \mathbf{r}_N, t_N)\Delta t_1 \ldots \Delta t_N = C_N \Delta t_1 \ldots \Delta t_N \langle \mathcal{T}:\hat{I}(\mathbf{r}_1, t_1) \ldots \hat{I}(\mathbf{r}_N, t_N):\rangle$$

$$= C_N \Delta t_1 \ldots \Delta t_N \langle I(\mathbf{r}_1, t_1) \ldots I(\mathbf{r}_N, t_N) \rangle_\phi, \tag{12.2–14}$$

with the last line again holding for a free field. This expression is similar to the semiclassical expression given by Eq. (9.6–4) when the electromagnetic field is not quantized. When the phase space functional $\phi(\{v\})$ that enters into the calculation of expectation values has the character of a classical probability functional, there is no difference between the expressions given by quantum electrodynamics and by semiclassical treatments. We shall see in Section 14.5 that Eqs. (12.2–13) and (12.2–14) can be derived by treating the interaction of the field with the detector explicitly.

The normal ordering symbol $::$ used in Eqs. (12.2–13) and (12.2–14) should not be interpreted as a linear operator, but simply as an element of notation. Although $::$ has the linear property $:\hat{A} + \hat{B}: = :\hat{A}: + :\hat{B}:$, the operation represented by $::$ may violate an equality, in that $\hat{A} = \hat{B}$ does not imply $:\hat{A}: = :\hat{B}:$. For example, from the commutation rules for the creation and annihilation operators, we have

$$\hat{a}\hat{a}^\dagger = \hat{a}^\dagger\hat{a} + 1,$$

but

$$:\hat{a}\hat{a}^\dagger: \neq :\hat{a}^\dagger\hat{a} + 1:,$$

because $:\hat{a}\hat{a}^\dagger: = \hat{a}^\dagger\hat{a} \neq \hat{a}^\dagger\hat{a} + 1$.

However, it is possible to introduce a linear operator that operates on a c-number functional of the set of complex numbers $\{v\}$, and results in a normally ordered operator in which each v is replaced by an $\hat{a}$ and each v^* by an $\hat{a}^\dagger$ [Agarwal and Wolf (1970a,b,c)]. We denote the linear normal ordering operator by $\hat{\Omega}^{(\mathrm{N})}$ and define it by the relation

$$\hat{\Omega}^{(\mathrm{N})} f(\{v\}, \{v^*\}) = f^{(\mathrm{N})}(\{\hat{a}\}, \{\hat{a}^\dagger\}), \tag{12.2–15}$$

where $f^{(\mathrm{N})}(\{\hat{a}\}, \{\hat{a}^\dagger\})$ is a normally ordered functional of the set of annihilation and creation operators. For example,

$$\hat{\Omega}^{(\mathrm{N})}(vv^*) = \hat{a}^\dagger\hat{a}.$$

We obtained an explicit expression for $\hat{\Omega}^{(\mathrm{N})}$ (as well as other ordering operators) in Section 11.10. With its help we can also write Eq. (12.2–14) in the form

$$P_N(\mathbf{r}_1, t_1, \ldots, \mathbf{r}_N, t_N)\Delta t_1 \ldots \Delta t_N$$
$$= C_N \Delta t_1 \ldots \Delta t_N \langle \mathcal{T} \hat{\Omega}^{(\mathrm{N})}[I(\mathbf{r}_1, t_1) \ldots I(\mathbf{r}_N, t_N)] \rangle. \tag{12.2–16}$$

12.3 Photon density operator

We have seen that it is convenient to associate an intensity operator $\hat{I}(\mathbf{r}, t) = \hat{\mathbf{F}}^{(-)}(\mathbf{r}, t) \cdot \hat{\mathbf{F}}^{(+)}(\mathbf{r}, t)$ with every configuration space annihilation operator $\hat{\mathbf{F}}^{(+)}(\mathbf{r}, t)$. Of the different field operators $\hat{\mathbf{F}}^{(+)}(\mathbf{r}, t)$ and $\hat{\mathbf{F}}^{(-)}(\mathbf{r}, t)$ that may be defined by the plane-wave expansions (12.2–1) and (12.2–2), the ones for which

$l(k) = 1$ have a special significance; the corresponding intensity plays the role of photon density and it is often simply related to photoelectric measurements.

Let us define the positive frequency operator $\hat{\mathbf{V}}(\mathbf{r}, t)$ by the expansion

$$\hat{\mathbf{V}}(\mathbf{r}, t) \equiv \frac{1}{L^{3/2}} \sum_{\mathbf{k},s} \hat{a}_{\mathbf{k}s} \boldsymbol{\varepsilon}_{\mathbf{k}s}\, e^{i(\mathbf{k}\cdot\mathbf{r}-\omega t)}, \quad (\omega = ck) \tag{12.3–1}$$

and consider the integral of the corresponding intensity $\hat{I}(\mathbf{r}, t) = \hat{\mathbf{V}}^\dagger(\mathbf{r}, t) \cdot \hat{\mathbf{V}}(\mathbf{r}, t)$ over all space. Then we have

$$\int_{L^3} \hat{I}(\mathbf{r}, t)\, d^3r = \frac{1}{L^3} \int_{L^3} \sum_{\mathbf{k},s} \sum_{\mathbf{k}',s'} \hat{a}^\dagger_{\mathbf{k}s} \hat{a}_{\mathbf{k}'s'} (\boldsymbol{\varepsilon}^*_{\mathbf{k}s} \cdot \boldsymbol{\varepsilon}_{\mathbf{k}'s'})\, e^{i[(\mathbf{k}'-\mathbf{k})\cdot\mathbf{r}-(\omega'-\omega)t]}\, d^3r.$$

After formally interchanging the order of summation and integration, we may carry out the integration over space and obtain $L^3 \delta^3_{\mathbf{k}\mathbf{k}'}$. We therefore have

$$\int_{L^3} \hat{I}(\mathbf{r}, t)\, d^3r = \sum_{\mathbf{k},s} \sum_{s'} \hat{a}^\dagger_{\mathbf{k}s} \hat{a}_{\mathbf{k}s'} (\boldsymbol{\varepsilon}^*_{\mathbf{k}s} \cdot \boldsymbol{\varepsilon}_{\mathbf{k}s'})$$

and, because of the orthonormality of the base vectors, this reduces to

$$\int_{L^3} \hat{I}(\mathbf{r}, t)\, d^3r = \sum_{\mathbf{k},s} \hat{a}^\dagger_{\mathbf{k}s} \hat{a}_{\mathbf{k}s} \equiv \hat{n}, \tag{12.3–2}$$

where $\hat{n}$ is the total photon number operator. Therefore, $\hat{I}(\mathbf{r}, t)$ plays the role of the photon number per unit volume, or the photon density.

We shall see in Section 12.11 that the identification of $\hat{I}(\mathbf{r}, t)$ with the photon density, although not precise, is more than purely formal, in that the integral of $\hat{I}(\mathbf{r}, t)$ over some volume smaller than L^3 (but of dimensions much larger than the wavelength) to some extent acts as a configuration space or local photon number operator. However, $\hat{I}(\mathbf{r}, t)$ should not be looked on as an operator that localizes photons precisely at position $\mathbf{r}$ at time t.

In many situations in optics the fields with which we are concerned are quasi-monochromatic, with a frequency spread that is small compared with the midfrequency. Then the function $l(k)$ in the mode expansion [Eqs. (12.2–1) to (12.2–3)] does not vary much over the range of occupied modes of the field, and behaves substantially as a constant under the summation over the occupied modes. Under these circumstances, normally ordered correlations of the $\hat{\mathbf{V}}(\mathbf{r}, t)$ operators given by Eq. (12.3–1) and their conjugates, differ only by a constant from the normally ordered correlations of the more general operators given by Eqs. (12.2–1) and (12.2–2). The reason is that the unoccupied modes of the field make no contribution to the normally ordered correlations. Similar remarks also apply to the expectation value of the light intensity $\langle \hat{I}(\mathbf{r}, t)\rangle$, which is a special case of a normally ordered correlation. For a quasi-monochromatic field we may therefore use the mean photon density as a measure of the mean field intensity.

This choice of units for $\langle \hat{I}(\mathbf{r}, t)\rangle$ is especially convenient for the description of photoelectric measurements of quasi-monochromatic light, for it leads to a very simple interpretation of the constant C_1 in Eqs. (12.2–10) or (12.2–11) for the differential detection probability. Let us consider the typical experimental situation shown in Fig. 12.2. The photosensitive surface of the photodetector is a very thin layer in a plane that is normal to the incident field. The exposed area S of the sensitive surface is made small enough to ensure that the field at the detector looks like a plane wave over S. Under these circumstances each detector electron of the exposed photocathode sees essentially the same field, and the point $\mathbf{r}$ in Eq. (12.2–11) may be taken to be any point in the exposed photocathode. The constant C_1 in Eq. (12.2–11) is then expected to be proportional to the number of exposed detector electrons, or to the exposed cathode area S. If we write for the differential detection probability

$$P_1(\mathbf{r}, t)\Delta t = \alpha c S \Delta t \langle \hat{I}(\mathbf{r}, t) \rangle, \tag{12.3–3}$$

where α is some other constant and $\langle \hat{I}(\mathbf{r}, t) \rangle$ is the expectation value of the photon density, we see that α has to be a dimensionless number. The term, $cS\Delta t \langle \hat{I}(\mathbf{r}, t) \rangle$, is the mean number of photons contained in a cylinder whose base is the area S and whose height is the distance $c\Delta t$ traveled by the light in the short measurement time Δt (see Fig. 12.2). But this is just the number of photons that would be expected to strike the exposed part of the photocathode on the average, if the photons were viewed as particles moving normally to the photocathode with velocity c. The constant α is clearly the probability that any one of these photons is detected. It is usually known as the *quantum efficiency* of the photodetector and is expressed in units of photoelectrons per photon (on the understanding that most emissions are single electron emissions).

In the same way, when we are dealing with a number of normally illuminated detectors, we may surmise that the constant C_N in Eq. (12.2–13) is given by

$$C_N = \alpha_1 \alpha_2 \ldots \alpha_N S_1 S_2 \ldots S_N c^N, \tag{12.3–4}$$

where $\alpha_1, \ldots, \alpha_N, S_1, \ldots, S_N$ are the quantum efficiencies and exposed surface areas of the detectors, provided the $\hat{I}(\mathbf{r}, t)$ operators are photon densities.

However, when the field is polychromatic or of wide bandwidth, the local correspondence between the number of photons in some region and the probability of detection breaks down, and one encounters non-local relationships. This situation will be discussed in Section 12.11.

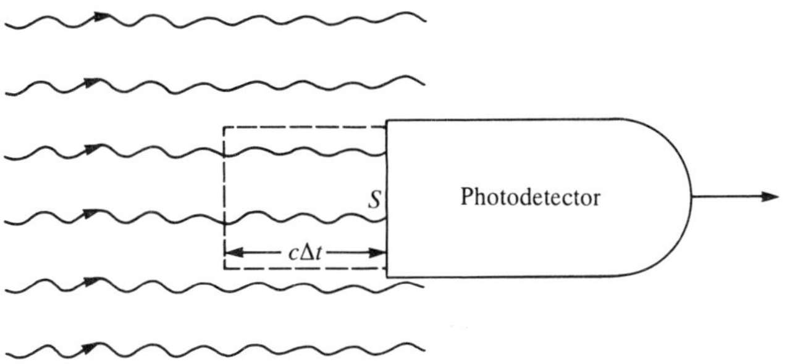

Fig. 12.2 The geometry of a photodetector of surface area S that is normal to the incident field and is exposed for a time Δt. The detector measures the number of photons in the cylinder of volume $cS\Delta t$ as shown.

12.4 Interference experiments; second-order correlation functions

The correlation functions that arise in connection with photoelectric measurements of the field are of a particular kind. They are normally ordered, have equal numbers of creation and annihilation operators, and each argument is repeated and appears once with a creation operator and once with an annihilation operator. However, in interference experiments in which two or more beams are superposed and the resultant light intensity is measured, we encounter second-order correlation functions with different arguments, exactly as in the classical treatment of Section 4.3.

Let us consider the simple interference experiment illustrated in Fig. 12.3, in which the optical field at some point $\mathbf{r}$ contains contributions from two beams emerging from two small apertures at points $\mathbf{r}_1$ and $\mathbf{r}_2$. If each of these beams can be blocked out in turn, without disturbance of the other one, we can describe the total field by a joint phase space functional $\phi_{12}(\{v_1\}, \{v_2\})$ (see Section 11.8). Then the integral over one or the other set of complex numbers $\{v_2\}$ or $\{v_1\}$ generates the phase space functional $\phi_1(\{v_1\})$ or $\phi_2(\{v_2\})$ for the field associated with one or the other beam separately. Thus,

$$\phi_1(\{v_1\}) = \int \phi_{12}(\{v_1\}, \{v_2\})\,\mathrm{d}\{v_2\}, \tag{12.4–1}$$

$$\phi_2(\{v_2\}) = \int \phi_{12}(\{v_1\}, \{v_2\})\,\mathrm{d}\{v_1\}, \tag{12.4–2}$$

while the phase space functional of the total, superposed field is given by the expression [cf. Eq. (11.8–24)]

$$\phi(\{v\}) = \int \phi_{12}(\{v_1\}, \{v - v_1\})\,\mathrm{d}\{v_1\}. \tag{12.4–3}$$

Now suppose that we wish to measure the light intensity $\hat{I}(\mathbf{r}, t)$ at the point $\mathbf{r}$ at time t. Its expectation value is given by

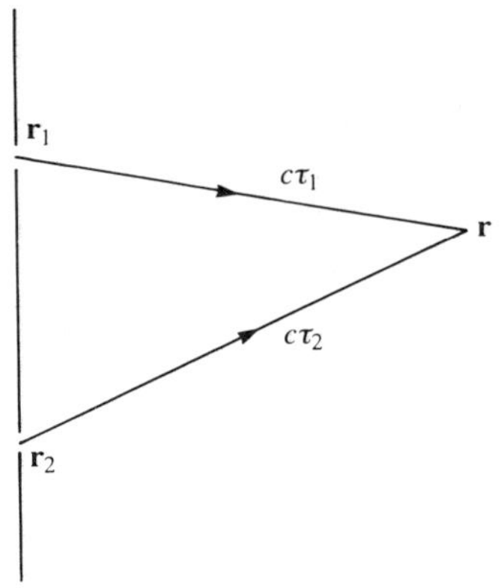

Fig. 12.3 Illustrating a simple interference experiment in which light waves from $\mathbf{r}_1$ and $\mathbf{r}_2$ are superposed at $\mathbf{r}$.

$$\langle \hat{I}(\mathbf{r}, t)\rangle = \mathrm{Tr}\,[\hat{\rho}\hat{\mathbf{F}}^{(-)}(\mathbf{r}, t)\cdot\hat{\mathbf{F}}^{(+)}(\mathbf{r}, t)]$$

$$= \int \phi\{v\}\mathbf{F}^{(-)}(\mathbf{r}, t)\cdot\mathbf{F}^{(+)}(\mathbf{r}, t)\,\mathrm{d}\{v\}, \qquad (12.4\text{–}4)$$

by the optical equivalence theorem, where $\mathbf{F}^{(+)}(\mathbf{r}, t)$ and $\mathbf{F}^{(-)}(\mathbf{r}, t)$ are the right and left eigenvalues belonging to the coherent state $|\{v\}\rangle$ of $\hat{\mathbf{F}}^{(+)}(\mathbf{r}, t)$ and $\hat{\mathbf{F}}^{(-)}(\mathbf{r}, t)$, respectively. On substituting for $\phi(\{v\})$ from Eq. (12.4–3) in Eq. (12.4–4) and putting $v - v_1 = v_2$, we obtain

$$\begin{aligned}\langle \hat{I}(\mathbf{r}, t)\rangle &= \int \phi_{12}(\{v_1\}, \{v_2\})[\mathbf{F}_1^{(-)}(\mathbf{r}, t) + \mathbf{F}_2^{(-)}(\mathbf{r}, t)]\\ &\quad\cdot[\mathbf{F}_1^{(+)}(\mathbf{r}, t) + \mathbf{F}_2^{(+)}(\mathbf{r}, t)]\,\mathrm{d}\{v_1\}\,\mathrm{d}\{v_2\}\\ &= \langle\mathbf{F}_1^{(-)}(\mathbf{r}, t)\cdot\mathbf{F}_1^{(+)}(\mathbf{r}, t)\rangle_{\phi_{12}} + \langle\mathbf{F}_2^{(-)}(\mathbf{r}, t)\cdot\mathbf{F}_2^{(+)}(\mathbf{r}, t)\rangle_{\phi_{12}}\\ &\quad+ \langle\mathbf{F}_1^{(-)}(\mathbf{r}, t)\cdot\mathbf{F}_2^{(+)}(\mathbf{r}, t)\rangle_{\phi_{12}} + \langle\mathbf{F}_2^{(-)}(\mathbf{r}, t)\cdot\mathbf{F}_1^{(+)}(\mathbf{r}, t)\rangle_{\phi_{12}}.\end{aligned} \qquad (12.4\text{–}5)$$

Here $\mathbf{F}_1^{(+)}(\mathbf{r}, t)$ and $\mathbf{F}_2^{(+)}(\mathbf{r}, t)$ are the eigenvalues of $\hat{\mathbf{F}}^{(+)}(\mathbf{r}, t)$ belonging to the coherent states $|\{v_1\}\rangle$ and $|\{v_2\}\rangle$, i.e. they are the complex functions obtained when we make the substitutions $v = v_1$ or $v = v_2$ in the expansion of $\mathbf{F}^{(+)}(\mathbf{r}, t)$. Evidently $\mathbf{F}_1^{(+)}(\mathbf{r}, t)$ and $\mathbf{F}_2^{(+)}(\mathbf{r}, t)$ refer to the fields produced by each beam separately.

As before, the first two terms on the right of Eq. (12.4–5) describe the mean light intensities at $\mathbf{r}$, t due to each beam separately, while the term $\langle\mathbf{F}_1^{(-)}(\mathbf{r}, t)\cdot\mathbf{F}_2^{(+)}(\mathbf{r}, t)\rangle_{\phi_{12}}$ describes the correlation between the two fields at $\mathbf{r}$, t.

By the optical propagation laws that follow directly from Maxwell's equations, $\mathbf{F}_1^{(+)}(\mathbf{r}, t)$ and $\mathbf{F}_2^{(+)}(\mathbf{r}, t)$ are simply related to the fields $\mathbf{F}^{(+)}(\mathbf{r}_1, t - \tau_1)$ and $\mathbf{F}^{(+)}(\mathbf{r}_2, t - \tau_2)$ at $\mathbf{r}_1$ and $\mathbf{r}_2$ at earlier times, where $c\tau_1 = |\mathbf{r} - \mathbf{r}_1|$ and $c\tau_2 = |\mathbf{r} - \mathbf{r}_2|$. For approximately quasi-monochromatic and plane-wave fields

$$\mathbf{F}_1^{(+)}(\mathbf{r}, t) \approx K_1\mathbf{F}^{(+)}(\mathbf{r}_1, t - \tau_1), \qquad (12.4\text{–}6)$$

$$\mathbf{F}_2^{(+)}(\mathbf{r}, t) \approx K_2\mathbf{F}^{(+)}(\mathbf{r}_2, t - \tau_2), \qquad (12.4\text{–}7)$$

by the same argument as in Section 4.3. The cross-correlation function appearing in Eq. (12.4–5) is therefore of the form

$$\left.\begin{aligned}\Gamma &\equiv \langle\mathbf{F}^{(-)}(\mathbf{r}_1, t - \tau_1)\cdot\mathbf{F}^{(+)}(\mathbf{r}_2, t - \tau_2)\rangle_{\phi_{12}}\\ &\equiv \langle\hat{\mathbf{F}}^{(-)}(\mathbf{r}_1, t - \tau_1)\cdot\hat{\mathbf{F}}^{(+)}(\mathbf{r}_2, t - \tau_2)\rangle,\end{aligned}\right\} \qquad (12.4\text{–}8)$$

which is a second-order, normally ordered correlation function, in which the space and time arguments are different in general, exactly as in the classical theory.

With the help of Eqs. (12.4–6) and (12.4–7) the equation (12.4–5) for the expectation value of the total intensity then takes the form

$$\begin{aligned}\langle\hat{I}(\mathbf{r}, t)\rangle &= |K_1|^2\langle\hat{I}(\mathbf{r}_1, t - \tau_1)\rangle + |K_2|^2\langle\hat{I}(\mathbf{r}_2, t - \tau_2)\rangle\\ &\quad+ 2\,\mathrm{Re}\,[K_1^*K_2\langle\hat{\mathbf{F}}^{(-)}(\mathbf{r}_1, t - \tau_1)\hat{\mathbf{F}}^{(+)}(\mathbf{r}_2, t - \tau_2)\rangle].\end{aligned} \qquad (12.4\text{–}9)$$

The similarity of Eq. (12.4–9) to the corresponding classical result [Eq. (4.3–5)] is at once apparent.

For a small variation of τ_1 and τ_2, Eq. (12.4–9) predicts a sinusoidal modulation with respect to τ_1 and τ_2, of relative modulation amplitude

$$\frac{2}{[\langle \hat{I}(\mathbf{r}_1, t-\tau_1)\rangle/\langle \hat{I}(\mathbf{r}_2, t-\tau_2)\rangle]^{1/2} + [\langle \hat{I}(\mathbf{r}_2, t-\tau_2)\rangle/\langle \hat{I}(\mathbf{r}_1, t-\tau_1)\rangle]^{1/2}} \times \frac{|\langle \hat{\mathbf{F}}^{(-)}(\mathbf{r}_1, t-\tau_1)\cdot \hat{\mathbf{F}}^{(+)}(\mathbf{r}_2, t-\tau_2)\rangle|}{[\langle \hat{I}(\mathbf{r}_1, t-\tau_1)\rangle\langle \hat{I}(\mathbf{r}_2, t-\tau_2)\rangle]^{1/2}},$$

as does the corresponding equation in the classical case. The second factor represents the degree of second-order coherence. For a stationary field, the mean light intensities are of course independent of time, and the correlation function depends only on the difference $\tau_1 - \tau_2$ between the time arguments.

12.5 Correlation functions and cross-spectral densities of arbitrary order

Let us now combine the results of the previous section with the conclusions reached in Section 12.2 regarding multiple photoelectric detections. It is not difficult to see that the description of a photoelectric correlation experiment involving two photodetectors, in which the field at each detector results from the superposition of two or more light beams, will lead to terms of the form

$$\langle \hat{F}_i^{(-)}(\mathbf{r}_1, t_1)\hat{F}_j^{(-)}(\mathbf{r}_2, t_2)\hat{F}_j^{(+)}(\mathbf{r}_3, t_3)\hat{F}_i^{(+)}(\mathbf{r}_4, t_4)\rangle.$$

This is a normally ordered correlation function of the fourth order, in which the arguments are all different. Similarly, a correlation measurement based on N photodetectors will involve a $2N$'th-order, normally ordered correlation function with (possibly) different arguments. Therefore, in a general treatment of correlation functions we have to allow the arguments to be arbitrary.

One more generalization is desirable. All the normally ordered correlations we have encountered so far are of even order, i.e. the number of creation operators in the expression equals the number of annihilation operators. This restriction also has to be removed when we encounter nonlinear dielectric media, in which optical harmonics are produced.

A general nonlinear dielectric is characterized by a set of susceptibility tensors $\boldsymbol{\chi}$, such that the induced polarization $\mathbf{P}$ is related to the incident electric field $\mathbf{E}$ by the equation[‡]

$$P_i = \varepsilon_0[\chi_{ij}E_j + \chi_{ijk}E_jE_k + \chi_{ijkl}E_jE_kE_l + \ldots], \tag{12.5–1}$$

in which the terms beyond the first represent departures from linearity. As a result, the field emerging from the dielectric is related to the incident field by a somewhat similar non-linear, tensorial relation. If we divide the real field into positive frequency and negative frequency components, as before, the expression

‡ For a discussion of nonlinear media and their propagation characteristics, see, for example, Bloembergen (1965), Shen (1984), and Boyd (1992).

for the quantum expectation value of the intensity $\hat{I}(\mathbf{r}, t)$ will involve contributions from terms of the form

$$\langle \hat{F}_i^{(-)}(\mathbf{r}, t)\hat{F}_j^{(+)}(\mathbf{r}, t)\hat{F}_k^{(+)}(\mathbf{r}, t)\rangle,$$
$$\langle \hat{F}_i^{(-)}(\mathbf{r}, t)\hat{F}_j^{(-)}(\mathbf{r}, t)\hat{F}_l^{(+)}(\mathbf{r}, t)\hat{F}_m^{(+)}(\mathbf{r}, t)\hat{F}_n^{(+)}(\mathbf{r}, t)\rangle$$

among others. These are correlation functions of odd order, in which the number of creation operators is not equal to the number of annihilation operators, and in which the Cartesian indices may all be different.

It should now be clear that the most general type of correlation function that may be encountered in measurements with photodetectors, when both interference effects and tensorial non-linearities may be present, is the normally ordered correlation of order (N, M),

$$\Gamma^{(N,M)}_{i_1,\ldots,i_N;j_M,\ldots,j_1}(\mathbf{r}_1, t_1, \ldots, \mathbf{r}_N, t_N; \mathbf{r}'_M, t'_M, \ldots, \mathbf{r}'_1, t'_1) \equiv$$
$$\langle \hat{F}_{i_1}^{(-)}(\mathbf{r}_1, t_1)\hat{F}_{i_2}^{(-)}(\mathbf{r}_2, t_2)\ldots \hat{F}_{i_N}^{(-)}(\mathbf{r}_N, t_N)\hat{F}_{j_M}^{(+)}(\mathbf{r}'_M, t'_M)\ldots \hat{F}_{j_2}^{(+)}(\mathbf{r}'_2, t'_2)\hat{F}_{j_1}^{(+)}(\mathbf{r}'_1, t'_1)\rangle. \tag{12.5–2}$$

This is a function of $N + M$ space-time points and $N + M$ indices.

It is sometimes convenient to abbreviate the notation, and to let the symbol x_l, for example, stand for the space coordinate $\mathbf{r}_l$, the time t_l, and the Cartesian component i_l, collectively. Then Eq. (12.5–2) can be written more compactly as

$$\Gamma^{(N,M)}(x_1, x_2, \ldots, x_N; y_M, \ldots, y_2, y_1) \equiv$$
$$\langle \hat{F}^{(-)}(x_1)\hat{F}^{(-)}(x_2)\ldots \hat{F}^{(-)}(x_N)\hat{F}^{(+)}(y_M)\ldots \hat{F}^{(+)}(y_2)\hat{F}^{(+)}(y_1)\rangle. \tag{12.5–3}$$

12.5.1 Properties of correlation functions

We now examine some of the properties of the normally ordered correlation functions in complete analogy with what was done in Chapter 8 for the classical correlations.‡ Indeed many of the properties of the quantum and classical correlations are similar. By inspection of Eq. (12.5–3) we have

$$\Gamma^{(N,M)*}(x_1, \ldots, x_N; y_M, \ldots, y_1) = \langle \hat{F}^{(-)}(y_1)\ldots \hat{F}^{(-)}(y_M)\hat{F}^{(+)}(x_N)\ldots \hat{F}^{(+)}(x_1)\rangle$$
$$= \Gamma^{(M,N)}(y_1, \ldots, y_M; x_N, \ldots, x_1). \tag{12.5–4}$$

Since the creation operators commute among themselves and the annihilation operators commute among themselves for a free field, it follows that

$$\Gamma^{(N,M)}(P[x_1, \ldots, x_N]; P[y_M, \ldots, y_1]) = \Gamma^{(N,M)}(x_1, \ldots, x_N; y_M, \ldots, y_1) \tag{12.5–5}$$

where $P[x_1, \ldots, x_N]$ stands for any permutation of the ordered set $x_1, \ldots, x_N$.

When $N = M$ and the collective coordinates are repeated, so that $x_i = y_i$, $i = 1, 2, \ldots$, etc., we have already seen that the resulting correlation function has the character of a joint detection probability. It follows that it must be real and

‡ Various properties of correlation functions are discussed by Glauber (1963a,b); Mehta (1966, 1967); Mehta and Mandel (1967); Mandel and Mehta (1969).

non-negative, i.e.

$$\Gamma^{(N,N)}(x_1, \ldots, x_N; x_N, \ldots, x_1) \geqslant 0. \tag{12.5–6}$$

From the Schwarz inequality applied to the operator product in Eq. (12.5–3) we obtain the inequality

$$|\Gamma^{(N,M)}(x_1, \ldots, x_N; y_M, \ldots, y_1)|^2 \leqslant \Gamma^{(N,N)}(x_1, \ldots, x_N; x_N, \ldots, x_1) \times \Gamma^{(M,M)}(y_1, \ldots, y_M; y_M, \ldots, y_1) \tag{12.5–7}$$

which reduces, for $N = M = 1$, to the familiar result

$$|\Gamma^{(1,1)}(x; y)|^2 \leqslant \langle \hat{I}(x)\rangle\langle \hat{I}(y)\rangle. \tag{12.5–8}$$

The second-order correlation functions also obey a non-negative definiteness condition more general than Eq. (12.5–8). If we set

$$\hat{O} \equiv \sum_{i=1}^{r} \lambda_i \hat{F}^{(+)}(x_i), \tag{12.5–9}$$

where the λ_i ($i = 1, 2, \ldots, r$) are arbitrary complex numbers, and x_i ($i = 1, 2, \ldots, r$) are arbitrary parameters, and note that $\hat{O}^\dagger\hat{O}$ is a non-negative definite operator, we find that

$$0 \leqslant \langle \hat{O}^\dagger \hat{O}\rangle = \sum_{i=1}^{r}\sum_{j=1}^{r} \lambda_i^* \lambda_j \Gamma^{(1,1)}(x_i; x_j). \tag{12.5–10}$$

The right side is a non-negative definite quadratic form in the λ's. The determinant formed by the coefficients $\Gamma^{(1,1)}(x_i; x_j)$ must therefore be non-negative definite also, and we may write

$$\det[\Gamma^{(1,1)}(x_i; x_j)] \geqslant 0. \tag{12.5–11}$$

For $r = 1$, Eq. (12.5–11) is a special case of Eq. (12.5–6), and for $r = 2$ it reduces to Eq. (12.5–8). When $r = 3$, Eq. (12.5–11) implies the relation

$$\begin{aligned}\langle \hat{I}(x_1)\rangle\langle \hat{I}(x_2)\rangle\langle \hat{I}(x_3)\rangle &- \langle \hat{I}(x_1)\rangle|\Gamma^{(1,1)}(x_2; x_3)|^2 - \langle \hat{I}(x_2)\rangle|\Gamma^{(1,1)}(x_3; x_1)|^2 \\ &- \langle \hat{I}(x_3)\rangle|\Gamma^{(1,1)}(x_1; x_2)|^2 \\ &+ \Gamma^{(1,1)}(x_1; x_2)\Gamma^{(1,1)}(x_2; x_3)\Gamma^{(1,1)}(x_3; x_1) + \text{c.c.} \geqslant 0,\end{aligned} \tag{12.5–12}$$

and the implications become rapidly more complicated for larger values of r. A non-negative definiteness condition of the form (12.5–10) must also be satisfied by the higher-order correlation functions of even order. If we put

$$\hat{O} \equiv \sum_{i=1}^{r} \lambda_i \hat{F}^{(+)}(x_1^{(i)}) \ldots \hat{F}^{(+)}(x_N^{(i)}),$$

where $x_n^{(i)}$ ($n = 1, 2, \ldots, N$; $i = 1, 2, \ldots, r$) are arbitrary arguments, the condition $\langle \hat{O}^\dagger \hat{O} \rangle \geqslant 0$ implies that

$$\sum_{i=1}^{r} \sum_{j=1}^{r} \lambda_i^* \lambda_j \Gamma^{(N,N)}(x_1^{(i)}, \ldots, x_N^{(i)}; x_N^{(j)}, \ldots, x_1^{(j)}) \geqslant 0. \tag{12.5–13}$$

For $r = 1$ and $r = 2$ this equation reproduces the conditions (12.5–6) and (12.5–7) with $N = M$, respectively. For $r = 3$, Eq. (12.5–13) leads to a natural generalization of Eq. (12.5–12), and the implications are increasingly less transparent for larger values of r. These relations among the quantum correlation functions do not reflect any special quantum mechanical features of the field; they are all equivalent to the corresponding relations of classical coherence theory.

The odd-order correlation functions for $N \neq M$ generally play a much less significant role than the even-order ones. We have seen that they may arise in connection with nonlinear media. We shall show in Section 12.8 that when the electromagnetic field is stationary and quasi-monochromatic, as it often is in practice, the odd-order correlations must vanish unless both N and M are very large.‡ A proof for the corresponding classical correlation functions has already been given in Section 8.3. In both the classical and the quantum mechanical cases the odd-order correlations vanish for a bandlimited electromagnetic field unless

$$\frac{|N - M|}{N + M} \leqslant \frac{1}{2}\frac{\Delta\omega}{\omega_0}, \tag{12.5–14}$$

where ω_0 is the midfrequency and $\Delta\omega$ the bandwidth of the field. In particular, for $N = 0$ or $M = 0$ we therefore have the results that

$$\langle \hat{F}^{(+)}(x_1)\hat{F}^{(+)}(x_2) \ldots \hat{F}^{(+)}(x_M)\rangle = 0, \tag{12.5–15a}$$

and

$$\langle \hat{F}^{(-)}(x_1)\hat{F}^{(-)}(x_2) \ldots \hat{F}^{(-)}(x_N)\rangle = 0, \tag{12.5–15b}$$

for a stationary field. The quantum implications of stationarity will be discussed further in Section 12.8. For the moment we may simply think of stationarity as the condition that all correlations are invariant under translations of the origin of time.

Although many properties of the quantum correlation functions correspond closely to those of the classical theory, there are certain features of the quantum correlations that have no classical analogy. Since the operators are normally ordered, the repeated application of the annihilation operators on the right, or of the creation operators on the left, will cause the correlation to vanish if the state has fewer photons than there are operators. More precisely,

$$\left.\begin{aligned} &\Gamma^{(N,M)}(x_1, \ldots, x_N; y_M, \ldots, y_1) = 0, \\ \text{if} \\ &\langle m|\hat{\rho}|n\rangle = 0 \quad \text{for all } n \geqslant N, \\ &\qquad\qquad\qquad\qquad m \geqslant M, \end{aligned}\right\} \tag{12.5–16}$$

where $|n\rangle$ is the state of total photon number n.

‡ Another proof and discussion of this point are given in the paper by Mandel and Mehta (1969).

12.5.2 *Cross-spectral densities of arbitrary order*

It is sometimes convenient also to introduce the normally ordered cross-spectral densities of order (N, M). They are defined by the formula

$$\Theta^{(N,M)}(\mathbf{k}_1, s_1, \ldots, \mathbf{k}_N, s_N; \mathbf{k}'_M, s'_M, \ldots, \mathbf{k}'_1, s'_1) \equiv \langle \hat{a}^\dagger_{\mathbf{k}_1 s_1} \ldots \hat{a}^\dagger_{\mathbf{k}_N s_N} \hat{a}_{\mathbf{k}'_M s'_M} \ldots \hat{a}_{\mathbf{k}'_1 s'_1} \rangle, \quad (12.5\text{–}17)$$

and they are scalar functions of $N + M$ wave vectors and $N + M$ polarization indices. For certain states of the field $\Theta^{(N+M)}$ may not be finite, and a number of products of delta functions of the difference of two wave vectors can sometimes be factored out of $\Theta^{(N+M)}$. These higher-order spectral densities satisfy several conditions that are analogous to those satisfied by the correlation functions $\Gamma^{(N,M)}$, such as the conjugation condition

$$\Theta^{(N,M)*}(\mathbf{k}_1, s_1, \ldots, \mathbf{k}_N, s_N; \mathbf{k}'_M, s'_M, \ldots, \mathbf{k}'_1, s'_1) \equiv \Theta^{(M,N)}(\mathbf{k}'_1, s'_1, \ldots, \mathbf{k}'_M, s'_M; \mathbf{k}_N, s_N, \ldots, \mathbf{k}_1, s_1), \quad (12.5\text{–}18)$$

the non-negativeness condition

$$\Theta^{(N,N)}(\mathbf{k}_1, s_1, \ldots, \mathbf{k}_N, s_N; \mathbf{k}_N, s_N, \ldots, \mathbf{k}_1, s_1) \geq 0, \quad (12.5\text{–}19)$$

and the inequality

$$|\Theta^{(N,M)}(\mathbf{k}_1, s_1, \ldots, \mathbf{k}_N, s_N; \mathbf{k}'_M, s'_M, \ldots, \mathbf{k}'_1, s'_1)|^2 \leq \Theta^{(N,N)}(\mathbf{k}_1, s_1, \ldots, \mathbf{k}_N, s_N; \mathbf{k}_N, s_N, \ldots, \mathbf{k}_1, s_1) \times \Theta^{(M,M)}(\mathbf{k}'_1, s'_1, \ldots, \mathbf{k}'_M, s'_M; \mathbf{k}'_M, s'_M, \ldots, \mathbf{k}'_1, s'_1). \quad (12.5\text{–}20)$$

From the expansions (12.2–1) and (12.2–2) of the $\hat{\mathbf{F}}^{(+)}(\mathbf{r}, t)$ and $\hat{\mathbf{F}}^{(-)}(\mathbf{r}, t)$ operators in terms of $\hat{a}_{\mathbf{k}s}$ and $\hat{a}^\dagger_{\mathbf{k}s}$ operators, we may express the correlation $\Gamma^{(N,M)}$ as a multiple Fourier transform of the spectral density $\Theta^{(N,M)}$ (in the limit where the Fourier series may be replaced by an integral), and vice versa. For example, by inversion of Eq. (12.2–1) we have

$$\hat{a}_{\mathbf{k}s} = \frac{1}{L^{3/2}} \frac{1}{l(k)} \int \boldsymbol{\varepsilon}^*_{\mathbf{k}s} \cdot \hat{\mathbf{F}}^{(+)}(\mathbf{r}, t)\, \mathrm{e}^{-\mathrm{i}(\mathbf{k}\cdot\mathbf{r}-\omega t)}\, \mathrm{d}^3 r \quad (12.5\text{–}21)$$

and when this result is used in the definition (12.5–17) of $\Theta^{(N,M)}$, we obtain the relation

$$\Theta^{(N,M)}(\mathbf{k}_1, s_1, \ldots, \mathbf{k}_N, s_N; \mathbf{k}'_M, s'_M, \ldots, \mathbf{k}'_1, s'_1) = \left(\frac{1}{L^{3/2}}\right)^{N+M} \int \Gamma^{(N,M)}(\mathbf{r}_1, t_1, i_1, \ldots, \mathbf{r}_N, t_N, i_N; \mathbf{r}'_M, t'_M, j'_M, \ldots, \mathbf{r}'_1, t'_1, j'_1) \times \prod_{n=1}^{N} \prod_{m=1}^{M} \left[\frac{1}{l^*(k_n)} \frac{1}{l(k'_m)} \times (\boldsymbol{\varepsilon}_{\mathbf{k}_n s_n})_{i_n} (\boldsymbol{\varepsilon}^*_{\mathbf{k}'_m s'_m})_{j_m} \exp \mathrm{i}(\mathbf{k}_n \cdot \mathbf{r}_n - \mathbf{k}'_m \cdot \mathbf{r}'_m - \omega_n t_n + \omega'_m t'_m)\, \mathrm{d}^3 r_n\, \mathrm{d}^3 r'_m \right]. \quad (12.5\text{–}22)$$

It follows immediately that $\Theta^{(N,M)}$ vanishes whenever $\Gamma^{(N,M)}$ vanishes, and vice versa.

12.5.3 Phase-independent phase space functions

A particularly interesting situation arises when $\Theta^{(N,M)}(\mathbf{k}_1, s_1, \ldots, \mathbf{k}_N, s_N; \mathbf{k}'_M, s'_M, \ldots, \mathbf{k}'_1, s'_1)$ is non-zero only for repeated indices, i.e. for $N = M$, and $\mathbf{k}_i = \mathbf{k}'_i$, $s_i = s'_i$, $i = 1, 2, \ldots, N$. It is not difficult to show that this has certain implications regarding the state of the field, which are most readily expressed as a condition on the phase space functional $\phi(\{v\})$ in the diagonal coherent state representation of the density operator. Another proof and discussion of this point have been given by Mandel and Mehta (1969).

Consider the normally ordered characteristic functional $C_{\mathrm{N}}(\{u\})$ defined by Eq. (11.11–13). By repeated differentiation of $C_{\mathrm{N}}(\{u\})$ we may generate a spectral density $\Theta^{(N,M)}$ of arbitrary order,‡

$$\Theta^{(N,M)}(\mathbf{k}_1, s_1, \ldots, \mathbf{k}_N, s_N; \mathbf{k}'_M, s'_M, \ldots, \mathbf{k}'_1, s'_1) = \prod_{n=1}^{N}\left(\frac{\partial}{\partial u_{\mathbf{k}_n s_n}}\right)\prod_{m=1}^{M}\left(-\frac{\partial}{\partial u^*_{\mathbf{k}'_m s'_m}}\right)C_{\mathrm{N}}(\{u\})\Bigg|_{\substack{\{u\}=0\\ \{u^*\}=0}}, \qquad (12.5\text{–}23)$$

so that $C_{\mathrm{N}}(\{u\})$ is a generating functional for normally ordered spectral densities. Alternatively, by expanding the exponential terms in the definition of $C_{\mathrm{N}}(\{u\})$, we may express $C_{\mathrm{N}}(\{u\})$ as a series involving the spectral densities. Thus,

$$C_{\mathrm{N}}(\{u\}) = \sum_{\{n\}}\sum_{\{m\}}\left\langle :\prod_{\mathbf{k},s}\frac{u_{\mathbf{k}s}^{n_{\mathbf{k}s}}(-u^*_{\mathbf{k}s})^{m_{\mathbf{k}s}}}{n_{\mathbf{k}s}!\,m_{\mathbf{k}s}!}(\hat{a}^\dagger_{\mathbf{k}s})^{n_{\mathbf{k}s}}(\hat{a}_{\mathbf{k}s})^{m_{\mathbf{k}s}}:\right\rangle. \qquad (12.5\text{–}24)$$

If the spectral densities all vanish except for repeated indices, the double summation reduces to a single summation, and we have

$$C_{\mathrm{N}}(\{u\}) = \sum_{\{n\}}\left\langle :\prod_{\mathbf{k},s}\frac{(-|u_{\mathbf{k}s}|^2)^{n_{\mathbf{k}s}}}{(n_{\mathbf{k}s}!)^2}(\hat{a}^\dagger_{\mathbf{k}s})^{n_{\mathbf{k}}}(\hat{a}_{\mathbf{k}s})^{n_{\mathbf{k}s}}:\right\rangle. \qquad (12.5\text{–}25)$$

This shows that the characteristic functional $C_{\mathrm{N}}(\{u\})$ is actually a functional only of the set of moduli $\{|u|\}$, and is independent of the phases of $\{u\}$. But since the phase space functional $\phi(\{v\})$ is expressible as a multiple Fourier transform of $C_{\mathrm{N}}(\{u\})$ according to Eq. (11.11–14), it follows that $\phi(\{v\})$ similarly is a functional only of the set of moduli $\{|v|\}$. For

$$\phi(\{v\}) = \int C_{\mathrm{N}}(\{u\})\prod_{\mathbf{k},s}\left\{\frac{1}{\pi^2}\exp\left(u^*_{\mathbf{k}s}v_{\mathbf{k}s} - u_{\mathbf{k}s}v^*_{\mathbf{k}s}\right)\mathrm{d}^2u_{\mathbf{k}s}\right\}$$

$$= \int_{\{r\}=0}^{\infty}\int_{\{\theta\}=0}^{2\pi} C_{\mathrm{N}}(\{r\})\prod_{\mathbf{k},s}\left\{\frac{1}{\pi^2}\exp\left[2\mathrm{i}r_{\mathbf{k}s}|v_{\mathbf{k}s}|\sin\left(\arg v_{\mathbf{k}s} - \theta_{\mathbf{k}s}\right)\right]r_{\mathbf{k}s}\,\mathrm{d}r_{\mathbf{k}s}\,\mathrm{d}\theta_{\mathbf{k}s}\right\},$$

$$(12.5\text{–}26)$$

‡ The suffix N of $C_{\mathrm{N}}(\{u\})$ stands for 'normal ordering', and has nothing to do with the indices N, M of $\Theta^{(N,M)}$.

where $u_{\mathbf{k}s} = r_{\mathbf{k}s}\,\mathrm{e}^{\mathrm{i}\theta_{\mathbf{k}s}}$. By carrying out the integration over the $\{\theta\}$ we see at once that $\phi(\{v\})$ does not depend on the phases of $\{v\}$. The converse of this result is also true, and the spectral densities without repeated indices all vanish when $\phi(\{v\})$ is a functional only of the set of moduli $\{|v|\}$. This can be seen most easily if the spectral densities are expressed as c-number integrals with the help of the optical equivalence theorem. In particular, both the spectral densities $\Theta^{(N,M)}$ and the correlation functions $\Gamma^{(N,M)}$ with $N \neq M$ must vanish if ϕ is of the form $\phi(\{|v|\})$.

We shall find in Section 12.8 that a phase space functional of the form $\phi(\{|v|\})$ implies that the electromagnetic field is both stationary and homogeneous. The vanishing of the spectral densities, except for repeated indices, is therefore sufficient to ensure both stationarity and homogeneity. Of course, in practice, these conditions are not likely to be satisfied exactly to all orders. However, from the nature of the expansion (12.5–24), which involves the factors $n_{\mathbf{k}s}!m_{\mathbf{k}s}!$ in the denominator, we might expect the contributions of the high-order terms to be less than those of the low-order terms. The low-order spectral densities will therefore tend to dominate the behavior of $C_{\mathrm{N}}(\{u\})$, and $C_{\mathrm{N}}(\{u\})$ will be almost phase independent if the low-order spectral densities vanish except for repeated indices.

12.6 Degree and order of coherence

In the discussion in Sections 4.3 and 12.4 of interference between two light beams we were led naturally to a normalized second-order correlation function

$$\frac{\Gamma^{(1,1)}(x, y)}{[\Gamma^{(1,1)}(x, x)\Gamma^{(1,1)}(y, y)]^{1/2}} \equiv \gamma^{(1,1)}(x, y), \qquad (12.6\text{–}1)$$

provided $\Gamma^{(1,1)}(x,x)$ and $\Gamma^{(1,1)}(y,y)$ do not vanish. $|\gamma^{(1,1)}(x,y)|$ provides a natural measure of the degree of coherence of certain Cartesian components of the optical field at space-time points x and y, in the context of second-order correlations, that is simply related to the visibility of the interference fringes formed. By virtue of the inequality (12.5–7), $\gamma^{(1,1)}(x, y)$ is automatically normalized so that

$$0 \leqslant |\gamma^{(1,1)}(x, y)| \leqslant 1. \qquad (12.6\text{–}2)$$

Moreover, as we have already found from our classical treatment of coherence in Section 4.5, and as we show in Section 12.7, the condition $|\gamma^{(1,1)}(x, y)| = 1$ implies the factorization of $\Gamma^{(1,1)}(x, y)$ into a product of the form $U^*(x)U(y)$, and therefore agrees with intuitive notions of what is implied by coherence.

Several attempts have been made to introduce normalized correlation functions of higher order, and, by implication, to define a degree of coherence of arbitrary order (N, M). When the normalized function is unimodular, this would imply that the field has unit degree of coherence to order (N, M). None of these normalized correlation functions has as meaningful an interpretation as the second-order one given above. Let us examine some of them.

For even-order correlation functions of type $\Gamma^{(N,N)}$, Glauber (1963a) has extended the definition of $\gamma^{(1,1)}$ given by Eq. (12.6–1). He introduced the normalized quantity‡

$$g^{(N,N)}(x_1, \ldots, x_N; y_N, \ldots, y_1) \equiv \frac{\Gamma^{(N,N)}(x_1, \ldots, x_N; y_N, \ldots, y_1)}{\prod_{r=1}^{N} [\Gamma^{(1,1)}(x_r; x_r)\Gamma^{(1,1)}(y_r; y_r)]^{1/2}}. \quad (12.6\text{–}3)$$

Leaving aside the mathematical significance of this quantity for the moment, we can see that if $g^{(N,N)}$ were unity with $x_r = y_r$, $r = 1, 2, \ldots, N$, then $\Gamma^{(N,N)}(x_1, \ldots, x_N; x_N, \ldots, x_1)$ would factorize into the product $\prod_{r=1}^{N}\langle \hat{I}(x_r)\rangle$. But $\Gamma^{(N,N)}(x_1, \ldots, x_N; x_N, \ldots, x_1)$ is the normally ordered intensity correlation function, which is proportional to the joint probability of photodetection at N space-time points [cf. Eq. (12.2–14)]. The factorization therefore implies that the joint N-fold detection probability reduces to the product of the N separate detection probabilities, so that the photodetection events are independent. In other words, the condition $g^{(N,N)}(x_1, \ldots, x_N; x_N, \ldots, x_1) = 1$ ensures that no N'th-order intensity or photoelectric correlations exist at the N space-time points.

It seems tempting to try and interpret $|g^{(N,N)}(x_1, \ldots, x_N; y_N, \ldots, y_1)|$ as the *degree of coherence of order* (N, N), by analogy with $|\gamma^{(1,1)}|$. A field would then be said to be coherent up to order (N, N), if $|g^{(r,r)}(x_1, \ldots, x_r; y_r, \ldots, y_1)|$ $(r = 1, 2, \ldots)$ were unity for all $r \leq N$, in the sense that no intensity correlations up to the N'th order are present. Coherence to any order (N, N) would then automatically imply coherence to any lower order (r, r), with $r < N$. Unfortunately, $|g^{(N,N)}|$ does not in general satisfy an inequality of the type (12.6–2), unless $N = 1$, so that it is not very meaningful to interpret $|g^{(N,N)}|$ as a degree of (N, N)'th-order coherence. Moreover, the condition $|g^{(N,N)}| = 1$ does not in general imply the factorization of the correlation function $\Gamma^{(N,N)}$ into a product of complex factors, as it does for $N = 1$, nor does $g^{(N,N)}(x, \ldots, x; x, \ldots, x)$ in general equal unity. We can however adopt the factorization of $\Gamma^{(r,r)}$ in the form

$$\Gamma^{(r,r)}(x_1, \ldots, x_r; y_r, \ldots, y_1) = U^*(x_1) \ldots U^*(x_r)U(y_r) \ldots U(y_1), \text{ for all } r \leq N, \quad (12.6\text{–}4)$$

to be the condition for coherence up to order (N, N), and this automatically implies that $|g^{(N,N)}| = 1$.

It is shown in Section 13.3 (and a corresponding classical result can be obtained – see Section 8.4) that for polarized light from all thermal sources,

$$\Gamma^{(N,N)}(x, \ldots, x; x, \ldots, x) \equiv \langle{:}\hat{I}(x)^N{:}\rangle = N!\langle \hat{I}(x)\rangle^N, \quad (12.6\text{–}5)$$

so that

$$g^{(N,N)}(x, \ldots, x; x, \ldots, x) = N! \quad (12.6\text{–}6)$$

It follows that, on the basis of this definition of (N, N)'th-order coherence, no thermal radiation field can be coherent beyond the second order.

A different normalized correlation function for arbitrary order (N, M) was introduced by Mehta, (1966, 1967), in an attempt to satisfy the same inequality as

‡ Glauber (1963a) considered only even-order correlations of type $\Gamma^{(N,N)}$ to which he referred as 'N'th order correlations'.

$\gamma^{(1,1)}$. He defined

$$\gamma^{(N,M)}(x_1, \ldots, x_N; y_M, \ldots, y_1) \equiv \frac{\Gamma^{(N,M)}(x_1, \ldots, x_N; y_M, \ldots, y_1)}{\prod_{r=1}^{N}[\Gamma^{(N,N)}(x_r, \ldots, x_r; x_r, \ldots, x_r)]^{1/2N} \prod_{r=1}^{M}[\Gamma^{(M,M)}(y_r, \ldots, y_r; y_r, \ldots, y_r)]^{1/2M}}, \tag{12.6–7}$$

provided the correlation functions in the denominator are non-zero. This function evidently satisfies the relation

$$\gamma^{(N,N)}(x, \ldots, x; x, \ldots, x) = 1. \tag{12.6–8}$$

Moreover, when $|\gamma^{(N,N)}(x_1, \ldots, x_N; y_N, \ldots, y_1)| = 1$, the correlation function $\Gamma^{(N,N)}(x_1, \ldots, x_N; y_N, \ldots, y_1)$ may be shown to factorize in the form (12.6–4), so that the condition $|\gamma^{(N,N)}| = 1$ appears to be a condition for (N, N)'th-order coherence in an intuitively acceptable sense. A proof of the factorization property is given in Section 12.7, based on the argument of Titulaer and Glauber (1965). For fields for which the phase space functional $\phi(\{v\})$ in the diagonal coherent-state representation is non-negative and has the character of a classical probability density, it may also be shown that

$$0 \leq |\gamma^{(N,M)}(x_1, \ldots, x_N; y_M, \ldots, y_1)| \leq 1, \tag{12.6–9}$$

so that $|\gamma^{(N,M)}|$ might be interpreted as a degree of (N, M)'th-order coherence in a certain sense.

Unfortunately, this inequality is not satisfied in general for a quantized field, for which there appears to be no upper bound for $|\gamma^{(N,M)}|$. Moreover, coherence to any even order (N, N) automatically implies coherence to all higher even orders when there is no bound to the number of photons, and to higher even orders up to (M, M), with $M \leq n$, when there are n photons present (in which case $\Gamma^{(r,r)}$ vanishes and $\gamma^{(r,r)}$ is not defined for $r > n$). On this definition of higher-order coherence, second-order coherence is the strongest form of coherence, with consequences for the higher-order coherence, rather than the other way around. As we show in the next section second-order coherence does indeed carry implications for the higher-order correlation functions.

Yet another normalized correlation function was discussed by Sudarshan (1969; see also Klauder and Sudarshan, 1968, Chap. 8, Eq. (8–84)), who defined

$$s^{(N,M)}(x_1, \ldots, x_N; y_M, \ldots, y_1) \equiv \frac{\Gamma^{(N,M)}(x_1, \ldots, x_N; y_M, \ldots, y_1)}{[\Gamma^{(N,N)}(x_1, \ldots, x_N; x_N, \ldots, x_1)\Gamma^{(M,M)}(y_1, \ldots, y_M; y_M, \ldots, y_1)]^{1/2}} \tag{12.6–10}$$

provided the correlations in the denominator are non-vanishing. These functions satisfy the condition

$$s^{(N,N)}(x, \ldots, x; x, \ldots, x) = 1, \tag{12.6–11}$$

and the inequality

$$0 \leqslant |s^{(N,M)}(x_1, \ldots, x_N; y_M, \ldots, y_1)| \leqslant 1 \tag{12.6–12}$$

for all fields. In addition, they share the property of $|\gamma^{(N,M)}|$ that $|s^{(1,1)}(x; y)| = 1$ implies $|s^{(N,N)}(x_1, \ldots, x_N; y_N, \ldots, y_1)| = 1$ for all N, provided the number of photons is not bounded. However, $s^{(N,N)}(x_1, \ldots, x_N; x_N, , \ldots, x_1) = 1$ of course, which makes the interpretations of $s^{(N,N)}$ and $\gamma^{(N,N)}$ altogether different.

Although there is no question that the higher-order correlation functions are important for the complete description of the electromagnetic field and its coherence properties, the usefulness of the concept of higher-order coherence, and of the degree of higher-order coherence, is doubtful. Admittedly, a coherent state of the field is coherent to all orders, and the normalized correlation function is unimodular on just about all definitions. However, no examples are known of fields that are coherent to some orders (except the second) but not others. It is questionable whether such fields are ever encountered in practice, or whether they exist at all. It may be that the concept of second-order coherence, which was the starting point for the development of coherence theory, is after all the only really meaningful one.

12.7 Implications of second-order coherence

In this section we shall examine some of the consequences of the condition of second-order coherence

$$|\gamma^{(1,1)}(x; y)| = 1, \tag{12.7–1}$$

and show that the condition has certain implications for the state of the electromagnetic field, and for the higher-order correlation functions, as was first pointed out by Titulaer and Glauber (1965, 1966). Not only $\Gamma^{(1,1)}(x, y)$, but all the correlation functions $\Gamma^{(N,M)}(x_1, \ldots, x_N; y_M, \ldots, y_1)$ will be shown to factorize into a product of functions associated with each space-time component x_r ($r = 1, 2, \ldots, N$) and y_r ($r = 1, 2, \ldots, M$). Some of the implications of complete coherence will be found to be analogous to those encountered in Sections 4.7 and 8.5.2 for classical fields.

12.7.1 Factorization of the correlation functions‡

We start from the general operator Schwarz inequality

$$|\langle \hat{A}^\dagger \hat{B} \rangle|^2 \leqslant \langle \hat{A}^\dagger \hat{A} \rangle \langle \hat{B}^\dagger \hat{B} \rangle, \tag{12.7–2}$$

holding for any two operators $\hat{A}$ and $\hat{B}$ for which the various expectation values exist. In particular, if

$$\langle \hat{A}^\dagger \hat{A} \rangle = 0,$$

then

$$\left.\begin{aligned} &\langle \hat{A}^\dagger \hat{A} \rangle = 0, \\ &\langle \hat{A}^\dagger \hat{B} \rangle = 0 = \langle \hat{B}^\dagger \hat{A} \rangle, \end{aligned}\right\} \tag{12.7–3}$$

‡ In this section we follow essentially the argument of Titulaer and Glauber (1965).

for any operator $\hat{B}$. The expectation values can be written out explicitly in terms of the density operator $\hat{\rho}$, so that we have

$$\mathrm{Tr}\,(\hat{\rho}\hat{A}^\dagger\hat{B}) = 0 = \mathrm{Tr}\,(\hat{\rho}\hat{B}^\dagger\hat{A}) = \mathrm{Tr}\,(\hat{A}\hat{\rho}\hat{B}^\dagger),$$

and since $\hat{B}$ is completely arbitrary, this requires that

$$\hat{\rho}\hat{A}^\dagger = 0 = \hat{A}\hat{\rho}, \quad \text{if } \mathrm{Tr}\,(\hat{\rho}\hat{A}^\dagger\hat{A}) = 0. \tag{12.7–4}$$

An example of an operator $\hat{A}$ satisfying the condition $\langle\hat{A}^\dagger\hat{A}\rangle = 0$ is the following:

$$\hat{A} \rightarrow \hat{F}^{(+)}(x) - \frac{\Gamma^{(1,1)*}(x;\, y)}{\Gamma^{(1,1)}(y;\, y)}\hat{F}^{(+)}(y), \tag{12.7–5}$$

provided $\Gamma^{(1,1)}(y;\, y) \neq 0$, and provided that Eq. (12.7–1) holds, so that

$$|\Gamma^{(1,1)}(x;\, y)|^2 = \Gamma^{(1,1)}(x;\, x)\Gamma^{(1,1)}(y;\, y), \tag{12.7–6}$$

as may readily be verified. With this choice of $\hat{A}$, the relations (12.7–4) become

$$\hat{F}^{(+)}(x)\hat{\rho} = \frac{\Gamma^{(1,1)*}(x;\, y)}{\Gamma^{(1,1)}(y;\, y)}\hat{F}^{(+)}(y)\hat{\rho}, \tag{12.7–7}$$

and

$$\hat{\rho}\hat{F}^{(-)}(x) = \frac{\Gamma^{(1,1)}(x;\, y)}{\Gamma^{(1,1)}(y;\, y)}\hat{\rho}\hat{F}^{(-)}(y). \tag{12.7–8}$$

We now apply these equations to the calculation of the correlation function $\Gamma^{(1,1)}(x;\, y)$. We first suppose that there exists a third parameter z for which $\Gamma^{(1,1)}(z;\, z) \neq 0$, such that the degree of coherence is unity between any two of the parameters x, y, z. We may then replace x or y by z in any of the foregoing equations. From the definition,

$$\Gamma^{(1,1)}(x;\, y) \equiv \mathrm{Tr}\,[\hat{\rho}\hat{F}^{(-)}(x)\hat{F}^{(+)}(y)],$$

and from Eq. (12.7–8) with y replaced by z we have

$$\begin{aligned}\Gamma^{(1,1)}(x;\, y) &= \frac{\Gamma^{(1,1)}(x;\, z)}{\Gamma^{(1,1)}(z;\, z)}\mathrm{Tr}\,[\hat{\rho}\hat{F}^{(-)}(z)\hat{F}^{(+)}(y)] \\ &= \Gamma^{(1,1)}(x;\, z)\Gamma^{(1,1)}(z;\, y)/\Gamma^{(1,1)}(z;\, z).\end{aligned} \tag{12.7–9}$$

If z is a fixed point, this shows that the correlation function has factorized into the product of a function of x alone and a function of y alone. Of course, if x and y were fixed points this conclusion would not have much significance, but we may suppose that there exists at least a neighborhood of points x and y for which Eq. (12.7–9) holds. By defining

$$U(y) \equiv \frac{\Gamma^{(1,1)}(z;\, y)}{[\Gamma^{(1,1)}(z;\, z)]^{1/2}}, \tag{12.7–10}$$

we may re-express this result in the compact form

$$\Gamma^{(1,1)}(x;\, y) = U^*(x)U(y), \tag{12.7–11}$$

just as for classical fields having second-order coherence (see Section 4.5). However, we must emphasize that the functions $U^*(x)$, $U(y)$ are not in general the expectation values of $\hat{F}^{(-)}(x)$ and $\hat{F}^{(+)}(y)$, as they would be if the field were in a coherent state. The coherent state is a rather special example of a field having second-order coherence.

The function $U(x)$ appears to depend on the choice of the fixed point z in such a way that a different fixed point z' results in a different function $U'(x)$. However, as we must have

$$U^*(x)U(y) = U'^*(x)U'(y), \tag{12.7–12}$$

for a range of values of x and y, $U(x)$ and $U'(x)$ must be related through a constant unimodular factor, that is not of much significance.

12.7.2 Correlations of arbitrary order

We may apply a similar argument to calculate a correlation function $\Gamma^{(N,M)}$ of arbitrary order for a free field,

$$\Gamma^{(N,M)}(x_1, \ldots, x_N; y_M, \ldots, y_1) = \\ \mathrm{Tr}[\hat{\rho}\hat{F}^{(-)}(x_1) \ldots \hat{F}^{(-)}(x_M)\hat{F}^{(+)}(y_M) \ldots \hat{F}^{(+)}(y_1)]. \tag{12.7–13}$$

Because the creation operators commute with one another, as do the annihilation operators, and the trace is invariant under cyclic permutation of operators, each operator in Eq. (12.7–13) may be brought adjacent to $\hat{\rho}$ in turn. If the field is coherent to the second order between any of the points $x_1, \ldots, x_N, y_1, \ldots, y_M$ and the reference point z, we simply apply the relation (12.7–8) N times with $y = z$, and $x = x_1, x_2, \ldots, x_N$ in turn, and the relation (12.7–7) M times with $y = z$ and $x = y_1, y_2, \ldots, y_M$ in turn. We then obtain the formula

$$\begin{aligned}
&\Gamma^{(N,M)}(x_1, \ldots, x_N; y_M, \ldots, y_1) \\
&\quad = \prod_{r=1}^{N}\left[\frac{\Gamma^{(1,1)}(x_r; z)}{\Gamma^{(1,1)}(z; z)}\right]\prod_{r=1}^{M}\left[\frac{\Gamma^{(1,1)*}(y_r; z)}{\Gamma^{(1,1)}(z; z)}\right]\mathrm{Tr}[\hat{\rho}(\hat{F}^{(-)}(z))^N(\hat{F}^{(+)}(z))^M] \\
&\quad = \frac{\Gamma^{(N,M)}(z, \ldots, z; z, \ldots, z)}{[\Gamma^{(1,1)}(z; z)]^{(N+M)/2}}U^*(x_1) \ldots U^*(x_N)U(y_M) \ldots U(y_1) \\
&\quad = g^{(N,M)}U^*(x_1) \ldots U^*(x_N)U(y_M) \ldots U(y_1),
\end{aligned} \tag{12.7–14}$$

when we introduce the functions U given by Eq. (12.7–10) and let

$$g^{(N,M)} \equiv \Gamma^{(N,M)}(z, \ldots, z; z, \ldots, z)/[\Gamma^{(1,1)}(z; z)]^{(N+M)/2}. \tag{12.7–15}$$

Once again we see that $\Gamma^{(N,M)}$ factorizes into a product of functions of one parameter, in the manner that we might intuitively expect for a coherent field, although $U(x)$ and $U^*(x)$ are not, in general, expectation values of $\hat{F}^{(+)}(x)$ and $\hat{F}^{(-)}(x)$. It is really rather remarkable that the second-order coherence condition (12.7–1) has such a profound consequence for the higher-order correlation functions. However, the (generally) complex quantity $g^{(N,M)}$, which is independent of $x_1, \ldots, x_N, y_1, \ldots, y_M$, depends on the order (N, M). Although $g^{(N,M)}$ is a constant, its modulus generally departs from unity. In a sense $|g^{(N,M)}|$ is a measure of the departure of the field from the completely coherent state, for which $|g^{(N,M)}| = 1$ and for which the factorization of $\Gamma^{(N,M)}$ follows from first principles. For such a coherent state $|\{v\}\rangle$, $U(y)$ in Eq. (12.7–10) reduces to

$$U(y) = \exp[-\mathrm{i} \arg F^{(+)}(z)] F^{(+)}(y), \tag{12.7–16}$$

where $F^{(+)}(y)$ and $F^{(+)}(z)$ are the eigenvalues of $\hat{F}^{(+)}(y)$ and $\hat{F}^{(+)}(z)$ belonging to the coherent state $|\{v\}\rangle$.

12.7.3 Density operator of the field

We now examine some of the implications of the second-order coherence condition (12.7–1) for the state of the electromagnetic field. We start with some simple examples of fields having second-order coherence, and then show that a generalization of the examples leads to an interesting characterization of the field as having only one type of excitation present (cf. also Sections 4.7 and 8.5.2). We follow the outline of the argument by Titulaer and Glauber (1966).

Consider a state of the field such as‡ $|0, \ldots, 0, 1_{\mathbf{k}s}, 0, \ldots, 0\rangle$, having only one photon of type $\mathbf{k}, s$. By using the expansions (12.2–1) and (12.2–2) for $\hat{\mathbf{F}}^{(+)}(\mathbf{r}, t)$ and $\hat{\mathbf{F}}^{(-)}(\mathbf{r}, t)$ we readily find that

$$\begin{aligned} \Gamma^{(1,1)}_{i;j}(\mathbf{r}_1, t_1; \mathbf{r}_2, t_2) &= \langle 0, \ldots, 0, 1_{\mathbf{k}s}, 0, \ldots, 0|\hat{F}^{(-)}_i(\mathbf{r}_1, t_1)\hat{F}^{(+)}_j(\mathbf{r}_2, t_2)|0, \ldots, 0, 1_{\mathbf{k}s}, 0, \ldots, 0\rangle \\ &= \frac{1}{L^3}|l(k)|^2(\boldsymbol{\varepsilon}^*_{\mathbf{k},s})_i(\boldsymbol{\varepsilon}_{\mathbf{k}s})_j\, \mathrm{e}^{-\mathrm{i}(\mathbf{k}\cdot\mathbf{r}_1-\omega t_1)}\, \mathrm{e}^{\mathrm{i}(\mathbf{k}\cdot\mathbf{r}_2-\omega t_2)}. \end{aligned} \tag{12.7–17}$$

This shows that $\Gamma^{(1,1)}$ factorizes into a product of a function of $\mathbf{r}_1$, t_1, i and a function of $\mathbf{r}_2$, t_2, j, as required in order to satisfy the condition for second-order coherence. A field that is empty except for one photon in one mode is therefore coherent to the second order.

Two different generalizations of this result can be made immediately. In the first place, we consider an arbitrary one-photon state, in which the photon is not necessarily associated with just one mode $\mathbf{k}, s$. The density operator for a pure one-photon state of this type is given by

$$\hat{\rho} = \sum_{\mathbf{k},s}\sum_{\mathbf{k}',s'} \alpha(\mathbf{k}, s)\alpha^*(\mathbf{k}', s')|0, \ldots, 0, 1_{\mathbf{k}s}, 0, \ldots, 0\rangle\langle 0, \ldots, 0, 1_{\mathbf{k}'s'}, 0, \ldots, 0|, \tag{12.7–18a}$$

where $\alpha(\mathbf{k}, s)$ is some complex function of $\mathbf{k}, s$ satisfying the condition

$$\sum_{\mathbf{k},s} |\alpha(\mathbf{k}, s)|^2 = 1. \tag{12.7–18b}$$

For this state

‡ We emphasize the fact, which may be obscured by the notation for the state, that the number of modes is actually unbounded.

$$\Gamma^{(1,1)}_{i;j}(\mathbf{r}_1, t_1; \mathbf{r}_2, t_2) = \sum_{\mathbf{k},s}\sum_{\mathbf{k}',s'} \alpha(\mathbf{k}, s)\alpha^*(\mathbf{k}', s')\langle 0, \dots, 0, 1_{\mathbf{k}'s'}, 0, \dots, 0| \times \hat{F}^{(-)}_i(\mathbf{r}_1, t_1)\hat{F}^{(+)}_j(\mathbf{r}_2, t_2)|0, \dots, 0, 1_{\mathbf{k}s}, 0, \dots, 0\rangle. \tag{12.7–19}$$

Now evidently the effect of the annihilation operator $\hat{F}^{(+)}_j(\mathbf{r}_2, t_2)$ on the one-photon state $|0, \dots, 0, 1_{\mathbf{k}s}, 0, \dots, 0\rangle$ is to produce the vacuum state $|\text{vac}\rangle$, apart from a multiplicative factor, and similarly for $\hat{F}^{(-)}_i(\mathbf{r}_1, t_1)$ operating to the left on $\langle 0, \dots, 0, 1_{\mathbf{k}'s'}, 0, \dots, 0|$. If we define

$$\sum_{\mathbf{k},s} \alpha(\mathbf{k}, s)\hat{F}^{(+)}_j(\mathbf{r}_2, t_2)|0, \dots, 0, 1_{\mathbf{k}s}, 0, \dots, 0\rangle \equiv U_j(\mathbf{r}_2, t_2)|\text{vac}\rangle, \tag{12.7–20}$$

we can re-write Eq. (12.7–19) in the form

$$\Gamma^{(1,1)}_{i;j}(\mathbf{r}_1, t_1; \mathbf{r}_2, t_2) = U^*_i(\mathbf{r}_1, t_1)U_j(\mathbf{r}_2, t_2), \tag{12.7–21}$$

which shows that any pure one-photon state satisfies the condition of second-order coherence. The restriction to a pure state is necessary here.

We can make an alternative generalization of Eq. (12.7–17) by allowing one mode $\mathbf{k}'$, s' to be excited in a completely arbitrary manner, with all other modes remaining empty. The density operator for such a state has the factored form

$$\hat{\rho} = \hat{\rho}_{\mathbf{k}'s'} \prod_{\mathbf{k},s\neq\mathbf{k}',s'} |0_{\mathbf{k}s}\rangle\langle 0_{\mathbf{k}s}|, \tag{12.7–22}$$

where $\hat{\rho}_{\mathbf{k}'s'}$ is the arbitrary density operator for the occupied $\mathbf{k}'$, s' mode alone. With this choice of $\hat{\rho}$ we find for the correlation function

$$\Gamma^{(1,1)}_{i;j}(\mathbf{r}_1, t_1; \mathbf{r}_2, t_2) = \frac{1}{L^3}|l(k')|^2(\boldsymbol{\varepsilon}^*_{\mathbf{k}'s'})_i(\boldsymbol{\varepsilon}_{\mathbf{k}'s'})_j\, e^{-i(\mathbf{k}'\cdot\mathbf{r}_1-\omega' t_1)}\, e^{i(\mathbf{k}'\cdot\mathbf{r}_2-\omega' t_2)} \times \operatorname{Tr}(\hat{\rho}_{\mathbf{k}'s'}\hat{a}^\dagger_{\mathbf{k}'s'}\hat{a}_{\mathbf{k}'s'}). \tag{12.7–23}$$

This again factorizes into a product of the form given by Eq. (12.7–21), as required for second-order coherence, if we put

$$U_i(\mathbf{r}, t) \equiv \frac{1}{L^{3/2}} l(k')(\boldsymbol{\varepsilon}_{\mathbf{k}'s'})_i[\Theta^{(1,1)}(\mathbf{k}', s'; \mathbf{k}', s')]^{1/2}\, e^{i(\mathbf{k}'\cdot\mathbf{r}-\omega' t)}. \tag{12.7–24}$$

Indeed, it is not difficult to see that in this case all correlation functions of arbitrary order must factorize in a similar manner, so that we have

$$\begin{aligned}\Gamma^{(N,M)}_{i_1,\dots,i_N;j_M,\dots,j_1}&(\mathbf{r}_1, t_1, \dots, \mathbf{r}_N, t_N; \mathbf{r}'_M, t'_M, \dots, \mathbf{r}'_1, t'_1)\\ &= \Theta^{(N,M)}(\mathbf{k}', s', \dots, \mathbf{k}', s'; \mathbf{k}', s', \dots, \mathbf{k}', s')\\ &\quad\times W^*_{i_1}(\mathbf{r}_1, t_1)\dots W^*_{i_N}(\mathbf{r}_N, t_N)W_{j_M}(\mathbf{r}'_M, t'_M)\dots W_{j_1}(\mathbf{r}'_1, t'_1),\end{aligned} \tag{12.7–25}$$

if we put

$$W_i(\mathbf{r}, t) \equiv \frac{1}{L^{3/2}} l(k')(\boldsymbol{\varepsilon}_{\mathbf{k}'s'})_i\, e^{i(\mathbf{k}'\cdot\mathbf{r}-\omega' t)}. \tag{12.7–26}$$

A field having only one excited mode therefore satisfies the factorization condition, although the factors $\Theta^{(N,M)}$ depend on the order $N + M$ in general.

12.7.4 Wave packets as modes

The modes we have been considering are plane waves of definite wave vector **k** and definite polarization s. However, we can use linear combinations of plane-wave modes to form wave packets of different forms, all of which will be solutions of the wave equation. If these different wave packets form a complete set, they form a basis for the representation of field vectors, and may be regarded as an alternative set of modes of the electromagnetic field. We shall see that it is true once again that a field having only one excited mode of this new kind is coherent to all orders.

Consider the wave packets formed by the linear combination of plane-wave modes

$$\mathbf{u}_m(\mathbf{r}, t) \equiv \frac{1}{L^{3/2}} \sum_\lambda \alpha_{m\lambda} l(k_\lambda) \boldsymbol{\varepsilon}_\lambda \, \mathrm{e}^{\mathrm{i}(\mathbf{k}_\lambda \cdot \mathbf{r} - \omega_\lambda t)}, \qquad (12.7\text{–}27)$$

where we use the index λ to stand for the combination **k**, s of wave vector and polarization indices. The coefficients $\alpha_{m\lambda}$ are chosen to be the elements of a unitary matrix, so that

$$\sum_m \alpha^*_{m\lambda} \alpha_{m\lambda'} = \delta_{\lambda\lambda'} = \sum_m \alpha_{\lambda m} \alpha^*_{\lambda' m}. \qquad (12.7\text{–}28)$$

The factor $l(k_\lambda)$ in Eq. (12.7–27) is the same coefficient that enters in the expansion (11.12–3) of the complex field vector $\hat{\mathbf{F}}^{(+)}(\mathbf{r}, t)$, and is different for different vectors. The functions $\mathbf{u}_m(\mathbf{r}, t)$, are therefore intimately connected with the choice of a particular field vector $\hat{\mathbf{F}}^{(+)}(\mathbf{r}, t)$. However, like the plane-wave modes, they satisfy the Helmholtz equation throughout the space, and therefore constitute a set of modes of the field. We can expand $\hat{\mathbf{F}}^{(+)}(\mathbf{r}, t)$ in terms of them by using Eq. (12.7–27) and (12.7–28) and write

$$\begin{aligned}\hat{\mathbf{F}}^{(+)}(\mathbf{r}, t) &= \frac{1}{L^{3/2}} \sum_\lambda \hat{a}_\lambda l(k_\lambda) \boldsymbol{\varepsilon}_\lambda \, \mathrm{e}^{\mathrm{i}(\mathbf{k}_\lambda \cdot \mathbf{r} - \omega_\lambda t)} \\ &= \frac{1}{L^{3/2}} \sum_m \sum_{\lambda'} \sum_\lambda \hat{a}_{\lambda'} \alpha^*_{m\lambda'} \alpha_{m\lambda} l(k_\lambda) \boldsymbol{\varepsilon}_\lambda \, \mathrm{e}^{\mathrm{i}(\mathbf{k}_\lambda \cdot \mathbf{r} - \omega_\lambda t)} \\ &= \sum_m \sum_{\lambda'} \hat{a}_{\lambda'} \alpha^*_{m\lambda'} \mathbf{u}_m(\mathbf{r}, t). \end{aligned} \qquad (12.7\text{–}29)$$

If we define a new operator $\hat{c}_m$ by the linear combination of annihilation operators

$$\hat{c}_m \equiv \sum_\lambda \alpha^*_{m\lambda} \hat{a}_\lambda, \qquad (12.7\text{–}30)$$

we may express $\hat{\mathbf{F}}^{(+)}(\mathbf{r}, t)$ in the compact form

$$\hat{\mathbf{F}}^{(+)}(\mathbf{r}, t) = \sum_m \hat{c}_m \mathbf{u}_m(\mathbf{r}, t). \qquad (12.7\text{–}31)$$

It is not difficult to see that the operators $\hat{c}_m$, and their adjoints $\hat{c}^\dagger_m$, play the role of boson annihilation and creation operators for excitations characterized by the mode functions $\mathbf{u}_m(\mathbf{r}, t)$. From the commutation relations (10.3–9)–(10.3–11) for the $\hat{a}_\lambda$, $\hat{a}_\lambda^\dagger$ operators, it follows at once that the $\hat{c}_m$, $\hat{c}_m^\dagger$ operators obey the commutation relations

$$[\hat{c}_m, \hat{c}_n] = 0 = [\hat{c}_m^\dagger, \hat{c}_n^\dagger], \tag{12.7–32}$$

and

$$[\hat{c}_m, \hat{c}_n^\dagger] = \sum_\lambda \alpha_{\lambda m}\alpha_{\lambda n}^* = \delta_{mn}, \tag{12.7–33}$$

which are the usual rules for boson annihilation and creation operators. Because of these relations, excitations of the field corresponding to different mode functions $\mathbf{u}_m(\mathbf{r}, t)$ can be produced independently.‡

Now consider a field in which only one of the modes $\mathbf{u}_n(\mathbf{r}, t)$ is excited, while all the others are empty. Then the density operator $\hat{\rho}$ of the field must be of the factored form

$$\hat{\rho} = \hat{\rho}_n \prod_{m\neq n} |0_m\rangle\langle 0_m|. \tag{12.7–34}$$

As $\hat{c}_m|0_m\rangle = 0$ for $m \neq n$, it follows from the expansion (12.7–31) that the correlation function $\Gamma_{i;j}^{(1,1)}(\mathbf{r}_1, t_1; \mathbf{r}_2, t_2)$ is given by

$$\Gamma_{i;j}^{(1,1)}(\mathbf{r}_1, t_1; \mathbf{r}_2, t_2) = [u_n^*(\mathbf{r}_1, t_1)]_i[u_n(\mathbf{r}_2, t_2)]_j \operatorname{Tr}(\hat{\rho}_n\hat{c}_n^\dagger\hat{c}_n), \tag{12.7–35}$$

which is a product of a function of $\mathbf{r}_1$, t_1, i and a function of $\mathbf{r}_2$, t_2, j. More generally, for a normally ordered correlation function of any order (N, M) we have

$$\Gamma^{(N,M)}(x_1, \ldots, x_N; y_M, \ldots, y_1) = u_n^*(x_1)\ldots u_n^*(x_N)u_n(y_M)\ldots u_n(y_1)\operatorname{Tr}[\rho_n\hat{c}_n^{\dagger N}\hat{c}_n^M], \tag{12.7–36}$$

where we use the abbreviated notation of Section 12.5. This shows that the correlation functions of all orders satisfy the factorization condition, and that the degree of second-order coherence is unity when only one mode of the field is excited, even when this mode is one of the more general kind. If we make use of the definition (12.6–7) for $\gamma^{(N,M)}$, then the degree of coherence $|\gamma^{(N,N)}|$ of any even order (N, N) is also unity in that case. Since the form of $\hat{\rho}_n$, the density operator associated with the one excitation n, has been left completely arbitrary, it is clear that there exists a great variety of different states satisfying the condition for coherence.

The converse of this result is also true. A field satisfying the condition (12.7–1) of second-order coherence must have a density operator of the form (12.7–34), corresponding to the excitation of only one mode of arbitrary form.§ In order to

‡ By leaving out the coefficient $l(k_\lambda)$ in the expansion (12.7–27) we could, of course, generate mode functions that are independent of the choice of field operator, and form a complete orthonormal set. Then any operator could be expanded in terms of the mode functions in the form of Eq. (12.7–31). However, the resulting operators $\hat{c}_m$, $\hat{c}_m^\dagger$ do not obey such a simple set of commutation rules as Eqs. (12.7–32) and (12.7–33).

§ See Titulaer and Glauber (1966). We follow the essence of their argument. Some similarities between a single-mode quantized field and a single-mode classical field, briefly discussed in Section 8.5.2, should be noted.

show this we start from the relation (12.7–7), that holds for any field having second-order coherence, which we write in the form

$$\hat{F}_i^{(+)}(\mathbf{r}, t)\hat{\rho} = \frac{\Gamma_{j;i}^{(1,1)}(\mathbf{r}_0, t_0; \mathbf{r}, t)}{\Gamma_{j;j}^{(1,1)}(\mathbf{r}_0, t_0; \mathbf{r}_0, t_0)}\hat{F}_j^{(+)}(\mathbf{r}_0, t_0)\hat{\rho}, \tag{12.7–37}$$

where $\mathbf{r}_0$, t_0 is some fixed space-time point and j is some fixed Cartesian index. No summation is implied in this equation. We now multiply both sides of the equation by

$$\frac{1}{L^{3/2}}\frac{1}{l(k)}(\boldsymbol{\varepsilon}_{\mathbf{k}s}^*)_i\, \mathrm{e}^{-\mathrm{i}(\mathbf{k}\cdot\mathbf{r}-\omega t)},$$

sum on i and integrate with respect to $\mathbf{r}$ over all space. With the help of Eq. (12.5–21) the result may then be written as

$$\hat{a}_\lambda\hat{\rho} = \beta_\lambda\hat{F}^{(+)}(z)\hat{\rho}, \tag{12.7–38}$$

where z stands collectively for $(\mathbf{r}_0, t_0, j)$, λ stands collectively for $\mathbf{k}, s$, and β_λ is given by the formula

$$\beta_\lambda \equiv \frac{1}{\Gamma^{(1,1)}(z, z)}\frac{1}{L^{3/2}}\frac{1}{l(k)}\sum_i\int \Gamma_{j;i}^{(1,1)}(\mathbf{r}_0, t_0; \mathbf{r}, t)(\boldsymbol{\varepsilon}_{\mathbf{k}s}^*)_i\,\mathrm{e}^{-\mathrm{i}(\mathbf{k}\cdot\mathbf{r}-\omega t)}\,\mathrm{d}^3r. \tag{12.7–39}$$

The t-dependence disappears in the integration, so that β_λ is a constant.

Now consider the density operator $\hat{\rho}$ in the Fock representation. This may be written as

$$\begin{aligned}\hat{\rho} &= \sum_{\{n\}}\sum_{\{m\}}|\{n\}\rangle\langle\{n\}|\hat{\rho}|\{m\}\rangle\langle\{m\}| \\ &= \sum_{\{n\}}\sum_{\{m\}}\prod_\lambda\left[\frac{\hat{a}_\lambda^{\dagger n_\lambda}}{\sqrt{(n_\lambda!)}}\right]|\mathrm{vac}\rangle\langle\mathrm{vac}|\prod_\lambda\left[\frac{\hat{a}_\lambda^{n_\lambda}}{\sqrt{(n_\lambda!)}}\right]\hat{\rho}\prod_\lambda\left[\frac{\hat{a}_\lambda^{\dagger m_\lambda}}{\sqrt{(m_\lambda!)}}\right]|\mathrm{vac}\rangle\langle\mathrm{vac}|\prod_\lambda\left[\frac{\hat{a}_\lambda^{m_\lambda}}{\sqrt{(m_\lambda!)}}\right],\end{aligned} \tag{12.7–40}$$

when we make use of the representation (10.4–16) of the Fock states. With the help of Eq. (12.7–38) and its Hermitian conjugate we can simplify Eq. (12.7–40) and express it as

$$\hat{\rho} = \sum_{\{n\}}\sum_{\{m\}}\prod_\lambda\left[\frac{(\hat{a}_\lambda^\dagger\beta_\lambda)^{n_\lambda}}{n_\lambda!}\right]|\mathrm{vac}\rangle\langle\mathrm{vac}|[\hat{F}^{(+)}(z)]^{\mathrm{n}}\hat{\rho}[\hat{F}^{(-)}(z)]^{\mathrm{m}}|\mathrm{vac}\rangle\langle\mathrm{vac}|\prod_\lambda\left[\frac{(\hat{a}_\lambda\beta_\lambda^*)^{m_\lambda}}{m_\lambda!}\right], \tag{12.7–41}$$

where we have written

$$\left.\begin{aligned} n &\equiv \sum_\lambda n_\lambda, \\ m &\equiv \sum_\lambda m_\lambda. \end{aligned}\right\} \tag{12.7–42}$$

We now make use of the multinomial theorem in the form

$$\sum_{\{n\}}\prod_{\lambda}\left[\frac{(\hat{a}_{\lambda}^{\dagger}\beta_{\lambda})^{n_{\lambda}}}{n_{\lambda}!}\right]\delta_{nn'} = \frac{1}{n'!}\left(\sum_{\lambda}\hat{a}_{\lambda}^{\dagger}\beta_{\lambda}\right)^{n'},$$

and carry out the partial summations in Eq. (12.7–41) with n and m fixed. We then obtain the formula

$$\hat{\rho} = \sum_{n}\sum_{m}\frac{\left(\sum_{\lambda}\hat{a}_{\lambda}^{\dagger}\beta_{\lambda}\right)^{n}}{n!}|\text{vac}\rangle\langle\text{vac}|[\hat{F}^{(+)}(z)]^{\text{n}}\hat{\rho}[\hat{F}^{(-)}(z)]^{\text{m}}|\text{vac}\rangle\langle\text{vac}|\frac{\left(\sum_{\lambda}\hat{a}_{\lambda}\beta_{\lambda}^{*}\right)^{m}}{m!}. \tag{12.7–43}$$

The operator $\sum_{\lambda}\hat{a}_{\lambda}^{\dagger}\beta_{\lambda}$ appearing in this equation defines a new creation operator for a new non-monochromatic mode of the field. If we choose $\alpha_{\lambda m}$ in Eq. (12.7–30) to be given by

$$\left.\begin{aligned}\alpha_{\lambda r} &= \frac{\beta_{\lambda}^{*}}{\sum_{\lambda}|\beta_{\lambda}|^{2}} \equiv \frac{\beta_{\lambda}^{*}}{B}, \quad \text{for } r = 1\\ &= 0, \qquad\qquad\qquad \text{otherwise}\end{aligned}\right\} \tag{12.7–44}$$

where B is finite so long as the mean number of photons in the field is finite, we may re-write Eq. (12.7–43) in the form

$$\hat{\rho} = \sum_{n}\sum_{m}\frac{\hat{c}_{1}^{\dagger n}}{n!}|\text{vac}\rangle\langle\text{vac}|\frac{\hat{c}_{1}^{m}}{m!}B^{n+m}\langle\text{vac}|[\hat{F}^{(+)}(z)]^{n}\hat{\rho}[\hat{F}^{(-)}(z)]^{m}|\text{vac}\rangle. \tag{12.7–45}$$

Inspection of this equation shows that it corresponds to a field in which only the first of the new set of modes is excited, while all other modes are unoccupied. We have therefore shown that the condition of second-order coherence [Eq. (12.7–1)] implies that the only type of excitation – or photon – present can be associated with one mode of the field. The nature of the excitation is however quite arbitrary. The state may be one of discrete excitation, i.e. a Fock state, or a coherent state, or some arbitrary impure state. In the special case in which the state corresponds to a coherent excitation of one mode, i.e. to an eigenstate of $\hat{c}_1$, it may be shown that the state is also a coherent state in the basis formed by any other set of modes (Titulaer and Glauber, 1966).

12.8 Stationarity, homogeneity, isotropy

12.8.1 Stationarity

We have already encountered the concept of a stationary field in our discussion of the classical coherence functions. Roughly speaking, a stationary field is one whose statistical properties do not change in time, and this is reflected in the fact that the expectation value of any function or functional of the field operators is invariant under translation of the origin of time. The same criterion of stationarity is immediately applicable to the quantized field, whose state is described by the density operator $\hat{\rho}$. In the Schrödinger picture, $\hat{\rho}$ obeys the equation of motion

$$\frac{\mathrm{d}\hat{\rho}}{\mathrm{d}t} = \frac{1}{\mathrm{i}\hbar}[\hat{H}, \hat{\rho}], \tag{12.8–1}$$

where $\hat{H}$ is the total energy of the field. Stationarity requires that the density operator $\hat{\rho}$ be constant in time in the Schrödinger picture, in order that the expectation value of any dynamical variable be time independent. The Eq. (12.8–1) therefore implies that for a stationary field

$$[\hat{H}, \hat{\rho}] = 0. \tag{12.8–2}$$

But if the commutator vanishes in the Schrödinger picture, it vanishes in the Heisenberg picture and in any other picture also. We may therefore take Eq. (12.8–2) to be a defining condition for a stationary field.‡

Now $\hat{H}$ is also the generator of time translation, for we have, quite generally, from the operator expansion theorem (10.7–7), for any field operator $\hat{O}(t)$ in the Heisenberg picture,

$$\begin{aligned}\exp(\mathrm{i}\hat{H}\tau/\hbar)\hat{O}(t)\exp(-\mathrm{i}\hat{H}\tau/\hbar) &= \hat{O}(t) + \frac{\mathrm{i}\tau}{\hbar}[\hat{H}, \hat{O}(t)] + \left(\frac{\mathrm{i}\tau}{\hbar}\right)^2 \frac{1}{2!}[\hat{H}, [\hat{H}, \hat{O}(t)]] \\ &\quad + \ldots \\ &= \hat{O}(t) + \tau\frac{\mathrm{d}\hat{O}(t)}{\mathrm{d}t} + \frac{\tau^2}{2!}\frac{\mathrm{d}^2\hat{O}(t)}{\mathrm{d}t^2} + \ldots \\ &= \hat{O}(t+\tau).\end{aligned} \tag{12.8–3}$$

This property puts the condition (12.8–2) for stationarity in perspective and illustrates an important general result. *The state of the field is invariant under some transformation (in this case time translation), if the density operator $\hat{\rho}$ commutes with the generator of that transformation.*

As an example let us consider the correlation function of order (N, M)

$$\Gamma^{(N,M)} \equiv \langle \hat{F}^{(-)}_{i1}(\mathbf{r}_1, t_1+\tau) \ldots \hat{F}^{(-)}_{iN}(\mathbf{r}_N, t_N+\tau)\hat{F}^{(+)}_{jM}(\mathbf{r}'_M, t'_M+\tau) \ldots \hat{F}^{(+)}_{j1}(\mathbf{r}'_1, t'_1+\tau)\rangle.$$

By making use of the time-translation property (12.8–3) we may express this in the form

$$\begin{aligned}\Gamma^{(N,M)} &= \mathrm{Tr}\,[\hat{\rho}\exp(\mathrm{i}\hat{H}\tau/\hbar)\hat{F}^{(-)}_{i1}(\mathbf{r}_1, t_1)\exp(-\mathrm{i}\hat{H}\tau/\hbar) \ldots \\ &\quad \times \exp(\mathrm{i}\hat{H}\tau/\hbar)\hat{F}^{(-)}_{iN}(\mathbf{r}_N, t_N)\exp(-\mathrm{i}\hat{H}\tau/\hbar) \\ &\quad \times \exp(\mathrm{i}\hat{H}\tau/\hbar)\hat{F}^{(+)}_{jM}(\mathbf{r}'_M, t'_M)\exp(-\mathrm{i}\hat{H}\tau/\hbar) \ldots \\ &\quad \times \exp(\mathrm{i}\hat{H}\tau/\hbar)\hat{F}^{(+)}_{j1}(\mathbf{r}'_1, t'_1)\exp(-\mathrm{i}\hat{H}\tau/\hbar)] \\ &= \mathrm{Tr}\,[\exp(-\mathrm{i}\hat{H}\tau/\hbar)\hat{\rho}\exp(\mathrm{i}\hat{H}\tau/\hbar)\hat{F}^{(-)}_{i1}(\mathbf{r}_1, t_1) \ldots \\ &\quad \times \hat{F}^{(-)}_{iN}(\mathbf{r}_N, t_N)\hat{F}^{(+)}_{jM}(\mathbf{r}'_M, t'_M) \ldots \hat{F}^{(+)}_{j1}(\mathbf{r}'_1, t'_1)].\end{aligned} \tag{12.8–4}$$

‡ Implications of stationarity for a quantum field have been discussed by Kano (1964), Eberly and Kujawski (1967), Dialetis and Mehta (1968), and Mandel and Mehta (1969). The last two papers contain some discussion of homogeneity also, and the second paper includes a treatment of isotropy.

The last line is obtained from the previous one by moving the operator $\exp(-\mathrm{i}\hat{H}\tau/\hbar)$ from the rear to the front, and this is permissible because the trace is invariant under cyclic permutation of operators. If the field is stationary it follows from Eq. (12.8–2) that

$$\exp(-\mathrm{i}\hat{H}\tau/\hbar)\hat{\rho}\exp(\mathrm{i}\hat{H}\tau/\hbar) = \hat{\rho}, \tag{12.8–5}$$

so that

$$\langle \hat{F}^{(-)}_{i_1}(\mathbf{r}_1, t_1 + \tau) \ldots \hat{F}^{(-)}_{i_N}(\mathbf{r}_N, t_N + \tau)\hat{F}^{(+)}_{j_M}(\mathbf{r}'_M, t'_M + \tau) \ldots \hat{F}^{(+)}_{j_1}(\mathbf{r}'_1, t'_1 + \tau)\rangle = \langle \hat{F}^{(-)}_{i_1}(\mathbf{r}_1, t_1) \ldots \hat{F}^{(-)}_{i_N}(\mathbf{r}_N, t_N)\hat{F}^{(+)}_{j_M}(\mathbf{r}'_M, t'_M) \ldots \hat{F}^{(+)}_{j_1}(\mathbf{r}'_1, t'_1)\rangle. \tag{12.8–6}$$

The correlation function is therefore invariant under any time translation τ, as expected. Moreover, on putting $\tau = -t_1$ in Eq. (12.8–6) we obtain for $\Gamma^{(N,M)}$ the expression

$$\Gamma^{(N,M)} = \langle \hat{F}^{(-)}_{i_1}(\mathbf{r}_1, 0)\hat{F}^{(-)}_{i_2}(\mathbf{r}_2, t_2 - t_1) \ldots \hat{F}^{(-)}_{i_N}(\mathbf{r}'_N, t_N - t_1)\hat{F}^{(+)}_{j_M}(\mathbf{r}'_M, t'_M - t_1) \ldots \hat{F}^{(+)}_{j_1}(\mathbf{r}'_1, t'_1 - t_1)\rangle, \tag{12.8–7}$$

which shows that, when the field is stationary, $\Gamma^{(N,M)}$ is really a function of the $(N + M - 1)$ differences between the various time parameters. We can write down similar and equivalent relations by choosing $\tau = -t_r$ $(r = 2, 3, \ldots, N)$ or $\tau = -t'_r$ $(r = 1, 2, \ldots, M)$, instead.

The converse of this result is also true. If Eq. (12.8–6) holds for all values of τ and for any correlation function $\Gamma^{(N,M)}$ then the field must be stationary. In order to show this we return to Eq. (12.8–4) and make explicit use of the operator expansion theorem (10.7–7). We then have

$$\begin{aligned}\langle \hat{F}^{(-)}_{i_1}(\mathbf{r}_1, t_1 + \tau) \ldots \hat{F}^{(-)}_{i_N}(\mathbf{r}_N, t_N + \tau)\hat{F}^{(+)}_{j_M}(\mathbf{r}'_M, t'_M + \tau) \ldots \hat{F}^{(+)}_{j_1}(\mathbf{r}'_1, t'_1 + \tau)\rangle \\ = \mathrm{Tr}\left\{\left(\hat{\rho} + \frac{-\mathrm{i}\tau}{\hbar}[\hat{H}, \hat{\rho}] + \left(\frac{-\mathrm{i}\tau}{\hbar}\right)^2 \frac{1}{2!}[\hat{H}, [\hat{H}, \hat{\rho}]] + \ldots\right) \right. \\ \left. \times\, \hat{F}^{(-)}_{i_1}(\mathbf{r}_1, t_1) \ldots \hat{F}^{(-)}_{i_N}(\mathbf{r}_N, t_N)\hat{F}^{(+)}_{j_M}(\mathbf{r}'_M, t'_M) \ldots \hat{F}^{(+)}_{j_1}(\mathbf{r}'_1, t'_1)\right\}.\end{aligned} \tag{12.8–8}$$

As the right side of this equation is a power series in τ, it will be independent of τ for an arbitrary correlation function only if the coefficient of τ^r vanishes for $r = 1, 2, \ldots$, i.e. if Eq. (12.8–2) holds.

As each of the factors in the definition of $\Gamma^{(N,M)}$, being a positive or negative frequency part of a field operator, obeys the wave equation, it is clear that $\Gamma^{(N,M)}$ itself obeys a set of $N + M$ wave equations in the $N + M$ different sets of space-time variables,

$$(\nabla_i^2 - \partial^2/\partial t_i^2)\Gamma^{(N,M)} = 0, \quad i = 1, 2, \ldots, N, \tag{12.8–9}$$

and

$$(\nabla_i'^2 - \partial^2/\partial t_i'^2)\Gamma^{(N,M)} = 0, \quad i = 1, 2, \ldots, M, \tag{12.8–10}$$

where $\boldsymbol{\nabla}_i$ and $\boldsymbol{\nabla}'_i$ denote the vector gradients with respect to the coordinates $\mathbf{r}_i$ and $\mathbf{r}'_i$, respectively. These equations imply that, in a certain sense, the correlation functions themselves propagate through space like the electromagnetic field.

12.8.2 *Condition on the density operator*

It is often an advantage to be able to recognize at once whether a field described by a given density operator $\hat{\rho}$ is stationary or not. For a stationary field $\hat{\rho}$ is evidently diagonal in the energy representation, but it is a trivial matter to translate this into a condition on the matrix elements of $\hat{\rho}$ in the Fock representation. If we put

$$\hat{\rho} = \sum_{\{n\}}\sum_{\{n'\}} \rho_{\{n\}\{n'\}} |\{n\}\rangle\langle\{n'\}|$$

and denote the energy eigenvalues belonging to the states $|\{n\}\rangle$ and $|\{n'\}\rangle$ by E and E' respectively, we may write

$$[\hat{H}, \hat{\rho}] = 0 = \sum_{\{n\}}\sum_{\{n'\}} (E - E')\rho_{\{n\}\{n'\}} |\{n\}\rangle\langle\{n'\}|. \qquad (12.8\text{–}11)$$

As the Fock states are orthogonal, each coefficient in this expansion must vanish separately, and we have the condition

$$(E - E')\rho_{\{n\}\{n'\}} = 0. \qquad (12.8\text{–}12)$$

It follows that all matrix elements $\rho_{\{n\}\{n'\}}$ associated with Fock states of unequal energies must vanish for a stationary state of the field. An immediate corollary of this result is that a density operator which is diagonal in the Fock representation corresponds to a stationary field.

The condition for stationarity is less easily expressed as a condition on the phase space functional $\phi(\{v\})$ in the diagonal coherent-state representation. However, there is one form of $\phi(\{v\})$, which is quite frequently encountered in practice, that is sufficient to ensure stationarity. Consider a field for which the phase space functional $\phi(\{|v|\})$ depends only on the set of moduli $\{|v|\}$ and not on the phases of the set of complex numbers $\{v\}$. We have already shown in Section 12.5 that the spectral densities associated with such a field vanish except for repeated indices. Let us now evaluate the commutator of $\hat{\rho}$ with the number operator $\hat{n}_{\mathbf{k}'s'}$ for some mode $\mathbf{k}'s'$. On making a diagonal representation of $\hat{\rho}$ and then expanding the coherent state projector $|v_{\mathbf{k}'s}\rangle\langle v_{\mathbf{k}'s'}|$ in terms of Fock states, we find that

$$[\hat{\rho}, \hat{n}_{\mathbf{k}'s'}] = \int \phi(\{|v|\})\, \mathrm{d}^2 v_{\mathbf{k}'s'} \prod_{\mathbf{k}s \neq \mathbf{k}'s'} [|v_{\mathbf{k}s}\rangle\langle v_{\mathbf{k}s}|\, \mathrm{d}^2 v_{\mathbf{k}s}]$$

$$\times\, \mathrm{e}^{-|v_{\mathbf{k}'s'}|^2} \sum_{n_{\mathbf{k}'s'}} \sum_{m_{\mathbf{k}'s'}} \frac{(v_{\mathbf{k}'s'})^{n_{\mathbf{k}'s'}}\,(v^*_{\mathbf{k}'s'})^{m_{\mathbf{k}'s'}}}{\sqrt{(n_{\mathbf{k}'s'}!)}\,\sqrt{(m_{\mathbf{k}'s'}!)}} [|n_{\mathbf{k}'s'}\rangle\langle m_{\mathbf{k}'s'}|, \hat{n}_{\mathbf{k}'s'}]. \qquad (12.8\text{–}13)$$

The commutator under the integral vanishes when $n_{\mathbf{k}'s'} = m_{\mathbf{k}'s'}$. On the other hand, on putting $v_{\mathbf{k}'s'} \equiv r\,\mathrm{e}^{\mathrm{i}\theta}$, $\mathrm{d}^2 v_{\mathbf{k}'s'} = r\,\mathrm{d}r\,\theta$, and carrying out the integration over θ, we see at once that the θ integral vanishes when $n_{\mathbf{k}'s'} \neq m_{\mathbf{k}'s'}$. The right-hand side is therefore zero always, and it follows that

$$[\hat{\rho}, \hat{n}_{\mathbf{k}'s'}] = 0, \qquad (12.8\text{–}14)$$

and that $\hat{\rho}$ is diagonal in the Fock representation. The density operator $\hat{\rho}$ therefore represents a stationary field whenever $\phi(\{v\})$ depends only on the set of moduli $\{|v|\}$. We have therefore found a condition that is sufficient, although not necessary, for stationarity. We shall see in the next chapter that density operators having this general form are not uncommon.

12.8.3 Properties of cross-spectral densities for stationary fields

Let us now examine some further implications of the stationarity condition, as it affects the cross-spectral densities. From the commutation rules (10.4–1) and (10.4–2) it follows that

$$[\hat{a}_{\mathbf{k}s}, \hat{H}] = \hbar\omega\hat{a}_{\mathbf{k}s}, \tag{12.8–15}$$

and that

$$[\hat{a}^{\dagger}_{\mathbf{k}s}, \hat{H}] = -\hbar\omega\hat{a}^{\dagger}_{\mathbf{k}s}, \quad (\omega = ck). \tag{12.8–16}$$

We now use these relations to transform the general cross-spectral density of order $N+M$ defined by Eq. (12.5–17). By substituting for $\hat{a}^{\dagger}_{\mathbf{k}_1 s_1}$ from Eq. (12.8–16) we obtain the relation

$$\begin{aligned}
&\Theta^{(N,M)}(\mathbf{k}_1, s_1, \ldots, \mathbf{k}_N, s_N; \mathbf{k}'_M, s'_M, \ldots, \mathbf{k}'_1, s'_1)\\
&\quad\equiv \mathrm{Tr}\,[\hat{\rho}\hat{a}^{\dagger}_{\mathbf{k}_1 s_1}\ldots\hat{a}^{\dagger}_{\mathbf{k}_N s_N}\hat{a}_{\mathbf{k}'_M s'_M}\ldots\hat{a}_{\mathbf{k}'_1 s'_1}]\\
&\quad= \mathrm{Tr}\left\{\hat{\rho}\left(-\frac{1}{\hbar\omega_1}\right)[\hat{a}^{\dagger}_{\mathbf{k}_1 s_1}, \hat{H}]\hat{a}^{\dagger}_{\mathbf{k}_2 s_2}\ldots\hat{a}^{\dagger}_{\mathbf{k}_N s_N}\hat{a}_{\mathbf{k}'_M s'_M}\ldots\hat{a}_{\mathbf{k}'_1 s'_1}\right\}\\
&\quad= \frac{1}{\hbar\omega_1}\mathrm{Tr}\,\{\hat{\rho}\hat{H}\hat{a}^{\dagger}_{\mathbf{k}_1 s_1}\ldots\hat{a}^{\dagger}_{\mathbf{k}_N s_N}\hat{a}_{\mathbf{k}'_M s'_M}\ldots\hat{a}_{\mathbf{k}'_1 s'_1}\}\\
&\quad\quad - \frac{1}{\hbar\omega_1}\mathrm{Tr}\,\{\hat{\rho}\hat{a}^{\dagger}_{\mathbf{k}_1 s_1}(\hat{a}^{\dagger}_{\mathbf{k}_2 s_2}\hat{H} + \hbar\omega_2\hat{a}^{\dagger}_{\mathbf{k}_2 s_2})\hat{a}^{\dagger}_{\mathbf{k}_3 s_3}\ldots\hat{a}^{\dagger}_{\mathbf{k}_N s_N}\hat{a}_{\mathbf{k}'_M s'_M}\ldots\hat{a}_{\mathbf{k}'_1 s'_1}\}.
\end{aligned} \tag{12.8–17}$$

By repeated application of Eqs. (12.8–15) and (12.8–16) to the last term, we may move the $\hat{H}$ operator to the right of all the creation and annihilation operators in successive steps. Each such step generates an additional term of the type

$$\mp\frac{\hbar\omega_r}{\hbar\omega_1}\Theta^{(N,M)}(\mathbf{k}_1, s_1, \ldots, \mathbf{k}_N, s_N; \mathbf{k}'_M, s'_M, \ldots, \mathbf{k}'_1, s'_1), \quad r = 2, 3, \ldots,$$

as is evident by inspection of Eq. (12.8–17). The upper sign applies when $\hat{H}$ is moved past a creation operator and the lower sign when $\hat{H}$ is moved past an annihilation operator. Since the trace is invariant under cyclic permutation of operators, the term in which $\hat{H}$ eventually appears on the extreme right may be combined with the first term on the right of Eq. (12.8–17) to yield finally

$$\begin{aligned}
&\Theta^{(N,M)}(\mathbf{k}_1, s_1, \ldots, \mathbf{k}_N, s_N; \mathbf{k}'_M, s'_M, \ldots, \mathbf{k}'_1, s'_1)\\
&\quad= \frac{1}{\hbar\omega_1}\mathrm{Tr}\,\{[\hat{\rho}, \hat{H}]\hat{a}^{\dagger}_{\mathbf{k}_1 s_1}\ldots\hat{a}^{\dagger}_{\mathbf{k}_N s_N}\hat{a}_{\mathbf{k}'_M s'_M}\ldots\hat{a}_{\mathbf{k}'_1 s'_1}\}\\
&\quad\quad + (-\hbar\omega_2 - \hbar\omega_3 - \ldots - \hbar\omega_N + \hbar\omega'_M + \ldots + \hbar\omega'_1)\frac{1}{\hbar\omega_1}\\
&\quad\quad \times \Theta^{(N,M)}(\mathbf{k}_1, s_1, \ldots, \mathbf{k}_N, s_N; \mathbf{k}'_M, s'_M, \ldots, \mathbf{k}'_1, s'_1),
\end{aligned}$$

i.e.

$$\hbar(\omega_1 + \omega_2 + \ldots + \omega_N - \omega'_M - \ldots - \omega'_1) \times \Theta^{(N,M)}(\mathbf{k}_1, s_1, \ldots, \mathbf{k}_N, s_N; \mathbf{k}'_M, s'_M, \ldots, \mathbf{k}'_1, s'_1) = \mathrm{Tr}\{[\hat{\rho}, \hat{H}]\hat{a}^\dagger_{\mathbf{k}_1 s_1} \ldots \hat{a}^\dagger_{\mathbf{k}_N s_N}\hat{a}_{\mathbf{k}'_M s'_M} \ldots \hat{a}_{\mathbf{k}'_1 s'_1}\}. \quad (12.8\text{–}18)$$

This is a useful general relation. We conclude at once that when the field is stationary, so that the commutator on the right vanishes, we must have

$$(\omega_1 + \ldots + \omega_N - \omega'_M - \ldots - \omega'_1) \times \Theta^{(N,M)}(\mathbf{k}_1, s_1, \ldots, \mathbf{k}_N, s_N; \mathbf{k}'_M, s'_M, \ldots, \mathbf{k}'_1, s'_1) = 0. \quad (12.8\text{–}19)$$

A number of relations follow from Eq. (12.8–19). We see that the cross-spectral density of a stationary field can be non-zero only if

$$(\omega_1 + \ldots + \omega_N - \omega'_M - \ldots - \omega'_1) = 0, \quad (12.8\text{–}20)$$

which is the condition that is also found from a classical analysis [cf. Section 8.3, Eq. (8.3–10)]. When the cross-spectral density $\Theta^{(N,M)}$ is of even order with $N = M$, Eq. (12.8–20) can readily be satisfied by choice of repeated arguments, i.e. by making $\omega_r = \omega'_r$, $r = 1, 2, \ldots, N$, although this is not the only way to satisfy the equation. However, when $N \neq M$, the number of positive terms on the left of Eq. (12.8–20) does not equal the number of negative terms, and it is much harder to satisfy the equation. If the field is bandlimited, and the cross-spectral densities vanish unless each frequency ω lies within the range $\omega_0 - \frac{1}{2}\Delta\omega \leqslant \omega \leqslant \omega_0 + \frac{1}{2}\Delta\omega$, then Eq. (12.8–20) can only be satisfied if

$$|N - M|\omega_0 - \tfrac{1}{2}(N + M)\Delta\omega \leqslant 0 \leqslant |N - M|\omega_0 + \tfrac{1}{2}(N + M)\Delta\omega,$$

i.e. [cf. Eq. (8.3–17)]

$$\frac{|N - M|}{(N + M)} \leqslant \frac{1}{2}\frac{\Delta\omega}{\omega_0}. \quad (12.8\text{–}21)$$

If the field is quasi-monochromatic, so that $\Delta\omega \ll \omega_0$, this condition can only be met when $N = M$, or for unequal N and M if both N and M are very large, which means that all low-order cross-spectral densities of order $N \neq M$ must vanish. Since the correlation functions are expressible as multiple Fourier transforms of the spectral densities, the corresponding odd-order correlations must vanish also. The same conclusion can be drawn from a classical analysis of a stationary field (cf. Section 8.3).

In the special case in which the frequency arguments of $\Theta^{(N,M)}$ are all equal, it is clear that Eq. (12.8–20) cannot be satisfied at all when $N \neq M$, so that the corresponding cross-spectral density vanishes for $N \neq M$ in a stationary field. As a corollary of this result we have

$$\langle(\hat{a}^\dagger_{\mathbf{k}s})^N(\hat{a}_{\mathbf{k}s'})^M\rangle = 0, \quad \text{for } N \neq M, \quad (12.8\text{–}22)$$

in a stationary field. In particular, if either $N = 0$ or $M = 0$, we arrive at the conclusion that

$$\langle(\hat{a}^\dagger_{\mathbf{k}s})^N\rangle = 0 \quad (12.8\text{–}23)$$

and

$$\langle (\hat{a}_{\mathbf{k}s})^N \rangle = 0 \tag{12.8–24}$$

for any positive integer N in a stationary field. As the various field operators are expressible as linear combinations of annihilation and/or creation operators, we have

$$\langle \hat{\mathbf{F}}^{(+)}(\mathbf{r}, t) \rangle = 0 = \langle \hat{\mathbf{F}}^{(-)}(\mathbf{r}, t) \rangle, \tag{12.8–25}$$

and the expectation value of any Hermitian field operator $\hat{\mathbf{F}}(\mathbf{r}, t)$ vanishes also when the field is stationary.

Another special case of Eq. (12.8–19) arises when $\omega_1 = \omega_2 = \ldots = \omega_N = \omega$ and $\omega_1' = \omega_2' = \ldots = \omega_M' = \omega'$, but $\omega \neq \omega'$ and $N = M$. In that case Eq. (12.8–20) once again cannot be satisfied, and the corresponding cross-spectral density vanishes. As a corollary we have the conclusion that

$$\langle (\hat{a}^\dagger_{\mathbf{k}s})^N (\hat{a}_{\mathbf{k}'s'})^N \rangle = 0, \quad \text{for } \omega \neq \omega' \ (\omega = c|\mathbf{k}|), \tag{12.8–26}$$

in a stationary field, which may be interpreted to signify that the Fourier amplitudes of the field associated with different frequencies are uncorrelated to all orders.

Let us now examine some examples. A Fock state of the field, being an eigenstate of the energy, obviously corresponds to a stationary state. A coherent state $|\{v\}\rangle$ has a density operator whose matrix elements in the Fock representation are [cf. Eq. (11.2–11)]

$$\rho_{\{n\}\{n'\}} = \langle \{n\}|\{v\}\rangle\langle\{v\}|\{n'\}\rangle = \prod_{\mathbf{k},s}\left[e^{-|v_{\mathbf{k}s}|^2} \frac{v_{\mathbf{k}s}^{n_{\mathbf{k}s}}}{(n_{\mathbf{k}s}!)} \frac{(v_{\mathbf{k}s}^{n_{\mathbf{k}s}})^*}{(n'_{\mathbf{k}s}!)}\right]. \tag{12.8–27}$$

These matrix elements do not satisfy the condition (12.8–12), unless $v_{\mathbf{k}s} = 0$ for all $\mathbf{k}$, s, and the coherent state is generally a non-stationary state of the field. The sole exception is the vacuum state $|\text{vac}\rangle$, which happens to be simultaneously an eigenstate both of the energy and of the annihilation operator. As we have already seen (cf. Eq. 11.11–8), any normally ordered correlation function factorizes when the field is in a coherent state, but each factor depends explicitly on one of the time arguments of the correlation function, which is therefore not invariant under translation of the origin of time. On the other hand, the phase-averaged mixture of coherent states given by Eq. (11.11–18), which is sometimes used to represent the field of a single mode laser, is a functional only of the moduli $\{|v|\}$ and therefore, by the foregoing results, represents a stationary field.‡

12.8.4 Homogeneity

The concept of statistical homogeneity has the same implication for the variation of the electromagnetic field over space, as the concept of stationarity has for variation in time. We describe a field as statistically homogeneous if the expectation value of any operator that is a function of position is invariant under translation of the origin of space.

‡ It is worth noting that the differential phase space probability $\delta^2(v_{\mathbf{k}s})\,\mathrm{d}^2v_{\mathbf{k}s}$ is formally equivalent to $(1/2\pi|v_{\mathbf{k}s}|)\delta(|v_{\mathbf{k}s}|)|v_{\mathbf{k}s}|\,\mathrm{d}|v_{\mathbf{k}s}|\,\mathrm{d}\theta_{\mathbf{k}s}$ where $v_{\mathbf{k}s} = |v_{\mathbf{k}s}|\,e^{i\theta_{\mathbf{k}s}}$.

Now the generator of spatial translation is the total momentum of the field $\hat{\mathbf{P}}$. If $\hat{\mathbf{F}}^{(+)}(\mathbf{r}, t)$ is some arbitrary non-Hermitian field operator given by Eq. (12.2–1), and $\mathbf{r}_0$ is some arbitrary displacement, we have from the operator expansion theorem (10.7–7)

$$\exp(\mathrm{i}\mathbf{r}_0 \cdot \hat{\mathbf{P}}/\hbar)\hat{\mathbf{F}}^{(+)}(\mathbf{r}, t)\exp(-\mathrm{i}\mathbf{r}_0 \cdot \hat{\mathbf{P}}/\hbar) =$$

$$\hat{\mathbf{F}}^{(+)}(\mathbf{r}, t) + \frac{\mathrm{i}}{\hbar}[\mathbf{r}_0 \cdot \hat{\mathbf{P}}, \hat{\mathbf{F}}^{(+)}(\mathbf{r}, t)] + \left(\frac{\mathrm{i}}{\hbar}\right)^2 \frac{1}{2!}[\mathbf{r}_0 \cdot \hat{\mathbf{P}}, [\mathbf{r}_0 \cdot \hat{\mathbf{P}}, \hat{\mathbf{F}}^{(+)}(\mathbf{r}, t)]] + \ldots . \tag{12.8–28}$$

From the expansions of $\hat{\mathbf{P}}$ and $\hat{\mathbf{F}}^{(+)}(\mathbf{r}, t)$, and the commutation rules (10.4–1) and (10.4–2) it follows at once that

$$\frac{\mathrm{i}}{\hbar}[\mathbf{r}_0 \cdot \hat{\mathbf{P}}, \hat{\mathbf{F}}^{(+)}(\mathbf{r}, t)] = \frac{1}{L^{3/2}}\sum_{\mathbf{k},s} l(k)(\mathrm{i}\mathbf{r}_0 \cdot \mathbf{k})\hat{a}_{\mathbf{k}s}\boldsymbol{\varepsilon}_{\mathbf{k}s}\,\mathrm{e}^{\mathrm{i}(\mathbf{k}\cdot\mathbf{r}-\omega t)} \quad (\omega = ck),$$

and similarly that the n'th order commutator

$$\left(\frac{\mathrm{i}}{\hbar}\right)^n [\mathbf{r}_0 \cdot \hat{\mathbf{P}}, [\mathbf{r}_0 \cdot \hat{\mathbf{P}}, [\ldots [\mathbf{r}_0 \cdot \hat{\mathbf{P}}, \hat{\mathbf{F}}^{(+)}(\mathbf{r}, t)]] \ldots]] =$$

$$\frac{1}{L^{3/2}}\sum_{\mathbf{k},s} l(k)(\mathrm{i}\mathbf{r}_0 \cdot \mathbf{k})^n\hat{a}_{\mathbf{k}s}\boldsymbol{\varepsilon}_{\mathbf{k}s}\,\mathrm{e}^{\mathrm{i}(\mathbf{k}\cdot\mathbf{r}-\omega t)}. \tag{12.8–29}$$

On substituting from Eq. (12.8–29) in Eq. (12.8–28) we find that

$$\begin{aligned}\exp(\mathrm{i}\mathbf{r}_0 \cdot \hat{\mathbf{P}}/\hbar)\hat{\mathbf{F}}^{(+)}(\mathbf{r}, t)\exp(-\mathrm{i}\mathbf{r}_0 \cdot \hat{\mathbf{P}}/\hbar) &= \frac{1}{L^{3/2}}\sum_{\mathbf{k},s} l(k)\hat{a}_{\mathbf{k}s}\boldsymbol{\varepsilon}_{\mathbf{k}s}\sum_{n=0}^{\infty}\frac{(\mathrm{i}\mathbf{r}_0 \cdot \mathbf{k})^n}{n!}\mathrm{e}^{\mathrm{i}(\mathbf{k}\cdot\mathbf{r}-\omega t)} \\ &= \frac{1}{L^{3/2}}\sum_{\mathbf{k},s} l(k)\hat{a}_{\mathbf{k}s}\boldsymbol{\varepsilon}_{\mathbf{k}s}\,\mathrm{e}^{\mathrm{i}[\mathbf{k}\cdot(\mathbf{r}+\mathbf{r}_0)-\omega t]} \\ &= \hat{\mathbf{F}}^{(+)}(\mathbf{r} + \mathbf{r}_0, t),\end{aligned} \tag{12.8–30}$$

which confirms the role of $\hat{\mathbf{P}}$ as the generator of spatial translation.

It now follows that, if the expectation value of some arbitrary product of creation and annihilation operators $\hat{F}_{i_1}^{(-)}(\mathbf{r}_1, t)\hat{F}_{i_2}^{(-)}(\mathbf{r}_2, t) \ldots$, such as occurs in the correlation functions $\Gamma^{(N,M)}$ that we have been discussing, is to be invariant under the translation $\mathbf{r}_0$, we must have

$$\begin{aligned}&\mathrm{Tr}[\hat{\rho}\hat{F}_{i_1}^{(-)}(\mathbf{r}_1, t_1)\hat{F}_{i_2}^{(-)}(\mathbf{r}_2, t_2) \ldots] \\ &\quad = \mathrm{Tr}[\hat{\rho}\hat{F}_{i_1}^{(-)}(\mathbf{r}_1 + \mathbf{r}_0, t_1)\hat{F}_{i_2}^{(-)}(\mathbf{r}_2 + \mathbf{r}_0, t_2) \ldots] \\ &\quad = \mathrm{Tr}[\hat{\rho}\exp(\mathrm{i}\mathbf{r}_0 \cdot \hat{\mathbf{P}}/\hbar)\hat{F}_{i_1}^{(-)}(\mathbf{r}_1, t_1)\hat{F}_{i_2}^{(-)}(\mathbf{r}_2, t_2) \ldots \exp(-\mathrm{i}\mathbf{r}_0 \cdot \hat{\mathbf{P}}/\hbar)] \\ &\quad = \mathrm{Tr}[\exp(-\mathrm{i}\mathbf{r}_0 \cdot \hat{\mathbf{P}}/\hbar)\hat{\rho}\exp(\mathrm{i}\mathbf{r}_0 \cdot \hat{\mathbf{P}}/\hbar)\hat{F}_{i_1}^{(-)}(\mathbf{r}_1, t_1)\hat{F}_{i_2}^{(-)}(\mathbf{r}_2, t_2) \ldots],\end{aligned} \tag{12.8–31}$$

when we recall that the trace is invariant under cyclic permutation of operators. This equation will hold for an arbitrary correlation function and an arbitrary displacement $\mathbf{r}_0$ only if $\hat{\rho}$ commutes with $\hat{\mathbf{P}}$. We may therefore take the condition

$$[\hat{\rho}, \hat{\mathbf{P}}] = 0, \tag{12.8–32}$$

to be the condition for homogeneity of the field. By putting $\mathbf{r}_0 = -\mathbf{r}_1$, or $-\mathbf{r}_2$, etc.

in Eq. (12.8–31) we see that the expectation value of $\hat{F}^{(-)}_{i_1}(\mathbf{r}_1, t_1)\hat{F}^{(-)}_{i_2}(\mathbf{r}_2, t_2) \ldots$ is actually a function only of the differences between the position variables $\mathbf{r}_1$, $\mathbf{r}_2$, etc. Condition (12.8–32) is completely analogous to the condition (12.8–2) for stationarity. Indeed, it should be evident from the foregoing argument that the expectation values of operators will generally be invariant under some transformation if the density operator commutes with the generator of that transformation.

From Eq. (10.5–7), we have

$$\hat{\mathbf{P}} = \sum_{\mathbf{k},s} \hbar\mathbf{k}\hat{n}_{\mathbf{k}s},$$

so that $\hat{\mathbf{P}}$ commutes with the total energy $\hat{H}$, and the conditions under which $\hat{\rho}$ satisfies Eq. (12.8–32) generally correspond rather closely to those we have found for $\hat{\rho}$ to satisfy Eq. (12.8–2). Corresponding to Eq. (12.8–12), the condition for homogeneity may be expressed in terms of the matrix elements $\rho_{\{n\}\{n'\}}$ of $\hat{\rho}$ in the Fock representation in the form

$$(\mathbf{P} - \mathbf{P}')\rho_{\{n\}\{n'\}} = 0, \tag{12.8–33}$$

where $\mathbf{P}$ and $\mathbf{P}'$ are the expectation values of the momentum in the Fock states $|\{n\}\rangle$ and $|\{n'\}\rangle$, respectively. Evidently, a Fock state corresponds to a field that is both homogeneous and stationary. By the same analysis that was used to derive Eq. (12.8–14), we see that a state whose phase space functional in the diagonal coherent state representation depends only on the set of moduli $\{|v|\}$ is both homogeneous and stationary.

By an argument similar to that leading to Eq. (12.8–19), we may show that the condition for homogeneity on the cross-spectral density of order (N, M) is

$$(\mathbf{k}_1 + \ldots + \mathbf{k}_N - \mathbf{k}'_M - \ldots - \mathbf{k}'_1)$$
$$\Theta^{(N,M)}(\mathbf{k}_1, s_1, \ldots, \mathbf{k}_N, s_N; \mathbf{k}'_M, s'_M, \ldots, \mathbf{k}'_1, s'_1) = 0. \tag{12.8–34}$$

Evidently the cross-spectral density $\Theta^{(N,M)}$ can be non-zero for a homogeneous field only if

$$\mathbf{k}_1 + \ldots + \mathbf{k}_N - \mathbf{k}'_M - \ldots - \mathbf{k}'_1 = 0. \tag{12.8–35}$$

As this is a vector equation, it is possible to satisfy it for a great variety of combinations of $\mathbf{k}_1, \ldots, \mathbf{k}_N, \mathbf{k}'_M, \ldots, \mathbf{k}'_1$, even when $N \neq M$ and when N and M are not too large. On putting $\mathbf{k}_1 = \mathbf{k}_2 = \ldots = \mathbf{k}_N = \mathbf{k} = \mathbf{k}'_1 = \mathbf{k}'_2 = \ldots = \mathbf{k}'_M$ we find that Eq. (12.8–22) must be satisfied also for a homogeneous field, and similarly for Eqs. (12.8–23), (12.8–24) and (12.8–25). The relation (12.8–26) for a stationary field is replaced by the more general relation

$$\langle(\hat{a}^\dagger_{\mathbf{k}s})^N(\hat{a}_{\mathbf{k}'s'})^M\rangle = 0, \quad \text{if } \mathbf{k} \text{ is not parallel to } \mathbf{k}', \tag{12.8–36}$$

for a homogeneous field.

12.8.5 Isotropy

The concept of statistical isotropy bears the same relation to spatial rotations as the concept of statistical homogeneity does to spatial translations. Loosely speaking, in an isotropic field expectations of field operators do not change under rotation about some point. The center of isotropy may be a single preferred point in space, just as there may exist a preferred direction in a homogeneous field. However, if the field is also homogeneous, then it is necessarily isotropic about all points in space if it is isotropic about one point.

Let us make these considerations a little more quantitative. If a field is isotropic about the point $\mathbf{r}_0$, then the expectation value of any field vector $\hat{\mathbf{F}}$ at a point in space obtained by rotating the position vector $\mathbf{r}$ about $\mathbf{r}_0$, equals the result of making the same rotation of the expectation of the vector $\hat{\mathbf{F}}$ at the point $\mathbf{r}$ (see Fig. 12.4). If R is the rotation matrix that performs the desired rotation about $\mathbf{r}_0$, we can express this by writing

$$\langle \hat{\mathbf{F}}(\mathsf{R}\mathbf{r})\rangle = \mathsf{R}\langle \hat{\mathbf{F}}(\mathbf{r})\rangle,$$

or

$$\mathsf{R}\langle \hat{\mathbf{F}}(\mathsf{R}^{-1}\mathbf{r})\rangle = \langle \hat{\mathbf{F}}(\mathbf{r})\rangle. \tag{12.8–37}$$

More generally, if $\Gamma^{(N,M)}_{i_1,\ldots,i_N;j_M,\ldots,j_1}(\mathbf{r}_1,\ldots,\mathbf{r}_N;\mathbf{r}'_M,\ldots,\mathbf{r}'_1)$ is a correlation tensor of order N, M, [cf. Eq. (12.5–2)], the field is isotropic if

$$\begin{aligned}\mathsf{R}_{i_1p_1}\ldots\mathsf{R}_{i_Np_N}\mathsf{R}_{j_Mq_M}\ldots\mathsf{R}_{j_1q_1}\Gamma^{(N,M)}_{i_1,\ldots,i_N;j_M,\ldots,j_1}(\mathsf{R}^{-1}\mathbf{r}_1,\ldots,\mathsf{R}^{-1}\mathbf{r}_N;\mathsf{R}^{-1}\mathbf{r}'_M,\ldots,\mathsf{R}^{-1}\mathbf{r}'_1)\\ = \Gamma^{(\mathrm{N,M})}_{p_1,\ldots,p_N;q_M,\ldots,q_1}(\mathbf{r}_1,\ldots,\mathbf{r}_N;\mathbf{r}'_M,\ldots,\mathbf{r}'_1).\end{aligned} \tag{12.8–38}$$

The time arguments have been suppressed because they play no significant role here.

In order to translate this symmetry condition into a condition on the density operator $\hat{\rho}$ of the field, as before, we need the generator of rotation about $\mathbf{r}_0$. It may be shown (Simmons and Guttmann, 1970; Lenstra and Mandel, 1982) that the total angular moment $\mathbf{J}(\mathbf{r}_0)$ given by Eq. (10.6–2) is the required operator. In order to demonstrate this, we start from the following commutation relation (Lenstra and Mandel, 1982), which we shall not prove, between the positive frequency part $\hat{\mathbf{F}}^{(+)}(\mathbf{r},t)$ of any field vector and the total angular momentum $\hat{\mathbf{J}}(\mathbf{r}_0)$,

$$[\hat{F}^{(+)}_m(\mathbf{r},t),\hat{J}_n(\mathbf{r}_0)] = \mathrm{i}\hbar\left[\epsilon_{mnp}\hat{F}^{(+)}_p(\mathbf{r},t) - \epsilon_{nqp}(r_q - r_{0q})\frac{\partial}{\partial r_p}\hat{F}^{(+)}_m(\mathbf{r},t)\right]. \tag{12.8–39}$$

We now perform an infinitesimal unitary transformation on the vector operator $\hat{\mathbf{F}}^{(+)}(\mathbf{r},t)$, in which $\boldsymbol{\delta\theta}$ is a small vector angle of rotation, i.e.

$$\hat{\mathbf{F}}^{(+)\prime} = \exp(\mathrm{i}\hat{\mathbf{J}}(\mathbf{r}_0)\cdot\boldsymbol{\delta\theta}/\hbar)\hat{\mathbf{F}}^{(+)}(\mathbf{r},t)\exp(-\mathrm{i}\hat{\mathbf{J}}(\mathbf{r}_0)\cdot\boldsymbol{\delta\theta}/\hbar).$$

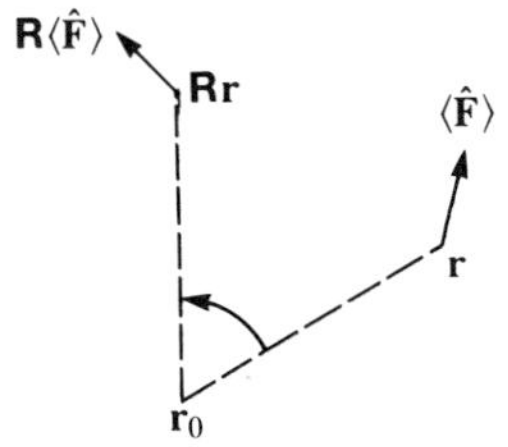

Fig. 12.4 Illustrating the concept of isotropy about $\mathbf{r}_0$.

Then, to the first order in $\boldsymbol{\delta\theta}$,

$$\hat{F}_m^{(+)\prime} = \hat{F}_m^{(+)}(\mathbf{r}, t) + \frac{i\delta\theta_n}{\hbar}[\hat{J}_n(r_0), \hat{F}_m^{(+)}(\mathbf{r}, t)]$$

and with the help of Eq. (12.8–39), after making a cyclic permutation of indices in the last term, we have

$$\hat{F}_m^{(+)\prime} = \hat{F}_m^{(+)}(\mathbf{r}, t) + \epsilon_{mnp}\delta\theta_n\hat{F}_p^{(+)}(\mathbf{r}, t) - \epsilon_{pnq}\delta\theta_n(r_q - r_{0q})\frac{\partial}{\partial r_p}\hat{F}_m^{(+)}(\mathbf{r}, t).$$

To the first order in $\delta\theta$ this can be written as a vector relation involving the simultaneous rotation of $\hat{\mathbf{F}}^{(+)}$ and the coordinate $\mathbf{r}$ of $\hat{\mathbf{F}}^{(+)}$,

$$\hat{\mathbf{F}}^{(+)\prime} = (1 + \boldsymbol{\delta\theta}\times)\hat{\mathbf{F}}^{(+)}(\mathbf{r} - \boldsymbol{\delta\theta} \times (\mathbf{r} - \mathbf{r}_0), t), \tag{12.8–40}$$

and it may also be expressed symbolically as

$$\hat{\mathbf{F}}^{(+)\prime} = \mathbf{R}\hat{\mathbf{F}}^{(+)}(\mathbf{R}^{-1}\mathbf{r}, t), \tag{12.8–41}$$

when we associate $\mathbf{R}$ with the rotation $\boldsymbol{\delta\theta}$ about $\mathbf{r}_0$. It follows that $\hat{\mathbf{J}}(\mathbf{r}_0)$ is the generator of rotation about $\mathbf{r}_0$, and by the same argument as was used to derive Eq. (12.8–2) this imposes a condition on the density operator $\hat{\rho}$. From Eq. (12.8–6), for example, the condition (12.8–38) for isotropy requires that $\hat{\mathbf{J}}(\mathbf{r}_0)$ commutes with the density operator, or

$$[\hat{\mathbf{J}}(\mathbf{r}_0), \hat{\rho}] = 0. \tag{12.8–42}$$

The implications of isotropy and of this condition for the structure of $\hat{\rho}$ are not as intuitively obvious as those of the conditions for stationarity and homogeneity that we encountered earlier. We shall therefore not pursue the subject here. However, in the next chapter we shall encounter an example of a radiation field (blackbody radiation) that is simultaneously stationary, homogeneous, and isotropic about all points in space.

12.9 Anti-normally ordered correlations

So far our discussion of quantum correlation functions has been confined almost entirely to normally ordered products of creation and annihilation operators. This emphasis on normal ordering is perfectly reasonable, for we showed in Section 12.2 that measurements with optical detectors that function by absorption of photons lead naturally to normally ordered correlation functions. If there existed measuring instruments that functioned instead by emission of photons, the results of measurements performed with these devices would be describable naturally by anti-normally ordered correlation functions. Although they are not generally encountered in the laboratory, in principle such devices could certainly be constructed. We shall refer to them as *quantum counters*,‡ and examine some of

‡ Atomic counting devices that were also called quantum counters were described earlier by Bloembergen (1959) and by Basov, Krokhin and Popov (1960). However, their mode of operation is rather different from the devices we are envisaging here. We have adopted the name quantum counter to distinguish devices functioning by emission of photons from photodetectors functioning by absorption of photons.

their properties in this section. We shall find that quantum counters suffer from certain fundamental disadvantages as compared with photoelectric detectors, which make them unattractive as practical devices. Nevertheless, an examination of some of their features will prove to be rewarding in affording us a little more insight into the character of quantum correlations. Much of the material of this section is based on a paper by Mandel (1966b).

12.9.1 The quantum counter

Consider an atomic system having the energy level structure indicated in Fig 12.5. Here a represents the ground state of the system, or some other terminal energy level, and b is a metastable level that is radiatively coupled to a broad band c, corresponding to a very short-lived state. We suppose that the atomic system makes spontaneous radiative transitions from c to the terminal level a, and that the energy jump from c to a is significantly greater than that from b to c, or

$$E_c - E_a \gg E_b - E_c. \tag{12.9–1}$$

In order to operate as a quantum counter, the system first has to be prepared in the state b. The method of preparation does not really concern us here, but if there exists a broad energy band d above level b, which is non-radiatively coupled to b, we could prepare the state b by optical pumping with an intense light source to level d, from which the system relaxes spontaneously and non-radiatively to level b. Some features of this proposed energy level scheme for the quantum counter will be seen to resemble the energy level structure of the ruby laser.

Now suppose that we have such an atomic system prepared in the metastable state b at position $\mathbf{r}$ at time t. Under the influence of an external electromagnetic field having occupied modes within the frequency range $(E_b - E_{c\max})/\hbar$ and $(E_b - E_{c\min})/\hbar$, the atomic system may be induced to make a downward transition to level c, with the stimulated emission of a photon.‡ The system therefore responds to the external field by emitting – rather than by absorbing – a photon.

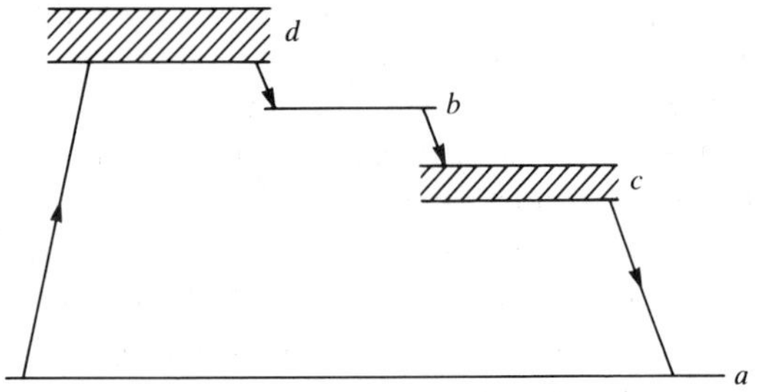

Fig. 12.5 Example of a possible energy level scheme for a quantum counter. The counter responds to frequencies within the range $(E_b - E_{c\min})/\hbar$ to $(E_b - E_{c\max})/\hbar$. (Reproduced from Mandel, 1966b.)

‡ The subject of stimulated and spontaneous emission from atoms is treated more fully in Sections 15.4 and 15.5.

However, for the device to qualify as a quantum counter, this emission must be registered in some way. That is the purpose of the second downward transition, because state c, being extremely short-lived, will immediately decay spontaneously to level a, with the emission of a second, higher-frequency photon. Since the second photon is clearly distinguishable from the first, we may imagine that our atomic system has an adjacent thin photoelectric layer that has a sufficiently high photoelectric threshold to be transparent to frequencies in the range $(E_b - E_{c\max})/\hbar$ to $(E_b - E_{c\min})/\hbar$, but to act as a photoemissive absorber for the second photon. The combination of the atomic system, or a large number of such atomic systems, with this photodetector then acts as a quantum counter that functions by stimulated emission of photons and registers the emission time of each photon. The photodetector here only plays the auxiliary role of registering the secondary photon; the external field is actually measured by means of the induced transition.

An argument similar to that given in Section 12.2, which led us to identify the configuration space absorption operator $\hat{\mathbf{F}}^{(+)}(\mathbf{r}, t)$ as the operator corresponding most closely to a measurement with a photodetector, now suggests that the photon creation operator $\hat{\mathbf{F}}^{(-)}(\mathbf{r}, t)$ given by Eq. (12.2–2) should correspond to a measurement with the quantum counter at $\mathbf{r}$, t. However, there are some significant distinctions between the two cases. One is connected with the fact that the response of the quantum counter is limited to frequencies within a finite range determined by the particular energy level structure, as shown in Fig. 12.5. In a finite normalization volume L^3, the quantum counter is therefore coupled only to a finite number of modes of the field.[‡] Second, photon emission, unlike absorption, can occur spontaneously, or in the presence of a vacuum field, and this is reflected in the fact that, when the $\hat{\mathbf{F}}^{(-)}(\mathbf{r}, t)$ operator is applied to the vacuum state $|\text{vac}\rangle$, it leads to a state in which all modes contributing to the expansion of $\hat{\mathbf{F}}^{(-)}(\mathbf{r}, t)$ are partly occupied. For this reason, the appropriate photon creation operator $\hat{\mathbf{F}}^{(-)}(\mathbf{r}, t)$ must be so defined that it contains contributions only from those modes of the field that are coupled to, and can be excited by, the quantum counter. In addition, we shall find it convenient to let $\hat{\mathbf{F}}^{(-)}(\mathbf{r}, t)\cdot\hat{\mathbf{F}}^{(+)}(\mathbf{r}, t)$ have the dimensions of photon density (see Section 12.3). We therefore take the operator representing a measurement of the field by the quantum counter to be

$$\hat{\mathbf{V}}^{\dagger}(\mathbf{r}, t) \equiv \frac{1}{L^{3/2}} \sum_{[\mathbf{k},s]} \hat{a}^{\dagger}_{\mathbf{k}s} \boldsymbol{\varepsilon}^{*}_{\mathbf{k}s}\, \mathrm{e}^{-\mathrm{i}(\mathbf{k}\cdot\mathbf{r}-\omega t)}, \tag{12.9–2}$$

where the symbol $[\mathbf{k}, s]$ stands for the finite set of modes that can be excited by the quantum counter. The other modes of the electromagnetic field play no role in the measurement, and we may suppose that they are unoccupied.

We now make use of the same argument as in Section 12.2 to derive an expression for the differential probability $P_1(\mathbf{r}, t)\Delta t$ that a count is registered at $\mathbf{r}$, t within some short time interval Δt. Let the quantum counter be exposed to

[‡] The frequency response of the photodetector is often supposed to have a lower but no upper bound. Actually this is not so, and there is also an upper bound connected with the finite size of the potential well binding the photoelectron. This leads to a frequency cutoff of a few multiples of the threshold frequency, as was shown by Mandel and Meltzer (1969).

an electromagnetic field described by a density operator $\hat{\rho}$, such that

$$\hat{\rho} = \sum_{\psi_1} p(\psi_1)|\psi_1\rangle\langle\psi_1|, \tag{12.9–3}$$

where the states $|\psi_1\rangle$ are arbitrary and $p(\psi_1)$ is the probability of the state $|\psi_1\rangle$. In the interaction picture we then have, on expressing P_1 in terms of the transition probability amplitude,

$$P_1(\mathbf{r}, t)\Delta t = C_2\Delta t \sum_{\psi_1} p(\psi_1) \sum_{\text{all } \psi_2} |\langle\psi_2|\hat{\mathbf{V}}^\dagger(\mathbf{r}, t)|\psi_1\rangle|^2, \tag{12.9–4}$$

where C_2 is some constant characteristic of the detailed structure of the quantum counter. The matrix element $\langle\psi_2|\hat{\mathbf{V}}^\dagger(\mathbf{r}, t)|\psi_1\rangle$ is the probability amplitude for the transition from the state $|\psi_1\rangle$ to some final state $|\psi_2\rangle$ under the influence of the photon emission $\hat{\mathbf{V}}^\dagger(\mathbf{r}, t)$, and the sum over a complete set of final states $|\psi_2\rangle$ makes $P_1(\mathbf{r}, t)$ independent of the final state. On expanding the square and performing the summation over ψ_2 in Eq. (12.9–4) we arrive at

$$\begin{aligned} P_1(\mathbf{r}, t)\Delta t &= C_2\Delta t \sum_{\psi_1} p(\psi_1)\langle\psi_1|\hat{\mathbf{V}}(\mathbf{r}, t)\cdot\hat{\mathbf{V}}^\dagger(\mathbf{r}, t)|\psi_1\rangle \\ &= C_2\Delta t\, \mathrm{Tr}\,[\hat{\rho}\hat{\mathbf{V}}(\mathbf{r}, t)\cdot\hat{\mathbf{V}}^\dagger(\mathbf{r}, t)] \\ &= C_2\Delta t\langle\hat{\mathbf{V}}(\mathbf{r}, t)\cdot\hat{\mathbf{V}}^\dagger(\mathbf{r}, t)\rangle, \end{aligned} \tag{12.9–5}$$

which should be compared with Eq. (12.2–7) for the probability of photodetection. It will be seen that the expressions are similar, except that the operator on the right now appears in anti-normal order. This order is characteristic of a measurement based on the emission of photons with the help of the quantum counter. It is true that the expansion for $\hat{\mathbf{V}}^\dagger(\mathbf{r}, t)$ in Eq. (12.2–7) has contributions from all modes of the field, rather than from the limited set $[\mathbf{k}, s]$ in Eq. (12.9–2), but this is of no importance when only the limited set of modes is occupied, because the unoccupied modes make no contribution to the expectation value of a normally ordered product of operators. If the sensitive surface of the quantum counter is normal to the incident light and has a surface area S, we may suppose that C_2 is proportional to S, and write, as in Eq. (12.3–3),

$$P_1(r, t) = c\beta S\langle\hat{\mathbf{V}}(\mathbf{r}, t)\cdot\hat{\mathbf{V}}^\dagger(\mathbf{r}, t)\rangle, \tag{12.9–6}$$

where β is some dimensionless number that may be identified as the quantum efficiency of the quantum counter.

In order to compare the effectiveness of the quantum counter with that of the photodetector under similar conditions, we need the scalar product commutator of $\hat{\mathbf{V}}(\mathbf{r}, t)$ and $\hat{\mathbf{V}}^\dagger(\mathbf{r}, t)$. From Eq. (12.9–2) and its conjugate we have

$$[\hat{V}_i(\mathbf{r}, t), \hat{V}_i^\dagger(\mathbf{r}, t)] = \frac{1}{L^3}\sum_{[\mathbf{k},s]}\sum_{[\mathbf{k}',s']}[\hat{a}_{\mathbf{k}s}, \hat{a}^\dagger_{\mathbf{k}'s'}]\boldsymbol{\varepsilon}_{\mathbf{k}s}\cdot\boldsymbol{\varepsilon}^*_{\mathbf{k}'s'}\,e^{i[(\mathbf{k}-\mathbf{k}')\cdot\mathbf{r}-(\omega-\omega')t]},$$

and with the help of Eqs. (10.2–16) and (10.3–9) we obtain

$$[\hat{V}_i(\mathbf{r}, t), \hat{V}_i^\dagger(\mathbf{r}, t)] = \frac{1}{L^3}\sum_{[\mathbf{k},s]} 1 = \frac{\mu}{L^3}, \tag{12.9–7}$$

where we denote by μ the total number of modes contributing to the finite set $[\mathbf{k}, s]$. Equation (12.9–6) can therefore be re-expressed in the form

$$P_1(\mathbf{r}, t) = c\beta S\left[\langle \hat{\mathbf{V}}^\dagger(\mathbf{r}, t)\cdot\hat{\mathbf{V}}(\mathbf{r}, t)\rangle + \frac{\mu}{L^3}\right]$$
$$= c\beta S\left[\langle \hat{I}(\mathbf{r}, t)\rangle + \frac{\mu}{L^3}\right], \qquad (12.9\text{–}8)$$

where $\hat{I}(\mathbf{r}, t) \equiv \hat{\mathbf{V}}^\dagger(\mathbf{r}, t)\cdot\hat{\mathbf{V}}(\mathbf{r}, t)$ is the photon density as before. Comparison of this equation with Eq. (12.3–3) now shows that, for the same state of the field with all modes outside the set $[\mathbf{k}, s]$ unoccupied, and for similar quantum efficiencies α and β, the quantum counter will count at a higher rate than the photoelectric detector. The difference between the two counting rates $c\beta S\mu/L^3$ is independent of the state of the field. As the number of modes μ is proportional to L^3, $c\beta S\mu/L^3$ is actually independent of the normalization volume L^3, as it must be if it is a quantity of physical significance. From Eq. (12.9–8), $c\beta S\mu/L^3$ is also the rate of counting of the quantum counter in the vacuum, so that it is evidently the contribution attributable to photon emission in the spontaneous decay of the atomic system from level b to level c. It is this spontaneous counting in the vacuum which distinguishes the quantum counter most from the photoelectric detector.

The importance of the spontaneous counting rate obviously depends on the relative magnitudes of $\langle \hat{I}(\mathbf{r}, t)\rangle$ and μ/L^3. As we have seen in Section 12.3, $\langle \hat{I}(\mathbf{r}, t)\rangle$ is the mean number of photons per unit volume. Since μ/L^3 is the number of modes per unit volume, the ratio of the first term to the second is the average number of photons per mode within the response band of the detector. When this average occupation number is large, then the behavior of the quantized field approaches that of a classical field in the sense of the correspondence principle, and the spontaneous counting rate is negligible compared with the stimulated counting rate. Under these circumstances there is little to choose between the quantum counter and the photoelectric detector as regards effectiveness in the measurements of the field. Under these circumstances, also, the expectation values of the normally and anti-normally ordered products in Eqs. (12.2–7) and (12.9–5) are nearly equal, and we are justified in treating the field vectors as c-numbers. On the other hand, when the average photon occupation number per mode is small, then $\langle \hat{I}(\mathbf{r}, t)\rangle \ll \mu/L^3$ and the output of the quantum counter is dominated by the spontaneous counts. Indeed, its counting rate is then almost constant and independent of the state of the field. Evidently such a device is not expected to be in great demand as a laboratory measuring instrument. Unfortunately, it so happens that for light emitted by all the familiar thermal sources, the average photon occupation number per mode is much less than unity (see Section 13.3). Not until the source temperature is far in excess of about 100 000 K does the mean photon occupation number become appreciably greater than unity. The quantum counter is therefore quite inappropriate for measuring the intensity of thermal light. However, the situation is very different for the field produced by a laser or parametric oscillator. Here the average photon occupation number per mode can be very large indeed, so that the field should be describable in classical terms and measurable both by the quantum counter and by the photoelectric detector.

Equation (12.9–6) may readily be generalized for N different quantum counters. It is not difficult to see that an argument similar to that leading to Eq. (12.2–12) now gives

$$P_N(\mathbf{r}_1, t_1, \ldots, \mathbf{r}_N, t_N) = \beta_1 \ldots \beta_N S_1 \ldots S_N c^N \langle \hat{V}_{i_1}(\mathbf{r}_1, t_1) \ldots \times \hat{V}_{i_N}(\mathbf{r}_N, t_N) \hat{V}^\dagger_{i_N}(\mathbf{r}_N, t_N) \ldots \hat{V}^\dagger_{i_1}(\mathbf{r}_1, t_1) \rangle \quad (12.9\text{–}9)$$

for the N-fold joint probability density that N counts are registered by quantum counters at $\mathbf{r}_1, t_1, \ldots, \mathbf{r}_N, t_N$, respectively. $\beta_1, \ldots, \beta_N$ and $S_1, \ldots, S_N$ are the quantum efficiencies and the illuminated surface areas of the N quantum counters, which need not all be different. Evidently measurements with quantum counters correspond to anti-normally ordered correlation functions, just as measurements with photoelectric detectors correspond to normally ordered ones. The importance of the latter as compared with the former is simply a reflection of the greater utility of the photodetector.

As in Section 12.2, it is often convenient to introduce a notation that allows us to write the correlation function in Eq. (12.9–9) in abbreviated form. In Section 12.2 we encountered both an ordering symbol and an ordering operator that served the purpose of normal ordering, and we shall introduce an analogous notation here. We shall use the symbol "$\hat{O}$" to denote the arrangement of the operator $\hat{O}$ in anti-normal order, without regard for commutation relations. For example, "$\hat{a}^\dagger\hat{a}$" $= \hat{a}\hat{a}^\dagger$, "$\hat{n}^2$" $= \hat{a}^2\hat{a}^{\dagger 2}$. As before, the symbol " " is not to be regarded as an operator, but simply as an element of notation. With its help we may re-write Eq. (12.9–9) in the form

$$P_N(\mathbf{r}_1, t_1, \ldots, \mathbf{r}_N, t_N) = \beta_1 \ldots \beta_N S_1 \ldots S_N \langle \text{“}\hat{I}(\mathbf{r}_1, t_1) \ldots \hat{I}(\mathbf{r}_N, t_N)\text{”} \rangle. \quad (12.9\text{–}10)$$

Alternatively, we may introduce a linear ordering operator $\hat{\Omega}^{(A)}$ (cf. Section 11.10) that has the effect of converting a c-number functional into an anti-normally ordered functional of creation and annihilation operators according to the rule

$$\hat{\Omega}^{(A)} f(\{v\}, \{v^*\}) = f^{(A)}(\{\hat{a}\}, \{\hat{a}^\dagger\}), \quad (12.9\text{–}11)$$

where $f^{(A)}(\{\hat{a}\}, \{\hat{a}^\dagger\})$ is an anti-normally ordered functional. The operator $\hat{\Omega}^{(A)}$ allows us to express Eq. (12.9–9) in the form

$$P_N(\mathbf{r}_1, t_1, \ldots, \mathbf{r}_N, t_N) = \beta_1 \ldots \beta_N S_1 \ldots S_N c^N \langle \hat{\Omega}^{(A)}[I(\mathbf{r}_1, t_1) \ldots I(\mathbf{r}_N, t_N)] \rangle, \quad (12.9\text{–}12)$$

where $I(\mathbf{r}_1, t_1) \equiv \mathbf{V}^*(\mathbf{r}_1, t_1) \cdot \mathbf{V}(\mathbf{r}_1, t_1)$, etc. are c-number intensities in which $\mathbf{V}(\mathbf{r}_1, t)$ and $\mathbf{V}^*(\mathbf{r}_1, t)$ are right and left eigenvalues of $\hat{\mathbf{V}}(\mathbf{r}_1, t)$ and $\hat{\mathbf{V}}^\dagger(\mathbf{r}_1, t)$.

Let us now turn to the problem of evaluating the anti-normally ordered correlation functions. We have already seen that for normally ordered correlations the problem can be reduced to the evaluation of an integral involving only c-number functions, and we can show that the same is true in this case, as is expected from the general relation (11.10–9). However, the diagonal coherent-state representation of the density operator, which has proved so valuable for normally ordered correlations, will not be of direct help here. It allows us to write

$$\langle \hat{V}(x_1) \ldots \hat{V}(x_N)\hat{V}^\dagger(y_M) \ldots \hat{V}^\dagger(y_1)\rangle = \mathrm{Tr}\int \phi(\{v\})\hat{V}(x_1) \ldots \hat{V}(x_N)\hat{V}^\dagger(y_M) \ldots \times \hat{V}^\dagger(y_1)|\{v\}\rangle\langle\{v\}|\,\mathrm{d}\{v\}, \qquad (12.9\text{–}13)$$

where $x_1, \ldots,$ etc., stand collectively for $\mathbf{r}_1$, t_1, $i_1, \ldots,$ etc., as before, but as none of the operators on the right of this equation stands next to its own left or right eigenstate, even after cyclic permutation, the operators cannot be eliminated directly. We could of course re-order the operators by repeated application of the commutation rules, and then make use of the optical equivalence theorem, but the resulting expression would generally be rather complicated.

12.9.2 Substitution of differential operators

An alternative procedure permits the creation and annihilation operators under the integral in Eq. (12.9–13) to be replaced by differential operators (Louisell, 1969). We first observe that the field operator product under the integral is of the general form

$$f^{(\mathrm{A})}(\{\hat{a}\}, \{\hat{a}^\dagger\}) \equiv \sum_{\lambda_1} \cdots \sum_{\lambda_N}\sum_{\lambda'_1} \cdots \sum_{\lambda'_M} c_{1\lambda_1} \ldots c_{N\lambda_N}c'_{M\lambda'_M} \ldots c'_{1\lambda'_1}\hat{a}_{\lambda_1} \ldots \hat{a}_{\lambda_N}\hat{a}^\dagger_{\lambda'_M} \ldots \hat{a}^\dagger_{\lambda'_1}, \qquad (12.9\text{–}14)$$

where $\lambda_r (r = 1, 2, \ldots, N)$ and λ'_r $(r = 1, 2, \ldots, M)$ are labels for the modes $\mathbf{k}$, s and the $c_{r\lambda_r}(r = 1, 2, \ldots, N)$ and $c'_{r\lambda'_r}(r = 1, 2, \ldots, M)$ are coefficients that generally depend on space and time parameters. The superscript (A) emphasizes that the functional $f^{(\mathrm{A})}$ is in anti-normal order. We also recall that the coherent state may be regarded as a displaced vacuum state according to Eq. (11.3–1), so that we may express the coherent state projector $|\{v\}\rangle\langle\{v\}|$ as

$$|\{v\}\rangle\langle\{v\}| = \exp(-|v_{\lambda_r}|^2)\exp(v_{\lambda_r}\hat{a}^\dagger_{\lambda_r})|0_{\lambda_r}\rangle\langle 0_{\lambda_r}|\exp(v^*_{\lambda_r}\hat{a}_{\lambda_r})\prod_{m\neq r}|v_{\lambda_m}\rangle\langle v_{\lambda_m}|. \qquad (12.9\text{–}15)$$

The product of $\hat{a}^\dagger_{\lambda_r}$ and $|\{v\}\rangle\langle\{v\}|$ may therefore be written formally

$$\begin{aligned}\hat{a}^\dagger_{\lambda_r}|\{v\}\rangle\langle\{v\}| &= \exp(-|v_{\lambda_r}|^2)\frac{\partial}{\partial v_{\lambda_r}}\exp(v_{\lambda_r}\hat{a}^\dagger_{\lambda_r})|0_{\lambda_r}\rangle\langle 0_{\lambda_r}|\exp(v^*_{\lambda_r}\hat{a}_{\lambda_r})\prod_{m\neq r}|v_{\lambda_m}\rangle\langle v_{\lambda_m}| \\ &= \exp(-|v_{\lambda_r}|^2)\frac{\partial}{\partial v_{\lambda_r}}[\exp(|v_{\lambda_r}|^2)|\{v\}\rangle\langle\{v\}|] \\ &= \left(v^*_{\lambda_r} + \frac{\partial}{\partial v_{\lambda_r}}\right)|\{v\}\rangle\langle\{v\}|, \qquad (12.9\text{–}16)\end{aligned}$$

where $\partial/\partial v_{\lambda_r}$ denotes partial differentiation with respect to v_{λ_r}, with $v^*_{\lambda_r}$ regarded as an independent variable. Equation (12.9–16) shows that the Hilbert space operator $\hat{a}_\lambda^\dagger$ may be eliminated and replaced by the differential operator $(v_\lambda^* + \partial/\partial v_\lambda)$ when it stands to the left of the projector $|\{v\}\rangle\langle\{v\}|$. In the same way we may show that

$$\begin{aligned}|\{v\}\rangle\langle\{v\}|\hat{a}_\lambda &= \exp(-|v_\lambda|^2)\frac{\partial}{\partial v_\lambda^*}[\exp(|v_\lambda|^2)|\{v\}\rangle\langle\{v\}|] \\ &= \left(v_\lambda + \frac{\partial}{\partial v_\lambda^*}\right)|\{v\}\rangle\langle\{v\}|,\end{aligned} \tag{12.9–17}$$

which implies that the annihilation operator $\hat{a}_\lambda$ standing to the right of $|\{v\}\rangle\langle\{v\}|$ may be replaced by the differential operator $(v_\lambda + \partial/\partial v_\lambda^*)$. These results may readily be combined, so that we have for example,

$$\hat{a}_\lambda^\dagger|\{v\}\rangle\langle\{v\}|\hat{a}_\lambda = \left(v_\lambda^* + \frac{\partial}{\partial v_\lambda}\right)\left(v + \frac{\partial}{\partial v_\lambda^*}\right)|\{v\}\rangle\langle\{v\}|, \tag{12.9–18}$$

$$\hat{a}_\lambda^\dagger\hat{a}_\lambda|\{v\}\rangle\langle\{v\}| = \hat{a}_\lambda^\dagger|\{v\}\rangle\langle\{v\}|v_\lambda = v_\lambda\left(v_\lambda^* + \frac{\partial}{\partial v_\lambda}\right)|\{v\}\rangle\langle\{v\}|, \tag{12.9–19}$$

$$|\{v\}\rangle\langle\{v\}|\hat{a}^\dagger\hat{a} = v^*|\{v\}\rangle\langle\{v\}|\hat{a} = v_\lambda^*\left(v_\lambda + \frac{\partial}{\partial v_\lambda^*}\right)|\{v\}\rangle\langle\{v\}|. \tag{12.9–20}$$

By making use of the relations (12.9–16) and (12.9–17) to determine the expectation value of $f^{(A)}(\{\hat{a}\}, \{\hat{a}^\dagger\})$ given by Eq. (12.9–14) we arrive at

$$\begin{aligned}\langle f^{(A)}(\{\hat{a}\}\ \{\hat{a}^\dagger\})\rangle &= \mathrm{Tr}\int \phi(\{v\})f^{(A)}(\{\hat{a}\}, \{\hat{a}^\dagger\})|\{v\}\rangle\langle\{v\}|\,\mathrm{d}\{v\} \\ &= \mathrm{Tr}\int \phi(\{v\})f^{(A)}\left(\{\hat{a}\}, \left\{v^* + \frac{\partial}{\partial v}\right\}\right)|\{v\}\rangle\langle\{v\}|\,\mathrm{d}\{v\} \\ &= \mathrm{Tr}\int \phi(\{v\})f^{(A)}\left(\left\{v + \frac{\partial}{\partial v^*}\right\}, \left\{v^* + \frac{\partial}{\partial v}\right\}\right)|\{v\}\rangle\langle\{v\}|\,\mathrm{d}\{v\},\end{aligned} \tag{12.9–21}$$

where the last line follows from the previous one after cyclic permutation of operators under the trace. We have therefore succeeded in eliminating all Hilbert space operators (other than the projector $|\{v\}\rangle\langle\{v\}|$) under the integral. By carrying out the integration by parts, we can usually move the differential operators from the projector $|\{v\}\rangle\langle\{v\}|$ to the phase space functional, whereupon the trace reduces to an ordinary integral of c-number functions. This procedure is rarely used directly to evaluate the correlation function. However, we shall later find it valuable in reducing the operator master equation of motion of an interacting field to the form of a differential equation (see Section 18.5).

12.9.3 Phase space functional for anti-normally ordered correlations

It is possible to reduce the expectation value of an anti-normally ordered operator product directly to an integral over c-number functions, in a manner that is strictly analogous to the procedure we adopted in Sections 11.9 and 11.10. The phase space functional for anti-normal ordering has been discussed by Kano

(1965) and by Mehta and Sudarshan (1965). More comprehensive recent treatments have been given by Cahill and Glauber (1969a,b) and by Agarwal and Wolf (1970a,b,c). We make use of the expansion (11.10–8) of the unit operator in terms of coherent-state projectors, except that we limit the product over modes to the set $[\mathbf{k}, s]$. Moreover, we shall assume that modes outside the set $[\mathbf{k}, s]$ are empty and that the density operator $\hat{\rho}$ has already been traced over these unoccupied modes, which are irrelevant to our present discussion. We insert the unit operator between the annihilation and creation operators in the definition of the correlation function, and obtain the formula

$$\langle \hat{V}(x_1) \ldots \hat{V}(x_N)\hat{V}^\dagger(y_M) \ldots \hat{V}^\dagger(y_1)\rangle =$$
$$\mathrm{Tr}\int \hat{\rho}\hat{V}(x_1) \ldots \hat{V}(x_N)\prod_{[\mathbf{k},s]}[|v_{\mathbf{k}s}\rangle\langle v_{\mathbf{k}s}|]\hat{V}^\dagger(y_M) \ldots \hat{V}^\dagger(y_1)\prod_{[\mathbf{k},s]}\frac{\mathrm{d}^2 v_{\mathbf{k}s}}{\pi}. \qquad (12.9\text{–}22)$$

We now observe that each of the creation and annihilation operators under the integral stands to the left or to the right of its respective eigenstate. The operators may therefore be replaced by their eigenvalues. If the trace is then evaluated under the integral we arrive at

$$\begin{aligned}\langle \hat{V}(x_1) \ldots \hat{V}(x_N)\hat{V}^\dagger(y_M) \ldots \hat{V}^\dagger(y_1)\rangle &= \int \langle\{v\}|\hat{\rho}|\{v\}\rangle \prod_{[\mathbf{k},s]}\left[\frac{1}{\pi}\right] V(x_1) \ldots \\ &\quad \times V(x_N)V^*(y_M) \ldots V^*(y_1)\,\mathrm{d}\{v\} \\ &= \int Q(\{v\})V(x_1) \ldots V(x_N)V^*(y_M) \ldots \\ &\quad \times V^*(y_1)\,\mathrm{d}\{v\} \\ &= \langle V(x_1) \ldots V(x_N)V^*(y_M) \ldots V^*(y_1)\rangle_Q,\end{aligned}$$
$$(12.9\text{–}23)$$

where

$$Q(\{v\}) \equiv \langle\{v\}|\hat{\rho}|\{v\}\rangle \prod_{[\mathbf{k},s]}\left[\frac{1}{\pi}\right]. \qquad (12.9\text{–}24)$$

We have therefore succeeded in expressing the anti-normally ordered correlation function as a weighted average of the corresponding c-number function, exactly as we did earlier for normally ordered operator products. It should be evident that a similar result must hold for any anti-normally ordered functional $f^{(\mathrm{A})}(\{\hat{a}\}, \{\hat{a}^\dagger\})$ of creation and annihilation operators having a power series expansion, so that

$$\langle f^{(\mathrm{A})}(\{\hat{a}\}, \{\hat{a}^\dagger\})\rangle = \int Q(\{v\})f^{(\mathrm{A})}(\{v\}, \{v^*\})\,\mathrm{d}\{v\} \equiv \langle f^{(\mathrm{A})}(\{v\}, \{v^*\})\rangle_Q.$$
$$(12.9\text{–}25)$$

The weighting, or phase space, functional this time is the functional $Q(\{v\})$ given by Eq. (12.9–24), [known as the $Q(\{v\})$ functional], rather than the functional

$\phi(\{v\})$ [or the $P(\{v\})$ functional] appearing in the diagonal coherent-state representation of $\hat{\rho}$. The difference between the expectation values of normally ordered and anti-normally ordered operators must therefore reside in the character of the two functionals $\phi(\{v\})$ and $Q(\{v\})$. We have already investigated some properties of the phase space functional $\phi(\{v\})$, and shown that, although it has some features of a probability functional, it may at times be more singular than a classical probability functional.

Let us now examine the corresponding properties of $Q(\{v\})$. We summarize the four principal properties of $Q(\{v\})$ below:

(a) $Q(\{v\})$ is real.
Since $Q(\{v\})$ is equal to the expectation value of a Hermitian operator, apart from a real proportionality factor, $Q(\{v\})$ must be real, i.e.

$$Q^*(\{v\}) = Q(\{v\}). \tag{12.9–26}$$

(b) $Q(\{v\})$ is normalized to unity.
By retracing some of the steps used to derive Eq. (12.9–23), we have

$$\begin{aligned}\int Q(\{v\})\,\mathrm{d}\{v\} &= \int \langle\{v\}|\hat{\rho}\prod_{[\mathbf{k},s]}\left[\frac{1}{\pi}|v_{\mathbf{k}s}\rangle\,\mathrm{d}^2v_{\mathbf{k}s}\right] \\ &= \mathrm{Tr}\,\hat{\rho}\int\prod_{[\mathbf{k},s]}\left[\frac{1}{\pi}|v_{\mathbf{k}s}\rangle\langle v_{\mathbf{k}s}|\,\mathrm{d}^2v_{\mathbf{k}s}\right] \\ &= \mathrm{Tr}\,\hat{\rho} = 1,\end{aligned} \tag{12.9–27}$$

which shows that $Q(\{v\})$ is normalized to unity.

Properties (a) and (b) are shared by $\phi(\{v\})$ also. However, $\phi(\{v\})$ in general does not possess the following properties.

(c) $Q(\{v\})$ is non-negative.
Since $Q(\{v\})$ is proportional to the expectation value of the non-negative operator $\hat{\rho}$, it must itself be non-negative. If we express $\hat{\rho}$ in the form

$$\hat{\rho} = \sum_i p_i|s_i\rangle\langle s_i|, \quad \sum_i p_i = 1, \tag{12.9–28}$$

where $|s_i\rangle$ is any state of the field and p_i is the probability of that state, we have

$$Q(\{v\}) = \sum_i p_i|\langle s_i|\{v\}\rangle|^2\prod_{[\mathbf{k},s]}\left(\frac{1}{\pi}\right) \geqslant 0, \tag{12.9–29}$$

since $|\langle s_i|\{v\}\rangle|^2 \geqslant 0$ and $p_i \geqslant 0$ for all i.

Properties (a), (b) and (c) are the essential properties of all classical probability densities, so that $Q(\{v\})$, unlike $\phi(\{v\})$, falls within the class of probability densities. The fourth property (d) given below establishes $Q(\{v\})$ as a special subclass of probability densities.

(d) $Q(\{v\})$ is bounded from above.

From Eq. (12.9–29) we have

$$Q(\{v\}) = \sum_i p_i |\langle s_i|\{v\}\rangle|^2 \prod_{[\mathbf{k},s]}\left(\frac{1}{\pi}\right) \leqslant \prod_{[\mathbf{k},s]}\left(\frac{1}{\pi}\right), \tag{12.9–30}$$

since $|\langle s_i|\{v\}\rangle|^2 \leqslant 1$ and $\sum_i p_i = 1$, which shows that $Q(\{v\})$ is bounded both from above and from below. In this respect $Q(\{v\})$ is even better behaved than a classical phase space weighting function or probability density, which can be as singular as a delta function.

Despite this excellent mathematical behavior of $Q(\{v\})$, and the fact that it is a subclass of classical probability densities, $Q(\{v\})$ appears to be remote from a probability density in significance. In this respect it is quite unlike $\phi(\{v\})$, which generally coincides with the classical probability density when a classical analog of the state exists. Some examples will illustrate the situation.

If the state is the coherent state $|\{v'\}\rangle$, then

$$\begin{aligned} Q(\{v\}) &= \mathrm{Tr} \prod_{[\mathbf{k},s]} \frac{1}{\pi} |\langle v_{\mathbf{k}s}|v'_{\mathbf{k}s}\rangle|^2 \\ &= \prod_{[\mathbf{k},s]} \frac{1}{\pi} \mathrm{e}^{-|v_{\mathbf{k}s}-v'_{\mathbf{k}s}|^2}, \end{aligned} \tag{12.9–31}$$

which is a Gaussian distribution in $\{v\}$ centered on $\{v'\}$, of unit variance. This should be compared with the corresponding form of $\phi(\{v\})$,

$$\phi(\{v\}) = \prod_{[\mathbf{k},s]} \delta^2(v_{\mathbf{k}s} - v'_{\mathbf{k}s}), \tag{12.9–32}$$

for the same state. It is evident that, despite its good mathematical behavior, $Q(\{v\})$ does not correspond to a classical phase space function for an electromagnetic field of well-defined complex amplitude, whereas $\phi(\{v\})$ does. The difference between $Q(\{v\})$ and $\phi(\{v\})$ is very great for small $v'_{\mathbf{k}s}$, and becomes most pronounced for the vacuum state for which $v'_{\mathbf{k}s} = 0$ for all $\mathbf{k}$, s. However, when the real and imaginary parts of $v'_{\mathbf{k}s}$ are all large numbers, $Q(\{v\})$ given by Eq. (12.9–31) is strongly peaked for $\{v\}$ in the neighborhood of $\{v'\}$, and begins to take on some of the features of a delta function.

As another example, we consider the Fock state $|\{n\}\rangle$, for which

$$\begin{aligned} Q(\{v\}) &= \prod_{[\mathbf{k},s]}\left[\frac{1}{\pi}|\langle v_{\mathbf{k}s}|n_{\mathbf{k}s}\rangle|^2\right] \\ &= \prod_{[\mathbf{k},s]}\left[\frac{1}{\pi}\frac{|v_{\mathbf{k}s}|^{2n_{\mathbf{k}s}}\mathrm{e}^{-|v_{\mathbf{k}s}|^2}}{n_{\mathbf{k}s}!}\right]. \end{aligned} \tag{12.9–33}$$

This is a product of γ-distributions in $v_{\mathbf{k}s}$ (see Section 1.5), which are well-behaved functions, that are approximately Gaussian in form for large $n_{\mathbf{k}s}$ and are again very different from the corresponding $\phi(\{v\})$ given by Eq. (11.8–20). In this case no classical analog of the state exists, and this is reflected in the very singular form of $\phi(\{v\})$. However, it would be difficult to judge from the form of $Q(\{v\})$ given by Eq. (12.9–33) that $Q(\{v\})$ represents a non-classical state.

It is not difficult to obtain an explicit relation between the functionals $Q(\{v\})$ and $\phi(\{v\})$. A connection between them is already implicit in the Fourier expansion (11.9–16) of the characteristic function for anti-normally ordered operators, after an obvious generalization for the multimode case. But it is simpler to proceed directly from the definition (12.9–24). We make a diagonal coherent-state representation of $\hat{\rho}$, and find, with the help of Eq. (11.11–5), that

$$\begin{aligned} Q(\{v\}) &= \int \phi(\{v'\}) \prod_{[\mathbf{k},s]} \left[\frac{1}{\pi} |\langle v_{\mathbf{k}s} | v'_{\mathbf{k}s} \rangle|^2 \, \mathrm{d}^2 v'_{\mathbf{k}s} \right] \\ &= \int \phi(\{v'\}) \prod_{[\mathbf{k},s]} \left[\frac{1}{\pi} \mathrm{e}^{-|v_{\mathbf{k}s} - v'_{\mathbf{k}s}|^2} \, \mathrm{d}^2 v'_{\mathbf{k}s} \right]. \end{aligned} \qquad (12.9\text{–}34)$$

The functional $Q(\{v\})$ is therefore a kind of smoothed form of $\phi(\{v\})$, in which the smoothing functional is a product over the modes of Gaussian functions. We have already seen that, for large $v_{\mathbf{k}s}$, $\exp(-|v_{\mathbf{k}s} - v'_{\mathbf{k}s}|^2)$ has some of the character of the $\delta^2(v_{\mathbf{k}s} - v'_{\mathbf{k}s})$ function. If $\phi(\{v'\})$ is itself a smooth and well-behaved function, we may treat it as a test function under the integral in Eq. (12.9–34), and conclude that for large $\{v\}$,

$$Q(\{v\}) \approx \int \phi(\{v'\}) \prod_{[\mathbf{k},s]} [\delta^2(v_{\mathbf{k}s} - v'_{\mathbf{k}s}) \, \mathrm{d}^2 v'_{\mathbf{k}s}] = \phi(\{v\}). \qquad (12.9\text{–}35)$$

We have therefore shown that for states for which a classical limit exists, the phase space functionals $Q(\{v\})$ and $\phi(\{v\})$ tend to become equal in the classical limit. This implies that the distinction between normally ordered and anti-normally ordered correlations tends to disappear in the classical limit, when the field vectors may be treated as c-numbers. The distinction between the results of measurements with photoelectric detectors and quantum counters then tends to vanish also.

The phase space or quasi-probability functionals $\phi(\{v\})$ and $Q(\{v\})$ are only two examples of a wide class of possible functionals that can be associated with different operator orderings (cf. Section 11.10).

12.10 Photon statistics

In Section 10.4 we encountered the Fock state of the field, which is a state of definite photon occupation number for each mode. We showed that Fock states are also eigenstates of the energy, of the momentum and, for modes of circular polarization, of the helicity. On the other hand, the field operators have zero expectations but no definite values in a Fock state, which is extremely non-classical in character. Later we encountered the coherent states of the field, which are more nearly analogous to classical states, in which the positive and negative frequency parts of the field operators are well defined, but in which the photon numbers have no definite values. Now the states of the electromagnetic field that we encounter in practice are almost always states of indefinite photon number, in which the number of photons can be given only a statistical description. We now turn our attention to the problem of the photon statistics. We shall later find that the photon fluctuations are quite well reflected in the fluctuations of photoelectric counts registered by photodetectors placed in the field.

12.10.1 Probabilities

If $\{n\}$ denotes some definite set of photon occupation numbers, then the probability $p(\{n\})$ of finding the set $\{n\}$, when the electromagnetic field is in some arbitrary state described by the density operator $\hat{\rho}$, is simply the expectation value of the Fock state projector $|\{n\}\rangle\langle\{n\}|$, or

$$p(\{n\}) = \mathrm{Tr}\,[\hat{\rho}|\{n\}\rangle\langle\{n\}|]. \tag{12.10–1}$$

With the help of the diagonal coherent-state representation (11.11–11) of $\hat{\rho}$, and the previously derived scalar product of the coherent state with the Fock state [Eq. (11.2–12)], we find that

$$\begin{aligned} p(\{n\}) &= \mathrm{Tr}\int \phi(\{v\})|\{v\}\rangle\langle\{v\}|\{n\}\rangle\langle\{n\}|\,\mathrm{d}\{v\} \\ &= \int \phi(\{v\})|\langle\{n\}|\{v\}\rangle|^2\,\mathrm{d}\{v\} \\ &= \int \phi(\{v\})\prod_{\mathbf{k},s}\left[\frac{|v_{\mathbf{k}s}|^{2n_{\mathbf{k}s}}\,\mathrm{e}^{-|v_{\mathbf{k}s}|^2}}{n_{\mathbf{k}s}!}\right]\mathrm{d}\{v\} \\ &= \left\langle \prod_{\mathbf{k},s}\left[\frac{|v_{\mathbf{k}s}|^{2n_{\mathbf{k}s}}\,\mathrm{e}^{-|v_{\mathbf{k}s}|^2}}{n_{\mathbf{k}s}!}\right]\right\rangle_{\phi} && (12.10\text{–}2\mathrm{a}) \\ &= \left\langle :\prod_{\mathbf{k},s}\left[\frac{\hat{n}_{\mathbf{k}s}^{n_{\mathbf{k}s}}\,\mathrm{e}^{-\hat{n}_{\mathbf{k}s}}}{n_{\mathbf{k}s}!}\right]:\right\rangle. && (12.10\text{–}2\mathrm{b}) \end{aligned}$$

The last expression for $p(\{n\})$ follows from the preceding one by the optical equivalence theorem. The relation (12.10–2a) has an interesting structure; it expresses $p(\{n\})$ as a product of Poisson distributions in the occupation numbers $n_{\mathbf{k}s}$, which is then to be averaged with respect to the phase space functional $\phi(\{v\})$ (Ghielmetti, 1964). However, the operation of averaging with respect to $\phi(\{v\})$ in general will not preserve the structure of $p(\{n\})$ as a product of Poisson distributions. The resulting expression may look entirely different, and may no longer be in the form of a product over modes.

Of course, in special cases $p(\{n\})$ may have the form of a Poisson product. For example, if the state is the coherent state $|\{v'\}\rangle$, then $\phi(\{v\})$ is given by the relation

$$\phi(\{v\}) = \prod_{\mathbf{k},s}\delta^2(v_{\mathbf{k}s} - v'_{\mathbf{k}s}), \tag{12.10–3}$$

and from Eq. (12.10–2) we have

$$p(\{n\}) = \prod_{\mathbf{k},s}\left[\frac{|v'_{\mathbf{k}s}|^{2n_{\mathbf{k}s}}\,\mathrm{e}^{-|v'_{\mathbf{k}s}|^2}}{n_{\mathbf{k}s}!}\right], \tag{12.10–4}$$

which is a product of the Poisson distribution for each mode, as expected. For the phase space functional $\phi(\{v\})$ given by Eq. (11.11–18), corresponding to the randomly phased single-mode laser, we find that

$$p(\{n\}) = \frac{r^{2n_{\mathbf{k}'s'}}\,\mathrm{e}^{-r^2}}{n_{\mathbf{k}'s'}} \prod_{\mathbf{k},s\neq\mathbf{k}',s'} \delta_{n_{\mathbf{k}s}0}, \tag{12.10–5}$$

which shows that the number of photons in the $\mathbf{k}'$, s' mode has a Poisson distribution, while all other modes are empty. In Section 13.2 we shall show that for the thermal equilibrium state of the field, for which $\phi(\{v\})$ is Gaussian, Eq. (12.10–2) leads to a product of Bose–Einstein distributions.

Sometimes we are less interested in the probability distribution of the set of occupation number $\{n\}$, than in the probability distribution $P(n)$ of the total number of photons n, where

$$n \equiv \sum_{\mathbf{k},s} n_{\mathbf{k}s}. \tag{12.10–6}$$

We can readily derive $P(m)$ from $p(\{n\})$ by summing $p(\{n\})$ over all combinations of $\{n\}$ for which the total number $n = m$. With the help of Eq. (12.10–2) we then have

$$\begin{aligned} P(m) &= \sum_{\{n\}} p(\{n\})\,\delta_{nm} \\ &= \sum_{\{n\}} \left\langle \prod_{\mathbf{k},s}\left[\frac{|v_{\mathbf{k}s}|^{2n_{\mathbf{k}s}}\,\mathrm{e}^{-|v_{\mathbf{k}s}|^2}}{n_{\mathbf{k}s}!}\right]\right\rangle_\phi \delta_{nm}. \end{aligned}$$

By interchanging the orders of summation, multiplication and phase space averaging, and making use of the multinomial theorem in the form

$$\left[\sum_{\mathbf{k},s}|v_{\mathbf{k}s}|^2\right]^m = \sum_{\{n\}}\prod_{\mathbf{k},s}\left[\frac{|v_{\mathbf{k}s}|^{2n_{\mathbf{k}s}}}{n_{\mathbf{k}s}!}\right] m!\,\delta_{nm}, \tag{12.10–7}$$

we arrive at

$$P(m) = \left\langle \frac{1}{m!}\left[\sum_{\mathbf{k},s}|v_{\mathbf{k},s}|^2\right]^m \exp\left[-\sum_{\mathbf{k},s}|v_{\mathbf{k},s}|^2\right]\right\rangle_\phi$$

$$\left.\begin{aligned} &= \left\langle \frac{U^m\,\mathrm{e}^{-U}}{m!}\right\rangle_\phi \\ &= \left\langle :\frac{\hat{n}^m\,\mathrm{e}^{-\hat{n}}}{m!}:\right\rangle, \end{aligned}\right\} \tag{12.10–8}$$

where

$$U \equiv \sum_{\mathbf{k},s}|v_{\mathbf{k},s}|^2 = \int_{L^3} \mathbf{V}^*(\mathbf{r},t)\cdot\mathbf{V}(\mathbf{r},t)\,\mathrm{d}^3x, \tag{12.10–9}$$

and $\mathbf{V}^*\cdot\mathbf{V}$ is the photon density. The last form of Eq. (12.10–8) again follows from the previous one by the optical equivalence theorem. We see that $P(m)$ too can be written formally as an average over Poisson distributions, with the phase space functional $\phi(\{v\})$ playing the role of weighting functional. For the coherent state $|\{v'\}\rangle$ with $\phi(\{v\})$ given by Eq. (12.10–3), and for the single-mode laser with $\phi(\{v\})$ given by Eq. (11.11–18), $P(m)$ is actually a Poisson distribution in m. But, in general, the operation of phase space averaging may lead to an entirely different form of $P(m)$.

The parameter U in Eq. (12.10–9) is the c-number light intensity integrated over all space, and, in a sense, is the classical analog of the total photon number $\hat{n}$. The right side of Eq. (12.10–8) may also be interpreted to be the average of $U^m \mathrm{e}^{-U}/m!$ with respect to some random variable U with quasi-probability density $\mathcal{P}(U)$ given by the expression

$$\mathcal{P}(U') = \int \phi(\{v\})\delta(U' - U)\,\mathrm{d}\{v\}. \tag{12.10–10}$$

In order to see this we observe that Eq. (12.10–8) may also be written as

$$\begin{aligned} P(m) &= \int \phi(\{v\}) \frac{U^m \mathrm{e}^{-U}}{m!} \mathrm{d}\{v\} \\ &= \iint \phi(\{v\})\delta(U - U') \frac{U^m \mathrm{e}^{-U}}{m!} \mathrm{d}U'\,\mathrm{d}\{v\} \\ &= \iint \phi(\{v\})\delta(U - U') \frac{U'^m \mathrm{e}^{-U'}}{m!} \mathrm{d}U'\,\mathrm{d}\{v\} \\ &= \int \mathcal{P}(U') \frac{U'^m \mathrm{e}^{-U'}}{m!} \mathrm{d}U', \end{aligned} \tag{12.10–11}$$

where the right-hand side has the usual form of a classical average. However, when $\phi(\{v\})$ lies outside the domain of a classical phase space density, $\mathcal{P}(U')$ given by Eq. (12.10–10) may also not have the character of a classical probability density.

12.10.2 Tests for non-classical states

It was already pointed out in Section 11.8 that the photon counting probability $P(m)$ cannot vanish for any state – other than the vacuum – for which a classical analog exists, i.e. for which the phase space density $\phi(\{v\})$ is non-negative. This is apparent by inspection of the form of Eq. (12.10–11). It is possible to express the sufficient condition for a non-classical state of the field in other ways (Hillery, 1985). Let us denote by P_e and P_o the probabilities for an even and an odd number of photons, respectively. Then from Eq. (12.10–11), we have on interchanging summation and integration,

$$\begin{aligned} P_\mathrm{e} &= \sum_{m=0}^{\infty} P(2m) = \int_0^\infty \mathcal{P}(U')\mathrm{e}^{-U'} \sum_{m=0}^{\infty} \frac{U'^{2m}}{(2m)!} \mathrm{d}U' \\ &= \int_0^\infty \mathcal{P}(U')\mathrm{e}^{-U'} \cosh U'\,\mathrm{d}U' = \tfrac{1}{2}\int_0^\infty \mathcal{P}(U')(1 + \mathrm{e}^{-2U'})\,\mathrm{d}U', \\ P_\mathrm{o} &= \sum_{m=0}^{\infty} P(2m+1) = \int_0^\infty \mathcal{P}(U')\mathrm{e}^{-U'} \sum_{m=0}^{\infty} \frac{U'^{2m+1}}{(2m+1)!} \mathrm{d}U' \\ &= \int_0^\infty \mathcal{P}(U')\mathrm{e}^{-U'} \sinh U'\,\mathrm{d}U' = \tfrac{1}{2}\int_0^\infty \mathcal{P}(U')(1 - \mathrm{e}^{-2U'})\,\mathrm{d}U'. \end{aligned}$$

It follows by inspection that P_o must be less than 1/2 and P_e greater than 1/2 for any state having a classical analog for which $P(U') \geqslant 0$. Conversely, if $P_o > 1/2$ we have a quantum state without classical analog, and since $P_o + P_e = 1$, the same is true if $P_e < 1/2$.

12.10.3 *Moments of $\hat{n}$*

With the help of Eq. (12.10–8) or Eq. (12.10–11) we may readily relate the moments of the total photon number $\hat{n}$ and the moments of U. On calculating the factorial moment generating function $F(\xi)$ of $P(m)$, we find that

$$\begin{aligned} F(\xi) \equiv \langle (1-\xi)^{\hat{n}} \rangle &= \sum_{m=0}^{\infty} P(m)(1-\xi)^m \\ &= \left\langle \sum_{m=0}^{\infty} \frac{U^m(1-\xi)^m}{m!} e^{-U} \right\rangle_\phi \\ &= \langle e^{-\xi U} \rangle_\phi \\ &= \langle : e^{-\xi \hat{n}} : \rangle. \end{aligned} \qquad (12.10\text{–}12)$$

Expansion of $F(\xi)$ in a power series then yields the factorial moments of $\hat{n}$ and we obtain, by comparing coefficients of $\xi^r/r!$ on both sides of the equation,

$$\begin{aligned} \langle \hat{n}^{(r)} \rangle \equiv \langle \hat{n}(\hat{n}-1) \dots (\hat{n}-r+1) \rangle &= \langle U^r \rangle_\phi \\ &= \langle : \hat{n}^r : \rangle. \end{aligned} \qquad (12.10\text{–}13a)$$

This result is valid for any quantum state of the field, so that the operator equality

$$\hat{n}^{(r)} = :\hat{n}^r: \qquad (12.10\text{–}13b)$$

is actually implied by Eq. (12.10–13a). In particular, when $r = 1$ or 2 we have

$$\langle \hat{n} \rangle = \langle U \rangle_\phi, \qquad (12.10\text{–}14)$$

and

$$\begin{aligned} \langle (\Delta \hat{n})^2 \rangle &= \langle \hat{n}(\hat{n}-1) \rangle + \langle \hat{n} \rangle - \langle \hat{n} \rangle^2 \\ &= \langle \hat{n} \rangle + \langle (\Delta U)^2 \rangle_\phi. \end{aligned} \qquad (12.10\text{–}15)$$

The last relation is particularly interesting, for it is reminiscent of a well-known formula derived by Einstein (1909) for the fluctuation of energy in blackbody radiation (see also Mandel, Sudarshan and Wolf, 1964). He showed that the expression for the variance is separable into the sum of two terms, that can be interpreted as due to the fluctuations of classical particles and classical waves, respectively. According to Einstein, electromagnetic radiation therefore appears as a strange hybrid of particles and waves. It is interesting to note that a similar interpretation may be given to Eq. (12.10–15), which holds quite generally for any state of the field, because the first term on the right is the variance of randomly fluctuating particles, while the second term is the variance of fluctuating waves. Such an interpretation must not be carried too far, for $\langle (\Delta U)^2 \rangle_\phi$ is not necessarily a positive quantity, as it would be for classical waves. As an example, we need only consider a Fock state of the field, for which the photon number is definite, so that

$$\langle (\Delta \hat{n})^2 \rangle = 0,$$

and from Eq. (12.10–15) we find that

$$\begin{aligned} -\langle \hat{n} \rangle &= \langle (\Delta U)^2 \rangle_\phi \\ &= \int \mathcal{P}(U')(\Delta U')^2 \, \mathrm{d}U'. \end{aligned}$$

Evidently $\mathcal{P}(U')$ given by Eq. (12.10–10) is not a non-negative function for a Fock state. More generally, it is clear from Eq. (12.10–15) that whenever the variance of the photon number is less than the mean, so that the photon statistics are sub-Poissonian, $\phi(\{v\})$ cannot be a probability functional; the state of the field is then purely quantum mechanical, without classical analog. A convenient way to characterize the non-classical state is via the parameter Q defined by the formula (Mandel, 1979)

$$Q \equiv \frac{\langle (\Delta \hat{n})^2 \rangle - \langle \hat{n} \rangle}{\langle \hat{n} \rangle}, \qquad (12.10\text{–}16)$$

which is negative whenever the statistics are sub-Poissonian. The parameter Q takes on its greatest possible negative value of -1 for a Fock state.

Although the total photon number in an infinite domain is not accessible to direct measurement, it may be shown that relations rather similar to the foregoing hold also for the number of photons in a finite volume (Mandel, 1966a), provided its linear dimensions are large compared with the wavelengths of occupied modes. This problem will be discussed briefly in Section 12.11. When we come to examine the problem of photoelectric counting in detail in Chapter 14, we shall find that the photoelectric counts are also governed by a probability distribution very similar to Eq. (12.10–8). It seems therefore that we are dealing with relations of rather general validity. Sub-Poissonian statistics were first encountered in practice in photoelectric counting experiments on the fluorescence from a single atom (Short and Mandel, 1983).

12.10.4 Generating functions for photon numbers in normal and anti-normal order

We may obtain the same equations connecting the moments of $\hat{n}$ and of U by calculating the characteristic function $C(\xi)$ of $P(m)$. We then have from Eq. (12.10–8)

$$\begin{aligned} C(\xi) \equiv \langle \mathrm{e}^{\mathrm{i}\xi\hat{n}} \rangle &= \sum_{m=0}^{\infty} P(m)\, \mathrm{e}^{\mathrm{i}\xi m} \\ &= \langle \mathrm{e}^{U(\mathrm{e}^{\mathrm{i}\xi}-1)} \rangle_\phi \\ &= \langle :\mathrm{e}^{\hat{n}(\mathrm{e}^{\mathrm{i}\xi}-1)}: \rangle. \end{aligned} \qquad (12.10\text{–}17)$$

By expanding both sides of this equation as a power series in ξ and comparing coefficients, we can relate the moments of $\hat{n}$ and U. It is obvious that the results must be the same as before, for Eq. (12.10–17) is identical with Eq. (12.10–12) when we make the substitution $e^{i\xi} - 1 \to -\xi$. Both these equations may also be regarded as ordering formulas, because they allow us to relate the normally ordered products of $\hat{n}$ to the direct products, as in Eq. (12.10–13).

It is sometimes useful to have available a corresponding relation for anti-normally ordered products of the number operator $\hat{n}$. Although the problem of operator ordering will be treated more generally in Chapter 14, we can readily obtain the required relation by starting from the definition of the r'th order anti-normally ordered product of $\hat{n}$,

$$\text{“}\hat{n}^r\text{”} \equiv \sum_{\lambda_1} \ldots \sum_{\lambda_r} \hat{a}_{\lambda_1} \ldots \hat{a}_{\lambda_r} \hat{a}^\dagger_{\lambda_r} \ldots \hat{a}^\dagger_{\lambda_1} \qquad (12.10\text{–}18)$$

where “ ” denotes anti-normal ordering as before, and $\lambda_1, \lambda_2, \ldots, \lambda_r$ are collective mode labels standing for $\mathbf{k}_1, s_1, \mathbf{k}_2, s_2, \ldots, \mathbf{k}_r, s_r$. Now from the commutation relation (10.3–9) it follows that

$$\text{“}\hat{n}\text{”} \equiv \sum_\lambda \hat{a}_\lambda \hat{a}^\dagger_\lambda = \sum_\lambda (\hat{a}^\dagger_\lambda \hat{a}_\lambda + 1) = \hat{n} + \mu, \qquad (12.10\text{–}19)$$

where μ is the total number of modes contributing to the sum. As in our previous discussion of anti-normally ordered correlations in Section 12.9, we must take this number to be finite for the same reasons as before, and we must be prepared to see it appear explicitly in the equations. With the help of Eq. (12.10–19), Eq. (12.10–18) can be written

$$\text{“}\hat{n}^r\text{”} = \sum_{\lambda_1} \ldots \sum_{\lambda_{r-1}} \hat{a}_{\lambda_1} \ldots \hat{a}_{\lambda_{r-1}} (\hat{n} + \mu) \hat{a}^\dagger_{\lambda_{r-1}} \ldots \hat{a}^\dagger_{\lambda_1}. \qquad (12.10\text{–}20)$$

We now make use of the commutation relation (10.4–2), in the form

$$[\hat{a}^\dagger_\lambda, \hat{n}] = -\hat{a}^\dagger_\lambda, \qquad (12.10\text{–}21)$$

to move the $\hat{n}$ operator in Eq. (12.10–20) repeatedly to the right. We then obtain the formula

$$\begin{aligned} \text{“}\hat{n}^r\text{”} &= \sum_{\lambda_1} \ldots \sum_{\lambda_{r-1}} \hat{a}_{\lambda_1} \ldots \hat{a}_{\lambda_{r-1}} \hat{a}^\dagger_{\lambda_{r-1}} \ldots \hat{a}^\dagger_{\lambda_1} (\hat{n} + \mu + r - 1) \\ &= \text{“}\hat{n}^{r-1}\text{”}(\hat{n} + \mu + r - 1), \end{aligned} \qquad (12.10\text{–}22)$$

which is a recursion relation. Repeated application immediately leads to the result

$$\text{“}\hat{n}^r\text{”} = (\hat{n} + \mu)(\hat{n} + \mu + 1) \ldots (\hat{n} + \mu + r - 1), \qquad (12.10\text{–}23a)$$

or

$$\langle \text{“}\hat{n}^r\text{”} \rangle = \langle (\hat{n} + \mu)(\hat{n} + \mu + 1) \ldots (\hat{n} + \mu + r - 1) \rangle, \qquad (12.10\text{–}23b)$$

when we take the expectation values of both sides. We have therefore arrived at an expansion for the anti-normally ordered moments of the number operator. The resulting Eqs. (12.10–23) should be compared with Eqs. (12.10–13) for the normally ordered moment. We see that the structure of the expressions is related to the underlying physical situation in which the field may be measured with photodetectors or quantum counters. We recall that normal ordering corresponds to successive photon absorptions, whereas anti-normal ordering corresponds to successive photon emissions. It is interesting to note also that, according to Eq. (12.10–23b), $\langle``\hat{n}^r"\rangle$ cannot vanish for any state of the field, whereas $\langle:\hat{n}^r:\rangle$ vanishes whenever the field has fewer than r photons.

From Eq. (12.10–23b) we can immediately derive the characteristic generating function for anti-normally ordered moments,

$$\begin{aligned}\langle``e^{i\xi\hat{n}}"\rangle &= \left\langle \sum_r \frac{(i\xi)^r}{r!} ``\hat{n}^r" \right\rangle \\ &= \left\langle \sum_r \frac{(i\xi)^r}{r!} (\hat{n}+\mu)(\hat{n}+\mu+1)\ldots(\hat{n}+\mu+r-1) \right\rangle \\ &= \langle (1-i\xi)^{-(\hat{n}+\mu)} \rangle, \end{aligned} \tag{12.10–24}$$

which can be compared with Eq. (12.10–12). By substituting $(1-i\xi) = e^{-ix}$ in Eq. (12.10–24) and $1-\xi = e^{ix}$ in Eq. (12.10–12), we can combine both results and write

$$\langle:\exp[(e^{ix}-1)\hat{n}]:\rangle = \langle e^{ix\hat{n}} \rangle = ``\exp[(1-e^{-ix})\hat{n}]"\, e^{-i\mu x}, \tag{12.10–25}$$

which relates the normally ordered and anti-normally ordered moments directly.

12.11 The problem of localizing photons‡

The number operator $\hat{n}$ with which we have dealt so far refers to the total photon number in all space. It is therefore not expected to be accessible to direct measurement. In order to make contact with experiment, it would seem more appropriate to introduce a description in which the photons are localized. However, in attempting to localize photons in space we encounter some fundamental difficulties. No position operator exists for the photon (Newton and Wigner, 1949); the maximum precise localization appears to be in the form of a wavefront (Acharya and Sudarshan, 1960). Although in Section 12.3 we encountered the intensity operator $\hat{I}(\mathbf{r}, t) = \hat{\mathbf{V}}^{\dagger}(\mathbf{r}, t)\cdot\hat{\mathbf{V}}(\mathbf{r}, t)$ having the dimensions of photons per unit volume, attempts to interpret this as an operator that localizes photons at the space-time point $\mathbf{r}$, t in a *precise* sense lead to contradictions. For example, the intensities $\hat{I}(\mathbf{r}_1, t_1)$ and $\hat{I}(\mathbf{r}_2, t_2)$ associated with two different space-time points $\mathbf{r}_1$, t_1 and $\mathbf{r}_2$, t_2 do not commute in general. Nevertheless, the temptation to interpret the electronic signal registered by a photodetector as due to a photon that is localized in some sense is quite strong. In this section we shall look a little more carefully at the extent to which this can be justified, at least approximately.

As we indicated in Section 12.3, the counts registered by a detector whose surface is normal to the incident field and exposed for some finite time Δt, are interpreted most naturally as a measurement of the number of photons in a cylindrical volume whose base is the sensitive surface of the detector and whose height is $c\Delta t$ (see Fig. 12.2). In the following we shall show that the integral of

‡ The treatment in this section is based on the paper by Mandel (1966a).

the intensity $\hat{I}(\mathbf{r}, t)$ over such a volume can be interpreted as a configuration space number operator, at least in an approximate sense, provided the linear dimensions of the volume are all large compared with the wavelengths of contributing modes.‡

12.11.1 Configuration space photon number operator

We start by introducing a photon absorption operator $\hat{\mathbf{V}}(\mathbf{r}, t)$ as in Section 12.3, defined by the expansion

$$\hat{\mathbf{V}}(\mathbf{r}, t) = \frac{1}{L^{3/2}} \sum_{[\mathbf{k},s]} \hat{a}_{\mathbf{k}s} \boldsymbol{\varepsilon}_{\mathbf{k}s}\, e^{i(\mathbf{k}\cdot\mathbf{r}-\omega t)}, \tag{12.11–1}$$

except that it is convenient to explicitly limit the sum to the finite set of modes $[\mathbf{k}, s]$ to which the detector responds. If we are dealing with an optical detector, it is reasonable to suppose that it responds only to optical frequencies in some range and to certain polarizations. Other frequencies play no role in the analysis, and we may assume that the corresponding modes are empty. If we take the conjugate $\hat{\mathbf{V}}^\dagger(\mathbf{r}, t)$ of $\hat{\mathbf{V}}(\mathbf{r}, t)$ defined by Eq. (12.11–1), then the scalar product intensity $\hat{\mathbf{V}}^\dagger(\mathbf{r}, t)\cdot\hat{\mathbf{V}}(\mathbf{r}, t)$ has the dimensions of photons per unit volume. We now define the configuration space number operator $\hat{n}(\mathcal{V}, t)$ by the relation

$$\hat{n}(\mathcal{V}, t) = \int_{\mathcal{V}} \hat{\mathbf{V}}^\dagger(\mathbf{r}, t)\cdot\hat{\mathbf{V}}(\mathbf{r}, t)\, d^3r, \tag{12.11–2}$$

where the integral is taken over some finite volume $\mathcal{V}$, such as a rectangular box of sides l_1, l_2, l_3. It is easy to show that $\hat{n}(\mathcal{V}, t)$ commutes with the total number operator $\hat{n}$, in which the corresponding integral is taken over all space. In order to see this we expand $\hat{\mathbf{V}}^\dagger(\mathbf{r}, t)\cdot\hat{\mathbf{V}}(\mathbf{r}, t)$ under the integral in Eq. (12.11–2) and observe, with the help of Eqs. (10.4–1) and (10.4–2), that

$$[\hat{a}^\dagger_{\mathbf{k}'s'}\hat{a}_{\mathbf{k}''s''}, \hat{n}] = 0. \tag{12.11–3}$$

In the following we shall examine some of the properties of $\hat{n}(\mathcal{V}, t)$ with the help of certain approximations. Therefore many of the relationships derived below hold only in an approximate sense.

12.11.2 Commutation relations

In order to evaluate the commutator of $\hat{n}(\mathcal{V}_1, t)$ and $\hat{n}(\mathcal{V}_2, t)$, where $\mathcal{V}_1$ and $\mathcal{V}_2$ are two different regions of space, it is useful to begin by examining the equal-time commutator of $\hat{\mathbf{V}}(\mathbf{r}, t)$ with $\hat{n}(\mathcal{V}, t)$. For simplicity we limit ourselves to equal times. The more general commutator is evaluated in the paper by Mandel (1966a). From Eqs. (12.11–1) and (12.11–2) we have

$$[\hat{V}_i(\mathbf{r}, t), \hat{n}(\mathcal{V}, t)] = \int_{\mathcal{V}} [\hat{V}_i(\mathbf{r}, t), \hat{V}_j^\dagger(\mathbf{r}', t)\hat{V}_j(\mathbf{r}', t)]\, d^3r'$$

‡ A slightly different configuration space number operator, but one that has certain features in common with $\hat{n}(\mathcal{V}, t)$ defined in Eq. (12.11–2), was introduced by Cook (1982) (see also Cook, 1984). The problem of photon localization has been treated more rigorously by Jauch and Piron (1967), by Amrein (1969) and by Pike and Sarkar (1986).

$$= \int_{\mathscr{V}} [\hat{V}_i(\mathbf{r}, t), \hat{V}_j^\dagger(\mathbf{r}', t)]\hat{V}_j(\mathbf{r}', t)\,\mathrm{d}^3r'. \qquad (12.11\text{–}4)$$

We now make use of the commutation relation (11.12–6), which we write in the form

$$[\hat{V}_i(\mathbf{r}, t), \hat{V}_j^\dagger(\mathbf{r}', t)] = \frac{1}{L^3}\sum_{[\mathbf{k}]}\left(\delta_{ij} - \frac{k_ik_j}{k^2}\right)\mathrm{e}^{\mathrm{i}\mathbf{k}\cdot(\mathbf{r}-\mathbf{r}')}, \qquad (12.11\text{–}5)$$

to transform the integral in Eq. (12.11–4). We then obtain the relation

$$\begin{aligned}
&[\hat{V}_i(\mathbf{r}, t), \hat{n}(\mathscr{V}, t)] \\
&= \frac{1}{L^{9/2}}\sum_{[\mathbf{k}]}\left(\delta_{ij} - \frac{k_ik_j}{k^2}\right)\sum_{[\mathbf{k}',s']}\hat{a}_{\mathbf{k}'s'}(\boldsymbol{\varepsilon}_{\mathbf{k}'s'})_j\,\mathrm{e}^{\mathrm{i}(\mathbf{k}\cdot\mathbf{r}-\omega' t)}\int_{\mathscr{V}}\mathrm{e}^{\mathrm{i}(\mathbf{k}'-\mathbf{k})\cdot\mathbf{r}'}\,\mathrm{d}^3r' \\
&= \frac{1}{L^3}\sum_{[\mathbf{k}]}\left(\delta_{ij} - \frac{k_ik_j}{k^2}\right)\frac{1}{L^{3/2}}\sum_{[\mathbf{k}',s']}\hat{a}_{\mathbf{k}'s'}(\boldsymbol{\varepsilon}_{\mathbf{k}'s'})_j\,\mathrm{e}^{\mathrm{i}(\mathbf{k}\cdot\mathbf{r}-\omega' t)}\,\mathrm{e}^{\mathrm{i}(\mathbf{k}'-\mathbf{k})\cdot\mathbf{r}_0} \\
&\quad\times\prod_{m=1}^{3}\left[\frac{\sin\frac{1}{2}(k'_m - k_m)l_m}{\frac{1}{2}(k'_m - k_m)}\right] \\
&= \frac{1}{L^{3/2}}\sum_{[\mathbf{k}',s']}\hat{a}_{\mathbf{k}'s'}(\boldsymbol{\varepsilon}_{\mathbf{k}'s'})_j\,\mathrm{e}^{\mathrm{i}(\mathbf{k}'\cdot\mathbf{r}-\omega' t)}\frac{1}{L^3}\sum_{[\mathbf{k}]}\left(\delta_{ij} - \frac{k_ik_j}{k^2}\right)\mathrm{e}^{\mathrm{i}(\mathbf{k}-\mathbf{k}')\cdot(\mathbf{r}-\mathbf{r}_0)} \\
&\quad\times\prod_{m=1}^{3}\left[\frac{\sin\frac{1}{2}(k'_m - k_m)l_m}{\frac{1}{2}(k'_m - k_m)}\right],
\end{aligned} \qquad (12.11\text{–}6)$$

where $\mathbf{r}_0$ is the midpoint of the volume $\mathscr{V}$. Let us now consider the summation over $[\mathbf{k}]$. From the appearance of the product term in Eq. (12.11–6) it is clear that the principal contribution to the sum will come from values of $\mathbf{k}$ such that

$$\left.\begin{aligned}
|k_1 - k'_1| &\lesssim 1/l_1, \\
|k_2 - k'_2| &\lesssim 1/l_2, \\
|k_3 - k'_3| &\lesssim 1/l_3.
\end{aligned}\right\} \qquad (12.11\text{–}7)$$

Both $\mathbf{k}$ and $\mathbf{k}'$ are wave vectors corresponding to optical or still higher frequencies. If the lengths l_1, l_2, l_3 determining the volume of integration are all large compared with optical wavelengths, then Eqs. (12.11–7) imply equality between $\mathbf{k}$ and $\mathbf{k}'$ to a good approximation. The principal contribution to the sum over $[\mathbf{k}]$ in Eq. (12.11–6) then comes from values of $\mathbf{k}$ close to $\mathbf{k}'$. If we put $\mathbf{k} = \mathbf{k}'$ in the term k_ik_j/k^2 that varies slowly with $\mathbf{k}$ under the summation, we see that this term makes a small contribution, since $k'_j(\boldsymbol{\varepsilon}_{\mathbf{k}'s'})_j = 0$. The equation may therefore be simplified to read

$$\begin{aligned}
[\hat{V}_i(\mathbf{r}, t), \hat{n}(\mathscr{V}, t)] \approx{}& \frac{1}{L^{3/2}}\sum_{[\mathbf{k}',s']}\hat{a}_{\mathbf{k}'s'}(\boldsymbol{\varepsilon}_{\mathbf{k}'s'})_i\,\mathrm{e}^{\mathrm{i}(\mathbf{k}'\cdot\mathbf{r}-\omega' t)} \\
&\times\frac{1}{L^3}\sum_{[\mathbf{k}]}\mathrm{e}^{\mathrm{i}(\mathbf{k}-\mathbf{k}')\cdot(\mathbf{r}-\mathbf{r}_0)}\prod_{m=1}^{3}\left[\frac{\sin\frac{1}{2}(k_m - k'_m)l_m}{\frac{1}{2}(k_m - k'_m)}\right].
\end{aligned} \qquad (12.11\text{–}8)$$

At this stage it is convenient to replace the summation over [**k**] by an integral, according to the usual rule

$$\frac{1}{L^3}\sum_{[\mathbf{k}]} \to \frac{1}{(2\pi)^3}\int_{[\mathbf{k}]} \mathrm{d}^3k,$$

and to introduce the new variable of integration $\mathbf{k}'' = \mathbf{k} - \mathbf{k}'$. Because of the presence of the product factor in Eq. (12.11–8), which severely limits the contributions to the integral to small values of $\mathbf{k}''$, we may effectively allow the limits of integration over $\mathbf{k}''$ to become infinite. We then have

$$[\hat{V}_i(\mathbf{r}, t), \hat{n}(\mathcal{V}, t)] \approx \hat{V}_i(\mathbf{r}, t)\frac{1}{(2\pi)^3}\int \mathrm{e}^{\mathrm{i}\mathbf{k}''\cdot(\mathbf{r}-\mathbf{r}_0)} \prod_{m=1}^{3}\left[\frac{\sin(\frac{1}{2}k''_m l_m)}{\frac{1}{2}k''_m}\right]\mathrm{d}^3k''. \quad (12.11\text{–}9)$$

The integral is the three-dimensional version of the well-known Dirichlet discontinuous integral (Bracewell, 1978, p. 99, 129)

$$\left.\begin{aligned}\frac{1}{2\pi}\int_{-\infty}^{\infty} \mathrm{e}^{\mathrm{i}kx}\frac{\sin\frac{1}{2}kl}{\frac{1}{2}k}\mathrm{d}k &= 1, \quad \text{if } |x| \leqslant \tfrac{1}{2}l \\ &= 0, \quad \text{otherwise,}\end{aligned}\right\} \quad (12.11\text{–}10)$$

so that Eq.(12.11–9) can be written as

$$\left.\begin{aligned}[\hat{V}_i(\mathbf{r}, t), \hat{n}(\mathcal{V}, t)] &\approx \hat{V}_i(\mathbf{r}, t), \quad \text{if } \mathbf{r} \in \mathcal{V}, \\ &\approx 0, \quad \text{otherwise,}\end{aligned}\right\} \quad (12.11\text{–}11)$$

where $\mathbf{r} \in \mathcal{V}$ signifies that $\mathbf{r}$ lies inside the volume $\mathcal{V}$.

This result is entirely analogous to the corresponding $\mathbf{k}$, s space commutation relation

$$\left.\begin{aligned}\left[\hat{a}_{\mathbf{k}s}, \sum_{[\mathbf{k},s]}\hat{n}_{\mathbf{k}s}\right] &= \hat{a}_{\mathbf{k}s}, \quad \text{if } \mathbf{k}, s \in [\mathbf{k}, s], \\ &= 0, \quad \text{otherwise,}\end{aligned}\right\} \quad (12.11\text{–}12)$$

and suggests that $\hat{n}(\mathcal{V}, t)$ plays the role of a number operator associated with the volume $\mathcal{V}$ and the time t. However, we must not lose sight of the fact that approximations were introduced in the derivation of Eq. (12.11–11). In particular, the distinction between points inside and outside the volume $\mathcal{V}$ becomes hazy for points lying close to the boundary of $\mathcal{V}$, within distances of order an optical wavelength.‡ By taking the Hermitian conjugate of Eq. (12.11–11) we obtain the result

$$\left.\begin{aligned}[\hat{V}_i^\dagger(\mathbf{r}, t), \hat{n}(\mathcal{V}, t)] &\approx -\hat{V}_i^\dagger(\mathbf{r}, t), \quad \text{if } \mathbf{r} \in \mathcal{V}, \\ &\approx 0, \quad \text{otherwise.}\end{aligned}\right\} \quad (12.11\text{–}13)$$

With the help of these results we may readily evaluate the commutator of two

‡ The commutation relation corresponding to Eq. (12.11–11) for unequal times is less simple. When the two space-time regions $\mathbf{r}$, t and $\mathcal{V}$, t' are disjoint, the commutator vanishes, and when they overlap appreciably, the commutator equals $\hat{V}_i(\mathbf{r}, t)$. But in intermediate regions the commutator has no simple interpretation (see Mandel, 1966a).

configuration space number operators associated with two different volumes $\mathcal{V}$, $\mathcal{V}'$. We find that

$$\begin{aligned}[\hat{n}(\mathcal{V}, t), \hat{n}(\mathcal{V}', t)] &\approx \int_{\mathcal{V}} [\hat{V}_i^\dagger(\mathbf{r}, t)\hat{V}_i(\mathbf{r}, t)\hat{n}(\mathcal{V}', t) - \hat{n}(\mathcal{V}', t)\hat{V}_i^\dagger(\mathbf{r}, t)\hat{V}_i(\mathbf{r}, t)]\, \mathrm{d}^3 r \\ &= \int_{\mathcal{V}} \{\hat{V}_i^\dagger(\mathbf{r}, t)[\hat{n}(\mathcal{V}', t) + U(\mathbf{r}, \mathcal{V}')]\hat{V}_i(\mathbf{r}, t) \\ &\quad - \hat{V}_i^\dagger(\mathbf{r}, t)[\hat{n}(\mathcal{V}', t) + U(\mathbf{r}, \mathcal{V}')]\hat{V}_i(\mathbf{r}, t)\}\, \mathrm{d}^3 r \\ &= 0, \end{aligned} \tag{12.11–14}$$

where $U(\mathbf{r}, \mathcal{V})$ is a function that is unity or zero according as $\mathbf{r}$ lies inside or outside $\mathcal{V}$, respectively. The commutator of two configuration space number operators therefore vanishes at least approximately, in complete analogy with the corresponding result in $\mathbf{k}$, s space, where we have

$$\left[\sum_{[\mathbf{k},s]} \hat{n}_{\mathbf{k}s}, \sum_{[\mathbf{k},s]'} \hat{n}_{\mathbf{k}s}\right] = 0, \tag{12.11–15}$$

irrespective of whether the two sets of modes $[\mathbf{k}, s]$ and $[\mathbf{k}, s]'$ overlap or not.

12.11.3 *Eigenstates of $\hat{n}(\mathcal{V}, t)$*

Let us now examine the eigenstates of the $\hat{n}(\mathcal{V}, t)$ operator, which should correspond to localized excitations of the electromagnetic field in some sense. Consider the state $\hat{V}_i^\dagger(\mathbf{r}, t)|\text{vac}\rangle$ produced when the creation operator $\hat{V}_i^\dagger(\mathbf{r}, t)$ acts on the vacuum. With the help of Eq. (12.11–13) we have

$$\begin{aligned}\hat{n}(\mathcal{V}, t)\hat{V}_i^\dagger(\mathbf{r}, t)|\text{vac}\rangle &= \hat{V}_i^\dagger(\mathbf{r}, t)[\hat{n}(\mathcal{V}, t) + U(\mathbf{r}, \mathcal{V})]|\text{vac}\rangle \\ &= U(\mathbf{r}, \mathcal{V})\hat{V}_i^\dagger(\mathbf{r}, t)|\text{vac}\rangle, \end{aligned} \tag{12.11–16}$$

since $\hat{n}(\mathcal{V}, t)|\text{vac}\rangle = 0$. This equation shows that $\hat{V}_i^\dagger(\mathbf{r}, t)|\text{vac}\rangle$ is an eigenstate of $\hat{n}(\mathcal{V}, t)$ belonging to the eigenvalue $U(\mathbf{r}, \mathcal{V})$, which is unity if $\mathbf{r}$ lies inside the volume $\mathcal{V}$ and zero otherwise.

Similarly it may be shown that

$$\begin{aligned}&\hat{n}(\mathcal{V}, t)\hat{V}_{i_1}^\dagger(\mathbf{r}_1, t) \ldots \hat{V}_{i_N}^\dagger(\mathbf{r}_N, t)|\text{vac}\rangle = \\ &\qquad [U(\mathbf{r}_1, \mathcal{V}) + \ldots + U(\mathbf{r}_N, \mathcal{V})]\hat{V}_{i_1}^\dagger(\mathbf{r}_1, t) \ldots \hat{V}_{i_N}^\dagger(\mathbf{r}_N, t)|\text{vac}\rangle, \end{aligned} \tag{12.11–17}$$

so that $\hat{V}_{i_1}^\dagger(\mathbf{r}_1, t) \ldots \hat{V}_{i_N}^\dagger(\mathbf{r}_N, t)|\text{vac}\rangle$ is also an eigenstate of $\hat{n}(\mathcal{V}, t)$, with eigenvalue $U(\mathbf{r}_1, \mathcal{V}) + \ldots + U(\mathbf{r}_N, \mathcal{V})$ equal to the number of points $\mathbf{r}_1, \ldots, \mathbf{r}_N$ that lie within the volume $\mathcal{V}$. The spectrum of $\hat{n}(\mathcal{V}, t)$, like that of $\hat{n}$, is therefore the set of numbers 0, 1, 2, ... It seems natural to interpret the eigenstate in Eq. (12.11–17) as one in which photons are localized approximately at $(\mathbf{r}_1, t), \ldots, (\mathbf{r}_N, t)$, provided we do not attempt to define the positions to better than a few optical wavelengths.

Although states of the type $\hat{V}_{i_1}^\dagger(\mathbf{r}_1, t) \ldots \hat{V}_{i_N}^\dagger(\mathbf{r}_N, t)|\text{vac}\rangle$ are not normalized to unity and are not strictly orthogonal, they form a complete set. Two different states

$$\left.\begin{aligned}|\psi\rangle &\equiv \hat{V}^{\dagger}_{i_1}(\mathbf{r}_1, t)\ldots \hat{V}^{\dagger}_{i_N}(\mathbf{r}_N, t)|\text{vac}\rangle\\ \text{and}\qquad & \\ |\psi'\rangle &\equiv \hat{V}^{\dagger}_{j_1}(\mathbf{r}'_1, t)\ldots \hat{V}^{\dagger}_{j_M}(\mathbf{r}'_M, t)|\text{vac}\rangle\end{aligned}\right\} \tag{12.11–18}$$

are, however, approximately orthogonal even when $N = M$, provided $\mathbf{r}_p$ is not close to $\mathbf{r}'_p$, for every integer p.‡ We may then construct a box of volume $\mathcal{V}$, of linear dimensions much larger than a wavelength, which is chosen so that the number of points inside $\mathcal{V}$ associated with the state $|\psi\rangle$ is different from the number associated with state $|\psi'\rangle$. The situation is illustrated in Fig. 12.6 for two states $|\psi\rangle$ and $|\psi'\rangle$ having $N = 3$, $M = 3$ or three photons each, for which $\mathbf{r}_3$ and $\mathbf{r}'_3$ are appreciably far apart. The volume $\mathcal{V}$ illustrated by the broken line encloses two points associated with state $|\psi\rangle$ but only one point associated with state $|\psi'\rangle$. It follows that the state $|\psi\rangle$ belongs to the eigenvalue 2 of the operator $\hat{n}(\mathcal{V}, t)$, while the state $|\psi'\rangle$ belongs to the eigenvalue 1 of the same operator. As both $|\psi\rangle$ and $|\psi'\rangle$ are eigenstates of the same Hermitian operator belonging to different eigenvalues, they must be (at least approximately) orthogonal.

In order to show that states of the type $|\psi\rangle$ above form a complete set, we observe that, by Fourier inversion of the expansion for $\hat{\mathbf{V}}^{\dagger}(\mathbf{r}, t)$ [the conjugate of Eq. (12.11–1)], we have,

$$\hat{a}^{\dagger}_{\mathbf{k}s} = \frac{1}{L^{3/2}}\int_{L^3} \hat{\mathbf{V}}^{\dagger}(\mathbf{r}, t)\cdot \boldsymbol{\varepsilon}_{\mathbf{k}s}\, \mathrm{e}^{\mathrm{i}(\mathbf{k}\cdot\mathbf{r}-\omega t)}\, \mathrm{d}^3 r. \tag{12.11–19}$$

Now any Fock state $|\{n\}\rangle$ with occupied modes belonging to the set $[\mathbf{k}, s]$ may be expressed in the form

$$|\{n\}\rangle = \prod_{[\mathbf{k},s]}\left[\frac{(\hat{a}^{\dagger}_{\mathbf{k}s})^{n_{\mathbf{k}s}}}{\sqrt{(n_{\mathbf{k}s}!)}}|0_{\mathbf{k}s}\rangle\right]. \tag{12.11–20}$$

By substituting for $\hat{a}^{\dagger}_{\mathbf{k}s}$ from Eq. (12.11–19) we see at once that we arrive at an

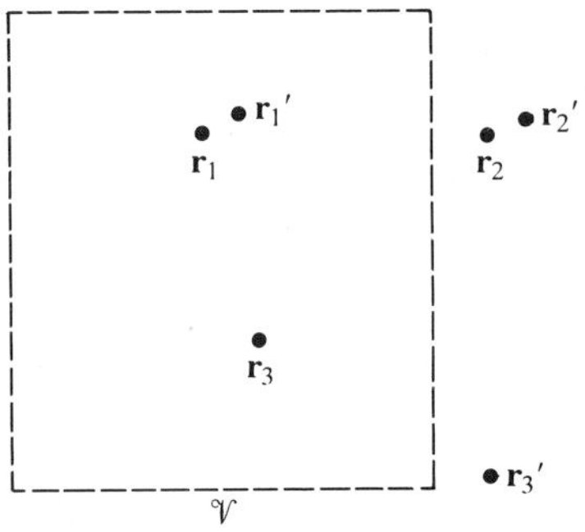

Fig. 12.6 Illustrating that a volume $\mathcal{V}$ may be chosen that contains two excitations associated with state $|\psi\rangle$ and one excitation associated with state $|\psi'\rangle$.

‡ It is important to bear in mind that the positions $\mathbf{r}_1, \ldots, \mathbf{r}_N$ in the definition of the state are not meant to be defined to an accuracy better than a few wavelengths.

expansion of the Fock state $|\{n\}\rangle$ in terms of states of the type $|\psi\rangle$ given by Eq. (12.11–18). As the Fock states form a complete set, so do the eigenstates of $\hat{n}(\mathcal{V}, t)$.

12.11.4 *Photon statistics in a finite volume*

As the spectrum of $\hat{n}(\mathcal{V}, t)$ is the set of integers 0, 1, 2, . . . , approximately, and as $\hat{n}(\mathcal{V}, t)$ behaves as a number operator in certain important respects, we might expect to find relations for the normally ordered products of $\hat{n}(\mathcal{V}, t)$ and for the distribution of eigenvalues of $\hat{n}(\mathcal{V}, t)$ that are analogous to the corresponding relations for the number operator $\hat{n}$. This is indeed the case. It may be shown (Mandel, 1966a) that,

$$\langle :[\hat{n}(\mathcal{V}, t)]^N: \rangle = \langle \hat{n}(\mathcal{V}, t)[\hat{n}(\mathcal{V}, t) - 1] \ldots [\hat{n}(\mathcal{V}, t) - N + 1] \rangle, \quad (12.11\text{–}21)$$

in complete analogy with Eq. (12.10–13). On the other hand, if the volumes $\mathcal{V}_1, \mathcal{V}_2, \ldots, \mathcal{V}_N$ are non-overlapping we find that

$$\langle :\hat{n}(\mathcal{V}_1, t)\hat{n}(\mathcal{V}_2, t) \ldots \hat{n}(\mathcal{V}_N, t): \rangle = \langle \hat{n}(\mathcal{V}_1, t)\hat{n}(\mathcal{V}_2, t) \ldots \hat{n}(\mathcal{V}_N, t) \rangle, \quad (12.11\text{–}22)$$

in view of Eq. (12.11–11). The probability distribution $p(n)$ of the eigenvalues n of $\hat{n}(\mathcal{V}, t)$ may be expressed in the form

$$p(n) = \left\langle \frac{\mathcal{N}^n \mathrm{e}^{-\mathcal{N}}}{n!} \right\rangle_\phi, \quad (12.11\text{–}23)$$

where

$$\mathcal{N} \equiv \int_\mathcal{V} \mathbf{V}^*(\mathbf{r}, t) \cdot \mathbf{V}(\mathbf{r}, t)\, \mathrm{d}^3 r, \quad (12.11\text{–}24)$$

and $\mathbf{V}(\mathbf{r}, t)$ and $\mathbf{V}^*(\mathbf{r}, t)$ are right and left eigenvalues of $\hat{\mathbf{V}}(\mathbf{r}, t)$ and $\hat{\mathbf{V}}^\dagger(\mathbf{r}, t)$, respectively. This relation is again similar to Eq. (12.10–8) for the distribution of eigenvalues of the total number operator $\hat{n}$. The generator of moments of $\hat{n}(\mathcal{V}, t)$ has the form

$$\langle \exp[\mathrm{i}x\hat{n}(\mathcal{V}, t)] \rangle = \langle \exp[\mathcal{N}(\mathrm{e}^{\mathrm{i}x} - 1)] \rangle_\phi, \quad (12.11\text{–}25)$$

in analogy with Eq. (12.10–17), and so on.

We see therefore that, provided we do not insist on localizing the excitation too precisely, we can introduce states of localized excitations or photons, and we can define a configuration space number operator that measures the number of photons in a finite volume. However, it is important to bear in mind that the procedure is meaningful only because the wavelengths of optical photons are so small on a laboratory scale.

12.11.5 *Polychromatic photons and non-locality*

We have seen that the expectation value of the light intensity $\hat{I}(\mathbf{r}, t) = \hat{\mathbf{V}}^\dagger(\mathbf{r}, t) \cdot \hat{\mathbf{V}}(\mathbf{r}, t)$, with $\hat{\mathbf{V}}$ given by Eq. (12.11–1), is closely related to the expected response of a photodetector at $\mathbf{r}$ at time t. But this intimate relationship holds only for a quasi-monochromatic field. As soon as the field becomes polychromatic, the presence or absence of the factor $l(k)$ [see Eqs. (12.2–1) and (12.2–2)] in the mode expansion of the field has non-trivial consequences, with the result that the positive frequency part of the electric field, for example, is no longer locally connected to the detection operator $\hat{\mathbf{V}}(\mathbf{r}, t)$. We illustrate the problem by considering a simple one-photon state.

The one-photon state $\hat{a}^{\dagger}_{\mathbf{k}s}|\text{vac}\rangle$, obtained when the photon creation operator $\hat{a}^{\dagger}_{\mathbf{k}s}$ acts on the vacuum state $|\text{vac}\rangle$, corresponds to a photon that is associated with a plane wave and is therefore distributed over all space. However, a linear superposition of such states of the form

$$|\phi\rangle = \frac{1}{L^{3/2}}\sum_{\mathbf{k},s}\phi(\mathbf{k}, s)\hat{a}^{\dagger}_{\mathbf{k}s}|\text{vac}\rangle, \tag{12.11–26}$$

in which $\phi(\mathbf{k}, s)$ is any weight function satisfying the condition

$$\langle\phi|\phi\rangle = \frac{1}{L^3}\sum_{\mathbf{k},s}|\phi(\mathbf{k}, s)|^2 \to \sum_{s}\frac{1}{(2\pi)^3}\int|\phi(\mathbf{k}, s)|^2\,\mathrm{d}^3k = 1, \tag{12.11–27}$$

corresponds to a photon that is at least partially localized in space. The vector function

$$\boldsymbol{\Phi}(\mathbf{r}, t) = \frac{1}{L^3}\sum_{\mathbf{k},s}\phi(\mathbf{k}, s)\boldsymbol{\varepsilon}_{\mathbf{k}s}\,\mathrm{e}^{\mathrm{i}(\mathbf{k}\cdot\mathbf{r}-\omega t)} \to \sum_{s}\frac{1}{(2\pi)^3}\int\phi(\mathbf{k}, s)\boldsymbol{\varepsilon}_{\mathbf{k}s}\,\mathrm{e}^{\mathrm{i}(\mathbf{k}\cdot\mathbf{r}-\omega t)}\,\mathrm{d}^3k \tag{12.11–28}$$

then represents the corresponding position space wave function of the photon in state $|\phi\rangle$. However, by virtue of the linear superposition in Eq. (12.11–26), this photon does not have a definite momentum or a definite energy.

In order to show that $|\boldsymbol{\Phi}(\mathbf{r}, t)|^2$ gives the photon probability density we let the local detection operator

$$\hat{\mathbf{V}}(\mathbf{r}, t) = \frac{1}{L^{3/2}}\sum_{\mathbf{k},s}\hat{a}_{\mathbf{k}s}\boldsymbol{\varepsilon}_{\mathbf{k}s}\,\mathrm{e}^{\mathrm{i}(\mathbf{k}\cdot\mathbf{r}-\omega t)}$$

act on the state $|\phi\rangle$. With the help of the commutation relation between $\hat{a}_{\mathbf{k}s}$, $\hat{a}^{\dagger}_{\mathbf{k}'s'}$, we then obtain

$$\begin{aligned}\hat{\mathbf{V}}(\mathbf{r}, t)|\phi\rangle &= \frac{1}{L^3}\sum_{\mathbf{k},s}\sum_{\mathbf{k}',s'}\phi(\mathbf{k}', s')\hat{a}_{\mathbf{k}s}\hat{a}^{\dagger}_{\mathbf{k}'s'}|\text{vac}\rangle\boldsymbol{\varepsilon}_{\mathbf{k}s}\,\mathrm{e}^{\mathrm{i}(\mathbf{k}\cdot\mathbf{r}-\omega t)}\\ &= \frac{1}{L^3}\sum_{\mathbf{k},s}\phi(\mathbf{k}, s)\boldsymbol{\varepsilon}_{\mathbf{k}s}\,\mathrm{e}^{\mathrm{i}(\mathbf{k}\cdot\mathbf{r}-\omega t)}|\text{vac}\rangle\\ &= \boldsymbol{\Phi}(\mathbf{r}, t)|\text{vac}\rangle,\end{aligned} \tag{12.11–29}$$

from which it is at once apparent that $\boldsymbol{\Phi}(\mathbf{r}, t)$ is just the projection of the state $|\phi\rangle$ on to the localized one-photon state $\hat{\mathbf{V}}^{\dagger}(\mathbf{r}, t)|\text{vac}\rangle$. Now the probability of finding the photon in state $|\phi\rangle$ in the volume $\mathcal{V}$ at time t is given by the expectation of the configuration space number operator $\hat{n}(\mathcal{V}, t)$. From Eqs. (12.11–2) and

(12.11–29) we then have

$$\langle\phi|\hat{n}(\mathcal{V}, t)|\phi\rangle = \int_{\mathcal{V}} \langle\phi|\hat{\mathbf{V}}^{\dagger}(\mathbf{r}, t)\cdot\hat{\mathbf{V}}(\mathbf{r}, t)|\phi\rangle\, \mathrm{d}^3 r$$

$$= \int_{\mathcal{V}} |\boldsymbol{\Phi}(\mathbf{r}, t)|^2\, \mathrm{d}^3 r, \tag{12.11–30}$$

which is the usual expression for the probability of locating a particle with wave function $\boldsymbol{\Phi}(\mathbf{r}, t)$ within the volume $\mathcal{V}$. In this case the integral is meaningful only when the linear dimensions of $\mathcal{V}$ are large compared with optical wavelengths.

However, it is not difficult to see that neither the energy of the photon nor the photoelectric detection probability are localized in the same place. The average photon energy is

$$\langle\phi|\hat{H}|\phi\rangle = \sum_{\mathbf{k},s} \hbar\omega\langle\phi|\hat{n}_{\mathbf{k}s}|\phi\rangle$$

and with the help of Eq. (12.11–26) this becomes

$$\langle\phi|\hat{H}|\phi\rangle = \frac{1}{L^3}\sum_{\mathbf{k},s}\hbar\omega|\phi(\mathbf{k}, s)|^2 \to \sum_{s}\frac{1}{(2\pi)^3}\int \hbar\omega|\phi(\mathbf{k}, s)|^2\, \mathrm{d}^3 k. \tag{12.11–31}$$

By introducing the function

$$\boldsymbol{\Psi}(\mathbf{r}, t) = \frac{1}{L^3}\sum_{\mathbf{k},s}(\hbar\omega)^{1/2}\phi(\mathbf{k}, s)\boldsymbol{\varepsilon}_{\mathbf{k}s}\, \mathrm{e}^{\mathrm{i}(\mathbf{k}\cdot\mathbf{r}-\omega t)}, \tag{12.11–32}$$

which might be called the energy wave function to distinguish it from $\boldsymbol{\Phi}(\mathbf{r}, t)$, and differs from $\boldsymbol{\Phi}(\mathbf{r}, t)$ only in having the factor $(\hbar\omega)^{1/2}$ in the expansion, we readily find that

$$\langle\phi|\hat{H}|\phi\rangle = \int |\boldsymbol{\Psi}(\mathbf{r}, t)|^2\, \mathrm{d}^3 r. \tag{12.11–33}$$

$|\boldsymbol{\Psi}(\mathbf{r}, t)|^2$ therefore plays the role of energy density. But this energy density is not locally connected with the photon density $|\boldsymbol{\Phi}(\mathbf{r}, t)|^2$. Indeed, from the Fourier expansions of $\boldsymbol{\Phi}(\mathbf{r}, t)$ and $\boldsymbol{\Psi}(\mathbf{r}, t)$ we find with the help of the convolution theorem that they are connected through the spatial convolution

$$\boldsymbol{\Psi}(\mathbf{r}, t) = \int G(\mathbf{r} - \mathbf{r}')\boldsymbol{\Phi}(\mathbf{r}', t)\, \mathrm{d}^3 r', \tag{12.11–34}$$

where the spread function $G(\mathbf{r})$ is the three-dimensional Fourier transform of $(\hbar\omega)^{1/2}$, or

$$G(\mathbf{r}) = \frac{(\hbar c)^{1/2}}{(2\pi)^3}\int k^{1/2}\, \mathrm{e}^{\mathrm{i}\mathbf{k}\cdot\mathbf{r}}\, \mathrm{d}^3 k. \tag{12.11–35}$$

Because of the spread associated with $G(\mathbf{r})$, $\boldsymbol{\Psi}(\mathbf{r}, t)$ can be non-zero at positions where $\boldsymbol{\Phi}(\mathbf{r}, t)$ is zero.

Strictly speaking the integral in Eq. (12.11–35) diverges, but it can be regularized by introducing an exponential factor and defining the function [Amrein,

1969]

$$
\begin{aligned}
G_\varepsilon(\mathbf{r}) &= \frac{(\hbar c)^{1/2}}{(2\pi)^3}\int k^{1/2}\,\mathrm{e}^{\mathrm{i}\mathbf{k}\cdot\mathbf{r}}\,\mathrm{e}^{-\varepsilon k}\,\mathrm{d}^3k \\
&= \frac{(\hbar c)^{1/2}}{(2\pi)^3}\int_0^\infty \mathrm{d}k\,k^{5/2}\,\mathrm{e}^{-\varepsilon k}\int_0^\pi \mathrm{d}\theta\,\mathrm{e}^{\mathrm{i}kr\cos\theta}\sin\theta\int_0^{2\pi}\mathrm{d}\phi \\
&= \frac{(\hbar c)^{1/2}}{2\pi^2 r}\int_0^\infty k^{3/2}\,\mathrm{e}^{-\varepsilon k}\sin kr\,\mathrm{d}k \\
&= \frac{3(\hbar c)^{1/2}}{8\pi^{3/2}}\frac{1}{r(r^2+\varepsilon^2)^{5/4}}\sin\left(\tfrac{5}{2}\arctan r/\varepsilon\right). \qquad (12.11\text{–}36)
\end{aligned}
$$

When $\varepsilon \to 0$, or more generally whenever $\varepsilon \ll r$, this reduces to

$$
G(\mathbf{r}) = \pm\frac{3(\hbar c)^{1/2}}{8\sqrt{2}\,\pi^{3/2}r^{7/2}}. \qquad (12.11\text{–}37)
$$

It follows from Eqs. (12.11–34) and (12.11–37) that even when the position wave function $\boldsymbol{\Phi}(\mathbf{r}, t)$ is strongly concentrated near the origin, the energy wave function $\boldsymbol{\Psi}(\mathbf{r}, t)$ is spread out over space asymptotically like $r^{-7/2}$. Alternatively, we may say that even when the probability distribution of the photon is strongly localized near the origin, the energy distribution extends over large distances and falls off as r^{-7} (see also Hegerfeldt, 1974; Hegerfeldt and Ruijsenaar, 1980).

Let us now turn to the problem of detecting the localized photon with a photoelectric detector. If the detector electrons interact with the light via the electric field $\hat{\mathbf{E}}(\mathbf{r}, t)$, one would expect the probability of photoelectric detection to be proportional to the expectation of the electric field intensity $\hat{I}_j(\mathbf{r}, t) \equiv \hat{\mathbf{E}}^{(-)}(\mathbf{r}, t)\cdot\hat{\mathbf{E}}^{(+)}(\mathbf{r}, t)$ (cf. Section 12.2), where

$$
\hat{\mathbf{E}}^{(+)}(\mathbf{r}, t) = \frac{1}{L^{3/2}}\sum_{\mathbf{k},s}\mathrm{i}\left(\frac{\hbar\omega}{2\varepsilon_0}\right)^{1/2}\hat{a}_{\mathbf{k}s}\boldsymbol{\varepsilon}_{\mathbf{k}s}\,\mathrm{e}^{\mathrm{i}(\mathbf{k}\cdot\mathbf{r}-\omega t)}. \qquad (12.11\text{–}38)
$$

Actually, as we show in Chapter 14, the situation is usually more complicated in practice. Nevertheless, for simplicity, we shall suppose here that the detection probability is proportional to $\langle\phi|\hat{\mathbf{E}}^{(-)}(\mathbf{r}, t)\cdot\hat{\mathbf{E}}^{(+)}(\mathbf{r}, t)|\phi\rangle$. Now from Eq. (12.11–26)

$$
\begin{aligned}
\hat{\mathbf{E}}^{(+)}(\mathbf{r}, t)|\phi\rangle &= \frac{1}{L^3}\sum_{\mathbf{k},s}\sum_{\mathbf{k}',s'}\mathrm{i}\left(\frac{\hbar\omega}{2\varepsilon_0}\right)^{1/2}\phi(\mathbf{k}', s')\hat{a}_{\mathbf{k}s}\hat{a}^\dagger_{\mathbf{k}',s'}|\mathrm{vac}\rangle\boldsymbol{\varepsilon}_{\mathbf{k}s}\,\mathrm{e}^{\mathrm{i}(\mathbf{k}\cdot\mathbf{r}-\omega t)} \\
&= \mathrm{i}\left(\frac{1}{2\varepsilon_0}\right)^{1/2}\frac{1}{L^3}\sum_{\mathbf{k},s}(\hbar\omega)^{1/2}\phi(\mathbf{k}, s)\boldsymbol{\varepsilon}_{\mathbf{k}s}\,\mathrm{e}^{\mathrm{i}(\mathbf{k}\cdot\mathbf{r}-\omega t)}|\mathrm{vac}\rangle \\
&= \frac{\mathrm{i}}{(2\varepsilon_0)^{1/2}}\boldsymbol{\Psi}(\mathbf{r}, t)|\mathrm{vac}\rangle, \qquad (12.11\text{–}39)
\end{aligned}
$$

where $\boldsymbol{\Psi}(\mathbf{r}, t)$ is again given by Eq. (12.11–32), so that the probability of photodetection is proportional to

$$\langle\phi|\hat{\mathbf{E}}^{(-)}(\mathbf{r}, t)\cdot\hat{\mathbf{E}}^{(+)}(\mathbf{r}, t)|\phi\rangle = \frac{1}{2\varepsilon_0}|\boldsymbol{\Psi}(\mathbf{r}, t)|^2. \tag{12.11–40}$$

It follows that this quantity bears the same non-local relation to the photon probability distribution $|\boldsymbol{\Phi}(\mathbf{r}, t)|^2$ as does the energy distribution. Once again we find that, for a photon which is strongly localized close to the origin, there is a non-vanishing probability falling off as r^{-7} that it will be detected by a photoelectric detector at a distance r. From these considerations it is apparent that the concept of the photon as a localized particle traveling with velocity c can be quite inappropriate and misleading under some circumstances, even though it works in other cases.

12.12 Effect of an attenuator or beam splitter on the quantum field

In optics it is often necessary to attenuate a beam of light for one reason or another, perhaps in order to protect a measuring instrument. This can be achieved by inserting a filter of some kind into the beam, and a convenient form of filter is a (possibly multilayered) dielectric. Such a partial reflector, when inserted at an angle as shown in Fig. 12.7, also serves as a beam splitter, and it gives rise to a reflected and a transmitted beam, whose intensities depend on the reflectivity and transmissivity of the beam splitter. Because it can serve as an attenuator as well, we shall take the beam splitter shown in Fig. 12.7 as the prototype of an attenuator or beam splitter. Because the reflectivity r and transmissivity t of the beam splitter may depend on the frequency, the polarization and perhaps also the direction of the incident light, it is desirable to make a mode decomposition of the incident, the reflected, and the transmitted fields, and to relate the corresponding mode amplitudes, rather than the total fields.

For simplicity we assume that there is no absorption within the beam splitter itself. For a single dielectric layer the phases of the reflected and transmitted waves always differ by $\pm\pi/2$ when the same medium is in contact with both sides of the beam splitter, and this holds more generally for any symmetric stratified beam splitter. But an unsymmetric beam splitter can have a different reflectivity r' and transmissivity t' for light incident from the other side (see Fig. 12.8). The four quantities r, t, r', t' must however, satisfy the following reciprocity relations due to Stokes (1849) (see also Born and Wolf, 1980, Sec. 1.6; Vašíček, 1960;

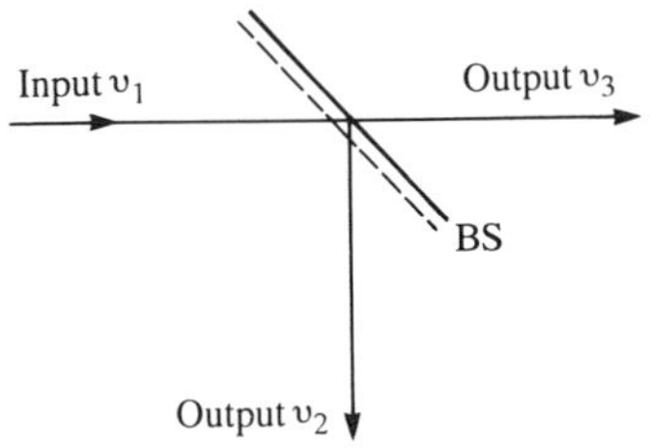

Fig. 12.7 The beam splitter BS with classical fields at input and output.

Friberg and Drummond 1983a,b; Nieto-Vesperinas and Wolf, 1986; Ou and Mandel, 1989):

$$\left.\begin{aligned} |r'| &= |r| \\ |t'| &= |t| \\ |r|^2 + |t|^2 &= 1 \\ r^*t' + r't^* &= 0 \\ r^*t + r't'^* &= 0, \end{aligned}\right\} \tag{12.12–1}$$

which can be derived, for example, with the help of an energy balance argument. If we think of r, t, r', t' as the four elements of the 2×2 'scattering matrix'

$$\mathbf{S} \equiv \begin{bmatrix} t' & r \\ r' & t \end{bmatrix} \tag{12.12–2}$$

for the beam splitter, that connects the two inputs and the two outputs [see Eqs. (12.12–4) below], then Eqs. (12.12–1) just express the unitarity of the matrix $\mathbf{S}$. We have already seen in Section 10.9.5 that, whereas any unused input port can be ignored in a classical treatment, this is not so in a quantum treatment of the beam splitter.

12.12.1 Operator relations

If the field is quantized, the classical complex mode amplitudes v_1, v_2, v_3 shown in Fig. 12.7 have to be replaced by photon annihilation operators $\hat{a}_1$, $\hat{a}_2$, $\hat{a}_3$, which obey the canonical commutation relations

$$\left.\begin{aligned} [\hat{a}_1, \hat{a}_1^\dagger] = [\hat{a}_2, \hat{a}_2^\dagger] = [\hat{a}_3, \hat{a}_3^\dagger] = 1 \\ [\hat{a}_2, \hat{a}_3^\dagger] = 0. \end{aligned}\right\} \tag{12.12–3}$$

It is then impossible for the two output operators $\hat{a}_2$, $\hat{a}_3$ to be proportional to the input $\hat{a}_1$ like the corresponding classical output amplitudes, because this would violate the commutation relations (12.12–3), as was already shown in Section 10.9.5. In order to reconcile the known effect of the beam splitter on the incident light with quantum mechanics, we have to incorporate the vacuum field at the unused input port. This problem has been treated in a number of different ways (Yuen and Shapiro, 1980; Ley and Loudon, 1985; Fearn and Loudon, 1987; Ou, Hong and Mandel, 1987; Campos, Saleh and Teich, 1989).

Let $\hat{a}_0$ be the complex amplitude of the field, at the inport port 0 (see Fig. 12–8). We now construct the output dynamical variables $\hat{a}_2$, $\hat{a}_3$ from the input variables $\hat{a}_0$, $\hat{a}_1$ by reference to Fig. 12.8, allowing for the possibility that the beam splitter is asymmetric. Thus

$$\left.\begin{aligned} \hat{a}_2 &= t'\hat{a}_0 + r\hat{a}_1 \\ \hat{a}_3 &= r'\hat{a}_0 + t\hat{a}_1. \end{aligned}\right\} \tag{12.12–4}$$

Then if $\hat{a}_0$, $\hat{a}_1$ obey the commutation relations

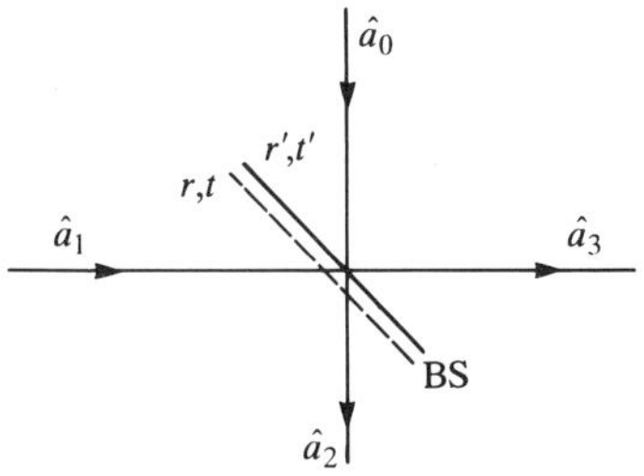

Fig. 12.8 The beam splitter with quantum fields.

$$\left.\begin{aligned}[\hat{a}_0, \hat{a}_0^\dagger] &= 1 = [\hat{a}_1, \hat{a}_1^\dagger] \\ [\hat{a}_0, \hat{a}_1^\dagger] &= 0,\end{aligned}\right\} \tag{12.12-5}$$

we obtain from Eq. (12.12–4)

$$\begin{aligned}[\hat{a}_2, \hat{a}_2^\dagger] &= [t'\hat{a}_0 + r\hat{a}_1, t'^*\hat{a}_0^\dagger + r^*\hat{a}_1^\dagger] \\ &= |t|^2 + |r|^2 \\ &= 1,\end{aligned} \tag{12.12-6}$$

where the last line follows with the help of the reciprocity relations (12.12–1). Similarly we find that

$$[\hat{a}_3, \hat{a}_3^\dagger] = 1,$$

and

$$\begin{aligned}[\hat{a}_2, \hat{a}_3^\dagger] &= [t'\hat{a}_0 + r\hat{a}_1, r'^*\hat{a}_0^\dagger + t^*\hat{a}_1^\dagger] \\ &= t'r'^* + rt^* \\ &= 0\end{aligned} \tag{12.12-7}$$

by use of Eq. (12.12–1). Hence Eqs. (12.12–4) satisfy all the required commutation relations, and we take them to be the proper input–output relations.

We can readily construct the photon number operators from Eqs. (12.12–4), and we obtain the formulas

$$\left.\begin{aligned}\hat{n}_2 &= \hat{a}_2^\dagger\hat{a}_2 = (t'^*\hat{a}_0^\dagger + r^*\hat{a}_1^\dagger)(t'\hat{a}_0 + r\hat{a}_1) \\ &= |t|^2\hat{n}_0 + |r|^2\hat{n}_1 + t'^*r\hat{a}_0^\dagger\hat{a}_1 + r^*t'\hat{a}_1^\dagger\hat{a}_0 \\ \hat{n}_3 &= |r|^2\hat{n}_0 + |t|^2\hat{n}_1 + r'^*t\hat{a}_0^\dagger\hat{a}_1 + t^*r'\hat{a}_1^\dagger\hat{a}_0.\end{aligned}\right\} \tag{12.12-8}$$

From these it follows immediately with the help of the reciprocity relations that

$$\hat{n}_2 + \hat{n}_3 = \hat{n}_0 + \hat{n}_1, \tag{12.12-9}$$

which expresses the conservation of photons between the inputs and outputs.

12.12.2 Photon correlations

All the relations we have obtained so far are relations among dynamical variables. The physics enters more explicitly when we calculate expectation values in some quantum state $\hat{\rho}$. Let us now assume that mode 0 is in the vacuum state. Then the effect of $\hat{a}_0$ acting on the input density operator $\hat{\rho}$ is expressed by the property

$$\hat{a}_0\hat{\rho} = 0 = \hat{\rho}\hat{a}_0^\dagger. \tag{12.12–10}$$

The expectation values of $\hat{n}_2$ and $\hat{n}_3$ are given by the expressions

$$\langle \hat{n}_2 \rangle = \mathrm{Tr}\,(\hat{n}_2\hat{\rho})$$
$$\langle \hat{n}_3 \rangle = \mathrm{Tr}\,(\hat{n}_3\hat{\rho}),$$

and with the help of Eqs. (12.12–8) and (12.12–10) we readily obtain for the average number of photons in the reflected and the transmitted beams

$$\left.\begin{aligned} \langle \hat{n}_2 \rangle &= |r|^2 \langle \hat{n}_1 \rangle \\ \langle \hat{n}_3 \rangle &= |t|^2 \langle \hat{n}_1 \rangle. \end{aligned}\right\} \tag{12.12–11}$$

All other terms have zero expectations by virtue of Eq. (12.12–10). These conclusions are similar to what would be obtained from the classical input–output relations.

A more interesting non-classical result is obtained if we calculate the correlation between output beams 2 and 3. Suppose that photodetectors are placed at the two exit ports 2 and 3. From Section 12.2 we know that the joint probability of detecting a photon in both the reflected and transmitted beams is proportional to

$$\begin{aligned} P_{23} = \langle \hat{a}_2^\dagger \hat{a}_3^\dagger \hat{a}_3 \hat{a}_2 \rangle &= \langle {:}\hat{n}_2\hat{n}_3{:} \rangle \\ &= \langle \hat{n}_2\hat{n}_3 \rangle, \end{aligned} \tag{12.12–12}$$

where the last expression follows by virtue of Eqs. (12.12–3). Let us now use Eqs. (12.12–8) for $\hat{n}_2$ and $\hat{n}_3$ to calculate $\langle \hat{n}_2\hat{n}_3 \rangle$. With the input $\hat{a}_0$ in the vacuum state, we readily obtain with the help of Eq. (12.12–10) that

$$\begin{aligned} \langle \hat{n}_2\hat{n}_3 \rangle &= |r|^2|t|^2 \langle \hat{n}_1^2 \rangle - |r|^2|t|^2 \langle \hat{a}_1^\dagger \hat{a}_0 \hat{a}_0^\dagger \hat{a}_1 \rangle \\ &= |r|^2|t|^2 [\langle \hat{n}_1^2 \rangle - \langle \hat{n}_1(\hat{a}_0^\dagger \hat{a}_0 + 1) \rangle] \\ &= |r|^2|t|^2 [\langle \hat{n}_1^2 \rangle - \langle \hat{n}_1 \rangle]. \end{aligned} \tag{12.12–13}$$

The second line follows from the first by use of the commutation relations (12.12–5).

An interesting consequence of this relation is the conclusion that $\langle \hat{n}_2\hat{n}_3 \rangle = 0$ when the incident field is in a one-photon state, for which $\langle \hat{n}_1^2 \rangle = \langle \hat{n}_1 \rangle$. The probability of detecting a photon both in the reflected and in the transmitted beam is then zero. Needless to say, this conclusion is without analogy for a classical field, for which the analogous quantity $\langle |v_2|^2 |v_3|^2 \rangle$ cannot vanish unless the field vanishes identically. It is also of interest to evaluate the correlation

$$\langle \Delta\hat{n}_2 \Delta\hat{n}_3 \rangle = \langle \hat{n}_2\hat{n}_3 \rangle - \langle \hat{n}_2 \rangle \langle \hat{n}_3 \rangle.$$

With the help of Eqs. (12.12–11) and (12.12–13) we immediately find that

$$\begin{aligned} \langle \Delta\hat{n}_2 \Delta\hat{n}_3 \rangle &= |r|^2|t|^2 [\langle \hat{n}_1^2 \rangle - \langle \hat{n}_1 \rangle - \langle \hat{n}_1 \rangle^2] \\ &= |r|^2|t|^2 [\langle (\Delta\hat{n}_1)^2 \rangle - \langle \hat{n}_1 \rangle]. \end{aligned} \tag{12.12–14}$$

This shows that the equal-time correlation between the fluctuations of the reflected and transmitted beams is positive or negative according as the fluctuations of the incident photons are super-Poissonian or sub-Poissonian. For any field having a classical description we must have $\langle(\Delta\hat{n}_1)^2\rangle \geqslant \langle\hat{n}_1\rangle$, with equality for a completely coherent state. Hence there are no correlations between the beam splitter outputs 2 and 3 for a coherent field, because there are no intensity fluctuations. For a field that is classical but not coherent the correlation between the outputs 2 and 3 is positive, and the phenomenon is then known as the Hanbury Brown–Twiss effect (in this connection see Sections 8.4 and 12.2). We have already seen that sub-Poissonian statistics are possible only for a non-classical field (cf. Section 12.10), and the same is therefore true also for negative correlations between the outputs of the beam splitter.

Finally we note that the operator relations (12.12–4) are equally valid when the field $\hat{a}_0$ entering at port 0 is not in the vacuum state. As the two light beams represented by $\hat{a}_1$ and $\hat{a}_0$ interfere, the same operator equations describe an interference or homodyne experiment.

As an example, we again consider the joint probability $P_{23} = \langle:\hat{n}_2\hat{n}_3:\rangle$ for detecting a photon at each of the two output ports of the beam splitter shown in Fig. 12.8, but this time we shall assume that one photon enters at each of the input ports 0 and 1. The input state is then the Fock state $|1_0, 1_1\rangle$. If we make use of Eqs. (12.12–8) for the number operator and take expectation values in the state $|1_0, 1_1\rangle$, we readily obtain the result

$$\begin{aligned} P_{23} &= \langle\hat{n}_2\hat{n}_3\rangle \\ &= (|t|^2 - |r|^2)^2. \end{aligned} \qquad (12.12\text{–}15)$$

This vanishes for a 50%:50% beam splitter for which $|t|^2 = 0.5 = |r|^2$. Therefore when two similar photons enter a 50%:50% beam splitter, one at each input port, we will never encounter one photon exiting at each output port; either both photons exit at port 2 or both exit at port 3. This is an example of quantum mechanical interference of the probability amplitudes for a photon pair.

The effect can be understood as follows. There are two different ways in which the situation with one photon exiting at port 2 and one exiting at port 3 can arise. Either both incident photons are transmitted through the beam splitter or both are reflected from it. As the detectors observing the output cannot distinguish between these possibilities, the corresponding two-photon probability amplitudes must be added before squaring to determine the probability. But because of the phase shifts on reflection and transmission, the two probability amplitudes are exactly 180° out of phase, and they add to give zero for a 50%:50% beam splitter.

12.12.3 Michelson interferometer

Next we consider the Michelson interferometer illustrated in Fig. 12.9. In the figure $\hat{a}_0$ and $\hat{a}_1$ represent the two inputs, and the beams emerging from the beam splitter, represented by $\hat{a}_2$ and $\hat{a}_3$, fall on two perpendicular, perfectly reflecting mirrors. They are reflected back to the beam splitter, where they are represented by $\hat{a}_2 e^{i\phi_2}$ and $\hat{a}_3 e^{i\phi_3}$. These now serve as inputs to the beam splitter, and they

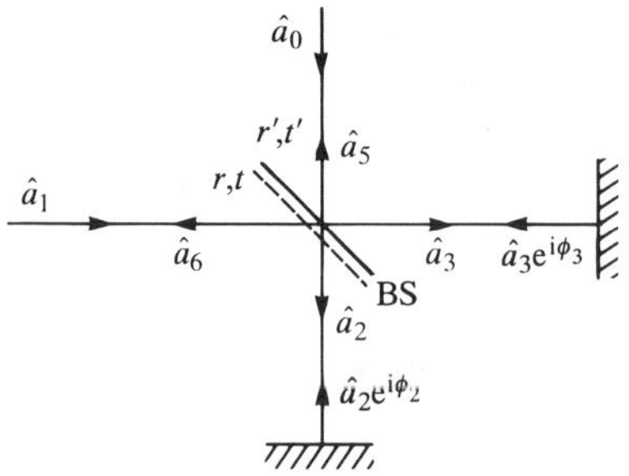

Fig. 12.9 Illustrating the modes of a Michelson interferometer.

generate two new outputs $\hat{a}_5$ and $\hat{a}_6$, which are related to $\hat{a}_2\,e^{i\phi_2}$ and $\hat{a}_3\,e^{i\phi_3}$ by

$$\left.\begin{aligned}\hat{a}_5 &= t\hat{a}_2\,e^{i\phi_2} + r'\hat{a}_3\,e^{i\phi_3}\\ \hat{a}_6 &= t'\hat{a}_3\,e^{i\phi_3} + r\hat{a}_2\,e^{i\phi_2}.\end{aligned}\right\} \qquad (12.12\text{–}16)$$

Hence we have with the help of Eqs. (12.12–4)

$$\left.\begin{aligned}\hat{a}_5 &= \hat{a}_1[tr\,e^{i\phi_2} + r't\,e^{i\phi_3}] + \hat{a}_0[tt'\,e^{i\phi_2} + r'^2\,e^{i\phi_3}]\\ \hat{a}_6 &= \hat{a}_1[tt'\,e^{i\phi_3} + r^2\,e^{i\phi_2}] + \hat{a}_0[t'r'\,e^{i\phi_3} + rt'\,e^{i\phi_2}].\end{aligned}\right\} \qquad (12.12\text{–}17)$$

If $\hat{a}_0$ represents a vacuum mode, we obtain with the help of Eq. (12.12–10) for the average number of photons in each output beam

$$\left.\begin{aligned}\langle\hat{n}_5\rangle &= \langle\hat{n}_1\rangle 2|r|^2|t|^2[1 + \cos(\phi_2 - \phi_3 + \arg r - \arg r')]\\ \langle\hat{n}_6\rangle &= \langle\hat{n}_1\rangle[|t|^4 + |r|^4 - 2|r|^2|t|^2\cos(\phi_2 - \phi_3 + \arg r - \arg r')].\end{aligned}\right\} \qquad (12.12\text{–}18)$$

Hence $\langle\hat{n}_5\rangle$ and $\langle\hat{n}_6\rangle$ are in antiphase, and the mean photon numbers $\langle\hat{n}_5\rangle$ and $\langle\hat{n}_6\rangle$ can be used to determine the phase difference $\phi_2 - \phi_3$. Also

$$\langle\hat{n}_5\rangle + \langle\hat{n}_6\rangle = \langle\hat{n}_1\rangle,$$

so that photon numbers are conserved, as expected. We may readily show from Eqs. (12.12–17), for a symmetric 50%:50% beam splitter, that the cross-correlation between the two output beams is given by the relation

$$\langle\Delta\hat{n}_5\Delta\hat{n}_6\rangle = \tfrac{1}{4}[\langle(\Delta\hat{n}_1)^2\rangle - \langle\hat{n}_1\rangle]\sin^2(\phi_2 - \phi_3), \qquad (12.12\text{–}19)$$

and this is positive or negative according as the input photon statistics are super- or sub-Poissonian. Similarly, the fluctuations of $\hat{n}_5$ and of $\hat{n}_6$ reflect the super- or sub-Poisson character of the input field.

12.12.4 Relationship between input and output states for the beam splitter

We have seen that Eqs. (12.12–4) connect the dynamical variables of the input and output fields of the beam splitter. If $f(\hat{a}_2, \hat{a}_3)$ is an operator function of the output variables and their conjugates, we can use these equations to calculate the expectation value $\langle f(\hat{a}_2, \hat{a}_3)\rangle$, by transforming from $\hat{a}_3$, $\hat{a}_2$ to $\hat{a}_1$, $\hat{a}_0$. However, an alternative approach is to specify the state on the output side of the beam splitter directly in terms of the input state. We will now derive a general procedure for doing this (Ou, Hong and Mandel, 1987).

Let us assume that the density operator $\hat{\rho}_{\text{in}}$ at the same beam splitter input shown in Fig. 12.8 has been given a diagonal coherent-state representation in the two complex input mode amplitudes v_0, v_1, in the usual form

$$\hat{\rho}_{\text{in}} = \int \phi_{\text{in}}(v_0, v_1)|v_0, v_1\rangle\langle v_0, v_1|\,\mathrm{d}^2v_0\,\mathrm{d}^2v_1. \tag{12.12–20}$$

Similarly, we suppose that the density operator $\hat{\rho}_{\text{out}}$ at the output can be written in the diagonal form in terms of the two complex output mode amplitudes v_2, v_3,

$$\hat{\rho}_{\text{out}} = \int \phi_{\text{out}}(v_2, v_3)|v_2, v_3\rangle\langle v_2, v_3|\,\mathrm{d}^2v_2\,\mathrm{d}^2v_3. \tag{12.12–21}$$

Then the state $\hat{\rho}_{\text{out}}$ corresponds correctly to the state $\hat{\rho}_{\text{in}}$ if, for any function $f(\hat{a}_2, \hat{a}_3)$ of $\hat{a}_2$, $\hat{a}_3$ and their conjugates

$$\operatorname{Tr}[f(\hat{a}_2, \hat{a}_3)\hat{\rho}_{\text{out}}] = \operatorname{Tr}[f(r\hat{a}_1 + t'\hat{a}_0, t\hat{a}_1 + r'\hat{a}_0)\hat{\rho}_{\text{in}}]. \tag{12.12–22}$$

As any operator can, in principle, be put into normal order, we shall limit ourselves to the case in which $f(\hat{a}_2, \hat{a}_3)$ is a normally ordered operator function $f^{(\mathrm{N})}(\hat{a}_2, \hat{a}_3)$. Then, because of the form of the operator relations (12.12–4), it follows that when $\hat{a}_2$, $\hat{a}_3$ are expressed in terms of $\hat{a}_0$, $\hat{a}_1$, the operator $f^{(\mathrm{N})}$ is again in normal order.

We now make use of the optical equivalence theorem for normally ordered operators (see Section 11.9), according to which

$$\operatorname{Tr}[f^{(\mathrm{N})}(\hat{a}_2, \hat{a}_3)\hat{\rho}_{\text{out}}] = \int f^{(\mathrm{N})}(v_2, v_3)\phi_{\text{out}}(v_2, v_3)\,\mathrm{d}^2v_2\,\mathrm{d}^2v_3, \tag{12.12–23}$$

and

$$\operatorname{Tr}[f^{(\mathrm{N})}(r\hat{a}_1 + t'\hat{a}_0, t\hat{a}_1 + r'\hat{a}_0)\hat{\rho}_{\text{in}}] =$$
$$\int f^{(\mathrm{N})}(rv_1 + t'v_0, tv_1 + r'v_0)\phi_{\text{in}}(v_0, v_1)\,\mathrm{d}^2v_0\,\mathrm{d}^2v_1. \tag{12.12–24}$$

Because of Eq. (12.12–22) the integrals on the right of these two relations must be equal. By making the change of variables

$$\left.\begin{aligned} v_2 &= rv_1 + t'v_0 \\ v_3 &= tv_1 + r'v_0 \end{aligned}\right\} \tag{12.12–25}$$

or

$$\left.\begin{aligned} v_0 &= (-tv_2 + rv_3)F \\ v_1 &= (r'v_2 - t'v_3)F, \quad F \equiv (rr' - tt')^{-1} \end{aligned}\right\} \tag{12.12–26}$$

in the last integral, and noting that

$$\mathrm{d}^2v_2\,\mathrm{d}^2v_3 = \mathrm{d}^2v_0\,\mathrm{d}^2v_1,$$

we are immediately led to the equation

$$\phi_{\text{out}}(v_2, v_3) = \phi_{\text{in}}(-Ftv_2 + Frv_3, Fr'v_2 - Ft'v_3). \tag{12.12–27}$$

This determines the output state in terms of the input state.

In particular, if the input field is in the pure coherent state $|v_0', v_1'\rangle$, so that

$$\hat{\rho}_{\text{in}} = |v_0', v_1'\rangle\langle v_0', v_1'|$$

and

$$\phi_{\text{in}}(v_0, v_1) = \delta^2(v_0 - v_0')\delta^2(v_1 - v_1'),$$

then

$$\phi_{\text{out}}(v_2, v_3) = \delta^2(-Ftv_2 + Frv_3 - v_0')\delta^2(Fr'v_2 - Ft'v_3 - v_1').$$

Substituting for $\phi_{\text{out}}(v_2, v_3)$ in Eq. (12.12–21), and making the change of variables (12.12–25) shows immediately that the output state is the two-mode coherent state

$$\hat{\rho}_{\text{out}} = |rv_1' + t'v_0', tv_1' + r'v_0'\rangle\langle rv_1' + t'v_0', tv_1' + r'v_0'|. \quad (12.12\text{–}28)$$

When the input state $|\Psi_{\text{in}}\rangle$ is a pure state in the form of a linear superposition of coherent states

$$|\Psi_{\text{in}}\rangle = c'|v_0', v_1'\rangle + c''|v_0'', v_1''\rangle, \quad (12.12\text{–}29)$$

the associated phase space density $\phi_{\text{in}}(v_0, v_1)$ becomes highly singular, as we saw in Section 11.8. By using Eq. (12.12–22) we may readily show, however, that the associated output state $|\Psi_{\text{out}}\rangle$ is just the corresponding linear superposition of output coherent states

$$|\Psi_{\text{out}}\rangle = c'|rv_1' + t'v_0', tv_1' + r'v_0'\rangle + c''|rv_1'' + t'v_0'', tv_1'' + r'v_0''\rangle. \quad (12.12\text{–}30)$$

By observing that a complex factor t is always associated with the amplitude transmission of a photon through the beam splitter, and a facter r is associated with the reflection of a photon from it, we can often write down the output state directly in terms of the input state. Thus, let us suppose that the beam splitter is symmetric and that the input is the two-photon state $|1_0, 1_1\rangle$ corresponding to one photon entering at port 0 and one entering at port 1. There are four possibilities. If both photons are transmitted, this results in the output state $|1_2, 1_3\rangle$ with complex amplitude t^2, which we take to be real. If both photons are reflected, we again encounter the output state $|1_2, 1_3\rangle$ with complex amplitude $\text{i}|r|i|r| = -|r|^2$. If one photon is reflected and one is transmitted, we have either the output state $|2_2, 0_3\rangle$ with complex amplitude $\sqrt{2}\,\text{i}|rt|$, or the output state $|0_2, 2_3\rangle$ with complex amplitude $\sqrt{2}\,\text{i}|rt|$. It follows that the output $|\Psi_{\text{out}}\rangle$ corresponding to the input $|1_0, 1_1\rangle$ is the linear superposition state

$$|\psi_{\text{out}}\rangle = (|t|^2 - |r|^2)|1_2, 1_3\rangle + \text{i}\sqrt{2}\,|rt|(|2_2, 0_3\rangle + |0_2, 2_3\rangle). \quad (12.12\text{–}31)$$

The $\sqrt{2}$ factor, which is the result of interference, is obviously necessary to ensure that $|\psi_{\text{out}}\rangle$ is normalized to unity. Note that when $|t|^2 = |r|^2 = \frac{1}{2}$ the two photons always emerge together either at port 2 or at port 3, never one at port 2 and one at port 3, in confirmation of Eq. (12.12–15). This principle has been used to measure the time separation between two similar input photon wave packets with femtosecond accuracy (Hong, Ou and Mandel, 1987).

12.13 Effect of a polarizer on the field

In discussing the effect of a beam splitter or linear attenuator, we resolved the total field into modes and dealt with one mode at a time. But when the polarization properties are of interest, it is often advantageous to deal with two modes, corresponding to two orthogonal polarization components, at once. The electromagnetic field corresponding to one wave vector $\mathbf{k}$, that we take to define the z-axis, can then be represented by the vector amplitude $\hat{\mathscr{A}}(\mathbf{k})$ such that

$$\hat{\mathscr{A}}(\mathbf{k}) = \hat{a}_x \boldsymbol{\varepsilon}_x + \hat{a}_y \boldsymbol{\varepsilon}_y. \tag{12.13–1}$$

Here $\hat{a}_x$, $\hat{a}_y$ are photon annihilation operators corresponding to orthogonal polarization components in the x- and y-directions, and $\boldsymbol{\varepsilon}_x$, $\boldsymbol{\varepsilon}_y$ are unit polarization vectors. Alternatively, we can resolve the vector field amplitude $\hat{\mathscr{A}}$ into two orthogonal elliptic polarization components, as discussed in Section 10.2. For simplicity we shall take $\boldsymbol{\varepsilon}_x$, $\boldsymbol{\varepsilon}_y$ to be real vectors for the moment.

We now suppose that the light falls on a polarizer. The effect of an elliptic polarizer on the polarization matrix $\mathsf{J}(\mathbf{k})$ is represented by the 2×2 transmission matrix [cf. Eqs (6.4–9a) and (6.4–20)]

$$\mathsf{T}(\theta, \delta) = \begin{bmatrix} \cos^2\theta & \cos\theta\sin\theta\,\mathrm{e}^{-\mathrm{i}\delta} \\ \cos\theta\sin\theta\,\mathrm{e}^{\mathrm{i}\delta} & \sin^2\theta \end{bmatrix}, \tag{12.13–2}$$

where θ is the angle between the major axis of the ellipse and the x-axis, and δ determines the eccentricity. In particular, when $\delta = 0$, T represents a linear polarizer whose polarizing direction is inclined at an angle θ to the x-axis.

In order to determine the vector field amplitude $\hat{\mathscr{A}}'(\mathbf{k})$ behind the polarizer, we use the transformation law

$$\hat{\mathscr{A}}' = T_{ij}\hat{a}_i \boldsymbol{\varepsilon}_j, \quad (i, j = x, y), \tag{12.13–3}$$

where summation on repeated indices is understood. On writing out the transformation we arrive at the formula

$$\begin{aligned} \hat{\mathscr{A}}' &= (\cos^2\theta\,\hat{a}_x + \cos\theta\sin\theta\,\mathrm{e}^{\mathrm{i}\delta}\hat{a}_y)\boldsymbol{\varepsilon}_x + (\cos\theta\sin\theta\,\mathrm{e}^{-\mathrm{i}\delta}\hat{a}_x + \sin^2\theta\,\hat{a}_y)\boldsymbol{\varepsilon}_y \\ &= (\cos\theta\,\hat{a}_x + \sin\theta\,\mathrm{e}^{\mathrm{i}\delta}\hat{a}_y)(\cos\theta\,\boldsymbol{\varepsilon}_x + \sin\theta\,\mathrm{e}^{-\mathrm{i}\delta}\boldsymbol{\varepsilon}_y) \\ &= \hat{a}'\boldsymbol{\varepsilon}'. \end{aligned} \tag{12.13–4}$$

But $\cos\theta\,\boldsymbol{\varepsilon}_x + \sin\theta\,\mathrm{e}^{-\mathrm{i}\delta}\boldsymbol{\varepsilon}_y$ is just the complex unit polarization vector $\boldsymbol{\varepsilon}'$ (cf. Section 10.2) that characterizes the polarizer, so that the coefficient of $\boldsymbol{\varepsilon}'$ is the field amplitude $\hat{a}'$ behind the polarizer, or

$$\hat{a}' = \cos\theta\,\hat{a}_x + \sin\theta\,\mathrm{e}^{\mathrm{i}\delta}\,\hat{a}_y. \tag{12.13–5}$$

This formula can be used to calculate expectations of the field behind the polarizer in terms of the dynamical variables at the input side. We note that $\hat{a}'$ satisfies the same commutation relation as $\hat{a}_x$, $\hat{a}_y$. In particular, the average number of photons behind the polarizer is given by

$$\langle \hat{a}'^{\dagger}\hat{a}' \rangle = \cos^2\theta \langle \hat{a}_x^{\dagger}\hat{a}_x \rangle + \sin^2\theta \langle \hat{a}_y^{\dagger}\hat{a}_y \rangle + \cos\theta \sin\theta (e^{i\delta}\langle \hat{a}_x^{\dagger}\hat{a}_y \rangle + e^{-i\delta}\langle \hat{a}_y^{\dagger}\hat{a}_x \rangle), \tag{12.13–6}$$

and this depends on some of the correlation properties of the incident light.

12.14 Einstein locality and photon correlations

It was first pointed out by Einstein, Podolsky and Rosen (EPR) in a classic paper (1935) that, according to the usual interpretation of quantum mechanics, there exist certain two-particle states with the property that a measurement of one chosen variable of particle 1 completely determines the outcome of a measurement of the corresponding variable of particle 2. At the time of measurement, the two particles may be so far apart that no influence resulting from one measurement can possibly propagate to the other particle in the available time. Such a situation may arise when both particles are emitted from a common source in some entangled (non-factorizable) quantum state

$$|\psi\rangle = \frac{1}{\sqrt{2}}[|\phi\rangle_1|\chi\rangle_2 - |\chi\rangle_1|\phi\rangle_2].$$

According to Einstein, Podolsky and Rosen (1935) when the outcome of a measurement of some particle variable can be predicted with certainty, without disturbing the particle, then '. . . there exists an element of physical reality corresponding to this physical quantity . . .'. In other words, then *particle 2 really has this value of the variable*, irrespective of whether it is actually measured or not. This must be contrasted with the quantum point of view, according to which the measurement creates the reality, in a sense. On the other hand, suppose that a different variable, say one that is canonically conjugate to the previous one, is measured for particle 1. Then this predetermines the value of the conjugate variable for particle 2, and by the foregoing arguments particle 2 really *has* this value of the conjugate variable. But, if the two variables are canonical conjugates, then according to quantum mechanics they do not commute and they cannot both have definite values at the same time. Now the decision whether to measure one or the other conjugate variable of particle 1 can be made when the two particles are far apart and cannot communicate in the available time, yet it influences the state of particle 2. We appear to have a contradiction, which led Einstein, Podolsky and Rosen to conclude that quantum mechanics is 'incomplete'. Such counter-intuitive non-local correlations have, however, been observed experimentally. The phenomenon is sometimes referred to as a *violation of Einstein locality*, and its implications have been widely discussed (Bohr, 1935; Bohm, 1952a,b; Bell, 1964, 1966; Clauser, Horne, Shimony and Holt, 1969; Wigner, 1970; Clauser and Horne, 1974; Clauser and Shimony, 1978; d'Espagnat 1979; Mermin, 1981, 1985).

To illustrate the paradox more explicitly, let us consider a source that emits two spin $\frac{1}{2}$ particles in opposite directions with total spin angular momentum zero. If we were to measure the spin $\hat{S}_x^{(1)}$ of particle 1 in the x-direction, for example, and obtain the value $\hbar/2$, we would know that particle 2 is in an eigenstate of its spin $\hat{S}_x^{(2)}$ with eigenvalue $-\hbar/2$. On the other hand, had we chosen instead to measure the spin $\hat{S}_y^{(1)}$ of particle 1 in the y-direction and obtained $\hbar/2$, this would have determined the eigenvalue of $\hat{S}_y^{(2)}$ for particle 2 to be $-\hbar/2$. In both cases, a measurement of particle 1 determines the outcome of a certain measurement of particle 2 with probability unity, and according to Einstein, Podolsky and Rosen the values of $\hat{S}_x^{(2)}$ and $\hat{S}_y^{(2)}$ should therefore have 'an element of physical reality'. But $\hat{S}_x^{(2)}$ and $\hat{S}_y^{(2)}$ do not commute and according to quantum mechanics they cannot both have definite values. The paradox arises because we tend intuitively to think in classical terms, i.e. to associate an objective physical reality with each particle and its variables, whereas in quantum mechanics a dynamical variable does not actually have a value until it is measured. In a sense, the measurement creates the physical reality. Attempts have been made to account for the predicted (and later observed) correlations between two particles in terms of *hidden variables*, or unmeasurable parameters that are supposed to determine the outcome of an experiment (Bohm, 1952a,b). But it was later shown by Bell (1964, 1966) and others (Bohm and Aharanov, 1957; Clauser, Horne, Shimony and Holt, 1969; Clauser and Horne, 1974; Clauser and Shimony, 1978) that such non-local effects are fundamentally quantum mechanical, and that no realistic local theory can account for the correlations quantitatively.

12.14.1 The Einstein–Podolsky–Rosen paradox for an entangled two-photon state

To illustrate the paradox within the optical domain, let us consider a two-photon state of zero angular momentum, such as might be created in the cascade decay of an atom making a two-stage transition of type $\Delta J = 0$. Let us suppose that the photons 1 and 2 leave the atom in opposite directions along the z-axis in the singlet state (see Fig. 12.10)

$$|\psi\rangle = \frac{1}{\sqrt{2}}(|1_{1x}, 0_{1y}, 0_{2x}, 1_{2y}\rangle - |0_{1x}, 1_{1y}, 1_{2x}, 0_{2y}\rangle), \qquad (12.14\text{–}1)$$

in which the two photons are polarized orthogonally. We have denoted the state in which photon j ($j = 1, 2$) is linearly polarized in the x-direction by $|1_{jx}\rangle$, etc. Then it is easy to see that each of the two photons considered separately is unpolarized. For from Eq. (12.14–1) we obtain for the polarization matrix $\mathbf{J}_j$ ($j = 1, 2$) (cf. Section 6.2) of photon j, after taking the trace over the other variables,

$$\mathbf{J}_j = \begin{bmatrix} \langle \hat{a}_{jx}^{+}\hat{a}_{jx}\rangle & \langle \hat{a}_{jx}^{+}\hat{a}_{jy}\rangle \\ \langle \hat{a}_{jy}^{+}\hat{a}_{jx}\rangle & \langle \hat{a}_{jy}^{+}\hat{a}_{jy}\rangle \end{bmatrix} = \frac{1}{2}\begin{bmatrix} 1 & 0 \\ 0 & 1 \end{bmatrix} \quad (j = 1, 2), \qquad (12.14\text{–}2)$$

and this represents unpolarized light. However, the polarization of the two photons is strongly coupled.

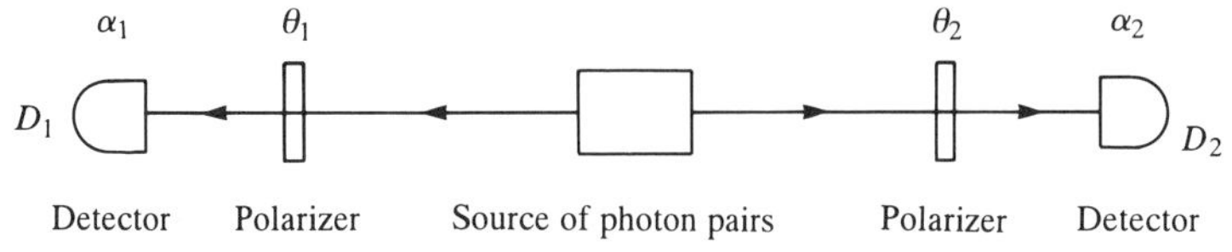

Fig. 12.10 Outline of an experiment for exhibiting polarization correlations between two photons.

Let us now suppose that linear polarizers inclined at angles θ_1 and θ_2 to the x-direction, respectively, are inserted in the paths of the two photons, which then fall on two photodetectors D_1, D_2, as shown in Fig. 12.10. Let $P_j(\theta_j)$ be the probability that the photon in arm j is detected, and let $P_{12}(\theta_1, \theta_2)$ be the joint probability that both photons are detected by their respective detectors when both polarizers are in position. Let α_1, α_2 be the quantum efficiencies of the two detectors. In order to relate the dynamical variables $\hat{a}_1$, $\hat{a}_2$ of the field behind the polarizers to those $\hat{a}_{1x}$, $\hat{a}_{1y}$, $\hat{a}_{2x}$, $\hat{a}_{2y}$ before the polarizers we shall use the relation derived in Section 12.13 above [cf. Eq. (12.13–5)],

$$\hat{a}_j = \hat{a}_{jx}\cos\theta_j + \hat{a}_{jy}\sin\theta_j, \quad (j = 1, 2), \tag{12.14–3}$$

which takes account of the projective property of each polarizer and also preserves the commutation relations. Then from Eqs. (12.14–1) and (12.14–3) we obtain

$$\left.\begin{aligned} P_1(\theta_1) &= \alpha_1\langle\psi|\hat{a}_1^+\hat{a}_1|\psi\rangle = \tfrac{1}{2}\alpha_1 \\ P_2(\theta_2) &= \alpha_2\langle\psi|\hat{a}_2^+\hat{a}_2|\psi\rangle = \tfrac{1}{2}\alpha_2, \end{aligned}\right\} \tag{12.14–4}$$

while the joint probability $P_{12}(\theta_1, \theta_2)$ is given by

$$\begin{aligned} P_{12}(\theta_1, \theta_2) &= \alpha_1\alpha_2\langle\psi|\hat{a}_1^\dagger\hat{a}_2^\dagger\hat{a}_2\hat{a}_1|\psi\rangle \\ &= \tfrac{1}{2}\alpha_1\alpha_2[\sin^2\theta_1\cos^2\theta_2 + \cos^2\theta_1\sin^2\theta_2 - 2\sin\theta_1\cos\theta_2\cos\theta_1\sin\theta_2] \\ &= \tfrac{1}{2}\alpha_1\alpha_2\sin^2(\theta_1 - \theta_2). \end{aligned} \tag{12.14–5}$$

This last probability depends on the settings of both polarizers. The conditional probability $P_c(\theta_2|\theta_1)$ of detecting photon 2, given the detection of photon 1 (cf. Section 1.2), is therefore

$$P_c(\theta_2|\theta_1) = P_{12}(\theta_1, \theta_2)/P_1(\theta_1) = \alpha_2\sin^2(\theta_1 - \theta_2). \tag{12.14–6}$$

This probability can be unity for a perfect detector when $\theta_1 - \theta_2 = \pm\pi/2$, and zero when $\theta_1 = \theta_2$, showing that photon 2 is definitely polarized at right angles to photon 1; yet the polarization of the latter was chosen at will by the orientation of polarizer 1. The outcome of the measurement on photon 2 therefore appears to be influenced by the orientation of the polarizer in arm 1, even though the two photons may be well separated at the time of the measurement. This is the non-locality paradox.

However, it does not follow that an observer located near D_1 can influence the outcome of a measurement made near D_2 by adjusting the angle θ_1 of polarizer 1. This is because setting the polarizer angle θ_1 in general does not guarantee the polarization of the photon in arm 1, except in the special case when the photon 1 emerges from the polarizer θ_1.

The joint probability that photon 2 emerges from polarizer θ_2 and that photon 1 emerges from polarizer θ_1 is given by Eq. (12.14–5) with $\alpha_1 = 1 = \alpha_2$. If + denotes the emergence of the photon from the polarizer and − denotes failure to emerge, then, since the latter probability can be obtained from the former by incrementing θ_1, θ_2 by $\pi/2$, we have

$$P(+, \theta_2, +, \theta_1) = \tfrac{1}{2}\sin^2(\theta_1 - \theta_2) = P(-, \theta_2, -, \theta_1). \tag{12.14–7}$$

Similarly, we find that the joint probability that photon 2 emerges from θ_2 and that photon 1 does not emerge from polarizer θ_1 is

$$P(+, \theta_2, -, \theta_1) = \tfrac{1}{2}\cos^2(\theta_1 - \theta_2) = P(-, \theta_2, +, \theta_1). \qquad (12.14\text{–}8)$$

This last result can be obtained from Eq. (12.14–7) simply by incrementing θ_1 by $\pi/2$.

Now, obviously, setting the angle θ_1 of polarizer 1 does not guarantee either the outcome $+$, θ_1 or the outcome $-$, θ_1. We obtain the probability $P(+, \theta_2|\theta_1)$ that photon 2 emerges from polarizer 2 when polarizer 1 is set to the angle θ_1 by adding $P(+, \theta_2, +, \theta_1)$ and $P(+, \theta_2, -, \theta_1)$. With the help of Eqs. (12.14–7) and (12.14–8) this becomes

$$\begin{aligned} P(+, \theta_2|\theta_1) &= P(+, \theta_2, +, \theta_1) + P(+, \theta_2, -, \theta_1) \\ &= \tfrac{1}{2}. \end{aligned} \qquad (12.14\text{–}9)$$

As this result is independent of θ_1, it follows that setting the angle of polarizer 1 has no influence on the outcome of the measurement in arm 2. Causality is therefore preserved despite the non-locality of the system.

In order to show that these quantum mechanical predictions are not compatible with a local probabilistic theory of the measurement involving hidden variables, we shall now derive one of the so-called Bell inequalities (Bell, 1964, 1966; Bohm and Aharonov, 1957; Clauser, Horne, Shimony, and Holt, 1969; Clauser and Horne, 1974; Clauser and Shimony, 1978) that such a theory must satisfy.

12.14.2 Bell's inequality

Let $A(a)$ and $B(b)$ be two so-called dichotomic observables, characterized by parameters a, b, whose measurement can have only two possible outcomes, labeled $+1$ and -1. For example, $A(a) = +1$ might represent the emergence of a photon from the output port of polarizer θ_1 in arm 1 of the apparatus in Fig. 12.10, and $A(a) = -1$ might represent the failure of the photon to emerge from there. Let

$$C(a, b) \equiv \langle A(a)B(b)\rangle \qquad (12.14\text{–}10)$$

be the correlation between the two observables averaged over the ensemble of all possible outcomes. According to the reality criterion of Einstein, Podolsky and Rosen, there are elements of reality that determine the outcome of the measurement. Let these be characterized by some hidden variable λ, having a probability density $\rho(\lambda)$, such that $\int\rho(\lambda)\,\mathrm{d}\lambda = 1$. Then the correlation between the observables $A(a, \lambda)$ and $B(b, \lambda)$ can be expressed in the form

$$C(a, b) = \int A(a, \lambda)B(b, \lambda)\rho(\lambda)\,\mathrm{d}\lambda. \qquad (12.14\text{–}11)$$

This incorporates the assumption of locality, in that A depends only on a and B only on b. Then since $|A(a, \lambda)| = 1$,

$$|C(a, b) - C(a, b')| \leq \int |A(a, \lambda)[B(b, \lambda) - B(b', \lambda)]|\rho(\lambda)\,\mathrm{d}\lambda$$

$$|C(a, b) - C(a, b')| \leq \int |B(b, \lambda) - B(b', \lambda)|\rho(\lambda)\,\mathrm{d}\lambda \qquad (12.14\text{–}12)$$

and

$$|C(a', b) + C(a', b')| \leq \int |A(a', \lambda)[B(b, \lambda) + B(b', \lambda)]|\rho(\lambda)\,\mathrm{d}\lambda$$

$$|C(a', b) + C(a', b')| \leq \int |B(b, \lambda) + B(b', \lambda)|\rho(\lambda)\,\mathrm{d}\lambda. \qquad (12.14\text{–}13)$$

We now add Eqs. (12.14–12) and (12.14–13). Then

$$|C(a, b) - C(a, b')| + |C(a', b) + C(a', b')|$$
$$\leq \int [|B(b, \lambda) - B(b', \lambda)| + |B(b, \lambda) + B(b', \lambda)|]\rho(\lambda)\,\mathrm{d}\lambda. \qquad (12.14\text{–}14)$$

But because each B takes on only the values ± 1, it follows that

$$|B(b, \lambda) - B(b', \lambda)| + |B(b, \lambda) + B(b', \lambda)| = 2, \qquad (12.14\text{–}15)$$

and when this result is used in Eq. (12.14–14) and we make use of the normalization of $\rho(\lambda)$,

$$\int \rho(\lambda)\,\mathrm{d}\lambda = 1,$$

we arrive at the *Bell inequality* (Bell, 1964, 1966)

$$|C(a, b) - C(a, b')| + |C(a', b) + C(a', b')| \leq 2. \qquad (12.14\text{–}16)$$

Let us now apply the inequality to the experiment shown in Fig. 12.10. We identify a with the polarizer angle θ_1 and b with polarizer angle θ_2, and we associate the outcome $+1$ with the emergence of a photon from the polarizer and -1 with the failure of the photon to emerge. Then from Eqs. (12.14–7) and (12.14–8) we have for the correlation

$$\begin{aligned} C(\theta_1, \theta_2) &= P(+, \theta_2, +, \theta_1) + P(-, \theta_2, -, \theta_1) - P(+, \theta_2, -, \theta_1) \\ &\quad - P(-, \theta_2, +, \theta_1) \\ &= \sin^2(\theta_1 - \theta_2) - \cos^2(\theta_1 - \theta_2) \\ &= -\cos 2(\theta_1 - \theta_2). \end{aligned} \qquad (12.14\text{–}17)$$

This is unity when $\theta_2 - \theta_1 = \pi/2$, as expected for completely correlated events, and -1 when $\theta_1 = \theta_2$, in which case we have complete anti-correlation.

With the particular choice of polarizing angles,

$$\begin{aligned} \theta_1 &= 0 \\ \theta_2 &= 3\pi/8 \\ \theta_1' &= -\pi/4 \\ \theta_2' &= \pi/8 \end{aligned}$$

for example, we obtain the result

$$|C(\theta_1, \theta_2) - C(\theta_1, \theta_2')| + |C(\theta_1', \theta_2) + C(\theta_1', \theta_2')| = \left|-\frac{1}{\sqrt{2}} - \frac{1}{\sqrt{2}}\right| + \left|\frac{1}{\sqrt{2}} + \frac{1}{\sqrt{2}}\right| = 2\sqrt{2} \approx 2.828, \quad (12.14\text{–}18)$$

and this evidently violates the Bell inequality (12.14–16). It follows that the quantum mechanical two-photon system in a singlet state cannot be described in terms of a local, realistic probabilistic theory.

However, in deriving $C(\theta_1, \theta_2)$ we have deliberately excluded the photodetectors, or we have effectively treated the quantum efficiencies α_1, α_2 as being unity. Had the quantum efficiencies been incorporated in $C(\theta_1, \theta_2)$, as they were in Eq. (12.14–5) for example, with the (*fair sampling*) hypothesis that the subensemble of detected photons is typical of the whole ensemble, then the Bell inequality would not have been violated for small values of α_1, α_2. We shall therefore examine another classical inequality that is violated by the two-photon singlet state, which does not directly depend on the magnitudes of α_1, α_2.

12.14.3 Clauser–Horne form of Bell's inequality

We again consider the experimental situation illustrated in Fig. 12.10. A source S emits two photons in two different directions labeled 1, 2, which fall on polarizers in the form of Wollaston prisms that divide the incident light into two orthogonally polarized beams. We assume that the sum of the probabilities for an incident photon to emerge from one or the other exit port of each polarizer is unity, so that there is no absorption. The component linearly polarized in some direction θ_1 is transmitted by the Wollaston prism and allowed to fall on detector D_1, and similarly for the other arm. Let us suppose that each photon pair is characterized by some hidden variable λ, whose value is unknown, and that λ has some probability density $\rho(\lambda)$. Let $p_j(\theta_j, \lambda)$ be the probability that the photon in arm j reaches the detector D_j when the linear polarizer in arm j is set to transmit at the polarization angle θ_j and when the hidden variable is λ. The locality assumption is contained in the statement that $p_j(\theta_j, \lambda)$ depends only on θ_j and not on the polarizer setting in the other arm. Let $p_j(-, \lambda)$ be the corresponding probability when there is no polarizer in arm j. Then since $p_j(-, \lambda)$ cannot depend on θ_j, we have

$$p_j(\theta_j, \lambda) \leq p_j(-, \lambda). \quad (12.14\text{–}19)$$

This relation is sometimes described as the 'no enhancement assumption' (Clauser and Horne, 1974), but it is a requirement here (Reid and Walls, 1986) given the properties of the polarizer.

The joint probability $P_{12}(\theta_1, \theta_2)$ for detections by both D_1 and D_2 when the linear polarizers are set at angles θ_1, θ_2, and when the detectors are not polarization sensitive, is then obtained by multiplying $p_1(\theta_1, \lambda)$ and $p_2(\theta_2, \lambda)$ and summing over all values λ of the hidden variable in the form

$$P_{12}(\theta_1, \theta_2) = \alpha_1\alpha_2 \int p_1(\theta_1, \lambda)p_2(\theta_2, \lambda)\rho(\lambda)\,d\lambda. \quad (12.14\text{–}20)$$

α_1, α_2 are the quantum efficiencies of the two detectors, which are assumed to be polarization independent, and fair sampling is again assumed. Similarly, if $P_{12}(\theta_1, -)$, $P_{12}(-, \theta_2)$ denote the joint probability of detection of the two photons by both D_1 and D_2 when one or the other linear polarizer is removed, we have

$$\left.\begin{aligned} P_{12}(\theta_1, -) &= \alpha_1\alpha_2\int p_1(\theta_1, \lambda)p_2(-, \lambda)\rho(\lambda)\,\mathrm{d}\lambda \\ P_{12}(-, \theta_2) &= \alpha_1\alpha_2\int p_1(-, \lambda)p_2(\theta_2, \lambda)\rho(\lambda)\,\mathrm{d}\lambda. \end{aligned}\right\} \qquad (12.14\text{–}21)$$

We now make use of the algebraic inequality (Clauser and Horne, 1974, Appendix A)

$$-1 \leqslant xy - xy' + x'y + x'y' - x' - y \leqslant 0, \qquad (12.14\text{–}22)$$

which holds for all $0 \leqslant x, y, x', y' \leqslant 1$. With the help of relation (12.14–19) we identify

$$\left.\begin{aligned} x &= p_1(\theta_1, \lambda)/p_1(-, \lambda) \\ y &= p_2(\theta_2, \lambda)/p_2(-, \lambda) \\ x' &= p_1(\theta_1', \lambda)/p_1(-, \lambda) \\ y' &= p_2(\theta_2', \lambda)/p_2(-, \lambda). \end{aligned}\right\} \qquad (12.14\text{–}23)$$

After multiplying each term in the inequality (12.14–22) by the expression $\alpha_1\alpha_2 p_1(-, \lambda)p_2(-, \lambda)\rho(\lambda)$, integrating over all λ, and making use of the definitions (12.14–20), (12.14–21), we arrive at the following Bell-type inequality in the Clauser–Horne form (1974):

$$S \equiv P_{12}(\theta_1, \theta_2) - P_{12}(\theta_1, \theta_2') + P_{12}(\theta_1', \theta_2) + P_{12}(\theta_1', \theta_2') - P_{12}(\theta_1', -) - P_{12}(-, \theta_2) \leqslant 0. \qquad (12.14\text{–}24)$$

Any local realistic theory is expected to obey this relation, which involves only readily measured quantities. As every term is proportional to α_1, α_2, we can divide through by $\alpha_1\alpha_2$ and obtain a relation that does not depend on the quantum efficiencies of the detectors.

We now show that the quantum mechanical equations (12.14–4) and (12.14–5) violate this inequality. For this purpose we again adopt the choice of angles

$$\begin{aligned} \theta_1 &= 0 \\ \theta_2 &= 3\pi/8 \\ \theta_1' &= -\pi/4 \\ \theta_2' &= \pi/8, \end{aligned}$$

and use Eqs. (12.14–4) and (12.14–5) to calculate the probabilities. We then obtain for S defined by Eq. (12.14–24)

$$S/\alpha_1\alpha_2 = \frac{\sqrt{2} - 1}{2} \approx 0.207. \qquad (12.14\text{–}25)$$

This is positive and clearly violates the inequality (12.14–24).

12.14.4 Experimental confirmation

Several experiments confirming such locality violations have been carried out (Freedman and Clauser, 1972; Clauser, 1976; Fry and Thompson, 1976; Aspect, Grangier and Roger, 1981, 1982; Aspect, Dalibard and Roger, 1982; Ou and Mandel, 1988; Shih and Alley, 1988). Figure 12.11 shows an outline of the experiment of Aspect, Dalibard and Roger (1982). The photon pair is derived from the two-photon cascade decay $J = 0 \rightarrow J = 1 \rightarrow J = 0$ of calcium atoms. C_I and C_{II} are accousto-optical switches that direct each photon to one of two different polarizers followed by a detector, depending on the setting of the switch. Switching occurs several times while the photons are traveling between the source and the detectors, so that the polarization direction is effectively chosen at random while the photon is on the way. Strong quantum correlations between the polarizations of the two photons were nevertheless observed, as predicted by Eq. (12.14–5), and the Bell inequality was found to be violated by five standard deviations.

Similar polarization correlations also occur in the process of two-photon down-conversion in a nonlinear crystal (see Section 22.4), when the polarizations of the two photons are arranged to be orthogonal and when the photons are mixed at a beam splitter. Figure 12.12 shows the outline of such an experiment. Down-converted pairs of photons, polarized orthogonally, are fed to the two input ports of a 50%:50% beam splitter, while linear polarizers oriented at angles θ_1 and θ_2 followed by detectors are placed near the two output ports. When the path lengths are equal, the joint detection probability $P_{12}(\theta_1, \theta_2)$ for each emitted photon pair is given by the formula

$$P_{12}(\theta_1, \theta_2) = \frac{\alpha_1 \alpha_2}{4} \sin^2(\theta_1 + \theta_2), \qquad (12.14\text{–}26)$$

which may be compared with Eq. (12.14–5). It is found that an inequality similar to Eq. (12.14–24), which must be satisfied by any local hidden variable theory, is violated by six standard deviations, and that classical wave optics is violated also. It follows that in this experiment the light cannot be described either by classical particles or by classical waves. Still larger violations have been observed in more recent experiments (Kiess, Shih, Sergienko and Alley, 1994).

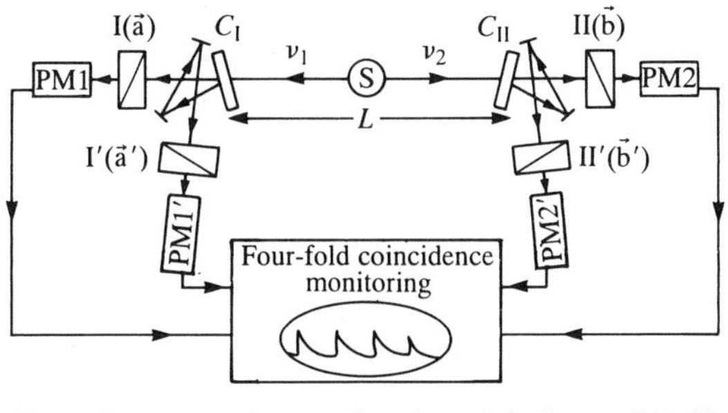

Fig. 12.11 Outline of an experiment showing violations of Bell's inequality in polarization correlation measurements. Optical switches C_I and C_{II} direct each photon into one of two possible paths through a polarizer to a detector. (Reproduced from Aspect, Dalibard and Roger, 1982.)

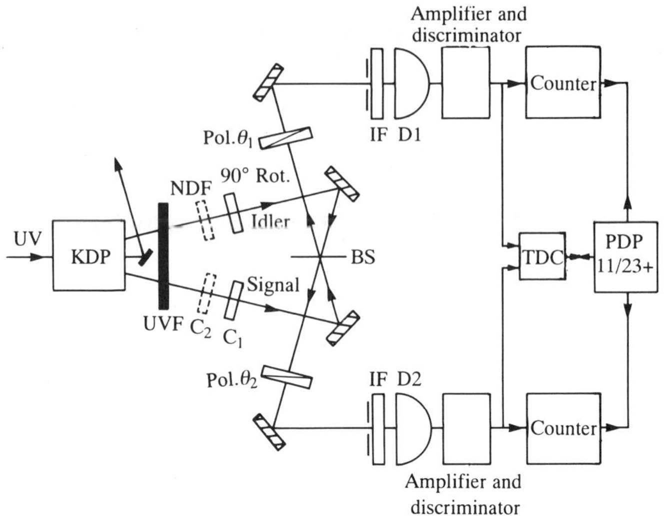

Fig. 12.12 Experimental set up for demonstrating the violation of locality in the measurement of polarization correlations between mixed, down-converted signal and idler photons. (Reproduced from Ou and Mandel, 1988.)

12.14.5 Non-classical states and Bell inequalities

Finally, let us briefly return to the integral relation (12.14–20) for the joint probability $P_{12}(\theta_1, \theta_2)$ of detecting the photons behind the polarizers θ_1, θ_2 with the two detectors, which was derived by applying the rules of classical probability to a local hidden variable theory. In order to understand what distinguishes this from the corresponding quantum mechanical expression, let us recall from Section 12.2 that quantum mechanically we can write (for perfect detectors)

$$P_{12}(\theta_1, \theta_2) = \langle{:}\hat{n}_1\hat{n}_2{:}\rangle = \mathrm{Tr}\,(\hat{a}_1^\dagger\hat{a}_2^\dagger\hat{a}_2\hat{a}_1\hat{\rho}), \qquad (12.14\text{–}27)$$

where $\hat{n}_1 = \hat{a}_1^\dagger\hat{a}_1$, $\hat{n}_2 = \hat{a}_2^\dagger\hat{a}_2$ are photon number operators for the fields behind the polarizers at the detectors, and $\hat{\rho}$ is the density operator. We now invoke the so-called optical equivalence theorem (cf. Section 11.8) for normally ordered operators, which allows us to express the expectation in Eq. (12.14–5) as an integral over phase space,

$$P_{12}(\theta_1, \theta_2) = \int |v_1|^2|v_2|^2\phi(v_1, v_2)\,\mathrm{d}^2v_1\,\mathrm{d}^2v_2. \qquad (12.14\text{–}28)$$

Here $\phi(v_1, v_2)$ is the diagonal coherent-state representation of $\hat{\rho}$. We may interpret $|v_1|^2$ and $|v_2|^2$ as measures of the probabilities (apart from a real positive factor) $\mathscr{P}_1(\theta_1|v_1)$ and $\mathscr{P}_2(\theta_2|v_2)$ respectively for photons to be detected behind the polarizers when the complex field amplitude has a given value. Then Eq. (12.14–28) has a formal structure rather like Eq. (12.14–20). But whereas $\rho(\lambda)$ is a probability density, we have seen that quantum mechanics admits of states for which $\phi(v_1, v_2)$ is not a true probability density. It is because of the existence of these non-classical states, for which $\phi(v_1, v_2)$ may be negative and very singular, that the quantum mechanical probabilities $P_{12}(\theta_1, \theta_2)$ are not constrained by the Bell inequality (12.14–24) and other similar inequalities (Reid and Walls, 1986; Su and Wódkiewicz, 1991).

Problems

12.1 Calculate the normalized, normally ordered intensity autocorrelation function $\lambda(\tau) = \langle:\Delta\hat{I}(t)\Delta\hat{I}(t+\tau):\rangle/\langle\hat{I}\rangle^2$ for a plane electromagnetic wave traveling in one direction which is in a Fock state $|\{n\}\rangle$. Take $\hat{I}(t)$ to be the photon density.

12.2 In the previous problem if the mode spacing $\delta\omega$ tends to zero and the total photon number $n \to \infty$ in such a way that the ratio $n_{\mathbf{k}s}/n \to \phi(\omega)\delta\omega$, where $\phi(\omega)$ is a continuous, normalized spectral density, find the limiting form of $\lambda(\tau)$.

12.3 Express the operator $\exp(\hat{n}_{\mathbf{k}s}x)$ in normal and in antinormal order.

12.4 Starting from expressions for the joint differential probabilities for photoelectric detections, derive a general formula for the probability density $P(\tau)$ that two successive photoelectric detections by a photodetector illuminated by a stationary light beam are separated by a time interval τ. Evaluate $P(\tau)$ for a randomly phased single-mode laser when the average photon counting rate is R.

12.5 A 45° beam splitter has input ports labeled 0, 1 and output ports labeled 2, 3. Let r, t be the amplitude reflectivity and transmissivity from port 0 and r', t' from port 1, respectively, and assume that $|r| = |r'|$, $|t| = |t'|$, $|r|^2 + |t|^2 = 1$, $r't^* + r^*t' = 0$. A single-mode field in the coherent state $|v\rangle$ enters at port 0, and a photon of the same frequency enters at port 1. Determine under what conditions the cross-correlation $\langle:\Delta\hat{n}_2\Delta\hat{n}_3:\rangle$ between the two outputs is negative.

12.6 Consider a single-mode electromagnetic field. With the help of the expression for the probability $P(m)$ of m photons, calculate the expectation of $\cos(\hat{n}\theta)$ in the coherent state $|v\rangle$.

12.7 Use the same procedure as in Problem 12.6 to evaluate the expectation $\langle v|(\hat{n}^2 + 5\hat{n} + 2)^{-1}|v\rangle$.

12.8 Show that for a single-mode field in a state with P-representation $\phi(v)$

$$\langle\hat{n}^3\rangle = \int \phi(v)[|v|^6 + 3|v|^4 + |v|^2]\,\mathrm{d}^2v.$$

12.9 Calculate the probability $p(n)$ for n photons in a single-mode optical field when the state:

(a) is an equally weighted superposition of coherent states $|v'\rangle$ and $|v''\rangle$;

(b) is an equally weighted incoherent mixture of coherent states $|v'\rangle$ and $|v''\rangle$;

(c) results from the interference of two electromagnetic fields in states $|v'\rangle$ and $|v''\rangle$.

13

Radiation from thermal equilibrium sources

We have now derived a number of general properties of a quantized electromagnatic field, and have encountered some useful formalisms for treating certain problems in quantum optics. We have introduced the correlation functions of the field, and we have seen in a general way how they are related to measurements. In selecting examples for illustration we have tended to focus our attention largely on certain idealized quantum states of the field, such as Fock states and coherent states. However, there exists an important class of optical fields with simple properties, the so-called thermal fields, which includes most fields commonly encountered in practice, that has not yet been discussed. These fields are produced by sources in thermal equilibrium, and they exhibit many features that can be treated almost exactly in our formalism. We now turn our attention to such fields.

13.1 Blackbody radiation

13.1.1 The density operator

Blackbody radiation is the name given to an electromagnetic field in thermal equilibrium with a large thermal reservoir or heat bath at some temperature T. By definition, such a field is assumed to be coupled to the heat bath, and it is therefore not a strictly free field in the sense of the previous chapters. However, the coupling can be as weak as we wish, and it is well known from the general theory of statistical thermodynamics that the properties of a system with many degrees of freedom in thermal equilibrium (described by a canonical ensemble) are often similar to those of an equivalent isolated system (described by a microcanonical ensemble). In general, a system of total energy $\hat{H}$, which is in thermal equilibrium at temperature T, has a density operator $\hat{\rho}$ given by the Boltzmann distribution

$$\hat{\rho} = \frac{\exp(-\hat{H}/k_B T)}{\operatorname{Tr}[\exp(-\hat{H}/k_B T)]}, \qquad (13.1\text{–}1)$$

where k_B is Boltzmann's constant ($k_B \approx 1.38 \times 10^{-23}$ joule/kelvin). Equation (13.1–1) describes the well-known canonical ensemble, in which the probability of any state varies exponentially with its energy, such that, for each degree of freedom, energies much in excess of $k_B T$ are very improbable.

Let us apply Eq. (13.1–1) to a quantized optical field contained in a cubical cavity of side L. Then from Eq. (10.3–15)

$$\hat{H} = \sum_{\mathbf{k},s}(\hat{n}_{\mathbf{k}s} + \tfrac{1}{2})\hbar\omega,$$

and the density operator of the field becomes

$$\hat{\rho} = \frac{\exp(-\sum_{\mathbf{k},s}\hat{n}_{\mathbf{k}s}\hbar\omega/k_{\mathrm{B}}T)}{\mathrm{Tr}[\exp(-\sum_{\mathbf{k},s}\hat{n}_{\mathbf{k}s}\hbar\omega/k_{\mathrm{B}}T)]}. \tag{13.1–2}$$

It is worth noting that the troublesome zero point energy has dropped out automatically in the normalization. The dimensionless ratio $\hbar\omega/k_{\mathrm{B}}T$ is encountered very frequently, and will be denoted by

$$\beta \equiv \hbar\omega/k_{\mathrm{B}}T. \tag{13.1–3}$$

On observing that

$$\mathrm{Tr}\, f(\hat{n}_{\mathbf{k}s}) = \sum_{n_{\mathbf{k}s}}\langle n_{\mathbf{k}s}|f(\hat{n}_{\mathbf{k}s})|n_{\mathbf{k}s}\rangle = \sum_{n_{\mathbf{k}s}} f(n_{\mathbf{k}s}),$$

so that

$$\mathrm{Tr}\left[\exp\left(\sum_{\mathbf{k},s}\hat{n}_{\mathbf{k}s}\beta\right)\right] = \mathrm{Tr}\prod_{\mathbf{k},s}(\mathrm{e}^{-\hat{n}_{\mathbf{k}s}\beta}) = \prod_{\mathbf{k},s}\sum_{n_{\mathbf{k}s}}\mathrm{e}^{-\beta n_{\mathbf{k}s}}$$
$$= \prod_{\mathbf{k},s}\left(\frac{1}{1-\mathrm{e}^{-\beta}}\right),$$

and substituting in Eq. (13.1–2), we obtain the formula

$$\hat{\rho} = \prod_{\mathbf{k},s}(1-\mathrm{e}^{-\beta})\,\mathrm{e}^{-\beta\hat{n}_{\mathbf{k}s}} \equiv \prod_{\mathbf{k},s}\hat{\rho}_{\mathbf{k}s}. \tag{13.1–4}$$

We are allowed to interchange the order of the Tr and $\prod_{\mathbf{k},s}$ operations because operators associated with different modes operate in different subspaces of the Hilbert space.

Equation (13.1–4) shows that the density operator factorizes into a product of density operators $\hat{\rho}_{\mathbf{k},s}$ for each mode, so that all modes of the electromagnetic field are statistically independent of each other. The fact that the only parameter in Eq. (13.1–4), β, is independent of the direction of $\mathbf{k}$ or of the polarization s for each $\mathbf{k}$, s mode, already indicates that the field is isotropic and unpolarized, as we show more explicitly below. Moreover, because $\hat{\rho}$ is a functional of the set of photon number operators $\hat{n}_{\mathbf{k},s}$, it describes a field that is both stationary and spatially homogeneous (cf. Section 12.8).

13.1.2 Photon statistics

From Eq. (13.1–4) $\hat{\rho}$ is clearly diagonal in the Fock representation $|\{n\}\rangle$. On multiplying each factor in Eq. (13.1–4) by unity in the form

$$1 = \sum_{n_{\mathbf{k}s}}|n_{\mathbf{k}s}\rangle\langle n_{\mathbf{k}s}|,$$

and making use of the fact that $|n_{\mathbf{k}s}\rangle$ is the eigenstate of $\hat{n}_{\mathbf{k}s}$, we have

$$\begin{aligned}\hat{\rho} &= \prod_{\mathbf{k},s}(1 - \mathrm{e}^{-\beta})\sum_{n_{\mathbf{k}s}} \mathrm{e}^{-\beta\hat{n}_{\mathbf{k}s}}|n_{\mathbf{k}s}\rangle\langle n_{\mathbf{k}s}| \\ &= \sum_{\{n\}}\prod_{\mathbf{k},s}(1 - \mathrm{e}^{-\beta})\,\mathrm{e}^{-\beta n_{\mathbf{k}s}}|n_{\mathbf{k}s}\rangle\langle n_{\mathbf{k}s}|. \qquad (13.1\text{–}5)\end{aligned}$$

The coefficient of $|n_{\mathbf{k}s}\rangle\langle n_{\mathbf{k}s}|$ is the probability $p(n_{\mathbf{k}s})$ for $n_{\mathbf{k}s}$ photons in the mode $\mathbf{k}, s$. We then have for the joint probability $p(\{n\})$ of the set of photon occupation numbers $\{n\}$,

$$p(\{n\}) = \prod_{\mathbf{k},s} p(n_{\mathbf{k}s}),$$

with

$$p(n_{\mathbf{k}s}) = (1 - \mathrm{e}^{-\beta})\,\mathrm{e}^{-\beta n_{\mathbf{k}s}}. \qquad (13.1\text{–}6)$$

The occupation numbers for different modes are statistically independent and they are distributed exponentially in $n_{\mathbf{k}s}$.

Having obtained the joint probability $p(\{n\})$, we can readily calculate all the moments of $\hat{n}_{\mathbf{k}s}$. Instead of determining the moments individually, we can more easily determine them all at once from the factorial moment generating function

$$F(\xi) \equiv \langle (1 + \xi)^{\hat{n}_{\mathbf{k}s}}\rangle = \sum_{r=0}^{\infty}\frac{\xi^r}{r!}\langle \hat{n}_{\mathbf{k}s}^{(r)}\rangle.$$

The coefficient of $\xi^r/r!$ in the power series expansion of $F(\xi)$ is the r'th factorial moment

$$\langle \hat{n}_{\mathbf{k}s}^{(r)}\rangle \equiv \langle \hat{n}_{\mathbf{k}s}(\hat{n}_{\mathbf{k}s} - 1)\ldots(\hat{n}_{\mathbf{k}s} - r + 1)\rangle.$$

With the help of Eq. (13.1–6) we find immediately that

$$\begin{aligned}F(\xi) &= \sum_{n_{\mathbf{k}s}=0}^{\infty}(1 + \xi)^{n_{\mathbf{k}s}}(1 - \mathrm{e}^{-\beta})\,\mathrm{e}^{-\beta n_{\mathbf{k}s}} \\ &= \frac{1 - \mathrm{e}^{-\beta}}{1 - (1 + \xi)\,\mathrm{e}^{-\beta}} \\ &= \frac{1}{1 - \xi/(\mathrm{e}^{\beta} - 1)} \\ &= 1 + \frac{\xi}{\mathrm{e}^{\beta} - 1} + \frac{\xi^2}{(\mathrm{e}^{\beta} - 1)^2} + \ldots\end{aligned}$$

and it follows from this by inspection that the coefficient of $\xi^r/r!$ is

$$\langle \hat{n}_{\mathbf{k}s}^{(r)}\rangle = \frac{r!}{(\mathrm{e}^{\beta} - 1)^r}. \qquad (13.1\text{–}7)$$

Hence by putting $r = 1$, we obtain for the mean photon number of mode $\mathbf{k}, s$

$$\langle \hat{n}_{\mathbf{k}s}\rangle = \frac{1}{\mathrm{e}^{\beta} - 1} = \frac{1}{\mathrm{e}^{\hbar\omega/k_{\mathrm{B}}T} - 1}. \qquad (13.1\text{–}8)$$

When $r = 2$, we have

$$\langle \hat{n}_{\mathbf{k}s}^{(2)} \rangle = 2\langle \hat{n}_{\mathbf{k}s} \rangle^2,$$

so that

$$\langle \hat{n}_{\mathbf{k}s}^2 \rangle = \langle \hat{n}_{\mathbf{k}s} \rangle + 2\langle \hat{n}_{\mathbf{k}s} \rangle^2$$

and

$$\langle (\Delta \hat{n}_{\mathbf{k}s})^2 \rangle = \langle \hat{n}_{\mathbf{k}s}^2 \rangle - \langle \hat{n}_{\mathbf{k}s} \rangle^2 = \langle \hat{n}_{\mathbf{k}s} \rangle (1 + \langle \hat{n}_{\mathbf{k}s} \rangle). \tag{13.1–9}$$

By making use of Eq. (13.1–8) we can replace e^{β} by $1 + 1/\langle n_{\mathbf{k}s} \rangle$ in Eq. (13.1–6), and express the probability $p(\{n\})$ directly in terms of the average occupation number $\langle n_{\mathbf{k}s} \rangle = \langle \hat{n}_{\mathbf{k}s} \rangle$. We then obtain the result

$$p(\{n\}) = \prod_{\mathbf{k},s} \frac{1}{[1 + \langle n_{\mathbf{k}s} \rangle][1 + 1/\langle n_{\mathbf{k}s} \rangle]^{n_{\mathbf{k}s}}}, \tag{13.1–10}$$

which has the standard form of a product of Bose–Einstein distributions (see Section 1.5.3). Thermal equilibrium photons therefore obey Bose–Einstein statistics.

Equation (13.1–8) shows that the average occupation number of each mode depends only on the frequency ω and on the temperature T, and that all modes associated with different directions of the wave vector $\mathbf{k}$ and with different polarizations are equally occupied on the average. Note also that the size of the quantization volume L^3 does not enter into the average number $\langle n_{\mathbf{k}s} \rangle$. Although the average total photon number $\sum_{\mathbf{k},s} \langle n_{\mathbf{k}s} \rangle$ is proportional to L^3, the density of modes varies as $1/L^3$, so that $\langle n_{\mathbf{k}s} \rangle$ is independent of geometry. It is not difficult to make an order of magnitude estimate of $\langle n_{\mathbf{k}s} \rangle$ in the optical domain for a typical radiation temperature T. Since optical frequencies cover a relatively small range, $\hbar\omega$ is always in the range 1.5 to 2.5 eV. At a temperature of $T \sim 3000$ K, which is characteristic of the radiation produced by incandescent lamps, $k_B T \sim 0.3$ eV, and therefore

$$\langle n_{\mathbf{k}s} \rangle \sim \frac{1}{e^{2/0.3} - 1} \sim 10^{-3},$$

whereas at the surface of the sun, where $T \sim 5000$ K and $k_B T \sim 0.5$ eV,

$$\langle n_{\mathbf{k}s} \rangle \sim 10^{-2}.$$

The corresponding numbers are of course smaller at lower temperatures. We see that the average photon occupation numbers are very small for the thermal optical fields usually encountered in practice, and that most of the modes are generally empty. Since the different modes are independent, the optical photons produced by familiar thermal equilibrium sources are also independent to a first approximation. This is also indicated by Eq. (13.1–9), which gives $\langle (\Delta n_{\mathbf{k}s})^2 \rangle \sim \langle n_{\mathbf{k}s} \rangle$ under these conditions, which is characteristic of Poisson statistics. Nevertheless, the small departures from strict Poisson statistics can lead to important observable effects, as we shall see in Chapter 14. Of course, the situation changes at sufficiently high temperatures, when $\langle n_{\mathbf{k}s} \rangle$ becomes proportional to T, as can be seen from Eq. (13.1–8). However, extraordinarily high temperatures, such as those encountered in thermonuclear fusion reactions, are needed before the optical photon occupation numbers become large. The number $\langle n_{\mathbf{k}s}\rangle$ is still only of order unity at optical wavelengths at a temperature of about 30 000 K.

13.1.3 Polarization

Let us examine the state of polarization of the field in a little more detail. For any plane wave with wave vector **k** we can define a Hermitian 2 × 2 coherence (or polarization) matrix **J**(**k**) (cf. Section 6.2) by

$$\mathbf{J}(\mathbf{k}) = \begin{bmatrix} \langle \hat{a}^\dagger_{\mathbf{k}1}\hat{a}_{\mathbf{k}1}\rangle & \langle \hat{a}^\dagger_{\mathbf{k}1}\hat{a}_{\mathbf{k}2}\rangle \\ \langle \hat{a}^\dagger_{\mathbf{k}2}\hat{a}_{\mathbf{k}1}\rangle & \langle \hat{a}^\dagger_{\mathbf{k}2}\hat{a}_{\mathbf{k}2}\rangle \end{bmatrix},$$

which describes the polarization and is an obvious extension of the usual definition to a quantum field. By virtue of the factorization of the density operator $\hat{\rho}$ into factors corresponding to the constituent modes, it follows that the modes are all independent and that

$$\langle \hat{a}^\dagger_{\mathbf{k}1}\hat{a}_{\mathbf{k}2}\rangle = \langle \hat{a}^\dagger_{\mathbf{k}1}\rangle\langle \hat{a}_{\mathbf{k}2}\rangle$$
$$\langle \hat{a}^\dagger_{\mathbf{k}2}\hat{a}_{\mathbf{k}1}\rangle = \langle \hat{a}^\dagger_{\mathbf{k}2}\rangle\langle \hat{a}_{\mathbf{k}1}\rangle,$$

and since $\hat{\rho}$ is diagonal in $\hat{n}_{\mathbf{k}s}$,

$$\langle \hat{a}_{\mathbf{k}1}\rangle = 0 = \langle \hat{a}_{\mathbf{k}2}\rangle \tag{13.1–11}$$

irrespective of the nature of the polarization basis.

From Eq. (13.1–11) the expectation value of the positive and negative frequency parts of the electric field, the magnetic field, the vector potential, or of any homogeneous linear transformation of these fields, must vanish for blackbody radiation,

$$\left.\begin{aligned} \langle \hat{\mathbf{E}}^{(+)}(\mathbf{r}, t)\rangle &= 0 \\ \langle \hat{\mathbf{B}}^{(+)}(\mathbf{r}, t)\rangle &= 0 \\ \langle \hat{\mathbf{A}}^{(+)}(\mathbf{r}, t)\rangle &= 0. \end{aligned}\right\} \tag{13.1–12}$$

With the help of Eq. (13.1–8), we then have

$$\langle \hat{a}^\dagger_{\mathbf{k}s}\hat{a}_{\mathbf{k}s'}\rangle = \delta_{ss'}\left(\frac{1}{e^\beta - 1}\right), \tag{13.1–13}$$

so that the matrix **J**(**k**) is diagonal and reduces to the unit matrix

$$\mathbf{J}(\mathbf{k}) = \left(\frac{1}{e^\beta - 1}\right)\begin{bmatrix} 1 & 0 \\ 0 & 1 \end{bmatrix}. \tag{13.1–14}$$

As we have seen in Section 6.3.1, this form of **J** is characteristic of unpolarized light, and it is independent of the polarization basis adopted for the representation. Each **k** component of blackbody radiation is therefore unpolarized.

Next let us consider the total field. Let $\hat{\mathbf{F}}^{(+)}(\mathbf{r}, t)$ be the positive frequency part of any one of the electromagnetic field vectors like $\hat{\mathbf{E}}$, $\hat{\mathbf{A}}$, $\hat{\mathbf{B}}$ having the plane-wave mode expansion

$$\hat{\mathbf{F}}^{(+)}(\mathbf{r}, t) = \frac{1}{L^{3/2}}\sum_{\mathbf{k},s} l(\omega)\hat{a}_{\mathbf{k}s}\boldsymbol{\varepsilon}_{\mathbf{k}s}\, \mathrm{e}^{\mathrm{i}(\mathbf{k}\cdot\mathbf{r}-\omega t)}.$$

Then we have for the correlation between Cartesian components,

$$\langle \hat{F}_i^{(-)}(\mathbf{r}, t)\hat{F}_j^{(+)}(\mathbf{r}, t)\rangle =$$

$$\frac{1}{L^3}\sum_{\mathbf{k},s}\sum_{\mathbf{k}',s'} l^*(\omega)l(\omega')\langle \hat{a}^\dagger_{\mathbf{k}s}\hat{a}_{\mathbf{k}'s'}\rangle(\boldsymbol{\varepsilon}^*_{\mathbf{k}s})_i(\boldsymbol{\varepsilon}_{\mathbf{k}'s'})_j\, \mathrm{e}^{\mathrm{i}[(\mathbf{k}'-\mathbf{k})\cdot\mathbf{r}-(\omega'-\omega)t)]}.$$

In view of the independence of the different plane-wave modes, with the help of Eqs. (13.1–11) and (13.1–13) we have

$$\langle \hat{a}^\dagger_{\mathbf{k}s}\hat{a}_{\mathbf{k}'s'}\rangle = \delta^3_{\mathbf{k}\mathbf{k}'}\delta_{ss'}\left(\frac{1}{\mathrm{e}^\beta - 1}\right),$$

so that

$$\langle \hat{F}_i^{(-)}(\mathbf{r}, t)\hat{F}_j^{(+)}(\mathbf{r}, t)\rangle = \frac{1}{L^3}\sum_{\mathbf{k},s}\frac{|l(\omega)|^2}{(\mathrm{e}^\beta - 1)}(\boldsymbol{\varepsilon}^*_{\mathbf{k}s})_i(\boldsymbol{\varepsilon}_{\mathbf{k}'s'})_j$$

$$\Rightarrow \frac{1}{(2\pi)^3}\int\frac{|l(\omega)|^2}{(\mathrm{e}^\beta - 1)}\left(\delta_{ij} - \frac{k_i k_j}{k^2}\right)\mathrm{d}^3k,$$

after we make use of the tensor expansion Eq. (10.2–19c) and go to the continuum limit. If $i = j$ the answer is independent of i by symmetry. If $i \neq j$ $\delta_{ij} = 0$, and the term $k_i k_j$ integrates to zero by symmetry. It follows that the coherence or polarization matrix $\langle \hat{F}_i^{(-)}\hat{F}_j^{(+)}\rangle$ is again proportional to the unit matrix, as in Eq. (13.1–14).

13.1.4 Spectral distributions

With the help of Eq. (13.1–8) we can derive a spectral distribution function $\Phi(\omega)$, such that $\Phi(\omega)\,\mathrm{d}\omega$ gives the density of photons within the frequency interval $\mathrm{d}\omega$. By making use of the density of modes $(L/2\pi)^3$, we obtain the following equation for the density of photons in the differential interval d^3k,

$$\begin{array}{r}\text{Density of photons within the}\\ \text{differential interval } \mathrm{d}^3k\end{array} = \left(\frac{\langle n_{\mathbf{k}1}\rangle + \langle n_{\mathbf{k}2}\rangle}{L^3}\right)\left(\frac{L}{2\pi}\right)^3\mathrm{d}^3k$$

$$= \frac{1}{4\pi^3(\mathrm{e}^{\hbar\omega/k_\mathrm{B}T} - 1)}\mathrm{d}^3k,$$

and after putting $\mathrm{d}^3k = (\omega^2/c^3)\,\mathrm{d}\omega\,\mathrm{d}\Omega$, where $\mathrm{d}\Omega$ is an element of solid angle, and integrating over all directions, we have immediately

$$\Phi(\omega)\,\mathrm{d}\omega = \frac{\omega^2}{\pi^2c^3(\mathrm{e}^{\hbar\omega/k_\mathrm{B}T} - 1)}\mathrm{d}\omega. \tag{13.1–15}$$

Alternatively, since each photon of frequency ω carries an energy $\hbar\omega$, we can write a corresponding equation for the energy density $u(\omega)\,d\omega$ within the frequency interval $d\omega$,

$$u(\omega)\,d\omega = \hbar\omega\Phi(\omega)\,d\omega = \frac{\hbar\omega^3}{\pi^2 c^3(e^{\hbar\omega/k_B T} - 1)}\,d\omega. \qquad (13.1\text{–}16)$$

This energy distribution is known as Planck's distribution law, and is illustrated in Fig. 13.1 as a function of wavelength. It was first derived by Planck on the basis of empirical data, and was then used by him to establish the energy quantization of the electromagnetic field (Planck, 1901a,b). For small values of ω the distribution $u(\omega)$ is proportional to ω^2, which is characteristic of Rayleigh's law, but it reaches a maximum for a frequency ω of order $3k_B T/\hbar$, and then falls back towards zero. The position of the maximum is given by Wien's displacement law. Finally, by integrating over all frequencies, we arrive at the total energy density u of the electromagnetic field at temperature T,

$$\begin{aligned} u &= \int_0^\infty \frac{\hbar\omega^3}{\pi^2 c^3(e^{\hbar\omega/k_B T} - 1)}\,d\omega \\ &= \frac{(k_B T)^4}{\pi^2 c^3 \hbar^3}\int_0^\infty \frac{x^3}{e^x - 1}\,dx \\ &= \frac{(k_B T)^4}{\pi^2 c^3 \hbar^3} 3!\,\zeta(4), \end{aligned} \qquad (13.1\text{–}17a)$$

where

$$\zeta(r) \equiv \frac{1}{\Gamma(r)}\int_0^\infty \frac{x^{r-1}}{e^x - 1}\,dx = \sum_{n=1}^\infty \frac{1}{n^r} \qquad (r > 1)$$

is the Riemann zeta function. It can be shown that $\zeta(4) = \pi^4/90$. The fourth power law dependence of u on temperature T given by Eq. (13.1–17a) is known as the *Stefan–Boltzmann law*.

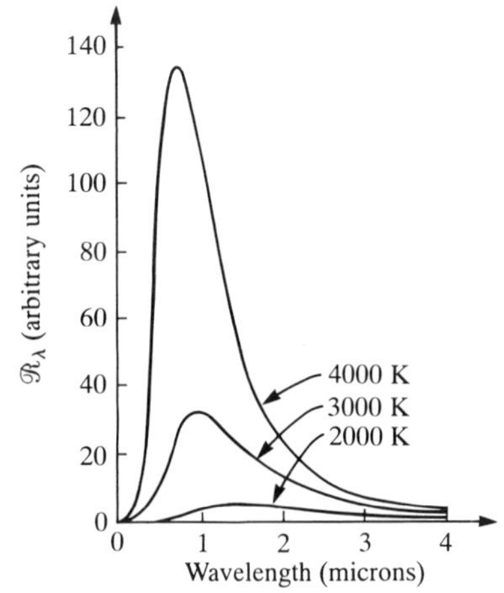

Fig. 13.1 The distribution of radiant energy with wavelength according to Planck's law at three different temperatures. (Reproduced from Halliday and Resnick, 1970, p. 759.)

In a completely analogous manner, we may integrate $\Phi(\omega)$ given by Eq. (13.1–15) over all frequencies to obtain the average density of photons at temperature T, and we find

$$\text{mean photon density} = \int_0^\infty \Phi(\omega)\,\mathrm{d}\omega = \frac{(k_B T)^3}{\pi^2 c^3 \hbar^3} 2\zeta(3). \qquad (13.1\text{–}17\text{b})$$

As we show below, the range of field correlations in blackbody radiation is always of order $\hbar c/k_B T$, which makes $(\hbar c/k_B T)^3$ of the order of the coherence volume. Equation (13.1–17b) then expresses the interesting result that the average number of photons within a coherence volume of blackbody radiation is always of order unity (Mandel, 1979).

13.1.5 The diagonal coherent-state representation of $\hat{\rho}$

Although we have found the density operator $\hat{\rho}$ of the field for blackbody radiation, which is diagonal in the Fock representation, we recall from Section 11.8 that it should also be possible to cast the same density operator into a diagonal form in the coherent-state representation. Because of the general usefulness and power of this representation, we now turn to the problem of expressing $\hat{\rho}$ in terms of coherent states.

In Section 11.8 we encountered some general procedures for deriving the diagonal representation $\phi(v)$ for a single-mode optical field. In particular, we saw from Eq. (11.8–11) that $\phi(v)$ can be derived by Fourier inversion of the matrix element $\exp(|u|^2)\langle -u|\hat{\rho}|u\rangle$, where u is a complex number and both $|u\rangle$ and $|-u\rangle$ are coherent states. By an obvious generalization to a multimode field, we have the following integral representation of the density operator,

$$\hat{\rho} = \int \phi(\{v\})|\{v\}\rangle\langle\{v\}|\,\mathrm{d}\{v\}, \qquad (13.1\text{–}18\text{a})$$

with the phase-space functional given by

$$\phi(\{v\}) = \int \langle -\{u\}|\hat{\rho}|\{u\}\rangle \prod_{\mathbf{k},s}\left[\mathrm{e}^{|v_{\mathbf{k}s}|^2+|u_{\mathbf{k}s}|^2}\,\mathrm{e}^{u^*_{\mathbf{k}s}v_{\mathbf{k}s}-u_{\mathbf{k}s}v^*_{\mathbf{k}s}}\frac{\mathrm{d}^2u_{\mathbf{k}s}}{\pi^2}\right]. \qquad (13.1\text{–}18\text{b})$$

We can make use of Eqs. (13.1–5) and (13.1–10) to evaluate the matrix element and we find that

$$\begin{aligned}
\langle -\{u\}|\hat{\rho}|\{u\}\rangle &= \prod_{\mathbf{k},s}\sum_{n_{\mathbf{k}s}} \frac{1}{[1+\langle n_{\mathbf{k}s}\rangle][1+1/\langle n_{\mathbf{k}s}\rangle]^{n_{\mathbf{k}s}}}\langle -u_{\mathbf{k}s}|n_{\mathbf{k}s}\rangle\langle n_{\mathbf{k}s}|u_{\mathbf{k}s}\rangle \\
&= \prod_{\mathbf{k},s}\sum_{n_{\mathbf{k}s}} \frac{1}{[1+\langle n_{\mathbf{k}s}\rangle][1+1/\langle n_{\mathbf{k}s}\rangle]^{n_{\mathbf{k}s}}}\frac{(-|u_{\mathbf{k}s}|^2)^{n_{\mathbf{k}s}}\,\mathrm{e}^{-|u_{\mathbf{k}s}|^2}}{n_{\mathbf{k}s}!} \\
&= \prod_{\mathbf{k},s}\frac{\exp\left[-|u_{\mathbf{k}s}|^2\left(1+\dfrac{1}{1+1/\langle n_{\mathbf{k}s}\rangle}\right)\right]}{[1+\langle n_{\mathbf{k}s}\rangle]},
\end{aligned} \qquad (13.1\text{–}19)$$

where we have introduced Eq. (11.2–11) for the scalar product of the Fock state with the coherent state. We now use Eq. (13.1–19) in the integral in Eq. (13.1–18b) and obtain the formula

$$\phi(\{v\}) = \prod_{\mathbf{k},s} \frac{e^{|v_{\mathbf{k}s}|^2}}{[1+\langle n_{\mathbf{k}s}\rangle]} \int \exp\left(\frac{-|u_{\mathbf{k}s}|^2}{1+1/\langle n_{\mathbf{k}s}\rangle}\right) e^{u^*_{\mathbf{k}s}v_{\mathbf{k}s}-u_{\mathbf{k}s}v^*_{\mathbf{k}s}} \frac{d^2u_{\mathbf{k}s}}{\pi^2}.$$

The integral is the characteristic function (or the two-dimensional Fourier transform) of a complex Gaussian variate $u_{\mathbf{k}s}$, and from Section 1.5.5 we have, after carrying out the integration,

$$\phi(\{v\}) = \prod_{\mathbf{k},s} \frac{e^{|v_{\mathbf{k}s}|^2}}{[1+\langle n_{\mathbf{k}s}\rangle]} \frac{[1+1/\langle n_{\mathbf{k}s}\rangle]}{\pi} \exp[-|v_{\mathbf{k}s}|^2(1+1/\langle n_{\mathbf{k}s}\rangle)]$$

$$= \prod_{\mathbf{k},s} \frac{1}{\pi\langle n_{\mathbf{k}s}\rangle} e^{-|v_{\mathbf{k}s}|^2/\langle n_{\mathbf{k}s}\rangle}. \tag{13.1–20}$$

The average occupation number $\langle n_{\mathbf{k}s}\rangle$ is, of course, given by Eq. (13.1–8). Hence the phase space functional is simply a product of Gaussian functions in each of the complex variables $v_{\mathbf{k}s}$. Equation (13.1–20) is another remarkable illustration of the power of the diagonal coherent-state representation, for the states $|\{n\}\rangle$ and $|\{v\}\rangle$ are eigenstates of non-commuting operators, and yet $\hat{\rho}$ is diagonal in both $|\{n\}\rangle$ and $|\{v\}\rangle$, with well-behaved representations in both.

It is worth noting also that, in addition to being well behaved mathematically, $\phi(\{v\})$ has the same form as the probability functional that one would use to describe a blackbody radiation field classically. Since many independent radiators contribute to the total field and to the excitation of each mode, the distribution of the complex amplitude $v_{\mathbf{k}s}$ for each mode is expected to be Gaussian by the central limit theorem. Once the spectral distribution is given, blackbody radiation also has a good classical description, no matter how small the photon occupation numbers may be, despite the fact that $\hat{\rho}$ is diagonal in Fock states and that these states are non-classical.

13.1.6 Correlation functions of blackbody radiation

We can use the density operator to evaluate certain second-order correlation functions of the optical field. For historical reasons, we concentrate on the correlations associated with the electric and magnetic fields, for which the positive frequency parts of the field operators are given by the first terms in Eqs. (10.4–39) and (10.4–40),

$$\hat{\mathbf{E}}^{(+)}(\mathbf{r}, t) = \frac{i}{L^{3/2}} \sum_{\mathbf{k},s} \left(\frac{\hbar\omega}{2\varepsilon_0}\right)^{1/2} \hat{a}_{\mathbf{k}s} \boldsymbol{\varepsilon}_{\mathbf{k}s} e^{i(\mathbf{k}\cdot\mathbf{r}-\omega t)} \tag{13.1–21}$$

$$\hat{\mathbf{B}}^{(+)}(\mathbf{r}, t) = \frac{i}{cL^{3/2}} \sum_{\mathbf{k},s} \left(\frac{\hbar\omega}{2\varepsilon_0}\right)^{1/2} \hat{a}_{\mathbf{k}s}(\boldsymbol{\kappa}\times\boldsymbol{\varepsilon}_{\mathbf{k}s}) e^{i(\mathbf{k}\cdot\mathbf{r}-\omega t)}, \tag{13.1–22}$$

with $\boldsymbol{\kappa} \equiv \mathbf{k}/k$. Hence the second-order normally ordered correlation tensor of the electric field is given by

$$\begin{aligned}\mathscr{E}_{ij}(\mathbf{r}_1, t_1; \mathbf{r}_2, t_2) &\equiv \langle \hat{E}_i^{(-)}(\mathbf{r}_1, t_1)\hat{E}_j^{(+)}(\mathbf{r}_2, t_2)\rangle \\ &= \frac{1}{L^3}\sum_{\mathbf{k},s}\sum_{\mathbf{k}',s'}\left(\frac{\hbar\omega}{2\varepsilon_0}\right)^{1/2}\left(\frac{\hbar\omega'}{2\varepsilon_0}\right)^{1/2}\langle \hat{a}_{\mathbf{k}s}^{\dagger}\hat{a}_{\mathbf{k}'s'}\rangle(\boldsymbol{\varepsilon}_{\mathbf{k}s}^{*})_i(\boldsymbol{\varepsilon}_{\mathbf{k}'s'})_j \\ &\quad\times \mathrm{e}^{\mathrm{i}(\mathbf{k}'\cdot\mathbf{r}_2-\mathbf{k}\cdot\mathbf{r}_1-\omega' t_2+\omega t_1)}.\end{aligned}$$

As all modes are independent, and $\langle \hat{a}_{\mathbf{k}s}\rangle = 0$ according to Eq. (13.1–11), this reduces with the help of Eq. (13.1–13) to (Bourret, 1960; Sarfatt, 1963; Mehta and Wolf, 1964 a,b)

$$\mathscr{E}_{ij}(\mathbf{r}_1, t_1; \mathbf{r}_2, t_2) = \frac{1}{L^3}\sum_{\mathbf{k},s}\left(\frac{\hbar\omega}{2\varepsilon_0}\right)\frac{1}{(\mathrm{e}^{\hbar\omega/k_\mathrm{B}T}-1)}(\boldsymbol{\varepsilon}_{\mathbf{k}s}^{*})_i(\boldsymbol{\varepsilon}_{\mathbf{k}s})_j\,\mathrm{e}^{\mathrm{i}[\mathbf{k}\cdot(\mathbf{r}_2-\mathbf{r}_1)-\omega(t_2-t_1)]}.$$

On making use of the tensor relation (10.8–2), and replacing the sum over $\mathbf{k}$ by an integral according to Eq. (10.8–4), we obtain

$$\mathscr{E}_{ij}(\mathbf{r}_1, t_1; \mathbf{r}_2, t_2) = \frac{1}{(2\pi)^3}\left(\frac{\hbar}{2\varepsilon_0}\right)\int\frac{\omega}{(\mathrm{e}^{\hbar\omega/k_\mathrm{B}T}-1)}\left(\delta_{ij}-\frac{k_ik_j}{k^2}\right)\mathrm{e}^{\mathrm{i}[\mathbf{k}\cdot(\mathbf{r}_2-\mathbf{r}_1)-\omega(t_2-t_1)]}\,\mathrm{d}^3k. \tag{13.1–23}$$

We observe that the correlation function depends only on the difference between both the space and the time arguments, which shows that the field is both spatially homogeneous and stationary, at least in the wide sense. As we show in Section 13.1.8, the integral in Eq. (13.1–23) also has the necessary symmetry properties for the field to be isotropic (cf. Section 12.8). Similarly, it follows from Eq. (13.1–22) for the magnetic field that

$$\begin{aligned}&\mathscr{B}_{ij}(\mathbf{r}_1, t_1; \mathbf{r}_2, t_2) \\ &\quad\equiv \langle \hat{B}_i^{(-)}(\mathbf{r}_1, t_1)\hat{B}_j^{(+)}(\mathbf{r}_2, t_2)\rangle \\ &\quad= \frac{1}{(2\pi)^3}\left(\frac{\hbar}{2\varepsilon_0c^2}\right)\int\frac{\omega}{(\mathrm{e}^{\hbar\omega/k_\mathrm{B}T}-1)}\left(\delta_{ij}-\frac{k_ik_j}{k^2}\right)\mathrm{e}^{\mathrm{i}[\mathbf{k}\cdot(\mathbf{r}_2-\mathbf{r}_1)-\omega(t_2-t_1)]}\,\mathrm{d}^3k,\end{aligned} \tag{13.1–24}$$

and for the mixed correlation tensor

$$\begin{aligned}\mathscr{G}_{ij}(\mathbf{r}_1, t_1; \mathbf{r}_2, t_2) &\equiv \langle \hat{E}_i^{(-)}(\mathbf{r}_1, t_1)\hat{B}_j^{(+)}(\mathbf{r}_2, t_2)\rangle \\ &= \frac{\epsilon_{ijl}}{(2\pi)^3}\left(\frac{\hbar}{2\varepsilon_0c}\right)\int\frac{k_l}{(\mathrm{e}^{\hbar\omega/k_\mathrm{B}T}-1)}\mathrm{e}^{\mathrm{i}[\mathbf{k}\cdot(\mathbf{r}_2-\mathbf{r}_1)-\omega(t_2-t_1)]}\,\mathrm{d}^3k.\end{aligned} \tag{13.1–25}$$

The integrals have not been evaluated in closed form, although they can be developed into infinite series. The behavior of $\mathscr{E}_{ij}$ and $\mathscr{B}_{ij}$ is illustrated in Fig. 13.2, where some of the properties of the normalized correlation tensor

$$\gamma_{ij}(\mathbf{r}, \tau) \equiv \frac{\mathscr{E}_{ij}(\mathbf{r}_1, t_1; \mathbf{r}_1+\mathbf{r}, t_1+\tau)}{\underset{\text{(no summation)}}{\mathscr{E}_{ii}(\mathbf{r}_1, t_1; \mathbf{r}_1, t_1)}} = \frac{\mathscr{B}_{ij}(\mathbf{r}_1, t_1; \mathbf{r}_1+\mathbf{r}, t_1+\tau)}{\underset{\text{(no summation)}}{\mathscr{B}_{ii}(\mathbf{r}_1, t_1; \mathbf{r}_1, t_1)}} \tag{13.1–26}$$

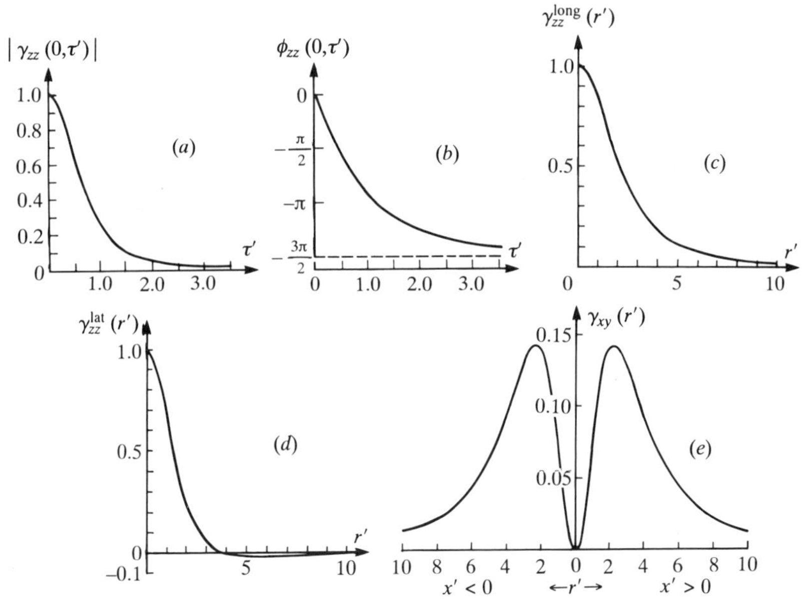

Fig. 13.2 Normalized correlation functions of electric or magnetic fields of blackbody radiation. (*a*) The modulus and (*b*) the phase of a typical diagonal component as a function of time $\tau' = (k_B T/\hbar)\tau$. (*c*) The variation of γ_{zz} with distance in the direction of the z-axis, with $r' = (\pi k_B T/\hbar c)z$. (*d*) The variation of γ_{zz} with distance in the direction of the x-axis, with $r' = (\pi k_B T/\hbar c)x$. (*e*) The variation of γ_{xy} with distance in the direction $x = y$ in the z-plane, with $r' = (\pi k_B T/\hbar c)(x^2 + y^2)^{1/2}$. [Figs. (*a*), (*b*) reproduced from Kano and Wolf, 1962. Figs. (*c*), (*d*), (*e*), reproduced from Mehta and Wolf, 1964a.]

are shown as functions of their arguments. In time $|\gamma_{ij}(\mathbf{r}, \tau)|$ has a range of order $\hbar/k_B T$, and in distance it has a range of order $\hbar c/k_B T$, corresponding roughly to the period and to the wavelength at which the Planck spectral distribution has a maximum. But we note that orthogonal components of $\hat{\mathbf{E}}^{(+)}$ and orthogonal components of $\hat{\mathbf{B}}^{(+)}$ at the same space-time point are uncorrelated, as both integrals in Eq. (13.1–23) and Eq. (13.1–24) then become proportional to δ_{ij}. Indeed, when the space-time points coincide, $\mathscr{E}_{ij}$ and $\mathscr{B}_{ij}$ become independent of both position and time arguments and reduce to integrals that can be evaluated, and give

$$\left.\begin{aligned} \mathscr{E}_{ij} &= \delta_{ij}\frac{\hbar}{3\varepsilon_0}\frac{1}{(2\pi)^3}\int\frac{\omega}{(e^{\hbar\omega/k_B T} - 1)}\,\mathrm{d}^3k \\ &= \delta_{ij}\frac{\pi^2 k_B^4 T^4}{90\varepsilon_0\hbar^3 c^3} = \tfrac{1}{3}\langle \hat{I}_E\rangle\delta_{ij} \\ \text{and} \qquad & \\ \mathscr{B}_{ij} &= \delta_{ij}\frac{\pi^2 k_B^4 T^4}{90\varepsilon_0\hbar^3 c^5} = \tfrac{1}{3}\langle \hat{I}_B\rangle\delta_{ij} \end{aligned}\right\} \qquad (13.1\text{–}27)$$

where $\langle \hat{I}_E\rangle$ and $\langle \hat{I}_B\rangle$ are the average intensities associated with the electric and magnetic fields.

Figure 13.3 illustrates some features of the normalized mixed correlation tensor (no summation on i, j)

$$\sigma_{ij}(\mathbf{r}, \tau) \equiv \frac{\mathcal{G}_{ij}(\mathbf{r}_1, t_1; \mathbf{r}_1 + \mathbf{r}, t_1 + \tau)}{[\mathcal{E}_{ii}(\mathbf{r}_1, t_1; \mathbf{r}_1, t_1)\mathcal{B}_{jj}(\mathbf{r}_1 + \mathbf{r}, t_1 + \tau; \mathbf{r}_1 + \mathbf{r}, t_1 + \tau)]^{1/2}}. \quad (13.1\text{–}28)$$

Because of the ϵ_{ijl} factor in Eq. (13.1–25), the diagonal components of $\sigma_{ij}(\mathbf{r}, \tau)$ all vanish, which means that parallel electric and magnetic field components are always uncorrelated. Moreover, when $\mathbf{r} = 0$, or $\mathbf{r}_2 = \mathbf{r}_1$ in Eq. (13.1–25), we see immediately that the three-dimensional $\mathbf{k}$-integral vanishes by symmetry. Hence, there is no temporal coherence between electric and magnetic field components at the same point in space.

13.1.7 Higher-order correlations

Because of the Gaussian form of the phase space functional $\phi(\{v\})$ given by Eq. (13.1–20), we can easily express all higher-order correlations in normal order in terms of second-order ones. Thus, for the correlation of order (N, M) we have, with the help of the optical equivalence theorem [Eq. (11.9–4)],

$$\begin{aligned}\Gamma^{(N,M)}_{i_1\ldots i_N;j_M\ldots j_1}&(\mathbf{r}_1, t_1, \ldots, \mathbf{r}_N, t_N; \mathbf{r}'_M, t'_M, \ldots, \mathbf{r}'_1, t'_1)\\ &= \langle \hat{F}^{(-)}_{i_1}(\mathbf{r}_1, t_1) \ldots \hat{F}^{(-)}_{i_N}(\mathbf{r}_N, t_N)\hat{F}^{(+)}_{j_M}(\mathbf{r}'_M, t'_M) \ldots \hat{F}^{(+)}_{j_1}(\mathbf{r}'_1, t'_1)\rangle\\ &= \langle F^{(-)}_{i_1}(\mathbf{r}_1, t_1) \ldots F^{(-)}_{i_N}(\mathbf{r}_N, t_N)F^{(+)}_{j_M}(\mathbf{r}'_M, t'_M) \ldots F^{(+)}_{j_1}(\mathbf{r}'_1, t'_1)\rangle_\phi,\end{aligned}$$

where $\hat{\mathbf{F}}^{(+)}(\mathbf{r}, t)$ is the positive frequency part of any one of the field operators, and $\mathbf{F}^{(+)}(\mathbf{r}, t)$ is its right eigenvalue. Since $\phi(\{v\})$ is a Gaussian functional, all the $F^{(+)}_i$ and $F^{(-)}_i$ are Gaussian variates, and we can use the Gaussian moment theorem (cf. Section 1.6.1) to express the answer in terms of second-order correlations. We obtain, just as in the derivation of Eqs. (8.4–2),

$$\begin{aligned}\Gamma^{(N,M)}_{i_1\ldots i_N;j_M\ldots j_1}&(\mathbf{r}_1, t_1, \ldots, \mathbf{r}_N, t_N; \mathbf{r}'_M, t'_M, \ldots, \mathbf{r}'_1, t'_1)\\ &= \delta_{NM}\sum_P \langle F^{(-)}_{i_1}(\mathbf{r}_1, t_1)F^{(+)}_{j_1}(\mathbf{r}'_1, t'_1)\rangle_\phi \ldots \langle F^{(-)}_{i_N}(\mathbf{r}_N, t_N)F^{(+)}_{j_N}(\mathbf{r}'_N, t'_N)\rangle_\phi. \quad (13.1\text{–}29)\end{aligned}$$

The sum is to be taken over all $N!$ pairings of the positive with negative frequency fields. As an obvious application of Eq. (13.1–29) we note that

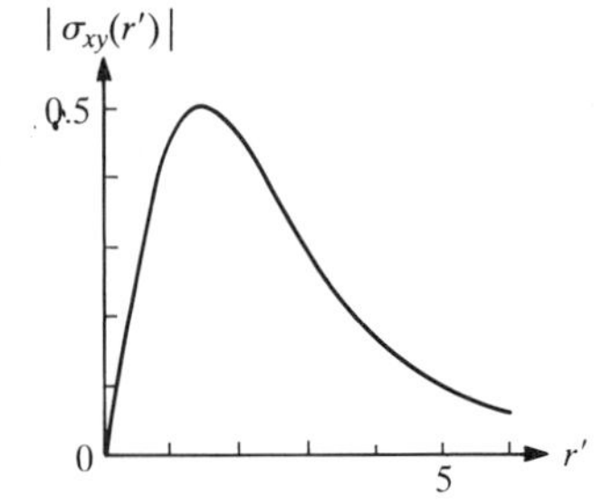

Fig. 13.3 The variation of the normalized mixed correlation $|\sigma_{xy}|$ of electric and magnetic fields of blackbody radiation with distance, in the direction of the z-axis, with $r' = (\pi k_B T/\hbar c)z$. (Reproduced from Mehta and Wolf, 1964a.)

$$\langle [\hat{F}_i^{(-)}(\mathbf{r}_1, t_1)]^N [\hat{F}_j^{(+)}(\mathbf{r}_2, t_2)]^M \rangle = \delta_{NM} N! \langle F_i^{(-)}(\mathbf{r}_1, t_1) F_j^{(+)}(\mathbf{r}_2, t_2) \rangle_\phi^N . \quad (13.1\text{–}30)$$

13.1.8 Isotropy of blackbody radiation

In Section 12.8 we encountered a general criterion for the directional isotropy of a radiation field [cf. Eq. (12.8–38)]. If $\hat{\mathbf{F}}^{(+)}(\mathbf{r}, t)$ is the positive frequency part of any one of the field vectors like $\hat{\mathbf{E}}(\mathbf{r}, t)$ and $\hat{\mathbf{B}}(\mathbf{r}, t)$, and if, for simplicity, we limit our attention to second-order correlation functions, then this criterion takes the form

$$\mathsf{R}_{ip}\mathsf{R}_{jq}\langle \hat{F}_p^{(-)}(\mathbf{R}^{-1}\mathbf{r}_1, t_1)\hat{F}_q^{(+)}(\mathbf{R}^{-1}\mathbf{r}_2, t_2)\rangle = \langle \hat{F}_i^{(-)}(\mathbf{r}_1, t_1)\hat{F}_j^{(+)}(\mathbf{r}_2, t_2)\rangle .$$

Here $\mathbf{R}$ is a 3×3 rotation matrix and $\mathbf{R}^{-1}$ is its inverse. Let us see if the correlation tensors for blackbody radiation given by Eqs. (13.1–23) and (13.1–24) obey this condition.

We first note that both tensors are of the general form

$$\langle \hat{F}_p^{(-)}(\mathbf{r}_1, t_1)\hat{F}_q^{(+)}(\mathbf{r}_2, t_2)\rangle = \frac{G}{(2\pi)^3}\int \frac{\omega}{(e^{\hbar\omega/k_B T} - 1)}\left(\delta_{pq} - \frac{k_p k_q}{k^2}\right) e^{i[\mathbf{k}\cdot(\mathbf{r}_2-\mathbf{r}_1)-\omega(t_2-t_1)]}\, d^3k, \quad (13.1\text{–}31)$$

and differ only in the constant G. Then

$$\mathsf{R}_{ip}\mathsf{R}_{jq}\langle \hat{F}_p^{(-)}(\mathbf{R}^{-1}\mathbf{r}_1, t_1)\hat{F}_q^{(+)}(\mathbf{R}^{-1}\mathbf{r}_2, t_2)\rangle = \mathsf{R}_{ip}\mathsf{R}_{jq}\frac{G}{(2\pi)^3}\int \frac{\omega}{(e^{\hbar\omega/k_B T} - 1)}\left(\delta_{pq} - \frac{k_p k_q}{k^2}\right) \times e^{i[\mathbf{k}\cdot\mathbf{R}^{-1}(\mathbf{r}_2-\mathbf{r}_1)-\omega(t_2-t_1)]}\, d^3k. \quad (13.1\text{–}32)$$

We now make the change of variable $k_p = \mathsf{R}_{pi}^{-1} k_i'$ and observe that, because of the orthogonality of the rotation matrices,

$$\begin{aligned}\mathbf{k}\cdot\mathbf{R}^{-1}(\mathbf{r}_2 - \mathbf{r}_1) &= k_p \mathsf{R}_{pl}^{-1}(\mathbf{r}_2 - \mathbf{r}_1)_l \\ &= \mathsf{R}_{pm}^{-1} k_m' \mathsf{R}_{pl}^{-1}(\mathbf{r}_2 - \mathbf{r}_1)_l \\ &= \delta_{ml} k_m' (\mathbf{r}_2 - \mathbf{r}_1)_l \\ &= \mathbf{k}'\cdot(\mathbf{r}_2 - \mathbf{r}_1)\end{aligned}$$

and

$$\mathsf{R}_{ip}\mathsf{R}_{jq} k_p k_q = \mathsf{R}_{ip}\mathsf{R}_{jq}\mathsf{R}_{pl}^{-1}\mathsf{R}_{qm}^{-1} k_l' k_m' = \delta_{il}\delta_{jm} k_l' k_m' = k_i' k_j'$$
$$k^2 = k_p k_p = \mathsf{R}_{pi}^{-1}\mathsf{R}_{pj}^{-1} k_i' k_j' = \delta_{ij} k_i' k_j' = k'^2.$$

Moreover, $d^3k = d^3k'$, because the rotational transformation does not change the scale. When these results are used on the right of Eq. (13.1–32) we obtain the formula

$$\begin{aligned}&\mathsf{R}_{ip}\mathsf{R}_{jq}\langle \hat{F}_p^{(-)}(\mathbf{R}^{-1}\mathbf{r}_1, t_1)\hat{F}_q^{(+)}(\mathbf{R}^{-1}\mathbf{r}_2, t_2)\rangle \\ &\quad = \frac{G}{(2\pi)^3}\int \frac{\omega'}{(e^{\hbar\omega'/k_B T} - 1)}\left(\delta_{ij} - \frac{k_i' k_j'}{k'^2}\right) e^{i[\mathbf{k}'\cdot(\mathbf{r}_2-\mathbf{r}_1)-\omega'(t_2-t_1)]}\, d^3k' \\ &\quad = \langle \hat{F}_i^{(-)}(\mathbf{r}_1, t_1)\hat{F}_j^{(+)}(\mathbf{r}_2, t_2)\rangle . \end{aligned} \quad (13.1\text{–}33)$$

Hence the correlation tensors satisfy the condition for isotropy, and blackbody radiation is isotropic about all points in space.

13.1.9 Intensity fluctuations of blackbody radiation

By virtue of the fact that the phase space functional $\phi(\{v\})$ describing the state of the field is an ordinary probability functional, we can also think of the blackbody radiation field in classical terms involving the fluctuating complex field amplitude $\mathbf{F}^{(+)}(\mathbf{r}, t)$ and the fluctuating instantaneous light intensity $I(\mathbf{r}, t) \equiv |\mathbf{F}^{(+)}(\mathbf{r}, t)|^2$. The amplitude $\mathbf{F}^{(+)}(\mathbf{r}, t)$ is the positive frequency part, or the analytic signal representation, of any of the field vectors like $\mathbf{E}(\mathbf{r}, t)$, $\mathbf{B}(\mathbf{r}, t)$, $\mathbf{A}(\mathbf{r}, t)$, and it can be given a Fourier expansion in terms of the set of mode amplitudes $\{v\}$,

$$\mathbf{F}^{(+)}(\mathbf{r}, t) = \frac{1}{L^{3/2}} \sum_{\mathbf{k},s} l(\omega) v_{\mathbf{k},s} \boldsymbol{\varepsilon}_{\mathbf{k}s}\, \mathrm{e}^{\mathrm{i}(\mathbf{k}\cdot\mathbf{r}-\omega t)}. \tag{13.1–34}$$

$l(\omega)$ is a simple slowly varying function of ω, which takes the form $\mathrm{i}(\hbar\omega/2\varepsilon_0)^{1/2}$ for the electric field, $(\hbar/2\omega\varepsilon_0)^{1/2}$ for the vector potential, etc. As all the Fourier amplitudes $v_{\mathbf{k}s}$ have a joint Gaussian distribution, and as $\mathbf{F}^{(+)}(\mathbf{r}, t)$ is linear in the $v_{\mathbf{k}s}$, it follows that $\mathbf{F}^{(+)}(\mathbf{r}, t)$ is a complex vector Gaussian random process.

Let us resolve $\mathbf{F}^{(+)}(\mathbf{r}, t)$ into its three Cartesian components $F_x^{(+)}$, $F_y^{(+)}$, $F_z^{(+)}$, which are complex, scalar Gaussian random variables that, by virtue of Eq. (13.1–27), are uncorrelated and statistically independent. The light intensities $I_x \equiv |F_x^{(+)}|^2$, $I_y \equiv |F_y^{(+)}|^2$, $I_z \equiv |F_z^{(+)}|^2$ associated with these three scalar processes therefore have exponential probability distributions (cf. Eq. (13.3–7) below)

$$\mathcal{P}_x(I_x) = \frac{1}{\langle I_x \rangle} \mathrm{e}^{-I_x/\langle I_x \rangle}, \tag{13.1–35}$$

with corresponding expressions for $\mathcal{P}_y(I_y)$ and $\mathcal{P}_z(I_z)$, and all three I_x, I_y, I_z are statistically independent of each other. But the total light intensity is given by

$$I = I_x + I_y + I_z, \tag{13.1–36}$$

and the expectation values satisfy the relations

$$\langle I_x \rangle = \langle I_y \rangle = \langle I_z \rangle = \tfrac{1}{3}\langle I \rangle, \tag{13.1–37}$$

by virtue of the isotropy of the field. For electric or magnetic fields $\langle I \rangle$ is obtainable from Eqs. (13.1–27). It follows that the total light intensity I has a probability density $\mathbb{P}(I)$ given by the formula

$$\mathbb{P}(I) = \iiint_0^\infty \mathrm{d}I_x\, \mathrm{d}I_y\, \mathrm{d}I_z \mathcal{P}_x(I_x)\mathcal{P}_y(I_y)\mathcal{P}_z(I_z)\delta(I - I_x - I_y - I_z),$$

and with the help of Eqs. (13.1–36) and (13.1–37) this immediately reduces to

$$\begin{aligned}\mathbb{P}(I) &= \frac{27}{\langle I \rangle^3} \int_0^I \mathrm{d}I_x\, \mathrm{e}^{-3I_x/\langle I \rangle} \int_0^{I-I_x} \mathrm{d}I_y\, \mathrm{e}^{-3I_y/\langle I \rangle}\, \mathrm{e}^{-3(I-I_x-I_y)/\langle I \rangle} \\ &= \frac{27}{2} \frac{I^2}{\langle I \rangle^3} \mathrm{e}^{-3I/\langle I \rangle}. \end{aligned} \tag{13.1–38}$$

This probability distribution, which has the form of a $\gamma - 3$ distribution (cf. Section 1.5.7), is illustrated in Fig. 13.4. The most probable value of I is $\frac{2}{3}\langle I \rangle$, rather than 0, as for the components I_x, I_y, I_z. The same distribution can of course be derived directly from the phase space functional $\phi(\{v\})$ with the definition

$$\mathbb{P}(I') = \int \delta(I' - I)\phi(\{v\})\,\mathrm{d}\{v\}, \qquad (13.1\text{–}39)$$

where $I \equiv |\mathbf{F}^{(+)}|^2$, but the calculation is less simple despite the formal simplicity of Eq. (13.1–39).

The moments of the light intensity follow immediately from Eq. (13.1–38), and we have

$$\begin{aligned}\langle I^n \rangle &= \int_0^\infty \frac{27}{2}\frac{I^{n+2}}{\langle I \rangle^3}\,\mathrm{e}^{-3I/\langle I \rangle}\,\mathrm{d}I \\ &= \frac{(n+2)!}{2}\left(\frac{\langle I \rangle}{3}\right)^n, \end{aligned} \qquad (13.1\text{–}40)$$

so that the variance

$$\begin{aligned}\langle (\Delta I)^2 \rangle &= \langle I^2 \rangle - \langle I \rangle^2 \\ &= \tfrac{1}{3}\langle I \rangle^2. \end{aligned} \qquad (13.1\text{–}41)$$

The characteristic function $C(\xi)$ of $\mathbb{P}(I)$ is given by

$$C(\xi) = \int_0^\infty \mathrm{e}^{\mathrm{i}\xi I}\,\mathbb{P}(I)\,\mathrm{d}I,$$

and this reduces to

$$C(\xi) = \frac{1}{(1 - \mathrm{i}\xi\langle I \rangle/3)^3}, \qquad (13.1\text{–}42)$$

when the integral is evaluated, which is of the expected form for a $\gamma - 3$ distribution.

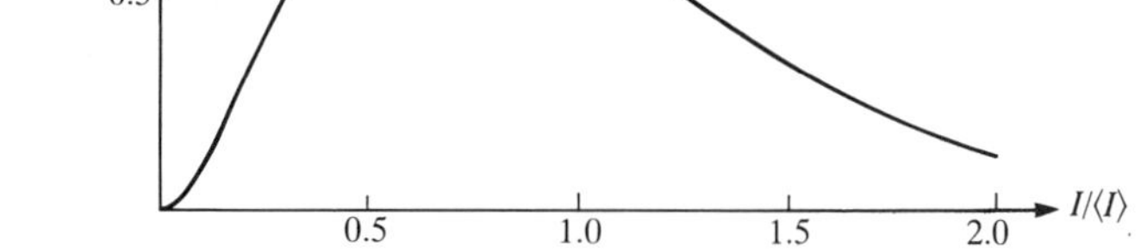

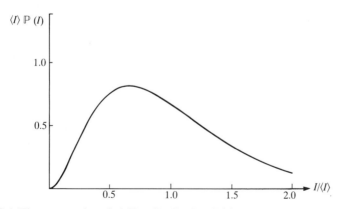

Fig. 13.4 The expected probability distribution $\mathbb{P}(I)$ of the light intensity I for blackbody radiation.

13.2 Thermal light

Many familar light sources do not emit true blackbody radiation, even though the sources are approximately in thermal equilibrium. The emitted light may be directional rather than isotropic, partially polarized rather than unpolarized, spatially inhomogeneous, and it may have a spectral distribution different from that given by the Planck distribution. Such light is generally obtained when blackbody radiation is passed through some filter. We shall call radiation that is derivable from blackbody radiation by any linear filtering process *thermal radiation*. It has sometimes also been called *chaotic radiation*. Typical linear filters include apertures, gratings, mirrors, lenses, polarizers, spectral filters, polarization rotators, phase plates, or any combination thereof. All the important symmetry properties of blackbody radiation, including stationarity if the filters are time dependent, may be lost in the filtering, and yet certain statistical features are preserved.

If the effect of the filter is represented by a homogeneous linear transformation of the field, then the Gaussian statistics of the Fourier amplitudes of blackbody radiation will be retained by thermal radiation, and the phase space functional $\phi(\{v\})$ describing the state will again be Gaussian. However, $\phi(\{v\})$ is now of the general multivariate Gaussian form (cf. Section 1.6)

$$\phi(\{v\}) = \frac{1}{\det(\pi\boldsymbol{\mu})} \mathrm{e}^{-\mathbf{v}^{\dagger}\boldsymbol{\mu}^{-1}\mathbf{v}}, \tag{13.2–1}$$

rather than of the particular form of Eq. (13.1–20). Here $\mathbf{v}$ stands for the column matrix formed by the set of complex amplitudes $\{v\}$, $\mathbf{v}^{\dagger}$ is the Hermitian conjugate, and $\boldsymbol{\mu}$ is the covariance matrix with elements $\mu_{\mathbf{k}s,\mathbf{k}'s'} = \langle v_{\mathbf{k}s}^{*} v_{\mathbf{k}'s'}\rangle_{\phi}$. The denominator is the determinant of the matrix $\pi\boldsymbol{\mu}$. The exponent in Eq. (13.2–1) is a general bilinear functional of the set $\{v\}$, and it can be written as

$$\mathbf{v}^{\dagger}\boldsymbol{\mu}^{-1}\mathbf{v} = \sum_{\mathbf{k},s}\sum_{\mathbf{k}',s'} v_{\mathbf{k}s}^{*}\mu_{\mathbf{k}s,\mathbf{k}'s'}^{-1} v_{\mathbf{k}'s'}. \tag{13.2–2}$$

Since $\boldsymbol{\mu}^{-1}$ is not necessarily diagonal, the phase space functional $\phi(\{v\})$ no longer factorizes into a product of separate mode distributions, and the different mode amplitudes $v_{\mathbf{k}s}$ are not statistically independent in general. However, if we integrate $\phi(\{v\})$ over all modes except one, say $\mathbf{k}'$, s', in order to derive the phase space distribution of the $\mathbf{k}'$, s' mode, we find that

$$\phi(v_{\mathbf{k}'s'}) = \int \phi(\{v\}) \prod_{\mathbf{k},s\neq\mathbf{k}',s'} \mathrm{d}^2 v_{\mathbf{k}s} = \frac{1}{\pi\langle n_{\mathbf{k}'s'}\rangle} \mathrm{e}^{-|v_{\mathbf{k}'s'}|^2/\langle n_{\mathbf{k}'s'}\rangle}, \tag{13.2–3}$$

which has the same form as that for blackbody radiation, except that the average occupation numbers $\langle n_{\mathbf{k}'s'}\rangle$ are no longer given by Eq. (13.1–8). If the filters are passive $\langle n_{\mathbf{k}'s'}\rangle$ will, in general, be no larger than for blackbody radiation, as given by Eq. (13.1–8), but even this restriction does not apply to active filters. We may therefore regard the average occupation numbers in Eq. (13.2–3) as completely arbitrary. When $\langle n_{\mathbf{k}'s'}\rangle = 0$, the corresponding phase space function $\phi(v_{\mathbf{k}'s'})$ has to be interpreted as a $\delta^2(v_{\mathbf{k}'s'})$ function.

Certain features of blackbody radiation are retained by thermal fields. Thus it follows from Eq. (13.2–3), with the help of Eq. (12.10–2), that the probability distribution $p(n_{\mathbf{k}s})$ of the photon occupation number $n_{\mathbf{k}s}$ for any mode again has the Bose–Einstein form

$$p(n_{\mathbf{k}s}) = \frac{1}{[1 + \langle n_{\mathbf{k}s}\rangle][1 + 1/\langle n_{\mathbf{k}s}\rangle]^{n_{\mathbf{k}s}}}, \qquad (13.2\text{–}4)$$

although the joint distribution $p(\{n\})$ is not a product of factors of the same form. Also, since $\phi(\{v\})$ is Gaussian, the Gaussian moment theorem is still applicable, and relations like Eqs. (13.1–29) and (13.1–30) also hold for thermal light. However, in its most general form thermal light does not have the symmetries characteristic of blackbody radiation.

13.3 Stationary, thermal light beams

The light derived from thermal equilibrum sources is often filtered in such a way that an approximate plane wave or a beam traveling in a well–defined direction is produced. Figure 13.5 illustrates one way in which such a beam might be obtained. It can be shown quite generally that, when the optical field is thermal, has a well-defined direction and is stationary and cross-spectrally pure with respect to its polarization (cf. Section 4.5.1), the phase space functional $\phi(\{v\})$ can be simplified.

We have seen in Section 12.8 that for any stationary field

$$\langle \hat{a}^\dagger_{\mathbf{k}s}\hat{a}_{\mathbf{k}'s'}\rangle = \langle v^*_{\mathbf{k}s}v_{\mathbf{k}'s'}\rangle_\phi = 0 \text{ unless } \omega = \omega'. \qquad (13.3\text{–}1)$$

If the field has a well-defined direction of propagation characterized by the unit wave vector $\boldsymbol{\kappa}_0$, all modes except those having a wave vector pointing in the direction $\boldsymbol{\kappa}_0$ are empty, and therefore $\langle \hat{a}^\dagger_{\mathbf{k}s}\hat{a}_{\mathbf{k}'s'}\rangle = 0$ unless $\mathbf{k} = \mathbf{k}'$. Let us choose the polarization basis so that the polarization matrix $\mathbf{J}(\mathbf{k})$ is diagonal, i.e.

$$\langle \hat{a}^\dagger_{\mathbf{k}s}\hat{a}_{\mathbf{k}s'}\rangle = \langle v^*_{\mathbf{k}s}v_{\mathbf{k}s'}\rangle_\phi = \langle n_{\mathbf{k}s}\rangle\delta_{ss'}. \qquad (13.3\text{–}2)$$

If the cross-spectral purity condition holds for polarizations, this basis will be the same for all occupied modes. It follows from Eqs. (13.3–1) and (13.3–2) that for such a unidirectional field

$$\langle \hat{a}^\dagger_{\mathbf{k}s}\hat{a}_{\mathbf{k}'s'}\rangle = \langle v^*_{\mathbf{k}s}v_{\mathbf{k}'s'}\rangle_\phi = \langle n_{\mathbf{k}s}\rangle\delta^3_{\mathbf{k}\mathbf{k}'}\delta_{ss'}. \qquad (13.3\text{–}3)$$

Hence the complex amplitudes $v_{\mathbf{k}s}$ associated with different modes are uncorrelated Gaussian variates, and therefore they are also statistically independent, and

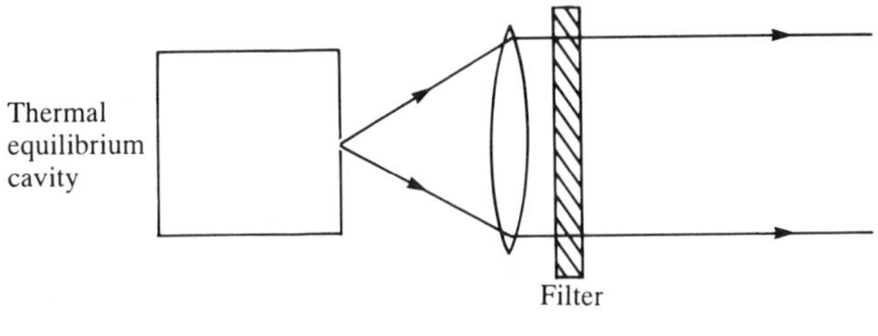

Fig. 13.5 Illustrating the formation of a thermal light beam from blackbody radiation.

the phase space functional $\phi(\{v\})$ then reduces from the general form (13.2–1) to

$$\phi(\{v\}) = \prod_{\mathbf{k},s} \frac{1}{\pi\langle n_{\mathbf{k}s}\rangle} \mathrm{e}^{-|v_{\mathbf{k}s}|^2/\langle n_{\mathbf{k}s}\rangle}. \tag{13.3–4}$$

This has exactly the same structure as the phase space functional given by Eq. (13.1–20) for blackbody radiation. The only difference is that the mean photon occupation numbers $\langle n_{\mathbf{k}s}\rangle$ are all zero except for $\mathbf{k}$ pointing in the direction $\boldsymbol{\kappa}_0$, and for those wave vectors $\langle n_{\mathbf{k}s}\rangle$ is arbitrary. Of course it is not possible to produce a *strictly* unidirectional field or light beam by the method illustrated in Fig. 13.5, but Eq. (13.3–4) often provides a good approximation to the description of the state. Needless to say, a substantial simplification is achieved whenever Eq. (13.3–4) can be used instead of Eq. (13.2–1).

13.3.1 Intensity fluctuations of a thermal light beam

Let us consider a thermal light beam whose state is represented by the phase space density given by Eq. (13.3–4), and let us calculate the probability distribution $\mathbb{P}(I)$ of the light intensity I. As $\phi(\{v\})$ has the same form as the phase space functional for blackbody radiation, it might seem at first that the probability distribution $\mathbb{P}(I)$ would be given by Eq. (13.1–38). However we have to remember that a thermal light beam does not have all the symmetry properties of a blackbody radiation field, and that the occupation numbers $\langle n_{\mathbf{k}s}\rangle$ are arbitrary.

If all Fourier components of the light beam have similar polarization properties, then the polarization matrix $\mathbf{J}(\mathbf{k})$ divided by $\langle n_{\mathbf{k}}\rangle \equiv \langle n_{\mathbf{k}1}\rangle + \langle n_{\mathbf{k}2}\rangle$ must be the same for all occupied $\mathbf{k}$-modes of the field. Accordingly, we can resolve the complex vector field amplitude $\mathbf{F}^{(+)}(\mathbf{r}, t)$ [$\mathbf{F}^{(+)}(\mathbf{r}, t)$ is the eigenvalue of the positive frequency part of the electric field, or the magnetic field, etc.] into two components proportional to the orthogonal unit polarization vectors $\boldsymbol{\varepsilon}_{\mathbf{k}1}$, $\boldsymbol{\varepsilon}_{\mathbf{k}2}$ and write

$$\mathbf{F}^{(+)}(\mathbf{r}, t) = F_1^{(+)}(\mathbf{r}, t)\boldsymbol{\varepsilon}_{\mathbf{k}1} + F_2^{(+)}(\mathbf{r}, t)\boldsymbol{\varepsilon}_{\mathbf{k}2}. \tag{13.3–5}$$

The Gaussian form of the phase space functional $\phi(\{v\})$ ensures that $F_1^{(+)}$, $F_2^{(+)}$ are complex Gaussian scalar variates, and as the basis vectors $\boldsymbol{\varepsilon}_{\mathbf{k}1}$, $\boldsymbol{\varepsilon}_{\mathbf{k}2}$ in Eq. (13.3–5) were chosen so as to diagonalize the polarization matrix, $F_1^{(+)}$ and $F_2^{(+)}$ are uncorrelated and therefore also statistically independent. From Eq. (13.3–5) it follows that

$$I \equiv \mathbf{F}^{(+)*}\cdot\mathbf{F}^{(+)} = |F_1^{(+)}|^2 + |F_2^{(+)}|^2 = I_1 + I_2, \tag{13.3–6}$$

where $I_1 \equiv |F_1^{(+)}|^2$ and $I_2 \equiv |F_2^{(+)}|^2$ are also statistically independent. Now since $F_1^{(+)}$ and $F_2^{(+)}$ are complex Gaussian variates, they have probability distributions of the form

$$p_1(F_1^{(+)})\,\mathrm{d}^2F_1^{(+)} = \frac{1}{\pi\langle I_1\rangle}\exp\left(-|F_1^{(+)}|^2/\langle I_1\rangle\right)\mathrm{d}^2F_1^{(+)}$$

$$p_2(F_2^{(+)})\,\mathrm{d}^2F_2^{(+)} = \frac{1}{\pi\langle I_2\rangle}\exp\left(-|F_2^{(+)}|^2/\langle I_2\rangle\right)\mathrm{d}^2F_2^{(+)},$$

where $\langle I_1\rangle$, $\langle I_2\rangle$ are expectation values of $|F_1^{(+)}|^2$ and $|F_2^{(+)}|^2$ evaluated with the help of the phase space functional in Eq. (13.3–4). By writing $F_1^{(+)} = \sqrt{(I_1)}\exp \mathrm{i}\theta_1$, $F_2^{(+)} = \sqrt{(I_2)}\exp \mathrm{i}\theta_2$, observing that $\mathrm{d}^2F_1^{(+)} = \frac{1}{2}\mathrm{d}I_1\,\mathrm{d}\theta_1$, $\mathrm{d}^2F_2^{(+)} = \frac{1}{2}\mathrm{d}I_2\,\mathrm{d}\theta_2$, and integrating over the phase angles θ_1, θ_2, we immediately find for the probability densities of I_1 and I_2 the expressions,

$$\left.\begin{aligned}\mathcal{P}_1(I_1) &= \frac{1}{\langle I_1\rangle}\mathrm{e}^{-I_1/\langle I_1\rangle}\\ \mathcal{P}_2(I_2) &= \frac{1}{\langle I_2\rangle}\mathrm{e}^{-I_2/\langle I_2\rangle}.\end{aligned}\right\} \tag{13.3–7}$$

From Eqs. (13.3–6) and (13.3–7) we then have for the probability density of I,

$$\begin{aligned}\mathbb{P}(I) &= \iint_0^\infty \mathcal{P}(I_1)\mathcal{P}(I_2)\delta(I - I_1 - I_2)\,\mathrm{d}I_1\,\mathrm{d}I_2\\ &= \int_0^I \mathcal{P}_1(I_1)\mathcal{P}_2(I - I_1)\,\mathrm{d}I_1\\ &= \frac{1}{\langle I_1\rangle - \langle I_2\rangle}[\mathrm{e}^{-I/\langle I_1\rangle} - \mathrm{e}^{-I/\langle I_2\rangle}].\end{aligned} \tag{13.3–8}$$

We can express $\langle I_1\rangle$, $\langle I_2\rangle$ in terms of the mean total light intensity $\langle I\rangle$ and the degree of polarization P of the light. From Eq. (13.3–6) it follows that

$$\langle I\rangle = \langle I_1\rangle + \langle I_2\rangle. \tag{13.3–9}$$

The degree of polarization P of a plane wave of wave vector $\mathbf{k}$ is related to the eigenvalues of the polarization matrix $\mathbf{J}(\mathbf{k})$ by (cf. Eq. (6.3–31))

$$P = \frac{|\langle|v_{\mathbf{k}1}|^2\rangle - \langle|v_{\mathbf{k}2}|^2\rangle|}{\langle|v_{\mathbf{k}1}|^2\rangle + \langle|v_{\mathbf{k}2}|^2\rangle}, \tag{13.3–10}$$

and, by virtue of the fact that all occupied $\mathbf{k}$-modes have the same polarization, this must also be expressible as

$$P = \frac{\langle I_1\rangle - \langle I_2\rangle}{\langle I_1\rangle + \langle I_2\rangle}, \quad \text{if } \langle I_1\rangle \geqslant \langle I_2\rangle. \tag{13.3–11}$$

From Eqs. (13.3–9) and (13.3–11) it follows that

$$\left.\begin{aligned}\langle I_1\rangle &= \tfrac{1}{2}\langle I\rangle(1 + P)\\ \langle I_2\rangle &= \tfrac{1}{2}\langle I\rangle(1 - P),\end{aligned}\right\} \tag{13.3–12}$$

and when these relations are substituted in Eq. (13.3–8), we finally obtain the expression for $\mathbb{P}(I)$ (Mandel, 1963)

$$\mathbb{P}(I) = \frac{1}{P\langle I\rangle}\left\{\exp\left[\frac{-2I}{\langle I\rangle(1 + P)}\right] - \exp\left[\frac{-2I}{\langle I\rangle(1 - P)}\right]\right\}. \tag{13.3–13}$$

Some examples of the probability distribution $\mathbb{P}(I)$ are illustrated in Fig. 13.6. It is worth noting that the case $P = 1$, corresponding to a completely polarized light beam, is somewhat singular, in that the most probable value of the light intensity is zero, whereas $\mathbb{P}(0) = 0$ in all other cases. When the light beam is

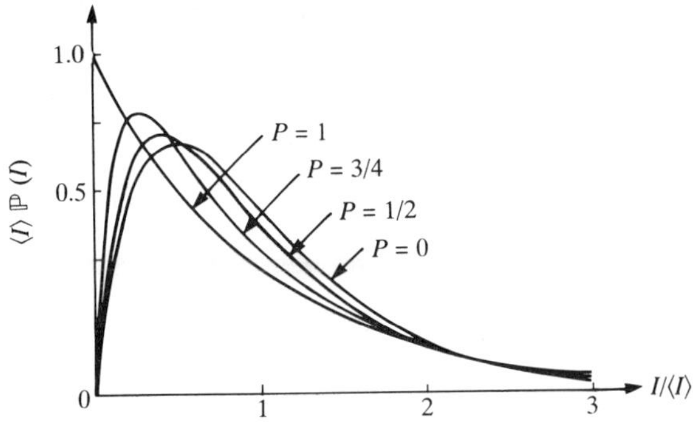

Fig. 13.6 Some examples of the probability distribution $\mathbb{P}(I)$ for a thermal light beam, for several different degrees of polarization P. (Reproduced from Mandel, 1963.)

completely unpolarized and $P = 0$, we obtain the corresponding probability distribution $\mathbb{P}(I)$ by taking the limit of the righthand side of Eq. (13.3–13) as $P \to 0$. We readily find by expansion of the exponentials (Hurwitz, 1945),

$$\mathbb{P}(I) = \frac{4I}{\langle I \rangle^2} \mathrm{e}^{-2I/\langle I \rangle}. \tag{13.3–14}$$

The moments of the light intensity corresponding to Eq. (13.3–13) are easily evaluated, with the result

$$\begin{aligned} \langle I^n \rangle &= \int_0^\infty I^n \mathbb{P}(I)\,\mathrm{d}I \\ &= \frac{n!\langle I \rangle^n}{2^{n+1}P}[(1 + P)^{n+1} - (1 - P)^{n+1}]. \end{aligned} \tag{13.3–15}$$

In particular, the variance is given by

$$\begin{aligned} \langle (\Delta I)^2 \rangle &= \langle I^2 \rangle - \langle I \rangle^2 \\ &= \tfrac{1}{2}(1 + P^2)\langle I \rangle^2. \end{aligned} \tag{13.3–16}$$

The variance therefore ranges between the relatively narrow limits of $\langle I \rangle^2$ for a fully polarized light beam to $\frac{1}{2}\langle I \rangle^2$ for an unpolarized beam.

Finally let us examine the two-time intensity correlation function of such a thermal light beam. As the field is stationary, we express the normally ordered correlation function $\Gamma^{(2,2)}$ in the form

$$\Gamma^{(2,2)} \equiv \langle : \hat{I}(\mathbf{r}_1, t)\hat{I}(\mathbf{r}_2, t + \tau): \rangle$$

and with the help of the optical equivalence theorem,

$$\langle : \hat{I}(\mathbf{r}_1, t)\hat{I}(\mathbf{r}_2, t + \tau): \rangle = \langle I(\mathbf{r}_1, t)I(\mathbf{r}_2, t + \tau) \rangle_\phi.$$

Let us again decompose the complex vector field amplitude $\mathbf{F}^{(+)}(\mathbf{r}, t)$ into two orthogonal, statistically independent, complex scalar components $F_1^{(+)}(\mathbf{r}, t)$ and $F_2^{(+)}(\mathbf{r}, t)$, which are Gaussian processes. Then we have from Eq. (13.3–6)

$$\langle : \hat{I}(\mathbf{r}_1, t)\hat{I}(\mathbf{r}_2, t + \tau): \rangle = \langle I_1(\mathbf{r}_1, t)I_1(\mathbf{r}_2, t + \tau) \rangle_\phi + \langle I_1(\mathbf{r}_1, t)I_2(\mathbf{r}_2, t + \tau) \rangle_\phi$$

$$+ \langle I_2(\mathbf{r}_1, t)I_1(\mathbf{r}_2, t+\tau)\rangle_\phi + \langle I_2(\mathbf{r}_1, t)I_2(\mathbf{r}_2, t+\tau)\rangle_\phi.$$

With the help of the Gaussian moment theorem and the homogeneity of the field, which implies that

$$\langle I_i(\mathbf{r}_1, t)\rangle = \langle I_i(\mathbf{r}_2, t)\rangle \equiv \langle I_i\rangle \quad (i = 1, 2), \tag{13.3–17}$$

and with the statistical independence of I_1 and I_2, we obtain

$$\langle :\hat{I}(\mathbf{r}_1, t)\hat{I}(\mathbf{r}_2, t+\tau): \rangle = (\langle I_1\rangle + \langle I_2\rangle)^2 + |\langle F_1^{(-)}(\mathbf{r}_1, t)F_1^{(+)}(\mathbf{r}_2, t+\tau)\rangle|^2 + |\langle F_2^{(-)}(\mathbf{r}_1, t)F_2^{(+)}(\mathbf{r}_2, t+\tau)\rangle|^2. \tag{13.3–18}$$

It is understood here that the expectation values of all c-number functions are to be evaluated with the help of the phase space functional $\phi(\{v\})$ given by Eq. (13.3–4). We now introduce the normalized second-order correlation functions

$$\left.\begin{aligned} \gamma_{11}(\mathbf{r}_1, \mathbf{r}_2, \tau) &\equiv \frac{\langle F_1^{(-)}(\mathbf{r}_1, t)F_1^{(+)}(\mathbf{r}_2, t+\tau)\rangle}{[\langle I_1(\mathbf{r}_1)\rangle\langle I_1(\mathbf{r}_2)\rangle]^{1/2}} \\ \gamma_{22}(\mathbf{r}_1, \mathbf{r}_2, \tau) &\equiv \frac{\langle F_2^{(-)}(\mathbf{r}_1, t)F_2^{(+)}(\mathbf{r}_2, t+\tau)\rangle}{[\langle I_2(\mathbf{r}_1)\rangle\langle I_2(\mathbf{r}_2)\rangle]^{1/2}}, \end{aligned}\right\} \tag{13.3–19}$$

and make use of the cross-spectral purity condition with respect to polarization, which ensures that

$$\gamma_{11}(\mathbf{r}_1, \mathbf{r}_2, \tau) = \gamma_{22}(\mathbf{r}_1, \mathbf{r}_2, \tau) \equiv \gamma(\mathbf{r}_1, \mathbf{r}_2, \tau). \tag{13.3–20}$$

Then Eq. (13.3–18) reduces to

$$\langle :\hat{I}(\mathbf{r}_1, t)\hat{I}(\mathbf{r}_2, t+\tau): \rangle = (\langle I_1\rangle + \langle I_2\rangle)^2 + (\langle I_1\rangle^2 + \langle I_2\rangle^2)|\gamma(\mathbf{r}_1, \mathbf{r}_2, \tau)|^2. \tag{13.3–21}$$

Finally we express $\langle I_1\rangle$, $\langle I_2\rangle$ in terms of the total mean light intensity $\langle I\rangle$ and the degree of polarization P by making use of Eqs. (13.3–12). Then we arrive at

$$\langle :\hat{I}(\mathbf{r}_1, t)\hat{I}(\mathbf{r}_2, t+\tau): \rangle = \langle I\rangle^2[1 + \tfrac{1}{2}(1 + P^2)|\gamma(\mathbf{r}_1, \mathbf{r}_2, \tau)|^2], \tag{13.3–22}$$

which implies that the normalized intensity correlation function

$$\lambda(\mathbf{r}_1, \mathbf{r}_2, \tau) \equiv \frac{\langle :\Delta\hat{I}(\mathbf{r}_1, t)\Delta\hat{I}(\mathbf{r}_2, t+\tau): \rangle}{\langle \hat{I}(\mathbf{r}_1, t)\rangle\langle \hat{I}(\mathbf{r}_2, t+\tau)\rangle}$$

is given by

$$\lambda(\mathbf{r}_1, \mathbf{r}_2, \tau) = \tfrac{1}{2}(1 + P^2)|\gamma(\mathbf{r}_1, \mathbf{r}_2, \tau)|^2. \tag{13.3–23}$$

Because of the homogeneity of the field, each side of this equation actually depends only on the difference $\mathbf{r}_1 - \mathbf{r}_2$ between the two space arguments.

13.3.2 Photon statistics with equal average occupation numbers

As the phase space functional for a thermal light beam given by Eq. (13.3–4) has the same structure as that for blackbody radiation, it follows immediately that the joint probability distribution $p(\{n\})$ of the set of photon occupation numbers $\{n\}$

of the field must also be of the form of Eq. (13.1–10), namely

$$p(\{n\}) = \prod_{\mathbf{k},s} \frac{1}{[1 + \langle n_{\mathbf{k}s} \rangle][1 + 1/\langle n_{\mathbf{k}s} \rangle]^{n_{\mathbf{k}s}}}, \tag{13.3–24}$$

except that the average occupation numbers $\langle n_{\mathbf{k}s} \rangle$ are arbitrary. For any mode for which $\langle n_{\mathbf{k}s} \rangle = 0$ the corresponding factor must be interpreted as $\delta_{n_{\mathbf{k}s}0}$. In the following we shall suppose that only a subset $[\mathbf{k}, s]$ consisting of μ modes of the field is actually occupied, and we shall restrict our attention to this subset of the modes. If n is the total number of photons,

$$n = \sum_{[\mathbf{k},s]} n_{\mathbf{k}s}, \tag{13.3–25}$$

and $P(n)$ is the probability distribution of n, then evidently

$$P(m) = \sum_{\{n\}} p(\{n\})\delta_{nm}. \tag{13.3–26}$$

Although $P(m)$ is difficult to evaluate in general, the evaluation becomes easy in the special case in which the average occupation numbers $\langle n_{\mathbf{k}s} \rangle$ of all the μ occupied modes become equal. This corresponds to a thermal light beam with a rectangular spectral density, which is either fully polarized or fully unpolarized. In that case

$$\langle n \rangle = \sum_{[\mathbf{k},s]} \langle n_{\mathbf{k}s} \rangle = \mu \langle n_{\mathbf{k}s} \rangle, \tag{13.3–27}$$

and from Eqs. (13.3–24) and (13.3–27) the joint probability distribution of the occupied modes becomes

$$p(\{n\}) = \frac{1}{[1 + \langle n \rangle/\mu]^{\mu}[1 + \mu/\langle n \rangle]^{n}}. \tag{13.3–28}$$

It will be seen that this depends only on the total number of photons n, and not at all on the manner in which they are actually distributed among the μ modes. Because of this fact, every non-vanishing term in the summation in Eq. (13.3–26) has the same value, and the required probability $P(m)$ is simply $p(\{n\})$ given by Eq. (13.3–28) multiplied by the number of different ways $\mathcal{N}$ of distributing n photons among μ modes.

The combinatorial factor $\mathcal{N}$ is well known in quantum statistics. We can derive it by imagining that the n photons are indistinguishable balls to be distributed among μ boxes. We arrive at different distributions by arranging the boxes and balls at random on a line, with the convention that the balls lying to the immediate right of a box belong in that box. That requires the left extremity of the line to be occupied by a box, and leads to $\mu(n + \mu - 1)!$ different possible arrangements. But as the n photons are indistinguishable, and the actual locations of the $\mu - 1$ boxes have no physical significance, we must divide this number by $n!(\mu - 1)!$ to arrive at $\mathcal{N}$. Thus we have finally, (Mandel, 1959)

$$P(n) = \frac{(n + \mu - 1)!}{(\mu - 1)!n!} \frac{1}{[1 + \langle n \rangle/\mu]^{\mu}[1 + \mu/\langle n \rangle]^{n}}. \tag{13.3–29}$$

The moments of $P(n)$ are evaluated most easily with the help of the factorial moment generating function

$$F(\xi) \equiv \sum_{n=0}^{\infty}(1+\xi)^n P(n) = \sum_{r=0}^{\infty}\frac{\xi^r}{r!}\langle n^{(r)}\rangle, \qquad (13.3\text{–}30)$$

which generates the factorial moments

$$\langle n^{(r)}\rangle \equiv \langle n(n-1)(n-2)\ldots(n-r+1)\rangle$$

by a power series expansion in ξ. If we make use of the algebraic identity

$$\frac{(n+\mu-1)!}{(\mu-1)!n!} = \binom{-\mu}{n}(-1)^n,$$

we can express $F(\xi)$ as a binomial series in $-(1+\xi)/(1+\mu/\langle n\rangle)$, the so-called negative binomial, which is easily summed, with the result that

$$\begin{aligned} F(\xi) &= \frac{1}{(1+\langle n\rangle/\mu)^{\mu}}\sum_{n=0}^{\infty}\binom{-\mu}{n}\left[\frac{-(1+\xi)}{1+\mu/\langle n\rangle}\right]^n \\ &= \left[1+\frac{\langle n\rangle}{\mu}\right]^{-\mu}\left[1-\frac{1+\xi}{1+\mu/\langle n\rangle}\right]^{-\mu} \\ &= \left[1-\frac{\xi\langle n\rangle}{\mu}\right]^{-\mu} \\ &= 1+\xi\langle n\rangle + \frac{\xi^2}{2!}\frac{\mu+1}{\mu}\langle n\rangle^2 + \ldots. \end{aligned} \qquad (13.3\text{–}31)$$

By inspection, the mean $\langle n\rangle$ is the coefficient of ξ, and the second factorial moment is the coefficient of $\xi^2/2!$, so that

$$\langle n(n-1)\rangle = \left(\frac{\mu+1}{\mu}\right)\langle n\rangle^2,$$

and the variance is given by

$$\langle(\Delta n)^2\rangle = \langle n\rangle\left(1+\frac{\langle n\rangle}{\mu}\right). \qquad (13.3\text{–}32)$$

In the special case $\mu = 1$, Eqs. (13.3–31) and (13.3–32) reduce to the results obtained earlier for the single-mode Bose–Einstein distribution, as required. However, when μ becomes very large for a fixed $\langle n\rangle$, Eq. (13.3–31) leads to

$$F(\xi) \to e^{\xi\langle n\rangle}, \qquad (13.3\text{–}33)$$

which is the factorial moment generating function of a Poisson distribution, and $\langle(\Delta n)^2\rangle \to \langle n\rangle$. In this limit the photons become almost independent, and obey the statistics of classical particles.

We shall have occasion to refer to some of these results in connection with photoelectric counting statistics.

Problems

13.1 A quasi-monochromatic, polarized light beam from a stationary thermal source is incident normally on a photodetector. Show that if the normalized spectral distribution $\phi(\omega)$ of the light is symmetric about some known frequency ω_0, then $\phi(\omega)$ can be completely determined from measurements of the rate at which pairs of photoelectric pulses separated by an interval τ are being detected as a function of τ. Express $\phi(\omega)$ in terms of this rate.

13.2 Calculate the normally ordered dispersion $\langle:(\Delta\hat{I}^2):\rangle$ of the photon density $\hat{I}$ of blackbody radiation at temperature T.

13.3 Consider a thermal light beam with the same average photon occupation numbers n_0 for μ occupied modes and with all other modes empty. Calculate the characteristic function $C(x) = \langle\exp(\mathrm{i}x\hat{n})\rangle$ of the total photon number $\hat{n}$. Hence, derive the probability distribution of the total photon number.

13.4 Express the operator $\exp(\hat{n}_{\mathbf{k},s}x)$ in antinormal order. Hence show how the phase space density $\phi(v_{\mathbf{k},s})$ for one $\mathbf{k}, s$ mode of black body radiation can be readily obtained from the expression $\hat{\rho}_{\mathbf{k},s} = (1 - \mathrm{e}^{-\beta})\exp(-\beta\hat{n}_{\mathbf{k},s})$ for the density operator of the mode.

13.5 A monochromatic light beam of frequency ω_1 and mean intensity $\langle I_1\rangle$ produced by a randomly phased single-mode laser, and a parallel quasi-monochromatic thermal light beam of mean intensity $\langle I_2\rangle$, are superposed. Both light beams are similarly polarized. The superposed light is allowed to impinge normally on a photodetector of quantum efficiency α and photocathode area S for a time interval T. If T is long compared with the coherence time of the thermal light, calculate the variance of the number of photoelectric counts registered by the detector in terms of $\langle I_1\rangle$, $\langle I_2\rangle$, ω_1 and the normalized spectral density $\phi(\omega)$ of the thermal light beam.

13.6 In the previous problem, how would the answer change if the counting time T were short compared with the coherence time of the thermal light beam? Under what circumstances would the excess fluctuations over a Poisson distribution be independent of the intensity of the laser beam?

13.7 Calculate the average density of photons within a coherence volume of blackbody radiation. Show that it is independent of temperature.

13.8 A polarized quasi-monochromatic, thermal light beam has a Lorentzian spectrum of width $\Delta\omega$ centered on frequency ω_0. Calculate the normalized, normally ordered intensity correlation function $\langle:\Delta\hat{I}_\mathrm{E}(t)\Delta\hat{I}_\mathrm{E}(t+\tau):\rangle/\langle\hat{I}_\mathrm{E}\rangle^2$, where $\hat{I}_\mathrm{E} = \hat{\mathbf{E}}^{(-)}\cdot\hat{\mathbf{E}}^{(+)}$ is the electric field intensity.

14

Quantum theory of photoelectric detection of light

14.1 Interactions of a quantized electromagnetic field

In the preceding chapters we have studied the quantum properties of the electromagnetic field, but we have treated it as a free or non-interacting quantum system until now. However, both the emission and the absorption of light imply interactions with other quantum systems, and we now turn to the treatment of such interaction problems.

It is true that, to a limited extent, we have already succeeded in treating some interactions of light in previous chapters, without invoking the full apparatus of the quantum theory of interacting systems. For example, in Chapter 9 the electromagnetic field was treated as a classical potential acting on an atomic quantum system, and in Sections 12.2 and 12.9 we invoked simple heuristic arguments to describe the operation of certain optical detectors. But these are limited applications, and, in any case, the validity of results obtained in this way needs to be confirmed.

To describe the state of a quantized electromagnetic field (F) in interaction with some other quantum system (A), we evidently require an enlarged, or product Hilbert space, which encompasses both F and A, of which the Hilbert spaces of F and A are subspaces. We shall find it convenient to refer to the other system A as an 'atomic system', simply to give a name, without restricting the nature of A. The dynamical variables of the field $\hat{O}_F(t)$ and of the atomic system $\hat{O}_A(t)$ commute at the same time t, and at the beginning, when the interaction between them is assumed to be turned on, each operator acts only on the state vectors within its respective Hilbert space. In the Heisenberg picture, the equation of motion of any dynamical variable $\hat{O}(t)$ takes the usual form

$$\frac{\mathrm{d}\hat{O}(t)}{\mathrm{d}t} = \frac{1}{\mathrm{i}\hbar}[\hat{O}(t), \hat{H}] + \frac{\partial \hat{O}(t)}{\partial t}, \qquad (14.1\text{–}1)$$

in which the last term implies differentiation with respect to an explicitly time-dependent variable. The total energy $\hat{H}$ of the coupled system can always be decomposed into three parts,

$$\hat{H} = \hat{H}_A + \hat{H}_F + \hat{H}_I. \qquad (14.1\text{–}2)$$

Here $\hat{H}_A$ and $\hat{H}_F$ are the energies of the free, or non-interacting, atomic system and the quantum field, respectively, and $\hat{H}_I$ is the interaction energy between them. Even when $\hat{H}$ and $\hat{O}$ are not explicitly time dependent, and Eq. (14.1–1) can be formally integrated to give

$$\hat{O}(t) = \exp(\mathrm{i}\hat{H}t/\hbar)\hat{O}(0)\exp(-\mathrm{i}\hat{H}t/\hbar), \tag{14.1–3}$$

but this formula rarely leads to a closed form expression for $\hat{O}(t)$; approximations are usually required.

In the Schrödinger picture, the state of the coupled system is described by a density operator $\hat{\rho}(t)$ that spans the entire Hilbert space, and obeys the Schrödinger equation of motion

$$\frac{\mathrm{d}\hat{\rho}(t)}{\mathrm{d}t} = \frac{1}{\mathrm{i}\hbar}[\hat{H}, \hat{\rho}(t)]. \tag{14.1–4}$$

The states of the separate atomic and field components $\hat{\rho}_{\mathrm{A}}(t)$ and $\hat{\rho}_{\mathrm{F}}(t)$ can be extracted from $\hat{\rho}(t)$ by tracing over the unwanted portions of the total Hilbert space, so that

$$\left.\begin{aligned} \hat{\rho}_{\mathrm{A}}(t) &= \mathrm{Tr}_{\mathrm{F}}\,\hat{\rho}(t) \\ \hat{\rho}_{\mathrm{F}}(t) &= \mathrm{Tr}_{\mathrm{A}}\,\hat{\rho}(t). \end{aligned}\right\} \tag{14.1–5}$$

When $\hat{H}$ is not explicitly time dependent, Eq. (14.1–4) is readily integrated to yield

$$\hat{\rho}(t) = \exp(-\mathrm{i}\hat{H}t/\hbar)\hat{\rho}(0)\exp(\mathrm{i}\hat{H}t/\hbar), \tag{14.1–6}$$

but the transformation implied by this equation can rarely be carried out exactly. Moreover, even when $\hat{\rho}(t)$ can be found, this still does not allow multi-time correlation functions to be evaluated without further assumptions. Occasionally, the equation of motion for the component density operator $\hat{\rho}_{\mathrm{A}}(t)$ or $\hat{\rho}_{\mathrm{F}}(t)$, usually known as the master equation, is simpler than that for the total system, and it can sometimes be further simplified by Markovian – or short memory – assumptions about the evolution of the system. Alternatively, when the interaction lasts only a short time, we can make use of time-dependent perturbation theory to solve Eq. (14.1–4).

14.1.1 Perturbative solution in the interaction picture

For perturbative solutions of the Schrödinger equation it is often more convenient to work in the interaction picture (I) rather than in the Schrödinger picture (S). In this picture, the dynamical variables evolve in time according to the non-interacting part $\hat{H}_0 \equiv \hat{H}_{\mathrm{A}} + \hat{H}_{\mathrm{F}}$ of the total energy, i.e. as they would in the Heisenberg picture in the absence of any interaction, and the states do not change except for the interaction $\hat{H}_{\mathrm{I}}$. If we suppose that the different pictures coincide at time t_0, so that

$$\hat{O}^{(\mathrm{S})}(t_0) = \hat{O}^{(\mathrm{I})}(t_0) = \hat{O}^{(\mathrm{H})}(t_0),$$

where the labels identify the picture, then at other times

$$\begin{aligned} \hat{O}^{(\mathrm{I})}(t) &= \exp[\mathrm{i}\hat{H}_0(t - t_0)/\hbar]\hat{O}^{(\mathrm{S})}(t)\exp[-\mathrm{i}\hat{H}_0(t - t_0)/\hbar] \\ &= \exp[\mathrm{i}\hat{H}_0(t - t_0)/\hbar]\hat{O}^{(\mathrm{S})}(t_0)\exp[-\mathrm{i}\hat{H}_0(t - t_0)/\hbar], \end{aligned} \tag{14.1–7}$$

and

$$\hat{\rho}^{(\mathrm{I})}(t) = \exp[\mathrm{i}\hat{H}_0(t-t_0)/\hbar]\hat{\rho}^{(\mathrm{S})}(t)\exp[-\mathrm{i}\hat{H}_0(t-t_0)/\hbar]. \quad (14.1\text{–}8)$$

The Schrödinger equation of motion (14.1–4), when expressed in terms of states and operators in the interaction picture, takes the simpler form

$$\frac{\mathrm{d}\hat{\rho}^{(\mathrm{I})}(t)}{\mathrm{d}t} = \frac{1}{\mathrm{i}\hbar}[\hat{H}_\mathrm{I}^{(\mathrm{I})}(t), \hat{\rho}^{(\mathrm{I})}(t)], \quad (14.1\text{–}9)$$

which shows clearly that the evolution of the state is governed by the interaction energy $\hat{H}_\mathrm{I}$ alone, and this usually simplifies the calculation. In the following we shall take it for granted that we are in the interaction picture, and the superscript label (I) that identifies the picture will be suppressed from now on.

Equation (14.1–9) may be formally integrated with respect to time to give

$$\hat{\rho}(t) = \hat{\rho}(t_0) + \frac{1}{\mathrm{i}\hbar}\int_{t_0}^{t}[\hat{H}_\mathrm{I}(t_1), \hat{\rho}(t_1)]\,\mathrm{d}t_1, \quad (14.1\text{–}10)$$

which is a Volterra-type integral equation that can be solved by iteration. Let us regard $\hat{\rho}(t_0)$ as a zero'th-order approximation to $\hat{\rho}(t)$, and write

$$\hat{\rho}^{(0)}(t) = \hat{\rho}(t_0).$$

We now substitute this approximation for $\hat{\rho}(t)$ under the integral in Eq. (14.1–10), to yield a first-order approximation $\hat{\rho}^{(1)}(t)$ to $\hat{\rho}(t)$. Thus we write

$$\hat{\rho}^{(1)}(t) = \hat{\rho}(t_0) + \frac{1}{\mathrm{i}\hbar}\int_{t_0}^{t}[\hat{H}_\mathrm{I}(t_1), \hat{\rho}(t_0)]\,\mathrm{d}t_1,$$

and when this approximation is substituted for $\hat{\rho}(t)$ under the integral in Eq. (14.1–10), we obtain the second-order approximation

$$\hat{\rho}^{(2)}(t) = \hat{\rho}(t_0) + \frac{1}{\mathrm{i}\hbar}\int_{t_0}^{t}\mathrm{d}t_1[\hat{H}_\mathrm{I}(t_1), \hat{\rho}(t_0)]$$
$$+ \frac{1}{(\mathrm{i}\hbar)^2}\int_{t_0}^{t}\mathrm{d}t_1\int_{t_0}^{t_1}\mathrm{d}t_2[\hat{H}_\mathrm{I}(t_1), [\hat{H}_\mathrm{I}(t_2), \hat{\rho}(t_0)]].$$

Repetition of the same procedure yields the following infinite series (cf. Section 9.2), that may be regarded as an exact, explicit solution for $\hat{\rho}(t)$, provided the series converges:

$$\hat{\rho}(t) = \hat{\rho}(t_0) + \sum_{r=1}^{\infty}\frac{1}{(\mathrm{i}\hbar)^r}\int_{t_0}^{t}\mathrm{d}t_1\int_{t_0}^{t_1}\mathrm{d}t_2$$
$$\times \ldots \int_{t_0}^{t_{r-1}}\mathrm{d}t_r[\hat{H}_\mathrm{I}(t_1), [\hat{H}_\mathrm{I}(t_2), \ldots, [\hat{H}_\mathrm{I}(t_r), \hat{\rho}(t_0)]]\ldots]. \quad (14.1\text{–}11)$$

In practice, the infinite series is rarely, if ever, used; instead the series is terminated at the point where it yields a sufficiently good approximation to the solution. For many applications the first non-vanishing contribution suffices, and the series is terminated there.

If the states of the atomic system and of the field are known separately at time t_0 when the interaction is assumed to commence, then the initial density operator

$\hat{\rho}(t_0)$ of the coupled system factorizes into a product of two independent density operators,

$$\hat{\rho}(t_0) = \hat{\rho}_A(t_0) \otimes \hat{\rho}_F(t_0), \tag{14.1–12}$$

each of which spans its own Hilbert space.

In a number of problems we shall be interested in the probability $P(t_0 + T)$ that, after a time interval T following the start of the interaction, the coupled system can be found in some quantum state $|\Psi_F\rangle$, with density operator $\hat{\rho}_F \equiv |\Psi_F\rangle\langle\Psi_F|$, which is orthogonal to the initial state $\hat{\rho}(t_0)$, i.e.

$$\hat{\rho}_F\hat{\rho}(t_0) = 0 = \hat{\rho}(t_0)\hat{\rho}_F. \tag{14.1–13}$$

An example of such a problem occurs in the photoelectric detection of light, where we start with bound electrons and are interested in the probability that one or more of them becomes a free electron under the influence of the electromagnetic interaction. To obtain the probability $P(t_0 + T)$, we have to project the final state on to the state of interest, so that we obtain

$$P(t_0 + T) = \mathrm{Tr}\,[\hat{\rho}_F\hat{\rho}(t_0 + T)]$$

and from Eq. (14.1–11)

$$\begin{aligned} P(t_0 + T) = {} & \mathrm{Tr}\,\hat{\rho}_F\hat{\rho}(t_0) + \frac{1}{i\hbar}\mathrm{Tr}\,\hat{\rho}_F\int_{t_0}^{t_0+T} \mathrm{d}t_1[\hat{H}_I(t_1), \hat{\rho}(t_0)] \\ & + \frac{1}{(i\hbar)^2}\mathrm{Tr}\,\hat{\rho}_F\int_{t_0}^{t_0+T}\mathrm{d}t_1\int_{t_0}^{t_1}\mathrm{d}t_2[\hat{H}_I(t_1), [\hat{H}_I(t_2), \hat{\rho}(t_0)]] \\ & + \text{etc.} \end{aligned}$$

By virtue of the orthogonality expressed by Eq. (14.1–13) the first two terms in the expansion vanish, and we have to the lowest non-vanishing order,

$$P(t_0 + T) = \frac{1}{\hbar^2}\int_{t_0}^{t_0+T}\mathrm{d}t_1\int_{t_0}^{t_1}\mathrm{d}t_2\langle\Psi_F|\hat{H}_I(t_1)\hat{\rho}(t_0)\hat{H}_I(t_2)|\Psi_F\rangle + \text{c.c.} \tag{14.1–14}$$

We shall have occasion to use this simple expression in several problems below.

14.1.2 The electromagnetic interaction between fields and charges

We have already shown in Chapter 10 that the energy $\hat{H}_F$ of the free, quantized electromagnetic field may be expressed in the form (in SI units)

$$\hat{H}_F = \tfrac{1}{2}\int\left[\varepsilon_0\hat{\mathbf{E}}^2(\mathbf{r}, t) + \frac{1}{\mu_0}\hat{\mathbf{B}}^2(\mathbf{r}, t)\right]\mathrm{d}^3r = \sum_{\mathbf{k},s}(\hat{n}_{\mathbf{k}s} + \tfrac{1}{2})\hbar\omega.$$

Here $\hat{\mathbf{E}}$, $\hat{\mathbf{B}}$, $\hat{n}_{\mathbf{k}s}$ are operators in the Hilbert space of the field. On the other hand, the energy of a particle of charge e, of mass m, and momentum $\hat{\mathbf{p}}$ located at position $\mathbf{r}$ in some external potential $U(\mathbf{r})$ is given by

$$\hat{H}_A = \frac{\hat{\mathbf{p}}^2}{2m} + eU(\hat{\mathbf{r}}),$$

in which $\hat{\mathbf{p}}$ and $\hat{\mathbf{r}}$ must be regarded as operators in the Hilbert space of the particle.

In order to obtain an expression for the interaction between the charged particle, such as an electron, and the field, we shall allow ourselves to be guided by classical electromagnetic theory. According to this theory the canonical momentum of a system consisting of a charge e in an electromagnetic field characterized by the vector potential $\mathbf{A}(\mathbf{r}, t)$ is obtained by the substitution

$$\mathbf{p} \to \mathbf{p} - e\mathbf{A}(\mathbf{r}, t),$$

where $\mathbf{r}$ is the position of the charge. The same procedure also holds in quantum mechanics, except that $\hat{\mathbf{p}}$ and $\hat{\mathbf{r}}$ become operators in the Hilbert space of the charged particle, while $\hat{\mathbf{A}}$ becomes an operator in the Hilbert space of the field. If we combine this with the expression for $\hat{H}_{\mathrm{F}}$ and $\hat{H}_{\mathrm{A}}$, we obtain for the total energy $\hat{H}$ of the coupled system

$$\hat{H} = \tfrac{1}{2}\int\left[\varepsilon_0\hat{\mathbf{E}}^2(\mathbf{r}, t) + \frac{1}{\mu_0}\hat{\mathbf{B}}^2(\mathbf{r}, t)\right]\mathrm{d}^3r + \frac{1}{2m}[\hat{\mathbf{p}} - e\hat{\mathbf{A}}(\hat{\mathbf{r}}, t)]^2 + eU(\hat{\mathbf{r}}) \quad (14.1\text{–}15a)$$

$$= \sum_{\mathbf{k},s}(\hat{n}_{\mathbf{k}s} + \tfrac{1}{2})\hbar\omega + \left[\frac{\hat{\mathbf{p}}^2}{2m} + eU(\hat{\mathbf{r}})\right]$$

$$- \frac{e}{2m}[\hat{\mathbf{p}}\cdot\hat{\mathbf{A}}(\hat{\mathbf{r}}, t) + \hat{\mathbf{A}}(\hat{\mathbf{r}}, t)\cdot\hat{\mathbf{p}}] + \frac{e^2}{2m}\hat{\mathbf{A}}^2(\hat{\mathbf{r}}, t). \quad (14.1\text{–}15b)$$

The first and second terms on the right of this equation are, of course, $\hat{H}_{\mathrm{F}}$ and $\hat{H}_{\mathrm{A}}$, respectively, whereas the remaining terms can be identified as the interaction $\hat{H}_{\mathrm{I}}(t)$. This form of $\hat{H}$ is known as the *minimal coupling form* of the Hamiltonian.

At first sight it may seem strange that the interaction energy $\hat{H}_{\mathrm{I}}$ should depend on the potential $\hat{\mathbf{A}}$, which in turn depends on the chosen gauge. So far we have been working in the Coulomb gauge, in which $\hat{\mathbf{A}}$ is a transverse vector field, and it should be understood that $\hat{\mathbf{A}}$ in Eqs. (14.1–15) is in the Coulomb gauge. However, we may replace $\hat{\mathbf{A}}$ by the transverse part of $\hat{\mathbf{A}}_{\mathrm{T}}$ of the vector potential $\hat{\mathbf{A}}$ to obtain a gauge-invariant expression for $\hat{H}_{\mathrm{I}}$. By using the Coulomb gauge, in which $\hat{\mathbf{A}}_{\mathrm{T}}$ and $\hat{\mathbf{A}}$ coincide, we simply avoid the need to distinguish between them.

A number of simplifications are often made in the expression for $\hat{H}_{\mathrm{I}}(t)$. When the vector potential $\hat{\mathbf{A}}(\hat{\mathbf{r}}, t)$ of the field is expressed in the Coulomb gauge, it is possible to treat $\hat{\mathbf{p}}$ and $\hat{\mathbf{A}}(\hat{\mathbf{r}}, t)$ as commuting variables, even though $\hat{\mathbf{p}}$ and $\hat{\mathbf{r}}$ do not commute. To see this, we let $|\Psi\rangle$ be any quantum state of the atomic system and let $|\mathbf{r}\rangle$ be any position state so that

$$\langle\mathbf{r}|\Psi\rangle \equiv \Psi(\mathbf{r}).$$

Then we find for the following matrix element of the scalar product commutator

$$\langle\mathbf{r}|[\hat{\mathbf{p}}\cdot\hat{\mathbf{A}}(\hat{\mathbf{r}}, t)]|\Psi\rangle = \frac{\hbar}{\mathrm{i}}[\nabla\cdot\hat{\mathbf{A}}(\mathbf{r}, t)\Psi(\mathbf{r}) - \hat{\mathbf{A}}(\mathbf{r}, t)\cdot\nabla\Psi(\mathbf{r})]$$

$$= \frac{\hbar}{\mathrm{i}}[\nabla\cdot\hat{\mathbf{A}}(\mathbf{r}, t)]\Psi(\mathbf{r})$$

$$= 0,$$

by virtue of the fact that $\nabla \cdot \hat{\mathbf{A}}(\mathbf{r}, t) = 0$ in the Coulomb gauge. As this holds for any state $|\Psi\rangle$, it follows that we need not distinguish between $\hat{\mathbf{p}} \cdot \hat{\mathbf{A}}(\hat{\mathbf{r}}, t)$ and $\hat{\mathbf{A}}(\hat{\mathbf{r}}, t) \cdot \hat{\mathbf{p}}$ in the Coulomb gauge, even though $\hat{\mathbf{p}}$ and $\hat{\mathbf{r}}$ do not commute.

A further simplification is possible when the atomic state $|\Psi\rangle$ or $\langle \Psi|$ on which $\hat{\mathbf{A}}(\hat{\mathbf{r}}, t)$ acts, either to the right or to the left, is a bound state such that the wave function $\Psi(\mathbf{r})$ vanishes for $\mathbf{r}$ outside some small region surrounding a point $\mathbf{r}_0$. To show this we make use of the plane-wave mode expansion of the vector potential together with the completeness of the states $|\mathbf{r}\rangle$, and write

$$
\begin{aligned}
&\hat{\mathbf{A}}(\hat{\mathbf{r}}, t)|\Psi\rangle \\
&= \frac{1}{L^{3/2}} \sum_{\mathbf{k},s} \left(\frac{\hbar}{2\omega\varepsilon_0}\right)^{1/2} [\hat{a}_{\mathbf{k}s}(t)\boldsymbol{\varepsilon}_{\mathbf{k}s}\, \mathrm{e}^{\mathrm{i}\mathbf{k}\cdot\hat{\mathbf{r}}} + \mathrm{h.c.}]|\Psi\rangle \\
&= \frac{1}{L^{3/2}} \sum_{\mathbf{k},s} \left(\frac{\hbar}{2\omega\varepsilon_0}\right)^{1/2} [\hat{a}_{\mathbf{k}s}(t)\boldsymbol{\varepsilon}_{\mathbf{k}s}\, \mathrm{e}^{\mathrm{i}\mathbf{k}\cdot\hat{\mathbf{r}}} + \mathrm{h.c.}]\int |\mathbf{r}\rangle\langle\mathbf{r}|\Psi\rangle\, \mathrm{d}\mathbf{r} \\
&= \frac{1}{L^{3/2}} \sum_{\mathbf{k},s} \left(\frac{\hbar}{2\omega\varepsilon_0}\right)^{1/2} \left[\hat{a}_{\mathbf{k}s}(t)\boldsymbol{\varepsilon}_{\mathbf{k}s}\int \mathrm{e}^{\mathrm{i}\mathbf{k}\cdot\mathbf{r}}\,\Psi(\mathbf{r})\,\mathrm{d}\mathbf{r} + \hat{a}^\dagger_{\mathbf{k}s}(t)\boldsymbol{\varepsilon}^*_{\mathbf{k}s}\int \mathrm{e}^{-\mathrm{i}\mathbf{k}\cdot\mathbf{r}}\,\Psi(\mathbf{r})\,\mathrm{d}\mathbf{r}\right]|\mathbf{r}\rangle
\end{aligned}
$$

If the spread of the wave function $\Psi(\mathbf{r})$ about $\mathbf{r}_0$ in the direction $\mathbf{k}$ is much smaller than the wavelength $2\pi/k$ of any occupied mode $\mathbf{k}$, s of the electromagnetic field, we can evidently replace the factors $\mathrm{e}^{\pm\mathrm{i}\mathbf{k}\cdot\mathbf{r}}$ by $\mathrm{e}^{\pm\mathrm{i}\mathbf{k}\cdot\mathbf{r}_0}$ under the integrals to a good degree of approximation. Hence

$$\hat{\mathbf{A}}(\hat{\mathbf{r}}, t)|\Psi\rangle \approx \hat{\mathbf{A}}(\mathbf{r}_0, t)|\Psi\rangle,$$

and the position operator $\hat{\mathbf{r}}$ in the vector potential can be replaced by its average value and treated as a constant. This is usually known as the *dipole approximation*. It has immediate application to the interaction between light and an atomic electron for example, as the size of an atom is much smaller than any optical wavelength. When this replacement is applicable the Hamiltonian can be written in the simpler form

$$\hat{H} = \tfrac{1}{2}\int\left[\varepsilon_0\hat{\mathbf{E}}^2(\mathbf{r}, t) + \frac{1}{\mu_0}\hat{\mathbf{B}}^2(\mathbf{r}, t)\right]\mathrm{d}^3r + \frac{1}{2m}[\hat{\mathbf{p}} - e\hat{\mathbf{A}}(\mathbf{r}_0, t)]^2 + eU(\hat{\mathbf{r}}) \tag{14.1–16a}$$

$$= \sum_{\mathbf{k},s}\hbar\omega(\hat{n}_{\mathbf{k}s} + \tfrac{1}{2}) + \frac{\hat{\mathbf{p}}^2}{2m} + eU(\hat{\mathbf{r}}) - \frac{e}{m}\hat{\mathbf{p}}\cdot\hat{\mathbf{A}}(\mathbf{r}_0, t) + \frac{e^2}{2m}\hat{\mathbf{A}}^2(\mathbf{r}_0, t). \tag{14.1–16b}$$

Finally, for all but the most intense optical fields, we may ignore the interaction term $e^2\hat{\mathbf{A}}^2/2m$ compared with the term $e\hat{\mathbf{p}}\cdot\hat{\mathbf{A}}/m$. To see this we note that the c-number ratio

$$\mathcal{R} = \frac{e^2A^2/m}{epA/m} = \frac{eA}{p}$$

can be re-expressed in the approximate form

$$\mathcal{R} \sim \frac{eE}{\omega p} \sim \frac{eE\lambda}{pc} \sim \frac{v}{c}\frac{eE\lambda}{pv} \sim \left(\frac{v}{c}\right)\left(\frac{\text{work done on charge } e \text{ by the field in distance } \lambda}{\text{kinetic energy of charge } e}\right)$$

where E is the electric field, ω and λ are its frequency and wavelength, and v is the electron velocity. The kinetic energies of electrons encountered in most optical interactions are of the order of a few electron volts. Even if we are dealing with the light from a laser of 10^6 W power concentrated in a 1 mm^2 cross-section, for which $E \sim 2 \times 10^7$ V/m at a typical wavelength of 0.6 μm, the ratio $\mathcal{R}$ is only about 0.02. It is apparent therefore that for all weak or moderately intense light beams the $e^2\mathbf{A}^2/2m$ term in the interactions can be neglected. Only at the highest light powers, where multiphoton interactions dominate, does this term become important. We shall therefore further simplify the expression for the interaction energy and write

$$\hat{H}_{\text{I}}(t) \approx -\frac{e}{m}\hat{\mathbf{p}}(t) \cdot \hat{\mathbf{A}}(\mathbf{r}_0, t). \tag{14.1–17}$$

14.1.3 Multipolar Hamiltonian

Besides the minimal coupling form of the Hamiltonian given by Eqs. (14.1–16) or (14.1–17), another form, the so-called *multipolar Hamiltonian* is sometimes found to be more convenient for treating the interaction between an atomic electron and a field (Lamb, 1952; Power and Zienau, 1959). This other form is derivable from Eq. (14.1–16) by the unitary transformation (Ackerhalt and Milonni, 1984; we largely follow their treatment; see also Milonni, Cook and Ackerhalt, 1989),

$$\hat{U} \equiv \exp[-ie\hat{\mathbf{r}} \cdot \hat{\mathbf{A}}(\mathbf{r}_0, t)/\hbar], \tag{14.1–18}$$

where $\mathbf{r}_0$ may be taken to be the coordinate of the atomic nucleus for example. We shall label the new operators resulting from the transformation $\hat{U}$ by the superscript (N). Evidently

$$\left.\begin{aligned} \hat{\mathbf{r}}^{(\text{N})} &= \hat{U}^\dagger \hat{\mathbf{r}}\, \hat{U} = \hat{\mathbf{r}} \\ \hat{\mathbf{A}}^{(\text{N})}(\mathbf{r}, t) &= \hat{U}^\dagger \hat{\mathbf{A}}(\mathbf{r}, t)\hat{U} = \hat{\mathbf{A}}(\mathbf{r}, t) \\ \hat{\mathbf{B}}^{(\text{N})}(\mathbf{r}, t) &= \hat{U}^\dagger \hat{\mathbf{B}}(\mathbf{r}, t)\hat{U} = \hat{\mathbf{B}}(\mathbf{r}, t), \end{aligned}\right\} \tag{14.1–19}$$

the last because $\hat{\mathbf{A}}(\mathbf{r}_0, t)$ commutes with $\hat{\mathbf{B}}(\mathbf{r}, t)$. However, both the momentum $\hat{\mathbf{p}}$ and the electric field $\hat{\mathbf{E}}(\mathbf{r}, t)$ change under the unitary transformation. We obtain with the help of the operator expansion theorem [Eq. (10.11–1)]

$$\begin{aligned} \hat{\mathbf{p}}^{(\text{N})} &= \exp[ie\hat{\mathbf{r}} \cdot \hat{\mathbf{A}}(\mathbf{r}_0, t)/\hbar]\hat{\mathbf{p}} \exp[-ie\hat{\mathbf{r}} \cdot \hat{\mathbf{A}}(\mathbf{r}_0, t)/\hbar] \\ &= \hat{\mathbf{p}} + \frac{ie}{\hbar}[\hat{\mathbf{r}} \cdot \hat{\mathbf{A}}(\mathbf{r}_0, t), \hat{\mathbf{p}}] \\ &= \hat{\mathbf{p}} - e\hat{\mathbf{A}}(\mathbf{r}_0, t), \end{aligned} \tag{14.1–20}$$

and

$$\hat{E}_i^{(\mathrm{N})}(\mathbf{r}, t) = \exp[ie\hat{\mathbf{r}}\cdot\hat{\mathbf{A}}(\mathbf{r}_0, t)/\hbar]\hat{E}_i(\mathbf{r}, t)\exp[-ie\hat{\mathbf{r}}\cdot\hat{\mathbf{A}}(\mathbf{r}_0, t)/\hbar]$$
$$= \hat{E}_i(\mathbf{r}, t) + \frac{ie}{\hbar}\hat{r}_j[\hat{A}_j(\mathbf{r}_0, t), \hat{E}_i(\mathbf{r}, t)]$$
$$= \hat{E}_i(\mathbf{r}, t) + \frac{e\hat{r}_j}{\varepsilon_0}\delta_{ji}^T(\mathbf{r} - \mathbf{r}_0), \tag{14.1–21}$$

when we make use of the commutation relation (10.8–15), with summation on repeated indices implied. Now

$$e\hat{r}_j\delta_{ji}^T(\mathbf{r} - \mathbf{r}_0) \equiv \hat{P}_i(\mathbf{r}, t) \tag{14.1–22}$$

is the transverse polarization produced by the point charge e in both the old and the transformed variables. It follows that Eq. (14.1–21) may be written

$$\hat{\mathbf{E}}^{(\mathrm{N})}(\mathbf{r}, t) = \hat{\mathbf{E}}(\mathbf{r}, t) + \frac{1}{\varepsilon_0}\hat{\mathbf{P}}(\mathbf{r}, t) = \frac{1}{\varepsilon_0}\hat{\mathbf{D}}(\mathbf{r}, t), \tag{14.1–23}$$

where $\hat{\mathbf{D}}(\mathbf{r}, t)$ is the transverse dielectric displacement vector.

If we now substitute for $\hat{\mathbf{r}}$, $\hat{\mathbf{p}}$, $\hat{\mathbf{A}}$, $\hat{\mathbf{E}}$, $\hat{\mathbf{B}}$ from Eqs. (14.1–19), (14.1–20) and (14.1–23) into Eq. (14.1–16a), we obtain the Hamiltonian in terms of the transformed variables in the form

$$\hat{H} = \tfrac{1}{2}\int\left[\varepsilon_0\hat{\mathbf{E}}^{(\mathrm{N})2}(\mathbf{r}, t) + \frac{1}{\mu_0}\hat{\mathbf{B}}^{(\mathrm{N})2}(\mathbf{r}, t)\right]\mathrm{d}^3r + \left[\frac{\hat{\mathbf{p}}^{(\mathrm{N})2}}{2m} + eU(\hat{\mathbf{r}}^{(\mathrm{N})})\right]$$
$$- \int\hat{\mathbf{E}}^{(\mathrm{N})}(\mathbf{r}, t)\cdot\hat{\mathbf{P}}(\mathbf{r}, t)\,\mathrm{d}^3r + \frac{1}{2\varepsilon_0}\int\hat{\mathbf{P}}^2(\mathbf{r}, t)\,\mathrm{d}^3r. \tag{14.1–24}$$

If the first two terms are interpreted as the non-interacting parts of the Hamiltonian in the new variables, then the other two terms represent the interaction. The last term is usually important only in level shift calculation, and if we discard it and make use of Eq. (14.1–22) for $\hat{\mathbf{P}}(\mathbf{r}, t)$ and Eq. (14.1–23) for $\hat{\mathbf{D}}(\mathbf{r}, t)$, the interaction takes the form

$$\hat{H}_\mathrm{I} = -e\int\hat{r}_j\delta_{ji}^T(\mathbf{r} - \mathbf{r}_0)\hat{E}_i^{(\mathrm{N})}(\mathbf{r}, t)\,\mathrm{d}^3r$$
$$= -e\hat{\mathbf{r}}\cdot\hat{\mathbf{E}}^{(\mathrm{N})}(\mathbf{r}_0, t)$$
$$= -\frac{e}{\varepsilon_0}\hat{\mathbf{r}}\cdot\hat{\mathbf{D}}(\mathbf{r}_0, t). \tag{14.1–25}$$

As $e\hat{\mathbf{r}}$ is the dipole moment of the charge e, this is known as the *electric dipole* form of the interaction, expressed in terms of the original variables. However, it must be emphasized that the 'non-interacting' part of the total energy is not the same as that in Eqs. (14.1–16) so that the interaction $\hat{H}_\mathrm{I}$ has been re-defined. It differs from the more familiar expression $-e\hat{\mathbf{r}}\cdot\hat{\mathbf{E}}(\mathbf{r}_0, t)$ which is sometimes used in place of Eq. (14.1–25) by the added polarization term. Discarding the polarization can lead to inconsistencies at times. This question has been the subject of some controversy (Woolley, 1971; Fried, 1973; Kobe, 1978, 1979; Carter and Kelley, 1979; Mandel, 1979; Power and Thirunamachandran, 1980,

1982; Healy, 1980, 1982; Haller, 1982; Ackerhalt and Milonni, 1984; Milonni, Cook and Ackerhalt, 1989).

In cases where we are dealing with a bound electron interacting with an oscillating electromagnetic field, a much simpler, approximate argument can be given for the multipolar form of the electromagnetic interaction. We first note that the so-called 'kinetic momentum' of the electron, defined by the formula

$$\hat{\mathbf{p}}^{(\mathrm{Kin})} = m\frac{\mathrm{d}\hat{\mathbf{r}}}{\mathrm{d}t} = \frac{m}{\mathrm{i}\hbar}[\hat{\mathbf{r}}, \hat{H}],$$

can be expressed with the help of Eq. (14.1–16a) and the summation convention as

$$\begin{aligned}\hat{p}_j^{(\mathrm{Kin})} &= \frac{m}{\mathrm{i}\hbar}\left\{\left[\hat{r}_j, \frac{\hat{p}_i\hat{p}_i}{2m}\right] - \frac{e}{m}[\hat{r}_j, \hat{p}_k]\hat{A}_k(\mathbf{r}_0, t)\right\} \\ &= \hat{p}_j - e\hat{A}_j(\mathbf{r}_0, t). \qquad (14.1\text{–}26)\end{aligned}$$

We now take the scalar product of $\hat{\mathbf{p}}^{(\mathrm{Kin})}$ with $(-e/m)\hat{\mathbf{A}}(\mathbf{r}_0, t)$, and find that

$$\begin{aligned}-\frac{e}{m}\hat{\mathbf{p}}^{(\mathrm{Kin})}\cdot\hat{\mathbf{A}}(\mathbf{r}_0, t) &= -\frac{e}{m}\hat{\mathbf{p}}\cdot\hat{\mathbf{A}}(\mathbf{r}_0, t) + \frac{e^2}{m}\hat{\mathbf{A}}^2(\mathbf{r}_0, t) \\ &= -e\frac{\partial}{\partial t}[\hat{\mathbf{r}}\cdot\hat{\mathbf{A}}(\mathbf{r}_0, t)] + e\hat{\mathbf{r}}\cdot\frac{\partial\hat{\mathbf{A}}(\mathbf{r}_0, t)}{\partial t} \\ &= -e\frac{\partial}{\partial t}[\hat{\mathbf{r}}\cdot\hat{\mathbf{A}}(\mathbf{r}_0, t)] - e\hat{\mathbf{r}}\cdot\hat{\mathbf{E}}(\mathbf{r}_0, t). \qquad (14.1\text{–}27)\end{aligned}$$

We now average each term over one cycle of oscillation. If $\ddot{\hat{\mathbf{r}}}$ and $\hat{\mathbf{E}}$ are nearly in phase, then so are $\hat{\mathbf{r}}$ and $\hat{\mathbf{E}}$ and so are $\hat{\mathbf{p}}$ and $\hat{\mathbf{A}}(\mathbf{r}_0, t)$. On the other hand, $\hat{\mathbf{r}}$ and $\hat{\mathbf{A}}(\mathbf{r}_0, t)$ will be approximately $\pi/2$ out of phase, so that the first term on the right of Eq. (14.1–27) has an average close to zero. Hence

$$-\frac{e}{m}\hat{\mathbf{p}}^{(\mathrm{Kin})}\cdot\hat{\mathbf{A}}(\mathbf{r}_0, t) \approx -e\hat{\mathbf{r}}\cdot\hat{\mathbf{E}}(\mathbf{r}_0, t) = -\frac{e}{m}\hat{\mathbf{p}}\cdot\hat{\mathbf{A}}(\mathbf{r}_0, t) + \frac{e^2}{m}\hat{\mathbf{A}}^2(\mathbf{r}_0, t) \qquad (14.1\text{–}28)$$

14.2 The one-electron photodetection probability

We have already noted in Chapter 12 that most measurements of light are based on the absorption of photons via the photoelectric effect. The situation is quite different and more complicated at longer wavelengths, in the radio frequency domain for example, where the electric field itself is easily measurable, and the interaction with the measuring instrument involves both emissions and absorptions. But optical periods are sufficiently short that in any measurable time interval many oscillations of the field occur, and these allow the final state energy to be well defined if it was well defined initially.

In order to tackle the detection problem, we consider the somewhat idealized photodetector illustrated in Fig. 14.1. It has a conducting photosensitive surface, the photocathode, of area S and of thickness much smaller than any optical wavelength, and it is illuminated at normal incidence by a light beam. For simplicity we assume that over the surface S the field looks like a plane wave. The

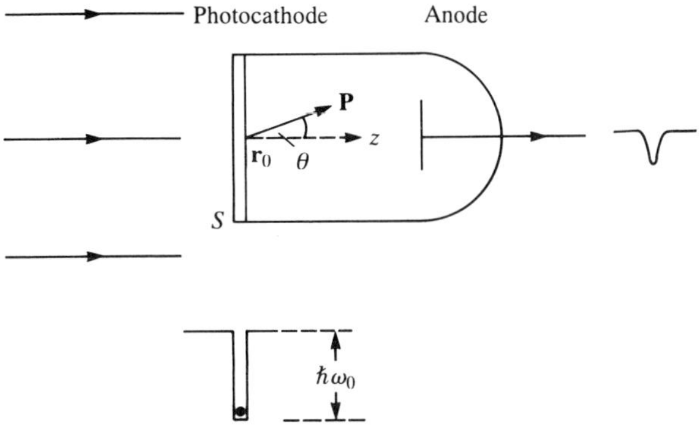

Fig. 14.1 An idealized photodetector with a bound electron in a potential well.

cathode contains many electrons in bound states, some of which may be emitted under the influence of the incident light. Any emitted electron is rapidly attracted towards an anode, where it produces a detectable electric pulse (possibly after further amplification).

We begin by focusing on a single electron located at position $\mathbf{r}_0$, which is initially in some bound state $|\Phi_0\rangle$ with negative energy eigenvalue $E_0 = -\hbar\omega_0$ of the non-interacting electron Hamiltonian, so that

$$\hat{H}_A|\Phi_0\rangle = E_0|\Phi_0\rangle = -\hbar\omega_0|\Phi_0\rangle. \tag{14.2–1}$$

We shall assume that the binding energy $|E_0|$ is of order 1 or 2 eV, as is typical for an optical photodetector, so that ω_0 is a frequency of order $10^{15}\,\mathrm{s}^{-1}$. We now suppose that this electron is acted on by an optical field which is in some quantum state characterized by the density operator $\hat{\rho}_F(t_0)$ at the initial time t_0 when the interaction commences. The density operator $\hat{\rho}(t_0)$ at time t_0 of the combined system of electron and field then factorizes into the direct product

$$\hat{\rho}(t_0) = |\Phi_0\rangle\langle\Phi_0| \otimes \hat{\rho}_F(t_0). \tag{14.2–2}$$

We now proceed to calculate the probability $P(r_0, t_0, \Delta t)$ that, under the influence of the electromagnetic field, the electron makes a transition to an unbound or positive energy state $|\Phi\rangle$ within some short time interval Δt that is, however, very long compared with the optical period. The implication is that a free electron will be rapidly attracted to the anode and give rise to an electric pulse that registers the photoelectric emission at time $t_0 + \Delta t$. We shall see that the solution to this problem lends itself to generalization.

We take the interaction energy $\hat{H}_I(t)$ of the coupled system to be given by Eq. (14.1–17), with both the electron momentum $\hat{\mathbf{p}}(t)$ and the vector potential $\hat{\mathbf{A}}(\mathbf{r}_0, t)$ of the field expressed in the interaction picture. Then

$$\hat{\mathbf{p}}(t) = \exp[\mathrm{i}\hat{H}_A(t - t_0)/\hbar]\hat{\mathbf{p}}\exp[-\mathrm{i}\hat{H}_A(t - t_0)/\hbar], \tag{14.2–3}$$

and

$$\hat{\mathbf{A}}(\mathbf{r}_0, t) = \frac{1}{L^{3/2}}\sum_{\mathbf{k},s}\left(\frac{\hbar}{2\omega\varepsilon_0}\right)^{1/2}[\hat{a}_{\mathbf{k}s}\boldsymbol{\varepsilon}_{\mathbf{k}s}\,\mathrm{e}^{\mathrm{i}[\mathbf{k}\cdot\mathbf{r}_0-\omega(t-t_0)]} + \mathrm{h.c.}]$$

$$\equiv \hat{\mathbf{A}}^{(+)}(\mathbf{r}_0, t) + \hat{\mathbf{A}}^{(-)}(\mathbf{r}_0, t), \tag{14.2–4}$$

where $\hat{\mathbf{p}}$ and $\hat{a}_{\mathbf{k}s}$ without time argument denote the operators at time t_0 when the interaction commences.

Let $|\,\Phi\rangle$ be a free electron energy state with positive eigenvalue E, so that

$$\hat{H}_{\mathrm{A}}|\Phi\rangle = E|\Phi\rangle, \quad E > 0, \tag{14.2–5}$$

and let $|\chi\rangle$ be an arbitrary final state of the electromagnetic field. As we are not usually interested in the final state of the field, we shall eventually sum over all possible states $|\chi\rangle$. Then the probability that the system makes a transition within time Δt from the initial state to the final state $|\Phi\rangle|\chi\rangle$ is given by the expression $\langle\chi|\langle\Phi|\hat{\rho}(t_0 + \Delta t)|\Phi\rangle|\chi\rangle$. Because the final state is orthogonal to the initial state, we can make use of the simplified relation (14.1–14) obtained from perturbation theory, and write:

Probability of transition to state $|\Phi\rangle|\chi\rangle$ at time $t_0 + \Delta t$

$$= \frac{1}{\hbar^2}\int_{t_0}^{t_0+\Delta t} \mathrm{d}t_1 \int_{t_0}^{t_1} \mathrm{d}t_2 \langle\chi|\langle\Phi|\hat{H}_{\mathrm{I}}(t_1)\hat{\rho}(t_0)\hat{H}_{\mathrm{I}}(t_2)|\Phi\rangle|\chi\rangle + \mathrm{c.c.}$$

We now substitute for $\hat{\rho}(t_0)$ and $\hat{H}_{\mathrm{I}}(t)$ from Eqs. (14.2–2) and (14.1–17), and use Eqs. (14.2–3) and (14.2–4) under the integral. As electron and field operators commute at the same time, the matrix element under the integral may be decomposed into a product of two matrix elements, one for the electron and one for the field. The operators $\exp[\pm\mathrm{i}\hat{H}_{\mathrm{A}}(t - t_0)/\hbar]$ are all adjacent to their own eigenstates, and after replacing $\hat{H}_{\mathrm{A}}$ by E or E_0 with the help of Eqs. (14.2–1) and (14.2–5), we obtain:

Prob. of transition to state $|\Phi\rangle|\chi\rangle$ at time $t_0 + \Delta t$

$$= \left(\frac{e}{m\hbar}\right)^2 \langle\Phi|\hat{p}_i|\Phi_0\rangle\langle\Phi_0|\hat{p}_j|\Phi\rangle \int_{t_0}^{t_0+\Delta t} \mathrm{d}t_1 \int_{t_0}^{t_1} \mathrm{d}t_2\, \mathrm{e}^{\mathrm{i}(E-E_0)(t_1-t_2)/\hbar}$$
$$\times \langle\chi|\hat{A}_i(\mathbf{r}_0, t_1)\hat{\rho}_{\mathrm{F}}(t_0)\hat{A}_j(\mathbf{r}_0, t_2)|\chi\rangle + \mathrm{c.c.} \tag{14.2–6}$$

Generally we have no interest in, and make no attempt to determine, the final state $|\chi\rangle$ of the field. The probability of a transition to a free electron state, irrespective of the final state of the field, is then obtained by summing over all possible states $|\chi\rangle$.

Let us now make a diagonal coherent-state representation of $\hat{\rho}_{\mathrm{F}}(t_0)$ (see Section 11.8) by putting

$$\hat{\rho}_{\mathrm{F}}(t_0) = \int \phi(\{v\}, t_0)|\{v\}\rangle\langle\{v\}|\,\mathrm{d}\{v\},$$

and substitute this in Eq. (14.2–6). On interchanging the order of the two matrix elements $\langle\chi|\hat{A}_i(\mathbf{r}_0, t_1)|\{v\}\rangle$ and $\langle\{v\}|\hat{A}_j(\mathbf{r}_0, t_2)|\chi\rangle$ under the integral, and summing over a complete set of all final states $|\chi\rangle$ of the field, we arrive at the probability of an electron transition irrespective of the final field state. With the help of the identity

$$\sum_{\mathrm{all}\,\chi} |\chi\rangle\langle\chi| = 1,$$

we obtain

Prob. of transition to electron state $|\Phi\rangle$ at time $t_0 + \Delta t$,
irrespective of the final state of the field

$$= \left(\frac{e}{m\hbar}\right)^2 \langle \Phi|\hat{p}_i|\Phi_0\rangle\langle \Phi_0|\hat{p}_j|\Phi\rangle \int_{t_0}^{t_0+\Delta t} dt_1 \int_{t_0}^{t_1} dt_2\, e^{i(E-E_0)(t_1-t_2)/\hbar}$$
$$\times \langle \hat{A}_j(\mathbf{r}_0, t_2)\hat{A}_i(\mathbf{r}_0, t_1)\rangle + \text{c.c.}, \tag{14.2–7}$$

where the factor involving the expectation of the field stands for

$$\langle \hat{A}_j(\mathbf{r}_0, t_2)\hat{A}_i(\mathbf{r}_0, t_1)\rangle = \int \phi(\{v\}, t_0)\langle\{v\}|\hat{A}_j(\mathbf{r}_0, t_2)\hat{A}_i(\mathbf{r}_0, t_1)|\{v\}\rangle\, d\{v\}.$$

Finally, we sum the probability in Eq. (14.2–7) over all final electron states $|\Phi\rangle$ of positive energy E, in order to arrive at the photoelectric detection probability. In the course of performing this sum we can also make allowance for the possibility that electrons in different final states $|\Phi\rangle$ have different probabilities g of being collected and registered by the detector. In order to make this more explicit we shall suppose that the final electron state $|\Phi(E, \Omega)\rangle$ is characterized by the electron energy E [cf. Eq. (14.2–5)] and possibly by other variables represented collectively by Ω, and that the probability $g(E, \Omega)$ also depends on these variables. We express the sums over E and Ω as integrals with the help of the density of states $\sigma(E, \Omega)$, where $\sigma(E, \Omega)\, dE\, d\Omega$ is the number of final electron states within $dE\, d\Omega$. Hence we have from Eq. (14.2–7) (Kimble and Mandel, 1984)

Prob. of photoelectric detection at time $t_0 + \Delta t$

$$= \int_{t_0}^{t_0+\Delta t} dt_1 \int_{t_0}^{t_1} dt_2 k_{ij}(t_1 - t_2)\langle \hat{A}_j(\mathbf{r}_0, t_2)\hat{A}_i(\mathbf{r}_0, t_1)\rangle + \text{c.c.}, \tag{14.2–8}$$

where we have written

$$k_{ij}(\tau) \equiv \left(\frac{e}{m\hbar}\right)^2 \int_0^\infty dE \int d\Omega\, \sigma(E, \Omega) g(E, \Omega)\, e^{i(E-E_0)\tau/\hbar} \langle \Phi(E, \Omega)|\hat{p}_i|\Phi_0\rangle$$
$$\times \langle \Phi_0|\hat{p}_j|\Phi(E, \Omega)\rangle \tag{14.2–9}$$

for the effective response function of the detector.

Inspection of Eq. (14.2–9) shows that $k_{ij}(\tau)$ obeys the symmetry condition

$$k_{ji}(-\tau) = k_{ij}^*(\tau). \tag{14.2–10}$$

If we make the change of variable $E - E_0 = \hbar\omega$, so that the energy integral in Eq. (14.2–9) becomes a frequency integral, then the Fourier transform $K_{ij}(\omega)$ of $k_{ij}(\tau)$, which is the frequency response function of the photodetector, is evidently given by

$$K_{ij}(\omega) = 2\pi\hbar\left(\frac{e}{m\hbar}\right)^2 \int g(E_0 + \hbar\omega, \Omega)\sigma(E_0 + \hbar\omega, \Omega)\langle \Phi(E_0 + \hbar\omega, \Omega)|\hat{p}_i|\Phi_0\rangle$$
$$\times \langle \Phi_0|\hat{p}_j|\Phi(E_0 + \hbar\omega, \Omega)\rangle \Theta(\omega - \omega_0)\, d\Omega. \tag{14.2–11}$$

Here $\Theta(\omega)$ is the Heaviside unit step function defined by $\Theta(\omega) = 0$ for $\omega < 0$ and

$\Theta(\omega) = 1$ for $\omega > 0$, and ω_0 is given by Eq. (14.2–1). Because of the presence of the step function $\Theta(\omega - \omega_0)$, we have for any positive frequency ω, by Fourier inversion of $k_{ij}(\tau)$,

$$\left.\begin{aligned} \int_{-\infty}^{\infty} k_{ij}(\tau)\,\mathrm{e}^{-\mathrm{i}\omega\tau}\,\mathrm{d}\tau &= K_{ij}(\omega) \\ \int_{-\infty}^{\infty} k_{ij}(\tau)\,\mathrm{e}^{\mathrm{i}\omega\tau}\,\mathrm{d}\tau &= K_{ij}(-\omega) = 0. \end{aligned}\right\} \tag{14.2–12}$$

Now it is typical of most photodetectors that the frequency response $K_{ij}(\omega)$ has a large bandwidth, often of the same order as the optical frequency ω_0. Under these conditions the range in time of $k_{ij}(\tau)$ is extremely short, of the order of the optical period. The infinite limits of integration in Eq. (14.2–12) can then, to a good approximation, be replaced by finite limits $-\tau_1$, τ_2, provided τ_1 and τ_2 are at least as long as several optical periods, and we may write

$$\left.\begin{aligned} \int_{-\tau_1}^{\tau_2} k_{ij}(\tau)\,\mathrm{e}^{-\mathrm{i}\omega\tau}\,\mathrm{d}\tau &\approx K_{ij}(\omega) \\ \int_{-\tau_1}^{\tau_2} k_{ij}(\tau)\,\mathrm{e}^{\mathrm{i}\omega\tau}\,\mathrm{d}\tau &\approx 0. \end{aligned}\right\} \tag{14.2–13}$$

It is convenient to express the operator products under the integral in Eq. (14.2–8) in normal order by decomposing $\hat{\mathbf{A}}(\mathbf{r}, t)$ into its positive and negative frequency parts

$$\hat{\mathbf{A}}(\mathbf{r}, t) = \hat{\mathbf{A}}^{(+)}(\mathbf{r}, t) + \hat{\mathbf{A}}^{(-)}(\mathbf{r}, t),$$

which play the role of configuration space annihilation and creation operators (cf. Section 11.11). Thus we have (Yurke, 1985)

$$\begin{aligned} \hat{A}_j(\mathbf{r}, t_2)\hat{A}_i(\mathbf{r}, t_1) &= \hat{A}_j^{(+)}(\mathbf{r}, t_2)\hat{A}_i^{(+)}(\mathbf{r}, t_1) + \hat{A}_j^{(-)}(\mathbf{r}, t_2)\hat{A}_i^{(-)}(\mathbf{r}, t_1) \\ &\quad + \hat{A}_j^{(-)}(\mathbf{r}, t_2)\hat{A}_i^{(+)}(\mathbf{r}, t_1) + \hat{A}_i^{(-)}(\mathbf{r}, t_1)\hat{A}_j^{(+)}(\mathbf{r}, t_2) \\ &\quad + [\hat{A}_j^{(+)}(\mathbf{r}, t_2), \hat{A}_i^{(-)}(\mathbf{r}, t_1)] \\ &= :\hat{A}_j(\mathbf{r}, t_2)\hat{A}_i(\mathbf{r}, t_1): + [\hat{A}_j^{(+)}(\mathbf{r}, t_2), \hat{A}_i^{(-)}(\mathbf{r}, t_1)], \end{aligned} \tag{14.2–14}$$

where the pair of colons : : denotes normal ordering. The commutator is readily evaluated with the help of the usual mode expansion (14.2–4) [see also Eq. (10.4–38)], and we find

$$\begin{aligned} [\hat{A}_j^{(+)}(\mathbf{r}, t_2), \hat{A}_i^{(-)}(\mathbf{r}, t_1)] &= \frac{1}{L^3}\sum_{\mathbf{k},s}\sum_{\mathbf{k}'s'}\left(\frac{\hbar}{2\omega\varepsilon_0}\right)^{1/2}\left(\frac{\hbar}{2\omega'\varepsilon_0}\right)^{1/2} \\ &\quad \times [\hat{a}_{\mathbf{k}s}, \hat{a}^{\dagger}_{\mathbf{k}'s'}](\boldsymbol{\varepsilon}_{\mathbf{k}s})_j(\boldsymbol{\varepsilon}^{*}_{\mathbf{k}'s'})_i\,\mathrm{e}^{\mathrm{i}[(\mathbf{k}-\mathbf{k}')\cdot\mathbf{r}-\omega(t_2-t_0)+\omega'(t_1-t_0)]} \\ &= \frac{1}{L^3}\sum_{\mathbf{k}}\left(\frac{\hbar}{2\omega\varepsilon_0}\right)\left(\delta_{ij} - \frac{k_ik_j}{k^2}\right)\mathrm{e}^{\mathrm{i}\omega(t_1-t_2)} \\ &\to \frac{1}{(2\pi)^3}\int\left(\frac{\hbar}{2\omega\varepsilon_0}\right)\left(\delta_{ij} - \frac{k_ik_j}{k^2}\right)\mathrm{e}^{\mathrm{i}\omega(t_1-t_2)}\,\mathrm{d}^3k. \end{aligned} \tag{14.2–15}$$

In deriving this result we have made use of the commutation relation (10.3–9), the tensor equality (10.2–19c), and have gone to the continuum limit $L \to \infty$. With the help of Eqs. (14.2–14) and (14.2–15) Eq. (14.2–8) then becomes:

Prob. of photoelectric detection at time $t_0 + \Delta t$

$$= \int_{t_0}^{t_0+\Delta t} \mathrm{d}t_1 \int_{t_0}^{t_1} \mathrm{d}t_2 k_{ij}(t_1 - t_2)[\langle : \hat{A}_j(\mathbf{r}_0, t_2)\hat{A}_i(\mathbf{r}_0, t_1) : \rangle + \frac{1}{(2\pi)^3} \int \frac{\hbar}{2\omega\varepsilon_0}\left(\delta_{ij} - \frac{k_i k_j}{k^2}\right) \mathrm{e}^{\mathrm{i}\omega(t_1 - t_2)}\, \mathrm{d}^3 k] + \text{c.c.} \quad (14.2\text{–}16)$$

Let us examine the contribution of the second term, coming from the commutator in Eq. (14.2–15), in more detail. We may call this the 'vacuum contribution' to the probability, because it is independent of the state of the incoming optical field. As the integrand for the t_1, t_2 integration depends only on the difference between the two time arguments, we can reduce the double time integral to a single time integral with respect to the difference $t_1 - t_2$. We put $t_1 - t_0 = t'$, $t_2 - t_0 = t''$, $t' - t'' = \tau$, and make use of the symmetry property (14.2–10). Then we obtain, after interchanging the order of integrations,

Vacuum contribution to the detection probability

$$\begin{aligned} &= \frac{1}{(2\pi)^3} \int \mathrm{d}^3 k \left(\frac{\hbar}{2\omega\varepsilon_0}\right)\left(\delta_{ij} - \frac{k_i k_j}{k^2}\right) \int_0^{\Delta t} \mathrm{d}t' \int_0^{t'} \mathrm{d}t''[k_{ij}(t' - t'')\, \mathrm{e}^{\mathrm{i}\omega(t' - t'')} \\ &\quad + k_{ij}(t'' - t')\, \mathrm{e}^{\mathrm{i}\omega(t'' - t')}] \\ &= \frac{1}{(2\pi)^3} \int \mathrm{d}^3 k \left(\frac{\hbar}{2\omega\varepsilon_0}\right)\left(\delta_{ij} - \frac{k_i k_j}{k^2}\right) \int_{-\Delta t}^{\Delta t} \mathrm{d}\tau (\Delta t - |\tau|) k_{ij}(\tau)\, \mathrm{e}^{\mathrm{i}\omega\tau} \\ &= 0, \end{aligned} \quad (14.2\text{–}17)$$

by virtue of Eqs. (14.2–13), provided $\Delta t \gg 1/\omega_0$, because the factor $\Delta t - |\tau|$ can be replaced by Δt to a very good approximation. [The case $\Delta t \not\gg 1/\omega_0$ was treated by Mandel and Meltzer (1969).] The second equation follows from the first when the integration is carried out along strips inclined at 45° to the t' and t'' axes, of area $(\Delta t - |\tau|)\,\mathrm{d}\tau$. [A similar transformation from a double to a single time integral occurs in Section 14.9, in connection with Eq. (14.9–11), and is discussed there in somewhat more detail.] With the help of Eq. (14.2–17), Eq. (14.2–16) now simplifies to

Prob. of photoelectric detection at time $t_0 + \Delta t$

$$= P(\mathbf{r}_0, t_0, \Delta t) = \int_{t_0}^{t_0+\Delta t} \mathrm{d}t_1 \int_{t_0}^{t_1} \mathrm{d}t_2 k_{ij}(t_1 - t_2)\langle : \hat{A}_j(\mathbf{r}_0, t_2)\hat{A}_i(\mathbf{r}_0, t_1) : \rangle + \text{c.c.} \quad (14.2\text{–}18)$$

14.2.1 Application to the pure coherent state

We now apply this result to the special case in which the incident optical field is in the pure coherent state $|\{v\}\rangle$. As $|\{v\}\rangle$ is the right eigenstate of $\hat{\mathbf{A}}^{(+)}(\mathbf{r}_0, t)$ and $\langle\{v\}|$ is the left eigenstate of $\hat{\mathbf{A}}^{(-)}(\mathbf{r}_0, t)$, we can immediately evaluate the normally ordered expectation in Eq. (14.2–18) and obtain the relation

$$P(\mathbf{r}_0, t_0, \Delta t) = \int_{t_0}^{t_0+\Delta t} \mathrm{d}t_1 \int_{t_0}^{t_1} \mathrm{d}t_2 k_{ij}(t_1 - t_2) A_j(\mathbf{r}_0, t_2) A_i(\mathbf{r}_0, t_1) + \text{c.c.}, \quad (14.2\text{–}19)$$

where $\mathbf{A}(\mathbf{r}_0, t)$ is the c-number vector potential corresponding to the coherent state $|\{v\}\rangle$.

It is interesting to compare this result for the special case of the coherent state $|\{v\}\rangle$ with the answer given directly by Eq. (14.2–6), if we put $\hat{\rho}_F(t_0) = |\{v\}\rangle\langle\{v\}|$ and we take the final state $|\chi\rangle$ to be the same coherent state $|\{v\}\rangle$. Since

$$\langle\{v\}|\hat{\mathbf{A}}(\mathbf{r}_0, t)|\{v\}\rangle = \mathbf{A}(\mathbf{r}_0, t),$$

Eq. (14.2–6) yields exactly the same answer for the detection probability as Eq. (14.2–19). In other words, in a coherent state, the detection probability, subject to the state of the field remaining unchanged, equals the detection probability irrespective of the final state of the field. This implies that the initial state remains $|\{v\}\rangle$ after the detection with probability unity, to the degree of approximation to which we are working here.‡

It is not difficult to see why this should be. In discarding the vacuum contribution and retaining only the normally ordered terms, as in Eq. (14.2–18), we have effectively put the emphasis on the energy-conserving terms, corresponding to the emission of the electron accompanied by the absorption of a photon. But as we know from our study of the coherent state in Chapter 11, the absorption of a photon from the coherent state leaves the state unchanged. That is why the process of photoelectric detection in a coherent state in a sufficiently long time interval Δt also leaves the state unchanged. It is this feature of the coherent state that makes its choice as initial state attractive in any calculation involving absorption of photons.

In order to carry out the time integrations in Eq. (14.2–19) we now make a mode expansion of each of the two vector potentials [cf. Eq. (14.2–4)], which yields oscillatory factors like $\exp[\pm \mathrm{i}\omega(t_1 - t_0) \pm \mathrm{i}\omega'(t_2 - t_0)]$, and we use the integral representation of $k_{ij}(\tau)$ given by Eq. (14.2–9). We are then led to four different double time integrals of the form

$$\int_{t_0}^{t_0+\Delta t} \mathrm{d}t_1 \int_{t_0}^{t_1} \mathrm{d}t_2\, \mathrm{e}^{\mathrm{i}[(E-E_0)/\hbar \pm \omega](t_1 - t_0)}\, \mathrm{e}^{-\mathrm{i}[(E-E_0)/\hbar \pm \omega'](t_2 - t_0)}$$
$$= \int_0^{\Delta t} \mathrm{d}t' \int_0^{t'} \mathrm{d}t''\, \mathrm{e}^{\mathrm{i}[(E-E_0)/\hbar \pm \omega]t'}\, \mathrm{e}^{-\mathrm{i}[(E-E_0)/\hbar \pm \omega']t''}. \quad (14.2\text{–}20)$$

Since $E > 0$, $-E_0/\hbar = \omega_0$, ω, ω' are all optical frequencies, and the measurement time interval Δt is typically thousands of optical periods long, it is clear that the

‡ This is an example of the theorem that if A, B are two events, and if the joint probability

$$p(A, B) = \sum_{\text{all } B} p(A, B) = P(A),$$

where $P(A)$ is the probability of A irrespective of B, then the conditional probability of B given A is

$$P_c(B|A) = \frac{p(A, B)}{P(A)} = 1.$$

integrands oscillate rapidly at double an optical frequency, and therefore make a small contribution to the integral, unless there is a minus in front of both ω and ω' in the exponents. It follows that the only significant contribution to the time integrals comes from the term $A_j^{(-)}(\mathbf{r}_0, t_2)A_i^{(+)}(\mathbf{r}_0, t_1)$ in $A_j(\mathbf{r}_0, t_2)A_i(\mathbf{r}_0, t_1)$, and we may write in place of Eq. (14.2–19)

$$P(\mathbf{r}_0, t_0, \Delta t) = \int_{t_0}^{t_0+\Delta t} \mathrm{d}t_1 \int_{t_0}^{t_1} \mathrm{d}t_2 k_{ij}(t_1 - t_2) A_j^{(-)}(\mathbf{r}_0, t_2) A_i^{(+)}(\mathbf{r}_0, t_1) + \text{c.c.}$$

or, on substituting $t_1 = t_0 + t'$, $t_2 = t_0 + t''$,

$$P(\mathbf{r}_0, t_0, \Delta t) = \int_0^{\Delta t} \mathrm{d}t' \int_0^{t'} \mathrm{d}t'' k_{ij}(t' - t'') A_j^{(-)}(\mathbf{r}_0, t_0 + t'') A_i^{(+)}(\mathbf{r}_0, t_0 + t') + \text{c.c.} \tag{14.2–21}$$

14.2.2 Evaluation of the probability in the quasi-monochromatic approximation

We can simplify the answer further by making the quasi-monochromatic approximation, according to which the only contributions to the mode expansions of $A_j^{(-)}(\mathbf{r}_0, t_2)$ and $A_i^{(+)}(\mathbf{r}_0, t_1)$ come from frequencies in a narrow range within $\omega_1 \pm \frac{1}{2}\Delta\omega$ centered on some optical frequency ω_1, with $\Delta\omega \ll \omega_1$ (see Fig. 14.2.) [For the treatment of photoelectric detection of polychromatic light see Kimble and Mandel (1984).] If we now take Δt to satisfy the inequalities

$$1/\omega_1 \ll \Delta t \ll 1/\Delta\omega, \tag{14.2–22}$$

then in the mode expansion of $A_j^{(+)}(\mathbf{r}_0, t_0 + t')$ we may use the approximation

$$\mathrm{e}^{-\mathrm{i}\omega(t_0+t')} = \mathrm{e}^{-\mathrm{i}\omega t_0} \mathrm{e}^{-\mathrm{i}\omega_1 t'} \mathrm{e}^{-\mathrm{i}(\omega-\omega_1)t'} \approx \mathrm{e}^{-\mathrm{i}\omega t_0} \mathrm{e}^{-\mathrm{i}\omega_1 t'}, \tag{14.2–23}$$

because $|\omega - \omega_1|t' < \frac{1}{2}\Delta\omega t' < \frac{1}{2}\Delta\omega\Delta t \ll 1$. It follows that

$$A_i^{(+)}(\mathbf{r}_0, t_0 + t') \approx A_i^{(+)}(\mathbf{r}_0, t_0)\,\mathrm{e}^{-\mathrm{i}\omega_1 t'},$$

and similarly

$$A_j^{(-)}(\mathbf{r}_0, t_0 + t'') \approx A_j^{(-)}(\mathbf{r}_0, t_0)\,\mathrm{e}^{\mathrm{i}\omega_1 t''},$$

to a good approximation. The field amplitudes can then be taken out of the time

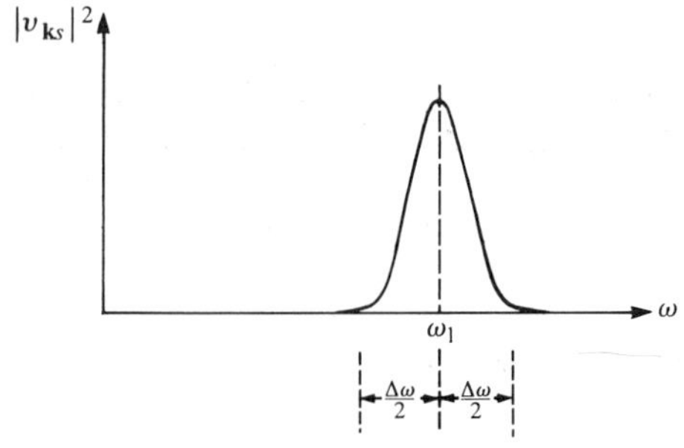

Fig. 14.2 A quasi-monochromatic spectral distribution centered on frequency ω_1.

integrals, and Eq. (14.2–21) becomes

$$
\begin{aligned}
P(\mathbf{r}_0, t_0, \Delta t) &= A_j^{(-)}(\mathbf{r}_0, t_0)A_i^{(+)}(\mathbf{r}_0, t_0)\int_0^{\Delta t} \mathrm{d}t' \int_0^{t'} \mathrm{d}t'' k_{ij}(t' - t'')\,\mathrm{e}^{-\mathrm{i}\omega_1(t'-t'')} + \text{c.c.} \\
&= A_j^{(-)}(\mathbf{r}_0, t_0)A_i^{(+)}(\mathbf{r}_0, t_0)\int_0^{\Delta t} \mathrm{d}t' \int_0^{t'} \mathrm{d}\tau k_{ij}(\tau)\,\mathrm{e}^{-\mathrm{i}\omega_1\tau} + \text{c.c.} \\
&= A_j^{(-)}(\mathbf{r}_0, t_0)A_i^{(+)}(\mathbf{r}_0, t_0)\int_0^{\Delta t} \mathrm{d}t' \int_{-t'}^{t'} \mathrm{d}\tau k_{ij}(\tau)\,\mathrm{e}^{-\mathrm{i}\omega_1\tau},
\end{aligned}
$$

when we put $t' - t'' = \tau$ and make use of the symmetry property (14.2–10). With the understanding that $\Delta t \gg 1/\omega_0$, so that the τ-integral spans virtually the whole support of $k_{ij}(\tau)$ for almost all t', we can use Eq. (14.2–13) and identify the τ-integral with $K_{ij}(\omega_1)$. Hence Eq. (14.2–21) leads to the relation

$$
\left.
\begin{aligned}
P(\mathbf{r}_0, t_0, \Delta t) &= A_j^{(-)}(\mathbf{r}_0, t_0)A_i^{(+)}(\mathbf{r}_0, t_0)\Delta t K_{ij}(\omega_1) \\
&= 2\pi\hbar\Delta t\left(\frac{e}{m\hbar}\right)^2 \int \sigma(E_0 + \hbar\omega_1, \Omega)g(E_0 + \hbar\omega_1, \Omega) \\
&\quad \times |\langle \Phi(E_0 + \hbar\omega_1, \Omega)|\hat{\mathbf{p}}|\Phi_0\rangle \cdot \mathbf{A}^{(+)}(\mathbf{r}_0, t_0)|^2 \\
&\quad \times \Theta(\omega_1 - \omega_0)\,\mathrm{d}\Omega,
\end{aligned}
\right\} \qquad (14.2\text{–}24)
$$

when we make use of the expression for $K_{ij}(\omega)$ given by Eq. (14.2–11). By virtue of the factor $\Theta(\omega_1 - \omega_0)$ this is non-zero only when $\omega_1 > \omega_0$, i.e. when the energy $\hbar\omega_1$ of the absorbed photon is large enough to exceed the electron binding energy $\hbar\omega_0$.

Four important features of the result embodied in Eq. (14.2–24) are worth noting:

(a) The photoelectric detection probability $P(\mathbf{r}_0, t_0, \Delta t)$ is proportional to Δt for small Δt, and will sometimes be written as $P(\mathbf{r}_0, t_0)\Delta t$, with the understanding that $P(\mathbf{r}_0, t_0)$ is a probability density.

(b) The analytic signal $\mathbf{A}^{(+)}(\mathbf{r}_0, t)$ for the vector potential representing the state of the optical field appears naturally, because it is the eigenvalue of the photon absorption operator belonging to the coherent state $|\{v\}\rangle$, and the measurement corresponds to a photon absorption.

(c) The frequency of the optical field must satisfy the Einstein photoelectric condition $\hbar\omega_1 > -E_0$ for photoemission to occur.

(d) To a good approximation, the final state of the optical field is the same coherent state as the initial state.

We now attempt to evaluate the detector frequency response function $K_{ij}(\omega)$ in Eq. (14.2–24) with the help of some reasonable approximations.

14.2.3 Evaluation of the electron matrix element

Let us assume that the final electron state $|\Phi(E_0 + \hbar\omega_1, \Omega)\rangle = |\Phi(\hbar\omega_1 - \hbar\omega_0, \Omega)\rangle$ may be taken as almost a free particle state. For non-relativistic free electrons, internal degrees of freedom like spin play a very small role, and the state is then almost completely determined by the electron energy E and by its direction of motion, which we now identify with the variable Ω. Let the photocathode be the xy-plane, with the z-axis pointing in the direction of the detector axis, and let θ, ϕ be the polar and azimuthal angles of the electron momentum $\mathbf{P}_e$ in the state $|\Phi(\hbar\omega_1 - \hbar\omega_0, \Omega)\rangle$ (see Fig. 14.1). Then the Schrödinger wave functions corresponding to the initial and final electron states are

$$\left.\begin{aligned} \Phi_0(\mathbf{r}) &= \langle \mathbf{r}|\Phi_0\rangle \\ \Phi(\mathbf{r}) &= \langle \mathbf{r}|\Phi\rangle \approx \frac{1}{(2\pi\hbar)^{3/2}}\,\mathrm{e}^{\mathrm{i}\mathbf{P}_e\cdot\mathbf{r}/\hbar}, \end{aligned}\right\} \tag{14.2–25}$$

with $P_e^2/2m = \hbar(\omega_1 - \omega_0)$. Of course, strictly speaking, $|\Phi\rangle$ is not really a free electron state of definite energy and momentum, because of the presence of the potential well, but this form of the wave function $\Phi(\mathbf{r})$ will serve as a rough approximation. We use it to evaluate the matrix element of the electron,

$$\begin{aligned} \langle \Phi(\hbar\omega_1 - \hbar\omega_0, \Omega)|\hat{\mathbf{p}}|\Phi_0\rangle &= \int \langle \Phi(\hbar\omega_1 - \hbar\omega_0, \Omega)|\hat{\mathbf{p}}|\mathbf{r}\rangle\langle \mathbf{r}|\Phi_0\rangle\,\mathrm{d}^3r \\ &\approx -\frac{\hbar}{\mathrm{i}(2\pi\hbar)^{3/2}}\int [\nabla\,\mathrm{e}^{-\mathrm{i}\mathbf{P}_e\cdot\mathbf{r}/\hbar}]\Phi_0(\mathbf{r})\,\mathrm{d}^3r \\ &= \mathbf{P}_e\frac{1}{(2\pi\hbar)^{3/2}}\int \mathrm{e}^{-\mathrm{i}\mathbf{P}_e\cdot\mathbf{r}/\hbar}\Phi_0(\mathbf{r})\,\mathrm{d}^3r \\ &= \mathbf{P}_e\,\tilde{\Phi}_0(\mathbf{P}_e), \end{aligned} \tag{14.2–26}$$

where $\tilde{\Phi}_0(\mathbf{P}_e)$ is the three-dimensional Fourier transform of the initial state wave function $\Phi_0(\mathbf{r})$. Hence the scalar product

$$\langle \Phi(\hbar\omega_1 - \hbar\omega_0, \Omega)|\hat{\mathbf{p}}|\Phi_0\rangle \cdot \mathbf{A}^{(+)}(\mathbf{r}_0, t_0) = \tilde{\Phi}_0(\mathbf{P}_e)\mathbf{P}_e \cdot \mathbf{A}^{(+)}(\mathbf{r}_0, t_0), \tag{14.2–27}$$

and as $\mathbf{A}$ is transverse this equation shows that final states in which the emitted electron travels parallel to the direction of the incident light make no contribution to the emission probability. Evidently the absorbed photon does not simply pass its momentum on to the photoelectron.

14.2.4 Evaluation of the detection probability for an axially symmetric detector

We now make two further assumptions in order to simplify the integral in Eq. (14.2–24). We assume that the detector has approximate cylindrical symmetry about the z-axis, so that both the weight function g and the Fourier transform $\tilde{\Phi}_0$ of the initial state wave function depend only on the polar angle θ and not on ϕ. Hence we may write

$$\left.\begin{aligned} g(\hbar\omega_1 - \hbar\omega_0, \Omega) &= g(\hbar\omega_1 - \hbar\omega_0, \theta) \\ \tilde{\Phi}_0(\mathbf{P}_e) &= \tilde{\Phi}_0(P_e, \theta), \end{aligned}\right\} \tag{14.2–28}$$

where the magnitude of the electron momentum $\mathbf{P}_e$ is given by $P_e^2/2m = \hbar(\omega_1 - \omega_0)$. Also we assume that the density of states σ is isotropic, independent of direction θ, ϕ, or

$$\sigma(\hbar\omega_1 - \hbar\omega_0, \Omega) = \sigma(\hbar\omega_1 - \hbar\omega_0). \tag{14.2–29}$$

From Eq. (14.2–27), since $\mathbf{P}_e \cdot \mathbf{A}^{(+)} = P_e \sin\theta(A_x^{(+)}\cos\phi + A_y^{(+)}\sin\phi)$, we then have

$$|\langle \Phi(\hbar\omega_1 - \hbar\omega_0, \Omega)|\hat{\mathbf{p}}|\Phi_0\rangle \cdot \mathbf{A}^{(+)}(\mathbf{r}_0, t_0)|^2 = |\widetilde{\Phi}_0(P_e, \theta)|^2 2m\hbar(\omega_1 - \omega_0)$$
$$\times \sin^2\theta[|A_x^{(+)}|^2\cos^2\phi + |A_y^{(+)}|^2\sin^2\phi + A_x^{(-)}A_y^{(+)}\cos\phi\sin\phi + \text{c.c.}], \quad (14.2\text{–}30)$$

and with the help of Eqs. (14.2–28) to (14.2–30) we obtain from Eq. (14.2–24), on putting $\mathrm{d}\Omega = \sin\theta\,\mathrm{d}\theta\,\mathrm{d}\phi$,

$$P(\mathbf{r}_0, t_0)\Delta t$$
$$= \frac{4\pi e^2}{m}(\omega_1 - \omega_0)\sigma(\hbar\omega_1 - \hbar\omega_0)\Delta t\int_0^\pi \mathrm{d}\theta\, g(\hbar\omega_1 - \hbar\omega_0, \theta)|\widetilde{\Phi}_0(P_e, \theta)|^2 \sin^3\theta$$
$$\times \int_0^{2\pi} \mathrm{d}\phi[|A_x^{(+)}|^2\cos^2\phi + |A_y^{(+)}|^2\sin^2\phi + A_x^{(-)}A_y^{(+)}\cos\phi\sin\phi + \text{c.c.}]\Theta(\omega_1 - \omega_0)$$
$$= \frac{4\pi^2 e^2}{m}(\omega_1 - \omega_0)\sigma(\hbar\omega_1 - \hbar\omega_0)\Delta t\int_0^\pi g(\hbar\omega_1 - \hbar\omega_0, \theta)|\widetilde{\Phi}_0(P_e, \theta)|^2 \sin^3\theta$$
$$\times\, \Theta(\omega_1 - \omega_0)(|A_x^{(+)}|^2 + |A_y^{(+)}|^2)\,\mathrm{d}\theta, \quad (14.2\text{–}31)$$

after we carry out the ϕ-integration.

Let us introduce the parameter

$$\eta_A(\omega_1) \equiv \frac{4\pi^2 e^2}{m}(\omega_1 - \omega_0)\sigma(\hbar\omega_1 - \hbar\omega_0)\int_0^\pi g(\hbar\omega_1 - \hbar\omega_0, \theta')|\widetilde{\Phi}_0(P_e, \theta')|^2 \sin^3\theta'$$
$$\times\, \Theta(\omega_1 - \omega_0)\,\mathrm{d}\theta' \quad (14.2\text{–}32)$$

characterizing the detection efficiency of the detector for light of frequency ω_1. Calling $|A_x^{(+)}|^2 + |A_y^{(+)}|^2 \equiv I_A(\mathbf{r}_0, t_0)$, the intensity associated with the vector potential of the incident field, we can re-write Eq. (14.2–31) in the very compact form

$$P(\mathbf{r}_0, t_0)\Delta t = \eta_A I_A(\mathbf{r}_0, t_0)\Delta t. \quad (14.2\text{–}33)$$

This should be compared with the semiclassical result given by Eq. (9.3–10). The detection efficiency is independent of the polarization of light under the assumed conditions.

The fact that the light intensity $I_A(\mathbf{r}_0, t_0)$ appearing in Eq. (14.2–33) is associated with the vector potential $\mathbf{A}(\mathbf{r}_0, t_0)$ is of course a reflection of the fact that the interaction was taken to be proportional to $\hat{\mathbf{p}} \cdot \hat{\mathbf{A}}(\mathbf{r}_0, t)$. However, because of the assumed quasi-monochromaticity, all other measures of the light intensity are proportional to $I_A(\mathbf{r}_0, t_0)$. For example, if we wish to express the probability in terms of the photon density $I(\mathbf{r}_0, t_0)$ (see Section 12.3), we simply note that for a quasi-monochromatic field,

$$I_A(\mathbf{r}_0, t) = \frac{\hbar}{2\omega_1\varepsilon_0} I(\mathbf{r}_0, t),$$

so that

$$P(\mathbf{r}_0, t)\Delta t = \eta_A(\omega_1)\left(\frac{\hbar}{2\omega_1\varepsilon_0}\right) I(\mathbf{r}_0, t)\Delta t \equiv \eta(\omega_1) I(\mathbf{r}_0, t)\Delta t. \quad (14.2\text{–}34)$$

Although the expressions (14.2–33) and (14.2–34) for the photoemission probability have a very simple structure, we must not forget that they are of limited validity. They apply to the detection of light in a pure coherent state by a single bound electron. Fortunately, the expressions readily lend themselves to several generalizations, as we now show.

14.3 An N-electron photodetector

Let us now consider a photodetector similar to the one shown in Fig. 14.1 that contains a large number N of bound electrons within the photocathode. For simplicity we suppose that light falls on the detector at right angles, and that its photosensitive area S is small enough for the field to look like a plane wave over the area S. If the electric potential binding any one electron is not significantly influenced by the emission of other electrons, then the interaction of the light with one electron can be regarded as proceeding almost independently of the interaction with another.

Now the probability of detection of light in a coherent state by one electron has been shown to be $\eta I(\mathbf{r}, t)\Delta t$, and if all bound electrons see the same field and have the same probability of being released and counted, then the probability $P(\mathbf{r}, t, \Delta t)$ of photodetection by one or more of the N bound electrons is given by the binomial sum

$$\begin{aligned} P(\mathbf{r}, t, \Delta t) &= \sum_{n=1}^{N} \binom{N}{n} [\eta I(\mathbf{r}, t)\Delta t]^n [1 - \eta I(\mathbf{r}, t)\Delta t]^{N-n} \\ &= 1 - [1 - \eta I(\mathbf{r}, t)\Delta t]^N \\ &= N\eta I(\mathbf{r}, t)\Delta t - \frac{N(N-1)}{2!}[\eta I(\mathbf{r}, t)\Delta t]^2 + \dots . \end{aligned} \qquad (14.3\text{–}1)$$

If $N\eta I(\mathbf{r}, t)\Delta t \ll 1$, then the dominant term in this series is the first, and we can write

$$P(\mathbf{r}, t, \Delta t) \simeq N\eta I(\mathbf{r}, t)\Delta t, \qquad (14.3\text{–}2)$$

with the understanding that the $P(\mathbf{r}, t, \Delta t)$ is still a small differential probability. The detection probability for an N-electron detector is then just N times as great as for a one-electron detector.

More generally the quantity η that measures the likelihood of detection by one electron may not be the same for all electrons, but may depend on the position $\mathbf{r}$ of the bound electron within the photocathode. Moreover, the light intensity $I(\mathbf{r}, t)$ could vary with position $\mathbf{r}$, if the field is not uniform over the cathode and if the light is attenuated on propagation. We should then replace Eq. (14.3–2) for the detection probability by a more general integral relation

$$P(\mathbf{r}_0, t, \Delta t) = \int \mathcal{N}(\mathbf{r})\eta(\mathbf{r})I(\mathbf{r}, t)\Delta t \, \mathrm{d}^3 r, \qquad (14.3\text{–}3)$$

in which $\mathcal{N}(\mathbf{r})$ is the bound electron density and $\mathbf{r}_0$ is some reference point characterizing the detector, like the midpoint of the photocathode. In the special case when $\mathcal{N}$ is constant over the photocathode, and η and I vary with position only in the z-direction, $P(\mathbf{r}_0, t, \Delta t)$ reduces to the simpler expression

$$P(\mathbf{r}_0, t, \Delta t) = S\mathcal{N}\int_0^{\delta z} \eta(z) I(z, t)\, \mathrm{d}z,$$

where the integral extends over the thickness δz of the photocathode. Now $I(z, t)$ is proportional to $I(0, t) = I(\mathbf{r}_0, t)$, so that we may write $I(z, t) = I(\mathbf{r}_0, t) f(z)$. The detection probability then takes the form

$$P(\mathbf{r}_0, t, \Delta t) = \alpha c S I(\mathbf{r}_0, t)\Delta t, \tag{14.3–4}$$

where we have written

$$\alpha \equiv (\mathcal{N}/c)\int_0^{\delta z} \eta(z) f(z)\, \mathrm{d}z \tag{14.3–5}$$

for the dimensionless quantum efficiency of the detector. Equation (14.3–4) will be seen to be a special case of Eq. (12.3–3), which was obtained earlier by a simple heuristic argument, when the incident field is in a coherent state.

14.4 The multiple photoelectric detection probability

Although the probability of photoelectron detection at a certain place and time is an important quantity characterizing the optical field, it refers to the simplest possible measurement. More sophisticated measurements involve correlations between photodetections at several different space-time points. Let us consider the situation illustrated in Fig. 14.3, in which there are two photodetectors of quantum efficiencies α_1, α_2 with photocathodes of areas S_1, S_2 centered at points $\mathbf{r}_1$, $\mathbf{r}_2$, which are exposed to the optical field in such a way that the light is incident normally on both. Once again we assume, for simplicity, that the field over each photocathode looks like a plane wave. We wish to determine the joint probability $P_2(\mathbf{r}_1, t_1, \Delta t_1; \mathbf{r}_2, t_2, \Delta t_2)$ that photodetections will be registered by one detector at time t_1 within Δt_1, and by the other detector at time t_2 within Δt_2, with $t_2 > t_1$, when the field is initially in a coherent state $|\{v\}\rangle$. In principle, we can approach this problem by a perturbative analysis as before, but we have to carry the perturbation expansion through to the fourth order in the interaction energy (Glauber, 1965, p. 84 *et seq.*). However, the special features of the coherent state make it possible to short-circuit the calculation, and to use our previous results for the single detection probability.

We shall evaluate the joint probability in two stages. We first calculate the probability for the first photodetection to occur, which has been shown to be

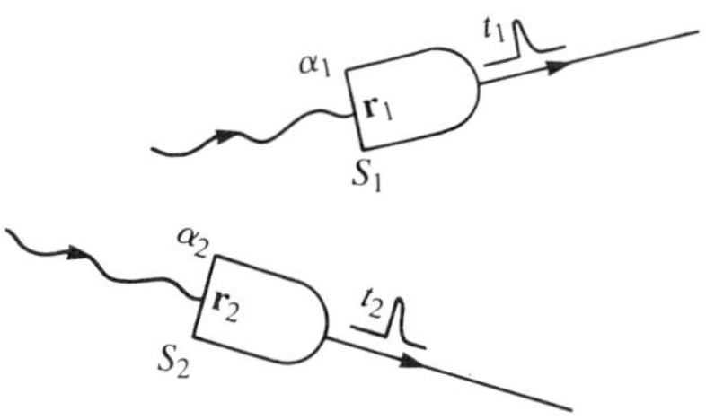

Fig. 14.3 Illustrating multiple photodetections with two detectors.

$\alpha_1 c S_1 I(\mathbf{r}_1, t_1)\Delta t_1$ [cf. Eq. (14.3–4)], and we determine the state of the electromagnetic field following the measurement. We take this state to be the initial state of the field for the second measurement, and then calculate the probability for the second detection at time t_2, given this initial state. These two probabilities are then multiplied together to yield $P_2(\mathbf{r}_1, t_1, \Delta t_1; \mathbf{r}_2, t_2, \Delta t_2)$. But the state of the field, before any measurement takes place, was taken to be a pure coherent state, and we have shown that this state remains unchanged, to a very good approximation, following a photodetection. The same coherent state is therefore also the appropriate initial state of the field for the second photodetection at time t_2, which has probability $\alpha_2 c S_2 I(\mathbf{r}_2, t_2)\Delta t_2$, where $I(\mathbf{r}_2, t_2)$ is the light intensity (photon density) at $\mathbf{r}_2$, t_2 in the given coherent state. Hence we have finally

$$P_2(\mathbf{r}_1, t_1, \Delta t_1; \mathbf{r}_2, t_2, \Delta t_2) = \alpha_1 c S_1 I(\mathbf{r}_1, t_1)\Delta t_1 \alpha_2 c S_2 I(\mathbf{r}_2, t_2)\Delta t_2. \quad (14.4\text{–}1)$$

Needless to say, this equation has an obvious generalization, and a similar argument can be used to derive the joint probability $P_N(\mathbf{r}_1, t_1, \Delta t_1; \ldots; \mathbf{r}_N, t_N, \Delta t_N)$ for photodetections at N different space-time points,

$$P_N(\mathbf{r}_1, t_1\, \Delta t_1; \ldots; \mathbf{r}_N, t_N, \Delta t_N) = \prod_{l=1}^{N} \alpha_l c S_l I(\mathbf{r}_l, t_l)\Delta t_l, \quad (14.4\text{–}2)$$

$$t_N \geqslant t_{N-1} \geqslant \ldots \geqslant t_1, \quad N = 2, 3, \ldots$$

The different photodetections need not necessarily refer to different detectors, but a single photodetector could be used. The quantum efficiencies α_1, α_2, ... and the surface areas S_1, S_2, ... are then necessarily equal, and the c-number light intensities I differ only in their time arguments.

As each joint probability $P_N(\mathbf{r}_1, t_1, \Delta t_1; \ldots; \mathbf{r}_N, t_N, \Delta t_N)$ is proportional to $\Delta t_1 \ldots \Delta t_N$, this probability is sometimes written in the alternative form

$$P_N(\mathbf{r}_1, t_1; \ldots; \mathbf{r}_N, t_N)\Delta t_1 \ldots \Delta t_N,$$

with the understanding that P_N is a *probability density* whenever the differential time intervals $\Delta t_1, \Delta t_2, \ldots$ do not appear as explicit arguments of P_N.

14.5 The multiple detection probability for an arbitrary initial state of the field

The simplicity of the result embodied in Eqs. (14.4–1) and (14.4–2) is of course a reflection of the fact that we have chosen a pure coherent state $|\{v\}\rangle$ of the electromagnetic field to be the initial state. At first sight it might seem that our previous treatment is of no use at all if the detection probability is to be calculated for another initial state. However, at the cost of a certain amount of mathematical rigor, we can immediately generalize Eq. (14.4–2) for an arbitrary initial field state.

Let us suppose that, instead of an initial pure coherent state $|\{v\}\rangle$, we have to deal with an ensemble of coherent states. For each realization of the ensemble we use the previous argument to derive the joint detection probability $P_N \Delta t_1 \ldots \Delta t_N$, and we arrive at an expression of the form (14.4–2). If each realization $|\{v\}\rangle$ of the ensemble is characterized by a weight $\phi(\{v\})$, so that the initial density operator $\hat{\rho}_F$ of the field is

$$\hat{\rho}_F = \int \phi(\{v\}) |\{v\}\rangle\langle\{v\}| \, d\{v\}, \tag{14.5-1}$$

then the corresponding probability given by Eq. (14.4–2) must also be given the same weight $\phi(\{v\})$. The correct joint detection probability is then obtained by averaging with $\phi(\{v\})$ the probability derived for each realization, so that

$$\begin{aligned} P_N(\mathbf{r}_1, t_1, \ldots, \mathbf{r}_N, t_N)\Delta t_1 \ldots \Delta t_N &= \int \phi(\{v\}) \prod_{l=1}^{N} [\alpha_l c S_l I(\mathbf{r}_l, t_l)\Delta t_l] \, d\{v\} \\ &= \left\langle \prod_{l=1}^{N} [\alpha_l c S_l I(\mathbf{r}_l, t_l)\Delta t_l] \right\rangle_\phi . \end{aligned} \tag{14.5-2}$$

The angular brackets denote the ensemble average with respect to $\phi\{v\}$. But we have seen in Section 11.8 that there is a sense in which any state of the field $\hat{\rho}_F$ can be regarded as a generalized 'ensemble' of coherent states and be given an expansion (14.5–1), provided the weight functionals $\phi(\{v\})$ include rather singular generalized functionals also. In this wider sense, Eq. (14.5–2) should then apply to any initial state of the electromagnetic field, with $\phi(\{v\})$ given by the diagonal coherent-state representation of the density operator of the field. The result can be compared with the semiclassical Eq. (9.6–4).

We now make use of the so-called optical equivalence theorem described in Section 11.9, and given by Eq. (11.9–4), to rewrite Eq. (14.5–2). According to the theorem, the expectation value of any normally ordered field operator can be written as an ensemble average, in which the creation and annihilation operators are replaced by their left and right eigenvalues and the average is to be taken with the phase space functional $\phi(\{v\})$ used as weighting functional. Let us apply the theorem in reverse to Eq. (14.5–2). The eigenvalues $\mathbf{V}^*(\mathbf{r}_l, t_l)$, $\mathbf{V}(\mathbf{r}_l, t_l)$, of which the intensities $I(\mathbf{r}_l, t_l) = \mathbf{V}^*(\mathbf{r}_l, t_l) \cdot \mathbf{V}(\mathbf{r}_l, t_l)$ are composed, are then to be replaced by the corresponding creation and annihilation operators $\hat{\mathbf{V}}^\dagger(\mathbf{r}_l, t_l)$, $\hat{\mathbf{V}}(\mathbf{r}_l, t_l)$ in normal order, and the quantum expectation is to be calculated. As we have treated the optical field as effectively free, or uncoupled from any source, the creation and annihilation operators commute among themselves, and their order in time is not important. But in the more general situation in which the field is interacting with a source, the time order of the operators is significant also (cf. Section 12.2). In invoking the optical equivalence theorem in reverse we shall therefore write the expression for P_N in the time-ordered form (Glauber, 1963, 1965, p. 83)

$$\begin{aligned} &P_N(\mathbf{r}_1, t_1; \ldots; \mathbf{r}_N, t_N)\Delta t_1 \ldots \Delta t_N \\ &\quad = \prod_{l=1}^{N} [\alpha_l c S_l \Delta t_l] \langle \hat{V}_{i_1}^\dagger(\mathbf{r}_1, t_1) \ldots \hat{V}_{i_N}^\dagger(\mathbf{r}_N, t_N) \hat{V}_{i_N}(\mathbf{r}_N, t_N) \ldots \hat{V}_{i_1}(\mathbf{r}_1, t_1)\rangle \\ &\quad = \prod_{l=1}^{N} [\alpha_l c S_l \Delta t_l] \langle \mathcal{T}\!: \hat{I}(\mathbf{r}_1, t_1) \ldots \hat{I}(\mathbf{r}_N, t_N) :\rangle \\ &\quad = \prod_{l=1}^{N} [\alpha_l c S_l \Delta t_l] \Gamma^{(N,N)}_{i_1 \ldots i_N; i_N \ldots i_1}(\mathbf{r}_1, t_1, \ldots, \mathbf{r}_N, t_N; \mathbf{r}_N, t_N, \ldots, \mathbf{r}_1, t_1) \\ &\qquad\qquad t_N \geqslant t_{N-1} \geqslant \ldots \geqslant t_1, \end{aligned} \tag{14.5-3}$$

where $\Gamma^{(N,N)}$ is the $2N$'th-order, normally ordered, time-ordered correlation function in the notation of Chapter 12. Here $::$ and $\mathcal{T}$ are the normal ordering and time-ordering symbols and $\langle\ \rangle$ denotes the quantum expectation. In this form the equation for the joint N-fold detection probability applies to any quantum state of the field, and represents the appropriate generalization of Eq. (14.4–2). It is usually most convenient to regard the field operators on the right of Eq. (14.5–3) as being in the Heisenberg picture, so that the field may be taken to be in its initial state. We note that the answer is precisely of the form obtained earlier in Section 12.2 from simple heuristic arguments. It is also identical with the result given by semiclassical theory (Mandel, Sudarshan and Wolf, 1964) (cf. Section 9.4) if $\Gamma^{(N,N)}$ in Eq. (14.5–3) is a classical field correlation function, except that the quantum field theory also allows us to treat states having no classical counterparts.

As an immediate application of Eq. (14.5–3) to a non-classical situation, let us calculate the joint two-fold photodetection probability $P_2(\mathbf{r}_1, t_1; \mathbf{r}_2, t_2)\Delta t_1 \Delta t_2$ for a field in the one-photon Fock state $|1_{\mathbf{k}s}, \{0\}\rangle$. As two photon annihilation operators acting to the right on this state give zero, it follows that

$$\langle 1_{\mathbf{k}s}, \{0\}|\mathcal{T}\colon \hat{I}(\mathbf{r}_1, t_1)\hat{I}(\mathbf{r}_2, t_2)\colon |1_{\mathbf{k}s}, \{0\}\rangle = 0$$

and

$$P_2(\mathbf{r}_1, t_1; \mathbf{r}_2, t_2)\Delta t_1 \Delta t_2 = 0,$$

as is to be expected, because each detection involves the absorption of a photon.

It should be emphasized that the probabilities given by Eq. (14.5–3) are differential probabilities, and they cannot be normalized to unity by integration over the times. In order to make this clear, let us focus attention on the special case of a single detector. The joint detection probability then reduces to

$$P_2(t, t+\tau)\Delta t \Delta\tau = (\alpha c S)^2 \langle \mathcal{T}\colon \hat{I}(t)\hat{I}(t+\tau)\colon\rangle \Delta t \Delta\tau, \qquad (14.5\text{–}4)$$

where we have denoted by τ the time interval between the two photoelectric counts and have suppressed the space arguments. If we divide both sides of the equation by $P_1(t)\Delta t$, we arrive at the conditional probability $P_c(t+\tau|t)\Delta\tau$ that, following a count at time t, there is another count at time $t+\tau$ within $\Delta\tau$,

$$P_c(t+\tau|t)\Delta\tau = \frac{P_2(t, t+\tau)\Delta t \Delta\tau}{P_1(t)\Delta t} = \frac{\alpha c S \langle \mathcal{T}\colon \hat{I}(t)\hat{I}(t+\tau)\colon\rangle \Delta\tau}{\langle \hat{I}(t)\rangle}. \qquad (14.5\text{–}5)$$

For a stationary field, each average in this equation is independent of t and depends only on τ. However, the probability $P_c(t+\tau|t)\Delta\tau$ must not be confused with the probability distribution $\mathcal{P}(\tau)\Delta\tau$ of the time interval τ between *successive* photoelectric counts, which refers to mutually exclusive events and is normalized to unity. If τ is the time interval between successive counts, then, by definition, there are no other counts between t and $t+\tau$, whereas this possibility is not excluded in Eq. (14.5–5). We shall show in Section 14.7 below that $\mathcal{P}(\tau)$ is given by the formula

$$\mathcal{P}(\tau) = \alpha c S \left\langle \mathcal{T}\colon \exp\left[-\alpha c S \int_t^{t+\tau} \hat{I}(t')\,\mathrm{d}t'\right] \hat{I}(t)\hat{I}(t+\tau)\colon \right\rangle \Big/ \langle \hat{I}(t)\rangle, \qquad (14.5\text{–}6)$$

which is correctly normalized to unity so that

$$\int_0^\infty \mathcal{P}(\tau)\,\mathrm{d}\tau = 1.$$

14.6 Photoelectric correlations

It is an immediate consequence of Eq. (14.5–3) for the joint detection probability that if

$$\langle \mathcal{T}\colon \hat{I}(\mathbf{r}_1, t_1)\hat{I}(\mathbf{r}_2, t_2)\colon\rangle \neq \langle \hat{I}(\mathbf{r}_1, t_1)\rangle\langle \hat{I}(\mathbf{r}_2, t_2)\rangle, \tag{14.6–1}$$

then

$$P_2(\mathbf{r}_1, t_1; \mathbf{r}_2, t_2) \neq P_1(\mathbf{r}_1, t_1)P_1(\mathbf{r}_2, t_2), \tag{14.6–2}$$

where $P_1(\mathbf{r}, t)$ is the single or one-fold probability density for photodetection. This implies that the two photodetections at $\mathbf{r}_1$, t_1 and $\mathbf{r}_2$, t_2 are not independent but are correlated. This conclusion is exactly the same as was reached in Section 9.6 from semiclassical arguments. Only for special states of the field does the inequality in (14.6–1) have to be replaced by an equality, and then the joint probability factorizes into a product of the separate probabilities, because the photoelectric detections are independent. An example of that is provided by a pure coherent state $|\{v\}\rangle$, or by the randomly phased single-mode laser model described by Eq. (11.11–18), for which

$$\langle \mathcal{T}\colon \hat{I}(\mathbf{r}_1, t_1)\hat{I}(\mathbf{r}_2, t_2)\colon\rangle = \langle \hat{I}(\mathbf{r}_1, t_1)\rangle\langle \hat{I}(\mathbf{r}_2, t_2)\rangle.$$

In general, however, correlations between successive photoelectric pulses are to be expected. This is true even if one photoemission cannot physically influence another one, because the events $\mathbf{r}_1$, t_1 and $\mathbf{r}_2$, t_2 have a space-like separation, for example. The reason is that each photodetection yields information about the field that influences our estimate of the probability of another photodetection. We can express this in another way, that invokes a more classical picture of the optical field, by saying that the intensity fluctuations cause the photodetections at two or more space-time points to be correlated. Only in the absence of any intensity fluctuations are successive photodetections strictly independent.

By introducing the normalized intensity correlation function $\lambda(\mathbf{r}_1, t_1; \mathbf{r}_2, t_2)$ defined by the formula

$$\lambda(\mathbf{r}_1, t_1; \mathbf{r}_2, t_2) \equiv \frac{\langle \mathcal{T}\colon \hat{I}(\mathbf{r}_1, t_1)\hat{I}(\mathbf{r}_2, t_2)\colon\rangle}{\langle \hat{I}(\mathbf{r}_1, t_1)\rangle\langle \hat{I}(\mathbf{r}_2, t_2)\rangle} - 1, \tag{14.6–3}$$

we can re-express the joint detection probability in the form

$$\begin{aligned} P_2(\mathbf{r}_1, t_1; \mathbf{r}_2, t_2)\Delta t_1,\Delta t_2 \\ &= \alpha_1 c S_1\langle \hat{I}(\mathbf{r}_1, t_1)\rangle\Delta t_1\alpha_2 c S_2\langle \hat{I}(\mathbf{r}_2, t_2)\rangle\Delta t_2[1 + \lambda(\mathbf{r}_1, t_1; \mathbf{r}_2, t_2)] \\ &= P_1(\mathbf{r}_1, t_1)\Delta t_1 P_1(\mathbf{r}_2, t_2)\Delta t_2[1 + \lambda(\mathbf{r}_1, t_1; \mathbf{r}_2, t_2)]. \end{aligned} \tag{14.6–4}$$

This shows clearly that the normalized correlation function $\lambda(\mathbf{r}_1, t_1;\ \mathbf{r}_2, t_2)$ provides a measure of the lack of statistical independence of the photoelectric pulses. Only for those states of the field for which $\lambda = 0$ do the pulses become independent.

14.6.1 The Hanbury Brown–Twiss effect (quantum treatment)

For states for which photoelectric correlations exist, the photoelectric pulses produced by illuminated photodetectors evidently do not occur strictly at random, and by studying their correlations in space and time one should be able to obtain information about the nature of the optical field. As was already mentioned in Section 9.9, the first experimental proof of this, and the first evidence for the existence of photoelectric correlations, was provided by the experiments performed in the 1950s by Brown and Twiss (1956, 1957) with the apparatus shown in Fig. 9.6(*a*). The light beam from a mercury arc was divided into two by a beam splitter, and the two beams fell on two photomultipliers, one of which could be translated across the field. The individual photoelectric pulses were not resolved, but instead the two photoelectric currents $J_1(t)$ and $J_2(t)$ were amplified and fed to a correlator that provided a signal proportional to the correlation $\langle J_1(t)J_2(t)\rangle$ (with $\langle J_1\rangle = 0 = \langle J_2\rangle$). This correlation was measured for various displacements of the movable photodetector with the results shown in Fig. 9.6(*b*).

Let us attempt to account for these results from Eq. (14.6–4). As the light is produced by a stationary, thermal source, the average intensity does not depend on time and the normalized correlation function $\lambda(\mathbf{r}_1, t_1; \mathbf{r}_2, t_2)$ depends only on the time difference $t_2 - t_1 = \tau$ and can be denoted by $\lambda(\mathbf{r}_1, \mathbf{r}_2, \tau)$. Moreover, as we have shown in Chapter 13 [see Eq. (13.3–23)], for a thermal state of the field the normalized intensity correlation function $\lambda(\mathbf{r}_1, \mathbf{r}_2, \tau)$ can be simply related to the normalized second-order correlation function defined by

$$\gamma^{(1,1)}(\mathbf{r}_1, t_1; \mathbf{r}_2, t_2) \equiv \frac{\langle \hat{\mathbf{V}}^{\dagger}(\mathbf{r}_1, t_1) \cdot \hat{\mathbf{V}}(\mathbf{r}_2, t_2)\rangle}{[\langle \hat{I}(\mathbf{r}_1, t_1)\rangle \langle \hat{I}(\mathbf{r}_2, t_2)\rangle]^{1/2}}. \tag{14.6–5}$$

For stationary, polarized light this relation takes the simple form (cf. Eq. (8.4–4))

$$\lambda(\mathbf{r}_1, \mathbf{r}_2, \tau) = |\gamma(\mathbf{r}_1, \mathbf{r}_2, \tau)|^2. \tag{14.6–6}$$

We have abbreviated $\gamma^{(1,1)}(\mathbf{r}_1, t_1;\ \mathbf{r}_2, t_1 + \tau)$ to $\gamma(\mathbf{r}_1, \mathbf{r}_2, \tau)$. Equation (14.6–4) therefore becomes

$$P_2(\mathbf{r}_1, t_1; \mathbf{r}_2, t_1 + \tau)\Delta t \Delta \tau = \alpha_1 \alpha_2 c^2 S_1 S_2 \langle \hat{I}(\mathbf{r}_1)\rangle \langle \hat{I}(\mathbf{r}_2)\rangle \Delta t \Delta \tau [1 + |\gamma(\mathbf{r}_1, \mathbf{r}_2, \tau)|^2], \tag{14.6–7}$$

which shows that photoelectric correlations are expected at $\mathbf{r}_1$, $\mathbf{r}_2$ so long as the degree of coherence is not zero at these two points. This is confirmed by the experimental results shown in Fig. 9.6(*b*). However the equation still does not adequately describe a measurement in which the individual photoelectric pulses are not resolved. This problem has already been tackled in Section 9.8 within the context of the semiclassical theory of photoelectric detection. Here we will briefly repeat the calculation for a quantum field.

14.6.2 Photocurrent correlations

Let us suppose, as before, that when a photoemission occurs at time t_0, each photodetector delivers a standard, short current pulse of the form shown in Fig. 9.5 after amplification, which will be denoted by $k(t - t_0)$. The area under each pulse is the total charge Q delivered per pulse,

$$Q = \int_0^\infty k(t)\,dt,$$

which is actually zero for an amplifier that does not pass a direct current, as in the Hanbury Brown–Twiss experiment. The length T_r of the pulse is of the order of the reciprocal pass bandwidth of the amplifier. Then the currents $J_1(t)$, $J_2(t)$ emerging from the two detectors/amplifiers can be expressed in the form

$$\left.\begin{aligned} J_1(t) &= \sum_i k(t - t_i) \\ J_2(t) &= \sum_j k(t - t_j), \end{aligned}\right\} \tag{14.6–8}$$

where the sum is to be taken over the various times t_i, t_j at which photoemissions occur. We can now use our expressions for the differential detection probabilities $P_1(\mathbf{r}, t)\Delta t$ and $P_2(\mathbf{r}_1, t_1; \mathbf{r}_2, t_2)\Delta t_1 \Delta t_2$ to calculate the averages of $J_1(t)$, $J_2(t)$ and their correlations. Thus, in the stationary state,

$$\begin{aligned}\langle J_1(t)\rangle = \left\langle \sum_i k(t - t_i)\right\rangle = \int_{-\infty}^{\infty} k(t - t_i) P_1(\mathbf{r}_1, t_i)\,dt_i &= \alpha_1 c S_1 \langle \hat{I}(\mathbf{r}_1)\rangle \int_0^\infty k(t')\,dt' \\ &= 0, \end{aligned} \tag{14.6–9}$$

because $\int_0^\infty k(t')\,dt' = 0$, and similarly for $\langle J_2(t)\rangle$, while

$$\begin{aligned} &\langle J_1(t) J_2(t)\rangle \\ &= \left\langle \sum_i \sum_j k(t - t_i) k(t - t_j)\right\rangle \\ &= \int_{-\infty}^{\infty} dt_i \int_{-\infty}^{\infty} dt_j k(t - t_i) k(t - t_j) P_2(\mathbf{r}_1, t_i; \mathbf{r}_2, t_j) \\ &= \alpha_1 \alpha_2 c^2 S_1 S_2 \langle \hat{I}(\mathbf{r}_1)\rangle \langle \hat{I}(\mathbf{r}_2)\rangle \iint_{-\infty}^{\infty} dt_i\, dt_j k(t - t_i) k(t - t_j)[1 + \lambda(\mathbf{r}_1, \mathbf{r}_2, t_j - t_i)], \end{aligned} \tag{14.6–10}$$

with $\lambda(\mathbf{r}_1, \mathbf{r}_2, \tau)$ given by Eq. (14.6–6). This result is similar to that obtained by the semi classical theory (Section 9.8). For light that is cross-spectrally pure (see Section 4.5.1), $\gamma(\mathbf{r}_1, \mathbf{r}_2, \tau)$ may be factorized in the form

$$\gamma(\mathbf{r}_1, \mathbf{r}_2, \tau) = \gamma(\mathbf{r}_1, \mathbf{r}_2, 0)\gamma(\tau), \tag{14.6–11}$$

where $\gamma(\tau)$ is the normalized autocorrelation function of the field, or the Fourier transform of the normalized spectral density. The field had a bandwidth of many thousand MHz in the experiment. The range of $\gamma(\tau)$, which is the coherence time T_c of the light, is therefore very short compared with the range of $k(t)$ in Fig. 9.5 and $\lambda(\mathbf{r}_1, \mathbf{r}_2, \tau)$ may be treated almost as a delta function under the integral in Eq. (14.6–10). We then find with the help of Eqs. (14.6–9) and (14.6–11) that

$$\begin{aligned}\langle J_1(t)J_2(t)\rangle &\simeq \alpha_1\alpha_2 c^2 S_1 S_2 \langle \hat{I}(\mathbf{r}_1)\rangle\langle \hat{I}(\mathbf{r}_2)\rangle |\gamma(\mathbf{r}_1, \mathbf{r}_2, 0)|^2 \int_{-\infty}^{\infty} |\gamma(\tau)|^2\,\mathrm{d}\tau \int_0^{\infty} k^2(t')\,\mathrm{d}t' \\ &= \alpha_1\alpha_2 c^2 S_1 S_2 \langle \hat{I}(\mathbf{r}_1)\rangle\langle \hat{I}(\mathbf{r}_2)\rangle |\gamma(\mathbf{r}_1, \mathbf{r}_2, 0)|^2 T_c \int_0^{\infty} k^2(t')\,\mathrm{d}t', \qquad (14.6\text{–}12)\end{aligned}$$

where we have made use of the fact that the integral

$$\int_{-\infty}^{\infty} |\gamma(\tau)|^2\,\mathrm{d}\tau = T_c \qquad (14.6\text{–}13)$$

provides a natural measure of the coherence time T_c of the light, because $|\gamma(\tau)|^2$ is unity for $\tau = 0$ and has a range of order T_c, as shown in Fig. 14.4. The same quantity has already been encountered in Eq. (2.4–29) and Eq. (4.3–82). Equation (14.6–12) shows that the current correlation is expected to be proportional to the square of the degree of coherence, as was actually observed in the experiment.

Of more interest than the value of $\langle J_1(t)J_2(t)\rangle$ is the normalized cross-correlation coefficient

$$\sigma_{12} \equiv \frac{\langle \Delta J_1(t)\Delta J_2(t)\rangle}{[\langle(\Delta J_1)^2\rangle\langle(\Delta J_2)^2\rangle]^{1/2}}, \qquad (14.6\text{–}14)$$

for which we require the variance of each current $J_1(t)$, $J_2(t)$ as well as the cross-correlation. For the sake of somewhat greater generality we shall calculate the autocorrelation function $\langle J_1(t)J_1(t+\tau)\rangle$. The calculation is very similar to the foregoing one, with one significant difference. We obtain from Eqs. (14.6–8)

$$\begin{aligned}\langle J_1(t)J_1(t+\tau)\rangle &= \left\langle \sum_i\sum_j k(t-t_i)k(t+\tau-t_j)\right\rangle \\ &= \left\langle \sum_i k(t-t_i)k(t+\tau-t_i)\right\rangle + \left\langle \sum\sum_{i\neq j} k(t-t_i)k(t+\tau-t_j)\right\rangle,\end{aligned}$$

where we have divided the sum into contributions that involve only single

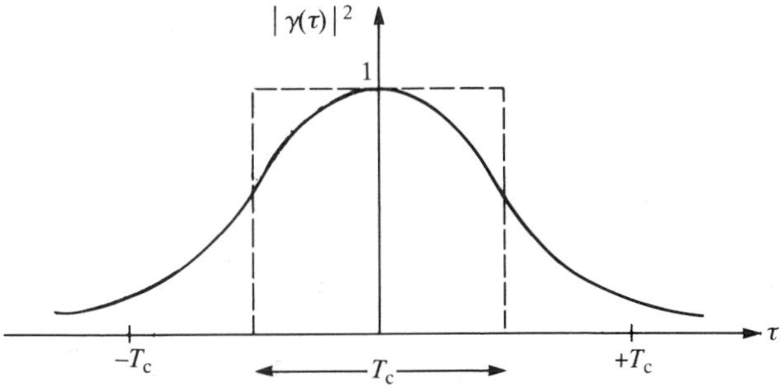

Fig. 14.4 Illustrating a typical form of $|\gamma(\tau)|^2$.

photoemissions, and those involving pairs of photoemissions, as in Section 9.8. The first term has no counterpart in the cross-correlation function. The averages are calculated as before, with the help of the differential probabilities $P_1(t_i)\,\mathrm{d}t_i$ and $P_2(t_i, t_j)\,\mathrm{d}t_i\,\mathrm{d}t_j$, and we find that (cf. Section 9.8 of the semiclassical treatment)

$$\langle J_1(t)J_1(t+\tau)\rangle = \alpha_1 c S_1\langle \hat{I}(\mathbf{r}_1)\rangle \int_{-\infty}^{\infty} k(t-t_i)k(t+\tau-t_i)\,\mathrm{d}t_i$$
$$+ (\alpha_1 c S_1\langle \hat{I}(\mathbf{r}_1)\rangle)^2 \iint_{-\infty}^{\infty} k(t-t_i)k(t+\tau-t_j)\lambda(\mathbf{r}_1, \mathbf{r}_1, t_j - t_i)\,\mathrm{d}t_i\,\mathrm{d}t_j. \quad (14.6\text{–}15)$$

When $\tau = 0$, with $\lambda(\mathbf{r}_1, \mathbf{r}_1, \tau)$ given by Eq. (14.6–6), and with the same approximations as before, this reduces to

$$\langle(\Delta J_1)^2\rangle = \langle J_1^2\rangle = \alpha_1 c S_1\langle \hat{I}(\mathbf{r}_1)\rangle[1 + \alpha_1 c S_1\langle \hat{I}(\mathbf{r}_1)\rangle T_\mathrm{c}]\int_0^{\infty} k^2(t')\,\mathrm{d}t'. \quad (14.6\text{–}16)$$

Equations (14.6–15) and (14.6–16) show that the current fluctuations can be expressed as the sum of two terms, of which the first one is independent of the coherence properties of the incident light, whereas the second one depends on $\lambda(\mathbf{r}_1, \mathbf{r}_1, \tau)$ or T_c. The first term is attributable to the shot noise of the photoelectric current, while the second term depends on the fluctuation properties of the optical field. Together with a similar expression for $\langle(\Delta J_2)^2\rangle$ and with Eq. (14.6–12) we then find for the correlation coefficient σ_{12} defined by Eq. (14.6–14)

$$\sigma_{12} = \frac{[\alpha_1\alpha_2 c^2 S_1 S_2\langle \hat{I}(\mathbf{r}_1)\rangle\langle \hat{I}(\mathbf{r}_2)\rangle]^{1/2} T_\mathrm{c}}{[1 + \alpha_1 c S_1\langle \hat{I}(\mathbf{r}_1)\rangle T_\mathrm{c}]^{1/2}[1 + \alpha_2 c S_2\langle \hat{I}(\mathbf{r}_2)\rangle T_\mathrm{c}]^{1/2}}|\gamma(\mathbf{r}_1, \mathbf{r}_2, 0)|^2,$$

which can be further simplified to

$$\sigma_{12} = \frac{\delta}{1+\delta}|\gamma(\mathbf{r}_1, \mathbf{r}_2, 0)|^2, \quad (14.6\text{–}17)$$

where

$$\alpha_1 c S_1\langle \hat{I}(\mathbf{r}_1)\rangle T_\mathrm{c} \equiv \delta = \alpha_2 c S_2\langle \hat{I}(\mathbf{r}_2)\rangle T_\mathrm{c}. \quad (14.6\text{–}18)$$

The important parameter δ that appears naturally is a measure of the average number of photoelectric emissions produced at each detector in a time equal to the coherence time T_c by a spatially coherent light beam, and it evidently determines the size of the correlation coefficient σ_{12}. When δ is a large number, $\sigma_{12} \approx |\gamma(\mathbf{r}_1, \mathbf{r}_2, 0)|^2$, but when δ is small, σ_{12} becomes much smaller than $|\gamma(\mathbf{r}_1, \mathbf{r}_2, 0)|^2$. It is because δ was a very small number in the Hanbury Brown–Twiss experiment, and because the current fluctuations were dominated by shot noise according to Eq. (14.6–18), that the correlation measurement was so difficult and the error bars in Fig. 9.6(*b*) were so large.

We can see why this was to be expected, by noting that δ/α is the average number of photons falling on the photodetector in a time equal to the coherence time. We have assumed that the size of the photocathode is small enough for the field to look like a plane wave, so that it is smaller than the coherence area. The number δ/α is therefore smaller than the number of photons within a coherence volume, the so-called degeneracy parameter and, as we have seen in Section 13.1, for thermal light produced by a source at temperatures below about 10^5 K, the photon occupation number per mode is less than one. We therefore expect δ/α to be much smaller than unity and δ to be very much smaller than unity, as was the case in the experiment.

Finally, after taking the Fourier transform of each term in Eq. (14.6–15) we obtain an expression for the spectral density $X(\omega)$ of the photoelectric current. This leads to the relation

$$X(\omega) = (\alpha_1 c S_1 \langle \hat{I}(\mathbf{r}_1)\rangle)|K(\omega)|^2[1 + (\alpha_1 c S_1 \langle \hat{I}(\mathbf{r}_1)\rangle)\psi(\omega)], \quad (14.6\text{–}19)$$

in which $K(\omega)$ is the frequency response of the photodetector–amplifier combination,

$$K(\omega) = \int_0^\infty k(\tau)\, e^{i\omega\tau}\, d\tau, \quad (14.6\text{–}20)$$

and $\psi(\omega)$ is the spectral density of the normalized intensity correlation function

$$\psi(\omega) = \int_0^\infty \lambda(\tau)\, e^{i\omega\tau}\, d\tau. \quad (14.6\text{–}21)$$

For a classical field $\psi(\omega) \geqslant 0$, and the second term in Eq. (14.6–19) represents the additional photocurrent fluctuation over the shot noise of the photocurrent due to the presence of the fluctuating electromagnetic field. However, as we shall see in Chapter 21, it is a remarkable feature of the quantum field that there exist states with $\psi(\omega) < 0$, which are known as *squeezed states*. Turning on a beam of squeezed light falling on the detector would have the effect of *lowering* the photocurrent fluctuations below the shot noise.

14.7 Bunching and antibunching

We have already seen from Eq. (14.6–2) that photoelectric detections within an optical field are correlated, in general, both in space and in time. One way of exhibiting this correlation is by effectively multiplying the output currents of two photodetectors, as in the experiment shown in Fig. 9.6(*a*). Another procedure is to measure the joint detection probability $P_2(\mathbf{r}_1, t_1, \Delta t_1; \mathbf{r}_2, t_2, \Delta t_2)$ as a function of the separations $|\mathbf{r}_2 - \mathbf{r}_1|$ in space and $|t_2 - t_1|$ in time. This involves measurements in which the individual photoelectric pulses are registered. As is shown by Eq. (14.6–4), the joint probability may be larger or smaller than the product of the two separate one-photon detection probabilities, depending on the sign of the normalized correlation function $\lambda(\mathbf{r}_1, t_1; \mathbf{r}_2, t_2)$.

Let us concentrate on two detections as a function of the time separation $t_2 - t_1 = \tau$, and ignore the space coordinates for the moment. For a stationary field λ is then a function only of the time difference τ. If $\lambda(0) > \lambda(\tau)$, any two photoelectric pulses are more likely to occur close together in time than farther apart, and we speak of *bunching* of photoelectric detections (Mandel and Wolf, 1965, Sec. 6.3). If, on the other hand, $\lambda(0) < \lambda(\tau)$, so that two photoelectric detections are less likely to appear close together than further apart, we speak of *antibunching*. It can be shown from the Schwarz inequality that the laws of classical probability require $\lambda(0) \geqslant \lambda(\tau)$, so that antibunching is a purely quantum mechanical phenomenon. It should be emphasized that the sign of $\lambda(0)$ by itself does not determine whether bunching or antibunching is present in general, but only the relationship between $\lambda(0)$ and $\lambda(\tau)$. For example, a state with $\lambda(0) < 0$ but $\lambda(0) = \lambda(\tau)$ for all τ does not exhibit antibunching, although it is a quantum state without classical analogy.

14.7.1 Coincidence detection

One procedure for gathering information about the joint probability $P_2(\mathbf{r}_1, t, \Delta t; \mathbf{r}_2, t+\tau, \Delta\tau)$ that was historically important is to feed the photoelectric pulses from the two detectors to a coincidence counter, as shown in Fig. 14.5, and to measure the rate at which pulses from the two detectors arrive 'in coincidence'. Several such experiments were performed in a search for correlations in the 1950s, some of which succeeded (Twiss, Little and Brown, 1957; Rebka and Pound, 1957; Brannen, Ferguson and Wehlau, 1958) and some of which failed (Adam, Janossy and Varga, 1955a,b; Brannen and Ferguson, 1956). Let us examine the results to be expected in such an experiment more quantitatively.

Loosely speaking, a coincidence counter delivers an output pulse whenever pulses appear at its two inputs simultaneously or in coincidence, and not otherwise. However, the phrase 'in coincidence' must be interpreted by reference to the intrinsic time resolution T_r of the circuit, which is limited by the width or rise time of the pulses. By definition, two pulses at the two inputs that begin at times t and $t+\tau$, respectively, will be judged to be 'in coincidence' if $|\tau| \leqslant \frac{1}{2}T_r$ (see Fig. 14.5). If $P_2(\mathbf{r}_1, t; \mathbf{r}_2, t+\tau)\Delta t\Delta\tau$ is the joint probability that one pulse appears at one input at time t within Δt and another one at the other input at time $t+\tau$ within $\Delta\tau$, the average rate $\mathscr{R}_c$ at which the coincidence circuit delivers output pulses will be

$$\mathscr{R}_c = \int_{-\frac{1}{2}T_r}^{\frac{1}{2}T_r} P_2(\mathbf{r}_1, t; \mathbf{r}_2, t+\tau)\,d\tau.$$

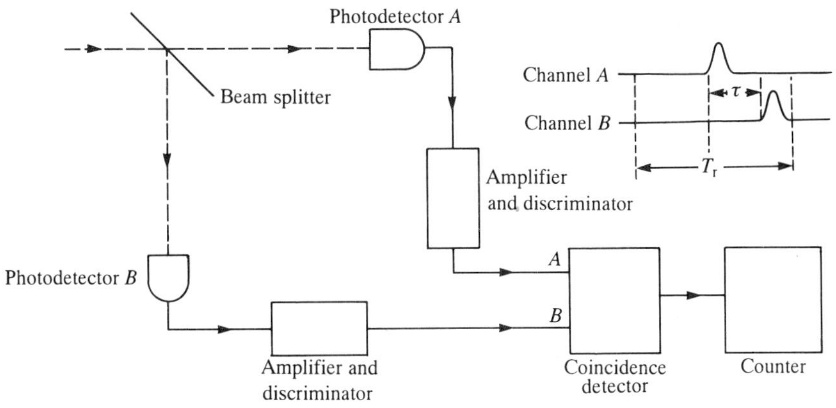

Fig. 14.5 Outline of correlation experiment based on coincidence detection.

With the help of Eq. (14.6–4) this becomes

$$\mathcal{R}_c = R_1 R_2 T_r\left[1 + \frac{1}{T_r}\int_{-\frac{1}{2}T_r}^{\frac{1}{2}T_r} \lambda(\mathbf{r}_1, t; \mathbf{r}_2, t+\tau)\,d\tau\right], \tag{14.7–1}$$

in which $R_1 \equiv \alpha_1 c S_1 \langle \hat{I}(\mathbf{r}_1)\rangle$ and $R_2 \equiv \alpha_2 c S_2 \langle \hat{I}(\mathbf{r}_2)\rangle$ are the average rates at which pulses arrive in the two input channels. The first term on the right of this equation represents the random or accidental contribution to the coincidence rate $\mathcal{R}_c$, resulting from purely random overlap of input pulses. The second term represents a correction or excess contribution attributable to the light intensity fluctuations, which can be either positive or negative, in principle. For light derived from stationary thermal sources, as in all experiments mentioned above, we may express $\lambda(\mathbf{r}_1, t;\ \mathbf{r}_2, t+\tau)$ in terms of $\gamma(\mathbf{r}_1, \mathbf{r}_2, \tau)$, and if the light is polarized and cross-spectrally pure we can use Eqs. (14.6–6) and (14.6–11) and replace $\lambda(\mathbf{r}_1, t;\ \mathbf{r}_2, t+\tau)$ by $|\gamma(\mathbf{r}_1, \mathbf{r}_2, 0)|^2 |\gamma(\tau)|^2$. The excess term in Eq. (14.7–1) is then necessarily positive. Now $|\gamma(\tau)|^2$ has the general structure shown in Fig. 14.4 (its detailed shape is, of course, dependent on the spectral distribution). It is unity for $\tau \approx 0$, and it becomes very small once $|\tau|$ appreciably exceeds the correlation time T_c. In fact, T_r greatly exceeded T_c in all the above-mentioned coincidence counting experiments, so that we may replace the limits in the integral in Eq. (14.7–1) by $\pm\,\infty$ to a reasonable approximation, and write

$$\int_{-\frac{1}{2}T_r}^{\frac{1}{2}T_r} |\gamma(\tau)|^2\,d\tau \approx \int_{-\infty}^{\infty} |\gamma(\tau)|^2\,d\tau = T_c, \tag{14.7–2}$$

where T_c is a natural measure of the coherence time of the light, as we show below. Equation (14.7–1) then becomes

$$\mathcal{R}_c = R_1 R_2 T_r[1 + (T_c/T_r)|\gamma(\mathbf{r}_1, \mathbf{r}_2, 0)|^2]. \tag{14.7–3}$$

It is clear that the effect of thermal light fluctuations is to produce an excess contribution to the coincidence rate $\mathcal{R}_c$, over the accidental rate $R_1 R_2 T_r$, but that the feasibility of detecting this excess depends strongly on the ratio T_c/T_r. In all the unsuccessful experiments the light source had an appreciable spectral width, with coherence time $T_c \lesssim 10^{-12}$ s. Even with the fastest electronics, the ratio T_c/T_r was therefore extremely small, and it is not surprising that the excess proved to be undetectable. In all the successful experiments the light source was a low pressure gas discharge producing spectral lines with coherence times $T_c \sim 10^{-9}$ s. When this is combined with a coincidence circuit of resolving time $T_r \sim 10^{-8}$ s, the excess contribution becomes detectable without difficulty. The reasons for the successes and failures of certain experiments are therefore readily understandable.

14.7.2 Two-time pulse correlation measurements

Successful time-resolved experiments with thermal light, in which the joint detection probability $P_2(\mathbf{r}_1, t;\ \mathbf{r}_2, t+\tau)\Delta t \Delta\tau$ is directly measured as a function of τ, were not reported until the late 1960s. This is hardly surprising, for the rise times of photomultiplier pulses are themselves of the same order (~1 ns) as the coherence times of the most monochromatic thermal light beams. Figure 14.6(a)

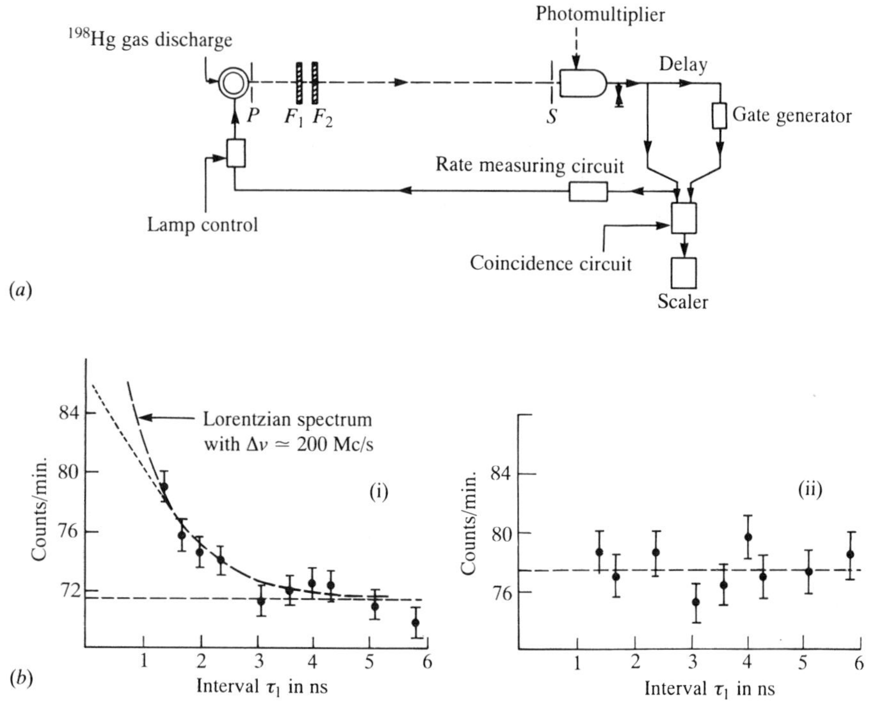

Fig. 14.6 (*a*) Outline of the apparatus used to demonstrate photoelectric bunching. (*b*) Results of the measurement, showing the pulse pair counting rate as a function of time separation τ_1; (i) for the ^{198}Hg vapor light source, (ii) for a tungsten light source. (Reproduced from Morgan and Mandel, 1966.)

shows an outline of the first successful time-resolved experiment (Morgan and Mandel, 1966). The light source was a low-pressure gas discharge in a single ^{198}Hg isotope, that produced a single spectral line of coherence time $T_c \sim 2$ ns with the help of an interference filter. The light fell on a photomultiplier tube, whose output pulses were sent to a high-resolution coincidence detector via two separate paths, one of which contained an adjustable delay τ. Under these conditions a single photomultiplier pulse produces no output from the coincidence circuit, because the two inputs do not arrive simultaneously. However, two pulses with time separation τ produce an output, because the first pulse taking the long path arrives in coincidence with the second pulse taking the short path. It was therefore possible to obtain $P_2(\mathbf{r}, t; \mathbf{r}, t + \tau)$ by measuring the coincidence counting rate for various values of the delay τ. Figure 14.6(*b*) shows the rate at which pairs of photoelectric pulses with time separation τ were being detected as a function of τ. For τ greater than 5 or 6 ns, the pulse pair counting rate was approximately constant and given by $R_1 R_2 T_r$ [cf. Eq. (14.7–1)], which shows that pulses with such large separations were arriving independently. But for pulse time separations τ of order of less than the 2 ns coherence time of the light, the counting rate increased in accordance with Eq. (14.6–4). This shows that the photoelectric pulses tend to be emitted in bunches rather than at random, for the arrival of one pulse enhances the probability for the appearance of another one immediately afterwards. No such bunching was observed with white light from a tungsten lamp, for which the coherence time was thousands of times shorter. When the resolving time is sufficiently short, the shape of the experimentally derived curve for $P_2(\mathbf{r}, t{:}\, \mathbf{r}, t + \tau)$ directly yields the intensity correlation function $\lambda(\mathbf{r}, t{:}\, \mathbf{r}, t + \tau) \equiv \lambda(\tau)$, and therefore $|\gamma(\tau)|$ for thermal light. The measurement of the joint probability density $P_2(\mathbf{r}, t{:}\, \mathbf{r}, t + \tau)$ therefore offers another opportunity for making spectroscopic measurements of thermal light in the time domain. Such measurements become progressively easier as the bandwidth of the light is reduced. It is more common now to divide the light beam into two parts with the aid of a beam splitter, and to use two photodetectors (Scarl, 1966; Phillips, Kleiman, and Davis, 1967; Davidson and Mandel, 1968; Davidson, 1969). This allows measurements to be extended down to zero time separation, and avoids some problems connected with spurious multiple pulse emissions from some detectors.

A variety of different electronic methods has been developed for measuring the number of events in which two pulses from the two detectors are separated by a delay τ. Sometimes the pulses are fed to the 'start' and 'stop' inputs of a time-to-amplitude converter (TAC), that generates a ramp voltage whose height is proportional to the time interval τ between the two pulses (see Fig. 14.7). This height is then digitized and stored in a digital memory in such a way that the value of τ is treated as an address, and the number stored at the address τ is incremented by unity by each detection with delay τ (Davidson and Mandel, 1968; Davidson, 1969). After many such events, the number accumulated at address τ is proportional to the probability density for a time interval τ. This method is most suitable for the shortest correlation times, because the digitizing can be performed slowly. However, strictly speaking, it yields, not the probability density $P_2(\mathbf{r}_1, t{:}\, \mathbf{r}_2, t + \tau)$ that a stop pulse follows a start pulse after a time delay τ, but the more complicated probability that the *first stop pulse* following the start pulse occurs after a time τ. The difference between the two is discussed in more detail below. At low pulse counting rates the two probabilities are rather similar, and it is possible to relate one to the other (Davidson and Mandel, 1968; Davidson, 1969), but they become increasingly different as the counting rate is increased. However, the method has been used to make more accurate measurements of $\lambda(\mathbf{r}_1, t{:}$

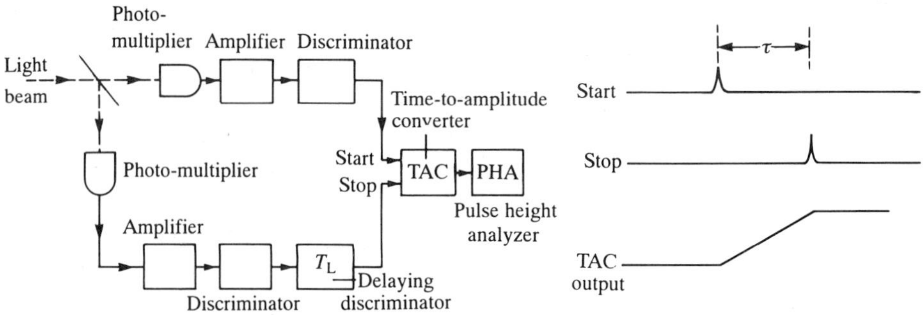

Fig. 14.7 Outline of an apparatus for photoelectric correlation measurements based on time-to-amplitude conversion. The time interval between successive start and stop pulses is recorded. (Reproduced from Davidson and Mandel, 1968.)

$\mathbf{r}_2, t + \tau$) for thermal light (Scarl, 1966; Phillips, Kleiman and Davis, 1967). Figure 14.8 shows the results of an experiment in which a time-to-amplitude converter and pulse height analyzer were used to measure the probability distribution of the time interval τ for positive and negative τ. The spectral width of the light was narrowed interferometrically in successive measurements, as a result of which the bunching extended over progressively longer time intervals. This clearly shows the spectroscopic application of the technique.

Figure 14.9 shows the schematic outline of another electronic technique for making correlation measurements that yields $P_2(\mathbf{r}_1, t; \mathbf{r}_2, t + \tau)$. This time the time interval between 'start' and 'stop' pulses from two photodetectors is digitized in steps of T with the help of delays and AND gates, without regard for

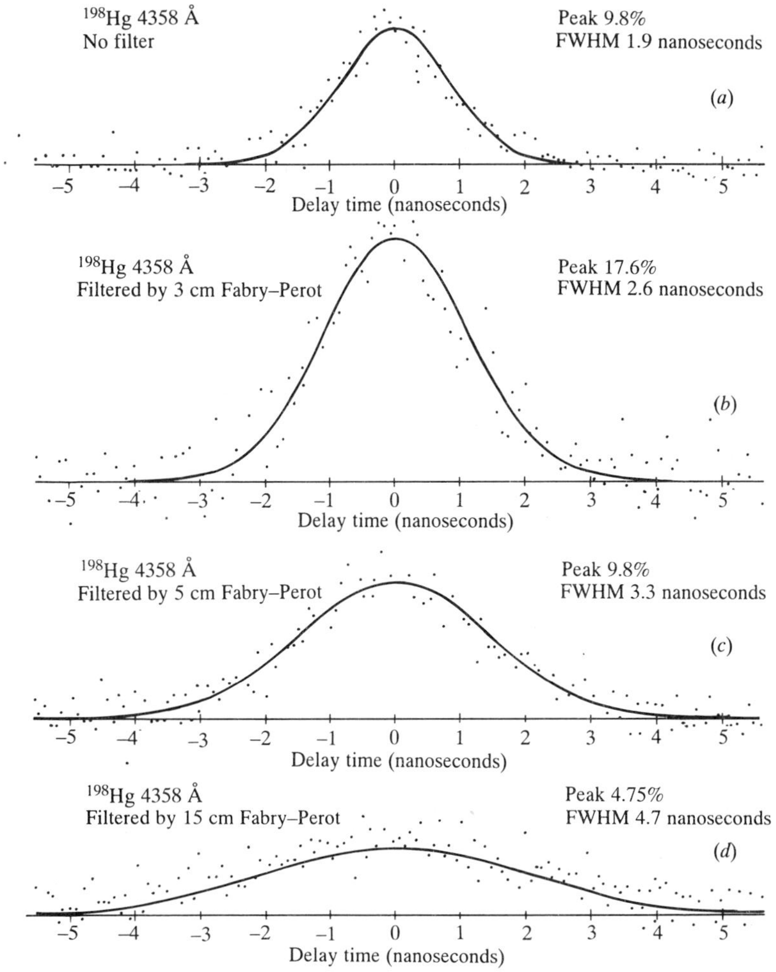

Fig. 14.8 Results of two-time correlation measurements with a time-to-amplitude converter. The number of photoelectric pulse pairs is plotted against the time delay for the light from a ^{198}Hg lamp; (*a*) without a filter, (*b*) behind a 3 cm Fabry–Perot filter, (*c*) behind a 5 cm Fabry–Perot filter, (*d*) behind a 15 cm Fabry–Perot filter. (Reproduced from Phillips, Kleiman and Davis, 1967.)

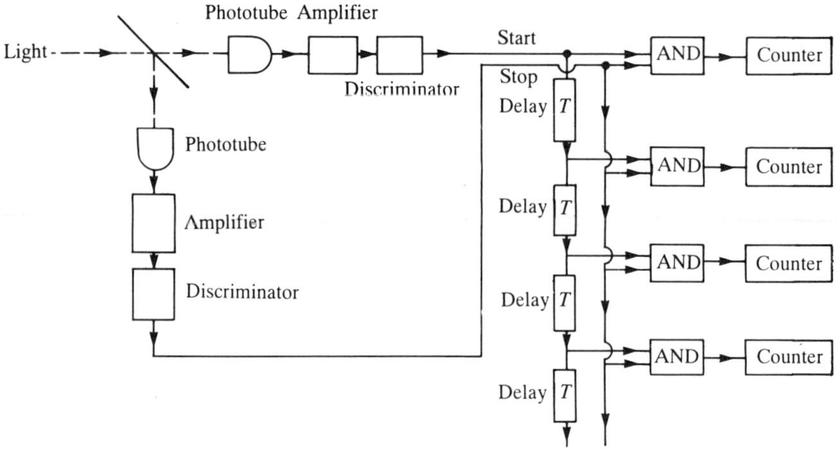

Fig. 14.9 Outline of an apparatus for photoelectric correlation measurements. The time interval between start and stop pulses is digitized in steps of T, without regard for intervening pulses.

intervening pulses. The number of pulse pairs corresponding to a delay NT is then registered on counter N ($N = 1, 2, \ldots$), and this is proportional to $P_2(\mathbf{r}_1, t; \mathbf{r}_2, t + NT)$. Finally, we should mention that much more sophisticated correlators, some of which make use of fast shift registers, have also been developed and used to study the statistical properties of light (Cummins and Pike, 1977; Swinney, 1983; Pike, 1986).

It is possible to give general arguments as to why bunching of photoelectric pulses is to be expected when thermal light falls on a photodetector. If we adopt a classical picture of the optical field, we can argue that thermal fluctuations of the source cause the light intensity to fluctuate. But more photoemissions occur when the instantaneous light intensity is high than when the intensity is low, so that the photoelectric pulses are not distributed strictly at random, as they would be under constant intensity illumination. The bunching therefore appears as a manifestation of fluctuating electromagnetic waves, and is seen to have a semiclassical explanation.

However, we can also adopt a point of view in which the thermal light beam is regarded as a random stream of particles, with overlapping wave functions that interfere. When we symmetrize the quantum state of the system, because the particles are intrinsically indistinguishable, we find that the interference increases the probability of detecting one particle close to another one. The bunching now appears as a manifestation of the Bose–Einstein statistics obeyed by thermal photons (Purcell, 1956). Despite the differences between these two points of view, they are, in a sense, equivalent descriptions of the same phenomenon.

Time-resolved pulse correlation measurements are not, of course, limited to thermal light. They have been applied to scattered light in order to yield information about the nature of the scatterers (Cummins and Swinney, 1970; Chu, 1974; Crosignani, DiPorto and Bertolotti, 1975), and they have been used to investigate correlation properties of laser light, particularly in the neighborhood of the threshold of oscillation (Arecchi, Gatti and Sona, 1966; Davidson and

Mandel, 1967; Chopra and Mandel, 1973a; Corti, Degiorgio and Arecchi, 1973). Several different digital methods have been developed for accumulating the information about the number of photoelectric pulse pairs with time separation τ. Sometimes the time intervals between pulses are digitized and recorded directly (Chopra and Mandel, 1972), and sometimes the pulses are sorted into different channels according to their delays, and the numbers corresponding to each delay are registered (Jakeman and Pike, 1969; Jakeman, 1970). Figure 14.10 shows the results of an experiment (Abate, Kimble and Mandel, 1976) in which the correlation properties of light emitted from a dye laser in the neighborhood of the oscillation threshold were investigated. The normalized intensity correlation function $\lambda(\tau)$ exhibits two characteristic time constants of order 10 μs and 100 μs. Finally, the third-order joint detection probability density $P_3(\mathbf{r}_1, t; \mathbf{r}_2, t+\tau; \mathbf{r}_3, t+\tau')$, which depends on two time intervals τ, τ', has also been measured for a laser near threshold. (Chopra and Mandel, 1973b; Corti and Degiorgio, 1976). It yields information about the sixth-order correlation function $\Gamma^{(3,3)}$ of the optical field. It has been pointed out that the phase of the second order correlation function can be determined from measurements of P_3 (Gamo, 1963a,b).

14.7.3 Antibunching

In all the foregoing experiments the photoelectric pulses either exhibit bunching or they occur strictly at random, as shown by the fact that the joint detection probability always satisfies the inequality

$$P_2(\mathbf{r}, t; \mathbf{r}, t+\tau) \leq P_2(\mathbf{r}, t; \mathbf{r}, t).$$

According to Eqs. (14.6–3) and (14.6–4) this implies that

$$\langle \mathcal{T}\!: \hat{I}(\mathbf{r}, t)\hat{I}(\mathbf{r}, t+\tau)\!:\rangle \leq \langle : I^2(\mathbf{r}, t) : \rangle, \tag{14.7–4a}$$

or, in terms of the phase space functional $\phi(\{v\})$ of the field and c-number light

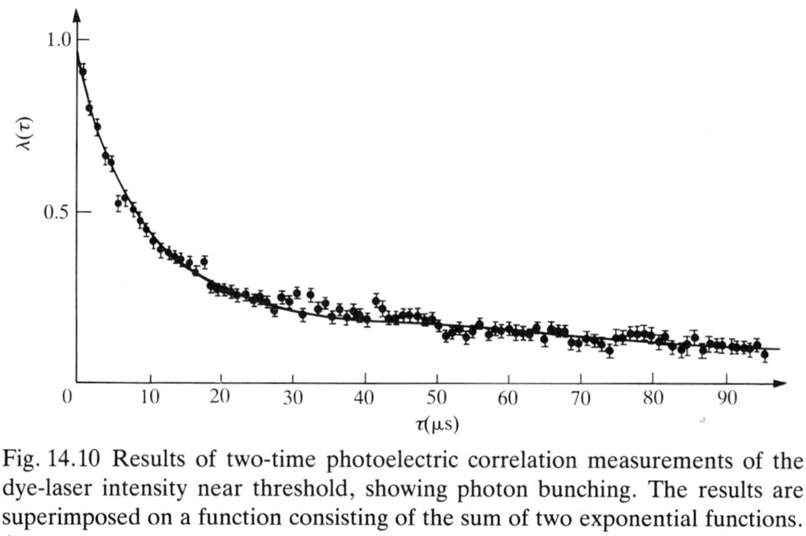

Fig. 14.10 Results of two-time photoelectric correlation measurements of the dye-laser intensity near threshold, showing photon bunching. The results are superimposed on a function consisting of the sum of two exponential functions. (Reproduced from Abate, Kimble and Mandel, 1976.)

intensities, that

$$\langle I(\mathbf{r}, t)I(\mathbf{r}, t+\tau)\rangle_\phi \leq \langle I^2(\mathbf{r}, t)\rangle_\phi. \qquad (14.7\text{–}4\text{b})$$

In other words, the two-time intensity autocorrelation function either falls from its initial value at $\tau = 0$ or remains constant. Now when $\phi(\{v\})$ is a true, non-negative, probability function, one can show from the Schwarz inequality that Eq. (14.7–4b) must hold quite generally for a stationary field. It follows that for such states the photoelectric pulses must exhibit either bunching or complete randomness; antibunching, which implies that $P_2(\mathbf{r}, t;\ \mathbf{r}, t) < P_2(\mathbf{r}, t;\ \mathbf{r}, t+\tau)$ and violates (14.7–4), is impossible under these circumstances. If antibunching occurs, so that

$$\langle I(\mathbf{r}, t)I(\mathbf{r}, t+\tau)\rangle_\phi > \langle I^2(\mathbf{r}, t)\rangle_\phi, \qquad (14.7\text{–}5)$$

i.e. the correlation function rises instead of falling from its initial value at $\tau = 0$, then $\phi(\{v\})$ cannot be a probability functional. Now states of the electromagnetic field for which $\phi(\{v\})$ is negative and more singular than a delta function exist, but, as we know from Section 11.8, they are decidedly non-classical. Both thermal light and laser light have to be excluded. Observation of photoelectric antibunching therefore provides evidence for an explicitly quantum mechanical state of the optical field, whereas bunching carries no such implication.

Figure 14.11 shows the results of experiments in which antibunching was observed (Kimble, Dagenais and Mandel, 1977; Dagenais and Mandel, 1978). The light source consisted of single, coherently excited atoms in an atomic beam; the atomic fluorescence was collected by a microscope objective and projected on to two photomultipliers, whose output pulses provided the start and stop inputs of a time-to-digital converter. As Fig. 14.11 shows, more photoelectric pulse pairs were observed with time separation τ up to about 25 ns than with zero time separation. This is in agreement with the predictions of the quantum theory of resonance fluorescence, which gives $\lambda(\tau) > \lambda(0)$, as we shall see in Section 15.6. Antibunching has also been observed in other more recent experiments (Cresser, Häger, Leuchs, Rateike and Walther, 1982; Walker and Jakeman, 1985; Grangier, Roger and Aspect, 1986; Grangier, Roger and Aspect, Heidmann and Reynaud, 1986).

14.7.4 The time interval distribution of photoelectric pulses

As we have seen, when light falls on a photodetector the joint probability of detecting two photoelectric pulses at time t within Δt and at time $t+\tau$ within $\Delta\tau$ is given by

$$P_2(t, t+\tau)\Delta t\Delta\tau = (\alpha cS)^2\Delta t\Delta\tau\langle\mathcal{T}{:}\hat{I}(t)\hat{I}(t+\tau){:}\rangle, \qquad (14.7\text{–}6)$$

in which the space coordinates have been suppressed. It might seem at first that we may look on the conditional probability

$$P_c(t+\tau|t)\Delta\tau = \frac{P_2(t, t+\tau)\Delta t\Delta\tau}{P_1(t)\Delta t} \qquad (14.7\text{–}7)$$

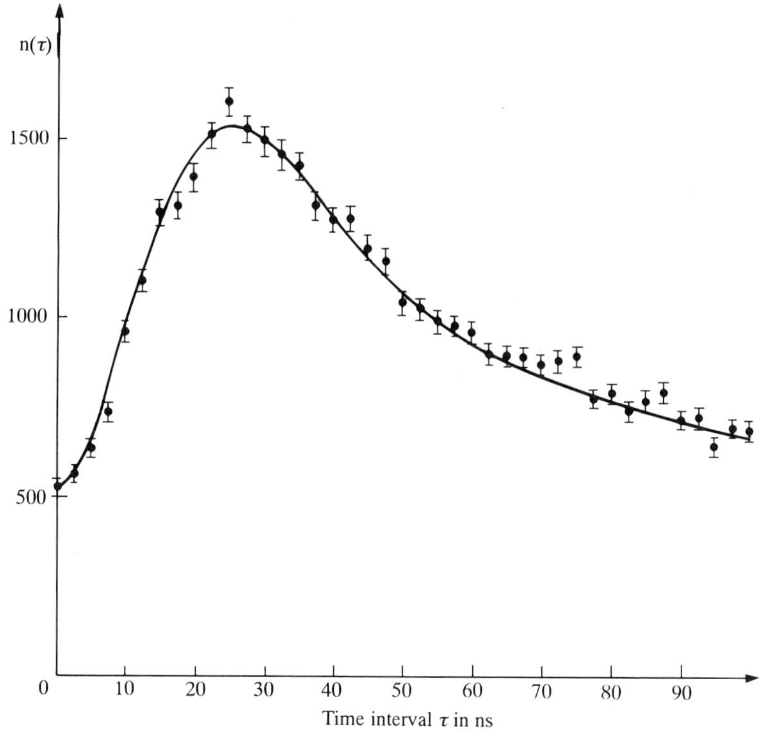

Fig. 14.11 Results of two-time correlation measurements of the light from one Na atom, showing antibunching. (Reproduced from Dagenais and Mandel, 1978.)

of registering a detection at time $t + \tau$ within $\Delta\tau$, given a detection at time t, as an expression for the distribution of the time interval τ between successive photoelectric pulses. However, $P_2(t, t + \tau)$ does not refer to pulses that are necessarily consecutive, nor does it tend to zero as $\tau \to \infty$. It gives the differential probability that one photoelectric pulse occurs at time t and one occurs at time $t + \tau$, irrespective of what other pulses may occur in between. We have already referred to several different experimental techniques for measuring photoelectric correlations, and we have pointed out that in some cases, such as that shown in Fig. 14.7, what is measured is not $P_2(t, t + \tau)$ or $P_c(t + \tau|t)$ but the probability density that the *first* stop pulse following a start pulse at time t occurs at time $t + \tau$. Let us now derive this probability density, which is just the probability distribution $\mathcal{P}(\tau)$ of the time interval τ between successive pulses.

In any differential time interval of width Δt the probability of no detection is given by $\langle[1 - \alpha c S\hat{I}(t)\Delta t]\rangle$. As might be expected, the probability for no detections to occur in the finite time interval from t to $t + \tau$ involves the product of many similar factors in time order and normal order, of the form

$$\mathcal{T}{:}\prod_{n=1}^{\tau/\Delta t}[1 - \alpha c S\hat{I}(t + n\Delta t)]{:},$$

and in the limit $\Delta t \to 0$ this can be shown (see Section 14.8 below) to reduce to

$$\mathcal{T}{:}\exp\left[-\alpha c S \int_t^{t+\tau} \hat{I}(t')\,\mathrm{d}t'\right]{:}.$$

Combination of this factor with the expression for the conditional probability density that a detection occurs at time $t + \tau$ given an earlier detection at time t, leads to the following formula for $\mathcal{P}(\tau)$:

$$\mathcal{P}(\tau) = \frac{\alpha c S}{\langle \hat{I}(t) \rangle} \left\langle \mathcal{T}{:}\hat{I}(t) \exp\left[-\alpha c S \int_t^{t+\tau} \hat{I}(t')\,\mathrm{d}t'\right] \hat{I}(t+\tau){:} \right\rangle. \quad (14.7\text{–}8)$$

For a stationary field $\langle \hat{I}(t) \rangle$ and $\mathcal{P}(\tau)$ are of course independent of t. As the negative exponent increases with τ, $\mathcal{P}(\tau)$ tends to zero as $\tau \to \infty$, so that $\mathcal{P}(\tau)$, unlike $P_c(t + \tau|t)$ given by Eq. (14.7–7), is normalizable. It is easy to confirm by direct integration that

$$\int_0^\infty \mathcal{P}(\tau)\,\mathrm{d}\tau = -\frac{1}{\langle \hat{I}(t) \rangle} \left[\left\langle \mathcal{T}{:}\hat{I}(t) \exp\left[-\alpha c S \int_t^{t+\tau} \hat{I}(t')\,\mathrm{d}t'\right]{:} \right\rangle \right]_{\tau=0}^{\tau=\infty} = 1, \quad (14.7\text{–}9)$$

as required. As $\alpha c S \langle \hat{I}(t) \rangle \equiv R$ is the average counting rate of the detector, it is clear that $\mathcal{P}(\tau)$ differs from $P_c(t + \tau|t)$ significantly only when $R\tau \not\ll 1$. When $R\tau$ is small it is possible to relate $\mathcal{P}(\tau)$ and $P_c(t + \tau|t)$ by making a power series expansion of the exponential factor (Davidson and Mandel, 1968; Davidson, 1969). In the particularly simple special case in which the field is in a coherent state, with the light intensity $I(t)$ constant in time, it follows from Eq. (14.7–8) that

$$\mathcal{P}(\tau) = R\mathrm{e}^{-R\tau} \qquad (0 \leqslant \tau). \quad (14.7\text{–}10)$$

More generally, $\mathcal{P}(\tau)$ is expected to vary approximately exponentially with τ for time intervals τ that are sufficiently long that intensity correlations have died out, provided the process is ergodic.

It is possible to formally derive both the probability densities $\mathcal{P}(\tau)$ and $P_c(t + \tau|t)$ from the generating function (Glauber, 1967; Mandel, 1967)

$$G(t, \tau, \xi) \equiv \frac{1}{\langle \hat{I}(t) \rangle} \left\langle \mathcal{T}{:}\hat{I}(t) \exp\left[-\xi \int_t^{t+\tau} \alpha c S \hat{I}(t')\,\mathrm{d}t'\right]{:} \right\rangle, \quad (14.7\text{–}11)$$

which, in principle, involves correlations of all orders. For we obtain by partial differentiation that

$$P_c(t + \tau|t) = \left[-\frac{1}{\xi}\frac{\partial G(t, \tau, \xi)}{\partial \tau}\right]_{\xi=0} \quad (14.7\text{–}12)$$

and

$$\mathcal{P}(\tau) = \left[-\frac{\partial G(t, \tau, \xi)}{\partial \tau}\right]_{\xi=1}. \quad (14.7\text{–}13)$$

We point out once again that when τ is sufficiently long, or when the photoelectric counting rate is sufficiently high, a distinction needs to be made between $P_c(t + \tau|t)$ and $\mathcal{P}(\tau)$. Then $\mathcal{P}(\tau)$ would be measured with an apparatus such as that shown in Fig. 14.7, whereas $P_c(t+\tau|t)$ would be measured with the apparatus illustrated in Fig. 14.9.

Finally we mention that the so-called waiting time distribution $W(\tau)$, which is the probability density for encountering a time interval τ from an arbitrary instant t until the next pulse, is given by

$$W(\tau) = \alpha cS\left\langle \mathcal{T}\text{:exp}\left[-\alpha cS\int_t^{t+\tau}\hat{I}(t')\,\mathrm{d}t'\right]\hat{I}(t+\tau)\text{:}\right\rangle. \quad (14.7\text{–}14)$$

14.8 Photoelectric counting statistics

So far we have dealt only with the differential probabilities of photoelectric detection in various short time intervals Δt. The interval Δt is not a true infinitesimal, because it was assumed to be long compared with the optical period, but it is sufficiently short that the differential probability is much less than unity. In practice, measurements are often made in a finite time interval, say from t to $t+T$, such that there is an appreciable probability $p(n, t, t+T)$ for n detections to occur within the interval T. We now turn our attention to the calculation of this integral counting probability $p(n, t, t+T)$ for an open system.

At first sight it might seem that the previous results would not be applicable to this situation at all, because they were obtained by the method of short time perturbation theory. A moment's thought will show, however, that when a beam of light falls on a photodetector, each element of the optical field generally interacts locally with the detector only very briefly, rather than for the whole measurement time T. Moreover, because the response time of the detector is so short and the energy of the electron state is well defined after an interaction time Δt, each electron is either free or still bound after that time Δt. We may therefore think of the measurement in the finite time interval T as being equivalent to a large number of successive measurements, each in a short time interval Δt, so that the previous results for the differential detection probability should be applicable. The situation would be different for a closed system, which we do not consider here.

We start by dividing the time interval T into $T/\delta t$ short intervals of duration δt, labeled $t_1, t_2, t_3, \ldots, t_{T/\delta t}$, respectively. Let $P_n(t_{i_1}, t_{i_2}, \ldots, t_{i_n})$ denote the joint probability of making n detections in the n time intervals $t_{i_1}, t_{i_2}, \ldots, t_{i_n}$, which is given by

$$P_n(t_{i_1}, t_{i_2}, \ldots, t_{i_n}) = \langle \mathcal{T}\text{:}[\alpha cS\hat{I}(t_{i_1})\delta t]\ldots[\alpha cS\hat{I}(t_{i_n})\delta t]\text{:}\rangle, \quad (14.8\text{–}1)$$

according to Eq. (14.5–3). We have dropped the position label $\mathbf{r}$ for the intensity $\hat{I}$ in order to simplify the notation. We shall also need to make use of the probability $\tilde{P}_n(t_{i_1}, t_{i_2}, \ldots, t_{i_n}; t_{j_1}, t_{j_2}, \ldots, t_{j_n})$ for making n detections in the time intervals $t_{i_1}, t_{i_2}, \ldots, t_{i_n}$ and *no detections* in the intervals $t_{j_1}, t_{j_2}, \ldots, t_{j_n}$ which can be calculated by the application of the unitarity principle for probabilities. For example, by summing the joint probability for one detection at t_i together with zero, or one, or two, etc. detections at t_j, we arrive at $P_1(t_i)$. Thus

$$P_1(t_i) = \tilde{P}_1(t_i; t_j) + P_2(t_i, t_j) + P_3(t_i, t_j, t_j) + \ldots,$$

so that

$$\begin{aligned}\tilde{P}_1(t_i;\, t_j) &= P_1(t_i) - P_2(t_i,\, t_j) - P_3(t_i,\, t_j,\, t_j) - \ldots \\ &= \langle \mathcal{T}{:}[\alpha c S \hat{I}(t_i)\delta t][1 - \alpha c S \hat{I}(t_j)\delta t - (\alpha c S \hat{I}(t_j)\delta t)^2 - \ldots]{:}\rangle. \qquad (14.8\text{–}2)\end{aligned}$$

More generally, we can show by a similar argument that the expression for the joint probability $\tilde{P}_n(t_{i_1}, \ldots, t_{i_n};\, t_{j_1}, t_{j_2}, \ldots, t_{j_n})$ has a factor of the form $[\alpha c S \hat{I}(t_i)\delta t]$ for every interval δt in which a detection occurs, and a factor of the form $[1 - \alpha c S \hat{I}(t_j)\delta t - (\alpha c S \hat{I}(t_j)\delta t)^2 - \ldots]$ for every interval δt in which no detection occurs. Hence the probability for exactly n detections in the intervals $t_{i_1}, t_{i_2}, \ldots, t_{i_n}$ is given by

$$\tilde{P}_n(t_{i_1}, \ldots, t_{i_n};\ \text{all other intervals})$$
$$= \left\langle \mathcal{T}{:}\prod_{s=1}^{n}[\alpha c S \hat{I}(t_{i_s})\delta t] \prod_{\substack{r=1 \\ r \neq i_1, i_2, \ldots, i_n}}^{T/\delta t} [1 - \alpha c S \hat{I}(t_r)\delta t - (\alpha c S \hat{I}(t_r)\delta t)^2 - \ldots]{:} \right\rangle. \qquad (14.8\text{–}3)$$

14.8.1 *The integral detection probability*

We can now use Eq. (14.8–3) to determine the integral counting probability $p(n, t, t+T)$ by summing $\tilde{P}_n$ over all possible times when the n detections can occur, and letting $\delta t \to 0$. But if we allow each index $i_1, i_2, \ldots, i_n$ to range from 1 to $T/\delta t$, then the probability for every n-detection event is repeated $n!$ times in the sum. We therefore need to divide by $n!$ to obtain the probability $p(n, t, t+T)$, namely

$$\begin{aligned}p(n, t, t+T) &= \lim_{\delta t \to 0} \frac{1}{n!} \sum_{i_1=1}^{T/\delta t} \cdots \sum_{i_n=1}^{T/\delta t} \tilde{P}_n(i_1, i_2, \ldots, i_n;\ \text{all other intervals}) \\ &= \lim_{\delta t \to 0} \frac{1}{n!} \Bigg\langle \mathcal{T}{:} \sum_{i_1=1}^{T/\delta t} \cdots \sum_{i_n=1}^{T/\delta t} [\alpha c S \hat{I}(t_{i_1})\delta t] \ldots [\alpha c S \hat{I}(t_{i_n})\delta t] \\ &\quad \times \prod_{r=1}^{T/\delta t} [1 - \alpha c S \hat{I}(t_r)\delta t - (\alpha c S \hat{I}(t_r)\delta t)^2 - \ldots]{:} \Bigg\rangle. \qquad (14.8\text{–}4)\end{aligned}$$

We have interchanged the order of summation and averaging in the last equation, and we have simplified the product at the end by inserting the n factors for $r = i_1, i_2, \ldots, i_n$ that were missing in Eq. (14.8–3). This is legitimate because each of these n factors tends to unity as $\delta t \to 0$ and therefore does not affect the answer. Needless to say, such an argument cannot be applied to *all* the $T/\delta t$ factors, because their number tends to infinity as $\delta t \to 0$ for a given n.

Now the product factor in Eq. (14.8–4) can be expressed in the following way,

$$\prod_{r=1}^{T/\delta t} [1 - \alpha c S \hat{I}(t_r)\delta t - (\alpha c S \hat{I}(t_r)\delta t)^2 - \ldots]$$
$$= 1 - \sum_{i=1}^{T/\delta t} [\alpha c S \hat{I}(t_i)\delta t + (\alpha c S \hat{I}(t_i)\delta t)^2 + \ldots]$$

$$+ \frac{1}{2!} \sum_{\substack{i=1 \\ i \neq j}}^{T/\delta t} \sum_{j=1}^{T/\delta t} [\alpha c S \hat{I}(t_i)\delta t + (\alpha c S \hat{I}(t_i)\delta t)^2 + \ldots]$$

$$\times [\alpha c S \hat{I}(t_j)\delta t + (\alpha c S \hat{I}(t_j)\delta t)^2 + \ldots]$$

$$- \frac{1}{3!} \sum_{\substack{i=1 \\ i \neq j \neq k \neq i}}^{T/\delta t} \sum_{j=1}^{T/\delta t} \sum_{k=1}^{T/\delta t} [\alpha c S \hat{I}(t_i)\delta t + (\alpha c S \hat{I}(t_i)\delta t)^2 + \ldots]$$

$$\times [\alpha c S \hat{I}(t_j)\delta t + (\alpha c S \hat{I}(t_j)\delta t)^2 + \ldots]$$

$$\times [\alpha c S \hat{I}(t_k)\delta t + (\alpha c S \hat{I}(t_k)\delta t)^2 + \ldots]$$

$$+ \text{etc.}$$

$$= 1 - \sum_{i=1}^{T/\delta t} \alpha c S \hat{I}(t_i)\delta t + \frac{1}{2!}\left[\sum_{i=1}^{T/\delta t} \alpha c S \hat{I}(t_i)\delta t\right]^2 - \frac{1}{3!}\left[\sum_{i=1}^{T/\delta t} \alpha c S \hat{I}(t_i)\delta t\right]^3$$

$$+ \ldots + 0(\delta t)$$

$$= \exp\left[-\sum_{i=1}^{T/\delta t} \alpha c S \hat{I}(t_i)\delta t\right] + 0(\delta t). \tag{14.8–5}$$

On substituting Eq. (14.8–5) in Eq. (14.8–4) and discarding terms of order δt or smaller, we obtain the formula

$$p(n, t, t+T) = \operatorname*{Lim}_{\delta t \to 0} \frac{1}{n!}\left\langle \mathcal{T}\!:\!\left[\sum_{i=1}^{T/\delta t} \alpha c S \hat{I}(t_i)\delta t\right]^n \exp\left[-\sum_{i=1}^{T/\delta t} \alpha c S \hat{I}(t_i)\delta t\right]\!:\right\rangle.$$

As $\delta t \to 0$ the sums become time integrals, so that

$$\sum_{i=1}^{T/\delta t} \alpha c S \hat{I}(t_i)\delta t \to \int_t^{t+T} \alpha c S \hat{I}(t')\,\mathrm{d}t' \equiv \hat{W}. \tag{14.8–6}$$

Hence, we have finally,

$$p(n, t, t+T) = \left\langle \mathcal{T}\!:\!\frac{\hat{W}^n \exp(-\hat{W})}{n!}\!:\right\rangle, \tag{14.8–7}$$

which shows that the photoelectric counting probability is given by the expectation value of a certain Poisson distribution. Although, strictly speaking, δt should always be much longer than an optical period, in practice the limit $\delta t \to 0$ yields a very good approximation.

When the time ordering is not important, as for a free field, we can re-express this result in another form with the help of the optical equivalence theorem (11.9–4) for normally ordered operators. We then have

$$p(n, t, t+T) = \int \phi(\{v\}) \frac{W^n \mathrm{e}^{-W}}{n!} \mathrm{d}\{v\} = \left\langle \frac{W^n \mathrm{e}^{-W}}{n!} \right\rangle_{\phi}, \tag{14.8–8a}$$

where $\phi(\{v\})$ is the phase space functional, or the diagonal coherent-state representation, or the P-representation, of the density operator for the field, and

$$W = \alpha c S \int_t^{t+T} I(t')\,\mathrm{d}t' \tag{14.8–8b}$$

is the c-number integrated intensity, which is obtained from $\hat{W}$ by replacing creation and annihilation operators by their left and right eigenvalues. We can re-express this result in yet another way, by writing

$$p(n, t, t+T) = \iint \phi(\{v\})\delta(W' - W)\frac{W'^n \mathrm{e}^{-W'}}{n!}\mathrm{d}W'\,\mathrm{d}\{v\}$$
$$= \int_0^\infty \mathcal{P}(W')\frac{W'^n \mathrm{e}^{-W'}}{n!}\mathrm{d}W', \tag{14.8–9}$$

where

$$\mathcal{P}(W') \equiv \int \phi(\{v\})\delta(W - W')\,\mathrm{d}\{v\} \tag{14.8–10}$$

can be regarded as the 'probability density' for the random variable W'. But because $\phi(\{v\})$ is not always a true probability density, neither is $\mathcal{P}(W')$, although it is properly normalized to unity.

Equation (14.8–9) was actually first derived from semiclassical arguments (Mandel, 1958, 1959, 1963a) and indeed is superficially identical with Eq. (9.7–3), which was also obtained without field quantization. The formula was later derived more rigorously for a quantized field (Kelley and Kleiner, 1964; Glauber, 1965, p. 181). However, it should be emphasized that despite the formal similarity of Eqs. (9.7–3) and (14.8–9), the equations are not completely identical, because $\mathcal{P}(W')$ in Eq. (9.7–3) is a probability density, whereas there exist quantum states without classical analogy for which $\mathcal{P}(W')$ in Eq. (14.8–9) is not a true probability density. We should also note that insofar as our derivation was based on the differential probability (14.8–1), the various expressions for $p(n, t, t+T)$ are applicable to situations in which the electromagnetic field interacts with the detector briefly and unabsorbed photons propagate away, i.e. to an open system. This is the usual experimental situation when a light beam falls on a detector. The various forms of $p(n, t, t+T)$ are not applicable to a closed system, e.g. one in which the electromagnetic field and the detector are contained within a resonant cavity. The light then interacts continuously with the detector, and the field intensity decays to zero in time as a result of measurement. This situation has been treated by Mollow (1968) and Srinivas and Davis (1981). It is important not to confuse the two cases (Mandel, 1981).

It is interesting to compare Eq. (14.8–8a) for the photoelectric counting probability $p(n, t, t+T)$ with Eq. (12.10–8) for the probability $P(n)$ of n photons in the quantum field, namely

$$P(n) = \langle U^n \mathrm{e}^{-U}/n!\rangle_\phi,$$

where

$$U \equiv \int_{L^3} I(\mathbf{r}, t)\,\mathrm{d}^3\mathbf{r},$$

is the light intensity integrated over all space. The similarity of the two expressions strongly suggests that the statistics of the photoelectric detections mirror the statistics of the photons. Indeed it is possible to look on W defined by Eq. (14.8–8b) as α times a volume integral of the light intensity taken over a cylinder of height cT whose base is the photocathode area S. As $I(\mathbf{r}, t)$ is also the photon density, we see that $cS\int_t^{t+T} I(\mathbf{r}, t')\,\mathrm{d}t'$ is the expected number of photons in the cylindrical volume, when the state of the field is the coherent state $|\{v\}\rangle$.

When the counting time interval T is very short compared with the correlation time, or the reciprocal bandwidth, of the light, Eq. (14.8–8b) can be simplified somewhat, for T is then too short for $I(\mathbf{r}, t')$ to vary appreciably under the integral. Accordingly we can pull $I(\mathbf{r}, t)$ out of the time integral and put

$$W \approx \alpha cSIT. \tag{14.8–11}$$

This allows us to transform Eq. (14.8–9) by writing

$$\begin{aligned} p(n, t, t+T) &= \iint \phi(\{v\})\delta(I' - I)\frac{(\alpha cSI'T)^n}{n!}\mathrm{e}^{-\alpha cSI'T}\,\mathrm{d}I'\,\mathrm{d}\{v\} \\ &= \int_0^\infty \mathbb{P}(I')\frac{(\alpha cSI'T)^n}{n!}\mathrm{e}^{-\alpha cSI'T}\,\mathrm{d}I' \end{aligned} \tag{14.8–12}$$

with

$$\mathbb{P}(I') \equiv \int \phi(\{v\})\delta(I' - I)\,\mathrm{d}\{v\}. \tag{14.8–13}$$

The probability $p(n, t, t+T)$ is therefore determined when the 'probability density' $\mathbb{P}(I)$ of the instantaneous light intensity I is known, although we emphasize again, that for certain states of the field $\mathbb{P}(I)$ will not have the character of a true probability density.

In recent years the measurement of photoelectric counting statistics has found important application as a means of investigating the nature of several kinds of optical fields, and for testing such things as laser theories, for example. The subject has been comprehensively reviewed (Arecchi, 1969; Mehta, 1970; Peřina, 1970; Saleh, 1978; Teich and Saleh, 1988). We now consider several examples of the application of Eqs. (14.8–8) or (14.8–9). We shall find that despite the formal Poisson structure of the expressions for $p(n, t, t+T)$, this probability can be very different from Poissonian for certain states of the field.

14.8.2 *Examples of the detection probability*

(a) A quasimonochromatic coherent state $|\{v'\}\rangle$

In this case the phase space functional $\phi(\{v\})$ is a product of delta functions of the form

$$\phi(\{v\}) = \prod_{\mathbf{k},s} \delta^2(v_{\mathbf{k},s} - v'_{\mathbf{k},s}).$$

When this expression is substituted in Eq. (14.8–8a), we obtain at once

$$p(n, t, t+T) = \frac{W'^n\,\mathrm{e}^{-W'}}{n!}, \tag{14.8–14}$$

where $W' \equiv \alpha cS\int_t^{t+T} \mathbf{V}'^*(\mathbf{r}, t')\cdot\mathbf{V}'(\mathbf{r}, t')\,\mathrm{d}t'$, and $\mathbf{V}'(\mathbf{r}, t)$ is the right eigenvalue of $\hat{\mathbf{V}}(\mathbf{r}, t)$ belonging to the coherent state $|\{v'\}\rangle$. The photoelectric pulses therefore obey Poisson statistics, just like the photons of the electromagnetic field.

(b) *A randomly phased single-mode laser*

If the only excited mode is $\mathbf{k}'$, s' and the absolute value of the mode amplitude is $|v'|$, then the phase space functional $\phi(\{v\})$ for this rather idealized laser model is given by Eq. (11.11–18). When this is inserted under the integral in Eq. (14.8–8a), we again find that

$$p(n, t, t+T) = \frac{W'^n \mathrm{e}^{-W'}}{n!}, \tag{14.8–15}$$

where $W' \equiv \alpha cS\int_t^{t+T} I'(\mathbf{r}, t')\,\mathrm{d}t'$, and $I'(\mathbf{r}, t') = |v'|^2/L^3$. The distribution is again Poissonian, because the light intensity has a well-defined value.

(c) *A single-mode Fock state* $|m_{\mathbf{k}',s'}, \{0\}\rangle$

In this case all modes are empty other than the $\mathbf{k}'$, s' mode, which contains m photons. We can proceed either directly from Eq. (14.8–7) and express the normally ordered operator in terms of photon number operators, or we can make use of the expression (11.8–20) found earlier for the phase space function of the occupied mode. Since the other modes are empty, we have

$$\phi(\{v\}) = \frac{\mathrm{e}^{|v_{\mathbf{k}',s'}|^2}}{m!}\left[\frac{\partial^{2m}}{\partial v^{*m}_{\mathbf{k}'s'}\partial v^m_{\mathbf{k}'s'}}\delta^2(v_{\mathbf{k}'s'})\right]\prod_{\mathbf{k},s\neq\mathbf{k}',s'}\delta^2(v_{\mathbf{k}s}).$$

Although this phase space functional is highly singular, we can use it without difficulty in Eq. (14.8–8a), because the 'test function' $W^n \mathrm{e}^{-W}$ under the integral is infinitely differentiable. As only one mode is occupied, we find that W can be reduced to $\alpha cS|v_{\mathbf{k}'s'}|^2 T/L^3$, so that

$$\begin{aligned}
p(n, t, t+T) \\
&= \int \frac{\mathrm{e}^{|v|^2}}{n!m!}\left[\frac{\partial^{2m}}{\partial v^{*m}\partial v^m}\delta^2(v)\right]\left(\frac{\alpha cS|v|^2 T}{L^3}\right)^n \exp(-\alpha cS|v|^2T/L^3)\,\mathrm{d}^2v \\
&= \left(\frac{\alpha cST}{L^3}\right)^n \frac{1}{n!m!}\int \delta^2(v)\left[\frac{\partial^{2m}}{\partial v^{*m}\partial v^m}(vv^*)^n \exp[vv^*(1-\alpha cST/L^3)]\right]\mathrm{d}^2v,
\end{aligned}$$

where the last line follows from the previous one by integration by parts. We need to impose the condition $cST < L^3$ to ensure that the volume of space sampled by the detector always lies within the quantization volume. The integrand can be expanded with the help of the Leibnitz chain rule, and the integral is found to have the value

$$\begin{aligned}
&m!n!\binom{m}{n}\left(1-\frac{\alpha cST}{L^3}\right)^{m-n}, &&\text{if } m \geqslant n, \\
&0, &&\text{otherwise},
\end{aligned}$$

so that finally

$$\left.\begin{aligned} p(n, t, t+T) &= \binom{m}{n}\left(\frac{\alpha cST}{L^3}\right)^n\left(1-\frac{\alpha cST}{L^3}\right)^{m-n}, &&\text{if } n \leqslant m, \\ &= 0, &&\text{otherwise.} \end{aligned}\right\} \tag{14.8–16}$$

This is a Bernoulli or binomial distribution in n, and its form has a natural physical interpretation. As the photons are distributed uniformly over the quantization volume L^3, the probability that any one photon is contained in the cylinder of base area S and height cT, which is the volume of space effectively sampled by the detector in time T, is just cTS/L^3 ($cTS < L^3$), and the probability that the photon is detected is therefore $\alpha cST/L^3$. This is the so-called 'success' parameter of the distribution, and the probability of registering n 'successes' when there are m photons is just of the form of Eq. (14.8–16). Had we started from a Fock state with multimode occupation, we could have encountered a much more complicated success parameter.

(d) *A polarized thermal light beam and short counting time T*

We have shown in Section 13.2 that the phase space functional for a stationary field of this general type can always be reduced to the form

$$\phi(\{v\}) = \prod_{\mathbf{k},s}\frac{1}{\pi\langle n_{\mathbf{k}s}\rangle}\exp\left(-\frac{|v_{\mathbf{k}s}|^2}{\langle n_{\mathbf{k}s}\rangle}\right), \tag{14.8–17}$$

in which the $\langle n_{\mathbf{k}s}\rangle$ are arbitrary occupation numbers of the different modes. For any mode for which $\langle n_{\mathbf{k}s}\rangle$ is zero, the corresponding factor in the product must be interpreted as a $\delta^2(v_{\mathbf{k}s})$ function. Hence the phase space function for each mode is an independent Gaussian distribution, and the total complex field amplitude $\mathbf{V}(\mathbf{r}, t)$ also has a Gaussian distribution at each space-time point $\mathbf{r}$, t. For a polarized field that is cross-spectrally pure (Section 4.6) with respect to polarization we can write $\mathbf{V} = V\boldsymbol{\varepsilon}$, where $\boldsymbol{\varepsilon}$ is a unit polarization vector, and the instantaneous intensity $I = V^*V$ then obeys the probability distribution [cf. Eqs. (13.3–7)

$$\mathbb{P}(I) = \frac{1}{\langle I\rangle}\mathrm{e}^{-I/\langle I\rangle}. \tag{14.8–18}$$

Despite the simplicity of the phase space functional given by Eq. (14.8–17), no exact closed form expression for either $\mathcal{P}(W)$ or $p(n, t, t+T)$ has been found in general. However, when the measurement interval T is sufficiently short compared with the coherent time or reciprocal bandwidth $1/\Delta\omega$, so that Eq. (14.8–12) applies, we can make use of Eq. (14.8–18) for $\mathbb{P}(I)$, and obtain the result

$$\begin{aligned} p(n, t, t+T) &= \int_0^\infty \frac{\mathrm{e}^{-I/\langle I\rangle}}{\langle I\rangle}\frac{(\alpha cSIT)^n\,\mathrm{e}^{-\alpha cSIT}}{n!}\,\mathrm{d}I \\ &= \frac{1}{[1+\alpha cS\langle I\rangle T][1+1/\alpha cS\langle I\rangle T]^n}. \end{aligned} \tag{14.8–19}$$

This is a Bose–Einstein distribution in n with parameter $\langle n\rangle = \alpha cS\langle I\rangle T$. It has the same structure as the distribution of photon occupation numbers for one mode of a thermal field [see Eq. (13.2–4)]. Although the value of $\langle n \rangle$ was to be expected, the reason for the single-mode Bose–Einstein form of $p(n, t, t+T)$ is less obvious, because we are dealing with a multimode field.

In order to understand this, we first recall that the cylindrical volume of space cTS effectively sampled by the detector in a time T is smaller than a coherence volume, because S is smaller than the coherence area, and cT is shorter than the coherence length of the field. But the coherence volume is also the projection onto three-dimensional space of the unit cell of phase space, and can be used to define a 'mode' of the field; the sampled volume therefore lies within the corresponding cell of phase space. It is in this sense that the measurement is effectively limited to sampling one mode of the field. Another way of expressing the same thing is to observe that a photon of frequency spread $\Delta\omega$ has a fundamental time uncertainty of $1/\Delta\omega$, and that our measurement interval T lies entirely within this fundamental time uncertainty. Similar remarks can also be made about the position uncertainty. Figure 14.12 shows the results of measurements of $p(n, t, t+T)$ in a short time interval for polarized thermal light and for single-mode laser light, in agreement with the predictions of Eqs. (14.8–15) and (14.8–19).

(e) A partially polarized thermal light beam

If a light beam is cross-spectrally pure with respect to polarization (cf. Sections 4.5 and 13.3), we can always resolve the total field into the sum of two orthogonal,

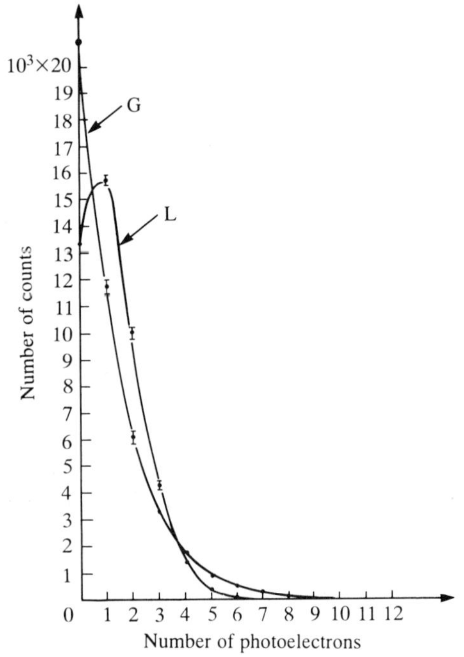

Fig. 14.12 Results of short-time photoelectric counting measurements of $p(n, t, t+T)$ for a laser beam (L) and for a polarized thermal beam with Gaussian statistics (G). (Adapted from Arecchi, 1965.)

fully polarized components, which are jointly Gaussian, with light intensities I_1, I_2. If we choose the polarization basis such that the polarization matrix is diagonal (cf. Section 6.2), then the two field components are uncorrelated, and since they are Gaussian, they are also statistically independent. Hence the intensities I_1, I_2 of the two polarized components both obey probability distributions of the form (14.8–18), in which the corresponding mean light intensities $\langle I_1\rangle$, $\langle I_2\rangle$ are related to the total mean light intensity $\langle I\rangle$ and to the degree of polarization P by Eqs. (13.3–12). We have already shown that under these conditions the probability density $\mathbb{P}(I)$ of the total light intensity is given by [see Eq. (13.3–13)]

$$\mathbb{P}(I) = \frac{1}{P\langle I\rangle}\left\{\exp\left[\frac{-2I}{(1+P)\langle I\rangle}\right] - \exp\left[\frac{-2I}{(1-P)\langle I\rangle}\right]\right\}. \qquad (14.8\text{–}20)$$

When this result is inserted in Eq. (14.8–12), it leads to the following counting probability (Mandel, 1963b):

$$\begin{aligned} &p(n, t, t+T) \\ &= \frac{1}{P\langle I\rangle n!}\int_0^\infty (\alpha cSIT)^n \exp(-\alpha cSIT)\left\{\exp\left[\frac{-2I}{\langle I\rangle(1+P)}\right] - \exp\left[\frac{-2I}{\langle I\rangle(1-P)}\right]\right\}\mathrm{d}I \\ &= \frac{1}{Pn!}\left\{\frac{1}{\left[1+\dfrac{2}{(1+P)\langle n\rangle}\right]^{n+1}} - \frac{1}{\left[1+\dfrac{2}{(1-P)\langle n\rangle}\right]^{n+1}}\right\}, \qquad (14.8\text{–}21) \end{aligned}$$

where $\langle n\rangle = \alpha cS\langle I\rangle T$, provided the time T is sufficiently short.

(f) A polarized thermal light beam and arbitrary counting time T

We can use the interpretation of the coherence length as the length of the unit cell of phase space in order to arrive at an approximate expression for the detection probability $p(n, t, t+T)$ when T is not necessarily short. Let T_c be the coherence time of the light beam (to be defined precisely below), and suppose that for long T

$$T/T_c = \mu, \qquad (14.8\text{–}22)$$

so that the measurement time interval contains μ coherence times. Then we may expect μ, which is not necessarily an integer, to play the role of the number of equally occupied cells of phase space with respect to the measurement. As $p(n, t, t+T)$ has the form of a single-mode Bose–Einstein distribution for a very short measurement time interval T, we may reasonably expect it to have the form of a multimode Bose–Einstein distribution for longer times T, corresponding to n photons distributed over μ equally occupied modes. The form of such a probability distribution was derived in Section 13.3, and it is given by Eq. (13.3–29). Accordingly we now make the ansatz that for an arbitrary time interval T, the probability $p(n, t, t+T)$ is well approximated by

$$p(n, t, t+T) = \frac{(n+\mu-1)!}{n!(\mu-1)!}\,\frac{1}{[1+\langle n\rangle/\mu]^\mu[1+\mu/\langle n\rangle]^n}, \qquad (14.8\text{–}23)$$

where $\langle n \rangle$ is chosen to be the correct mean and where the precise value of μ remains to be specified. It then follows that the variance of n is given by

$$\langle(\Delta n)^2\rangle = \sum_{n=0}^{\infty}(n - \langle n \rangle)^2 \frac{(n + \mu - 1)!}{n!(\mu - 1)!} \frac{1}{[1 + \langle n \rangle/\mu]^{\mu}[1 + \mu/\langle n \rangle]^n},$$

which has been shown [cf. Eq. (13.3–32)] to reduce to

$$\langle(\Delta n)^2\rangle = \langle n \rangle\left[1 + \frac{\langle n \rangle}{\mu}\right]. \qquad (14.8\text{–}24)$$

But, as we show below, the variance of n can be found quite generally from Eqs. (14.8–7) or (14.8–8), and it can be expressed in the form [see Eq. (14.9–11)]

$$\langle(\Delta n)^2\rangle = \langle n \rangle\left[1 + \langle n \rangle\frac{\theta(T)}{T}\right], \qquad (14.8\text{–}25)$$

where $\theta(T)$ is a time interval depending on T that is given by

$$\theta(T) \equiv \int_{-T}^{T}\left(1 - \frac{|\tau|}{T}\right)\lambda(\tau)\,\mathrm{d}\tau. \qquad (14.8\text{–}26)$$

For polarized thermal light we may identify the normalized intensity correlation function $\lambda(\tau)$ with $|\gamma(\tau)|^2$, the square of the normalized second-order field amplitude correlation, and write

$$\theta(T) = \int_{-T}^{T}\left(1 - \frac{|\tau|}{T}\right)|\gamma(\tau)|^2\,\mathrm{d}\tau. \qquad (14.8\text{–}27)$$

If we now compare Eq. (14.8–24) obtained from the ansatz (14.8–23), with the exact Eq. (14.8–25) obtained directly from the formula (14.8–7), we observe that they can be made identical if we identify the number μ by

$$\mu = \frac{T}{\theta(T)}. \qquad (14.8\text{–}28)$$

With this identification, Eq. (14.8–23) for $p(n, t, t + T)$ automatically gives the first two moments of n correctly under all circumstances. Moreover, for sufficiently short times T we can replace $|\gamma(\tau)|^2$ by 1 under the integral in Eq. (14.8–27), so that $\theta(T) \approx T$ and $\mu = 1$, and Eq. (14.8–23) is automatically correct. We also expect it to hold for very long times T because μ then becomes a large number and can be well approximated by an integer. It follows that Eq. (14.8–23), with μ defined by Eq. (14.8–28), should give a good approximation to the exact probability $p(n, t, t + T)$ for polarized thermal light under a wide range of conditions. The formula was first proposed from heuristic arguments (Mandel, 1959, 1963a), and its accuracy has been confirmed for several different spectral distributions (Bédard, Chang and Mandel, 1967).

A formula of the form of Eq. (14.8–23) should also be a good approximation to the photoelectric counting probability when the light falling on the photodetector is not spatially coherent. The number of coherence areas $\mathscr{A}_c$ falling within the photocathode area S plays a role similar to that of the number of coherence times T_c falling within the counting interval T. When $S \gg \mathscr{A}_c$ and $T \gg T_c$, the effective number μ of equally occupied cells of phase space in Eq. (14.8–23) is given by

$(S/\mathscr{A}_c)(T/T_c)$ approximately (Bures, Delisle and Zardecki, 1971, 1972; Zardecki and Delisle, 1973).

14.8.3 A natural measure of coherence time

Comparison of Eqs. (14.8–22) and (14.8–28) for very long times T allows us to arrive at a precise, natural measure of coherence time T_c, at least for a thermal light beam. With the help of Eq. (14.8–27) we then obtain

$$\begin{aligned} T_c &= \lim_{T\to\infty} \theta(T) \\ &= \lim_{T\to\infty} \int_{-T}^{T} \left(1 - \frac{|\tau|}{T}\right) |\gamma(\tau)|^2 \, d\tau \\ &= \int_{-\infty}^{\infty} |\gamma(\tau)|^2 \, d\tau. \end{aligned} \qquad (14.8\text{–}29)$$

This is an expression that we have already encountered and regarded as an approximation to the hitherto ill-defined coherence time. Our discussion of photoelectric counting statistics shows that the limiting value of $\theta(T)$ as $T \to \infty$ is also the natural measure of the length of the unit cell of phase space and therefore of the coherence time of the light. Figure 14.13 illustrates the variation of $\theta(T)$ with T. The effective bandwidth $\Delta\nu$ of a thermal optical field can always be defined as the reciprocal of the coherence time T_c.

For non-thermal light, for which the intensity correlation function $\lambda(\tau)$ is not generally expressible in terms of $\gamma(\tau)$, we could adopt the more general definition

$$T_c = \frac{\int_{-\infty}^{\infty} |\lambda(\tau)| \, d\tau}{|\lambda(0)|}, \qquad (14.8\text{–}30)$$

for the correlation time of the intensity fluctuations, and this reduces to Eq.

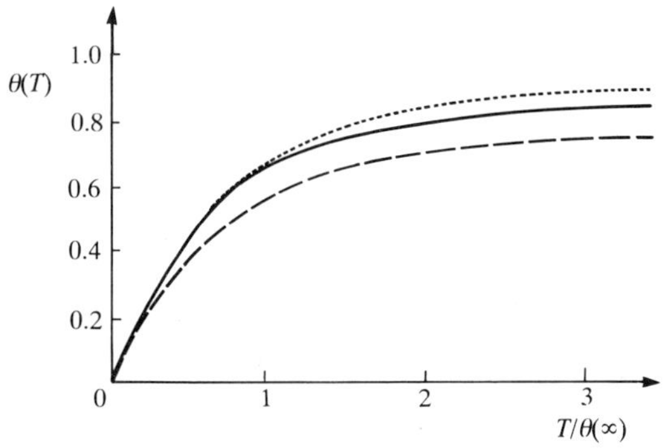

Fig. 14.13 The variation of $\theta(T)$ with T for three different spectral distributions of similar spectral width (the full curve corresponds to a rectangular profile, the dotted curve to a Gaussian profile and the broken curve to a rectangular doublet). (Reproduced from a paper by Mandel, 1963a.)

(14.8–29) when $\lambda(\tau) = |\gamma(\tau)|^2$. However, for non-thermal light there is, in general, no single time that characterizes the range of all the correlation functions. There may exist a whole hierarchy of correlation times, characterizing correlations of different orders.

14.9 Properties of the detection probability $p(n, t, t + T)$

14.9.1 Generating functions and the inversion problem

Having examined some special cases for which the detection probability $p(n, t, t + T)$ takes on a simple form, we now return to the general formula (14.8–7) or (14.8–8), and examine some of its properties. Because of the Poisson structure of the distribution, several generating functions are easily evaluated. Thus we find for the characteristic function of the number of photoelectric pulses n,

$$C_n(\xi) \equiv \langle e^{i\xi n}\rangle = \left\langle :\mathcal{T}\sum_{n=0}^{\infty} e^{i\xi n}\frac{\hat{W}^n e^{-\hat{W}}}{n!}:\right\rangle$$

$$= \langle \mathcal{T}: \exp[\hat{W}(e^{i\xi}-1)]:\rangle \tag{14.9–1a}$$

$$\left.\begin{aligned} &= \langle \exp[W(e^{i\xi} - 1)]\rangle_\phi \\ &= C_W\left(\frac{e^{i\xi} - 1}{i}\right), \end{aligned}\right\} \tag{14.9–1b}$$

where we have made use of the optical equivalence theorem (Section 11.9), and we have denoted the characteristic function of the detected integrated light intensity W by C_W. Of course, in invoking a classical description we assume that the field is free, so that time-ordering is unnecessary. For those quantum states of the optical field for which a classical analog exists, in the sense that the phase space functional $\phi(\{v\})$ is a probability functional, we can make use of Eqs. (14.9–1b) to derive the probability density $\mathcal{P}(W)$ of W from $C_n(\xi)$ by Fourier inversion of the characteristic function (Mandel, 1959; Wolf and Mehta, 1964). Thus wc have

$$\mathcal{P}(W) = \frac{1}{2\pi}\int_{-\infty}^{\infty} C_W(x)\, e^{-iWx}\, dx$$

$$= \frac{1}{2\pi}\int_{-\infty}^{\infty} C_n(\xi)\, e^{-iWx}\, dx \quad \text{with } i\xi = \ln(1 + ix). \tag{14.9–2a}$$

For real x, this requires that $\xi = \xi_r + i\xi_i$ is a complex number with

$$\left.\begin{aligned} \xi_r &= \tan^{-1} x \\ \xi_i &= \ln(\cos\xi_r). \end{aligned}\right\} \tag{14.9–2b}$$

Thus the probability density $\mathcal{P}(W)$ can, in principle, be determined from $p(n, t, t + T)$ through the intermediary of the characteristic functions (Mandel, 1959; Wolf and Mehta, 1964). The inversion may however be so sensitive to small variations of $p(n, t, t + T)$ that it is unstable and of no use in practice.

As an obvious, if somewhat trivial, application of Eqs. (14.9–2) we note that if the photoelectric counting probability $p(n, t, t + T)$ is Poissonian, then

$C_n(\xi) = \exp[\langle n\rangle(e^{i\xi} - 1)]$ and therefore

$$\begin{aligned}\mathcal{P}(W) &= \frac{1}{2\pi}\int_{-\infty}^{\infty} e^{-i(\langle n\rangle - W)x}\,dx \\ &= \delta(W - \langle n\rangle).\end{aligned}$$

Hence $\langle \hat{W}\rangle = \langle n\rangle$ and the integrated light intensity has the definite value $\langle n\rangle/\alpha$ (α is the quantum efficiency) and is free from fluctuations.

Instead of the characteristic function (or moment generating function if $i\xi$ is replaced by ξ), we may make use of the factorial moment generating function $F_n(\xi)$ of n (see Section 1.4.1), which becomes especially simple for a probability distribution with Poisson structure. Thus we have from the definition and from Eq. (14.8–7),

$$\left.\begin{aligned}F_n(\xi) = \langle(1-\xi)^n\rangle &= \left\langle \mathcal{T}\colon \sum_{n=0}^{\infty}(1-\xi)^n \frac{\hat{W}^n \exp(-\hat{W})}{n!}\colon\right\rangle \\ &= \langle \mathcal{T}\colon \exp(-\hat{W}\xi)\colon\rangle \\ &= \langle \exp(-W\xi)\rangle_\phi.\end{aligned}\right\} \tag{14.9–3}$$

Since $F_n(\xi)$ generates the factorial moments $\langle n^{(r)}\rangle$ of n when we make a power series expansion in ξ, we have, on comparing coefficients of ξ^r on both sides of this equation [cf. Eq. (12.10–13)],

$$\langle n^{(r)}\rangle = \langle \mathcal{T}\colon \hat{W}^r\colon\rangle = \langle W^r\rangle_\phi, \quad r = 1, 2, \ldots \tag{14.9–4}$$

We note once again that the photoelectric counts stand in the same statistical relationship to the integrated light intensity as the photons of the field.

Alternatively, we can calculate the cumulant generating function (see Section 1.4) $K_n(\xi)$, which is the logarithm of $C_n(\xi/i)$. With the help of Eq. (14.9–1b) we find (Mandel, 1959; Mandel and Wolf, 1965) that

$$\begin{aligned}K_n(\xi) = \ln\langle e^{\xi n}\rangle &= \ln\langle \exp[W(e^{\xi} - 1)]\rangle_\phi \\ &= K_W(e^{\xi} - 1),\end{aligned} \tag{14.9–5}$$

where K_W is the cumulant generating function of W. Expansion of each side of Eq. (14.9–5) in a power series allows us to relate the cumulants $\kappa_r^{(n)}$, $\kappa_r^{(W)}$ of n and W. Thus we obtain the formula

$$\sum_{r=1}^{\infty}\frac{\xi^r}{r!}\kappa_r^{(n)} = \sum_{r=1}^{\infty}\frac{(e^{\xi}-1)^r}{r!}\kappa_r^{(W)}, \tag{14.9–6}$$

and on comparing coefficients of ξ^r ($r = 1, 2, \ldots$) on both sides of the equation, we find that

$$\left.\begin{aligned}\kappa_1^{(n)} &= \kappa_1^{(W)} \\ \kappa_2^{(n)} &= \kappa_1^{(W)} + \kappa_2^{(W)} \\ \kappa_3^{(n)} &= \kappa_1^{(W)} + 3\kappa_2^{(W)} + \kappa_3^{(W)} \\ \kappa_4^{(n)} &= \kappa_1^{(W)} + 7\kappa_2^{(W)} + 6\kappa_3^{(W)} + \kappa_4^{(W)}\end{aligned}\right\} \tag{14.9–7}$$

etc.

At very high light intensities, the right-hand side of each of these equations is dominated by the last term, so that $\kappa_r^{(n)} \to \kappa_r^{(W)}$, $r = 1, 2, \ldots$ and the probability distributions of n and W tend to become identical, as might be expected in the classical limit. On the other hand, when the light intensity is very low, the first term on the right of each equation is the dominant term, and we find that $\kappa_r^{(n)} \to \kappa_1^{(W)} = \kappa_1^{(n)} = \langle n \rangle$. Hence all the cumulants tend to become identical to the first, which is characteristic of a Poisson distribution. In this very weak field limit the photoelectric pulses tend to become independent of each other.

14.9.2 The second moment and sub-Poissonian counting statistics

The cumulant relations (14.9–7) are completely equivalent to the moment relations (14.9–4). Let us examine the first two in a little more detail. They may also be written as

$$\langle n \rangle = \langle \hat{W} \rangle = \alpha \int_t^{t+T} cS\langle \hat{I}(\mathbf{r}, t')\rangle \, \mathrm{d}t' = \alpha c S \langle \hat{I} \rangle T \tag{14.9–8}$$

and

$$\left.\begin{aligned} \langle (\Delta n)^2 \rangle &= \langle \hat{W} \rangle + \langle \mathcal{T}{:}(\Delta \hat{W})^2{:}\rangle \\ &= \langle n \rangle + \langle (\Delta W)^2 \rangle_\phi . \end{aligned}\right\} \tag{14.9–9}$$

Eq. (14.9–8) makes the evident statement that the mean number of photoelectric counts is given by the product of the detector quantum efficiency α and the integrated photon density. Eq. (14.9–9) shows that the mean squared fluctuation of the photoelectric pulses is expressible as the sum of contributions from fluctuations of independent particles and from fluctuations of integrated wave fields. This result may be regarded as a generalization of a well-known formula first derived by Einstein for blackbody radiation (Einstein, 1909; Fürth, 1928a,b; Mandel, Sudarshan and Wolf, 1964). However, Eq. (14.9–9) does not actually imply that $\langle (\Delta n)^2 \rangle$ always exceeds $\langle n \rangle$, because $\langle \mathcal{T}{:}(\Delta \hat{W})^2{:}\rangle$ may be negative for certain states of the electromagnetic field. The counting statistics then become sub-Poissonian, and this is a characteristic feature of a non-classical state. As an example, we note that $\langle \mathcal{T}{:}(\Delta \hat{W})^2{:}\rangle$ is always negative for a Fock state of the field.

If we make use of the definition (14.8–6) of $\hat{W}$, we can re-write Eq. (14.9–9) in the form

$$\begin{aligned} \langle (\Delta n)^2 \rangle &= \langle n \rangle + (\alpha c S)^2 \iint_t^{t+T} \mathrm{d}t' \, \mathrm{d}t'' \langle \mathcal{T}{:}\Delta \hat{I}(\mathbf{r}, t')\Delta \hat{I}(\mathbf{r}, t''){:}\rangle \\ &= \langle n \rangle + (\alpha c S \langle \hat{I}(\mathbf{r})\rangle)^2 \iint_t^{t+T} \mathrm{d}t' \, \mathrm{d}t'' \lambda(t', t''), \end{aligned} \tag{14.9–10}$$

where we have again introduced the normalized intensity correlation function λ defined by Eq. (14.6–3), with the space coordinates suppressed. For a stationary field, $\lambda(t', t'')$ is a function only of the time difference $t'' - t'$, and can be written as $\lambda(t'' - t')$. If we substitute $t' = t + t_1$, $t'' = t + t_2$ in the integral, the limits become 0 and T. Moreover, the double integral can be reduced to a single integral with the help of the transformation $\tau = t_2 - t_1$, if we integrate over the 45° diagonal strips shown in Fig. 14.14, just as in Section 2.2.2, because λ depends only on τ. Each strip has a length $(T - |\tau|)/\sqrt{2}$ and a width $(\sqrt{2})\,d\tau$, so that we can write

$$\begin{aligned}\langle(\Delta n)^2\rangle &= \langle n\rangle + (\alpha c S\langle \hat{I}(\mathbf{r})\rangle)^2 \int_{-T}^{T} (T - |\tau|)\lambda(\tau)\,d\tau \\ &= \langle n\rangle\left[1 + \langle n\rangle \frac{\theta(T)}{T}\right], \end{aligned} \tag{14.9–11}$$

where we have put

$$\theta(T) \equiv \int_{-T}^{T}\left(1 - \frac{|\tau|}{T}\right)\lambda(\tau)\,d\tau. \tag{14.9–12}$$

The function $\theta(T)$ has the dimension of time, and is the same function that was illustrated in Fig. 14.13 for the special case of thermal light. Evidently the term $\langle n\rangle^2\theta(T)/T$ in Eq. (14.9–11) measures the departure of the photoelectric counting statistics from the Poisson form, for which $\langle(\Delta n)^2\rangle = \langle n\rangle$.

From Eq. (14.9–11) the fractional deviation of the variance $\langle(\Delta n)^2\rangle$ from the Poisson variance $\langle n\rangle$ for independent photoelectric pulses, may be expressed by the parameter

$$Q \equiv \frac{\langle(\Delta n)^2\rangle - \langle n\rangle}{\langle n\rangle} = \langle n\rangle\frac{\theta(T)}{T} = R\theta(T), \tag{14.9–13}$$

where $R = \alpha c S\langle \hat{I}(\mathbf{r})\rangle$ is the average counting rate of the illuminated photodetector. Hence the variance is greater or less than the Poisson variance according as $\theta(T)$ is positive or negative. In particular, for a counting interval T that is much shorter than the correlation time T_c, we can replace $\lambda(\tau)$ by $\lambda(0)$ under the integral in Eq. (14.9–12) to a good approximation. Then

$$\theta(T) \approx \lambda(0)T, \tag{14.9–14}$$

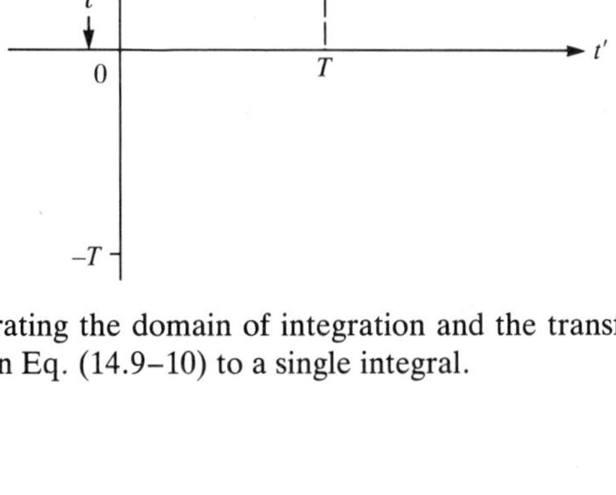

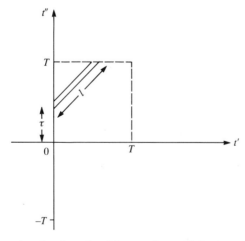

Fig. 14.14 Illustrating the domain of integration and the transformation of the double integral in Eq. (14.9–10) to a single integral.

and this is negative whenever $\lambda(0)$ is negative. It follows that $\lambda(0) < 0$ is the condition for sub-Poissonian counting statistics in a short counting time T.

As an example of an optical field that exhibits a sub-Poissonian variance of the photoelectric counts, we mention a single-mode Fock state, for which the photoelectric counting probability $p(n, t, t+T)$ is given by the Bernoulli distribution in Eq. (14.8–16). We readily find that for this distribution

$$Q \equiv \frac{\langle(\Delta n)^2\rangle - \langle n\rangle}{\langle n\rangle} = -\left(\frac{\alpha cST}{L^3}\right), \tag{14.9–15}$$

which is always negative, and implies a negative value of $\theta(T)$. Another, perhaps more physically realizable, example is provided by the fluorescent light radiated by a two-level atom in the presence of a coherent exciting field on resonance (see Section 15.6). For such a field one can show that (Carmichael and Walls, 1976a,b; Kimble and Mandel, 1976; Cook, 1981; Lenstra, 1982; Singh, 1983)

$$\lambda(\tau) = -e^{-3\beta\tau/2}\left(\cos q\beta\tau + \frac{3}{2q}\sin q\beta\tau\right), \tag{14.9–16}$$

with $q \equiv (\Omega^2/\beta^2 - \frac{1}{4})^{1/2}$. The parameter 2β is the spontaneous radiation rate of the undisturbed atom, or the Einstein A-coefficient (see Section 15.4), and Ω is the so-called Rabi frequency, which is proportional to the field strength of the exciting light. As $\lambda(0) = -1$, the short time counting statistics are sub-Poissonian. But in addition we also find from Eqs. (14.9–12) and (14.9–16) that

$$\theta(\infty) = -\frac{3/\beta}{(\Omega^2/2\beta^2) + 1}, \tag{14.9–17}$$

so that the counts in a long time interval also fluctuate in a sub-Poissonian manner. This has been confirmed experimentally (Short and Mandel, 1983, 1984). Other procedures for generating sub-Poissonian counting statistics have also been reported (Teich, Saleh and Periña, 1984; Saleh and Teich, 1985; Teich and Saleh, 1985; Hong and Mandel, 1986; Machida and Yamamoto, 1986; Rarity, Tapster and Jakeman, 1987).

14.9.3 Deriving correlation functions from counting statistics

Equation (14.9–11) shows that the second moment of the photoelectric counts contains information about the normalized intensity correlation function $\lambda(\tau)$ of the light. This suggests that $\lambda(\tau)$ might be derivable from measurements of the photoelectric counting probability $p(n, t, t+T)$ for different counting intervals T. To see this, we observe that if

$$G(T) \equiv \langle(\Delta n)^2\rangle - \langle n\rangle = R^2 T\theta(T), \tag{14.9–18}$$

then

$$\frac{dG(T)}{dT} = 2R^2\int_0^T \lambda(\tau)\,d\tau,$$

when we make use of the fact that $\lambda(-\tau) = \lambda(\tau)$, and

$$\frac{\mathrm{d}^2 G(T)}{\mathrm{d}T^2} = 2R^2\lambda(T). \tag{14.9–19}$$

Hence $\lambda(T)$ is derivable by double differentiation of $G(T)$, which is obtainable from photoelectric counting measurements for various times T.

Of course, the requirements for double differentiation imposes severe constraints on the accuracy with which $G(T)$ has to be determined. Still, very accurate measurements of the counting probability are possible, in principle. The method has been used to derive $\lambda(\tau)$ for a laser operating near its threshold of oscillation (Meltzer and Mandel, 1970).

Problems

14.1 Consider an electromagnetic field in which a single plane wave mode $\mathbf{k}_0$, s_0 is in an m-photon Fock state within the quantization volume L^3, while all other modes are in the vacuum state. A photodetector of surface area S and quantum efficiency α is arranged so that the photoelectric surface is perpendicular to $\mathbf{k}_0$. Calculate the probability $p(n, t, t+T)$ of registering n photodetections in the time interval from t to $t+T$.

14.2 A laser is oscillating simultaneously with equal amplitudes in two longitudinal modes of frequencies ω_1 and ω_2. Both modes are randomly phased and independent of each other. Calculate the probability density $p(\tau)$ that one photoelectric detection occurs at time t and one at time $t+\tau$ when the laser light is incident on a photodetector.

14.3 A photodetector has quantum efficiency α for light of frequency ω. Show that the probability $P(m)$ of detecting m photons of this frequency in some time interval T is the same as for a perfect detector with 100% quantum efficiency, which has a beam splitter of amplitude transmissivity $\sqrt{\alpha}$ placed immediately in front of it.

14.4 A light beam from a randomly phased single-mode laser passes through a lossless beam splitter BS of unknown transmissivity and reflectivity and is then incident on photodetector D_T of 100% quantum efficiency. An experimenter repeatedly measures the number n of photons counted by D_T in some time interval T in order to derive the photon emission rate R and the photon statistics of the laser. But because of the attenuation introduced by BS, $\langle n\rangle/T$ will not give the correct value of R, and neither will the higher moments of n be correct for the laser.

The experimenter has available a second photodetector D_R of 100% quantum efficiency with which to measure the light reflected from BS, and argues that by retaining only those counting data obtained with D_T for which D_R registers no photons within the same counting time interval T, the attenuation of the beam splitter can effectively be overcome. Is this argument correct? How will the moments of n obtained in this way compare with the values that would be obtained in the absence of the beam splitter?

14.5 A radiation field in the form of a plane wave produced by a thermal source has μ modes with equal average occupation number n_0, while all other modes are empty. Let $\mathbf{k}_1$, s_1 and $\mathbf{k}_2$, s_2 label two modes of the occupied set. Calculate the correlation $\langle \hat{n}^{(2)}_{\mathbf{k}_1 s_1} \hat{n}^{(2)}_{\mathbf{k}_2 s_2} \rangle$.

14.6 A polarized thermal beam of light of mean intensity $\langle \hat{I} \rangle$ falls at right angles on a photodetector of surface area S and quantum efficiency α. The light is quasi-monochromatic with a Lorentzian normalized spectral density given by $2B/[(\omega - \omega_0)^2 + B^2]$. Calculate the conditional probability density that, given one photodetection at time t, there will be another one τ seconds later, and sketch its τ-dependence.

14.7 A stationary thermal beam of light is linearly polarized in a certain direction characterized by the unit polarization vector $\boldsymbol{\varepsilon}$. Consider a Hilbert space restricted only to the occupied modes of the field. Calculate the characteristic function $\langle \exp (i\boldsymbol{\varepsilon} \cdot \hat{\mathbf{E}} x) \rangle$ of the electric field $\hat{\mathbf{E}}$, when $\boldsymbol{\varepsilon}$ is a real vector in the direction of the electric field.

14.8 A stationary, polarized light beam with normalized intensity correlation function $\gamma(\tau) = \exp(-\beta|\tau|) - 2\exp(-2\beta|\tau|)$ is incident normally on a photodetector for a time interval T. Calculate the variance of the number of photoelectric counts n received in the time interval T. Comment on the significance of the answer.

15

Interaction between light and a two-level atom

Light is both radiated and absorbed by atoms, and the interaction between the quantized electromagnetic field and an atom represents one of the most fundamental problems in quantum optics. However, real atoms are complicated systems, and even the simplest real atom, the hydrogen atom, has a non-trivial energy level structure. It is therefore often necessary or desirable to approximate the behavior of a real atom by that of a much simpler quantum system. For many purposes only two atomic energy levels play a significant role in the interaction with the electromagnetic field, so that it has become customary in many theoretical treatments to represent the atom by a quantum system with only two energy eigenstates. This is the most basic of all quantum systems, and it generally simplifies the treatment substantially.

In a real atom the selection rules limit the allowed transitions between states, so that, in some cases, a certain state may couple to only one other. Moreover, optical pumping techniques have been developed that allow such preferred states to be prepared in the laboratory, and they have been successfully used in experiments (Abate, 1974). The two-level atom approximation is therefore close to the truth and not merely a mathematical convenience in some experimental situations. In the following we begin by developing the algebra for such a two-level atom.‡

15.1 Dynamical variables for a two-level atom

We consider an atomic quantum system with the two energy levels shown in Fig. 15.1. The levels are separated in energy by $\hbar\omega_0$, and they are centered around the energy E_0. The quantum states $|1\rangle$ and $|2\rangle$ are therefore eigenstates of the non-interacting atomic Hamiltonian $\hat{H}_A$, with energy eigenvalues $E_0 \pm \frac{1}{2}\hbar\omega_0$, i.e.

$$\left.\begin{aligned} \hat{H}_A|2\rangle &= (E_0 + \tfrac{1}{2}\hbar\omega_0)|2\rangle \\ \hat{H}_A|1\rangle &= (E_0 - \tfrac{1}{2}\hbar\omega_0)|1\rangle, \end{aligned}\right\} \tag{15.1–1}$$

and they form a complete orthonormal set for the representation of atomic states,

$$\langle\lambda|\lambda'\rangle = \delta_{\lambda\lambda'}, \tag{15.1–2}$$

‡ The properties of two-level atoms are reviewed in some detail in the book by Allen and Eberly (1975).

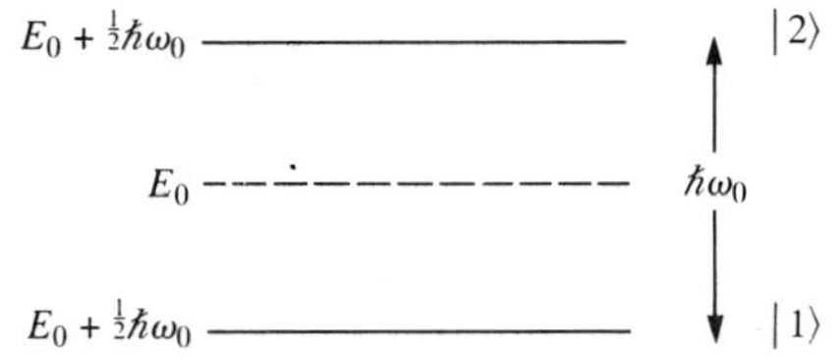

Fig. 15.1 The energy level structure of the two-level atom.

$$\sum_{\lambda=1}^{2} |\lambda\rangle\langle\lambda| = 1, \quad \lambda, \lambda' = 1, 2. \tag{15.1–3}$$

Now a quantum system with only two possible energy levels is mathematically equivalent to a spin-$\frac{1}{2}$ particle in a magnetic field, which also has just two energy levels. The formalism that has been developed for treating the dynamics of such spins (Rabi, 1937; Bloch, 1946) is therefore immediately applicable to the two-level atom. In treating the quantized electromagnetic field we found it convenient to introduce non-Hermitian operators $\hat{a}_{\mathbf{k}s}$ and $\hat{a}^\dagger_{\mathbf{k}s}$, that play the role of lowering and raising the excitation of the field by $\hbar\omega$ ($\omega = ck$). In the same way we now introduce two atomic operators $\hat{b}$ and $\hat{b}^\dagger$, that lower and raise the excitation of the atom by $\hbar\omega_0$. As the atomic excitation has both a lower and an upper bound, the effect of $\hat{b}$ acting on the lower state $|1\rangle$ and of $\hat{b}^\dagger$ acting on the upper state $|2\rangle$ must result in zero. We therefore have

$$\left.\begin{aligned} \hat{b}|2\rangle &= |1\rangle, & \hat{b}^\dagger|2\rangle &= 0 \\ \hat{b}|1\rangle &= 0, & \hat{b}^\dagger|1\rangle &= |2\rangle. \end{aligned}\right\} \tag{15.1–4}$$

By repeated application of these operators we immediately find that

$$\left.\begin{aligned} \hat{b}\hat{b}^\dagger|2\rangle &= 0, & \hat{b}^\dagger\hat{b}|2\rangle &= |2\rangle \\ \hat{b}\hat{b}^\dagger|1\rangle &= |1\rangle, & \hat{b}^\dagger\hat{b}|1\rangle &= 0, \end{aligned}\right\} \tag{15.1–5}$$

so that $\hat{b}\hat{b}^\dagger$ and $\hat{b}^\dagger\hat{b}$ play the role of 'number operators' with eigenvalues 0, 1 for the lower state and upper state excitations, respectively. Moreover, the repeated application of $\hat{b}$ or of $\hat{b}^\dagger$ to any state yields zero, so that

$$\hat{b}^2 = 0 = \hat{b}^{\dagger 2}. \tag{15.1–6}$$

We can summarize these properties by the anti-commutation rules

$$\left.\begin{aligned} \{\hat{b}, \hat{b}\} = 0 &= \{\hat{b}^\dagger, \hat{b}^\dagger\}, \\ \{\hat{b}, \hat{b}^\dagger\} &= 1, \end{aligned}\right\} \tag{15.1–7}$$

in which $\{\hat{A}, \hat{B}\} \equiv \hat{A}\hat{B} + \hat{B}\hat{A}$ denotes the anti-commutator of $\hat{A}$, $\hat{B}$. These relations are characteristic of a fermion algebra, and should be compared with Eqs. (10.3–9) to (10.3–11) for a single-mode electromagnetic field, which is a boson field. It will be seen that Eqs. (15.1–7) differ from Eqs. (10.3–9) to (10.3–11) only by the replacement of the commutators by anti-commutators.

Although all the internal atomic operators are expressible in terms of $\hat{b}$ and $\hat{b}^\dagger$, for the purpose of making closer contact with the physical observables that describe the atom, it is often convenient to introduce Hermitian dynamical variables. We shall make use of the set of three traceless Pauli spin operators defined by the equations‡

$$\left.\begin{aligned} \hat{R}_1 &= \tfrac{1}{2}(\hat{b}^\dagger + \hat{b}) \\ \hat{R}_2 &= \frac{1}{2\mathrm{i}}(\hat{b}^\dagger - \hat{b}) \\ \hat{R}_3 &= \tfrac{1}{2}(\hat{b}^\dagger\hat{b} - \hat{b}\hat{b}^\dagger), \end{aligned}\right\} \tag{15.1–8}$$

which is sometimes augmented by the introduction of a fourth element

$$\hat{R}_0 = \tfrac{1}{2}\hat{1}. \tag{15.1–9}$$

These four operators form a complete set of linearly independent, Hermitian observables in the two-dimensional Hilbert space of the atom, so that for any atomic operator $\hat{O}$, we can always write

$$\hat{O} = \sum_{\alpha=0}^{3} g_\alpha \hat{R}_\alpha, \tag{15.1–10}$$

where the g_α are coefficients determined by $\hat{O}$. We may easily show from the definitions that the operators obey the commutation and anti-commutation relations

$$\left.\begin{aligned} [\hat{R}_l, \hat{R}_m] &= \mathrm{i}\epsilon_{lmn}\hat{R}_n \\ \{\hat{R}_l, \hat{R}_m\} &= \tfrac{1}{2}\delta_{lm}, \quad (l, m, n = 1, 2, 3) \end{aligned}\right\} \tag{15.1–11}$$

and that

$$\left.\begin{aligned} \hat{R}_\alpha^2 &= \tfrac{1}{4}, \quad (\alpha = 0, 1, 2, 3) \\ \sum_{\alpha=0}^{3} \hat{R}_\alpha^2 &= 1. \end{aligned}\right\} \tag{15.1–12}$$

Moreover, when the $\hat{b}$'s and $\hat{R}$'s occur together, we have the relations

$$\left.\begin{aligned} [\hat{b}, \hat{R}_1] &= -\hat{R}_3 \\ [\hat{b}, \hat{R}_2] &= \mathrm{i}\hat{R}_3 \\ [\hat{b}, \hat{R}_3] &= \hat{b}, \end{aligned}\right\} \tag{15.1–13}$$

which are sometimes useful.

We can readily make a representation of the various operators in terms of the complete set of states $|1\rangle$, $|2\rangle$ by multiplying each operator by unity in the form given by Eq. (15.1–3) on the right and on the left, and then using the properties

‡ Our definitions of these operators may differ by a factor 2 and by the order of the labels 1, 2, 3 from those used by some other workers, but they are the same as those employed by Dicke (1954), and by Allen and Eberly (1975).

(15.1–4) and (15.1–5). We then obtain the relations

$$\left.\begin{aligned}\hat{b} &= |1\rangle\langle 2| \\ \hat{b}^{\dagger} &= |2\rangle\langle 1| \\ \hat{b}\hat{b}^{\dagger} &= |1\rangle\langle 1| \\ \hat{b}^{\dagger}\hat{b} &= |2\rangle\langle 2|,\end{aligned}\right\} \tag{15.1–14}$$

and it follows immediately that

$$\left.\begin{aligned}\hat{R}_3|2\rangle &= \tfrac{1}{2}|2\rangle \\ \hat{R}_3|1\rangle &= -\tfrac{1}{2}|1\rangle,\end{aligned}\right\} \tag{15.1–15}$$

so that the states $|2\rangle$, $|1\rangle$ are eigenstates of $\hat{R}_3$, which can be regarded as a measure of the atomic inversion.

15.1.1 Atomic energy and atomic dipole moment

If we make a representation of the atomic energy $\hat{H}_A$ in terms of the states $|2\rangle$, $|1\rangle$ in the same way and use Eqs. (15.1–1) we immediately obtain the formula

$$\hat{H}_A = E_0 + \tfrac{1}{2}\hbar\omega_0(|2\rangle\langle 2| - |1\rangle\langle 1|),$$

which, with the help of Eqs. (15.1–14), gives

$$\hat{H}_A = E_0 + \hbar\omega_0\hat{R}_3. \tag{15.1–16}$$

Hence, $\hbar\omega_0$ times $\hat{R}_3$ measures the energy of the atom relative to the reference level E_0. If the lower state is the atomic ground state then $E_0 = \frac{1}{2}\hbar\omega_0$, and $\hat{H}_A = \hbar\omega_0(\hat{R}_3 + \frac{1}{2}) = \hbar\omega_0\hat{b}^{\dagger}\hat{b}$. For some purposes the value of E_0 is immaterial and can even be put equal to zero.

The significance of the other Hermitian variables $\hat{R}_1$, $\hat{R}_2$ becomes clearer if we examine the atomic dipole moment $\hat{\boldsymbol{\mu}}$. For a real atom $\hat{\boldsymbol{\mu}}$ can be defined as $\sum_i e\hat{\mathbf{r}}_i$, where $\hat{\mathbf{r}}_i$ is the position operator of the i'th atomic electron of charge e. Our model two-level atom has no physical structure, but we can always make a representation of $\hat{\boldsymbol{\mu}}$ by multiplying by unity on the left and on the right with the help of Eq. (15.1–3), and writing

$$\begin{aligned}\hat{\boldsymbol{\mu}} &= (|2\rangle\langle 2| + |1\rangle\langle 1|)\hat{\boldsymbol{\mu}}(|2\rangle\langle 2| + |1\rangle\langle 1|) \\ &= \boldsymbol{\mu}_{22}\hat{b}^{\dagger}\hat{b} + \boldsymbol{\mu}_{11}\hat{b}\hat{b}^{\dagger} + \boldsymbol{\mu}_{12}\hat{b} + \boldsymbol{\mu}_{21}\hat{b}^{\dagger},\end{aligned}$$

where $\boldsymbol{\mu}_{ij}$ is the matrix element $\langle i|\hat{\boldsymbol{\mu}}|j\rangle$, $(i, j = 1, 2)$. In fact $\boldsymbol{\mu}_{11}$ and $\boldsymbol{\mu}_{22}$, which are expectation values of the dipole moment $\hat{\boldsymbol{\mu}}$ in the lower and upper states, must vanish from considerations of symmetry for states of definite parity, because the dipole moment has odd parity. Hence we have

$$\hat{\boldsymbol{\mu}} = \boldsymbol{\mu}_{12}\hat{b} + \boldsymbol{\mu}_{12}^{*}\hat{b}^{\dagger}, \tag{15.1–17}$$

and $\hat{\boldsymbol{\mu}}$ is entirely off-diagonal in the states $|1\rangle$, $|2\rangle$. If the transition from state $|2\rangle$ to state $|1\rangle$ corresponds to a $\Delta m = 0$ transition of a real atom, we may take $\boldsymbol{\mu}_{12}$ to be a real vector. On the other hand, for an atomic $\Delta m = \pm 1$ transition, such as might be induced by circularly polarized light, $\boldsymbol{\mu}_{12}$ is necessarily a complex vector.

To illustrate this we consider two states of atomic hydrogen as an example. If the states $|1\rangle$, $|2\rangle$ correspond respectively to the s and p states $|n=1,\ l=0,\ m=0\rangle$, $|n=2,\ l=1,\ m=0\rangle$ of atomic hydrogen, we are dealing with a $\Delta m = 0$ transition, and if the quantization axis is the z-axis, as usual, we find from the wave functions of the hydrogen atom that (cf. Allen and Eberly, 1975)

$$\boldsymbol{\mu}_{12} = \langle 1|\hat{\boldsymbol{\mu}}|2\rangle = \frac{128\sqrt{2}}{243} e a_0 \mathbf{z}_1. \qquad (15.1\text{–}18)$$

Here a_0 is the Bohr radius and $\mathbf{x}_1$, $\mathbf{y}_1$, $\mathbf{z}_1$ are unit vectors in the directions of the three axes. Hence $\boldsymbol{\mu}_{12}$ can be treated as a real vector. On the other hand, if the upper state $|2\rangle$ corresponds to the state $|n=2,\ l=1,\ m=1\rangle$ of atomic hydrogen, then we have a $\Delta m = \pm 1$ transition, and we find that

$$\langle 1|\hat{\boldsymbol{\mu}}|2\rangle = -\frac{128}{243} e a_0 (\mathbf{x}_1 + \mathrm{i}\mathbf{y}_1), \qquad (15.1\text{–}19)$$

so that $\boldsymbol{\mu}_{12}$ is necessarily complex. When $\boldsymbol{\mu}_{12}$ is real, we can use Eqs. (15.1–17) and (15.1–8) to write

$$\hat{\boldsymbol{\mu}} = \boldsymbol{\mu}_{12}(\hat{b} + \hat{b}^\dagger) = 2\boldsymbol{\mu}_{12}\hat{R}_1, \qquad (15.1\text{–}20)$$

and when $\boldsymbol{\mu}_{12}$ is complex we can always re-write Eq. (15.1–17) in the form

$$\begin{aligned}\hat{\boldsymbol{\mu}} &= \boldsymbol{\mu}_{12}(\hat{R}_1 - \mathrm{i}\hat{R}_2) + \boldsymbol{\mu}_{12}^*(\hat{R}_1 + \mathrm{i}\hat{R}_2) \\ &= 2\,\mathrm{Re}\,(\boldsymbol{\mu}_{12})\hat{R}_1 + 2\,\mathrm{Im}\,(\boldsymbol{\mu}_{12})\hat{R}_2. \end{aligned} \qquad (15.1\text{–}21)$$

The $\hat{R}_1$, $\hat{R}_2$ operators are therefore intimately related to the dipole moment $\hat{\boldsymbol{\mu}}$.

Occasionally we also need the rate of change of the dipole moment $\hat{\boldsymbol{\mu}}$, which corresponds to e times the electron velocity $\hat{\mathbf{v}}$ in a real atom. From the Heisenberg equation of motion

$$\frac{\mathrm{d}\hat{\boldsymbol{\mu}}}{\mathrm{d}t} = \frac{1}{\mathrm{i}\hbar}[\hat{\boldsymbol{\mu}}, \hat{H}_\mathrm{A}],$$

and with the help of Eqs. (15.1–16) and (15.1–17) and the commutation relations (15.1–11) or (15.1–13), we obtain the equation

$$\begin{aligned} e\hat{\mathbf{v}} = \frac{\mathrm{d}\hat{\boldsymbol{\mu}}}{\mathrm{d}t} &= \frac{1}{\mathrm{i}\hbar}[\boldsymbol{\mu}_{12}\hat{b} + \boldsymbol{\mu}_{12}^*\hat{b}^\dagger, \hbar\omega_0\hat{R}_3 + E_0] \\ &= -\mathrm{i}\omega_0(\boldsymbol{\mu}_{12}\hat{b} - \boldsymbol{\mu}_{12}^*\hat{b}^\dagger). \end{aligned} \qquad (15.1\text{–}22)$$

Of course, in the interaction picture operators such as $\hat{b}$ and $\hat{b}^\dagger$ evolve in time according to the rule

$$\hat{b}(t) = \exp[\mathrm{i}\hat{H}_\mathrm{A}(t-t_0)/\hbar]\hat{b}(t_0)\exp[-\mathrm{i}\hat{H}_\mathrm{A}(t-t_0)/\hbar], \quad t \geqslant t_0.$$

If we make use of Eq. (15.1–16) and the operator expansion theorem, together with the commutation rules (15.1–13), we readily find that

$$\left.\begin{aligned} \hat{b}(t) &= \hat{b}(t_0)\,\mathrm{e}^{-\mathrm{i}\omega_0(t-t_0)} \\ \hat{b}^\dagger(t) &= \hat{b}^\dagger(t_0)\,\mathrm{e}^{\mathrm{i}\omega_0(t-t_0)}, \end{aligned}\right\} \qquad (15.1\text{–}23)$$

so that

$$\left.\begin{aligned}\hat{\boldsymbol{\mu}}(t) &= 2\,\mathrm{Re}\,(\boldsymbol{\mu}_{12})\hat{R}_1(t) + 2\,\mathrm{Im}\,(\boldsymbol{\mu}_{12})\hat{R}_2(t)\\ &= \boldsymbol{\mu}_{12}\hat{b}(t_0)\,\mathrm{e}^{-\mathrm{i}\omega_0(t-t_0)} + \boldsymbol{\mu}_{12}^*\hat{b}^\dagger(t_0)\,\mathrm{e}^{\mathrm{i}\omega_0(t-t_0)}.\end{aligned}\right\} \tag{15.1–24}$$

15.2 Bloch-representation of the state

Any pure state $|\psi\rangle$ of the two-level atom can always be written as a linear superposition

$$|\psi\rangle = c_1|1\rangle + c_2|2\rangle, \tag{15.2–1}$$

with

$$|c_1|^2 + |c_2|^2 = 1.$$

More generally, when the state is not necessarily pure, but has to be described by an ensemble, we can represent it by an atomic density operator $\hat{\rho}^{(\mathrm{A})}$,

$$\hat{\rho}^{(\mathrm{A})} = \rho_{11}|1\rangle\langle 1| + \rho_{22}|2\rangle\langle 2| + \rho_{12}|1\rangle\langle 2| + \rho_{21}|2\rangle\langle 1|,$$

in which ρ_{ij} is the ensemble average,

$$\rho_{ij} = \langle c_i c_j^*\rangle, \quad i, j = 1, 2. \tag{15.2–2}$$

Hence $\hat{\rho}^{(\mathrm{A})}$ has a representation in terms of a two-dimensional, Hermitian, covariance matrix. There is however a simple geometric representation of the state in terms of a real three-dimensional vector $\mathbf{r}$ with components r_1, r_2, r_3 that is often intuitively attractive. This is known as the *Bloch-representation* of the atomic state (Bloch, 1946), and it is analogous to the representation of the state of polarization of a light beam by the Stokes vector (see Chapter 6). The correspondence between the density matrix and the Bloch-vector representations reflects a well-known symmetry property, which is encountered repeatedly in physics, namely the correspondence between the special unitary group SU2 and the real orthogonal group O3. The components of the Bloch-vector $\mathbf{r}$ representing the state in the Schrödinger picture are defined by the expressions

$$\left.\begin{aligned} r_1 &= 2\,\mathrm{Re}\,(\rho_{12})\\ r_2 &= 2\,\mathrm{Im}\,(\rho_{12})\\ r_3 &= \rho_{22} - \rho_{11},\end{aligned}\right\} \tag{15.2–3}$$

and these three components are sometimes augmented by a fourth one given by

$$r_0 = \rho_{22} + \rho_{11} = 1. \tag{15.2–4}$$

The pure unexcited state $|1\rangle$ has the unit Bloch-vector $(0, 0, -1)$ pointing straight down, whereas the pure excited state $|2\rangle$ has the unit Bloch-vector $(0, 0, 1)$ pointing straight up. Intermediate states have Bloch-vectors pointing in various directions, and any state that is an equal mixture of upper and lower states ($\rho_{22} = \rho_{11}$) has a horizontal Bloch-vector (see Fig. 15.2).

It is easy to show generally that for any mixed or impure state the length of the Bloch-vector is less than unity. From Eqs. (15.2–3) it follows that

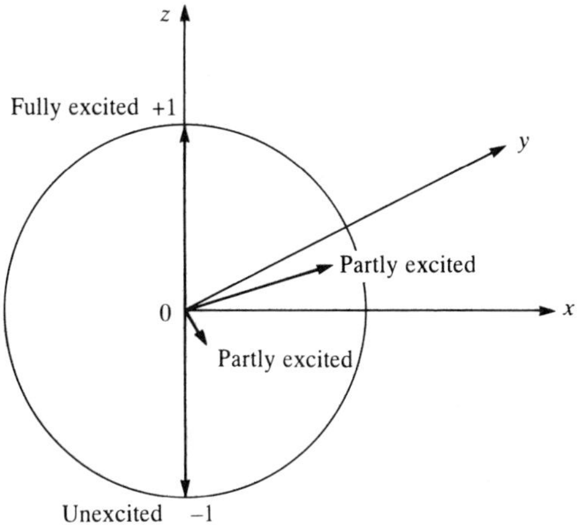

Fig. 15.2 The representation of the state of a two-level atom by a Bloch-vector. The vectors can vary in length between zero and one, but are always of unit length for a pure quantum state.

$$r_1^2 + r_2^2 + r_3^2 = 4|\rho_{12}|^2 + (\rho_{22} - \rho_{11})^2$$
$$= 1 - 4(\rho_{22}\rho_{11} - |\rho_{12}|^2). \qquad (15.2\text{–}5)$$

Now from the Schwarz inequality we have

$$\rho_{22}\rho_{11} - |\rho_{12}|^2 = \langle |c_2|^2 \rangle \langle |c_1|^2 \rangle - |\langle c_1 c_2^* \rangle|^2 \geqslant 0, \qquad (15.2\text{–}6)$$

with equality holding only when the ensemble degenerates to a single realization, i.e. for a pure state. Hence

$$r_1^2 + r_2^2 + r_3^2 \leqslant 1, \qquad (15.2\text{–}7)$$

with equality only for a pure quantum state. A randomly phased, equally weighted mixture of upper and lower states would correspond to a zero Bloch-vector.

We have already seen that the augmented set of spin operators $\hat{R}_0$, $\hat{R}_1$, $\hat{R}_2$, $\hat{R}_3$ forms a complete set for the representation of any operator in the two-dimensional Hilbert space of the atom. It follows from Eq. (15.1–10) that the density operator $\hat{\rho}^{(A)}$ may be expanded in the form

$$\hat{\rho}^{(A)} = \sum_{\alpha=0}^{3} g_\alpha \hat{R}_\alpha,$$

in which the $\hat{R}_\alpha$ are time-independent spin operators, and the time dependence of $\hat{\rho}^{(A)}$ is carried by the coefficients g_α. By comparing matrix elements $\langle i|\hat{\rho}^{(A)}|j\rangle$ (i, $j = 1, 2$) on both sides of this equation, and making use of Eqs. (15.2–3) and (15.2–4), we readily find that $g_\alpha = r_\alpha$, so that

$$\hat{\rho}^{(A)} = \sum_{\alpha=0}^{3} r_\alpha \hat{R}_\alpha. \qquad (15.2\text{–}8)$$

The components of the Bloch-vector are therefore the coefficients in the expansion of the density operator in the Schrödinger picture in terms of the spin operators.

The Bloch-vector is sometimes more conveniently described by spherical polar coordinates (r, θ, ϕ) rather than by the Cartesian coordinates (r_1, r_2, r_3). From Eq. (15.2–5) we have

$$r = [1 - 4(\rho_{22}\rho_{11} - |\rho_{12}|^2)]^{1/2}. \tag{15.2–9}$$

Also from Eqs. (15.2–3) and (15.2–4)

$$\left.\begin{aligned} \rho_{22} &= \tfrac{1}{2}(1 + r_3) = \tfrac{1}{2}(1 + r\cos\theta) \\ \rho_{11} &= \tfrac{1}{2}(1 - r_3) = \tfrac{1}{2}(1 - r\cos\theta) \\ \rho_{12} &= \tfrac{1}{2}(r_1 + ir_2) = \tfrac{1}{2}r\sin\theta\, \mathrm{e}^{\mathrm{i}\phi}, \end{aligned}\right\} \tag{15.2–10}$$

which gives

$$\cos\theta = (\rho_{22} - \rho_{11})/r$$

and

$$\tan\phi = \frac{\mathrm{Im}\,(\rho_{12})}{\mathrm{Re}\,(\rho_{12})}. \tag{15.2–11}$$

In the special case of a pure quantum state, $r = 1$ and the relations simplify. Then $|c_2| = \sqrt{\rho_{22}} = \cos\frac{1}{2}\theta$, $|c_1| = \sqrt{\rho_{11}} = \sin\frac{1}{2}\theta$, $\arg c_1 - \arg c_2 = \phi$, and, according to Eq. (15.2–1), we can represent an arbitrary pure atomic state $|\psi\rangle$ at time $t = 0$ by

$$|\psi\rangle = \sin\tfrac{1}{2}\theta\, \mathrm{e}^{\mathrm{i}\phi/2}|1\rangle + \cos\tfrac{1}{2}\theta\, \mathrm{e}^{-\mathrm{i}\phi/2}|2\rangle, \tag{15.2–12}$$

up to a phase factor.

In the Schrödinger picture the upper and lower states $|2\rangle$ and $|1\rangle$ evolve in time via the phase factors $\exp[-\mathrm{i}(E_0/\hbar + \frac{1}{2}\omega_0)t]$ and $\exp[-\mathrm{i}(E_0/\hbar - \frac{1}{2}\omega_0)t]$, respectively. When these factors are inserted in Eq. (15.2–12) we obtain for the state $|\psi(t)\rangle$ at time t, the result

$$|\psi(t)\rangle = \sin\tfrac{1}{2}\theta\, \mathrm{e}^{\mathrm{i}(\phi+\omega_0 t)/2}|1\rangle + \cos\tfrac{1}{2}\theta\, \mathrm{e}^{-\mathrm{i}(\phi+\omega_0 t)/2}|2\rangle, \tag{15.2–13}$$

again up to a phase factor. It follows that the azimuthal angle $\phi + \omega_0 t$ increases linearly with time, and that the Bloch-vector $\mathbf{r}(t)$ in the Schrödinger pictures rotates continuously about the z-axis with angular velocity ω_0.

15.2.1 Expectation values of spin operators

We have already seen from Eq. (15.2–8) that there is a connection between the spin operators $\hat{R}_0$, $\hat{R}_1$, $\hat{R}_2$, $\hat{R}_3$ and the four components r_0, r_1, r_2, r_3 of the Bloch-representation. We now establish a further connection by calculating the expectation values of the $\hat{R}$'s in an arbitrary quantum state described by the density operator $\hat{\rho}^{(\mathrm{A})}$. We obtain

$$\langle \hat{R}_\alpha \rangle = \mathrm{Tr}\,\hat{\rho}^{(\mathrm{A})}\hat{R}_\alpha, \quad \alpha = 0, 1, 2, 3$$

and with the help of Eq. (15.2–8) for the density operator in the Schrödinger picture we have

$$\langle \hat{R}_\alpha \rangle = \sum_{\beta=0}^{3} \operatorname{Tr} r_\beta \hat{R}_\beta \hat{R}_\alpha.$$

Now the product of two different $\hat{R}$ operators is another operator $\hat{R}_1$, or $\hat{R}_2$, or $\hat{R}_3$, each of which is traceless. It follows that the only contribution to the sum comes from the term $\beta = \alpha$, and with the help of Eq. (15.1–12) we find that

$$\langle \hat{R}_\alpha \rangle = \operatorname{Tr}(\tfrac{1}{4} r_\alpha \hat{1}) = \tfrac{1}{2} r_\alpha, \quad \alpha = 0, 1, 2, 3, \tag{15.2–14}$$

since the trace of the unit operator is 2 in the two-dimensional Hilbert space. The elements of the Bloch-representation can therefore also be defined in terms of the expectation values of the spin operators.

Some caution is in order here because the relations (15.2–3) and (15.2–14) are not completely equivalent. The expectation $\langle \hat{R}_\alpha \rangle$ is independent of the picture in which it is evaluated, whereas the density matrix ρ_{ij} depends on the picture. The two definitions (15.2–3) and (15.2–14) coincide only in the Schrödinger picture, because the representation of the quantum state would be a fixed vector in the Heisenberg picture, and a rotating vector in the Schrödinger picture. It is therefore preferable to regard (15.2–14) as the more fundamental definition, because it relates $\mathbf{r}$ to the expectation values of certain physical observables. The Bloch-vector then depends on time in general, and represents the state properly only in the Schrödinger picture.

With the help of Eq. (15.1–16) we now have for the expectation value of the energy,

$$\langle \hat{H}_A \rangle = E_0 + \tfrac{1}{2}\hbar\omega_0 r_3, \tag{15.2–15}$$

and from Eq. (15.1–21) the expectation value of the dipole moment is given by

$$\langle \hat{\boldsymbol{\mu}} \rangle = \operatorname{Re}(\boldsymbol{\mu}_{12}) r_1 + \operatorname{Im}(\boldsymbol{\mu}_{12}) r_2. \tag{15.2–16}$$

The z-component of the Bloch vector is therefore a measure of the energy of the atom, while the x- and y-components are related to the dipole moment.

For atoms without a permanent dipole moment the expectation value of $\hat{\boldsymbol{\mu}}$ vanishes both in the lower and upper atomic states, because $\hat{\boldsymbol{\mu}}$ has no well-defined value in these states and fluctuates. From Eqs. (15.1–17) or (15.1–21) we readily find that

$$\hat{\boldsymbol{\mu}}^2 = |\boldsymbol{\mu}_{12}|^2 \hat{1}, \tag{15.2–17}$$

so that the variance of the dipole moment

$$\langle 1|(\Delta\hat{\boldsymbol{\mu}})^2|1\rangle = |\boldsymbol{\mu}_{12}|^2 = \langle 2|(\Delta\hat{\boldsymbol{\mu}})^2|2\rangle \tag{15.2–18}$$

has its greatest value in these states. Because of the fluctuations of $\hat{\boldsymbol{\mu}}$ the atom can interact with an electromagnetic field even when it is in the quantum state $|1\rangle$ or $|2\rangle$, for which $\langle \hat{\boldsymbol{\mu}} \rangle = 0$. We now turn to a discussion of such interactions.

15.3 Interaction of an atom with a classical field

We suppose that the atom is located at a fixed point in an electromagnetic field that is describable classically, for example by its electric field $\mathbf{E}(t)$. If we regard the atom as an electric point dipole with moment $\hat{\boldsymbol{\mu}}(t)$, then the interaction energy $\hat{H}_{\mathrm{I}}(t)$ may be taken as the usual expression for the potential energy of a dipole in a field, and from Eq. (15.1–21) we have

$$\hat{H}_{\mathrm{I}}(t) = -\hat{\boldsymbol{\mu}}(t)\cdot\mathbf{E}(t) = -2[\mathrm{Re}\,(\boldsymbol{\mu}_{12})\hat{R}_1 + \mathrm{Im}\,(\boldsymbol{\mu}_{12})\hat{R}_2]\cdot\mathbf{E}(t). \qquad (15.3\text{–}1)$$

On the other hand, in a real atom the important interaction is that between the atomic electrons and the field, which is given by Eq. (14.1–17). If we discard the $e^2\mathbf{A}^2/2m$ term for reasons that were pointed out in Section 14.1, and if we identify the electron canonical momentum $\hat{\mathbf{p}}$ with $m\dot{\hat{\boldsymbol{\mu}}}/e$, then we have with the help of Eq. (15.1–22)

$$\hat{H}_{\mathrm{I}}(t) = -\dot{\hat{\boldsymbol{\mu}}}(t)\cdot\mathbf{A}(t) = \mathrm{i}\omega_0[\boldsymbol{\mu}_{12}\hat{b}(t) - \boldsymbol{\mu}_{12}^*\hat{b}^\dagger(t)]\cdot\mathbf{A}(t). \qquad (15.3\text{–}2)$$

Which of the two interactions is the more appropriate in a given situation has been the subject of some discussion (see Section 14.1). Both expressions have been used successfully, and in most cases they lead to similar conclusions. The total energy $\hat{H}$ of the atom in the field is of course the sum $\hat{H} = \hat{H}_{\mathrm{A}} + \hat{H}_{\mathrm{I}}$, in which $\hat{H}_{\mathrm{A}}$ is diagonal in the states $|1\rangle$, $|2\rangle$, whereas $\hat{H}_{\mathrm{I}}$ is entirely off-diagonal.

The time evolution of the atomic density operator $\hat{\rho}^{(\mathrm{A})}(t)$ in the Schrödinger picture is governed by the Schrödinger equation

$$\frac{\mathrm{d}\hat{\rho}(t)}{\mathrm{d}t} = \frac{1}{\mathrm{i}\hbar}[\hat{H}_{\mathrm{A}} + \hat{H}_{\mathrm{I}}, \hat{\rho}^{(\mathrm{A})}(t)]. \qquad (15.3\text{–}3)$$

15.3.1 Bloch equations

On taking matrix elements of both sides of this equation, inserting unity between $\hat{H}$ and $\hat{\rho}^{\mathrm{A}}(t)$ in the form of Eq. (15.1–3) and using the aforementioned properties of $\hat{H}_{\mathrm{A}}$ and $\hat{H}_{\mathrm{I}}$, we obtain the following equations:

$$\left.\begin{aligned}
\dot{\rho}_{11} &= \frac{1}{\mathrm{i}\hbar}[\langle 1|\hat{H}_{\mathrm{I}}(t)|2\rangle\rho_{21} - \mathrm{c.c.}] \\
\dot{\rho}_{22} &= -\frac{1}{\mathrm{i}\hbar}[\langle 1|\hat{H}_{\mathrm{I}}(t)|2\rangle\rho_{21} - \mathrm{c.c}] \\
\dot{\rho}_{12} &= \frac{1}{\mathrm{i}\hbar}[-\hbar\omega_0\rho_{12} + \langle 1|\hat{H}_{\mathrm{I}}(t)|2\rangle(\rho_{22} - \rho_{11})] \\
\dot{\rho}_{21} &= \frac{1}{\mathrm{i}\hbar}[\hbar\omega_0\rho_{21} + \langle 2|\hat{H}_{\mathrm{I}}(t)|1\rangle(\rho_{11} - \rho_{22})].
\end{aligned}\right\} \qquad (15.3\text{–}4)$$

As these equations of motion are written in the Schrödinger picture, we may use Eqs. (15.2–3) to identify the components of the Bloch-vector $\mathbf{r}$. We then arrive at the following equations of motion for the three components r_1, r_2, r_3:

$$\dot{r}_1 = \frac{1}{\hbar}2\,\mathrm{Im}\,[\langle 1|\hat{H}_{\mathrm{I}}|2\rangle]r_3 - \omega_0 r_2 \qquad (15.3\text{–}5a)$$

$$\dot{r}_2 = -\frac{1}{\hbar}2\,\mathrm{Re}\,[\langle 1|\hat{H}_{\mathrm{I}}|2\rangle]r_3 + \omega_0 r_1 \qquad (15.3\text{–}5b)$$

$$\dot{r}_3 = -\frac{2}{\hbar}\,\mathrm{Im}\,[\langle 1|\hat{H}_{\mathrm{I}}|2\rangle]r_1 + \frac{2}{\hbar}\,\mathrm{Re}\,[\langle 1|\hat{H}_{\mathrm{I}}|2\rangle]r_2. \tag{15.3–5c}$$

These are sometimes known as *Bloch equations*, describing the time evolution of the atom under the influence of the interaction $\hat{H}_{\mathrm{I}}$. They were first used in connection with problems in nuclear magnetic resonance by Bloch (1946). We note that, in the absence of an external field, r_3 remains constant. If we multiply the first equation by r_1, the second by r_2, and the third by r_3 and add the result, we readily find that

$$\frac{\mathrm{d}}{\mathrm{d}t}(r_1^2 + r_2^2 + r_3^2) = 0. \tag{15.3–6}$$

The length of the Bloch-vector therefore remains constant in the presence of a classical field. In particular, this implies that an atom that starts in a pure quantum state remains in a pure state, and one that starts in a mixed quantum state remains in a mixed state. The assumption of a pure state is often convenient and sufficient in many cases.

The equations (15.3–5) have an interesting geometric interpretation, that was found by Feynman, Vernon and Hellwarth (1957). To see this we introduce a new vector $\mathbf{Q}(t)$ with components defined by

$$\left.\begin{aligned} Q_1 &\equiv \frac{2}{\hbar}\,\mathrm{Re}\,[\langle 1|\hat{H}_{\mathrm{I}}|2\rangle] \\ Q_2 &\equiv \frac{2}{\hbar}\,\mathrm{Im}\,[\langle 1|\hat{H}_{\mathrm{I}}|2\rangle] \\ Q_3 &\equiv \omega_0. \end{aligned}\right\} \tag{15.3–7}$$

The three equations (15.3–5) are then equivalent to the vector equation

$$\frac{\mathrm{d}}{\mathrm{d}t}\mathbf{r} = \mathbf{Q} \times \mathbf{r}, \tag{15.3–8}$$

which shows that the motion of the Bloch-vector is a precession about the $\mathbf{Q}$-vector, at a rate determined by the magnitude of $\mathbf{Q}$. Of course, in general, when $\mathbf{Q}$ itself varies with time, the precession may become very complicated.

If we make use of Eq. (15.3–1) or Eq. (15.3–2) we can evaluate the matrix element $\langle 1|\hat{H}_{\mathrm{I}}|2\rangle$ explicitly. From Eq. (15.3–1) we obtain the result

$$\langle 1|\hat{H}_{\mathrm{I}}|2\rangle = -\boldsymbol{\mu}_{12}\cdot\mathbf{E}(t).$$

We shall suppose that the oscillations of the electric field are centered on some frequency ω_1 close to ω_0, so that we may write

$$\begin{aligned} \mathbf{E}(t) &= \boldsymbol{\varepsilon}\mathscr{E}(t)\,\mathrm{e}^{-\mathrm{i}\omega_1 t} + \text{c.c.} \\ &= 2|\mathscr{E}(t)|\{\mathrm{Re}\,(\boldsymbol{\varepsilon})\cos\,[\omega_1 t - \phi(t)] + \mathrm{Im}\,(\boldsymbol{\varepsilon})\sin\,[\omega_1 t - \phi(t)]\}, \end{aligned} \tag{15.3–9}$$

where $\mathscr{E}(t) = |\mathscr{E}(t)|\,\mathrm{e}^{\mathrm{i}\phi(t)}$ is a slowly varying complex amplitude, and $\boldsymbol{\varepsilon}$ is a unit polarization vector. Then

$$\langle 1|\hat{H}_{\mathrm{I}}|2\rangle = -|\mathscr{E}(t)|\boldsymbol{\mu}_{12}\cdot[\boldsymbol{\varepsilon}\,\mathrm{e}^{-\mathrm{i}[\omega_1 t - \phi(t)]} + \text{c.c.}] \tag{15.3–10}$$

and this expression should be substituted in the Bloch equations (15.3–5).

The equations simplify in certain cases. For example, if we are dealing with an atomic $\Delta m = \pm 1$ transition, so that we may set $\boldsymbol{\mu}_{12} = |\boldsymbol{\mu}_{12}|(\mathbf{x}_1 + \mathrm{i}\mathbf{y}_1)/\sqrt{2}$ ($\mathbf{x}_1$, $\mathbf{y}_1$ are unit vectors), and if the external field is circularly polarized and propagating in the z-direction, so that $\boldsymbol{\varepsilon} = (\mathbf{x}_1 + \mathrm{i}\mathbf{y}_1)/\sqrt{2}$, then $\boldsymbol{\mu}_{12} \cdot \boldsymbol{\varepsilon} = 0$ and $\boldsymbol{\mu}_{12} \cdot \boldsymbol{\varepsilon}^* = |\boldsymbol{\mu}_{12}|$. The equations (15.3–5) then reduce to

$$\left.\begin{aligned} \dot{r}_1 &= -\Omega(t) r_3 \sin[\omega_1 t - \phi(t)] - \omega_0 r_2 \\ \dot{r}_2 &= \Omega(t) r_3 \cos[\omega_1 t - \phi(t)] + \omega_0 r_1 \\ \dot{r}_3 &= \Omega(t) r_1 \sin[\omega_1 t - \phi(t)] - \Omega(t) r_2 \cos[\omega_1 t - \phi(t)], \end{aligned}\right\} \tag{15.3–11}$$

where we have introduced the abbreviation

$$\Omega(t) \equiv 2\boldsymbol{\mu}_{12} \cdot \boldsymbol{\varepsilon}^* |\mathscr{E}(t)|/\hbar. \tag{15.3–12}$$

The parameter Ω is known as the *Rabi frequency*. It is a measure of the strength of the time-varying external field. The components of the **Q**-vector defined by Eq. (15.3–7) then become

$$\left.\begin{aligned} Q_1(t) &= -\Omega(t) \cos[\omega_1 t - \phi(t)] \\ Q_2(t) &= -\Omega(t) \sin[\omega_1 t - \phi(t)] \\ Q_3(t) &= \omega_0. \end{aligned}\right\} \tag{15.3–13}$$

On the other hand, if we are dealing with an atomic $\Delta m = 0$ transition, when $\boldsymbol{\mu}_{12}$ may be taken to be a real vector, and if the incident light is linearly polarized so that $\boldsymbol{\varepsilon}$ is also real, then the equations (15.3–5) become

$$\left.\begin{aligned} \dot{r}_1 &= -\omega_0 r_2 \\ \dot{r}_2 &= 2\Omega r_3 \cos[\omega_1 t - \phi(t)] + \omega_0 r_1 \\ \dot{r}_3 &= -2\Omega r_2 \cos[\omega_1 t - \phi(t)]. \end{aligned}\right\} \tag{15.3–14}$$

The corresponding **Q**-vector, when the equations of motion are written in the form of Eq. (15.3–8), is given by

$$\left.\begin{aligned} Q_1(t) &= -2\Omega(t) \cos[\omega_1 t - \phi(t)] \\ Q_2(t) &= 0 \\ Q_3(t) &= \omega_0. \end{aligned}\right\} \tag{15.3–15}$$

Although, at first sight, Eqs. (15.3–14) and (15.3–15) appear to be very different from Eqs. (15.3–11) and (15.3–13), respectively, they actually differ only by the contributions of certain anti-resonant terms.‡ We may look on the **Q**-vector defined by Eq. (15.3–15) as the sum of two **Q**-vectors, of which the first is given by Eq. (15.3–13) as before, while the second, or auxiliary, **Q**-vector has components $-\Omega(t)\cos[\omega_1 t - \phi(t)]$, $+\Omega(t)\sin[\omega_1 t - \phi(t)]$, 0. This represents a vector rotating about the z-axis at frequency $\omega_1 - \dot{\phi}(t)$, whereas the Bloch-vector rotates in the opposite direction around the z-axis at frequency ω_0. Relative to the

‡ See Section 2.4 of Allen and Eberly (1975) for a good discussion of this point.

Bloch-vector $\mathbf{r}$, the auxiliary $\mathbf{Q}$-vector rotates at frequency $\omega_1 + \omega_0 - \dot{\phi}(t)$, and when we integrate the equations of motion over any measurable time interval, its effect is expected to be very small. We may therefore neglect this auxiliary $\mathbf{Q}$-vector to a good approximation, in which case $\mathbf{Q}$ is again given by Eq. (15.3–13), and the equations of motion again become Eqs. (15.3–11). This procedure is known as the *rotating wave approximation*, and it allows us to use the same set of equations in both circumstances. Of course there is no approximation in the former case ($\Delta m = \pm 1$).

15.3.2 Bloch equations in the rotating frame

The Bloch vector described by the equations of motion (15.3–11) still rotates at an optical frequency about the z-axis. For this reason it is sometimes preferable to describe the motion in a reference frame that is itself rotating, in which the $\mathbf{r}$-vector moves much more slowly. Inspection of Eqs. (15.3–11) suggests that we should let this frame rotate at the atomic frequency ω_0. However, because the atomic frequency may vary from one atom to another in a real medium, and because it is desirable to refer the time development of different atoms to the same reference frame, it is customary to let the frame rotate at the frequency ω_1 of the applied field. The coordinates r_1', r_2', r_3' of the Bloch-vector in the rotating frame are then related to the coordinates r_1, r_2, r_3 in the stationary frame by the transformation

$$\mathbf{r}' = \mathbf{\Theta}\mathbf{r}, \tag{15.3–16}$$

in which $\mathbf{r}$, $\mathbf{r}'$ are 3×1 column matrices of the coordinates r_1, r_2, r_3 and r_1', r_2', r_3', respectively, and $\mathbf{\Theta}$ is the 3×3 orthogonal rotation matrix

$$\mathbf{\Theta} \equiv \begin{bmatrix} \cos\omega_1 t & \sin\omega_1 t & 0 \\ -\sin\omega_1 t & \cos\omega_1 t & 0 \\ 0 & 0 & 1 \end{bmatrix}. \tag{15.3–17a}$$

This transformation leaves the z-component r_3 unchanged, and it connects r_1, r_2 and r_1', r_2' so that

$$r_1' + \mathrm{i}r_2' = (r_1 + \mathrm{i}r_2)\,\mathrm{e}^{-\mathrm{i}\omega_1 t}. \tag{15.3–17b}$$

In order to transform the Bloch equations (15.3–11) to the rotating frame, we write them in matrix form

$$\dot{\mathbf{r}} = \mathbf{C}\mathbf{r},$$

where $\mathbf{C}$ is the 3×3 matrix of the coefficients, and then use Eq. (15.3–16). Thus we have

$$\begin{aligned} \dot{\mathbf{r}}' &= \dot{\mathbf{\Theta}}\mathbf{r} + \mathbf{\Theta}\dot{\mathbf{r}} \\ &= \dot{\mathbf{\Theta}}\mathbf{\Theta}^{-1}\mathbf{\Theta}\mathbf{r} + \mathbf{\Theta}\mathbf{C}\mathbf{\Theta}^{-1}\mathbf{\Theta}\mathbf{r} \\ &= (\dot{\mathbf{\Theta}}\mathbf{\Theta}^{-1} + \mathbf{\Theta}\mathbf{C}\mathbf{\Theta}^{-1})\mathbf{r}', \end{aligned} \tag{15.3–18}$$

and insertion of the explicit expressions for $\mathbf{\Theta}$ and $\mathbf{C}$ from Eqs. (15.3–17a) and

(15.3–11) leads to the equations

$$\left.\begin{aligned}\dot{r}_1' &= (\omega_1 - \omega_0)r_2' + \Omega\sin\phi r_3' \\ \dot{r}_2' &= (\omega_0 - \omega_1)r_1' + \Omega\cos\phi r_3' \\ \dot{r}_3' &= -\Omega\sin\phi r_1' - \Omega\cos\phi r_2'.\end{aligned}\right\} \qquad (15.3\text{–}19)$$

These are the Bloch equations in the rotating frame. They may again be expressed in the form

$$\frac{\mathrm{d}\mathbf{r}'}{\mathrm{d}t} = \mathbf{Q}' \times \mathbf{r}', \qquad (15.3\text{–}20)$$

in which the $\mathbf{Q}'$-vector in the rotating frame has components $-\Omega\cos\phi$, $\Omega\sin\phi$, $\omega_0 - \omega_1$. Hence the Bloch-vector $\mathbf{r}'$ precesses about the vector $\mathbf{Q}'$ at a rate determined by the magnitude $[\Omega^2 + (\omega_1 - \omega_0)^2]^{1/2}$ and the orientation of $\mathbf{Q}'$. However, if the detuning $|\omega_0 - \omega_1|$ is much larger than the Rabi frequency Ω and the rate of change of $\phi(t)$, then $\mathbf{Q}'$ points approximately down or up, and its motion is relatively simple in the rotating frame. If the Bloch-vector $\mathbf{r}'$ initially points in a direction that is approximately parallel to $\mathbf{Q}'$, then $\mathbf{r}'$ precesses about $\mathbf{Q}'$ in a cone of small semi-angle, and this cone tends to follow slow variations of $\mathbf{Q}'$. The Bloch-vector is then said to follow the $\mathbf{Q}'$-vector adiabatically, and this *adiabatic following* phenomenon can be exploited to prepare the atom in a certain quantum state (Treacy and DeMaria, 1969; Grischkowsky, 1970; Grischkowsky, Courtens and Armstrong, 1973; Loy, 1974). For example, if the frequency of the exciting light changes so that $\omega_1 - \omega_0$ goes from a large positive value to a large negative value, then the $\mathbf{Q}'$-vector rotates almost through 180°, as does the cone of precession, and an atom starting in the ground state can be brought near to the excited state. It is important to note that the actual magnitude $|\omega_0 - \omega_1|$ of the detuning is not very important. The technique can therefore be applied to an inhomogeneously broadened medium like a gas, in which different atoms moving with different velocities have different natural frequencies in the laboratory frame, because of the associated Doppler shifts.

15.3.3 The Rabi problem

In the special case in which the exciting field is strictly sinusoidal, so that the complex amplitude $\mathscr{E}$ is constant and the phase ϕ may be made zero by proper choice of the origin of time, the equations simplify. The solution to the problem of how an atom responds to the field was first given by Rabi (1937) in connection with the behavior of a spin $\frac{1}{2}$ in a magnetic field. If the atom starts in the lower state $|1\rangle$ at time $t = 0$, so that $r_3(0) = -1$ and $r_1(0) = 0 = r_2(0)$, the solution to the Bloch equations in the rotating frame takes the form

$$\left.\begin{aligned}r_1'(t) &= \frac{(\omega_0 - \omega_1)\Omega}{\Omega^2 + (\omega_0 - \omega_1)^2}\{1 - \cos[\Omega^2 + (\omega_0 - \omega_1)^2]^{1/2}t\} \\ r_2'(t) &= \frac{-\Omega}{[\Omega^2 + (\omega_0 - \omega_1)^2]^{1/2}}\sin[\Omega^2 + (\omega_0 - \omega_1)^2]^{1/2}t \\ r_3'(t) &= -\frac{(\omega_0 - \omega_1)^2 + \Omega^2\cos[\Omega^2 + (\omega_0 - \omega_1)^2]^{1/2}t}{\Omega^2 + (\omega_0 - \omega_1)^2}.\end{aligned}\right\} \qquad (15.3\text{–}21)$$

This corresponds to a rather complicated rotation. Perhaps the most interesting aspect of the solution is the oscillatory behavior of the atomic excitation as measured by r_3', which oscillates about the mid-value $-(\omega_0 - \omega_1)^2/[\Omega^2 + (\omega_0 - \omega_1)^2]$ with frequency $[\Omega^2 + (\omega_0 - \omega_1)^2]^{1/2}$ and amplitude $\Omega^2/[\Omega^2 + (\omega_0 - \omega_1)^2]$. This phenomenon is known as *Rabi oscillation* or *optical nutation*. If the exciting field is sufficiently strong or the detuning $|\omega_0 - \omega_1|$ is sufficiently small, then r_3' oscillates almost between the values ± 1, and almost at the frequency Ω. The applied field therefore has the effect of repeatedly exciting and de-exciting the atom. When the same problem is treated in a fully quantum mechanical manner with a quantized field, the effects of spontaneous atomic emission cause $r_3'(t)$ to damp out gradually in time, even though the Rabi oscillations continue. This phenomenon has been observed experimentally (Walther, 1977; Dagenais and Mandel, 1978). Figure 15.3 shows the observed time variation of the fluorescence emitted by an atom in a coherent exciting field that is detuned from resonance. The oscillatory excitation or Rabi oscillation has the effect of modulating the amplitude of the atomic fluorescence in time, and this amplitude modulation is reflected in the appearance of sidebands in the spectrum of the atomic fluorescence centered at frequencies $\omega_1 \pm [\Omega^2 + (\omega_0 - \omega_1)^2]^{1/2}$. This can also be seen more directly if we use Eqs. (15.3–21) to calculate the expectation value of the dipole moment $\langle \hat{\boldsymbol{\mu}}(t) \rangle$. From Eq. (15.2–16)

$$\langle \hat{\boldsymbol{\mu}}(t) \rangle = \mathrm{Re}\,(\boldsymbol{\mu}_{12}) r_1 + \mathrm{Im}\,(\boldsymbol{\mu}_{12}) r_2,$$

and, if we take $\boldsymbol{\mu}_{12}$ to be real, we obtain with the help of Eqs. (15.3–17) and (15.3–21) the relation

$$\begin{aligned}\langle \hat{\boldsymbol{\mu}}(t) \rangle &= \boldsymbol{\mu}_{12}[\mathbf{r}_1' \cos \omega_1 t - \mathbf{r}_2' \sin \omega_1 t] \\ &= \frac{\boldsymbol{\mu}_{12}\Omega}{\Omega^2 + (\omega_0 - \omega_1)^2}[(\omega_0 - \omega_1) \cos \omega_1 t \\ &\quad + \tfrac{1}{2}\{[\Omega^2 + (\omega_0 - \omega_1)^2]^{1/2} - (\omega_0 - \omega_1)\} \cos\{\omega_1 - [\Omega^2 + (\omega_0 - \omega_1)^2]^{1/2}\}t \\ &\quad - \tfrac{1}{2}\{[\Omega^2 + (\omega_0 - \omega_1)^2]^{1/2} + (\omega_0 - \omega_1)\} \cos\{\omega_1 + [\Omega^2 + (\omega_0 - \omega_1)^2]^{1/2}\}t].\end{aligned} \qquad (15.3\text{–}22)$$

This equation shows explicitly that the dipole moment oscillates at the three frequencies ω_1, $\omega_1 \pm [\Omega^2 + (\omega_0 - \omega_1)^2]^{1/2}$. A three-peaked spectrum has been observed in resonance fluorescence experiments (Schuda, Stroud and Herscher, 1974; Wu, Grove and Ezekiel, 1975; Hartig, Rasmussen, Schieder and Walther, 1976), as shown in Fig. 15.4, although the three lines are not sharp. We shall return to this subject in Section 15.6.

The Bloch equations (15.3–11) or (15.3–19) are sometimes augmented by the addition of phenomenological damping terms, which are introduced in an ad hoc manner to account for the observed damping effects. The damped oscillatory solutions of the modified equations were first obtained by Torrey (1949). We shall not pursue this phenomenological approach here, but rather return to the problem in connection with the fully quantum mechanical treatment of the phenomenon of resonance fluorescence. However, even when damping effects are present, Eqs. (15.3–11), or their counterparts (15.3–19) in the rotating frame, are

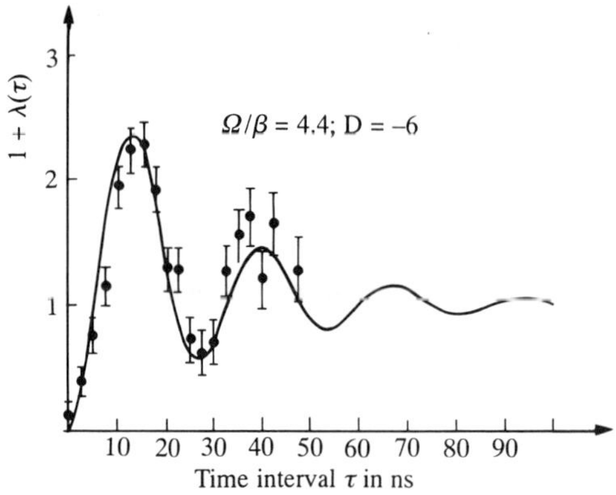

Fig. 15.3 Observed time variation of the fluorescence emitted by an atom in a coherent exciting field with Rabi frequency $\Omega = 2.2$ linewidths and detuning $\omega_0 - \omega_1 = 3$ linewidths. The full line is the theoretically expected curve. (Reproduced from Dagenais and Mandel, 1978.)

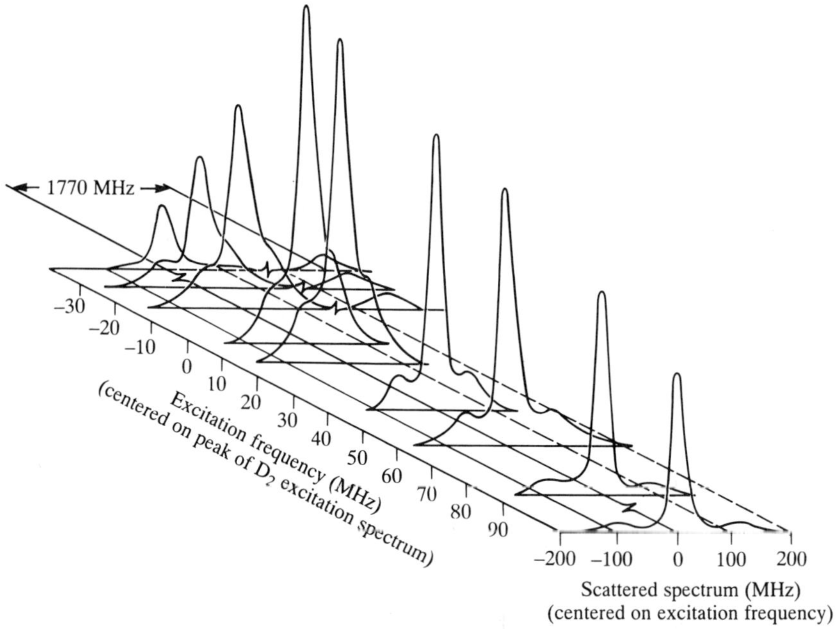

Fig. 15.4 The spectrum of the fluorescence emitted by an atom in a coherent exciting field at various excitation frequencies. (Reproduced from Schuda, Stroud and Herscher, 1974.)

adequate to describe the atomic motion for time intervals short compared with the damping time.

15.3.4 The response of the atom to a light pulse

In the foregoing we have examined how a two-level atom responds to a constant monochromatic exciting field. Actually, Eqs. (15.3–21) can also be used when the atom is subjected to a rectangular exciting pulse for the duration of the pulse. However, the general solution for an arbitrary exciting field is much more complicated. Instead we shall simplify the problem by supposing that the time-dependent phase $\phi(t)$ in Eq. (15.3–19) may be put equal to zero, and that the exciting frequency ω_1 coincides with the atomic frequency ω_0. Then the Bloch equations (15.3–19) in the rotating frame simplify to

$$\left.\begin{aligned} \dot{r}_1' &= 0 \\ \dot{r}_2' &= \Omega r_3' \\ \dot{r}_3' &= -\Omega r_2'. \end{aligned}\right\} \qquad (15.3\text{–}23)$$

The first of these equations ensures that the Bloch-vector moves entirely in the y', z'-plane, and that $r_1'(t) = r_1'(-\infty)$. We shall suppose that the initial quantum state is a pure state, so that $\mathbf{r}'$ has unit length throughout and merely rotates in the y', z'-plane. Let us put

$$r_2'(t) = [1 - r_1'^2(t)]^{1/2} \sin[\Theta(t) + K],$$

where $\Theta(t)$ is a new variable and K is a constant. Then

$$r_3'(t) = [1 - r_1'^2(t)]^{1/2} \cos[\Theta(t) + K],$$

in order to ensure that $r_1'^2 + r_2'^2 + r_3'^2 = 1$ at all times. Substitution in the Bloch equations (15.3–23) then leads to the equation

$$\dot{\Theta}(t) = \Omega(t),$$

or

$$\Theta(t) = \int_{-\infty}^{t} \Omega(t')\,\mathrm{d}t', \qquad (15.3\text{–}24)$$

and, when K is expressed in terms of initial values, we have the solution

$$\left.\begin{aligned} r_1'(t) &= r_1'(-\infty) \\ r_2'(t) &= r_2'(-\infty)\cos\Theta(t) + r_3'(-\infty)\sin\Theta(t) \\ r_3'(t) &= -r_2'(-\infty)\sin\Theta(t) + r_3'(-\infty)\cos\Theta(t). \end{aligned}\right\} \qquad (15.3\text{–}25)$$

$r_1'(-\infty)$, $r_2'(-\infty)$, $r_3'(-\infty)$ are the initial coordinates of the Bloch-vector at time $t \to -\infty$, when we may suppose that the exciting field has not yet been turned on. This solution represents a rotation of the Bloch-vector in the y', z'-plane about the x'-axis through an angle $\Theta(t)$, which is sometimes known as the *tipping angle* associated with the time-dependent optical field. If the light is in the form of a pulse whose amplitude is non-zero for a limited time only, then at the end of the pulse the Bloch-vector has been rotated through the angle

$$A \equiv \Theta(\infty) = \int_{-\infty}^{\infty} \Omega(t')\,\mathrm{d}t'. \qquad (15.3\text{–}26)$$

The angle A is sometimes referred to as the *area* of the pulse, as it is proportional to the area under the pulse envelope in the time domain. For example, if $A = \pi$, and if the atom starts off in the lower state $|1\rangle$, it ends up in the excited state $|2\rangle$

at the end of the pulse. Such a light pulse is known as a π-pulse. If the pulse area $A = 2\pi$, the Bloch-vector makes one complete revolution in the y', z'-plane, and the atom ends up back in the state $|1\rangle$. Note that the rotation is determined entirely by the area A of the light pulse given by the integral in Eq. (15.3–26), and it is independent of the actual pulse shape.

15.3.5 *Hyperbolic secant 2π-pulse*

If the frequency ω_1 of the exciting light is detuned from the atomic frequency ω_0, the situation is more complicated, and the single variable $\Theta(t)$ no longer determines the motion of the Bloch-vector. The position in which an atom that starts in the lower state $|1\rangle$ is left at the end of the pulse depends both on the pulse shape and on the detuning $\omega_0 - \omega_1$, in general. There is, however, one very special form of 2π-pulse that has the property of leaving a lower-state atom in the lower state, irrespective of the detuning. In order to find the form of this pulse we shall start with a plausible assumption (Torrey, 1949). As we have seen from Eqs. (15.3–23) with $r_1'(-\infty) = 0 = r_2'(-\infty)$ and $r_3'(-\infty) = -1$, on resonance, the z'-component of the Bloch-vector has the form

$$r_3'(t) = -\cos\Theta(t) = -1 + 2\sin^2[\tfrac{1}{2}\Theta(t)],$$

which starts from -1 at $t = -\infty$, increases to $+1$, and returns to -1 again after a 2π pulse. It seems plausible that the excitation of an atom off-resonance, that starts and ends in the ground state, undergoes a somewhat smaller amplitude cycle with

$$r_3' = -1 + 2B\sin^2(\tfrac{1}{2}\Theta), \tag{15.3–27}$$

where $B \leqslant 1$ and depends on the detuning $\omega_0 - \omega_1$ in such a way that $B = 1$ when $\omega_0 = \omega_1$. If r_3' is given by Eq. (15.3–27), then from the last two Eqs. (15.3–19), with $\phi = 0$, we have

$$r_2' = -\dot{r}_3'/\dot{\Theta} = -B\sin\Theta \tag{15.3–28}$$

and

$$r_1' = \frac{1-B}{\omega_0 - \omega_1}\dot{\Theta}. \tag{15.3–29}$$

If we substitute these values of r_1', r_2', r_3' in the first Eq. (15.3–19), we find that Θ must satisfy the differential equation

$$\ddot{\Theta} = \frac{B(\omega_0 - \omega_1)^2}{1-B}\sin\Theta, \tag{15.3–30a}$$

or equivalently

$$\dot{\Theta} = 2\left(\frac{B}{1-B}\right)^{1/2}(\omega_0 - \omega_1)\sin(\tfrac{1}{2}\Theta), \tag{15.3–30b}$$

which determines the shape of the 2π-pulse. The coefficient of $\sin\Theta$ in Eq. (15.3–30a) has the dimensions of reciprocal time squared and should not depend on frequency. If we denote it by $1/T^2$, we obtain for B the expression

$$B = \frac{1}{1 + (\omega_0 - \omega_1)^2 T^2}, \tag{15.3–31}$$

which satisfies the required condition on B for any value of T. Equation (15.3–30b), which describes a simple pendulum, is directly integrable and gives

$$\int_\pi^\Theta \frac{d\Theta}{2\sin(\frac{1}{2}\Theta)} = \frac{1}{T}\int_{t_0}^t dt,$$

where t_0 is the time when $\Theta = \pi$ and is arbitrary. Hence

$$\ln|\tan\tfrac{1}{4}\Theta| = \frac{t - t_0}{T},$$

or

$$\Theta(t) = 4\arctan\left[\exp\left(\frac{t - t_0}{T}\right)\right]. \tag{15.3–32}$$

With this expression for $\Theta(t)$ we find that

$$\Omega(t) = \dot{\Theta}(t) = \frac{2}{T}\operatorname{sech}\left(\frac{t - t_0}{T}\right), \tag{15.3–33}$$

so that the electric field amplitude $|\mathscr{E}(t)|$ of the applied light pulse has the shape of a hyperbolic secant. This is illustrated in Fig. 15.5. The pulse has a maximum at time $t = t_0$ and an effective width of order T, although it is, strictly speaking, of infinite duration. With the help of the triangle relations in Fig. 15.6 we obtain the formulas

$$\sin(\tfrac{1}{2}\Theta) = \operatorname{sech}\left(\frac{t - t_0}{T}\right)$$

$$\sin\Theta = 2\operatorname{sech}\left(\frac{t - t_0}{T}\right)\tanh\left(\frac{t - t_0}{T}\right),$$

so that the solutions of the Bloch equations in the rotating frame, when the atom

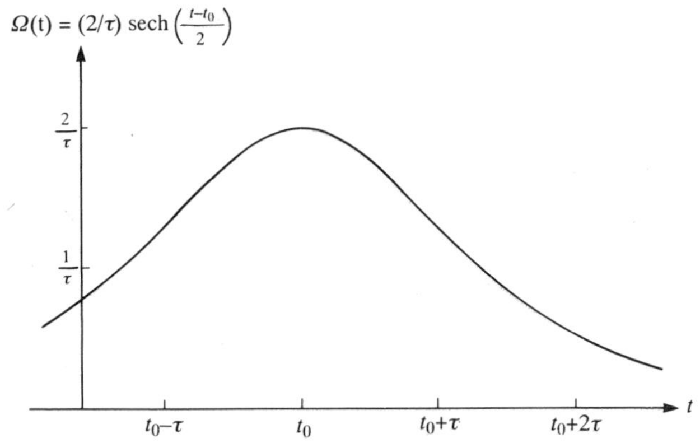

Fig. 15.5 Illustrating the form of a hyperbolic secant light pulse $\Omega(t) = (2/\tau)\operatorname{sech}[(t - t_0)/\tau]$.

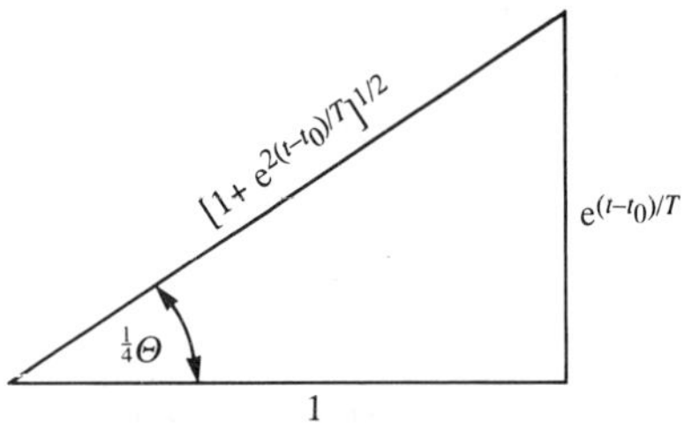

Fig. 15.6 Illustrating the relation between the tipping angle Θ and time t.

starts in the ground state, take the form

$$\left.\begin{aligned} r_1'(t) &= \frac{2(\omega_0-\omega_1)T}{1+(\omega_0-\omega_1)^2T^2}\operatorname{sech}\left(\frac{t-t_0}{T}\right) \\ r_2'(t) &= -\frac{2}{1+(\omega_0-\omega_1)^2T^2}\operatorname{sech}\left(\frac{t-t_0}{T}\right)\tanh\left(\frac{t-t_0}{T}\right) \\ r_3'(t) &= -1+\frac{2}{1+(\omega_0-\omega_1)^2T^2}\operatorname{sech}^2\left(\frac{t-t_0}{T}\right). \end{aligned}\right\} \quad (15.3\text{–}34)$$

These solutions were first obtained by McCall and Hahn (1967, 1969), and the corresponding motion of the Bloch-vector for various detunings $\omega_0 - \omega_1$ is illustrated in Fig. 15.7. It should be noted that although an atom that is appreciably detuned from the exciting field undergoes a much smaller excitation cycle than one near resonance, the atom ends up back in the lower state $|1\rangle$ at the end of the pulse in every case.

This result has an important implication for an inhomogeneously broadened material medium or group of atoms, in which different atoms have different resonant frequencies ω_0. If these atoms are subjected to the hyperbolic secant

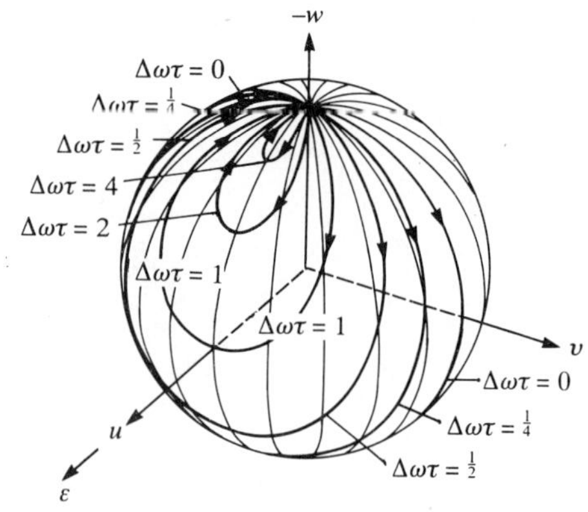

Fig. 15.7 The motion of the atomic Bloch-vector in response to a hyperbolic secant 2π-pulse for various detunings $\Delta\omega$. The atom starts and ends in the lower state $|1\rangle$. The coordinates u, v, w correspond to our r_1', r_2', r_3', but the w-axis is drawn with negative values up. (Reproduced from McCall and Hahn, 1969.)

2π-pulse given by Eq. (15.3–33) at frequency ω_1, each one undergoes a different cycle of excitation, but in such a way that all of them are back in the ground state when the pulse is over. As a result, no energy is absorbed out of the exciting pulse, despite the fact that the frequency of the light is close to the atomic resonance, where the medium is expected to be highly absorbing. As we shall see below, this is the basis of the phenomenon of *self-induced transparency* that was discovered by McCall and Hahn (1967, 1969).

15.4 Interaction between an atom and a quantum field – perturbative treatment

In all the foregoing problems we have regarded the electromagnetic field as an external field that is unaffected by the atom and may be treated classically. If the field is a strong, coherent field, such as might be produced by a laser, for example, this is an adequate approximation in many cases. However, an atom can also radiate spontaneously when it is partly excited, and such processes lie outside the domain of the semiclassical treatment given in Sections 15.1 to 15.3.

A number of attempts have been made to incorporate spontaneous emission into the framework of an essentially semiclassical theory. One approach has been to suppose that there exists a universal, fluctuating electromagnetic background that induces certain processes.‡ Another approach was adopted by Jaynes and his co-workers, (Jaynes and Cummings, 1963; Crisp and Jaynes, 1969; Stroud and Jaynes, 1970; Jaynes, 1973) who succeeded in showing that, when the radiation reaction of the field on the atom is taken into account, then a partially excited atom radiates even when the electromagnetic field is treated classically. However the detailed predictions of this 'neoclassical' theory have been contradicted by experiments (Gibbs, 1972, 1973; Wessner, Anderson and Robiscoe, 1972; Freedman and Clauser, 1972; Schuda, Herscher and Stroud, 1973). The only satisfactory treatment of the interaction between an atom and light is one in which both the atom and the field are quantized.

By combining Eq. (14.1–2) for the total energy of the coupled system of the two-level atom and the field

$$\hat{H} = \hat{H}_A + \hat{H}_F + \hat{H}_I,$$

with Eq. (15.1–16) for $\hat{H}_A$, Eq. (10.3–16) for $\hat{H}_F$, and Eq. (15.3–2) for $\hat{H}_I$, we arrive at

$$\hat{H} = \hbar\omega_0(\hat{R}_3 + \tfrac{1}{2}) + \sum_{\mathbf{k},s}\hbar\omega(\hat{n}_{\mathbf{k},s} + \tfrac{1}{2}) + i\omega_0[(\boldsymbol{\mu}_{12})\hat{b}(t) - (\boldsymbol{\mu}_{12}^*)\hat{b}^\dagger(t)]\cdot\hat{\mathbf{A}}(r_0, t), \tag{15.4–1}$$

where $\mathbf{r}_0$ is the position of the atom, which we take to be fixed. We have chosen the $-(e/m)\hat{\mathbf{p}}\cdot\hat{\mathbf{A}} \to -\dot{\hat{\boldsymbol{\mu}}}\cdot\hat{\mathbf{A}}$ form over the $-\hat{\boldsymbol{\mu}}\cdot\hat{\mathbf{E}}$ form of the interaction Hamiltonian for later convenience, although both forms work equally well for the perturbative treatment below, and we have taken the lower atomic state to be the ground state.

‡ See for example the papers by Marshall (1963, 1965) and Boyer (1969a,b, 1970a,b, 1973, 1974).

15.4.1 Absorption and emission of photons

As usual, we assume that the coupled system of the atom and the field is in a product state initially at time $t = 0$. However, in order to emphasize certain elementary quantum features of the process, we shall take the field to be in a Fock state this time rather than in the more commonly encountered coherent state. Then the initial density operator $\hat{\rho}_0$ of the coupled system has the form

$$\hat{\rho}_0 = |\{n\}\rangle\langle\{n\}| \otimes |\psi\rangle\langle\psi|, \tag{15.4–2}$$

in which the atomic state $|\psi\rangle$ is an arbitrary pure state, and the field has $n_{\mathbf{k}s}$ photons in the mode $\mathbf{k}$, s. We are interested in the probability of a photon of type $\mathbf{k}_1$, s_1 being either emitted or absorbed after a short time Δt, and this can be obtained by a simple perturbative treatment, as in Section 14.1.

As before, we shall find it convenient to work in the interaction picture, in which the evolution proceeds from time $t = 0$, and the interaction Hamiltonian is given by

$$\begin{aligned}\hat{H}_{\mathrm{I}}(t) &= \mathrm{i}\omega_0[\boldsymbol{\mu}_{12}\hat{b}\,\mathrm{e}^{-\mathrm{i}\omega_0 t} - \boldsymbol{\mu}_{12}^*\hat{b}^\dagger\,\mathrm{e}^{\mathrm{i}\omega_0 t}]\cdot\left[\frac{1}{L^{3/2}}\sum_{\mathbf{k},s}\left(\frac{\hbar}{2\omega\varepsilon_0}\right)^{1/2}\hat{a}_{\mathbf{k}s}\boldsymbol{\varepsilon}_{\mathbf{k}s}\,\mathrm{e}^{\mathrm{i}(\mathbf{k}\cdot\mathbf{r}_0-\omega t)} + \text{h.c.}\right]\\ &= \frac{\mathrm{i}\omega_0}{L^{3/2}}\sum_{\mathbf{k},s}\left(\frac{\hbar}{2\omega\varepsilon_0}\right)^{1/2}[\boldsymbol{\mu}_{12}\cdot\boldsymbol{\varepsilon}_{\mathbf{k}s}^*\hat{b}\hat{a}_{\mathbf{k}s}^\dagger\,\mathrm{e}^{-\mathrm{i}\mathbf{k}\cdot\mathbf{r}_0}\,\mathrm{e}^{\mathrm{i}(\omega-\omega_0)t} - \text{h.c.}].\end{aligned} \tag{15.4–3}$$

Here all operators without explicit time labels refer to time $t = 0$. We have discarded terms oscillating at optical or higher frequencies in the last line, because they make negligible contributions in any time interval Δt that is many optical periods long. As we are interested in the probability for the emission or absorption of a photon, the final state is necessarily orthogonal to the initial state, and we can make use of the general relation (14.1–14) for the transition probability in a short time Δt. With the help of Eqs. (15.4–2) and (15.4–3), and after summing over all possible final atomic states, we find that

Probability for the emission (or absorption) of a photon of type $\mathbf{k}_1$, s_1 in time Δt

$$\begin{aligned}&= \frac{-\omega_0^2}{\hbar^2 L^3}\sum_{j=1}^{2}\sum_{\mathbf{k},s}\sum_{\mathbf{k}',s'}\left(\frac{\hbar}{2\omega\varepsilon_0}\right)^{1/2}\left(\frac{\hbar}{2\omega'\varepsilon_0}\right)^{1/2}\\
&\times\Big\{\langle j|\hat{b}|\psi\rangle\langle\psi|\hat{b}|j\rangle(\boldsymbol{\mu}_{12}\cdot\boldsymbol{\varepsilon}_{\mathbf{k}s}^*)(\boldsymbol{\mu}_{12}\cdot\boldsymbol{\varepsilon}_{\mathbf{k}'s'}^*)\\
&\quad\times\mathrm{e}^{-\mathrm{i}(\mathbf{k}+\mathbf{k}')\cdot\mathbf{r}_0}\langle\{n\}, n_{\mathbf{k}_1s_1}\pm 1|\hat{a}_{\mathbf{k}s}^\dagger|\{n\}\rangle\langle\{n\}|\hat{a}_{\mathbf{k}'s'}^\dagger|\{n\}, n_{\mathbf{k}_1s_1}\pm 1\rangle\\
&\quad\times\int_0^{\Delta t}\mathrm{d}t_1\int_0^{t_1}\mathrm{d}t_2\,\mathrm{e}^{\mathrm{i}(\omega-\omega_0)t_1}\,\mathrm{e}^{\mathrm{i}(\omega'-\omega_0)t_2} + \text{c.c.}\\
&\quad - \langle j|\hat{b}|\psi\rangle\langle\psi|\hat{b}^\dagger|j\rangle(\boldsymbol{\mu}_{12}\cdot\boldsymbol{\varepsilon}_{\mathbf{k}s}^*)(\boldsymbol{\mu}_{12}^*\cdot\boldsymbol{\varepsilon}_{\mathbf{k}'s'})\\
&\quad\times\mathrm{e}^{\mathrm{i}(\mathbf{k}'-\mathbf{k})\cdot\mathbf{r}_0}\langle\{n\}, n_{\mathbf{k}_1s_1}\pm 1|\hat{a}_{\mathbf{k}s}^\dagger|\{n\}\rangle\langle\{n\}|\hat{a}_{\mathbf{k}'s'}|\{n\}, n_{\mathbf{k}_1s_1}\pm 1\rangle\\
&\quad\times\int_0^{\Delta t}\mathrm{d}t_1\int_0^{t_1}\mathrm{d}t_2\,\mathrm{e}^{\mathrm{i}(\omega-\omega_0)t_1}\,\mathrm{e}^{-\mathrm{i}(\omega'-\omega_0)t_2}\\
&\quad - \langle j|\hat{b}^\dagger|\psi\rangle\langle\psi|\hat{b}|j\rangle(\boldsymbol{\mu}_{12}^*\cdot\boldsymbol{\varepsilon}_{\mathbf{k}s})(\boldsymbol{\mu}_{12}\cdot\boldsymbol{\varepsilon}_{\mathbf{k}'s'}^*)\\
&\quad\times\mathrm{e}^{\mathrm{i}(\mathbf{k}-\mathbf{k}')\cdot\mathbf{r}_0}\langle\{n\}, n_{\mathbf{k}_1s_1}\pm 1|\hat{a}_{\mathbf{k}s}|\{n\}\rangle\langle\{n\}|\hat{a}_{\mathbf{k}'s'}^\dagger|\{n\}, n_{\mathbf{k}_1s_1}\pm 1\rangle\end{aligned}$$

$$\times \int_0^{\Delta t} dt_1 \int_0^{t_1} dt_2 \, e^{-i(\omega-\omega_0)t_1} e^{i(\omega'-\omega_0)t_2} \Bigg\} + \text{c.c.} \tag{15.4–4}$$

Of the four terms within the { } brackets, only the fourth makes an appreciable contribution if we are interested in the absorption of a photon, and only the third makes an appreciable contribution if we are interested in the emission of a photon. Also from Eqs. (10.4–5) and (10.4–8),

$$\left.\begin{aligned} \langle\{n\}, n_{\mathbf{k}_1 s_1} - 1|\hat{a}_{\mathbf{k}s}|\{n\}\rangle &= (n_{\mathbf{k}_1 s_1})^{1/2}\delta^3_{\mathbf{k}\mathbf{k}_1}\delta_{ss_1} \\ \langle\{n\}, n_{\mathbf{k}_1 s_1} + 1|\hat{a}^\dagger_{\mathbf{k}s}|\{n\}\rangle &= (n_{\mathbf{k}_1 s_1} + 1)^{1/2}\delta^3_{\mathbf{k}\mathbf{k}_1}\delta_{ss_1}. \end{aligned}\right\} \tag{15.4–5}$$

The time integrals are straightforward and are easily evaluated, and we find with the help of Eqs. (15.4–5) that

Probability for absorption of a photon of type $\mathbf{k}_1$, s_1 in time Δt

$$= \frac{\omega_0^2}{2\hbar\omega_1\varepsilon_0 L^3}|\boldsymbol{\mu}^*_{12}\cdot\boldsymbol{\varepsilon}_{\mathbf{k}_1 s_1}|^2\langle\psi|\hat{b}\hat{b}^\dagger|\psi\rangle n_{\mathbf{k}_1 s_1}\left[\frac{\sin\frac{1}{2}(\omega_1-\omega_0)\Delta t}{\frac{1}{2}(\omega_1-\omega_0)}\right]^2 \tag{15.4–6}$$

and that

Probability for emission of a photon of type $\mathbf{k}_1$, s_1 in time Δt

$$= \frac{\omega_0^2}{2\hbar\omega_1\varepsilon_0 L^3}|\boldsymbol{\mu}^*_{12}\cdot\boldsymbol{\varepsilon}_{\mathbf{k}_1 s_1}|^2\langle\psi|\hat{b}^\dagger\hat{b}|\psi\rangle (n_{\mathbf{k}_1 s_1}+1)\left[\frac{\sin\frac{1}{2}(\omega_1-\omega_0)\Delta t}{\frac{1}{2}(\omega_1-\omega_0)}\right]^2. \tag{15.4–7}$$

A number of interesting general conclusions follow from these expressions. In the first place, both probabilities are very small unless the frequency ω_1 of the photon in question is fairly close to the atomic frequency ω_0, so that the frequency difference $|\omega_1 - \omega_0|$ is of order $1/\Delta t$ or less. Next, the dipole moment vector must not be orthogonal to the polarization vector of the photon in question, or the probabilities vanish. For example, if the dipole moment $\boldsymbol{\mu}_{12}$ is real and points in the z-direction, the atom can neither emit nor absorb a photon whose wave vector points in the z-direction, because its polarization $\boldsymbol{\varepsilon}_{\mathbf{k}_1 s_1}$ would be orthogonal to $\boldsymbol{\mu}_{12}$. This is simply a reflection of the well-known properties of the dipole radiation pattern, which always vanishes in the direction of the dipole moment vector. On the other hand, if we are dealing with an atomic $\Delta m = -1$ transition, and we put $\boldsymbol{\mu}_{12} = (|\boldsymbol{\mu}_{12}|/\sqrt{2})(\mathbf{x}_1 + i\mathbf{y}_1)$, then we find from Eq. (15.4–7) that the photon emission probability vanishes for left circularly polarized photons with $\boldsymbol{\varepsilon}_{\mathbf{k}_1 s_1} = (i\mathbf{x}_1 + \mathbf{y}_1)/\sqrt{2}$, and is greatest for right circularly polarized photons with $\boldsymbol{\varepsilon}_{\mathbf{k}_1 s_1} = (\mathbf{x}_1 + i\mathbf{y}_1)/\sqrt{2}$. This is a consequence of the requirement for angular momentum conservation. For a non-vanishing absorption probability, the atom must not start off in the fully excited state $|\psi\rangle = |2\rangle$, and for a non-vanishing emission probability the atom must not start off in the ground state $|\psi\rangle = |1\rangle$. However, the atom can absorb or emit a photon from any other quantum state.

The absorption probability for a photon of type $\mathbf{k}_1$, s_1 is proportional to the number of photons $n_{\mathbf{k}_1 s_1}$ of that type initially present in the field, as might be expected. However, the emission probability for a photon of type $\mathbf{k}_1, s_1$ is proportional to $n_{\mathbf{k}_1 s_1} + 1$, and therefore has two contributions. The term proportional to $n_{\mathbf{k}_1 s_1}$ is known as the *stimulated emission* or *induced emission* probability. It is proportional to the intensity of the field associated with the mode $\mathbf{k}_1$, s_1 just like the absorption probability. The other term, on the other hand, is independent of the photon occupation number, and is non-zero even in the vacuum state. It is known as the *spontaneous emission* probability. Whereas stimulated emission and absorption processes are encountered even with a classical field, as was shown implicitly in Section 15.3, spontaneous emission is characteristic of a quantum field. The spontaneous process is occasionally described as induced by the vacuum fluctuations of the quantum field.

By summing the spontaneous emission probability given by Eq. (15.4–7) over all modes of the field, we arrive at the total probability for spontaneous photon emission in time Δt. To perform the sum over polarizations we make use of the tensor relation (10.2–19c) and write

$$\begin{aligned}\sum_s |\boldsymbol{\mu}_{12}^* \cdot \boldsymbol{\varepsilon}_{\mathbf{k}s}|^2 &= (\boldsymbol{\mu}_{12}^*)_i(\boldsymbol{\mu}_{12})_j \sum_s (\boldsymbol{\varepsilon}_{\mathbf{k}s})_i(\boldsymbol{\varepsilon}_{\mathbf{k}s}^*)_j \\ &= (\boldsymbol{\mu}_{12}^*)_i(\boldsymbol{\mu}_{12})_j\left(\delta_{ij} - \frac{k_i k_j}{k^2}\right) \\ &= |\boldsymbol{\mu}_{12}|^2 - \frac{|\mathbf{k}\cdot\boldsymbol{\mu}_{12}|^2}{k^2}. \end{aligned} \qquad (15.4\text{–}8)$$

We replace the sum over all wave vectors $\mathbf{k}$ by an integral according to the usual rule $(1/L^3)\sum_{\mathbf{k}} \to 1/(2\pi)^3 \int \mathrm{d}^3 k$. Let θ, ϕ be the polar and azimuthal angles of the $\mathbf{k}$-vector. If we suppose that $\boldsymbol{\mu}_{12}$ lies in the x, y-plane, we can put $\boldsymbol{\mu}_{12} = |\boldsymbol{\mu}_{12}|\mathbf{x}_1$ for a real dipole moment, and $\boldsymbol{\mu}_{12} = (|\boldsymbol{\mu}_{12}|/\sqrt{2})(\mathbf{x}_1 + \mathrm{i}\mathbf{y}_1)$ for a complex dipole moment. Then $|\boldsymbol{\mu}_{12}|^2 - |\mathbf{k}\cdot\boldsymbol{\mu}_{12}|^2/k^2$ becomes $|\boldsymbol{\mu}_{12}|^2(1 - \sin^2\theta\cos^2\phi)$ in the former case and $|\boldsymbol{\mu}_{12}|^2(1 - \frac{1}{2}\sin^2\theta)$ in the latter case. On integrating over all azimuthal angles ϕ we arrive at $2\pi|\boldsymbol{\mu}_{12}|^2(1 - \frac{1}{2}\sin^2\theta)$ in both cases, so that finally, from Eqs. (15.4–7) and (15.4–8) we obtain

Probability of spontaneous emission in time Δt

$$= \frac{\omega_0^2|\boldsymbol{\mu}_{12}|^2}{8\pi^2\varepsilon_0\hbar c^3}\int_0^\pi (1 - \tfrac{1}{2}\sin^2\theta)\sin\theta\,\mathrm{d}\theta \int_0^\infty \omega_1\left[\frac{\sin\frac{1}{2}(\omega_1 - \omega_0)\Delta t}{\frac{1}{2}(\omega_1 - \omega_0)}\right]^2 \langle\psi|\hat{b}^\dagger\hat{b}|\psi\rangle\,\mathrm{d}\omega_1.$$

The ω_1-integrand is strongly peaked near $\omega_1 = \omega_0$ for $\Delta t \gg 1/\omega_0$, so that the integral over ω_1 gives $2\pi\omega_0\Delta t$ to a good approximation. The θ-integral gives 4/3, so that finally, on dividing both sides by Δt, we have the result

$$\text{Rate of spontaneous emission} = \frac{1}{4\pi\varepsilon_0}\left(\frac{4|\boldsymbol{\mu}_{12}|^2\omega_0^3}{3\hbar c^3}\right)\langle\psi|\hat{b}^\dagger\hat{b}|\psi\rangle. \qquad (15.4\text{–}9)$$

If we express the state ψ in terms of upper and lower states with the help of Eq. (15.2–12), we immediately find that

$$\langle\psi|\hat{b}^\dagger\hat{b}|\psi\rangle = \langle\psi|(\hat{R}_3 + \tfrac{1}{2})|\psi\rangle = \cos^2\tfrac{1}{2}\theta = \tfrac{1}{2}(r_3 + 1), \qquad (15.4\text{–}10)$$

where θ is the polar angle of the atomic Bloch vector and r_3 is its z-component. This factor is, of course, a maximum when $\theta = 0$, i.e. when the atom is fully excited. The rate of spontaneous photon emission is then known as the *Einstein A-coefficient* (it is sometimes also denoted by 2β), and we have the result

$$A = 2\beta = \frac{1}{4\pi\varepsilon_0}\left(\frac{4}{3}\frac{|\boldsymbol{\mu}_{12}|^2\omega_0^3}{\hbar c^3}\right). \tag{15.4–11}$$

The reciprocal of this rate is a measure of the lifetime T_1 of the excited atomic state.

Although our perturbative treatment is strictly appropriate only for short interaction times Δt, we can use a simple energy balance argument to obtain information about the time development of the spontaneous emission process. As the emitted photons are peaked in frequency near $\omega_1 = \omega_0$, we can convert Eq. (15.4–9) into an approximate expression for the rate of energy emission by multiplying by $\hbar\omega_0$, with the result that

$$\text{Rate of energy emission by the atom} = \hbar\omega_0 A\tfrac{1}{2}(r_3 + 1). \tag{15.4–12}$$

If we equate this to the rate at which the atom is losing energy, we obtain an equation of motion for r_3. Thus, with the help of Eqs. (15.1–16) and (15.2–14) we have

$$\hbar\omega_0 A\tfrac{1}{2}(r_3 + 1) = -\frac{\mathrm{d}}{\mathrm{d}t}[\hbar\omega_0\tfrac{1}{2}(r_3 + 1)],$$

and this equation can be immediately integrated to give

$$\langle \hat{H}_\mathrm{A}(t)\rangle = \tfrac{1}{2}\hbar\omega_0[r_3(t) + 1] = \langle \hat{H}_\mathrm{A}(0)\rangle\, \mathrm{e}^{-At}. \tag{15.4–13}$$

The perturbative expression (15.4–12) for the rate of emission of energy holds, even though the field is not strictly in the vacuum state at all times, because most modes of the field remain empty throughout the atomic decay. The average atomic energy therefore decays exponentially in time at the rate A by spontaneous emission from the initial state. However, the smooth behavior of the expectation value for the energy conceals the quantum jumps associated with the photon emission process. This phenomenon is examined more explicitly in Section 15.6.

15.5 Interaction between an atom and a quantum field – non-perturbative treatment

15.5.1 Heisenberg equations of motion

In the preceding section we made use of time-dependent perturbation theory to calculate the transition rates for certain processes. Such calculations are necessarily limited to short interaction times, and they do not allow us to investigate time-dependent features of the process in general, even though we succeeded in obtaining some information about spontaneous atomic decay with the help of a simple energy balance argument. In order to treat the interaction process more generally as a function of time, and to calculate certain multi-time correlation functions, we shall find it convenient to make use of the Heisenberg picture. We start by deriving the Heisenberg equations of motion

$$\frac{d}{dt}\hat{O}(t) = \frac{1}{i\hbar}[\hat{O}(t), \hat{H}] \tag{15.5–1}$$

for certain key dynamical variables $\hat{O}(t)$, and integrate them over time. Expectation values can then be calculated for any assumed initial state.‡

We start with the system Hamiltonian $\hat{H}$ given by Eq. (15.4–1) for a field and an atom at a fixed position that we take to be the origin. The vector potential has the mode expansion

$$\hat{\mathbf{A}}(\mathbf{r}, t) = \frac{1}{L^{3/2}}\sum_{\mathbf{k},s}\left(\frac{\hbar}{2\omega\varepsilon_0}\right)^{1/2}[\hat{a}_{\mathbf{k}s}(t)\boldsymbol{\varepsilon}_{\mathbf{k}s}\,e^{i\mathbf{k}\cdot\mathbf{r}} + \text{h.c.}]. \tag{15.5–2}$$

With this choice of Hamiltonian, the Heisenberg equations of motion (15.5–1) for the atomic operators $\hat{b}(t)$ and $\hat{R}_3(t)$ take the forms

$$\frac{d}{dt}\hat{b}(t) = -i\omega_0\hat{b}(t) + \frac{2\omega_0}{\hbar}\boldsymbol{\mu}_{12}^*\cdot\hat{\mathbf{A}}(0, t)\hat{R}_3(t) \tag{15.5–3}$$

$$\frac{d}{dt}\hat{R}_3(t) = -\frac{\omega_0}{\hbar}[\boldsymbol{\mu}_{12}\hat{b}(t) + \boldsymbol{\mu}_{12}^*\hat{b}^\dagger(t)]\cdot\hat{\mathbf{A}}(0, t), \tag{15.5–4}$$

while the equations of motion for the electric and magnetic field operators reduce to Maxwell's equations,

$$\left.\begin{aligned}\nabla\times\hat{\mathbf{E}}(\mathbf{r}, t) &= -\frac{\partial}{\partial t}\hat{\mathbf{B}}(\mathbf{r}, t)\\ \nabla\times\hat{\mathbf{B}}(\mathbf{r}, t) &= \frac{1}{c^2}\frac{\partial}{\partial t}\hat{\mathbf{E}}(\mathbf{r}, t) + \mu_0\hat{\mathbf{j}}(\mathbf{r}, t).\end{aligned}\right\} \tag{15.5–5}$$

The effective current density $\hat{\mathbf{j}}(\mathbf{r}, t)$ is proportional to the transverse delta function and is given by (Kimble and Mandel, 1976)

$$\left.\begin{aligned}&\hat{\mathrm{j}}_l(\mathbf{r}, t) = -i\omega_0\delta_{lm}^T(\mathbf{r})(\boldsymbol{\mu}_{12})_m\hat{b}(t) + \text{h.c.}\\ &\text{i.e.}\\ &\hat{\mathbf{j}}(\mathbf{r}, t) = -i\omega_0\left[\frac{2}{3}\boldsymbol{\mu}_{12}\delta^3(\mathbf{r}) - \frac{\boldsymbol{\mu}_{12}}{4\pi r^3} + \frac{3(\boldsymbol{\mu}_{12}\cdot\mathbf{r})\mathbf{r}}{4\pi r^5}\right]\hat{b}(t) + \text{h.c.}\end{aligned}\right\} \tag{15.5–6}$$

As is well known, the Maxwell equations imply that the vector potential $\hat{\mathbf{A}}(\mathbf{r}, t)$ obeys the wave equation

$$\nabla^2\hat{\mathbf{A}}(\mathbf{r}, t) - \frac{1}{c^2}\frac{\partial^2\hat{\mathbf{A}}(\mathbf{r}, t)}{\partial t^2} = -\mu_0\hat{\mathbf{j}}(\mathbf{r}, t),$$

whose solution has the so-called retarded potential form

‡ The calculations in this section and in Section 15.6 are largely based on the papers by Ackerhalt, Knight and Eberly (1973); Wódkiewicz and Eberly (1976); Kimble and Mandel (1976). See also Allen and Eberly (1975), Chapter 7. In particular, the procedure of writing products of mixed operators in normal order was introduced by Wódkiewicz and Eberly (1976).

$$\hat{\mathbf{A}}(\mathbf{r}, t) = \frac{\mu_0}{4\pi}\int \frac{\hat{\mathbf{j}}(\mathbf{r}', t - |\mathbf{r} - \mathbf{r}'|/c)}{|\mathbf{r} - \mathbf{r}'|}\Theta(t - |\mathbf{r} - \mathbf{r}'|/c)\,\mathrm{d}^3r' + \hat{\mathbf{A}}_{\text{free}}(\mathbf{r}, t). \tag{15.5–7}$$

Here $\Theta(\tau)$ is the unit step function that vanishes for $\tau < 0$, and $\hat{\mathbf{A}}_{\text{free}}(\mathbf{r}, t)$ is the solution of the homogeneous or free-field wave equation. As always, the total field is obtained by adding the contribution produced by the source to the free-field solution.

With the help of Eq. (15.5–7), the electric or magnetic field at any space-time point $\mathbf{r}$, t may be expressed in terms of the atomic current $\hat{\mathbf{j}}(\mathbf{r}, t)$ or in terms of $\hat{b}$ and $\hat{b}^\dagger$. For example, for a point in the far field of the atom, we may show that, to the lowest order in $1/r$, and for $t > r/c$,

$$\hat{\mathbf{E}}(\mathbf{r}, t) = \frac{\omega_0^2}{4\pi\varepsilon_0 c^2}\left[\frac{\boldsymbol{\mu}_{12}}{r} - \frac{(\boldsymbol{\mu}_{12}\cdot\mathbf{r})\mathbf{r}}{r^3}\right]\hat{b}\left(t - \frac{r}{c}\right) + \text{h.c.} + \hat{\mathbf{E}}_{\text{free}}(\mathbf{r}, t). \tag{15.5–8}$$

This is similar to the well-known expression for the far-field of an oscillating classical dipole (cf. Born and Wolf, 1980, Sec. 2.2.3). In the derivation of Eq. (15.5–8) we have used the approximation of replacing the time derivative of $\hat{b}(t)$ by the dominant term $-\mathrm{i}\omega_0\hat{b}(t)$ given by Eq. (15.5–3). Equation (15.5–8) breaks naturally into two separate equations, corresponding to the positive frequency and the negative frequency parts of the field. However, it applies only to the far field, whereas Eqs. (15.5–3) and (15.5–4) involve the field at $\mathbf{r} = 0$.

In order to determine the field at the position of the atom, it is easier to start with the Heisenberg equation of motion for the single-mode operator $\hat{a}_{\mathbf{k}s}(t)$, namely

$$\frac{\mathrm{d}}{\mathrm{d}t}\hat{a}_{\mathbf{k}s}(t) = -\mathrm{i}\omega\hat{a}_{\mathbf{k}s}(t) + \frac{\omega_0}{\hbar}\frac{1}{L^{3/2}}\left(\frac{\hbar}{2\omega\varepsilon_0}\right)^{1/2}[\hat{b}(t)\boldsymbol{\mu}_{12} - \hat{b}^\dagger(t)\boldsymbol{\mu}_{12}^*]\cdot\boldsymbol{\varepsilon}_{\mathbf{k}s}^*, \tag{15.5–9}$$

and to synthesize the appropriate field operators from $\hat{a}_{\mathbf{k}s}(t)$. Before we attempt to integrate the equation, it is convenient to replace the atomic operator $\hat{b}(t)$, which oscillates close to the atomic frequency ω_0, according to Eq. (15.5–3), by a more slowly varying dynamical variable $\hat{b}_{\mathrm{s}}(t)$, defined by

$$\hat{b}_{\mathrm{s}}(t) \equiv \hat{b}(t)\,\mathrm{e}^{\mathrm{i}\omega_0 t}. \tag{15.5–10}$$

From Eq. (15.5–3), $\hat{b}_{\mathrm{s}}(t)$ obeys the equation of motion

$$\frac{\mathrm{d}}{\mathrm{d}t}\hat{b}_{\mathrm{s}}(t) = \frac{2\omega_0}{\hbar}[\hat{R}_3(t)\boldsymbol{\mu}_{12}^*\cdot\hat{\mathbf{A}}^{(+)}(0, t) + \boldsymbol{\mu}_{12}^*\cdot\hat{\mathbf{A}}^{(-)}(0, t)\hat{R}_3(t)]\,\mathrm{e}^{\mathrm{i}\omega_0 t}, \tag{15.5–11}$$

in which we have decomposed $\hat{\mathbf{A}}(\mathbf{r}, t)$ into its positive and negative frequency parts $\hat{\mathbf{A}}^{(+)}(\mathbf{r}, t)$ and $\hat{\mathbf{A}}^{(-)}(\mathbf{r}, t)$. Because equal-time atomic and field operators commute, the order of the operators in each term may be chosen at will. We have adopted a normal order for later convenience. When Eq. (15.5–10) is substituted in Eq. (15.5–9) and the equation is formally integrated, one obtains

$$\hat{a}_{\mathbf{k}s}(t) = \hat{a}_{\mathbf{k}s}(0)\,\mathrm{e}^{-\mathrm{i}\omega t} + \left(\frac{\omega_0}{\hbar}\right)\frac{1}{L^{3/2}}\left(\frac{\hbar}{2\omega\varepsilon_0}\right)^{1/2}\mathrm{e}^{-\mathrm{i}\omega t}$$
$$\times\ \boldsymbol{\varepsilon}_{\mathbf{k}s}^*\cdot\int_0^t[\boldsymbol{\mu}_{12}\hat{b}_{\mathrm{s}}(t')\,\mathrm{e}^{\mathrm{i}(\omega-\omega_0)t'} - \boldsymbol{\mu}_{12}^*\hat{b}_{\mathrm{s}}^\dagger(t')\,\mathrm{e}^{\mathrm{i}(\omega+\omega_0)t'}]\,\mathrm{d}t'. \tag{15.5–12}$$

This formula expresses the mode amplitude $\hat{a}_{\mathbf{k}s}(t)$ as the sum of a free-field amplitude and a contribution from the atomic source. By summing over all modes, we can derive the positive or negative frequency parts of $\hat{\mathbf{A}}(\mathbf{r}, t)$ or of any other field operator. Thus we have for example,

$$\begin{aligned}
&\boldsymbol{\mu}_{12}^* \cdot \hat{\mathbf{A}}^{(+)}(0, t) \\
&= \boldsymbol{\mu}_{12}^* \cdot \frac{1}{L^{3/2}} \sum_{\mathbf{k},s} \left(\frac{\hbar}{2\omega\varepsilon_0}\right)^{1/2} \hat{a}_{\mathbf{k}s}(t) \boldsymbol{\varepsilon}_{\mathbf{k}s} \\
&= \boldsymbol{\mu}_{12}^* \cdot \hat{\mathbf{A}}_{\text{free}}^{(+)}(0, t) + \frac{\omega_0}{\hbar} \frac{1}{L^3} \sum_{\mathbf{k},s} \left(\frac{\hbar}{2\omega\varepsilon_0}\right) \boldsymbol{\mu}_{12}^* \cdot \boldsymbol{\varepsilon}_{\mathbf{k}s} \\
&\quad \times \boldsymbol{\varepsilon}_{\mathbf{k}s}^* \cdot \int_0^t [\boldsymbol{\mu}_{12} \hat{b}_s(t') \, \mathrm{e}^{\mathrm{i}(\omega-\omega_0)(t'-t)} \, \mathrm{e}^{-\mathrm{i}\omega_0 t} - \boldsymbol{\mu}_{12}^* \hat{b}_s^\dagger(t') \, \mathrm{e}^{\mathrm{i}(\omega+\omega_0)(t'-t)} \, \mathrm{e}^{\mathrm{i}\omega_0 t}] \, \mathrm{d}t'.
\end{aligned} \tag{15.5–13}$$

We have written

$$\hat{\mathbf{A}}_{\text{free}}^{(+)}(0, t) \equiv \frac{1}{L^{3/2}} \sum_{\mathbf{k},s} \left(\frac{\hbar}{2\omega\varepsilon_0}\right)^{1/2} \hat{a}_{\mathbf{k}s}(0) \boldsymbol{\varepsilon}_{\mathbf{k}s} \, \mathrm{e}^{-\mathrm{i}\omega t} \tag{15.5–14}$$

for the homogeneous or free-field part of the vector potential operator in the absence of the atom. The sums over the polarization indices s yield, with the help of the tensorial relation for the polarization vectors [cf. Eq. (10.2–19c)], the results

$$\sum_s \boldsymbol{\mu}_{12}^* \cdot \boldsymbol{\varepsilon}_{\mathbf{k}s} \boldsymbol{\varepsilon}_{\mathbf{k}s}^* \cdot \boldsymbol{\mu}_{12} = |\boldsymbol{\mu}_{12}|^2 - |\boldsymbol{\mu}_{12} \cdot \mathbf{k}|^2/k^2$$

$$\sum_s \boldsymbol{\mu}_{12}^* \cdot \boldsymbol{\varepsilon}_{\mathbf{k}s} \boldsymbol{\varepsilon}_{\mathbf{k}s}^* \cdot \boldsymbol{\mu}_{12}^* = \boldsymbol{\mu}_{12}^{*2} - (\boldsymbol{\mu}_{12}^* \cdot \mathbf{k})^2/k^2,$$

and the sums over $\mathbf{k}$ can be replaced by an integral according to the usual rule

$$\frac{1}{L^3} \sum_{\mathbf{k}} \rightarrow \frac{1}{(2\pi)^3} \int \mathrm{d}^3 k \rightarrow \frac{1}{(2\pi)^3 c^3} \int_0^\infty \mathrm{d}\omega \, \omega^2 \int_0^\pi \mathrm{d}\theta \sin\theta \int_0^{2\pi} \mathrm{d}\phi.$$

The θ and ϕ integrals are easily evaluated as in Section 15.4, and we find that

$$\frac{1}{L^3} \sum_{\mathbf{k},s} \left(\frac{\hbar}{2\omega\varepsilon_0}\right) \boldsymbol{\mu}_{12}^* \cdot \boldsymbol{\varepsilon}_{\mathbf{k}s} \boldsymbol{\varepsilon}_{\mathbf{k}s}^* \cdot \boldsymbol{\mu}_{12} f(\omega) \rightarrow \frac{4}{3} \frac{|\boldsymbol{\mu}_{12}|^2}{4\pi^2 c^3} \int_0^\infty \left(\frac{\hbar\omega}{2\varepsilon_0}\right) f(\omega) \, \mathrm{d}\omega, \tag{15.5–15a}$$

whether $\boldsymbol{\mu}_{12}$ is real, as for a $\Delta m = 0$ transition, or complex, as for a $\Delta m = \pm 1$ transition. On the other hand, the second term under the summation in Eq. (15.5–13) yields

$$\left.\begin{aligned}
\frac{1}{L^3} \sum_{\mathbf{k},s} \left(\frac{\hbar}{2\omega\varepsilon_0}\right) \boldsymbol{\mu}_{12}^* \cdot \boldsymbol{\varepsilon}_{ks} \boldsymbol{\varepsilon}_{ks}^* \cdot \boldsymbol{\mu}_{12}^* f(\omega) &\rightarrow \frac{4}{3} \frac{|\boldsymbol{\mu}_{12}|^2}{4\pi^2 c^3} \int_0^\infty \left(\frac{\hbar\omega}{2\varepsilon_0}\right) f(\omega) \, \mathrm{d}\omega \\
&\qquad \text{for a } \Delta m = 0 \text{ transition,} \\
&\rightarrow 0 \quad \text{for a } \Delta m = \pm 1 \text{ transition.}
\end{aligned}\right\} \tag{15.5–15b}$$

For simplicity, let us therefore treat the case of a $\Delta m = \pm 1$ transition. In that case Eq. (15.5–13) reduces to

$$\boldsymbol{\mu}_{12}^* \cdot \hat{\mathbf{A}}^{(+)}(0, t) = \boldsymbol{\mu}_{12}^* \cdot \hat{\mathbf{A}}_{\text{free}}^{(+)}(0, t) + \frac{\hbar}{\omega_0}\left(\frac{2|\boldsymbol{\mu}_{12}|^2\omega_0^2}{4\pi^2\varepsilon_0 3\hbar c^3}\right) e^{-i\omega_0 t}\int_0^\infty d\omega \int_0^t dt'\,\omega\, e^{-i(\omega-\omega_0)(t-t')}\hat{b}_s(t'), \tag{15.5–16}$$

and an analogous expression for $\boldsymbol{\mu}_{12}^* \cdot \hat{\mathbf{A}}^{(-)}(0, t)$ may be derived in a similar manner:

$$\boldsymbol{\mu}_{12}^* \cdot \hat{\mathbf{A}}^{(-)}(0, t) = \boldsymbol{\mu}_{12}^* \cdot \hat{\mathbf{A}}_{\text{free}}^{(-)}(0, t) - \frac{\hbar}{\omega_0}\left(\frac{2|\boldsymbol{\mu}_{12}|^2\omega_0^2}{4\pi^2\varepsilon_0 3\hbar c^3}\right) e^{-i\omega_0 t}\int_0^\infty d\omega \int_0^t dt'\,\omega\, e^{i(\omega+\omega_0)(t-t')}\hat{b}_s(t'). \tag{15.5–17}$$

When these results are substituted in Eq. (15.5–11), we arrive at the equation

$$\begin{aligned}\frac{d}{dt}\hat{b}_s(t) &= \frac{2\omega_0}{\hbar}[\hat{R}_3(t)\boldsymbol{\mu}_{12}^* \cdot \hat{\mathbf{A}}_{\text{free}}^{(+)}(0, t) + \boldsymbol{\mu}_{12}^* \cdot \hat{\mathbf{A}}_{\text{free}}^{(-)}(0, t)\hat{R}_3(t)]\,e^{i\omega_0 t} \\ &+ \frac{4}{3}\frac{|\boldsymbol{\mu}_{12}|^2\omega_0^2}{4\pi^2\varepsilon_0\hbar c^3}\int_0^\infty d\omega\int_0^t dt'\,\omega[e^{-i(\omega-\omega_0)(t-t')}\hat{R}_3(t)\hat{b}_s(t') - e^{i(\omega+\omega_0)(t-t')}\hat{b}_s(t')\hat{R}_3(t)].\end{aligned} \tag{15.5–18}$$

This is an integro-differential equation for $\hat{b}_s(t)$ in which the only field operators are free-field or zero-time operators. The unknown field has therefore been effectively eliminated.

15.5.2 *Approximate solution–the Einstein A-coefficient and the Lamb shift*

Up to now the equations of motion follow directly from the Hamiltonian (15.4–1) without approximation. However, at this stage some approximations are desirable in order to simplify the calculation [see footnote in Section 15.5.1]. Let us expand $\hat{b}_s(t')$ under the integral in Eq. (15.5–18) in a Taylor series about $t' = t$,

$$\hat{b}_s(t') = \hat{b}_s(t) + (t' - t)\dot{\hat{b}}_s(t) + \dots, \tag{15.5–19}$$

and then substitute in Eq. (15.5–18). We shall only retain terms up to the first order in the fine structure constant $\alpha = e^2/4\pi\varepsilon_0\hbar c$ in the equation of motion for $\hat{b}_s(t)$. If we express $\boldsymbol{\mu}_{12}$ in the form $e\mathbf{r}_{12}$, where the vector $\mathbf{r}_{12}$ has a length of the order of an atomic electron orbit, then the factor $|\boldsymbol{\mu}_{12}|^2\omega_0^2/4\pi^2\varepsilon_0\hbar c^3 \approx 4\pi\alpha(r_{12}^2/\lambda_0^2)$, where λ_0 is the wavelength at the atomic resonance frequency ω_0. Terms of first order in the fine structure constant α are therefore of second order in $|\boldsymbol{\mu}_{12}|$. As $\hat{b}_s(t)$ is itself of order $|\boldsymbol{\mu}_{12}|$ or higher, only the first term in the expansion has to be retained, and, with the help of the relations

$$\hat{R}_3(t)\hat{b}_s(t) = -\tfrac{1}{2}\hat{b}_s(t) = -\hat{b}_s(t)\hat{R}_3(t),$$

Eq. (15.5–18) reduces to

$$\begin{aligned}\frac{d}{dt}\hat{b}_s(t) &= \frac{2\omega_0}{\hbar}[\hat{R}_3(t)\boldsymbol{\mu}_{12}^* \cdot \hat{\mathbf{A}}_{\text{free}}^{(+)}(0, t) + \boldsymbol{\mu}_{12}^* \cdot \hat{\mathbf{A}}_{\text{free}}^{(-)}(0, t)\hat{R}_3(t)]\,e^{i\omega_0 t} \\ &\quad - \frac{2|\boldsymbol{\mu}_{12}|^2\omega_0^2}{12\pi^2\varepsilon_0\hbar c^3}\hat{b}_s(t)\int_0^\infty d\omega\int_0^t d\tau\,\omega[e^{-i(\omega-\omega_0)\tau} + e^{i(\omega+\omega_0)\tau}]. \qquad (15.5\text{–}20)\end{aligned}$$

The net effect of the approximation we have just made is therefore to replace $\hat{b}_s(t')$ under the integral in Eq. (15.5–18) by its most recent value $\hat{b}_s(t)$, and this also reduces the integro-differential equation (15.5–18) to an ordinary differential equation. For this reason the procedure is sometimes known as the short memory or *Markovian* approximation. However, we emphasize that we have been led to Eq. (15.5–20) not by assuming short memory for the atomic system a priori, but by restricting our calculation to terms of order $|\boldsymbol{\mu}_{12}|^2$.

To evaluate the double integral we proceed in the following somewhat heuristic manner. We express the τ-integral as the difference between two integrals, by writing

$$\begin{aligned}\mathcal{I} \equiv \int_0^\infty d\omega\int_0^t d\tau\omega[e^{-i(\omega-\omega_0)\tau} + e^{i(\omega+\omega_0)\tau}] &= \int_0^\infty d\omega\int_0^\infty d\tau\omega[e^{-i(\omega-\omega_0)\tau} + e^{i(\omega+\omega_0)\tau}] \\ &\quad - \int_0^\infty d\omega\int_t^\infty d\tau\omega[e^{-i(\omega-\omega_0)\tau} + e^{i(\omega+\omega_0)\tau}].\end{aligned}$$

The first two τ-integrals yield the functions $2\pi\delta_+(\omega - \omega_0)$ and $2\pi\delta_-(\omega + \omega_0)$, respectively, that can be written explicitly in the form (see Section 3.1.1 and Appendix A4.1)

$$2\pi\delta_\pm(\omega) = \mp iP\left(\frac{1}{\omega}\right) + \pi\delta(\omega),$$

where P denotes the Cauchy principal part. The third double integral can be written as

$$\int_0^\infty d\omega\int_t^\infty d\tau\omega e^{-i(\omega-\omega_0)\tau} = \int_t^\infty d\tau\int_{-\omega_0}^\infty d\omega'\omega' e^{-i\omega'\tau} + \omega_0\int_t^\infty d\tau\int_{-\omega_0}^\infty d\omega' e^{-i\omega'\tau},$$

and, provided that $\omega_0 t \gg 1$, this yields

$$\begin{aligned}&\approx \int_t^\infty 2\pi i\delta'(\tau)\,d\tau + \omega_0\int_t^\infty 2\pi\delta(\tau)\,d\tau \\ &\approx 0.\end{aligned}$$

The last double integral in the expression for $\mathcal{I}$ can be treated in a similar manner and shown to yield a very small value also. Hence, to a good approximation,

$$\begin{aligned}\mathcal{I} &= \int_0^\infty \omega\left\{\pi\delta(\omega-\omega_0) + \pi\delta(\omega+\omega_0) - iP\left[\frac{1}{(\omega-\omega_0)} - \frac{1}{(\omega+\omega_0)}\right]\right\}d\omega \\ &= \pi\omega_0 - i\int_0^\infty \omega\left[P\frac{1}{(\omega-\omega_0)} - P\frac{1}{(\omega+\omega_0)}\right]d\omega,\end{aligned}$$

and substitution in Eq. (15.5–20) finally gives the equation of motion

$$\frac{\mathrm{d}}{\mathrm{d}t}\hat{b}_s(t) = \frac{2\omega_0}{\hbar}[\hat{R}_3(t)\boldsymbol{\mu}_{12}^* \cdot \hat{\mathbf{A}}_{\text{free}}^{(+)}(0, t) + \boldsymbol{\mu}_{12}^* \cdot \hat{\mathbf{A}}_{\text{free}}^{(-)}(0, t)\hat{R}_3(t)]\,\mathrm{e}^{\mathrm{i}\omega_0 t} + (-\beta + \mathrm{i}\gamma)\hat{b}_s(t). \tag{15.5–21}$$

Here

$$\beta \equiv \frac{2}{3}\frac{|\boldsymbol{\mu}_{12}|^2\omega_0^3}{4\pi\varepsilon_0\hbar c^3}$$

is the same decay constant that was encountered in Section 15.4 and γ is a frequency shift parameter, known as the *Lamb shift*, given by the formula

$$\gamma \equiv \frac{2}{3}\frac{|\boldsymbol{\mu}_{12}|^2\omega_0^2}{4\pi^2\varepsilon_0\hbar c^3}\int_0^\infty \omega P\left[\frac{1}{\omega - \omega_0} - \frac{1}{\omega + \omega_0}\right]\mathrm{d}\omega. \tag{15.2–22a}$$

The physical significance of γ will become clear when we examine the spectrum of the light radiated by the atom. The integral diverges logarithmically at the high frequency end, and therefore needs to be provided with a frequency cut-off to yield a finite value. Such a cut-off appears automatically in a relativistic calculation, where the upper limit is of order $mc^2/\hbar$ (m = electron mass). The expression (15.2–22a) is similar to one first calculated by Bethe (1947) for a real hydrogen atom, except that he obtained a series of terms of the type given by Eq. (15.2–22a) over all allowed transitions between one given level and all the remaining levels. Bethe calculated the Lamb shift between the 2s and 2p levels of atomic hydrogen from the series to be ~1040 MHz, in good agreement with experimental measurements (Lamb and Retherford, 1947). If we provide the integral in Eq. (15.5–22a) with an upper cut-off at $\omega = mc^2/\hbar$, it can be evaluated and gives the value $2\omega_0 \ln(mc^2/\hbar\omega_0)$ very nearly, in which case

$$\gamma \approx \beta\frac{2}{\pi}\ln\left(\frac{mc^2}{\hbar\omega_0}\right). \tag{15.5–22b}$$

However, real atoms have more than two energy levels, so that this value of γ, although finite, is a poor estimate of the Lamb shift for a real atom.

15.5.3 Integral equations of motion

The equation of motion for $\hat{R}_3(t)$ can now be treated in a similar manner. If we decompose $\hat{\mathbf{A}}(0, t)$ in Eq. (15.5–4) into its positive and negative frequency parts, arrange terms in normal order, then substitute for $\boldsymbol{\mu}_{12} \cdot \hat{\mathbf{A}}^{(+)}(0, t)$, etc. and make the same approximations as above, we arrive at the equation

$$\frac{\mathrm{d}}{\mathrm{d}t}\hat{R}_3(t) = -2\beta[\hat{R}_3(t) + \tfrac{1}{2}] - \frac{\omega_0}{\hbar}[\hat{b}_s^\dagger(t)\,\mathrm{e}^{\mathrm{i}\omega_0 t}\boldsymbol{\mu}_{12}^* \cdot \hat{\mathbf{A}}_{\text{free}}^{(+)}(0, t) + \text{h.c.} + \hat{b}_s(t)\,\mathrm{e}^{-\mathrm{i}\omega_0 t}\boldsymbol{\mu}_{12} \cdot \hat{\mathbf{A}}_{\text{free}}^{(+)}(0, t) + \text{h.c.}]. \tag{15.5–23}$$

The last two terms in this equation are anti-resonant and make a very small contribution when we integrate with respect to time. If we formally integrate Eqs. (15.5–21) and (15.5–23) over some finite time t that is many optical periods long, we can discard the contributions of the anti-resonant terms and we finally obtain the formulas (Kimble and Mandel, 1976)

$$\hat{b}_s(t) = \hat{b}_s(0)\,e^{(-\beta+i\gamma)t} + \frac{2\omega_0}{\hbar}\,e^{(-\beta+i\gamma)t}\int_0^t \hat{R}_3(t')\boldsymbol{\mu}_{12}^*\cdot\hat{\mathbf{A}}_{\text{free}}^{(+)}(0,t')\,e^{(\beta-i\gamma+i\omega_0)t'}\,dt' \tag{15.5–24}$$

$$\hat{R}_3(t) + \tfrac{1}{2} = [\hat{R}_3(0) + \tfrac{1}{2}]\,e^{-2\beta t} - \frac{\omega_0}{\hbar}\,e^{-2\beta t}\int_0^t [\hat{b}_s^\dagger(t')\boldsymbol{\mu}_{12}^*\cdot\hat{\mathbf{A}}_{\text{free}}^{(+)}(0,t')\,e^{(i\omega_0+2\beta)t'} + \text{h.c.}]\,dt'. \tag{15.5–25}$$

We note that the unknown electromagnetic field variables have been eliminated from these equations of motion, because $\hat{\mathbf{A}}_{\text{free}}^{(+)}(0,t)$ is actually made up of zero-time operators $\hat{a}_{\mathbf{k}s}(0)$ which behave as known quantities.

15.5.4 Spontaneous emission

Up to now we have derived equations of motion connecting various dynamical variables. In order to draw physical conclusions from these equations we have to calculate expectation values of the appropriate variables in the state under consideration. As an example, let us consider the time evolution of a system consisting of an atom in some quantum state described by the density operator $\hat{\rho}^{(A)}$, and a field in the vacuum state $|\text{vac}\rangle$. This is the problem of spontaneous emission. The density operator $\hat{\rho}$ of the coupled system then factorizes in the form

$$\hat{\rho} = \hat{\rho}^{(A)} \otimes |\text{vac}\rangle\langle\text{vac}|. \tag{15.5–26}$$

We also have the relations

$$\hat{a}_{\mathbf{k}s}(0)|\text{vac}\rangle = 0 = \boldsymbol{\mu}_{12}^*\cdot\hat{\mathbf{A}}_{\text{free}}^{(+)}(0,t)|\text{vac}\rangle, \tag{15.5–27}$$

because $\hat{\mathbf{A}}_{\text{free}}^{(+)}(0,t)$ is actually a function of the $\hat{a}_{\mathbf{k}s}(0)$ operators. If we use this relation to calculate expectation values of every term in Eq. (15.5–25), we find that the contribution of the integral vanishes, and we obtain the result

$$\langle\hat{R}_3(t)\rangle + \tfrac{1}{2} = [\langle\hat{R}_3(0)\rangle + \tfrac{1}{2}]\,e^{-2\beta t},$$

or from Eq. (15.1–16)

$$\langle\hat{H}_A(t)\rangle = \langle\hat{H}_A(0)\rangle\,e^{-2\beta t}. \tag{15.5–28}$$

The mean energy of the atom therefore decays exponentially in time at the rate $2\beta = A$, in agreement with Eq. (15.4–13).

In order to examine the time evolution of the field radiated by the atom we now calculate $\langle\hat{\mathbf{E}}^{(+)}(\mathbf{r},t)\rangle$ at some point $\mathbf{r}$ in the far field. If we decompose Eq. (15.5–8) into its positive and negative frequency parts we find at once that

$$\langle\hat{\mathbf{E}}^{(+)}(\mathbf{r},t)\rangle = \frac{\omega_0^2}{4\pi\varepsilon_0 c^2 r}\left[\boldsymbol{\mu}_{12} - \frac{(\boldsymbol{\mu}_{12}\cdot\mathbf{r})\mathbf{r}}{\mathbf{r}^2}\right]\left\langle\hat{b}\left(t-\frac{r}{c}\right)\right\rangle,$$

since $\hat{\mathbf{E}}^{(+)}_{\text{free}}(\mathbf{r}, t)|\text{vac}\rangle = 0$. After taking expectation values of both sides of Eq. (15.5–24), with the help of Eq. (15.5–27), and noting that the integral in Eq. (15.5–24) again does not contribute, we obtain the relation

$$\langle \hat{\mathbf{E}}^{(+)}(\mathbf{r}, t + r/c)\rangle = \langle \hat{\mathbf{E}}^{(+)}(\mathbf{r}, r/c)\rangle\, e^{-\beta t}\, e^{-i(\omega_0-\gamma)t}, \quad (t \geqslant 0). \quad (15.5\text{–}29)$$

r/c is of course the propagation time of the light from the atom to point $\mathbf{r}$. The mean electric field amplitude therefore decays exponentially at the rate β, but the frequency of the field is not centered on the frequency ω_0, but is displaced from ω_0 by γ, the Lamb shift. Evidence for this frequency shift was first provided by the celebrated experiments of Lamb and Retherford (1947), who measured a frequency shift of the order of 1000 MHz between the 2s and 2p levels of atomic hydrogen.

Of course, strictly speaking, we should not rely on Eq. (15.5–29) to yield the spectrum of the emitted light, because $\langle \hat{\mathbf{E}}^{(+)}(\mathbf{r}, r/c)\rangle$ may vanish and, in any case, the mean electric field is not usually measured in optical experiments. We can however derive the spectrum from the second-order correlation function of the field. If we again decompose Eq. (15.5–8) into its positive and negative frequency parts, we find that

$$\langle \hat{\mathbf{E}}^{(-)}(\mathbf{r}, t)\cdot \hat{\mathbf{E}}^{(+)}(\mathbf{r}, t+\tau)\rangle =$$

$$\left(\frac{\omega_0^2|\boldsymbol{\mu}_{12}|}{4\pi\varepsilon_0 c^2 r}\right)^2 (1 - \tfrac{1}{2}\sin^2\theta)\left\langle \hat{b}^\dagger\left(t - \frac{r}{c}\right)\hat{b}\left(t - \frac{r}{c} + \tau\right)\right\rangle, \quad t > \frac{r}{c}, \tau \geqslant 0, \quad (15.5\text{–}30)$$

where θ is the polar angle of the $\mathbf{r}$-vector and the complex dipole moment $\boldsymbol{\mu}_{12}$ lies in the x, y-plane. Now from Eq. (15.5–24) we immediately obtain, when we multiply $\hat{b}_s^\dagger(t)$ by $\hat{b}_s(t+\tau)$ and take expectation values, the result

$$\begin{aligned}\langle \hat{b}^\dagger(t)\hat{b}(t+\tau)\rangle &= \langle \hat{b}^\dagger(0)\hat{b}(0)\rangle\, e^{-2\beta t}\, e^{-\beta\tau}\, e^{-i(\omega_0-\gamma)\tau}\\ &= [\langle \hat{R}_3(0)\rangle + \tfrac{1}{2}]\, e^{-2\beta t}\, e^{-\beta\tau}\, e^{-i(\omega_0-\gamma)\tau}, \quad \tau \geqslant 0. \quad (15.5\text{–}31)\end{aligned}$$

The correlation function therefore decays exponentially with τ and it is shifted with respect to the atomic frequency ω_0 by the Lamb shift γ. Of course, the correlation function also depends on t, so that the radiation process is not stationary. However, if we average over many different atoms that are in various states of excitation at various times, the resulting field may be stationary. The spectral density $\Phi(\mathbf{r}, \omega)$ corresponding to the exponential correlation function given by Eqs. (15.5–30) and (15.5–31) is Lorentzian,

$$\Phi(\mathbf{r}, \omega) \propto \frac{1}{\beta^2 + (\omega - \omega_0 + \gamma)^2},$$

with a full width at half-maximum of 2β. Also, on putting $\tau = 0$ in Eqs. (15.5–30) and (15.5–31) we obtain for the mean light intensity the expression

$$\begin{aligned}\langle \hat{I}(\mathbf{r}, t)\rangle &= \langle \hat{\mathbf{E}}^{(-)}(\mathbf{r}, t)\cdot \hat{\mathbf{E}}^{(+)}(\mathbf{r}, t)\rangle\\ &= \left(\frac{\omega_0^2|\boldsymbol{\mu}_{12}|}{4\pi\varepsilon_0 c^2 r}\right)^2 (1 - \tfrac{1}{2}\sin^2\theta)[\langle \hat{R}_3(0)\rangle + \tfrac{1}{2}]\, e^{-2\beta(t-r/c)} \quad (t \geqslant r/c).\end{aligned} \quad (15.5\text{–}32)$$

If the atom is fully excited initially, then $\langle \hat{R}_3(0) \rangle + \frac{1}{2} = 1$, whereas the same factor has a value between 0 and 1 in all other cases. The predicted variation of the fluorescent light intensity with the excitation $\langle \hat{R}_3(0) \rangle + \frac{1}{2}$ has been confirmed experimentally (Gibbs, 1973), and so has the exponential decay law, over several orders of magnitude, in measurements of the Lyman α radiation from atomic hydrogen (Wessner, Anderson and Robiscoe, 1972). The angular distribution implied by the factor $(1 - \frac{1}{2}\sin^2\theta)$ is similar to the radiation pattern produced by a classical oscillating dipole.

15.6 Resonance fluorescence

The physical consequences of the coupled equations (15.5–24) and (15.5–25) become considerably more varied and interesting when the atom is in the presence of an external electromagnetic field, whose frequency is near the atomic resonance frequency. Such a situation occurs, for example, when the atom is exposed to the spontaneous radiation from a group of similar atoms, or to a laser beam of similar frequency, and the resulting phenomenon is known as *resonance fluorescence*. With the development of tunable lasers in recent years, it has become relatively easy to excite a given atomic resonance by an almost monochromatic, coherent optical field, and to study the fluorescence. Let us now examine how the atom responds in this situation. In a sense, this is the Rabi problem again, which was treated semiclassically in section 15.3, but which is now to be analyzed in the framework of a fully quantized theory.

We consider an atom located at the origin, with the complex transition dipole moment $\boldsymbol{\mu}_{12} = |\boldsymbol{\mu}_{12}|(\mathbf{x}_1 + \mathrm{i}\mathbf{y}_1)/\sqrt{2}$ lying in the x,y-plane, where $\mathbf{x}_1$, $\mathbf{y}_1$ are unit vectors (see Fig. 15.8). Let $\mathbf{r}$ be an arbitrary point in the far field of the atomic dipole, with polar coordinates (r, θ, ϕ). We shall again take the initial state of the coupled system to be a product state, but the optical field is now assumed to be in a coherent state $|\{v\}\rangle$, such as might be produced by an ideal laser. From Eq. (11.11–2) this state is a right eigenstate of $\hat{\mathbf{A}}^{(+)}_{\text{free}}(\mathbf{r}, t)$, and for an ideal monochromatic laser we take

$$\hat{\mathbf{A}}^{(+)}_{\text{free}}(0, t)|\{v\}\rangle = \boldsymbol{\varepsilon}\mathscr{A}\,\mathrm{e}^{-\mathrm{i}(\omega_1 t - \phi)}|\{v\}\rangle, \tag{15.6–1}$$

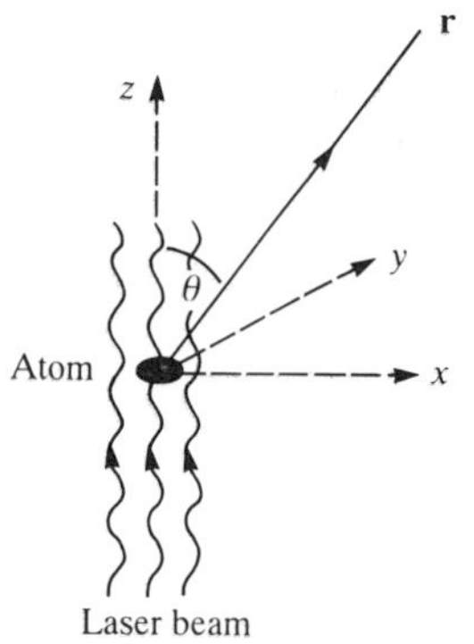

Fig. 15.8 The geometry for the resonance fluorescence problem.

where ω_1 is the laser frequency, $\mathscr{A}$ is a real amplitude, ϕ is a phase and $\boldsymbol{\varepsilon}$ is a complex unit polarization vector that represents circular polarization in the x,y-plane. To make the situation more explicit we shall assume that the atom is initially in the ground state $|1\rangle$, so that the state of the coupled system is the direct product

$$|\psi\rangle = |1\rangle|\{v\}\rangle. \tag{15.6–2}$$

15.6.1 Time development of the light intensity

If we decompose Eq. (15.5–8) into its positive and negative frequency parts, as before, and form the scalar product in normal order, we find for the average electric field intensity at some space-time point $\mathbf{r}$, t in the far field of the radiating atom, the expression

$$\begin{aligned}\langle \hat{I}(\mathbf{r}, t)\rangle &= \langle \hat{\mathbf{E}}^{(-)}(\mathbf{r}, t)\cdot\hat{\mathbf{E}}^{(+)}(\mathbf{r}, t)\rangle \\ &= \left(\frac{\omega_0^2|\boldsymbol{\mu}_{12}|}{4\pi\varepsilon_0 c^2 r}\right)^2(1 - \tfrac{1}{2}\sin^2\theta)\left\langle \hat{b}^\dagger\left(t - \frac{r}{c}\right)\hat{b}\left(t - \frac{r}{c}\right)\right\rangle \\ &= \left(\frac{\omega_0^2|\boldsymbol{\mu}_{12}|}{4\pi\varepsilon_0 c^2 r}\right)^2(1 - \tfrac{1}{2}\sin^2\theta)\left[\left\langle \hat{R}_3\left(t - \frac{r}{c}\right)\right\rangle + \tfrac{1}{2}\right],\end{aligned} \tag{15.6–3}$$

provided the point $\mathbf{r}$ is chosen so that the external or applied field vanishes there, i.e.

$$\hat{\mathbf{E}}^{(+)}_{\text{free}}(\mathbf{r}, t)|\{v\}\rangle = 0. \tag{15.6–4}$$

Equation (15.6–3) is equivalent to Eq. (15.5–30) with $\tau = 0$, because both equations depend only on the requirement, expressed by Eq. (15.6–4), that there be zero external field at $\mathbf{r}$, t. Apart from a scale factor, the mean fluorescent light intensity is therefore given by $\langle \hat{R}_3(t - r/c)\rangle + \frac{1}{2}$, which is also proportional to the mean atomic energy. Of course, r/c represents the propagation time of the light from the atom to the point $\mathbf{r}$.

If we take expectation values with respect to the state $|\psi\rangle$ of each term in Eqs. (15.5–24) and (15.5–25), we find that all the field operators $\hat{\mathbf{A}}^{(+)}_{\text{free}}(0, t)$ and $\hat{\mathbf{A}}^{(-)}_{\text{free}}(0, t)$ under the integral signs can be replaced by their right and left eigenvalues, because of the normal ordering. The two equations remain coupled, but they can be easily uncoupled, for example, by substituting for $\langle \hat{b}_s(t)\rangle$ and $\langle \hat{b}_s^\dagger(t)\rangle$ from Eq. (15.5–24) and its conjugate in Eq. (15.5–25). After integrating the double integral by parts, we arrive at the following integral equation for $\langle \hat{R}_3(t)\rangle$:

$$\langle \hat{R}_3(\tau)\rangle = y(\tau) + \int_0^\tau K(\tau - t')\langle \hat{R}_3(t')\rangle\,\mathrm{d}t'. \tag{15.6–5}$$

Here $y(\tau) = -\frac{1}{2}$ and the kernel $K(\tau)$ is given by the expression

$$K(\tau) = \frac{\Omega^2}{\beta(1 + D^2)}[\mathrm{e}^{-2\beta\tau} - \mathrm{e}^{-\beta\tau}(\cos\beta D\tau + D\sin\beta D\tau)], \tag{15.6–6}$$

with

$$\left.\begin{aligned}\Omega &\equiv \frac{2\omega_0|\boldsymbol{\mu}_{12}^*\cdot\boldsymbol{\varepsilon}|\mathscr{A}}{\hbar}; \quad D \equiv \frac{\omega_1-\omega_0+\gamma}{\beta}\\ \beta &\equiv \frac{1}{4\pi\varepsilon_0}\frac{2|\boldsymbol{\mu}_{12}|^2\omega_0^3}{3\hbar c^3}.\end{aligned}\right\} \quad (15.6\text{–}7)$$

The parameter Ω will be recognized as the Rabi frequency, and it is completely equivalent to the parameter defined by Eq. (15.3–12) in the semiclassical treatment. D is a measure of the detuning of the applied field from the Lamb-shifted resonance frequency $\omega_0 - \gamma$ in units of β.

The integral equation (15.6–5) is of the Volterra type, with a kernel that depends only on the difference between the two time arguments. Such an equation can be easily solved by a Laplace transform method. If we denote the Laplace transforms of $\langle \hat{R}_3(\tau)\rangle$ and $K(\tau)$ by $\widetilde{R}_3(p)$ and $\widetilde{K}(p)$, respectively, i.e.

$$\left.\begin{aligned}\widetilde{R}_3(p) &\equiv \int_0^\infty \langle \hat{R}_3(\tau)\rangle\, \mathrm{e}^{-p\tau}\,\mathrm{d}\tau\\ \widetilde{K}(p) &\equiv \int_0^\infty K(\tau)\,\mathrm{e}^{-p\tau}\,\mathrm{d}\tau,\end{aligned}\right\} \quad (15.6\text{–}8)$$

and take the Laplace transform of each term in Eq. (15.6–5), we obtain the equation

$$\widetilde{R}_3(p) = -\frac{1}{2p} + \widetilde{K}(p)\widetilde{R}_3(p),$$

which can be solved for $\widetilde{R}_3(p)$ to give

$$\widetilde{R}_3(p) = \frac{1}{2p[\widetilde{K}(p)-1]}. \quad (15.6\text{–}9)$$

The inverse Laplace transformation then yields the solution

$$\langle \hat{R}_3(\tau)\rangle = \frac{1}{2\pi\mathrm{i}}\int_{-\mathrm{i}\infty+p_r}^{\mathrm{i}\infty+p_r} \frac{\mathrm{e}^{p\tau}}{2p[\widetilde{K}(p)-1]}\,\mathrm{d}p, \quad \tau \geqslant 0, \quad (15.6\text{–}10)$$

where p_r is a constant chosen so that all the singularities of the integrand lie to the left of the line $\mathrm{Re}\,(p) = p_r$ in the complex p-plane. Despite the relative simplicity of this integral and of the kernel $\widetilde{K}(p)$, the expression for $\langle \hat{R}_3(t)\rangle$ is fairly complicated in general, and we only state the result. It may be shown that (Kimble and Mandel, 1976)

$$\langle \hat{R}_3(t)\rangle + \tfrac{1}{2} = \frac{\tfrac{1}{4}\Omega^2/\beta^2}{(\tfrac{1}{2}\Omega^2/\beta^2)+1+D^2} - \sum_{\substack{i=1\\ j\neq i\neq k}}^{3} \frac{(2\beta+p_i)[(\beta+p_i)^2+\beta^2D^2]\,\mathrm{e}^{p_it}}{2p_i(p_i-p_j)(p_i-p_k)}, \quad (15.6\text{–}11a)$$

where p_1, p_2, p_3 are the roots of the cubic equation

$$p^3 + 4\beta p^2 + (5\beta^2 + \beta^2D^2 + \Omega^2)p + 2\beta^3 + 2\beta^3D^2 + \Omega^2\beta = 0. \quad (15.6\text{–}11b)$$

We shall carry out the detailed calculation only for the simpler case of zero detuning, when the three roots effectively reduce to two. On putting $D = 0$ in Eq. (15.6–6) and taking the Laplace transform we immediately obtain the expression

$$\begin{aligned}\widetilde{K}(p) &= \left(\frac{\Omega^2}{\beta}\right)\left[\frac{1}{p+2\beta} - \frac{1}{p+\beta}\right] \\ &= \frac{-\Omega^2}{(p+\beta)(p+2\beta)},\end{aligned}$$

so that

$$\frac{1}{\widetilde{K}(p)-1} = \frac{-(p+\beta)(p+2\beta)}{(p+\beta)(p+2\beta)+\Omega^2} = \frac{-(p+\beta)(p+2\beta)}{(p+\frac{3}{2}\beta+\mathrm{i}\Omega')(p+\frac{3}{2}\beta-\mathrm{i}\Omega')}, \tag{15.6–12}$$

where $\Omega' \equiv (\Omega^2 - \frac{1}{4}\beta^2)^{1/2}$. After substituting this result in Eq. (15.6–10), and applying the Cauchy residue theorem to the integral we obtain the formula

$$\begin{aligned}\langle \hat{R}_3(t)\rangle &= \frac{-2\beta^2}{2(2\beta^2+\Omega^2)} + \left[\frac{(\frac{1}{2}\beta+\mathrm{i}\Omega')(\frac{1}{2}\beta-\mathrm{i}\Omega')}{2(-\frac{3}{2}\beta-\mathrm{i}\Omega')(-2\mathrm{i}\Omega')}\,\mathrm{e}^{(-3\beta/2-\mathrm{i}\Omega')t} + \text{c.c.}\right] \\ &= \left[\frac{-\frac{1}{2}}{(\frac{1}{2}\Omega^2/\beta^2)+1}\right]\left[1 + (\tfrac{1}{2}\Omega^2/\beta^2)\,\mathrm{e}^{-3\beta t/2}\left(\cos\Omega' t + \frac{3\beta}{2\Omega'}\sin\Omega' t\right)\right],\end{aligned}$$

so that the average energy of the atom at time t is

$$\begin{aligned}\langle \hat{H}_{\mathrm{A}}(t)\rangle &= \hbar\omega_0[\langle \hat{R}_3(t)\rangle + \tfrac{1}{2}] \\ &= \hbar\omega_0\left[\frac{\frac{1}{4}\Omega^2/\beta^2}{(\frac{1}{2}\Omega^2/\beta^2)+1}\right]\left[1 - \mathrm{e}^{-3\beta t/2}\left(\cos\Omega' t + \frac{3\beta}{2\Omega'}\sin\Omega' t\right)\right].\end{aligned} \tag{15.6–13}$$

Once $\langle \hat{R}_3(t)\rangle + \frac{1}{2}$ is known, the average fluorescent light intensity $\langle \hat{I}(\mathbf{r}, t + r/c)\rangle$ at any point $\mathbf{r}$ in the far field, which is also proportional to $\langle \hat{R}_3(t)\rangle + \frac{1}{2}$, is obtainable immediately from Eq. (15.6–3).

$\langle \hat{R}_3(t)\rangle + \frac{1}{2}$, and also the light intensity, starts from zero at time $t = 0$, as is to be expected for an initially unexcited atom, and then grows towards its steady-state value. The detailed time development is illustrated in Fig. 15.9 for several different exciting field strengths and detunings, and this behavior has been confirmed experimentally (see Fig. 15.12) below. On resonance, in a relatively weak exciting field, the approach to the steady state is approximately exponential, but it becomes damped oscillatory in a strong field with $\Omega/\beta \gg 1$. This is a reflection of the fact that the atom undergoes *Rabi oscillations*, as we found from semiclassical arguments in Section 15.3. The oscillatory behavior becomes more pronounced as the detuning is increased, although the excitation is then smaller. However, the quantum fluctuations of the field always cause damping on the average, so that the expectation value $\langle \hat{R}_3(t)\rangle + \frac{1}{2}$ tends towards a constant after a long time, with an upper bound of $\frac{1}{2}$ in a very strong field.

The implications of this statement should not be misinterpreted. Although the atomic energy and the fluorescent light intensity become constant in time *on the average*, this is the result of dephasing of the Rabi oscillations. Actually the atomic excitation continues to undergo Rabi oscillation indefinitely. This is illustrated more clearly by the spectral density of the fluorescence, which exhibits

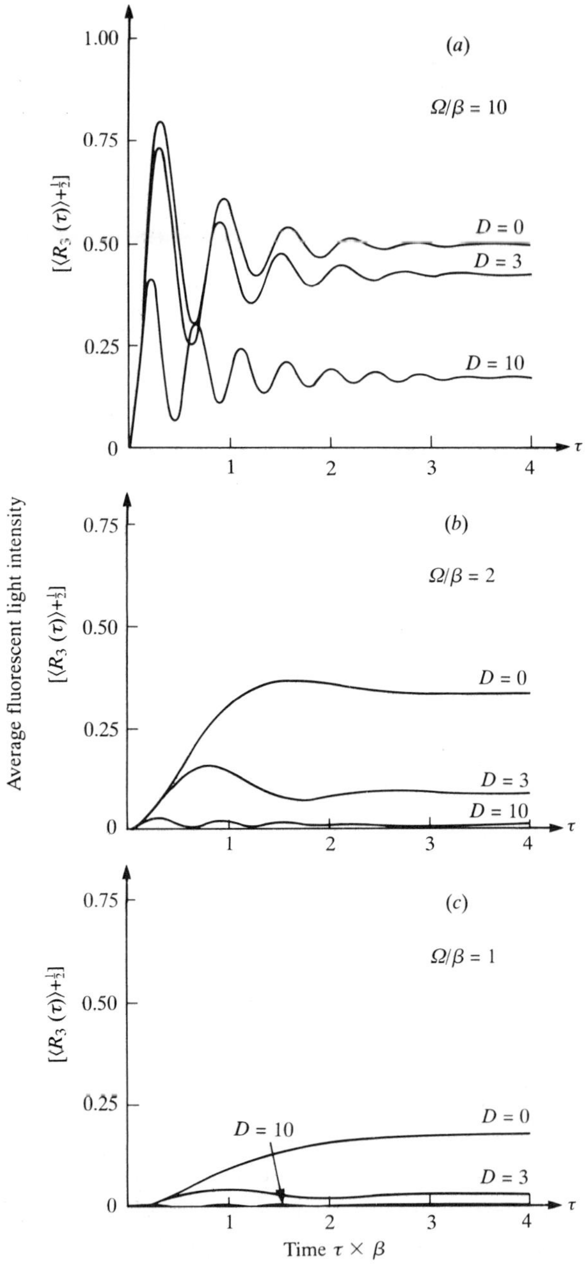

Fig. 15.9 Time development of the fluorescent light intensity in units of $1/\beta$ produced by a driven two-level atom, for various strengths Ω/β of the exciting field and for various detunings D. (Figures (*a*) and (*b*) are reproduced from Kimble and Mandel, 1976.)

symmetric sidebands in the steady state, reflecting the periodic modulation of the excitation, as we show more explicitly below. The steady-state value of the atomic energy $\hbar\omega_0[\langle \hat{R}_3(\infty)\rangle + \frac{1}{2}]$ tends to $\frac{1}{2}\hbar\omega_0$ in a strong field, corresponding to a half-excited atom, while the atomic excitation actually oscillates between 0 and 1.

It is possible to derive a damped oscillatory solution of the form (15.6–11a) directly from the semiclassical Bloch equations (15.3–19), when phenomenological damping terms are added in an ad hoc manner. This was first accomplished by Torrey (1949).

Finally, we point out that when t is very great and $\langle \hat{R}_3(t)\rangle$ takes on its asymptotic value, $\langle \hat{E}^{(+)}(\mathbf{r}, t)\rangle$, which is proportional to $\langle \hat{b}(t - r/c)\rangle$, also takes on a simple form. From Eq. (15.5–24) we find immediately with the help of Eq. (15.6–1), in the long-time limit ($t \to \infty$),

$$\langle \hat{b}(t)\rangle \to -\frac{\frac{1}{2}(\Omega/\beta)(1 + \mathrm{i}D)}{(\frac{1}{2}\Omega^2/\beta^2) + 1 + D^2}\,\mathrm{e}^{-\mathrm{i}(\omega_1 t - \phi)}. \tag{15.6–14}$$

This formula shows that the expectation value of the fluorescent field is directly proportional to the eigenvalue of the applied field given by Eq. (15.6–1), and that it follows the oscillations of the applied field. The fluorescent field is therefore not strictly stationary even in the long-time limit, although it has some of the properties of a stationary field. We shall refer to it as *quasi-stationary*. However, the average field does not tell the full story. An important feature of the phenomenon is that the fluctuations of the fluorescent field do not die out in the long-time limit. If we calculate the dimensionless ratio

$$[\langle \hat{\mathbf{E}}^{(-)}\cdot\hat{\mathbf{E}}^{(+)}\rangle - \langle \hat{\mathbf{E}}^{(-)}\rangle\cdot\langle \hat{\mathbf{E}}^{(+)}\rangle]/\langle \hat{\mathbf{E}}^{(-)}\cdot\hat{\mathbf{E}}^{(+)}\rangle,$$

we find that it is non-zero and tends to $\frac{1}{2}\Omega^2/\beta^2[(\frac{1}{2}\Omega^2/\beta^2) + 1 + D^2]$ in the long-time limit, which is close to unity for a sufficiently strong applied field.

The fluorescent field is therefore not coherent in the steady state, but is subject to quantum fluctuations. These quantum fluctuations are perhaps most evident in the spectrum of the fluorescence, which is not monochromatic in the steady state, despite the monochromaticity of the driving field.

15.6.2 The atomic scattering cross-section

From Eqs. (15.6–3) and (15.6–13) it is clear that, after a long time, in the steady state, the fluorescent light intensity of the electric field becomes

$$\langle \hat{I}(\mathbf{r}, \infty)\rangle = \left(\frac{\omega_0^2|\boldsymbol{\mu}_{12}|}{4\pi\varepsilon_0 c^2 r}\right)^2 (1 - \tfrac{1}{2}\sin^2\theta)\frac{\frac{1}{4}\Omega^2/\beta^2}{(\frac{1}{2}\Omega^2/\beta^2) + 1 + D^2}.$$

We can convert this into an expression for the rate of emission of photons, by recalling that $2\varepsilon_0\langle \hat{\mathbf{E}}^{(-)}(\mathbf{r}, t)\cdot\hat{\mathbf{E}}^{(+)}(\mathbf{r}, t)\rangle = 2\varepsilon_0\langle \hat{I}(\mathbf{r}, t)\rangle$ is the energy density at the space-time point $\mathbf{r}$, t. By multiplying $2\varepsilon_0\langle \hat{I}(\mathbf{r}, \infty)\rangle$ by $c/\hbar\omega_0$ we arrive at the flux density of fluorescent photons in the steady state,

$$\begin{aligned}\text{fluorescent photon flux density at } \mathbf{r} &= \frac{2\varepsilon_0\langle \hat{I}(\mathbf{r}, \infty)\rangle c}{\hbar\omega_0}\\ &= \frac{3\beta}{4\pi r^2}\left[\frac{\frac{1}{4}\Omega^2/\beta^2}{(\frac{1}{2}\Omega^2/\beta^2) + 1 + D^2}\right][1 - \tfrac{1}{2}\sin^2\theta],\end{aligned} \tag{15.6–15}$$

where we have made use of the definition (15.6–7) of β. Integration of this flux density over a sphere of radius r then yields the total photon flux, i.e. the rate of photon emission,

$$\begin{aligned}\text{rate of emission of fluorescent photons} &= \frac{3\beta}{4\pi}\left[\frac{\frac{1}{4}\Omega^2/\beta^2}{(\frac{1}{2}\Omega^2/\beta^2)+1+D^2}\right]\int_0^{2\pi}\mathrm{d}\phi\int_0^{\pi}\mathrm{d}\theta\sin\theta[1-\tfrac{1}{2}\sin^2\theta] \\ &= \beta\left[\frac{\frac{1}{2}\Omega^2/\beta^2}{(\frac{1}{2}\Omega^2/\beta^2)+1+D^2}\right]. \qquad (15.6\text{–}16)\end{aligned}$$

This expression tends towards $\beta = A/2$ in a sufficiently strong exciting field, as expected for an atom that oscillates between the excited state and the lower state.

We can obtain an expression for the effective scattering cross-section of the atom for the incident photons, by comparing Eq. (15.6–15) for the fluorescent flux density with the corresponding expression for the photon flux density incident on the atom. This gives

$$\text{incident photon flux density} = \frac{2\varepsilon_0 c}{\hbar\omega_0}\langle\{v\}|\hat{\mathbf{E}}^{(-)}_{\text{free}}(0,t)\cdot\hat{\mathbf{E}}^{(+)}_{\text{free}}(0,t)|\{v\}\rangle,$$

and with the help of Eq. (15.6–1) this reduces to

$$= \frac{2\varepsilon_0 c}{\hbar\omega_0}\mathscr{A}^2\omega_0^2.$$

By making use of the definitions (15.6–7) of the Rabi frequency parameter Ω and the decay constant β, we can express the result in the form

$$\text{incident photon flux density} = \frac{\pi\beta}{3\lambda^2}\frac{\Omega^2}{\beta^2},$$

where λ is the wavelength of the incident light. The ratio of the fluorescent to the incident flux density, multiplied by r^2, gives the effective scattering cross-section $\sigma(\theta)$ of the atom for the incident photons. We therefore have, for scattering through an angle θ,

$$\sigma(\theta) = \frac{9\lambda^2}{16\pi^2}\frac{1-\frac{1}{2}\sin^2\theta}{(\frac{1}{2}\Omega^2/\beta^2)+1+D^2}. \qquad (15.6\text{–}17)$$

The cross-section is greatest when $\theta = 0$, but it is non-zero even for $\theta = 90°$ in the scattering of circularly polarized light by an atom undergoing a $\Delta m = \pm 1$ transition.

Finally, on integrating $\sigma(\theta)$ over all angles we arrive at the total scattering cross-section σ_{T}:

$$\begin{aligned}\sigma_{\mathrm{T}} &= \frac{9\lambda^2}{16\pi^2}\left[\frac{1}{(\frac{1}{2}\Omega^2/\beta^2)+1+D^2}\right]\int_0^{2\pi}\mathrm{d}\phi\int_0^{\pi}\mathrm{d}\theta\sin\theta(1-\tfrac{1}{2}\sin^2\theta) \\ &= \frac{3\lambda^2}{2\pi}\left[\frac{1}{(\frac{1}{2}\Omega^2/\beta^2)+1+D^2}\right]. \qquad (15.6\text{–}18)\end{aligned}$$

The scattering cross-section is seen to be of order λ^2 in a weak exciting field ($\Omega/\beta \ll 1$) near resonance. However, it falls off as $1/\Omega^2$ in a sufficiently strong exciting field, because of the saturation of the fluorescence that we have already noted. Since the atom has been modeled as a point dipole in this calculation, its size cannot of course enter directly here. In a real atom the atomic size would provide a lower limit for the scattering cross-section, but other effects like atomic ionization would come into play long before that point, so that Eq. (15.6–18) would cease to apply.

15.6.3 Spectrum of the fluorescence

Perhaps the easiest way to derive the spectral density of the fluorescence emitted by the driven atom is to calculate the second-order autocorrelation $\Gamma^{(1,1)}$ of the electric field, say. According to Eq. (15.5–8) this is given by the formula [see Eq. (15.5–30)]

$$\langle \hat{\mathbf{E}}^{(-)}(\mathbf{r}, t) \cdot \hat{\mathbf{E}}^{(+)}(\mathbf{r}, t+\tau)\rangle = \left(\frac{\omega_0^2|\boldsymbol{\mu}_{12}|}{4\pi\varepsilon_0 c^2 r}\right)^2 (1 - \tfrac{1}{2}\sin^2\theta)\left\langle \hat{b}^\dagger\left(t - \frac{r}{c}\right)\hat{b}\left(t - \frac{r}{c} + \tau\right)\right\rangle, \tag{15.6–19}$$

provided Eq. (15.6–4) holds as before. We can obtain an expression for the spectral density by taking the Fourier transform with respect to τ in the steady state. The spectrum was first calculated by Mollow (1969), who found that it exhibits three peaks in the presence of a strong exciting field. We shall make use of our coupled equations of motion (15.5–24) and (15.5–25) to solve simultaneously for the three quantities

$$g(t, \tau) \equiv \langle \hat{b}_s^\dagger(t)\hat{b}_s(t+\tau)\rangle\, e^{i(\omega_1-\omega_0)\tau}$$

$$f(t, \tau) \equiv \langle \hat{b}_s^\dagger(t)\hat{b}_s^\dagger(t+\tau)\rangle\, e^{i(\omega_0-\omega_1)(2t+\tau)}\, e^{2i\phi}$$

$$h(t, \tau) \equiv \langle \hat{b}_s^\dagger(t)\hat{R}_3(t+\tau)\rangle\, e^{i(\omega_0-\omega_1)t}\, e^{i\phi},$$

which are closely related to three atomic correlation functions first introduced by Milonni (1974). These quantities are defined in such a way that they become independent of time t and phase ϕ after a sufficiently long time t. After forming the required products of the quantities $\hat{b}_s$, $\hat{b}_s^\dagger$, $\hat{R}_3$ from Eqs. (15.5–24) and (15.5–25), and taking expectation values with the help of the eigenvalue relation (15.6–1), one obtains the following three coupled integral equations (Kimble and Mandel, 1976)

$$g(t, \tau) = [\langle \hat{R}_3(t)\rangle + \tfrac{1}{2}]\, e^{-\beta(1-iD)\tau} + \Omega\int_0^\tau h(t, t')\, e^{\beta(1-iD)(t'-\tau)}\, dt' \tag{15.6–20a}$$

$$f(t, \tau) = \Omega\int_0^\tau h(t, t')\, e^{\beta(1+iD)(t'-\tau)}\, dt' \tag{15.6–20b}$$

$$h(t, \tau) = -\tfrac{1}{2}\langle \hat{b}^\dagger(t)\rangle\, e^{-i(\omega_1 t-\phi)} - \tfrac{1}{2}\Omega\int_0^\tau [f(t, t') + g(t, t')]\, e^{2\beta(t'-\tau)}\, dt'. \tag{15.6–20c}$$

In arriving at these equations we have made use of the relations

$$\hat{b}_s^\dagger(t)\hat{b}_s^\dagger(t) = 0 = \hat{b}_s^\dagger(t)[\hat{R}_3(t) + \tfrac{1}{2}] \tag{15.6–21a}$$

and

$$[\hat{b}_s(t), \boldsymbol{\mu}_{12}^* \cdot \hat{\mathbf{A}}_{\text{free}}^{(+)}(0, t+\tau)] = 0 = [\hat{b}_s(t), \boldsymbol{\mu}_{12} \cdot \hat{\mathbf{A}}_{\text{free}}^{(-)}(0, t+\tau)]. \tag{15.6–21b}$$

Equations (15.6–21a) follow immediately from Eqs. (15.1–7) and (15.1–8). Equations (15.6–21b) express the fact that the free field is not influenced by the behavior of the atom at earlier times (Mollow, 1975; Renaud, Whitley and Stroud, 1976; Agarwal, 1977). By adding Eqs. (15.6–20a) and (15.6–20b) and then substituting for $h(t, t')$ under the integral from Eq. (15.6–20c), we obtain a Volterra-type integral equation for the sum $g(t, \tau) + f(t, \tau)$ that is of the same form as Eq. (15.6–5) for $\langle \hat{R}_3(\tau) \rangle$. Moreover, the integral kernel $K(\tau)$ is given by Eq. (15.6–6) as before, although the inhomogeneous term $y(\tau)$ is more complicated and depends both on time t and on the variable τ. It takes the form

$$y(\tau, t) = [\langle \hat{R}_3(t) \rangle + \tfrac{1}{2}]\, e^{-\beta(1-iD)\tau} - \langle \hat{b}^\dagger(t) \rangle\, e^{-i(\omega_1 t - \phi)} \frac{\Omega}{\beta(1 + D^2)} [1 - e^{-\beta\tau}(\cos \beta D\tau - D \sin \beta D\tau)]. \tag{15.6–22}$$

Let us concentrate on the long-time limit, in which $\langle \hat{R}_3(t) \rangle$ and $\langle \hat{b}^\dagger(t) \rangle$ are obtainable from Eqs. (15.6–11) and (15.6–14). In this limit $y(\tau, t)$ depends only on τ and not on t or ϕ, and so does $g(t, \tau) + f(t, \tau)$. Equations (15.6–20a) to (15.6–20c) then show that all three quantities $f(t, \tau)$, $g(t, \tau)$, $h(t, \tau)$ become independent of t and ϕ in the long-time limit, so that a quasi-stationary state is reached. This quasi-stationarity is a consequence of the quantum fluctuations.

The integral equation for $g(t, \tau) + f(t, \tau)$ can be solved by a Laplace transform method exactly as before. Once again we confine our attention to the on-resonance case ($D = 0$) in the long-time limit $t \to \infty$. In that case, with the help of Eqs. (15.6–13) and (15.6–14), the inhomogeneous term given by Eq. (15.6–22) reduces to

$$y(\tau) = \frac{\tfrac{1}{4}\Omega^2/\beta^2}{(\tfrac{1}{2}\Omega^2/\beta^2) + 1} e^{-\beta\tau} + \frac{\tfrac{1}{2}(\Omega/\beta)^2}{(\tfrac{1}{2}\Omega^2/\beta^2) + 1}(1 - e^{-\beta\tau}).$$

The Laplace transform $\tilde{y}(p)$ of $y(\tau)$ is given by

$$\tilde{y}(p) = \frac{\tfrac{1}{4}\Omega^2/\beta^2}{(\tfrac{1}{2}\Omega^2/\beta^2) + 1} \frac{p + 2\beta}{p(p + \beta)}.$$

This solution of the integral equation for $g(t, \tau) + f(t, \tau)$ is then obtained by contour integration, in the same way as Eq. (15.6–10) was in connection with the solution for $\langle \hat{R}_3(\tau) \rangle$. We find that in the long-time limit,

$$\begin{aligned} g(\infty, \tau) &+ f(\infty, \tau) \\ &= \frac{1}{2\pi i} \int_{-i\infty + p_r}^{i\infty + p_r} \frac{\tilde{y}(p)}{1 - \tilde{K}(p)} e^{p\tau}\, dp \\ &= \left\{ \frac{\tfrac{1}{4}\Omega^2/\beta^2}{[(\tfrac{1}{2}\Omega^2/\beta^2) + 1]^2} \right\} \Big[2 + \frac{e^{-3\beta\tau/2}}{-4\beta^2 i\Omega'} \{ (\tfrac{1}{2}\beta - i\Omega')^2 (-3\beta/2 + i\Omega')\, e^{-i\Omega'\tau} \\ &\qquad - (\tfrac{1}{2}\beta + i\Omega')^2(-3\beta/2 - i\Omega')\, e^{i\Omega'\tau} \} \Big]. \end{aligned} \tag{15.6–23a}$$

Also it follows from Eqs. (15.6–20a) and (15.6–20b) that when $D = 0$,

$$g(t, \tau) - f(t, \tau) = [\langle \hat{R}_3(t) \rangle + \tfrac{1}{2}]\, e^{-\beta\tau},$$

which becomes, with the help of Eq. (15.6–13), in the long-time limit,

$$g(\infty, \tau) - f(\infty, \tau) = \frac{\tfrac{1}{4}\Omega^2/\beta^2}{(\tfrac{1}{2}\Omega^2/\beta^2) + 1}\, e^{-\beta\tau}. \qquad (15.6\text{–}23b)$$

Hence, by adding Eqs. (15.6–23a) and (15.6–23b) we arrive at the result for $g(\infty, \tau)$:

$$g(\infty, \tau) = \frac{\tfrac{1}{8}\Omega^2/\beta^2}{[(\tfrac{1}{2}\Omega^2/\beta^2) + 1]^2} \Bigg\{ 2 + [(\tfrac{1}{2}\Omega^2/\beta^2) + 1]\, e^{-\beta\tau} + \left[\frac{e^{-3\beta\tau/2}}{4\beta^2 i\Omega'} (\tfrac{1}{2}\beta + i\Omega')^2 (-\tfrac{3}{2}\beta - i\Omega')\, e^{i\Omega'\tau} + \text{c.c.} \right] \Bigg\}$$

$$\tau \geqslant 0. \qquad (15.6\text{–}24)$$

For negative τ we make use of the symmetry relation $g(\infty, -\tau) = g^*(\infty, \tau)$.

The Fourier transform of the field correlation function $\langle \hat{\mathbf{E}}^{(-)}(\mathbf{r}, t) \cdot \hat{\mathbf{E}}^{(+)}(\mathbf{r}, t + \tau) \rangle$ with respect to τ, which is proportional to $g(t, \tau)$ according to Eq. (15.6–19), yields the spectral density $\Phi(\mathbf{r}, \omega)$ of the fluorescent light at $\mathbf{r}$. Thus we have from Eq. (15.6–24) in the long-time limit, and for a sufficiently strong exciting field so that $\Omega > \tfrac{1}{2}\beta$ and Ω' is real,

$$\begin{aligned}
\Phi(\mathbf{r}, \omega) &= \int_{-\infty}^{\infty} \langle \hat{\mathbf{E}}^{(-)}(\mathbf{r}, t) \cdot \hat{\mathbf{E}}^{(+)}(\mathbf{r}, t + \tau) \rangle\, e^{i\omega\tau}\, d\tau \\
&= \left(\frac{\omega_0^2 |\boldsymbol{\mu}_{12}|}{4\pi\varepsilon_0 c^2 r} \right)^2 (1 - \tfrac{1}{2}\sin^2\theta) 2 \int_0^{\infty} g(\infty, \tau)\, e^{i(\omega-\omega_1)\tau}\, d\tau \\
&= \left(\frac{\omega_0^2 |\boldsymbol{\mu}_{12}|}{4\pi\varepsilon_0 c^2 r} \right)^2 (1 - \tfrac{1}{2}\sin^2\theta) \left(\frac{\tfrac{1}{4}\Omega^2/\beta^2}{(\tfrac{1}{2}\Omega^2/\beta^2) + 1} \right) \\
&\quad \times \left[\frac{2\pi\delta(\omega - \omega_1)}{(\tfrac{1}{2}\Omega^2/\beta^2) + 1} + \frac{\beta}{\beta^2 + (\omega - \omega_1)^2} \right. \\
&\quad + \left(\frac{\tfrac{1}{2}\beta}{\Omega^2 + 2\beta^2} \right) \frac{4\Omega^2 - 4\beta^2 + (5\Omega^2/2 - \beta^2)(\omega - \omega_1)/\Omega'}{9\beta^2/4 + (\omega - \omega_1 + \Omega')^2} \\
&\quad \left. + \left(\frac{\tfrac{1}{2}\beta}{\Omega^2 + 2\beta^2} \right) \frac{4\Omega^2 - 4\beta^2 - (5\Omega^2/2 - \beta^2)(\omega - \omega_1)/\Omega'}{9\beta^2/4 + (\omega - \omega_1 - \Omega')^2} \right]. \qquad (15.6\text{–}25a)
\end{aligned}$$

In the special case when $\Omega \gg \beta$, the spectral density can be expressed in the simpler and more recognizable form

$$\begin{aligned}
\Phi(\mathbf{r}, \omega) &= \left(\frac{\omega_0^2 |\boldsymbol{\mu}_{12}|}{4\pi\varepsilon_0 c^2 r} \right)^2 (1 - \tfrac{1}{2}\sin^2\theta) \left[\frac{2\pi\delta(\omega - \omega_1)}{\Omega^2/\beta^2} + \frac{\beta/2}{\beta^2 + (\omega - \omega_1)^2} \right. \\
&\quad \left. + \frac{3\beta/8}{9\beta^2/4 + (\omega - \omega_1 + \Omega')^2} + \frac{3\beta/8}{9\beta^2/4 + (\omega - \omega_1 - \Omega')^2} \right]. \qquad (15.6\text{–}25b)
\end{aligned}$$

We note that this expression has split naturally into the sum of four contributions. The first term, proportional to $\delta(\omega - \omega_1)$, corresponds to the light from the external field that is elastically scattered by the atom, and therefore has the same spectral distribution as the applied field. Although this term is infinite when $\omega = \omega_1$ under monochromatic excitation, after integration over all frequencies the total elastically scattered contribution actually becomes proportionately small when the exciting field is strong ($\Omega/\beta \gg 1$). The remaining three terms in Eq. (15.6–25b) represent inelastically scattered light, and the corresponding spectral distribution differs from that of the applied field in general. It exhibits three peaks, each of which has a Lorentzian form. The central component has a peak at the atomic or driving frequency ω_1 and a halfwidth β, whereas the two sidebands are displaced from ω_1 by the modified Rabi frequency Ω' and have halfwidths $\frac{3}{2}\beta$. These last two shifted components, whose heights are $\frac{1}{3}$ the height of the inelastically scattered central peak, are a reflection of the Rabi oscillations of the atomic excitation. The splitting of the spectral line into three components is sometimes also known as the *a.c. Stark effect*. The phenomenon was first predicted by Mollow (1969) and was first observed in resonance fluorescence measurements of an atomic beam of sodium by Schuda, Stroud, and Herscher (1974). The atoms were excited by a light beam from a tunable dye laser that was incident at right angles to the atomic beam, and the fluorescent light was analyzed by a scanning Fabry–Perot interferometer. The use of an atomic beam is necessary to avoid Doppler broadening of the spectral line. Figure 15.10 shows the results of later and somewhat more accurate measurements. Very good agreement with Eqs. (15.6–25) was obtained when the instrumental width of the interferometer was taken into account. In all these measurements the exciting field was relatively strong ($\Omega/\beta \geqslant 1$).

If the exciting field is so weak that $\Omega < (1/2)\beta$, then Ω' is imaginary, and $g(\infty, \tau) + f(\infty, \tau)$ given by Eq. (15.6–23a) consists of a constant plus two purely exponential contributions. As a result $g(\infty, \tau)$ is exponential also, apart from a constant. The corresponding spectral density $\Phi(\mathbf{r}, \omega)$ then has only a single peak at $\omega = \omega_1$, in addition to the $\delta(\omega - \omega_1)$ contribution. The reason for this becomes obvious if we refer to Fig. 15.9; in a weak exciting field there are no atomic Rabi oscillations, and it is these Rabi oscillations which are responsible for the two side peaks in the spectral density at $\omega_1 \pm \Omega'$ exhibited in Fig. 15.10. Also, in a weak exciting field, with $\Omega/\beta \ll 1$, the elastically scattered component in Eqs. (15.6–25) becomes relatively more important, and results in a narrowing of the fluorescence spectrum as compared with the natural line. This effect has been observed (Gibbs and Venkatesan, 1976). When the finite bandwidth of the exciting laser beam is significant and is taken into account, the spectral density may show appreciable asymmetries (Agarwal, 1976; Eberly, 1976; Kimble and Mandel, 1977).

Finally, we mention that the inelastic contribution to the steady-state spectrum of the fluorescence given by Eqs. (15.6–25) has a physical significance that goes beyond the significance of the corresponding spontaneous emission spectrum. In a weak exciting field the former reduces to a single Lorentzian peak of width 2β, just like the latter. But in the spontaneous emission from an initially excited atom we are dealing with the radiation of a finite amount of energy. The radiation

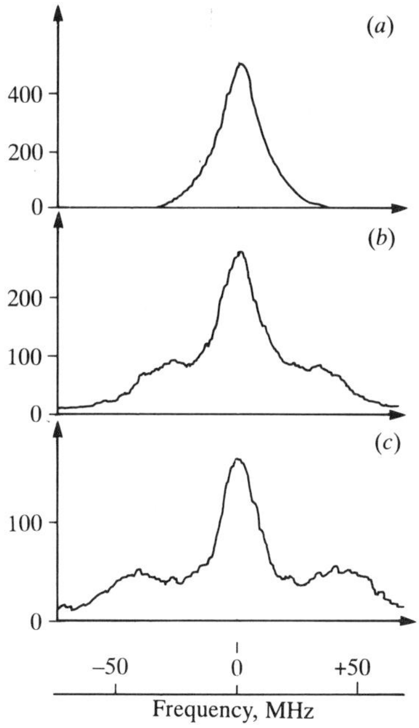

Fig. 15.10 The measured spectrum of the fluorescence produced by a driven sodium atom near resonance for various exciting fields strengths. (*a*) 85 mW/cm^2; (*b*) 490 mW/cm^2; (*c*) 920 mW/cm^2. (Reproduced from Wu, Grove and Ezekiel, 1975.)

process must therefore die out, and the associated linewidth has to be non-zero. This is so even in a semiclassical theory. On the other hand, in the process of resonance fluorescence the atom is continually excited, in our calculation by a strictly monochromatic field, and yet it radiates a field of non-zero linewidth. This may be regarded as a reflection of the quantum fluctuations that are inherent in the coupled atom-plus-field system. These quantum fluctuations are manifest in the most striking way in the intensity correlations of the fluorescence.

15.6.4 Intensity correlation of the fluorescence

As we have seen in Chapter 14, the joint probability that a photodetector exposed to the atomic fluorescence registers two photons at times t and $t+\tau$ is proportional to the fourth-order correlation function in normal order and time order,

$$\begin{aligned}\Gamma^{(2,2)}(\mathbf{r}; t, t+\tau) &\equiv \langle \hat{E}_i^{(-)}(\mathbf{r}, t)\hat{E}_j^{(-)}(\mathbf{r}, t+\tau)\hat{E}_j^{(+)}(\mathbf{r}, t+\tau)\hat{E}_i^{(+)}(\mathbf{r}, t)\rangle \\ &= \langle \mathcal{T}\!:\!\hat{I}(\mathbf{r}, t)\hat{I}(\mathbf{r}, t+\tau)\!:\rangle. \end{aligned} \tag{15.6–26}$$

As this probability can be measured directly, we turn our attention to the calculation of the intensity correlation function $\Gamma^{(2,2)}(\mathbf{r}; t, t+\tau)$. With the help of the positive and negative frequency parts of $\hat{E}(\mathbf{r}, t)$ given by Eq. (15.5–8), and the commutation relations (15.6–21), together with Eq. (15.6–4), we find that

$$\begin{aligned}
\Gamma^{(2,2)}&(\mathbf{r}; t, t, + \tau) \\
&= \left(\frac{\omega_0^2|\boldsymbol{\mu}_{12}|}{4\pi\varepsilon_0 c^2 r}\right)^4 (1 - \tfrac{1}{2}\sin^2\theta)^2 \\
&\quad \times \left\langle \hat{b}_s^\dagger\left(t - \frac{r}{c}\right)\hat{b}_s^\dagger\left(t - \frac{r}{c} + \tau\right)\hat{b}_s\left(t - \frac{r}{c} + \tau\right)\hat{b}_s\left(t - \frac{r}{c}\right)\right\rangle \\
&= \left(\frac{\omega_0^2|\boldsymbol{\mu}_{12}|}{4\pi\varepsilon_0 c^2 r}\right)^4 (1 - \tfrac{1}{2}\sin^2\theta)^2 \\
&\quad \times \left[\left\langle \hat{b}_s^\dagger\left(t - \frac{r}{c}\right)\hat{R}_3\left(t - \frac{r}{c} + \tau\right)\hat{b}_s\left(t - \frac{r}{c}\right)\right\rangle + \frac{1}{2}\left\langle \hat{R}_3\left(t - \frac{r}{c}\right)\right\rangle + \frac{1}{4}\right], \\
&\qquad \tau \geqslant 0. \qquad (15.6\text{–}27)
\end{aligned}$$

We have already evaluated the second term, which is given by Eq. (15.6–11a). In order to evaluate the first term it is convenient to define two atomic correlation functions (cf. Kimble and Mandel, 1976),

$$\mathcal{H}(t, \tau) \equiv \langle \hat{b}_s^\dagger(t)\hat{R}_3(t + \tau)\hat{b}_s(t)\rangle$$

$$\mathcal{F}(t, \tau) \equiv \langle \hat{b}_s^\dagger(t)\hat{b}_s^\dagger(t + \tau)\hat{b}_s(t)\rangle\, \mathrm{e}^{\mathrm{i}(\omega_0 - \omega_1)(t+\tau)}\, \mathrm{e}^{\mathrm{i}\phi},$$

which become independent of t and ϕ in the long-time limit $t \to \infty$. These correlations can be formed from Eqs. (15.5–24) and (15.5–25), with the help of Eqs. (15.6–1) and (15.6–21), and they lead to the coupled integral equations

$$\mathcal{H}(t, \tau) = -\tfrac{1}{2}[\langle \hat{R}_3(t)\rangle + \tfrac{1}{2}] - \tfrac{1}{2}\Omega\int_0^\tau [\mathcal{F}(t, t') + \mathcal{F}^*(t, t')]\,\mathrm{e}^{-2\beta(\tau - t')}\,\mathrm{d}t' \qquad (15.6\text{–}28)$$

$$\mathcal{F}(t, \tau) = \Omega\int_0^\tau \mathcal{H}(t, t')\,\mathrm{e}^{-\beta(1+\mathrm{i}D)(\tau - t')}\,\mathrm{d}t'. \qquad (15.6\text{–}29)$$

By substituting for $\mathcal{F}(t, t')$ and $\mathcal{F}^*(t, t')$ from the last equation in Eq. (15.6–28) and integrating by parts, we arrive at an integral equation for $\mathcal{H}(t, \tau)$ that is once again of the form of Eq. (15.6–5), with the same kernel $K(\tau)$ as before, but with the inhomogeneous term

$$y(\tau, t) = -\tfrac{1}{2}[\langle \hat{R}_3(t)\rangle + \tfrac{1}{2}], \qquad (15.6\text{–}30)$$

rather than $y(\tau) = -\frac{1}{2}$, as it was in the equation for $\langle R_3(\tau)\rangle_G$, the average atomic inversion. We added the suffix G to signify that the atom starts in the ground state at $\tau = 0$. This inhomogeneous term is also constant with respect to the variable τ of the integral equation, although it depends on t. As the solution of the integral equation is directly proportional to the inhomogeneous term when it is constant, we can make use of our previous results to write down the answer immediately. We find by inspection that

$$\mathcal{H}(t, \tau) = [\langle \hat{R}_3(t)\rangle + \tfrac{1}{2}]\langle \hat{R}_3(\tau)\rangle_G. \qquad (15.6\text{–}31)$$

From Eqs. (15.6–27) and (15.6–31) we then have

$$\Gamma^{(2,2)}(\mathbf{r}; t, t+\tau) = \left(\frac{\omega_0^2|\boldsymbol{\mu}_{12}|}{4\pi\varepsilon_0 c^2 r}\right)^4 (1 - \tfrac{1}{2}\sin^2\theta)^2\left[\left\langle \hat{R}_3\left(t - \frac{r}{c}\right)\right\rangle + \frac{1}{2}\right][\langle \hat{R}_3(\tau)\rangle_{\mathrm{G}} + \tfrac{1}{2}],$$

$$\tau \geqslant 0, \quad (15.6\text{–}32)$$

or, when the transit time r/c can be neglected,

$$\langle \mathcal{T}{:}\hat{I}(\mathbf{r}, t)\hat{I}(\mathbf{r}, t+\tau){:}\rangle = \langle \hat{I}(\mathbf{r}, t)\rangle\langle \hat{I}(\mathbf{r}, \tau)\rangle_{\mathrm{G}}. \quad (15.6\text{–}33)$$

This remarkably simple result has a natural physical interpretation. The first factor on the right $\langle \hat{I}(\mathbf{r}, t)\rangle$ is proportional to the probability that one photon is emitted or detected at time t, and it becomes constant in the steady state. But in the process of emission at time t the atom makes a quantum jump to the ground state, and the second factor $\langle \hat{I}(\mathbf{r}, \tau)\rangle_{\mathrm{G}}$ is proportional to the probability that another photon is emitted after a subsequent time delay τ, given that the atom is in the ground state at $\tau = 0$. The average light intensity $\langle \hat{I}(\mathbf{r}, \tau)\rangle_{\mathrm{G}}$ is given by Eqs (15.6–3) and (15.6–13) and its τ-dependence is illustrated in Fig. 15.9. It always starts from zero at time $\tau = 0$, and then grows with increasing τ, as shown in Fig. 15.9. This is quite unlike a classical correlation function, which can fall below its initial value at $\tau = 0$, but can never rise above it. The photoelectric pulses produced when the fluorescent light falls on a photodetector should therefore exhibit antibunching (Carmichael & Walls, 1976a,b; Kimble & Mandel, 1976), as was discussed in more detail in Section 14.7.

15.6.5 Measurements of photon antibunching

Photon antibunching was first observed in two-time photoelectric correlation measurements of the fluorescence produced by single sodium atoms in an atomic beam (Kimble, Dagenais and Mandel, 1977, 1978; Dagenais and Mandel, 1978). Figure 15.11 shows the apparatus used in the experiment. Sodium atoms produced from an oven in the form of a 100 μm wide atomic beam are excited by the perpendicular (circularly polarized) light beam from a tunable dye laser, and the fluorescence is collected by a microscope objective and imaged on two photodetectors. The photoelectric pulses from the two detectors are fed to the start and stop inputs of a time-to-digital converter (TDC), that registers their time separation τ. The number of pulse pairs $n(\tau)$ recorded with time separation τ is a direct measure of the joint photodetection probability and therefore of $\Gamma^{(2,2)}$ $(\mathbf{r}; t, t+\tau)$. Actually, the atoms pass successively through two light beams, the first of which is used merely to prepare a certain initial quantum state (the $3^2S_{1/2}$, $F = 2$, $m_F = 2$ state) by optical pumping. When the exciting light beam is circularly polarized and tuned to the $3^2S_{1/2}$, $F = 2$ to $3^2P_{3/2}$, $F = 3$ transition in sodium, the dipole selection rules allow a transition from the $3^2S_{1/2}$, $F = 2$, $m_F = 2$ magnetic sublevel only to the $3^2P_{3/2}$, $F = 3$, $m_F = 3$ state. The atoms therefore behave like two-level quantum systems to a good approximation, and the foregoing theory applies.

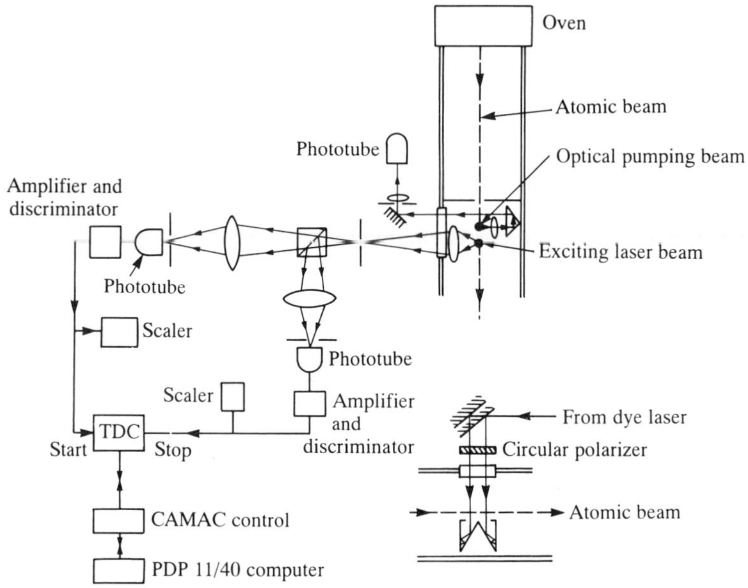

Fig. 15.11 Outline of the apparatus used for two-time photon correlation measurements. (Reproduced from Dagenais and Mandel, 1978.)

Figure 15.12 shows the results obtained for the normalized intensity correlation function in the steady state or long-time limit $t \to \infty$, namely

$$1 + \lambda(\tau) = \frac{\langle \mathcal{T}:\hat{I}(\mathbf{r}, t)\hat{I}(\mathbf{r}, t + \tau):\rangle}{\langle \hat{I}(\mathbf{r}, \infty)\rangle^2}, \tag{15.6–34}$$

after some corrections were applied to the data. The full curves are theoretical and are based on Eq. (15.6–32). The experiment provided the first example of photon antibunching, and demonstrates the quantum character of the electromagnetic field. The fact that all the curves start from zero at $\tau = 0$ also provides experimental evidence that the atom undergoes a quantum jump to the ground state at the instant of emission because of its inability to emit a second photon immediately after the first. Such quantum jumps are inconsistent with any classical picture of radiation.

15.6.6 Sub-Poissonian photon statistics

One can see from Eq. (15.6–33) that if the number n of detected fluorescent photons emitted by the atom in a short time interval T is counted, then the photon statistics must be sub-Poissonian. It was shown in Section 14.9 [see Eqs. (14.9–11) and (14.9–12)] that the variance of n is given by

$$\langle(\Delta n)^2\rangle = \langle n\rangle(1 + \langle n\rangle\theta(T)/T), \tag{15.6–35}$$

where the function $\theta(T)$ is defined by the integral

$$\theta(T) \equiv \int_{-T}^{T} (1 - |\tau|/T)\lambda(\tau)\,d\tau.$$

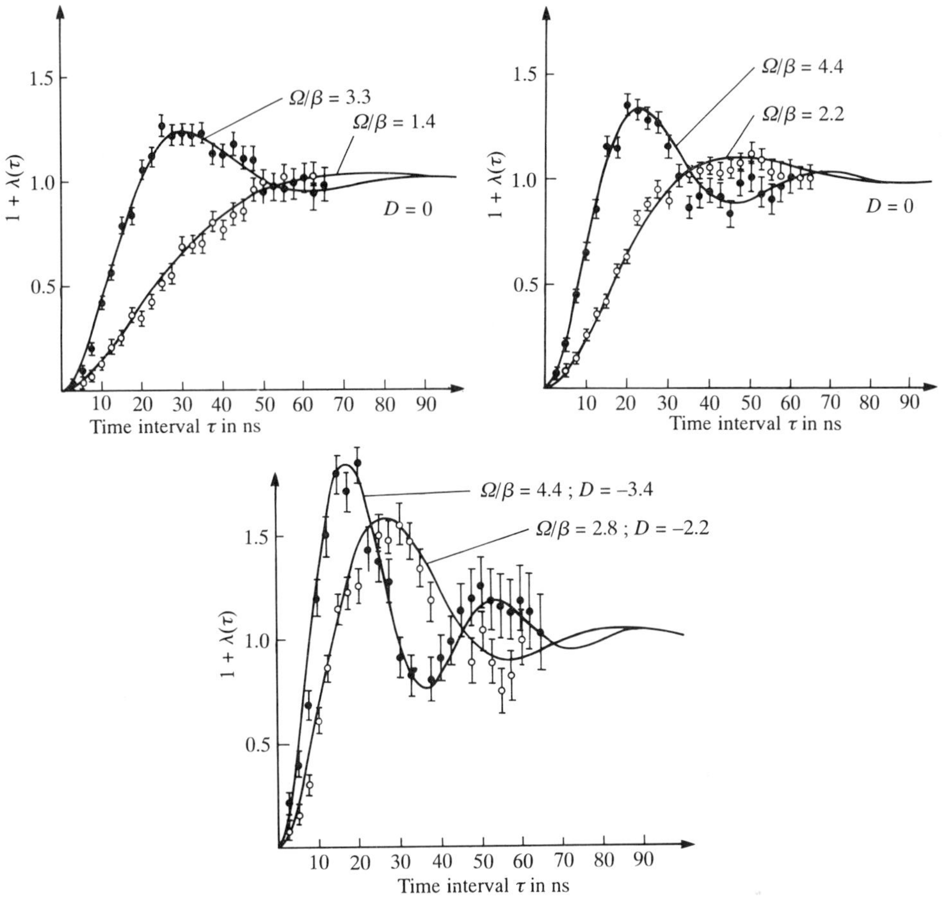

Fig. 15.12 Results of two-time photon correlation measurements for a single driven atom, for various exciting field strengths Ω/β and detunings D. The full curves are theoretical and they all exhibit antibunching. (Reproduced from Dagenais and Mandel, 1978.)

For a sufficiently short time interval $T \ll 1/\beta$ and $T \ll 1/\Omega$, $\lambda(\tau)$ may be effectively replaced by $\lambda(0)$ under the integral, so that

$$\theta(T) \approx \lambda(0)\int_{-T}^{T}(1 - |\tau|/T)\,\mathrm{d}\tau = \lambda(0)T. \tag{15.6–36}$$

Hence the variance becomes

$$\langle(\Delta n)^2\rangle = \langle n\rangle(1 + \lambda(0)\langle n\rangle). \tag{15.6–37}$$

But from Eqs. (15.6–33) and (15.6–34), $1 + \lambda(0) = 0$, or

$$\lambda(0) = -1, \tag{15.6–38}$$

so that

$$\langle(\Delta n)^2\rangle = \langle n\rangle(1 - \langle n\rangle). \tag{15.6–39}$$

Hence the detected photons are sub-Poissonian, although the departure from

Poisson statistics will be very small when $\langle n\rangle \ll 1$. The Q-parameter characterizing the departure from Poisson statistics is given by [cf. Eq. (14.9–13)]

$$Q = [\langle(\Delta n)^2\rangle - \langle n\rangle]/\langle n\rangle = -\langle n\rangle. \tag{15.6–40}$$

Similarly, it is easy to show that the photon statistics are sub-Poissonian in the long-time limit $T \to \infty$, for which

$$\theta(T) \to 2\int_0^\infty \lambda(\tau)\,d\tau. \tag{15.6–41}$$

With the help of Eq. (15.6–33) we readily find that in the steady state

$$\lambda(\tau) = \frac{\langle \hat{I}(\mathbf{r}, \tau)\rangle_G}{\langle \hat{I}(\mathbf{r})\rangle} - 1, \quad (\tau \geqslant 0),$$

so that

$$\theta(T) \Rightarrow \frac{2}{\langle \hat{I}(\mathbf{r})\rangle}\int_0^\infty \langle \hat{I}(\mathbf{r}, \tau)\rangle_G\,d\tau. \tag{15.6–42}$$

As may be shown with the help of Eq. (15.6–13) (Mandel, 1979a),

$$\theta(T) \to -\frac{3}{\beta}\,\frac{1}{(\Omega^2/2\beta^2) + 1}, \tag{15.6–43}$$

so that in the long-time limit,

$$Q = -\frac{3\langle n\rangle}{\beta T}\,\frac{1}{(\Omega^2/2\beta^2) + 1}, \tag{15.6–44}$$

which is always negative, but may be numerically very small when few of the emitted photons are counted.

The existence of sub-Poissonian photon statistics in resonance fluorescence was first confirmed experimentally (Short and Mandel, 1983a,b) with Q as small as -0.002. Indeed, resonance fluorescence provided the first demonstration of non-classical counting statistics.[‡]

Atomic resonance fluorescence is an example of a phenomenon that cannot be adequately treated by semiclassical theory, no matter how intense or 'classical' the exciting field may be, because the fluorescence always exhibits quantum features.

15.7 Deflection of atoms by light

In the foregoing treatment of the interaction between an atom and a near resonant beam of light, any motion of the atom was completely ignored. In fact, both in the processes of absorption and emission of a photon some momentum has to be transferred to the atom, which may result in a deflection of the atomic trajectory. In addition, this momentum transfer may lead to a Doppler shift of the exciting light as seen in the frame of the moving atom that brings the optical frequency either closer to or further from resonance. The inclusion of the atomic recoil substantially complicates the treatment of resonance fluorescence, and we

‡ For a review of later experiments exhibiting non-classical statistics, see Teich and Saleh (1988, 1990).

shall not discuss it here (Baklanov and Dubetskii, 1976; Lam and Berman, 1976; Agarwal and Saxena, 1978; Stenholm, 1978; Cook and Bernhardt, 1978). However, we can readily use some of the results of the previous sections to calculate the recoil when the Doppler shifts are small. When an optical photon of energy ~ 2 eV, is absorbed by an atom of sodium, say, the atom acquires a recoil velocity of about 3 cm/s, which results in a maximum Doppler shift of about 50 000 Hz. This is very small compared with the natural linewidth, so that the Doppler shifts resulting from the interaction can safely be neglected so long as we are dealing with a small number of absorptions and emissions. The situation is, of course, quite different in the prolonged interaction of atoms with a sufficiently strong field. In that case the atoms can be both cooled and trapped by the light.

Two fundamentally different kinds of forces may be exerted on the atom by the incident light; the distinction depends on whether the re-emission by the atom of the absorbed photon is stimulated or spontaneous. As spontaneous emission can occur in almost any direction at random, the average momentum transfer to the atom by spontaneous emission is zero. Hence the net transfer by absorption and re-emission is just the average momentum transfer by absorption alone. On the other hand, if the re-emission is stimulated by the presence of a strong resonant electromagnetic field, then there is a definite direction for the momentum transfer on absorption and a definite direction for the transfer on emission. For a plane wave these directions are of course equal, and there is no net momentum transfer or force. But if the electromagnetic field is not uniform, then in general the moving atom will find the directions of the Poynting vector to be different on absorption and emission, with the result that there is a net momentum transfer in an inhomogeneous electromagnetic field. The resulting force is referred to as the *dipole or gradient force*, and it can be understood in terms of the interaction of the induced atomic dipole moment with the oscillating field, without the need for field quantization. This force was first investigated by Letokhov (1968) and Ashkin (1978), and was observed in atomic beam focusing experiments by Bjorkholm, Freeman, Ashkin and Pearson (1978).

The subject of the effect of near resonant light on moving atoms has mushroomed and generated a vast literature in recent years, and we shall merely touch on some of the important topics in the following sections (see for example Cohen-Tannoudji, 1992). We start by calculating the momentum transfer or the force associated with spontaneous atomic emissions.

15.7.1 Momentum transfer after n spontaneous emissions

To simplify the problem let us suppose that the two-level atom is initially either at rest, or moving exactly at right angles to a beam of light of wave vector $\mathbf{k}_0$, whose frequency is close to the atomic frequency ω_0. We shall follow the approach of Mandel (1979b). Every time the atom absorbs a photon from the field and becomes excited, a momentum $\hbar\mathbf{k}_0$ is transferred to it, and whenever the atom emits a photon of wave vector $\mathbf{k}_1$ and returns to the ground state, it recoils and acquires an additional momentum $-\hbar\mathbf{k}_1$. If the photon emission is stimulated, then $\mathbf{k}_1 = \mathbf{k}_0$ and there is no net momentum transfer to the atom. However, after each spontaneous emission there is a net momentum transfer of $\hbar(\mathbf{k}_0 - \mathbf{k}_1)$. After

n spontaneous emissions resulting in photons with wave vectors $\mathbf{k}_1, \mathbf{k}_2, \ldots, \mathbf{k}_n$ the atom acquires an additional momentum

$$\mathbf{p} = \hbar(n\mathbf{k}_0 - \mathbf{k}_1 - \mathbf{k}_2 - \ldots - \mathbf{k}_n). \tag{15.7–1}$$

From Sections 15.5 and 15.6 we know that the probability for the detection of a spontaneously emitted photon at position $\mathbf{r}$ in the far field of an atom located at the origin is proportional to

$$|\boldsymbol{\mu}_{12}|^2 - \frac{|\boldsymbol{\mu}_{12} \cdot \mathbf{r}|^2}{r^2},$$

and we may interpret the detection at $\mathbf{r}$ as being associated with a photon of wave vector $\mathbf{k}$ such that $\mathbf{r}/r = \mathbf{k}/k$. The frequency of the emitted photon may of course differ from the frequency of the exciting photon, but the fractional frequency difference is generally very small, and relative to the momentum transfer can be neglected. We may therefore write, to a good approximation, for the probability $P_1(\mathbf{k})\,\mathrm{d}^3k$ that the spontaneously emitted photon has wave vector $\mathbf{k}$ within d^3k,

$$P_1(\mathbf{k})\,\mathrm{d}^3k = C\left(|\boldsymbol{\mu}_{12}|^2 - \frac{|\boldsymbol{\mu}_{12} \cdot \mathbf{k}|^2}{k^2}\right)\delta(k - k_0)\,\mathrm{d}^3k, \tag{15.7–2}$$

where C is a normalization constant. Because the atom returns to the ground state after each spontaneous emission, the wave vectors of successively emitted photons are independent, and the joint probability density $P_n(\mathbf{k}_1, \ldots, \mathbf{k}_n)$ for the wave vectors of n spontaneously emitted photons to be $\mathbf{k}_1, \mathbf{k}_2, \ldots, \mathbf{k}_n$ is given by

$$P_n(\mathbf{k}_1, \ldots, \mathbf{k}_n)\,\mathrm{d}^3k_1 \ldots \mathrm{d}^3k_n = C^n \prod_{r=1}^{n}\left[\left(|\boldsymbol{\mu}_{12}|^2 - \frac{|\boldsymbol{\mu}_{12} \cdot \mathbf{k}_r|^2}{k_r^2}\right)\delta(k_r - k_0)\,\mathrm{d}^3k_r\right]. \tag{15.7–3}$$

With the help of Eq. (15.7–1) it then follows at once that the probability $\mathcal{P}_n(\mathbf{p})\,\mathrm{d}^3p$ for the atom to acquire a momentum $\mathbf{p}$ within d^3p after n absorptions and spontaneous emissions is

$$\begin{aligned}\mathcal{P}_n(\mathbf{p})\,\mathrm{d}^3p &= \int \ldots \int P_n(\mathbf{k}_1, \ldots, \mathbf{k}_n)\delta^3(\mathbf{p} - n\hbar\mathbf{k}_0 + \hbar\mathbf{k}_1 + \ldots + \hbar\mathbf{k}_n)\,\mathrm{d}^3k_1 \ldots \mathrm{d}^3k_n \\ &= \int \mathcal{P}_{n-1}(\mathbf{p}')P_1(\mathbf{p} - \mathbf{p}')\,\mathrm{d}^3p', \end{aligned} \tag{15.7–4}$$

with P_n given by Eq. (15.7–3).

A number of general properties of the moments of the momentum transfer $\mathbf{p}$ follow immediately from this relation. For example, the average momentum transfer $\langle\mathbf{p}\rangle_n$ to the atom after n absorptions and emissions is given by the expression

$$\begin{aligned}\langle\mathbf{p}\rangle_n &= \int \mathbf{p}\mathcal{P}_n(\mathbf{p})\,\mathrm{d}^3p \\ &= \int \ldots \int (\mathbf{p}' + n\hbar\mathbf{k}_0 - \hbar\mathbf{k}_1 - \ldots - \hbar\mathbf{k}_n)P_1(\mathbf{k}_1) \ldots P_1(\mathbf{k}_n)\delta^3(\mathbf{p}')\,\mathrm{d}^3k_1\end{aligned}$$

$$\ldots \mathrm{d}^3 k_n \, \mathrm{d}^3 p'$$
$$= n\hbar\mathbf{k}_0, \tag{15.7–5}$$

and equals the momentum of the absorbed photons, because the momentum of the spontaneously emitted photons averages to zero. The second moment of $\mathbf{p}$ after n absorptions and emissions is similarly obtained, and we find that

$$\langle \mathbf{p}^2 \rangle_n = \int \mathbf{p}^2 \mathscr{P}_n(\mathbf{p}) \, \mathrm{d}^3 p$$
$$= \int \ldots \int (\mathbf{p}' + n\hbar\mathbf{k}_0 - \hbar\mathbf{k}_1 - \ldots - \hbar\mathbf{k}_n)^2 P_1(\mathbf{k}_1) \ldots P_1(\mathbf{k}_n) \delta^3(\mathbf{p}') \, \mathrm{d}^3 k_1$$
$$\ldots \mathrm{d}^3 k_n \, \mathrm{d}^3 p'$$
$$= n^2 (\hbar\mathbf{k}_0)^2 + n\hbar^2 \langle \mathbf{k}^2 \rangle_1,$$

so that

$$\langle (\Delta\mathbf{p})^2 \rangle_n = n \langle (\Delta\mathbf{p})^2 \rangle_1. \tag{15.7–6}$$

The momentum dispersion therefore increases linearly with the number of absorptions and emissions. Whereas, according to Eq. (15.7–5), the average momentum transfer increases only in the direction of the incident light beam, the dispersion of $\mathbf{p}$ increases in all directions. More generally, it may be shown that the r'th cumulant of the momentum transfer after n spontaneous emissions is n times as great as the r'th cumulant after one emission (Mandel, 1979b). When n becomes large, it follows from the central limit theorem and from the fact that the directions of the successively emitted photons are random and independent, that $\mathscr{P}_n(\mathbf{p})$ tends to a Gaussian form. Moreover, we may readily show from Eq. (15.7–4) that

$$\mathscr{P}_n(2n\hbar\mathbf{k}_0 - \mathbf{p}) = \mathscr{P}_n(\mathbf{p}), \tag{15.7–7}$$

so that the distribution $\mathscr{P}_n(\mathbf{p})$ is symmetric about $n\hbar\mathbf{k}_0$.

Let us now calculate $\mathscr{P}_1(\mathbf{p})$ explicitly. If we are dealing with an atomic $\Delta m = \pm 1$ transition, and the incident field is circularly polarized and propagates in the z-direction, then

$$\boldsymbol{\mu}_{12} = \frac{|\boldsymbol{\mu}_{12}|}{\sqrt{2}} (\mathbf{x}_1 + \mathrm{i}\mathbf{y}_1),$$

where $\mathbf{x}_1$, $\mathbf{y}_1$ are unit vectors in the directions of the x,y-axes. Then

$$\left(|\boldsymbol{\mu}_{12}|^2 - \frac{|\boldsymbol{\mu}_{12} \cdot \mathbf{k}|^2}{k^2} \right) \delta(k - k_0) = \tfrac{1}{2} |\boldsymbol{\mu}_{12}|^2 \left(1 + \frac{k_z^2}{k_0^2} \right) \delta(k - k_0),$$

and Eq. (15.7–2) yields

$$P_1(\mathbf{k}) \, \mathrm{d}^3 k = \frac{3}{16\pi k_0^2} \left(1 + \frac{k_z^2}{k_0^2} \right) \delta(k - k_0) \, \mathrm{d}^3 k, \tag{15.7–8}$$

after we evaluate and insert the normalization constant. After integrating over two of the variables and using Eq. (15.7–4), we obtain for the atomic momentum distributions the results

$$\left.\begin{aligned}\mathcal{P}_1(p_z) &= \frac{3}{8\hbar k_0}\left[1+\left(1-\frac{p_z}{\hbar k_0}\right)^2\right], \quad \text{if } 0 \leqslant p_z \leqslant 2\hbar k_0, \\ &= 0, \quad \text{otherwise,}\end{aligned}\right\} \tag{15.7–9}$$

and

$$\left.\begin{aligned}\mathcal{P}_1(p_y) &= \frac{3}{8\hbar k_0}\left[\frac{3}{2}-\frac{1}{2}\left(\frac{p_y}{\hbar k_0}\right)^2\right], \quad \text{if } |p_y| \leqslant \hbar k_0, \\ &= 0, \quad \text{otherwise,}\end{aligned}\right\} \tag{15.7–10}$$

and similarly for $\mathcal{P}_1(p_x)$. The higher-order probabilities then follow from these expressions with the help of the convolution integral (15.7–4).

A few of the momentum transfer probabilities are illustrated in Fig. 15.13. It should be noted that $\mathcal{P}_1(p_y)$ is symmetric about the origin because $\langle p_y \rangle_1 = 0$, whereas $\mathcal{P}_1(p_z)$, $\mathcal{P}_2(p_z)$, $\mathcal{P}_3(p_z)$ are symmetric about $\hbar k_0$, $2\hbar k_0$, $3\hbar k_0$, respectively, and illustrate the progressive momentum transfer to the atom. Also $\mathcal{P}_1(p_y)$, $\mathcal{P}_1(p_z)$ have discontinuities and $\mathcal{P}_2(p_z)$ has a discontinuous derivative, whereas the higher-order probabilities are smoother. Indeed $\mathcal{P}_3(p_z)$ is already fairly close to the asymptotic (Gaussian) form.

The momentum transfer to an atom has been studied experimentally in measurements of the deflection of an atomic beam (Frisch, 1933; Picqué and Vialle, 1972; Schieder, Walther and Wöste, 1972), and also in the presence of a standing wave (Arimondo, Lew and Oka, 1979). Figure 15.14 shows the results of measurements in which a beam of sodium atoms was illuminated at right angles with the light from a sodium lamp, and the deflection was studied with the help of a hot wire ionization detector. It is estimated that the average deflection of about 50 μm was produced by one absorbed photon. It should be noted that the atomic beam suffers a net deflection and broadening in the direction of the incident light, but only a broadening in the perpendicular direction, as expected. The measured

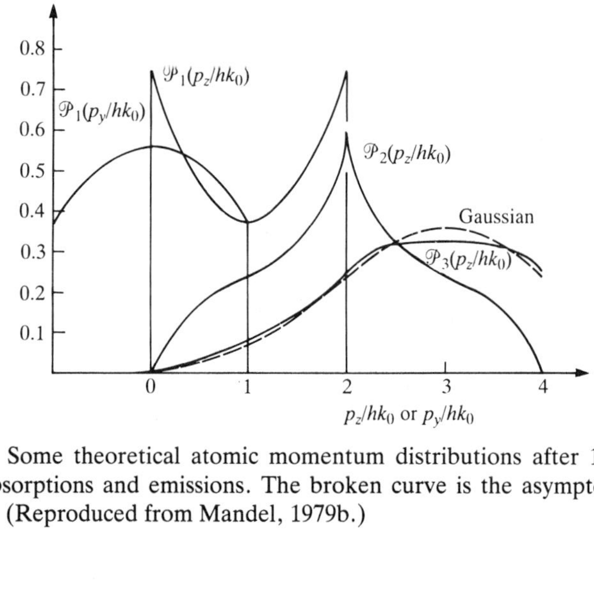

Fig. 15.13 Some theoretical atomic momentum distributions after 1, 2 and 3 photon absorptions and emissions. The broken curve is the asymptotic Gaussian form. (Reproduced from Mandel, 1979b.)

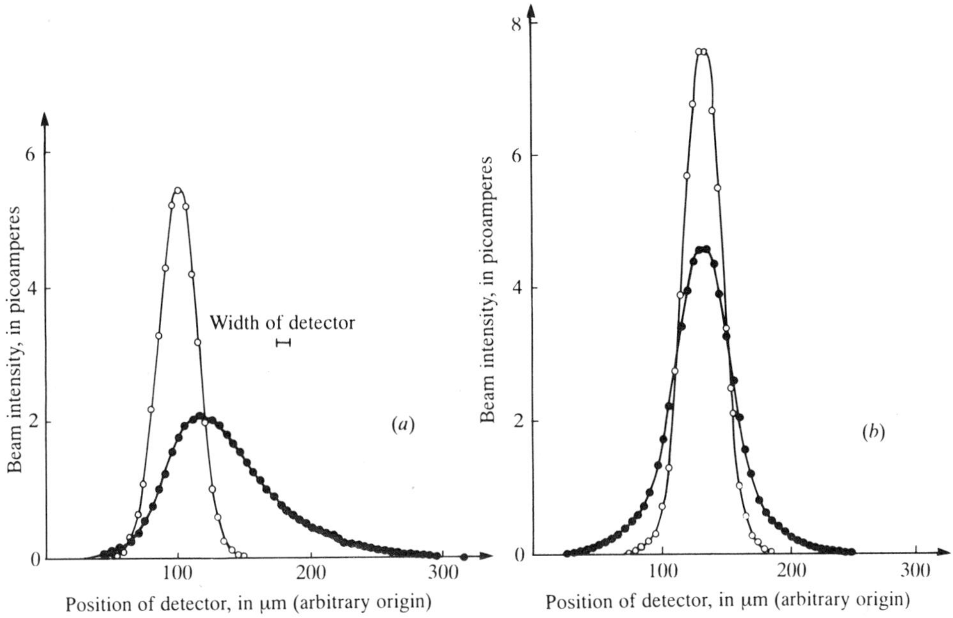

Fig. 15.14 Measured atomic beam shapes with and without resonance radiation (*a*) Parallel to the direction of the incident light, (*b*) perpendicular to the direction of the incident light. The open and the black circles correspond to measurements without the light and with the light, respectively. (Reproduced from Picqué and Vialle, 1972.)

distributions are, however, very different from the corresponding curves in Fig. 15.13. The difference is largely attributable to the velocity spread of the incident atoms.

15.7.2 Momentum transfer by stimulated emission or gradient forces in a strong field

When the field is sufficiently strong, stimulated emissions dominate over spontaneous emissions, and, in a time interval short compared with the natural lifetime, the spontaneous emission can be largely ignored. We are then justified in treating the field essentially in classical terms. As we have already pointed out, in a plane-wave field there is no net momentum transfer to the atom when a photon absorption is followed by a stimulated photon emission, because the two momentum exchanges just cancel. However, the situation is quite different in a non-uniform electromagnetic field, in which the net forces on the induced atomic dipole can be very large. This subject has been investigated by Letokhov and Pavlik (1976); Letokhov, Minogin and Pavlik (1976, 1977); Ashkin (1978); Bjorkholm, Freeman, Ashkin and Pearson (1978, 1980); Cook (1978, 1979, 1980a,b); Cook and Bernhardt (1978); Kazantsev (1978); Balykin, Letokhov and Mushin (1979); Ashkin and Gordon (1979); Letokhov and Minogin (1979); Salomon, Dalibard, Aspect, Metcalf and Cohen-Tannoudji (1987); Westbrook,

Watts, Tanner, Rolston, Phillips, Lett and Gould (1990); Bigelow and Prentiss (1990).

As an example, let us consider a standing wave field of the form

$$\mathbf{E}(z, t) = 2\boldsymbol{\varepsilon} E_0 \cos kz \cos \omega t,$$

which can, of course, be regarded as a superposition of two plane waves propagating in opposite directions along the z-axis. An atom travelling in the x-direction, say, may absorb a photon from one of the plane waves, and it may then be induced to emit a photon by the other plane wave, so as to acquire a net momentum component $\pm 2\hbar k$ in the z-direction. This process could be repeated any number of times, and if we ignore spontaneous emissions, would result in the deflection of an atomic beam travelling in the direction of the x-axis in several sharply defined directions. In effect, the standing wave pattern of light acts as a diffraction grating for the resonant atoms (Cook and Bernhardt, 1978; Arimondo, Lew and Oka, 1979; Martin, Gould, Oldaker, Miklich and Pritchard, 1987).

More generally, let us consider the equations of motion for a two-level atom in a classical electromagnetic field of the form

$$\mathbf{E}(\mathbf{r}, t) = \boldsymbol{\varepsilon}\mathscr{E}(\mathbf{r})\, \mathrm{e}^{\mathrm{i}[\phi(\mathbf{r}) - \omega t]} + \text{c.c.}$$

The total energy $\hat{H}$ of the atom of mass M and momentum $\hat{\mathbf{p}}$ is then

$$\hat{H} = (E_0 + \hbar\omega \hat{R}_3) + (\hat{\mathbf{p}}^2/2M) - \hat{\boldsymbol{\mu}} \cdot \mathbf{E}(\hat{\mathbf{r}}, t), \tag{15.7–11}$$

in which the first term represents the internal atomic energy, the second term the kinetic energy and the third term the dipole interaction energy. As the atomic position $\hat{\mathbf{r}}$ is now one of the dynamical variables, it has to be treated as a Hilbert space operator.

The rate of change of the atomic momentum is the force $\hat{\mathbf{F}}$ on the atom, and from Eq. (15.7–11) and the Heisenberg equation of motion for $\hat{\mathbf{p}}$ this force is given by

$$\hat{\mathbf{F}} = \dot{\hat{\mathbf{p}}} = \frac{\mathrm{i}}{\hbar}[\hat{\mathbf{p}}, \hat{\boldsymbol{\mu}} \cdot \mathbf{E}(\hat{\mathbf{r}}, t)].$$

In this equation $\hat{\mathbf{p}}$ and $\hat{\mathbf{r}}$ operate on the state of the atom that characterizes the motion, while $\hat{\boldsymbol{\mu}}$ operates on the internal state that describes the atomic excitation. If the initial state is one in which the atomic position is sufficiently well defined that $\mathbf{E}(\mathbf{r}, t)$ does not change appreciably over this range of $\mathbf{r}$, we can replace the position operator $\hat{\mathbf{r}}$ by its expectation value $\mathbf{r}$ at time t [cf. Eq. (14.1–19)]. When the momentum operator $\hat{\mathbf{p}}$ is expressed in the differential form $(\hbar/i)\boldsymbol{\nabla}$, the equation of motion becomes

$$\hat{\mathbf{F}} = \dot{\hat{\mathbf{p}}} = \boldsymbol{\nabla}\hat{\boldsymbol{\mu}} \cdot \mathbf{E}(\mathbf{r}, t). \tag{15.7–12}$$

We can now take expectation values of both sides of this equation, which allows us to write the following classical equation of motion for the atomic trajectory:

$$\langle \hat{\mathbf{F}} \rangle = M\ddot{\mathbf{r}} = \boldsymbol{\nabla}\langle \hat{\boldsymbol{\mu}} \rangle \cdot \mathbf{E}(\mathbf{r}, t). \tag{15.7–13}$$

If $\langle \hat{\boldsymbol{\mu}} \rangle$ does not change rapidly with position, in a standing wavefield the force may change sign at the nodes and cause the atom to be effectively trapped

(Kazantsev, 1976; Ashkin 1978). However, strictly speaking, $\langle \hat{\boldsymbol{\mu}} \rangle$, which is given by Eq. (15.2–16), must be regarded as a function of atomic position and time. In order to determine the atomic trajectory, it is necessary to solve Eq. (15.7–13) and the Bloch equations (15.3–11) [or Eqs. (15.3–19) in the rotating frame] simultaneously. In general this is a non-trivial problem, especially when we recall that both the atomic Rabi frequency Ω and the phase ϕ in Eqs. (15.3–11) or Eq. (15.3–19) are not constant in the frame of the moving atom, even though they may be constant in the laboratory frame (Cook, 1980b; Gordon and Ashkin, 1980). In general we have the relations

$$\dot{\phi} = \dot{\mathbf{r}} \cdot \nabla \phi(\mathbf{r})$$
$$\dot{\Omega} = \dot{\mathbf{r}} \cdot \nabla \Omega(\mathbf{r}).$$

The problem of solving the coupled equations has been discussed by Cook (1978, 1979, 1980a,b), who showed that under certain circumstances, the force on the atom may lead to focusing and defocusing of the atomic trajectory. Here we follow his treatment.

The stimulated emission force on the atom can contribute to the momentum fluctuations, even when the average force is zero, as in a plane wave (in the absence of spontaneous emissions). This is because any momentum gained during the absorption of a photon results in a perturbation of the atomic motion, even though this momentum is again lost in the process of stimulated emission. The fluctuation of $\hat{\mathbf{F}}$ is readily calculated from Eq. (15.7–12). If we make use of Eq. (15.1–17) for the dipole moment $\hat{\boldsymbol{\mu}}$, and introduce the Rabi frequency $\Omega(\mathbf{r})$ through Eq. (15.3–12) as usual, we obtain from Eq. (15.7–12)

$$\hat{\mathbf{F}} = \tfrac{1}{2}\hbar[\hat{b}\, e^{i\omega t}\nabla(\Omega e^{-i\phi}) + \text{h.c.}],$$

when terms that oscillate at double the optical frequency are discarded. With the help of the anticommutation relations (15.1–7) for the atomic variables, this immediately leads to the result

$$\langle \hat{\mathbf{F}}^2 \rangle = \tfrac{1}{4}\hbar^2[(\nabla\Omega)^2 + \Omega^2(\nabla\phi)^2] \tag{15.7–14}$$

for any quantum state. If follows that an atom experiences a fluctuating stimulated emission force even in a plane wave for which $\phi(\mathbf{r}) = \mathbf{k} \cdot \mathbf{r}$, $\nabla\phi(\mathbf{r}) = \mathbf{k}$ and $\Omega(\mathbf{r})$ is constant.

With the help of some simplifying assumptions, the motion of the two-level atom can be treated quantum mechanically, without the introduction of expectation values from the beginning, as in Eq. (15.7–13). If the optical frequency ω coincides with the atomic frequency ω_0 and the phase $\phi(\mathbf{r}) = 0$, and if we are dealing with linearly polarized light and a $\Delta m = 0$ transition, so that the dipole moment $\hat{\boldsymbol{\mu}} = 2\boldsymbol{\mu}_{12}\hat{R}_1$, then the total energy (15.7–11) reduces to

$$\hat{H} = (E_0 + \hbar\omega\hat{R}_3) + \hat{\mathbf{p}}^2/2M - 2\boldsymbol{\mu}_{12} \cdot \boldsymbol{\varepsilon}\mathscr{E}(\mathbf{r})\hat{R}_1 \cos\omega t.$$

The state $|\psi(t)\rangle$ of the atom at any time satisfies the Schrödinger equation

$$i\hbar\frac{\partial}{\partial t}|\psi(t)\rangle = \hat{H}|\psi(t)\rangle, \tag{15.7–15}$$

which governs both the external and the internal motion. If $|\psi(t)\rangle$ is expanded in terms of atomic states $|n, \mathbf{r}\rangle$ of definite excitation $n = 1, 2$, and definite position $\mathbf{r}$, in the form

$$|\psi(t)\rangle = \sum_{n=1}^{2} \int \psi_n(\mathbf{r}, t)\, \mathrm{e}^{-\mathrm{i}E_n t/\hbar} |n, \mathbf{r}\rangle\, \mathrm{d}^3 r,$$

with $E_1 = E_0 - \frac{1}{2}\hbar\omega_0$, $E_2 = E_0 + \frac{1}{2}\hbar\omega_0$, and this expansion is substituted in Eq. (15.7–15), then $\psi_1(\mathbf{r}, t)$ and $\psi_2(\mathbf{r}, t)$ are found to obey the two coupled equations of motion

$$\left.\begin{aligned} -\frac{\hbar^2}{2m}\nabla^2\psi_1(\mathbf{r}, t) - \tfrac{1}{2}\boldsymbol{\mu}_{12}\cdot\boldsymbol{\varepsilon}\mathscr{E}(\mathbf{r})\psi_2(\mathbf{r}, t) &= \mathrm{i}\hbar\frac{\partial}{\partial t}\psi_1(\mathbf{r}, t) \\ -\frac{\hbar^2}{2m}\nabla^2\psi_2(\mathbf{r}, t) - \tfrac{1}{2}\boldsymbol{\mu}_{12}\cdot\boldsymbol{\varepsilon}\mathscr{E}(\mathbf{r})\psi_1(\mathbf{r}, t) &= \mathrm{i}\hbar\frac{\partial}{\partial t}\psi_2(\mathbf{r}, t), \end{aligned}\right\} \tag{15.7–16}$$

after terms oscillating at twice the optical frequency are discarded. The two equations can be uncoupled by introducing orthogonal, symmetric and antisymmetric wave functions

$$\left.\begin{aligned} \phi_1(\mathbf{r}, t) &= \frac{1}{\sqrt{2}}[\psi_1(\mathbf{r}, t) + \psi_2(\mathbf{r}, t)] \\ \phi_2(\mathbf{r}, t) &= \frac{1}{\sqrt{2}}[\psi_1(\mathbf{r}, t) - \psi_2(\mathbf{r}, t)], \end{aligned}\right\} \tag{15.7–17}$$

which satisfy the equations of motion

$$\left.\begin{aligned} -\frac{\hbar^2}{2m}\nabla^2\phi_1(\mathbf{r}, t) - \tfrac{1}{2}\boldsymbol{\mu}_{12}\cdot\boldsymbol{\varepsilon}\mathscr{E}(\mathbf{r})\phi_1(\mathbf{r}, t) &= \mathrm{i}\hbar\frac{\partial}{\partial t}\phi_1(\mathbf{r}, t) \\ -\frac{\hbar^2}{2m}\nabla^2\phi_2(\mathbf{r}, t) + \tfrac{1}{2}\boldsymbol{\mu}_{12}\cdot\boldsymbol{\varepsilon}\mathscr{E}(\mathbf{r})\phi_2(\mathbf{r}, t) &= \mathrm{i}\hbar\frac{\partial}{\partial t}\phi_2(\mathbf{r}, t). \end{aligned}\right\} \tag{15.7–18}$$

The two wave functions $\phi_1(\mathbf{r}, t)$ and $\phi_2(\mathbf{r}, t)$ therefore propagate independently of each other, and, because they are orthogonal, they do not interfere. An atom that is initially in the ground state has equal amplitudes $|\phi_1(\mathbf{r}, 0)|$ and $|\phi_2(\mathbf{r}, 0)|$. Because the potential energies in the two Schrödinger equations (15.7–18) appear with opposite sign, the electromagnetic forces acting on an atom in the two states $[|1, \mathbf{r}\rangle + |2, \mathbf{r}\rangle]/\sqrt{2}$ and $[|1, \mathbf{r}\rangle - |2, \mathbf{r}\rangle]/\sqrt{2}$ are opposite. An atomic beam of atoms initially in the ground state should therefore split into two components. Cook (1978, 1979) has shown that a perpendicular standing wave-field should act somewhat like a converging lens for atoms described by $\phi_1(\mathbf{r}, t)$ and like a diverging lens for atoms described by $\phi_2(\mathbf{r}, t)$. Figure 15.15 shows the expected distribution of the atomic flux associated with the $\phi_1(\mathbf{r}, t)$ and $\phi_2(\mathbf{r}, t)$ waves at a distance from the interaction region where the focusing is most pronounced. The peaks are the result of converging and diverging trajectories. The observed distribution when the incident atoms are in the ground state should be the sum $|\phi_1(\mathbf{r}, t)|^2 + |\phi_2(\mathbf{r}, t)|^2$ of the two illustrated distributions.

Finally, we mention that the resonant interaction responsible for the deflection of atoms, may, under certain circumstances, also cause deflection of light beams

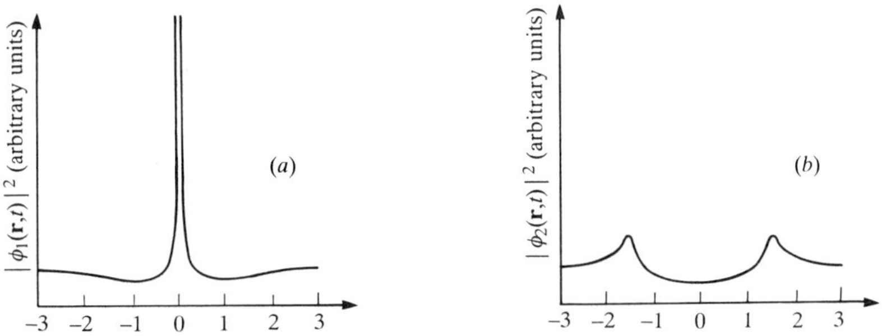

Fig. 15.15 The atomic density (*a*) $|\phi_1(\mathbf{r}, t)|^2$, (*b*) $|\phi_2(\mathbf{r}, t)|^2$ as a function of position transverse to the atomic beam, in the focal plane. (*a*) shows focusing and (*b*) defocusing of the incident atoms. (Reproduced from Cook, 1978.)

traveling through a gas of resonant atoms. This has been demonstrated in experiments by Tam and Happer (1977), who observed that two initially parallel, circularly polarized light beams with equal polarizations passing through alkali vapor 'attract' each other, while beams of opposite polarizations 'repel'. The effects are attributable to the exchange of polarized atoms between the two light beams.

15.8 Cooling and trapping of atoms

When the number of photon absorptions and spontaneous emissions becomes large, the resonant interaction between an atom and light can cause the atom to be slowed down and cooled, and eventually to be trapped by the light. It was suggested by Hänsch and Schawlow (1975) that neutral atoms might be cooled to temperatures of the order of a millikelvin by the action of a near resonant laser beam that excites the atoms, followed by spontaneous emission, and a similar suggestion was made independently by Wineland and Dehmelt (1975) for trapped ions. Although the dipole forces on atoms associated with stimulated rather than spontaneous emissions can also be used to cool and trap atoms (Ashkin, 1978; Ashkin and Gordon, 1979; Gordon and Ashkin, 1980; Dalibard and Cohen-Tannoudji, 1985; Kazantsev, Ryabenko, Surdutovich and Yakovlev, 1985; Aspect, Dalibard, Heidmann, Salomon and Cohen-Tannoudji, 1986; Chu, Bjorkholm, Ashkin and Cable, 1986; Prentiss and Cable, 1989), we shall here confine our attention to the first mechanism.

In order to understand the principle, let us consider the situation shown in Fig. 15.16, in which a two-level atom of midfrequency ω_0, mass M and velocity $\mathbf{v}$ encounters a quasi-monochromatic laser beam of wave vector $\mathbf{k}_L$ and frequency ω_L. Suppose that the laser linewidth is much narrower than both the natural line

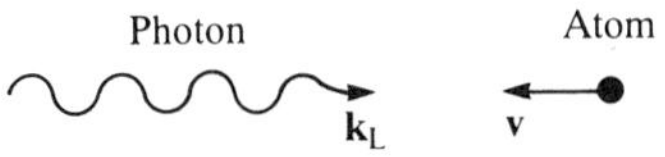

Fig. 15.16 Illustration of a situation in which the atom sees a Doppler-shifted photon of frequency $\omega_L - \mathbf{k}_L \cdot \mathbf{v} = \omega_L\,(1 + \omega v/c)$.

and the Doppler-broadened atomic line, and that ω_L is detuned so as to lie below the atomic midfrequency ω_0 by some amount D. If the vectors $\mathbf{v}$ and $\mathbf{k}_L$ point in opposite directions, then in the rest frame of the moving atom the light is recognized as having the Doppler-shifted frequency

$$\omega_L - \mathbf{k}_L \cdot \mathbf{v} = \omega_L(1 + v/c).$$

If $\omega_L v/c$ is of order D, the incoming photons are strongly resonant with the atomic transition, and absorption of one such photon results in a momentum transfer of $\hbar \mathbf{k}_L$ to the atom in the laboratory frame, which reduces its momentum $M\mathbf{v}$ to $M(\mathbf{v} + \Delta\mathbf{v})$. The change of velocity $\Delta\mathbf{v}$ is given by the relation

$$M\Delta\mathbf{v} = \hbar\mathbf{k}_L, \tag{15.8–1}$$

and it is in a direction opposite to $\mathbf{v}$, so that the atom slows down. When the excited atom spontaneously emits a photon of wave vector $\mathbf{k}$, its momentum changes again by $-\hbar\mathbf{k}$. But since $\langle \hbar\mathbf{k} \rangle = 0$ it follows that the mean momentum transfer to the atom in one absorption and one emission is given by Eq. (15.8–1) and results in the slowing of the atom. Needless to say, if the vectors $\mathbf{v}$ and $\mathbf{k}_L$ pointed in the same direction, then the Doppler-shifted frequency would be $\omega_L(1 - v/c)$, which is sufficiently far below the atomic midfrequency ω_0 to make absorption of a photon unlikely.

Atoms moving towards the light are therefore slowed down by repeated absorption and spontaneous re-emission, whereas those moving away are not. By arranging for the atoms to be illuminated from all sides, so that some $\mathbf{k}_L$ always has a component opposing the atomic velocity, we can ensure that virtually all atoms are slowed down. Equation (15.8–1) yields $\Delta v \approx 6$ cm/s for magnesium atoms exposed to light below the resonance frequency ($\lambda = 2851.1$ Å). As the r.m.s. velocity $v = (3k_B T/M)^{1/2}$, where k_B is the Boltzmann constant, at $T = 600$ K is about 800 m/s, approximately $v/\Delta v = 13\,000$ cycles of photon absorption and emission should reduce the velocity to near zero. In this case the atomic linewidth is reduced from the Doppler-broadened value close to the natural value. If the pump light intensity is high enough to saturate the atomic transition, with a radiative lifetime T_1 of 2 ns for the upper state, the cycling time for photon absorption and spontaneous emission is close to 4 ns. It will therefore take a minimum of $(v/\Delta v)T_1 \approx 60$ μs for an atom to be cooled to near zero temperature and brought virtually to rest. Hänsch and Schawlow (1975), who proposed this idea, suggested that six laser beams arranged parallel and antiparallel to the three coordinate axes should make it possible to implement the scheme in practice, and these beams also have the effect of concentrating the slow atoms. Effective temperatures of order 70 mK were realized in experiments with neutral sodium atoms by Phillips and Metcalf (1982) and by Prodan, Phillips and Metcalf (1982). The first neutral atomic trap based on spontaneous radiation forces was reported by Raab, Prentiss, Cable, Chu and Pritchard (1987).

15.8.1 Optical molasses

Although Eq. (15.8–1) and the accompanying argument accounts in general terms for the observation of atomic slowing and trapping, it is important to remember that it refers only to the average momentum transfer, averaged over many cycles of absorption and emission. Because the spontaneous emissions are randomly directed they introduce large fluctuations into the atomic motion, and in any given time interval the number of fluorescence cycles is itself a random variable. The result is that the motion of the atom acquires some of the character of Brownian motion, and the r.m.s. distance traveled by the atom in a given time interval t is proportional to $\sqrt{t}$ rather than to t. The diffusive atomic motion produced by the light is reminiscent of motion in a viscous fluid, and it has been labeled *optical molasses*. The phenomenon was first observed by Chu, Hollberg, Bjorkholm, Cable and Ashkin (1985), and it was also studied by Phillips, Prodan and Metcalf (1985).

Figure 15.17 shows an outline of an apparatus used to investigate cooling and viscous confinement of sodium atoms in optical molasses. The atomic beam of sodium travels along the axis of a tapered solenoid, that compensates for the changing Doppler shift with a changing Zeeman shift, where it is cooled continuously by a counterpropagating laser beam. Atoms that are not stopped completely by the time they reach the end of the solenoid escape into the main experimental chamber, where some are trapped at the common intersection of six laser beams in the molasses. The maximum light power density for the molasses was of order 60 mW/cm^2 in the experiment.

15.8.2 An estimate of achievable low temperature based on energy balance

Let us briefly examine how the final atomic temperature T is expected to depend on the frequency detuning $D \equiv \omega_0 - \omega_L$ of the laser below resonance and on the light intensity I. An atom of mass M traveling in the pseudo-viscous fluid is subject to a damping force $-\alpha v$ proportional to the mean velocity v, that causes it to lose energy (to be cooled) at the average rate

$$\left(\frac{dE}{dt}\right)_{\text{cool}} = -\alpha v^2. \tag{15.8–2}$$

The damping coefficient α is approximately given by (Lett, Phillips, Rolston, Tanner, Watts and Westbrook, 1989)

$$\alpha = -4\hbar k^2 \frac{I}{I_0} \frac{2D/\Gamma}{[1 + (2D/\Gamma)^2]^2} \quad \text{when } kv \ll \Gamma,\ D, \quad (I \ll I_0), \tag{15.8–3}$$

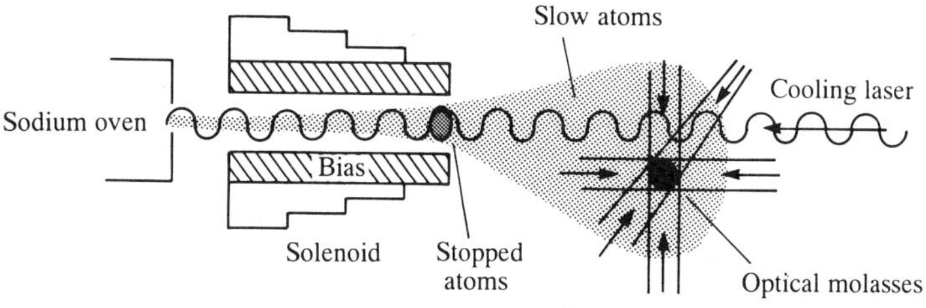

Fig. 15.17 Outline of the apparatus used to investigate cooling and trapping of atoms in optical molasses. (Reproduced from Phillips, Gould and Lett, 1988.)

where I_0 is the saturation light intensity, Γ is the natural linewidth and $k = \omega/c$. Equation (15.8–3) is obtainable from the theory of resonance fluorescence of a two-level atom (Section 15.6). At the same time the atom gains energy at the rate $(\mathrm{d}E/\mathrm{d}t)_{\mathrm{heat}}$ from the heating or dissipation that always accompanies Brownian motion, due to the random photon emissions and absorptions (see the fluctuation–dissipation theorem treated in Section 17.2). In one dimension the heating rate can be expressed in the form (Lett, Phillips, Rolston, Tanner, Watts and Westbrook, 1989)

$$\left(\frac{\mathrm{d}E}{\mathrm{d}t}\right)_{\mathrm{heat}} = \frac{\hbar^2 k^2 \Gamma}{M} \frac{I/I_0}{1 + (2D/\Gamma)^2}. \qquad (15.8\text{–}4)$$

In the steady state the energy loss and gain rates must be equal, so that

$$\langle v \rangle^2 = \frac{\hbar \Gamma}{4M} \frac{1 + (2D/\Gamma)^2}{2D/\Gamma}. \qquad (15.8\text{–}5)$$

If we now identify the mean squared velocity in one dimension multipled by $\frac{1}{2}M$ in the usual way with $\frac{1}{2}k_{\mathrm{B}}T$, then we obtain the standard relation for the achievable low temperature T

$$k_{\mathrm{B}}T = \frac{\hbar \Gamma}{4} \frac{1 + (2D/\Gamma)^2}{2D/\Gamma}. \qquad (15.8\text{–}6)$$

This expression has its minimum value when $2D/\Gamma = 1$, in which case $k_{\mathrm{B}}T_{\mathrm{min}} = \hbar\Gamma/2$. The formula defines the so-called Doppler limit for cooling atoms by this technique. For sodium atoms Eq. (15.8–6) implies an achievable low temperature of the order of 1 mK.

Experiments conducted at NIST (the National Institute for Science and Technology) and elsewhere during the late 1980s (Lett, Watts, Westbrook, Phillips, Gould and Metcalf, 1988; Lett, Phillips, Rolston, Tanner, Watts and Westbrook, 1989; Weiss, Riis, Shevy, Ungar and Chu, 1989) demonstrated, however, that substantially lower temperatures can be realized by laser polarization gradient cooling. Figure 15.18 shows the results of temperature measurements for sodium

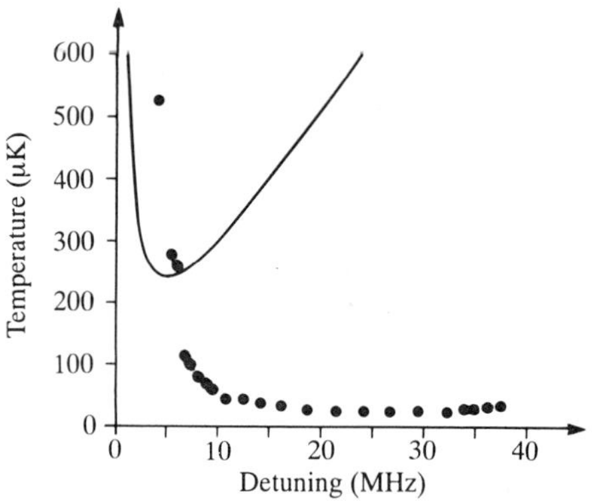

Fig. 15.18 Results of measurements of atom temperature versus frequency detuning of the laser below the atomic line center. The solid curve is the predicted variation according to Eq. (15.8–6). (Reproduced from Lett, Phillips, Rolston, Tanner, Watts and Westbrook, 1989.)

optical molasses as a function of frequency detuning D below resonance. Evidently low temperatures of order some tens of microkelvins can be reached by the technique, and this is more than an order of magnitude lower than the limit given by Eq. (15.8–6). The full curve in Fig. 15.18 corresponds to Eq. (15.8–6).

These and other results obtained at Bell Laboratories (Chu, Prentiss, Cable and Bjorkholm, 1987) demonstrated the inadequacy of the simple treatment given above and the need for a newer and better theory. Such a theory was developed by Dalibard and Cohen-Tannoudji (1989) and by Ungar, Weiss, Riis and Chu, (1989) shortly thereafter (see also Cohen-Tannoudji and Phillips, 1990). The new approach is much more complicated and includes a more realistic, multilevel treatment of the atoms, the role of optical pumping and light shifts and the role of polarization gradients in the laser beam. The new theories exhibit much improved agreement with experiment, but they are beyond the scope of this chapter.

Laser cooling and trapping of atoms became the subject of a great deal of research activity in the 1980s and 1990s, with an ever expanding literature (see for example Dalibard, Raimond and Zinn-Justin, 1992; Arimondo, Philips and Strumia, 1992).

Problems

15.1 A two-level atom which is held fixed at the origin interacts with a quantized electromagnetic field according to the interaction Hamiltonian

$$\hat{H}_I(t) = -\hat{\boldsymbol{\mu}}(t) \cdot \hat{\mathbf{A}}(0, t).$$

Assume that the quantum numbers of the two atomic states differ by $\Delta m = \pm 1$, so that the transition dipole moment $\boldsymbol{\mu}_{12}$ is a complex vector. Show that the equations of motion of the field vectors $\hat{\mathbf{E}}(\mathbf{r}, t)$ and $\hat{\mathbf{B}}(\mathbf{r}, t)$ in the Heisenberg picture are Maxwell's equations with current density $\hat{\mathbf{j}}(\mathbf{r}, t)$ given by the formula

$$\hat{\mathbf{j}}_l(\mathbf{r}, t) = -\mathrm{i}\omega_0 \delta^{\mathrm{T}}_{lm}(\mathbf{r})(\boldsymbol{\mu}_{12})_m \hat{b}(t) + \text{h.c.},$$

where $\delta^{\mathrm{T}}_{lm}(\mathbf{r})$ is the transverse delta function.

15.2 A two-level atom of transition dipole moment $\boldsymbol{\mu}_{12}$ is located at the origin. Show that, to order $1/r$, the electric field $\hat{\mathbf{E}}^{(+)}(\mathbf{r}, t)$ at a point $\mathbf{r}$ in the far field of the atom is given by the formula

$$\hat{\mathbf{E}}^{(+)}(\mathbf{r}, t) = \frac{\omega_0^2}{4\pi\varepsilon_0 c^2 r}\left[\boldsymbol{\mu}_{12} - \frac{(\boldsymbol{\mu}_{12} \cdot \mathbf{r})\mathbf{r}}{r^2}\right]\hat{b}(t - r/c) + \hat{\mathbf{E}}^{(+)}_{\text{free}}(\mathbf{r}, t), \quad (t \leqslant r/c).$$

15.3 A two-level atom is undergoing resonance fluorescence in a coherent exciting field of constant amplitude that is turned on at time $t = 0$. Calculate the time development of the average light intensity $\langle \hat{I}(t) \rangle$ radiated by the atom, when the exciting field is detuned from the atomic resonance.

15.4 Under the circumstances described in the previous problem, explain why the cross-section for scattering by the atom in the steady state decreases with increasing pump-field intensity.

15.5 Under the conditions described in Problem 15.3, show that the expectation of any atomic operator expressible in terms of the atomic inversion $\hat{R}_3(t)$ becomes independent of the initial state of the atom in the long-time limit $t \to \infty$.

15.6 Calculate the spectrum of the light radiated by an atom undergoing resonance fluorescence in the steady state, when the exciting field is off resonance and weak, and when the Rabi frequency is much smaller than the natural line width.

16

Collective atomic interactions

In the previous chapter we studied the interaction between a single two-level atom and the electromagnetic field, both when the field is treated classically and when it is quantized. We encountered some interesting phenomena such as Rabi oscillations, the a.c. Stark effect, and photon antibunching, all of which have been observed. These phenomena are essentially single-atom effects, in the sense that either they require a single atom for the effect to be observed, as in the last case, or at least they do not require more than one, although a group of atoms may be used in practice.

In this chapter we shall turn to a discussion of some effects that depend in an essential way on the presence of a group of atoms. In some cases we shall find that the group or collective behavior of the atomic system is relatively trivial, in the sense that we can account for the phenomenon by summing the contributions of the individual atoms to the total field, and treating each of them as if it acts almost independently of the others. This is the situation in *free induction decay* and in the *photon echo*. In other cases it is essential to include the effect of each atom on all the other atoms, because this modifies the behavior of each in a significant way. These phenomena, such as *self-induced transparency* and *superradiance*, are collective effects in a deeper sense. They are sometimes called *cooperative effects*.

16.1 Optical free induction decay

This is a simple dephasing effect resulting from the spread of atomic frequencies. Let us consider a group of N inhomogeneously broadened two-level atoms distributed over a region of space that is small compared with the wavelength. The restriction to a small region simplifies the problem by allowing us to ignore propagation delays. By an *inhomogeneously broadened system* we mean one in which different atoms have different resonant frequencies ω, with some distribution $g(\omega)$ centered on the midfrequency ω_0 normalized so that

$$\frac{1}{2\pi}\int_0^\infty g(\omega)\,\mathrm{d}\omega = 1. \tag{16.1–1}$$

The spectrum of the fluorescent light emitted by the group of atoms is therefore broader than the natural linewidth, by an amount that depends on the width of

$g(\omega)$. This width is usually denoted by $1/T_2^*$, where T_2^* is the *inhomogeneous lifetime*. In many cases this inhomogeneous width is much larger than the natural linewidth $1/T_1$, where T_1 is the *natural lifetime*. In a solid, the inhomogeneous broadening is generally the result of variations in the field of the crystal lattice, which affects different atoms differently. In a gas, inhomogeneous broadening arises from the different Doppler shifts associated with atoms moving with different velocities. In addition, *homogeneous broadening* effects that affect all atoms equally, and that are characterized by a certain homogeneous lifetime T_2', with $T_2' \leqslant T_1$, may also be present. The total optical linewidth $1/T_2$ is the sum of the homogeneous and inhomogeneous linewidths, i.e.

$$\frac{1}{T_2} = \frac{1}{T_2'} + \frac{1}{T_2^*}. \tag{16.1–2}$$

However, in the following we shall suppose that $T_2^* \ll T_2'$, so that the total linewidth is dominated by inhomogeneous broadening. The Fourier transform of the spectral distribution $g(\omega)$ is a certain autocorrelation function, that we denote by $G(\tau)\,\mathrm{e}^{-\mathrm{i}\omega_0\tau}$,

$$\mathrm{e}^{-\mathrm{i}\omega_0\tau}G(\tau) = \frac{1}{2\pi}\int_0^\infty g(\omega)\,\mathrm{e}^{-\mathrm{i}\omega\tau}\,\mathrm{d}\omega$$

so that

$$G(\tau) = \frac{1}{2\pi}\int_{-\omega_0}^\infty g(\omega_0 + \omega')\,\mathrm{e}^{-\mathrm{i}\omega'\tau}\,\mathrm{d}\omega', \tag{16.1–3}$$

where $G(\tau)$ is a slowly varying function with a range of order T_2^*. For example, if $g(\omega)$ is Gaussian, i.e. if

$$g(\omega) = \sqrt{(2\pi)}T_2^*\exp\left[-\tfrac{1}{2}(\omega - \omega_0)^2 T_2^{*2}\right], \tag{16.1–4a}$$

then

$$G(\tau) = \exp\left[-\tfrac{1}{2}\tau^2/T_2^{*2}\right]. \tag{16.1–4b}$$

For time intervals short compared with T_2', damping or atomic decay can be neglected, and, as we know from the previous chapter, each atomic Bloch-vector $\mathbf{r}(\omega, t)$ rotates at frequency ω about the z-axis, and contributes a mean dipole moment given by the expression

$$\mathrm{Re}\,(\boldsymbol{\mu}_{12})r_1(\omega, t) + \mathrm{Im}\,(\boldsymbol{\mu}_{12})r_2(\omega, t) = \tfrac{1}{2}\boldsymbol{\mu}_{12}[r_1(\omega, t) - \mathrm{i}r_2(\omega, t)] + \mathrm{c.c.}$$

If the transition dipole moment $\boldsymbol{\mu}_{12}$ is the same for all the atoms, then the total mean dipole moment $\langle\hat{\boldsymbol{\mu}}(t)\rangle$ produced by all N atoms is

$$\langle\hat{\boldsymbol{\mu}}(t)\rangle = \frac{N}{4\pi}\int_0^\infty \boldsymbol{\mu}_{12}[r_1(\omega, t) - \mathrm{i}r_2(\omega, t)]g(\omega)\,\mathrm{d}\omega + \mathrm{c.c.} \tag{16.1–5}$$

Provided the atoms are not all in the ground state, $\hat{\boldsymbol{\mu}}(t)$ oscillates in time even in the absence of any external field, and this causes electromagnetic radiation to be emitted. Let us calculate this radiation field semiclassically for a time short compared with T_2', in which damping can be neglected and very little of the atomic energy is lost.

We suppose that the atoms start off in the ground state, and that they are subjected to a constant, coherent optical excitation of frequency ω_1 for some time T that is short compared with T_2', as discussed in Section 15.3. At the end of the exciting pulse the components of the Bloch-vector in the rotating frame are given by Eqs. (15.3–21) with $t = T$. From this time on each Bloch-vector rotates uniformly about the z-axis with angular velocity ω, so that at a later time t,

$$r_1(\omega, t) - \mathrm{i}r_2(\omega, t) = [r_1(\omega, T) - \mathrm{i}r_2(\omega, T)]\,\mathrm{e}^{-\mathrm{i}\omega(t-T)}$$

and, with the help of Eq. (15.3–17b),

$$= [r_1'(\omega, T) - \mathrm{i}r_2'(\omega, T)]\,\mathrm{e}^{-\mathrm{i}\omega_1 T}\,\mathrm{e}^{-\mathrm{i}\omega(t-T)}$$

in terms of components in the rotating frame. With the help of Eqs. (15.3–21) we then have

$$r_1(\omega, t) - \mathrm{i}r_2(\omega, t) = \frac{\Omega}{\Omega'}\left[\frac{\omega - \omega_1}{\Omega'}(1 - \cos \Omega' T) + \mathrm{i}\sin \Omega' T\right]\mathrm{e}^{-\mathrm{i}\omega_1 T}\,\mathrm{e}^{-\mathrm{i}\omega(t-T)},$$
$$t \geqslant T, \quad (16.1\text{–}6)$$

where we have used the abbreviation

$$\Omega' \equiv [\Omega^2 + (\omega - \omega_1)^2]^{1/2}. \quad (16.1\text{–}7)$$

We now substitute from Eq. (16.1–6) into Eq. (16.1–5) and obtain the expression

$$\langle \hat{\boldsymbol{\mu}}(t)\rangle = \frac{N\boldsymbol{\mu}_{12}}{4\pi}\mathrm{e}^{-\mathrm{i}\omega_1 T}\,\mathrm{e}^{-\mathrm{i}\omega_0(t-T)}\int_{-\omega_0}^{\infty}\frac{\Omega}{\Omega'}\left[\frac{\omega'}{\Omega'}(1 - \cos \Omega' T) + \mathrm{i}\sin \Omega' T\right]$$
$$\times\, g(\omega_0 + \omega')\,\mathrm{e}^{-\mathrm{i}\omega'(t-T)}\,\mathrm{d}\omega' + \text{c.c.}, \quad t \geqslant T.$$

If the exciting field is sufficiently strong and is tuned to the center ω_0 of the atomic line, then Ω' depends only weakly on the atomic frequency ω over the range of $g(\omega)$. If we neglect this weak frequency dependence of Ω' under the integral, we have, with the help of Eq. (16.1–3),

$$\langle \hat{\boldsymbol{\mu}}(t)\rangle = \tfrac{1}{2}\mathrm{i}N\boldsymbol{\mu}_{12}\,\mathrm{e}^{-\mathrm{i}\omega_0 t}\left[G(t - T)\sin \Omega T + \dot{G}(t - T)\frac{(1 - \cos \Omega T)}{\Omega}\right] + \text{c.c.},\ t \geqslant T.$$
$$(16.1\text{–}8a)$$

If the dipole moment $\boldsymbol{\mu}_{12}$ is real and if the spectral distribution $g(\omega)$ is symmetric about ω_0, so that $G(t)$ is a real function, this can be written

$$\langle \hat{\boldsymbol{\mu}}(t)\rangle = N\boldsymbol{\mu}_{12}\sin \omega_0 t\left[G(t - T)\sin \Omega T + \dot{G}(t - T)\frac{(1 - \cos \Omega T)}{\Omega}\right],\ t \geqslant T.$$
$$(16.1\text{–}8b)$$

This shows that the macroscopic dipole moment of the collective atomic system oscillates at the center frequency ω_0, but that it falls to zero following the excitation pulse in a time of the order of the inhomogeneous lifetime T_2^*, when $G(t - T)$ and $\dot{G}(t - T)$ become very small. This decay of $\langle \hat{\boldsymbol{\mu}}(t)\rangle$, of course, has nothing to do with the decay of the atomic excitation, but is a consequence of the dephasing that occurs among atomic dipoles of different frequencies. Actually, the assumption we made at the beginning that all atoms are concentrated within a region of space that is small compared with a wavelength, has implications that go beyond the subject matter of this section. As we shall see in Section 16.4, the excitation of a collection of closely spaced excited atoms decays to zero much more rapidly than that of a single atom, so that it is not always possible to ignore this decay. Still, for sufficiently short inhomogeneous lifetimes T_2^*, Eqs. (16.1–8) give a good approximation to the macroscopic dipole moment.

As the far field of an oscillating dipole is proportional to the second time derivative of the dipole moment,‡ it follows that the intensity of the light radiated by the atomic system is proportional to

$$\omega_0^4 N^2 \left[G(t-T) \sin \Omega T + \frac{\dot{G}(t-T)}{\Omega}(1 - \cos \Omega T) \right]^2,$$

to a good approximation, and also decays to zero in a time of order T_2^* following the excitation. The light pulse radiated by the system is known as the *free induction pulse* and the phenomenon is called *free induction decay*. Of course the radiation process implies that the atoms must lose some energy, and the z-components of their Bloch-vectors cannot remain strictly constant in time, although they may be constant to a good approximation.

16.1.1 Experiments

Free induction decay was first observed in nuclear magnetic resonance experiments by Hahn, (1950a), and has since been studied in the optical domain also, particularly by Brewer and Shoemaker (1972). Figure 16.1 shows the free induction pulse at 10.6 μm wavelength that was observed by them in a sample of NH_2D. The excitation was produced by switching the frequency rather than the amplitude of the exciting light beam.

It should be emphasized that dependence of the free induction pulse intensity on the square of the number of atoms, is merely a reflection of the fact that the fields radiated by the different atoms remain in phase for a brief time of order T_2^*.

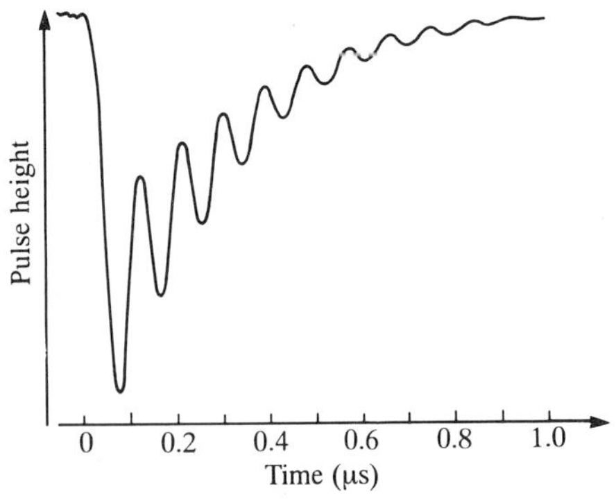

Fig. 16.1 The optical free induction pulse from a sample of NH_2D. (Reproduced from Brewer and Shoemaker, 1972.)

‡ For the derivation of the field produced by a classical oscillating point dipole see for example Born and Wolf (1980), Sec. 2.2.3.

This does not necessarily imply that the atoms are radiating cooperatively (cf. Section 16.5), although such cooperative effects may be present also. If the atoms were sufficiently far apart that they could not influence each other, we would still observe the free induction pulse at some distant point where all the fields interfere.

16.2 Photon echo

This phenomenon is a striking illustration of the rephasing that can occur in an inhomogeneously broadened atomic system, when the motion of the Bloch-vectors is reversed following the initial dephasing of the atomic dipoles. The effect is most easily understood by reference to Fig. 16.2. We consider an inhomogeneously broadened group of atoms whose Bloch-vectors in the rotating frame initially point in the direction of the y'-axis. If the rotating frame rotates at an angular velocity ω_0, which is the mid-frequency of the inhomogeneous line, then the Bloch-vectors of atoms whose natural frequencies exceed ω_0 will precess clockwise around the z'-axis, while those belonging to atoms with frequencies below ω_0 precess in the counter-clockwise direction. Initially the atomic system has a macroscopic dipole moment $\langle \hat{\boldsymbol{\mu}} \rangle$, but as time goes on a dephasing occurs, because the different Bloch-vectors precess at different rates, until the atomic Bloch-vectors are distributed almost uniformly around the x', y'-plane. The macroscopic dipole moment is then zero, and the system does not radiate, as we showed in the previous section, at least not in times short compared with the lifetime T_2'.

If the motion of the different Bloch-vectors at some time τ could be reversed, then the vectors would again become aligned along the y'-axis after a time 2τ, when each one has retraced its path. The atomic system would therefore radiate at this time. In fact, such a reversal can almost be brought about by the application of a short, intense π-pulse, because this causes a reversal of the y'-coordinate, or more generally, a rotation through π of the projection of the Bloch-vector in the y', z'-plane, as we have seen (see Fig. 16.2). The atomic system therefore delivers an echo pulse at time 2τ following the application of a π-pulse at time τ. This is the basic principle underlying both the *spin echo* phenomenon, that was discovered in nuclear spins by Hahn (1950b), and its optical analog the *photon echo*, that was first observed by Kurnit, Abella and Hartmann (1964), and Abella, Kurnit and Hartmann (1966). Let us examine the process more explicitly.

As in the previous section, we consider an inhomogeneously broadened group of N atoms close together, and initially in the ground state. As before we assume that $T_2^* \ll T_2'$, we take ω_0 as the center of the inhomogeneous line, and we confine our attention to time intervals short compared with T_2'. In order to prepare the state in which nearly all Bloch-vectors point along the y'-axis, we subject the atomic system to a rectangular, coherent $\pi/2$-pulse of frequency ω_0 for a short time T (Fig. 16.3), such that $T \ll T_2^*$ and the pulse area

$$A_1 = \frac{\pi}{2} = \Omega_1 T. \tag{16.2-1}$$

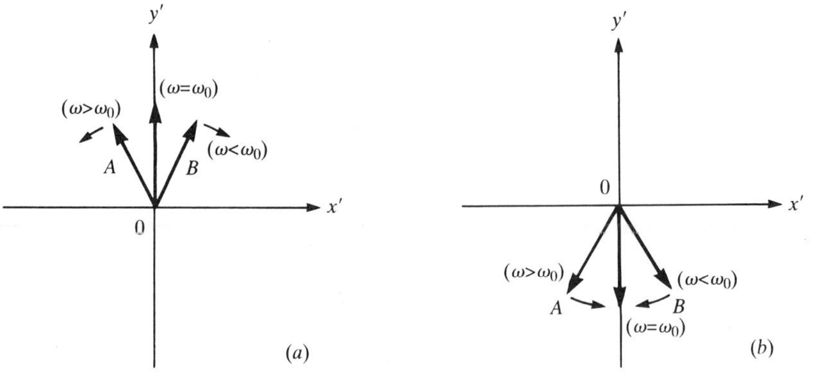

Fig. 16.2 Illustrating the precession of Bloch-vectors in the x', y'-plane and the reversal brought about by a π-pulse: (a) before application of the π-pulse; (b) after application of the π-pulse.

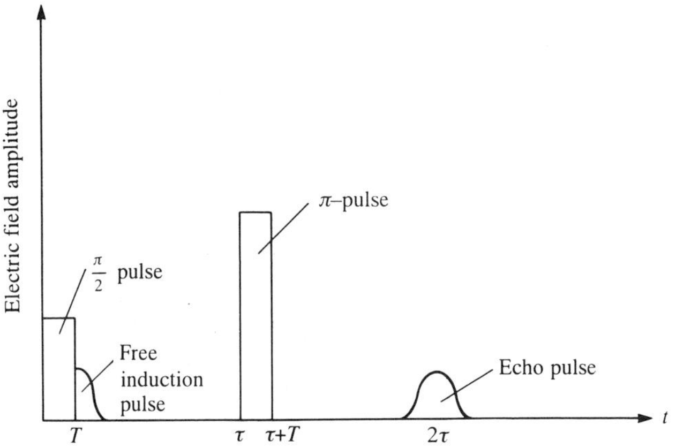

Fig. 16.3 The sequence of pulses leading to the photon echo.

At the end of this pulse the Bloch-vector components in the rotating frame are given by Eqs. (15.3–21) with $t = T$ [but note that ω_0 is the atomic frequency and ω_1 the optical frequency in Eqs. (15.3–21)]. If the pulse is sufficiently intense that the Rabi frequency Ω_1 is large compared with the frequency deviation $|\omega - \omega_0|$ of most atoms from the line center, so that terms of order $(\omega - \omega_0)^2/\Omega_1^2$ can be neglected, then the Bloch-vector components are approximately given by the relations

$$\left.\begin{aligned} r_1'(\omega, T) &\approx \frac{\omega - \omega_0}{\Omega_1} \\ r_2'(\omega, T) &\approx -1 \\ r_3'(\omega, T) &\approx 0. \end{aligned}\right\} \tag{16.2–2}$$

Following the application of this short pulse, the different Bloch-vectors precess freely, such that at a later time τ that is still short compared with T_2',

$$r_1'(\omega, \tau) - \mathrm{i}r_2'(\omega, \tau) = \mathrm{e}^{-\mathrm{i}(\omega-\omega_0)(\tau-T)}[r_1'(\omega, T) - \mathrm{i}r_2'(\omega, T)]$$

$$= \mathrm{e}^{-\mathrm{i}(\omega-\omega_0)(\tau-T)}\left[\frac{\omega-\omega_0}{\Omega_1} + \mathrm{i}\right], \quad \tau \geqslant T. \qquad (16.2\text{–}3)$$

At time τ we subject the atomic system to a second short light pulse of duration T and area [see Fig. 16.3]

$$A_2 = \pi = \Omega_2 T, \qquad (16.2\text{–}4)$$

which has twice the amplitude of the preceding one. In order to determine its effect on the atom, we need the solutions of the Bloch equations (15.3–19), with $\phi = 0$, for a more general initial state than the ground state. We may readily show that with the initial values $r_1'(\omega, 0)$, $r_2'(\omega, 0)$, 0 the components $r_1'(\omega, t)$, $r_2'(\omega, t)$, $r_3'(\omega, t)$ are given by the equations

$$\left.\begin{aligned} r_1'(\omega, t) &\approx r_1'(\omega, 0) - \left[\frac{\omega-\omega_0}{\Omega}\sin\Omega t\right] r_2'(\omega, 0) \\ r_2'(\omega, t) &\approx \left[\frac{(\omega-\omega_0)}{\Omega}\sin\Omega t\right] r_1'(\omega, 0) + [\cos\Omega t] r_2'(\omega, 0) \\ r_3'(\omega, t) &\approx \left[-\frac{(\omega-\omega_0)}{\Omega}(1-\cos\Omega t)\right] r_1'(\omega, 0) - [\sin\Omega t] r_2'(\omega, 0), \end{aligned}\right\} \qquad (16.2\text{–}5)$$

when we make the same strong field approximation as before. If we take the initial values in Eqs. (16.2–5) to be those given by Eq. (16.2–3) for $r_1'(\omega, \tau)$ and $r_2'(\omega, \tau)$, $\Omega = \Omega_2$ and the time interval $t = T$, then following the application of the π-pulse, we obtain the formulas

$$\left.\begin{aligned} r_1'(\omega, \tau+T) &\approx \frac{\omega-\omega_0}{\Omega_1}\cos(\omega-\omega_0)(\tau-T) + \sin(\omega-\omega_0)(\tau-T) \\ r_2'(\omega, \tau+T) &\approx -\frac{\omega-\omega_0}{\Omega_1}\sin(\omega-\omega_0)(\tau-T) + \cos(\omega-\omega_0)(\tau-T) \\ r_3'(\omega, \tau+T) &\approx -\frac{\omega-\omega_0}{\Omega_1}\sin(\omega-\omega_0)(\tau-T), \end{aligned}\right\} \qquad (16.2\text{–}6)$$

so that

$$r_1'(\omega, \tau+T) - \mathrm{i}r_2'(\omega, \tau+T) \approx \left(-\mathrm{i} + \frac{\omega-\omega_0}{\Omega_1}\right)\mathrm{e}^{\mathrm{i}(\omega-\omega_0)(\tau-T)}, \quad \tau > T. \qquad (16.2\text{–}7)$$

From this time on the Bloch-vector again precesses freely, and at any later time $t > \tau + T$,

$$r_1'(\omega, t) - r_2'(\omega, t) = \mathrm{e}^{-\mathrm{i}(\omega-\omega_0)(t-\tau-T)}[r_1'(\omega, \tau+T) - \mathrm{i}r_2'(\omega, \tau+T)]$$

$$= \mathrm{e}^{-\mathrm{i}(\omega-\omega_0)(t-2\tau)}\left[-\mathrm{i} + \frac{\omega-\omega_0}{\Omega_1}\right]. \qquad (16.2\text{–}8)$$

With the help of Eq. (16.1–5), and the transformation (15.3–17b) to the non-rotating frame, the macroscopic dipole moment $\langle\hat{\boldsymbol{\mu}}(t)\rangle$ is therefore given by the relation

$$\langle \hat{\boldsymbol{\mu}}(t) \rangle = \frac{N}{4\pi} \int_0^\infty \left[\boldsymbol{\mu}_{12} \left(-\mathrm{i} + \frac{\omega - \omega_0}{\Omega_1} \right) \mathrm{e}^{-\mathrm{i}(\omega-\omega_0)(t-2\tau)} \, \mathrm{e}^{-\mathrm{i}\omega_0 t} + \text{c.c.} \right] g(\omega)\, \mathrm{d}\omega,$$

and from Eq. (16.1–3),

$$\langle \hat{\boldsymbol{\mu}}(t) \rangle = -\mathrm{i}\frac{N\boldsymbol{\mu}_{12}}{2} \mathrm{e}^{-\mathrm{i}\omega_0 t} \left[G(t-2\tau) - \frac{1}{\Omega_1} \dot{G}(t-2\tau) \right] + \text{c.c.}, \quad t > \tau + T. \tag{16.2–9}$$

For a symmetric spectral distribution, for which $G(\tau)$ is real, and for a real dipole moment, this simplifies further to

$$\langle \hat{\boldsymbol{\mu}}(t) \rangle = -N\boldsymbol{\mu}_{12} \sin \omega_0 t \left[G(t-2\tau) - \frac{1}{\Omega_1} \dot{G}(t-2\tau) \right], \quad t > \tau + T, \tag{16.2–10a}$$

whereas for a complex dipole moment $\boldsymbol{\mu}_{12} = (|\boldsymbol{\mu}_{12}|/\surd 2)(\mathbf{x}_1 + \mathrm{i}\mathbf{y}_1)$, as for a $\Delta m = \pm 1$ transition, we obtain

$$\langle \hat{\boldsymbol{\mu}}(t) \rangle = \frac{N|\boldsymbol{\mu}_{12}|}{\surd 2} [-\mathbf{x}_1 \sin \omega_0 t + \mathbf{y}_1 \cos \omega_0 t] \left[G(t-2\tau) - \frac{1}{\Omega_1} \dot{G}(t-2\tau) \right], t > \tau + T. \tag{16.2–10b}$$

Now both the functions $G(t)$ and $\dot{G}(t)$ are very small except in a range $|t| \lessapprox T_2^*$, when they are of order unity. It follows from Eqs. (16.2–10) that at time $t = 2\tau$ a macroscopic dipole moment appears, whose duration is of order T_2^* (see Fig. 16.3), and this causes a pulse of light to be radiated by the atomic system. This is the *echo pulse*, which follows the π-pulse after a delay equal to the delay between the two exciting pulses. At time $t = 2\tau$ $G(t - 2\tau)$ takes on its greatest value of unity, and $\dot{G}(t - 2\tau)$ is usually zero. The greatest value of $\dot{G}/\Omega_1$ is generally of order $1/\Omega_1 T_2^*$, so that this term can be neglected in Eqs. (16.2–10) when $\Omega_1 T_2^* \gg 1$. For example, if $G(\tau)$ is given by Eq. (16.1–4), as for a Gaussian inhomogeneously broadened line, then Eq. (16.2–10a) becomes

$$\langle \hat{\boldsymbol{\mu}}(t) \rangle = -N\boldsymbol{\mu}_{12} \sin \omega_0 t \exp [-\tfrac{1}{2}(t-2\tau)^2/T_2^{*2}]. \tag{16.2–11}$$

Figure 16.3 illustrates the sequence of exciting pulses and the echo pulse. The free induction pulse that follows the first excitation pulse is also shown. The echo phenomenon is a simple illustration of the fact that the oscillations of each atomic dipole moment remain phase coherent for times that are short compared with T_2', but these times may be long compared with T_2^*. Despite the name *photon echo*, the phenomenon can be readily understood in terms of the classical electromagnetic field radiated by a group of point dipoles.

16.2.1 Echo experiments

Photon echoes were discovered and first observed in a cooled sample of ruby by Kurnit, Abella and Hartmann (1964), who made use of a ruby laser to generate the two exciting pulses. Figure 16.4 shows the results of an experiment in which the time interval τ was 137 ns. The laser was divided into two parts with the aid of

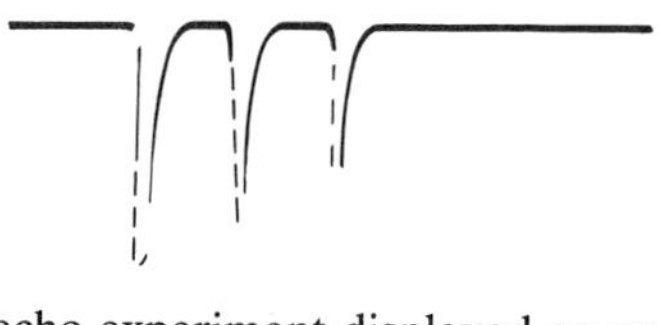

Fig. 16.4 Results of an echo experiment displayed on an oscilloscope. The first two pulses, which are separated by 137 ns, are the excitation pulses, and the third pulse is the echo. The first two photoelectric pulses have been attenuated to make all three of them comparable in size. (Reproduced from Abella, Kurnit and Hartmann, 1966.)

a beam splitter, and the second or π-pulse was delayed relative to the first one by an optical delay line. The light beams delivering the two pulses were actually deliberately misaligned by about 3°. It was found that the emerging echo pulse made an angle of about 6° with the first light pulse and about 3° with the second one, which makes it much easier to resolve the weak echo pulse (Abella, Kurnit and Hartmann, 1966). These directional features are the result of spatial phasing effects, which are somewhat analogous to the temporal phasing we have discussed. They do not show up in a calculation in which all the atoms are taken as concentrated within a space region of the order of a wavelength.

As in the preceding section, we emphasize that, in calculating the macroscopic dipole moment and deriving the form of the echo pulse radiated by the atomic system, we have ignored the effects of the atomic dipoles on each other. Such effects are not necessarily negligible, and we shall discuss them in Section 16.5 in connection with superradiance. However, they do not play an essential role in the formation of either the free induction pulse or the echo pulse. The light intensities of both these pulses are proportional to the square of the number of atoms N, because the fields radiated by the atoms interfere constructively for a short time, and not because the atoms necessarily influence each other. It is not difficult to visualize a situation in which the different atoms are so far apart that they cannot influence each other during the formation of the pulse, but their fields can still interfere constructively at some point in space. An observer at this point would see the induction and echo pulses despite the fact that cooperative atomic effects are ruled out.

16.3 Self-induced transparency

In the previous two sections we have encountered some optical phenomena that depend on the excitation of a group of atomic dipoles by an external field, and on the superposition or interference of the resulting fields radiated by each of these dipoles. Of course, the atomic dipole fields in turn modify the exciting fields acting on each atom, but this process does not play an essential role in the formation of the free induction pulse or of the photon echo, and was ignored in our treatment. However, there are circumstances when the field acting on each atom that is produced by all other atomic dipoles is as important as, or even more important than, the external field, and cannot be ignored. The propagation of a pulse of light through a near resonant material medium provides one example. We now turn our attention to the treatment of this problem.

Let us consider the one-dimensional problem of a classical electromagnetic wave propagating in the z-direction through a material medium, in which there exists a macroscopic polarization $\mathbf{P}(z,t)$. In a solid, the polarizable atoms may be imbedded in a host medium with some refractive index n. If damping is neglected, the electric field $\mathbf{E}(z,t)$ in the medium then satisfies the inhomogeneous wave equation

$$\left(\frac{\partial^2}{\partial z^2} - \frac{n^2}{c^2}\frac{\partial^2}{\partial t^2}\right)\mathbf{E}(z,t) = \frac{1}{\varepsilon_0 c^2}\frac{\partial^2 \mathbf{P}(z,t)}{\partial t^2}, \tag{16.3–1}$$

in which the polarization $\mathbf{P}(z,t)$ plays the role of the source. Now the electric field $\mathbf{E}(z,t)$ acts on the atomic dipoles in accordance with the Bloch equations, and induces a certain dipole moment, such that the sum of all the dipole moments per unit volume in the neighborhood of z constitutes the macroscopic polarization $\mathbf{P}(z,t)$. We now attempt to find a self-consistent solution of the wave equation (16.3–1) when it is coupled to the Bloch equations (15.3–11) or (15.3–19). This formulation of the pulse propagation problem, which was given by McCall and Hahn (1967, 1969), is analogous to the laser problem as formulated by Lamb (1964) that we shall discuss in Chapter 18. Because damping effects and spontaneous emission are omitted from the analysis, it is implicit in our calculation that we are dealing with time intervals short compared with T_2'.

16.3.1 Equations of motion of the pulse envelope

Let us consider a polarized plane wave of midfrequency ω_0 centered on the inhomogeneous spectral line, and wave number $k = n\omega_0/c$, propagating in the z-direction. We make an envelope representation of $\mathbf{E}(z,t)$ by writing

$$\mathbf{E}(z,t) = \mathscr{E}(z,t)\boldsymbol{\varepsilon}\mathrm{e}^{-\mathrm{i}[\omega_0 t - kz - \phi(z)]} + \text{c.c.}, \tag{16.3–2}$$

where $\boldsymbol{\varepsilon}$ is a unit polarization vector, $\phi(z)$ and $\mathscr{E}(z,t)$ are slowly varying phase and envelope functions that we take to be real. In taking $\mathscr{E}(z,t)$ to be real and $\phi(z)$ to be time independent, we are discarding possible phase modulation effects in order to simplify the problem. However, the z-dependent phase $\phi(z)$ allows for the possibility that the phase velocity of waves in the active medium may differ from the phase velocity c/n in the host medium. We assume that $\mathscr{E}(z,t)$ and $\phi(z)$ vary sufficiently slowly with distance z and time t that

$$\left.\begin{aligned} \left|\frac{\partial \mathscr{E}}{\partial z}\right| &\ll k|\mathscr{E}| \\ \left|\frac{\partial \mathscr{E}}{\partial t}\right| &\ll \omega_0|\mathscr{E}| \\ \left|\frac{\partial \phi}{\partial z}\right| &\ll k. \end{aligned}\right\} \tag{16.3–3}$$

As the electromagnetic wave or pulse propagates through the medium it excites the atoms, and the Bloch-vector $\mathbf{r}(\omega,z,t)$ in general depends on the atomic frequency ω, the position z and the time t. From Eq. (16.1–5) we then find that the polarization $\mathbf{P}(z, t)$ at position z and time t is given by

$$\mathbf{P}(z, t) = \frac{\mathcal{N}}{4\pi}\int_0^\infty \boldsymbol{\mu}_{12}[r_1(\omega, z, t) - \mathrm{i}r_2(\omega, z, t)]g(\omega)\,\mathrm{d}\omega + \text{c.c.}, \qquad (16.3\text{–}4)$$

where $\mathcal{N}$ is the density of atomic dipoles, and $g(\omega)$ is the inhomogeneous spectral profile, as before. As usual, we can express $r_1 - \mathrm{i}r_2$ in the integrand in terms of the Bloch-vector components r_1', r_2', r_3' in the rotating frame, and we set

$$r_1(\omega, z, t) - \mathrm{i}r_2(\omega, z, t) = \mathrm{e}^{-\mathrm{i}[\omega_0 t - kz - \phi(z)]}[r_1'(\omega, z, t) - \mathrm{i}r_2'(\omega, z, t)]. \qquad (16.3\text{–}5)$$

It is convenient to introduce the z-dependent phase factor $\mathrm{e}^{\mathrm{i}kz+\phi}$ in the transformation in anticipation of the fact that the polarization $\mathbf{P}(z, t)$ propagates just like $\mathbf{E}(z, t)$, and that $r_1' - \mathrm{i}r_2'$ varies slowly in space and time. This makes the phase of the rotating frame dependent on position z, but its frequency of rotation is always ω_0. The Bloch equations in the rotating frame are then Eqs. (15.3–19) with the phase angle put equal to zero.

We now substitute the envelope representation for $\mathbf{E}(z, t)$ from Eq. (16.3–2) in the wave equation (16.3–1), use the inequalities (16.3–3) to discard terms that are negligible, and make use of Eqs. (16.3–4) and (16.3–5). This yields the equation

$$\left[2\mathrm{i}k\left(\frac{\partial\mathscr{E}}{\partial z} + \frac{n}{c}\frac{\partial\mathscr{E}}{\partial t}\right) - 2k\mathscr{E}\frac{\partial\phi}{\partial z}\right]\boldsymbol{\varepsilon}\,\mathrm{e}^{-\mathrm{i}(\omega_0 t - kz - \phi)} + \text{c.c.}$$
$$= -\frac{\omega_0^2\mathcal{N}}{4\pi\varepsilon_0 c^2}\boldsymbol{\mu}_{12}\,\mathrm{e}^{-\mathrm{i}(\omega_0 t - kz - \phi)}\int_0^\infty [r_1'(\omega, z, t) - \mathrm{i}r_2'(\omega, z, t)]g(\omega)\,\mathrm{d}\omega + \text{c.c.}$$

On comparing coefficients of $\mathrm{e}^{-\mathrm{i}(\omega_0 t - kz - \phi)}$ on both sides of the equation, and taking the scalar product with $\boldsymbol{\varepsilon}^*$, we obtain the equation

$$2\mathrm{i}k\left(\frac{\partial\mathscr{E}}{\partial z} + \frac{n}{c}\frac{\partial\mathscr{E}}{\partial t}\right) - 2k\mathscr{E}\frac{\partial\phi}{\partial z} =$$
$$-\frac{\omega_0^2\mathcal{N}(\boldsymbol{\mu}_{12}\cdot\boldsymbol{\varepsilon}^*)}{4\pi\varepsilon_0 c^2}\int_0^\infty [r_1'(\omega, z, t) - \mathrm{i}r_2'(\omega, z, t)]g(\omega)\,\mathrm{d}\omega. \qquad (16.3\text{–}6)$$

Up to now we have not specified the nature of the atomic transition or the polarization of the light wave. If we are dealing with a $\Delta m = 0$ atomic transition and $\boldsymbol{\mu}_{12}$ is real, and if we are using linearly polarized light, then $\boldsymbol{\mu}_{12}\cdot\boldsymbol{\varepsilon}^*$ is also real. On the other hand, if the atomic transition is of the $\Delta m = \pm 1$ type, with $\boldsymbol{\mu}_{12} = (|\boldsymbol{\mu}_{12}|\sqrt{2})(\mathbf{x}_1 + \mathrm{i}\mathbf{y}_1)$, and if the light is circularly polarized, then $\boldsymbol{\mu}_{12}\cdot\boldsymbol{\varepsilon}^* = |\boldsymbol{\mu}_{12}|$ or 0, according to the sense of the circular polarization. It follows that the factor $\boldsymbol{\mu}_{12}\cdot\boldsymbol{\varepsilon}^*$ may be taken as real in both cases. If we now compare real and imaginary parts on both sides of Eq. (16.3–6), we arrive at the two equations

$$\frac{\partial\mathscr{E}}{\partial z} + \frac{n}{c}\frac{\partial\mathscr{E}}{\partial t} = \frac{\omega_0\mathcal{N}\boldsymbol{\mu}_{12}\cdot\boldsymbol{\varepsilon}^*}{8\pi\varepsilon_0 cn}\int_0^\infty r_2'(\omega, z, t)g(\omega)\,\mathrm{d}\omega \qquad (16.3\text{–}7)$$

$$\mathscr{E}\frac{\partial\phi}{\partial z} = \frac{\omega_0\mathcal{N}\boldsymbol{\mu}_{12}\cdot\boldsymbol{\varepsilon}^*}{8\pi\varepsilon_0 cn}\int_0^\infty r_1'(\omega, z, t)g(\omega)\,\mathrm{d}\omega, \qquad (16.3\text{–}8)$$

which describe the motion of the light pulse through the medium.

16.3.2 The area theorem of McCall and Hahn

We now introduce the 'area' of the pulse, as in Section 15.3. We define the Rabi frequency $\Omega(z, t)$ as before by the relation [cf. Eq. (15.3–12)]

$$\Omega(z, t) = 2\boldsymbol{\mu}_{12} \cdot \boldsymbol{\varepsilon}^* \mathscr{E}(z, t)/\hbar.$$

Then the tipping angle $\Theta(z, t)$ is given by [cf. Eq. (15.3–24)] the pulse area $A(z, t)$, i.e.

$$A(z, t) = \Theta(z, t) = \int_{-\infty}^{t} \Omega(z, t')\, \mathrm{d}t'.$$

On multiplying both sides of Eq. (16.3–7) by $2\boldsymbol{\mu}_{12} \cdot \boldsymbol{\varepsilon}^*/\hbar$, and integrating over time from $t = -\infty$ before the pulse begins to some long time t after the pulse has died out and $\Omega(t) = 0$, we obtain the equation

$$\frac{\partial \Theta(z, t)}{\partial z} = \frac{\omega_0 \mathcal{N}(\boldsymbol{\mu}_{12} \cdot \boldsymbol{\varepsilon}^*)^2}{4\pi\varepsilon_0 \hbar c n} \int_{-\infty}^{t} \mathrm{d}t' \int_0^{\infty} \mathrm{d}\omega\, r_2'(\omega, z, t') g(\omega). \qquad (16.3\text{–}9)$$

In order to perform the time integration we shall make use of the first Bloch equation in the rotating frame, corresponding to the transformation (16.3–5), which takes the form [cf. Eqs. (15.3–19)]

$$\dot{r}_1'(\omega, z, t) = (\omega_0 - \omega) r_2'(\omega, z, t),$$

so that

$$\int_{-\infty}^{t} r_2'(\omega, z, t')\, \mathrm{d}t' = \frac{r_1'(\omega, z, t)}{\omega_0 - \omega}, \qquad (16.3\text{–}10)$$

provided $\omega \neq \omega_0$, and provided the atoms start in the ground state at $t = -\infty$. The case $\omega = \omega_0$ has to be approached as a limiting case. But if t is so large that the pulse is over, then the Bloch-vector $\mathbf{r}'(\omega, z, t)$ precesses freely, and $r_1'(\omega, z, t)$ oscillates at the frequency $\omega - \omega_0$. More explicitly, if $\Omega(t)$ is already zero at some earlier time t_0, then from the Bloch equations we find at the later time

$$r_1'(\omega, z, t) = r_1'(\omega, z, t_0) \cos(\omega - \omega_0)(t - t_0) - r_2'(\omega, z, t_0) \sin(\omega - \omega_0)(t - t_0). \qquad (16.3\text{–}11)$$

If we use Eqs. (16.3–10) and (16.3–11) in Eq. (16.3–9), and put $\omega - \omega_0 = \omega'$, we obtain the equation

$$\frac{\partial \Theta(z, t)}{\partial z} = \frac{\omega_0 \mathcal{N}(\boldsymbol{\mu}_{12} \cdot \boldsymbol{\varepsilon}^*)^2}{4\pi\varepsilon_0 \hbar c n} \int_{-\omega_0}^{\infty} \left[-\frac{r_1'(\omega_0 + \omega', z, t_0) \cos \omega'(t - t_0)}{\omega'} + \frac{r_2'(\omega_0 + \omega', z, t_0) \sin \omega'(t - t_0)}{\omega'} \right] g(\omega_0 + \omega')\, \mathrm{d}\omega'.$$

Now, as $t \to \infty$, the function

$$\frac{\sin \omega'(t - t_0)}{\omega'}$$

becomes one of the representations of the $\delta(\omega')$ function, so that the second integral reduces to $\pi r_2'(\omega_0, z, t_0) g(\omega_0)$. Also, as $t \to \infty$,

$$\frac{1 - \cos \omega'(t - t_0)}{\omega'}$$

becomes a representation of the Cauchy principal part $P(1/\omega')$ (see Heitler, 1954, p. 69), so that

$$-\int_{-\omega_0}^{\infty} r_1'(\omega_0 + \omega', z, t_0)\frac{\cos \omega'(t - t_0)}{\omega'}g(\omega_0 + \omega')\,d\omega' \rightarrow$$

$$\int_{-\omega_0}^{\infty} r_1'(\omega_0 + \omega', z, t_0)\left[P\left(\frac{1}{\omega'}\right) - \frac{1}{\omega'}\right]g(\omega_0 + \omega')\,d\omega' = 0,$$

because $r_1'(\omega_0 + \omega', z, t_0)g(\omega_0 + \omega')$ is a smooth function of ω' near $\omega' = 0$. It follows that

$$\frac{\partial \Theta(z, t)}{\partial z} = \frac{\omega_0 \mathcal{N}(\boldsymbol{\mu}_{12} \cdot \boldsymbol{\varepsilon}^*)^2 g(\omega_0)}{4\varepsilon_0 \hbar c n} r_2'(\omega_0, z, t_0). \qquad (16.3\text{–}12)$$

$r_2'(\omega_0, z, t_0)$ is the y'-component of the Bloch-vector of an atom that is resonant with the applied field. But from Eqs. (15.3–25) this is directly expressible in terms of the tipping angle $\Theta(z, t_0)$ of the light pulse and the initial values of r_1', r_2', r_3'. If the atoms all start in the ground state at time $t = -\infty$, then from Eqs. (15.3–25)

$$r_2'(\omega_0, z, t_0) = -\sin \Theta(z, t_0) = -\sin A(z), \qquad (16.3\text{–}13)$$

where $A(z)$ is the total *area* of the pulse. Hence for sufficiently long times t, Eq. (16.3–12) becomes

$$\frac{dA(z)}{dz} = -\tfrac{1}{2}\alpha \sin A(z), \qquad (16.3\text{–}14)$$

where we have written

$$\alpha \equiv \frac{\omega_0 \mathcal{N}(\boldsymbol{\mu}_{12} \cdot \boldsymbol{\varepsilon}^*)^2 g(\omega_0)}{2\varepsilon_0 \hbar c n}. \qquad (16.3\text{–}15)$$

Equation (10.3–14) represents the so-called *area theorem* of McCall and Hahn (1967, 1969), which determines the propagation of the pulse area through the medium.

The significance of the parameter α can readily be found if we consider the propagation of a weak light pulse of small area A. Then $\sin A$ can be replaced by A to a good approximation, and we obtain the equation

$$\frac{dA(z)}{dz} = -\tfrac{1}{2}\alpha A(z), \qquad (16.3\text{–}16a)$$

or

$$A(z) = A(0)\,e^{-\alpha z/2}. \qquad (16.3\text{–}16b)$$

This is Beer's law [cf. Eq. (5.7–86) for the special case of homogeneous, isotropic media]. The parameter $\alpha/2$ is therefore the attenuation constant of the electric field amplitude, or α is the *absorption coefficient* of the light intensity. The reciprocal $1/\alpha$ is also known as the *absorption length*. It is important to note that a weak pulse is gradually absorbed as it progresses through the medium, even in the absence of any dissipation mechanism. The pulse energy is simply transferred to the atoms, which remain close to the unexcited state.

However, the situation is very different for an intense light pulse that produces appreciable atomic excitation. In particular, if the pulse area $A = 2n\pi$, $n = 1, 2, 3, \ldots$, then from Eq. (16.3–14) $\mathrm{d}A/\mathrm{d}z = 0$ and the pulse propagates through the medium without any attenuation of its area. The medium therefore appears to be totally transparent to the light pulse, despite the fact that its midfrequency coincides with the atomic resonance. McCall and Hahn have called this phenomenon *self-induced transparency*. Of course the area theorem alone tells us nothing about the shape of the pulse. We shall see below that the pulse is reshaped in general, and may even be broken up into several component pulses.

Equation (16.3–14) is immediately integrable and we find that

$$\int_{A_0}^{A} \frac{\mathrm{d}A}{\sin A} = -\tfrac{1}{2}\alpha \int_0^z \mathrm{d}z,$$

$$\ln\left|\frac{\tan\frac{1}{2}A(z)}{\tan\frac{1}{2}A_0}\right| = -\tfrac{1}{2}\alpha z,$$

or

$$\tan\tfrac{1}{2}A(z) = \tan\left(\tfrac{1}{2}A_0\right)\mathrm{e}^{-\alpha z/2}. \qquad (16.3\text{–}17)$$

Figure 16.5 shows how the pulse area varies with distance of propagation z for various initial values of the pulse area (not necessarily at $z = 0$). Pulses of area less than π are attenuated to zero, whereas those between π and 2π grow in area towards 2π, and similarly for larger pulses. Pulses of area 2π, 4π, etc., remain constant in area as they propagate. Although it appears from Eqs. (16.3–14) or (16.3–17) that pulses whose areas are an odd multiple of π should also remain unattenuated, these pulse areas are actually unstable against small perturbations, as is indicated by Fig. 16.5.

The transparency phenomenon implied by the area theorem was first observed by McCall and Hahn (1967) with pulses from a liquid nitrogen cooled ruby laser traversing a ruby rod cooled to liquid helium temperature. Figure 16.6 shows the results of some later measurements of transmittance through ruby. A fairly sharp threshold for the onset of transparency is observed.

16.3.3 *The pulse shape*

Equations (16.3–14) or (16.3–17) refer only to the pulse area, and tell us nothing about the shape of the pulse. It is possible for the area of the pulse to remain

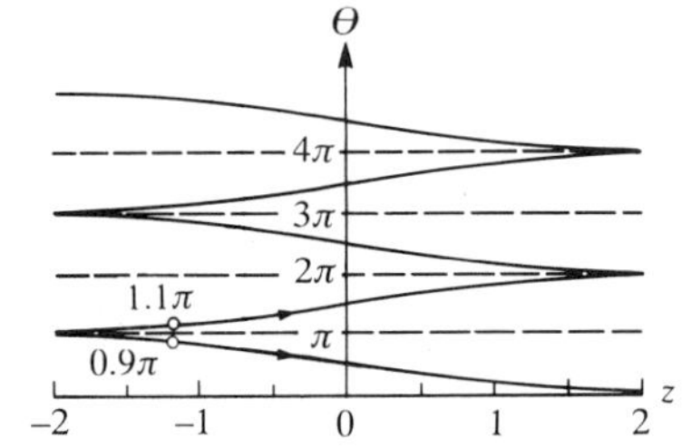

Fig. 16.5 Expected variation of pulse area or tipping angle Θ with propagation distance z. (Reproduced from McCall and Hahn, 1967.)

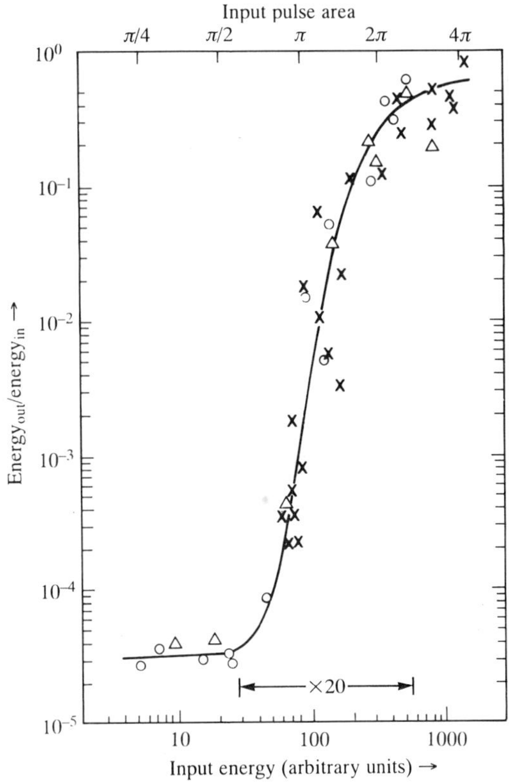

Fig. 16.6 Results of transmittance measurements for the propagation of light pulses through a ruby rod. (Reproduced from Asher and Scully, 1971.)

constant as it propagates, while its shape changes, and indeed this happens in general. For information about the pulse shape we have to return to the equations of motion (16.3–7) and (16.3–8).

A light pulse that propagates in one dimension with some velocity V and preserves its shape must be of the general form

$$\mathscr{E}(z, t) = f\left(t - \frac{z}{V}\right). \tag{16.3–18}$$

The pulse velocity V may of course be different from the phase velocity of waves in the medium. If we substitute this form for $\mathscr{E}(z, t)$ in Eq. (16.3–7), we obtain the equation

$$\left(\frac{\eta}{c} - \frac{1}{V}\right)\dot{f}\left(t - \frac{z}{V}\right) = \frac{\omega_0 \mathcal{N} \boldsymbol{\mu}_{12} \cdot \boldsymbol{\varepsilon}^*}{8\pi\varepsilon_0 cn} \int_0^\infty r_2'(\omega, z, t) g(\omega)\, \mathrm{d}\omega. \tag{16.3–19}$$

In discussing the response of an atom to a pulse of light in Section 15.3, we encountered a rather special form of 2π-pulse, that has the property of taking any ground state atom, irrespective of its frequency ω, through a cycle, so that it ends up back in the ground state. This is the hyperbolic secant pulse given by Eq.

(15.3–33). If such a pulse propagates through the medium with velocity V, then

$$\Omega(z, t) = \frac{2\boldsymbol{\mu}_{12} \cdot \boldsymbol{\varepsilon}^* \mathscr{E}(z, t)}{\hbar} = \frac{2}{T} \operatorname{sech}\left(\frac{t - z/V}{T}\right), \tag{16.3–20}$$

where T is a measure of the pulse duration, and the origin has been chosen so that $\Omega(z, t)$ is maximum at time $t = z/V$. Let us see if this form of pulse is shape preserving as it propagates through the resonant medium. In that case the function in Eq. (16.3–18) is given by

$$f\left(t - \frac{z}{V}\right) = \frac{\hbar}{2\boldsymbol{\mu}_{12} \cdot \boldsymbol{\varepsilon}^* T} \operatorname{sech}\left(\frac{t - z/V}{T}\right). \tag{16.3–21}$$

From Eqs. (15.3–34) an atom of frequency ω responds to the pulse in such a way that

$$\left.\begin{aligned} r_1'(\omega, z, t) &= \frac{2(\omega_0 - \omega)T}{1 + (\omega - \omega_0)^2 T^2} \operatorname{sech}\left(\frac{t - z/V}{T}\right) \\ r_2'(\omega, z, t) &= -\frac{2}{1 + (\omega - \omega_0)^2 T^2} \operatorname{sech}\left(\frac{t - z/V}{T}\right) \tanh\left(\frac{t - z/V}{T}\right). \end{aligned}\right\} \tag{16.3–22}$$

If we substitute this form of $f(t - z/V)$ and $r_2'(\omega, z, t)$ in the equation of motion (16.3–19), we find that

$$\begin{aligned} &\frac{\hbar}{\boldsymbol{\mu}_{12} \cdot \boldsymbol{\varepsilon}^* T^2}\left(\frac{\eta}{c} - \frac{1}{V}\right) \operatorname{sech}\left(\frac{t - z/V}{T}\right) \tanh\left(\frac{t - z/V}{T}\right) = \\ &\quad -\frac{\omega_0 \mathscr{N}(\boldsymbol{\mu}_{12} \cdot \boldsymbol{\varepsilon}^*)}{4\pi\varepsilon_0 cn} \operatorname{sech}\left(\frac{t - z/V}{T}\right) \tanh\left(\frac{t - z/V}{T}\right) \int_0^\infty \frac{g(\omega)}{1 + (\omega - \omega_0)^2 T^2}\, \mathrm{d}\omega, \end{aligned}$$

so that the equation of motion is satisfied provided that

$$\frac{1}{V} - \frac{n}{c} = \frac{\alpha}{2\pi g(\omega_0)} \int_0^\infty \frac{g(\omega)}{(\omega - \omega_0)^2 + 1/T^2}\, \mathrm{d}\omega. \tag{16.3–23}$$

It follows that the hyperbolic secant 2π-pulse given by Eq. (16.3–20) is the solution of the equations of motion that preserves its shape as it propagates without attenuation. The medium is perfectly transparent to this form of 2π-pulse. This was of course to be expected from the special property of the pulse, namely that it leaves an atom that starts in the ground state back in the ground state. The hyperbolic secant solution to the pulse propagation problem was first obtained by McCall and Hahn (1967) and it appears to be the only stable solution of the equations. Figure 16.7 shows some computer solutions of the equations of motion, that illustrate how hyperbolic secant 2π-pulses are formed from pulses of various initial areas and shapes. When the initial pulse area is some multiple of 2π the pulse tends to break up into several smaller 2π-pulses. This phenomenon has also been observed experimentally. Figure 16.8 shows the results of some measurements by Slusher and Gibbs (1972) of pulse propagation in rubidium vapor. The tendency of the larger pulses to break up is clearly visible. The whole subject of self-induced transparency has been reviewed by Slusher (1974).

When $r_1'(\omega, z, t)$ and $r_2'(\omega, z, t)$ are given by Eqs. (16.3–22), we obtain from

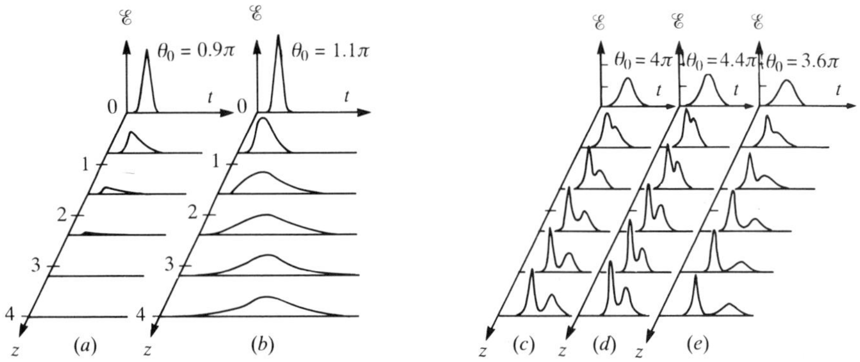

Fig. 16.7 Computer solutions showing the propagation of pulses of various areas. Distance z is measured in units of π/α. In the first example (*a*) the pulse is absorbed, and in example (*b*) it is reshaped to a hyperbolic secant form. In (*c*), (*d*) and (*e*) the pulse breaks up into two separate 2π hyperbolic secant pulses. (Reproduced from McCall and Hahn, 1969.)

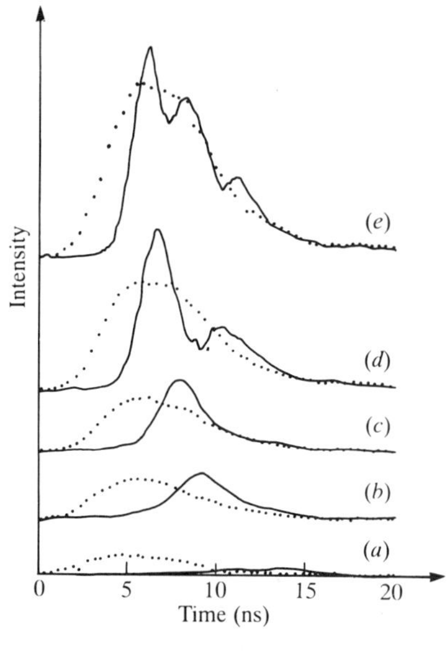

Fig. 16.8 Results of pulse propagation measurements in rubidium vapor, showing pulse breakup. Input pulses are shown by dotted lines and output pulses by solid lines. The input pulse areas increase progressively from (*a*) to (*e*). (Reproduced from Slusher and Gibbs, 1972.)

Eq. (16.3–8)

$$\frac{\partial \phi}{\partial z} = \frac{-\alpha}{2\pi g(\omega_0)} \int_0^\infty \frac{(\omega - \omega_0)}{(\omega - \omega_0)^2 + 1/T^2} g(\omega)\, d\omega. \qquad (16.3\text{–}24)$$

The right-hand side is independent of z and, according to Eq. (16.3–2), represents a change in the wave number of the electromagnetic wave from k to $k + \partial\phi/\partial z$, and therefore of the effective phase velocity. However, for any spectral distribution $g(\omega)$ that is symmetric about $\omega = \omega_0$ the integral vanishes, and the phase velocity in the active medium coincides with the phase velocity c/n in the host medium.

16.3.4 The pulse velocity

From Eq. (16.3–23) the velocity V of pulse propagation is smaller than the phase velocity c/n in an absorbing medium. At first sight this is a puzzling aspect of the phenomenon, because the hyperbolic secant pulse propagates without change. Actually the leading portion of the pulse is continually absorbed by the atoms, which draw energy from the pulse, and they then reconstitute the pulse by induced emission. The pulse is therefore continually re-formed as it travels through the medium, and it is this process and the associated energy storage which slows down the propagation. We can estimate the velocity of propagation from Eq. (16.3–23). If the pulse length T is much shorter than the inhomogeneous lifetime T_2^*, then the denominator under the integral varies very slowly with frequency compared with $g(\omega)$, and we obtain the formula

$$\frac{1}{V} - \frac{n}{c} \approx \frac{\alpha T^2}{g(\omega_0)} \frac{1}{2\pi} \int_0^\infty g(\omega)\,\mathrm{d}\omega = \frac{\alpha T^2}{g(\omega_0)} \sim \frac{\alpha T^2}{T_2^*}.$$

On the other hand, if the pulse is very long compared with T_2^*, as is usually the case, then $g(\omega)$ varies slowly with ω as compared with the denominator, and $g(\omega)$ can be approximated by $g(\omega_0)$ under the integral. Then

$$\frac{1}{V} - \frac{n}{c} \approx \frac{\alpha}{2\pi} \int_0^\infty \frac{\mathrm{d}\omega}{(\omega - \omega_0)^2 + 1/T^2} \approx \frac{\alpha T}{2}. \qquad (16.3\text{–}25)$$

For sufficiently long pulses, the difference between V and n/c can be substantial, and the pulse can travel much more slowly than c/n. The tendency of long pulses to travel more slowly than short ones can also be seen from the examples in Fig. 16.7 (*c*) to (*e*).

Equation (16.3–23) has a curious consequence if we consider an amplifying medium, in which the atoms start in the excited state, because α then becomes negative. Of course, the hyperbolic secant 2π pulse is no longer a stable solution under these circumstances, and the area theroem predicts that a π-pulse should be stable. According to Eq. (16.3–23) the pulse velocity then exceeds the phase velocity in the medium, and apparently can even exceed c. However, as the pulse extends to infinity in time it cannot be used to transmit information. Nor does it carry energy, for the energy is already stored at each point in the medium before the pulse arrives.

16.4 Optical bistability

Optical bistability is a remarkble phenomenon that exploits both the cooperative nature of the interaction between a group of atoms and a field and its strong nonlinearity. It was first predicted by Szöke, Daneu, Goldhar and Kurnit, (1969) and was treated in more detail by McCall (1974). The phenomenon was later observed in experiments in sodium vapor by Gibbs, McCall and Venkatesan

(1976). Under certain circumstances the propagation of light through a resonant medium may exhibit both bistable behavior and hysteresis, so that the medium appears as either transparent or absorbent. The possibility of applications of this effect to an optical switch, or 'optical transistor', has given rise to a large number of different treatments, both semiclassical and fully quantum mechanical. On the whole, the semiclassical treatments account for the behavior of the mean field quite well, but a quantized field treatment is required if such features as the fluctuations and photon statistics are to be described. In the following, we shall make use of a simplified semiclassical treatment, partly in order to emphasize the connection between optical bistability and light propagation through a resonant medium, as in self-induced transparency. The essential element in optical bistability is the nonlinear relation between the applied electromagnetic field and the electromagnetic field that is radiated by the atomic dipoles, when the actions of all the other dipoles are taken into account. Cooperative radiation therefore plays a key role in bistability.

We can see this from the following over-simplified, heuristic argument, in which we ignore all spatial variations and think of the N atoms as concentrated at a point. The effective field amplitude $\mathscr{E}_{\text{eff}}$ acting on each atom is the sum of the external or applied field amplitude $\mathscr{E}_{\text{app}}$ and the field amplitudes produced by all other atomic dipoles. From Eqs. (15.5–8) and (15.6–16) the dipole field on resonance is proportional to

$$\frac{-\Omega_{\text{eff}}}{(\frac{1}{2}\Omega_{\text{eff}}^2/\beta^2) + 1},$$

where $\Omega_{\text{eff}} \equiv 2(\boldsymbol{\mu}_{12} \cdot \boldsymbol{\varepsilon}^*/\hbar)\mathscr{E}_{\text{eff}}$ is the Rabi frequency associated with the field amplitude $\mathscr{E}_{\text{eff}}$, and 2β is the Einstein A-coefficient. Hence we can write

$$\mathscr{E}_{\text{eff}} = \mathscr{E}_{\text{app}} - \frac{K(N-1)\Omega_{\text{eff}}}{(\frac{1}{2}\Omega_{\text{eff}}^2/\beta^2) + 1},$$

where K is a constant, or, on putting $\Omega_{\text{app}} \equiv 2(\boldsymbol{\mu}_{12} \cdot \boldsymbol{\varepsilon}^*/\hbar)\mathscr{E}_{\text{app}}$, we have

$$\frac{\Omega_{\text{app}}}{\beta} = \frac{\Omega_{\text{eff}}}{\beta} + C\frac{\Omega_{\text{eff}}/\beta}{(\frac{1}{2}\Omega_{\text{eff}}^2/\beta^2) + 1}, \tag{16.4–1}$$

where C is a dimensionless constant that can be much greater than unity. From Eq. (16.4–1) we find that for weak fields ($\Omega_{\text{eff}}/\beta \ll 1$),

$$\frac{\Omega_{\text{app}}}{\beta} \approx \frac{\Omega_{\text{eff}}}{\beta}(1 + C),$$

so that $\Omega_{\text{eff}} \ll \Omega_{\text{app}}$, whereas for strong fields ($\Omega_{\text{eff}}/\beta \gg 1$),

$$\frac{\Omega_{\text{app}}}{\beta} \approx \frac{\Omega_{\text{eff}}}{\beta}.$$

However, in the intermediate region we have a cubic equation in Ω_{eff}, and when $C > 8$, for a given applied field there may be three different effective fields acting on the atoms, two of which represent stable solutions. In general, the atomic dipole fields, corresponding to cooperative atomic radiation, are more important when the applied field is weak, whereas the applied field dominates when it is strong. The relationship between Ω_{app} and Ω_{eff} when $C = 20$ is illustrated by the curve in Fig. 16.9. It is the multivaluedness of the response that is the essential feature of optical bistability, and we note that cooperative atomic radiation plays a fundamental role in the phenomenon.

The effects can be enhanced if the active medium is placed within a resonant optical cavity. We shall now discuss a somewhat more realistic model of absorptive bistability, in which the atoms are not all concentrated at a point, and propagation effects are taken into account. This will enable us to make use of some of the equations of motion that we derived in Section 16.3. Our treatment is similar to one given by Bonifacio and Lugiato (1978, a,b,c).

16.4.1 Absorptive bistability in a ring cavity

In order to demonstrate absorptive optical bistability, a resonant medium is placed between the mirrors of a Fabry–Perot interferometer tuned to the atomic transition, and an external, resonant light beam is passed through the interferometer. The system therefore bears a resemblance to a laser with an externally applied field, except that there is no auxiliary optical pumping mechanism. This was the arrangement used in the earliest successful experiments (Gibbs, McCall and Venkatesan, 1976), except that these relied more on the dispersive properties of the medium than on the resonant absorption. We shall briefly discuss dispersive bistability below, but, for the moment, we concentrate on the absorption mechanism. Moreover, for simplicity, we shall consider, not the Fabry–Perot resonator, but the ring laser cavity illustrated in Fig. 16.10, in which there is an active medium of length L in one arm, and the light traverses the medium in one direction only. A monochromatic plane wave of atomic frequency ω_0 and of amplitude $\mathscr{E}_i$ is incident on the system from one side as shown, and a wave of amplitude $\mathscr{E}_0$ emerges from the other side. We suppose that the input and output mirrors have amplitude reflectance $\sqrt{R}$ and amplitude transmittance $\sqrt{(1 - R)}$, whereas the other two mirrors are perfectly reflecting. Some of the light emerging

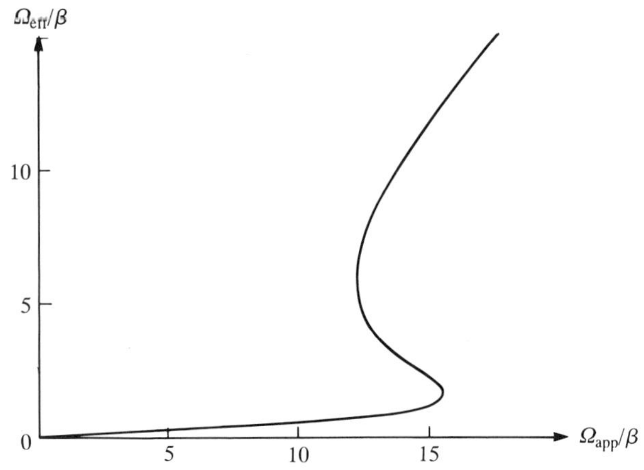

Fig. 16.9 The theoretical relationship between Ω_{eff} and Ω_{app}.

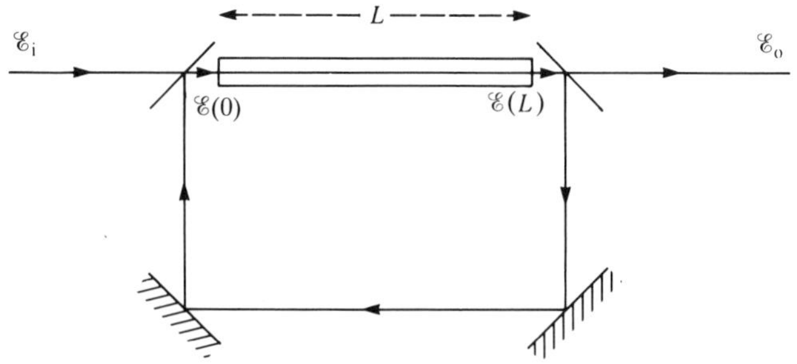

Fig. 16.10 The geometry for a system exhibiting optical bistability.

from the medium is fed back by the mirrors and combined with the incident light beam, as shown.

Within the active medium the propagation of the light is described by the same Eqs. (16.3–7) and (16.3–8) for the amplitude $\mathscr{E}(z, t)$ and the phase $\phi(z, t)$ of the field that we encountered in connection with self-induced transparency. These equations for $\mathscr{E}$ and ϕ have to be coupled with the Bloch equations (15.3–19) for the internal motion of the atoms in the field. We shall make two modifications to these equations for the purpose of treating this problem. In the first place, in order to simplify the calculation, we shall suppose that any inhomogeneous broadening is small, so that $g(\omega)$ describes the natural, Lorentzian line, which is a narrow spectral distribution centered on ω_0. Then Eqs. (16.3–7) and (16.3–8) reduce to

$$\frac{\partial \mathscr{E}(z, t)}{\partial z} + \frac{n}{c}\frac{\partial \mathscr{E}(z, t)}{\partial t} = \frac{\omega_0 \mathscr{N} \boldsymbol{\mu}_{12} \cdot \boldsymbol{\varepsilon}^*}{4\varepsilon_0 cn} r_2'(z, t) \tag{16.4–2}$$

$$\mathscr{E}(z, t)\frac{\partial \phi(z, t)}{\partial z} = \frac{\omega_0 \mathscr{N} \boldsymbol{\mu}_{12} \cdot \boldsymbol{\varepsilon}^*}{4\varepsilon_0 cn} r_1'(z, t), \tag{16.4–3}$$

in which all the symbols have the same meaning as in Section 16.3; η is the refractive index of the host medium and $\mathscr{N}$ is the density of active two-level atoms. In the second place, we shall have to augment the Bloch equations (15.3–19) by introducing phenomenological damping terms. We have seen in Chapter 15 that the semiclassical Bloch equations do not account for spontaneous atomic radiation in the absence of an external field. However, when the field is quantized, it is found that r_1 and r_2, which are related to the dipole moment, decay exponentially to zero at the rate β (half the Einstein A-coefficient), whereas $r_3 + 1$, which is proportional to the atomic energy, decays exponentially to zero at the rate 2β [cf. Eq. (15.5–28) and (15.5–29)]. These features of the atomic motion are sometimes introduced into the semiclassical treatment in an ad hoc manner, by the addition of phenomenological damping terms to the Bloch equations. Thus $-\beta r_1'$ and $-\beta r_2'$ are added to the first two Eqs. (15.3–19), respectively, and $-2\beta(r_3' + 1)$ is added to the third. The augmented Bloch equations then become, on resonance

$$\dot{r}_1'(z, t) = \Omega(z, t) r_3'(z, t) \sin\phi(z, t) - \beta r_1'(z, t) \tag{16.4–4}$$

$$\dot{r}_2'(z, t) = \Omega(z, t) r_3'(z, t) \cos\phi(z, t) - \beta r_2'(z, t) \tag{16.4–5}$$

$$\dot{r}'_3(z, t) = -\Omega(z, t) r'_1(z, t) \sin\phi(z, t) - \Omega(z, t) r'_2(z, t) \cos\phi(z, t) - 2\beta[r'_3(z, t) + 1], \qquad (16.4\text{–}6)$$

in which the field at position z at time t is represented by the atomic Rabi frequency

$$\Omega(z, t) = \frac{2\boldsymbol{\mu}_{12} \cdot \boldsymbol{\varepsilon}^*}{\hbar} \mathscr{E}(z, t)$$

and by the phase $\phi(z, t)$. The Bloch-vector components r'_1, r'_2, r'_3 in the rotating frame are, of course, also functions of position z and time t. In any time interval short compared with the decay time $1/\beta$, the damping terms do not exert much influence. That is why we were able to treat the propagation of short pulses through the medium, at least in an approximate way, without the introduction of damping. However, in the description of the steady-state behavior of the system, the damping cannot be ignored.

In the following we shall not attempt to solve the coupled Maxwell–Bloch equations in general, but we merely look for the steady-state solution. In that case all time derivatives in Eq. (16.4–2) to (16.4–6) can be put equal to zero, and $\mathscr{E}$, Ω, ϕ, r'_1, r'_2, r'_3 become functions of z only. Equations (16.4–4) to (16.4–6) have the steady-state solutions

$$r'_1(z) = \frac{-(\Omega/\beta)\sin\phi}{(\frac{1}{2}\Omega^2/\beta^2) + 1} \qquad (16.4\text{–}7)$$

$$r'_2(z) = \frac{-(\Omega/\beta)\cos\phi}{(\frac{1}{2}\Omega^2/\beta^2) + 1} \qquad (16.4\text{–}8)$$

$$r'_3(z) = \frac{-1}{(\frac{1}{2}\Omega^2/\beta^2) + 1}, \qquad (16.4\text{–}9)$$

which show incidentally that the atomic Bloch-vectors are no longer unit vectors for the coupled quantum system, because the atomic states are no longer pure quantum states. Also from Eq. (16.4–2) we have in the steady state, in terms of the Rabi frequency $\Omega(z)$,

$$\frac{\mathrm{d}\Omega}{\mathrm{d}z} = \frac{\alpha}{g(\omega_0)} r'_2,$$

where α is the absorption coefficient for the light intensity given by Eq. (16.3–15). If we substitute for r'_2 from Eq. (16.4–8) this becomes

$$\frac{\mathrm{d}\Omega}{\mathrm{d}z} = -\frac{\alpha}{g(\omega_0)} \frac{(\Omega/\beta)\cos\phi}{(\frac{1}{2}\Omega^2/\beta^2) + 1}. \qquad (16.4\text{–}10)$$

Equation (16.4–3) yields in the steady state

$$\Omega\frac{\mathrm{d}\phi}{\mathrm{d}z} = \frac{\alpha}{g(\omega_0)} r'_1,$$

and with the help of Eq. (16.4–7) we obtain

$$\Omega\frac{\mathrm{d}\phi}{\mathrm{d}z} = -\frac{\alpha}{g(\omega_0)} \frac{(\Omega/\beta)\sin\phi}{(\frac{1}{2}\Omega^2/\beta^2) + 1}. \qquad (16.4\text{–}11)$$

Equations (16.4–10) and (16.4–11) can be combined to give

$$\Omega\frac{d\phi}{dz}\bigg/\frac{d\Omega}{dz} = \tan\phi,$$

and this formula can be written as a differential equation connecting Ω and ϕ,

$$\Omega\frac{d\phi}{d\Omega} = \tan\phi.$$

The equation can be integrated immediately, and gives

$$\Omega(z)\sin\phi(0) = \Omega(0)\sin\phi(z), \qquad (16.4\text{–}12)$$

from which it follows that in the special case $\phi(0) = 0$ and $\Omega(0) \neq 0$, $\phi(z) = 0$ for all z. We shall henceforth confine our attention to this special case. Also, if $g(\omega)$ is the natural, Lorentzian lineshape function, we can replace $g(\omega_0)$ by $2/\beta$.

When $\phi = 0$, Eq. (16.4–10) can be integrated directly with respect to z and gives

$$\int_{\Omega(0)}^{\Omega(z)} [(\tfrac{1}{2}\Omega^2/\beta^2) + 1]\frac{d\Omega}{\Omega} = -\tfrac{1}{2}\alpha\int_0^z dz$$

or

$$\ln\frac{\Omega(z)}{\Omega(0)} + \frac{1}{4}\left[\frac{\Omega^2(z)}{\beta^2} - \frac{\Omega^2(0)}{\beta^2}\right] = -\tfrac{1}{2}\alpha z. \qquad (16.4\text{–}13)$$

It should be noted that both sides of the equation are negative, because the medium is an absorber. In a strong field, with $\Omega/\beta \gg 1$, the logarithmic term in Ω will be numerically small compared with the quadratic term, and the opposite is true in a sufficiently weak field with $\Omega/\beta \ll 1$.

Finally, we need to relate $\Omega(0)$ and $\Omega(L)$ to the input and output amplitudes $\mathscr{E}_i$ and $\mathscr{E}_0$ of the electric field. It is convenient to represent the electric fields by the corresponding Rabi frequencies

$$\left.\begin{aligned} \Omega_i &\equiv \frac{2\boldsymbol{\mu}_{12}\cdot\boldsymbol{\varepsilon}^*}{\hbar}\mathscr{E}_i \\ \Omega_0 &\equiv \frac{2\boldsymbol{\mu}_{12}\cdot\boldsymbol{\varepsilon}^*}{\hbar}\mathscr{E}_0. \end{aligned}\right\} \qquad (16.4\text{–}14)$$

It can be seen by inspection of Fig. 16.10 that

$$\Omega(0) = \sqrt{(1-R)}\Omega_i + R\Omega(L)$$
$$\Omega_0 = \sqrt{(1-R)}\Omega(L),$$

so that

$$\Omega(0) = \sqrt{(1-R)}\Omega_i + \frac{R}{\sqrt{(1-R)}}\Omega_0 \qquad (16.4\text{–}15)$$

and

$$\Omega(L) = \frac{1}{\sqrt{(1-R)}}\Omega_0. \qquad (16.4\text{–}16)$$

We now put $z = L$ in Eq. (16.4–13) and substitute for $\Omega(0)$ and $\Omega(L)$ from Eqs. (16.4–15) and (16.4–16). We then obtain the following relation between the input and output field amplitudes Ω_i and Ω_0:

$$\ln\left[R + (1-R)\frac{\Omega_i}{\Omega_0}\right] + \tfrac{1}{4}\left[(1-R)\frac{\Omega_i^2}{\beta^2} - (1+R)\frac{\Omega_0^2}{\beta^2} + 2R\frac{\Omega_0\Omega_i}{\beta^2}\right] = \tfrac{1}{2}\alpha L. \tag{16.4 17}$$

The relation (16.4–17) is illustrated graphically in Fig. 16.11 for the case $R = 0.95$, and for several different values of the absorption as measured by αL. It will be seen that so long as αL is not too small, a moderately small input signal produces a very small output, because the medium is strongly absorbing. On the other hand, a sufficiently large input signal produces an output of comparable amplitude, because the absorbing medium eventually becomes saturated. The most interesting behavior is to be found in the intermediate region, where the slope $d\Omega_0/d\Omega_i$ may become negative if αL is not too small, and there are several possible values of output Ω_0 for a given input Ω_i. In practice the regions of negative slope are unstable, so that the system switches discontinuously from one branch of the curve to another. For example, let us suppose that we start with a very small light input Ω_i, and that this is gradually increased. When the system reaches point A in Fig. 16.11, the output jumps to the larger value represented by point B, and the system thereafter follows the upper branch of the curve as the input continues to increase. If the input is now gradually decreased, the system follows the upper branch of the curve past B to point C, at which point the output jumps to the lower value represented by point D. Further decrease of the input causes the system to follow the lower branch of the curve. The system therefore

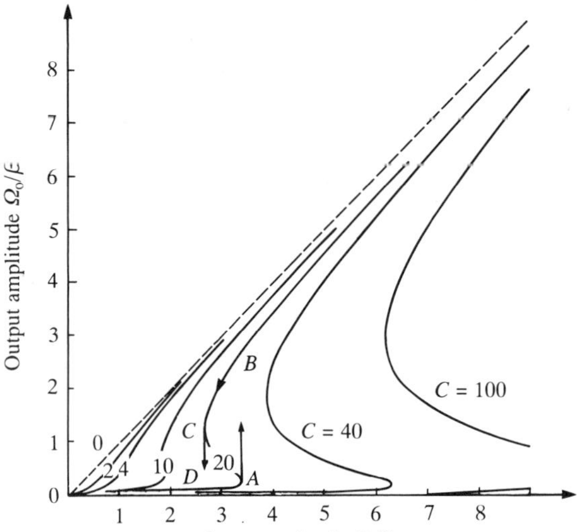

Fig. 16.11 The relation between the input and the output amplitudes given by Eq. (16.4–17). The reflectivity R is fixed at 0.95, and each graph corresponds to a different value of $\alpha L = 2CT$.

exhibits *hysteresis* and *bistability*, and this behavior has been demonstrated experimentally (McCall, 1974).

It should be noted that both the resonantly absorbing medium and the optical feedback are needed for bistability in this case. If we dispense with the medium by putting $\alpha L = 0$, then Eq. (16.4–17) leads to $\Omega_0 = \Omega_i$, which is the relationship illustrated by the broken line in Fig. 16.11. On the other hand, if we dispense with the feedback mirrors and put $R = 0$, then Eq. (16.4–17) reduces to an equation like Eq. (16.4–13), which implies a monotonic relation between Ω_0 and Ω_i, and again there is no bistability. This is to be contrasted with the situation in which the absorber is concentrated at a point, when the response is apparently bistable without the need for a resonant cavity (Walls, Drummond, Hassan and Carmichael, 1978; Bowden and Sung, 1979). Generally when $R \neq 0$, there is a minimum value of αL, of order unity when R is 0.95, below which no bistability occurs. Under certain conditions the upper region of the curve between points B and C in Fig. 16.11 may also be unstable in a different sense, in that self-pulsing of the output may occur (Bonifacio and Lugiato, 1978a,b,c).

When the transmittance $T = 1 - R$ of the mirrors is sufficiently small that we can expand the logarithm to the first order in T, and if αL is also small and we put $\frac{1}{2}\alpha L/T \equiv C$, then Eq. (16.4–17) simplifies to

$$\left(\frac{\Omega_i}{\Omega_0} - 1\right) + \frac{1}{2}\frac{\Omega_0}{\beta T}\left(\frac{\Omega_i}{\beta} - \frac{\Omega_0}{\beta}\right) \approx C$$

or

$$\frac{\Omega_i}{\beta} \approx \frac{\Omega_0}{\beta} + C\frac{\Omega_0/\beta}{(\frac{1}{2}\Omega_0^2/\beta^2) + 1}, \tag{16.4–18}$$

which is of the same general form as Eq. (16.4–1). This equation also exhibits bistability, so long as C exceeds 8. C is sometimes known as the *cooperativity parameter*.

This simple semiclassical analysis ignores the quantum fluctuations of the system, which become particularly important in the neighborhood of the bistability. Indeed they will generally cause the system to switch somewhere between points D and A and between point C and B, rather than right at the turning points shown in Fig. 16.11. It can be shown from a fully quantized treatment (Bonifacio and Lugiato, 1978a; Narducci, Gilmore, Da Hsuan Feng and Agarwal, 1978) that large field fluctuations appear in the critical switching region, and that the output light has an incoherent component whose spectral density changes drastically as we progress around the bistability cycle. The spectrum is much narrower than the natural line at point D in Fig. 16.11; it broadens as we move towards point A, and becomes three-peaked at point B; it again becomes single-peaked but much broader than the natural line at point C. These changes of the spectrum reflect fundamental changes in the radiation mechanism. Near point B and above, where the absorbing medium is close to saturation, each atom radiates more or less independently in the resonant driving field. We therefore expect to find the three-peaked spectrum that is characteristic of spontaneous radiation in a coherent field (cf. Fig. 15.10). However, near point C the driving field is not strong enough to overcome the mutual interaction between the atoms, and the system radiates in a cooperative or superradiant manner, to be discussed in more detail in Section 16.6. This is characterized by a broadening of the emitted line. In the region D to A the strong absorption of the resonant medium causes a narrowing of the spectrum of the transmitted light as compared with the natural line. This shows once again that the switching that occurs in the region of bistability represents transitions between cooperative and non-cooperative radiation processes.

One way of testing the theory of absorptive bistability is to measure the light intensities Y_1, Y_2, corresponding to points A and D in Fig. 16.11, where switching from one state to the other occurs. This has been done in systematic measurements by Orozco, Kimble and Rosenberger (1987), who studied Y_1, Y_2 as a function of the cooperativity parameter C. Some of their results are shown in Fig. 16.12, superimposed on the theoretically predicted curves, and they exhibit very good agreement.

16.4.2 Dispersive bistability

The phenomenon we have discussed so far depends on the saturation of the atomic absorption, and is known as *absorptive bistability*, to distinguish it from

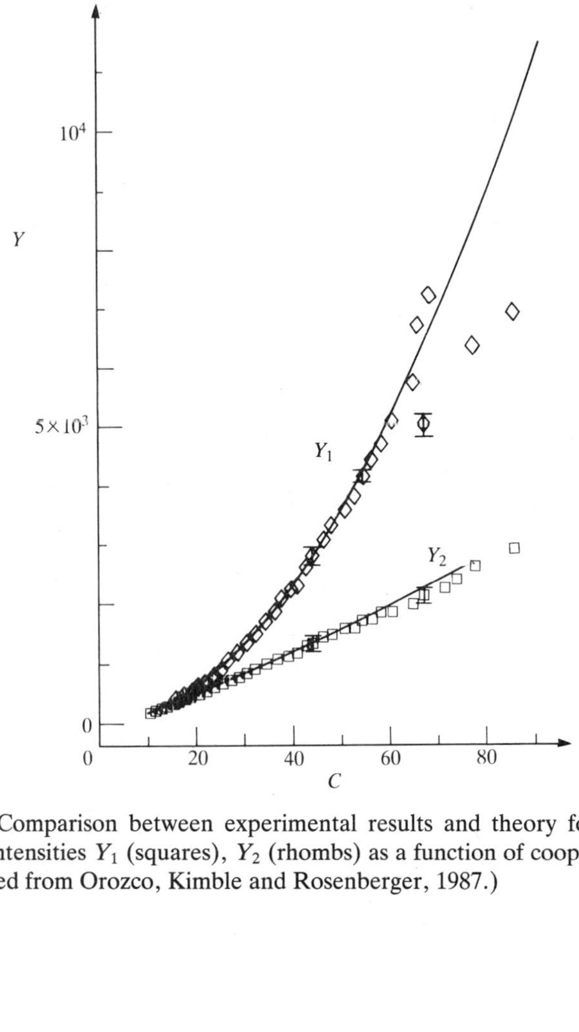

Fig. 16.12 Comparison between experimental results and theory for the two switching intensities Y_1 (squares), Y_2 (rhombs) as a function of cooperativity C (Reproduced from Orozco, Kimble and Rosenberger, 1987.)

dispersive bistability, which arises under similar circumstances when the dispersive, rather than the absorptive, properties of the medium play the more important role. We shall not treat dispersive bistability in any detail, but instead introduce a simplified heuristic argument (Gibbs, McCall and Venkatesan, 1978).

Let us suppose that the field propagating through the medium is not attenuated by the medium, but experiences an intensity-dependent phase shift ϕ, so that

$$\Omega(L) = \Omega(0)\,\mathrm{e}^{-\mathrm{i}\phi}, \tag{16.4–19}$$

with

$$\phi = \phi_0 - k|\Omega_0|^2. \tag{16.4–20}$$

This might be the situation if the refractive index has the sort of non-linearity that is encountered in the optical Kerr effect, or when the system of two-level atoms is very far from resonance. The coefficient k is a positive constant and ϕ_0 is an adjustable parameter that depends on the detuning of the cavity from the atomic resonance. Equation (16.4–19) now replaces Eq. (16.4–13) as the relation between $\Omega(0)$ and $\Omega(L)$, and, together with the boundary conditions (16.4–15) and (16.4–16), we now obtain the result

$$\Omega_0 = [(1-R)\Omega_\mathrm{i} + R\Omega_0]\,\mathrm{e}^{-\mathrm{i}(\phi_0 - k|\Omega_0|^2)}. \tag{16.4–21}$$

If we denote $|\Omega_\mathrm{i}|^2$ by I_i and $|\Omega_0|^2$ by I_0, where I_i, I_0 play the role of input and output intensities, and if we take the square modulus of both sides of this equation, and assume that the phase angle ϕ is sufficiently small that we can make the approximation

$$\cos(\phi_0 - kI_0) \approx 1 - \tfrac{1}{2}(\phi_0 - kI_0)^2,$$

then Eq. (16.4–21) reduces to

$$I_\mathrm{i} = I_0\left[1 + \frac{R\phi_0^2}{(1-R)^2} - \frac{2k\phi_0 R}{(1-R)^2}I_0 + \frac{k^2R}{(1-R)^2}I_0^2\right]. \tag{16.4–22}$$

This cubic equation in I_0 again exhibits bistability behavior, in the sense that $\mathrm{d}I_\mathrm{i}/\mathrm{d}I_0$ can be made to vanish for two real values of I_0, provided that

$$\phi_0^2 > 3T^2/R. \tag{16.4–23}$$

This time the asymptotic relation between I_i and I_0 is not linear, as for the case of absorptive bistability. However, Eq. (16.4–22) comes closer than Eq. (16.4–17) to describing the first experiments on optical bistability performed by Gibbs, McCall and Venkatesan (1976). They made use of a Fabry–Perot interferometer filled with sodium vapor and operated far from resonance, so that dispersion was the principal non-linearity. Figure 16.13 shows the results of measurements of the transmitted power as a function of input power, both when the cell was empty and when it was filled with sodium vapor. The hysteresis in the transmittance can be seen very clearly.

16.4.3 Chaos in optical bistability

It was pointed out by Ikeda (1979) [see also Ikeda, Daido and Akimoto (1980)] that when the time taken by the light to propagate from the output end ($z = L$)

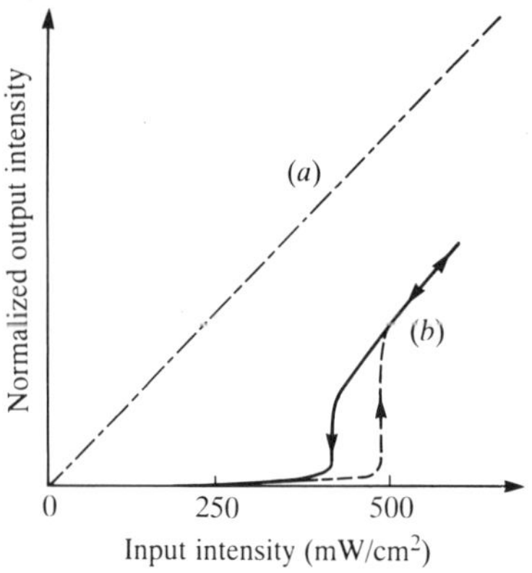

Fig. 16.13 Results of measurements of transmitted power versus input power for a Fabry–Perot interferometer (*a*) without and (*b*) with sodium vapor. (Adapted from Gibbs, McCall and Venkatesan, 1976.)

back to the input end ($z = 0$) of the nonlinear medium (see Fig. 16.10) is substantially greater than the relaxation time of the medium, entirely new types of behavior may appear. Under these conditions the differential equations governing the time evolution of the light become time delay difference equations, and can be reduced to a discrete mapping. This mapping, although completely deterministic, may nevertheless result in a time development that resembles random fluctuations and is best described as *chaotic*. It is characterized by a spectrum that consists of a continuous distribution rather than discrete frequencies, and the system exhibits very sensitive dependence on initial conditions. This behavior has been observed experimentally (Gibbs, Hopf, Kaplan and Shoemaker, 1981; Nakatsuka, Asaka, Itoh, Ikeda and Matsuoka, 1983). The subject of chaos in optical bistability has already spawned a large literature, but we shall not go into it here. However, chaotic behavior in the laser is treated briefly in Section 18.8.

16.5 Collective atomic states and collective dynamical variables

We have now encountered several optical phenomena that depend on collective atomic effects, although so far we have succeeded in treating them in the framework of an essentially single-atom theory. The effects produced by the different atoms were simply added together. However, for some problems it is desirable to develop the atomic formalism so that it is directly applicable to a group of atoms. As we shall see, this leads not only to some new dynamical variables, but also to some collective atomic states that exhibit interesting new features.

We consider a group of N identical two-level atoms described by spin operators $\hat{R}_1^{(j)}$, $\hat{R}_2^{(j)}$, $\hat{R}_3^{(j)}$ ($j = 1, 2, \ldots, N$), or by the atomic lowering and raising operators $\hat{b}^{(j)}$, $\hat{b}^{(j)\dagger}$. As the atoms are distinct, all dynamical variables asssociated with different atoms at the same time commute. In order to describe the system collectively it is sometimes convenient to introduce the collective atomic spin operators

$$\left.\begin{aligned}\hat{\mathcal{R}}_1 &\equiv \sum_{j=1}^{N} \hat{R}_1^{(j)} \\ \hat{\mathcal{R}}_2 &\equiv \sum_{j=1}^{N} \hat{R}_2^{(j)} \\ \hat{\mathcal{R}}_3 &\equiv \sum_{j=1}^{N} \hat{R}_3^{(j)},\end{aligned}\right\} \tag{16.5–1}$$

or the collective lowering and raising operators

$$\left.\begin{aligned}\hat{\mathcal{R}}^{(-)} &\equiv \sum_{j=1}^{N} \hat{b}^{(j)} = \hat{\mathcal{R}}_1 - \mathrm{i}\hat{\mathcal{R}}_2 \\ \hat{\mathcal{R}}^{(+)} &\equiv \sum_{j=1}^{N} \hat{b}^{(j)\dagger} = \hat{\mathcal{R}}_1 + \mathrm{i}\hat{\mathcal{R}}_2.\end{aligned}\right\} \tag{16.5–2}$$

We readily find from the commutation relations (15.1–11) and (15.1–13) obeyed by the single-atom operators that

$$\begin{aligned}[\hat{\mathcal{R}}_l, \hat{\mathcal{R}}_m] = \sum_{j=1}^{N}\sum_{k=1}^{N}[\hat{R}_l^{(j)}, \hat{R}_m^{(k)}] &= \sum_{j=1}^{N}[\hat{R}_l^{(j)}, \hat{R}_m^{(j)}] \\ &= \mathrm{i}\epsilon_{lmn}\hat{\mathcal{R}}_n, \quad (l, m, n = 1, 2, 3),\end{aligned} \tag{16.5–3}$$

$$\left.\begin{aligned}[\hat{\mathcal{R}}^{(+)}, \hat{\mathcal{R}}^{(-)}] &= 2\hat{\mathcal{R}}_3 \\ [\hat{\mathcal{R}}_3, \hat{\mathcal{R}}^{(\pm)}] &= \pm\hat{\mathcal{R}}^{(\pm)}.\end{aligned}\right\} \tag{16.5–4}$$

Just as $\hat{b}^{(j)\dagger}$, $\hat{b}^{(j)}$ raise and lower the excitation of the j'th atom by unity, so the $\hat{\mathcal{R}}^{(+)}$ and $\hat{\mathcal{R}}^{(-)}$ operators raise and lower the excitation of the collective atomic system by unity, but in a manner such that this excitation is distributed over all atoms. As an example, let us suppose that each of the N atoms is in the lower state $|1\rangle$, so that the collective atomic system is in the product state

$$|\{1\}\rangle \equiv \prod_{j=1}^{N} |1_j\rangle.$$

Then

$$\hat{\mathcal{R}}^{(+)}|\{1\}\rangle = \sum_{k=1}^{N} |2_k\rangle \prod_{j\neq k} |1_j\rangle,$$

which is a superposition state in which one excitation is distributed with equal weight over all atoms.

In addition to the collective operators $\hat{\mathcal{R}}_1$, $\hat{\mathcal{R}}_2$, $\hat{\mathcal{R}}_3$ we shall find it useful to introduce the operator

$$\begin{aligned}\hat{\mathcal{R}}^2 \equiv \hat{\mathcal{R}}_1^2 + \hat{\mathcal{R}}_2^2 + \hat{\mathcal{R}}_3^2 &= \tfrac{3}{4}N + \sum\sum_{j\neq k}[\hat{R}_1^{(j)}\hat{R}_1^{(k)} + \hat{R}_2^{(j)}\hat{R}_2^{(k)} + \hat{R}_3^{(j)}\hat{R}_3^{(k)}] \\ &= \tfrac{3}{4}N + \sum\sum_{j\neq k}[\hat{b}^{(j)}\hat{b}^{(k)\dagger} + \hat{R}_3^{(j)}\hat{R}_3^{(k)}].\end{aligned} \tag{16.5–5}$$

It follows from the commutation relations that

$$[\hat{\mathcal{R}}_l, \hat{\mathcal{R}}^2] = 0, \quad l = 1, 2, 3, \tag{16.5–6}$$

so that the collective $\hat{\mathcal{R}}$ operators obey the same algebra as the angular momentum. We can therefore find atomic states that are simultaneous eigenstates of $\hat{\mathcal{R}}_3$ and $\hat{\mathcal{R}}^2$, for example.

In terms of the collective operators, the total energy of the system of N atoms when the lower state is the ground state, is

$$\hat{H}_A = \hbar\omega_0(\hat{\mathcal{R}}_3 + \tfrac{1}{2}N), \tag{16.5–7}$$

and the total dipole moment $\hat{\boldsymbol{\mu}}$ is

$$\left.\begin{aligned}\hat{\boldsymbol{\mu}} &= 2\,\mathrm{Re}\,(\boldsymbol{\mu}_{12})\hat{\mathcal{R}}_1 + 2\,\mathrm{Im}\,(\boldsymbol{\mu}_{12})\hat{\mathcal{R}}_2 \\ &= \boldsymbol{\mu}_{12}\hat{\mathcal{R}}^{(-)} + \boldsymbol{\mu}_{12}^*\hat{\mathcal{R}}^{(+)}.\end{aligned}\right\} \tag{16.5–8}$$

16.5.1 Dicke states

Let us consider an atomic product state $|\Phi\rangle$ in which N_1 of the N atoms are in the lower state $|1\rangle$ and N_2 are in the excited state $|2\rangle$, where

$$\left.\begin{aligned}N &= N_1 + N_2 \\ m &\equiv \tfrac{1}{2}(N_2 - N_1),\end{aligned}\right\} \tag{16.5–9a}$$

or

$$\left.\begin{aligned}N_2 &= \tfrac{1}{2}N + m \\ N_1 &= \tfrac{1}{2}N - m.\end{aligned}\right\} \tag{16.5–9b}$$

Evidently m is a measure of total atomic inversion, and it is either an integer or a half-integer. The state $|\Phi\rangle$ is of the form

$$|\Phi\rangle = |1_1\rangle|1_2\rangle|2_3\rangle|1_4\rangle \ldots \quad \ldots |2_N\rangle,$$

and from the definition of $\hat{\mathcal{R}}_3$ we can readily see that it is an eigenstate of $\hat{\mathcal{R}}_3$ with eigenvalue m,

$$\hat{\mathcal{R}}_3|\Phi\rangle = m|\Phi\rangle, \quad -\tfrac{1}{2}N \leqslant m \leqslant \tfrac{1}{2}N. \tag{16.5–10}$$

$|\Phi\rangle$ is therefore an energy eigenstate of the atomic system with energy eigenvalue $\hbar\omega_o(m + \frac{1}{2}N)$, and this eigenvalue is independent of the manner in which the N_2 excitations are actually distributed over the N atoms. In general, the eigenvalue m has the degeneracy

$$d_m = \frac{N!}{N_2!N_1!} = \frac{N!}{(\frac{1}{2}N + m)!(\frac{1}{2}N - m)!} \tag{16.5–11}$$

under permutations among the N atoms, which is greatest when $m = 0$. There is no degeneracy at all when $m = \pm\frac{1}{2}N$, in which case all the atoms are either fully excited or fully de-excited.

The degeneracy of the energy eigenvalue can be reduced substantially if we consider states satisfying Eq. (16.5–10) that are also eigenstates of $\hat{\mathcal{R}}^2$. Because

$\hat{\mathcal{R}}_3$ and $\hat{\mathcal{R}}^2$ commute, according to Eq. (16.5–6), there exist states which are simultaneous eigenstates of both $\hat{\mathcal{R}}_3$ and $\hat{\mathcal{R}}^2$. If we denote the eigenvalue of $\hat{\mathcal{R}}^2$ by $l(l+1)$, and label the corresponding eigenstates $|l, m\rangle$, we can write

$$\left.\begin{aligned} \hat{\mathcal{R}}_3|l, m\rangle &= m|l, m\rangle \\ \hat{\mathcal{R}}^2|l, m\rangle &= l(l+1)|l, m\rangle. \end{aligned}\right\} \tag{16.5–12}$$

The definition (16.5–5) of $\hat{\mathcal{R}}^2$ and the obvious analogy between the $\hat{\mathcal{R}}_l$ operators ($l = 1, 2, 3$) and angular momentum operators, allows us to identify the constraints on l immediately, and we find that l can be an integer or half-integer, with

$$|m| \leqslant l \leqslant \tfrac{1}{2}N. \tag{16.5–13}$$

The states $|l, m\rangle$ were first introduced by Dicke (1954) in connection with the treatment of radiation by a group of N atoms interacting through the electromagnetic field. He called the number l the *cooperation number*, because it plays a key role in determining the rate of cooperative radiation by the atomic system. We shall return to a discussion of this problem in the next section.

From the commutation relation (16.5–4) and from Eq. (16.5–12) it follows immediately that

$$\begin{aligned} \hat{\mathcal{R}}_3\hat{\mathcal{R}}^{(\pm)}|l, m\rangle &= \hat{\mathcal{R}}^{(\pm)}(\hat{\mathcal{R}}_3 \pm 1)|l, m\rangle \\ &= (m \pm 1)\hat{\mathcal{R}}^{(\pm)}|l, m\rangle, \end{aligned} \tag{16.5–14}$$

so that the effect of $\hat{\mathcal{R}}^{(\pm)}$ on a Dicke state is to increase or decrease the eigenvalue m by unity, while l remains unchanged. Also, from the well-known properties of angular momentum operators it follows that (e.g. see Dicke and Wittke, 1960, Chapt. 9)

$$\left.\begin{aligned} \hat{\mathcal{R}}^{(-)}|l, m\rangle &= [(l+m)(l-m+1)]^{1/2}|l, m-1\rangle \\ \hat{\mathcal{R}}^{(+)}|l, m\rangle &= [(l-m)(l+m+1)]^{1/2}|l, m+1\rangle. \end{aligned}\right\} \tag{16.5–15}$$

As the interaction between the atomic system and the field is through the collective dipole moment $\hat{\boldsymbol{\mu}}$ given by Eq. (16.5–8), which is expressible in terms of $\hat{\mathcal{R}}^{(+)}$ and $\hat{\mathcal{R}}^{(-)}$ operators, any dipole interaction with the atomic system in a Dicke state will be governed by the first-order selection rules

$$\left.\begin{aligned} \Delta m &= \pm 1 \\ \Delta l &= 0. \end{aligned}\right\} \tag{16.5–16}$$

For any given cooperation number l, the 'ground state' of the atomic system is represented by the state $|l, -l\rangle$, for which m has the lowest possible value. More highly excited Dicke states $|l, m\rangle$ with $m > -l$ can be generated from the ground state by repeated application of the collective raising operator $\hat{\mathcal{R}}^{(+)}$, so that from Eqs. (16.5–15) we can write

$$|l, m\rangle = \left(\frac{(l-m)!}{(2l)!(l+m)!}\right)^{1/2} (\hat{\mathcal{R}}^{(+)})^{l+m}|l, -l\rangle. \tag{16.5–17}$$

We can illustrate the formation of the $|l, m\rangle$ states most easily if we first confine our attention to a system of $N = 2$ two-level atoms. The fully excited state $|2_1\rangle|2_2\rangle$ has $m = \frac{1}{2}N = 1$, and from Eq. (16.5–13) it follows that $l = 1$ also. Hence we can identify the fully excited state $|2_1\rangle|2_2\rangle$ with the Dicke state $|l = 1, m = 1\rangle$, or

$$|1, 1\rangle = |2_1\rangle|2_2\rangle. \tag{16.5–18}$$

In order to generate other Dicke states from $|1, 1\rangle$ we apply the lowering operator $\hat{\mathscr{R}}^{(-)}$, and make use of the definition (16.5–2) and Eq. (16.5–15). We then find that

$$(\sqrt{2})|1, 0\rangle = \hat{\mathscr{R}}^{(-1)}|1, 1\rangle = |2_1\rangle|1_2\rangle + |1_1\rangle|2_2\rangle$$

so that

$$|1, 0\rangle = \frac{1}{\sqrt{2}}[|2_1\rangle|1_2\rangle + |1_1\rangle|2_2\rangle], \tag{16.5–19}$$

and another application of $\hat{\mathscr{R}}^{(-)}$ leads to the formula

$$(\sqrt{2})|1, -1\rangle = \hat{\mathscr{R}}^{(-)}|1, 0\rangle = (\sqrt{2})|1_1\rangle|1_2\rangle,$$

so that

$$|1, -1\rangle = |1_1\rangle|1_2\rangle. \tag{16.5–20}$$

All these states correspond to cooperation number $l = 1$. To see which states correspond to $l = 0$, we make use of Eq. (16.5–5) to find the eigenstates of $\hat{\mathscr{R}}^2$. Then we readily find for any complex numbers c_1, c_2

$$\hat{\mathscr{R}}^2[c_1|2_1\rangle|1_2\rangle + c_2|1_1\rangle|2_2\rangle] = (c_1 + c_2)[|2_1\rangle|1_2\rangle + |1_1\rangle|2_2\rangle].$$

The state $c_1|2_1\rangle|1_2\rangle + c_2|1_1\rangle|2_2\rangle$ is therefore an eigenstate of $\hat{\mathscr{R}}^2$ with eigenvalue λ if

$$c_1 + c_2 = \lambda c_1 = \lambda c_2,$$

which requires that either

$$c_1 = c_2, \quad \lambda = 2,$$

or

$$c_1 + c_2 = 0, \quad \lambda = 0.$$

The former conditions lead to the state $|1, 0\rangle$ given by Eq. (16.5–19), that we have already found. The latter conditions clearly imply $l = 0$, and lead to the state

$$|0, 0\rangle = \frac{1}{\sqrt{2}}[|2_1\rangle|1_2\rangle - |1_1\rangle|2_2\rangle] \tag{16.5–21}$$

after normalization. The four possible states $|l, m\rangle$ when $N = 2$ are illustrated in Fig. 16.14. $l = 0$ is a singlet state, whereas $l = 1$ corresponds to a triplet.

In one sense the example with $N = 2$ is perhaps oversimple, because each of the four Dicke states $|l, m\rangle$ is expressible unambiguously in terms of the four states $|1_1\rangle|1_2\rangle$, $|2_1\rangle|1_2\rangle$, $|1_1\rangle|2_2\rangle$, $|2_1\rangle|2_2\rangle$, and there is no degeneracy. When N exceeds 2 the situation is a little more complicated.

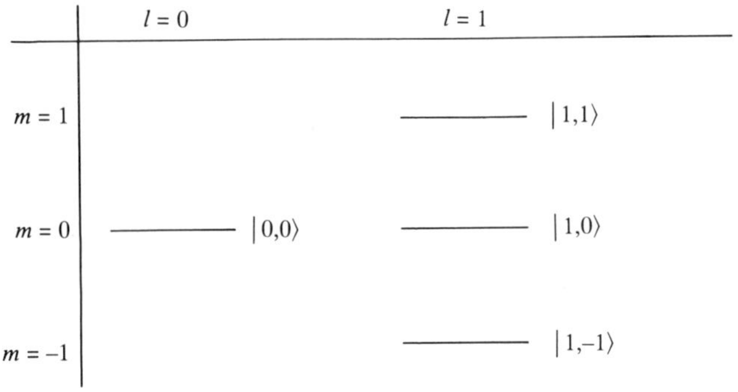

Fig. 16.14 Energy levels of Dicke states for the two-atom system.

Thus, when $N = 3$ we can readily identify the ground state of the three-atom system with the state

$$|\tfrac{3}{2}, -\tfrac{3}{2}\rangle = |1_1\rangle|1_2\rangle|1_3\rangle, \tag{16.5–22}$$

and by repeated application of the raising operator we obtain the result

$$\begin{aligned} |\tfrac{3}{2}, -\tfrac{1}{2}\rangle &= \frac{1}{\sqrt{3}}\hat{\mathscr{R}}^{(+)}|1_1\rangle|1_2\rangle|1_3\rangle \\ &= \frac{1}{\sqrt{3}}[|2_1\rangle|1_2\rangle|1_3\rangle + |1_1\rangle|2_2\rangle|1_3\rangle + |1_1\rangle|1_2\rangle|2_3\rangle] \end{aligned} \tag{16.5–23}$$

$$|\tfrac{3}{2}, \tfrac{1}{2}\rangle = \frac{1}{\sqrt{3}}[|2_1\rangle|2_2\rangle|1_3\rangle + |2_1\rangle|1_2\rangle|2_3\rangle + |1_1\rangle|2_2\rangle|2_3\rangle] \tag{16.5–24}$$

$$|\tfrac{3}{2}, \tfrac{3}{2}\rangle = |2_1\rangle|2_2\rangle|2_3\rangle. \tag{16.5–25}$$

So far each Dicke state is identifiable unambiguously. However, when we come to the states $|\tfrac{1}{2}, -\tfrac{1}{2}\rangle$ and $|\tfrac{1}{2}, \tfrac{1}{2}\rangle$, both of which are eigenstates of $\hat{\mathscr{R}}^2$ with eigenvalue $\tfrac{3}{4}$, we readily find that any state of the form

$$(|c_1|^2 + |c_2|^2 + |c_3|^2)^{-1/2}[c_1|2_1\rangle|1_2\rangle|1_3\rangle + c_2|1_1\rangle|2_2\rangle|1_3\rangle + c_3|1_1\rangle|1_2\rangle|2_3\rangle]$$

with $c_1 + c_2 + c_3 = 0$ is a state of type $|\tfrac{1}{2}, -\tfrac{1}{2}\rangle$, and similarly any state of the form

$$(|c_1|^2 + |c_2|^2 + |c_3|^2)^{-1/2}[c_1|1_1\rangle|2_2\rangle|2_3\rangle + c_2|2_1\rangle|1_2\rangle|2_3\rangle + c_3|2_1\rangle|2_2\rangle|1_3\rangle]$$

with $c_1 + c_2 + c_3 = 0$ is a state of type $|\tfrac{1}{2}, \tfrac{1}{2}\rangle$. The states $|\tfrac{1}{2}, -\tfrac{1}{2}\rangle$ and $|\tfrac{1}{2}, \tfrac{1}{2}\rangle$ are therefore not uniquely determined by Eqs. (16.5–12) and by the two eigenvalues l, m. However, we may easily show that all states of type $|\tfrac{1}{2}, -\tfrac{1}{2}\rangle$ are expressible as linear combinations of the two linearly independent states

$$|\tfrac{1}{2}, -\tfrac{1}{2}, a\rangle = \frac{1}{\sqrt{2}}[|2_1\rangle|1_2\rangle|1_3\rangle - |1_1\rangle|2_2\rangle|1_3\rangle] \tag{16.5–26}$$

$$|\tfrac{1}{2}, -\tfrac{1}{2}, b\rangle = \frac{1}{\sqrt{6}}[|2_1\rangle|1_2\rangle|1_3\rangle + |1_1\rangle|2_2\rangle|1_3\rangle - 2|1_1\rangle|1_2\rangle|2_3\rangle]. \tag{16.5–27}$$

Similarly all states of the type $|\tfrac{1}{2}, \tfrac{1}{2}\rangle$ can be constructed from the two linearly independent states

$$|\tfrac{1}{2}, \tfrac{1}{2}, a\rangle = \frac{1}{\sqrt{2}}[|1_1\rangle|2_2\rangle|2_3\rangle - |2_1\rangle|1_2\rangle|2_3\rangle] \qquad (16.5\text{–}28)$$

$$|\tfrac{1}{2}, \tfrac{1}{2}, b\rangle = \frac{1}{\sqrt{6}}[|1_1\rangle|2_2\rangle|2_3\rangle + |2_1\rangle|1_2\rangle|2_3\rangle - 2|2_1\rangle|2_2\rangle|1_3\rangle]. \qquad (16.5\text{–}29)$$

There are therefore two states of type $|\tfrac{1}{2}, -\tfrac{1}{2}\rangle$ and two of type $|\tfrac{1}{2}, \tfrac{1}{2}\rangle$, with slightly different symmetry properties, and any dipole transitions between them will only couple states with the same symmetry. The system of energy levels is illustrated in Fig. 16.15.

However, if we require that all three states of one excitation $|2_1\rangle|1_2\rangle|1_3\rangle$, $|1_1\rangle|2_2\rangle|1_3\rangle$, $|1_1\rangle|1_2\rangle|2_3\rangle$ contribute equally to the superposition state $|\tfrac{1}{2}, -\tfrac{1}{2}\rangle$, so that $|c_1| = |c_2| = |c_3|$, then the only possible Dicke states of type $|\tfrac{1}{2}, -\tfrac{1}{2}\rangle$ are of the form

$$|\tfrac{1}{2}, -\tfrac{1}{2}, \alpha\rangle = \frac{1}{\sqrt{3}}[|2_1\rangle|1_2\rangle|1_3\rangle + e^{2\pi i/3}|1_1\rangle|2_2\rangle|1_3\rangle + e^{4\pi i/3}|1_1\rangle|1_2\rangle|2_3\rangle] \qquad (16.5\text{–}30)$$

and

$$|\tfrac{1}{2}, -\tfrac{1}{2}, \beta\rangle = \frac{1}{\sqrt{3}}[|2_1\rangle|1_2\rangle|1_3\rangle + e^{-2\pi i/3}|1_1\rangle|2_2\rangle|1_3\rangle + e^{-4\pi i/3}|1_1\rangle|1_2\rangle|2_3\rangle]. \qquad (16.5\text{–}31)$$

These two states are orthogonal. Both $|\tfrac{1}{2}, -\tfrac{1}{2}, \alpha\rangle$ and $|\tfrac{1}{2}, -\tfrac{1}{2}, \beta\rangle$ remain unchanged under cyclic permutation of atoms, but one turns into the other under the exchange of two atoms. The labels α, β stand for two different symmetries. Similarly the only possible Dicke states of type $|\tfrac{1}{2}, \tfrac{1}{2}\rangle$ with $|c_1| = |c_2| = |c_3|$ are of the form

$$|\tfrac{1}{2}, \tfrac{1}{2}, \alpha\rangle = \frac{1}{\sqrt{3}}[|1_1\rangle|2_2\rangle|2_3\rangle + e^{2\pi i/3}|2_1\rangle|1_2\rangle|2_3\rangle + e^{4\pi i/3}|2_1\rangle|2_2\rangle|1_3\rangle] \qquad (16.5\text{–}32)$$

$$|\tfrac{1}{2}, \tfrac{1}{2}, \beta\rangle = \frac{1}{\sqrt{3}}[|1_1\rangle|2_2\rangle|2_3\rangle + e^{-2\pi i/3}|2_1\rangle|1_2\rangle|2_3\rangle + e^{-4\pi i/3}|2_1\rangle|2_2\rangle|1_3\rangle], \qquad (16.5\ 33)$$

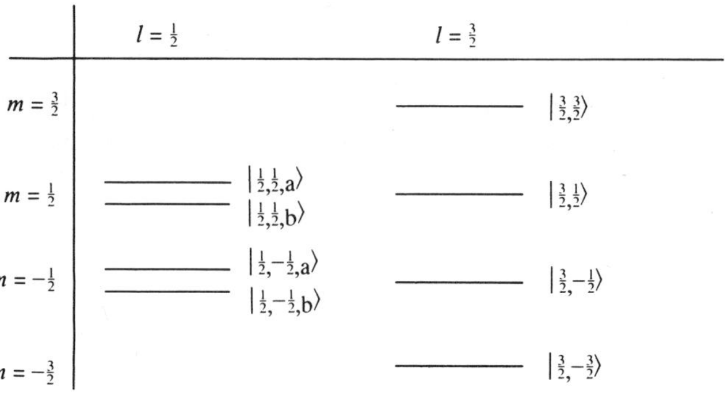

Fig. 16.15 Energy levels of Dicke states for the three-atom system. The energies at $l = \tfrac{1}{2}$, $m = \pm\tfrac{1}{2}$ are degenerate.

which again differ only in their symmetry. Under a dipole transition only states with the same symmetry α or β transform into each other.

16.5.2 *Degeneracy of Dicke states*

We can arrive at a general expression for the degeneracy of the Dicke state $|l, m\rangle$ in the following manner. Let us suppose that the state $|l, l\rangle$ has a degeneracy D_l. In particular, $D_{N/2} = 1$. By applying the lowering operator $\hat{\mathscr{R}}^{(-)}$ to $|l, l\rangle$ repeatedly, we can successively generate all the states of type $|l, m\rangle$ with $|m| \leqslant l$. It follows that each of them has the same degeneracy D_l as the original state $|l, l\rangle$, and that this degeneracy depends only on the quantum number l.

Now let us consider all the states with $m = l - 1$. From Eq. (16.5–11) there are d_{l-1} such states. Of these d_{l-1} different states, D_{l-1} are of the type $|l-1, l-1\rangle$, as we have just seen. Because of the condition $|m| \leqslant l$, all the other ones must be of the type $|l, l-1\rangle, |l+1, l-1\rangle, \ldots, |\frac{1}{2}N, l-1\rangle$. Hence

$$d_{l-1} = D_{l-1} + D_l + \ldots + D_{N/2-1} + D_{N/2}.$$

When l is replaced by $l + 1$, this becomes

$$d_l = D_l + \ldots + D_{N/2},$$

and by taking the difference we find that

$$D_{l-1} = d_{l-1} - d_l. \tag{16.5–34}$$

With the help of Eq. (16.5–11) this result immediately leads to the expression

$$D_l = \frac{(2l+1)N!}{(\frac{1}{2}N + l + 1)!(\frac{1}{2}N - l)!}, \quad l \leqslant \tfrac{1}{2}N. \tag{16.5–35}$$

In particular, we note that when $N = 3$ and $l = \frac{1}{2}$, $D_{1/2} = 2$, as we found directly by constructing the states $|\frac{1}{2}, \frac{1}{2}\rangle$, $|\frac{1}{2}, -\frac{1}{2}\rangle$.

16.6 Cooperative atomic radiation

The concept of cooperative spontaneous radiation, or *superradiance*, was introduced by Dicke in 1954. He showed that as a result of the mutual coupling between atoms through the electromagnetic field, the rate at which any one excited atom radiates is significantly influenced by the presence of all the other atoms. If several atoms are very close together, each one has to work against the radiation reaction produced not only by its own field, but also by the fields of its neighbors. As a result, each atom may work harder and radiate at an enhanced rate, so that it loses energy more rapidly than it would on its own. This would be manifest in the appearance of a short, intense pulse of light from the atomic system.

In order to make these considerations more quantitative, let us consider a group of N identical atoms within a region of space whose linear dimensions are smaller than the wavelength $2\pi c/\omega_0$ of the natural resonance frequency ω_0. Under these conditions we may assume that each atom interacts with the same electromagnetic field, and with the help of Eq. (16.5–8) we may write for the total interaction energy

$$\hat{H}_I(t) = -\hat{\boldsymbol{\mu}}(t)\cdot\hat{\mathbf{E}}(t)$$
$$= -\boldsymbol{\mu}_{12}\cdot\hat{\mathbf{E}}(t)\hat{\mathscr{R}}^{(-)}(t) - \boldsymbol{\mu}_{12}^*\cdot\hat{\mathbf{E}}(t)\hat{\mathscr{R}}^{(+)}(t). \qquad (16.6\text{–}1)$$

16.6.1 Dicke superradiance

Let us suppose that the atomic system is initially in the Dicke state $|l, m\rangle$, and that the field is initially in the vacuum state $|\text{vac}\rangle$. The probability amplitude for a transition from this state to some final state $|\psi\rangle$ is given by the matrix element

$$-\langle\psi|\boldsymbol{\mu}_{12}\cdot\hat{\mathbf{E}}(t)\hat{\mathscr{R}}^{(-)}(t) + \boldsymbol{\mu}_{12}^*\cdot\hat{\mathbf{E}}(t)\hat{\mathscr{R}}^{(+)}(t)|l, m\rangle|\text{vac}\rangle.$$

If we are interested in transitions which conserve energy and in which a photon is emitted, then the only appreciable contribution to the probability amplitude comes from the term $\langle\psi|\boldsymbol{\mu}_{12}\cdot\hat{\mathbf{E}}^{(-)}(t)\hat{\mathscr{R}}^{(-)}(t)|l, m\rangle|\text{vac}\rangle$, in which $\hat{\mathbf{E}}^{(-)}(t)$ is the negative frequency part of the electric field $\hat{\mathbf{E}}(t)$. This corresponds to a lowering of the excitation of the atomic system through $\hat{\mathscr{R}}^{(-)}$ and the creation of a photon through $\hat{\mathbf{E}}^{(-)}(t)$. By squaring the probability amplitude and summing over all possible final states, we arrive at an expression for the rate of emission of photons. This leads to the formula

$$\text{Rate of photon emission} \propto \sum_{\text{all }\psi}|\langle\psi|\boldsymbol{\mu}_{12}\cdot\hat{\mathbf{E}}^{(-)}(t)\hat{\mathscr{R}}^{(-)}(t)|l, m\rangle|\text{vac}\rangle|^2$$
$$= \langle l, m|\hat{\mathscr{R}}^{(+)}(t)\hat{\mathscr{R}}^{(-)}(t)|l, m\rangle\langle\text{vac}|\boldsymbol{\mu}_{12}\cdot\hat{\mathbf{E}}^{(+)}(t)\boldsymbol{\mu}_{12}^*\cdot\hat{\mathbf{E}}^{(-)}(t)|\text{vac}\rangle, \qquad (16.6\text{–}2)$$

when we separate matrix elements of the atomic system and of the field. With the help of Eq. (16.5–15) this equation gives immediately

$$\text{Rate of photon emission} = (l + m)(l - m + 1)A, \qquad (16.6\text{–}3)$$

where A is independent of the state of the atomic system. We can readily identify the physical significance of the factor A by applying Eq. (16.6–3) to a single excited atom. In that case $N = 1$, and the inversion $m = \frac{1}{2}$, and from the inequality

$$|m| \leqslant l \leqslant \tfrac{1}{2}N \qquad (16.6\text{–}4)$$

we find $l = \frac{1}{2}$. Hence the rate of photon emission is A, and we see that A is simply the Einstein A-coefficient for each atom.

Let us now explore some of the implications of Eq. (16.6–3). If all N atoms are in the ground state, then $m = -\frac{1}{2}N$, and from the inequality (16.6–4) we must have $l = \frac{1}{2}N$. Hence the radiation rate is zero, as expected. If all N atoms are in the excited state, then $m = \frac{1}{2}N$ and we again have $l = \frac{1}{2}N$. Hence the rate of photon emission is NA, which is just what one would expect from a group of N independently radiating atoms. However, the situation is quite different if the initial state is not the fully excited state. Let us consider a state in which half the atoms are excited and half are not, so that $m = 0$. Then

$$\text{Rate of photon emission} = l(l + 1)A, \qquad (16.6\text{–}5)$$

and l can have any value between 0 and $\frac{1}{2}N$. The larger the value of l, the larger is the collective rate of radiation of the atomic system, which led Dicke to call l the *cooperation number*. In particular, if $l = \frac{1}{2}N$, then

$$\text{Rate of photon emission} = \tfrac{1}{2}N(\tfrac{1}{2}N + 1)A, \tag{16.6–6}$$

and for large N this rate is proportional to the square of the number of atoms, which can be a very large number. Such an N^2-dependent rate is characteristic of a cooperative radiation process, which Dicke labeled *superradiance*. It follows that the radiation rate of a partially excited atomic system can be much higher than that of a fully excited system, even though the expectation value of the total dipole moment $\hat{\boldsymbol{\mu}}$ always vanishes in a Dicke state. However from Eqs. (16.5–8) and (16.5–15), we note that the mean squared value of the dipole moment is

$$\langle l, m|\hat{\boldsymbol{\mu}}^2|l, m\rangle = 2|\boldsymbol{\mu}_{12}|^2[l(l+1) - m^2], \tag{16.6–7}$$

and this is greatest when $m = 0$ and the atomic system is half-excited.

We should emphasize, however, that cooperative processes do not necessarily imply an enhanced radiation rate. In particular, if $l = 0$ in Eq. (16.6–5), then the rate of photon emission from the atomic system is zero, even though the system is half-excited. This would correspond to the unusual, but not impossible, situation in which the atomic dipoles are oppositely phased in pairs, so that the field radiated by one atom is absorbed by another. The reduction of the spontaneous emission rate, and the corresponding lengthening of the lifetime, that is associated with atomic radiators in antiphase has been observed in experiments with atoms close to a metallic mirror (Drexhage, 1970). The mirror image of one atom behaves as a second radiator in antiphase with the first. An example of such an *anti-superradiant* state for two atoms is the anti-symmetric state

$$|l = 0, m = 0\rangle = \frac{1}{\sqrt{2}}(|2_1\rangle|1_1\rangle - |1_2\rangle|2_2\rangle),$$

for which the probability amplitude coupling this state to the atomic ground state $|1_1\rangle|1_1\rangle$ via the interaction $\boldsymbol{\mu}_{12} \cdot \hat{\mathbf{E}}^{(-)}(t)\hat{\mathcal{R}}^{(-)}(t)$ vanishes.

Although the Dicke states provide a convenient illustration of the idea of superradiance, they are not easily realized experimentally. The dynamical features of cooperative atomic radiation can be illustrated more simply in terms of atomic product states (Agarwal, Brown, Narducci and Vetri, 1977).

16.6.2 Cooperative radiation in an atomic product state

We suppose that the initial state of the N-atom system is given by the product

$$|\Phi\rangle \equiv \prod_{j=1}^{N}[c_1|1_j\rangle + c_2|2_j\rangle],$$

in which c_1, c_2 are arbitrary complex numbers, with $|c_1|^2 + |c_2|^2 = 1$, that are the same for every atom. Each atom of the system is therefore assumed to be in the same pure quantum state. We take the initial state of the field to be the vacuum state, so that the initial density operator of the coupled system at some time t, say, is

$$\hat{\rho}(t) = |\Phi\rangle\langle\Phi| \otimes |\text{vac}\rangle\langle\text{vac}|. \tag{16.6–8}$$

We are interested in transitions within a short time Δt from this state to a state in which a photon of wave vector $\mathbf{k}'$ and polarization s' is emitted. The probability for the emission of such a photon is given by

$$\begin{matrix}\text{Probability of emission of} \\ \mathbf{k}',\, s' \text{ photon in time } \Delta t\end{matrix} = \sum_{\text{all } A} \langle \Psi_A | \langle 1_{\mathbf{k}'s'}, \text{vac} | \hat{\rho}(t + \Delta t) | 1_{\mathbf{k}'s'}, \text{vac}\rangle | \Psi_A \rangle, \tag{16.6–9}$$

where $\hat{\rho}(t + \Delta t)$ is the density operator of the coupled system at the later time $t + \Delta t$, and the sum is to be taken over a complete set of final atomic states $|\Psi_A\rangle$. As the final one-photon state is orthogonal to the initial state of the system, we can make use of Eq. (14.1–14) for the probability to the lowest non-vanishing order. After we insert Eq. (16.6–1) for the interaction $\hat{H}_{\text{I}}(t)$ this leads to

Probability of emission of $\mathbf{k}'$, s' photon in time Δt

$$\begin{aligned}
&= \sum_{\text{all } A} \frac{1}{\hbar^2} \int_t^{t+\Delta t} \mathrm{d}t_1 \int_t^{t_1} \mathrm{d}t_2 \langle \Psi_A | \sum_{m=1}^{N} [(\boldsymbol{\mu}_{12})_i \hat{b}^{(m)} \, \mathrm{e}^{-\mathrm{i}\omega_0 t_1} + \text{h.c.}] | \Phi \rangle \\
&\qquad \times \langle \Phi | \sum_{n=1}^{N} [(\boldsymbol{\mu}_{12})_j \hat{b}^{(n)} \, \mathrm{e}^{-\mathrm{i}\omega_0 t_2} + \text{h.c.}] | \Psi_{\text{A}} \rangle \\
&\qquad \times \langle 1_{\mathbf{k}'s'}, \text{vac} | \frac{1}{L^{3/2}} \sum_{\mathbf{k},s} \left(\frac{\hbar\omega}{2\varepsilon_0}\right)^{1/2} [\mathrm{i}\hat{a}_{\mathbf{k}s} (\boldsymbol{\varepsilon}_{\mathbf{k}s})_i \, \mathrm{e}^{-\mathrm{i}\omega t_1} + \text{h.c.}] | \text{vac} \rangle \\
&\qquad \times \langle \text{vac} | \frac{1}{L^{3/2}} \sum_{\mathbf{k},s} \left(\frac{\hbar\omega}{2\varepsilon_0}\right)^{1/2} [\mathrm{i}\hat{a}_{\mathbf{k}s} (\boldsymbol{\varepsilon}_{\mathbf{k}s})_j \, \mathrm{e}^{-\mathrm{i}\omega t_2} + \text{h.c.}] | 1_{\mathbf{k}'s'}, \text{vac} \rangle + \text{h.c.} \\
&\approx |\boldsymbol{\mu}_{12} \cdot \boldsymbol{\varepsilon}^*_{\mathbf{k}'s'}|^2 \frac{\omega'}{L^3 \hbar \varepsilon_0} \int_0^{\Delta t} \mathrm{d}t_1 \int_0^{t_1} \mathrm{d}t_2 \cos[(\omega' - \omega_0)(t_1 - t_2)] \\
&\qquad \times \langle \Phi | \sum_{m=1}^{N} \sum_{n=1}^{N} b^{(m)\dagger} b^{(n)} | \Phi \rangle \\
&= |\boldsymbol{\mu}_{12} \cdot \boldsymbol{\varepsilon}^*_{\mathbf{k}'s'}|^2 \frac{\omega'}{L^3 \hbar \varepsilon_0} \int_0^{\Delta t} \mathrm{d}t_1 \int_0^{t_1} \mathrm{d}t_2 \cos[(\omega' - \omega_0)(t_1 - t_2)] \\
&\qquad \times [N|c_2|^2 + N(N-1)|c_2|^2|c_1|^2],
\end{aligned} \tag{16.6–10}$$

in which we have discarded any highly oscillatory contributions to the time integrals. Finally, we sum over all $\mathbf{k}'$, s' corresponding to all possible final one-photon states, as in Section 15.4, and arrive at the following expression for the probability $P(t)\Delta t$ that a photon is emitted in time Δt:

$$P(t)\Delta t = A\Delta t[N|c_2|^2 + N(N-1)|c_2|^2|c_1|^2]. \tag{16.6–11}$$

Here $A = 2\beta$ is the usual Einstein A-coefficient.

It can be seen that there is a term in this expression proportional to the number of atoms N, as would be expected for independently radiating atoms. If the initial state is the fully excited state, so that $c_1 = 0$, then this is the only non-vanishing term on the right of the equation. The other term proportional to $N(N-1)$ is clearly associated with a cooperative radiation process when $c_1 \neq 0$, because it vanishes unless $N \geqslant 2$. In particular, when N is a large number, this term will become proportional to N^2 and will tend to dominate the emission probability, unless c_1 is close to zero and the atoms are almost fully excited. It follows therefore that the rate of radiation of N interacting atoms in a product state can be much greater than for N independently radiating atoms, provided the atoms are not fully excited. This is the phenomenon of superradiance, for which the radiation rate is proportional to N^2. The rate has its greatest value when $|c_1|^2|c_2|^2$ is greatest, which requires that $|c_1| = 1/\sqrt{2} = |c_2|$, or that the initial excitation of the atomic system is exactly half its maximum value. The rate of photon emission is then very nearly $\frac{1}{4}N^2A$, which is similar to the radiation rate in the Dicke state given by Eq. (16.6–6). In this respect the atomic system behaves very much in the same way when it is in a half-excited product state and when it is in a half-excited Dicke state. However, we must not lose sight of the fact that the expectation value of the collective atomic dipole moment is greatest in the former case and zero in the latter case, so that the former case bears a much closer analogy with a set of classical dipoles. The term *superfluorescence* has sometimes been used to distinguish superradiance phenomena in which there is a vanishing net atomic dipole moment from those in which the dipole moment is non-zero.

16.6.3 Time development of superradiance

Our perturbative calculation of the photon emission probability in a short time Δt does not allow us to determine the time evolution of the system for arbitrary times. However, with the help of a simple energy balance argument (Rehler and Eberly, 1971), we can readily obtain some information about how the radiation rate varies with time. A more precise quantum treatment would of course go beyond first-order perturbation theory (Bonifacio, Schwendimann and Haake, 1971a,b). If we multiply both sides of Eq. (16.6–11) by the excitation energy $\hbar\omega_0$ of each atom and divide by Δt, we obtain the mean rate of emission of energy in the form of photons. This rate must equal the rate at which the atomic system is losing energy on the average, which, according to Eq. (16.5–7) is given by

$$\begin{aligned}\text{Rate of loss of energy} &= -\frac{\mathrm{d}}{\mathrm{d}t}\hbar\omega_0[\langle\hat{\mathcal{R}}_3(t)\rangle + \tfrac{1}{2}N] \\ &= -\frac{\mathrm{d}}{\mathrm{d}t}\hbar\omega_0 N|c_2(t)|^2. \qquad (16.6\text{–}12)\end{aligned}$$

By equating the rate of radiation of energy to the rate of loss, we arrive at the equation of motion

$$-\frac{\mathrm{d}}{\mathrm{d}t}|c_2(t)|^2 = A|c_2(t)|^2[1 + (N-1)|c_1(t)|^2] \qquad (16.6\text{–}13)$$

for the time evolution of the system. The presumption that the atomic system remains in a pure product state has of course been built into the equation. We are also assuming that most modes of the field are empty, and that, from the point of view of the atomic system, the fraction of occupied modes of the electromagnetic field remains sufficiently small that the state of the field can continue to be approximated by the vacuum state. When $N = 1$ the second term on the right of Eq. (16.6–13) drops out and we recover the exponential decay rate that is characteristic of spontaneous atomic radiation. However, in general, the solution is much more complicated.

In order to solve Eq. (16.6–13) we put $|c_2(t)|^2 = y$ and $|c_1(t)|^2 = 1 - y$. The differential equation then becomes

$$\frac{\mathrm{d}y}{\mathrm{d}t} = -ANy\left[1 - \left(1 - \frac{1}{N}\right)y\right]. \tag{16.6–14}$$

If the atomic system starts from the fully inverted state with $|c_2(0)|^2 = 1$, we obtain after integrating from time 0 to time t,

$$\int_1^y \frac{\mathrm{d}y}{y[1 - (1 - 1/N)y]} = -ANt.$$

After decomposing the integrand in the form

$$\frac{1}{y[1 - (1 - 1/N)y]} = \frac{1}{y} + \frac{1 - 1/N}{1 - (1 - 1/N)y},$$

and carrying out the integration we obtain the formula

$$\ln\left(\frac{1}{[N - (N - 1)y]}\right) = -ANt$$

or

$$|c_2(t)|^2 = y(t) = \frac{N}{N - 1 + \mathrm{e}^{ANt}}. \tag{16.6–15}$$

From Eq. (16.6–11) the rate of emission of photons from the atomic system is then given by the equation

$$\begin{aligned} R(t) &= AN|c_2(t)|^2[1 + (N - 1)|c_1(t)|^2] \\ &= \frac{AN^3\,\mathrm{e}^{ANt}}{(N - 1 + \mathrm{e}^{ANt})^2}, \end{aligned} \tag{16.6–16}$$

while the average atomic energy is

$$\langle \hat{H}_\mathrm{A}(t) \rangle = \hbar\omega_0 N|c_2(t)|^2 = \hbar\omega_0\left(\frac{N^2}{N - 1 + \mathrm{e}^{ANt}}\right). \tag{16.6–17}$$

When $t = 0$, we readily find that $R(0) = AN$, and $\langle \hat{H}_\mathrm{A}(0) \rangle = \hbar\omega_0 N$, which are the relations expected for N non-interacting atoms. However, at later times t when cooperative effects come into play the rate of emission $R(t)$ can be very much greater. Figure 16.16 illustrates the radiation rate given by Eq. (16.6–16) for a system of $N = 20$ atoms.

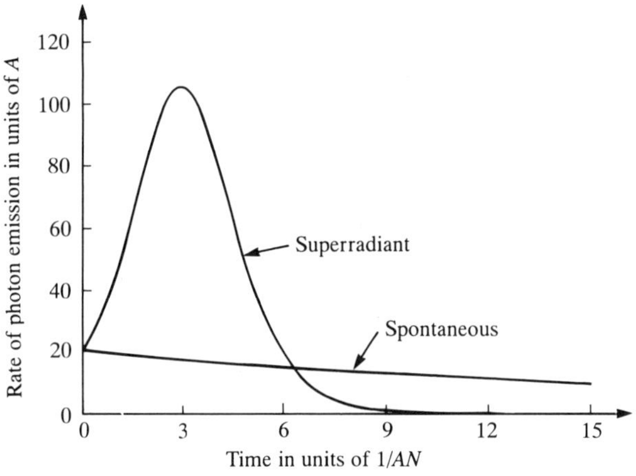

Fig. 16.16 Comparison of the superradiant radiation rate given by Eq. (16.6–16) with normal spontaneous emission for a system of $N = 20$ atoms. (Courtesy of D. James.)

When $N \gg 1$ the superradiant pulse tends to become symmetric and it is useful to re-express these quantities in another way. Let us introduce the polar angle $\theta(t)$ of the atomic Bloch-vector for the atoms, by putting

$$\left.\begin{aligned} |c_2(t)| &= \cos\left[\tfrac{1}{2}\theta(t)\right] \\ |c_1(t)| &= \sin\left[\tfrac{1}{2}\theta(t)\right]. \end{aligned}\right\} \tag{16.6–18}$$

Let t_0 be the time when the atomic Bloch-vector is horizontal and the atomic system is half excited. Then $\theta(t_0) = \pi/2$ and $|c_2(t_0)|^2 = \frac{1}{2} = |c_1(t_0)|^2$. From Eq. (16.6–15) it follows that

$$t_0 = \frac{\ln(N+1)}{2\beta N}, \tag{16.6–19}$$

where $2\beta \equiv A$ as before. Then from Eq. (16.6–18) we have

$$\tan(\theta/2) = \left|\frac{c_1(t)}{c_2(t)}\right| = \left(\frac{e^{2\beta N t} - 1}{N}\right)^{1/2}. \tag{16.6–20}$$

With the help of Eqs. (16.6–19) and (16.6–20) we can write this result in the form

$$e^{N\beta(t-t_0)} = \left(\frac{N\tan^2(\theta/2) + 1}{N+1}\right)^{1/2}. \tag{16.6–21}$$

So long as N is a large number and the angle θ is large enough that $N\tan^2(\theta/2) \gg 1$, and from Eq. (16.6–20) this is true at all times for which $t \gg 1/2\beta N$, the 1's in the numerator and denominator in Eq. (16.6–21) are negligible compared with the terms in N. We then have, to a very good approximation,

$$e^{N\beta(t-t_0)} \approx \tan(\theta/2). \tag{16.6–22}$$

Only during a very short time at the beginning of the radiation process, before cooperative effects become significant and before the Bloch-vector has moved significantly, is this approximation invalid.

We now use Eqs. (16.6–16) to (16.6–18) to express the photon radiation rate $R(t)$ and the mean atomic energy $\langle \hat{H}_A(t)\rangle$ in terms of the polar angle $\theta(t)$. Then so long as $t \gg 1/2\beta N$ and the cooperative effects are dominant, we may write

$$R(t) = \tfrac{1}{2}\beta N^2 \sin^2\theta(t) \tag{16.6–23}$$

$$\langle \hat{H}_A(t)\rangle = \tfrac{1}{2}\hbar\omega_0 N[1 + \cos\theta(t)]. \tag{16.6–24}$$

From Fig. 16.17 we readily obtain

$$\sin\theta(t) = 2\sin(\theta/2)\cos(\theta/2) = \frac{2\,e^{N\beta(t-t_0)}}{1 + e^{2N\beta(t-t_0)}} = \operatorname{sech} N\beta(t - t_0),$$

and

$$\begin{aligned} 1 + \cos\theta(t) &= 2\cos^2(\theta/2) \\ &= 1 + \frac{1 - e^{N\beta(t-t_0)}}{1 + e^{N\beta(t-t_0)}} \\ &= 1 - \tanh N\beta(t - t_0), \end{aligned}$$

so that finally

$$R(t) = \tfrac{1}{2}\beta N^2 \operatorname{sech}^2 N\beta(t - t_0) \tag{16.6–25}$$

and

$$\langle \hat{H}_A(t)\rangle = \tfrac{1}{2}\hbar\omega_0 N[1 - \tanh N\beta(t - t_0)]. \tag{16.6–26}$$

The superradiant light intensity therefore has the form of a hyperbolic secant squared pulse, and the amplitude of this superradiant pulse has the same form as the one we encountered previously in Fig. 15.5. When the radiation process is over, $\theta(\infty) = \pi$, and then the Bloch-vector points straight down.

Figure 16.18 illustrates the light intensity as given by Eq. (16.6–25) as a function of time. The peak intensity occurs at time $t = t_0$ and it is proportional to N^2; the width of the pulse is of order $1/N\beta$, which can be very much shorter than the natural atomic lifetime. The onset of the peak is delayed by $t_0 = \ln(N + 1)/2N\beta$. These are the characteristic features of superradiance, and they have all been observed experimentally, (Skribanowitz, Herman, MacGillivray and Feld, 1973; Gross, Fabre, Pillet and Haroche, 1976; Flusberg, Mossberg and

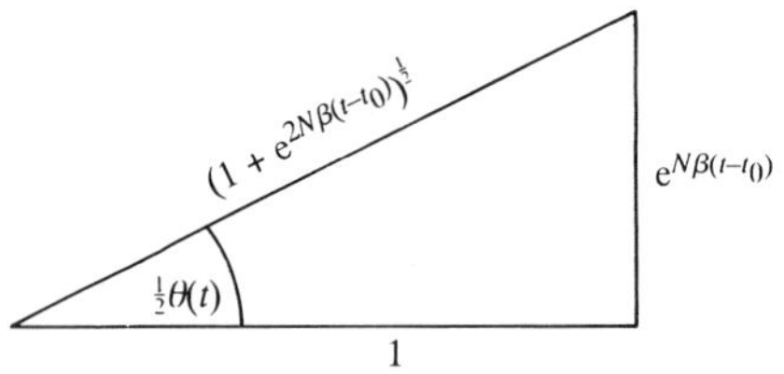

Fig. 16.17 Illustrating the time dependence of $\theta(t)$.

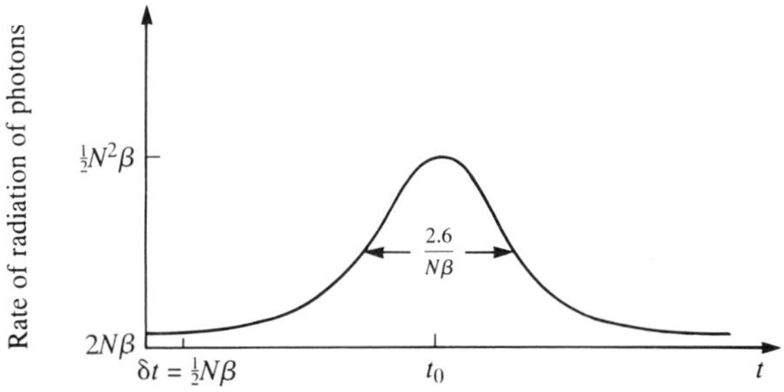

Fig. 16.18 The form of the superradiant light pulse.

Hartmann, 1976; Gibbs, Vrehen and Hikspoors, 1977; Raimond, Goy, Gross, Fabre and Haroche, 1982), although the experiments cannot, of course, be carried out under the idealized conditions that were assumed for our analysis.

Figure 16.19 shows the time development of the average atomic energy. It will be seen that the decay of the energy proceeds very rapidly after an initial slow start, and that at time $t = t_0$ the mean energy has dropped to half its initial value.

16.6.4 Some additional complications

One of the significant limitations of our treatment is the localization of all the atoms at a point, whereas excited atomic systems always have finite size. It has been pointed out that the mutual interaction between the atomic dipoles in a microscopic sample could cause dephasing and effectively quench the cooperative radiation process at high densities (Friedberg and Hartmann, 1974). In practice, pulse propagation effects may also come into play, and they may lead to multiple pulses or ringing, which have been discussed, for example, by Bonifacio and Lugiato (1975a,b). The effects can be made small by choosing a sample of sufficiently low atomic density, as in the experiments by Gibbs, Vrehen and Hikspoors (1977). Figure 16.20 shows some results of those experiments, in which

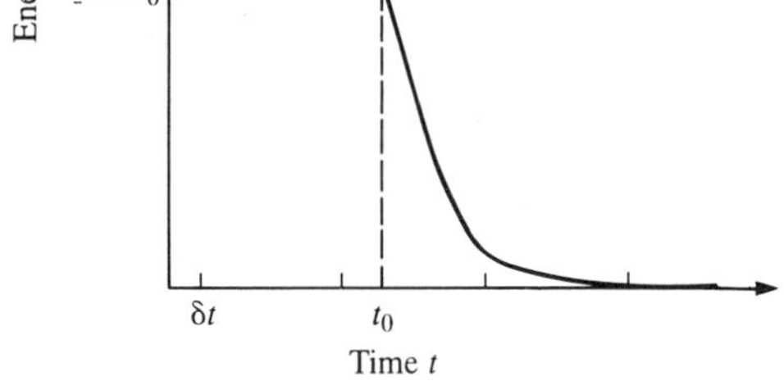

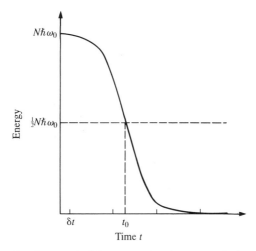

Fig. 16.19 Time development of the mean atomic energy during superradiance.

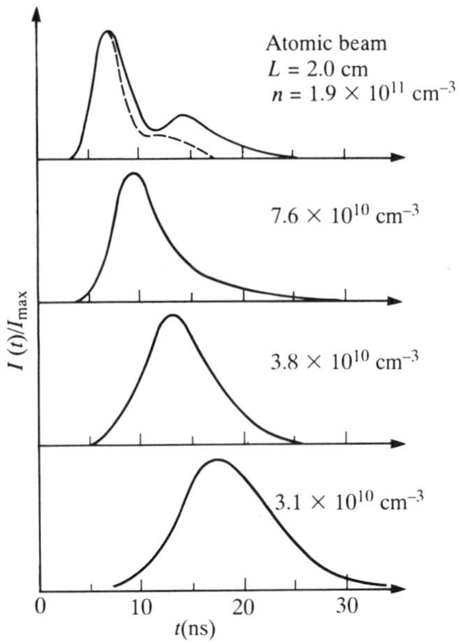

Fig. 16.20 Measured single shot superradiant pulse shapes for several different atomic densities (Reproduced from Gibbs, Vrehen and Hikspoors, 1977.)

cesium atoms in an atomic beam of varying density were excited by a short 2 ns pulse of light from a dye laser. Of course, the finite size of the atomic sample imposes additional constraints, because atoms separated by a distance greater than about $c/N\beta$ cannot cooperate in a time of order $1/N\beta$, and the actual shape of the sample also comes into play in determining the radiation rate. These effects have been discussed by Arecchi and Courtens (1970) and by Rehler and Eberly (1971), respectively.

16.6.5 More general cooperative radiation

The simplifying assumptions from which we started, that all atoms are initially in the same state and that they see essentially the same field, are actually rather strong. However, the main features of our solution are effectively preserved even if the different atoms are initially in different quantum states. It is only necessary to treat the vector sum of the individual atomic Bloch-vectors as a new super Bloch-vector (Stroud, Eberly, Lama and Mandel, 1972). This can be seen in the following simple way. When the number of atoms is very large, the radiated field can be regarded almost as a classical field, except during initial times of order of or less than $1/2N\beta$. Now as we showed in Section 15.3, the motion of an atom with Bloch-vector $\mathbf{r}$ in a classical field can be described by the equation of motion

$$\frac{\mathrm{d}\mathbf{r}}{\mathrm{d}t} = \boldsymbol{\Omega} \times \mathbf{r},$$

where the vector $\boldsymbol{\Omega}$ characterizes the applied field and has a magnitude of the order of the Rabi frequency. If two atoms of Bloch-vectors $\mathbf{r}_1$ and $\mathbf{r}_2$ are exposed to the same field, they both obey a similar equation of motion, and we have immediately

$$\begin{aligned}\frac{\mathrm{d}}{\mathrm{d}t}(\mathbf{r}_1\cdot\mathbf{r}_2) &= (\boldsymbol{\Omega}\times\mathbf{r}_1)\cdot\mathbf{r}_2+\mathbf{r}_1\cdot(\boldsymbol{\Omega}\times\mathbf{r}_2)\\ &= \boldsymbol{\Omega}\cdot[\mathbf{r}_1\times\mathbf{r}_2+\mathbf{r}_2\times\mathbf{r}_1]\\ &= 0.\end{aligned} \tag{16.6–27}$$

It follows that the angles between the Bloch-vectors do not change during the cooperative radiation process. The internal motion of the collection of N atoms can be described by the motion of a rigidly connected set of N Bloch-vectors, whose vector sum $\mathbf{R}$ obeys the equation of motion (16.6–14), but with N replaced by the length R of this new vector (Stroud, Eberly, Lama and Mandel, 1972). At the end of the superradiant pulse the vector $\mathbf{R}$ points down. As a consequence of the constraint on the motion of the individual Bloch-vectors, these vectors may not point down at the end, and energy may remain trapped in the atomic system, to be dissipated ultimately by non-cooperative processes. When inhomogeneous broadening of the atomic system is present and taken into account (Eberly, 1971; Agarwal, 1971; Ressayre and Tallet, 1973; Jodoin and Mandel, 1974a,b), the effect is generally also to limit the amount of cooperative radiation, and to force some of the energy of excitation to be radiated non-cooperatively. Figure 16.21 shows that the superradiant pulse is expected to be reduced and delayed as the inhomogeneous lifetime T_2^* is progressively shortened.

Finally, we should mention an important limitation of the mean field treatment we have given, namely that it tells us nothing about the fluctuations associated with a succession of superradiant pulses. For information on fluctuations we will need to investigate the higher-order moments of the light intensity. The delay t_0 of the superradiant pulse is actually very sensitive to the initial state of excitation of the atomic system. The initial radiation rate is governed largely by non-cooperative or single-atom spontaneous emission. However, as we well know from Chapter 15, spontaneous emissions are subject to very large quantum

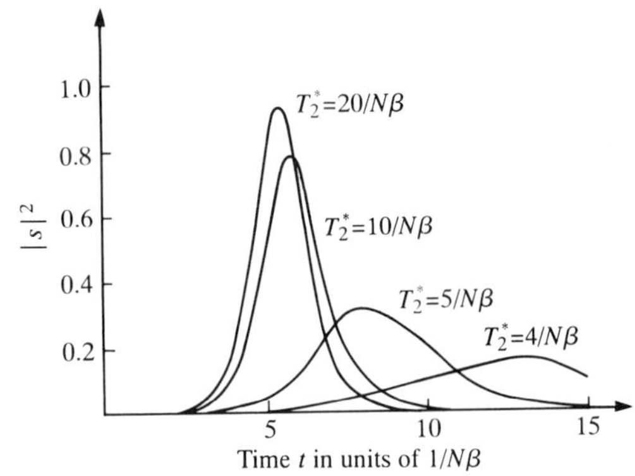

Fig. 16.21 The predicted dependence of the superradiant pulse on the inhomogeneous lifetime T_2^*. When $T_2^* \gg 1/N\beta$ the pulse shape is given by Eq. (16.6–24). (Reproduced from Jodoin and Mandel, 1974a.)

fluctuations. It is therefore to be expected that in practice the delay t_0 may vary widely from pulse to pulse. This subject has been treated in detail by several authors (Haake, King, Schröder, Haus and Glauber, 1979; Polder, Schuurmans and Vrehen, 1979; Haake, Haus, King, Schröder and Glauber, 1980), who also found that substantial fluctuations of pulse height and shape are to be expected. Experimental evidence for large fluctuations of superradiant pulses has been found (Gibbs, Vrehen and Hikspoors, 1977).

16.6.6 Superradiant classical oscillations

As we have already pointed out, superradiance was first predicted for certain collective quantum states of an atomic system for which the expectation value of the dipole moment is zero (Dicke, 1954). Later work showed, and we have emphasized this in our treatment, that superradiant effects should also appear for atomic product states in which there is a macroscopic atomic dipole moment. In that case the atoms can be regarded as oscillating dipoles, and the system can be visualized in classical terms to a certain limited extent. Nevertheless, it is still a quantum system, and our treatment may leave the reader with the impression that there are always quantum mechanical aspects to superradiance. We therefore wish to show that a completely classical system of coupled harmonic oscillators will also exhibit superradiant effects.

The displacement x of a classical damped harmonic oscillator of natural undamped frequency ω_0 and damping constant β, obeys the equation of motion

$$\ddot{x} + 2\beta\dot{x} + \omega_0^2 x = 0, \tag{16.6–28}$$

which has the solution

$$x(t) = x(0)\,\mathrm{e}^{-\beta t}\cos(\omega t + \phi), \tag{16.6–29}$$

with $\omega = (\omega_0^2 - \beta^2)^{1/2}$. Let us suppose that the oscillator emits radiation, sound waves for example, and that the velocity damping responsible for the exponential decay of the oscillation is largely radiative damping. Now suppose that N similar oscillators are placed close together, such that their separations are smaller than the radiation wavelength. Then each oscillator is damped not only by its own radiation reaction field, but by the fields of all the other oscillators, which it experiences with equal amplitude. Alternatively, we may look on the fields produced by all the other oscillators as sources driving the given oscillator i ($i = 1, 2, \ldots, N$), and express the equation of motion in the form

$$\ddot{x}_i + 2\beta\dot{x}_i + \omega_0^2 x_i = \sum_{\substack{j=1 \\ j\neq i}}^{N} -2\beta\dot{x}_j, \quad i = 1, 2, \ldots, N. \tag{16.6–30}$$

By summing each term of this equation on i, and denoting

$$\sum_{i=1}^{N} x_i \equiv X,$$

we immediately obtain the equation of motion

$$\ddot{X} + 2N\beta\dot{X} + \omega_0^2 X = 0. \tag{16.6–31}$$

This equation has a solution of the form

$$X(t) = X(0)\,\mathrm{e}^{-N\beta t}\cos(\omega t + \Phi), \tag{16.6–32}$$

with $\omega = (\omega_0^2 - N^2\beta^2)^{1/2}$, which should be compared with Eq. (16.6–29). If all the oscillators are in phase, and $x_i(0) = x(0)$, $i = 1, 2, \ldots, N$, then the amplitude of the radiated field is N times as great as the field of a single oscillator, or the intensity is N^2 times as great, and the field decays to zero N times as rapidly as the field of a single oscillator. These are both characteristic features of superradiance. The effect has been demonstrated experimentally in the decay of two radiatively coupled tuning forks mounted close together (Lama, Jodoin and Mandel, 1972). Some results are shown in Fig. 16.22. When the tuning forks are excited in phase initially, their oscillations are damped out more rapidly than for one oscillating alone. By contrast, the oscillation dies away less rapidly when the tuning forks are excited in antiphase initially. This state is somewhat analogous to a Dicke state with $l = 0$, $m = 0$. The lifetime is not exactly halved in the fomer case, or made infinite in the latter case, mainly because the damping of the tuning forks is not purely radiative, and the propagation delay between them is not

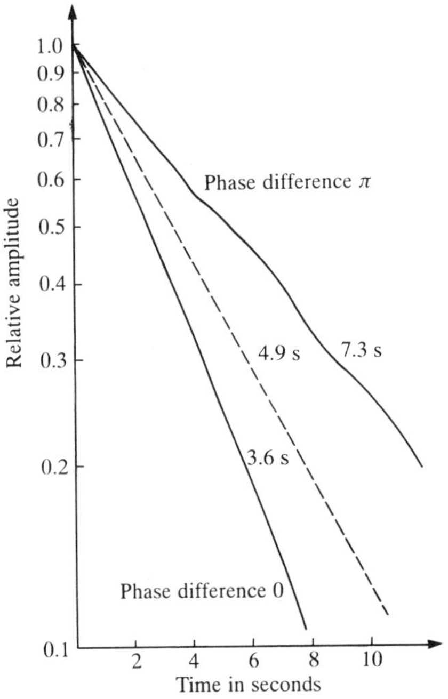

Fig. 16.22 Observed time developments of the sound amplitude from two spontaneously radiating tuning forks. The broken curve shows the behavior of each fork separately. The lower curve corresponds to the two forks oscillating in phase and the upper one in antiphase. (Reproduced from Lama, Jodoin and Mandel, 1972.)

strictly zero. Nevertheless, the experiment shows clearly that cooperative radiation is not confined to atomic or quantum systems. Superradiant damping effects have been encountered in the piano, in which some musical tones are produced by a group of two or three identical strings stretched side by side that are struck together. It is known that the sound decays away too rapidly if the tuning of the group of strings is too perfect, and a small amount of detuning is sometimes introduced deliberately to reduce the superradiant damping.

16.7 Atomic coherent states

The Dicke states, as we have seen, are states of definite atomic excitation, in which the excitation is, however, distributed among the different atoms. In a sense, the Dicke states of the N-atom system are somewhat analogous to the Fock states of the electromagnetic field. We shall now show that there exist collective atomic states that are analogous to the coherent states of the electromagnetic field (Radcliffe, 1971; Arecchi, Courtens, Gilmore and Thomas, 1972).

We saw in Chapter 11 [see. Eqs. (11.3–1) and (11.3–2)] that the coherent states of the field can be generated from the vacuum or ground state by application of the displacement operator. In an analogous manner, we now introduce the collective atomic *displacement operator*

$$\hat{D} \equiv \exp(\zeta\hat{\mathcal{R}}^{(+)} - \zeta^*\hat{\mathcal{R}}^{(-)}) \tag{16.7–1}$$

for collective atomic states, where ζ is some complex number. This operator can be factorized with the help of the Campbell–Baker–Hausdorff theorem, although the factorization is somewhat more complicated than for the corresponding boson displacement operator. It can be shown that (Arecchi, Courtens, Gilmore and Thomas, 1972)

$$\left.\begin{aligned} \exp(\zeta\hat{\mathcal{R}}^{(+)} - \zeta^*\hat{\mathcal{R}}^{(-)}) &= \exp(z\hat{\mathcal{R}}^{(+)})\exp[\hat{\mathcal{R}}_3\ln(1+|z|^2)]\exp(-z^*\hat{\mathcal{R}}^{(-)}) \\ &\equiv \hat{D}(z), \\ \text{where } z \text{ and } \zeta \text{ are related by the formula} \\ z &= \tan|\zeta|\,\mathrm{e}^{\mathrm{i}\arg\zeta}. \end{aligned}\right\} \tag{16.7–2}$$

Then within the subspace of cooperation number l, we define the *atomic coherent state* $|l, z\rangle$, labeled by some complex number z, by applying the displacement operator $\hat{D}(z)$ to the ground state $|l, -l\rangle$, or the Dicke state of lowest excitation,

$$|l, z\rangle = \hat{D}(z)|l, -l\rangle. \tag{16.7–3}$$

With the help of Eq. (16.7–2) this result reduces to

$$\begin{aligned} |l, z\rangle &= \exp(z\hat{\mathcal{R}}^{(+)})\exp[\hat{\mathcal{R}}_3\ln(1+|z|^2)]\exp(-z^*\hat{\mathcal{R}}^{(-)})|l, -l\rangle \\ &= \frac{1}{(1+|z|^2)^l}\exp(z\hat{\mathcal{R}}^{(+)})|l, -l\rangle, \end{aligned} \tag{16.7–4}$$

if we make use of the fact that $\exp(-z^*\hat{\mathcal{R}}^{(-)})$ acting on $|l, -l\rangle$ merely reproduces the state $|l, -l\rangle$, and that $|l, -l\rangle$ is an eigenstate of $\hat{\mathcal{R}}_3$ with eigenvalue $-l$. The atomic coherent state, like the coherent state of the electromagnetic field, may therefore be looked on as a displaced ground state.

If we expand the operator $\exp(z\hat{\mathcal{R}}^{(+)})$ as a power series, and make use of Eqs. (16.5–15), we obtain the following series expansion of $|l, z\rangle$ in terms of Dicke states:

$$|l, z\rangle = \frac{1}{(1+|z|^2)^l}\sum_{n=0}^{2l}\frac{z^n}{n!}\left[\frac{n!(2l)!}{(2l-n)!}\right]^{1/2}|l, -l+n\rangle,$$

or, if we put $-l + n = m$,

$$|l, z\rangle = \frac{1}{(1+|z|^2)^l}\sum_{m=-l}^{l}\binom{2l}{l+m}^{1/2} z^{l+m}|l, m\rangle. \qquad (16.7\text{–}5)$$

This expansion may be regarded as the analog of the expansion (11.2–9) of the coherent state of a boson field in terms of Fock states. Moreover, when $z = 0$, the atomic coherent state coincides with the Dicke state of lowest excitation. In the following we shall see that the analogy carries over to many other properties, and that, like the coherent state of the field, the atomic coherent state has the physical significance of being produced by a classical source.

As usual the cooperation number l is bounded by zero and $\frac{1}{2}N$,

$$0 \leqslant l \leqslant \tfrac{1}{2}N.$$

In the special case $l = \frac{1}{2}N$, the lowest energy state $|l, -l\rangle$ becomes the absolute ground state $|\frac{1}{2}N, -\frac{1}{2}N\rangle = |1\rangle_1|1\rangle_2 \ldots |1\rangle_N$ of the atomic system, and Eq. (16.7–4) reduces to

$$|\tfrac{1}{2}N, z\rangle = \prod_{i=1}^{N}\frac{1}{(1+|z|^2)^{1/2}}\,e^{z\hat{b}_i^\dagger}|1\rangle_i = \prod_{i=1}^{N}\frac{1}{(1+|z|^2)^{1/2}}(|1\rangle_i + z|2\rangle_i). \quad (16.7\text{–}6)$$

This is simply an atomic product state, in which each atom is in the same pure state, which can be characterized by the Bloch-vector (θ, ϕ) with

$$z = \tan\tfrac{1}{2}\theta\, e^{-i\phi}.$$

In this special case the atomic coherent state coincides with a product state. Other atomic coherent states are simply generalizations, applicable to the more general situation when the cooperation number is smaller than $\frac{1}{2}N$.

In one important respect the two kinds of coherent state are different however. Whereas the coherent state of the field is an eigenstate of the photon absorption or lowering operator, the atomic coherent state $|l, z\rangle$ is not an eigenstate of $\hat{\mathcal{R}}^{(-)}$. Instead we have from Eqs. (16.5–15) and (16.7–5)

$$\hat{\mathcal{R}}^{(-)}|l, z\rangle = \frac{1}{(1+|z|^2)^l}\sum_{m=-l+1}^{l}\binom{2l}{l+m}^{1/2}(l+m)^{1/2}(l-m+1)^{1/2}z^{l+m}|l, m-1\rangle$$

$$= \frac{z}{(1+|z|^2)^l}\sum_{m'=-l}^{l-1}\left[\frac{(2l)!}{(l+m')!(l-m')!}\right]^{1/2}(l-m')z^{l+m'}|l, m'\rangle.$$

Hence the expectation value of $\hat{\mathcal{R}}^{(-)}$ is

$$\langle l, z|\hat{\mathcal{R}}^{(-)}|l, z\rangle = \frac{2lz}{(1+|z|^2)^{2l}} \sum_{m=-l}^{l-1} \binom{2l-1}{l+m} |z|^{2(l+m)}$$
$$= \frac{2lz}{(1+|z|^2)^{2l}}(1+|z|^2)^{2l-1}$$
$$= \frac{2lz}{1+|z|^2}, \qquad (16.7\text{–}7a)$$

and similarly

$$\langle l, z|\hat{\mathcal{R}}^{(+)}|l, z\rangle = \frac{2lz^*}{1+|z|^2}. \qquad (16.7\text{–}7b)$$

Apart from the factor $2l/(1+|z|^2)$, the expectation values of $\hat{\mathcal{R}}^{(-)}$ and $\hat{\mathcal{R}}^{(+)}$ in the atomic coherent state are proportional to z and z^*, respectively. This allows us to write down the expectations of $\hat{\mathcal{R}}_1$ and $\hat{\mathcal{R}}_2$ immediately, namely

$$\langle l, z|\hat{\mathcal{R}}_1|l, z\rangle = \frac{2l\,\mathrm{Re}\,(z)}{1+|z|^2} \qquad (16.7\text{–}8)$$

$$\langle l, z|\hat{\mathcal{R}}_2|l, z\rangle = \frac{-2l\,\mathrm{Im}\,(z)}{1+|z|^2}. \qquad (16.7\text{–}9)$$

Also, with the help of Eq. (16.7–5) we have

$$\langle l, z|\hat{\mathcal{R}}_3|l, z\rangle = \frac{1}{(1+|z|^2)^{2l}} \sum_{m=-l}^{l} \binom{2l}{l+m} m|z|^{2(l+m)}$$

and if $m = r - l$,

$$\langle l, z|\hat{\mathcal{R}}_3|l, z\rangle = \frac{1}{(1+|z|^2)^{2l}}\left[\sum_{r=1}^{2l}\frac{(2l)!|z|^{2r}}{(r-1)!(2l-r)!} - l\sum_{r=0}^{2l}\frac{(2l)!|z|^{2r}}{r!(2l-r)!}\right]$$
$$= \frac{1}{(1+|z|^2)^{2l}}[2l|z|^2(1+|z|^2)^{2l-1} - l(1+|z|^2)^{2l}]$$
$$= \frac{l(|z|^2-1)}{1+|z|^2}. \qquad (16.7\text{–}10)$$

16.7.1 Bloch-representation of the atomic coherent state

Just as we found it convenient to represent the state of a single two-level atom by a real vector in a symbolic three-dimensional Bloch space, so it is sometimes convenient to represent the atomic coherent state of the collection of atoms by a real vector. The atomic coherent states are defined within a subspace of definite cooperation number l, and within this subspace the state is completely determined by the complex number z, which can be mapped onto the direction of a vector on a sphere by a projective transformation. For this purpose we introduce polar and azimuthal angles θ, ϕ such that

$$z = \cot(\tfrac{1}{2}\theta)\, e^{-i\phi}, \tag{16.7–11}$$

which defines a direction for each complex number z. Then

$$\operatorname{Re}(z) = \cot(\tfrac{1}{2}\theta)\cos\phi$$

$$\operatorname{Im}(z) = \cot(\tfrac{1}{2}\theta)\sin\phi,$$

and we have immediately from Eqs. (16.7–8) to (16.7–10)

$$\left.\begin{aligned} \langle l, z|\hat{\mathcal{R}}_1|l, z\rangle &= l\sin\theta\cos\phi \\ \langle l, z|\hat{\mathcal{R}}_2|l, z\rangle &= l\sin\theta\sin\phi \\ \langle l, z|\hat{\mathcal{R}}_3|l, z\rangle &= l\cos\theta. \end{aligned}\right\} \tag{16.7–12}$$

These relations are analogous to Eqs. (15.2–14) for a single atom and suggest that the atomic coherent state is a certain symmetrized generalization of the single-atom state. It follows that if we represent the atomic coherent state by a three-dimensional Bloch-vector of length l, then the expectation values of $\hat{\mathcal{R}}_1$, $\hat{\mathcal{R}}_2$, $\hat{\mathcal{R}}_3$ are the three Cartesian components of this vector. The ground state $|l, z = 0\rangle$ has a Bloch-vector pointing straight down. Because of this Bloch-representation, the atomic coherent state is occasionally also called a *Bloch state*.

Although the state $|l, z\rangle$ is not an eigenstate of any simple operator like $\hat{\mathcal{R}}^{(-)}$, it can be shown that (Arecchi, Courtens, Gilmore and Thomas, 1972)

$$(\hat{\mathcal{R}}^{(-)} e^{i\phi}\sin\tfrac{1}{2}\theta + \hat{\mathcal{R}}_3\cos\tfrac{1}{2}\theta)|l, z\rangle = l\cos\tfrac{1}{2}\theta|l, z\rangle.$$

This could be used as the defining equation for atomic coherent states, except that it has no simple, intuitive interpretation.

16.7.2 Non-orthogonality and over-completeness

Just as in Section 11.6 the coherent states of the electromagnetic field were found to be non-orthogonal and not linearly independent, we shall find that the atomic coherent states have analogous properties. We start by calculating the scalar product of two such states. With the help of Eq. (16.7–5) we find

$$\begin{aligned} \langle l, z|l, z'\rangle &= \frac{1}{(1+|z|^2)^l(1+|z'|^2)^l}\sum_{m=-l}^{l}\binom{2l}{l+m}(z^*z')^{l+m} \\ &= \frac{(1+z^*z')^{2l}}{(1+|z|^2)^l(1+|z'|^2)^l}, \end{aligned} \tag{16.7–13}$$

which shows that the atomic coherent states are properly normalized, and that they are *non-orthogonal in general*. However, for any given atomic coherent state $|l, z\rangle$ with $z \neq 0$ there exists an orthogonal atomic coherent state $|l, z'\rangle$, having $z' = -1/z^*$. In this respect the atomic coherent states differ from the coherent states of the field, for which no two states are ever strictly orthogonal.

Equation (16.7–13) may be re-expressed in another form that brings out the geometric relationship between the Bloch-vectors of orthogonal atomic coherent states. When we introduce the polar coordinates l, θ, ϕ of the corresponding Bloch-vectors we have from Eq. (16.7–13) with the help of Eq. (16.7–11)

$$\begin{aligned}
|\langle l, z|l, z'\rangle|^2 &= \frac{|1+\cot\frac{1}{2}\theta\cot\frac{1}{2}\theta'\,\mathrm{e}^{\mathrm{i}(\phi-\phi')}|^{4l}}{(\operatorname{cosec}\frac{1}{2}\theta)^{4l}\operatorname{cosec}(\frac{1}{2}\theta')^{4l}} \\
&= |\sin\tfrac{1}{2}\theta\sin\tfrac{1}{2}\theta' + \cos\tfrac{1}{2}\theta\cos\tfrac{1}{2}\theta'\,\mathrm{e}^{\mathrm{i}(\phi-\phi')}|^{4l} \\
&= [\tfrac{1}{2} + \tfrac{1}{2}\cos\theta\cos\theta' + \tfrac{1}{2}\sin\theta\cos\phi\sin\theta'\cos\phi' \\
&\quad + \tfrac{1}{2}\sin\theta\sin\phi\sin\theta'\sin\phi']^{2l} \\
&= [\tfrac{1}{2}(1+\cos\Theta)]^{2l} \\
&= [\cos\tfrac{1}{2}\Theta]^{4l},
\end{aligned}$$

where Θ is the angle between the two Bloch vectors (l, θ, ϕ), (l, θ', ϕ'). Hence

$$|\langle l, z|l, z'\rangle| = \cos^{2l}(\tfrac{1}{2}\Theta), \tag{16.7–14}$$

which shows at once that orthogonal coherent states correspond to diametrically opposite vectors on the Bloch sphere. We note also that if l is a very large number, then the scalar product of two atomic coherent states that differ only slightly ($\Theta \ll 1$) can be very small. In this sense different atomic coherent states are almost orthogonal and behave somewhat like the coherent states of the field.

In order to test the completeness of the set of states $|l, z\rangle$ within the subspace of the Hilbert space corresponding to all Dicke states with cooperation number l, we integrate the projector $|l, z\rangle\langle l, z|$ over the entire phase space. Then we have from Eqs. (16.7–5) and (16.7–11).

$$\begin{aligned}
&\int_0^\pi \mathrm{d}\theta\sin\theta\int_0^{2\pi}\mathrm{d}\phi|l, z\rangle\langle l, z| \\
&= \sum_{m=-l}^{l}\sum_{m'=-l}^{l}\binom{2l}{l+m}^{1/2}\binom{2l}{l+m'}^{1/2}|l, m\rangle\langle l, m'| \\
&\quad\times\int_0^\pi \mathrm{d}\theta\sin\theta(\sin\tfrac{1}{2}\theta)^{4l}(\cot\tfrac{1}{2}\theta)^{2l+m+m'}\int_0^{2\pi}\mathrm{d}\phi\,\mathrm{e}^{\mathrm{i}(m'-m)\phi} \\
&= 4\pi\sum_{m=-l}^{l}\binom{2l}{l+m}\int_0^\pi(\sin\tfrac{1}{2}\theta)^{2l-2m+1}(\cos\tfrac{1}{2}\theta)^{2l+2m+1}|l, m\rangle\langle l, m|\,\mathrm{d}\theta.
\end{aligned}$$

The θ-integral yields the $\mathscr{B}$-function

$$\mathscr{B}(l+m+1, l-m+1) = \frac{(l+m)!(l-m)!}{(2l+1)!}.$$

Hence finally

$$\int_0^\pi \mathrm{d}\theta\sin\theta\int_0^{2\pi}\mathrm{d}\phi|l, z\rangle\langle l, z| = \frac{4\pi}{2l+1}\sum_{m=-l}^{l}|l, m\rangle\langle l, m|,$$

or, since the sum over m within the subspace of all Dicke states with definite cooperation number l must be unity,

$$\frac{2l+1}{4\pi}\int |l, z\rangle\langle l, z| \, \mathrm{d}\Omega = 1, \tag{16.7–15}$$

where $\mathrm{d}\Omega$ stands for the element of solid angle.

It follows from this that the atomic coherent states form a complete set for the representation of any atomic state $|\Psi\rangle$ that is expressible as a linear combination of Dicke states. We merely multiply $|\Psi\rangle$ on the left by unity in the form given by Eq.(16.7–15) and obtain

$$|\Psi\rangle = \frac{2l+1}{4\pi}\int \langle l, z|\Psi\rangle |l, z\rangle \, \mathrm{d}\Omega. \tag{16.7–16}$$

In particular, if $|\Psi\rangle$ is one of the atomic coherent states $|l, z'\rangle$, because two different states $|l, z\rangle$ and $|l, z'\rangle$ are not orthogonal in general, we find that we have a representation of one atomic coherent state in terms of the complete set. It follows that, except for those states which are orthogonal to each other, the different atomic coherent states are not linearly independent, and the set is over-complete. Nevertheless, as with the coherent states of the field, it often forms a convenient basis. It may be shown that because of the over-completeness, there also exists a diagonal atomic coherent-state representation of the density operator for the atomic system, and that the representative weighting function is a quasi-probability density in general, as for the electromagnetic field (Arecchi, Courtens, Gilmore and Thomas, 1972).

16.7.3 The atomic state produced by a classical field

In Section 11.12 we showed that the coherent states of the electromagnetic field are more than just convenient mathematical constructs, in that these states are actually produced when a classical current interacts with a quantum field that is initially in the ground state. In a similar manner we shall now show that the atomic coherent states are physically realizable when a classical field interacts with an N-atom system of sufficiently small dimensions that is initially in the ground state.

The electric dipole interaction $\hat{H}_{\mathrm{I}}(t)$ between the system of N two-level atoms and the electromagnetic field can be written in the form of Eq. (16.6–1). When the field is classical and $\mathbf{E}(t)$ is a c-number, $\hat{H}_{\mathrm{I}}(t)$ becomes in the interaction picture

$$\hat{H}_{\mathrm{I}}(t) = -\boldsymbol{\mu}_{12}^{*}\cdot\mathbf{E}(t)\,\mathrm{e}^{\mathrm{i}\omega_0 t}\hat{\mathcal{R}}^{(+)} - \boldsymbol{\mu}_{12}\cdot\mathbf{E}(t)\,\mathrm{e}^{-\mathrm{i}\omega_0 t}\hat{\mathcal{R}}^{(-)}, \tag{16.7–17}$$

where we have explicitly displayed the time dependence of the operators, and unlabeled operators refer to time $t = 0$. The infinitesimal time evolution operator in the interaction picture is then

$$\begin{aligned}\hat{U}(t, t+\delta t) &= \exp[-\mathrm{i}\hat{H}_{\mathrm{I}}(t)\delta t/\hbar] \\ &= \exp[\zeta(t)\hat{\mathcal{R}}^{(+)} - \zeta^{*}(t)\hat{\mathcal{R}}^{(-)}],\end{aligned} \tag{16.7–18}$$

where we have written

$$\zeta(t) \equiv \mathrm{i}\frac{\boldsymbol{\mu}_{12}^{*}\cdot\mathbf{E}(t)}{\hbar}\mathrm{e}^{\mathrm{i}\omega_0 t}\delta t. \tag{16.7–19}$$

When the classical field acts on the N-atom system which is initially in the lowest energy Dicke state $|l, -l\rangle$, we obtain a new state $|\Psi(\delta t)\rangle$ given by the formula

$$\begin{aligned}|\Psi(\delta t)\rangle &= \hat{U}(0, \delta t)|l, -l\rangle \\ &= \exp[\zeta(0)\hat{\mathcal{R}}^{(+)} - \zeta^*(0)\hat{\mathcal{R}}^{(-)}]|l, -l\rangle.\end{aligned} \tag{16.7–20}$$

More generally, after a finite time interval t, which can be divided into $t/\delta t$ infinitesimal intervals δt labeled $t_1, t_2, \ldots,$ we obtain the state

$$|\Psi(t)\rangle = \prod_{r=1}^{t/\delta t}[\exp(\zeta(t_r)\hat{\mathcal{R}}^{(+)} - \zeta^*(t_r)\hat{\mathcal{R}}^{(-)})]|l, -l\rangle. \tag{16.7–21}$$

In order to be able to evaluate the operator product and identify the state $|\Psi(t)\rangle$ we need to make use of the following operator relation, that can be proved with the help of the Baker–Haussdorff theorem (Arecchi, Courtens, Gilmore and Thomas, 1972),

$$\exp(\zeta\hat{\mathcal{R}}^{(+)} - \zeta^*\hat{\mathcal{R}}^{(-)})\exp(\zeta'\hat{\mathcal{R}}^{(+)} - \zeta'^*\hat{\mathcal{R}}^{(-)}) = \exp(\zeta''\hat{\mathcal{R}}^{(+)} - \zeta''^*\hat{\mathcal{R}}^{(-)})\exp(-\mathrm{i}\phi\hat{\mathcal{R}}_3), \tag{16.7–22}$$

where ζ, ζ', ζ'' and ϕ are related by the matrix equation

$$\frac{1}{(1+|z|^2)^{1/2}(1+|z'|^2)^{1/2}}\begin{bmatrix} 1 - z^*z' & z + z' \\ -z^* - z'^* & 1 - zz'^* \end{bmatrix} = \frac{1}{(1+|z''|^2)^{1/2}}\begin{bmatrix} \mathrm{e}^{-\mathrm{i}\phi/2} & z''\mathrm{e}^{\mathrm{i}\phi/2} \\ -z''^*\mathrm{e}^{-\mathrm{i}\phi/2} & \mathrm{e}^{\mathrm{i}\phi/2} \end{bmatrix} \tag{16.7–23}$$

and

$$\left.\begin{aligned} z' &= \tan|\zeta'|\exp(\mathrm{i}\arg\zeta') \\ z'' &= \tan|\zeta''|\exp(\mathrm{i}\arg\zeta''). \end{aligned}\right\} \tag{16.7–24}$$

Equation (16.7–22) may be regarded as a statement of the reproducing property for atomic displacement operators, in analogy with Eq. (11.3–14) for the boson field, although the reproducing property is less direct in that it involves $\hat{\mathcal{R}}_3$ also. However, as the state $|l, -l\rangle$ is an eigenstate of $\hat{\mathcal{R}}_3$, it follows immediately from Eq. (16.7–22) that the state $|\Psi(t)\rangle$ given by Eq. (16.7–21) can be expressed as a single displacement operator acting on $|l, -l\rangle$, apart from a phase factor. According to Eq. (16.7–3), such a state is an atomic coherent state. We have therefore shown that the action of a classical field on a collection of N sufficiently close two-level atoms that are initially in the ground state is an atomic coherent state.

Evidently there is a fundamental correspondence between coherent states of the field and atomic coherent states, that carries over to many other properties. These have been discussed by Arecchi, Courtens, Gilmore and Thomas (1972).

Problems

16.1 Consider a system of N identical two-level atoms close to each other. Compare the rate of photon emission in (a) an atomic product state with each atom in the same state, and (b) a Dicke state with the same average excitation as in (a).

16.2 For the system of identical atoms in the preceeding problem, compare the expectation value of the total atomic energy in the atomic coherent state $|\theta, \phi\rangle$ with that in the product state in which the state of each atom is given by the same Bloch vector $|\theta, \phi\rangle$.

16.3 For the system of identical atoms in Problem 16.1, compare the expectation values of the total atomic dipole moment in the atomic coherent state $|\theta, \phi\rangle$ and in the product state in which the state of each atom is given by the same Bloch vector $|\theta, \phi\rangle$.

16.4 A two-level atom with energy spacing $\hbar\omega$ is in its lower state. It is subjected to a brief interaction of duration δt with a classical electromagnetic field at time $t = 0$, and to an identical interaction at a later time τ. Show that, subsequently, the expectation value of the dipole moment at time t contains a contribution varying with time as $\cos[\omega(t - 2\tau)]$.

Discuss the implication of this result for a large number of atoms having slightly different resonant frequencies. What physically significant effect would be expected to occur under these circumstances?

16.5 Consider N similar, partly excited two-level atoms distributed in some manner in space. The intensity of the electric field at a distant point is measured and it is found to be N^2 times as large as that for one atom. Discuss the question whether the atoms are radiating cooperatively.

17

Some general techniques for treating interacting systems

In the preceding chapters we have solved a number of specific problems in which electromagnetic fields interact with charges, atoms, or molecules, and these have been approached in several different ways. For example, for the problem of photoelectric detection, which involves short interaction times, we found it convenient to use a perturbative method, whereas the resonance fluorescence problem was treated by solving the Heisenberg equations of motion. In the following sections we shall encounter a number of general methods for tackling interaction problems that can often simplify the problems substantially when they are applicable. We shall illustrate the utility of these methods by recalculating a number of results that were obtained in a different manner before.

17.1 The quantum regression theorem

It was shown by Lax (1963; see also Louisell, 1973, Sec. 6.6) that, with the help of a certain factorization assumption, it is often possible to express multi-time correlation functions of certain quantum mechanical operators in terms of single-time expectations. The result is now known as the *regression theorem*. As multi-time correlation functions play a rather important role in quantum optics, the theorem is often of great utility, and it can drastically simplify certain calculations. In the following we largely adopt the procedure given by Lax.

We consider two coupled quantum systems, to which, for the sake of convenience, we shall refer as the *system* (S) and the *reservoir* (R). We suppose that S has a complete orthonormal set of eigenstates $|\phi_i\rangle$, $i = 1, 2, \ldots$ forming a basis, which may be discrete or continuous, but will here be treated as discrete for simplicity. The density operator $\hat{\rho}(t)$ of the coupled system in the Schrödinger picture spans the Hilbert spaces of S and R, and we can derive the reduced density operator $\hat{\rho}_S$ for S or $\hat{\rho}_R$ for R by tracing over the opposite variables, so that

$$\left.\begin{aligned} \hat{\rho}_S(t) &= \mathrm{Tr}_R\, \hat{\rho}(t) \\ \hat{\rho}_R(t) &= \mathrm{Tr}_S\, \hat{\rho}(t). \end{aligned}\right\} \qquad (17.1\text{–}1)$$

In particular, $\hat{\rho}_S(t)$ can be given a representation in terms of the basis $|\phi_i\rangle$ in the form

$$\hat{\rho}_S(t) = \sum_{ij} \rho_{ij}(t)|\phi_i\rangle\langle\phi_j|, \tag{17.1–2}$$

where

$$\begin{aligned}\rho_{ij}(t) &= \mathrm{Tr}\,[\hat{\rho}(t)|\phi_j\rangle\langle\phi_i|] \\ &= \langle\phi_i|\hat{\rho}_S(t)|\phi_j\rangle. \end{aligned}\tag{17.1–3}$$

The time evolution of the density operator $\hat{\rho}(t)$ for the coupled system is governed by the unitary time evolution operator $\hat{U}(t, t_0)(t \geqslant t_0)$, such that

$$\hat{\rho}(t) = \hat{U}(t, t_0)\hat{\rho}(t_0)\hat{U}^\dagger(t, t_0), \tag{17.1–4}$$

and we shall make the key assumption that S and R are uncoupled at the 'initial' time t_0, so that

$$\hat{\rho}(t_0) = \hat{\rho}_R(t_0) \times \hat{\rho}_S(t_0). \tag{17.1–5}$$

We can now determine the time evolution of $\rho_{ij}(t)$ by using Eqs. (17.1–3), (17.1–4) and (17.1–5) and writing

$$\begin{aligned}\rho_{ji}(t) &= \mathrm{Tr}\,[\hat{U}(t, t_0)\hat{\rho}_R(t_0)\hat{\rho}_S(t_0)\hat{U}^\dagger(t, t_0)|\phi_i\rangle\langle\phi_j|] \\ &= \mathrm{Tr}\,[\hat{U}^\dagger(t, t_0)|\phi_i\rangle\langle\phi_j|\hat{U}(t, t_0)\hat{\rho}_R(t_0)\hat{\rho}_S(t_0)]. \end{aligned}\tag{17.1–6}$$

The last line is obtained from the previous one with the help of a cyclic permutation of operators under the trace. We now make a representation of the operator $\hat{U}^\dagger(t, t_0)|\phi_i\rangle\langle\phi_j|\hat{U}(t, t_0)$ in the basis $|\phi_p\rangle$ by writing

$$\hat{U}^\dagger(t, t_0)|\phi_i\rangle\langle\phi_j|\hat{U}(t, t_0) = \sum_{p,q} \hat{C}^{(R)}_{ijpq}(t, t_0)|\phi_p\rangle\langle\phi_q|, \tag{17.1–7}$$

in which the coefficients $\hat{C}^{(R)}_{ijpq}(t, t_0)$ are c-numbers with respect to the Hilbert space S, but operators in the Hilbert space R. This is indicated by the caret ^ and the superscript (R). If we substitute Eq. (17.1–7) in Eq. (17.1–6), and also use the representation (17.1–2) for $\hat{\rho}_S(t_0)$, we obtain the relation

$$\begin{aligned}\rho_{ji}(t) &= \mathrm{Tr} \sum_{p,q}\sum_{l,m} \hat{C}^{(R)}_{ijpq}(t, t_0)\hat{\rho}_R(t_0)|\phi_p\rangle\langle\phi_q|\phi_l\rangle\langle\phi_m|\rho_{lm}(t_0) \\ &= \mathrm{Tr}_R \sum_{p,q} \hat{C}^{(R)}_{ijpq}(t, t_0)\hat{\rho}_R(t_0)\rho_{qp}(t_0) \\ &= \sum_{p,q} G_{ijpq}(t, t_0)\rho_{qp}(t_0), \end{aligned}\tag{17.1–8}$$

where we have written

$$G_{ijpq}(t, t_0) \equiv \mathrm{Tr}_R \hat{C}^{(R)}_{ijpq}(t, t_0)\hat{\rho}_R(t_0). \tag{17.1–9}$$

Equation (17.1–8) shows that we may look on $G_{ijpq}(t, t_0)$ as the Green function for the time evolution of the system. In particular, if $t = t_0$,

$$G_{ijpq}(t_0, t_0) = \delta_{ip}\delta_{jq}. \tag{17.1–10}$$

17.1.1 *Single-time expectation values*

It is easy to show that the Green function $G(t, t_0)$ can immediately be used to calculate the time evolution of the expectation value of any dynamical variable $\hat{M}$ belonging to S. Thus we have

$$\begin{aligned}\langle \hat{M}(t)\rangle &= \mathrm{Tr}[\hat{M}\hat{\rho}(t)] \\ &= \mathrm{Tr}_S[\hat{M}\hat{\rho}_S(t)],\end{aligned}$$

and if we insert the unit projectors $\sum_i|\phi_i\rangle\langle\phi_i|$ and $\sum_j|\phi_j\rangle\langle\phi_j|$ in front of $\hat{M}$ and behind $\hat{M}$, respectively, and take the trace, we obtain the result

$$\langle \hat{M}(t)\rangle = \sum_{ij} M_{ij}\rho_{ji}(t), \tag{17.1–11}$$

where we have written $\langle\phi_i|\hat{M}|\phi_j\rangle \equiv M_{ij}$ for the matrix element of $\hat{M}$. Substitution of the right-hand side of Eq. (17.1–8) for $\rho_{ji}(t)$ now gives immediately

$$\langle \hat{M}(t)\rangle = \sum_{ij}\sum_{pq} M_{ij}G_{ijpq}(t, t_0)\rho_{qp}(t_0). \tag{17.1–12}$$

The Green function $G_{ijpq}(t, t_0)$ therefore allows the expectation value of $\hat{M}$ at any time t to be determined.

17.1.2 *Multi-time expectation values*

Although, at first sight, it might appear that the same procedure cannot be used to calculate two-time expectation values, because $\hat{\rho}_S(t)$ is different at different times, the difficulty disappears if one of the two times is t_0, the time at which $\hat{\rho}(t_0)$ factorizes according to Eq. (17.1–5).

Let $\hat{L}$, $\hat{M}$, $\hat{N}$ be three dynamical variables belonging to the system S, and suppose that we wish to evaluate the expectation $\langle \hat{L}(t_0)\hat{M}(t)\hat{N}(t_0)\rangle$, with $t \geqslant t_0$. Then since

$$\hat{M}(t) = \hat{U}^\dagger(t, t_0)\hat{M}(t_0)\hat{U}(t, t_0) \tag{17.1–13}$$

in the Heisenberg picture, it follows that

$$\langle \hat{L}(t_0)\hat{M}(t)\hat{N}(t_0)\rangle = \mathrm{Tr}[\hat{L}(t_0)\hat{U}^\dagger(t, t_0)\hat{M}(t_0)\hat{U}(t, t_0)\hat{N}(t_0)\hat{\rho}_R(t_0)\hat{\rho}_S(t_0)]. \tag{17.1–14}$$

We now insert unit projectors in the form $\sum_i|\phi_i\rangle\langle\phi_i|$ after $\hat{L}$, $\hat{U}^\dagger$, $\hat{M}$, $\hat{U}$, $\hat{N}$, $\hat{\rho}_S$ in Eq. (17.1–14), and obtain

$$\begin{aligned}\langle \hat{L}(t_0)\hat{M}(t)\hat{N}(t_0)\rangle = \mathrm{Tr}\sum_{ij}\sum_{kl}\sum_{mn} &\hat{L}(t_0)|\phi_k\rangle\langle\phi_k|\hat{U}^\dagger(t, t_0)|\phi_i\rangle \\ &\times \langle\phi_j|\hat{U}(t, t_0)|\phi_l\rangle\langle\phi_i|\hat{M}(t_0)|\phi_j\rangle\langle\phi_l|\hat{N}(t_0)|\phi_m\rangle \\ &\times \hat{\rho}_R(t_0)\langle\phi_m|\hat{\rho}_S(t_0)|\phi_n\rangle\langle\phi_n|.\end{aligned}$$

It is convenient to introduce the notation

$$\begin{aligned}\langle\phi_n|\hat{L}(t_0)|\phi_k\rangle &\equiv L_{nk} \\ \langle\phi_i|\hat{M}(t_0)|\phi_j\rangle &\equiv M_{ij}\end{aligned}$$

$$\langle \phi_l|\hat{N}(t_0)|\phi_m\rangle \equiv N_{lm}$$

for the matrix elements of the dynamical variables in the Schrödinger picture. After taking the trace over S we then have

$$\langle \hat{L}(t_0)\hat{M}(t)\hat{N}(t_0)\rangle = \mathrm{Tr}_R \sum_{ij}\sum_{kl}\sum_{mn} L_{nk}\langle \phi_k|\hat{U}^\dagger(t, t_0)|\phi_i\rangle\langle \phi_j|\hat{U}(t, t_0)|\phi_l\rangle$$

$$\times\ M_{ij}N_{lm}\hat{\rho}_R(t_0)\rho_{mn}(t_0). \qquad (17.1\text{–}15)$$

Finally, we use the expansion (17.1–7) for $\hat{U}^\dagger(t, t_0)|\phi_i\rangle\langle\phi_j|\hat{U}(t, t_0)$ and take the trace over R as in Eq. (17.1–9). This leads to the equation

$$\langle \hat{L}(t_0)\hat{M}(t)\hat{N}(t_0)\rangle = \sum_{ij}\sum_{kl}\sum_{mn}\sum_{pq} L_{nk}\langle \phi_k|\phi_p\rangle G_{ijpq}(t, t_0)\langle\phi_q|\phi_l\rangle M_{ij}N_{lm}\rho_{mn}(t_0)$$

$$= \sum_{ij}\sum_{kl}\sum_{mn} G_{ijkl}(t, t_0)L_{nk}M_{ij}N_{lm}\rho_{mn}(t_0), \qquad (17.1\text{–}16)$$

and shows that the same Green function as before can be used to evaluate the expectation of a multi-time operator product, once the matrix elements have been determined. The result, which is known as the *quantum regression theorem*, implies that the fluctuations regress in time like the macroscopic averages. Equation (17.1–16) holds exactly, but the factorization of the density operator $\hat{\rho}(t_0)$ at time t_0 plays an essential role in the derivation.

If the interaction is effectively turned on at time t_0, or if the coupled system is measured at that time, so that the states $\hat{\rho}_S(t_0)$ and $\hat{\rho}_R(t_0)$ are known separately, the regression theorem is applicable, and Eq. (17.1–16) can be used to evaluate the correlation function. However, in practice, the theorem is often applied in an approximate sense when no exact factorization of $\hat{\rho}(t_0)$ is to be expected, but when the coupled system is Markovian. The implication is that, even though R and S are coupled, the effect on the time evolution of S is sufficiently small that it makes little difference whether the initial state is uncoupled or not. The assumption may be appropriate when the reservoir R has many degrees of freedom, and the states associated with most of the degrees of freedom remain largely unaffected by the system S. We now illustrate the application of the regression theorem by a few examples.

17.1.3 Spontaneous atomic emission

We consider a quantum system S in the form of a two-level atom coupled to an electromagnetic field that is initially in the vacuum state and is to be regarded as playing the role of the reservoir R. The two energy eigenstates $|2\rangle$ and $|1\rangle$ of the atom form a basis, and the important dynamical variables are the atomic lowering and raising operators $\hat{b}$ and $\hat{b}^\dagger$, and the excitation operator $\hat{R}_3 + \frac{1}{2}$. These have matrix elements (cf. Section 15.1)

$$\left.\begin{aligned} \langle i|\hat{b}|j\rangle &= \delta_{i1}\delta_{j2} \\ \langle i|\hat{b}^\dagger|j\rangle &= \delta_{i2}\delta_{j1} \\ \langle i|\hat{R}_3 + \tfrac{1}{2}|j\rangle &= \delta_{i2}\delta_{j2}, \end{aligned}\right\} \qquad (17.1\text{–}17)$$

and initial time expectation values

$$\left.\begin{aligned}\langle\hat{b}(t_0)\rangle &= \sum_{ij} b_{ij}\rho_{ji}(t_0) = \rho_{21}(t_0)\\ \langle\hat{b}^\dagger(t_0)\rangle &= \rho_{12}(t_0)\\ \langle\hat{R}_3(t_0)+\tfrac{1}{2}\rangle &= \rho_{22}(t_0).\end{aligned}\right\} \tag{17.1–18}$$

At a later time t we have from Eq. (17.1–12), on identifying $\hat{M}$ with $\hat{b}$,

$$\begin{aligned}\langle\hat{b}(t)\rangle &= \sum_{ij}\sum_{pq} G_{ijpq}(t, t_0)\delta_{i1}\delta_{j2}\rho_{qp}(t_0)\\ &= \sum_{pq} G_{12pq}(t, t_0)\rho_{qp}(t_0).\end{aligned} \tag{17.1–19}$$

However, from our discussion in Section 15.5 we already know that [cf. Eq. (15.5–29)]

$$\begin{aligned}\langle\hat{b}(t)\rangle &= \mathrm{e}^{-\beta(t-t_0)}\,\mathrm{e}^{-\mathrm{i}(\omega_0-\gamma)(t-t_0)}\langle\hat{b}(t_0)\rangle\\ &= \mathrm{e}^{-\beta(t-t_0)}\,\mathrm{e}^{-\mathrm{i}(\omega_0-\gamma)(t-t_0)}\rho_{21}(t_0),\end{aligned} \tag{17.1–20}$$

and comparison of Eqs. (17.1–19) and (17.1–20) allows us to identify four elements of the Green function,

$$G_{12pq}(t, t_0) = \delta_{p1}\delta_{q2}\,\mathrm{e}^{-\beta(t-t_0)}\,\mathrm{e}^{-\mathrm{i}(\omega_0-\gamma)(t-t_0)}. \tag{17.1–21}$$

Let us now use these results in order to calculate the two-time atomic correlation function $\langle\hat{b}^\dagger(t)\hat{b}(t+\tau)\rangle$ from the regression theorem. If the factorization assumption is appropriate, we have from Eq. (17.1–16) on putting $\hat{L}=\hat{b}^\dagger$, $\hat{M}=\hat{b}$, $\hat{N}=1$,

$$\begin{aligned}\langle\hat{b}^\dagger(t)\hat{b}(t+\tau)\rangle &= \sum_{ij}\sum_{kl}\sum_{mn} G_{ijkl}(t+\tau, t)\delta_{n2}\delta_{k1}\delta_{i1}\delta_{j2}\delta_{lm}\rho_{mn}(t)\\ &= \sum_m G_{121m}\rho_{m2}(t),\end{aligned}$$

and with the help of Eq. (17.1–21)

$$\begin{aligned}&= \mathrm{e}^{-\beta\tau}\,\mathrm{e}^{-\mathrm{i}(\omega_0-\gamma)\tau}\rho_{22}(t)\\ &= \mathrm{e}^{-\beta\tau}\,\mathrm{e}^{-\mathrm{i}(\omega_0-\gamma)\tau}\langle\hat{b}^\dagger(t)\hat{b}(t)\rangle.\end{aligned} \tag{17.1–22}$$

This result is in agreement with the result obtained in Section 15.5. It should be noted that although, strictly speaking, the atom and the electromagnetic field are not decoupled at time t, the state of the field is nevertheless only weakly affected by the atom, in the sense that almost all the modes remain unoccupied at all times. So long as our interest is focused on the atom and not on the emitted photon, we may regard the state of the field as remaining almost unchanged from the vacuum state.

17.1.4 Resonance fluorescence of a two-level atom

As an example of a slightly more complicated situation, we now consider the problem of an atom interacting with an electromagnetic field that is initially in a coherent state $|\{v\}\rangle$. We wish to evaluate the two-time atomic correlation function

$$\langle \hat{b}^\dagger(t)\hat{b}^\dagger(t+\tau)\hat{b}(t+\tau)\hat{b}(t)\rangle = \langle \hat{b}^\dagger(t)[\hat{R}_3(t+\tau)+\tfrac{1}{2}]\hat{b}(t)\rangle$$

that arises in photoelectric correlation measurements [cf. Eq. (15.6–27)]. From the regression theorem, when we identify $\hat{L}$ with $\hat{b}^\dagger$, $\hat{M}$ with $\hat{R}_3+\frac{1}{2}$ and $\hat{N}$ with $\hat{b}$ in Eq. (17.1–16), we obtain the relation

$$\begin{aligned}\langle \hat{b}^\dagger(t)[\hat{R}_3(t+\tau)+\tfrac{1}{2}]\hat{b}(t)\rangle &= \sum_{ij}\sum_{kl}\sum_{mn} G_{ijkl}(t+\tau, t)\delta_{n2}\delta_{k1}\delta_{i2}\delta_{j2}\delta_{l1}\delta_{m2}\rho_{mn}(t)\\ &= G_{2211}(t+\tau, t)\rho_{22}(t). \qquad (17.1\text{–}23)\end{aligned}$$

Let us suppose that the problem of calculating the time evolution of $\langle \hat{R}_3(t)+\frac{1}{2}\rangle$ from some arbitrary initial state at time t_0 is already solved, and that the answer is expressible in the form

$$\langle \hat{R}_3(t)+\tfrac{1}{2}\rangle = \sum_{pq} D_{pq}(t-t_0)\rho_{qp}(t_0). \qquad (17.1\text{–}24)$$

Explicit expressions for $D_{pq}(\tau)$ have been given (Kimble and Mandel, 1976). In particular, $D_{11}(\tau)$ is given by the right-hand side of Eq. (15.6–11a) when τ replaces t, which happens to be the expectation value of $\langle \hat{R}_3(\tau)+\frac{1}{2}\rangle$ for an atom that starts in the ground state $|1\rangle$ at $\tau=0$. We may therefore write

$$D_{11}(\tau) = \langle \hat{R}_3(\tau)+\tfrac{1}{2}\rangle_{\mathrm{G}}, \qquad (17.1\text{–}25)$$

where the G serves to remind us that the initial state is the ground state.

But from Eq. (17.1–12) we also have the expression

$$\begin{aligned}\langle \hat{R}_3(t)+\tfrac{1}{2}\rangle &= \sum_{ij}\sum_{pq} G_{ijpq}(t, t_0)\delta_{i2}\delta_{j2}\rho_{qp}(t_0)\\ &= \sum_{pq} G_{22pq}(t, t_0)\rho_{qp}(t_0), \qquad (17.1\text{–}26)\end{aligned}$$

and comparison with Eq. (17.1–24) allows us to identify the following elements of the Green function:

$$G_{22pq}(t, t_0) = D_{pq}(t-t_0). \qquad (17.1\text{–}27)$$

If we substitute this in Eq. (17.1–23), and recognize $\rho_{22}(t)$ as $\langle \hat{R}_3(t)+\frac{1}{2}\rangle$, we obtain at once the result

$$\langle \hat{b}^\dagger(t)[\hat{R}_3(t+\tau)+\tfrac{1}{2}]\hat{b}(t)\rangle = \langle \hat{R}_3(t)+\tfrac{1}{2}\rangle D_{11}(\tau), \qquad (17.1\text{–}28)$$

and with the help of Eq. (17.1–25),

$$\langle \hat{b}^\dagger(t)[\hat{R}_3(t+\tau)+\tfrac{1}{2}]\hat{b}(t)\rangle = \langle \hat{R}_3(t)+\tfrac{1}{2}\rangle\langle \hat{R}_3(\tau)+\tfrac{1}{2}\rangle_{\mathrm{G}}. \qquad (17.1\text{–}29)$$

This is the interesting factorization theorem, expressed in atomic variables, that was derived for the light intensity correlation function in Section 15.6. Its physical implications, which are related to the atomic quantum jumps associated with each photon emission, have already been discussed there. The validity of the factorization assumption in this problem again rests on the fact that, from the point of view of the atomic behavior, the state of the field does not change very much. Those modes of the field that were heavily populated initially remain almost unchanged, whereas almost all other modes remain empty.

17.1.5 Quantum regression theorem for normally ordered field operators

The quantum regression theorem can be expressed in a particularly simple form for certain normally ordered operators of the quantum field when it is interacting with a 'reservoir' (Louisell, 1969). In Section 11.9 we saw that the expectation value of a normally ordered field operator can be calculated just like a classical c-number average with the help of the diagonal coherent-state representation of the density operator $\hat{\rho}_F(t)$. For a single-mode of the electromagnetic field this takes the form

$$\hat{\rho}_F(t) = \int \phi(v, t)|v\rangle\langle v|\, d^2v, \qquad (17.1\text{–}30)$$

in which $|v\rangle$ is the coherent state, and $\phi(v, t)$ is the generalized phase space density. Then we can calculate $\langle \hat{a}^\dagger(t)\hat{a}(t)\rangle$, for example, from (Sudarshan, 1963; Klauder, 1966; Klauder and Sudarshan, 1968) the formula

$$\langle \hat{a}^\dagger(t)\hat{a}(t)\rangle = \int v^* v\phi(v, t)\, d^2v \qquad (17.1\text{–}31)$$

where we used the optical equivalence theorem (see Section 11.9). At a later time $t + \tau$, $\phi(v, t + \tau)$ is given by the relation

$$\phi(v, t + \tau) = \int G(v, t + \tau|v', t)\phi(v', t)\, d^2v', \qquad (17.1\text{–}32)$$

where $G(v, t + \tau|v', t)$ is the Green function for the time evolution of $\phi(v, t)$. We can therefore write

$$\langle \hat{a}^\dagger(t + \tau)\hat{a}(t + \tau)\rangle = \int v^* vG(v, t + \tau|v', t)\phi(v', t)\, d^2v'\, d^2v. \qquad (17.1\text{–}33)$$

This equation has the structure of Eq. (17.1–12) except that sums have been replaced by integrals over the complex plane.

The quantum regression theorem allows us to write down the expectation value of a two-time average like $\langle \hat{a}^\dagger(t + \tau)\hat{a}(t)\rangle$ directly for an interacting electromagnetic field by combining the Green function with the optical equivalence theorem. It yields the result

$$\langle \hat{a}^\dagger(t + \tau)\hat{a}(t)\rangle = \int v^* v' G(v, t + \tau|v', t)\phi(v', t)\, d^2v\, d^2v'. \qquad (17.1\text{–}34)$$

Similarly, if we are interested in two-time correlations of the light intensity, we have by application of the same theorem

$$\langle \hat{a}^\dagger(t)\hat{a}^\dagger(t + \tau)\hat{a}(t + \tau)\hat{a}(t)\rangle = \int |v|^2|v'|^2 G(v, t + \tau|v', t)\phi(v', t)\, d^2v\, d^2v'. \qquad (17.1\text{–}35)$$

Once again we note that the same Green function that governs the time evolution of expectation values, as in Eq. (17.1–33), allows us to calculate multi-time correlations also. The general theory of the evolution of multi-time correlations and multi-time distribution functions has been studied by Srinivas and Wolf (1977).

17.2 The fluctuation–dissipation theorem

It has been known since the earliest work of Einstein on Brownian motion and molecular diffusion (Einstein, 1905), and the work of Nyquist on thermal noise in resistors (Nyquist, 1928), that there exists a connection in some physical systems between thermal fluctuations and the energy dissipation produced by an external disturbance. This connection was generalized and given a quantum mechanical foundation in a paper by Callen and Welton (1951), and it has become known as the *fluctuation–dissipation theorem*. As we shall see, the theorem allows the spectral density of the thermal fluctuations in some simple physical systems to be determined with a minimum of calculation, and it has led to further developments (Kubo, 1966).

In order to understand the physical basis of the theorem, we consider the example of a galvanometer mirror suspended in air. The mirror is subjected to a large number of random molecular impacts that causes its angular displacement to fluctuate and to perform a Brownian motion in time. The random motion can be observed and recorded with a sufficiently sensitive instrument. At the same time, the molecular impacts give rise to a frictional or dissipative force that resists the motion of the mirror. In macroscopic terms this resistance is interpreted as viscosity, and it causes any periodic oscillations of the galvanometer mirror to damp out in time. It is clear that the same mechanism – the molecular impacts – is responsible both for the fluctuations of the mirror and for its damping, and a relationship between fluctuations and dissipation is therefore to be expected.

17.2.1 A simple classical linear dissipative system

In order to make this more quantitative, let us consider a physical system subject to random perturbations whose motion can be described by a first-order differential equation of the Langevin type. An example would be the galvanometer mirror to which we have just referred, or a particle undergoing Brownian motion in a fluid, for which the equation of motion is of the form

$$m\frac{\mathrm{d}\mathbf{v}}{\mathrm{d}t} = -\frac{1}{B}\mathbf{v} + \mathbf{f}(t). \tag{17.2-1}$$

Here $\mathbf{v}$ is the velocity of the particle and m is its mass. The constant B is sometimes known as the mobility. The terms $-\mathbf{v}/B$ and $\mathbf{f}(t)$ together give the total fluctuating force exerted on the particle by the surrounding fluid. We have divided this force into its mean $-\mathbf{v}/B$, which is usually interpreted as the macroscopic viscous drag on the particle, and a fluctuating force $\mathbf{f}(t)$ representing the departure from the mean, which is attributable to molecular impacts. By definition

$$\langle \mathbf{f}(t) \rangle = 0.$$

Another example of a physically different, but mathematically equivalent, system is illustrated in Fig. 17.1. Here an electric current $I(t)$ flows through a resistance R and an inductance L as a result of a fluctuating e.m.f. $v(t)$ induced by the thermal motion of the charges in the resistor. This is sometimes known as *Johnson noise*. The differential equation governing the current flow is

$$L\frac{\mathrm{d}I}{\mathrm{d}t} = -RI + v(t), \tag{17.2–2}$$

with $\langle v(t)\rangle = 0$. If we restrict the motion of the particle considered above to one dimension, the two problems are clearly equivalent mathematically. Equation (17.2–2) can be integrated directly, and we obtain

$$I(t) = I(0)\,\mathrm{e}^{-Rt/L} + \frac{1}{L}\mathrm{e}^{-Rt/L}\int_0^t \mathrm{e}^{Rt'/L} v(t')\,\mathrm{d}t'. \tag{17.2–3}$$

If we calculate the mean of each term, and make use of the fact that $\langle v(t')\rangle = 0$, we immediately find that

$$\langle I(t)\rangle = \langle I(0)\rangle\,\mathrm{e}^{-Rt/L}. \tag{17.2–4}$$

The average current therefore tends to zero, no matter what its initial value may be. However, the mean squared current does not tend to zero, as we now show. In squaring $I(t)$ and calculating the average, we shall make use of the fact that the current $I(0)$ at time $t = 0$ is not correlated with the fluctuating e.m.f. $v(t)$ at a later time t, so that

$$\langle I(0)v(t)\rangle = \langle I(0)\rangle\langle v(t)\rangle = 0 \quad \text{if } t > 0.$$

We then have

$$\langle I^2(t)\rangle = \langle I^2(0)\rangle\,\mathrm{e}^{-2Rt/L} + \frac{1}{L^2}\mathrm{e}^{-2Rt/L}\int_0^t\!\!\int_0^t \mathrm{e}^{R(t'+t'')/L}\langle v(t')v(t'')\rangle\,\mathrm{d}t'\,\mathrm{d}t''.$$

Now $\langle v(t')v(t'')\rangle$ is the correlation function of the Johnson noise, which may be assumed to be stationary at a given temperature. Therefore, as usual for a stationary process, $\langle v(t')v(t'')\rangle$ is a function of $t'' - t'$, and we may write

$$\langle v(t')v(t'')\rangle = \Gamma(t'' - t'). \tag{17.2–5}$$

As the integrand contains only functions of $t'' - t'$ and of $t'' + t'$, the integral can be simplified as in Sections 2.2 and 14.9 by the substitution

$$\tfrac{1}{2}(t' + t'') = t_1$$

$$t'' - t' = t_2.$$

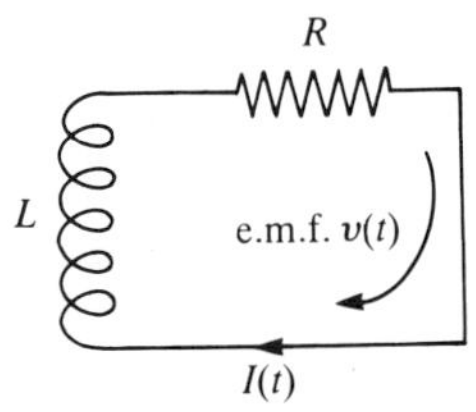

Fig. 17.1 A simple linear dissipative system subject to fluctuations.

The Jacobian of this transformation is unity, and the limits on the new variables t_1, t_2 are

$$\text{for} \quad 0 \leqslant t_1 \leqslant \tfrac{1}{2}t, \quad -2t_1 \leqslant t_2 \leqslant 2t_1$$
$$\text{for} \quad \tfrac{1}{2}t \leqslant t_1 \leqslant t, \quad -2(t-t_1) \leqslant t_2 \leqslant 2(t-t_1),$$

as can be seen by reference to Figs. 2.3 and 14.9. We then obtain the following formula:

$$\langle I^2(t)\rangle = \langle I^2(0)\rangle\, \mathrm{e}^{-2Rt/L}$$
$$+ \frac{1}{L^2}\mathrm{e}^{-2Rt/L}\left[\int_0^{t/2} \mathrm{d}t_1 \int_{-2t_1}^{2t_1} \mathrm{d}t_2\, \mathrm{e}^{2Rt_1/L}\Gamma(t_2) + \int_{t/2}^{t} \mathrm{d}t_1 \int_{-2(t-t_1)}^{2(t-t_1)} \mathrm{d}t_2\, \mathrm{e}^{2Rt_1/L}\Gamma(t_2)\right].$$

Now the correlation function $\Gamma(t_2)$ always has a limited range governed by the bandwidth of the random noise $v(t)$. If we are particularly interested in the steady-state value of $\langle I^2(t)\rangle$, we shall naturally allow t to be large. It then follows that for almost all values of t_1, both the integrals over t_2 can be well approximated by the relation

$$\int_{-\infty}^{\infty} \Gamma(t_2)\,\mathrm{d}t_2 = \Phi(0),$$

where

$$\Phi(\omega) = \int_{-\infty}^{\infty} \Gamma(t_2)\,\mathrm{e}^{\mathrm{i}\omega t_2}\,\mathrm{d}t_2 \tag{17.2–6}$$

is the spectral density of the Johnson noise. The equation then simplifies to

$$\langle I^2(t)\rangle = \langle I^2(0)\rangle\,\mathrm{e}^{-2Rt/L} + \frac{1}{L^2}\mathrm{e}^{-2Rt/L}\int_0^t \mathrm{e}^{2Rt_1/L}\,\Phi(0)\,\mathrm{d}t_1$$
$$= \langle I^2(0)\rangle\,\mathrm{e}^{-2Rt/L} + \frac{1}{2RL}(1 - \mathrm{e}^{-2Rt/L})\Phi(0), \tag{17.2–7}$$

and as $t \to \infty$ this further reduces to

$$\langle I^2\rangle = \frac{1}{2RL}\Phi(0) \tag{17.2–8}$$

in the steady state.

Now $\frac{1}{2}L\langle I^2\rangle$ is the average energy stored in the inductance through the magnetic field, and in thermal equilibrium we expect this to equal $\frac{1}{2}k_{\mathrm{B}}T$ from the equipartition law, where k_{B} is Boltzmann's constant. This allows us to re-write Eq. (17.2–8) in the form

$$\Phi(0) = 2k_{\mathrm{B}}TR, \tag{17.2–9}$$

which is a relation connecting the magnitude of the voltage fluctuations, represented by the spectral density at zero frequency $\Phi(0)$, with the dissipation in the system represented by the resistance R. It is perhaps the simplest example of a *fluctuation–dissipation theorem*. It was shown by Nyquist (1928), with the help of simple physical arguments, that the spectral density $\Phi(\omega)$ of the thermal noise is constant (at least up to very high frequencies), so that Eq. (17.2–9) can be generalized to

$$\Phi(\omega) = 2k_B TR, \qquad (17.2\text{–}10)$$

and a further generalization for a frequency-dependent resistance $R(\omega)$ is straightforward also.

Both the arguments we have used, and the conclusions to which they lead, turn out to be of rather general validity. They hold for a wide class of linear systems and even for quantum mechanical ones. We now turn our attention to the derivation of the relationship between fluctuations and dissipation in a quantum mechanical linear dissipative system. We shall follow the basic argument of Callen and Welton (1951). Another example of a quantum system that obeys the fluctuation–dissipation theorem will be encountered in Section 17.4 below in connection with the quantum Langevin process.

17.2.2 Quantum mechanical linear dissipative system

We consider a quantum system of unperturbed energy $\hat{H}_0$ interacting with some external perturbation $V(t)$, that will be treated classically, through a dynamical variable $\hat{Q}$ of the quantum system. Then the total energy $\hat{H}$ of the interacting system is given by the equation

$$\hat{H} = \hat{H}_0 + V(t)\hat{Q}. \qquad (17.2\text{–}11)$$

We suppose that $\hat{H}_0$ has a complete set of eigenstates $|E_n\rangle$, $n = 1, 2, \ldots$, that may be discrete or continuous, but will here be treated as discrete for simplicity, and we take the lowest energy eigenvalue to be zero. We further assume that the variable $\hat{Q}$ is off-diagonal in the states $|E_n\rangle$. We shall take the perturbation $V(t)$ to be oscillatory at frequency ω, and of the form

$$V(t) = \mathrm{Re}\,[V(\omega)\,\mathrm{e}^{-\mathrm{i}\omega t}]. \qquad (17.2\text{–}12)$$

An example might be a two-level atom having $\hat{H}_0 = \hbar\omega_0(\hat{R}_3 + \tfrac{1}{2})$, interacting with a classical electromagnetic field $V(t) = \mathrm{Re}\,[\mathscr{E}(t)\,\mathrm{e}^{-\mathrm{i}\omega t}]$ through its electric dipole moment $\hat{Q} = 2\mu_{12}\hat{R}_1$ (cf. Section 15.1).

If the quantum system is linear in the sense that it responds linearly to the perturbation, as we assume, then the expectation value $\langle\hat{Q}(t)\rangle$ will also oscillate at frequency ω, and so will the rate of change of $\langle\hat{Q}(t)\rangle$ or the 'current' $\langle\dot{\hat{Q}}(t)\rangle$. We express this fact by the formula

$$\langle\dot{\hat{Q}}(t)\rangle = \mathrm{Re}\,[I(\omega)\,\mathrm{e}^{-\mathrm{i}\omega t}], \qquad (17.2\text{–}13)$$

where $I(\omega)$ is in general complex. The linearity of the system makes $I(\omega)$ proportional to $V(\omega)$, so that we may put

$$Z(\omega) \equiv \frac{V(\omega)}{I(\omega)}, \qquad (17.2\text{–}14)$$

where $Z(\omega)$ is known as the *generalized impedance* of the system at frequency ω. It is clear that there is an analogy here with the behavior of a linear electric circuit, and the notation and the terminology have been chosen with the analogy in mind. In line with this terminology, we decompose $Z(\omega)$ into its real and imaginary parts, namely

$$Z(\omega) \equiv R(\omega) + \mathrm{i}X(\omega), \tag{17.2–15}$$

and we refer to them as the *generalized resistance* and *generalized reactance*, respectively.

It is easy to see from Eq. (17.2–11) that the disturbance $V(t)$ produces a net power dissipation in the system given by $V(t)\langle\dot{\hat{Q}}(t)\rangle$. If we discard terms oscillating at frequency 2ω, that average to zero, we obtain from Eqs. (17.2–12) and (17.2–13) for the average power dissipation $P(\omega)$ at frequency ω,

$$P(\omega) = \tfrac{1}{2}\mathrm{Re}\,[V^*(\omega)I(\omega)].$$

With the help of Eqs. (17.2–14) and (17.2–15) this formula reduces to

$$\begin{aligned} P(\omega) &= \frac{1}{2}\mathrm{Re}\left[\frac{|V(\omega)|^2}{Z(\omega)}\right] \\ &= \frac{1}{2}\frac{R(\omega)}{|Z(\omega)|^2}|V(\omega)|^2. \end{aligned} \tag{17.2–16}$$

We note that, despite the fact that $V(t)$ is oscillatory, there is a net dissipation when the resistance $R(\omega) \neq 0$, which implies that the current $\langle\dot{\hat{Q}}(t)\rangle$ has a component in phase with the perturbation $V(t)$. On the other hand, when $R(\omega) = 0$ the external perturbation does no net work on the system. We now calculate the power dissipation more explicitly for a system in thermal equilibrium.

17.2.3 Power dissipation

If the system is initially in one of the eigenstates $|E_n\rangle$ of $\hat{H}_0$, then, under the influence of the perturbation $V(t)$ given by Eq. (17.2–12), it may be induced to make a transition to another energy eigenstate $|E_n + \hbar\omega\rangle$ or $|E_n - \hbar\omega\rangle$ (if the eigenstates are sufficiently densely packed). The rate of transition $\mathcal{R}_\pm$ from $|E_n\rangle$ to one or the other of these states, averaged over several periods $2\pi/\omega$, can be calculated by perturbation theory in the usual way and is given by the Fermi Golden Rule in the form

$$\mathcal{R}_\pm = \frac{2\pi}{\hbar}|\langle E_n \pm \hbar\omega|\hat{H}_1|E_n\rangle|^2\sigma(E_n \pm \hbar\omega). \tag{17.2–17}$$

$\hat{H}_1$ is the interaction energy $V(t)\hat{Q}$, and $\sigma(E)$ is the density of states of the system at energy E. We can allow for the possibility that $E_n - \hbar\omega$ may turn out to be negative by adopting the convention that the state $|E\rangle$ is zero when $E < 0$. In Eq. (17.2–17) $\mathcal{R}_+$ refers to absorption of energy from the disturbance, because the system ends up in a higher energy state, whereas $\mathcal{R}_-$ refers to energy emission for the opposite reason. By taking the difference $\mathcal{R}_+ - \mathcal{R}_-$ and multiplying by $\hbar\omega$ we arrive at the net rate of energy absorption from the perturbation in the initial state $|E_n\rangle$. With the help of Eq. (17.2–12) for $V(t)$ we then obtain the result

$$\begin{aligned} &\text{Net rate of energy absorption in state } |E_n\rangle \text{ from disturbance} = \pi\omega|V(\omega)|^2[|\langle E_n + \hbar\omega|\hat{Q}|E_n\rangle|^2\sigma(E_n + \hbar\omega) \\ &\qquad - |\langle E_n - \hbar\omega|\hat{Q}|E_n\rangle|^2\sigma(E_n - \hbar\omega)]. \end{aligned} \tag{17.2–18}$$

If the system starts from a state of thermal equilibrium at temperature T, then all the different energy eigenstates $|E_n\rangle$ are populated initially with probability $p(E_n)$ given by the normalized Boltzmann distribution

$$p(E_n) = C\,\mathrm{e}^{-E_n/k_\mathrm{B}T}, \tag{17.2–19}$$

in which C is a normalization constant. The net rate of energy absorption from the disturbance at frequency ω, or the power dissipation $P(\omega)$, is then obtained by multiplying the rate given by Eq. (17.2–18) by $p(E_n)$ and summing over all E_n. As usual we may replace the sum by an integral with the help of the density of states and write

$$\begin{aligned}
P(\omega) &= \pi\omega|V(\omega)|^2\sum_{E_n}p(E_n)\{|\langle E_n+\hbar\omega|\hat{Q}|E_n\rangle|^2\sigma(E_n+\hbar\omega)\\
&\qquad - |\langle E_n-\hbar\omega|\hat{Q}|E_n\rangle|^2\sigma(E_n-\hbar\omega)\}\\
&\to \pi\omega|V(\omega)|^2\int_0^\infty\sigma(E)p(E)\{|\langle E+\hbar\omega|\hat{Q}|E\rangle|^2\sigma(E+\hbar\omega)\\
&\qquad - |\langle E-\hbar\omega|\hat{Q}|E\rangle|^2\sigma(E-\hbar\omega)\}\,\mathrm{d}E\\
&= \pi\omega|V(\omega)|^2\int_0^\infty[p(E)-p(E+\hbar\omega)]\sigma(E)\sigma(E+\hbar\omega)|\langle E+\hbar\omega|\hat{Q}|E\rangle|^2\,\mathrm{d}E.
\end{aligned} \tag{17.2–20}$$

The last line follows from the previous one with the help of the substitution $E-\hbar\omega\to E$ in the second integral. On comparing this result with Eq. (17.2–16) for the power dissipation $P(\omega)$, we can immediately make the identification

$$\frac{R(\omega)}{|Z(\omega)|^2} = 2\pi\omega\int_0^\infty[p(E)-p(E+\hbar\omega)]\sigma(E)\sigma(E+\hbar\omega)|\langle E|\hat{Q}|E+\hbar\omega\rangle|^2\,\mathrm{d}E,$$

and with the help of Eq. (17.2–19) we have,

$$\frac{R(\omega)}{|Z(\omega)|^2} = 2\pi\omega[1-\mathrm{e}^{-\hbar\omega/k_\mathrm{B}T}]\int_0^\infty p(E)\sigma(E)\sigma(E+\hbar\omega)|\langle E|\hat{Q}|E+\hbar\omega\rangle|^2\,\mathrm{d}E. \tag{17.2–21}$$

This expresses the relevant dissipative part of the impedance in terms of the characteristics of the system. It is worth noting that the disturbance $V(t)$ or $V(\omega)$ no longer appears. We introduced it merely in order to help us calculate the response of the system, but it can now be removed, so that the system remains in thermal equilibrium.

17.2.4 Current fluctuations in thermal equilibrium

As the dynamical variable $\hat{Q}$ was assumed to be off-diagonal in the eigenstates $|E_n\rangle$ of $\hat{H}_0$, it is clear that $\hat{Q}$ does not have a definite value in the state $|E_n\rangle$, but rather some probability distribution with mean $\langle E_n|\hat{Q}|E_n\rangle = 0$ and dispersion $\langle E_n|\hat{Q}^2|E_n\rangle$, and the same is true for the current $\dot{\hat{Q}}$. From the Heisenberg equation of motion for $\hat{Q}$ and from Eq. (17.2–11) it follows that

$$\begin{aligned}\dot{\hat{Q}} &= \frac{1}{\mathrm{i}\hbar}[\hat{Q}, \hat{H}] \\ &= \frac{1}{\mathrm{i}\hbar}[\hat{Q}, \hat{H}_0], \end{aligned} \tag{17.2–22}$$

so that the matrix element

$$\langle E_n|\dot{\hat{Q}}|E_m\rangle = \frac{1}{\mathrm{i}\hbar}(E_m - E_n)\langle E_n|\hat{Q}|E_m\rangle. \tag{17.2–23}$$

When $n = m$ this expression vanishes, so that the expectation of $\dot{\hat{Q}}$ is zero in the state $|E_n\rangle$.

Let us now calculate the dispersion of the current $\dot{\hat{Q}}$ in the state $|E_n\rangle$, which is given by

$$\langle E_n|(\Delta\dot{\hat{Q}})^2|E_n\rangle = \langle E_n|\dot{\hat{Q}}^2|E_n\rangle.$$

If we insert the unit operator in the form $\sum_{E_m}|E_m\rangle\langle E_m|$ between the two factors $\dot{\hat{Q}}$, and make use of Eq. (17.2–23), we immediately obtain the relation

$$\begin{aligned}\langle E_n|(\Delta\dot{\hat{Q}})^2|E_n\rangle &= \sum_{E_m}\langle E_n|\dot{\hat{Q}}|E_m\rangle\langle E_m|\dot{\hat{Q}}|E_n\rangle \\ &= \frac{1}{\hbar^2}\sum_{E_m}(E_m - E_n)^2|\langle E_n|\hat{Q}|E_m\rangle|^2 \\ &= \frac{1}{\hbar^2}\left[\sum_{E_m\geq E_n} + \sum_{E_m<E_n}\right](E_m - E_n)^2|\langle E_n|\hat{Q}|E_m\rangle|^2,\end{aligned}$$

and if we replace sums by integrals by introducing the density of states, as before, this becomes

$$\begin{aligned}\langle E_n|(\Delta\dot{\hat{Q}})^2|E_n\rangle &= \frac{1}{\hbar^2}\int_{E_n}^{\infty}\sigma(E)(E - E_n)^2|\langle E_n|\hat{Q}|E\rangle|^2\,\mathrm{d}E \\ &\quad + \frac{1}{\hbar^2}\int_0^{E_n}\sigma(E)(E - E_n)^2|\langle E_n|\hat{Q}|E\rangle|^2\,\mathrm{d}E \\ &= \hbar\int_0^{\infty}\omega^2\sigma(E_n + \hbar\omega)|\langle E_n|\hat{Q}|E_n + \hbar\omega\rangle|^2\,\mathrm{d}\omega \\ &\quad + \hbar\int_0^{E_n/\hbar}\omega^2\sigma(E_n - \hbar\omega)|\langle E_n|\hat{Q}|E_n - \hbar\omega\rangle|^2\,\mathrm{d}\omega. \end{aligned} \tag{17.2–24}$$

The last equation follows from the previous one with the help of the substitutions $E - E_n = \hbar\omega$ and $E - E_n = -\hbar\omega$ in the first and second integrals, respectively. Because of our convention that $|E\rangle = 0$ if $E < 0$, the upper limit in the last integral can be extended to infinity.

Equation (17.2–24) give the dispersion of $\dot{\hat{Q}}$ in the state $|E_n\rangle$. By considering a state which is a mixture of states $|E_n\rangle$ with weights $p(E_n)$ given by Eq. (17.2–19), we immediately obtain the dispersion of $\dot{\hat{Q}}$ for a system in thermal equilibrium.

Once again we perform the average over E_n by replacing the sum by an integral, and find that

$$\begin{aligned}\langle(\Delta\dot{\hat{Q}})^2\rangle &= \int_0^\infty \sigma(E)p(E)\langle E|(\Delta\dot{\hat{Q}})^2|E\rangle\,\mathrm{d}E \\ &= \hbar\int_0^\infty \mathrm{d}E\,\sigma(E)p(E)\int_0^\infty \mathrm{d}\omega\,\omega^2[\sigma(E+\hbar\omega)|\langle E|\hat{Q}|E+\hbar\omega\rangle|^2 \\ &\quad + \sigma(E-\hbar\omega)|\langle E|\hat{Q}|E-\hbar\omega\rangle|^2] \\ &= \hbar\int_0^\infty \mathrm{d}\omega\,\omega^2[1+\mathrm{e}^{-\hbar\omega/k_\mathrm{B}T}]\int_0^\infty \mathrm{d}E\,p(E)\sigma(E)\sigma(E+\hbar\omega)|\langle E|\hat{Q}|E+\hbar\omega\rangle|^2.\end{aligned} \tag{17.2–25}$$

The last line follows from the previous one by interchanging the E and ω integrals, by putting $E-\hbar\omega \rightarrow E$ in the second integral, and by expressing $p(E+\hbar\omega)$ in terms of $p(E)$ with the help of Eq. (17.2–19).

17.2.5 *Spectral density of the fluctuations*

We have now expressed the total dispersion of the fluctuating current $\dot{\hat{Q}}$ as an integral over all frequencies. If we re-write Eq. (17.2–25) in the form

$$\langle(\Delta\dot{\hat{Q}})^2\rangle = \frac{1}{2\pi}\int_0^\infty \Phi_{\dot{Q}}(\omega)\,\mathrm{d}\omega, \tag{17.2–26}$$

it is clear that the non-negative function

$$\Phi_{\dot{Q}}(\omega) \equiv 2\pi\hbar\omega^2[1+\mathrm{e}^{-\hbar\omega/k_\mathrm{B}T}]\int_0^\infty p(E)\sigma(E)\sigma(E+\hbar\omega)|\langle E|\hat{Q}|E+\hbar\omega\rangle|^2\,\mathrm{d}E \tag{17.2–27}$$

is the *spectral density* of the current fluctuations. Because of the linearity of the system expressed by Eq. (17.2–14), according to which changes in current and changes in V are proportional to each other, the current fluctuations are equivalent to the fluctuations produced by an applied disturbance V of zero mean, with spectral density

$$\begin{aligned}\Phi_V(\omega) &= |Z(\omega)|^2\Phi_{\dot{Q}}(\omega) \\ &= 2\pi\hbar\omega^2|Z(\omega)|^2[1+\mathrm{e}^{-\hbar\omega/k_\mathrm{B}T}]\int_0^\infty p(E)\sigma(E)\sigma(E+\hbar\omega)|\langle E|\hat{Q}|E+\hbar\omega\rangle|^2\,\mathrm{d}E.\end{aligned} \tag{17.2–28}$$

The integral of $\Phi_V(\omega)$ over all positive frequencies gives the total mean squared fluctuation $\langle(\Delta V)^2\rangle$ of the external disturbance that would give rise to the current fluctuation $\langle(\Delta\dot{\hat{Q}})^2\rangle$ given by Eq. (17.2–25).

If we compare the result given by Eq. (17.2–28) with Eq. (17.2–21), we observe that they are proportional to each other and contain the same integral. We may therefore relate $R(\omega)$ and $\Phi_V(\omega)$ and write

$$\Phi_V(\omega) = \hbar\omega\left[\frac{1 + e^{-\hbar\omega/k_B T}}{1 - e^{-\hbar\omega/k_B T}}\right]R(\omega). \qquad (17.2\text{–}29)$$

The factor on the right can be re-expressed in the form

$$\begin{aligned}\mathscr{E}(\omega, T) &\equiv \frac{1}{2}\hbar\omega\left[\frac{1 + e^{-\hbar\omega/k_B T}}{1 - e^{-\hbar\omega/k_B T}}\right] \\ &= \hbar\omega\left[\frac{1}{2} + \frac{1}{e^{\hbar\omega/k_B T} - 1}\right],\end{aligned} \qquad (17.2\text{–}30)$$

which will be recognized as the average energy of a harmonic oscillator in thermal equilibrium at temperature T. It is also the average energy of a mode of thermal equilibrium radiation [cf. Eq. (13.1–8)], for which $(e^{\hbar\omega/k_B T} - 1)^{-1}$ is the average photon occupation number, when the zero point energy $\frac{1}{2}\hbar\omega$ is added.

With the help of Eq. (17.2–30) we can now re-write Eq. (17.2–29) in the compact form

$$\Phi_V(\omega) = 2\mathscr{E}(\omega, T)R(\omega), \qquad (17.2\text{–}31)$$

which is the usual statement of the *fluctuation–dissipation theorem*. By using the relationship between $\Phi_V(\omega)$ and $\Phi_{\dot{Q}}(\omega)$ we can also express it in the equivalent form

$$\Phi_{\dot{Q}}(\omega) = 2\mathscr{E}(\omega, T)\frac{R(\omega)}{|Z(\omega)|^2}. \qquad (17.2\text{–}32)$$

The theorem relates the thermal fluctuations at frequency ω on the left with the dissipation by the system at frequency ω on the right, and the proportionality of $\Phi_V(\omega)$ and $R(\omega)$ brings out the close connection between fluctuations and dissipation. When the frequency ω is sufficiently low, or the temperature T is sufficiently high that $\hbar\omega \ll k_B T$, we can approximate $\mathscr{E}(\omega)$ by its classical limit $k_B T$ and write

$$\Phi_V(\omega) \approx 2k_B T R(\omega), \qquad (17.2\text{–}33)$$

which is a relationship that can also be derived by classical arguments (Nyquist, 1928) as we showed earlier. At the other extreme $\hbar\omega \gg k_B T$, $\mathscr{E}(\omega, T) \approx \frac{1}{2}\hbar\omega$, and we have

$$\Phi_V(\omega) \approx \hbar\omega R(\omega). \qquad (17.2\text{–}34)$$

However, in calculating the total fluctuation $\langle(\Delta(V)^2\rangle$, which requires integration of $\Phi_V(\omega)$ over all positive frequencies, we generally need to use the full expression (17.2–30) for $\mathscr{E}(\omega, T)$. When it is desirable to include both positive and negative frequencies in the spectral density, so that

$$\langle(\Delta V)^2\rangle = \frac{1}{2\pi}\int_{-\infty}^{\infty} \Phi_V(\omega)\,d\omega,$$

we re-write Eq. (17.2–31) in the form

$$\Phi_V(\omega) = \mathscr{E}(\omega, T)R(\omega), \qquad (17.2\text{–}35)$$

with the understanding that $\mathscr{E}(-\omega, T) = \mathscr{E}(\omega, T)$, $R(-\omega) = R(\omega)$.

In order to demonstrate the utility of the fluctuation–dissipation theorem, we now consider two simple applications.

17.2.6 *Brownian motion of a particle*

A small particle suspended in a fluid experiences a fluctuating force F as a result of numerous collisions with the molecules. These collisions give rise to the erratic movement of the particle known as *Brownian motion*. At the same time, the repeated molecular impacts result in a resistance to the motion of the particle under the influence of an external force, such as gravity. We now calculate the spectral density of the fluctuating force.

If we identify the disturbance $V(t)$ in our formalism with the applied force F, then clearly the dynamical variable Q in Eq. (17.2–11) has to be identified with the displacement x in one-dimensional motion. The current $\dot{Q}$ then becomes the particle velocity v, and at low frequencies the generalized Ohm's law (17.2–14) can be written

$$F = Rv, \tag{17.2–36}$$

in which the generalized impedance is purely resistive and real. Equation (17.2–36) is just Stokes' law for the motion of a particle through a viscous fluid. For a spherical particle of radius a the resistance R has the form

$$R = 6\pi a\eta, \tag{17.2–37}$$

where η is the viscosity. The fluctuation–dissipation theorem now enables us to write down the spectral density $\Phi_F(\omega)$ of the force fluctuations immediately. From Eqs. (17.2–31) and (17.2–37) we have, at low frequencies,

$$\Phi_F(\omega) = 2k_B TR = 12\pi a\eta k_B T, \tag{17.2–38}$$

and from Eq. (17.2–32) we have for the spectral density $\Phi_V(\omega)$ of the velocity fluctuations

$$\Phi_V(\omega) = \frac{2k_B T}{R} = \frac{k_B T}{3\pi a\eta}. \tag{17.2–39}$$

Of course, the spectral densities cannot remain constant at all frequencies. At higher frequencies phase shifts are to be expected.

17.2.7 *Field fluctuations in blackbody radiation*

We consider a blackbody radiation field in thermal equilibrium at temperature T. Because of the interaction with the thermal reservoir, the electric field $\mathbf{E}$ fluctuates in time, and we wish to calculate the spectral density $\Phi_E(\omega)$ of these fluctuations. In order to make use of the foregoing results, it is convenient to introduce a small electric point charge e of mass m, that experiences a fluctuating force $e\mathbf{E}$ with spectral density $e^2\Phi_E(\omega)$.

If we identify the disturbance $V(t)$ in Eq. (17.2–11) with the x-component of the force $eE_x(t)$ on the charge, and the variable Q with its displacement $x(t)$ in the same direction, then the current $\hat{Q}$ becomes the velocity $v_x(t)$. We suppose as usual that

$$eE_x(t) = \mathrm{Re}\,[e\tilde{E}_x(\omega)\,\mathrm{e}^{-\mathrm{i}\omega t}], \tag{17.2–40}$$

and that the velocity responds linearly to the force, so that

$$v_x(t) = \mathrm{Re}\,[\tilde{v}_x(\omega)\,\mathrm{e}^{-\mathrm{i}\omega t}]. \tag{17.2–41}$$

In order to determine the generalized impedance in this case, we start from the equation of motion for a charged particle in an electromagnetic field

$$eE_x(t) + \frac{1}{4\pi\varepsilon_0}\frac{2}{3}\frac{e^2}{c^3}\ddot{v}_x(t) = m\dot{v}_x, \tag{17.2–42}$$

in which the second term on the left arises from radiative damping, and we substitute Eqs. (17.2–40) and (17.2–41) in Eq. (17.2–42). We then obtain

$$e\tilde{E}_x(\omega) = \left[\frac{1}{4\pi\varepsilon_0}\frac{2}{3}\frac{e^2}{c^3}\omega^2 - \mathrm{i}m\omega\right]\tilde{v}_x(\omega),$$

from which the generalized impedance can be identified at once as the factor in front of $v_x(\omega)$. Its resistive part is

$$R(\omega) = \frac{1}{4\pi\varepsilon_0}\frac{2e^2\omega^2}{3c^3}. \tag{17.2–43}$$

We can now apply the fluctuation–dissipation theorem in the form of Eq. (17.2–31), to obtain the spectral density of the x-component of the fluctuating force on the charge, with the result that

$$\begin{aligned} e^2\Phi_{E_x}(\omega) &= 2\mathscr{E}(\omega, T)R(\omega) \\ &= 2\hbar\omega\left[\frac{1}{2} + \frac{1}{\mathrm{e}^{\hbar\omega/k_\mathrm{B}T} - 1}\right]\frac{1}{4\pi\varepsilon_0}\frac{2}{3}\frac{e^2\omega^2}{c^3}. \end{aligned}$$

When we add similar expressions for the fluctuations in the y- and z-directions, we arrive at the following result for the spectral density of the total electric field:

$$\Phi_E(\omega) = \frac{\hbar\omega^3}{\pi\varepsilon_0 c^3}\left[\frac{1}{2} + \frac{1}{\mathrm{e}^{\hbar\omega/k_\mathrm{B}T} - 1}\right]. \tag{17.2–44}$$

It is worth noting that all characteristics of the charged particle have dropped out in this expression. The zero-point contribution associated with the first term on the right leads to a divergence when integrated over all frequencies, as we have already seen in Chapter 10. This represents the vacuum fluctuations of the quantized electromagnetic field. The second term, when multiplied by ε_0 and integrated, yields the usual expression for the energy density of blackbody radiation [cf. Eq. (13.1–16)].

These simple examples show how easily the spectral densities of various fluctuating quantities can be obtained from the fluctuation–dissipation theorem, once the equivalent generalized resistance of the system has been identified.

17.3 Master equations

In the Schrödinger picture, the fundamental problem in the study of the time evolution of interacting quantum systems is to determine how the state, or the density operator, of the system on which our interest is focused evolves in time.

Even the solution of this problem does not allow us to evaluate multi-time correlation functions without additional information about the Green function of the system, but we have seen that this problem can sometimes be by-passed with the help of the quantum regression theorem. However, the equation of motion either for the total, or for the reduced, density operator has to be solved as a function of time. Such an equation is known as the *master equation* for the system, although the same name is sometimes also applied to equations of motion for various probability distributions. A number of powerful and rather general master equations have been developed over the years (Pauli, 1928; Van Hove, 1955, 1957, 1962; Zwanzig, 1961; Montroll, 1961; Prigogine, 1962; Agarwal, 1973, 1974; Oppenheim, Shuler and Weiss, 1977). In the following we shall only consider two special forms of master equation and indicate how these can be derived.

17.3.1 The Pauli master equation

One of the first forms of master equation was obtained by Pauli (1928) [see also Van Hove (1955, 1957, 1962)]. We will derive it first for a classical system and then examine under what conditions a quantum system obeys the same equation of motion.

We consider the state of a classical system or discrete random process (cf. Chapter 2) to be characterized by some integer n $(0, 1, 2, \ldots)$, so that the value n occurs with probability $p(n, t)$. If the transition rate $\mathscr{A}_{mn}$ from state n to state m depends only on n and m, so that the process is Markovian, then the rate of change of $p(n, t)$ must be the difference between the rate at which the population of state n increases due to transitions from other states, and the rate at which it decreases due to transitions to other states. Thus, we have

$$\frac{\partial}{\partial t}p(n, t) = \sum_{m\neq n} \mathscr{A}_{nm}p(m, t) - \sum_{m\neq n} \mathscr{A}_{mn}p(n, t). \qquad (17.3\text{–}1)$$

This is the Pauli master equation for the probability $p(n, t)$, [cf. Eq. (2.7–11)] and it has the form of a rate equation. In principle, it can be solved for $p(n, t)$ if the transition rates $\mathscr{A}_{nm}$ are all known. In the special case in which the system obeys the *principle of detailed balance*, we obtain the steady-state solution

$$\mathscr{A}_{nm}p(m) = \mathscr{A}_{mn}p(n), \qquad (17.3\text{–}2)$$

which can be regarded as a recursion relation and yields all the $p(n)$ in terms of $p(0)$, say.

When we treat a quantum system whose time evolution is governed by quantum mechanics, we find that an equation of motion like Eq. (17.3–1) holds only under special circumstances, despite the seemingly universal character of the equation. We consider a complete orthonormal set of states $|\phi_n\rangle$ that are eigenstates of the non-interacting energy $\hat{H}_0$. Let $\hat{\rho}(t)$ be the density operator of the system at time t in the interaction picture. Under the influence of some perturbation $\hat{H}_\mathrm{I}(t)$ the density operator evolves according to the equation

$$\hat{\rho}(t + \Delta t) = \hat{U}(t + \Delta t, t)\hat{\rho}(t)\hat{U}^\dagger(t + \Delta t, t), \qquad (17.3\text{–}3)$$

where the unitary time evolution operator $\hat{U}(t+\Delta t, t)$ for the time differential Δt is given by the formula

$$\hat{U}(t+\Delta t, t) = \exp(-i\hat{H}_I \Delta t/\hbar). \tag{17.3–4}$$

The transition probability from state $|m\rangle$ at time t to state $|n\rangle$ at time $t+\Delta t$ is then given by the relation

$$\mathcal{P}(n, t+\Delta t|m, t) = |\langle n|\hat{U}(t+\Delta t, t)|m\rangle|^2 \equiv |U_{mn}|^2, \tag{17.3–5}$$

and this obeys the normalization conditions

$$\sum_n |U_{nm}|^2 = 1 = \sum_m |U_{nm}|^2. \tag{17.3–6}$$

If the energy spectrum is continuous or densely packed, then, as is well known, the transition probability from a state of energy E_m to one of energy E_n is proportional to Δt and is given by the Fermi Golden Rule:

$$\mathcal{P}(n, t+\Delta t|m, t) = |U_{nm}|^2 \approx \frac{2\pi}{\hbar}\Delta t\,\delta(E_n - E_m)|\langle\phi_n|\hat{H}_I|\phi_m\rangle|^2.$$

The transition rate $\mathcal{A}_{nm}$ for $n \neq m$ is therefore

$$\mathcal{A}_{nm} = \frac{\mathcal{P}(n, t+\Delta t|m, t)}{\Delta t} = \frac{|U_{nm}|^2}{\Delta t} = \frac{2\pi}{\hbar}\delta(E_n - E_m)|\langle\phi_n|\hat{H}_I|\phi_m\rangle|^2. \tag{17.3–7}$$

We now use these results to calculate the rate of change of the probability $p(n, t)$, by writing

$$\frac{\partial p(n, t)}{\partial t} = \frac{\partial \rho_{nn}(t)}{\partial t} = \lim_{\Delta t\to 0}\frac{p(n, t+\Delta t) - p(n, t)}{\Delta t}$$
$$= \lim_{\Delta t\to 0}\frac{\langle\phi_n|\hat{\rho}(t+\Delta t) - \hat{\rho}(t)|\phi_n\rangle}{\Delta t}.$$

From Eq. (17.3–3) we have

$$\frac{\partial \rho_{nn}(t)}{\partial t} = \lim_{\Delta t\to 0}\frac{1}{\Delta t}\{\langle\phi_n|\hat{U}(t+\Delta t, t)\hat{\rho}(t)\hat{U}^\dagger(t+\Delta t, t)|\phi_n\rangle - \rho_{nn}(t)\}$$
$$= \lim_{\Delta t\to 0}\frac{1}{\Delta t}\left\{\sum_m\sum_r[U_{nm}\rho_{mr}(t)U^*_{rn}] - \rho_{nn}(t)\right\}, \tag{17.3–8}$$

where the last line follows from the preceding one after we insert the unit operators $\sum_m|\phi_m\rangle\langle\phi_m|$ and $\sum_r|\phi_r\rangle\langle\phi_r|$ between $\hat{U}$ and $\hat{\rho}(t)$ and between $\hat{\rho}(t)$ and $\hat{U}^\dagger$, respectively.

This is the basic equation of motion for $\rho_{nn}(t)$. It is not a rate equation in general, and it is not reducible to Eq. (17.3–1) without further assumption. The terms with $m \neq r$ generally give oscillating contributions that are less important than those with $m = r$, but strictly speaking, these terms cannot be discarded. However, if we make the so-called *random phase assumption*, and take the density operator $\hat{\rho}(t)$ at time t to be diagonal in $|\phi_n\rangle$, so that

$$\rho_{mr}(t) = \delta_{mr}\rho_{mm}(t),$$

then, with the help of Eq. (17.3–6), we obtain

$$\frac{\partial p(n,t)}{\partial t} = \lim_{\Delta t \to 0} \frac{1}{\Delta t} \sum_{m \neq n} [|U_{nm}|^2 \rho_{mm}(t) - |U_{mn}|^2 \rho_{nn}(t)], \qquad (17.3\text{–}8a)$$

and this equation becomes identical with Eq. (17.3–1) when Eq. (17.3–7) is used. A quantum system therefore obeys the Pauli master equation only when the density operator $\hat{\rho}(t)$ is diagonal in $|\phi_n\rangle$. The reason for this is, of course, that when $\hat{\rho}(t)$ is not diagonal, the system is not necessarily in one of the states $|\phi_n\rangle$ with some probability $p(n, t)$, but it may be in a superposition of these states. Actually, even if the density operator $\hat{\rho}(t)$ is diagonal at time t, it cannot remain strictly diagonal at later times in the presence of the interaction $\hat{H}_\mathrm{I}$. The random phase assumption and the Pauli master equation therefore hold only in a certain limiting sense, that can be made precise (Van Hove, 1955, 1957, 1962; Zwanzig, 1961; Montroll, 1961; Prigogine, 1962; Agarwal, 1973, 1974). Nevertheless, the equation is useful in the treatment of some problems.

17.3.2 Zwanzig's generalized master equation

Whereas the Pauli master equation holds under restricted circumstances and describes only the behavior of the diagonal elements of the density operator $\hat{\rho}(t)$, more general equations of motion for $\hat{\rho}(t)$ have been obtained. We shall briefly describe an interesting method developed by Zwanzig (1961) that makes use of projection operators to project out of the density operator that part on which our interest is focused. We follow the derivation given by Agarwal (1973).

We start from the general Schrödinger equation of motion for the density operator $\hat{\rho}(t)$ of the system, namely

$$\frac{\partial \hat{\rho}(t)}{\partial t} = \frac{1}{\mathrm{i}\hbar}[\hat{H}, \hat{\rho}(t)] = -\mathrm{i}\mathcal{L}\hat{\rho}(t), \qquad (17.3\text{–}9)$$

where $\mathcal{L} \equiv (1/\hbar)[\hat{H}, \ldots]$ is the Liouville operator for the system. The advantage of using the Liouvillian form is that the equation of motion (17.3–9) applies equally to a classical system having some density distribution $\rho(t)$. If we are working in the interaction picture, $\hat{H}$ can be replaced by the interaction energy $\hat{H}_\mathrm{I}(t)$ in Eq. (17.3–9), but the Liouville operator $\mathcal{L}(t)$ then becomes time-dependent. We now introduce a time-independent projection operator $\hat{P}$:

$$\hat{P}^2 = \hat{P}, \qquad (17.3\text{–}10)$$

that is chosen so as to project out the most relevant or interesting part of $\hat{\rho}$. Then $(1 - \hat{P})\hat{\rho}(t)$ represents the part of the density operator in which we are not especially interested, and evidently

$$\hat{\rho}(t) = \hat{P}\hat{\rho}(t) + (1 - \hat{P})\hat{\rho}(t). \qquad (17.3\text{–}11)$$

$\hat{P}$ can have many different forms. For example, if we are particularly interested in the diagonal part of $\hat{\rho}$ in some representation $|\phi_n\rangle$, we can choose

$$\hat{P} \ldots = \sum_n |\phi_n\rangle\langle\phi_n| \operatorname{Tr}[|\phi_n\rangle\langle\phi_n| \ldots]. \qquad (17.3\text{–}12a)$$

Then if $\hat{\rho} = \sum_n \sum_m \rho_{nm} |\phi_n\rangle\langle\phi_m|$,

$$\hat{P}\hat{\rho} = \sum_n \rho_{nn} |\phi_n\rangle\langle\phi_n|. \tag{17.3–12b}$$

On the other hand, if we are dealing with a quantum system S coupled to a reservoir R, and we are interested mainly in S, we can focus on the reduced density operator

$$\hat{\rho}_S(t) = \mathrm{Tr}_R\, \hat{\rho}(t)$$

by choosing

$$\hat{P}\ldots = \hat{\rho}_R(0)\,\mathrm{Tr}_R\ldots. \tag{17.3–13}$$

On multiplying both sides of Eq. (17.3–9) by $\hat{P}$ from the left and then using Eq. (17.3–11) to substitute for $\hat{\rho}(t)$ on the right, we have

$$\hat{P}\frac{\partial\hat{\rho}}{\partial t} = -\mathrm{i}\hat{P}\mathscr{L}(t)\hat{P}\hat{\rho}(t) - \mathrm{i}\hat{P}\mathscr{L}(t)(1-\hat{P})\hat{\rho}(t). \tag{17.3–14}$$

Similarly, after multiplying Eq. (17.3–9) by $(1-\hat{P})$ from the left, we obtain

$$(1-\hat{P})\frac{\partial\hat{\rho}}{\partial t} = -\mathrm{i}(1-\hat{P})\mathscr{L}(t)\hat{P}\hat{\rho}(t) - \mathrm{i}(1-\hat{P})\mathscr{L}(t)(1-\hat{P})\hat{\rho}(t). \tag{17.3–15}$$

We now formally integrate the first-order differential Eq. (17.3–15) and solve for $(1-\hat{P})\hat{\rho}(t)$. We find that

$$\begin{aligned}(1-\hat{P})\hat{\rho}(t) &= \exp\left[-\mathrm{i}(1-\hat{P})\int_0^t \mathscr{L}(t')\,\mathrm{d}t'\right](1-\hat{P})\hat{\rho}(0)\\ &\quad - \mathrm{i}\int_0^t \exp\left[-\mathrm{i}(1-\hat{P})\int_\tau^t \mathscr{L}(t')\,\mathrm{d}t'\right](1-\hat{P})\mathscr{L}(\tau)\hat{P}\hat{\rho}(\tau)\,\mathrm{d}\tau,\end{aligned}$$

or, if $t-\tau=\tau'$,

$$\begin{aligned}&= \exp\left[-\mathrm{i}(1-\hat{P})\int_0^t \mathscr{L}(t')\,\mathrm{d}t'\right](1-\hat{P})\hat{\rho}(0) - \mathrm{i}\int_0^t \exp\left[-\mathrm{i}(1-\hat{P})\int_{t-\tau'}^t \mathscr{L}(t')\,\mathrm{d}t'\right]\\ &\quad \times (1-\hat{P})\mathscr{L}(t-\tau')\hat{P}\hat{\rho}(t-\tau')\,\mathrm{d}\tau'.\end{aligned}$$

After substituting for $(1-\hat{P})\hat{\rho}(t)$ in Eq. (17.3–14) we arrive at

$$\begin{aligned}\hat{P}\frac{\partial\hat{\rho}}{\partial t} &= -\mathrm{i}\hat{P}\mathscr{L}(t)\hat{P}\hat{\rho}(t) - \mathrm{i}\hat{P}\mathscr{L}(t)\exp\left[-\mathrm{i}(1-\hat{P})\int_0^t \mathscr{L}(t')\,\mathrm{d}t'\right](1-\hat{P})\hat{\rho}(0)\\ &\quad - \hat{P}\mathscr{L}(t)\int_0^t \exp\left[-\mathrm{i}(1-\hat{P})\int_\tau^t \mathscr{L}(t')\,\mathrm{d}t'\right](1-\hat{P})\mathscr{L}(\tau)\hat{P}\hat{\rho}(\tau)\,\mathrm{d}\tau,\end{aligned}$$

or on setting

$$\exp\left[-\mathrm{i}(1-\hat{P})\int_\tau^t \mathscr{L}(t')\,\mathrm{d}t'\right] \equiv \hat{U}(t,\tau),$$

$$\begin{aligned}\hat{P}\frac{\partial\hat{\rho}}{\partial t} &= -\mathrm{i}\hat{P}\mathscr{L}(t)\hat{P}\hat{\rho}(t) - \mathrm{i}\hat{P}\mathscr{L}(t)\hat{U}(t,0)(1-\hat{P})\hat{\rho}(0)\\ &\quad - \hat{P}\mathscr{L}(t)\int_0^t \hat{U}(t,\tau)(1-\hat{P})\mathscr{L}(\tau)\hat{P}\hat{\rho}(\tau)\,\mathrm{d}\tau.\end{aligned} \tag{17.3–16}$$

This is *Zwanzig's generalized master equation*. In this form it holds exactly, but the equation is not easy to apply in practice. Its utility rests partly on the fact that it is an equation of motion for $\hat{P}\hat{\rho}(t)$ in which the 'irrelevant' part $(1 - \hat{P})\hat{\rho}(t)$ appears only at the initial time $t = 0$. Sometimes the initial conditions make $(1 - \hat{P})\hat{\rho}(0) = 0$, in which case the second term on the right vanishes.

17.3.3 Application to the Pauli equation

It is possible to derive the Pauli master equation (17.3–1) from the more general Eq. (17.3–16), but the problem is not trivial and several approximations and limits have to be introduced. For the Pauli equation we focus attention on the diagonal elements of the density operator $\hat{\rho}(t)$ by choosing $\hat{P}$ as in Eq. (17.3–12a). If the initial state is diagonal, then

$$(1 - \hat{P})\hat{\rho}(0) = 0, \qquad (17.3\text{–}17)$$

which makes the second term on the right of Eq. (17.3–16) vanish. By recalling that $\mathscr{L} \ldots \equiv (1/\hbar)[\hat{H}_0 + \hat{H}_{\mathrm{I}}, \ldots]$, taking $\hat{H}_{\mathrm{I}}$ to be off-diagonal in the states $|\phi_n\rangle$ and time independent for simplicity and using Eq. (17.3–12b), one may readily show that

$$\hat{P}\mathscr{L}\hat{P}\hat{\rho}(t) = 0, \qquad (17.3\text{–}18)$$

so that the first term on the right of Eq. (17.3–16) also vanishes. It also follows immediately from the same argument that

$$\mathscr{L}\hat{P}\hat{\rho}(t) = (1/\hbar)[\hat{H}_{\mathrm{I}}, \hat{P}\hat{\rho}(t)] \equiv \mathscr{L}_{\mathrm{I}}\hat{P}\hat{\rho}(t). \qquad (17.3\text{–}19)$$

Equation (17.3–16) can therefore be reduced to

$$\frac{\partial}{\partial t}[\hat{P}\hat{\rho}(t)] = -\int_0^t \hat{K}(\tau)[\hat{P}\hat{\rho}(t - \tau)]\,\mathrm{d}\tau, \qquad (17.3\text{–}20)$$

in which the operator $\hat{K}(\tau)$ is given by the formula

$$\hat{K}(t) \equiv \hat{P}\mathscr{L}_{\mathrm{I}}(t)\exp[-\mathrm{i}(1 - \hat{P})\mathscr{L}(\tau)](1 - \hat{P})\mathscr{L}_{\mathrm{I}}(t - \tau). \qquad (17.3\text{–}21)$$

By making a representation of the operators in Eq. (17.3–20) in the basis $|\phi_n\rangle$, and comparing coefficients of $|\phi_n\rangle\langle\phi_n|$ on both sides of the equation, one finds that

$$\frac{\partial}{\partial t}\rho_{nn}(t) = -\sum_m \int_0^t k_{nm}(\tau)\rho_{mm}(t - \tau)\,\mathrm{d}\tau, \qquad (17.3\text{–}22)$$

in which the kernel $k_{nm}(\tau)$ is given by the relation

$$k_{nm}(\tau) \equiv \frac{1}{\hbar^2}\langle\phi_n|\Bigg[\hat{H}_{\mathrm{I}}, \exp\left\{\frac{-\mathrm{i}\tau}{\hbar}[\hat{H}_0 + (1 - \hat{P})\hat{H}_{\mathrm{I}}(1 - \hat{P})]\right\}$$
$$\times [\hat{H}_{\mathrm{I}}, |\phi_m\rangle\langle\phi_m|]\exp\left\{\frac{\mathrm{i}\tau}{\hbar}[\hat{H}_0 + (1 - \hat{P})\hat{H}_{\mathrm{I}}(1 - \hat{P})]\right\}\Bigg]|\phi_n\rangle.$$

The kernel $k_{nm}(\tau)$ has the property

$$\sum_m k_{nm}(\tau) = 0, \qquad (17.3\text{–}23)$$

which follows immediately from the fact that

$$\left[\hat{H}_{\mathrm{I}}, \sum_m |\phi_m\rangle\langle\phi_m|\right] = 0,$$

because $\hat{H}_{\mathrm{I}}$ is off-diagonal. By using Eq. (17.3–23) we can formally re-write Eq. (17.3–22) in the form

$$\frac{\partial \rho_{nn}(t)}{\partial t} = \sum_{m \neq n} \int_0^t [k_{nm}(\tau)\rho_{nn}(t-\tau) - k_{nm}(\tau)\rho_{mm}(t-\tau)]\,\mathrm{d}\tau, \quad (17.3\text{–}24)$$

which resembles the Pauli equation (17.3–1) except for the integral over τ. However, if the interaction $\hat{H}_{\mathrm{I}}$ is sufficiently weak and the time τ sufficiently long, and if the number of degrees of freedom of the system is sufficiently large, it is possible to replace $\rho_{nn}(t-\tau)$ under the integral by $\rho_{nn}(t)$ and extend the upper limit to infinity. The precise conditions have been given (Van Hove, 1955, 1957, 1962; Zwanzig, 1961; Montroll, 1961; Prigogine, 1962; Agarwal, 1973, 1974). Equation (17.3–24) then reduces to the form of Eq. (17.3–1). Loosely speaking, the approximation implies that memory effects persist only for very brief times, so that the system is approximately Markovian. It should be clear from this brief discussion that the Pauli master equation holds only under rather special circumstances, even though it may have the appearance of universal validity when formulated in classical terms. The Zwanzig master equation (17.3–16), on the other hand, holds very generally, but it is so complicated that one usually has to introduce several approximations in order to apply it in practice.

17.3.4 Application to the Dicke problem

As a further example, we will consider the application of Eq. (17.3–16) to the Dicke problem that was treated in Chapter 16, i.e. to the interaction of a group of N closely spaced, identical two-level atoms with the electromagnetic field (Agarwal, 1973, 1974). We shall work in the interaction picture and assume that the atoms are at the origin. For simplicity, we also make the rotating wave approximation and drop counter-rotating terms right from the beginning, so that the interaction Hamiltonian is taken to be of the form [cf. Eq. (15.4–3)]

$$\hat{H}_{\mathrm{I}}(t) = \sum_{\mathbf{k}s} \hbar g_{\mathbf{k}s} \hat{a}^{\dagger}_{\mathbf{k}s} \hat{\mathscr{R}}^{(-)}\, \mathrm{e}^{\mathrm{i}(\omega-\omega_0)t} + \mathrm{h.c.}, \quad (17.3\text{–}25)$$

where $\hat{\mathscr{R}}^{(-)}$ is the collective atomic lowering operator [see Eq. (16.5–2)], and the coupling constant $g_{\mathbf{k}s}$ is given by the formula

$$g_{\mathbf{k}s} = \frac{\mathrm{i}\omega_0}{\hbar L^{3/2}} \left(\frac{\hbar}{2\omega\varepsilon_0}\right)^{1/2} \boldsymbol{\mu}_{12} \cdot \boldsymbol{\varepsilon}^*_{\mathbf{k}s}. \quad (17.3\text{–}26)$$

All the symbols have their usual meanings, as in Chapter 15. Because of the premature use of the rotating wave approximation, we shall find that the calculation does not give the correct frequency shifts, although the dynamics are correct.

We take the initial state of the coupled system of atoms and field to be the direct product state with the field in the vacuum state,

$$\left.\begin{aligned}\hat{\rho}(0) &= \hat{\rho}_A(0) \otimes \hat{\rho}_F(0),\\ \hat{\rho}_F(0) &= |\text{vac}\rangle\langle\text{vac}|.\end{aligned}\right\} \tag{17.3–27}$$

In the following we shall focus our attention on the atomic system A and identify the 'reservoir' with the field F, so that the projection operator $\hat{P}$ is given by

$$\hat{P}\ldots = \hat{\rho}_F(0)\,\mathrm{Tr}_F\ldots. \tag{17.3–28}$$

Under these conditions the term $\hat{P}\mathscr{L}(t)\hat{P}\hat{\rho}(t)$ in Eq. (17.3–16) is given by the formula

$$\begin{aligned}\hat{P}\mathscr{L}(t)\hat{P}\hat{\rho}(t) &= \hat{\rho}_F(0)\,\mathrm{Tr}_F\frac{1}{\hbar}[\hat{H}_I(t),\,\hat{\rho}_F(0)\hat{\rho}_A(t)]\\ &= \hat{\rho}_F(0)\left[\sum_{\mathbf{k}s} g_{\mathbf{k}s}\langle\text{vac}|\hat{a}^\dagger_{\mathbf{k}s}|\text{vac}\rangle\hat{\mathscr{R}}^{(-)}\,\mathrm{e}^{\mathrm{i}(\omega-\omega_0)t}\hat{\rho}_A(t) - \text{h.c.}\right]\\ &= 0,\end{aligned} \tag{17.3–29}$$

because

$$\langle\text{vac}|\hat{a}^\dagger_{\mathbf{k}s}|\text{vac}\rangle = 0 = \langle\text{vac}|\hat{a}_{\mathbf{k}s}|\text{vac}\rangle. \tag{17.3–30}$$

Moreover, we also have

$$(1-\hat{P})\hat{\rho}(0) = [1-\hat{\rho}_F(0)\,\mathrm{Tr}_F]\hat{\rho}_A(0)\hat{\rho}_F(0) = 0, \tag{17.3–31}$$

so that both the first and second terms on the right of Eq. (17.3–16) vanish.

With the help of Eq. (17.3–29) and the approximation $\hat{U}(t,\tau)\approx 1$ under the integral, which is sometimes known as the Born approximation, Eq. (17.3–16) then reduces to

$$\hat{P}\frac{\partial\hat{\rho}}{\partial t} = -\hat{P}\mathscr{L}(t)\int_0^t \mathscr{L}(\tau)\hat{P}\hat{\rho}(\tau)\,\mathrm{d}\tau,$$

and after we substitute for $\hat{P}$ from Eq. (17.3–28), we have

$$\hat{\rho}_F(0)\frac{\partial\hat{\rho}_A(t)}{\partial t} = -\hat{\rho}_F(0)\,\mathrm{Tr}_F\,\mathscr{L}(t)\int_0^t \mathscr{L}(\tau)\hat{\rho}_F(0)\hat{\rho}_A(\tau)\,\mathrm{d}\tau.$$

Substituting $\tau = t-\tau'$ in the last integral yields the equation

$$\frac{\partial\hat{\rho}_A(t)}{\partial t} = -\mathrm{Tr}_F\,\mathscr{L}(t)\int_0^t \mathscr{L}(t-\tau)\hat{\rho}_F(0)\hat{\rho}_A(t-\tau)\,\mathrm{d}\tau. \tag{17.3–32}$$

We now make the further approximation of replacing $\hat{\rho}_A(t-\tau)$ under the integral by its most recent value $\hat{\rho}_A(t)$, which is often known as the Markov or short memory approximation [cf. also the previous calculation leading to Eq. (15.5–20)]. Then we finally have the relation

$$\begin{aligned}\frac{\partial\hat{\rho}_A(t)}{\partial t} &= -\mathrm{Tr}_F\,\mathscr{L}(t)\int_0^t \mathscr{L}(t-\tau)\hat{\rho}_F(0)\hat{\rho}_A(t)\,\mathrm{d}\tau\\ &= -\frac{1}{\hbar^2}\mathrm{Tr}_F\int_0^t [\hat{H}_I(t),\,[\hat{H}_I(t-\tau),\,\hat{\rho}_F(0)\hat{\rho}_A(t)]]\,\mathrm{d}\tau\end{aligned}$$

$$= -\frac{1}{\hbar^2}\mathrm{Tr_F}\int_0^\infty [\hat{H}_\mathrm{I}(t)\hat{H}_\mathrm{I}(t-\tau)\hat{\rho}_\mathrm{F}(0)\hat{\rho}_\mathrm{A}(t) - \hat{H}_\mathrm{I}(t)\hat{\rho}_\mathrm{F}(0)\hat{\rho}_\mathrm{A}(t)\hat{H}_\mathrm{I}(t-\tau)]\,\mathrm{d}\tau$$
$$+ \text{h.c.} \tag{17.3–33}$$

After substituting for $\hat{H}_\mathrm{I}$ from Eq. (17.3–25) we find that each term under the integral breaks up into the sum of four terms involving $\langle\mathrm{vac}|\hat{a}_{\mathbf{k}s}\hat{a}_{\mathbf{k}'s'}|\mathrm{vac}\rangle$, $\langle\mathrm{vac}|\hat{a}^\dagger_{\mathbf{k}s}\hat{a}^\dagger_{\mathbf{k}'s'}|\mathrm{vac}\rangle$, $\langle\mathrm{vac}|\hat{a}^\dagger_{\mathbf{k}s}\hat{a}_{\mathbf{k}'s'}|\mathrm{vac}\rangle$ and $\langle\mathrm{vac}|\hat{a}_{\mathbf{k}s}\hat{a}^\dagger_{\mathbf{k}'s'}|\mathrm{vac}\rangle$. Of these only the last is non-zero, namely

$$\langle\mathrm{vac}|\hat{a}_{\mathbf{k}s}\hat{a}^\dagger_{\mathbf{k}'s'}|\mathrm{vac}\rangle = \delta^3_{\mathbf{kk}'}\delta_{ss'},$$

so that we obtain the equation

$$\frac{\partial\hat{\rho}_\mathrm{A}(t)}{\partial t} = -\int_0^t \sum_{\mathbf{k},s}|g_{\mathbf{k}s}|^2[\mathrm{e}^{\mathrm{i}(\omega_0-\omega)\tau}\hat{\mathscr{R}}^{(+)}\hat{\mathscr{R}}^{(-)}\hat{\rho}_\mathrm{A}(t)\,\mathrm{d}\tau + \text{h.c.}$$
$$- 2\cos(\omega-\omega_0)\tau\hat{\mathscr{R}}^{(-)}\hat{\rho}_\mathrm{A}(t)\hat{\mathscr{R}}^{(+)}]\,\mathrm{d}\tau. \tag{17.3–34}$$

In the limit $L\to\infty$, the summation on $\mathbf{k}$ becomes an integral in the usual way. Indeed, the evaluation of the expression

$$\int_0^t \sum_{\mathbf{k},s}|g_{\mathbf{k},s}|^2\,\mathrm{e}^{\mathrm{i}(\omega_0-\omega)\tau}\,\mathrm{d}\tau$$

is very similar to the calculation that was carried out previously in Section 15.5 in the course of deriving Eq. (15.5–21), except that a counter-rotating contribution is missing here because of the rotating wave approximation. With the understanding that $t \gg 1/\omega_0$ we put

$$\int_0^t \sum_{\mathbf{k},s}|g_{\mathbf{k}s}|^2\,\mathrm{e}^{\mathrm{i}(\omega_0-\omega)\tau}\,\mathrm{d}\tau \equiv \beta + \mathrm{i}\gamma, \tag{17.3–35}$$

as before, where β is half the Einstein A-coefficient and γ is a frequency shift. Then Eq. (17.3–34) becomes

$$\frac{\partial\hat{\rho}_\mathrm{A}(t)}{\partial t} = \beta[-\hat{\mathscr{R}}^{(+)}\hat{\mathscr{R}}^{(-)}\hat{\rho}_\mathrm{A}(t) + 2\hat{\mathscr{R}}^{(-)}\hat{\rho}_\mathrm{A}(t)\hat{\mathscr{R}}^{(+)} - \hat{\rho}_\mathrm{A}(t)\hat{\mathscr{R}}^{(+)}\hat{\mathscr{R}}^{(-)}]$$
$$- \mathrm{i}\gamma[\hat{\mathscr{R}}^{(+)}\hat{\mathscr{R}}^{(-)}, \hat{\rho}_\mathrm{A}(t)]. \tag{17.3–36}$$

Equation (17.3–36) is the appropriate master equation describing spontaneous emission from a collective atomic system. It can be shown that in the presence of a classical field whose amplitude is characterized by the atomic Rabi frequency Ω (cf. Section 15.3), there is an additional term $-\mathrm{i}\Omega[\hat{\mathscr{R}}_1, \hat{\rho}_\mathrm{A}(t)]$ on the right of Eq. (17.3–36). The resulting equation has been used to treat the problem of resonance fluorescence from a collection of atoms (Bonifacio, Schwendimann and Haake, 1971; Agarwal, Brown, Narducci and Vetri, 1977; Agarwal, Feng, Narducci, Gilmore and Tuft, 1979) and optical bistability (Carmichael and Walls, 1977; Narducci, Gilmore, Feng and Agarwal, 1978).

As a simple application of Eq. (17.3–36) we use the equation to calculate the rate of loss of energy of the atomic system. As the rate of loss of energy is proportional to the rate of change of $\langle\hat{\mathscr{R}}_3\rangle$, we multiply each term in Eq.

(17.3–36) by $\hat{\mathscr{R}}_3$ on the right and take the trace. We then find

$$\frac{\partial}{\partial t}\langle\hat{\mathscr{R}}_3\rangle = \beta[-\langle\hat{\mathscr{R}}_3\hat{\mathscr{R}}^{(+)}\hat{\mathscr{R}}^{(-)}\rangle + 2\langle\hat{\mathscr{R}}^{(+)}\hat{\mathscr{R}}_3\hat{\mathscr{R}}^{(-)}\rangle - \langle\hat{\mathscr{R}}^{(+)}\hat{\mathscr{R}}^{(-)}\hat{\mathscr{R}}_3\rangle]$$
$$- i\gamma[\langle\hat{\mathscr{R}}_3\hat{\mathscr{R}}^{(+)}\hat{\mathscr{R}}^{(-)}\rangle - \langle\hat{\mathscr{R}}^{(+)}\hat{\mathscr{R}}^{(-)}\hat{\mathscr{R}}_3\rangle]. \tag{17.3–37}$$

We now make use of the relations

$$[\hat{\mathscr{R}}^{(+)}, \hat{\mathscr{R}}_3] = -\hat{\mathscr{R}}^{(+)} \tag{17.3–38}$$

$$\hat{\mathscr{R}}^{(+)}\hat{\mathscr{R}}^{(-)} = \hat{\mathscr{R}}_3 + \tfrac{1}{2}N + \sum_{i\neq j}^{N}\sum^{N}\hat{b}^{(i)\dagger}\hat{b}^{(j)} \tag{17.3–39}$$

that follow from the definitions of the collective operators [see Eq. (16.5–2)], together with

$$[\hat{\mathscr{R}}^{(+)}\hat{\mathscr{R}}^{(-)}, \hat{\mathscr{R}}_3] = \sum_{i\neq j}^{N}\sum^{N}[\hat{b}^{(i)\dagger}\hat{b}^{(j)}, \hat{\mathscr{R}}_3]$$
$$= \sum_{i\neq j}^{N}\sum^{N}[\hat{b}^{(i)\dagger}\hat{b}^{(j)} - \hat{b}^{(i)\dagger}\hat{b}^{(j)}]$$
$$= 0, \tag{17.3–40}$$

which follows from Eq. (17.3–39). This last result allows us to simplify Eq. (17.3–37), and one finds that

$$\frac{\partial\langle\hat{\mathscr{R}}_3\rangle}{\partial t} = 2\beta[-\langle\hat{\mathscr{R}}_3\hat{\mathscr{R}}^{(+)}\hat{\mathscr{R}}^{(-)}\rangle + \langle\hat{\mathscr{R}}^{(+)}\hat{\mathscr{R}}_3\hat{\mathscr{R}}^{(-)}\rangle],$$

and with the help of Eq. (17.3–38), this leads to the equation

$$\frac{\partial\langle\hat{\mathscr{R}}_3\rangle}{\partial t} = 2\beta[-\langle\hat{\mathscr{R}}_3\hat{\mathscr{R}}^{(+)}\hat{\mathscr{R}}^{(-)}\rangle + \langle(\hat{\mathscr{R}}_3\hat{\mathscr{R}}^{(+)} - \hat{\mathscr{R}}^{(+)})\hat{\mathscr{R}}^{(-)}\rangle]$$
$$= -2\beta\langle\hat{\mathscr{R}}^{(+)}\hat{\mathscr{R}}^{(-)}\rangle. \tag{17.3–41}$$

If we now suppose that the atomic system is in the Dicke state $|l, m\rangle$, in which m is half the atomic inversion and l is the cooperation number (see Section 16.5), with $|m| \leqslant l \leqslant \frac{1}{2}N$, then with the help of Eqs. (16.5–15) we immediately obtain the equation

$$\frac{\partial\langle\hat{\mathscr{R}}_3\rangle}{\partial t} = -2\beta(l + m)(l - m + 1). \tag{17.3–42}$$

This is the same result that was obtained perturbatively in Section 16.6 [cf. Eq. (16.6–3)]. The rate of emission has the greatest numerical value $\beta N(\frac{1}{2}N + 1)$ when the atomic system is half inverted, with $m = 0$ and $l = \frac{1}{2}N$, and it is in the superfluorescent state that was treated previously.

17.3.5 Linear damping of off-diagonal matrix elements

Eq. (17.3–36) giving the rate of change of the density operator $\hat{\rho}$ of a collection of atoms actually holds much more generally for systems with linear damping. For example, it applies to a single-mode electromagnetic field subject to linear damping by some loss mechanism, such as a heat bath of harmonic oscillators, to which energy can be transferred. In that case we may identify $\hat{\mathscr{R}}^{(+)}$ and $\hat{\mathscr{R}}^{(-)}$ with the creation and annihilation operators $\hat{a}^\dagger$, $\hat{a}$ for the field, and obtain the equation of motion

$$\frac{\partial\hat{\rho}(t)}{\partial t} = -\beta[\hat{a}^\dagger\hat{a}\hat{\rho}(t) + \hat{\rho}(t)\hat{a}^\dagger\hat{a} - 2\hat{a}\hat{\rho}(t)\hat{a}^\dagger] - \mathrm{i}\gamma[\hat{a}^\dagger\hat{a}\hat{\rho}(t) - \hat{\rho}(t)\hat{a}^\dagger\hat{a}]. \tag{17.3–43}$$

This is another master equation for a linearly damped system when $\beta > 0$.

Let us use the equation to show that off-diagonal matrix elements of $\hat{\rho}(t)$ in the photon number basis damp out rapidly in time. If we multiply Eq. (17.3–43) by $\langle m|$ on the left and by $|n\rangle$ on the right and put $\langle m|\hat{\rho}(t)|n\rangle \equiv \rho_{mn}(t)$, we obtain the relation

$$\frac{\partial\rho_{mn}(t)}{\partial t} = -\beta[m\rho_{mn}(t) + n\rho_{mn}(t) - 2[(m+1)(n+1)]^{1/2}\rho_{m+1\;n+1}(t)] \\ - \mathrm{i}\gamma(m-n)\rho_{mn}(t). \tag{17.3–44}$$

Now if m, n are reasonable large numbers and $\rho_{mn}(t)$ changes only slowly with m, n we may, to a good approximation, replace $m+1$ by m, $n+1$ by n and $\rho_{m+1\;n+1}(t)$ by $\rho_{mn}(t)$. We then have

$$\begin{aligned}\frac{\partial\rho_{mn}(t)}{\partial t} &= \{-\beta[m+n-2(mn)^{1/2}] - \mathrm{i}\gamma(m-n)\}\rho_{mn}(t) \\ &= [-\beta(\sqrt{m}-\sqrt{n})^2 - \mathrm{i}\gamma(m-n)]\rho_{mn}(t).\end{aligned} \tag{17.3–45}$$

This simple linear differential equation has the solution,

$$\rho_{mn}(t) = \rho_{mn}(0)\,\mathrm{e}^{[-\beta(\sqrt{m}-\sqrt{n})^2-\mathrm{i}\gamma(m-n)]t}, \tag{17.3–46}$$

which shows at once that the off-diagonal matrix elements $\rho_{mn}(t)$ decay to zero in time. Moreover, the greater the off-diagonality, as measured by $\sqrt{m}-\sqrt{n}$, the greater is the damping rate. After a long time, the density operator $\hat{\rho}(t)$ therefore tends to become diagonal in the basis of photon number states (Walls and Milburn, 1985).

17.4 Quantum noise sources and quantum Langevin equations

17.4.1 Introduction

We have seen that the Heisenberg equations of motion for certain dynamical variables of a quantum system often bear a close resemblance to the equations of motion for the corresponding classical system. An obvious example is provided by the Maxwell equations for the free electromagnetic field, which are identical for a classical and a quantum field (cf. Chapter 10). However, as soon as damping or loss comes into the picture, the quantum mechanical situation becomes more complicated, because the average energy of a fully quantized system has to be conserved in the long-time limit. In practice, the problem is sometimes solved by the introduction of a large number of loss or reservoir systems, whose function is to soak up energy without being very much affected. But as a result of being coupled to the original quantum system under investigation, the loss systems not only introduce losses but also fluctuations, which are connected by the fluctuation–dissipation theorem. We shall now examine the equations of motion that govern this behavior of a quantum system (Senitzky, 1960, 1961; Weidlich and Haake, 1965a,b; Lax, 1966, 1967; Haken, 1970; Louisell, 1973, Sec. 6.6).

We start with an example from classical physics. Let $a(t)$ be the complex amplitude of a classical harmonic oscillator of frequency ω_0. In the absence of damping, $a(t)$ obeys the equation of motion

$$\dot{a}(t) = -\mathrm{i}\omega_0 a(t). \tag{17.4–1}$$

If the oscillator is damped at some rate κ, the term $-\kappa a(t)$ has to be added on the right-hand side, so that the equation becomes

$$\dot{a}(t) = -\mathrm{i}\omega_0 a(t) - \kappa a(t), \tag{17.4–2}$$

and this is an acceptable equation in classical physics. However, strictly speaking, the mechanism responsible for the damping will also disturb the oscillatory motion in microscopic ways, which can be modeled by the introduction of Langevin forces $F(t)$. Equation (17.4–2) then should be replaced by

$$\dot{a}(t) = -(\mathrm{i}\omega_0 + \kappa)a(t) + F(t). \tag{17.4–3}$$

The random force $F(t)$ is usually taken to have a zero mean, and it may or may not be strictly δ-correlated in time. However, its memory is always short, so that $F(t)$ is generally uncorrelated with the oscillator amplitude $a(t')$ at earlier times t', so that

$$\langle a(t')F(t)\rangle = 0, \quad \text{if } t > t'. \tag{17.4–4}$$

When we turn to a quantum mechanical oscillator we find that the loss-less oscillator obeys an equation of motion just like Eq. (17.4–1), but with $a(t)$ replaced by the Hilbert space operator $\hat{a}(t)$. On the other hand, a damped quantum oscillator cannot be described by an equation like Eq. (17.4–2), because the solution of Eq. (17.4–2) cannot satisfy the commutation relations at all times. However, Eq. (17.4–3) is valid also for a damped quantum oscillator, provided $\hat{F}(t)$ is a Hilbert space operator like $\hat{a}(t)$. This is illustrated by the discussion below.

17.4.2 Equations of motion of the quantum system

We consider a simple quantum system in the form of a harmonic oscillator at frequency ω_0, that we refer to as the *system S*. In order to subject the system to loss or damping, we suppose that it is coupled to a large number of loss or reservoir oscillators at various frequencies ω, that we collectively designate as the *reservoir R*. The reservoir is sometimes visualized as a heat bath in thermal equilibrium at some temperature T, but that is not always necessary. All that is necessary is that the reservoir be so large that its state is not very much affected by the coupling to the system. By taking every element of the combination $S + R$

to be a harmonic oscillator, we have constructed the simplest possible model of damping in a quantum system.

We take the energy of the coupled system and reservoir to be

$$\hat{H} = \hbar\omega_0(\hat{a}^\dagger\hat{a} + \tfrac{1}{2}) + \sum_\omega \hbar\omega[\hat{A}^\dagger(\omega)\hat{A}(\omega) + \tfrac{1}{2}] + \sum_\omega \hbar[g(\omega)\hat{a}^\dagger\hat{A}(\omega) + g^*(\omega)\hat{A}^\dagger(\omega)\hat{a}], \tag{17.4–5}$$

in which $\hat{a}$, $\hat{A}(\omega)$ and their conjugates are annihilation and creation operators for the system and reservoir oscillator modes, respectively, which are time dependent in the Heisenberg picture, and $g(\omega)$ is a frequency dependent coupling constant. All operators belonging to different modes commute at the same time, but not necessarily at different times, and we have of course

$$[\hat{a}(t), \hat{a}^\dagger(t)] = 1 = [\hat{A}(\omega, t), \hat{A}^\dagger(\omega, t)] \tag{17.4–6}$$

for all times t.

The time evolution of $\hat{a}(t)$ and $\hat{A}(\omega, t)$ is governed by the Heisenberg equations of motion

$$\begin{aligned}\dot{\hat{a}}(t) &= \frac{1}{i\hbar}[\hat{a}(t), \hat{H}] \\ &= -i\omega_0\hat{a}(t) - i\sum_\omega g(\omega)\hat{A}(\omega, t),\end{aligned} \tag{17.4–7}$$

and

$$\begin{aligned}\dot{\hat{A}}(\omega, t) &= \frac{1}{i\hbar}[\hat{A}(\omega, t), \hat{H}] \\ &= -i\omega\hat{A}(\omega, t) - ig^*(\omega)\hat{a}(t).\end{aligned} \tag{17.4–8}$$

The last equation is easily integrated with respect to time and yields

$$\hat{A}(\omega, t) = \hat{A}(\omega, 0)\,e^{-i\omega t} - ig^*(\omega)\,e^{-i\omega t}\int_0^t \hat{a}(t')\,e^{i\omega t'}\,dt', \tag{17.4–9}$$

and when this expression is inserted in Eq. (17.4–7) it leads to

$$\dot{\hat{a}}(t) = -i\omega_0\hat{a}(t) - i\sum_\omega g(\omega)\hat{A}(\omega, 0)\,e^{-i\omega t} - \int_0^t \sum_\omega |g(\omega)|^2\,e^{i\omega(t'-t)}\hat{a}(t')\,dt'. \tag{17.4–10}$$

Let us examine the last term more carefully. If the loss or reservoir oscillators are densely distributed, we can approximate the sum on ω by an integral. Let $\eta(\omega)\delta\omega$ be the number of oscillators lying within the frequency interval $\delta\omega$. Then we may write

$$\begin{aligned}\int_0^t \sum_\omega |g(\omega)|^2\,e^{i\omega(t'-t)}\hat{a}(t')\,dt' &\to \int_0^t dt' \int_0^\infty d\omega\,\eta(\omega)|g(\omega)|^2\,e^{i\omega(t'-t)}\hat{a}(t') \\ &= \int_0^t d\tau \int_{-\omega_0}^\infty d\omega'\,\eta(\omega_0 + \omega')|g(\omega_0 + \omega')|^2\,e^{-i\omega'\tau}\hat{a}(t - \tau)\,e^{-i\omega_0\tau}.\end{aligned} \tag{17.4–11}$$

The last line follows from the previous one by the substitutions $\omega = \omega_0 + \omega'$ and $t - t' = \tau$, which we have introduced because $\hat{a}(t)$ oscillates at frequency ω_0 to a first approximation. The combination $\hat{a}(t-\tau)\,\mathrm{e}^{-\mathrm{i}\omega_0\tau}$ is therefore a slowly varying function of τ. Now we may regard the non-negative function $\eta(\omega_0+\omega')|g(\omega_0+\omega')|^2$ as a spectral density for the loss oscillators, whose total frequency range is covered by the range of integration. By the Wiener–Khintchine theorem [Eq. (2.4–16)] the ω'-integral then gives the autocorrelation function

$$\frac{1}{2\pi}\int_{-\omega_0}^{\infty}\eta(\omega_0+\omega')|g(\omega_0+\omega')|^2\,\mathrm{e}^{-\mathrm{i}\omega'\tau}\,\mathrm{d}\omega' \equiv \Gamma(\tau), \tag{17.4–12}$$

which plays the role of a memory function for the reservoir. The memory or correlation time is determined by the spectral width of $\eta(\omega_0+\omega')|g(\omega_0+\omega')|^2$, and this memory is very short if the frequency spread is great. Indeed if $\eta(\omega_0+\omega')|g(\omega_0+\omega')|^2$ varies slowly with ω' and has a large frequency spread, and if ω_0 is near the middle of the frequency range, it may be an adequate approximation to take this term out of the integral and write

$$\begin{aligned}\Gamma(\tau) &= \frac{1}{2\pi}\int_{-\omega_0}^{\infty}\eta(\omega_0+\omega')|g(\omega_0+\omega')|^2\,\mathrm{e}^{-\mathrm{i}\omega'\tau}\,\mathrm{d}\omega' \\ &\approx \eta(\omega_0)|g(\omega_0)|^2\frac{1}{2\pi}\int_{-\omega_0}^{\infty}\mathrm{e}^{-\mathrm{i}\omega'\tau}\,\mathrm{d}\omega' \\ &\approx \eta(\omega_0)|g(\omega_0)|^2\delta(\tau),\end{aligned} \tag{17.4–13}$$

so that $\Gamma(\tau)$ reduces to a $\delta(\tau)$ function. The δ-correlated form of the reservoir is the ideal form, in the sense that the reservoir then becomes strictly Markovian without memory, and this is frequently assumed (Senitzky, 1960, 1961; Weidlich and Haake, 1965a,b; Lax, 1966, 1967; Haken, 1970; Louisell, 1973, Sec. 6.6). However, for now we shall retain the more general expression for $\Gamma(\tau)$ given by Eq. (17.4–12).

With the help of Eq. (17.4–12), Eq. (17.4–11) becomes

$$\int_0^t\sum_{\omega}|g(\omega)|^2\,\mathrm{e}^{\mathrm{i}\omega(t'-t)}\hat{a}(t')\,\mathrm{d}t' = 2\pi\int_0^t\Gamma(\tau)\hat{a}(t-\tau)\,\mathrm{e}^{-\mathrm{i}\omega_0\tau}\,\mathrm{d}\tau,$$

in which $\Gamma(\tau)$ is a strongly peaked or $\delta(\tau)$-like function, whereas $\hat{a}(t-\tau)\,\mathrm{e}^{-\mathrm{i}\omega_0\tau}$ varies slowly with τ. We may therefore look on the latter as the 'test function' on which $\Gamma(\tau)$ acts under the integral. Whether $\Gamma(\tau)$ is treated as a true $\delta(\tau)$ function or not, it is clear that by virtue of its strongly peaked form we have, to a good approximation,

$$\begin{aligned}\int_0^t\sum_{\omega}|g(\omega)|^2\,\mathrm{e}^{\mathrm{i}\omega(t'-t)}\hat{a}(t')\,\mathrm{d}t' &\approx 2\pi\hat{a}(t)\int_0^t\Gamma(\tau)\,\mathrm{d}\tau \\ &\approx \pi\hat{a}(t)\eta(\omega_0)|g(\omega_0)|^2\end{aligned} \tag{17.4–14}$$

provided t is large compared with the very short range of $\Gamma(\tau)$. The last line follows from the previous one by Fourier inversion of Eq. (17.4–12). It is convenient to set

$$\pi\eta(\omega_0)|g(\omega_0)|^2 \equiv \kappa, \tag{17.4–15}$$

which has the dimension of frequency or rate. Equation (17.4–10) then becomes, with the help of Eq. (17.4–14),

$$\dot{\hat{a}}(t) = (-\mathrm{i}\omega_0 - \kappa)\hat{a}(t) - \hat{F}(t), \tag{17.4–16}$$

where

$$\hat{F}(t) \equiv \mathrm{i}\sum_{\omega} g(\omega)\hat{A}(\omega, 0)\,\mathrm{e}^{-\mathrm{i}\omega t}. \tag{17.4–17}$$

This equation of motion for $\hat{a}(t)$ has precisely the form of the classical Langevin equation (17.4–3) with damping κ, except that the noise term $\hat{F}(t)$ is now an operator. Therefore $\hat{F}(t)$ describes a form of quantum noise. From Eq. (17.4–15) the spectral density of the quantum noise determines the damping of the system, which is another example of the *fluctuation–dissipation* theorem (see Section 17.2). Apart from the operator character of $\hat{F}(t)$, the analogy with the classical Langevin noise $F(t)$ is very close, as we shall now show.

17.4.3 Commutation relations

By definition $\hat{F}(t)$ is expressible entirely in terms of zero time $\hat{A}(\omega, 0)$ operators, so that it commutes with $\hat{a}(0)$, $\hat{a}^\dagger(0)$,

$$[\hat{F}(t), \hat{a}(0)] = 0 = [\hat{F}(t), \hat{a}^\dagger(0)], \quad (t > 0), \tag{17.4–18}$$

although it does not necessarily commute with both $\hat{a}(t)$ and $\hat{a}^\dagger(t)$, as we show below. The operator character of $\hat{F}(t)$ is in fact essential in order to preserve the commutation relation between $\hat{a}(t)$, $\hat{a}^\dagger(t)$.

Let us first examine the commutators of $\hat{F}(t)$ and $\hat{F}^\dagger(t)$. From the definition (17.4–17) it follows immediately that

$$[\hat{F}(t), \hat{F}(t')] = 0 = [\hat{F}^\dagger(t), \hat{F}^\dagger(t')] \tag{17.4–19}$$

for all t, t'. However, $\hat{F}(t)$ and $\hat{F}^\dagger(t')$ do not commute in general. It is convenient to start with the more slowly varying variables $\hat{F}(t)\,\mathrm{e}^{\mathrm{i}\omega_0 t}$ and $\hat{F}^\dagger(t)\,\mathrm{e}^{-\mathrm{i}\omega_0 t}$, that have had their oscillatory behavior canceled out, for which the commutator is given by the expression

$$\begin{aligned}
&[\hat{F}(t)\,\mathrm{e}^{\mathrm{i}\omega_0 t}, \hat{F}^\dagger(t')\,\mathrm{e}^{-\mathrm{i}\omega_0 t'}] \\
&\quad = \sum_{\omega}\sum_{\omega'} g(\omega)g^*(\omega')[\hat{A}(\omega, 0), \hat{A}^\dagger(\omega', 0)]\,\mathrm{e}^{-\mathrm{i}(\omega-\omega_0)t}\,\mathrm{e}^{\mathrm{i}(\omega'-\omega_0)t'} \\
&\quad = \sum_{\omega} |g(\omega)|^2\,\mathrm{e}^{\mathrm{i}(\omega-\omega_0)(t'-t)}.
\end{aligned}$$

We now convert the sum to an integral and put $\omega = \omega_0 + \omega'$, exactly as in the calculation leading to Eq. (17.4–11) above. Hence

$$[\hat{F}(t)\,\mathrm{e}^{\mathrm{i}\omega_0 t}, \hat{F}^\dagger(t')\,\mathrm{e}^{-\mathrm{i}\omega_0 t'}] = \int_{-\omega_0}^{\infty} \eta(\omega_0 + \omega')|g(\omega_0 + \omega')|^2\,\mathrm{e}^{-\mathrm{i}\omega'(t-t')}\,\mathrm{d}\omega',$$

or, with the help of Eq. (17.4–12),

$$[\hat{F}(t), \hat{F}^\dagger(t')] = 2\pi\Gamma(t - t')\,\mathrm{e}^{-\mathrm{i}\omega_0(t-t')}. \qquad (17.4\text{–}20)$$

The commutator therefore falls off very rapidly with increasing $|t - t'|$. To the extent that $\Gamma(\tau)$ can be approximated by a $\delta(\tau)$-function, as in Eq. (17.4–13), we can also write

$$[\hat{F}(t), \hat{F}^\dagger(t')] \approx 2\kappa\delta(t - t'), \qquad (17.4\text{–}21)$$

with κ given by Eq. (17.4–15) as before. Once again, the delta function reflects the dense nature and the large spectral range of the reservoir oscillators.

It is not difficult to show that the relation (17.4–20) or (17.4–21) ensures the proper commutation relation between $\hat{a}(t)$ and $\hat{a}^\dagger(t)$. We start by integrating Eq. (17.4–16) with respect to time, i.e.

$$\hat{a}(t) = \hat{a}(0)\,\mathrm{e}^{(-\mathrm{i}\omega_0-\kappa)t} - \mathrm{e}^{(-\mathrm{i}\omega_0-\kappa)t}\int_0^t \hat{F}(t')\,\mathrm{e}^{(\mathrm{i}\omega_0+\kappa)t'}\,\mathrm{d}t', \qquad (17.4\text{–}22)$$

and use this to construct the commutation relation

$$[\hat{a}(t), \hat{a}^\dagger(t)] = \mathrm{e}^{-2\kappa t} + \mathrm{e}^{-2\kappa t}\iint_0^t [\hat{F}(t'), \hat{F}^\dagger(t'')]\,\mathrm{e}^{\mathrm{i}\omega_0(t'-t'')}\,\mathrm{e}^{\kappa(t'+t'')}\,\mathrm{d}t'\,\mathrm{d}t''.$$

With the help of the change of variables

$$\tfrac{1}{2}(t'' + t') = t_1$$

$$t'' - t' = t_2,$$

together with Eq. (17.4–20) we obtain, as in Section 17.2,

$$[\hat{a}(t), \hat{a}^\dagger(t)] =$$

$$\mathrm{e}^{-2\kappa t} + \mathrm{e}^{-2\kappa t}\left[\int_0^{t/2}\mathrm{d}t_1\int_{-2t_1}^{2t_1}\mathrm{d}t_2 2\pi\Gamma(-t_2)\,\mathrm{e}^{2\kappa t_1} + \int_{t/2}^{t}\mathrm{d}t_1\int_{-2(t-t_1)}^{2(t-t_1)}\mathrm{d}t_2 2\pi\Gamma(-t_2)\,\mathrm{e}^{2\kappa t_1}\right].$$

Now $\Gamma(\tau)$ is a function of very short range that can be approximated by a $\delta(\tau)$-function, so that for almost all t_1 we have by Fourier inversion of Eq. (17.4–12),

$$\int_{-2t_1}^{2t_1}\Gamma(-t_2)\,\mathrm{d}t_2 \approx \int_{-\infty}^{\infty}\Gamma(t_2)\,\mathrm{d}t_2 = \eta(\omega_0)|g(\omega_0)|^2 = \kappa/\pi$$

$$\int_{-2(t-t_1)}^{2(t-t_1)}\Gamma(-t_2)\,\mathrm{d}t_2 \approx \int_{-\infty}^{\infty}\Gamma(t_2)\,\mathrm{d}t_2 = \eta(\omega_0)|g(\omega_0)|^2 = \kappa/\pi.$$

Hence

$$[\hat{a}(t), \hat{a}^\dagger(t)] = \mathrm{e}^{-2\kappa t} + \mathrm{e}^{-2\kappa t}\int_0^t 2\kappa\,\mathrm{e}^{2\kappa t_1}\,\mathrm{d}t_1 = 1, \qquad (17.4\text{–}23)$$

for all times t that are long compared with the range of $\Gamma(\tau)$. The lack of commutativity of $\hat{F}(t)$ and $\hat{F}^\dagger(t)$ is therefore essential to preserve the commutation relation connecting $\hat{a}(t)$ and $\hat{a}^\dagger(t)$, and the Langevin Eq. (17.4–16) describes our quantum system S correctly only if the reservoir R is quantum mechanical also.

Finally, let us examine some mixed commutators involving both $\hat{a}(t)$ [or $\hat{a}^\dagger(t)$] and $\hat{F}(t)$ [or $\hat{F}^\dagger(t)$]. From Eqs. (17.4–22) and (17.4–17) $\hat{a}(t)$ is expressible in terms of $\hat{a}(0)$ and $\hat{A}(\omega, 0)$, from which it follows that

$$[\hat{a}(t), \hat{F}(t')] = 0 = [\hat{a}^\dagger(t), \hat{F}^\dagger(t')] \quad \text{for all } t, t', \tag{17.4–24}$$

which is a generalization of Eq. (17.4–18). We now use Eq. (17.4–22) to calculate $[\hat{a}(t), \hat{F}^\dagger(t')]$, and we find that

$$[\hat{a}(t), \hat{F}^\dagger(t')] = -\mathrm{e}^{(-\mathrm{i}\omega_0-\kappa)t}\int_0^t [\hat{F}(t''), \hat{F}^\dagger(t')]\,\mathrm{e}^{(\mathrm{i}\omega_0+\kappa)t''}\,\mathrm{d}t'',$$

and with the help of Eq. (17.4–20)

$$\begin{aligned}[\hat{a}(t), \hat{F}^\dagger(t')] &= -2\pi\,\mathrm{e}^{(-\mathrm{i}\omega_0-\kappa)t}\int_0^t \Gamma(t''-t')\,\mathrm{e}^{\kappa t''+\mathrm{i}\omega_0 t'}\,\mathrm{d}t'' \\ &= -2\pi\,\mathrm{e}^{(\mathrm{i}\omega_0+\kappa)(t'-t)}\int_{-t'}^{t-t'} \Gamma(\tau)\,\mathrm{e}^{\kappa\tau}\,\mathrm{d}\tau.\end{aligned}$$

Now $\Gamma(\tau)$ is a very short-range function that can be approximated by a $\delta(\tau)$-function. Provided $t > t'$, the τ-integral is well approximated for almost all t, t' by the relation

$$\int_{-t'}^{t-t'} \Gamma(\tau)\,\mathrm{e}^{\kappa\tau}\,\mathrm{d}\tau \approx \int_{-\infty}^{\infty} \Gamma(\tau)\,\mathrm{d}\tau = \kappa/\pi,$$

whereas

$$\int_{-t'}^{t-t'} \Gamma(\tau)\,\mathrm{e}^{\kappa\tau}\,\mathrm{d}\tau \approx 0 \quad \text{if } t < t',$$

because the peak of the $\Gamma(\tau)$ function then lies outside the range of integration. When $t = t'$ it is to be expected that the integral covers half the area under $\Gamma(\tau)$. We then obtain the result

$$\left.\begin{aligned}[\hat{a}(t), \hat{F}^\dagger(t')] &= -2\kappa\,\mathrm{e}^{-(\mathrm{i}\omega_0+\kappa)(t-t')} && \text{if } t > t', \\ &= -\kappa && \text{if } t = t', \\ &= 0 && \text{if } t < t'.\end{aligned}\right\} \tag{17.4–25}$$

Similarly it may be shown that

$$\left.\begin{aligned}[\hat{a}^\dagger(t), \hat{F}(t')] &= 2\kappa\,\mathrm{e}^{(\mathrm{i}\omega_0-\kappa)(t-t')} && \text{if } t > t', \\ &= \kappa && \text{if } t = t', \\ &= 0 && \text{if } t < t'.\end{aligned}\right\} \tag{17.4–26}$$

As a zero value of a commutator generally implies that the two dynamical variables can be measured without influencing each other, we conclude that the system is influenced by the reservoir at earlier times but not at later times. In this respect the quantum Langevin equation behaves very much like the corresponding classical equation.

17.4.4 Two-time correlation functions

So far we have examined relations among the various dynamical variables without introducing the quantum states. However, the physics is ultimately described by quantum mechanical expectation values, for which the state has to be specified. Let us then assume that the state of system plus reservoir is expressible as a direct product of states characterizing the system S and the reservoir R, either or both of which can be in a mixed state represented by a density operator. In addition, we assume that the reservoir oscillators are in states that satisfy the relations

$$\left.\begin{aligned} \langle \hat{A}(\omega, 0) \rangle &= 0 \\ \langle \hat{A}(\omega, 0)\hat{A}(\omega', 0) \rangle &= 0 \\ \langle \hat{A}^\dagger(\omega, 0)\hat{A}(\omega', 0) \rangle &= \delta_{\omega\omega'} N(\omega), \end{aligned}\right\} \qquad (17.4\text{–}27)$$

where $N(\omega)$ is the average number of the reservoir oscillators at frequency ω. These conditions are satisfied if all the reservoir modes are independent, and if the density operator for each mode is diagonal in the number of excitations. Sometimes it is even assumed that the reservoir oscillators are in thermal equilibrium at temperature T, in which case $N(\omega) = [\exp(\hbar\omega/k_B T) - 1]^{-1}$. However, these are merely examples of states that satisfy the conditions (17.4–27), which are sometimes characterized as the *random phase assumption*.

From Eqs. (17.4–17) and (17.4–27) it then follows immediately that

$$\langle \hat{F}(t) \rangle = 0, \qquad (17.4\text{–}28)$$

and

$$\langle \hat{F}(t)\hat{F}(t') \rangle = 0 = \langle \hat{F}^\dagger(t)\hat{F}^\dagger(t') \rangle, \qquad (17.4\text{–}29)$$

just as for classical Langevin forces. Of more interest is the two-time correlation function

$$\begin{aligned} \langle \hat{F}^\dagger(t)\hat{F}(t') \rangle &= \sum_{\omega}\sum_{\omega'} g^*(\omega) g(\omega') \langle \hat{A}^\dagger(\omega, 0)\hat{A}(\omega', 0) \rangle \, e^{i(\omega t - \omega' t')} \\ &= \sum_{\omega} |g(\omega)|^2 N(\omega) \, e^{i\omega(t-t')}, \end{aligned}$$

where we have used Eq. (17.4–27). We now replace the sum by an integral by introducing the density $\eta(\omega)$, and put $\omega = \omega_0 + \omega'$, as before. We then find that

$$\langle \hat{F}^\dagger(t)\hat{F}(t') \rangle = e^{i\omega_0(t-t')} \int_{-\omega_0}^{\infty} N(\omega_0 + \omega') |g(\omega_0 + \omega')|^2 \eta(\omega_0 + \omega') \, e^{i\omega'(t-t')}.$$

The integrand is dominated by the oscillatory factor $e^{i\omega'(t-t')}$, and the integral has the same form as that in Eq. (17.4–12), except for the factor $N(\omega_0 + \omega')$. If this is a slowly varying function of ω' in the neighborhood of $\omega' = 0$, as we shall assume, we are justified in pulling the factor $N(\omega_0)$ outside the integral, which then reduces to that in Eq. (17.4–12), and yields the result

$$\langle \hat{F}^\dagger(t)\hat{F}(t') \rangle = 2\pi \, e^{i\omega_0(t-t')} N(\omega_0) \Gamma(t' - t). \qquad (17.4\text{–}30a)$$

The autocorrelation function of the Langevin forces therefore has the same structure as their commutator given by Eq. (17.4–20), and it also has a very short range in time. To the extent that it is permissible to approximate $\Gamma(\tau)$ by a $\delta(\tau)$-function as in Eq. (17.4–13), we also have the relation

$$\langle \hat{F}^\dagger(t)\hat{F}(t') \rangle \approx 2\kappa N(\omega_0) \delta(t - t'). \qquad (17.4\text{–}30b)$$

The random forces therefore have a short memory and are Markovian. Once again we note that the same constant κ that governs the damping of the system also governs the fluctuations of the quantum Langevin forces, as required by the fluctuation–dissipation theorem. It follows immediately with the help of Eq. (17.4–20), that when the operators are expressed in antinormal order we obtain, instead of Eqs. (17.4–30),

$$\langle \hat{F}(t)\hat{F}^\dagger(t')\rangle = 2\pi\, e^{i\omega_0(t'-t)}[N(\omega_0)+1]\Gamma(t-t') \tag{17.4–31a}$$

$$\approx 2\kappa[N(\omega_0)+1]\delta(t-t'). \tag{17.4–31b}$$

Higher-order correlations are also readily calculated from Eq. (17.4–17) if it is assumed that the reservoir is in thermal equilibrium at temperature T. For example, for normally ordered correlations one may show that

$$\langle \hat{F}^\dagger(t_1)\hat{F}^\dagger(t_2)\ldots\hat{F}^\dagger(t_N)\hat{F}(t'_M)\ldots\hat{F}(t'_2)\hat{F}(t'_1)\rangle =$$

$$\delta_{NM}\sum_{\mathbb{P}}\langle \hat{F}^\dagger(t_1)\hat{F}(t'_1)\rangle\langle \hat{F}^\dagger(t_2)\hat{F}(t'_2)\rangle\ldots\langle \hat{F}^\dagger(t_N)\hat{F}(t'_N)\rangle,$$

where $\sum_{\mathbb{P}}$ stands for the sum over all $N!$ pairings of the $\hat{F}^\dagger$ and $\hat{F}$ operators. This equation will be recognized as the Gaussian moment theorem [Eq. (1.6–33)], and it suggests that the quantum noise $\hat{F}(t)$ may be regarded as a Gaussian process. From Eqs. (17.4–28) and onward it is apparent that expectations of the reservoir operators $\hat{F}(t)$ are independent of the state of the system oscillator S, which is consistent with the assumption that the reservoir R is large and barely influenced by the system S.

Next we shall examine some correlations between system and reservoir operators. From Eqs. (17.4–17) and (17.4–22) we have

$$\langle \hat{a}(t)\hat{F}(t')\rangle = i\sum_{\omega} g(\omega)\langle \hat{a}(0)\hat{A}(\omega,0)\rangle\, e^{-i(\omega_0 t+\omega t')-\kappa t}$$

$$-\int_0^t \langle \hat{F}(t'')\hat{F}(t')\rangle\, e^{(i\omega_0+\kappa)(t''-t)}\,dt''.$$

The integral vanishes by virtue of Eq. (17.4–29), and, because of the assumed product state, the expectation $\langle \hat{a}(0)\hat{A}(\omega,0)\rangle$ factorizes into the product of two expectations and vanishes by virtue of Eq. (17.4–27). We therefore have

$$\langle \hat{a}(t)\hat{F}(t')\rangle = 0 = \langle \hat{a}^\dagger(t)\hat{F}^\dagger(t')\rangle. \tag{17.4–32}$$

On the other hand, with the help of Eq. (17.4–30a) the same procedure gives

$$\langle \hat{a}^\dagger(t)\hat{F}(t')\rangle = i\sum_{\omega} g(\omega)\langle \hat{a}^\dagger(0)\hat{A}(\omega,0)\rangle\, e^{i(\omega_0 t-\omega t')-\kappa t}$$

$$-\int_0^t \langle \hat{F}^\dagger(t'')\hat{F}(t')\rangle\, e^{(-i\omega_0+\kappa)(t''-t)}\,dt''$$

$$= -2\pi N(\omega_0)\, e^{i\omega_0(t-t')}\int_0^t e^{\kappa(t''-t)}\Gamma(t'-t'')\,dt''$$

$$= -2\pi N(\omega_0)\, e^{(i\omega_0-\kappa)(t-t')}\int_{t'-t}^{t'} \Gamma(\tau)\, e^{-\kappa\tau}\,d\tau.$$

Because $\Gamma(\tau)$ behaves almost like a delta function, we obtain different answers according to the sign of $t' - t$. If $t > t'$, then $t' - t < 0$, and for almost all values of t, t' we may approximate the integral as before, namely

$$\int_{t'-t}^{t'} \Gamma(\tau)\,\mathrm{e}^{-\kappa\tau}\,\mathrm{d}\tau \approx \int_{-\infty}^{\infty} \Gamma(\tau)\,\mathrm{d}\tau = \kappa/\pi.$$

On the other hand, if $t' > t$, then the range of integration does not include the peak of $\Gamma(\tau)$, and for almost all t, t' the integral is vanishingly small. When $t = t'$, the integral may be assumed to capture half the area under the $\Gamma(\tau)$ function. Hence we have finally,

$$\left.\begin{aligned} \langle \hat{a}^\dagger(t)\hat{F}(t')\rangle &= -2\kappa N(\omega_0)\,\mathrm{e}^{(\mathrm{i}\omega_0-\kappa)(t-t')} && \text{if } t > t', \\ &= -\kappa N(\omega_0) && \text{if } t = t', \\ &= 0 && \text{if } t < t', \end{aligned}\right\} \tag{17.4–33}$$

and it may similarly be shown that

$$\left.\begin{aligned} \langle \hat{a}(t)\hat{F}^\dagger(t')\rangle &= -2\kappa[N(\omega_0)+1]\,\mathrm{e}^{(-\mathrm{i}\omega_0-\kappa)(t-t')} && \text{if } t > t', \\ &= -\kappa[N(\omega_0)+1] && \text{if } t = t', \\ &= 0 && \text{if } t < t'. \end{aligned}\right\} \tag{17.4–34}$$

These relations imply that the system variables $\hat{a}(t)$, $\hat{a}^\dagger(t)$ are always uncorrelated with the quantum Langevin forces at times greater than t, although not at earlier times. In this respect also the Langevin forces behave like the corresponding classical ones.

These conclusions are not limited to $\hat{a}(t)$ and $\hat{F}(t)$ and their conjugates. For example, by using Eq. (17.4–22) to construct $\hat{n}(t) = \hat{a}^\dagger(t)\hat{a}(t)$, one may also show that

$$\left.\begin{aligned} \langle \hat{n}(t)\hat{F}(t')\rangle &= -2\kappa N(\omega_0)\langle \hat{a}(0)\rangle\,\mathrm{e}^{\kappa(t'-2t)}\,\mathrm{e}^{-\mathrm{i}\omega_0 t'} && \text{if } t > t', \\ &= -\kappa N(\omega_0)\langle \hat{a}(0)\rangle\,\mathrm{e}^{-\kappa t}\,\mathrm{e}^{-\mathrm{i}\omega_0 t} && \text{if } t = t', \\ &= 0 && \text{if } t < t', \end{aligned}\right\} \tag{17.4–35}$$

provided all triple correlations of the type $\langle \hat{F}^\dagger(t')\hat{F}(t'')\hat{F}(t''')\rangle$ vanish.

17.4.5 Langevin equation for the excitation of the system oscillator

In the foregoing we have demonstrated that the Heisenberg equation of motion for the system variable $\hat{a}(t)$ can be cast into the form of the Langevin equation (17.4–16), with the reservoir variables providing the quantum mechanical Langevin forces. However, lest it be thought that the approach is limited to the variables $\hat{a}(t)$, $\hat{a}^\dagger(t)$, we now show that an equation of motion of the Langevin type may readily be obtained for other operators, such as the Hermitian operator $\hat{n}(t) = \hat{a}^\dagger(t)\hat{a}(t)$, also. From the definition and Eq. (17.4–16) we have

$$\begin{aligned} \dot{\hat{n}}(t) &= \dot{\hat{a}}^\dagger(t)\hat{a}(t) + \hat{a}^\dagger(t)\dot{\hat{a}}(t) \\ &= -2\kappa\hat{n}(t) - \hat{a}^\dagger(t)\hat{F}(t) - \hat{F}^\dagger(t)\hat{a}(t). \end{aligned} \tag{17.4–36}$$

Let us introduce a new 'quantum noise' $\hat{f}(t)$ of zero expectation by writing

$$\hat{f}(t) \equiv \hat{a}^\dagger(t)\hat{F}(t) + \hat{F}^\dagger(t)\hat{a}(t) - \langle \hat{a}^\dagger(t)\hat{F}(t) + \hat{F}^\dagger(t)\hat{a}(t)\rangle. \quad (17.4\text{–}37)$$

We make use of Eq. (17.4–22) to evaluate the last term and find, with the help of exactly the same argument as that leading to Eq. (17.4–33),

$$\begin{aligned}\langle \hat{a}^\dagger(t)\hat{F}(t)\rangle + \text{c.c.} &= -\mathrm{e}^{(\mathrm{i}\omega_0-\kappa)t}\int_0^t \langle \hat{F}^\dagger(t')\hat{F}(t)\rangle\, \mathrm{e}^{(-\mathrm{i}\omega_0+\kappa)t'}\,\mathrm{d}t' + \text{c.c.} \\ &= -2\kappa N(\omega_0). \end{aligned} \quad (17.4\text{–}38)$$

Equation (17.4–36) then becomes

$$\dot{\hat{n}}(t) = -2\kappa[\hat{n}(t) - N(\omega_0)] - \hat{f}(t), \quad (17.4\text{–}39)$$

which is in the form of a Langevin equation for $\hat{n}(t) - N(\omega_0)$. On integrating with respect to time and taking expectation values, we immediately have the solution

$$\langle \hat{n}(t)\rangle - N(\omega_0) = [\langle \hat{n}(0)\rangle - N(\omega_0)]\,\mathrm{e}^{-2\kappa t}, \quad (17.4\text{–}40)$$

which shows that $\langle \hat{n}(t)\rangle \to N(\omega_0)$ as $t \to \infty$, when the system comes into equilibrium with the reservoir oscillators. With the help of Eq. (17.4–22) $\hat{f}(t)$ may be expressed in terms of the $\hat{F}(t)$ Langevin operators in the form

$$\hat{f}(t) = \hat{a}^\dagger(0)\hat{F}(t)\,\mathrm{e}^{(\mathrm{i}\omega_0-\kappa)t} - \mathrm{e}^{(\mathrm{i}\omega_0-\kappa)t}\int_0^t \hat{F}^\dagger(t')\hat{F}(t)\,\mathrm{e}^{(-\mathrm{i}\omega_0+\kappa)t'}\,\mathrm{d}t' + \text{h.c.} + 2\kappa N(\omega_0), \quad (17.4\text{–}41)$$

from which it is apparent that two-time correlations between the $\hat{f}$'s involve four-time correlations among the $\hat{F}$-operators. The higher-order correlations are readily evaluated if it is assumed that the reservoir oscillators are in thermal equilibrium for example, so that the normally ordered moments of $\hat{F}(t)$ obey the Gaussian moment theorem. To the extent that $\hat{F}(t)$ is delta-correlated in time, one then finds that $\hat{f}(t)$ is delta-correlated also.

17.4.6 *Irreversibility and the arrow of time*

We started by considering a simple quantum oscillator S coupled to a quantum reservoir R, with total energy $\hat{H}$ given by Eq. (17.4–5). According to quantum mechanics, such a coupled system undergoes Hamiltonian evolution, with the time evolution operator

$$\hat{U}(t) = \exp(-\mathrm{i}\hat{H}t/\hbar). \quad (17.4\text{–}42)$$

Then the state $|\psi(t)\rangle$ at time t in the Schrödinger picture takes the form

$$|\psi(t)\rangle = \hat{U}(t)|\psi(0)\rangle, \quad (17.4\text{–}43)$$

and the dynamical variable $\hat{Q}(t)$ in the Heisenberg picture becomes

$$\hat{Q}(t) = \hat{U}^\dagger(t)\hat{Q}(0)\hat{U}(t). \quad (17.4\text{–}44)$$

Since $\hat{U}(t)$ is unitary, the time development of the expectation of any dynamical variable then should be strictly reversible in time. Yet it is clear from Eqs. (17.4–22) or (17.4–33) or (17.4–40) that the oscillator S is damped in time, and that irreversibility has crept into the solution. How has this come about, and what determines the direction of the arrow of time?

A key element in the calculation is the so-called *random phase assumption* at time $t = 0$, which is embodied in Eqs. (17.4–27) and in the higher-order moments of $\hat{A}(\omega, 0)$. It implies that the different reservoir oscillators are independent, both of each other and of the system S. However, because of the coupling between the various modes introduced by the Hamiltonian $\hat{H}$, correlations must gradually develop as time goes on. It is easy to verify that relations like (17.4–27) cannot strictly hold at later times t if they hold at $t = 0$. The random phase assumption is therefore the key element governing the direction of time. If at some time t the direction of time were to be reversed, the initial (i.e. $t = 0$) conditions would be recovered only if the loss oscillators were correlated at time t. This may suggest that the instant $t = 0$ plays a unique role. Actually, if the reservoir is sufficiently large and contains many modes, the correlations that are introduced after some time t can be so small that the state of the reservoir at time t in the Schrödinger picture need not differ significantly from its state at $t = 0$. Formally, this can be achieved by taking the reservoir to be in thermal equilibrium at all times, which, in turn, implies an infinite reservoir. The infinite number of degrees of freedom give a direction to the arrow of time.

Problems

17.1 The fluctuating real electromotive force $V(t)$ across the ends of a resistance R at temperature T has the n'th order correlation function

$$\langle V(t_1)V(t_2)\ldots V(t_n)\rangle = 0, \qquad \text{when } n \text{ is odd,}$$

$$= \sum (2kTR)^{n/2}\delta(t_1 - t_2)\delta(t_3 - t_4)\ldots\delta(t_{n-1} - t_n),$$

$$\text{when } n \text{ is even,}$$

where the sum is taken over all possible pairings of the n different times. The resistance R is connected across an inductance L. Using only the information provided, calculate the probability distribution and the spectral density of the current flowing in the R, L circuit in the steady state.

17.2 A quantum harmonic oscillator of frequency ω_0 and amplitude $\hat{a}(t)$ is coupled to a set of quantum loss oscillators of various frequencies ω and amplitudes $\hat{A}(\omega, t)$. The total energy of the system is given by

$$\hat{H} = \hbar\omega_0(\hat{a}^\dagger\hat{a} + \tfrac{1}{2}) + \sum_{\omega}\hbar\omega[\hat{A}^\dagger(\omega)\hat{A}(\omega) + \tfrac{1}{2}] + \sum_{\omega}\hbar[g(\omega)\hat{a}^\dagger\hat{A}(\omega) + \text{h.c.}].$$

Show that the equation of motion of $\hat{a}(t)$ can be expressed in the form of a Langevin equation

$$\dot{\hat{a}}(t) = (-\mathrm{i}\omega_0 - \kappa)\hat{a}(t) - \hat{F}(t),$$

with quantum noise $\hat{F}(t)$, and evaluate the commutator $[\hat{F}(t), \hat{F}^\dagger(t')]$.

17.3 A quantum system is undergoing Hamiltonian evolution. The density matrix at time t in the Schrödinger picture is $\rho_{\mathrm{nm}}(t)$. Under what conditions does the probability $p_n(t)$ of being in the state $|n\rangle$ at time t satisfy the equation

$$p_n(t) = \sum_m \mathcal{P}(n, t|m, t_0) p_m(t_0), \qquad (t \geqslant t_0),$$

in which $\mathcal{P}(n, t|m, t_0)$ is the transition probability?

17.4 For the system in Problem 17.3, show that the probability amplitude $\langle n|\exp(-\mathrm{i}\hat{H}t/\hbar)|m\rangle$ satisfies a relation of the Smoluchowski–Chapman–Kolmogorov type.

17.5 If the quantum system in Problem 17.3 is a harmonic oscillator with energy eigenstates $|n\rangle$, and it is subject to some perturbation $\hat{V}(t)$ which is off-diagonal in $|n\rangle$, calculate the transition probability $\mathcal{P}(n, t+\Delta t|m, t)$ to the second order in Δt.

18

The single-mode laser

18.1 Introduction

The idea of making use of the phenomenon of stimulated emission from an atom or molecule for amplification of the electromagnetic field, and then combining the amplifier with a resonator to make an oscillator, is due to Townes and his co-workers (Gordon, Zeiger and Townes, 1954, 1955) and independently to Basov and Prokhorov (1954, 1955). The former constructed the first MASER, or *microwave amplifier by stimulated emission of radiation*. In 1958 Schawlow and Townes proposed an application of the same principle to the optical domain (Schawlow and Townes, 1958), in which a two-mirror Fabry–Perot interferometer serves as the optical resonator and an excited group of atoms as the gain medium. The first LASER, or *light amplifier by stimulated emission of radiation*,‡ was constructed by Maiman (1960). The gain medium was ruby, which was excited by a bright flash of light from a gas discharge tube, whereupon the laser delivered a short optical pulse. The first continuously operating – or CW – laser was developed by Javan and his co-workers (Javan, Bennett and Herriott, 1961); it made use of a He:Ne gas mixture, that was continuously excited by a discharge, as the gain medium. This type of laser is still widely used today. Since that time many different kinds of laser have been developed, ranging in wavelength from the infrared to the ultraviolet. Some produce light at several different frequencies at once, and some are tunable over a wide range. The radiation that lasers emit can be highly coherent, and, except for the wavelength, of the same general character as the radiation from a radio frequency oscillator. Indeed, the quantum state of the laser field can be close to the coherent state that we discussed in Chapter 11. Moreover, as the electromagnetic energy is generally concentrated in one mode, or in a small number of modes, the photon occupation number per mode can be exceedingly large (Mandel, 1961). As a result, the radiation field produced by a laser can come closer to being classical than that from just about any other light source.

We can identify four common elements in nearly all lasers:

(a) an optical resonator, generally formed by two or more mirrors;

‡ The A in MASER and LASER is sometimes taken to stand for 'amplification' rather than 'amplifier', in which case MASER and LASER become processes rather than devices.

(b) a gain medium in which an inverted atomic population between the laser energy levels is established;‡
(c) an optical pump or energy source to excite the gain medium;
(d) a loss mechanism by which energy is dissipated or dispersed.

Figure 18.1 shows a typical form of laser, in which the resonator is a Fabry–Perot interferometer, and the amplifier is a gas plasma tube wherein a discharge is maintained. To reduce reflection losses from the plasma tube, its end windows are generally arranged at the Brewster angle for linearly polarized light at the laser frequency. The end mirrors are usually provided with multi-layer dielectric coatings to make them highly reflecting. Of course, the output mirror needs to have less than 100% reflectivity, so that it is commonly the main source of the energy loss that has to be compensated by the gain medium. The cavity mirrors play the important role of feeding photons belonging to the laser modes back into the laser cavity. Most of the spontaneously emitted photons traveling in various other directions are lost. However, photons associated with a cavity resonance mode interact repeatedly with the atoms of the gain medium, and their number grows through stimulated emission, as illustrated in Fig. 18.2. Once one mode is sufficiently populated, the probability for stimulated emission into that mode exceeds the spontaneous emission probability (cf. Section 15.4).

In general, when the rate at which photons are fed into the optical cavity mode exceeds the rate at which they are lost from the cavity by the loss mechanism, the amplitude of the laser field starts to grow until a steady state is reached. At that point the rate of radiation by the laser equals the net rate at which energy is supplied. It is easy to see that this is achievable by an inverted population between the two atomic laser levels, with more atoms in the upper laser state than in the lower. For if N_2, N_1 are the upper state and lower state populations, the rate of absorption of laser photons by the atomic system is proportional to N_1,

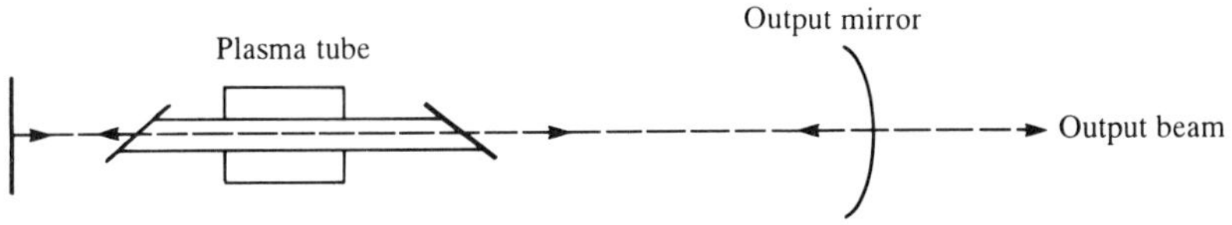

Fig. 18.1 A simple form of laser.

Fig. 18.2 Illustrating the growth of the stimulated emission probability with the intensity of the mode.

‡ Lasers that operate without population inversion have also been proposed and investigated more recently (Kocharovskaya and Khanin, 1988; Harris, 1989; Scully, Zhu and Gavrielides, 1989; Imamoglu, 1989; Imamoglu and Harris, 1989; Agarwal, Ravi and Cooper, 1990; Narducci, Doss, Ru, Scully, Zhu and Keitel, 1991). We shall not consider them here.

and the rate of stimulated emission of laser photons by the system is proportional to N_2, with the same constant of proportionality for both (cf. Section 15.4). If N_2 exceeds N_1 sufficiently all the radiation losses can be made good by the atomic system.

18.1.1 Condition for laser action

It is not difficult to arrive at a more quantitative, but still very approximate, condition for laser oscillation with an inverted population. If there are N_2 excited atoms, the total rate of spontaneous photon emission is AN_2 from Eq. (15.4–9), where A is the Einstein A-coefficient for the transition being considered. These emissions span M distinct cavity modes, all of which fall within the natural linewidth A. From the known density of cavity modes $V/4\pi^3$ for both polarizations (cf. Chapter 10), in which V is the cavity volume, it follows that [cf. the calculation leading to Eq. (13.1–15)]

$$M \approx \frac{V}{4\pi^3}\left(\frac{\omega_0^2 A}{c^3}\right)4\pi = \frac{\omega_0^2 VA}{\pi^2 c^3}, \tag{18.1–1}$$

where ω_0 is the frequency of the atomic transition. Actually the modes of an open Fabry–Perot type of cavity are not the plane-wave modes that were used in Chapter 10. The Fabry–Perot modes have a more complicated structure and were treated by Fox and Li (1961) and others (Boyd and Gordon, 1961; Kogelnik and Li, 1966). For the moment we shall ignore this complication.

The rate of spontaneous photon emission into one cavity mode is therefore

$$\frac{AN_2}{M} = \frac{\pi^2 c^3}{\omega_0^2 V} N_2.$$

The rate of stimulated emission into the same mode equals the rate of spontaneous emission multiplied by the photon occupation number n of that mode (cf. Section 15.4),

$$\begin{matrix}\text{rate of stimulated emission by excited} \\ \text{atoms into the given cavity mode}\end{matrix} = \frac{\pi^2 c^3 N_2 n}{\omega_0^2 V}. \tag{18.1–2}$$

We can immediately write a similar expression for the rate of absorption of photons from the cavity mode by the lower-state laser atoms, by replacing N_2 in Eq. (18.1–2) by N_1; the difference between the two rates is the net rate of photon emission. For laser oscillation to take place, this rate must exceed or equal the rate n/T_L at which photons are lost from the given cavity mode, so that

$$\frac{n\pi^2 c^3 (N_2 - N_1)}{\omega_0^2 V} \geqslant \frac{n}{T_L}. \tag{18.1–3}$$

T_L is the average lifetime of a photon in the cavity mode, which can be estimated from the cavity length l and the reflectivity R of the laser output mirror, when transmission through the mirror is the principal source of photon loss. We may then regard $(1 - R)$ as the probability that a photon is lost whenever it impinges on the output mirror, and when this probability is multiplied by $c/2l$, the rate at which a cavity photon strikes the mirror, we obtain the rate of emission $1/T_L$ per photon,

$$1/T_L = (c/2l)(1 - R). \tag{18.1–4}$$

From Eqs. (18.1–3) and (18.1–4) we arrive at the following condition for laser oscillation in one cavity mode,

$$N_2 - N_1 \geqslant \frac{\omega_0^2 \mathscr{A}}{2\pi^2 c^2}(1 - R) = \frac{2\mathscr{A}}{\lambda^2}(1 - R), \tag{18.1–5}$$

where $\mathscr{A}$ is the cross-sectional area of the laser and λ is the wavelength. It is clear that the greater the transmissivity $1 - R$ of the output mirror, the greater is the atomic inversion $N_2 - N_1$ required for laser action to take place. An equivalent form of Eq. (18.1–5) that is sometimes useful makes use of the so-called *cavity Q-factor* defined by the formula

$$Q \equiv \frac{2\pi l/\lambda}{1 - R}. \tag{18.1–6}$$

Roughly speaking, Q is a measure of the total phase of a damped oscillation that elapses before the oscillation is reduced e times (2.718 times) in amplitude. If we introduce the densities ρ_2, ρ_1 of upper- and lower-state atoms given by $\rho_i = N_i/\mathscr{A}l (i = 1, 2)$, we can re-write Eq. (18.1–5) in the equivalent form

$$(\rho_2 - \rho_1)\lambda^3 \geqslant 4\pi/Q, \tag{18.1–7}$$

which shows that the atomic inversion within a wavelength cube must exceed 4π divided by the cavity Q-factor. These conditions make it possible to estimate the rate of optical pumping that is required for laser action, but they should not be regarded as precise.

The inequalities (18.1–5) and (18.1–7) of course tell us nothing about how the laser field develops. The most important early contributions to the dynamics of the laser were made by Lamb (1964) and Haken (1964); see also Haken and Sauermann (1963a,b). We begin by outlining the approach used by Lamb, in which the radiation field is treated classically throughout.‡

18.2 Semiclassical theory of the laser

In the Lamb theory (Lamb, 1964), the laser field is introduced in a self-consistent manner, and an equation of motion for the field amplitude is obtained. The argument can be roughly summarized as follows:

(a) It is assumed that there are atoms, with energy separation $\hbar\omega_0$ between two laser levels, contained within an optical cavity that has been tuned to a resonance frequency close to ω_0, and that the atomic population has been inverted with respect to these two levels.

(b) It is assumed that there exists a classical laser field $\mathbf{E}$ of frequency ω_0, or close to ω_0, within the cavity.

‡ For a good general introduction to lasers see the books by Siegman (1986) and by Milonni and Eberly (1988), which also contain a wealth of detail on specific lasers.

(c) The field acts on each excited atom and induces a dipole moment through the quantum mechanical equations of motion.
(d) The sum of all the elementary atomic dipole moments constitutes a macroscopic polarization **P** of the gain medium.
(e) The polarization **P** acts as the source term in Maxwell's equations for the electromagnetic field and gives rise to the laser **E**-field.
(f) This **E**-field is equated to the field that was assumed to exist in (b).

In a realistic treatment of the laser problem it is necessary to take into account the natural atomic lineshape, and any motion of the atoms that results in inhomogeneous broadening. Although this brings out some interesting features of the laser, such as the *Lamb dip* in the spectrum (Lamb, 1964), it complicates the calculation, and detracts somewhat from an understanding of the basic laser mechanism. We shall therefore discard these complications and concentrate on the essence of the laser process, by treating the atoms as identical, sharply resonant, two-level quantum systems.

In a semiclassical theory, the losses of the electromagnetic field from the cavity are modeled most easily by the artifice of introducing a conducting medium, that naturally causes attenuation of the field within the cavity in time. If σ is the ohmic conductivity of this artificial medium, then the current density $\mathbf{j}$ in the cavity can be replaced by $\sigma\mathbf{E}$, and the Maxwell equations for the field take the form, in SI units,

$$\left.\begin{aligned} \boldsymbol{\nabla}\cdot\mathbf{D} &= 0 \\ \boldsymbol{\nabla}\cdot\mathbf{B} &= 0 \\ \boldsymbol{\nabla}\times\mathbf{E} &= -\frac{\partial\mathbf{B}}{\partial t} \\ \boldsymbol{\nabla}\times\mathbf{B} &= \mu_0\sigma\mathbf{E} + \mu_0\frac{\partial\mathbf{D}}{\partial t}, \end{aligned}\right\} \tag{18.2–1}$$

where $\mathbf{D} = \varepsilon_0\mathbf{E} + \mathbf{P}$.

If we take the curl of the third of these equations, and then use the fourth one to eliminate **B** in the usual way, we arrive at the following equation of motion for the laser field **E**,

$$\boldsymbol{\nabla}\times(\boldsymbol{\nabla}\times\mathbf{E}) + \mu_0\sigma\frac{\partial\mathbf{E}}{\partial t} + \frac{1}{c^2}\frac{\partial^2\mathbf{E}}{\partial t^2} + \mu_0\frac{\partial^2\mathbf{P}}{\partial t^2} = 0. \tag{18.2–2}$$

A number of simplifications are now generally made. We first suppose that the field is polarized, as it usually is in practice, and that both **E** and **P** can be treated as scalars. Moreover, the principal spatial variation of E will be in the direction of the laser axis, which we take to be the z-axis, so that $\boldsymbol{\nabla}\times(\boldsymbol{\nabla}\times\mathbf{E})$ can be approximated by $-\partial^2E/\partial z^2$. Then Eq. (18.2–2) becomes

$$-\frac{\partial^2E}{\partial z^2} + \mu_0\sigma\frac{\partial E}{\partial t} + \frac{1}{c^2}\frac{\partial^2E}{\partial t^2} + \mu_0\frac{\partial^2P}{\partial t^2} = 0. \tag{18.2–3}$$

Next we expand $E(\mathbf{r}, t)$ in terms of the normal modes of the optical cavity, and make a slowly varying amplitude approximation.

18.2.1 Normal modes of a cavity

The *normal modes* of an empty cavity can be defined as the solutions of the Helmholtz equation

$$(\nabla^2 + \omega^2/c^2)u(\mathbf{r}) = 0$$

satisfying appropriate boundary conditions. In general there exists a discrete set of eigenfrequencies ω_n and an orthogonal set of eigenfunctions $u_n(\mathbf{r})$, which are conveniently labeled by some index n. We can then expand any electric field $E(\mathbf{r}, t)$ in the cavity in normal modes by writing

$$E(\mathbf{r}, t) = \sum_n \mathscr{E}_n(t)u_n(\mathbf{r})\,\mathrm{e}^{-\mathrm{i}\omega_n t} + \text{c.c.}, \tag{18.2-4}$$

in which the coefficients $\mathscr{E}_n(t)$ are complex mode amplitudes. By explicitly introducing the oscillatory factor $\mathrm{e}^{-\mathrm{i}\omega_n t}$ we ensure that $\mathscr{E}_n(t)$ is a relatively slowly varying amplitude. In general n stands collectively for three integers.

However, most laser cavities are open; they are not bounded by closed surfaces, but by two mirrors with less than perfect reflectivities, and the formal definition of a mode is then not strictly applicable. Nevertheless it is customary to speak of cavity modes – or quasi-modes – in the sense that, after many reflections, additional reflections have very little effect on the field distribution in the cavity of a quasi-mode. This condition has been used to derive the approximate form of the cavity mode functions for a two-mirror cavity (Fox and Li, 1961; Boyd and Gordon, 1961; cf also Section 7.4).

When the geometry is simple, it is easy to find approximate forms of $u_n(\mathbf{r})$. For example, for an optical cavity formed by plane mirrors located at positions $z = 0$ and $z = l$, an obvious approximation to a set of frequencies ω_n and mode functions $u_n(\mathbf{r})$ would be

$$\left.\begin{aligned} \omega_n &= n\pi c/l \\ u_n(\mathbf{r}) &= \sin(\pi n z/l). \end{aligned}\right\} \tag{18.2-5}$$

$$n = 1, 2, 3, \ldots$$

$u_n(\mathbf{r})$ vanishes at each mirror surface, and the integer n is normally a very large number. However, when the mode is not approximated well by a plane wave traveling along the axis, the situation is more complicated. It has been shown by Boyd and Gordon (1961) and by Kogelnik and Li (1966), that for a confocal cavity formed by two spherical mirrors of radius l separated by an axial distance l, when certain paraxial approximations are made, the mode functions $u(\mathbf{r})$ in general have to be labeled by three integers n, p, q, and they are given by the formula

$$u_{npq}(x, y, z) = KH_p\left[\left(\frac{2kl}{l^2 + 4z^2}\right)^{1/2} x\right] H_q\left[\left(\frac{2kl}{l^2 + 4z^2}\right)^{1/2} y\right] \exp\left[-\frac{(x^2 + y^2)kl}{l^2 + 4z^2}\right] \mathrm{e}^{-\mathrm{i}\psi(x,y,z)}. \tag{18.2-6}$$

K is a normalizing constant. The origin of the coordinate system is the midpoint of the confocal cavity, and the z-axis coincides with the cavity axis. $H_p(X)$ is the p'th order Hermite polynomial, and the phase function $\psi(x, y, z)$, whose sine vanishes at the two mirrors, is given by the expression

$$\psi(x, y, z) = k\left[z + \frac{1}{2}l + \frac{2z(x^2 + y^2)}{l^2 + 4z^2}\right] - (1 + p + q)\left[\frac{1}{2}\pi - \arctan\left(\frac{1 - 2z/l}{1 + 2z/l}\right)\right]. \tag{18.2–7}$$

k is the wave number of the cavity mode n, p, q, and the condition for resonance leads to the equation

$$k = (\pi/2l)(2n + 1 + p + q). \tag{18.2–8}$$

The lowest order or *axial modes* have $p = 0 = q$, in which case the corresponding mode functions $u_{n00}(\mathbf{r})$ become Gaussian distributions with axial symmetry. It should be noted that, whereas the off-axis mode numbers p, q are generally zero or small integers, the axial mode number n is a very large number that measures the number of antinodes of the standing wave within the laser cavity. The frequency separation $\Delta\nu$ between two successive axial modes is $c/2l$, and because of their Gaussian form they are sometimes known as *Gaussian modes*. It can be shown that the different mode functions given by Eq. (18.2–6) form a complete orthogonal set, and we shall choose the constant K in Eq. (18.2–6) so that they are normalized to the cavity volume V, which makes the $u(\mathbf{r})$ function dimensionless, i.e.

$$\int u^*_{npq}(\mathbf{r})u_{n'p'q'}(\mathbf{r})\,d\mathbf{r} = \delta_{nn'}\delta_{pp'}\delta_{qq'}V. \tag{18.2–9}$$

The volume of integration extends over all x and y, and over a range of z from one mirror to the other. The modes given by Eqs. (18.2–5) are normalizable in three dimensions only if the standing waves have a finite cross-sectional area $\mathscr{A}$, in which case the constant K becomes $(2/\mathscr{A}l)^{1/2}$.

The modes have been generalized (Fox and Li, 1961; Boyd and Gordon, 1961; Kogelnik and Li, 1966) for a cavity of length l formed by two mirrors with arbitrary radii of curvature R_1 and R_2. They have the same general structure as those given by Eq. (18.2–6), except that the origin of coordinates has to be located at the beam waist, which may not coincide with the midpoint of the cavity. An additional constraint is that the inequality

$$0 \leqslant \left(1 - \frac{l}{R_1}\right)\left(1 - \frac{l}{R_2}\right) \leqslant 1 \tag{18.2–10}$$

must hold. This is known as the condition for a stable cavity, and when it is violated there are no normal modes in the usual sense. Nevertheless, 'unstable' resonators (see for example, Siegman, 1971) are occasionally used in applications of high-power pulsed lasers, because they tend to spread the optical field over a larger volume of the active medium. Figure 18.3 shows some off-axis mode patterns that were obtained by photographing the end face of a gas laser. It should be noted that the number of bright spots in the x- or y-direction exceeds the mode number p or q by one.

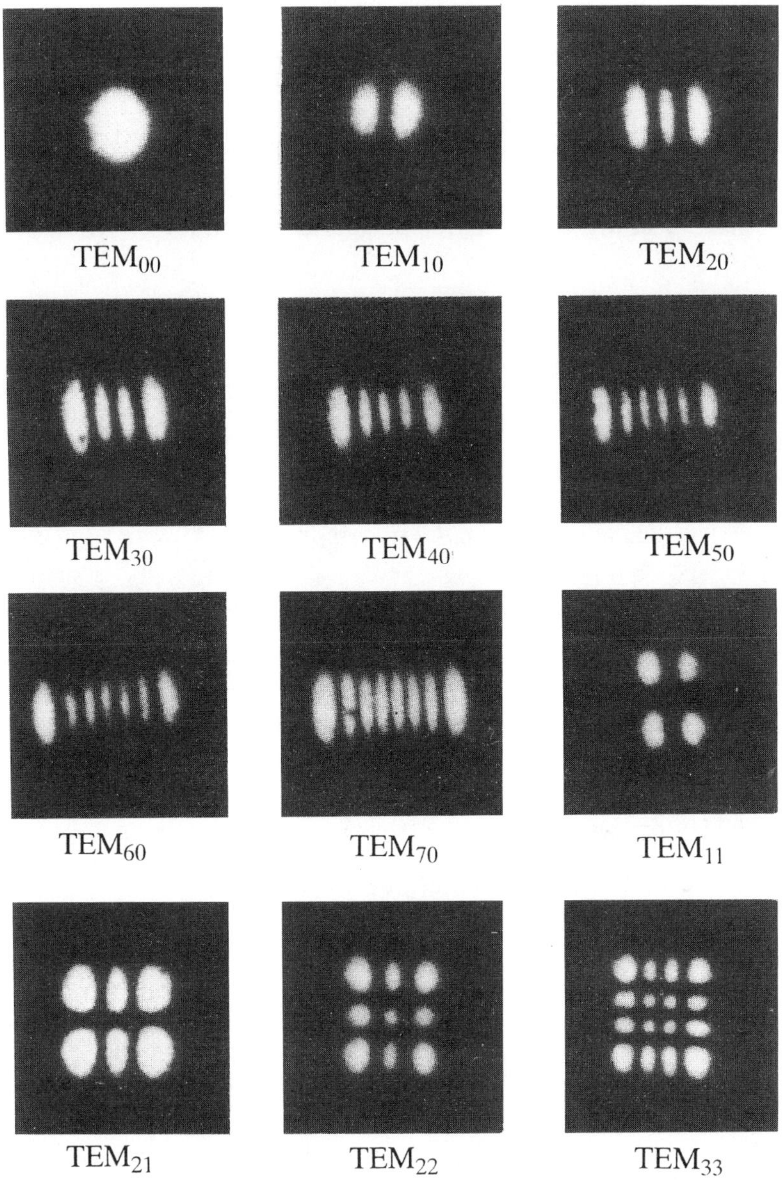

Fig. 18.3 The intensity distributions in various transverse electric modes (TEM). The suffices correspond to p and q in our notation. (Reproduced from Kogelnik and Rigrod, 1962.)

In this chapter we shall be concerned with lasers in which only a single cavity mode is excited, so that the mode labels can be dropped, and we can replace Eq. (18.2–4) by

$$E(\mathbf{r}, t) = \mathscr{E}(t)\,\mathrm{e}^{-\mathrm{i}\omega t}\,u(\mathbf{r}) + \text{c.c.} \tag{18.2–11}$$

The form of the mode function $u(\mathbf{r})$ plays no major role in the equation of motion of the laser field, and for many purposes need not be specified explicitly. When an explicit form of $u(\mathbf{r})$ is needed, we shall generally choose the axial mode function $u_{n00}(\mathbf{r})$, or even the still simpler form given by Eq. (18.2–5).

18.2.2 Equation of motion of the laser field

We now introduce the expression (18.2–11) into the equation of motion (18.2–3), and then make a slowly varying envelope approximation for $\mathscr{E}(t)$, which implies that

$$\left|\frac{\mathrm{d}\mathscr{E}(t)}{\mathrm{d}t}\right| \ll \omega|\mathscr{E}(t)|$$
$$\left|\frac{\mathrm{d}^2\mathscr{E}(t)}{\mathrm{d}t^2}\right| \ll \omega\left|\frac{\mathrm{d}\mathscr{E}(t)}{\mathrm{d}t}\right|.$$

This allows us to simplify the time derivatives

$$\frac{\partial E}{\partial t} = \left[\frac{\mathrm{d}\mathscr{E}(t)}{\mathrm{d}t} - \mathrm{i}\omega\mathscr{E}(t)\right]\mathrm{e}^{-\mathrm{i}\omega t}\,u(\mathbf{r}) + \text{c.c.}$$
$$\frac{\partial^2 E}{\partial t^2} = \left[\frac{\mathrm{d}^2\mathscr{E}(t)}{\mathrm{d}t^2} - 2\mathrm{i}\omega\frac{\mathrm{d}\mathscr{E}(t)}{\mathrm{d}t} - \omega^2\mathscr{E}(t)\right]\mathrm{e}^{-\mathrm{i}\omega t}\,u(\mathbf{r}) + \text{c.c.}$$

by dropping the first term in each expression on the right. We will also assume that the polarization $P(\mathbf{r}, t)$ is nearly oscillatory at the same frequency ω as the electric field, so that $\partial^2 P/\partial t^2$ can be approximated by $-\omega^2 P$. Finally, the z-variation of $u(\mathbf{r})$ is dominated by the $\sin kz$ term in Eq. (18.2–5) or Eqs. (18.2–6) and (18.2–7), so that

$$\frac{\partial^2 u(\mathbf{r})}{\partial z^2} \approx -k^2 u(\mathbf{r}) = -\frac{\omega^2}{c^2}u(\mathbf{r}).$$

With these approximations, Eq. (18.2–3) reduces to

$$-\mathrm{i}\omega\left(2\frac{\mathrm{d}\mathscr{E}}{\mathrm{d}t} + \frac{\sigma}{\varepsilon_0}\mathscr{E}\right)\mathrm{e}^{-\mathrm{i}\omega t}\,u(\mathbf{r}) + \text{c.c.} = \frac{\omega^2}{\varepsilon_0}P(\mathbf{r}, t). \qquad (18.2\text{–}12)$$

This equation may not appear to be strictly consistent internally, because the solution for $\mathscr{E}(t)$ depends on $\mathbf{r}$, but it becomes consistent as soon as we integrate over the cavity volume. The reason is that the distribution of the active medium within the cavity has to be specified before the mode and its amplitude are determined. If we decompose the polarization $P(\mathbf{r}, t)$ into its positive frequency and negative frequency parts by writing

$$P(\mathbf{r}, t) = P^{(+)}(\mathbf{r}, t) + P^{(-)}(\mathbf{r}, t) = P^{(+)}(\mathbf{r}, t) + \text{c.c.},$$

in which $P^{(+)}(\mathbf{r}, t)$ varies as $\mathrm{e}^{-\mathrm{i}\omega t}$ and $P^{(-)}(r, t)$ varies as $\mathrm{e}^{\mathrm{i}\omega t}$ to a first approximation, then clearly

$$-\mathrm{i}\omega\left(2\frac{\mathrm{d}\mathscr{E}}{\mathrm{d}t} + \frac{\sigma}{\varepsilon_0}\mathscr{E}\right)\mathrm{e}^{-\mathrm{i}\omega t}\,u(\mathbf{r}) = \frac{\omega^2}{\varepsilon_0}P^{(+)}(\mathbf{r}, t).$$

We now multiply each term by $u^*(\mathbf{r})$ and integrate over the cavity volume. With the help of the orthonormality relation (18.2–9) we then obtain the equation

$$-\mathrm{i}\omega\left(2\frac{\mathrm{d}\mathscr{E}}{\mathrm{d}t}+\frac{\sigma}{\varepsilon_0}\mathscr{E}\right)\mathrm{e}^{-\mathrm{i}\omega t}=\frac{1}{V}\frac{\omega^2}{\varepsilon_0}\int P^{(+)}(\mathbf{r},t)u^*(\mathbf{r})\,\mathrm{d}^3r. \qquad (18.2\text{–}13)$$

If the active laser atoms are distributed throughout the cavity with some density $\eta(\mathbf{r})$, and the average dipole moment of an atom at position $\mathbf{r}$ at time t is $\langle\hat{\boldsymbol{\mu}}(\mathbf{r},t)\rangle$ in some well-defined direction, then the macroscopic polarization $\mathbf{P}(\mathbf{r},t)$ is given by the relation

$$\mathbf{P}(\mathbf{r},t)=\eta(\mathbf{r})\langle\hat{\boldsymbol{\mu}}(\mathbf{r},t)\rangle. \qquad (18.2\text{–}14)$$

It is at this point that quantum mechanics enters into the calculation, through the evaluation of the average atomic dipole moment $\langle\hat{\boldsymbol{\mu}}(\mathbf{r},t)\rangle$.

In order to calculate $\langle\hat{\boldsymbol{\mu}}(\mathbf{r},t)\rangle$ for a typical two-level atom, we shall assume that by incoherent optical pumping the atom is initially prepared in a mixed state with density operator

$$\hat{\rho}_{\mathrm{A}}(0)=p_1|1\rangle\langle 1|+p_2|2\rangle\langle 2|, \qquad (18.2\text{–}15)$$

so that there is a probability p_2 that the atom is excited and p_1 that it is not excited. We have already seen that p_2 should exceed p_1, on the average, for laser action to take place. We now allow this atom to interact with the classical field $\mathbf{E}(\mathbf{r},t)$ for a short time Δt, after which time it may be supposed that the state has decayed by radiative or non-radiative means or by collision.

In Section 15.3 we discussed the classical Rabi problem, in which a two-level atom interacts with a classical electromagnetic field of constant amplitude. In this laser problem the amplitude $\mathscr{E}(t)$ of the field is itself varying in time. However, if the time scale Δt on which the atom interacts with the field is very short compared with the time scale on which the laser field $\mathscr{E}(t)$ itself is changing, we may treat $\mathscr{E}(t)$ as effectively constant during the interaction. The solutions given by Eqs. (15.3–21) and (15.3–22) therefore apply, and if the detuning between the atomic and field frequencies is negligible, Eq. (15.3–22) reduces to the following expression for the average atomic dipole moment $\langle\hat{\boldsymbol{\mu}}(\mathbf{r},t)\rangle$ at time t, when the atom starts in the lower state at time $t-\Delta t$:

$$\langle\hat{\boldsymbol{\mu}}(\mathbf{r},t)\rangle=\boldsymbol{\mu}_{12}\sin(\Omega\Delta t)\sin\omega t.$$

As usual $\boldsymbol{\mu}_{12}$ is the transition dipole moment, which we take to be real, and Ω is the Rabi frequency. The time Δt is the time for which the interaction persists, and it plays the role of the lifetime of the laser atom. We may suppose that after a time Δt the atom makes a transition to some other state in which it no longer interacts with the laser field. It may easily be shown that a similar expression, but with opposite sign, is obtained when the initial atomic state is the excited state. When the initial state is mixed, and upper and lower states occur with probabilities p_2 and p_1, respectively, as in Eq. (18.2–15), the expectation value of the dipole moment becomes

$$\langle\hat{\boldsymbol{\mu}}(\mathbf{r},t)\rangle=-\tfrac{1}{2}\mathrm{i}(p_2-p_1)\boldsymbol{\mu}_{12}\sin(\Omega\Delta t)\,\mathrm{e}^{-\mathrm{i}\omega t}+\mathrm{c.c.} \qquad (18.2\text{–}16)$$

In practice, there is an approximately exponential distribution of lifetimes Δt, and one needs to average over Δt. If T is the average natural lifetime, for simplicity assumed to be similar for the upper and lower states, then Eq.

(18.2–16) has to be replaced by the equation

$$\begin{aligned}\langle \hat{\boldsymbol{\mu}}(\mathbf{r}, t)\rangle &= -\tfrac{1}{2}\mathrm{i}(p_2 - p_1)\boldsymbol{\mu}_{12}\,\mathrm{e}^{-\mathrm{i}\omega t}\int_0^\infty \frac{1}{T}\exp\left(-\frac{\Delta t}{T}\right)\sin(\Omega\Delta t)\,\mathrm{d}\Delta t + \text{c.c.}\\ &= -\tfrac{1}{2}\mathrm{i}(p_2 - p_1)\boldsymbol{\mu}_{12}\frac{\Omega T}{1+\Omega^2T^2}\mathrm{e}^{-\mathrm{i}\omega t} + \text{c.c.}\\ &= -\tfrac{1}{2}\mathrm{i}(p_2 - p_1)\boldsymbol{\mu}_{12}\Omega T(1 - \Omega^2T^2 + \ldots)\,\mathrm{c}^{-\mathrm{i}\omega t} + \text{c.c.}\end{aligned} \qquad (18.2\text{–}17)$$

If ΩT is small, it is generally sufficient to retain only the first two terms in the expansion of $(1 + \Omega^2T^2)^{-1}$ in powers of ΩT; this implies that we exclude very strong laser fields. The Rabi frequency Ω in our present notation is given by [cf. Eq. (15.3–12)]

$$\Omega = 2|\mu_{12}|\,|\mathscr{E}(t)|\,|u(\mathbf{r})|/\hbar, \qquad (18.2\text{–}18)$$

provided the dipole moment vector $\boldsymbol{\mu}_{12}$ and the polarization vector of the field are parallel. But it should be noted that, on resonance, the phase of the dipole moment $\langle\hat{\boldsymbol{\mu}}(\mathbf{r}, t)\rangle$ is related to the phase of the field, and when the sign of $\mathscr{E}(t)$ changes, the direction of the dipole moment changes also. Accordingly we may re-write Eq. (18.2–17) more explicitly as a scalar equation:

$$\langle|\hat{\mu}(\mathbf{r}, t)|\rangle = \frac{-\mathrm{i}(p_2 - p_1)\mu_{12}^2 T}{\hbar}|\mathscr{E}(t)|\,\mathrm{e}^{-\mathrm{i}\omega t}\,u(\mathbf{r})\left[1 - \frac{4\mu_{12}^2T^2}{\hbar^2}|\mathscr{E}(t)|^2|u(\mathbf{r})|^2 + \ldots\right] + \text{c.c.} \qquad (18.2\text{–}19)$$

By multiplying by the atomic density $\eta(\mathbf{r})$ we can immediately identify the positive frequency part $P^{(+)}(\mathbf{r}, t)$ of the polarization $P(\mathbf{r}, t)$. When this result is combined with Eq. (18.2–13) we obtain the relation

$$\frac{\mathrm{d}\mathscr{E}}{\mathrm{d}t} + \frac{\sigma}{2\varepsilon_0}\mathscr{E} = \frac{(p_2 - p_1)\mu_{12}^2\omega T}{2\varepsilon_0\hbar V}\mathscr{E}(t)\left[\int\eta(\mathbf{r})|u(\mathbf{r})|^2\,\mathrm{d}^3r - \frac{4\mu_{12}^2T^2}{\hbar^2}|\mathscr{E}(t)|^2\int\eta(\mathbf{r})|u(\mathbf{r})|^4\,\mathrm{d}^3r + \ldots\right]. \qquad (18.2\text{–}20)$$

This can be written a little more compactly if we introduce the mean atomic density

$$\bar{\eta} \equiv \frac{1}{V}\int\eta(\mathbf{r})|u(\mathbf{r})|^2\,\mathrm{d}^3r \qquad (18.2\text{–}21\text{a})$$

and the quantity

$$F \equiv \frac{1}{V}\int\eta(\mathbf{r})|u(\mathbf{r})|^4\,\mathrm{d}^3r. \qquad (18.2\text{–}21\text{b})$$

We then arrive at the equation of motion

$$\frac{\mathrm{d}\mathscr{E}}{\mathrm{d}t} = \left[\frac{\omega\bar{\eta}(p_2 - p_1)\mu_{12}^2T}{2\varepsilon_0\hbar} - \frac{\sigma}{2\varepsilon_0}\right]\mathscr{E} - \frac{2\omega F\mu_{12}^4(p_2 - p_1)T^3}{\varepsilon_0\hbar^3}|\mathscr{E}|^2\mathscr{E}. \qquad (18.2\text{–}22)$$

This is similar in structure to the equation first derived by van der Pol (1922) for the behavior of an electronic oscillator, when the slowly varying amplitude approximation is made. The equation applies to a homogeneously broadened gain medium, in which all the atoms see the field on resonance. When the medium is inhomogeneously broadened, Eq. (18.2–22) still holds approximately, provided the atomic density $\eta(\mathbf{r})$ is taken to refer to those atoms whose frequencies lie within a natural linewidth of the laser frequency.

Let us briefly examine the implications of the equation. When the field amplitude $\mathscr{E}$ is sufficiently small, we can ignore the cubic term in $\mathscr{E}$ to a first approximation, and the equation then describes either exponential growth or exponential decay, accordingly as

$$\frac{\omega\bar{\eta}(p_2 - p_1)\mu_{12}^2 T}{2\varepsilon_0\hbar} \gtrless \frac{\sigma}{2\varepsilon_0}.$$

The term on the left, which is proportional to the population inversion, clearly represents the *gain* of the system (provided $p_2 > p_1$), and the term on the right, which is proportional to the conductivity σ, clearly represents the *loss*. Indeed, σ can be related to the Q-factor of the cavity mode by $\sigma/2\varepsilon_0 = \omega/Q$. When exact equality holds, the gain just equals the loss, and we are at the *threshold* of oscillation. Threshold therefore requires a minimum atomic inversion density given by the relation

$$\bar{\eta}(p_2 - p_1)_{\text{threshold}} = \frac{\hbar\sigma}{\omega\mu_{12}^2 T}. \tag{18.2–23}$$

Above threshold, when the gain exceeds the loss, the amplitude $\mathscr{E}(t)$ grows exponentially until the nonlinear or saturation term begins to assert itself and to inhibit further growth. However, it should be noted that the initial growth depends on the presence of a non-zero field at the beginning, and in its absence the field remains zero for all time, no matter how high the population inversion may be. This is a fundamental weakness of the semi-classical treatment. With the introduction of the more compact notation‡

$$\left.\begin{aligned} \mathscr{A} &\equiv \omega\bar{\eta}(p_2 - p_1)\mu_{12}^2 T/2\varepsilon_0\hbar \\ \mathscr{C} &\equiv \sigma/2\varepsilon_0 \\ \mathscr{B} &\equiv 2\omega F(p_2 - p_1)\mu_{12}^4 T^3/\varepsilon_0\hbar^3 \end{aligned}\right\} \tag{18.2–24}$$

for the coefficients of the gain, the loss, and the nonlinear terms, Eq. (18.2–22) can be simplified to

$$\frac{\mathrm{d}\mathscr{E}}{\mathrm{d}t} = \mathscr{B}\left[\frac{\mathscr{A} - \mathscr{C}}{\mathscr{B}} - |\mathscr{E}|^2\right]\mathscr{E}. \tag{18.2–25}$$

It is sometimes convenient to eliminate one or more of the three parameters $\mathscr{A}$, $\mathscr{C}$, $\mathscr{B}$ by the introduction of new dimensionless variables. For example, we could

‡ $\mathscr{A}$ should not be confused with the laser beam cross-sectional area that was also denoted by $\mathscr{A}$ in Section 18.1.1.

define a dimensionless measure $\tilde{\mathscr{E}}$ of the field amplitude by putting

$$\tilde{\mathscr{E}}(t) \equiv (\mathscr{B}/\mathscr{A})^{1/2}\mathscr{E}(t), \tag{18.2–26}$$

and a dimensionless *pump parameter* a, by normalizing the difference between the gain and loss coefficients,

$$a \equiv \frac{\mathscr{A} - \mathscr{C}}{\mathscr{A}}.$$

From the definition, a is positive above threshold and negative below threshold. The equation of motion for the laser field amplitude $\tilde{\mathscr{E}}(t)$ then takes the more compact form

$$\frac{\mathrm{d}\tilde{\mathscr{E}}}{\mathrm{d}t} - \mathscr{A}[a - |\tilde{\mathscr{E}}|^2]\tilde{\mathscr{E}} = 0. \tag{18.2–27}$$

The parameter $\mathscr{A}$ could also be eliminated by the introduction of a dimensionless measure of time $\tilde{t} = \mathscr{A}t$. However, we will leave $\mathscr{A}$ in place for the moment, because a somewhat different scaling for the variables will prove to be more convenient later.

The steady-state solution is obtained by putting $\mathrm{d}\tilde{\mathscr{E}}/\mathrm{d}t = 0$. Below threshold the only steady-state solution is the one for which $\tilde{\mathscr{E}} = 0$, whereas there is a non-zero steady-state solution above threshold, and we have in the long-time limit,

$$\left.\begin{aligned} \tilde{\mathscr{E}} &= 0, && \text{for } a \leqslant 0 \\ |\tilde{\mathscr{E}}|^2 &= a, && \text{for } a \geqslant 0. \end{aligned}\right\} \tag{18.2–28}$$

Above threshold, whenever the light intensity $|\tilde{\mathscr{E}}(t)|^2$ is smaller than a, the amplitude grows in time until $|\tilde{\mathscr{E}}(t)|^2 = a$, and whenever $|\tilde{\mathscr{E}}(t)|^2$ exceeds a, the amplitude falls until $|\tilde{\mathscr{E}}(t)|^2 = a$. Evidently, in our dimensionless units a gives the intensity of the laser field in the steady state above threshold. Figure 18.4 shows plots of the laser amplitude and intensity in the steady state against the pump

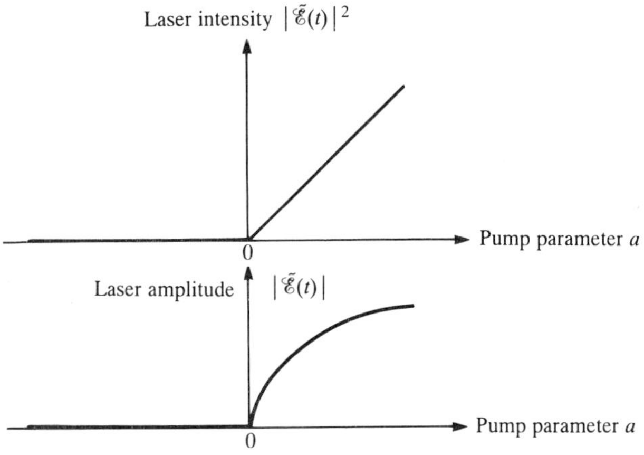

Fig. 18.4 Variation of the steady-state laser intensity and absolute amplitude with pump parameter.

parameter a, which is itself a linear function of the rate of excitation. It will be seen that once the population inversion exceeds the critical value given by Eq. (18.2–23), the steady-state light intensity grows in proportion to a. Of course, when the pump parameter is very large and the light intensity is sufficiently high, the cubic approximation in Eq. (18.2–17) may cease to be adequate. This generally happens when the gain exceeds the loss by more than a few percent, but this already corresponds to quite a high excitation. In practice the cubic approximation holds over quite a wide range of laser operation.

Despite the simplicity of Eq. (18.2–27), its general solution is not very simple, because of the nonlinear character of the equation. We shall discuss the time-dependent properties of the laser field a little later in Section 18.6, after introducing spontaneous emission noise into the problem.

18.2.3 The phase transition analogy

We can express the steady state solution of the equation of motion (18.2–27) in the form

$$(\mathscr{A} - \mathscr{C})\tilde{\mathscr{E}} - \mathscr{A}|\tilde{\mathscr{E}}|^2\tilde{\mathscr{E}} = 0. \tag{18.2–29}$$

It has been observed by a number of scientists working on the laser problem, that this equation bears a striking resemblance to the equations encountered in describing *phase transitions* in certain physical systems (Degiorgio and Scully, 1970; Graham and Haken, 1970; Scully, 1973; Graham, 1975; Haken, 1975a,b, 1977, Chap. 6, 1985). For example, the magnetization M of a ferromagnetic material in the neighborhood of the Curie temperature T_C (the temperature at which a phase transition occurs) varies with temperature T according to the equation (Weiss' law)

$$c(T - T_C)M + gTM^3 = 0, \tag{18.2–30}$$

in which c, g are positive constants. So long as the temperature is above the Curie temperature, the only solution of the equation is $M = 0$. However, once $T < T_C$, Eq. (18.2–30) gives a non-zero solution for the magnetization and we have

$$\left.\begin{aligned} M &= 0, && \text{for } T \geqslant T_C \\ M^2 &= \frac{c}{g}\left(\frac{T_C - T}{T}\right), && \text{for } T \leqslant T_C. \end{aligned}\right\} \tag{18.2–31}$$

If we associate the absolute laser field amplitude $|\tilde{\mathscr{E}}|$ with the magnetization M, the loss rate $\mathscr{C}$ with the Curie temperature T_C, and the gain rate $\mathscr{A}$ with the temperature T, we observe that there is an almost complete correspondence between Eqs. (18.2–29) and (18.2–30), and the solution (18.2–31) corresponds to the solution (18.2–28). In the ferromagnet the appearance of a spontaneous magnetization M below the Curie temperature is associated with a certain ordering of the magnetic dipoles, and in the laser the appearance of a macroscopic field $\tilde{\mathscr{E}}$ when the gain exceeds the loss, is associated with a certain ordering or phasing of the atomic oscillators. In both cases a fundamental change in order of the system occurs when $T = T_C$, or when $\mathscr{A} = \mathscr{C}$, which is known as a *phase transition*. The magnetization M and the field amplitude $|\tilde{\mathscr{E}}|$ are known as *order parameters* for the corresponding phase transitions. In Fig. 18.5 the dependence of $\tilde{\mathscr{E}}$ on $\mathscr{A}$ and the dependence of M on T are compared. In both cases the order parameter vanishes on the disordered side of the phase transition, and grows on the well-ordered side. Both order parameters are macroscopic variables that may be regarded as manifestations of the amount of microscopic order in the system and, at the same time, as the macroscopic forces tending to drive or *slave* the microscopic subsystems in a certain direction. The order parameters represent the collective action of all the microscopic subsystems on any one of them; they give instructions to all the subsystems and keep them in order. Order parameters may therefore be regarded as the generals commanding and issuing marching orders to the troops (Graham, 1975; Haken, 1975a, 1977).

The analogy between the laser and the ferromagnet can be extended even further. For example, if we impose an external magnetizing field H on the ferromagnetic material, then the right-hand side of Eq. (18.2–30) has to be replaced by H, and the equation becomes

$$c(T - T_C)M + gTM^3 = H. \tag{18.2–32}$$

If we apply a resonant external electromagnetic field of amplitude $\tilde{\mathscr{E}}_0$ to the laser, Eq. (18.2–29) has to be replaced by

$$(\mathscr{A} - \mathscr{C})\tilde{\mathscr{E}} - \mathscr{A}|\tilde{\mathscr{E}}|^2\tilde{\mathscr{E}} = -\tilde{\mathscr{E}}_0. \tag{18.2–33}$$

We now introduce the *susceptibility* $\chi = \partial M/\partial H$ of the ferromagnet, and, in the limit $H \to 0$, we find with the help of Eqs. (18.2–31) that this takes the form

$$\left.\begin{aligned} \chi = \left(\frac{\partial M}{\partial H}\right)_{H=0} &= \frac{1}{c(T - T_C)} \qquad \text{if } T > T_C \\ &= \frac{1}{2c(T_C - T)} \qquad \text{if } T < T_C. \end{aligned}\right\} \tag{18.2–34}$$

Evidently the susceptibility diverges at the phase transition, and for $T < T_C$ $\chi \propto M^{-2}$. Similarly (Degiorgio and Scully, 1970), we may define a *responsivity* or susceptibility for the laser by $\chi_L = \partial|\tilde{\mathscr{E}}|/\partial|\tilde{\mathscr{E}}_0|$, and from Eq. (18.2–33) in the limit $\tilde{\mathscr{E}}_0 \to 0$ the susceptibility becomes

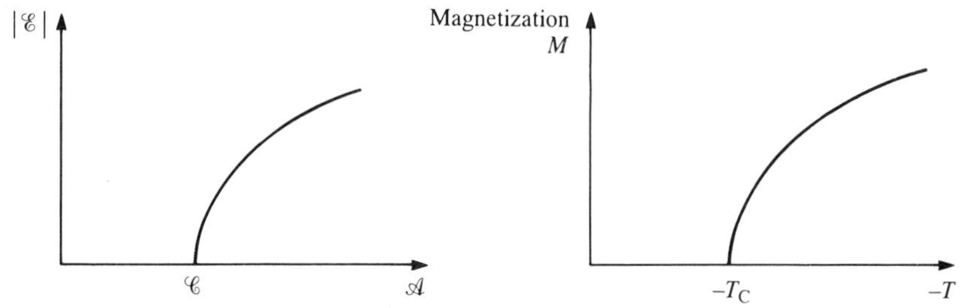

Fig. 18.5 Comparison of the laser and the ferromagnetic phase transition. $|\tilde{\mathscr{E}}|$ is an order parameter in the former case and M in the latter case. A phase transition occurs when $\mathscr{A} = \mathscr{C}$ or when $T = T_C$.

$$\left.\begin{aligned} \chi_{\mathrm{L}} = \left(\frac{\partial|\tilde{\mathscr{E}}|}{\partial|\tilde{\mathscr{E}}_0|}\right)_{\tilde{\mathscr{E}}_0=0} &= \frac{1}{\mathscr{C} - \mathscr{A}} \qquad \text{if } \mathscr{C} > \mathscr{A} \\ &= \frac{1}{2(\mathscr{A} - \mathscr{C})} \qquad \text{if } \mathscr{A} > \mathscr{C}. \end{aligned}\right\} \tag{18.2-35}$$

χ_{L} diverges at the laser threshold, and with the same critical exponent, $\chi_{\mathrm{L}} \propto |\tilde{\mathscr{E}}|^{-2}$ when $\mathscr{A} > \mathscr{C}$, as does χ at the Curie temperature. The reason is that a very small change of the applied field $\tilde{\mathscr{E}}_0$ can give rise to a large change of the laser field $\tilde{\mathscr{E}}$ at the threshold. The behavior of the laser near threshold therefore bears quite a close analogy with an order–disorder type of phase transition.

Actually, more phase transitions are encountered when the pump parameter is increased further, although this is not apparent from the equation of motion (18.2–27), because this equation is limited to a single laser mode and to the lowest order nonlinearity. As the excitation of the laser is progressively increased, more and more modes can become excited. If their phases are correlated or locked, the resulting laser field exhibits regular periodic pulsing or spiking, whereas if the phases are not locked, the field becomes essentially chaotic, even though it continues to be described by a deterministic equation of motion (Uspenskii, 1963; Korobkin and Uspenskii, 1963; Risken and Nummedal, 1968a,b; Graham, 1975; Haken, 1975a,b, 1977, Chap. 6; Haken and Ohno, 1976a,b; Casperson, 1978; Mayr, Risken and Vollmer, 1981).

When the effects of fluctuations are included in the theory, we shall find that the sharp discontinuities illustrated in Fig. 18.4 in the response of the system disappear. However, the phase transition concept remains applicable, and the analogy between phase transitions in lasers and in other systems is preserved. Although the light amplitude $\mathscr{E}$ is no longer zero below threshold when quantum fluctuations are included, the phase transition is manifest in a change of the statistical properties of the laser field.

18.3 Semiclassical laser theory with spontaneous emission noise

Although the equation of motion (18.2–27) describes the dynamical behavior of the laser field, it does so in a completely deterministic manner. The optical field is free from all fluctuations, and therefore any questions about the coherence properties of laser light, which differ in important ways from the properties of other types of light, are outside the domain of the theory. For example, Eq. (18.2–27) can tell us nothing about the spectral width of the laser field, which is here regarded as completely monochromatic, although a finite spectral width is always observed. Moreover, the discontinuous onset of radiation at threshold described by the equation, and illustrated in Fig. 18.4, is not actually observed in practice. The key to an understanding of these questions lies in the fluctuations of the optical field, which are brought about by random spontaneous atomic emissions. Spontaneous emissions are of course included automatically in any fully quantized treatment of the laser problem, such as we give below in Section 18.4. However, they do not appear naturally in deterministic semiclassical theories.

Nevertheless, attempts have been made to include the effects of spontaneous emissions in a semiclassical laser theory by the device of adding a noise term to the equation of motion (18.2–27) (Risken, 1965, 1966; Risken and Vollmer, 1967a,b; Hempstead and Lax, 1967; Risken, 1970), which then takes the form

$$\frac{\mathrm{d}\tilde{\mathscr{E}}(t)}{\mathrm{d}t} - \mathscr{A}[a - |\tilde{\mathscr{E}}(t)|^2]\tilde{\mathscr{E}}(t) = \zeta(t). \tag{18.3–1}$$

Here $\zeta(t)$ is taken to be a complex Gaussian random function of zero mean, with a correlation time that is of the order of the lifetime for spontaneous atomic emission. However, as this lifetime is generally very short compared with the time in which the laser amplitude $\tilde{\mathscr{E}}(t)$ changes, we can simplify the problem by treating $\zeta(t)$ as a complex *δ-correlated noise or white noise* satisfying the relations

$$\left.\begin{aligned} \langle\zeta(t)\rangle &= 0 \\ \langle\zeta(t)\zeta(t')\rangle &= 0 \\ \langle\zeta^*(t)\zeta(t')\rangle &= 4Q_s\delta(t-t'). \end{aligned}\right\} \tag{18.3–2}$$

The noise $\zeta(t)$ is therefore *Markovian* (see Section 2.6). The value of Q_s is chosen appropriately to represent the magnitude of the spontaneous emission fluctuations.

The new equation of motion (18.3–1) is a *Langevin* type of equation, in which the field amplitude $\tilde{\mathscr{E}}(t)$ is no longer deterministic. Instead, $\tilde{\mathscr{E}}(t)$ is a random process, with a certain probability density $p(\tilde{\mathscr{E}}, t)$ determined by the statistical properties of the noise $\zeta(t)$. We can now ask questions about the moments and the correlation functions of $\tilde{\mathscr{E}}(t)$, and arrive at meaningful answers that were not obtainable from the deterministic equation.

18.3.1 Fokker–Planck equation

As was shown in Section 2.9, corresponding to the Langevin equation of motion with Markovian noise, there exists an equation of motion for the probability density $p(\tilde{\mathscr{E}}, t)$, known as the *Fokker–Planck equation*. It is convenient to replace the complex amplitude $\tilde{\mathscr{E}}$ by a two-dimensional real vector $\tilde{\boldsymbol{\mathscr{E}}}$ with components $\tilde{\mathscr{E}}_1$, $\tilde{\mathscr{E}}_2$ that are the real and imaginary parts of $\tilde{\mathscr{E}}$. The Fokker–Planck equation corresponding to Eq. (18.3–1) then takes the form (see Section 2.9)

$$\frac{\partial p(\tilde{\boldsymbol{\mathscr{E}}}, t)}{\partial t} = -\sum_{i=1}^{2}\frac{\partial}{\partial\tilde{\mathscr{E}}_i}[\mathscr{A}(a - \tilde{\boldsymbol{\mathscr{E}}}^2)\tilde{\mathscr{E}}_i p(\tilde{\boldsymbol{\mathscr{E}}}, t)] + \sum_{i=1}^{2}\frac{\partial^2}{\partial\tilde{\mathscr{E}}_i^2}Q_s p(\tilde{\boldsymbol{\mathscr{E}}}, t). \tag{18.3–3}$$

The solution of this equation allows various averages of $\tilde{\boldsymbol{\mathscr{E}}}$ to be calculated. The equation contains three parameters, two of which can be eliminated by a re-scaling of the amplitude $\tilde{\boldsymbol{\mathscr{E}}}$, the time t, and the pump parameter a. If we introduce a new dimensionless amplitude $\boldsymbol{\mathscr{E}}'$ by the transformation

$$\tilde{\mathscr{E}}_i \to \mathscr{E}_i'(Q_s/\mathscr{A})^{1/4},$$

a new pump parameter a' by

$$a \to a'(Q_s/\mathscr{A})^{1/2},$$

and a new dimensionless time t' by

$$t \to t'(1/Q_s\mathscr{A})^{1/2},$$

we arrive at the following simplified equation

$$\frac{\partial p(\boldsymbol{\mathscr{E}}, t)}{\partial t} = -\sum_{i=1}^{2}\frac{\partial}{\partial \mathscr{E}_i}[(a - \boldsymbol{\mathscr{E}}^2)\mathscr{E}_i p(\boldsymbol{\mathscr{E}}, t)] + \sum_{i=1}^{2}\frac{\partial^2 p(\boldsymbol{\mathscr{E}}, t)}{\partial \mathscr{E}_i^2}, \quad (18.3\text{–}4)$$

in which the primes have been dropped for simplicity. The equation contains just a single parameter, the pump parameter a, and should have universal validity for any laser in which the gain medium consists of two-level atoms, provided the light intensity is not too high. The solution can of course be re-scaled to a dimensional form for the particular laser under consideration. We shall investigate the general time-dependent solution of Eq. (18.3–4) in Sections 18.6 and 18.7 below.

Once the equation has been integrated, and the solution $p(\boldsymbol{\mathscr{E}}, t)$ for some given initial state $p(\boldsymbol{\mathscr{E}}, 0)$ has been obtained, we can immediately evaluate any single-time expectation of the laser field, such as

$$\langle \mathscr{E}_i^r(t)\rangle = \int \mathscr{E}_i^r p(\boldsymbol{\mathscr{E}}, t)\,\mathrm{d}^2\mathscr{E}, \; r = 1, 2, \ldots. \quad (18.3\text{–}5)$$

However $p(\boldsymbol{\mathscr{E}}, t)$ still does not allow the evaluation of multi-time correlation functions, which calls for additional information, such as the joint probability density $p_2(\boldsymbol{\mathscr{E}}^{(2)}, t_2; \boldsymbol{\mathscr{E}}^{(1)}, t_1)$ for the field amplitudes $\boldsymbol{\mathscr{E}}^{(1)}$, $\boldsymbol{\mathscr{E}}^{(2)}$ at two different times t_1, t_2. Fortunately $p_2(\boldsymbol{\mathscr{E}}^{(2)}, t_2; \boldsymbol{\mathscr{E}}^{(1)}, t_1)$ is expressible in terms of the conditional probability density $P(\boldsymbol{\mathscr{E}}^{(2)}, t_2|\boldsymbol{\mathscr{E}}^{(1)}, t_1)$ for the field amplitude $\boldsymbol{\mathscr{E}}^{(2)}$ at time t_2, given the amplitude $\boldsymbol{\mathscr{E}}^{(1)}$ at time t_1 ($t_2 \geqslant t_1$), together with $p(\boldsymbol{\mathscr{E}}^{(1)}, t_1)$, via the relation

$$p_2(\boldsymbol{\mathscr{E}}^{(2)}, t_2; \boldsymbol{\mathscr{E}}^{(1)}, t_1) = P(\boldsymbol{\mathscr{E}}^{(2)}, t_2|\boldsymbol{\mathscr{E}}^{(1)}, t_1)p(\boldsymbol{\mathscr{E}}^{(1)}, t_1). \quad (18.3\text{–}6)$$

Now the conditional probability density $P(\boldsymbol{\mathscr{E}}^{(2)}, t_2|\boldsymbol{\mathscr{E}}^{(1)}, t_1)$ is also a solution of the Fokker–Planck equation, with a particular boundary condition. By definition, when the two times are equal,

$$P(\boldsymbol{\mathscr{E}}^{(2)}, t|\boldsymbol{\mathscr{E}}^{(1)}, t) = \delta^2(\boldsymbol{\mathscr{E}}^{(2)} - \boldsymbol{\mathscr{E}}^{(1)}), \quad (18.3\text{–}7)$$

which implies that $P(\boldsymbol{\mathscr{E}}^{(2)}, t_2|\boldsymbol{\mathscr{E}}^{(1)}, t_1)$ is the Green function of the partial differential Eq. (18.3–4). Once $p(\boldsymbol{\mathscr{E}}, t)$ and the Green function $P(\boldsymbol{\mathscr{E}}^{(2)}, t_2|\boldsymbol{\mathscr{E}}^{(1)}, t_1)$ have been found, we can calculate any two-time tensor correlation function, such as $\langle \mathscr{E}_i(t)\mathscr{E}_j(t')\rangle$, from the equation

$$\langle \mathscr{E}_i(t)\mathscr{E}_j(t')\rangle = \iint \mathscr{E}_i'\mathscr{E}_j'' P(\boldsymbol{\mathscr{E}}'', t'|\boldsymbol{\mathscr{E}}', t)p(\boldsymbol{\mathscr{E}}', t)\,\mathrm{d}^2\mathscr{E}'\,\mathrm{d}^2\mathscr{E}'', \quad t' > t. \quad (18.3\text{–}8)$$

Moreover, because of the Markovian character of the fluctuations, one can express higher-order joint probability densities in terms of products of Green's functions in the form

$$\begin{aligned} p_n(\boldsymbol{\mathscr{E}}^{(n)}, t_n; \boldsymbol{\mathscr{E}}^{(n-1)}, t_{n-1}; \ldots, \boldsymbol{\mathscr{E}}^{(1)}, t_1) \\ = P(\boldsymbol{\mathscr{E}}^{(n)}, t_n|\boldsymbol{\mathscr{E}}^{(n-1)}, t_{n-1})P(\boldsymbol{\mathscr{E}}^{(n-1)}, t_{n-1}|\boldsymbol{\mathscr{E}}^{(n-2)}, t_{n-2}) \\ \times \ldots P(\boldsymbol{\mathscr{E}}^{(2)}, t_2|\boldsymbol{\mathscr{E}}^{(1)}, t_1)p(\boldsymbol{\mathscr{E}}^{(1)}, t_1), \end{aligned} \quad (18.3\text{–}9)$$

and this formula allows higher-order correlations to be calculated also. For example,

$$\langle \mathscr{E}_l(t_4)\mathscr{E}_k(t_3)\mathscr{E}_j(t_2)\mathscr{E}_i(t_1)\rangle = \iiiint \mathscr{E}_l''''\mathscr{E}_k'''\mathscr{E}_j''\mathscr{E}_i' p_4(\boldsymbol{\mathscr{E}}'''', t_4; \boldsymbol{\mathscr{E}}''', t_3; \boldsymbol{\mathscr{E}}'', t_2; \boldsymbol{\mathscr{E}}', t_1) \times d^2\mathscr{E}'\, d^2\mathscr{E}''\, d^2\mathscr{E}'''\, d^2\mathscr{E}''''. \tag{18.3–10}$$

The problem of how the laser field behaves can therefore be regarded as solved, in principle, once the general solution of Eq. (18.3–4) has been obtained.

18.3.2 *Steady-state solution*

Before attempting to obtain the time-dependent solution of Eq. (18.3–4), which will be treated in Section 18.6, we shall examine the solution in the steady state, which is reached after a sufficiently long time has elapsed. In that case $p(\boldsymbol{\mathscr{E}}, t)$ no longer depends on time t, and the left-hand side of the Fokker–Planck equation vanishes. The equation may therefore be rewritten in the form

$$\sum_{i=1}^{2} \frac{\partial J_i}{\partial \mathscr{E}_i} = 0,$$

where **J** is the *probability current* given by

$$J_i \equiv (a - \boldsymbol{\mathscr{E}}^2)\mathscr{E}_i p_s - \frac{\partial p_s}{\partial \mathscr{E}_i}, \tag{18.3–11}$$

and $p_s(\boldsymbol{\mathscr{E}})$ is the steady-state distribution of the laser field $\boldsymbol{\mathscr{E}}$. The vector $(a - \boldsymbol{\mathscr{E}}^2)\boldsymbol{\mathscr{E}} \equiv \mathbf{A}$ is known as the *drift vector*, and in the special case in which the drift vector obeys the potential condition (cf. Section 2.9)

$$\frac{\partial A_i}{\partial \mathscr{E}_j} = \frac{\partial A_j}{\partial \mathscr{E}_i},$$

as it does in this case, the solution of the Fokker–Planck equation is obtained by equating the probability current **J** to zero. This yields the differential equation

$$(a - \boldsymbol{\mathscr{E}}^2)\mathscr{E}_i p_s - \frac{\partial p_s}{\partial \mathscr{E}_i} = 0,$$

and this equation can be integrated directly to give

$$\begin{aligned} \ln p_s &= \int_0^{\boldsymbol{\mathscr{E}}} (a - \boldsymbol{\mathscr{E}}^2)\boldsymbol{\mathscr{E}} \cdot d\boldsymbol{\mathscr{E}} + \text{const.} \\ &= \tfrac{1}{2}a\boldsymbol{\mathscr{E}}^2 - \tfrac{1}{4}\boldsymbol{\mathscr{E}}^4 + \text{const.}, \end{aligned}$$

or

$$p_s(\boldsymbol{\mathscr{E}}) = K \exp\left(\tfrac{1}{2}a\boldsymbol{\mathscr{E}}^2 - \tfrac{1}{4}\boldsymbol{\mathscr{E}}^4\right). \tag{18.3–12}$$

K is a constant that ensures the normalization of $p_s(\boldsymbol{\mathscr{E}})$. It is sometimes convenient to introduce the *potential* function

$$U(\mathscr{E}) \equiv -\tfrac{1}{2}a\mathscr{E}^2 + \tfrac{1}{4}\mathscr{E}^4, \tag{18.3–13}$$

which allows us to re-write the steady-state solution in the form

$$p_s(\mathscr{E}) = K\,\mathrm{e}^{-U(\mathscr{E})}, \tag{18.3–14}$$

in analogy with the probability and the potential in thermodynamics. We should emphasize, however, that the analogy is a purely formal one; the laser is not a system in thermodynamic equilibrium in general.

As the steady-state probability distribution of the field amplitude $\mathscr{E}$ involves only the square of $\mathscr{E}$, it is more convenient to introduce the light intensity I and the phase ϕ of the field, by writing

$$\left.\begin{aligned} \mathscr{E}_1 &= \sqrt{I}\cos\phi \\ \mathscr{E}_2 &= \sqrt{I}\sin\phi. \end{aligned}\right\} \tag{18.3–15}$$

Then the joint probability density $\mathscr{P}'(I, \phi)$ of I and ϕ is related to $p_s(\mathscr{E})$ by the transformation

$$\mathscr{P}'(I, \phi)\,\mathrm{d}I\,\mathrm{d}\phi = p_s(\mathscr{E})\,\mathrm{d}\mathscr{E}_1\,\mathrm{d}\mathscr{E}_2,$$

or, when $\mathscr{E}^2$ is replaced by I, and $\mathrm{d}I\,\mathrm{d}\phi$ by $2\,\mathrm{d}\mathscr{E}_1\,\mathrm{d}\mathscr{E}_2$, we obtain the formula

$$\mathscr{P}'(I, \phi) = \tfrac{1}{2}K\exp\left(\tfrac{1}{2}aI - \tfrac{1}{4}I^2\right). \tag{18.3–16}$$

As the phase ϕ does not appear at all on the right of this equation, it is evidently distributed uniformly over all angles from 0 to 2π in the steady state. On integrating over ϕ from 0 to 2π we arrive at the probability density $\mathscr{P}(I)$ of the light intensity I,

$$\begin{aligned} \mathscr{P}(I) &= \pi K\exp\left(\tfrac{1}{2}aI - \tfrac{1}{4}I^2\right) \\ &= C\exp\left[-\tfrac{1}{4}(I-a)^2\right], \end{aligned} \tag{18.3–17}$$

where C is another normalization constant. The probability density $\mathscr{P}(I)$ therefore has the structure of a Gaussian distribution in I, except that the Gaussian is truncated at $I = 0$ because the light intensity I is necessarily non-negative.

As the operating regime of the laser is varied from below threshold to above threshold, and the pump parameter a changes from negative to positive values, the Gaussian function moves to the right along the I-axis (see Fig. 18.6). It also changes in height, because of the need to preserve the normalization

$$\int_0^\infty \mathscr{P}(I)\,\mathrm{d}I = 1. \tag{18.3–18}$$

However, once a exceeds 1 or 2, further increases in the pump parameter a merely lead to a translation of the Gaussian distribution, with a negligible change of shape, and the variance of I remains equal to 2 approximately in our dimensionless units. Of course, it should be emphasized that this conclusion, which was derived within the framework of the third-order theory, should not be expected to hold very far above threshold. However, it applies typically up to mean light intensities that are 100 times above the threshold intensity.

By inspection of Eq. (18.3–17) or Fig. 18.6, we can see immediately that when

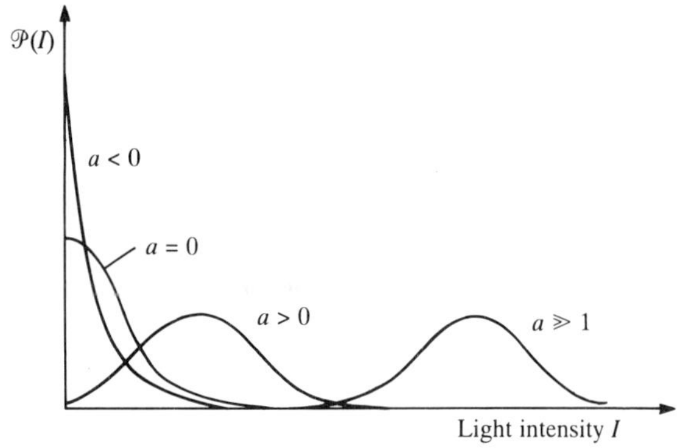

Fig. 18.6 Probability distribution of the laser light intensity I in the steady state for various pump parameters a. The value $a = 0$ defines the laser threshold.

the laser is operating well above threshold, and $a \gg 1$, then

$$\left.\begin{aligned} \langle I \rangle &\approx a \\ \langle (\Delta I)^2 \rangle &\approx 2 \\ \langle (\Delta I)^2 \rangle^{1/2} / \langle I \rangle &\approx (\sqrt{2})/a. \end{aligned}\right\} \tag{18.3–19}$$

It follows that although the fluctuations of the light intensity do not decrease in an absolute sense above threshold, the relative fluctuations tend to zero as the pump parameter increases. On the other hand, it can be shown from Eq. (18.3–12) that the dispersion of the absolute amplitude $|\mathscr{E}|$ tends towards zero above threshold [cf. Eqs. (18.5–15) below]. The laser field therefore acquires the character of an oscillation of constant amplitude, as shown in Fig. 18.7(a), which is however still modulated in frequency and has a random phase. Moreover, as we show later, the bandwidth becomes increasingly narrow well above threshold. It is in the sense of having a constant amplitude and a very narrow bandwidth that the laser field is often described as being *coherent* well above threshold.

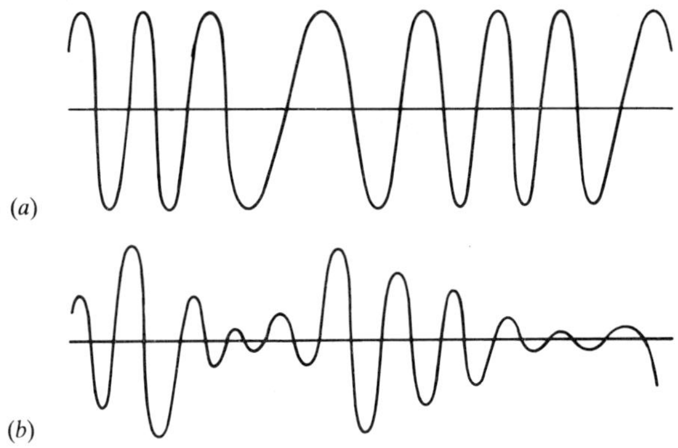

Fig. 18.7 Oscillation of the laser field as function of time (a) well above threshold, (b) below threshold. The bandwidth shown in the figure is greatly exaggerated.

We now turn to the other extreme, in which the laser is operating far below threshold, and the pump parameter a is negative and numerically large. In that case we see immediately from Eq. (18.3–17) that

$$\mathcal{P}(I) \propto \exp(-\tfrac{1}{2}|a|I - \tfrac{1}{4}I^2) \approx \exp(-\tfrac{1}{2}|a|I), \qquad (18.3\text{–}20)$$

because only small values of I have appreciable probability. The probability distribution is therefore very nearly exponential in I, as illustrated in Fig. 18.6. As was shown in Section 13.3, this probability distribution is characteristic of polarized thermal light, and we have from Eq. (18.3–20)

$$\left.\begin{aligned} \langle I\rangle &\approx 2/|a| \\ \langle(\Delta I)^2\rangle^{1/2}/\langle I\rangle &\approx 1. \end{aligned}\right\} \qquad (18.3\text{–}21)$$

The field itself has a Gaussian distribution, and it fluctuates both in amplitude and in phase, as illustrated in Fig. 18.7(b).

Finally, right at the laser threshold, where the pump parameter $a = 0$, $\mathcal{P}(I)$ has the character of a half-truncated Gaussian distribution, and the slope of $\mathcal{P}(I)$ at $I = 0$ is zero. Very small values of the light intensity therefore have approximately equal probability.

The change in the form of $\mathcal{P}(I)$ as the pump parameter is varied has been tested in photoelectric counting measurements. In these experiments the laser beam is allowed to fall on a photodetector, and one records repeatedly the number n of photoelectric pulses produced in a short time interval T, so as to build up a probability distribution $p(n, T)$. If T is short compared with the natural intensity correlation time of the light, then $p(n, T)$ is related to the probability distribution $\mathcal{P}(I)$ of the light intensity I by the integral relation (14.8–12). This allows the theoretical form of $\mathcal{P}(I)$ to be tested, and good agreement with Eq. (18.3–17) has been obtained (Arecchi, Berné and Burlamacchi, 1966; Freed and Haus, 1966a,b; Smith and Armstrong, 1966; Meltzer, Davis and Mandel, 1970). An example of such a measurement is shown in Fig. 18.8.

Alternatively, we can make use of the factorial moments of the photoelectric counts derived from $p(n, T)$, and their relationship to the moments of I [cf. Eq. (14.9–4)], to test the theory. We shall find that $\langle I\rangle$ varies smoothly with pump parameter a, and that, because of the quantum fluctuations, there is no sharp change of $\langle I\rangle$ at threshold as there was in the deterministic case described by Eq. (18.2–29). Nevertheless, an order parameter, which vanishes below threshold on the disordered side of the phase transition and is non-zero above threshold, can again be identified.

18.3.3 The phase transition analogy for a fluctuating laser field

Although the probability distribution $\mathcal{P}(I)$ of the laser light intensity changes in a continuous manner as the operating point of the laser is varied from below threshold to above threshold, the electromagnetic field still undergoes a phase transition at threshold. However, in contrast to the situation we encountered in Section 18.2, neither the mean field amplitude $\langle|\mathcal{E}|\rangle$ nor the mean intensity $\langle I\rangle$ is now a suitable order parameter, because both are non-zero for all values of the pump parameter a.

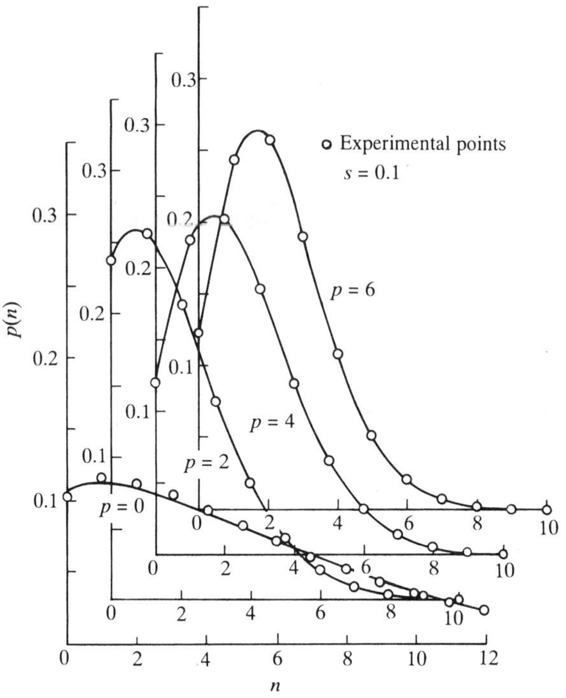

Fig. 18.8 Comparison of the measured photoelectric counting probabilities $P(n)$ with the theoretical forms having the same mean, shown as continuous curves, for four different pump parameters (here called p). The ratio of the photon counting time to the coherence time $s = 0.1$. (Reproduced from Meltzer, Davis and Mandel, 1970.)

In order to identify a suitable order parameter, it is convenient to examine the potential function $U(\mathscr{E})$ given by Eq. (18.3–13) which was introduced to describe $p_s(\mathscr{E})$. The potential is illustrated as a function of $|\mathscr{E}|$ in Fig. 18.9(a) for negative, zero, and positive values of the pump parameter a. It is clear from Eq. (18.3–14) that there exists a one-to-one relationship between the probability $p_s(\mathscr{E})$ and the potential $U(\mathscr{E})$, such that the most probable value of $|\mathscr{E}|$ is always associated with the lowest value of the potential $U(\mathscr{E})$. It is then at once obvious from Fig. 18.9(a) that, so long as the laser is operating below or at threshold, and $a \leqslant 0$, the most probable field amplitude $|\mathscr{E}|$ is zero. However, once the threshold is exceeded, and $a > 0$, the most probable value of $|\mathscr{E}|$ becomes $\sqrt{a}$. The most probable value of the field amplitude $|\mathscr{E}|$ therefore changes from zero to non-zero at the laser threshold, and it is a natural order parameter for the phase transition.

We can give an equally valid description of the phase transition in terms of the light intensity I. Let us introduce a potential $U(I)$ by writing Eq. (18.3–17) in the form

$$\left.\begin{aligned} \mathscr{P}(I) &= \pi K\,\mathrm{e}^{-U(I)}, \\ \text{with}\qquad U(I) &= -\tfrac{1}{2}aI + \tfrac{1}{4}I^2. \end{aligned}\right\} \qquad (18.3\text{–}22)$$

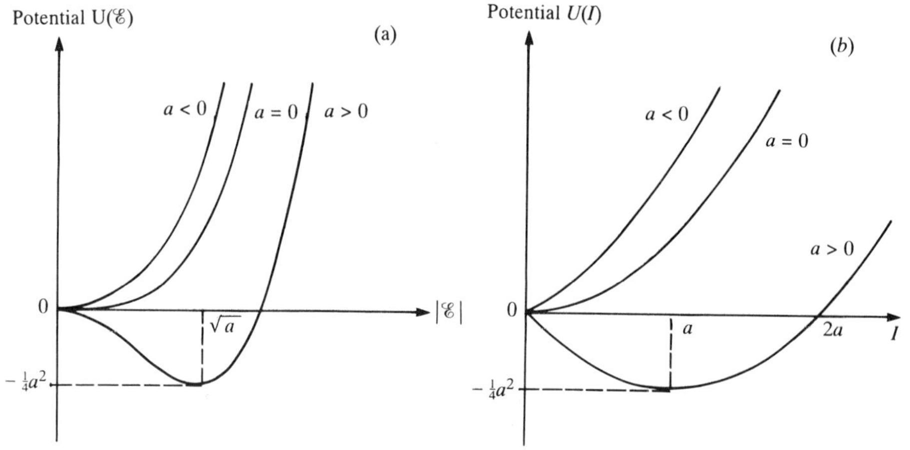

Fig. 18.9 The variation of the potential with (a) absolute field amplitude $|\mathscr{E}|$, (b) intensity I, for several different pump parameters below threshold ($a < 1$), at threshold ($a = 1$), and above threshold ($a > 1$).

The form of the potential $U(I)$ is illustrated in Fig. 18.9(b) for three different values of the pump parameter a. The most probable value of the light intensity is again associated with the lowest value of the potential, and this most probable value is seen to be zero at and below the laser threshold and non-zero and equal to a only above threshold. The most probable light intensity therefore also serves as a possible order parameter for the phase transition.

In the Landau theory of phase transitions it is customary to make an expansion of the thermodynamic potential in a power series in the order parameter near the phase transition (Haken, 1977, Chap. 6; Landau and Lifshitz, 1980). The expansion coefficients in general are functions of the temperature, and symmetry considerations may cause some of them to vanish. If one inserts the proper value of the order parameter at each temperature into the potential, one generally encounters discontinuities in some of the derivatives of the potential with respect to temperature. These discontinuities determine the order of the phase transition.

We can apply an analogous procedure to the potential given by Eq. (18.3–22). We regard $U(I, a)$ as analogous to the thermodynamic potential, and the negative of the pump parameter a, which determines the degree of excitation of the system, as being analogous to the temperature. The most probable value $I_{\mathrm{m}}(a)$ of the light intensity is the order parameter, as we have just seen, and it is given by the relations

$$\left.\begin{aligned} I_{\mathrm{m}}(a) &= 0 \quad \text{if } a \leqslant 0 \\ &= a \quad \text{if } a \geqslant 0. \end{aligned}\right\} \tag{18.3–23}$$

If we substitute $I_{\mathrm{m}}(a)$ for I in $U(I, a)$ we obtain the most probable potential $U_{\mathrm{m}}(a)$,

$$\left.\begin{aligned} U_{\mathrm{m}}(a) &= 0 \qquad\quad \text{if } a \leqslant 0 \\ &= -\tfrac{1}{4}a^2 \quad \text{if } a \geqslant 0, \end{aligned}\right\} \tag{18.3–24}$$

whose form is illustrated in Fig. 18.10. It has a continuous first derivative, but its second derivative $d^2U_m(a)/da^2$ is discontinuous at the laser threshold $a = 0$. We can also construct a quantity that is the analog of the *thermodynamic entropy* S, even though the laser is not a thermal equilibrium system, by writing

$$S(a) = \left.\frac{\partial U(I, a)}{\partial a}\right|_{I=I_m}. \tag{18.3–25}$$

With the help of Eqs. (18.3–22) and (18.3–23) we find immediately that

$$\left.\begin{aligned} S(a) &= 0 && \text{if } a \leqslant 0 \\ &= -\tfrac{1}{2}a && \text{if } a \geqslant 0, \end{aligned}\right\} \tag{18.3–26}$$

which happens to be the derivative of $U_m(a)$ in this case. The form of $S(a)$ is also illustrated in Fig. 18.10. It is continuous at the phase transition, but its derivative has a discontinuity. The laser is therefore described as undergoing a *continuous* or *second-order phase transition* at the threshold. A first-order, or discontinuous, phase transition would be associated with a discontinuity of the entropy, and with a less smooth most probable potential. We shall see in Chapter 19 that a two-mode laser can, under certain circumstances, exhibit a first-order phase transition, because of competition effects between the modes.

The potential $U(I, a)$ or $U(\mathscr{E}, a)$ is sometimes defined so that the normalization constant for the probability distribution is included within it. This clearly does not affect its variation with light intensity I, or the positions of maxima or minima or points of discontinuity. The entropy $S(a)$ is affected to the extent of an additive, smooth function of a. On the other hand, discontinuities of the entropy or its derivatives remain unchanged.

18.3.4 Moments of the light intensity

By making use of the probability distribution (18.3–17), we can readily evaluate the moments of the light intensity I as a function of the pump parameter a (Risken, 1965, 1966, 1970; Risken and Vollmer, 1967a,b; Hempstead and Lax,

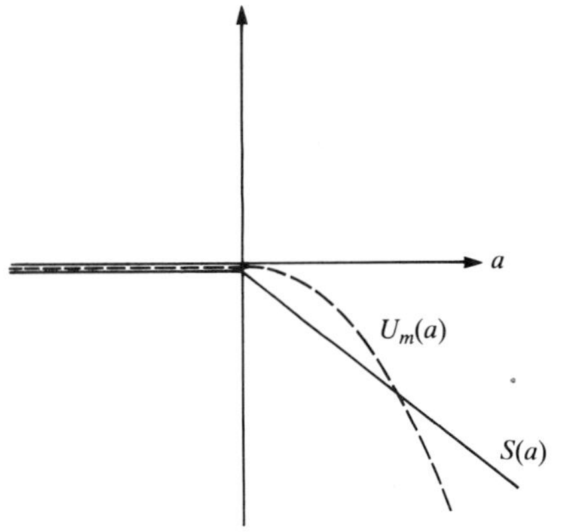

Fig. 18.10 Variation of the most probable potential $U_m(a)$ and the entropy $S(a)$ with pump parameter a.

1967). For example, for the mean intensity we have

$$\langle I\rangle = \int_0^\infty I\,\mathrm{e}^{-(I-a)^2/4}\,\mathrm{d}I \bigg/ \int_0^\infty \mathrm{e}^{-(I-a)^2/4}\,\mathrm{d}I.$$

With the help of the change of variable $(I-a)/2 = x$, this takes the form

$$\begin{aligned}\langle I\rangle &= a + 2\int_{-a/2}^\infty x\,\mathrm{e}^{-x^2}\,\mathrm{d}x \bigg/ \int_{-a/2}^\infty \mathrm{e}^{-x^2}\,\mathrm{d}x \\ &= a + \mathrm{e}^{-a^2/4} \bigg/ \int_{-a/2}^\infty \mathrm{e}^{-x^2}\,\mathrm{d}x.\end{aligned}$$

We now introduce the *Gaussian error function* in the usual way, by putting

$$\operatorname{erf} y \equiv \frac{2}{\sqrt{\pi}}\int_0^y \mathrm{e}^{-x^2}\,\mathrm{d}x.$$

This function has the properties $\operatorname{erf}(-y) = -\operatorname{erf} y$ and $\operatorname{erf}(\infty) = 1$, so that we can immediately express the mean light intensity in the form

$$\langle I(a)\rangle = a + \frac{2\,\mathrm{e}^{-a^2/4}}{(\sqrt{\pi})[1 + \operatorname{erf}(\tfrac{1}{2}a)]}. \tag{18.3–27}$$

For large pump parameters a, the second term is very small, and it becomes a small correction to the solution $\langle I\rangle \approx a$ that we gave in Eq. (18.3–19). For large negative pump parameters a, we recover the result in Eq. (18.3–21) after making an asymptotic expansion of the error function. Right at the laser threshold, where $a = 0$, we obtain $\langle I(0)\rangle = 2/\sqrt{\pi}$ in our dimensionless units. Figure 18.11(a) shows a plot of the mean light intensity $\langle I\rangle$ against the pump parameter a. The curve should be compared with Fig. 18.4, which was based on the deterministic equation of motion of the laser. The main difference is that the average light intensity is not zero below threshold, because spontaneous emission causes some light to be emitted under all conditions, so that there is no discontinuity at the threshold. Well below threshold ($-a \gg 1$) an asymptotic expansion of the error function shows that $\langle I\rangle \to 2/|a|$. These conclusions have been confirmed experimentally, as shown in Fig. 18.11(b) and (c).

We can evaluate the higher moments of the light intensity in a similar manner from the formula

$$\langle I^r\rangle = \int_0^\infty I^r\,\mathrm{e}^{-(I-a)^2/4}\,\mathrm{d}I \bigg/ \int_0^\infty \mathrm{e}^{-(I-a)^2/4}\,\mathrm{d}I, \quad r = 1, 2, 3, \ldots.$$

By introducing the same change of variable as before, we may readily show that the variance is given by the equation

$$\begin{aligned}\langle(\Delta I)^2\rangle &= \langle I^2\rangle - \langle I\rangle^2 \\ &= 2\left[1 - \frac{a\,\mathrm{e}^{-a^2/4}}{(\sqrt{\pi})[1 + \operatorname{erf}(\tfrac{1}{2}a)]} - \frac{2\,\mathrm{e}^{-a^2/2}}{\pi[1 + \operatorname{erf}(\tfrac{1}{2}a)]^2}\right],\end{aligned} \tag{18.3–28}$$

and this expression tends to 2 for large a, in agreement with Eq. (18.3–19). At the laser threshold $a = 0$, $\langle(\Delta I)^2\rangle = 2 - 4/\pi$, and the dimensionless ratio $\langle(\Delta I)^2\rangle/\langle I\rangle^2 = \pi/2 - 1 \approx 0.57$. This number provides a convenient means of identifying the laser threshold in practice. Equations (18.3–27) and (18.3–28)

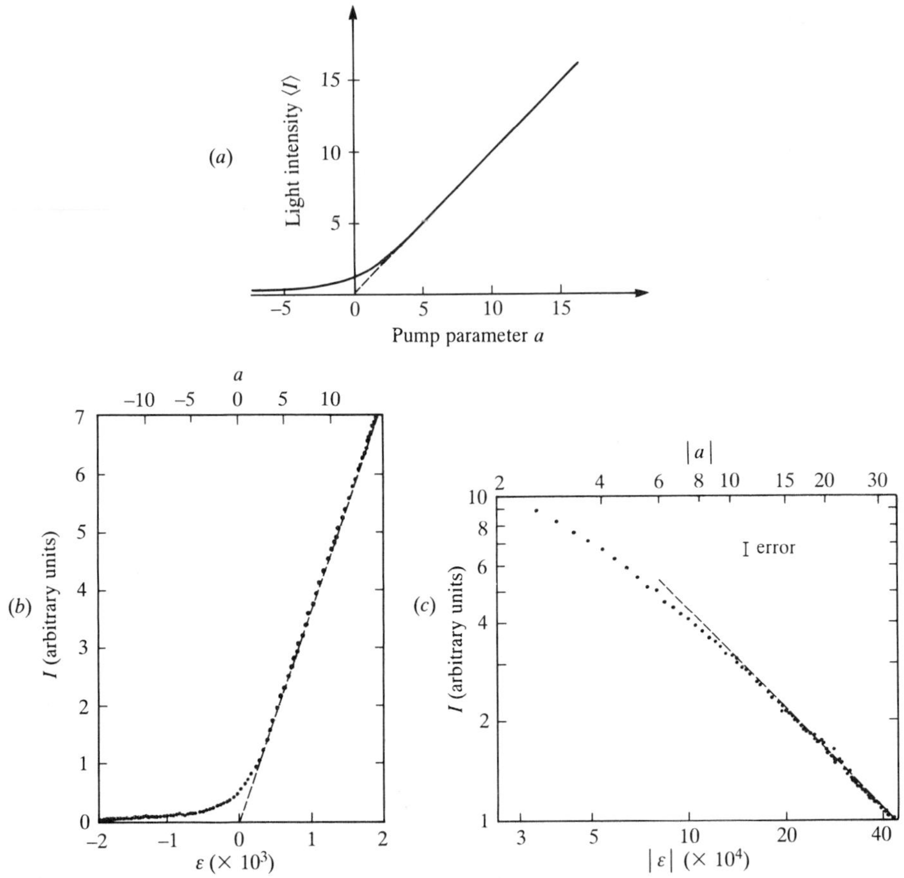

Fig. 18.11 (*a*) Predicted variation of the mean laser light intensity with pump parameter. (*b*) and (*c*) Measured variation of the mean laser light intensity with relative population inversion ε above the critical value and with pump parameter a, (*b*) above and below threshold; (*c*) well below threshold. (Reproduced from Corti and Degiorgio, 1976b.)

allow us to plot the ratio $\langle(\Delta I)^2\rangle/\langle I\rangle^2$ against the mean light intensity $\langle I\rangle$. The result is shown by the continuous curve in Fig. 18.12, and it has been confirmed experimentally by photoelectric counting and correlation measurements (Freed and Haus, 1966a,b; Smith and Armstrong, 1966; Arecchi, Rodari and Sona, 1967; Davidson and Mandel, 1967; Chang, Korenman, Alley and Detenbeck, 1969). We recall, with the help of Eq. (14.9–4), that the number of photoelectric counts n registered by an illuminated photodetector in a time that is very short compared with the intensity correlation time, has factorial moments which are simply related to the moments of the light intensity, namely

$$\frac{\langle n(n-1)(n-2)\ldots(n-r+1)\rangle}{\langle n\rangle^r} = \frac{\langle I^r\rangle}{\langle I\rangle^r}, \quad \mathrm{r} = 1, 2, 3, \ldots . \tag{18.3–29}$$

The moments of I are therefore easily determined from photoelectric counting measurements.

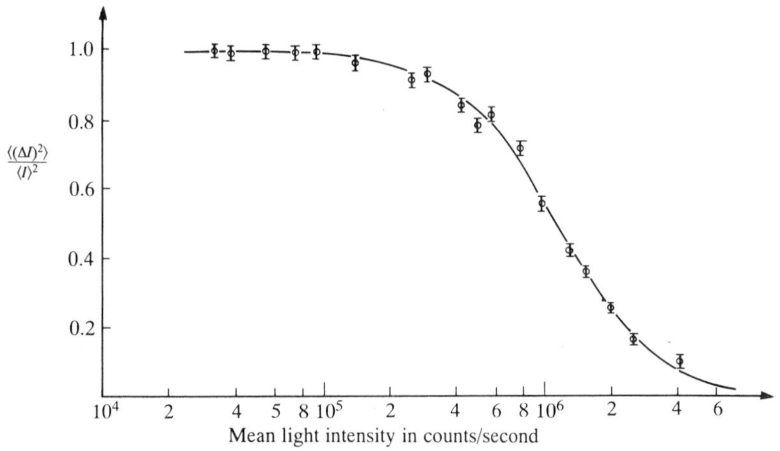

Fig. 18.12 Variation of the relative mean squared light intensity fluctuation of the laser with mean intensity. The full curve is based on Eqs. (18.3–27) and (18.3–28), with the scale factor for the logarithmic intensity scale chosen arbitrarily for best agreement with the experimental results shown. (Reproduced from Davidson and Mandel, 1967.)

Alternatively, one can plot $\langle(\Delta I)^2\rangle/\langle I\rangle^2$ directly against the pump parameter a, and arrive at the curve shown in Fig. 18.13. However, this is less easy to confirm directly by experiment, because of the difficulty of determining the pump parameter. Indeed, in practice, a is often derived from measurements of $\langle(\Delta I)^2\rangle/\langle I\rangle^2$ with the help of Eqs. (18.3–27) and (18.3–28).

We shall not calculate higher-order moments of the light intensity directly, because it turns out to be simpler to calculate the cumulants (cf. Section 1.4), which are related to the moments, and can be derived from each other by successive differentiation (Risken, 1965, 1966, 1970). In order to show this, let us define

$$F(a) \equiv \int_0^\infty \exp\left(\tfrac{1}{2}aI - \tfrac{1}{4}I^2\right) \mathrm{d}I.$$

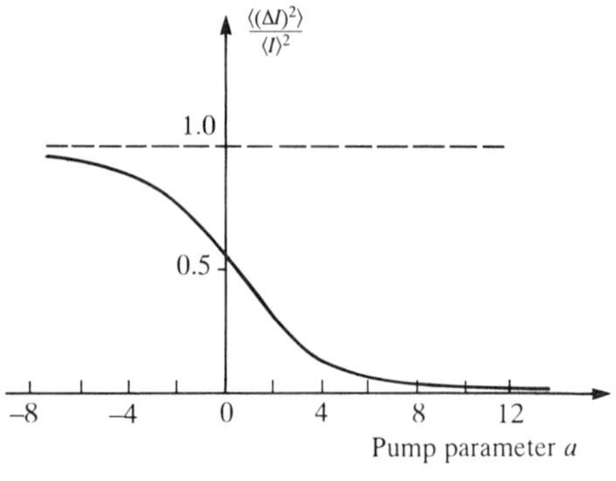

Fig. 18.13 Expected variation of the relative mean squared light intensity fluctuations with pump parameter a.

Then the moment generating function $M(s, a)$ of I (cf. Section 1.4) is given by the relation

$$M(s, a) = \int_0^\infty e^{sI} \exp\left(\tfrac{1}{2}aI - \tfrac{1}{4}I^2\right) dI \Big/ \int_0^\infty \exp\left(\tfrac{1}{2}aI - \tfrac{1}{4}I^2\right) dI$$
$$= F(a + 2s)/F(a), \qquad (18.3\text{–}30)$$

and the moments of I can be obtained by successive differentiation of $M(s, a)$. For example, the mean intensity can be written as

$$\langle I \rangle = \frac{1}{F(a)}\left[\frac{d}{ds}F(a + 2s)\right]_{s=0} = \left[\frac{d}{ds}\ln F(a + 2s)\right]_{s=0}. \qquad (18.3\text{–}31)$$

Now the cumulants κ_n $(n = 1, 2, \ldots)$ of the probability density $\mathcal{P}(I)$ are defined to be the coefficients of $s^n/n!$ in the expansion of $\ln[M(s, a)]$ as a power series in s (cf. Section 1.4), so that

$$\kappa_{n+1}(a) = \left[\frac{d^n}{ds^n}\frac{d}{ds}\ln F(a + 2s)\right]_{s=0}$$
$$= 2^n \frac{d^n}{da^n}\left[\frac{d}{ds}\ln F(a + 2s)\right]_{s=0}$$
$$= 2^n \frac{d^n}{da^n}\langle I(a)\rangle. \qquad (18.3\text{–}32)$$

The last line follows immediately from Eq. (18.3–31). The first cumulant $\kappa_1(a)$ is just the mean light intensity $\langle I(a)\rangle$, and successive cumulants can be obtained by successive differentiations. The behavior of the first four cumulants is illustrated in Fig. 18.14.

In practice, it is often preferable to work with normalized cumulants

$$c_n(a) \equiv \kappa_n(a)/\langle I(a)\rangle^n. \qquad (18.3\text{–}33)$$

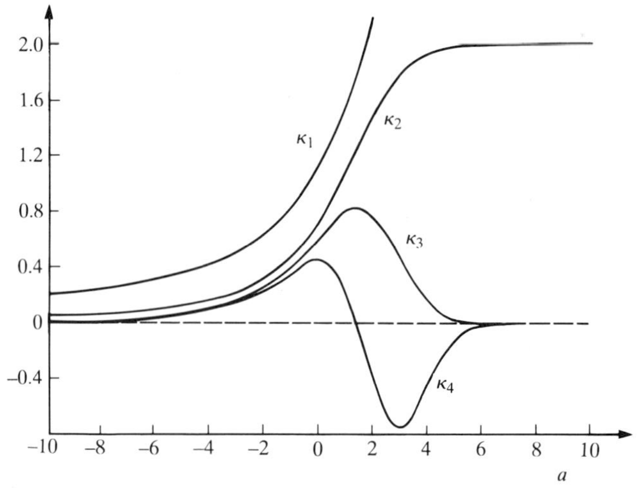

Fig. 18.14 The first four cumulants of the laser light intensity as a function of pump parameter a. (Reproduced from Risken, 1970.)

The variation of $c_n(a)$ with the normalized mean light intensity $\langle I(a)\rangle/\langle I(0)\rangle$ of the laser has been tested up to $n = 4$ by Chang, Korenman, Alley and Detenbeck (1969) in photoelectric counting experiments. The results are shown in Fig. 18.15 and demonstrate good agreement between theory and experiment. As the n'th cumulant measures intrinsic fluctuations of the n'th order (or intrinsic n-photon correlations in a certain sense), these measurements represent quite a searching test of the steady-state solution given by Eq. (18.3–17). The steady state laser theory can therefore be considered as being very well confirmed.

Before returning to the time-dependent equation of motion (18.3–4), and investigating its general solution, we shall examine the laser problem within the framework of the quantum theory of radiation. We shall see in Section 18.5, with the help of the diagonal coherent-state representation that was discussed in Section 11.8 and with some approximations, that the quantum theory leads to the same Fokker–Planck equation (18.3–4) as the semiclassical theory with added noise.

18.4 Quantum theory of the laser

As we have seen, the semiclassical theory of the laser leads to a deterministic equation of motion for the field amplitude, and questions about linewidths and fluctuations are, strictly speaking, outside the domain of the theory. Fluctuations were incorporated at a later stage by the introduction of Langevin noise sources, but this is an ad hoc procedure that cannot be easily justified in a rigorous

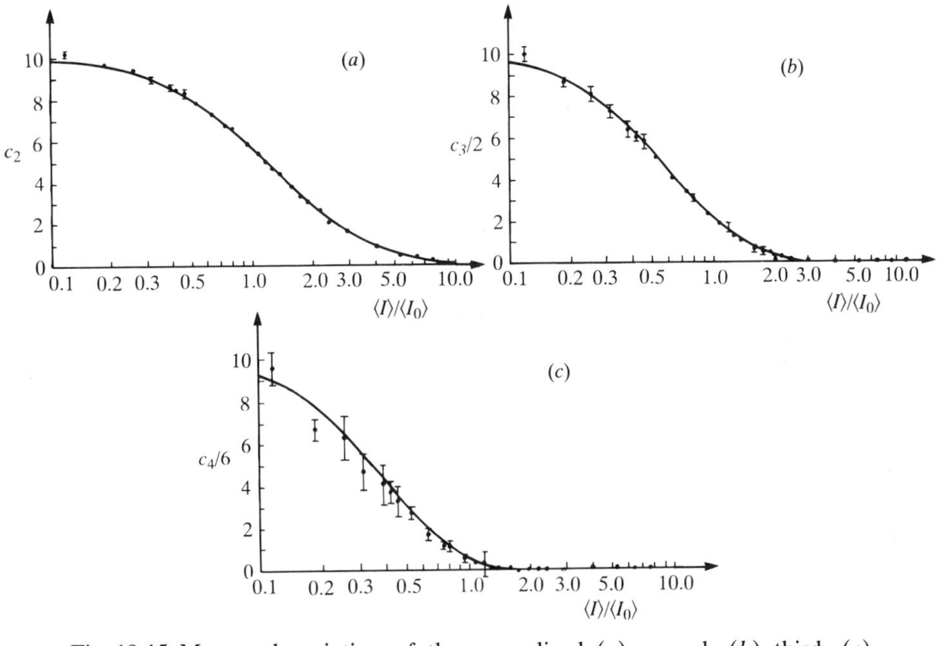

Fig. 18.15 Measured variation of the normalized (*a*) second, (*b*) third, (*c*) fourth cumulants of the laser light intensity with the mean intensity. The full curves are theoretical. (Reproduced from Chang, Korenman, Alley and Detenbeck, 1969.)

manner, whose validity rests largely on the agreement obtained with other approaches. The quantized field treatment provides a sounder and more consistent foundation for the laser problem. In addition, it is able to answer certain questions that are not meaningful at all within the framework of a semiclassical theory, such as how many photons are present in the laser cavity at threshold, and what is their probability distribution.

The laser was treated in a fully quantum mechanical manner by several workers (Haken and Sauermann, 1963a,b; Haken, 1964, 1970; see also Lax and Louisell, 1967). In the following treatment we shall largely make use of the approach of Scully and Lamb (1967, 1969, 1970; Sargent, Scully and Lamb, 1974). However, for simplicity, we replace the Weisskopf–Wigner procedure used by them with a simpler perturbative procedure that leads to essentially similar dynamics, but with slightly different coefficients (Scully, 1969). We shall make other simplifying approximations, in parallel with the semiclassical treatment given earlier. For example, we shall neglect any motion, or inhomogeneous broadening, of the atomic systems, and we take the cavity resonator to be very nearly on resonance with the atomic frequency ω_0.

Once again we consider a set of identical two-level atoms or atomic dipoles located at various positions within the laser cavity, interacting with the field of a single cavity mode. If the atomic population is inverted, the coupling between atoms and the field may cause the number of photons in the cavity mode to grow in time. We start by examining the effect of a single excited atom on the state of the laser field.

The energy of the coupled system consisting of a single-mode electromagnetic field of frequency ω and a two-level atom of resonance frequency ω_0 located at position $\mathbf{r}$ may be taken to have the form [cf. Eq. (15.3–1)]

$$\hat{H} = E_0 + \hbar\omega_0\hat{R}_3 + \hbar\omega(\hat{n} + \tfrac{1}{2}) - \boldsymbol{\mu}_{12}\cdot\hat{\mathbf{E}}(\mathbf{r}, t)[\hat{b}(t) + \hat{b}^\dagger(t)]. \quad (18.4\text{–}1)$$

Here E_0 is the average energy of the two laser levels, $\boldsymbol{\mu}_{12}$ is the transition dipole moment of the atom, which is here assumed to be real, as for a $\Delta m = 0$ transition, and $\hat{n}$ is the photon number operator, as usual. It will be convenient to work in the interaction picture (cf. Section 14.1), in which the atomic lowering and raising operators evolve in time according to the relations

$$\hat{b}(t) = \hat{b}(t_0)\,\mathrm{e}^{-\mathrm{i}\omega_0(t-t_0)}$$
$$\hat{b}^\dagger(t) = \hat{b}^\dagger(t_0)\,\mathrm{e}^{\mathrm{i}\omega_0(t-t_0)},$$

where t_0 is the initial time from which evolution is assumed to proceed. The single-mode electric field $\hat{\mathbf{E}}(\mathbf{r}, t)$ in the interaction picture may then be expressed in the form [cf. Eq. (18.2–4)]

$$\hat{\mathbf{E}}(\mathbf{r}, t) = \frac{\boldsymbol{\varepsilon}}{\sqrt{V}}\left(\frac{\hbar\omega}{2\varepsilon_0}\right)^{1/2}[\hat{a}(t_0)u(\mathbf{r})\,\mathrm{e}^{-\mathrm{i}\omega(t-t_0)} + \mathrm{h.c.}], \quad (18.4\text{–}2)$$

where $u(\mathbf{r})$ is the normalized cavity mode function discussed in Section 18.2, V is the cavity volume and $\boldsymbol{\varepsilon}$ is a real unit polarization vector representing a linearly polarized wave. The interaction term in Eq. (18.4–1) becomes

$$\hat{H}_I(t) = \hbar g[\hat{a}\hat{b}^\dagger u(\mathbf{r})\,e^{i(\omega_0-\omega)(t-t_0)} + \text{h.c.} + \hat{a}\hat{b}u(\mathbf{r})\,e^{-i(\omega_0+\omega)(t-t_0)} + \text{h.c.}]. \tag{18.4–3}$$

We have suppressed the time argument in the operators $\hat{a}$, $\hat{a}^\dagger$, $\hat{b}$, $\hat{b}^\dagger$ with the understanding that it is always t_0 unless shown otherwise, and we define

$$g \equiv -\left(\frac{\omega}{2\hbar\varepsilon_0 V}\right)^{1/2} \boldsymbol{\mu}_{12}\cdot\boldsymbol{\varepsilon}. \tag{18.4–4}$$

Here g is the coupling constant characterizing the strength of the interaction.

Let us suppose that at time t_0, when the interaction is assumed to be turned on, the state of the coupled system is factorizable, so that we can express the density operator in the form

$$\hat{\rho}(t_0) = \hat{\rho}_A(t_0) \otimes \hat{\rho}_F(t_0), \tag{18.4–5}$$

where $\hat{\rho}_A(t_0)$, $\hat{\rho}_F(t_0)$ are reduced density operators of the atom and the field, respectively. As a result of the interaction, $\hat{\rho}$ evolves in time in the interaction picture, and its form at a later time t can be given the perturbation expansion (14.1–11). By taking the trace over the atomic variables, we can express the effect of the interaction on the state of the laser field at time t, in the form

$$\hat{\rho}_F(t) = \hat{\rho}_F(t_0) + \text{tr}_A \sum_{r=1}^{\infty} \frac{1}{(i\hbar)^r} \int_{t_0}^{t} dt_1 \int_{t_0}^{t_1} dt_2 \ldots \int_{t_0}^{t_{r-1}} dt_r$$
$$\times [\hat{H}_I(t_1), [\hat{H}_I(t_2), [\ldots [\hat{H}_I(t_r), \hat{\rho}(t_0)]] \ldots]]. \tag{18.4–6}$$

18.4.1 Master equation of the laser field

In order to determine the effect of a single excited atom on the laser field, we take the initial atomic state to be the upper state

$$\hat{\rho}_A(t_0) = |2\rangle\langle 2|, \tag{18.4–7}$$

and we allow the interaction to proceed for a time that is of the order of the lifetime of the upper state. As indicated in Fig. 18.16, it is assumed that the two laser levels $|1\rangle$ and $|2\rangle$ decay to other states with average lifetimes T_1 and T_2, respectively. The effect of an ensemble of such excited atoms on the field is therefore obtainable, at least approximately, by choosing $t - t_0 = \Delta t$ in Eq.

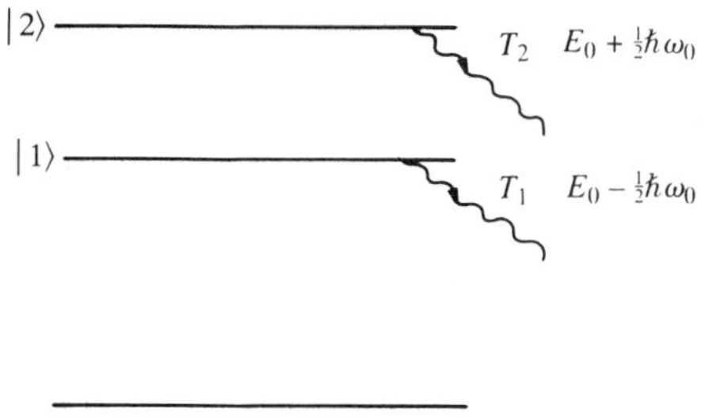

Fig. 18.16 Energy level scheme for the laser atom.

(18.4–6) to be the atomic lifetime, and averaging over the ensemble of Δt with an exponential probability distribution of average T_2, as in Section 18.2. Of course, it would be somewhat more realistic to assume that the excited state population decays continuously exponentially in time, as is done in the Weisskopf–Wigner procedure used by Scully and Lamb (1967, 1969, 1970) (see also Sargent, Scully and Lamb, 1974, Chap. 17). However, the perturbative approach leads to a substantial simplification and essentially to the same answer (Scully, 1969), besides making contact with the semiclassical treatment given earlier in Section 18.2. Moreover, as before, we shall limit ourselves to the first two non-vanishing contributions in the expansion (18.4–6).

We now substitute Eqs. (18.4–3) and (18.4–5) in Eq. (18.4–6). It is at once clear that if the lifetime $\Delta t = t - t_0$ is very long compared with the optical period $2\pi/\omega_0$, then the terms in $e^{\pm i(\omega+\omega_0)t}$ in Eq. (18.4–3) will integrate almost to zero and can be neglected. On the other hand, the contributions made by the terms in $e^{\pm i(\omega-\omega_0)t}$ depend strongly on the magnitude of the detuning $\omega - \omega_0$. For simplicity we shall assume that the detuning is so small that

$$|\omega - \omega_0|\Delta t \ll 1,$$

which allows us to neglect the oscillatory factors altogether, and to treat $\hat{H}_I$ as independent of time under the integral in Eq. (18.4–6). We then arrive at the equation

$$\hat{\rho}_F(t_0 + \Delta t) = \hat{\rho}_F(t_0) + \mathrm{tr}_A\left\{\frac{\Delta t}{i\hbar}[\hat{H}_I, \hat{\rho}(t_0)] + \left(\frac{\Delta t}{i\hbar}\right)^2 \frac{1}{2!}[\hat{H}_I, [\hat{H}_I, \hat{\rho}(t_0)]] + \ldots\right\}. \tag{18.4–8}$$

The problem therefore reduces to the evaluation of commutators of successively higher orders.

These are easily calculated with the help of Eqs. (15.1–4) and (15.1–5). Thus we find

$$\begin{aligned}[\hat{H}_I, \hat{\rho}(t_0)] &= \hbar g[\hat{a}\hat{b}^\dagger u(\mathbf{r}) + \text{h.c.}, |2\rangle\langle 2| \otimes \hat{\rho}_F(t_0)] \\ &= \hbar g[\hat{b}\hat{a}^\dagger \hat{\rho}_F(t_0) u^*(\mathbf{r}) - \hat{b}^\dagger \hat{\rho}_F(t_0)\hat{a} u(\mathbf{r})],\end{aligned} \tag{18.4–9}$$

when we make use of the fact that atomic and field operators commute at the same time, and that $|1\rangle\langle 2|$ and $|2\rangle\langle 1|$ are representations of the lowering and raising operators $\hat{b}$ and $\hat{b}^\dagger$, respectively. As $\hat{b}$ and $\hat{b}^\dagger$ are traceless, it follows immediately from Eq. (18.4–9) that

$$\mathrm{tr}_A[\hat{H}_I, \hat{\rho}(t_0)] = 0, \tag{18.4–10}$$

so that the first term under the trace in the perturbation expansion (18.4–8) vanishes. With the help of Eq. (18.4–9) we now calculate the second-order commutator, and we find

$$\begin{aligned}[\hat{H}_I, [\hat{H}_I, \hat{\rho}(t_0)]] &= (\hbar g)^2[\hat{a}\hat{b}^\dagger u(\mathbf{r}) + \text{h.c.}, \hat{b}\hat{a}^\dagger \hat{\rho}_F(t_0) u^*(\mathbf{r}) - \hat{b}^\dagger \hat{\rho}_F(t_0)\hat{a} u(\mathbf{r})] \\ &= (\hbar g)^2 |u(\mathbf{r})|^2 [\hat{b}^\dagger \hat{b}(\hat{a}\hat{a}^\dagger \hat{\rho}_F(t_0) + \hat{\rho}_F(t_0)\hat{a}\hat{a}^\dagger) - 2\hat{b}\hat{b}^\dagger \hat{a}^\dagger \hat{\rho}_F(t_0)\hat{a}].\end{aligned} \tag{18.4–11}$$

Because

$$\mathrm{tr}_A\,\hat{b}\hat{b}^\dagger = 1 = \mathrm{tr}_A\,\hat{b}^\dagger\hat{b},$$

it follows that

$$\mathrm{tr}_A\,[\hat{H}_I, [\hat{H}_I, \hat{\rho}(t_0)]] = (\hbar g)^2|u(\mathbf{r})|^2[\hat{a}\hat{a}^\dagger\hat{\rho}_F(t_0) - \hat{a}^\dagger\hat{\rho}_F(t_0)\hat{a} + \text{h.c.}], \quad (18.4\text{–}12)$$

which is evidently the lowest-order non-vanishing contribution in the perturbation expansion. It is of the second order in the amplitude of the field. By a similar argument we readily find that

$$[\hat{H}_I, [\hat{H}_I, [\hat{H}_I, \hat{\rho}(t_0)]]] = (\hbar g)^3|u(\mathbf{r})|^2[u^*(\mathbf{r})\hat{b}(\hat{a}^\dagger\hat{a}\hat{a}^\dagger\hat{\rho}_F(t_0) + 3\hat{a}^\dagger\hat{\rho}_F(t_0)\hat{a}\hat{a}^\dagger) - \text{h.c.}], \quad (18.4\text{–}13)$$

which again vanishes when traced over atomic variables, whereas

$$\begin{aligned}[\hat{H}_I, [\hat{H}_I, [\hat{H}_I, [\hat{H}_I, \hat{\rho}(t_0)]]]] &= (\hbar g)^4|u(\mathbf{r})|^4[\hat{b}^\dagger\hat{b}(\hat{a}\hat{a}^\dagger\hat{a}\hat{a}^\dagger\hat{\rho}_F(t_0) + 3\hat{a}\hat{a}^\dagger\hat{\rho}_F(t_0)\hat{a}\hat{a}^\dagger + \text{h.c.}) \\ &\quad - 4\hat{b}\hat{b}^\dagger(\hat{a}^\dagger\hat{a}\hat{a}^\dagger\hat{\rho}_F(t_0)\hat{a} + \text{h.c.})], \quad (18.4\text{–}14)\end{aligned}$$

and this result leads to the formula

$$\mathrm{tr}_A\,[\hat{H}_I, [\hat{H}_I, [\hat{H}_I, [\hat{H}_I, \hat{\rho}(t_0)]]]] = (\hbar g)^4|u(\mathbf{r})|^4[\hat{a}\hat{a}^\dagger\hat{a}\hat{a}^\dagger\hat{\rho}_F(t_0) + 3\hat{a}\hat{a}^\dagger\hat{\rho}_F(t_0)\hat{a}\hat{a}^\dagger - 4\hat{a}^\dagger\hat{a}\hat{a}^\dagger\hat{\rho}_F(t_0)\hat{a} + \text{h.c.}]. \quad (18.4\text{–}15)$$

Finally, after averaging over the lifetimes Δt with the exponential probability distribution

$$P(\Delta t) = (1/T_2)\,\mathrm{e}^{-\Delta t/T_2},$$

we obtain from Eq. (18.4–8), with the help of Eqs. (18.4–12) and (18.4–15),

$$\begin{aligned}\Delta\hat{\rho}_F(t_0) = &-(gT_2)^2|u(\mathbf{r})|^2[\hat{a}\hat{a}^\dagger\hat{\rho}_F(t_0) - \hat{a}^\dagger\hat{\rho}_F(t_0)\hat{a} + \text{h.c.}] \\ &+ (gT_2)^4|u(\mathbf{r})|^4[\hat{a}\hat{a}^\dagger\hat{a}\hat{a}^\dagger\hat{\rho}_F(t_0) + 3\hat{a}\hat{a}^\dagger\hat{\rho}_F(t_0)\hat{a}\hat{a}^\dagger - 4\hat{a}^\dagger\hat{a}\hat{a}^\dagger\hat{\rho}_F(t_0)\hat{a} + \text{h.c.}]\end{aligned} \quad (18.4\text{–}16)$$

for the average change in the state of the laser field brought about by the interaction with a single excited atom at $\mathbf{r}$. As in the semiclassical approach given in Section 18.2, the calculation has been carried through to the second non-vanishing contribution from the perturbation expansion. The first term in Eq. (18.4–16) corresponds to a single photon emission by the excited atom, whereas the second represents the contribution of two cycles of emission and re-excitation by the atom.

A real laser of course has many excited atoms or molecules distributed throughout the active medium, whose excitations are replenished by some pumping mechanism at a certain rate R_2. The contribution to the change of $\hat{\rho}_F$ made by any one of them is therefore small, and if the laser field evolves slowly we may look on the change $\Delta\hat{\rho}_F(t_0)$ given by Eq. (18.4–16) as an approximate differential. When $\Delta\hat{\rho}_F(t_0)$ is multiplied by R_2, and averaged over the different atomic positions $\mathbf{r}$, it gives the average rate of change of the density operator of the laser field brought about by pumping, or by the gain mechanism, provided that the state of the field is changing much more slowly than that of each atom. The result is

$$\left(\frac{\partial \hat{\rho}_F}{\partial t}\right)_{\text{gain}} =$$
$$-\frac{1}{N}R_2(gT_2)^2\int \eta(\mathbf{r})|u(\mathbf{r})|^2[\hat{a}\hat{a}^\dagger\hat{\rho}_F - \hat{a}^\dagger\hat{\rho}_F\hat{a} + \text{h.c.}]\,\mathrm{d}^3r$$
$$+\frac{1}{N}R_2(gT_2)^4\int \eta(\mathbf{r})|u(\mathbf{r})|^4[\hat{a}\hat{a}^\dagger\hat{a}\hat{a}^\dagger\hat{\rho}_F + 3\hat{a}\hat{a}^\dagger\hat{\rho}_F\hat{a}\hat{a}^\dagger - 4\hat{a}^\dagger\hat{a}\hat{a}^\dagger\hat{\rho}_F\hat{a} + \text{h.c.}]\,\mathrm{d}^3r. \tag{18.4–17}$$

Here $\eta(\mathbf{r})$ is the density of active laser atoms, as before, N is their total number, and the integral is to be taken over the entire cavity volume, as in Eq. (18.2–9). The average rate of change of the field derived in this manner is sometimes known as the *coarse-grained derivative*. It is an essential feature of the derivation that the state of the laser field changes much more slowly than the state of the atom, so that the former varies very little during the average interaction time T_2 with the atom. This assumption, which is borne out well in practice for a so-called 'good' cavity, allows us to make an *adiabatic elimination* of the atomic variables from the equation of motion for the field. It should be emphasized that even though we have calculated the contributions brought about by the atoms one at a time, mutual atomic interactions through the intermediary of the laser field are included implicitly in the treatment.

So far our laser model contains no loss mechanism to account for the loss of laser photons from the cavity. In the semiclassical Lamb theory of the laser given in Section 18.2, loss was introduced artificially by the device of allowing the gain medium to be slightly conducting or dissipative. In the fully quantized theory an artifice for loss is again needed, because photons can be lost only through their coupling to another quantum system. We therefore introduce a set of loss resonators in the form of 'atoms' with the same energy level structure as shown in Fig. 18.16, which are assumed to be in the lower state $|1\rangle$ initially. The purpose of these 'atoms' is to absorb photons of the laser field at a rate proportional to the field intensity, but without re-radiation.

We can calculate the effect of a single lower-state loss atom by the same procedure that was used to derive Eq. (18.4–16), except that we replace Eq. (18.4–7) for the initial atomic state by the equation

$$\hat{\rho}_A(t_0) = |1\rangle\langle 1|,$$

and we terminate the perturbation expansion at the lowest non-vanishing term. This ensures that the loss is proportional to the laser intensity, and that re-radiation by the loss atom is excluded. If T_1 is the lifetime of the state $|1\rangle$ for which the interaction persists, we obtain in place of Eq. (18.4–16), for the change of state produced by a loss atom at position $\mathbf{r}$,

$$\Delta\hat{\rho}_F(t_0) = -\,(gT_1)^2|u(\mathbf{r})|^2[\hat{a}^\dagger\hat{a}\hat{\rho}_F(t_0) - \hat{a}\hat{\rho}_F(t_0)\hat{a}^\dagger + \text{h.c.}]. \tag{18.4–18}$$

If we multiply this small change by the rate R_1 at which loss atoms are introduced in the state $|1\rangle$, we obtain the average rate of change of $\hat{\rho}_F$, as before, provided the field is changing much more slowly than the atoms. We also average over position $\mathbf{r}$, assuming that the N_1 loss atoms are distributed with some density $\eta_1(\mathbf{r})$. This leads to the equation

$$\left(\frac{\partial \hat{\rho}_F}{\partial t}\right)_{\text{loss}} = -\frac{1}{N_1} R_1 (gT_1)^2 \int \eta_1(\mathbf{r}) |u(\mathbf{r})|^2 [\hat{a}^\dagger \hat{a} \hat{\rho}_F - \hat{a} \hat{\rho}_F \hat{a}^\dagger + \text{h.c.}]\, d^3 r, \quad (18.4\text{–}19)$$

in which the rate of change is again a coarse-grained derivative. Despite the unphysical nature of the loss model, its effect on the field closely parallels the action of the real loss mechanism, because it results in photon absorptions.

Finally, we combine the effects of the gain and the loss by adding the contributions given by Eqs. (18.4–17) and (18.4–19). This leads to the following *master equation* for the density operator $\hat{\rho}_F$ of the laser field:

$$\begin{aligned}\frac{\partial \hat{\rho}_F}{\partial t} = &-\tfrac{1}{2} A [\hat{a}\hat{a}^\dagger \hat{\rho}_F - \hat{a}^\dagger \hat{\rho}_F \hat{a} + \text{h.c.}] - \tfrac{1}{2} C [\hat{a}^\dagger \hat{a} \hat{\rho}_F - \hat{a} \hat{\rho}_F \hat{a}^\dagger + \text{h.c.}] \\ &+ \tfrac{1}{8} B [\hat{a}\hat{a}^\dagger \hat{a}\hat{a}^\dagger \hat{\rho}_F + 3\hat{a}\hat{a}^\dagger \hat{\rho}_F \hat{a}\hat{a}^\dagger - 4\hat{a}^\dagger \hat{a}\hat{a}^\dagger \hat{\rho}_F \hat{a} + \text{h.c.}]. \quad (18.4\text{–}20)\end{aligned}$$

We have introduced the abbreviations‡

$$\left.\begin{aligned} A &\equiv 2(R_2/N)(gT_2)^2 \int \eta(\mathbf{r}) |u(\mathbf{r})|^2 \, d^3 r \\ C &\equiv 2(R_1/N_1)(gT_1)^2 \int \eta_1(\mathbf{r}) |u(\mathbf{r})|^2 \, d^3 r \\ B &\equiv 8(R_2/N)(gT_2)^4 \int \eta(\mathbf{r}) |u(\mathbf{r})|^4 \, d^3 r \end{aligned}\right\} \quad (18.4\text{–}21)$$

for the coefficients characterizing the gain, the loss, and the nonlinearity or the saturation of the laser, respectively. In practice, the loss rate would be identified with the actual rate at which photons are lost from the laser cavity, which is largely determined by the reflectivity of the mirrors. Moreover, in practice, the gain and loss rates A and C are usually approximately comparable, even well above and well below threshold. On the other hand, the ratio of B to A is usually extremely small. It is given by the relation

$$\frac{B}{A} = 4(gT_2)^2 \frac{\int \eta(\mathbf{r}) |u(\mathbf{r})|^4 \, d^3 r}{\int \eta(\mathbf{r}) |u(\mathbf{r})|^2 \, d^3 r} \sim (gT_2)^2 F, \quad (18.4\text{–}22)$$

where F is a dimensionless factor of order unity. We can make a rough numerical estimate of B/A, by first expressing g in terms of the Einstein A-coefficient $1/T$ for the $|2\rangle$ to $|1\rangle$ atomic transition, as given by Eq. (15.4–11). Then

$$\frac{B}{A} \sim (gT_2)^2 \approx \frac{3}{8\pi} \frac{cT_2^2 \lambda^2}{TV}.$$

‡ The symbol A has also been used for the Einstein A-coefficient, and for some other quantities. There should be no confusion with the symbols in Eqs. (18.4–21), which are standard notation and widely used.

If we take $T_2 \sim T \sim 10^{-8}$ s, $\lambda \sim 6 \times 10^{-5}$ cm, $V \sim 10^{-1}$ to 1 cm^3, which are typical values for some lasers, we find that

$$B/A \sim 10^{-6} \text{ to } 10^{-7}.$$

In general, the smaller the saturation coefficient B, the stronger will be the laser field in the cavity required to establish a steady state.

18.4.2 Photon statistics

The operator equation of motion (18.4–20) allows us to derive equations for the expectations or probability distributions of various dynamical variables of the field. As an illustration, we shall now obtain the equation of motion for the probability $p(n, t)$ that there are n photons in the laser cavity at time t. This probability is given by taking the expectation value of $\hat{\rho}_F(t)$ in the Fock state $|n\rangle$,

$$p(n, t) = \langle n|\hat{\rho}_F(t)|n\rangle. \tag{18.4–23}$$

We multiply each term in Eq. (18.4–20) by $|n\rangle$ on the right and $\langle n|$ on the left, and make use of the relations (10.4–5) and (10.4–8). We then find that

$$\begin{aligned}\frac{\partial p(n, t)}{\partial t} &= -A[(n+1)p(n, t) - np(n-1, t)] \\ &\quad - C[np(n, t) - (n+1)p(n+1, t)] \\ &\quad + B[(n+1)^2 p(n, t) - n^2 p(n-1, t)] \\ &= -A(n+1)\left[1 - \frac{B}{A}(n+1)\right]p(n, t) + An\left[1 - \frac{B}{A}n\right]p(n-1, t) \\ &\quad + C(n+1)p(n+1, t) - Cnp(n, t). \end{aligned} \tag{18.4–24}$$

If we restrict our discussion to photon numbers n for which the terms $(B/A)n$ are much smaller than unity, we may re-write this equation in the form

$$\begin{aligned}\frac{\partial p(n, t)}{\partial t} &= -\frac{A(n+1)}{1 + (B/A)(n+1)}p(n, t) + \frac{An}{1 + (B/A)n}p(n-1, t) \\ &\quad + C(n+1)p(n+1, t) - Cnp(n, t). \end{aligned} \tag{18.4–25}$$

This formula is precisely the form of the equation given directly by the Weisskopf–Wigner procedure (Haken, 1964, 1970; Scully and Lamb, 1967, 1969), which places no restriction on the strength of the laser field. We note that the equation of motion (18.4–24) that we derived by the perturbation expansion follows from the more general Eq. (18.4–25) by expanding the two denominators and retaining only the first two terms in the expansion. This illustrates the relationship between the two methods and emphasizes that the laser must not be operating too far above threshold, so that $nB/A \ll 1$, if Eq. (18.4–24) is to hold. The coefficients A, B obtained from the Weisskopf–Wigner procedure are also somewhat different; they take the form (Scully and Lamb, 1967, 1969)

$$\left.\begin{aligned} A &= \frac{2R_2 g^2}{\gamma_2 \gamma_{12}} \\ B &= \frac{8R_2 g^4}{\gamma_1 \gamma_2^2 \gamma_{12}}, \end{aligned}\right\} \tag{18.4–26}$$

where $\gamma_2 \approx 1/T_2$ is the decay rate of the upper state $|2\rangle$, γ_1 is the decay rate of the lower state $|1\rangle$, and γ_{12} is the mean $\frac{1}{2}(\gamma_1 + \gamma_2)$.

The master equation (18.4–25) has the character of a rate equation, in which the change in the photon population of state $|n\rangle$ is attributable to transitions from and to neighboring states $|n-1\rangle$ and $|n+1\rangle$. This is illustrated in Fig. 18.17. The first term on the right of Eq. (18.4–25) evidently describes the decrease of the probability $p(n)$ resulting from stimulated atomic emission, with the field going from the state $|n\rangle$ to the state $|n+1\rangle$, while the second term describes the increase of $p(n)$ resulting from stimulated emission, with the field going from state $|n-1\rangle$ to state $|n\rangle$. Similarly, the third and fourth terms describe increases and decreases of $p(n)$ associated with absorption of laser radiation.

18.4.3 Steady-state probability

Let us examine the steady-state solution of Eq. (18.4–25). In the steady state $\partial p(n,t)/\partial t = 0$, and we expect detailed balance to hold between neighboring states, so that (see Fig. 18.17)

$$\frac{A(n+1)}{1+(B/A)(n+1)} p(n) = C(n+1)p(n+1)$$

$$\frac{An}{1+(B/A)n} p(n-1) = Cnp(n).$$

It is at once apparent that these two equations are equivalent, and that they require

$$p(n) = \frac{A/C}{1+(B/A)n} p(n-1). \tag{18.4–27}$$

This formula is a recursion relation, and repeated application of it leads to the result

$$p(n) = p(0)\prod_{r=1}^{n}\left(\frac{A/C}{1+r(B/A)}\right). \tag{18.4–28}$$

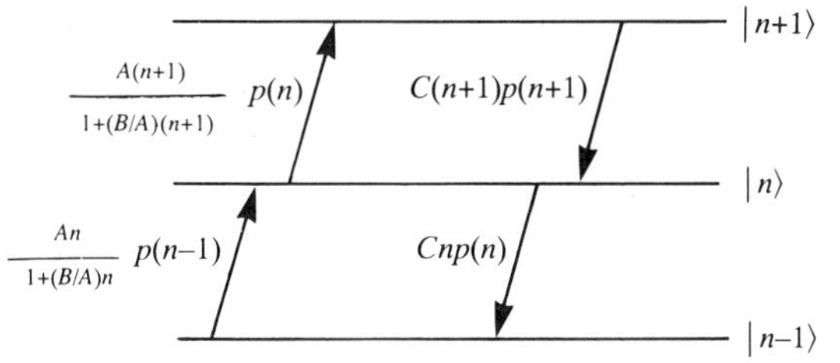

Fig. 18.17 Illustrating the flow of probability between neighboring states of the laser field. (Adapted from Scully and Lamb, 1967.)

Had we started from Eq. (18.4–24), which is the equation obtained by terminating the perturbation expansion at the second non-vanishing term, we would have been led to the result

$$p(n) = p(0)\prod_{r=1}^{n}\left[\frac{A}{C}\left(1 - r\frac{B}{A}\right)\right], \tag{18.4–29}$$

which is equivalent to Eq. (18.4–28) when nB/A is sufficiently small. Only when the laser is operating far above threshold do we need to distinguish between the two expressions and to use Eq. (18.4–28).

Equation (18.4–28) can be written in the equivalent form

$$p(n) = \text{const.} \times \frac{(A^2/BC)^n}{\Gamma[(A/B) + n + 1]}. \tag{18.4–30}$$

This tends to the Poisson form when the laser is operated so far above threshold that the photon number n appreciably exceeds A/B. However, this implies that $A \gg C$, and this condition is not always achievable in a steady-state laser. Frequently the gain A and the loss C are comparable, and the photon number n is much smaller than A/B. It is not difficult to show that Eq. (18.4–30) reduces to an analytically simpler form under these conditions.

We start by replacing the Γ-function by the Stirling approximation (Korn and Korn, 1961, p. 822)

$$\begin{aligned}\Gamma\left(\frac{A}{B} + n + 1\right) &= \sqrt{(2\pi)}\left(n + \frac{A}{B}\right)^{n+(A/B)+1/2} \exp\left(-n - \frac{A}{B}\right)\\ &= \sqrt{(2\pi)}\left(\frac{A}{B}\right)^{n+(A/B)+1/2}\left(1 + n\frac{B}{A}\right)^{n+(A/B)+1/2} \exp\left(-n - \frac{A}{B}\right),\end{aligned} \tag{18.4–31}$$

and then substitute this expression in Eq. (18.4–30). We then obtain

$$\begin{aligned}p(n) &= \text{const.}\left(\frac{A}{C}\right)^n e^n \exp\left[-\left(n + \frac{A}{B} + 1/2\right)\ln\left(1 + n\frac{B}{A}\right)\right]\\ &= \text{const.}\left(\frac{A}{C}\right)^n e^n \exp\left[-\left(n + \frac{A}{B} + \frac{1}{2}\right)\left(n\frac{B}{A} - \tfrac{1}{2}n^2\frac{B^2}{A^2} + \ldots\right)\right]\end{aligned}$$

when we expand the logarithm in a power series in the small quantity nB/A. This leads to the relation

$$p(n) = \text{const.}\left(\frac{A}{C}\right)^n \exp\left[-\tfrac{1}{2}n^2\frac{B}{A}\right]\exp\left[O\left(n\frac{B}{A}\right)\right], \tag{18.4–32}$$

where O denotes the order of magnitude.

With the help of the substitutions

$$n_0 \equiv (A/B)^{1/2} \tag{18.4–33}$$

$$\exp\left[a/(\sqrt{2}n_0)\right] \equiv A/C, \tag{18.4–34}$$

Eq. (18.4–32) simplifies to

$$p(n) = \text{const.} \times \exp\left[-\frac{1}{4}\left(\frac{\sqrt{2}n}{n_0} - a\right)^2\right], \tag{18.4–35}$$

and the same result is obtained if we start from Eq. (18.4–29). Except for the fact that n is a discrete rather than a continuous variable, this is a truncated Gaussian distribution in n/n_0. The result can be compared with Eq. (18.3–17) giving the probability distribution of the classical light intensity I. The two distributions are seen to be identical if we treat n/n_0 as almost continuous and identify $(\sqrt{2})n/n_0$ with the light intensity I. The dimensionless parameter a introduced in Eq. (18.4–34) is the pump parameter of the laser. It is positive above threshold, when $A > C$, and negative below threshold, when $A < C$. Figure 18.6, which illustrates the shape of the probability distribution $\mathcal{P}(I)$ of the classical light intensity I for various pump parameters, therefore also illustrates the forms of $p(n)$. Moreover, the same calculation that was carried out in Section 18.3 to determine the first two moments of the light intensity now yields for the moments of n,

$$\langle n \rangle = \frac{n_0}{\sqrt{2}}\left[a + \frac{2\,\mathrm{e}^{-a^2/4}}{\sqrt{\pi}[1 + \operatorname{erf}(\frac{1}{2}a)]}\right], \tag{18.4–36}$$

and

$$\langle(\Delta n)^2\rangle = n_0^2\left[1 - \frac{a\,\mathrm{e}^{-a^2/4}}{\sqrt{\pi}[1 + \operatorname{erf}(\frac{1}{2}a)]} - \frac{2\,\mathrm{e}^{-a^2/2}}{\pi[1 + \operatorname{erf}(\frac{1}{2}a)]^2}\right]. \tag{18.4–37}$$

Although Eq. (18.4–35) is rather similar to the semiclassical equation (18.3–17), Eq. (18.4–35) yields information on the absolute number of photons, unlike Eq. (18.3–17).

Let us first interpret the physical significance of n_0. From Eq. (18.4–36) it follows that when the laser is at threshold and $a = 0$, then $\langle n \rangle = n_0\sqrt{(2/\pi)}$. Hence, apart from a factor of order unity, n_0 is the average number of photons present in the laser cavity at threshold. From the definition (18.4–33) and the approximate value of B/A, one finds that for a typical laser the number n_0 is of order a few thousand photons. The condition $nB/A \ll 1$ that we invoked in the derivation leading to Eq. (18.4–35) is therefore equivalent to the inequality

$$n \ll n_0^2. \tag{18.4–38}$$

Equation (18.4–34) then gives

$$a/(\sqrt{2}n_0) \approx (A - C)/C \approx (A - C)/A$$

when $A \approx C$, and

$$a \approx \frac{A - C}{(AB/2)^{1/2}}, \tag{18.4–39}$$

which is an alternative definition of the pump parameter a. Well above threshold, when $a \gg 1$, Eq. (18.4–36) yields the result

$$\langle n \rangle \approx an_0/\sqrt{2} \approx (A - C)/B, \tag{18.4–40}$$

so that the average photon number is the difference between the gain and the loss divided by the saturation.

The condition $n \ll n_0^2$ that was implied in the derivation of Eq. (18.4–35), can now be translated into a condition on the pump parameter a, and with the help of Eq. (18.4–40) it becomes

$$a \ll n_0. \tag{18.4–41}$$

As n_0 is usually at least several thousand, the dimensionless pump parameter a can certainly be made as large as 100 without contradicting the requirement that $nB/A \ll 1$ and $A \approx C$. It is therefore possible for a laser to be operating well above threshold, and yet to be well described by equations of motion for which the perturbation expansion is terminated at the second non-vanishing contribution. However, it should be emphasized that this is not the limit of large pump parameter, and that at this level of excitation $p(n)$ is still far from Poissonian. Figure 18.18 shows a comparison of the form of $p(n)$ given by Eq. (18.4–30), with $\langle n \rangle = 10^6$ and pump parameter about 500, corresponding to $(A - C)/C = 0.2$, and a Poissonian distribution with the same mean. The latter would be characteristic of a field in a coherent state.

The probability distribution $p(n)$, like the distribution $\mathcal{P}(I)$, can readily be related to photoelectric counting measurements, if we suppose that each photon in the laser cavity has the same small probability β of escaping from the cavity and being counted by the detector within a certain time interval. If n photons are present, the probability that m of them ($m \leqslant n$) will be counted in an experiment is then given by the Bernoulli distribution

$$\binom{n}{m}\beta^m(1-\beta)^{n-m}.$$

More generally, if n photons have a probability distribution $p(n)$, the probability $P(m)$ of registering m photoelectric counts is given by the convolution

$$P(m) = \sum_{n=m}^{\infty} p(n)\binom{n}{m}\beta^m(1-\beta)^{n-m}. \tag{18.4–42}$$

From this relation it follows at once that the factorial moments of the photoelectric counts m are proportional to the corresponding factorial moments of the number of photons n. For we have

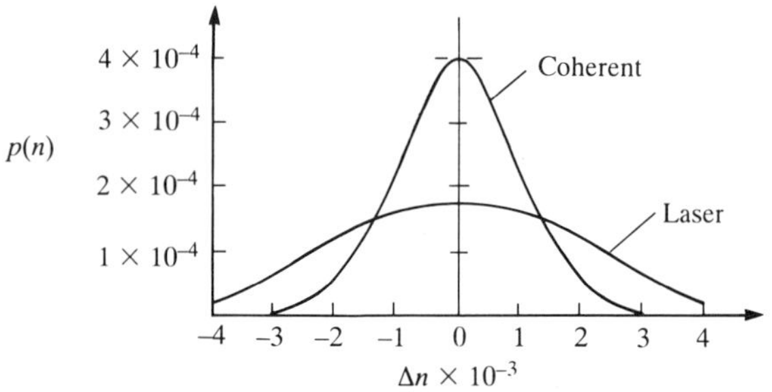

Fig. 18.18 Comparison of the probability distribution $p(n)$ for the photon number given by Eq. (18.4–30), for $A/C = 1.2$ and $\langle n \rangle = 10^6$, with a Poisson distribution having the same mean. (Reproduced from Scully and Lamb, 1967.)

$$\begin{aligned}\langle m^{(r)}\rangle &= \langle m(m-1)\ldots(m-r+1)\rangle \\ &= \sum_{m=r}^{\infty}\sum_{n=m}^{\infty}\frac{m!}{(m-r)!}p(n)\binom{n}{m}\beta^m(1-\beta)^{n-m} \\ &= \sum_{n=r}^{\infty}\sum_{s=0}^{\infty}p(n)\frac{n!}{(n-r)!}\beta^r\binom{n-r}{s}\beta^s(1-\beta)^{n-r-s} \\ &= \sum_{n=r}^{\infty}p(n)\frac{n!}{(n-r)!}\beta^r \\ &= \langle n^{(r)}\rangle\beta^r. \end{aligned} \tag{18.4–43}$$

The third line follows from the second if we exchange the order of summation and put $s = m - r$. Measurements of the factorial moments of the photoelectric counts can therefore be related very easily to the calculated photon statistics. However, because of the close correspondence between Eqs. (18.4–35) and (18.3–17), the predictions for photoelectric counting of the quantum theory of the laser are found to be essentially indistinguishable from those of the semiclassical theory with additive quantum noise.

18.5 Relationship between quantum and semiclassical laser theories

We have seen that the quantum theory of the laser provides information on the absolute strength – or the absolute number of photons – of the laser field. However, the distribution of the light intensity in the steady state is no different from that given by semiclassical theory with additive Langevin noise. This suggests a deeper and more general connection between the two different treatments of the laser problem, which we now explore. We shall find, with the help of the diagonal coherent-state representation of the laser field that was introduced in Section 11.8, that the operator master equation (18.4–20) can be cast into the form of the Fokker–Planck equation (18.3–4) (Lax and Louisell, 1969; Louisell, 1969; Wang and Lamb 1973). This shows the essential equivalence of the two approaches for the calculation of expectation values of normally ordered operators.

18.5.1 Coherent-state representation of the laser field

According to Eq. (11.8–1), it is possible to represent the density operator $\hat{\rho}_F(t)$ of the single-mode laser field in the diagonal form

$$\hat{\rho}_F(t) = \int \phi(v, t)|v\rangle\langle v|\,d^2v, \tag{18.5–1}$$

in which the integral is to be taken over the entire complex v-plane. Here $|v\rangle$ is the coherent state with complex amplitude v, and $\phi(v, t)$ is a real weight function or phase space density that need not be a true probability density. We now substitute this representation for $\hat{\rho}_F(t)$ in the master equation (18.4–20) and attempt to eliminate the operators $\hat{a}$, $\hat{a}^\dagger$. For certain operator products such as

$$\hat{a}|v\rangle\langle v| = v|v\rangle\langle v|$$
$$\hat{a}|v\rangle\langle v|\hat{a}^\dagger = |v|^2|v\rangle\langle v|, \tag{18.5–2}$$

the elimination is trivial, when we recall that $|v\rangle$ is the right eigenstate of $\hat{a}$. For other products like $\hat{a}^\dagger|v\rangle\langle v|$ and $|v\rangle\langle v|\hat{a}$ the problem is less trivial, and we shall make use of the procedure developed by Louisell (1969) for replacing $\hat{a}$ and $\hat{a}^\dagger$ by differential operators that was described in Section 12.9. It was shown there that

$$\hat{a}^\dagger|v\rangle\langle v| = \left(v^* + \frac{\partial}{\partial v}\right)|v\rangle\langle v|$$
$$|v\rangle\langle v|\hat{a} = \left(v + \frac{\partial}{\partial v^*}\right)|v\rangle\langle v|,$$

which allows the $\hat{a}$, $\hat{a}^\dagger$ operators to be eliminated. In addition to the relations (12.9–18) to (12.9–20) derived in Section 12, we shall require the following results:

$$\begin{aligned}\hat{a}\hat{a}^\dagger\hat{a}\hat{a}^\dagger|v\rangle\langle v| &= (\hat{a}^{\dagger 2}\hat{a}^2 + 3\hat{a}^\dagger\hat{a} + 1)|v\rangle\langle v| \\ &= \left[v^2\left(v^* + \frac{\partial}{\partial v}\right)^2 + 3v\left(v^* + \frac{\partial}{\partial v}\right) + 1\right]|v\rangle\langle v| \\ &= \left[|v|^4 + 2v^2v^*\frac{\partial}{\partial v} + v^2\frac{\partial^2}{\partial v^2} + 3v\frac{\partial}{\partial v} + 3|v|^2 + 1\right]|v\rangle\langle v|\end{aligned} \tag{18.5–3}$$

$$\begin{aligned}\hat{a}\hat{a}^\dagger|v\rangle\langle v|\hat{a}\hat{a}^\dagger &= (\hat{a}^\dagger\hat{a} + 1)|v\rangle\langle v|(\hat{a}^\dagger\hat{a} + 1) \\ &= \left[v\left(v^* + \frac{\partial}{\partial v}\right) + 1\right]\left[v^*\left(v + \frac{\partial}{\partial v^*}\right) + 1\right]|v\rangle\langle v| \\ &= \left[|v|^4 + 3|v|^2 + (|v|^2 + 1)\left(v\frac{\partial}{\partial v} + v^*\frac{\partial}{\partial v^*}\right)\right. \\ &\quad \left. + |v|^2\frac{\partial^2}{\partial v\partial v^*} + 1\right]|v\rangle\langle v|\end{aligned} \tag{18.5–4}$$

$$\begin{aligned}\hat{a}^\dagger|v\rangle\langle v|\hat{a}\hat{a}^\dagger\hat{a} &= \hat{a}^\dagger|v\rangle\langle v|(\hat{a}^\dagger\hat{a}^2 + \hat{a}) \\ &= \left[v^*\left(v^* + \frac{\partial}{\partial v}\right)\left(v + \frac{\partial}{\partial v^*}\right)^2 + \left(v^* + \frac{\partial}{\partial v}\right)\left(v + \frac{\partial}{\partial v^*}\right)\right]|v\rangle\langle v| \\ &= \left[|v|^4 + v^*(2|v|^2 + 3)\frac{\partial}{\partial v^*} + v(|v|^2 + 1)\frac{\partial}{\partial v} + (2|v|^2 + 1)\frac{\partial^2}{\partial v\partial v^*}\right. \\ &\quad \left. + v^{*2}\frac{\partial^2}{\partial v^{*2}} + v^*\frac{\partial^3}{\partial v\partial v^{*2}} + 3|v|^2 + 1\right]|v\rangle\langle v|.\end{aligned} \tag{18.5–5}$$

We now insert the representation (18.5–1) of the density operator $\hat{\rho}_F(t)$ into the master equation (18.4–20), and we eliminate the $\hat{a}$, $\hat{a}^\dagger$ operators with the help of these relations. After some rearrangement of terms, we arrive at the equation of motion

$$\int \frac{\partial \phi(v,t)}{\partial t} |v\rangle\langle v| \, d^2v$$
$$= \int \phi(v,t) \left\{ \frac{1}{2} A \left[v\frac{\partial}{\partial v} + v^* \frac{\partial}{\partial v^*} + \frac{2\partial^2}{\partial v \partial v^*} \right] - \frac{1}{2} C \left[v\frac{\partial}{\partial v} + v^* \frac{\partial}{\partial v^*} \right] \right.$$
$$+ B \left[-\frac{1}{2}|v|^2 \left(v\frac{\partial}{\partial v} + v^*\frac{\partial}{\partial v^*} \right) - \frac{3}{8}\left(v^2\frac{\partial^2}{\partial v^2} + v^{*2}\frac{\partial^2}{\partial v^{*2}} \right) - \frac{5}{4}|v|^2 \frac{\partial^2}{\partial v \partial v^*} \right.$$
$$\left. \left. - \frac{7}{8}\left(v\frac{\partial}{\partial v} + v^*\frac{\partial}{\partial v^*} \right) - \frac{\partial^2}{\partial v \partial v^*} - \frac{1}{2}\left(v\frac{\partial^3}{\partial v^* \partial v^2} + v^* \frac{\partial^3}{\partial v \partial v^{*2}} \right) \right] \right\} |v\rangle\langle v| \, d^2v, \tag{18.5–6}$$

in which the differential operators under the integral on the right operate on the coherent-state projector $|v\rangle\langle v|$. By formally integrating by parts, with the assumption that $\phi(v, t)$ vanishes at infinity faster than any power of v, v^*, we can convert the integrand on the right into a product of $|v\rangle\langle v|$ and a c-number function of v, v^*, and we obtain the formula

$$\int |v\rangle\langle v| \frac{\partial \phi(v,t)}{\partial t} \, d^2v$$
$$= \int |v\rangle\langle v| \left\{ -\frac{1}{2}(A - C)\left(\frac{\partial}{\partial v} v + \frac{\partial}{\partial v^*} v^* \right) + A \frac{\partial^2}{\partial v \partial v^*} \right.$$
$$+ B \left[\frac{1}{2}\left(\frac{\partial}{\partial v} v + \frac{\partial}{\partial v^*} v^* \right)|v|^2 - \frac{3}{8}\left(\frac{\partial^2}{\partial v^2} v^2 + \frac{\partial^2}{\partial v^{*2}} v^{*2} \right) - \frac{5}{4} \frac{\partial^2}{\partial v \partial v^*} |v|^2 \right.$$
$$\left. \left. + \frac{7}{8}\left(\frac{\partial}{\partial v} v + \frac{\partial}{\partial v^*} v^* \right) - \frac{\partial^2}{\partial v \partial v^*} + \frac{1}{2}\left(\frac{\partial^3}{\partial v^* \partial v^2} v + \frac{\partial^3}{\partial v \partial v^{*2}} v^* \right) \right] \right\} \phi(v,t) \, d^2v. \tag{18.5–7}$$

Comparison of the coefficients of $|v\rangle\langle v|$ on both sides of the equation leads to a partial differential equation for $\phi(v, t)$, from which all Hilbert space operators have disappeared. We shall simplify this equation by discarding certain terms. We first recall that B is a very small coefficient compared with A or C, with B/A typically of the order of 10^{-6} or less. We therefore retain only the most important terms in B. On the other hand, when the laser is operating anywhere near the steady state, $|v|$ is a large number, and $|v|^2$ is of the order of the average number of photons. It follows that among the terms in B containing first derivatives,

$$\frac{1}{2}\left(\frac{\partial}{\partial v} v + \frac{\partial}{\partial v^*} v^* \right) |v|^2 \phi(v,t)$$

is the most important, and that among the second derivative terms in B

$$-\frac{3}{8}\left(\frac{\partial^2}{\partial v^2} v^2 + \frac{\partial^2}{\partial v^{*2}} v^{*2} \right) \phi(v,t) - \frac{5}{4} \frac{\partial^2}{\partial v \partial v^*} |v|^2 \phi(v,t)$$

are the most important terms. But unless the laser is operating so far above threshold that $|v|^2$ is 1000 times, say, greater than its threshold value (which implies that the pump parameter is of the order of 1000 or higher), these last terms are still small compared with the second derivative term in A. Similarly, it can be argued that the terms in B involving third derivatives are small compared with those involving first derivatives. With the understanding that the laser is not operating too far above threshold, we shall therefore retain only the first term in B. The differential equation for $\phi(v, t)$ then takes the much simpler form

$$\frac{\partial\phi(v, t)}{\partial t} = \frac{1}{2}\frac{\partial}{\partial v}[B|v|^2 - (A - C)]v\phi(v, t) + \frac{1}{2}\frac{\partial}{\partial v^*}[B|v|^2 - (A - C)]v^*\phi(v, t) + A\frac{\partial^2\phi(v, t)}{\partial v\partial v^*}, \tag{18.5–8}$$

which is a Fokker–Planck equation for the phase space density $\phi(v, t)$.

It is often convenient to replace the complex amplitude v by a real two-dimensional vector $\mathbf{x}$, such that

$$v = x_1 + \mathrm{i}x_2.$$

Then

$$x_1 = \frac{1}{2}(v + v^*)$$

$$x_2 = \frac{1}{2\mathrm{i}}(v - v^*)$$

and

$$\frac{\partial}{\partial v} \to \frac{1}{2}\left(\frac{\partial}{\partial x_1} - \mathrm{i}\frac{\partial}{\partial x_2}\right)$$

$$\frac{\partial}{\partial v^*} \to \frac{1}{2}\left(\frac{\partial}{\partial x_1} + \mathrm{i}\frac{\partial}{\partial x_2}\right),$$

and the Fokker–Planck equation becomes

$$\frac{\partial\phi(\mathbf{x}, t)}{\partial t} = -\sum_{i=1}^{2}\frac{1}{2}\frac{\partial}{\partial x_i}(A - C - B\mathbf{x}^2)x_i\phi(\mathbf{x}, t) + \frac{1}{4}A\sum_{i=1}^{2}\frac{\partial^2}{\partial x_i^2}\phi(\mathbf{x}, t). \tag{18.5–9}$$

Finally, we may re-scale the equation by putting

$$\left.\begin{aligned}(\tfrac{1}{8}AB)^{1/2}t &\equiv t' \\ (2B/A)^{1/4}\mathbf{x} &\equiv \mathbf{x}',\end{aligned}\right\} \tag{18.5–10}$$

and if we define

$$\frac{A - C}{(\mathrm{AB}/2)^{1/2}} \equiv a, \tag{18.5–11}$$

we arrive at the equation

$$\frac{\partial\phi(\mathbf{x}, t)}{\partial t} = -\sum_{i=1}^{2}\frac{\partial}{\partial x_i}[(a - \mathbf{x}^2)x_i\phi(\mathbf{x}, t)] + \sum_{i=1}^{2}\frac{\partial^2\phi(\mathbf{x}, t)}{\partial x_i^2}. \tag{18.5–12}$$

The primes that were introduced in Eqs. (18.5–10) have been dropped again for simplicity. This equation of motion is identical to the scaled equation (18.3–4) that was obtained from the semiclassical theory with additive noise, in which a is the dimensionless pump parameter. It follows that, within the limits of the approximations that were made, we may identify $\mathbf{x}$ with the scaled amplitude of the laser electric field, and $\phi(\mathbf{x}, t)$ with its probability density. However, the quantized field treatment has the advantage over the semiclassical one that no ad hoc assumption about the strength of the quantum fluctuations is needed. All scaling factors are well-defined in terms of the fundamental gain, loss and saturation coefficients A, C, B of the laser.

18.5.2 Steady-state solution of the master equation

As was shown in Section 18.3, and particularly by Eq. (18.3–12), the steady-state solution of the Fokker–Planck equation can be written as

$$\phi_s(v) = \text{const.} \times \exp[-\tfrac{1}{4}(a - |v|^2)^2]. \tag{18.5–13}$$

For large pump parameter a this has the form of a Gaussian distribution in $|v|^2$. The phase of v is completely random and is distributed uniformly over the range 0 to 2π, while, for large a, the first two moments of $|v|$ are given approximately by the relations

$$\left.\begin{aligned} \langle |v| \rangle &\approx (\sqrt{a})(1 - 1/4a^2) \\ \langle |v|^2 \rangle &\approx a. \end{aligned}\right\} \tag{18.5–14}$$

Hence

$$\left.\begin{aligned} \langle (\Delta|v|)^2 \rangle &\approx \frac{1}{2a} \\ \frac{\langle (\Delta|v|)^2 \rangle^{1/2}}{\langle |v| \rangle} &\approx \frac{1}{(\sqrt{2})a}. \end{aligned}\right\} \tag{18.5–15}$$

The phase space density $\phi_s(v)$, regarded as a function of $|v|$, therefore becomes narrower and narrower, both in an absolute and in a relative sense as the pump parameter a increases, while the phase of v remains random. For large a, $\phi_s(v)$ is sometimes approximated by a delta function in $|v|$, and we may write [cf. Eq. (11.8–14)]

$$\phi_s(v) \to \frac{1}{2\pi\sqrt{a}}\delta(|v| - \sqrt{a}). \tag{18.5–16}$$

It is in the sense that $\phi_s(v)$ tends toward a delta function, that the laser field well above threshold is sometimes described as being in a coherent state. More precisely, it tends towards a randomly phased mixture of coherent states.

18.6 Solution of the time-dependent equation of motion

So far we have derived and discussed only the steady-state behavior of the laser field. But having demonstrated that the field is describable by a Fokker–Planck equation within the framework of both the fully quantized and the semiclassical theory, we now turn our attention to the general solution of the time-dependent equation of motion (18.3–4) or (18.5–12) (Risken, 1965, 1966, Risken and Vollmer 1967a,b; Hempstead and Lax, 1967)

$$\frac{\partial p(\mathscr{E}, t)}{\partial t} = -\sum_{i=1}^{2} \frac{\partial}{\partial \mathscr{E}_i}[(a - \mathscr{E}^2)\mathscr{E}_i p(\mathscr{E}, t)] + \sum_{i=1}^{2} \frac{\partial^2 p(\mathscr{E}, t)}{\partial \mathscr{E}_i^2}. \quad (18.6\text{–}1)$$

It is possible to obtain a solution by separation of variables. Let us attempt to express the time-dependent probability density $p(\mathscr{E}, t)$ in the form

$$p(\mathscr{E}, t) = f(\mathscr{E})k(t) \quad (18.6\text{–}2)$$

and substitute in Eq. (18.6–1). We then obtain the equation

$$-\frac{1}{k(t)}\frac{\mathrm{d}k(t)}{\mathrm{d}t} = \frac{1}{f(\mathscr{E})}\left[\sum_{i=1}^{2} \frac{\partial}{\partial \mathscr{E}_i}(a - \mathscr{E}^2)\mathscr{E}_i f(\mathscr{E}) - \sum_{i=1}^{2} \frac{\partial^2 f(\mathscr{E})}{\partial \mathscr{E}_i^2}\right].$$

As the left-hand side of the equation contains only the t-variable, and the right-hand side only the $\mathscr{E}$-variable, each side must be a constant λ. We are therefore led to two separate equations for $k(t)$ and $f(\mathscr{E})$,

$$\frac{\mathrm{d}k(t)}{\mathrm{d}t} = -\lambda k(t)$$

with solution

$$k(t) = k(0)\,\mathrm{e}^{-\lambda t}, \quad (18.6\text{–}3)$$

and

$$\sum_{i=1}^{2} \frac{\partial}{\partial \mathscr{E}_i}(a - \mathscr{E}^2)\mathscr{E}_i f(\mathscr{E}) - \sum_{i=1}^{2} \frac{\partial^2 f(\mathscr{E})}{\partial \mathscr{E}_i^2} = \lambda f(\mathscr{E}). \quad (18.6\text{–}4)$$

Equation (18.6–4) is an eigenvalue equation in a two-dimensional vector space, and the corresponding eigenvalues λ and eigenfunctions $f(\mathscr{E})$ form a two-dimensional array, conveniently labeled by integral subscripts m, n. The general solution $p(\mathscr{E}, t)$ is therefore expressible in the form

$$p(\mathscr{E}, t) = \sum_{m,n} c_{mn} f_{mn}(\mathscr{E})\,\mathrm{e}^{-\lambda_{mn} t}, \quad (18.6\text{–}5)$$

with the constants c_{mn} to be determined by the initial conditions. Because a steady-state solution $p_s(\mathscr{E})$ given by Eq. (18.3–12) exists, one of the eigenvalues must vanish, and we shall take this to be $\lambda_{00} = 0$. The associated eigenfunction $f_{00}(\mathscr{E})$ is therefore $p_s(\mathscr{E})$.

With the help of the transformation

$$f_{mn}(\mathscr{E}) = [p_s(\mathscr{E})]^{1/2} g_{mn}(\mathscr{E}), \quad (18.6\text{–}6)$$

we can convert the differential equation (18.6–4) into a Sturm–Liouville equation

$$\left[-\sum_{i=1}^{2} \frac{\partial^2}{\partial \mathscr{E}_i^2} + a - 2\mathscr{E}^2 + \frac{1}{4}\mathscr{E}^2(a - \mathscr{E}^2)^2\right] g_{mn}(\mathscr{E}) = \lambda_{mn} g_{mn}(\mathscr{E}), \quad (18.6\text{–}7)$$

with

$$\left.\begin{aligned} g_{00}(\mathscr{E}) &= [p_s(\mathscr{E})]^{1/2} = \text{const.} \times \exp[-\tfrac{1}{8}(\mathscr{E}^2 - a)^2] \\ \lambda_{00} &= 0. \end{aligned}\right\} \tag{18.6–8}$$

From the general properties of the solution of such an equation (Ross, 1984, Chap. 12) it can be shown that the different eigenfunctions are orthogonal, and we shall take them to be normalized also, so that

$$\int g^*_{mn}(\mathscr{E}) g_{m'n'}(\mathscr{E})\, d^2\mathscr{E} = \delta_{mm'}\delta_{nn'}. \tag{18.6–9}$$

Moreover, if the eigenvalues λ_{mn} have a lower bound, which is the case here, because the lowest eigenvalue must be zero if the solution (18.6–5) is not to diverge as $t \to \infty$, the eigenfunctions form a complete set,

$$\sum_{m,n} g_{mn}(\mathscr{E}) g^*_{mn}(\mathscr{E}') = \delta^{(2)}(\mathscr{E} - \mathscr{E}'). \tag{18.6–10}$$

The equation (18.6–7) is more easily solved in polar coordinates r, θ. We shall put

$$\left.\begin{aligned} \mathscr{E}_1 &= r\cos\theta \\ \mathscr{E}_2 &= r\sin\theta, \end{aligned}\right\} \tag{18.6–11}$$

so that

$$\sum_{i=1}^{2} \frac{\partial^2}{\partial \mathscr{E}_i^2} = \frac{\partial^2}{\partial r^2} + \frac{1}{r}\frac{\partial}{\partial r} + \frac{1}{r^2}\frac{\partial^2}{\partial \theta^2}, \tag{18.6–12}$$

and we try another separation of variables by writing

$$g_{mn}(r, \theta) = \frac{1}{\sqrt{r}} \psi_{mn}(r) \chi_n(\theta). \tag{18.6–13}$$

Substitution in Eq. (18.6–7) and division by $g_{mn}(r, \theta) r^{-2}$ then yields the equation

$$\frac{1}{\psi_{mn}(r)}\left[-r^2\frac{\partial^2}{\partial r^2} + \frac{1}{4}r^4(a - r^2)^2 + r^2(a - 2r^2) - \lambda_{mn}r^2 - \frac{1}{4}\right]\psi_{mn}(r) = \frac{1}{\chi_n(\theta)}\frac{\partial^2 \chi_n(\theta)}{\partial \theta^2}.$$

Once again we note that the left-hand side of the equation depends only on r and the right-hand side only on θ, so that both sides must be constant and independent of both r and θ, although the constant in general will depend on n. As we are looking for solutions that are periodic in θ, it is natural to take the constant to be $-n^2$. This leads to the differential equation in θ,

$$\frac{\partial^2 \chi_n(\theta)}{\partial \theta^2} = -n^2 \chi_n(\theta), \tag{18.6–14}$$

with normalized solution

$$\chi_n(\theta) = \frac{1}{\sqrt{(2\pi)}} e^{-in\theta}, \tag{18.6–15}$$

and to the equation in r,

$$-\frac{\partial^2 \psi_{mn}(r)}{\partial r^2} + V_n(r)\psi_{mn}(r) = \lambda_{mn}\psi_{mn}(r), \qquad m = 0, 1, 2, \ldots \tag{18.6–16}$$

with

$$V_n(r) \equiv (n^2 - \tfrac{1}{4})\frac{1}{r^2} + a + (\tfrac{1}{4}a^2 - 2)r^2 - \tfrac{1}{2}ar^4 + \tfrac{1}{4}r^6. \tag{18.6–17}$$

This will be recognized as a one-dimensional Schrödinger equation for the motion of a particle in the 'potential' $V_n(r)$. The laser problem has therefore been reduced to the problem of finding the eigenfunctions $\psi_{mn}(r)$ and eigenvalues λ_{mn} of this Schrödinger-type equation. The form of the potential $V_0(r)$ is illustrated in Fig. 18.19 for several values of the pump parameter a. The functions $V_n(r)$ with $|n| \geqslant 1$ are qualitatively similar, except that they rise near $r = 0$ instead of falling.

It is clear from an inspection of Fig. 18.19 that the solutions $\psi_{mn}(r)$ correspond to bound states, and both the eigenfunctions and eigenvalues are obviously identical for n and $-n$. It has so far not proved possible to obtain analytic solutions of Eq. (18.6–16), but numerical solutions for both $\psi_{mn}(r)$ and λ_{mn} have been given (Risken, 1965, 1966; Risken and Vollmer, 1967a,b; Hempstead and Lax, 1967). Figure 18.20 shows a plot of the first four eigenvalues of type λ_{m0} as a function of the laser pump parameter a. We note that the eigenvalues all exhibit a minimum somewhat above threshold. The full curve in Fig. 18.21 shows the behavior of the eigenvalue λ_{01} as a function of pump parameter a. The eigenvalue λ_{01} tends asymptotically to zero like $1/a$ for large a, but the higher-order eigenvalues of type λ_{0n} exhibit minima, just like λ_{m0}. The eigenvalue λ_{00} is of course zero, as we have seen, and from Eqs. (18.6–8), (18.6–13) and (18.6–15) the zero-order eigenfunction is given by

$$\psi_{00}(r) = [2\pi r p_s(r)]^{1/2} = \text{const.} \times (\sqrt{r})\,e^{-(r^2-a)^2/8}, \tag{18.6–18}$$

with the constant appropriately chosen to ensure the normalization of $\psi_{00}(r)$. Because of the orthonormality relations (18.6–9), all the eigenfunctions $\psi_{mn}(r)$

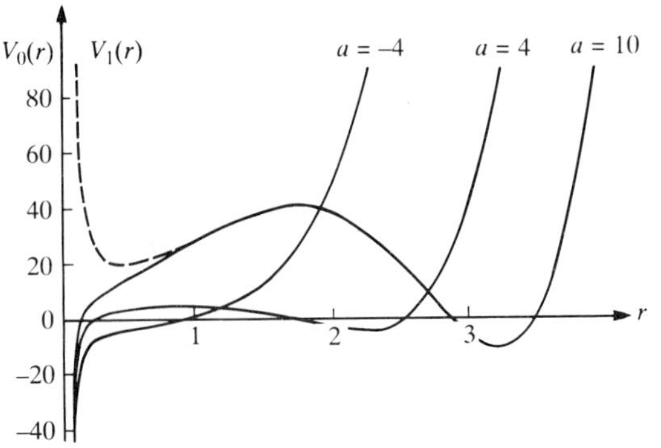

Fig. 18.19 The form of the potential $V_0(r)$ for three different values of the pump parameter a. The potentials $V_n(r)$ $n = 1, 2, \ldots$ rise to ∞ at $r = 0$. (Reproduced from Risken and Vollmer, 1967a.)

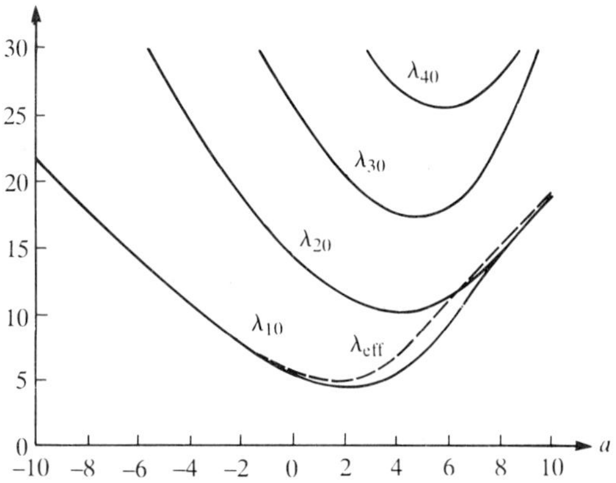

Fig. 18.20 Variation of the first four eigenvalues λ_{m0} with laser pump parameter a. The dashed curve shows the variation of $\lambda_{\text{eff}}^{-1}$, the effective mean of the reciprocals $1/\lambda_{m0}$. (Reproduced from Risken and Vollmer, 1967a.)

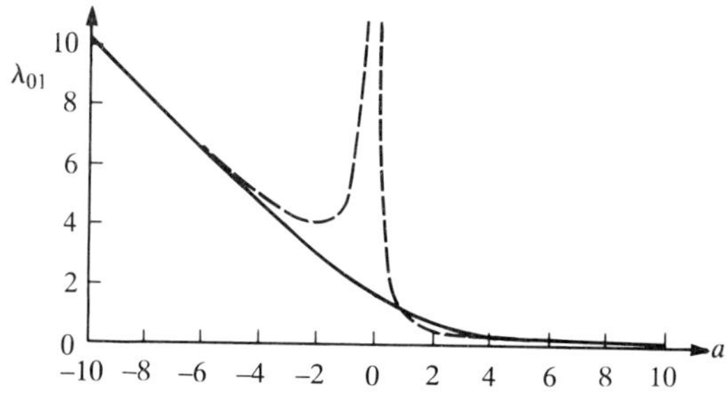

Fig. 18.21 Variation of the eigenvalue λ_{01} with laser pump parameter a. The broken line shows the approximations provided by the asymptotic forms $|a| + 4/|a|$ for a negative and $|a| \gg 1$ and $1/a$ for $a \gg 1$. (Reproduced from Risken, 1970.)

are normalized so that

$$\begin{aligned}\delta_{mm'}\delta_{nn'} &= \int_0^\infty \mathrm{d}r \int_0^{2\pi} \mathrm{d}\theta\, \psi^*_{mn}(r)\psi_{m'n'}(r)\frac{1}{2\pi}\,\mathrm{e}^{\mathrm{i}(n-n')\theta} \\ &= \delta_{nn'}\int_0^\infty \psi^*_{mn}(r)\psi_{m'n'}(r)\,\mathrm{d}r, \end{aligned} \qquad (18.6\text{–}19)$$

whereas the completeness relation (18.6–10) implies that

$$\sum_{m=0}^{\infty}\sum_{n=-\infty}^{\infty}\frac{1}{2\pi}\,\mathrm{e}^{in(\theta'-\theta)}\psi_{mn}(r)\psi^*_{mn}(r') = \delta(r-r')\delta(\theta-\theta'). \qquad (18.6\text{–}20)$$

With the help of Eqs. (18.6–5), (18.6–6), (18.6–15) and (18.6–18) the general solution of the Fokker–Planck equation (18.6–1) can now be expressed in the form

$$p(r,\theta,t) = \frac{1}{2\pi r}\sum_{m=0}^{\infty}\sum_{n=-\infty}^{\infty} c_{mn}\psi_{00}(r)\psi_{mn}(r)\,\mathrm{e}^{-in\theta}\,\mathrm{e}^{-\lambda_{mn}t}, \quad t \geqslant 0, \qquad (18.6\text{–}21)$$

with the constants c_{mn} chosen to ensure that the boundary conditions are satisfied at $t = 0$, and with the normalization

$$\int_0^\infty \mathrm{d}r r \int_0^{2\pi} \mathrm{d}\theta p(r, \theta, t) = 1. \tag{18.6–22}$$

18.6.1 *Growth of the laser field from the vacuum state*

As an example, we consider the solution when the laser field starts from the vacuum state. If the field amplitude is zero initially at $t = 0$, then the coefficients c_{mn} are given by

$$c_{mn} = \delta_{n0}[\psi^*_{m0}(0)/\psi_{00}(0)]. \tag{18.6–23}$$

The ratio on the right-hand side has to be interpreted as the limit of $\psi^*_{m0}(r)/\psi_{00}(r)$ as $r \to 0$, because $\psi_{m0}(0)$ vanishes for $m = 0, 1, 2, \ldots$. The solution then becomes

$$p(r, \theta, t) = \frac{1}{2\pi r}\frac{\psi_{00}(r)}{\psi_{00}(0)} \sum_{m=0}^{\infty} \psi^*_{m0}(0)\psi_{m0}(r)\, \mathrm{e}^{-\lambda_{m0}t}, \; t \geqslant 0. \tag{18.6–24}$$

By putting $t = 0$ and using the completeness relation for the eigenfunctions, we can see immediately that

$$p(r, \theta, 0) = \frac{1}{2\pi r}\delta(r),$$

which is the correct representation of the initial vaccum state. As $p(r, \theta, t)$ given by Eq. (18.6–24) satisfies the Fokker–Planck equation (18.6–1) and the correct boundary condition, it must be the solution we are seeking. It is worth noting that all phases in the range 0 to 2π are equally likely at time t, just as they are at time $t = 0$. The probability density $\mathcal{P}(I, t)$ of the light intensity $I = r^2$ at time t can immediately be obtained from $p(r, \theta, t)$:

$$\begin{aligned}\mathcal{P}(I, t) &= \frac{1}{2}\int_0^{2\pi} p(r = \sqrt{I}, \theta, t)\, \mathrm{d}\theta \\ &= \frac{1}{2\sqrt{I}}\frac{\psi_{00}(\sqrt{I})}{\psi_{00}(0)} \sum_{m=0}^{\infty} \psi^*_{m0}(0)\psi_{m0}(\sqrt{I})\, \mathrm{e}^{-\lambda_{m0}t}, \; t \geqslant 0.\end{aligned} \tag{18.6–25}$$

The growth of the probability distribution predicted by Eq. (18.6–25) has been evaluated numerically and the results are shown in Fig. 18.22 for the case $a = 8$. The probability density $\mathcal{P}(I, t)$ starts off as a delta function centered at $I = 0$ at time $t = 0$, turns into an exponential form as for thermal light, then develops a maximum, and finally ends up in the steady state as the Gaussian function that was derived in Section 18.3 [cf. Eq. (18.3–17)]. The solution $\mathcal{P}(I, t)$ can of course be used to calculate the moments of the light intensity as a function of time t, from the relation

$$\langle I^r(t)\rangle = \int_0^\infty I^r \mathcal{P}(I, t)\, \mathrm{d}I. \tag{18.6–26}$$

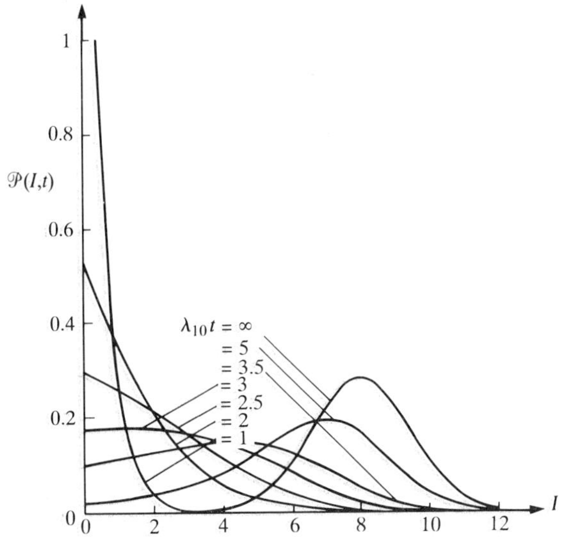

Fig. 18.22 Time development of the probability distribution $\mathcal{P}(I, t)$ given by Eq. (18.6–25) with pump parameter $a = 8$. The different curves correspond to different times t ranging from $1/\lambda_{10}$ to ∞, with $\lambda_{10} = 14.651$. (Reproduced from Risken, 1970.)

The results of a calculation of $\langle I(t) \rangle$ are illustrated by the full curves in Fig. 18.23 for three different laser pump parameters. It is worth noting that the initial growth of $\langle I(t) \rangle$ with t is independent of pump parameter a, whereas the steady-state values depend strongly on a. The full curves in Fig. 18.24 show the calculated time development of the normalized relative intensity fluctuations for the same three pump parameters. We observe that initially the relative intensity fluctuations are unity, just as for thermal light. However, as time goes on, and as

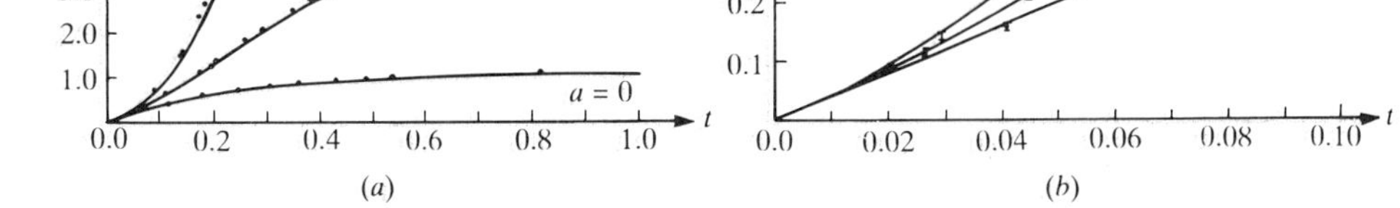

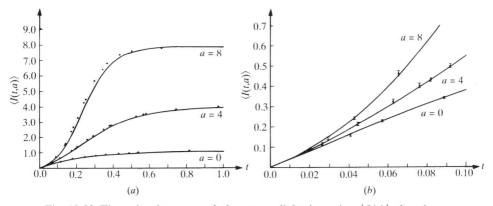

Fig. 18.23 Time development of the mean light intensity $\langle I(t) \rangle$ for three different pump parameters a. The full curves are derived from Eq. (18.6–25). The experimental points are obtained from photoelectric counting measurements. Figure (b) is an expansion of Fig. (a) near the origin. (Reproduced from Meltzer and Mandel, 1970.)

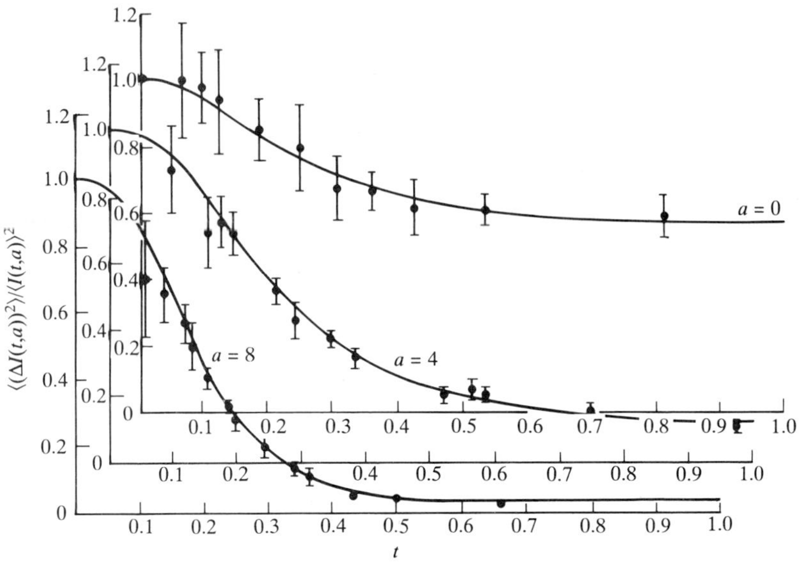

Fig. 18.24 Time development of relative intensity fluctuations $\langle(\Delta I(t))^2\rangle/\langle I(t)\rangle^2$ for three different pump parameters a. The full curves are derived from Eqs. (18.6–25) and (18.6–26), and the experimental points are obtained from photoelectric counting measurements. (Reproduced from Meltzer and Mandel, 1970.)

the characteristics of the laser field develop, the relative intensity fluctuations fall, and eventually they become very small for large pump parameters.

Actually, for short times t, it is possible to obtain a much simpler solution of the Fokker–Planck equation directly. For if the initial state corresponds to the vacuum state, or to a field of zero amplitude, then the field will still be weak after a short time t, and we are justified in dropping the nonlinear term in the Fokker–Planck equation (18.6–1). The equation is then simple enough to be integrated directly, and we may readily confirm that the time-dependent probability density

$$\mathcal{P}(I, t) = \frac{a}{2(e^{2at} - 1)} \exp\left[\frac{-aI}{2(e^{2at} - 1)}\right], \tag{18.6–27}$$

satisfies this simpler Fokker–Planck equation. The same $\mathcal{P}(I, t)$ should also be a solution of the nonlinear equation (18.6–1) for short times t, before the light intensity has grown appreciably. The exponential form of the probability distribution is characteristic of thermal radiation (cf. Section 13.2), and the mean intensity is given by

$$\begin{aligned}\langle I(t)\rangle &= (2/a)(e^{2at} - 1)\\ &\approx 4t\end{aligned} \tag{18.6–28}$$

for short times t. The initial growth of $\langle I(t)\rangle$ is therefore expected to be independent of the pump parameter a, as shown in Fig. 18.23.

18.6.2 Experimental investigations

Several of the foregoing predicted features of the laser dynamics have been tested in photoelectric counting experiments, in which the laser was turned on under various conditions of excitation (Arecchi, Degiorgio and Querzola, 1967; Meltzer and Mandel, 1970, 1971; Arecchi and Degiorgio, 1971). Figure 18.25 shows the apparatus used by Meltzer and Mandel. The active laser medium was a He:Ne plasma tube placed between cavity mirrors M_1 and M_2. The main output light beam, after attenuation, was directed to a counting photomultiplier, whose output pulses, after amplification and pulse shaping, were counted by a scaler for a time interval of the order of 1 μs, which is short compared with the natural rise-time of the laser field. Mirror M_2 was mounted on a piezoelectric crystal, which allowed the cavity length to be varied over a range of the order of one or two wavelengths. As the gain medium is inhomogeneously broadened, it was possible to vary the number of atoms participating in laser action, and therefore the laser gain parameter, by varying the tuning of the cavity. With the help of a monitor photomultiplier and amplifier arranged in a feedback loop, the working point of the laser in the steady state could be maintained anywhere from below to well above threshold.

In order to switch the system on and off rapidly, an external mirror M_3 was introduced along the laser axis, and positioned so that the reflected beam was sufficiently dephased from the standing wave in the cavity to extinguish the laser.

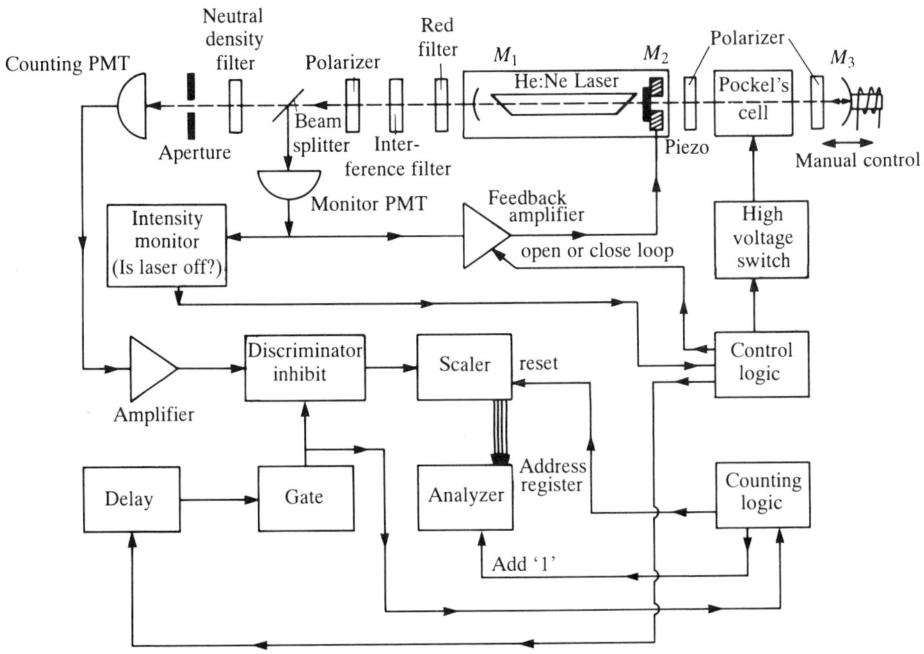

Fig. 18.25 Outline of the apparatus used to study the time evolution of a He:Ne laser field from the instant the laser is turned on. (Reproduced from Meltzer and Mandel, 1971.)

A Pockels cell and two polarizers placed between the mirror M_3 and the laser made it possible to effectively block the return beam within a few nanoseconds by a pulse applied to the Pockels cell; this blocking caused the laser to turn on. The pulse initiating the switch-on was also sent through a variable delay τ to an electronic gate that controlled the position of the 1 μs photon counting interval relative to the initiation pulse. The number of counts n registered by the scaler at the end of the counting interval was used to increment the number stored at address n in a digital memory. The laser was then extinguished again, the scaler was cleared, and the cycle was repeated. After many thousands of similar counting cycles, the number stored at address n provided a measure of the probability $p(n, \tau)$ that n photons were counted after a time delay τ following the switch-on pulse.

As we know from Chapter 14, for a short counting interval T, $p(n, \tau)$ is related to the probability density $\mathcal{P}(I, \tau)$ by the relation [cf. Eq. (14.8–12)]

$$p(n, \tau) = \frac{1}{n!}\int_0^\infty \mathcal{P}(I, \tau)(\alpha I T)^n \,\mathrm{e}^{-\alpha I T}\,\mathrm{d}I, \tag{18.6–29}$$

where α is a scale factor that converts the average light intensity $\langle I \rangle$ in our dimensionless units to the average photoelectric counting rate per second during the counting interval T. This equation makes it possible to test the predicted form of $\mathcal{P}(I, \tau)$ by photoelectric counting measurements. In particular, the factorial moments of n are proportional to the moments of the light intensity I [cf. Eq. (14.9–4)], so that

$$\left.\begin{aligned} \langle n(\tau)\rangle &= \alpha\langle I(\tau)\rangle T \\ \langle n(\tau)(n(\tau)-1)\rangle &= \alpha^2\langle I^2(\tau)\rangle T^2, \end{aligned}\right\} \tag{18.6–30}$$

which allows the theoretical curves in Figs. 18.23 and 18.24 to be subjected to direct experimental test.

The results of the measurements are shown by the points on these figures, and the agreement confirms the dynamical predictions of the laser theory. It is less easy to test the curves in Fig. 18.22 directly, because of the difficulty of inverting Eq. (18.6–29) and extracting $\mathcal{P}(I, \tau)$ with any accuracy. However, one can use Eq. (18.6–29) to compare the measured probabilities $p(n, \tau)$ with those expected from the theoretical form of $\mathcal{P}(I, \tau)$ given by Eq. (18.6–25). The results are shown in Fig. 18.26 for three different steady-state pump parameters a. The experimental points are again found to be in very good agreement with the full curves based on the laser theory.

18.7 Correlation functions

We have now obtained the steady-state solution of the Fokker–Planck Eq. (18.6–1) for the laser and also the transient or time-dependent solution.

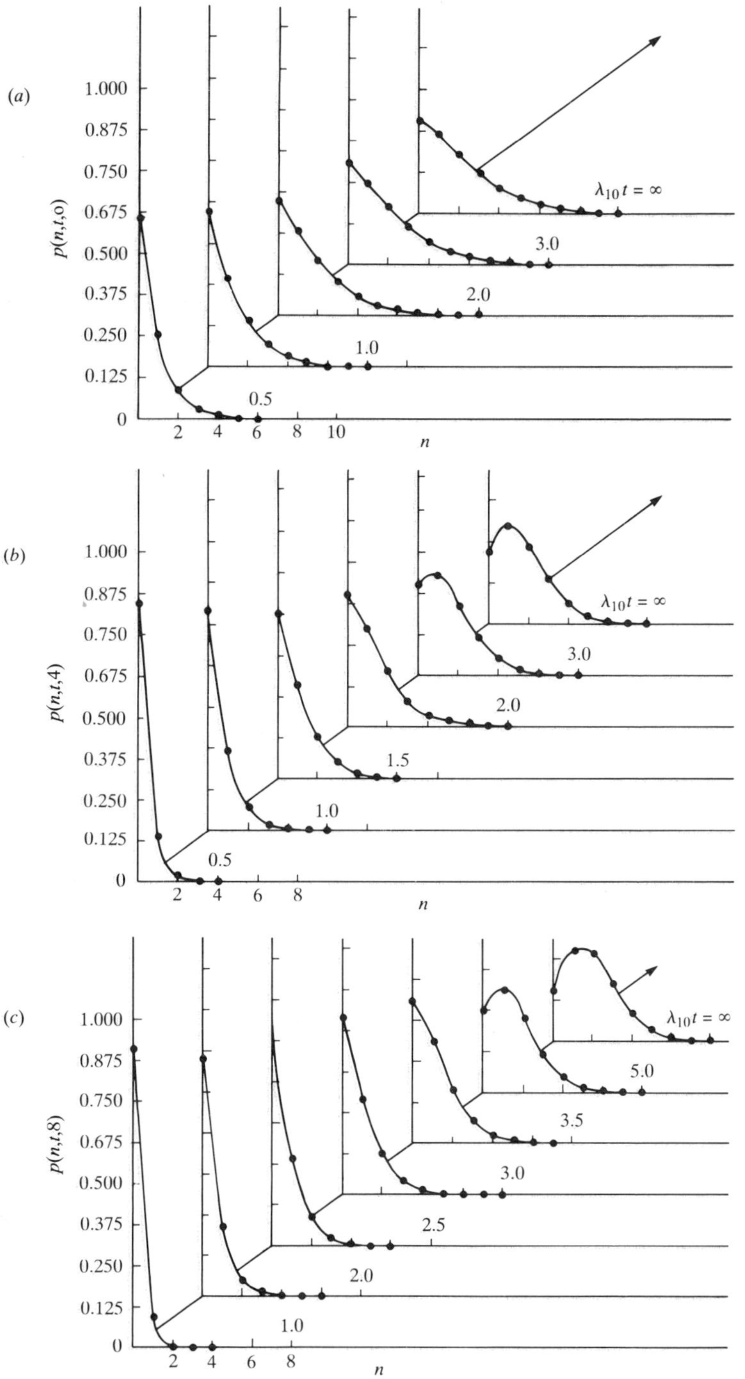

Fig. 18.26 Comparison of experimentally measured probabilities $p(n, t, a)$ with theoretical distributions given by Eqs. (18.6–25) to (18.6–27) for various time delays t expressed in units of $1/\lambda_{10}$, and for steady-state pump parameters (a) $a = 0$; (b) $a = 4$; (c) $a = 8$. (Reproduced from Meltzer and Mandel, 1971.)

However, as was pointed out in Section 18.3, one more quantity is needed before multi-time correlation functions of the laser field can be calculated. This is the Green function or conditional probability density $P(\mathscr{E}_2, t_2|\mathscr{E}_1, t_1)$ that the field has amplitude $\mathscr{E}_2$ at time t_2, given that it has amplitude $\mathscr{E}_1$ at the earlier time t_1.

18.7.1 Green's functions

As the Green function is itself a probability density, it is also a solution of Eq. (18.6–1). In polar coordinates, it must therefore be expressible in the form of Eq. (18.6–21) with r, θ replaced by r_2, θ_2, the time interval t replaced by $t_2 - t_1$, and with the constants c_{mn} determined by the initial values r_1, θ_1. Moreover, when $t_2 = t_1$, the Green function must reduce to

$$P(r_2, \theta_2, t_1|r_1, \theta_1, t_1) = \delta^{(2)}(\mathscr{E}_2 - \mathscr{E}_1) = \frac{1}{r_1}\delta(r_2 - r_1)\delta(\theta_2 - \theta_1). \quad (18.7\text{–}1)$$

Let us put

$$c_{mn} = \frac{\psi^*_{mn}(r_1)\,\mathrm{e}^{in\theta_1}}{\psi_{00}(r_1)} \quad (18.7\text{–}2)$$

in Eq. (18.6–21); then the resulting probability density becomes the Green function

$$P(r_2, \theta_2, t_2|r_1, \theta_1, t_1) = \frac{1}{2\pi r_2}\frac{\psi_{00}(r_2)}{\psi_{00}(r_1)}\sum_{m=0}^{\infty}\sum_{n=-\infty}^{\infty}\psi_{mn}(r_2)\psi^*_{mn}(r_1)\,\mathrm{e}^{in(\theta_1-\theta_2)}\,\mathrm{e}^{-\lambda_{mn}(t_2-t_1)}.$$

$$t_2 \geqslant t_1. \quad (18.7\text{–}3)$$

In order to check this we put $t_2 = t_1$ and make use of Eq. (18.6–20), and we see at once that the right-hand side reduces to $(1/r_1)\delta(r_2 - r_1)\delta(\theta_2 - \theta_1)$, as required. We observe that the Green function depends only on the difference between the two time arguments, as expected if a stationary state is to be reached.

By multiplying $P(r_2, \theta_2, t_2|r_1, \theta_1, t_1)$ by the steady-state probability distribution $p_s(r_1, \theta_1)$ given by Eq. (18.6–18), we obtain the joint probability density $p_2(r_2, \theta_2, t_2; r_1, \theta_1, t_1)$ of the field at two times t_1, t_2 in the steady state:

$$\begin{aligned}
p_2(r_2, \theta_2, t_2; r_1, \theta_1, t_1)\\
&= P(r_2, \theta_2, t_2|r_1, \theta_1, t_1)p_s(r_1, \theta_1)\\
&= \frac{\psi_{00}(r_2)\psi_{00}(r_1)}{2\pi r_2 2\pi r_1}\sum_{m=0}^{\infty}\sum_{n=-\infty}^{\infty}\psi_{mn}(r_2)\psi^*_{mn}(r_1)\,\mathrm{e}^{in(\theta_1-\theta_2)}\,\mathrm{e}^{-\lambda_{mn}(t_2-t_1)}, \quad t_2 \geqslant t_1.
\end{aligned}$$

(18.7–4)

18.7.2 Intensity correlation

We are now in a position to calculate any two-time autocorrelation function, such as the intensity correlation $\langle I(t_1)I(t_2)\rangle$ with $t_2 \geqslant t_1$, from the relation

$$\langle I(t_1)I(t_2)\rangle = \iint_0^\infty dr_1\, dr_2 r_1 r_2 \iint_0^{2\pi} d\theta_1\, d\theta_2\, r_1^2 r_2^2 p_2(r_2, \theta_2, t_2; r_1, \theta_1, t_1). \quad (18.7\text{–}5)$$

With the help of Eq. (18.7–4) this becomes

$$\begin{aligned}\langle I(t_1)I(t_2)\rangle &= \frac{1}{(2\pi)^2}\iint_0^\infty dr_1\, dr_2\, r_1^2 r_2^2 \psi_{00}(r_2)\psi_{00}(r_1)\sum_{m=0}^\infty \sum_{n=-\infty}^\infty \psi_{mn}(r_2)\psi_{mn}^*(r_1)\\ &\quad\times \iint_0^{2\pi} d\theta_1\, d\theta_2\, e^{in(\theta_1-\theta_2)}\, e^{-\lambda_{mn}(t_2-t_1)}\\ &= \iint_0^\infty dr_1\, dr_2\, r_1^2 r_2^2 \psi_{00}(r_2)\psi_{00}(r_1)\sum_{m=0}^\infty \psi_{m0}(r_2)\psi_{m0}^*(r_1)\, e^{-\lambda_{m0}(t_2-t_1)},\end{aligned}$$

which can be expressed as the series

$$\langle I(t_1)I(t_2)\rangle = \langle(\Delta I)^2\rangle \sum_{m=0}^\infty M_m\, e^{-\lambda_{m0}|t_2-t_1|}. \quad (18.7\text{–}6)$$

The coefficients M_m are given by the equation

$$M_m \equiv \left|\int_0^\infty dr\, r^2 \psi_{00}(r)\psi_{m0}(r)\right|^2 \Big/ \langle(\Delta I)^2\rangle, \quad (18.7\text{–}7)$$

and $\langle(\Delta I)^2\rangle$ is obtainable from Eq. (18.3–28). We have replaced $t_2 - t_1$ in the exponent in Eq. (18.7–6) by $|t_2 - t_1|$ in order to ensure that $\langle I(t_1)I(t_2)\rangle$ is symmetric in $t_2 - t_1$, as required for a stationary field. The first term corresponding to $m = 0$ in the series (18.7–6) is the only one that survives when $|t_2 - t_1| \to \infty$, and this term must therefore equal $\langle I\rangle^2$. If we subtract $\langle I\rangle^2$ from both sides of Eq. (18.7–6) and divide by $\langle(\Delta I)^2\rangle$, we arrive at the equation

$$\frac{\langle \Delta I(t)\Delta I(t+\tau)\rangle}{\langle(\Delta I)^2\rangle} = \sum_{m=1}^\infty M_m\, e^{-\lambda_{m0}|\tau|}, \quad (18.7\text{–}8)$$

which gives the time dependence of the normalized intensity correlation function. It is clear from Eqs. (18.7–6) and (18.7–8) that the coefficients M_m are normalized so that

$$\sum_{m=1}^\infty M_m = 1 = \frac{\langle(\Delta I)^2\rangle}{\langle I^2\rangle}\sum_{m=0}^\infty M_m. \quad (18.7\text{–}9)$$

We observe that only eigenfunctions and eigenvalues of the type $\psi_{m0}(r)$ and λ_{m0} are needed for the evaluation of the correlation function.

We have already seen how the eigenvalues $\lambda_{m0}(a)$ vary with the laser pump parameter a (see Fig. 18.20). Near threshold, the first few eigenvalues are generally of the same order of magnitude, although they gradually increase. However, to determine how many terms in Eq. (18.7–8) contribute appreciably to the series requires evaluation of the coefficients $M_m(a)$, which has been carried out by Hempstead and Lax (1967), Risken and Vollmer (1967a,b) and Risken (1970). Figure 18.27 shows the first four coefficients $M_m(a)$ as a function of pump parameter a. It will be seen that below threshold, up to about $a = -4$, the series

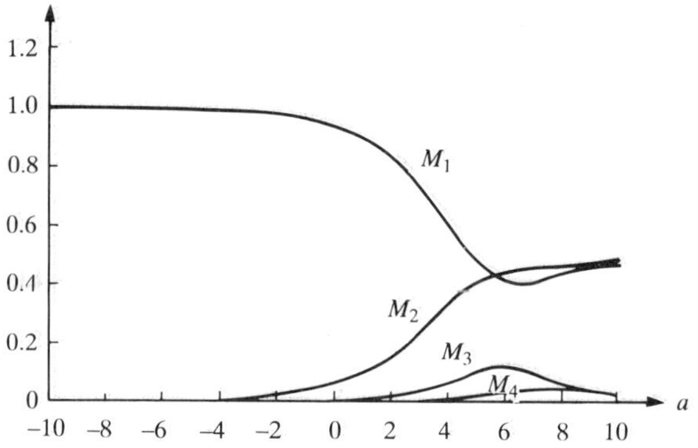

Fig. 18.27 Variation of the first four coefficients M_m in the series expansion of the intensity correlation function with laser pump parameter a. (Reproduced from Risken and Vollmer, 1967a.)

in Eq. (18.7–9) is dominated by the first term, and the intensity correlation function is very nearly of exponential form with correlation time $1/\lambda_{10}$. However, a little above threshold, other terms in the series also contribute, and the correlation function is no longer of purely exponential form. Of course, well above threshold, the relative intensity fluctuations become very small, and the form of the correlation function is then moot.

Equation (18.7–8) has been tested in photoelectric correlation measurements (Chopra and Mandel, 1972, 1973a; Corti, Degiorgio and Arecchi, 1973), in which the form of the intensity correlation function was accurately determined. Figure 18.28 shows some experimental results obtained by Corti, Degiorgio and Arecchi (1973) for the first four eigenvalues λ_{m0} ($m = 1, 2, 3, 4$) as a function of average laser light intensity, superimposed on the theoretically predicted curves. In other experiments the average intensity correlation time T_c corresponding to these eigenvalues was determined. Figure 18.29 shows some results of Singh, Friberg and Mandel (1983), who also investigated the behavior of the laser when it was

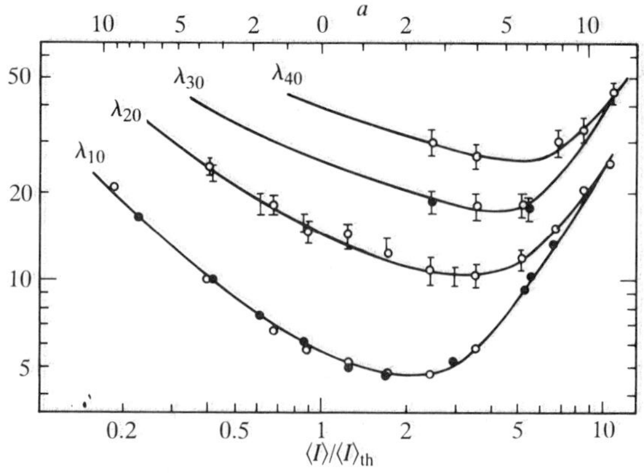

Fig. 18.28 The measured eigenvalues λ_{m0} as a function of mean laser intensity $\langle I \rangle$ relative to the threshold intensity $\langle I_{\text{th}} \rangle$, superimposed on the theoretical curves. (Reproduced from Corti, Degiorgio and Arecchi, 1973.)

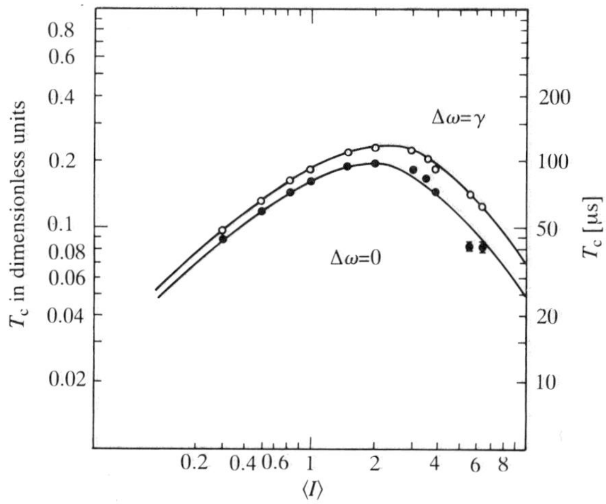

Fig. 18.29 Measured values of the average intensity correlation time T_c as a function of mean light intensity $\langle I \rangle$, superimposed on the theoretical curves. The two curves correspond to a laser on resonance ($\Delta\omega = 0$) and one whose cavity is detuned one natural linewidth ($\Delta\omega = \gamma$). (Reproduced from Singh, Friberg and Mandel, 1983.)

detuned from resonance. Once again we note that the theory is well confirmed by the measurements.

18.7.3 Field amplitude correlation function and spectral density

Let us now examine the second-order amplitude correlation function of the complex laser field $\mathscr{E}(t)$, which is given by the formula

$$\langle \mathscr{E}^*(t_1)\mathscr{E}(t_2)\rangle\, \mathrm{e}^{-\mathrm{i}\omega_1(t_2-t_1)} = \langle r(t_1)r(t_2)\,\mathrm{e}^{\mathrm{i}[\theta(t_2)-\theta(t_1)]}\rangle\, \mathrm{e}^{-\mathrm{i}\omega_1(t_2-t_1)} \quad (18.7\text{–}10)$$

in our notation, where ω_1 is the frequency to which the cavity is tuned. With the help of Eq. (18.7–4) we have in the long-time limit

$$\begin{aligned}\langle \mathscr{E}^*(t_1)\mathscr{E}(t_2)\rangle &= \iint_0^\infty \mathrm{d}r_1\,\mathrm{d}r_2\, r_1^2 r_2^2 \iint_0^{2\pi} \mathrm{d}\theta_1\,\mathrm{d}\theta_2\, \mathrm{e}^{\mathrm{i}(\theta_2-\theta_1)} p_2(r_2, \theta_2, t_2; r_1, \theta_1, t_1)\\
&= \frac{1}{(2\pi)^2}\iint_0^\infty \mathrm{d}r_1\,\mathrm{d}r_2 r_1 r_2 \psi_{00}(r_2)\psi_{00}(r_1)\sum_{m=0}^{\infty}\sum_{n=-\infty}^{\infty} \psi_{mn}(r_2)\psi^*_{mn}(r_1)\\
&\quad\times \iint_0^{2\pi} \mathrm{d}\theta_1\,\mathrm{d}\theta_2\, \mathrm{e}^{\mathrm{i}(n-1)(\theta_1-\theta_2)}\,\mathrm{e}^{-\lambda_{mn}(t_2-t_1)}\\
&= \iint_0^\infty \mathrm{d}r_1\,\mathrm{d}r_2 r_1 r_2 \psi_{00}(r_2)\psi_{00}(r_1)\sum_{m=0}^{\infty} \psi_{m1}(r_2)\psi^*_{m1}(r_1)\,\mathrm{e}^{-\lambda_{m1}(t_2-t_1)},\end{aligned}$$

$$t_2 \geqslant t_1. \quad (18.7\text{–}11)$$

The resulting correlation function can be expressed in the form

$$\Gamma(\tau) = \langle \mathscr{E}^*(t)\mathscr{E}(t+\tau)\rangle\, \mathrm{e}^{-\mathrm{i}\omega_1\tau} = \langle I\rangle \sum_{m=0}^{\infty} V_m\, \mathrm{e}^{-\lambda_{m1}|\tau|}\, \mathrm{e}^{-\mathrm{i}\omega_1\tau}, \qquad (18.7\text{–}12)$$

where we have written

$$V_m \equiv \left|\int_0^\infty \mathrm{d}r\, r\, \psi_{00}(r)\psi_{m1}(r)\right|^2 \Big/ \langle I\rangle, \qquad (18.7\text{–}13)$$

and $\langle I\rangle$ is given by Eq. (18.3–27).

As $\Gamma(\tau)\,\mathrm{e}^{\mathrm{i}\omega_1\tau}$ turns out to be real, and $\Gamma(-\tau) = \Gamma^*(\tau)$ for a stationary field, it follows that the exponents $-\lambda_{m1}\tau$ in Eq. (18.7–12) must depend only on the modulus of τ. If we put $\tau = 0$ on both sides of the equation, it is at once evident that the coefficients V_m are normalized so that

$$\sum_{m=0}^{\infty} V_m = 1. \qquad (18.7\text{–}14)$$

Moreover, the normalized second-order correlation function (complex degree of self-coherence) of the laser field is given by the formula

$$\gamma(\tau) = \langle \mathscr{E}^*(t)\mathscr{E}(t+\tau)\rangle\, \mathrm{e}^{-\mathrm{i}\omega_1\tau}/\langle I\rangle = \sum_{m=0}^{\infty} V_m\, \mathrm{e}^{-\lambda_{m1}|\tau|}\, \mathrm{e}^{-\mathrm{i}\omega_1\tau}, \qquad (18.7\text{–}15)$$

and the Fourier transform of $\gamma(\tau)$ yields the normalized spectral density $\phi(\omega)$, as usual. In this case

$$\begin{aligned}\phi(\omega) &= \int_{-\infty}^{\infty} \gamma(\tau)\, \mathrm{e}^{\mathrm{i}\omega\tau}\, \mathrm{d}\tau \\ &= \sum_{m=0}^{\infty} V_m \left[\frac{2\lambda_{m1}}{\lambda_{m1}^2 + (\omega - \omega_1)^2}\right]. \end{aligned} \qquad (18.7\text{–}16)$$

The spectral density of the laser field is therefore a sum of Lorentzian spectral densities, all centered on the frequency ω_1, but with various widths λ_{m1}. The importance of the higher terms in the series (18.7–15) is again determined by the magnitude of the coefficients V_m. Calculations show that V_0 is by far the largest coefficient in the expansion; the sum of all other coefficients is never more than 2% of V_0, and substantially less away from threshold. The spectral density is therefore approximated well by a single Lorentzian function of width λ_{01}.

The form of λ_{01} is illustrated in Fig. 18.21 as a function of pump parameter a, and it is seen to fall steadily with increasing a. To a first approximation, λ_{01} is found to vary inversely with the mean light intensity $\langle I\rangle$, and we may write

$$\lambda_{01}(a) = \frac{\alpha_{\mathrm{L}}(a)}{\langle \mathrm{I}(a)\rangle}, \qquad (18.7\text{–}17)$$

where $\alpha_{\mathrm{L}}(a)$ is of order unity and does not vary much. This emphasizes that the linewidth of the laser becomes narrower and narrower as the excitation and the light intensity are increased. In the limit, the field becomes strictly monochromatic. However, this limit is not attainable, and, in practice, one often finds that, well above threshold, the laser bandwidth is limited by cavity vibrations rather than by these more fundamental considerations. Nevertheless, bandwidths below 1 kHz are attainable when great care is taken, which means that the frequency can be defined to better than 1 part in 10^{12}.

The form of $\alpha_L(a)$ given by Eq. (18.7–17) as a function of a is illustrated in Fig. 18.30, and we observe that it varies between the values 2 below threshold and 1 above threshold. The halving of the linewidth above threshold may be regarded as a reflection of the fact that only phase fluctuations remain in this region and intensity fluctuations die out. The latter therefore cease to contribute to the laser linewidth above threshold, whereas both intensity and phase fluctuations contribute equally to the linewidth below threshold.

These predictions of the theory have been tested in interference and heterodyne experiments (Siegman, Daino and Manes, 1967; Siegman and Arrathoon, 1968; Manes and Siegman, 1971; Gerhardt, Welling and Güttner, 1972; Güttner, Welling, Gericke and Seifert, 1978). Although the correlation functions are nearly exponential, the small predicted departures from the exponential form have been observed, and the values of the first few time constants in the expansion (18.7–15) have been determined. Figure 18.31 shows the results of measurements by Güttner, Welling, Gericke and Seifert (1978) for the first four eigenvalues λ_{m1} as a function of pump parameter a plotted on a logarithmic scale. It will be seen that the change implied by Fig. 18.30 is observed in λ_{01}, and that there is generally good agreement between theory and experiment. Figure 18.32 shows the values of the coefficients V_m ($m = 1, 2, 3$) extracted from the measurements. Because of the smallness of the V_m's and the difficulty of extracting their values from the data, the results exhibit much more scatter, but they are in general agreement with the theoretical predictions.

18.7.4 Higher-order correlations

The Green function given by Eq. (18.7–3) can readily be used to determine higher-order correlations also. For example, in order to calculate a three-time intensity correlation function of the light intensity, we start from the three-fold joint probability density $p_3(r_3, \theta_3, t_3; r_2, \theta_2, t_2; r_1, \theta_1, t_1)$ of the laser field at three different times. Because of the Markovian nature of the system, the Green function suffices to determine all joint probability densities, as was shown in Eq. (18.3–9), and in the steady state, for $t_3 \geqslant t_2 \geqslant t_1$, we have with the help of Eqs.

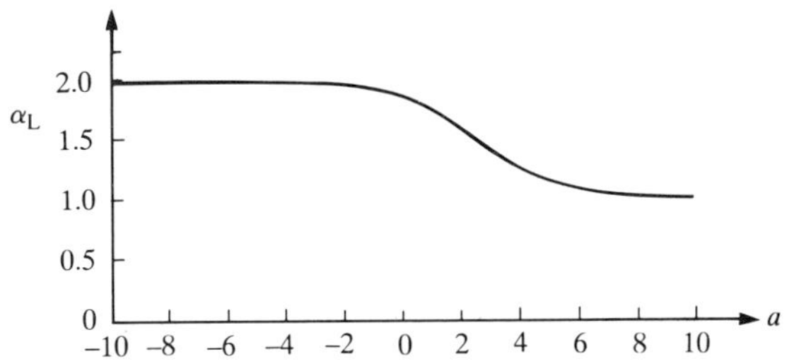

Fig. 18.30 Variation of the linewidth parameter α_L with laser pump parameter a. (Reproduced from Risken, 1970.)

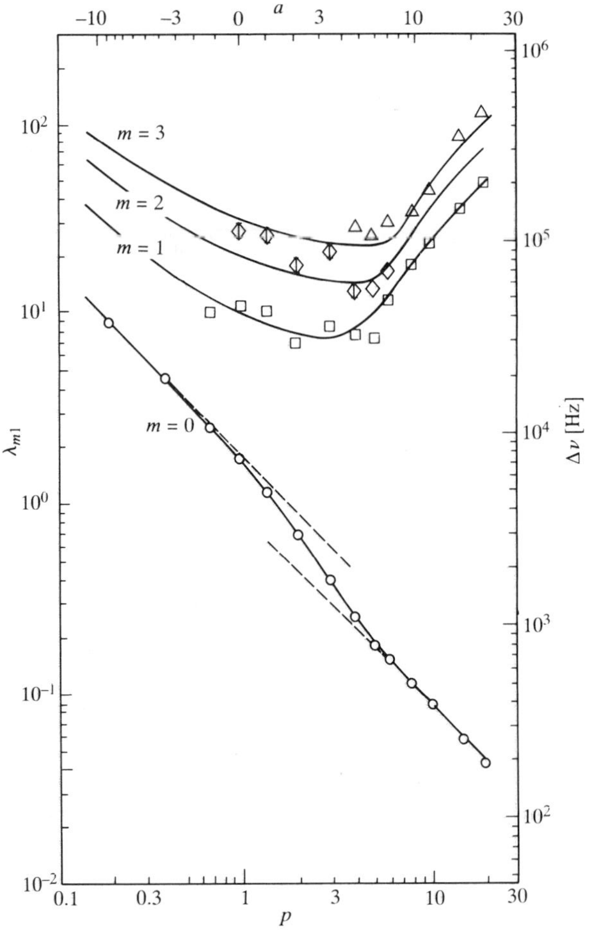

Fig. 18.31 Measured variation of the first four eigenvalues λ_{m1} with pump parameter (see the top scale), superimposed on the theoretical curves. (Reproduced from Güttner, Welling, Gericke and Seifert, 1978.)

(18.6–18) and (18.7–3),

$$\begin{aligned}
&p_3(r_3, \theta_3, t_3; r_2, \theta_2, t_2; r_1, \theta_1, t_1) \\
&\quad = P(r_3, \theta_3, t_3 | r_2, \theta_2, t_2) P(r_2, \theta_2, t_2 | r_1, \theta_1, t_1) p_s(r_1, \theta_1) \\
&\quad = \frac{\psi_{00}(r_3)\psi_{00}(r_1)}{(2\pi)^3 r_1 r_2 r_3} \sum_{m=0}^{\infty} \sum_{n=-\infty}^{\infty} \sum_{m'=0}^{\infty} \sum_{n'=-\infty}^{\infty} \psi_{mn}(r_3)\psi^*_{mn}(r_2) \\
&\qquad \times \psi_{m'n'}(r_2)\psi^*_{m'n'}(r_1)\, e^{in(\theta_2-\theta_3)}\, e^{in'(\theta_1-\theta_2)}\, e^{-\lambda_{mn}(t_3-t_2)}\, e^{-\lambda_{m'n'}(t_2-t_1)}. \qquad (18.7\text{–}18)
\end{aligned}$$

The three-time intensity correlation function is therefore expressible in the form

$$\begin{aligned}
\langle I(t)I(t+\tau_1)I(t+\tau_1+\tau_2)\rangle = &\iiint_0^{\infty} dr_1\, dr_2\, dr_3 \iiint_0^{2\pi} d\theta_1\, d\theta_2\, d\theta_3\, r_1^3 r_2^3 r_3^3 \\
&\times p_3(r_3, \theta_3, t+\tau_1+\tau_2; r_2, \theta_2, t+\tau_1; r_1, \theta_1, t)
\end{aligned}$$

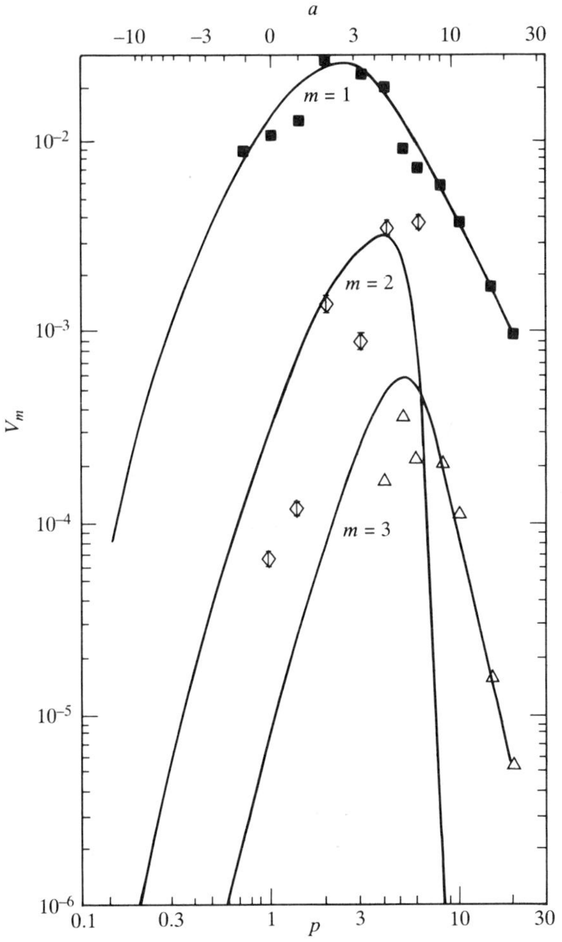

Fig. 18.32 Measured variation of the coefficients V_m given by Eq. (18.7–13) with pump parameters $m = 1, 2, 3$ (see the top scale) superimposed on the theoretical curves. (Reproduced from Güttner, Welling, Gericke, and Seifert, 1978.)

$$= \langle I^3 \rangle \sum_{m=0}^{\infty} \sum_{m'=0}^{\infty} N_{mm'} \, e^{-\lambda_{m0}\tau_2} e^{-\lambda_{m'0}\tau_1}, \quad \tau_1, \tau_2 \geqslant 0, \tag{18.7–19}$$

where the coefficients N_{mm} are normalized so that

$$\sum_{m=0}^{\infty} \sum_{m'=0}^{\infty} N_{mm'} = 1 \tag{18.7–20}$$

and are given by

$$N_{mm'} = \frac{1}{\langle I^3 \rangle} \iiint_0^{\infty} dr_1 \, dr_2 \, dr_3 r_1^2 r_2^2 r_3^2 \psi_{00}(r_1) \psi^*_{m'0}(r_1) \psi^*_{m0}(r_2) \psi_{m'0}(r_2) \psi_{00}(r_3) \psi_{m0}(r_3). \tag{18.7–21}$$

As usual, it is convenient to introduce a normalized intensity correlation function, defined by the relation

$$\lambda^{(3)}(\tau_1, \tau_2) \equiv \langle \Delta I(t)\Delta I(t+\tau_1)\Delta I(t+\tau_1+\tau_2)\rangle / \langle I \rangle^3, \qquad (18.7\text{–}22)$$

which is a function of two time arguments τ_1, τ_2.

The form of $\lambda^{(3)}(\tau_1, \tau_2)$ has been calculated for several combinations of τ_1, τ_2, and pump parameter a, and some of the results are illustrated in Fig. 18.33. Close to or below threshold, the variation with τ_1 or τ_2 is approximately exponential. Above threshold, $\lambda^{(3)}(\tau_1, \tau_2)$ can become negative and develop a minimum, although the absolute values are small. Still higher-order correlations have been calculated in a similar manner (Cantrell and Smith, 1971; Cantrell, Lax and Smith, 1973a,b).

These theoretical predictions have been tested in photoelectric correlation measurements involving three photon detections. As we know from Chapter 14, the three-fold joint probability of registering photoelectric detections at times t, $t+\tau_1$ and $t+\tau_1+\tau_2$ is proportional to $\langle I(t)I(t+\tau_1)I(t+\tau_1+\tau_2)\rangle$, so that measurement of the rate of arrival of photoelectric pulse triplets as a function of the intervals τ_1 and τ_2 between pulse pairs yields the required correlation function. Such measurements have been carried out (Davidson and Mandel, 1968; Davidson, 1969; Chopra and Mandel, 1973b; Corti and Degiorgio, 1976a,b). Figures 18.34 (a) and (b) show some measured values of $\lambda^{(3)}(\tau_1, \tau_2)$ below and above threshold superimposed on the theoretical curves calculated from Eqs. (18.7–19) and (18.7–21).

It is apparent that there is generally good agreement between theory and experiment. The higher-order correlation measurements probably represent the most searching test of the laser theory to date. Despite the simplifications in the theory that were introduced in the beginning, it has evidently proved to be very successful.

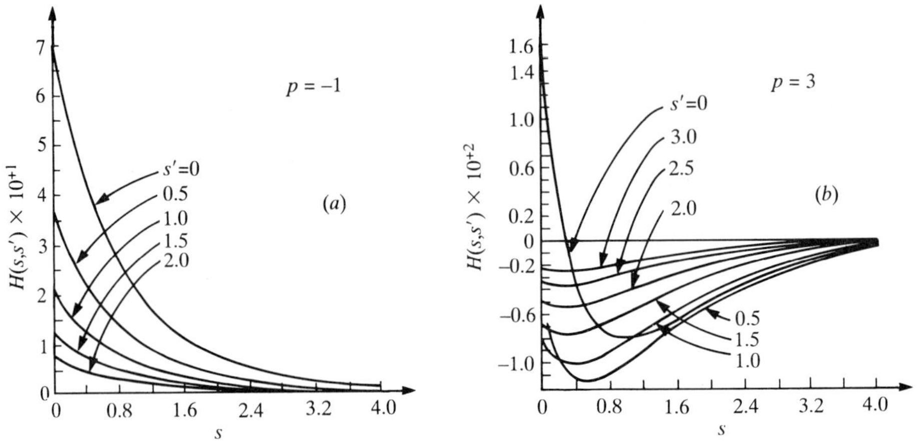

Fig. 18.33 The calculated form of $\lambda^{(3)}(\tau_1, \tau_2)$ [here called $H(s, s')$] as function of τ_1, τ_2 [denoted here as s, s'] (a) below threshold with pump parameter -1; (b) above threshold with pump parameter 3. (Reproduced from Cantrell, Lax and Smith, 1973a.)

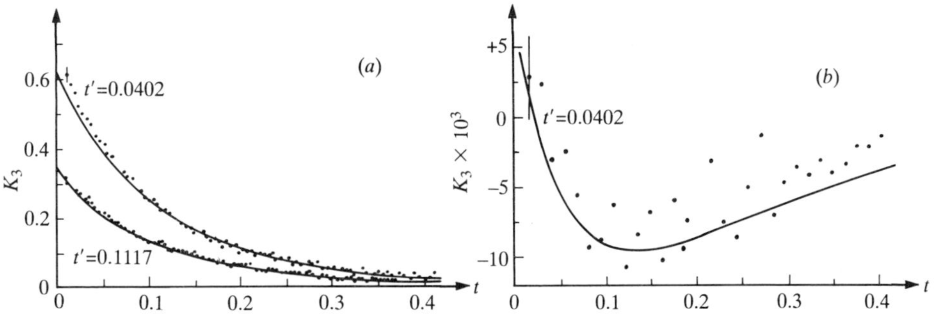

Fig. 18.34 Some measured values of the normalized third-order intensity correlation function $\lambda^{(3)}(t, t')$ [here denoted as $K_3(t, t')$], superimposed on the theoretical curves (*a*) for a laser with pump parameter $a = -2$; (*b*) for a laser with pump parameter $a = 2.8$. (Reproduced from Corti and Degiorgio, 1974.)

18.8 Laser instabilities and chaos

In the foregoing sections we have explored various aspects of the laser field, both classically and quantum mechanically. The quantum treatment automatically includes fluctuations, and introduces stochastic features into the description. Sometimes these statistical features are modelled semiclassically through the ad hoc introduction of Langevin forces into the equations of motion, but their origin is nevertheless quantum mechanical. In the absence of quantum effects, we saw that the behavior of the laser is deterministic and that a steady state is always reached after a sufficiently long time.

However, it has been discovered that, even when the laser is treated semiclassically by deterministic equations, the equations are capable of exhibiting quasi-periodic and even chaotic behavior, but this has nothing to do with quantum fluctuations (Uspenskii, 1963; Buley and Cummings, 1964; Risken and Nummedal, 1968a,b). Instead the behavior is reminiscent of the turbulence displayed by the classical equations of fluid dynamics (Lorenz, 1963; Haken, 1975a,b, 1985). However, in order to encounter these features one has to go beyond some of the simplifying approximations that were introduced in the semiclassical treatment in Section 18.2. In particular, one needs to avoid the adiabatic elimination of atomic variables that makes it possible to relate the expectation of the atomic dipole moment at time t to the electric field at the same time, when the field varies relatively slowly [cf. Eq. (18.2–16)].

In order to demonstrate the possibility of chaotic behavior, we start from the coupled equations of motion for a system of identical two-level atoms interacting with a single-mode field in the laser cavity on resonance. The equation of motion for the amplitude $\mathscr{E}(t)$ in the rotating wave approximation is again Eq. (18.2–13), but we shall simplify the calculation by supposing that we are dealing with a traveling plane wavefield with mode function

$$u(\mathbf{r}) = e^{ikz} \tag{18.8–1}$$

in a cavity of length L and cross-sectional area A, and that the atomic density η is constant. As before the atomic polarization $P(t)$ is given by the expression

$$P(t) = \eta\langle\hat{\mu}(t)\rangle,$$

and with the help of Eq. (15.2–16) we have

$$\begin{aligned}P(t) &= \eta[\mathrm{Re}\,(\mu_{12})r_1(t) + \mathrm{Im}\,(\mu_{12})r_2(t)]\\ &= \tfrac{1}{2}\eta\mu_{12}[r_1(t) - \mathrm{i}r_2(t)] + \text{c.c.},\end{aligned}$$

where $r_1(t)$, $r_2(t)$ are components of the atomic Bloch-vector. We transform to the rotating frame as described in Section 15.3.2 with the help of Eq. (15.3–17b) and find that

$$P(t) = \tfrac{1}{2}\eta\mu_{12}[r_1'(t) - \mathrm{i}r_2'(t)]\mathrm{e}^{-\mathrm{i}\omega t} + \text{c.c.}, \tag{18.8–2}$$

where $r_1'(t)$, $r_2'(t)$ are slowly varying Bloch-vector components, from which the positive frequency part $P^{(+)}(t)$ can be obtained by inspection. Equation (18.2–13) then yields the following differential equation:

$$-\mathrm{i}\omega\left(2\frac{\mathrm{d}\mathscr{E}}{\mathrm{d}t} + \frac{\sigma}{\varepsilon_0}\mathscr{E}\right) = \frac{1}{2}\eta\frac{\omega^2}{\varepsilon_0}\mu_{12}[r_1'(t) - \mathrm{i}r_2'(t)]\left[\frac{\sin(kL/2)}{(kL/2)}\right]. \tag{18.8–3}$$

In the special case in which the transition dipole moment μ_{12} and the amplitude $\mathscr{E}(t)$ are real, the Bloch-vector moves in the y', z'-plane and $r_1'(t) = 0$. The equation of motion for the field then simplifies to

$$\frac{\mathrm{d}\Omega}{\mathrm{d}t} + \frac{\sigma}{2\varepsilon_0}\Omega = \frac{\mu_{12}^2\omega\eta}{2\varepsilon_0\hbar}\left[\frac{\sin(kL/2)}{(kL/2)}\right]r_2'$$

or

$$\frac{\mathrm{d}\Omega}{\mathrm{d}t} + \gamma_{\mathrm{c}}\Omega = \omega g r_2'. \tag{18.8–4}$$

Here we have used Eq. (18.2–18) to represent the field amplitude $\mathscr{E}(t)$ by the Rabi frequency $\Omega(t)$, with the understanding that it can be positive or negative, and we have written

$$\left.\begin{aligned}\gamma_{\mathrm{c}} &\equiv \sigma/2\varepsilon_0\\ g &\equiv \frac{\mu_{12}^2\eta}{2\varepsilon_0\hbar}\left[\frac{\sin(kL/2)}{(kL/2}\right]\end{aligned}\right\} \tag{18.8–5}$$

for the cavity damping rate and for the coupling rate between the cavity field and the atom.

Next we need the equations of motion for the atomic Bloch-vector to accompany Eq. (18.8–4). These are given by the Bloch equations (15.3–19) in the rotating frame, with the simplification $\phi = 0$, $r_1' = 0$ in the present situation, and with $\omega_0 = \omega_1 = \omega$. However, we augment the Bloch equations in one respect by introducing phenomenological damping terms $\gamma_\perp r_2'$ and $\gamma_\parallel r_3'$ for both r_2' and the atomic inversion r_3'. The three coupled equations of motion for the laser then take the form

$$\left.\begin{aligned}\frac{dr_2'}{dt} &= -\gamma_\perp r_2' + \Omega r_3' \\ \frac{dr_3'}{dt} &= -\gamma_\parallel(r_3' - r_e') - \Omega r_2' \\ \frac{d\Omega}{dt} &= -\gamma_c\Omega + \omega g r_2',\end{aligned}\right\} \tag{18.8–6}$$

in which r_e' is the equilibrium value of r_3' in the absence of the field.

Although these equations look quite different from the van der Pol oscillator type Eq. (18.2–25) that we derived earlier, they actually reduce to this form when the atomic variables are eliminated adiabatically. The usual laser threshold, in which the gain rate equals the loss rate, corresponds to the condition

$$\frac{\omega g r_e'}{\gamma_\perp} = \gamma_c. \tag{18.8–6a}$$

However, in the process of reducing the problem from three variables to one variable we also eliminate the possibility of encountering certain instabilities that one-dimensional equations cannot exhibit.

18.8.1 *Relationship to the Lorenz model*

In 1963 Lorenz (Lorenz, 1963) developed a simplified model for the convective flow of a fluid that takes the form of the following set of coupled equations in the three variables x, y, z:

$$\left.\begin{aligned}\frac{dx}{d\tau} &= -\sigma(x - y) \\ \frac{dy}{d\tau} &= -y - xz + \rho x \\ \frac{dz}{d\tau} &= xy - bz.\end{aligned}\right\} \quad \sigma, \rho, b > 0. \tag{18.8–7}$$

Although the subject of fluid flow does not directly concern us here, we should take note of the fact that these equations can describe turbulent motion and *chaos* when $\sigma > b + 1$. The term chaos has come to be applied to behavior governed by deterministic equations that is nevertheless irregular and seemingly random, with the solution after a long time t exhibiting very sensitive dependence on initial conditions. The irregularity of the motion $x(t)$, $y(t)$, $z(t)$ is also reflected in the fact that the spectra, or the Fourier transforms of $x(t)$, $y(t)$, $z(t)$, are continuous functions of frequency.

It was first observed by Haken (1975a,b) that the laser equations (18.8–6) are actually isomorphic to the Lorenz equations, so that a laser governed by these equations is capable of exhibiting the same chaotic behavior. To establish the correspondence we make the substitutions

$$\left.\begin{aligned} r_2' &\to \left(\frac{\gamma_c\gamma_\perp}{\omega g}\right)y \\ r_3' - r_e' &\to -\left(\frac{\gamma_c\gamma_\perp}{\omega g}\right)z \\ \Omega &\to \gamma_\perp x \\ t &\to \tau/\gamma_\perp \\ \frac{\omega g r_e'}{\gamma_c\gamma_\perp} &\to \rho \\ \gamma_\parallel/\gamma_\perp &\to b \\ \gamma_c/\gamma_\perp &\to \sigma \end{aligned}\right\} \tag{18.8–8}$$

in Eqs. (18.8–6), which then become identical to Eqs. (18.8–7). The condition $\sigma > b + 1$ required for chaos then translates into the following condition on the damping constants:

$$\gamma_c > \gamma_\parallel + \gamma_\perp. \tag{18.8–9}$$

This implies that the cavity damping rate must exceed the atomic decay rates, which is exactly the opposite of the usual condition that was assumed in Section 18.2 in order to make an adiabatic elimination of the atomic variables. It is now apparent why chaos could not show up in our previous deterministic treatment of the laser problem; it requires a 'bad cavity', in which the spectral width of the cavity field exceeds the atomic homogeneous linewidth. Moreover, a three-dimensional system of equations such as Eqs. (18.8–6) is the smallest dimensional system capable of exhibiting chaos.

18.8.2 Linear stability analysis

Steady-state solutions of Eqs. (18.8–6) can be obtained by putting the time derivatives equal to zero. We then find the following solutions in which the field is non-zero,

$$\left.\begin{aligned} r_{3ss}' &= \gamma_c\gamma_\perp/\omega g \\ r_{2ss}' &= (\gamma_\parallel\gamma_c)^{1/2}(-\gamma_c\gamma_\perp + \omega g r_e')^{1/2}/\omega g \\ \Omega_{ss} &= (\gamma_\parallel/\gamma_c)^{1/2}(-\gamma_c\gamma_\perp + \omega g r_e')^{1/2}, \end{aligned}\right\} \tag{18.8–10}$$

provided that $\omega g r_e' \geqslant \gamma_c\gamma_\perp$. It is convenient to introduce normalized variables in terms of these steady-state values, as was done by Risken and Nummedal (1968a):

$$\left.\begin{aligned} \tilde{r}_2 &= r_2'/r_{2ss}' \\ \tilde{r}_3 &= r_3'/r_{3ss}' \\ \tilde{\Omega} &= \Omega/\Omega_{ss}. \end{aligned}\right\} \tag{18.8–11}$$

The equations of motion in these normalized variables then take the form

$$\left.\begin{aligned}\frac{\mathrm{d}\tilde{r}_2}{\mathrm{d}t} &= -\gamma_\perp \tilde{r}_2 + \gamma_\perp \tilde{\Omega}\tilde{r}_3\\ \frac{\mathrm{d}\tilde{r}_3}{\mathrm{d}t} &= \gamma_\parallel\left(-\tilde{r}_3 + \frac{\omega g r_e'}{\gamma_c\gamma_\perp}\right) - \gamma_\parallel\left(-1 + \frac{\omega g r_e'}{\gamma_c\gamma_\perp}\right)\tilde{\Omega}\tilde{r}_2\\ \frac{\mathrm{d}\tilde{\Omega}}{\mathrm{d}t} &= -\gamma_c\tilde{\Omega} + \gamma_c\tilde{r}_2.\end{aligned}\right\} \tag{18.8–12}$$

In order to test whether the steady-state solution for the fixed point given by Eqs. (18.8–10) represents a stable state, we examine the equations of motion in the neighborhood of the fixed point with the help of the linearizing approximations

$$\left.\begin{aligned}\tilde{r}_2 &= 1 + h_2\\ \tilde{r}_3 &= 1 + h_3\\ \tilde{\Omega} &= 1 + h_4.\end{aligned}\right\} \tag{18.8–13}$$

On substituting these into Eqs. (18.8–12) and retaining terms only of the first order in the small quantities h_1, h_2, h_3, we obtain the set of simultaneous differential equations

$$\left.\begin{aligned}\dot{h}_2 &= -\gamma_\perp(h_2 - h_3 - h_4)\\ \dot{h}_3 &= -\gamma_\parallel\left[\left(\frac{\omega g r_e'}{\gamma_c\gamma_\perp} - 1\right)(h_2 + h_4) + h_3\right]\\ \dot{h}_4 &= -\gamma_c(h_4 - h_2).\end{aligned}\right\} \tag{18.8–14}$$

These equations can be solved by the 'ansatz'

$$h_j(t) = h_j(0)\,\mathrm{e}^{-\lambda t}, \quad (j = 2, 3, 4), \tag{18.8–15}$$

which leads to a non-trivial solution only if

$$\begin{vmatrix} \lambda - \gamma_\perp & \gamma_\perp & \gamma_\perp \\ -\gamma_\parallel\left(\dfrac{\omega g r_e'}{\gamma_c\gamma_\perp} - 1\right) & \lambda - \gamma_\parallel & -\gamma_\parallel\left(\dfrac{\omega g r_e'}{\gamma_c\gamma_\perp} - 1\right) \\ \gamma_c & 0 & \lambda - \gamma_c \end{vmatrix} = 0. \tag{18.8–16}$$

This condition yields a cubic equation in the decay constant λ. It is evident that if the fixed point is to be stable, then the real parts of all three roots of λ must be positive.

A detailed analysis of the roots shows that an instability is encountered when

$$\gamma_c > \gamma_\parallel + \gamma_\perp \tag{18.8–17}$$

and when

$$\frac{\omega g r_e'}{\gamma_c\gamma_\perp} > \frac{\gamma_c}{\gamma_\perp}\left(\frac{\gamma_c + \gamma_\parallel + 3\gamma_\perp}{\gamma_c - \gamma_\parallel - \gamma_\perp}\right). \tag{18.8–18}$$

The first inequality requires that the cavity loss rate exceeds the atomic decay rates, and therefore calls for a 'bad cavity'. The second inequality places a lower bound on the excitation or pumping level of the laser. From Eq. (18.8–6a) the parameter $\omega g r'_e/\gamma_\perp$ represents the 'gain rate' of the laser, which equals the loss rate γ_c at the laser threshold. The ratio $\omega g r'_e/\gamma_c\gamma_\perp$ is therefore a measure of how many times the gain exceeds the loss. Let us see how large this has to be for an instability to be encountered.

The right-hand side of Eq. (18.8–18) considered as a function of $\gamma_c/\gamma_\perp$ has a minimum when

$$\gamma_c/\gamma_\perp = \gamma_\parallel/\gamma_\perp + 1 + [2(\gamma_\parallel/\gamma_\perp + 1)(\gamma_\parallel/\gamma_\perp + 2)]^{1/2},$$

i.e. when

$$\frac{\gamma_c}{\gamma_\perp + \gamma_\parallel} = 1 + \left[\frac{2(\gamma_\parallel/\gamma_\perp + 2)}{(\gamma_\parallel/\gamma_\perp + 1)}\right]^{1/2}, \tag{18.8–19}$$

in which case it takes on the value

$$\text{minimum gain-to-loss ratio} = 2\left[2\left(\frac{\gamma_\parallel}{\gamma_\perp} + 1\right)\left(\frac{\gamma_\parallel}{\gamma_\perp} + 2\right)\right]^{1/2} + 3\left(\frac{\gamma_\parallel}{\gamma_\perp} + 1\right) + 2. \tag{18.8–20}$$

This minimum value of the gain-to-loss ratio, which is sometimes referred to as the second laser threshold, depends on $\gamma_\parallel/\gamma_\perp$, but always exceeds 9. It equals 15 approximately when $\gamma_\parallel/\gamma_\perp = 1$, and it equals 21 approximately when $\gamma_\parallel/\gamma_\perp = 2$. In the latter case, according to Eq. (18.8–19), the minimum gain-to-loss ratio then requires $\gamma_c/(\gamma_\perp + \gamma_\parallel) \sim 2.6$. Hence the laser gain has to exceed the loss by more than an order of magnitude for the instability to be encountered. In terms of its output intensity, such a laser is very far above the first threshold.

18.8.3 Examples of laser instabilities

The equations of motion (18.8–6) can be integrated numerically. Figures 18.35 (*a*) and (*b*) show solutions (Ackerhalt, Milonni and Shih, 1985) for the atomic variable $r'_2(t)$ for two different values of the excitation parameter $\omega g r'_e/\gamma_c\gamma_\perp$, with the initial state $r'_2(0)\omega g/\gamma_c\gamma_\perp = 1$, $(r'_3(0) - r'_e)\omega g/\gamma_c\gamma_\perp = -1$, $\Omega(0)/\gamma_\perp = 1$, and with $\gamma_\parallel/\gamma_\perp = 1$, $\gamma_c/\gamma_\perp = 3$, so that condition (18.8–17) is satisfied. It is apparent that $r'_2(t)$ has the form of a damped oscillation when $\omega g r'_e/\gamma_c\gamma_\perp = 4$, but becomes chaotic when $\omega g r'_e/\gamma_c\gamma_\perp$ is increased to 22. Evidently there is a fundamental change in the behavior of the laser system when $\omega g r'_e/\gamma_c\gamma_\perp$ exceeds a certain threshold value.

From the point of view of experiment, the instability threshold is generally too high to be readily observable in a homogeneously broadened single-mode laser. However, by making use of a very high gain single-mode NH_3 ring laser operating in the far infrared, Weiss, Klische, Ering and Cooper (1985) succeeded in observing a sequence of pulsations of the light output, including successive period doubling and eventually chaos. The chaotic state is most readily recognized by the fact that the spectrum changes from a set of discrete frequencies to a continuous

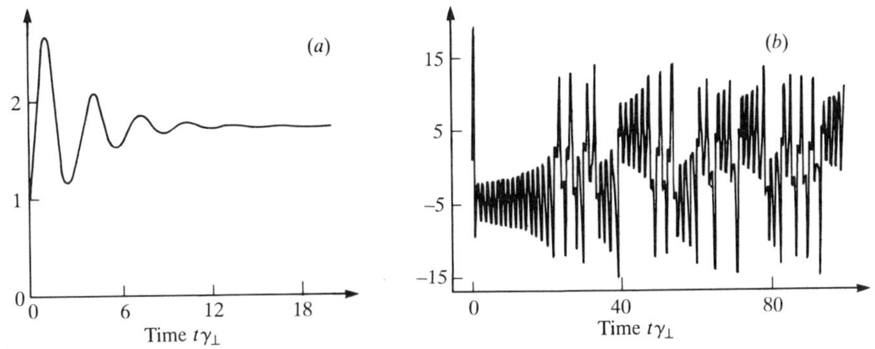

Fig. 18.35 $r_2'(t)\omega g/\gamma_c\gamma_\perp$ given by Eqs. (18.8–6) as a function of time, with the excitation parameter (*a*) $\omega g r_e'/\gamma_c\gamma_\perp = 4$. (*b*) $\omega g r_e'/\gamma_c\gamma_\perp = 22$. (Reproduced from Ackerhalt, Milonni and Shih, 1985.)

form. Figure 18.36 shows a sequence of spectra recorded as the NH_3 ring laser was gradually tuned towards line center. The progression from a line spectrum to a continuous spectrum is apparent. The scenario that allows a deterministic system to pass through a succession of period doublings to the chaotic state was first outlined by M. J. Feigenbaum (1978, 1979).

It was shown by Risken and Nummedal (1968a) that when multimode operation is allowed, a self-pulsing instability can appear when $\omega g r_e'/\gamma_c\gamma_\perp$ is greater than about 9. Self-pulsing in a laser operating far above threshold was first observed experimentally in inhomogeneously broadened Xe, He:Xe and He:Ne lasers (Casperson 1978, 1983; Maeda and Abraham, 1982; Bentley and Abraham, 1982; Weiss and King, 1982; Weiss, Godone and Olafsson, 1983; Gioggia and Abraham, 1983) that are described by somewhat more complicated equations than Eqs. (18.8–6). In some cases period doubling of the pulsations and chaos were also seen.

Although the chaotic state of a deterministic system bears some resemblance to random noise, and perhaps even to quantum fluctuations, we should emphasize

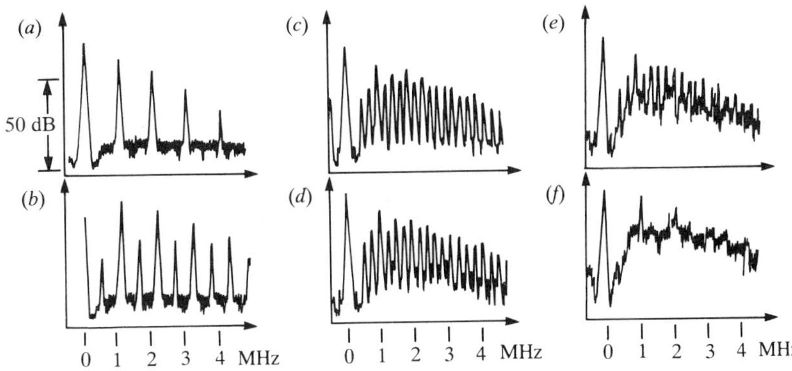

Fig. 18.36 Observed sequence of instabilities in a NH_3 ring laser, showing period doublings (*a*)–(*d*) and chaos (*e*), (*f*). (Reproduced from Weiss, Klische, Ering and Cooper, 1985.)

that it is nevertheless recognizably different. Whereas quantum noise is generally stationary and ergodic, the chaotic state is usually neither. Moreover, the chaotic state after a long time is usually strongly dependent on the initial state, instead of being independent of it. More explicitly if $\mathbf{x}(t)$ represents some vector chaotic process, and we change $\mathbf{x}(0)$ by a small amount $\delta\mathbf{x}(0)$, then the corresponding change $\delta\mathbf{x}(t)$ from $\mathbf{x}(t)$ after a long time t diverges exponentially in time, and we may write

$$\delta x_i(t) = \sum_j c_{ij}(t)\delta x_j(0)\,\mathrm{e}^{\lambda_j t}. \tag{18.8–21}$$

λ is known as the *Lyapunov exponent*. It is characteristic of a deterministically chaotic process that it has at least one positive Lyapunov exponent, which makes $\mathbf{x}(t)$ extremely sensitive to small changes of initial conditions.

18.8.4 A test for deterministic chaos

As chaotic and stochastic processes have certain similarities, it is useful to have a practical procedure for testing whether a recorded set of data is deterministically chaotic or stochastic. Such a test has been developed by Grassberger and Procaccia (1983) and Ben-Mizrachi, Procaccia and Grassberger (1984). The procedure has been applied to certain laser fields (Albano, Abounadi, Chyba, Searle, Yong, Gioggia and Abraham, 1985; Chyba, Christian, Gage, Lett and Mandel, 1986). We illustrate the approach that was used in the last experiment.

Let us suppose that a time series, such as the laser light intensity $I(t)$, has been recorded at regular time intervals $t = t_0 + n\tau$ $(n = 1, 2, 3, \ldots)$. Let us denote $I(t_0 + n\tau)$ by I_n. From the numbers $I_1, I_2, I_3, \ldots$ we construct d-dimensional $(d = 1, 2, \ldots)$ difference vectors defined by

$$\boldsymbol{\Delta}_{nm} \equiv (I_n - I_m, I_{n+1} - I_{m+1}, \ldots, I_{n+d-1} - I_{m+d-1}), \quad (m > n = 1, 2, 3, \ldots).$$

d is usually known as the *embedding dimension*. Let $N^d(\varepsilon)$ be the number of difference vectors whose lengths $|\boldsymbol{\Delta}_{nm}|$ fall below ε. We make a plot of $\ln N^d(\varepsilon)$ against $\log \varepsilon$ for increasing embedding dimensions d. The resulting curves usually have a linear region for small ε, except for possible disturbing effects caused by extraneous noise. For large ε the curves usually flatten out because of the finite size of the sample. Let us concentrate on the linear portion of the curves for small ε and denote the slopes of the curves by $D_2^d(\varepsilon)$. For a truly random process one would expect the representative points in phase space to be distributed uniformly within a d-dimensional sphere of sufficiently small radius ε, so that $N^d(\varepsilon)$ is proportional to ε^d. When $\varepsilon \to 0$, the slope $D_2^d(\varepsilon)$ then coincides with the embedding dimension d, and it increases without limit as $d \to \infty$ for a stochastic process. On the other hand, for a deterministic system that exhibits chaotic behavior, for which the representative points are not uniformly distributed on a small scale, the slope tends towards a finite value as d increases. The limit

$$\operatorname*{Lim}_{\varepsilon\to 0}\operatorname*{Lim}_{d\to\infty} D_2^d(\varepsilon) = D_2, \tag{18.8–22}$$

if it exists, is known as the *order 2 information dimension* of the data set, and it is characteristic of chaotic behavior.

Figure 18.37 illustrates a plot of $\ln N^d(\varepsilon)$ versus $\log \varepsilon$ for several different d that was derived from periodic measurements of the light intensity of a dye ring laser. A plot of the slope of the straight portions in Fig. 18.37 versus embedding dimension d reveals a steady increase without limit, and strongly suggests that this laser is stochastic and not chaotic.

Another useful test for deterministic chaos is based on the ratio of neighboring curves $N^d(\varepsilon)$. In order to see how this ratio behaves for a truly random process, we need to examine the structure of $N^d(\varepsilon)$ a little more carefully. When the representative points are uniformly distributed in phase space over a small region, then $N^d(\varepsilon)$ is the product of the density ρ of points and the volume V of a hypersphere of radius ε. If a is the average separation between $I_1, I_2, I_3, \ldots$ then

$$\rho = \frac{1}{a^d},$$

and in d-dimensions the hypersphere has a volume

$$V = \frac{\pi^{(1/2)d} \varepsilon^d}{\Gamma(1 + \frac{1}{2}d)}.$$

It follows that

$$N^d(\varepsilon) = \left[\frac{(\sqrt{\pi})\varepsilon}{a}\right]^d \frac{1}{\Gamma(1 + \frac{1}{2}d)}, \tag{18.8–23}$$

so that

$$\frac{N^d(\varepsilon)}{N^{d+1}(\varepsilon)} = \left[\frac{a}{(\sqrt{\pi})\varepsilon}\right] \frac{\Gamma(\frac{3}{2} + \frac{1}{2}d)}{\Gamma(1 + \frac{1}{2}d)}. \tag{18.8–24}$$

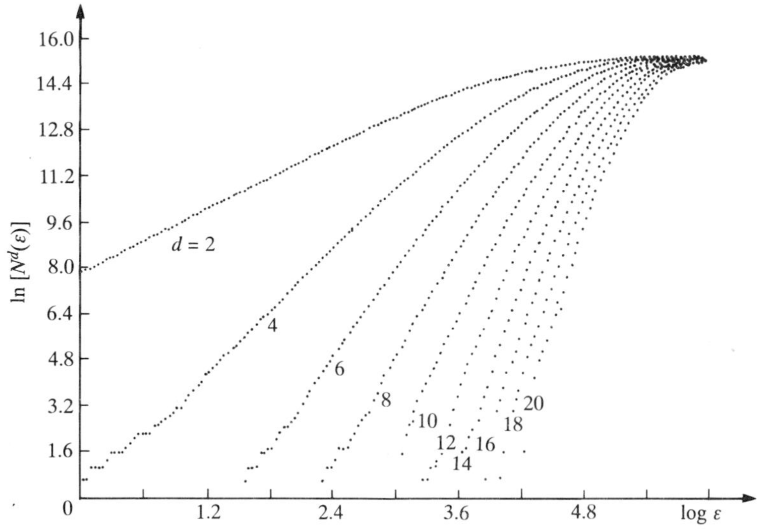

Fig. 18.37 An example of plots of $\ln[N^d(\varepsilon)]$ versus $\log(\varepsilon)$ for several different values of d from 2 to 20 derived from experiment. (Reproduced from Chyba, 1990.)

For large d we may use the Stirling approximation to the Γ-function which leads to the result

$$\frac{N^d(\varepsilon)}{N^{d+1}(\varepsilon)} = \frac{a}{\sqrt{(2\pi \mathrm{e})}\varepsilon} d^{1/2}[1 + (1/d)]^{1+(1/2)d}$$

$$= \frac{a}{\sqrt{(2\pi \mathrm{e})}\varepsilon} d^{1/2} \exp[(1 + \tfrac{1}{2}d)\ln(1 + (1/d)]$$

$$= \frac{a}{\sqrt{(2\pi)}\varepsilon} d^{1/2}\, \mathrm{e}^{3/4d - 1/3d^2 + \ldots} \tag{18.8–25}$$

when we make a power series expansion of the logarithm in the exponent.

Let

$$K_2^d(\varepsilon) = \frac{1}{\tau}\ln\left[\frac{N^d(\varepsilon)}{N^{d+1}(\varepsilon)}\right]$$

$$= \frac{1}{\tau}\left[-\frac{1}{2}\ln 2\pi - \ln\left(\frac{\varepsilon}{a}\right) + \frac{1}{2}\ln d + \frac{3}{4d} - \frac{1}{3d^2} + \ldots\right]. \tag{18.8–26}$$

Then it is clear that $K_2^d(\varepsilon)$ grows logarithmically without limit as d increases for this completely random process. On the other hand, for a deterministic process, $K_2^d(\varepsilon)$ may tend to a limit as d increases, for small ε. When the limit exists, we write

$$\lim_{\varepsilon\to 0}\lim_{d\to\infty} K_2^d(\varepsilon) = K_2. \tag{18.8–27}$$

K_2 is known as the *order 2 Kolmogorov entropy* of the system and it is characteristic of chaotic behavior. The data in fig. 18.37 reveal a steady increase of $K_2^d(\mathrm{s})$ with d, showing that this particular laser is fluctuating stochastically and not chaotically.

Problems

18.1 A single-mode He:Ne laser delivers 1 mW of power in a 1 mm wide diffraction-limited beam of light, with a spectral width of 30 kHz. Calculate the average photon occupation number per unit cell of phase space in the laser beam.

18.2 A single-mode laser has an output mirror of intensity transmissivity $\mathcal{T}$, which can be varied from a fraction of one percent to several percent. The loss of photons through this mirror is the principal loss of the laser. If the laser is operating well above threshold, but with gain exceeding loss only slightly, determine how the emerging light intensity varies with $\mathcal{T}$.

18.3 Consider a single-mode laser whose gain medium occupies only a small fraction of the cavity length. Show that, well above threshold, the mean laser light intensity $\langle I \rangle$ is independent of the cavity length, whereas the variance $\langle (\Delta I)^2 \rangle$ is inversely proportional to the cavity length. Explain this behavior in physical terms. How would $\langle I \rangle$ and $\langle (\Delta I)^2 \rangle$ vary with cavity length well below threshold?

18.4 $\mathscr{E}_1$, $\mathscr{E}_2$ are the real and imaginary parts of the complex electric field amplitude $\mathscr{E}$ of a single-mode laser. Show that

$$\frac{\partial}{\partial t}\langle \Delta\mathscr{E}_1(t)\Delta\mathscr{E}_2(t)\rangle =$$
$$\langle \Delta\mathscr{E}_1(t)\Delta[\mathscr{E}_2(t)(a - |\mathscr{E}(t)|^2)]\rangle + \langle \Delta\mathscr{E}_2(t)\Delta[\mathscr{E}_1(t)(a - |\mathscr{E}(t)|^2)]\rangle$$

where a is the dimensionless pump parameter.

19

The two-mode ring laser

When more than one mode of a laser is excited, we have to return to the mode expansion (18.2–4) of the electromagnetic field within the optical cavity. The laser field is now described by several mode amplitudes $\mathscr{E}_1(t)$, $\mathscr{E}_2(t)$, etc., and it therefore has more degrees of freedom than before. Although it is straightforward to repeat the calculations given in Sections 18.2 and 18.3 and to derive equations of motion, the additional degrees of freedom have a profound effect on the behavior of the laser, which is capable of exhibiting some entirely new phenomena. For this reason we devote a separate chapter to the discussion of the two-mode laser.

Among the various forms of two-mode laser, one that is particularly interesting and convenient for experimental work is the ring laser, one possible form of which is illustrated in Fig. 19.1. Whereas the optical cavity of a conventional laser has two mirrors, and the normal modes of the cavity are standing waves (cf. Section 18.2), a *ring laser* makes use of an open cavity formed by three or more mirrors arranged at the corners of a polygon or in a 'ring', and its normal modes are traveling waves. Some waves travel in the clockwise direction around the cavity, and some in the counterclockwise direction. Although it is possible for the clockwise and counterclockwise travelling waves to be of the same frequency, they are nevertheless distinct cavity modes, which can be excited separately. Moreover, as can be seen from Fig. 19.1 the light associated with these modes that escapes from the cavity emerges in different directions, which makes it easy to investigate the excitations of the two modes separately. By contrast, two longitudinal modes of a standing-wave laser can usually only be separated with an interferometer.

In this chapter we shall concentrate on the two-mode ring laser, in which waves propagate in each direction around the ring. In place of the simple standing-wave mode functions given by Eq. (18.2–5), we represent the two traveling-wave modes along one arm of the ring by the plane wave modes

$$\left.\begin{aligned} u_1(\mathbf{r}) &= K\,\mathrm{e}^{\mathrm{i}\mathbf{k}_1\cdot\mathbf{r}} \\ u_2(\mathbf{r}) &= K\,\mathrm{e}^{-\mathrm{i}\mathbf{k}_2\cdot\mathbf{r}}, \end{aligned}\right\} \tag{19.1–1}$$

with the vectors $\mathbf{k}_1$, $\mathbf{k}_2$ pointing in the same direction. The lengths of these vectors are given by $\pi n_i/l$, where n_i is an integer, and l is the path length around the ring. Normalization of these mode functions is only possible if the plane waves are

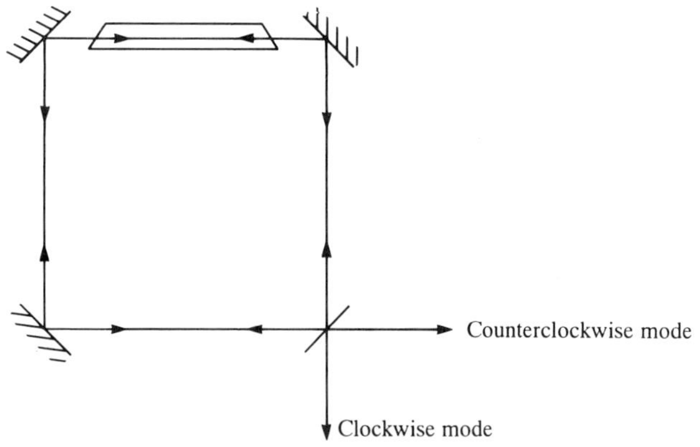

Fig. 19.1 A possible geometry for a ring laser.

assumed to have a cross-sectional area $\mathscr{A}$, in which case we can take the normalization constant K to be $(\mathscr{A}l)^{-1/2}$. Alternatively, for a more realistic description of the modes of a confocal ring laser, we could use mode functions like those given by Eq. (18.2–6), except that the factor $\sin\psi$ would be replaced by $e^{\pm i\psi}$. Because the counterpropagating waves of a ring laser share the same gain medium and the same optical cavity, the gains and losses of the two modes are usually rather similar. Nevertheless, because the laser geometry may not be symmetric, and because reflection and diffraction losses may differ somewhat in the two directions, we shall require two pump parameters a_1, a_2 to describe the excitations of the system, although the two parameters usually do not differ very much.

As long as the two-mode ring laser is at rest, the two counterpropagating traveling-wave modes are generally of the same frequency. However, when the system is rotated about an axis perpendicular to the plane of the ring, the frequency of one mode is increased and that of the other one is decreased. The frequency separation is proportional to the angular velocity of rotation, and this makes it possible to use the ring laser as an inertial gyro for the absolute measurement of rotation (Lamb, 1964; Aronowitz, 1965, 1971; Menegozzi and Lamb, 1973; Sargent, Scully and Lamb, 1974).

19.1 Equations of motion

For a semiclassical treatment of a two-mode laser, we shall adopt the same general procedure that was used in Section 18.2. Equation (18.2–11) for the field has to be replaced by the more general mode expansion (18.2–4), and the equation of motion (18.2–12) now becomes

$$\sum_{n=1}^{2} -i\omega_n\left[2\frac{d\mathscr{E}_n}{dt}+\frac{\sigma_n}{\varepsilon_0}\mathscr{E}_n\right]e^{-i\omega_n t}u_n(\mathbf{r})+\text{c.c.}=\frac{\bar{\omega}^2}{\varepsilon_0}P(\mathbf{r},t), \quad (19.1\text{–}2)$$

where ω_1, ω_2 are the frequencies of the two laser modes and $\bar{\omega}$ is their average. We have allowed the 'conductivities' σ_n to be different for the two modes, in order to allow for the possibility that the losses may be different. Also Eq.

(18.2–19) for the expectation value of the dipole moment now has to be replaced by

$$\langle\hat{\mu}(\mathbf{r},t)\rangle =$$

$$\frac{-\mathrm{i}(p_2-p_1)\mu_{12}^2 T}{\hbar}\sum_{n=1}^{2}\mathscr{E}_n(t)\,\mathrm{e}^{-\mathrm{i}\omega_n t}u_n(\mathbf{r})\left[1-\frac{2\mu_{12}^2T^2}{3\hbar^2}\left|\sum_{n=1}^{2}\mathscr{E}_n(t)\,\mathrm{e}^{-\mathrm{i}\omega_n t}u_n(\mathbf{r})\right|^2+\ldots\right]$$

$$+\ \text{c.c.} \qquad (19.1\text{–}3)$$

For simplicity, we shall henceforth take the two mode frequencies to be equal and on resonance with the atomic frequency ω_0, so that the equations describe a non-rotating ring laser.

We now proceed to single out one mode amplitude on the left of Eq. (19.1–2), such as $\mathscr{E}_1$, by multiplying both sides of the equation by $u_1^*(\mathbf{r})$ and integrating over the cavity volume, as before. The orthonormality property (18.2–9) causes the terms in $\mathscr{E}_2$ on the left to vanish, and, on making use of Eq. (19.1–3) to express the polarization $P(\mathbf{r})$ in terms of the atomic density $\eta(\mathbf{r})$, we obtain

$$\begin{aligned}\frac{\mathrm{d}\mathscr{E}_1}{\mathrm{d}t}&+\frac{\sigma_1}{2\varepsilon_0}\mathscr{E}_1\\ &=\frac{(p_2-p_1)\mu_{12}^2\omega_0 T}{2\varepsilon_0\hbar}\bigg\{\mathscr{E}_1\int\eta(\mathbf{r})|u_1(\mathbf{r})|^2\,\mathrm{d}^3r+\mathscr{E}_2\int\eta(\mathbf{r})u_2(\mathbf{r})u_1^*(\mathbf{r})\,\mathrm{d}^3r\\ &\quad-\frac{2}{3}\left(\frac{\mu_{12}T}{\hbar}\right)^2\bigg[\mathscr{E}_1|\mathscr{E}_1|^2\int\eta(\mathbf{r})|u_1(\mathbf{r})|^4\,\mathrm{d}^3r+2\mathscr{E}_1|\mathscr{E}_2|^2\int\eta(\mathbf{r})|u_1(\mathbf{r})|^2|u_2(\mathbf{r})|^2\,\mathrm{d}^3r\\ &\quad+2\mathscr{E}_2|\mathscr{E}_1|^2\int\eta(\mathbf{r})|u_1(\mathbf{r})|^2u_1^*(\mathbf{r})u_2(\mathbf{r})\,\mathrm{d}^3r+\mathscr{E}_2|\mathscr{E}_2|^2\int\eta(\mathbf{r})|u_2(\mathbf{r})|^2u_1^*(\mathbf{r})u_2(\mathbf{r})\,\mathrm{d}^3r\\ &\quad+\mathscr{E}_2^*|\mathscr{E}_1|^2\int\eta(\mathbf{r})|u_1(\mathbf{r})|^2u_1^*(\mathbf{r})u_2^*(\mathbf{r})\,\mathrm{d}^3r+\mathscr{E}_1^*\mathscr{E}_2^2\int\eta(\mathbf{r})u_1^{*2}(\mathbf{r})u_2^2(\mathbf{r})\,\mathrm{d}^3r\bigg]\bigg\}.\end{aligned}$$

$$(19.1\text{–}4)$$

If the mode function $u_1(\mathbf{r})$ and $u_2(\mathbf{r})$ can be approximated by Eqs. (19.1–1), corresponding to traveling waves within the gain medium, then several of the integrands become highly oscillatory, and the corresponding integrals make a negligible contribution. Of the terms linear in $\mathscr{E}$, only the first survives, and of the terms nonlinear in $\mathscr{E}$, only the first two are not negligible. If we define

$$\bar{\eta}\equiv\int\eta(\mathbf{r})|u_i(\mathbf{r})|^2\,\mathrm{d}^3r,\quad i=1,2 \qquad (19.1\text{–5a})$$

$$F\equiv\int\eta(\mathbf{r})|u_i(\mathbf{r})|^4\,\mathrm{d}^3r=\int\eta(\mathbf{r})|u_1(\mathbf{r})|^2|u_2(\mathbf{r})|^2\,\mathrm{d}^3r,\quad i=1,2 \qquad (19.1\text{–5b})$$

as in Section 18.2, Eq. (19.1–4) simplifies to

$$\frac{\mathrm{d}\mathscr{E}_1}{\mathrm{d}t}=\left[\frac{\omega_0\bar{\eta}(p_2-p_1)\mu_{12}^2T}{2\varepsilon_0\hbar}-\frac{\sigma_1}{2\varepsilon_0}\right]\mathscr{E}_1-\frac{\omega_0F(p_2-p_1)\mu_{12}^4T^3}{3\varepsilon_0\hbar^3}(|\mathscr{E}_1|^2+2|\mathscr{E}_2|^2)\mathscr{E}_1$$

$$(19.1\text{–}6)$$

With the introduction of the parameters $\mathscr{A}$, $\mathscr{C}_i$, $\mathscr{B}$ representing the gain, the loss, and the nonlinearity, as before,

$$\left.\begin{aligned}\mathscr{A} &\equiv \omega_0\bar{\eta}(p_2 - p_1)\mu_{12}^2 T/2\varepsilon_0\hbar \\ \mathscr{C}_i &\equiv \sigma_i/2\varepsilon_0, \quad i = 1, 2 \\ \mathscr{B} &\equiv \omega_0 F(p_2 - p_1)\mu_{12}^4 T^3/3\varepsilon_0\hbar^3,\end{aligned}\right\} \tag{19.1–7}$$

Eq. (19.1–6) reduces to

$$\frac{\mathrm{d}\mathscr{E}_1}{\mathrm{d}t} = [\mathscr{A} - \mathscr{C}_1 - \mathscr{B}(|\mathscr{E}_1|^2 + 2|\mathscr{E}_2|^2)]\mathscr{E}_1. \tag{19.1–8}$$

In a similar manner, by multiplying each term in Eq. (19.1–2) by $u_2^*(\mathbf{r})$ and integrating as before, we obtain the equation

$$\frac{\mathrm{d}\mathscr{E}_2}{\mathrm{d}t} = [\mathscr{A} - \mathscr{C}_2 - \mathscr{B}(|\mathscr{E}_2|^2 + 2|\mathscr{E}_1|^2)]\mathscr{E}_2 \tag{19.1–9}$$

for the equation of motion of the other mode amplitude. Except for the cross-terms $2\mathscr{B}|\mathscr{E}_2|^2\mathscr{E}_1$ and $2\mathscr{B}|\mathscr{E}_1|^2\mathscr{E}_2$, these equations are similar to the equation (18.2–25) governing the behavior of the single-mode laser, that we derived in the previous chapter. However, because of the cross-terms, the two equations of motion are coupled, and they have to be solved simultaneously. The coupling gives rise to some new phenomena, as we shall see. It should be noted that, when the two mode intensities $|\mathscr{E}_1|^2$ and $|\mathscr{E}_2|^2$ are approximately equal, the nonlinearity introduced by the cross-terms is twice as great as that contributed by the self-terms $\mathscr{B}|\mathscr{E}_1|^2\mathscr{E}_1$ or $\mathscr{B}|\mathscr{E}_2|^2\mathscr{E}_2$, because there are twice as many ways in which the cross-coupling contributes to the nonlinear polarization.

However, this is true only so long as the gain medium is homogeneously broadened, and all atoms contribute equally to the laser field. When inhomogeneous broadening dominates, because of the Doppler shifts introduced by atomic motion, the coupling is reduced by a factor 2 at line center; in general it depends on the detuning $\Delta\omega$ of the cavity frequency from the center of the inhomogeneous line. It can be shown that the mode coupling constant, which is represented by the factor $2\mathscr{B}$ in Eqs. (19.1–8) and (19.1–9), then has to be replaced by $1/[1 + (\Delta\omega T_1)^2]$, where T_1 is the natural lifetime of the laser transition (Smirnov and Zhelnov, 1969; M-Tehrani and Mandel, 1976, 1977). In addition the gain and saturation coefficients $\mathscr{A}$ and $\mathscr{B}$ become functions of the detuning $\Delta\omega$. We shall not go into these complications here, but in order to cover several different cases we express Eqs. (19.1–8) and (19.1–9) in the more general form

$$\left.\begin{aligned}\frac{\mathrm{d}\mathscr{E}_1}{\mathrm{d}t} &= [\mathscr{A} - \mathscr{C}_1 - \mathscr{B}(|\mathscr{E}_1|^2 + \xi|\mathscr{E}_2|^2)]\mathscr{E}_1 \\ \frac{\mathrm{d}\mathscr{E}_2}{\mathrm{d}t} &= [\mathscr{A} - \mathscr{C}_2 - \mathscr{B}(|\mathscr{E}_2|^2 + \xi|\mathscr{E}_1|^2)]\mathscr{E}_2,\end{aligned}\right\} \tag{19.1–10}$$

with the understanding that the dimensionless coupling constant ξ is given by

$$\left.\begin{aligned}&\xi = 2 \text{ for a homogeneously broadened gain medium;}\\&\xi = 1 \text{ for an inhomogeneously broadened gain medium at line center;}\\&\xi < 1 \text{ for a detuned, inhomogeneously broadened laser.}\end{aligned}\right\} \quad (19.1\text{–}11)$$

When the cavity detuning $\Delta\omega$ is sufficiently large, the coupling constant $\xi \ll 1$, and the two equations (19.1–10) then describe two almost independent single-mode lasers.

In order to understand the reason for the frequency dependence of ξ, consider an atom moving with velocity component v in the direction of cavity mode no. 1 (see Fig. 19.2). This atom will interact with mode 1 only if the mode frequency ω_1 exceeds the atomic resonance frequency ω_0 so that $\omega_1 = \omega_0 + \omega_1 v/c$ ($\pm$ the natural linewidth). The reason is that in the rest frame of the moving atom the frequency ω_1 is seen Doppler-shifted down to $\omega_1 - \omega_1 v/c$. Similarly, this atom will interact with mode 2 only if the mode frequency $\omega_2 = \omega_0 - \omega_2 v/c$ ($\pm$ the natural linewidth), because the frequency ω_2 is seen by the atom as Doppler-shifted up. If ω_1 actually equals ω_2, but differs from ω_0 by more than the natural linewidth, we see that the two modes cannot interact with the same atom, but only with different atoms, and they are virtually uncoupled. Only when ω_1, ω_2 are within a natural linewidth of ω_0, is there an appreciable amount of mode coupling. It can be shown (Sargent, Scully and Lamb, 1974; Singh, 1980) that, for certain atomic transitions in a two-mode Zeeman laser, ξ can have values between 1 and 2.

Although the emphasis in this chapter is on the ring laser, under some conditions Eqs. (19.1–10) with $\xi = 2$ can also describe a homogeneously broadened two-mode standing-wave laser to a first approximation, provided the frequency difference between the two modes is small compared with the natural atomic linewidth.

19.1.1 Inclusion of spontaneous emission fluctuations

As was shown in Sections 18.3 and 18.5, we may include the effects of spontaneous atomic emission by adding noise terms to the equation of motion, and arrive at results that are almost indistinguishable from those given by a fully quantized field theory. As the semiclassical theory is so much simpler, we shall adopt the same procedure here, and replace Eqs. (19.1–10) by

$$\left.\begin{aligned}\frac{\mathrm{d}\mathscr{E}_1}{\mathrm{d}t} &= [\mathscr{A} - \mathscr{C}_1 - \mathscr{B}(|\mathscr{E}_1|^2 + \xi|\mathscr{E}_2|^2)]\mathscr{E}_1 + n_1(t),\\ \frac{\mathrm{d}\mathscr{E}_2}{\mathrm{d}t} &= [\mathscr{A} - \mathscr{C}_2 - \mathscr{B}(|\mathscr{E}_2|^2 + \xi|\mathscr{E}_1|^2)]\mathscr{E}_2 + n_2(t),\end{aligned}\right\} \quad (19.1\text{–}12)$$

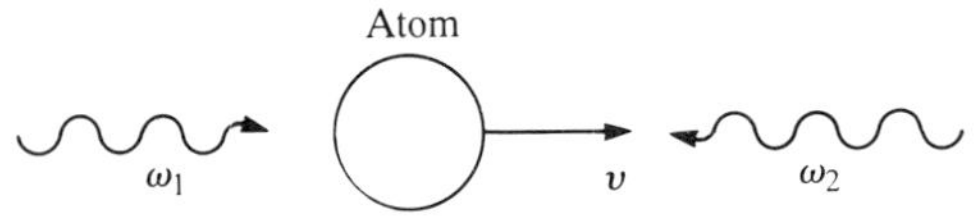

Fig. 19.2 A moving atom in the presence of two counter-propagating waves.

in which $n_1(t)$ and $n_2(t)$ represent two independent, complex, Gaussian, white noises of equal average strength, with

$$\left.\begin{aligned} \langle n_1^*(t)n_1(t')\rangle &= 2S\delta(t-t') = \langle n_2^*(t)n_2(t')\rangle \\ \langle n_1^*(t)n_2(t')\rangle &= 0. \end{aligned}\right\} \tag{19.1–13}$$

We require two independent noise terms because spontaneous emission takes place almost independently into each cavity mode. As before, the delta correlated form of $n_i(t)$ $(i = 1, 2)$ is an approximation, reflecting the fact that the time scale for spontaneous emission is usually short compared with the times in which the laser field amplitudes $\mathscr{E}_1(t)$, $\mathscr{E}_2(t)$ are changing.

As in Section 18.3, it is convenient to replace the various quantities by dimensionless scaled variables. If we consider a cavity on resonance with the atomic transition, and put

$$\left.\begin{aligned} \mathscr{E}_i &= (S/\mathscr{B})^{1/4}\tilde{\mathscr{E}}_i, \quad && i = 1, 2 \\ \mathscr{A} - \mathscr{C}_i &= (S\mathscr{B})^{1/2}a_i, && i = 1, 2 \\ n_i &= S^{3/4}\mathscr{B}^{1/4}q_i, && i = 1, 2 \\ t &= (S\mathscr{B})^{-1/2}\tilde{t} \end{aligned}\right\} \tag{19.1–14}$$

then the scaled noise functions $q_i(\tilde{t})$ have the correlations

$$\left.\begin{aligned} \langle q_1^*(\tilde{t})q_1(\tilde{t}')\rangle &= 2\delta(\tilde{t}-\tilde{t}') = \langle q_2^*(\tilde{t})q_2(\tilde{t}')\rangle \\ \langle q_1^*(\tilde{t})q_2(\tilde{t}')\rangle &= 0, \end{aligned}\right\} \tag{19.1–15}$$

and the scaled field amplitudes $\tilde{\mathscr{E}}_i$ obey the equations of motion

$$\left.\begin{aligned} \frac{d\tilde{\mathscr{E}}_1}{d\tilde{t}} &= [a_1 - (|\tilde{\mathscr{E}}_1|^2 + \xi|\tilde{\mathscr{E}}_2|^2)]\tilde{\mathscr{E}}_1 + q_1(\tilde{t}) \\ \frac{d\tilde{\mathscr{E}}_2}{d\tilde{t}} &= [a_2 - (|\tilde{\mathscr{E}}_2|^2 + \xi|\tilde{\mathscr{E}}_1|^2)]\tilde{\mathscr{E}}_2 + q_2(\tilde{t}). \end{aligned}\right\} \tag{19.1–16}$$

The parameters a_1, a_2 are dimensionless pump parameters of the same form as the parameter a introduced in Section 18.3. Because of their somewhat simpler structure, we shall work with the scaled equations from now on. However, in order to simplify the notation, we hereafter discard the tilde ~, with the understanding that the scale factors used in Eqs. (19.1–14) have to be put back into the equations whenever dimensional quantities are needed.

Because of the quantum noise terms, the complex mode amplitudes $\mathscr{E}_1(t)$ and $\mathscr{E}_2(t)$ are now random processes with a certain probability density $p(\mathscr{E}_1, \mathscr{E}_2, t)$. Whereas $\mathscr{E}_1(t)$ and $\mathscr{E}_2(t)$ obey the stochastic equations of motion [Eqs. (19.1–16)], $p(\mathscr{E}_1, \mathscr{E}_2, t)$ obeys a four-dimensional Fokker–Planck equation of the general form (Singh, 1984, Sec. 2.1)

$$\frac{\partial p(\mathbf{x}, t)}{\partial t} = -\sum_{i=1}^{4}\frac{\partial}{\partial x_i}[A_i p(\mathbf{x}, t)] + \frac{1}{2}\sum_{i=1}^{4}\sum_{j=1}^{4}\frac{\partial^2}{\partial x_i \partial x_j}[D_{ij}p(\mathbf{x}, t)]. \tag{19.1–17}$$

The two complex mode amplitudes $\mathscr{E}_1$, $\mathscr{E}_2$ have been replaced by a real, four-dimensional vector amplitude $\mathbf{x}$, whose components, which are related to $\mathscr{E}_1$ and $\mathscr{E}_2$

$$\left.\begin{aligned} \mathscr{E}_1 &\equiv x_1 + \mathrm{i}x_2 \\ \mathscr{E}_2 &\equiv x_3 + \mathrm{i}x_4, \end{aligned}\right\} \tag{19.1–18}$$

characterize the state of the laser field at any time. The drift vector $\mathbf{A}$ has components

$$\left.\begin{aligned} A_1 &= [a_1 - (x_1^2 + x_2^2) - \xi(x_3^2 + x_4^2)]x_1 \\ A_2 &= [a_1 - (x_1^2 + x_2^2) - \xi(x_3^2 + x_4^2)]x_2 \\ A_3 &= [a_2 - (x_3^2 + x_4^2) - \xi(x_1^2 + x_2^2)]x_3 \\ A_4 &= [a_2 - (x_3^2 + x_4^2) - \xi(x_1^2 + x_2^2)]x_4, \end{aligned}\right\} \tag{19.1–19}$$

and because of Eq. (19.1–15) the diffusion tensor D_{ij} has the simple diagonal form

$$D_{ij} = 2\delta_{ij}, \quad i, j = 1, 2, 3, 4. \tag{19.1–20}$$

The general solution of the Fokker–Planck equation allows us to calculate any average of interest, including multi-time correlation functions, as we showed in Section 18.3. However, the four-dimensional equation presents more formidable problems than did the corresponding two-dimensional equation for the single-mode laser that we treated in Chapter 18.

19.2 The steady-state solution

It is not difficult to find the steady-state solution $p_s(\mathbf{x})$ of the Fokker–Planck equation (19.1–17). In the steady state the left-hand side vanishes and the equation can be written in the form

$$\sum_{i=1}^{4} \frac{\partial j_i}{\partial x_i} = 0, \tag{19.2–1}$$

where

$$j_i \equiv A_i p_s - \partial p_s / \partial x_i.$$

$\mathbf{j}$ is the four-dimensional *probability current density*, which vanishes when the drift vector $\mathbf{A}$ satisfies the condition (called the potential condition) (Stratonovich, 1963, Chap. 4)

$$\frac{\partial A_i}{\partial x_j} = \frac{\partial A_j}{\partial x_i}, \tag{19.2–2}$$

and we may readily confirm that the components given by Eqs. (19.1–19) do indeed obey this condition. The drift vector $\mathbf{A}$ is then expressible as the gradient of a scalar potential function $U(\mathbf{x})$,

$$A_i(\mathbf{x}) = -\frac{\partial U(\mathbf{x})}{\partial x_i}. \tag{19.2–3}$$

By integration of this equation, for example along a straight line from the origin to point $\mathbf{x}$, we obtain, apart from an additive constant (M-Tehrani and Mandel, 1977, 1978a,b; Singh and Mandel, 1979),

$$\begin{aligned} U(\mathbf{x}) &= -\int_0^{\mathbf{x}} \mathbf{A}(\mathbf{x}') \cdot \mathrm{d}\mathbf{x}' \\ &= -\tfrac{1}{2}a_1(x_1^2 + x_2^2) - \tfrac{1}{2}a_2(x_3^2 + x_4^2) \\ &\quad + \tfrac{1}{4}(x_1^4 + x_2^4 + 2x_1^2x_2^2 + x_3^4 + x_4^4 + 2x_3^2x_4^2) \\ &\quad + \tfrac{1}{2}\xi(x_1^2 + x_2^2)(x_3^2 + x_4^2). \end{aligned} \tag{19.2–4}$$

Finally, by equating the probability current density $\mathbf{j}$ to zero and integrating the differential equation, we find for the probability density $p_s(\mathbf{x})$ in the steady state the result

$$\begin{aligned} p_s(\mathbf{x}) &= \exp\left[\int_0^{\mathbf{x}} \mathbf{A}(\mathbf{x}) \cdot \mathrm{d}\mathbf{x} + \text{const.}\right] \\ &= B\,\mathrm{e}^{-U(\mathbf{x})}, \end{aligned} \tag{19.2–5}$$

where B is a normalizing constant. Evidently $U(\mathbf{x})$ plays the same role as the potential in Section 18.3.

At this point it is more convenient to introduce the two laser mode intensities I_1, I_2 given by

$$\left.\begin{aligned} I_1 &\equiv |\mathscr{E}_1|^2 = x_1^2 + x_2^2 \\ I_2 &\equiv |\mathscr{E}_2|^2 = x_3^2 + x_4^2, \end{aligned}\right\} \tag{19.2–6}$$

because the potential U is expressible entirely in terms of I_1 and I_2, in the form

$$U = -\tfrac{1}{2}a_1I_1 - \tfrac{1}{2}a_2I_2 + \tfrac{1}{4}I_1^2 + \tfrac{1}{4}I_2^2 + \tfrac{1}{2}\xi I_1I_2. \tag{19.2–7}$$

The phases of the two modes do not appear in the steady-state solution, and they must therefore be completely random. From Eq. (19.2–5) we now obtain for the joint probability density $\mathscr{P}(\mathrm{I}_1, \mathrm{I}_2)$ of the two mode intensities in the steady state the expression

$$\mathscr{P}(I_1, I_2) = K^{-1}\exp\left(\tfrac{1}{2}a_1I_1 + \tfrac{1}{2}a_2I_2 - \tfrac{1}{4}I_1^2 - \tfrac{1}{4}I_2^2 - \tfrac{1}{2}\xi I_1I_2\right). \tag{19.2–8}$$

The constant $K(a_1, a_2, \xi)$ is determined from the normalization condition

$$\iint_0^\infty \mathscr{P}(I_1, I_2)\,\mathrm{d}I_1\,\mathrm{d}I_2 = 1,$$

giving

$$K = \iint_0^\infty \exp\left(\tfrac{1}{2}a_1I_1 + \tfrac{1}{2}a_2I_2 - \tfrac{1}{4}I_1^2 - \tfrac{1}{4}I_2^2 - \tfrac{1}{2}\xi I_1I_2\right)\mathrm{d}I_1\,\mathrm{d}I_2. \tag{19.2–9}$$

We shall see shortly that K plays the role of a generating function for the moments of I_1 and I_2.

Certain limiting forms of $\mathscr{P}(I_1, I_2)$ can be recognized at once. For example, if the laser is operating well below threshold, so that the two pump parameters a_1, a_2 are both negative and numerically large, then only very small light intensities

I_1, I_2 have appreciable probability. Equation (19.2–8) then reduces to

$$\mathcal{P}(I_1, I_2) \approx \tfrac{1}{4}|a_1 a_2|\,\mathrm{e}^{-|a_1|I_1/2}\,\mathrm{e}^{-|a_2|I_2/2}, \tag{19.2–10}$$

which is a product of two exponential distributions, as for two independent, polarized thermal light beams [cf. Eq. (13.3–7)]. On the other hand, well above threshold, when a_1, a_2 are both positive and large, and when $\xi \leqslant 1$, $\mathcal{P}(I_1, I_2)$ reduces approximately to a Gaussian distribution in the two variables I_1, I_2. However, the situation becomes more complicated when the coupling constant ξ exceeds unity.

We may integrate the joint probability density $\mathcal{P}(I_1, I_2)$ with respect to one variable, say I_2, in order to obtain the probability density $\mathcal{P}_1(I_1)$ of the other mode intensity. We readily find from Eq. (19.2–8) that (Singh and Mandel, 1979; see also Singh, 1981)

$$\begin{aligned}\mathcal{P}_1(I_1) &= \int_0^\infty \mathcal{P}(I_1, I_2)\,\mathrm{d}I_2 \\ &= K^{-1}(\sqrt{\pi})\exp[\tfrac{1}{4}(\xi^2-1)I_1^2 - \tfrac{1}{2}(a_2\xi - a_1)I_1 + \tfrac{1}{4}a_2^2][1 - \operatorname{erf}(\tfrac{1}{2}\xi I_1 - \tfrac{1}{2}a_2)].\end{aligned} \tag{19.2–11}$$

This probability distribution has different forms for $\xi < 1$ and for $\xi > 1$, corresponding to quite different kinds of behavior of the ring laser.

19.2.1 Moments of the light intensity

Once the steady-state probability density $\mathcal{P}(I_1, I_2)$ has been found, we can readily calculate any moment of the light intensities I_1, I_2 from the formula

$$\begin{aligned}\langle I_1^n I_2^m\rangle &= K^{-1}\iint_0^\infty I_1^n I_2^m \exp(\tfrac{1}{2}a_1 I_1 + \tfrac{1}{2}a_2 I_2 - \tfrac{1}{4}I_1^2 - \tfrac{1}{4}I_2^2 - \tfrac{1}{2}\xi I_1 I_2)\,\mathrm{d}I_1\,\mathrm{d}I_2 \\ &= \frac{2^{n+m}}{K(a_1, a_2, \xi)}\frac{\partial^n}{\partial a_1^n}\frac{\partial^m}{\partial a_2^m}\iint_0^\infty \exp(\tfrac{1}{2}a_1 I_1 + \tfrac{1}{2}a_2 I_2 - \tfrac{1}{4}I_1^2 - \tfrac{1}{4}I_2^2 - \tfrac{1}{2}\xi I_1 I_2)\,\mathrm{d}I_1\,\mathrm{d}I_2 \\ &= \frac{2^{n+m}}{K(a_1, a_2, \xi)}\frac{\partial^{n+m}}{\partial a_1^n \partial a_2^m}K(a_1, a_2, \xi).\end{aligned} \tag{19.2–12}$$

The mean light intensities, the variances and the cross-correlations are all special cases of Eq. (19.2–12), so that $K(a_1, a_2, \xi)$ acts as a *generating function* for all the moments of I_1, I_2. It can be shown, after some manipulation of terms, that (Grossmann and Richter, 1971; M-Tehrani and Mandel, 1978a,b)

$$\langle I_1\rangle = \frac{a_1 - \xi a_2}{1-\xi^2} + \frac{2\sqrt{\pi}}{(1-\xi^2)K}[\mathrm{e}^{a_2^2/4}(1 + \operatorname{erf}\tfrac{1}{2}a_2) - \xi\,\mathrm{e}^{a_1^2/4}(1 + \operatorname{erf}\tfrac{1}{2}a_1)], \tag{19.2–13}$$

$$\begin{aligned}\langle(\Delta I_1)^2\rangle = \frac{2}{1-\xi^2} - \left(\langle I_1\rangle - \frac{a_1 - \xi a_2}{1-\xi^2}\right)^2 \\ - \frac{2\sqrt{\pi}}{(1-\xi^2)^2 K}\Big[\mathrm{e}^{a_2^2/4}(a_1 - \xi a_2)(1 + \operatorname{erf}\tfrac{1}{2}a_2)\end{aligned}$$

$$+ \xi^2 e^{a_1^2/4}(a_2 - \xi a_1)(1 + \operatorname{erf}\tfrac{1}{2}a_1) + \frac{2\xi}{\sqrt{\pi}}(1 - \xi^2)\Bigg], \tag{19.2–14}$$

and similar expressions with a_1, a_2 interchanged hold for $\langle I_2\rangle$ and $\langle(\Delta I_2)^2\rangle$. Also the cross-correlation is given by the equation

$$\langle \Delta I_1 \Delta I_2\rangle = - \frac{2\xi}{1-\xi^2} - \left(\langle I_1\rangle - \frac{a_1 - \xi a_2}{1-\xi^2}\right)\left(\langle I_2\rangle - \frac{a_2 - \xi a_1}{1-\xi^2}\right) + \frac{2\xi\sqrt{\pi}}{(1-\xi^2)^2 K}$$
$$\times \left[e^{a_1^2/4}(a_2 - \xi a_1)(1 + \operatorname{erf}\tfrac{1}{2}a_1) + e^{a_2^2/4}(a_1 - \xi a_2)(1 + \operatorname{erf}\tfrac{1}{2}a_2) + \frac{2(1-\xi^2)}{\xi\sqrt{\pi}}\right]. \tag{19.2–15}$$

In order to see the implications of these relations it is convenient to examine separately the cases $\xi \leqslant 1$ corresponding to an inhomogeneously broadened gain medium, and $\xi = 2$ corresponding to a homogeneously broadened one.

Let us first consider an inhomogeneously broadened ring laser operating at line center, so that $\xi = 1$. Figure 19.3 shows plots of the average light intensities $\langle I_1\rangle$ and $\langle I_2\rangle$ as a function of a_1 for a modest asymmetry $\Delta a = a_1 - a_2 = 0.8$. In practice, for most ring lasers Δa is generally small but not zero, and it does not change very much as the excitation a_1 of the laser is varied. The more unusual

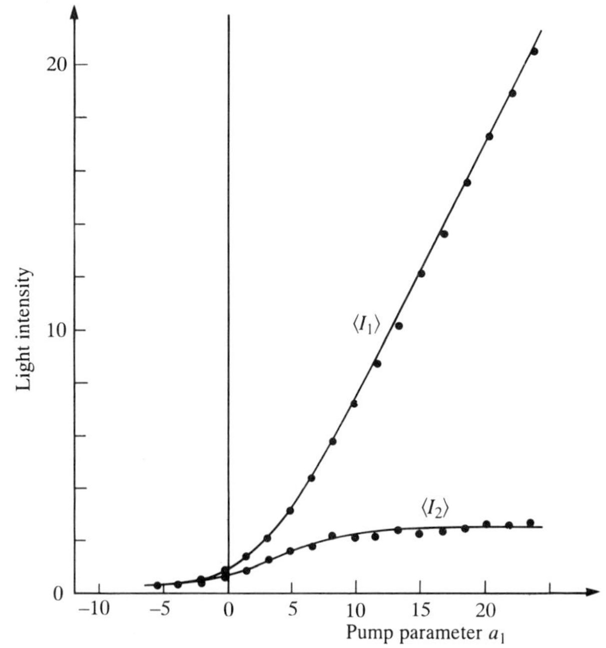

Fig. 19.3 Results of measurements of the average light intensity of each ring laser mode as a function of pump parameter a_1 at line center, superimposed on the theoretical curves with $\Delta a = 0.8$. (Reproduced from M-Tehrani and Mandel, 1978b.)

behavior is exhibited by the more lossy mode 2 (the mode subject to greater loss). As a_1 increases, $\langle I_1 \rangle$ grows steadily with a_1 above threshold, as for a single-mode laser, but the growth of $\langle I_2 \rangle$ is reduced above threshold and $\langle I_2 \rangle$ reaches an asymptotic value. In other words, the less lossy mode 1 suppresses the growth of the more lossy mode 2 with increasing a_1. Even more interesting behavior is exhibited by the relative intensity fluctuations $\langle (\Delta I)^2 \rangle / \langle I \rangle^2$ of the two modes, shown in Fig. 19.4, for $\Delta a = a_1 - a_2 = 0.8$. Whereas the relative fluctuations of mode 1 tend to zero with increasing a_1, as is typical of a laser above threshold, at line center the relative fluctuations of the more lossy mode 2 reach a minimum value just above threshold, and then return to unity, which is the value characteristic of thermal, or incoherent light (cf. Section 13.3). The mode competition therefore results in a drastic asymmetry not only of the intensities of the two modes, but also of their fluctuation properties. Finally, Fig. 19.4 also shows the relative cross-correlation of the two mode intensities. It is always negative, because the modes compete for photons from a common atomic population, and it reaches its largest numerical value a little above threshold. A careful analysis of Eqs. (19.2–13) to (19.2–15) at line center ($\xi = 1$) shows that in the asymptotic limit as $a_1 \to \infty$,

$$\left.\begin{aligned} \langle I_1 \rangle &\sim a_1 - \frac{2}{\Delta a} \\ \langle I_2 \rangle &\sim \frac{2}{\Delta a} \\ \frac{\langle (\Delta I_1)^2 \rangle}{\langle I_1 \rangle^2} &\sim \frac{2 + 4/(\Delta a)^2}{a_1^2} \\ \frac{\langle (\Delta I_2)^2 \rangle}{\langle I_2 \rangle^2} &\sim 1 \\ \frac{\langle \Delta I_1 \Delta I_2 \rangle}{\langle I_1 \rangle \langle I_2 \rangle} &\sim -\frac{2}{a_1 \Delta a}. \end{aligned}\right\} \qquad (19.2\text{–}16)$$

19.2.2 Comparison with experiment

Many of these theoretical predictions have been confirmed by photoelectric counting measurements of the two modes of a He:Ne ring laser, which is inhomogeneously broadened. Figure 19.5 shows the apparatus used in one set of measurements. The ring laser had two plane and two concave mirrors arranged at the corners of a square of side 12.5 cm, with a He:Ne plasma tube in one arm. One of the mirrors was mounted on a piezoelectric crystal, which allowed the cavity frequency to be varied. A knife-edge, activated by another piezoelectric driver and inserted in one arm, acted as a variable loss element and allowed the working point of the laser to be varied from below to above threshold. The mean light intensity was held constant against drifts by a negative feedback arrangement, with the help of a monitor phototube. The light leaking out from the two counterpropagating laser modes fell on two photon counting photomultipliers, whose output pulses, after amplification and shaping, were ultimately counted by

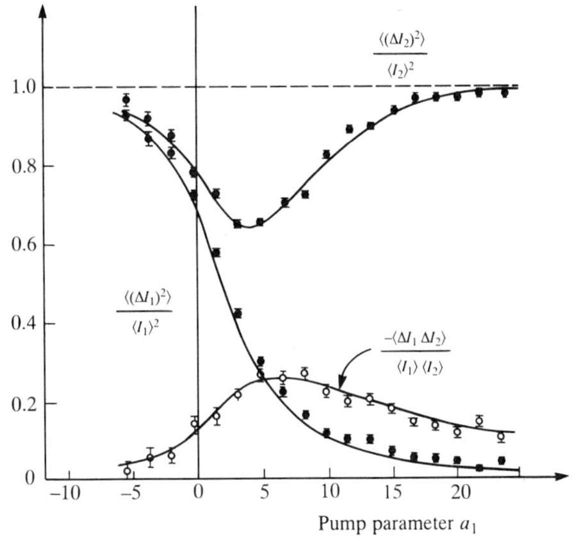

Fig. 19.4 Results of measurements of the relative intensity fluctuations and cross-correlations of the two ring laser modes as a function of pump parameter a_1 at line center, superimposed on the theoretical curves with $\Delta a = 0.8$. (Reproduced from M-Tehrani and Mandel, 1978b.)

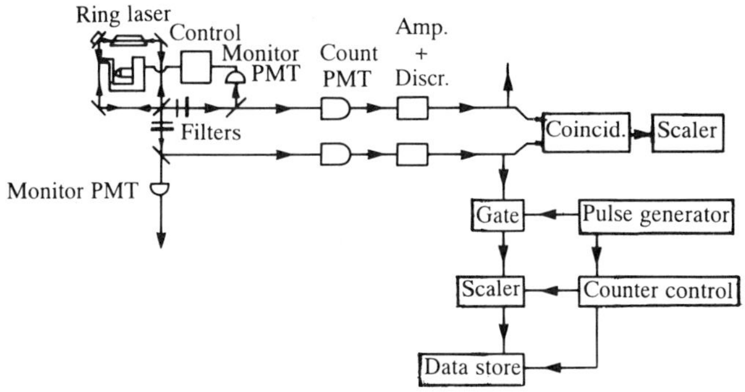

Fig. 19.5 Outline of the apparatus used for photoelectric counting measurements of a He:Ne ring laser. (Reproduced from M-Tehrani and Mandel, 1978b.)

a scaler during 1 μs time intervals. This yielded the average light intensity $\langle I \rangle$ and also the relative intensity fluctuation $\langle(\Delta I)^2\rangle/\langle I\rangle^2$, as described in Section 14.9. At the same time, the rate of arrival of pulses in coincidence from both channels provided a measure of the cross-correlation $\langle \Delta I_1 \Delta I_2 \rangle$.

Figures 19.3 and 19.4 show results of the measurements superimposed on the theoretical curves derived from Eqs. (19.2–13) to (19.2–15). It will be seen that the predicted effects are indeed observed. The growth of mode 2 is suppressed by mode 1, and the relative intensity fluctuations of mode 2 become close to unity well above threshold. For large pump parameter a_1, the electromagnetic waves propagating one way around the ring acquire the characteristics of coherent laser light, while those propagating in the opposite direction have the character of thermal or incoherent light. We have here a very striking manifestation of mode competition in a laser, and we note that the large asymmetry in the properties of the two modes has been produced by a relatively small asymmetry in their respective losses. The mode competition was observed to become less pronounced with detuning, and far from line center ($\Delta\omega T_1 \gg 1$) the two modes behaved essentially like two uncoupled single-mode lasers (M-Tehrani and Mandel, 1977, 1978a,b).

Equations (19.2–13) to (19.2–15) predict still stronger competition effects in a homogeneously broadened ring laser, for which $\xi = 2$. Figure 19.6 shows the mean values of the two mode intensities as a function of a_1 and Δa when $\xi = 2$. It is assumed that the cavity resonance lies entirely within the natural atomic line, in which case cavity detuning has no effect. It will be seen that whenever there is an asymmetry and $\Delta a \neq 0$, mode 1 not only suppresses the growth of mode 2 with increasing a_1, but actually forces it towards zero intensity. Figure 19.7 shows plots

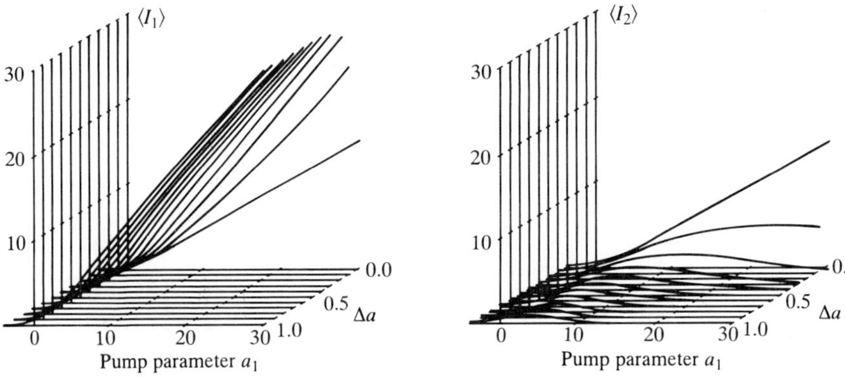

Fig. 19.6 Predicted variation of the mean light intensities of the two ring laser modes with pump parameter a_1, for various values of Δa, with $\xi = 2$. (Reproduced from Singh and Mandel, 1979.)

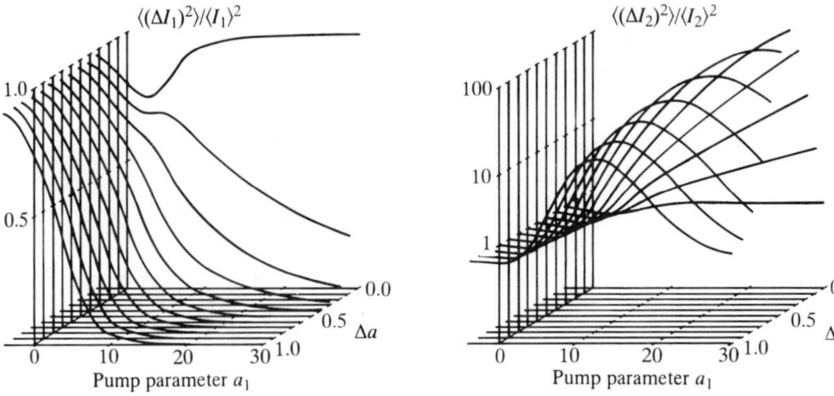

Fig. 19.7 Predicted variation of the relative intensity fluctuations of the two ring laser modes with pump parameter a_1, for various values of Δa, with $\xi = 2$. (Reproduced from Singh and Mandel, 1979.)

of the relative intensity fluctuations of each mode as a function of a_1. While the fluctuations of mode 1 tend to zero, as the excitation is increased as before, for $\Delta a > 0$, those of mode 2 can greatly exceed unity, although they tend asymptotically to unity as $a_1 \to \infty$. Large relative intensity fluctuations are often associated with on–off switching, and we shall see below, that such switching can indeed occur in the two-mode laser. Figure 19.8 shows the relative cross-correlation, which again is always negative, as a function of a_1 and Δa. When $\Delta a \gtrsim 1$, this behavior is qualitatively similar to that exhibited by the inhomogeneously broadened ring laser. However, when $\Delta a = 0$, the correlation $\langle I_1 I_2 \rangle / \langle I_1 \rangle \langle I_2 \rangle$ tends to zero asymptotically with increasing a, despite the fact that $I_1/\langle I_1 \rangle$ and $I_2/\langle I_2 \rangle$ are both non-negative, and each has a unit mean. The correlation can vanish only if one mode intensity is zero whenever the other one is non-zero, which implies switching of the excitation between the two counterpropagating modes. We shall return to a discussion of this phenomenon in Section 19.4.

19.3 The phase transition analogy

As in the previous chapter, it is convenient to make an analogy between the behavior of the laser field and certain kinds of phase transition (see for example Graham, 1975; Haken, 1975, 1977, Chapts. 6 and 7). For simplicity, we shall start with the deterministic equations of motion, before the terms representing quantum noise are added. In the steady state, when the time derivatives vanish, Eqs. (19.1–16) with $q_1 = 0 = q_2$ imply that

$$\left.\begin{aligned}(a_1 - I_1 - \xi I_2)I_1 &= 0\\ (a_2 - I_2 - \xi I_1)I_2 &= 0.\end{aligned}\right\} \tag{19.3–1}$$

The solution of these two coupled equations has three distinct branches in general, that we label (a), (b) and (c). They are (Mandel, Roy and Singh, 1981)

$$\text{(a)} \qquad \begin{aligned} I_1 = 0, \quad I_2 = a_2 = a - \tfrac{1}{2}\Delta a, &\quad \text{if } a_2 > 0\\ I_2 = 0, &\quad \text{if } a_2 < 0; \end{aligned} \tag{19.3–2}$$

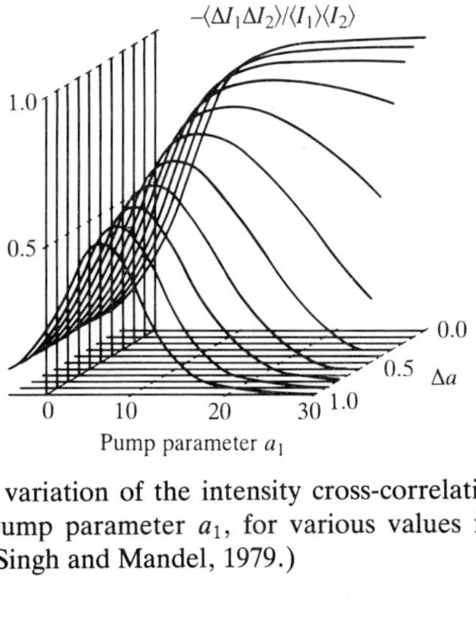

Fig. 19.8 Predicted variation of the intensity cross-correlation of the two ring laser modes with pump parameter a_1, for various values for Δa with $\xi = 2$. (Reproduced from Singh and Mandel, 1979.)

(b)
$$I_2 = 0, \quad I_1 = a_1 = a + \tfrac{1}{2}\Delta a, \quad \text{if } a_1 > 0 \qquad (19.3\text{–}3)$$
$$I_1 = 0, \qquad \text{if } a_1 < 0;$$

(c)
$$I_1 = \frac{a_2\xi - a_1}{\xi^2 - 1} = \frac{a}{\xi + 1} - \frac{\tfrac{1}{2}\Delta a}{\xi - 1}, \quad I_2 = \frac{a_1\xi - a_2}{\xi^2 - 1} = \frac{a}{\xi + 1} + \frac{\tfrac{1}{2}\Delta a}{\xi - 1}, \qquad (19.3\text{–}4)$$

provided
$$-\frac{2a}{\xi + 1} \leqslant \frac{\Delta a}{\xi - 1} \leqslant \frac{2a}{\xi + 1},$$

where
$$\left.\begin{aligned} a &\equiv \tfrac{1}{2}(a_1 + a_2) \\ \Delta a &\equiv a_1 - a_2 \end{aligned}\right\} \qquad (19.3\text{–}5)$$

represent the mean of and the difference between the two pump parameters. Figure 19.9 shows plots of the mode intensity I_1 as a function of the asymmetry parameter Δa for fixed a and for several different values of the coupling constant ξ. The value of I_1 is always zero for large negative Δa, and it is always equal to $a_1 = a + \frac{1}{2}\Delta a$ for large positive Δa. But these two branches of the solution are joined by a line whose slope depends on ξ. The slope is positive when $\xi < 1$, negative when $\xi > 1$, and infinite when $\xi = 1$. A negative slope corresponds to a multivalued, and generally unstable, solution and a system described by a graph such as $DACB$ would be expected to exhibit hysteresis, with a jump from A to B as Δa increases beyond $2a/3$, and a jump from C to D as Δa decreases below $-2a/3$. The solution given by Eq. (19.3–4), in which both mode intensities are non-zero, is therefore unstable when the coupling constant ξ exceeds 1. It follows that the value $\xi = 1$ divides two domains of rather different laser behavior. Figure 19.10 illustrates the solution of the coupled time-dependent equations of motion (without quantum noise) when $a_1 = a_2$ and $\xi = 2$. It should be noted that in the steady state one or the other mode intensity vanishes, depending on the initial conditions.

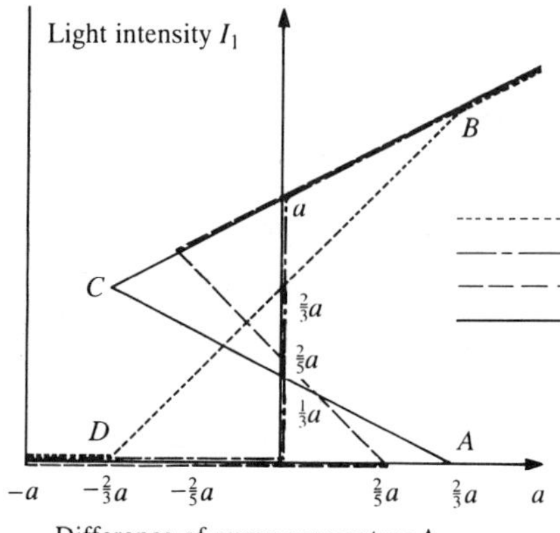

Fig. 19.9 The variation of one laser mode intensity I_1 with Δa for several different coupling constants ξ, in the steady state, according to the deterministic equations of motion. (Reproduced from Mandel, Roy and Singh, 1981.)

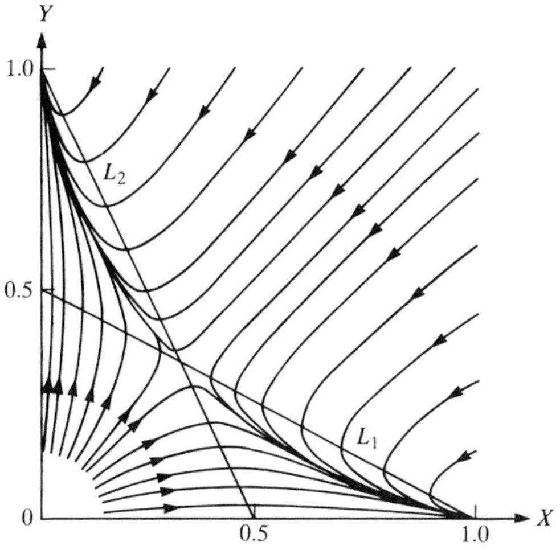

Fig. 19.10 Curves illustrating the solutions of the deterministic equations of motion for the two laser mode intensities I_1 and I_2 (here called X, Y), with $a_1 = a_2 = 1$, $\xi = 2$. (Reproduced from Lamb, 1964.)

19.3.1 Minima of the potential

These conclusions are based on the deterministic equations of motion, and they should be re-examined in the framework of the statistical solution given by Eqs. (19.2–7) or (19.2–8). Let us therefore find the most probable values of the two mode intensities I_1, I_2 by looking for the minimum of the potential U. Once again we encounter several different branches of the solution, that we shall discuss separately.

(a) $a_1 < 0$, $a_2 < 0$

It is obvious from Eq. (19.2–7) that below threshold, when both a_1 and a_2 are negative, the potential U has its smallest possible value when

$$I_1 = 0 = I_2, \tag{19.3–6}$$

and this represents the most probable state of the laser. We shall call this phase I.

(b) $a_1 < 0$, $a_2 > 0$

From Eq. (19.2–7) the potential U is smallest when $I_1 = 0$ and $I_2 \neq 0$, and the optimum value of I_2 is given by $\partial U/\partial I_2 = 0$, or

$$-a_2 + I_2 + \xi I_1 = 0$$

so that

$$\left.\begin{aligned} I_1 &= 0 \\ I_2 &= a_2. \end{aligned}\right\} \tag{19.3–7}$$

(c) $a_2 < 0,\ a_1 > 0$

A similar argument shows that the most probable state is given by

$$\left.\begin{aligned} I_1 &= a_1 \\ I_2 &= 0. \end{aligned}\right\} \qquad (19.3\text{–}8)$$

The region in which a_1 and a_2 are both positive requires further subdivision.

(d) $a_1 > 0,\ a_2 > 0,\ a_2\xi > a_1$

By putting $\partial U/\partial I_2 = 0$, and making I_1 very small we readily confirm that the state

$$\left.\begin{aligned} I_1 &= 0 \\ I_2 &= a_2 \end{aligned}\right\} \qquad (19.3\text{–}9)$$

corresponds to a local minimum. This is the same solution as is given by Eq. (19.3–7), and it shows that (b) and (d) can be combined into one phase, characterized by $a_2 > 0,\ a_2\xi > a_1$, that we shall call phase II.

(e) $a_1 > 0,\ a_2 > 0,\ a_1\xi > a_2$

Similarly, we can show that the state

$$\left.\begin{aligned} I_1 &= a_1 \\ I_2 &= 0 \end{aligned}\right\} \qquad (19.3\text{–}10)$$

represents a local minimum. Hence (c) and (e) can also be combined into one phase characterized by $a_1 > 0,\ a_1\xi > a_2$, that we shall call phase III.

(f) $a_1 > 0,\ a_2 > 0,\ a_2\xi < a_1,\ a_1\xi < a_2$

By equating both partial derivatives $\partial U/\partial I_1$ and $\partial U/\partial I_2$ to zero, we obtain the formulas

$$-a_1 + I_1 + \xi I_2 = 0$$

$$-a_2 + I_2 + \xi I_1 = 0,$$

and these simultaneous equations have the non-zero solutions given by Eqs. (19.3–4) above, namely

$$\left.\begin{aligned} I_1 &= \frac{a_2\xi - a_1}{\xi^2 - 1} \\ I_2 &= \frac{a_1\xi - a_2}{\xi^2 - 1}. \end{aligned}\right\} \qquad (19.3\text{–}11a)$$

However, these solutions represent a local minimum of the potential U only if

$$\left(\frac{\partial^2 U}{\partial I_1 \partial I_2}\right)^2 < \frac{\partial^2 U}{\partial I_1^2}\frac{\partial^2 U}{\partial I_2^2},$$

i.e. if

$$\xi < 1, \qquad (19.3\text{–}11b)$$

in which case Eqs. (19.3–11a) obviously require the constraints $a_2\xi < a_1$ and $a_1\xi < a_2$. We shall denote this state as phase IV. On the other hand, when $\xi > 1$ there is no minimum of the potential at which both I_1 and I_2 are non-zero, and phase IV does not exist. It can be shown that in this case the coordinates given by Eqs. (19.3–11a) correspond to a saddle point of the potential. We shall examine the situation when $\xi > 1$ more carefully later.

19.3.2 The case coupling constant $\xi < 1$: second-order phase transition

The most probable intensities I_1 and I_2 play the roles of order parameters for the phase transition of the laser field, as for the single-mode laser. But as a result of having two order parameters instead of one, we now encounter four distinguishable phases when $\xi < 1$. These are illustrated by the phase diagram in Fig. 19.11(*a*), which follows the scheme of Agarwal and Dattagupta (1982). The heavy lines separate the different stable phases. In phase I, both order parameters vanish. In phases II and III one of the two order parameters is zero while the other one is non-zero. In phase IV both order parameters are non-zero. All phases meet at the point C which is the critical point. If we substitute the most probable values of the intensities into the potential U, we obtain the most probable potential $U_\mathrm{m}(a_1, a_2, \xi)$ (cf. Section 18.3), which is to some extent analogous to a thermodynamic potential like the free energy. With the help of Eqs. (19.2–7) and (19.3–6) to (19.3–11) we find for the four different phases,

$$\left.\begin{aligned} U_m &= 0 && \text{in phase I} \\ U_m &= -\tfrac{1}{4}a_2^2 && \text{in phase II} \\ U_m &= -\tfrac{1}{4}a_1^2 && \text{in phase III} \\ U_m &= -[\tfrac{1}{4}(a_1^2 + a_2^2) - \tfrac{1}{2}a_1a_2\xi]/(1-\xi^2) && \text{in phase IV.} \end{aligned}\right\} \qquad (19.3\text{–}12)$$

Alternatively, as in Section 18.3, we can define quantities that are analogous to

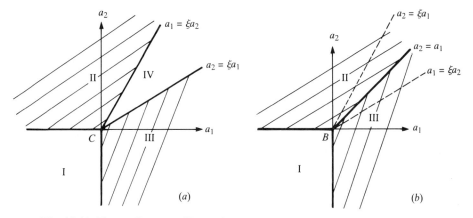

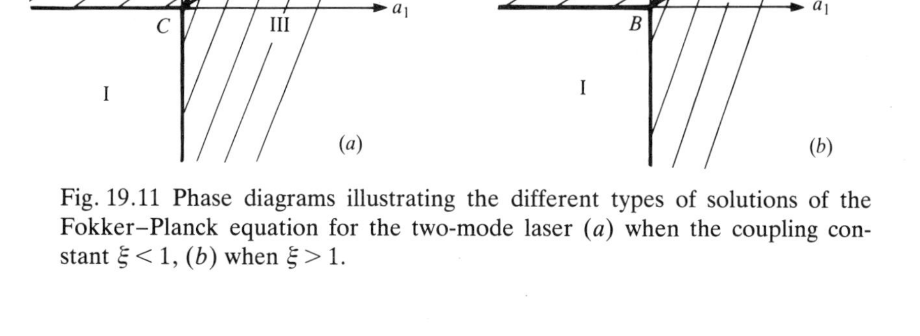

Fig. 19.11 Phase diagrams illustrating the different types of solutions of the Fokker–Planck equation for the two-mode laser (*a*) when the coupling constant $\xi < 1$, (*b*) when $\xi > 1$.

the entropy

$$S_1 = \left(\frac{\partial U}{\partial a_1}\right)_{\substack{I_1=I_{m1},\\ I_2=I_{m2}}} \quad S_2 = \left(\frac{\partial U}{\partial a_2}\right)_{\substack{I_1=I_{m1},\\ I_2=I_{m2}}}$$

where I_{m1}, I_{m2} stand for the most probable values of the light intensities I_1, I_2 in the corresponding phase. We then obtain with the help of Eqs. (19.2–7), (19.3–6) to (19.3–11) the relations

$$\left.\begin{aligned} S_1 &= 0 \\ S_2 &= 0 \end{aligned}\right\} \quad \text{in phase I}$$

$$\left.\begin{aligned} S_1 &= 0 \\ S_2 &= -\tfrac{1}{2}a_2 = -\tfrac{1}{2}a + \tfrac{1}{4}\Delta a \end{aligned}\right\} \quad \text{in phase II}$$

$$\left.\begin{aligned} S_1 &= -\tfrac{1}{2}a_1 = -\tfrac{1}{2}a - \tfrac{1}{4}\Delta a \\ S_2 &= 0 \end{aligned}\right\} \quad \text{in phase III} \qquad (19.3\text{–}13)$$

$$\left.\begin{aligned} S_1 &= \frac{-a_1 + a_2\xi}{2(1-\xi^2)} \\ S_2 &= \frac{-a_2 + a_1\xi}{2(1-\xi^2)} \end{aligned}\right\} \quad \text{in phase IV,}$$

and it is easy to confirm that these entropies are all continuous across the phase boundaries. It follows that, so long as $\xi < 1$, the ring laser exhibits a *continuous* or *second-order phase transition*, just like the single-mode laser (cf. Section 18.3).

19.3.3 The case coupling constant $\xi > 1$: first-order phase transition

When $\xi > 1$, as for a homogeneously broadened two-mode laser, the phase diagram changes (Hambenne and Sargent, 1975). We have already seen that phase IV disappears in this case. However, in addition, it follows from the definition that what we previously called phases II and III now overlap partially within the region $a_1/\xi < a_2 < a_1\xi$, because there are positive values of the pump parameters a_1, a_2 for which both inequalities $a_2\xi > a_1$ and $a_1\xi > a_2$ are satisfied simultaneously. But as the minimum potential is $-\frac{1}{4}a_2^2$ in phase II and $-\frac{1}{4}a_1^2$ in phase III, the state we previously referred to as phase II is actually more probable than the state referred to as phase III in the region $a_1 < a_2$, whereas state III is actually more probable than state II in the region $a_2 < a_1$. Clearly the line $a_1 = a_2$ separates these two regions. If the most probable state is taken to define the phase, then in the case $\xi > 1$ we must redefine phase II as the region $a_2 > 0$, $a_2 > a_1$, and phase III as the region $a_1 > 0$, $a_1 > a_2$. The corresponding phase diagram is illustrated in Fig. 19.11(*b*). The most probable values of the potential for the three phases are

$$\left.\begin{aligned} U_m &= 0 && \text{in phase I,} && a_1 < 0,\, a_2 < 0 \\ U_m &= -\tfrac{1}{4}a_2^2 && \text{in phase II,} && a_2 > 0,\, a_2 > a_1 \\ U_m &= -\tfrac{1}{4}a_1^2 && \text{in phase III,} && a_1 > 0,\, a_1 > a_2. \end{aligned}\right\} \qquad (19.3\text{–}14)$$

For the corresponding entropies we find

$$\left.\begin{aligned} S_1 &= 0 \\ S_2 &= 0 \end{aligned}\right\} \text{in phase I}$$
$$\left.\begin{aligned} S_1 &= 0 \\ S_2 &= -\tfrac{1}{2}a_2 = -\tfrac{1}{2}(a - \tfrac{1}{2}\Delta a) \end{aligned}\right\} \text{in phase II} \qquad (19.3\text{–}15)$$
$$\left.\begin{aligned} S_1 &= -\tfrac{1}{2}a_1 = -\tfrac{1}{2}(a + \tfrac{1}{2}\Delta a) \\ S_2 &= 0 \end{aligned}\right\} \text{in phase III,}$$

and we observe that both S_1 and S_2 have jump discontinuities on the line $a_1 = a_2$ separating phases II and III. The phase transition across the line $a_1 = a_2$ in Fig. 19.11(b) is therefore a *first-order or discontinuous phase transition*. The point B where all the lines meet is the critical point for the two-mode laser. The phase diagram in the neighborhood of B is analogous to that for a weakly anisotropic antiferromagnet (Agarwal and Dattagupta, 1982), when the two pump parameters a_1, a_2 characterizing the state of the system are replaced by the applied magnetic field and the temperature. The paramagnetic phase is separated from the antiferromagnetic phase and from the so-called spin-flop phase by lines of second-order phase transitions, while the antiferromagnetic and spin-flop phases are separated by a line of first-order phase transition.

Figure 19.12 shows three-dimensional plots of the potential $U(I_1, I_2)$ as a function of I_1, I_2 when $a_1 = 15$, $a_2 = 14.5$, both for the case $\xi = 0.5$ and for $\xi = 2$. The former case corresponds to phase IV of the phase diagram 19.11(a), whereas the latter corresponds to phase III of the phase diagram 19.11(b). It is clear from this that the situations $\xi = 0.5$ and $\xi = 2$ are fundamentally different. In the first case U has one minimum, whereas in the second case U has two local minima, one at $I_1 \approx 15$, $I_2 = 0$, and one at $I_1 = 0$, $I_2 \approx 14.5$, and these minima are separated by a saddle. Both minima represent highly probable states of the laser, but the first minimum is somewhat lower and therefore corresponds to the most probable state. The three-dimensional graphs make it possible to visualize how the change from a second-order to a first-order phase transition comes about as ξ increases above unity.

Sometimes it is helpful to integrate over one of the variables and to deal with a single mode of the laser at one time (Mandel, Roy and Singh, 1981). For this purpose we may make use of Eq. (19.2–11), which gives the probability distribution $\mathcal{P}_1(I_1)$ of the intensity of mode 1 in the steady state, expressed in the form

$$\mathcal{P}_1(I_1) = \text{const.} \times e^{-U_1(I_1)}, \qquad (19.3\text{–}16a)$$

with the one-mode potential $U_1(I_1)$ given by the expression

$$U_1(I_1) = -\tfrac{1}{4}(\xi^2 - 1)I_1^2 + \tfrac{1}{2}(a_2\xi - a_1)I_1 - \tfrac{1}{4}a_2^2 - \ln[1 - \operatorname{erf}(\tfrac{1}{2}\xi I_1 - \tfrac{1}{2}a_2)].$$
$$(19.3\text{–}16b)$$

Figure 19.13 shows plots of the potential $U_1(I_1)$ for fixed values of the pump parameters but for several different coupling constants ξ. Whereas $U_1(I_1)$ has a single minimum somewhat below $I_1 = a_1$ so long as $\xi \leqslant 1$, the potential begins to

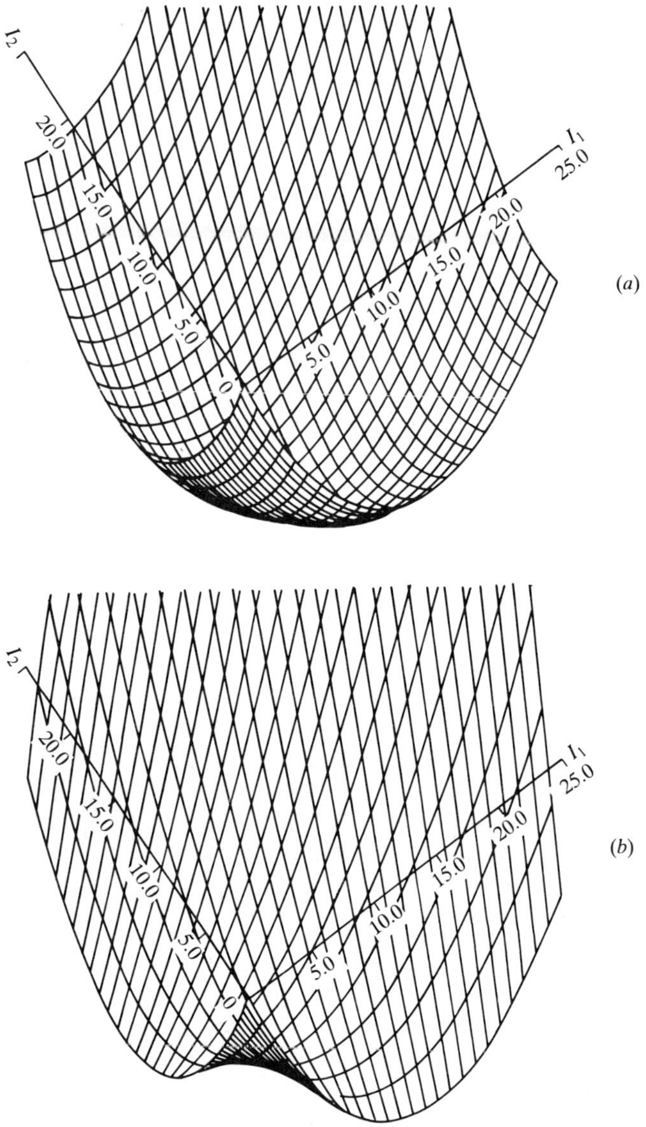

Fig. 19.12 Plots illustrating the form of the potential $U(I_1, I_2)$ when $a_1 = 15$, $a_2 = 14.5$ for (*a*) $\xi = 0.5$, (*b*) $\xi = 2$. (Reproduced from Lett, Christian, Singh and Mandel, 1981.)

exhibit a maximum as well as a minimum once ξ exceeds unity sufficiently. The maximum, which is quite pronounced when $\xi = 2$, separates two local minima, near $I_1 = 0$ and $I_1 = a_1$, which both correspond to very probable states. Of course I_2 is non-zero whenever $I_1 = 0$, and vice versa. Although it might appear that the states would be stable, both states are actually only metastable because of quantum fluctuations. Whatever state the system may be in at one time, it will sooner or later encounter a sufficiently large fluctuation that causes a transition from one metastable state to the other. The lifetime of the state depends on the

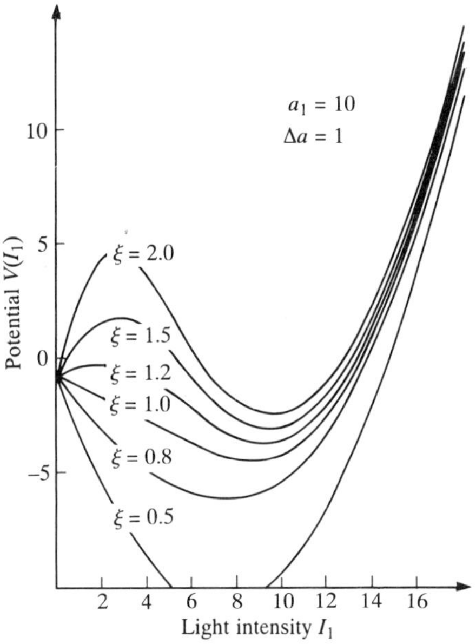

Fig. 19.13 The form of the potential for one mode [here called $V(I_1)$] of a two-mode ring laser with $a_1 = 10$, $\Delta a = 1$, for several different coupling constants ξ. (Reproduced from Mandel, Roy and Singh, 1981.)

height of the potential maximum, and it can be quite long, as we shall see shortly. In some respects the laser in this state exhibits features that one associates with quantum mechanical tunneling through a potential barrier.

In order to see more clearly why the transition between these two metastable states corresponds to a first-order phase transition, let us plot the potential $U_1(I_1)$ once again, but this time with fixed $\xi = 2$ for several different pump parameters a_1 (with $\Delta a \equiv a_1 - a_2 = 1$). Remembering that the lowest potential corresponds to the most probable value I_m of the mode intensity I_1, we see from Fig. 19.14 that the most probable intensity is zero up to some value of a_1 between 4 and 6, and then jumps to a non-zero value for larger a_1. The broken curve in Fig. 19.14 is the locus of the most probable intensity I_m, which is an order parameter of the phase transition. As usual, the order parameter vanishes on the disordered side of the phase transition, but we note that it has a jump discontinuity at the phase transition. As a result, the most probable potential U_m, obtained by substituting the most probable intensity I_m in $U_1(I_1, a_1, a_2, \xi)$, has a discontinuous first derivative. This is clearly illustrated in Fig. 19.15, where the form of the most probable potential U_m is shown as a function of a_1, for several different values of the coupling constant ξ. It will be seen from this figure that the discontinuity of the derivative is associated with coupling constants ξ greater than 1.

We have already seen from Eqs. (19.3–9) and (19.3–10) that zero and a_1 are the two most probable values I_m of the light intensity I_1. The same conclusion holds also when the one-dimensional potential $U_1(I_1, a_1, a_2, \xi)$ in Eq. (19.3–16b) is used. For convenience we let $a \equiv \frac{1}{2}(a_1 + a_2)$ and $\Delta a \equiv a_1 - a_2$. For small a, inspection of Eq. (19.3–16b) or Fig. 19.14 shows clearly that $I_m = 0$. To determine the other value of I_m we differentiate $U_1(I_1, a, \Delta a, \xi)$ with respect to I_1.

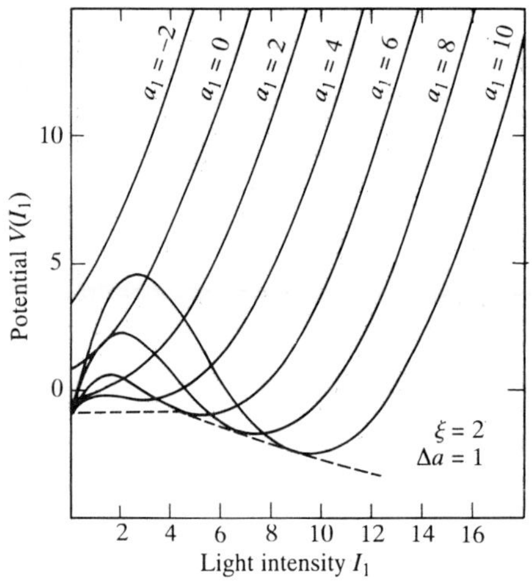

Fig. 19.14 The variation of the potential with intensity I_1 for one mode [here called $V(I_1)$] of a two-mode ring laser for several different pump parameters a_1, with $\Delta a = 1$, $\xi = 2$. The broken curve shows the most probable light intensity I_1. (Reproduced from Mandel, Roy and Singh, 1981.)

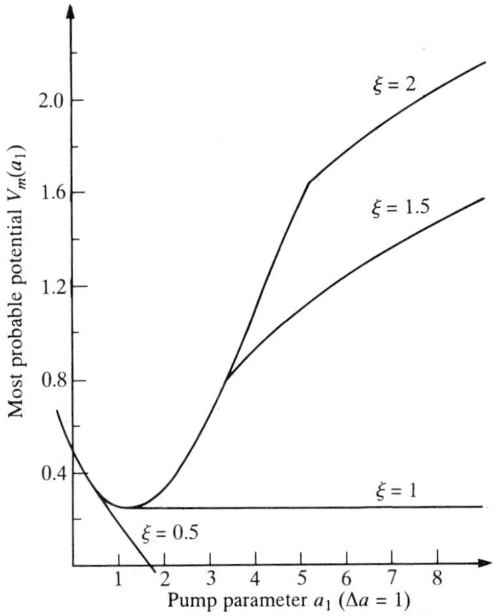

Fig. 19.15 The form of the most probable potential for one mode of a two-mode laser as a function of pump parameter a_1, for several different coupling constants ξ, with $\Delta a = 1$. (Reproduced from Mandel, Roy and Singh, 1981.)

We readily find (Mandel, Roy and Singh, 1981) that the non-zero value of I_m is given by the intersection of the two curves

$$y = [(\xi^2 - 1)I_1 - a(\xi - 1) + \tfrac{1}{2}\Delta a(\xi + 1)][1 - \text{erf}\,(\tfrac{1}{2}\xi I_1 - \tfrac{1}{2}a + \tfrac{1}{4}\Delta a)]$$

$$y = \frac{2\xi}{\sqrt{\pi}} \exp -\tfrac{1}{4}(\xi I_1 - a + \tfrac{1}{2}\Delta a)^2,$$

which occurs in the high-intensity tail of the Gaussian function. Provided a is not too small, we may make use of the asymptotic expansion of the error function (Abramowitz and Stegun, 1965, Sect. 7)

$$1 - \operatorname{erf} z \sim \frac{1}{z\sqrt{\pi}} e^{-z^2}\left(1 - \frac{1}{2z^2} + \ldots\right) \quad \text{as } z \to \infty, \; |\arg z| < \frac{3\pi}{4}. \qquad (19.3\text{–}17)$$

It follows from this expansion that I_m is the solution of the equation

$$(\xi^2 - 1)I_m - a(\xi - 1) + \tfrac{1}{2}\Delta a(\xi + 1) = \xi(\xi I_m - a + \tfrac{1}{2}\Delta a)$$

i.e.

$$I_m = a + \tfrac{1}{2}\Delta a. \qquad (19.3\text{–}18)$$

The two most probable values of I_m are therefore 0 and $a_1 = a + \tfrac{1}{2}\Delta a$. In order to determine which one is the more probable, we have to calculate the difference between the potentials $U_1(0, a, \Delta a)$ and $U_1(a_1, a, \Delta a)$. With the same approximation as before we find from Eq. (19.3–16b) that

$$U_1(0, a, \Delta a) - U_1(a_1, a, \Delta a) = \tfrac{1}{2}a\Delta a - \ln\{\sqrt{\pi}[(\xi - 1)a + \tfrac{1}{2}(\xi + 1)\Delta a]\}.$$

It follows that the cross-over from one minimum of U_1 to the other occurs at the value $a = a_0$ which is the solution of the transcendental equation

$$\tfrac{1}{2}a_0\Delta a = \ln\{\sqrt{\pi}[(\xi - 1)a_0 + \tfrac{1}{2}(\xi + 1)\Delta a]\}. \qquad (19.3\text{–}19)$$

When $\xi = 2$ and Δa is very small, an order of magnitude approximation to the solution is given by

$$a_0 \sim \frac{10}{\Delta a}. \qquad (19.3\text{–}20)$$

The most probable value I_m of the light intensity I_1 is therefore

$$\left.\begin{aligned} I_m &= 0 && \text{when } a < a_0 \\ I_m &= a + \tfrac{1}{2}\Delta a && \text{when } a > a_0. \end{aligned}\right\} \qquad (19.3\text{–}21)$$

On recalling that the negative of the pump parameter a is the analog of temperature for a thermodynamic system, we may now introduce an entropy $S_1(a, \Delta a)$ for the system, by taking the derivative of the potential with respect to the pump parameter, i.e.

$$S_1(a, \Delta a) = \left[\frac{\partial U_1(I_1, a, \Delta a)}{\partial a}\right]_{I_1 = I_m} \qquad (19.3\text{–}22)$$

From Eq. (19.3–16b) it follows that,

$$\left[\frac{\partial U_1(I_1, a, \Delta a)}{\partial a}\right]_{I_1 = I_m} = \tfrac{1}{2}(\xi - 1)I_m - \tfrac{1}{2}(a - \tfrac{1}{2}\Delta a) - \frac{\exp[-\tfrac{1}{4}(\xi I_m - a + \tfrac{1}{2}\Delta a)^2]}{\sqrt{\pi}[1 - \operatorname{erf}(\tfrac{1}{2}\xi I_m - \tfrac{1}{2}a + \tfrac{1}{4}\Delta a)]},$$

and we may use Eq. (19.3–21) to substitute the appropriate values of I_m. If a exceeds about 3 and $I_m = a + \frac{1}{2}\Delta a$, we may introduce the approximation $\operatorname{erf}(-\frac{1}{2}a + \frac{1}{4}\Delta a) \approx -1$, and the asymptotic expansion (19.3–17) for $1 - \operatorname{erf}(\frac{1}{2}\xi I_m - \frac{1}{2}a + \frac{1}{4}\Delta a)$. This finally leads to the result

$$\left.\begin{aligned} S_1(a, \Delta a) & \\ &= -\tfrac{1}{2}(a - \tfrac{1}{2}\Delta a) - \frac{1}{2\sqrt{\pi}}\exp[-\tfrac{1}{4}(a - \tfrac{1}{2}\Delta a)^2], \quad \text{when } a < a_0 \\ &= -\tfrac{1}{2}(a - \tfrac{1}{2}\Delta a) - \tfrac{1}{2}\Delta a - \frac{1}{(\xi - 1)a + \frac{1}{2}(\xi + 1)\Delta a}, \quad \text{when } a > a_0. \end{aligned}\right\} \quad (19.3\text{–}23)$$

The entropy has a discontinuity at $a = a_0$, which shows that the laser exhibits a discontinuous, or first-order, phase transition at this point. This result is attributable to the bistable form of the potential. Indeed, potentials of the general form (19.2–4) have been used as a model to describe first-order phase transitions in other systems. (For a discussion of first-order phase transitions see, for example, Patashinskiĭ and Pokrovskiĭ, 1979, Chap. 8.) An example of the entropy given by Eq. (19.3–23) is shown in Fig. 19.16.

19.3.4 Latent heat of the phase transition

For large a_0 the discontinuity ΔS_1 of the entropy is given by the relation

$$\Delta S_1 \equiv S_1(a_0-, \Delta a) - S_1(a_0+, \Delta a) \approx \tfrac{1}{2}\Delta a + \frac{1}{(\xi - 1)a_0 + \frac{1}{2}(\xi + 1)\Delta a}. \quad (19.3\text{–}24)$$

If we multiply this expression by a_0, we obtain a quantity L that is the analog of the *latent heat* associated with the phase transition (Mandel, 1982), namely

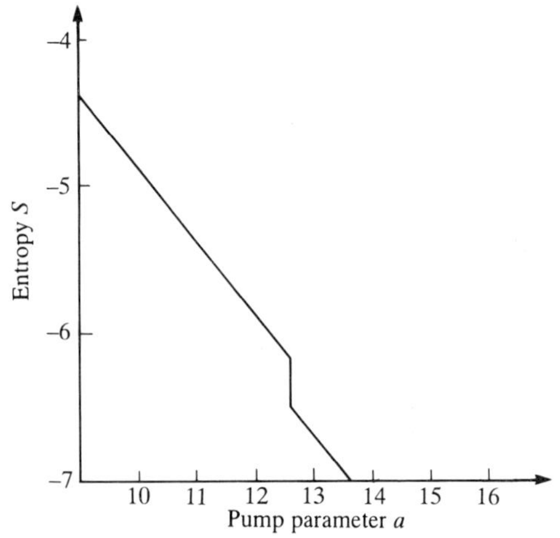

Fig. 19.16 The variation of entropy S given by Eq. (19.3–23) of a two-mode ring laser with pump parameter a, for $\Delta a = 1/2$. (Reproduced from Mandel, 1982.)

$$L = a_0[S_1(a_0-, \Delta a) - S_1(a_0+, \Delta a)] \approx \tfrac{1}{2}a_0\Delta a + \frac{a_0}{(\xi - 1)a_0 + \frac{1}{2}(\xi + 1)\Delta a}. \tag{19.3–25}$$

When $\xi = 2$, as for a homogeneously broadened ring laser, and with the help of the order-of-magnitude relation (19.3–20), we find the interesting, surprisingly general result that $L \sim 6$ in our dimensionless units.

The bistable form of the potential $U_1(I_1)$ when $\xi = 2$ is of course reflected in the double-peaked form of the probability distribution $\mathcal{P}_1(I_1)$ given by Eq. (19.2–11). Figure 19.17 shows a plot of $\mathcal{P}_1(I_1)$ for three different values of Δa with $a_1 = 15$ (Singh and Mandel, 1979). Each probability density consists of two almost disconnected branches, one which is peaked at $I_1 = 0$ and the other one near $I_1 = 15$. When $\Delta a \equiv a_1 - a_2 = 0$ the areas under each of these two branches are exactly equal, because it is equally probable that mode 1 is excited and mode 2 is off, and that mode 2 is excited while mode 1 is off. However, when the two pump parameters are not equal, one of the two states is more probable than the other.

The double-peaked form of $\mathcal{P}_1(I_1)$ for a ring laser with a homogeneously broadened gain medium (so that $\xi = 2$) has been demonstrated experimentally (Roy and Mandel, 1980; Lett, Christian, Singh and Mandel, 1981). In these measurements the light from one mode of a dye ring laser was allowed to fall on a photodetector, and the output of the photodetector was sampled electronically at regular intervals for a short time. The samples of varying amplitude were sorted out by a pulse height analyzer according to their amplitudes, and this yielded $\mathcal{P}_1(I_1)$. Figure 19.18 shows the results of one measurement together with the corresponding theoretical prediction given by Eq. (19.2–11). It will be seen that a large sharp peak occurs at $I_1 = 0$ and a broader one close to $I_1 = a + \frac{1}{2}\Delta a$, as the ring laser spontaneously switches from the clockwise to the counterclockwise

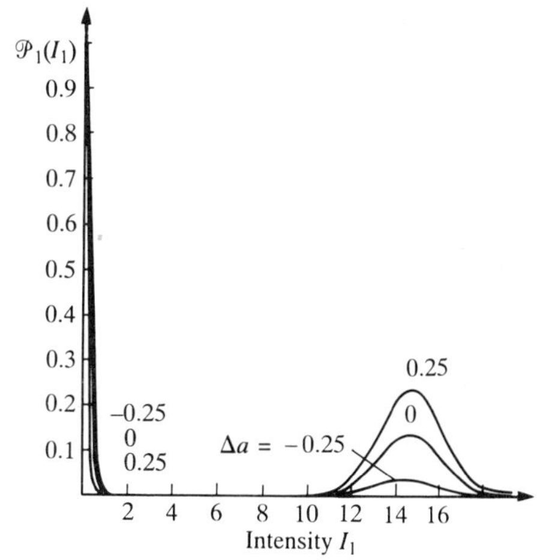

Fig. 19.17 Predicted form of the probability distribution $\mathcal{P}_1(I_1)$ of the intensity I_1 for one ring laser mode, with $a_1 = 15$, $\xi = 2$, for three different values of Δa. (Reproduced from Singh and Mandel, 1979.)

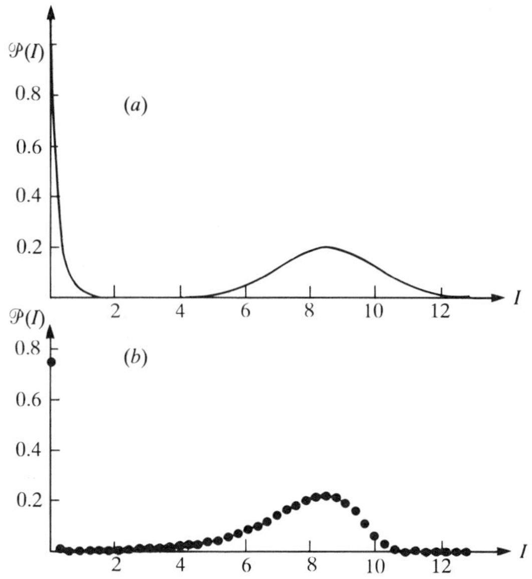

Fig. 19.18 Comparison between (*a*) the theoretical form and (*b*) the measured form of the probability distribution $\mathcal{P}(I)$. (Reproduced from Lett, Christian, Singh and Mandel, 1981.)

mode and back again. The measured distribution falls more rapidly from its peak values than expected. This may be connected with backscattering of light from one laser mode to the other. However, the existence of two highly probable metastable states is clearly confirmed.

19.4 Time-dependent solution when mode coupling $\xi = 1$

As the Fokker–Planck equation (19.1–17) describing the two-mode ring laser is a four-dimensional partial differential equation, its general solution presents more formidable problems than does the solution of the corresponding equation for the single-mode laser. However, in certain special cases the situation simplifies. In particular, when the mode coupling constant ξ is unity, and when there is symmetry so that $a_1 = a_2$, the two-mode laser problem can be solved as completely as that for the single-mode laser (cf. Section 18.6). For small asymmetries $|\Delta a| \ll a$ and $\xi = 1$, a perturbative argument makes it possible to express the solution in terms of the solutions for the symmetric laser (Hioe, Singh and Mandel, 1979). Alternatively, when the laser is operating well above threshold, it is possible to obtain solutions by making quadratic approximations to the minima of the steady-state potential (Hioe and Singh, 1981). In this section we shall illustrate the procedures for finding time-dependent solutions for the case $\xi = 1$ and $a_1 = a = a_2$, because the procedure closely parallels that used earlier in Section 18.6 for the single-mode laser. We shall follow the method described by M-Tehrani and Mandel (1978a).

19.4.1 Solution by separation of variables when $\xi = 1$, $a_1 = a_2$

If we put

$$p(\mathbf{x}, t) = f(\mathbf{x})k(t) \tag{19.4–1}$$

in the Fokker–Planck equation (19.1–17), use Eq. (19.1–20), and divide by $p(\mathbf{x}, t)$, we obtain the equation

$$\frac{1}{k(t)}\frac{\mathrm{d}k(t)}{\mathrm{d}t} = -\frac{1}{f(\mathbf{x})}\sum_{i=1}^{4}\left[\frac{\partial A_i f(\mathbf{x})}{\partial x_i} - \frac{\partial^2 f(\mathbf{x})}{\partial x_i^2}\right],$$

where $\mathbf{A}$ is the drift vector given by Eqs. (19.1–19). As the left-hand side of this equation depends only on t, while the right-hand side depends only on $\mathbf{x}$, each side must equal the same constant λ. This gives the two equations

$$\frac{\mathrm{d}k(t)}{\mathrm{d}t} = -\lambda k(t),$$

which has the solution

$$k(t) = k(0)\,\mathrm{e}^{-\lambda t}, \tag{19.4–2}$$

and

$$-\sum_{i=1}^{4}\left[\frac{\partial^2 f(\mathbf{x})}{\partial x_i^2} - \frac{\partial A_i f(\mathbf{x})}{\partial x_i}\right] = \lambda f(\mathbf{x}). \tag{19.4–3}$$

Because of the four dimensions, we must expect λ and f to consist of four-dimensional arrays and we label them by four suffices, λ_{lmnp} and $f_{lmnp}(\mathbf{x})$.

The general time-dependent solution of Eq. (19.1–17) is then of the form

$$p(\mathbf{x}, t) = \sum_{l,m,n,p} c_{lmnp} f_{lmnp}(\mathbf{x})\,\mathrm{e}^{-\lambda_{lmnp} t}, \tag{19.4–4}$$

with the coefficients c_{lmnp} to be determined by the initial conditions. As before, it is possible to convert the differential equation (19.4–3) into a Sturm–Liouville equation with the help of the transformation

$$\left.\begin{aligned} f_{lmnp}(\mathbf{x}) &= [p_{\mathrm{s}}(x)]^{1/2} g_{lmnp}(\mathbf{x}) \\ &= \sqrt{B}\,\mathrm{e}^{-U(x)/2} g_{lmnp}(\mathbf{x}), \end{aligned}\right\} \tag{19.4–5}$$

where $p_{\mathrm{s}}(\mathbf{x})$ is the steady-state solution (19.2–5). The function $g_{lmnp}(\mathbf{x})$ satisfies the self-adjoint equation

$$\sum_{i=1}^{4}\left[-\frac{\partial^2}{\partial x_i^2} + \frac{1}{4}A_i^2 + \tfrac{1}{2}\left(\frac{\partial A_i}{\partial x_i}\right)\right] g_{lmnp}(\mathbf{x}) = \lambda_{lmnp} g_{lmnp}(\mathbf{x}), \tag{19.4–6}$$

whose solutions are orthogonal and complete, and can be normalized so that

$$\int g^{*}_{lmnp}(\mathbf{x}) g_{l'm'n'p'}(\mathbf{x})\,\mathrm{d}^4x = \delta_{ll'}\delta_{mm'}\delta_{nn'}\delta_{pp'} \tag{19.4–7}$$

and

$$\sum_{l,m,n,p} g_{lmnp}(\mathbf{x}) g^*_{lmnp}(\mathbf{x}') = \delta^4(\mathbf{x} - \mathbf{x}'). \qquad (19.4\text{–}8)$$

Up to this point the procedure has been identical to that used in Chapter 18 to solve the equation for the single-mode laser.

In order to further simplify the partial differential equation, we now make a change of variables, first to polar coordinates by writing

$$\left.\begin{aligned} x_1 + \mathrm{i}x_2 &= \surd(\mathrm{I}_1)\,\mathrm{e}^{\mathrm{i}\phi_1}, \quad 0 \leqslant \phi_1 < 2\pi \\ x_3 + \mathrm{i}x_4 &= \surd(\mathrm{I}_2)\,\mathrm{e}^{\mathrm{i}\phi_2}, \quad 0 \leqslant \phi_2 < 2\pi, \end{aligned}\right\} \qquad (19.4\text{–}9)$$

and then to new variables

$$\left.\begin{aligned} u &= I_1 + I_2, \quad u \geqslant 0 \\ v &= (I_1 - I_2)/(I_1 + I_2), \quad -1 \leqslant v \leqslant 1 \end{aligned}\right\} \qquad (19.4\text{–}10\text{a})$$

by putting

$$\left.\begin{aligned} I_1 &= \tfrac{1}{2}u(1 + v) \\ I_2 &= \tfrac{1}{2}u(1 - v). \end{aligned}\right\} \qquad (19.4\text{–}10\text{b})$$

When the explicit form of **A**, given by Eq. (19.1–19), is substituted in Eq. (19.4–6) the function $g_{lmnp}(u, v, \phi_1, \phi_2)$ is found to satisfy the equation (M-Tehrani and Mandel, 1978a,b)

$$\left[u^2\frac{\partial^2}{\partial u^2} + 2u\frac{\partial}{\partial u} + \tfrac{1}{4}u(-2a + 3u - \tfrac{1}{4}(a - u)^2 u + \lambda_{lmnp}) + (1 - v^2)\frac{\partial^2}{\partial v^2} - 2v\frac{\partial}{\partial v}\right.$$
$$\left. + \frac{1}{2(1 + v)}\frac{\partial^2}{\partial \phi_1^2} + \frac{1}{2(1 - v)}\frac{\partial^2}{\partial \phi_2^2}\right] g_{lmnp} = 0. \qquad (19.4\text{–}11)$$

We note that the differential operators in u and v have separated, and this suggests that we try a further separation of variables in the form

$$g_{lmnp}(u, v, \phi_1, \phi_2) = S_{lnp}(v) R_{lmnp}(u) V_n(\phi_1) W_p(\phi_2), \qquad (19.4\text{–}12)$$

in which each of the four functions is assumed to be normalized to unity. Substitution of Eq. (19.4–12) in Eq. (19.4–11) immediately suggests that there are oscillatory solutions for V_n, W_p

$$\left.\begin{aligned} V_n(\phi_1) &= \frac{1}{\surd(2\pi)}\mathrm{e}^{-\mathrm{i}n\phi_1}, \quad n = 0, \pm 1, \pm 2, \ldots \\ W_p(\phi_2) &= \frac{1}{\surd(2\pi)}\mathrm{e}^{-\mathrm{i}p\phi_2}, \quad p = 0, \pm 1, \pm 2, \ldots \end{aligned}\right\} \qquad (19.4\text{–}13)$$

while the equations for $S_{lnp}(v)$ and $R_{lmnp}(u)$ become uncoupled and reduce to

$$\left[-(1 - v^2)\frac{\mathrm{d}^2}{\mathrm{d}v^2} + 2v\frac{\mathrm{d}}{\mathrm{d}v} + \frac{1}{2}\left(\frac{n^2}{1 + v} + \frac{p^2}{1 - v}\right)\right] S_{lnp}(v) = \beta_{lnp} S_{lnp}(v), \qquad (19.4\text{–}14)$$

$$\left[u^2\frac{d^2}{du^2} + 2u\frac{d}{du} + \frac{1}{4}u(-2a + 3u - \tfrac{1}{4}u(a-u)^2 + \lambda_{lmnp})\right]R_{lmnp}(u)$$
$$= \beta_{lnp}R_{lmnp}(u). \quad (19.4\text{–}15)$$

Here β_{lnp} is another eigenvalue or parameter to be determined. With the help of the transformation

$$S_{lnp} = \sqrt{(M_{lnp})}(1+v)^{n/2}(1-v)^{p/2}P_l^{(p,n)}(v), \quad (19.4\text{–}16)$$

we may show that the differential equation obeyed by $P_l^{(p,n)}(v)$ is

$$\left\{(1-v^2)\frac{d^2}{dv^2} + [n - p - (n+p+2)v]\frac{d}{dv} + \beta_{lnp} - \frac{1}{4}(n+p)(n+p+2)\right\}$$
$$\times P_l^{(p,n)}(v) = 0, \quad (19.4\text{–}17)$$

which is the equation satisfied by the Jacobi polynomials $P_l^{(p,n)}(v)$ (see, for example, Gradshteyn and Ryzhik (1980), p. 1036, 8.964) with

$$\beta_{lnp} = l(l+n+p+1) + \tfrac{1}{4}(n+p)(n+p+2), \quad l = 0, 1, 2, \ldots. \quad (19.4\text{–}18)$$

The functions $S_{lnp}(v)$ are then orthogonal with unit weight if the normalization constant M_{lnp} is chosen so that

$$S_{lnp}(v) = \left[\frac{(2l+n+p+1)}{2^{n+p+1}}\frac{l!(l+n+p)!}{(l+n)!(l+p)!}\right]^{1/2}(1+v)^{n/2}(1-v)^{p/2}P_l^{(p,n)}(v).$$
$$(19.4\text{–}19)$$

The Jacobi polynomials are usually defined for the non-negative integers n, p, but, with some restriction, they can be generalized to negative integer values of n, p. However, for the purpose of calculating the most important correlation functions of the laser field, we shall require only the combinations $n = 0 = p$, $n = 1$, $p = 0$, $n = 0$, $p = 1$, for which the Jacobi polynomials reduce to Legendre and associated Legendre functions.

Finally, we turn to the differential equation (19.4–15) obeyed by $R_{lmnp}(u)$. With the help of the transformations

$$\left.\begin{aligned} u &= y^2, \quad y \geqslant 0 \\ R_{lmnp}(y^2) &= 2y^{-3/2}\psi_{lmnp}(y), \end{aligned}\right\} \quad (19.4\text{–}20)$$

we may readily show that $\psi_{lmnp}(y)$ is the solution of the one-dimensional Schrödinger equation

$$-\frac{d^2\psi_{lmnp}(y)}{dy^2} + V_{lnp}(y)\psi_{lmnp}(y) = \lambda_{lmnp}\psi_{lmnp}(y), \quad m = 0, 1, 2, \ldots, \quad (19.4\text{–}21)$$

with the 'potential' $V_{lnp}(y)$ given by the relation

$$V_{lnp}(y) = [4l(l+n+p+1) + (n+p)(n+p+2) + \tfrac{3}{4}]/y^2 + 2a$$
$$+ (\tfrac{1}{4}a^2 - 3)y^2 - \tfrac{1}{2}ay^4 + \tfrac{1}{4}y^6, \quad (19.4\text{–}22)$$

and λ_{lmnp} is the corresponding eigenvalue. Although all three suffices l, n, p appear in Eq. (19.4–22), certain combinations of them lead to the same potential.

If we denote by L the combination

$$L \equiv 2l + n + p, \tag{19.4–23}$$

the potential is completely determined by L alone, and it can be written as

$$V_L(y) = [L(L+2) + \tfrac{3}{4}]/y^2 + 2a + (\tfrac{1}{4}a^2 - 3)y^2 - \tfrac{1}{2}ay^4 + \tfrac{1}{4}y^6. \tag{19.4–24}$$

We could therefore replace the three suffices l, n, p by L, and write $\psi_{Lm}(y)$, $R_{Lm}(y^2)$, λ_{Lm} for the eigenfunctions and eigenvalues. However, we will keep the longer notation for the sake of the functions $S_{lnp}(v)$, $V_n(\phi_1)$, $W_p(\phi_2)$.

With the solution of the Schrödinger equation (19.4–21), the problem of finding the general time-dependent solution of the Fokker–Planck equation is solved, in principle. It is interesting to note that the Schrödinger equation is again one-dimensional, as in the single-mode laser problem, so that the eigenfunctions may be taken as real, and that the potential is remarkably similar in the two cases [cf. Eq. (18.6–17)]. Unfortunately, it has not so far proved possible to solve the equation analytically, although numerical solutions have been obtained. Figure 19.19 shows graphs of some eigenvalues of the type λ_{0m00}, λ_{1m00}, λ_{1m10} as a function of a. It should be noted that λ_{1000} and λ_{1010}, unlike the other eigenvalues, tend to zero with increasing pump parameter a. As always, the lowest eigenvalue

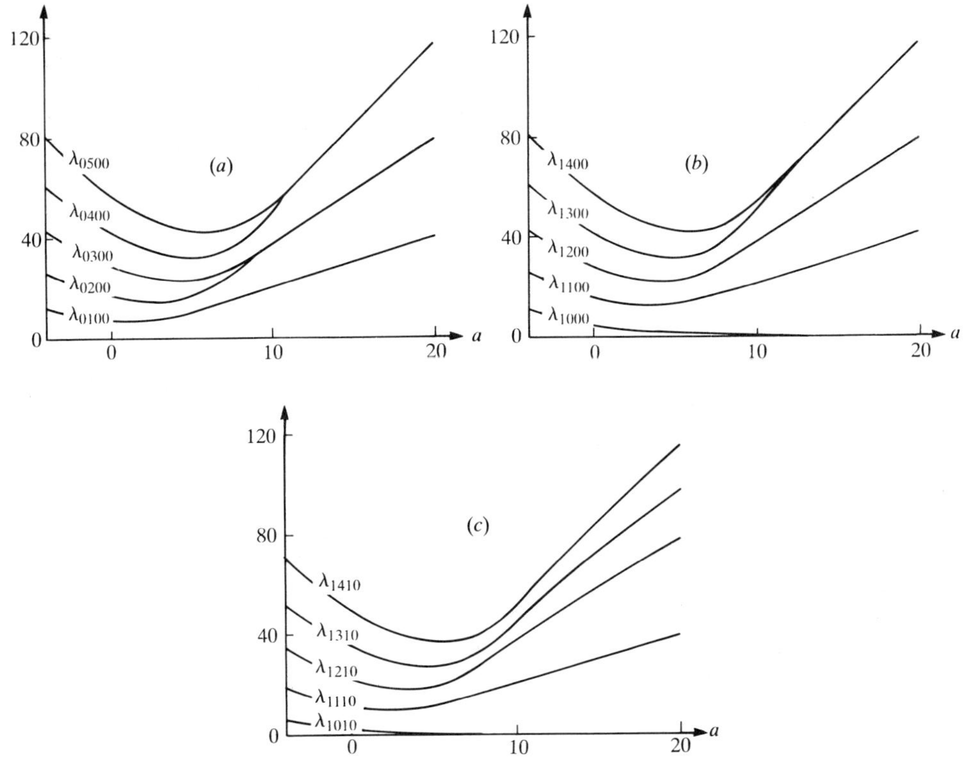

Fig. 19.19 Predicted variation of the eigenvalues (a) λ_{0m00}; (b) λ_{1m00}; (c) λ_{1m10} with pump parameter a for $\xi = 1$ and $\Delta a = 0$. (Reproduced from M-Tehrani and Mandel, 1978a.)

$$\lambda_{0000} = 0, \tag{19.4–25}$$

and the corresponding eigenfunction

$$R_{0000}(u) = 2\pi(2B)^{1/2} \exp\left(\tfrac{1}{4}au - \tfrac{1}{8}u^2\right), \tag{19.4–26}$$

must yield the steady-state solution of the Fokker–Planck equation. This conclusion is obvious from Eq. (19.4–4). The most general solution of the equation, expressed in the variables u, v, ϕ_1, ϕ_2, therefore takes the form

$$p(u, v, \phi_1, \phi_2) = \sum_{l,m,n,p} c_{lmnp}[p_s(u, v)]^{1/2} S_{lnp}(v) R_{lmnp}(u) \frac{1}{2\pi} e^{-in\phi_1 - ip\phi_2}. \tag{19.4–27}$$

From the transformations (19.4–9) and (19.4–10) it follows that the differential element

$$d^4x = \tfrac{1}{8}u \, du \, dv \, d\phi_1 \, d\phi_2, \tag{19.4–28}$$

and that the eigenfunctions $S_{lnp}(v)$ and $R_{lmnp}(u)$ are normalized so that

$$\left.\begin{aligned} \int_{-1}^{1} |S_{lnp}(v)|^2 \, dv &= 1 \\ \frac{1}{8}\int_0^\infty |R_{lmnp}(u)|^2 u \, du &= 1. \end{aligned}\right\} \tag{19.4–29}$$

19.4.2 *Green's function*

For the purpose of calculating various two-time correlation functions of the laser field we also need to know the Green function, or the conditional probability density $G(\mathbf{x}, t + \tau|\mathbf{x}_0, t)$ that $\mathbf{x}$ characterizes the state at time $t + \tau$ if it was $\mathbf{x}_0$ at time t. In the steady state $G(\mathbf{x}, t + \tau|\mathbf{x}_0, t)$ depends only on τ and not on t, and, being a probability density, it must satisfy the same Fokker–Planck equation as $p(\mathbf{x}, t)$. It follows that $G(\mathbf{x}, t + \tau|\mathbf{x}_0, t) = G(\mathbf{x}, \tau|\mathbf{x}_0, 0)$ must also be expressible in the form of Eq. (19.4–4), with the special restriction that it reduces to $\delta^4(\mathbf{x} - \mathbf{x}_0)$ when $\tau = 0$.

Inspection of Eq. (19.4–4) shows that the choice

$$c_{lmnp} = \frac{f^*_{lmnp}(\mathbf{x}_0)}{p_s(\mathbf{x}_0)} = \frac{g^*_{lmnp}(\mathbf{x}_0)}{[p_s(\mathbf{x}_0)]^{1/2}}$$

leads to the correct Green function

$$G(\mathbf{x}, \tau|\mathbf{x}_0, 0) = \sum_{l,m,n,p} \left(\frac{p_s(\mathbf{x})}{p_s(\mathbf{x}_0)}\right)^{1/2} g^*_{lmnp}(\mathbf{x}_0) g_{lmnp}(\mathbf{x}) \, e^{-\lambda_{lmnp}\tau}, \tag{19.4–30}$$

because it reduces to $\delta^4(\mathbf{x} - \mathbf{x}_0)$ when $\tau = 0$. The joint probability density of $\mathbf{x}_0$ at time t and $\mathbf{x}$ at time $t + \tau$ $(\tau \geqslant 0)$ in the steady state is therefore given by the formula

$$\begin{aligned} p(\mathbf{x}, t + \tau; \mathbf{x}_0, t) &= G(\mathbf{x}, \tau|\mathbf{x}_0, 0) p_s(\mathbf{x}_0) \\ &= [p_s(\mathbf{x}) p_s(\mathbf{x}_0)]^{1/2} \sum_{l,m,n,p} g^*_{lmnp}(\mathbf{x}_0) g_{lmnp}(\mathbf{x}) \, e^{-\lambda_{lmnp}\tau}. \end{aligned} \tag{19.4–31}$$

In terms of the new variables u, v, ϕ_1, ϕ_2 this equation becomes

$$p(u, v, \phi_1, \phi_2, t+\tau; u_0, v_0, \phi_{10}, \phi_{20}, t) = B\exp\left(\tfrac{1}{4}au - \tfrac{1}{8}u^2 + \tfrac{1}{4}au_0 - \tfrac{1}{8}u_0^2\right)$$

$$\times \sum_{l,m,n,p} S^*_{lnp}(v_0)S_{lnp}(v)R^*_{lmnp}(u_0)R_{lmnp}(u)\frac{1}{(2\pi)^2}\,\mathrm{e}^{\mathrm{i}n(\phi_{10}-\phi_1)}\,\mathrm{e}^{\mathrm{i}p(\phi_{20}-\phi_2)}\,\mathrm{e}^{-\lambda_{lmnp}\tau},$$

$$(\tau \geqslant 0). \quad (19.4\text{–}32)$$

19.4.3 *Correlation functions*

The calculation of any two-time correlation function is now straightforward, and we have immediately from Eqs. (19.4–10), with $s, s' = 1, 2$,

$$\langle I_{s'}(t)I_s(t+\tau)\rangle = \int_0^\infty \mathrm{d}u_0\,\mathrm{d}u\frac{1}{64}u_0u\int_{-1}^1 \mathrm{d}v_0\,\mathrm{d}v\frac{1}{4}u_0u[1-(-1)^{s'}v_0][1-(-1)^s v]$$

$$\times\int_0^{2\pi}\mathrm{d}\phi_1\,\mathrm{d}\phi_2\,\mathrm{d}\phi_{10}\,\mathrm{d}\phi_{20}\,p(u, v, \phi_1, \phi_2, t+\tau; u_0, v_0, \phi_{10}, \phi_{20}, t)$$

$$= \frac{B}{64}\int_0^\infty \mathrm{d}u_0\,\mathrm{d}u\int_{-1}^1 \mathrm{d}v_0\,\mathrm{d}v\exp\left(\tfrac{1}{4}au - \tfrac{1}{8}u^2 + \tfrac{1}{4}au_0 - \tfrac{1}{8}u_0^2\right)$$

$$\times\tfrac{1}{4}u_0^2u^2[1-(-1)^{s'}v_0][1-(-1)^s v]$$

$$\times\sum_{l,m,n,p}R^*_{lmnp}(u_0)R_{lmnp}(u)S^*_{lnp}(v_0)S_{lnp}(v)$$

$$\times\frac{1}{(2\pi)^2}\int_0^{2\pi}\mathrm{d}\phi_{10}\,\mathrm{d}\phi_{20}\,\mathrm{d}\phi_1\,\mathrm{d}\phi_2\,\mathrm{e}^{\mathrm{i}n(\phi_{10}-\phi_1)}\,\mathrm{e}^{\mathrm{i}p(\phi_{20}-\phi_2)}\,\mathrm{e}^{-\lambda_{lmnp}\tau}$$

$$= \frac{B\pi^2}{64}\sum_{l=0}^\infty\sum_{m=0}^\infty\left|\int_0^\infty \mathrm{d}u\,u^2\exp\left(\tfrac{1}{4}\mathrm{au} - \tfrac{1}{8}\mathrm{u}^2\right)R_{lm00}(u)\right|^2$$

$$\times\int_{-1}^1 \mathrm{d}v_0\,\mathrm{d}v[1-(-1)^{s'}v_0][1-(-1)^s v]S_{l00}(v_0)S_{l00}(v)\,\mathrm{e}^{-\lambda_{lm00}\tau}.$$

Using the properties of the Jacobi polynomials one can show that

$$\int_{-1}^1[1-(-1)^s v]S_{l00}(v)\,\mathrm{d}v = \frac{1}{(l+\frac{1}{2})^{1/2}}[\delta_{l0} - \tfrac{2}{3}(-1)^s\delta_{l1}],$$

so that finally,

$$\langle I_{s'}(t)I_s(t+\tau)\rangle = \frac{\pi^2 B}{32}\sum_{m=0}^\infty\left\{\left|\int_0^\infty u^2\exp\left(\tfrac{1}{4}au - \tfrac{1}{8}u^2\right)R_{0m00}(u)\,\mathrm{d}u\right|^2\mathrm{e}^{-\lambda_{0m00}\tau}\right.$$

$$\left.+\frac{(-1)^{s+s'}}{3}\left|\int_0^\infty u^2\exp\left(\tfrac{1}{4}\mathrm{au} - \tfrac{1}{8}\mathrm{u}^2\right)R_{1m00}(u)\,\mathrm{d}u\right|^2\mathrm{e}^{-\lambda_{1m00}\tau}\right\},\quad \tau\geqslant 0.$$

$$(19.4\text{–}33)$$

Some examples of intensity correlation functions calculated from this formula are given in Fig. 19.20. In the threshold region the autocorrelations are very nearly exponential in form, but contributions from additional exponential terms become noticeable above threshold. Because the eigenvalue λ_{1000} tends to zero with increasing a, as shown in Fig. 19.19, the correlation functions tend to be dominated by the $\exp(-\lambda_{1000}\tau)$ term for large τ, and the correlation times increase steadily with pump parameter above threshold. In this respect the two-mode ring laser is quite different from the single-mode laser.

Figure 19.21 shows the results of some measurements of the decay rate of the intensity correlation function in a ring laser with $\xi = 1$, when there was a small asymmetry $\Delta a = 0.1$ between the two modes. The average decay time $\bar{T}$ defined by

$$\bar{T} = \sum_{l=0}^{1}\sum_{m=0}^{\infty}\frac{c_{lm}}{\lambda_{lm00}}\Bigg/\sum_{l=0}^{1}\sum_{m=0}^{\infty}c_{lm} \qquad (19.4\text{–}34)$$

is plotted as a function of pump parameter a, with the coefficients c_{lm} given by the corresponding coefficients of $\exp(-\lambda_{lm00}\tau)$ in Eq. (19.4–33). Because of the asymmetry, a small difference between the decay constants for the two modes is to be expected, although it is really too small to be resolved experimentally. It should be noted that the observed behavior is generally of the form suggested by the eigenvalue λ_{1000} [see Fig. 19.19(b)], over a range of pump parameters from -12 to 2. Above $a = 2$ there are increasing departures from the theoretical predictions, that are believed to be due to backscattering of light from one mode into the other.

We can calculate second-order field correlations of the type $\langle \mathscr{E}^*(t)\mathscr{E}(t+\tau)\rangle$ in a similar manner. Because of the statistical independence of the two phases, $\langle \mathscr{E}_1^*(t)\mathscr{E}_2(t+\tau)\rangle = 0$, and by symmetry

$$\langle \mathscr{E}_1^*(t)\mathscr{E}_1(t+\tau)\rangle = \langle \mathscr{E}_2^*(t)\mathscr{E}_2(t+\tau)\rangle. \qquad (19.4\text{–}35)$$

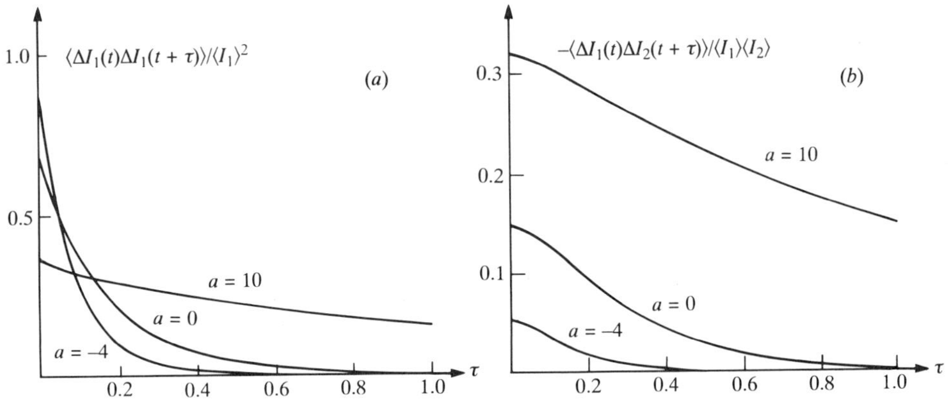

Fig. 19.20 Normalized intensity correlation functions calculated from Eq. (19.4–33) for various pump parameters a; (a) autocorrelation; (b) cross-correlation. (Reproduced from M-Tehrani and Mandel, 1978a.)

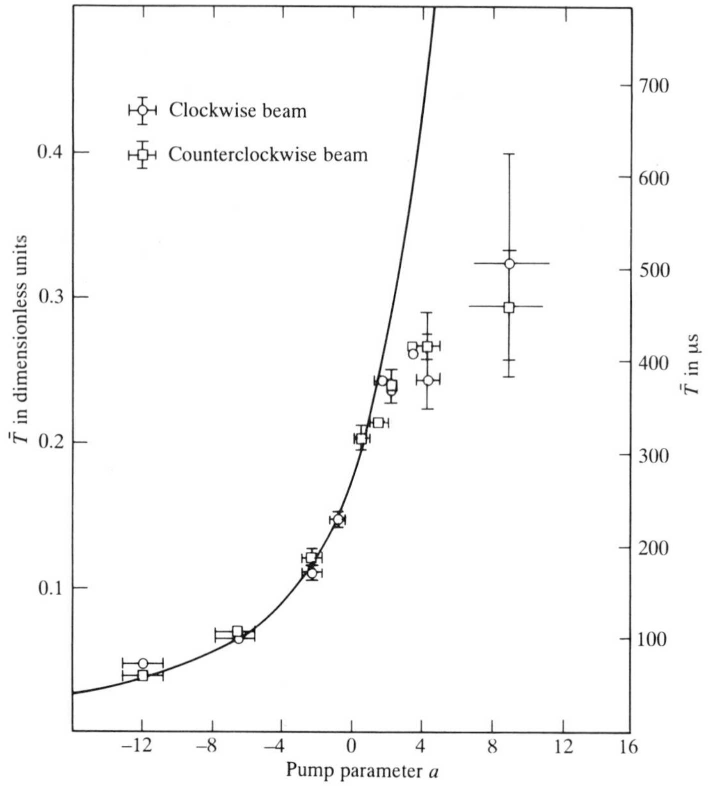

Fig. 19.21 Comparison between theory and experiment for the variation of the average intensity autocorrelation time $\bar{T}$ of a ring laser with pump parameter a. The full curve is theoretical and calculated from Eq. (19.4–33). (Reproduced from Singh and Mandel, 1981.)

Now

$$
\begin{aligned}
\langle \mathscr{E}_1^*(t)\mathscr{E}_1(t+\tau)\rangle &= \int_0^\infty \mathrm{d}u_0\,\mathrm{d}u \frac{1}{64} u_0 u \int_{-1}^1 \mathrm{d}v_0\,\mathrm{d}v [\tfrac{1}{4}u_0 u(1+v_0)(1+v)]^{1/2} \\
&\quad \times \int_0^{2\pi} \mathrm{d}\phi_{10}\,\mathrm{d}\phi_{20}\,\mathrm{d}\phi_1\,\mathrm{d}\phi_2 \\
&\quad \times \mathrm{e}^{\mathrm{i}(\phi_1-\phi_{10})} p(u, v, \phi_1, \phi_2, t+\tau; u_0, v_0, \phi_{10}, \phi_{20}, t) \\
&= \frac{B}{64}\int_0^\infty \mathrm{d}u_0\,\mathrm{d}u \int_{-1}^1 \mathrm{d}v_0\,\mathrm{d}v \exp(\tfrac{1}{4}au - \tfrac{1}{8}u^2 + \tfrac{1}{4}au_0 - \tfrac{1}{8}u_0^2) \\
&\quad \times [\tfrac{1}{4}u^3 u_0^3(1+v_0)(1+v)]^{1/2} \\
&\quad \times \sum_{l,m,n,p} R^*_{lmnp}(u_0) R_{lmnp}(u) S^*_{lnp}(v_0) S_{lnp}(v) \\
&\quad \times \frac{1}{(2\pi)^2}\int_0^{2\pi} \mathrm{d}\phi_{10}\,\mathrm{d}\phi_{20}\,\mathrm{d}\phi_1\,\mathrm{d}\phi_2\, \mathrm{e}^{\mathrm{i}(n-1)(\phi_{10}-\phi_1)}\, \mathrm{e}^{\mathrm{i}p(\phi_{20}-\phi_2)}\, \mathrm{e}^{-\lambda_{lmnp}\tau}
\end{aligned}
$$

$$= \frac{\pi^2 B}{32} \sum_{l=0}^{\infty} \sum_{m=0}^{\infty} \left| \int_0^{\infty} u^{3/2} \exp\left(\tfrac{1}{4}au - \tfrac{1}{8}u^2\right) R_{lm10}(u)\, du \right|^2$$

$$\times \left| \int_{-1}^{1} (1+v)^{1/2} S_{l10}(v)\, dv \right|^2 e^{-\lambda_{lm10}\tau}.$$

From the properties of the Jacobi polynomials it follows that

$$\int_{-1}^{1} (1+v)^{1/2} S_{l10}(v)\, dv = (\sqrt{2})\delta_{l1},$$

so that finally

$$\langle \mathscr{E}_1^*(t)\mathscr{E}_1(t+\tau)\rangle\, e^{-i\omega_0\tau} =$$

$$\frac{\pi^2 B}{16} \sum_{m=0}^{\infty} \left| \int_0^{\infty} u^{3/2} \exp\left(\tfrac{1}{4}au - \tfrac{1}{8}u^2\right) R_{1m10}(u)\, du \right|^2 e^{-\lambda_{1m10}\tau}\, e^{-i\omega_0\tau}, \quad \tau \geqslant 0. \quad (19.4\text{–}36)$$

The spectral density of the laser field is therefore a sum of Lorentzian functions of various widths centered on frequency ω_0. However, because the eigenvalue λ_{1010}, unlike the others, falls towards zero with increasing a, we expect this eigenvalue to dominate the behavior of the correlation function as time τ increases. Figure 19.22 shows some examples of second-order correlations calculated from Eq. (19.4–35) for various values of a. They are seen to be nearly exponential in form, so that the spectral density should be almost Lorentzian.

19.5 Time-dependent behavior in the more general case of mode coupling $\xi \neq 1$

Up to now, in dealing with the time-dependent solution of the laser problem, we have confined our attention to the case in which $\xi = 1$ and there is symmetry $a_1 = a_2$. This corresponds to a symmetric, inhomogeneously broadened ring laser on resonance. We shall now briefly consider the more general case, in which ξ can be larger or smaller than unity and there is no symmetry. Although the self-adjoint differential equation Eq. (19.4–6) still holds, the separate equations

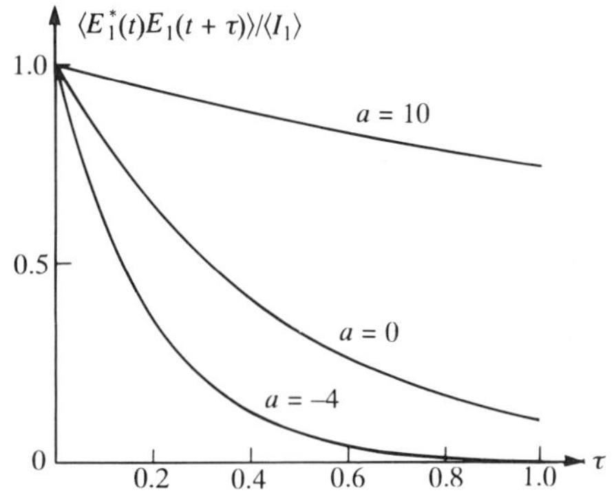

Fig. 19.22 Predicted form of the normalized second-order correlation function of the complex mode amplitude $\mathscr{E}_1(t)$ of a ring laser, calculated from Eq. (19.4–36), for various pump parameters a with $\Delta a = 0$. (Reproduced from M-Tehrani and Mandel, 1978a.)

(19.4–14) and (19.4–15) derived from it are no longer valid. However, with the introduction of polar coordinates for the two complex mode amplitudes,

$$x_1 = r_1 \cos\phi_1, \quad x_2 = r_1 \sin\phi_1$$

$$x_3 = r_2 \cos\phi_2, \quad x_4 = r_2 \sin\phi_2,$$

and with the assumed factorization of $g_{lmnp}(\mathbf{x})$, as before, in the form

$$g_{lmnp}(\mathbf{x}) = \chi_{lmnp}(r_1, r_2)\frac{1}{\sqrt{(2\pi)}}\,\mathrm{e}^{-\mathrm{i}n\phi_1}\frac{1}{\sqrt{(2\pi)}}\,\mathrm{e}^{-\mathrm{i}p\phi_2}, \tag{19.5–1}$$

the radial part of the solution can be shown to satisfy the equation (Hioe and Singh, 1981)

$$\left[\left(-\frac{1}{r_1}\frac{\partial}{\partial r_1}r_1\frac{\partial}{\partial r_1} - \frac{1}{r_2}\frac{\partial}{\partial r_2}r_2\frac{\partial}{\partial r_2}\right) + \frac{n^2}{r_1^2} + \frac{p^2}{r_2^2} + V(r_1, r_2)\right]\chi_{lmnp}(r_1, r_2) = \lambda_{lmnp}\chi_{lmnp}(r_1, r_2). \tag{19.5–2}$$

The 'potential' $V(r_1, r_2)$ is given by the expression

$$V(r_1, r_2) = a_1 + a_2 - (2 + \xi)(r_1^2 + r_2^2) + \tfrac{1}{4}r_1^2(a_1 - r_1^2 - \xi r_2^2)^2 + \tfrac{1}{4}r_2^2(a_2 - r_2^2 - \xi r_1^2)^2. \tag{19.5–3}$$

This again has the character of a Schrödinger-type equation, although it is in two variables. As usual, the possible solutions of the equation are dominated by the behavior of the potential $V(r_1, r_2)$ near its minima. If the potential can be well approximated by a quadratic function in the neighborhood of each minimum, then analytic solutions can be found, as has been shown by Hioe and Singh (1981). They succeeded in obtaining approximate analytic expressions both for the eigenvalues and the eigenfunctions of Eq. (19.5–2) that hold well below and well above threshold. We shall not investigate the details of their solution here.

19.5.1 Mode switching and first passage times

The most interesting features of the solution are encountered when the coupling constant $\xi > 1$, in particular when $\xi = 2$, as for a homogeneously broadened two-mode laser. In that case, provided $\xi a_2 > a_1$ and $\xi a_1 > a_2$, the Schrödinger potential $V(r_1, r_2)$ exhibits four minima, two of which are associated with zero eigenvalues, and therefore with the steady-state solution. They lead to double-peaked steady-state probability distributions $\mathcal{P}_1(I_1)$ and $\mathcal{P}_2(I_2)$ for each mode intensity, as we have already seen in Section 19.2. Each peak corresponds to a metastable state, and from time to time the laser switches from one metastable state, in which one mode intensity is zero, to the other one in which the other mode intensity vanishes. The rate of switching is determined by the lowest two non-zero eigenvalues λ_{1000} and λ_{0100}, which are much smaller than the other ones. Whereas the higher eigenvalues govern the approach of the system to one or the other local, metastable equilibrium state, the lowest two non-zero eigenvalues govern the rate of transition or switching between these metastable states. Actually, when there is symmetry, $\Delta a = 0$, all the higher eigenvalues are split by the same amount that separates the lowest two eigenvalues. The same splitting of energy levels is well known in quantum mechanics in the presence of a double potential well. One finds that for $a_1 = a_2 = a$

$$\lambda_{1000} = \lambda_{0100} = \frac{a^2[(1+\xi)^{1/2} - 1]}{[\pi(1+\xi)]^{1/2}} \exp\left[-\frac{a^2[(1+\xi)^{1/2} - 1]^2}{1+\xi}\right], \quad (19.5\text{–}4)$$

and this eigenvalue governs the rate at which the laser switches between its metastable states.

Such mode switching in homogeneously broadened two-mode lasers has been observed experimentally (Rigrod and Bridges, 1965). Figure 19.23 shows an oscilloscope photograph of the time development of the light intensities for the two modes of a dye ring laser. It will be seen that the light intensity of each mode switches on and off randomly and that, whenever one mode turns on, the other one turns off, and vice versa. Clearly the two mode intensities are strongly anticorrelated, as was already suggested by the curves in Fig. 19.8.

We can estimate the rate of mode switching in another way through the formalism of the first passage time in phase space (see for example Stratonovich, 1963; see also Weiss, 1977) which leads to values similar to those given by Eq. (19.5–4) (Singh and Mandel, 1979; Lenstra and Singh, 1983; Shenoy and Agarwal, 1984). From the form of the steady-state solution, as illustrated by the potential $U(I_1, I_2)$ in Fig. 19.12, it is apparent that the representative point in phase space characterizing the instantaneous state spends most of its time near one or the other potential minimum, either at $I_1 = a_1$, $I_2 = 0$ or at $I_1 = 0$, $I_2 = a_2$. However, from time to time a particularly large fluctuation causes the representative point to make a transition from the neighborhood of one minimum to the other, and this is manifest in the switching exhibited in Fig. 19.23. Although the exact path between the minima in phase space is not predictable, it is most likely to follow the valley in $U(I_1, I_2)$ and to pass through the saddle point. We may therefore look on the motion in phase space as one-dimensional to a first approximation. This allows us to obtain a rough estimate of the average time taken to move from one metastable state to the other with the help of the formalism for first passage times in one dimension.

In order to understand this, let us focus on the curve labeled $\xi = 2$ in Fig. 19.13, which gives the form of the potential $U(I_1)$ associated with the steady-state

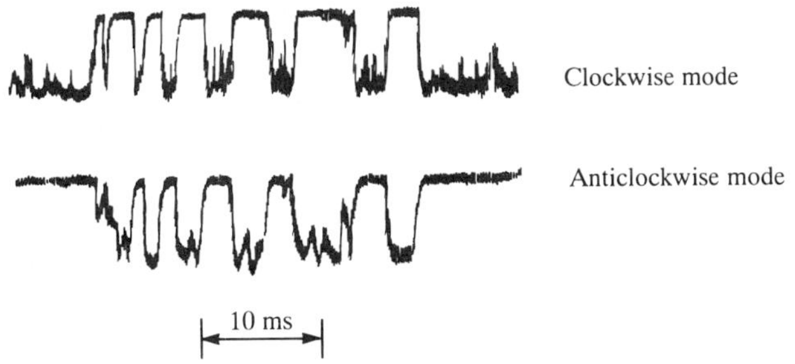

Fig. 19.23 Sketch of an oscilloscope photograph showing simultaneous variations of the two mode intensities of a dye ring laser. (After Mandel, Roy and Singh, 1981.)

solution for one mode of the laser. If the representative point lies near the minimum at $I_1 = 0$, so that this mode intensity is off, we can deem a first passage to the on-state to have occurred when the representative point first crosses the maximum of the potential, after which it tends to drop quickly into the valley at $I_1 = a_1$.

More generally, let us suppose that the value $I = I_A$ divides the one-dimensional phase space into two regions Ω and $\bar{\Omega}$, and that a first passage occurs when the representative point, starting at $I_0 \in \Omega$ at time $t = 0$, first leaves Ω and crosses I_A. Let T be the time taken for the first passage to occur. In general, T is a random variable with some probability distribution $\eta(T|I_0)$ that depends on I_0. We will now show that the integral of $\eta(T|I_0)$ is connected with the Green function $G(I, t|I_0, 0)$ of the one-dimensional random process $I(t)$, and that it obeys the backward Fokker–Planck or Kolmogorov equation.

In order to demonstrate this we introduce the integral first passage time distribution

$$\Phi(T|I_0) = \int_0^T \eta(T'|I_0)\,\mathrm{d}T', \tag{19.5–5}$$

giving the probability that the first passage time is less that T, and we observe that, by definition, $\Phi(T|I_0)$ obeys the composition law

$$\Phi(T + \Delta T|I_0) = \int_\Omega G(I', \Delta T|I_0, 0)\Phi(T|I')\,\mathrm{d}I'. \tag{19.5–6}$$

This merely expresses the fact that $\Phi(T + \Delta T|I_0)$ is the probability of going from I_0 to a neighboring point I' in some short interval ΔT, multiplied by the probability that the first passage time from I' does not exceed T, integrated over all I' within Ω. We now make a change of variable by putting $I' - I_0 = \Delta I$ and expand $\Phi(T|I_0 + \Delta I)$ in a Taylor series in ΔI about I_0. After subtracting $\Phi(T|I_0)$ from both sides of the equation, dividing by ΔT and going to the limit $\Delta T \to 0$, we arrive at the equation

$$\frac{\partial \Phi(T|I)}{\partial T} = A(I)\frac{\partial \Phi(T|I)}{\partial I} + \tfrac{1}{2}D(I)\frac{\partial^2 \Phi(T|I)}{\partial I^2}, \tag{19.5–7}$$

in which $A(I)$ and $D(I)$ are the usual drift and diffusion coefficients, defined by the formulas

$$A(I) = \operatorname*{Lim}_{\Delta T \to 0} \int \Delta I G(I + \Delta I, \Delta T|I, 0)\,\mathrm{d}(\Delta I)$$

$$D(I) = \operatorname*{Lim}_{\Delta T \to 0} \int (\Delta I)^2 G(I + \Delta I, \Delta T|I, 0)\,\mathrm{d}(\Delta I).$$

As usual, we assume that the higher transition moments vanish. The differential equation is the adjoint of the usual Fokker–Planck equation, and it is known as the *backward Fokker–Planck* or *Kolmogorov* equation. As $\partial \Phi(T|I)/\partial T$ is the probability distribution $\eta(T|I)$, we can obtain an equation for the average first passage time

$$\langle T(I) \rangle \equiv \mu_1(I) = \int_0^\infty T\eta(T|I)\,\mathrm{d}T$$

by differentiating each term in Eq. (19.5–7) with respect to T, multiplying by T and integrating over all T. We then find

$$\int_0^\infty T \frac{\partial^2 \Phi(T|I)}{\partial T^2} \mathrm{d}T = A(I)\frac{\mathrm{d}\mu_1(I)}{\mathrm{d}I} + \tfrac{1}{2}D(I)\frac{\mathrm{d}^2\mu_1(I)}{\mathrm{d}I^2}.$$

After integration by parts the left-hand side reduces to

$$-\int_0^\infty \frac{\partial \Phi(T|I)}{\partial T} \mathrm{d}T = -1,$$

so that finally

$$-1 = A(I)\frac{\mathrm{d}\mu_1(I)}{\mathrm{d}I} + \tfrac{1}{2}D(I)\frac{\mathrm{d}^2\mu_1(I)}{\mathrm{d}I^2}. \tag{19.5–8}$$

This is an ordinary differential equation for $\mu_1(I)$, which can be integrated. The general solution has the form

$$\mu_1(I) = -2\int_0^I \mathrm{d}I'\, \mathrm{e}^{U(I')}\int_0^{I'} \mathrm{d}I'' \frac{\mathrm{e}^{-U(I'')}}{D(I'')} + C_1\int_0^I \mathrm{e}^{U(I')}\,\mathrm{d}I' + C_2, \tag{19.5–9}$$

where C_1, C_2 are constants and

$$U(I) = -2\int_0^I \frac{A(I')}{D(I')}\mathrm{d}I' \tag{19.5–10}$$

is the same potential that enters into the steady-state solution $\mathcal{P}(I)$ for one laser mode intensity I [see Eqs. (19.3–16)]. By integrating the two-mode Fokker–Planck equation (19.1–17) over all variables except I_1, we can readily show (Singh and Mandel, 1979) that the one-dimensional diffusion constant $D(I_1) = 8I_1$. The boundary conditions

$$\mu_1(I_\mathrm{A}) = 0,$$

and

$$\mu_1(0) \geqslant \mu_1(I), \quad \text{if } I < I_\mathrm{A}$$

$$\frac{\mathrm{d}\mu_1(I)}{\mathrm{d}I} \to 0 \text{ as } I \to \infty, \quad \text{if } I > I_\mathrm{A},$$

allow the constants C_1, C_2 in Eq. (19.5–9) to be evaluated. We finally obtain for the average first passage time $\langle T_\mathrm{off}\rangle$ for passing from the off-state $I_1 = 0$ to the on-state the expression

$$\langle T_\mathrm{off}\rangle = \int_0^{I_\mathrm{A}} \mathrm{d}I' \frac{2}{\mathcal{P}(I')D(I')}\int_0^{I'} \mathrm{d}I''\mathcal{P}(I''). \tag{19.5–11}$$

Similarly, if the system is initially in the on-state with $I_1 \sim a_1$, the average first passage time is given by

$$\langle T_\mathrm{on}\rangle = \int_{I_\mathrm{A}}^{a_1} \mathrm{d}I' \frac{2}{\mathcal{P}(I')D(I')}\int_{I'}^{\infty} \mathrm{d}I''\mathcal{P}(I''). \tag{19.5–12}$$

The evaluation of the integrals is very much simplified by the double-peaked form of $\mathscr{P}(I)$ above threshold. Within the domain $I < I_A$, $\mathscr{P}(I)$ is appreciable only near $I = 0$, and in the domain $I > I_A$, $\mathscr{P}(I)$ is appreciable only near $I = a_1$. In both cases $\mathscr{P}(I)$ is very small close to the boundary $I = I_A$. It follows that the dominant contribution to the I'-integrals in Eqs. (19.5–11) and (19.5–12) comes from values of I' close to I_A, in which case the I''-integrals are almost independent of I', and may be replaced by

$$\left.\begin{aligned} \int_0^{I_A} \mathscr{P}(I'')\,\mathrm{d}I'' &\equiv P_L \\ \int_{I_A}^{\infty} \mathscr{P}(I'')\,\mathrm{d}I'' &\equiv P_H. \end{aligned}\right\} \tag{19.5–13}$$

Here P_L, P_H are the probabilities of finding the light intensity 'low' or 'high', with

$$P_L + P_H = 1.$$

By making use of Eqs. (19.3–16) for $\mathscr{P}(I)$, and introducing the asymptotic form for large argument of the error function, we can show that

$$P_{\substack{H\\L}} \sim \left[1 + \mathrm{e}^{\mp a\Delta a/2} \frac{(\xi - 1)a \pm \frac{1}{2}(\xi + 1)\Delta a}{(\xi - 1)a \mp \frac{1}{2}(\xi + 1)\Delta a}\right]^{-1}, \tag{19.5–14a}$$

and for small asymmetries

$$P_{\substack{H\\L}} \approx \{1 + \exp[\mp\tfrac{1}{2}a\Delta a \pm (\xi + 1)\Delta a/(\xi - 1)a]\}^{-1}, \tag{19.5–14b}$$

where $a \equiv \frac{1}{2}(a_1 + a_2)$, $\Delta a \equiv a_1 - a_2$. The I'-integrals in Eqs. (19.5–11) and (19.5–12) can now be evaluated approximately by noting that near $I' = I_A$, $1/\mathscr{P}(I')$ behaves almost as a Gaussian distribution in I' centered at $I' = I_A$. If we also approximate $1/D(I')$ by its value $1/D(I_A) = 1/8I_A$ at the Gaussian peak, the I'-integrals can be written down at once. We finally arrive at the result (Roy, Short, Durnin and Mandel, 1980)

$$\langle T_{\text{off}} \rangle \approx \frac{(\sqrt{\pi}) \exp[\frac{1}{4}(\xi^2 - 1)I_A^2]}{2I_A^2(\xi^2 - 1)^{3/2}}, \tag{19.5–15}$$

in which I_A is the position of the maximum of the potential $U(I)$. By differentiation of Eq. (19.3–16b) for the potential one can show that

$$I_A \approx \frac{a}{\xi + 1} - \frac{\Delta a/2}{\xi - 1}. \tag{19.5–16}$$

The expression for $\langle T_{\text{on}} \rangle$ is similar, but with Δa replaced by $-\Delta a$. In the special case $\xi = 2$, $\Delta a = 0$, which holds for a homogeneously broadened, symmetric ring laser, we obtain approximately

$$\langle T_{\text{off}} \rangle = \langle T_{\text{on}} \rangle \approx \frac{1.5}{a^2} \mathrm{e}^{a^2/12}. \tag{19.5–17}$$

This expression can be compared with the answer given by Eq. (19.5–4) for the lowest non-zero eigenvalue when $\xi = 2$,

$$\frac{1}{\lambda_{1000}} \approx \frac{4}{a^2} e^{a^2/5.6}.$$

The two expressions have a similar structure, although they differ in detail.

It has been pointed out that the average time given by Eq. (19.5–17) should really be doubled (Lenstra and Singh, 1983), because once the representative point reaches the boundary, it is about as likely to fall back into the first potentital well as to move on to the other well. Moreover, we have arrived at our answer by considering the time development of one mode intensity only, whereas the representative point is really moving on the three-dimensional surface of Fig. 19.12. A more realistic treatment of the first passage time problem (Shenoy and Agarwal, 1984) leads to a value that is six times as large as that given by Eq. (19.5–17).

Because of the exponential factor in Eq. (19.5–17) the average first passage time tends to become very long when the pump parameter is made large. The full curve in Fig. 19.24 is a plot of $\langle T_{\text{off}} \rangle$ as a function of a with $\Delta a \ll 1$. It will be seen that $\langle T_{\text{off}} \rangle$ increases by four orders of magnitude as a increases modestly from the value 4 to 12. The reason is that the height of the potential maximum at I_A becomes larger and larger as a increases, so that ever larger fluctuations are needed to cause a transition from one metastable state to the other. Needless to say, the larger the fluctuations the less frequently do they occur, and at sufficiently high pump parameters $\langle T_{\text{on}} \rangle$ and $\langle T_{\text{off}} \rangle$ become so long that the laser may appear to be stable. These switching or first passage times have been measured for a homogeneously broadened two-mode dye laser (Roy, Short, Durnin and Mandel, 1980), and as the circles in Fig. 19.24 show, after scaling, reasonable agreement with the theory was obtained.

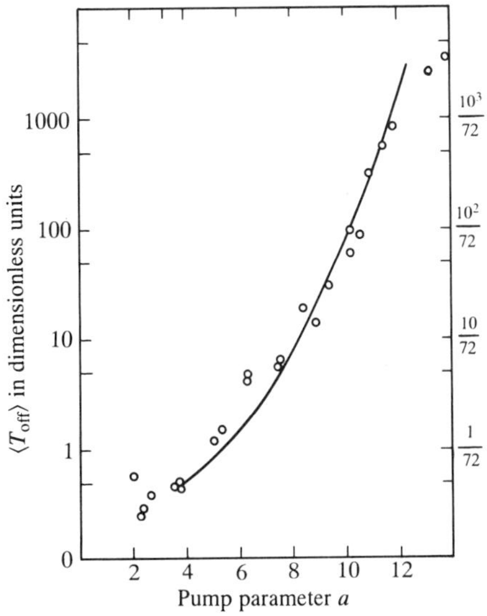

Fig. 19.24 Comparison of theoretical and experimental values of the mean first passage time $\langle T_{\text{off}} \rangle$ as a function of pump parameter a. The scale on the right gives the measured time in seconds. (Reproduced from Roy, Short, Durnin and Mandel, 1980.)

19.5.2 First passage time distributions

The foregoing formalism also allows higher-order moments of the first passage times T_{off} and T_{on} to be calculated. By multiplying each term in Eq. (19.5–7) by T^r ($r = 2, 3, \ldots$), and proceeding as in the derivation of $\mu_1(I)$, we can readily obtain a differential equation for the r'th moment of T,

$$\langle T^r(I)\rangle \equiv \mu_r(I),$$

in the form

$$-r\mu_{r-1}(I) = A(I)\frac{\mathrm{d}\mu_r(I)}{\mathrm{d}I} + \tfrac{1}{2}D(I)\frac{\mathrm{d}^2\mu_r(I)}{\mathrm{d}I^2}. \tag{19.5–18}$$

The higher-order moments of the first passage time T can therefore be derived recursively from the lower-order ones. By integrating Eq. (19.5–18) we can express $\mu_r(I)$ as the multiple integral

$$\begin{aligned}\langle T^r_{\text{off}}\rangle = r!\int_0^{I_A}\mathrm{d}I_{2r}\frac{2}{\mathcal{P}(I_{2r})D(I_{2r})}\int_0^{I_{2r}}\mathrm{d}I_{2r-1}\mathcal{P}(I_{2r-1})\int_{I_{2r-1}}^{I_A}\mathrm{d}I_{2r-2}\ldots\int_0^{I_4}\mathrm{d}I_3\mathcal{P}(I_3)\\ \times\int_{I_3}^{I_A}\mathrm{d}I_2\frac{2}{\mathcal{P}(I_2)D(I_2)}\int_0^{I_2}\mathrm{d}I_1\mathcal{P}(I_1),\quad r = 2, 3, \ldots\end{aligned} \tag{19.5–19}$$

and

$$\begin{aligned}\langle T^r_{\text{on}}\rangle = r!\int_{I_A}^{a_1}\mathrm{d}I_{2r}\frac{2}{\mathcal{P}(I_{2r})D(I_{2r})}\int_{I_{2r}}^{\infty}\mathrm{d}I_{2r-1}\mathcal{P}(I_{2r-1})\int_{I_A}^{I_{2r-1}}\mathrm{d}I_{2r-2}\ldots\int_{I_4}^{\infty}\mathrm{d}I_3\mathcal{P}(I_3)\\ \times\int_{I_A}^{I_3}\mathrm{d}I_2\frac{2}{\mathcal{P}(I_2)D(I_2)}\int_{I_2}^{\infty}\mathrm{d}I_1\mathcal{P}(I_1),\quad r = 2, 3, \ldots.\end{aligned} \tag{19.5–20}$$

The integrals can be evaluated with the help of the same approximations that were used in connection with Eqs. (19.5–11) and (19.5–12), and the result is (Roy, Short, Durnin and Mandel, 1980)

$$\left.\begin{aligned}\langle T^r_{\text{off}}\rangle &\approx r!\langle T_{\text{off}}\rangle^r\\ \langle T^r_{\text{on}}\rangle &\approx r!\langle T_{\text{on}}\rangle^r.\end{aligned}\right\} \tag{19.5–21}$$

This implies that the times T_{off} and T_{on} are distributed in an approximately exponential manner, with the shortest times the most probable. Actually, as the probability of making a first passage in zero time has to be zero, this answer cannot be strictly correct, but the probability distribution can rise very rapidly from zero. Figure 19.25 shows the results of some measurements of the distribution of switching times T_{on} in a two-mode dye laser, for two different pump parameters. The distributions are approximately exponential in form, but a rapid rise in the first millisecond to the most probable value is observed when $\langle T_{\text{on}}\rangle \approx 6.1$ ms. At the longer switching times, such as $\langle T_{\text{on}}\rangle \approx 0.38$ s, this rise is not resolvable, but it is presumed to be still there, and its shape is derivable from a more accurate and realistic treatment of the problem (Lenstra and Singh, 1983). In view of the oversimplification that was made in treating the first passage problem as one-dimensional, the general agreement between theory and experiment is quite reasonable.

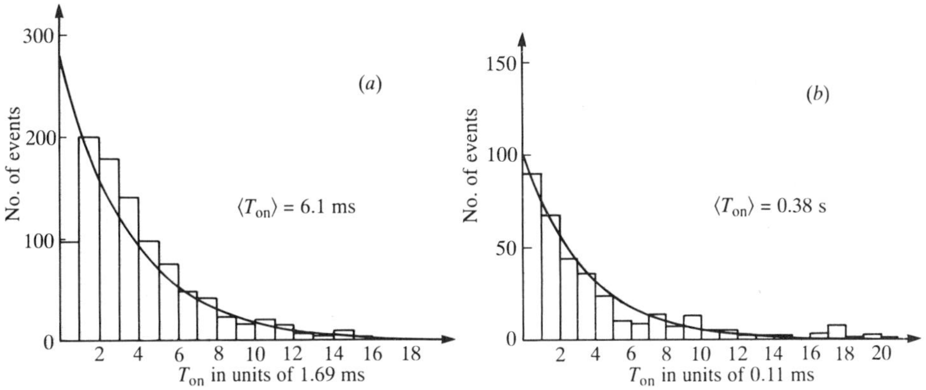

Fig. 19.25 Histograms of the measured probability distributions of the first passage time T_{on} (*a*) for a pump parameter 3.5; (*b*) for a pump parameter 9. The full curves are exponential distributions having the same mean $\langle T_{on} \rangle$. (Reproduced from Roy, Short, Durnin and Mandel, 1980.)

One interesting aspect of the mode-switching phenomenon is that it is a macroscopic manifestation of microscopic quantum fluctuations. A single spontaneously emitted photon can initiate the switching process in which one quite intense laser beam is turned on and another one is turned off. There are not many phenomena in physics in which quantum fluctuations are so strikingly evident.

In this chapter we have encountered a number of different phenomena, such as coherence and incoherence, metabistability and mode switching, first- and second-order phase transitions, which are exhibited by ring lasers under different conditions. To some extent they are all consequences of the competition for photons between the laser modes. However, our treatment of the ring laser by no means exhausts its possible forms of behavior. Under suitable conditions the equations of motion can lead to self-pulsing (Risken and Nummedal, 1968; Graham and Haken, 1968), periodic breathing of pulses, and chaotic behavior even in the absence of spontaneous emission fluctuations (Mayr, Risken and Vollmer, 1981). The ring laser is evidently a system that is extremely rich in physical properties.

Problems

19.1 Starting from the equation of motion for the joint probability density of the two-mode field of a ring laser, show that the equation of motion for the probability density $\mathcal{P}(I_1)$ of the light intensity I_1 of one mode can be cast in the form of a Fokker–Planck equation provided that the conditional mean intensity $\langle I_2 \rangle_{I_1}$ of mode 2, given I_1, can be replaced by the unconditional mean $\langle I_2 \rangle$. Hence determine the diffusion rate of I_1 in dimensionless units.

19.2 Use the steady-state solutions for the moments $\langle I_j \rangle$, $\langle (\Delta I_j)^2 \rangle$, $(j = 1, 2)$, and $\langle \Delta I_1 \Delta I_2 \rangle$ of a two-mode ring laser to show that, with a homogeneously broadened gain medium, and with the laser tuned to line center, with pump parameters a_1, a_2 and with the difference $\Delta a \equiv a_1 - a_2$ positive and constant, the moments have the following asymptotic forms as $a_1 \to \infty$:

$$\langle I_1 \rangle \sim a_1 - 2/\Delta a,$$
$$\langle I_2 \rangle \sim 2/\Delta a,$$
$$\langle (\Delta I_1)^2 \rangle / \langle I_1 \rangle^2 \sim [2 + 4/\Delta a)^2]/a_1^2,$$
$$\langle (\Delta I_2)^2 \rangle / \langle I_2 \rangle^2 \sim 1,$$
$$\langle \Delta I_1 \Delta I_2 \rangle / \langle I_1 \rangle \langle I_2 \rangle \sim -2/(a_1 \Delta a).$$

19.3 In Problem 19.2 show that, when $\Delta a = 0$ and $a_1 = a = a_2$, the asymptotic values of the moments for large a are given by:

$$\langle I_j \rangle \sim a/2, \qquad (j = 1, 2)$$
$$\langle (\Delta I_j)^2 \rangle / \langle I_j \rangle^2 \sim 1/3, \qquad (j = 1, 2)$$
$$\frac{\langle \Delta I_1 \Delta I_2 \rangle}{\langle I_1 \rangle \langle I_2 \rangle} \sim -1/3.$$

19.4 A two-mode ring laser with mode-coupling constant ξ $(0 \leqslant \xi < 1)$ is operating at threshold $(a_1 = 0 = a_2)$. Show that the following relations among the light intensities hold:

$$\langle I_j \rangle = \sqrt{\pi} \left(\frac{1 - \xi}{1 + \xi} \right)^{1/2} (\arccos \xi)^{-1}, \qquad (j = 1, 2)$$
$$\frac{\langle (\Delta I_j)^2 \rangle}{\langle I_j \rangle^2} = \frac{2(\arccos \xi)^2}{\pi(1 - \xi)^2} - \frac{2\xi(1 + \xi) \arccos \xi}{\pi(1 - \xi)(1 - \xi^2)^{1/2}} - 1, \qquad (j = 1, 2)$$
$$\frac{\langle \Delta I_1 \Delta I_2 \rangle}{\langle I_1 \rangle \langle I_2 \rangle} = \frac{2 \arccos \xi}{\pi(1 - \xi^2)^{3/2}} - \frac{2\xi(\arccos \xi)^2}{\pi(1 - \xi^2)^2} - 1.$$

20

The linear light amplifier

20.1 Introduction

As we have seen in Chapters 18 and 19, laser action results from the combination of a light amplifier with an optical resonator and a pumping source. It could therefore be argued that the treatment of the light amplifier is already included in Chapters 18 and 19. However, saturation effects and nonlinearities play an essential role in the laser, whereas they are absent by definition in the linear light amplifier. It is therefore more convenient to discuss the linear light amplifier on its own, in order to focus on its distinctive features.

It has occasionally been argued that the light amplifier, when inserted into an interferometer arm, is potentially capable of violating the uncertainty principle, because it allows one to register the path of the photon while interference is taking place. Consider the Mach–Zehnder interferometer shown in Fig. 20.1, which has a linear light amplifier in each arm followed by a beam splitter and a detector. Suppose that a photon is incident at the input of the interferometer. Because the light emerging from the light amplifier is phase coherent with the input, it might appear that interference fringes would be observable at the output of the interferometer, even while detector A or detector B tells us which path the

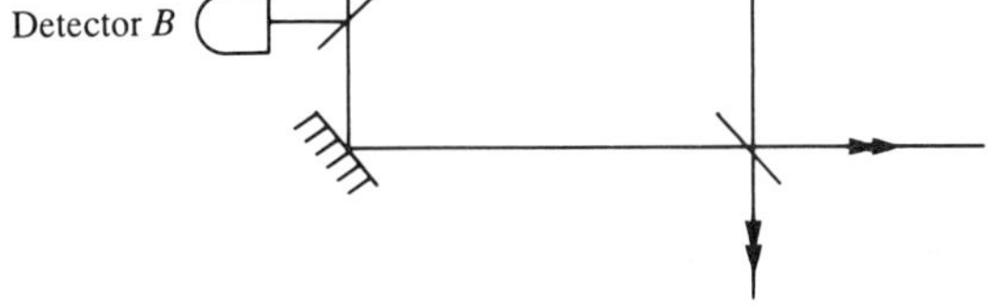

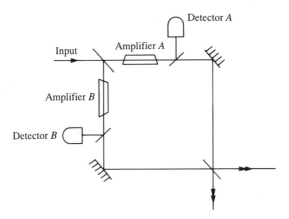

Fig. 20.1 Mach–Zehnder interferometer with a light amplifier in each arm, which is supposed to allow the path chosen by an incident photon to be identified, according as detector A or detector B registers an output.

photon followed through the interferometer. This would of course violate the principle that interference is always associated with the intrinsic indistinguishability of the two paths. But the argument is fallacious; the fallacy results from ignoring the spontaneous emission emerging from the light amplifier. The problem has been carefully analyzed by Glauber (1986). It is because of the spontaneous emission that no linear amplifier is capable of strictly cloning an incident photon, as was shown earlier by a rather general argument (Wooters and Zurek, 1982; Mandel, 1983).

In this chapter we shall concentrate on the simplest kind of amplifier, consisting of a partly inverted population of two-level atoms that effectively all see the same optical field. The population inversion is assumed to be maintained by some optical pumping scheme, so that it behaves like a reservoir. We shall examine how the amplifier responds to various kinds of incident fields in both classical and non-classical states, and we shall find that as the amplifier gain increases any non-classical features of the light are lost (Friberg and Mandel 1983, 1984; Hong, Friberg and Mandel, 1985). A rather general treatment of fluctuations in linear amplifiers has been given by Caves (1982), which applies to both phase sensitive and phase insensitive situations.

20.2 Master equation for the amplifier field

We consider a system of N identical two-level atoms, of which N_2 are excited and N_1 are unexcited, interacting with a single-mode quantum field. In Chapter 18 we discussed a similar problem in connection with the laser, and we took the single field mode to be an eigenmode of a cavity. Although we could of course choose to do the same here, the transmissivities of the cavity mirrors would then play an important role, because the light amplifier has both an input and an output. In order to make the problem as simple as possible, we shall suppose that we are dealing with an eigenmode of a free field, and that all atoms see essentially the same field. We assume that the field frequency is resonant with the atomic frequency, and we ignore any possible direct dipole interaction between atoms. We shall further suppose that the numbers N_2, N_1 are maintained approximately constant in time by some pump and loss mechanism, without considering the details. Let $\hat{\rho}$ be the reduced density operator of the electromagnetic field. We wish to derive a master equation of motion for $\hat{\rho}(t)$ in the interaction picture.

This problem is of course similar to the one we addressed in Section 18.4 in the quantum treatment of the laser. The main difference is that this time we focus entirely on the linear regime of the laser, and we exclude all non-linear atomic interactions involving more than one photon emission. We then arrive, by the same argument as before, at the master equation (18.4–20), except that we now put the coefficient B describing the saturation or non-linearity of the laser equal to zero:

$$\frac{\partial\hat{\rho}}{\partial t} = -\tfrac{1}{2}A(\hat{a}\hat{a}^\dagger\hat{\rho} - \hat{a}^\dagger\hat{\rho}\hat{a} + \text{h.c.}) - \tfrac{1}{2}C(\hat{a}^\dagger\hat{a}\hat{\rho} - \hat{a}\hat{\rho}\hat{a}^\dagger + \text{h.c.}). \quad (20.2\text{–}1)$$

As before A represents the gain rate that is associated with the excited atomic population, and C represents the loss rate that is associated with the unexcited atoms. For simplicity we shall take the gain and loss rates to be proportional to the populations N_2, N_1, so that

$$\left.\begin{aligned} A &= 2\lambda N_2 \\ C &= 2\lambda N_1, \end{aligned}\right\} \tag{20.2–2}$$

where λ is a rate that is of the order of the atomic linewidth.

In Section 18.5 we showed that, by making a diagonal coherent-state representation of the density operator $\hat{\rho}(t)$, namely

$$\hat{\rho}(t) = \int \phi(v, t)|v\rangle\langle v|\,\mathrm{d}^2 v, \tag{20.2–3}$$

substituting this representation into the master equation and replacing $\hat{a}$, $\hat{a}^\dagger$ by differential operators, one can convert the operator equation for $\hat{\rho}(t)$ into a c-number Fokker–Planck equation for $\phi(v, t)$. We may then look at the initial phase space density $\phi(v, 0)$ as representing the input field to the light amplifier whose output, after a time t, is $\phi(v, t)$. On putting $B = 0$ and using Eqs. (20.2–2) we have from Eq. (18.5–8),

$$\frac{1}{\lambda}\frac{\partial \phi(v, t)}{\partial t} = -(N_2 - N_1)\left[\frac{\partial}{\partial v}(v\phi(v, t)) + \frac{\partial}{\partial v^*}(v^*\phi(v, t))\right] + 2N_2\frac{\partial^2 \phi(v, t)}{\partial v \partial v^*}. \tag{20.2–4}$$

We shall take this as the equation of motion describing the linear light amplifier with input $\phi(v, 0)$ and output $\phi(v, t)$.

20.3 Solution of the master equation

In Section 18.6 we described a procedure for solving the Fokker–Planck equation for the laser based on the Sturm–Liouville approach. This led to a solution in the form of an infinite series, which could not be summed analytically, but was usable for numerical calculations. However, the Fokker–Planck equation (20.2–4) for the linear amplifier is significantly simpler, and it is possible to solve it analytically (Carusotto, 1975; Abraham and Smith, 1977; Rockower, Abraham and Smith, 1978). We shall not derive the solution explicitly here, but merely present the answer, and leave it to the reader to check that it satisfies Eq. (20.2–4). It may be shown that $\phi(v, t)$ representing the amplifier field after a time t, which we shall regard as the output field, is expressible as a simple convolution of the input or zero-time phase space density $\phi_0(v) \equiv \phi(v, 0)$ with the phase space density $\phi_s(v, t)$ of a thermal field [cf. Eq. (13.2–3)] that is associated with spontaneous emission. Thus

$$\phi(v, t) = \int \phi_0(v')\phi_s(v - G(t)v', t)\,\mathrm{d}^2 v', \tag{20.3–1}$$

where

$$\phi_s(v, t) \equiv \frac{1}{\pi m(t)}\mathrm{e}^{-|v|^2/m(t)} \tag{20.3–2}$$

$$m(t) \equiv \left(\frac{N_2}{N_2 - N_1}\right)[|G(t)|^2 - 1] \qquad (20.3\text{–}3)$$

$$G(t) \equiv e^{(N_2 - N_1)\lambda t}\, e^{-i\omega t}. \qquad (20.3\text{–}4)$$

It is evident from the form of Eq. (20.3–2) that $m(t)$ represents the average photon number of the thermal or spontaneous emission field. We shall see shortly that $G(t)$ represents the amplifier gain after a time t.

If we make the change of variable $v - G(t)v' = v''$ in Eq. (20.3–1) and write

$$\phi(v, t) = \int \phi_0\left(\frac{v - v''}{G}\right)\phi_s(v'')\, d^2v'', \qquad (20.3\text{–}5)$$

then it is apparent from the convolution structure that the output of the light amplifier can be regarded as resulting from the interference of the amplified input field with the spontaneous emission field [cf. the discussion leading to Eq. (11.8–26)]. For example, if we calculate the average output field amplitude $\langle \hat{a}(t)\rangle$ at time t from Eq. (20.3–1), we have

$$\begin{aligned}\langle \hat{a}(t)\rangle = \langle v\rangle_t &= \int v\phi(v, t)\, d^2v \\ &= \iint (v - G(t)v')\phi_s(v - G(t)v', t)\phi_0(v')\, d^2v'\, d^2v \\ &\quad + G(t)\iint v'\phi_0(v')\phi_s(v - G(t)v', t)\, d^2v'\, d^2v.\end{aligned}$$

The first integral on the right yields zero because the average value of the thermal field is zero, i.e.

$$\int v\phi_s(v)\, d^2v = 0,$$

whereas the second term gives

$$G(t)\int v'\phi_0(v')\, d^2v' = G(t)\langle v'\rangle_0,$$

because the integral over v yields unity. Finally

$$\langle \hat{a}(t)\rangle = \langle v\rangle_t = G(t)\langle v\rangle_0 = G(t)\langle \hat{a}(0)\rangle. \qquad (20.3\text{–}6)$$

Hence the average output field equals the average input field multiplied by $G(t)$, which is the amplifier gain, because the average spontaneous emission field is zero. By a similar argument we may show that

$$\langle \hat{a}^r(t)\rangle = G^r(t)\langle \hat{a}^r(0)\rangle. \qquad (20.3\text{–}7)$$

The spontaneous emission field does however contribute to the average photon number $\langle \hat{n}(t)\rangle$ at the output. Thus we have from Eq. (20.3–1) with the help of the optical equivalence theorem (see Section 11.9),

$$\begin{aligned}\langle \hat{n}(t)\rangle = \langle \hat{a}^\dagger(t)\hat{a}(t)\rangle &= \langle |v|^2\rangle_t \\ &= \int |v|^2\phi(v, t)\, d^2v\end{aligned}$$

$$= \iint |v|^2 \phi_s(v, t)\phi_0(v')\, d^2v'\, d^2v$$

$$+ |G(t)|^2 \iint |v'|^2 \phi_0(v')\phi_s(v, t)\, d^2v'\, d^2v$$

$$+ G(t)\iint v^* v' \phi_0(v')\phi_s(v, t)\, d^2v'\, d^2v + \text{c.c.}$$

$$= m(t) + |G(t)|^2\langle \hat{n}(0)\rangle, \qquad (20.3\text{–}8)$$

because the last two integrals on the right vanish. Thus the amplification contributes $|G(t)|^2\langle \hat{n}(0)\rangle$ photons and the spontaneous emission contributes $m(t)$ photons to the output field on the average.

20.3.1 Input–output correlations

In order to show that, except for the spontaneous emission, the amplifier output field is coherent with the input field, we now calculate the input–output cross-correlation function $\langle \hat{a}^\dagger(0)\hat{a}(t)\rangle$. For this purpose we need to construct the Green function $\mathcal{G}(v, t|v_0, 0)$, or the conditional phase space density, from Eq. (20.3–1) by putting $\phi_0(v') = \delta^2(v' - v_0)$. We then obtain the equation

$$\mathcal{G}(v, t|v_0, 0) = \phi_s(v - G(t)v_0, t) = \frac{1}{\pi m(t)} e^{-|v-G(t)v_0|^2/m^2(t)}, \qquad (20.3\text{–}9)$$

and it follows that

$$\langle \hat{a}^\dagger(0)\hat{a}(t)\rangle = \iint v_0^* v \mathcal{G}(v, t|v_0, 0)\phi_0(v_0)\, d^2v_0\, d^2v$$

$$= \iint v_0^*(v' + G(t)v_0)\frac{1}{\pi m(t)} e^{-|v'|^2/m^2(t)}\phi_0(v_0)\, d^2v_0\, d^2v',$$

where we have used Eq. (20.3–9) and have put $v = v' + G(t)v_0$. Hence

$$\left.\begin{aligned} \langle \hat{a}^\dagger(0)\hat{a}(t)\rangle &= G(t)\langle \hat{n}(0)\rangle, \\ \text{or} \qquad & \\ \langle v^*(0)v(t)\rangle &= G(t)\langle v^*(0)v(0)\rangle, \end{aligned}\right\} \qquad (20.3\text{–}10)$$

because the integral involving the term $v_0^* v'$ vanishes. The input–output cross-correlation therefore coincides with the input autocorrelation except for the amplifier gain factor $G(t)$.

20.4 Photon statistics

After obtaining a general expression (20.3–1) for the output field in terms of the input field, we can readily relate various moments of the photon number at the amplifier output to those at the input, or more generally we can connect the output probability distribution $p(n, t)$ to the input probability $p(n, 0)$ (Friberg and Mandel, 1983, 1984).

Let us consider the second factorial moment of the photon number. With the help of the general relation (12.10–13) and the optical equivalence theorem for normally ordered operators, we have, for the second factorial moment of the number of output photons,

$$\begin{aligned}\langle \hat{n}^{(2)}(t)\rangle &= \langle \hat{a}^{\dagger 2}(t)\hat{a}^{2}(t)\rangle \\ &= \int |v|^4 \phi(v, t)\, \mathrm{d}^2 v,\end{aligned}$$

and after using Eq. (20.3–1) and putting $v = G(t)v' + v''$, we find that

$$\langle \hat{n}^{(2)}(t)\rangle = \iint [|v''|^4 + |G(t)|^4|v'|^4 + 4|G(t)|^2|v'|^2|v''|^2]\phi_s(v'')\phi_0(v')\,\mathrm{d}^2v'\,\mathrm{d}^2v'',$$

because all other terms integrate to zero. After the insertion of the Gaussian moments of $\phi_s(v'')$, we obtain the formula

$$\langle \hat{n}^{(2)}(t)\rangle = 2m^2(t) + |G(t)|^4\langle \hat{n}^{(2)}(0)\rangle + 4|G(t)|^2 m(t)\langle \hat{n}(0)\rangle. \quad (20.4\text{–}1)$$

By combining Eq. (20.3–8) with Eq. (20.4–1) we readily find that

$$\begin{aligned}\langle \hat{n}^{(2)}(t)\rangle - \langle \hat{n}(t)\rangle^2 &= m^2(t) + |G(t)|^4[\langle \hat{n}^{(2)}(0)\rangle - \langle \hat{n}(0)\rangle^2] \\ &\quad + 2|G(t)|^2 m(t)\langle \hat{n}(0)\rangle. \quad (20.4\text{–}2)\end{aligned}$$

At this point we recall that the difference $\langle \hat{n}^{(2)}(t)\rangle - \langle \hat{n}(t)\rangle^2$ represents the departure from Poisson statistics, and it is positive or negative according as the fluctuations are super- or sub-Poissonian. In the latter case the state of the field has no classical description. Now it is apparent by inspection of Eq. (20.4–2) that the photons at the amplifier output cannot be sub-Poissonian unless the input photons are sub-Poissonian. In general the output statistics are sub-Poissonian if at the input

$$\langle \hat{n}^{(2)}(0)\rangle - \langle \hat{n}(0)\rangle^2 < -\frac{m(t)}{|G(t)|^2}\left[2\langle \hat{n}(0)\rangle + \frac{m(t)}{|G(t)|^2}\right]. \quad (20.4\text{–}3)$$

Let us denote the ratio

$$\frac{\langle \hat{n}^{(2)}(t)\rangle - \langle \hat{n}(t)\rangle^2}{\langle \hat{n}(t)\rangle} \equiv Q_t, \quad (20.4\text{–}4)$$

where $-1 \leq Q_t < 0$ when the statistics are sub-Poissonian. Then from the inequality (20.4–3), sub-Poisson statistics at the amplifier output require

$$\left[\frac{m(t)}{|G(t)|^2}\right]^2 + 2\langle \hat{n}(0)\rangle\frac{m(t)}{|G(t)|^2} - \langle \hat{n}(0)\rangle|Q_0| < 0,$$

and because of the quadratic structure of the left-hand side, this requires

$$\frac{m(t)}{|G(t)|^2} < [\langle \hat{n}(0)\rangle^2 + \langle \hat{n}(0)\rangle|Q_0|]^{1/2} - \langle \hat{n}(0)\rangle. \quad (20.4\text{–}5)$$

But from the definition (20.3–3) it follows that

$$\frac{m(t)}{|G(t)|^2} = \frac{1 - 1/|G(t)|^2}{1 - N_1/N_2}. \quad (20.4\text{–}6)$$

When this is substituted on the left of inequality (20.4–5), it imposes an upper bound on $|G(t)|^2$, namely

$$|G(t)|^2 < \frac{1}{1 - (1 - N_1/N_2)\{[\langle \hat{n}(0)\rangle^2 + \langle \hat{n}(0)\rangle |Q_0|]^{1/2} - \langle \hat{n}(0)\rangle\}}. \quad (20.4\text{–}7)$$

Now $[\langle \hat{n}(0)\rangle^2 + \langle \hat{n}(0)\rangle |Q_0|]^{1/2} - \langle \hat{n}(0)\rangle$ increases monotonically with $\langle \hat{n}(0)\rangle$ and has an upper bound of $\frac{1}{2}|Q_0|$, because

$$\langle \hat{n}(0)\rangle^2 + \langle \hat{n}(0)\rangle |Q_0| \leqslant [\langle \hat{n}(0)\rangle + \tfrac{1}{2}|Q_0|]^2.$$

Hence the inequality (20.4–7) implies that

$$|G(t)|^2 < \frac{1}{1 - (|Q_0|/2)(1 - N_1/N_2)}. \quad (20.4\text{–}8)$$

But $|Q_0|$ cannot exceed 1 when the statistics are sub-Poissonian, so that

$$|G(t)|^2 < \frac{1}{1 - \frac{1}{2}(1 - N_1/N_2)} \leqslant 2, \quad (20.4\text{–}9)$$

with equality possible only when $N_1 = 0$, in which case the atomic system is fully inverted. We have therefore shown that if the photons at the amplifier output are to be sub-Poissonian, the amplifier intensity gain $|G(t)|^2$ must not exceed 2. Once $|G(t)|^2$ exceeds 2 the non-classical photon statistics are lost, no matter how non-classical the input to the amplifier may be.

The gain $|G(t)|^2 = 2$ corresponds to the so-called 'cloning' value of the amplifier gain, for which one photon at the amplifier input results in two similar photons at the output, except for the spontaneous emission contribution. But the average photon number $m(t)$ contributed by spontaneous emission is not negligible. From Eq. (20.4–6) it follows that

$$m(t) = [|G(t)|^2 - 1]/(1 - N_1/N_2) \geqslant |G(t)|^2 - 1 \quad (20.4\text{–}10)$$

so that $m(t) \geqslant 1$ when $|G(t)|^2 = 2$. There is therefore at least one spontaneously emitted photon, on the average, for each amplified photon.

20.4.1 Probability distributions

We may readily use Eq. (20.3–1) to calculate the probability $p(n, t)$ that there are n photons at the amplifier output, and relate it to the corresponding input probability $p(n, 0)$. The calculation is straightforward in principle, but a little complicated, and we shall only outline the procedure (Friberg and Mandel, 1983, 1984). From the general relationship [cf. Eq. (12.10–2)] between $p(n, t)$ and the phase space density $\phi(v, t)$, namely

$$p(n, t) = \frac{1}{n!}\int |v|^{2n} e^{-|v|^2} \phi(v, t)\, d^2v, \quad (20.4\text{–}11)$$

we have

$$p(n, t) = \frac{1}{n!}\int\int |v|^{2n} e^{-|v|^2} \frac{1}{\pi m(t)} e^{-|v - G(t)v'|^2/m^2(t)} \times \phi_0(v')\, d^2v'\, d^2v. \quad (20.4\text{–}12)$$

On putting $v = (\sqrt{x})\,\mathrm{e}^{\mathrm{i}\phi}$, $\mathrm{d}^2 v = \frac{1}{2}\mathrm{d}x\,\mathrm{d}\phi$ and carrying out the integrations first over ϕ and then over x, we find that

$$p(n, t) = \int \mathrm{d}^2 v' \int_0^\infty \mathrm{d}x\, x^n\, \mathrm{e}^{-x(1+1/m(t))} I_0\left[\frac{2|G(t)|\,|v'|\sqrt{x}}{m(t)}\right] \phi_0(v') \frac{1}{n!\,m(t)} \times \mathrm{e}^{-|G(t)|^2|v'|^2/m(t)}$$

$$= \frac{1}{[1 + m(t)][1 + 1/m(t)]^n} \int \phi_0(v')\, \mathrm{e}^{-|G(t)|^2|v'|^2/m(t)} \times {}_1F_1(n + 1, 1, |G(t)|^2|v'|^2/m(t)[m(t) + 1])\, \mathrm{d}^2 v', \tag{20.4–13}$$

where $I_0(z)$ is the zero-order modified Bessel function and ${}_1F_1(a, b, z)$ is the confluent hypergeometric function.

If we now make a power series expansion of ${}_1F_1$ and integrate term by term with respect to v', we obtain a series expansion for $p(n, t)$,

$$p(n, t) = \frac{1}{[1 + m(t)][1 + 1/m(t)]^n} \sum_{r=0}^\infty \frac{(n + r)!}{n!\,r!\,(m + 1)^r} \mathscr{P}_0(r, |G(t)|^2/m(t)), \tag{20.4–14}$$

with

$$\mathscr{P}_0(r, \alpha) = \frac{1}{r!} \int \phi_0(v')(\alpha|v'|^2)^r\, \mathrm{e}^{-\alpha|v'|^2}\, \mathrm{d}^2 v'. \tag{20.4–15}$$

Except for the parameter $\alpha \equiv |G(t)|^2/m(t)$, $\mathscr{P}_0(n, \alpha)$ has the structure of the input probability $p(n, 0)$, as comparison with Eq. (20.4–11) shows. Indeed, when we recall the treatment of photoelectric detection given in Section 14.8, we recognize that $\mathscr{P}_0(n, \alpha)$ has the form of the photon detection probability for a detector whose input field is described by $\phi_0(v)$ and whose effective quantum efficiency is α. This interpretation of α allows us to express $\mathscr{P}_0(n, \alpha)$ in the alternative form of a Bernoulli convolution between $p(m, 0)$ and the probability for detecting n out of the m incident photons, i.e.

$$\mathscr{P}_0(n, \alpha) = \sum_{m=n}^\infty \binom{m}{n} \alpha^n (1 - \alpha)^{m-n} p(m, 0). \tag{20.4–16}$$

Actually the interpretation of $\alpha = |G(t)|^2/m(t)$ as a quantum efficiency is meaningful only if $|G(t)|^2/m(t) \leq 1$, so that α is a probability. However Eqs. (20.4–15) and (20.4–16) are formally equivalent whether or not α is a probability.

By substituting for $\mathscr{P}_0$ from Eq. (20.4–16) in Eq. (20.4–14) and interchanging the order of summations we finally arrive at the formula

$$p(n, t) = \frac{1}{[1 + m(t)][1 + 1/m(t)]^n} \sum_{l=0}^\infty \sum_{r=0}^l \binom{-n - 1}{r} \binom{l}{r} \times \left[\frac{-|G(t)|^2/m(t)(m(t) + 1)}{1 - |G(t)|^2/m(t)}\right]^r \left[1 - \frac{|G(t)|^2}{m(t)}\right]^l p(l, 0), \tag{20.4–17}$$

which expresses the photon probability $p(n, t)$ at the amplifier output in terms of the probability $p(l, 0)$ at the input. Needless to say, $p(n, t)$ must be a probability whether or not $|G(t)|^2/m(t)$ is a probability.

Let us examine the form of $|G(t)|^2/m(t)$ a little more closely. From Eq. (20.4–6) it follows at once that

$$|G(t)|^2/m(t) < 1 \quad \text{only if} \quad |G(t)|^2 > N_2/N_1. \tag{20.4–18}$$

Unless the intensity gain exceeds N_2/N_1, the interpretation of $|G(t)|^2/m(t)$ as a detection probability breaks down. In particular, if the light amplifier has a fully inverted atomic system, so that $N_1 = 0$, then this interpretation always fails.

20.4.2 The fully inverted light amplifier

Let us examine this limiting case in more detail. When $N_1 = 0$, it follows from the definition (20.3–3) that

$$-\frac{|G(t)|^2/m(t)(m(t)+1)}{1-|G(t)|^2/m(t)} = 1, \tag{20.4–19}$$

and

$$1-|G(t)|^2/m(t) = -1/[|G(t)|^2-1]. \tag{20.4–20}$$

When these results are substituted in Eq. (20.4–17), the summation on r can be carried out. With the help of the algebraic relations

$$\left.\begin{aligned} \sum_{r=0}^{l}\binom{-n-1}{r}\binom{l}{r} &= \binom{n}{l}(-1)^l \quad && \text{if } l \leqslant n \\ &= 0 && \text{if } l > n, \end{aligned}\right\} \tag{20.4–21}$$

we arrive at the formula

$$p(n, t) = \frac{1}{[1+m(t)][1+1/m(t)]^n}\sum_{l=0}^{n}\binom{n}{l}\left[\frac{1}{|G(t)|^2-1}\right]^l p(l, 0). \tag{20.4–22}$$

The cut-off at $l = n$ is the result of the fact that the fully inverted light amplifier can only increase the photon number and not decrease it.

The series in Eq. (20.4–22) can be summed in special cases. If $p(l, 0)$ is a Bose–Einstein distribution, as for thermal light at the amplifier input, then so is $p(n, t)$. If the amplifier input is in a Fock state $|n_0\rangle$, then $p(l, 0) = \delta_{ln_0}$, and the sum in Eq. (20.4–22) reduces to a single term. With the help of the substitution $n = n_0 + r$ we can then express $p(n, t)$ as a negative binomial distribution, i.e.

$$p(n_0+r, t) = \frac{1}{|G(t)|^{2(n_0+1)}}\binom{-n_0-1}{r}[1/|G(t)|^2-1]^r. \tag{20.4–23}$$

Despite the appearance of the negative factor $1/|G(t)|^2 - 1$, $p(n_0 + r, t)$ is of course positive for all $r \geqslant 0$.

Figure 20.2 shows the form of the probability $p(n, t)$ for various input photon

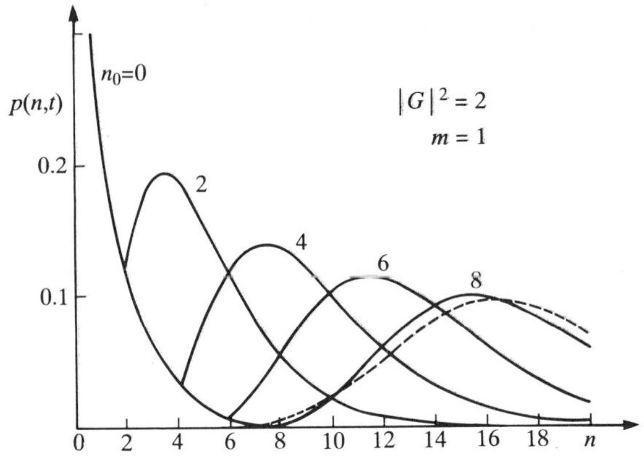

Fig. 20.2 The probability $p(n, t)$ of the photon number n at the amplifier output (shown as continuous curves for clarity) for various photon numbers n_0 at the input, with $|G(t)|^2 = 2$, $N_1 = 0$. The broken curve is a Poisson distribution with the same mean as for $n_0 = 8$. (Reproduced from Friberg and Mandel, 1984.)

numbers n_0, for the special case when the intensity gain of the amplifier has the cloning value $|G(t)|^2 = 2$. Although the photons at the amplifier input are of course sub-Poissonian, those at the output are super-Poissonian, as expected from Eq. (20.4–23), or directly from Eq. (20.4–2), because of the spontaneous emission contributed by the amplifier. When $|G|^2 = 2$, $p(n, t)$ tends towards the Poisson form as $n_0 \to \infty$, as is apparent from Fig. 20.2.

20.5 Squeezed light

If the field at the amplifier input is in a (quadrature) squeezed state (see Section 21.1), there exists some phase angle θ such that the real quadrature amplitude

$$\hat{Q}(0) \equiv \hat{a}(0)\,e^{-i\theta} + \hat{a}^\dagger(0)\,e^{i\theta} \tag{20.5–1}$$

has a dispersion below the vacuum value 1, i.e.

$$\langle(\Delta\hat{Q}(0))^2\rangle < 1. \tag{20.5–2}$$

The quadrature component $\hat{Q}$ then fluctuates less than it would in a vacuum field, at the cost of a corresponding increase of the fluctuations of the conjugate quadrature amplitude. Let us examine under what circumstances the slowly varying amplifier output field $\hat{Q}(t)$ defined by the formula

$$\hat{Q}(t) \equiv \hat{a}(t)\,e^{i(\omega t-\theta)} + \hat{a}^\dagger(t)\,e^{-i(\omega t-\theta)} \tag{20.5–3}$$

exhibits squeezing when the input field is squeezed (Friberg and Mandel, 1983; Hong, Friberg and Mandel, 1985).

From Eq. (20.5–3) and the commutation relations we have

$$\langle(\Delta\hat{Q}(t))^2\rangle = \langle(\Delta\hat{a}(t))^2\rangle\,e^{2i(\omega t-\theta)} + \langle(\Delta\hat{a}^\dagger(t))^2\rangle\,e^{-2i(\omega t-\theta)} + 2\langle\Delta\hat{a}^\dagger(t)\Delta\hat{a}(t)\rangle + 1, \tag{20.5–4}$$

and with the help of Eqs. (20.3–7) and (20.3–8) we obtain the relation

$$\langle(\Delta\hat{Q}(t))^2\rangle = |G(t)|^2[\langle[\Delta\hat{a}(0)]^2\rangle\, e^{-2i\theta} + \langle(\Delta\hat{a}^\dagger(0))^2\rangle\, e^{2i\theta} + 2\langle\Delta\hat{a}^\dagger(0)\Delta\hat{a}(0)\rangle] + 2m(t) + 1. \quad (20.5\text{–}5)$$

If we now introduce $\langle(\Delta\hat{Q}(0))^2\rangle$ given by Eq. (20.5–1) and use the relation (20.3–3), we can express this result in the form

$$\begin{aligned}\langle(\Delta\hat{Q}(t))^2\rangle &= |G(t)|^2\langle(\Delta\hat{Q}(0))^2\rangle + 2m(t) + 1 - |G(t)|^2 \\ &= |G(t)|^2\langle(\Delta\hat{Q}(0))^2\rangle + [|G(t)|^2 - 1](N_2 + N_1)/(N_2 - N_1).\end{aligned} \quad (20.5\text{–}6)$$

If $\hat{Q}(t)$ is to exhibit squeezing so that $\langle(\Delta\hat{Q}(t))^2\rangle$ is less than unity, we require that

$$|G(t)|^2 < \frac{2}{(1 + N_1/N_2) + (1 - N_1/N_2)\langle(\Delta\hat{Q}(0))^2\rangle}. \quad (20.5\text{–}7)$$

It is readily seen that this condition can be satisfied only if $\langle(\Delta\hat{Q}(0))^2\rangle < 1$, i.e. if the input to the amplifier is squeezed. Moreover, since $\langle(\Delta\hat{Q}(0))^2\rangle \geqslant 0$, condition (20.5–7) requires that

$$|G(t)|^2 < \frac{2}{(1 + N_1/N_2)} \leqslant 2. \quad (20.5\text{–}8)$$

Hence we find once again that the non-classical features, this time associated with squeezing at the output, are all lost if the intensity gain of the light amplifier exceeds the so-called cloning value $|G(t)|^2 = 2$.

20.6 Condition for the amplifier output field to be classical

Although we have shown that two non-classical features of the amplifier output, namely sub-Poissonian statistics and squeezing, are lost when the intensity gain $|G(t)|^2$ exceeds the value 2, the possibility that some other non-classical properties may exist still remains open. However, it is not difficult to show by a more general argument that all non-classical features of the light must disappear at the amplifier output when the gain is sufficiently high, although this general argument generally leads to a much weaker condition on $|G(t)|^2$.

We start by re-writing Eq. (20.3–1) for the output of the amplifier in the form

$$\phi(v, t) = \frac{1}{\pi m(t)}\int \phi_0(v')\exp[-|v' - v/G(t)|^2/(m(t)/|G(t)|^2)]\, d^2v'. \quad (20.6\text{–}1)$$

Now it follows immediately from the diagonal representation (20.2–3) that the matrix element $\langle v/G(t)|\hat{\rho}(0)|v/G(t)\rangle$, where $|v/G(t)\rangle$ is a coherent state, is given by the formula

$$\frac{1}{\pi m(t)}\langle v/G(t)|\hat{\rho}(0)|v/G(t)\rangle = \frac{1}{\pi m(t)}\int \phi_0(v')|\langle v/G(t)|v'\rangle|^2\,\mathrm{d}^2v'$$
$$= \frac{1}{\pi m(t)}\int \phi_0(v')\,\mathrm{e}^{-|v'-v/G(t)|^2}\,\mathrm{d}^2v', \qquad (20.6\text{–}2)$$

when we make use of the scalar product relation for two coherent states. Now the matrix element $\langle v/G|\hat{\rho}(0)|v/G\rangle$, being the expectation value of a Hermitian, non-negative definite operator, is of course real and non-negative, so that

$$\frac{1}{\pi m(t)}\int \phi_0(v')\,\mathrm{e}^{-|v'-v/G(t)|^2}\,\mathrm{d}^2v' \geqslant 0 \qquad (20.6\text{–}3)$$

for any $\phi_0(v')$. Let us put $v/G(t) = v_0$ in the integral, multiply by $\exp(-|v_0 - v''|^2/\beta)$, with β real and positive, and integrate with respect to v_0 over all complex v_0. Then we obtain from relation (20.6–3), for any v'',

$$\frac{\pi}{1+1/\beta}\int \phi_0(v')\,\mathrm{e}^{-|v'-v''|^2/(1+\beta)}\,\mathrm{d}^2v' \geqslant 0. \qquad (20.6\text{–}4)$$

When this relation is compared with Eq. (20.6–1), it is evident that if $(1+\beta)$ is identified with $m(t)/|G(t)|^2$, then $\phi(v, t)$ is real and

$$\phi(v, t) \geqslant 0. \qquad (20.6\text{–}5)$$

This inequality implies that the phase space density $\phi(v, t)$ is a probability density, and that the corresponding quantum state at the amplifier output has a classical descripion. In other words, under these conditions all quantum features of the output field are absent. However, in order to be able to identify $1+\beta$ with $m(t)/|G(t)|^2$ one obviously requires that

$$m(t)/|G(t)|^2 \geqslant 1, \qquad (20.6\text{–}6)$$

and from Eqs. (20.3–3) this, in turn, requires that

$$|G(t)|^2 \geqslant N_2/N_1. \qquad (20.6\text{–}7)$$

We have therefore arrived at a bound on the amplifier gain $|G(t)|^2$ that is sufficient to make the output field classical. However, condition (20.6–7) is not necessary for quantum features to be absent. In particular, when $N_2 \gg N_1$ the condition (20.6–7) is hard to satisfy, and conditions (20.4–9) and (20.5–8) are more useful. The domain lying in between the conditions (20.4–9) or (20.5–8) and (20.6–7) has been explored by Agarwal and Tara (1992). They find that the real part of the complex amplifier output can also exhibit non-classical features, but these are all lost when $|G|^2 > 2(1 + N_1/N_2)^{-1}$.

Problems

20.1 Can a linear light amplifier 'amplify' an incident photon so that the output is independent of the photon polarization? Consider an amplifier consisting of two two-level atoms that are both exposed to the same incident optical field.

20.2 Consider a light amplifier consisting of N_2 excited and N_1 unexcited two-level atoms. Show that the condition $|G|^2 \geqslant N_2/N_1$ on the gain G is not necessary for ensuring that the output field of the amplifier is in a classical state. Give an example in which the condition is violated while the amplifier output is in a classical state.

21

Squeezed states of light

In Section 11.5 we showed that, although the complex amplitude of the electromagnetic field has a well-defined value in any coherent state, yet the real and imaginary (Hermitian and anti-Hermitian) parts of the field fluctuate with equal dispersions. The phenomenon of vacuum fluctuations is a manifestation of this effect, because the vacuum state is an example of a particular coherent state. This behavior is quite different from that of an ordinary, classical field. In a squeezed state, which is even more non-classical, as we shall see, one part of the field fluctuates less and another part fluctuates more than in the vacuum state. In general, a squeezed state is one in which the distribution of canonical variables over the phase space has been distorted or 'squeezed' in such a way that the dispersion of one variable is reduced at the cost of an increase in the dispersion of the other variable. In the following we shall examine the properties of squeezed states when the two canonical variables are two quadratures of the electromagnetic field. Although the squeezing terminology is sometimes applied to variables other than the two field quadratures, it is less meaningful in those cases. A number of review articles on squeezing have been published and can be consulted for more details (Walls, 1983; Schumaker, 1986; Loudon and Knight, 1987; Teich and Saleh, 1989, 1990; Kimble, 1992).

21.1 Definition of quadrature squeezing

We start by considering squeezing of a single-mode field. Two-mode squeezing exhibits additional features and has been treated by Caves (1981, 1982), Milburn (1984), Caves and Schumaker (1985), Schumaker and Caves (1985), Fan, Zaidi and Klauder (1987), and Fan (1990). We first define two dimensionless variables $\hat{Q}'$, $\hat{P}'$ representing the real and imaginary parts of the complex amplitude in terms of annihilation and creation operators, by the formulas

$$\left.\begin{aligned} \hat{Q}' &= \hat{a}^\dagger + \hat{a} \\ \hat{P}' &= \mathrm{i}(\hat{a}^\dagger - \hat{a}). \end{aligned}\right\} \qquad (21.1\text{–}1)$$

Then we readily find from the commutation relations obeyed by $\hat{a}$, $\hat{a}^\dagger$ that

$$[\hat{Q}', \hat{P}'] = 2\mathrm{i}, \qquad (21.2\text{–}2)$$

so that $\hat{Q}'$ and $\hat{P}'$ behave like dimensionless canonical conjugates. They therefore obey the uncertainty relations

$$\langle(\Delta\hat{Q}')^2\rangle^{1/2}\langle(\Delta\hat{P}'^2)\rangle^{1/2} \geqslant 1, \tag{21.1–3}$$

no matter what the quantum state may be. Another interpretation of $\hat{Q}'$, $\hat{P}'$ is suggested if we express the field vector $\hat{\mathbf{E}}(\mathbf{r}, t)$ for a single-mode, linearly polarized field in the form

$$\hat{\mathbf{E}}(\mathbf{r}, t) = l(\omega)\boldsymbol{\varepsilon}[\hat{a}\,\mathrm{e}^{\mathrm{i}(\mathbf{k}\cdot\mathbf{r}-\omega t)} + \hat{a}^\dagger\,\mathrm{e}^{-\mathrm{i}(\mathbf{k}\cdot\mathbf{r}-\omega t)}], \tag{21.1–4}$$

where $l(\omega)$ is some real function of frequency. Then we find, on substituting for $\hat{a}$, $\hat{a}^\dagger$ in terms of $\hat{Q}'$, $\hat{P}'$, that

$$\hat{\mathbf{E}}(\mathbf{r}, t) = l(\omega)\boldsymbol{\varepsilon}[\hat{Q}'\cos(\mathbf{k}\cdot\mathbf{r} - \omega t) - \hat{P}'\sin(\mathbf{k}\cdot\mathbf{r} - \omega t)]. \tag{21.1–5}$$

Hence the canonical variables $\hat{Q}'$, $\hat{P}'$ are also the amplitudes of the quadratures into which the oscillating field can be decomposed.

It is characteristic of the coherent state, including the vacuum state, that the dispersions of the dimensionless quadrature amplitudes $\hat{Q}'$, $\hat{P}'$ are equal, with

$$\left.\begin{aligned}\langle(\Delta\hat{Q}')^2\rangle &= 1\\ \langle(\Delta\hat{P}')^2\rangle &= 1,\end{aligned}\right\} \tag{21.1–6}$$

so that the uncertainty product has its minimum value. For this state we can find a phase space distribution of Q', P' which has circular symmetry, with the unit cell a circle (see Fig. 21.1). If there exists a state for which either $\hat{Q}'$ or $\hat{P}'$ has a dispersion below unity, i.e. below the vacuum level, at the cost of a corresponding increase in the dispersion of the other variable, then the corresponding phase space distribution takes on an elliptic shape, and we call the corresponding state a *squeezed state* (see Fig. 21.1).

Squeezing need not be confined to the $\hat{Q}'$ or $\hat{P}'$ variables defined by Eq. (21.1–1) and, as illustrated in Fig. 21.1, the major axis of the ellipse may point in a direction other than the Q', P' axes. We can easily accommodate this possibility in the definition of the squeezed state by defining more general variables $\hat{Q}$, $\hat{P}$ for any angle β, by the relations

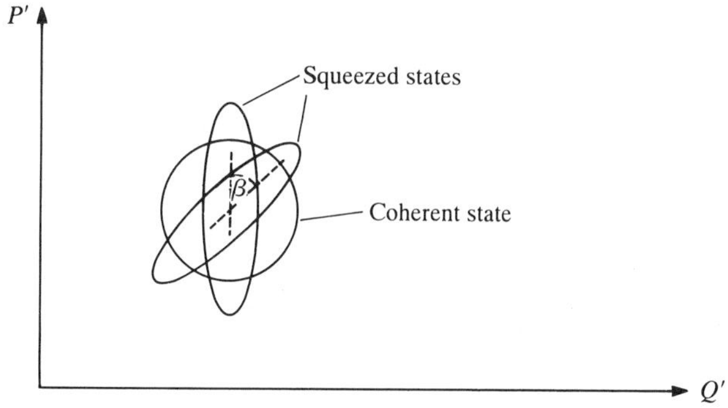

Fig. 21.1 Illustrating a squeezed (elliptical) phase space distribution in comparison with a coherent (circular) one.

any angle β, by the relations

$$\left.\begin{aligned}\hat{Q} &= \hat{a}^\dagger e^{i\beta} + \hat{a}\, e^{-i\beta}\\ \hat{P} &= \hat{a}^\dagger e^{i(\beta+\pi/2)} + \hat{a}\, e^{-i(\beta+\pi/2)}.\end{aligned}\right\} \tag{21.1–7}$$

These obey the same commutation and uncertainty relations as $\hat{Q}'$, $\hat{P}'$, and their dispersions are both unity in the vacuum state. In terms of $\hat{Q}$ and $\hat{P}$, $\hat{\mathbf{E}}(\mathbf{r}, t)$ is given by

$$\hat{\mathbf{E}}(\mathbf{r}, t) = l(\omega)\boldsymbol{\varepsilon}[\hat{Q}\cos(\mathbf{k}\cdot\mathbf{r} - \omega t + \beta) - \hat{P}\sin(\mathbf{k}\cdot\mathbf{r} - \omega t + \beta)]. \tag{21.1–8}$$

From the definition it is apparent that $\hat{P}$ differs from $\hat{Q}$ only in that the angle β is incremented by $\pi/2$. The deviations $\Delta\hat{Q}$, $\Delta\hat{P}$ are related to $\Delta\hat{Q}'$, $\Delta\hat{P}'$ by a rotation through β, i.e.

$$\left.\begin{aligned}\Delta\hat{Q} &= \Delta\hat{Q}'\cos\beta + \Delta\hat{P}'\sin\beta\\ \Delta\hat{P} &= -\Delta Q'\sin\beta + \Delta\hat{P}\cos\beta.\end{aligned}\right\} \tag{21.1–9}$$

We then define a squeezed state more generally by the condition that there exists an angle β for which the dispersion $\langle(\Delta\hat{Q})^2\rangle$ is smaller than in the vacuum state, i.e. $\langle(\Delta\hat{Q})^2\rangle < 1$. From the uncertainty relation (21.1–3) it then follows that in this state $\langle(\Delta\hat{P})^2\rangle > 1$ necessarily. The uncertainty product may exceed its minimum value unity in a squeezed state.

It is instructive to illustrate the effect of squeezing on the fluctuations of the electric field $\mathbf{E}$, if these could be measured and displayed on an oscilloscope. The results might look as shown in Fig. 21.2. Here (*a*) corresponds to the vacuum state; (*b*) corresponds to a coherent state, for which the uncertainties of $\mathbf{E}$ are distributed uniformly over the whole cycle; (*c*) corresponds to a squeezed state with reduced phase uncertainty and increased amplitude uncertainty; (*d*) corresponds to a squeezed state with reduced amplitude uncertainty and increased phase uncertainty.

The squeezing terminology is occasionally applied to canonical variables other than the two quadrature amplitudes we have considered above, even when the two variables are quite different and have different dimensions. This can result in ambiguities in the interpretation. We shall limit our discussion to the case for which the canonical variables are field quadratures. The squeezing phenomenon is then sometimes referred to as *quadrature squeezing*. Such squeezed states were introduced and studied by Stoler (1970, 1971), Yuen (1976), Caves (1981), Caves and Schumaker (1985), Schumaker and Caves (1985), and others.

21.2 Quantum nature of the squeezed state

At first sight it might appear that a quantum state of the field in which the fluctuations are reduced below those for the coherent state would look even more classical than the coherent state. But, of course, in a squeezed state only one field quadrature has reduced fluctuations, while the other quadrature fluctuates more. Although it is possible to construct a classical phase space distribution with features such as those shown in Figs. 21.2 (*c*) and (*d*), the squeezed state is always

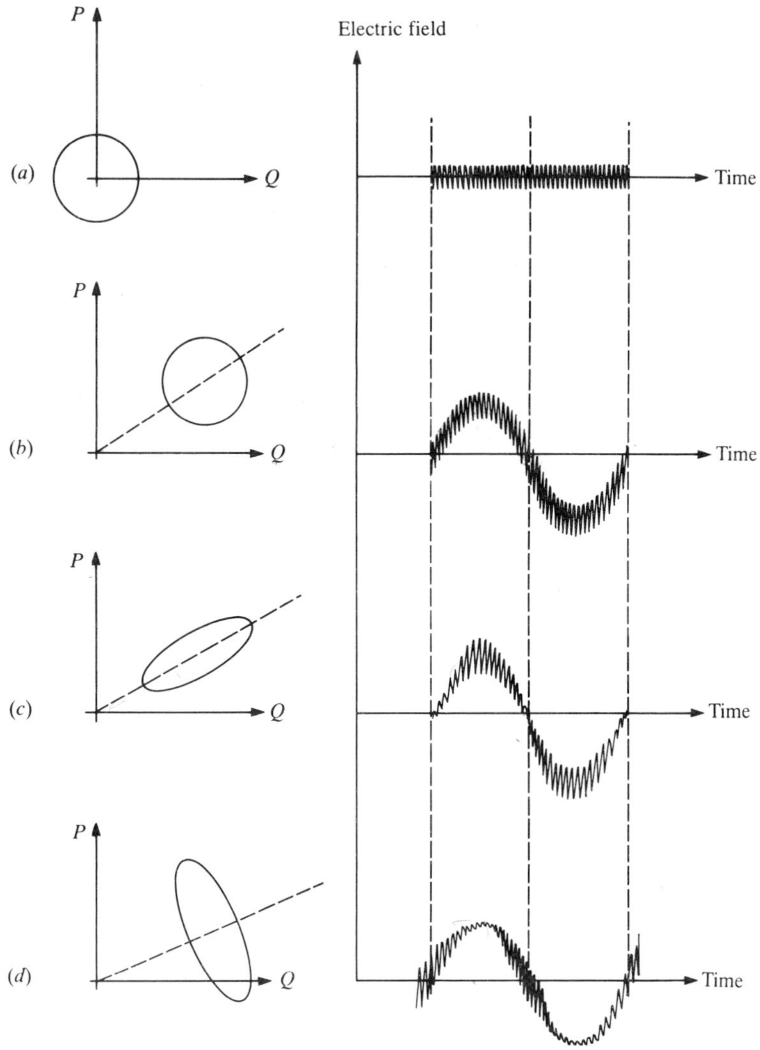

Fig. 21.2 Illustrating the relation between the phase space distribution and the fluctuations of the electric field for (*a*) the vacuum state; (*b*) the coherent state; (*c*) squeezed state with reduced phase uncertainty; (*d*) squeezed state with reduced amplitude uncertainty.

non-classical in the usual sense; i.e. the diagonal coherent-state representation of the density operator is not a classical probability density (cf. Section 11.8).

We define the squeezed state by the property that

$$\langle(\Delta\hat{Q})^2\rangle < 1 \quad \text{for some } \beta, \tag{21.2–1}$$

and use Eqs. (21.1–7) to relate $\langle\hat{Q}^2\rangle$ to the normally ordered expectation $\langle:\hat{Q}^2:\rangle$. We then obtain the relations

$$\langle:\hat{Q}^2:\rangle = \langle\hat{a}^{\dagger 2}\rangle\, e^{2i\beta} + \langle\hat{a}^2\rangle\, e^{-2i\beta} + 2\langle\hat{a}^\dagger\hat{a}\rangle$$
$$\langle\hat{Q}^2\rangle = \langle\hat{a}^{\dagger 2}\rangle\, e^{2i\beta} + \langle\hat{a}^2\rangle\, e^{-2i\beta} + \langle\hat{a}^\dagger\hat{a}\rangle + \langle\hat{a}\hat{a}^\dagger\rangle.$$

Since $[\hat{a}, \hat{a}^\dagger] = 1$ we find immediately that

$$\langle :\hat{Q}^2: \rangle = \langle \hat{Q}^2 \rangle - 1$$

so that

$$\langle :(\Delta\hat{Q})^2: \rangle = \langle (\Delta\hat{Q})^2 \rangle - 1. \tag{21.2–2}$$

It follows from the definition (21.2–1) that for a squeezed state

$$\langle :(\Delta\hat{Q})^2: \rangle < 0 \quad \text{for some } \beta, \tag{21.2–3}$$

and this inequality may be regarded as an alternative but equivalent definition of squeezing.

We now invoke the optical equivalence theorem for expectations of normally ordered operators (cf. Section 11.9), which involves expressing the density operator $\hat{\rho}$ in the diagonal coherent-state representation, i.e.

$$\hat{\rho} = \int \phi(v)|v\rangle\langle v|\mathrm{d}^2 v. \tag{21.2–4}$$

Then we have the relation

$$\langle :(\Delta\hat{Q})^2: \rangle = \int (\Delta Q)^2 \phi(v)\,\mathrm{d}^2 v. \tag{21.2–5}$$

Here $(\Delta Q)^2$ is the c-number corresponding to $:(\Delta\hat{Q})^2:$, which is obtained by replacing each $\hat{a}$ by v and each $\hat{a}^\dagger$ by v^*. From Eq. (21.1–7) we see that $\langle (\Delta\hat{Q})^2 \rangle$ is given by

$$(\Delta Q)^2 = (\Delta v^* \,\mathrm{e}^{\mathrm{i}\beta} + \Delta v\,\mathrm{e}^{-\mathrm{i}\beta})^2. \tag{21.2–6}$$

The condition (21.2–3) for squeezing can now be expressed in the form

$$\langle (\Delta Q)^2 \rangle_\phi = \int \phi(v)(\Delta v^* \,\mathrm{e}^{\mathrm{i}\beta} + \Delta v\,\mathrm{e}^{-\mathrm{i}\beta})^2\,\mathrm{d}^2 v < 0 \text{ for some } \beta. \tag{21.2–7}$$

As $(\Delta Q)^2$ is real and non-negative, it is now obvious by inspection that $\phi(v)$ cannot be a classical probability density if the inequality (21.2–7) is to be satisfied. Classical states are characterized by having a phase space density $\phi(v)$ which is a true probability density. It follows that a squeezed state is always quantum mechanical and has no analog in classical electromagnetic theory.

21.3 The unitary squeeze operator

It is possible to generate a squeezed single-mode state from an unsqueezed one by the action of the following simple unitary operator (Stoler, 1970, 1971; Yuen, 1976; Hollenhorst, 1979; Caves, 1981; Fisher, Nieto and Sandberg, 1984; Caves and Schumaker, 1985; Schumaker and Caves, 1985),

$$\hat{S}(z) = \exp\tfrac{1}{2}(z^*\hat{a}^2 - z\hat{a}^{\dagger 2}), \quad z = r\,\mathrm{e}^{\mathrm{i}\theta}, \tag{21.3–1}$$

which is known as the *squeeze operator*. This is a more general form of the operator we encountered in Section 11.5. With the help of the operator expansion theorem [Eq. (10.11–1)] we readily find for the unitary transformation of the operator $\hat{a}$ by $\hat{S}(z)$,

$$\begin{aligned}\hat{A}(z) &\equiv \hat{S}(z)\hat{a}\hat{S}^\dagger(z)\\ &= \hat{a} + z\hat{a}^\dagger + \frac{|z|^2\hat{a}}{2!} + \frac{z|z|^2\hat{a}^\dagger}{3!} + \ldots\\ &= \hat{a}\cosh r + \hat{a}^\dagger e^{i\theta}\sinh r\\ &= \mu\hat{a} + \nu\hat{a}^\dagger, \end{aligned} \tag{21.3–2}$$

where we have put

$$\left.\begin{aligned}\mu &\equiv \cosh r\\ \nu &\equiv e^{i\theta}\sinh r.\end{aligned}\right\} \tag{21.3–3}$$

Similarly, we obtain the equation

$$\begin{aligned}\hat{A}^\dagger(z) &= \hat{S}(z)\hat{a}^\dagger\hat{S}^\dagger(z)\\ &= \mu\hat{a}^\dagger + \nu^*\hat{a}.\end{aligned} \tag{21.3–4}$$

In order to abbreviate the notation we shall sometimes write $\hat{A}$ for $\hat{A}(z)$, but it is important to bear in mind that $\hat{A}$ depends on z. Because

$$|\mu|^2 - |\nu|^2 = 1 \tag{21.3–5}$$

by definition, it follows immediately that

$$[\hat{A}, \hat{A}^\dagger] = 1, \tag{21.3–6}$$

so that the $\hat{A}$, $\hat{A}^\dagger$ are pseudo-annihilation and creation operators, similar in some respects to $\hat{a}$, $\hat{a}^\dagger$. By inversion of Eqs. (21.3–2) and (21.3–4) we immediately obtain the relations

$$\left.\begin{aligned}\hat{a} &= \mu\hat{A} - \nu\hat{A}^\dagger\\ \hat{a}^\dagger &= \mu\hat{A}^\dagger - \nu^*\hat{A}.\end{aligned}\right\} \tag{21.3–7}$$

The state obtained by letting the squeeze operator $\hat{S}(z)$ act on the coherent state $|v\rangle$ has been studied at some length by Yuen (1976), who labeled it the *two-photon coherent state*. We shall use the notation

$$|[z, v]\rangle \equiv \hat{S}(z)|v\rangle \tag{21.3–8}$$

for the two-photon coherent state. From the definitions we readily find that

$$\begin{aligned}\hat{A}(z)|[z, v]\rangle &= \hat{S}(z)\hat{a}\hat{S}^\dagger(z)\hat{S}(z)|v\rangle\\ &= \hat{S}(z)\hat{a}|v\rangle\\ &= v\hat{S}(z)|v\rangle\\ &= v|[z, v]\rangle,\end{aligned} \tag{21.3–9}$$

and by Hermitian conjugation

$$\langle[z, v]|\hat{A}^\dagger(z) = v^*\langle[z, v]|. \tag{21.3–10}$$

Hence $\hat{A}$, $\hat{A}^\dagger$ bear the same relation and have the same eigenvalues with respect to $|[z, v]\rangle$ as do $\hat{a}$, $\hat{a}^\dagger$ with respect to the coherent state $|v\rangle$.

21.3.1 *Squeezing of the two-photon coherent state*

In order to show that the state $|[z, v]\rangle$ is indeed squeezed we now calculate $\langle\hat{Q}\rangle$ and the dispersion $\langle(\Delta\hat{Q})^2\rangle$ in the two-photon coherent state. From the definition (21.1–7) and Eqs. (21.3–7) to (21.3–10), we obtain the expressions

$$\begin{aligned}\langle[z, v]|\hat{Q}|[z, v]\rangle &= \langle[z, v]|(\mu\hat{A}^\dagger - \nu^*\hat{A})\,e^{i\beta} + (\mu\hat{A} - \nu\hat{A}^\dagger)\,e^{-i\beta}|[z, v]\rangle \\ &= (\mu v^* - \nu^* v)\,e^{i\beta} + (\mu v - \nu v^*)\,e^{-i\beta}, \qquad (21.3\text{–}11)\end{aligned}$$

and

$$\begin{aligned}\langle[z, v]|\hat{Q}^2|[z, v]\rangle &= \langle[z, v]|[(\mu\hat{A}^\dagger - \nu^*\hat{A})^2\,e^{2i\beta} + (\mu\hat{A} - \nu\hat{A}^\dagger)^2\,e^{-2i\beta} \\ &\quad + 2(\mu\hat{A}^\dagger - \nu^*\hat{A})(\mu\hat{A} - \nu\hat{A}^\dagger) + 1]|[z, v]\rangle \\ &= [(\mu v^* - \nu^* v)^2 - \mu\nu^*]\,e^{2i\beta} + [(\mu v - \nu v^*)^2 - \mu\nu]\,e^{-2i\beta} \\ &\quad + 2|\mu v^* - \nu^* v|^2 + 2|\nu|^2 + 1, \qquad (21.3\text{–}12)\end{aligned}$$

so that

$$\begin{aligned}\langle(\Delta\hat{Q})^2\rangle &= \langle\hat{Q}^2\rangle - \langle\hat{Q}\rangle^2 \\ &= 2|\nu|^2 + 1 - (\mu\nu^*\,e^{2i\beta} + \mu\nu\,e^{-2i\beta}) \\ &= \cosh 2r - \sinh 2r\,\cos(\theta - 2\beta). \qquad (21.3\text{–}13)\end{aligned}$$

It follows that, when $\beta = \theta/2$, $\langle(\Delta\hat{Q})^2\rangle$ takes on its smallest value,

$$\langle[z, v]|(\Delta\hat{Q})^2|[z, v]\rangle = \cosh 2r - \sinh 2r = e^{-2r}, \qquad (21.3\text{–}14)$$

which is less than unity for all $z \neq 0$. The two-photon coherent state $|[z, v]\rangle$ is therefore a squeezed state according to our definition. With this choice of β we obtain from Eqs. (21.3–11) and (21.3–12) for the first two moments of $\hat{Q}$:

$$\langle[z, v]|\hat{Q}|[z, v]\rangle = 2|v|\cos(\arg v - \theta/2)\,e^{-r} \qquad (21.3\text{–}15)$$

$$\langle[z, v]|\hat{Q}^2|[z, v]\rangle = [1 + 4|v|^2\cos^2(\arg v - \theta/2)]\,e^{-2r}. \qquad (21.3\text{–}16)$$

In a similar manner we may readily show that for the same β the dispersion $\langle(\Delta\hat{P})^2\rangle = e^{2r}$. The fluctuations of $\hat{P}$ therefore exceed those in the vacuum state, while those of $\hat{Q}$ are below the vacuum level. However, the uncertainty product remains unity and has the minimum value. The phase space distribution is elliptical (see Fig. 21.1), with eccentricity given by $(1 - e^{-4r})^{1/2}$. This is close to unity when $r \gg 1$, in which case the ellipse degenerates almost to a straight line. The parameter r is sometimes referred to as the *squeeze parameter*.

21.3.2 *Action of the squeeze operator on any state*

It is not difficult to show that, with the same choice of β, the squeeze operator $\hat{S}(z)$ acting on any quantum state reduces the dispersion of $\hat{Q}$ by the same factor

e^{-2r}. Consider a single-mode electromagnetic field in a state described by the density operator $\hat{\rho}$. We make a diagonal coherent-state representation of $\hat{\rho}$ (see Section 11.8) in the form

$$\hat{\rho} = \int \phi(v)|v\rangle\langle v|\,d^2v. \tag{21.3–17}$$

The first two moments of $\hat{Q}$ in this state are

$$\begin{aligned}\langle\hat{Q}\rangle &= \mathrm{Tr}\,(\hat{Q}\hat{\rho}) = \mathrm{Tr}\int\phi(v)(\hat{a}\,e^{-i\beta} + \hat{a}^\dagger e^{i\beta})|v\rangle\langle v|\,d^2v \\ &= \int\phi(v)2|v|\cos(\arg v - \beta)\,d^2v \\ \langle\hat{Q}^2\rangle &= \mathrm{Tr}\,(\hat{Q}^2\hat{\rho}) = \mathrm{Tr}\int\phi(v)(\hat{a}\,e^{-i\beta} + \hat{a}^\dagger e^{i\beta})^2|v\rangle\langle v|\,d^2v \\ &= \int\phi(v)[4|v|^2\cos^2(\arg v - \beta) + 1]\,d^2v,\end{aligned}$$

so that

$$\langle(\Delta\hat{Q})^2\rangle = 1 + \langle 4|v|^2\cos^2(\arg v - \beta)\rangle_\phi - \langle 2|v|\cos(\arg v - \beta)\rangle_\phi^2. \tag{21.3–18}$$

The notation $\langle\ \rangle_\phi$ denotes an averaging operation with respect to the phase space density $\phi(v)$, whether $\phi(v)$ is a true probability density or not.

Under the action of the squeeze operator $\hat{S}(z)$ the initial state $\hat{\rho}$ becomes

$$\hat{\rho}'' = \hat{S}(z)\hat{\rho}\hat{S}^\dagger(z). \tag{21.3–19}$$

The first moment of $\hat{Q}$ in the new state is therefore given by

$$\begin{aligned}\langle\hat{Q}''\rangle &= \mathrm{Tr}\,[\hat{Q}\hat{\rho}''] \\ &= \mathrm{Tr}\int\phi(v)\hat{Q}\hat{S}(z)|v\rangle\langle v|\hat{S}^\dagger(z)\,d^2v \\ &= \int\phi(v)\langle[z, v]|\hat{Q}|[z, v]\rangle\,d^2v,\end{aligned}$$

and with the help of Eq. (21.3–15) we obtain the formula

$$\langle\hat{Q}''\rangle = e^{-r}\int\phi(v)2|v|\cos(\arg v - \theta/2)\,d^2v. \tag{21.3–20}$$

Similarly we find from Eq. (21.3–16) that

$$\langle\hat{Q}''^2\rangle = e^{-2r}\int\phi(v)[1 + 4|v|^2\cos^2(\arg v - \theta/2)]\,d^2v. \tag{21.3–21}$$

From this it follows that the dispersion of $\hat{Q}$ in the new state $\hat{\rho}''$ is given by

$$\langle(\Delta\hat{Q}'')^2\rangle = e^{-2r}[1 + \langle 4|v|^2\cos^2(\arg v - \theta/2)\rangle_\phi - \langle 2|v|\cos(\arg v - \theta/2)\rangle_\phi^2]. \tag{21.3–22}$$

Comparison between Eqs. (21.3–18) and (21.3–22) shows immediately that if we choose $\beta = \theta/2$, then

$$\langle(\Delta\hat{Q}'')^2\rangle = e^{-2r}\langle(\Delta\hat{Q})^2\rangle, \tag{21.3–23}$$

which is what we set out to prove.

21.4 Ideal squeezed states

We have already encountered the two-photon coherent state $|[z, v]\rangle$ (Yuen, 1976), which is obtained by letting the squeezed operator $\hat{S}(z)$ act on the coherent state $|v\rangle$. The coherent state $|v\rangle$ can be obtained by letting the unitary displacement operator (Glauber, 1963) $D(v) \equiv \exp(v\hat{a}^\dagger - v^*\hat{a})$ act on the vacuum state (cf. Section 11.3), so that

$$|[z, v]\rangle \equiv \hat{S}(z)\hat{D}(v)|\text{vac}\rangle. \tag{21.4–1}$$

Alternatively, we can produce a squeezed state by allowing the two unitary operators $\hat{S}(z)$ and $\hat{D}(v)$ to act in the reverse order. The resulting state

$$|(v, z)\rangle \equiv \hat{D}(v)\hat{S}(z)|\text{vac}\rangle \tag{21.4–2}$$

has been called the *ideal squeezed state* (Caves, 1981) because of its simple properties. As $\hat{S}(z)$ and $\hat{D}(v)$ do not commute, the two states $|[z, v]\rangle$ and $|(v, z)\rangle$ are different. The relation between the two states can be obtained from the following operator identities (Fisher, Nieto and Sandberg, 1984) which we shall not prove here,

$$\left.\begin{aligned} \hat{D}(v)\hat{S}(z) &= \hat{S}(z)\hat{D}(\tilde{v}_+), \\ \hat{S}(z)\hat{D}(v) &= \hat{D}(\tilde{v}_-)\hat{S}(z), \end{aligned}\right\} \tag{21.4–3}$$

where

$$\tilde{v}_\pm(z) = \mu v \pm \nu v^*, \tag{21.4–4}$$

and μ, ν are given by Eqs. (21.3–3) above. To simplify the notation we shall frequently write $\tilde{v}_\pm$ for $\tilde{v}_\pm(z)$, but its dependence on z is still implied. This leads to the following relation between the ideal squeezed state and the two-photon coherent state:

$$\left.\begin{aligned} |(v, z)\rangle &= |[z, \tilde{v}_+]\rangle \\ |[z, v]\rangle &= |(\tilde{v}_-, z)\rangle. \end{aligned}\right\} \tag{21.4–5}$$

It follows that each ideal squeezed state is also an example of a two-photon coherent state, and vice versa. Because the latter state is squeezed, as we have seen, the former is squeezed also. It should be emphasized that z and v in the equations above are independent variables, whereas $\tilde{v}_\pm$ given by Eq. (21.4–4) are actually functions of both v and z.

Just as the two-photon coherent state is a right eigenstate of the pseudo-annihilation operator $\hat{A}$, so is the ideal squeezed state, for we have from Eqs. (21.3–9) and (21.4–5)

$$\hat{A}|(v, z)\rangle = \hat{A}|[z, \tilde{v}_+]\rangle$$

$$= \tilde{v}_+ |[z, \tilde{v}_+]\rangle$$
$$= \tilde{v}_+ |(v, z)\rangle, \tag{21.4–6}$$

and similarly, after taking the conjugate, we obtain the equation

$$\langle (v, z)| \tilde{v}_+^* = \langle (v, z)| \hat{A}^\dagger. \tag{21.4–7}$$

With the help of Eqs. (21.3–7), (21.4–4), (21.4–6) and (21.4–7), we find that for the expectation of $\hat{a}$ in the ideal squeezed state is

$$\begin{aligned}\langle (v, z)|\hat{a}|(v, z)\rangle &= \langle (v, z)|\mu\hat{A} - \nu\hat{A}^\dagger|(v, z)\rangle \\ &= \mu\tilde{v}_+ - \nu\tilde{v}_+^* \\ &= (|\mu|^2 - |\nu|^2)v + (\mu\nu - \mu\nu)v^* \\ &= v,\end{aligned} \tag{21.4–8}$$

whereas the corresponding expectation in the two-photon coherent state is

$$\begin{aligned}\langle [z, v]|\hat{a}|[z, v]\rangle &= \langle [z, v]|\mu\hat{A} - \nu\hat{A}^\dagger|[z, v]\rangle \\ &= \mu v - \nu v^* \\ &= \tilde{v}_-.\end{aligned} \tag{21.4–9}$$

Now, loosely speaking, $\langle \hat{a} \rangle$ identifies the center of the phase space distribution for the state. Therefore the ellipse characterizing the ideal squeezed state $|(v, z)\rangle$ is centered at v, whereas that for the two-photon coherent state $|[z, v]\rangle$ is centered at $\tilde{v}_-$ (cf. Fig. 21.3).

Figure 21.3 illustrates the formation of the states $|(v, z)\rangle$ and $|[z, v]\rangle$ from the vacuum by squeezing and displacement. The vacuum state is represented by a circle centered at the origin in phase space. If the squeeze operator is applied first, the circle becomes an ellipse and the corresponding state is sometimes called

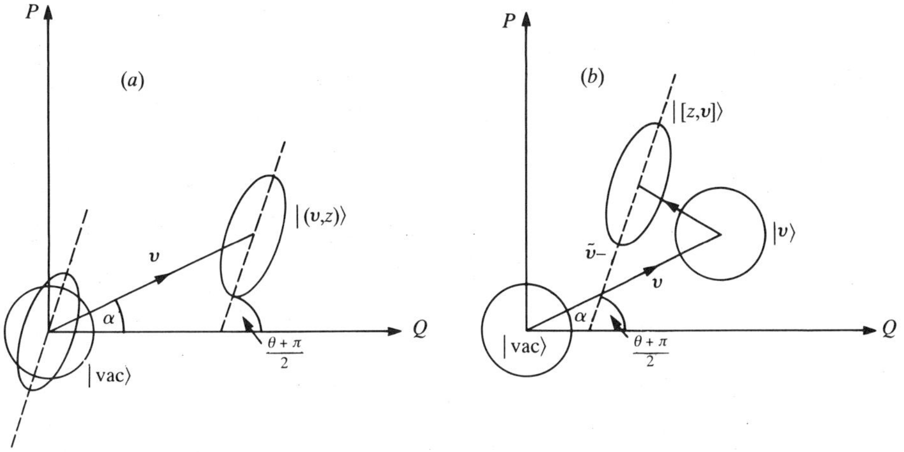

Fig. 21.3 Illustrating the generation from the vacuum state of (a) the ideal squeezed state $|(v, z)\rangle$; (b) the two-photon coherent state $|[z, v]\rangle$. The angle θ is negative in the illustration.

the *squeezed vacuum state*, although it is not the vacuum as we show below [see Fig. 21.3(a)]. The action of the displacement operator $\hat{D}(v)$ is then to translate the ellipse by v from the origin [see Fig. 21.3(a)]. On the other hand, if we start by applying the displacement operator $\hat{D}(v)$ first, the vacuum state becomes the coherent state $|v\rangle$ as the circle is translated by v [see Fig. 21.3(b)]. The subsequent action of the squeeze operator $\hat{S}(z)$ turns the circle into an ellipse, but it also produces an additional translation to the point $\tilde{v}_-$ in the plane. It is the absence of such an interference effect between squeezing and displacement in Fig. 21.3(a) that makes the ideal squeezed state particularly simple.

21.4.1 Photon statistics

The mean number of photons in the ideal squeezed state $|(v, z)\rangle$ is given by the formula

$$\begin{aligned}\langle (v, z)|\hat{a}^\dagger \hat{a}|(v, z)\rangle &= \langle (v, z)|(\mu\hat{A}^\dagger - \nu^*\hat{A})(\mu\hat{A} - \nu\hat{A}^\dagger)|(v, z)\rangle \\ &= \langle (v, z)|\,|\mu|^2\hat{A}^\dagger\hat{A} - \mu\nu\hat{A}^{\dagger 2} - \mu\nu^*\hat{A}^2 + |\nu|^2(\hat{A}^\dagger\hat{A} + 1)|(v, z)\rangle \\ &= |\mu|^2|\tilde{v}_+|^2 - \mu\nu\tilde{v}_+^{*2} - \mu\nu^*\tilde{v}_+^2 + |\nu|^2(|\tilde{v}_+|^2 + 1).\end{aligned}$$

On substituting for $\tilde{v}_+$ from Eq. (21.4–4) we find after some algebra that

$$\langle (v, z)|\hat{n}|(v, z)\rangle = |v|^2 + |\nu|^2. \tag{21.4–10}$$

Hence the squeezing contributes $|\nu|^2$ photons on the average, while the displacement contributes $|v|^2$ photons. In particular, for the so-called 'squeezed vacuum' state $|(0, z)\rangle$, we have

$$\langle (0, z)|\hat{n}|(0, z)\rangle = |\nu|^2, \tag{21.4–11}$$

from which it is apparent that this state is not a vacuum state at all. By contrast, we find with the help of Eq. (21.4–5) that for the two-photon coherent state

$$\begin{aligned}\langle [z, v]|\hat{n}|[z, v]\rangle &= \langle (\tilde{v}_-, z)|\hat{n}|(\tilde{v}_-, z)\rangle \\ &= |\tilde{v}_-|^2 + |\nu|^2.\end{aligned} \tag{21.4–12}$$

Next, we consider the photon fluctuations in the ideal squeezed state. With the help of Eqs. (21.3–7) we can write

$$\begin{aligned}\langle (v, z)|\hat{n}(\hat{n} - 1)|(v, z)\rangle &= \langle (v, z)|\hat{a}^{\dagger 2}\hat{a}^2|(v, z)\rangle \\ &= \langle (v, z)|(\mu\hat{A}^\dagger - \nu^*\hat{A})^2(\mu\hat{A} - \nu\hat{A}^\dagger)^2|(v, z)\rangle.\end{aligned}$$

After putting the operator into normal order, making use of the eigenfunction relations (21.4–6) and (21.4–7), and collecting terms, we obtain after some algebraic manipulation,

$$\begin{aligned}\langle (v, z)|\hat{n}(\hat{n} - 1)|(v, z)\rangle &= |\tilde{v}_+|^4(\mu^4 + |\nu|^4 + 4\mu^2|\nu|^2) + (\tilde{v}_+^4\mu^2\nu^{*2} + \text{c.c.}) \\ &\quad - 2|\tilde{v}_+|^2[\tilde{v}_+^2\mu\nu^*(|\nu|^2 + \mu^2) + \text{c.c.}] \\ &\quad + |\tilde{v}_+|^2(4|\nu|^4 + 8\mu^2|\nu|^2) \\ &\quad - [\tilde{v}_+^2\mu\nu^*(5|\nu|^2 + \mu^2) + \text{c.c.}] + 2|\nu|^4 + \mu^2|\nu|^2.\end{aligned}$$

After combining this with Eq. (21.4–10) for $\langle \hat{n} \rangle$ and using Eq. (21.3–3), we obtain for the departure from Poisson statistics the formula

$$\begin{aligned}\langle \hat{n}(\hat{n}-1)\rangle - \langle \hat{n}\rangle^2 \equiv \langle (\Delta\hat{n})^2\rangle - \langle \hat{n}\rangle &= |\tilde{v}_+|^2(\cosh 2r - 1)(2\cosh 2r + 1)\\ &\quad - (\cosh r)(\sinh r)(2\cosh 2r - 1)(\tilde{v}_+^2\,\mathrm{e}^{-\mathrm{i}\theta} + \text{c.c.})\\ &\quad + \sinh^2 r\cosh 2r. \qquad (21.4\text{–}13)\end{aligned}$$

Finally, we use the definition (21.4–4) to obtain the relations ($v = |v|\,\mathrm{e}^{\mathrm{i}\alpha}$)

$$\left.\begin{aligned}|\tilde{v}_+|^2 &= |v|^2[\cosh 2r + \sinh 2r\cos(2\alpha - \theta)]\\ \tilde{v}_+^2\,\mathrm{e}^{-\mathrm{i}\theta} &= |v|^2[\mathrm{i}\sin(2\alpha - \theta) + \cosh 2r\cos(2\alpha - \theta) + \sinh 2r],\end{aligned}\right\} \qquad (21.4\text{–}14)$$

and then substitute these in Eq. (21.4–13). After rearranging terms we eventually arrive at the result

$$\langle (\Delta\hat{n})^2\rangle - \langle \hat{n}\rangle = |v|^2[-1 + \cosh 2r - \sinh 2r\cos(2\alpha - \theta)] + \sinh^2 r\cosh 2r. \qquad (21.4\text{–}15)$$

Depending on the phase angle α, this can be positive or negative, so that the photon statistics of the ideal squeezed state can be super- or sub-Poissonian. Reference to Fig. 21.3(a) shows that when $\alpha = \theta/2$, the minor axis of the ellipse lies along the radius vector from the origin to the center, and the intensity or photon number fluctuations are then smallest. We have seen (cf. Section 12.10.3) that a convenient way to characterize the departure from Poisson statistics is by the parameter

$$\mathcal{Q} = \frac{\langle (\Delta\hat{n})^2\rangle - \langle \hat{n}\rangle}{\langle \hat{n}\rangle},$$

which is negative for sub-Poisson statistics, and is equal to -1 for a Fock state. With the help of Eqs. (21.4–15) and (21.4–10) we find that when $\alpha = \theta/2$

$$\mathcal{Q} = \frac{|v|^2(-1 + \mathrm{e}^{-2r}) + \sinh^2 r\cosh 2r}{|v|^2 + \sinh^2 r}. \qquad (21.4\text{–}16)$$

Let us suppose that the squeeze parameter $r \gg 1$ and the displacement is sufficiently large that $|v|^2 \gg \mathrm{e}^{2r}$. Then Eq. (21.4–16) yields

$$\mathcal{Q} \approx -1 + \frac{\sinh^2 \cosh 2r}{|v|^2}. \qquad (21.4\text{–}17)$$

If $|v|^2$ is large compared even with e^{4r}, then $\mathcal{Q} \approx -1$, which corresponds to complete absence of photon number fluctuations. A squeezed state with $\mathcal{Q} < 0$ is sometimes referred to as a *photon number squeezed state*.

By contrast when $\alpha = \theta/2 + \pi/2$, the ellipse is rotated through 90° and the major axis of the ellipse lies along the radius vector. Then $\mathcal{Q} \approx (-1 + \mathrm{e}^{2r}) + (\sinh^2 r\cosh 2r)/|v|^2$ which is large and positive, corresponding to large super-Poisson photon number fluctuations, while the phase fluctuations of the field are reduced.

21.5 Two-photon coherent states

As we have seen, the two-photon coherent state $|[z, v]\rangle$ is one form of ideal squeezed state, and the two states are connected by Eqs. (21.4–5). Because Yuen (1976) studied the former states at considerable length, more properties are known about two-photon coherent states than about ideal squeezed states, even though the two are intimately connected. In the following we shall therefore discuss the two-photon coherent states in a little more detail.

Following Yuen, we first generalize the definition slightly. Let $\hat{A}(\mu, \nu)$, $\hat{A}^\dagger(\mu, \nu)$ be pseudo-annihilation and pseudo-creation operators defined by the relations [cf. Eq. (21.3–2)]

$$\left.\begin{aligned} \hat{A}(\mu, \nu) &= \mu\hat{a} + \nu\hat{a}^\dagger \\ \hat{A}^\dagger(\mu, \nu) &= \mu^*\hat{a}^\dagger + \nu^*\hat{a} \end{aligned}\right\} \tag{21.5–1a}$$

for two complex numbers μ, ν satisfying the condition [cf. Eq. (21.3–5)]

$$|\mu|^2 - |\nu|^2 = 1, \tag{21.5–1b}$$

Let $\hat{U}(\mu, \nu)$ be any unitary transformation that generates $\hat{A}$, $\hat{A}^\dagger$ from $\hat{a}$, $\hat{a}^\dagger$. Evidently $\hat{U}(\mu, \nu) = \hat{S}(z)$ [Eq. (21.3–1)] is an example of such a transformation, but the more general unitary transformation

$$\hat{U} = \exp \mathrm{i}(K\hat{a}^\dagger\hat{a} + k\hat{a}^2 + k^*\hat{a}^{\dagger 2}), \tag{21.5–2}$$

with K real and k is a constant, has this property. The exponent will be recognized as having the structure of a quadratic Hamiltonian, corresponding to a 'two-photon' interaction with a classical current, and it has three free parameters, which are determined by the complex numbers μ, ν.

The generalized *two-photon coherent state* $|[\mu, \nu; v]\rangle$ can be defined to be the state resulting from the operation of $\hat{U}(\mu, \nu)$ on the coherent state $|v\rangle$, i.e.

$$|[\mu, \nu; v]\rangle = \hat{U}(\mu, \nu)|v\rangle, \tag{21.5–3}$$

which obviously corresponds to Eq. (21.3–8). The same argument as before leads to the eigenvalue relation

$$\begin{aligned} \hat{A}(\mu, \nu)|[\mu, \nu; v]\rangle &= \hat{U}(\mu, \nu)\hat{a}\hat{U}^\dagger(\mu, \nu)\hat{U}(\mu, \nu)|v\rangle \\ &= v\hat{U}(\mu, \nu)|v\rangle \\ &= v|[\mu, \nu; v]\rangle, \end{aligned} \tag{21.5–4}$$

and this equation provides an alternative definition of $|[\mu, \nu; v]\rangle$. The name 'two-photon coherent state' is of course connected with the form of the exponent in Eq. (21.5–2).

These states exhibit some of the features of the coherent states. For example, just as the expectation value of a normally ordered product of $\hat{a}$, $\hat{a}^\dagger$ factorizes in the coherent state $|v\rangle$, so does the expectation of a normally ordered product of $\hat{A}$, $\hat{A}^\dagger$ operators factorize in the state $|[\mu, \nu; v]\rangle$. Under the influence of the Hamiltonian

$$\hat{H} = \hbar\omega\hat{a}^\dagger\hat{a} + \hbar(g\hat{a} + g^*\hat{a}^\dagger) + \hbar(f\hat{a}^2 + f^*\hat{a}^{\dagger 2}), \tag{21.5–5}$$

a field in a two-photon coherent state $|[\mu, \nu; v]\rangle$ can be shown (Yuen, 1976) to evolve into other two-photon coherent states, but with different v, μ, ν that generally change in time. As we saw in Section 11.5, coherent states evolve into other coherent states when $f = 0$ in the Hamiltonian. For this reason it has been suggested that, while coherent states are produced by lasers in which the atomic source undergoes a one-photon interaction, two-photon coherent states might be produced by lasers in which the atoms undergo two-photon interactions (Yuen, 1976).

However, unlike the coherent states $|v\rangle$, the states $|[\mu, \nu; v]\rangle$ are not states of definite complex amplitude. With the help of the inverse of Eqs. (21.5–1a) [cf. Eqs. (21.3–7)] we obtain the relations

$$\begin{aligned}\langle[\mu, \nu; v]|\hat{a}|[\mu, \nu; v]\rangle &= \langle[\mu, \nu; v]|\mu^*\hat{A} - \nu\hat{A}^\dagger|[\mu, \nu; v]\rangle \\ &= \mu^* v - \nu v^*,\end{aligned}$$

and

$$\begin{aligned}\langle[\mu, \nu; v]|\hat{a}^2|[\mu, \nu; v]\rangle &= \langle[\mu, \nu; v]|\mu^{*2}\hat{A}^2 + \nu^2\hat{A}^{\dagger 2} - 2\mu^*\nu\hat{A}^\dagger\hat{A} - \mu^*\nu|[\mu, \nu; v]\rangle \\ &= (\mu^* v - \nu v^*)^2 - \mu^*\nu.\end{aligned}$$

Hence the dispersion of $\hat{a}$ is given by the formula

$$\langle[\mu, \nu; v]|(\Delta\hat{a})^2|[\mu, \nu; v]\rangle = -\mu^*\nu, \tag{21.5–6}$$

and is always non-zero when $\nu \neq 0$.

21.5.1 Transformed Fock states

Just as the state $|[\mu, \nu; v]\rangle$ was derived from $|v\rangle$ by application of the unitary transformation $\hat{U}(\mu, \nu)$, it is possible to define a generalization $|[\mu, \nu; n]\rangle$ of the Fock state $|n\rangle$ by applying the same transformation to $|n\rangle$ (Yuen, 1976). We then have

$$|[\mu, \nu; n]\rangle \equiv \hat{U}(\mu, \nu)|n\rangle. \tag{21.5–7}$$

It is easy to show that the newly defined states are eigenstates of the operator $\hat{N} = \hat{A}^\dagger\hat{A}$, just as the Fock states $|n\rangle$ are eigenstates of $\hat{n} = \hat{a}^\dagger\hat{a}$. For we have

$$\begin{aligned}\hat{N}|[\mu, \nu; n]\rangle &= \hat{U}\hat{a}^\dagger\hat{U}^\dagger\hat{U}\hat{a}\hat{U}^\dagger\hat{U}|n\rangle \\ &= \hat{U}\hat{a}^\dagger\hat{a}|n\rangle \\ &= n|[\mu, \nu; n]\rangle,\end{aligned} \tag{21.5–8}$$

so that the spectrum of $\hat{N}$ also consists of the integers 0, 1, 2, ... For this reason $\hat{N}$ has been called the *quasi-photon number operator* (Yuen 1976). Moreover, the action of $\hat{A}$ or $\hat{A}^\dagger$ on $|[\mu, \nu; n]\rangle$ is similar to the action of $\hat{a}$ or $\hat{a}^\dagger$ on $|n\rangle$, and we find that

$$\begin{aligned}\hat{A}|[\mu, \nu; n]\rangle &= \hat{U}\hat{a}\hat{U}^\dagger\hat{U}|n\rangle \\ &= \hat{U}\hat{a}|n\rangle \\ &= \left.\begin{aligned}&n^{1/2}|[\mu, \nu; n-1]\rangle \quad &&\text{if } n = 1, 2, 3, \ldots \\ &0 &&\text{if } n = 0,\end{aligned}\right\}\end{aligned} \tag{21.5–9}$$

and similarly

$$\hat{A}^\dagger|[\mu, \nu; n]\rangle = (n+1)^{1/2}|[\mu, \nu; n+1]\rangle. \tag{21.5–10}$$

It follows from this that the states $|[\mu, \nu; n]\rangle$ can also be written as

$$|[\mu, \nu; n]\rangle = \frac{\hat{A}^{\dagger n}}{\sqrt{(n!)}}|[\mu, \nu; 0]\rangle, \tag{21.5–11}$$

rather like the states $|n\rangle$.

Moreover, we may readily show that the states $|[\mu, \nu; n]\rangle$ form a complete orthonormal basis, just like the Fock states, because

$$\langle[\mu, \nu; n]|[\mu, \nu; n']\rangle = \langle n|\hat{U}^\dagger\hat{U}|n'\rangle = \delta_{nn'}, \tag{21.5–12}$$

$$\sum_{n=0}^{\infty}|[\mu, \nu; n]\rangle\langle[\mu, \nu; n]| = \hat{U}\sum_{n=0}^{\infty}|n\rangle\langle n|\hat{U}^\dagger = 1. \tag{21.5–13}$$

We can therefore represent any state in the basis $|[\mu, \nu; n]\rangle$. In particular, we may write

$$|[\mu, \nu; v]\rangle = \sum_n C_n|[\mu, \nu; n]\rangle,$$

and application of the same procedure that was used in Section 11.2 to expand the coherent state $|v\rangle$ in terms of $|n\rangle$, leads to the result

$$|[\mu, \nu; v]\rangle = \mathrm{e}^{-|v|^2/2}\sum_{n=0}^{\infty}\frac{v^n}{\sqrt{n!}}|[\mu, \nu; n]\rangle. \tag{21.5–14}$$

Combination of this with Eq. (21.5–11) and the Campbell–Baker–Hausdorff theorem [Eq. (10.11–26)] allows us to express $|[\mu, \nu; v]\rangle$ in the form

$$|[\mu, \nu; v]\rangle = \mathrm{e}^{v\hat{A}^\dagger - v^*\hat{A}}|[\mu, \nu; 0]\rangle, \tag{21.5–15}$$

in complete analogy with Eq. (11.3–6). We can also use the expansion (21.5–14) to show, as in Section 11.6, that the two-photon coherent states satisfy the relation

$$\langle[\mu, \nu; v]|[\mu, \nu; v']\rangle = \mathrm{e}^{(v^*v' - vv'^*)/2}\,\mathrm{e}^{-|v-v'|^2/2}, \tag{21.5–16}$$

and that

$$\frac{1}{\pi}\int|[\mu, \nu; v]\rangle\langle[\mu, \nu; v]|\,\mathrm{d}^2v = 1. \tag{21.5–17}$$

The two photon coherent states therefore form an over-complete basis for the representation of states and operators. With the same restriction as before, a diagonal two-photon coherent-state representation of the density operator exists.

21.5.2 Coherent-state representation of the two-photon coherent state

We showed in Section 11.7 that the projection $\langle v|\psi\rangle$ on the coherent state $\langle v|$ forms a unique representation of any state $|\psi\rangle$. Accordingly, we now derive the coherent-state representation of the two-photon coherent state $|[\mu, \nu; w]\rangle$. Following Yuen we first derive a differential equation for $\langle v|[\mu, \nu; w]\rangle$ by using Eq. (21.5–4) to obtain

$$\langle v|\hat{A}|[\mu, \nu; w]\rangle = \langle v|\mu\hat{a} + \nu\hat{a}^\dagger|[\mu, \nu; w]\rangle = w\langle v|[\mu, \nu; w]\rangle. \quad (21.5\text{–}18)$$

We now express $\langle v|\hat{a}|\psi\rangle$ in differential form with the help of Eq. (11.3–1), by noting that

$$\begin{aligned}\langle v|\hat{a}|\psi\rangle &= \langle \text{vac}|\text{e}^{-|v|^2/2+v^*\hat{a}}\hat{a}|\psi\rangle \\ &= \left(\frac{\partial}{\partial v^*} + \frac{1}{2}v\right)\langle \text{vac}|\text{e}^{-|v|^2/2+v^*\hat{a}}|\psi\rangle \\ &= \left(\frac{\partial}{\partial v^*} + \frac{1}{2}v\right)\langle v|\psi\rangle. \end{aligned} \quad (21.5\text{–}19)$$

If we substitute this result in Eq. (21.5–18), we obtain

$$\left[\mu\left(\frac{\partial}{\partial v^*} + \frac{1}{2}v\right) + \nu v^*\right]\langle v|[\mu, \nu; w]\rangle = w\langle v|[\mu, \nu; w]\rangle$$

or

$$\frac{\partial}{\partial v^*}\langle v|[\mu, \nu; w]\rangle = \left(\frac{w}{\mu} - \frac{1}{2}v - \frac{\nu v^*}{\mu}\right)\langle v|[\mu, \nu; w]\rangle.$$

This equation can be integrated immediately with respect to v^* to give

$$\langle v|[\mu, \nu; w]\rangle = \exp\left[\frac{wv^*}{\mu} - \frac{1}{2}|v|^2 - \frac{1}{2}\frac{\nu v^{*2}}{\mu} + f(w, w^*)\right]. \quad (21.5\text{–}20)$$

The function $f(w, w^*)$ is arbitrary, but will be chosen so as to ensure the normalization

$$\frac{1}{\pi}\int|\langle v|[\mu, \nu; w]\rangle|^2\,\text{d}^2v = 1.$$

Equation (21.5–20) then yields, apart from an arbitrary phase factor

$$\langle v|[\mu, \nu; w]\rangle = \frac{1}{\sqrt{\mu}}\exp\left[-\frac{1}{2}|v|^2 - \frac{1}{2}|w|^2 - \frac{1}{2}\frac{\nu}{\mu}v^{*2} + \frac{w}{\mu}v^* + \frac{1}{2}\frac{\nu^*}{\mu}w^2\right]. \quad (21.5\text{–}21)$$

If we decompose the complex variable v into its real and imaginary parts by writing $v = x + \text{i}y$, it follows immediately that $|\langle v|[\mu, \nu; w]\rangle|^2$, which can be interpreted as a probability density in a certain limited sense, has the form of a bivariate Gaussian distribution in x and y. However, the variables x and y are not independent in general, nor are their variances equal unless ν is zero.

21.5.3 Photon statistics of the two-photon coherent state

Although the two-photon coherent states have the same expansion in the states $|[\mu, \nu; n]\rangle$ as do the coherent states in $|n\rangle$, the photon statistics of the former are much more complicated than Poissonian (Yuen, 1976; Yuen and Shapiro, 1980; Shapiro, Yuen and Machado Mata, 1979). As a first step we now calculate the projection $\langle n|[\mu, \nu; w]\rangle$. It turns out to be convenient to make use of the coherent-state representation (21.5–21), by writing

$$\frac{1}{\sqrt{\mu}}\exp\left[-\frac{1}{2}|v|^2-\frac{1}{2}|w|^2-\frac{1}{2}\frac{\nu}{\mu}v^{*2}+\frac{w}{\mu}v^*+\frac{1}{2}\frac{\nu^*}{\mu}w^2\right]$$

$$=\langle v|[\mu, \nu; w]\rangle$$

$$=\sum_{n=0}^{\infty}\langle v|n\rangle\langle n|[\mu, \nu; w]\rangle$$

$$=\sum_{n=0}^{\infty}e^{-|v|^2/2}\frac{v^{*n}}{\sqrt{n!}}\langle n|[\mu, \nu; w]\rangle,$$

or

$$\exp\left(-\frac{1}{2}\frac{\nu}{\mu}v^{*2}+\frac{w}{\mu}v^*\right)=(\sqrt{\mu})\exp\left(\frac{1}{2}|w|^2-\frac{1}{2}\frac{\nu^*}{\mu}w^2\right)\sum_{n=0}^{\infty}\frac{v^{*n}}{\sqrt{n!}}\langle n|[\mu, \nu; w]\rangle. \tag{21.5–22}$$

We now introduce the generating function for the Hermite polynomials $H_n(z)$ (see for example Abramowitz and Stegun, 1965, Sec. 22.9; Rainville, 1960, Chap. 11)

$$e^{-z^2+2sz}=\sum_{n=0}^{\infty}\frac{1}{n!}H_n(s)z^n,$$

and use this result to expand the left-hand side of Eq. (21.5–22), after substituting

$$z=v^*\left(\frac{\nu}{2\mu}\right)^{1/2}$$

$$s=\frac{w}{(2\mu\nu)^{1/2}}.$$

We then obtain the formula

$$\sum_{n=0}^{\infty}\frac{1}{n!}H_n\left[\frac{w}{(2\mu\nu)^{1/2}}\right]v^{*n}\left(\frac{\nu}{2\mu}\right)^{n/2}=(\sqrt{\mu})\exp\left(\frac{1}{2}|w|^2-\frac{1}{2}\frac{\nu^*}{\mu}w^2\right)\sum_{n=0}^{\infty}\frac{v^{*n}}{\sqrt{n!}}\langle n|[\mu, \nu; w]\rangle, \tag{21.5–23}$$

and comparison of the coefficients of v^{*n} shows that

$$\langle n|[\mu, \nu; w]\rangle=\frac{1}{\sqrt{(\mu n!)}}\left(\frac{\nu}{2\mu}\right)^{n/2}\exp\left(-\frac{1}{2}|w|^2+\frac{1}{2}\frac{\nu^*}{\mu}w^2\right)H_n\left[\frac{w}{(2\mu\nu)^{1/2}}\right]. \tag{21.5–24}$$

The squared modulus of this gives the probability $p(n)$ that there are n photons in the two-photon coherent state:

$$p(n)=e^{-|w|^2}\exp\left(\frac{1}{2}\frac{\nu^*}{\mu}w^2+\frac{1}{2}\frac{\nu}{\mu^*}w^{*2}\right)\frac{1}{|\mu n!|}\left|\frac{\nu}{2\mu}\right|^n\left|H_n\left[\frac{w}{(2\mu\nu)^{1/2}}\right]\right|^2. \tag{21.5–25}$$

Evidently $p(n)$ is rather different from Poissonian, and its rather complicated structure makes it difficult to use in calculations.

As before the first two moments of the photon number $\hat{n}$ are derivable directly from the definitions. For the mean we obtain the relation

$$\begin{aligned}\langle \hat{n} \rangle &= \langle [\mu, v; w] | \hat{a}^\dagger \hat{a} | [\mu, v; w] \rangle \\ &= \langle [\mu, v; w] | (\mu \hat{A}^\dagger - v^* \hat{A})(\mu^* \hat{A} - v \hat{A}^\dagger) | [\mu, v; w] \rangle \\ &= \langle [\mu, v; w] | \, |\mu|^2 \hat{A}^\dagger \hat{A} + |v|^2 (\hat{A}^\dagger \hat{A} + 1) - \mu v \hat{A}^{\dagger 2} - \mu^* v^* \hat{A}^2 | [\mu, v; w] \rangle \\ &= |\mu w^* - v^* w|^2 + |v^2|. \end{aligned} \tag{21.5–26}$$

Also with the help of the commutation relations for $\hat{a}$, $\hat{a}^\dagger$ we find

$$\begin{aligned}\langle \hat{n}(\hat{n} - 1) \rangle &= \langle [\mu, v; w] | \hat{a}^{\dagger 2} \hat{a}^2 | [\mu, v; w] \rangle \\ &= \langle [\mu, v; w] | (\mu \hat{A}^\dagger - v^* \hat{A})^2 (\mu^* \hat{A} - v \hat{A}^\dagger)^2 | [\mu, v; w] \rangle \\ &= |(\mu w^* - v^* w)^2 - \mu v^*|^2 + |v|^2 [4|\mu|^2 |w|^2 + 4|v|^2 |w|^2 + 2|v|^2 \\ &\quad - 4\mu^* v^* w^2 - 4\mu v w^{*2}], \end{aligned} \tag{21.5–27}$$

so that

$$\begin{aligned}\langle (\Delta \hat{n})^2 \rangle - \langle \hat{n} \rangle &= \langle \hat{n}(\hat{n} - 1) \rangle - \langle \hat{n} \rangle^2 \\ &= |v|^2 [8|v|^2 |w|^2 + 6|w|^2 + |\mu|^2 + |v|^2] - (4|v|^2 + 1)[\mu v w^{*2} + \text{c.c.}]. \end{aligned} \tag{21.5–28}$$

The difference $\langle (\Delta \hat{n})^2 \rangle - \langle \hat{n} \rangle$, measuring the departure from Poisson statistics, can be either positive, zero, or negative, depending on the values of μ, v, w. It always vanishes when $v = 0$, in which case we recover the coherent state, but it can be negative for certain combinations of μ, v, w, as we showed in Section 21.4 for the ideal squeezed states.

21.6 Detection of squeezing by homodyning with coherent light

We have seen that in a quadrature squeezed state a light beam has one field quadrature that fluctuates less than it does in the vacuum state, while the other one fluctuates more. If information could be impressed on and extracted from the former quadrature, this would offer obvious advantages for measurements of weak signals. So far our treatment of squeezing has been entirely theoretical; we now turn briefly to the question how squeezing might be recognized in practice.

Let us consider the experimental situation illustrated in Fig. 21.4, in which the incoming single-mode squeezed light is mixed by a beam splitter with the coherent light of a strong local oscillator of the same optical frequency, whose phase θ can be varied. In order to simplify the problem as much as possible we suppose that the transmissivity of the beam splitter is close to unity, so that the squeezed light beam is hardly affected by it, while its reflectivity $\mathcal{R}$ is very small. The complex amplitude V of the local oscillator is assumed to be sufficiently large that $\mathcal{R}V = v$ is still large and may be treated classically, while the amplitude $\hat{a}$ of the incoming squeezed light is treated as an operator. The combined light of amplitude $\hat{a} + v$

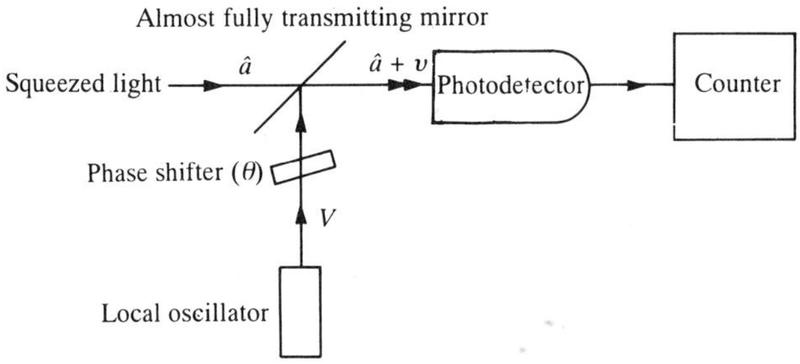

Fig. 21.4 Detection of a squeezed state by interference or homodyning with the coherent light beam of a local oscillator.

then falls on a photodetector that counts the number n of photons detected in some time interval short compared with the coherence time. The moments of n can be determined by repeating the measurement many times. The object of the experiment is to vary the local oscillator phase θ so as to identify the squeezed field quadrature, which manifests itself in reduced fluctuations of the combined light intensity.

This interference or homodyne problem has been treated several times (Yuen and Shapiro, 1980; Shapiro, Yuen and Machado Mata, 1979; Mandel, 1982a). Here we shall largely follow the last treatment. The field amplitude at the detector is proportional to $\hat{a} + v$. Hence the average number of counts $\langle \hat{n} \rangle$ registered in some short time interval is given by the expression

$$\langle \hat{n} \rangle = \eta \langle (\hat{a}^\dagger + v^*)(\hat{a} + v) \rangle, \tag{21.6–1}$$

and for the second factorial moment of the number $\hat{n}$ we have, since $\hat{n}^{(2)} = :\hat{n}^2:$,

$$\langle \hat{n}(\hat{n} - 1) \rangle = \eta^2 \langle (\hat{a}^\dagger + v^*)^2 (\hat{a} + v)^2 \rangle. \tag{21.6–2}$$

The parameter η is a measure of the detector efficiency and of the counting time. We first use Eqs. (21.1–7) to express $\hat{a}$, $\hat{a}^\dagger$ in terms of the quadrature amplitudes $\hat{Q}$, $\hat{P}$,

$$\left.\begin{aligned} \hat{a} &= \tfrac{1}{2}(\hat{Q} + \mathrm{i}\hat{P})\,\mathrm{e}^{\mathrm{i}\beta} \\ \hat{a}^\dagger &= \tfrac{1}{2}(\hat{Q} - \mathrm{i}\hat{P})\,\mathrm{e}^{-\mathrm{i}\beta}. \end{aligned}\right\} \tag{21.6–3}$$

It follows that

$$\begin{aligned} \langle \hat{n} \rangle &= \eta \langle [\tfrac{1}{2}(\hat{Q} - \mathrm{i}\hat{P})\,\mathrm{e}^{-\mathrm{i}\beta} + v^*][\tfrac{1}{2}(\hat{Q} + \mathrm{i}\hat{P})\,\mathrm{e}^{\mathrm{i}\beta} + v] \rangle \\ &= \eta \{ \tfrac{1}{4}[\langle \hat{Q}^2 \rangle + \langle \hat{P}^2 \rangle + \mathrm{i}\langle \hat{Q}\hat{P} - \hat{P}\hat{Q} \rangle] + [\tfrac{1}{2}(\langle \hat{Q} \rangle - \mathrm{i}\langle \hat{P} \rangle) v\,\mathrm{e}^{-\mathrm{i}\beta} + \mathrm{c.c.}] + |v|^2 \}, \end{aligned}$$

and with the help of the commutation relation $[\hat{Q}, \hat{P}] = 2\mathrm{i}$, this becomes, if $v = |v|\,\mathrm{e}^{\mathrm{i}\theta}$,

$$\langle \hat{n} \rangle = \eta \{ \tfrac{1}{4}(\langle \hat{Q}^2 \rangle + \langle \hat{P}^2 \rangle - 2) + |v|[\langle \hat{Q} \rangle \cos(\theta - \beta) + \langle \hat{P} \rangle \sin(\theta - \beta)] + |v|^2 \}. \tag{21.6–4}$$

Similarly we may evaluate the second factorial moment $\langle \hat{n}(\hat{n} - 1) \rangle$. Finally we combine the results to determine the departure from Poisson statistics and we obtain, after some rearrangement of terms,

$$
\begin{aligned}
\langle(\Delta\hat{n})^2\rangle - \langle\hat{n}\rangle &= \langle\hat{n}(\hat{n}-1)\rangle - \langle\hat{n}\rangle^2 \\
&= \eta^2|v|^2[\langle:(\Delta\hat{Q})^2:\rangle\cos^2(\theta-\beta) + \langle:(\Delta\hat{P})^2:\rangle\sin^2(\theta-\beta) \\
&\quad + \tfrac{1}{2}(\langle\Delta\hat{P}\Delta\hat{Q}\rangle + \langle\Delta\hat{Q}\Delta\hat{P}\rangle)\sin 2(\theta-\beta)] \\
&\quad + \eta^2[\langle\hat{a}^{\dagger 2}\hat{a}^2\rangle - \langle\hat{a}^\dagger\hat{a}\rangle^2 + 2v\langle\Delta\hat{a}^\dagger\Delta\hat{n}_a\rangle + 2v^*\langle\Delta\hat{n}_a\Delta\hat{a}\rangle],
\end{aligned}
\tag{21.6–5}
$$

where $\hat{n}_a \equiv \hat{a}^\dagger\hat{a}$. We have explicitly introduced the phase of v by writing

$$v = |v|e^{i\theta}, \tag{21.6–6}$$

in order to show that the difference $\langle(\Delta\hat{n})^2\rangle - \langle\hat{n}\rangle$ can be made to change with the phase θ of the coherent local oscillator field. If the latter is made sufficiently intense, then the dominant term in (21.6–5) is the term proportional to $|v|^2$, and the other terms can be discarded.

By choosing $\theta - \beta = 0$ and $\theta - \beta = \pi/2$ in turn, we readily see that

$$
\left.
\begin{aligned}
\langle(\Delta\hat{n})^2\rangle - \langle\hat{n}\rangle &\approx \eta^2|v|^2\langle:(\Delta\hat{Q})^2:\rangle, \quad \text{if } \theta-\beta = 0, \\
&\approx \eta^2|v|^2\langle:(\Delta\hat{P})^2:\rangle, \quad \text{if } \theta-\beta = \pi/2.
\end{aligned}
\right\}
\tag{21.6–7}
$$

It follows that if the state of the field is a squeezed state, so that $\langle:(\Delta\hat{Q})^2:\rangle < 0$ for some angle β [see (Eq. (21.2–3)], then $\langle(\Delta\hat{n})^2\rangle - \langle\hat{n}\rangle < 0$ for that choice of phase angle. The photon counting fluctuations are therefore narrower than Poissonian for some value of the phase angle β, and this confirms that the state is intrinsically quantum mechanical. We see that interference or homodyning of squeezed light converts the squeezing into sub-Poissonian photon statistics, which shows up easily in a measurement. Therefore the act of turning on the squeezed light in Fig. 21.4 actually *lowers* the fluctuations of the photoelectric counts. This is of course a reflection of the fact that the fluctuations of the field in a squeezed state are smaller than in the vacuum state. As β is varied, the numerically largest, most negative value of the dimensionless ratio

$$\mathcal{Q} = [\langle(\Delta\hat{n})^2\rangle - \langle\hat{n}\rangle]/\langle\hat{n}\rangle, \quad \mathcal{Q} \geqslant -1,$$

provides a convenient measure of the amount of squeezing achieved. The ideal squeezed state $|(v, z)\rangle$ with $z \neq 0$ is an example of a state for which $\mathcal{Q}$ is always negative, and when $\eta = 1$, $\mathcal{Q}$ tends towards its limiting value $\mathcal{Q} = -1$ as $|z| \to \infty$.

21.7 Squeezing produced in practice: Degenerate parametric down-conversion

So far we have only discussed squeezed states that were constructed mathematically, although the treatment of the two-photon coherent states in Section 21.5 suggested a possible mechanism for realizing the conditions under which squeezing might be achieved in practice. Actually the theory indicates that squeezing should show up in numerous situations in which light interacts with a nonlinear medium. Examples are resonance fluorescence (Walls and Zoller, 1981), harmonic generation (Mandel, 1982b), four-wave mixing (Yuen and Shapiro, 1979; Reid and Walls, 1985, Yurke, 1985a,b), parametric amplification and down-conversion (Milburn and Walls, 1981; Lugiato and Strini, 1982a,b,c; Friberg and

Mandel, 1984; Gardiner and Savage, 1984; Collett and Gardiner, 1984; Collett and Walls, 1985), and optical bistability (Lugiato and Strini, 1982c).

Squeezing was first demonstrated in the laboratory by Slusher, Hollberg, Yurke, Mertz and Valley (1985) in the process of four-wave mixing in a beam of sodium (see Chapter 22). Their apparatus is shown in Fig. 21.5 (*a*), and the results are reproduced in (*b*). The light emerging from the resonant cavity via SM2 is mixed with the light from the local oscillator (the dye laser) by a beam splitter BS2, and the combined beams fall on detectors D_A and D_B. The difference between the two photocurrent fluctuations is measured in order to cancel the effects of local oscillator noise (Yuen and Chan, 1983). As the phase ϕ_{LO} of the local oscillator is varied, the measured noise difference fluctuations sometimes dip a few percent below the corresponding level with the cavity field blocked (the vacuum level). This is the signature of the squeezed state.

Much larger squeezing effects have been observed in the process of parametric down-conversion by Wu, Kimble, Hall and Wu (1986). Their results are illustrated in Fig. 21.6(*a*) and (*b*), which also shows the measured r.m.s. noise voltage as a function of local oscillator phase. The phase space distribution in (*a*) to the left of the experimental curve in (*b*) is meant to illustrate the magnitude of the observed squeezing effect by the eccentricity of the ellipse. Because of the large size of the effect, we shall briefly examine the theory of squeezed state production in the process of degenerate down-conversion.

21.7.1 Production of a squeezed state in degenerate parametric down-conversion

As an example of a squeezed state we now consider the process of degenerate parametric down-conversion, in which a coherent beam of light at frequency ω_0 interacts with a nonlinear crystal so as to produce light at the sub-harmonic

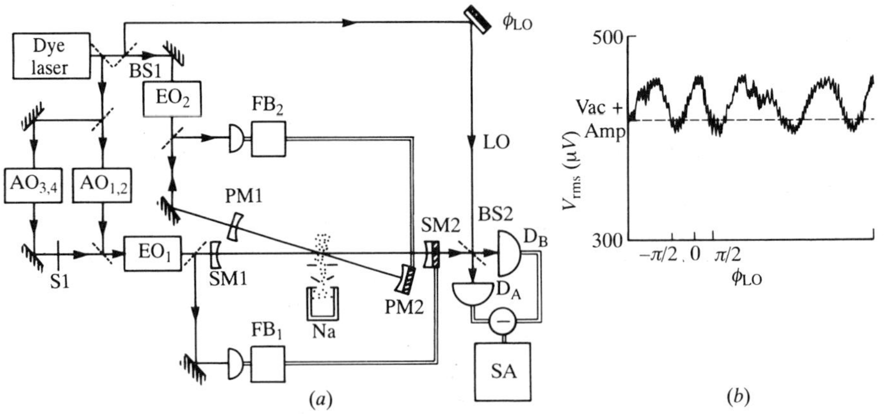

Fig. 21.5 (*a*) Outline of the apparatus used in the first successful demonstration of squeezing. LO stands for local oscillator and ϕ_{LO} is the local oscillator phase shifter. The difference between the signals from the two photon detectors D_A and D_B is fed to the spectrum analyzer SA. (*b*) Observed r.m.s. noise as a function of local oscillator phase. The thin horizontal line gives the results with the squeezed light turned off. (Reproduced from Slusher, Hollberg, Yurke, Mertz and Valley, 1985.)

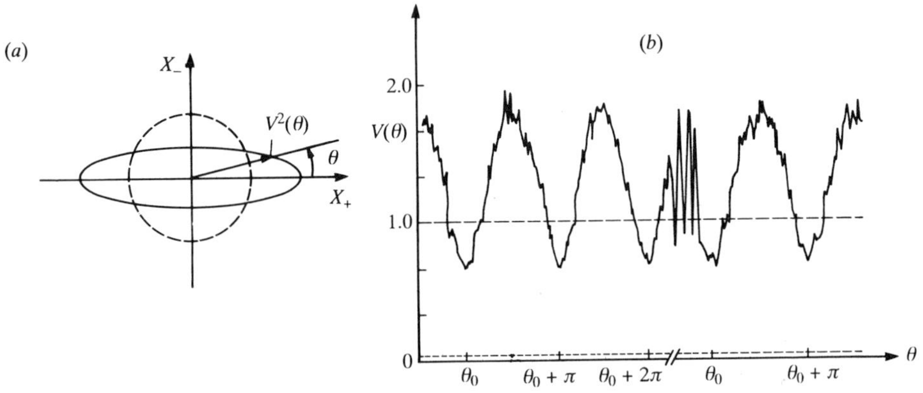

Fig. 21.6 (*a*) Phase space distribution corresponding to the experimental results shown in (*b*); (*b*) r.m.s. noise as a function of local oscillator phase observed in parametric down-conversion experiments. (Reproduced from Wu, Kimble, Hall and Wu, 1986.)

frequency $\frac{1}{2}\omega_0$. A photon of energy $\hbar\omega_0$ in effect splits into two photons of frequency $\omega_1 = \frac{1}{2}\omega_0$. If the incident field is very intense, we may be justified in treating it classically, although the down-converted field has to be quantized.

The simplest single-mode Hamiltonian that describes this interaction has the form

$$\hat{H} = \hbar\omega_1 \hat{a}_1^\dagger \hat{a}_1 + \hbar g(\hat{a}_1^{\dagger 2} v_0 \mathrm{e}^{-2\mathrm{i}\omega_1 t} + v_0^* \hat{a}_1^2 \mathrm{e}^{2\mathrm{i}\omega_1 t}), \tag{21.7–1}$$

in which v_0 is the complex amplitude of the incident light beam, and g is a real mode coupling constant that depends on the nonlinear susceptibility of the medium. The interaction term corresponds to the simultaneous creation or annihilation of two photons of mode 1. The Heisenberg equation of motion for $\hat{a}_1(t)$ then becomes

$$\dot{\hat{a}}_1 = \frac{1}{\mathrm{i}\hbar}[\hat{a}_1, \hat{H}] = -\mathrm{i}\omega_1 \hat{a}_1 - 2\mathrm{i}g\hat{a}_1^\dagger v_0 \mathrm{e}^{-2\mathrm{i}\omega_1 t}, \tag{21.7–2}$$

and this has the general solution (Mollow, 1973; for a general discussion see Yariv, 1967, Sec. 22.3)

$$\hat{a}_1(t) = \hat{a}_1(0)\cosh(2g|v_0|t)\,\mathrm{e}^{-\mathrm{i}\omega_1 t} - \mathrm{i}\frac{v_0}{|v_0|}\hat{a}_1^\dagger(0)\sinh(2g|v_0|t)\,\mathrm{e}^{-\mathrm{i}\omega_1 t}, \tag{21.7–3}$$

as may be confirmed by substitution back into the differential equation.

We can use this result to calculate any expectation of interest at time t, on the assumption that the field starts in the vacuum state, with $\hat{a}_1(0)|\mathrm{vac}\rangle = 0 = \langle\mathrm{vac}|\hat{a}_1^\dagger(0)$. Thus, we find immediately that $\langle\hat{a}_1(t)\rangle = 0$, whereas

$$\begin{aligned}\langle\hat{a}_1^2(t)\rangle &= -\mathrm{i}\frac{v_0}{|v_0|}\sinh(2g|v_0|t)\cosh(2g|v_0|t)\,\mathrm{e}^{-2\mathrm{i}\omega_1 t}\langle\hat{a}_1(0)\hat{a}_1^\dagger(0)\rangle \\ &= -\frac{1}{2}\mathrm{i}\frac{v_0}{|v_0|}\sinh(4g|v_0|t)\,\mathrm{e}^{-2\mathrm{i}\omega_1 t},\end{aligned} \tag{21.7–4}$$

and

$$\langle \hat{n}_1(t) \rangle = \langle \hat{a}_1^\dagger(t)\hat{a}_1(t) \rangle = \sinh^2(2g|v_0|t)\langle \hat{a}_1(0)\hat{a}_1^\dagger(0) \rangle$$
$$= \sinh^2(2g|v_0|t). \tag{21.7–5}$$

If we now construct the Hermitian operator

$$\hat{Q} = \hat{a}_1\,\mathrm{e}^{\mathrm{i}(\omega_1 t-\beta)} + \hat{a}_1^\dagger\,\mathrm{e}^{-\mathrm{i}(\omega_1 t-\beta)}, \tag{21.7–6}$$

(which becomes $\hat{P}$ when β is replaced by $\beta + \pi/2$) we find for the dispersion of $\hat{Q}$,

$$\langle [\Delta Q(t)]^2 \rangle = \langle \hat{a}_1^2(t) \rangle\,\mathrm{e}^{2\mathrm{i}(\omega_1 t-\beta)} + \langle \hat{a}_1^{\dagger 2}(t) \rangle\,\mathrm{e}^{-2\mathrm{i}(\omega_1 t-\beta)} + 2\langle \hat{a}_1^\dagger(t)\hat{a}_1(t) \rangle + 1,$$

and from Eqs. (21.7–4) and (21.7–5) it follows that

$$\langle [\Delta Q(t)]^2 \rangle = -\sinh(4g|v_0|t)\sin(2\beta - \arg v) + \cosh(4g|v_0|t). \tag{21.7–7}$$

If we choose the phase angle β so that $2\beta - \arg v = +\pi/2$, then $\langle (\Delta\hat{Q}(t))^2 \rangle = \mathrm{e}^{-4g|v_0|t}$, which is less than unity for all $t > 0$. Hence the field that results from the degenerate parametric down-conversion or frequency splitting of a coherent light beam is always squeezed, and this squeezing can be substantial (Wu, Kimble, Hall and Wu, 1986).

21.8 Broadband squeezed light

Up to now we have discussed squeezing within the context of a single-mode field. Certain processes, such as non-degenerate parametric down-conversion, naturally generate two photons simultaneously in two conjugate modes whose frequencies are simply related. The corresponding state can be generated from the vacuum by the action of the *two-mode squeeze operator* [cf. Eq. (21.3–1)]

$$\hat{S}_{12}(z) = \mathrm{e}^{z^*\hat{a}_1\hat{a}_2 - z\hat{a}_1^\dagger\hat{a}_2^\dagger}. \tag{21.8–1}$$

This may be followed by displacements in each mode represented by the operators

$$\hat{D}_1(v_1) = \mathrm{e}^{v_1\hat{a}_1^\dagger - v_1^*\hat{a}_1}$$
$$\hat{D}_2(v_2) = \mathrm{e}^{v_2\hat{a}_2^\dagger - v_2^*\hat{a}_2}.$$

More generally, one has to deal with a field having a continuum of modes, some of which may exhibit squeezing while some do not. Experimentally it is possible to make a spectral analysis of the photoelectric current produced by homodyning or heterodyning, as shown in Fig. 21.7, and to measure the fluctuations of different Fourier components, rather than of the total photoelectric signal. Squeezing may be present only for some Fourier components and not for others, or for different Fourier components to different degrees.

We shall now briefly analyze this situation, which has been treated many times in different ways (Yuen and Shapiro, 1980; Walls and Zoller, 1981; Yuen and Chan, 1983; Collett, Walls and Zoller, 1984; Collett and Gardiner, 1984; Friberg and Mandel, 1984; Gardiner and Savage, 1984; Heidmann, Reynaud and Cohen-Tannoudji, 1984; Loudon, 1984; Richter, 1984; Schumaker, 1984; Caves and Schumaker, 1985; Schumaker and Caves, 1985; Collett and Walls, 1985; Gardiner

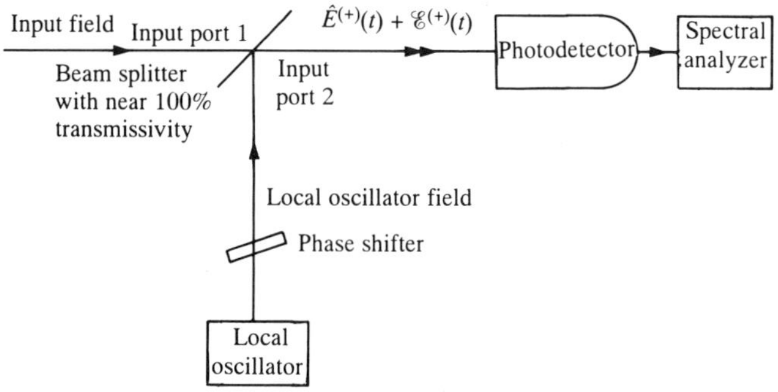

Fig. 21.7 Outline of the homodyne experiment for studying the spectral density of the field fluctuations.

and Collett, 1985; Shapiro, 1985; Yurke, 1985a,b; Ou, Hong and Mandel, 1987). Our approach is based on the method of the last named authors. To simplify the calculations, we consider a polarized, quasi-monochromatic quantum field of midfrequency ω_1, whose positive frequency part is represented by the scalar field operator $\hat{E}^{(+)}(\mathbf{r}, t)$. We begin by decomposing $\hat{E}^{(+)}(\mathbf{r}, t)$ into discrete, normal, plane-wave modes in the form [cf. Eq. (10.4–39)]

$$\hat{E}^{(+)}(\mathbf{r}, t) = \frac{1}{\sqrt{\mathcal{V}}}\sum_{[\mathbf{k}]} l(\omega)\hat{a}_{\mathbf{k}}\, e^{i(\mathbf{k}\cdot\mathbf{r}-\omega t)}, \tag{21.8–2}$$

in which $\mathcal{V}$ is the quantization volume, $l(\omega)$ is some simple frequency-dependent factor [such as $i(\hbar\omega/2\varepsilon_0)^{1/2}$ for the electric field], and $[\mathbf{k}]$ stands for the set of plane-wave modes to which we confine our attention, and to which the photodetector responds. At a later state of the calculation we shall let $\mathcal{V} \to \infty$, in which case the sum becomes an integral over the continuum. Henceforth we suppress the position variable $\mathbf{r}$ to simplify the notation. The term $\hat{E}^{(+)}(t)$ and its conjugate $\hat{E}^{(-)}(t)$ are annihilation and creation operators that obey the equal-time commutation relation

$$[\hat{E}^{(+)}(t), \hat{E}^{(-)}(t)] = \frac{1}{\mathcal{V}}\sum_{[\mathbf{k}]} |l(\omega)|^2 \equiv C. \tag{21.8–3}$$

It will be useful to adopt units in which the light intensity $\hat{I} = \hat{E}^{(-)}\hat{E}^{(+)}$ is expressed in photons per second. Then C is expressed in units of photons per second also.

In analogy with Eqs. (21.1–7) we now construct two Hermitian quadrature operators $\hat{E}_1(t)$, $\hat{E}_2(t)$, namely

$$\left.\begin{aligned}\hat{E}_1(t) &\equiv \hat{E}^{(+)}(t)\, e^{i(\omega_1 t-\beta)} + \hat{E}^{(-)}(t)\, e^{-i(\omega_1 t-\beta)}\\ \hat{E}_2(t) &\equiv \hat{E}^{(+)}(t)\, e^{i(\omega_1 t-\beta-\pi/2)} + \hat{E}^{(-)}(t)\, e^{-i(\omega_1 t-\beta-\pi/2)},\end{aligned}\right\} \tag{21.8–4}$$

where β is arbitrary for now. The oscillatory factors are introduced in order to make $\hat{E}_1(t)$, $\hat{E}_2(t)$ slowly varying functions of t. This allows us to express the total field $\hat{E}(t)$ in the form

$$\hat{E}(t) = \hat{E}_1(t)\cos(\omega_1 t - \beta) - \hat{E}_2(t)\sin(\omega_1 t - \beta). \tag{21.8–5}$$

From Eq. (21.8–3) $\hat{E}_1(t)$ and $\hat{E}_2(t)$ obey the equal-time commutation relations

$$[\hat{E}_1(t),\ \hat{E}_2(t)] = 2\mathrm{i}C, \tag{21.8–6a}$$

which shows that they are canonical conjugates satisfying the uncertainty relations

$$[\langle(\Delta\hat{E}_1)^2\rangle\langle(\Delta\hat{E}_2)^2\rangle]^{1/2} \geqslant C. \tag{21.8–6b}$$

It is easy to show that in the vacuum state $\langle(\Delta\hat{E}_1)^2\rangle = C = \langle(\Delta\hat{E}_2)^2\rangle$. Accordingly, the state is a squeezed state if for some angle β (cf. Section 21.1)

$$\langle(\Delta\hat{E}_1)^2\rangle < C, \tag{21.8–7}$$

and we shall refer to this condition as *squeezing in the full sense*, because we are dealing with the fluctuations of the total field.

21.8.1 Homodyning and correlation functions

The following treatment is based on the approach of Ou, Hong and Mandel (1987). Let us suppose that the local oscillator in Fig. 21.7 generates a strong, monochromatic field at frequency ω_1 in a coherent state, that we represent classically by the c-number $\mathscr{E}^{(+)}(t)/\mathscr{R}$ before the beam splitter, with

$$\mathscr{E}^{(+)}(t) = |\mathscr{E}|\,\mathrm{e}^{-\mathrm{i}(\omega_1 t-\theta)}, \quad (\mathscr{E}^{(-)}\mathscr{E}^{(+)} = |\mathscr{E}|^2), \tag{21.8–8}$$

where $\mathscr{R}$ is the reflectivity of the beam splitter. If $|\mathscr{R}| \ll 1$ and the beam splitter transmissivity is close to unity, then the total field at the detector becomes $\hat{E}^{(+)}(t) + \mathscr{E}^{(+)}(t)$. The expected value of the light intensity at the detector is therefore of the form

$$\begin{aligned}\langle\hat{I}(t)\rangle &= \langle[\hat{E}^{(-)}(t) + \mathscr{E}^{(-)}(t)][\hat{E}^{(+)}(t) + \mathscr{E}^{(+)}(t)]\rangle \\ &= |\mathscr{E}|^2 + \mathscr{E}^{(-)}(t)\langle\hat{E}^{(+)}(t)\rangle + \langle\hat{E}^{(-)}(t)\rangle\mathscr{E}^{(+)}(t) + \langle\hat{E}^{(-)}(t)\hat{E}^{(+)}(t)\rangle\end{aligned} \tag{21.8–9}$$

and the normally ordered, time-ordered two-time intensity correlation function (cf. Section 12.2) for $\tau \geqslant 0$ is given by

$$\begin{aligned}\Gamma^{(2,2)}(\tau,\theta) &\equiv \langle[\hat{E}^{(-)}(t) + \mathscr{E}^{(-)}(t)][\hat{E}^{(-)}(t+\tau) + \mathscr{E}^{(-)}(t+\tau)] \\ &\quad\times[\hat{E}^{(+)}(t+\tau) + \mathscr{E}^{(+)}(t+\tau)][\hat{E}^{(+)}(t) + \mathscr{E}^{(+)}(t)]\rangle \\ &= |\mathscr{E}|^4 + |\mathscr{E}|^3[\langle\hat{E}^{(+)}(t)\rangle\,\mathrm{e}^{\mathrm{i}(\omega_1 t-\theta)} + \langle\hat{E}^{(+)}(t+\tau)\rangle\,\mathrm{e}^{\mathrm{i}(\omega_1(t+\tau)-\theta)} + \mathrm{c.c.}] \\ &\quad + |\mathscr{E}|^2\{\langle\hat{E}^{(-)}(t)\hat{E}^{(+)}(t)\rangle + \langle\hat{E}^{(-)}(t+\tau)\hat{E}^{(+)}(t+\tau)\rangle \\ &\qquad + [\langle\hat{E}^{(-)}(t)\hat{E}^{(+)}(t+\tau)\rangle\,\mathrm{e}^{\mathrm{i}\omega_1\tau} + \mathrm{c.c.}] \\ &\qquad + [\langle\hat{E}^{(-)}(t)\hat{E}^{(-)}(t+\tau)\rangle\,\mathrm{e}^{-\mathrm{i}(2\omega_1 t+\omega_1\tau-2\theta)} + \mathrm{c.c.}]\} \\ &\quad + O[|\mathscr{E}|].\end{aligned} \tag{21.8–10}$$

If the local oscillator is strong and $|\mathscr{E}|^2 \gg 1$, the higher-order terms in $|\mathscr{E}|$ dominate the lower-order ones. From Eqs. (21.8–9) and (21.8–10) we obtain for the correlation of the intensity fluctuations

$$\begin{aligned}\langle \mathscr{T}{:}\Delta\hat{I}(t)\Delta\hat{I}(t+\tau){:}\rangle &= \Gamma^{(2,2)}(\tau,\theta) - \langle \hat{I}(t)\rangle\langle \hat{I}(t+\tau)\rangle \\ &= |\mathscr{E}|^2\{\Gamma^{(1,1)}(\tau) + \Gamma^{(2,0)}(\tau)\,\mathrm{e}^{2\mathrm{i}\theta} + \text{c.c.}\} + O(|\mathscr{E}|).\end{aligned} \tag{21.8–11}$$

Here we have introduced the notation for second-order correlation functions‡:

$$\left.\begin{aligned}\langle \Delta\hat{E}^{(-)}(t)\Delta\hat{E}^{(+)}(t+\tau)\rangle\,\mathrm{e}^{\mathrm{i}\omega_1\tau} &\equiv \Gamma^{(1,1)}(\tau)\\ \langle \Delta\hat{E}^{(-)}(t)\Delta\hat{E}^{(-)}(t+\tau)\rangle\,\mathrm{e}^{-\mathrm{i}\omega_1(2t+\tau)} &\equiv \Gamma^{(2,0)}(\tau)\quad (\Delta\hat{E} \equiv \hat{E} - \langle \hat{E}\rangle).\end{aligned}\right\} \tag{21.8–12}$$

Correlations of the type $\Gamma^{(2,0)}(\tau)$ vanish for strictly stationary fields, but because the field fluctuations in the squeezed state are assumed to depend on the phase relative to some nearly monochromatic carrier, the field is not strictly stationary. Moreover, $\langle \Delta\hat{E}^{(-)}(t)\Delta\hat{E}^{(-)}(t+\tau)\rangle$ still depends on t, but $\Gamma^{(2,0)}(\tau)$ does not, and $\langle \hat{E}^{(+)}(t)\rangle\,\mathrm{e}^{\mathrm{i}\omega_1 t}$ is not zero but is independent of t. The right-hand side of Eq. (21.8–11) is then independent of t also. We shall refer to this field as *quasi-stationary*. It has been assumed so far that $\tau \geqslant 0$, but with the help of the symmetry property

$$\left.\begin{aligned}\Gamma^{(1,1)}(-\tau) &= \Gamma^{(1,1)*}(\tau)\\ \Gamma^{(2,0)}(-\tau) &= \Gamma^{(2,0)}(\tau),\end{aligned}\right\} \tag{21.8–13}$$

Eq. (21.8–11) holds for both positive and negative τ.

Finally, we introduce the normalized intensity correlation function

$$\lambda(\tau,\theta) \equiv \langle \mathscr{T}{:}\Delta\hat{I}(t)\Delta\hat{I}(t+\tau){:}\rangle / \langle \hat{I}(t)\rangle\langle \hat{I}(t+\tau)\rangle. \tag{21.8–14}$$

Then, since $\langle \hat{I}(t)\rangle \approx |\mathscr{E}|^2$ from Eq. (21.8–9) to a good approximation, it follows that

$$\lambda(\tau,\theta) \approx \frac{1}{|\mathscr{E}|^2}[\Gamma^{(1,1)}(\tau) + \Gamma^{(2,0)}(\tau)\,\mathrm{e}^{2\mathrm{i}\theta} + \text{c.c.}]. \tag{21.8–15}$$

Because $\Gamma^{(2,2)}(\tau,\theta) \geqslant 0$ by definition, it follows immediately that $\lambda(\tau,\theta) \geqslant -1$. We shall see shortly that $\lambda(\tau,\theta) < 0$ in the squeezed state, and that the limit $\lambda(\tau,\theta) = -1$ corresponds to perfect squeezing.

We now need to relate $\lambda(\tau,\theta)$ to the homodyne measurement illustrated in Fig. 21.7. When the light falls on the photoelectric detector, it causes photoelectric emissions at certain times $t_1, t_2, \ldots$. If $k(t)$ is the current output pulse produced by a photoemission at time $t = 0$, then the total photoelectric current $J(t)$ can be represented by the sum of pulses over all times t_j, namely

$$J(t) = \sum_j k(t - t_j). \tag{21.8–16}$$

From a similar kind of analysis as was given in Section 14.6, we find for the mean and the autocorrelation function of the current $J(t)$,

$$\langle J(t)\rangle = \eta|\mathscr{E}|^2\int_{-\infty}^{\infty} k(t-t')\,\mathrm{d}t' \equiv \eta|\mathscr{E}|^2 q \tag{21.8–17}$$

‡ The correlation functions defined by Eqs. (21.8–12) differ from the correlation functions $\Gamma^{(1,1)}$ and $\Gamma^{(2,2)}$ introduced earlier, in that the fields $\hat{E}^{(-)}$ and $\hat{E}^{(+)}$ are now replaced by $\Delta\hat{E}^{(-)}$ and $\Delta\hat{E}^{(+)}$.

$$\langle \Delta J(t)\Delta J(t+\tau)\rangle = \eta|\mathscr{E}|^2\int_0^\infty \mathrm{d}t' k(t')k(t'+\tau) + \eta^2|\mathscr{E}|^4\iint_0^\infty \mathrm{d}t'\,\mathrm{d}t'' k(t')k(t'')\lambda(\tau+t'-t'',\theta), \qquad (21.8\text{–}18)$$

where η is characteristic of the photodetector and q is the total electric charge delivered by the current pulse resulting from one photoelectron. As $|\mathscr{E}|^2$ like $\hat{E}^{(-)}\hat{E}^{(+)}$ is in units of photons per second, $J(t)$ has units of charge per second, as required, and η is evidently the probability that one photon gives rise to a photoelectric pulse.

Let us examine Eq. (21.8–18) a little more closely. For an incident field which is completely coherent, $\lambda(\tau, \theta) = 0$ and the second term on the right vanishes. The first term on the right then accounts for the current fluctuations, which are sometimes referred to as *shot noise*. But if the incident field at port 1 in Fig. 21.7 is blocked completely, so that only the 'vacuum field' enters at input port 1, then the second term is again zero and only the first term remains. This time we might be inclined to interpret the current fluctuations as a consequence of the vacuum field fluctuations entering at input port 1. Clearly these two interpretations are equivalent. In general, when an incoherent field enters at port 1 the current fluctuations increase over the vacuum or shot noise level. But if the input field is in a squeezed state, then for some phase of the local oscillator the current fluctuations are reduced below the vacuum level. This is the signature of squeezing in the full sense. We shall now show that this interpretation agrees with the usual definition of squeezing, by expressing $\lambda(\tau, \theta)$ in terms of the field quadratures.

21.8.2 *Quadrature correlations*

Let us first write $\hat{E}^{(+)}(t)$ and $\hat{E}^{(-)}(t)$ in terms of the quadrature amplitudes $\hat{E}_1(t)$, $\hat{E}_2(t)$ given by Eqs. (21.8–4). It is convenient to introduce the following normally ordered, time-ordered correlation functions of the fluctuations $\Delta\hat{E}_j(t)$ $(j = 1, 2)$,

$$\langle \mathscr{T}{:}\Delta\hat{E}_i(t)\Delta\hat{E}_j(t+\tau){:}\rangle \equiv \Gamma_{ij}(\tau), \quad (i, j = 1, 2). \qquad (21.8\text{–}19)$$

The time-ordering symbol $\mathscr{T}$ time orders the $\hat{E}^{(+)}(t)$ and $\hat{E}^{(-)}(t)$ operators among themselves, so that

$$\mathscr{T}\hat{E}^{(-)}(t_2)\hat{E}^{(-)}(t_1) \equiv \hat{E}^{(-)}(t_1)\hat{E}^{(-)}(t_2) \qquad \text{if } t_1 < t_2$$
$$\mathscr{T}\hat{E}^{(+)}(t_1)\hat{E}^{(+)}(t_2) \equiv \hat{E}^{(+)}(t_2)\hat{E}^{(+)}(t_1) \qquad \text{if } t_1 < t_2.$$

Then we readily find from the definitions (21.8–4) that

$$\left.\begin{aligned} \mathrm{Re}\,\Gamma^{(1,1)}(\tau) &= \tfrac{1}{4}[\Gamma_{11}(\tau) + \Gamma_{22}(\tau)] \\ \mathrm{Im}\,\Gamma^{(1,1)}(\tau) &= \tfrac{1}{4}[\Gamma_{12}(\tau) - \Gamma_{21}(\tau)] \\ \mathrm{Re}\,[\Gamma^{(2,0)}(\tau)\,\mathrm{e}^{2\mathrm{i}\beta}] &= \tfrac{1}{4}[\Gamma_{11}(\tau) - \Gamma_{22}(\tau)] \\ \mathrm{Im}\,[\Gamma^{(2,0)}(\tau)\,\mathrm{e}^{2\mathrm{i}\beta}] &= \tfrac{1}{4}[\Gamma_{12}(\tau) + \Gamma_{21}(\tau)]. \end{aligned}\right\} \qquad (21.8\text{–}20)$$

These relations immediately allow us to express $\lambda(\tau, \theta)$ given by Eq. (21.8–15) in terms of $\Gamma_{ij}(\tau)$, and we obtain

$$\lambda(\tau, \theta) = \frac{1}{2|\mathscr{E}|^2}\{\Gamma_{11}(\tau)[1 + \cos 2(\theta - \beta)] + \Gamma_{22}(\tau)[1 - \cos 2(\theta - \beta)] + [\Gamma_{12}(\tau) + \Gamma_{21}(\tau)] \sin 2(\theta - \beta)\}. \tag{21.8–21}$$

In particular, when $\theta = \beta$,

$$\lambda(\tau, \beta) = \Gamma_{11}(\tau)/|\mathscr{E}|^2 \tag{21.8–22}$$

and when $\theta = \beta \pm \pi/2$,

$$\lambda(\tau, \beta \pm \pi/2) = \Gamma_{22}(\tau)/|\mathscr{E}|^2. \tag{21.8–23}$$

Now for a state that is squeezed in the full sense, $\langle:(\Delta \hat{E}_1)^2:\rangle < 0$ for some β [cf. Eq. (21.2–3)] so that $\langle:(\Delta \hat{E}_1)^2:\rangle = \Gamma_{11}(0) < 0$ for some β. Hence for such a state

$$\lambda(0, \beta) = \frac{\langle:(\Delta \hat{E}_1)^2:\rangle}{|\mathscr{E}|^2} < 0. \tag{21.8–24}$$

In order to appreciate the implications of this result, let us focus on the situation in which the detector response is fast enough to follow the field fluctuations. Then the pulses $k(t)$ have a range in time that is much shorter than the coherence time of $\Gamma_{11}(\tau)$, and $k(t)$ can be approximated as a delta function under the double integral in Eq. (21.8–18). Hence, on putting $\tau = 0$, and using Eq. (21.8–24) we obtain the formula

$$\langle(\Delta J)^2\rangle = \eta|\mathscr{E}|^2\int_0^\infty k^2(t')\,\mathrm{d}t' + \eta^2|\mathscr{E}|^2 q^2\langle:(\Delta \hat{E}_1)^2:\rangle. \tag{21.8–25}$$

It is at once apparent from this equation that the second term on the right vanishes for a vacuum field, i.e. when the field at input port 1 in Fig. 21.7 is blocked. Hence the first term provides a measure of the vacuum fluctuations, and $\langle(\Delta J)^2\rangle$ drops below the vacuum level whenever the input field is squeezed in the full sense.

It is possible, however, that the field is not squeezed in the full sense, although certain Fourier components are squeezed. Evidently, this would not be revealed by a measurement of the total current fluctuations $\langle(\Delta J)^2\rangle$, but only after the current is first passed through a spectral analyzer.

21.8.3 Spectral correlations

We now examine the situation shown in Fig. 21.7, where the photocurrent is fed to a spectral analyzer, which provides a measure of the spectral density $\chi(\omega)$ of the photocurrent fluctuations at various frequencies ω. Thus by definition,

$$\chi(\omega) = \int_{-\infty}^{\infty} \langle \Delta J(t)\Delta J(t + \tau)\rangle\, \mathrm{e}^{\mathrm{i}\omega\tau}\,\mathrm{d}\tau. \tag{21.8–26}$$

Let $K(\omega)$ be the Fourier transform of the current pulse $k(t)$, i.e.

$$K(\omega) \equiv \int_0^\infty k(\tau)\,\mathrm{e}^{\mathrm{i}\omega\tau}\,\mathrm{d}\tau, \tag{21.8–27}$$

which may be interpreted as the frequency response of the photodetector, and let $\psi(\omega, \theta)$ be the Fourier transform of $\lambda(\tau, \theta)$, i.e.

$$\psi(\omega, \theta) \equiv \int_{-\infty}^{\infty} \lambda(\tau, \theta)\, e^{i\omega\tau}\, d\tau. \tag{21.8–28}$$

Then $\psi(\omega, \theta)$ is the normalized spectral density of the light intensity fluctuations at the photodetector. After taking the Fourier transform with respect to τ of each term in Eq. (21.8–18), and invoking the convolution theorem, we obtain

$$\chi(\omega) = \eta|\mathscr{E}|^2|K(\omega)|^2[1 + \eta|\mathscr{E}|^2\psi(\omega, \theta)]. \tag{21.8–29}$$

Once again, the first term on the right corresponds to the vacuum noise, or shot noise, that is present when the light at input port 1 is blocked and $\psi(\omega, \theta)$ vanishes. We shall say that the light is squeezed, this time at frequency $\omega_1 + \omega$, whenever the act of opening input port 1 causes the measured fluctuations represented by $\chi(\omega)$ to drop below the vacuum level, i.e. when $\psi(\omega, \theta) < 0$ for some phase angle θ.

This result may be expressed in another way, if we make use of the definition of $\psi(\omega, \theta)$ given by Eq. (21.8–28) together with Eq. (21.8–21) for $\lambda(\tau, \theta)$. It is convenient to introduce the Fourier transforms of $\Gamma^{(N,M)}(\tau)$ and $\Gamma_{ij}(\tau)$ with respect to τ, which are of course functions of frequency ω,

$$\left.\begin{aligned} \Phi^{(N,M)}(\omega) &\equiv \int_{-\infty}^{\infty} \Gamma^{(N,M)}(\tau)\, e^{i\omega\tau}\, d\tau \qquad (N, M = 0, 1, 2) \\ \Phi_{ij}(\omega) &\equiv \int_{-\infty}^{\infty} \Gamma_{ij}(\tau)\, e^{i\omega\tau}\, d\tau \qquad (i, j = 1, 2). \end{aligned}\right\} \tag{21.8–30}$$

Then after taking the Fourier transform of each term, we obtain from Eq. (21.8–21) the result

$$\begin{aligned} \psi(\omega, \theta) = \frac{1}{2|\mathscr{E}|^2}\{&\Phi_{11}(\omega)[1 + \cos 2(\theta - \beta)] + \Phi_{22}(\omega)[1 - \cos 2(\theta - \beta)] \\ &+ [\Phi_{12}(\omega) + \Phi_{21}(\omega)]\sin 2(\theta - \beta)\}. \end{aligned} \tag{21.8–31}$$

When this expression for $\psi(\omega, \theta)$ is inserted in Eq. (21.8–29), we see that in the two special cases

(a) $\theta = \beta$

$$\chi(\omega) = \eta|\mathscr{E}|^2|K(\omega)|^2[1 + \eta\Phi_{11}(\omega)], \tag{21.8–32}$$

(b) $\theta = \beta + \pi/2$

$$\chi(\omega) = \eta|\mathscr{E}|^2|K(\omega)|^2[1 + \eta\Phi_{22}(\omega)]. \tag{21.8–33}$$

It follows that if the measured spectral density $\chi(\omega)$ is to fall below the vacuum level (represented by the first term) for some θ, say for $\theta = \beta$, then we must have

$$\Phi_{11}(\omega) < 0. \tag{21.8–34}$$

This is the condition for squeezing the spectral component of the light at frequency $\omega_1 + \omega$. The spectral function $\Phi_{11}(\omega)$ is sometimes called the *spectrum of squeezing*.

21.8.4 *Spectral component squeezing and degree of squeezing*

It is useful to have distinguishing labels for the different situations. If the field is not squeezed in the full sense with $\langle:(\Delta\hat{E}_1)^2:\rangle < 0$, but $\Phi_{11}(\omega) < 0$ for some frequencies ω, we shall refer to this as *spectral component squeezing*. Irrespective of whether the squeezing is in the full sense or not, if $\Phi_{11}(\omega) < 0$ for some frequencies but $\Phi_{11}(\omega) > 0$ for others, so that the field is squeezed only at some frequencies, we shall refer to the squeezing as *inhomogeneous*. On the other hand, if $\Phi_{11}(\omega) < 0$ for all frequencies, then the squeezing is *homogeneous*, and in that case the field is also squeezed in the full sense, because we have, by Fourier inversion of the second Eq. (21.8–30),

$$\langle:(\Delta\hat{E}_1)^2:\rangle = \Gamma_{11}(0) = \frac{1}{2\pi}\int_{-\infty}^{\infty}\Phi_{11}(\omega)\,\mathrm{d}\omega. \tag{21.8–35}$$

It is convenient to have a measure of the extent to which the field is squeezed at frequency $\omega_1 + \omega$ that we call the *degree of squeezing* $Q(\omega)$, which is defined in analogy with the $\mathcal{Q}$-parameter introduced in Section 21.6. Squeezing is associated with $Q(\omega) < 0$, and it is complete $[Q(\omega) = -1]$ if the spectral density $\chi(\omega) = 0$ in Eq. (21.8–32) with a perfect detector of quantum efficiency $\alpha = 1$. With the understanding that the photodetector collects all the light falling on it, we define $Q(\omega)$ by

$$Q(\omega) \equiv \frac{\eta\Phi_{11}(\omega)}{\alpha}. \tag{21.8–36}$$

In the special case in which $\hat{E}^2$ and $|\mathscr{E}|^2$ are expressed in units of photons per second and $\Phi_{11}(\omega)$ is dimensionless, we may take $\eta = \alpha$. In any case $Q(\omega)$ is independent of the photodetector efficiency, and $Q(\omega) < 0$ whenever the field component at frequency $\omega_1 + \omega$ is squeezed. Its most negative value $Q(\omega) = -1$ corresponds to perfect squeezing.

21.8.5 *Examples of the degree of squeezing $Q(\omega)$*

The parameter $Q(\omega)$ has been determined for a number of different situations in which squeezing occurs. For a two-level atom undergoing resonance fluorescence in a coherent exciting field on resonance, one finds that the squeezing is homogeneous whenever $\Omega/\beta < 1$, where Ω is the atomic Rabi frequency and β is half the Einstein A-coefficient (Walls and Zoller, 1981; Collett, Walls and Zoller, 1984; Loudon, 1984; Ou, Hong and Mandel, 1987). Figure 21.8 shows plots of $Q(\omega)$ versus frequency for several different values of Ω/β. Evidently the squeezing is inhomogeneous for values $\Omega/\beta > 1$. It can be shown that the field is squeezed in the full sense whenever $\Omega/\beta < \sqrt{2}$. When $\Omega/\beta > \sqrt{2}$, there is still spectral component squeezing for frequencies for which

$$\omega > 2(\Omega^2 - \beta^2)^{1/2}.$$

For the problem of parametric down-conversion in a cavity, the squeezing is always homogeneous and $Q(\omega)$ can approach -1 at the resonance frequency $\omega = 0$. Figure 21.9 shows plots of $Q(\omega)$ versus frequency for several different

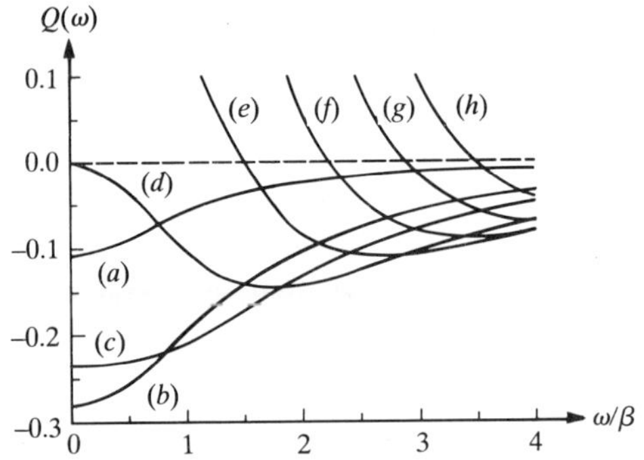

Fig. 21.8 Predicted variation of the squeeze parameter $Q(\omega)$ with frequency ω in resonance fluorescence from a two-level atom for several different values of Ω/β: (*a*) $\Omega/\beta = 0.25$; (*b*) $\Omega/\beta = (3 - \sqrt{7})^{1/2} \approx 0.595$; (*c*) $\Omega/\beta = 0.75$; (*d*) $\Omega/\beta = 1$; (*e*) $\Omega/\beta = 1.25$; (*f*) $\Omega/\beta = 1.5$; (*g*) $\Omega/\beta = 1.75$; (*h*) $\Omega/\beta = 2$. (Reproduced from Ou, Hong and Mandel, 1987.)

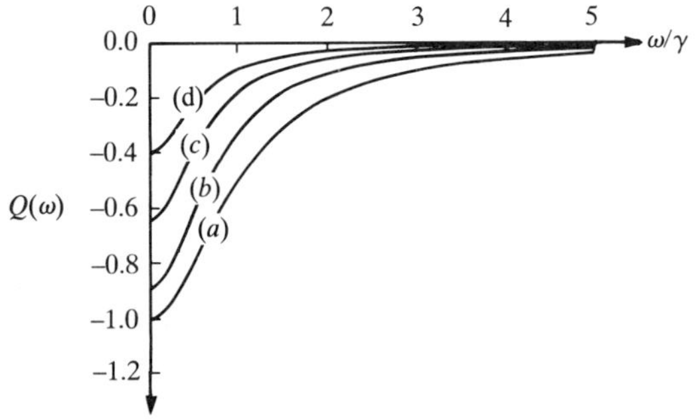

Fig. 21.9 Predicted variation of the squeeze parameter $Q(\omega)$ with frequency ω in parametric down-conversion for several different values of $\gamma/2\varepsilon$: (*a*) $\gamma/2\varepsilon = 1$; (*b*) $\gamma/2\varepsilon = 2$; (*c*) $\gamma/2\varepsilon = 4$; (*d*) $\gamma/2\varepsilon = 8$. (Reproduced from Ou, Hong and Mandel, 1987.)

values of the parameter $\gamma/2\varepsilon$, where γ is the cavity damping rate $[\gamma \equiv (1 - \mathcal{R})c/l]$ for a cavity of length l and mirror reflectivity $\mathcal{R}$, and ε is related to the dimensionless mode coupling constant g by $\varepsilon = (g - g^{-1})c/4l$. When $\gamma = 2\varepsilon$, $Q(\omega)$ approaches the limiting value -1 as $\omega \to 0$.

Finally, for the case of four-wave mixing in a nonlinear medium (Yariv and Pepper, 1977; Yuen and Shapiro, 1979; Reid and Walls, 1985; Yurke, 1985a; Ou, Hong and Mandel, 1987), in which the light emerging from opposite ends of the medium is combined by a beam splitter of reflectivity $\mathcal{R}$, which was the basis of the first successful demonstration of squeezing (Slusher, Hollberg, Yurke, Mertz and Valley, 1985), one again finds that the squeezing is inhomogeneous. However $Q(\omega)$ can approach its limiting value -1 as $\omega \to 0$ for mirror reflectivity $\mathcal{R} = 1/2$. On the other hand when $\mathcal{R} \neq 1/2$, the largest (but not perfect) squeezing occurs for non-zero ω.

21.9 Higher-order squeezing

Our definition and the treatment of squeezing given above all relate to the second moments of the squeezed field. The phenomenon we have been discussing might therefore be appropriately called second-order squeezing. However, the squeezing concept can readily be generalized so as to relate to the higher moments of the field. We shall now briefly consider this generalization, and examine the definition of higher-order squeezing as treated by Hong and Mandel (1985a,b). We then examine one simple example. For simplicity we again confine ourselves to a single-mode field.

We have seen that, based on the definition (21.1–7) of the field quadrature components $\hat{Q}$ and $\hat{P}$, the squeezed state is generally characterized by the property that, for some phase angle β, $\langle(\Delta\hat{Q})^2\rangle$ is smaller than in the vacuum state. Let $\hat{E}_1$, $\hat{E}_2$ be two field quadratures given by Eqs. (21.8–4) which are not necessarily dimensionless and which are proportional to $\hat{Q}$, $\hat{P}$. Then a similar statement can be made about $\langle(\Delta\hat{E}_1)^2\rangle$. This has a natural generalization to any even power N. We shall say that a field exhibits N'th order squeezing for even N if $\langle(\Delta\hat{E}_1)^N\rangle$ is smaller than the corresponding N'th moment in the vacuum state, for some phase angle β. However, the alternative definition (21.2–3), that $\langle:(\Delta\hat{E}_1)^2:\rangle < 0$ in the squeezed state, has to be generalized with caution, because the condition $\langle:(\Delta\hat{E}_1)^N:\rangle < 0$ is not equivalent to the foregoing definition of N'th order squeezing when $N > 2$.

Just as before, we can relate $\langle(\Delta\hat{E}_1)^N\rangle$ to the normally ordered moments of the field, although the relationship is more complicated. With the help of the Campbell–Baker–Hausdorff identity [see Eq. (10.11–26)] one can readily show that for any x,

$$\langle\exp(x\Delta\hat{E}_1)\rangle = \langle:\exp(x\Delta\hat{E}_1):\rangle\, e^{x^2C/2}, \tag{21.9–1}$$

where : : is the usual normal ordering symbol, and C stands for the commutator $[\hat{E}^{(+)}, \hat{E}^{(-)}]$. By expanding both sides in a power series in x and equating the coefficients of $x^N/N!$ we arrive at the following relation for any even number N:

$$\langle(\Delta\hat{E}_1)^N\rangle = \langle:(\Delta\hat{E}_1)^N:\rangle + \frac{N^{(2)}}{1!}(\tfrac{1}{2}C)\langle:(\Delta\hat{E}_1)^{N-2}:\rangle + \frac{N^{(4)}}{2!}(\tfrac{1}{2}C)^2\langle:(\Delta\hat{E}_1)^{N-4}:\rangle + \ldots + (N-1)!!C^{N/2} \quad (N \text{ even}), \tag{21.9–2}$$

where $N^{(r)} = N(N-1)\ldots(N-r+1)$. Now the normally ordered moments $\langle:(\Delta\hat{E}_1)^r:\rangle$ all vanish in the vacuum state, because an annihilation operator acting to the right on the vacuum gives zero. It follows that the last term $(N-1)!!C^{N/2}$ on the right of the equation is the N'th moment of $\Delta\hat{E}_1$ in the vacuum state. Therefore for N'th-order (N even) squeezing to occur, one or more of the normally ordered moments has to be negative.

It is worth noting, however, that not all the normally ordered moments need be negative for N'th-order squeezing. Indeed, it might suffice for just one term in the series of moments in Eq. (21.9–2) to be negative, if this negative term dominates the other moments. This leads us to the definition of *intrinsic N'th-order squeezing*. We shall say that a field is squeezed intrinsically to order N

(even) if for some β

$$\langle :(\Delta \hat{E}_1)^N: \rangle < 0. \tag{21.9–3}$$

When $N = 2$, the definition of N'th-order squeezing and intrinsic N'th-order squeezing coincide, but not so when $N = 4, 6, 8, \ldots$. A field may be intrinsically squeezed only to the second order and yet exhibit higher-order squeezing.

It can be shown that this is indeed the situation in resonance fluorescence from an atom. The field produced by N'th harmonic generation is squeezed intrinsically to order N when N is even, and to order $N - 1$ when N is odd (Kozierowski, 1986). In multiphoton absorption N'th-order squeezing is intrinsic except for $N = 6$ (García-Fernández, Sainz de los Terreros, Bermejo and Santoro, 1986). As an example of higher-order squeezing we now examine the two-photon coherent state.

21.9.1 Nth-order squeezing of the two-photon coherent state

Let us calculate $\langle (\Delta \hat{E}_1)^N \rangle$ for the two-photon coherent state $|[z, v]\rangle$. We first express $\hat{E}_1$ in the form

$$\hat{E}_1 = g\hat{a} + g^*\hat{a}^\dagger, \tag{21.9–4}$$

where g is some complex coefficient which includes the phase factor $e^{-i\beta}$, so that the commutator

$$C = |g|^2. \tag{21.9–5}$$

Then with the help of Eq. (21.3–7) we can express $\hat{E}_1$ in terms of the $\hat{A}(z)$, $\hat{A}^\dagger(z)$ operators in the form

$$\hat{E}_1 = f\hat{A} + f^*\hat{A}^\dagger, \tag{21.9–6}$$

where

$$f \equiv g \cosh r - g^* e^{i\theta} \sinh r \quad (z = r e^{i\theta}). \tag{21.9–7}$$

We now start from the operator identity

$$\exp(x\Delta \hat{E}_1) = ::\exp(x\Delta \hat{E}_1):: e^{x^2|f|^2/2}, \tag{21.9–8}$$

which is similar to Eq. (21.9–1), in which $::\hat{O}::$ denotes normal ordering of $\hat{O}$ with respect to $\hat{A}$, $\hat{A}^\dagger$ instead of $\hat{a}$, $\hat{a}^\dagger$. In place of Eq. (21.9–2), we then have the relation, after expanding both sides in a power series in x,

$$\begin{aligned}\langle (\Delta \hat{E}_1)^N \rangle &= \langle ::(\Delta \hat{E}_1)^N:: \rangle + \frac{N^{(2)}}{1!}(|f|^2/2)\langle ::(\Delta \hat{E}_1)^{N-2}:: \rangle \\ &+ \frac{N^{(4)}}{2!}(|f|^2/2)^2 \langle ::(\Delta \hat{E}_1)^{N-4}:: \rangle + \ldots \\ &+ (N-1)!!|f|^N \quad (N \text{ even}).\end{aligned} \tag{21.9–9}$$

As the state $|[z, v]\rangle$ is the right eigenstate of $\hat{A}$ with eigenvalue v [cf. Eq. (21.3–9)] it follows that

$$\langle ::(\Delta \hat{E}_1)^r:: \rangle = 0, \quad r = 1, 2, \ldots, \tag{21.9–10}$$

so that in this state

$$\langle(\Delta\hat{E}_1)^N\rangle = (N-1)!!|f|^N$$
$$= (N-1)!!|g|^N[\cosh 2r - \sinh 2r\cos(\theta + 2\arg g)]^{N/2} \quad (N \text{ even}). \tag{21.9–11}$$

Now the factor $(N-1)!!|g|^N$, obtained by putting $r = 0$ in Eq. (21.9–11), is the value of $\langle(\Delta\hat{E}_1)^N\rangle$ in the vacuum state. By choosing $\arg(g)$ so that $\cos[\theta + 2\arg(g)] = 1$, we can make the last factor equal to e^{-Nr}, which makes $\langle(\Delta\hat{E}_1)^N\rangle$ smaller than in the vacuum state for any even N when $z \neq 0$. Hence the field is squeezed not only to the second order, but to all even orders in the state $|[z, v]\rangle$, and similarly for the state $|(v, z)\rangle$.

There remains the question whether the N'th-order squeezing is intrinsic in the sense of the relation (21.9–3) or not. By making use of the ordering relation (21.9–1) in reverse, one can show that

$$\langle:(\Delta\hat{E}_1)^N:\rangle = (-1)^{N/2}(N-1)!!|g|^N(1-e^{-2r})^{N/2}, \quad (N \text{ even}), \tag{21.9–12}$$

which is negative for $N = 2, 6, 10, \ldots$. Hence even N'th-order squeezing is intrinsic in the two-photon coherent state whenever $N/2$ is an odd number.

The presence of higher-order squeezing can also be detected in a photoelectric homodyne experiment of the kind discussed in Section 21.6, but it requires measurements of the higher-order moments (Mandel and Hong, 1986). Such measurements are likely to be more difficult than the corresponding second-order ones. Still, in principle, it is possible to achieve even greater squeezing or noise reduction with respect to the higher moments of the field than with respect to the second moments.

21.9.2 Amplitude-squared squeezing

An entirely different definition of higher-order squeezing was introduced by Hillery (1987a,b), who labelled it *amplitude-squared squeezing*. Whereas we have treated squeezing of a single-mode field in terms of the quadrature operators

$$\hat{E}_1 = \hat{a}^\dagger e^{-i(\omega t-\beta)} + \hat{a}\, e^{i(\omega t-\beta)}$$
$$\hat{E}_2 = i[\hat{a}^\dagger e^{-i(\omega t-\beta)} - \hat{a}\, e^{i(\omega t-\beta)}],$$

Hillery has introduced the two dimensionless variables defined in terms of $\hat{a}^2$, $\hat{a}^{\dagger 2}$, by the formulas

$$\left.\begin{aligned} \hat{Y}_1 &= \tfrac{1}{2}[\hat{a}^{\dagger 2} e^{-2i\omega t} + \hat{a}^2 e^{2i\omega t}] \\ \hat{Y}_2 &= \frac{i}{2}[\hat{a}^{\dagger 2} e^{-2i\omega t} - \hat{a}^2 e^{2i\omega t}], \end{aligned}\right\} \tag{21.9–13}$$

which obey the commutation relation

$$[\hat{Y}_1, \hat{Y}_2] = 2i(\hat{n} + \tfrac{1}{2}), \tag{21.9–14}$$

and therefore satisfy the uncertainty relation

$$[\langle(\Delta\hat{Y}_1)^2\rangle\langle(\Delta\hat{Y}_2)^2\rangle]^{1/2} \geqslant \langle\hat{n} + \tfrac{1}{2}\rangle. \tag{21.9–15}$$

In a coherent state each dispersion $\langle(\Delta\hat{Y}_1)^2\rangle$ or $\langle(\Delta\hat{Y}_2)^2\rangle$ is separately equal to $\langle\hat{n}\rangle+\frac{1}{2}$. The state is then defined as being amplitude-squared squeezed in the variable $\hat{Y}_1$ if

$$\langle(\Delta\hat{Y}_1)^2\rangle < \langle\hat{n}+\tfrac{1}{2}\rangle. \tag{21.9–16}$$

It can be shown that, like quadrature squeezing, this form of squeezing exists only in a non-classical state, and that it arises in second harmonic generation, in degenerate parametric down-conversion and in two-photon absorption. However, the fluctuations of $\hat{Y}_1$ in an amplitude-squared squeezed state are not smaller than those in the vacuum state, which makes this form of squeezing fundamentally different.

Problems

21.1 Consider a single-mode quantum field in the superposition state $|\psi\rangle = \alpha|0\rangle + \beta|1\rangle$, $(|\alpha|^2 + |\beta|^2 = 1)$. Determine the values of α, β for which the state exhibits quadrature squeezing.

21.2 A single-mode field is in the quantum state $|\psi\rangle = (1/\sqrt{2})(|n\rangle + |n+2\rangle)$ which is a superposition of Fock states. For what values of n is the state (a) quadrature squeezed; (b) one with sub-Poissonian statistics? If we consider the more general state $|\psi\rangle = (1/\sqrt{2})(|n\rangle + |n+r\rangle)$, $(r = 1, 2, 3, \ldots)$ is quadrature squeezing easier to achieve?

21.3 Consider the nonlinear interaction in the interaction picture between two single-mode fields with mode labels 1 and 2 of the form

$$\hat{H}_{\mathrm{I}}(t) = \hbar g[\hat{a}_1^2(t)\hat{a}_2^\dagger(t) + \text{h.c.}], \qquad (g \text{ real}).$$

When mode 2 has twice the frequency of mode 1 this interaction is sometimes taken to represent the process of second-harmonic generation. Assume that initially mode 2 is in the vacuum state, while mode 1 is in the coherent state $|v_1\rangle$. Show, to order $(gt)^2$, that after a short interaction time t mode 1 exhibits quadrature squeezing.

21.4 In the previous problem show that, to order $(gt)^2$, there is no intrinsic N'th order squeezing of the fundamental mode beyond $N = 2$, even though the N'th (even) order deviation of one field quadrature may be smaller than in the vacuum state.

22

Some quantum effects in nonlinear optics

22.1 Introduction

We have already encountered situations in which light of one frequency falling on an atomic system gives rise to light of different frequencies (cf. Section 15.6). The atom may here be considered to play the role of a (noisy) nonlinear transducer for the incident field. Even stronger nonlinear effects arise when one is dealing with a large number of atoms, or a *nonlinear medium*. Under these circumstances it is sometimes permissible to ignore the atomic structure and to treat the medium as a continuum, as in Maxwell's electromagnetic theory. The subject of the interaction of the incident field with the nonlinear medium is usually known as *nonlinear* optics.

The subject had its beginnings in an experiment in which a strong beam of red light (wavelength 6943 Å) from a ruby laser was allowed to fall on a quartz crystal, and a faint beam of blue light at a wavelength of 3472 Å, the first harmonic of the red, was produced (Franken, Hill, Peters and Weinreich, 1961). The development of the subject of nonlinear optics into a mature field is due largely to the work of Bloembergen and his collaborators. In the following we shall consider only a few simple illustrative examples of phenomena in nonlinear optics. More topics and more detail can be found in the books by Bloembergen and others (Bloembergen, 1965; Yariv, 1967, Chap. 21; Shen, 1984; Schubert and Wilhelmi, 1986; Butcher and Cotter, 1990; Boyd, 1992).

22.2 Energy of the field in a dielectric

In classical electromagnetic theory, the energy of the electromagnetic field within a non-magnetic medium in SI units is given by the expression

$$H = \int \frac{1}{2\mu_0} \mathbf{B}^2(\mathbf{r}, t)\, \mathrm{d}^3 r + \int \mathrm{d}^3 r \int_0^{\mathbf{D}(\mathbf{r},t)} \mathbf{E}(\mathbf{r}, t) \cdot \mathrm{d}\mathbf{D}(\mathbf{r}, t). \qquad (22.2\text{–}1)$$

Here $\mathbf{D}(\mathbf{r}, t)$ is the electric displacement vector, and strictly speaking, the integration with respect to $\mathbf{D}$ cannot be carried out, because in general there exists no simple relationship between $\mathbf{E}$ and $\mathbf{D}$. Of course, in empty space, or in a linear isotropic medium, $\mathbf{D}$ is proportional to $\mathbf{E}$ and the $\mathbf{D}$-integral reduces to $\frac{1}{2}\mathbf{E}(\mathbf{r}, t) \cdot \mathbf{D}(\mathbf{r}, t)$, which is the usual expression for the electric energy density. However in a nonlinear medium no comparable simplification is possible, although one can express $\mathbf{D}(\mathbf{r}, t)$ (in SI units) in the form

$$\mathbf{D}(\mathbf{r}, t) = \varepsilon_0 \mathbf{E}(\mathbf{r}, t) + \mathbf{P}(\mathbf{r}, t), \tag{22.2–2}$$

and it is sometimes possible to expand the polarization $\mathbf{P}(\mathbf{r}, t)$ induced in the medium in a power series in $\mathbf{E}$, in the form

$$P_i = \chi_{ij}^{(1)} E_j + \chi_{ijk}^{(2)} E_j E_k + \chi_{ijkl}^{(3)} E_j E_k E_l + \ldots . \tag{22.2–3}$$

Here $\boldsymbol{\chi}^{(n)}$ is a susceptibility tensor of rank $n + 1$, and summation on repeated indices is understood. This equation is appropriate for a non-dispersive medium, or under circumstances when the effective frequencies of the field are not too close to the resonance frequencies of the medium. When the susceptibility is strongly frequency-dependent, it is more natural to make a Fourier decomposition of both P_i and E_i and to relate the Fourier components $P_i(\omega)$ and $E_i(\omega)$ via a power series. As $\boldsymbol{\chi}^{(n)}$ in general involves n different frequencies, we write in place of Eq. (22.2–3)

$$\begin{aligned} P_i(\omega_1) = {} & \chi_{ij}^{(1)}(\omega_1; \omega_1) E_j(\omega_1) + \chi_{ijk}^{(2)}(\omega_1; \omega_1 - \omega_2, \omega_2) E_j(\omega_1 - \omega_2) E_k(\omega_2) \\ & + \chi_{ijkl}^{(3)}(\omega_1; \omega_1 - \omega_2 - \omega_3, \omega_2, \omega_3) E_j(\omega_1 - \omega_2 - \omega_3) E_k(\omega_2) E_l(\omega_3) \\ & + \ldots . \end{aligned} \tag{22.2–4}$$

We can now use Eqs. (22.2–2) to (22.2–4) in Eq. (22.2–1) for the energy, which leads to an expression of the form

$$H = \int \left[\frac{1}{2\mu_0} \mathbf{B}^2(\mathbf{r}, t) + \tfrac{1}{2}\varepsilon_0 \mathbf{E}^2(\mathbf{r}, t) + X_1(\mathbf{r}) + X_2(\mathbf{r}) + \ldots \right] \mathrm{d}^3 r, \tag{22.2–5}$$

where

$$X_1(\mathbf{r}) \equiv \tfrac{1}{2} \int\!\!\int \mathrm{d}\omega\, \mathrm{d}\omega' \chi_{ij}^{(1)}(\omega, \omega') E_i(\mathbf{r}, \omega') E_j(\mathbf{r}, \omega)$$

and

$$X_2(\mathbf{r}) \equiv \tfrac{1}{3} \int\!\!\int\!\!\int \mathrm{d}\omega\, \mathrm{d}\omega'\, \mathrm{d}\omega'' \chi_{ijk}^{(2)}(\omega''; \omega - \omega', \omega') E_i(\mathbf{r}, \omega'') E_j(\mathbf{r}, \omega - \omega') E_k(\mathbf{r}, \omega')$$

The term in $\chi_{ijk}^{(2)}$ represents the lowest-order nonlinear contribution to the energy.

The canonical quantization of the macroscopic field in a nonlinear medium is a non-trivial problem (see for example the books and papers by Shen, 1967, 1984; Hillery and Mlodinow, 1984; Schubert and Wilhelmi, 1986; Drummond, 1990; Peřina, 1991), which will not be attempted here. Approximations are usually introduced that hold under various conditions. So long as the nonlinearities are small, it is natural to try replacing the $\mathbf{E}(\mathbf{r}, t)$ and $\mathbf{B}(\mathbf{r}, t)$ vectors above by the corresponding free-field Hilbert space operators, provided the resulting energy is Hermitian. When attention is focused on a particular nonlinear process, one sometimes picks out terms from the expansion that characterize the interaction and discards all the others. We shall illustrate this procedure in one or two simple cases.

22.3 Optical harmonic generation

Harmonic generation is the oldest and best known example of a process in nonlinear optics (Franken, Hill, Peters and Weinreich, 1961; Armstrong, Bloembergen, Ducuing and Pershan, 1962). A monochromatic light beam of frequency ω_1 incident on the non-linear medium generates a field at the harmonic frequency $\omega_2 = 2\omega_1$. The process is mediated by combinations of three creation and annihilation operators, and is therefore described by terms in $\boldsymbol{\chi}^{(2)}$ in the operator form of Eq. (22.2–5). It has been treated by numerous workers. We shall largely adopt the approach of Kielich and his collaborators (Kozierowski and Tanas, 1977; Kielich, Kozierowski and Tanas, 1978), in which the Hilbert space is limited to only two modes at frequencies ω_1 and ω_2, and a Taylor expansion is used to describe the time evolution of the dynamical variables.

We express the energy of the two-mode field in the abbreviated form

$$\hat{H} = \sum_{i=1}^{2} \hbar\omega_i(\hat{n}_i + \tfrac{1}{2}) + \hbar g[\hat{a}_2^\dagger \hat{a}_1^2 + \hat{a}_1^{\dagger 2}\hat{a}_2], \tag{22.3–1}$$

in which suffices 1, 2 refer to the fundamental and the harmonic mode, respectively. The real mode coupling constant g contains the nonlinear susceptibility $\boldsymbol{\chi}^{(2)}$. The Hamiltonian $\hat{H}$ describes a process in which two photons at frequency ω_1 are absorbed and give rise to a new photon at the harmonic frequency $\omega_2 = 2\omega_1$, together with the inverse process. We immediately find from Eq. (22.3–1), with the help of the commutation relations among the $\hat{a}$, $\hat{a}^\dagger$ and $\hat{n}$ operators, that

$$[\hat{n}_1 + 2\hat{n}_2, \hat{H}] = 0, \tag{22.3–2}$$

so that the sum $\hat{n}_1(t) + 2\hat{n}_2(t)$ is a constant of the motion. Two photons of the fundamental mode are therefore absorbed for every harmonic photon emitted.

It is somewhat more convenient to replace $\hat{a}_1$, $\hat{a}_2$ by the more slowly varying operators $\hat{\mathscr{A}}_1$, $\hat{\mathscr{A}}_2$ defined by

$$\left.\begin{aligned} \hat{\mathscr{A}}_1 &= \hat{a}_1 \mathrm{e}^{\mathrm{i}\omega_1 t} \\ \hat{\mathscr{A}}_2 &= \hat{a}_2 \mathrm{e}^{\mathrm{i}\omega_2 t}, \end{aligned}\right\} \tag{22.3–3}$$

which obey the same commutation rules as the $\hat{a}$'s but do not oscillate at optical frequencies. Then the Heisenberg equations of motion for $\hat{\mathscr{A}}_1$ and $\hat{\mathscr{A}}_2$ take the form

$$\begin{aligned} \dot{\hat{\mathscr{A}}}_1 &= \frac{1}{\mathrm{i}\hbar}[\hat{\mathscr{A}}_1, \hat{H}] + \frac{\partial \hat{\mathscr{A}}_1}{\partial t} \\ &= -2\mathrm{i}g\hat{\mathscr{A}}_1^\dagger \hat{\mathscr{A}}_2, \end{aligned} \tag{22.3–4}$$

$$\begin{aligned} \dot{\hat{\mathscr{A}}}_2 &= \frac{1}{\mathrm{i}\hbar}[\hat{\mathscr{A}}_2, \hat{H}] + \frac{\partial \hat{\mathscr{A}}_2}{\partial t} \\ &= -\mathrm{i}g\hat{\mathscr{A}}_1^2. \end{aligned} \tag{22.3–5}$$

Similarly, we may calculate the second derivatives, and obtain

$$\begin{aligned} \ddot{\hat{\mathscr{A}}}_1 &= -2\mathrm{i}g(\dot{\hat{\mathscr{A}}}_1^\dagger \hat{\mathscr{A}}_2 + \hat{\mathscr{A}}_1^\dagger \dot{\hat{\mathscr{A}}}_2) \\ &= 4g^2(\hat{n}_2 - \hat{n}_1/2)\hat{\mathscr{A}}_1 \end{aligned} \tag{22.3–6}$$

and

$$\begin{aligned}\ddot{\hat{\mathscr{A}}}_2 &= -ig(\dot{\hat{\mathscr{A}}}_1\hat{\mathscr{A}}_1 + \hat{\mathscr{A}}_1\dot{\hat{\mathscr{A}}}_1)\\ &= -4g^2(\hat{n}_1 + \tfrac{1}{2})\hat{\mathscr{A}}_2.\end{aligned} \tag{22.3–7}$$

When a plane wave of a certain frequency propagates through the medium in a certain direction, distance and time are proportional to each other and can be used more or less interchangeably. Because the interaction time t is effectively the propagation time through the medium, if this is sufficiently short, one can approximate $\hat{\mathscr{A}}_1(t)$ and $\hat{\mathscr{A}}_2(t)$ by a Taylor expansion about $t = 0$. This leads to the formulas

$$\begin{aligned}\hat{\mathscr{A}}_1(t) &= \hat{\mathscr{A}}_1(0) + t\dot{\hat{\mathscr{A}}}_1(0) + \frac{t^2}{2!}\ddot{\hat{\mathscr{A}}}_1(0) + \ldots\\ &= \hat{\mathscr{A}}_1(0) - 2igt\hat{\mathscr{A}}_1^\dagger(0)\hat{\mathscr{A}}_2(0) + 2g^2t^2[\hat{n}_2(0) - \tfrac{1}{2}\hat{n}_1(0)]\hat{\mathscr{A}}_1(0) + \ldots\end{aligned} \tag{22.3–8}$$

$$\begin{aligned}\hat{\mathscr{A}}_2(t) &= \hat{\mathscr{A}}_2(0) + t\dot{\hat{\mathscr{A}}}_2(0) + \frac{t^2}{2!}\ddot{\hat{\mathscr{A}}}_2(0) + \ldots\\ &= \hat{\mathscr{A}}_2(0) - igt\hat{\mathscr{A}}_1^2(0) - 2g^2t^2[\hat{n}_1(0) + \tfrac{1}{2}]\hat{\mathscr{A}}_2(0) + \ldots\end{aligned} \tag{22.3–9}$$

up to terms of order $(gt)^2$. The expansions are valid as long as $\langle\hat{n}_1(0)\rangle(gt)^2 \ll 1$.

These equations can now be used to construct other operators like the photon numbers $\hat{n}_1(t)$ and $\hat{n}_2(t)$ and to calculate their moments. Let us assume that the state $|\psi\rangle$ of the field at the beginning (time $t = 0$) is the coherent state with complex amplitude v for mode 1 and the vacuum for mode 2. Then

$$\left.\begin{aligned}\hat{\mathscr{A}}_1(0)|\psi\rangle &= v|\psi\rangle\\ \hat{\mathscr{A}}_2(0)|\psi\rangle &= 0,\end{aligned}\right\} \tag{22.3–10}$$

and we readily find from Eqs. (22.3–8) and (22.3–9), after multiplying out, expressing all operators in normal order and taking expectation values, that

$$\begin{aligned}\langle\hat{n}_1(t)\rangle &= \langle\psi|\hat{\mathscr{A}}_1^\dagger(t)\hat{\mathscr{A}}_1(t)|\psi\rangle\\ &= |v|^2 - 2(gt)^2|v|^4 + \ldots\end{aligned} \tag{22.3–11}$$

$$\begin{aligned}\langle\hat{n}_2(t)\rangle &= \langle\psi|\hat{\mathscr{A}}_2^\dagger(t)\hat{\mathscr{A}}_2(t)|\psi\rangle\\ &= (gt)^2|v|^4 + \ldots.\end{aligned} \tag{22.3–12}$$

The intensity of the harmonic component therefore grows with the square of the propagation time and the square of the fundamental intensity.

The fluctuations of the photon numbers may be calculated in a similar manner, and one finds with the help of Eqs. (22.3–8) and (22.3–9) for the difference between the variance and the mean (Kozierowski and Tanas, 1977; Kielich, Kozierowski and Tanas, 1978) the expressions

$$\langle(\Delta\hat{n}_1(t))^2\rangle - \langle\hat{n}_1(t)\rangle = -2(gt)^2|v|^4 + \ldots \tag{22.3–13}$$

$$\langle(\Delta\hat{n}_2(t))^2\rangle - \langle\hat{n}_2(t)\rangle = 0(gt)^6, \tag{23.3–14}$$

where $O(x)$ denotes the order of magnitude. The statistics of the harmonic photons are therefore close to Poissonian. For the fundamental mode, on the other hand, one finds a narrowing of the photon distribution, or sub-Poissonian statistics, although the effect is small for small $\langle \hat{n}_2(t) \rangle$.

22.3.1 *Squeezing in harmonic generation*

It is not difficult to show that the fundamental mode also becomes *squeezed* (cf. Chapter 21) as the light propagates through the nonlinear medium (Mandel, 1982). For this purpose we construct the two Hermitian operators

$$\left.\begin{aligned} \hat{Q}_1(t) &= \hat{\mathscr{A}}_1(t)\,\mathrm{e}^{-\mathrm{i}\phi} + \hat{\mathscr{A}}_1^\dagger(t)\,\mathrm{e}^{\mathrm{i}\phi} \\ \hat{P}_1(t) &= \hat{\mathscr{A}}_1(t)\,\mathrm{e}^{-\mathrm{i}(\phi+\pi/2)} + \hat{\mathscr{A}}_1^\dagger\,\mathrm{e}^{\mathrm{i}(\phi+\pi/2)}, \end{aligned}\right\} \tag{22.3–15}$$

which are canonical conjugates and correspond to the amplitudes of two quadrature components of the fundamental mode. The arbitrary phase angle ϕ may be chosen at will, and it is apparent that $\hat{P}_1(t)$ differs from $\hat{Q}_1(t)$ only by having ϕ incremented by $\pi/2$. We have a squeezed state if the dispersion of either $\hat{Q}_1(t)$ or $\hat{P}_1(t)$ for some ϕ is less than unity, which is the value of the dispersion in the vacuum state. Now we readily find from Eq. (22.3–8) that

$$\langle \hat{Q}_1(t) \rangle = (v\,\mathrm{e}^{-\mathrm{i}\phi} + v^*\,\mathrm{e}^{\mathrm{i}\phi})(1 - g^2t^2|v|^2) + O(gt)^3 \tag{22.3–16}$$

and

$$\begin{aligned} \langle \hat{Q}_1^2(t) \rangle = {} & [v^2\,\mathrm{e}^{-2\mathrm{i}\phi}(1 - 2g^2t^2|v|^2) + \mathrm{c.c.}] + 2|v|^2(1 - 2g^2t^2|v|^2) + 1 \\ & - [g^2t^2v^2\,\mathrm{e}^{-2\mathrm{i}\phi} + \mathrm{c.c.}] + 0(gt)^3, \end{aligned} \tag{22.3–17}$$

so that

$$\langle [\Delta \hat{Q}_1(t)]^2 \rangle = 1 - 2g^2t^2|v|^2\cos 2(\theta - \phi) + O(gt)^3 \quad (v = |v|\,\mathrm{e}^{\mathrm{i}\theta}). \tag{22.3–18}$$

If we choose $\phi = \theta$, then the dispersion of $\hat{Q}_1(t)$ is less than unity, so that the fundamental mode becomes squeezed, at least to a small extent. The amount of squeezing, as measured by the departure of $\langle [\Delta \hat{Q}_1(t)]^2 \rangle$ from unity, is given by the ratio $2\langle \hat{n}_2(t) \rangle / \langle \hat{n}_1(0) \rangle$. We see therefore that the process of second harmonic generation is not describable completely classically, because it is accompanied by the generation of sub-Poissonian statistics and squeezing, both of which are purely quantum mechanical phenomena.

In practice the situation is always more complicated, because one is not dealing with just two well-defined modes. Moreover, for processes having appreciable probability one expects the momentum as well as the energy to be conserved in the interaction. Actually, it is possible to satisfy the conditions for energy and momentum conservation simultaneously in certain crystals by appropriate choice of the directions of propagation and of the polarization components of the light beams relative to the crystal axes. The technique is known as *phase matching* (Armstrong, Bloembergen, Ducuing and Pershan, 1962, Giordmaine, 1962; Maker, Terhune, Nisenoff and Savage, 1962; Yariv, 1967, Chap. 21; Shen, 1984; Boyd, 1992).

22.4 Parametric down-conversion

For our second example of a nonlinear interaction we consider the process of parametric down-conversion, which is, in a sense, the inverse of harmonic generation. Whereas in the latter case two incident photons generate one photon of double the original frequency, in down-conversion one photon incident on the dielectric having a $\chi^{(2)}$ nonlinearity breaks up into two new photons of lower frequencies (see Fig. 22.1). For historical reasons they are known as the *signal photon* and the *idler photon*. If the two new photons are indistinguishable, it is possible to describe the process by the same Hamiltonian (22.3–1) as before. However, we shall treat the slightly more general situation in which incident photons of frequency ω_0 break up into two photons of lower frequencies that differ either in the directions or the magnitudes of their wave vectors $\mathbf{k}_1$, $\mathbf{k}_2$ or both. In the steady state we always have

$$\omega_0 = \omega_1 + \omega_2, \tag{22.4–1}$$

where ω_0 is known as the *pump frequency* of the parametric process, and ω_1, ω_2 are known as the *signal and idler frequencies*. The process of spontaneous parametric down-conversion in a nonlinear crystal was first investigated theoretically by Klyshko (1968) and experimentally by Burnham and Weinberg (1970), who showed that the signal and idler photons appeared 'simultaneously' within the resolving time of the detectors and the associated electronics. There is a substantial amount of literature going back to the 1960s on the theory of parametric amplification and up- and down-conversion. (For the early work on parametric amplification and down-conversion see Louisell, 1960; Yariv, 1967; Mollow, 1967, 1973; Mollow and Glauber, 1967a,b; Kleinman, 1968; Zel'dovich and Klyshko, 1969; Tucker and Walls, 1969). When the phase matching condition is satisfied, the wave vectors $\mathbf{k}_0$, $\mathbf{k}_1$, $\mathbf{k}_2$ of the pump, signal and idler photons are related by

$$\mathbf{k}_0 = \mathbf{k}_1 + \mathbf{k}_2, \tag{22.4–2}$$

which will be recognized as the condition for momentum conservation.

The Hamiltonian for the three-mode parametric process can be written in the following form, which is an obvious generalization of Eq. (22.3–1),

$$\hat{H} = \sum_{i=0}^{2} \hbar\omega_i(\hat{n}_i + \tfrac{1}{2}) + \hbar g[\hat{a}_1^\dagger \hat{a}_2^\dagger \hat{a}_0 + \text{h.c.}]. \tag{22.4–3}$$

We may readily confirm that

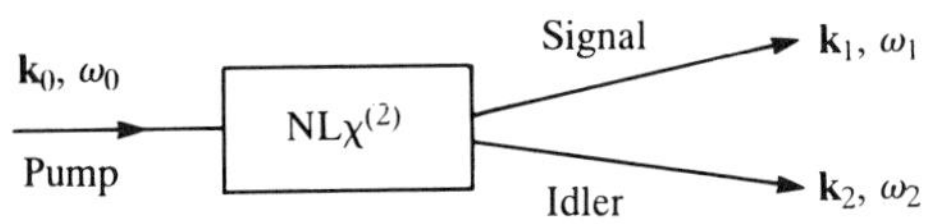

Fig. 22.1 Illustrating the process of parametric down-conversion in a nonlinear crystal NL with susceptibility $\chi^{(2)}$.

$$[\hat{n}_1 + \hat{n}_2 + 2\hat{n}_0, \hat{H}] = 0,$$

so that $\hat{n}_1 + \hat{n}_2 + 2\hat{n}_0$ is a constant of the motion, which reflects the fission of one pump photon into one signal and one idler photon. By making a Taylor expansion of $\hat{a}_j(t)$ $(j = 1, 2, 3)$ and using the Heisenberg equations of motion to construct the time derivatives, we can obtain short-time solutions for $\hat{a}_j(t)$ as before. However, with the help of a simplification of $\hat{H}$, the problem can be reduced to one with a simple analytic solution (Graham, 1984).

22.4.1 *Solution of the equations of motion*

Let us suppose that the incident pump field is intense and that the pump mode $\hat{a}_0$ can be treated classically as a field of complex amplitude $a_0 = v_0\,\mathrm{e}^{-\mathrm{i}\omega_0 t}$. Then the Hamiltonian in Eq. (22.4–3) has only two quantized field modes corresponding to the signal 1 and idler 2; the contribution from the classical pump no longer appears. However, we must bear in mind that, because the pump amplitude v_0 is now treated as constant, the solution will cease to be valid once appreciable down-conversion, and consequently appreciable depletion of the pump field, occurs. We shall therefore restrict our calculation to the case $\langle \hat{n}_1(t)\rangle$, $\langle \hat{n}_2(t)\rangle \ll |v_0|^2$.

We now have from the Hamiltonian

$$\hat{H} = \sum_{i=1}^{2} \hbar\omega_i(\hat{n}_i + \tfrac{1}{2}) + \hbar g[\hat{a}_1^\dagger \hat{a}_2^\dagger v_0\,\mathrm{e}^{-\mathrm{i}\omega_0 t} + \mathrm{h.c.}], \tag{22.4–4}$$

$[\hat{n}_1 - \hat{n}_2, \hat{H}] = 0$, so that $\hat{n}_1(t) - \hat{n}_2(t)$ is a constant of the motion, and

$$\hat{n}_1(t) - \hat{n}_2(t) = \hat{n}_1(0) - \hat{n}_2(0). \tag{22.4–5}$$

This relation expresses the fact that signal and idler photons are always created together.

The Heisenberg equation of motion for $\hat{a}_1(t)$ takes the form

$$\begin{aligned}\dot{\hat{a}}_1(t) &= \frac{1}{\mathrm{i}\hbar}[\hat{a}_1(t), \hat{H}] \\ &= -\mathrm{i}\omega_1\hat{a}_1(t) - \mathrm{i}g\hat{a}_2^\dagger(t)v_0\,\mathrm{e}^{-\mathrm{i}\omega_0 t}.\end{aligned}$$

As before it is convenient to introduce slowly varying complex mode amplitudes as before, defined by the relations

$$\begin{aligned}\hat{\mathscr{A}}_1(t) &= \hat{a}_1(t)\,\mathrm{e}^{\mathrm{i}\omega_1 t} \\ \hat{\mathscr{A}}_2(t) &= \hat{a}_2(t)\,\mathrm{e}^{\mathrm{i}\omega_2 t}.\end{aligned}$$

Then $\hat{\mathscr{A}}_1(t)$ obeys the simpler equation of motion

$$\frac{\mathrm{d}\hat{\mathscr{A}}_1(t)}{\mathrm{d}t} = -\mathrm{i}gv_0\hat{\mathscr{A}}_2^\dagger(t)\,\mathrm{e}^{\mathrm{i}(\omega_1+\omega_2-\omega_0)t}.$$

If $\omega_1 + \omega_2 = \omega_0$, we have

$$\left.\begin{aligned} \frac{\mathrm{d}\hat{\mathscr{A}}_1(t)}{\mathrm{d}t} &= -\mathrm{i}gv_0\hat{\mathscr{A}}_2^\dagger(t), \\ \text{and similarly} \qquad & \\ \frac{\mathrm{d}\hat{\mathscr{A}}_2(t)}{\mathrm{d}t} &= -\mathrm{i}gv_0\hat{\mathscr{A}}_1^\dagger(t). \end{aligned}\right\} \tag{22.4–6}$$

By differentiating a second time and substituting for $\mathrm{d}\mathscr{A}_2^\dagger(t)/\mathrm{d}t$ from the second equation in the first, we obtain the uncoupled equations for real g,

$$\frac{\mathrm{d}^2\hat{\mathscr{A}}_1(t)}{\mathrm{d}t^2} = g^2|v_0|^2\hat{\mathscr{A}}_1(t),$$

and similarly

$$\frac{\mathrm{d}^2\hat{\mathscr{A}}_2(t)}{\mathrm{d}t^2} = g^2|v_0|^2\hat{\mathscr{A}}_2(t).$$

The general solutions of these equations can be obtained at once, and when the appropriate boundary conditions are imposed, they take the form

$$\hat{\mathscr{A}}_1(t) = \hat{\mathscr{A}}_1(0)\cosh(g|v_0|t) - \mathrm{i}\,\mathrm{e}^{\mathrm{i}\theta}\hat{\mathscr{A}}_2^\dagger(0)\sinh(g|v_0|t) \tag{22.4–7}$$

$$\hat{\mathscr{A}}_2(t) = \hat{\mathscr{A}}_2(0)\cosh(g|v_0|t) - \mathrm{i}\,\mathrm{e}^{\mathrm{i}\theta}\hat{\mathscr{A}}_1^\dagger(0)\sinh(g|v_0|t), \tag{22.4–8}$$

where we have written

$$v_0 = |v_0|\,\mathrm{e}^{\mathrm{i}\theta}.$$

It is easy to confirm that these solutions satisfy Eq. (22.4–5).

22.4.2 *Photon statistics*

In order to calculate expectation values, we shall take the initial state of both signal and idler modes to be the vacuum state $|\text{vac}\rangle_{1,2}$. Recalling that $\hat{\mathscr{A}}_j(0)|\text{vac}\rangle_{1,2} = 0$ $(j = 1, 2)$, we can readily calculate both the moments and the generating function for the moments of the photon numbers $\hat{n}_1(t)$ and $\hat{n}_2(t)$. From Eq. (22.4–7) we obtain for the r'th moment of $\hat{n}_1(t)$ in normal order, which is also the r'th factorial moment, the expression

$$\begin{aligned} \langle\hat{n}_1^{(r)}(t)\rangle &= \langle:\hat{n}_1^r(t):\rangle = {}_{1,2}\langle\text{vac}|\hat{\mathscr{A}}_1^{\dagger r}(t)\hat{\mathscr{A}}_1^r(t)|\text{vac}\rangle_{1,2} \\ &= {}_{1,2}\langle\text{vac}|[\hat{\mathscr{A}}_1^\dagger(0)\cosh(g|v_0|t) + \mathrm{i}\,\mathrm{e}^{-\mathrm{i}\theta}\hat{\mathscr{A}}_2(0)\sinh(g|v_0|t)]^r \\ &\quad\times[\hat{\mathscr{A}}_1(0)\cosh(g|v_0|t) - \mathrm{i}\,\mathrm{e}^{\mathrm{i}\theta}\hat{\mathscr{A}}_2^\dagger(0)\sinh(g|v_0|t)]^r|\text{vac}\rangle_{1,2} \\ &= |(-\mathrm{i}\,\mathrm{e}^{\mathrm{i}\theta})^r\sinh^r(g|v_0|t)\hat{\mathscr{A}}_2^{\dagger r}(0)|\text{vac}\rangle_{1,2}|^2 \\ &= \sinh^{2r}(g|v_0|t)\,{}_{1,2}\langle\text{vac}|\hat{\mathscr{A}}_2^r(0)\hat{\mathscr{A}}_2^{\dagger r}(0)|\text{vac}\rangle_{1,2} \\ &= \sinh^{2r}(g|v_0|t)\,{}_{1,2}\langle\text{vac}|(\hat{n}_2(0)+1)(\hat{n}_2(0)+2)\ldots(\hat{n}_2(0)+r)|\text{vac}\rangle_{1,2} \\ &= r!\sinh^{2r}(g|v_0|t), \end{aligned} \tag{22.4–9}$$

and an identical result for $\langle:\hat{n}_2^r(t):\rangle$ follows from Eq. (22.4–8). For the step from the fifth to the sixth line of Eq. (22.4–9) we have made use of the anti-normal ordering theorem [cf. Eq. (12.10–23a)]

$$``\hat{n}_2^r" = (\hat{n}_2 + 1)(\hat{n}_2 + 2) \ldots (\hat{n}_2 + r).$$

These moments are typical of photons that are governed by the Bose–Einstein probability distribution [cf. Eq. (13.1–7)], which applies to photons emitted by a source in thermal equilibrium.

In particular, when $r = 1$, we obtain the results

$$\langle \hat{n}_1(t) \rangle = \sinh^2(g|v_0|t) = \langle \hat{n}_2(t) \rangle, \tag{22.4–10}$$

and when $r = 2$,

$$\langle :\hat{n}_1^2(t): \rangle = 2\sinh^4(g|v_0|t) = \langle :\hat{n}_2^2(t): \rangle, \tag{22.4–11}$$

so that

$$\begin{aligned}\langle (\Delta\hat{n}_1(t))^2 \rangle &= \langle :\hat{n}_1^2(t): \rangle - \langle \hat{n}_1(t) \rangle^2 + \langle \hat{n}_1(t) \rangle \\ &= \langle \hat{n}_1(t) \rangle [1 + \langle \hat{n}_1(t) \rangle] = \langle (\Delta\hat{n}_2(t))^2 \rangle. \end{aligned} \tag{22.4–12}$$

From Eq. (22.4–10) the average number of down-converted photons grows quadratically in time by spontaneous emission so long as $g|v_0|t \ll 1$. But once $g|v_0|t$ exceeds unity, stimulated emission dominates, and $\langle \hat{n}_j(t) \rangle$ then grows exponentially in time. However, it should be noted again that the assumption of a constant pump amplitude v_0 cannot be justified once $\langle \hat{n}_j(t) \rangle$ becomes sufficiently large. In practice, the interaction time t may be taken to be the propagation time through the nonlinear medium, which is usually very short, so that under conditions of steady-state pumping generally $g|v_0|t \ll 1$.

Next we make use of Eqs. (22.4–7) and (22.4–8) to calculate the cross-correlation $\langle :\hat{n}_1(t)\hat{n}_2(t): \rangle$. We find that

$$\begin{aligned}\langle :\hat{n}_1(t)\hat{n}_2(t): \rangle &= {}_{1,2}\langle \text{vac}|\hat{\mathscr{A}}_1^\dagger(t)\hat{\mathscr{A}}_2^\dagger(t)\hat{\mathscr{A}}_2(t)\hat{\mathscr{A}}_1(t)|\text{vac}\rangle_{1,2} \\ &= |[\hat{\mathscr{A}}_2(0)\cosh(g|v_0|t) - \mathrm{i}\,\mathrm{e}^{\mathrm{i}\theta}\hat{\mathscr{A}}_1^\dagger(0)\sinh(g|v_0|t)] \\ &\quad \times [\hat{\mathscr{A}}_1(0)\cosh(g|v_0|t) - \mathrm{i}\,\mathrm{e}^{\mathrm{i}\theta}\hat{\mathscr{A}}_2^\dagger(0)\sinh(g|v_0|t)]|\text{vac}\rangle_{1,2}|^2 \\ &= |[-\mathrm{i}\,\mathrm{e}^{\mathrm{i}\theta}(\hat{\mathscr{A}}_2^\dagger(0)\hat{\mathscr{A}}_2(0) + 1)\cosh(g|v_0|t)\sinh(g|v_0|t) \\ &\quad - \mathrm{e}^{2\mathrm{i}\theta}\hat{\mathscr{A}}_1^\dagger(0)\hat{\mathscr{A}}_2^\dagger(0)\sinh^2(g|v_0|t)]|\text{vac}\rangle_{1,2}|^2 \\ &= \cosh^2(g|v_0|t)\sinh^2(g|v_0|t) + \sinh^4(g|v_0|t) \\ &= \sinh^2(g|v_0|t)[1 + 2\sinh^2(g|v_0|t)] \\ &= \langle \hat{n}_j(t) \rangle + 2\langle \hat{n}_j(t) \rangle^2 \quad (j = 1, 2) \\ &= \langle \hat{n}_j(t) \rangle + \langle :\hat{n}_j^2(t): \rangle \quad (j = 1, 2). \end{aligned} \tag{22.4–13}$$

The cross-correlation of the photon number fluctuations is therefore given by the relation

$$\begin{aligned}\langle :\Delta\hat{n}_1(t)\Delta\hat{n}_2(t): \rangle &= \langle :\hat{n}_1(t)\hat{n}_2(t): \rangle - \langle \hat{n}_1(t) \rangle \langle \hat{n}_2(t) \rangle \\ &= \langle \hat{n}_j(t) \rangle (1 + \langle \hat{n}_j(t) \rangle), \end{aligned} \tag{22.4–14}$$

and from this result, together with Eq. (22.4–12) we obtain for the normalized cross-correlation coefficient between signal and idler photon numbers,

$$\sigma_{12} \equiv \frac{\langle:\Delta\hat{n}_1(t)\Delta\hat{n}_2(t):\rangle}{[\langle(\Delta\hat{n}_1(t))^2\rangle\langle(\Delta\hat{n}_2(t))^2\rangle]^{1/2}} = 1. \qquad (22.4\text{–}15)$$

Hence signal and idler are completely correlated, and for every increase in signal photons there is an equal increase in idler photons. Another way to interpret the same conclusion is to note from Eq. (22.4–13) that when $\langle\hat{n}_j(t)\rangle \ll 1$,

$$\langle:\hat{n}_1(t)\hat{n}_2(t):\rangle \approx \langle\hat{n}_j(t)\rangle \quad (j = 1, 2). \qquad (22.4\text{–}16)$$

Now when $\langle\hat{n}_j(t)\rangle \ll 1$ each side of this equation can be interpreted as a probability. The right-hand side $\langle\hat{n}_j(t)\rangle$ is a measure of the probability that a signal or idler photon is detected by a perfect detector, and the left-hand side $\langle:\hat{n}_1(t)\hat{n}_2(t):\rangle$ measures the joint probability of detecting both a signal and an idler photon (cf. Section 12.2). Equation (22.4–16) therefore expresses the fact that the two down-converted photons are always produced together, because the joint two-photon probability equals the single photon probability. Needless to say, this interpretation breaks down when $\langle\hat{n}_j(t)\rangle \gtrsim 1$.

Once again we must point out that the situation is always more complicated in reality, because more than just two modes of the field become excited, so that a more general multimode treatment is needed (Mollow, 1973; Hong and Mandel, 1985; Ou, Wang and Mandel, 1989). One can then also address questions about the time intervals between signal and idler photons. Such a treatment will be given in Section 22.4.4 below.

It is worth noting that if we introduce a normalized intensity correlation function, namely

$$\lambda_{12} = \frac{\langle:\Delta\hat{n}_1(t)\Delta\hat{n}_2(t):\rangle}{\langle\hat{n}_1(t)\rangle\langle\hat{n}_2(t)\rangle},$$

then from Eq. (22.4–14)

$$\lambda_{12} = 1 + \frac{1}{\langle\hat{n}_j(t)\rangle} \approx \frac{1}{\langle\hat{n}_j(t)\rangle} \gg 1, \quad (j = 1, 2).$$

Thus λ_{12} should be inversely proportional to the down-converted light intensity. This conclusion has been confirmed experimentally. Figure 22.2 shows the linear relationship between $1/\lambda_{12}$ and $\langle\hat{n}_1(t)\rangle$ obtained from experiment.

22.4.3 Proof of non-classical behavior

The non-classical character of the down-converted field can be demonstrated explicitly by use of the relation (22.4–13) in the form (Graham, 1984)

$$\langle:\hat{n}_1(t)\hat{n}_2(t):\rangle = \langle:\hat{n}_j^2(t):\rangle + \langle\hat{n}_j(t)\rangle, \quad (j = 1, 2).$$

By using the optical equivalence theorem for normally ordered operators (see Section 11.9), we can express this in the form of an inequality

$$\langle|v_1|^2|v_2|^2\rangle_\phi > \langle|v_j|^4\rangle_\phi = [\langle|v_1|^4\rangle_\phi\langle|v_2|^4\rangle_\phi]^{1/2}, \qquad (22.4\text{–}17)$$

where $\phi(v)$ is the diagonal coherent-state representation of the density operator. It is evident that this violates the Schwarz inequality

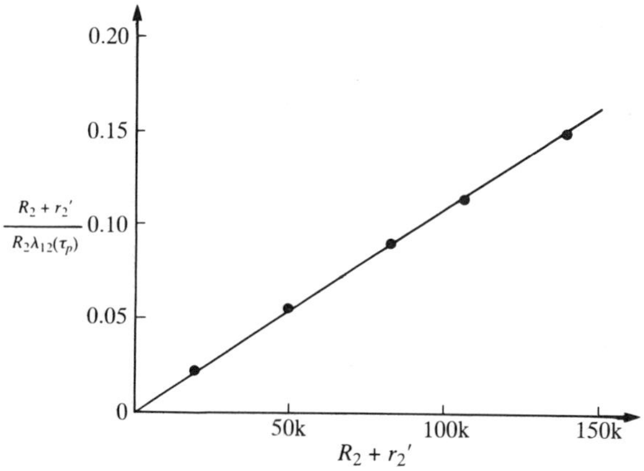

Fig. 22.2 Results of measurements illustrating the reciprocal relationship between λ_{12} and the mean light intensity in parametric down-conversion. $R_2 + r_2'$ is proportional to the light intensity and the ordinate is proportional to $1/\lambda_{12}$. The continuous curve is a straight line passing through the origin. (Reproduced from Friberg, Hong and Mandel, 1985a.)

$$\langle |v_1|^2 |v_2|^2 \rangle \leqslant [\langle |v_1|^4 \rangle \langle |v_2|^4 \rangle]^{1/2}$$

for classical fields, and it shows explicitly that the down-converted light has no classical description. Indeed, in practice $\langle |v_1|^2 |v_2|^2 \rangle_\phi \gg \langle |v_1|^4 \rangle_\phi$ when $\langle \hat{n}_1(t) \rangle \ll 1$, so that the violation is very large. A more recent experiment (Zou, Wang and Mandel, 1991a) in which the two-photon coincidence detection rates R_{12} between signal and idler, R_{11} between two signals and R_{22} between two idlers, were compared, led to the conclusion that, after subtraction of accidentals, R_{12} exceeds $(R_{11} + R_{22})$ by about 600 standard deviations. Needless to say there are no classical fields with the property that the joint probability of photodetection at two different points in space where the mean intensities are equal is much larger than that for two detections at the same point.

The phenomenon of spontaneous parametric down-conversion is capable of providing us with a very close approximation to an ideal one-photon state, that can be used as a probe or input for other processes, by virtue of the fact that when a signal photon is detected, we know that it must be accompanied by a sister idler photon. Figure 22.3 shows the results of photon counting measurements of the probability $p(n)$ for the appearance of n idler photons at a certain place and time, conditioned on the detection of a signal photon at the corresponding conjugate position within a certain time interval. Evidently there is one idler photon for every detected signal photon.

22.4.4 Multimode perturbative treatment of parametric down-conversion

Although the Hamiltonian with two quantized field modes given by Eq. (22.4–3) describes some features of the down-conversion process, it fails in other cases, because the down-converted light can be broadband and far from monochromatic. Even if the sum of the signal and idler frequencies has a sharp value, each signal photon and each idler photon can have a wide bandwidth, so that it behaves more like a short wave packet than a monochromatic wave.

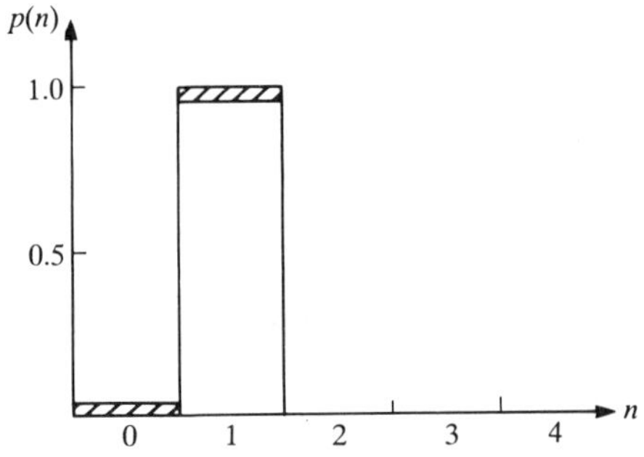

Fig. 22.3 The measured probability $p(n)$ of detecting n idler photons conditioned on the detection of a signal photon. (Reproduced from Hong and Mandel, 1986.)

We can allow for this possibility by making a plane-wave mode expansion of each field vector and expressing the interaction Hamiltonian $\hat{H}_I$ in the form (Ou, Wang and Mandel, 1989)

$$\hat{H}_I(t) = \frac{1}{L^3}\sum_{\mathbf{k}',s'}\sum_{\mathbf{k}'',s''} V_l \chi^{(2)}_{lij}(\omega_0, \omega', \omega'')(\boldsymbol{\varepsilon}^*_{\mathbf{k}'s'})_i(\boldsymbol{\varepsilon}^*_{\mathbf{k}''s''})_j$$

$$\times \int_{\mathscr{V}} e^{i(\mathbf{k}_0-\mathbf{k}'-\mathbf{k}'')\cdot\mathbf{r}}\, e^{i(\omega'+\omega''-\omega_0)t}\hat{a}^\dagger_{\mathbf{k}'s'}\hat{a}^\dagger_{\mathbf{k}''s''}\, d^3r + \text{h.c.} \quad (22.4\text{–}18)$$

$\mathbf{k}_0$ is the wave vector and ω_0 is the frequency of the monochromatic pump wave of vector amplitude $\mathbf{V}$, which is again treated classically, and the volume integral is to be taken over the active region $\mathscr{V}$ of the nonlinear medium. In order to avoid complications associated with refraction at the dielectric–air interface, we may suppose that the nonlinear medium is embedded in a passive linear dielectric of the same refractive index.

If $|\Psi(0)\rangle$ is the state of the field at time $t = 0$ in the interaction picture, then the state $|\Psi(t)\rangle$ at a later time t is given by

$$|\Psi(t)\rangle = \exp\left[\frac{1}{i\hbar}\int_0^t \hat{H}_I(t')\right]|\Psi(0)\rangle\, dt'. \quad (22.4\text{–}19)$$

In the special case in which the initial state of the down-converted light is the vacuum state of the signal and idler $|\Psi(0)\rangle = |\text{vac}\rangle_s|\text{vac}\rangle_i$, and for times t which are short compared with the average time interval between successive down-conversions, we have by expansion of the exponential,

$$|\Psi(t)\rangle = |\text{vac}\rangle_s|\text{vac}\rangle_i + L^{-3}\frac{1}{i\hbar}\sum_{[\mathbf{k}'s']_s}\sum_{[\mathbf{k}''s'']_i} V_l\chi^{(2)}_{lij}(\omega_0, \omega', \omega'')(\boldsymbol{\varepsilon}^*_{\mathbf{k}'s'})_i(\boldsymbol{\varepsilon}^*_{\mathbf{k}''s''})_j$$

$$\times \prod_{m=1}^{3}\left[\frac{\sin\frac{1}{2}(\mathbf{k}_0-\mathbf{k}'-\mathbf{k}'')_m l_m}{\frac{1}{2}(\mathbf{k}_0-\mathbf{k}'-\mathbf{k}'')_m}\right] e^{i(\omega'+\omega''-\omega_0)t/2}\frac{\sin\frac{1}{2}(\omega'+\omega''-\omega_0)t}{\frac{1}{2}(\omega'+\omega''-\omega_0)}|\mathbf{k}', s'\rangle_s|\mathbf{k}'', s''\rangle_i$$

$$+ \ldots \quad (22.4\text{–}20)$$

We take the nonlinear dielectric to be a parallelepiped with sides l_1, l_2, l_3 whose center is at the origin. We denote the set of signal and idler modes by $[\mathbf{k}', s']_s$ and $[\mathbf{k}'', s'']_i$, respectively, and we assume that they do not overlap.

Because the expression for $|\Psi(t)\rangle$ is rather complicated, we now simplify it in several respects. Let us suppose that signal and idler waves have similar polarizations and that their directions are well defined by apertures, etc. We may then treat s', s'' and the directions of $\mathbf{k}'$, $\mathbf{k}''$ as fixed and sum only over signal and idler frequencies. In place of Eq. (22.4–20) we then write the simpler relation

$$|\Psi(t)\rangle = M|\text{vac}\rangle_s|\text{vac}\rangle_i + \frac{\eta V\delta\omega}{2\pi}\sum_{\omega'}\sum_{\omega''}\phi(\omega', \omega'')\frac{\sin\frac{1}{2}(\omega' + \omega'' - \omega_0)t}{\frac{1}{2}(\omega' + \omega'' - \omega_0)} \times e^{i(\omega'+\omega''-\omega_0)t/2}|\omega'\rangle_s|\omega''\rangle_i + \ldots . \qquad (22.4\text{–}21)$$

The spectral function $\phi(\omega', \omega'')$, which we take to be symmetric with respect to ω', ω'', incorporates the frequency dependence of the various factors under the sum in Eq. (22.4–20). It is peaked at $\omega' = \omega_0/2 = \omega''$. The mode spacing $\delta\omega$ may be taken to tend to zero at the end of the calculation, when sums are replaced by integrals. For convenience we take $\phi(\omega', \omega'')$ to be normalized so that

$$\left(\frac{1}{2\pi}\right)\delta\omega\sum_{\omega}|\phi(\omega, \omega_0 - \omega)|^2 \Rightarrow \left(\frac{1}{2\pi}\right)\int_0^\infty|\phi(\omega, \omega_0 - \omega)|^2\,d\omega = 1, \qquad (22.4\text{–}22)$$

and we represent the magnitude of the perturbative contribution by the parameter η. If the pump intensity $|V|^2$ is expressed in units of photons per second, then η is dimensionless. The constant M in Eq. (22.4–21) is very close to unity because photon pairs are emitted only rarely, although it cannot be unity.

Normalization of $|\Psi(t)\rangle$ given by Eq. (22.4–21) requires that when $\delta\omega \to 0$,

$$|M|^2 + \frac{|\eta V|^2}{(2\pi)^2}\iint_0^\infty d\omega'\,d\omega''|\phi(\omega', \omega'')|^2\left[\frac{\sin\frac{1}{2}(\omega' + \omega'' - \omega_0)t}{\frac{1}{2}(\omega' + \omega'' - \omega_0)}\right]^2 + \ldots \to 1,$$

so that $|M|$ must be slightly less than unity. Let us make the substitution $\omega'' + \omega' - \omega_0 = \Omega''$, and consider the steady-state limit of long t. Then

$$1 = |M|^2 + \frac{|\eta V|^2}{(2\pi)^2}\int_0^\infty d\omega'\int_{\omega'-\omega_0}^\infty d\Omega''|\phi(\omega', \omega_0 - \omega' + \Omega'')|^2\left[\frac{\sin\frac{1}{2}\Omega''t}{\frac{1}{2}\Omega''}\right]^2. \qquad (22.4\text{–}23)$$

Now the dominant contributions to the Ω''-integral come from small values of Ω'' such that $\Omega''t \lesssim 1$. If $|\phi(\omega', \omega'')|^2$ is a sufficiently slowly varying function of ω'', we can make the approximation of replacing $|\phi(\omega', \omega_0 - \omega' + \Omega'')|^2$ by $|\phi(\omega', \omega_0 - \omega')|^2$ under the integral. The integrations with respect to ω' and Ω'' then become separable, and we may write

$$\int_{\omega'-\omega_0}^\infty\left[\frac{\sin\frac{1}{2}\Omega''t}{\frac{1}{2}\Omega''}\right]^2 d\Omega'' \approx \int_{-\infty}^\infty\left[\frac{\sin\frac{1}{2}\Omega''t}{\frac{1}{2}\Omega''}\right]^2 d\Omega'' = 2\pi t. \qquad (22.4\text{–}24)$$

With the help of Eq. (22.4–22) we then obtain from Eq. (22.4–23) the relation

$$1 = |M|^2 + |\eta V|^2 t.$$

The validity of the perturbative approximation we have made then depends on the condition

$$|\eta V|^2 t \ll 1. \tag{22.4–25}$$

At first sight this seems to contradict the long-time t assumption that was used to derive Eq. (22.4–24). However, condition (22.4–25) only requires t to be short compared with the average time interval between down-conversions, and this time can be several orders of magnitude longer than any correlation time which t has to exceed.

22.4.5 Entangled quantum state

The quantum state represented by Eq. (22.4–21)), describing the down-converted photon pair, exhibits some interesting and counter-intuitive features. Quite apart from the additive contribution of the vacuum state $|\text{vac}\rangle_s|\text{vac}\rangle_i$, which allows the down-converted photons to carry information about the phase of the pump, we see that the sum over frequencies does not allow factorization into a product of signal and idler states. The signal and idler photons are said to be *entangled* with each other in the frequency domain.

This has some unusual consequences. Suppose that near the nonlinear crystal we introduce a filter into the path of the signal photons that transmits only the frequency ω_F. This is obviously an oversimplification as regards real filters, but it is sufficient to illustrate the point we wish to emphasize. Let $\hat{P} \equiv |\omega_F\rangle_{s\,s}\langle\omega_F|$ be the projector that represents the effect of the filter. Then the state of the emerging down-converted field can be represented by the density operator

$$\hat{\rho} = K\hat{P}|\Psi(t)\rangle\langle\Psi(t)|,$$

where $|\Psi(t)\rangle$ is given by Eq. (22.4–21) and K is a normalization factor. The state of the idler alone, given the presence of the filter, is obtained by tracing $\hat{\rho}$ over the signal sub-space, or

$$\hat{\rho}_i = K\,\text{Tr}_s\,\hat{P}|\Psi(t)\rangle\langle\Psi(t)|.$$

In the long-time limit we find with the help of a similar approximation as in Section 22.4.4 that

$$\hat{\rho}_i \to K|\eta V|^2|\phi(\omega_F, \omega_0 - \omega_F)|^2|\omega_0 - \omega_F\rangle_{i\,i}\langle\omega_0 - \omega_F|. \tag{22.4–26}$$

In other words, as a consequence of choosing the frequency ω_F of the signal photon of the entangled pair we have found the idler photon to be of definite frequency also, such that the two frequencies add up to ω_0. Had we chosen a filter that transmits frequencies over a certain passband, which would be represented by a sum over projectors P, perhaps with some weight function, then the state of the corresponding idler field would be given by a sum over ω_F of expressions like those on the right of Eq. (22.4–26). Therefore the state of the idler photon is governed by observations made on the signal photon, even though the two photons may be so far apart at the time the measurement is made, that there is no time for them to communicate with each other. This is an example of the curious *non-locality* that is characteristic of entangled quantum states. At first sight it appears to violate causality and the principle of relativity, although closer examination shows that it does not. Such arguments led Einstein, Podolsky and

Rosen (1935) to conclude that quantum mechanics must be incomplete, (cf. Section 12.14).

The non-local behavior has been observed many times and was discussed in more detail in Section 12.14. The effects represented by Eq. (22.4–26) have also been observed, perhaps most closely in the experiments of Rarity and Tapster (1990) and Kwiat, Steinberg and Chiao (1992).

22.4.6 Rate of down-conversion

Let us now use the perturbative formalism to calculate the rate of down-conversion. Let a photodetector be located in the path of the signal beam at a distance $c\tau_s$ from the nonlinear medium. Then the signal field at the detector at time t can be represented by the mode expansion

$$\hat{E}_s^{(+)}(t) = \left(\frac{\delta\omega}{2\pi}\right)^{1/2} \sum_{\omega} \hat{a}_s(\omega)\, e^{-i\omega(t-\tau_s)}, \tag{22.4–27}$$

which has been normalized so that $\hat{E}_s^{(-)}\hat{E}_s^{(+)}$ is in units of photons per second. The term $\hat{a}_s(\omega)$ is the photon annihilation operator for the signal mode of frequency ω.

The average rate R_s at which the detector registers signal photons is now given by the expression

$$R_s = \alpha_s \langle \Psi(t)|\hat{E}_s^{(-)}(t)\hat{E}_s^{(+)}(t)|\Psi(t)\rangle,$$

where α_s is the quantum efficiency of the detector, and with the help of Eqs. (22.4–21) and (22.4–27) this formula gives

$$\begin{aligned}R_s = \alpha_s \left(\frac{\delta\omega}{2\pi}\right)^3 |\eta V|^2 \sum_{\omega'}\sum_{\omega''} e^{i(\omega'-\omega'')(t-\tau_s)} \sum_{\omega_1}\sum_{\omega_2}\sum_{\omega_3}\sum_{\omega_4} \phi^*(\omega_1, \omega_2)\phi(\omega_3, \omega_4)\\ \times \frac{\sin\frac{1}{2}(\omega_1+\omega_2-\omega_0)t}{\frac{1}{2}(\omega_1+\omega_2-\omega_0)} \frac{\sin\frac{1}{2}(\omega_3+\omega_4-\omega_0)t}{\frac{1}{2}(\omega_3+\omega_4-\omega_0)} e^{i(\omega_3+\omega_4-\omega_1-\omega_2)t/2}\\ \times {}_i\langle\omega_2|_s\langle\omega_1|\hat{a}_s^\dagger(\omega')\hat{a}_s(\omega'')|\omega_3\rangle_s|\omega_4\rangle_i.\end{aligned}$$

Non-vanishing contributions to the sum require $\omega_1 = \omega'$, $\omega_2 = \omega_4$, $\omega_3 = \omega''$. After proceeding to the limit $\delta\omega \to 0$ we obtain

$$R_s = \alpha_s |\eta V|^2 \frac{1}{2\pi} \int_0^\infty |F(\omega; t)|^2\, d\omega \tag{22.4–28}$$

where

$$F(\omega; t) = \int_0^\infty \phi(\omega', \omega) \frac{\sin\frac{1}{2}(\omega'+\omega-\omega_0)t}{\frac{1}{2}(\omega'+\omega-\omega_0)} e^{-i\omega'(t/2-\tau_s)}\, d\omega'. \tag{22.4–29}$$

Once again we consider the long-time limit and put $\omega' + \omega - \omega_0 = \Omega'$, so that

$$F(\omega; t) = \int_{\omega-\omega_0}^\infty \phi(\omega_0 - \omega + \Omega', \omega) \frac{\sin\frac{1}{2}\Omega' t}{\frac{1}{2}\Omega'} e^{-i\Omega'(t/2-\tau_s)}\, e^{-i(\omega-\omega_0)(t/2-\tau_s)}\, d\Omega'.$$

When t is large the dominant contributions to the Ω'-integral come from small Ω'. If $\phi(\omega_0 - \omega + \Omega', \omega)$ varies sufficiently slowly with Ω', we can replace

$\phi(\omega_0 - \omega + \Omega', \omega)$ by $\phi(\omega_0 - \omega, \omega)$ to a good approximation. Moreover, as $\omega_0 - \omega$ is an optical frequency, we may also replace the lower limit $\omega - \omega_0$ of the integral by $-\infty$ to a good approximation. The Ω'-integral then reduces to the well-known Dirichlet discontinuous integral [see Eq. (12.11–10)],

$$\left.\begin{aligned} \frac{1}{2\pi}\int_{-\infty}^{\infty} \frac{\sin\frac{1}{2}\Omega' t}{\frac{1}{2}\Omega'} \mathrm{e}^{-\mathrm{i}\Omega'(t/2-\tau_s)}\,\mathrm{d}\Omega' &= 1 \quad \text{if } |t/2 - \tau_s| < t/2 \\ &= 0 \quad \text{otherwise.} \end{aligned}\right\} \tag{22.4–30}$$

Hence, provided $0 < \tau_s < t$, we obtain for large t

$$\begin{aligned} R_s &= \alpha_s |\eta V|^2 \frac{1}{2\pi}\int_0^{\infty} |\phi(\omega_0 - \omega, \omega)|^2\,\mathrm{d}\omega \\ &= \alpha_s |\eta V|^2. \end{aligned} \tag{22.4–31}$$

The last line follows with the help of Eq. (22.4–22). As $|V|^2$ gives the rate at which pump photons fall on the nonlinear medium, it is apparent that $|\eta|^2$ gives the fraction of incident pump photons that convert to signal–idler pairs.

22.4.7 *Time separation between signal and idler photons*

We have already seen that signal and idler photons appear together in the process of parametric down-conversion. An obvious question concerns the time scale T_c within which signal and idler photons may be regarded as 'simultaneous'. It was found by Burnham and Weinberg (1970) in a coincidence counting experiment that the two photons are separated by no more than a few nanoseconds. However they recognized that this measured correlation time T_c was probably determined by the time resolution of their photodetectors, rather than being intrinsic to the down-conversion process. Although later measurements with faster detectors led to lower values of T_c (Friberg, Hong and Mandel, 1985b), the intrinsic correlation time determined by the reciprocal bandwidth $\Delta\omega$ of the down-converted light was known to be in the picosecond or sub-picosecond range. This time is so many orders of magnitude below the time resolution of the detector/electronics that it might appear to be unobservable. However, by using an interference technique rather than direct detection, Hong, Ou and Mandel (1987) were able to measure T_c for the down-converted photons.

In order to understand the principle of the experiment, let us consider the beam splitter BS illustrated in Fig. 22.4 having two inputs 1 and 2 and two outputs 3 and 4. Suppose that a signal photon enters port 1 after traveling a distance $c\tau_1$ from the down-converter and the conjugate idler photon enters port 2 after travelling a distance $c\tau_2$. Let us calculate the joint probability for detecting a pair of photons in coincidence at the output ports 3 and 4. The fields at the exit ports can be expressed in the form

$$\left.\begin{aligned} \hat{E}_3^{(+)}(t) &= \left(\frac{\delta\omega}{2\pi}\right)^{1/2} \sum_{\omega} \mathcal{T}\hat{a}_s(\omega)\,\mathrm{e}^{-\mathrm{i}\omega(t-\tau_1)} + \mathcal{R}\hat{a}_i(\omega)\,\mathrm{e}^{-\mathrm{i}\omega(t-\tau_2)} \\ \hat{E}_4^{(+)}(t) &= \left(\frac{\delta\omega}{2\pi}\right)^{1/2} \sum_{\omega} \mathcal{R}\hat{a}_s(\omega)\,\mathrm{e}^{-\mathrm{i}\omega(t-\tau_1)} + \mathcal{T}\hat{a}_i(\omega)\,\mathrm{e}^{-\mathrm{i}\omega(t-\tau_2)}, \end{aligned}\right\} \tag{22.4–32}$$

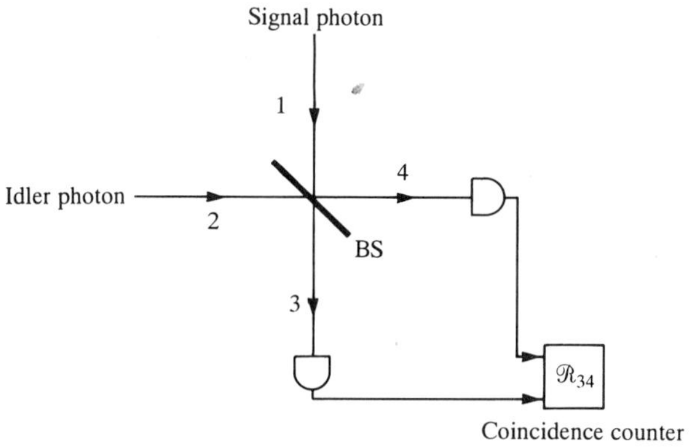

Fig. 22.4 Illustrating the principle of determining the time separation between two photons by interference at a beam splitter (BS).

where $\mathscr{R}$, $\mathscr{T}$ are the complex amplitude reflectivity and transmissivity of the beam splitter. The joint probability density that a photon is detected at port 3 at time t and another one at port 4 at time $t + \tau$ is proportional to

$$P_{34}(t, t+\tau) = \alpha_3\alpha_4\langle \Psi(t)|\hat{E}_3^{(-)}(t)\hat{E}_4^{(-)}(t+\tau)\hat{E}_4^{(+)}(t+\tau)\hat{E}_3^{(+)}(t)|\Psi(t)\rangle, \tag{22.4–33}$$

with $|\Psi(t)\rangle$ given by Eq. (22.4–21). Here, α_3, α_4 are the quantum efficiencies of the detectors at ports 3 and 4, respectively.

We now substitute for $\hat{E}_3^{(+)}(t)$, $\hat{E}_4^{(+)}(t)$ and $|\Psi(t)\rangle$ and evaluate the expectation for large t, just as in the calculation leading to Eq. (22.4–31) above. After a somewhat lengthy but straightforward calculation we obtain the result that

$$\begin{aligned} &P_{34}(t, t+\tau) \\ &= \alpha_3\alpha_4|\eta V|^2|G(0)|^2[|\mathscr{T}|^4|g(\tau_2-\tau_1+\tau)|^2 + |\mathscr{R}|^4|g(\tau_2-\tau_1-\tau)|^2 \\ &\quad - \mathscr{R}^2\mathscr{T}^{*2}g^*(\tau_2-\tau_1-\tau)g(\tau_2-\tau_1+\tau) - \mathscr{R}^{*2}\mathscr{T}^2g^*(\tau_2-\tau_1+\tau)g(\tau_2-\tau_1-\tau)]. \end{aligned} \tag{22.4–34}$$

Here

$$G(\tau) \equiv \int_0^\infty \phi(\tfrac{1}{2}\omega_0 + \omega, \tfrac{1}{2}\omega_0 - \omega)\,\mathrm{e}^{-\mathrm{i}\omega\tau}\,\mathrm{d}\omega, \tag{22.4–35}$$

is the Fourier transform of the spectral function ϕ, which is the autocorrelation function of the down-converted light, and

$$g(\tau) \equiv G(\tau)/G(0) \tag{22.4–36}$$

is the normalized autocorrelation. It has a range in τ of order $1/\Delta\omega$, and it becomes very small for $\tau \gg 1/\Delta\omega$.

In practice one usually measures the coincidence counting rate R_{34}, which is the rate at which detections are registered at ports 3 and 4 within the resolving time

T_R of the detectors and the electronics. Thus

$$R_{34} = \int_{-T_R/2}^{T_R/2} P_{34}(t, t+\tau)\, d\tau. \qquad (22.4\text{–}37)$$

After substituting for $P_{34}(t, t+\tau)$ under the integral, carrying out the integrations, and remembering that T_R is typically much longer than the coherence time $1/\Delta\omega$, so that the limits in Eq. (22.4–37) may be taken as effectively $\pm\infty$, we arrive at the relation (Hong, Ou and Mandel, 1987)

$$R_{34} = \alpha_3\alpha_4|\eta V|^2|G(0)|^2 \times \left\{|\mathcal{T}|^4 + |\mathcal{R}|^4 - \left[\mathcal{R}^{*2}\mathcal{T}^2\int_{-\infty}^{\infty} g^*(\tau_2 - \tau_1 + \tau)g(\tau_2 - \tau_1 - \tau)\, d\tau + \text{c.c.}\right]\right\}. \qquad (22.4\text{–}38)$$

Let us examine the form of R_{34} as a function of $\tau_2 - \tau_1$ in the special case when $|\mathcal{R}|^2 = |\mathcal{T}|^2 = \frac{1}{2}$, and the spectral function ϕ is Gaussian, so that

$$g(\tau) = e^{-(\tau\Delta\omega)^2/2}. \qquad (22.4\text{–}39)$$

Then we readily find from Eq. (22.4–38) that

$$R_{34} = \alpha_3\alpha_4|\eta V|^2|G(0)|^2[1 - e^{-\Delta\omega^2(\tau_2-\tau_1)^2}]. \qquad (22.4\text{–}40)$$

This is zero when $\tau_2 - \tau_1 = 0$ and rises with increasing $|\tau_2 - \tau_1|$ to $\alpha_3\alpha_4|\eta V|^2$ when $|\tau_2 - \tau_1| \gg 1/\Delta\omega$. In other words, by varying the time delay $\tau_2 - \tau_1$ introduced between the signal and idler photons and measuring the two-photon coincidence counting rate at the output of the beam splitter as a function of $\tau_2 - \tau_1$, we can determine the correlation time between the two photons. Moreover, because this is an interferometric measurement, it should be possible to measure correlation times $1/\Delta\omega$ in the sub-picosecond time range that are much shorter than the resolving times of the detectors and the counting electronics.

Figure 22.5 shows the outline of an experiment to measure the distribution of time intervals between signal and idler photons produced in the down-conversion process. The two photons fall on the beam splitter BS from opposite sides; this results in two output beams with contributions from signal and idler photons, and the mixed signal and idler photons are detected by D1 and D2 both separately and in coincidence. The interference filters IF1 and IF2 placed in front of the detectors have passbands of about 10^{13} Hz, which means that the photons falling on the detectors should be regarded as wave packets of length ~ 100 fs. In order to introduce a differential time delay between signal and idler photons, BS is translated slightly as shown. This shortens the path for one photon relative to the other one.

Figure 22.6 shows the results of the measurement superimposed on the theoretical curve derived from Eqs. (22.4–38) or (22.4–40). From the distribution of the coincidence counts it appears that the two photons have a correlation time ~ 100 fs, as expected from the passband of IF1 and IF2. Note that the time resolution achieved in this interference experiment is almost one million times shorter than the resolving times of the detectors and electronics.

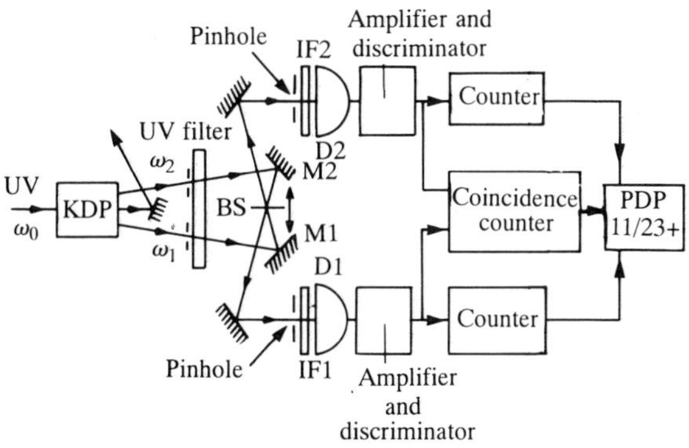

Fig. 22.5 Outline of an experiment to measure the time separation between two photons by interference at a beam splitter BS. KDP is a nonlinear crystal of potassium dihydrogen phosphate functioning as down-converter and PDP 11/23 is a computer. (Reproduced from Hong, Ou and Mandel, 1987.)

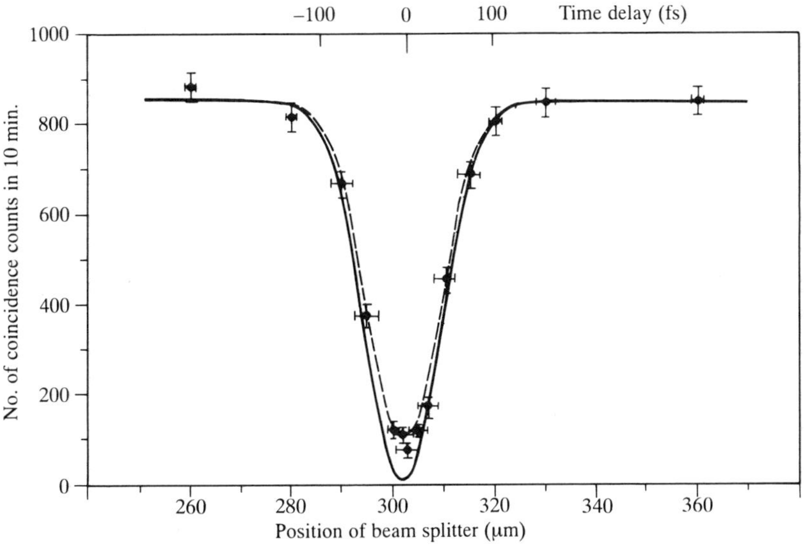

Fig. 22.6 Results of two-photon coincidence measurements as a function of differential time delay between the two photons superimposed on the theoretical (full) curve. (Reproduced from Hong, Ou and Mandel, 1987.)

Finally, one is tempted to ask if there is an intuitively simple way to understand this experiment. It has been shown (see Section 12.12) that when one photon enters a 50%:50% beam splitter at port 1 and a similar photon enters at port 2, destructive interference makes it impossible for one photon to emerge at port 3 and one at port 4 (cf. Fig. 22.4). Instead both photons emerge together either from port 3 or port 4. The coincidence rate is therefore zero for two identical, simultaneous photons. However, if one photon is delayed relative to the other one, so that the two wave packets no longer overlap completely, the destructive interference is no longer complete. The coincidence rate then rises with the delay, until, for long delays, it becomes constant and independent of the time delay. This accounts for the behavior exhibited in Fig. 22.6.

22.4.8 *Interference experiments with two down-converters*

Some interesting quantum effects show up in interference experiments utilizing two down-converters of the kind we have been discussing. For example, let us consider the experiment illustrated in Fig. 22.7, in which two similar nonlinear crystals NL1 and NL2, both functioning as down-converters, are both optically pumped by light of frequency ω_0 derived from the same laser beam. As a result, down-conversion can occur at NL1 with the simultaneous emission of a pair of signal s_1 and idler i_1 photons, or down-conversion can occur at NL2 with the emission of a pair of s_2 and i_2 photons (the probability of simultaneous emissions from both NL1 and NL2 may be regarded as negligibly small). Suppose now that we allow the signals s_1 and s_2 to come together and mix at the 50%:50% beam splitter BS_A, such that the combined field falls on detector D_A. Similarly, the idlers i_1, i_2 come together at BS_B and the combined idler field falls on detector D_B. We are interested to know whether the two signal fields s_1, s_2 and the two idler fields i_1, i_2 interfere under these conditions.

An experiment to answer this question has been carried out (Ou, Wang, Zou and Mandel, 1990), and the theory of the process has been examined in some detail (Ou, Wang and Mandel, 1989). The results are shown in Fig. 22.8. Figure 22.8(a) gives the observed results when the photon counting rates R_A and R_B of detectors D_A and D_B are plotted against the optical path difference or the differential phase shift between the pump beams reaching NL1 and NL2. It is apparent that both superposed light intensities are constant and independent of the path difference, so that s_1 and s_2 do not interfere, and neither do i_1 and i_2. Figure 22.8(b) shows the result of plotting the two-photon coincidence counting rate R_{AB} for detectors D_A and D_B together against the optical path difference. This time we find an unmistakable interference pattern of the expected periodicity, even though s_1 and s_2 are mutually incoherent as are i_1 and i_2.

We shall now attempt to account for these experimental results by an oversim-

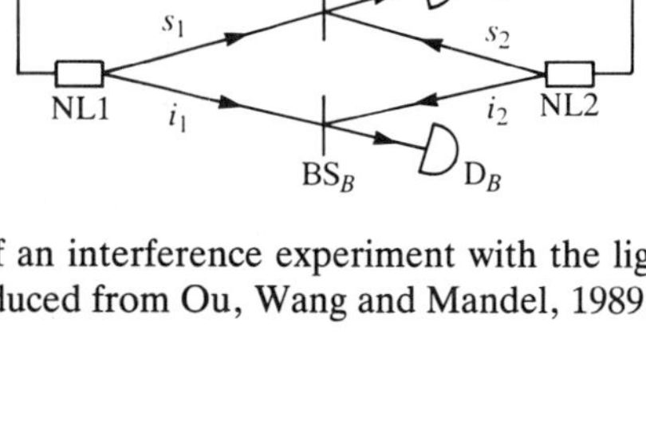

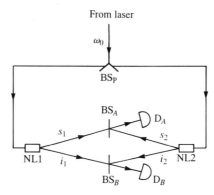

Fig. 22.7 Outline of an interference experiment with the light from two down-converters. (Reproduced from Ou, Wang and Mandel, 1989.)

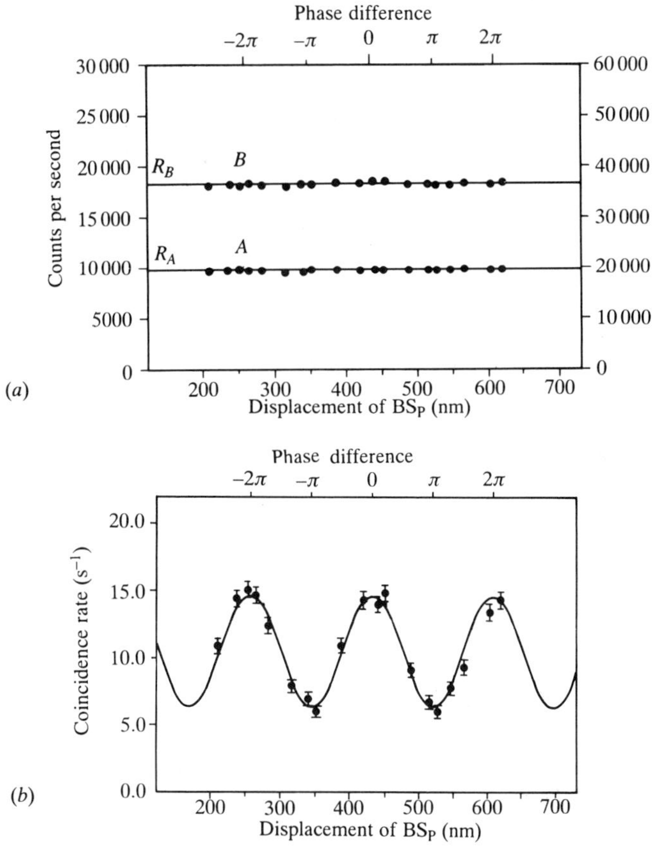

Fig. 22.8 Results of interference experiments with two down-converters; (*a*) one-photon rates R_A (left scale), R_B (right scale), (*b*) two-photon coincidence rate R_{AB}. (Reproduced from Ou, Wang, Zou and Mandel, 1990.)

plified treatment in which the down-converted fields are all treated as monochromatic. If $|\psi_1\rangle_1$ and $|\psi_2\rangle_2$ are the quantum states of the down-converted fields in the interaction picture produced by crystals NL1 and NL2, then by reference to Eq. (22.4–20), when the sum over modes can be discarded we may write

$$|\psi_1\rangle_1 = M_1|\text{vac}\rangle_{s1,i_1} + \eta_1 V_1 F_1|\omega\rangle_{s_1}|\omega'\rangle_{i_1} \qquad (22.4\text{–}41)$$

$$|\psi_2\rangle_2 = M_2|\text{vac}\rangle_{s2,i_2} + \eta_2 V_2 F_2|\omega\rangle_{s_2}|\omega'\rangle_{i_2}. \qquad (22.4\text{–}42)$$

Here ω, ω' are the signal and idler frequencies, respectively, V_1, V_2 are the complex classical pump amplitudes at NL1, NL2, η_1, η_2 are dimensionless factors such that $|\eta_1|^2$, $|\eta_2|^2$ give the down-conversion efficiences, F_1, F_2 are constants characterizing the two down-converters, and M_1, M_2 are complex coefficients which are very close to unity in practice. If $\hat{E}_A^{(+)}$, $\hat{E}_B^{(+)}$ are the positive frequency parts of the total electric fields at detectors D_A, D_B, then

$$\hat{E}_A^{(+)} = K(\hat{a}_{s_1} + \mathrm{i}\hat{a}_{s_2})\,\mathrm{e}^{-\mathrm{i}\omega t} \qquad (22.4\text{–}43)$$

$$\hat{E}_B^{(+)} = K(\hat{a}_{i_1} + \mathrm{i}\hat{a}_{i_2})\,\mathrm{e}^{-\mathrm{i}\omega' t}. \qquad (22.4\text{–}44)$$

Hence the photon counting rate of detector D_A is proportional to the expression

$$R_A = {}_1\langle\psi_1|{}_2\langle\psi_2|\hat{E}_A^{(-)}\hat{E}_A^{(+)}|\psi_2\rangle_2|\psi_1\rangle_1$$
$$= |K|^2[|\eta_1 V_1 F_1|^2 + |\eta_2 V_2 F_2|^2], \tag{22.4–45}$$

and similarly for detector D_B. Evidently these photon rates exhibit no interference.

We now turn to the calculation of the two-photon coincidence counting rate R_{AB},

$$R_{AB} = {}_1\langle\psi_1|{}_2\langle\psi_2|\hat{E}_A^{(-)}\hat{E}_B^{(-)}\hat{E}_B^{(+)}\hat{E}_A^{(+)}|\psi_2\rangle_2|\psi_1\rangle_1. \tag{22.4–46}$$

With the help of Eqs. (22.4–41) to (22.4–44) we immediately obtain the result

$$R_{AB} = |\eta_1 V_1 F_1 M_2 - \eta_2 V_2 F_2 M_1|^2$$
$$\approx |\eta_1 V_1 F_1|^2 + |\eta_2 V_2 F_2|^2 - 2|\eta_1 V_1 F_1|\,|\eta_2 V_2 F_2| \cos[(\arg V_1 - \arg V_2) + \text{const.}]. \tag{22.4–47}$$

This two-photon coincidence rate evidently exhibits interference as the phase difference ($\arg V_1 - \arg V_2$) is varied, and this was observed, although the observed visibility was lower than predicted by Eq. (22.4–47). It is not difficult to show that a more realistic multimode treatment of the problem leads to essentially the same result (Ou, Wang and Mandel, 1989).

We now turn to the question why the coincidence rate R_{AB} exhibits interference, whereas the single photon rates R_A and R_B do not. One way of approaching the problem is to invoke the relationship between mutual coherence and intrinsic indistinguishability of the photon path (Mandel, 1991). If, without disturbing the interference, it is impossible to determine whether the photons originate in NL1 or in NL2, then the corresponding probability amplitudes for the two paths have to be added in order to arrive at the detection probability, and then interference results. On the other hand, if there is a possibility, even in principle, of determining the source of the photons, then all interference is wiped out.

Now in the two-photon coincidence measurements there is indeed no way to determine the source of each photon pair without introducing disturbances. But suppose that we are interested only in the signal photons and whether they interfere. Because the idlers are of no interest we may remove beam splitter BS_B without disturbing the interference of the signal photons. But once BS_B has been removed, detections by D_A which are accompanied by detections by D_B occur only when both photons originate in NL1. By contrast, if D_B is a near 100% efficient detector, a photon detection by D_A which is not accompanied by a detection by D_B must be attributed to an s_2 photon emitted by NL2. It follows that the source of each detected signal photon can be identified in this way, and therefore all indistinguishability of the sources is lost, and so is all interference. It is not necessary, however, for the auxiliary measurement to be actually carried out. The possibility that it can be performed is sufficient to suppress the interference of the signal photons. A similar argument shows that the idlers do not interfere either.

Lest it be thought that in order to observe interference with two down-converters it is always necessary to measure two-photon coincidences, we now consider the experimental arrangement shown in Fig. 22.9. Once again we have two similar nonlinear crystals NL1 and NL2 functioning as down-converters, which are optically pumped by mutually coherent pump beams derived from the same laser. But this time the two crystals have been arranged so that the idler i_1 from NL1 passes through NL2 and lines up with the idler i_2. As before we allow the two signals s_1 and s_2 to come together, to be mixed by output beam splitter BS_0 and to fall on the signal detector D_s. Once again we are interested in the question whether the two signals exhibit interference, in which case the output of D_s will undergo a sinusoidal variation as BS_0 is translated in a direction normal to its face.

Such an experiment has been carried out (Zou, Wang and Mandel, 1991b; Wang, Zou and Mandel, 1991), with the results shown in Fig. 22.10. So long as the two idlers are colinear, the counting rate of D_s follows along curve *A* and clearly exhibits interference. But if i_1 and i_2 are misaligned, or if i_1 is blocked by insertion of a beam stop and prevented from reaching NL2, the experimental points fall on line *B* and all interference disappears. It is worth noting that the

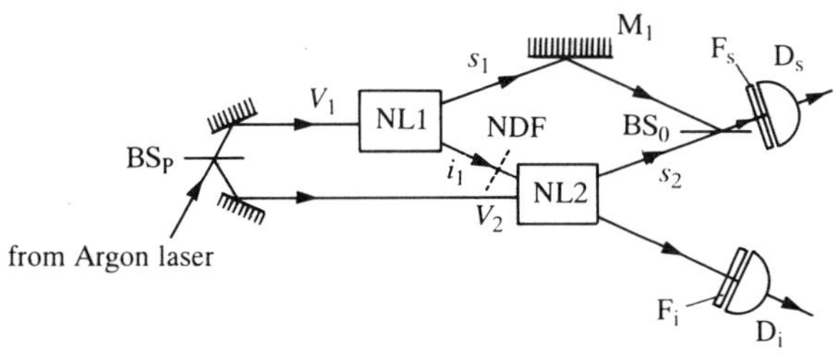

Fig. 22.9 Outline of another interference experiment with two down-converters. (Reproduced from Zou, Wang and Mandel, 1991b.)

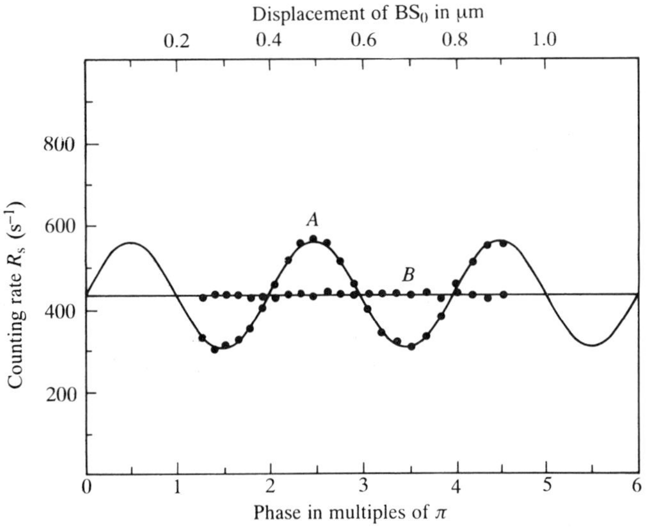

Fig. 22.10 Results of the interference experiment giving the photon counting rate as a function of the displacement of BS_0 (*a*) with idlers i_1 and i_2 aligned, (*b*) with idler i_1 blocked. (Reproduced from Zou, Wang and Mandel, 1991b.)

average rate of photon emission does not change; it is only the mutual coherence between s_1 and s_2 which is affected. If, instead of blocking i_1 altogether, we insert a filter of variable amplitude transmissivity $\mathcal{T}$ between NL1 and NL2, it is found that the visibility of the interference is directly proportional to $|\mathcal{T}|$, as shown in Fig. 22.11. If i_1 can be said to induce coherence between s_1 and s_2 in this experiment, then we are dealing with a new and unusual form of induced coherence that is not accompanied by induced emission. Evidently, control of $|\mathcal{T}|$ provides us with a new method for varying the degree of mutual coherence between two light beams that leaves the light intensities unchanged.

Once again we shall attempt to account for these observations by an oversimplified, but still useful, treatment in which the fields are all monochromatic, and we write

$$\hat{E}_s^{(+)} = K(\hat{a}_{s_1} e^{i\theta_1} + \hat{a}_{s_2} e^{i\theta_2}) e^{-i\omega t} \tag{22.4–48}$$

for the total field at the signal detector, θ_1 and θ_2 being phase shifts associated with the propagation of the two signal beams. Let us model the variable transmissivity filter $\mathcal{T}$ as a beam splitter, as shown in Fig. 22.12, with a vacuum input field mode $\hat{a}_v$, so that

$$\hat{a}_{i_2} = \mathcal{T}\hat{a}_{i_1} + \mathcal{R}\hat{a}_v, \tag{22.4–49}$$

where $\mathcal{R}$ is the beam splitter reflectivity. The quantum state $|\psi\rangle$ of the field in the interaction picture produced by both down-converters will be given by the

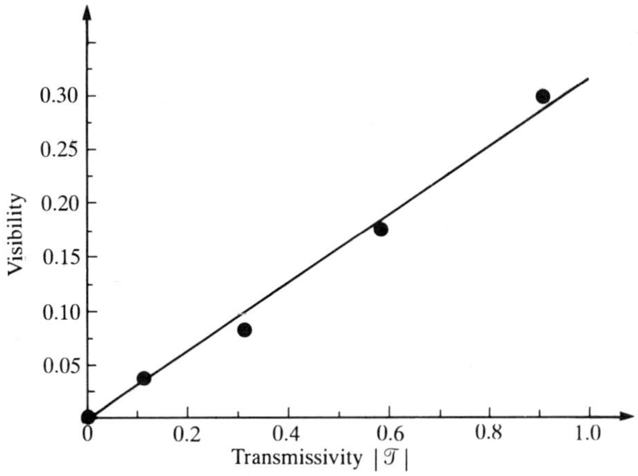

Fig. 22.11 Results of the interference experiment with a filter of transmissivity $\mathcal{T}$ introduced between NL1 and NL2. The visibility is shown as function of the transmissivity $|\mathcal{T}|$. (Reproduced from Zou, Wang and Mandel, 1991b.)

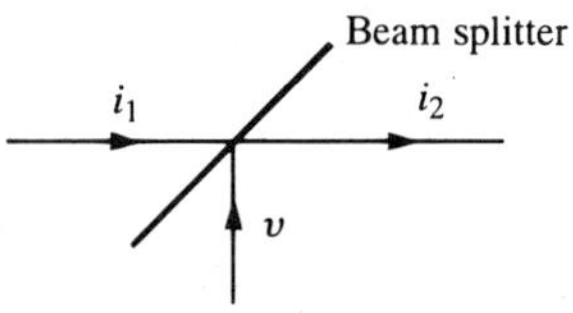

Fig. 22.12 Illustration of the relevant beam splitter modes.

equation [cf. Eqs. (22.4–41) and (22.4–42) above]

$$|\psi\rangle = |\text{vac}\rangle_{s1,s2,i1,v} + \eta_1 V_1(t) F_1 |\omega\rangle_{s1}|\omega'\rangle_{i1}|\text{vac}\rangle_{s2,v}$$
$$+ \eta_2 V_2(t+\tau_0) F_2 \, \mathrm{e}^{-\mathrm{i}\omega'\tau_0}[\mathcal{T}^*|\omega\rangle_{s2}|\omega'\rangle_{i1}|\text{vac}\rangle_{s1,v} + \mathcal{R}^*|\omega\rangle_{s2}|\omega'\rangle_{v}|\text{vac}\rangle_{s1,i1}]. \tag{22.4–50}$$

Here τ_0 is the propagation time of the idler i_1 between crystals NL1 and NL2.

The rate of photon counting by D_s is then expressible as

$$R_s = \langle\psi|\hat{E}_s^{(-)}\hat{E}_s^{(+)}|\psi\rangle$$

and with the help of Eqs. (22.4–48) and (22.4–50) this gives

$$R_s = |K|^2[|\eta_1|^2|F_1|^2\langle I_1\rangle + |\eta_2|^2|F_2|^2\langle I_2\rangle + 2|\eta_1\eta_2|\,|F_1|\,|F_2|(\langle I_1\rangle\langle I_2\rangle)^{1/2}$$
$$\times |\gamma_{12}|\,|\mathcal{T}|\cos(\omega'\tau_0 + \theta_2 - \theta_1 - \arg\mathcal{T} + \text{const})]. \tag{22.4–51}$$

Here $\gamma_{12} \equiv \langle V_1^*(t)V_2(t+\tau_0)\rangle/(\langle I_1\rangle\langle I_2\rangle)^{1/2}$ is the complex degree of coherence. Therefore detector D_s should register interference with visibility proportional to the filter transmissibility $|\mathcal{T}|$, as was indeed observed. Once again it may be shown that a fuller multimode treatment leads to essentially the same conclusion.

Let us now examine how these results can be understood in terms of the indistinguishability of the photon paths. If the idler i_1 path is blocked, it becomes possible to determine where each photon detected by D_s originated with the help of a simple auxiliary measurement that does not disturb the interference experiment in any way. For this purpose let us suppose that a near perfect photodetector D_i is positioned in the i_2 beam at the same distance from NL2 as D_s, as illustrated in Fig. 22.9. Let us note whether a photon detection by D_s is accompanied by a simultaneous detection by D_i or not. If it is, then both detected photons obviously originated in NL2. If it is not, then the photon detected by D_s must have originated in NL1. It follows that the source of each detected signal photon is identifiable with the help of D_i, and this destroys all indistinguishability and all interference. Once again we emphasize that the auxiliary measurement with D_i need not actually be carried out. It is the possibility that it could be carried out that destroys the interference. When $\mathcal{T} = 0$ the state of the signal photons can be shown to be diagonal in photon numbers, and a diagonal density operator reflects not so much what is known but what is knowable in principle.

Finally we emphasize that blocking i_1 wipes out the s_1, s_2 interference not because we have created a large uncontrollable disturbance of the system, nor for reasons connected with the uncertainty principle. The disturbance is actually of a rather subtle kind: it is the possibility of obtaining 'which path' information that destroys the interference.

22.5 Degenerate four-wave mixing

Degenerate four-wave mixing is a nonlinear process involving the mutual interaction of four different waves of the same frequency through the nonlinear medium (Hellwarth, 1977; Yariv and Pepper, 1977; Bloom and Bjorklund, 1977;

Jensen and Hellwarth, 1978). For simplicity we shall concentrate on the experimental situation shown in Fig. 22.13, in which the two intense pump waves 1 and 2 and the two signal and idler waves 3 and 4, are counter-propagating and perpendicular, as shown (Yuen and Shapiro, 1979). The fundamental interaction of interest is one in which two photons are absorbed from the pump beams and two photons are created in the signal and idler beams. Comparison with Eq. (22.4–3) suggests that we start from an interaction Hamiltonian of the general form $\boldsymbol{\chi}^{(3)}(\hat{a}_4^\dagger\hat{a}_3^\dagger\hat{a}_2\hat{a}_1 + \text{h.c.})$, but, as before, we shall simplify the problem by treating the pump modes classically. The four-wave mixing process is distinguished from the parametric process treated above in that we are dealing with four different directions of propagation, and this must be reflected in the structure of the Hamiltonian.

We start by expressing the interaction $\hat{H}_\text{I}$ as an integral over the volume $\mathcal{V}$ of the nonlinear medium, in the form

$$\hat{H}_\text{I} = \int_{\mathcal{V}} \chi^{(3)}_{ijlm}\mathbf{E}^{(1)}_i(\mathbf{r}, t)\mathbf{E}^{(2)}_j(\mathbf{r}, t)\hat{\mathbf{E}}^{(s)}_l(\mathbf{r}, t)\hat{\mathbf{E}}^{(i)}_m(\mathbf{r}, t)\,\mathrm{d}^3r, \qquad (22.5\text{–}1)$$

where $\mathbf{E}^{(1)}$, $\mathbf{E}^{(2)}$ are two classical pump waves and $\hat{\mathbf{E}}^{(s)}$, $\hat{\mathbf{E}}^{(i)}$ are the quantized signal and idler waves. If V is the phase velocity of the waves in the medium, we represent these four quasi-monochromatic, polarized plane waves in the form

$$\left.\begin{aligned}
\mathbf{E}^{(1)}(\mathbf{r}, t) &= v_1\boldsymbol{\varepsilon}_1\,\mathrm{e}^{\mathrm{i}\omega(x/V-t)} + \text{c.c.}\\
\mathbf{E}^{(2)}(\mathbf{r}, t) &= v_2\boldsymbol{\varepsilon}_2\,\mathrm{e}^{\mathrm{i}\omega(-x/V-t)} + \text{c.c.}\\
\hat{\mathbf{E}}^{(s)}(\mathbf{r}, t) &= \hat{\mathscr{A}}_3(z/V - t)\boldsymbol{\varepsilon}_3\,\mathrm{e}^{\mathrm{i}\omega(z/V-t)} + \text{h.c.}\\
\hat{\mathbf{E}}^{(i)}(\mathbf{r}, t) &= \hat{\mathscr{A}}_4(z/V + t)\boldsymbol{\varepsilon}_4\,\mathrm{e}^{\mathrm{i}\omega(-z/V-t)} + \text{h.c.},
\end{aligned}\right\} \qquad (22.5\text{–}2)$$

where $\boldsymbol{\varepsilon}_j$ $(j = 1, 2, 3, 4)$ are unit polarization vectors. Note that the complex amplitudes v_1, v_2 of the pump waves are treated as constant, whereas those of the signal and idler are slowly varying operator functions of $(z/V - t)$ and $(z/V + t)$, respectively. We now substitute these expressions in Eq. (22.5–1). If the linear dimensions of the volume $\mathcal{V}$ are very large compared with the wavelength $2\pi V/\omega$, then the principal contributions to the volume integral come from terms in which the oscillatory factors all cancel. This leads to the following simplified form of the interaction energy,

$$\begin{aligned}
\hat{H}_\text{I} &= \mathcal{V}\chi^{(3)}_{ijlm}v_1v_2(\boldsymbol{\varepsilon}_1)_i(\boldsymbol{\varepsilon}_2)_j(\boldsymbol{\varepsilon}_3^*)_l(\boldsymbol{\varepsilon}_4^*)_m\hat{\mathscr{A}}_3^\dagger\hat{\mathscr{A}}_4^\dagger + \text{h.c.}\\
&= \hbar g(v_1v_2\hat{\mathscr{A}}_3^\dagger\hat{\mathscr{A}}_4^\dagger + \text{h.c.}).
\end{aligned} \qquad (22.5\text{–}3)$$

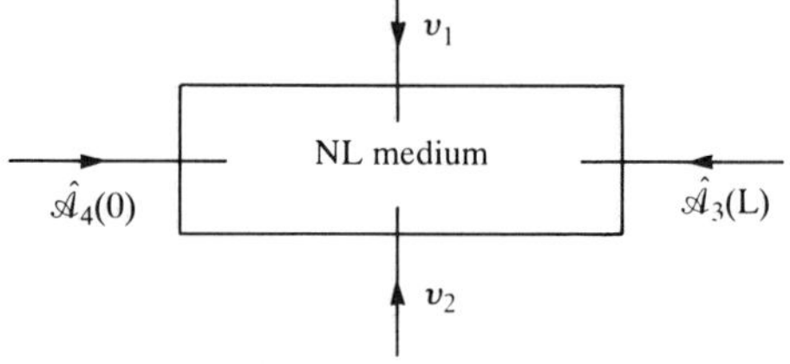

Fig. 22.13 Illustrating the geometry for the four-wave interaction.

We shall take v_1, v_2, $\hat{\mathscr{A}}_3$, $\hat{\mathscr{A}}_4$ to be dimensionless, so that g is a real frequency.

22.5.1 *Equations of motion*

From Eq. (22.5–3) for $\hat{H}_I$ we obtain the Heisenberg equations of motion for the operators $\hat{\mathscr{A}}_3$, $\hat{\mathscr{A}}_4$, in the form

$$\left.\begin{aligned} \frac{\partial \hat{\mathscr{A}}_3}{\partial t} &= -\mathrm{i}g v_1 v_2 \hat{\mathscr{A}}_4^\dagger \\ \frac{\partial \hat{\mathscr{A}}_4}{\partial t} &= -\mathrm{i}g v_1 v_2 \hat{\mathscr{A}}_3^\dagger . \end{aligned}\right\} \tag{22.5–4}$$

These equations are very similar to Eqs. (22.4–6) that we encountered in the process of down-conversion, and they can be solved in a similar manner.

However, in this problem it is more convenient to consider $\hat{\mathscr{A}}_3$, $\hat{\mathscr{A}}_4$ as functions of position z, and to solve the equations accordingly. Recalling that $\hat{\mathscr{A}}_3$ is actually a function of $(z/V - t)$ and $\hat{\mathscr{A}}_4$ is actually a function of $(z/V + t)$, we see that

$$\left.\begin{aligned} \frac{1}{V}\frac{\partial \hat{\mathscr{A}}_3}{\partial t} &= -\frac{\partial \hat{\mathscr{A}}_3}{\partial z} \\ \frac{1}{V}\frac{\partial \hat{\mathscr{A}}_4}{\partial t} &= \frac{\partial \hat{\mathscr{A}}_4}{\partial z}, \end{aligned}\right\} \tag{22.5–5a}$$

so that Eqs. (22.5–4) can also be written as

$$\left.\begin{aligned} \frac{\partial \hat{\mathscr{A}}_3}{\partial z} &= \frac{\mathrm{i}g}{V} v_1 v_2 \hat{\mathscr{A}}_4^\dagger \\ \frac{\partial \hat{\mathscr{A}}_4}{\partial z} &= -\frac{\mathrm{i}g}{V} v_1 v_2 \hat{\mathscr{A}}_3^\dagger . \end{aligned}\right\} \tag{22.5–5b}$$

After being differentiated once with respect to z the equations become uncoupled, and take the form of harmonic oscillator equations

$$\begin{aligned} \frac{\partial^2 \hat{\mathscr{A}}_3}{\partial z^2} &= -(g/V)^2 |v_1 v_2|^2 \hat{\mathscr{A}}_3 \\ \frac{\partial^2 \hat{\mathscr{A}}_4}{\partial z^2} &= -(g/V)^2 |v_1 v_2|^2 \hat{\mathscr{A}}_4 , \end{aligned}$$

whose solutions are expressible as functions of z in the form

$$\left.\begin{aligned} \hat{\mathscr{A}}_3(z) &= \hat{\mathscr{A}}_3(0)\cos Kz + \mathrm{i}\hat{\mathscr{A}}_4^\dagger(0)\,\mathrm{e}^{\mathrm{i}(\theta_1+\theta_2)}\sin Kz \\ \hat{\mathscr{A}}_4(z) &= \hat{\mathscr{A}}_4(0)\cos Kz - \mathrm{i}\hat{\mathscr{A}}_3^\dagger(0)\,\mathrm{e}^{\mathrm{i}(\theta_1+\theta_2)}\sin Kz . \end{aligned}\right\} \tag{22.5–6}$$

Here we have put $K \equiv g|v_1 v_2|/V$ and $v_j \equiv |v_j|\,\mathrm{e}^{\mathrm{i}\theta_j}$, $(j = 1, 2)$.

Let L be the length of the nonlinear medium, with the origin at $z = 0$. Then, as wave 3 is incident from the right and wave 4 from the left, we identify the input and output amplitudes by the relations

$$\left.\begin{aligned}\hat{\mathscr{A}}_{4\mathrm{in}} &= \hat{\mathscr{A}}_4(0)\\ \hat{\mathscr{A}}_{4\mathrm{out}} &= \hat{\mathscr{A}}_4(L)\\ \hat{\mathscr{A}}_{3\mathrm{in}} &= \hat{\mathscr{A}}_3(L)\\ \hat{\mathscr{A}}_{3\mathrm{out}} &= \hat{\mathscr{A}}_3(0).\end{aligned}\right\} \tag{22.5–7}$$

Hence, on putting $z = L$ we obtain the following input/output relations from Eqs. (22.5–6):

$$\left.\begin{aligned}\hat{\mathscr{A}}_{3\mathrm{out}} &= \hat{\mathscr{A}}_{3\mathrm{in}} \sec(KL) - \mathrm{i}\hat{\mathscr{A}}^\dagger_{4\mathrm{in}}\, \mathrm{e}^{\mathrm{i}(\theta_1+\theta_2)} \tan(KL)\\ \hat{\mathscr{A}}_{4\mathrm{out}} &= \hat{\mathscr{A}}_{4\mathrm{in}} \sec(KL) - \mathrm{i}\hat{\mathscr{A}}^\dagger_{3\mathrm{in}}\, \mathrm{e}^{\mathrm{i}(\theta_1+\theta_2)} \tan(KL).\end{aligned}\right\} \tag{22.5–8}$$

It is at once apparent that the approximation of treating the pump waves as constant, without depletion, ceases to be adequate when KL approaches $\pi/2$.

22.5.2 *Application to the coherent state*

Let us suppose that the state at the two inputs to the nonlinear medium is the coherent product state $|v_3\rangle|v_4\rangle$ with $v_j = |v_j|\,\mathrm{e}^{\mathrm{i}\theta_j}$ $(j = 3, 4)$. The expectations of the outputs $\hat{\mathscr{A}}_{3\mathrm{out}}$, $\hat{\mathscr{A}}_{4\mathrm{out}}$ given by Eqs. (22.5–8) are then easily evaluated, with the result

$$\left.\begin{aligned}\langle\hat{\mathscr{A}}_{3\mathrm{out}}\rangle &= v_3 \sec(KL) - \mathrm{i}v_4^*\, \mathrm{e}^{\mathrm{i}(\theta_1+\theta_2)} \tan(KL)\\ \langle\hat{\mathscr{A}}_{4\mathrm{out}}\rangle &= v_4 \sec(KL) - \mathrm{i}v_3^*\, \mathrm{e}^{\mathrm{i}(\theta_1+\theta_2)} \tan(KL).\end{aligned}\right\} \tag{22.5–9}$$

In particular, if either mode 3 or mode 4 starts from the vacuum state, that mode amplitude grows with L by spontaneous emission. We shall return to this problem below in connection with the topic of phase conjugation.

From Eqs. (22.5–8) we can also calculate the mean intensity or photon number at the output end of the medium, and we find that

$$\begin{aligned}\langle\hat{n}_{3\mathrm{out}}\rangle &= \langle\hat{n}_{3\mathrm{in}}\rangle \sec^2(KL) + (\langle\hat{n}_{4\mathrm{in}}\rangle + 1)\tan^2(KL)\\ &\quad + [\mathrm{i}\,\mathrm{e}^{-\mathrm{i}(\theta_1+\theta_2)}\langle\hat{\mathscr{A}}_{4\mathrm{in}}\hat{\mathscr{A}}_{3\mathrm{in}}\rangle + \mathrm{c.c.}]\sec(KL)\tan(KL)\\ &= |v_3|^2\sec^2(KL) + (|v_4|^2 + 1)\tan^2(KL)\\ &\quad + [\mathrm{i}v_3v_4\,\mathrm{e}^{-\mathrm{i}(\theta_1+\theta_2)} + \mathrm{c.c.}]\sec(KL)\tan(KL),\end{aligned} \tag{22.5–10}$$

and similarly for $\langle\hat{n}_{4\mathrm{out}}\rangle$. We note that, because of the term $\tan^2(KL)$, $\langle\hat{n}_{3\mathrm{out}}\rangle$ grows by spontaneous emission from zero with the length L, even when both v_3, v_4 are zero initially.

22.5.3 *Squeezing in four-wave mixing*

As nonlinear media are known to give rise to squeezed states, it is natural to inquire whether either modes 3 or 4 exhibit quadrature squeezing at the output end of the crystal. We therefore introduce a real variable

$$\hat{Q}_3 \equiv \hat{\mathscr{A}}_{3\mathrm{out}}\,\mathrm{e}^{-\mathrm{i}\phi} + \hat{\mathscr{A}}^\dagger_{3\mathrm{out}}\,\mathrm{e}^{\mathrm{i}\phi}, \tag{22.5–11}$$

where ϕ is some adjustable phase, and ask whether there exists some value of ϕ for which $\langle(\Delta\hat{Q}_3)^2\rangle \leq 1$, when the input state is the product state $|v_3\rangle|v_4\rangle$.

From Eq. (22.5–9) we have immediately

$$\langle\hat{Q}_3\rangle = [v_3 \sec(KL) - iv_4^* e^{i(\theta_1+\theta_2)} \tan(KL)] e^{-i\phi} + \text{c.c.}, \qquad (22.5\text{–}12)$$

whereas from Eq. (22.5–8)

$$\begin{aligned}\langle\hat{Q}_3^2\rangle &= [v_3 \sec(KL) - iv_4^* e^{i(\theta_1+\theta_2)} \tan(KL)]^2 e^{-2i\phi} + \text{c.c.} \\ &\quad + 2[|v_3|^2 \sec^2(KL) + (|v_4|^2 + 1) \tan^2(KL) \\ &\quad - iv_3^* v_4^* e^{i(\theta_1+\theta_2)} \sec(KL) \tan(KL) + \text{c.c.}] + 1.\end{aligned}$$

Hence

$$\begin{aligned}\langle(\Delta\hat{Q}_3)^2\rangle &= \langle\hat{Q}_3^2\rangle - \langle\hat{Q}_3\rangle^2 \\ &= 1 + 2\tan^2(KL),\end{aligned} \qquad (22.5\text{–}13)$$

and similarly for $\langle(\Delta\hat{Q}_4)^2\rangle$. As these dispersions never drop below unity, it follows that neither mode 3 nor mode 4 is squeezed at the output end of the nonlinear medium. Nevertheless, squeezing is achievable when the light emerging from the medium in modes 3 and 4 is mixed at a beam splitter (Yuen and Shapiro, 1979), as we shall now demonstrate.

Let us consider the situation illustrated in Fig. 22.14, in which the emerging light signals of amplitudes $\hat{\mathscr{A}}_{3\text{out}} e^{i\chi_3}$ and $\hat{\mathscr{A}}_{4\text{out}} e^{i\chi_4}$ are mixed at a beam splitter of complex amplitude transmissivity $\mathscr{T}$ and reflectivity $\mathscr{R}$. Let the outputs from the beam splitter have mode amplitudes $\hat{\mathscr{A}}_5$ and $\hat{\mathscr{A}}_6$. Then reference to Fig. 22.14 shows that

$$\left.\begin{aligned}\hat{\mathscr{A}}_5 &= \mathscr{T}\hat{\mathscr{A}}_{3\text{out}} e^{i\chi_3} + \mathscr{R}\hat{\mathscr{A}}_{4\text{out}} e^{i\chi_4} \\ \hat{\mathscr{A}}_6 &= \mathscr{R}\hat{\mathscr{A}}_{3\text{out}} e^{i\chi_3} + \mathscr{T}\hat{\mathscr{A}}_{4\text{out}} e^{i\chi_4},\end{aligned}\right\} \qquad (22.5\text{–}14)$$

and these relations ensure that $\hat{\mathscr{A}}_5$ and $\hat{\mathscr{A}}_6$ obey the usual canonical commutation relations for single-mode amplitudes, provided that $\hat{\mathscr{A}}_3$ and $\hat{\mathscr{A}}_4$ do also. We now form the field quadrature

$$\hat{Q}_5 \equiv \hat{\mathscr{A}}_5 e^{-i\phi} + \hat{\mathscr{A}}_5^\dagger e^{i\phi} \qquad (22.5\text{–}15)$$

in the usual way and calculate its dispersion. With the help of Eqs. (22.5–8) and

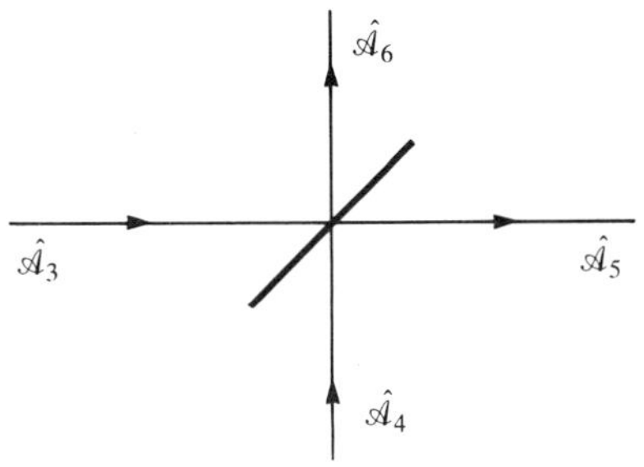

Fig. 22.14 Illustrating the use of a beam splitter to combine two fields $\hat{\mathscr{A}}_3$, $\hat{\mathscr{A}}_4$ so as to create two new fields $\hat{\mathscr{A}}_5$, $\hat{\mathscr{A}}_6$.

(22.5–14) we obtain the relation

$$\hat{Q}_5 = [(\mathcal{T}\,\mathrm{e}^{\mathrm{i}\chi_3}\hat{\mathscr{A}}_{3\mathrm{in}} + \mathcal{R}\,\mathrm{e}^{\mathrm{i}\chi_4}\hat{\mathscr{A}}_{4\mathrm{in}})\sec(KL) \\ - \mathrm{i}(\mathcal{T}\,\mathrm{e}^{\mathrm{i}(\theta_1+\theta_2+\chi_3)}\hat{\mathscr{A}}^\dagger_{4\mathrm{in}} + \mathcal{R}\,\mathrm{e}^{\mathrm{i}(\theta_1+\theta_2+\chi_4)}\hat{\mathscr{A}}^\dagger_{3\mathrm{in}})\tan(KL)]\,\mathrm{e}^{-\mathrm{i}\phi} + \mathrm{h.c.}, \quad (22.5\text{–}16)$$

and from this equation it follows, after a little algebra, that

$$\langle(\Delta\hat{Q}_5)^2\rangle = \sec^2(KL) + \tan^2(KL) \\ + 4\sec(KL)\tan(KL)|\mathcal{R}\mathcal{T}|\sin(\theta_1+\theta_2+\chi_3+\chi_4+\beta-2\phi), \quad (22.5\text{–}17)$$

in which $\beta = \arg(\mathcal{R}\mathcal{T})$. In particular, when $|\mathcal{R}|^2 = \frac{1}{2} = |\mathcal{T}|^2$ and $\theta_1 + \theta_2 + \chi_3 + \chi_4 + \beta - 2\phi = -\pi/2$,

$$\langle(\Delta Q_5)^2\rangle = [\sec(KL) - \tan(KL)]^2,$$

which is less than unity for almost all L and approaches zero as $KL \to \pi/2$. Therefore one can always find a phase angle ϕ such that $\langle(\Delta\hat{Q}_5)^2\rangle < 1$, which is the condition for squeezing. A similar conclusion holds also with respect to $\hat{Q}_6$. The effect of mixing the two outputs from the nonlinear medium is therefore to generate squeezed fields from unsqueezed ones.

Squeezing effects occurring in four-wave mixing in a sodium beam were first observed in experiments by Slusher, Hollberg, Yurke, Mertz and Valley (1985) at the AT&T Bell Laboratories. The authors measured the fluctuations of the field by heterodyning against a local oscillator, and they found that fluctuations below the shot-noise or Poisson limit are achievable somewhat off resonance. Figure 21.5(*b*) shows some of the measured fluctuations expressed as a r.m.s. noise voltage as a function of local oscillator phase. For certain phase angles the fluctuations dip about 10% below the shot-noise limit that one expects for a completely coherent field. This is the signature of the squeezed state.

22.5.4 Phase conjugation‡

Let us consider the situation in which a signal wave in the coherent state $|v_3\rangle$ is incident from the right on the nonlinear medium in Fig. 22.13, while the idler incident from the left is in the vacuum state. Then, as Eqs. (22.5–9) indicate, there is a return wave of mean amplitude $\langle\hat{\mathscr{A}}_{4\mathrm{out}}\rangle$ leaving the medium that looks like a reflection of the signal wave. On putting $v_4 = 0$ in Eqs. (22.5–9) we see immediately that the mean amplitude of the wave emerging to the right is given by the expression

$$\langle\hat{\mathscr{A}}_{4\mathrm{out}}\rangle = -\mathrm{i}v_3^*\,\mathrm{e}^{\mathrm{i}(\theta_1+\theta_2)}\tan(KL) = -\mathrm{i}\,\mathrm{e}^{\mathrm{i}(\theta_1+\theta_2)}\tan(KL)\langle\hat{\mathscr{A}}^\dagger_{3\mathrm{in}}\rangle, \quad (22.5\text{–}18)$$

and that it is proportional to v_3^* rather than to v_3, as it would be for the reflection

‡ There is a large amount of literature dealing with the subject of phase conjugation. See for example a review article by Pepper (1982) and books by Fisher (1983), and Zel'dovich, Pilipetsky and Shkunov (1985).

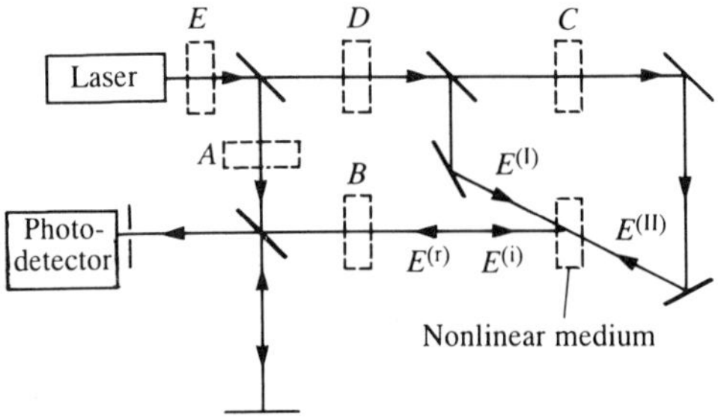

Fig. 22.15 Outline of the experiment to illustrate phase conjugation. The broken lines show positions of the movable air cell. (Reproduced from Boyd, Habashy, Jacobs, Mandel, Nieto-Vesperinas, Tompkin and Wolf, 1987.)

from an ordinary mirror. This unusual mirror is said to be a phase conjugate mirror and the phenomenon is known as *phase conjugation*.

Figure 22.15 shows the outline of an experiment (Boyd, Habashy, Jacobs, Mandel, Nieto-Vesperinas, Tompkin and Wolf, 1987) that illustrates the basic difference between a phase conjugate mirror, like that produced by four-wave mixing in a nonlinear medium, and an ordinary metallic mirror. The nonlinear medium is pumped by counter-propagating waves $E^{(I)}$ and $E^{(II)}$, and it is probed by the incident wave $E^{(i)}$. All three waves are derived by splitting the light from an argon ion laser oscillating at 488 nm, as shown. The wave $E^{(r)}$ reflected from the phase conjugate mirror interferes with the incident wave in a Michelson type interferometer. The interference pattern is probed by the photodetector as variable phase shifts θ are introduced by a pressure controlled glass cell placed at one of five different positions A, B, C, D, E. Figure 22.16 shows the results of some of the measurements.

With the cell placed at position A, the phase of the incident wave increases with θ, and the phase of the reflected wave decreases with θ, as predicted by Eq. (22.5–18), so that the phase difference grows linearly with θ. By contrast, the phase difference is constant when the phase conjugate mirror (PCM) is replaced by an ordinary mirror. With the cell placed at position B, so that the reflected wave passes through B also, the phase shift introduced by the phase conjugate mirror exactly compensates for any phase shift introduced into the incident wave. The result is phase shift cancellation, and the interference pattern does not depend on θ. By contrast, with an ordinary metallic mirror, the phase shifts suffered by the incident and reflected waves add up and the interference pattern exhibits the usual linear variation with θ.

The ability of a phase conjugate mirror to compensate for phase shifts suffered by an incident wave offers interesting possibilities for compensating for wave front distortion in scattering or for phase aberrations of an image forming system (see, for example, Giuliano, 1981; Agarwal, Friberg and Wolf, 1982, 1983). An example of such an application is illustrated in Fig. 22.17, which shows the image after the light emerging from a distorting glass plate is reflected by a phase conjugate mirror and allowed to traverse the plate again in the reverse direction. It is apparent that the phase distortion has been corrected by the double transit, as a consequence of the phase reversal on reflection.

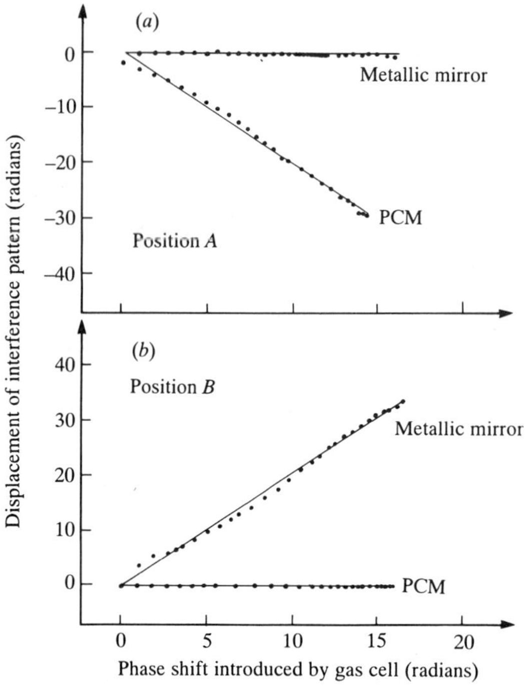

Fig. 22.16 Measured displacement of the interference pattern as a function of phase shift introduced by the air cell in Fig. 22.15 (*a*) at position *A*; (*b*) at position *B*. (Reproduced from Boyd, Habashy, Jacobs, Mandel, Nieto-Vesperinas, Tompkin and Wolf, 1987.)

22.6 Quantum non-demolition measurements

Most measurements of a quantum system disturb that system by introducing uncontrollable quantum fluctuations into it. As a consequence, a repeated measurement of the same variable of the system a short while later can result in an entirely different outcome. One of the obvious examples of this situation is the photoelectric measurement of an optical field, which generally results in the absorption of one or more photons. A later photoelectric measurement of the same optical field may therefore yield a different answer.

In principle, there exist dynamical variables of a quantum system which remain undisturbed by the measurement, although the measurement may introduce a disturbance in the conjugate variables of the system. Such a dynamical variable is known as a *quantum non-demolition* (QND) variable (Caves, Thorne, Drever, Sandberg and Zimmermann, 1980; Milburn, Lane and Walls, 1983; Milburn and Walls, 1983a,b; Imoto, Haus and Yamamoto, 1985; Braginsky and Khalil, 1992, Chapt. 4). In the Heisenberg picture $A(t)$ is a QND variable if

$$[\hat{A}(t'), \hat{A}(t)] = 0, \tag{22.6–1}$$

for all t, t'. In that case a repeated measurement of $\hat{A}(t)$ will yield the same

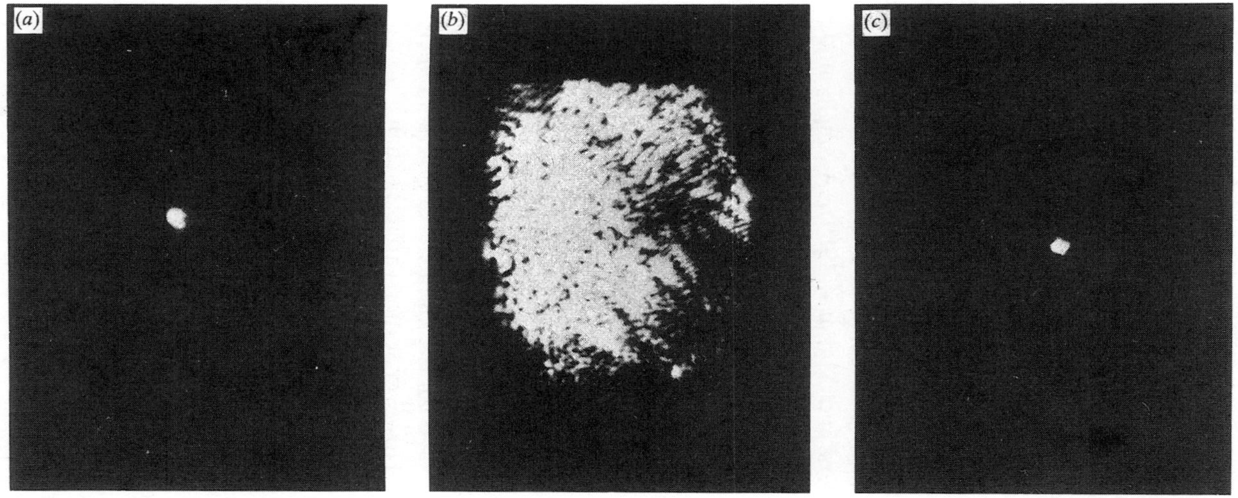

Fig. 22.17 Illustrating the ability of a phase conjugate mirror to correct aberrations: (*a*) photograph of the incident laser beam; (*b*) the distorted image resulting from the passage of the beam through an etched glass plate; (*c*) the restored image of the same beam after phase conjugate reflection and a second traversal of the plate. (Reproduced from Giuliano, 1981; see also Jain and Lind, 1983.)

outcome. Evidently this will be so if $\hat{A}$ commutes with the total system Hamiltonian $\hat{H}$, i.e. if

$$[\hat{A}, \hat{H}] = 0. \tag{22.6–2}$$

In the Schrödinger picture, a system satisfying Eq. (22.6–2) that starts in an eigenstate of $\hat{A}$ remains in an eigenstate of $\hat{A}$, so that $\hat{A}$ is a QND variable. A distinction is sometimes made between a QND variable that merely satisfies Eq. (22.6–1) and a *back-action evading* variable $\hat{A}$, with the property that the interaction Hamiltonian $\hat{H}_{\mathrm{I}}$ for the measurement depends only on the observable $\hat{A}$. Then $\hat{A}$ is unaffected by the interaction with the measurement apparatus.

22.6.1 The Kerr effect – an example of a QND variable

As an example of a QND and back-action evading measurement, we consider the parametric coupling between two optical field modes in a Kerr medium having a $\chi^{(3)}$ nonlinear susceptibility (Imoto, Haus and Yamamoto, 1985; Kitagawa and Yamamoto, 1986; Sanders and Milburn, 1989). A single-mode field in a Kerr medium is characterized by the Hamiltonian

$$\begin{aligned}\hat{H} &= \hbar\omega_0\hat{n} + \hbar\chi\hat{a}^{\dagger 2}\hat{a}^2\\ &= \hbar(\omega_0 - \chi)\hat{n} + \hbar\chi\hat{n}^2,\end{aligned} \tag{22.6–3}$$

when we recall that $\hat{a}^{\dagger 2}\hat{a}^2 = \hat{n}(\hat{n} - 1)$. The anharmonicity parameter χ is real and proportional to $\chi^{(3)}$. The refractive index of the medium is intensity-dependent with a coefficient that is proportional to χ. It follows from Eq. (22.6–3) that

$$[\hat{n}, \hat{H}] = 0, \tag{22.6–4}$$

so that $\hat{n}$ is a constant of the motion, and it is a QND variable that is also back-action evading. We can measure $\hat{n}$ and the moments of $\hat{n}$ without affecting subsequent measurements. As all the moments of $\hat{n}$ remain unchanged in time, so do the photon statistics. However, other variables change in time. For example, from Eq. (22.6–3) we have, for the Heisenberg equation of motion for the annihilation operator $\hat{a}(t)$,

$$\begin{aligned}\frac{\mathrm{d}\hat{a}(t)}{\mathrm{d}t} &= -\mathrm{i}(\omega_0 - \chi)\hat{a} - \mathrm{i}\chi(\hat{a}\hat{n} + \hat{n}\hat{a})\\ &= -\mathrm{i}[\omega_0 + \chi 2\hat{n}]\hat{a}(t).\end{aligned} \tag{22.6–5}$$

As $\hat{n}$ is a constant of the motion, this equation can be integrated at once to yield

$$\hat{a}(t) = \mathrm{e}^{-\mathrm{i}[\omega_0 + 2\chi\hat{n}]t}\hat{a}(0). \tag{22.6–6}$$

If the field starts off in the coherent state $|v\rangle$, with $|v|^2 \gg 1$ so that $\langle\hat{n}\rangle$ has a value close to $|v|^2$, then subsequently

$$\langle\hat{a}(t)\rangle \approx v\,\mathrm{e}^{-\mathrm{i}(\omega_0 + 2\chi|v|^2)t}. \tag{22.6–7}$$

Evidently, the phase of the field is affected by its propagation through the Kerr medium, but not its modulus.

We can characterize a two-mode field in a Kerr medium by the Hamiltonian

$$\hat{H} = \hbar\omega_{\mathrm{s}}\hat{n}_{\mathrm{s}} + \hbar\omega_{\mathrm{p}}\hat{n}_{\mathrm{p}} + \tfrac{1}{2}\hbar\chi_{\mathrm{s}}\hat{n}_{\mathrm{s}}^2 + \tfrac{1}{2}\hbar\chi_{\mathrm{p}}\hat{n}_{\mathrm{p}}^2 + 2\hbar\chi\hat{n}_{\mathrm{s}}\hat{n}_{\mathrm{p}}, \tag{22.6–8}$$

where we refer to one mode as the signal (s) and the other as the probe (p). The parameters χ_s, χ_p, χ are all elements of the $\boldsymbol{\chi}^{(3)}$ tensor susceptibility. In some media such as optical fibers, the three are equal, whereas χ_s, χ_p may vanish in other media. It is evident from Eq. (22.6–8) that $\hat{n}_s$ and $\hat{n}_p$ are both QND variables of the back-action evading type, which are constants of the motion. We therefore have, as in the derivation of Eq. (22.6–6), that

$$\hat{a}_s(t) = \exp[-i(\omega_s + \tfrac{1}{2}\chi_s + \chi_s\hat{n}_s + 2\chi\hat{n}_p)t]\hat{a}_s(0) \qquad (22.6\text{–}9)$$

$$\hat{a}_p(t) = \exp[-i(\omega_p + \tfrac{1}{2}\chi_p + \chi_p\hat{n}_p + 2\chi\hat{n}_s)t]\hat{a}_p(0). \qquad (22.6\text{–}10)$$

It follows that the interaction between the two modes affects the phases of the field, but not the photon numbers. Moreover, not only does the signal field affect the phase of the probe field, but the probe field causes an intensity-dependent phase shift itself, and vice versa. However, when $\hat{n}_p$ has a fairly well-defined value n_p, we can often choose the interaction time so that

$$\chi_p n_p t \approx 2\pi N \ (N = 1, 2, \ldots), \qquad (22.6\text{–}11)$$

and then the $\chi_p\hat{n}_p t$ can be discarded in the exponent in Eq. (22.6–10). Henceforth we shall assume that this simplification has been made.

Because the Kerr effect allows us to measure the intensity, or photon number, of a signal field without disturbing this number, it might seem that incorporating a Kerr probe into one arm of an interferometer would allow us to tell 'which way the photon went' through the interferometer, without disturbing the fringe pattern. This would, of course, imply a violation of the uncertainty principle, because an interference pattern is always a manifestation of the intrinsic indistinguishability of the different paths. An analysis of such an interference experiment is therefore of considerable pedagogic interest.

22.6.2 *Analysis of an interference experiment*

We shall follow the treatment of Sanders and Milburn (1989) (see also Kärtner and Haus, 1993) and consider the Mach–Zehnder interferometer arrangement shown in Fig. 22.18. The interferometer is formed by two 50%:50% beam splitters BS_A and BS_B, together with the mirrors M_A and M_B. The output field $\hat{b}(t)$ is measured by the photodetector PD. One interferometer arm contains the Kerr medium KM, which is also probed by probe wave $\hat{a}_p$, whose phase is eventually measured by some homodyne arrangement. Another interferometer arm contains a phase shifter PS that introduces a variable phase shift θ into that interferometer arm. The interference pattern is investigated by measuring the intensity registered by the photodetector PD as a function of θ.

We now insert one photon at the input beam splitter BS_A at the port 1 and attempt to determine with the help of KM which path it follows through the interferometer. This measurement leaves the photon number undisturbed. But from Eq. (22.6–10) it is apparent that the measurement introduces a phase shift into the probe wave that is different according to whether the signal wave contains a photon or not.

Let us denote by $\hat{a}_0$, $\hat{a}_1$ the input mode amplitudes shown in Fig. 22.18 and by

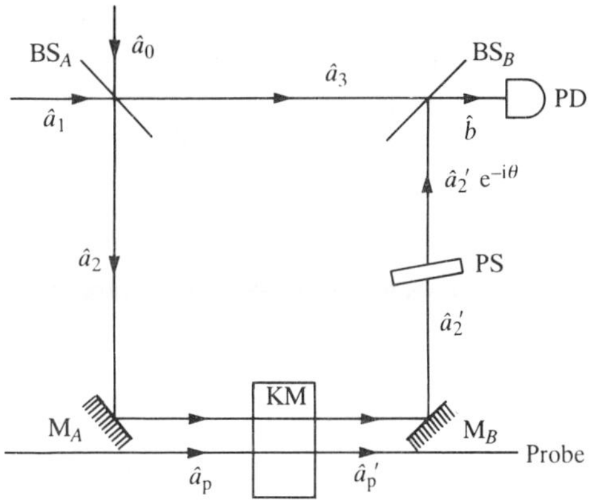

Fig. 22.18 Outline of an interference experiment in which the path taken by the photon can be inferred from a probe of the Kerr Medium KM with another light beam. (Adapted from Sanders and Milburn, 1989.)

$\hat{a}_2$, $\hat{a}_3$ the amplitudes inside the interferometer. Then for a 50%:50% beam splitter we have,

$$\hat{a}_3 = \frac{1}{\sqrt{2}}(\hat{a}_1 + i\hat{a}_0) \tag{22.6–12}$$

$$\hat{a}_2 = \frac{1}{\sqrt{2}}(i\hat{a}_1 + \hat{a}_0). \tag{22.6–13}$$

From Eq. (22.6–9) the output amplitude $\hat{a}_2'$ of the field from the Kerr medium is related to the input amplitude $\hat{a}_2$ by the formula

$$\hat{a}_2' = \exp[-i(\omega_2 + \tfrac{1}{2}\chi_2 + \chi_2\hat{n}_2 + 2\chi\hat{n}_p)T]\hat{a}_2, \tag{22.6–14}$$

where T is the propagation time inside KM. Finally, the field $\hat{b}$ emerging from the output beam splitter is given by

$$\hat{b} = \frac{1}{\sqrt{2}}(\hat{a}_3 + i\hat{a}_2' \, e^{-i\theta}), \tag{22.6–15}$$

where θ represents the effect of the phase shifter PS and the additional propagation time around the longer interferometer path.

From Eqs. (22.6–12) to (22.6–15) we readily obtain the relation

$$\hat{b} = \tfrac{1}{2}(\hat{a}_1 + i\hat{a}_0) + \tfrac{1}{2}i\,e^{-i\theta}\exp[-i(\omega_2 + \tfrac{1}{2}\chi_2 + \chi_2\hat{n}_2 + 2\chi\hat{n}_p)T](i\hat{a}_1 + \hat{a}_0), \tag{22.6–16}$$

from which it follows that

$$\begin{aligned}\hat{b}^\dagger\hat{b} = \tfrac{1}{2}[&\hat{a}_1^\dagger\hat{a}_1 + \hat{a}_0^\dagger\hat{a}_0 + \tfrac{1}{2}i\,e^{-i\theta}(\hat{a}_1^\dagger - i\hat{a}_0^\dagger) \\ &\times \exp\{-i[\omega_2 + \tfrac{1}{2}\chi_2 + 2\chi\hat{n}_p + \tfrac{1}{2}\chi_2(-i\hat{a}_1^\dagger + \hat{a}_0^\dagger)(i\hat{a}_1 + \hat{a}_0)]T\} \\ &\times (i\hat{a}_1 + \hat{a}_0) + \text{h.c.}].\end{aligned} \tag{22.6–17}$$

22.6.3 *Calculation of fringe visibility*

We now consider the initial quantum state $|\psi\rangle$ to be a product state in which mode 1 is in a one-photon state, mode 0 is in the vacuum state, and the pump mode P is in the coherent state $|v\rangle_P$. Then

$$|\psi\rangle = |\text{vac}\rangle_0 |1\rangle_1 |v\rangle_P. \tag{22.6–18}$$

We take expectations of Eq. (22.6–17) in the state $|\psi\rangle$ and obtain the result

$$\langle \hat{b}^\dagger \hat{b} \rangle = \tfrac{1}{2}\{1 - \tfrac{1}{2}\, {}_P\langle v| \exp[-\mathrm{i}(\omega_2 + \tfrac{1}{2}\chi_2 + 2\chi\hat{n}_\mathrm{p})T] |v\rangle_P \,\mathrm{e}^{-\mathrm{i}\theta} + \text{c.c.}\}. \tag{22.6–19}$$

Recalling that for the coherent state (cf. Section 11.5)

$$\mathrm{e}^{-\mathrm{i}2\chi\hat{n}_\mathrm{p}T}|v\rangle_P = |v\,\mathrm{e}^{-2\mathrm{i}\chi T}\rangle_P,$$

and making use of the scalar product of two coherent states [cf. Eq. (11.6–1)], we finally obtain from Eq. (22.6–19), the result

$$\langle \hat{b}^\dagger \hat{b} \rangle = \tfrac{1}{2}[1 - \mathrm{e}^{-2|v|^2 \sin^2 \chi T} \cos(\omega_2 T + \tfrac{1}{2}\chi_2 T + |v|^2 \sin 2\chi T + \theta)]. \tag{22.6–20}$$

Hence the visibility $\mathcal{V}$ of the interference fringes is given by

$$\mathcal{V} = \exp[-2|v|^2 \sin^2(\chi T)], \tag{22.6–21}$$

and it depends on the strength of the probe wave, on the Kerr coupling constant χ, and on the propagation time T.

22.6.4 *The probe wave phase shift*

We have already seen from Eq. (22.6–10) that the probe wave experiences a phase shift as a result of the interaction with the signal wave $\hat{a}_2$ that depends on the number of photons in the signal wave. Let $\hat{a}'_\mathrm{p}$ be the probe wave amplitude emerging from the Kerr medium KM after the interaction. Then

$$\hat{a}'_\mathrm{p} = \exp[-\mathrm{i}(\omega_\mathrm{p} + \tfrac{1}{2}\chi_\mathrm{p} + 2\chi\hat{n}_2)T]\hat{a}_\mathrm{p}, \tag{22.6–22}$$

where we have taken $\chi_\mathrm{p} n_\mathrm{p} T \approx 2N\pi$ for simplicity. Hence there is an additional phase shift $2\chi T$ introduced in the presence of a one-photon field of mode 2, as compared with a vacuum field. The phase shift can be measured by a homodyne technique. Let $\hat{Y}$ be a real variable defined by

$$\hat{Y} \equiv \tfrac{1}{2}(\hat{a}'_\mathrm{p}\,\mathrm{e}^{-\mathrm{i}\phi} + \text{h.c.}), \tag{22.6–23}$$

where ϕ is some phase angle. Then the expectation value of $\hat{Y}$, when the probe field is initially in the coherent state $|v\rangle_P$, is different according to whether the photon passed through arm 2 or arm 3 of the interferometer. In the first case $\langle \hat{Y} \rangle_{\text{arm}\,2}$ is given by Eqs. (22.6–22) and (22.6–23) with $\hat{n}_2 \to 1$, and it results in the expression

$$\begin{aligned}\langle \hat{Y} \rangle_{\text{arm}\,2} &= \tfrac{1}{2}\{v \exp[-\mathrm{i}(\omega_\mathrm{p} + \tfrac{1}{2}\chi_\mathrm{p} + 2\chi)T]\,\mathrm{e}^{-\mathrm{i}\phi} + \text{c.c.}\} \\ &= |v| \cos(\omega_\mathrm{p} T + \tfrac{1}{2}\chi_\mathrm{p} T + 2\chi T + \phi - \arg v),\end{aligned} \tag{22.6–24}$$

whereas in the second case we make the substitution $\hat{n}_2 \to 0$ and obtain

$$\langle \hat{Y} \rangle_{\text{arm}\,3} = \tfrac{1}{2}\{v \exp[-\mathrm{i}(\omega_\mathrm{p} + \tfrac{1}{2}\chi_\mathrm{p})T]\,\mathrm{e}^{-\mathrm{i}\phi} + \text{c.c.}\}$$
$$= |v| \cos(\omega_\mathrm{p} T + \tfrac{1}{2}\chi_\mathrm{p} T + \phi - \arg v). \qquad (22.6\text{–}25)$$

The difference in these two values of $\langle \hat{Y} \rangle$ due to the presence of the photon in arm 2 is

$$\Delta Y \equiv \langle \hat{Y} \rangle_{\text{arm}\,3} - \langle \hat{Y} \rangle_{\text{arm}\,2} \qquad (22.6\text{–}26)$$
$$= 2|v| \sin\chi T \sin(\omega_\mathrm{p} T + \tfrac{1}{2}\chi_\mathrm{p} T + \chi T + \phi - \arg v\}, \qquad (22.6\text{–}27)$$

and this has the largest possible value $2|v| \sin\chi T$ when ϕ is chosen appropriately.

Let us compare this change of $\langle \hat{Y} \rangle$ due to the presence of a photon in arm 2 with the natural quantum fluctuations of $\hat{Y}$. When the photon is in arm 3, we have

$$\langle \hat{Y}^2 \rangle_{\text{arm}\,3} = \tfrac{1}{4}[\langle \hat{a}_\mathrm{p}'^2 \rangle\,\mathrm{e}^{-2\mathrm{i}\phi} + \text{c.c.} + 2\langle \hat{n}_\mathrm{p}' \rangle + 1]$$
$$= \tfrac{1}{4}[v^2\,\mathrm{e}^{-\mathrm{i}(2\omega_\mathrm{p} T + \chi_\mathrm{p} T + 2\phi)} + \text{c.c.} + 2|v|^2 + 1]$$
$$= |v|^2 \cos^2(\omega_\mathrm{p} T + \tfrac{1}{2}\chi_\mathrm{p} T + \phi - \arg v) + \tfrac{1}{4}, \qquad (22.6\text{–}28)$$

and when this result is combined with Eq. (22.6–25), we readily obtain the formula

$$\langle (\Delta\hat{Y})^2 \rangle_{\text{arm}\,3} = \langle \hat{Y}^2 \rangle_{\text{arm}\,3} - \langle \hat{Y} \rangle^2_{\text{arm}\,3} = \tfrac{1}{4}. \qquad (22.6\text{–}29)$$

This allows us to associate a signal-to-noise ratio R_{SN} with the phase measurement that we define by

$$R_{\text{SN}} \equiv \frac{\Delta Y_{\max}}{\sqrt{\langle (\Delta\hat{Y})^2 \rangle_{\text{arm}\,3}}} = 4|v| \sin\chi T. \qquad (22.6\text{–}30)$$

When R_{SN} is large, the phase measurement is accurate, and we can clearly distinguish between paths 2 and 3 for the photon. On the other hand, when R_{SN} is of order 1 or smaller, the phase measurement is of poor accuracy, and it is difficult to tell which path the photon followed through the interferometer.

22.6.5 Complementarity

By combining Eq. (22.6–21) with Eq. (22.6–30) for the accuracy of the phase measurement, we obtain for the visibility $\mathcal{V}$, the expression

$$\mathcal{V} = \exp(-R^2_{\text{SN}}/8). \qquad (22.6\text{–}31)$$

It is apparent from this that the fringe visibility can be unity only when the signal-to-noise ratio is zero, in which case nothing at all is known regarding the path followed by the photon through the interferometer. Conversely, when $R_{\text{SN}} \gg 1$, and the path followed by the photon through the interferometer is well established, then $\mathcal{V} \ll 1$, and there is very little interference. The fringe visibility and the signal-to-noise ratio therefore obey a kind of uncertainty relationship, in the sense that they cannot both be large simultaneously. This conclusion reinforces the idea that interference is always a manifestation of the intrinsic indistinguishability of the path of the photon.

Problems

22.1 The signal and idler photons produced by down-conversion from an optically pumped nonlinear crystal are in the following entangled state:

$$|\psi\rangle = |\text{vac}\rangle_{\text{s,i}} + g\sum_{\omega'}\phi(\omega')|\omega'\rangle_{\text{s}}|\omega_0 - \omega'\rangle_{\text{i}}.$$

Here suffices s, i refer to signal and idler, $\phi(\omega)$ is some real function of frequency, and ω_0 is the optical pump frequency. Frequency filters with transmissivities $F_{\text{s}}(\omega)$ and $f_{\text{i}}(\omega)$ followed by photodetectors are inserted in the signal and the idler paths, respectively. The filter $f_{\text{i}}(\omega)$ has a very narrow passband, which is tunable over frequency. The rate of two photon coincidence detection is measured as the frequency of filter $f_{\text{i}}(\omega)$ is scanned, in order to determine the bandwidth of the idler. Show that the passband of $F_{\text{s}}(\omega)$ determines the idler bandwidth.

22.2 Starting from the multimode quantum state $|\psi\rangle$ for signal and idler photons given by Eq. (22.4–21), show that the experiment illustrated in Fig. 22.7 leads to interference when signal and idler photons are detected in coincidence, but not when they are detected separately.

22.3 Two similar nonlinear crystals functioning as parametric down-converters are pumped by a common source and arranged so that the idler from crystal 1 passes through crystal 2 and is colinear with the idler from crystal 2. The signal photon from each crystal falls on a photodetector, and the rate $\mathcal{R}_{\text{ss}}$ of two-photon coincidence counting is measured. Calculate $\mathcal{R}_{\text{ss}}$ and show that crystal 2 behaves as a photon amplifier for the idler photon from crystal 1.

22.4 Consider two nonlinear crystals NL1 and NL2 acting as parametric down-converters in the arrangement shown in the figure. Two mutually coherent

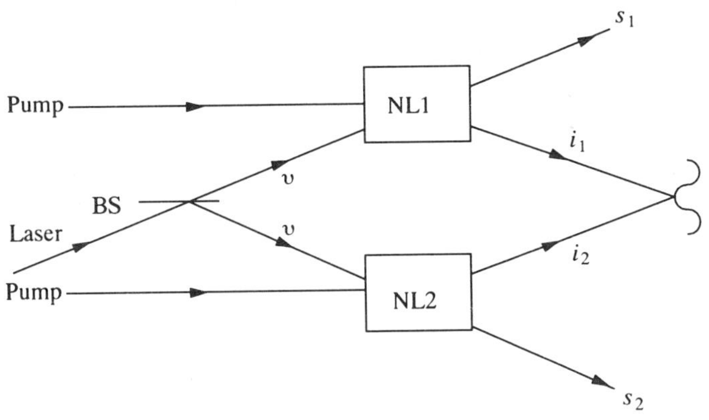

light beams of complex amplitude v emerging from a laser are made colinear with the signal 1 and signal 2 beams produced by the crystals, while idler 1 and idler 2 are superposed and interfere. Show by a two-mode treatment of each down-converter that the visibility of the interference is equal to $|v|^2/(|v|^2 + 1)$.

22.5 Use a multimode treatment of the previous problem to determine the visibility of the interference fringes.

References

Chapter 1

Bochner, S. (1959), *Lectures on Fourier Integrals* (Princeton University Press, Princeton, NJ).
Goldberg, R. R. (1961), *Fourier Transforms* (Cambridge University Press, Cambridge).
Lukacs, E. (1970), *Characteristic Functions*, 2nd edn. (Charles Griffin, London).
Kendall, M. G. (1952), *The Advanced Theory of Statistics*, Vol. 1, 5th edn. (Charles Griffin, London).
Mehta, C. L. (1965), in *Lectures in Theoretical Physics*, Vol. VIIC, ed. W. E. Brittin (University of Colorado Press, Boulder, CO), p. 345.
Yaglom, A. M. (1987), *IEEE ASSP Magazine* **4**, 7.

Chapter 2

Beran, M. J. (1968), *Statistical Continuum Theories* (Wiley, New York).
Bremerman, H. (1965), *Distributions, Complex Variables, and Fourier Transforms* (Addison-Wesley, Reading, MA).
Carter, W. H. and Wolf, E. (1975), *J. Opt. Soc. Am.* **65**, 1067.
Chapman, S. (1916), *Phil. Trans. Roy. Soc. (London) A* **216**, 279.
Davenport, W. B., Jr. and Root, W. L. (1958), *Random Signals and Noise* (McGraw-Hill, New York).
Davis, R. C. (1953), *Proc. Am. Math. Soc.* **4**, 564.
Einstein, A. (1906), *Ann. d. Physik* **19**, 371.
Einstein, A. (1914), *Archives de Sciences Physiques et Naturelles* **37**, 254.
Einstein, A. (1915), *Ann. d. Physik* **47**, 879.
Einstein, A. (1987), *IEEE ASSP Magazine* **4**, 6.
Einstein, A. and Hopf, L. (1910), *Ann. d. Physik* **33**, 1096.
Goldman, S. (1953), *Information Theory* (Prentice-Hall, New York).
Jones, D. S. (1982), *The Theory of Generalized Functions*, 2nd edn. (Cambridge University Press, Cambridge).
Kac, M. and Siegert, A. J. F. (1947), *Ann. Math. Statist.* **18**, 438.
Karhunen, K. (1946), *Ann. Acad. Sci. Fenn., Series A* **34**, 3.
Kestelman, H. (1960), *Modern Theories of Integration*, 2nd edn. (Dover, New York).
Khintchine, A. (1934), *Math. Ann.* **109**, 604.
Kolmogoroff, A. (1931), *Math. Ann.* **104**, 415.
Kramers, H. A. (1940), *Physica* **7**, 284.
Lax, M. (1966), in *Statistical Physics, Phase Transitions and Superfluidity*, eds. M. Chrétien, E. P. Gross and S. Deser (Gordon and Breach, New York), p. 269.
Lukacs, E. (1970), *Characteristic Functions*, 2nd edn. (Charles Griffin, London).
Mandel, L. and Wolf, E. (1976), *J. Opt. Soc. Am.* **66**, 529.

Mandel, L. and Wolf, E. (1981), *Opt. Commun.* **36**, 247.
Middleton, D. (1960), *An Introduction to Statistical Communication Theory* (McGraw-Hill, New York).
Morse, P. M. and Feshbach, H. (1953), *Methods of Theoretical Physics*, Part I (McGraw-Hill, New York).
Moyal, J. E. (1949), *J. Roy. Statist. Soc. B* **11**, 150.
Nussenzveig, H. M. (1972), *Causality and Dispersion Relations* (Academic Press, New York).
Oppenheim, I., Shuler, K. E. and Weiss, G. H. (1977), *Stochastic Processes in Chemical Physics: The Master Equation* (MIT Press, Cambridge, MA).
Pauli, W. (1928) in *Probleme der Modernen Physik, Arnold Sommerfeld zum 60. Geburtstage gewidmet von seinen Schülern*, ed. P. Debye (Hirzel Verlag, Leipzig). Reprinted in *Collected Scientific Papers by Wolfgang Pauli*, Vol. 1 (1964), eds. R. Kronig and V. F. Weisskopf (Interscience, New York).
Pogorzelski, W. (1966), *Integral Equations and Their Applications*, Vol. 1 (Pergamon Press, Oxford).
Rice, S. O. (1944), *Bell Syst. Tech. J.* **23**, 282; reprinted in *Selected Papers on Noise and Stochastic Processes*, ed. N. Wax (Dover, New York, 1954).
Riesz, F. and Sz.-Nagy, B. (1955), *Functional Analysis* (Ungar, New York).
Risken, H. (1984), *The Fokker–Planck Equation* (Springer, Berlin).
Root, W. L. and Pitcher, T. S. (1955), *Ann. Math. Statist.* **26**, 313.
Smithies, F. (1970), *Integral Equations* (Cambridge University Press, Cambridge).
Smoluchowski, M. V. (1906), *Ann. d. Physik* **21**, 756.
Srinivas, M. D. and Wolf, E. (1977), in *Statistical Mechanics and Statistical Methods in Theory and Applications*, ed. Uzi Landman (New York, Plenum), p. 219.
Titchmarsh, E. C. (1939), *The Theory of Functions*, 2nd edn. (Oxford University Press, London).
von Laue, M. (1915a), *Ann. d. Physik* **47**, 853.
von Laue, M. (1915b), *Ann. d. Physik* **48**, 668.
Wiener, N. (1923), *J. Math. and Phys. (Massachusetts Institute of Technology)*, **2**, 131.
Wiener, N. (1930), *Acta. Math.* **55**, 117.
Wolf, E. (1981), *Opt. Commun.* **38**, 3.
Wolf, E. (1982), *J. Opt. Soc. Am.* **72**, 343.
Wolf, E. (1986), *J. Opt. Soc. Am. A* **3**, 76.
Wolf, E. and Carter, W. H. (1975), *Opt. Commun.* **13**, 205.
Wolf, E. and Carter, W. H. (1976), *Opt. Commun.* **16**, 297.
Yaglom, A. M. (1962), *An Introduction to the Theory of Stationary Random Functions* (Prentice-Hall, Englewood Cliffs, NJ).
Yaglom, A. M. (1987), *IEEE ASSP Magazine* **4**, 7.

Chapter 3

Agarwal, G. S. and Wolf, E. (1972), *J. Math. Phys.* **13**, 1759.
Baker, B. B. and Copson, E. T. (1950), *The Mathematical Theory of Huygens' Principle* (Clarendon Press, Oxford).
Baños, A. (1966), *Dipole Radiation in the Presence of a Conducting Half-space* (Pergamon Press, Oxford).
Born, M. and Wolf, E. (1980), *Principles of Optics*, 6th edn. (Pergamon Press, Oxford).
Braun, G. (1956), *Acta Phys. Austriaca* **10**, 8.
Chako, N. (1965), *J. Inst. Maths. Applics.* **1**, 372.
Copson, E. T. (1935), *An Introduction to the Theory of Functions of a Complex Variable* (Oxford University Press, Oxford).

Copson, E. T. (1967), *Asymptotic Expansions* (Cambridge University Press, Cambridge).

Courant, R. and Hilbert, D. (1962), *Methods of Mathematical Physics* (Interscience, New York).

Davenport, W. B. and Root, W. L. (1958), *An Introduction to the Theory of Random Signals and Noise* (McGraw-Hill, New York).

Dennery, P. and Krzywicki, A. (1967), *Mathematics for Physicists* (Harper and Row, New York).

Dugundji, J. (1958), *I. R. E. Trans. on Information Theory* IT**4**, 53.

Eisenhart, L. P. (1947), *An Introduction to Differential Geometry* (Princeton University Press, Princeton, NJ).

Erdelyi, A. (1954), ed., *Tables of Integral Transforms*, Vol. II (McGraw-Hill, New York).

Erdelyi, A. (1955), *J. Soc. Ind. Appl. Math.* **3**, 17.

Erdelyi, A. (1956), *Asymptotic Expansions* (Dover, New York).

Focke, J. (1954), *Ber. Verh. Sächs. Akad. Wiss. (Leipzig)* **101**, Heft 3.

Gabor, D. (1946), *J. Inst. of Elect. Eng.* **93**, Pt. III, 429.

Gabor, D. (1961), in *Progress in Optics*, Vol. 1, ed. E. Wolf (North-Holland, Amsterdam), p. 109, Sec. 3).

Goodman, J. W. (1968), *Introduction to Fourier Optics* (McGraw-Hill, New York).

Gradshteyn, I. S. and Ryzhik, I. M. (1980), *Table of Integrals, Series and Products* (Academic Press, New York).

Heitler, W. (1954), *The Quantum Theory of Radiation*, 3rd edn. (Oxford University Press, Oxford).

Jones, D. S. and Kline, M. (1958), *J. Math. Phys.* **37**, 1.

Lalor, É. (1968), *J. Opt. Soc. Am.* **58**, 1235.

Luneburg, R. K. (1964), *Mathematical Theory of Optics* (University of California Press, Berkeley and Los Angeles).

Mandel, L. (1967), *J. Opt. Soc. Am.* **57**, 613.

Miyamoto, K. and Wolf, E. (1962), *J. Opt. Soc. Am.* **52**, 615, Appendix.

Morse, P. M. and Feshbach, H. (1953), *Methods of Theoretical Physics*, Part I (McGraw-Hill, New York).

Nussenzveig, H. M. (1972), *Causality and Dispersion Relations* (Academic Press, New York).

Oswald, J. R. V. (1956), *I. R. E. Trans.* CT **3**, 244.

Paley, A. R. E. A. C. and Wiener, N. (1934), *Fourier Transforms in the Complex Domain* (American Mathematical Society, New York).

Rayleigh, Lord (1897), *Phil. Mag.* **43**, 259.

Rice, S. O. (1944), *Bell Syst. Tech. J.* **23**, 282.

Sherman, G. C. (1969), *J. Opt. Soc. Am.* **59**, 697.

Sherman, G. C., Stamnes, J. J. and Lalor, É. (1976), *J. Math. Phys.* **17**, 760.

Stamnes, J. J. (1986), *Waves in Focal Regions* (Adam Hilger, Bristol and Boston).

Thompson, W. (1887), *Phil. Mag.*, **23**, 252; Reprinted in *Mathematical and Physical Papers by Sir William Thompson, Baron Kelvin*, Vol. IV (Cambridge University Press, 1910), p. 303.

Titchmarsh, E. C. (1948), *Introduction to the Theory of Fourier Integrals*, 2nd edn. (Oxford University Press, London and New York).

van der Corput, J. G. (1934–1935), *Compositio Mathematica*, **1**, 15.

van der Corput, J. G. (1936), *Compositio Mathematica* **3**, 328.

van Kampen, N. G. (1958), *Physica* **24**, 437.

Ville, J. (1948), *Cables et Trans.* **2**, 61.

Ville, J. (1950), *Cables et Trans.* **4**, 9.

Watson, G. N. (1944), *A Treatise on the Theory of Bessel Functions*, 2nd edn. (Cambridge University Press, Cambridge).

Weyl, H. (1919), *Ann. d. Physik* **60**, 481.

Whittaker, E. T. and Watson, G. N. (1940), *A Course of Modern Analysis*, 4th edn, reprinted (Cambridge University Press, Cambridge).

Chapter 4

Agarwal, G. S. and Wolf, E. (1993), *J. Mod. Opt.* **40**, 1489.
Bastiaans, M. J. (1977), *Opt. Acta* **24**, 261.
Blanc-Lapierre, A. and Dumontet, P. (1955), *Rev. Opt.* **34**, 1.
Born, M. and Wolf, E. (1980), *Principles of Optics*, 6th edn. (Pergamon Press, Oxford).
Bothe, W. (1927), *Z. Phys.* **41**, 345.
Brown, Hanbury R. and Twiss, R. Q. (1957), *Proc. Roy. Soc. (London) A* **242**, 300.
Einstein, A. (1912), in *La Théorie du Rayonnement et les Quanta* (Instituts Solvay, Brussels, 1911), ed. P. Langevin and L. de Broglie (Paris, Gauthier-Villars, 1912), pp. 407–435, discussion pp. 436–450.
Friberg, A. T. and Wolf, E. (1995), *Opt. Lett.* **20**, 623.
Fürth, R. (1928), *Z. Phys.* **50**, 310.
Goldberg, R. R. (1965), *Fourier Transforms* (Cambridge University Press, Cambridge).
Heiniger, F., Herden, A. and Tschudi, T. (1983), *Opt. Commun.* **48**, 237.
Heitler, W. (1984), *The Quantum Theory of Radiation*, 3rd edn. (Oxford University Press, Oxford).
Hopkins, H. H. (1951), *Proc. Roy. Soc. (London) A* **208**, 263.
Hopkins, H. H. (1953), *Proc. Roy. Soc. (London) A* **217**, 408.
Hopkins, H. H. (1957), *J. Opt. Soc. Am.* **47**, 508.
James, D. F. V., Kandpal, H. C. and Wolf, E. (1995), *Astrophys. J.* **445**, 406.
Kandpal, H. C., Saxena, K., Mehta, D. S., Vaishya, J. S. and Joshi, K. C. (1993), *Opt. Commun.* **99**, 157.
Kay, I. and Silverman, R. (1957), *Information and Control*, **1**, 64.
Kharkevich, A. A. (1960), *Spectra and Analysis*, translated from Russian, (New York, Consultants Bureau).
Mandel, L. (1959), *Proc. Phys. Soc. (London)* **74**, 233.
Mandel, L. (1961a), *J. Opt. Soc. Am.* **51**, 797.
Mandel, L. (1961b), *J. Opt. Soc. Am.* **51**, 1342.
Mandel, L. and Wolf, E. (1962), *Proc. Phys. Soc. (London)* **80**, 894.
Mandel, L. and Wolf, E. (1965), *Rev. Mod. Phys.* **37**, 231.
Mandel, L. and Wolf, E. (1976), *J. Opt. Soc. Am.* **66**, 529.
Mandel, L. and Wolf, E. (1981), *Opt. Commun.* **36**, 247.
Mehta, C. L. (1963), *Nuovo Cimento* **28**, 401.
Mehta, C. L., Wolf, E. and Balachandran, A. P. (1966), *J. Math. Phys.* **7**, 133.
Michelson, A. A. (1890), *Phil. Mag.* **30**, 1.
Michelson, A. A. (1891a), *Nature* **45**, 160.
Michelson, A. A. (1891b), *Phil. Mag.* **31**, 256.
Michelson, A. A. (1891c), *Phil. Mag.* **31**, 338.
Michelson, A. A. (1892), *Phil. Mag.* **34**, 280.
Michelson, A. A. (1920), *Astroph. J.* **51**, 257.
Pancharatnam, S. (1956), *Proc. Indian Acad. Sci. Sec. A* **44**, 398.
Pancharatnam, S. (1957), *Proc. Indian Acad. Sci. Sec. A* **46**, 1.
Pancharatnam, S. (1963a), *Proc. Indian Acad. Sci. Sec. A* **57**, 218.
Pancharatnam, S. (1963b), *Proc. Indian Acad. Sci. Sec. A* **57**, 231.
Pancharatnam, S. (1975), *Collected Works of S. Pancharatnam* (Oxford University Press, London).
Papas, C. H. (1965), *Theory of Electromagnetic Wave Propagation* (McGraw-Hill, New York).
Parrent, G. B. (1959), *J. Opt. Soc. Am.* **49**, 787.

Stokes, G. G. (1852), *Trans. Camb. Phil. Soc.* **9**, 399. Reprinted in his *Mathematical and Physical Papers*, Vol. III (Cambridge University Press, Cambridge, 1922), p. 233 and in W. Swindell (1975), p. 124.
Sudarshan, E. C. G. (1969), *J. Math. and Phys. Sci. (Madras)* **3**, 121.
Swindell, W. (1975), *Polarized Light* (Dowden, Hutchinson and Ross, Stroudsburg, PA).
Tricomi, F. G. (1957), *Integral Equations* (Interscience, New York; reprinted by Dover, New York, 1985).
van Cittert, P. H. (1934), *Physica* **1**, 201.
Verdet, E. (1865), *Ann. Scientif. l'École Normale Supérieure* **2**, 291.
Verdet, E. (1869), *Leçons d'Optique Physique* Vol. 1, p. 106. (L'Imprimerie Impériale, Paris).
von Laue, M. (1907a), *Ann. d. Physik* **23**, 1.
von Laue, M. (1907b), *Ann. d. Physik* **23**, 795.
Weyl, H. (1950), *The Theory of Groups and Quantum Mechanics*, translated from German (Dover, New York).
Wiener, N. (1927–1928), *J. Math. Phys. (M.I.T.)* **7**, 109.
Wiener, N. (1929), *J. Frankl. Inst.* **207**, 525.
Wiener, N. (1930), *Acta. Math. (Uppsala)*, **55**, 117.
Wolf, E. (1954a), *Proc. Roy. Soc. (London), A* **225**, 96.
Wolf, E. (1954b), *Nuovo Cimento* **12**, 884.
Wolf, E. (1955), *Proc. Roy. Soc. (London)* **230**, 246.
Wolf, E. (1958), *Proc. Phys. Soc. (London)* **71**, 257.
Wolf, E. (1959), *Nuovo Cimento* **13**, 1165.
Wolf, E. (1981a), *Optics in Four Dimensions* (ICO Ensenada, 1980), eds. M. A. Machado and L. M. Narducci (Conference Proceedings #65, American Institute of Physics, New York), p. 42.
Wolf, E. (1981b) *Opt. Commun.* **38**, 3.
Wolf, E. (1982), *J. Opt. Soc. Am.* **72**, 343.
Wolf, E. (1983), *Opt. Lett.* **8**, 250.
Wolf, E. (1986), *J. Opt. Soc. Am. A* **3**, 76.
Wolf, E. and Carter, W. H. (1975), *Opt. Commun.* **13**, 205.
Wolf, E. and Carter, W. H. (1976), *Opt. Commun.* **16**, 297.
Wolf, E. and Devaney, A. J. (1981), *Opt. Lett.* **6**, 168.
Wolf, E., Devaney, A. J. and Foley, J. T. (1981), *Optics in Four Dimensions* (ICO Ensenada, 1980), eds. M. A. Machado and L. M. Narducci (Conference Proceedings #65, American Institute of Physics, New York).
Wolf, E., Devaney, A. J. and Gori, F. (1983), *Opt. Commun.* **46**, 4.
Zernike, F. (1938), *Physica* **5**, 785.

Chapter 5

Agarwal, G. S., Foley, J. T. and Wolf, E. (1987), *Opt. Commun.* **62**, 67.
Antes, G., Baltes, H. P. and Steinle, B. (1976), *Helv. Phys. Acta.* **49**, 759.
Baltes, H. P. (1977), *Appl. Phys.* **12**, 221.
Baltes, H. P. and Steinle, B. (1977), *Nuovo Cimento* **41B**, 428.
Baltes, H. P., Steinle, B. and Antes, G. (1976), *Opt. Commun.* **18**, 242.
Baltes, H. P., Steinle, B. and Antes, G. (1978), in *Coherence and Quantum Optics IV*, eds. L. Mandel and E. Wolf (Plenum Press, New York), p. 431.
Bocko, M., Douglass, D. H. and Knox, R. S. (1987), *Phys. Rev. Lett.* **58**, 2649.
Born, M. and Wolf, E. (1980), *Principles of Optics*, 6th edn. (Pergamon Press, Oxford).
Carter, W. H. (1978), *Opt. Commun.* **26**, 1.
Carter, W. H. (1984), *J. Opt. Soc. Am. A* **1**, 716.
Carter, W. H. and Bertolotti, M. (1978), *J. Opt. Soc. Am.* **68**, 329.

Carter, W. H. and Wolf, E. (1975), *J. Opt. Soc. Am.* **65**, 1067.
Carter, W. H. and Wolf, E. (1977), *J. Opt. Soc. Am.* **67**, 785.
Carter, W. H. and Wolf, E. (1981a), *Opt. Acta* **28**, 227.
Carter, W. H. and Wolf, E. (1981b), *Opt. Acta* **28**, 245.
Chandrasekhar, S. (1960), *Radiative Transfer* (Dover, New York).
Collett, E. and Wolf, E. (1978), *Opt. Lett.* **2**, 27.
Collett, E. and Wolf, E. (1979), *J. Opt. Soc. Am.* **69**, 942.
Courjon, D. and Bulabois, J. (1979), in *Applications of Optical Coherence*, ed. W. H. Carter, *Proc. Soc. Photo-Opt. Instr. Eng.* **194**, 129.
Courjon, D., Bulabois, J. and Carter, W. H. (1981), *J. Opt. Soc. Am.* **71**, 469.
Dechamps, J., Courjon, D. and Bulabois, J. (1983), *J. Opt. Soc. Am.* **73**, 256.
De Santis, P., Gori, F. and Palma, C. (1979), *Opt. Commun.* **28**, 151.
De Santis, P., Gori, F., Guattari, G. and Palma, C. (1979), *Opt. Commun.* **29**, 256.
Faklis, D. and Morris, G. M. (1992), *J. Mod. Opt.* **39**, 941.
Farina, J. D., Narducci, L. M. and Collett, E. (1980), *Opt. Commun.* **32**, 203.
Foley, J. T. (1990), *Opt. Commun.* **75**, 347.
Foley, J. T. (1991), *J. Opt. Soc. Am. A* **8**, 1099.
Foley, J. T. and Wolf, E. (1985), *Opt. Commun.* **55**, 236.
Foley, J. T. and Wolf, E. (1991), *J. Mod. Opt.* **38**, 2053.
Foley, J. T. and Wolf, E. (1995), *J. Mod. Opt.* **42**, 787.
Foley, J. T. and Zubairy, M. S. (1978), *Opt. Commun.* **26**, 297.
Friberg, A. T. (1979), *J. Opt. Soc. Am.* **69**, 192.
Friberg, A. T. (1981), *Opt. Acta* **28**, 261.
Friberg, A. T. (ed.) (1993), *Selected Papers on Coherence and Radiometry* (SPIE Optical Engineering Press, Milestone Series, MS69, Bellingham, WA).
Friberg, A. T. and Sudol, R. J. (1982), *Opt. Commun.* **41**, 383.
Friberg, A. T. and Sudol, R. J. (1983), *Opt. Acta.* **30**, 1075.
Friberg, A. T. and Wolf, E. (1983), *Opt. Acta* **30**, 1417.
Friberg, A. T., Agarwal, G. S. Foley, J. T. and Wolf, E. (1992), *J. Opt. Soc. Am. B* **9**, 1386.
Gamliel, A. (1988), *Statistical Optics*, ed. G. M. Morris, *Proc. SPIE* **976**, 137.
Gamliel, A. and Wolf, E. (1988), *Opt. Commun.* **65**, 91.
Goodman, J. W. (1965), *Proc. IEEE* **53**, 1688.
Goodman, J. W. (1979) in *Applications of Optical Coherence*, ed. W. H. Carter, *Proc Soc. Photo–Opt. Instr. Eng.* **194**, 86.
Gori, F. (1980a), *Opt. Acta* **27**, 1025.
Gori, F. (1980b), *Opt. Commun.* **34**, 301.
Gori, F. (1981), *Opt. Commun.* **39**, 293.
Gori, F. (1983), *Opt. Commun.* **46**, 149.
Gori, F. (1994) in *International Trends in Optics*, Vol. II, ed. J. C. Dainty, (Academic Press, New York).
Gori, F. and Palma, C. (1978), *Opt. Commun.* **27**, 185.
Gori, F., Guattari, G., Palma, C. and Padovani, C. (1988), *Opt. Commun.* **67**, 1.
Gradshteyn, I. S. and Ryzhik, I. M. (1980), *Table of Integrals, Series & Products* (Academic Press, New York).
Hopf, E. (1934), *Mathematical Problems of Radiative Equilibrium* (Cambridge University Press, Cambridge).
Imai, Y. and Ohtsuka, Y. (1980), *Appl. Opt.* **19**, 542.
Imre, K., Özizmir, E., Rosenbaum, M. and Zweifel, P. F. (1967), *J. Math. Phys.* **8**, 1097.
Indebetouw, G. (1989), *J. Mod. Opt.* **36**, 251.
James, D. F. V. and Wolf, E. (1990), *Phys. Lett. A* **146**, 167.
James, D. F. V. and Wolf, E. (1991a), *Opt. Commun.* **81**, 150.
James, D. F. V. and Wolf, E. (1991b), *Phys. Lett. A* **157**, 6.

James, D. F. V. and Wolf, E. (1991c), *Radio Science* **26**, 1239.

James, D. F. V. and Wolf, E. (1994), *Phys. Lett.* A **188**, 239.

James, D. F. V., Savedoff, M. P. and Wolf, E. (1990), *Ap. J.* **359**, 67.

Kandpal, H. C. and Wolf, E. (1994), *Opt. Commun.* **110**, 255.

Kandpal, H. C., Vaishya, J. S. and Joshi, K. C. (1989), *Opt. Commun.* **73**, 169.

Kandpal, H. C., Vaishya, J. S., Chander, M., Saxena, K., Mehta, D. S. and Joshi, K. C. (1992), *Phys. Lett. A* **167**, 114.

Kim, K. and Wolf, E. (1987), *J. Opt. Soc. Am. A* **4**, 1233.

Kinzly, R. E. (1972), *J. Opt. Soc. Am.* **62**, 386.

Kogelnik, H. and Li, T. (1966), *Proc. IEEE* **54**, 1312.

Leader, J. C. (1978), *J. Opt. Soc. Am.* **68**, 1332.

Li, Y. and Wolf, E. (1982), *Opt. Lett.* **7** 256.

Lorentz, H. A. (1909), *The Theory of Electrons* (Teubner, Leipzig and Columbia University Press, New York; reprinted by Dover, New York).

Marchand, E. W. and Wolf, E. (1972), *J. Opt. Soc. Am.* **62**, 379.

Marchand, E. W. and Wolf, E. (1974a), *J. Opt. Soc. Am.* **64**, 1219.

Marchand, E. W. and Wolf, E. (1974b), *J. Opt. Soc. Am.* **64**, 1273.

Margenau, H. and Hill, R. N. (1961), in *Progr. Theoretical Phys. (Japan)* **26**, 722.

Martienssen, W. and Spiller, E. (1964), *Amer. J. Phys.* **32**, 919.

McGuire, D. (1979), *Opt. Commun.* **29**, 17.

Menzel, D. H. ed., (1966), *Selected Papers on the Transfer of Radiation* (Dover, New York).

Morris, G. M. and Faklis, D. (1987), *Opt. Commun.* **62**, 5.

Nyyssonen, D. (1977), *Appl. Opt.* **16**, 2223.

Nyyssonen, D. (1979), in *Applications of Optical Coherence*, ed. W. H. Carter, *Proc. Soc. Photo–Opt. Instr. Eng.* **194**, 34.

Ohtsuka, Y. and Imai, Y. (1979), *J. Opt. Soc. Am.* **69**, 684.

Oldham, W. G., Subramanian, S. and Neureuther, A. R. (1981), *Solid State Electr.* **24**, 975.

Planck, M. (1959), *The Theory of Heat Radiation* (Dover, New York).

Reynolds, G. O. and Smith, A. E. (1973), *Appl. Opt.* **12**, 1259.

Saleh, B. E. A. (1979), *Opt. Commun.* **30**, 135.

Saleh, B. E. A. and Irshid, M. I. (1982), *Opt. Lett.* **7**, 342.

Santarsiero, M. and Gori, F. (1992), *Phys. Lett. A* **167**, 123.

Schell, A. C. (1961), *The Multiple Plate Antenna* (Doctoral Dissertation. Massachusetts Institute of Technology), Sec. 7.5.

Schell, A. C. (1967), *IEEE Trans. Antennas and Propag.* AP-**15**, 187.

Scudieri, F., Bertolotti, M. and Bartolino, R. (1974), *Appl. Opt.* **13**, 181.

Siegman, A. E. (1971), *An Introduction to Lasers and Masers* (McGraw-Hill, New York).

Starikov, A. and Wolf, E. (1982), *J. Opt. Soc. Am.* **72**, 923.

Steinle, B. and Baltes, H. P. (1977), *J. Opt. Soc. Am.* **67**, 241.

Tervonen, E., Friberg, A. T. and Turunen, J. (1992), *J. Opt. Soc. Am. A* **9**, 796.

Trotter, A. P. (1919), *The Illuminating Engineer (London)* **12**, 243.

Turunen, J., Tervonen, E. and Friberg, A. T. (1990), *J. Appl. Phys.* **67**, 49.

Walther, A. (1968), *J. Opt. Soc. Am.* **58**, 1256.

Walther, A. (1973), *J. Opt. Soc. Am.* **63**, 1622.

Walther, A. (1974), *J. Opt. Soc. Am.* **64**, 1275.

Walther, A. (1978a), *J. Opt. Soc. Am.* **68**, 1606.

Walther, A. (1978b), *Opt. Lett.* **3**, 127.

Watson, G. N. (1966), *A Treatise on the Theory of Bessel Functions*, 2nd edn. (Cambridge University Press, Cambridge).

Wentzel, G. (1949), *Quantum Theory of Fields* (Interscience Publishers, New York).

Wigner, E. (1932), *Phys. Rev.* **40**, 749.

Wigner, E. (1971) in *Perspective in Quantum Theory*, ed. by W. Yourgrau and A. van der Merwe (M. I. T. Press, Cambridge, MA).
Wolf, E. (1978), *J. Opt. Soc. Am.* **68**, 6.
Wolf, E. (1986), *Phys. Rev. Lett.* **56**, 1370.
Wolf, E. (1987a), *Nature* **326**, 363.
Wolf, E. (1987b), *Opt. Commun.* **62**, 12.
Wolf, E. (1987c), *Phys. Rev. Lett.* **58**, 2646.
Wolf, E. (1989), *Phys. Rev. Lett.* **63**, 220.
Wolf, E. (1992), *J. Mod. Opt.* **22**, 9.
Wolf, E. (1994), *Opt. Lett.* **19**, 2024.
Wolf, E. and Carter, W. H. (1978a), in *Coherence and Quantum Optics IV*, eds. L. Mandel and E. Wolf, (Plenum Press, New York), p. 415.
Wolf, E. and Carter, W. H. (1978b), *J. Opt. Soc. Am.* **68**, 953.
Wolf, E. and Collett, E. (1978), *Opt. Commun.* **25**, 293.
Wolf, E. and Fineup, J. R. (1991), *Opt. Commun.* **82**, 209.
Wolf, E., Foley, J. T. and Gori, F. (1989), *J. Opt. Soc. Am. A* **6**, 1142, Appendix A.

Chapter 6

Aitken, A. C. (1944), *Determinants and Matrices* (Oliver and Boyd, Edinburgh).
Barakat, R. (1981), *Opt. Commun.* **38**, 159.
Beran, M. and Parrent, G. (1962), *J. Opt. Soc. Am.* **52**, 98.
Born, M. and Wolf, E. (1980), *Principles of Optics*, 6th edn. (Pergamon Press, Oxford).
Bourret, R. C. (1960), *Nuovo Cimento* **18**, 347.
Hurwitz, H. and Jones, R. C. (1941), *J. Opt. Soc. Am.* **31**, 493.
Jeffreys, H. (1931), *Cartesian Tensors* (Cambridge University Press, Cambridge).
Jones, R. C. (1941a), *J. Opt. Soc. Am.* **31**, 488.
Jones, R. C. (1941b), *J. Opt. Soc. Am.* **31**, 500.
Jones, R. C. (1942), *J. Opt. Soc. Am.* **32**, 486.
Jones, R. C. (1947a), *J. Opt. Soc. Am.* **37**, 107.
Jones, R. C. (1947b), *J. Opt. Soc. Am.* **37**, 110.
Jones, R. C. (1948), *J. Opt. Soc. Am.* **38**, 671.
Jones, R. C. (1956), *J. Opt. Soc. Am.* **46**, 126.
Kim, K., Mandel, L. and Wolf, E. (1987), *J. Opt. Soc. Am. A* **4**, 433.
Mehta, C. L. and Wolf, E. (1964a), *Phys. Rev. A* **134**, 1143.
Mehta, C. L. and Wolf, E. (1964b), *Phys. Rev. A* **134**, 1149.
Mehta, C. L. and Wolf, E. (1967a), *Phys. Rev.* **157**, 1183.
Mehta, C. L. and Wolf, E. (1967b), *Phys. Rev.* **157**, 1188.
Mehta, C. L. and Wolf, E. (1967c), *Phys. Rev.* **161**, 1328.
Mueller, H. (1948), *J. Opt. Soc. Am.* **38**, 661.
Parke, N. G., III (1949), *J. Math. Phys. (MIT)* **28**, 131.
Parrent, G. B. and Roman, P. (1960), *Nuovo Cimento* **15**, 370.
Perrin, F. (1942), *J. Chem. Phys.* **10**, 415.
Roman, P. (1961a), *Nuovo Cimento* **20**, 759.
Roman, P. (1961b), *Nuovo Cimento* **22**, 1005.
Roman, P. and Wolf, E. (1960a), *Nuovo Cimento* **17**, 462.
Roman, P. and Wolf, E. (1960b), *Nuovo Cimento* **17**, 477.
Shurcliff, W. A. (1962), *Polarized Light* (Harvard University Press, Cambridge, MA).
Simon, R. (1982), *Opt. Commun.* **42**, 293.
Soleillet, P. (1929), *Ann. de Physique* **12**, 23.
Stokes, G. G. (1852), *Trans. Cambr. Phil. Soc.* **9**, 399. Reprinted in his *Mathematical and Physical Papers*, Vol. III (Cambridge University Press, Cambridge, 1922), p. 233 and in W. Swindell (1975).

Swindell, W. (1975) *Polarized Light* (Dowden, Hutchinson and Ross, Stroudsburg, PA, 1975).
Wolf, E. (1954), *Nuovo Cimento* **12**, 884.
Wolf, E. (1956), in *Proc. Symposium on Astronomical Optics*, ed. Z. Kopal (North-Holland, Amsterdam).
Wolf, E. (1959), *Nuovo Cimento* **13**, 1165.

Chapter 7

Allen, L., Gatehouse, S. and Jones, D. G. C. (1971), *Opt. Commun.* **4**, 169.
Beard, T. D. (1969), *Appl. Phys. Lett.* **15**, 227.
Benedek, G. B. (1968), in *Statistical Physics, Phase Transitions and Superfluidity* (Brandeis University Summer Institute in Theoretical Physics, 1966), eds. M. Chretien, E. P. Gross and S. Desser (Gordon and Breach, New York), Vol. 2, pp. 1–98.
Berne, B. J. and Pecora, R. (1976), *Dynamic Light Scattering* (Wiley, New York).
Bertolotti, M. Daino, B., Gori, F. and Sette, D. (1965), *Nuovo Cimento* **38**, 1505.
Born, M. and Wolf, E. (1980), *Principles of Optics*, 6th edn. (Pergamon Press, Oxford).
Boyd, G. D. and Gordon, J. P. (1961), *Bell Tech. J.* **40**, 489.
Breckinridge, J. B. (1990) ed., *Amplitude and Intensity Spatial Interferometry* (SPIE Proceeding, Bellingham, WA), Vol. **1237**.
Brillouin, L. (1914), *Compt. Rend.* **158**, 1331.
Brillouin, L. (1922), *Ann. Phys. (Paris)* **17**, 88.
Burge, R. E., Fiddy, M. A., Greenaway, A. H. and Ross, G. (1976), *Proc. Roy. Soc. (London) A* **350**, 191.
Chu, B. (1974), *Laser Light Scattering* (Academic Press, New York).
Connes, J. (1961a), *Rev. d'Opt.* **40**, 45.
Connes, J. (1961b), *Rev. d'Opt.* **40**, 116.
Connes, J. (1961c), *Rev. d'Opt.* **40**, 171.
Connes, J. (1961d), *Rev. d'Opt.* **40**, 231.
Crosignani, B., Di Porto, P. and Bertolotti, M. (1975), *Statistical Properties of Scattered Light* (Academic Press, New York).
Cummins, H. Z. (1969), in *Quantum Optics* (Proc. Internat. School of Physics, 'Enrico Fermi', Course XLII, Varenna, 1967), ed. R. J. Glauber (Academic Press, New York), p. 247.
Cummins, H. Z. and Swinney, H. L. (1970), *Progress in Optics*, Vol. 8, ed. E. Wolf, (North-Holland, Amsterdam), p. 133.
Dainty, J. C. (1984), *Laser Speckle*, 2nd edn., ed. J. C. Dainty (Springer, Berlin), Chapt. 7.
Davis, J. and Tango, W. J. (1985), *Proc. Astr. Soc. Austr.* **6**, 34.
Davis, J. and Tango, W. J. (1986), *Nature* **323**, 234.
Debye, P. (1912), *Ann. d. Physik* **39**, 789.
Dialetis, D. (1967), *J. Math. Phys.* **8**, 1641.
Dialetis, D. and Wolf, E. (1967), *Nuovo Cimento* **47**, 113.
Einstein, A. (1910), *Ann. d. Physik* **33**, 1275; [An English translation of this paper was published in *Colloid Chemistry*, ed. J. Alexander (The Chemical Catalog Company, New York), Vol. 1, 1926), p. 323.]
Fabelinskii, I. L. (1968), *Molecular Scattering of Light* (Plenum Press, New York).
Fellgett, P. (1958a), *J. Phys. Rad.* **19**, 187.
Fellgett, P. (1958b), *J. Phys. Rad.* **19**, 237.
Ferwerda, H. A. (1978), in *Inverse Source Problems In Optics*, ed. H. P. Baltes, (Springer, New York), Chapt. 2.
Fleury, P. A. and Boon, J. P. (1969), *Phys. Rev.* **186**, 244.
Foley, J. T. and Wolf, E. (1989), *Phys. Rev. A* **40**, 588.

Forster, D. (1975), *Hydrodynamic Fluctuations, Broken Symmetry and Correlation Functions* (Benjamin, Reading, MA).
Fox, A. G. and Li, T. (1961), *Bell Tech. J.* **40**, 453.
Friberg, A. T. and Turunen, J. (1994), *J. Opt. Soc. Am. A* **11**, 227.
Friberg, A. T. and Wolf, E. (1983), *J. Opt. Soc. Am.* **73**, 26.
Gamo, H. (1963) in *Electromagnetic Theory and Antennas*, Part 2, ed. E. C. Jordan, (Macmillan, New York).
Gamo, H. (1963), *J. Appl. Phys.* **34**, 875.
Glauber, R. J. (1962) in *Lectures on Theoretical Physics IV* (Univ. of Colorado Summer Institute for Theoretical Physics), eds. W. E. Britten, B. W. Downs and J. Downs, (Interscience, New York), p. 571.
Gori, F. (1980), *Atti Fond. Giorgio Ronchi* **35**, 434.
Gross, E. (1930a), *Nature* **126**, 201.
Gross, E. (1930b), *Nature* **126**, 400.
Gross, E. (1930c), *Nature* **126**, 603.
Gross, E. (1932), *Nature* **129**, 722.
Jacquinot, P. (1958), *J. Phys. Rad.* **19**, 223.
Jacquinot, P. (1960), *Rep. Progr. Phys. (London)* **23**, 267.
James, D. F. V. and Wolf, E. (1990), *Phys. Lett. A* **146**, 167.
James, D. F. V. and Wolf, E. (1994), *Phys. Lett.* A. **188**, 239.
James, D. F. V., Savedoff, M. P. and Wolf, E. (1990), *Ap. J.* **359**, 67.
Kohler, D. and Mandel, L. (1973), *J. Opt. Soc. Am.* **63**, 126.
Komarov, L. I. and Fisher, I. Z. (1962), *Zh. Eksp. Teor. Fiz. (U.S.S.R.)* **43**, 1927–1933, [English translation in *Sov. Phys. JETP* **16**, 1358 (1963)].
Labeyrie, A. (1970), *Astron. & Astrophys.* **6**, 85.
Labeyrie, A. (1976) in *Progress in Optics*, Vol. 14, ed. E. Wolf (North-Holland, Amsterdam), p. 47.
Landau, L. D. and Lifshitz, E. M. (1960), *Electrodynamics of Continuous Media* (Pergamon Press, Oxford).
Landau, L. and Placzek, G. (1934), *Physik Z. Sovietunion* **5**, 172.
Mandel, L. (1969), *Phys. Rev.* **181**, 75.
Mandel, L. and Wolf, E. (1973), *Opt. Commun.* **8**, 95.
Marathay, A. S. and Roman, P. (1964), *Nuovo Cimento* **34**, 1821.
Mehta, C. L. (1965), *Nuovo Cimento* **36**, 202.
Mehta, C. L. (1968), *J. Opt. Soc. Am.* **58**, 1233.
Michelson, A. A. (1890), *Phil. Mag.* **30**, 1.
Michelson, A. A. (1891), *Phil. Mag.* **31**, 338.
Michelson, A. A. (1892), *Phil. Mag.* **34**, 280.
Michelson, A. A. (1920), *Astrophys. J.* **51**, 257.
Michelson, A. A. (1927), *Studies in Optics* (University of Chicago Press, Chicago, IL).
Michelson, A. A. and Pease, F. G. (1921), *Astrophys. J.* **53**, 249.
Millane, R. P. (1990), *J. Opt. Soc. Am.* **7**, 394.
Morse, P. M. and Feshbach, H. (1953), *Methods of Theoretical Physics*, Part I (McGraw-Hill, New York).
Mountain, R. D. (1966), *Rev. Mod. Phys.* **38**, 205.
Nussenzveig, H. M. (1967), *J. Math. Phys.* **8**, 561.
Page, C. H. (1955), *Physical Mathematics* (Van Nostrand, New York).
Paley, R. E. A. C. and Wiener, N. (1934), *Fourier Transforms in the Complex Domain* (American Mathematical Society, New York).
Papas, C. H. (1965), *Theory of Electromagnetic Wave Propagation* (McGraw-Hill, New York).
Pease, F. G. (1925), *Amour Eng.* **16**, 125.
Pease, F. G. (1930), *Sci. Amer.* **143**, 290.

Pease, F. G. (1931), *Ergeb. exact. Naturwiss.* **10**, 84.
Pecora, R. (1964), *J. Chem. Phys.* **40**, 1604.
Rayleigh, Lord (1892), *Phil. Mag.* **34**, 407.
Rohlfs, K. (1986), *Tools of Radio Astronomy* (Springer, Berlin).
Roman, P. and Marathay, A. S. (1963), *Nuovo Cimento* **30**, 1452.
Smithies, E. (1970), *Integral Equations* (Cambridge University Press, Cambridge).
Smoluchowski, M. (1908), *Ann. d. Physik*, **25**, 205.
Streifer, W. (1966), *J. Opt. Soc. Am.* **56**, 1481.
Strong, J. and Vanasse, G. A. (1959), *J. Opt. Soc. Am.* **49**, 844.
Thompson, A. R., Moran, J. M. and Swenson, G. W. (1986), *Inteferometry and Synthesis in Radio Astronomy* (Wiley, New York).
Toll, J. S. (1956), *Phys. Rev.* **104**, 1760.
Van Hove, L. (1954), *Phys. Rev.* **95**, 249.
Van Kampen, N. G. (1969) in *Quantum Optics* (Proc. Internat. School of Physics, 'Enrico Fermi', Course XLII, Varenna, 1967), ed. R. J. Glauber (Academic Press, New York), p. 235.
Vanasse, G. A. and Sakai, H. (1967), *Progress in Optics*, Vol. 6, ed. E. Wolf (North-Holland, Amsterdam), p. 259.
Weaver, R. L. and Pao, Y. (1981), *J. Math. Phys.* **22**, 1909.
Wolf, E. (1962), *Proc. Phys. Soc. (London)* **80**, 1269.
Wolf, E. (1963), *Phys. Lett.* **3**, 166.
Wolf, E. (1989), *Phys. Rev. Lett.* **63**, 2220.
Wolf, E. and Agarwal, G. S. (1984), *J. Opt. Soc. Am. A* **1**, 541.
Wolf, E. and Foley, J. T. (1989), *Phys. Rev. A* **40**, 579.
Wolf, E., Foley, J. T. and Gori, F. (1989), *J. Opt. Soc. Am. A* **6**, 1142; errata *ibid. A* **7**, 173 (1990).

Chapter 8

Agarwal, G. S. and Wolf, E. (1993), *J. Mod. Opt.* **40**, 1489.
Mandel, L. (1964), in *Quantum Electronics*, Proceedings of the Third International Congress, eds. N. Blombergen and P. Grivet (Dunod, Paris and Columbia University Press, New York), p. 101.
Mehta, C. L. and Mandel, L. (1967), in *Electromagnetic Wave Theory*, Proceeding of a Symposium held at Delft, Vol. 2, ed. T. Brown (Pergamon Press, Oxford), p. 1069.
Wolf, E. (1963), in *Proceeding of the Symposium on Optical Masers* (Brooklyn Polytechnic Press and Wiley, New York), ed. J. Fox, p. 29.
Wolf, E. (1964), in *Quantum Electronics*, Proceedings of the Third International Congress, eds. N. Blombergen and P. Grivet (Dunod, Paris and Columbia University Press, New York), p. 29.

Chapter 9

Arecchi, F. T., Gatti, E. and Sona, A. (1966), *Phys. Lett.* **20**, 27.
Brown, R. Hanbury (1974), *The Intensity Interferometer* (Taylor and Francis, London).
Brown, R. Hanbury and Twiss, R. Q. (1956a), *Nature* **177**, 27.
Brown, R. Hanbury and Twiss, R. Q. (1956b), *Nature* **178**, 1046.
Brown, R. Hanbury and Twiss, R. Q. (1957a), *Proc. Roy. Soc. (London) A* **242**, 300.
Brown, R. Hanbury and Twiss, R. Q. (1957b), *Proc. Roy. Soc. (London) A* **243**, 291.
Brown, R. Hanbury and Twiss, R. Q. (1958a), *Proc. Roy. Soc. (London) A* **248**, 199.
Brown, R. Hanbury and Twiss, R. Q. (1958b), *Proc. Roy. Soc. (London) A* **248**, 222.
Brown, R. Hanbury, Davis, J. and Allen, L. R. (1967a), *Mon. Not. R. Astron. Soc.* **137**, 375.

Brown, R. Hanbury, Davis, J. and Allen, L. R. (1967b), *Mon. Not. R. Astron. Soc.* **137**, 393.
Brown, R. Hanbury, Davis, J. and Allen, L. R. (1974), *Mon. Not. R. Astron. Soc.* **167**, 121.
Cohen-Tannoudji, C., Diu, B. and Laloë, F. (1977), *Quantum Mechanics*, Vols 1 and 2 (John Wiley, New York).
Cummins, H. Z., Knable, N. and Yeh, Y. (1964), *Phys. Rev. Lett.* **12**, 150.
Cummins, H. Z. and Swinney, H. L. (1970), in *Progress in Optics*, Vol. 8, ed. E. Wolf. (North-Holland, Amsterdam), p. 133.
Dicke, R. H. and Wittke, J. P. (1960), *Introduction to Quantum Mechanics* (Addison-Wesley, Reading, MA).
Einstein, A. (1905), *Ann. d. Physik* **17**, 132. [English translation in *Collected Papers of Albert Einstein*, Vol. 2 (1900–1909), Princeton University Press (Princeton, NJ), p. 186 (1989)].
Ford, N. C. and Benedek, G. B. (1965), *Phys. Rev. Lett.* **15**, 649.
Lamb, W. E., Jr. and Scully, M. O. (1969), in *Polarization: Matiere et Rayonnement.* (Presses Universitaires de France, Paris).
Lastovka, J. B. and Benedek, G. B. (1966), *Phys. Rev. Lett.* **17**, 1039.
Loudon, R. (1983), The *Quantum Theory of Light*, 2nd edn. (Clarendon Press, Oxford).
Mandel, L. (1958), *Proc. Phys. Soc. (London)* **72**, 1037.
Mandel, L. (1959), *Proc. Phys. Soc. (London)* **74**, 233.
Mandel, L. (1963) in *Progress in Optics*, Vol. 2, ed. E. Wolf. (North-Holland, Amsterdam), p. 181.
Mandel, L., Sudarshan, E. C. G. and Wolf, E. (1964), *Proc. Phys. Soc. (London)* **84**, 435.
Messiah, A. (1961), *Quantum Mechanics*, Vol. 1 (North-Holland, Amsterdam).
Messiah, A. (1962), *Quantum Mechanics*, Vol. 2 (North-Holland, Amsterdam).
Morgan, B. L. and Mandel, L. (1966), *Phys. Rev. Lett.* **16**, 1012.
Rebka, G. A. and Pound, R. V. (1957), *Nature* **180**, 1035.
Rice, S.O. (1944), *Bell Syst. Tech. J.* **23**, 282.
Scarl, D. B. (1966), *Phys. Rev. Lett.* **17**, 663.
Schiff, L. I. (1968), *Quantum Mechanics*, 3rd edn. (McGraw-Hill, New York).
Twiss, R. Q., Little, A. G. and Hanbury Brown, R. (1957), *Nature* **180**, 324.

Chapter 10

Ackerhalt, J. R., Knight, P. L. and Eberly, J. H. (1973), *Phys. Rev. Lett.* **30**, 456.
Amrein, W. O. (1969), *Helv. Phys. Acta* **42**, 149.
Baker, H. F. (1902), *Proc. London Math. Soc.* **34**, 347.
Baker, H. F. (1903), *Proc. London Math. Soc.* **35**, 333.
Bandilla, A., Paul, H. and Ritze, H.-H. (1991), *Quantum Opt.* **3**, 267.
Barnett, S. M. and Pegg, D. T. (1986), *J. Phys. A* **19**, 3849.
Barnett, S. M. and Pegg, D. T. (1989), *J. Mod. Opt.* **36**, 7.
Beth, R. A. (1936), *Phys. Rev.* **50**, 115.
Bethe, H. A. (1947), *Phys. Rev.* **72**, 339.
Campbell, J. E. (1898), *Proc. London Math. Soc.* **29**, 14.
Carruthers, P. and Nieto, M. M. (1965), *Phys. Rev. Lett.* **14**, 387.
Carruthers, P. and Nieto, M. M. (1968), *Rev. Mod. Phys.* **40**, 411.
Casimir, H. B. G. (1949), *J. de Chim. Physique* 46, 407.
Casimir, H. B. G. and Polder, D. (1948), *Phys. Rev.* **73**, 360.
Dalibard, J., Dupont-Roc, J. and Cohen-Tannoudji, C. (1982), *J. de Phys.* **43**, 1617.
Derjaguin, B. V., Abrikosova, I. I. and Lifshitz, E. M. (1956), *Quart. Rev. Chem. Soc.* **10**, 295.

Dirac, P. A. M. (1927), *Proc. Roy. Soc. (London) A* **114**, 243.

Dirac, P. A. M. (1958) *Principles of Quantum Mechanics*, 4th edn. (Oxford University Press, London).

Gottfried, K. (1966) *Quantum Mechanics*, Vol. 1. (Benjamin, New York).

Hausdorff, F. (1906), *Math. Naturwiss.* **58**, 19.

Healy, W. P. (1982) *Non-Relativistic Quantum Electrodynamics.* (Academic Press, London).

Heitler, W. (1954) *The Quantum Theory of Radiation*, 3rd edn. (Oxford University Press, London).

Jackson, J. D. (1975) *Classical Electrodynamics*, 2nd edn. (John Wiley, New York).

Kitchener, J. A. and Prosser, A. P. (1957), *Proc. Roy. Soc. (London) A* **242**, 403.

Lamb, W. E., Jr. and Retherford, R. C. (1947), *Phys. Rev.* **72**, 241.

Ledermann, W., (1944), *Proc. Roy. Soc. (London) A* **182**, 362.

Lenstra, D. and Mandel, L. (1982), *Phys. Rev. A* **26**, 3428.

Lerner, E. C. (1968), *Nuovo Cimento B* **56**, 183.

Louisell, W. H. (1963), *Phys. Lett.* **7**, 60.

Louisell, W. H. (1973), *Quantum Statistical Properties of Radiation* (Wiley, New York).

Lynch, R. (1986), *J. Opt. Soc. Am. B* **3**, 1006.

Lynch, R. (1987), *J. Opt. Soc. Am. B* **4**, 1723.

Messiah, A. (1961), *Quantum Mechanics*, Vol. 1. (North-Holland, Amsterdam).

Milonni, P. W. (1988), *Physica Scripta* **T21**, 102.

Milonni, P. W., Ackerhalt, J. R. and Smith, W. A. (1973), *Phys. Rev. Lett.* **31**, 958.

Newton, T. D. and Wigner, E. P. (1949), *Rev. Mod. Phys.* **21**, 400.

Noh, J. W., Fougères, A. and Mandel, L. (1991), *Phys. Rev. Lett.* **67**, 1426.

Noh, J. W., Fougères, A. and Mandel, L. (1992a), *Phys. Rev. A* **45**, 424.

Noh, J. W., Fougères, A. and Mandel, L. (1992b), *Phys. Rev. A* **46**, 2840.

Noh, J. W., Fougères, A. and Mandel, L. (1993), *Phys. Rev. Lett.* **71**, 2579.

Paul, H. (1974), *Forts. d. Physik* **22**, 657.

Pegg, D. T. and Barnett, S. M. (1988), *Europhys. Lett.* **6**, 483.

Pegg, D. T. and Barnett, S. M. (1989), *Phys. Rev. A* **39**, 1665.

Peierls, R. E. (1954), *Proc. Nat. Inst. of Science of India* **20**, 121.

Power, E. A. (1964), *Introductory Quantum Electrodynamics* (Longmans, Green and Co, London).

Sanders, B. C., Barnett, S. M. and Knight, P. L. (1986), *Opt. Commun.* **58**, 290.

Schleich, W., Horowicz, R. J. and Varro, S. (1989), *Phys. Rev. A* **40**, 7405.

Schubert, M. and Vogel, W. (1978), *Phys. Lett.* **68A**, 321.

Schweber, S. S. (1961), *An Introduction to Relativistic Quantum Field Theory* (Harper & Row, New York).

Senitzky, I. R. (1973), *Phys. Rev. Lett.* **31**, 955.

Series, G. W. (1957), *Spectrum of Atomic Hydrogen* (Oxford University Press, London).

Shapiro, J. H., Shepard, S. R. and Wong, N. C. (1989), *Phys. Rev. Lett.* **62**, 2377.

Simmons, J. W. and Guttmann, M. J. (1970), *States, Waves and Photons*, (Addison-Wesley, Reading, MA).

Stokes, G. G. (1849), *Cambridge & Dublin Math. J.* **4**, 1; Reprinted in *Mathematical and Physical Papers of G. G. Stokes*, Vol. II (Cambridge University Press, Cambridge 1883), pp. 89–103.

Susskind, L. and Glogower, J. (1964), *Physics* **1**, 49.

Turski, L. A. (1972), *Physica* **57**, 432.

Vogel, W. and Schleich, W. (1991), *Phys. Rev. A* **44**, 7642.

Vogel, W. and Welsch, D.-G. (1994), *Lectures on Quantum Optics* (Akademie Verlag, Berlin).

Welton, T. A. (1948), *Phys. Rev.* **74**, 1157.

Zak, J. (1969), *Phys. Rev.* **187**, 1803.

Chapter 11

Agarwal, G. S. and Wolf, E. (1968), *Phys. Rev. Lett.* **21**, 180.

Agarwal, G. S. and Wolf, E. (1970a), *Phys. Rev. D* **2**, 2161.

Agarwal, G. S. and Wolf, E. (1970b), *Phys. Rev. D* **2**, 2187.

Agarwal, G. S. and Wolf, E. (1970c), *Phys. Rev. D* **2**, 2206.

Bargmann, V. (1961), *Comm. Pure and Appl. Math.* **14**, 187.

Bargmann, V. (1962), *Proc. Nat. Acad. Science (U.S.)* **48**, 199.

Bremerman, H. (1965), *Distributions, Complex Variables and Fourier Transforms*, (Addison-Wesley, Reading, MA)

Cahill, K. E. (1965), *Phys. Rev.* **138**, B1566.

Cahill, K. E. and Glauber, R. J. (1969a), *Phys. Rev.* **177**, 1857.

Cahill, K. E. and Glauber, R. J. (1969b), *Phys. Rev.* **177**, 1882.

Glauber, R. J. (1963a), *Phys. Rev.* **130**, 2529.

Glauber, R. J. (1963b), *Phys. Rev.* **131**, 2766.

Glauber, R. J. (1963c), *Phys. Rev. Lett.* **10**, 84.

Glauber, R. J. (1965), in *Quantum Optics and Electronics*, (Les Houches Summer School of Theoretical Physics, University of Grenoble) eds. C. DeWitt, A. Blandin and C. Cohen-Tannoudji (Gordon and Breach, New York), p. 53.

Glauber, R. J. (1970), in *Quantum Optics*, eds. S. M. Kay and A. Maitland (Academic Press, New York), p. 70.

Jordan, T. F. (1964), *Phys. Lett.* **11**, 289.

Kibble, T. W. B. (1968), in *Cargèse Lectures in Physics*, Vol. 2, ed. M. Lèvy (Gordon and Breach, New York), p. 299.

Klauder, J. R. (1960), *Ann. of Phys.* **11**, 123.

Klauder, J. R. (1966), *Phys. Rev. Lett.* **16**, 534.

Klauder, J. R., McKenna, J. and Currie, D. G. (1965), *J. Math. Phys.* **6**, 734.

Klauder, J. R. and Sudarshan, E. C. G. (1968), *Fundamentals of Quantum Optics* (Benjamin, New York).

Lax, M. (1968), *Phys. Rev.* **172**, 350.

Mandel, L. (1965), *Phys. Rev.* **138**, B753.

Mehta, C. L. (1967), *Phys. Rev. Lett.* **18**, 752.

Mehta, C. L. and Sudarshan, E. C. G. (1965), *Phys. Rev.* **138**, B274.

Miller, M. M. and Mishkin, E. A. (1967), *Phys. Rev.* **164**, 1610.

Moyal, J. E. (1949), *Proc. Camb. Phil. Soc.* **45**, 99.

Nussenzveig, H. M. (1972), *Causality and Dispersion Relations* (Academic Press, New York).

Peřina, J. (1985), *Coherence of Light* (Second edition, Reidel, Dordrecht).

Peřina, J. (1991), *Quantum Statistics of Linear and Nonlinear Optical Phenomena*, 2nd edn. (Kluwer, Dordrecht).

Rocca, F. (1966), *Compt. Rend.* **262**, A547.

Schrödinger, E. (1926), *Naturwissenschaften* **14**, 664.

Stoler, D. (1970), *Phys. Rev. D* **1**, 3217.

Stoler, D. (1971), *Phys. Rev. D* **4**, 1925.

Sudarshan, E. C. G. (1963), *Phys. Rev. Lett.* **10**, 277.

Wigner, E. (1932), *Phys. Rev.* **40**, 749.

Wolf, E. (1963), in *Proceedings of Symposium of Optical Masers* (Brooklyn Polytechnic Press, New York, and Wiley, New York), p. 29.

Yuen, H. P. (1976), *Phys. Rev. A* **13**, 2226.

Chapter 12

Acharya, R. and Sudarshan, E. C. G. (1960), *J. Math. Phys.* **1**, 532.

Agarwal, G. S. and Wolf, E. (1970a), *Phys. Rev. D* **2**, 2161.

Agarwal, G. S. and Wolf, E. (1970b), *Phys. Rev. D* **2**, 2187.
Agarwal, G. S. and Wolf, E. (1970c), *Phys. Rev. D* **2**, 2206.
Amrein, W. O. (1969), *Helv. Phys. Acta.* **42**, 149.
Aspect, A., Dalibard, J. and Roger, G. (1982), *Phys. Rev. Lett.* **49**, 1804.
Aspect, A., Grangier, P. and Roger, G. (1981), *Phys. Rev. Lett.* **47**, 460.
Aspect, A., Grangier, P. and Roger, G. (1982), *Phys. Rev. Lett.* **49**, 91.
Basov, N. G., Krokhin, O. N. and Popov, Yu. M. (1960), *Usp. Fiz. Nauk.* **72**, 161, [English Translation in *Soviet Phys. Usp.* **3**, 702 (1961)].
Bell, J. S. (1964), *Physics* **1**, 195.
Bell, J. S. (1966), *Rev. Mod. Phys.* **38**, 447.
Bloembergen, N. (1959), *Phys. Rev. Lett.* **2**, 84.
Bloembergen, N. (1965), *Nonlinear Optics* (Benjamin, New York).
Bohr, N. (1935), *Phys. Rev.* **48**, 696.
Bohm, D. (1952a), *Phys. Rev.* **85**, 166.
Bohm, D. (1952b), *Phys. Rev.* **85**, 180.
Bohm, D. and Aharanov, Y. (1957), *Phys. Rev.* **108**, 1070.
Born, M. and Wolf, E. (1980), *Principles of Optics*, 6th edn (Pergamon Press, Oxford).
Boyd, R. W. (1992) *Nonlinear Optics* (Academic Press, Boston).
Bracewell, R. N. (1978), *The Fourier Transform and its Applications*, 2nd edn (McGraw Hill, New York).
Cahill, K. E. and Glauber, R. J. (1969a), *Phys. Rev.* **177**, 1857.
Cahill, K. E. and Glauber, R. J. (1969b), *Phys. Rev.* **177**, 1882.
Campos, R. A., Saleh, B. E. A. and Teich, M. C. (1989), *Phys. Rev. A* **40**, 1371.
Clauser, J. F. (1976), *Phys. Rev. Lett.* **36**, 1223.
Clauser, J. F. and Horne, M. A. (1974), *Phys. Rev. D* **10**, 526.
Clauser, J. F. and Shimony, A. (1978), *Rep. Progr. Phys. (London)* **41**, 1881.
Clauser, J. F., Horne, M. A., Shimony, A. and Holt, R. A. (1969), *Phys. Rev. Lett.* **23**, 880.
Cook, R. J. (1982), *Phys. Rev. A* **25**, 2164.
Cook, R. J. (1984), in *Coherence and Quantum Optics V*, eds. L. Mandel and E. Wolf (Plenum Press, New York) p. 539.
d'Espagnat, B. (1979), *Scientific American* (Nov.) **241**, 158.
Dialetis, D. and Mehta, C. L. (1968), *Nuovo Cimento* **56B**, 89.
Drummond, P. D. and Friberg, A. T. (1983), *J. Appl. Phys.* **54**, 5618.
Eberly, J. H. and Kujawski, A. (1967), *Phys. Lett.* **24A**, 426.
Einstein, A. (1909), *Physik Z.* **10**, 185; English translation *in The Collected Papers of Albert Einstein*, Vol. 2 (Princeton University Press, Princeton, NJ, 1989), p. 357.
Einstein, A., Podolsky, B. and Rosen, N. (1935), *Phys. Rev.* **47**, 777.
Fearn, H. and Loudon, R. (1987), *Opt. Commun.* **64**, 485.
Freedman, S. J. and Clauser, J. R. (1972), *Phys. Rev. Lett.* **28**, 938.
Friberg, A. T. and Drummond, P. D. (1983), *J. Opt. Soc. Am.* **73**, 1216.
Fry, E. S. and Thompson, R. C. (1976), *Phys. Rev. Lett.* **37**, 465.
Ghielmetti, F. (1964), *Phys. Lett.* **12**, 210.
Glauber, R. J. (1963a), *Phys. Rev.* **130**, 2529.
Glauber, R. J. (1963b), *Phys. Rev.* **131**, 2766.
Hegerfeldt, G. C. (1974), *Phys. Rev. D* **10**, 3320.
Hegerfeldt, G. C. and Ruijsenaar, S, N. M. (1980), *Phys. Rev. D* **22**, 377.
Hillery, M. (1985), *Phys. Rev. A* **31**, 338.
Hong, C. K., Ou, Z. Y. and Mandel, L. (1987), *Phys. Rev. Lett.* **59**, 2044.
Jauch, J. M. and Piron, C. (1967), *Helv. Phys. Acta* **40**, 559.
Kano, Y. (1964), *Ann. of Phys.* **30**, 127.
Kano, Y. (1965), *J. Math. Phys.* **6**, 1913.
Kiess, T. E., Shih, Y. H., Sergienko, A. V. and Alley, C. O. (1993), *Phys. Rev. Lett.* **71**, 3893.

Klauder, R. J. and Sudarshan, E. C. G. (1968), *Fundamentals of Quantum Optics* (Benjamin, New York).
Lenstra, D. and Mandel, L. (1982), *Phys. Rev. A* **26**, 3428.
Ley, M. and Loudon, R. (1985), *Opt. Commun.* **54**, 317.
Louisell, W. H. (1969), in *Quantum Optics*, Proc. Internat. School of Physics, 'Enrico Fermi', Course XLII, Varenna, ed. R. J. Glauber (Academic Press, New York), p. 680.
Mandel, L. (1966a), *Phys. Rev.* **144**, 1071.
Mandel, L. (1966b), *Phys. Rev.* **152**, 438.
Mandel, L. (1979), *Opt. Lett.* **4**, 205.
Mandel, L. and Mehta, C. L. (1969), *Nuovo Cimento* **61B**, 149.
Mandel, L. and Meltzer, D. (1969), *Phys. Rev.* **188**, 198.
Mandel, L., Sudarshan, E. C. G. and Wolf, E. (1964), *Proc. Phys. Soc. (London)* **84**, 435.
Mehta, C. L. (1966), *Nuovo Cimento* **45**, 280.
Mehta, C. L. (1967), *J. Math. Phys.* **8**, 1798.
Mehta, C. L. and Mandel, L. (1967), in *Electromagnetic Wave Theory*, Part 2, ed. J. Brown (Pergamon Press, Oxford), p. 1069.
Mehta, C. L. and Sudarshan, E. C. G. (1965), *Phys. Rev.* **138**, B274.
Mermin, N. D. (1981), *Am. J. Phys.* **49**, 940.
Mermin, N. D. (1985), *Physics Today* (April) **38**, 38.
Newton, T. D. and Wigner, E. P. (1949), *Rev. Mod. Phys.* **21**, 400.
Nieto-Vesperinas, M. and Wolf, E. (1986), *J. Opt. Soc. Am. A* **3**, 2038.
Ou, Z. Y. and Mandel, L. (1988), *Phys. Rev. Lett.* **61**, 50.
Ou, Z. Y. and Mandel, L. (1989), *Am. J. Phys.* **57**, 66.
Ou, Z. Y., Hong, C. K. and Mandel, L. (1987), *Opt. Commun.* **63**, 118.
Pike, E. R. and Sarkar, S. (1986), in *Frontiers in Quantum Optics*, eds. E. R. Pike and S. Sarkar (Adam Hilger, Bristol), p. 282.
Reid, M. D. and Walls, D. F. (1986), *Phys. Rev. A* **34**, 1260.
Shen, Y. R. (1984), *The Principles of Nonlinear Optics* (Wiley, New York).
Shih, Y. H. and Alley, C. O. (1988), *Phys. Rev. Lett.* **61**, 2921.
Short, R. and Mandel, L. (1983), *Phys. Rev. Lett.* **51**, 384.
Simmons, J. W. and Guttmann, M. J. (1970), *States, Waves and Photons* (Addison-Wesley, Reading, MA).
Stokes, G. G. (1849), *Cambridge & Dublin Math. J.* **4**, 1; Reprinted in *Mathematical and Physical Papers of G. G. Stokes*, Vol. II (Cambridge University Press, Cambridge, 1883), pp. 89–103.
Su, C. and Wódkiewicz, K. (1991), *Phys. Rev. A* **44**, 6097.
Sudarshan, E. C. G. (1969), *J. Math. and Phys. Science (Madras, India)* **3**, 121.
Titulaer, U. M. and Glauber, R. J. (1965), *Phys. Rev.* **140**, B676.
Titulaer, U. M. and Glauber, R. J. (1966), *Phys. Rev.* **145**, 1041.
Vašíček, A. (1960), *Optics of Thin Films* (North-Holland, Amsterdam).
Wigner, E. P. (1970), *Am. J. Phys.* **38**, 1005.
Yuen, H. P. and Shapiro, J. H. (1980), *IEEE Trans. Inf. Theory* **IT-26**, 78.

Chapter 13

Bourret, R. C. (1960), *Nuovo Cimento* **18**, 347.
Halliday, D. and Resnick, R. (1970), *Fundamentals of Physics*. (Wiley, New York).
Hurwitz, H. (1945), *J. Opt. Soc. Am.* **35**, 525.
Kano, Y. and Wolf, E. (1962), *Proc. Phys. Soc. (London)* **80**, 1273.
Mandel, L. (1959), *Proc. Phys. Soc. (London)* **74**, 233.
Mandel, L. (1963), *Proc. Phys. Soc. (London* **81**, 1104.
Mandel, L. (1979), *J. Opt. Soc. Am.* **69**, 1038.

Mehta, C. L. and Wolf, E. (1964a), *Phys. Rev.* **134**, A1143.
Mehta, C. L. and Wolf, E. (1964b), *Phys. Rev.* **134**, A1149.
Planck, M. (1901a), *Ann. d. Physik* **4**, 553.
Planck, M. (1901b), *Ann. d. Physik* **4**, 564.
Sarfatt, J. (1963), *Nuovo Cimento* **27**, 1119.

Chapter 14

Abate, J. A., Kimble, H. J. and Mandel, L. (1976), *Phys. Rev. A* **14**, 788.
Ackerhalt, J. R. and Milonni, P. W. (1984). *J. Opt. Soc. Am. B* **1**, 116.
Ádám, A., Jánossy, L. and Varga, P. (1955a), *Acta Phys. Acad. Sci. Hungarica* **4**, 301.
Ádám, A., Jánossy, L. and Varga, P. (1955b), *Ann. d. Physik* **16**, 408.
Arecchi, F. T. (1965), *Phys. Rev. Lett.* **15**, 912.
Arecchi, F. T. (1969), in *Quantum Optics*, Proc. Internat. School of Physics 'Enrico Fermi', Course XLII, Varenna, ed. R. J. Glauber (Academic Press, New York), p. 57.
Arecchi, F. T., Gatti, E. and Sona, A. (1966), *Phys. Lett.* **20**, 27.
Bédard, G., Chang, J. C. and Mandel, L. (1967), *Phys. Rev.* **160**, 1496.
Brannen, E. and Ferguson, H. I. S. (1956), *Nature* **178**, 481.
Brannen, E., Ferguson, H. I. S. and Wehlau, W. (1958), *Can J. Phys.* **36**, 871.
Brown, R. Hanbury and Twiss, R. Q. (1956), *Nature*, **177**, 27.
Brown, R. Hanbury and Twiss, R. Q. (1957), *Proc. Roy, Soc. (London) A* **243**, 291.
Bures, J., Delisle, C. and Zardecki, A. (1971), *Can J. Phys.* **49**, 3064.
Bures, J., Delisle, C. and Zardecki, A. (1972), *Can J. Phys.* **50**, 1307.
Carmichael, H. J. and Walls, D. F. (1976a), *J. Phys. B* **9**, L43.
Carmichael, H. J. and Walls, D. F. (1976b), *J. Phys. B* **9**, 1199.
Carter, S. L. and Kelley, H. P. (1979), *Phys. Rev. Lett.* **42**, 966.
Chopra, S. and Mandel, L. (1972), *Rev. Sci. Instrum.* **43**, 1489.
Chopra, S. and Mandel, L. (1973a), in *Coherence and Quantum Optics*, eds. L. Mandel and E. Wolf (Plenum Press, New York), p. 805.
Chopra, S. and Mandel, L. (1973b), *Phys. Rev. Lett.* **30**, 60.
Chu, B. (1974), *Laser Light Scattering* (Academic Press, New York).
Cook, R. J. (1981), *Phys. Rev. A* **23**, 1243.
Corti, M. and Degiorgio, V. (1976), *Phys. Rev. A* **14**, 1475.
Corti, M., Degiorgio, V. and Arecchi, F. T. (1973), *Opt. Commun.* **8**, 329.
Cresser, J. D., Häger, J., Leuchs, G., Rateike, M. and Walther, H. (1982), in *Dissipative Systems in Quantum Optics*, ed. R. Bonifacio (Springer, Berlin), p. 21.
Crosignani, B., DiPorto, P. and Bertolotti, M. (1975), *Statistical Properties of Scattered Light* (Academic Press, New York).
Cummins, H. Z. and Pike, E. R. (1977), *Photon Correlation Spectroscopy and Velocimetry* (Plenum Press, New York).
Cummins, H. Z. and Swinney, H. L. (1970), in *Progress in Optics*, Vol. 8, ed. E. Wolf (North-Holland, Amsterdam), p. 133.
Dagenais, M. and Mandel, L. (1978), *Phys. Rev. A* **18**, 2217.
Davidson, F. (1969), *Phys. Rev.* **185**, 446.
Davidson, F. and Mandel, L. (1967), *Phys. Rev.* **25 A**, 700.
Davidson, F. and Mandel, L. (1968), *J. Appl. Phys.* **39**, 62.
Einstein, A. (1909), *Zeits. f. Physik*, **10**, 185 [English translation in *The Collected Papers of Albert Einstein*, Vol. 2 (Princeton University Press, Princeton, NJ, 1989), p. 379].
Fried, Z. (1973), *Phys. Rev. A* **8**, 2835.
Fürth, R. (1928a), *Zeits. f. Physik* **48**, 323.
Fürth, R. (1928b), *Zeits. f. Physik* **50**, 310.
Gamo, H. (1963a), *J. Appl. Phys.* **34**, 875.

Gamo, H. (1963b), in *Electromagnetic Theory and Antennas*, Part 2, ed. E. C. Jordan (Pergamon Press, Oxford), p. 801.

Glauber, R. J. (1963), *Phys. Rev.* **130**, 2529.

Glauber, R. J. (1965), *Quantum Optics and Electronics*, (Les Houches Summer School of Theoretical Physics, University of Grenoble), eds. C. DeWitt, A. Blandin and C. Cohen-Tannoudji (Gordon and Breach, New York), p. 53.

Glauber, R. J. (1967), in *Modern Optics*, ed. J. Fox (Brooklyn Polytechnic Press and Interscience, New York), p. 1.

Grangier, P., Roger, G. and Aspect, A. (1986), *Europhys. Lett.* **1**, 173.

Grangier, P., Roger, G., Aspect, A., Heidmann, A. and Reynaud, S. (1986), *Phys. Rev. Lett.* **57**, 687.

Haller, K. (1982), *Phys. Rev. A* **26**, 1796.

Healy, W. P. (1980), *Phys. Rev. A* **22**, 2891.

Healy, W. P. (1982), *Phys. Rev. A* **26**, 1798.

Hong, C. K. and Mandel, L. (1986), *Phys. Rev. Lett.* **56**, 58.

Jakeman, E. (1970), *J. Phys. A* **3**, 201.

Jakeman, E. and Pike, E. R. (1969), *J. Phys. A* **2**, 411.

Kelley, P. L. and Kleiner, W. H. (1964), *Phys. Rev.* **136**, A316.

Kimble, H. J. and Mandel, L. (1976), *Phys. Rev. A* **13**, 2123.

Kimble, H. J. and Mandel, L. (1984), *Phys. Rev. A* **30**, 844.

Kimble, H. J., Dagenais, M. and Mandel, L. (1977), *Phys. Rev. Lett.* **39**, 691.

Kobe, D. H. (1978), *Phys. Rev. Lett.* **40**, 538.

Kobe, D. H. (1979), *Phys. Rev. A* **19**, 205.

Lamb, W. E., Jr. (1952), *Phys. Rev.* **85**, 259.

Lenstra, D. (1982), *Phys. Rev. A* **26**, 3369.

Machida, S. and Yamamoto, Y. (1986), *Opt. Commun.* **57**, 290.

Mandel, L. (1958), *Proc. Phys. Soc. (London)*, **72**, 1037.

Mandel, L. (1959), *Proc. Phys. Soc. (London)*, **74**, 233.

Mandel, L. (1963a), in *Progress in Optics*, Vol. 2, ed. E. Wolf (North-Holland, Amsterdam), p. 181.

Mandel, L. (1963b), *Proc. Phys. Soc. (London)*, **81**, 1104.

Mandel, L. (1967), in *Modern Optics*, ed. J. Fox (Brooklyn Polytechnic Press and Interscience, New York), p. 143.

Mandel, L. (1979), *Phys. Rev. A* **20**, 1590.

Mandel, L. (1981), *Opt. Acta.* **28**, 1447.

Mandel, L. and Meltzer, D. (1969), *Phys. Rev.* **188**, 198.

Mandel, L., Sudarshan, E. C. G. and Wolf, E. (1964), *Proc. Phys. Soc. (London)*, **84**, 435.

Mandel, L. and Wolf, E. (1965), *Rev. Mod. Phys.* **37**, 231.

Mehta, C. L. (1970), in *Progress in Optics*, Vol. 8, ed. E. Wolf (North-Holland, Amsterdam), p. 373.

Meltzer, D. and Mandel, L. (1970), *IEEE J. Quant. Electron.* **QE-6**, 661.

Milonni, P. W., Cook, R. J. and Ackerhalt, J. R. (1989), *Phys. Rev. A* **40**, 3764.

Mollow, B. R. (1968), *Phys. Rev.* **168**, 1896.

Morgan, B. L. and Mandel, L. (1966), *Phys. Rev. Lett.* **16**, 1012.

Peřina, J. (1970), in *Quantum Optics*, eds. S. M. Kay and A. Maitland (Academic Press, New York), p. 513.

Phillips, D. T., Kleiman, H. and Davis, S. P. (1967), *Phys. Rev.* **153**, 113.

Pike, E. R. (1986) in *Coherence, Cooperation and Fluctuations*, eds. F. Haake, L. M. Narducci and D. F. Walls (Cambridge University Press, Cambridge), p. 293.

Power, E. A. and Thirunamachandran, T. (1980), *Phys. Rev. A* **22**, 2894.

Power, E. A. and Thirunamachandran, T. (1982), *Phys. Rev. A* **26**, 1800.

Power, E. A. and Zienau, S. (1959), *Phil. Trans. Roy. Soc. London* **251**, 427.

Purcell, E. M. (1956), *Nature*, **178**, 1449.
Rarity, J. G., Tapster, P. R. and Jakeman, E. (1987), *Opt. Commun.* **62**, 201.
Rebka, G. A. and Pound, R. V. (1957), *Nature* **180**, 1035.
Saleh, B. (1978), *Photoelectron Statistics*, (Springer, Berlin).
Saleh, B. E. A. and Teich, M. C. (1985), *Opt. Commun.* **52**, 429.
Scarl, D. B. (1966), *Phys. Rev. Lett.* **17**, 663.
Singh, S. (1983), *Opt. Commun.* **44**, 254.
Short, R. and Mandel, L. (1983), *Phys. Rev. Lett.* **51**, 384.
Short, R. and Mandel, L. (1984), in *Coherence and Quantum Optics V*, eds. L. Mandel, and E. Wolf (Plenum Press, New York), p. 671.
Srinivas, M. D. and Davis, E. B. (1981), *Opt. Acta* **28**, 981.
Swinney, H. L. (1983), *Physica D* **7**, 3.
Teich, M. C. and Saleh, B. E. A. (1985), *J. Opt. Soc. Am. B* **2**, 275.
Teich, M. C. and Saleh, B. E. A. (1988), in *Progress in Optics*, Vol. 26, ed. E. Wolf, (North-Holland, Amsterdam), p. 1.
Teich, M. C., Saleh, B. E. A. and Peřina, J. (1984), *J. Opt. Soc. Am. B* **1**, 366.
Twiss, R. Q., Little, A. G. and Brown, R. Hanbury (1957), *Nature* **180**, 324.
Walker, J. G. and Jakeman, E. (1985), *Opt. Acta* **32**, 1303.
Wolf, E. and Mehta, C. L. (1964), *Phys. Rev. Lett.* **13**, 705.
Woolley, R. G. (1971), *Mol. Phys.* **22**, 1013.
Yurke, B. (1985), *Phys. Rev. A* **32**, 311.
Zardecki, A. and Delisle, C. (1973), *Can J. Phys.* **51**, 1017.

Chapter 15

Abate, J. A. (1974), *Opt. Commun.* **10**, 269.
Ackerhalt, J. R., Knight, P. L. and Eberly, J. H. (1973), *Phys. Rev. Lett.* **30**, 456.
Agarwal, G. S. (1976), *Phys. Rev. Lett.* **37**, 1383.
Agarwal, G. S. (1977), *Phys. Rev. A* **15**, 814.
Agarwal, G. S. and Saxena, R. (1978), *Opt. Commun.* **26**, 202.
Allen, L. and Eberly, J. H. (1975), *Optical Resonance and Two-Level Atoms* (Wiley, New York).
Arimondo, E., Lew, H. and Oka, T. (1979), *Phys. Rev. Lett.* **43**, 753.
Arimondo, E., Phillips, W. D. and Strumia, F. (1992) eds. *Laser Manipulation of Atoms and Ions*; Proc. Internat. School of Physics 'Enrico Fermi', Course CXVIII, Varenna (North-Holland, Amsterdam).
Ashkin, A. (1978), *Phys. Rev. Lett.* **40**, 729.
Ashkin, A. and Gordon, J. P. (1979), *Opt. Lett.* **4**, 161.
Aspect, A., Dalibard, J., Heidmann, A., Salomon, C. and Cohen-Tannoudji, C. (1986), *Phys. Rev. Lett.* **57**, 1688.
Baklanov, E. V. and Dubetskii, B. Ya. (1976), *Opt. Spectrosk.* **41**, 3 [English Translation in *Opt. and Spectroscopy* **41**, 1 (1976)].
Balykin, V. I., Letokhov, V. S. and Mushin, V. I. (1979), *Pis'ma Zh. Eksp. Teor. Fiz.* **29**, 614 [English translation in *JETP Lett.* **29**, 560 (1979)].
Bethe, H. A. (1947), *Phys. Rev.* **72**, 339.
Bigelow, N. P. and Prentiss, M. G. (1990). *Phys. Rev. Lett.* **65**, 29.
Bjorkholm, J. E., Freemen, R. R., Ashkin, A. and Pearson, D. B. (1978), *Phys. Rev. Lett.* **41**, 1361.
Bjorkholm, J. E., Freemen, R. R., Ashkin, A. and Pearson, D. B. (1980), *Opt. Lett.* **5**, 111.
Bloch, F. (1946), *Phys. Rev.* **70**, 460.
Born, M. and Wolf, E. (1980), *Principles of Optics*, 6th edn. (Pergamon Press, London).
Boyer, T. H. (1969a), *Phys. Rev.* **180**, 19.

Boyer, T. H. (1969b), *Phys. Rev.* **182**, 1374.
Boyer, T. H. (1970a), *Phys. Rev. D* **1**, 1526.
Boyer, T. H. (1970b), *Phys. Rev. D* **1**, 2257.
Boyer, T. H. (1973), *Phys. Rev. A* **7**, 1832.
Boyer, T. H. (1974), *Phys. Rev. A* **9**, 2078.
Carmichael, H. J. and Walls, D. F. (1976a), *J. Phys. B* **9**, L43.
Carmichael, H. J. and Walls, D. F. (1976b), *J. Phys. B* **9**, 1199.
Chu, S., Bjorkholm, J. E., Ashkin, A. and Cable, A. (1986), *Phys. Rev. Lett.* **57**, 314.
Chu, S., Hollberg, L., Bjorkholm, J., Cable, A. and Ashkin, A. (1985), *Phys. Rev. Lett.* **55**, 48.
Chu, S., Prentiss, M. G., Cable, A. E. and Bjorkholm, J. E. (1987), in *Laser Spectroscopy*, Vol. 8, eds. W. Persson and S. Svanberg (Springer, Berlin).
Cohen-Tannoudji, C. (1992), in *Fundamental Systems in Quantum Optics*, eds. J.-M. Dalibard, J.-M. Raimond, and J. Zinn-Justin (North-Holland, Amsterdam), Course 1.
Cohen-Tannoudji, C. and Phillips, W. D. (October, 1990), *Physics Today* **43**, October, 33.
Cook, R. J. (1978), *Phys. Rev. Lett.* **41**, 1788.
Cook, R. J. (1979), *Phys. Rev. A* **20**, 224.
Cook, R. J. (1980a), *Phys. Rev. Lett.* **44**, 976.
Cook, R. J. (1980b), *Phys. Rev. Lett.* **22**, 1078.
Cook, R. J. and Bernhardt, A. F. (1978), *Phys. Rev. A* **18**, 2533.
Crisp, M. D. and Jaynes, E. T. (1969), *Phys. Rev.* **179**, 1253.
Dagenais, M. and Mandel, L. (1978), *Phys. Rev. A* **18**, 2217.
Dalibard, J. and Cohen-Tannoudji, C. (1985), *J. Opt. Soc. Am. B* **2**, 1707.
Dalibard, J. and Cohen-Tannoudji, C. (1989), *J. Opt. Soc. Am. B* **6**, 2023.
Dalibard, J., Raimond, J. M. and Zinn-Justin, J. (1992), eds. *Fundamental Systems in Quantum Optics* (North-Holland, Amsterdam).
Dicke, R. H. (1954), *Phys. Rev.* **93**, 99.
Eberly, J. H. (1976), *Phys. Rev. Lett.* **37**, 1387.
Feynman, R. P., Vernon, F. L., Jr., and Hellwarth, R. W. (1957), *J. Appl. Phys.* **28**, 49.
Freedman, S. J. and Clauser, J. F. (1972), *Phys. Rev. Lett.* **28**, 938.
Frisch, R. (1933), *Zeits. f. Physik* **86**, 42.
Gibbs, H. M. (1972), *Phys. Rev. Lett.* **29**, 459.
Gibbs, H. M. (1973), *Phys. Rev. A* **8**, 456.
Gibbs, H. M. and Venkatesan, T. N. C. (1976), *Opt. Commun.* **17**, 87.
Gordon, J. P. and Ashkin, A. (1980), *Phys. Rev. A* **21**, 1606.
Grischkowsky, D. (1970), *Phys. Rev. Lett.* **24**, 866.
Grischkowsky, D., Courtens, E. and Armstrong, J. A. (1973), *Phys. Rev. Lett.* **31**, 422.
Hartig, W., Rasmussen, W., Schieder, R. and Walther, H. (1976), *Zeits. f. Physik A* **278**, 205.
Hänsch, T. W. and Schawlow, A. L. (1975), *Opt. Commun.* **13**, 68.
Jaynes, E. T. (1973) in *Coherence and Quantum Optics*, eds. L. Mandel and E. Wolf. (Plenum Press, New York), p. 35.
Jaynes, E. T. and Cummings, F. W. (1963), *Proc. IEEE* **51**, 89.
Kazantsev, A. P. (1976), *Opt. Commun.* **17**, 166.
Kazantsev, A. P. (1978), *Usp. Fiz. Nauk* **124**, 113 [English Translation in *Sov. Phys. Usp.* **21**, 58 (1978)].
Kazantsev, A. P., Ryabenko, G. A., Surdutovich, G. I., and Yakovlev, V. P. (1985), *Phys. Rep.* **129**, 76.
Kimble, H. J. and Mandel, L. (1976), *Phys. Rev. A* **13**, 2123.
Kimble, H. J. and Mandel, L. (1977), *Phys. Rev. A* **15**, 689.
Kimble, H. J., Dagenais, M. and Mandel, L. (1977), *Phys. Rev. Lett.* **39**, 691.
Kimble, H. J., Dagenais, M. and Mandel, L. (1978), *Phys. Rev. A* **18**, 201.
Lam, J. F. and Berman, P. R. (1976), *Phys. Rev. A* **14**, 1683.

Lamb, W. E., Jr., and Retherford, R. C. (1947), *Phys. Rev.* **72**, 241.
Letokhov, V. S. (1968), *Pis'ma Zh. Eksp. Teor. Fiz.* **7**, 348 [English Translation in *JETP Letters* **7**, 272 (1968)].
Letokhov, V. S. and Minogin, V. G. (1979), *J. Opt. Soc. Am.* **69**, 413.
Letokhov, V. S. and Pavlik, B. D. (1976), *Appl. Phys.* **9**, 229.
Letokhov, V. S., Minogin, V. G. and Pavlik, B. D. (1976), *Opt. Commun.* **19**, 72.
Letokhov, V. S., Minogin, V. G. and Pavlik, B. D. (1977), *Zh. Eksp. Teor. Fiz.* **72**, 1328 [English Translation in *Sov. Phys. JETP* **45**, 698 (1977)].
Lett, P. D., Phillips, W. D., Rolston, S. L., Tanner, C. E., Watts, R. N. and Westbrook, C. I. (1989), *J. Opt. Soc. Am. B* **6**, 2084.
Lett, P. D., Watts, R. N., Westbrook, C. I., Phillips, W. D., Gould, P. L. and Metcalf, H. J. (1988), *Phys. Rev. Lett.* **61**, 169.
Loy, M. M. T. (1974), *Phys. Rev. Lett.* **32**, 814.
Mandel, L. (1979a), *Opt. Lett.* **4**, 205.
Mandel, L. (1979b), *J. Optics (Paris)* **10**, 51.
Marshall, T. W. (1963), *Proc. Roy. Soc. (London) A* **276**, 475.
Marshall, T. W. (1965), *Nuovo Cimento* **38**, 206.
Martin, P. J., Gould, P. L., Oldaker, B. G., Miklich, A. H. and Pritchard, D. E. (1987), *Phys. Rev. A* **36**, 2495.
McCall, S. L. and Hahn, E. L. (1967), *Phys. Rev. Lett.* **18**, 908.
McCall, S. L. and Hahn, E. L. (1969), *Phys. Rev.* **183**, 457.
Milonni, P. W. (1974), *Theoretical Aspects of Spontaneous Emission from Atoms*, Ph.D. Thesis (University of Rochester, Rochester, NY).
Mollow, B. R. (1969), *Phys. Rev.* **188**, 1969.
Mollow, B. R. (1975), *J. Phys. A* **8**, L130.
Phillips, W. D. and Metcalf, H. (1982), *Phys. Rev. Lett.* **48**, 596.
Phillips, W. D., Gould, P. L. and Lett, P. D. (1988), *Science* **239**, 877.
Phillips, W. D., Prodan, J. V. and Metcalf, H. J. (1985), *J. Opt. Soc. Am. B* **2**, 1751.
Picqué, J.-L. and Vialle, J.-L. (1972), *Opt. Commun.* **5**, 402.
Prentiss, M. and Cable, A. (1989), *Phys. Rev. Lett.* **62**, 1354.
Prodan, J. V., Phillips, W. D. and Metcalf, H. (1982), *Phys. Rev. Lett.* **49**, 1149.
Raab, E. L., Prentiss, M., Cable, A., Chu, S. and Pritchard, D. E. (1987), *Phys. Rev. Lett.* **59**, 2631.
Rabi, I. I. (1937), *Phys. Rev.* **51**, 652.
Renaud, B., Whitley, R. M. and Stroud, C. R., Jr., (1976), *J. Phys. B* **9**, L19.
Salomon, C., Dalibard, J., Aspect, A., Metcalf, H. and Cohen-Tannoudji, C. (1987), *Phys. Rev. Lett.* **59**, 1659.
Schieder, R., Walther, H. and Wöste, L. (1972), *Opt. Commun.* **5**, 337.
Schuda, F., Hercher, M. and Stroud, C. R., Jr. (1973), *Appl. Phys. Lett.* **22**, 360.
Schuda, F., Stroud, C. R., Jr., and Hercher, M. (1974), *J. Phys. B* **7**, L198.
Short, R. and Mandel, L. (1983a), *Phys. Rev. Lett.* **51**, 384.
Short, R. and Mandel, L. (1983b), in *Coherence and Quantum Optics V*, eds. L. Mandel and E. Wolf. (Plenum Press, New York), p. 671.
Stenholm, S. (1978), *Appl. Phys.* **15**, 287.
Stroud, C. R., Jr. and Jaynes, E. T. (1970), *Phys. Rev. A* **1**, 106.
Tam, A. C. and Happer, W. (1977), *Phys. Rev. Lett.* **38**, 278.
Teich, M. C. and Saleh, B. E. A. (1988), in *Progress in Optics*, Vol. 26, ed. E. Wolf. (North-Holland, Amsterdam), p. 1.
Teich, M. C. and Saleh, B. E. A. (June, 1990), *Physics Today* **43**, 26.
Torrey, H. C. (1949), *Phys. Rev.* **76**, 1059.
Treacy, E. B. and DeMaria, A. J. (1969), *Phys. Lett.* **29A**, 369.
Ungar, P. J., Weiss, D. S., Riis, E. and Chu, S. (1989), *J. Opt. Soc. Am. B* **6**, 2058.
Walther, H. (1977), *Phys. Bl.* **33**, 653.

Weiss, D. S., Riis, E., Shevy, Y., Ungar, P. J. and Chu, S. (1989), *J. Opt. Soc. Am. B* **6**, 2072.
Wessner, J. M., Anderson, D. K. and Robiscoe, R. T. (1972), *Phys. Rev. Lett.* **29**, 1126.
Westbrook, C. I., Watts, R. N., Tanner, C. E., Rolston, S. L., Phillips, W. D., Lett, P. D. and Gould P. L. (1990), *Phys. Rev. Lett.* **65**, 33.
Wineland, D. and Dehmelt, H. (1975), *Bull. Am. Phys. Soc.* **20**, 637.
Wódkiewicz, K. and Eberly, J. H. (1976), *Ann. Phys.* **101**, 574.
Wu, F. Y., Grove, R. E. and Ezekiel, S. (1975), *Phys. Rev. Lett.* **35**, 1426.

Chapter 16

Abella, I. D., Kurnit, N. A. and Hartmann, S. R. (1966), *Phys. Rev.* **141**, 391.
Agarwal, G. S. (1971), *Phys. Rev. A* **4**, 1791.
Agarwal, G. S., Brown, A. C., Narducci, L. M. and Vetri, G. (1977), *Phys. Rev. A* **15**, 1613.
Arecchi, F. T. and Courtens, E. (1970), *Phys. Rev. A* **2**, 1730.
Arecchi, F. T., Courtens, E., Gilmore, R. and Thomas, H. (1972), *Phys. Rev. A* **6**, 2211.
Asher, I. M. and Scully, M. O. (1971), *Opt. Commun.* **3**, 395.
Bonifacio, R. and Lugiato, L. A. (1975a), *Phys. Rev. A* **11**, 1507.
Bonifacio, R. and Lugiato, L. A. (1975b), *Phys. Rev. A* **12**, 587.
Bonifacio, R. and Lugiato, L. A. (1978a), *Phys. Rev. Lett.* **40**, 1023.
Bonifacio, R. and Lugiato, L. A. (1978b), *Lett. Nuovo Cimento* **21**, 505.
Bonifacio, R. and Lugiato, L. A. (1978c), *Lett. Nuovo Cimento* **21**, 510.
Bonifacio, R., Gronchi, M. and Lugiato, L. A. (1978), *Phys. Rev. A* **18**, 2266.
Bonifacio, R., Schwendimann, P. and Haake, F. (1971a), *Phys. Rev. A* **4**, 302.
Bonifacio, R., Schwendimann, P. and Haake, F. (1971b), *Phys. Rev. A* **4**, 854.
Born, M. and Wolf, E. (1980), *Principles of Optics*, 6th edn. (Pergamon Press, Oxford).
Bowden, C. M. and Sung, C. C. (1979), *Phys. Rev. A* **19**, 2392.
Brewer, R. G. and Shoemaker, R. L. (1972), *Phys. Rev. A* **6**, 2001.
Dicke, R. H. (1954), *Phys. Rev.* **93**, 99.
Drexhage, K. H. (1970), *J. Luminescence* **1**, 693.
Eberly, J. H. (1971), *Lett. Nuovo Cimento* **1**, 182.
Flusberg, A., Mossberg, T. and Hartmann, S. R. (1976), *Phys. Lett.* **58A**, 373.
Friedberg, R. and Hartmann, S. R. (1974), *Phys. Rev. A* **10**, 1728.
Gibbs, H. M., McCall, S. L. and Venkatesan, T. N. C. (1976), *Phys. Rev. Lett.* **36**, 1135.
Gibbs, H. M., McCall, S. L. and Venkatesan, T. N. C. (1978), in *Coherence Spectroscopy and Modern Physics*, ed. F. T. Arecchi, R. Bonifacio, and M. O. Scully (Plenum Press, New York), p. 111.
Gibbs, H. M., Vrehen, Q. H. F. and Hikspoors, H. M. J. (1977), *Phys. Rev. Lett.* **39**, 547.
Gibbs, H. M., Hopf, F. A., Kaplan, D. L. and Shoemaker, R. L. (1981), *Phys. Rev. Lett.* **46**, 474.
Gross, M., Fabre, C., Pillet, P. and Haroche, S. (1976), *Phys. Rev. Lett.* **36**, 1035.
Haake, F., Haus, J., King, H., Schröder, G. and Glauber, R. J. (1980), *Phys. Rev. Lett.* **45**, 558.
Haake, F., King, H., Schröder, G., Haus, J. and Glauber, R. J. (1979), *Phys. Rev. A* **20**, 2047.
Hahn, E. L. (1950a), *Phys. Rev.* **77**, 297.
Hahn, E. L. (1950b), *Phys. Rev.* **80**, 580.
Heitler, W. (1954), *The Quantum Theory of Radiation*, 3rd edn. (Oxford University Press, London).
Ikeda, K. (1979), *Opt. Commun.* **30**, 257.
Ikeda, K., Daido, H. and Akimoto, O. (1980), *Phys. Rev. Lett.* **45**, 709.
Jodoin, R. and Mandel, L. (1974a) *Phys. Rev. A* **9**, 873.

Jodoin, R. and Mandel, L. (1974b) *Phys. Rev. A* **10**, 1898.
Kurnit, N. A., Abella, I. D. and Hartmann, S. R. (1964), *Phys. Rev. Lett.* **13**, 567.
Lama, W. L., Jodoin, R. and Mandel, L. (1972), *Am. J. Phys.* **40**, 32.
Lamb, W. E., Jr. (1964), *Phys. Rev.* **134**, A1429.
McCall, S. L. (1974), *Phys. Rev. A* **9**, 1515.
McCall, S. L. and Hahn, E. L. (1967), *Phys. Rev. Lett.* **18**, 908.
McCall, S. L. and Hahn, E. L. (1969), *Phys. Rev.* **183**, 457.
Nakatsuka, H., Asaka, S., Itoh, H., Ikeda, K. and Matsuoka, M. (1983), *Phys. Rev. Lett.* **50**, 109.
Narducci, L. M., Gilmore, R., Da Hsuan Feng and Agarwal, G. S. (1978), *Opt. Lett.* **2**, 88.
Orozco, L. A., Kimble, H. J. and Rosenberger, A. T. (1987), *Opt. Commun.* **62**, 54.
Polder, D., Schuurmans, M. F. H. and Vrehen, Q. H. F. (1979), *Phys. Rev. A* **19**, 1192.
Radcliffe, J. M. (1971), *J. Phys. A* **4**, 313.
Raimond, J. M., Goy, P., Gross, M., Fabre, C. and Haroche, S. (1982), *Phys. Rev. Lett.* **49**, 1924.
Rehler, N. E. and Eberly, J. H. (1971), *Phys. Rev. A* **3**, 1735.
Ressayre, E. and Tallet, A. (1973), *Phys. Rev. Lett.* **30**, 1239.
Skribanowitz, N., Herman, I. P., MacGillivray, J. C. and Feld, M. S. (1973), *Phys. Rev. Lett.* **30**, 309.
Slusher, R. E. (1974), in *Progress in Optics*, Vol. 12, ed. E. Wolf (North-Holland, Amsterdam), p. 53.
Slusher, R. E. and Gibbs, H. M. (1972), *Phys. Rev. A* **5**, 1634.
Stroud, Jr., C. R., Eberly, J. H., Lama, W. L. and Mandel, L. (1972), *Phys. Rev. A* **5**, 1094.
Szöke, A., Daneu, V., Goldhar, J. and Kurnit, N. A. (1969), *Appl. Phys. Lett.* **15**, 376.
Walls, D. F., Drummond, P. D., Hassan, S. S. and Carmichael, H. J. (1978), *Progr. Theor. Phys. (Japan) Suppl.* **64**, 307.

Chapter 17

Agarwal, G. S. (1973), in *Progress in Optics*, Vol. 11, ed. E. Wolf (North-Holland, Amsterdam), p. 3.
Agarwal, G. S. (1974), *Quantum Statistical Theories of Spontaneous Emission and their Relation to other Approaches. Springer Tracts in Modern Physics*, Vol. 70 (Springer, Berlin).
Agarwal, G. S., Brown, A. C., Narducci, L. M. and Vetri, G. (1977), *Phys. Rev. A* **15**, 1613.
Agarwal, G. S., Feng, D. H., Narducci, L. M., Gilmore, R. and Tuft, R. A. (1979), *Phys. Rev. A* **20**, 2040.
Bonifacio, R., Schwendimann, P. and Haake, F. (1971), *Phys. Rev. A* **4**, 302.
Callen, H. B. and Welton, T. A. (1951), *Phys. Rev.* **83**, 34.
Carmichael, H. J. and Walls, D. F. (1977), *J. Phys. B* **10**, L685.
Einstein, A. (1905), *Ann. Phys. (Leipz)*, **17**, 549 [English Translation in *The Collected Papers of Albert Einstein*, Vol. 2 (Princeton University Press, Princeton, NJ, 1989) p. 123].
Haken, H. (1970), *Laser Theory* in *Handbuch der Physik*, ed. S. Flügge; Vol. XXV/2c (Springer, Berlin).
Kimble, H. J. and Mandel, L. (1976), *Phys. Rev. A* **13**, 2123.
Klauder, J. R. (1966), *Phys. Rev. Lett.* **16**, 534.
Klauder, J. R. and Sudarshan, E. C. G. (1968), *Fundamentals of Quantum Optics* (Benjamin, New York).
Kubo, R. (1966), *Rep. Progr. in Phys.* **29**, (pt. 1), 255.

Lax, M. (1963), *Phys. Rev.* **129**, 2342.
Lax, M. (1966), *Phys. Rev.* **145**, 110.
Lax, M. (1967), *Phys. Rev.* **157**, 213.
Louisell, W. H. (1969), in *Quantum Optics*, Proc. Internat. School of Physics 'Enrico Fermi', course XLII, Varenna, ed. R. J. Glauber (Academic Press, New York), p. 680.
Louisell, W. H. (1973), *Quantum Statistical Properties of Radiation* (Wiley, New York).
Montroll, E. W. (1961), in *Lectures in Theoretical Physics*, Vol. 3, eds. W. E. Brittin, D. W. Downs and J. Downs (Interscience, New York) p. 221.
Narducci, L. M., Gilmore, R., Feng, D. H. and Agarwal, G. S. (1978), *Opt. Lett.* **2**, 88.
Nyquist, H. (1928), *Phys. Rev.* **32**, 110.
Oppenheim, I., Shuler, K. E. and Weiss, G. H. (1977), *Stochastic Processes in Chemical Physics: The Master Equation* (MIT Press, Cambridge, MA).
Pauli, W. (1928), in *Probleme der Modernen Physik, Arnold Sommerfeld zum 60. Geburtstage gewidmet von seinen Schülern* ed. P. Debye, (Hirzel Verlag, Leipzig) Vol. 1. Reprinted in *Collected Scientific Papers by Wolfgang Pauli* (1964), eds. R. Kronig and V. F. Weisskopf, (Interscience, New York) p. 549.
Prigogine, I. (1962), *Non-Equilibrium Statistical Mechanics*, (Interscience, New York).
Senitzky, I. R. (1960), *Phys. Rev.* **119**, 670.
Senitzky, I. R. (1961), *Phys. Rev.* **124**, 642.
Srinivas, M. D. and Wolf, E. (1977) in *Statistical Mechanics and Statistical Methods in Theory and Application*, ed. U. Landman, (Plenum Press, New York), p. 219.
Sudarshan, E. C. G. (1963), *Phys. Rev. Lett.* **10**, 277.
Van Hove, L. (1955), *Physica* **21**, 517.
Van Hove, L. (1957), *Physica* **23**, 441.
Van Hove, L. (1962), in *Fundamental Problems in Statistical Mechanics*, ed. E. G. D. Cohen (North-Holland, Amsterdam) p. 157.
Walls, D. F. and Milburn, G. J. (1985), *Phys. Rev. A* **31**, 2403.
Weidlich, W. and Haake, F. (1965a), *Zeits. f. Physik* **185**, 30.
Weidlich, W. and Haake, F. (1965b), *Zeits. f. Physik* **186**, 203.
Zwanzig, R. (1961), in *Lectures in Theoretical Physics*, Vol. 3, eds. W. E. Brittin, B. W. Downs and J. Downs (Interscience, New York) p. 106.

Chapter 18

Ackerhalt, J. R., Milonni, P. W. and Shih, M. L. (1985), *Physics Reports*, **128**, 205.
Agarwal, G. S., Ravi, S. and Cooper, J. (1990), *Phys. Rev. A* **41**, 4721.
Albano, A. M., Abounadi, J., Chyba, T. H., Searle, C. E., Yong, S., Gioggia, R. S. and Abraham, N. B. (1985), *J. Opt. Soc. Am. B* **2**, 47.
Arecchi, F. T. and Degiorgio, V. (1971), *Phys. Rev. A* **3**, 1108.
Arecchi, F. T., Berné, A. and Burlamacchi, P. (1966), *Phys. Rev. Lett.* **16**, 32.
Arecchi, F. T., Degiorgio, V. and Querzola, B. (1967), *Phys. Rev. Lett.* **19**, 1168.
Arecchi, F. T., Rodari, G. S. and Sona, A. (1967), *Phys. Lett.* **25A**, 59.
Basov, N. G., and Prokhorov, A. M. (1954), *Zh. Exsp. Teor. Fiz. USSR* **27**, 431.
Basov, N. G., and Prokhorov, A. M. (1955), *Zh. Exsp. Teor. Fiz. USSR* **28**, 249. [English translation in *Sov. Phys. JETP* **1**, 184 (1955)].
Ben-Mizrachi, A., Procaccia, I. and Grassberger, P. (1984), *Phys. Rev. A* **29**, 975.
Bentley, J. and Abraham, N. B. (1982), *Opt. Commun*, **41**, 52.
Boyd, G. D. and Gordon, J. P., (1961), *Bell Syst. Tech. J.* **40**, 489.
Buley, E. R. and Cummings, F. W. (1964), *Phys. Rev.* **134A**, 1454.
Cantrell, C. D. and Smith, W. A. (1971), *Phys. Lett.* **37A**, 167.
Cantrell, C. D., Lax, M. and Smith, W. A. (1973a), in *Coherence and Quantum Optics*, eds. L. Mandel and E. Wolf (Plenum Press, New York), p. 785.

Cantrell, C. D., Lax, M. and Smith, W. A. (1973b), *Phys. Rev. A* **7**, 175.
Casperson, L. W. (1978), *IEEE J. Quant. Electron.* **QE–14**, 756.
Casperson, L. W. (1983), in *Laser Physics*, eds. D. J. Harvey and D. F. Walls (Springer-Verlag, Berlin) p. 88.
Chang, R. F., Korenman, V., Alley, C. O. and Detenbeck, R. W. (1969), *Phys. Rev.* **178**, 612.
Chopra, S. and Mandel, L. (1972), *IEEE J. Quant. Electron.* **QE–8**, 324.
Chopra, S. and Mandel, L. (1973a) in *Coherence and Quantum Optics*, eds. L. Mandel and E. Wolf (Plenum Press, New York), p. 805.
Chopra, S. and Mandel, L. (1973b), *Phys. Rev. Lett.* **30**, 60.
Chyba, T. H. (1990), *Stochastic and Chaotic Effects in the Ring Laser*, Ph.D. Thesis, (University of Rochester, Rochester, NY), p. 69.
Chyba, T. H., Christian, W. R., Gage, E., Lett, P. and Mandel, L. (1986), in *Optical Instabilities*, eds. R. W. Boyd, M. G. Raymer and L. M. Narducci, (Cambridge University Press, Cambridge) p. 253.
Corti, M. and Degiorgio, V. (1974), *Opt. Commun.* **11**, 1.
Corti, M. and Degiorgio, V. (1976a), *Phys. Rev. A* **14**, 1475.
Corti, M. and Degiorgio, V. (1976b), *Phys. Rev. Lett.* **36**, 1173.
Corti, M., Degiorgio, V. and Arecchi, F. T. (1973), *Opt. Commun.* **8**, 329.
Davidson, F. (1969), *Phys. Rev.* **185**, 446.
Davidson, F. and Mandel, L. (1967), *Phys. Lett.* **25A**, 700.
Davidson, F. and Mandel, L. (1968), *Phys. Lett.* **27A**, 579.
Degiorgio V. and Scully, M. O. (1970), *Phys. Rev. A* **2**, 1170.
Feigenbaum, M. J. (1978), *J. Stat. Phys.* **19**, 25.
Feigenbaum, M. J. (1979), *J. Stat. Phys.* **21**, 669.
Fox, A. G. and Li, T. (1961), *Bell Syst. Tech. J.* **40**, 453.
Freed, C. and Haus, H. A. (1966a), *Phys. Rev.* **141**, 287.
Freed, C. and Haus, H. A. (1966b), *IEEE J. Quant. Electron* **QE–2**, 190.
Gerhardt, H., Welling, H. and Güttner, A. (1972), *Zeits. f. Physik* **253**, 113.
Gioggia, R. S. and Abraham, N. B. (1983), *Opt. Commun*, **47**, 278.
Gordon, J. P., Zeiger, H. J. and Townes, C. H. (1954), *Phys. Rev.* **95**, 282.
Gordon, J. P., Zeiger, H. J. and Townes, C. H. (1955), *Phys. Rev.* **99**, 1264.
Graham, R. (1975), in *Fluctuations, Instabilities, and Phase Transitions*, ed. T. Riste (Plenum Press, New York), p. 215.
Graham, R. and Haken, H. (1970), *Zeits. f. Physik* **237**, 31.
Grassberger, P. and Procaccia, I. (1983), *Physica* **9D**, 189.
Güttner, A., Welling, H., Gericke, K. H. and Seifert, W. (1978), *Phys. Rev. A* **18**, 1157.
Haken, H. (1964), *Zeits. f. Physik* **181**, 96.
Haken, H. (1970), *Laser Theory* in *Handbuch der Physik*, Vol. XXV/2c, ed. S. Flügge (Springer, Berlin).
Haken, H. (1975a), *Rev. Mod. Phys.* **47**, 67.
Haken, H. (1975b), *Phys. Lett.* **53A**, 77.
Haken, H. (1977), *Synergetics* (Springer, Berlin).
Haken, H. (1985), *Light*, Vol. 2. (North-Holland, Amsterdam).
Haken, H. and Ohno, H. (1976a), *Opt. Commun*, **16**, 205.
Haken, H. and Ohno, H. (1976b), *Phys. Lett.* **59A**, 261.
Haken, H. and Sauermann, H. (1963a). *Zeits. f. Physik* **173**, 261.
Haken, H. and Sauermann, H. (1963b). *Zeits. f. Physik* **176**, 47.
Harris, S. E. (1989), *Phys. Rev. Lett.* **62**, 1033.
Hempstead, R. D. and Lax, M. (1967), *Phys. Rev.* **161**, 350.
Imamoglu, A. (1989), *Phys. Rev. A* **40**, 2835.
Imamoglu, A. and Harris, S. E. (1989), *Opt. Lett.* **14**, 1344.
Javan, A., Bennett, W. R. and Herriott, D. R. (1961), *Phys. Rev. Lett.* **6**, 106.

Kocharovskaya, O. A. and Khanin, Ya I. (1988), *Pis'ma Zh. Eksp. Teor. Fiz.* **48**, 581 [English translation in *JETP Lett.* **48**, 630 (1988)].
Kogelnik, H. and Li, T. (1966), *Proc. IEEE* **54**, 1312.
Kogelnik, H. and Rigrod, W. W. (1962), *Proc. IRE* **50**, 220.
Korn, G. A. and Korn, T. M. *Mathematical Handbook for Engineers* (1961), (McGraw Hill, New York).
Korobkin, V. V. and Uspenskii, A. V. (1963), *Zh. Eksp. Teor. Fiz.* **45**, 1003 [English translation in *Sov. Phys. JETP* **18**, 693 (1964)].
Lamb, W. E. Jr. (1964), *Phys. Rev.* **134**, A1429.
Landau, L. D. and Lifshitz, E. H. (1980), *Statistical Physics*, 3rd edn. (Pergamon Press, Oxford).
Lax, M. and Louisell, W. H. (1967), *IEEE J. Quant. Electr. QE*—**3**, 47.
Lax, M. and Louisell, W. H. (1969), *Phys. Rev.* **185**, 568.
Lorenz, E. N. (1963), *J. Atmosph. Sci.* **20**, 130.
Louisell, W. H. (1969) in *Quantum Optics*, Proc. Internat. School of Physics 'Enrico Fermi', Course XLII, Varenna, ed. R. J. Glauber (Academic Press, New York) p. 680.
Maeda, M. and Abraham, N. B. (1982), *Phys. Rev. A* **26**, 3395.
Maiman, T. H. (1960), *Nature (London)* **187**, 493.
Mandel, L. (1961), *J. Opt. Soc. Am.* **51**, 797.
Manes, K. R. and Siegman, A. E. (1971), *Phys. Rev. A* **4**, 373.
Mayr, M., Risken, H. and Vollmer, H. D. (1981), *Opt. Commun.* **36**, 480.
Meltzer, D. and Mandel, L. (1970), *Phys. Rev. Lett.* **25**, 1151.
Meltzer, D. and Mandel, L. (1971), *Phys. Rev. A* **3**, 1763.
Meltzer, D., Davis, W. and Mandel, L. (1970), *Appl. Phys. Lett.* **17**, 242.
Milonni, P. W. and Eberly, J. H. (1988), *Lasers* (Wiley, New York).
Narducci, L. M., Doss, H. M., Ru, P., Scully, M. O., Zhu, S. Y. and Keitel, C. (1991), *Opt. Commun.* **81**, 379.
Risken, H. (1965), *Zeits. f. Physik* **186**, 85.
Risken, H. (1966), *Zeits. f. Physik* **191**, 302.
Risken, H. (1970), in *Progress in Optics*, Vol. 8, ed. E. Wolf (North-Holland, Amsterdam) p. 239.
Risken, H. and Nummedal, K. (1968a), *J. Appl. Phys.* **39**, 4662.
Risken, H. and Nummedal, K. (1968b), *Phys. Lett.* **26A**, 275.
Risken, H. and Vollmer, H. D. (1967a), *Zeits. f. Physik* **201**, 323.
Risken, H. and Vollmer, H. D. (1967b), *Zeits. f. Physik* **204**, 240.
Ross, S. L. (1984) *Differential Equations* (Wiley, New York).
Sargent, M., III, Scully, M. O. and Lamb, W. E., Jr. (1974), *Laser Physics* (Addison-Wesley, Reading, MA).
Schawlow, A. L., and Townes, C. H. (1958), *Phys. Rev.* **112**, 1940.
Scully, M. O. (1969), in *Quantum Optics*, Proc. Internat. School of Physics, 'Enrico Fermi', Course XLII, Varenna, ed. R. J. Glauber (Academic Press, New York) p. 586.
Scully, M. O. (1973), in *Coherence and Quantum Optics*, eds. L. Mandel and E. Wolf (Plenum Press, New York) p. 691.
Scully, M. O. and Lamb, W. E., Jr. (1967), *Phys. Rev.* **159**, 208.
Scully, M. O. and Lamb, W. E., Jr. (1969), *Phys. Rev.* **166**, 246.
Scully, M. O. and Lamb, W. E., Jr. (1970), *Phys. Rev.* **179**, 368.
Scully, M. O., Zhu, S. Y. and Gavrielides, A. (1989), *Phys. Rev. Lett.* **62**, 2813.
Siegman, A. E. (1971), *An Introduction to Lasers and Masers* (McGraw-Hill, New York).
Siegman, A. E. (1986), *Lasers* (University Science Books, Mill Valley, CA).
Siegman, A. E. and Arrathoon, R. (1968), *Phys. Rev. Lett.* **20**, 901.

Siegman, A. E., Daino, B. and Manes, K. R. (1967), *IEEE J. Quant. Electron.* **QE-3**, 180.
Singh, S., Friberg, S. and Mandel, L. (1983), *Phys. Rev. A* **27**, 381.
Smith, A. W. and Armstrong, J. A. (1966), *Phys. Rev. Lett.* **16**, 1169.
Uspenskii, A. V. (1963), *Radio Eng. Electron. Phys. (USSR)* **8**, 1145.
van der Pol, B. (1922), *Phil. Mag.* **43**, 700.
Wang, Y. K. and Lamb, W. E., Jr. (1973), *Phys. Rev. A* **8**, 866.
Weiss, C. O. and King, H. (1982), *Opt. Commun.* **44**, 59.
Weiss, C. O., Godone, A. and Olafsson, A. (1983), *Phys. Rev. A* **28**, 892.
Weiss, C. O., Klische, W., Ering, P. S. and Cooper, M. (1985), *Opt. Commun.* **52**, 405.

Chapter 19

Abramowitz, M. and Stegun, I. A. (1965), *Handbook of Mathematical Functions*, (Dover, New York).
Agarwal, G. S. and Dattagupta, S. (1982), *Phys. Rev. A* **26**, 880.
Aronowitz, F. (1965), *Phys. Rev.* **139**, A635.
Aronowitz, F. (1971) in *Laser Applications*, Vol. I, ed. M. Ross (Academic Press, New York), p. 133.
Gradshteyn, I. S. and Ryzhik, I. M. (1980), *Table of Integrals, Series and Products* (Academic Press, New York).
Graham, R. (1975), in *Fluctuations, Instabilities, and Phase Transitions*, ed. T. Riste (Plenum, New York), p. 215.
Graham, R. and Haken, H. (1968), *Zeits. f. Physik* **213**, 420.
Grossmann, S. and Richter, P. H. (1971), *Zeits. f. Physik* **249**, 43.
Haken, H. (1975), *Rev. Mod. Phys.* **47**, 67.
Haken, H. (1977), *Synergetics*, (Springer, Berlin).
Hambenne, J. B. and Sargent, M., III (1975), *IEEE J. Quant. Electron.* **QE-11**, 90.
Hioe, F. T. and Singh, S. (1981), *Phys. Rev. A* **24**, 2050.
Hioe, F. T., Singh, S. and Mandel, L. (1979), *Phys. Rev. A* **19**, 2036.
Lamb, W. E., Jr. (1964), *Phys. Rev.* **134**, A1429.
Lenstra, D. and Singh, S. (1983), *Phys. Rev. A* **28**, 2318.
Lett, P., Christian, W., Singh, S. and Mandel, L. (1981), *Phys. Rev. Lett.* **47**, 1892.
M-Tehrani, M. and Mandel, L. (1976), *Opt. Commun.* **16**, 16.
M-Tehrani, M. and Mandel, L. (1977), *Opt. Lett.* **1**, 196.
M-Tehrani, M. and Mandel, L. (1978a), *Phys. Rev. A* **17**, 677.
M-Tehrani, M. and Mandel, L. (1978b), *Phys. Rev. A* **17**, 694.
Mandel, L. (1982), *Opt. Commun.* **42**, 356.
Mandel, L., Roy, R. and Singh, S. (1981), in *Optical Bistability*, eds. C. M. Bowden, M. Ciftan and H. Robl (Plenum Press, New York), p. 127.
Mayr, M., Risken, H., and Vollmer, H. D. (1981), *Opt. Commun.* **36**, 480.
Menegozzi, L. N. and Lamb, W. E., Jr. (1973), *Phys. Rev. A* **8**, 2103.
Patashinskiĭ, A. Z. and Pokrovskiĭ, V. L. (1979), *Fluctuation Theory of Phase Transitions*, (Pergamon Press, Oxford).
Rigrod, W. W. and Bridges, T. J. (1965), *IEEE J. Quant. Electron.* **QE-1**, 298.
Risken, H. and Nummedal, H. (1968), *J. Appl. Phys.* **39**, 4662.
Roy, R. and Mandel, L. (1980), *Opt. Commun.* **34**, 133.
Roy, R., Short, R., Durnin, J. and Mandel, L. (1980), *Phys. Rev. Lett.* **45**, 1486.
Sargent M., III, Scully, M. O. and Lamb, W. E., Jr. (1974), *Laser Physics* (Addison-Wesley, Reading, MA).
Singh, S. (1980), *Opt. Commun.* **32**, 339.
Singh, S. (1981), *Phys. Rev. A* **23**, 837.

Singh, S. (1984), *Phys. Rep.* **108**, 217.
Singh, S. and Mandel, L. (1979), *Phys. Rev. A* **20**, 2459.
Singh, S. and Mandel, L. (1981), *Opt. Commun.* **40**, 139.
Shenoy, S. R. and Agarwal, G. S. (1984), *Phys. Rev. A* **29**, 1315.
Smirnov, V. S. and Zhelnov, B. L. (1969), *Zh. Eksp. Teor. Fiz.* **57**, 2043 [English translation in *Sov. Phys. JETP* (1970) **30**, 1108].
Stratonovich, R. L. (1963), *Topics in the Theory of Random Noise*, Vol. 1, (Gordon and Breach, New York).
Weiss, G. H. (1977), in *Stochastic Processes in Chemical Physics*, eds. I. Oppenheim, K. E. Shuler and G. H. Weiss, (MIT Press, Cambridge, MA), p. 361.

Chapter 20

Abraham, N. B. and Smith, S. R. (1977), *Phys. Rev. A* **15**, 421.
Agarwal, G. S. and Tara, K. (1992), *Phys. Rev. A* **46**, 485.
Carusotto, S. (1975), *Phys. Rev. A* **11**, 1629.
Caves, C. M. (1982), *Phys. Rev. D* **26**, 1817.
Friberg, S. and Mandel, L. (1983), *Opt. Commun.* **46**, 141.
Friberg, S. and Mandel, L. (1984), in *Coherence and Quantum Optics V*, eds. L. Mandel and E. Wolf. (Plenum Press, New York), p. 465.
Glauber, R. J. (1986), in *Frontiers in Quantum Optics*, eds. E. R. Pike and S. Sarkar (Adam Hilger, Bristol), p. 534.
Hong, C. K., Friberg, S. and Mandel, L. (1985), *J. Opt. Soc. Am. B* **2**, 494.
Mandel, L. (1983), *Nature* **304**, 188.
Rockower, E. B., Abraham, N. B. and Smith, S. R. (1978), *Phys. Rev. A* **17**, 1100.
Wooters, W. K. and Zurek, W. H. (1982), *Nature* **299**, 802.

Chapter 21

Abramowitz, M. and Stegun, I. A. (1965), *Handbook of Mathematical Functions*. (Dover, New York).
Caves, C. M. (1981), *Phys. Rev. D* **23**, 1693.
Caves, C. M. (1982), *Phys. Rev. D* **26**, 1817.
Caves, C. M. and Schumaker, B. L. (1985), *Phys. Rev. A* **31**, 3068.
Collett, M. J. and Gardiner, C. W. (1984), *Phys. Rev. A* **30**, 1386.
Collett, M. J. and Walls, D. F. (1985), *Phys. Rev. A* **32**, 2887.
Collett, M. J., Walls, D. F. and Zoller, P. (1984), *Opt. Commun.* **52**, 145.
Fan, H. Y. (1990), *Phys. Rev. A* **41**, 1526.
Fan, H. Y., Zaidi, H. R. and Klauder, J. R. (1987), *Phys. Rev. D* **35**, 1831.
Fisher, R. A., Nieto, M. M. and Sandberg, V. D. (1984), *Phys. Rev. D* **29**, 1107.
Friberg, S. and Mandel, L. (1984), *Opt. Commun.* **48**, 439.
Garcia-Fernandez, P., Sainz de los Terreros, L., Bermejo, F. J. and Santoro, J. (1986), *Phys. Lett. A* **118**, 400.
Gardiner, C. W. and Collett, M. J. (1985), *Phys. Rev. A* **31**, 3761.
Gardiner, C. W. and Savage, C. M. (1984), *Opt. Commun.* **50**, 173.
Glauber, R. J. (1963), *Phys. Rev.* **131**, 2766.
Heidmann, A., Reynaud, S. and Cohen-Tannoudji, C. (1984), *Opt. Commun.* **52**, 235.
Hillery, M. (1987a), *Opt. Commun.* **62**, 135.
Hillery, M. (1987b), *Phys. Rev. A* **36**, 3796.
Hollenhorst, J. N. (1979), *Phys. Rev. D* **19**, 1669.
Hong, C. K. and Mandel, L. (1985a), *Phys. Rev. Lett.* **54**, 323.
Hong, C. K. and Mandel, L. (1985b), *Phys. Rev. A* **32**, 974.
Kimble, H. J. (1992), in *Fundamental Systems in Quantum Optics*, eds. J. Dalibard, J. M. Raimond and J. Zinn-Justin (North-Holland, Amsterdam), Chapter 10.

Kozierowski, M. (1986), *Phys. Rev. A* **34**, 3474.
Loudon, R. (1984), *Opt. Commun.* **49**, 24.
Loudon, R. and Knight, P. L. (1987), *J. Mod. Opt.* **34**, 709.
Lugiato, L. A. and Strini, G. (1982a), *Opt. Commun.* **41**, 67.
Lugiato, L. A. and Strini, G. (1982b), *Opt. Commun.* **41**, 374.
Lugiato, L. A. and Strini, G. (1982c), *Opt. Commun.* **41**, 447.
Mandel, L. (1982a), *Phys. Rev. Lett.* **49**, 136.
Mandel, L. (1982b), *Opt. Commun.* **42**, 437.
Mandel, L. and Hong, C. K. (1986), in *Coherence, Cooperation and Fluctuations*, eds. F. Haake, L. M. Narducci and D. F. Walls. (Cambridge University Press, Cambridge), p. 254.
Milburn, G. J. (1984), *J. Phys. A* **17**, 737.
Milburn, G. and Walls, D. F. (1981), *Opt. Commun.* **39**, 401.
Mollow, B. R. (1973), *Phys. Rev. A* **8**, 2684.
Ou, Z. Y., Hong, C. K. and Mandel, L. (1987), *J. Opt. Soc. Am. B* **4**, 1574.
Rainville, E. D. (1960), *Special Functions* (Macmillan, New York).
Reid, M. D. and Walls, D. F. (1985), *Phys. Rev. A* **31**, 1622.
Richter, T. (1984), *Opt. Acta* **31**, 1045.
Schumaker, B. L. (1984), *Opt. Lett.* **9**, 189.
Schumaker, B. L. (1986), *Phys. Rep.* **135**, 317.
Schumaker, B. L. and Caves, C. M. (1985), *Phys. Rev. A* **31**, 3093.
Shapiro, J. H. (1985), *IEEE J. Quant. Electron.* **QE–21**, 237.
Shapiro, J. H., Yuen, H. P. and Machado Mata, J. A. (1979), *IEEE Trans. Inf. Theory* **IT–25**, 179.
Slusher, R. E., Hollberg, L. W., Yurke, B., Mertz, J. C. and Valley, J. F. (1985), *Phys. Rev. Lett.* **55**, 2409.
Stoler, D. (1970), *Phys. Rev. D* **1**, 3217.
Stoler, D. (1971), *Phys. Rev. D* **4**, 1925.
Teich, M. C. and Saleh, B. E. A. (1989), *Quant. Opt.* **1**, 153.
Teich, M. C. and Saleh, B. E. A. (1990), *Physics Today* (June) **43**, 26.
Walls, D. F. (1983), *Nature* **306**, 141.
Walls, D. F. and Zoller, P. (1981), *Phys. Rev. Lett.* **47**, 709.
Wu, L. A., Kimble, H. J., Hall, J. and Wu, H. (1986), *Phys. Rev. Lett.* **57**, 2520.
Yariv, A. (1967), *Quantum Electronics* (Wiley, New York).
Yariv, A. and Pepper, D. M. (1977), *Opt. Lett.* **1**, 16.
Yuen, H. P. (1976), *Phys. Rev. A* **13**, 2226.
Yuen, H. P. and Chan, V. W. S. (1983), *Opt. Lett.* **8**, 177.
Yuen, H. P. and Shapiro, J. H. (1979), *Opt. Lett.* **4**, 334.
Yuen, H. P. and Shapiro, J. H. (1980), *IEEE Trans. Inf. Theory* **IT–26**, 78.
Yurke, B. (1985a), *Phys. Rev. A* **32**, 300.
Yurke, B. (1985b), *Phys. Rev. A* **32**, 311.

Chapter 22

Agarwal, G. S., Friberg, A. T. and Wolf, E. (1982), *Opt. Commun.* **43**, 466.
Agarwal, G. S., Friberg, A. T. and Wolf, E. (1983), *J. Opt. Soc. Am.* **73**, 529.
Armstrong, J. A., Bloembergen, N., Ducuing, J. and Pershan, P. S. (1962) *Phys. Rev.* **127**, 1918.
Bloembergen, N. (1965), *Nonlinear Optics* (Benjamin, New York).
Bloom, D. M. and Bjorklund, G. C. (1977), *Appl. Phys. Lett.* **31**, 592.
Boyd, R. W. (1992), *Nonlinear Optics* (Academic Press, Boston).
Boyd, R. W., Habashy, T. M., Jacobs, A. A., Mandel, L., Nieto-Vesperinas, M., Tompkin, W. R. and Wolf, E. (1987), *Opt. Lett.* **12**, 42.

Braginsky, V. B. and Khalil, F. Y. (1992) *Quantum Measurements*, ed. K. S. Thorne (Cambridge University Press, Cambridge).
Burnham, D. C. and Weinberg, D. L. (1970), *Phys. Rev. Lett.* **25**, 84.
Butcher, P. N. and Cotter, D. (1990), *The Elements of Nonlinear Optics* (Cambridge University Press, Cambridge).
Caves, C. M., Thorne, K. S., Drever, R. W. P., Sandberg, V. D. and Zimmermann, M. (1980), *Rev. Mod. Phys.* **52**, 341.
Drummond, P. D. (1990), *Phys. Rev. A* **42**, 6845.
Einstein, A., Podolsky, B. and Rosen, N. (1935), *Phys. Rev.* **47**, 777.
Fisher, R. A. ed. (1983), *Optical Phase Conjugation* (Academic Press, New York).
Franken, P. A., Hill, A. E., Peters, C. W. and Weinreich, G. (1961), *Phys. Rev. Lett.* **7**, 118.
Friberg, S., Hong, C. K. and Mandel, L. (1985a), *Opt. Commun.* **54**, 311.
Friberg, S., Hong, C. K. and Mandel, L. (1985b), *Phys. Rev. Lett.* **54**, 2011.
Giordmaine, J. A. (1962), *Phys. Rev. Lett.* **8**, 19.
Giuliano, C. R. (1981), *Physics Today* (April) **34**, 27.
Graham, R. (1984), *Phys. Rev. Lett.* **52**, 117.
Hellwarth, R. W. (1977), *J. Opt. Soc. Am.* **67**, 1.
Hillery, M. and Mlodinow, L. D. (1984), *Phys. Rev. A* **30**, 1860.
Hong, C. K. and Mandel, L. (1985), *Phys. Rev. A* **31**, 2409.
Hong, C. K. and Mandel, L. (1986), *Phys. Rev. Lett.* **56**, 58.
Hong, C. K., Ou, Z. Y. and Mandel, L. (1987), *Phys. Rev. Lett.* **59**, 2044.
Imoto, N., Haus, H. A. and Yamamoto, Y. (1985), *Phys. Rev. A* **32**, 2287.
Jain, R. K. and Lind, R. C. (1983), *J. Opt. Soc. Am.* **73**, 647.
Jensen, S. M. and Hellwarth, R. W. (1978), *Appl. Phys. Lett.* **32**, 166.
Kärtner, F. X. and Haus, H. A. (1993), *Phys. Rev. A*. **47**, 4585.
Kielich, S., Kozierowski, M. and Tanás, R. (1978), in *Coherence and Quantum Optics IV*, eds. L. Mandel and E. Wolf. (Plenum Press, New York), p. 511.
Kitagawa, M. and Yamamoto, Y. (1986), *Phys. Rev. A* **34**, 3974.
Kleinman, D. A. (1968), *Phys. Rev.* **174**, 1027.
Klyshko, D. N. (1968), *Zh. Eksp. Teor. Fiz.* **55**, 1006 [English translation in *Sov. Phys. JETP* **28**, 522 (1969)].
Kozierowski, M. and Tanas, R. (1977), *Opt. Commun.* **21**, 229.
Kwiat, P. G., Steinberg, A. M. and Chiao, R. Y. (1992), in *Foundations of Quantum Mechanics*, eds. T. D. Black, M. M. Nieto, H. S. Pilloff, M. O. Scully and R. M. Sinclair (World Scientific, Singapore), p. 193.
Louisell, W. H. (1960), *Coupled Mode and Parametric Electronics* (Wiley, New York).
Maker, P. D., Terhune, R. W., Nisenoff, M. and Savage, C. M. (1962), *Phys. Rev. Lett.* **8**, 21.
Mandel, L. (1982), *Opt. Commun.* **42**, 437.
Mandel, L. (1991), *Opt. Lett.* **16**, 1882.
Milburn, G. J. and Walls, D. F. (1983a), *Phys. Rev. A* **28**, 2065.
Milburn, G. J. and Walls, D. F. (1983b), *Phys. Rev. A* **28**, 2646.
Milburn, G. J., Lane, A. S. and Walls, D. F. (1983), *Phys. Rev. A* **27**, 2804.
Mollow, B. R. (1967), *Phys. Rev.* **162**, 1256.
Mollow, B. R. (1973), *Phys. Rev. A* **8**, 2684.
Mollow, B. R. and Glauber, R. J. (1967a), *Phys. Rev.* **160**, 1076.
Mollow, B. R. and Glauber, R. J. (1967b), *Phys. Rev.* **160**, 1097.
Ou, Z. Y., Wang, L. J. and Mandel, L. (1989), *Phys. Rev. A* **40**, 1428.
Ou, Z. Y., Wang, L. J., Zou, X. Y. and Mandel, L. (1990), *Phys. Rev. A* **41**, 566.
Pepper, D. M. (1982), *Opt. Eng.* **21**, 156.
Peřina, J. (1991), *Quantum Statistics of Linear and Nonlinear Optical Phenomena*, 2nd edn. (Kluwer, Dordrecht).

Rarity, J. G. and Tapster, P. R. (1990), *Phys. Rev. Lett.* **64**, 2495.
Sanders, B. C. and Milburn, G. J. (1989), *Phys. Rev. A* **39**, 694.
Schubert, M. and Wilhelmi, B. (1986), *Nonlinear Optics and Quantum Electronics* (Wiley, New York).
Shen, Y. R. (1967), *Phys. Rev.* **155**, 921.
Shen, Y. R. (1984), *The Principles of Nonlinear Optics* (Wiley, New York).
Slusher, R. E., Hollberg, L. W., Yurke, B., Mertz, J. C. and Valley, J. F. (1985), *Phys. Rev. Lett.* **55**, 2409.
Tucker, J. and Walls, D. F. (1969), *Phys. Rev.* **178**, 2036.
Wang, L. J., Zou, X. Y. and Mandel, L. (1991), *Phys. Rev. A* **44**, 4614.
Yariv, A. (1967), *Quantum Electronics* (Wiley, New York).
Yariv, A. and Pepper, D. M. (1977), *Opt. Lett.* **1**, 16.
Yuen, H. P. and Shapiro, J. H. (1979), *Opt. Lett.* **4**, 334.
Zel'dovich, B. Ya and Klyshko, D. N. (1969), *Zh. Eksp. Teor. Fiz. Pis. Red.* **9**, 60 [English translation in *Soviet Phys. JETP Lett.* **9**, (1969) 40].
Zel'dovich, B. Ya, Pilipetsky, N. F. and Shkunov, V. V. (1985), *Principles of Phase Conjugation* (Springer, Berlin and New York).
Zou, X. Y., Wang, L. J. and Mandel, L. (1991a), *Opt. Commun.* **84**, 351.
Zou, X. Y., Wang, L. J. and Mandel, L. (1991b), *Phys. Rev. Lett.* **67**, 318.

Author index

Subject index